INTERNATIONAL ENCYCLOPEDIA OF INTEGRATED CIRCUITS

2nd Edition

Stan Gibilisco

SECOND EDITION
FIRST PRINTING

TAB Books is a division of McGraw-Hill, Inc.

Library of Congress Cataloging-in-Publication Data

Gibilisco, Stan.
International encyclopedia of integrated circuits / by Stan Gibilisco.—2nd ed.
p. cm.
Includes index.
ISBN 0-8306-3026-0
1. Integrated circuits—Catalogs. I. Title.
TK7874.G5 1991
621.381′5—dc20 91-697
CIP

TAB Books offers software for sale. For information and a catalog, please contact TAB Software Department, Blue Ridge Summit, PA 17294-0850.

Acquisitions Editor: Roland S. Phelps
Technical Editor: Andrew Yoder
Production: Katherine G. Brown
Book Design: Jaclyn J. Boone
Cover Design: Greg Schooley, Mars, PA

How To Use This Book

THIS ENCYCLOPEDIA is a substantially updated version of the first edition. This volume contains information on high-technology integrated circuits, according to the application(s) most often used. The application categories are abbreviated throughout, as follows:

CCT: Clocks, counters, timers
COM: Communications circuits
CON: Control circuits
DCP: Data-conversion and processing circuits
LOG: Logic circuits
MIC: Microcomputer peripherals
PTE: Power supplies and test equipment

Within each category, circuits are subcategorized by manufacturer in alphabetical order.

Information concerning any particular IC is provided just once. Often, circuits can be used for various purposes. Then, the IC is listed in the most applicable categories, with cross references guiding you to the data. You should use the *Integrated Circuits List* at the front of this book to find ICs according to application. If you know more specifically what you want, use the index in the back.

Microcomputer-related circuits form a category here, but it is impractical to provide a cross section of all currently available microcomputers. That information would require a separate encyclopedia.

This book is not a design manual, and it cannot begin to be a complete reference on all available ICs today (such a book would occupy a whole room). Rather, this book is an overview of the types of devices that are available. About two-thirds of the material in this book is updated or completely new, compared with the first edition, published just a few years ago. Technology in this field is rapidly advancing.

For those interested in circuit design, consult the manufacturers' data sheets and/or data books. Although this book is useful for generating design concepts, the detail for actual circuit construction and debugging is best found in the manufacturers' documents and, of course, by trial and error.

A philosophical difference between this edition and the first edition is the "advance, preliminary" data that is included in this volume. Such data is clearly labeled as such. The actual products might, in these cases, have specifications slightly different than those indicated here because of improvements made between the publication of the preliminary data and the availability of the IC.

The scope of the last category has been expanded to include test-equipment and instrumentation ICs. Suggestions for future editions are welcome.

EDITOR'S NOTE: During the publication process of this book, The Sprague Electric Company became Allegro Microsystems, Inc.

Integrated Circuits List

THIS LIST FEATURES all the ICs for which data appears in this book. You can use this list to locate a device based on the application that interests you. Within each application category, the ICs are listed in alphabetical order by manufacturer; within these subcategories, they are individually listed. Some devices have several different possible uses. In these cases, the description of the device and the page number are given in the application category for which the device would probably be used most often. Cross references are given when an IC can be used for a certain purpose, but the data is listed within some other category.

Acknowledgments

I THANK THE manufacturers who granted permissions for reproduction of their material in this book:

Analog Devices, Inc.
2 Technology Way
Norwood, MA 02062

GEC Plessey Semiconductors
Cheney Manor, Swindon, Wiltshire,
United Kingdom SN2 2QW

LSI Computer Systems, Inc.
1235 Walt Whitman Road
Melville, NY 11747

Mitel Corporation
350 Legget Drive
Kanata, Ontario, Canada K2K 1X3

National Semiconductor Corporation
2900 Semiconductor Drive
Santa Clara, CA 95052

Raytheon Company
350 Ellis Street
Mountain View, CA 94039

Rohm Corporation
8 Whatney
Irvine, CA 92718

Siliconix, Inc.
2201 Laurelwood Road
Santa Clara, CA 95054

Silicon Systems, Inc.
14351 Myford Road
Tustin, CA 92680

Sprague Electric Company
70 Pembroke Road
Concord, NH 03301

SECTION 1

Application: CCT 1

Manufacturer: Analog Devices, Inc.

Manufacturer: LSI Computer Systems, Inc.

Manufacturer: National Semiconductor Corporation

*If a listing appears in this column, the data for this IC is in the application category indicated by the abbreviation.

*If a listing appears in this column, the data for this IC is in the application category indicated by the abbreviation.

SECTION 2

Manufacturer: LSI Computer Systems, Inc.

Manufacturer: Mitel Corporation

Manufacturer: National Semiconductor Corporation

*If a listing appears in this column, the data for this IC is in the application category indicated by the abbreviation.

*If a listing appears in this column, the data for this IC is in the application category indicated by the abbreviation.

DEVICE DESIGNATOR	DEVICE FUNCTION	SEE*	PAGE
BU8320AF	"		270
BU8321	"		270
BU8321F	"		270
BU8322	"		270
BU8322F	"		270
BU8992	Pulse dialer		276
BU8302A	Tone/pulse dialer		277

Manufacturer: Siliconix, Inc.

DEVICE DESIGNATOR	DEVICE FUNCTION	SEE*	PAGE
DG200A		CON	
DG201A	Quad monolithic SPST CMOS analog switch		285
DG202	"		285
DG211	Low-cost 4-channel monolithic SPST CMOS analog switch		289
DG212	"		289
DG406	16-channel/dual 8-channel high-performance CMOS analog multiplexer		294
DG407	"		294
DG411		DCP	
DG412		DCP	
DG413		DCP	
DG421		DCP	
DG423		DCP	
DG425		DCP	
DG441		DCP	
DG442		DCP	
DG444		DCP	
DG445		DCP	
DG458	8-channel and dual 4-channel fault-protected CMOS analog multiplexer		296
DG459	"		296
DG485		CON	
DG534		MIC	
DG535		MIC	
DG536		MIC	
DG538		MIC	
DG540	Wideband/video "T" switch		280
DG541	"		280
DG542	"		280
DG884		MIC	

Manufacturer: Sprague Electric Company

DEVICE DESIGNATOR	DEVICE FUNCTION	SEE*	PAGE
ULN-2000A *series*	High-voltage, high-current Darlington arrays		298
ULN-2010A *series*	"		299
ULN-2020A *series*	"		300
ULN-2000L *series*	"		303
UDN-2522A	Quad bus transceiver		304

SECTION 3
Application: CON 307

Manufacturer: Analog Devices, Inc.

DEVICE DESIGNATOR	DEVICE FUNCTION	SEE*	PAGE
AD365	Programmable gain and T/H DAS amplifier		307
AD521	Precision instrumentation amplifier		311
AD522	High-accuracy data-acquisition instrumentation amplifier		314
AD524	Precision instrumentation amplifier		315
AD526	Software-programmable gain amplifier		319
AD537	Integrated circuit Voltage-to-frequency converter		346

*If a listing appears in this column, the data for this IC is in the application category indicated by the abbreviation.

*If a listing appears in this column, the data for this IC is in the application category indicated by the abbreviation.

DEVICE DESIGNATOR	DEVICE FUNCTION	SEE*	PAGE
LS7264	Four-phase brushless dc motor-speed controller		402
LS7331	Touch-sensitive light dimmer and ac motor-speed controller with computer control and monitoring		406

Manufacturer: Rohm Co., Ltd.

DEVICE DESIGNATOR	DEVICE FUNCTION	SEE*	PAGE
BA612	Large current driver		444
BA618	LED driver		453
BA718		COM	
BA728		COM	
BA728F		COM	
BA2266A	Radio servo controller		417
BA6208	Reversible motor driver		408
BA6218	"		410
BA6109	"		411
BA6209	"		412
BA6219	"		415
BA12001	High-voltage, high-current Darlington transistor array		448
BA12003	7-channel driver		449
BA12004	"		449
BR6116	2K × 8 high-speed CMOS SRAM		418
BR6116F	"		418
BR6264	8K × 8 CMOS SRAM		423
BR93C46	1024-bit serial electrically erasable PROM with 2V read capability		429
BR93CS46	1024-bit serial (3V and 5V) electrically erasable PROM		435
TA6270F		MIC	

Manufacturer: Silicon Systems, Inc.

DEVICE DESIGNATOR	DEVICE FUNCTION	SEE*	PAGE
32C260		MIC	
32C452		MIC	
32H4631		MIC	
32H6220		MIC	
32B451		MIC	

Manufacturer: Siliconix, Inc.

DEVICE DESIGNATOR	DEVICE FUNCTION	SEE*	PAGE
DG200A	Dual monolithic SPST CMOS analog switch		454
DG485	Low-power CMOS octal analog switch array		456
DG528	8-channel/dual 4-channel latchable multiplexers		462
DG529	"	CON	462

Manufacturer: Sprague Electric Company

DEVICE DESIGNATOR	DEVICE FUNCTION	SEE*	PAGE
UHP-400	Power and relay drivers		465
UHP-400-1	"		465
UHP-500	"		465
UHP-402	Quad 2-input OR power driver		468
UHP-402-1	"		468
UHP-502	"		468
UHP-403	Quad OR relay driver		469
UHP-403-1	"		469
UHP-503	"		469
UHP-406	Quad AND relay driver		470
UHP-406-1	"		470

*If a listing appears in this column, the data for this IC is in the application category indicated by the abbreviation.

*If a listing appears in this column, the data for this IC is in the application category indicated by the abbreviation.

SECTION 4

Application: DCP 509

Manufacturer: Analog Devices, Inc.

Manufacturer: GEC Plessey Semiconductors

Manufacturer: Mitel Corporation

*If a listing appears in this column, the data for this IC is in the application category indicated by the abbreviation.

DEVICE DESIGNATOR	DEVICE FUNCTION	SEE*	PAGE
MT8966		COM	
MT8967		COM	
MT8992		COM	
MT8993B		COM	
MT88630		COM	
MT88631		COM	
MT3530		COM	
MT8840		COM	
MT8880		MIC	
MT8880-1		MIC	
MT8880-2		MIC	
MT8952B		MIC	
MT8980D		MIC	
MT8981D		MIC	
MT8920		MIC	
MT8920-1		MIC	

Manufacturer: National Semiconductor Corporation

DEVICE DESIGNATOR	DEVICE FUNCTION	SEE*	PAGE
54LS138	Decoder/ demultiplexer		619
DM54LS138	"		619
DM74LS138	"		619
54LS139	"		619
DM54LS139	"		619
DM74LS139	"		619
54LS151	Data selector/ multiplexer		623
DM54LS151	"		623
DM74LS151	"		623
54LS152	8-input multiplexer		625
54LS153	Dual 4-line to 1-line data selector/ multiplexer		626
DM54LS153	"		626
DM74LS153	"		626
DM54LS154	4-line to 16-line decoder/ demultiplexer		628
DM74LS154	"		628
54LS155	Dual 2-line to 4-line decoder/ demultiplexer		630
DM54LS155	"		630
DM74LS155	"		630
54LS156	"		630
DM54LS156	"		630
DM74LS156	"		630
54LS157	Quad 2-line to 1-line data selector/ multiplexer		633
DM54LS157	"		633
DM74LS157	"		633
54LS158	"		633
DM54LS158	"		633
DM74LS158	"		633
DM54LS251	Tri-state data selector/ multiplexer		637
DM74LS251	"		637
54LS253	"		639
DM54LS253	"		639
DM74LS253	"		639
54LS257A	Tri-state Quad 2-data selector/ multiplexer		641
DM54LS257B	"		641
DM74LS257B	"		641
54LS258A	"		641
DM54LS258B	"		641
DM74LS258B	"		641
54LS322	8-bit serial/ parallel register with sign extend		645
DM74LS322	"		645
54LS323	8-bit universal shift/storage register with synchronous reset and common I/O pins		648
DM74LS323	"		648
54LS347	BCD to 7-segment decoder/driver		651
DM74LS347	"		651

*If a listing appears in this column, the data for this IC is in the application category indicated by the abbreviation.

DEVICE DESIGNATOR	DEVICE FUNCTION	SEE*	PAGE
54LS352	Dual 4-line to 1-line data selector/multiplexer		652
DM74LS352	″		652
54LS353	Dual 4-input multiplexer with Tri-state outputs		654
DM74LS353	″		654
DM54S138	Decoder/ demultiplexer		656
DM74S138	″		656
DM54S139	″		656
DM74S139	″		656
DM54S251	Tri-state 1 of 8-line data selector/ multiplexer		659
DM74S251	″		659
DM54S253	Dual Tri-state 1 of 4-line data selector/ multiplexer		661
DM74S253	″		661
DM54S257	Tri-state quad 1 of 2 data selector/ multiplexer		663
DM74S257	″		663
DM54S258	″		663
DM74S258	″		663
DM54S280	9-bit parity generator/ checker		666
DM74S280	″		666

Manufacturer: Rohm Co., Ltd.

DEVICE DESIGNATOR	DEVICE FUNCTION	SEE*	PAGE
BA820	8-bit serial-in, parallel-out driver		681
BA1610	FSK linear modem		669
BA6590S	Centronics interface		679
BA9101	8-bit successive comparison A/D converter		670
BA9101B	″		670
BA9101S	″		670
BA9201	8-bit D/A converter with latch		673
BA9201F	″		673
BA9211	10-bit D/A converter with internal reference voltage supply		674
BA9211F	″		674
BA9221	12-bit D/A converter		677
BA9221F	″		677

Manufacturer: Siliconix, Inc.

DEVICE DESIGNATOR	DEVICE FUNCTION	SEE*	PAGE
DG411	Precision monolithic quad SPST CMOS analog switch		685
DG412	″		685
DG413	″		685
DG421	Low-power, high-speed, latchable CMOS analog switch		688
DG423	″		688
DG425	″		688
DG441	Monolithic quad SPST CMOS analog switch		691
DG442	″		691
DG444	Low-cost quad SPST CMOS analog switch		696
DG445	″		696
DG485		CON	
DG528		CON	
DG529		CON	

Manufacturer: Silicon Systems, Inc.

DEVICE DESIGNATOR	DEVICE FUNCTION	SEE*	PAGE
32D441		MIC	

*If a listing appears in this column, the data for this IC is in the application category indicated by the abbreviation.

SECTION 5

Application: LOG 701

Manufacturer: Analog Devices, Inc.

Manufacturer: GEC Plessey Semiconductors

*If a listing appears in this column, the data for this IC is in the application category indicated by the abbreviation.

*If a listing appears in this column, the data for this IC is in the application category indicated by the abbreviation.

Manufacturer: Rohm Co., Ltd.

Manufacturer: Siliconix, Inc.

SECTION 6

Application: MIC 811

Manufacturer: Analog Devices, Inc.

*If a listing appears in this column, the data for this IC is in the application category indicated by the abbreviation.

Manufacturer: GEC Plessey Semiconductors

DEVICE DESIGNATOR	DEVICE FUNCTION	SEE*	PAGE
PDSP1601	Augmented arithmetic logic unit		815
PDSP1601A	"		815
NJ8821	Frequency synthesizer (microprocessor interface) with resettable counters		819
NJ8821A	"		822
NJ8821B	"		819
NJ8822	"		824
NJ8822A	"		827
NJ8822B	"		824
NJ8823	Frequency synthesizer (microprocessor interface) with non-resettable counters		829
NJ8823B	"		829
NJ8824	"		831
NJ8824B	"		831
NJ88C25	Frequency synthesizer		834

Manufacturer: LSI Computer Systems, Inc.

DEVICE DESIGNATOR	DEVICE FUNCTION	SEE*	PAGE
LS7066		CCT	
LS7270		CCT	

Manufacturer: Mitel Corporation

DEVICE DESIGNATOR	DEVICE FUNCTION	SEE*	PAGE
MT8880	Integrated DTMF transceiver		836
MT8880-1	"		836
MT8880-2	"		836
MT8952B	HDLC protocol controller		841
MT8980D	Digital switch		844
MT8981D	"		846
MT8920	ST-BUS parallel access circuit		849
MT8920-1	"		849
MSAN-119	Application note: MSAN-119		854

Manufacturer: National Semiconductor Corporation

DEVICE DESIGNATOR	DEVICE FUNCTION	SEE*	PAGE
DM74ALS242C	Quad Tri-state bidirectional bus driver		863
DM74ALS243A	"		863
DM54ALS244A	Octal Tri-state bus driver		865
DM74ALS244A	"		865
DM54ALS245A	Octal Tri-state bus transceiver		867
DM74ALS245A	"		867
DM74ALS465A	Octal Tri-state bidirectional bus driver		868
DM74ALS466A	"		868
DM74ALS467A	"		868
DM74ALS468A	"		868
DM74AS640	Tri-state octal bus transceiver		870
DM74AS645	"		872
DM74AS646	Octal bus transceiver and register		874
DM74AS648	"		874
DM74AS651	Bus transceiver and register		877
DM74AS652	"		877
54LS366A	Hex Tri-state inverting buffer		880
DM74LS366A	"		880
54LS367A	Hex Tri-state buffer		882
DM54LS367A	"		882
DM74LS367A	"		882
DM54S181	Arithmetic logic unit/function generator		883
DM74S181	"		883

*If a listing appears in this column, the data for this IC is in the application category indicated by the abbreviation.

*If a listing appears in this column, the data for this IC is in the application category indicated by the abbreviation.

SECTION 7

Application: PTE 979

Manufacturer: Analog Devices, Inc.

Manufacturer: GEC Plessey Semiconductors

Manufacturer: Raytheon Company

Manufacturer: Rohm Co., Ltd.

**Indicates applications data contained in this section; refer to the column "SEE" for specifications.

*If a listing appears in this column, the data for this IC is in the application category indicated by the abbreviation.

SECTION 1

CLOCKS, COUNTERS, TIMERS

This information is reproduced with permission of Analog Devices, Inc.

Analog Devices
AD9500
Digitally Programmable Delay Generator

- 10ps Delay resolution
- 2.5ns to 100μs + Full-scale range
- Fully differential inputs
- Separate trigger and rest inputs
- Low power dissipation — 310 mW

APPLICATIONS
- ATE
- Pulse deskewing
- Arbitrary waveform generators
- High-stability timing source
- Multiple phase clock generators

The AD9500 is a digitally programmable delay generator, which provides programmed delays, selected through an 8-bit digital code, in resolutions as small as 10ps. The AD9500 is constructed in a high-performance bipolar process, designed to provide high-speed operation for both digital and analog circuits.

The AD9500 employs differential TRIGGER and RESET inputs, which are designed primarily for ECL signal levels, but function with analog and TTL input levels. An on-board ECL reference midpoint allows both of the inputs to be driven by either single-ended or differential ECL circuits. The AD9500 output is a complementary ECL stage, which also provides a parallel Q_R output circuit to facilitate reset timing implementations.

The digital control data is passed to the AD9500 through a transparent latch controlled by the LATCH ENABLE signal. In the transparent mode, the internal DAC of the AD9500 will attempt to follow changes at the inputs. The LATCH ENABLE is otherwise used to strobe the digital data into the AD9500 latches.

The AD9500 is available as an industrial temperature-range device, −25 °C to +85 °C, and as an extended temperature-range device, −55 °C to 125 °C. Both grades are packaged in a 24-pin ceramic "skinny" DIP (0.3″ package width), as well as in 28-pin surface mount packages. Contact the factory for MIL-STD-883, revision C, qualified devices.

ABSOLUTE MAXIMUM RATINGS

PARAMETER	SYMBOL	VALUE
Positive supply voltage	$+V_S$	+7V
Negative supply voltage	$-V_S$	−7V
ECL COMMON to ground differential		−2.0V to +5.0V
Digital input voltage range		−3.5V to +5.0V
Trigger/reset input voltage range		±5.0V
Trigger/reset differential voltage		5.0V
Minimum R_{SET}		220Ω
Digital output current	Q and $\overline{Q}$	30mA
Digital output current	$\overline{Q_R}$	2mA
Offset adjust current (Sinking)		4mA

Power dissipation (+25 °C free air)	2.62W
Operating temperature range	
AD9500BP/BQ	−25 °C to +85 °C
AD9500TE/TQ	−55 °C to +125 °C
Storage temperature range	−65 °C to +150 °C
Junction temperature	+175 °C
Lead soldering temperature (10sec)	+300 °C

PIN NAME	FUNCTIONAL DESCRIPTION

D_4-D_6: One of eight digital inputs used to set the programmed delay.

D_7(MSB): One of eight digital inputs used to set the programmed delay. D_7(MSB) is the most significant bit of the digital input word.

ECL_{REF}: ECL midpoint reference, nominally −1.3V. Use of the ECL_{REF} allows either of the TRIGGER or the RESET inputs to be configured for single-ended ECL inputs.

OFFSET ADJUST: The OFFSET ADJUST is used to adjust the minimum propagation delay (t_{PD}) by pulling or pushing a small current out of or into the pin.

C_S: C_S allows the full-scale range to be extended by using an external timing capacitor. The value of C_{EXT}, connected between C_S and $+V_S$, can range from 0pF to 0.1μF+. See R_S ($C_{INTERNAL}=10$ pF).

$+V_S$: Positive supply terminal, nominally +50V.

TRIGGER: Noninverted input of the edge-sensitive differential trigger input stage. The output at Q will be delayed by the programmed delay, after the triggering event. The programmed delay is set by the digital input word. The $\overline{\text{TRIGGER}}$ input must be driven in conjunction with the TRIGGER input.

$\overline{\text{TRIGGER}}$: Inverted input of the edge-sensitive differential trigger input stage. The output at Q will be delayed by the programmed delay, after the triggering event. The programmed delay is set by the digital input word. The $\overline{\text{TRIGGER}}$ input must be driven in conjunction with the TRIGGER input.

$\overline{\text{RESET}}$: Inverted input of the level-sensitive differential reset input stage. The output at Q will be reset after a signal is received at the reset inputs. In the "minimum configuration" the minimum output pulse width will be equal to the "reset propagation delay," t_{RD}. The RESET input must be driven in conjunction with the $\overline{\text{RESET}}$ input.

RESET: Noninverted input of the level-sensitive differential reset input stage. The output at Q will be reset after a signal is received at the reset inputs. In the "minimum configuration," the minimum output pulse width will be equal to the "reset propagation delay," t_{RD}. The RESET input must be driven in conjunction with the $\overline{\text{RESET}}$ input.

Q: One of two complementary ECL outputs. A "triggering" event at the inputs will produce a logic HIGH on the Q output. A "resetting" event at the inputs will produce a logic LOW on the Q output.

$\overline{Q}$: One of two complementary ECL outputs. A "triggering" event at the inputs will produce a logic LOW on the $\overline{Q}$ output. A "resetting" event at the inputs will produce a logic HIGH on the $\overline{Q}$ output.

$\overline{Q_R}$: $\overline{Q_R}$ output is parallel to the $\overline{Q}$ output. The $\overline{Q_R}$ output is typically used to drive delaying circuits for extending output pulse widths. A "triggering" event at the inputs will produce a logic LOW on the $\overline{Q_R}$ output. A "resetting" event at the inputs will produce a logic HIGH on the $\overline{Q_R}$ output.

ECL COMMON: The collector common for the ECL output stage. The collector common can be tied to +5.0V, but normally it is tied to the circuit ground for standard ECL outputs.

$-V_S$: Negative supply terminal, nominally −5.2V.

R_S: R_S is the reference current setting terminal. An external setting resistor, R_{SET}, connected between R_S and $-V_S$, determines the internal reference current. See C_S ($250\Omega \leq R_{SET} \leq 50k\Omega$).

GROUND: The ground return for the TTL and analog inputs.

LATCH ENABLE: Transparent TTL latch control line. A logic HIGH on the LATCH ENABLE freezes the digital code at the logic inputs. A logic LOW on the LATCH ENABLE allows the internal current levels to be continuously updated through the logic inputs, D_0 thru D_7.

D_0(LSB): One of eight digital inputs used to set the programmed delay. D_0(LSB) is the least significant bit of the digital input word.

D_3-D_1: One of eight digital inputs used to set the programmed delay.

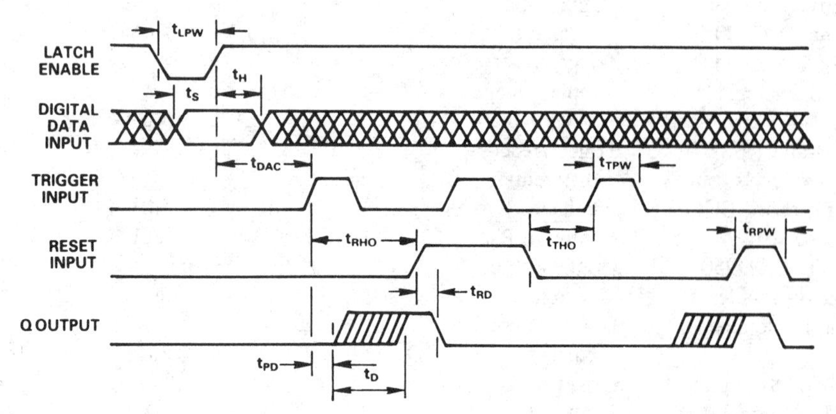

t_S - DIGITAL DATA SETUP TIME
t_H - DIGITAL DATA HOLD TIME
t_{LPW} - LATCH ENABLE PULSE WIDTH
t_{DAC} - INTERNAL DAC SETTLING TIME
t_{PD} - MINIMUM PROPAGATION DELAY
t_{RD} - RESET PROPAGATION DELAY
t_D - PROGRAMMED DELAY
t_{TPW} - TRIGGER PULSE WIDTH
t_{RPW} - RESET PULSE WIDTH
t_{THO} - RESET-TO-TRIGGER HOLDOFF
t_{RHO} - TRIGGER-TO-RESET HOLDOFF

NOTE
A TRIGGERING EVENT MAY OCCUR AT ANY TIME WHILE THE INTERNAL DAC (PROGRAMMED DELAY) IS BEING CHANGED. TRIGGERING EVENTS DURING THE INTERNAL DAC SETTLING TIME MAY NOT GENERATE AN ACCURATE PULSE DELAY.

System Timing Diagram

AD9500 FUNCTIONAL BLOCK DIAGRAM

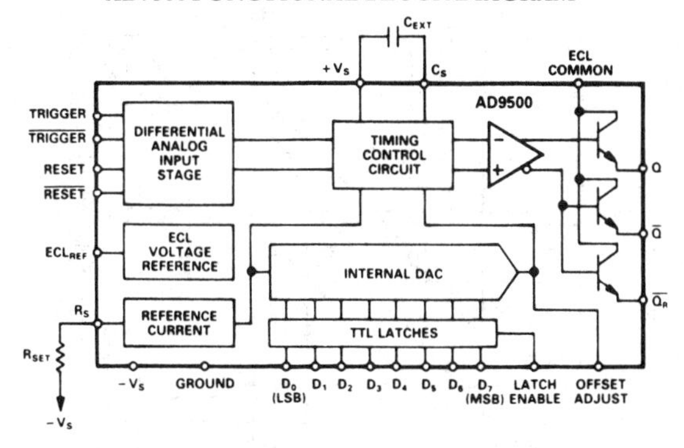

PIN CONFIGURATIONS

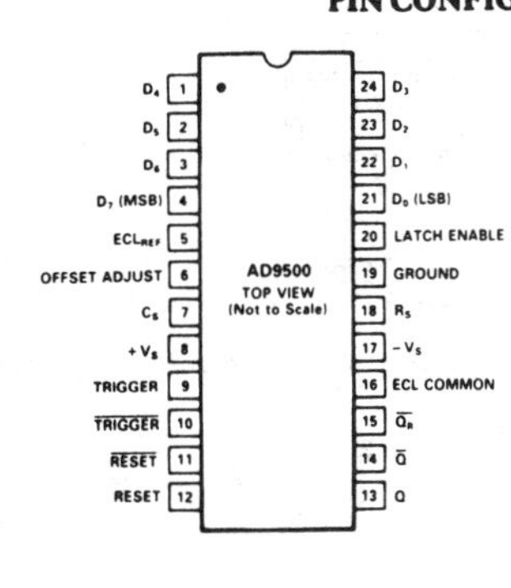

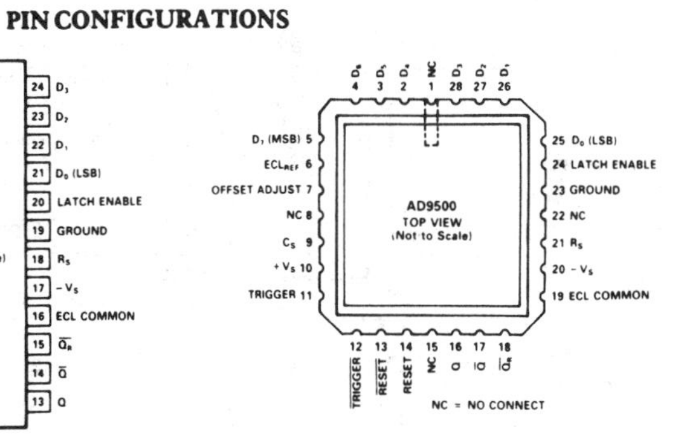

Input/Output Circuits

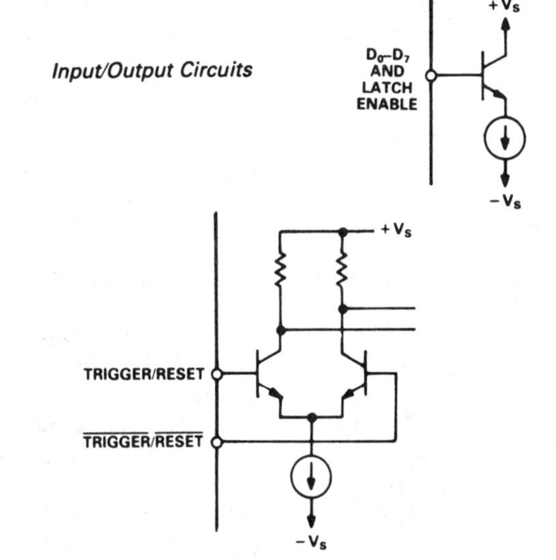

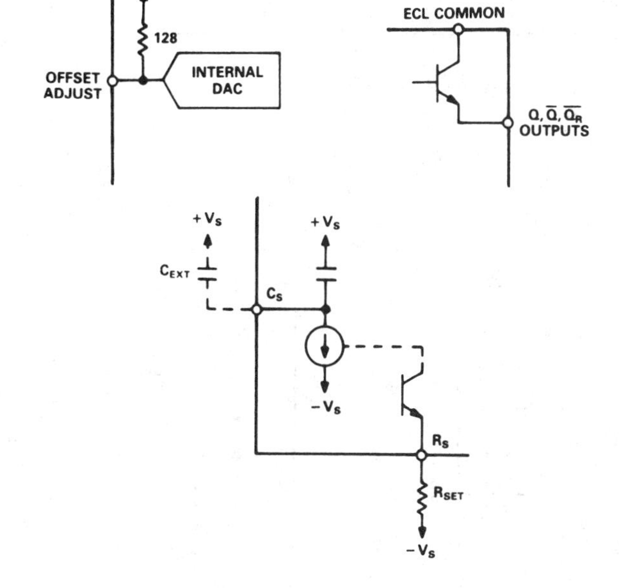

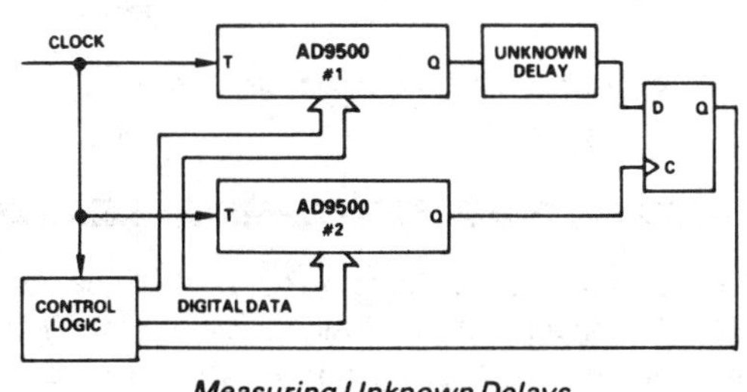

Measuring Unknown Delays

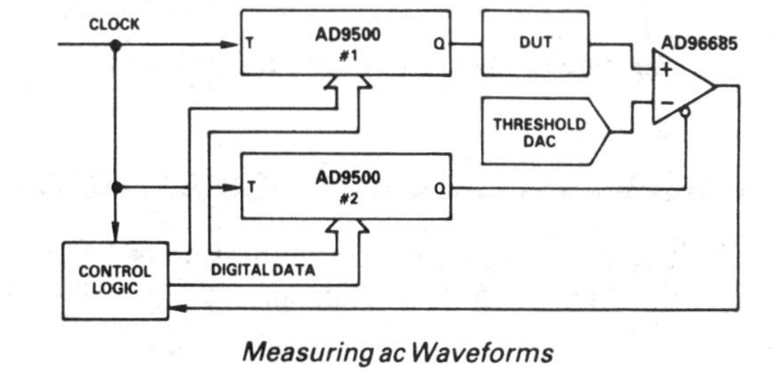

Measuring ac Waveforms

Burn-In Circuit

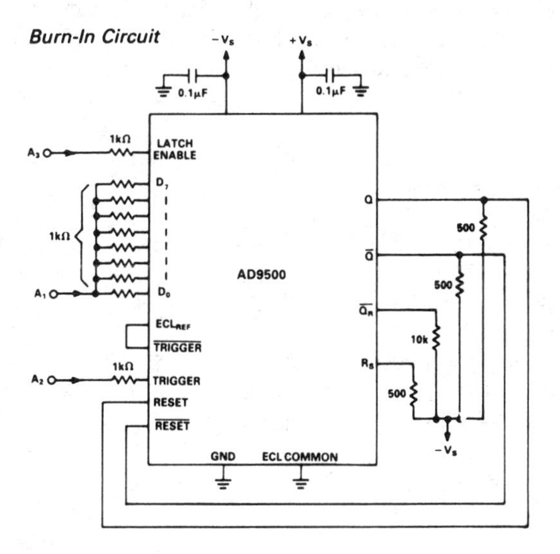

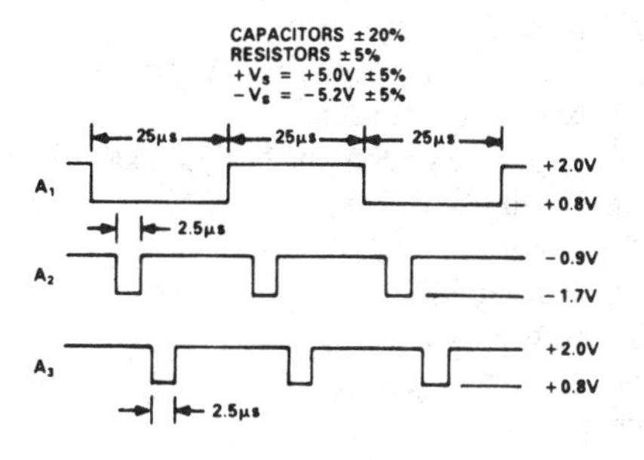

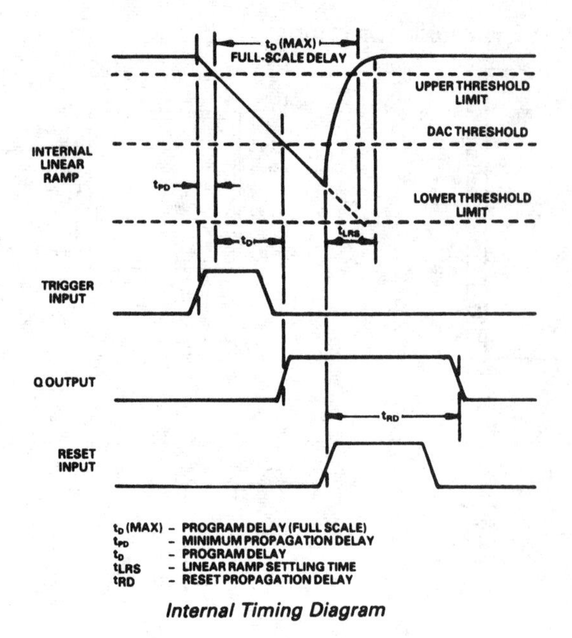

Internal Timing Diagram

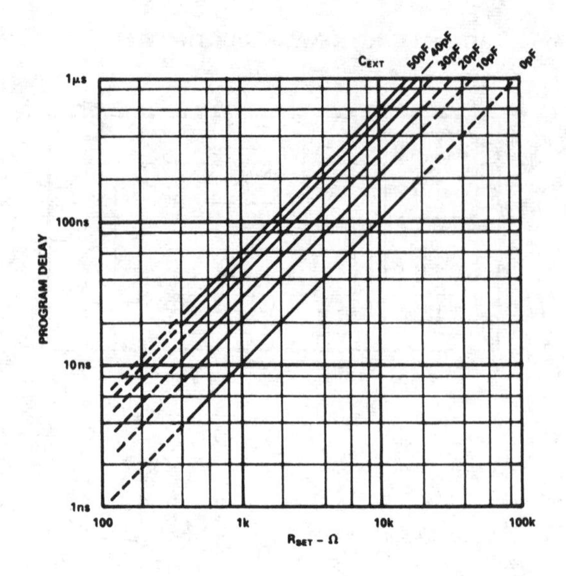

Typical Programmed Delay Ranges

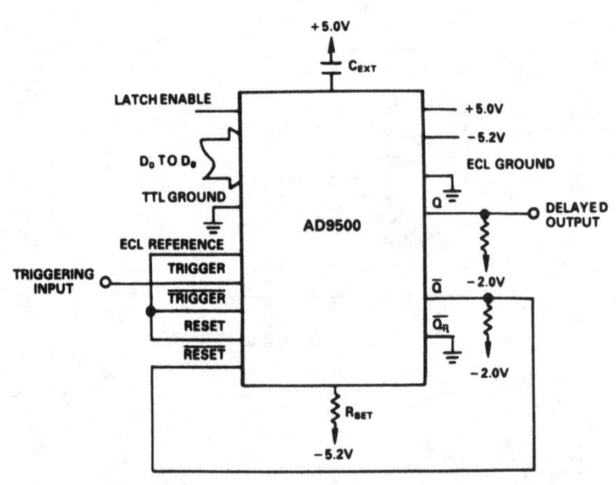

Single Input – Minimum Timing Configuration

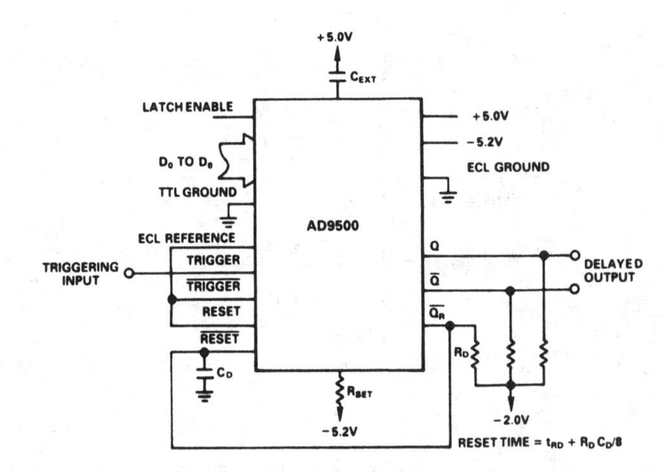

Extended Output Pulse Width Configuration

This information is reproduced with permission of LSI Computer Systems, Inc., 1235 Walt Whitman Road, Melville, NY 11747.

LSI
LS7030
Eight Decade Multiplexed MOS Up Counter

- DC to 7.5-MHz count frequency
- Multiplexed BCD and 7-segment outputs
- DC to 500-kHz scan frequency
- Single power supply operation, +4.75 VDC to 15 VDC
- Compatible with CMOS logic
- High input noise immunity
- Counter output latches
- Leading zero blanking
- Low power dissipation
- All inputs protected

The LS7030 is a monolithic, ion-implanted, 8-decade up counter. The circuit includes latches, multiplexer, leading zero blanking, BCD, and 7-segment data outputs.

The information included herein is believed to be accurate and reliable. However, LSI Computer Systems, Inc. assumes no responsibility for inaccuracies, nor for any infringements of patent rights of others which may result from its use.

The LS7030-1 is a selected higher count frequency version of the LS7030. The specification differences occur under dynamic electrical characteristics as follows:

PARAMETER	SYMBOL	MIN	MAX
Count and test count frequency			
(VSS = +5V ± 5%)	Fc, Fac	DC	10 MHz
(VSS = +10V)	Fc, Fac	DC	7.5 MHz
(VSS = +15V)	Fc, Fac	DC	6 MHz
Count pulse width			
(VSS = +5V ± 5%)	Tcpw	50ns	
(VSS = +10V)	Tcpw	62ns	
(VSS = +15V)	Tcpw	83ns	

Other specifications are unchanged.

MAXIMUM RATINGS

PARAMETER	SYMBOL	VALUE
Storage temperature	T_{STG}	−65 to +150°C
Operating temperature	T_A	−25 to +70°C
Voltage (any pin to VSS)	V_{MAX}	−30 to +0.5V

GUARDBANDED STROBE

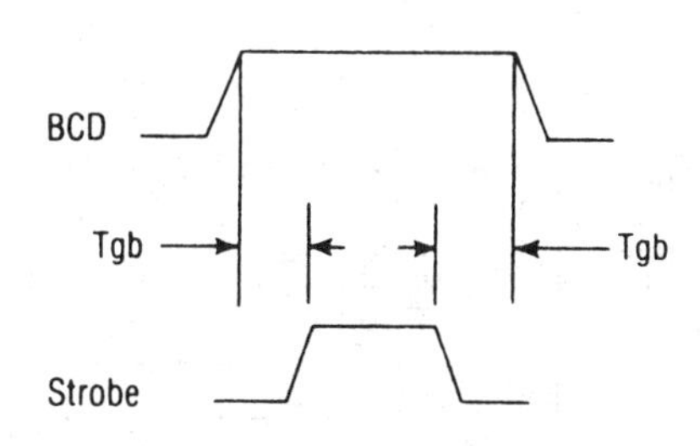

SEVEN-SEGMENT FONT

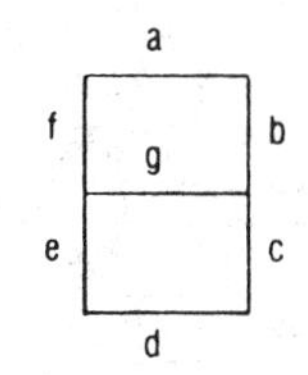

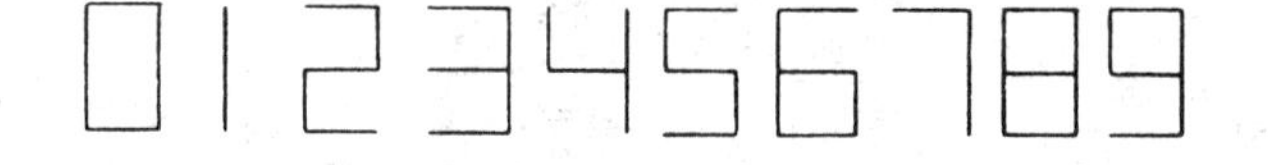

PACKAGE DIAGRAM

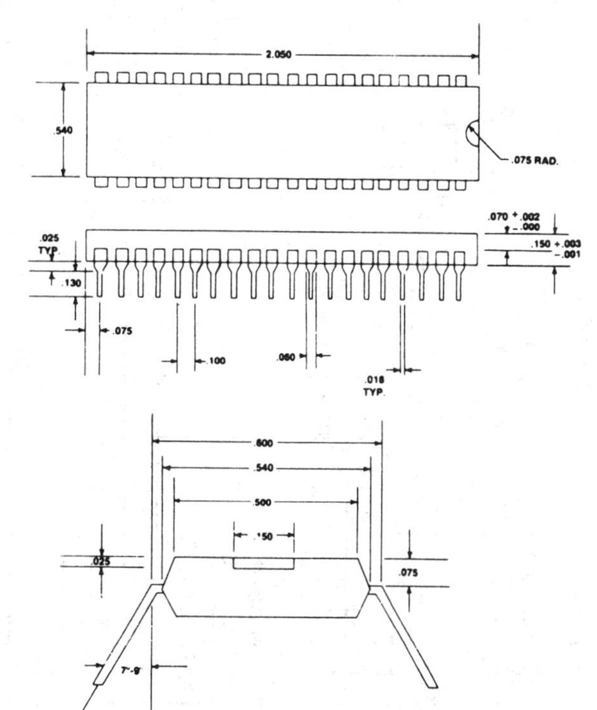

CONNECTION DIAGRAM

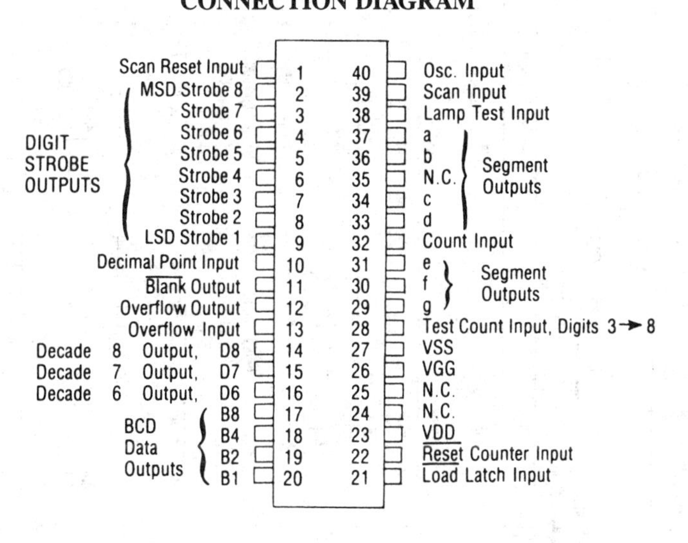

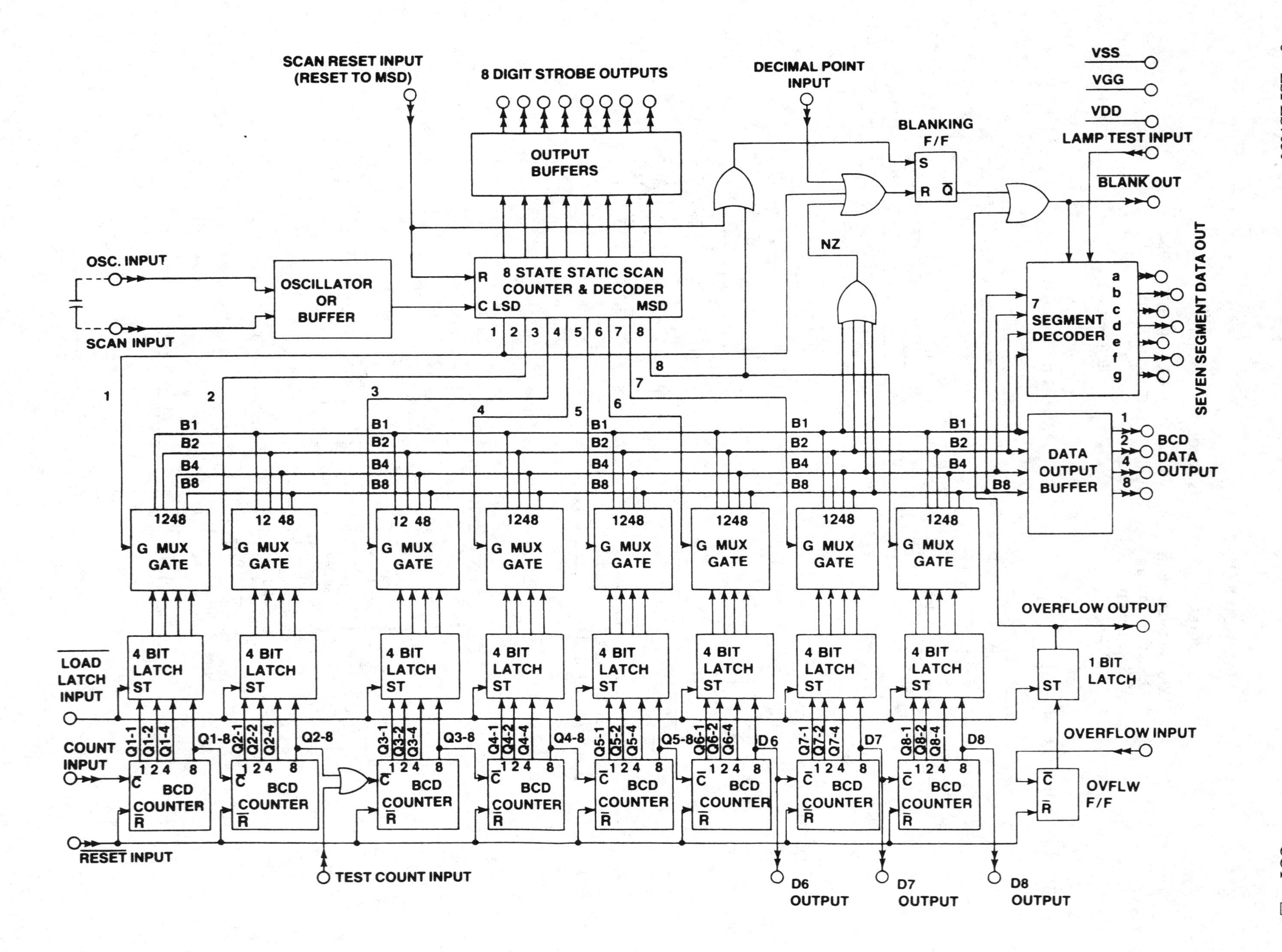

SCAN RESET INPUT (RESET TO MSD)
8 DIGIT STROBE OUTPUTS
DECIMAL POINT INPUT
VSS
VGG
VDD
BLANKING F/F
S R Q̄
LAMP TEST INPUT
BLANK OUT
OUTPUT BUFFERS
OSC. INPUT
SCAN INPUT
OSCILLATOR OR BUFFER
R 8 STATE STATIC SCAN COUNTER & DECODER
C LSD
MSD
1 2 3 4 5 6 7 8
NZ
7 SEGMENT DECODER
a b c d e f g
SEVEN SEGMENT DATA OUT
B1 B2 B4 B8
DATA OUTPUT BUFFER
1 2 4 8
BCD DATA OUTPUT
1248 G MUX GATE
12 48 G MUX GATE
4 BIT LATCH ST
OVERFLOW OUTPUT
1 BIT LATCH
ST
LOAD LATCH INPUT
COUNT INPUT
Q1-1 Q1-2 Q1-4 Q1-8
Q2-1 Q2-2 Q2-4 Q2-8
Q3-1 Q3-2 Q3-4 Q3-8
Q4-1 Q4-2 Q4-4 Q4-8
Q5-1 Q5-2 Q5-4 Q5-8
Q6-1 Q6-2 Q6-4
Q7-1 Q7-2 Q7-4
Q8-1 Q8-2 Q8-4
D6 D7 D8
OVERFLOW INPUT
BCD COUNTER
C̄ R̄
OVFLW F/F
RESET INPUT
TEST COUNT INPUT
D6 OUTPUT
D7 OUTPUT
D8 OUTPUT

LSI
LS7031
6-Decade MOS Up Counter with 8-Decade Latch and Multiplexer

- DC to 7.5-MHz count frequency
- Multiplexed BCD outputs
- DC to 500-kHz scan frequency
- Ability to latch external BCD data in the two LSD positions
- Leading zero blanking with decimal point and overflow controls
- Single power supply operation, +4.75 VDC to +15 VDC
- Compatible with CMOS logic
- High input noise immunity
- Low power dissipation
- All inputs protected

The LS7031 is a monolithic, ion-implanted MOS 6-decade up counter. The circuit includes latches, multiplexer, leading zero blanking, and BCD data outputs.

CONNECTION DIAGRAM

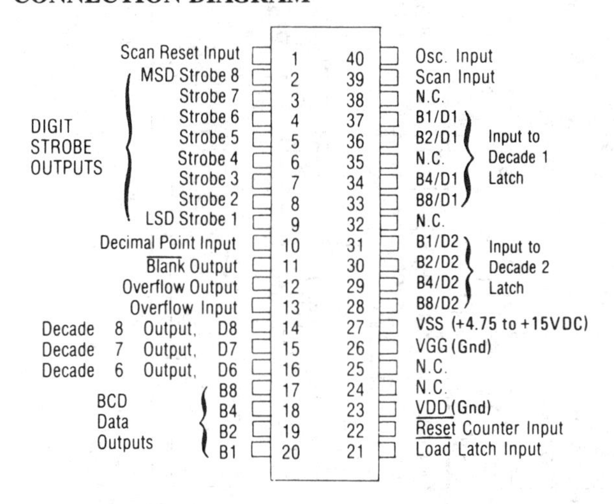

GUARDBANDED STROBE

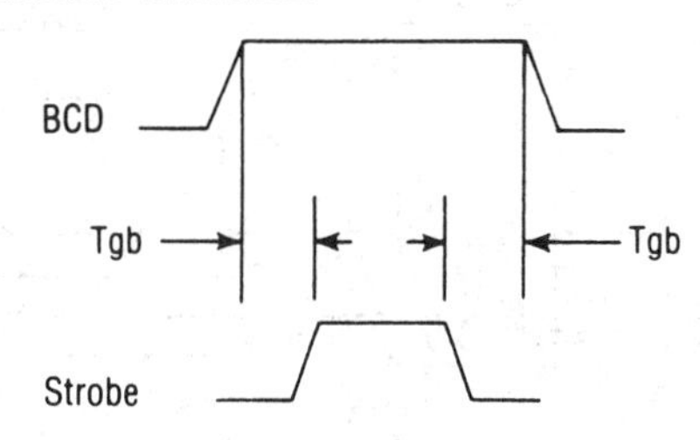

**Propagation Delay and Pulse Width

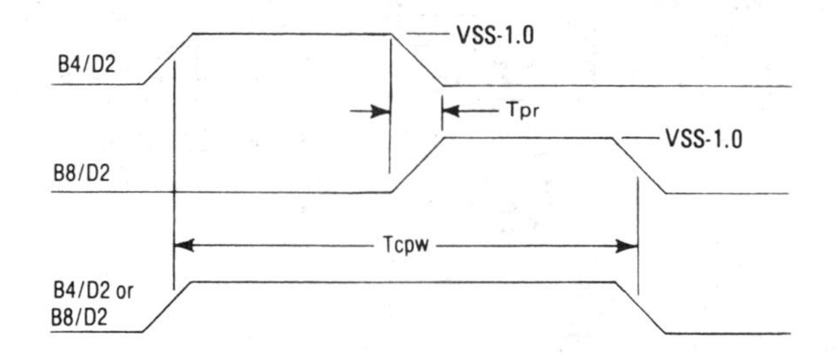

MAXIMUM RATINGS

PARAMETER	SYMBOL	VALUE
Storage temperature	T_{STG}	−65 to +150°C
Operating temperature	T_A	−25 to +70°C
Voltage (any pin to VSS)	V_{MAX}	−30 to +0.5V

The LS7031-1 is a selected higher count frequency version of the LS7031. The specification differences occur under dynamic electrical characteristics as follows:

PARAMETER	SYMBOL	MIN	MAX
Count frequency			
(VSS = +5V ± 5%)	Fc	DC	10 MHz
(VSS = +10V)	Fc	DC	7.5 MHz
(VSS = +15V)	Fc	DC	6 MHz
Count pulse width (Pulse applied to B4/D2 or B8/D2; 'OR' combination of B4/D2 and B8/D2)			
(VSS = +5V ± 5%)	Tcpw	50ns	
(VSS = +10V)	Tcpw	62ns	
(VSS = +15V)	Tcpw	83ns	

Other specifications are unchanged.

PACKAGE DIAGRAM

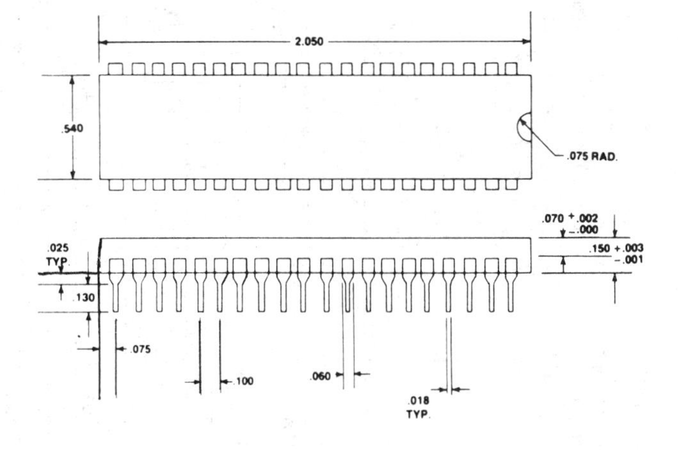

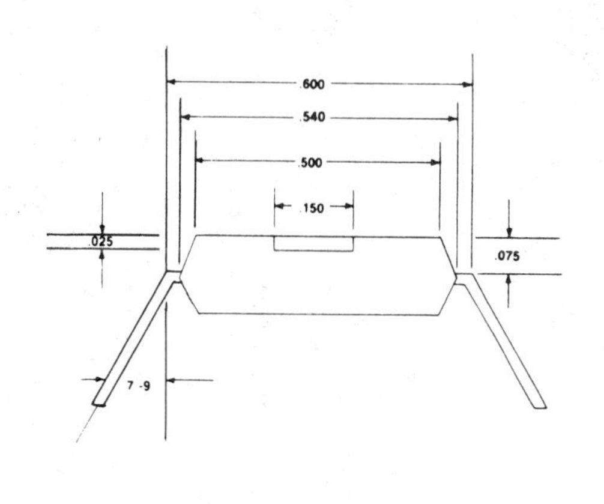

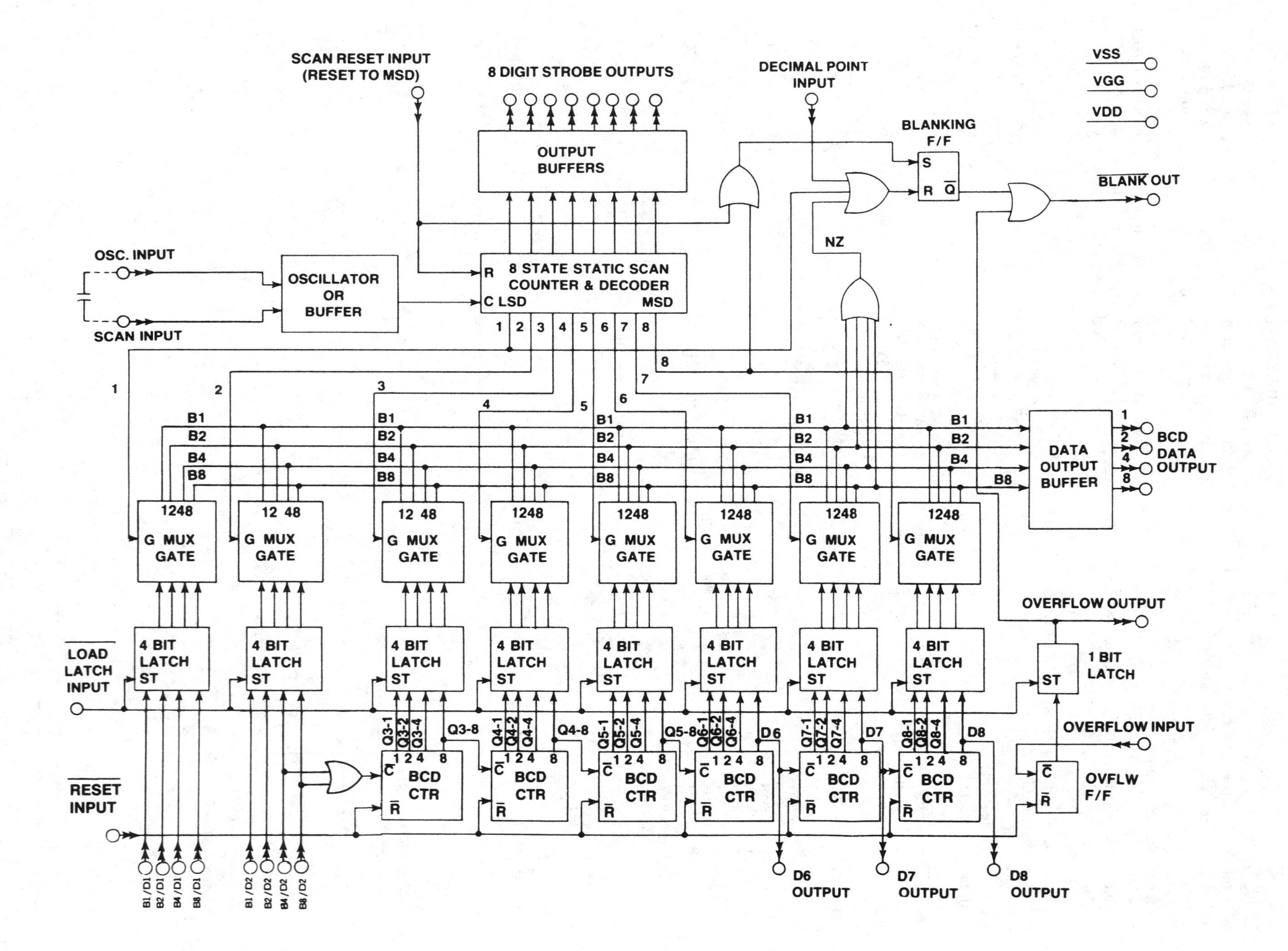
SCAN RESET INPUT (RESET TO MSD)
8 DIGIT STROBE OUTPUTS
DECIMAL POINT INPUT
VSS
VGG
VDD
BLANKING F/F
S
R
Q̄
BLANK OUT
OUTPUT BUFFERS
OSC. INPUT
SCAN INPUT
OSCILLATOR OR BUFFER
8 STATE STATIC SCAN COUNTER & DECODER
R
C LSD
MSD
NZ
B1
B2
B4
B8
1248
12 48
G MUX GATE
DATA OUTPUT BUFFER
BCD DATA OUTPUT
OVERFLOW OUTPUT
4 BIT LATCH ST
1 BIT LATCH
ST
LOAD LATCH INPUT
OVERFLOW INPUT
Q3-1 Q3-2 Q3-4 Q3-8
Q4-1 Q4-2 Q4-4 Q4-8
Q5-1 Q5-2 Q5-4 Q5-8
Q6-1 Q6-2 Q6-4
Q7-1 Q7-2 Q7-4
Q8-1 Q8-2 Q8-4
D6
D7
D8
C̄ 1 2 4 8 BCD CTR R̄
OVFLW F/F
RESET INPUT
B1/D1 B2/D1 B4/D1 B8/D1
B1/D2 B2/D2 B4/D2 B8/D2
D6 OUTPUT
D7 OUTPUT
D8 OUTPUT

TYPICAL APPLICATION

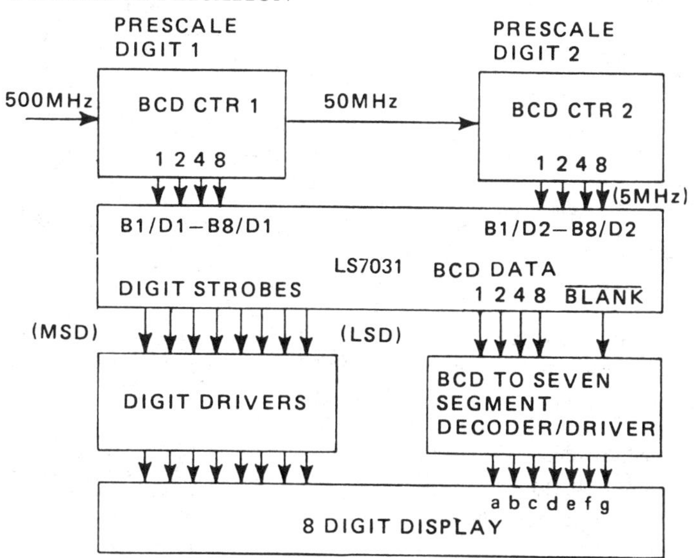

LSI
LS7040
Dual 3-Decade Up/Down Counter

- DC to 350-kHz count frequency at +5V operation
- Fully synchronous operation
- Cascadable
- Inputs CMOS, TTL, and DTL compatible at +5V operation
- Separate low-current-drain power supply for counter stages permits battery stand-by operation
- Reset
- Count enable
- Parallel BCD output data
- Power-on reset
- Count input can be applied to a regenerative circuit, which permits infinite rise and fall times
- Selectable as 6-decade or dual 3-decade up or down counter
- CMOS type noise immunity on all inputs
- Output latches
- Single power supply operation +4.75 VDC to +15 VDC

The LS7040 is a monolithic, ion implanted PMOS synchronous Dual 3-Decade or 6-Decade Up/Down Counter including latches and parallel BCD data outputs.

MAXIMUM RATINGS

PARAMETER	SYMBOL	VALUE
Storage temperature	T_{STG}	−65 to +150°C
Operating temperature	T_A	−25 to +70°C
Voltage (any pin to VSS)	V_{MAX}	−30 to +.5V

CONNECTION DIAGRAM: TOP VIEW
STANDARD 40 PIN PLASTIC DIP

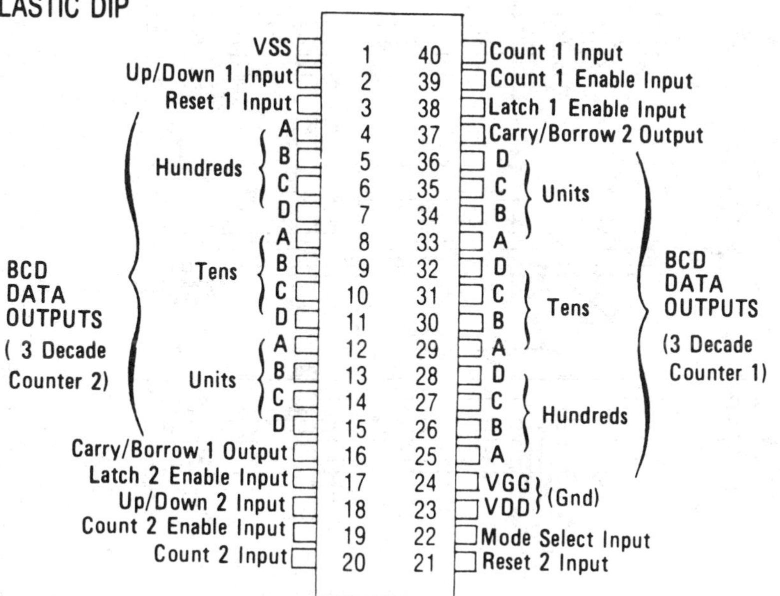

BATTERY STAND-BY SCHEMATICS

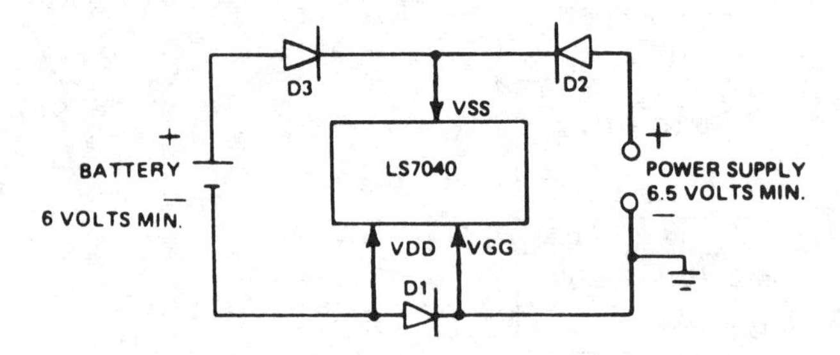

D3
VSS
+
BATTERY
—
6 VOLTS MIN.
LS7040
+
POWER SUPPLY
6.5 VOLTS MIN.
VDD
VGG
D1

Positive Supply System: The battery voltage is lower than the supply voltage and is used only in the power outage condition for counting. When the power supply is shut off, the battery supplies current only though VDD. Diode D1 is used to isolate VDD and VGG. Diode D2 is used to prevent VSS from going to ground potential when the power supply shuts off. Diode D3 is used to isolate the battery and the power supply.

Negative Supply System: Diode D2 is not needed since VGG going to ground potential will not affect standby operation.

The information included herein is believed to be accurate and reliable. However, LSI Computer Systems, Inc. assumes no responsibilities for inaccuracies, nor for any infringements of patent rights of others which may result from its use.

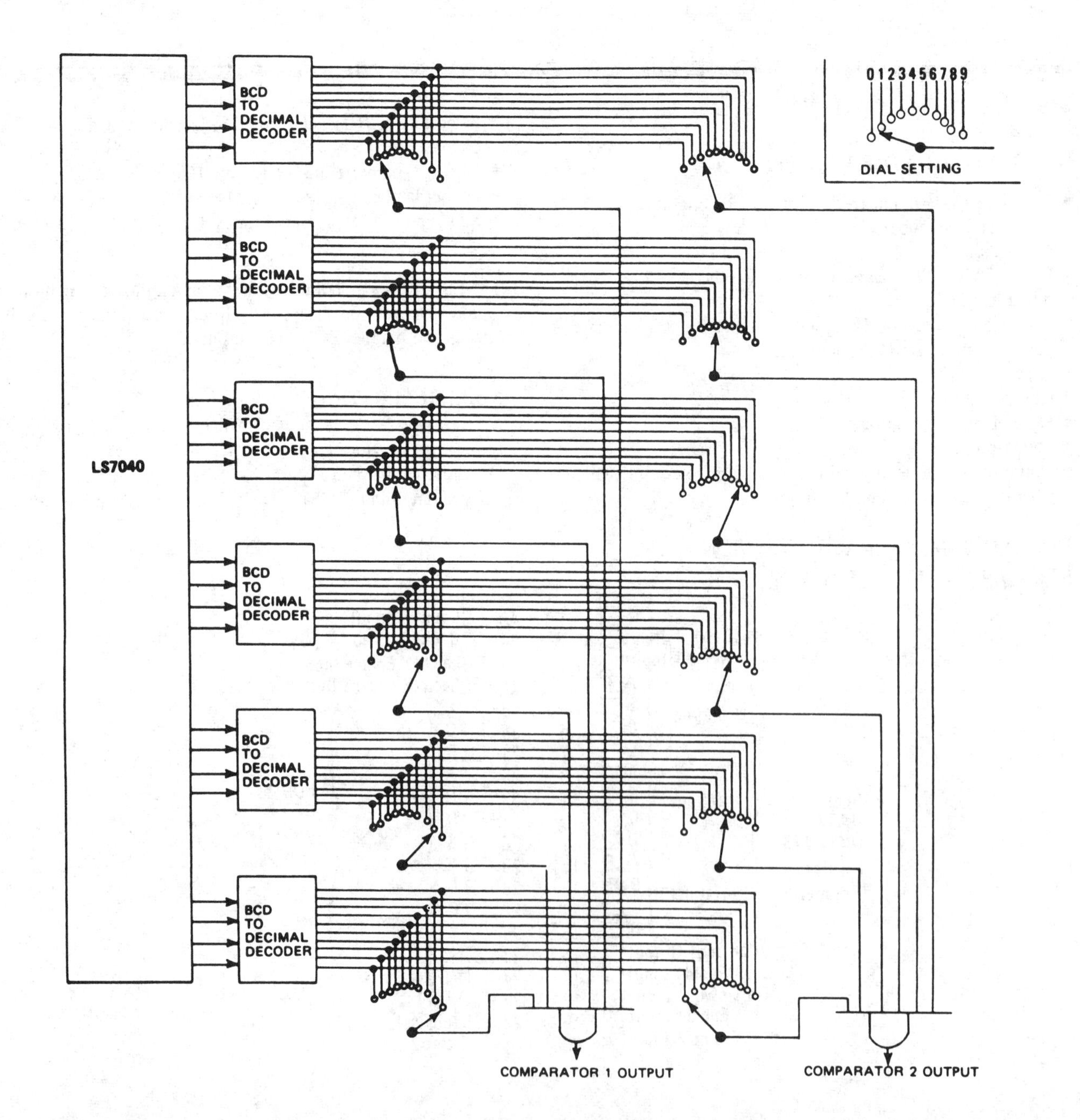

The unique feature of the LS7040 and its main advantage over multiplexed counters, is its parallel BCD outputs. These outputs can be applied to as many external preset comparators as desired with a minimum of hardware. The diagram here illustrates the circuitry for two 6-digit comparators. A BCD to decimal decoder and a 10-position switch is used for each of the decade outputs. The arms of the 6 switches are combined in an AND gate to provide the comparison output. The desired decimal number is selected and the AND gate produces a Logic 1 output when the LS7040 reaches that number. For each additional comparator, six 10-position switches and one AND gate are added. The number of comparators that can be used with one LS7040 and 6 BCD to decimal decoders are unlimited. Counter outputs can be displayed by applying the BCD outputs to a 7-segment decoder to drive LED displays.

Because of the cascadability of the LS7040, a 12-digit comparator scheme would use 2 LS7040's and 12 BCD to decade decoders. This scheme can be extended to as many digits as is desired.

The diagram here depicts an output that occurs at comparator 1 when the LS7040 reaches a count of 123789. Comparator 2 will produce an output at a count of 247650.

An additional advantage of the LS7040 over multiplexed counter outputs occurs when analog circuits and counter circuits are used together. The demultiplexing signals and associated hardware that are used by a multiplex counter can cause noise to interfere with analog signals. The use of the LS7040 in an analog application will negate the possibility of any noise generation.

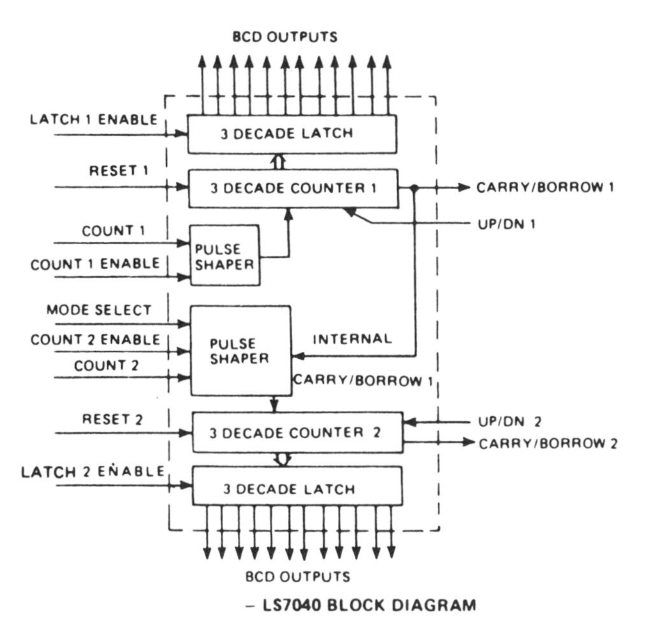

– LS7040 BLOCK DIAGRAM

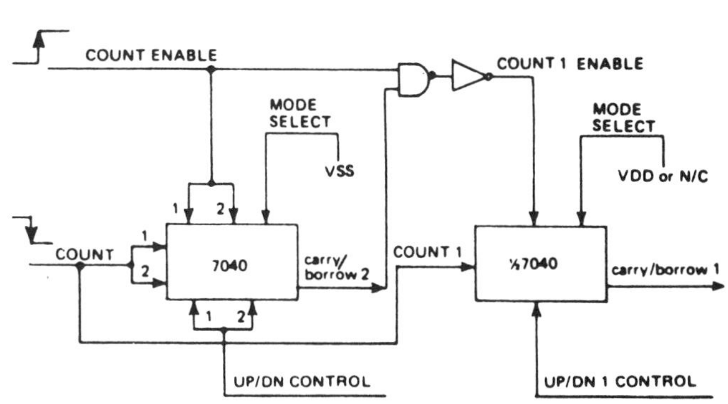

– SYNCHRONOUS 9 DECADE COUNTER

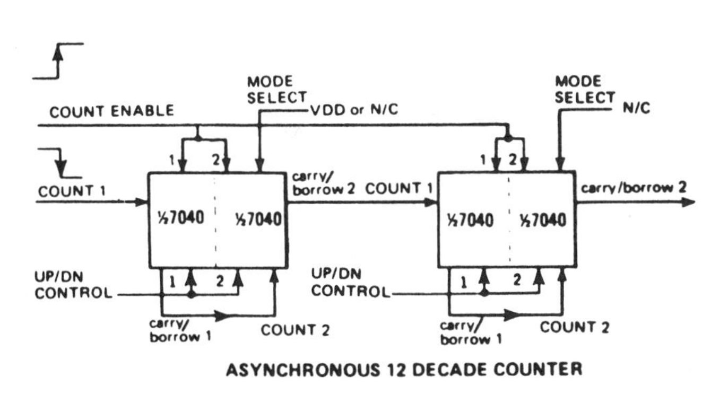

ASYNCHRONOUS 12 DECADE COUNTER

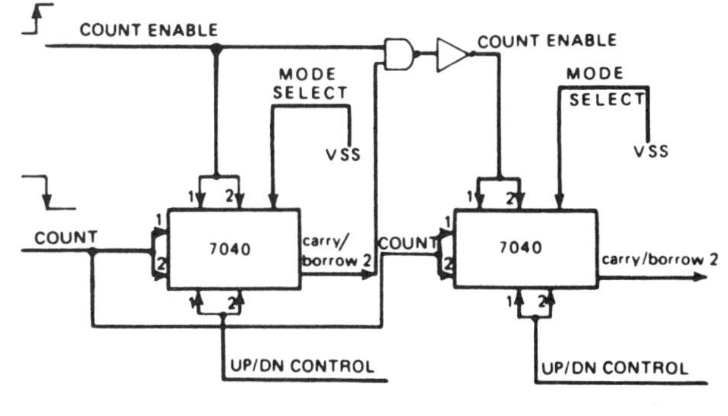

SYNCHRONOUS 12 DECADE COUNTER

LSI LS7060

32-Bit Binary Up Counter

- DC to 10-MHz count frequency
- 8-bit byte multiplexer
- DC to 1-MHz scan frequency
- Single power supply operation, +4.75 VDC to +5.25 VDC
- Three-state data outputs, bus and TTL compatible
- Inputs TTL, NMOS and CMOS compatible
- Unique cascade feature allows multiplexing of successive bytes of data in sequence in multiple counter systems
- Low power dissipation
- All inputs protected
- 18-pin DIP

The LS7060 is a monolithic, ion-implanted, N-channel MOS Silicon Gate, 32-bit up counter. The circuit includes latches, multiplexer, eight three-state binary data output drivers, and output cascading logic.

MAXIMUM RATINGS

PARAMETER	SYMBOL	VALUE
Storage temperature	T_{STG}	−55 to +150°C
Operating temperature	T_A	−0 to +70°C
Voltage (any pin to VSS)	V_{MAX}	+10 to −0.3V

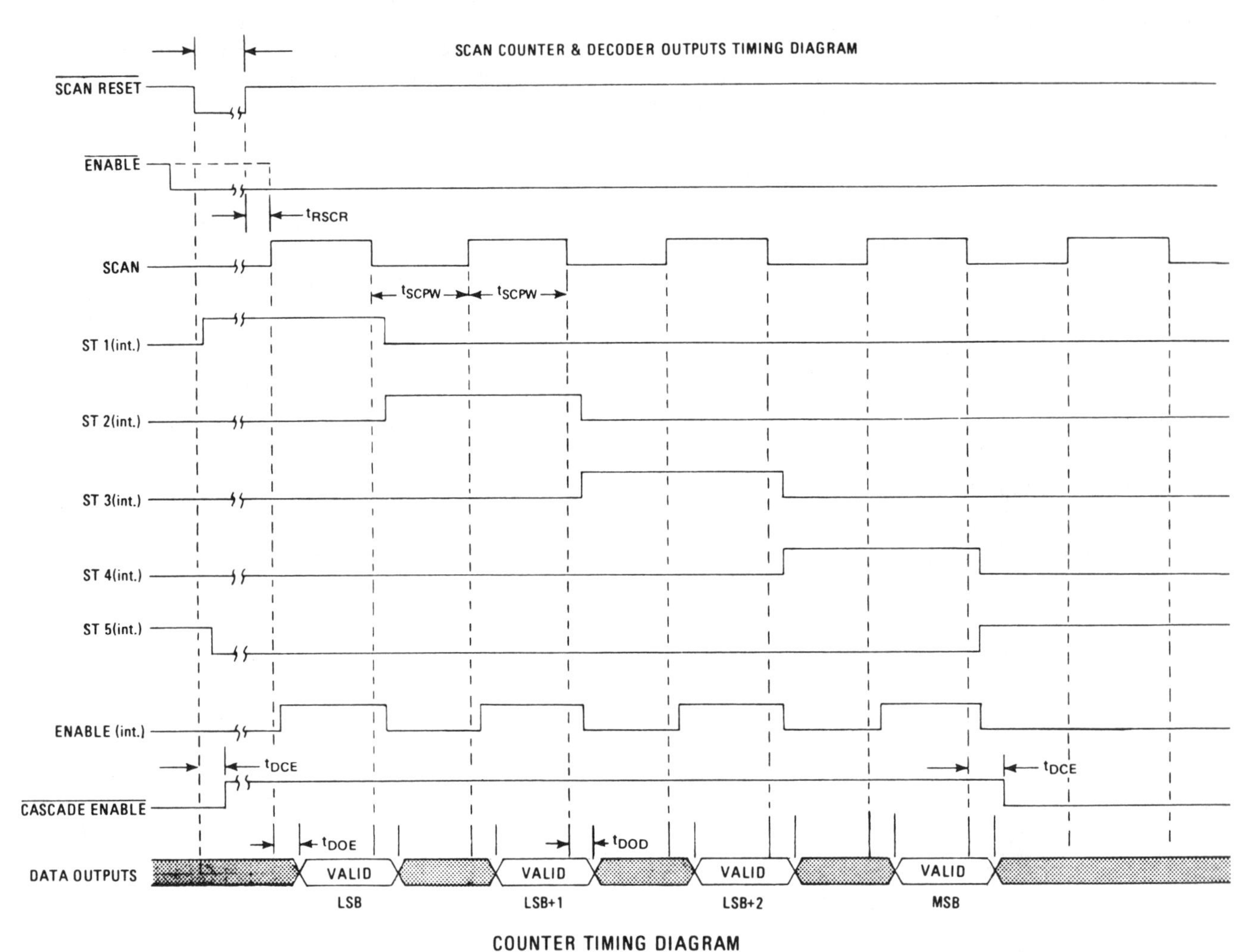

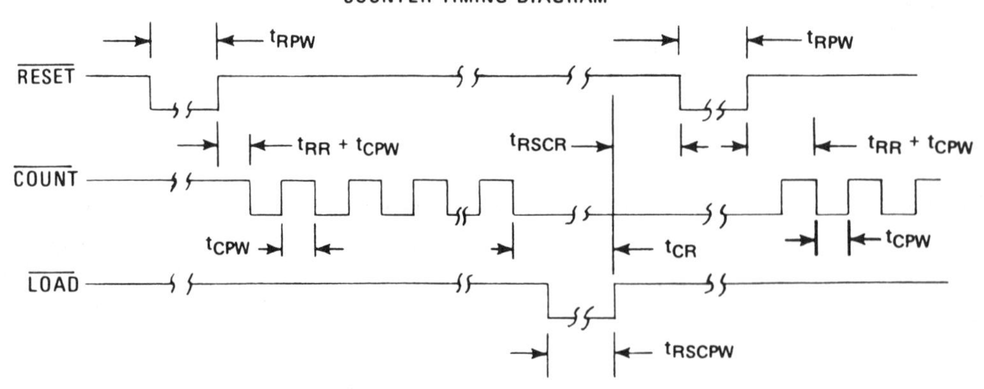

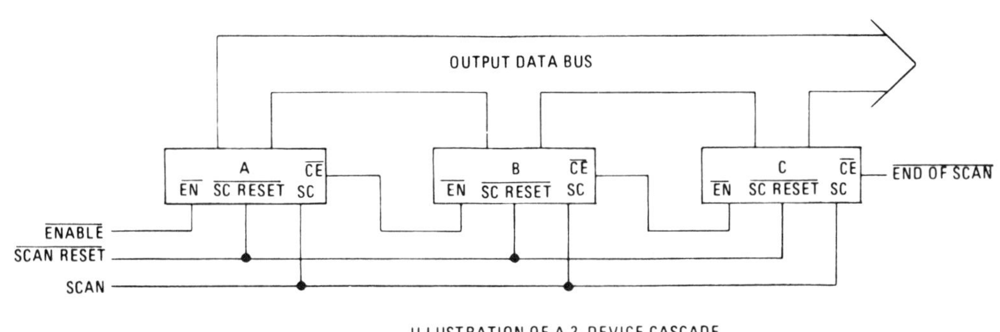

ILLUSTRATION OF A 3 DEVICE CASCADE

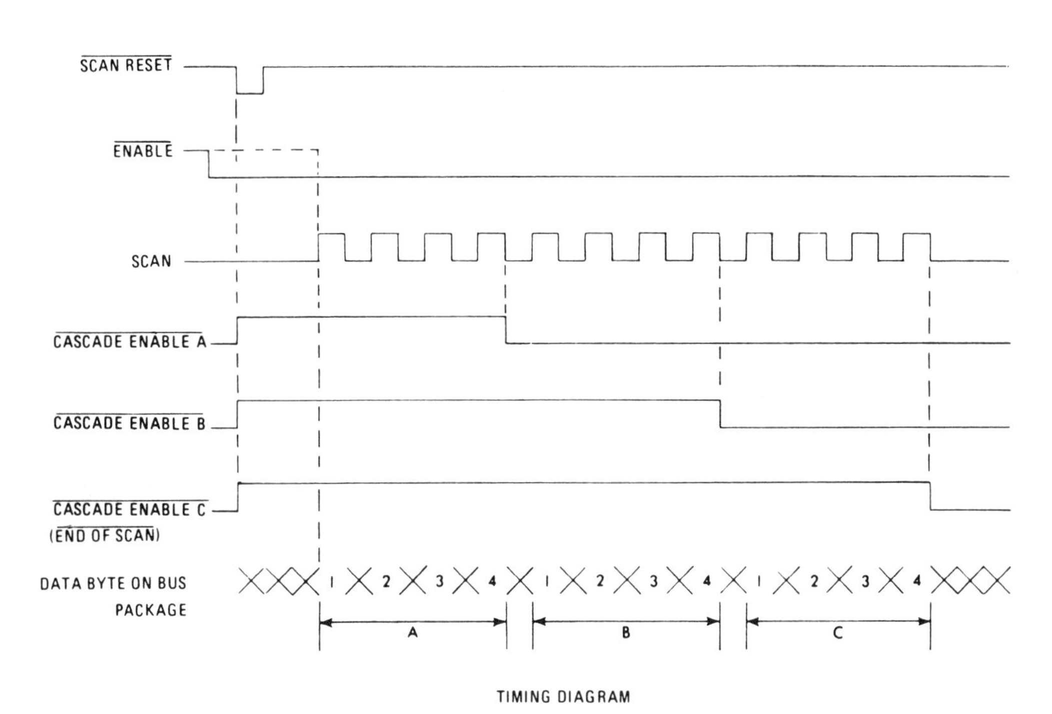

TIMING DIAGRAM

①

INHIBIT
S
Q
TO COUNT INPUT
R
COUNT
PULSES
(Same as input to Alt Count)

②

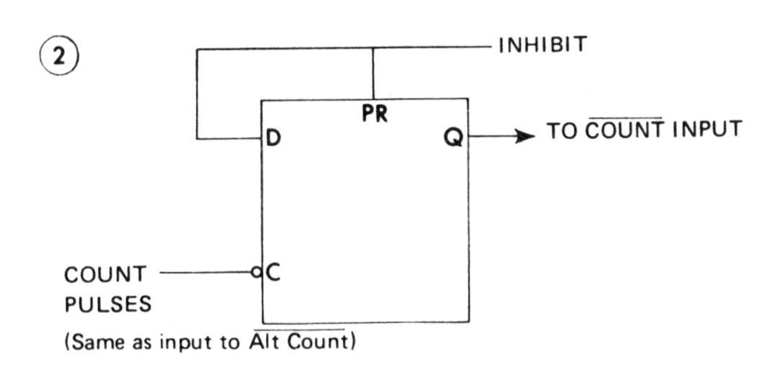

LS7060 — SYNCHRONIZING INHIBIT WITH C.P.

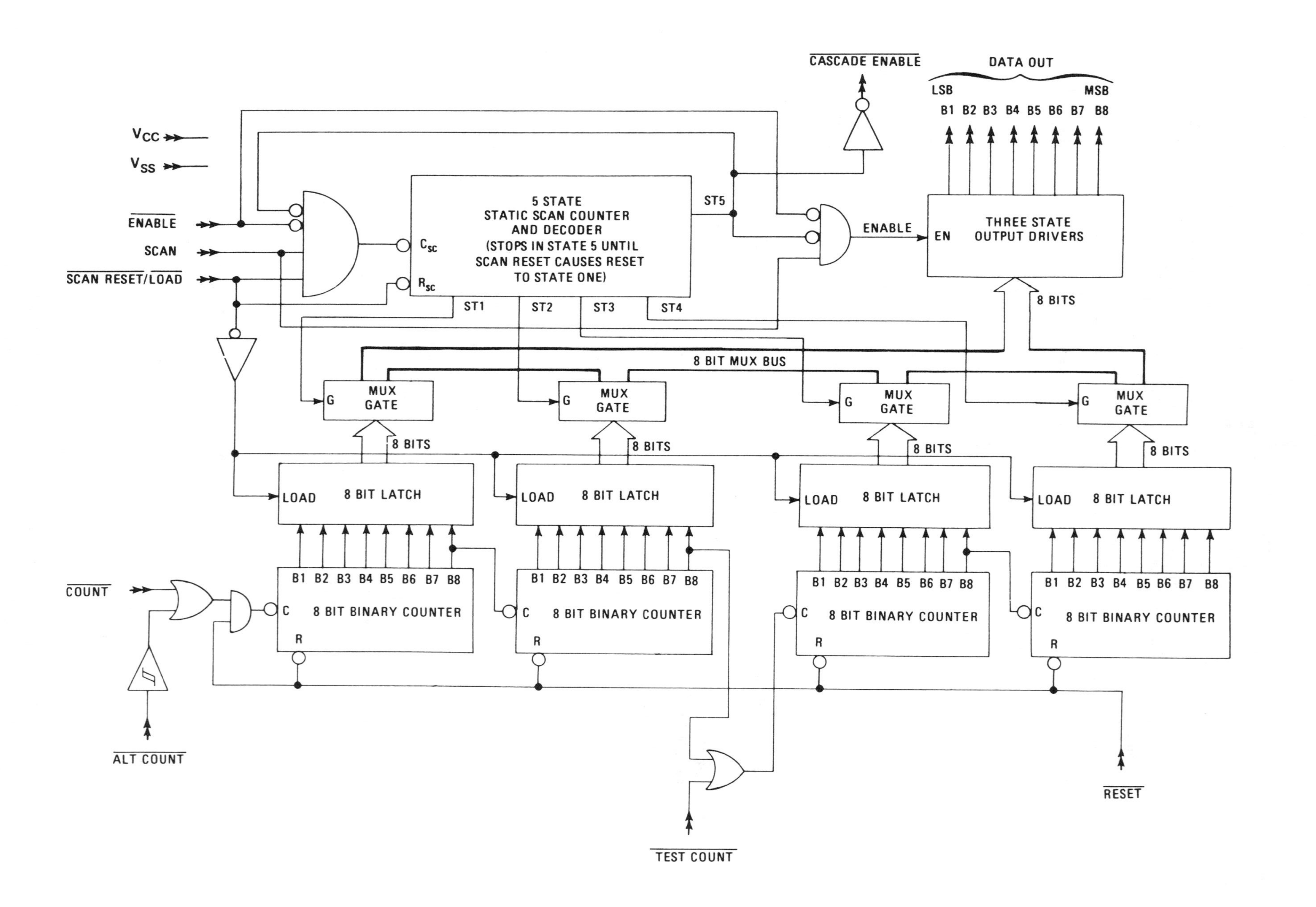
CASCADE ENABLE
DATA OUT
LSB
MSB
B1 B2 B3 B4 B5 B6 B7 B8
V_{CC}
V_{SS}
ENABLE
SCAN
SCAN RESET/LOAD
5 STATE
STATIC SCAN COUNTER
AND DECODER
(STOPS IN STATE 5 UNTIL
SCAN RESET CAUSES RESET
TO STATE ONE)
C_{SC}
R_{SC}
ST1
ST2
ST3
ST4
ST5
ENABLE
EN
THREE STATE
OUTPUT DRIVERS
8 BITS
8 BIT MUX BUS
G
MUX
GATE
8 BITS
LOAD
8 BIT LATCH
B1 B2 B3 B4 B5 B6 B7 B8
C
8 BIT BINARY COUNTER
R
COUNT
ALT COUNT
TEST COUNT
RESET

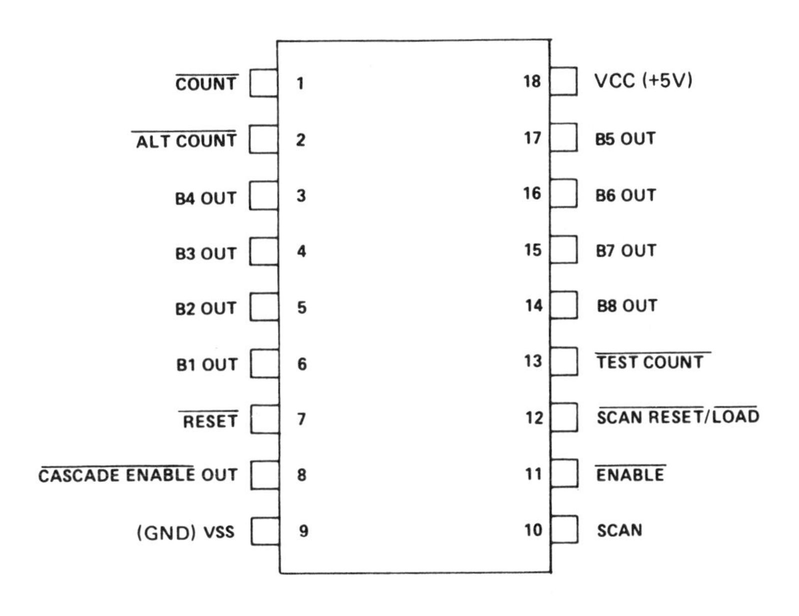

TOP VIEW
STANDARD 18 PIN DIP

LSI
LS7061
32-Bit Binary Up Counter

- DC to 10-MHz count frequency
- 8-bit byte multiplexer
- DC to 1-MHz scan frequency
- Ability to latch external 8 bits of high-speed external prescaler thereby extending count frequency to 2.56 GHz
- Single power supply operation, +4.75 VDC to +5.25 VDC
- Three-state data outputs, bus and TTL compatible
- Inputs TTL, NMOS, and CMOS compatible
- Unique cascade feature allows multiplexing of successive bytes of data in sequence in multiple counter systems
- Low power dissipation
- All inputs protected
- 24-pin DIP

The LS7061 is a monolithic, ion-implanted MOS Silicon Gate, 32-bit up counter. The circuit includes 40 latches, multiplexer, eight three-state binary data output drivers, and output cascading logic.

MAXIMUM RATINGS

PARAMETER	SYMBOL	VALUE
Storage temperature	T_{STG}	−55 to +150°C
Operating temperature	T_A	0 to +70°C
Voltage (any pin to VSS)	V_{MAX}	+10 to −0.3V

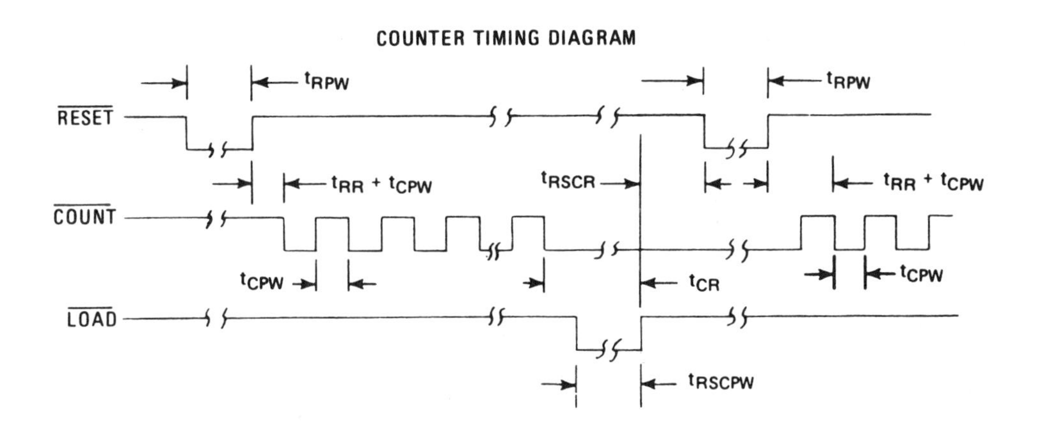

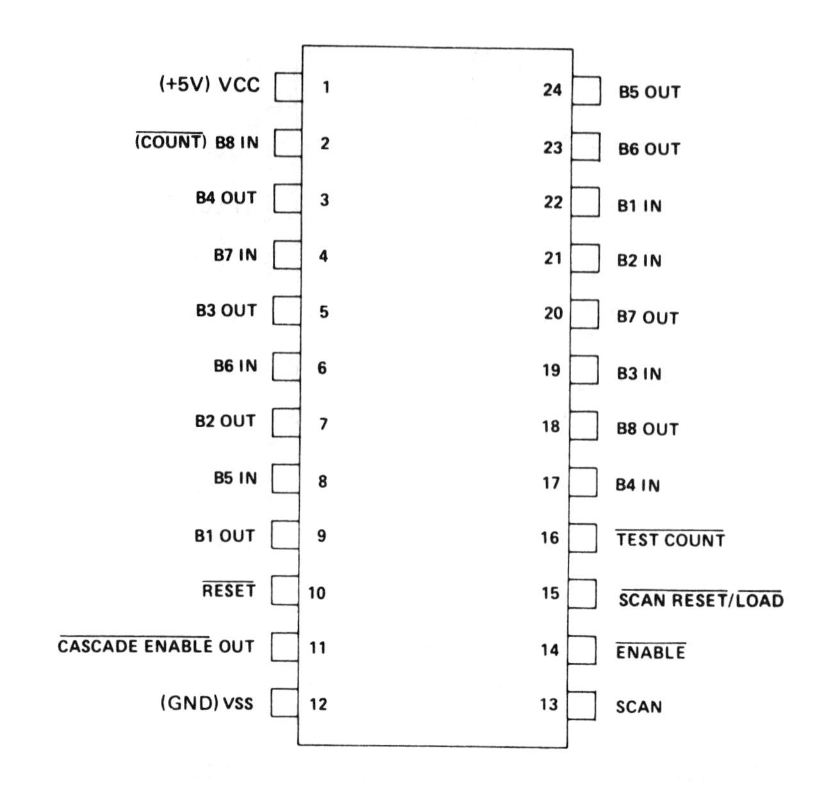

LS7061 PIN ASSIGNMENT

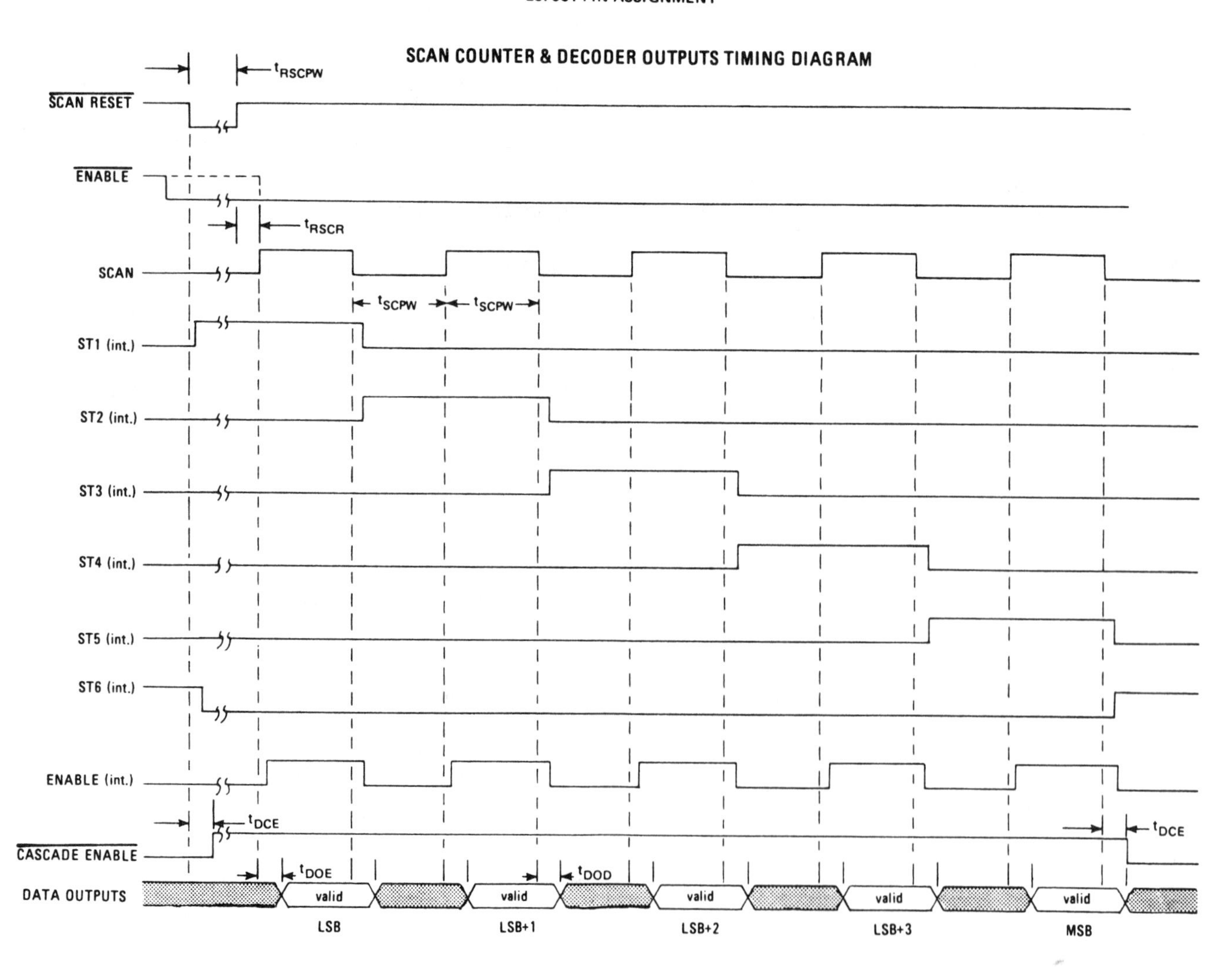

ILLUSTRATION OF A 3 DEVICE CASCADE

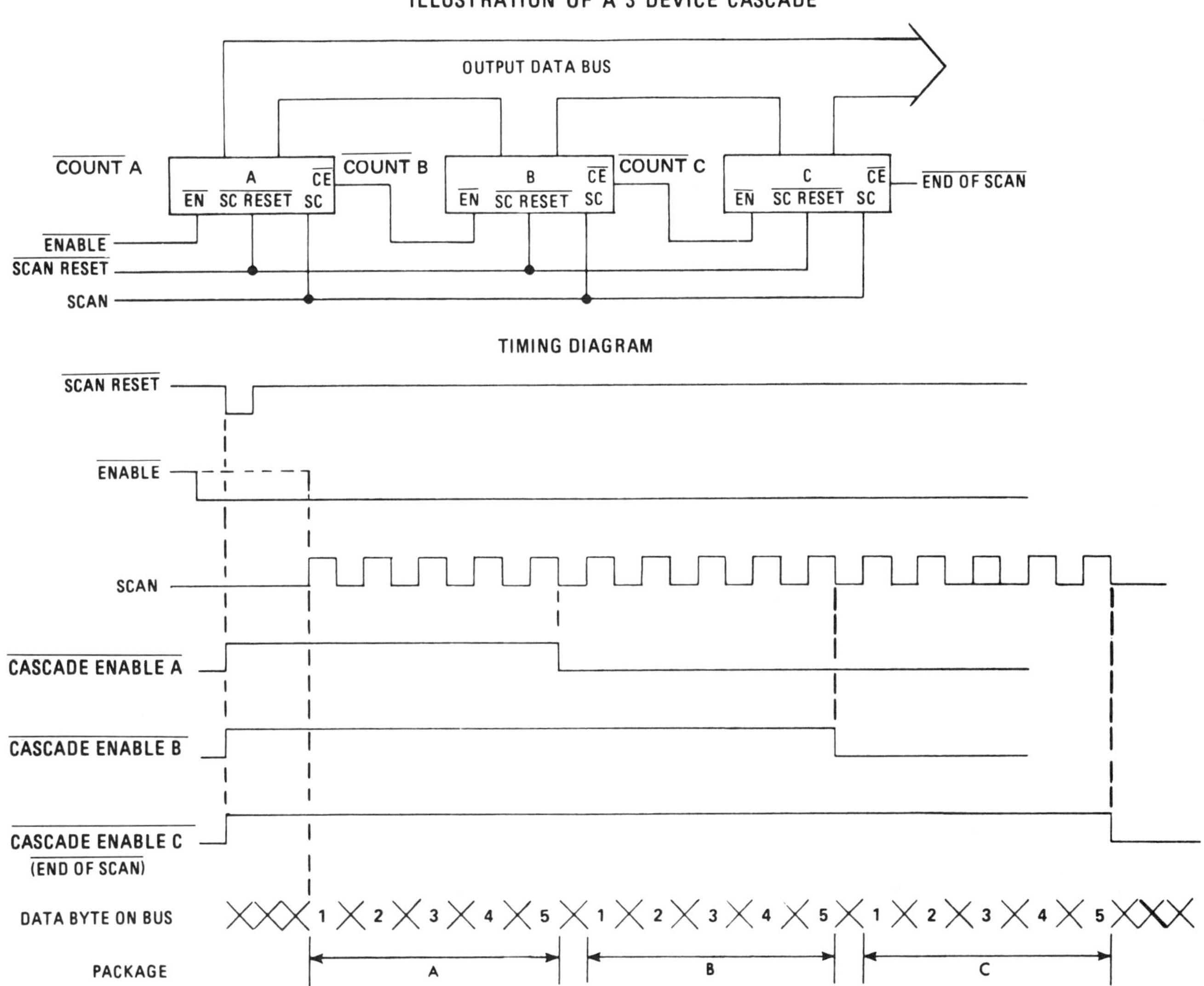

APPLICATION EXAMPLE: High Speed Differential Energy Analyzer

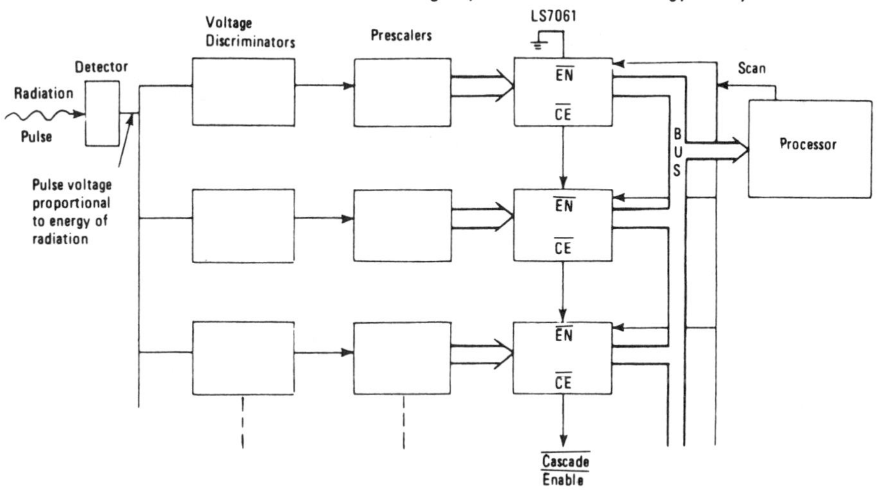

Note: The processor subtracts counts from successive counters to determine the differential energy spectrum.

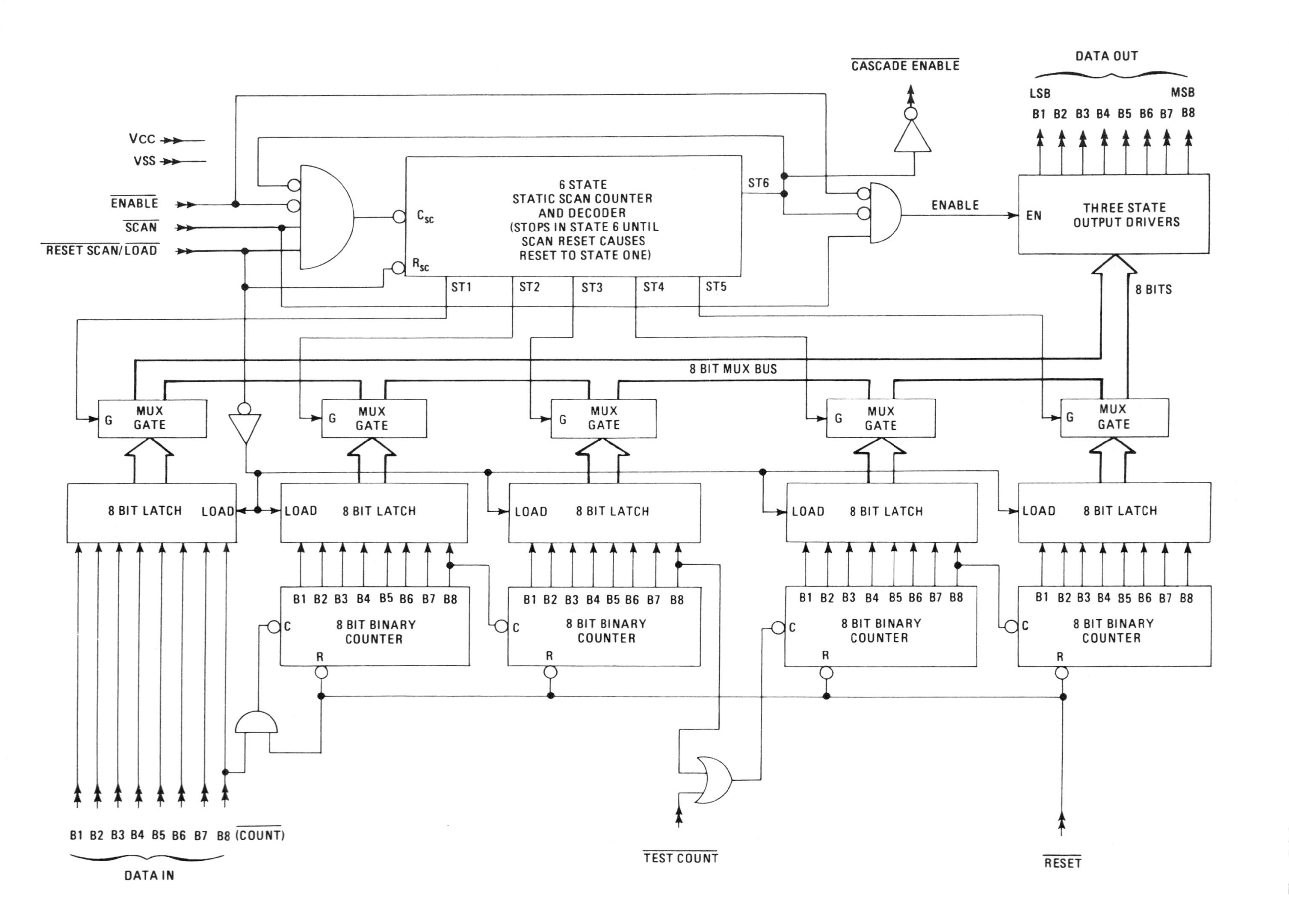
DATA OUT
LSB
MSB
B1 B2 B3 B4 B5 B6 B7 B8
CASCADE ENABLE
VCC
VSS
ENABLE
SCAN
RESET SCAN/LOAD
C_{SC}
R_{SC}
6 STATE
STATIC SCAN COUNTER
AND DECODER
(STOPS IN STATE 6 UNTIL
SCAN RESET CAUSES
RESET TO STATE ONE)
ST6
ST1
ST2
ST3
ST4
ST5
ENABLE
EN
THREE STATE
OUTPUT DRIVERS
8 BITS
8 BIT MUX BUS
G
MUX
GATE
8 BIT LATCH
LOAD
B1 B2 B3 B4 B5 B6 B7 B8
C
8 BIT BINARY
COUNTER
R
B1 B2 B3 B4 B5 B6 B7 B8 (COUNT)
DATA IN
TEST COUNT
RESET

LSI
LS7062
Dual 16-Bit Binary Up Counter

- DC to 10-MHz count frequency
- 8-bit byte multiplexer
- DC to 1-MHz scan frequency
- Single power supply operation, +4.75 VDC to + 5.25 VDC
- Three-state data outputs, bus, and TTL compatible
- Inputs TTL, NMOS, and CMOS compatible
- Unique cascade feature allows multiplexing of successive bytes of data in sequence in multiple counter systems.
- Low power dissipation
- All inputs protected
- 18-pin DIP

The LS7062 is a monolithic, ion-implanted MOS dual 16-bit up counter. The circuit includes latches, multiplexer, eight three-state binary data output drivers, and output cascading logic.

MAXIMUM RATINGS

PARAMETER	SYMBOL	VALUE
Storage temperature	T_{STG}	−55 to +150°C
Operating temperature	T_A	0 to +70°C
Voltage (any pin to VSS)	V_{MAX}	+10 to −0.3V

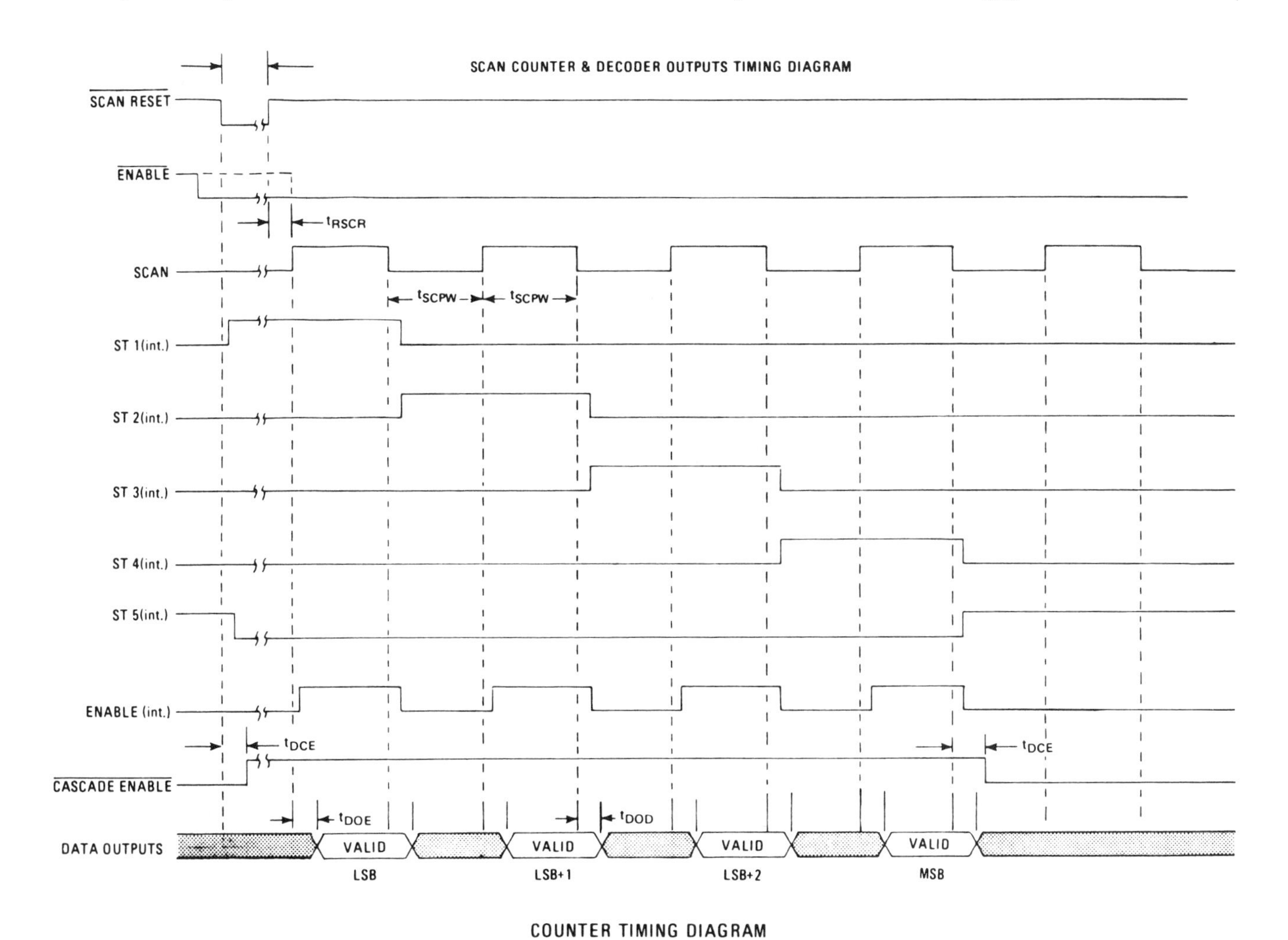

COUNTER TIMING DIAGRAM

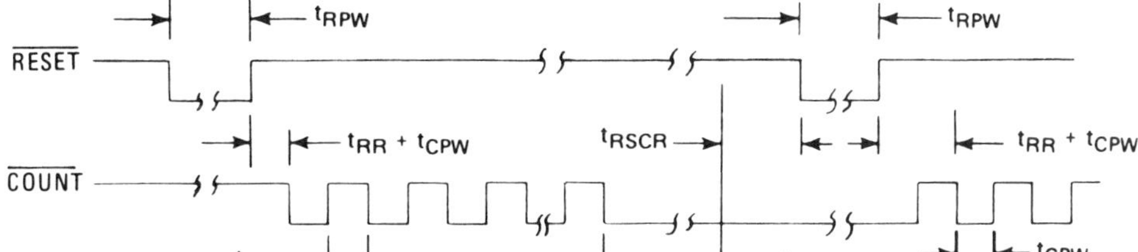

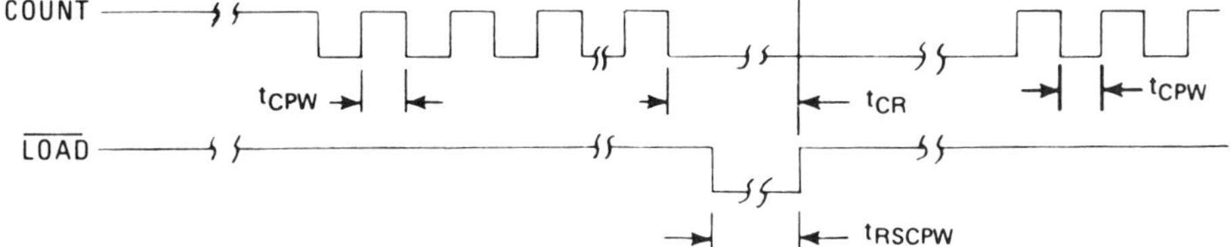

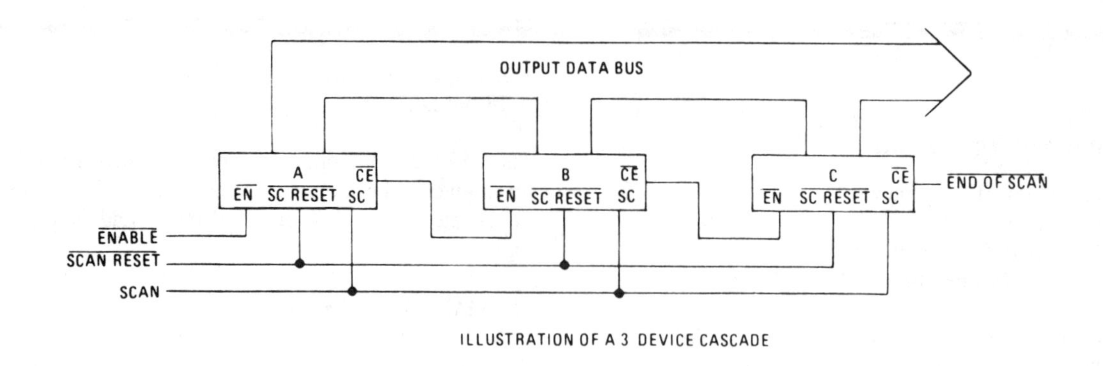

ILLUSTRATION OF A 3 DEVICE CASCADE

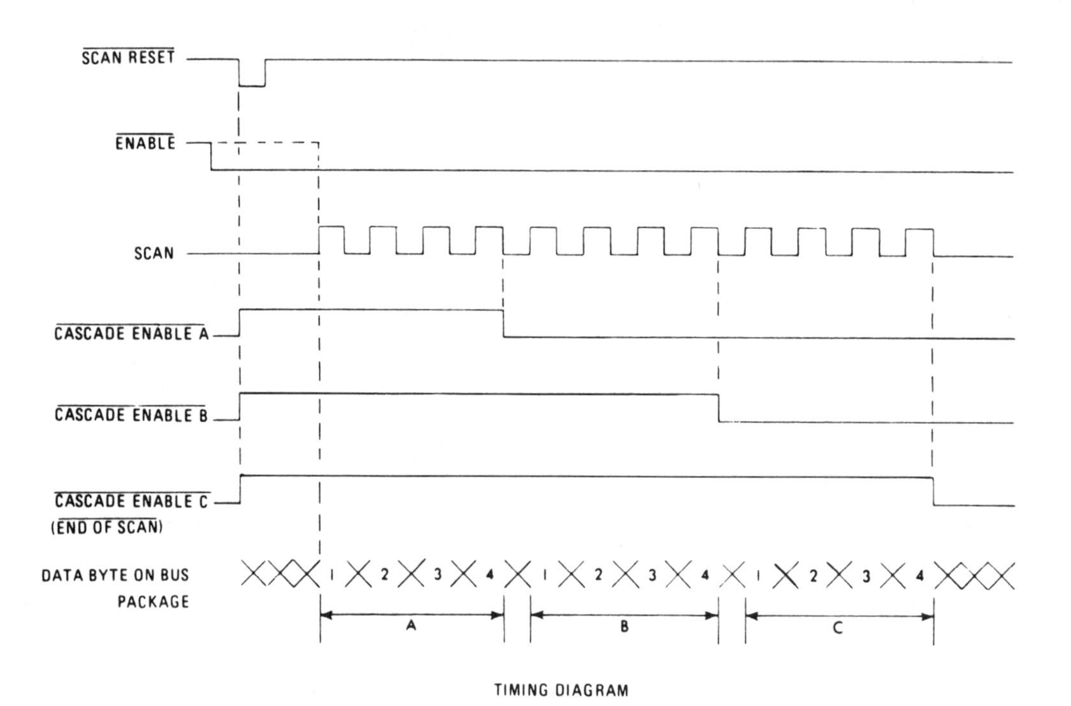

TIMING DIAGRAM

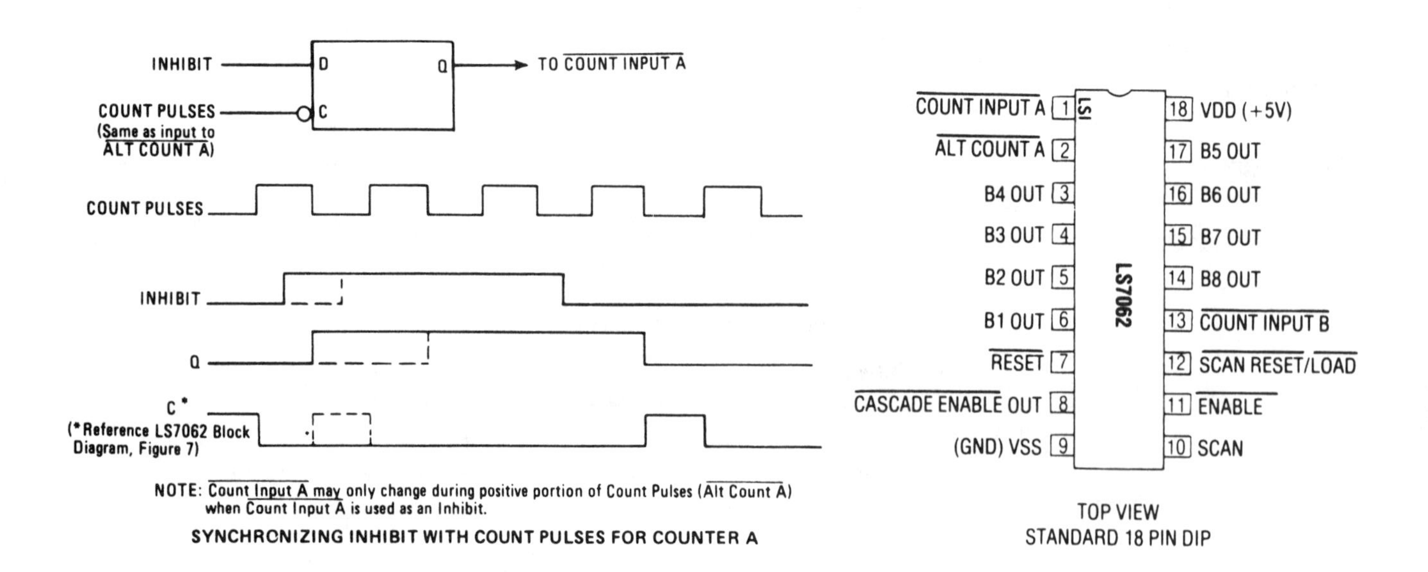

NOTE: Count Input A may only change during positive portion of Count Pulses (Alt Count A) when Count Input A is used as an Inhibit.

SYNCHRONIZING INHIBIT WITH COUNT PULSES FOR COUNTER A

TOP VIEW
STANDARD 18 PIN DIP

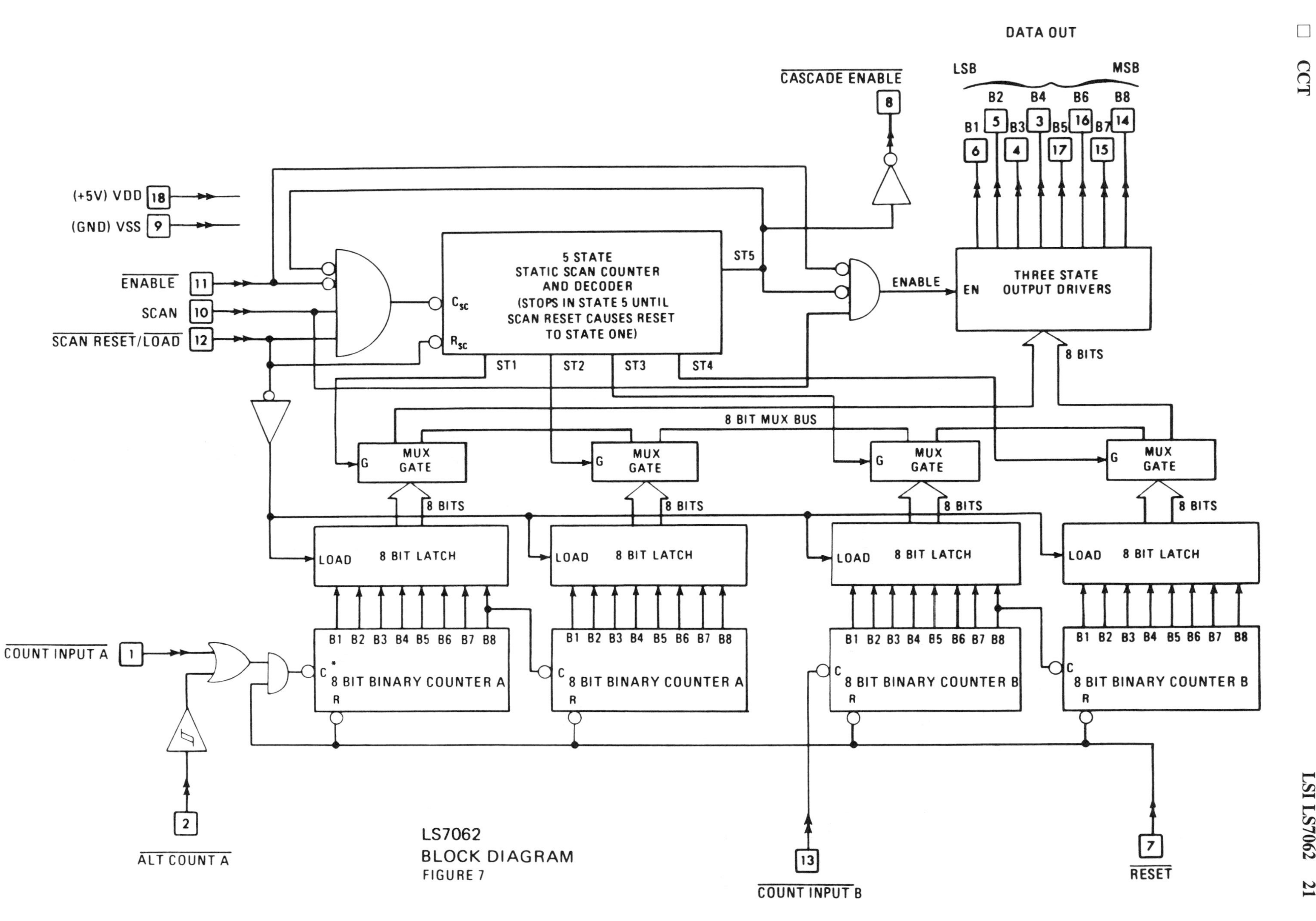

LS7062
BLOCK DIAGRAM
FIGURE 7

LSI
LS7066
24-Bit Multimode Counter

- Microprocessor compatible three state I/O bus
- Programmable modes are: Binary, BCD, 24-hour clock up, down, ÷ N, ×4 quadrature and single cycle. These modes can co-exist in different combinations.
- DC to 4 MHz
- 24-bit comparator for preset count comparison
- Readable status register
- Input/output TTL compatible
- Single +5 VDC power supply
- 20-pin plastic DIP

The LS7066 is a monolithic, ion-implanted MOS 24-bit counter that can be programmed to operate in several different modes. The operating mode is set up by writing control words into internal control registers. Three 6-bit and one 2-bit control registers set up the circuit functional characteristics. In addition to the control registers, the 5-bit output status register (OSR) indicates the current counter status. The LS7066 communicates with external circuits through an 8-bit three state I/O bus. Control and data words are written into the LS7066 through the bus. In addition to the I/O bus, a number of discrete inputs and outputs facilitate instantaneous hardware-based control functions and instantaneous status indication.

ABSOLUTE MAXIMUM RATINGS

PARAMETER	VALUE
Voltage at any pin with respect to VSS	−0.5 to 12V
Operating temperature	0 to +70°C
Storage temperature	−65 to +150°C

CONNECTION DIAGRAM — TOP VIEW
STANDARD 20 PIN PLASTIC DIP

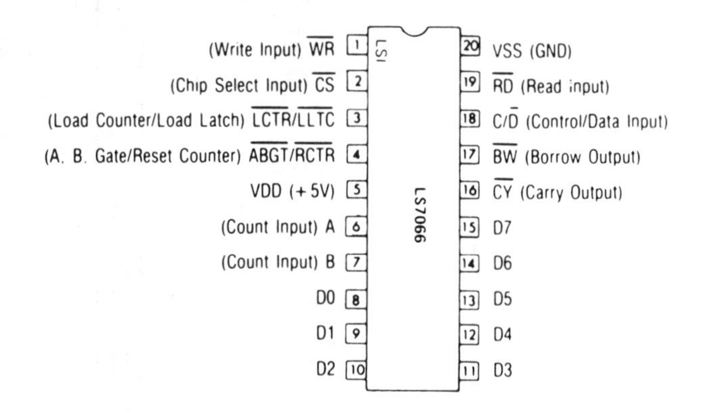

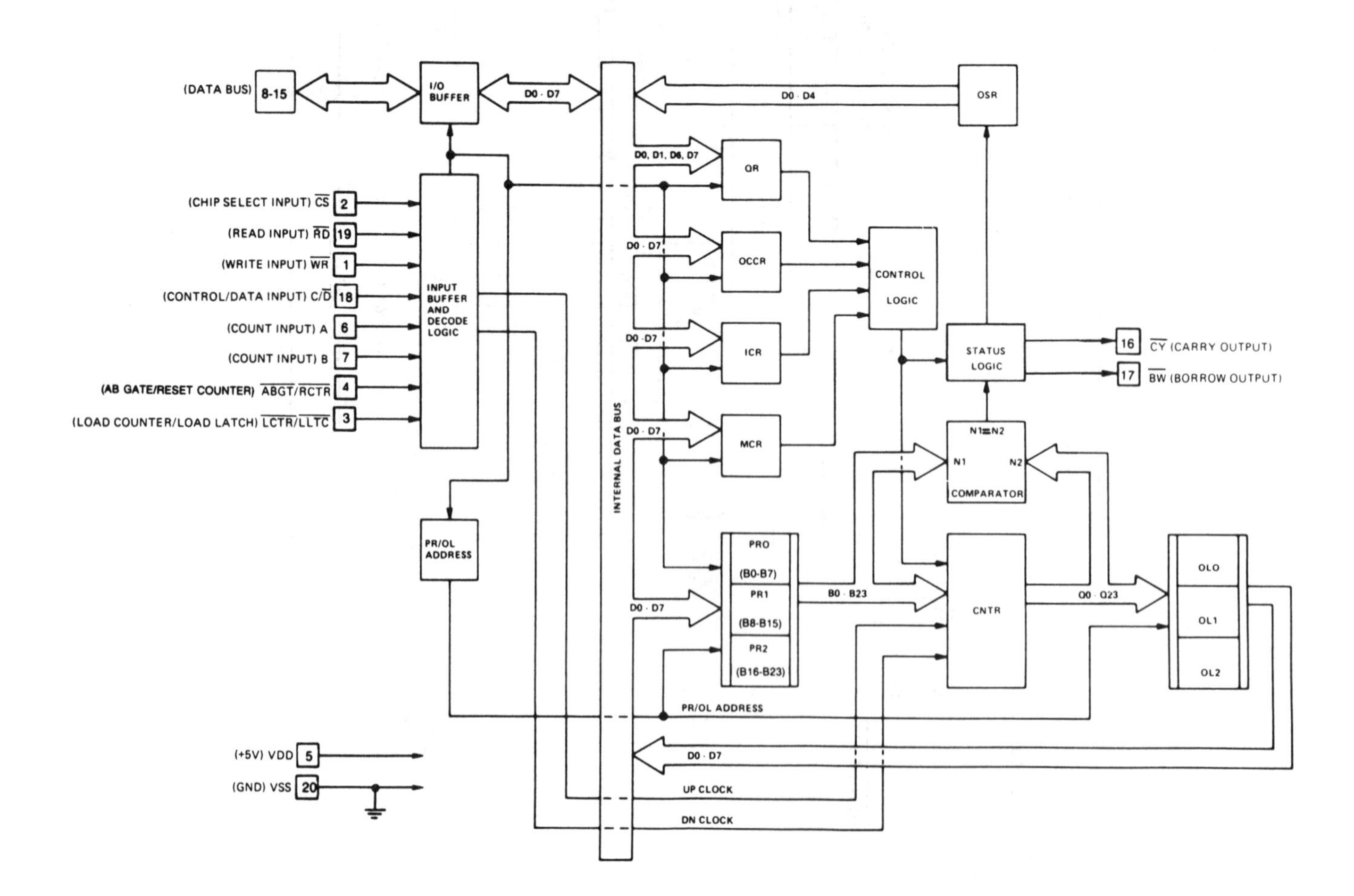

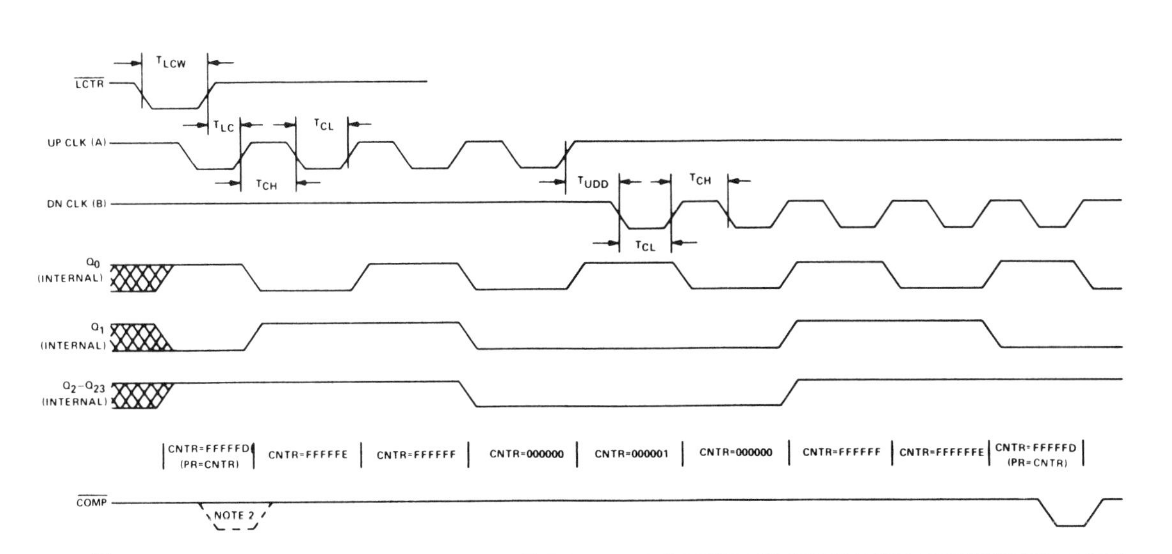

LOAD COUNTER UP CLOCK, DOWN CLOCK, COMPARE OUT, CARRY, BORROW

Note 1: The counter in this example is assumed to be operating in the binary mode.
Note 2: No COMP output is generated here, although PR=CNTR. COMP output is disabled with a counter load command and enabled with the rising edge of the next clock, thus eliminating invalid COMP outputs whenever the CNTR is loaded from the PR.
Note 3: When up Clock is active, the DN clock should be held HIGH and vice versa.

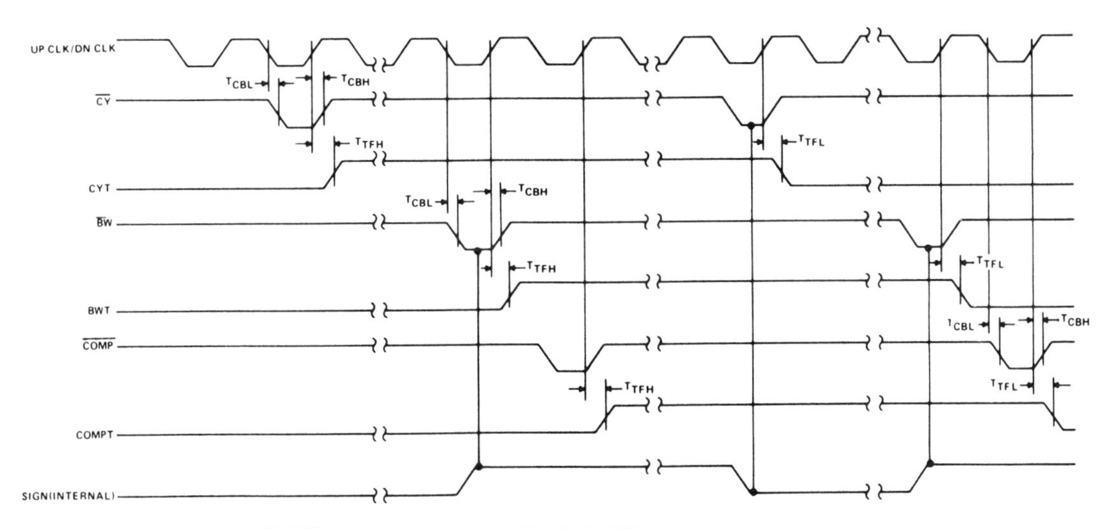

CLOCK TO CY/BW OUTPUT PROPAGATION DELAYS

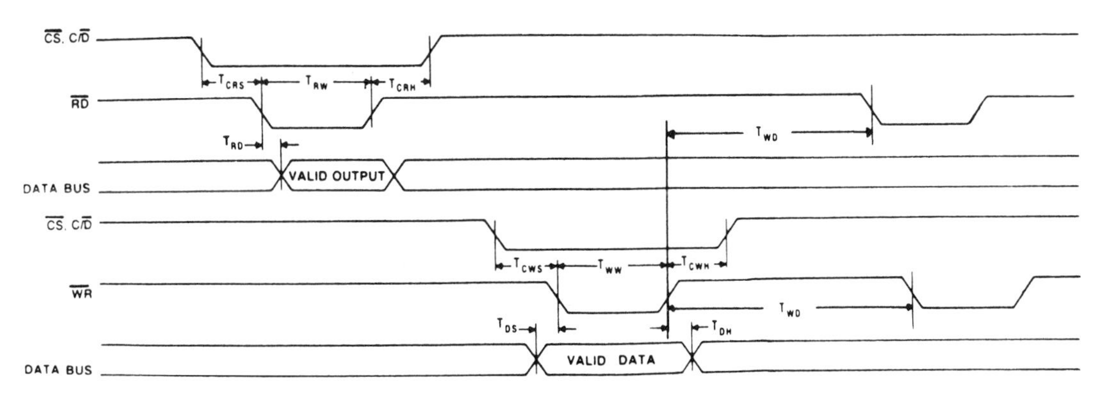

READ/WRITE CYCLES

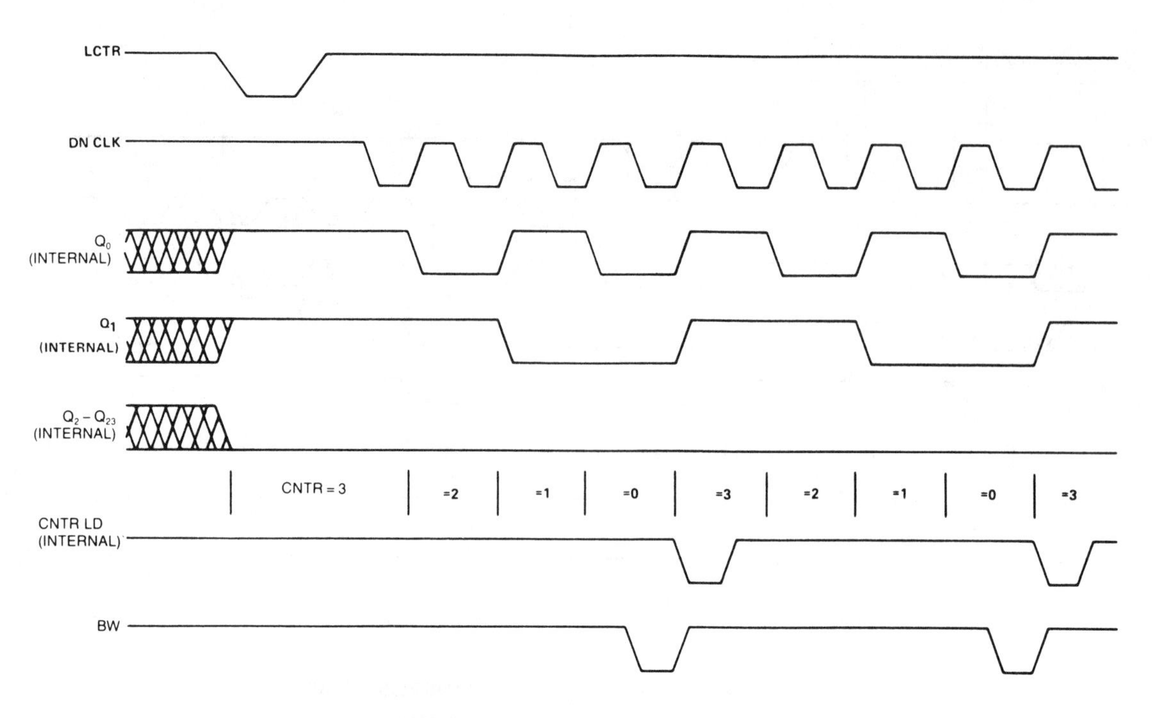

NOTE: EXAMPLE OF DIVIDE BY 4 IN DOWN COUNT MODE.

DIVIDE BY N MODE

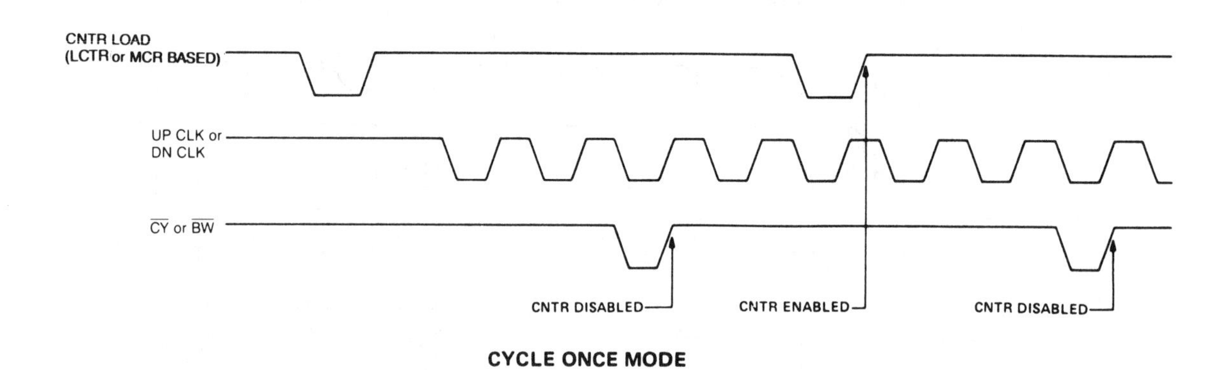

CYCLE ONCE MODE

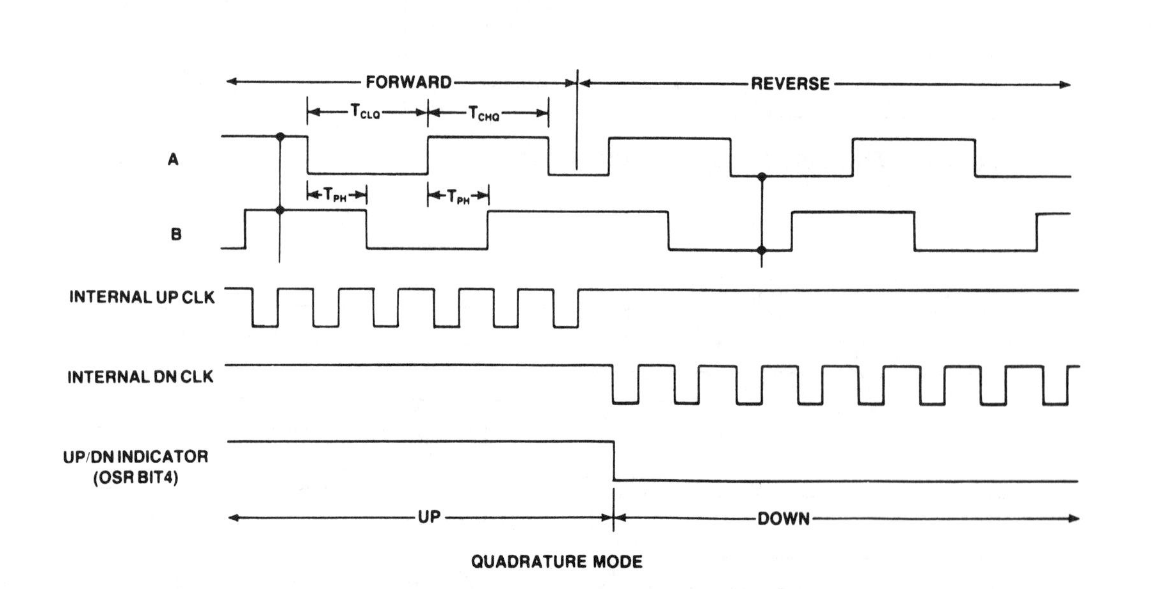

QUADRATURE MODE

LSI
LS7210
Programmable Digital Delay Timer

- Programmable delay from milliseconds to hours
- Can be cascaded for sequential events or extended delay
- Single power supply operation +4.75V to +15V
- On-chip oscillator
- Alternate clock input
- On-chip power-on reset
- Internal pull-ups on inputs
- Frequency range to 160 kHz
- CMOS-type noise immunity on all inputs
- All inputs are CMOS, PMOS and TTL compatible

The LS7210 is a monolithic, ion-implanted MOS programmable digital timer that can generate a delay in the range of 6ms to infinity. The delay is programmed by 5 binary weighted input bits, in combination with the oscillator provided. The chip can be operated into 4 different modes: delayed operate, delayed release, dual delay, and one-shot. These modes are selected by the control inputs A and B.

ABSOLUTE MAXIMUM RATINGS:
(All voltages referenced to V_{DD})

	SYMBOL	VALUE
DC supply voltage	V_{SS}	+18V
Voltage (any pin)	V_{IN}	0 to V_{SS}+.3V
Operating temperature	T_A	−25 to +70°C
Storage temperature	T_{STG}	−65 to +150°C

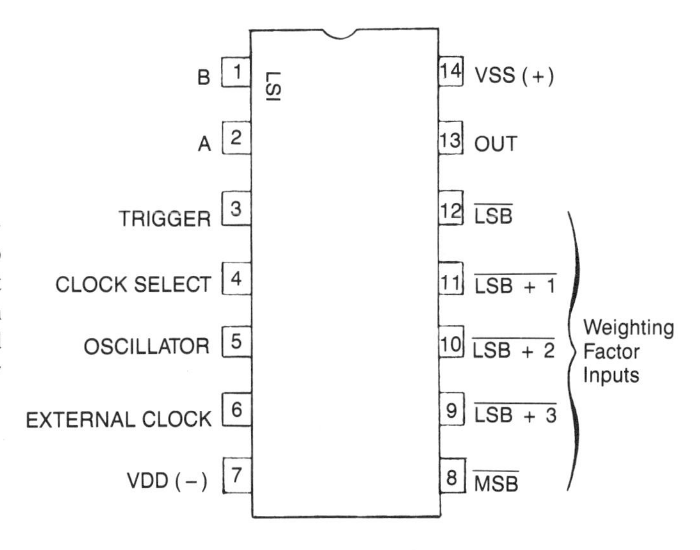

TOP VIEW
STANDARD 14 PIN DIP

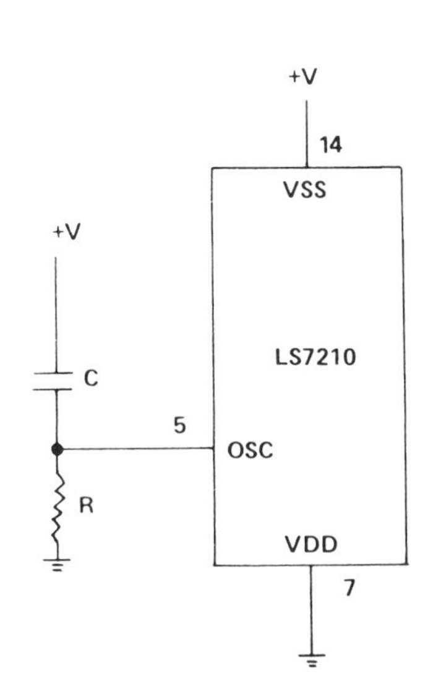

LS7210 OSCILLATOR CONNECTION

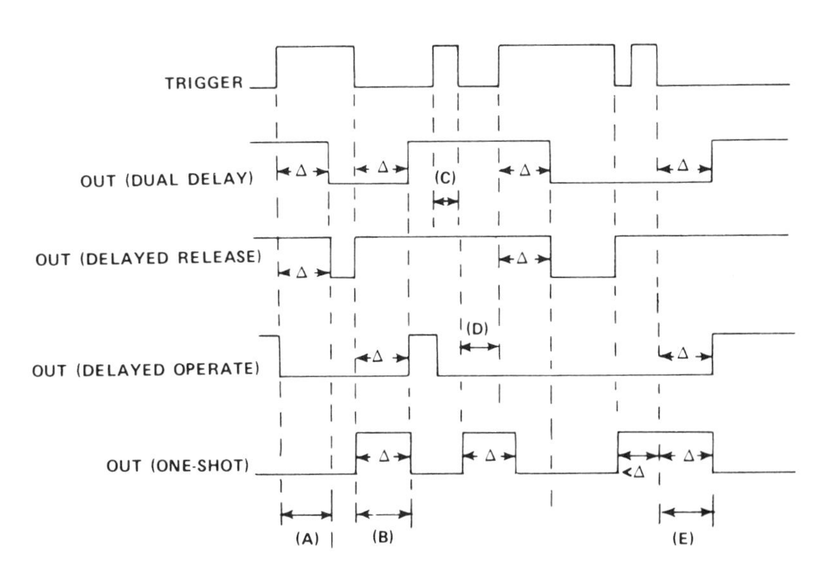

MODE DEFINITION TIMING DIAGRAM

A – Turn-off delay in "Dual Delay" and "Delayed Release" mode.
B – Turn-on delay in "Dual Delay" and "Delayed Operate" mode; one-shot period in "one-shot" mode.
C – Output remains on in "Delayed Release" and "Dual Delay" modes due to negative "trigger" transition before the turn-off delay is over.
D – Output remains off in "Delayed Operate" mode due to positive trigger transition before the turn-on delay is over.
E – One-Shot period extended by re-triggering.
NOTE: △ is the programmed delay.

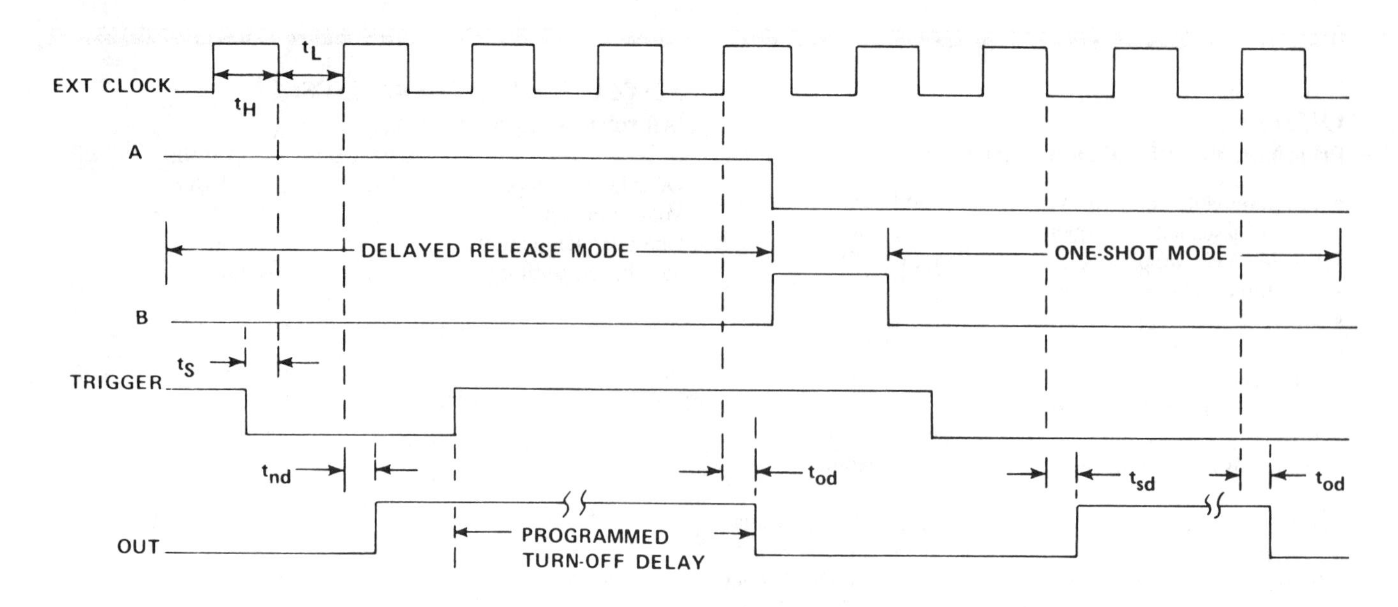

LS7210 TIMING DIAGRAM

Note 1. — A,B and trigger inputs are clocked into the input latches with the negative edge of the ext. clock.

Note 2. — In all modes except One-Shot, the output changes with the positive transition of the ext. clock. In One-Shot mode the output is turned on with the negative transition and turned off with the positive transition of the ext. clock.

APPLICATION EXAMPLES
SEQUENTIAL TURN ON

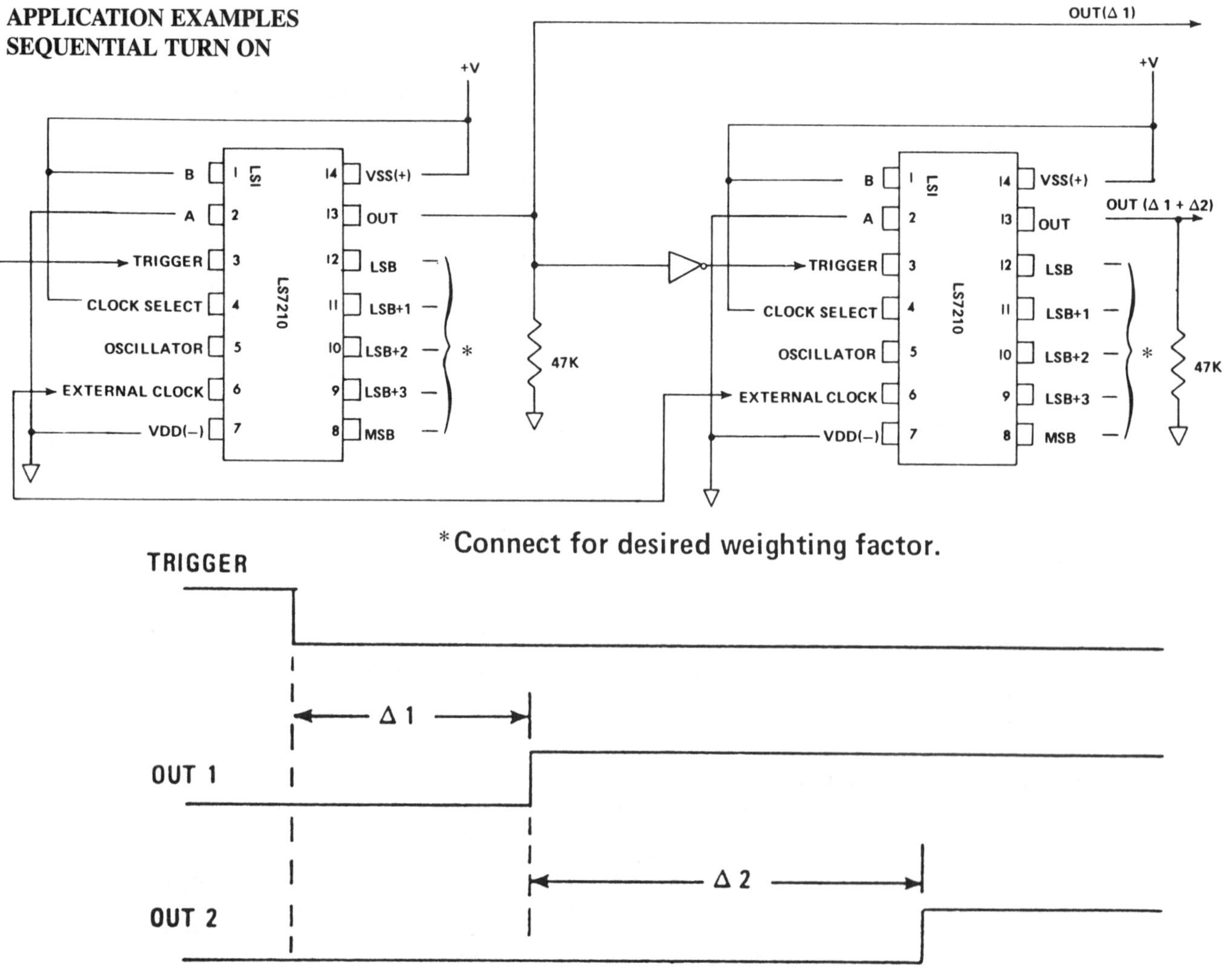

*NOTE:
Output of LS7210 is open drain FET. Some load to ground is required to cause output to go negative.

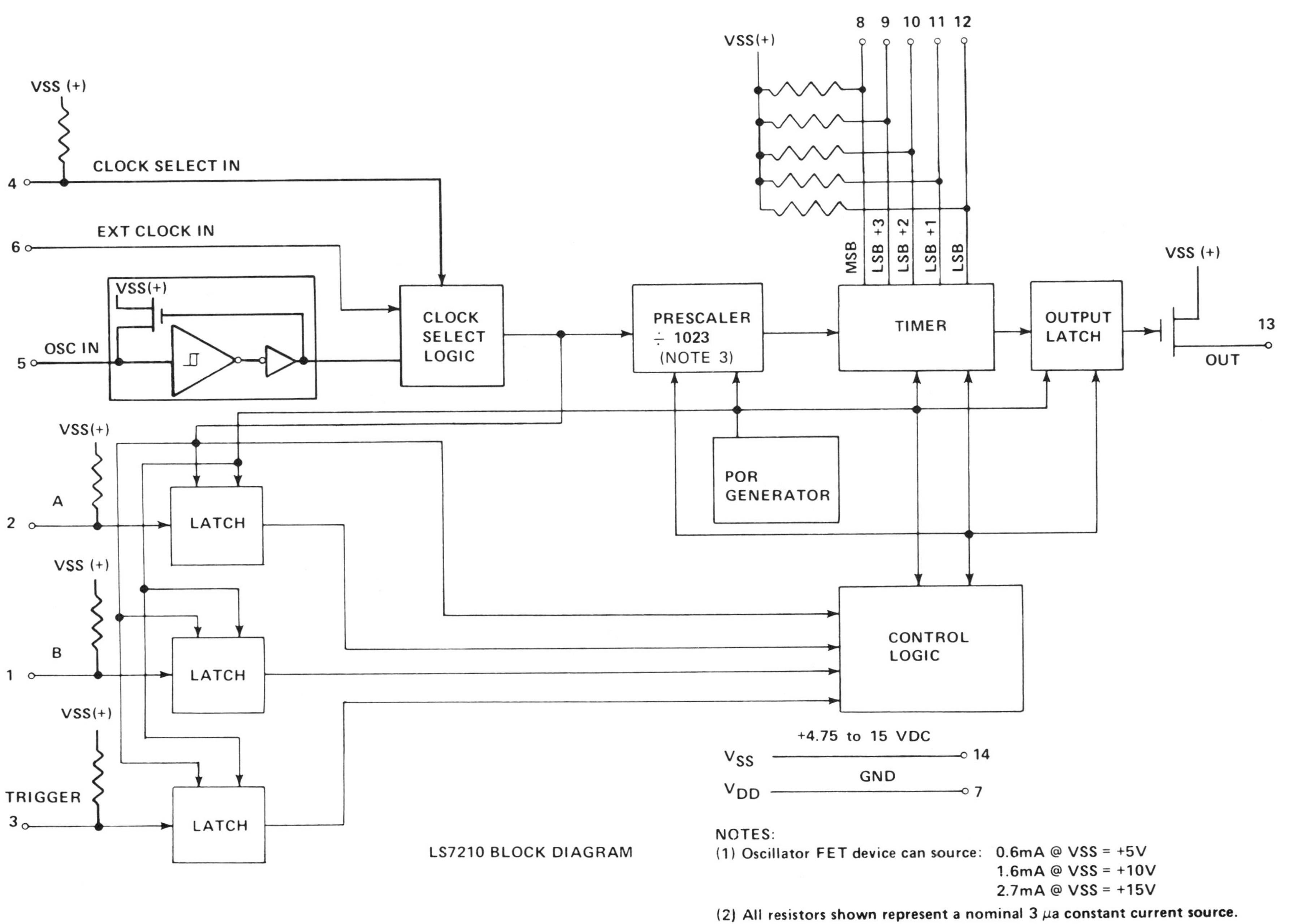

V_{SS} — +4.75 to 15 VDC — 14

V_{DD} — GND — 7

LS7210 BLOCK DIAGRAM

NOTES:

(1) Oscillator FET device can source: 0.6mA @ VSS = +5V
1.6mA @ VSS = +10V
2.7mA @ VSS = +15V

(2) All resistors shown represent a nominal 3 μa constant current source.

(3) ÷1023 is standard. Any number from 1 to 1022 can be mask programmed.

LS7210 IN DELAYED OPERATE MODE TO ACHIEVE ONE TO 31 MINUTE DELAY

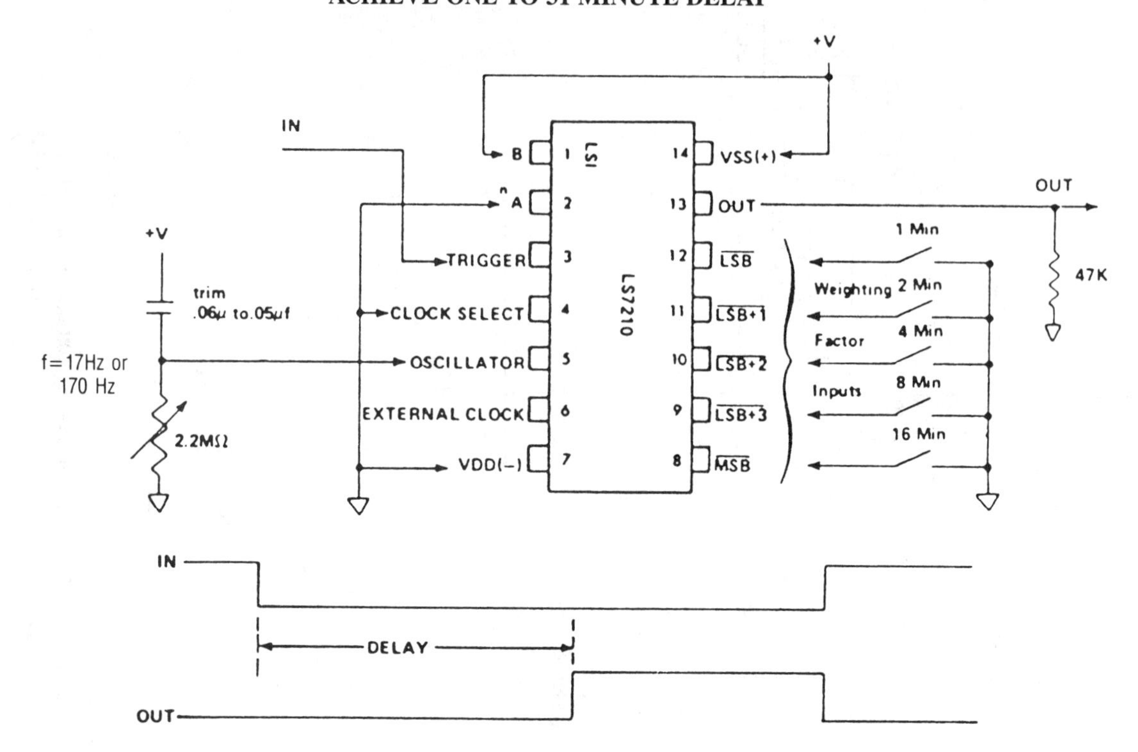

AUTO RESET WATCHDOG CIRCUIT

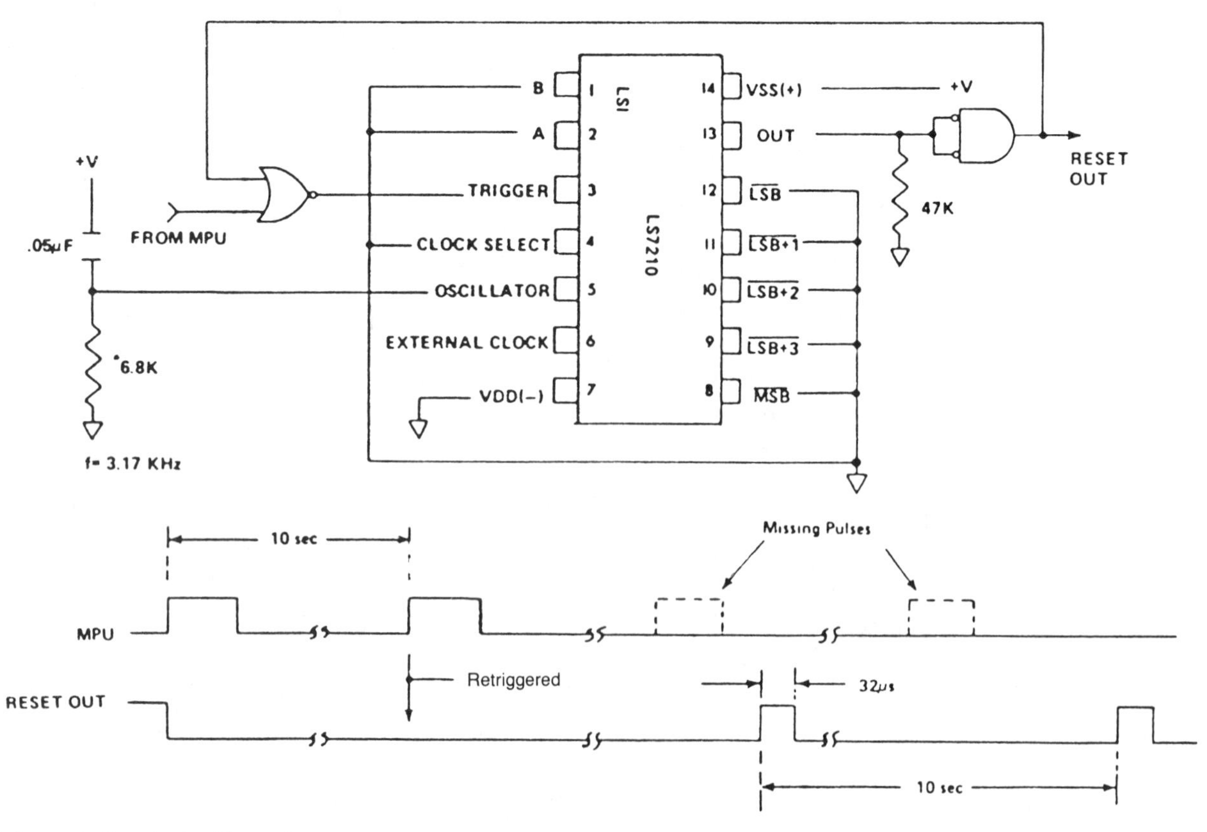

NOTE: Inputs A, B are in One Shot mode.
In this application an output is generated whenever the periodic sampling signal from the MPU is interrupted.

LSI
LS7270
Programmable Integrated Controller/Sequencer

- Hardware-oriented simple instruction set
- 4 on-chip 12-bit programmable down-counters
- 4 priority interrupt (JAM) inputs
- 12 discrete inputs
- 12 latched outputs
- 12 discrete memory bit registers
- Antibounce circuits on DI, CNT, and JAM inputs for direct interface with mechanical switches, keyboards, etc.
- Simple serial interface to external program memory (PROM or ROM)
- External program memory up to 2048 instructions
- On-chip clock generator
- Inputs TTL, NMOS, and CMOS compatible
- Outputs TTL, NMOS, and CMOS compatible
- Single power supply operation, +4.75 VDC to + 12 VDC
- 40-pin plastic DIP

The LS7270 is a monolithic, ion-implanted MOS logic controller/sequencer, designed to satisfy a wide variety of timing, sequencing, and controlling functions in small- to medium-sized systems requiring low-cost electronic control hardware. A "basic controller/sequencer" type machine can be thought of as a simple "black box" with inputs, outputs, and various chip support functions, such as power supply, oscillator, etc. As in any sequential logic machine, the present state of the machine is logically combined with the present state of the inputs to produce a new machine state with its corresponding outputs. Hence, as inputs change, the machine reacts, generating new outputs, depending on its previous state and the new inputs.

In a traditional hardwired logic machine, the sequence of the machine for all possible combinations of inputs is determined by the design of various random logic units—all permanently wired so that the results are not very flexible or amenable to change. The solution to this problem as implemented in the LS7270, is to utilize some form of computer- or microprocessor-type architecture that executes a series of instructions (the program steps) held in a memory (external to the chip) to perform the intended logical combinations of the inputs with the current machine state. In contrast to computers or microprocessors, however, the internal architecture of the LS7270 is geared to individual bit processing, Boolean processing, turn-on and turn-off functions, counting and timing operations, as opposed to numeric computations.

Broadly speaking, the LS7270 has discrete inputs (DI) that can be addressed and operated upon at the individual bit level, internal flags (T), and storage cells (M) also addressed and operated upon at the bit level, addressable internal counters that can be clocked by external sources and a group of individually addressable output registers (LO). Boolean processing is done by selecting and multiplexing various inputs into the Logic Unit (LU) along with the working Accumulator Flag (AF), thus sequentially performing the required Boolean expression and then outputting the result to the appropriate output. This is done under control of a sequence of instructions (a table of logical "0"s and "1"s) fetched from the external program memory.

In operation, the LS7270 serially shifts out a memory address when the chip is in the "shift cycle." An external shift register must be provided in which the address can be shifted and set up for addressing the memory. During the shift cycle, clocks are generated at the shift clock output, which are synchronized with the address bit changes at the address output. At the end of the shift cycle, instruction from the memory is loaded into the interface shift register. A new shift cycle begins, and the instruction from the Interface Shift Register is now shifted into the LS7270. Simultaneously, a new instruction address is shifted out into the interface shift register. The LS7270 continuously alternates between the "shift" and the "load" cycles executing the instruction in between whenever a complete instruction has been fetched. The address is automatically incremented by 1 in every shift cycle so that instructions from higher locations of a memory can be fetched sequentially. This general rule of address sequencing is broken only when an instruction involving an address jump is executed. When an instruction is executed one of the following events might take place:

1. Load 1 of 4 counters with a 12-bit number specified in the instruction field
2. Decrement one of the counters
3. Set or reset one of the internal registers
4. Load the AF with the true or complement value of one of the internal registers or discrete inputs
5. Combine AF with the true or complement value of any of the internal registers or discrete inputs in Boolean operation
6. Store the true or complement value of the AF in any of the internal registers or output latches
7. Branch out from normal addressing sequence and jump to an address specified in the instruction field

MAXIMUM RATINGS

PARAMETER	SYMBOL	VALUE
Storage temperature	T_{STG}	−55 to +150°C
Operating temperature	T_A	0 to +70°C
Voltage (any pin to VSS)	V_{MAX}	+15 to −0.3V

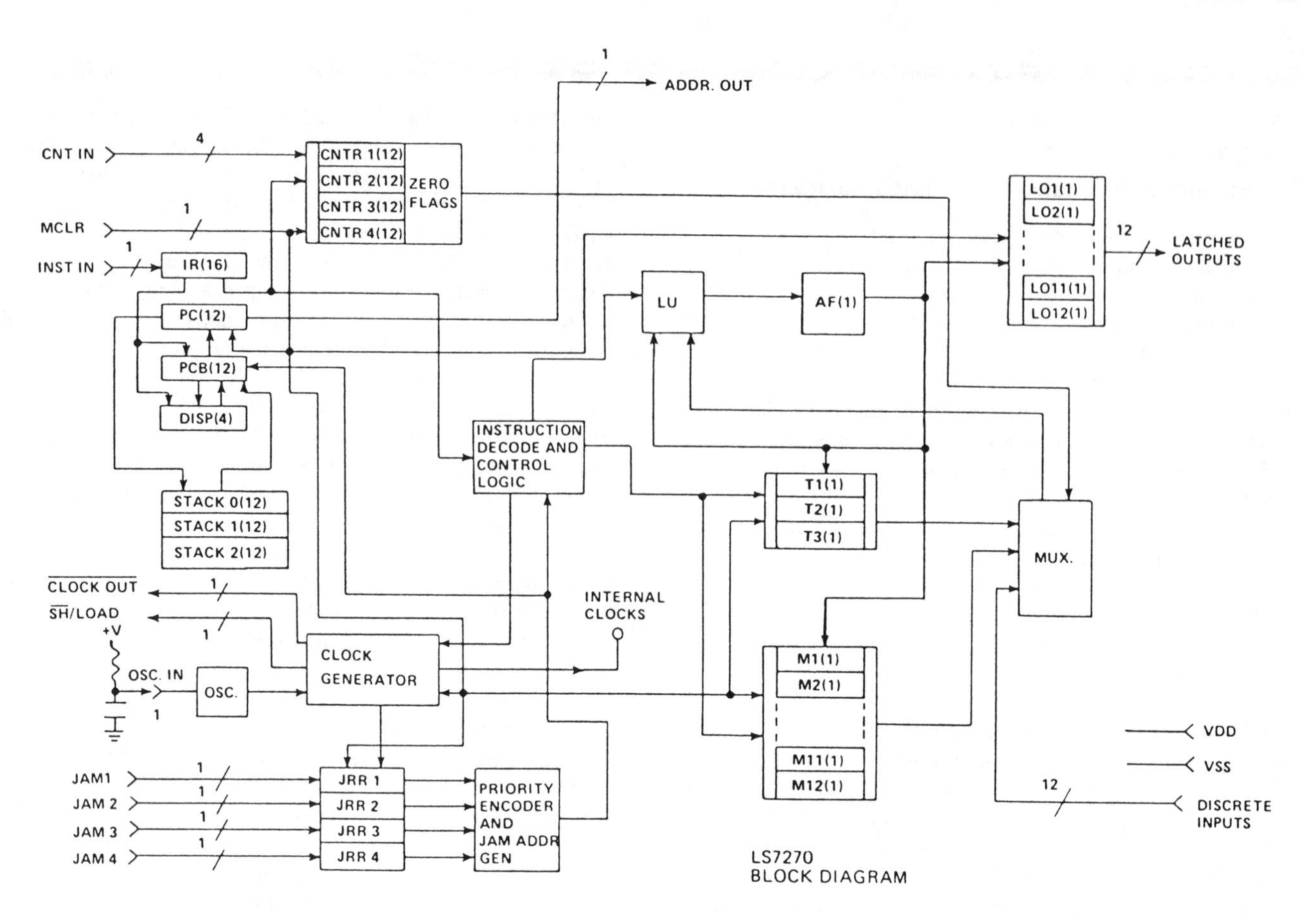

LS7270
BLOCK DIAGRAM

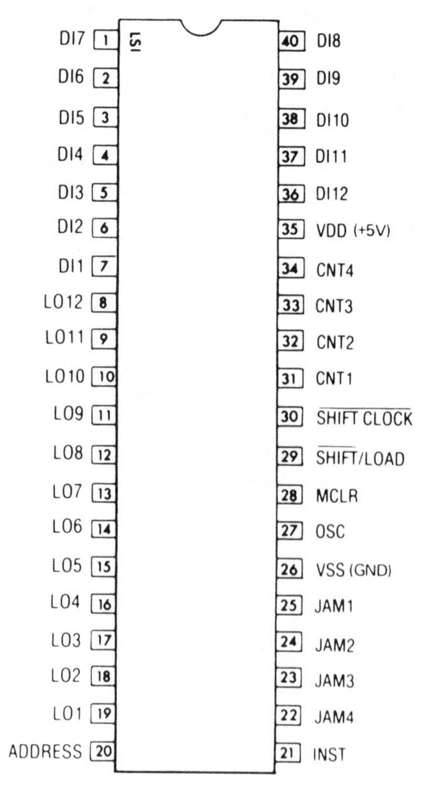

TOP VIEW

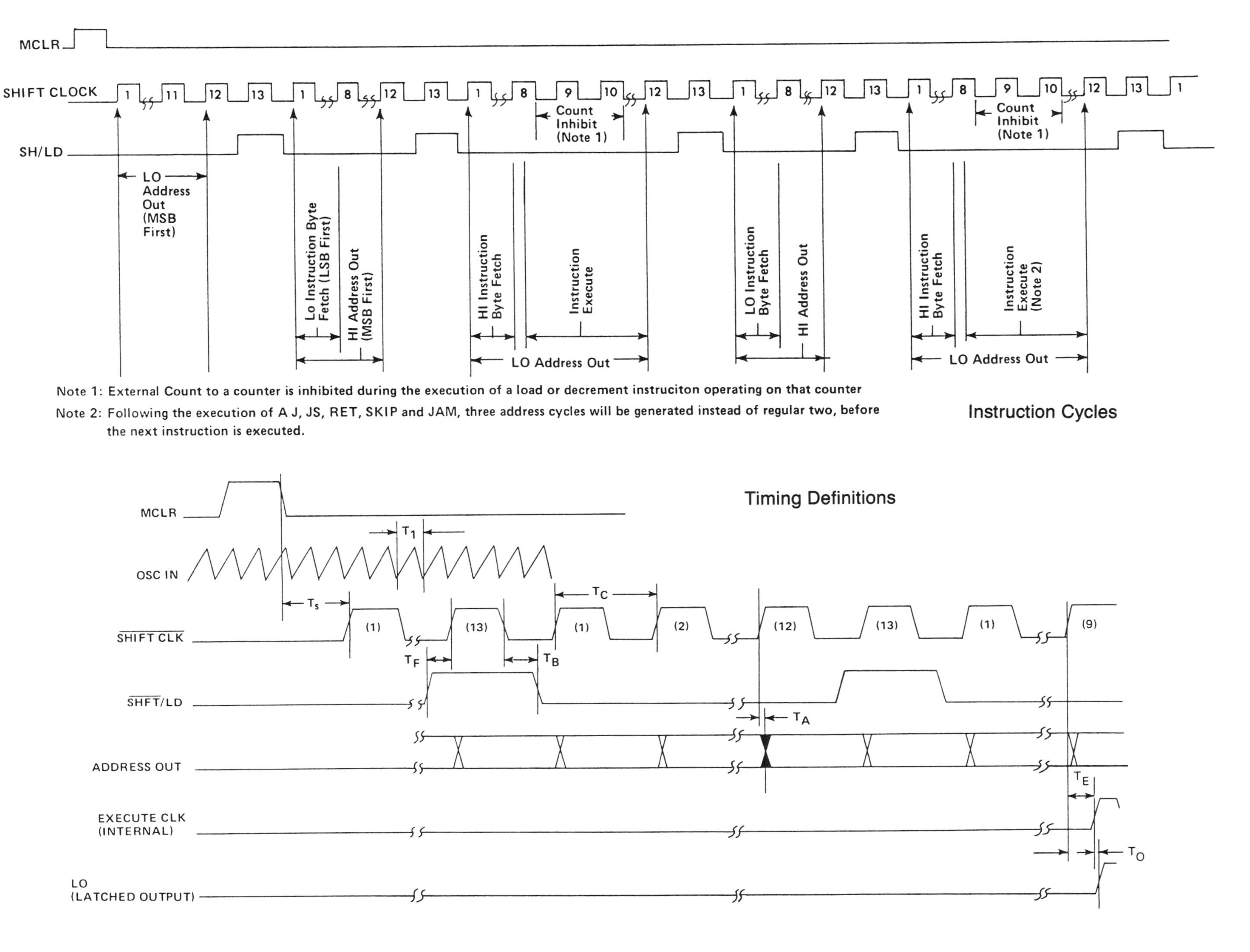

Note 1: External Count to a counter is inhibited during the execution of a load or decrement instruciton operating on that counter

Note 2: Following the execution of A J, JS, RET, SKIP and JAM, three address cycles will be generated instead of regular two, before the next instruction is executed.

Instruction Cycles

Timing Definitions

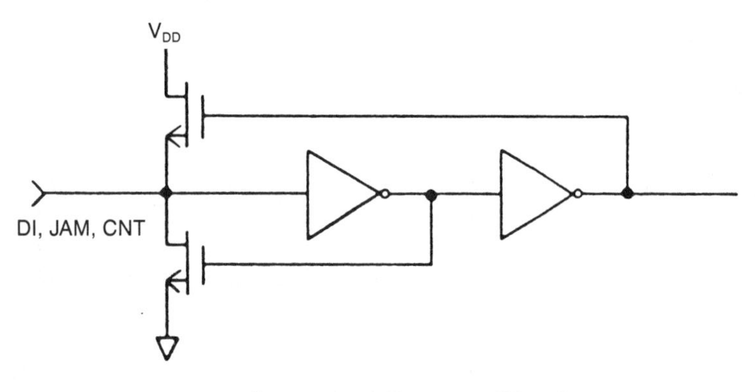

Input Anti-Bounce Circuit

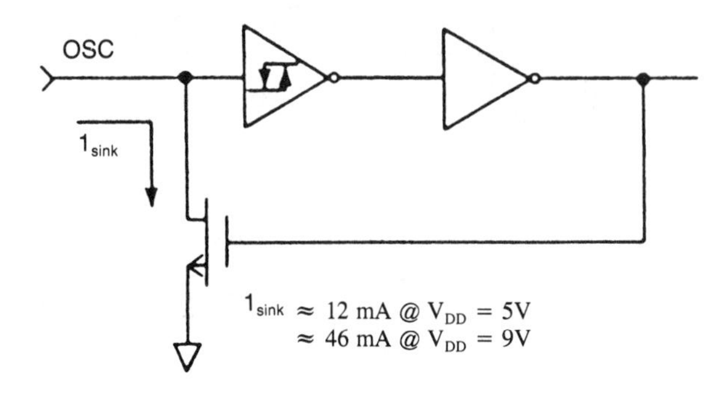

Internal Oscillator Circuit

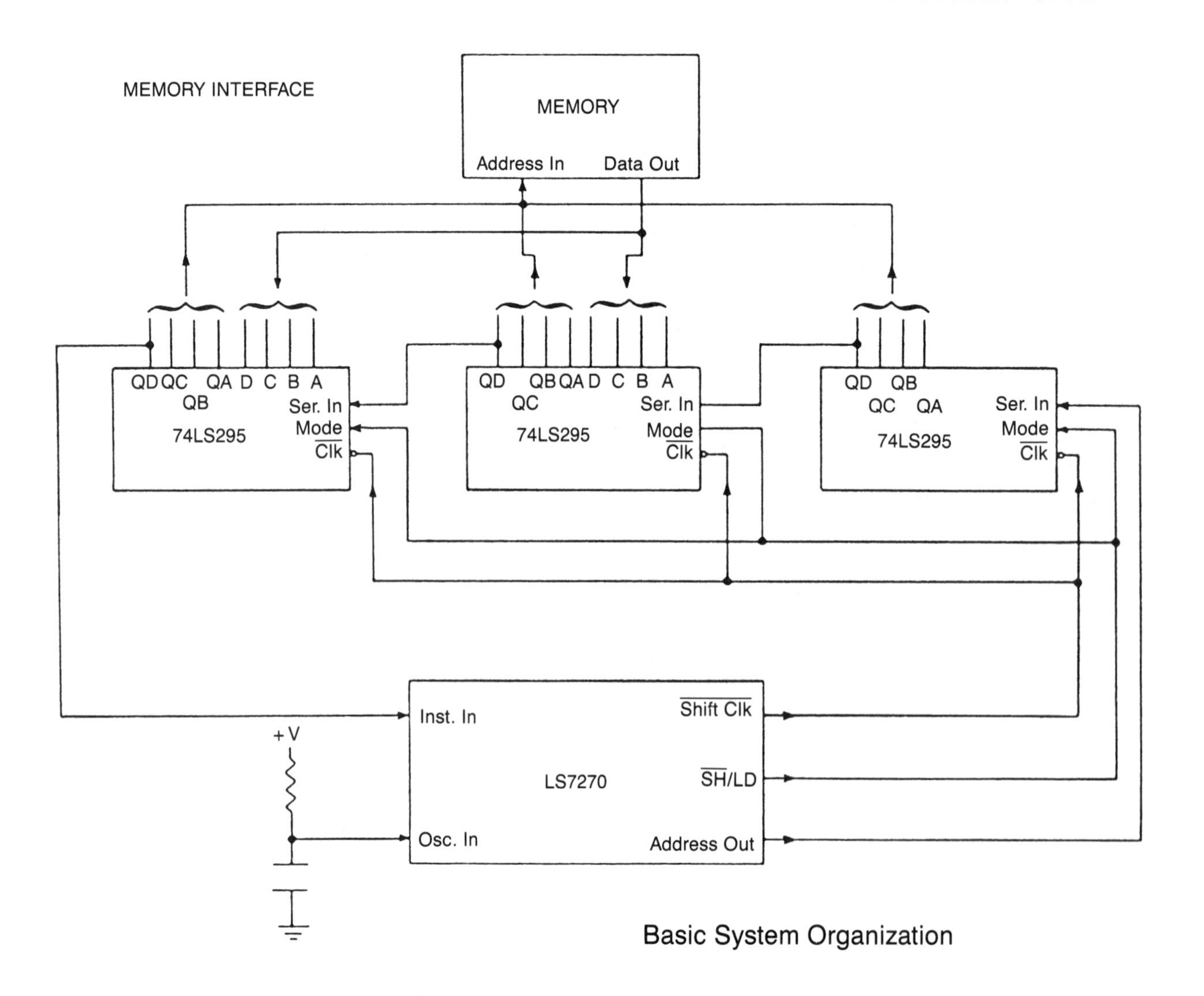

Basic System Organization

This information is reproduced with permission of National Semiconductor Corporation.

Life-Support Policy

NATIONAL'S PRODUCTS ARE NOT AUTHORIZED FOR USE AS CRITICAL COMPONENTS IN LIFE-SUPPORT DEVICES OR SYSTEMS WITHOUT THE EXPRESS WRITTEN APPROVAL OF THE PRESIDENT OF NATIONAL SEMICONDUCTOR CORPORATION.

1. Life-support devices or systems are devices or systems which, (a) are intended for surgical implant into the body, or (b) support or sustain life, and whose failure to perform, when properly used in accordance with instructions for use provided in the labeling, can be reasonably expected to result in a significant injury to the user.

2. A critical component is any component of a life-support device or system whose failure to perform can be reasonably expected to cause the failure of the life-support device or system, or to affect its safety or effectiveness.

National DM74LS90/DM74LS93 Decade and Binary Counters

ABSOLUTE MAXIMUM RATINGS (Note)

If Military/Aerospace specified devices are required, please contact the National Semiconductor Sales Office/Distributors for availability and specifications.

Supply Voltage	7V
Input Voltage (Reset)	7V
Input Voltage (A or B)	5.5V
Operating Free Air Temperature Range DM74LS	0°C to +70°C
Storage Temperature Range	−65°C to +150°C

Note: *The "Absolute Maximum Ratings" are those values beyond which the safety of the device cannot be guaranteed. The device should not be operated at these limits. The parametric values defined in the "Electrical Characteristics" table are not guaranteed at the absolute maximum ratings. The "Recommended Operating Conditions" table will define the conditions for actual device operation.*

RECOMMENDED OPERATING CONDITIONS

Symbol	Parameter		DM74LS90			Units
			Min	Nom	Max	
V_{CC}	Supply Voltage		4.75	5	5.25	V
V_{IH}	High Level Input Voltage		2			V
V_{IL}	Low Level Input Voltage				0.8	V
I_{OH}	High Level Output Current				−0.4	mA
I_{OL}	Low Level Output Current				8	mA
f_{CLK}	Clock Frequency (Note 1)	A to Q_A	0		32	MHz
		B to Q_B	0		16	
f_{CLK}	Clock Frequency (Note 2)	A to Q_A	0		20	MHz
		B to Q_B	0		10	
t_W	Pulse Width (Note 1)	A	15			ns
		B	30			
		Reset	15			
t_W	Pulse Width (Note 2)	A	25			ns
		B	50			
		Reset	25			
t_{REL}	Reset Release Time (Note 1)		25			ns
t_{REL}	Reset Release Time (Note 2)		35			ns
T_A	Free Air Operating Temperature		0		70	°C

Note 1: C_L = 15 pF, R_L = 2 kΩ, T_A = 25°C and V_{CC} = 5V.

Note 2: C_L = 50 pF, R_L = 2 kΩ, T_A = 25°C and V_{CC} = 5V.

'LS90 ELECTRICAL CHARACTERISTICS

over recommended operating free air temperature range (unless otherwise noted)

Symbol	Parameter	Conditions		Min	Typ (Note 1)	Max	Units
V_I	Input Clamp Voltage	V_{CC} = Min, I_I = −18 mA				−1.5	V
V_{OH}	High Level Output Voltage	V_{CC} = Min, I_{OH} = Max V_{IL} = Max, V_{IH} = Min		2.7	3.4		V
V_{OL}	Low Level Output Voltage	V_{CC} = Min, I_{OL} = Max V_{IL} = Max, V_{IH} = Min (Note 4)			0.35	0.5	V
		I_{OL} = 4 mA, V_{CC} = Min			0.25	0.4	
I_I	Input Current @ Max Input Voltage	V_{CC} = Max, V_I = 7V	Reset			0.1	mA
		V_{CC} = Max V_I = 5.5V	A			0.2	
			B			0.4	
I_{IH}	High Level Input Current	V_{CC} = Max, V_I = 2.7V	Reset			20	μA
			A			40	
			B			80	
I_{IL}	Low Level Input Current	V_{CC} = Max, V_I = 0.4V	Reset			−0.4	mA
			A			−2.4	
			B			−3.2	
I_{OS}	Short Circuit Output Current	V_{CC} = Max (Note 2)		−20		−100	mA
I_{CC}	Supply Current	V_{CC} = Max (Note 3)			9	15	mA

Note 1: All typicals are at V_{CC} = 5V, T_A = 25°C.

Note 2: Not more than one output should be shorted at a time, and the duration should not exceed one second.

Note 3: I_{CC} is measured with all outputs open, both RO inputs grounded following momentary connection to 4.5V and all other inputs grounded.

Note 4: Q_A outputs are tested at I_{OL} = Max plus the limit value of I_{IL} for the B input. This permits driving the B input while maintaining full fan-out capability.

'LS90 SWITCHING CHARACTERISTICS

at V_{CC} = 5V and T_A = 25°C (See Section 1 for Test Waveforms and Output Load)

Symbol	Parameter	From (Input) To (Output)	R_L = 2 kΩ				Units
			C_L = 15 pF		C_L = 50 pF		
			Min	Max	Min	Max	
f_{MAX}	Maximum Clock Frequency	A to Q_A	32		20		MHz
		B to Q_B	16		10		
t_{PLH}	Propagation Delay Time Low to High Level Output	A to Q_A		16		20	ns
t_{PHL}	Propagation Delay Time High to Low Level Output	A to Q_A		18		24	ns
t_{PLH}	Propagation Delay Time Low to High Level Output	A to Q_D		48		52	ns
t_{PHL}	Propagation Delay Time High to Low Level Output	A to Q_D		50		60	ns
t_{PLH}	Propagation Delay Time Low to High Level Output	B to Q_B		16		23	ns
t_{PHL}	Propagation Delay Time High to Low Level Output	B to Q_B		21		30	ns
t_{PLH}	Propagation Delay Time Low to High Level Output	B to Q_C		32		37	ns
t_{PHL}	Propagation Delay Time High to Low Level Output	B to Q_C		35		44	ns
t_{PLH}	Propagation Delay Time Low to High Level Output	B to Q_D		32		36	ns
t_{PHL}	Propagation Delay Time High to Low Level Output	B to Q_D		35		44	ns
t_{PLH}	Propagation Delay Time Low to High Level Output	SET-9 to Q_A, Q_D		30		35	ns
t_{PHL}	Propagation Delay Time High to Low Level Output	SET-9 to Q_B, Q_C		40		48	ns
t_{PHL}	Propagation Delay Time High to Low Level Output	SET-0 to Any Q		40		52	ns

RECOMMENDED OPERATING CONDITIONS

Symbol	Parameter		DM74LS93 Min	DM74LS93 Nom	DM74LS93 Max	Units
V_{CC}	Supply Voltage		4.75	5	5.25	V
V_{IH}	High Level Input Voltage		2			V
V_{IL}	Low Level Input Voltage				0.8	V
I_{OH}	High Level Output Current				−0.4	mA
I_{OL}	Low Level Output Current				8	mA
f_{CLK}	Clock Frequency (Note 1)	A to Q_A	0		32	MHz
		B to Q_B	0		16	MHz
f_{CLK}	Clock Frequency (Note 2)	A to Q_A	0		20	MHz
		B to Q_B	0		10	MHz
t_W	Pulse Width (Note 1)	A	15			ns
		B	30			ns
		Reset	15			ns
t_W	Pulse Width (Note 2)	A	25			ns
		B	50			ns
		Reset	25			ns
t_{REL}	Reset Release Time (Note 1)		25			ns
t_{REL}	Reset Release Time (Note 2)		35			ns
T_A	Free Air Operating Temperature		0		70	°C

Note 1: C_L = 15 pF, R_L = 2 kΩ, T_A = 25°C and V_{CC} = 5V.

Note 2: C_L = 50 pF, R_L = 2 kΩ, T_A = 25°C and V_{CC} = 5V.

'LS93 ELECTRICAL CHARACTERISTICS

over recommended operating free air temperature range (unless otherwise noted)

Symbol	Parameter	Conditions		Min	Typ (Note 1)	Max	Units
V_I	Input Clamp Voltage	V_{CC} = Min, I_I = −18 mA				−1.5	V
V_{OH}	High Level Output Voltage	V_{CC} = Min, I_{OH} = Max; V_{IL} = Max, V_{IH} = Min		2.7	3.4		V
V_{OL}	Low Level Output Voltage	V_{CC} = Min, I_{OL} = Max; V_{IL} = Max, V_{IH} = Min (Note 4)			0.35	0.5	V
		I_{OL} = 4 mA, V_{CC} = Min			0.25	0.4	V
I_I	Input Current @Max Input Voltage	V_{CC} = Max, V_I = 7V	Reset			0.1	mA
		V_{CC} = Max; V_I = 5.5V	A			0.2	mA
			B			0.4	mA
I_{IH}	High Level Input Current	V_{CC} = Max; V_I = 2.7V	Reset			20	μA
			A			40	μA
			B			80	μA
I_{IL}	Low Level Input Current	V_{CC} = Max, V_I = 0.4V	Reset			−0.4	mA
			A			−2.4	mA
			B			−1.6	mA
I_{OS}	Short Circuit Output Current	V_{CC} = Max (Note 2)		−20		−100	mA
I_{CC}	Supply Current	V_{CC} = Max (Note 3)			9	15	mA

Note 1: All typicals are at V_{CC} = 5V, T_A = 25°C.

Note 2: Not more than one output should be shorted at a time, and the duration should not exceed one second.

Note 3: I_{CC} is measured with all outputs open, both RO inputs grounded following momentary connection to 4.5V and all other inputs grounded.

Note 4: Q_A outputs are tested at I_{OL} = max plus the limit value of I_{IL} for the B input. This permits driving the B input while maintaining full fan-out capability.

'LS93 SWITCHING CHARACTERISTICS

at V_{CC} = 5V and T_A = 25°C (See Section 1 for Test Waveforms and Output Load)

Symbol	Parameter	From (Input) To (Output)	$R_L = 2\ k\Omega$				Units
			C_L = 15 pF		C_L = 50 pF		
			Min	Max	Min	Max	
f_{MAX}	Maximum Clock Frequency	A to Q_A	32		20		MHz
		B to Q_B	16		10		
t_{PLH}	Propagation Delay Time Low to High Level Output	A to Q_A		16		20	ns
t_{PHL}	Propagation Delay Time High to Low Level Output	A to Q_A		18		24	ns
t_{PLH}	Propagation Delay Time Low to High Level Output	A to Q_D		70		85	ns
t_{PHL}	Propagation Delay Time High to Low Level Output	A to Q_D		70		90	ns
t_{PLH}	Propagation Delay Time Low to High Level Output	B to Q_B		16		23	ns
t_{PHL}	Propagation Delay Time High to Low Level Output	B to Q_B		21		30	ns
t_{PLH}	Propagation Delay Time Low to High Level Output	B to Q_C		32		37	ns
t_{PHL}	Propagation Delay Time High to Low Level Output	B to Q_C		35		44	ns
t_{PLH}	Propagation Delay Time Low to High Level Output	B to Q_D		51		60	ns
t_{PHL}	Propagation Delay Time High to Low Level Output	B to Q_D		51		70	ns
t_{PHL}	Propagation Delay Time High to Low Level Output	SET-0 to Any Q		40		52	ns

CONNECTION DIAGRAMS (Dual-In-Line Packages)

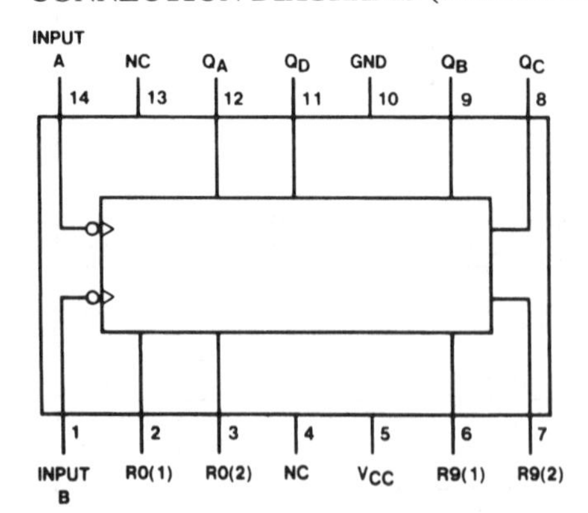

Order Number DM74LS90M or DM74LS90N
See NS Package Number M14A or N14A

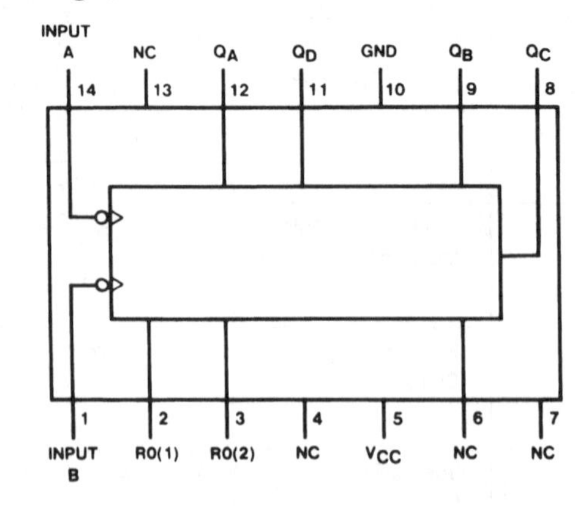

Order Number DM74LS93M or DM74LS93N
See NS Package Number M14A or N14A

LOGIC DIAGRAMS

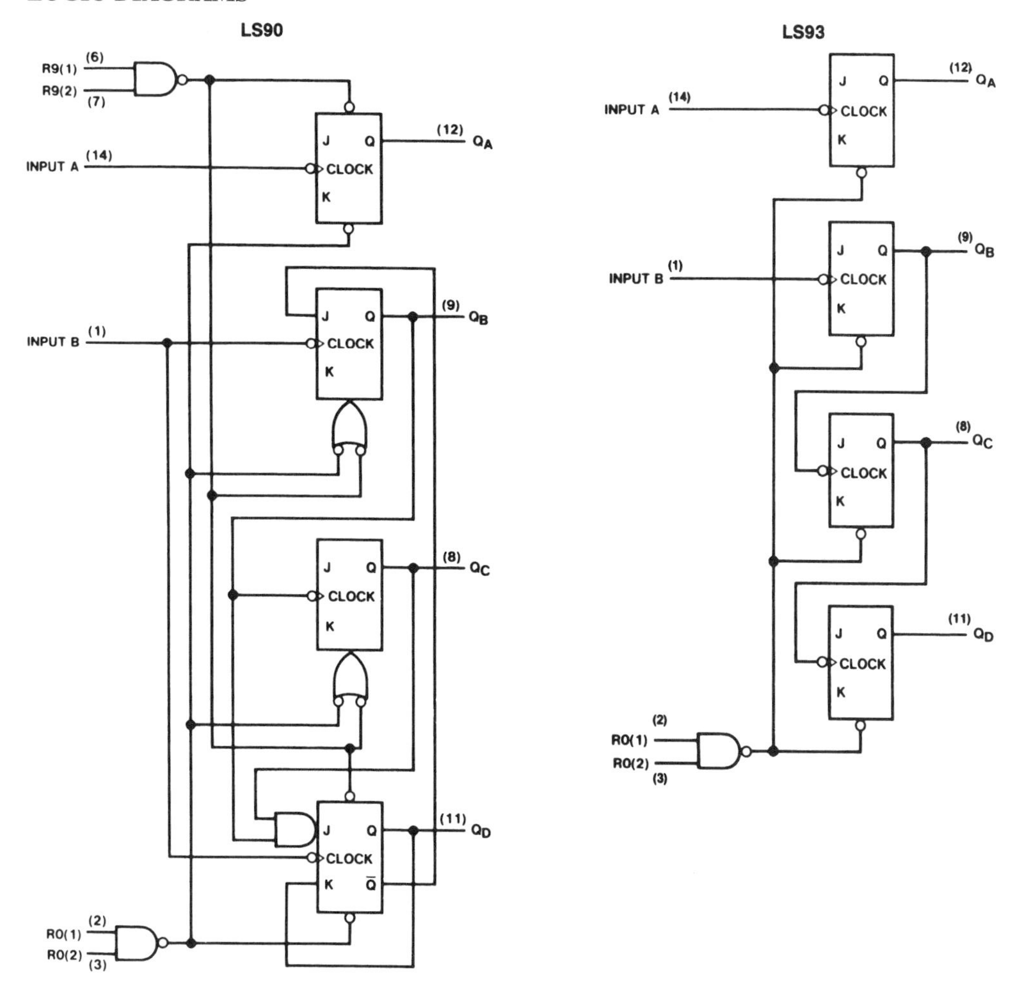

The J and K inputs shown without connection are for reference only and are functionally at a high level.

National 54LS160A/DM74LS160A, 54LS162A/DM74LS162A
Synchronous Presettable BCD Decade Counters

ABSOLUTE MAXIMUM RATINGS

If Military/Aerospace specified devices are required, please contact the National Semiconductor Sales Office/Distributors for availability and specifications.

Supply Voltage	7V
Input Voltage	7V
Operating Free Air Temperature Range	
54LS	−55°C to +125°C
DM74LS	0°C to +70°C
Storage Temperature Range	−65°C to +150°C

Note: *The "Absolute Maximum Ratings" are those values beyond which the safety of the device cannot be guaranteed. The device should not be operated at these limits. The parametric values defined in the "Electrical Characteristics" table are not guaranteed at the absolute maximum ratings. The "Recommended Operating Conditions" table will define the conditions for actual device operation.*

RECOMMENDED OPERATING CONDITIONS

Symbol	Parameter	54LS160A/162A			DM74LS160A/162A			Units
		Min	Nom	Max	Min	Nom	Max	
V_{CC}	Supply Voltage	4.5	5	5.5	4.75	5	5.25	V
V_{IH}	High Level Input Voltage	2			2			V
V_{IL}	Low Level Input Voltage			0.7			0.8	V
I_{OH}	High Level Output Current			−0.4			−0.4	mA
I_{OL}	Low Level Output Current			4			8	mA
T_A	Free Air Operating Temperature	−55		125	0		70	°C
$t_s(H)$ $t_s(L)$	Setup Time, HIGH or LOW P_n to CP	20 20			20 20			ns
$t_h(H)$ $t_h(L)$	Hold Time, HIGH or LOW P_n to CP	0.0 0.0			0.0 0.0			ns
$t_s(H)$ $t_s(L)$	Setup Time, HIGH or LOW $\overline{PE}$ to CP	20 20			20 20			ns
$t_h(H)$ $t_h(L)$	Hold Time, HIGH or LOW $\overline{PE}$ to CP	0 0			0 0			ns
$t_s(H)$ $t_s(L)$	Setup Time, HIGH or LOW CEP, CET or $\overline{SR}$ to CP	20 20			20 20			ns
$t_h(H)$ $t_h(L)$	Hold Time, HIGH or LOW CEP, CET or $\overline{SR}$ to CP	0 0			0 0			ns
$t_w(H)$ $t_w(L)$	CP Pulse Width, HIGH or LOW	15 25			15 25			ns
$t_w(L)$	$\overline{MR}$ Pulse Width LOW ('160)	15			15			ns
t_{rec}	Recovery Time $\overline{MR}$ to CP ('160)	20			20			ns

CONNECTION DIAGRAM

Dual-In-Line Package

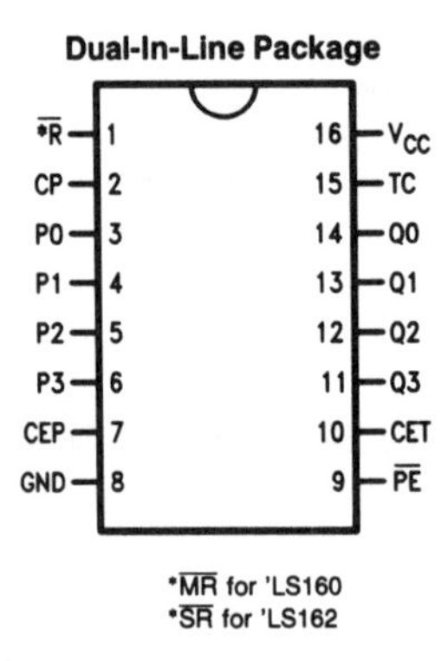

•$\overline{MR}$ for 'LS160
•$\overline{SR}$ for 'LS162

Order Number 54LS160ADMQB, 54LS160AFMQB, 54LS160ALMQB, 54LS162ADMQB, 54LS162AFMQB, 54LS162ALMQB, DM74LS160AM, DM74LS160AN, DM74LS162AM or DM74LS162AN
See NS Package Number E20A, J16A, M16A, N16E or W16A

Pin Names	Description
CEP	Count Enable Parallel Input
CET	Count Enable Trickle Input
CP	Clock Pulse Input (Active Rising Edge)
$\overline{MR}$ ('160)	Asynchronous Master Reset Input (Active LOW)
$\overline{SR}$ ('162)	Synchronous Reset Input (Active LOW)
P0–P3	Parallel Data Inputs
$\overline{PE}$	Parallel Enable Input (Active LOW)
Q0–Q3	Flip-Flop Outputs
TC	Terminal Count Output

LOGIC SYMBOL

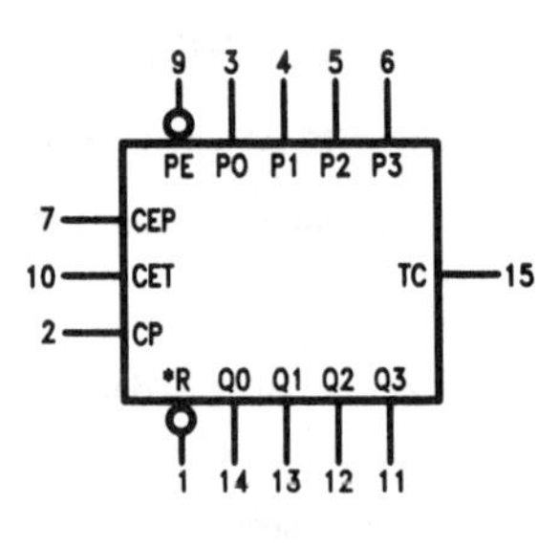

V_{CC} = Pin 16 •$\overline{MR}$ for 'LS160
GND = Pin 8 •$\overline{SR}$ for 'LS162

LOGIC DIAGRAMS

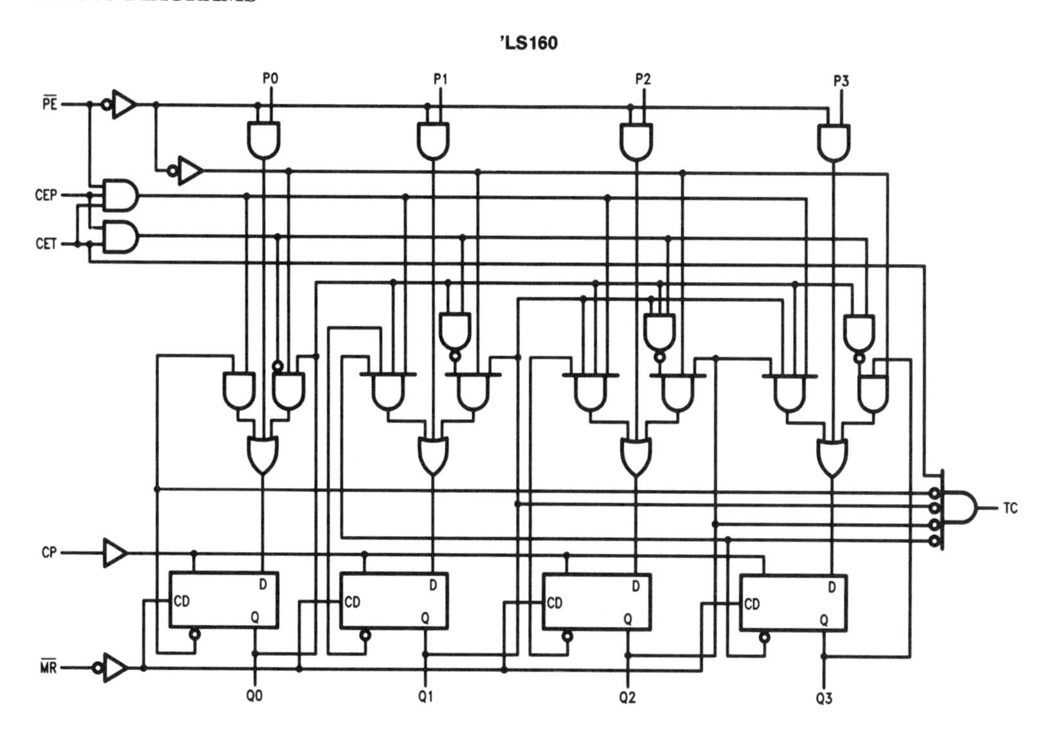

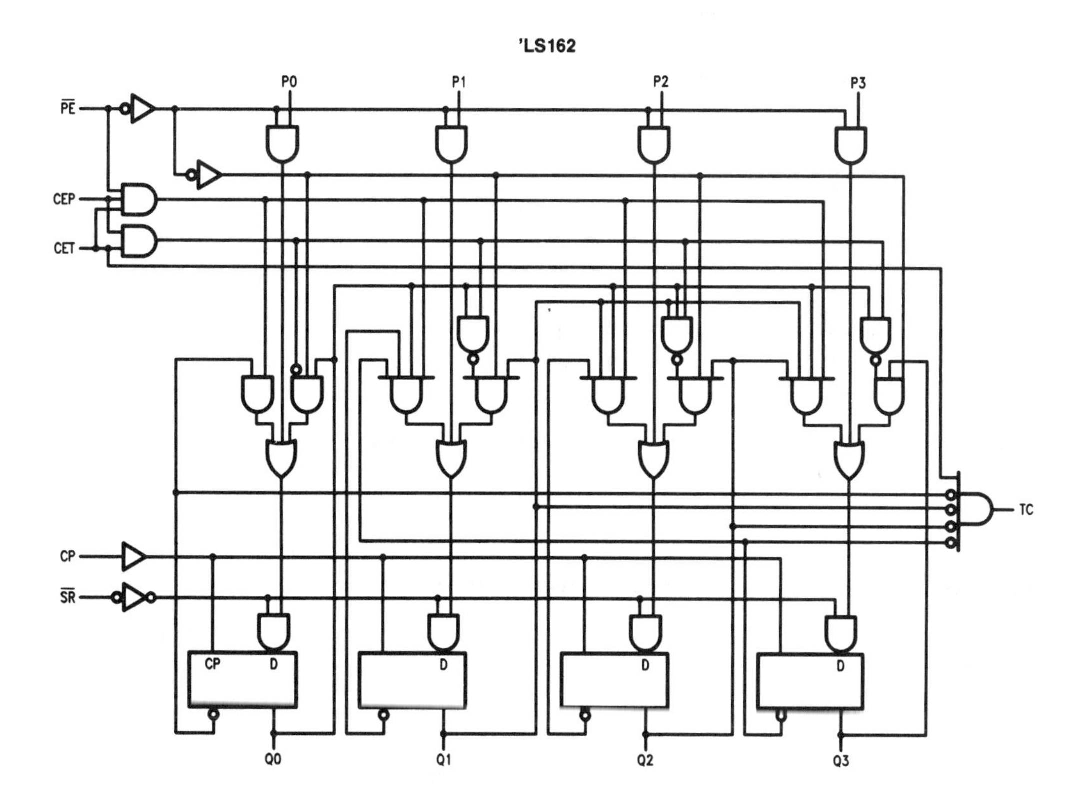

ELECTRICAL CHARACTERISTICS over recommended operating free air temperature range (unless otherwise noted)

Symbol	Parameter	Conditions		Min	Typ (Note 1)	Max	Units
V_I	Input Clamp Voltage	V_{CC} = Min, I_I = −18 mA				−1.5	V
V_{OH}	High Level Output Voltage	V_{CC} = Min, I_{OH} = Max, V_{IL} = Max	54LS	2.5			V
			DM74	2.7			
V_{OL}	Low Level Output Voltage	V_{CC} = Min, I_{OL} = Max, V_{IH} = Min	54LS			0.4	V
			DM74			0.5	
		I_{OL} = 4 mA, V_{CC} = Min	DM74			0.4	
I_I	Input Current @ Max Input Voltage	V_{CC} = Max, V_I = 10V Inputs $\overline{PE}$, CET Inputs				0.1 0.2	mA
I_{IH}	High Level Input Current	V_{CC} = Max, V_I = 2.7V Inputs $\overline{PE}$, CET Inputs				20 40	μA
I_{IL}	Low Level Input Current	V_{CC} = Max, V_I = 0.4V Inputs	54LS			−0.4	mA
			DM74			−1.6	
		$\overline{PE}$, CET Inputs				−0.8	mA
I_{OS}	Short Circuit Output Current	V_{CC} = Max (Note 2)	54LS	−20		−100	mA
			DM74	−20		−100	
I_{CCH}	Supply Current with Outputs HIGH	V_{CC} = Max, $\overline{PE}$ = GND CP = _/‾, Other Inputs = 4.5V				31	mA
I_{CCL}	Supply Current with Outputs LOW	V_{CC} = Max, V_{IN} = GND CP = _/‾				31	mA

SWITCHING CHARACTERISTICS V_{CC} = +5.0V, T_A = +25°C

Symbol	Parameter	R_L = 2 kΩ, C_L = 15 pF		Units
		Min	Max	
f_{max}	Maximum Clock Frequency	25		MHz
t_{PLH} t_{PHL}	Propagation Delay CP to TC		25 21	ns
t_{PLH} t_{PHL}	Propagation Delay CP to Q_n		20 27	ns
t_{PLH} t_{PHL}	Propagation Delay CP to Q_n		24 27	ns
t_{PLH} t_{PHL}	Propagation Delay CET to TC		14 14	ns
t_{PHL}	Propagation Delay $\overline{MR}$ to Q_n ('160)		28	ns

Note 1: All typicals are at V_{CC} = 5V, T_A = 25°C.

Note 2: Not more than one output should be shorted at a time, and the duration should not exceed one second.

National 54LS161A/DM54LS161A/DM74LS161A, 54LS163A/DM54LS163A/DM74LS163A

Synchronous 4-Bit Binary Counters

- Synchronously programmable
- Internal look-ahead for fast counting
- Carry output for n-bit cascading
- Synchronous counting
- Load control line
- Diode-clamped inputs
- Typical propagation time, clock to Q output 14 ns
- Typical clock frequency 32 MHz
- Typical power dissipation 93 mW

ABSOLUTE MAXIMUM RATINGS (Note)

If Military/Aerospace specified devices are required, please contact the National Semiconductor Sales Office/Distributors for availability and specifications.

Supply Voltage	7V
Input Voltage	7V
Operating Free Air Temperature Range	
DM54LS and 54LS	−55°C to +125°C
DM74LS	0°C to +70°C
Storage Temperature Range	−65°C to +150°C

Note: *The "Absolute Maximum Ratings" are those values beyond which the safety of the device cannot be guaranteed. The device should not be operated at these limits. The parametric values defined in the "Electrical Characteristics" table are not guaranteed at the absolute maximum ratings. The "Recommended Operating Conditions" table will define the conditions for actual device operation.*

RECOMMENDED OPERATING CONDITIONS

Symbol	Parameter		DM54LS161A			DM74LS161A			Units
			Min	Nom	Max	Min	Nom	Max	
V_{CC}	Supply Voltage		4.5	5	5.5	4.75	5	5.25	V
V_{IH}	High Level Input Voltage		2			2			V
V_{IL}	Low Level Input Voltage				0.7			0.8	V
I_{OH}	High Level Output Current				−0.4			−0.4	mA
I_{OL}	Low Level Output Current				4			8	mA
f_{CLK}	Clock Frequency (Note 1)		0		25	0		25	MHz
	Clock Frequency (Note 2)		0		20	0		20	MHz
t_W	Pulse Width (Note 1)	Clock	20	6		20	6		ns
		Clear	20	9		20	9		
	Pulse Width (Note 2)	Clock	25			25			ns
		Clear	25			25			
t_{SU}	Setup Time (Note 1)	Data	20	8		20	8		ns
		Enable P	25	17		25	17		
		Load	25	15		25	15		
	Setup Time (Note 2)	Data	20			20			ns
		Enable P	30			30			
		Load	30			30			
t_H	Hold Time (Note 1)	Data	0	−3		0	−3		ns
		Others	0	−3		0	−3		
	Hold Time (Note 2)	Data	5			5			ns
		Others	5			5			
t_{REL}	Clear Release Time (Note 1)		20			20			ns
	Clear Release Time (Note 2)		25			25			ns
T_A	Free Air Operating Temperature		−55		125	0		70	°C

Note 1: C_L = 15 pF, R_L = 2 kΩ, T_A = 25°C and V_{CC} = 5.5V.

Note 2: C_L = 50 pF, R_L = 2 kΩ, T_A = 25°C and V_{CC} = 5.5V.

CONNECTION DIAGRAM

Dual-In-Line Package

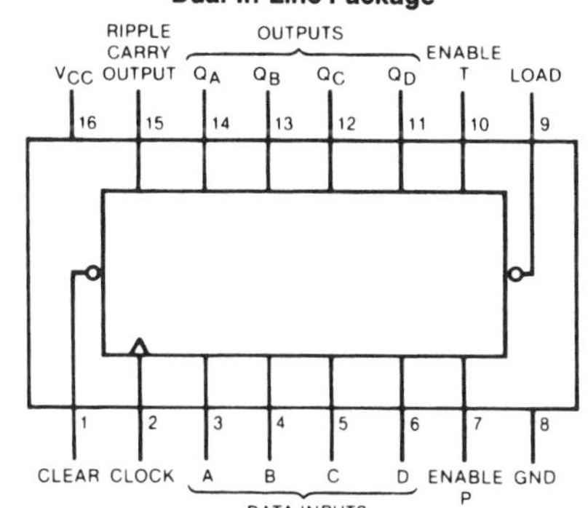

Order Numbers 54LS161ADMQB, 54LS161AFMQB, 54LS161ALMQB, 54LS163ADMQB, 54LS163AFMQB, 54LS163ALMQB, DM54LS161AJ, DM54LS161AW, DM54LS163AJ, DM54LS163AW, DM74LS161AM, DM74LS161AN, DM74LS163AM or DM74LS163AN See NS Package Number E20A, J16A, M16A, N16E or W16A

'LS161 ELECTRICAL CHARACTERISTICS

over recommended operating free air temperature range (unless otherwise noted)

Symbol	Parameter	Conditions		Min	Typ (Note 1)	Max	Units
V_I	Input Clamp Voltage	V_{CC} = Min, I_I = −18 mA				−1.5	V
V_{OH}	High Level Output Voltage	V_{CC} = Min, I_{OH} = Max, V_{IL} = Max, V_{IH} = Min	DM54	2.5	3.4		V
			DM74	2.7	3.4		
V_{OL}	Low Level Output Voltage	V_{CC} = Min, I_{OL} = Max, V_{IL} = Max, V_{IH} = Min	DM54		0.25	0.4	V
			DM74		0.35	0.5	
		I_{OL} = 4 mA, V_{CC} = Min	DM74		0.25	0.4	
I_I	Input Current @ Max Input Voltage	V_{CC} = Max, V_I = 7V	Enable T			0.2	mA
			Clock			0.2	
			Load			0.2	
			Others			0.1	
I_{IH}	High Level Input Current	V_{CC} = Max, V_I = 2.7V	Enable T			40	μA
			Clock			40	
			Load			40	
			Others			20	
I_{IL}	Low Level Input Current	V_{CC} = Max, V_I = 0.4V	Enable T			−0.8	mA
			Clock			−0.8	
			Load			−0.8	
			Others			−0.4	
I_{OS}	Short Circuit Output Current	V_{CC} = Max (Note 2)	DM54	−20		−100	mA
			DM74	−20		−100	
I_{CCH}	Supply Current with Outputs High	V_{CC} = Max (Note 3)			18	31	mA
I_{CCL}	Supply Current with Outputs Low	V_{CC} = Max (Note 4)			19	32	mA

Note 1: All typicals are at V_{CC} = 5V, T_A = 25°C.

Note 2: Not more than one output should be shorted at a time, and the duration should not exceed one second.

Note 3: I_{CCH} is measured with the load high, then again with the load low, with all other inputs high and all outputs open.

Note 4: I_{CCL} is measured with the clock input high, then again with the clock input low, with all other inputs low and all outputs open.

'LS161 SWITCHING CHARACTERISTICS

at V_{CC} = 5V and T_A = 25°C (See Section 1 for Test Waveforms and Output Load)

Symbol	Parameter	From (Input) To (Output)	R_L = 2 kΩ				Units
			C_L = 15 pF		C_L = 50 pF		
			Min	Max	Min	Max	
f_{MAX}	Maximum Clock Frequency		25		20		MHz
t_{PLH}	Propagation Delay Time Low to High Level Output	Clock to Ripple Carry		24		30	ns
t_{PHL}	Propagation Delay Time High to Low Level Output	Clock to Ripple Carry		30		38	ns
t_{PLH}	Propagation Delay Time Low to High Level Output	Clock to Any Q (Load High)		22		27	ns
t_{PHL}	Propagation Delay Time High to Low Level Output	Clock to Any Q (Load High)		27		38	ns
t_{PLH}	Propagation Delay Time Low to High Level Output	Clock to Any Q (Load Low)		24		30	ns
t_{PHL}	Propagation Delay Time High to Low Level Output	Clock to Any Q (Load Low)		29		38	ns
t_{PLH}	Propagation Delay Time Low to High Level Output	Enable T to Ripple Carry		18		27	ns
t_{PHL}	Propagation Delay Time High to Low Level Output	Enable T to Ripple Carry		15		27	ns
t_{PHL}	Propagation Delay Time High to Low Level Output	Clear to Any Q		35		45	ns

RECOMMENDED OPERATING CONDITIONS

Symbol	Parameter		DM54LS163A			DM74LS163A			Units
			Min	Nom	Max	Min	Nom	Max	
V_{CC}	Supply Voltage		4.5	5	5.5	4.75	5	5.25	V
V_{IH}	High Level Input Voltage		2			2			V
V_{IL}	Low Level Input Voltage				0.7			0.8	V
I_{OH}	High Level Output Current				−0.4			−0.4	mA
I_{OL}	Low Level Output Current				4			8	mA
f_{CLK}	Clock Frequency (Note 1)		0		25	0		25	MHz
	Clock Frequency (Note 2)		0		20	0		20	MHz
t_W	Pulse Width (Note 1)	Clock	20	6		20	6		ns
		Clear	20	9		20	9		
	Pulse Width (Note 2)	Clock	25			25			ns
		Clear	25			25			
t_{SU}	Setup Time (Note 1)	Data	20	8		20	8		ns
		Enable P	25	17		25	17		
		Load	25	15		25	15		
	Setup Time (Note 2)	Data	20			20			ns
		Enable P	30			30			
		Load	30			30			
t_H	Hold Time (Note 1)	Data	0	−3		0	−3		ns
		Others	0	−3		0	−3		
	Hold Time (Note 2)	Data	5			5			ns
		Others	5			5			
t_{REL}	Clear Release Time (Note 1)		20			20			ns
	Clear Release Time (Note 2)		25			25			ns
T_A	Free Air Operating Temperature		−55		125	0		70	°C

Note 1: C_L = 15 pF, R_L = 2 kΩ, T_A = 25°C and V_{CC} = 5V.

Note 2: C_L = 50 pF, R_L = 2 kΩ, T_A = 25°C and V_{CC} = 5V.

'LS163 SWITCHING CHARACTERISTICS

at V_{CC} = 5V and T_A = 25°C (See Section 1 for Test Waveforms and Output Load)

Symbol	Parameter	From (Input) To (Output)	R_L = 2 kΩ				Units
			C_L = 15 pF		C_L = 50 pF		
			Min	Max	Min	Max	
f_{MAX}	Maximum Clock Frequency		25		20		MHz
t_{PLH}	Propagation Delay Time Low to High Level Output	Clock to Ripple Carry		24		30	ns
t_{PHL}	Propagation Delay Time High to Low Level Output	Clock to Ripple Carry		30		38	ns
t_{PLH}	Propagation Delay Time Low to High Level Output	Clock to Any Q (Load High)		22		27	ns
t_{PHL}	Propagation Delay Time High to Low Level Output	Clock to Any Q (Load High)		27		38	ns
t_{PLH}	Propagation Delay Time Low to High Level Output	Clock to Any Q (Load Low)		24		30	ns
t_{PHL}	Propagation Delay Time High to Low Level Output	Clock to Any Q (Load Low)		29		38	ns
t_{PLH}	Propagation Delay Time Low to High Level Output	Enable T to Ripple Carry		18		27	ns
t_{PHL}	Propagation Delay Time High to Low Level Output	Enable T to Ripple Carry		15		27	ns
t_{PHL}	Propagation Delay Time High to Low Level Output	Clear to Any Q (Note 1)		35		45	ns

Note 1: The propagation delay clear to output is measured from the clock input transition.

'LS163 ELECTRICAL CHARACTERISTICS

over recommended operating free air temperature range (unless otherwise noted)

Symbol	Parameter	Conditions		Min	Typ (Note 1)	Max	Units
V_I	Input Clamp Voltage	V_{CC} = Min, I_I = −18 mA				−1.5	V
V_{OH}	High Level Output Voltage	V_{CC} = Min, I_{OH} = Max V_{IL} = Max, V_{IH} = Min	DM54	2.5	3.4		V
			DM74	2.7	3.4		
V_{OL}	Low Level Output Voltage	V_{CC} = Min, I_{OL} = Max V_{IL} = Max, V_{IH} = Min	DM54		0.25	0.4	V
			DM74		0.35	0.5	
		I_{OL} = 4 mA, V_{CC} = Min	DM74		0.25	0.4	
I_I	Input Current @ Max Input Voltage	V_{CC} = Max V_I = 7V	Enable T			0.2	mA
			Clock, Clear			0.2	
			Load			0.2	
			Others			0.1	
I_{IH}	High Level Input Current	V_{CC} = Max V_I = 2.7V	Enable T			40	μA
			Load			40	
			Clock, Clear			40	
			Others			20	
I_{IL}	Low Level Input Current	V_{CC} = Max V_I = 0.4V	Enable T			−0.8	mA
			Clock, Clear			−0.8	
			Load			−0.8	
			Others			−0.4	
I_{OS}	Short Circuit Output Current	V_{CC} = Max (Note 2)	DM54	−20		−100	mA
			DM74	−20		−100	
I_{CCH}	Supply Current with Outputs High	V_{CC} = Max (Note 3)			18	31	mA
I_{CCL}	Supply Current with Outputs Low	V_{CC} = Max (Note 4)			18	32	mA

Note 1: All typicals are at V_{CC} = 5V, T_A = 25°C.

Note 2: Not more than one output should be shorted at a time, and the duration should not exceed one second.

Note 3: I_{CCH} is measured with the load high, then again with the load low, with all other inputs high and all outputs open.

Note 4: I_{CCL} is measured with the clock input high, then again with the clock input low, with all other inputs low and all outputs open.

LOGIC DIAGRAM

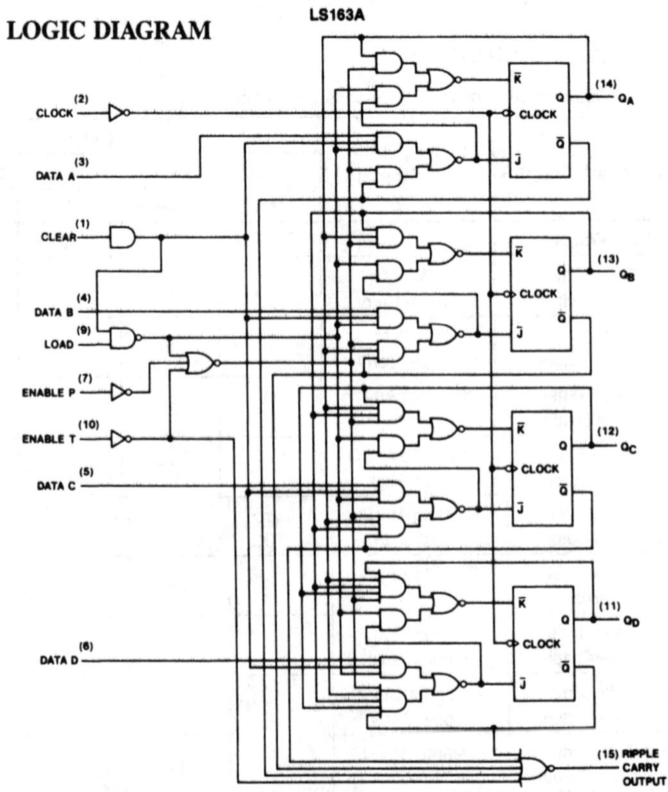

The LS161A is similar, however, the clear buffer is connected directly to the flip flops.

TIMING DIAGRAM

LS161A, LS163A Synchronous Binary Counters
Typical Clear, Preset, Count and Inhibit Sequences

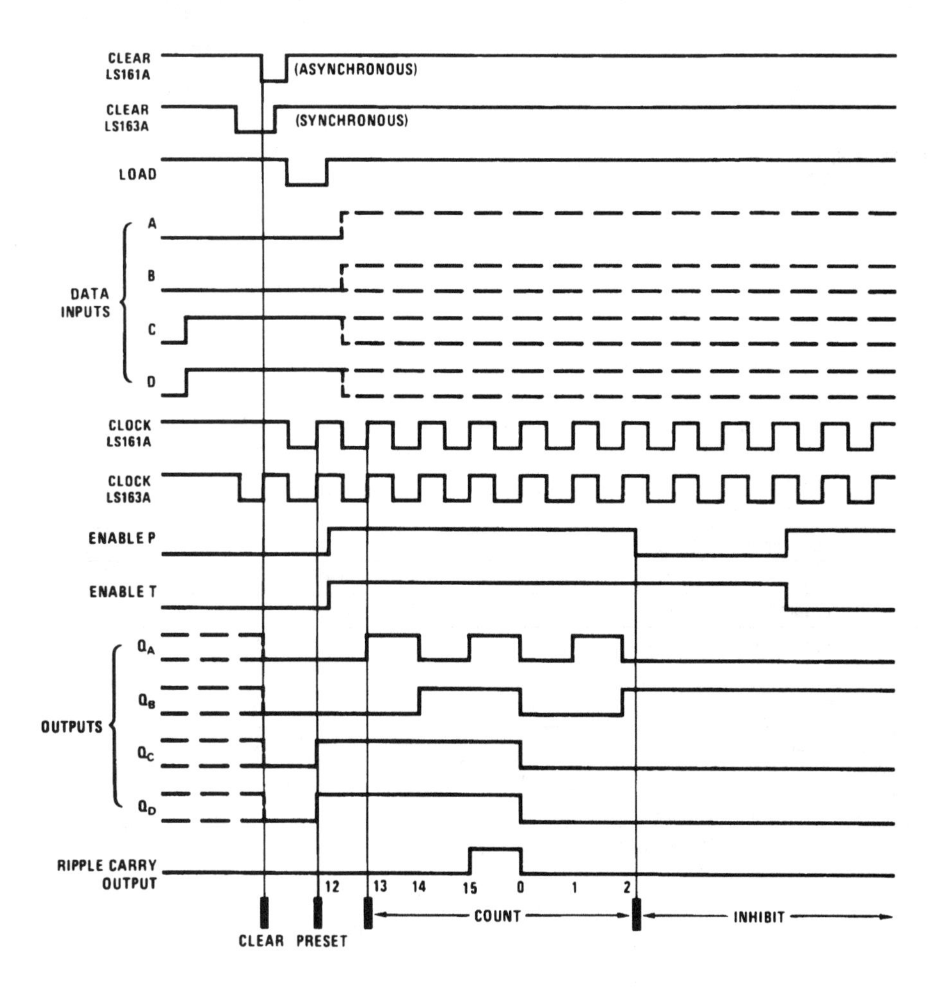

Sequence:

(1) Clear outputs to zero
(2) Preset to binary twelve
(3) Count to thirteen, fourteen, fifteen, zero, one, and two
(4) Inhibit

National DM54LS190/DM74LS190, DM54LS191/DM74LS191

Synchronous 4-Bit Up/Down Counters with Mode Control

- Counts 8-4-2-1 BCD or binary
- Single down/up count control line
- Count enable control input
- Ripple clock output for cascading
- Asynchronously presettable with load control
- Parallel outputs
- Cascadable for n-bit applications
- Average propagation delay 20 ns
- Typical clock frequency 25 MHz
- Typical power dissipation 100 mW

ABSOLUTE MAXIMUM RATINGS (Note)

If Military/Aerospace specified devices are required, please contact the National Semiconductor Sales Office/Distributors for availability and specifications.

Supply Voltage	7V
Input Voltage	7V
Operating Free Air Temperature Range	
DM54LS	−55°C to +125°C
DM74LS	0°C to +70°C
Storage Temperature Range	−65°C to +150°C

Note: *The "Absolute Maximum Ratings" are those values beyond which the safety of the device cannot be guaranteed. The device should not be operated at these limits. The parametric values defined in the "Electrical Characteristics" table are not guaranteed at the absolute maximum ratings. The "Recommended Operating Conditions" table will define the conditions for actual device operation.*

RECOMMENDED OPERATING CONDITIONS

Symbol	Parameter		DM54LS190, LS191			DM74LS190, LS191			Units
			Min	Nom	Max	Min	Nom	Max	
V_{CC}	Supply Voltage		4.5	5	5.5	4.75	5	5.25	V
V_{IH}	High Level Input Voltage		2			2			V
V_{IL}	Low Level Input Voltage				0.7			0.8	V
I_{OH}	High Level Output Current				−0.4			−0.4	mA
I_{OL}	Low Level Output Current				4			8	mA
f_{CLK}	Clock Frequency (Note 4)		0		20	0		20	MHz
t_W	Pulse Width (Note 4)	Clock	25			25			ns
		Load	35			35			ns
t_{SU}	Data Setup Time (Note 4)		20			20			ns
t_H	Data Hold Time (Note 4)		0			0			ns
t_{EN}	Enable Time to Clock (Note 4)		30			30			ns
T_A	Free Air Operating Temperature		−55		125	0		70	°C

'LS190 AND 'LS191 ELECTRICAL CHARACTERISTICS

over recommended operating free air temperature range (unless otherwise noted)

Symbol	Parameter	Conditions		Min	Typ (Note 1)	Max	Units
V_I	Input Clamp Voltage	V_{CC} = Min, I_I = −18 mA				−1.5	V
V_{OH}	High Level Output Voltage	V_{CC} = Min, I_{OH} = Max V_{IL} = Max, V_{IH} = Min	DM54	2.5	3.4		V
			DM74	2.7	3.4		
V_{OL}	Low Level Output Voltage	V_{CC} = Min, I_{OL} = Max V_{IL} = Max, V_{IH} = Min	DM54		0.25	0.4	V
			DM74		0.35	0.5	
		I_{OL} = 4 mA, V_{CC} = Min	DM74		0.25	0.4	
I_I	Input Current @ Max Input Voltage	V_{CC} = Max V_I = 7V	Enable			0.3	mA
			Others			0.1	
I_{IH}	High Level Input Current	V_{CC} = Max V_I = 2.7V	Enable			60	μA
			Others			20	
I_{IL}	Low Level Input Current	V_{CC} = Max V_I = 0.4V	Enable			−1.08	mA
			Others			−0.4	
I_{OS}	Short Circuit Output Current	V_{CC} = Max (Note 2)	DM54	−20		−100	mA
			DM74	−20		−100	
I_{CC}	Supply Current	V_{CC} = Max (Note 3)			20	35	mA

Note 1: All typicals are at V_{CC} = 5V, T_A = 25°C.

Note 2: Not more than one output should be shorted at a time, and the duration should not exceed one second.

Note 3: I_{CC} is measured with all inputs grounded and all outputs open.

Note 4: T_A = 25°C and V_{CC} = 5V.

'LS190 AND 'LS191 SWITCHING CHARACTERISTICS

at V_{CC} = 5V and T_A = 25°C (See Section 1 for Test Waveforms and Output Load)

Symbol	Parameter	From (Input) To (Output)	$R_L = 2\ k\Omega$				Units
			C_L = 15 pF		C_L = 50 pF		
			Min	Max	Min	Max	
f_{MAX}	Maximum Clock Frequency		20		20		MHz
t_{PLH}	Propagation Delay Time Low to High Level Output	Load to Any Q		33		43	ns
t_{PHL}	Propagation Delay Time High to Low Level Output	Load to Any Q		50		59	ns
t_{PLH}	Propagation Delay Time Low to High Level Output	Data to Any Q		22		26	ns
t_{PHL}	Propagation Delay Time High to Low Level Output	Data to Any Q		50		62	ns
t_{PLH}	Propagation Delay Time Low to High Level Output	Clock to Ripple Clock		20		24	ns
t_{PHL}	Propagation Delay Time High to Low Level Output	Clock to Ripple Clock		24		33	ns
t_{PLH}	Propagation Delay Time Low to High Level Output	Clock to Any Q		24		29	ns
t_{PHL}	Propagation Delay Time High to Low Level Output	Clock to Any Q		36		45	ns
t_{PLH}	Propagation Delay Time Low to High Level Output	Clock to Max/Min		42		47	ns
t_{PHL}	Propagation Delay Time High to Low Level Output	Clock to Max/Min		52		65	ns
t_{PLH}	Propagation Delay Time Low to High Level Output	Up/Down to Ripple Clock		45		50	ns
t_{PHL}	Propagation Delay Time High to Low Level Output	Up/Down to Ripple Clock		45		54	ns
t_{PLH}	Propagation Delay Time Low to High Level Output	Down/Up to Max/Min		33		36	ns
t_{PHL}	Propagation Delay Time High to Low Level Output	Down/Up to Max/Min		33		42	ns
t_{PLH}	Propagation Delay Time Low to High Level Output	Enable to Ripple Clock		33		36	ns
t_{PHL}	Propagation Delay Time High to Low Level Output	Enable to Ripple Clock		33		42	ns

CONNECTION DIAGRAM

Dual-In-Line-Package

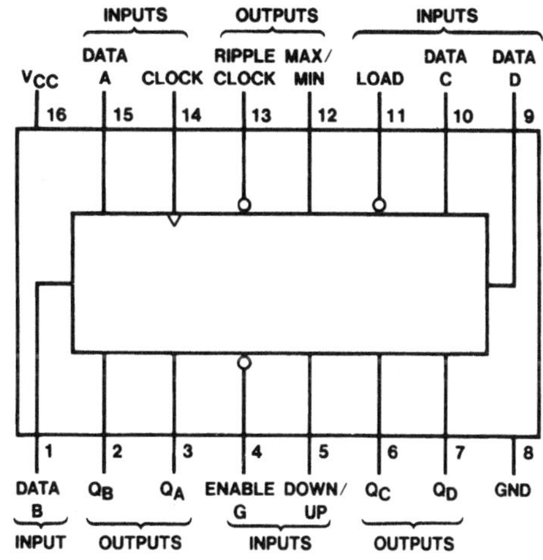

Order Number DM54LS190J, DM54LS191J, DM54LS190W, DM54LS191W, DM74LS190M, DM74LS191M, DM74LS190N, or DM74LS191N
See NS Package Number
J16A, M16A, N16A or W16A

LOGIC DIAGRAMS

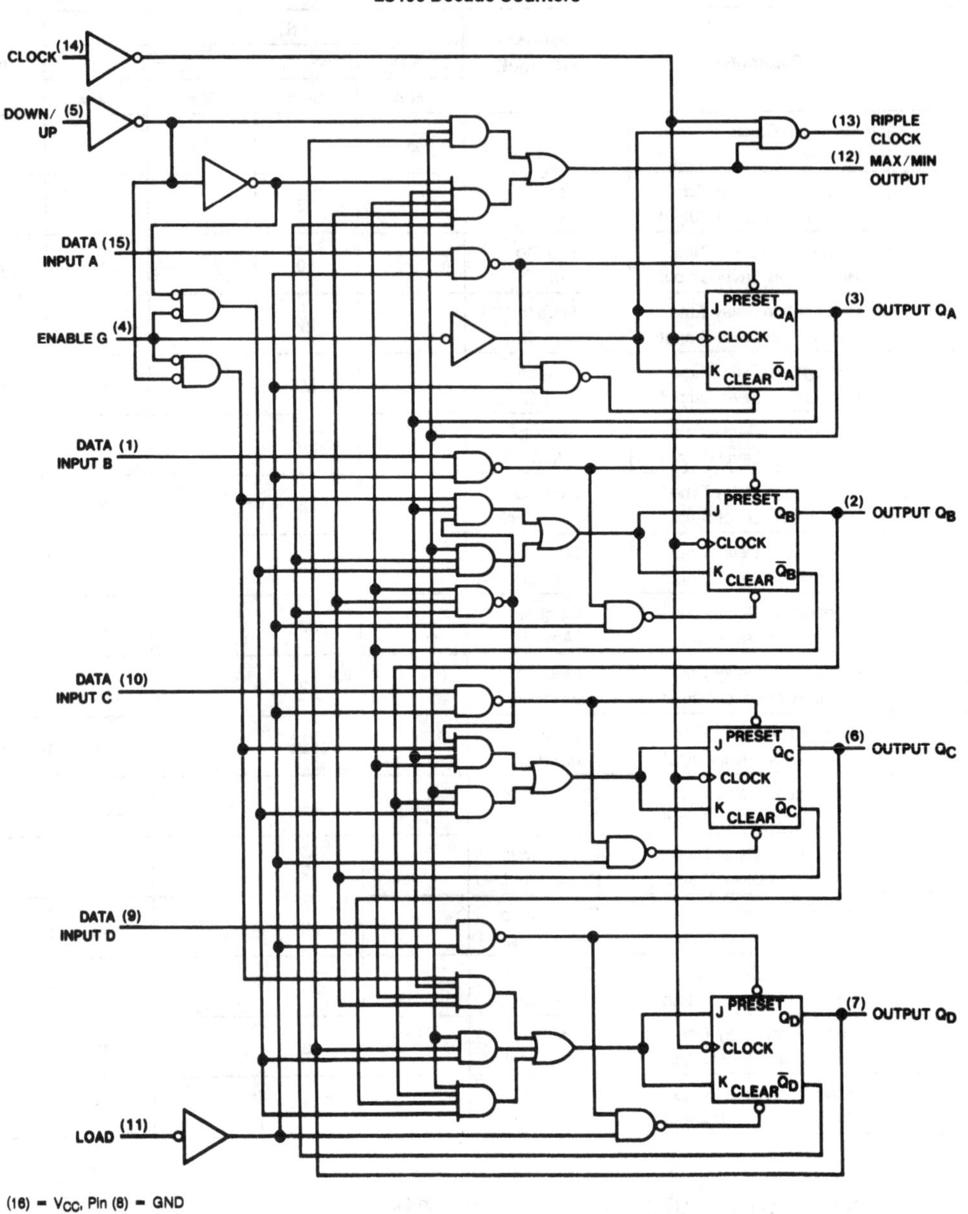

LOGIC DIAGRAMS (Continued)

LS191 Binary Counters

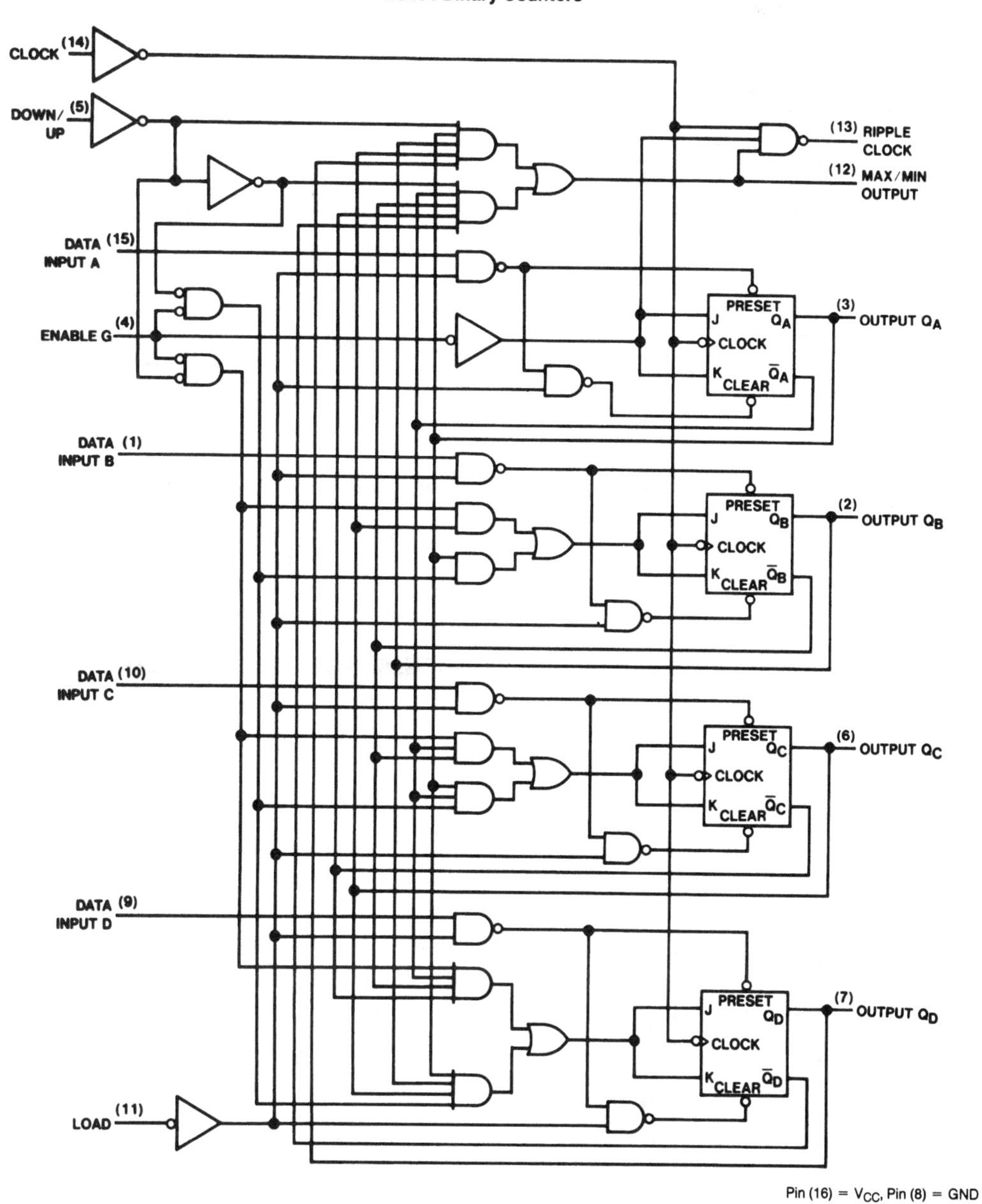

Pin (16) = V_{CC}, Pin (8) = GND

TIMING DIAGRAMS

LS190 Decade Counters
Typical Load, Count, and Inhibit Sequences

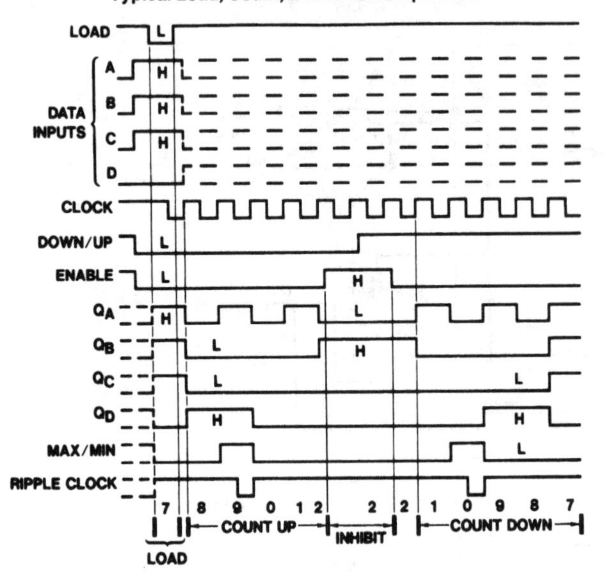

LS191 Binary Counters
Typical Load, Count, and Inhibit Sequences

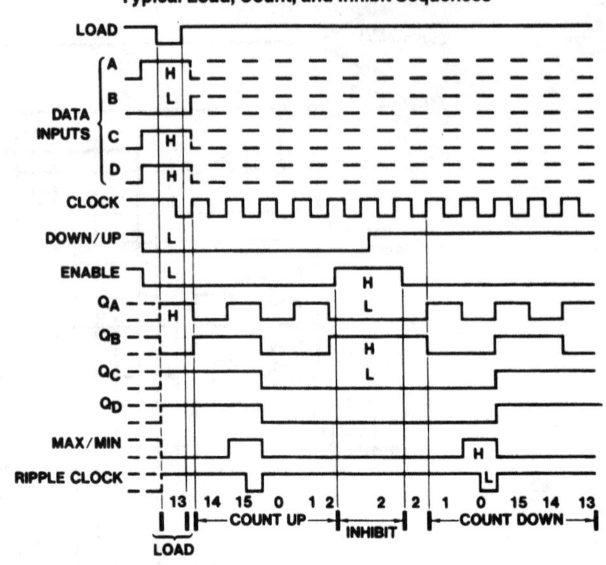

National 54LS192/DM74LS192
Up/Down Decade Counter with Separate Up/Down Clocks

CONNECTION DIAGRAM

Dual-In-Line Package

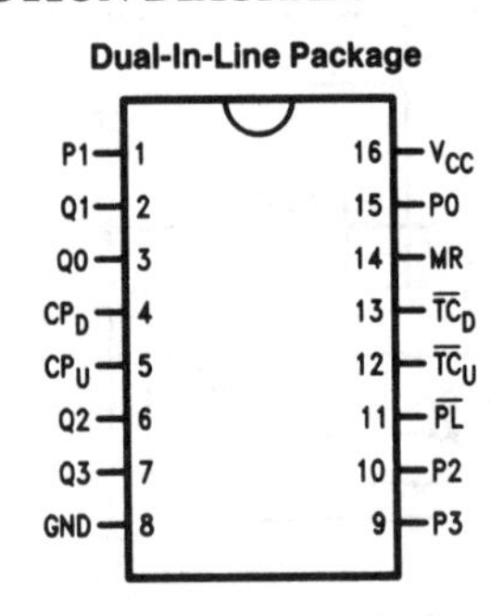

Order Number 54LS192DMQB, 54LS192FMQB, 54LS192LMQB, DM74LS192M or DM74LS192N
See NS Package Number E20A, J16A, M16A, N16E or W16A

LOGIC SYMBOL

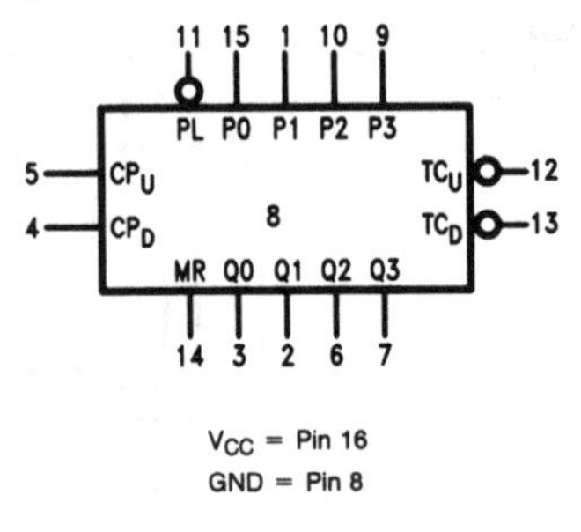

Pin Names	Description
CP_U	Count Up Clock Input (Active Rising Edge)
CP_D	Count Down Clock Input (Active Rising Edge)
MR	Asynchronous Master Reset Input (Active HIGH)
$\overline{PL}$	Asynchronous Parallel Load Input (Active LOW)
P0–P3	Parallel Data Inputs
Q0–Q3	Flip-Flop Outputs
$\overline{TC}_D$	Terminal Count Down (Borrow) Output (Active LOW)
$\overline{TC}_U$	Terminal Count Up (Carry) Output (Active LOW)

Mode Select Table

MR	$\overline{PL}$	CP_U	CP_D	Mode
H	X	X	X	Reset (Asyn.)
L	L	X	X	Preset (Asyn.)
L	H	H	H	No Change
L	H	⟋	H	Count Up
L	H	H	⟋	Count Down

H = HIGH Voltage Level
L = LOW Voltage Level
X = Immaterial

ABSOLUTE MAXIMUM RATINGS (Note)

If Military/Aerospace specified devices are required, please contact the National Semiconductor Sales Office/Distributors for availability and specifications.

Supply Voltage	7V
Input Voltage	7V
Operating Free Air Temperature Range	
54LS	−55°C to +125°C
DM74LS	0°C to +70°C
Storage Temperature Range	−65°C to +150°C

Note: *The "Absolute Maximum Ratings" are those values beyond which the safety of the device cannot be guaranteed. The device should not be operated at these limits. The parametric values defined in the "Electrical Characteristics" table are not guaranteed at the absolute maximum ratings. The "Recommended Operating Conditions" table will define the conditions for actual device operation.*

RECOMMENDED OPERATING CONDITIONS

Symbol	Parameter	54LS192			DM74LS192			Units
		Min	Nom	Max	Min	Nom	Max	
V_{CC}	Supply Voltage	4.5	5	5.5	4.75	5	5.25	V
V_{IH}	High Level Input Voltage	2			2			V
V_{IL}	Low Level Input Voltage			0.7			0.8	V
I_{OH}	High Level Output Voltage			−0.4			−0.4	mA
I_{OL}	Low Level Output Current			4			8	mA
T_A	Free Air Operating Temperature	−55		125	0		70	°C
t_s (H) t_s (L)	Setup Time HIGH or LOW Pn to $\overline{PL}$	20 20			20 20			ns
t_h (H) t_h (L)	Hold Time HIGH or LOW Pn to $\overline{PL}$	3 3			3 3			ns
t_w (L)	CP Pulse Width LOW	17			17			ns
t_w (L)	$\overline{PL}$ Pulse Width LOW	20			20			ns
t_w (H)	MR Pulse Width HIGH	15			15			ns
t_{rec}	Recovery Time, MR to CP	3			3			ns
t_{rec}	Recovery Time, $\overline{PL}$ to CP	10			10			ns

ELECTRICAL CHARACTERISTICS over recommended operating free air temperature range (unless otherwise noted)

Symbol	Parameter	Conditions		Min	Typ (Note 1)	Max	Units
V_I	Input Clamp Voltage	V_{CC} = Min, I_I = −18 mA				−1.5	V
V_{OH}	High Level Output Voltage	V_{CC} = Min, I_{OH} = Max, V_{IL} = Max	54LS	2.5			V
			DM74	2.7			
V_{OL}	Low Level Output Voltage	V_{CC} = Min, I_{OL} = Max,	54LS			0.4	V
		V_{IH} = Min	DM74			0.5	
		I_{OL} = 4 mA, V_{CC} = Min	DM74			0.4	
I_I	Input Current @ Max Input Voltage	V_{CC} = Max, V_I = 10V				0.1	mA
I_{IH}	High Level Input Current	V_{CC} = Max, V_I = 2.7V				20	μA
I_{IL}	Low Level Input Current	V_{CC} = Max, V_I = 0.4V				−0.4	mA
I_{OS}	Short Circuit Output Current	V_{CC} = Max (Note 2)	54LS	−20		−100	mA
			DM74	−20		−100	
I_{CC}	Supply Current	V_{CC} = Max, MR, $\overline{PL}$ = GND Other Inputs = 4.5V				31	mA

Note 1: All typicals are at V_{CC} = 5V, T_A = 25°C.

Note 2: Not more than one output should be shorted at a time, and the duration should not exceed one second.

SWITCHING CHARACTERISTICS

V_{CC} = +0.5V, T_A = +25°C (See Section 1 for waveforms and load configurations)

Symbol	Parameter	R_L = 2k, C_L = 15 pF Min	R_L = 2k, C_L = 15 pF Max	Units
f_{max}	Maximum Count Frequency	30		MHz
t_{PLH} t_{PHL}	Propagation Delay CP_U or CP_D to Q_n		31 28	ns
t_{PLH} t_{PHL}	Propagation Delay CP_U to $\overline{TC}_U$		16 21	ns
t_{PLH} t_{PHL}	Propagation Delay CP_D to $\overline{TC}_D$		16 24	
t_{PLH} t_{PHL}	Propagation Delay P_n to Q_n		20 30	ns
t_{PLH} t_{PHL}	Propagation Delay $\overline{PL}$ to Q_n		32 30	ns
t_{PHL}	Propagation Delay, MR to Q_n		25	

LOGIC DIAGRAM

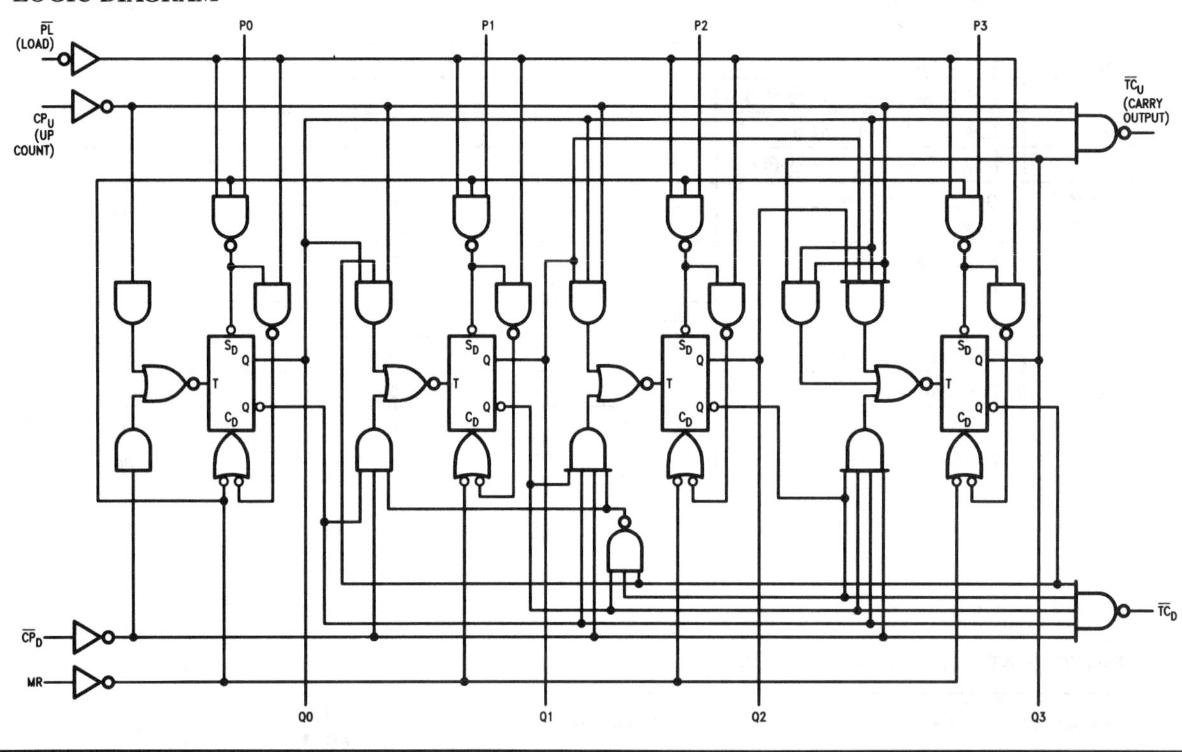

National 54LS193/DM54LS193/DM74LS193

Synchronous 4-Bit Up/Down Binary Counters with Dual Clock

- Fully independent clear input
- Synchronous operation
- Cascading circuitry provided internally
- Individual preset each flip-flop

CONNECTION DIAGRAM

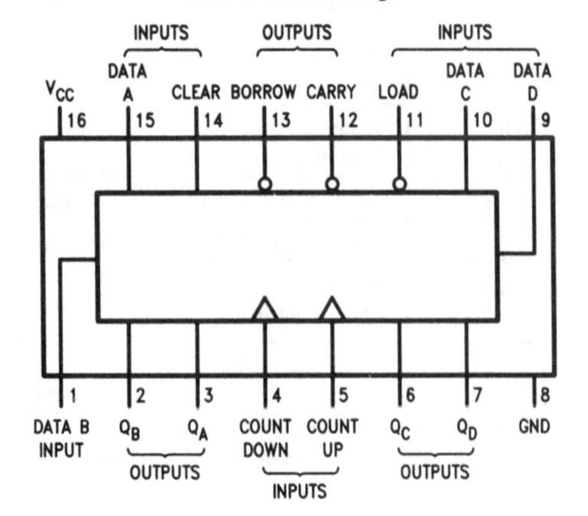

Order Number 54LS193DMQB, 54LS193FMQB, 54LS193LMQB, DM54LS193J, DM54LS193W, DM74LS193M or DM74LS193N
See NS Package Number E20A, J16A, M16A, N16E or W16A

ABSOLUTE MAXIMUM RATINGS (Note)

If Military/Aerospace specified devices are required, please contact the National Semiconductor Sales Office/Distributors for availability and specifications.

Supply Voltage	7V
Input Voltage	7V
Operating Free Air Temperature Range	
DM54LS and 54LS	−55°C to +125°C
DM74LS	0°C to +70°C
Storage Temperature Range	−65° C to +150° C

Note: *The "Absolute Maximum Ratings" are those values beyond which the safety of the device cannot be guaranteed. The device should not be operated at these limits. The parametric values defined in the "Electrical Characteristics" table are not guaranteed at the absolute maximum ratings. The "Recommended Operating Conditions" table will define the conditions for actual device operation.*

RECOMMENDED OPERATING CONDITIONS

Symbol	Parameter	DM54LS193			DM74LS193			Units
		Min	Nom	Max	Min	Nom	Max	
V_{CC}	Supply Voltage	4.5	5	5.5	4.75	5	5.25	V
V_{IH}	High Level Input Voltage	2			2			V
V_{IL}	Low Level Input Voltage			0.7			0.8	V
I_{OH}	High Level Output Current			−0.4			−0.4	mA
I_{OL}	Low Level Output Current			4			8	mA
f_{CLK}	Clock Frequency (Note 1)	0		25	0		25	MHz
	Clock Frequency (Note 2)	0		20	0		20	MHz
t_W	Pulse Width of Any Input (Note 6)	20			20			ns
t_{SU}	Data Setup Time (Note 6)	20			20			ns
t_H	Data Hold Time (Note 6)	0			0			ns
t_{REL}	Release Time (Note 6)	40			40			ns
T_A	Free Air Operating Temperature	−55		125	0		70	°C

ELECTRICAL CHARACTERISTICS over recommended operating free air temperature range (unless otherwise noted)

Symbol	Parameter	Conditions		Min	Typ (Note 3)	Max	Units
V_I	Input Clamp Voltage	V_{CC} = Min, I_I = −18 mA				−1.5	V
V_{OH}	High Level Output Voltage	V_{CC} = Min, I_{OH} = Max; V_{IL} = Max, V_{IH} = Min	DM54	2.5	3.4		V
			DM74	2.7	3.4		
V_{OL}	Low Level Output Voltage	V_{CC} = Min, I_{OL} = Max; V_{IL} = Max, V_{IH} = Min	DM54		0.25	0.4	V
			DM74		0.35	0.5	
		I_{OL} = 4 mA, V_{CC} = Min	DM74		0.25	0.4	
I_I	Input Current @ Max Input Voltage	V_{CC} = Max, V_I = 7V				0.1	mA
I_{IH}	High Level Input Current	V_{CC} = Max, V_I = 2.7V				20	µA
I_{IL}	Low Level Input Current	V_{CC} = Max, V_I = 0.4V				−0.4	mA
I_{OS}	Short Circuit Output Current	V_{CC} = Max (Note 4)	DM54	−20		−100	mA
			DM74	−20		−100	
I_{CC}	Supply Current	V_{CC} = Max (Note 5)			19	34	mA

Note 1: C_L = 15 pF, R_L = 2 kΩ, I_A = 25°C and V_{CC} = 5V.

Note 2: C_L = 50 pF, R_L = 2 kΩ, I_A = 25°C and V_{CC} = 5V.

Note 3: All typicals are at V_{CC} = 5V, T_A = 25°C.

Note 4: Not more than one output should be shorted at a time, and the duration should not exceed one second.

Note 5: I_{CC} is measured with all outputs open, CLEAR and LOAD inputs grounded, and all other inputs at 4.5V.

Note 6: T_A = 25°C and V_{CC} = 5V.

SWITCHING CHARACTERISTICS

Symbol	Parameter	From (Input) To (Output)	$R_L = 2\ k\Omega$				Units
			$C_L = 15\ pF$		$C_L = 50\ pF$		
			Min	Max	Min	Max	
f_{MAX}	Maximum Clock Frequency		25		20		MHz
t_{PLH}	Propagation Delay Time Low to High Level Output	Count Up to Carry		26		30	ns
t_{PHL}	Propagation Delay Time High to Low Level Output	Count Up to Carry		24		36	ns
t_{PLH}	Propagation Delay Time Low to High Level Output	Count Down to Borrow		24		29	ns
t_{PHL}	Propagation Delay Time High to Low Level Output	Count Down to Borrow		24		32	ns
t_{PLH}	Propagation Delay Time Low to High Level Output	Either Count to Any Q		38		45	ns
t_{PHL}	Propagation Delay Time High to Low Level Output	Either Count to Any Q		47		54	ns
t_{PLH}	Propagation Delay Time Low to High Level Output	Load to Any Q		40		41	ns
t_{PHL}	Propagation Delay Time High to Low Level Output	Load to Any Q		40		47	ns
t_{PHL}	Propagation Delay Time High to Low Level Output	Clear to Any Q		35		44	ns

TIMING DIAGRAMS

Typical Clear, Load, and Count Sequences

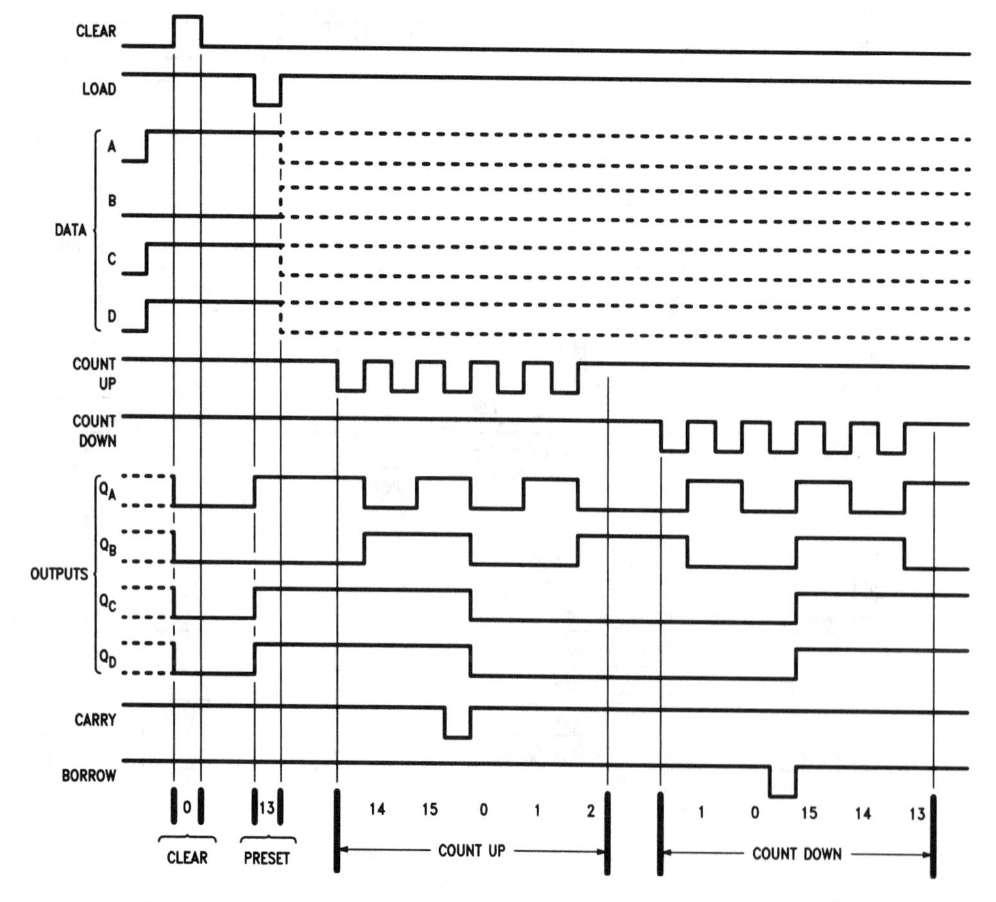

Note A: Clear overrides load, data, and count inputs.

Note B: When counting up, count-down input must be high; when counting down, count-up input must be high.

LOGIC DIAGRAM

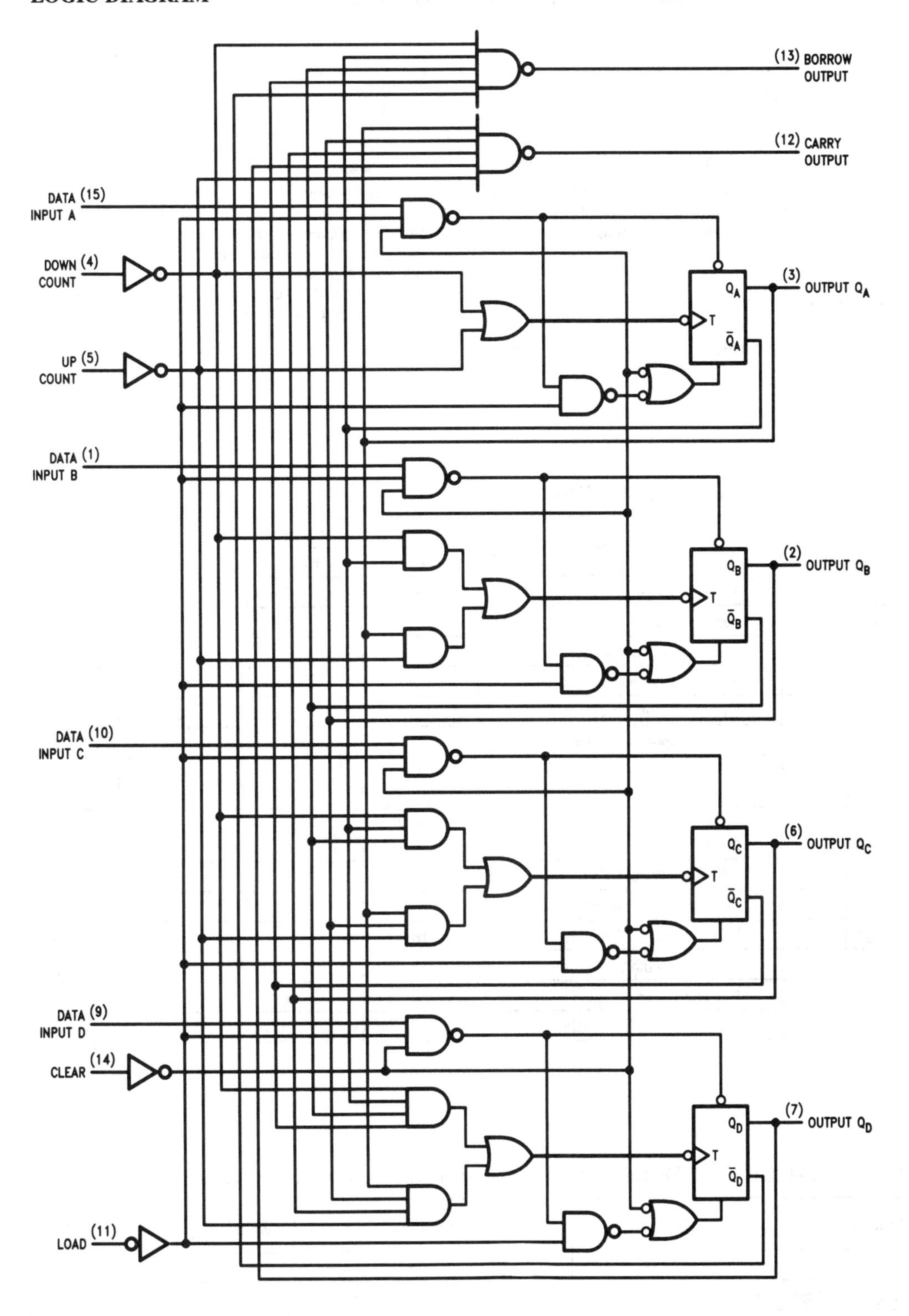

National DM74LS196 Presettable Decade Counter

ABSOLUTE MAXIMUM RATINGS (Note)

If Military/Aerospace specified devices are required, please contact the National Semiconductor Sales Office/Distributors for availability and specifications.

Supply Voltage	7V
Input Voltage	7V
Operating Free Air Temperature Range DM74LS	0°C to +70°C
Storage Temperature Range	−65°C to +150°C

Note: *The "Absolute Maximum Ratings" are those values beyond which the safety of the device cannot be guaranteed. The device should not be operated at these limits. The parametric values defined in the "Electrical Characteristics" table are not guaranteed at the absolute maximum ratings. The "Recommended Operating Conditions" table will define the conditions for actual device operation.*

RECOMMENDED OPERATING CONDITIONS

Symbol	Parameter	DM74LS196			Units
		Min	Nom	Max	
V_{CC}	Supply Voltage	4.75	5	5.25	V
V_{IH}	High Level Input Voltage	2			V
V_{IL}	Low Level Input Voltage			0.8	V
I_{OH}	High Level Output Current			−0.4	mA
I_{OL}	Low Level Output Current			8	mA
T_A	Free Air Operating Temperature	0		70	°C
t_s (H) t_s (L)	Setup Time HIGH or LOW Pn to $\overline{PL}$	8 12			ns
t_h (H) t_h (L)	Hold Time HIGH or LOW Pn to $\overline{PL}$	0 6			ns
t_w (H)	$\overline{CP}0$ Pulse Width HIGH	12			ns
t_w (H)	$\overline{CP}1$ Pulse Width HIGH	24			ns
t_w (L)	$\overline{PL}$ Pulse Width LOW	18			ns
t_w (L)	$\overline{MR}$ Pulse Width LOW	12			ns
t_{rec}	Recovery Time $\overline{PL}$ to $\overline{CP}n$	16			ns
t_{rec}	Recovery Time $\overline{MR}$ to $\overline{CP}n$	18			ns

ELECTRICAL CHARACTERISTICS over recommended operating free air temperature range (unless otherwise noted)

Symbol	Parameter	Conditions	Min	Typ (Note 1)	Max	Units
V_I	Input Clamp Voltage	V_{CC} = Min, I_I = −18 mA			−1.5	V
V_{OH}	High Level Output Voltage	V_{CC} = Min, I_{OH} = Max, V_{IL} = Max	2.7	3.4		V
V_{OL}	Low Level Output Voltage	V_{CC} = Min, I_{OL} = Max, V_{IH} = Min		0.35	0.5	V
		I_{OL} = 4 mA, V_{CC} = Min		0.25	0.4	
I_I	Input Current @ Max Input Voltage	V_{CC} = Max, V_I = 10V			0.1	mA
I_{IH}	High Level Input Current	V_{CC} = Max, V_I = 5.5V, $\overline{CP}1$			40	μA
I_{IL}	Low Level Input Current	V_{CC} = Max, V_I = 0.4V			−0.4	mA
I_{OS}	Short Circuit Output Current	V_{CC} = Max (Note 2)	−20		−100	mA
I_{CC}	Supply Current	V_{CC} = Max, V_{IN} = GND			20	mA

Note 1: All typicals are at V_{CC} = 5V, T_A = 25°C.

Note 2: Not more than one output should be shorted at a time, and the duration should not exceed one second.

SWITCHING CHARACTERISTICS

V_{CC} = +5.0V, T_A = +25°C (See Section 1 for waveforms and load configurations)

Symbol	Parameter	R_L = 2k, C_L = 15 pF		Units
		Min	Max	
f_{max}	Maximum Count Frequency at $\overline{CP0}$	45		MHz
f_{max}	Maximum Count Frequency at $\overline{CP1}$	22.5		MHz
t_{PLH} t_{PHL}	Propagation Delay $\overline{CP0}$ to Q0		15 15	ns
t_{PLH} t_{PHL}	Propagation Delay $\overline{CP1}$ to Q1		15 15	ns
t_{PLH} t_{PHL}	Propagation Delay $\overline{CP1}$ to Q2		34 34	ns
t_{PLH} t_{PHL}	Propagation Delay $\overline{CP1}$ to Q3		15 21	ns
t_{PLH} t_{PHL}	Propagation Delay Pn to Qn		25 35	ns
t_{PLH} t_{PHL}	Propagation Delay $\overline{PL}$ to Qn		31 37	ns
t_{PHL}	Propagation Delay $\overline{MR}$ to Qn		42	ns

CONNECTION DIAGRAM

Dual-In-Line Package

Pin	Name	Pin	Name
1	$\overline{PL}$	14	V_{CC}
2	Q2	13	$\overline{MR}$
3	P2	12	Q3
4	P0	11	P3
5	Q0	10	P1
6	$\overline{CP1}$	9	Q1
7	GND	8	$\overline{CP0}$

Order Number DM74LS196M or DM74LS196N
See NS Package Number M14A or N14A

LOGIC SYMBOL

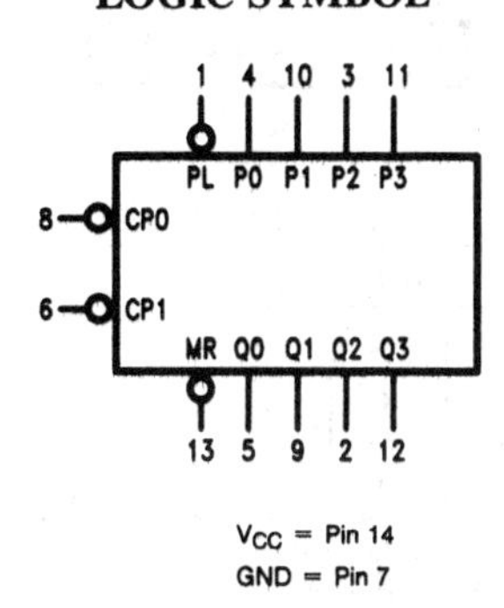

Pin Names	Description
$\overline{CP0}$	÷2 Section Clock Input (Active Falling Edge)
$\overline{CP1}$	÷5 Section Clock Input (Active Falling Edge)
$\overline{MR}$	Asynchronous Master Reset Input (Active LOW)
P0–P3	Parallel Data Inputs
$\overline{PL}$	Asynchronous Parallel Load Input (Active LOW)
Q0–Q3	Flip-Flop Outputs

LOGIC DIAGRAM

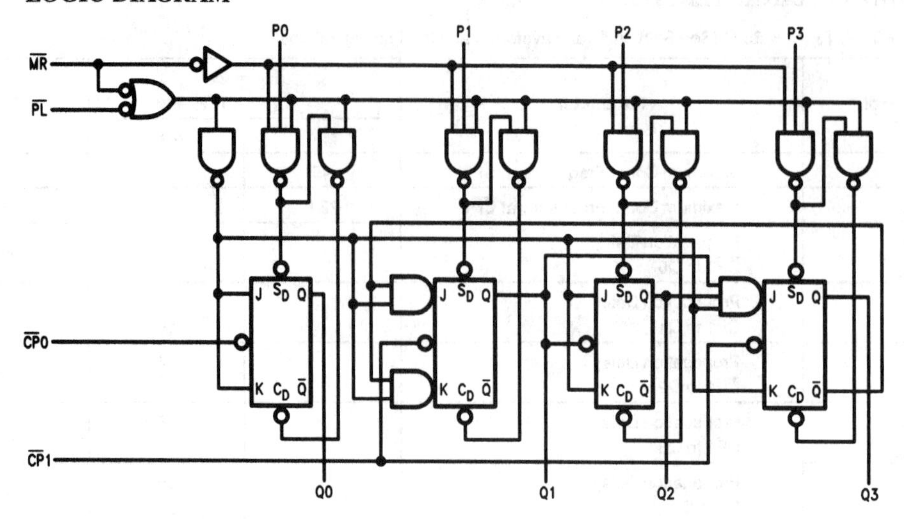

÷ 5 State Diagram

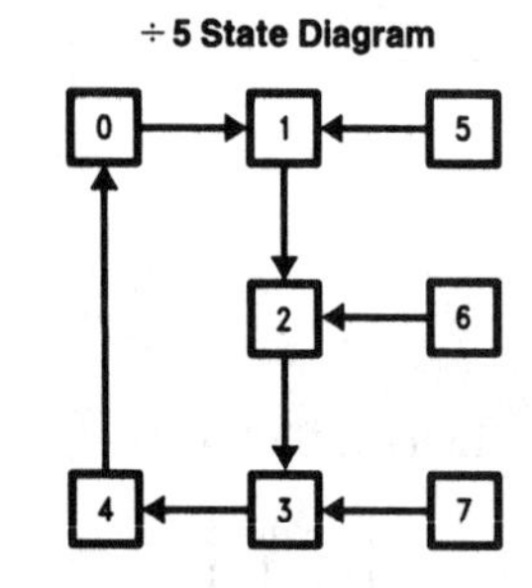

BCD State Diagram

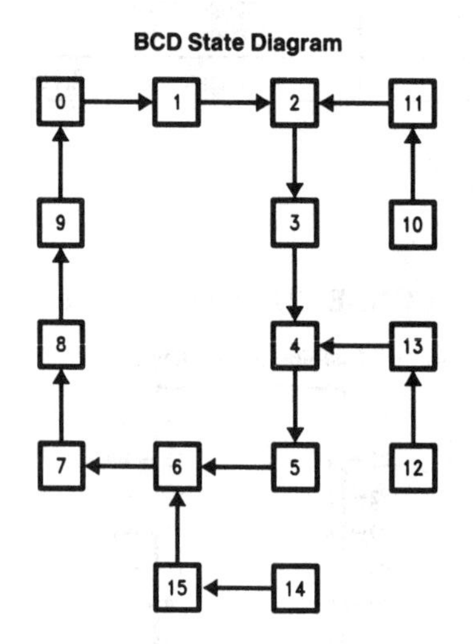

Mode Select Table

Inputs			Response
$\overline{MR}$	$\overline{PL}$	$\overline{CP}$	
L	X	X	Qn forced LOW
H	L	X	Pn → Qn
H	H	↘	Count Up

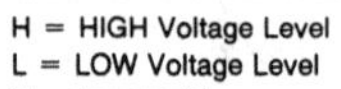

H = HIGH Voltage Level
L = LOW Voltage Level
X = Immaterial

National DM74LS197 Presettable Binary Counters

ABSOLUTE MAXIMUM RATINGS (Note)

If Military/Aerospace specified devices are required, please contact the National Semiconductor Sales Office/Distributors for availability and specifications.

Supply Voltage	7V
Input Voltage	7V
Operating Free Air Temperature Range	
DM74LS	0°C to +70°C
Storage Temperature Range	−65°C to +150°C

Note: *The "Absolute Maximum Ratings" are those values beyond which the safety of the device cannot be guaranteed. The device should not be operated at these limits. The parametric values defined in the "Electrical Characteristics" table are not guaranteed at the absolute maximum ratings. The "Recommended Operating Conditions" table will define the conditions for actual device operation.*

RECOMMENDED OPERATING CONDITIONS

Symbol	Parameter	74LS197			Units
		Min	Nom	Max	
V_{CC}	Supply Voltage	4.75	5	5.25	V
V_{IH}	High Level Input Voltage	2			V
V_{IL}	Low Level Input Voltage			0.8	V
I_{OH}	High Level Output Voltage			−0.4	mA
I_{OL}	Low Level Output Current			8	mA
T_A	Free Air Operating Temperature	0		70	°C

ELECTRICAL CHARACTERISTICS over recommended operating free air temperature range (unless otherwise noted)

Symbol	Parameter	Conditions	Min	Typ (Note 1)	Max	Units
V_I	Input Clamp Voltage	V_{CC} = Min, I_I = −18 mA			−1.5	V
V_{OH}	High Level Output Voltage	V_{CC} = Min, I_{OH} = Max, V_{IL} = Max	2.7	3.4		V
V_{OL}	Low Level Output Voltage	V_{CC} = Min, I_{OL} = Max, V_{IH} = Min		0.35	0.5	V
		I_{OL} = 4 mA, V_{CC} = Min		0.25	0.4	
I_I	Input Current @ Max Input Voltage	V_{CC} = Max, V_I = 10V			0.1	mA
I_{IH}	High Level Input Current	V_{CC} = Max, V_I = 2.7V			20	μA
I_{IL}	Low Level Input Current	V_{CC} = Max, V_I = 0.4V			−0.4	mA
I_{OS}	Short Circuit Output Current	V_{CC} = Max (Note 2)	−20		−100	mA
I_{CC}	Supply Current	V_{CC} = Max			27	mA

Note 1: All typicals are at V_{CC} = 5V, T_A = 25°C.

Note 2: Not more than one output should be shorted at a time, and the duration should not exceed one second.

SWITCHING CHARACTERISTICS

V_{CC} = +5.0V, T_A = +25°C (See Section 1 for Test Waveforms and Output Loads)

Symbol	Parameter	R_L = 2 kΩ, C_L = 15 pF		Units
		Min	Max	
f_{MAX}	Max CLK Frequency	55		MHz
t_{PLH} t_{PHL}	Propagation Delay $\overline{CP0}$ to Q0		15 15	ns
t_{PLH} t_{PHL}	Propagation Delay $\overline{CP1}$ to Q2		34 34	ns
t_{PLH} t_{PHL}	Propagation Delay P2 to Q2		27 44	ns
t_{PLH} t_{PHL}	Propagation Delay $\overline{PL}$ to Q2		39 45	ns
t_{PLH} t_{PHL}	Propagation Delay $\overline{CP1}$ to Q1		15 17	ns
t_{PLH} t_{PHL}	Propagation Delay $\overline{CP1}$ to Q3		55 63	ns
t_{PHL}	Propagation Delay $\overline{MR}$ to Q3		42	ns

LOGIC DIAGRAM

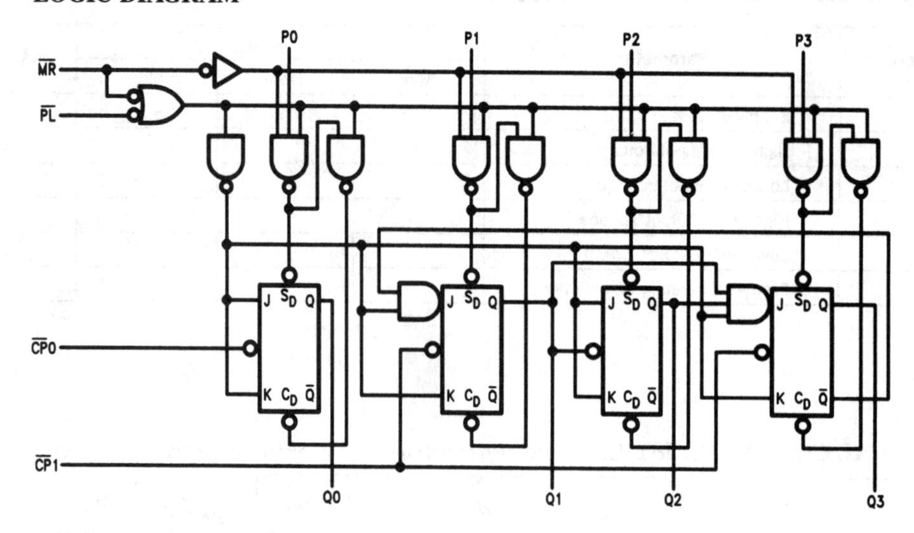

CONNECTION DIAGRAM

Dual-In-Line Package

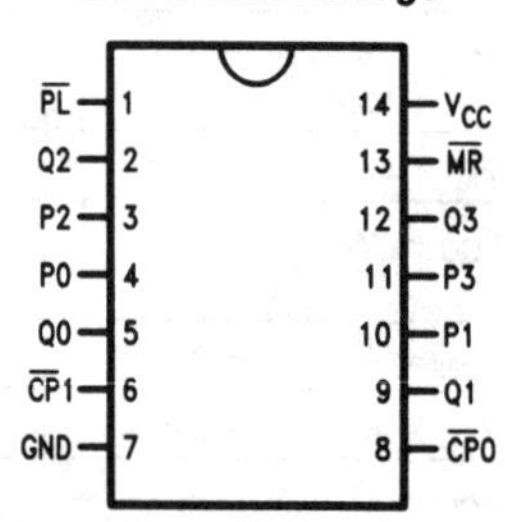

Order Number DM74LS197M or DM74LS197N
See NS Package Number M14A or N14A

Pin Names	Description
$\overline{CP}$0	÷2 Section Clock Input (Active Falling Edge)
$\overline{CP}$1	÷8 Section Clock Input (Active Falling Edge)
$\overline{MR}$	Asynchronous Master Reset Input (Active LOW)
P0–P3	Parallel Data Inputs
$\overline{PL}$	Asynchronous Parallel Load Input (Active LOW)
Q0	÷2 Section Output*
Q1–Q3	÷8 Section Outputs

*Q0 output is guaranteed to drive the full rated fan-out plus the $\overline{CP}$1 input.

Mode Select Table

Inputs			Response
$\overline{MR}$	$\overline{PL}$	$\overline{CP}$	
L	X	X	Qn Forced LOW
H	L	X	Pn → Qn
H	H	⌐_	Count Up

H = HIGH Voltage Level
L = LOW Voltage Level
X = Immaterial

National DM74LS290
4-Bit Decade Counter

ABSOLUTE MAXIMUM RATINGS

If Military/Aerospace specified devices are required, please contact the National Semiconductor Sales Office/Distributors for availability and specifications.

Supply Voltage	7V
Input Voltage	7V
Operating Free Air Temperature Range DM74LS	0°C to +70°C
Storage Temperature Range	−65°C to +150°C

Note: *The "Absolute Maximum Ratings" are those values beyond which the safety of the device cannot be guaranteed. The device should not be operated at these limits. The parametric values defined in the "Electrical Characteristics" table are not guaranteed at the absolute maximum ratings. The "Recommended Operating Conditions" table will define the conditions for actual device operation.*

RECOMMENDED OPERATING CONDITIONS

Symbol	Parameter		DM74LS290 Min	Nom	Max	Units
V_{CC}	Supply Voltage		4.75	5	5.25	V
V_{IH}	High Level Input Voltage		2			V
V_{IL}	Low Level Input Voltage				0.8	V
I_{OH}	High Level Output Current				−0.4	mA
I_{OL}	Low Level Output Current				8	mA
f_{CLK}	Clock Freq. (Note 1)	A to Q_A	0		32	MHz
		B to Q_B	0		16	
f_{CLK}	Clock Freq. (Note 2)	A to Q_A	0		20	MHz
		B to Q_B	0		10	
t_W	Pulse Width (Note 6)	A	15			ns
		B	30			
		Reset	15			
t_{REL}	Reset Release Time (Note 6)		25			ns
T_A	Free Air Operating Temperature		0		70	°C

SWITCHING CHARACTERISTICS at V_{CC} = 5V and T_A = 25°C (See Section 1 for Test Waveforms and Output Load)

Symbol	Parameter	From (Input) To (Output)	R_L = 2 kΩ				Units
			C_L = 15 pF		C_L = 50 pF		
			Min	Max	Min	Max	
f_{MAX}	Maximum Clock Frequency	A to Q_A	32		20		MHz
		B to Q_B	16		10		
t_{PLH}	Propagation Delay Time Low to High Level Output	A to Q_A		16		23	ns
t_{PHL}	Propagation Delay Time High to Low Level Output	A to Q_A		18		30	ns
t_{PLH}	Propagation Delay Time Low to High Level Output	A to Q_D		48		60	ns
t_{PHL}	Propagation Delay Time High to Low Level Output	A to Q_D		50		68	ns
t_{PLH}	Propagation Delay Time Low to High Level Output	B to Q_B		16		23	ns
t_{PHL}	Propagation Delay Time High to Low Level Output	B to Q_B		21		35	ns
t_{PLH}	Propagation Delay Time Low to High Level Output	B to Q_C		32		48	ns
t_{PHL}	Propagation Delay Time High to Low Level Output	B to Q_C		35		53	ns
t_{PLH}	Propagation Delay Time Low to High Level Output	B to Q_D		32		48	ns
t_{PHL}	Propagation Delay Time High to Low Level Output	B to Q_D		35		53	ns
t_{PLH}	Propagation Delay Time Low to High Level Output	SET-9 to Q_A, Q_D		30		38	ns
t_{PHL}	Propagation Delay Time High to Low Level Output	SET-9 to Q_B, Q_C		40		53	ns
t_{PHL}	Propagation Delay Time High to Low Level Output	SET-0 to Any Q		40		53	ns

ELECTRICAL CHARACTERISTICS over recommended operating free air temperature range (unless otherwise noted)

Symbol	Parameter	Conditions		Min	Typ (Note 3)	Max	Units
V_I	Input Clamp Voltage	V_{CC} = Min, I_I = −18 mA				−1.5	V
V_{OH}	High Level Output Voltage	V_{CC} = Min, I_{OH} = Max V_{IL} = Max, V_{IH} = Min		2.7	3.4		V
V_{OL}	Low Level Output Voltage	V_{CC} = Min, I_{OL} = Max V_{IL} = Max, V_{IH} = Min			0.35	0.5	V
		I_{OL} = 4 mA, V_{CC} = Min			0.25	0.4	
I_I	Input Current @ Max Input Voltage	V_{CC} = Max, V_I = 7V	Reset			0.1	mA
			A			0.2	
			B			0.4	
I_{IH}	High Level Input Current	V_{CC} = Max, V_I = 2.7V	Reset			20	μA
			A			40	
			B			80	
I_{IL}	Low Level Input Current	V_{CC} = Max V_I = 0.4V	Reset			−0.4	mA
			A			−2.4	
			B			−3.2	
I_{OS}	Short Circuit Output Current	V_{CC} = Max (Note 4)		−20		−100	mA
I_{CC}	Supply Current	V_{CC} = Max (Note 5)			9	15	mA

Note 1: C_L = 15 pF, R_L = 2 kΩ, T_A = 25°C and V_{CC} = 5V.

Note 2: C_L = 50 pF, R_L = 2 kΩ, T_A = 25°C and V_{CC} = 5V.

Note 3: All typicals are at V_{CC} = 5V, T_A = 25°C.

Note 4: Not more than one output should be shorted at a time, and the duration should not exceed one second.

Note 5: I_{CC} is measured with all outputs open, both RO inputs grounded following momentary connection to 4.5V and all other inputs grounded.

Note 6: T_A = 25°C and V_{CC} 5V.

FUNCTION TABLES

BCD Count Sequence (See Note A)

Count	Output Q_D	Q_C	Q_B	Q_A
0	L	L	L	L
1	L	L	L	H
2	L	L	H	L
3	L	L	H	H
4	L	H	L	L
5	L	H	L	H
6	L	H	H	L
7	L	H	H	H
8	H	L	L	L
9	H	L	L	H

Note A: Output Q_A is connected to input B for BCD count

H = High Logic Level

L = Low Logic Level

X = Either Low or High Logic Level

Bi-Quinary (5-2) (See Note B)

Count	Output Q_A	Q_B	Q_C	Q_D
0	L	L	L	L
1	L	L	L	H
2	L	L	H	L
3	L	L	H	H
4	L	H	L	L
5	H	L	L	L
6	H	L	L	H
7	H	L	H	L
8	H	L	H	H
9	H	H	L	L

Note B: Output Q_D is connected to input A for bi-quinary count.

Reset/Count Truth Table

Reset Inputs R0(1)	R0(2)	R9(1)	R9(2)	Outputs Q_D	Q_C	Q_B	Q_A
H	H	L	X	L	L	L	L
H	H	X	L	L	L	L	L
X	X	H	H	H	L	L	H
X	L	X	L	COUNT			
L	X	L	X	COUNT			
L	X	X	L	COUNT			
X	L	L	X	COUNT			

LOGIC DIAGRAM

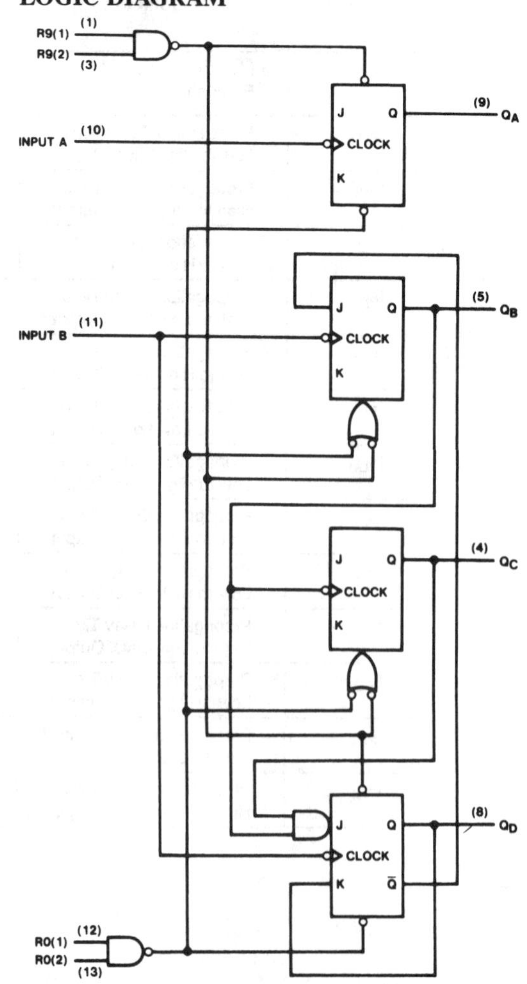

CONNECTION DIAGRAM

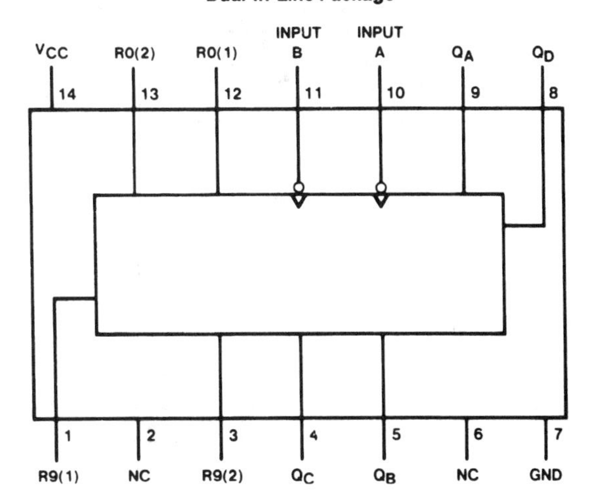

Order Number DM74LS290M or DM74LS290N
See NS Package Number M14A or N14A

National DM74LS293
4-Bit Binary Counter

ABSOLUTE MAXIMUM RATINGS (Note)

If Military/Aerospace specified devices are required, please contact the National Semiconductor Sales Office/Distributors for availability and specifications.

Supply Voltage	7V
Input Voltage	7V
Operating Free Air Temperature Range DM74LS	0°C to +70°C
Storage Temperature Range	−65°C to +150°C

Note: *The "Absolute Maximum Ratings" are those values beyond which the safety of the device cannot be guaranteed. The device should not be operated at these limits. The parametric values defined in the "Electrical Characteristics" table are not guaranteed at the absolute maximum ratings. The "Recommended Operating Conditions" table will define the conditions for actual device operation.*

RECOMMENDED OPERATING CONDITIONS

Symbol	Parameter		DM74LS293 Min	Nom	Max	Units
V_{CC}	Supply Voltage		4.75	5	5.25	V
V_{IH}	High Level Input Voltage		2			V
V_{IL}	Low Level Input Voltage				0.8	V
I_{OH}	High Level Output Current				−0.4	mA
I_{OL}	Low Level Output Current				8	mA
f_{CLK}	Clock Frequency (Note 1)	A to Q_A	0		32	MHz
		B to Q_B	0		16	
f_{CLK}	Clock Frequency (Note 2)	A to Q_A	0		20	MHz
		B to Q_B	0		10	
t_W	Pulse Width (Note 6)	A	15			ns
		B	30			
		Reset	15			
t_{REL}	Reset Release Time (Note 6)		25			ns
T_A	Free Air Operating Temperature		0		70	°C

ELECTRICAL CHARACTERISTICS over recommended operating free air temperature range (unless otherwise noted)

Symbol	Parameter	Conditions		Min	Typ (Note 3)	Max	Units
V_I	Input Clamp Voltage	V_{CC} = Min, I_I = −18 mA				−1.5	V
V_{OH}	High Level Output Voltage	V_{CC} = Min, I_{OH} = Max V_{IL} = Max, V_{IH} = Min		2.7	3.4		V
V_{OL}	Low Level Output Voltage	V_{CC} = Min, I_{OL} = Max V_{IL} = Max, V_{IH} = Min			0.35	0.5	V
		I_{OL} = 4 mA, V_{CC} = Min			0.25	0.4	
I_I	Input Current @ Max Input Voltage	V_{CC} = Max V_I = 7V	Reset			0.1	mA
			A			0.2	
			B			0.2	
I_{IH}	High Level Input Current	V_{CC} = Max V_I = 2.7V	Reset			20	μA
			A			40	
			B			40	
I_{IL}	Low Level Input Current	V_{CC} = Max V_I = 0.4V	Reset			−0.4	mA
			A			−2.4	
			B			−1.6	
I_{OS}	Short Circuit Output Current	V_{CC} = Max (Note 4)		−20		−100	mA
I_{CC}	Supply Current	V_{CC} = Max (Note 5)			9	15	mA

SWITCHING CHARACTERISTICS V_{CC} = 5V and T_A = 25°C (See Section 1 for Test Waveforms and Output Load)

Symbol	Parameter	From (Input) To (Output)	R_L = 2 kΩ				Units
			C_L = 15 pF		C_L = 50 pF		
			Min	Max	Min	Max	
t_{MAX}	Maximum Clock Frequency	A to Q_A	32		20		MHz
		B to Q_B	16		10		
t_{PLH}	Propagation Delay Time Low to High Level Output	A to Q_A		16		23	ns
t_{PHL}	Propagation Delay Time High to Low Level Output	A to Q_A		18		30	ns
t_{PLH}	Propagation Delay Time Low to High Level Output	A to Q_D		70		87	ns
t_{PHL}	Propagation Delay Time High to Low Level Output	A to Q_D		70		93	ns
t_{PLH}	Propagation Delay Time Low to High Level Output	B to Q_B		16		23	ns
t_{PHL}	Propagation Delay Time High to Low Level Output	B to Q_B		21		35	ns
t_{PLH}	Propagation Delay Time Low to High Level Output	B to Q_C		32		48	ns
t_{PHL}	Propagation Delay Time High to Low Level Output	B to Q_C		35		53	ns
t_{PLH}	Propagation Delay Time Low to High Level Output	B to Q_D		51		71	ns
t_{PHL}	Propagation Delay Time High to Low Level Output	B to Q_D		51		71	ns
t_{PHL}	Propagation Delay Time High to Low Level Output	SET-0 to Any Q		40		53	ns

Note 1: C_L = 15 pF, R_L = 2 kΩ, T_A = 25°C and V_{CC} = 5V.

Note 2: C_L = 50 pF, R_L = 2 kΩ, T_A = 25°C and V_{CC} = 5V.

Note 3: All typicals are at V_{CC} = 5V, T_A = 25°C.

Note 4: Not more than one output should be shorted at a time, and the duration should not exceed one second.

Note 5: I_{CC} is measured with all outputs open, both RO inputs grounded following momentary connection to 4.5V and all other inputs grounded.

Note 6: T_A = 25°C and V_{CC} = 5V.

CONNECTION DIAGRAM

Dual-In-Line Package

Pin	14	13	12	11	10	9	8
Signal	V_{CC}	R0(2)	R0(1)	INPUT B	INPUT A	Q_A	Q_D

Pin	1	2	3	4	5	6	7
Signal	NC	NC	NC	Q_C	Q_B	NC	GND

Order Number DM74LS293M or DM74LS293N
See NS Package Number M14A or N14A

FUNCTION TABLES

Count Sequence (See Note C)

Count	Outputs			
	Q_D	Q_C	Q_B	Q_A
0	L	L	L	L
1	L	L	L	H
2	L	L	H	L
3	L	L	H	H
4	L	H	L	L
5	L	H	L	H
6	L	H	H	L
7	L	H	H	H
8	H	L	L	L
9	H	L	L	H
10	H	L	H	L
11	H	L	H	H
12	H	H	L	L
13	H	H	L	H
14	H	H	H	L
15	H	H	H	H

Note C: Output Q_A is connected to input B.

Reset/Count Truth Table

Reset Inputs		Outputs			
R0(1)	R0(2)	Q_D	Q_C	Q_B	Q_A
H	H	L	L	L	L
L	X	COUNT			
X	L	COUNT			

H = High Level, L = Low Level, X = Don't Care.

LOGIC DIAGRAM

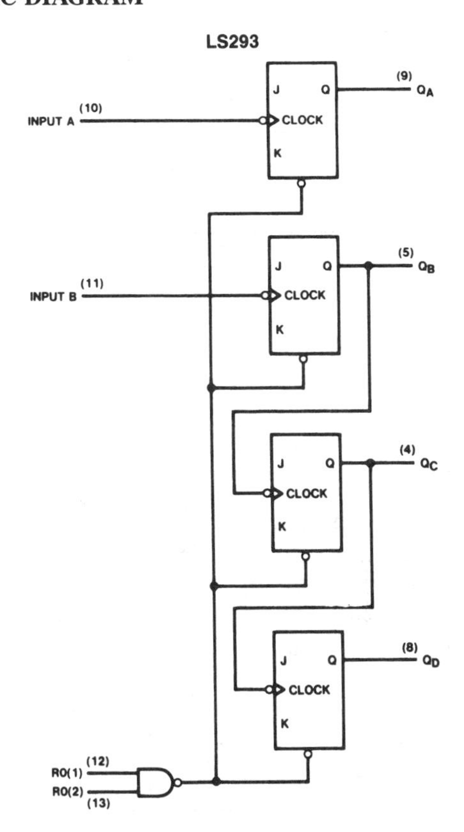

Note: The J and K inputs shown without connection are for reference only and are functionally at a high level.

National DM74LS390
Dual 4-Bit Decade Counter

- Dual version of the popular 'LS90
- Individual clocks for A and B flip-flops provide dual ÷ 2 and ÷ 5 counters
- Direct clear for each 4-bit counter
- Dual 4-bit version can significantly improve system densities by reducing counter package count by 50%
- Typical maximum count frequency: 35 MHz
- Buffered outputs reduce possibility of collector commutation

ABSOLUTE MAXIMUM RATINGS (Note)

If Military/Aerospace specified devices are required, please contact the National Semiconductor Sales Office/Distributors for availability and specifications.

Supply Voltage	7V
Input Voltage	
Clear	7V
A or B	5.5V
Operating Free Air Temperature Range	
DM74LS	0°C to +70°C
Storage Temperature Range	−65°C to +150°C

Note: *The "Absolute Maximum Ratings" are those values beyond which the safety of the device cannot be guaranteed. The device should not be operated at these limits. The parametric values defined in the "Electrical Characteristics" table are not guaranteed at the absolute maximum ratings. The "Recommended Operating Conditions" table will define the conditions for actual device operation.*

RECOMMENDED OPERATING CONDITIONS

Symbol	Parameter		DM74LS390 Min	Nom	Max	Units
V_{CC}	Supply Voltage		4.75	5	5.25	V
V_{IH}	High Level Input Voltage		2			V
V_{IL}	Low Level Input Voltage				0.8	V
I_{OH}	High Level Output Current				−0.4	mA
I_{OL}	Low Level Output Current				8	mA
f_{CLK}	Clock Frequency (Note 1)	A to Q_A	0		25	MHz
		B to Q_B	0		20	
f_{CLK}	Clock Frequency (Note 2)	A to Q_A	0		20	MHz
		B to Q_B	0		15	
t_W	Pulse Width (Note 1)	A	20			ns
		B	25			
		Clear High	20			
t_{REL}	Clear Release Time (Notes 3 & 4)		25↓			ns
T_A	Free Air Operating Temperature		0		70	°C

Note 1: C_L = 15 pF, R_L = 2 kΩ, T_A = 25°C and V_{CC} = 5V.

Note 2: C_L = 50 pF, R_L = 2 kΩ, T_A = 25°C and V_{CC} = 5V.

Note 3: The symbol (↓) indicates the falling edge of the clear pulse is used for reference.

Note 4: T_A = 25°C and V_{CC} = 5V.

ELECTRICAL CHARACTERISTICS over recommended operating free air temperature range (unless otherwise noted)

Symbol	Parameter	Conditions		Min	Typ (Note 1)	Max	Units
V_I	Input Clamp Voltage	V_{CC} = Min, I_I = −18 mA				−1.5	V
V_{OH}	High Level Output Voltage	V_{CC} = Min, I_{OH} = Max V_{IL} = Max, V_{IH} = Min		2.7	3.4		V
V_{OL}	Low Level Output Voltage	V_{CC} = Min, I_{OL} = Max V_{IL} = Max, V_{IH} = Min			0.35	0.5	V
		I_{OL} = 4 mA, V_{CC} = Min			0.25	0.4	
I_I	Input Current @ Max Input Voltage	V_{CC} = Max, V_I = 7V	Clear			0.1	mA
		V_{CC} = Max V_I = 5.5V	A			0.2	
			B			0.4	
I_{IH}	High Level Input Current	V_{CC} = Max V_I = 2.7V	Clear			20	μA
			A			40	
			B			80	
I_{IL}	Low Level Input Current	V_{CC} = Max, V_I = 0.4V	Clear			−0.4	mA
			A			−1.6	
			B			−2.4	
I_{OS}	Short Circuit Output Current	V_{CC} = Max (Note 2)	DM74	−20		−100	mA
I_{CC}	Supply Current	V_{CC} = Max (Note 3)			15	26	mA

Note 1: All typicals are at V_{CC} = 5V, T_A = 25°C.

Note 2: Not more than one output should be shorted at a time, and the duration should not exceed one second.

Note 3: I_{CC} is measured with all outputs open, both CLEAR inputs grounded following momentary connection to 4.5 and all other inputs grounded.

SWITCHING CHARACTERISTICS at V_{CC} = 5V and T_A = 25°C (See Section 1 for Test Waveforms and Output Load)

Symbol	Parameter	From (Input) To (Output)	$R_L = 2\,k\Omega$				Units
			C_L = 15 pF		C_L = 50 pF		
			Min	Max	Min	Max	
f_{MAX}	Maximum Clock Frequency	A to Q_A	25		20		MHz
		B to Q_B	20		15		
t_{PLH}	Propagation Delay Time Low to High Level Output	A to Q_A		20		24	ns
t_{PHL}	Propagation Delay Time High to Low Level Output	A to Q_A		20		30	ns
t_{PLH}	Propagation Delay Time Low to High Level Output	A to Q_C		60		81	ns
t_{PHL}	Propagation Delay Time High to Low Level Output	A to Q_C		60		81	ns
t_{PLH}	Propagation Delay Time Low to High Level Output	B to Q_B		21		27	ns
t_{PHL}	Propagation Delay Time High to Low Level Output	B to Q_B		21		33	ns
t_{PLH}	Propagation Delay Time Low to High Level Output	B to Q_C		39		51	ns
t_{PHL}	Propagation Delay Time High to Low Level Output	B to Q_C		39		54	ns
t_{PLH}	Propagation Delay Time Low to High Level Output	B to Q_D		21		27	ns
t_{PHL}	Propagation Delay Time High to Low Level Output	B to Q_D		21		33	ns
t_{PHL}	Propagation Delay Time High to Low Level Output	Clear to Any Q		39		45	ns

CONNECTION DIAGRAM

Dual-In-Line Package

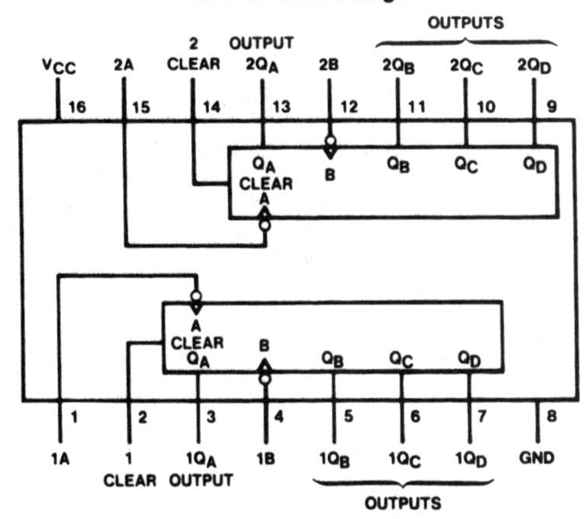

Order Number DM74LS390M or DM74LS390N
See NS Package Number M16A or N16E

FUNCTION TABLES

BCD Count Sequence (Each Counter) (See Note A)

Count	Outputs			
	Q_D	Q_C	Q_B	Q_A
0	L	L	L	L
1	L	L	L	H
2	L	L	H	L
3	L	L	H	H
4	L	H	L	L
5	L	H	L	H
6	L	H	H	L
7	L	H	H	H
8	H	L	L	L
9	H	L	L	H

Bi-Quinary (5-2) (Each Counter) (See Note B)

Count	Outputs			
	Q_A	Q_D	Q_C	Q_B
0	L	L	L	L
1	L	L	L	H
2	L	L	H	L
3	L	L	H	H
4	L	H	L	L
5	H	L	L	L
6	H	L	L	H
7	H	L	H	L
8	H	L	H	H
9	H	H	L	L

Note A: Output Q_A is connected to input B for BCD count.

Note B: Output Q_D is connected to input A for Bi-quinary count.

Note C: H = High Level, L = Low Level.

LOGIC DIAGRAM

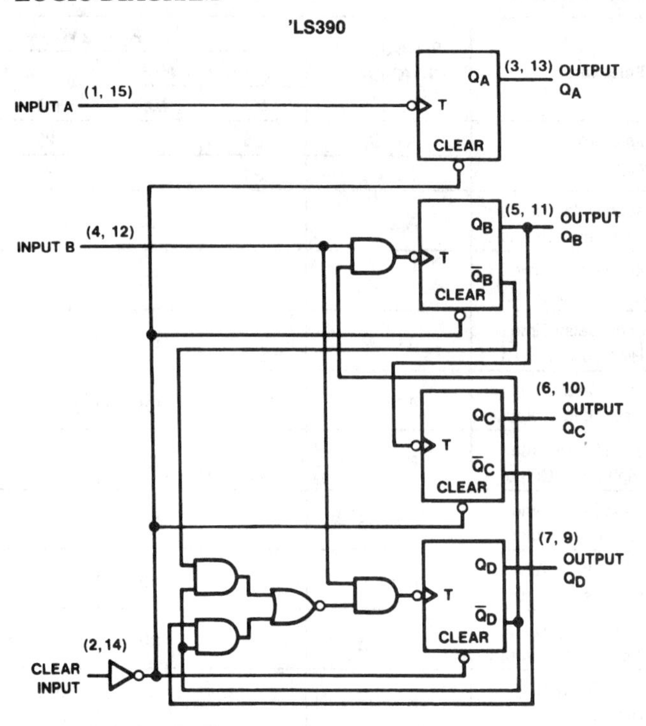

National DM74LS393
Dual 4-Bit Binary Counter

- Dual version of the popular 'LS93
- 'LS393 dual 4-bit binary counter with individual clocks
- Direct clear for each 4-bit counter
- Dual 4-bit versions can significantly improve system densities by reducing counter package count by 50%
- Typical maximum count frequency: 35 MHz
- Buffered outputs reduce possibility of collector commutation

ABSOLUTE MAXIMUM RATINGS (Note)

If Military/Aerospace specified devices are required, please contact the National Semiconductor Sales Office/Distributors for availability and specifications.

Supply Voltage	7V
Input Voltage	
Clear	7V
A	5.5V
Operating Free Air Temperature Range	
DM74LS	0°C to +70°C
Storage Temperature Range	−65°C to +150°C

Note: *The "Absolute Maximum Ratings" are those values beyond which the safety of the device cannot be guaranteed. The device should not be operated at these limits. The parametric values defined in the "Electrical Characteristics" table are not guaranteed at the absolute maximum ratings. The "Recommended Operating Conditions" table will define the conditions for actual device operation.*

RECOMMENDED OPERATING CONDITIONS

Symbol	Parameter		DM74LS393			Units
			Min	Nom	Max	
V_{CC}	Supply Voltage		4.75	5	5.25	V
V_{IH}	High Level Input Voltage		2			V
V_{IL}	Low Level Input Voltage				0.8	V
I_{OH}	High Level Output Current				−0.4	mA
I_{OL}	Low Level Output Current				8	mA
f_{CLK}	Clock Frequency (Note 1)		0		25	MHz
f_{CLK}	Clock Frequency (Note 2)		0		20	MHz
t_W	Pulse Width (Note 7)	A	20			ns
		Clear High	20			
t_{REL}	Clear Release Time (Notes 3 & 7)		25 ↓			ns
T_A	Free Air Operating Temperature		0		70	°C

ELECTRICAL CHARACTERISTICS over recommended operating free air temperature range (unless otherwise noted)

Symbol	Parameter	Conditions		Min	Typ (Note 4)	Max	Units
V_I	Input Clamp Voltage	V_{CC} = Min, I_I = −18 mA				−1.5	V
V_{OH}	High Level Output Voltage	V_{CC} = Min, I_{OH} = Max V_{IL} = Max, V_{IH} = Min		2.7	3.4		V
V_{OL}	Low Level Output Voltage	V_{CC} = Min, I_{OL} = Max V_{IL} = Max, V_{IH} = Min			0.35	0.5	V
		I_{OL} = 4 mA, V_{CC} = Min			0.25	0.4	
I_I	Input Current @ Max Input Voltage	V_{CC} = Max, V_I = 7V	Clear			0.1	mA
		V_{CC} = Max, V_I = 5.5V	A			0.2	
I_{IH}	High Level Input Current	V_{CC} = Max, V_I = 2.7V	Clear			20	µA
			A			40	
I_{IL}	Low Level Input Current	V_{CC} = Max, V_I = 0.4V	Clear			−0.4	mA
			A			−1.6	
I_{OS}	Short Circuit Output Current	V_{CC} = Max (Note 5)		−20		−100	mA
I_{CC}	Supply Current	V_{CC} = Max (Note 6)			15	26	mA

Note 1: C_L = 15 pF, R_L = 2 kΩ, T_A = 25°C and V_{CC} = 5V.

Note 2: C_L = 50 pF, R_L = 2 kΩ, T_A = 25°C and V_{CC} = 5V.

Note 3: The symbol (↓) indicates that the falling edge of the clear pulse is used for reference.

Note 4: All typicals are at V_{CC} = 5V, T_A = 25°C.

Note 5: Not more than one output should be shorted at a time, and the duration should not exceed one second.

Note 6: I_{CC} is measured with all outputs open, both CLEAR inputs grounded following momentary connection to 4.5V, and all other inputs grounded.

Note 7: T_A = 25°C, and V_{CC} = 5V.

SWITCHING CHARACTERISTICS at V_{CC} = 5V and T_A = 25°C (See Section 1 for Test Waveforms and Output Load)

Symbol	Parameter	From (Input) To (Output)	R_L = 2 kΩ				Units
			C_L = 15 pF		C_L = 50 pF		
			Min	Max	Min	Max	
f_{MAX}	Maximum Clock Frequency	A to Q_A	25		20		MHz
t_{PLH}	Propagation Delay Time Low to High Level Output	A to Q_A		20		24	ns
t_{PHL}	Propagation Delay Time High to Low Level Output	A to Q_A		20		30	ns
t_{PLH}	Propagation Delay Time Low to High Level Output	A to Q_D		60		87	ns
t_{PHL}	Propagation Delay Time High to Low Level Output	A to Q_D		60		87	ns
t_{PHL}	Propagation Delay Time High to Low Level Output	Clear to Any Q		39		45	ns

CONNECTION DIAGRAM

Dual-In-Line Package

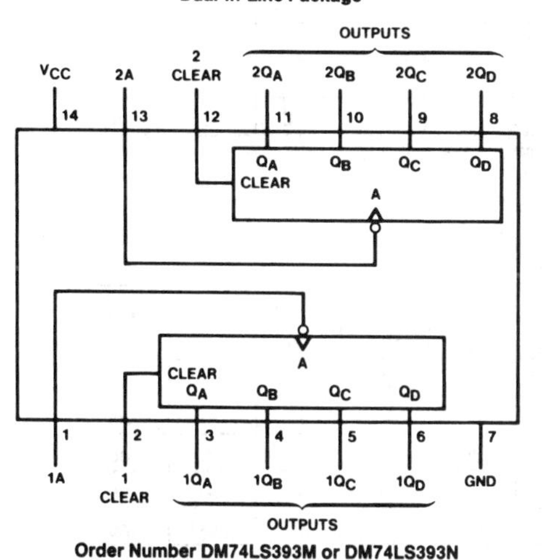

Order Number DM74LS393M or DM74LS393N
See NS Package Number M14A or N14A

FUNCTION TABLE

Count Sequence (Each Counter)

Count	Outputs			
	Q_D	Q_C	Q_B	Q_A
0	L	L	L	L
1	L	L	L	H
2	L	L	H	L
3	L	L	H	H
4	L	H	L	L
5	L	H	L	H
6	L	H	H	L
7	L	H	H	H
8	H	L	L	L
9	H	L	L	H
10	H	L	H	L
11	H	L	H	H
12	H	H	L	L
13	H	H	L	H
14	H	H	H	L
15	H	H	H	H

H = High Logic Level
L = Low Logic Level

LOGIC DIAGRAM

'LS393

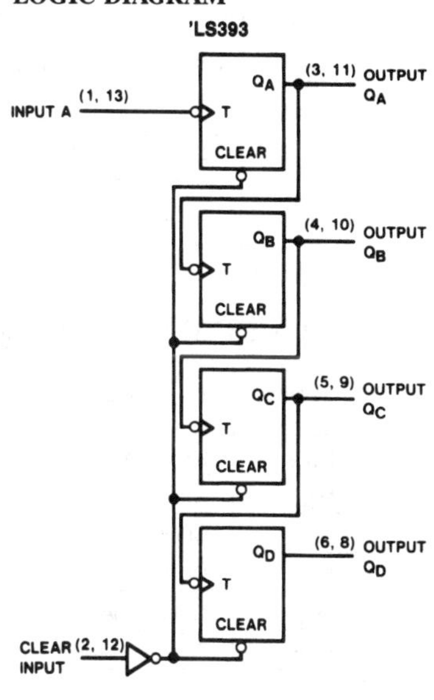

National DM54LS461/DM74LS461
Octal Counter

- Octal counter for microprogram-counter, DMA controller and general-purpose counting applications
- 8 bits match byte boundaries
- Bus-structured pinout
- 24-pin skinny DIP saves space
- TRI-STATE® outputs drive bus lines
- Low-current pnp inputs reduce loading
- Expandable in 8-bit increments

ABSOLUTE MAXIMUM RATINGS

If Military/Aerospace specified devices are required, please contact the National Semiconductor Sales Office/Distributors for availability and specifications.

Supply Voltage V_{CC}	7V
Input Voltage	5.5V
Off-State Output Voltage	5.5V
Storage Temperature	−65°C to +150°C

OPERATING CONDITIONS

Symbol	Parameter		Military			Commercial			Units
			Min	Typ	Max	Min	Typ	Max	
V_{CC}	Supply Voltage		4.5	5	5.5	4.75	5	5.25	V
T_A	Operating Free-Air Temperature		−55		125*	0		75	°C
t_W	Width of Clock	Low	40			35			ns
		High	30			25			
t_{SU}	Set Up Time		60			50			ns
t_h	Hold Time		0	−15		0	−15		

*Case Temperature

ELECTRICAL CHARACTERISTICS Over Operating Conditions

Symbol	Parameter	Test Conditions			Min	Typ†	Max	Units
V_{IL}	Low-Level Input Voltage						0.8	V
V_{IH}	High-Level Input Voltage				2			V
V_{IC}	Input Clamp Voltage	V_{CC}=MIN	I_I=−18 mA				−1.5	V
I_{IL}	Low-Level Input Current	V_{CC}=MAX	V_I=0.4V				−0.25	mA
I_{IH}	High-Level Input Current	V_{CC}=MAX	V_I=2.4V				25	μA
I_I	Maximum Input Current	V_{CC}=MAX	V_I=5.5V				1	mA
V_{OL}	Low-Level Output Voltage	V_{CC}=MIN V_{IL}=0.8V V_{IH}=2V	MIL	I_{OL}=12 mA			0.5	V
			COM	I_{OL}=24 mA				
V_{OH}	High-Level Output Voltage	V_{CC}=MIN V_{IL}=0.8V V_{IH}=2V	MIL	I_{OH}=−2 mA	2.4			V
			COM	I_{OH}=−3.2 mA				
I_{OZL}	Off-State Output Current	V_{CC}=MAX V_{IL}=0.8V V_{IH}=2V		V_O=0.4V			−100	μA
I_{OZH}				V_O=2.4V			100	μA
I_{OS}	Output Short-Circuit Current*	V_{CC}=5.0V		V_{CC}=0V	−30		−130	mA
I_{CC}	Supply Current	V_{CC}=MAX				120	180	mA

*No more than one output should be shorted at a time and duration of the short-circuit should not exceed one second

† All typical values are at V_{CC}=5V, T_A=25°C.

SWITCHING CHARACTERISTICS Over Operating Conditions

Symbol	Parameter	Test Conditions (See Test Load)	Military			Commercial			Units
			Min	Typ	Max	Min	Typ	Max	
f_{MAX}	Maximum Clock Frequency	C_L=50 pF R_1=200 Ω R_2=390 Ω	10.5			12.5			MHz
t_{PD}	$\overline{CBI}$ to $\overline{CBO}$ Delay			35	60		35	50	ns
t_{PD}	Clock to Q			20	35		20	30	ns
t_{PD}	Clock to $\overline{CO}$			55	95		55	80	ns
t_{PZX}	Output Enable Delay			35	55		35	45	ns
t_{PXZ}	Output Disable Delay			35	55		35	45	ns

LOGIC DIAGRAM

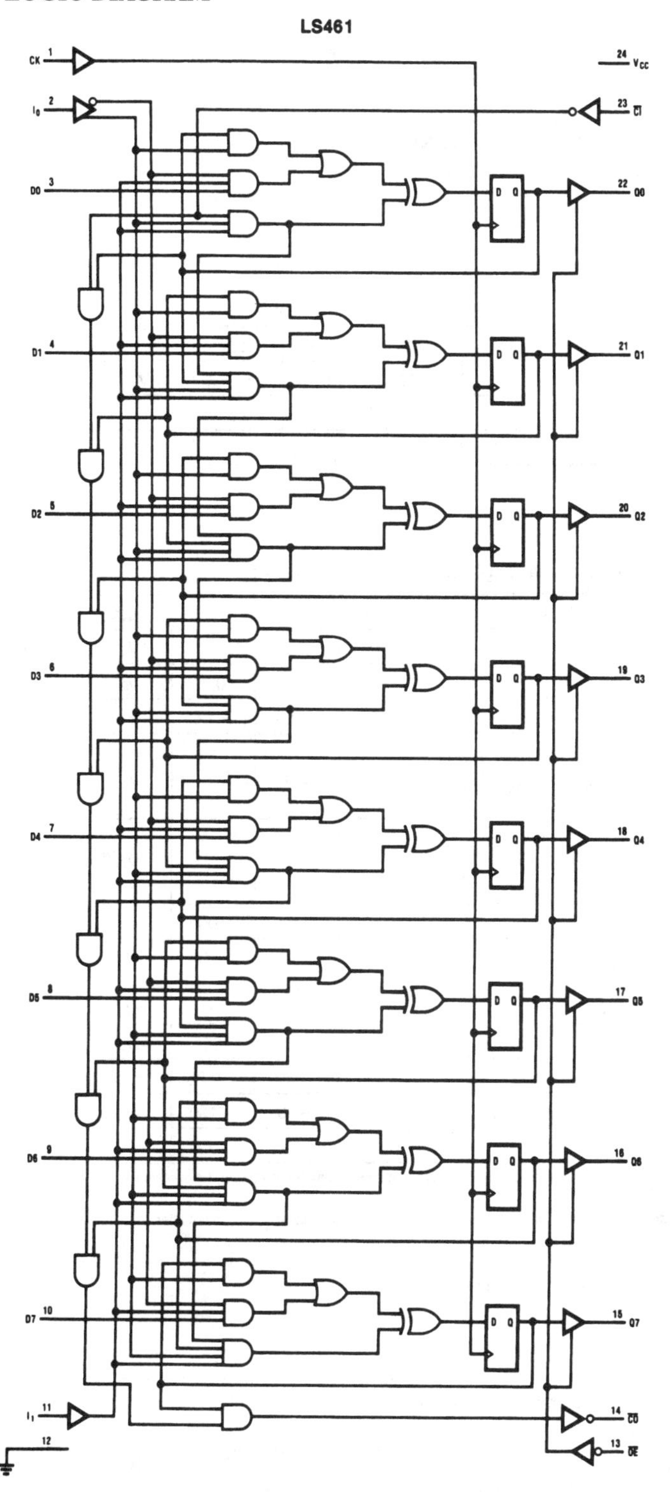

CONNECTION DIAGRAM

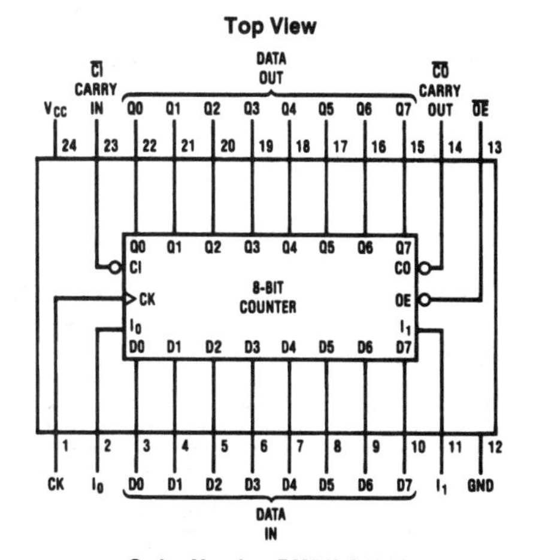

Order Number DM54LS461J, DM74LS461J or DM74LS461N
See NS Package Number J24F or N24C

FUNCTION TABLE

$\overline{OE}$	CK	I1	I0	$\overline{CI}$	D7–D0	Q7–Q0	Operation
H	X	X	X	X	X	Z	HI-Z
L	↑	L	L	X	X	L	CLEAR
L	↑	L	H	X	X	Q	HOLD
L	↑	H	L	X	D	D	LOAD
L	↑	H	H	H	X	Q	HOLD
L	↑	H	H	L	X	Q plus 1	INCREMENT

STANDARD TEST LOAD

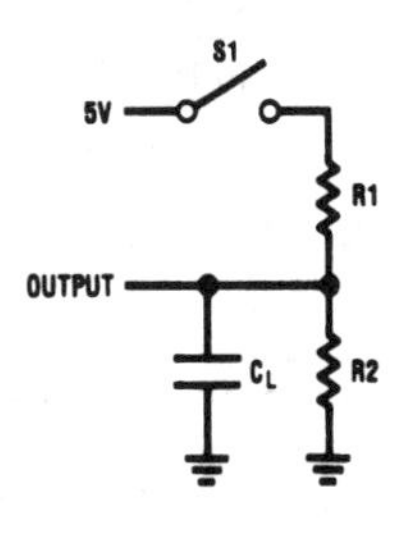

National DM54LS461A/DM74LS461A
Octal Counter

- Octal counter for microprogram-counter, DMA controller for general-purpose counting applications
- 8 bits match byte boundaries
- Low-current pnp inputs reduce loading
- Bus-structured pinout
- TRI-STATE® outputs drive bus lines
- 24-pin skinny DIP saves space
- Expandable in 8-bit increments

ABSOLUTE MAXIMUM RATINGS (Note 1)

If Military/Aerospace specified devices are required, please contact the National Semiconductor Sales Office/Distributors for availability and specifications.

V_{CC} Supply Voltage	7V
Input Voltage	5.5V
Off-State Output Voltage	5.5V
Storage Temperature	−65°C to +150°C
ESD Tolerance	>1000V
Czap = 100 pF	
Rzap = 150Ω	
Test Method: Human Body Model	
Test Specification: NSC SOP 5-028	

RECOMMENDED OPERATING CONDITIONS

Symbol	Parameter	Military			Commercial			Units
		Min	Typ	Max	Min	Typ	Max	
V_{CC}	Supply Voltage	4.5	5	5.5	4.75	5	5.25	V
T_A	Operating Free-Air Temperature	−55	25		0	25	75	°C
T_C	Operating Case Temperature			125				°C

ELECTRICAL CHARACTERISTICS Series 24A Over Recommended Operating Temperature Range

Symbol	Parameter	Test Conditions			Min	Typ	Max	Units
V_{IH}	High Level Input Voltage	(Note 2)			2			V
V_{IL}	Low Level Input Voltage	(Note 2)					0.8	V
V_{IC}	Input Clamp Voltage	V_{CC} = Min, II = −18 mA				−0.8	−1.5	V
V_{OH}	High Level Output Voltage	V_{CC} = Min V_{IL} = 0.8V V_{IH} = 2V	I_{OH} = −2 mA	MIL	2.4	2.9		V
			I_{OH} = −3.2 mA	COM				
V_{OL}	Low Level Output Voltage	V_{CC} = Min V_{IL} = 0.8V V_{IH} = 2V	I_{OL} = 12 mA	MIL		0.3	0.5	V
			I_{OL} = 24 mA	COM				
I_{OZH}	Off-State Output Current (Note 3)	V_{CC} = Max V_{IL} = 0.8V V_{IH} = 2V	V_O = 2.4V				100	μA
I_{OZL}			V_O = 0.4V				−100	μA
I_I	Maximum Input Current	V_{CC} = Max, V_I = 5.5V					1	mA
I_{IH}	High Level Input Current (Note 3)	V_{CC} = Max, V_I = 2.4V					25	μA
I_{IL}	Low Level Input Current (Note 3)	V_{CC} = Max, V_I = 0.4V				−0.04	−0.25	mA
I_{OS}	Output Short-Circuit Current	V_{CC} = 5V	V_O = 0V (Note 4)		−30	−70	−130	mA
I_{CC}	Supply Current	V_{CC} = Max				135	180	mA

Note 1: Absolute maximum ratings are those values beyond which the device may be permanently damaged. They do not mean that the device may be operated at these values.

Note 2: These are absolute voltages with respect to the ground pin on the device and include all overshoots due to system and/or tester noise. Do not attempt to test these values without suitable equipment.

Note 3: I/O leakage as the worst case of IOZX or IIX, e.g., I_{IL} and IOZL.

Note 4: During I_{OS} measurement, only one output at a time should be grounded. Permanent damage otherwise may result.

SWITCHING CHARACTERISTICS Over Recommended Operating Conditions

Symbol	Parameter		Test Conditions	Military			Commercial			Units
				Min	Typ	Max	Min	Typ	Max	
t_S	Set-Up Time from Input			40	20		30	20		ns
t_W	Width of Clock	High		20	7		15	7		ns
		Low		35	15		25	15		ns
T_{pd}	$\overline{CBI}$ to $\overline{CBO}$ Delay		C_L = 50 pF		23	35		23	30	ns
T_{clk}	Clock to Output		C_L = 50 pF		10	25		10	15	ns
T_{pzx}	Output Enable Delay		C_L = 50 pF		19	35		19	30	ns
T_{pzx}	Output Disable Delay		C_L = 5 pF		15	35		15	30	ns
t_H	Hold Time			0	−15		0	−15		ns
f_{max}	Maximum Frequency			15.3	32		22.2	32		MHz

LOGIC DIAGRAM

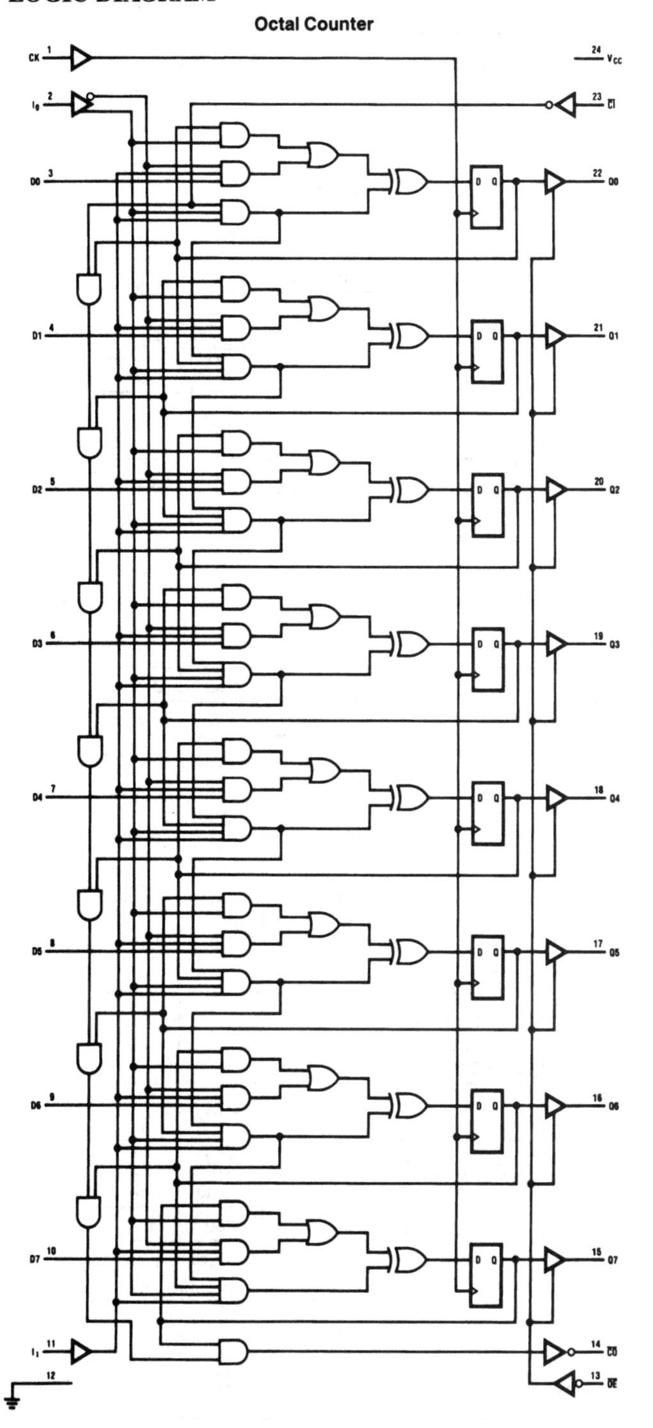

CONNECTION DIAGRAM

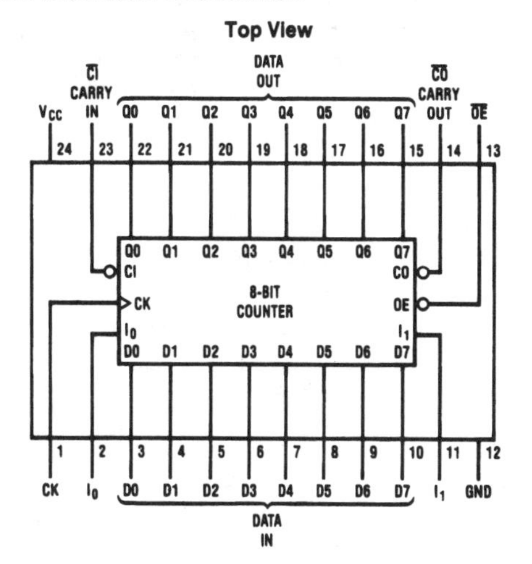

Order Number DM54LS461AJ, DM74LS461AJ, DM74LS461AN or DM74LS461AV
See NS Package Number J24F, N24C or V28A

FUNCTION TABLE

$\overline{OE}$	CK	I1	I0	$\overline{CI}$	D7–D0	Q7–Q0	Operation
H	X	X	X	X	X	Z	HI-Z
L	↑	L	L	X	X	L	CLEAR
L	↑	L	H	X	X	Q	HOLD
L	↑	H	L	X	D	D	LOAD
L	↑	H	H	H	X	Q	HOLD
L	↑	H	H	L	X	Q Plus 1	INCREMENT

National DM54LS469/DM74LS469

8-Bit Up/Down Counter

- 8-bit up/down counter for microprogram-counter, DMA controller and general-purpose counting applications
- 8 bits match byte boundaries
- Bus-structured pinout
- 24-pin skinny DIP saves space
- TRI-STATE® outputs drive bus lines
- Low-current pnp inputs reduce loading
- Expandable in 8-bit increments

ABSOLUTE MAXIMUM RATINGS

If Military/Aerospace specified devices are required, please contact the National Semiconductor Sales Office/Distributors for availability and specifications.

Supply Voltage V_{CC}	7V
Input Voltage	5.5V
Off-State Output Voltage	5.5V
Storage Temperature	−65°C to +150°C

OPERATING CONDITIONS

Symbol	Parameter		Military			Commercial			Units
			Min	Typ	Max	Min	Typ	Max	
V_{CC}	Supply Voltage		4.5	5	5.5	4.75	5	5.25	V
T_A	Operating Free-Air Temperature		−55		125*	0		75	°C
t_W	Width of Clock	Low	40			35	10		ns
		High	30			25			
t_{SU}	Set Up Time		60			50			ns
t_h	Hold Time		0	−15		0	−15		

*Case Temperature

ELECTRICAL CHARACTERISTICS Over Operating Conditions

Symbol	Parameter	Test Conditions			Min	Typ†	Max	Units
V_{IL}	Low-Level Input Voltage						0.8	V
V_{IH}	High-Level Input Voltage				2			V
V_{IC}	Input Clamp Voltage	V_{CC} = MIN		I_I = −18 mA			−1.5	V
I_{IL}	Low-Level Input Current	V_{CC} = MAX		V_I = 0.4V			−0.25	mA
I_{IH}	High-Level Input Current	V_{CC} = MAX		V_I = 2.4V			25	μA
I_I	Maximum Input Current	V_{CC} = MAX		V_I = 5.5V			1	mA
V_{OL}	Low-Level Output Voltage	V_{CC} = MIN V_{IL} = 0.8V V_{IH} = 2V	MIL	I_{OL} = 12 mA			0.5	V
			COM	I_{OL} = 24 mA				
V_{OH}	High-Level Output Voltage	V_{CC} = MIN V_{IL} = 0.8V V_{IH} = 2V	MIL	I_{OH} = −2 mA	2.4			V
			COM	I_{OH} = −3.2 mA				
I_{OZL}	Off-State Output Current	V_{CC} = MAX V_{IL} = 0.8V V_{IH} = 2V		V_O = 0.4V			−100	μA
I_{OZH}				V_O = 2.4V			100	μA
I_{OS}	Output Short-Circuit Current*	V_{CC} = 5.0V		V_O = 0V	−30		−130	mA
I_{CC}	Supply Current	V_{CC} = MAX				120	180	mA

*No more than one output should be shorted at a time and duration of the short-circuit should not exceed one second

† All typical values are V_{CC} = 5V, T_A = 25°C.

SWITCHING CHARACTERISTICS Over Operating Conditions

Symbol	Parameter	Test Conditions (See Test Load/Waveforms)	Military			Commercial			Units
			Min	Typ	Max	Min	Typ	Max	
f_{MAX}	Maximum Clock Frequency		10.5			12.5			MHz
t_{PD}	$\overline{CBI}$ to $\overline{CBO}$ Delay	C_L = 50 pF R_1 = 200Ω R_2 = 390Ω		35	60		35	50	ns
t_{PD}	Clock to Q			20	35		20	30	ns
t_{PD}	Clock to CBO			55	95		55	80	ns
t_{PZX}	Output Enable Delay			20	45		20	35	ns
t_{PXZ}	Output Disable Delay			20	45		20	35	ns

CONNECTION DIAGRAM

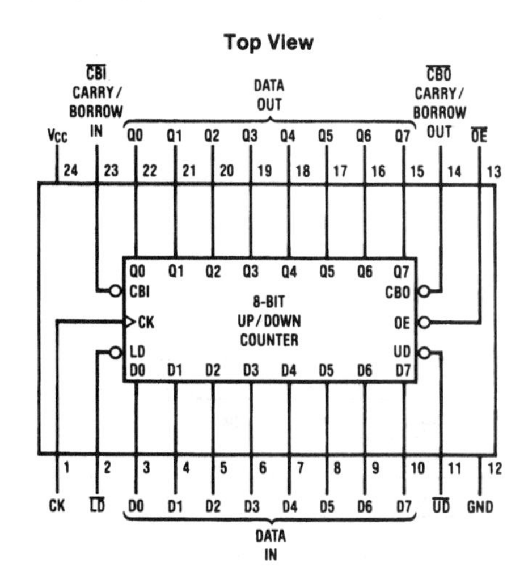

Order Number DM54LS469J, DM74LS469J or DM74LS469N
See NS Package Number J24F or N24C

LOGIC DIAGRAM

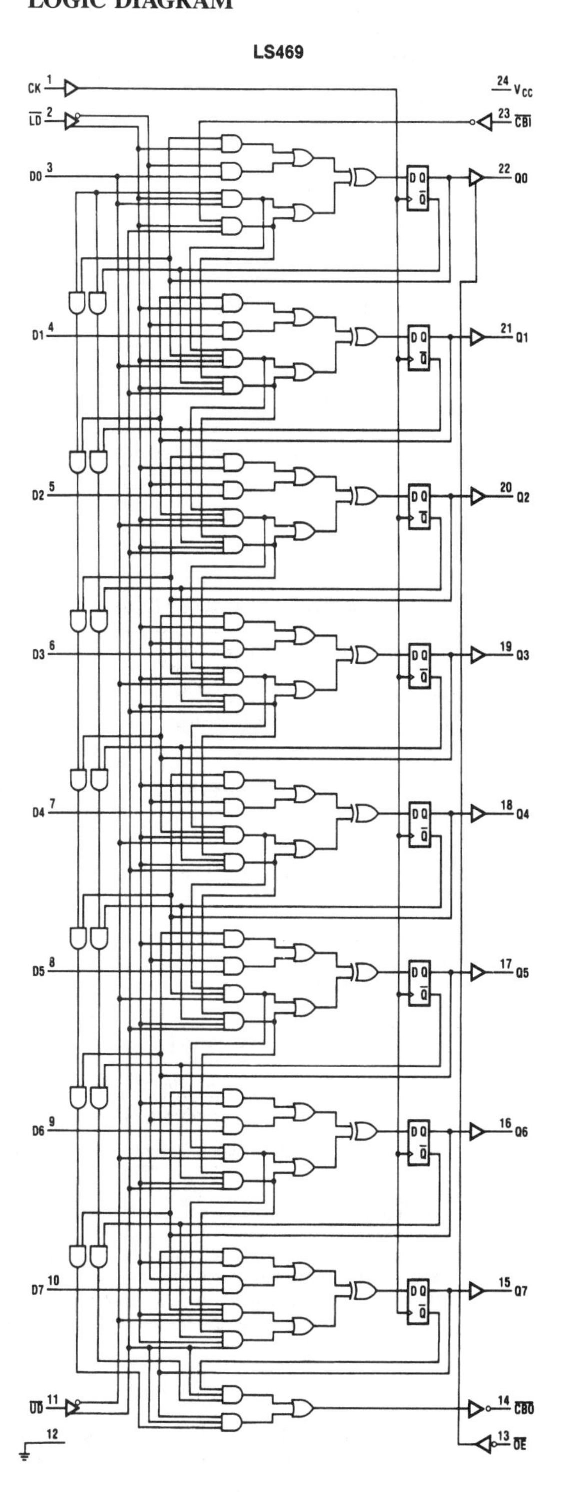

FUNCTION TABLE

$\overline{OE}$	CK	$\overline{LD}$	$\overline{UD}$	$\overline{CBI}$	D7–D0	Q7–Q0	Operation
H	X	X	X	X	X	Z	HI-Z
L	↑	L	X	X	D	D	LOAD
L	↑	H	L	H	X	Q	HOLD
L	↑	H	L	L	X	Q plus 1	INCREMENT
L	↑	H	H	H	X	Q	HOLD
L	↑	H	H	L	X	Q minus 1	DECREMENT

STANDARD TEST LOAD

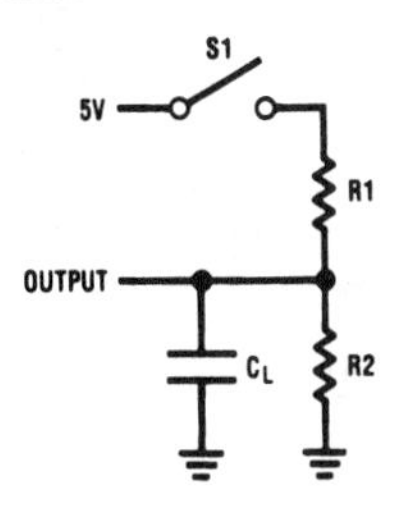

National DM54LS469A/DM74LS469A

8-Bit Up/Down Counter

- Octal register for general-purpose interfacing applications
- 8 bits match byte boundaries
- Low-current pnp inputs reduce loading
- Bus-structured pinout
- TRI-STATE® outputs
- 24-pin skinny dip saves space

ABSOLUTE MAXIMUM RATINGS (Note 1)

If Military/Aerospace specified devices are required, please contact the National Semiconductor Sales Office/Distributors for availability and specifications.

Supply Voltage (V_{CC})	7V
Input Voltage	5.5V
Off-State Output Voltage	5.5V
Storage Temperature	−65°C to +150°C

ESD Tolerance >1000V
Czap = 100 pF
Rzap = 150Ω
Test Method: Human Body Model
Test Specification: NSC SOP 5-028

RECOMMENDED OPERATING CONDITIONS

Symbol	Parameter	Military			Commercial			Units
		Min	Typ	Max	Min	Typ	Max	
V_{CC}	Supply Voltage	4.5	5	5.5	4.75	5	5.25	V
T_A	Operating Free-Air Temperature	−55	25		0	25	75	°C
T_C	Operating Case Temperature			125				°C

ELECTRICAL CHARACTERISTICS Series 24A Over Recommended Operating Temperature Range

Symbol	Parameter	Test Conditions			Min	Typ	Max	Units
V_{IH}	High Level Input Voltage	(Note 2)			2			V
V_{IL}	Low Level Input Voltage	(Note 2)					0.8	V
V_{IC}	Input Clamp Voltage	V_{CC} = Min, II = −18 mA				−0.8	−1.5	V
V_{OH}	High Level Output Voltage	V_{CC} = Min V_{IL} = 0.8V V_{IH} = 2V	I_{OH} = −2 mA	MIL	2.4	2.9		V
			I_{OH} = −3.2 mA	COM				
V_{OL}	Low Level Output Voltage	V_{CC} = Min V_{IL} = 0.8V V_{IH} = 2V	I_{OL} = 12 mA	MIL		0.3	0.5	V
			I_{OL} = 24 mA	COM				
I_{OZH}	Off-State Output Current (Note 3)	V_{CC} = Max V_{IL} = 0.8V V_{IH} = 2V	V_O = 2.4V				100	μA
I_{OZL}			V_O = 0.4V				−100	μA
I_I	Maximum Input Current	V_{CC} = Max, V_I = 5.5V					1	mA
I_{IH}	High Level Input Current (Note 3)	V_{CC} = Max, V_I = 2.4V					25	μA
I_{IL}	Low Level Input Current (Note 3)	V_{CC} = Max, V_I = 0.4V				−0.04	−0.25	mA
I_{OS}	Output Short-Circuit Current	V_{CC} = 5V	V_O = 0V (Note 4)		−30	−70	−130	mA
I_{CC}	Supply Current	V_{CC} = Max				135	180	mA

Note 1: Absolute maximum ratings are those values beyond which the device may be permanently damaged. They do not mean that the device maybe operated at these values.

Note 2: These are absolute voltages with respect to the ground pin on the device and include all overshoots due to system and/or tester noise. Do not attempt to test these values without suitable equipment.

Note 3: I/O leakage as the worst case of IOZX or IIX, e.g., I_{IL} and IOZL.

Note 4: During I_{OS} measurement, only one output at a time should be grounded. Permanent damage otherwise may result.

SWITCHING CHARACTERISTICS Over Recommended Operating Conditions

Symbol	Parameter		Test Conditions	Military			Commercial			Units
				Min	Typ	Max	Min	Typ	Max	
t_S	Set-Up Time from Input			40	20		30	20		ns
t_W	Width of Clock	High		20	7		15	7		ns
		Low		35	15		25	15		ns
t_{pd}	$\overline{CBI}$ to $\overline{CBO}$ Delay		C_L = 50 pF		23	35		23	30	ns
t_{clk}	Clock to Output		C_L = 50 pF		10	25		10	15	ns
t_{pzx}	Output Enable Delay		C_L = 50 pF		19	35		19	30	ns
t_{pzx}	Output Disable Delay		C_L = 5 pF		15	35		15	30	ns
t_H	Hold Time			0	−15		0	−15		ns
f_{max}	Maximum Frequency			15.3	32		22.2	32		MHz

CONNECTION DIAGRAM

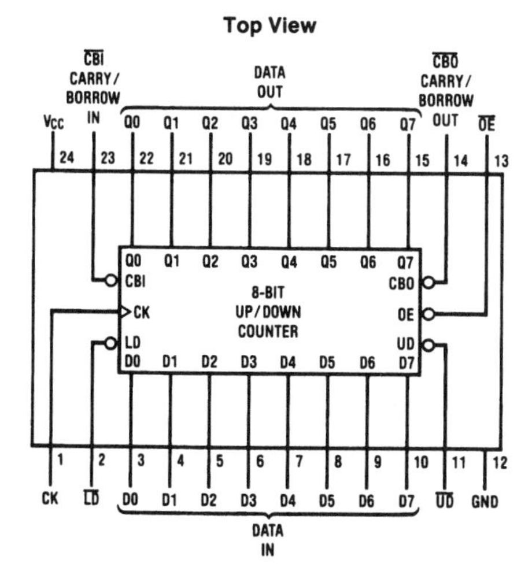

Order Number DM54LS469AJ, DM74LS469AJ, DM74LS469AN or DM74LS469AV
See NS Package Number J24F, N24C or V28A

LOGIC DIAGRAM

8-Bit Up/Down Counter

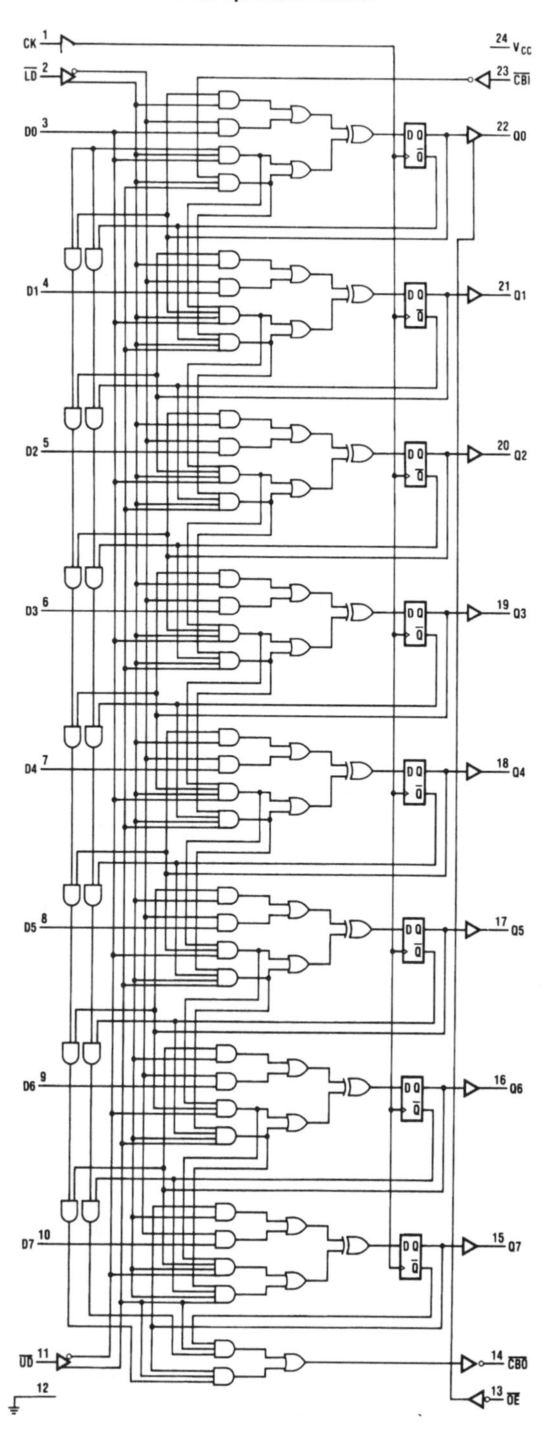

FUNCTION TABLE

OE	CK	LD	UD	CBI	D7–D0	Q7–Q0	Operation
H	X	X	X	X	X	Z	HI-Z
L	↑	L	X	X	D	D	LOAD
L	↑	H	L	H	X	Q	HOLD
L	↑	H	L	L	X	Q Plus 1	INCREMENT
L	↑	H	H	H	X	Q	HOLD
L	↑	H	H	L	X	Q Minus 1	DECREMENT

National 54LS490/DM74LS490
Dual Decade Counter

- Dual version 54LS/74LS90
- Individual asynchronous clear and preset to 9 for each counter
- Count frequency: typically 65 MHz
- Input clamp diodes limit high speed termination effects
- TTL and CMOS compatible

ABSOLUTE MAXIMUM RATINGS (Note)

If Military/Aerospace specified devices are required, please contact the National Semiconductor Sales Office/Distributors for availability and specifications.

Supply Voltage	7V
Input Voltage	7V
Operating Free Air Temperature Range	
54LS	−55°C to +125°C
DM74LS	0°C to +70°C
Storage Temperature Range	−65°C to +150°C

Note: *The "Absolute Maximum Ratings" are those values beyond which the safety of the device cannot be guaranteed. The device should not be operated at these limits. The parametric values defined in the "Electrical Characteristics" table are not guaranteed at the absolute maximum ratings. The "Recommended Operating Conditions" table will define the conditions for actual device operation.*

RECOMMENDED OPERATING CONDITIONS

Symbol	Parameter	54LS490			DM74LS490			Units
		Min	Nom	Max	Min	Nom	Max	
V_{CC}	Supply Voltage	4.5	5	5.5	4.75	5	5.25	V
V_{IH}	High Level Input Voltage	2			2			V
V_{IL}	Low Level Input Voltage			0.7			0.8	V
I_{OH}	High Level Output Current			−0.4			−0.4	mA
I_{OL}	Low Level Output Current			4			8	mA
T_A	Free Air Operating Temperature	−55		125	0		70	°C
t_w (L)	$\overline{CP}$ Pulse Width LOW	12.5			12.5			ns
t_w (H)	MR, MS Pulse Width HIGH	20			20			ns
t_{rec}	Recovery Time, MR or MS to $\overline{CP}$	15			15			ns

ELECTRICAL CHARACTERISTICS over recommended operating free air temperature range (unless otherwise noted)

Symbol	Parameter	Conditions		Min	Typ (Note 1)	Max	Units
V_I	Input Clamp Voltage	V_{CC} = Min, I_I = −18 mA				−1.5	V
V_{OH}	High Level Output Voltage	V_{CC} = Min, I_{OH} = Max, V_{IL} = Max	54LS	2.5			V
			DM74	2.7			
V_{OL}	Low Level Output Voltage	V_{CC} = Min, I_{OL} = Max, V_{IH} = Min	54LS			0.4	V
			DM74			0.5	
		I_{OL} = 4 mA, V_{CC} = Min	DM74			0.4	
I_I	Input Current @ Max Input Voltage	V_{CC} = Max, V_I = 10V	Inputs			100	μA
			$\overline{CP}$			200	
I_{IH}	High Level Input Current	V_{CC} = Max, V_I = 2.7V	Inputs			20	μA
			$\overline{CP}$			40	
I_{IL}	Low Level Input Current	V_{CC} = Max, V_I = 0.4V	Inputs	−0.03		−0.4	mA
			$\overline{CP}$	−0.18		−2.4	
I_{OS}	Short Circuit Output Current	V_{CC} = Max (Note 2)	54LS	−20		−100	mA
			DM74	−20		−100	
I_{CC}	Supply Current	V_{CC} = Max				26	mA

Note 1: All typicals are at V_{CC} = 5V, T_A = 25°C.

Note 2: Not more than one output should be shorted at a time, and the duration should not exceed one second.

SWITCHING CHARACTERISTICS

V_{CC} = +5.0V, T_A = +25°C (See Section 1 for test waveforms and output load)

Symbol	Parameter	R_L = 2 kΩ, C_L = 15 pF		Units
		Min	Max	
f_{max}	Maximum Clock Frequency	40		MHz
t_{PLH} t_{PHL}	Propagation Delay $\overline{CP}$ to Q0		15 15	ns
t_{PLH} t_{PHL}	Propagation Delay $\overline{CP}$ to Q1 or Q3		30 30	ns
t_{PLH} t_{PHL}	Propagation Delay $\overline{CP}$ to Q2		45 45	ns
t_{PLH} t_{PHL}	Propagation Delay MS to Qn		35 35	ns
t_{PHL}	Propagation Delay MR to Qn		39	ns

LOGIC DIAGRAM

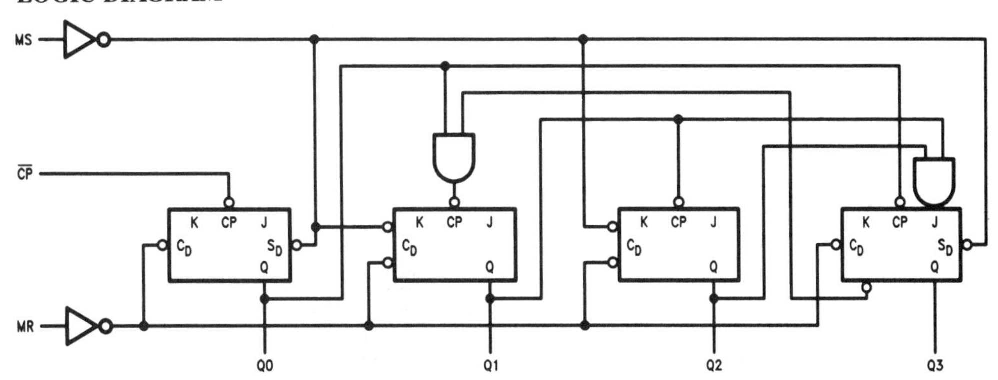

LOGIC SYMBOL

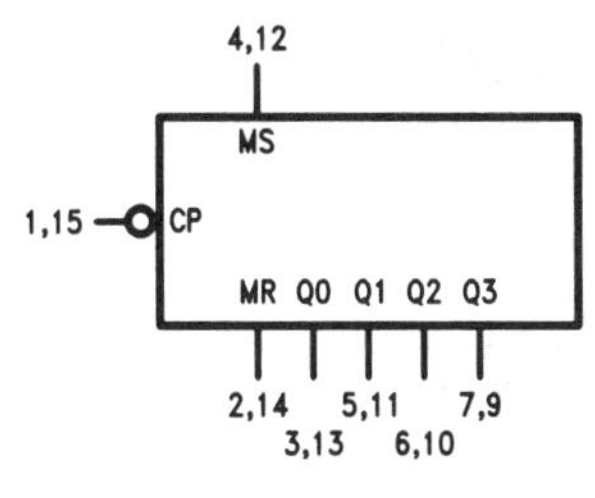

STATE DIAGRAM

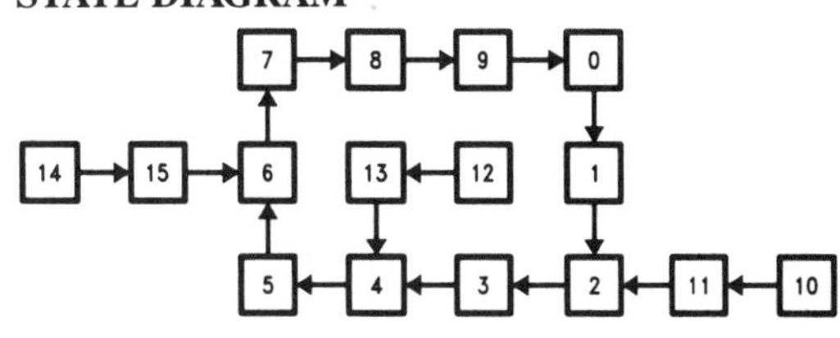

CONNECTION DIAGRAM

Dual-In-Line Package

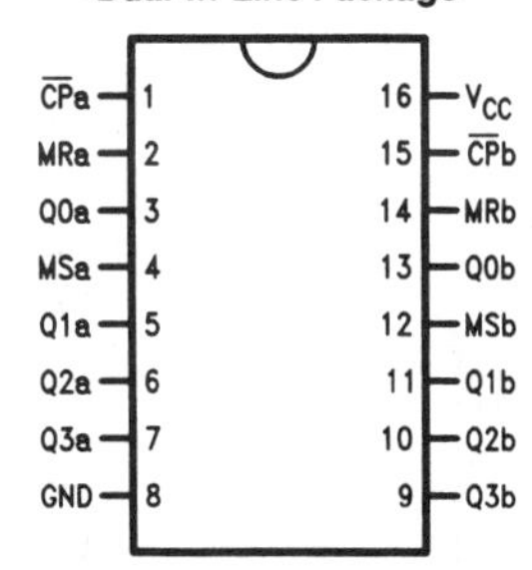

Order Number 54LS490DMQB, 54LS490FMQB, DM54LS490M or DM54LS490N
See NS Package Number J16A, M16A, N16E or W16A

Pin Names	Description
MS	Master Set (Set to 9) Input (Active HIGH)
MR	Master Reset Input (Active HIGH)
$\overline{CP}$	Clock Pulse Input (Active Falling Edge)
Q0–Q3	Counter Outputs

National DM54LS491/74LS491
10-Bit Counter

- CRT vertical and horizontal timing generation
- Bus-structured pinout
- 24-pin skinny DIP saves space
- TRI-STATE® outputs drive bus lines
- Low-current pnp inputs reduce loading

ABSOLUTE MAXIMUM RATINGS

If Military/Aerospace specified devices are required, please contact the National Semiconductor Sales Office/Distributors for availability and specifications.

Supply Voltage V_{CC}	7V
Input Voltage	5.5V
Off-State Output Voltage	5.5V
Storage Temperature	−65° to +150°C

OPERATING CONDITIONS

Symbol	Parameter		Military			Commercial			Units
			Min	Typ	Max	Min	Typ	Max	
V_{CC}	Supply Voltage		4.5	5	5.5	4.75	5	5.25	V
T_A	Operating Free-Air Temperature		−55		125*	0		75	°C
t_w	Width of Clock	High	40			40			ns
		Low	35			35			
t_{SU}	Set-Up Time		60			50			ns
t_h	Hold Time		0	−15		0	−15		

* Case temperature

ELECTRICAL CHARACTERISTICS Over Operating Conditions

Symbol	Parameter	Test Conditions			Min	Typ†	Max	Units
V_{IL}	Low-Level Input Voltage						0.8	V
V_{IH}	High-Level Input Voltage				2			V
V_{IC}	Input Clamp Voltage	V_{CC}=MIN	I_I=−18 mA				−1.5	V
I_{IL}	Low-Level Input Current	V_{CC}=MAX	V_I=0.4V				−0.25	mA
I_{IH}	High-Level Input Current	V_{CC}=MAX	V_I=2.4V				25	μA
I_I	Maximum Input Current	V_{CC}=MAX	V_I=5.5V				1	mA
V_{OL}	Low-Level Output Voltage	V_{CC}=MIN V_{IL}=0.8V V_{IH}=2V	MIL	I_{OL}=12 mA			0.5	V
			COM	I_{OL}=24 mA				
V_{OH}	High-Level Output Voltage	V_{CC}=MIN V_{IL}=0.8V V_{IH}=2V	MIL	I_{OH}=−2 mA	2.4			V
			COM	I_{OH}=3.2 mA				
I_{OZL}	Off-State Output Current	V_{CC}=MAX V_{IL}=0.8V V_{IH}=2V		V_O=0.4V			−100	μA
I_{OZH}				V_O=2.4V			100	μA
I_{OS}	Output Short-Circuit Current*	V_{CC}=5.0V		V_O=0V	−30		−130	mA
I_{CC}	Supply Current	V_{CC}=MAX				120	180	mA

* No more than one output should be shorted at a time and duration of the short-circuit should not exceed one second.

† All typical values are at V_{CC}=5V, T_A=25°C

SWITCHING CHARACTERISTICS Over Operating Conditions

Symbol	Parameter	Test Conditions (See Test Load)	Military			Commercial			Units
			Min	Typ	Max	Min	Typ	Max	
f_{MAX}	Maximum Clock Frequency	C_L=50 pF R_1=200Ω R_2=390Ω	10.5			12.5			MHz
t_{PD}	Clock to Q			20	35		20	30	ns
t_{PZX}	Output Enable Delay			35	55		35	45	ns
t_{PXZ}	Output Disable Delay			35	55		35	45	ns

CONNECTION DIAGRAM

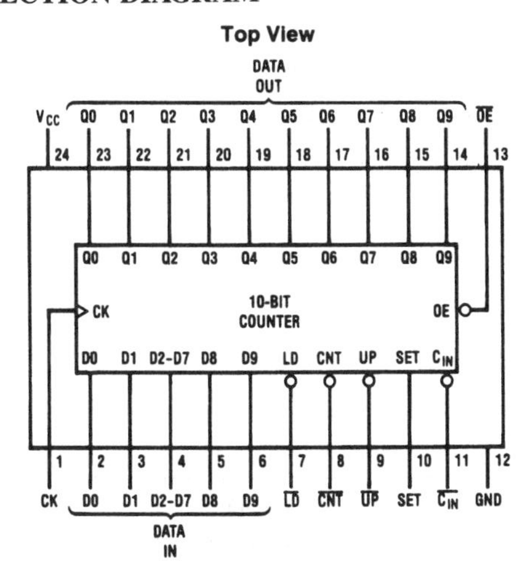

Order Number DM54LS491J, DM74LS491J or DM74LS491N
See NS Package Number J24F or N24C

LOGIC DIAGRAM

10-Bit Up/Down Counter

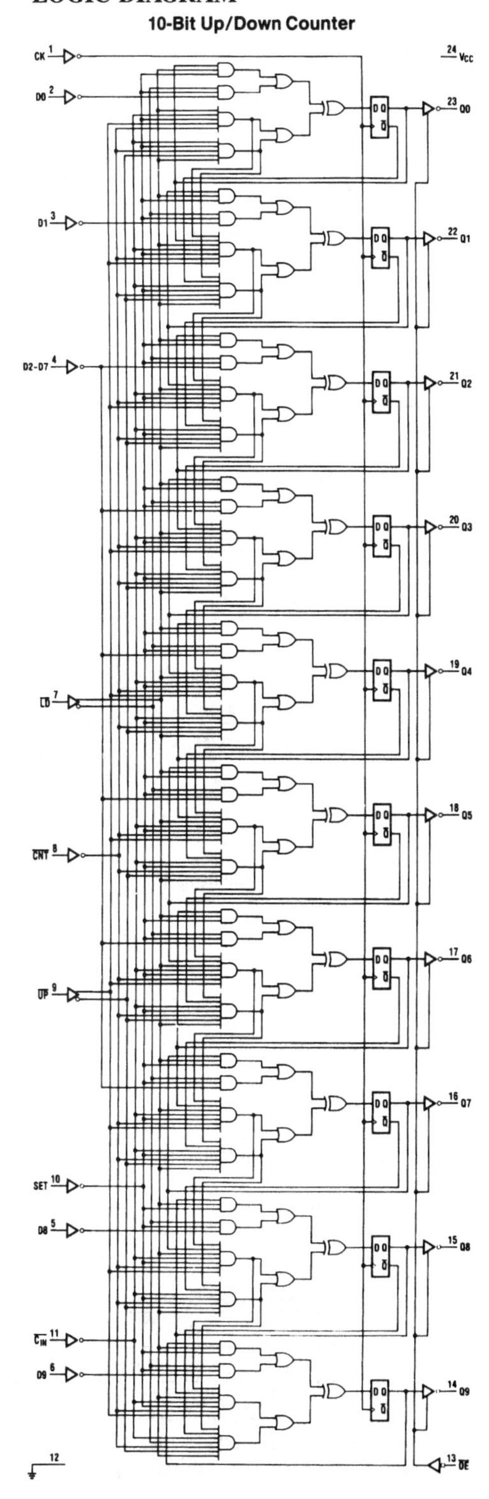

STANDARD TEST LOAD

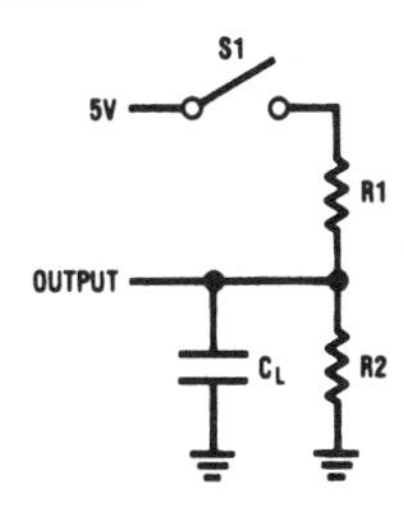

FUNCTION TABLE

$\overline{OE}$	CK	SET	LD	$\overline{CNT}$	$\overline{C_{IN}}$	$\overline{UP}$	D9-D0	Q9-Q0	Operation
H	X	X	X	X	X	X	X	Z	Hi-Z
L	↑	H	X	X	X	X	X	H	Set all HIGH
L	↑	L	L	X	X	X	D	D	LOAD D
L	↑	L	H	H	X	X	X	Q	HOLD
L	↑	L	H	L	H	X	X	Q	HOLD
L	↑	L	H	L	L	L	X	Q plus 1	Count UP
L	↑	L	H	L	L	H	X	Q minus 1	Count DN

National DM54S161/DM74S161, DM54S163/DM74S163

Synchronous 4-Bit Binary Counters

- Synchronously programmable
- Internal look-ahead for fast counting
- Carry output for n-bit cascading
- Synchronous counting
- Load control line
- Diode-clamped inputs

ABSOLUTE MAXIMUM RATINGS (Note)

If Military/Aerospace specified devices are required, please contact the National Semiconductor Sales Office/Distributors for availability and specifications.

Supply Voltage	7V
Input Voltage	5.5V
Operating Free Air Temperature Range	
DM54S	−55°C to +125°C
DM74S	0°C to +70°C
Storage Temperature Range	−65°C to +150°C

Note: *The "Absolute Maximum Ratings" are those values beyond which the safety of the device cannot be guaranteed. The device should not be operated at these limits. The parametric values defined in the "Electrical Characteristics" table are not guaranteed at the absolute maximum ratings. The "Recommended Operating Conditions" table will define the conditions for actual device operation.*

RECOMMENDED OPERATING CONDITIONS See Section 1 for Test Waveforms and Output Load

Symbol	Parameter		DM54S161/163			DM74S161/163			Units
			Min	Nom	Max	Min	Nom	Max	
V_{CC}	Supply Voltage		4.5	5	5.5	4.75	5	5.25	V
V_{IH}	High Level Input Voltage		2			2			V
V_{IL}	Low Level Input Voltage				0.8			0.8	V
I_{OH}	High Level Output Current				−1			−1	mA
I_{OL}	Low Level Output Current				20			20	mA
f_{CLK}	Clock Frequency (Note 1)		0		40	0		40	MHz
	Clock Frequency (Note 2)		0		35	0		35	
t_W	Pulse Width (Note 1)	Clock	10			10			ns
		Clear (Note 4)	10			10			
	Pulse Width (Note 2)	Clock	12			12			
		Clear (Note 4)	12			12			
t_{SU}	Setup Time (Note 1)	Data	4			4			ns
		Enable P or T	12			12			
		Load	14			14			
		Clear (Note 3)	14			14			
	Setup Time (Note 2)	Data	5			5			
		Enable P or T	14			14			
		Load	16			16			
		Clear (Note 3)	16			16			
t_H	Hold Time (Note 1)	Data	3			3			ns
		Others	0			0			
	Hold Time (Note 2)	Data	5			5			
		Others	2			2			
t_{REL}	Load or Clear Release Time (Note 1)		12			12			ns
	Load or Clear Release Time (Note 2)		14			14			
T_A	Free Air Operating Temperature		−55		125	0		70	°C

Note 1: C_L = 15 pF, R_L = 280Ω, T_A = 25°C and V_{CC} = 5V.

Note 2: C_L = 50 pF, R_L = 280Ω, T_A = 25°C and V_{CC} = 5V.

Note 3: Applies only to the 'S163 which has synchronous clear inputs.

Note 4: Applies only to the 'S161 which has asynchronous clear inputs.

ELECTRICAL CHARACTERISTICS over recommended operating free air temperature (unless otherwise noted)

Symbol	Parameter	Conditions		Min	Typ (Note 1)	Max	Units
V_I	Input Clamp Voltage	V_{CC} = Min, I_I = −18 mA				−1.2	V
V_{OH}	High Level Output Voltage	V_{CC} = Min, I_{OH} = Max, V_{IL} = Max, V_{IH} = Min	DM54	2.5	3.4		V
			DM74	2.7	3.4		
V_{OL}	Low Level Output Voltage	V_{CC} = Min, I_{OL} = Max, V_{IH} = Min, V_{IL} = Max				0.5	V
I_I	Input Current @ Max Input Voltage	V_{CC} = Max, V_I = 5.5V				1	mA
I_{IH}	Low Level Input Current	V_{CC} = Max, V_I = 2.7V	CLK, Data			50	μA
			Others	−10		−200	
I_{IL}	Low Level Input Current	V_{CC} = Max, V_I = 0.5V	Enable T			−4	mA
			Others			−2	
I_{OS}	Short Circuit Output Current	V_{CC} = Max (Note 2)	DM54	−40		−100	mA
			DM74	−40		−100	
I_{CC}	Supply Current	V_{CC} = Max			95	160	mA

SWITCHING CHARACTERISTICS at V_{CC} = 5V and T_A = 25°C (See Section 1 for Test Waveforms and Output Load)

Symbol	Parameter	From (Input) To (Output)	R_L = 280Ω				Units
			C_L = 15 pF		C_L = 50 pF		
			Min	Max	Min	Max	
f_{MAX}	Maximum Clock Frequency		40		35		MHz
t_{PLH}	Propagation Delay Time Low to High Level Output	Clock to Ripple Carry		25		25	ns
t_{PHL}	Propagation Delay Time High to Low Level Output	Clock to Ripple Carry		25		28	ns
t_{PLH}	Propagation Delay Time Low to High Level Output	Clock to Any Q		15		15	ns
t_{PHL}	Propagation Delay Time High to Low Level Output	Clock to Any Q		15		18	ns
t_{PLH}	Propagation Delay Time Low to High Level Output	Enable T to Ripple Carry		15		18	ns
t_{PHL}	Propagation Delay Time High to Low Level Output	Enable T to Ripple Carry		15		18	ns
t_{PHL}	Propagation Delay Time High to Low Level Output (Note 3)	Clear to Any Q		20		24	ns

Note 1: All typicals are at V_{CC} = 5V, T_A = 25°C.

Note 2: Not more than one output should be shorted at a time, and the duration should not exceed one second.

Note 3: Propagation delay for clearing is measured from clear input for the 'S161 and from the clock input transition for the 'S163.

CONNECTION DIAGRAM

Dual-In-Line Package

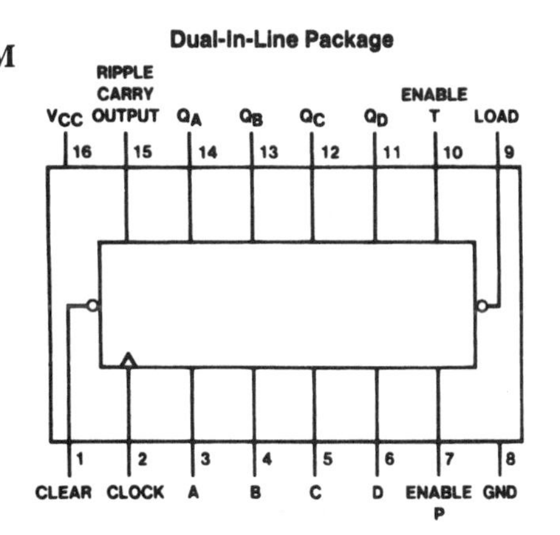

Order Number DM54S161J, DM54S163J, DM54S161W, DM74S161N, or DM74S163N
See NS Package Number J16A, N16E or W16A

LOGIC DIAGRAM

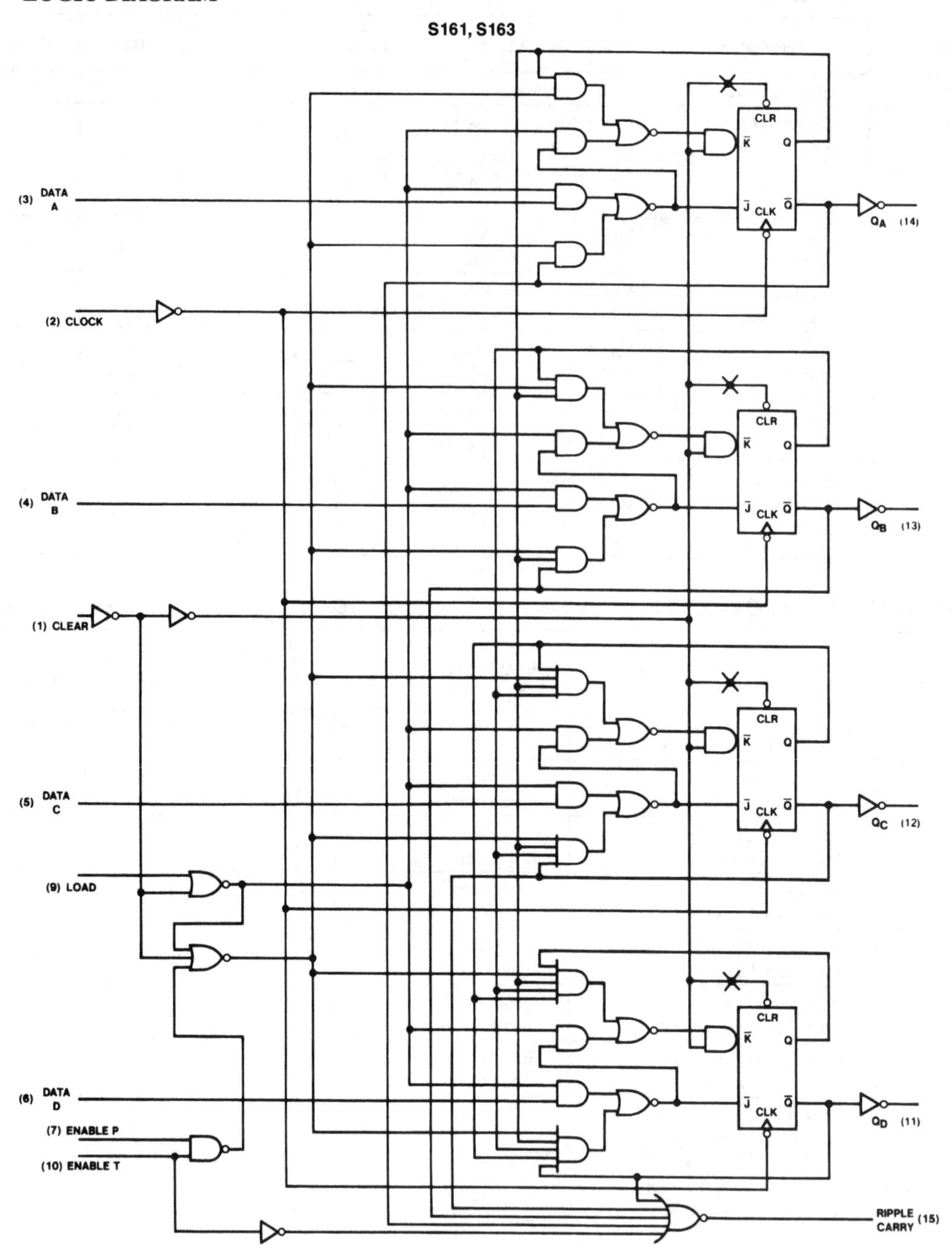

TIMING DIAGRAM

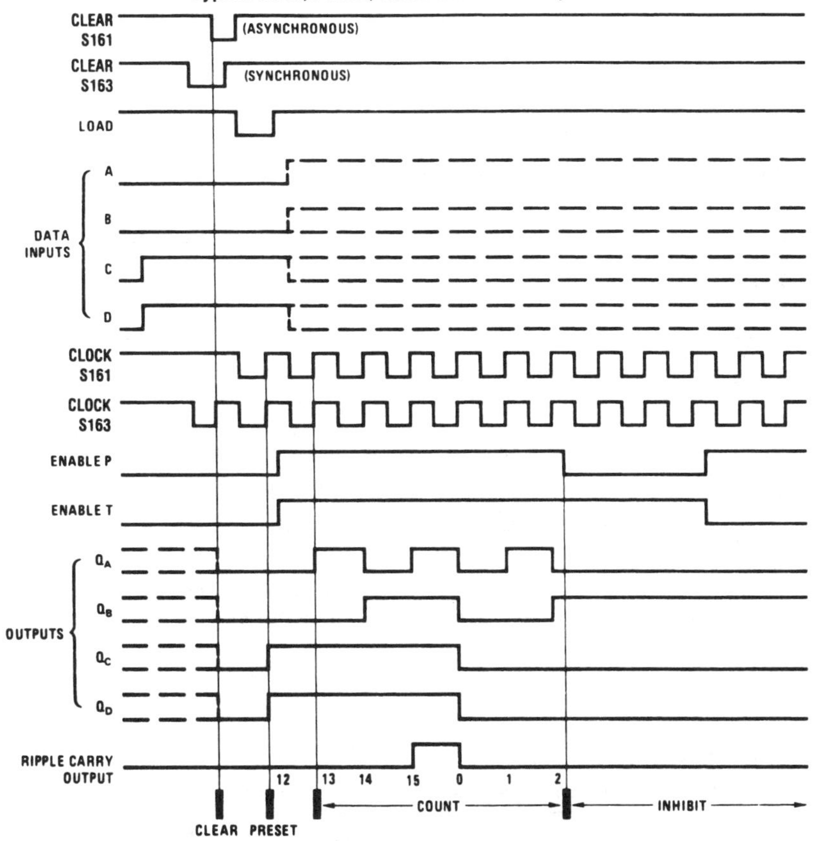

National 5490/DM5490A/DM7490A, DM5493A/DM7493A

Decade and Binary Counters

- Typical power dissipation
  - 90A 145 mW
  - 93A 130 mW
- Count frequency: 42 MHz
- Alternate Military/Aerospace device (5490) is available

ABSOLUTE MAXIMUM RATINGS (Note)

If Military/Aerospace specified devices are required, please contact the National Semiconductor Sales Office/Distributors for availability and specifications.

Supply Voltage	7V
Input Voltage	5.5V
Operating Free Air Temperature Range	
DM54 and 54	−55°C to +125°C
DM74	0°C to +70°C
Storage Temperature Range	−65°C to +150°C

Note: *The "Absolute Maximum Ratings" are those values beyond which the safety of the device cannot be guaranteed. The device should not be operated at these limits. The parametric values defined in the "Electrical Characteristics" table are not guaranteed at the absolute maximum ratings. The "Recommended Operating Conditions" table will define the conditions for actual device operation.*

RECOMMENDED OPERATING CONDITIONS

Symbol	Parameter		DM5490A			DM7490A			Units
			Min	Nom	Max	Min	Nom	Max	
V_{CC}	Supply Voltage		4.5	5	5.5	4.75	5	5.25	V
V_{IH}	High Level Input Voltage		2			2			V
V_{IL}	Low Level Input Voltage				0.8			0.8	V
I_{OH}	High Level Output Current				−0.8			−0.8	mA
I_{OL}	Low Level Output Current				16			16	mA
f_{CLK}	Clock Frequency (Note 5)	A	0		32	0		32	MHz
		B	0		16	0		16	
t_W	Pulse Width (Note 5)	A	15			15			ns
		B	30			30			
		Reset	15			15			
t_{REL}	Reset Release Time (Note 5)		25			25			ns
T_A	Free Air Operating Temperature		−55		125	0		70	°C

'90A ELECTRICAL CHARACTERISTICS

over recommended operating free air temperature range (unless otherwise noted)

Symbol	Parameter	Conditions		Min	Typ (Note 1)	Max	Units
V_I	Input Clamp Voltage	V_{CC} = Min, I_I = −12 mA				−1.5	V
V_{OH}	High Level Output Voltage	V_{CC} = Min, I_{OH} = Max V_{IL} = Max, V_{IH} = Min		2.4	3.4		V
V_{OL}	Low Level Output Voltage	V_{CC} = Min, I_{OL} = Max V_{IH} = Min, V_{IL} = Max (Note 4)			0.2	0.4	V
I_I	Input Current @ Max Input Voltage	V_{CC} = Max, V_I = 5.5V				1	mA
I_{IH}	High Level Input Current	V_{CC} = Max V_I = 2.7V	A			80	μA
			Reset			40	
			B			120	
I_{IL}	Low Level Input Current	V_{CC} = Max V_I = 0.4V	A			−3.2	mA
			Reset			−1.6	
			B			−4.8	
I_{OS}	Short Circuit Output Current	V_{CC} = Max (Note 2)	DM54	−20		−57	mA
			DM74	−18		−57	
I_{CC}	Supply Current	V_{CC} = Max (Note 3)			29	42	mA

Note 1: All typicals are at V_{CC} = 5V, T_A = 25°C.

Note 2: Not more than one output should be shorted at a time.

Note 3: I_{CC} is measured with all outputs open, both RO inputs grounded following momentary connection to 4.5V, and all other inputs grounded.

Note 4: Q_A outputs are tested at I_{OL} = Max plus the limit value of I_{IL} for the B input. This permits driving the B input while maintaining full fan-out capability.

Note 5: T_A = 25°C and V_{CC} = 5V.

'90A SWITCHING CHARACTERISTICS

at V_{CC} = 5V and T_A = 25°C (See Section 1 for Test Waveforms and Output Load)

Symbol	Parameter	From (Input) To (Output)	R_L = 400Ω, C_L = 15 pF		Units
			Min	Max	
f_{MAX}	Maximum Clock Frequency	A to Q_A	32		MHz
		B to Q_B	16		
t_{PLH}	Propagation Delay Time Low to High Level Output	A to Q_A		16	ns
t_{PHL}	Propagation Delay Time High to Low Level Output	A to Q_A		18	ns
t_{PLH}	Propagation Delay Time Low to High Level Output	A to Q_D		48	ns
t_{PHL}	Propagation Delay Time High to Low Level Output	A to Q_D		50	ns
t_{PLH}	Propagation Delay Time Low to High Level Output	B to Q_B		16	ns
t_{PHL}	Propagation Delay Time High to Low Level Output	B to Q_B		21	ns
t_{PLH}	Propagation Delay Time Low to High Level Output	B to Q_C		32	ns
t_{PHL}	Propagation Delay Time High to Low Level Output	B to Q_C		35	ns
t_{PLH}	Propagation Delay Time Low to High Level Output	B to Q_D		32	ns
t_{PHL}	Propagation Delay Time High to Low Level Output	B to Q_D		35	ns
t_{PLH}	Propagation Delay Time Low to High Level Output	SET-9 to Q_A, Q_D		30	ns
t_{PHL}	Propagation Delay Time High to Low Level Output	SET-9 to Q_B, Q_C		40	ns
t_{PHL}	Propagation Delay Time High to Low Level Output	SET-0 Any Q		40	ns

RECOMMENDED OPERATING CONDITIONS

Symbol	Parameter		DM5493A			DM7493A			Units
			Min	Nom	Max	Min	Nom	Max	
V_{CC}	Supply Voltage		4.5	5	5.5	4.75	5	5.25	V
V_{IH}	High Level Input Voltage		2			2			V
V_{IL}	Low Level Input Voltage				0.8			0.8	V
I_{OH}	High Level Output Current				−0.8			−0.8	mA
I_{OL}	Low Level Output Current				16			16	mA
f_{CLK}	Clock Frequency (Note 5)	A	0		32	0		32	MHz
		B	0		16	0		16	
t_W	Pulse Width (Note 5)	A	15			15			ns
		B	30			30			
		Reset	15			15			
t_{REL}	Reset Release Time (Note 5)		25			25			ns
T_A	Free Air Operating Temperature		−55		125	0		70	°C

'93A ELECTRICAL CHARACTERISTICS

over recommended operating free air temperature range (unless otherwise noted)

Symbol	Parameter	Conditions		Min	Typ (Note 1)	Max	Units
V_I	Input Clamp Voltage	V_{CC} = Min, I_I = −12 mA				−1.5	V
V_{OH}	High Level Output Voltage	V_{CC} = Min, I_{OH} = Max V_{IL} = Max, V_{IH} = Min		2.4	3.4		V
V_{OL}	Low Level Output Voltage	V_{CC} = Min, I_{OL} = Max V_{IH} = Min, V_{IL} = Max (Note 4)			0.2	0.4	V
I_I	Input Current @ Max Input Voltage	V_{CC} = Max, V_I = 5.5V				1	mA
I_{IH}	High Level Input Current	V_{CC} = Max V_I = 2.4V	Reset			40	μA
			A			80	
			B			80	
I_{IL}	Low Level Input Current	V_{CC} = Max V_I = 0.4V	Reset			−1.6	mA
			A			−3.2	
			B			−3.2	
I_{OS}	Short Circuit Output Current	V_{CC} = Max (Note 2)	DM54	−20		−57	mA
			DM74	−18		−57	
I_{CC}	Supply Current	V_{CC} = Max (Note 3)			26	39	mA

Note 1: All typicals are at V_{CC} = 5V, T_A = 25°C.

Note 2: Not more than one output should be shorted at a time.

Note 3: I_{CC} is measured with all outputs open, both R0 inputs grounded following momentary connection to 4.5V and all other inputs grounded.

Note 4: Q_A outputs are tested at I_{OL} = Max plus the limit value of I_{IL} for the B input. This permits driving the B input while maintaining full fan-out capability.

Note 5: T_A = 25°C and V_{CC} = 5V.

CONNECTION DIAGRAMS

Dual-In-Line Package

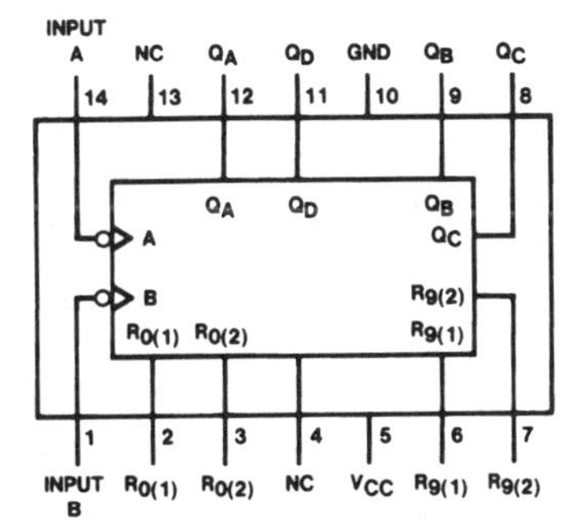

Order Number 5490DMQB, 5490FMQB, DM5490AJ, DM5490AW or DM7490AN
See NS Package Number J14A, N14A or W14B

Dual-In-Line Package

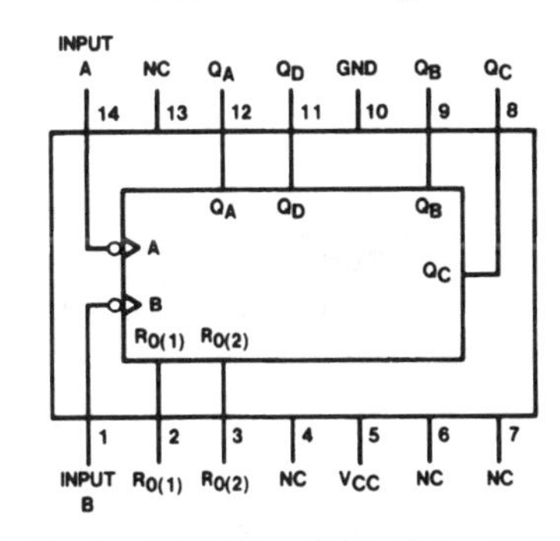

Order Number DM5493AJ, DM5493AW or DM7493AN
See NS Package Number J14A, N14A or W14B

'93A SWITCHING CHARACTERISTICS

at V_{CC} = 5V and T_A = 25°C (See Section 1 for Test Waveforms and Output Load)

Symbol	Parameter	From (Input) To (Output)	R_L = 400Ω C_L = 15 pF		Units
			Min	Max	
f_{MAX}	Maximum Clock Frequency	A to Q_A	32		MHz
		B to Q_B	16		
t_{PLH}	Propagation Delay Time Low to High Level Output	A to Q_A		16	ns
t_{PHL}	Propagation Delay Time High to Low Level Output	A to Q_A		18	ns
t_{PLH}	Propagation Delay Time Low to High Level Output	A to Q_D		70	ns
t_{PHL}	Propagation Delay Time High to Low Level Output	A to Q_D		70	ns
t_{PLH}	Propagation Delay Time Low to High Level Output	B to Q_B		16	ns
t_{PHL}	Propagation Delay Time High to Low Level Output	B to Q_B		21	ns
t_{PLH}	Propagation Delay Time Low to High Level Output	B to Q_C		32	ns
t_{PHL}	Propagation Delay Time High to Low Level Output	B to Q_C		35	ns
t_{PLH}	Propagation Delay Time Low to High Level Output	B to Q_D		51	ns
t_{PHL}	Propagation Delay Time High to Low Level Output	B to Q_D		51	ns
t_{PHL}	Propagation Delay Time High to Low Level Output	SET-0 to Any Q		40	ns

FUNCTION TABLES (Note D)

90A
BCD Count Sequence
(See Note A)

Count	Outputs			
	Q_D	Q_C	Q_B	Q_A
0	L	L	L	L
1	L	L	L	H
2	L	L	H	L
3	L	L	H	H
4	L	H	L	L
5	L	H	L	H
6	L	H	H	L
7	L	H	H	H
8	H	L	L	L
9	H	L	L	H

90A
BCD Bi-Quinary (5-2)
(See Note B)

Count	Outputs			
	Q_A	Q_D	Q_C	Q_B
0	L	L	L	L
1	L	L	L	H
2	L	L	H	L
3	L	L	H	H
4	L	H	L	L
5	H	L	L	L
6	H	L	L	H
7	H	L	H	L
8	H	L	H	H
9	H	H	L	L

93A
Count Sequence
(See Note C)

Count	Outputs			
	Q_D	Q_C	Q_B	Q_A
0	L	L	L	L
1	L	L	L	H
2	L	L	H	L
3	L	L	H	H
4	L	H	L	L
5	L	H	L	H
6	L	H	H	L
7	L	H	H	H
8	H	L	L	L
9	H	L	L	H
10	H	L	H	L
11	H	L	H	H
12	H	H	L	L
13	H	H	L	H
14	H	H	H	L
15	H	H	H	H

90A
Reset/Count Function Table

Reset Inputs				Outputs			
R0(1)	R0(2)	R9(1)	R9(2)	Q_D	Q_C	Q_B	Q_A
H	H	L	X	L	L	L	L
H	H	X	L	L	L	L	L
X	X	H	H	H	L	L	H
X	L	X	L	COUNT			
L	X	L	X	COUNT			
L	X	X	L	COUNT			
X	L	L	X	COUNT			

93A
Reset/Count Function Table

Reset Inputs		Outputs			
R0(1)	R0(2)	Q_D	Q_C	Q_B	Q_A
H	H	L	L	L	L
L	X	COUNT			
X	L	COUNT			

Note A: Output Q_A is connected to input B for BCD count.

Note B: Output Q_D is connected to input A for bi-quinary count.

Note C: Output Q_A is connected to input B.

Note D: H = High Level, L = Low Level, X = Don't Care.

LOGIC DIAGRAMS

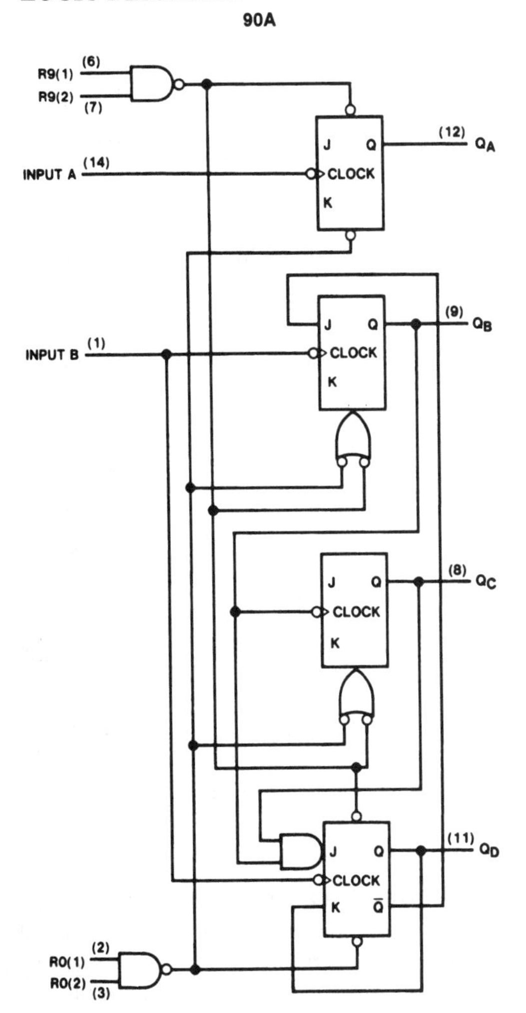

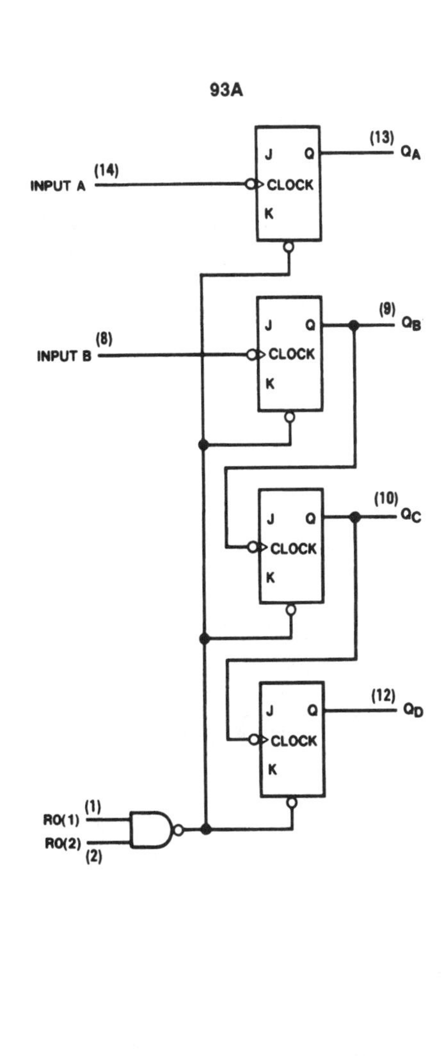

The J and K inputs shown without connection are for reference only and are functionally at a high level.

Raytheon RC555 Timer

- Times from microseconds through hours
- Operates in both astable and monostable modes
- Adjustable duty cycle
- Output drives TTL
- High-current output can source or sink 200 mA
- Temperature stability of 0.005%/°C
- Normally on and normally off output

The RC/RM555 monolithic timing circuit is a highly stable controller capable of producing accurate time delays or oscillation. In the time-delay mode, delay time is precisely controlled by only two external parts: a resistor and a capacitor. For operation as an oscillator, both the free-running frequency and the duty cycle are accurately controlled by two external resistors and a capacitor.

Terminals are provided for triggering and resetting. The circuit will trigger and reset on falling waveforms. The output can source or sink up to 200 mA or drive TTL circuits.

The information contained herein has been carefully compiled. However, it cannot, by implication or otherwise, become part of the terms and conditions of any subsequent sale. Raytheon's liability is determined solely by its standard terms and conditions of sale. No representation as to application or use or that the circuits are either licensed or free from patent infringement is intended or implied. Raytheon reserves the right to change the circuitry and other data at any time without notice and assumes no liability for inadvertent errors.

THERMAL CHARACTERISTICS

	8-Lead Plastic DIP	8-Lead Ceramic DIP	8-Lead TO-99 Metal Can
Max. Junction Temp.	125°C	175°C	175°C
Max. P_D T_A < 50°C	468mW	833mW	658mW
Therm. Res. θ_{JC}	—	45°C/W	50°C/W
Therm. Res. θ_{JA}	160°C/W	150°C/W	190°C/W
For T_A > 50°C Derate at	6.25mW per °C	8.33mW per °C	5.26mW per °C

+Vs (+5V to +15V)

Free-Running Operation

Free-Running Frequency vs. R_A, R_B and C

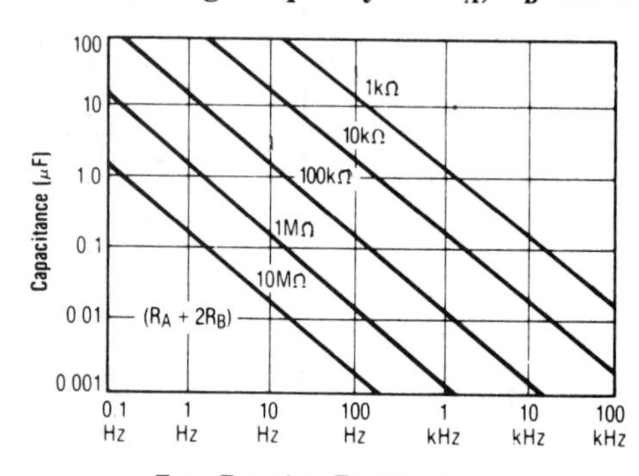

ORDERING INFORMATION

Part Number	Package	Operating Temperature Range
RC555DE	Ceramic	0°C to +70°C
RC555NB	Plastic	0°C to +70°C
RC555T	TO-99	0°C to +70°C
RV555NB	Plastic	-40°C to +85°C
RM555DE	Ceramic	-55°C to +125°C
RM555DE/883B*	Ceramic	-55°C to +125°C
RM555T	TO-99	-55°C to +125°C
RM555T/883B*	TO-99	-55°C to +125°C

FREE-RUNNING OPERATION (ASTABLE)

With the circuit connected as shown, it will trigger itself and free run as a multivibrator. The external capacitor charges through R_A and R_B and discharges through R_B only. Thus, the duty cycle is set by the ratio of these two resistors, and the capacitor charges and discharges between $1/3\ V_S$ and $2/3\ V_S$. Charge and discharge times, and therefore frequency, are independent of supply voltage. The free-running frequency versus R_A, R_B, and C is shown in the graph.

ABSOLUTE MAXIMUM RATINGS

Supply Voltage	+18V
Storage Temperature Range	-65°C to +150°C
Operating Temperature Range	
RC555	0°C to +70°C
RV555	-40°C to +85°C
RM555	-55°C to +125°C
Lead Soldering Temperature (60 Sec)	+300°C

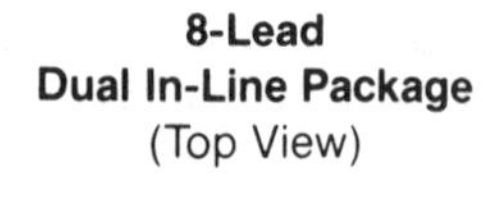

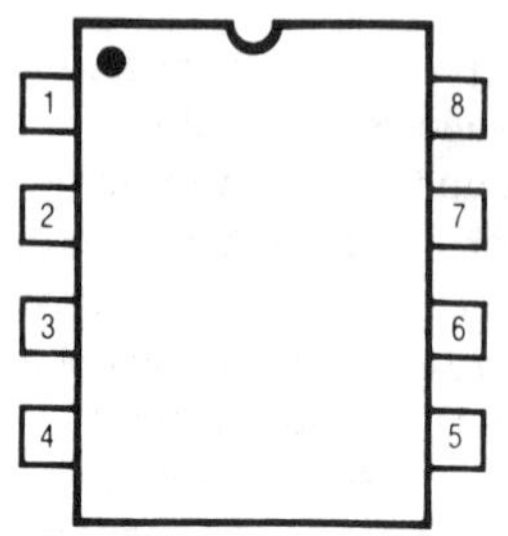

8-Lead
TO-99 Metal Can
(Top View)

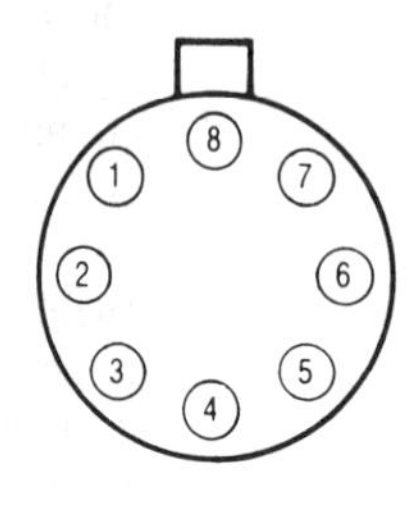

Pin	Function
1	Ground
2	Trigger
3	Output
4	Reset
5	Control Voltage
6	Threshold
7	Discharge
8	$+V_S$

ELECTRICAL CHARACTERISTICS (V_S = +5V to +15V and T_A = +25°C unless otherwise noted)

Parameters	Test Conditions	RM555			RV/RC555			Units
		Min	Typ	Max	Min	Typ	Max	
Supply Voltage		4.5		18	4.5		16	V
Supply Current[1]	V_S = +5V, R_L = ∞		3.0	5.0		4.0	6.0	mA
	V_S = +15V, R_L = ∞ Low State		10	12		10	15	mA
Timing Error[2] Initial Accuracy	R_A, R_B = 1kΩ to 100kΩ		0.5	2.0		1.0		%
V_S Temperature	C = 0.1µF		30	100		50		ppm/°C
V_S Supply Voltage			0.05	0.2		0.1		%/V
Threshold Voltage			2/3			2/3		x V_S
Trigger Voltage	V_S = +15V	4.8	5.0	5.2		5.0		V
	V_S = +5V	1.45	1.67	1.9		1.67		V
Trigger Current			0.5			0.5		µA
Reset Voltage		0.4	0.7	1.0	0.4	0.7	1.0	V
Reset Current			0.1			0.1		mA
Threshold Current[3]			0.1	0.25		0.1	0.25	µA
Control Voltage Level	V_S = +15V	9.6	10	10.4	9.0	10	11	V
	V_S = +5V	2.9	3.33	3.8	2.6	3.33	4.0	V
Output Voltage Drop (Low)	V_S = +15V, I_{SINK} = 10mA		0.1	0.15		0.1	0.25	V
	I_{SINK} = 50mA		0.4	0.5		0.4	0.75	V
	I_{SINK} = 100mA		2.0	2.2		2.0	2.5	V
	I_{SINK} = 200mA		2.5			2.5		V
	V_S = +5V, I_{SINK} = 8mA		0.1	0.25				V
	I_{SINK} = 5mA					0.25	0.35	V
Output Voltage Drop (High)	I_{SOURCE} = 200mA V_S = +15V		12.5			12.5		V
	I_{SOURCE} = 100mA V_S = +15V	13	13.3		12.75	13.3		V
	V_S = +5V	3.0	3.3		2.75	3.3		V
Rise Time of Output			100			100		nS
Fall Time of Output			100			100		nS

Notes: 1. Supply current when output high typically 1mA less.
2. Tested at V_S = +5V and V_S = +15V.
3. This will determine the maximum value of $R_A + R_B$. For +15V operation, the maximum total R = 20MΩ.

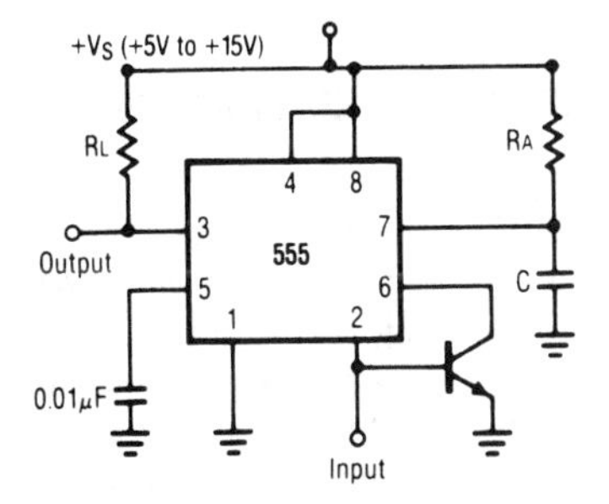

Missing Pulse Detector

MISSING PULSE DETECTOR

With the RC/RM555 connected as shown, the timing cycle will be continuously reset by the input pulse train. A change in frequency or a missing pulse allows the timing cycle to go to completion and change the output level. For proper operation, the time delay should be set slightly longer than the normal time between pulses.

SCHEMATIC DIAGRAM

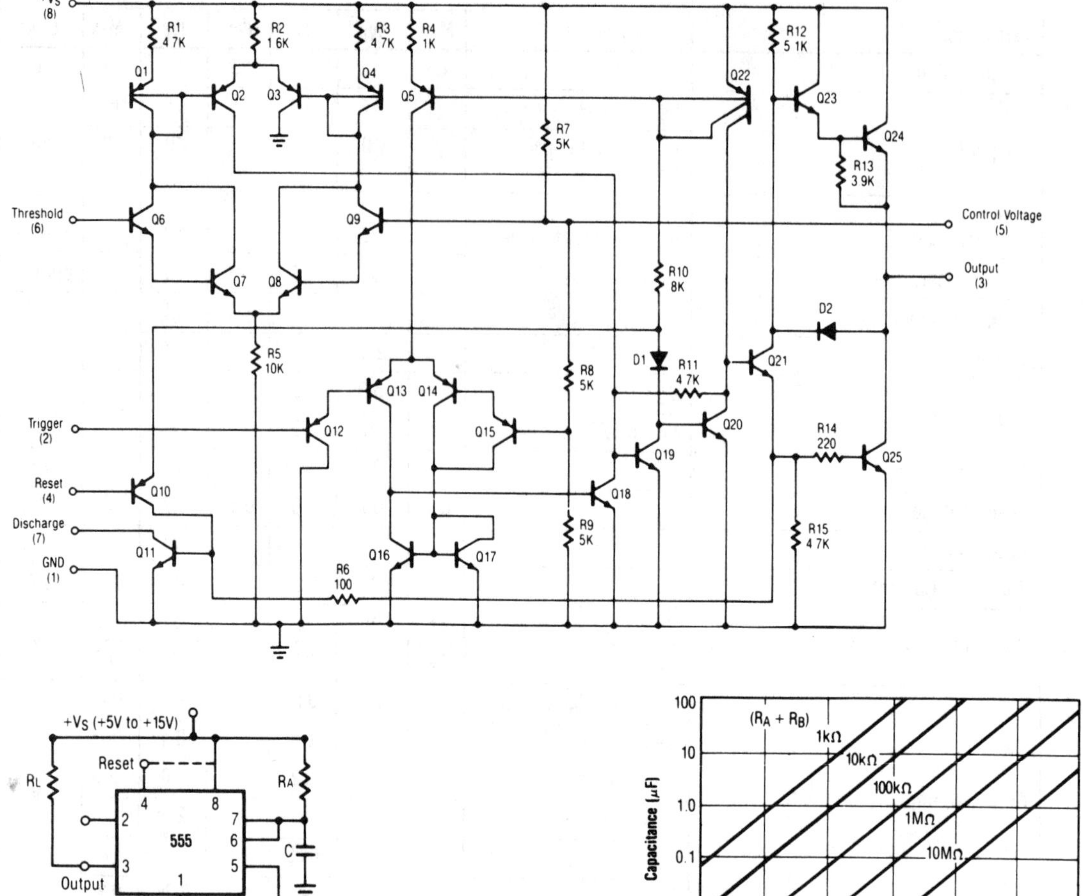

Time Delay vs. R_A, R_B and C

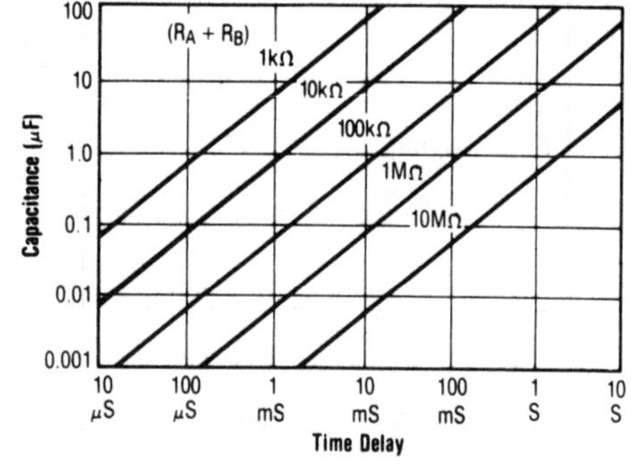

Monostable Operation

MONOSTABLE OPERATION

In this mode, the timer functions as a one-shot. The external capacitor is initially held discharged by a transistor internal to the timer. Applying a negative trigger pulse to pin 2 sets the flip-flop, driving the output high and releasing the short-circuit across the external capacitor. The voltage across the capacitor increases with time constant $r = R_AC$ to $2/3\ V_S$, where the comparator resets the flip-flop and discharges the external capacitor. The output is now in the low state.

Circuit triggering occurs when the negative-going trigger pulse reaches $1/3\ V_S$ and the circuit stays in the output high state until the set time elapses. The time the output remains in the high state is $1.1R_AC$ and it can be determined by the graph. A negative pulse applied to pin 4 (reset) during the timing cycle will discharge the external capacitor and start the cycle over again, beginning on the positive-going edge of the reset pulse. If the reset function is not used, pin 4 should be connected to V_S to avoid false resetting.

Raytheon RC556 Dual Timer

- Times from microseconds through hours
- Operates in both astable and monostable modes
- Adjustable duty cycle
- Output drives TTL
- High-current output can source or sink 200 mA
- Temperature stability of 0.005%/°C
- Normally on and normally off output

The RC556 dual monolithic timing circuit is a highly stable controller capable of producing accurate time delays or oscillation. In the time-delay mode, delay time is precisely controlled by only two external parts: a resistor and a capacitor. For operation as an oscillator, both the free-running frequency and the duty cycle are accurately controlled by two external resistors and a capacitor.

Terminals are provided for triggering and resetting. The circuit will trigger and reset on falling waveforms. The output can source or sink up to 200 mA or drive TTL circuits.

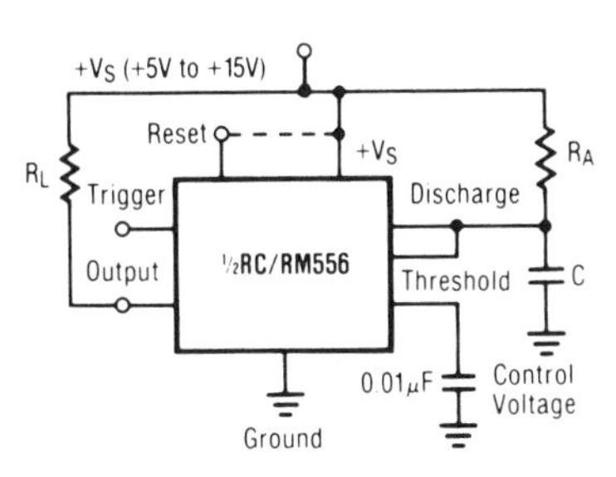

Monostable Operation

Time Delay vs. R_A, R_B and C

Capacitance (μF)

(R_A + R_B) 1kΩ 10kΩ 100kΩ 1MΩ 10MΩ

Time Delay

MONOSTABLE OPERATION

In this mode, the timer functions as a one-shot. The external capacitor is initially held discharged by a transistor internal to the timer. Applying a negative trigger pulse to pin 2 sets the flip-flop, driving the output high and releasing the short-circuit across the external capacitor. The voltage across the capacitor increases with time constant $r = R_AC$ to $2/3\ V_S$, where the comparator resets the flip-flop and discharges the external capacitor. The output is now in the low state.

Circuit triggering occurs when the negative-going trigger pulse reaches $1/3\ V_S$ and the circuit stays in the output high state until the set time elapses. The time the output remains in the high state is $1.1R_AC$, and it can be determined by the graph. A negative pulse applied to pin 4 (reset) during the timing cycle will discharge the external capacitor and start the cycle over again, beginning on the positive-going edge of the reset pulse. If the reset function is not used, pin 4 should be connected to V_S to avoid false resetting.

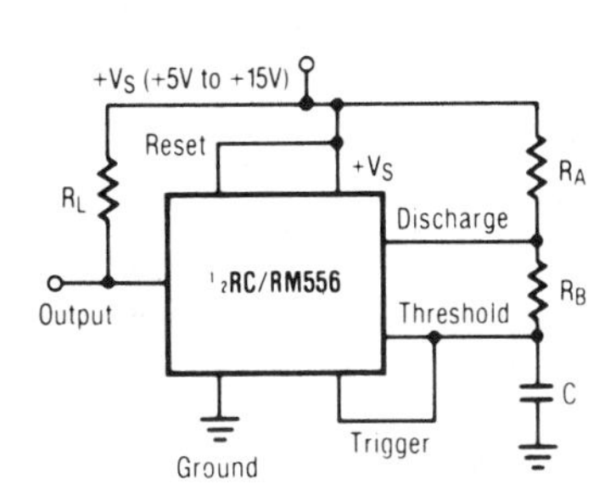

Free-Running Operation

Free-Running Frequency vs. R_A, R_B and C

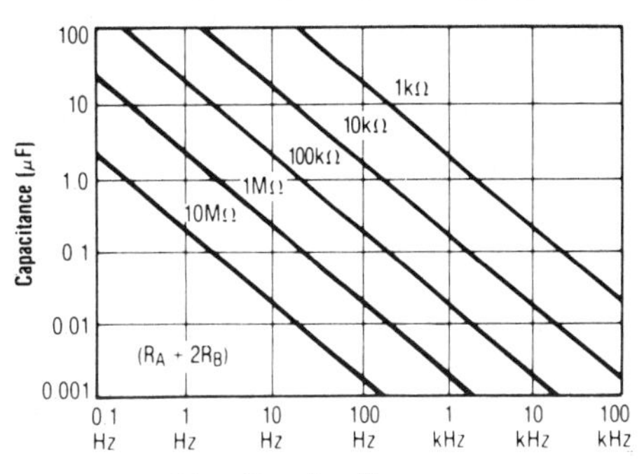

Free-Running Frequency

FREE-RUNNING OPERATION (ASTABLE)

With the circuit connected as shown in the figure, it will trigger itself and free run as a multivibrator. The external capacitor charges through R_A and R_B and discharges through R_B only. Thus, the duty cycle is set by the ratio of these two resistors, and the capacitor charges and discharges between $1/3\ V_S$ and $2/3\ V_S$. Charge and discharge times, and therefore frequency, are independent of supply voltage. The free-running frequency versus R_A, R_B, and C is shown in the graph.

14-Lead Dual In-Line Package (Top View)

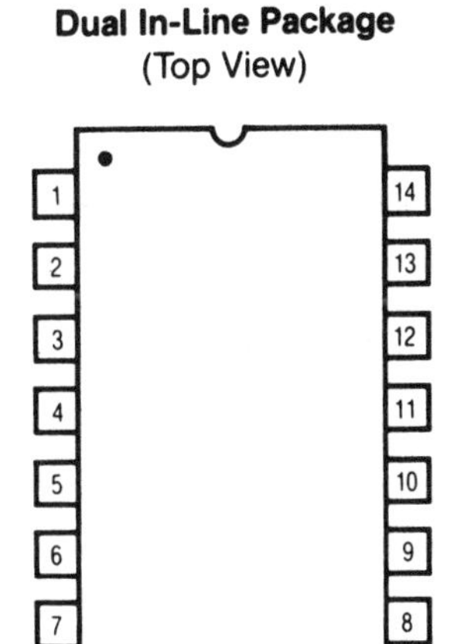

Pin	Function
1	Discharge A
2	Threshold A
3	Control A
4	Reset A
5	Output A
6	Trigger A
7	Ground
8	Trigger B
9	Output B
10	Reset B
11	Control B
12	Threshold B
13	Discharge B
14	$+V_S$

ELECTRICAL CHARACTERISTICS (V_S = +5V to +15V and T_A = +25°C unless otherwise noted)

Parameters	Test Conditions	RM556 Min	RM556 Typ	RM556 Max	RC/RV556 Min	RC/RV556 Typ	RC/RV556 Max	Units
Supply Voltage		4.5		18	4.5		16	V
Supply Current[1]	V_S = +5V, R_L = ∞		3.0	5.0		4.0	6.0	mA
	V_S = +15V, R_L = ∞ Low State		10	11		10	14	mA
Timing Error[2] (Free Running) Initial Accuracy	R_A, R_B = 2kΩ to 100kΩ		1.5			2.25		%
V_S Temperature	C = 0.1µF		90			150		ppm/°C
V_S Supply Voltage			0.15			0.2		%/V
Timing Error[2] (Monostable) Initial Accuracy	R_A, R_B = 2kΩ to 100kΩ		0.5	1.5		0.75		%
V_S Temperature	C = 0.1µF		30	100		50		ppm/°C
V_S Supply Voltage			0.05	0.2		0.1		%/V
Threshold Voltage			2/3			2/3		x V_S
Trigger Voltage	V_S = +15V	4.8	5.0	5.2		5.0		V
	V_S = +5V	1.45	1.67	1.9		1.67		V
Trigger Current			0.5			0.5		µA
Reset Voltage		0.4	0.7	1.0	0.4	0.7	1.0	V
Reset Current			0.1			0.1		mA
Threshold Current[3]			0.03	0.1		0.03	0.1	µA
Control Voltage Level	V_S = +15V	9.6	10	10.4	9.0	10	11	V
	V_S = +5V	2.9	3.33	3.8	2.6	3.33	4.0	V
Output Voltage Drop (Low)	V_S = +15V, I_{SINK} = 10mA		0.1	0.15		0.1	0.25	V
	I_{SINK} = 50mA		0.4	0.5		0.4	0.75	V
	I_{SINK} = 100mA		2.0	2.25		2.0	2.75	V
	I_{SINK} = 200mA		2.5			2.5		V
	V_S = +5V, I_{SINK} = 8mA		0.1	0.25				V
	I_{SINK} = 5mA					0.25	0.35	V
Output Voltage Drop (High)	I_{SOURCE} = 200mA V_S = +15V		12.5			12.5		V
	I_{SOURCE} = 100mA V_S = +15V	13	13.3		12.75	13.3		V
	V_S = +5V	3.0	3.3		2.75	3.3		V
Rise Time of Output			100			100		nS
Fall Time of Output			100			100		nS
Matching Characteristics Between Each Section Initial Timing Accuracy			0.3	0.6		0.5	1.0	%
V_S Temperature			±10			±10		ppm/°C
V_S Supply Voltage			0.1	0.2		0.2	0.5	%/V

Notes: 1. Supply current when output high typically 2mA less.
2. Tested at V_S = +5V and V_S = +15V.
3. This will determine the maximum value of $R_A + R_B$. For +15V operation, the maximum total R = 20MΩ.

SCHEMATIC DIAGRAM (½ SHOWN)

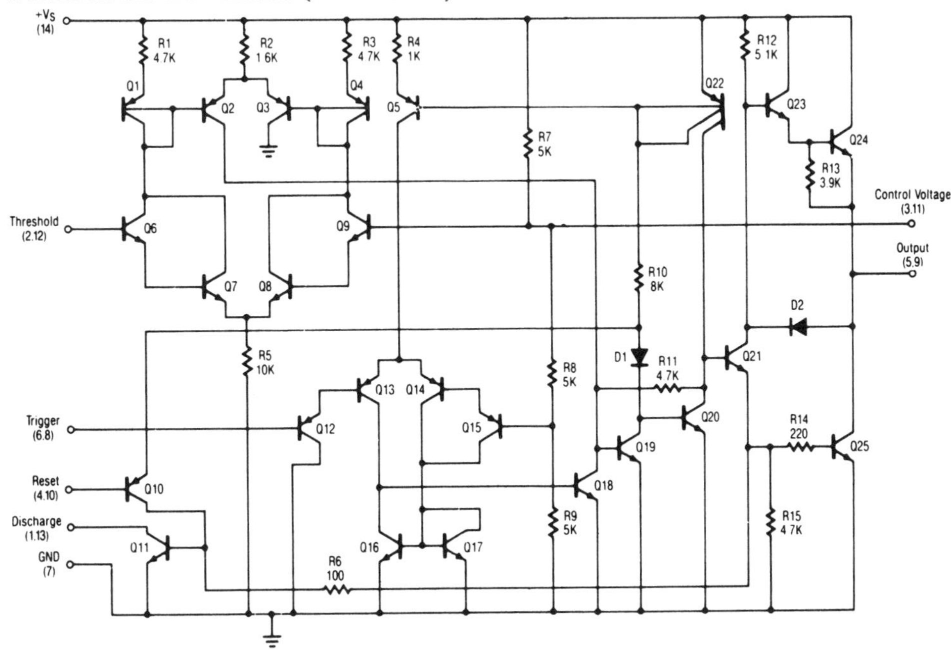

THERMAL CHARACTERISTICS

	14-Lead Plastic DIP	14-Lead Ceramic DIP
Max. Junction Temp.	125°C	175°C
Max. P_D T_A < 50°C	468mW	1042mW
Therm. Res. θ_{JC}	—	60°C/W
Therm. Res. θ_{JA}	160°C/W	120°C/W
For T_A > 50°C Derate at	6.25mW per °C	8.38mW per °C

ORDERING INFORMATION

Part Number	Package	Operating Temperature Range
RC556DB	Plastic	0°C to +70°C
RC556DC	Ceramic	0°C to +70°C
RV556DB	Plastic	-40°C to +85°C
RM556DC	Ceramic	-55°C to +125°C
RM556DC/883B*	Ceramic	-55°C to +125°C

*MIL-STD-883, Level B Processing

ABSOLUTE MAXIMUM RATINGS

Supply Voltage	+18V
Storage Temperature Range	-65°C to +150°C
Operating Temperature Range	
RC556	0°C to +70°C
RM556	-55°C to +125°C
RV556	-40°C to +85°C
Lead Soldering Temperature (60 Sec)	+300°C

Rohm BA222

Monolithic Timer

- Timer interval can be set from microseconds to several hours
- Load current: up to 200 mA
- Directly compatible with TTL and DTL devices
- Temperature stability: 50 ppm/°C (typ.)
- Supply voltage regulation: 0.1%/V (typ.)

The BA222 is a monolithic timer suitable for instrumentation, control and digital data processing applications with minimum external component requirements.

The timer interval can be set with an external RC time constant over a range from a few microseconds to several hours. The device has a wide range of applications, including monostable and astable multivibrators.

APPLICATIONS

- Delay timers
- Monostable multivibrators
- Astable multivibrators
- Pulse generators
- Frequency dividers
- Sequence timers

ABSOLUTE MAXIMUM RATINGS (Ta=25°C)

Parameter	Symbol	Limits	Unit
Supply voltage	V_{CC}	18	V
Power dissipation	Pd	500*	mW
Operating temperature range	Topr	−10~75	°C
Storage temperature range	Tstg	−55~125	°C

*Derating is done at 5.0mW/°C for operation above Ta=25°C.

ELECTRICAL CHARACTERISTICS (Ta=25°C, V_{CC}=+5V, +15V)

Parameter	Symbol	Min.	Typ.	Max.	Unit	Conditions	Test circuit
Supply voltage	V_{CC}	4.5	—	16.0	V	—	Fig. 13
Quiescent current	I_Q1	—	3	7	mA	V_{CC}=5V, $R_L=\infty$	Fig. 13
	I_Q2	—	10	15	mA	V_{CC}=15V, $R_L=\infty$	Fig. 13
Monostable operating timing	$T_{ERR}(M)$	—	1	—	%	R_A=1kΩ~100kΩ,C=0.1μF	Fig. 14
Monostable operating timing temperature regulation	$T_{DT}(M)$	—	50	—	ppm/°C	R_A=1kΩ~10kΩ, C=0.1μF	Fig. 14
Monostable operating timing power regulation	$T_{DS}(M)$	—	0.1	—	%/V	R_A=1kΩ~10kΩ, C=0.1μF	Fig. 14
Astable operating timing regulation	$T_{ERR}(A)$	—	2.5	—	%	$R_A=R_B$=1kΩ~100kΩ, C=0.1μF	Fig. 15
Astable operating timing temperature regulation	$T_{DT}(A)$	—	150	—	ppm/°C	$R_A=R_B$=1kΩ~10kΩ, C=0.1μF	Fig. 15
Astable operating timing power regulation	$T_{DS}(A)$	—	0.3	—	%/V	$R_A=R_B$=1kΩ~10kΩ, C=0.1μF	Fig. 15
Threshold voltage	V_{TH}	—	2/3 × V_{CC}	—	V	—	Fig. 13
Threshold current	I_{TH}	—	0.1	0.25	μA	—	Fig. 13
Trigger voltage	V_T	—	1/3 × V_{CC}	—	V	—	Fig. 13
Trigger current	I_T	—	0.5	—	μA	—	Fig. 13
Reset voltage	V_R	—	0.7	1.0	V	—	Fig. 13
Reset current	I_R	—	0.1	—	mA	—	Fig. 13
Control voltage	$V_{CRT}1$	2.60	3.33	4.00	V	—	Fig. 13
	$V_{CRT}2$	9.0	10.0	11.0	V	—	Fig. 13
Output low voltage	$V_{OL}1$	—	0.25	0.35	V	V_{CC}=5V, Isink=5mA	Fig. 13
	$V_{OL}2$	—	0.10	0.25	V	V_{CC}=15V, Isink=10mA	Fig. 13
	$V_{OL}3$	—	0.40	0.75	V	V_{CC}=15V, Isink=50mA	Fig. 13
	$V_{OL}4$	—	2.0	2.5	V	V_{CC}=15V, Isink=100mA	Fig. 13
	$V_{OL}5$	—	2.5	—	V	V_{CC}=15V, Isink=200mA	Fig. 13
Output high voltage	$V_{OH}1$	2.75	3.30	—	V	V_{CC}=5V, Isource=100mA	Fig. 13
	$V_{OH}2$	12.75	13.30	—	V	V_{CC}=15V, Isource=100mA	Fig. 13
	$V_{OH}3$	—	12.50	—	V	V_{CC}=15V, Isource=200mA	Fig. 13
Output rise time	tr	—	100	—	ns	—	Fig. 13
Output fall time	tf	—	100	—	ns	—	Fig. 13

CIRCUIT DIAGRAM

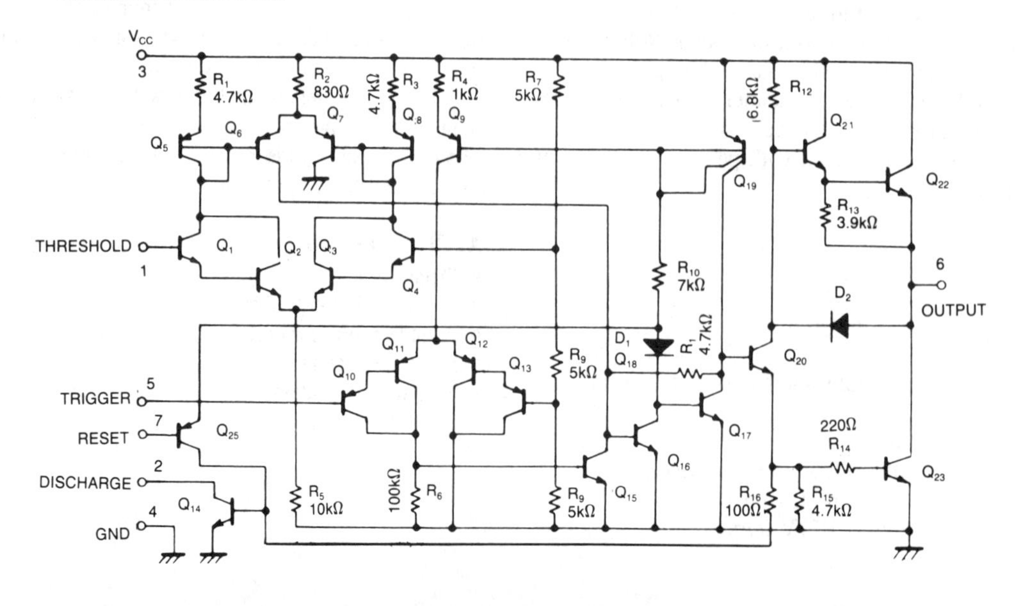

DIMENSIONS (Unit: mm)

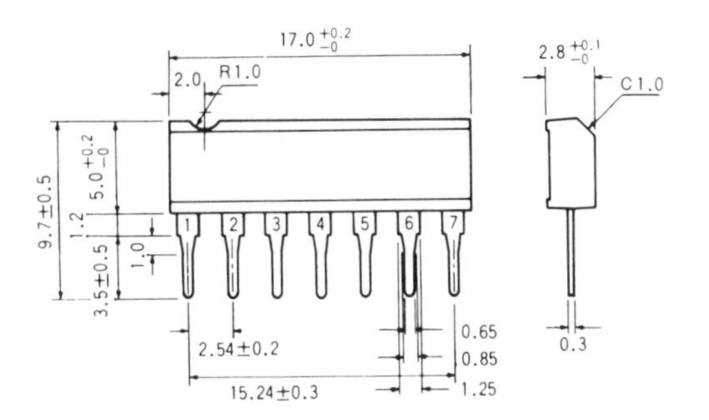

BLOCK DIAGRAM

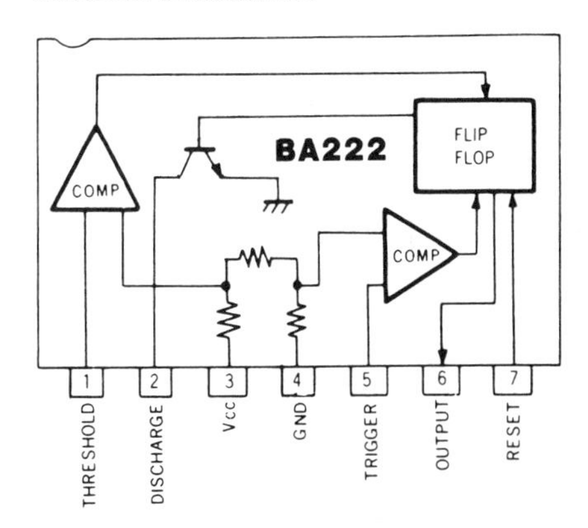

APPLICATION EXAMPLES

1. Monostable multivibrator

An application circuit and timing for the monostable multivibrator is shown.

When no trigger signal is applied, the output of the device is low with the timing capacitor C fully discharged. When a trigger signal is applied, the output is set high to start charging timing capacitor C. The charging time depends on the time constant, which is the product of the external timing resistor R_A and capacitor C. When the capacitor is charged to a potential of $2/3\ V_{CC}$, an internal flip-flop resets the output from high to low. At the same time, the capacitor is discharged to prepare for the next trigger input. The timer can be triggered by applying a voltage of $1/3\ V_{CC}$ or less to pin 5. Once the device is triggered, it ignores any subsequent trigger until the timer interval expires.

2. Astable multivibrator

An application as an astable multivibrator is shown. Timing capacitor C is charged from V_{CC} through external timing resistors R_A and R_B, and is discharged through R_B.

So, the output-duty ratio depends on the ratio of these timing resistors. Capacitor C repeatedly charges and discharges between potentials $1/3\ V_{CC}$ and $2/3\ V_{CC}$. When the capacitor's potential is at $1/3\ V_{CC}$, an internal flip-flop is set, which sets the output high and starts charging the capacitor. When the capacitor potential reaches $2/3\ V_{CC}$, the flip-flop is reset, which resets the output low. At the same time, capacitor C starts discharging through R_B.

TEST CIRCUITS

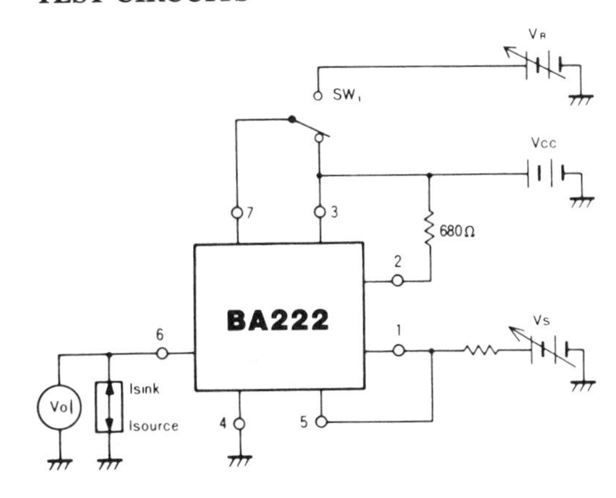

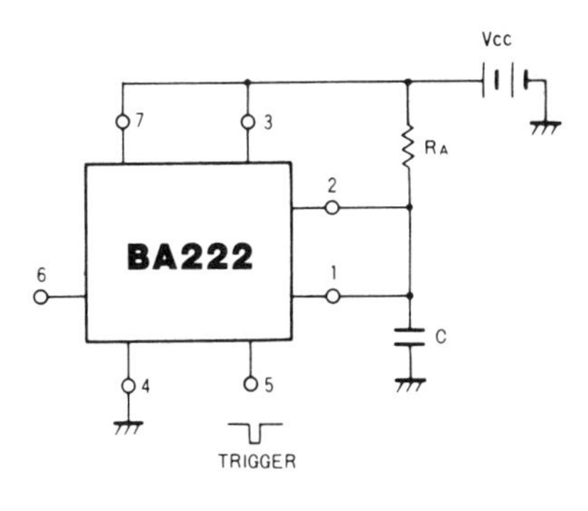

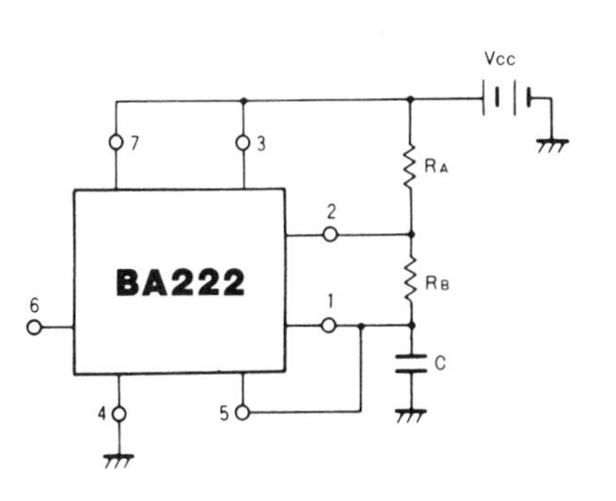

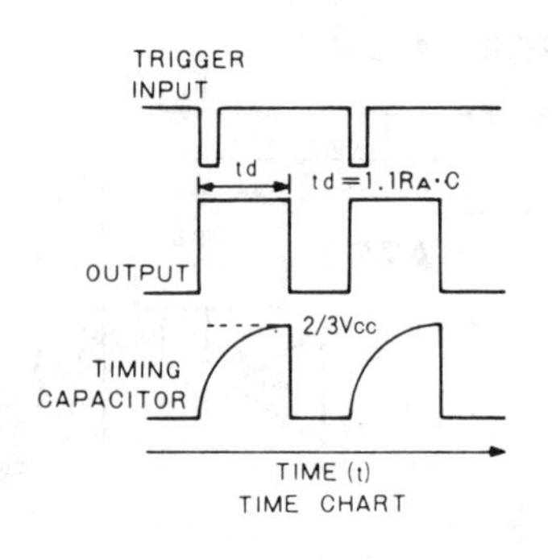

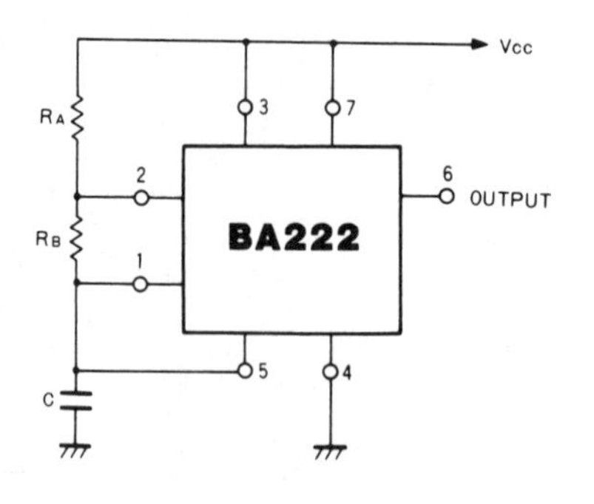

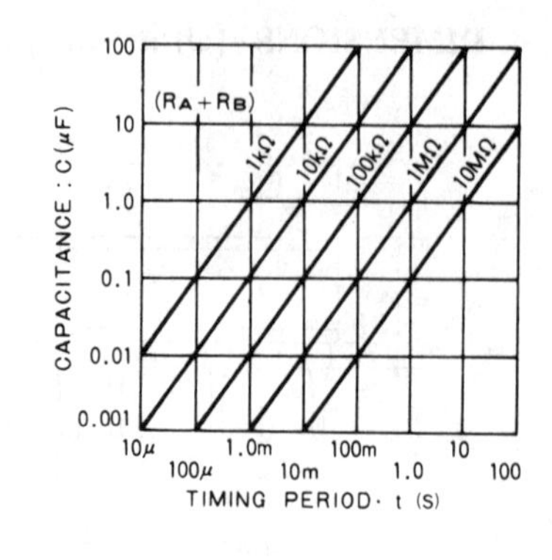

Monostable multivibrator

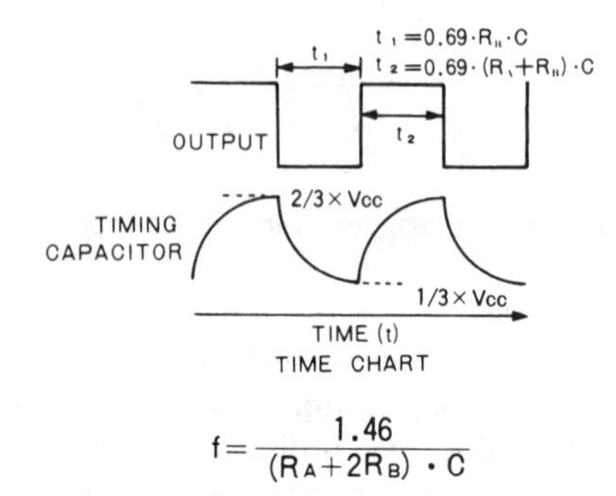

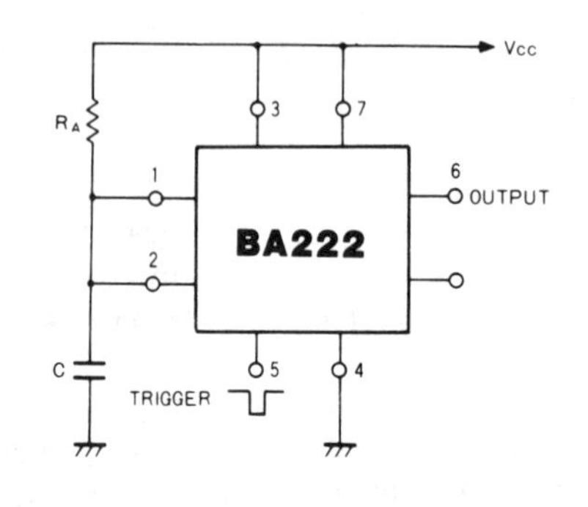

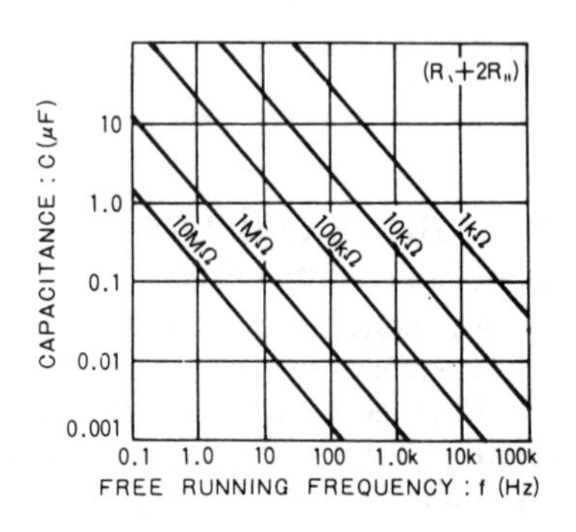

Astable multivibrator

ELECTRICAL CHARACTERISTIC CURVES

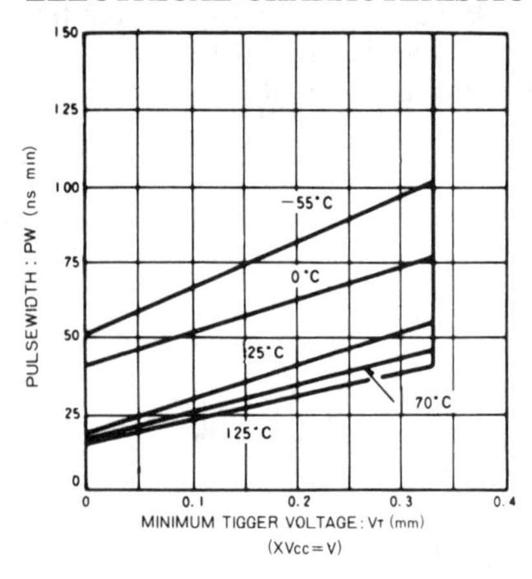

Pulse width vs. minimum trigger voltage

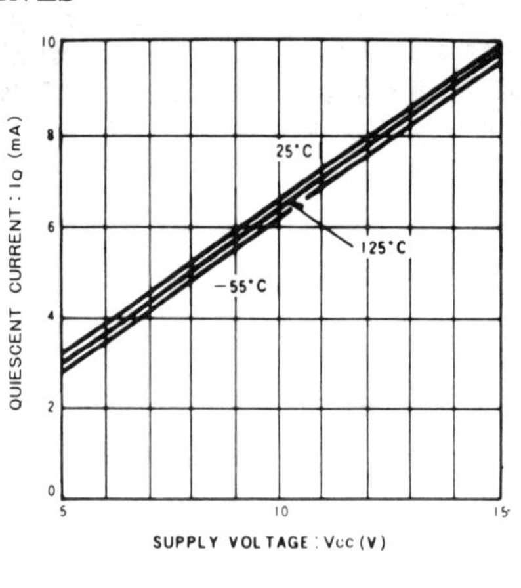

Quiescent current vs. supply voltage

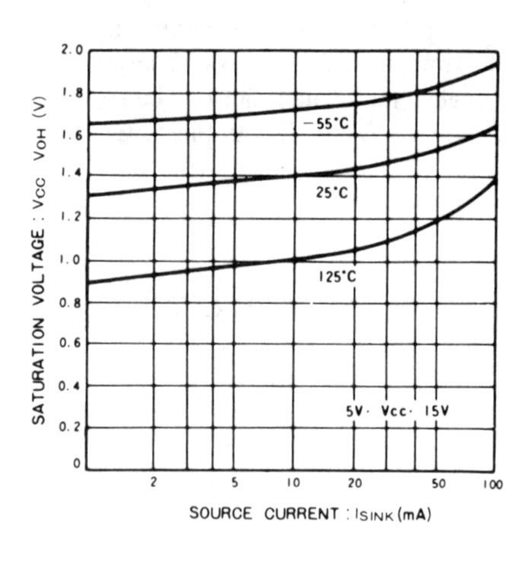

Saturation voltage vs. source current

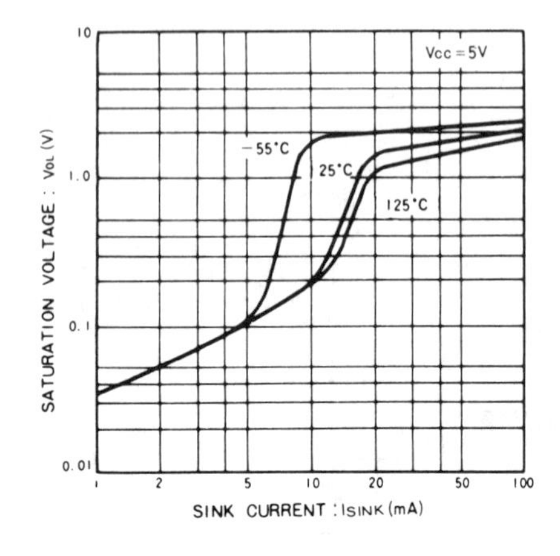

Saturation voltage vs. sink current

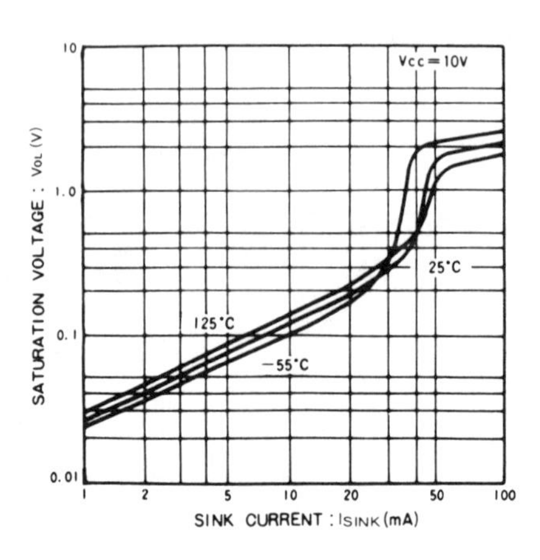

Saturation voltage vs. sink current

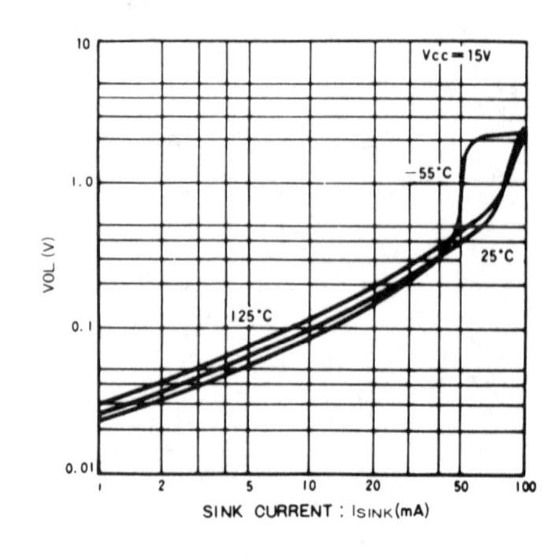

Voltage vs. sink current

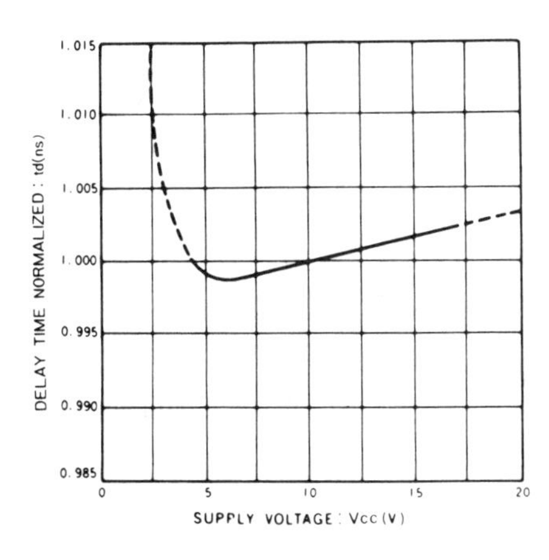

Delay time normalized vs. supply voltage

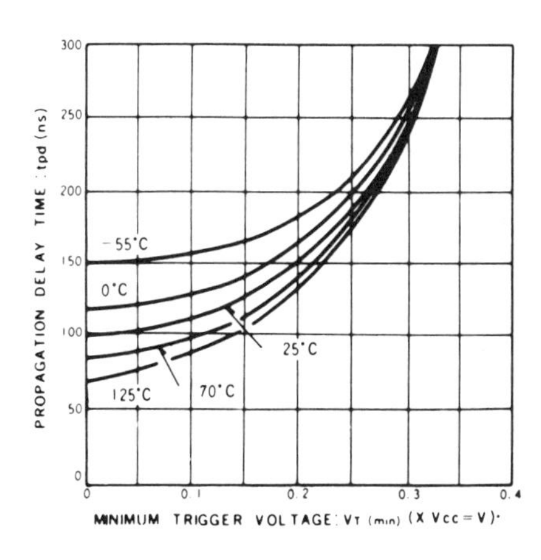

Propagation delay time vs. minimum trigger voltage

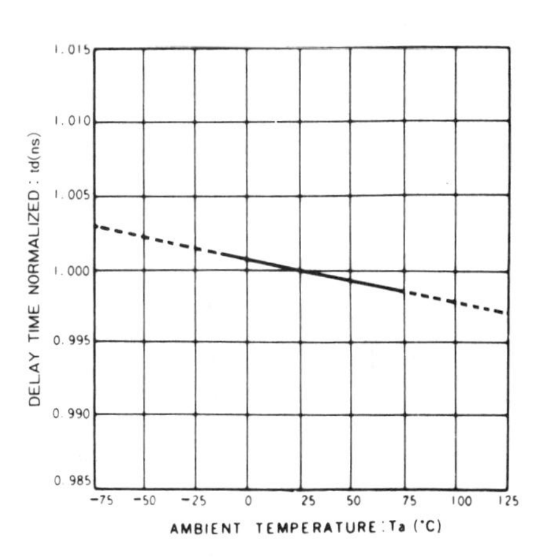

Delay time normalized vs. ambient temperature

Rohm BA223 Monolithic Timer

- Timer interval can be set from microseconds to several hours
- Wide range of applications, including monostable and astable multivibrators
- Load current up to 200 mA
- Temperature stability: 50 ppm/°C (typ.)
- Supply voltage regulation: 0.01%/V (typ.)
- Directly compatible with TTL and DTL devices

The BA223 is a monolithic timer that is suitable for instrumentation, control and digital data processing applications. The timer interval can be set with an external RC time constant over a range from a few microseconds to several hours. The device has a wide range of applications, including monostable and astable multivibrators.

With a load current of up to 200 mA, the device can directly drive TTL and DTL devices.

APPLICATIONS

- Delay timers
- Monostable multivibrators
- Astable multivibrators
- Pulse generators
- Frequency dividers
- Sequence timers

CIRCUIT DIAGRAM

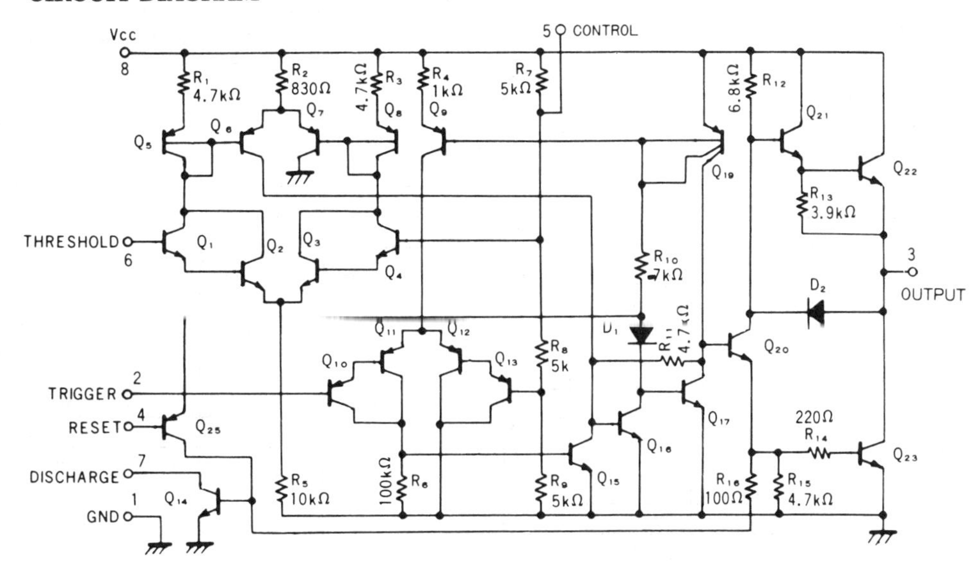

DIMENSIONS (Unit: mm)

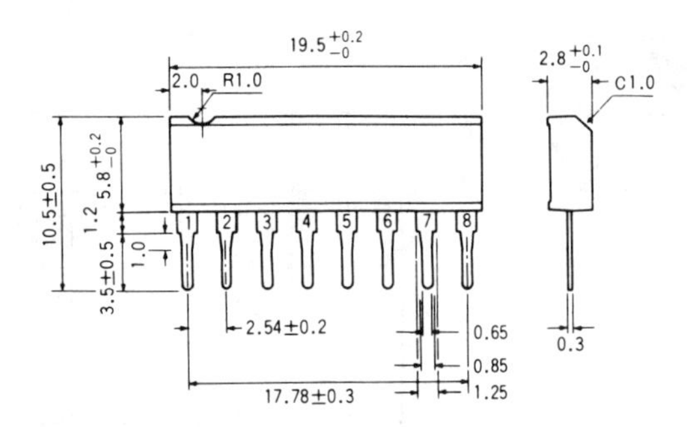

BLOCK DIAGRAM

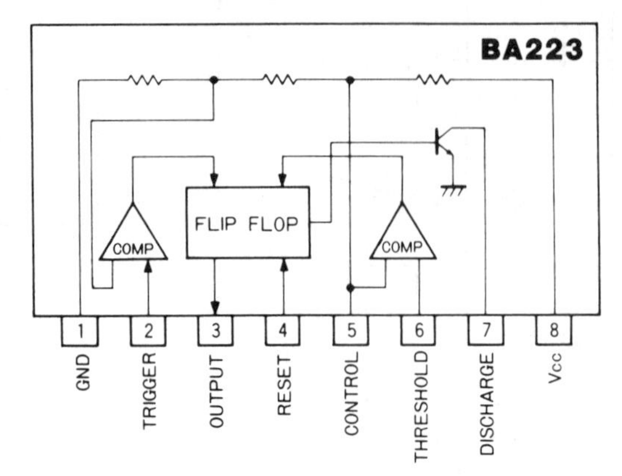

ABSOLUTE MAXIMUM RATINGS (Ta=25°C)

Parameter	Symbol	Limits	Unit
Supply voltage	V_{CC}	18	V
Power dissipation	Pd	550*	mW
Operating temperature range	Topr	−10~75	°C
Storage temperature range	Tstg	−55~125	°C

*Derating is done at 5.5mW/°C for operation above Ta=25°C.

ELECTRICAL CHARACTERISTICS (Ta=25°C, V_{CC}=+5V, +15V)

Parameter	Symbol	Min.	Typ.	Max.	Unit	Conditions	Test circuit
Supply voltage	V_{CC}	4.5	—	16.0	V	—	Fig. 13
Quiescent current	I_Q1	—	3	6	mA	V_{CC}=5V, $R_L=\infty$	Fig. 13
	I_Q2	—	10	15	mA	V_{CC}=15V, $R_L=\infty$	Fig. 13
Monostable operating timing	T_{ERR}(R)	—	1	—	%	R_A=1kΩ~100kΩ, C=0.1μF	Fig. 14
Monostable operating timing temperature regulation	T_{DT}(M)	—	50	—	ppm/°C	R_A=1kΩ~100kΩ, C=0.1μF	Fig. 14
Monostable operating timing power regulation	T_{DS}(M)	—	0.1	—	%/V	R_A=1kΩ~100kΩ, C=0.1μF	Fig. 14
Astable operating timing regulation	T_{ERR}(A)	—	2.5	—	%	$R_A=R_B$=1kΩ~100kΩ, C=0.1μF	Fig. 15
Astable operating timing temperature regulation	T_{DT}(A)	—	150	—	ppm/°C	$R_A=R_B$=1kΩ~100kΩ, C=0.1μF	Fig. 15
Astable operating timing power regulation	T_D(A)	—	0.3	—	%/V	$R_A=R_B$=1kΩ~100kΩ, C=0.1μF	Fig. 15
Threshold voltage	V_{TH}	—	2/3 x V_{CC}	—	V	—	Fig. 13
Threshold current	I_{TH}	—	0.1	0.25	μA	—	Fig. 13
Trigger voltage	V_T	—	1/3 x V_{CC}	—	V	—	Fig. 13
Trigger current	I_T	—	0.5	—	μA	—	Fig. 13
Reset voltage	V_R	—	0.7	1.0	V	—	Fig. 13
Reset current	I_R	—	0.1	—	mA	—	Fig. 13
Control voltage	$V_{CRT}1$	2.60	3.33	4.00	V	—	Fig. 13
	$V_{CRT}2$	9.0	10.0	11.0	V	—	Fig. 13
Output low voltage	$V_{OL}1$	—	0.25	0.35	V	V_{CC} =5V, Isink=5mA	Fig. 13
	$V_{OL}2$	—	0.10	0.25	V	V_{CC}=15V, Isink=10mA	Fig. 13
	$V_{OL}3$	—	0.40	0.75	V	V_{CC}=15V, Isink=50mA	Fig. 13
	$V_{OL}4$	—	2.0	2.5	V	V_{CC}=15V, Isink=100mA	Fig. 13
	$V_{OL}5$	—	2.5	—	V	V_{CC}=15V, Isink=200mA	Fig. 13
Output high voltage	$V_{OH}1$	2.75	3.30	—	V	V_{CC} =5V, Isource=100mA	Fig. 13
	$V_{OH}2$	12.75	13.30	—	V	V_{CC}15V, Isource=100mA	Fig. 13
	$V_{OH}3$	—	12.50	—	V	V_{CC}=15V, Isource=200mA	Fig. 13
Output rise time	tr	—	100	—	ns	—	Fig. 13
Output fall time	tf	—	100	—	ns	—	Fig. 13

ELECTRICAL CHARACTERISTIC CURVES

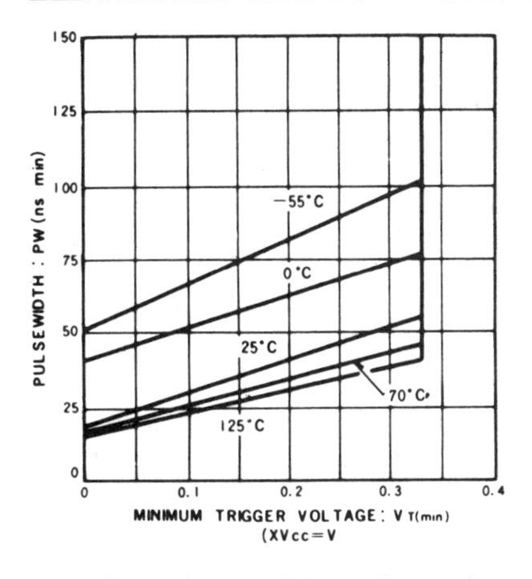

Pulse width vs. minimum trigger voltage

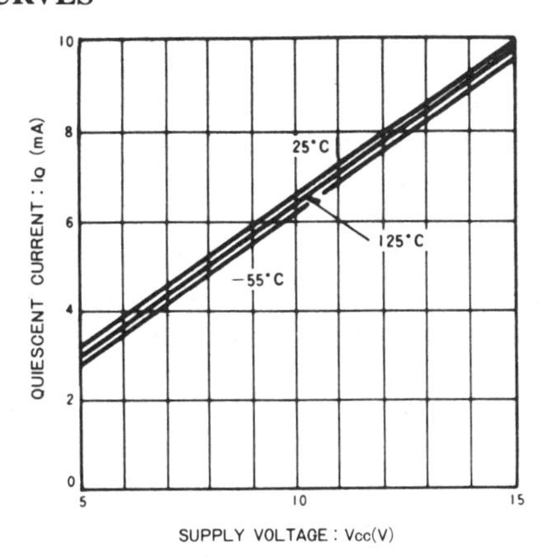

Quiescent current vs. supply voltage

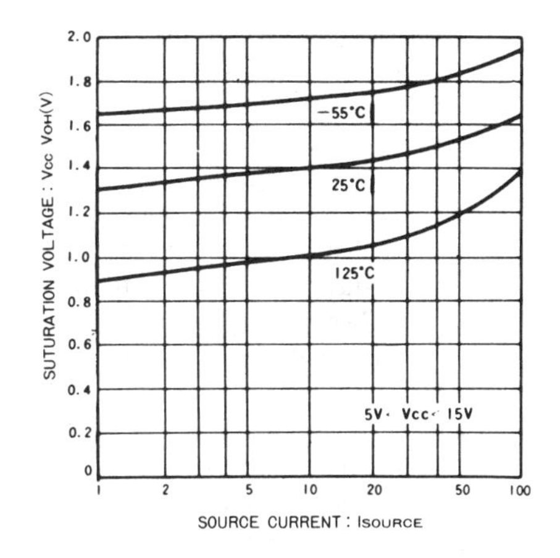

Saturation voltage vs. source current

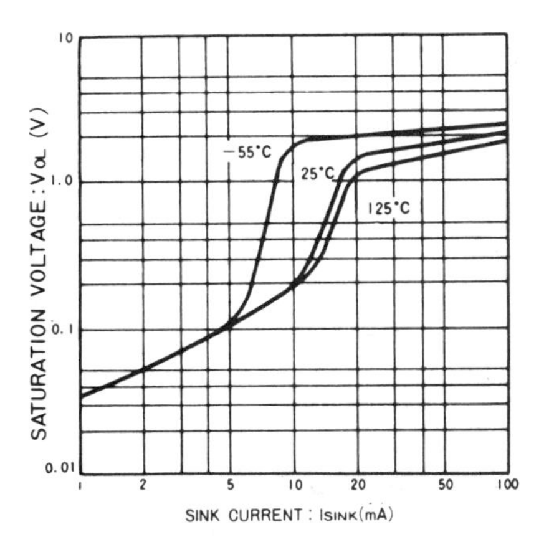

Saturation voltage vs. sink current

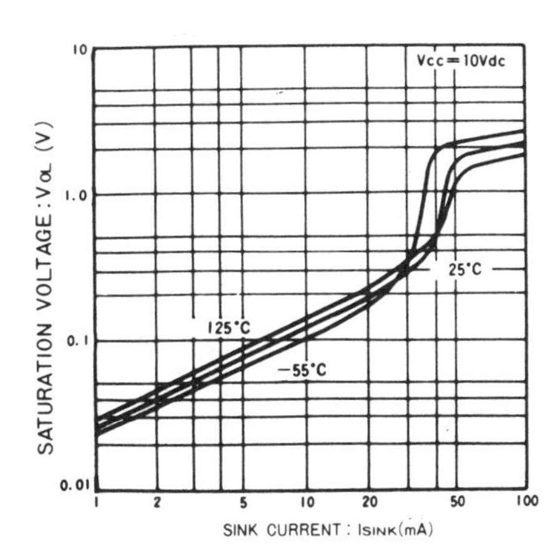

Saturation voltage vs. sink current

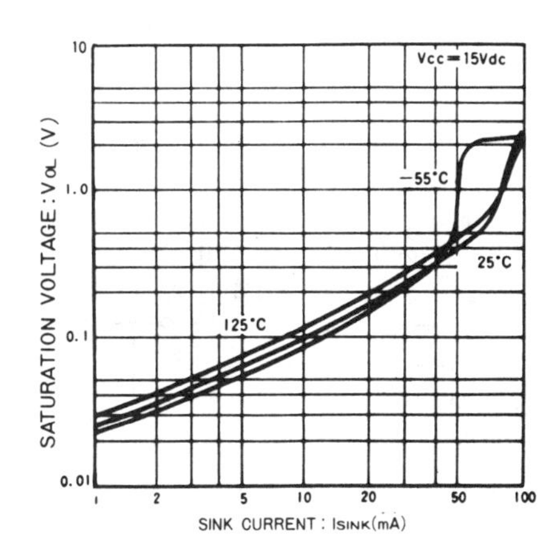

Saturation voltage vs. sink current

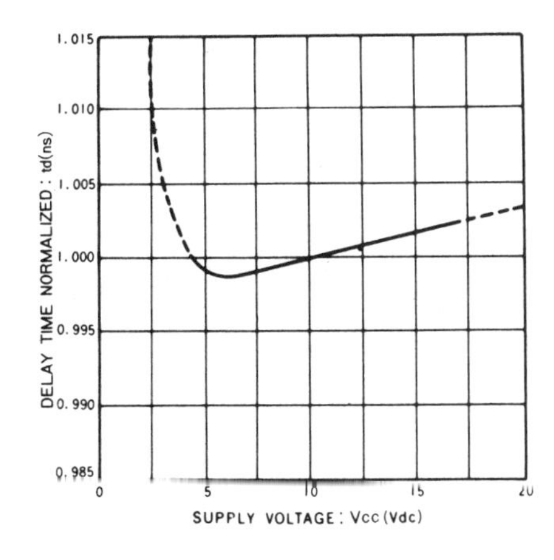

Delay time normalized vs. supply voltage

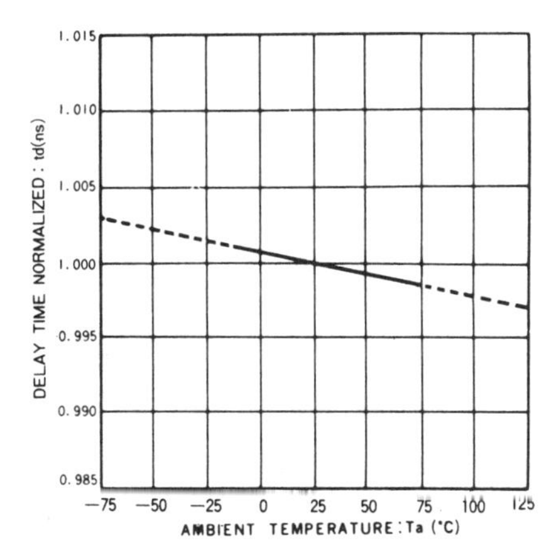

Delay time normalized vs. ambient temperature

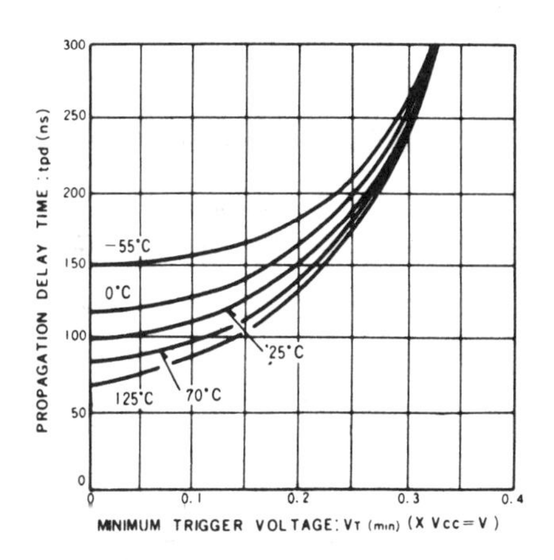

Propagation delay time vs. minimum trigger voltage

APPLICATION EXAMPLES

1. Monostable multivibrator

An application circuit for the monostable multivibrator is shown. When no trigger signal is applied, the output of the device is low and timing capacitor C is fully discharged. When a trigger signal is applied, the output is set high to start charging timing capacitor C.

The charging time depends on the time constant, which is the product of the external timing resistor R_A and capacitor C. When the capacitor is charged to a potential of 2/3 V_{CC}, an internal flip-flop resets the output from high to low. At the same time, the capacitor is discharged to prepare for the next trigger input.

The timer can be triggered by applying a voltage of 1/3 V_{CC} (or less) to pin 2. Once the device is triggered, it ignores any subsequent trigger until the timer interval expires. The capacitor (0.01 μF) connected to pin 5 is used to reduce the ac impedance for an internal comparison voltage.

2. Astable multivibrator

An application as an astable multivibrator is shown. Timing capacitor C is charged from V_{CC} through external timing resistors R_A and R_B, and is discharged through R_B.

So, the output-duty ratio depends on the ratio of these timing resistors. Capacitor C repeatedly charges and discharges between potentials 1/3 V_{CC} and 2/3 V_{CC}. When the capacitor's potential is at 1/3 V_{CC}, an internal flip-flop is set, which sets the output high and starts charging the capacitor.

When the capacitor potential reaches 2/3 V_{CC}, the flip-flop is reset, which resets the output low. At the same time, the capacitor starts discharging through R_B. The capacitor (0.01 μF) connected to pin 5 is used to reduce the ac impedance for an internal comparison voltage.

TEST CIRCUITS

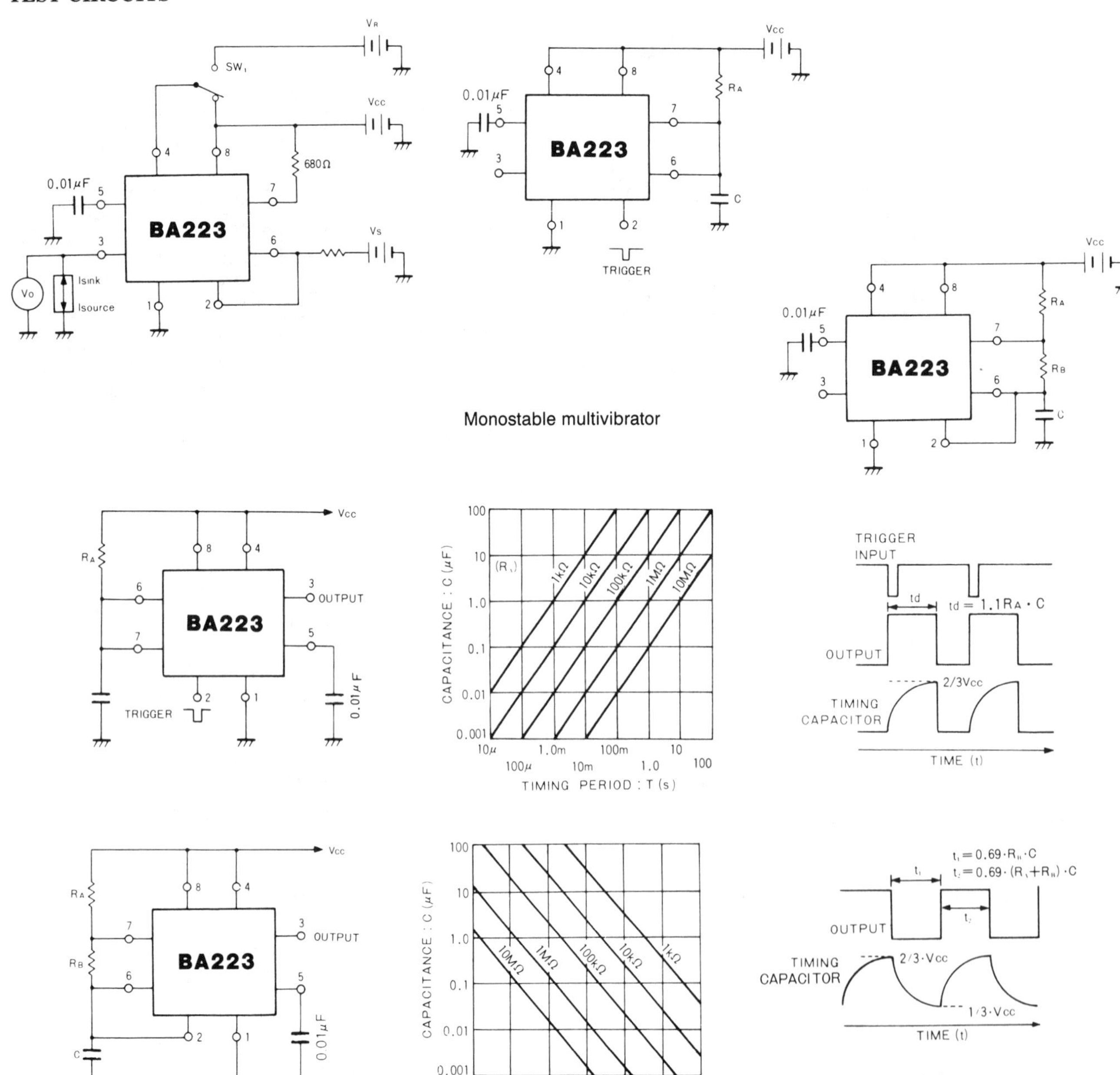

Monostable multivibrator

Astable multivibrator

Rohm BA225, BA225F, BA235 (BA226, BA226F, BA236)

Dual Monostable Multivibrators

- Leading-edge trigger input requires no external differentiation network
- Dual circuit configuration makes it possible to build a monolithic delay timer
- Minimum external component requirement
- Low current consumption: 0.75 mA per circuit
- Circuit current does not change between high and low output, which simplifies power supply circuit design
- Wide supply voltage range: 4.0 to 16.0 V
- BA235 and BA236 have symmetrical pin configuration allowing installation with opposite orientation
- Input hysteresis assures high noise immunity

The BA225, BA225F, and BA235 (BA226, BA226F, BA236) are monolithic dual independent monostable multivibrators, each with low current consumption (0.75 mA typ.). The timer interval can be set with an external RC time constant over a range from a few microseconds to a few minutes.

The BA225, BA225F, and BA235 are leading-edge trigger devices, and are suitable for instrumentation, control and digital data processing applications with minimum external component requirements. The BA226, BA226F, and BA236 are trailing-edge trigger devices.

APPLICATIONS

- Delay timers
- Monostable multivibrators (especially suited for VCR system control applications)
- Pulse generators, etc.

ABSOLUTE MAXIMUM RATINGS (Ta=25°C)

Parameter	Symbol	Limits	Unit
Supply voltage	V_{CC}	16	V
Power dissipation	Pd	450*	mW
Operating temperature range	Topr	−20~75	°C
Storage temperature range	Tstg	−55~125	°C

*Derating is done at 4.5mW/°C for operation above Ta=25°C.

ELECTRICAL CHARACTERISTICS (Ta=25°C, V_{CC}=5V)

Parameter	Symbol	Min.	Typ.	Max.	Unit	Conditions	Test circuit
Suppply voltage range	V_{CC}	4.0	5.0	16.0	V	—	Fig. 9
Quiescent current	I_Q	—	1.5	3.0	mA	—	Fig. 9
Timing	T_{ERR}	—	1	10	%	R=100kΩ=0.1μ *1	Fig. 9
Timing power regulation	T_{DS}	—	0.5	3.0	%/V	5V→16V	Fig. 9
Timing temperature regulation	T_{OT}	—	200	—	ppm/°C	—	Fig. 9
Trigger voltage	V_T	1.0	2.0	3.0	V	— *2	Fig. 9
Trigger current	I_T	—	70	200	μA	—	Fig. 9
Output low voltage	V_{OL}	—	0.5	1.0	V	Isink=5mA	Fig. 9
Output high voltage	V_{OH}	3.0	4.0	—	V	Isource=5mA	Fig. 9

*1 One-shot cycle T=0.5CR
*2 With input hysteresis (hysteresis width ≑ 200-600mV)

BLOCK DIAGRAMS

BA225/BA225F
BA226/BA226F

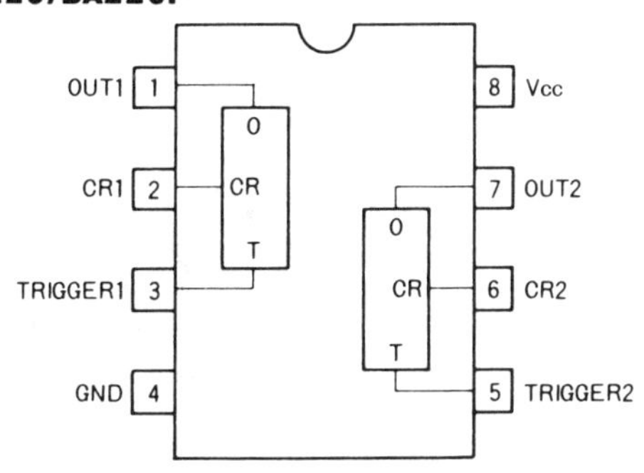

BA235/BA236

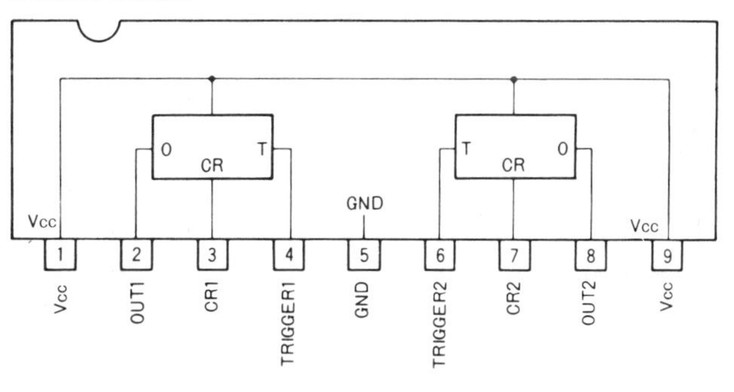

DIMENSIONS (Unit: mm)

BA225/BA226

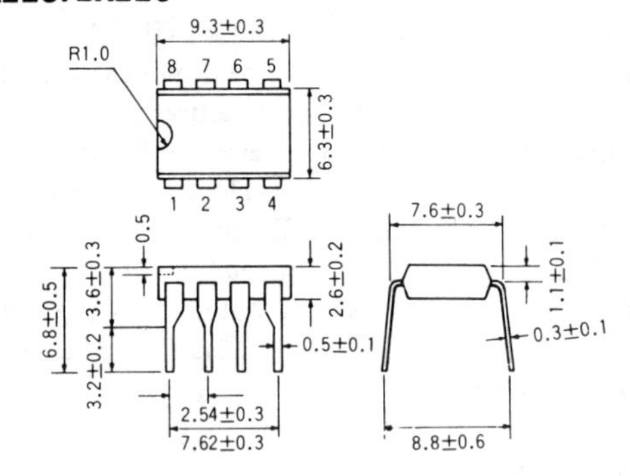

BA235/BA236

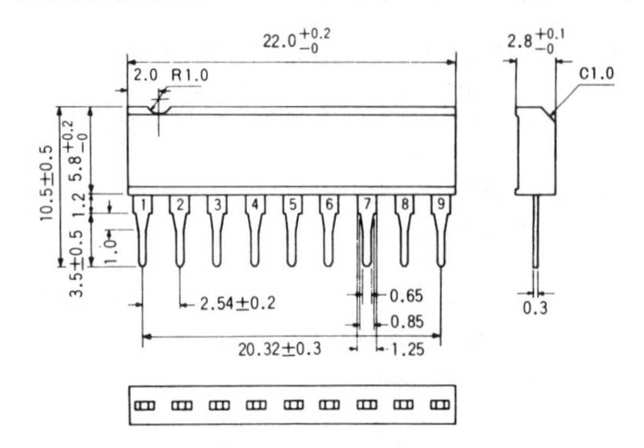

BA225F/BA226F

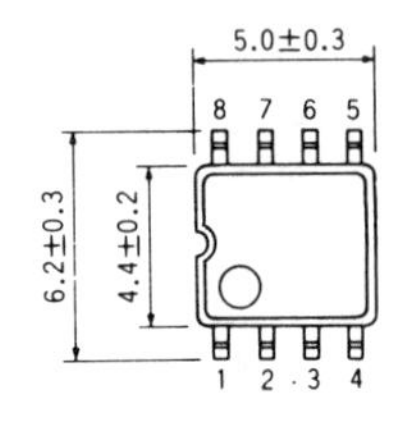

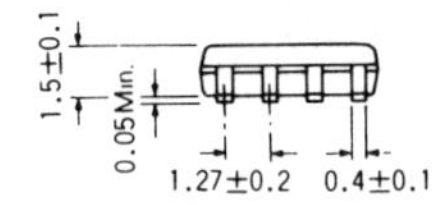

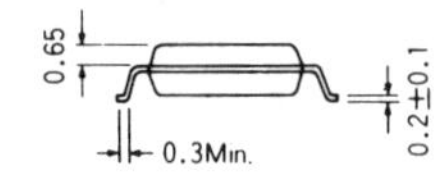

ELECTRICAL CHARACTERISTIC CURVES

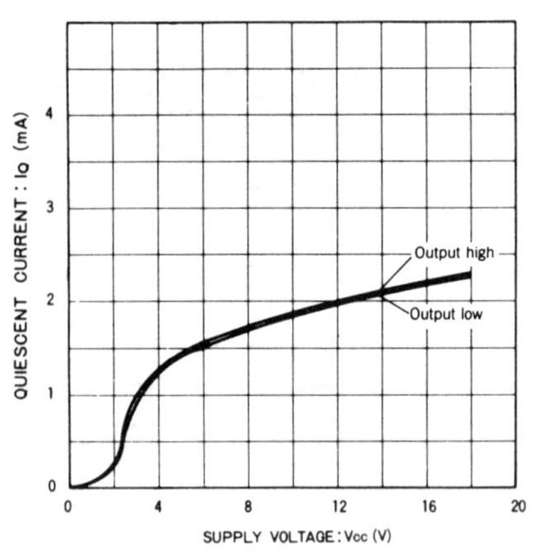

Quiescent current vs. supply voltage

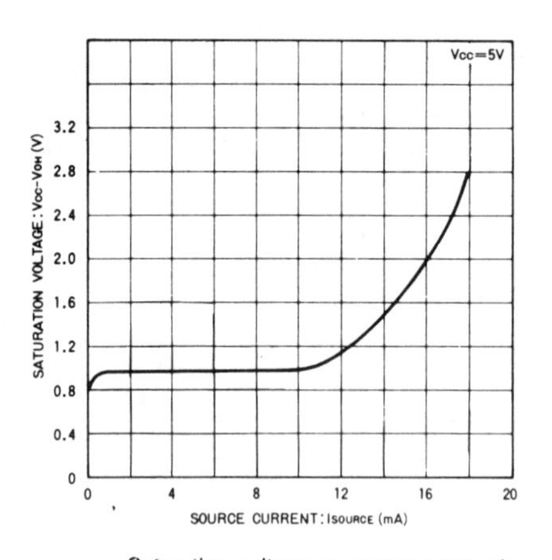

Saturation voltage vs. source current

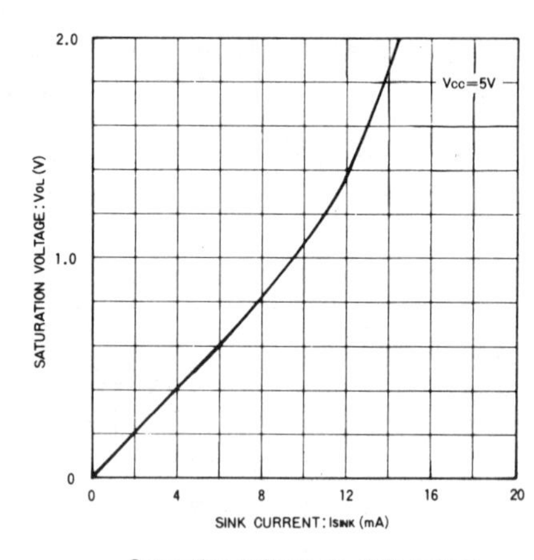

Saturation voltage vs. sink current

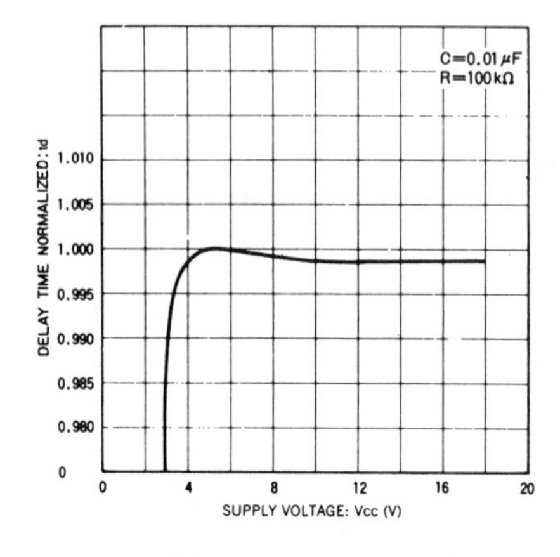

Delay time normalized vs. supply voltage

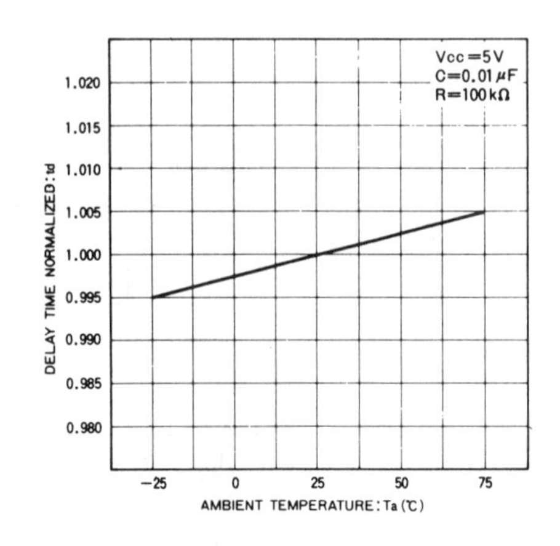

Delay time normalized vs. ambient temperature

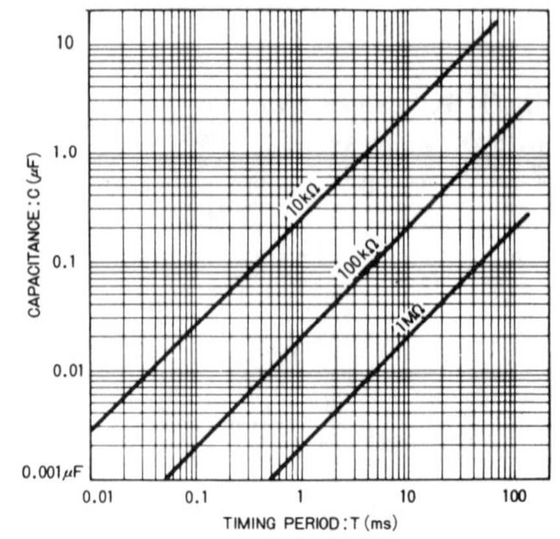

Capacitance vs. timing period

DESCRIPTION OF EXTERNAL COMPONENTS

1. **Timing capacitors (at pins 2 and 6)**
   These capacitors, along with timing resistors, determine the monostable time interval, $T = 1/2CR$. The other ends of the capacitors are grounded 1000 pF or more is recommended for these capacitors.
2. **Timing resistors (at pins 2 and 6)**
   These resistors, along with the timing capacitors, determine the monostable time interval. They are pulled up to V_{CC} so that the timing capacitors are charged through the resistors. A resistance of 10 kΩ to 1 MΩ is recommended for the timing resistors.
3. **Line capacitor (pin 8)**
   Ground the power supply pin of the device through a 0.02-μF capacitor, which will raise the dc line impedance and prevent malfunction.
4. **Loads (pins 1 and 7)**
   The device allows sink and drain load currents of up to 5 mA each.

PRECAUTIONS

1. **Operating voltage range**
   The recommended operating voltage range is 4.0 to 16.0 V. If the device is used at less than 3.0 V, a one-shot pulse with the same mode as the trigger input can be output at around 2.6 V.
2. **Input trigger**
   The threshold of the trigger input is 2.0 V. The device identifies a trigger input of 1.0 V (or less) as a low, and of 3.0 V or above as a high. The trigger pulse should have a rising and falling slope of 10 ms/V or less.
3. **Time constant RC**
   A capacitance of 1000 pF or more is recommended for the timing capacitors; a resistance of 10 kΩ to 1 MΩ is recommended for the timing resistors.

   A potential set by the ratio of an internal discharging resistor to the timing resistor remains at the RC pin. If the timing resistance is reduced below 5 kΩ, this potential will reach the input threshold voltage ($0.4 \times V_{CC}$) and cause the device to stop functioning. As the timing resistance is reduced, a potential remains at the RC pin, which causes the time constant to be reduced. If the timing resistance is increased to 2 MΩ or above, the device might stop functioning because the internal comparator might fail to operate.
4. **Pin configurations**
   The BA225/BA226 (BA225F/BA226F) and BA235/BA236 have different pin configurations.

   An application example and its basic timing sequences are shown. When no trigger signal is applied, the output remains low, with the timing capacitor fully discharged. When a trigger signal is applied, its leading edge sets the output high to start charging the timing capacitor.

   The charging time depends on the time constant, which is set by the external timing resistor and capacitor. When the capacitor is charged to a potential of 0.4 V_{CC}, an internal flip-flop resets the output from high to low. At the same time, the capacitor is discharged to prepare for the next trigger input.

TRAILING-EDGE TRIGGER

BA226, BA226F, and BA236 are trailing-edge trigger devices.

BA226: 8-pin DIP
BA226F: 8-pin MF
BA236: 9-pin SIP

The trailing-edge trigger devices have the same features, applications, absolute maximum ratings, electrical characteristics, internal circuitry, external components, precautions, and application examples as the leading-edge trigger devices. Basic timing sequences of the trailing-edge trigger devices are shown.

TEST CIRCUIT

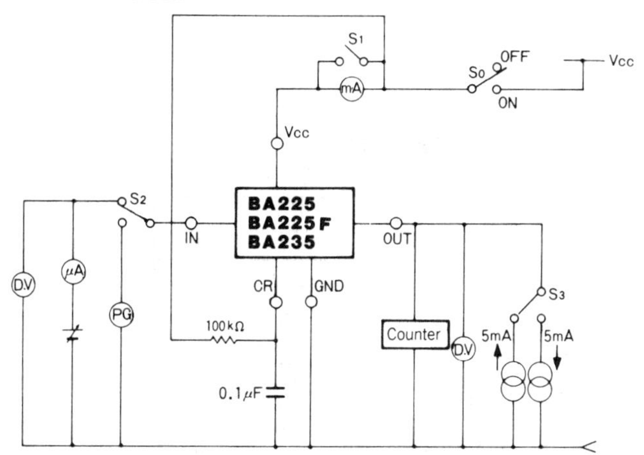

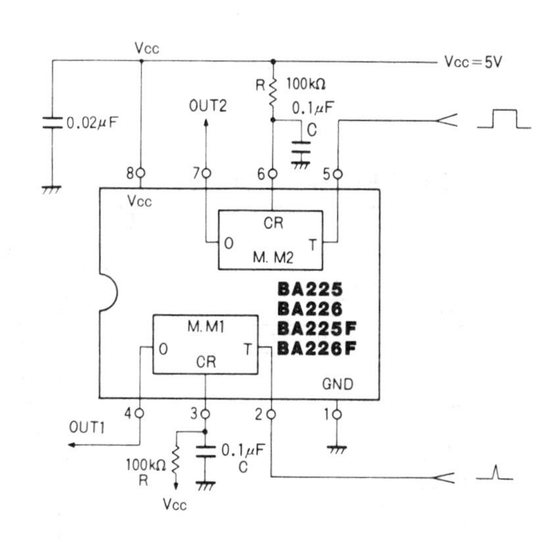

Trigger input
Leading-edge trigger
CR pin
T
0.4Vcc
Output
One-shot cycle: T≑0.5CR

Basic operating timing diagram
(Leading-edge trigger)

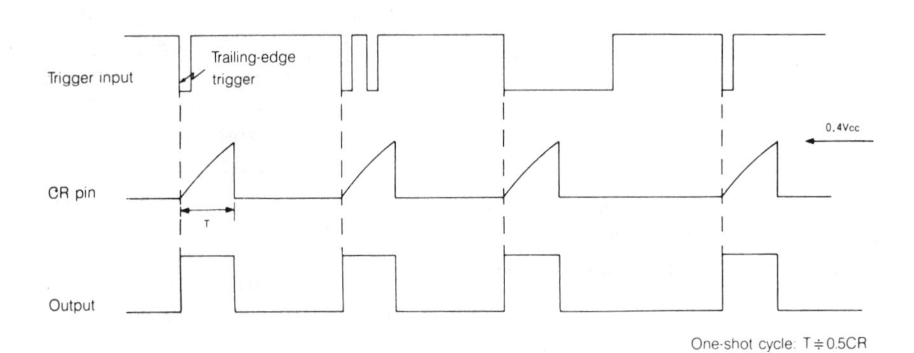

Basic operating timing diagram
(Trailing-edge trigger)

Sprague ULN-2430M Timer

- Microseconds to minutes
- Temperature compensated
- 400 mA output
- 8-pin dual in-line plastic package

ABSOLUTE MAXIMUM RATINGS

Regulator current, I_{REG}	15 mA
Latch current, I_4	3 mA
Output current, I_{OUT}	400 mA
Package power dissipation, P_D	330 mW*
Operating temperature range, T_A	−40 °C to + 85 °C
Storage temperature range, T_S	−65 °C to +150 °C

*Derate at the rate of 4.2 mW/ °C above T_A = +70 °C

ELECTRICAL CHARACTERISTICS at T_A = +25°C (unless otherwise noted)

Characteristic	Test Pin	Test Conditions	Limits Min.	Typ.	Max.	Units
Operating Voltage Range			10	—	16	V
Regulator Voltage	5		8.4	9.0	10.1	V
Output Breakdown Voltage	2	I_{LEAK} = 100 μA	30	—	—	V
Output Saturation Voltage	2	I_{OUT} = 400 mA	—	—	2.5	V
		I_{OUT} = 250 mA	—	—	1.3	V
Latch Voltage	4	Over Op. Temp. Range	5.5	7.0	8.0	V
Trigger Threshold	7	V_7/V_5	0.60	0.63	0.67	
Reference	8	V_8/V_5	0.58	0.63	0.68	
Temp. Coeff. of Trigger Threshold	7		−2.0	—	−4.0	mV/°C
Trigger Input Current	7		—	20	200	nA
Capacitor Discharge Time	7	C_1 = 220 μF, ±10%	—	—	2.0	s
Supply Current	5	V_{CC} = 16 V	—	—	10	mA

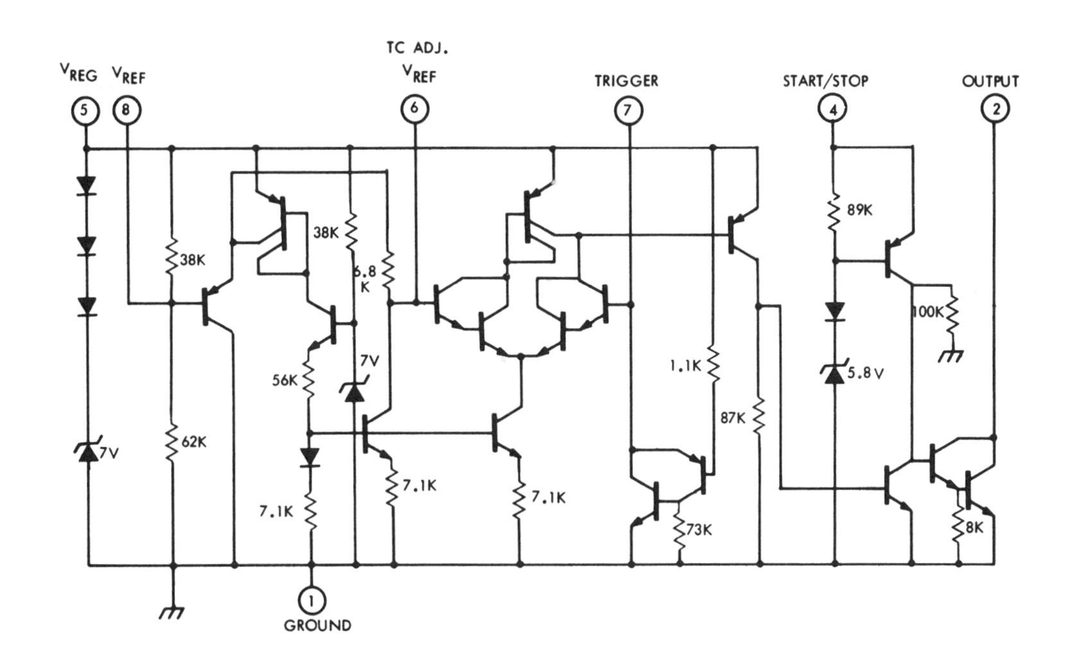

TYPICAL APPLICATION

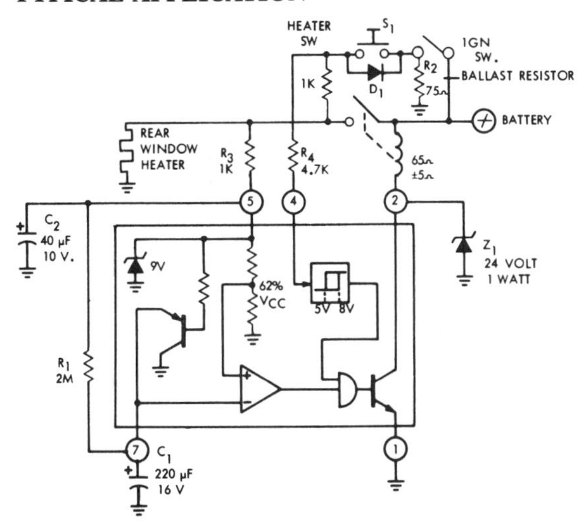

TIMER WAVEFORMS

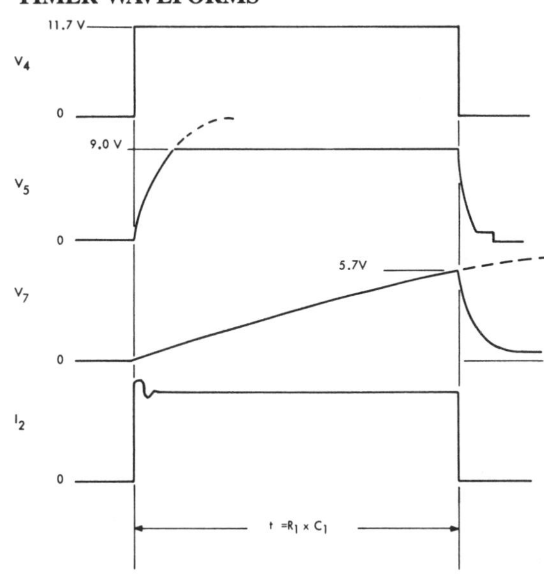

SECTION 2

COMMUNICATIONS CIRCUITS

This information is reproduced with permission of Analog Devices, Inc.

Analog Devices AD9685/AD9687 High-Speed Comparators

- 2.7ns Propagation delay
- 0.5ns Latch setup time
- 90dB CMRR
- +5V, −5.2V Supply voltages

APPLICATIONS

- High-speed triggers
- High-speed line receivers
- Peak detectors
- Threshold detectors

The AD9685 and the AD9687 are high-speed voltage comparators. The AD9685 and the AD9687 are manufactured in a high-performance bipolar process that allows improved speed and dc accuracy. The AD9685 is a single comparator with a 2.7ns propagation delay, and the AD9687 is a dual comparator of equal performance.

Both devices employ a high-precision differential input stage with a common-mode range of ±2.5V. The AD9685 and the AD9687 provide complementary digital outputs that are fully ECL compatible. The output stage is capable of driving 50Ω terminated transmission lines, given the 30mA output

PIN NAME	FUNCTIONAL DESCRIPTION
$+V_S$	– Positive supply terminal, nominally +5.0V.
NONINVERTING INPUT	– Noninverting analog input of the differential input stage. The NONINVERTING INPUT must be driven in conjunction with the INVERTING INPUT.
INVERTING INPUT	– Inverting analog input of the differential input stage. The INVERTING INPUT must be driven in conjunction with the NONINVERTING INPUT.
LATCH ENABLE	– In the "compare" mode (logic HIGH), the output will track changes at the input of the comparator. In the "latch" mode (logic LOW), the output will reflect the input state just prior to the comparator being placed in the "latch" mode. $\overline{\text{LATCH ENABLE}}$ must be driven in conjunction with LATCH ENABLE for the AD96687.
$\overline{\text{LATCH ENABLE}}$	– In the "compare" mode (logic LOW), the output will track changes at the input of the comparator. In the "latch" mode (logic HIGH), the output will reflect the input state just prior to the comparator being placed in the "latch" mode. LATCH ENABLE must be driven in conjunction with $\overline{\text{LATCH ENABLE}}$ for the AD96687.
$-V_S$	– Negative supply terminal, nominally −5.2V.
Q	– One of two complementary outputs. Q will be at logic HIGH if the analog voltage at the NONINVERTING INPUT is greater than the analog voltage at the INVERTING INPUT (provided the comparator is in the "compare" mode). See LATCH ENABLE and $\overline{\text{LATCH ENABLE}}$ (AD96687 only) for additional information.
$\overline{Q}$	– One of two complementary outputs. $\overline{Q}$ will be at logic LOW if the analog voltage at the NONINVERTING INPUT is greater than the analog voltage at the INVERTING INPUT (provided the comparator is in the "compare" mode). See LATCH ENABLE and $\overline{\text{LATCH ENABLE}}$ (AD96687 only) for additional information.
GROUND 1	– One of two grounds, but primarily associated with the digital ground. Both grounds should be connected together near the comparator.
GROUND 2	– One of two grounds, but primarily associated with the analog ground. Both grounds should be connected together near the comparator.

drive capacity. In addition, a latch enable input is provided, allowing operation in either a sample-hold mode or a track-hold mode.

The AD9685 and the AD9687 are both available as industrial-grade devices, −25°C to +85°C, and as extended temperature-range devices, −55°C to +125°C. The AD9685 is available in a 10-pin TO-100 metal can or a 16-pin ceramic package. The AD9687 is available in a 16-pin ceramic package.

ABSOLUTE MAXIMUM RATINGS

PARAMETER	SYMBOL	VALUE
Positive supply voltage	$+V_S$	+6V
Negative supply voltage	$-V_S$	−6V
Input voltage		±5V
Differential input voltage		5.5V
Latch enable voltage		$-V_S$ to 0V
Output current		30mA
Power dissipation		
AD9685		500mW
AD9687		600mW
Operating temperature range		
AD9685/87/BD/BH		−25°C to +85°C
AD9685/87/TD/TH		−55°C to +125°C
Storage temperature range		−55°C to +150°C
Junction temperature		+175°C
Lead soldering temperature (10sec)		+300°C

AD9685/AD9687 FUNCTIONAL BLOCK DIAGRAMS

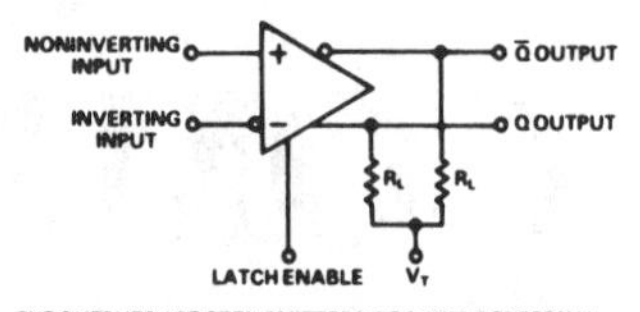

THE OUTPUTS ARE OPEN EMITTERS, REQUIRING EXTERNAL PULL-DOWN RESISTORS. THESE RESISTORS MAY BE IN THE RANGE OF 50Ω–200Ω CONNECTED TO −2.0V; OR 200Ω–2000Ω CONNECTED TO −5.2V.

AD9685

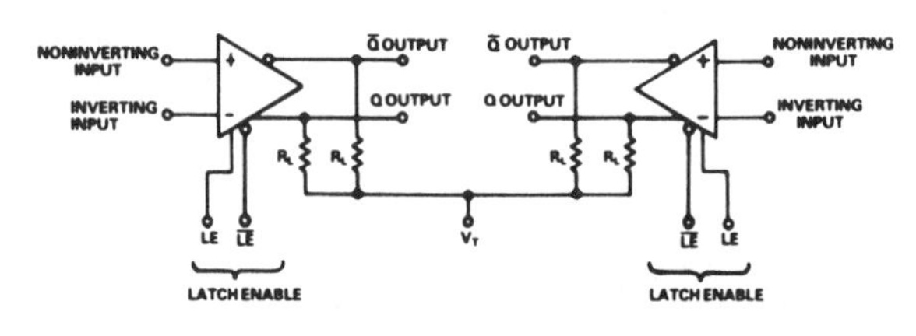

THE OUTPUTS ARE OPEN EMITTERS, REQUIRING EXTERNAL PULL-DOWN RESISTORS. THESE RESISTORS MAY BE IN THE RANGE OF 50Ω–200Ω CONNECTED TO −2.0V; OR 200Ω–2000Ω CONNECTED TO −5.2V.

AD9687

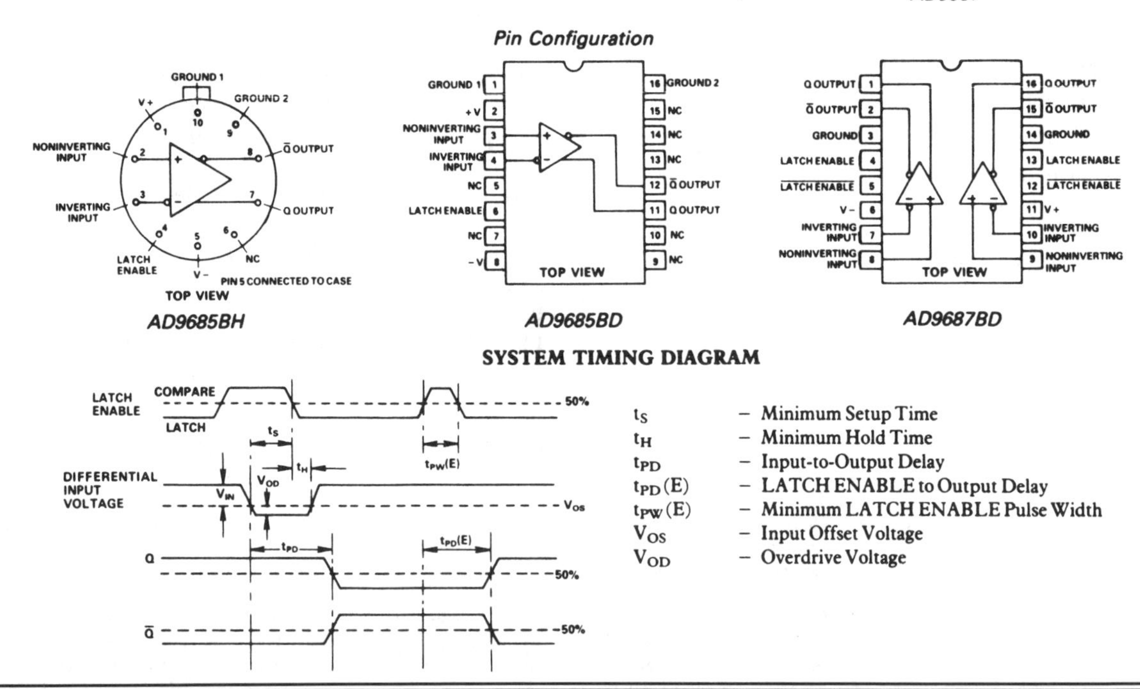

Analog Devices AD9686

High-Speed TTL Voltage Comparator

- 7ns Propagation delay
- Complementary TTL outputs
- 85dB CMRR
- +5V, −6V Supply voltages

APPLICATIONS

- High-speed triggers
- High-speed line receivers
- Peak detectors
- Threshold detectors

The AD9686 is a high-speed voltage comparator with complementary TTL outputs. The AD9686 is manufactured in a high-performance bipolar process that provides an excellent match between high-speed ac switching and dc accuracy. The AD9686 operates with a propagation delay of only 7ns.

The AD9686 incorporates a latch enable control line providing operation in either a sample-hold mode or a track-hold mode. The latch enable setup times are less than 2ns, which allows very high-speed voltage sampling.

The precision differential input stage has less than 2mV of offset voltage and requires an input bias current of only 4μA. This current, combined with the 85dB common-mode rejection ratio, makes the AD9686 especially well-suited for high-speed analog signal processing.

The AD9686 is useful as both an industrial temperature-range device, −25 °C to +85 °C, and as an extended temperature-range device, −55 °C to +125 °C. Both versions are available packaged in a TO-100 metal can and in a ceramic DIP. The extended temperature-range device is also available in a ceramic LCC package.

AD9686 FUNCTIONAL BLOCK DIAGRAM

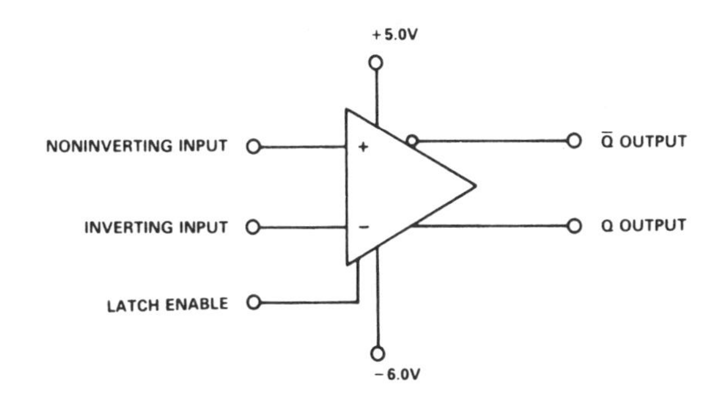

ABSOLUTE MAXIMUM RATINGS

PARAMETER	SYMBOL	VALUE
Positive supply voltage	$+V_S$	+7V
Negative supply voltage	$-V_S$	−7V
Input voltage range		±5V
Differential input voltage		6.0V
Latch enable voltage		0V to $+V_S$
Output current		
Sourcing		4mA
Sinking		14mA
Power dissipation		600mW
Operating temperature range		
AD9686/BH/BQ		−25°C to +85°C
AD9686TE/TH/TQ		−55°C to +125°C
Storage temperature range		−65°C to +150°C
Junction temperature		+175°C
Lead soldering temperature (10sec)		+300°C

PIN NAME	FUNCTIONAL DESCRIPTION
$+V_S$	– Positive supply terminal, nominally +5.0V.
NONINVERTING INPUT	– Noninverting analog input of the differential input stage. The NONINVERTING INPUT must be driven in conjunction with the INVERTING INPUT.
INVERTING INPUT	– Inverting analog input of the differential input stage. The INVERTING INPUT must be driven in conjunction with the NONINVERTING INPUT.
$-V_S$	– Negative supply terminal, nominally −6.0V.
LATCH ENABLE	– In the "compare" mode (logic LOW), the output will track changes at the input of the comparator. In the "latch" mode (logic HIGH), the output will reflect the input state just prior to the comparator being placed in the "latch" mode.
GROUND	– Analog and digital ground.
Q OUTPUT	– One of two complementary outputs. Q will be at logic HIGH if the analog voltage at the NON-INVERTING INPUT is greater than the analog voltage at the INVERTING INPUT (provided the comparator is in the "compare" mode). See LATCH ENABLE for additional information.
$\overline{Q}$ OUTPUT	– One of two complementary outputs. $\overline{Q}$ will be at logic LOW if the analog voltage at the NONIN-VERTING INPUT is greater than the analog voltage at the INVERTING INPUT (provided the comparator is in the "compare" mode). See LATCH ENABLE for additional information.
NC	– "NO CONNECT" pins are not internally connected.

PINOUT CONFIGURATIONS

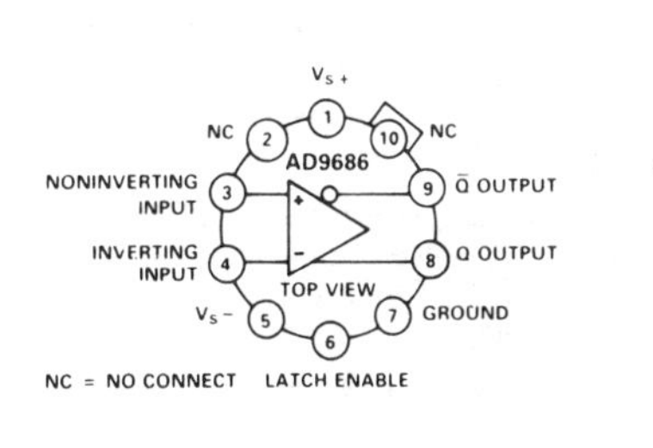

TO-100
10-Pin Can

NC 1 · 16 NC
NC 2 · 15 NC
V_S+ 3 · 14 $\overline{Q}$ OUTPUT
NONINVERTING INPUT 4 · 13 Q OUTPUT
INVERTING INPUT 5 · 12 GROUND
V_S− 6 · 11 LATCH ENABLE
NC 7 · 10 NC
NC 8 · 9 NC
AD9686
TOP VIEW
(Not to Scale)
NC = NO CONNECT

16-Pin DIP

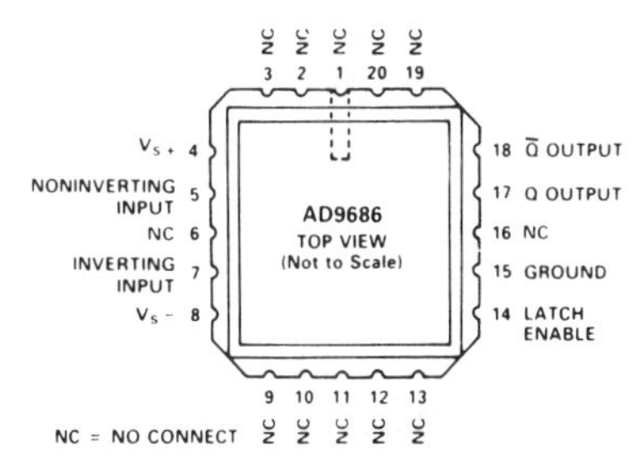

20-Pin LCC

SYSTEM TIMING DIAGRAM

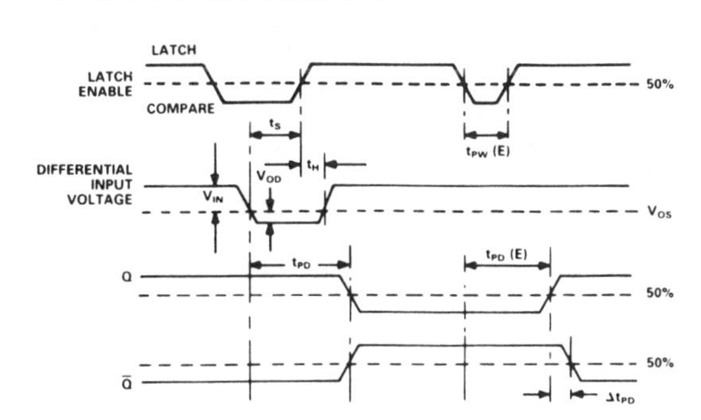

t_S	–	Minimum Setup Time
t_H	–	Minimum Hold Time
t_{PD}	–	Input to Output Delay
t_{PD}(E)	–	LATCH ENABLE to Output Delay
t_{PW}(E)	–	Minimum LATCH ENABLE Pulse Width
V_{OS}	–	Input Offset Voltage
V_{OD}	–	Overdrive Voltage
Δt_{PD}	–	Delta Delay Between Complementary Outputs

Analog Devices AD96685/AD96687
Ultrafast Comparators

- 2.5ns Propagation delay
- 0.5ns Latch setup time
- 90dB CMRR
- +5V, −5.2V Supply voltages

APPLICATIONS
- High-speed triggers
- High-speed line receivers
- Peak detectors
- Threshold detectors

The AD96685 and the AD96687 are ultrafast voltage comparators. The AD96685 and the AD96687 are manufactured in a high-performance bipolar process that allows improved speed and dc accuracy. The AD96685 is a single comparator with a 2.5ns propagation delay, 50ps dispersion; the AD96687 is an equally fast dual comparator.

Both devices employ a high-precision differential input stage with a common-mode range from −2.5V to +5.0V. The AD96685 and the AD96687 provide complementary digital outputs that are fully ECL compatible. The output stage is capable of driving 50Ω terminated transmission lines, given the 30mA output drive capacity. In addition, a latch enable input is provided, allowing operation in either a sample-hold mode or a track-hold mode.

The AD96685 and the AD96687 are both available as industrial temperature-range devices, −25 °C to +85 °C, and as extended temperature-range devices, −55 °C to +125 °C. The AD96685 is available in a 10-pin TO-100 metal can and in a 16-pin ceramic package. The AD96687 is available in a 16-pin ceramic package. Both comparators are also available in an extended-temperature range LCC package.

ABSOLUTE MAXIMUM RATINGS

PARAMETER	SYMBOL	VALUE
Positive supply voltage	$+V_S$	+6.5V
Negative supply voltage	$-V_S$	−6.5V
Input voltage range		±5V
Differential input voltage		5.5V
Latch enable voltage		$-V_S$ to 0V
Output current		30mA
Power dissipation		
AD96685		500mW
AD96687		600mW
Operating temperature range		
AD96685/87/BH/BQ		−25°C to +85°C
AD96685/87/TE/TH/TQ		−55°C to +125°C
Storage temperature range		−55°C to +150°C
Junction temperature		+175°C
Lead soldering temperature (10sec)		+300°C

PIN NAME	FUNCTIONAL DESCRIPTION
$+V_S$	– Positive supply terminal, nominally +5.0V.
NONINVERTING INPUT	– Noninverting analog input of the differential input stage. The NONINVERTING INPUT must be driven in conjunction with the INVERTING INPUT.
INVERTING INPUT	– Inverting analog input of the differential input stage. The INVERTING INPUT must be driven in conjunction with the NONINVERTING INPUT.
LATCH ENABLE	– In the "compare" mode (logic HIGH), the output will track changes at the input of the comparator. In the "latch" mode (logic LOW), the output will reflect the input state just prior to the comparator being placed in the "latch" mode. $\overline{\text{LATCH ENABLE}}$ must be driven in conjunction with LATCH ENABLE for the AD9687.
$\overline{\text{LATCH ENABLE}}$	– In the "compare" mode (logic LOW), the output will track changes at the input of the comparator. In the "latch" mode (logic HIGH), the output will reflect the input state just prior to the comparator being placed in the "latch" mode. $\overline{\text{LATCH ENABLE}}$ must be drive in conjunction with LATCH ENABLE for the AD9687.
$-V_S$	– Negative supply terminal, nominally −5.2V.
Q	– One of two complementary outputs, Q will be at logic HIGH, if the analog voltage at the NONINVERTING INPUT is greater than the analog voltage at the INVERTING INPUT (provided the comparator is in the "compare" mode). See LATCH ENABLE and $\overline{\text{LATCH ENABLE}}$ (AD9687 only) for additional information.
$\overline{Q}$	– One of two complementary outputs. $\overline{Q}$ will be at logic LOW, if the analog voltage at the NONINVERTING INPUT is greater than the analog voltage at the INVERTING INPUT (provided the comparator is in the "compare" mode). See LATCH ENABLE and $\overline{\text{LATCH ENABLE}}$ (AD9687 only) for additional information.
GROUND 1	– One of two grounds, but primarily associated with the digital ground. Both grounds should be connected together near the comparator.
GROUND 2	– One of two grounds, but primarily associated with the analog ground. Both grounds should be connected together near the comparator.

PIN DESIGNATIONS

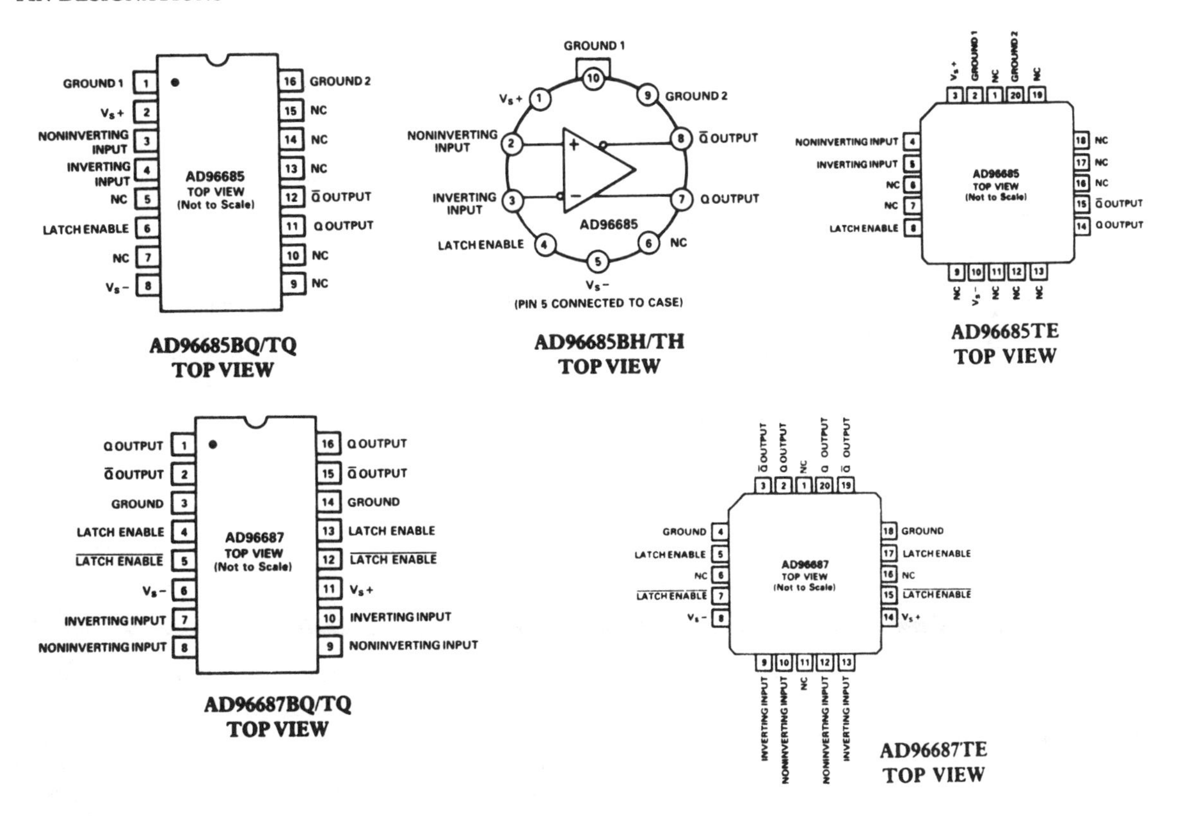

AD96685BQ/TQ TOP VIEW

AD96685BH/TH TOP VIEW

AD96685TE TOP VIEW

AD96687BQ/TQ TOP VIEW

AD96687TE TOP VIEW

AD96685/AD96687 FUNCTIONAL BLOCK DIAGRAMS

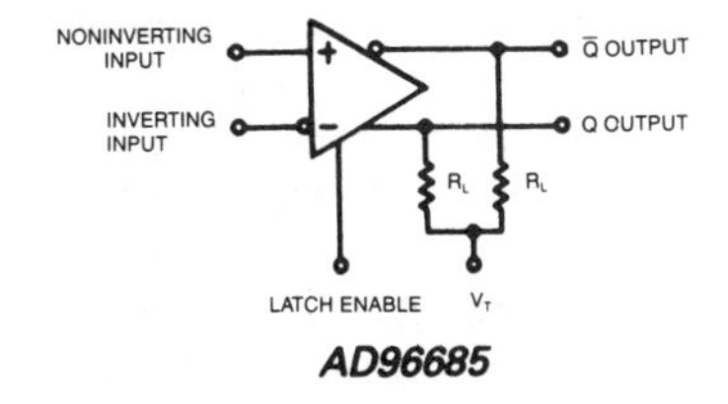

AD96685

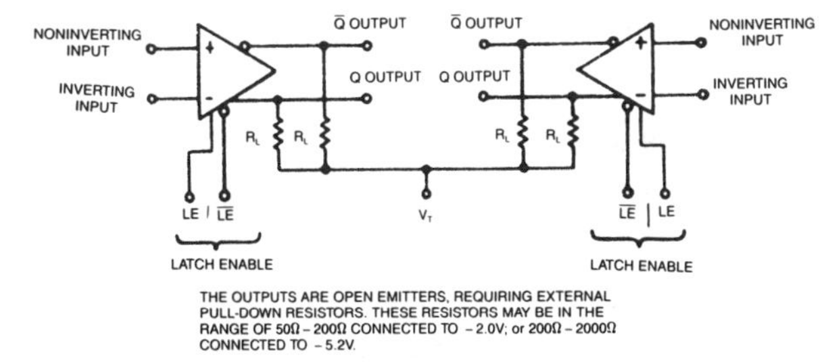

AD96687

SYSTEM TIMING DIAGRAM

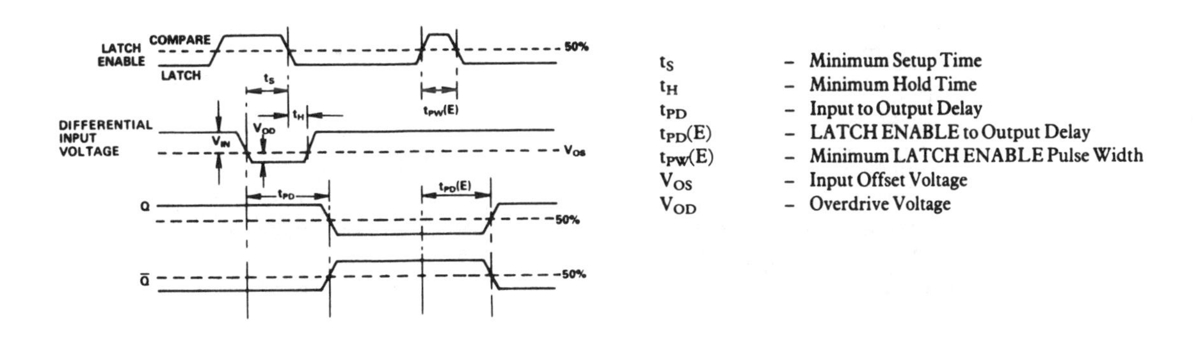

HIGH-SPEED SAMPLING CIRCUIT

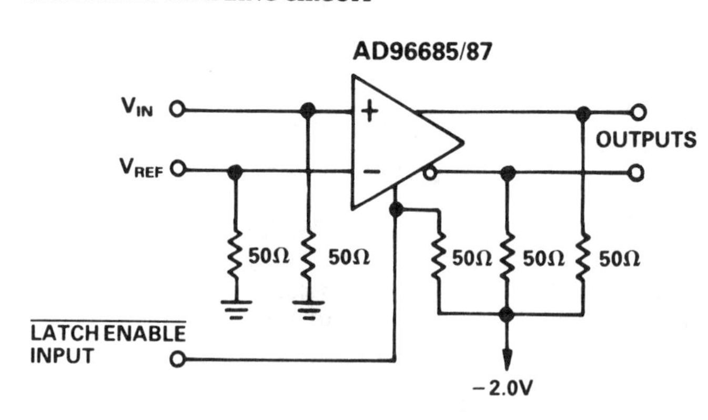

HIGH-SPEED WINDOW COMPARATOR

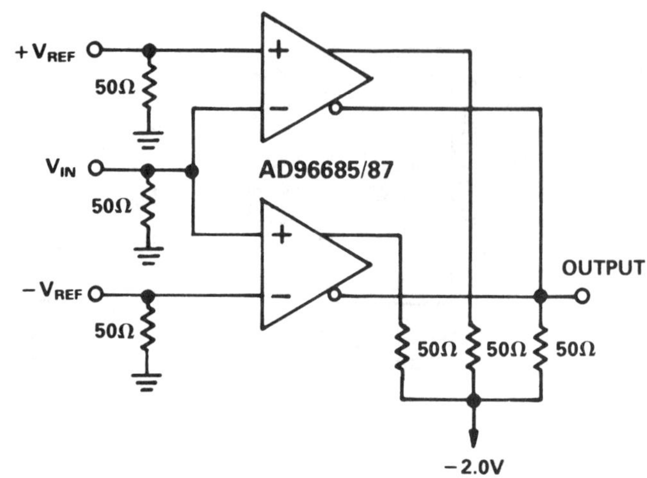

Analog Devices AD202/AD204

Low Cost, Miniature Isolation Amplifiers

- Small size: 4 channels/inch
- Low power: 35mW (AD204)
- High accuracy: ±0.025% max nonlinearity (K grade)
- High CMR: 130dB (gain = 100V/V)
- Wide bandwidth: 5kHz full-power (AD204)
- High CMV isolation: ±2000V pk continuous (K grade) (signal and power)
- Isolated power outputs
- Uncommitted input amplifier

APPLICATIONS

- Multichannel data acquisition
- Current shunt measurements
- Motor controls
- Process signal isolation
- High-voltage instrumentation amplifier

The AD202 and AD204 are members of a new generation of low-cost, high-performance isolation amplifiers. A new circuit design, transformer construction, and the use of surface-mounted components in an automated assembly process result in compact, economical isolators whose performance in many ways exceeds that previously available from very expensive devices. The primary distinction between the AD202 and AD204 is that the AD202 is powered directly from $+15V_{dc}$, and the AD204 is powered by an externally supplied clock (AD246).

The AD202 and AD204 employ transformer coupling and do not require the design compromises that must be made when optical isolators are used: each provides a complete isolation function, with both signal and power isolation internal to the module, and they exhibit no long-term parameter shifts under sustained common-mode stress. Power consumption, nonlinearity, and drift are each an order of magnitude lower than can be obtained from other isolation techniques, and these advantages are obtained without sacrifice of bandwidth or noise performance.

The design of the AD202 and AD204 emphasizes ease of use in a broad range of applications where signals must be measured or transmitted without a galvanic connection. In addition, the low cost and small size of these isolators makes component-level circuit applications of isolation practical for the first time.

PIN DESIGNATIONS

AD202/AD204 SIP PACKAGE

PIN	FUNCTION
1	+INPUT
2	INPUT/V_{ISO} COMMON
3	−INPUT
4	INPUT FEEDBACK
5	−V_{ISO} OUTPUT
6	+V_{ISO} OUTPUT
31	+15V POWER IN (AD202 ONLY)
32	CLOCK/POWER COMMON
33	CLOCK INPUT (AD204 ONLY)
37	OUTPUT LO
38	OUTPUT HI

AD202/AD204 DIP PACKAGE

PIN	FUNCTION
1	+INPUT
2	INPUT/V_{ISO} COMMON
3	−INPUT
18	OUTPUT LO
19	OUTPUT HI
20	+15V POWER IN (AD202 ONLY)
21	CLOCK INPUT (AD204 ONLY)
22	CLOCK/POWER COMMON
36	+V_{ISO} OUTPUT
37	−V_{ISO} OUTPUT
38	INPUT FEEDBACK

SPECIFICATIONS (typical @ +25°C and V_S = +15V unless otherwise noted)

Model	AD204J	AD204K	AD202J	AD202K
GAIN				
Range	1V/V-100V/V	*	*	*
Error	±0.5% typ (±4% max)	*	*	*
vs. Temperature	±20ppm/°C typ (±45ppm/°C max)	*	*	*
vs. Time	±50ppm/1000 Hours	*	*	*
vs. Supply Voltage	±0.001%/V	±0.001%/V	±0.01%/V	±0.01%/V
Nonlinearity (G = 1V/V)[1]	±0.05% max	±0.025% max	±0.05% max	±0.025% max
INPUT VOLTAGE RATINGS				
Linear Differential Range	±5V	*	*	*
Max CMV Input to Output				
ac, 60Hz, Continuous	750V rms	1500V rms	750V rms	1500V rms
Continuous (dc and ac)	±1000V peak	±2000V peak	±1000V peak	±2000V peak
Common-Mode Rejection (CMR), @ 60Hz				
R_S = ≤100Ω (HI & LO Inputs) G = 1	110dB	110dB	105dB	105dB
G = 100	130dB	*	*	*
R_S = ≤1kΩ (Input HI, LO, or Both) G = 1	104dB min	104dB min	100dB min	100dB min
G = 100	110dB min	*	*	*
Leakage Current Input to Output				
@240V rms, 60Hz	2μA rms max	*	*	*
INPUT IMPEDANCE				
Differential (G = 1V/V)	10^{12} Ω	*	*	*
Common Mode	2GΩ‖4.5pF	*	*	*
INPUT BIAS CURRENT				
Initial, @ +25°C	±30pA	*	*	*
vs. Temperature (0 to +70°C)	±10nA	*	*	*
INPUT DIFFERENCE CURRENT				
Initial, @ +25°C	±5pA	*	*	*
vs. Temperature (0 to +70°C)	±2nA	*	*	*
INPUT NOISE				
Voltage, 0.1 to 100Hz	4μV p-p	*	*	*
f>200Hz	$50nV/\sqrt{Hz}$	*	*	*
FREQUENCY RESPONSE				
Bandwidth (V_O ≤10V p-p, G = 1-50V/V)	5kHz	5kHz	2kHz	2kHz
Settling Time, to ±10mV (10V Step)	1ms	*	*	*
OFFSET VOLTAGE (RTI)				
Initial, @ +25°C Adjustable to Zero	(±15 ±15/G)mV max	(±5 ±5/G)mV max	(±15 ±15/G)mV max	(±5 ±5/G)mV max
vs. Temperature (0 to +70°C)	$\left(\pm 10 \pm \frac{10}{G}\right)\mu V/°C$	*	*	*
RATED OUTPUT				
Voltage (Out HI to Out LO)	±5V	*	*	*
Voltage at Out HI or Out LO (Ref. Pin 32)	±6.5V	*	*	*
Output Resistance	3kΩ	3kΩ	7kΩ	7kΩ
Output Ripple, 100kHz Bandwidth	10mV pk-pk	*	*	*
5kHz Bandwidth	0.5mV rms	*	*	*
ISOLATED POWER OUTPUT[2]				
Voltage, No Load	±7.5V	*	*	*
Accuracy	±10%	*	*	*
Current	2mA (Either Output)[3]	2mA (Either Output)[3]	400μA Total	400μA Total
Regulation, No Load to Full Load	5%	*	*	*
Ripple	100mV pk-pk	*	*	*
OSCILLATOR DRIVE INPUT				
Input Voltage	15V pk-pk nominal	15V pk-pk nominal	N/A	N/A
Input Frequency	25kHz nominal	25kHz nominal	N/A	N/A
POWER SUPPLY (AD202 Only)				
Voltage, Rated Performance	N/A	N/A	+15V ±5%	+15V ±5%
Voltage, Operating	N/A	N/A	+15V ±10%	+15V ±10%
Current, No Load (V_S = +15V)	N/A	N/A	5mA	5mA
TEMPERATURE RANGE				
Rated Performance	0 to +70°C	*	*	*
Operating	−40°C to +85°C	*	*	*
Storage	−40°C to +85°C	*	*	*
PACKAGE DIMENSIONS[4]				
SIP Package (Y)	2.08″ × 0.250″ × 0.625″	*	*	*
DIP Package (N)	2.10″ × 0.700″ × 0.350″	*	*	*

NOTES
*Specifications same as AD204J.
[1]Nonlinearity is specified as a % deviation from a best straight line.
[2]1.0μF min decoupling required (see text).
[3]3mA with one supply loaded.
[4]Width is 0.25″ typ, 0.26″ max.
Specifications subject to change without notice.

OUTLINE DIMENSIONS

Dimensions shown in inches and (mm).

AD202/AD204 SIP PACKAGE

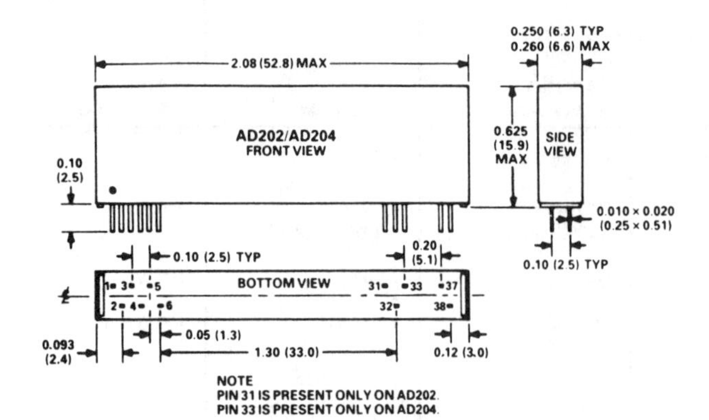

AD202/AD204 DIP PACKAGE

2.100 (53.3) MAX

AD202/AD204 SIDE VIEW

BOTTOM VIEW

NOTE
PIN 20 IS PRESENT ONLY ON AD202.
PIN 21 IS PRESENT ONLY ON AD204.

AC1058 MATING SOCKET

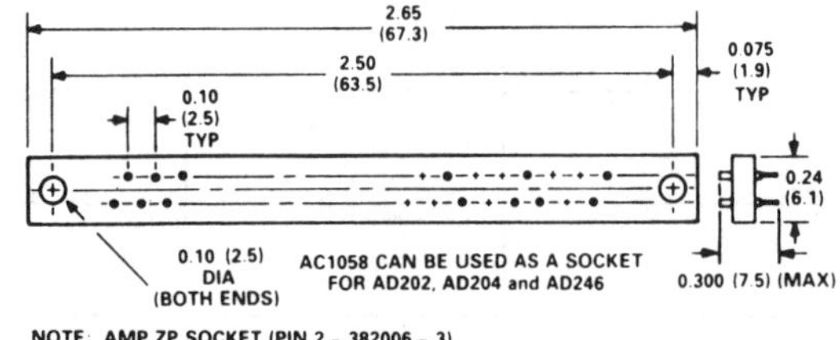

NOTE: AMP ZP SOCKET (PIN 2 - 382006 - 3)
MAY BE USED IN PLACE OF THE AC1058.

AC1060 MATING SOCKET

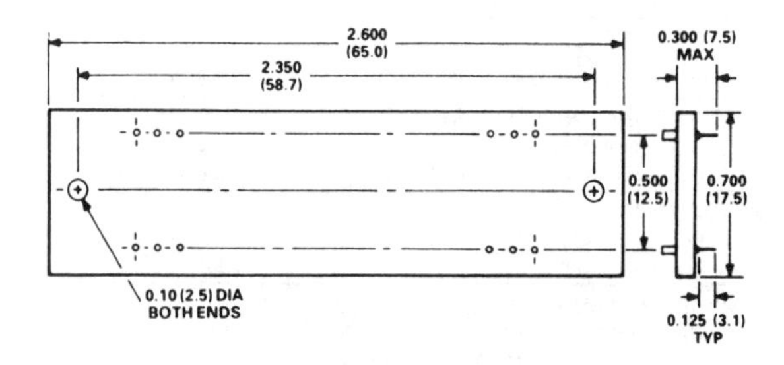

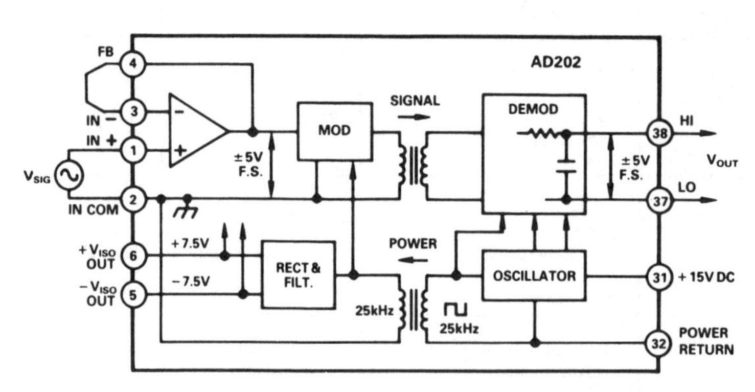

AD202 Functional Block Diagram

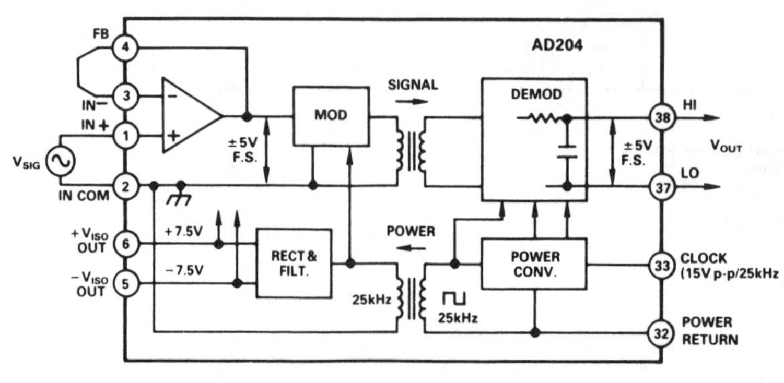

AD204 Functional Block Diagram

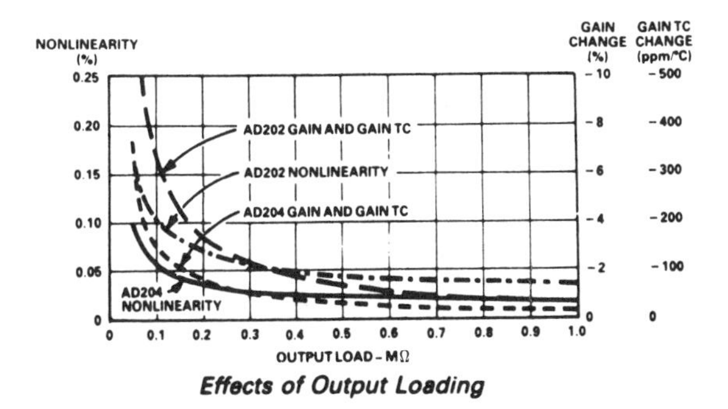

Effects of Output Loading

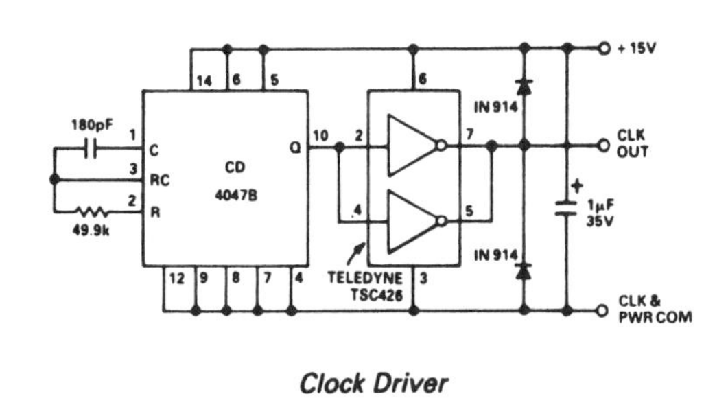

Clock Driver

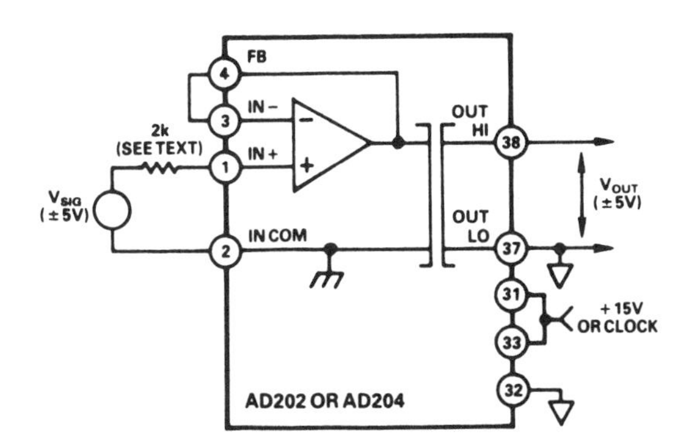

Basic Unity-Gain Application

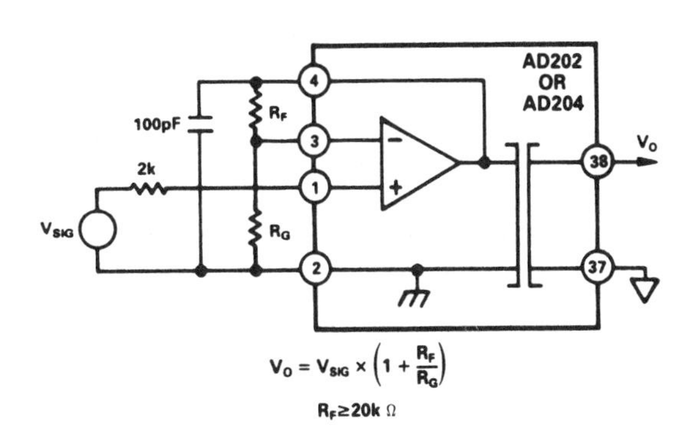

$$V_O = V_{SIG} \times \left(1 + \frac{R_F}{R_G}\right)$$

$R_F \geq 20k\ \Omega$

Input Connections for Gain > 1

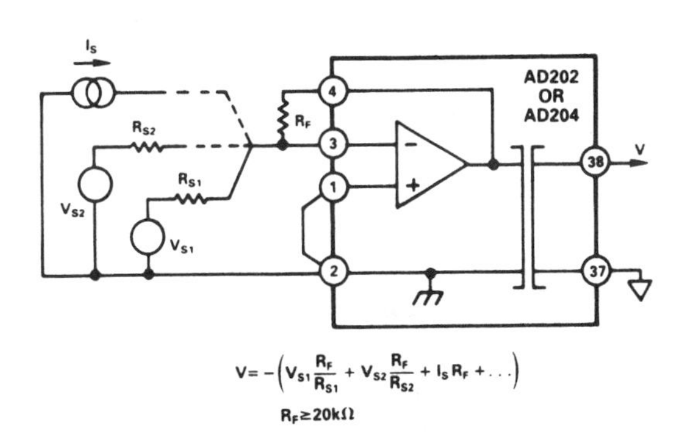

$$V = -\left(V_{S1}\frac{R_F}{R_{S1}} + V_{S2}\frac{R_F}{R_{S2}} + I_S R_F + \ldots\right)$$

$R_F \geq 20k\Omega$

Connections for Summing or Current Inputs

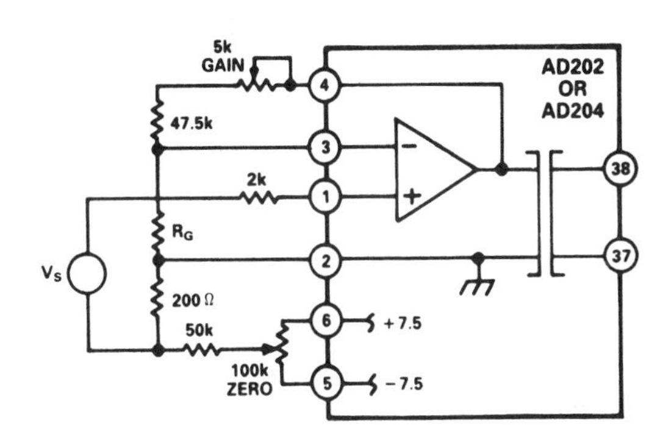

Adjustments for Noninverting Connection of Op Amp

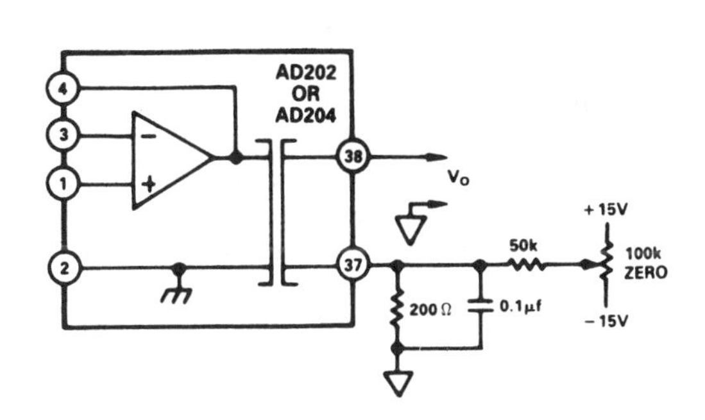

Output-Side Zero Adjustment

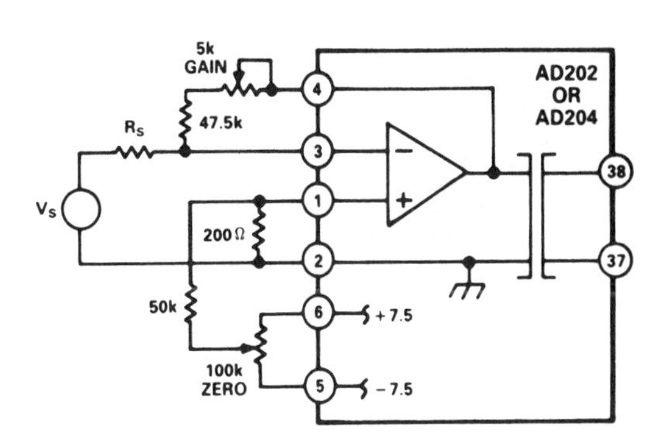

Adjustments for Summing or Current Input

Input Filter for Improved Step Response

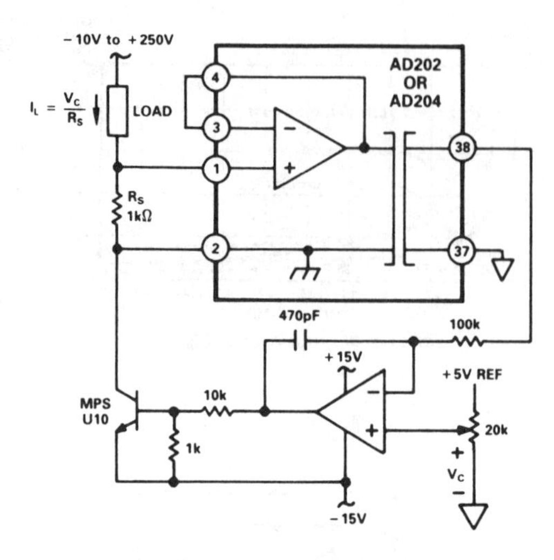

High-Compliance Current Source

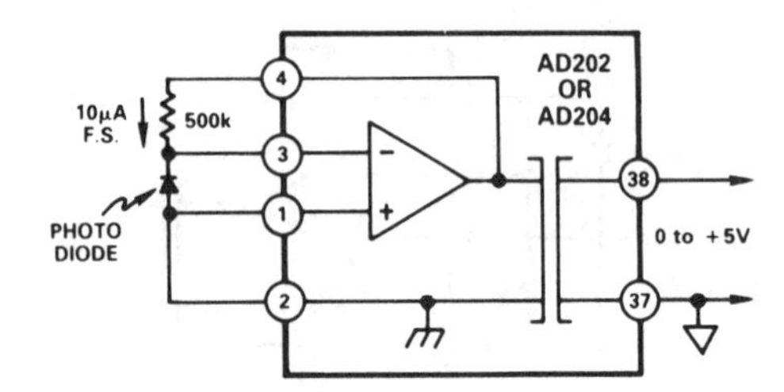

Photodiode Amplifier

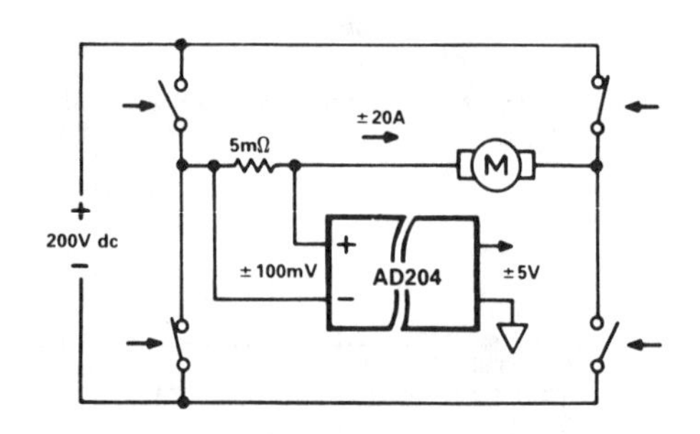

Motor Control Current Sensing

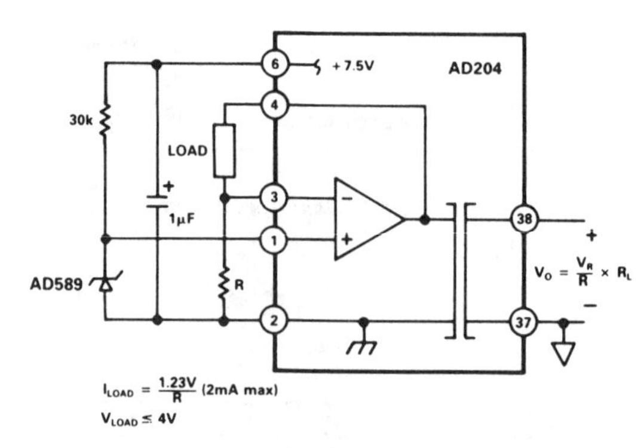

Floating Current Source

Analog Devices AD210

Precision, Wide Bandwidth, 3-Port Isolation Amplifier

- High CMV isolation: 2500V rms continuous
  ±3500V peak continuous
- Small size: 100″ × 2.10″ × 0.350″
- Three-port isolation: input, output, and power
- Low nonlinearity: ±0.012% max
- Wide bandwidth: 20kHz full-power (−3dB)
- Low-gain drift: ±25ppm/°C max
- High CMR: 120dB (G = 100V/V)
- Isolated power: ±15V @±5mA
- Uncommitted input amplifier

APPLICATIONS

- Multichannel data acquisition
- High-voltage instrumentation amplifier
- Current shunt measurements
- Process signal isolation

The AD210 is the latest member of a new generation of low-cost, high-performance isolation amplifiers. This three-port, wide bandwidth isolation amplifier is manufactured with surface-mounted components in an automated assembly process. The AD210 combines design expertise with state-of-the-art manufacturing technology to produce an extremely compact and economical isolator whose performance and abundant user features exceed those offered in more expensive devices.

The AD210 provides a complete isolation function with both signal and power isolation supplied via transformer coupling that is internal to the module. The AD210's functionally complete design, powered by a single +15V supply, eliminates the need for an external DC/DC converter, unlike optically coupled isolation devices. The true three-port design structure permits the AD210 to be applied as an input or output isolator, in single or multichannel applications. The AD210 will maintain its high performance under sustained common-mode stress.

Providing high accuracy and complete galvanic isolation, the AD210 interrupts ground loops and leakage paths, and rejects common-mode voltage and noise that may otherwise degrade measurement accuracy. In addition, the AD210 provides protection from fault conditions that might cause damage to other sections of a measurement system.

SPECIFICATIONS

(typical @ +25 °C, & V_s = +15V unless otherwise specified)

MODEL	AD210AN	AD210BN
GAIN		
Range	1V/V - 100V/V	*
Error	± 2% max	± 1% max
vs. Temperature (0 to +70°C)	± 25ppm/°C max	*
(−25°C to +85°C)	± 50ppm/°C max	*
vs. Supply Voltage	± 0.002%/V	*
Nonlinearity[1]	± 0.025% max	± 0.012% max
INPUT VOLTAGE RATINGS		
Linear Differential Range	± 10V	*
Maximum Safe Differential Input	± 15V	*
Max. CMV Input-to-Output		
ac, 60Hz, Continuous	2500V rms	*
dc, Continuous	± 3500V peak	*
Common-Mode Rejection		
60Hz, G = 100V/V		
$R_S \leq 500\Omega$ Impedance Imbalance	120dB	*
Leakage Current Input-to-Output		
@ 240Vrms, 60Hz	2μA rms max	*
INPUT IMPEDANCE		
Differential	$10^{12}\Omega$	*
Common Mode	5GΩ‖5pF	*
INPUT BIAS CURRENT		
Initial, @ +25°C	30pA typ (400pA max)	*
vs. Temperature (0 to +70°C)	10nA max	*
(−25°C to +85°C)	30nA max	*
INPUT DIFFERENCE CURRENT		
Initial, @ +25°C	5pA typ (200pA max)	*
vs. Temperature (0 to +70°C)	2nA max	*
(−25°C to +85°C)	10nA max	*
INPUT NOISE		
Voltage (1kHz)	$18nV/\sqrt{Hz}$	*
(10Hz to 10kHz)	4μV rms	*
Current (1kHz)	$0.01pA/\sqrt{Hz}$	*
FREQUENCY RESPONSE		
Bandwidth (−3dB)		
G = 1V/V	20kHz	*
G = 100V/V	15kHz	*
Settling Time (± 10mV, 20V Step)		
G = 1V/V	150μs	*
G = 100V/V	500μs	*
Slew Rate (G = 1V/V)	1V/μs	*
OFFSET VOLTAGE (RTI)[2]		
Initial, @ +25°C	(± 15 ± 45/G)mV max	(± 5 ± 15/G)mV max
vs. Temperature (0 to +70°C)	(± 10 ± 30/G)μV/°C	*
(−25°C to +85°C)	(± 10 ± 50/G)μV/°C	*
RATED OUTPUT[3]		
Voltage, 2kΩ Load	± 10V min	*
Impedance	1Ω max	*
Ripple, (Bandwidth = 100kHz)	10mV p-p max	*
ISOLATED POWER OUTPUTS[4]		
Voltage, No Load	± 15V	*
Accuracy	± 10%	*
Current	± 5mA	*
Regulation, No Load to Full Load	See Text	*
Ripple	See Text	*
POWER SUPPLY		
Voltage, Rated Performance	+ 15V dc ± 5%	*
Voltage, Operating	+ 15V dc ± 10%	*
Current, Quiescent	50mA	*
Current, Full Load – Full Signal	80mA	*
TEMPERATURE RANGE		
Rated Performance	−25°C to +85°C	*
Operating	−40°C to +85°C	*
Storage	−40°C to +85°C	*
PACKAGE DIMENSIONS		
Inches	1.00 × 2.10 × 0.350	*
Millimeters	25.4 × 53.3 × 8.9	*

NOTES
*Specifications same as AD210AN.
[1]Gain nonlinearity increases by ±0.002%/mA when the isolated power outputs are used.
[2]RTI – Referred to Input
[3]A reduced signal swing is recommended when both $\pm V_{ISS}$ and $\pm V_{OSS}$ supplies are fully loaded, due to supply voltage reduction.
[4]See text for detailed information.
Specifications subject to change without notice.

CAUTION

This device is ESD (ElectroStaticDischarge) sensitive. Permanent damage can occur on unconnected devices subject to high-energy electrostatic fields. Unused devices must be stored in conductive foam or shunts. The protective foam should be discharged to the destination socket before devices are removed.

OUTLINE DIMENSIONS

Dimensions shown in inches and (mm).

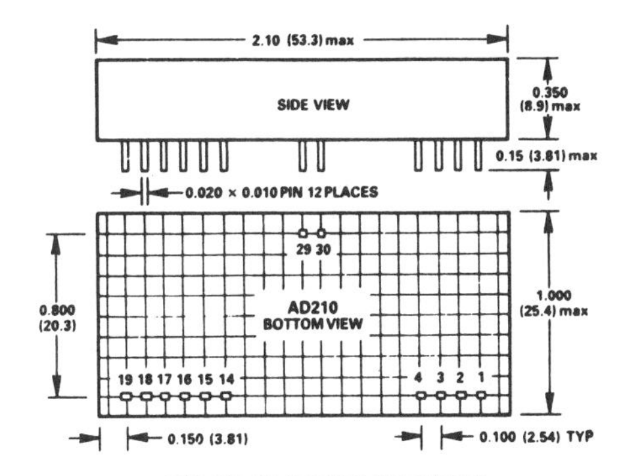

AC1059 MATING SOCKET

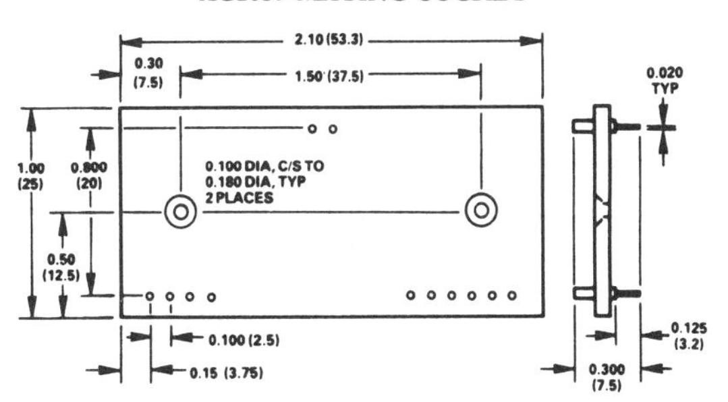

AD210 PIN DESIGNATIONS

PIN	DESIGNATION	FUNCTION
1	V_O	Output
2	O_{COM}	Output Common
3	$+V_{OSS}$	+Isolated Power @ Output
4	$-V_{OSS}$	−Isolated Power @ Output
14	$+V_{ISS}$	+Isolated Power @ Input
15	$-V_{ISS}$	−Isolated Power @ Input
16	FB	Input Feedback
17	−IN	−Input
18	I_{COM}	Input Common
19	+IN	+Input
29	Pwr Com	Power Common
30	Pwr	Power Input

AD210 FUNCTIONAL BLOCK DIAGRAM

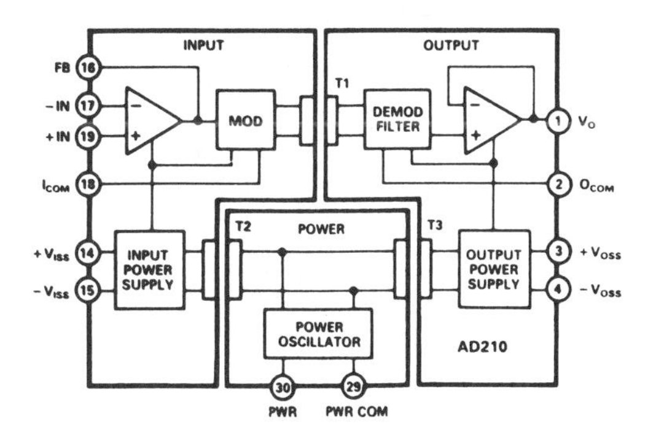

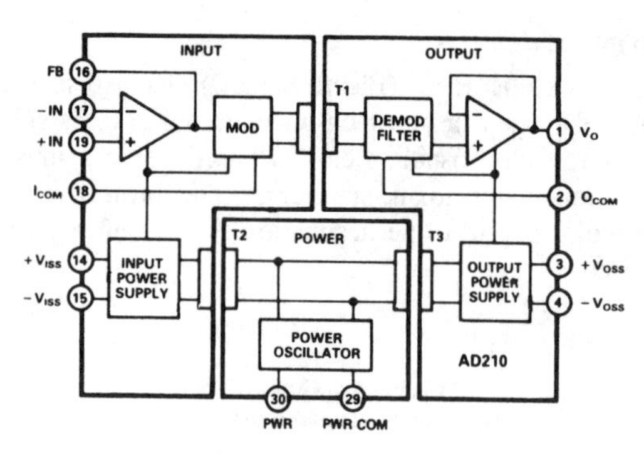

AD210 Block Diagram

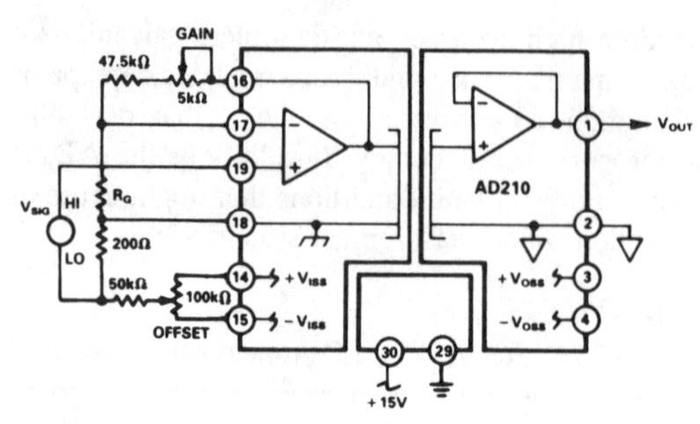

Adjustments for Noninverting Input

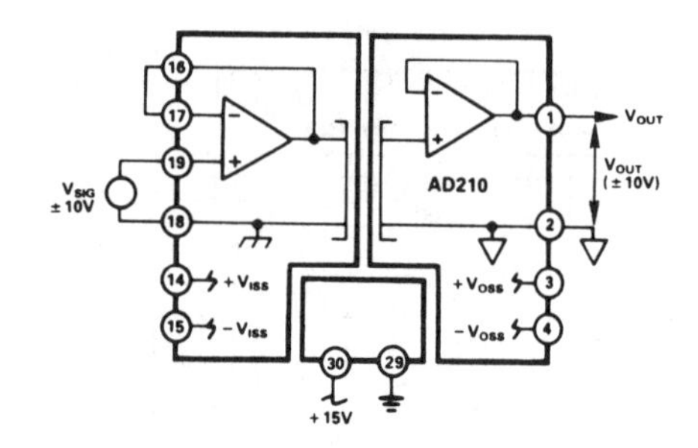

Basic Unity Gain Configuration

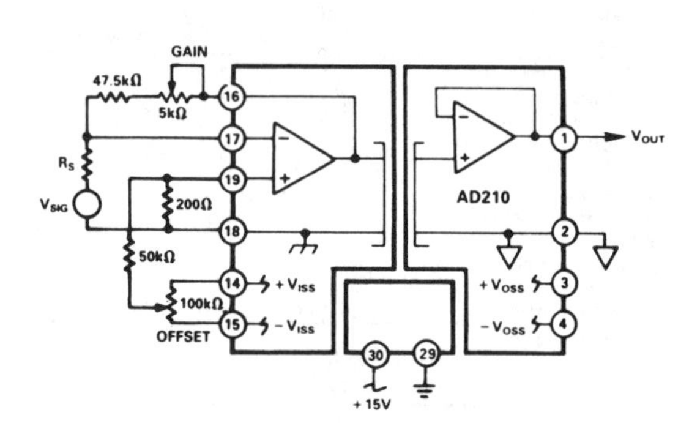

Adjustments for Inverting Input

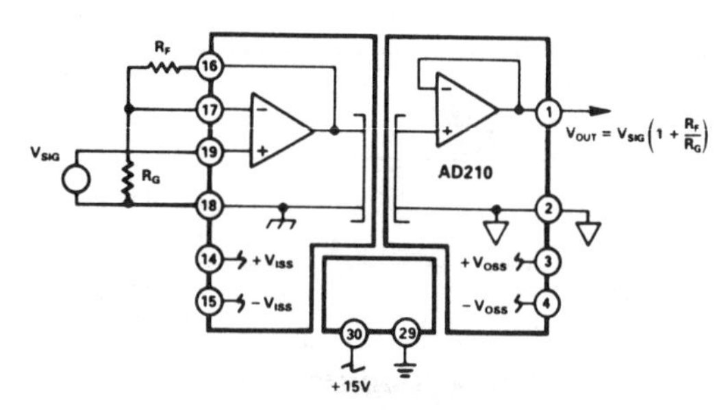

Input Configuration for G>1

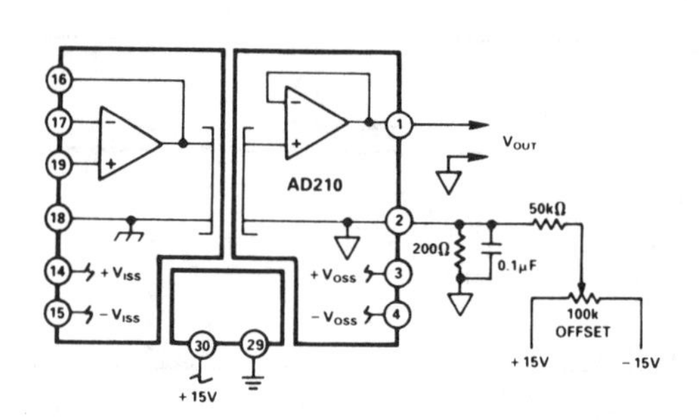

Output-Side Offset Adjustment

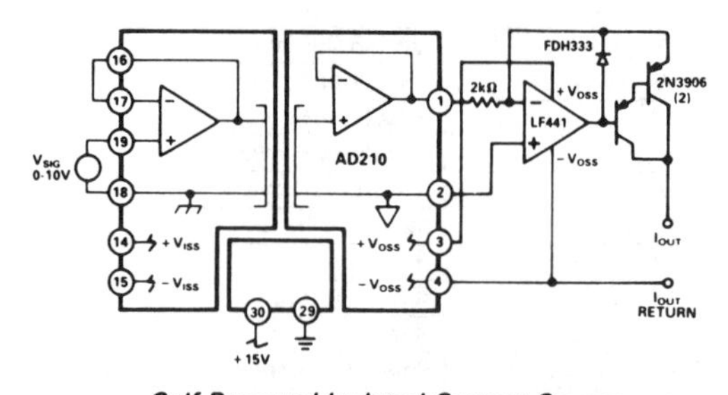

Self-Powered Isolated Current Source

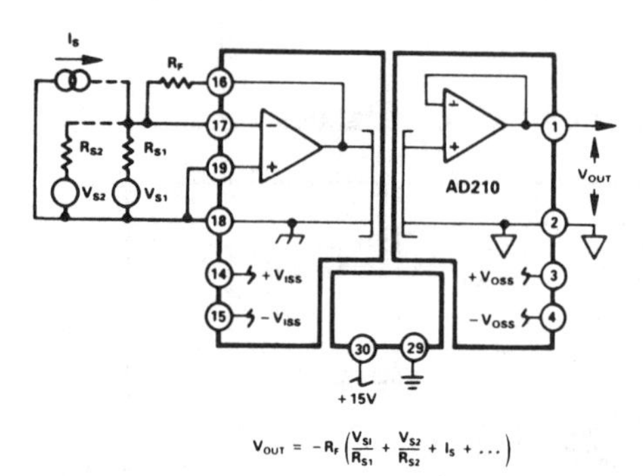

$$V_{OUT} = -R_F\left(\frac{V_{S1}}{R_{S1}} + \frac{V_{S2}}{R_{S2}} + I_S + \ldots\right)$$

Summing or Current Input Configuration

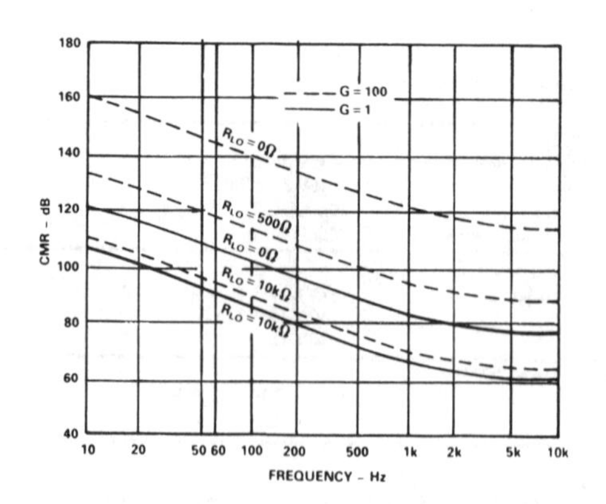

Common-Mode Rejection vs. Frequency

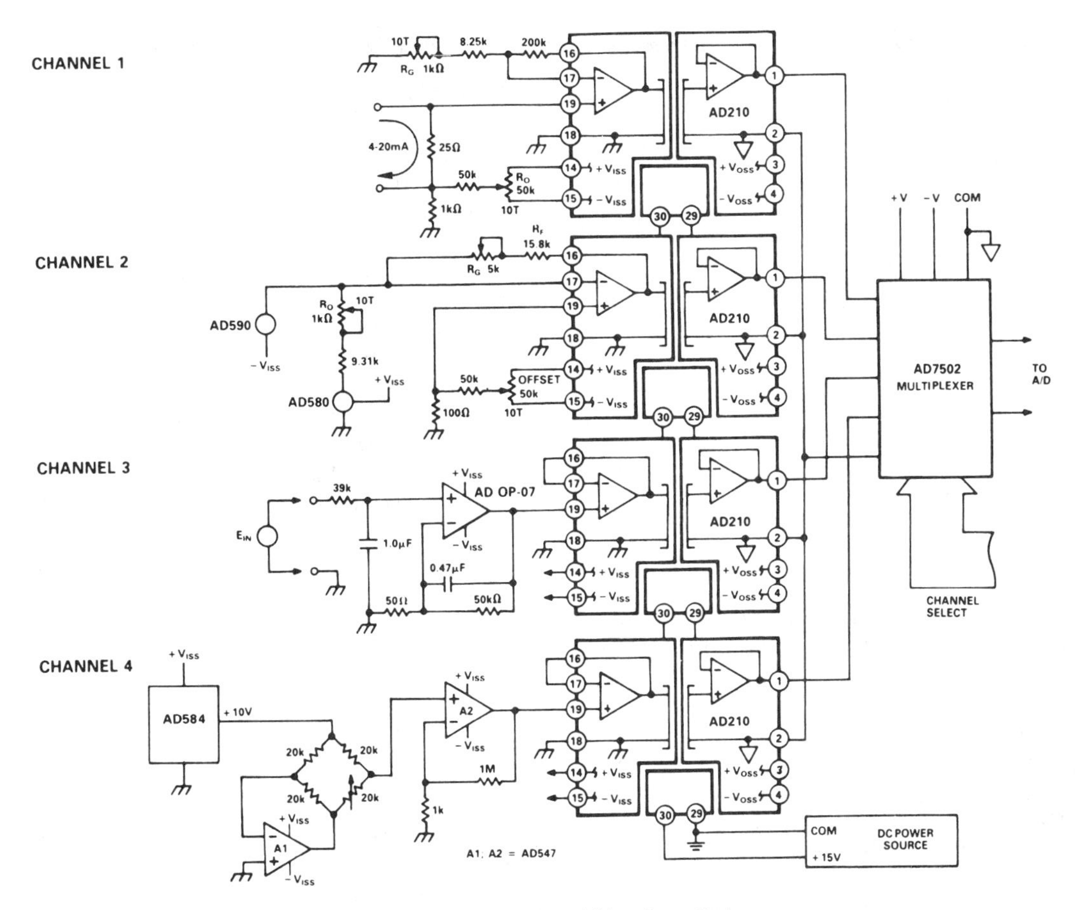

Multichannel Data Acquisition Front-End

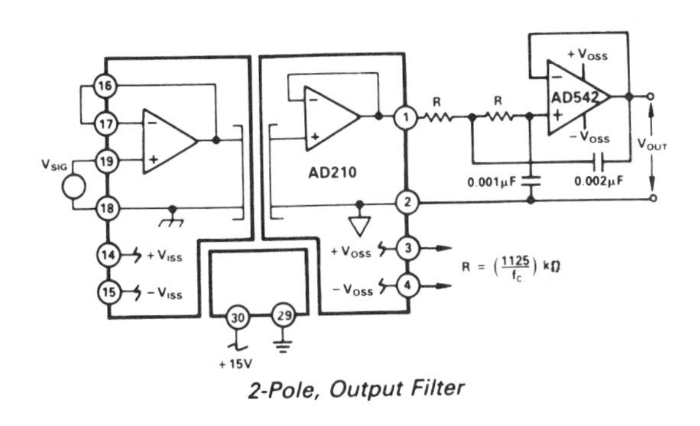

2-Pole, Output Filter

Isolated Voltage-to-Current Loop Converter

Isolated Thermocouple Amplifier

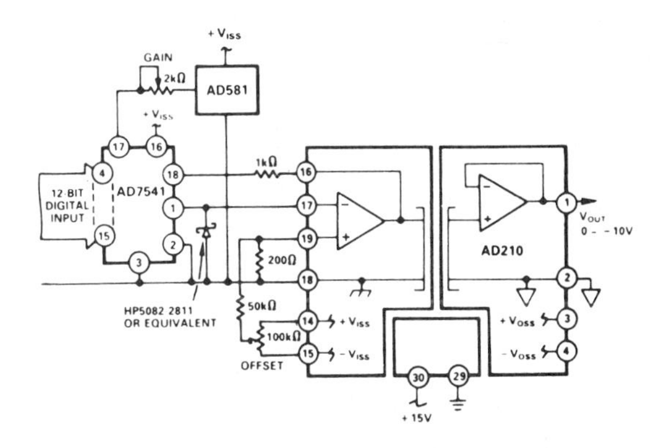

Precision Floating Programmable Reference

Analog Devices AD9300

4 × 1 Wideband Video Multiplexer

- 34 MHz Full power bandwidth
- ±0.1dB Gain flatness to 8 MHz
- 75dB Crosstalk rejection @ 10 MHz
- 0.05°/0.05% Differential phase/gain
- Cascadable for switch matrices

APPLICATIONS

- Video routing
- Medical imaging
- Electro-optics
- ECM systems
- Radar systems
- Data acquisition

The AD9300 is a monolithic high-speed video signal multiplexer useable in a wide variety of applications. Its four channels of video input signals can be randomly switched at megahertz rates to the single output. In addition, multiple devices can be configured in either parallel or cascade arrangements to form switch matrices. This flexibility in using the AD9300 is possible because the output of the device is in a high-impedance state when the chip is not enabled; when the chip is enabled, the unit acts as a buffer with a high-input impedance and low-output impedance.

An advanced bipolar process provides fast, wideband switching capabilities while maintaining crosstalk rejection of 75dB at 10 MHz. Full power bandwidth is a minimum 30 MHz. The device can be operated from +10V to ±15V power supplies. The AD9300KQ is packaged in a 16-pin ceramic DIP, and the AD9300KP is packaged in a 20-pin PLCC; both are designed to operate over the commercial temperature range of 0 to +70°C. For military temperatures of −55°C to +125°C, use part number AD9300TQ, which is also a 16-pin ceramic DIP. In addition, the AD9300 is available in a 20-pin LCC as the model AD9300TE, which operates over a temperature range of −55°C to +125°C.

ABSOLUTE MAXIMUM RATINGS[1]

Supply Voltages ($\pm V_S$) . . . . ±16V
Analog Input Voltage Each Input (IN_1 thru IN_4) . . . . ±3.5V
Differential Voltage Between Any Two Inputs (IN_1 thru IN_4) 5V
Digital Input Voltages (A_0, A_1, ENABLE) . −0.5V to +5.5V
Output Current
  Sinking 6.0mA
  Sourcing 6.0mA
Operating Temperature Range
  AD9300KQ/KP . . . . 0 to +70°C
  AD9300TQ/TE . . . . −55°C to +125°C
Storage Temperature Range . . . . −65°C to +150°C
Junction Temperature . . . . +175°C
Lead Soldering (10sec) . . . . +300°C

ELECTRICAL CHARACTERISTICS (± V_s = ±12V±5%; C_L = 10pF; R_L = 2kΩ, unless otherwise noted)

Parameter (Conditions)	Temp	Test Level	COMMERCIAL 0 to +70°C AD9300KQ/KP Min	Typ	Max	Military Subgroup[2]	MILITARY −55°C to +125°C AD9300TQ/TE Min	Typ	Max	Units
INPUT CHARACTERISTICS										
Input Offset Voltage	+25°C	I		3	10	1		3	10	mV
Input Offset Voltage	Full	VI			14	2, 3			18	mV
Input Offset Voltage Drift[3]	Full	V		75				83		μV/°C
Input Bias Current	+25°C	I		15	37	1		15	37	μA
Input Bias Current	Full	VI			55	2,3			55	μA
Input Resistance	+25°C	V		3.0				3.0		MΩ
Input Capacitance	+25°C	V		2				2		pF
Input Noise Voltage (dc to 8MHz)	+25°C	V		16				16		μV rms
TRANSFER CHARACTERISTICS										
Voltage Gain[4]	+25°C	I	0.990	0.994		1	0.990	0.994		V/V
Voltage Gain[4]	Full	VI	0.985			2,3	0.985			V/V
DC Linearity[5]	+25°C	V		0.01				0.01		%
Gain Tolerance (V_{IN} = ±1V)										
dc to 5MHz	+25°C	I		0.05	0.1	4		0.05	0.1	dB
5MHz to 8MHz	+25°C	I		0.1	0.3	4		0.1	0.3	dB
Small-Signal Bandwidth (V_{IN} = 100mV p-p)	+25°C	V		350				350		MHz
Full Power Bandwidth[6] (V_{IN} = 2V p-p)	+25°C	I	30	34		4	30	34		MHz
Output Swing	Full	VI	±2			1, 2, 3	±2			V
Output Current (Sinking @ = 25°C)	+25°C	V		5				5		mA
Output Resistance	+25°C	III		9	15	12		9	15	Ω
DYNAMIC CHARACTERISTICS										
Slew Rate[7]	+25°C	I	190	215		4	190	215		V/μs
Settling Time (to 0.1% on ±2V Output)	+25°C	III		70	100	12		70	100	ns

Parameter (Conditions)	Temp	Test Level	COMMERCIAL 0 to +70°C AD9300KQ/KP Min	Typ	Max	Military Subgroup[2]	MILITARY −55°C to +125°C AD9300TQ/TE Min	Typ	Max	Units
Overshoot										
To T-Step[8]	+25°C	V		<0.1				<0.1		%
To Pulse[9]	+25°C	V		<10				<10		%
Differential Phase[10]	+25°C	III		0.05	0.1	12		0.05	0.1	°
Differential Gain[10]	+25°C	III		0.05	0.1	12		0.05	0.1	%
Crosstalk Rejection										
Three Channels[11]	+25°C	IV	70	75			70	75		dB
One Channel[12]	+25°C	IV	78(75)	80			78(75)	80		dB
SWITCHING CHARACTERISTICS[13]										
A_X Input to Channel HIGH Time[14] (t_{HIGH})	+25°C	I		40	50	9		40	50	ns
A_X Input to Channel LOW Time[15] (t_{LOW})	+25°C	I		35	45	9		35	45	ns
Enable to Channel ON Time[16] (t_{ON})	+25°C	I		30	40	9		30	40	ns
Enable to Channel OFF Time[17] (t_{OFF})	+25°C	I		20	30	9		20	30	ns
Switching Transient[18]	+25°C	V		60				60		mV
DIGITAL INPUTS										
Logic "1" Voltage	Full	VI	2			1,2,3	2			V
Logic "0" Voltage	Full	VI			0.8	1,2,3			0.8	V
Logic "1" Current	Full	VI			5	1,2,3			5	μA
Logic "0" Current	Full	VI			1	1,2,3			1	μA
POWER SUPPLY										
Positive Supply Current (+12V)	+25°C	I		13	16	1		13	16	mA
Positive Supply Current (+12V)	Full	VI		13	16	2,3		13	16	mA
Negative Supply Current (−12V)	+25°C	I		12.5	15	1		12.5	15	mA
Negative Supply Current (−12V)	Full	VI		12.5	16	2,3		12.5	16	mA
Power Supply Rejection Ratio ($\pm V_S = \pm 12V \pm 5\%$)	Full	VI	67	75		1,2,3	67	75		dB
Power Dissipation (±12V)[19]	+25°C	V		306				306		mW

NOTES

For applications assistance, phone Computer Labs Division at (919) 668-9511

[1]Permanent damage may occur if any one absolute maximum rating is exceeded. Functional operation is not implied, and device reliability may be impaired by exposure to higher-than-recommended voltages for extended periods of time.

[2]Military Subgroups apply to military-qualified devices only.

[3]Measured at extremes of temperature range.

[4]Measured as slope of V_{OUT} versus V_{IN} with $V_{IN} = \pm 1V$.

[5]Measured as worst deviation from end-point fit with $V_{IN} = \pm 1V$.

[6]Full Power Bandwith (FPBW) based on Slew Rate (SR). $FPBW = SR/2\pi V_{PEAK}$

[7]Measured between 20% and 80% transition points of ±1V output.

[8]T-Step = Sin^2X Step, when Step between 0V and +700mV points has 10%-to-90% risetime = 125ns.

[9]Measured with a pulse input having slew rate >250V/μs.

[10]Measured at output between 0.28Vdc and 1.0Vdc with V_{IN} = 284mV p-p at 3.58MHz and 4.43MHz.

[11]This specification is critically dependent on circuit layout. Value shown is measured with selected channel grounded and 10MHz 2V p-p signal applied to remaining three channels. If selected channel is grounded through 75Ω, value is approximately 6dB higher.

[12]This specification is critically dependent on circuit layout. Value shown is measured with selected channel grounded and 10MHz 2V p-p signal applied to one other channel. If selected channel is grounded through 75Ω, value is approximately 6dB higher. Minimum specification in ( ) applies to DIPs

[13]Consult system timing diagram.

[14]Measured from address change to 90% point of −2V to +2V output LOW-to-HIGH transition.

[15]Measured from address change to 10% point of +2V to −2V output HIGH-to-LOW transition.

[16]Measured from 50% transition point of ENABLE input to 90% transition of 0V to −2V output.

[17]Measured from 50% transition point of ENABLE input to 10% transition of +2V to 0V output.

[18]Measured while switching between two grounded channels.

[19]Maximum power dissipation is a package-dependent parameter related to the following typical thermal impedances:

16-Pin Ceramic θ_{JA} = 87°C/W; θ_{JC} = 25°C/W

20-Pin LCC θ_{JA} = 74°C/W; θ_{JC} = 10°C/W

Specifications subject to change without notice.

ORDERING INFORMATION

Device	Temperature Range	Description	Package Options*
AD9300KQ	0 to +70°C	16-Pin Cerdip, Commercial	Q-16
AD9300TQ	−55°C to +125°C	16-Pin Cerdip, Military Temperature	Q-16
AD9300TE	−55°C to +125°C	20-Pin LCC, Military Temperature	E-20A
AD9300KP	0 to +70°C	20-Pin PLCC, Commercial	P-20A

MECHANICAL INFORMATION

Die Dimensions . . . . 84 × 104 × 18 (max) mils
Pad Dimensions . . . . 4 × 4 (min) mils
Metalization . . . . Aluminum
Backing . . . . None
Substrate Potential . . . . $-V_S$
Passivation . . . . Oxynitride
Die Attach . . . . Gold Eutectic
Bond Wire . . . . . 1.25 mil, Aluminum; Ultrasonic Bonding
or 1 mil, Gold; Gold Ball Bonding

AD9300 FUNCTIONAL BLOCK DIAGRAM
(Based on Cerdip)

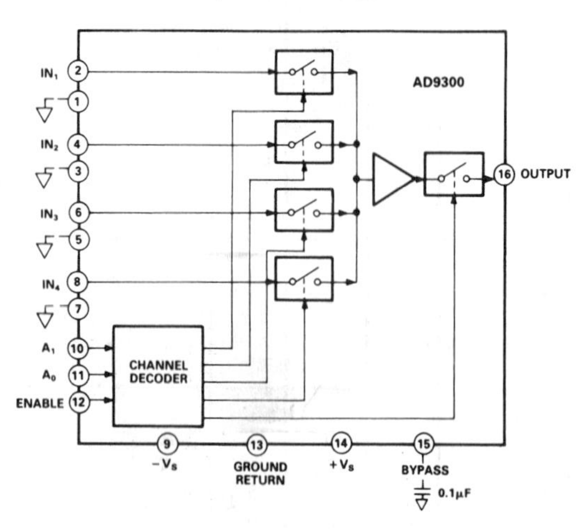

PIN DESIGNATIONS

DIP

Pin	Function	Pin	Function
1	GROUND	16	OUTPUT
2	IN_1	15	BYPASS
3	GROUND	14	$+V_S$
4	IN_2	13	GROUND RETURN
5	GROUND	12	ENABLE
6	IN_3	11	A_0
7	GROUND	10	A_1
8	IN_4	9	$-V_S$

AD9300 TOP VIEW (Not to Scale)

LCC

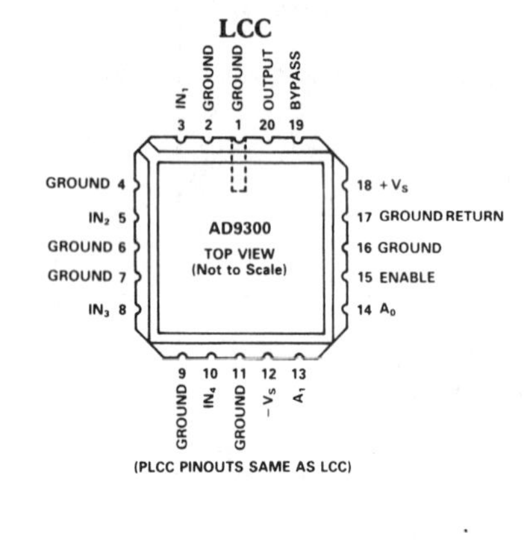

(PLCC PINOUTS SAME AS LCC)

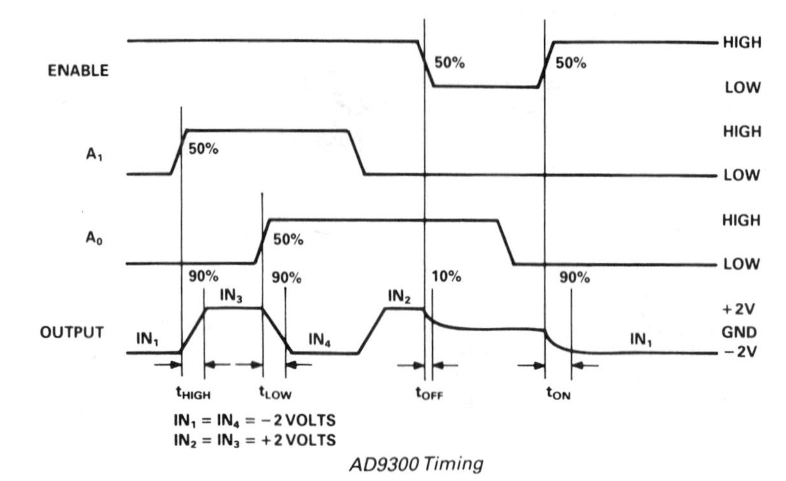

AD9300 Timing

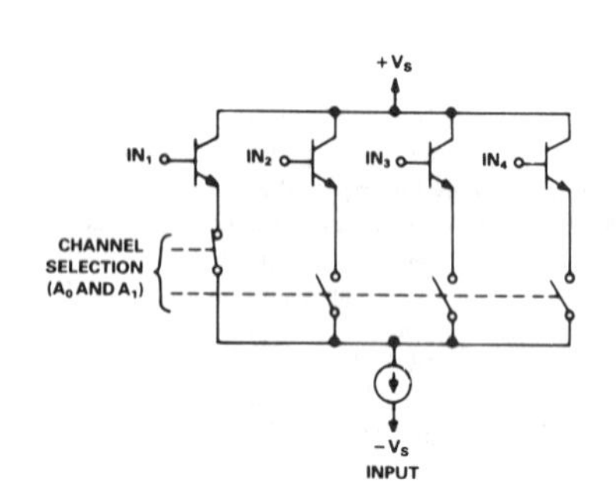

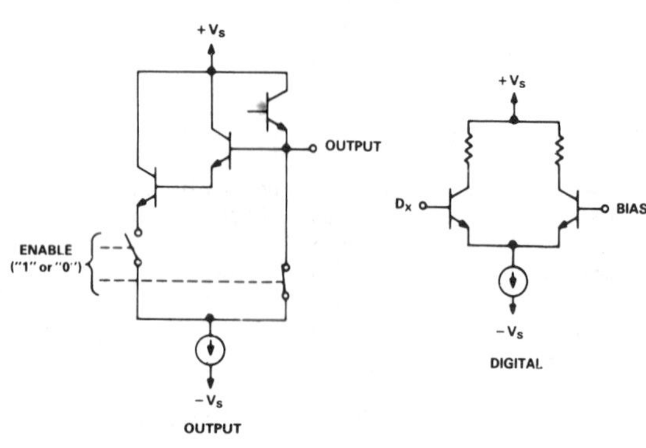

Input and Output Equivalent Circuits

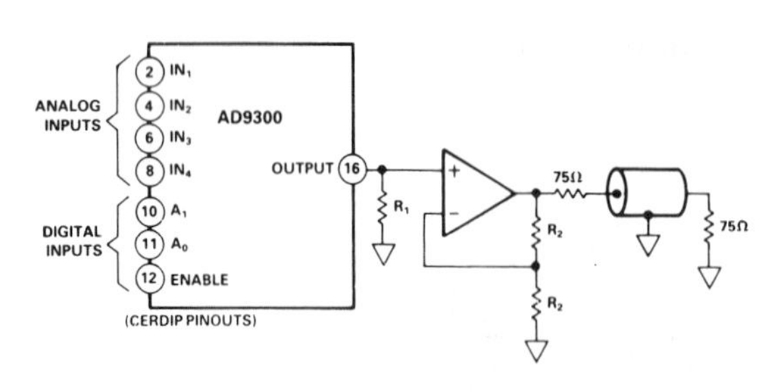

4 × 1 AD9300 Multiplexer with Buffered Output Driving 75Ω Coaxial Cable

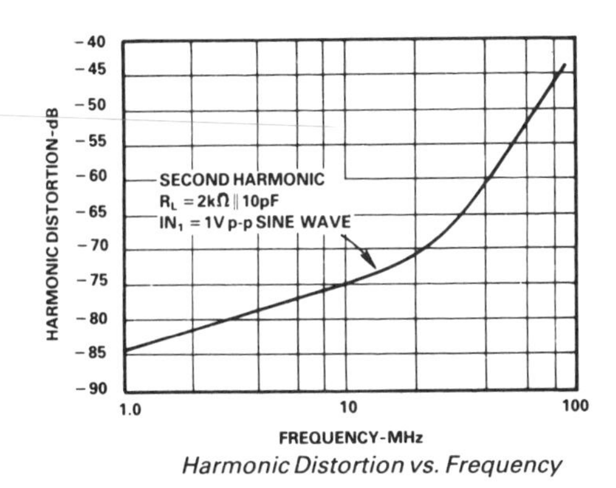

Harmonic Distortion vs. Frequency

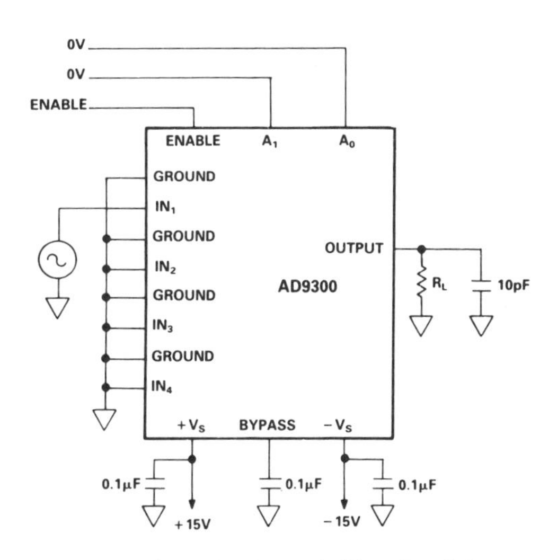

Test Circuit for Harmonic Distortion, Pulse Response, T-Step Response and Disable Characteristics

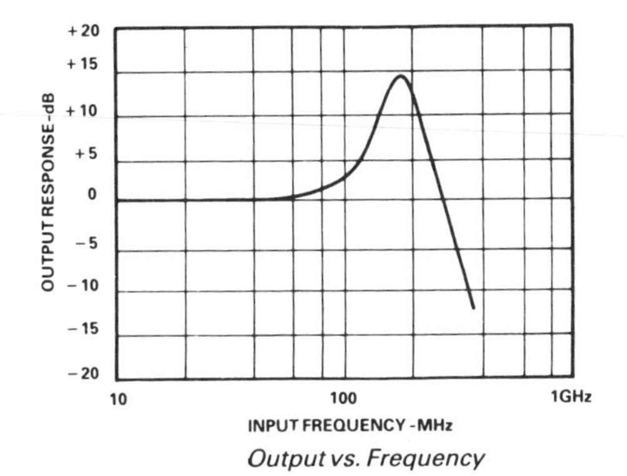

Output vs. Frequency

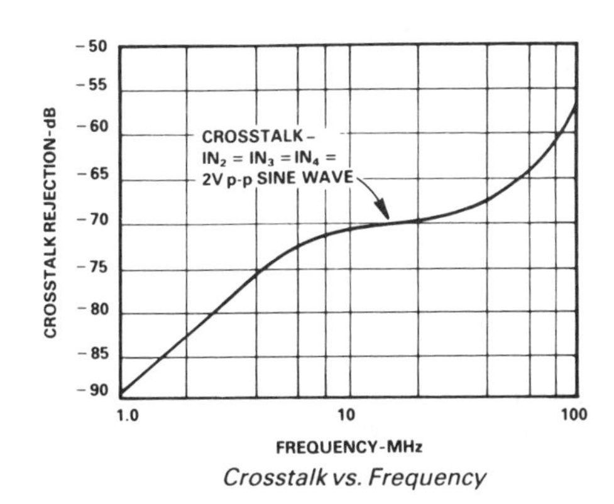

Crosstalk vs. Frequency

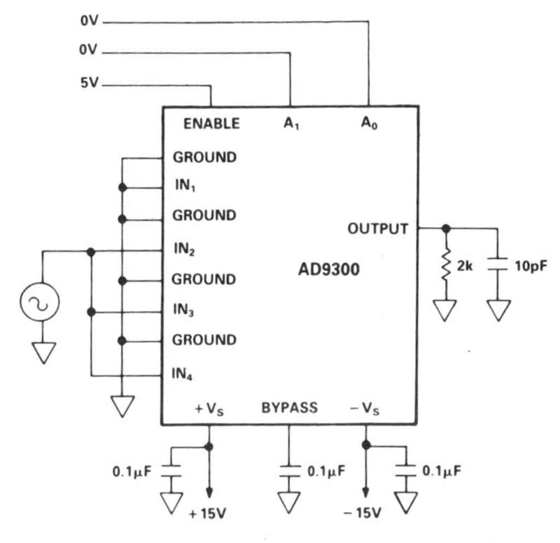

Crosstalk Rejection Test Circuit

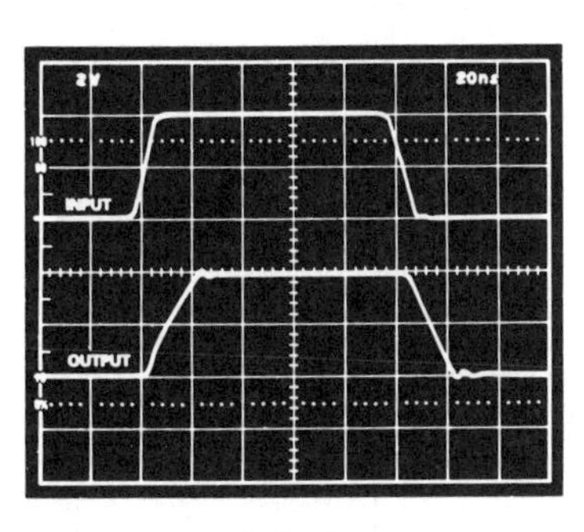

Pulse Response

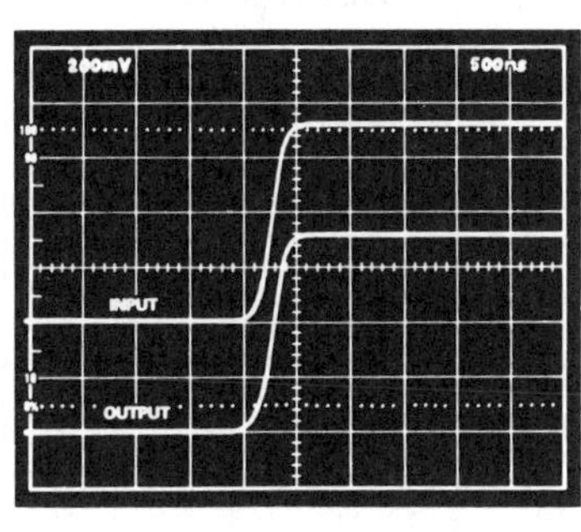

T-Step Response

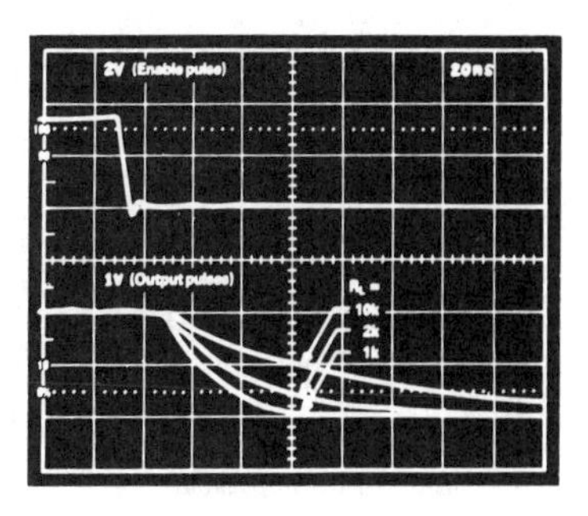

Enable to Channel "Off" Response

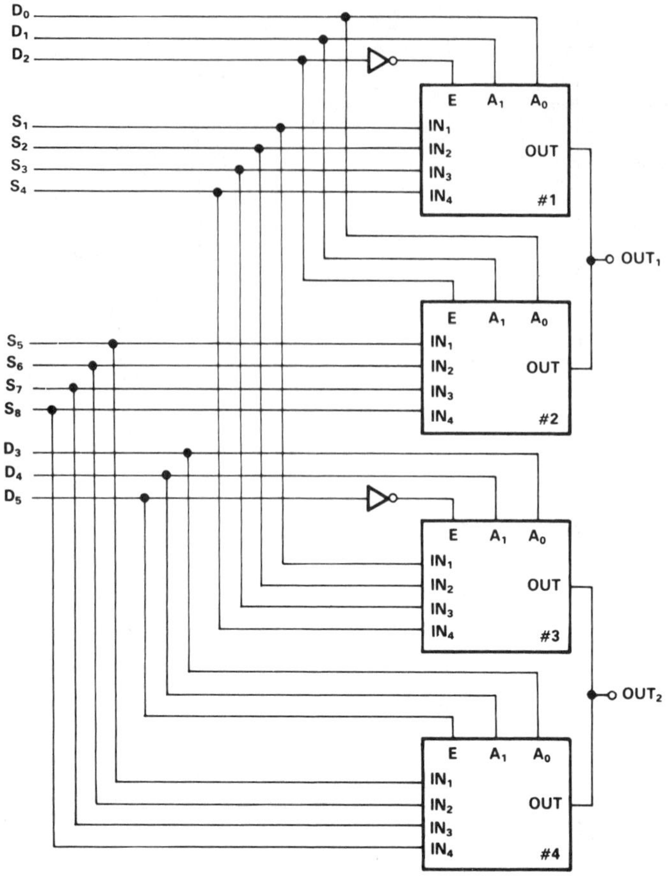

8 X 2 SIGNAL CROSSPOINT USING FOUR AD9300 MULTIPLEXERS

This information is reproduced with the permission of GEC Plessey Semiconductors.

Plessey MJ1444

PCM Synchronizing Word Generator

- $5V \pm 5\%$ supply—20mA typical
- Fully conforms to CCITT recommendation G732
- Outputs directly onto PCM data highway
- Provides both time slot 0 and time slot 16 channel pulses
- All inputs and outputs are TTL compatible

ABSOLUTE MAXIMUM RATINGS

The absolute maximum ratings are limiting values above which operating life might be shortened or specified parameters might be degraded.

Electrical Ratings

$+V_{CC}$	7V
Inputs	V_{CC} + 0.5V Gnd −0.3V
Outputs	V_{CC}, Gnd −0.3V

Thermal Ratings

Max Junction Temperature 175 °C

Thermal Resistance:	Chip to Case	Chip to Amb.
	35 °C/Watt	120 °C/Watt

ELECTRICAL CHARACTERISTICS

Test conditions (unless otherwise stated):
Supply voltage, $V_{CC} = 5V \pm 0.25V$
Ambient operating temperature −10°C to +70°C

Characteristic	Symbol	Pins	Value			Units	Conditions
			Min.	Typ.	Max.		
Low level input voltage	V_{IL}	1, 2, 3, 4, 5, 7, 11, 12, 13, 14.	−0.3		0.8	V	
Low level input current / High level input current	I_{IN}	11		1	50	μA	
High level input voltage	V_{IH}	11	2.4		V_{CC}		
Low level output voltage	V_{OL}	6, 9, 15			0.5	V	$I_{sink} = 2mA$
		10			0.7	V	$I_{sink} = 5mA$
High level output voltage	V_{OH}	6, 9, 15	2.8			V	$I_{source} = 200\mu A$
High level output leakage current	I_{OH}	10			20	μA	$V_{OUT} = V_{CC}$
Supply current	I_{CC}			20	40	mA	$V_{CC} = 5.25V$

DYNAMIC CHARACTERISTICS

Characteristic	Symbol	Value			Units	Conditions
		Min.	Typ.	Max.		
Max clock frequency	F_{max}	3			MHz	
Propagation delay, clock to TS0, TS0 $\overline{SF}$, TS16 and combined data outputs.	t_P	80		200	ns	See Figs.5 and 6
Set up time channel reset to clock	T_{S1}	100		450	ns	f_{clock} = 2.048 MHz
Hold time of channel reset input	t_{H1}	20		400	ns	
Set up time of bit 1 (SF) to datum B	t_{S2}	100			ns	
Hold time of bit 1 (SF) wrt datum B	t_{H2}	300			ns	
Set up time of bit 1 ($\overline{SF}$) and data bits 3—8 to datum B	t_{S2}	100			ns	
Hold time of bit 1 ($\overline{SF}$) and data bits 3—8 wrt datum B	t_{H2}	300			ns	

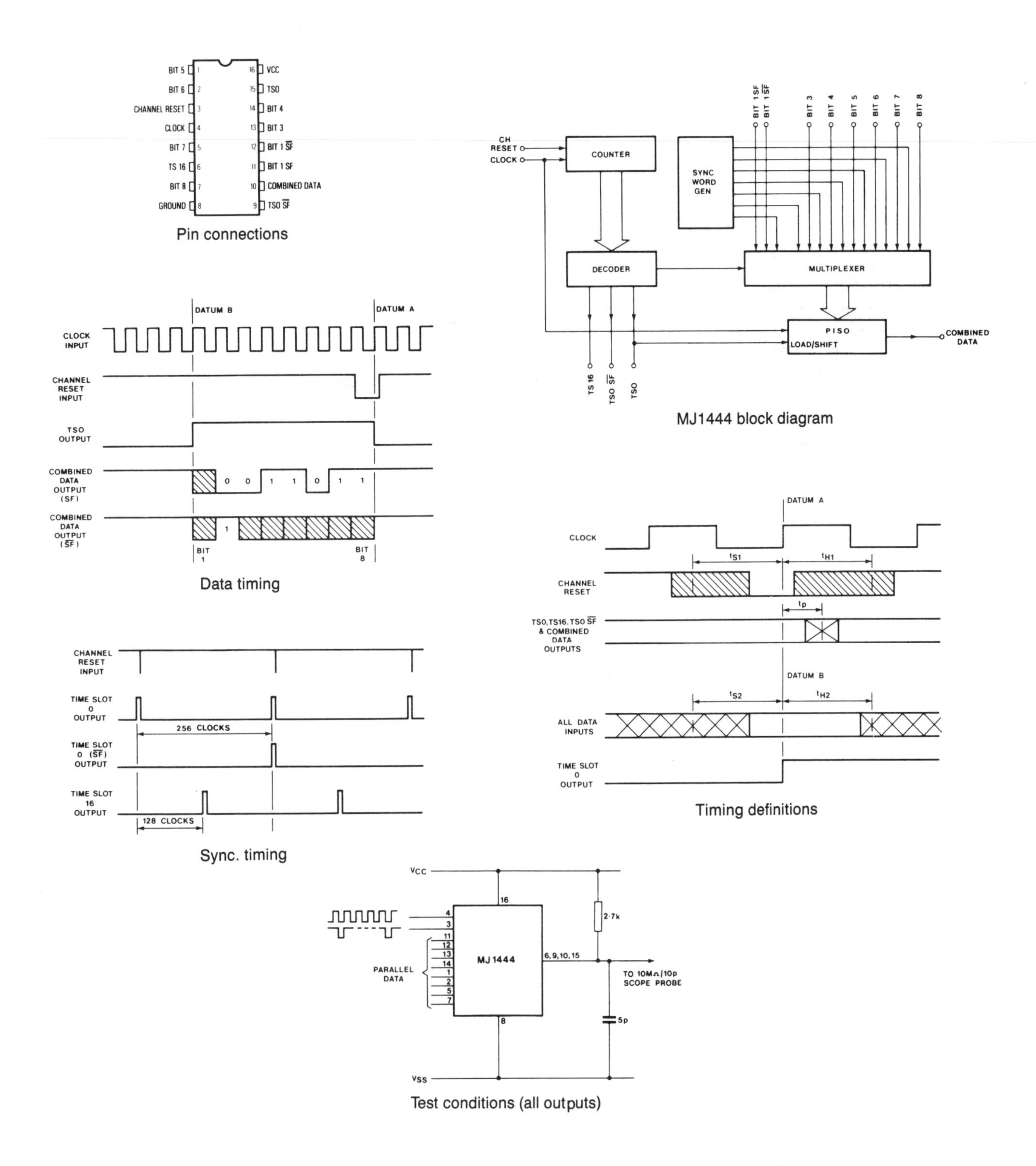

Pin connections

MJ1444 block diagram

Data timing

Timing definitions

Sync. timing

Test conditions (all outputs)

Plessey MJ1445
PCM Synchronizing Word Receiver

- 5V ±5% supply—20mA typical
- Conforms to CCITT recommendation G732
- Synchronizing word error monitor
- Out of sync alarm
- All inputs and outputs are TTL compatible

ABSOLUTE MAXIMUM RATINGS
The absolute maximum ratings are limiting values above which operating life might be shortened or specified parameters might be degraded.

Electrical Ratings

$+V_{CC}$	7V
Inputs	V_{CC} + 0.5V Gnd − 0.3V
Outputs	V_{CC}, Gnd − 0.3V

Thermal Ratings
Max Junction Temperature 175 °C

Thermal Resistance:	Chip to Case	Chip to Amb.
	35 °C/Watt	120 °C/Watt

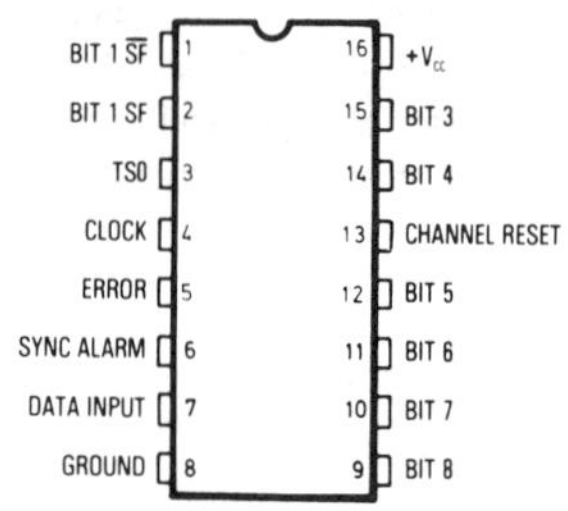

Pin connections

ELECTRICAL CHARACTERISTICS

Test conditions (unless otherwise stated):
Supply voltage, V_{CC} = 5V ±0.25V
Ambient temperature, T_{amb} = −10°C to +70°C

Static Characteristics

Characteristic	Symbol	Pins	Value Min.	Value Typ.	Value Max.	Units	Conditions
Low level input voltage	V_{IL}	4, 7	-0.3		0.8	V	
Low level input current / High level input current	I_{IN}	4, 7		1	50	μA	
High level input voltage	V_{IH}	4, 7	2.4		V_{CC}	V	
Low level output voltage	V_{OL}	1, 2, 3, 5, 6 9, 10, 11, 12 13, 14, 15			0.5	V	I_{sink} = 2mA
High level output voltage	V_{OH}		2.8				I_{source} = 200μA
Supply current	I_{CC}			20	40	mA	V_{CC} = 5.25V

Dynamic Characteristics

Characteristic	Symbol	Value Min.	Value Typ.	Value Max.	Units	Conditions
Max. clock frequency	f_{max}	2.2			MHz	
Input delay of data input	$t_{d\ data}$	20		200	ns	f_{clock} = 2.048MHz
Propagation delay, clock to TS0 output	$t_{d\ TS0}$	40		200	ns	
Propagation delay clock to error output, sync alarm and CH. Reset output high	t_d	50		400	ns	
Propagation delay, clock to CH. Reset output low (T — t_p)	t_p	100		450	ns	
Propagation delay clock to spare bits	$t_{d\ SB}$	50		300	ns	

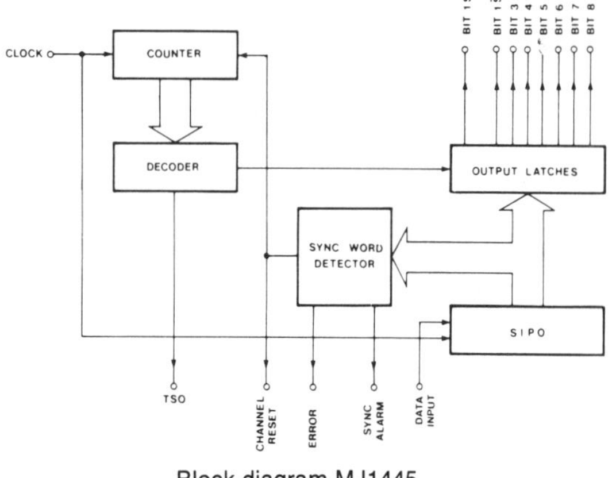

Block diagram MJ1445

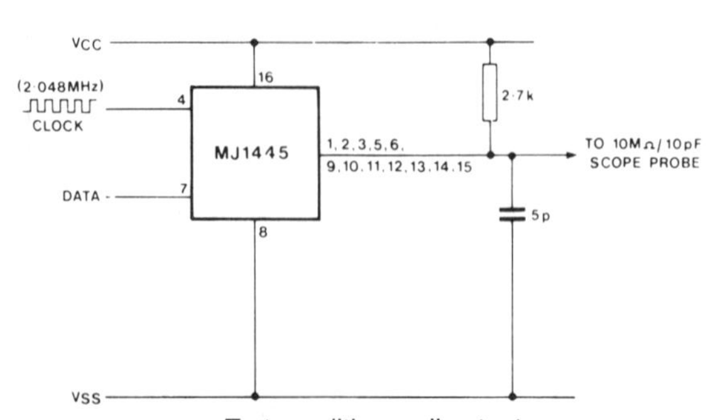

Test conditions, all outputs

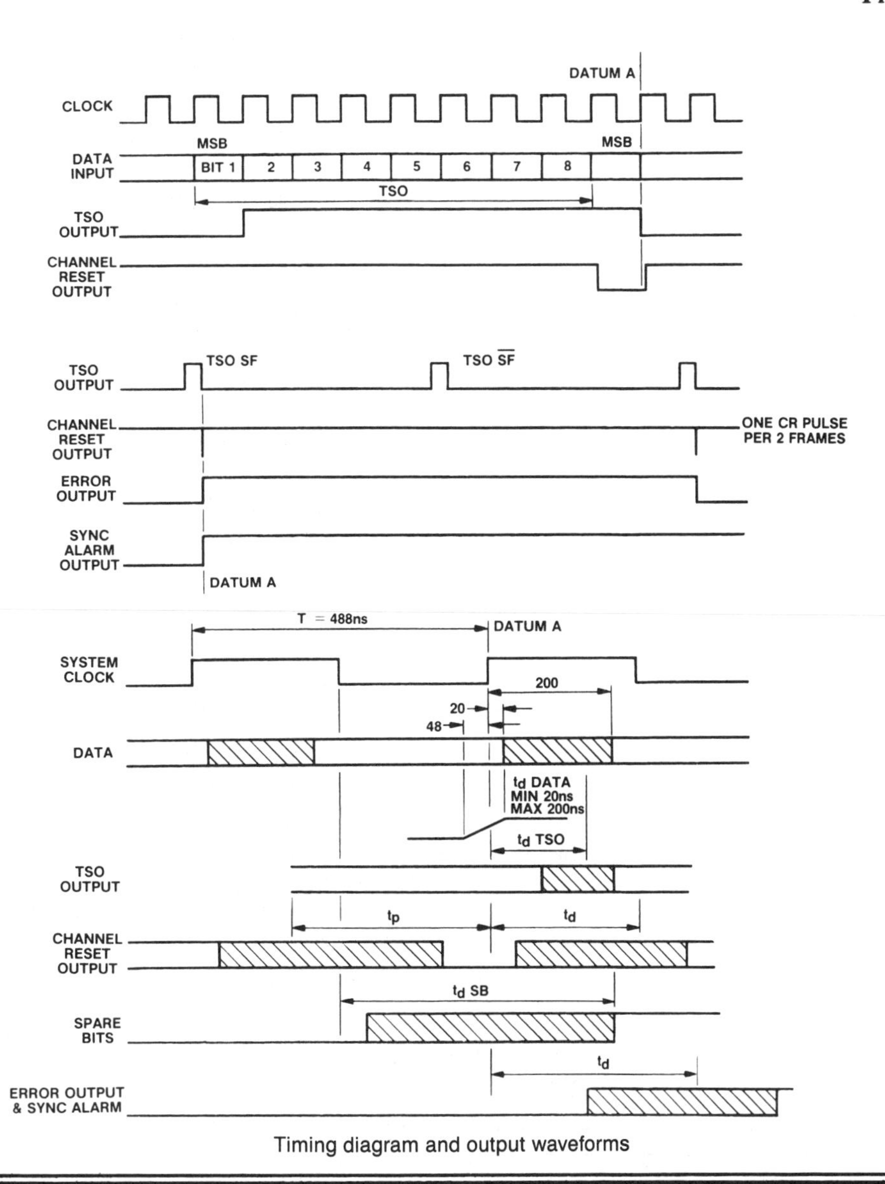

Timing diagram and output waveforms

Plessey MJ1446
Time Slot 16 Receiver and Transmitter

- 5V ± 5% supply—20mA typical
- Conforms to CCITT recommendations
- Provides both AMI and binary format data outputs
- Single-chip receive or transmit
- All inputs and outputs are TTL compatible

ELECTRICAL CHARACTERISTICS

Supply voltage V_{CC} = 5V ± 0.25V, Ambient temperature T_{amb} = -10°C to +70°C,

Static Characteristics

Characteristic	Symbol	Pins	Value			Units	Conditions
			Min.	Typ.	Max.		
Low level input voltage	V_{IL}	3, 4, 7, 9, 11, 12, 13, 14	-0.3		0.8	V	
Low level input current High level input current	I_{IN}	11		1	50	μA	
High level input voltage	V_{IH}	11	2.4		V_{CC}	V	
Low level output	V_{OL}	1, 2, 5, 6, 7, 9, 10, 11, 15			0.5	V	I_{sink} = 2mA
		12			0.5	V	I_{sink} = 5mA
High level output voltage	V_{OH}	1, 2, 10, 5, 6, 15	2.8			V	I_{source} = 200μA
High level output leakage current	I_{CH}	7, 9, 11, 12			20	μA	$V_{OUT} = V_{CC}$
Supply current	I_{CC}			20		mA	V_{CC} = 5.25V

DYNAMIC CHARACTERISTICS

Characteristic	Symbol	Value			Units
		Min.	Typ.	Max.	
Propogation delay clock to data out to digital highway	t_p	20		200	ns
Propogation delay clock to 64 kHz out	t_p	20		200	ns
Input delay, clock to digital highway access	$t_{d\,DATA}$	20		200	ns
Input delay, clock to time slot 16	$t_{d\,TS16}$	80		200	ns
Output delay 64 kHz to 16 kHz output	$t_{p\,16}$			70	ns
Output delay, 64 kHz to 8 kHz output	$t_{p\,8}$			170	ns
Output delay, 64 kHz to binary data output (64 kHz)	$t_{p\,BIN}$	20		450	ns
Output delay 64 kHz to AMI, $\overline{AMI}$, AMI data & $\overline{AMI}$ data o/p's	$t_{p\,AMI}$	20		400	ns
Input delay, 64 kHz to binary data in (64 kHz)	$t_{d\,BIN}$			100	ns

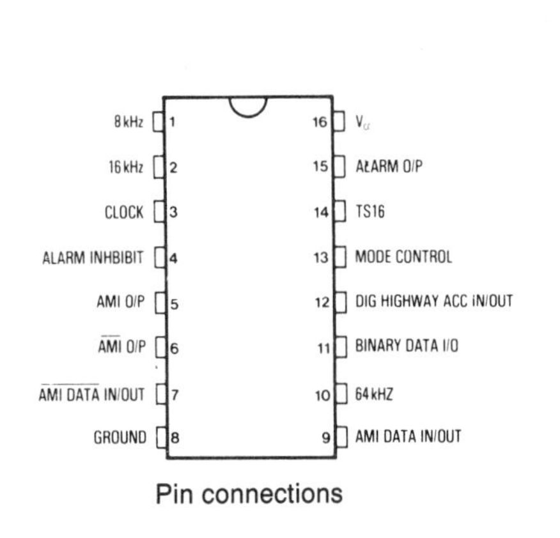

Pin connections

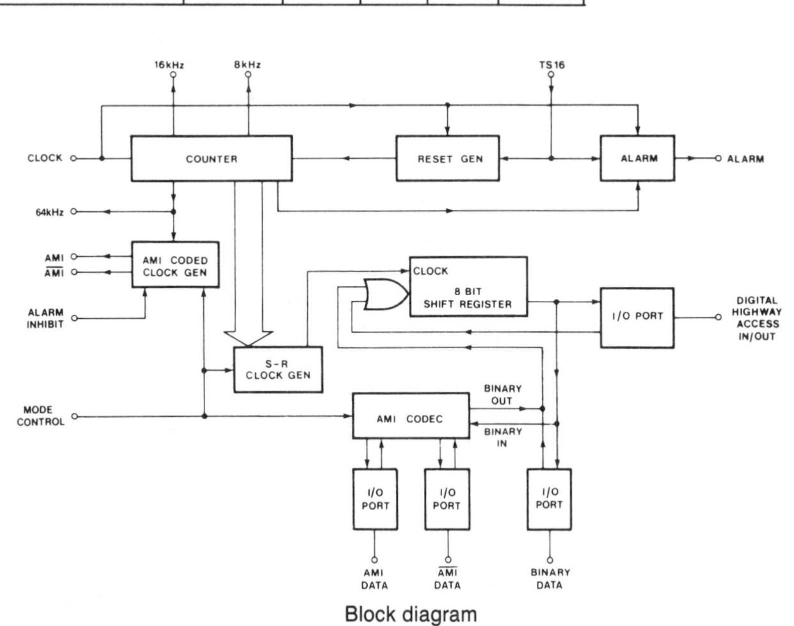

Block diagram

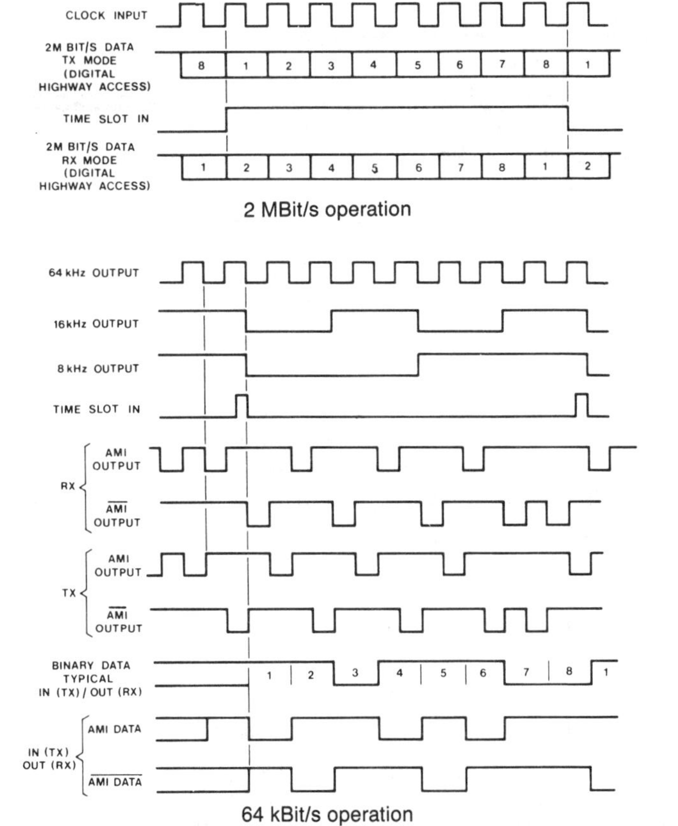

64 kBit/s operation

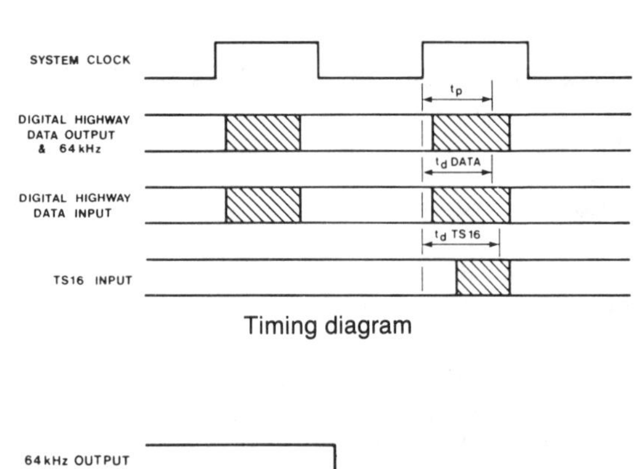

Timing diagram

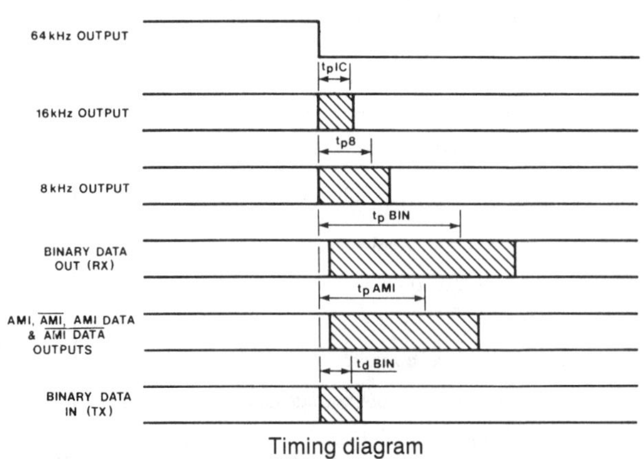

Timing diagram

Test conditions (transmit mode)

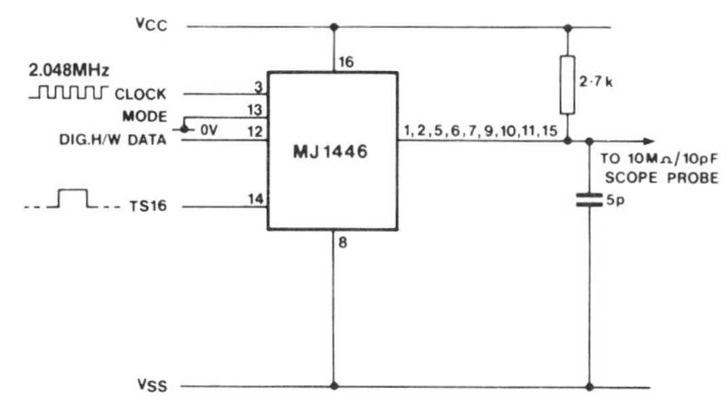

Test conditions (receive mode)

Plessey MS2014
Digital Filter and Detector (FAD)

- Linear 16-bit data
- 13-bit coefficient
- 2 MHz operating clock frequency
- Serial operation
- 448 bits of on-chip shift register data storage for 8th-order multiplex
- Nth-order multiplexing ($N \leq 8$)
- TTL compatible
- Single +5V supply

APPLICATIONS

- Low-cost digital filtering
- Level detection
- Spectral analysis
- Tone detectors (multifrequency receivers)
- Speech synthesis and analysis
- Data modems
- Group delay equalizers (all-pass networks)

ABSOLUTE MAXIMUM RATINGS

Supply voltage (V_{DD})	-0.5V to +7V
Input voltage	-0.5V to +7V
Maximum output voltage	+7V
Temperature: Storage	-65°C to 125°C
Operating	0°C to 70°C

NOTE
All voltages with respect to V_{SS}.

RECOMMENDED OPERATING CONDITIONS

Characteristic	Symbol	Value		Units	Conditions
		Min.	Max.		
Supply voltage	V_{DD}	4.75	5.25	V	
Input voltage (high state) except clock	V_{IH}	2.2	-	V	
Input voltage (low state) except clock	V_{IL}	-	0.7	V	
Input voltage (high state) clock	V_{IHC}	4.5	-	V	
Input voltage (low state) clock	V_{ILC}	-	0.5	V	
Clock rise and fall time	t_{cl}		30	ns	10% - 90% (Note 1)
Clock frequency	f_{cl}	0.5	2.048	MHz	
Operating temperature	T_{amb}	0	70	°C	

ELECTRICAL CHARACTERISTICS

Test conditions (unless otherwise stated):
V_{DD} = +5V T_{amb} = 25°C

Characteristic	Symbol	Value			Units	Conditions
		Min.	Typ.	Max.		
Supply current	I_{DD}		90	120	mA	
Output voltage, low	V_{OL}	-	-	0.5	V	I_{OL} = 0.4mA (Note 2)
Output voltage, high	V_{OH}	2.7	3.4	-	V	I_{OH} = -40μA (Note 2)
Input capacitance (except clock)	C_{in}		5	7	pF	
Input capacitance (clock)	C_{inc}		25		pF	
Input data set up time	t_{is}	50	-	-	ns	
Input data hold time	t_{ih}	150	-	-	ns	
Output data delay time	t_{os}	-	-	200	ns	

NOTES
1. An operating clock frequency of 2.048MHz is guaranteed over the supply voltage range and the full operating temperature range.
2. The output stage is designed to drive a standard TTL LS gate (74LS series).

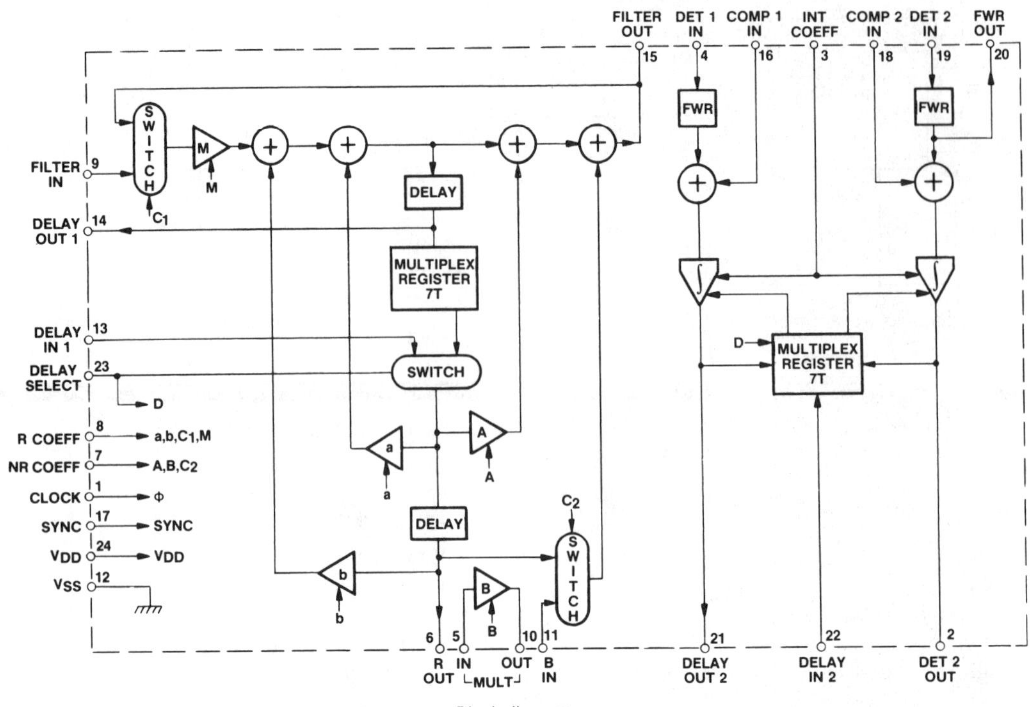

Block diagram

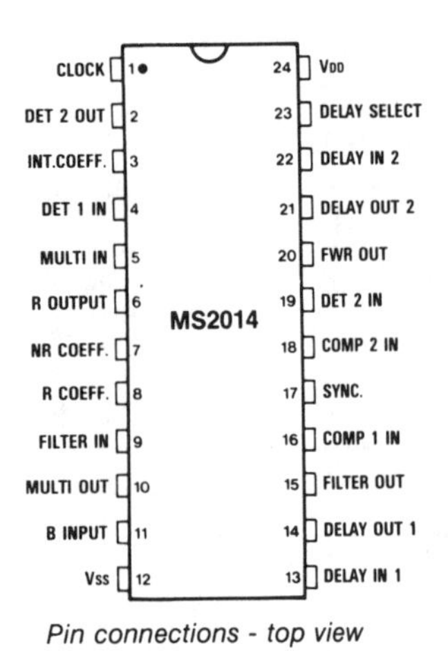

Pin connections - top view

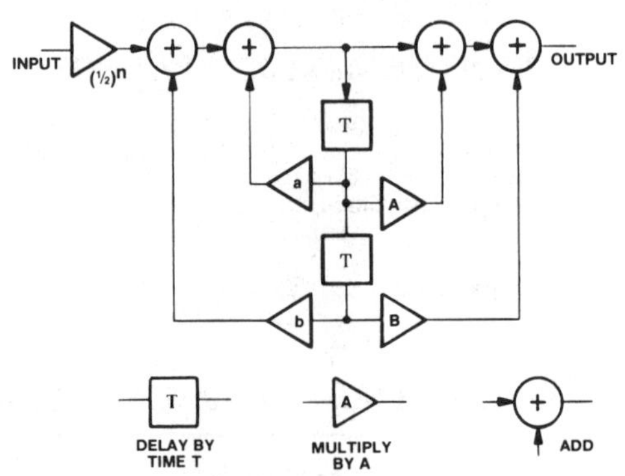

Basic 2nd order filter section

Leak factor	Rise time (0 to 90%)
1/2	3 Ts + T
3/4	8 Ts + T
7/8	17 Ts + T
15/16	35 Ts + T
31/32	72 Ts + T
63/64	146 Ts + T
127/128	293 Ts + T
255/256	588 Ts + T

Integrator rise times

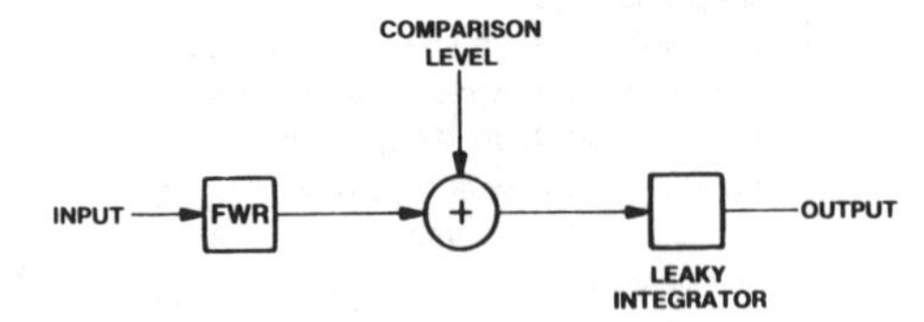

Simple level detector

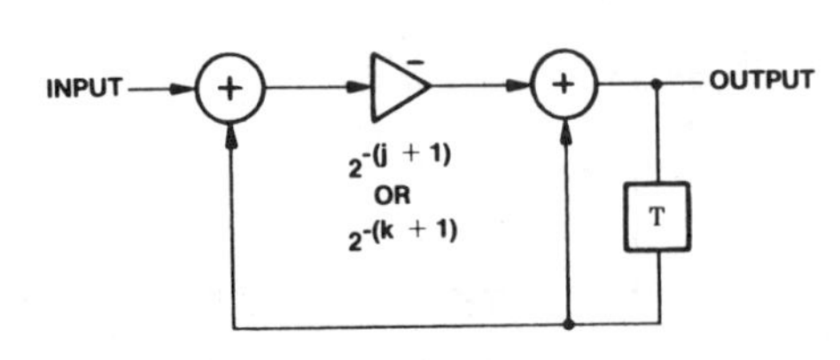

Leaky integrator

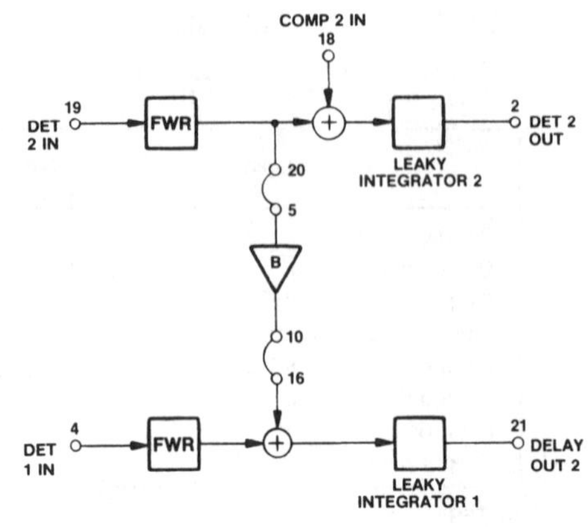

Relative level detection

32	31	30	29	28	27	26	25	24	23	22	21	20	19	18	17	16	15	14	13	12	11	10	9	8	7	6	5	4	3	2	1	Clock Pulse Number
C_1	M_1	M_2	M_3	M_4	b Coeff. msb.........lsb													a_s msb	a Coeff. (Recursive) ....lsb													
X	X	X	X	C_2	B Coeff. msb.........lsb													A_s msb	A Coeff. (Non-Recursive) ....lsb													
0.125					0													0.73217773437														
1	0	0	1	1	0	0	0	0	0	0	0	0	0	0	0	0	0	0	1	1	0	1	1	1	0	1	1	0	1	1	1	R
0	0	0	0	1	0	0	0	0	0	0	0	0	0	0	0	0	0	0	0	0	0	0	0	0	0	0	0	0	0	0	0	NR

Filter data format

Clock pulse number	32	31	30	29	28	27	26	25	24	23	22	21	20	19	18	17	16	15	14	13	12	11	10	9	8	7	6	5	4	3	2	1
Integrator coefficients	X	X	X	X	X	X	X	X	X	X	X	X	X	X	X	X	X	X	X	X	X	X	K_1	K_2	K_3	J_1	J_2	J_3	X	X	X	X
Comparison level 1 or 2	P_0	P_1	P_2	P_3	P_4	P_5	P_6	P_7	P_8	P_9	P_{10}	P_{11}	P_{12}	P_{13}	P_{14}	P_{15}	X	X	X	X	X	X	X	X	X	X	X	X	X	X	X	X

Detector data format

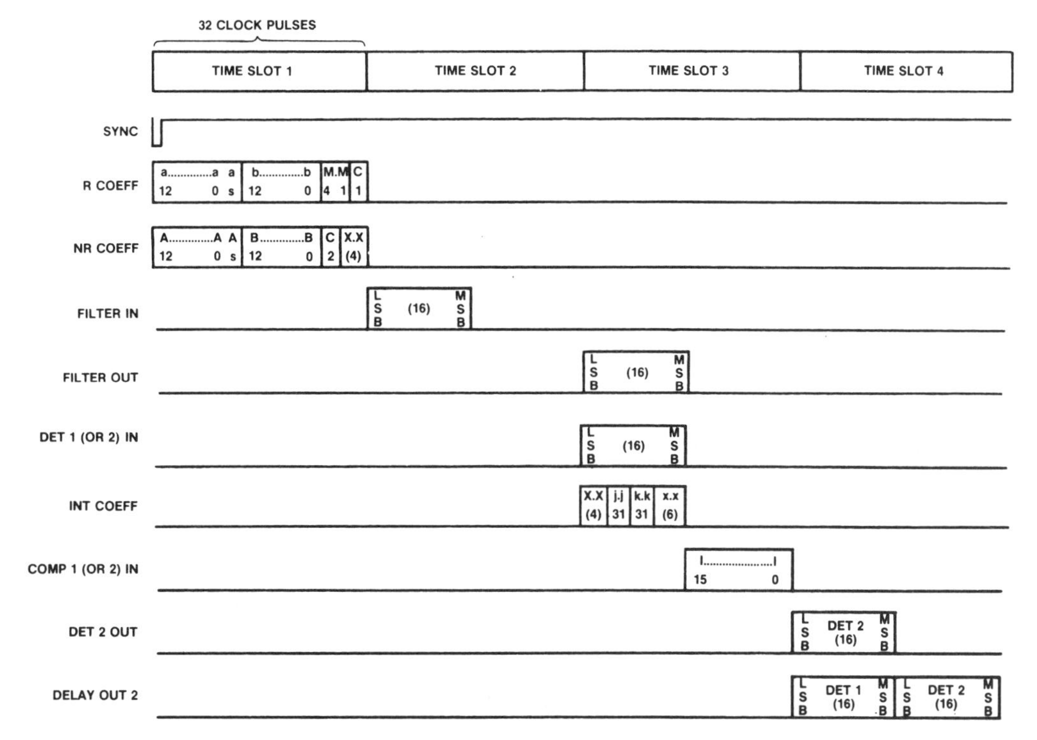

Timing diagram

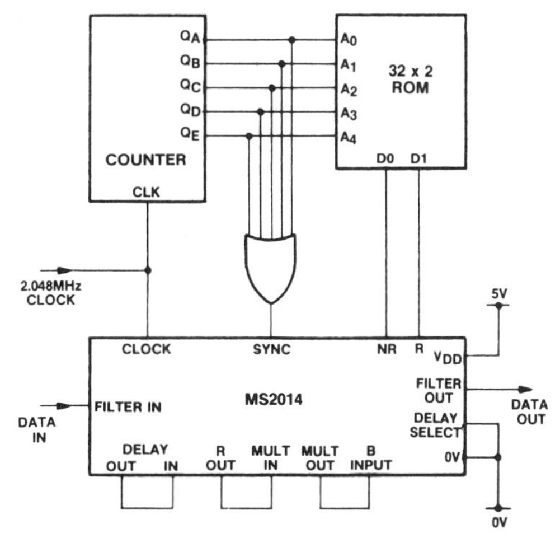

2nd order 32kHz bandwidth filter

A 16th order 4kHz bandwidth filter

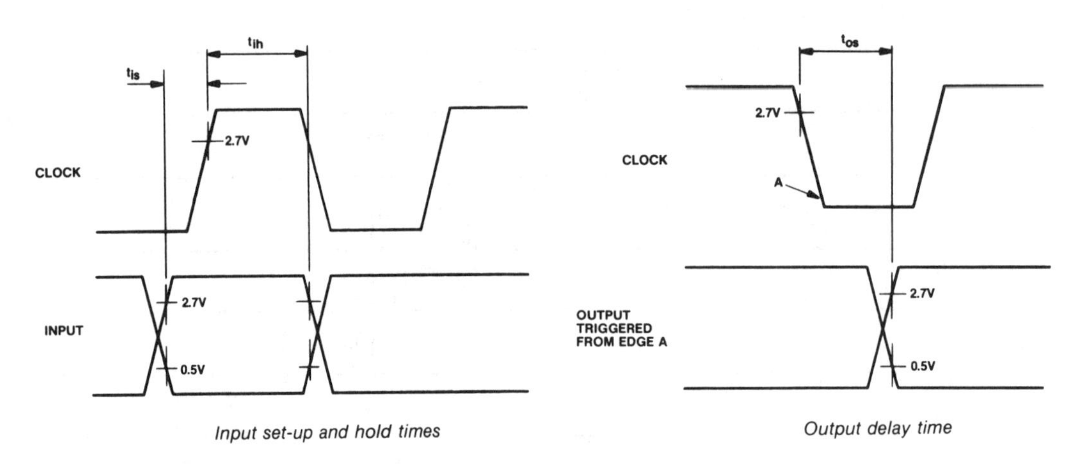

Input set-up and hold times

Output delay time

Input and output timing

Plessey MS2002 Digital Switch Module (DSM)

- Single 5V supply
- TTL compatible
- Interfaces directly with european-standard CCITT 32-channel format
- 256 Input/256 output channels
- Serial or parallel inputs and outputs
- Open-drain outputs for easy expansion
- One-system clock and one-frame sychronization pulse

APPLICATION

- Circuit-switched PCM or data systems

ABSOLUTE MAXIMUM RATINGS

Exceeding these ratings may cause permanent damage. Functional operation under these conditions is not implied.

Positive supply voltage, V_{DD}	-0.5V to +7V
Storage tempertaure, T_{ST}	-65°C to +150°C
Digital input voltage, V_{ID}	-0.3V to V_{DD} +0.3V
Clamp current (Sink or Source), I_C	50mA
Package power dissipation, P_P	800mW

PIN DESCRIPTIONS

Symbol	Pin No.	Pin name and description
DI0-DI7	1-8	**Data In 0 to 7 (Digital Inputs).** These are the inputs for the 256 incoming channels. The data presented at these pins is latched on the alternate negative edges of the CLK clock to those used by DO0-7. Frame synchronisation for these pins is established by the FS pulse.
C_{BB}	9	**Bias Decoupling Capacitor (Decoupling Node).** A bias decoupling capacitor of 1000pF should be connected between this pin and V_{SS}.
DO7-DO0	10-17	**Data Out 0 to 7 (Digital Pull-down Outputs).** These are the output pins for the 256 outgoing channels. Data is output at these pins on the alternate negative edges of the CLK clock to those used by DI0-7. Frame synchronisation for these pins is established by the FS pulse.
CDO	18	**Control Data Out (Digital Pull-down Output).** This pin outputs control data bytes. Bit synchronisation and frame synchronisation are stablished by the CLK and FS signal in a similar way as on the DO0-7 pins. It is high impedance for time slots which are not in use for control instructions. It is also high impedance for time slots corresponding to the 'write all ones' instruction. During time slots corresponding to other control instructions this pin outputs either the inverse of the 8 least significant bits at a control store location or the data at the speech store location selected by these 8 bits.
CWO	19	**Control Word Out (Digital Pull-down Output).** This pin outputs control word bytes. Bit synchronisation and frame synchronisation are established by the CLK and FS signals in a similar way as on the DO0-7 pins. It is high impedance for time slots which are not in use for control instructions and for time slots corresponding to the instruction 'write all ones'. During time slots corresponding to other instructions this pin outputs 4 bits which are the same as on CWI and 4 bits which indicate to the status of the chip.
V_{SS}	20	**Negative Supply Voltage (Power Input).** 0V.
CDI	21	**Control Data In (Digital Input).** The control data bytes are latched into the chip at this pin. Bit synchronisation and frame synchronisation are established by the CLK and FS signals in a similar way as on the DI0-7 pins. The bits in the input control data byte are inverted and written into the control store by the instruction 'write CWM bit and CI bits'.
CAI	22	**Column Address In (Digital Input).** This pin defines the column position of a chip in the control array.
RAI 1,0	23, 24	**Row Address In 1, 0 (Digital Inputs).** These pins defined the row position of a chip in the control array.
CWI	25	**Control Word In (Digital Input).** Control word bytes are latched into the chip at this pin. Bit synchronisation and frame synchronisation are established by the CLK and FS in a similar way as on the DI0-7 pins. The bits in the input control word byte control whether reads or writes occur, allow different chips in a control array to be addressed, and control whether connections are busy or free.
CLK	26	**Clock (Digital Input).** The system clock, nominally 4.096MHz, is input at this pin. It is used with the pulse on FS to establish bit synchronisation on the data and control inputs and outputs.
V_{DD}	27	**Positive Supply Voltage (Power Input).** 5V.
FS	28	**Frame Synchronisation (Digital Input).** The negative pulse input at this pin is used with the CLK clock to establish the frame synchronisation on the data and control inputs and outputs. The duration of the pulse determines the modes of the data input and output converters.

Duration (clock periods)	Data Inputs	Data Outputs
1	Serial	Serial
2	Serial	Parallel
3	Parallel	Serial
4	Parallel	Parallel

ELECTRICAL CHARACTERISTICS

Test Conditions - Voltages are with respect to ground (V_{SS}) unless otherwise stated

Characteristic	Symbol	Value			Units
		Min.	Typ.(1)	Max.	
Positive supply voltage	V_{DD}	4.75	5.0	5.25	V
Ambient temperature	T_{amb}	0		70	°C
Input low voltage	V_{IL}	0	0.4	0.8	V
Input high voltage	V_{IH}	2.0	2.4	V_{CC}	V
Output pullup resistor	R_{OP}	1000			Ω
Output load capacitor	C_{OP}	50			pF
Bias decoupling capacitor	C_{BB}	900	1000	1100	pF

Analog Characteristics - Voltages are with respect to ground (V_{SS}) unless otherwise stated

Characteristic	Symbol	Value			Units	Conditions
		Min.	Typ.(1)	Max.		
Pin capacitance	C_P		8	10	nF	Unloaded

Digital Static Characteristics - Voltages are with respect to ground (V_{SS}) unless otherwise stated

Characteristic	Symbol	Value			Units	Conditions
		Min.	Typ.(1)	Max.		
Supply current	I_{DD}		40	60	mA	Unloaded
Input leakage current	I_{LI}			50	μA	$0 < V < V_{CC}$
Output low voltage	V_{OL}	0		0.4	V	I_{OL} (Sink) = 2mA
Output low voltage	V_{OL}	0		2.0	V	I_{OL} (Sink) = 8mA
Output leakage current	I_{LO}			50	μA	$0 < V < V_{CC}$

Digital Switching Characteristics - Clock

Characteristic	Symbol	Value			Units	Conditions
		Min.	Typ.(1)	Max.		
Clock period	t_{CP}	225	244	275	ns	
Clock rise time	t_{CR}		50		ns	
Clock high period	t_{CH}	82			ns	
Clock fall time	t_{CF}		50		ns	
Clock low period	t_{CL}	82			ns	

Digital Switching Characteristics - Frame Synchronisation

Characteristic	Symbol	Value			Units	Conditions
		Min.	Typ.(1)	Max.		
Frame Synchronisation falling hold time	t_{FFH}	90	122		ns	
Frame synchronisation falling setup time	t_{FFS}	60	122		ns	
Frame synchronsiation rising hold time	t_{FRH}	90	122		ns	
Frame synchronisation rising setup time	t_{FFS}	60	122		ns	

Digital Switching Characteristics - Data and Control Inputs and Outputs

Characteristic	Symbol	Value			Units	Conditions
		Min.	Typ.(1)	Max.		
Input setup time	t_{IS}	60	244		ns	
Input hold time	t_{IH}	90	244		ns	
Output hold time	t_{OH}	5			ns	
Output delay	t_{OD}			150	ns	

NOTE
1. Typical figures are for design aid only. They are not guaranteed and not subject to production testing.

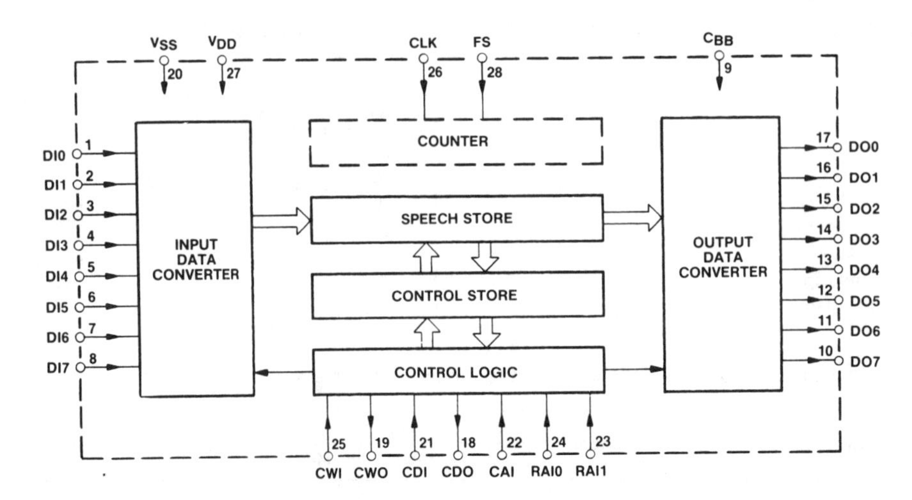

Functional block diagram

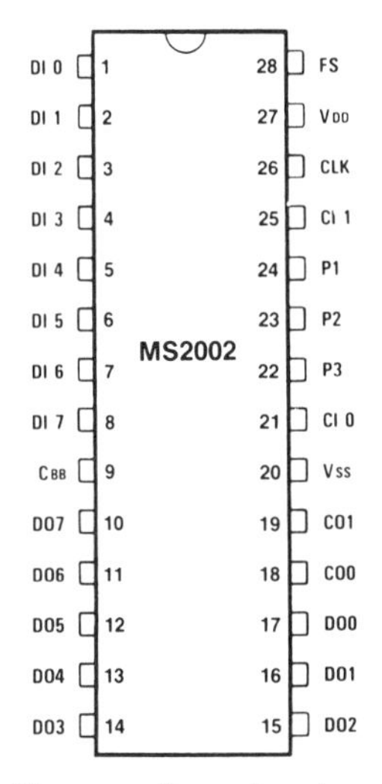

Pin connections - top view

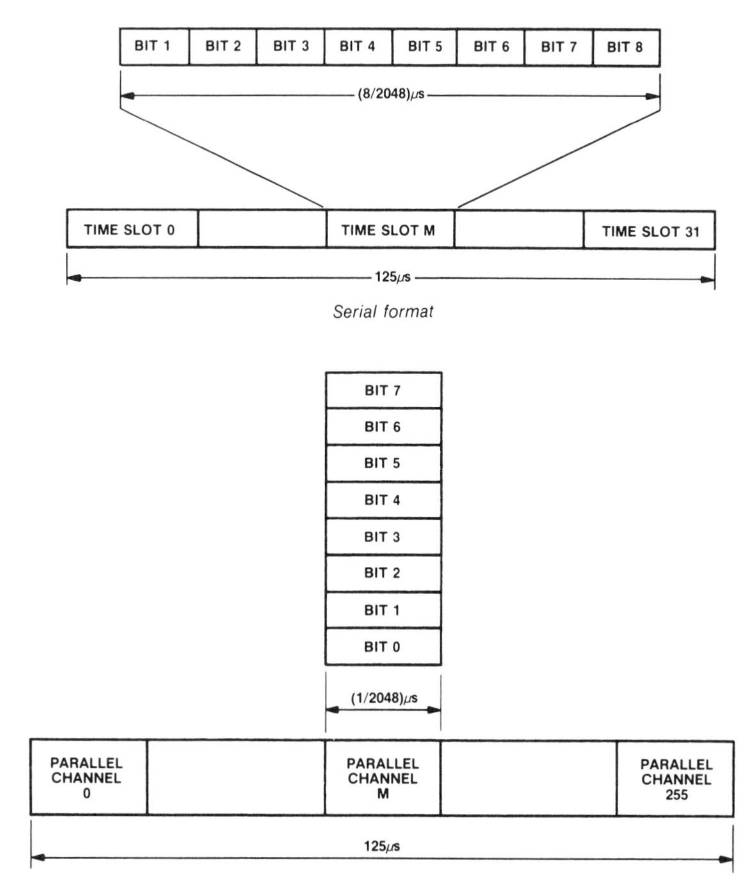

Serial format

Parallel format

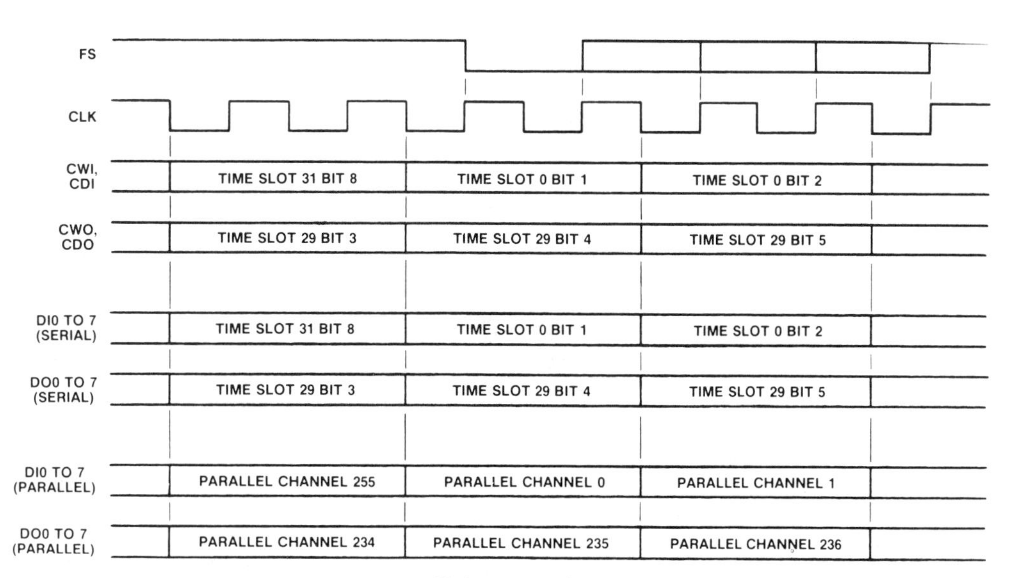

Timing - nominal

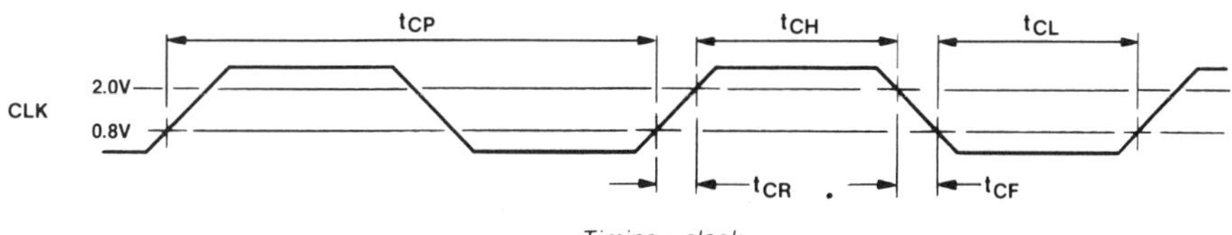

Timing - clock

Store Address (Speech or Control)	Serial Time Slot Address (Input or Output)	Serial Pin Address (Input or Outut)	Parallel Channel (Input or Output)
0	0	0	0
1	0	1	1
2	0	2	2
3	0	3	3
4	0	4	4
5	0	5	5
6	0	6	6
7	0	7	7
8	1	0	8
9	1	1	9
.	.	.	.
.	.	.	.
.	.	.	.
254	31	6	254
255	31	7	255

Relationship between inputs, outputs and stores

SPA2	SPA1	SPA0	STSA4	STSA3	STSA2	STSA1	STSA0	CM
1	2	3	4	5	6	7	8	9

Bit	Name	Description
1-3	SPA2-0	**Speech Pin Address 2 to 0.** These bits are the Serial Pin Address in the Speech Store. When used with the Speech Time Slot Address bits a unique Speech Store address is specified. This address corresponds to a Parallel Channel if parallel input is used.
4-8	STSA4-0	**Speech Time Slot Address 4 to 0.** These bits are the Serial Time Slot Address in the Speech Store. When used with the Speech Pin Address bits a unique Speech Store address is specified. This address corresponds to a Parallel Channel if parallel input is used.
9	CM	**Connection Mode.** This bit determines whether the connection is busy or free and also helps to control reads from the DSM. If this bit is 0 then the connection is busy. If it is 1 then the connection is free.

Bits at each control store address

Control Word In			CM Bit at Control Store Address	Instruction	Control Word Out			Control Data Out
R/W Bit	CWM Bit	Row Address			CAB Bit	CWM Bit	RAB1-0 Bits	
0	X	Matches	X	Write CWM bit + CDI bits	CAP Pin *†	Control Store Bit 9 (CM)†	RAP1-0 Pins *†	Control Store Bits 1-8†
0	X	Does Not Match	X	Write all 1s	High Impedance	High Impedance	High Impedance	High Impedance
1	0	X	X	Read Type 0	CAP Pin *†	Control Store Bit 9 (CM)	RAP1-0 Pins *	Control Store Bits 1-8
1	1	X	1	Read Type 1	High Impedance	Control Store Bit 9 (CM) = 1	High Impedance	Control Store Bits 1-8
1	1	X	0		CAP Pin†	Control Store Bit 9 (CM) = 0	RAP1-0 Pins	Speech Store Bits 1-8

Fig.11 The control operations

NB It is assumed that the Column Address matches, in which case $\overline{CPA2-0}$ and R/$\overline{W}$ are the same as on the control Time Slot on Control Word In. The control outputs are high impedance during the control Time Slot if the Column Address does not match.
* High Impedance if data at Control Store Address is all 1's.
† Should be identical to the data on the control inputs.

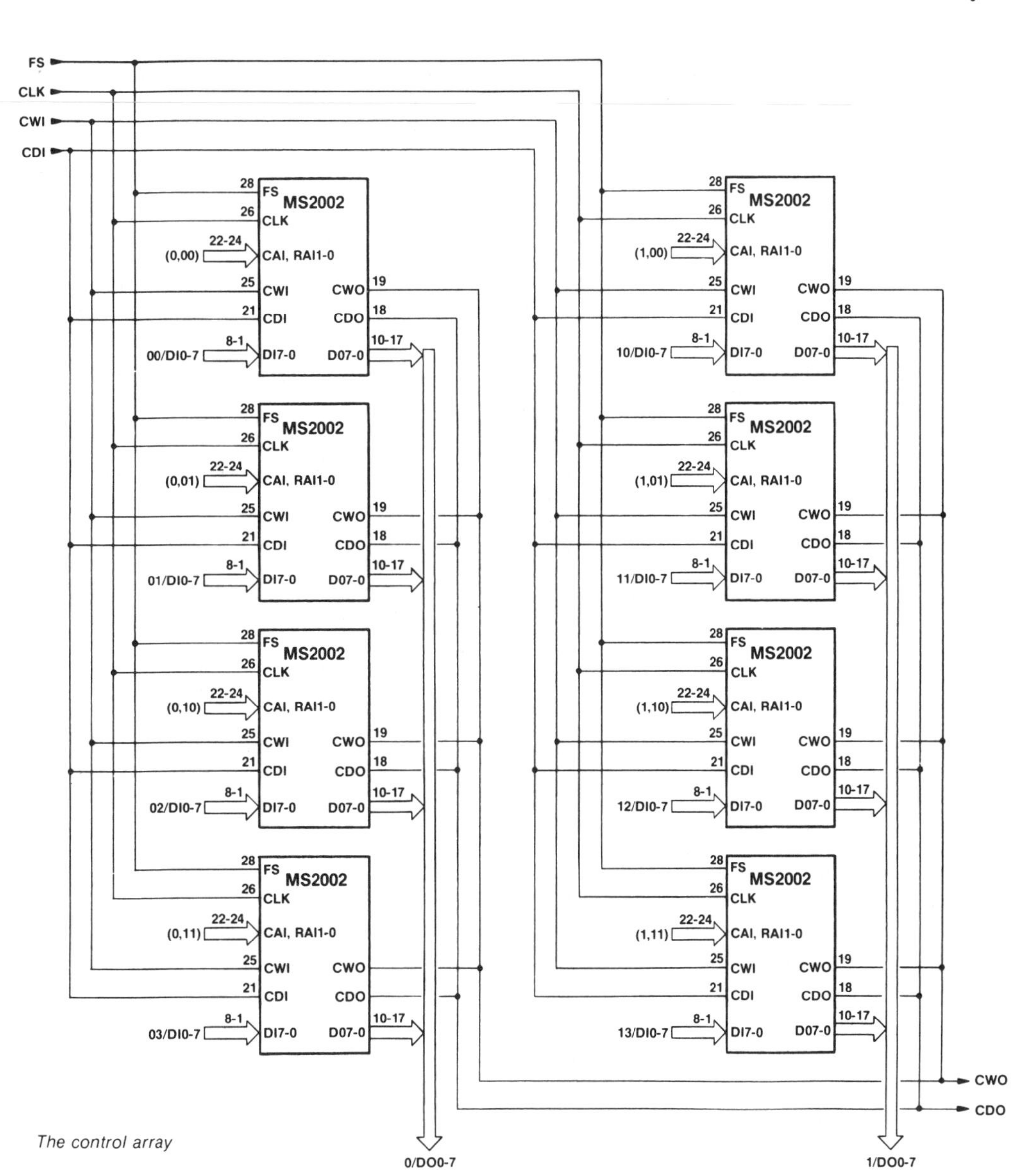

The control array

$\overline{SPA2}$	$\overline{SPA1}$	$\overline{SPA0}$	$\overline{STSA4}$	$\overline{STSA3}$	$\overline{STSA2}$	$\overline{STSA1}$	$\overline{STSA0}$
1	2	3	4	5	6	7	8

Bit	Name		Description
1-3	$\overline{SPA2-0}$		**Speech Pin Address 2 to 0**
		In	These bits replace the Speech Pin Address bits at the Control Store address during a write if the Row Addresses match (see Fig.10). **NB** These bits are inverted with respect to those at the Control Store address, i.e. if these are all 0 then they refer to Speech Pin Address 7 (see Fig.6).
		Out	These bits are high impedance unless a read or write occurs. During a read these bits can contain the Speech Pin Address bits at the Control Store address or bits 1 to 3 of the Speech Store location addressed by the bits at the Control Store address (see Fig.11). **NB** These bits are inverted with respect to the contents of the Control Store but not with respect to the contents of the Speech Store.
4-8	$\overline{STSA4-0}$		**Speech Time Slot Address 4 to 0**
		In	These bits replace the Speech Time Slot Address bits at the Control Store address during a write if the Row Addresses match (see Fig.11). **NB** These bits are inverted with respect to those at the Control Store address, i.e. if these are all 0 then they refer to Speech Time Slot Address 31 (see Fig.6).
		Out	These bits are high impedance unless a read or write occurs. During a read these bits can contain the Speech Time Slot Address bits at the Control Store address or bits 4 to 8 of the Speech Store location addressed by the bits at the Control Store address (see Fig.11). **NB** These bits are inverted with respect to the contents of the Control Store but not with respect to the contents of the Speech Store.

Control data bits (both input and output) **NB** *The Control Data Out bits are open-drain pulldown outputs. This means that output high is the same as output high impedance.*

CAB	$\overline{CPA2}$	$\overline{CPA1}$	$\overline{CPA0}$	R/$\overline{W}$	CWM	RAB1	RAB0
1	2	3	4	5	6	7	8

Bit	Name		Description
1	CAB		**Column Address Bit**
		In	If this bit matches the Column Address pin then the device will be written to or read from. If it does not match then Control Word Out and Control Data Out are high impedance during the output Time Slot.
		Out	This bit is set to the Column Address after a read from the Speech Store or after a write to the Control Store other than all 1's. It goes high impedance in all other cases.
2-4	$\overline{CPA2\text{-}0}$		**Control Pin Address 2 to 0**
		In	These bits are the Serial Pin Address at the Control Store (see Fig.6). The Time Slot Address is determined by the Time Slot on the Control Word In pin (see Fig.3). The Serial Pin and Time Slot Addresses define a unique address in the Control Store which corresponds to a Parallel channel if parallel output is used (Fig.6).
		Out	These bits are the same as those on the Time Slot on Control Word In if the Column Addresses match. They are high impedance otherwise.
5	R/$\overline{W}$		**Read or Write**
		In	This bit has no effect unless the Column Addresses match. If they do match then a read or write occurs depending whether it is 1 or 0 6 (see Fig.10).
		Out	This bit is the same as on the Time Slot on Control Word In if the Column Addresses match. It is high impedance otherwise.
6	CWM		**Control Word Mode**
		In	This bit has no effect unless the Column Addresses match. It can replace the Connection Mode bit at the Control Store address during writes or it can help to direct reads if the Column Addresses do match (see Fig.10).
		Out	This bit is the same as the Connection Mode bit at the Control Store address if the Column Addresses match. It is high impedance otherwise.
7-8	RAB1-0		**Row Address Bit 1 and 0**
		In	These bits have no effect unless the Column Addresses match. They help to control writes if the Column Addresses do match (see Fig.10). The Row Address Bits ensure that only one of the four MS2002s in a row of the Control Array can be active on the Data Out pins at any time.
		Out	These bits are set to the Row Address after certain operations and are high impedance otherwise. See Fig.11 for details.

Control word bits (both input and output) **NB** *The Control Word Out bits are open-drain pulldown outputs. This means that output high is the same as output high impedance.*

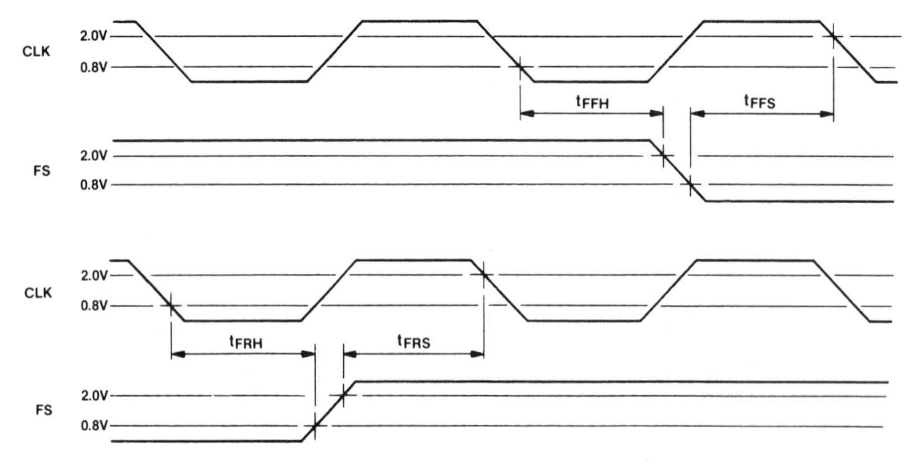

Timing - falling and rising edges of frame synchronisation

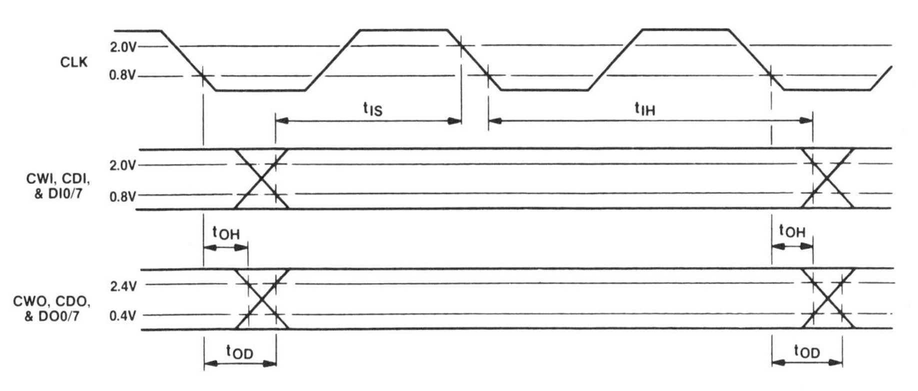

Timing - data and control inputs and outputs

Plessey MV1441 HDB3 Encoder/Decoder/Clock Regenerator

- On-chip digital clock regenerator
- HDB3 encoding and decoding to CCITT rec. G703
- Asynchronous operation
- Simultaneous encoding and decoding
- Clock recovery signal allows off-chip clock regeneration
- Loop-back control
- HDB3 error monitor
- "All ones" error monitor
- Loss of input alarm (all zeros detector)
- Decode data in NRZ form
- Low-power operation
- 2.048 MHz or 1.544 MHz operation

ABSOLUTE MAXIMUM RATINGS

The absolute maximum ratings are limiting values above which operating life might be shortened or specified parameters might be degraded.

Electrical Ratings

+Vcc	-0.5V to +7V
Inputs	Vcc +0.5V Gnd -0.3V
Outputs	Vcc, Gnd -0.3V

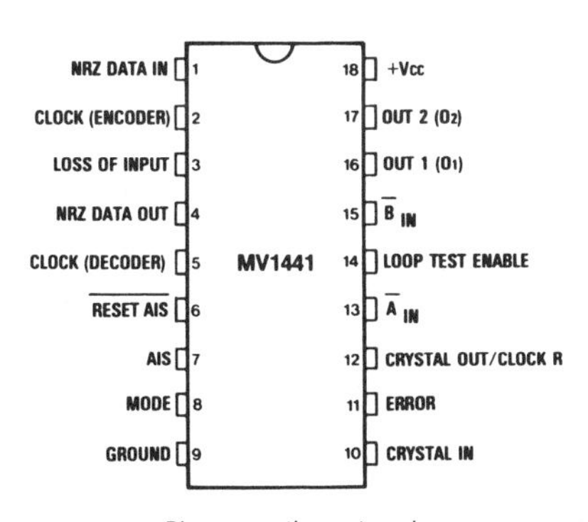

Pin connections - top view

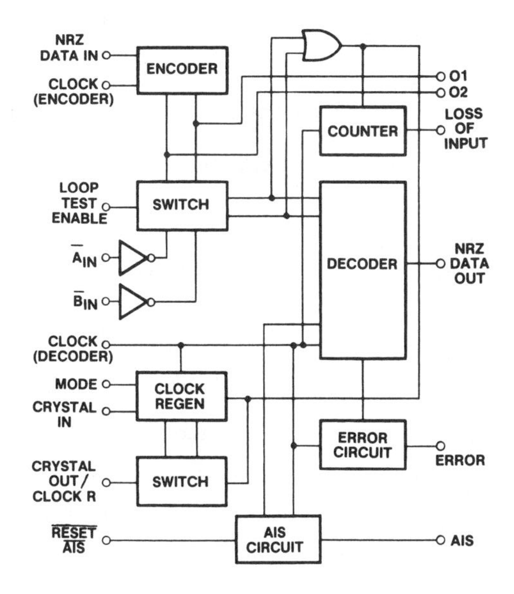

Block diagram

TYPICAL AIN BIN CLOCK R 3 CLOCK PERIODS CLOCK DECODER NRZ DATA OUT

Decode waveforms

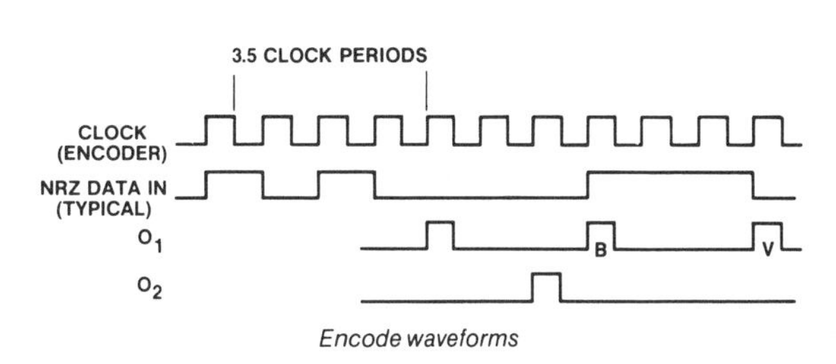

Encode waveforms

ELECTRICAL CHARACTERISTICS

Test conditions (unless otherwise stated):
Supply voltage V_{CC} = 5V ± 0.5V Ambient temperature T_{amb} = 0°C to +70°C

Static characteristics

Characteristic	Symbol	Pins	Value			Units	Conditions
			Min	Typ	Max		
Low level input voltage	V_{IL}	All inputs	-0.3		0.8	V	
Low level input current	I_{IL}				50	μA	V_{IL} = 0V
High level input voltage	V_{IH}		2.0		V_{CC}	V	
High level input current	I_{IH}				50	μA	V_{IH} = 5V
Low level output voltage	V_{OL}	All outputs			0.4	V	I_{sink} = 2.0mA
High level output voltage	V_{OH}		2.8			V	I_{source} = 2mA) both
			V_{CC}-0.75			V	I_{source} = 1mA) apply
Supply current	I_{CC}			2	4	mA	All inputs to 0V All outputs open circuit

Dynamic Characteristics

Characteristic	Symbol	Value			Units
		Min	Typ	Max	
Max. Clock (Encoder) frequency	f_{maxenc}	4.0	10		MHz
Max. Clock (Decoder) frequency	f_{maxdec}	2.2	5		MHz
Propagation Delay Clock (Encoder) to O_1, O_2	$t_{pd1A/B}$			100	ns
Rise and Fall times O_1, O_2				20	ns
t_{pd1A} - t_{pd1B} difference				20	ns
Propagation Delay Clock (Encoder) to Clock Regenerate	t_{pd3}			150	ns
Setup time of NRZ data in to Clock (Encoder)	t_{s3}	75			ns
Hold time of NRZ data in	t_{h3}	55			ns
Propagation delay $\overline{A}_{IN}$, $\overline{B}_{IN}$ to Clock Regenerate	t_{pd2}			150	ns
Propagation delay Clock (Decoder) to error	t_{pd4}			200	ns
Propagation delay $\overline{\text{Reset AIS}}$ falling edge to AIS output	t_{pd5}			200	ns
Propagation delay Clock (Decoder) to NRZ data out	t_{pd6}			150	ns
Setup time of $\overline{A}_{IN}$, $\overline{B}_{IN}$ to Clock (Decoder)	t_{sl}	75			ns
Hold time of $\overline{A}_{IN}$, $\overline{B}_{IN}$ to Clock (Decoder)	t_{h1}	5			ns
Hold time of $\overline{\text{Reset AIS}}$ = '0'	t_{h2}	30			ns
Setup time Clock (Decoder) to $\overline{\text{Reset AIS}}$	t_{s2}	100			ns
Setup time $\overline{\text{Reset AIS}}$ = 1 to Clock (Decoder)	t_{s2}	0			ns
Propagation Delay Clock (Decoder) to LIP				150	ns

NOTES
1. The Encoded ternary outputs (O_1, O_2) are delayed by 3.5 clock periods from $\overline{\text{NRZ Data}}$.
2. The decoded NRZ output is delayed by 3 clock periods from the HDB3 inputs ($\overline{A}_{IN}$, $\overline{B}_{IN}$).

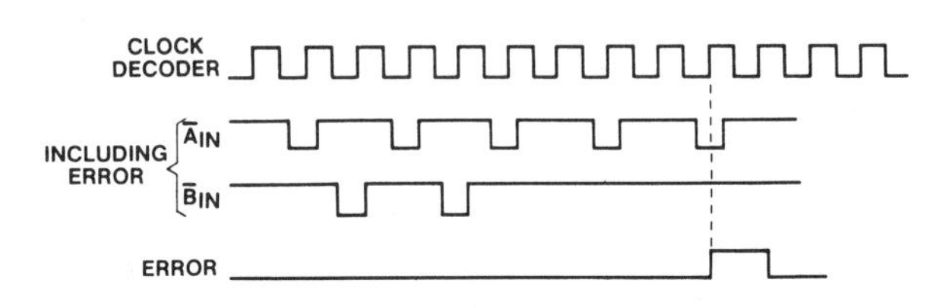

HDB3 error output waveforms

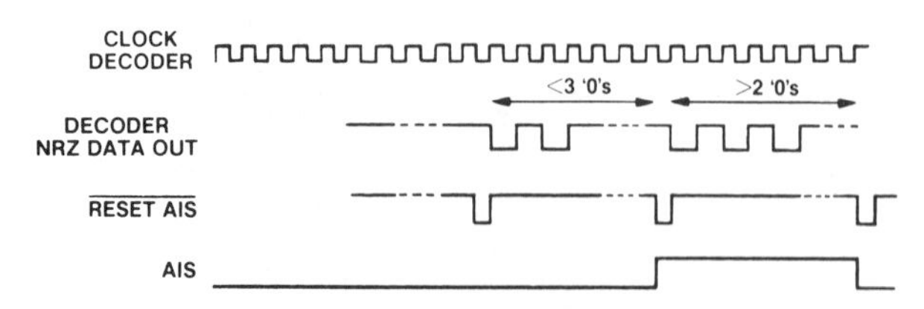

AIS error and Reset waveforms

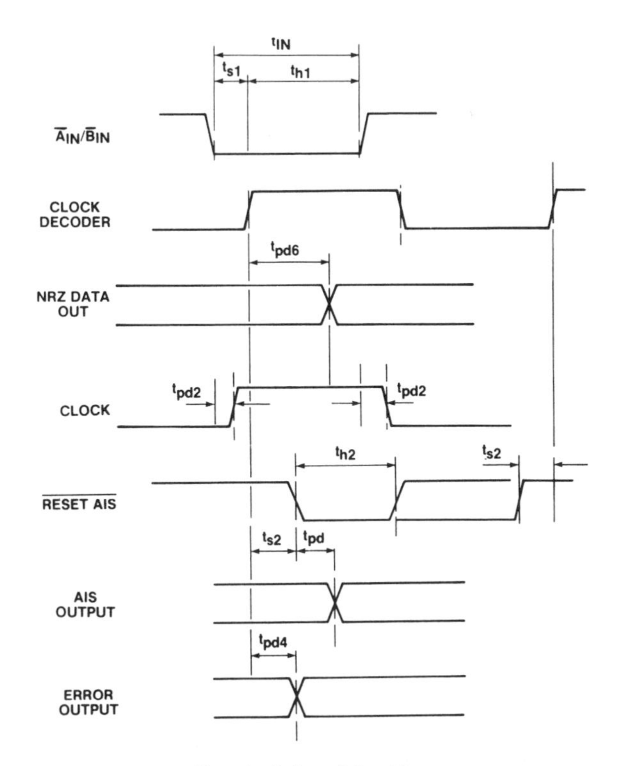

Decoder timing relationship

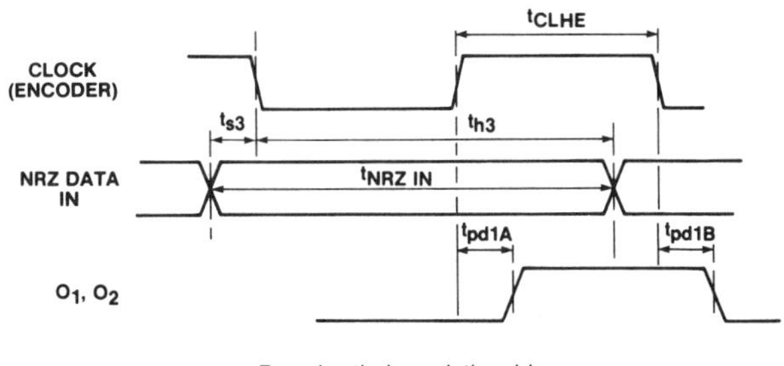

Encoder timing relationship

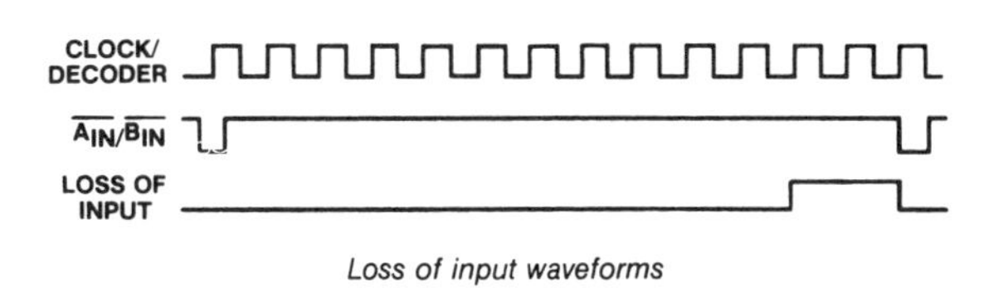

Loss of input waveforms

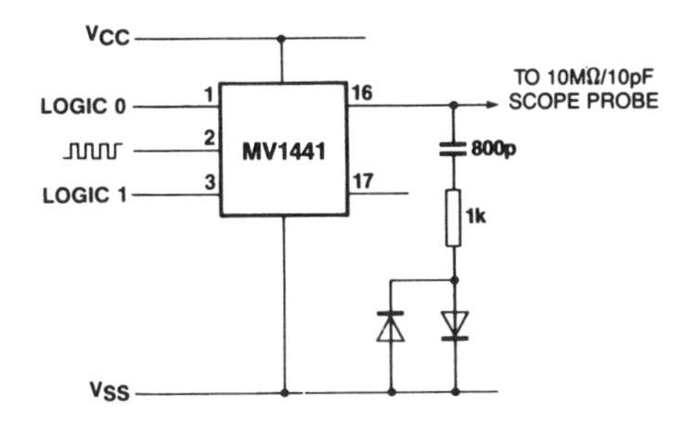

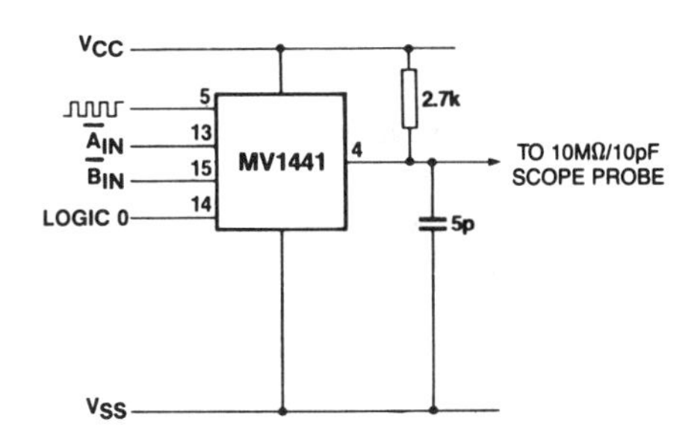

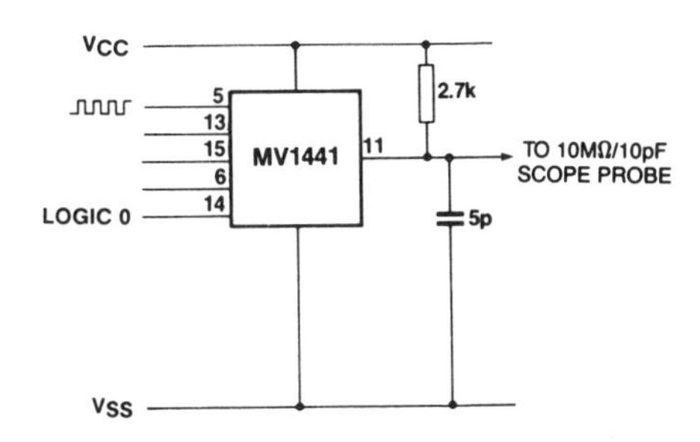

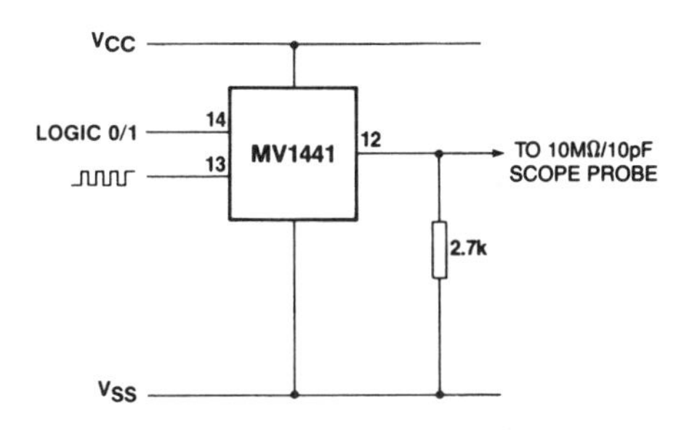

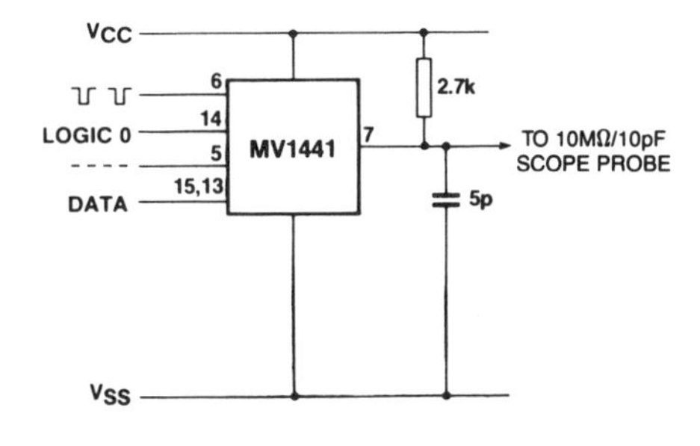

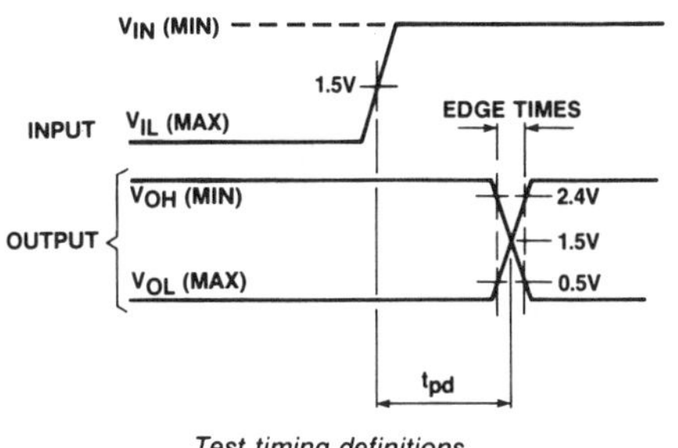

Test timing definitions

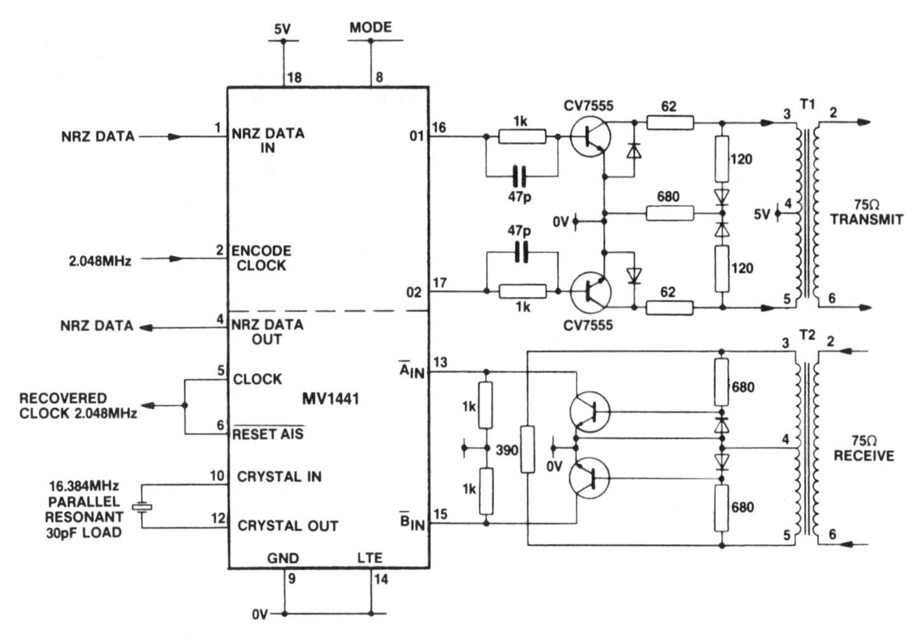

A typical application of the MV1441 with the interfacing to the transmission lines included

Plessey MV1448 HDB3 Encoder/Decoder

- HDB3 encoding and decoding to CCITT rec. G703
- Asynchronous operation
- Simultaneous encoding and decoding
- Clock recovery signal allows clock regeneration from incoming HDB3 data
- Loop-back control
- HDB3 error monitor
- "All ones" error monitor
- Loss of input alarm (all zeros detector)
- Decoded data in NRZ form
- Low-power operation
- 2.04 MHz or 8.544 MHz operation

ABSOLUTE MAXIMUM RATINGS

The absolute maximum ratings are limiting values above which operating life might be shortened or specified parameters might be degraded.

Electrical Ratings

+Vcc	-0.5V to +7V
Inputs	Vcc +0.5V to GND -0.3V
Outputs	Vcc to GND -0.3V

ELECTRICAL CHARACTERISTICS

Test conditions (unless otherwise stated):
Supply voltage V_{CC} = 5V ± 0.5V Ambient temperature T_{amb} = 0°C to +70°C

Static characteristics

Characteristic	Symbol	Pins	Value Min.	Value Typ.	Value Max.	Units	Conditions
Low level input voltage	V_{IL}	All inputs	-0.3		0.8	V	
Low level input current	I_{IL}				50	μA	V_{IL} = 0V
High level input voltage	V_{IH}		2.0		Vcc	V	
High level input current	I_{IH}				50	μA	V_{IH} = 5V
Low level output voltage	V_{OL}	All outputs			0.4	V	Isink = 2.0mA
High level output voltage	V_{OH}		2.8			V	Isource = 2mA } both apply
			Vcc-0.75			V	Isource = 1mA
Supply current	I_{CC}			2	4	mA	All inputs to 0V; All outputs open circuit

Dynamic characteristics

Characteristic	Symbol	Value			Units
		Min.	Typ.	Max.	
Max. clock (encoder) frequency	$F_{max\,enc}$	10			MHz
Max. clock (decoder) frequency	$F_{max\,dec}$	10			MHz
Propagation delay clock encoder to O_1, O_2	tpd1A/B		50		ns
Rise and fall times O_1, O_2				20	ns
tpd1A - tpd1B difference				20	ns
Propagation delay clock to clock regenerate (clock R)	tpd3		50		ns
Setup time of NRZ data in to clock (encoder)	ts3		40		ns
Hold time of NRZ data in	th3		40		ns
Propagation delay A_{IN}, B_{IN} to clock regenerate	tpd2				
	low-high			76	ns
	high-low			50	
Propagation delay clock (decoder) to error	tpd4			50	ns
Propagation delay $\overline{\text{Reset AIS}}$ falling edge to AIS output	tpd5		50	44	ns
Propagation delay clock (decoder) to NRZ data out	tpd6			50	ns
Setup time of $\bar{A}_{IN}$, $\bar{B}_{IN}$ to clock (decoder)	ts1	15	40		ns
Hold time of $\bar{A}_{IN}$, $\bar{B}_{IN}$ to clock (decoder)	th1	4	5		ns
Hold time of $\overline{\text{Reset AIS}}$ = '0'	th2	9			ns
Setup time clock (decoder) to $\overline{\text{Reset AIS}}$	ts2		50		ns
Setup time $\overline{\text{Reset AIS}}$ = '1' to clock (decoder)	ts2	31			ns
Propagation delay clock (decoder) to LIP			50	87	ns

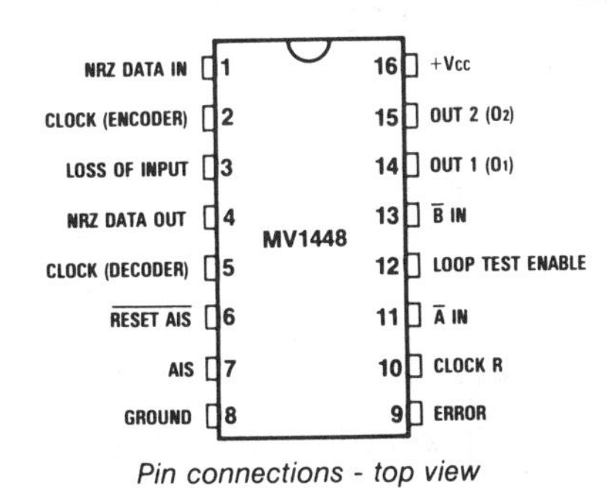

Pin connections - top view

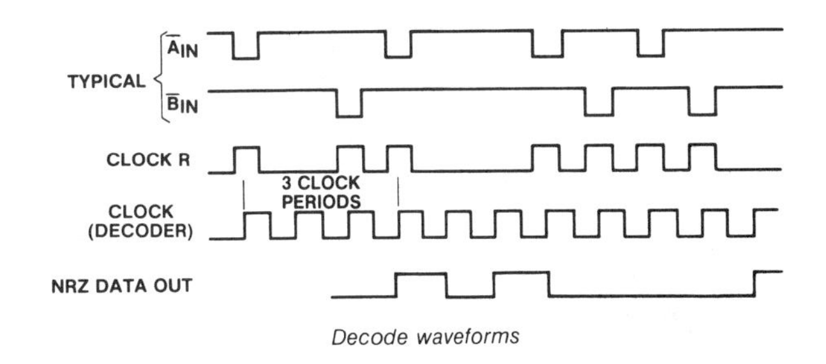

Decode waveforms

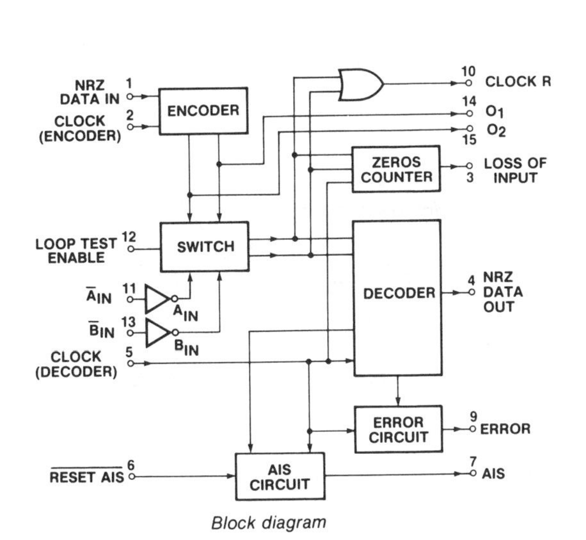

Block diagram

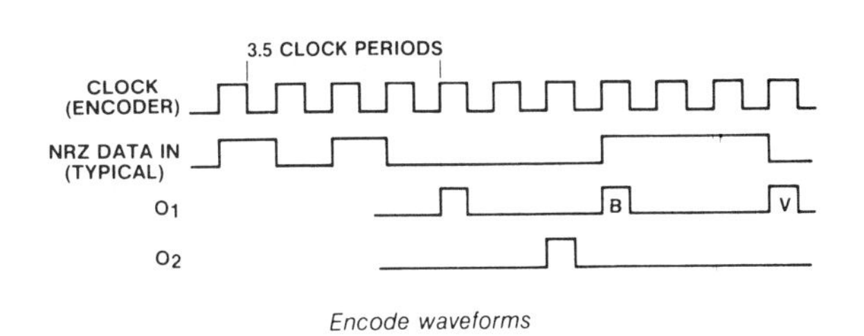

Encode waveforms

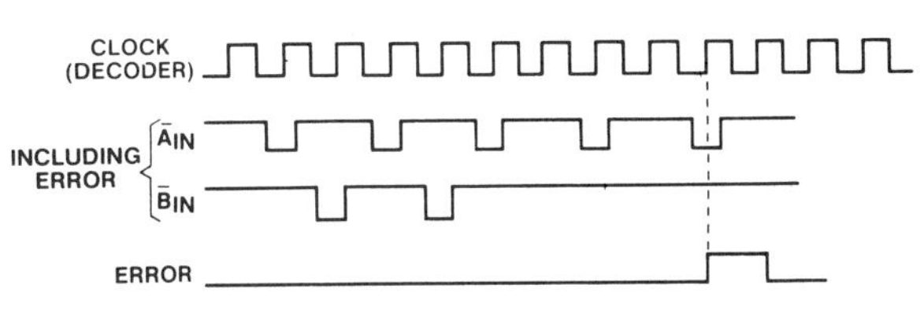

HDB3 error output waveforms

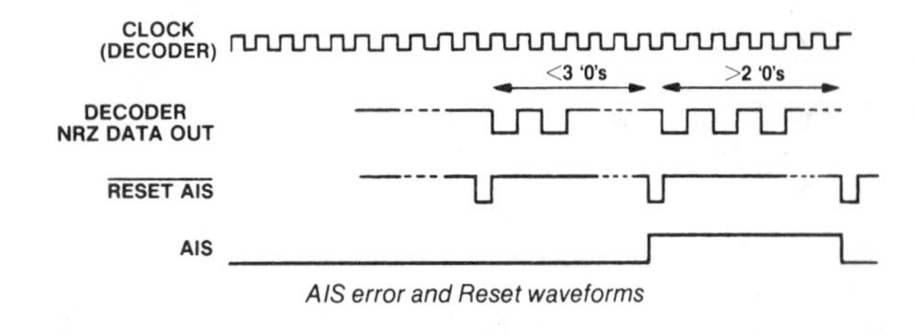

AIS error and Reset waveforms

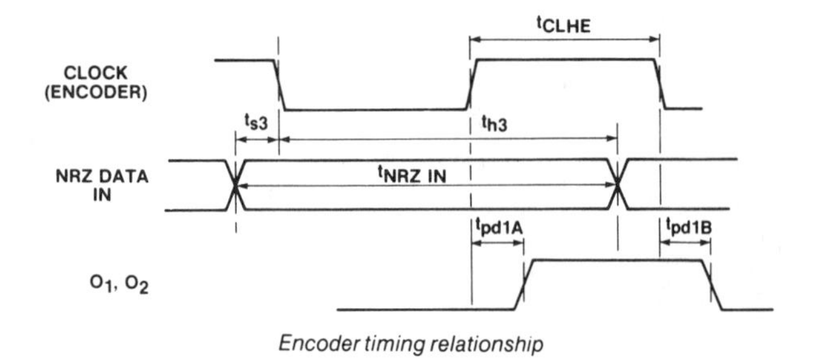

Encoder timing relationship

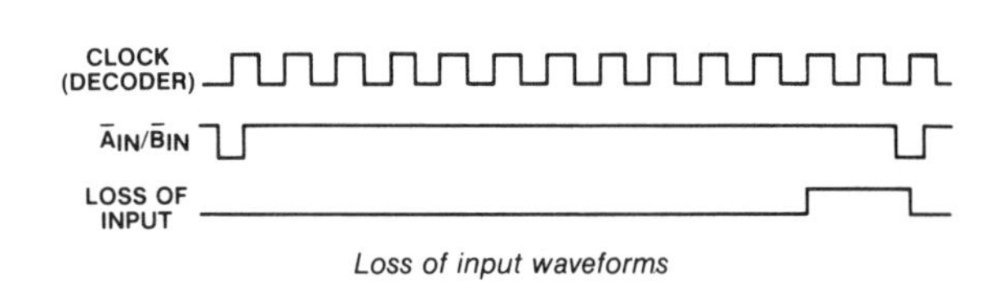

Loss of input waveforms

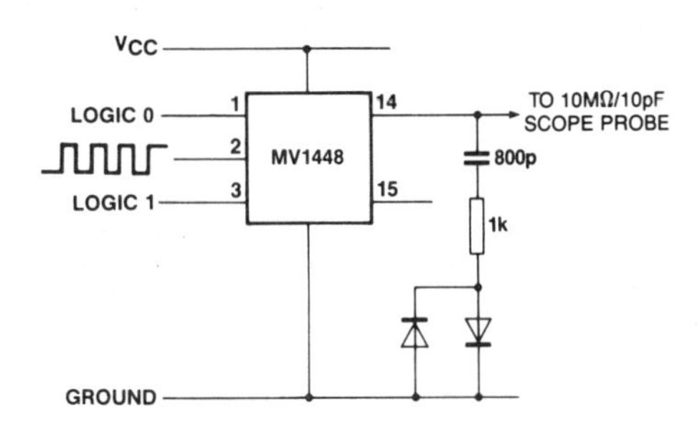

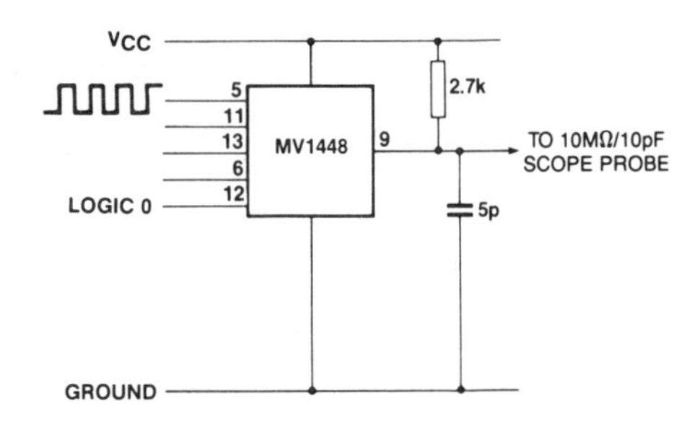

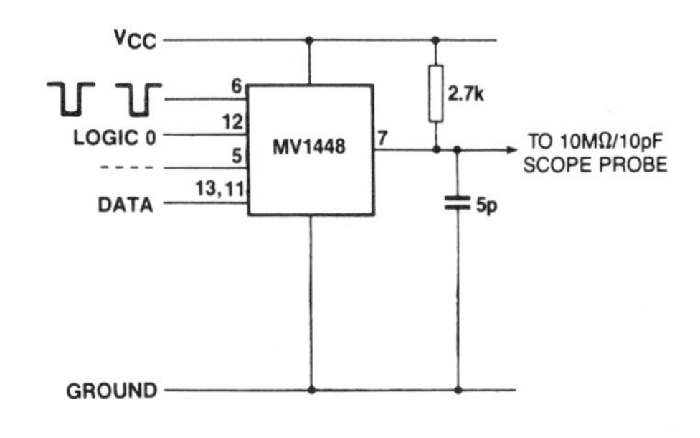

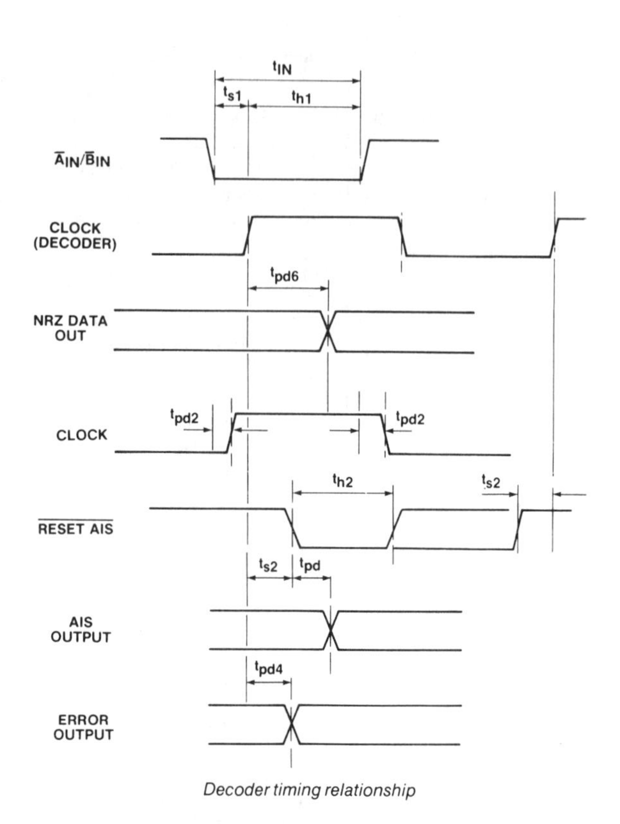

Decoder timing relationship

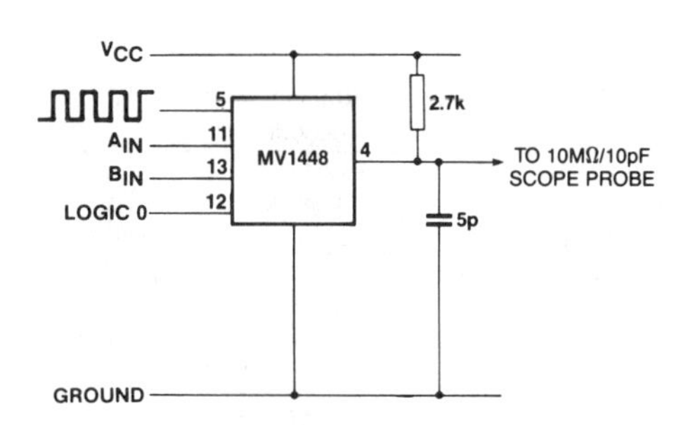

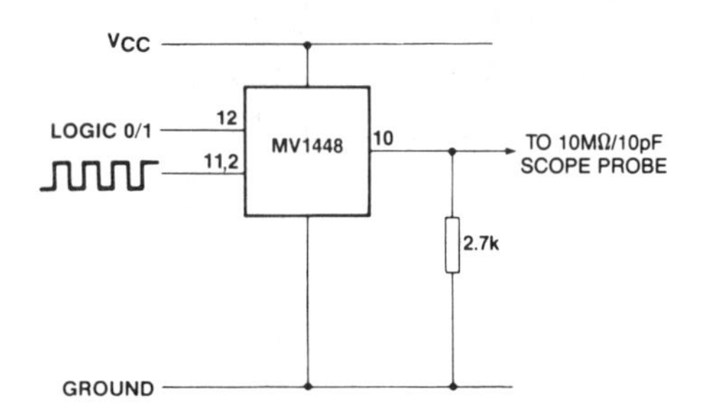

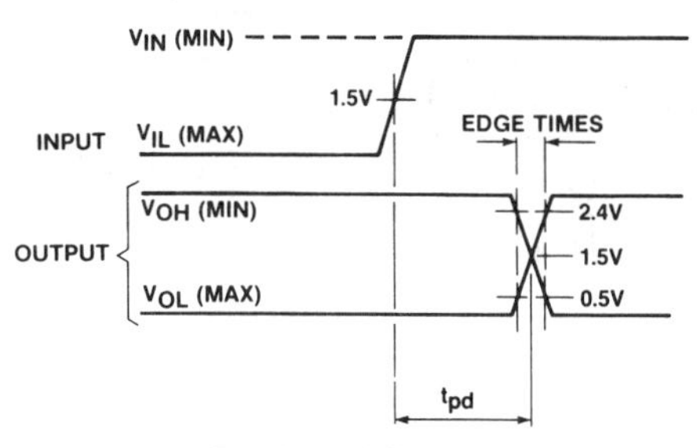

Test timing definitions

Plessey MV5087
DTMF Generator

- Pin-for-pin replacement for MK5087
- Low standby power
- Minimum external parts count
- 3.5V to 10V operation
- 2-of-8 keyboard or calculator-type single-contact (Form A) keyboard input
- On-chip regulation of output tone
- Mute and transmitter drivers on-chip
- High-accuracy tones provided by 3.58 MHz crystal oscillator
- Pin-selectable inhibit of single-tone generation

APPLICATIONS

- Telephone sets
- Mobile radio
- Remote control
- Point-of-sale and banking terminals
- Process control

ABSOLUTE MAXIMUM RATINGS

	MIN.	MAX.		MIN.	MAX.
V_{DD} - V_{SS}	-0.3V	10.5V	Power dissipation		850 mW
Voltage on any pin	V_{SS} - 0.3V	V_{DD} + 0.3V	Derate 16 mW/°C above 75°C		
Current on any pin		10 mA	(All leads soldered to PCB)		
Operating temperature	-40°C	+85°C			
Storage temperature	-65°C	+150°C			

DC ELECTRICAL CHARACTERISTICS

Test conditions (unless otherwise stated):
T_{amb} = +25°C, V_{DD} = 3.5V to 10V

	CHARACTERISTICS		SYMBOL	MIN	TYP	MAX	UNITS		
SUPPLY	Operating Supply Voltage		V_{DD}	3.5		10	V	Ref. to V_{SS}	
	Standby Supply Current		I_{DDS}		0.2	100	uA	V_{DD} = 3.5V	No Key Depressed
					0.5	200	uA	V_{DD} = 10V	All outputs Unloaded
	Operating Supply Current		I_{DD}		1.0	2.0	mA	V_{DD} = 3.5V	One Key Depressed
					5.0	10.0	mA	V_{DD} = 10V	All outputs Unloaded
INPUTS	$\overline{\text{SINGLE TONE INHIBIT}}$	INPUT HIGH VOLTAGE	V_{IH}	0.7V_{DD}		V_{DD}	V		
		INPUT LOW VOLTAGE	V_{IL}	0		0.3V_{DD}	V		
		INPUT RESISTANCE	R_{IN}		60		KΩ		
	ROW 1-4	INPUT HIGH VOLTAGE	V_{IH}	0.9V_{DD}			V		
		INPUT LOW VOLTAGE	V_{IL}			0.3V_{DD}	V		
	COLUMN 1-4	INPUT HIGH VOLTAGE	V_{IH}	0.7V_{DD}			V		
		INPUT LOW VOLTAGE	V_{IL}			0.1V_{DD}	V		
OUTPUTS	XMITR	SOURCE CURRENT	I_{OH}	-15	- 25		mA	V_{DD} = 3.5V, V_{OH} = 2.5V	No Keyboard Entry
				-50	- 100		mA	V_{DD} = 10V, V_{OH} = 8V	
		LEAKAGE CURRENT	I_{OZ}		0.1	10	uA	V_{DD} = 10V, V_{OH} = OV	Keyboard Entry
	MUTE	SINK CURRENT	I_{OL}	0.5			mA	V_{DD} = 3.5V, V_{OL} = 0.5V	No Keyboard Entry
				1.0			mA	V_{DD} = 10V, V_{OL} = 0.5V	
		SOURCE CURRENT	I_{OH}	-0.5			mA	V_{DD} = 3.5V, V_{OH} = 3.0V	Keyboard Entry
				-1.0			mA	V_{DD} = 10V, V_{OH} = 9.5V	

AC ELECTRICAL CHARACTERISTICS

Test conditions (unless otherwise stated):
T_{amb} = +25°C, V_{DD} = 3.5V to 10V

CHARACTERISTICS		SYMBOL	MIN	TYP	MAX	UNITS	
TONE OUT	ROW TONE OUTPUT VOLTAGE	V_{OR}	320	400	500	mV_{RMS}	Single Tone R_L = 1K Ω
	COLUMN TONE OUTPUT VOLTAGE	V_{OC}	400	500	630	mV_{RMS}	
	EXTERNAL LOAD IMPEDANCE	R_L	700			Ω	V_{DD} = 3.5V
			330			Ω	V_{DD} = 10V
OUTPUT DISTORTION					- 20	dB	Total out-of-band power relative to sum of row and column fundamental power
PRE EMPHASIS, High Band			1		3	dB	
Tone Output Rise Time		t_r		3	5	ms	

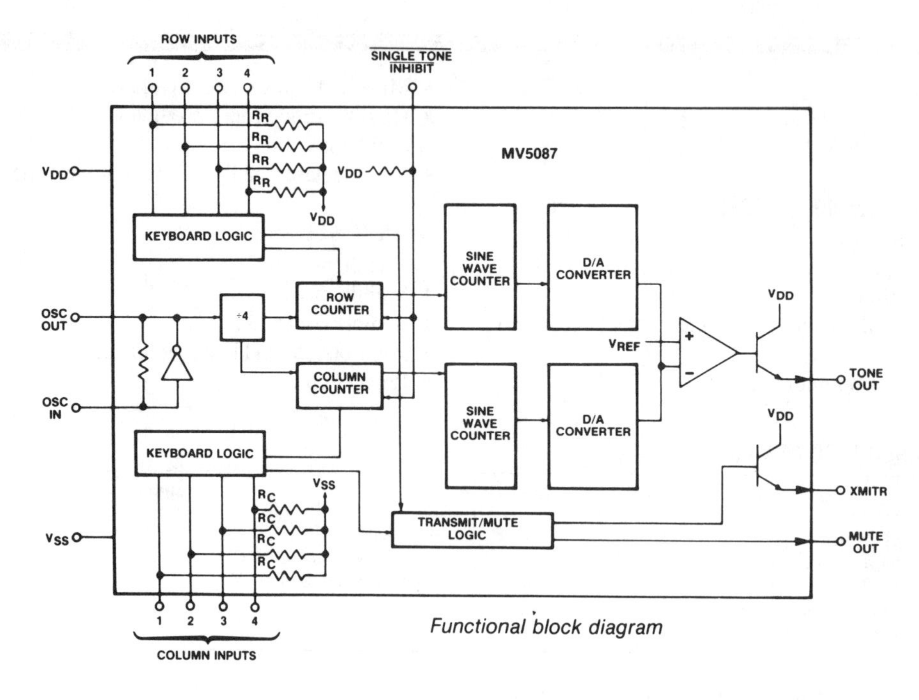

Functional block diagram

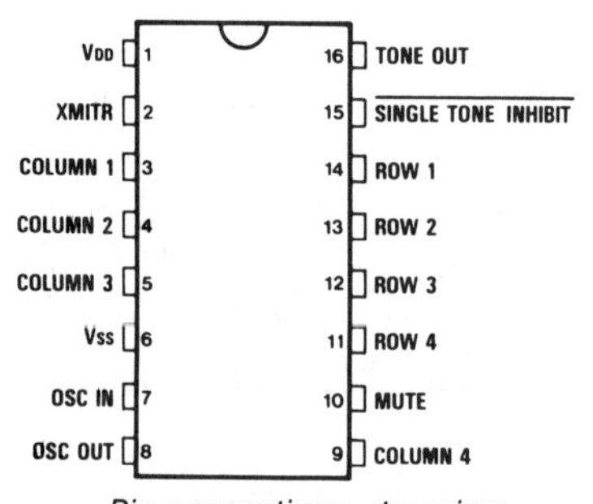

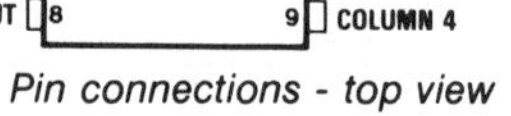

Pin connections - top view

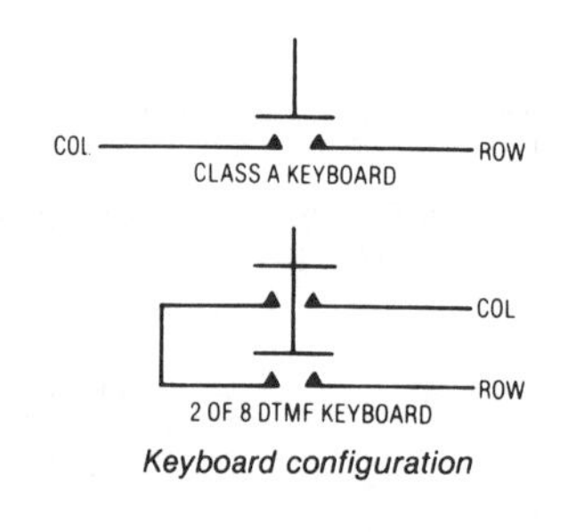

Keyboard configuration

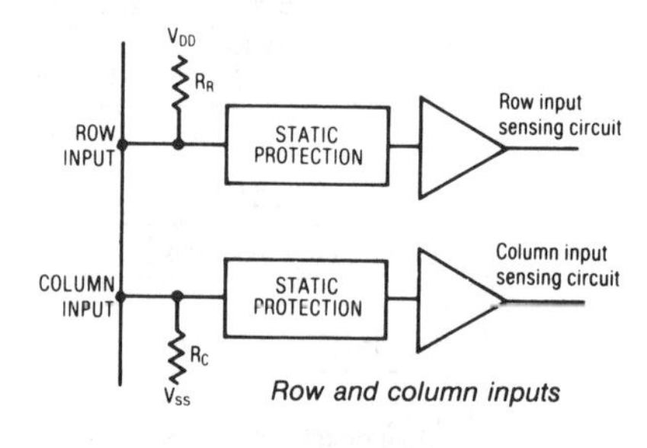

Row and column inputs

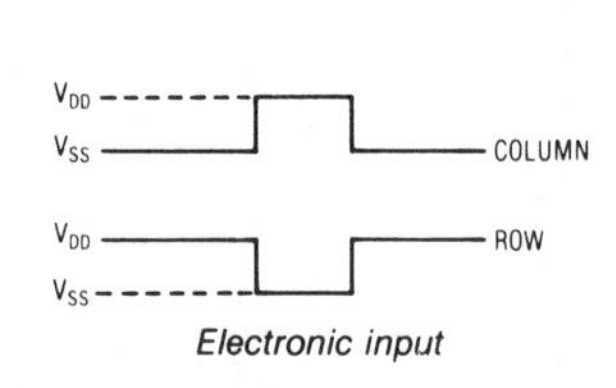

Electronic input

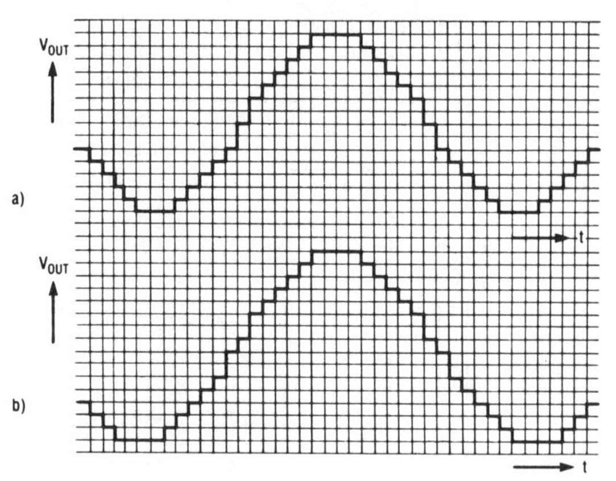

Typical sinewave output (a) Row tones (b) Column tones

		Standard DTMF (Hz)	Tone Output Frequency Using 3.579545 MHz Crystal	% Deviation From Standard	
Row	f_1	697	701.3	+0.62	Low Group
	f_2	770	771.4	+0.19	
	f_3	852	857.2	+0.61	
	f_4	941	935.1	−0.63	
Column	f_5	1209	1215.9	+0.57	High Group
	f_6	1336	1331.7	−0.32	
	f_7	1477	1471.9	−0.35	
	f_8	1633	1645.0	+0.73	

Output frequency deviation

V_{DD} SINGLE TONE INHIBIT COL 1 COL 2 COL 3 COL 4 ROW 1 ROW 2 ROW 3 ROW 4 MV5087 3.58 MHz XTAL XMITR TONE OUTPUT R_L V_{SS} MUTE

Connection diagram

Plessey MV8870-1
DTMF Receiver
Preliminary Information

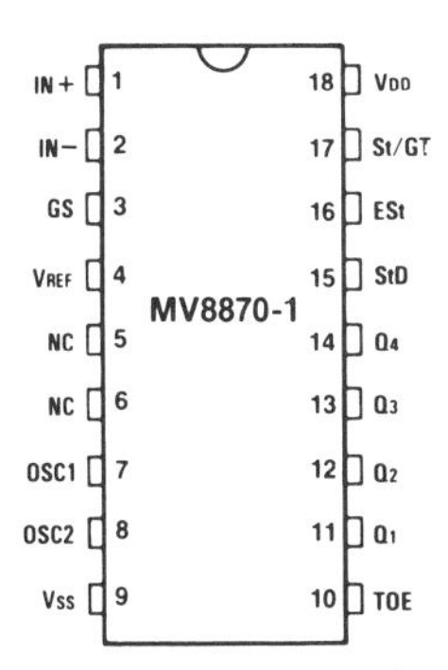

Pin connections - top view

- Complete DTMF receiver
- Low-power consumption
- Internal gain-setting amplifier
- Adjustable guard time
- Central office quality

APPLICATIONS

- Receiver systems for BT or CEPT specifications
- Paging systems
- Repeater systems/mobile radio
- Credit card systems
- Remote control

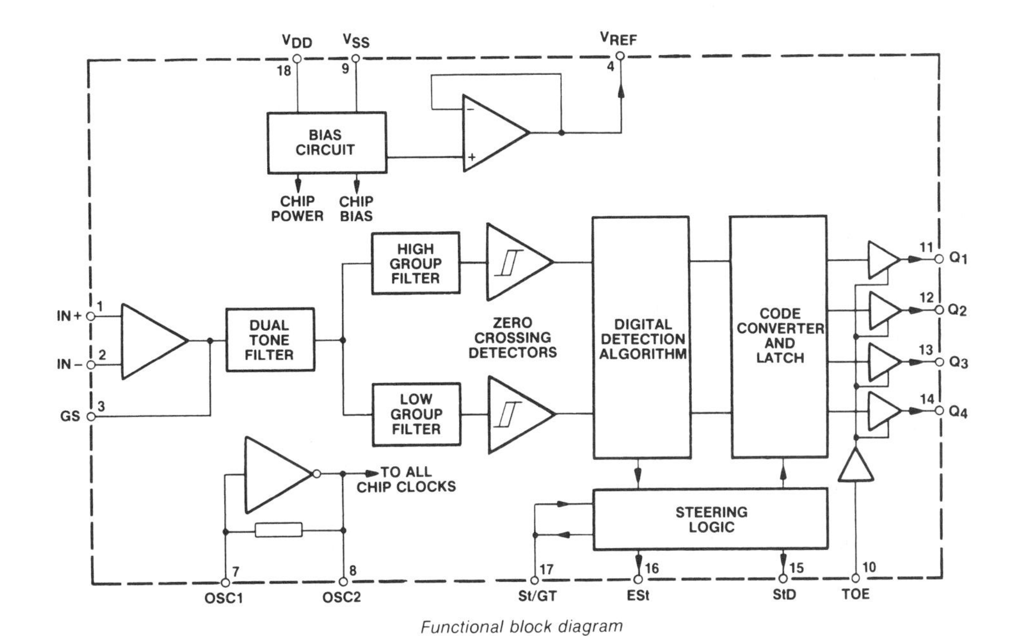

Functional block diagram

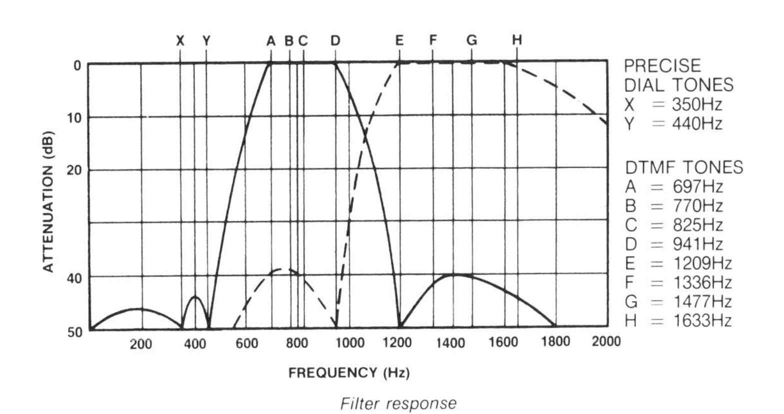

Filter response

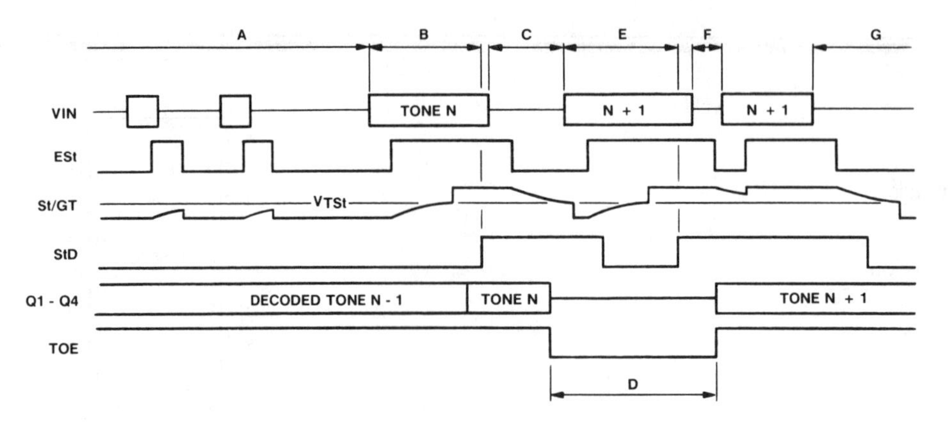

Explanation of Events

A. Tone bursts detected, but tone duration invalid and output latch unchanged.
B. Tone N detected, tone duration valid, output latch updated and new data signalled by StD.
C. End of tone N detected, tone absent duration valid, but output latch not updated until next valid tone.
D. Outputs switched to high impedance.
E. Tone N + 1 detected, tone duration valid, tone decoded, output latch updated (although outputs are currently high impedance) and new data signalled by StD.
F. Acceptable dropout of tone N + 1, tone absent duration invalid, StD and output latch unchanged.
G. End of tone N + 1 detected, tone absent duration valid, StD goes low but output latch not updated until next valid tone.

Timing diagram

F_{LOW}	F_{HIGH}	KEY	TOE	Q_4	Q_3	Q_2	Q_1
697	1209	1	H	0	0	0	1
697	1336	2	H	0	0	1	0
697	1477	3	H	0	0	1	1
770	1209	4	H	0	1	0	0
770	1336	5	H	0	1	0	1
770	1477	6	H	0	1	1	0
852	1209	7	H	0	1	1	1
852	1336	8	H	1	0	0	0
852	1477	9	H	1	0	0	1
941	1209	0	H	1	0	1	0
941	1336	*	H	1	0	1	1
941	1477	#	H	1	1	0	0
697	1633	A	H	1	1	0	1
770	1633	B	H	1	1	1	0
852	1633	C	H	1	1	1	1
941	1633	D	H	0	0	0	0
-	-	Any	L	Z	Z	Z	Z

Functional decode

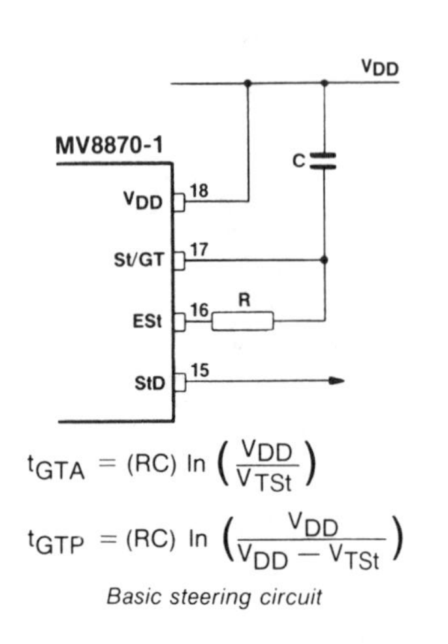

$$t_{GTA} = (RC)\ \ln\left(\frac{V_{DD}}{V_{TSt}}\right)$$

$$t_{GTP} = (RC)\ \ln\left(\frac{V_{DD}}{V_{DD} - V_{TSt}}\right)$$

Basic steering circuit

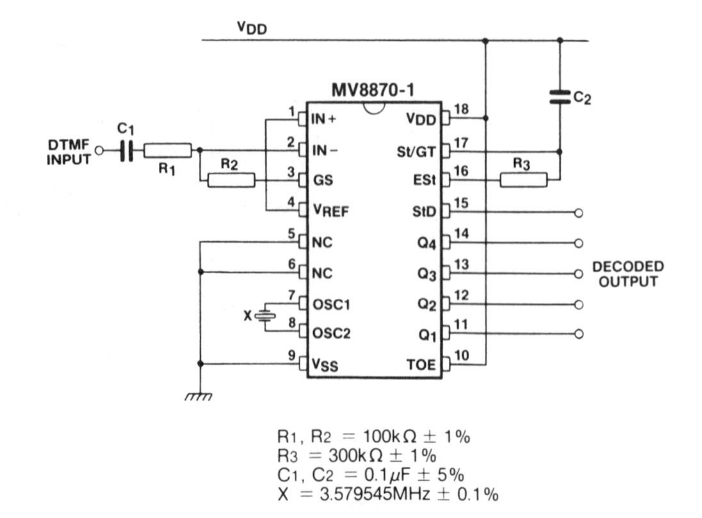

R_1, R_2 = 100kΩ ± 1%
R_3 = 300kΩ ± 1%
C_1, C_2 = 0.1μF ± 5%
X = 3.579545MHz ± 0.1%

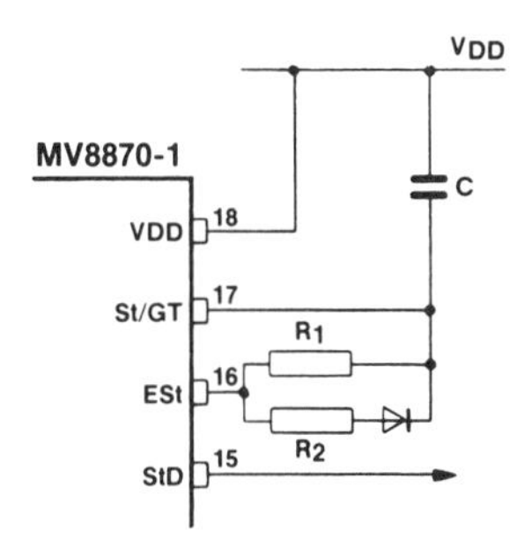

$$t_{GTA} = (R_p C)\ln \frac{V_{DD}}{V_{TSt}}$$

$$t_{GTP} = (R_1 C)\ln \frac{V_{DD}}{V_{DD} - V_{TSt}}$$

$$R_p = \frac{R_1 R_2}{R_1 + R_2}$$

Guard time adjustment ($t_{GTP} < t_{GTA}$)

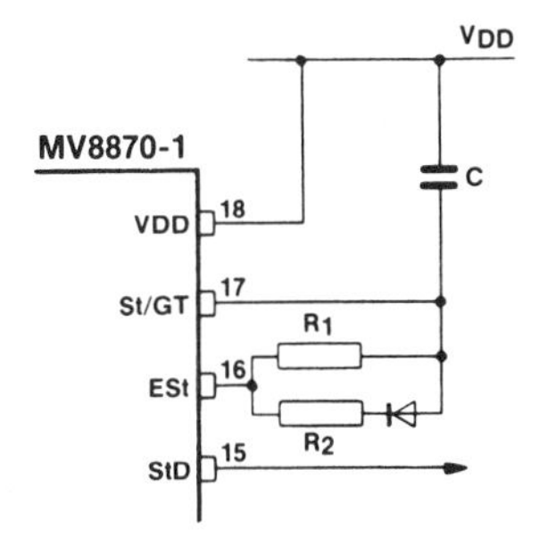

$$t_{GTA} = (R_1 C)\ln \frac{V_{DD}}{V_{TSt}}$$

$$t_{GTP} = (R_p C)\ln \frac{V_{DD}}{V_{DD} - V_{TSt}}$$

$$R_p = \frac{R_2 R_1}{R_2 + R_1}$$

Guard time adjustment ($t_{GTP} > t_{GTA}$)

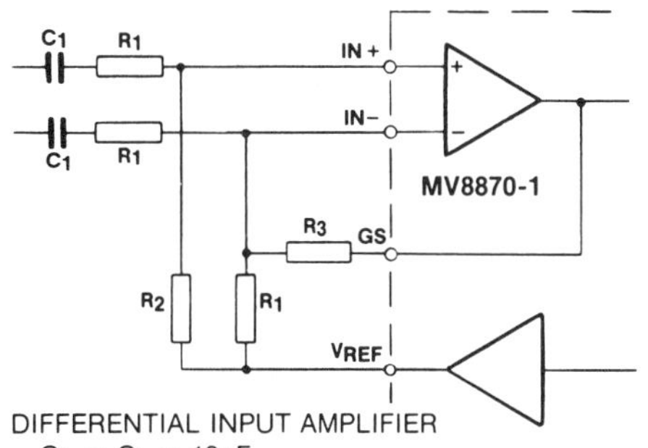

DIFFERENTIAL INPUT AMPLIFIER

$C_1 = C_1 = 10nF$ — All resistors ±1%

$R_1 = R_4 = R_1 = 100k\Omega$ — All capacitors ±5%

$R_2 = 60k\Omega$ $R_3 = 37.5k\Omega$

$$R_3 = \left(\frac{R_2 R_5}{R_2 + R_5}\right)$$

$$\text{VOLTAGE GAIN } (A_{V\,diff}) = \frac{R_5}{R_1}$$

INPUT IMPEDANCE

$$(Z_{IN\,diff}) = 2\sqrt{R_1^2 + \left(\frac{1}{\omega C}\right)^2}$$

Differential input configuration

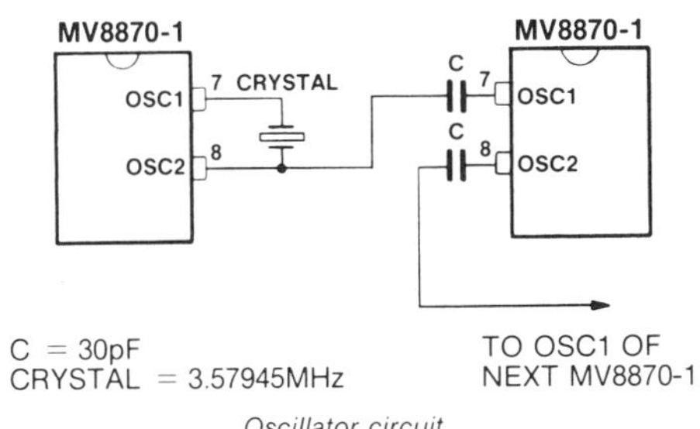

Oscillator circuit

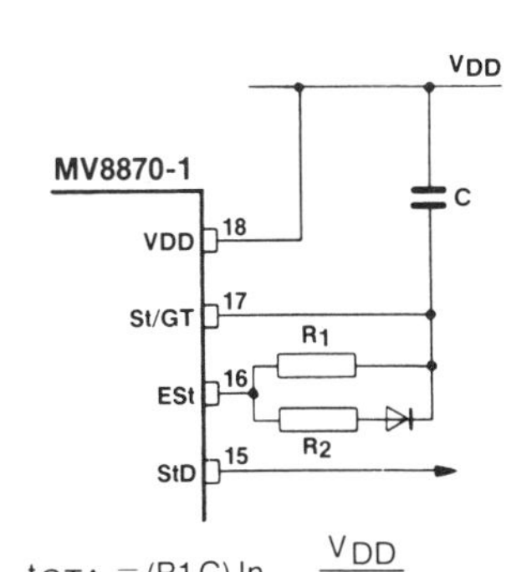

$$t_{GTA} = (R_1 C)\ln \frac{V_{DD}}{V_{TSt}}$$

$$t_{GTP} = (R_p C)\ln \frac{V_{DD}}{V_{DD} - V_{TSt}}$$

$$R_p = \frac{R_1 R_2}{R_1 + R_2}$$

$R_1 = 368k\Omega \pm 1\%$
$R_2 = 2.2M\Omega \pm \%$
$C = 0.1\mu F \pm 5\%$

Non-symmetric guard time circuit

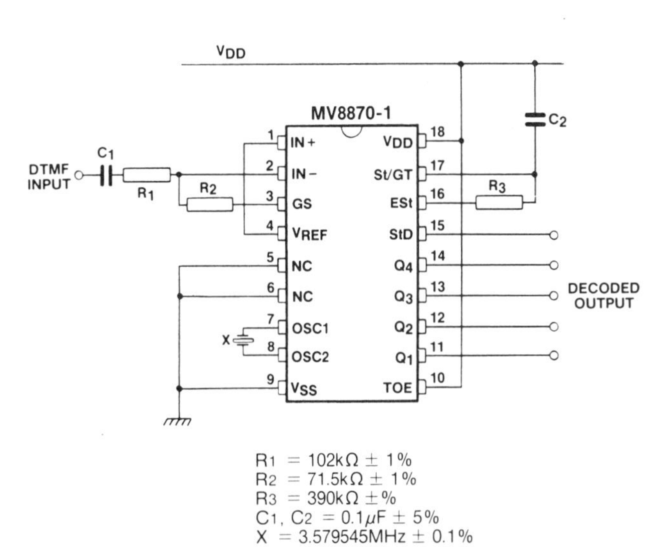

Single ended input configuration for BT or CEPT spec

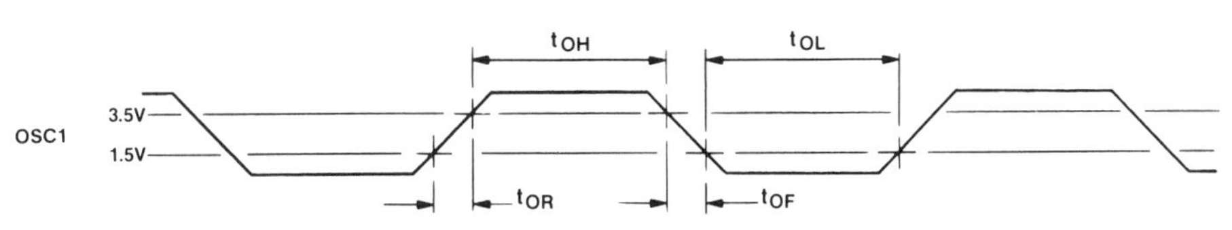

Timing - external oscillator input

ELECTRICAL CHARACTERISTICS

Test Conditions - Voltages are with respect to ground (V_{SS})

Parameter	Symbol	Value			Units
		Min.	Typ. (1)	Max.	
Positive supply voltage (V_{DD} pin)	V_{DD}	4.75	5	5.25	V
Ambient temperature	T_{amb}	-40		+85	°C
Op. amp. output capacitive load (GS pin)	C_{OUT}			100	pF
Op. amp. output resistive load (GS pin)	R_{OUT}	50			kΩ
Input high voltage (OSC1 and TOE pins)	V_{IH}	3.5		V_{DD}	V
Input low voltage (OSC1 and TOE pins)	V_{IL}	0		1.5	V
Oscillator frequency (OSC1 and OSC2 pins)	f_O	3.5759	3.579545	3.5831	MHz
Oscillator input rise time (OSC1 pin)	t_{OR}			110	ns
Oscillator input high time (OSC1 pin)	t_{OH}	110		170	ns
Oscillator input fall time (OSC1 pin)	t_{OF}			110	ns
Oscillator input low time (OSC1 pin)	t_{OL}	110		170	ns
Oscillator output load (OSC2 pin)	C_{LO}			30	pF

NOTE
1. Typical figures are for design aid only. They are not guaranteed and not subject to production testing.

Static Characteristics - Voltages are with respect to ground (V_{SS})

Characteristic	Symbol	Value			Units	Conditions
		Min.	Typ. (1)	Max.		
Power dissipation	P_D		15	37	mW	
Supply current (V_{DD} pin)	I_{DD}		3.0	7.0	mA	
Reference voltage (V_{REF} pin)	V_{REF}	2.4		2.8	V	
Reference output resistance (V_{REF} pin)	R_{REF}		10		kΩ	
Input leakage current (OSC1, IN+ and IN- pins)	I_I		100		nA	$0 \leqslant V_{PIN} \leqslant V_{DD}$
Internal pull-up current (TOE pin)	I_{PU}		7.5	15	A	$0 \leqslant V_{PIN} \leqslant V_{DD}$
Output low sink current (OSC2, Q_1 - Q_4, StD, ESt and St/GT pins)	I_{OL}	1	2.5		mA	$0.4V \leqslant V_{PIN} \leqslant V_{DD}$
Output high source current (OSC2, Q_1 - Q_4, StD, ESt and St/GT pins)	I_{OH}	0.4	0.8		mA	$0V \leqslant V_{PIN} \leqslant 4.6V$
Steering threshold voltage (St/GT pin)	V_{TSt}	2.2		2.5	V	
Pin capacitance	C_P		7	15	pF	Pin to supplies

NOTE
1. Typical figures are for design aid only. They are not guaranteed and not subject to production testing.

Dynamic Characteristics: Input Op. Amp. - Voltages are with respect to ground (V_{SS})

Characteristic	Symbol	Value			Units	Conditions
		Min.	Typ. (1)	Max.		
Input impedance (IN+ and IN- pins)	R_{IN}		10		MΩ	1kHz
Input offset voltage (IN+ and IN- pins)	V_{OS}		25		mV	
Power supply rejection	PSRR		60		dB	1kHz
Common mode range	V_{CM}		3.0		V p-p	No load
Common mode rejection	CMRR		60		dB	
DC open loop voltage gain	A_{VOL}		65		dB	
Open loop unit gain bandwidth	f_C		1.5		MHz	
Output voltage swing (GS pin)	V_O		4.5		V p-p	R_{OUT} to V_{SS} $\geqslant 100k\Omega$

NOTE
1. Typical figures are for design aid only. They are not guaranteed and not subject to production testing.

Dynamic Characteristics: Decoder - Voltages are with respect to ground V_{SS})

Characteristic	Symbol	Value			Units	Conditions
		Min.	Typ. (1)	Max.		
Propagation delay (St/GT to Q)	t_{PQ}		8	11	μs	TOE pin high
Propagation delay (St/GT to StD)	t_{PStD}		12		μs	
Output data set-up time (Q to StD)	t_{QStD}		3.4		μs	TOE pin high
Enable propagation delay (TOE to Q)	T_{PTE}		50		ns	R_L = 10kΩ (pulldown) C_L = 50pF
Disable propagation delay (TOE to Q)	t_{PTD}		300		ns	R_L = 10kΩ (pulldown) C_L = 50pF

NOTE
1. Typical figures are for design aid only. They are not guaranteed and not subject to production testing.

Dynamic Characteristics: Detector - Voltages are with respect to ground (V_{SS})

Characteristic	Symbol	Value			Units	Conditions
		Min.	Typ. (13)	Max.		
Valid input level (GS pin)	V_{VL}	61.7		2458	mV p-p	1,2,3,5,6,9
	P_{VL}	-31		1	dBm	
Invalid input level (GS pin)	V_{IL}			30.8	mV p-p	1,2,3,5,6,9
	P_{IL}			-37	dBm	
Acceptable positive twist	T_{AP}	6	10		dB	2,3,6,9
Acceptable negative twist	T_{AN}	6	10		dB	2,3,6,9
Acceptable frequency deviation	Δ_{FA}	-(1.5% +2Hz)		(1.5 % +2Hz)		2,3,5,9
Frequency deviation - rejected as too low	Δ_{FRL}		-5%	-3.5%		2,3,5,9
Frequency deviation - rejected as too high	Δ_{FRH}	3.5 %	5 %			2,3,5,9
Third tone tolerance	P_{TTT}	-18.5			dB	2,3,4,5,9,12
Noise tolerance	P_{NT}		-12		dB	2,3,4,5,7,9,10
Dial tone tolerance	P_{DTT}		22		dB	2,3,4,5,8,9,11
Tone present detect time	t_{DP}	5	11	14	ms	
Tone absent detect time	t_{DA}	0.5	4	8.5	ms	

NOTES
1. dBm = decibels above or below a reference power of 1mW into a 600 Ohm load.
2. Digit sequence consists of all DTMF tones.
3. Tone duration = 40ms. tone pause = 40ms.
4. Signal condition consists of nominal DTMF frequencies.
5. Both tone in composite signal have equal amplitudes.
6. Tone pair is deviated by ± (1.5% + 2Hz).
7. Bandwidth limited (3kHz) Gaussian Noise.
8. The precise dial tone frequencies are (350Hz and 440Hz) ±2%.
9. For an error rate of better than 1 in 10,000.
10. Referenced to lowest frequency component in DTMF signal.
11. Referenced to the minimum valid input level.
12. Referenced to Input DTMF Tone Level at -25dBm (-28dBm at GS pin). Interference Frequency Range is 480 to 3400Hz.
13. Typical figures are for design aid only. They are not guaranteed and not subject to production testing.

ABSOLUTE MAXIMUM RATINGS

Voltages are with respect to the negative power supply (V_{SS})

Exceeding these ratings may cause permanent damage. Functional operation under these conditions is not implied.

Positive supply voltage (pin 18), V_{DD}	6V
Voltage on any pin (other than supplies), V_{MAX}	−0.3V to V_{DD} +0.3V
Current at any pin (other than supplies), I_{MAX}	10mA
Storage temperature, T_{STG}	−65 °C to +150 °C
Package power dissipation, P_{DISS}	1000mW

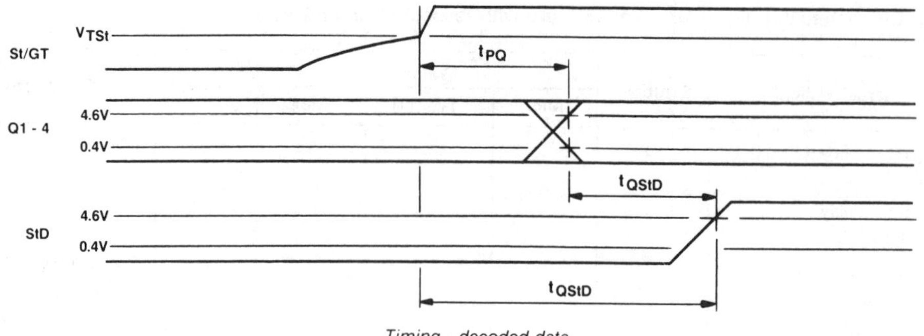

Timing - decoded data

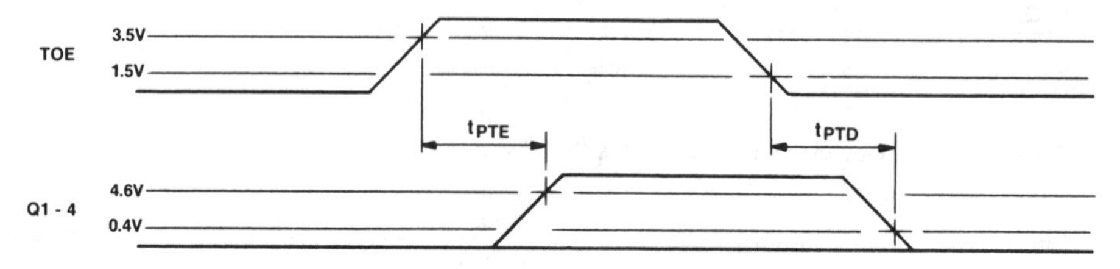

Timing - output enable and disable

Plessey SL376M

Metering Subscriber Line Interface Circuit Advance Product Information

- Low-power line feed via regulator
- Programmable constant-voltage resistive feed
- Programmable AC termination impedance
- Ground-key detection
- Programmable off-hook detection
- Ring relay driver
- Low-power standby mode
- Normal or reversed line polarity
- Supports 2.2V RMS metering pulses
- 75V rating to ease line-protector tolerancing

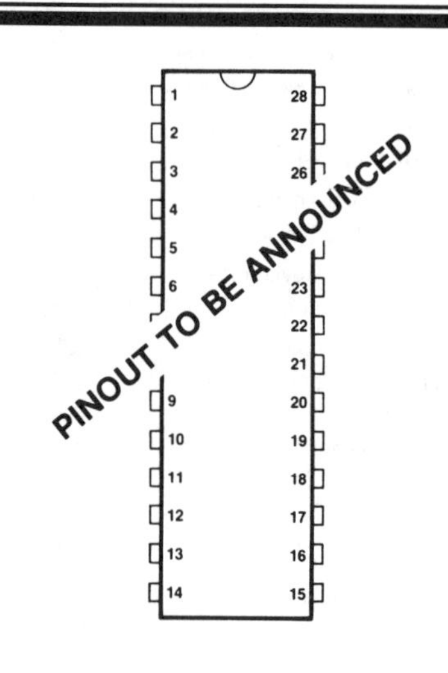

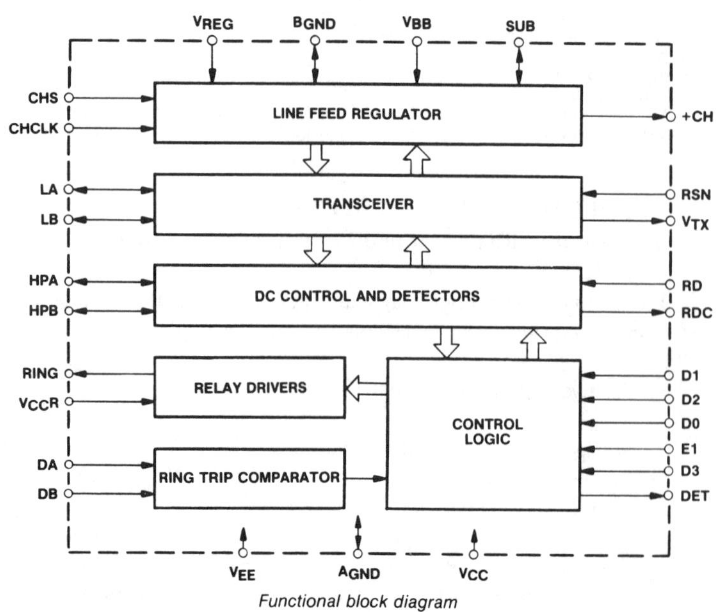

Functional block diagram

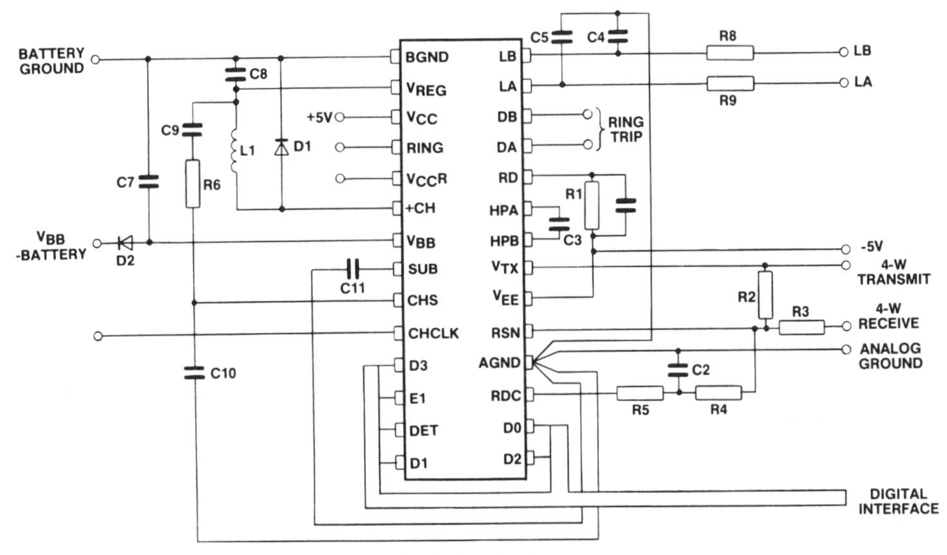

Typical application

Plessey SL9009

Adaptive Balance Circuit Preliminary Information

- Extracts the received signal
- Adapts automatically to line variations
- No microprocessor required
- Simple application circuit
- 40dB (Typ.) rejection of transmitted signal

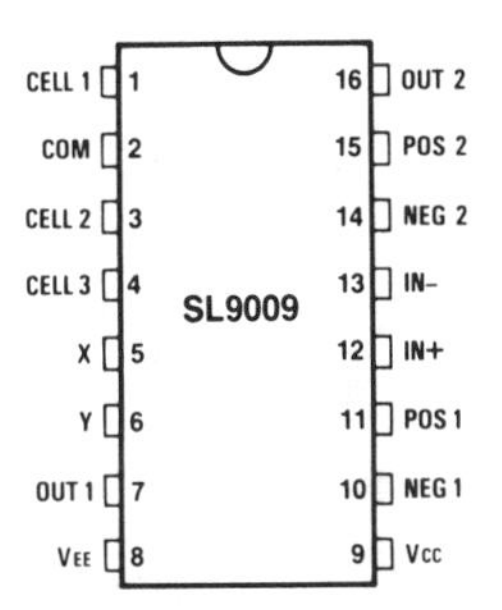

Pin connections - top view

APPLICATIONS

- Modems: extracting the received data
- Feature phones: extracting the received voice
- PBX/PABX/CO line cards: extracting the incoming signal from the telephone line

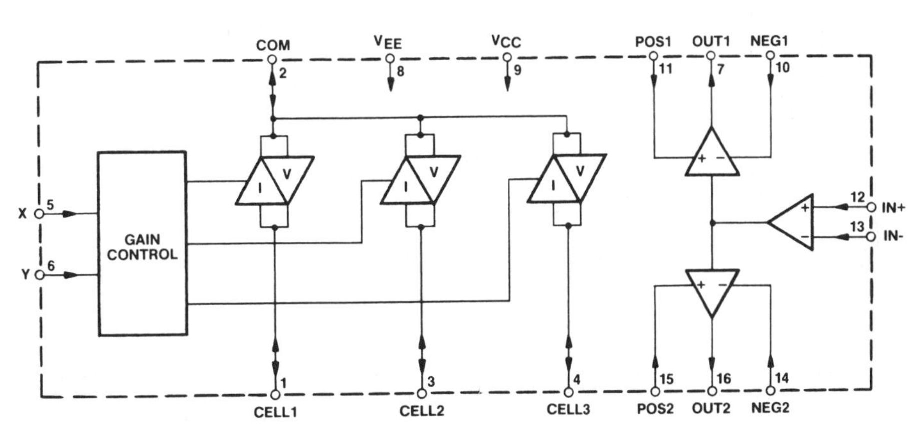

Functional block diagram

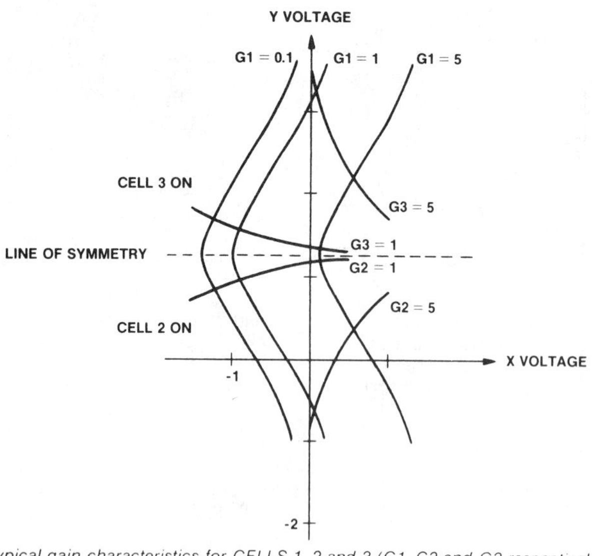

Typical gain characteristics for CELLS 1, 2 and 3 (G1, G2 and G3 respectively)

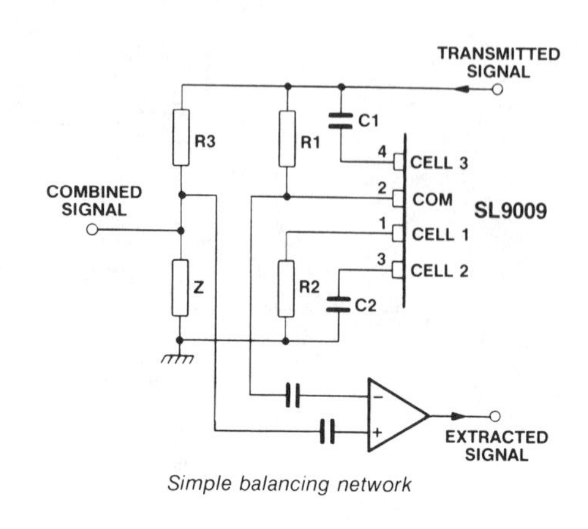

Simple balancing network

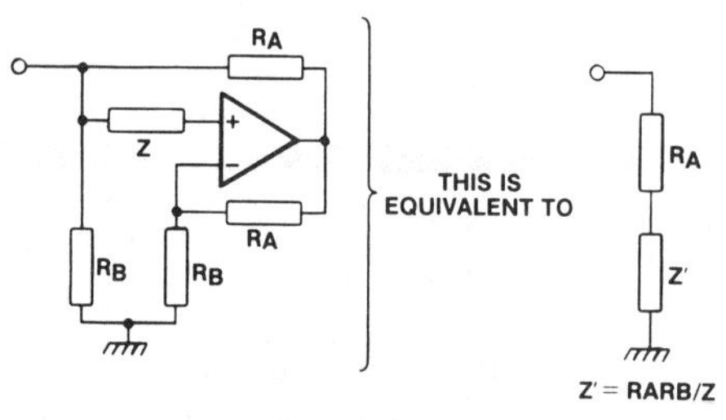

Gyrator circuit

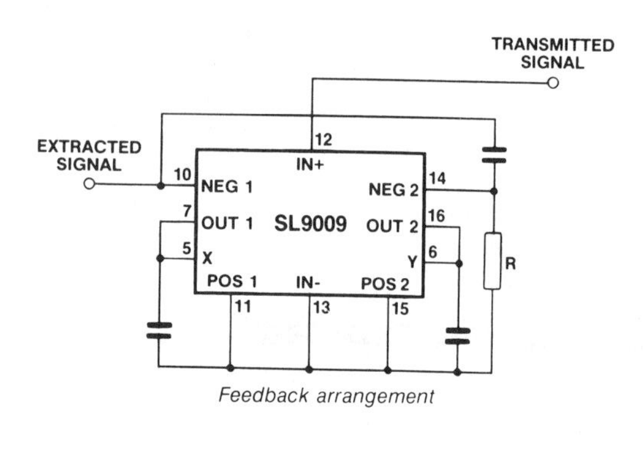

Feedback arrangement

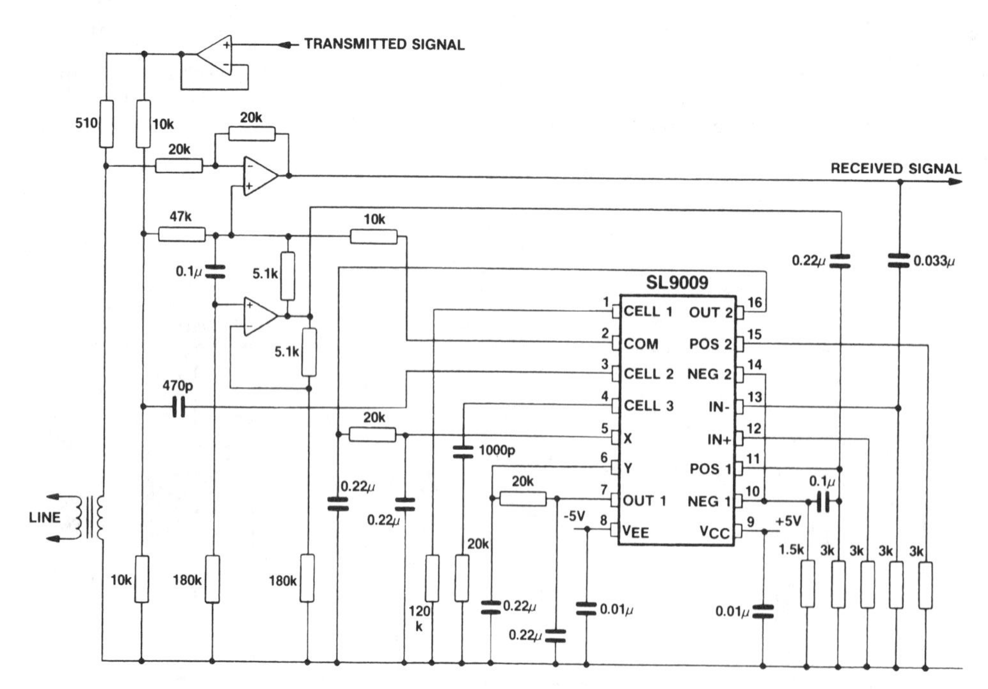

Line transformer:- Total primary + Secondary resistance = 100Ω; Inductance = 0.9H.

Complete circuit

TRANSMITTED SIGNAL

EXTRACTED SIGNAL

SIGN OF DERIVATIVE OF EXTRACTED SIGNAL

Signal relationship for 90° of lag on extracted signal

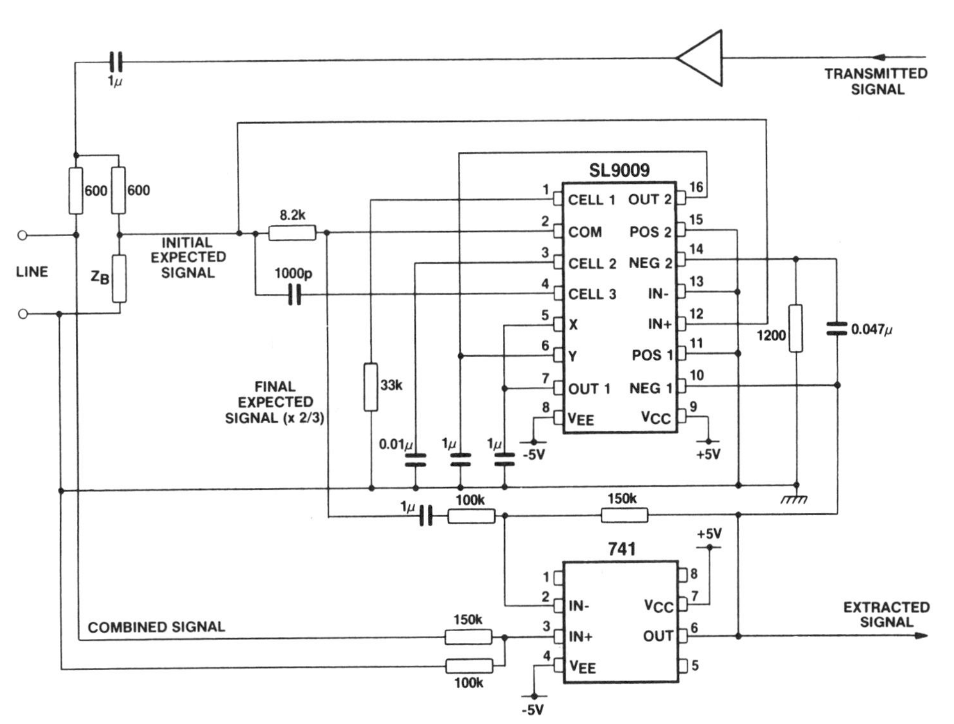

Circuit for improved line modelling

ABSOLUTE MAXIMUM RATINGS

Exceeding these ratings might cause permanent damage. Functional operation under these conditions is not implied.

Positive supply voltage (pin 9), V_{CC}	−0.3V to + 10V
Negative supply voltage (pin 9), V_{EE}	−10V to +0.3V
Input voltages (pins 2,5,6,10,11, 12,13,14 and 15), V_I	V_{EE} to V_{CC}
Output voltages (pins 1,3,4,7, 16), V_0	V_{EE} to V_{CC}
Cell voltage (pins 1,3,4) minus Common voltage (pin 2), V_C	−5V to +5
Storage temperature, T_{ST}	−10°C to +125°C

ELECTRICAL CHARACTERISTICS

Test conditions - Voltages are with respect to digital ground (V_{DGND} [V_{CC} - V_{EE}]/2)

Characteristic	Symbol	Value			Units
		Min.	Typ.(1)	Max.	
Positive supply voltage (pin 9)	V_{CC}	4.5	5	7.0	V
Negative supply voltage (pin 8)	V_{EE}	-7.0	-5	-4.5	V
Ambient temperature	T_{amb}	0		70	°C
Common cell pin voltage (pin 2)	V_{COM}	V_{EE} +2.7		V_{CC} -2	V
Cell input currents (pins 1,3 and 4)	I_{CELL}	-10		10	μA
X control voltage (pin 5)	V_X	V_{EE} +2.7		V_{EE} -6.0	V
Y control voltage (pin 6)	V_Y	-2.0		1.8	V
Detector input voltages (pins 12 and 13)	V_{IN}	V_{EE} +2.7		V_{CC} -2.7	V
Analysis input voltages (pins 10,11,14 and 15)	V_A	V_{EE} +2.7		V_{CC} -2.7	V
Detector output voltages (pins 7 and 16)	V_{OUT}	V_{EE} +2.5		V_{CC} -2	V

Operating Characteristics: General - Voltages are with respect to ground (V_{GND} = [V_{CC} - V_{EE}]/2)

Characteristic	Symbol	Value			Units	Conditions
		Min.	Typ.(1)	Max.		
Power dissipation	P_D			30	mW	
Supply current	I_{CC}		1.3	15	mA	
Pin capacitance	C_P		7	15	pF	Pin to supplies

Operating Characteristics: Cells - Voltages are with respect to ground (V_{GND} = [V_{CC} - V_{EE}]/2)

Characteristic	Symbol	Value			Units	Conditions
		Min.	Typ.(1)	Max.		
Internal resistance (pins 1, 3 and 4)	R_I	3		14	kΩ	
Control input leakage (pins 5 and 6)	I_C			0.12	μA	
Minimum cell gain	G_{MIN}		0.05			
Maximum cell gain	G_{MAX}		10			
DC bias current (pin 2)	I_{BDC}	-12	0	12	μA	
Residual impedance (pin 2)	Z_R	500			kΩ	Cell pins open circuit

Operating Characteristics: Detectors - Voltages are with respect to ground (V_{GND} = [V_{CC} - V_{EE}]/2)

Characteristic	Symbol	Value			Units	Conditions
		Min.	Typ.(1)	Max.		
Differential input offset (pins 10 to 15)	V_{DOFF}			13	mV	
Input offset current (pins 10 to 15)	I_{OFF}	-0.15		0.15	μs	
Input bias current (pins 10 to 15)	I_{IB}		0.1	0.7	μA	
Transconductance gain	G_T	250	500	1000	μΩ	Magnitude
Output offset current (pins 7 and 16)	I_{OFF}	-1.2		1.2	μA	
Maximum output current (pins 7 and 16)	I_{MAX}		50		μA	Sink or Source
Output impedance (pins 7 and 16)	Z_{OUT}		5		MΩ	

NOTE
1. Typical figures are for design aid only. They are not guaranteed and not subject to production testing.

This information is reproduced with permission of LSI Computer Systems, Inc., 1235 Walt Whitman Road, Melville, NY 11747

LSI
LS7501, LS7510
Tone Activated Line Isolation Device

- Low-power CMOS design
- On-chip oscillator (32,768Hz external crystal required)
- Tone input can be low-level sinusoid (as low as −30 dBM) or fully digital
- Mask programmable available frequencies: 11Hz to 4095Hz (in 1Hz steps)
- Sample interval −4.5 seconds (Mask programmable 0.5 to 8.0 seconds)

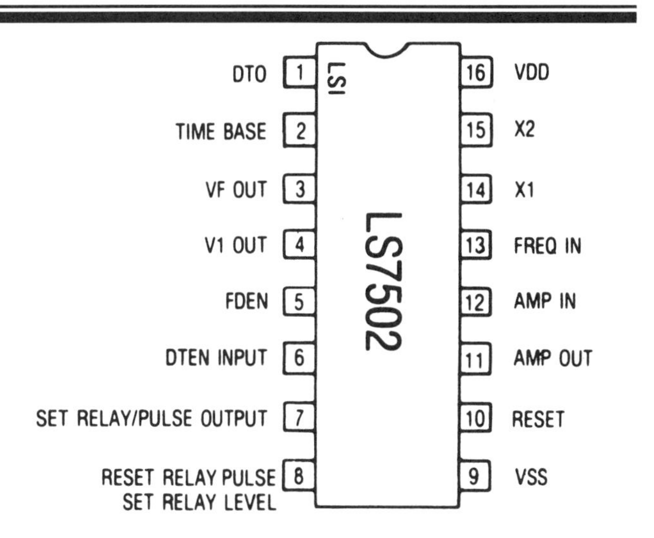

CIRCUIT BLOCK DIAGRAM

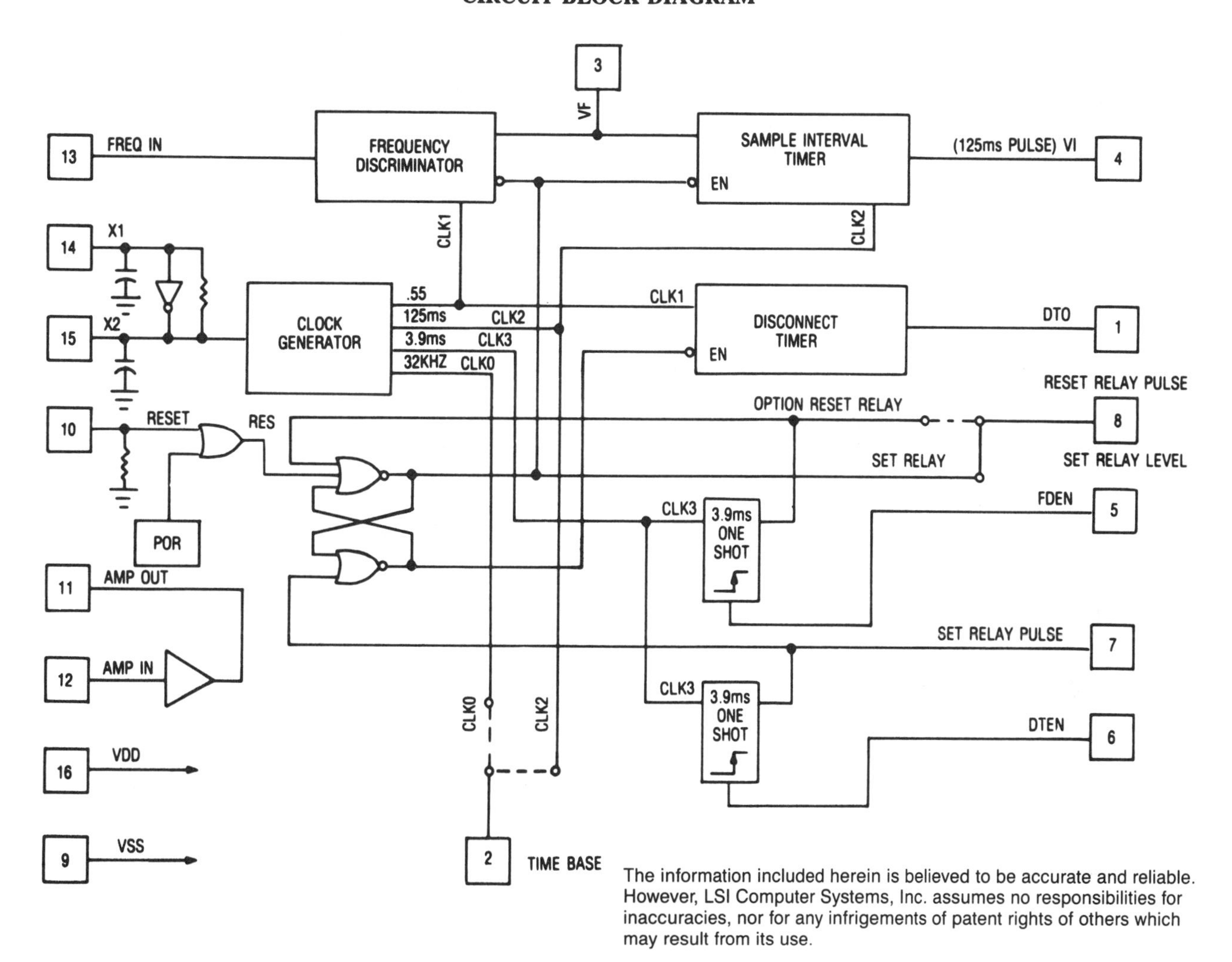

The information included herein is believed to be accurate and reliable. However, LSI Computer Systems, Inc. assumes no responsibilities for inaccuracies, nor for any infrigements of patent rights of others which may result from its use.

NOTE (1) All devices shown on the LS7501 through the LS7510 are configured with the set relay outpout on Pin 8. The reset option can be substituted by optional mask change.

NOTE (2) All devices shown with the exception of the LS7502 are configured with the clock-0, 32KHz output on Pin 2. The LS7502 is configured with the clock-2 time base output of 8Hz. These outputs may be changed with the same optional mask change referred to in Note 1.

The LS7501 — LS7510 are frequency discriminator circuits that respond to a standard frequency input if the input is maintained within ±10Hz during a 4.5 second continuous sample interval. During this interval, the input is being sampled every 0.5 seconds. If it is valid for the sample interval, then the circuit can be used to pulse a relay that disconnects the line to be tested. After 20 seconds of disconnect time, the relay is reset and the line is restored. Ten standard-frequency versions of this circuit are available.

PART NO.	FREQUENCY (HZ)
LS7501	2683
LS7502	2713
LS7503	2743
LS7504	2773
LS7505	2833
LS7506	2863
LS7507	2893
LS7508	2923
LS7509	2953
LS7510	2983

TELEPHONE LINE INTERROGATOR AND TONE BACK GENERATOR

This application indicates a method for interrogating a telephone line when a 2713Hz (±10Hz) tone is detected for a minimum of 4.5 seconds. (The LS7501 Circuit).

At the end of the 4.5 second sample period, an oscillator is energized and generates a tone back signal. This signal modulates the line at a voice level of −16dB or 3.5mV peak to peak.

Typical system input activation sensitivity is −30dBM. The unit should also be operational down to 6 volts at the tip/ring network terminals.

As shown, the differential op amp is connected to the telephone lines through 0.001μF coupling capacitors. This eliminates the DC component and acts as the first filter for 60Hz. The differential amplifier stage is followed by a bandpass filter centered around 2713Hz. This filter should be designed for high Q's (Q = 10) and yet utilize current efficient op amps.

The bandpass output is then squared and connected to the digital tone input (pin 13). The input signal is sampled by the digital discrimination section of the LS7502. If 2713Hz (±10Hz) is present for 4.5 seconds, a 125 millisecond pulse at pin 4 is applied to the DTEN input (pin 6), causing an internal flip-flop to set the relay output (pin 8) to go high, activating the tone back oscillator.

As the 10μF capacitor (C_4) builds up stored charge, it biases the FDEN input (pin 5) through R_7 until it is sufficient to reset the internal flip-flop and bring the circuit back to its idle state and turn the tone-back oscillator off. By varying the R_7 — C_4 network, the time constant for the tone-back duration can be varied.

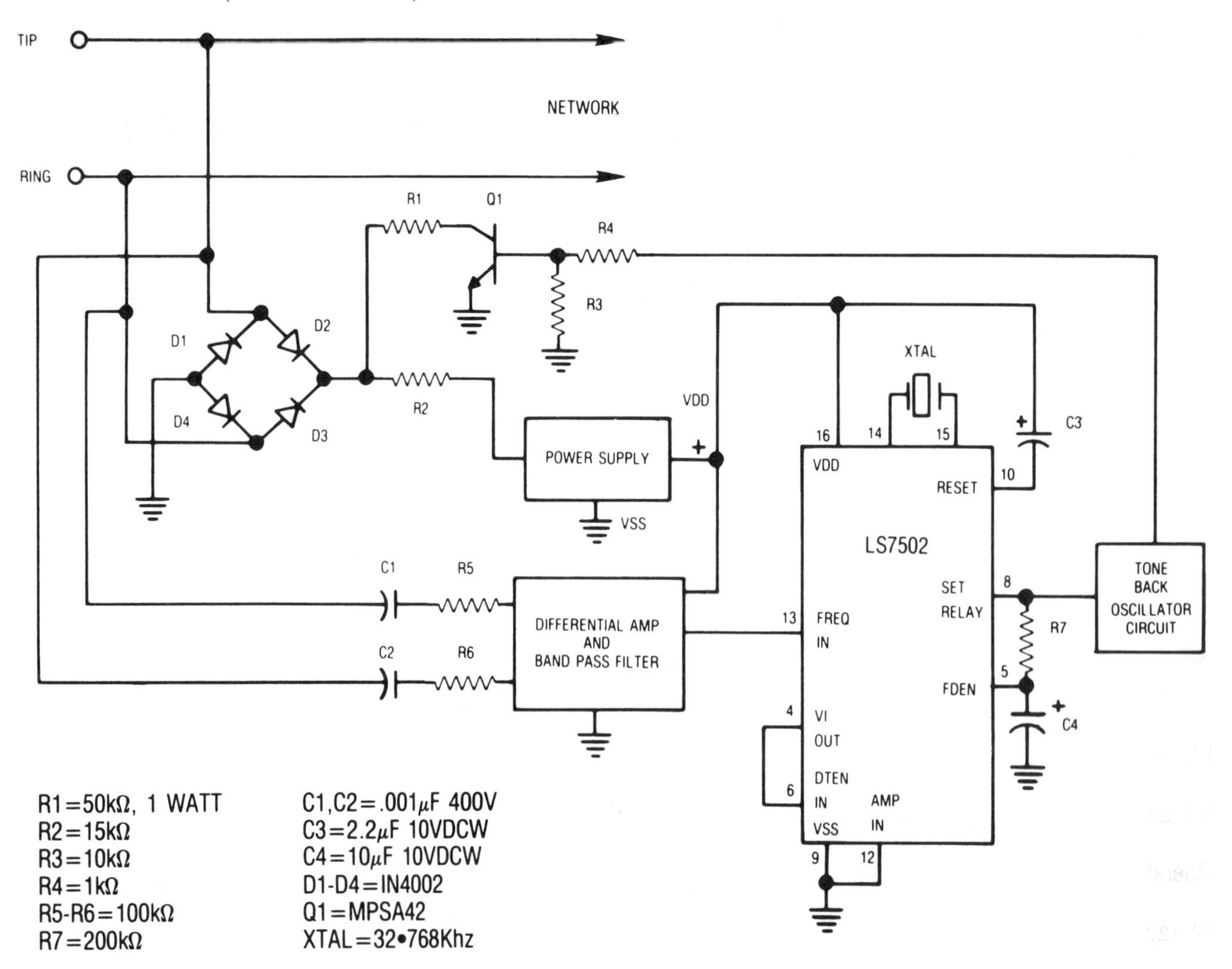

LSI
LS3404
High-Quality Melody Circuits

- Excellent pitch resolution
- Chime-like exponential envelope decay of each note
- Large ROM, 255 note capacity
- Wide variety of available fonts
- Mask programmable melody fonts
- Single or multiple melody capacity
- Auto-turn-off at end of play
- 4.5V to 15V operation
- Low standby current
- Direct drive of PIEZO transducers
- External drive of 8Ω dynamic speakers

The LS3404 Series are monolithic, ion-implanted MOS circuits designed for the generation of music. The circuit is mask programmable and can hold 255 notes in PROM.

The note pitch has an 0.8% resolution for notes up to 2kHz and 1.3% resolution for notes up to 3kHz. The note duration ranges from 125 milliseconds for a 1/16 note to 2.0 seconds for a full note. This is equivalent to 120 beats per minute.

The duration counter allows for 8-note durations out of a possible 16 durations to be programmed in each font.

The pitch counter allows for 15 different pitches out of a possible 511 pitches to be programmed in each font.

The pitch-counter ouput is conditioned by an external R/C envelope to provide proper envelope decay and applied to a pair of operational amplifiers which drive a piezoelectric speaker in a push-pull configuration. Only one output is used for driving an external transistor/dynamic speaker combination in a single-ended configuration.

The exponential decay envelope imposed on each note accounts for the uniquely realistic quality of sound exhibitied by the LS3404 series of circuits.

Upon application of supply V_{SS}, the chip will start to play after a small time delay caused by power on reset. Play will be terminated either by the V_{SS} being removed or by completion of the entire play. When being terminated by end of play (EOP), the circuit will continue to consume power at a reduced rate.

FONT NO.	SONG
3404-02	Christmas Medley
3404-03	"Somewhere My Love"
3404-04	"As Time Goes By"
3404-05	"Let Me Call You Sweetheart"
3404-08	"I'm In The Mood For Love"
3404-09	"Wedding March"
3404-10	"Happy Birthday I"
3404-11	"Zip-A-Dee-Doo-Dah"
3404-12	"Brahm's Lullabye"
3404-14	"Santa Claus Is Coming To Town"
3404-15	Christmas Angel Medley
3404-16	"We Wish You A Merry Christmas"
3404-17	"Walking In A Winter Wonderland"
3404-18	"Jingle Bells"
3404-19	"Joy To The World"
3404-20	"Love Makes The World Go Round"

FONT NO.	SONG
3404-21	"My Favorite Things"
3404-22	"What The World Needs Now"
3404-23	"I'd Do Anything"
3404-24	"Hail To The Chief"
3404-25	"Thanks For The Memories"
3404-26	"Gonna Fly Now" (Rocky)
3404-27	"Lazy Crazy Hazy Days of Summer"
3404-28	"For He's A Jolly Good Fellow"
3404-29	"Pomp & Circumstance"
3404-30	"More"
3404-31	"Ain't She Sweet"
3404-32	"You Are The Sunshine Of My Life"
3404-33	Nursery Rhyme Medley
3404-34	"Happy Birthday II"
3404-35	Brahms/Mozart Lullabye Medley

Listing of the 31 Presently Available Melodies

ABSOLUTE MAXIMUM RATINGS (all voltages referenced to VDD)

	SYMBOL	VALUE	UNIT
DC supply voltage	VSS	+18	Volts
Voltage (any pin)	VIN	0 to VSS+ .3	Volts
Operating Temperature	TA	0 to +70	°C
Storage Temperature	Tstg	-65 to +150	°C

ELECTRICAL CHARACTERISTICS

PARAMETER	SYMBOL	MIN	TYPICAL	MAX	UNITS	CONDITION
Supply Voltage	Vss	4.5		15	Volts	
Standby current At end of play	Iss			15 25	Microamperes Microamperes	VSS=6 volts VSS=10 volts } Duration Clock Resistor =2.2Meg Ω
Average operating current including Piezoelectric speaker			4 5		Milliamperes Milliamperes	VSS=6 VDC VSS=9 VDC
Pitch clock frequency		425	500	575	KHZ	R=15KΩ C=100 pF
Duration clock frequency			8		HZ	R=2.2 Meg Ω C=.1μF
Duration clock frequency range		2		30	HZ	
Envelope Resistance			3.3		MEG Ω	
Envelope Capacitance			.1		μF	
Minimum POR Capacitance		.01			μF	
Speaker output peak to peak voltage		4.5 8.0			Volts Volts	Vss=6volts Vss=10volts } pitch frequency =1KHZ
Tracking of output to the envelope		-2.0 -2.5		+.4 +.4	Volts Volts	Vss=6 volts Vss=10 volts
Composite output Peak to peak voltage		9 16			Volts Volts	Vss=6 volts Vss=10 Volts

INPUT/ OUTPUT	DESCRIPTION
SP-1, SP-2	Push-pull outputs for driving Piezoelectric capacitive type speaker. Typical speaker has a 27MM diameter with equivalent capacitance of approximately 20,000 PFD.
RCEN	R-C envelope input. External resistance-capacitance network for controlling the output envelope. The resistance is connected to the negative supply (VDD) and the capacitance is connected to the positive supply (VSS).
RCSS	R-C network for internal duration clock oscillator. The duration clock along with an internally programmed counter determines the time duration of each note.
RCHS	R-C network for internal pitch clock oscillator. The pitch clock generates the audio frequency output utilizing an internally programmed counter.
VDD	Negative voltage supply
VSS	Positive voltage supply
POR	Power-on-reset-external capacitor used for initializing circuit at the application of power.

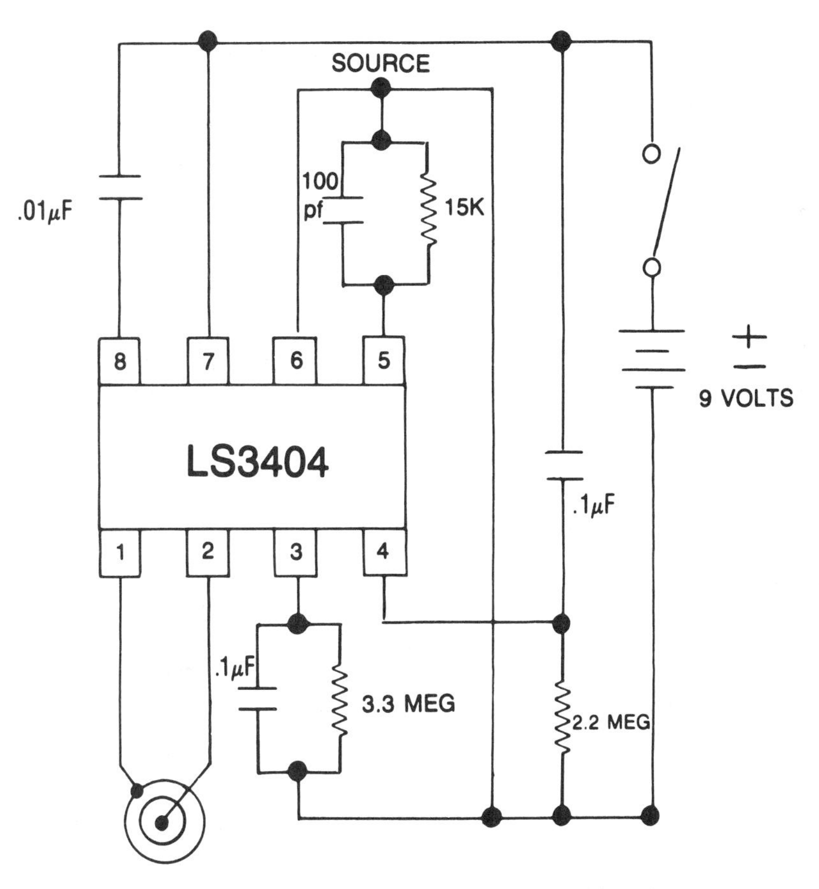

Typical Piezoelectric capacitance type speaker connection diagram.

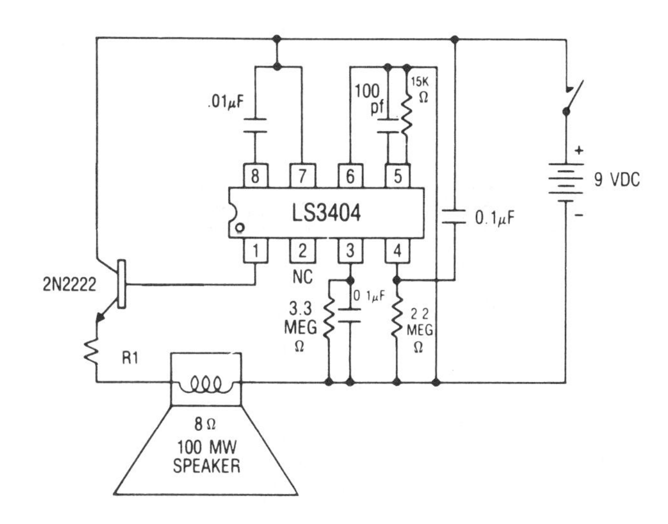

Typical 8 Ohm speaker connection. In this configuration only SP-1 is used to drive the external 8 ohm speaker in a single ended mode. Resistor R_1 is used as a volume control and can be omitted for maximum volume.

Standard 8 Pin Plastic Mini-DIP

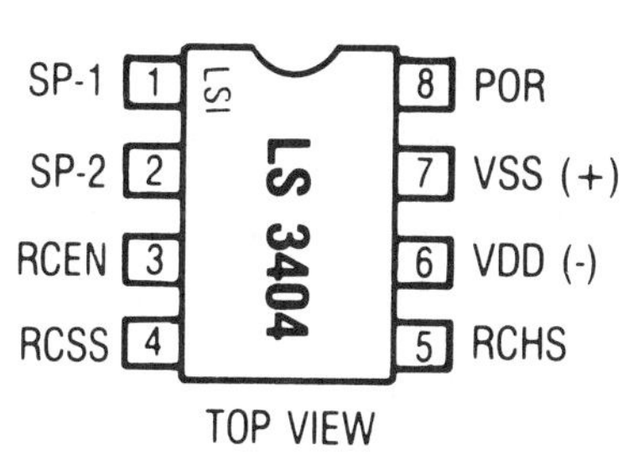

TOP VIEW

Mitel MT8971/72 Digital Network Interface Circuit

- Full duplex transmission over a single twisted pair
- Selectable 80 or 160 kbit/s line rate
- Adaptive echo cancellation
- Up to 3.7km at 80 kbit/s and 3.0km at 160 kbit/s (8971)
- Up to 5km at 80 kbit/s and 4km at 160 kbit/s (8972)
- ISDN-compatible (2B + D) data format
- Transparent modem capability
- Frame synchronization and clock extraction
- MITEL ST-BUS compatible
- Low-power (typically 50mW), single 5V supply

APPLICATIONS

- Digital subscriber lines
- High-speed data transmission over twisted wires
- Digital PABX line cards and telephone sets
- 80 or 160 kbit/s single-chip modem

MT8971/72 ISO^2-CMOS

ABSOLUTE MAXIMUM RATINGS** - Voltages are with respect to ground (V_{SS}) unless otherwise stated.

	Parameter	Symbol	Min	Max	Units
1	Supply Voltage	V_{DD}	-0.3	7	V
2	Voltage On Any Pin (other than supply)	V_{Max}	-0.3	$V_{DD}+0.3$	V
3	Current On Any Pin (other than supply)	I_{Max}		40	mA
4	Storage Temperature	T_{ST}	-65	+150	°C
5	Package Power Dissipation‡	P_{Diss}		750	mW

**Exceeding these values may cause permanent damage. Functional operation under these conditions is not implied.
‡Derate 16mW/°C above 75°C

RECOMMENDED OPERATING CONDITIONS† - Voltages are with respect to ground (V_{SS}) unless otherwise stated.

	Characteristics	Sym	Min	Typ*	Max	Units	Test Conditions
1	Operating Supply Voltage	V_{DD}	4.75	5.00	5.25	V	
2	Operating Temperature	T_{OP}	0		70	°C	
3	Input High Voltage (except OSC1)	V_{IH}	2.4		V_{DD}	V	for 400 mV noise margin
4	Input Low Voltage (except OSC1)	V_{IL}	0		0.4	V	for 400 mV noise margin

*Typical figures are at 25°C and are for design aid only: not guaranteed and not subject to production testing.
†Parameters over recommended temperature & power supply voltage ranges.

DC ELECTRICAL CHARACTERISTICS -Voltages are with respect to ground (V_{SS}) unless otherwise stated.

		Characteristics	Sym	Min	Typ*	Max	Units	Test Conditions
1	OUTPUTS	Operating Supply Current	I_{DD}		10	15	mA	
2		Output High Voltage (ex OSC2)	V_{OH}	2.4			V	$I_{OH}=10mA$
3		Output High Current (except OSC2)	I_{OH}	10	15		mA	Source current. $V_{OH}=2.4V$
4				8	12		mA	Source current. $V_{OH}=3.0V$
5		Output High Current - OSC2	I_{OH}	10			µA	Source current $V_{OH}=3.5V$
6		Output Low Voltage (ex OSC2)	V_{OL}			0.4	V	$I_{OL}=5mA$
7		Output Low Current (except OSC2)	I_{OL}	5	7.5		mA	Sink current. $V_{OL}=0.4V$
8				20	30		mA	Sink current. $V_{OL}=2.0V$
9		Output Low Current - OSC2	I_{OL}	10			µA	Sink current. $V_{OL}=1.5V$
10		High Imped. Output Leakage	I_{OZ}			10	µA	$V_{IN}=V_{SS}$ to V_{DD}
11		Output Voltage (V_{Ref})	V_O		V_{Bias}-1.8		V	
12		(V_{Bias})			$V_{DD}/2$		V	
13	INPUTS	Input High Voltage (ex OSC1)	V_{IH}	2.0			V	
14		Input Low Voltage (ex OSC1)	V_{IL}			0.8	V	
15		Input Leakage Current	I_{IL}			10	µA	$V_{IN}=V_{SS}$ to V_{DD}

*Typical figures are at 25°C and are for design aid only: not guaranteed and not subject to production testing.
†Parameters over recommended temperature & power supply voltage ranges.

AC ELECTRICAL CHARACTERISTICS - Voltages are with respect to ground (V_{SS}) unless otherwise stated.

		Characteristics	Sym	Min	Typ*	Max	Units	Test Conditions
1	INPUTS	Input Voltage (L_{IN})	V_{IN}			5.0	V_{pp}	
2		Input Current (L_{IN})	I_{IN}	-10		+10	μA	f_{Baud} = 160 kHz
3		Input Impedance (L_{IN})	Z_{IN}		50		kΩ	f_{Baud} = 160 kHz
4		Crystal/Clock Frequency	f_C		10.24		MHz	
5		Crystal/Clock Tolerance	T_C	-100	0	+100	ppm	
6a		Crystal/Clock Duty Cycle	DC_C	40	50	60	%	25°C
6b		Crystal/Clock Duty Cycle	DC_C	45	50	55	%	Recommended at max./ min. temp. & V_{DD}
7		Crystal/Clock Loading	C_L		33	50	pF	From OSC1 & OSC2 to V_{SS}.
8	OUTPUTS	Output Capacitance (L_{OUT})	C_o		8		pF	
9		Load Resistance (L_{OUT}) (V_{Bias}, V_{Ref})	$R_{L_{out}}$		500 100		Ω kΩ	
10		Load Capacitance (L_{OUT}) (V_{Bias}, V_{Ref})	$C_{L_{out}}$	 0.1		20	pF μF	Capacitance to V_{Bias}.
11		Output Voltage (L_{OUT})	V_o	3.8	4.3	4.8	V_{pp}	R_{Lout} = 500Ω, C_{Lout} = 20pF

AC ELECTRICAL CHARACTERISTICS† Clock Timing - DN Mode

		Sym	Min	Typ*	Max	Units	Test Conditions
1	$\overline{C4}$ Clock Period	t_{C4P}		244		ns	
2	$\overline{C4}$ Clock Width High or Low	t_{C4W}	90	122	150	ns	In Master Mode - Note 1
3	Frame Pulse Set Up Time	t_{F0S}	50			ns	
4	Frame Pulse Hold Time	t_{F0H}	50			ns	
5	Frame Pulse Width	t_{F0W}	172	244		ns	
6	10.24 MHz Clock Jitter (wrt $\overline{C4}$)	J_C	-15		+15	ns	Note 2

†Timing is over recommended temperature & power supply voltage ranges.
*Typical figures are at 25°C, for design aid only; not guaranteed and not subject to production testing.

Notes:
1) When operating as a SLAVE the $\overline{C4}$ clock has a 40% duty cycle.
2) If operating in MAS/ DN mode, the $\overline{C4}$ and Oscillator clocks must be externally frequency locked (i.e., $f_C = 2.5 \times f_{C4}$). The relative phase between these two clocks (φ in Fig. 4) is not critical and may vary from 0 ns. to t_{C4P}. However, the relative jitter must be less than J_C.(See Fig. 4).

PIN CONNECTIONS

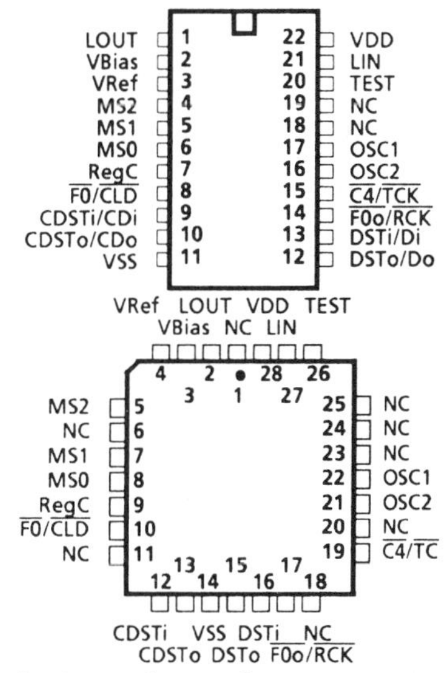

Ordering Information 0°C to +70°C

MT8972AC	22 Pin Cerdip
MT8971BE/MT8972AE	22 Pin Plastic DIP
MT8972AP	28 Pin Plastic LCC

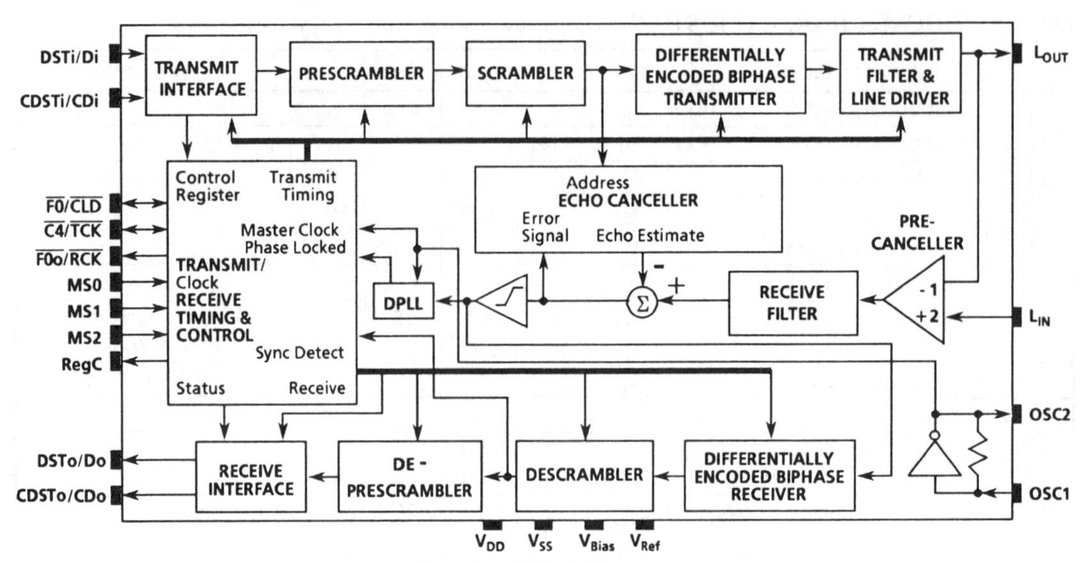

Functional Block Diagram

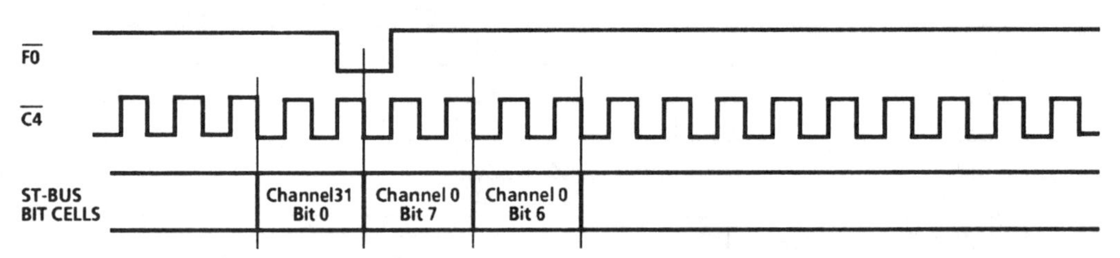

C4 Clock & Frame Pulse Alignment for ST-BUS Streams

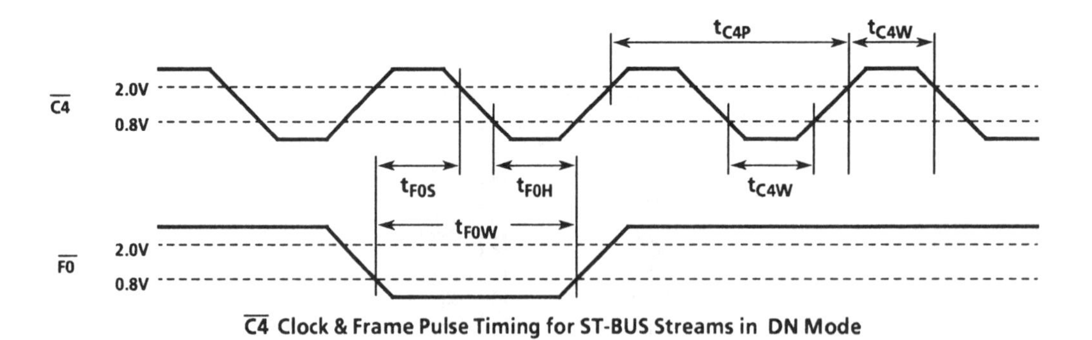

C4 Clock & Frame Pulse Timing for ST-BUS Streams in DN Mode

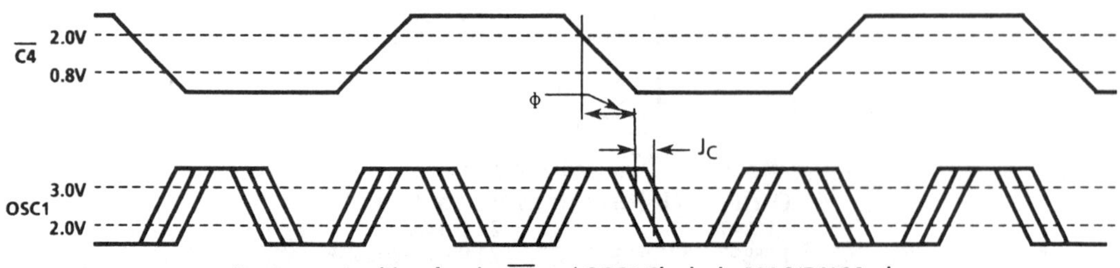

Frequency Locking for the C4 and OSC1 Clocks in MAS/DN Mode

AC ELECTRICAL CHARACTERISTICS† Clock Timing - MOD Mode

	Characteristics	Sym	80 kbit/s			160 kbit/s			Units	Test Conditions
			Min	Typ*	Max	Min	Typ*	Max		
1	$\overline{TCK}/\overline{RCK}$ Clock Period	t_{CP}		12.5			6.25		µs	
2	$\overline{TCK}/\overline{RCK}$ Clock Width	t_{CW}		6.25			3.125		µs	
3	$\overline{TCK}/\overline{RCK}$ Clock Transition Time	t_{CT}		20			20		ns	$C_L = 40pF$
4	$\overline{CLD}$ to $\overline{TCK}$ Setup Time	t_{CLDS}		3.125			1.56		µs	
5	$\overline{CLD}$ to $\overline{TCK}$ Hold Time	t_{CLDH}		3.125			1.56		µs	
6	$\overline{CLD}$ Width Low	t_{CLDW}		6.05			2.925		µs	
7	$\overline{CLD}$ Period	t_{CLDP}		$8xt_{CP}$			$8xt_{CP}$		µs	

†Timing is over recommended temperature & power supply voltage ranges.
* Typical figures are at 25°C, for design aid only: not guaranteed and not subject to production testing.

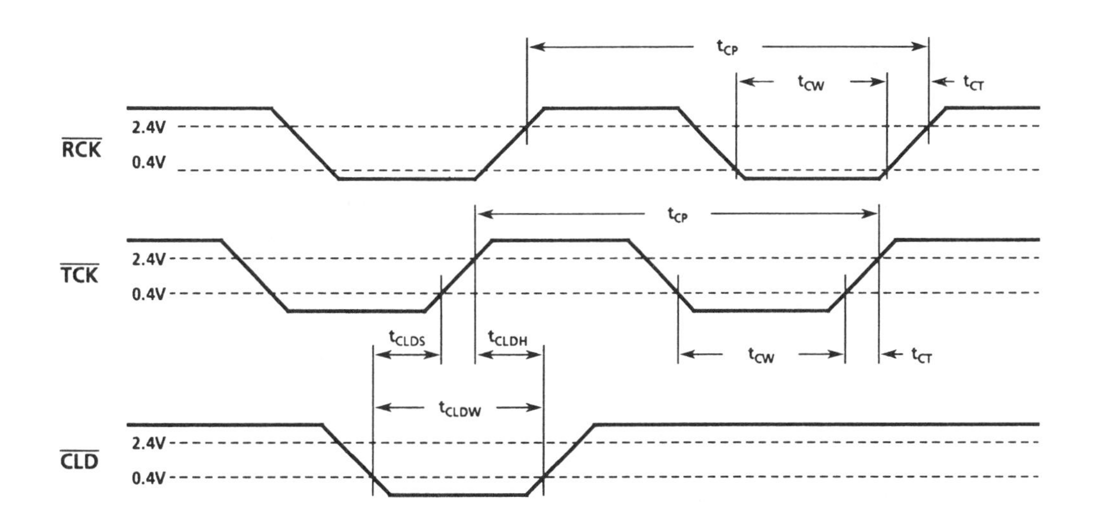

Note 1: $\overline{TCK}$ and $\overline{CLD}$ are generated on chip and provide the data clocks for the CD port and the transmit section of the DV port. $\overline{RCK}$, also generated on chip, is extracted from the receive data and only clocks out the data at the D_o output and may be skewed with respect to $\overline{TCK}$ due to end-to-end delay.

Note 2: At the slave end $\overline{TCK}$ is locked to, and in phase with, $\overline{RCK}$.

$\overline{RCK}$, $\overline{TCK}$ & $\overline{CLD}$ Timing For MOD Mode

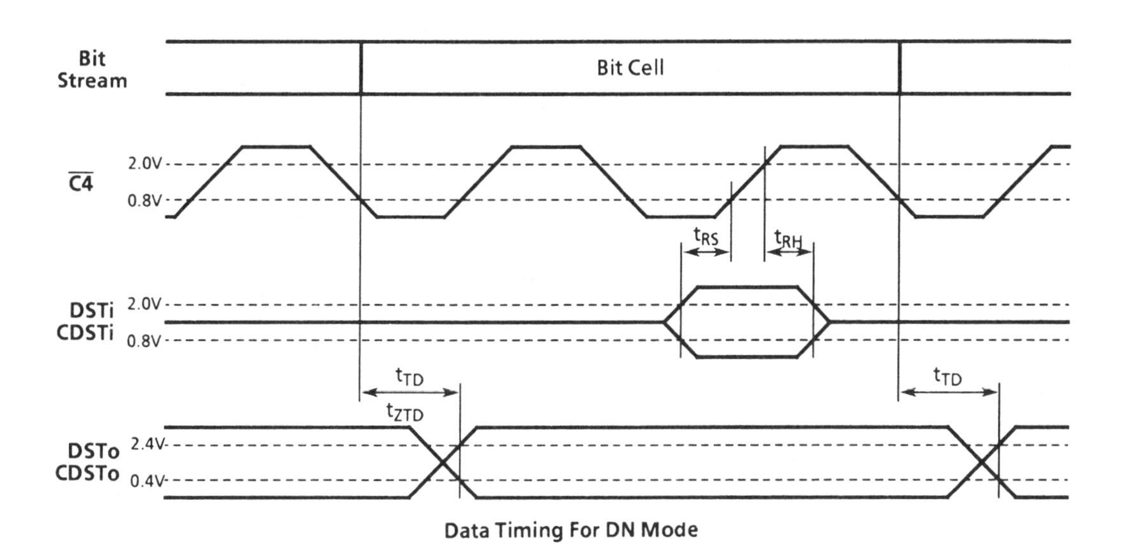

Data Timing For DN Mode

AC ELECTRICAL CHARACTERISTICS† Data Timing - DN Mode

	Characteristics	Sym	Min	Typ*	Max	Units	Test Conditions
1	DSTi/CDSTi Data Setup Time	t_{RS}	30		250	ns	
2	DSTi/CDSTi Data Hold Time	t_{RH}	50		122	ns	
3a	DSTo/CDSTo Data Delay	t_{TD}	20		120	ns	C_L=40pF
3b	DSTo/CDSTo High Z to Data Delay	t_{ZTD}		100		ns	C_L=40pF

†Timing is over recommended temperature & power supply voltage ranges.
* Typical figures are at 25°C, for design aid only: not guaranteed and not subject to production testing.

PERFORMANCE CHARACTERISTICS of the MT8972 DNIC

	Characteristics	Sym	Min	Typ*	Max	Units	Test Conditions
1	Allowable Attenuation for Bit Error Rate of 10^{-6} (Note 1)	A_{fb}	0	25		dB	SNR ≥ 16.5dB (300kHz bandlimited noise)
2	Line Length at 80 kbit/s -24 AWG -26 AWG	L_{80}		3.7 2.5		km	attenuation - 6.9 dB/km attenuation - 10.0 dB/km
3	Line Length at 160 kbit/s -24 AWG -26 AWG	L_{160}		3.0 2.2		km	attenuation - 8.0 dB/km attenuation - 11.5 dB/km

Note 1: Attenuation measured from Master L_{OUT} to Slave L_{IN} at $\frac{3}{4}$ baud frequency.
* Typical figures are at 25°C, for design aid only: not guaranteed and not subject to production testing.

PERFORMANCE CHARACTERISTICS of the MT8972 DNIC

	Characteristics	Sym	Min	Typ*	Max	Units	Test Conditions
1	Allowable Attenuation for Bit Error Rate of 10^{-6} (Note 1)	A_{fb}	0	40	33	dB	SNR ≥ 16.5dB (300kHz bandlimited noise)
2	Line Length at 80 kbit/s -24 AWG -26 AWG	L_{80}		5.0 3.4		km	attenuation - 6.9 dB/km attenuation - 10.0 dB/km
3	Line Length at 160 kbit/s -24 AWG -26 AWG	L_{160}		4.0 3.0		km	attenuation - 8.0 dB/km attenuation - 11.5 dB/km

Note 1: Attenuation measured from Master L_{OUT} to Slave L_{IN} at $\frac{3}{4}$ baud frequency.
* Typical figures are at 25°C, for design aid only: not guaranteed and not subject to production testing.

AC ELECTRICAL CHARACTERISTICS - Data Timing - MOD Mode

	Characteristics	Sym	80 kbit/s			160 kbit/s			Units	Test Conditions
			Min	Typ*	Max	Min	Typ*	Max		
1	Di/CDi Data Setup Time	t_{DS}		150			150		ns	
2	Di/CDi Data Hold Time	t_{DH}		4.5			2.5		µs	
3	Do Data Delay Time	t_{RD}			100			100	ns	C_L=40pF
4	CDo Data Delay Time	t_{TD}			100			100	ns	C_L=40pF

Note 1: Attenuation measured from Master L_{OUT} to Slave L_{IN} at $\frac{3}{4}$ baud frequency.
* Typical figures are at 25°C, for design aid only: not guaranteed and not subject to production testing.

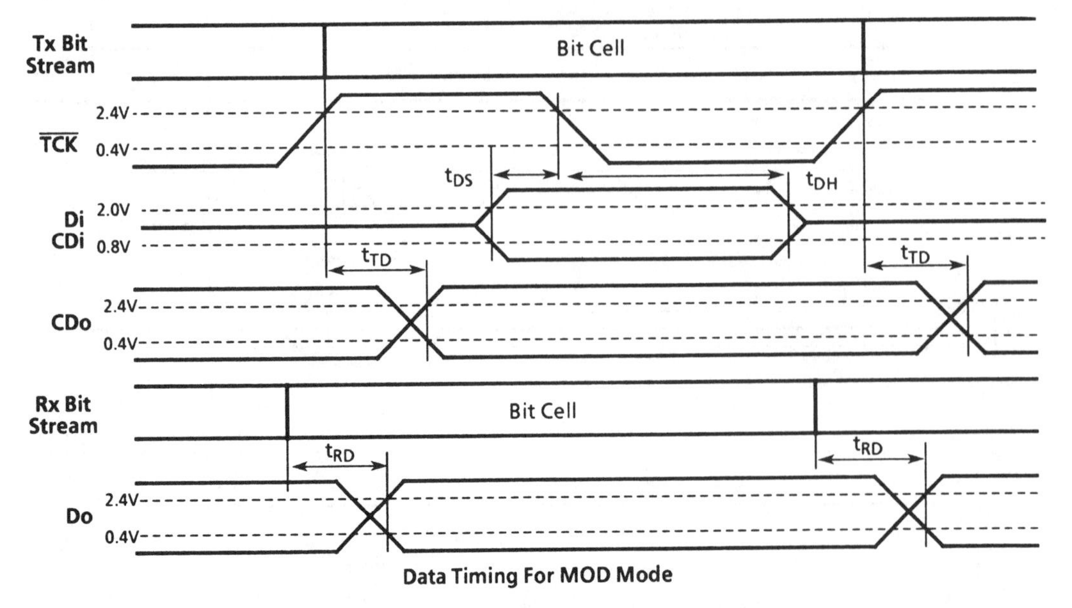

Data Timing For MOD Mode

Mode	Function
SLV	**SLAVE** - The chip timebase is extracted from the received line data and the external 10.24 MHz crystal is phase locked to it to provide clocks for the entire device and are output for the external system to synchronize to.
MAS	**MASTER** - The timebase is derived from the externally supplied data clocks and 10.24 MHz clock which must be frequency locked. The transmit data is synchronized to the system timing with the receive data recovered by a clock extracted from the receive data and resynchronized to the system timing.
DUAL	**DUAL PORT** - Both the CD and DV ports are active with the CD port transferring the C & D channels and the DV port transferring the B1& B2 channels.
SINGL	**SINGLE PORT** - The B1& B2, C and D channels are all transferred through the DV port. The CD port is disabled and CDSTi should be pulled high.
MOD	**MODEM** - Baseband operation at 80 or 160 kbits/s. The line data is received and transmitted through the DV port at the baud rate selected. The C-channel is transferred through CD port also at the baud rate and is synchronized to the $\overline{CLD}$ output.
DN	**DIGITAL NETWORK** - Intended for use in the digital network with the DV and CD ports operating at 2.048 Mbits/s and the line at 80 or 160 kbits/s configured according to the applicable ISDN recommendation.
D-C	**D BEFORE, C-CHANNEL** - The D-channel is transferred before the C channel following $\overline{F0}$.
C-D	**C BEFORE D-CHANNEL** - The C-channel is transferred before the D channel following $\overline{F0}$.
ODE	**OUTPUT DATA ENABLE** - When mode 7 is selected, the DV and CD ports are put in high impedance state. This is intended for power-up reset to avoid bus contention and possible damage to the device during the initial random state in a daisy chain configuration of DNICs. In all the other modes of operation DV and CD ports are enabled during the appropriate channel times.

Mode Definitions

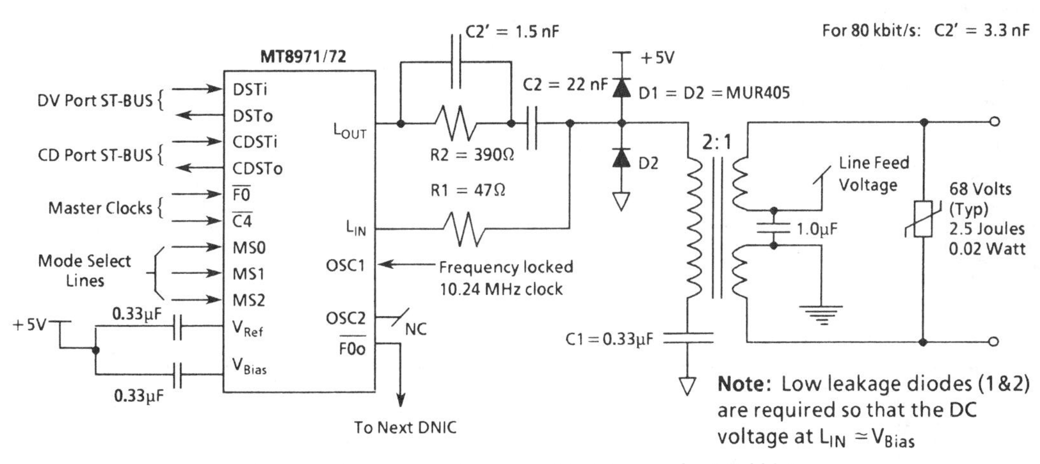

Typical Connection Diagram - MAS/DN Mode, 160 kbit/s

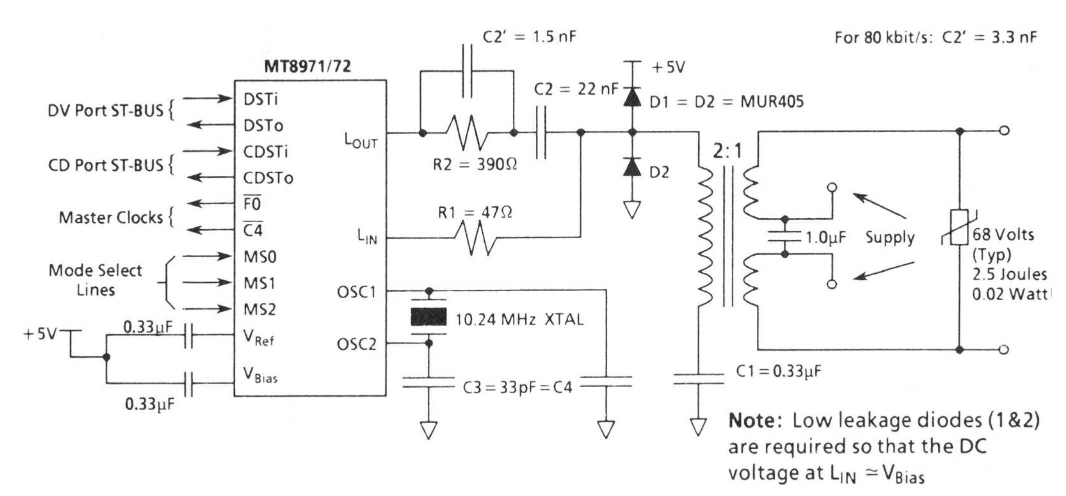

Typical Connection Diagram - SLV/DN Mode, 160 kbit/s

Mitel MT8972B
Digital Network Interface Circuit Preliminary Information

- Full duplex transmission over a single twisted pair
- Selectable 80 or 160 kbit/s line rate
- Adaptive echo cancellation
- Up to 5 km at 80 kbit/s and 4 km at 160 kbit/s
- ISDN-compatible (2B + D) data format
- Transparent modem capability
- Frame synchronization and clock extraction
- MITEL ST-BUS compatible
- Features unique to MT8972B
  - Fully backward compatible with MT8972A
  - Hardware disable of LOUT signal
  - Precanceller bypass option
  - Extended temperature range
- Low power consumption (typically 50mW)
- Single 5V power supply

APPLICATIONS
- Digital subscriber lines
- High-speed data transmission over twisted wires
- Digital PABX line cards and telephone sets
- 80 or 160 kbit/s single-chip modem

ABSOLUTE MAXIMUM RATINGS** - Voltages are with respect to ground (V_{SS}) unless otherwise stated.

	Parameter	Symbol	Min	Max	Units
1	Supply Voltage	V_{DD}	-0.3	7	V
2	Voltage On Any Pin (other than supply)	V_{Max}	-0.3	$V_{DD}+0.3$	V
3	Current On Any Pin (other than supply)	I_{Max}		40	mA
4	Storage Temperature	T_{ST}	-65	+150	°C
5	Package Power Dissipation (Derate 16mW/°C above 75°C)	P_{Diss}		750	mW

**Exceeding these values may cause permanent damage. Functional operation under these conditions is not implied.

RECOMMENDED OPERATING CONDITIONS - Voltages are with respect to ground (V_{SS}) unless otherwise stated.

	Characteristics	Sym	Min	Typ*	Max	Units	Test Conditions
1	Operating Supply Voltage	V_{DD}	4.75	5.00	5.25	V	
2	Operating Temperature	T_{OP}	−40		+85	°C	
3	Input High Voltage (except OSC1)	V_{IH}	2.4		V_{DD}	V	for 400 mV noise margin
4	Input Low Voltage (except OSC1)	V_{IL}	0		0.4	V	for 400 mV noise margin

*Typical figures are at 25°C and are for design aid only: not guaranteed and not subject to production testing.
†Parameters over recommended temperature & power supply voltage ranges.

DC ELECTRICAL CHARACTERISTICS† - Voltages are with respect to ground (V_{SS}) unless otherwise stated.

		Characteristics	Sym	Min	Typ*	Max	Units	Test Conditions
1	OUTPUTS	Operating Supply Current	I_{DD}		10	15	mA	
2		Output High Voltage (ex OSC2)	V_{OH}	2.4			V	$I_{OH}=10mA$
3		Output High Current (except OSC2)	I_{OH}	10	15		mA	Source current. $V_{OH}=2.4V$
4				8	12		mA	Source current. $V_{OH}=3.0V$
5		Output High Current - OSC2	I_{OH}	10			µA	Source current $V_{OH}=3.5V$
6		Output Low Voltage (ex OSC2)	V_{OL}			0.4	V	$I_{OL}=5mA$
7		Output Low Current (except OSC2)	I_{OL}	5	7.5		mA	Sink current. $V_{OL}=0.4V$
8				20	30		mA	Sink current. $V_{OL}=2.0V$
9		Output Low Current - OSC2	I_{OL}	10			µA	Sink current. $V_{OL}=1.5V$
10		High Imped. Output Leakage	I_{OZ}			10	µA	$V_{IN}=V_{SS}$ to V_{DD}
11		Output Voltage (V_{Ref})	V_O		V_{Bias}-1.8		V	
12		(V_{Bias})			$V_{DD}/2$		V	
13	INPUTS	Input High Voltage (ex OSC1)	V_{IH}	2.0			V	
14		Input Low Voltage (ex OSC1)	V_{IL}			0.8	V	
15		Input Leakage Current	I_{IL}			10	µA	$V_{IN}=V_{SS}$ to V_{DD}
16		Input Pulldown Impedance L_{OUT} DIS and Precan	Z_{PD}		50		kΩ	

*Typical figures are at 25°C and are for design aid only: not guaranteed and not subject to production testing.
†Parameters over recommended temperature & power supply voltage ranges.

AC ELECTRICAL CHARACTERISTICS - Voltages are with respect to ground (V_{SS}) unless otherwise stated.

		Characteristics		Sym	Min	Typ*	Max	Units	Test Conditions
1	INPUTS	Input Voltage	(L_{IN})	V_{IN}			5.0	V_{pp}	
2		Input Current	(L_{IN})	I_{IN}	-10		+10	μA	f_{Baud} = 160 kHz
3		Input Impedance	(L_{IN})	Z_{IN}	20		40	kΩ	f_{Baud} = 160 kHz
4		Crystal/Clock Frequency		f_C		10.24		MHz	
5		Crystal/Clock Tolerance		T_C	-100	0	+100	ppm	
6a		Crystal/Clock Duty Cycle		DC_C	40	50	60	%	Normal temp. & V_{DD}
6b		Crystal/Clock Duty Cycle		DC_C	45	50	55	%	Recommended at max./ min. temp. & V_{DD}
7		Crystal/Clock Loading		C_L		33	50	pF	From OSC1 & OSC2 to V_{SS}.
8	OUTPUTS	Output Capacitance	(L_{OUT})	C_o		8		pF	
9		Load Resistance	(L_{OUT}) (V_{Bias}, V_{Ref})	$R_{L_{out}}$		500 100		Ω kΩ	
10		Load Capacitance	(L_{OUT}) (V_{Bias}, V_{Ref})	$C_{L_{out}}$	 0.1		20	pF μF	Capacitance to V_{Bias}.
11		Output Voltage	(L_{OUT})	V_o	3.2	4.3	4.6	V_{pp}	R_{Lout} = 500Ω, C_{Lout} = 20pF

† Timing is over recommended temperature & power supply voltages

* Typical figures are at 25°C and are for design aid only: not guaranteed and not subject to production testing.

AC ELECTRICAL CHARACTERISTICS† - Clock Timing - DN Mode

		Characteristics	Sym	Min	Typ*	Max	Units	Test Conditions
1		$\overline{C4}$ Clock Period	t_{C4P}		244		ns	
2		$\overline{C4}$ Clock Width High or Low	t_{C4W}	90	122	150	ns	In Master Mode - Note 1
3		Frame Pulse Setup Time	t_{F0S}	50			ns	
4		Frame Pulse Hold Time	t_{F0H}	50			ns	
5		Frame Pulse Width	t_{F0W}	172	244		ns	
6		10.24 MHz Clock Jitter (wrt $\overline{C4}$)	J_C	−15		+15	ns	Note 2

† Timing is over recommended temperature & power supply voltages

* Typical figures are at 25°C and are for design aid only: not guaranteed and not subject to production testing.

Notes:
1) When operating as a SLAVE the C4 clock has a 40% duty cycle.
2) When operating in MAS/DN Mode, the $\overline{C4}$ and Oscillator clocks must be externally frequency-locked (i.e., $F_C = 2.5 \times f_{C4}$). The relative phase between these two clocks (Φ in Fig. 4) is not critical and may vary from 0 ns to t_{C4P}. However, the relative jitter must be less than J_C (see Figure 4).

PIN CONNECTIONS

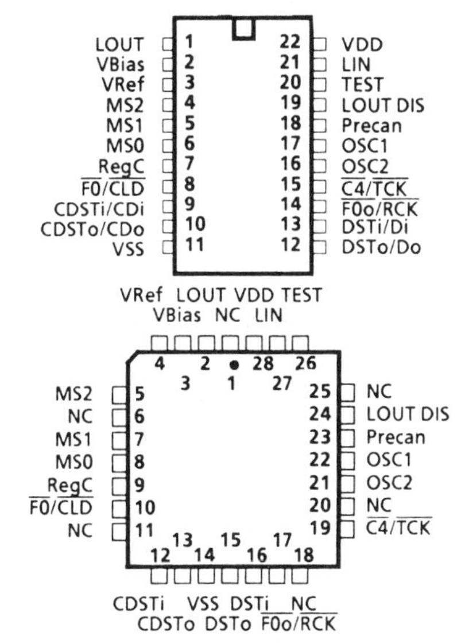

Ordering Information −40°C to +85°C

MT8972BC	22 Pin Cerdip
MT8972BE	22 Pin Plastic DIP
MT8972BP	28 Pin Plastic LCC

Functional Block Diagram

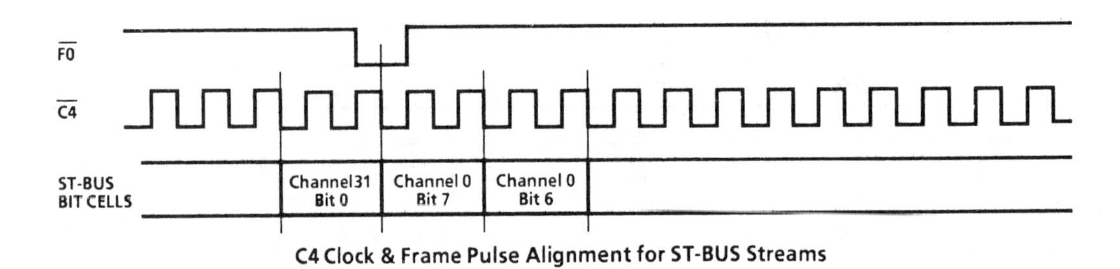

C4 Clock & Frame Pulse Alignment for ST-BUS Streams

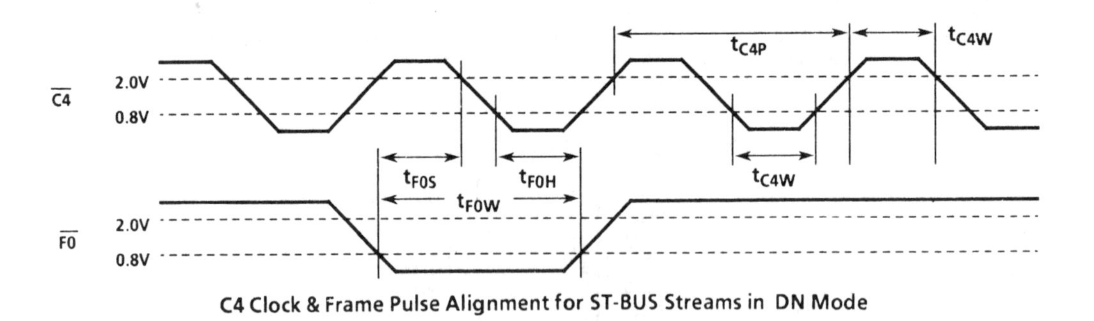

C4 Clock & Frame Pulse Alignment for ST-BUS Streams in DN Mode

AC ELECTRICAL CHARACTERISTICS† - Clock Timing - MOD Mode

	Characteristics	Sym	80 kbit/s			160 kbit/s			Units	Test Conditions
			Min	Typ*	Max	Min	Typ*	Max		
1	$\overline{TCK}$/$\overline{RCK}$ Clock Period	t_{CP}		12.5			6.25		µs	
2	$\overline{TCK}$/$\overline{RCK}$ Clock Width	t_{CW}		6.25			3.125		µs	
3	$\overline{TCK}$/$\overline{RCK}$ Clock Transition Time	t_{CT}		20			20		ns	$C_L = 40pF$
4	$\overline{CLD}$ to $\overline{TCK}$ Setup Time	t_{CLDS}		3.125			1.56		µs	
5	$\overline{CLD}$ to $\overline{TCK}$ Hold Time	t_{CLDH}		3.125			1.56		µs	
6	$\overline{CLD}$ Width Low	t_{CLDW}		6.05			2.925		µs	
7	$\overline{CLD}$ Period	t_{CLDP}		$8xt_{CP}$			$8xt_{CP}$		µs	

†Timing is over recommended temperature & power supply voltage ranges.
* Typical figures are at 25°C, for design aid only: not guaranteed and not subject to production testing.

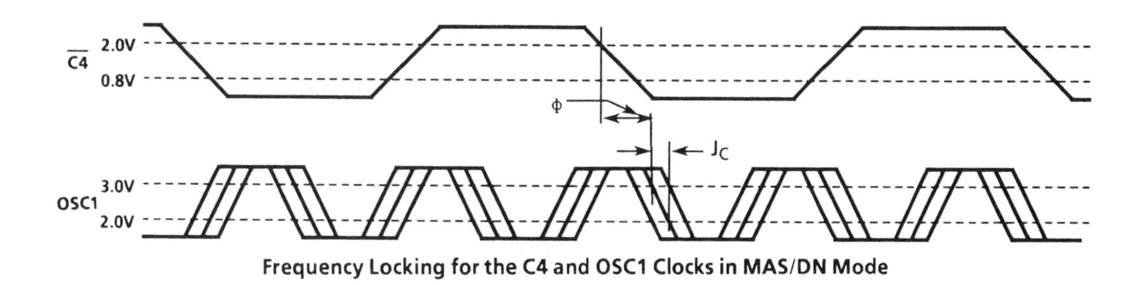

Frequency Locking for the C4 and OSC1 Clocks in MAS/DN Mode

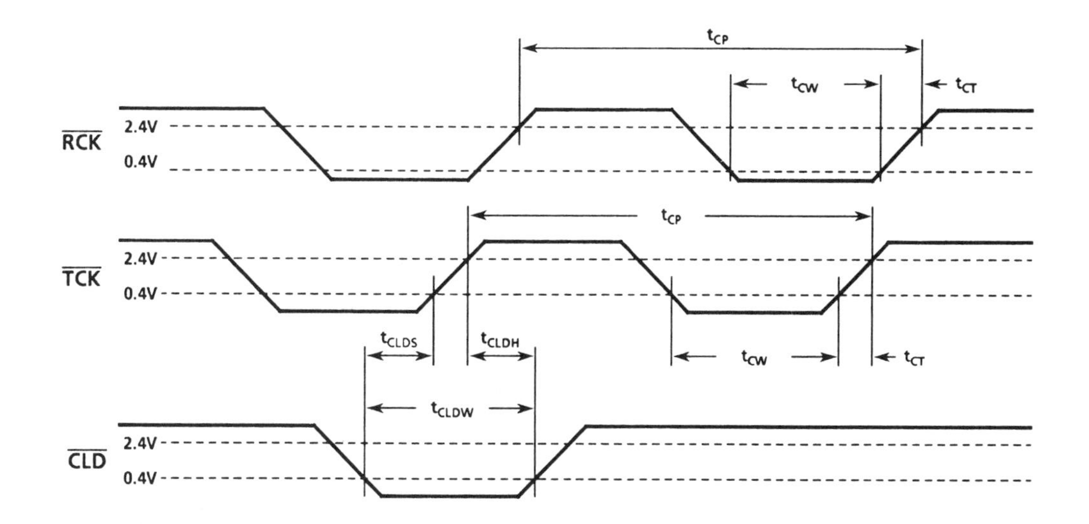

Note 1: TCK and CLD are generated on chip and provide the data clocks for the CD port and the transmit section of the DV port. RCK, also generated on chip, is extracted from the receive data and only clocks out the data at the D_o output and may be skewed with respect to TCK due to end-to-end delay.

Note 2: At the slave end TCK is locked to, and in phase with, RCK.

RCK, TCK & CLD Timing For MOD Mode

AC ELECTRICAL CHARACTERISTICS† - Data Timing - DN Mode

	Characteristics	Sym	Min	Typ*	Max	Units	Test Conditions
1	DSTi/CDSTi Data Setup Time	t_{RS}		100		ns	
2	DSTi/CDSTi Data Hold Time	t_{RH}		100		ns	
3a	DSTo/CDSTo Data Delay	t_{TD}		100		ns	$C_L = 40pF$
3b	DSTo/CDSTo High Z to Data Delay	t_{ZTD}		100		ns	$C_L = 40pF$

†Timing is over recommended temperature & power supply voltage ranges.
* Typical figures are at 25°C, for design aid only: not guaranteed and not subject to production testing.

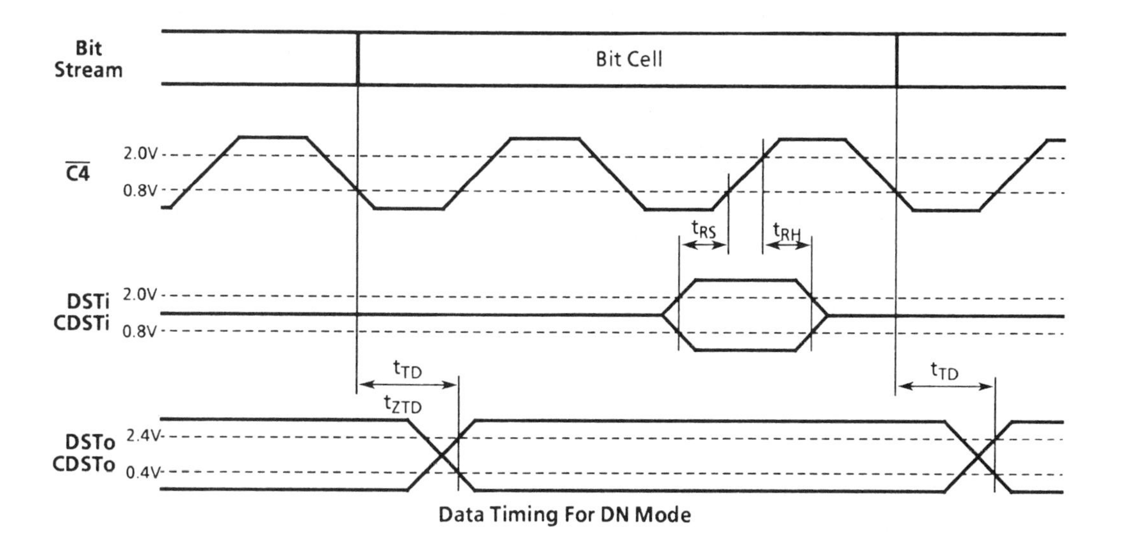

Data Timing For DN Mode

AC ELECTRICAL CHARACTERISTICS† - Data Timing - MOD Mode

	Characteristics	Sym	80 kbit/s			160 kbit/s			Units	Test Conditions
			Min	Typ*	Max	Min	Typ*	Max		
1	Di/CDi Data Setup Time	t_{DS}	150			150			ns	
2	Di/CDi Data Hold Time	t_{DH}	4.5			2.5			µs	
3	Do Data Delay Time	t_{RD}		100			100		ns	$C_L = 40pF$
4	CDo Data Delay Time	t_{TD}		100			100		ns	$C_L = 40pF$

†Timing is over recommended temperature & power supply voltage ranges.
* Typical figures are at 25°C, for design aid only: not guaranteed and not subject to production testing.

PERFORMANCE CHARACTERISTICS

	Characteristics	Sym	Min	Typ*	Max	Units	Test Conditions
1	Allowable Attenuation for Bit Error Rate of 10^{-6} (Note 1)	A_{fb}	0	40	33	dB	SNR ≥ 16.5dB (300kHz bandlimited noise)
2	Line Length at 80 kbit/s -24 AWG -26 AWG	L_{80}		5.0 3.4		km	attenuation - 6.9 dB/km attenuation - 10.0 dB/km
3	Line Length at 160 kbit/s -24 AWG -26 AWG	L_{160}		4.0 3.0		km	attenuation - 8.0 dB/km attenuation - 11.5 dB/km

Note 1: Attenuation measured from Master L_{OUT} to Slave L_{IN} at $\frac{1}{4}$ baud frequency.
* Typical figures are at 25°C, for design aid only: not guaranteed and not subject to production testing.

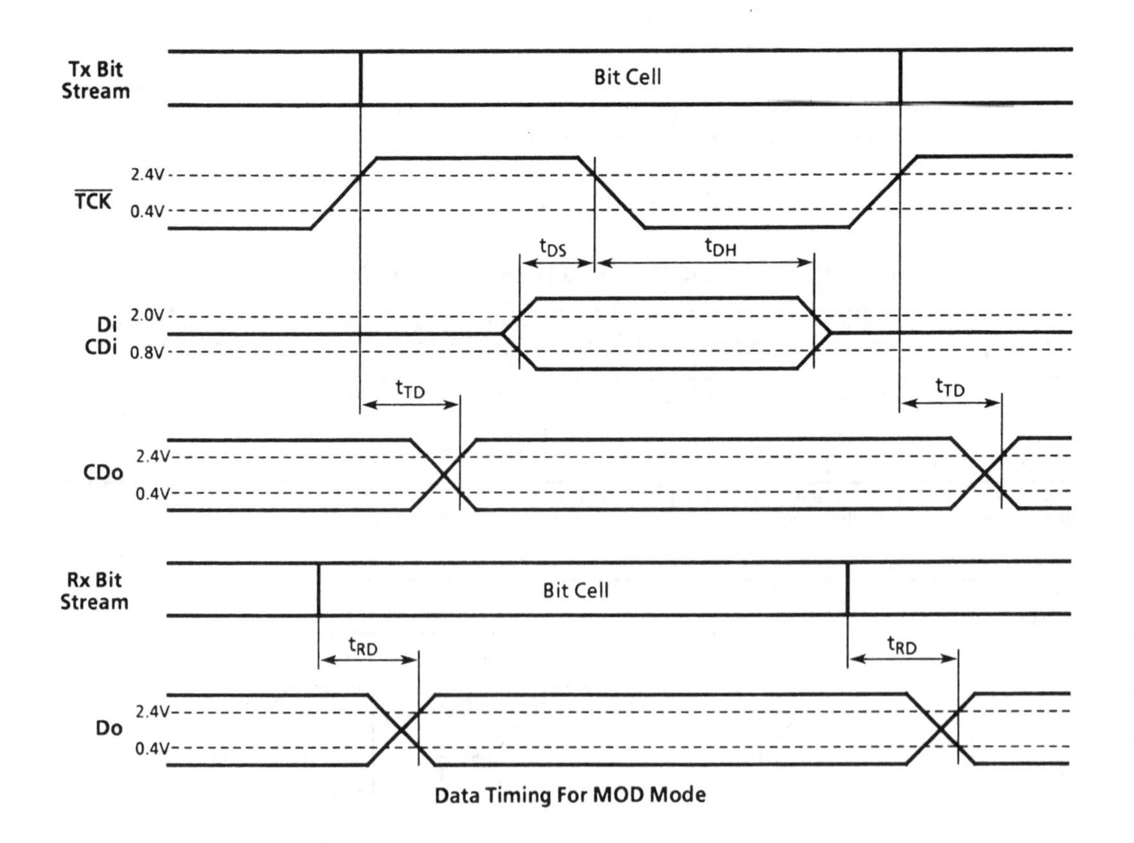

Data Timing For MOD Mode

Mode	Function
SLV	**SLAVE** - The chip timebase is extracted from the received line data and the external 10.24 MHz crystal is phase locked to it to provide clocks for the entire device and are output for the external system to synchronize to.
MAS	**MASTER** - The timebase is derived from the externally supplied data clocks and 10.24 MHz clock which must be frequency locked. The transmit data is synchronized to the system timing with the receive data recovered by a clock extracted from the receive data and resynchronized to the system timing.
DUAL	**DUAL PORT** - Both the CD and DV ports are active with the CD port transferring the C&D channels and the DV port transferring the B1& B2 channels.
SINGL	**SINGLE PORT** - The B1& B2, C and D channels are all transferred through the DV port. The CD port is disabled and CDSTi should be pulled high.
MOD	**MODEM** - Baseband operation at 80 or 160 kbits/s. The line data is received and transmitted through the DV port at the baud rate selected. The C-channel is transferred through the CD port also at the baud rate and is synchronized to the $\overline{CLD}$ output.
DN	**DIGITAL NETWORK** - Intended for use in the digital network with the DV and CD ports operating at 2.048 Mbits/s and the line at 80 or 160 kbits/s configured according to the applicable ISDN recommendation.
D-C	**D BEFORE C-CHANNEL** - The D-channel is transferred before the C-channel following $\overline{F0}$.
C-D	**C BEFORE D-CHANNEL** - The C-channel is transferred before the D-channel following $\overline{F0}$.
ODE	**OUTPUT DATA ENABLE** - When mode 7 is selected, the DV and CD ports are put in high impedance state. This is intended for power-up reset to avoid bus contention and possible damage to the device during the initial random state in a daisy chain configuration of DNICs. In all the other modes of operation DV and CD ports are enabled during the appropriate channel times.

Mode Definitions

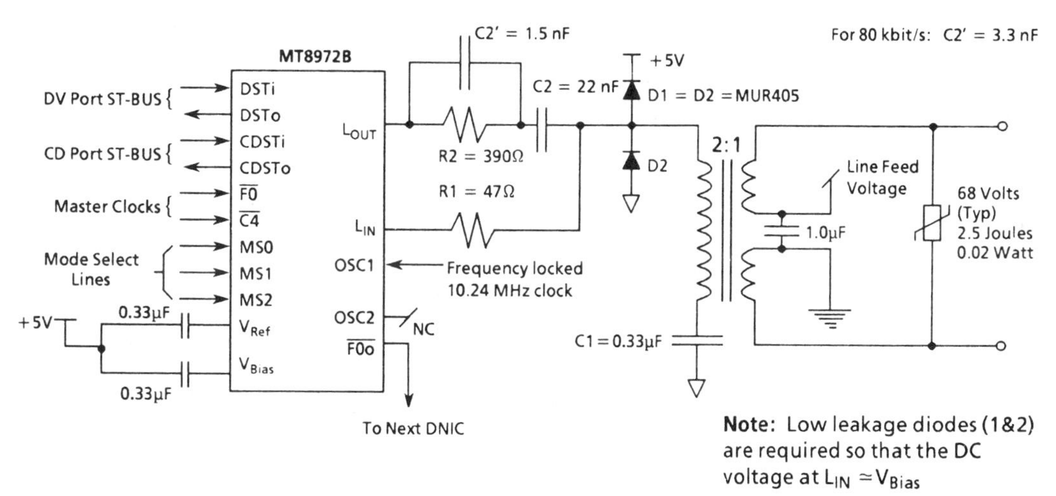

Note: Low leakage diodes (1&2) are required so that the DC voltage at $L_{IN} \simeq V_{Bias}$

Typical Connection Diagram - MAS/DN Mode, 160 kbit/s

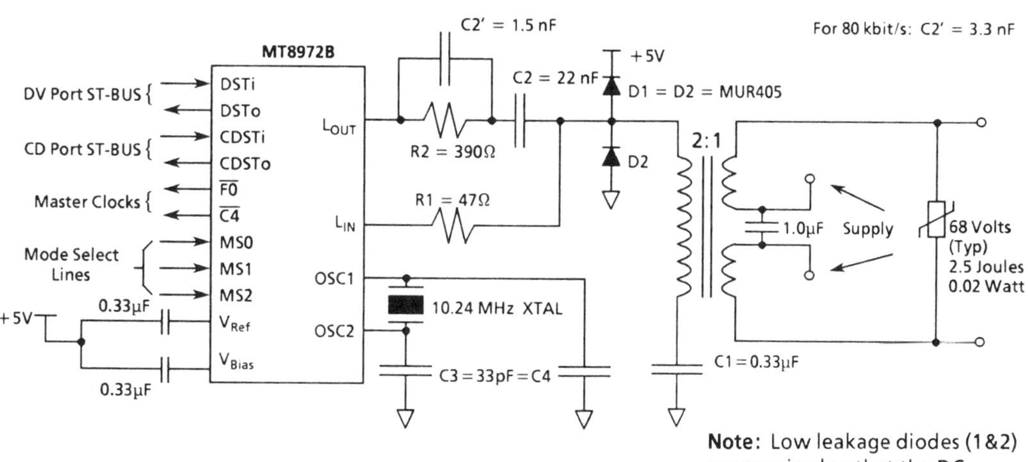

Note: Low leakage diodes (1&2) are required so that the DC voltage at $L_{IN} \simeq V_{Bias}$

Typical Connection Diagram - SLV/DN Mode, 160 kbit/s

Mitel MH89726/728
MT8972 Loop-Extender Circuits
Advance Information

- Operates from single or dual power supply
- MH89726 extends transmission performance for the MT8972 (DNIC) at 160 kbit/s line rate
- MH89728 extends transmission performance for the MT8972 (DNIC) at 80 kbit/s line rate
- MH89726 and MH89728 are pin for pin compatible with each other
- Compact SIL package
- Over 6km loop range on 14 AWG

APPLICATIONS
- Digital subscriber lines
- Digital PABX line cards and telephone sets
- High-speed limited-distance modem
- ISDN U-interface

PIN CONNECTIONS

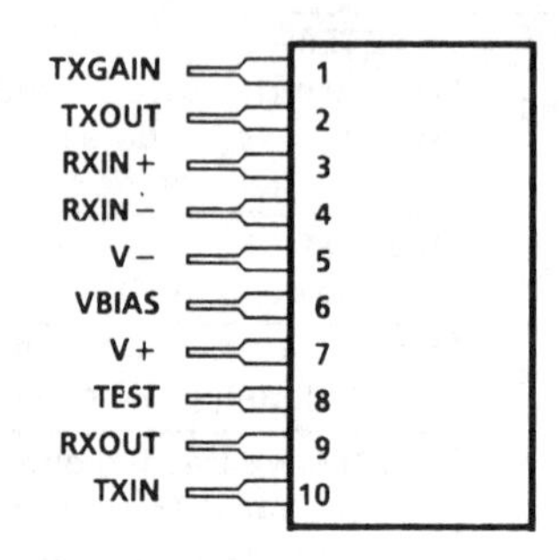

Ordering Information

MH89726/728 10 Pin SIL Package

0°C to 70C°

RECOMMENDED OPERATING CONDITIONS

	Characteristics	Sym	Min	Typ	Max	Units	Test Conditions
1	+15 V Supply	V_{DD}	13.5	15.0	16.5	V	Note 1
2	- 15 V Supply	V_{SS}	-16.5	-15.0	-13.5	V	Note 2
3	Standby Current	I_S		3.0		mA	

Note 1: A single voltage supply of +30V±10% can be used on V_{DD} with V_{Bias} grounded through a 0.33 µF capacitor and V_{SS} grounded.
Note 2: A single voltage supply of -30V±10% can be used on V_{SS} with V_{Bias} grounded through a 0.33 µF capacitor and V_{DD} grounded.

PIN DESCRIPTIONS

Pin #	Pin Name	Description
1	TX_{GAIN}	**Transmit Gain.** To be connected to Pin 2. An increase in gain can be achieved by connecting an external resistor R_{EXT} between Pin 1 and Pin 2. The resultant gain is calculated using; A= (R/24) + 2.8, where A=Gain and R= R_{EXT} (kΩ).
2	TX_{OUT}	**Transmit Output.** Connect to the termination network.
3	RX_{IN-}	**Negative Receive Signal.** Connect to TX_{OUT}, Pin 2.
4	RX_{IN+}	**Positive Receive Signal.** Connected to line transformer.
5	V -	**Negative power supply.**
6	V_{Bias}	**Internal Bias Voltage.** Connect to GND through 0.33µF for single power supplies. Connect to GND directly for split power supplies.
7	V +	**Positive power supply.**
8	TEST	**Test.** Used for production testing. Leave unconnected.
9	RX_{OUT}	**Receive Output.** Connect to L_{IN} (Pin 21) of MT8972.
10	TX_{IN}	**Transmit Input.** Connect to L_{OUT} (Pin 1) of MT8972.

AC ELECTRICAL CHARACTERISTICS

	Characteristics	Sym	Min	Typ	Max	Units	Test Conditions
1	Active Current	I_A		7.0		mA	@ 60 kHz for 80kbit/s @120 kHz for 160kbit/s
2	TX Gain	A_{TX}		9.0		dB	@ 60 kHz for 80kbit/s @120 kHz for 160kbit/s
3	TX Phase Angle	Φ_{TX}		-170.0		°	@ 60 kHz for 80kbit/s @120 kHz for 160kbit/s
4	RX Gain	A_{RX}		0.0		dB	@ 60 kHz for 80kbit/s @120 kHz for 160kbit/s
5	RX Phase Angle	Φ_{RX}		12.0		°	@ 60 kHz for 80kbit/s @120 kHz for 160kbit/s
6	Line Power			10.0		dBm	With 2:1 transformer

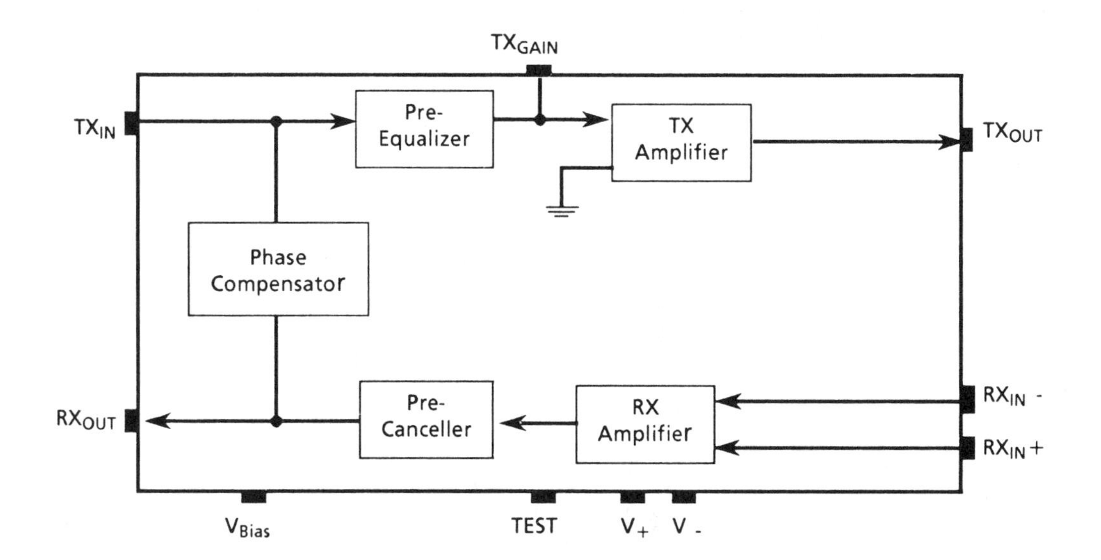

MH89726 and MH89728 Functional Block Diagram

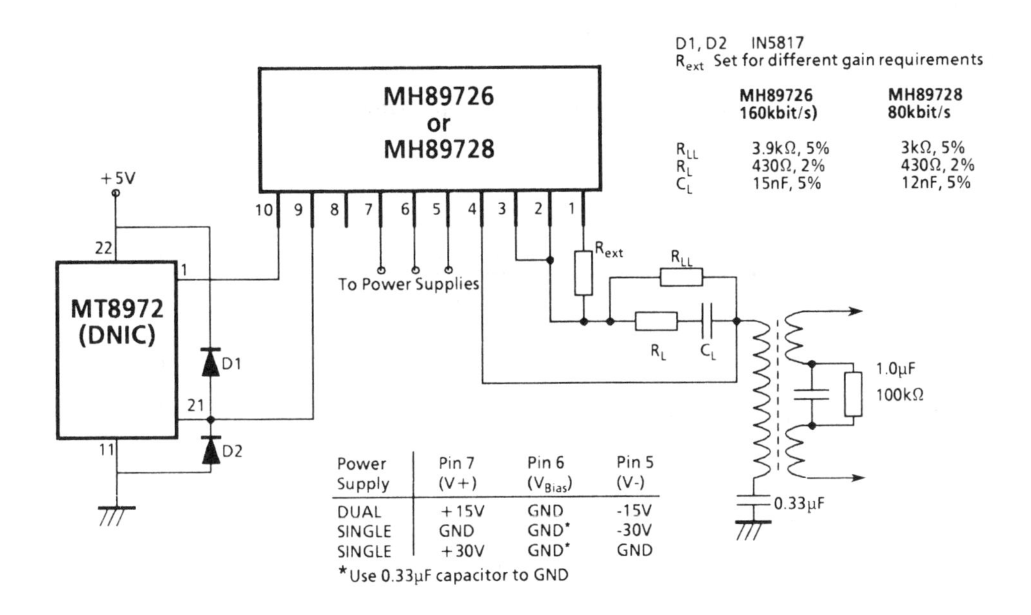

Typical Connection Diagram for MT8972 and MH89726/728

MH89726 (160 kbit/s)		
CONFIGURATION \ CABLE	24 AWG (8.0 dB/km at 120kHz)	26 AWG (11.5 dB/km at 120kHz)
DNIC to DNIC	0.0 to 4.1 km	0.0 to 3.0 km
DNIC with LEC to DNIC with LEC	0.65 to 5.1 km	0.5 to 3.6 km

Typical Transmission Performance at

MH89728 (80 kbit/s)		
CONFIGURATION \ CABLE	24 AWG (6.9 dB/km at 60kHz)(Note 1)	26 AWG (10.0 dB/km at 60kHz)
DNIC to DNIC	0.0 to 5.2 km	0.0 to 3.4 km
DNIC with LEC to DNIC and LEC	0.9 to 6.5 km	0.65 to 4.1 km

Typical Transmission Performance at

Note 1: The attenuation of the cable as specified by Bell System Technical Reference PUB 62411.

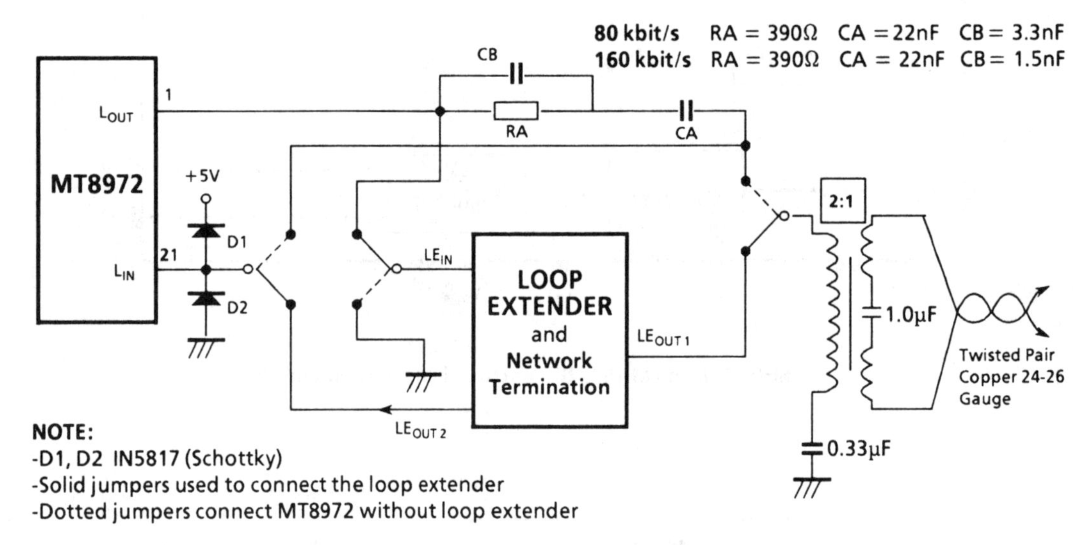

Typical Application of the Loop Extender

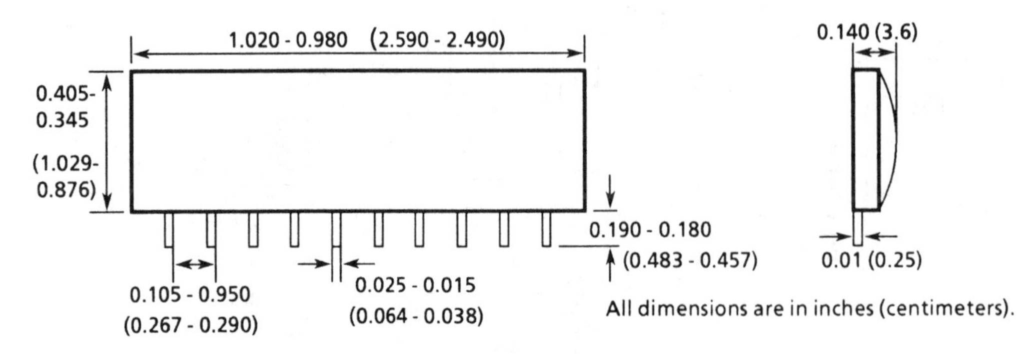

Mechanical Data

Mitel MT8960/61/62/63/64/65/66/67

Integrated PCM Filter/Codec

- ST-BUS compatible
- Transmit/Receive filters and PCM Codec in one IC
- Meets AT&T D3/D4 and CCITT G711 and G712
- μ-law: MT8960/62/64/66
  A-law: MT8961/63/65/67
- Low power consumption:
  Op.: 30mW typ.
  Stby.: 2.5mW typ
- Digital Coding Options:
  MT8964/65/66/67 CCITT Code
  MT8960/61/62/63 Alternative Code
- Digitally controlled gain adjust of both filters
- Analog and digital loopback
- Filters and codec independently user accessible for testing
- Powerdown mode available
- 2.048 MHz master clock input
- Up to six uncommitted control outputs
- $\pm$5V$\pm$5% power supply

ABSOLUTE MAXIMUM RATINGS*

	Parameter	Symbol	Min	Max	Units
1	DC Supply Voltages	V_{DD}-GNDD	-0.3	+6.0	V
		V_{EE}-GNDD	-6.0	+0.3	V
2	Reference Voltage	V_{Ref}	GNDA	V_{DD}	V
3	Analog Input	V_X	V_{EE}	V_{DD}	V
4	Digital Inputs	Except CA	GNDD-0.3	V_{DD}+0.3	V
		CA	V_{EE}-0.3	V_{DD}+0.3	V
5	Output Voltage	SD0-2	GNDD-0.3	V_{DD}+0.3	V
		SD3	V_{EE}-0.3	V_{DD}+0.3	V
		SD4-5	V_{EE}-0.3	V_{DD}+0.3	V
6	Current On Any Pin	I_I		20	mA
7	Storage Temperature	T_S	-55	+125	°C
8	Power Dissipation at 25°C (Derate 16 mW/°C above 75°C)	P_{Diss}		500	mW

*Exceeding these values may cause permanent damage. Functional operation under these conditions is not implied.

RECOMMENDED OPERATING CONDITIONS

- Voltages are with respect to GNDD unless otherwise stated

	Characteristics		Sym	Min	Typ*	Max	Units	Comments
1	Supply Voltage		V_{DD}	4.75	5.0	5.25	V	
			V_{EE}	-5.25	-5.0	-4.75	V	
			V_{Ref}		2.5		V	See Note 1
2	Voltage On Digital Ground		VGNDD	-0.1	0.0	+0.1	Vdc	Ref. to GNDA
				-0.4	0.0	+0.4	Vac	Ref. to GNDA 400ns max. duration in 125µs cycle
3	Operating Temperature		T_O	0		+70	°C	
4	Operating Current	V_{DD}	I_{DD}		3.0	4.0	mA	All digital inputs at V_{DD}
		V_{EE}	I_{EE}		3.0	4.0	mA	or GNDD (or V_{EE} for CA)
		V_{Ref}	I_{Ref}		2.0		µA	Mean current
5	Standby Current	V_{DD}	I_{DDO}		0.25	1.0	mA	All digital inputs at V_{DD}
		V_{EE}	I_{EEO}		0.25	1.0	mA	or GNDD (or V_{EE} for CA)

Note 1: Temperature coefficient of V_{Ref} should be better than 100 ppm/°C.

DC ELECTRICAL CHARACTERISTICS - Voltages are with respect to GNDD unless otherwise stated.

T_A=0 to 70°C, V_{DD}=5V±5%, V_{EE}=-5V±5%, V_{Ref}=2.5V±0.5%, GNDA=GNDD=0V, Clock Frequency =2.048MHz, Outputs unloaded unless otherwise specified

		Characteristics		Sym	Min	Typ*	Max	Units	Test Conditions
1	DIGITAL	Input Current	Except CA	I_I			10.0	µA	V_{IN} = GNDD to V_{DD}
			CA	I_{IC}			10.0	µA	V_{IN} = V_{EE} to V_{DD}
2		Input Low Voltage	Except CA	V_{IL}	0.0		0.8	V	
			CA	V_{ILC}	V_{EE}		V_{EE}+1.2	V	
3		Input High Voltage	All Inputs	V_{IH}	2.4		5.0	V	
4		Input Intermediate Voltage	CA	V_{IIC}	0.0		0.8	V	
5		Output Leakage Current (Tristate)	DSTo	I_{OZ}		±0.1		µA	Output High Impedence
			SD3-5				10.0	µA	

* Typical figures are at 25°C with nominal ±5V supplies. For design aid only: not guaranteed and not subject to production testing.

DC ELECTRICAL CHARACTERISTICS (cont'd)

		Characteristics		Sym	Min	Typ*	Max	Units	Test Conditions
6	DIGITAL	Output Low Voltage	DSTo	V_{OL}			0.4	V	I_{OUT} = 1.6 mA
			SD_{0-2}	V_{OL}			1.0	V	I_{OUT} = 1 mA
7		Output High Voltage	DSTo	V_{OH}	4.0			V	I_{OUT} = -100μA
			SD_{0-2}	V_{OH}	4.0			V	I_{OUT} = -1mA
8		Output Resistance	SD_{3-5}	R_{OUT}		1.0	2.0	KΩ	V_{OUT} = +1V
9		Output Capacitance	DSTo	C_{OUT}		4.0		pF	Output High Impedance
10	ANALOG	Input Current	V_X	I_{IN}			10.0	μA	$V_{EE} \leq V_{IN} \leq V_{CC}$
11		Input Resistance	V_X	R_{IN}		10.0		MΩ	
12		Input Capacitance	V_X	C_{IN}		30.0		pF	f_{IN} = 0 - 4 kHz
13		Input Offset Voltage	V_X	V_{OSIN}		+1.0		mV	See Note 2
14		Output Resistance	V_R	R_{OUT}			100	Ω	
15		Output Offset Voltage	V_R	V_{OSOUT}			100	mV	Digital Input= +0

Note 2: V_{OSIN} specifies the DC component of the digitally encoded PCM word.

AC ELECTRICAL CHARACTERISTICS - Voltages are with respect to GNDD unless otherwise stated

T_A=0 to 70°C, V_{DD}=5V±5%, V_{EE}=-5V±5%, V_{Ref}=2.5V±0.5%, GNDA=GNDD=0V, Clock Frequency=2.048 MHz, Outputs unloaded unless otherwise specified.

		Characteristics		Sym	Min	Typ*	Max	Units	Test Conditions
1	DIGITAL	Clock Frequency	C2i	f_C	2.046	2.048	2.05	MHz	See Note 3
2		Clock Rise Time	C2i	t_{CR}			50	ns	
3		Clock Fall Time	C2i	t_{CF}			50	ns	
4		Clock Duty Cycle	C2i		40	50	60	%	
5		Chip Enable Rise Time	$\overline{F1i}$	t_{ER}			100	ns	
6		Chip Enable Fall Time	$\overline{F1i}$	t_{EF}			100	ns	
7		Chip Enable Setup Time	$\overline{F1i}$	t_{ES}	50			ns	See Note 4
8		Chip Enable Hold Time	$\overline{F1i}$	t_{EH}	25			ns	See Note 4
9		Output Rise Time	DSTo	t_{OR}			100	ns	
10		Output Fall Time	DSTo	t_{OF}			100	ns	
11		Propagation Delay Clock to Output Enable	DSTo	t_{PZL}			122	ns	R_L = 10KΩ to V_{CC}
				t_{PZH}			122	ns	C_L = 100 pF
12		Propagation Delay Clock to Output	DSTo	t_{PLH}			100	ns	
				t_{PHL}			100	ns	
13		Input Rise Time	CSTi	t_{IR}			100	ns	
			DSTi				100	ns	
14		Input Fall Time	CSTi	t_{IF}			100	ns	
			DSTi				100	ns	
15		Input Setup Time	CSTi	t_{ISH}	25			ns	
			DSTi	t_{ISL}	0			ns	
16		Input Hold Time	CSTi	t_{IH}	60			ns	
			DSTi		60			ns	
17	DIGITAL	Propagation Delay Clock to SD Output	SD	t_{PCS}			400	ns	CL = 100 pF
18		SD Output Fall Time	SD	t_{SF}			200	ns	CL=20 pF
19		SD Output Rise Time	SD	t_{SR}			400	ns	
20		Digital Loopback Time DSTi to DSTo		t_{DL}			122	ns	

* Typical figures are at 25°C with nominal ±5V supplies. For design aid only: not guaranteed and not subject to production testing.

Note 3: The filter characteristics are totally dependent upon the accuracy of the clock frequency providing $\overline{F1i}$ is synchronized to C2i, the codec function is unaffected by changes in the clock frequency.

Note 4: This gives a 75 ns period, 50 ns before and 25 ns after the 50% point of C2i rising edge, when change in $\overline{F1i}$ will give an undetermined state to to the internally synchronized enable signal.

PIN CONNECTIONS

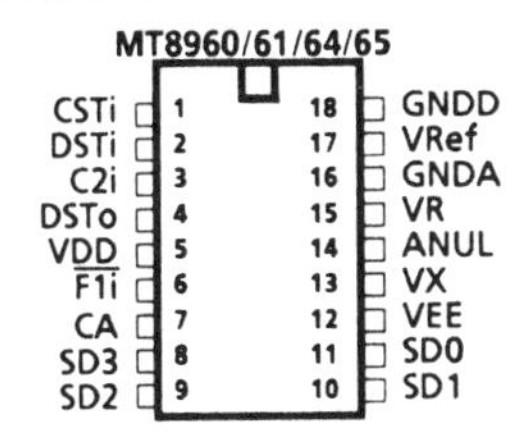

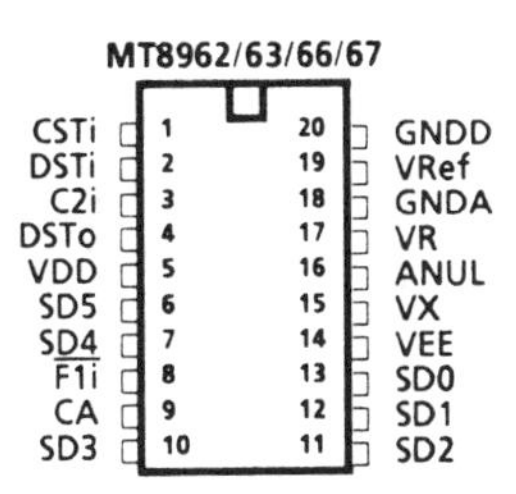

Ordering Information 0°C to +70°C

MT8960/61/64/65AC	18 Pin Cerdip
MT8960/61/64/65AE	18 Pin Plastic DIP
MT8962/63AC	20 Pin Cerdip
MT8962/63AE	20 Pin Plastic DIP
MT8962/63/66/67AS	20 Pin SOIC

AC ELECTRICAL CHARACTERISTICS

Transmit (A/D) Path - Voltages are with respect to GNDD unless otherwise stated.
T_A=0 to 70°C, V_{DD}=5V±5%, V_{EE}=-5V±5%, V_{Ref}=2.5V±0.5%, GNDA=GNDD=0V, Clock Frequency = 2.048MHz, Filter Gain Setting = 0dB, Outputs unloaded unless otherwise specified.

		Characteristics		Sym	Min	Typ*	Max	Units	Test Conditions
1	ANALOG	Analog Input at V_X equivalent to the overload decision level at the codec		V_{IN}					Level at codec:
						4.829		V_{PP}	μ-Law: 3.17 dBm0
						5.000		V_{PP}	A-Law: 3.14 dBm0
									See Note 6
2		Absolute Gain		G_{AX}	-0.25		+0.25	dB	0 dBm0 @ 1004 Hz
3		Deviation of Gain Adjustment			-0.1		+0.1	dB	From Nominal Value
4		Gain Variation	With Temp	G_{AXT}		0.01		dB	T_A=0°C to 70°C
			With Supplies	G_{AXS}		0.04		dB/V	
5		Gain Tracking (See Figure 5)	CCITT G712 Method 1)	GT_{X1}					Sinusoidal Level:
					-0.25		+0.25	dB	+3 to -20 dBm0
									Noise Signal Level:
					-0.25		+0.25	dB	-10 to -55 dBm0
					-0.50		+0.50	dB	-55 to -60 dBm0
			CCITT G712 (Method 2) AT&T	GT_{X2}					Sinusoidal Level:
					-0.25		+0.25	dB	+3 to -40 dBm0
					-0.50		+0.50	dB	-40 to -50 dBm0
					-1.50		+1.50	dB	-50 to -55 dBm0
6		Quantization Distortion (See Figure 6)	CCITT G712 (Method 1)	D_{QX1}					Noise Signal Level:
					28.00			dB	-3 dBm0
					35.60			dB	-6 to -27 dBm0
					33.90			dB	-34 dBm0
					29.30			dB	-40 dBm0
					14.20			dB	-55 dBm0

* Typical figures are at 25°C with nominal ±5V supplies. For design aid only: not guaranteed and not subject to production testing.

TRANSMIT (A/D) PATH (cont'd)

		Characteristics		Sym	Min	Typ*	Max	Units	Test Conditions
	A	Quantization	CCITT G712	D_{QX2}					Sinusoidal Input Level:
	N	Distortion	(Method 2)		35.30			dB	0 to -30 dBm0
	A	(cont'd)	AT&T		29.30			dB	-40 dBm0
	L				24.30			dB	-45 dBm0
7	O	Idle Channel	C-message	N_{CX}			18	dBrnC0	µ-Law Only
	G	Noise	Psophometric	N_{PX}			-67	dBm0p	CCITT G712
8		Single Frequency Noise		N_{SFX}			-56	dBm0	CCITT G712
9		Harmonic Distortion (2nd or 3rd Harmonic)					-46	dB	Input Signal: 0 dBm0 @ 1.02 kHz
10		Envelope Delay		D_{AX}			270	µs	@ 1004 Hz
11		Envelope Delay	1000-2600 Hz	D_{DX}		60		µs	Input Signal:
		Variation With	600-3000 Hz			150		µs	400-3200 Hz Sinewave
		Frequency	400-3200 Hz			250		µs	at 0 dBm0
12		Intermodulation Distortion	CCITT G712 50/60 Hz	IMD_{X1}			-55	dB	50/60 Hz @ -23 dBm0 and any signal within 300-3400 Hz at -9 dBm0
			CCITT G712 2 tone	IMD_{X2}			-41	dB	740 Hz and 1255 Hz @ -4 to -21 dBm0. Equal Input Levels
			AT&T	IMD_{X3}			-47	dB	2nd order products
			4 tone	IMD_{X4}			-49	dB	3rd order products
13		Gain Relative to	≤50 Hz	G_{RX}			-25	dB	0 dBm0 Input Signal
		Gain @ 1004 Hz	60 Hz				-30	dB	
			200 Hz		-1.8		0.00	dB	Transmit
			300-3000 Hz		-0.125		0.125	dB	Filter
			3200 Hz		-0.275		0.125	dB	Response
			3300 Hz		-0.350		0.030	dB	
			3400 Hz		-0.80		-0.100	dB	
			4000 Hz				-14	dB	
			≥4600 Hz				-32	dB	
14		Crosstalk D/A to A/D		CT_{RT}			-70	dB	0 dBm0 @ 1.02 kHz in D/A
15		Power Supply	V_{DD}	$PSSR_1$	33			dB	Input 50 mV_{RMS} at
		Rejection	V_{EE}	$PSSR_2$	35			dB	1.02 kHz
16		Overload Distortion							Input frequency = 1.02kHz

* Typical figures are at 25°C with nominal ±5V supplies. For design aid only: not guaranteed and not subject to production testing.

Note 6: 0dBm0 = 1.185 V_{RMS} for the µ-Law codec.
0dBm0 = 1.231 V_{RMS} for the A-Law codec.

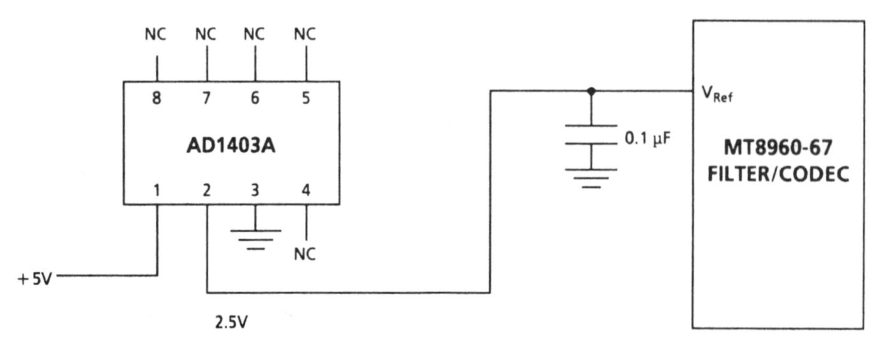

Typical Voltage Reference Circuit

AC ELECTRICAL CHARACTERISTICS

Receive (D/A) Path - Voltages are with respect to GNDD unless otherwise stated.

T_A=0 to 70°C, V_{DD}=5V±5%, V_{EE}=-5V±5%, V_{Ref}=2.5V±0.5%, GNDA=GNDD=0V, Clock Frequency = 2.048MHz, Filter Gain Setting = 0dB, Outputs unloaded unless otherwise specified.

		Characteristics		Sym	Min	Typ*	Max	Units	Test Conditions
1	ANALOG	Analog output at V_R equivalent to the overload decision level at codec		V_{OUT}		4.829 5.000		V_{pp} V_{pp}	Level at codec: µ-Law: 3.17 dBm0 A-Law: 3.14 dBm0 R_L=10 KΩ See Note 7
2		Absolute Gain		G_{AR}	-0.25		+0.25	dB	0 dBm0 @ 1004Hz
3		Deviation of Attenuation Adjustment			-0.10		+0.10	dB	From Nominal Value
4		Gain Variation	With Temp.	G_{ART}		0.01		dB	T_A=0°C to 70°C
			With Supplies	G_{ARS}		0.04		dB/V	
5		Gain Tracking	CCITT G712 (Method 1)	GT_{R1}	-0.25		+0.25	dB	Sinusoidal Level: +3 to -10 dBm0
					-0.25		+0.25	dB	Noise Signal Level: -10 to -55 dBm0
					-0.50		+0.50	dB	-55 to -60 dBm0
			CCITT G712 (Method 2) AT & T	GT_{R2}	-0.25		+0.25	dB	Sinusoidal Level: +3 to -40 dBm0
					-0.50		+0.50	dB	-40 to -50 dBm0
					-1.50		+1.50	dB	-50 to -55 dBm0
6		Quantization Distortion	CCITT G712 (Method 1)	D_{QR1}	28.00			dB	Noise Signal Level: -3 dBm0
					35.60			dB	-6 to -27 dBm0
					33.90			dB	-34 dBm0
					29.30			dB	-40 dBm0
					14.30			dB	-55 dBm0
			CCITT G712 (Method 2) AT & T	D_{QR2}	36.40			dB	Sinusoidal Input Level: 0 to -30 dBm0
					30.40			dB	-40 dBm0
					25.40			dB	-45 dBm0
7		Idle Channel Noise	C-message	N_{CR}			12	dBrnC0	µ-Law Only
			Psophometric	N_{PR}			-75	dBm0p	CCITT G712
8		Single Frequency Noise		N_{SFR}			-56	dBm0	CCITT G712
9		Harmonic Distortion (2nd or 3rd Harmonic)					-46	dB	Input Signal 0 dBm0 at 1.02 kHz
10		Intermodulation Distortion	CCITT G712 2 tone	IMD_{R2}			-41	dB	
			AT & T	IMD_{R3}			-47	dB	2nd order products
			4 tone	IMD_{R4}			-49	dB	3rd order products

* Typical figures are at 25°C with nominal ±5V supplies. For design aid only: not guaranteed and not subject to production testing.

RECEIVE (D/A) PATH (cont'd)

		Characteristics		Sym	Min	Typ*	Max	Units	Test Conditions
11	A N A L O G	Envelope Delay		D_{AR}			210	µs	@ 1004 Hz
12		Envelope Delay Variation with Frequency	1000-2600 Hz	D_{DR}		90		µs	Input Signal:
			600-3000 Hz			170		µs	400 - 3200 Hz digital
			400-3200 Hz			265		µs	sinewave at 0 dBm0
13		Gain Relative to Gain @ 1004 Hz	<200 Hz	G_{RR}			0.125	dB	0 dBm0 Input Signal
			200 Hz		-0.5		0.125	dB	
			300-3000 Hz		-0.125		0.125	dB	Receive
			3300 Hz		-0.350		0.030	dB	Filter
			3400 Hz		-0.80		-0.100	dB	Response
			4000 Hz				-14.0	dB	
			≥4600 Hz				-28.0	dB	
14		Crosstalk A/D to D/A		CT_{TR}			-70	dB	0 dBm0 @ 1.02 KHz in A/D
15		Power Supply Rejection	V_{DD}	$PSRR_3$	33			dB	Input 50 mV_{RMS} at
			V_{EE}	$PSRR_4$	35			dB	1.02 kHz
16		Overload Distortion							Input frequency = 1.02kHz

* Typical figures are at 25°C with nominal ±5V supplies. For design aid only: not guaranteed and not subject to production testing.

Note 7: 0dBm0 = 1.185 V_{RMS} for µ-Law codec and 0dBm0 = 1.231 V_{RMS} for A-Law codec.

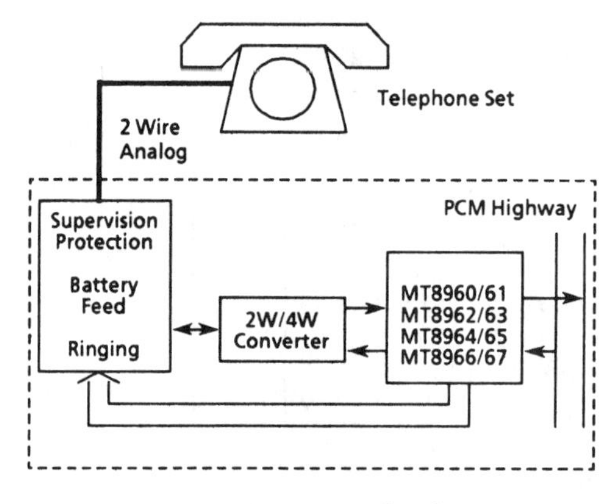

Typical Line Termination

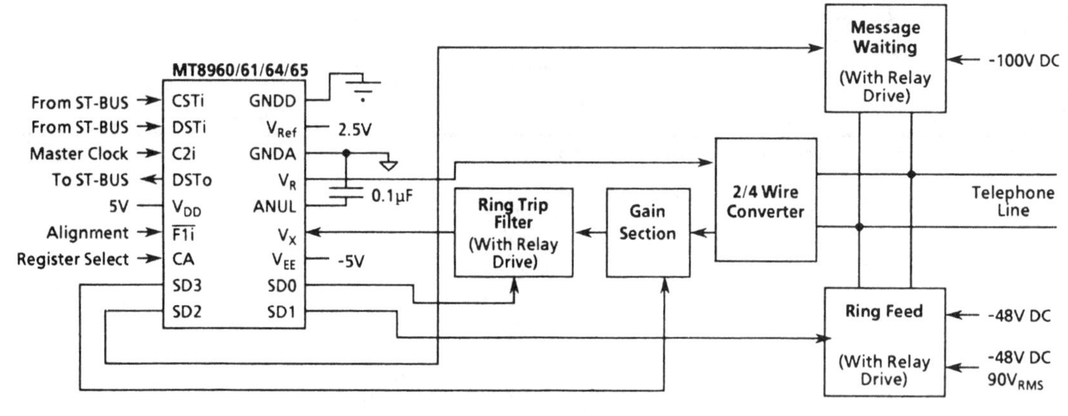

Typical Use of the Special Drive Outputs

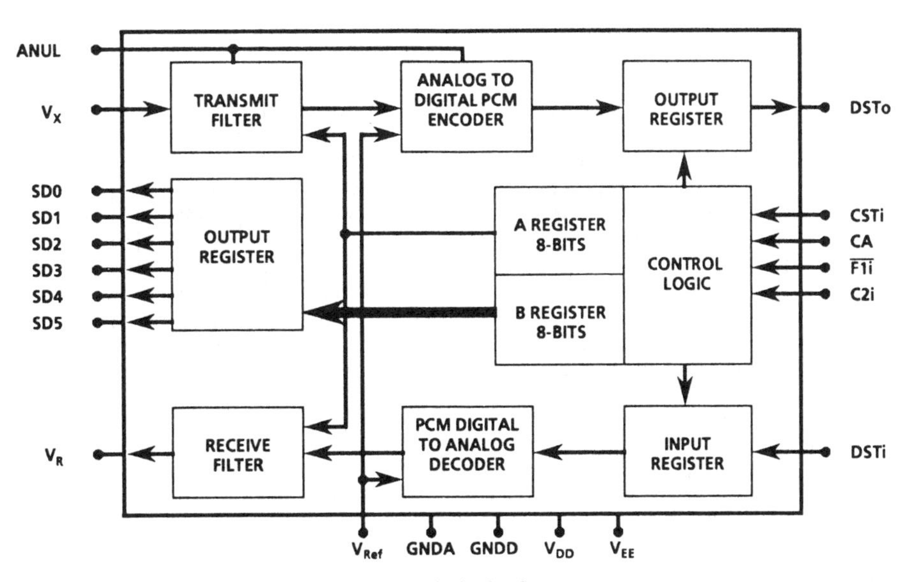

Functional Block Diagram

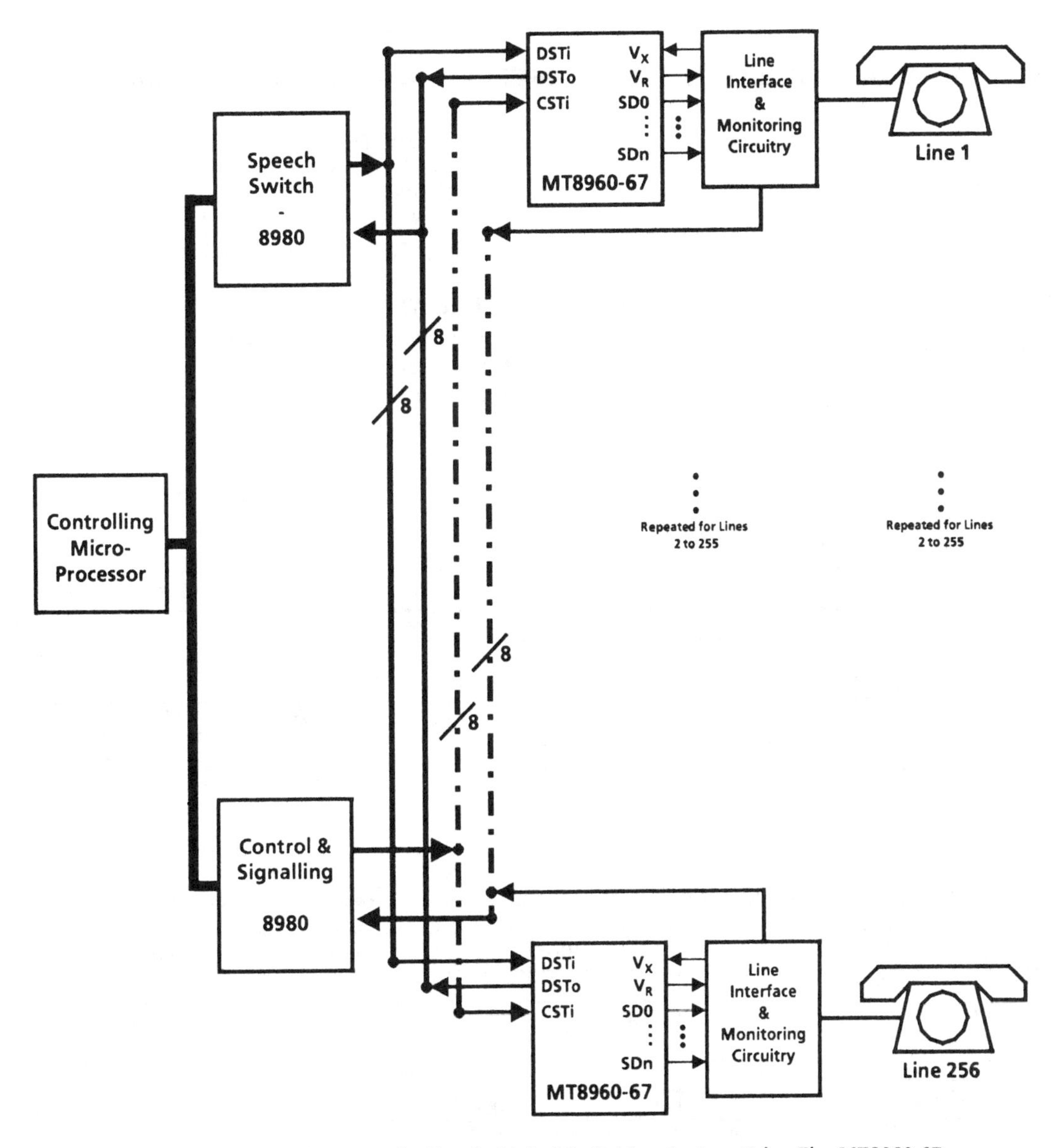

Example Architecture of a Simple Digital Switching System Using The MT8960-67

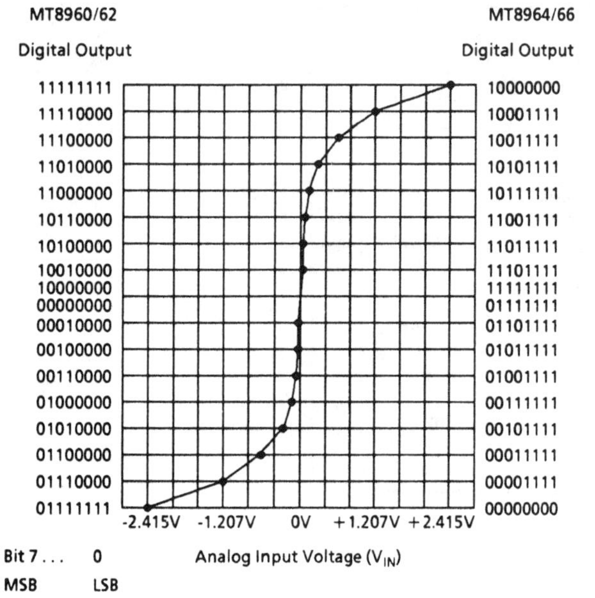

MT8960/62 Digital Output — MT8964/66 Digital Output

MT8960/62	MT8964/66
11111111	10000000
11110000	10001111
11100000	10011111
11010000	10101111
11000000	10111111
10110000	11001111
10100000	11011111
10010000	11101111
10000000	11111111
00000000	01111111
00010000	01101111
00100000	01011111
00110000	01001111
01000000	00111111
01010000	00101111
01100000	00011111
01110000	00001111
01111111	00000000

-2.415V -1.207V 0V +1.207V +2.415V

Bit 7 . . . 0 MSB LSB — Analog Input Voltage (V_{IN})

µ-Law Encoder Transfer Characteristic

MT8961/6 Digital Output — MT8965/67 Digital Output

MT8961/6	MT8965/67
11111111	10101010
11110000	10100101
11100000	10110101
11010000	10000101
11000000	10010101
10110000	11100101
10100000	11110101
10010000	11000101
10000000	11010101
00000000	01010101
00010000	01000101
00100000	01110101
00110000	01100101
01000000	00010101
01010000	00000101
01100000	00110101
01110000	00100101
01111111	00101010

-2.5V -1.25V 0V +1.25V +2.5V

Bit 7 . . . 0 MSB LSB — Analog Input Voltage (V_{IN})

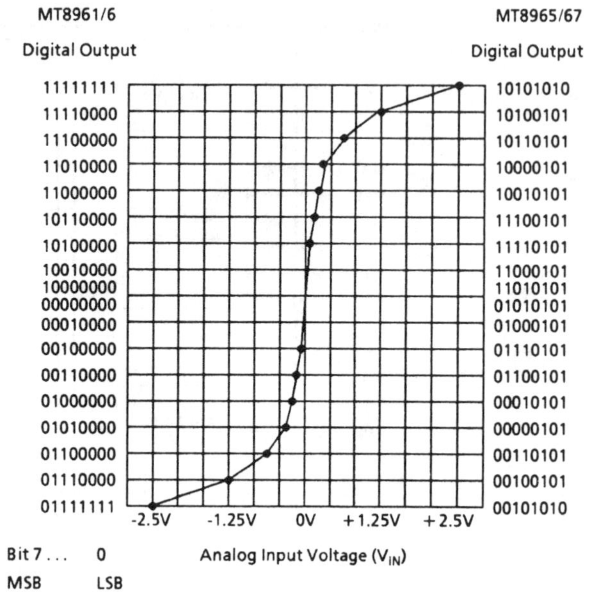

A-Law Encoder Transfer Characteristic

Mitel MT8992/3B Digital Telephone with HDLC (H-Phone)

- Integrated digital telephone circuit
- µ-Law/A-Law codec and filters
- Programmable receive gain
- DTMF/Tone generator and tone ringer
- Speakerphone operation
- Interface to standard telephony transducers
- Sense/drive ports
- Single 5V power supply
- Intel/Motorola bus interface
- ST-BUS compatible
- X.25 (CCITT) Level 2 HDLC data formatting

APPLICATIONS

- Featured digital telephone sets
- Voice/data terminals
- Cellular radio sets

ABSOLUTE MAXIMUM RATINGS*

	Parameter		Symbol	Min	Max	Units
1	Supply Voltage		$V_{DD} - V_{SS}$	−0.3	7	V
2	Voltage on any I/O pin		V_I / V_O	$V_{SS} - 0.3$	$V_{DD} + 0.3$	V
3	Current on any I/O pin		I_I / I_O		±20	mA
4	Storage Temperature		T_S	−65	+150	°C
5	Power Dissipation (package)	Ceramic	P_D		1.0	W

*Exceeding these values may cause permanent damage. Functional operation under these conditions is not implied.

RECOMMENDED OPERATING CONDITIONS - Voltages are with respect to V_{SS} unless otherwise stated

		Characteristics	Sym	Min	Typ‡	Max	Units	Test Conditions
1		Supply Voltage	V_{DD}	4.75	5	5.25	V	
2		Input Voltage (high)	V_{IH}	2.4		V_{DD}	V	Noise margin = 400 mV
3		Input Voltage (low)	V_{IL}	V_{SS}		0.4	V	Noise margin = 400 mV
4		Operating Temperature	T_O	0		+70	°C	
5		Clock Frequency	f_{CLK}		4096		kHz	

‡ Typical figures are at 25 °C and are for design aid only: not guaranteed and not subject to production testing.

DC ELECTRICAL CHARACTERISTICS† - Voltages are with respect to ground (V_{SS}) unless otherwise stated.

	Characteristics	Sym	Min	Typ‡	Max	Units	Test Conditions
1	Supply Current (clock disabled)	I_{DDC1}		1		mA	Outputs unloaded
	(clock enabled)	I_{DDC2}		2		mA	
2	Supply Current (handset speaker enabled) No Load	I_{DDH1}		6		mA	No signal
	Load = 150Ω	I_{DDH2}		7		mA	−20 dBm0, 1020 Hz
		I_{DDH3}		8		mA	0 dBm0, 1020 Hz
3	Supply Current (speaker driver enabled) No Load	I_{DDS1}		9		mA	No signal
	Load = 40Ω	I_{DDS2}		12		mA	−20 dBm0, 1020 Hz
		I_{DDS3}		45		mA	0 dBm0, 1020 Hz
4	Input HIGH Voltage TTL inputs	V_{IH}	2.0			V	
5	Input LOW Voltage TTL inputs	V_{IL}			0.8	V	
6	Input Voltage	V_{Bias}		2.5		V	
7	Input Leakage Current①	I_{IZ}		0.1	2.0	µA	$V_{IN}=V_{DD}$ to V_{SS}
8	Positive Going Threshold Voltage (PWRST only)	V_{T+}		1.7			$V_{DD}=5$ V
	Negative Going Threshold Voltage (PWRST only)	V_{T-}		1.2			$V_{DD}=5$ V
9	Output LOW Voltage TTL O/P	V_{OL}			0.4	V	$I_{OL}=2.8$ mA
10	Output HIGH Voltage TTL O/P	V_{OH}	2.4			V	$I_{OH}=-7$ mA
11	Output HIGH Current TTL O/P	I_{OH}		−15		mA	$V_{OH}=2.4$ V
12	Output LOW Current TTL O/P	I_{OL}		5		mA	$V_{OL}=0.4$ V
	SD0-6			3			$V_{OL}=0.4$ V
13	Output Voltage	V_{Ref}		0.5		V	No load
14	Output Leakage Current①	I_{OZ}		0.5		µA	$V_{OUT}=V_{DD}$ and V_{SS}

† DC Electrical Characteristics are over recommended temperature range & recommended power supply voltages.
‡ Typical figures are at 25 °C and are for design aid only: not guaranteed and not subject to production testing.
① TTL compatible and SD0-SD6 pins only.

AC ELECTRICAL CHARACTERISTICS†

	Characteristics	Sym	Min	Typ‡	Max	Units	Test Conditions
1	Input Pin Capacitance	C_I		8		pF	
2	Output Pin Capacitance	C_O		8		pF	

† Timing is over recommended temperature range & recommended power supply voltages.
‡ Typical figures are at 25°C and are for design aid only: not guaranteed and not subject to production testing.

AC ELECTRICAL CHARACTERISTICS - Microprocessor Bus Timing†

	Characteristics	Sym	Min	Typ‡	Max	Units
1	Cycle Time	t_{CYC}		667		ns
2	Pulse Width, DS Low or RD, WR High	PW_{EL}	240			ns
3	Pulse Width, DS High or RD, WR Low	PW_{EH}	225			ns
4	Input Rise and Fall Time	t_r, t_f			30	ns
5	R/W Hold Time	t_{RWH}		10		ns
6	R/W Setup Time Before DS Rise	t_{RWS}		20		ns
7	Chip Select Setup Time Before DS rise, RD, WR Fall	t_{CS}	50			ns
8	Chip Select Hold Time	t_{CH}	0			ns
9	Read Data Hold Time	t_{DHR}		10		ns
10	Write Data Hold Time	t_{DHW}	15			ns
11	Muxed Address Valid Time to AS, ALE Fall	t_{ASL}	40			ns
12	Muxed Address Hold Time	t_{AHL}	15			ns
13	Delay Time DS or RD or WR to AS, ALE Rise	t_{ASD}	50			ns
14	Pulse Width, AS, ALE High	PW_{ASH}	100			ns
15	Delay Time AS, ALE to DS Rise, RD, WR Fall	t_{ASED}	90			ns
16	Peripheral Output Data Delay Time from DS or RD	t_{DDR}			220	ns
17	Peripheral Data Setup Time	t_{DSW}	100			ns

† Timing is over recommended temperature range & recommended power supply voltages. All values shown assume a 50 pF load on the data pins. Read means read from the H-Phone and Write means write to the H-Phone.
‡ Typical figures are at 25°C and are for design aid only: not guaranteed and not subject to production testing.

AC ELECTRICAL CHARACTERISTICS†

	Characteristics	Sym	Min	Typ‡	Max	Units	Test Conditions
1	$\overline{C4i}$ Clock Period	t_{C4P}		244		ns	
2	$\overline{C4i}$ Clock High period	t_{C4H}		122		ns	
3	$\overline{C4i}$ Clock Low period	t_{C4L}		122		ns	
4	$\overline{C4i}$ Clock Transition Time	t_T		20		ns	
5	$\overline{F0i}$ Frame Pulse Setup Time	t_{F0iS}		50		ns	
6	$\overline{F0i}$ Frame Pulse Hold Time	t_{F0iH}		50		ns	
7	$\overline{F0i}$ Frame Pulse Width Low	t_{F0iW}		150		ns	
8	DSTo Delay	t_{DSToD}		125	200	ns	$C_L = 50$ pF
9	DSTi Setup Time	t_{DSTiS}		30		ns	
10	DSTi Hold Time	t_{DSTiH}		50		ns	

† Timing is over recommended temperature range & recommended power supply voltages.
‡ Typical figures are at 25°C and are for design aid only: not guaranteed and not subject to production testing.

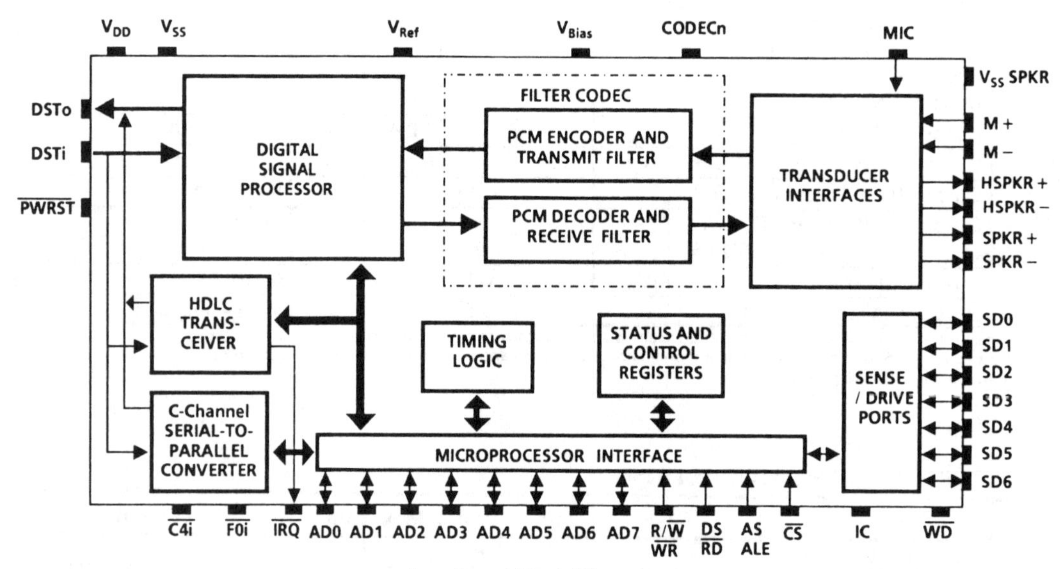

Functional Block Diagram

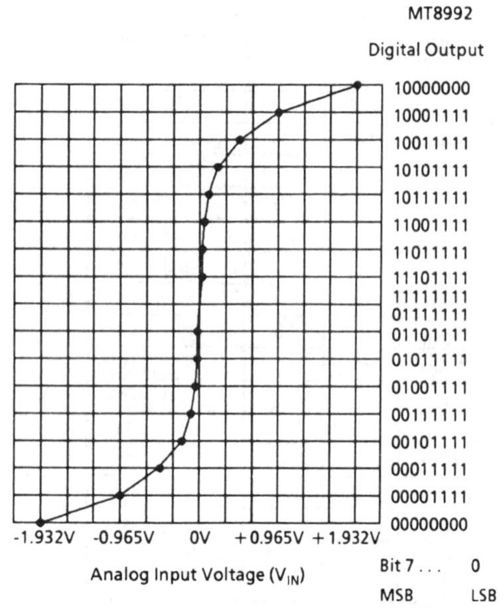

µ-Law Encoder Transfer Characteristic

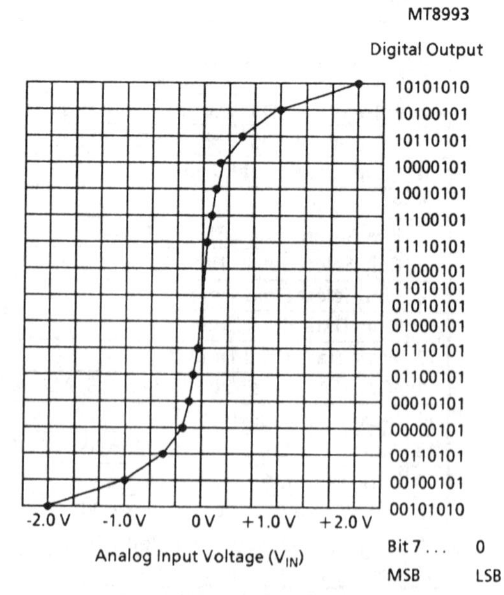

A-Law Encoder Transfer Characteristic

PIN CONNECTIONS

Name	Pin	Pin	Name
SPKR+	1	40	VDD
HSPKR+	2	39	SPKR−
CODECn	3	38	HSPKR−
PWRST	4	37	VSS SPKR
CS	5	36	VRef
AS,ALE	6	35	VBias
DS,RD	7	34	M+
R/W,WR	8	33	M−
AD7	9	32	MIC
AD6	10	31	F0i
AD5	11	30	DSTi
AD4	12	29	DSTo
AD3	13	28	C4i
AD2	14	27	SD6
AD1	15	26	SD5
AD0	16	25	SD4
IRQ	17	24	SD3
WD	18	23	SD2
IC	19	22	SD1
VSS	20	21	SD0

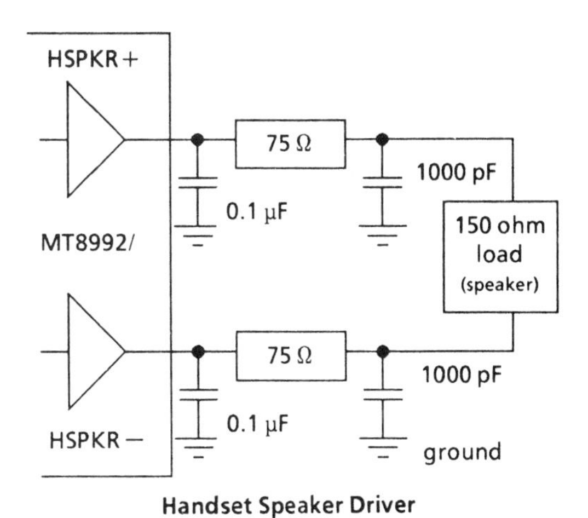

Handset Speaker Driver

Ordering Information

MT8992/3BC µ-Law/A-Law 40 Pin Cerdip
0 °C to +70 °C

RECOMMENDED OPERATING CONDITIONS - Speakerphone†

	Characteristics	Sym	Min	Typ‡	Max	Units	Test Conditions
1	Acoustic Separation			40		dB	
2	SPKR+ and SPKR− capacitive load allowed per pin				150	pF	direct speaker connection

† Parameters 1 and 2 are for design aid only and will give optimum results for speakerphone (handsfree) operation.
‡ Typical figures are at 25°C and are for design aid only: not guaranteed and not subject to production testing.

RECOMMENDED OPERATING CONDITIONS - Handset Microphone

	Characteristics	Sym	Min	Typ‡	Max	Units	Test Conditions
1	Sensitivity	S		−40		dBV	Acoustic input of 0 dBPa at 7.6 mm from lip ring at 1000 Hz
2	Output Impedance	Z_{OUT}		1		kΩ	At 1000 Hz
3	Harmonic Distortion	THD		1		%	Between 300 and 3000 Hz with an input level of 0 dBPa
4	External Interface Gain µ-Law ①A-Law	G_{EX}		7.3 16.7		dB dB	

‡ Typical figures are at 25 °C and are for design aid only. Refer to IEEE 269 for artificial mouth calibration procedure.
① Reflects need for higher audio levels typically found in networks for A-Law PCM coding.

RECOMMENDED OPERATING CONDITIONS - Speakerphone Microphone

	Characteristics	Sym	Min	Typ‡	Max	Units	Test Conditions
1	Sensitivity	S		−64		dB	(0 dB = 1 V/µbar) input of 1 µbar at 1000 Hz
2	Output Impedance	Z_{OUT}		2.2		kΩ	At 1000 Hz
3	Harmonic Distortion	THD		1		%	Between 300 and 3000 Hz with an input level of 0 dBPa

‡ Typical figures are at 25 °C and are for design aid only.

RECOMMENDED OPERATING CONDITIONS - Handset Speaker

	Characteristics	Sym	Min	Typ‡	Max	Units	Test Conditions
1	Efficiency			96.5		dBSPL	Reference frequency 1000 Hz, input signal 81.4 mV_{RMS} (open circuit) and closed loop generator impedance 150Ω
2	Impedance	Z_{HSPK}		150		Ω	Reference1000 Hz
3	Harmonic Distortion	T_{HD}		1		%	Between 300 and 3000 Hz

‡Typical figures are at 25 °C and are for design aid only.

RECOMMENDED OPERATING CONDITIONS - Speakerphone Speaker

	Characteristics	Sym	Min	Typ	Max	Units	Test Conditions
1	Impedance	Z_{SPK}		40		Ω	At 1000 Hz
2	Harmonic Distortion	T_{HD}		1		%	Between 300 and 3000 Hz

‡Typical figures are at 25 °C and are for design aid only.

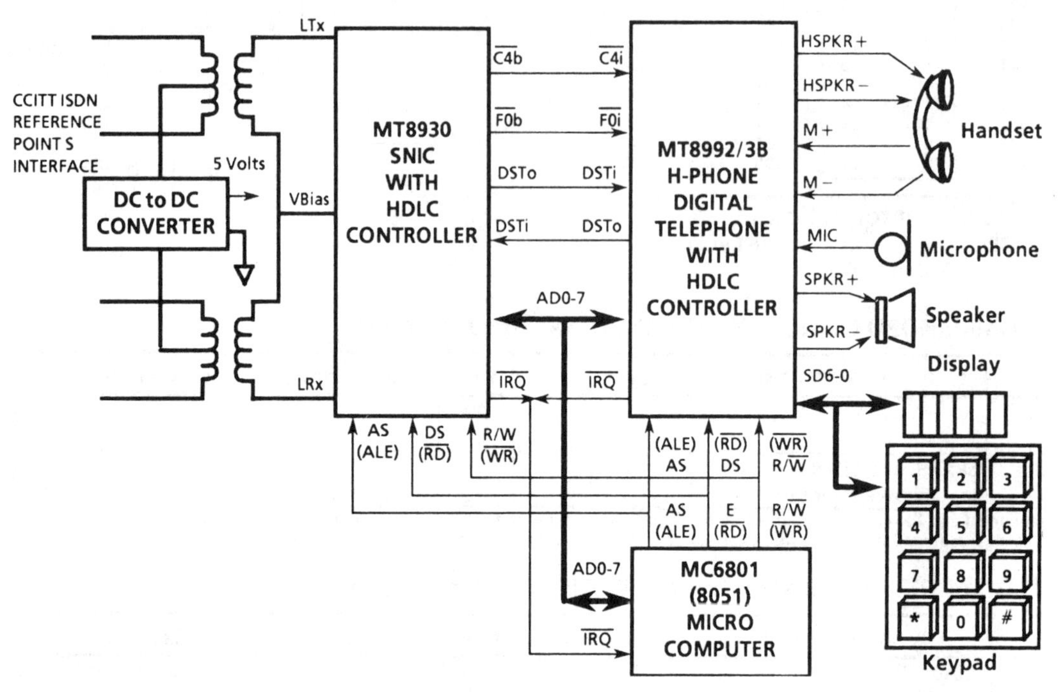

CCITT ISDN Voice/Data Terminal Equipment - TE1

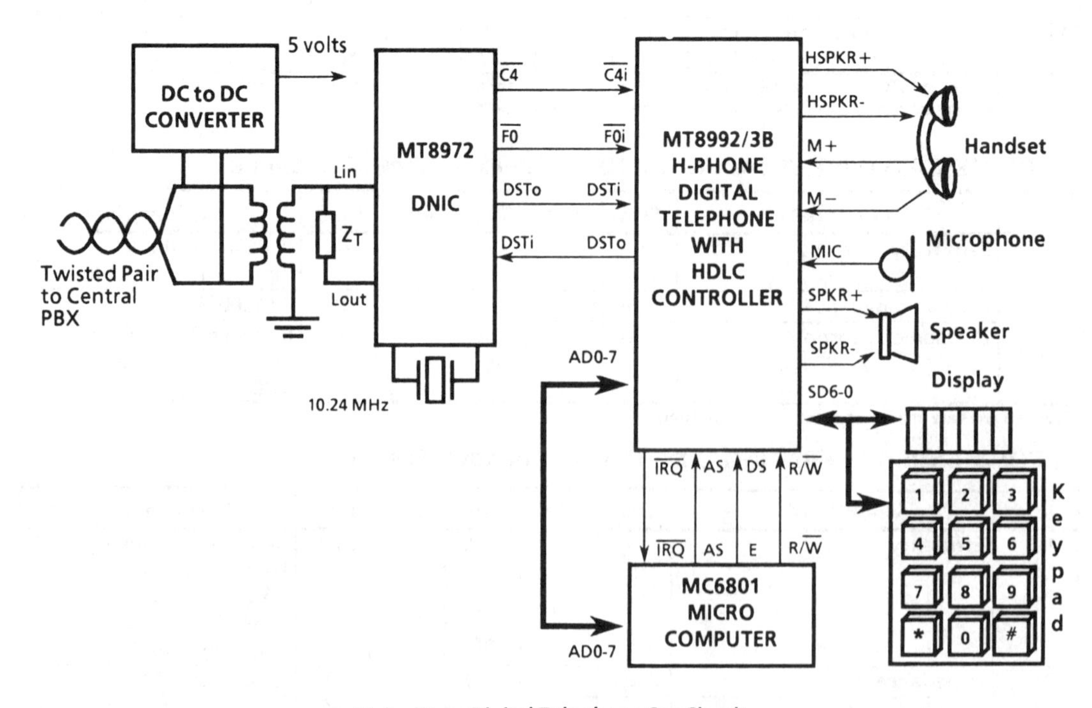

Voice/Data Digital Telephone Set Circuit

Note: Standard resistor values chosen may alter nominal gain.

Recommended Handset Microphone Interface Circuit

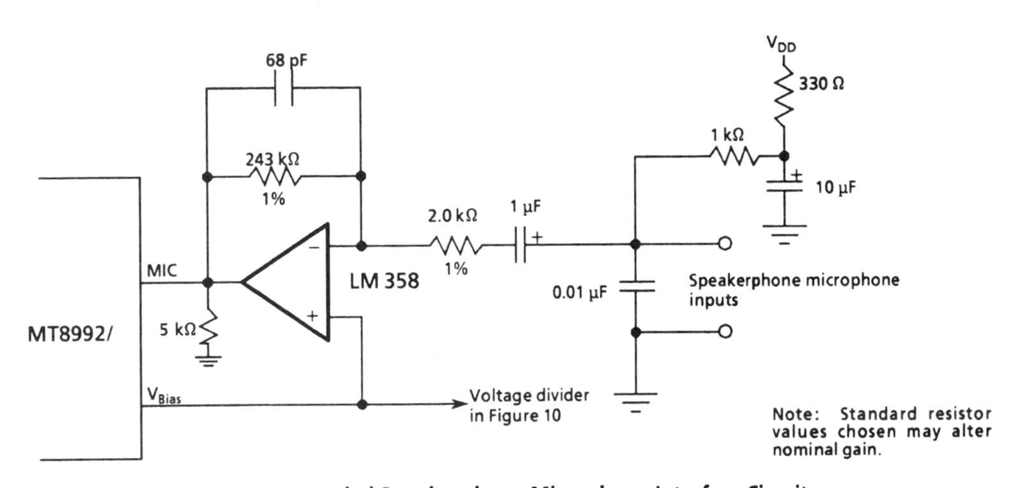

Recommended Speakerphone Microphone Interface Circuit

Mitel MH88630
Central Office Interface (LS/GS)
Advance Information

- Transformerless 2W to 4W conversion
- Line-state detection outputs:
  - ○ forward current
  - ○ reverse current
  - ○ ring ground
  - ○ tip ground
  - ○ ringing voltage
- Programmable audio transmit and receive gain
- Loop-start or ground-start termination
- Selectable 600Ω or AT&T compromise balance network

APPLICATIONS

- PBX interface to central office
- Channel bank
- Intercom
- Key system

PIN CONNECTIONS

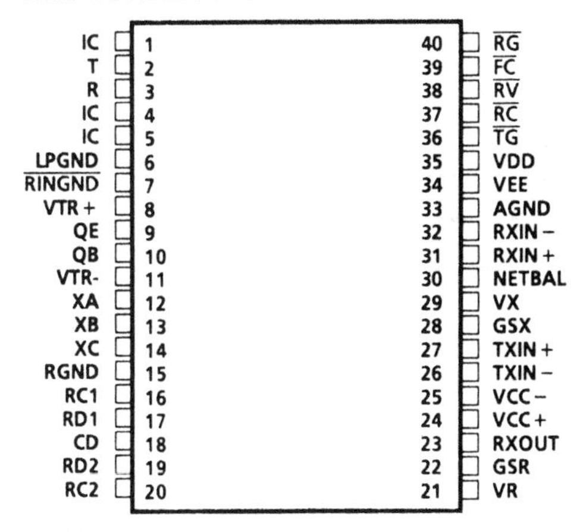

Ordering Information

MH88630 40 Pin DIL Package
0° C to 70° C

ABSOLUTE MAXIMUM RATINGS*

	Parameter	Symbol	Min	Max	Units
1	DC Supply Voltages	V_{DD}-GNDD V_{EE} - GNDD V_{CC+} - GNDA V_{CC-} - GNDA	-0.3 -6.0 -18.0	+6.0 +0.3 +18.0 	V V V V
2	Storage Temperature	T_{STG}	-55	+125	°C

*Exceeding these values may cause permanent damage. Functional operation under these conditions is not implied.

RECOMMENDED OPERATING CONDITIONS - Voltages are with respect to GNDD unless otherwise stated.

	Characteristics		Sym	Min	Typ*	Max	Units	Comments
1	Operating Supply Voltage		V_{DD}	4.75	5.0	5.25	V	
			V_{EE}	-5.25	-5.0	-4.75	V	
			V_{cc+}	11.4	12.0	12.6	V	
			V_{cc-}	-11.4	-12.0	-12.6	V	
			V_{BAT}		-48.0		V	
2	Operating Current	V_{DD}	I_{DD}			6.0	mA	$\overline{\text{RINGND}}$ High
		V_{EE}	I_{EE}			-6.0	mA	
		V_{cc+}	I_{cc+}			8.0	mA	
		V_{cc-}	I_{cc-}			-8.0	mA	
3	Power Consumption		P_C			265	mW	
4	Operating Temperature		T_O	0		70	°C	

* Typical figures are at 25°C. For design aid only: not guaranteed and not subject to production testing.

CONTROL INPUTS STATE TABLE

	Parameter	Active	Idle
1	RC1	Logic High	Logic Low
2	RC2	Logic High	Logic Low
3	$\overline{\text{RINGND}}$	Logic Low	Logic High
4	NETBAL AT&T compromise network (350Ω + 1 kΩ shunted by 0.21μF) 600 Ω network	 AGND Open (no connection)	

DC ELECTRICAL CHARACTERISTICS - control inputs

	Characteristics		Sym	Min	Typ*	Max	Units	Test Conditions
1	Input high voltage	RC1, RC2 $\overline{\text{RINGND}}$	V_{IH}	2.7 4.5			V V	
2	Input high current	RC1, RC2 $\overline{\text{RINGND}}$	I_{IH}	2.5		5.0 -100	mA μA	
3	Input low voltage	RC1, RC2 $\overline{\text{RINGND}}$	V_{IL}			0.7	V	
4	Input low current	RC1, RC2 $\overline{\text{RINGND}}$	I_{IL}			1.0 1.1	μA mA	

DC ELECTRICAL CHARACTERISTICS - TIP/RING Line State Outputs

	Characteristics	Sym	Min	Typ*	Max	Units	Test Conditions
1	Output High Voltage ($\overline{TG}$, $\overline{RC}$, $\overline{RV}$, $\overline{FC}$, $\overline{RG}$)			4.75		V	No Load on output
2	Output High Current ($\overline{TG}$, $\overline{RC}$, $\overline{RV}$, $\overline{FC}$, $\overline{RG}$)			0.17		mA	$V_{OH} = 2.7V_{DC}$
3	Output Low Voltage ($\overline{TG}$, $\overline{RC}$, $\overline{RV}$, $\overline{FC}$, $\overline{RG}$)			-0.30		V	No Load on output
4	Output Low Sink Current ($\overline{TG}$, $\overline{RC}$, $\overline{RV}$, $\overline{FC}$, $\overline{RG}$)			-0.40		mA	$V_{OL} = 0.4V_{DC}$

* Typical figures are at 25°C with nominal V_{DD} and V_{EE} supplies. For design aid only: not guaranteed and not subject to production

AC ELECTRICAL CHARACTERISTICS - Audio Transmission

	Characteristics	Sym	Min	Typ*	Max	Units	Test Conditions
1	Ringing Voltage	V_R	40	90	130	V_{RMS}	
2	Ringing Frequency			20		H_Z	Ringing Type A
3	Operating Loop Current	I_L	18		70	mA	
4	Off-Hook DC Resistance	R_T			300	Ω	@ 18mA
5	Operating Loop Resistance	R_L			2300	Ω	@ 18mA
6	On-Hook Leakage Current			10		μA	$\overline{RINGND} = 5.0\ V_{DC}$
7	Ring Ground Sink Current	I_{RG}			100	mA	$-48V_{DC}$ with 200Ω in series on Ring lead
8	Tip and Ring AC Impedance			600		Ω	with 10kΩ + 1.0μF in parallel with Tip and Ring
9	Longitudinal Balance metallic to longitudinal longitudinal to metallic		 60 40 58 53			 dB dB dB dB	 200 - 1000 Hz 1000 - 4000 Hz 200 - 1020 Hz 1020 - 3020 Hz
10	Return Loss Trunk to Line		20 26 30			dB dB dB	200 - 500 Hz 500 -1000 Hz 1000 - 3400 Hz
11	Transhybrid Loss (single frequency) into 600Ω	THL		18.5 34 30		dB dB dB	200 Hz 1000 Hz 3000 Hz
12	Transhybrid Loss (single frequency) into AT & T Compromise	THL		18 21		dB dB	200 - 1000 Hz 1000 - 4000 Hz
13	Frequency response		-0.05 -0.05 -0.05 -0.05		0.05 0.10 0.10 0.15	dB dB dB dB	200 Hz 300 Hz 3000 Hz 3400 Hz
14	Idle channel noise			8		dBrnc0	C-Message
15	Power supply rejection ratio	PSRR		40		dB	
16	Analog signal overload level (adjustable gain)				6	dBm	1 kHz, 0dBm = 0.775V_{RMS} into 600Ω

* Typical figures at 25°C are for design aid only and not guaranteed or subject to production testing.

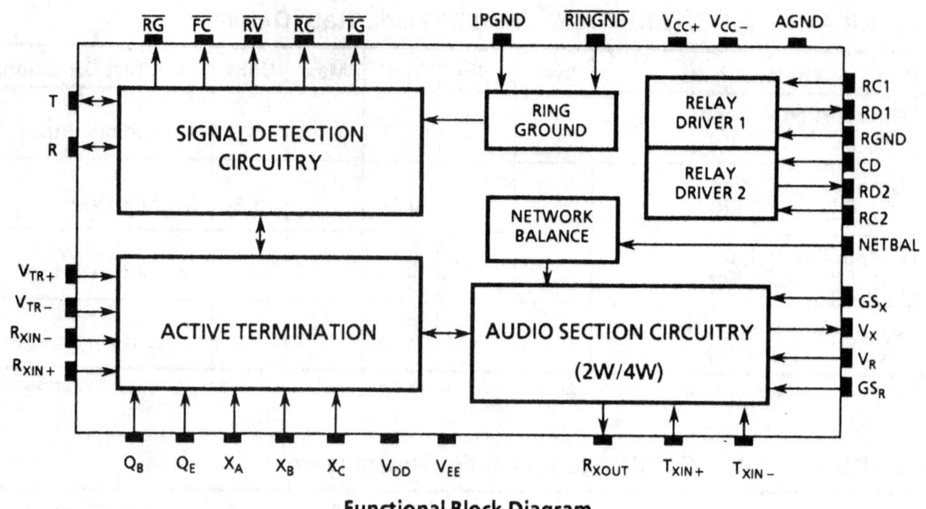

Functional Block Diagram

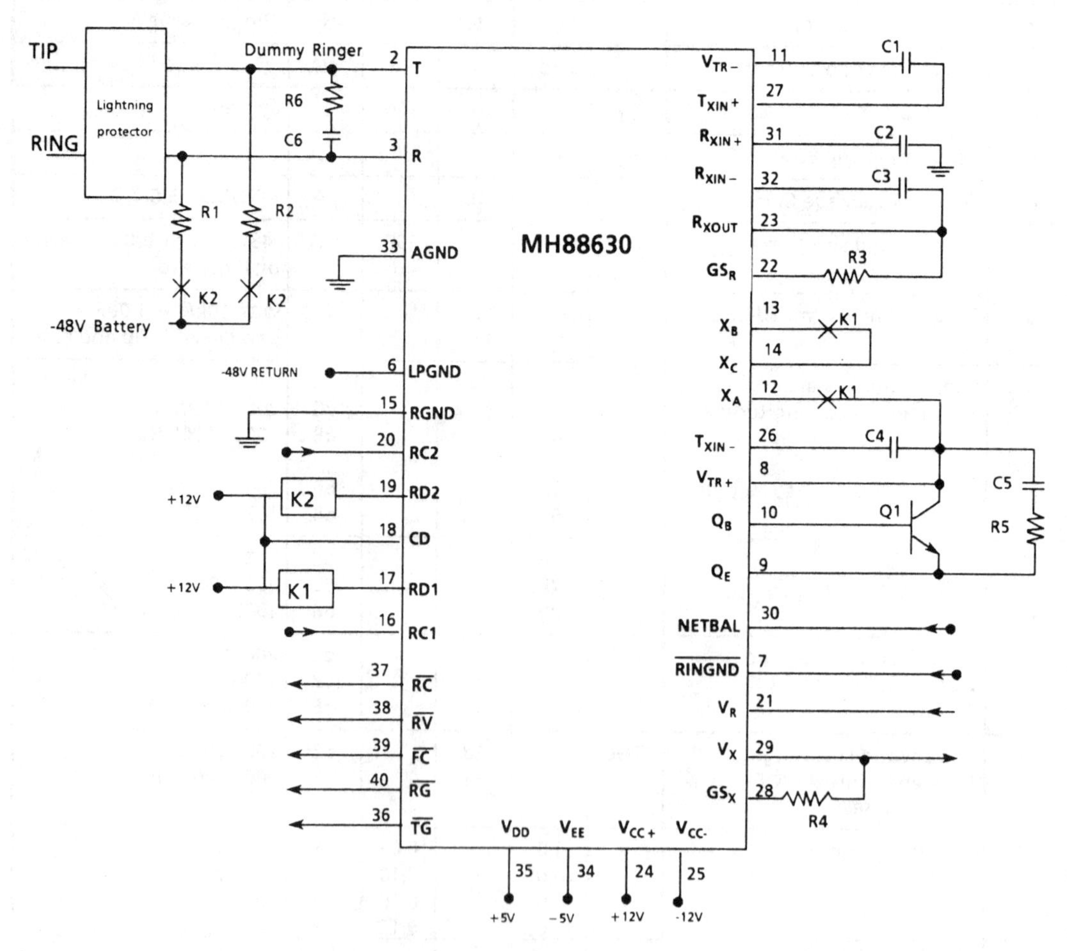

Application Circuit

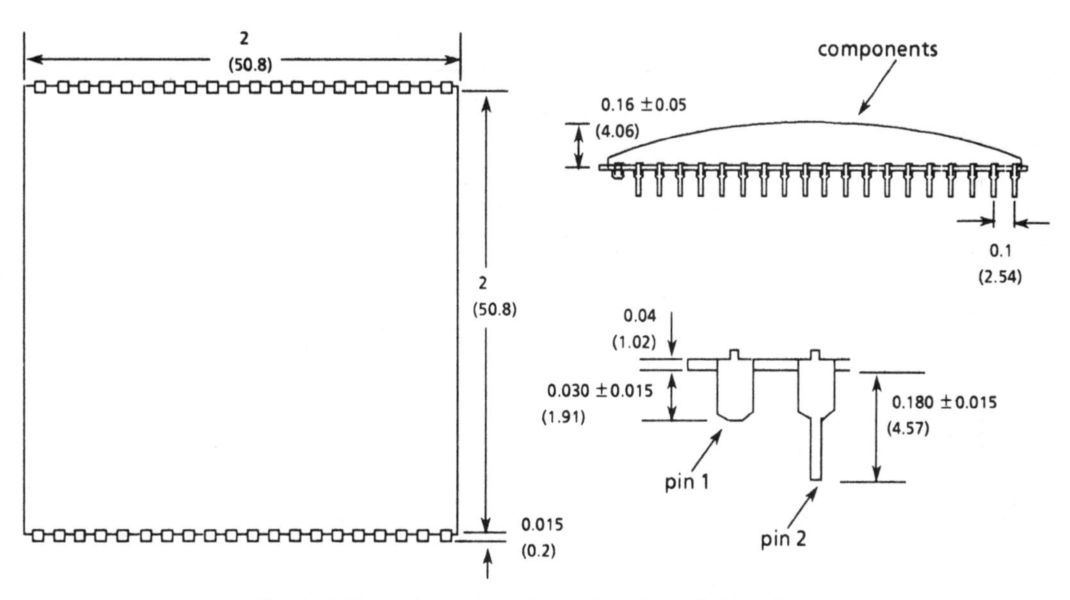

Physical Dimensions of 40 Pin Dual In Line Hybrid Package

Mitel MH88631
Central Office Interface (LS/GS)
Advance Information

- Transformerless 2W to 4W conversion
- Line-state detection outpus:
  - forward current
  - reverse current
  - ring ground
  - tip ground
  - ringing voltage
- Programmable audio transmit and receive gain
- Loop-start or ground-start termination
- Selectable 600Ω or AT&T compromise balance network

APPLICATIONS

- PBX interface to central office
- Channel bank
- Intercom
- Key system

PIN CONNECTIONS

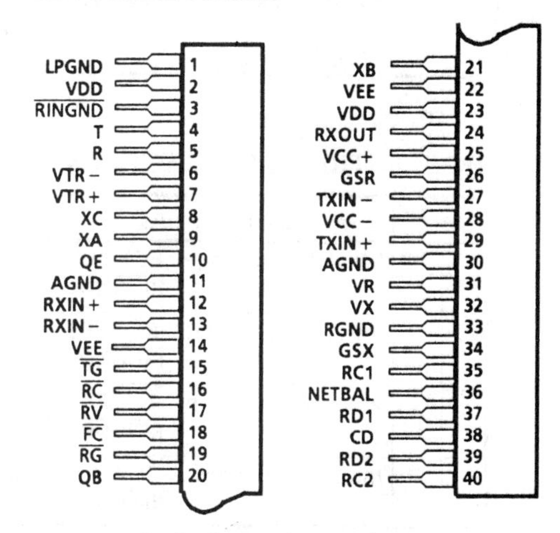

Ordering Information

MH88631 40 Pin SIL Package
0° C to 70° C

ABSOLUTE MAXIMUM RATINGS*

	Parameter	Symbol	Min	Max	Units
1	DC Supply Voltages	V_{DD}-GNDD	-0.3	+6.0	V
		V_{EE} - GNDD	-6.0	+0.3	V
		V_{CC+} - GNDA		+18.0	V
		V_{CC-} - GNDA	-18.0		V
2	Storage Temperature	T_{STG}	-55	+125	°C

*Exceeding these values may cause permanent damage. Functional operation under these conditions is not implied.

RECOMMENDED OPERATING CONDITIONS - Voltages are with respect to GNDD unless otherwise stated.

	Characteristics	Sym	Min	Typ*	Max	Units	Comments
1	Operating Supply Voltage	V_{DD}	4.75	5.0	5.25	V	
		V_{EE}	-5.25	-5.0	-4.75	V	
		V_{cc+}	11.4	12.0	12.6	V	
		V_{cc-}	-11.4	-12.0	-12.6	V	
		V_{BAT}		-48.0		V	
2	Operating Current V_{DD}	I_{DD}			6.0	mA	$\overline{RINGND}$ High
	V_{EE}	I_{EE}			-6.0	mA	
	V_{cc+}	I_{cc+}			8.0	mA	
	V_{cc-}	I_{cc-}			-8.0	mA	
3	Power Consumption	P_C			265	mW	
4	Operating Temperature	T_O	0		70	°C	

* Typical figures are at 25°C. For design aid only: not guaranteed and not subject to production testing.

CONTROL INPUT STATE TABLE

	Parameter	Active	Idle
1	RC1	Logic High	Logic Low
2	RC2	Logic High	Logic Low
3	$\overline{RINGND}$	Logic Low	Logic High
4	NETBAL AT&T compromise network (350Ω + 1 kΩ shunted by 0.21µF) 600 Ω network	AGND Open (no connection)	

DC ELECTRICAL CHARACTERISTICS -Control Inputs

	Characteristics	Sym	Min	Typ*	Max	Units	Test Conditions
1	Input high voltage RC1, RC2 $\overline{RINGND}$	V_{IH}	2.7 4.5			V V	
2	Input high current RC1, RC2 $\overline{RINGND}$	I_{IH}	2.5		5.0 -100	mA µA	
3	Input low voltage RC1, RC2 $\overline{RINGND}$	V_{IL}			0.7	V	
4	Input low current RC1, RC2 $\overline{RINGND}$	I_{IL}			1.0 1.1	µA mA	

DC ELECTRICAL CHARACTERISTICS - TIP/RING Line State Outputs

	Characteristics	Sym	Min	Typ*	Max	Units	Test Conditions
1	Output High Voltage ($\overline{TG}$, $\overline{RC}$, $\overline{RV}$, $\overline{FC}$, $\overline{RG}$)			4.75		V	No Load on output
2	Output High Current ($\overline{TG}$, $\overline{RC}$, $\overline{RV}$, $\overline{FC}$, $\overline{RG}$)			0.17		mA	$V_{OH} = 2.7V_{DC}$
3	Output Low Voltage ($\overline{TG}$, $\overline{RC}$, $\overline{RV}$, $\overline{FC}$, $\overline{RG}$)			-0.30		V	No Load on output
4	Output Low Sink Current ($\overline{TG}$, $\overline{RC}$, $\overline{RV}$, $\overline{FC}$, $\overline{RG}$)			-0.40		mA	$V_{OL} = 0.4V_{DC}$

* Typical figures are at 25°C with nominal V_{DD} and V_{EE} supplies. For design aid only: not guaranteed and not subject to production

AC ELECTRICAL CHARACTERISTICS - Audio Transmission

	Characteristics	Sym	Min	Typ*	Max	Units	Test Conditions
1	Ringing Voltage	V_R	40	90	130	V_{RMS}	
2	Ringing Frequency			20		Hz	Ringing Type A
3	Operating Loop Current	I_L	18		70	mA	
4	Off-Hook DC Resistance	R_T			300	Ω	@ 18mA
5	Operating Loop Resistance	R_L			2300	Ω	@ 18mA
6	On-Hook Leakage Current			10		µA	$\overline{RINGND}$ = 5.0V_{DC}
7	Ring Ground Sink Current	I_{RG}			100	mA	-48V_{DC} with 200Ω in series on Ring lead
8	Tip and Ring AC Impedance			600		Ω	with 10kΩ + 1.0µF in parallel with Tip and Ring
9	Longitudinal Balance metallic to longitudinal longitudinal to metallic		60 40 58 53			dB dB dB dB	200 - 1000 Hz 1000 - 4000 Hz 200 -1020 Hz 1020-3020 Hz
10	Return Loss Trunk to Line		20 26 30			dB dB dB	200-500 Hz 500 -1000 Hz 1000 -3400 Hz
11	Transhybrid Loss (single frequency) into 600Ω	THL		18.5 34 30		dB dB dB	200 Hz 1000 Hz 3000 Hz
12	Transhybrid Loss (single frequency) into AT & T Compromise	THL		18 21		dB dB	200-1000 Hz 1000-4000 Hz
13	Frequency response		-0.05 -0.05 -0.05 -0.05		0.05 0.10 0.10 0.15	dB dB dB dB	200 Hz 300 Hz 3000 Hz 3400 Hz
14	Idle channel noise			8		dBrnc0	C-Message
15	Power supply rejection ratio	PSRR		40		dB	
16	Analog signal overload level (adjustable gain)				6	dBm	1 kHz, 0dBm=0.775V_{RMS} into 600Ω

* Typical figures at 25°C are for design aid only and not guaranteed or subject to production testing.

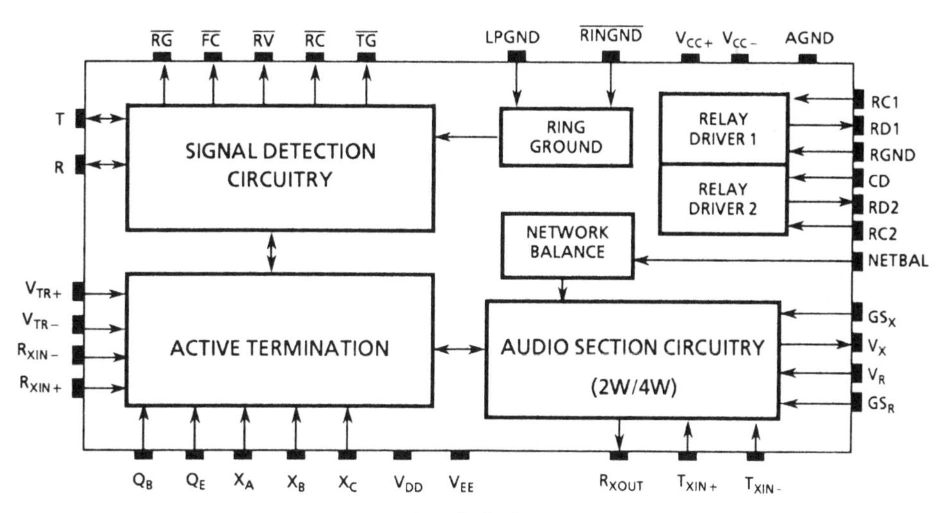

Functional Block Diagram

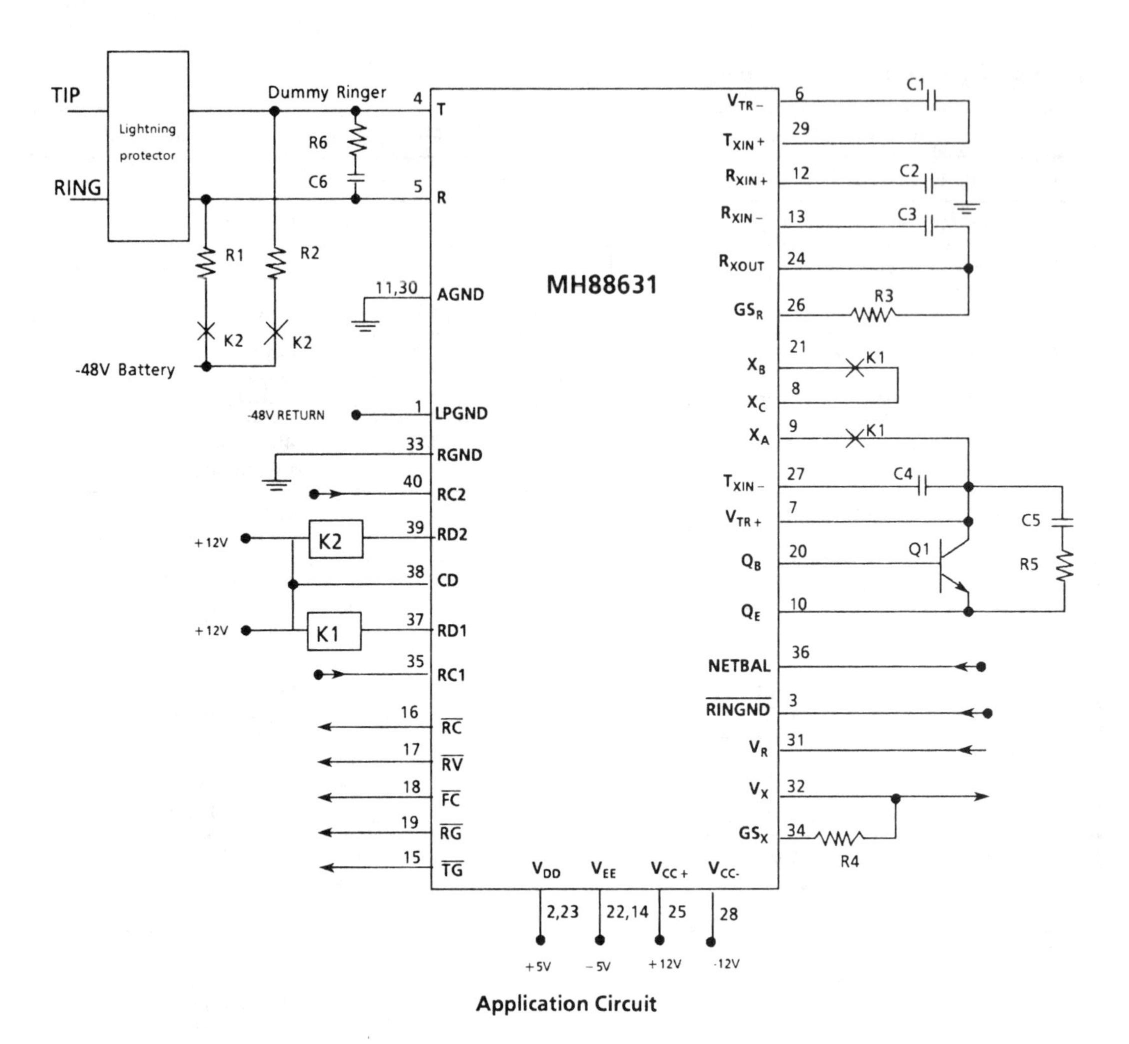

Application Circuit

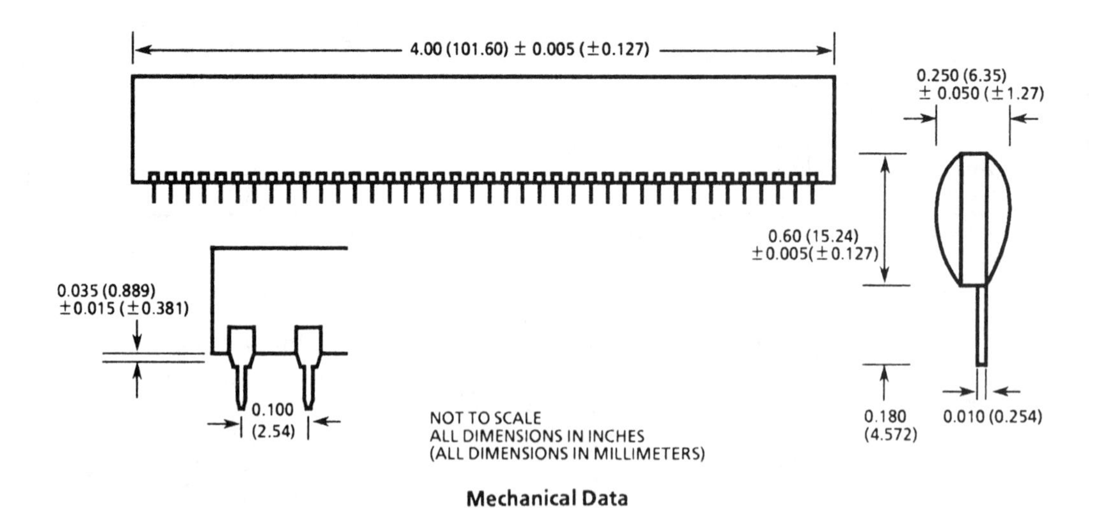

Mechanical Data

Mitel MT35212A
Bell 212A/CCITT V.22 Modem Filter

- Bell 212A and CCIT V.22 compatible
- Usable for Bell 103 and V.22 Bis applications
- Guard tone notch filters for V.22 applications
- High and low bandfilters with compromise group delay equalizers and smoothing filters
- Answer/originate operating modes
- Detection of call progress tones
- Choice of clocking frequencies: 2.4576 MHz, 1.2288 MHz, or 153.6 kHz
- Analog loopback test capability
- Two uncommitted operational amplifiers
- Pin compatible with AMI S35212A

PIN CONNECTIONS

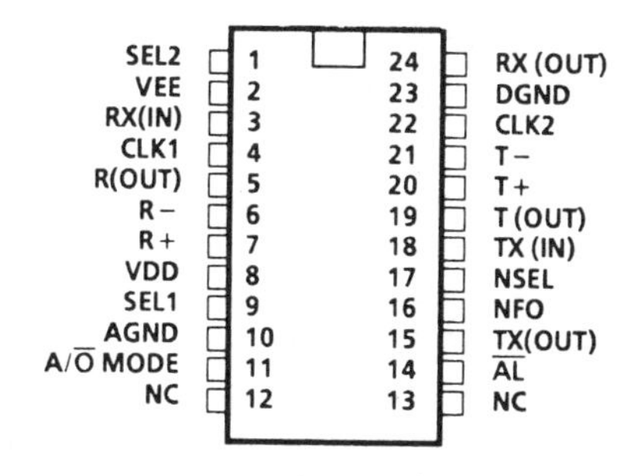

Ordering Information 0°C to 70°C

MT35212AE Plastic DIP
MT35212AC CERDIP

APPLICATIONS

- Modem filter/equalizer for 1200 bps full-duplex modem implementation
- Detection of tones in the call progress band by selecting filters

ABSOLUTE MAXIMUM RATINGS*

	Parameter	Symbol	Min	Max	Units
1	Positive Supply Voltage	V_{DD}		6.75	V
2	Negative Supply Voltage	V_{EE}		−6.75	V
3	Storage Temperature Range	T_{STG}	−55	+125	°C
4	Analog Input	V	V_{EE} −0.3	V_{DD} +0.3	V

*Exceeding these values may cause permanent damage. Functional operation under these conditions is not implied.

RECOMMENDED OPERATING CONDITIONS - Voltages are with respect to ground unless otherwise stated.

		Characteristics	Sym	Min	Typ‡	Max	Units	Test Conditions
1		Positive Supply Voltage	V_{DD}	4.75	5	5.25	V	DGND = AGND = 0 V
2		Negative Supply Voltage	V_{EE}	−4.75	−5	−5.25	V	DGND = AGND = 0 V
3		Operating Temperature Range	T_O	0	25	70	°C	

‡ Typical figures are at 25°C and are for design aid only: not guaranteed and not subject to production testing.

DC ELECTRICAL OPERATING CONDITIONS - T_O = 0°C to 70°C.

		Characteristics	Sym	Min	Typ‡	Max	Units	Test Conditions
1		Positive Supply Voltage	V_{DD}	+4.75	5.0	5.25	V	
2		Negative Supply Voltage	V_{EE}	−4.75	−5.0	−5.25	V	
3		Power Consumption	P_C		75	150	mW	V_{DD} = 5.25V; V_{EE} = −5.25V

‡ Typical figures are at 25°C and are for design aid only: not guaranteed and not subject to production testing.

DC ELECTRICAL CHARACTERISTICS - V_{DD} = +5 V ± 5%, V_{EE} = −5 V ± 5%, AGND = DGND = 0 V, T_O = 0°C to 70°C.

		Characteristics	Sym	Min	Typ‡	Max	Units	Test Conditions
1	INPUTS	High Level Logic	V_{IH}	4		V_{DD}	V	Pins 1, 9, 11,14, 17.
2		High Level Logic	V_{IH}	2.0		V_{DD}	V	Pins 4, 22.
3		Low Level Logic	V_{IL}	V_{EE}		0.8	V	Pins 1, 4, 9, 11,14, 17,22.
4		Resistance	R_{IN}		5		MΩ	Pins 3, 18.
5		Capacitance	C_{IN}		10		pF	Pins 3, 18.

‡ Typical figures are at 25°C and are for design aid only: not guaranteed and not subject to production testing.

AC ELECTRICAL CHARACTERISTICS - $V_{DD}=+5\ V \pm 5\%$, $V_{EE}=-5\ V \pm 5\%$, AGND = DGND = 0 V, $T_O=0°C$ to 70°C.

	Characteristics	Sym	Min	Typ‡	Max	Units	Test Conditions
1	Reference Signal Level Input	V_{REF}		1		V_{RMS}	
2	Maximum Signal Level Input	V_{MAX}			1.4125	V_{RMS}	
3	Bandwidth (both bands)	BW		960		Hz	
4	Gain at Center Frequencies	A_{FO}	−1.0	0	+1.0	dB	
5	Idle Channel Noise - Low Band Filter High Band Filter			 23 22	 33 33	 dBrnC0 dBrnC0	No load.
6	Harmonic Distortion	THD		−55		dB	
7	Clock Feed Through with respect to signal level	CLK_{FT}		−23 −60		dB dB	T_X (clock feedthrough R_X frequency is 76.8 kHz)

‡ Typical figures are at 25°C and are for design aid only: not guaranteed and not subject to production testing.

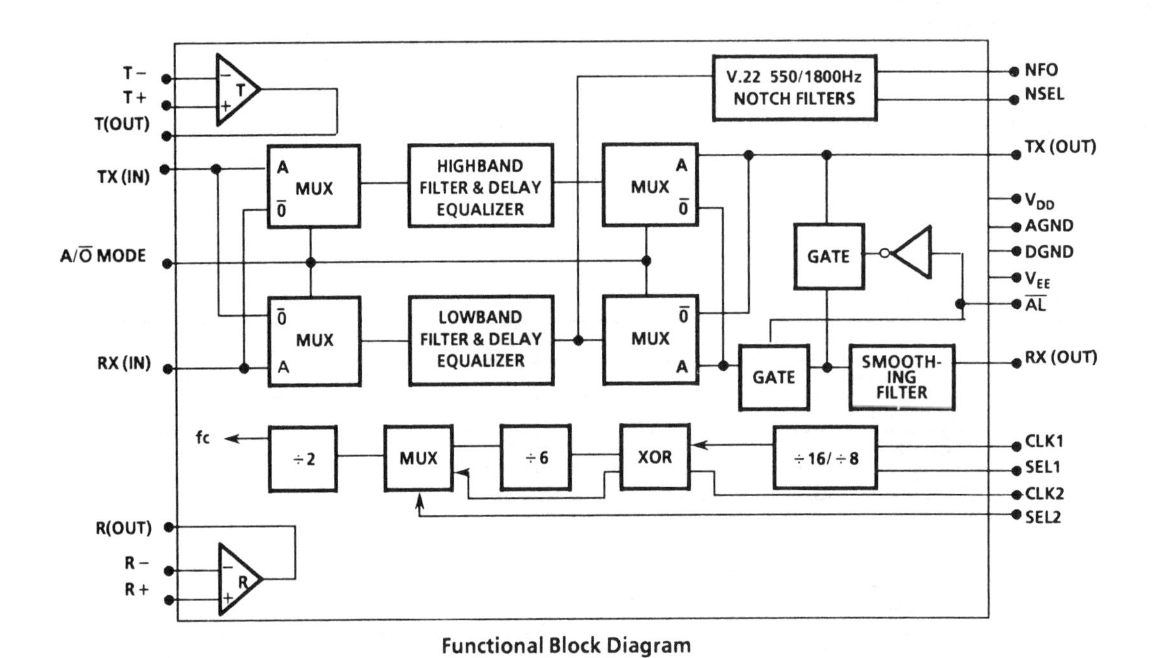

Functional Block Diagram

Typical Amplitude vs. Frequency Plot

FREQUENCY (Hz)		RELATIVE GAIN MIN.	RELATIVE GAIN MAX.
Lowband	400		−35
	800	−1	+1
	1200	−1	+1
	1600	−1.5	+1
	1800		−18
	2000		−48
	2400		−55
	2800		−50
Highband	800		−50
	1200		−53
	1600		−50
	2000	−2.5	+0.5
	2400	−1	+1
	2800	0	+2.5
	3200		−10
	3500		−20

Amplitude vs. Frequency Response

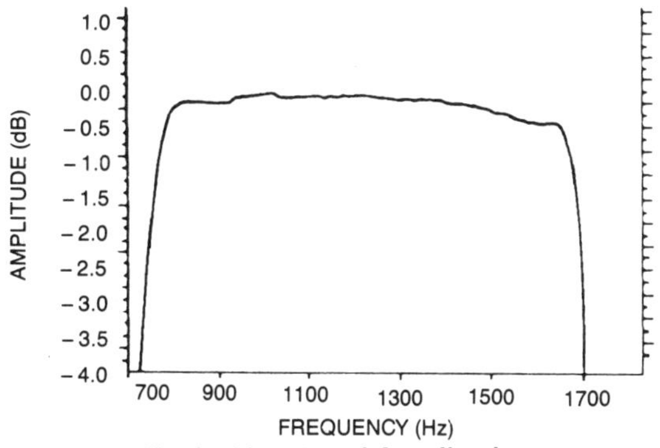

Typical Lowband Amplitude vs. Frequency Plot

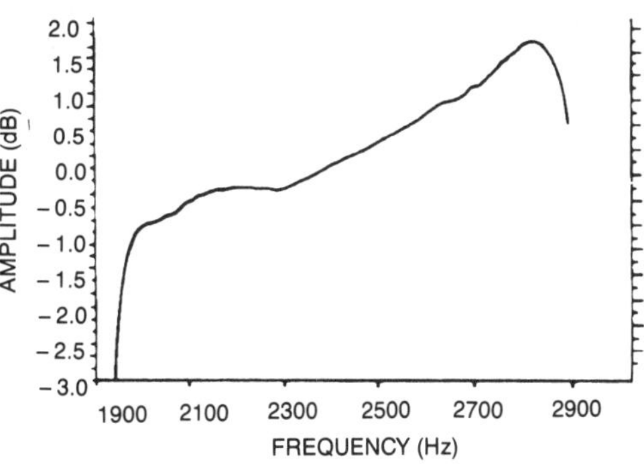

Typical Highband Amplitude vs. Frequency Plot

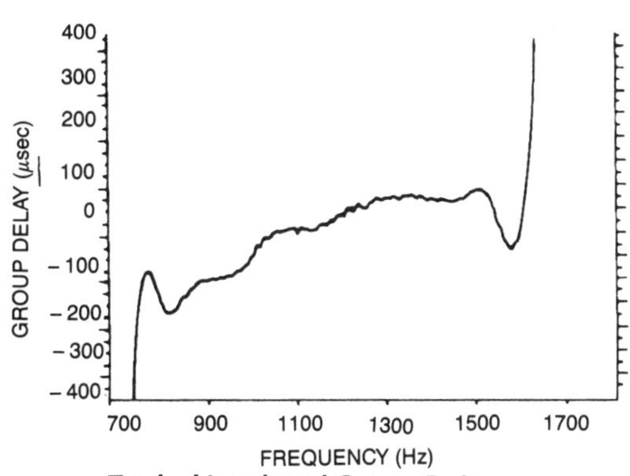

Typical Lowband Group Delay vs. Frequency Plot

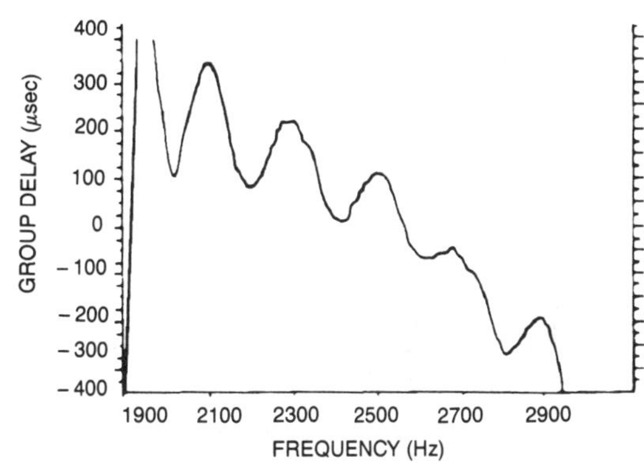

Typical Highband Group Delay vs. Frequency Plot

Mitel MT3530

Bell 103/V.21 Single-Chip Modem

- Single-chip 300 bps, full-duplex asynchronous FSK modem
- Bell 103/113 and CCITT V.21 selectable
- Auto answer/originate operating modes
- Manual mode
- Phase continuous transmit carrier switching
- Digital and analog loopback modes
- CCITT V.25 tone generation
- UART clock output
- Passthru mode for protocol independence
- No external filtering required
- DTE Interface—Functionally: RS-232C Compatible (CCITT V.24)
  Electrically: TTL-level compatible

APPLICATIONS

- Stand-alone RS-232C interface modem
- Add-on modem for personal computers and microprocessor systems

ABSOLUTE MAXIMUM RATINGS*

	Parameter	Symbol	Min	Max	Units
1	DC Supply Voltage	V_{DD}-V_{EE}		+12.0	V
2	Storage Temperature Range	T_{STG}	−65	+150	°C
3	Input Voltage, All Pins	V_{IN}	$V_{EE}-0.3$	$V_{DD}+0.3$	V

*Exceeding these values may cause permanent damage. Functional operation under these conditions is not implied.

DC ELECTRICAL OPERATING CONDITIONS - $T_O = 0°C$ to $+70°C$.

		Characteristics	Sym	Min	Typ‡	Max	Units	Test Conditions
1		Positive Supply Voltage	V_{DD}	+4.75	+5.0	+5.25	V	DGND = AGND = 0 Volt
2		Negative Supply Voltage	V_{EE}	−4.75	−5.0	−5.25	V	DGND = AGND = 0 Volt
3		Power Consumption	P_C		110	200	mW	$V_{DD}=5.0$ V; $V_{EE}=-5.0$ V

‡ Typical figures are at 25°C and are for design aid only: not guaranteed and not subject to production testing.

RECOMMENDED OPERATING CONDITIONS - Voltages are with respect to ground unless otherwise stated.

		Parameter	Sym	Min	Typ‡	Max	Units	Test Conditions
1	I	Positive Supply Voltage	V_{DD}		+5		V	DGND = AGND = 0 Volt
2	N	Negative Supply Voltage	V_{EE}		−5		V	DGND = AGND = 0 Volt
3	P U	Oscillator Clock Frequency	fosc		3.579545		MHz	
4	T	Oscillator Frequency Tolerance	Δfosc		±0.02		%	
5	S	Operating Temperature Range	T_O	0		70	°C	

‡ Typical figures are at 25°C and are for design aid only: not guaranteed and not subject to production testing.

ANALOG SIGNAL PARAMETERS T_O = 0°C to +70°C, ±5 Vdc, fosc = 3.579545 MHz.

		Parameter	Sym	Min	Typ‡	Max	Units	Test Conditions
1		Oscillator Clock Frequency Oscillator Frequency Tolerance Transmit Frequency Tolerance	fosc Δfosc Δft		3.579545 ±0.02 ±1.2	 ±3	MHz % Hz	
2		Transmit 2nd Harmonic Attenuation with respect to carrier level	T_{HD}		50		dB	
3		Transmit Output Level	T_{OUT}	−9	−8	−7	dBm	Load 10 kΩ 25 pF Max.
4		Carrier Input Range		−50		0	dBm	CDT open
5		Dynamic Range	DNR		50		dB	CDT open
6		Carrier Detect: On Level Off Level On/Off level Hysterisis	CD_{ON} CD_{OFF} CD_H	 −50 2.5	−43 −48 5	−41	dBm dBm dB	
7		Bit Jitter Bit Bias (Mark and Space) Bias Distortion			100 1 3		μs % %	Input = −30dBm

‡ Typical figures are at 25°C and are for design aid only: not guaranteed and not subject to production testing.

DC ELECTRICAL CHARACTERISTICS - V_{DD} = 5± 5% Vdc, V_{EE} = −5± 5% Vdc, AGND = DGND = 0V, T_O = 0°C to 70°C.

		Characteristics	Sym	Min	Typ‡	Max	Units	Test Conditions
1	I	CMOS Inputs Voltage High Voltage Low	V_{IH} V_{IL}	3		 −3	V V	Note 1
2	N P	TTL Inputs Voltage High Voltage Low	V_{IH} VIL	2		 0.8	V V	Note 2
3	U T	Input Resistance	R_{IN}	8			MΩ	All Inputs
4	S	Input Capacitance	C_{IN}			15	pF	All Inputs
1	O U T	LSTTL Outputs Voltage High Voltage Low	 V_{OH} V_{OL}	 2.4		 0.4	 V V	Note 3 I_{OL} = 0.4mA
2	P U T S	TTL Output Voltage High Voltage Low	 V_{OH} V_{OL}	 2.4		 0.4	 V V	Note 4 I_{OL} = 1.6mA

‡ Typical figures are at 25°C and are for design aid only: not guaranteed and not subject to production testing.

Note 1. Include $\overline{SH}$, $\overline{RI}$, TEST0, TEST1
Note 2. Include RTS, TD, DTR, AL, DL, SL
Note 3. Include $\overline{OH}$, $\overline{CLK}$, $\overline{CD}$, $\overline{DSR}$
Note 4. Include RD, $\overline{CTS}$
Note 5. Test conducted using Passthru mode

PIN CONNECTIONS

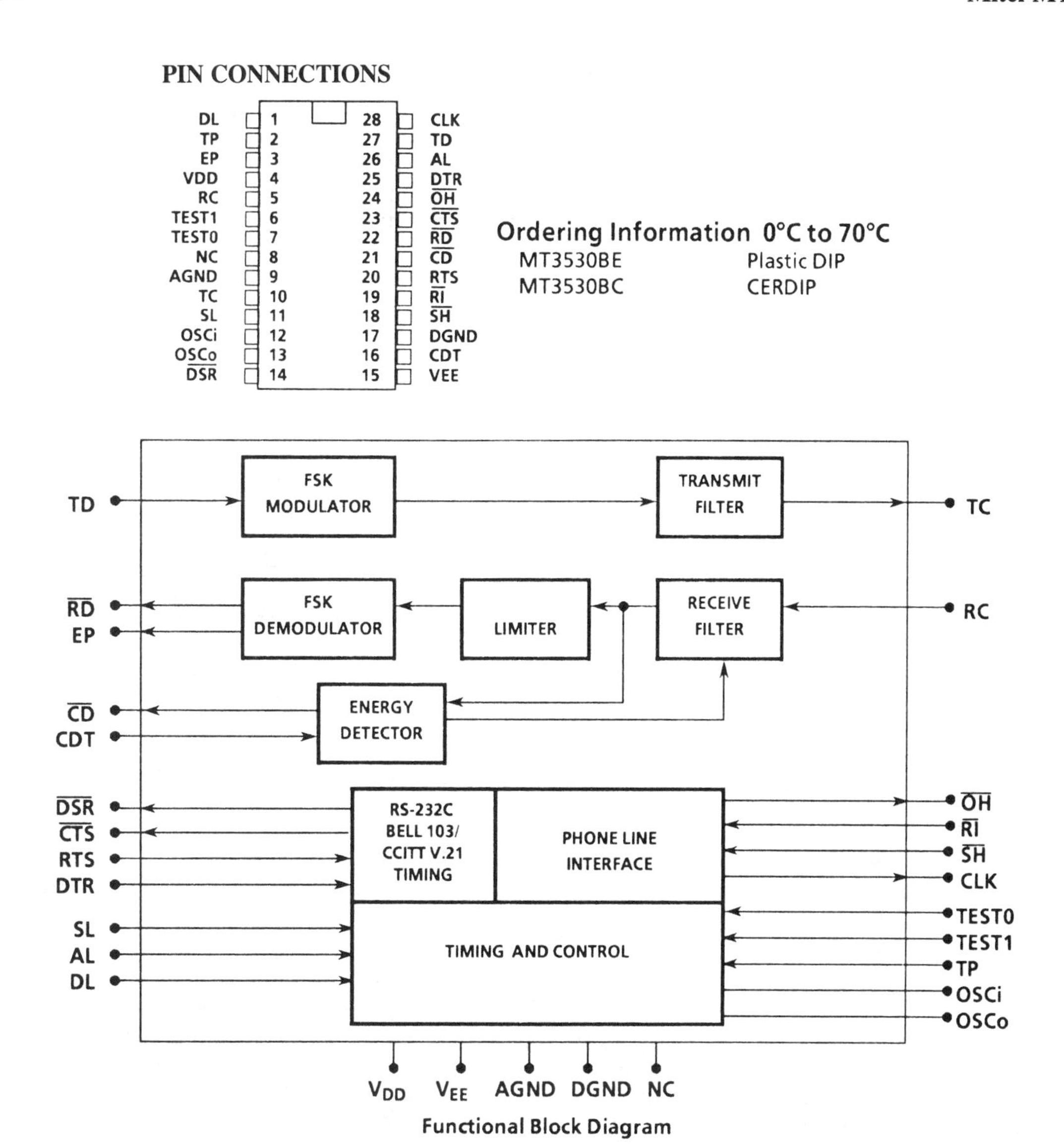

Functional Block Diagram

Pin Name	Pin No.	Input	Output	Voltage Level Low	Voltage Level High	Logic Family	I_{OL} mA
$\overline{SH}$	18	X		−3	+3	CMOS	
$\overline{RI}$	19	X		−3	+3	CMOS	
TEST0	7	X		−3	+3	CMOS	
TEST1	6	X		−3	+3	CMOS	
$\overline{OH}$	24		X	+0.4	+2.4	LSTTL	0.4
CLK	28		X	+0.4	+2.4	LSTTL	0.4
$\overline{CD}$	21		X	+0.4	+2.4	LSTTL	0.4
$\overline{RD}$	22		X	+0.4	+2.4	TTL	1.6
$\overline{CTS}$	23		X	+0.4	+2.4	TTL	1.6
$\overline{DSR}$	14		X	+0.4	+2.4	LSTTL	0.4
RTS	20	X		+0.8	+2.0	TTL	
TD	27	X		+0.8	+2.0	TTL	
DTR	25	X		+0.8	+2.0	TTL	
AL	26	X		+0.8	+2.0	TTL	
DL	1	X		+0.8	+2.0	TTL	
SL	11	X		+0.8	+2.0	TTL	

Signal Input and Output Compatibility

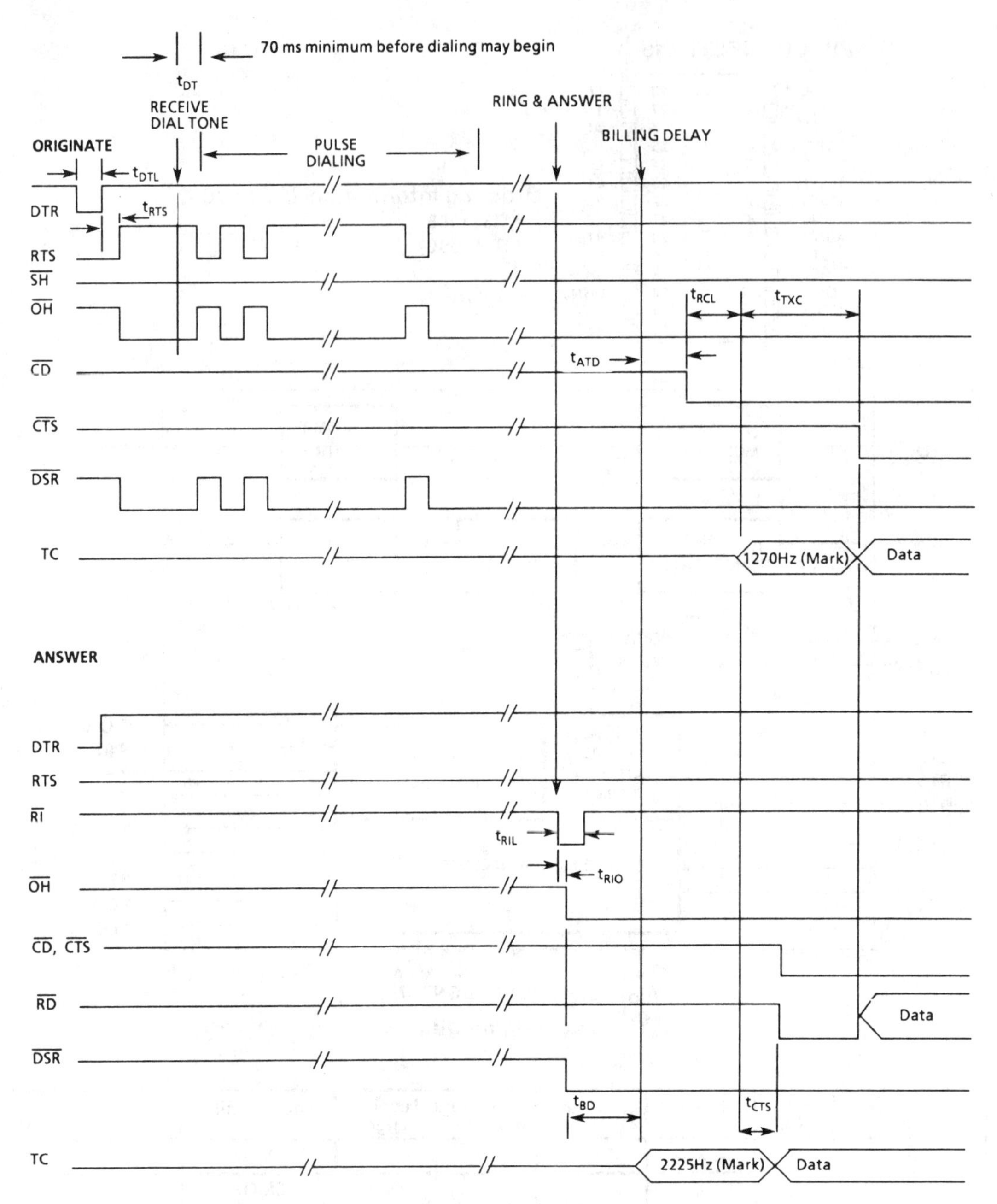

MT3530 Modem Timing Chart for Bell 103 Operating Mode

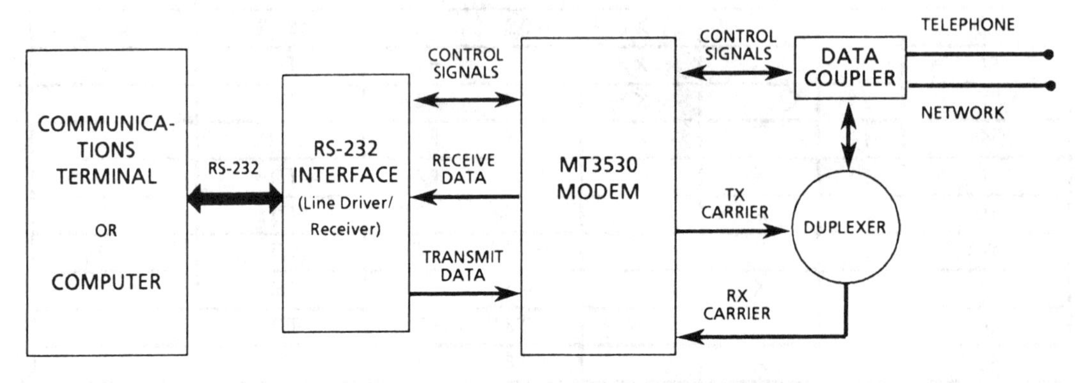

Serial Interface System Configuration for MT3530

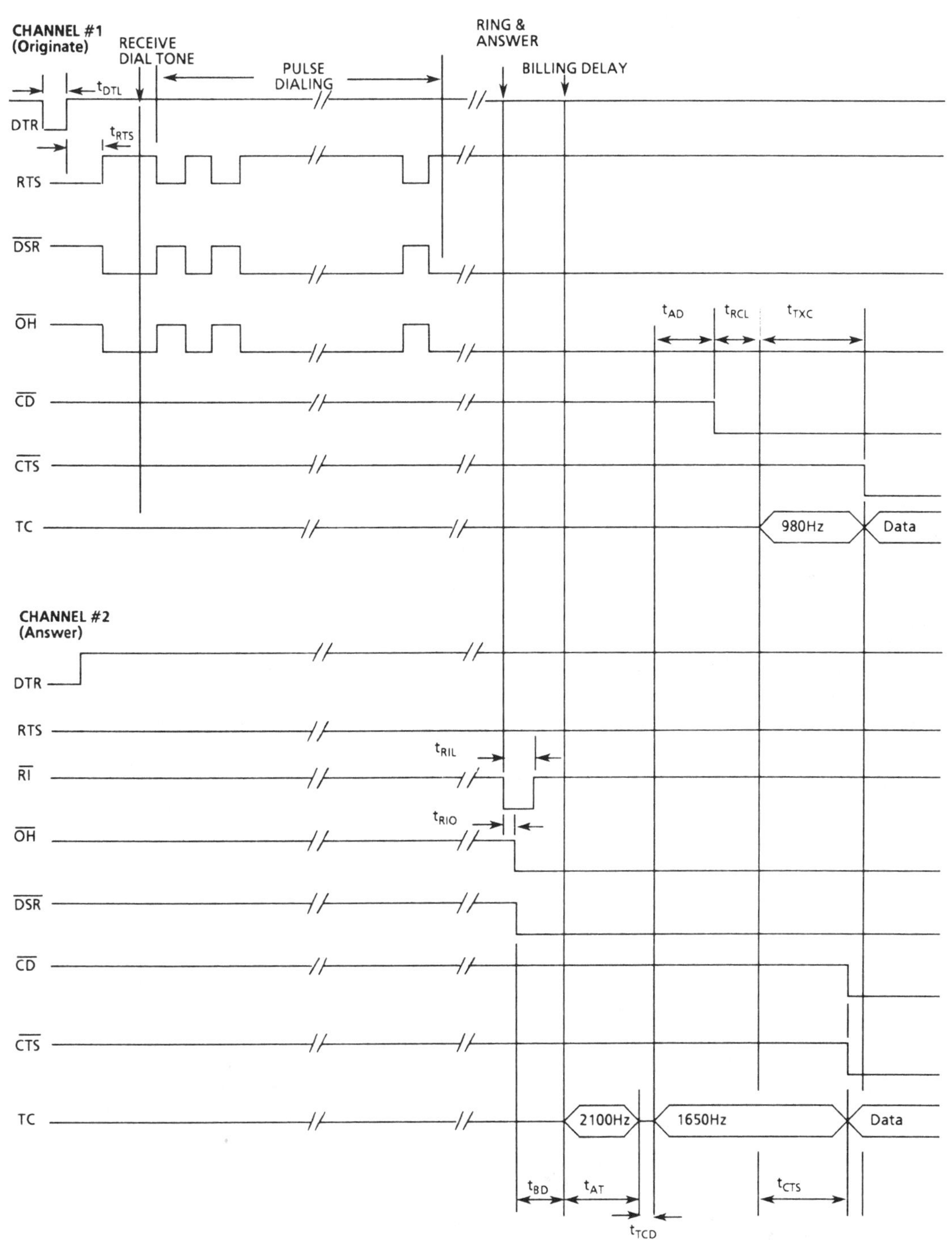

MT3530 Modem Timing Chart for CCITT V.21 Operating Mode

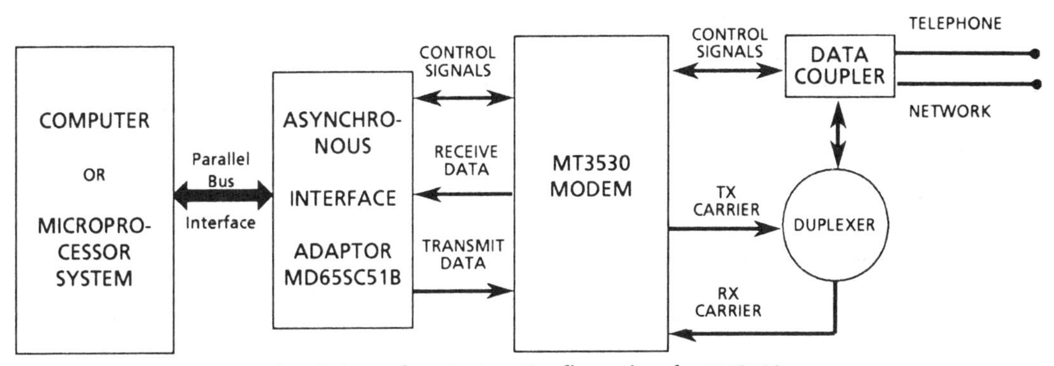

Parallel Interface System Configuration for MT3530

Mitel MT8840
Data-Over-Voice Modem

- Perform ASK (amplitude shift keyed) modulation and demodulation
- 32-kHz carrier frequency
- Up to 2 kbit/s full-duplex data-transfer rate
- On-chip oscillator
- On-chip tone caller for alerting functions
- Adjustable tone caller frequencies
- Selectable self-loop test mode
- 5V/2.5mA power supply
- ISO^2-CMOS and switched capacitor technologies
- 18-pin DIP

APPLICATIONS

- Simultaneous data and voice communication in PABXs
- 2 kbit/s data modem
- "Smart" telephone sets

PIN CONNECTIONS

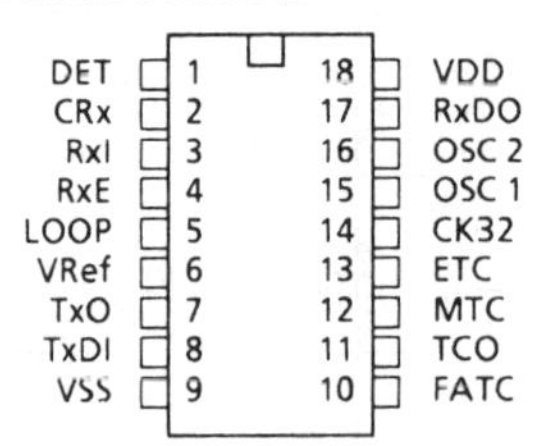

Ordering Information 0°C to +85°C

MT8840AC 18 Pin Ceramic Package

ABSOLUTE MAXIMUM RATINGS*

	Parameter	Symbol	Min	Max	Unit
1	Supply Voltage	V_{DD}-V_{SS}	-0.3	+7.0	V
2	Voltage On Any Pin	V_{Max}	V_{SS}-0.3	V_{DD}+0.3	V
3	Current On Any Pin	I_{Max}		20	mA
4	Storage Temperature	T_S	-65	+150	°C
5	Package Power Dissipation	P_{Diss}		850	mW

*Exceeding these ratings may cause permanent damage. Functional operation under these conditions is not implied.

RECOMMENDED OPERATING CONDITIONS

	Parameter	Symbol	Min	Typ	Max	Unit
1	Operating Supply Voltages	V_{DD}	4.75	5	5.25	V
2		V_{Ref}		$0.4V_{DD}$		V
3	Operating Supply Currents	I_{DD}		2.5	5.0	mA
4		I_{Ref}			200	µA
5	Operating Temperature	T_O	0		+85	°C
6	Load Capacitance (TxO)	C_L			50	pF
7	Load Resistance (TxO)	R_L	10			KΩ

DC CHARACTERISTICS V_{DD} = 5.0 V ± 5% V_{SS} = 0V T = 0 - 85°C (All voltages are referenced to V_{SS}/GND)

		Characteristics	Sym	Min	Typ	Max	Unit	Test Conditions
1	DIGITAL I/O	Input Current	I_{IN}			±10	µA	V_{IN}=0 to V_{DD}
2		Input Low Voltage	V_{IL}	0		1.5	V	
3		Input High Voltage	V_{IH}	3.5		5.0	V	
4		Output Low Voltage	V_{OL}			0.4	V	I_{OL}=0.4mA
5		Output High Voltage	V_{OH}	4.6			V	I_{OH}=0.4mA
6		Output Drive Current						
7		N Channel Sink (Except OSC2)	I_{OL}	0.4			mA	V_{OL}=0.4V
8		OSC2		0.1			mA	
9		P Channel Source (Except OSC2)	I_{OH}	0.4			mA	V_{OH}=4.6V
10		OSC2		0.1			mA	
11	ANALOG I/O	Input Current (RxI, FATC)	I_{IN}			±10	µA	V_{IN}=0 to 5.0V
12		Input Resistance (FATC)	R_{IN}	500			KΩ	
13		(DET to V_{DD})			170		KΩ	
14		(DET to V_{Ref})			23		KΩ	
15		Input Capacitance (RxI)	C_{IN}		50		pF	
16		(FATC)			10		pF	
17		Any Digital Input			5.0	7.5	pF	
18		Output Resistance (TxO)	R_O		100		Ω	
19		(TCO)			3		KΩ	MTC=0
20		(TCO)			30		KΩ	MTC=1
21		Output Offset Voltage (TxO)	V_O		±25	±200	mV	
22		Output Voltage (DET)	V_O	2.20	2.36	2.55	V	See Note 1

Notes: 1. Voltage specified is generated internally and measured with no external components connected to DET

AC CHARACTERISTICS - V_{DD}=5.0V±5% V_{SS}=0V T=0 - 85°C (All voltages are referenced to V_{SS}/GND)

		Characteristics	Sym	Min	Typ	Max	Unit	Test Conditions
1	DIGITAL I/O	Crystal/Clock Frequency	f_C	3.5759	3.5795	3.5831	MHz	OSC 1, OSC 2
2		Clock Input (OSC 1)						
3		Rise Time	t_{LHCI}			100	ns	10% - 90% of (V_{DD} - V_{SS})
4		Fall Time	t_{HLCI}			100	ns	
5		Duty Cycle	DC_{CI}	40	50	60	%	
6		Clock Output (OSC 2)						
7		Rise Time	t_{LHCO}		100		ns	C_L=30pF, 3.58MHz ext. clock to OSC1
8		Fall Time	t_{HLCO}		100		ns	
9		Duty Cycle	DC_{CO}		50		%	
10		Capacitive Load	C_{LCO}			30	pF	
11		Clock Output (CK32)	F_{C32}	32508	32541	32574	Hz	fc=3.5795MHz
12		Rise Time	t_{LH32}		100		ns	10% - 90% of (V_{DD} - V_{SS}) C_L=100pF
13		Fall Time	t_{HL32}		100		ns	
14		Duty Cycle	DC_{32}		50		%	
15		Capacitive Load	C_{L32}			100	pF	
16	TONE CALLER	Warbler Frequency (TCO)	f_W	7.935	7.945	7.955	Hz	fc=3.5795MHz±0.1%
17		Low Tone Frequency	f_{LT}	352	390	428	Hz	FATC=0, f_c=3.5795MHz
18				1036	1148	1260	Hz	FATC=V_{DD},f_c=3.5795MHz
19		High Tone Frequency	f_{HT}	440	487	535	Hz	FATC=0, f_c=3.5795MHz
20				1295	1434	1574	Hz	FATC=V_{DD},f_c=3.5795MHz
21		Harmonic Relationship	f_{HT}/f_{LT}		1.25			
22		Warbler Output (TCO)						
23		Rise Time	t_{LHWO}		500		ns	100KΩ load to V_{Ref} C_L=30pF, MTC=0
24		Fall Time	t_{HLWO}		500		ns	
25		Duty Cycle	DC_{WO}		50		%	
26		Output Level (TCO)	V_{TCC}		V_{DD}		V_{pp}	MTC=0
27					0.625		V_{pp}	MTC=1 (100KΩ load to V_{Ref})
28	MODULATOR	Modulated Frequency	f_{MOD}		32541		Hz	
29		Output Level (TxO)	V_{TxO}	225	250	270	mV_{pp}	V_{DD} = 5V
30		Output Level (TxO)						
31		variation vs. V_{DD}	V_{TxO}		100		%	
32		Transmit Data Input (TxDI)						
33		Rise Time	t_{LHTxDI}			100	ns	
34		Fall Time	t_{HLTxDI}			100	ns	
35		Data Rate (TxDI)	f_{Data}		2		kbit/s	See Note 1
36	DEMODULATOR	Input Impedance (RxI)	Z_{IN}		50		KΩ	32 kHz Input Frequency
37		Valid Input Level - Data (RxI)	V_{RxI}	40		400	mV_{pp}	See Note 2
38		Valid Input Level - Data + Voice	V_{RxI}			3.0	V_{pp}	
39		Receive Data Output (RxDO)	f_{Data}		2		kbit/s	
40		Rise Time			100		ns	10% - 90% of (V_{DD} - V_{SS}) C_L = 100pF
41		Fall Time			100		ns	
42		Capacitive Load				100	pF	
43		Duty Cycle		40	50	60	%	
44	DEMOD	Inband Noise Rejection (S/N)		12			dB	Input Sig. (RxI)=400mV_{pp}
45		Attenuation to Voice Signals		40			dB	f_{in} = 0 - 5KHz
46		Detect Filter Q	Q		3.8			
47		Detector Center Frequency			32		kHz	

Notes: 1. All A.C. parameters are based on a typical data rate of 2 kbit/s.
2. Measured with no external resistor to DET input. Detection level internally set to 2.36V typical

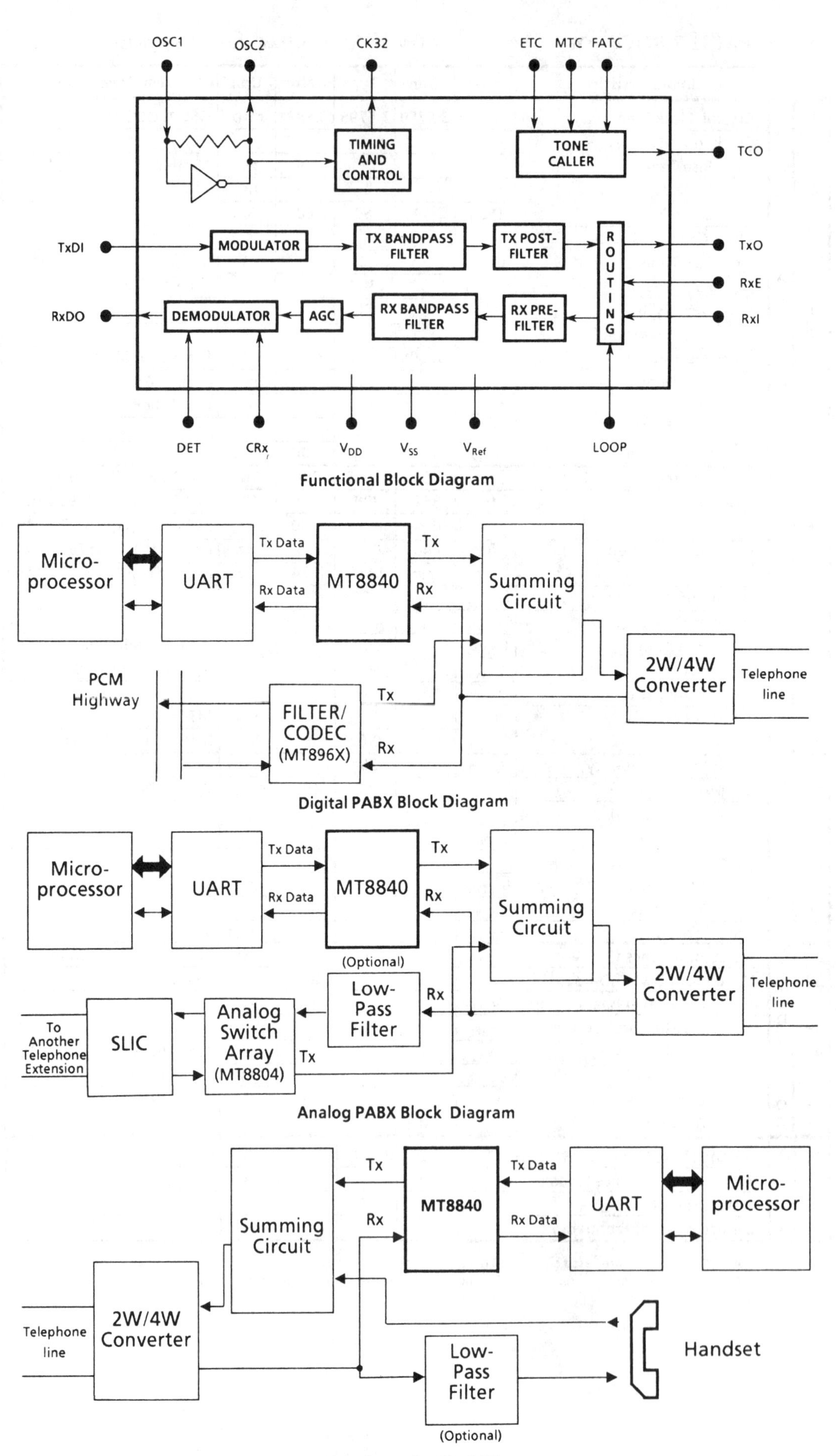

Functional Block Diagram

Digital PABX Block Diagram

Analog PABX Block Diagram

Smart Telephone Set Block Diagram

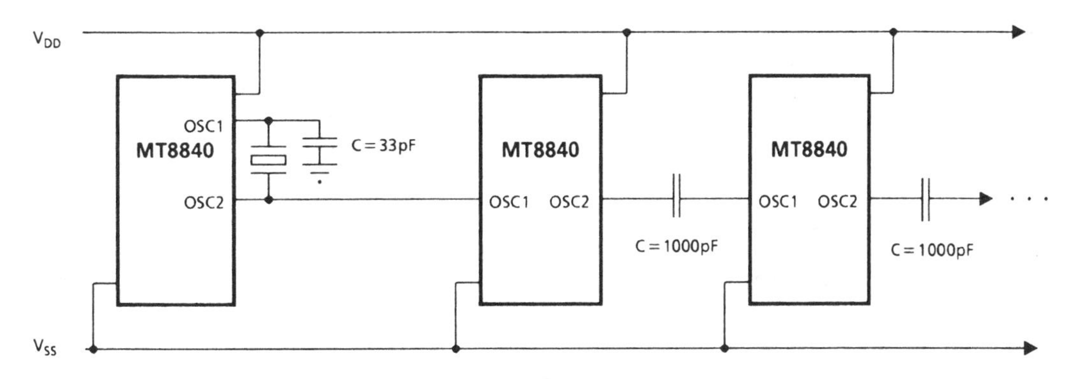

When a single crystal is shared among a number of MT8840 devices, OSC1 and OSC2 should be a.c. coupled with a 1000 pF capacitor as shown above. This capacitor is not needed between the device with the crystal and the first driven device. A capacitor should be used in the first stage whenever such a chain of devices is driven from a clock instead of a crystal. A 33pF capacitor should be connectced between OSC1 and V_{ss} to compensate for the load on OSC2.

Crystal Oscillator Connections for Driving Multiple MT8840's

This information is reproduced with permission of National Semiconductor Corporation.

Life-Support Policy

NATIONAL'S PRODUCTS ARE NOT AUTHORIZED FOR USE AS CRITICAL COMPONENTS IN LIFE-SUPPORT DEVICES OR SYSTEMS WITHOUT THE EXPRESS WRITTEN APPROVAL OF THE PRESIDENT OF NATIONAL SEMICONDUCTOR CORPORATION.

1. Life-support devices or systems are devices or systems which, (a) are intended for surgical implant into the body, or (b) support or sustain life, and whose failure to perform, when properly used in accordance with instructions for use provided in the labeling, can be reasonably expected to result in a significant injury to the user.

2. A critical component is any component of a life-support device or system whose failure to perform can be reasonably expected to cause the failure of the life-support device or system, or to affect its safety or effectiveness.

National 54LS240/DM54LS240/DM74LS240, 54LS241/DM54LS241/DM74LS241

Octal Tri-State® Buffers/Line Drivers/Line Receivers

- Tri-State outputs drive bus lines directly
- Pnp inputs reduce DC loading on bus lines
- Hysteresis at data inputs improves noise margins
- Typical I_{OL} (sink current)
  54LS 12 mA
  74LS 24 mA
- Typical I_{OH} (source current)
  54LS −12 mA
  74LS −15 mA
- Typical propagation delay times
  Inverting 10.5 ns
  Noninverting 12 ns
- Typical enable/disable time; 18 ns
- Typical power dissipation (enabled)
  Inverting 130 mW
  Noninverting 135 mW

ABSOLUTE MAXIMUM RATINGS (Note)

If Military/Aerospace specified devices are required, please contact the National Semiconductor Sales Office/Distributors for availability and specifications.

Supply Voltage	7V
Input Voltage	7V
Operating Free Air Temperature Range	
DM54LS, 54LS	−55°C to +125°C
DM74LS	0°C to +70°C
Storage Temperature Range	−65°C to +150°C

Note: *The "Absolute Maximum Ratings" are those values beyond which the safety of the device cannot be guaranteed. The device should not be operated at these limits. The parametric values defined in the "Electrical Characteristics" table are not guaranteed at the absolute maximum ratings. The "Recommended Operating Conditions" table will define the conditions for actual device operation.*

RECOMMENDED OPERATING CONDITIONS

Symbol	Parameter	DM54LS240, 241			DM74LS240, 241			Units
		Min	Nom	Max	Min	Nom	Max	
V_{CC}	Supply Voltage	4.5	5	5.5	4.75	5	5.25	V
V_{IH}	High Level Input Voltage	2			2			V
V_{IL}	Low Level Input Voltage			0.7			0.8	V
I_{OH}	High Level Output Current			−12			−15	mA
I_{OL}	Low Level Output Current			12			24	mA
T_A	Free Air Operating Temperature	−55		125	0		70	°C

ELECTRICAL CHARACTERISTICS over recommended operating free air temperature range (unless otherwise noted)

Symbol	Parameter	Conditions			Min	Typ (Note 1)	Max	Units
V_I	Input Clamp Voltage	V_{CC} = Min, I_I = −18 mA					−1.5	V
HYS	Hysteresis ($V_{T+} - V_{T-}$) Data Inputs Only	V_{CC} = Min			0.2	0.4		V
V_{OH}	High Level Output Voltage	V_{CC} = Min, V_{IH} = Min, V_{IL} = Max, I_{OH} = −1 mA		DM74	2.7			V
		V_{CC} = Min, V_{IH} = Min, V_{IL} = Max, I_{OH} = −3 mA		DM54/DM74	2.4	3.4		
		V_{CC} = Min, V_{IH} = Min, V_{IL} = 0.5V, I_{OH} = Max		DM54/DM74	2			
V_{OL}	Low Level Output Voltage	V_{CC} = Min, V_{IL} = Max, V_{IH} = Min	I_{OL} = 12 mA	DM74			0.4	V
			I_{OL} = Max	DM54			0.4	
				DM74			0.5	
I_{OZH}	Off-State Output Current, High Level Voltage Applied	V_{CC} = Max, V_{IL} = Max, V_{IH} = Min	V_O = 2.7V				20	μA
I_{OZL}	Off-State Output Current, Low Level Voltage Applied		V_O = 0.4V				−20	μA
I_I	Input Current at Maximum Input Voltage	V_{CC} = Max, V_I = 7V (DM74), V_I = 10V (DM54)					0.1	mA
I_{IH}	High Level Input Current	V_{CC} = Max, V_I = 2.7V					20	μA
I_{IL}	Low Level Input Current	V_{CC} = Max, V_I = 0.4V					−0.2	mA
I_{OS}	Short Circuit Output Current	V_{CC} = Max (Note 2)			−40		−225	mA
I_{CC}	Supply Current	V_{CC} = Max, Outputs Open	Outputs High	LS240, LS241		13	23	mA
			Outputs Low	LS240		26	44	
				LS241		27	46	
			Outputs Disabled	LS240		29	50	
				LS241		32	54	

Note 1: All typicals are at V_{CC} = 5V, T_A = 25°C.

Note 2: Not more than one output should be shorted at a time, and the duration should not exceed one second.

CONNECTION DIAGRAMS

Dual-In-Line Package

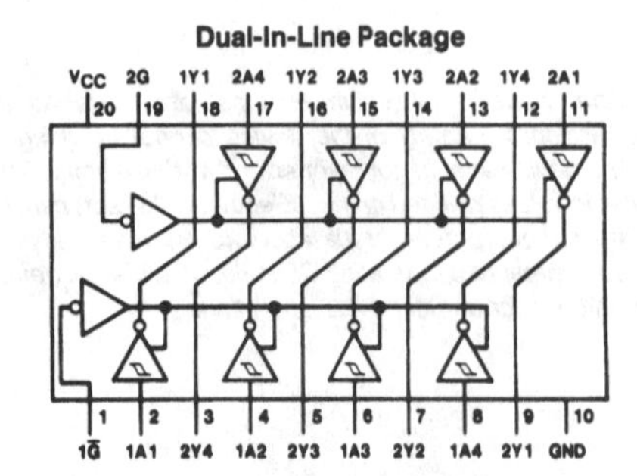

Order Number 54LS240DMQB, 54LS240FMQB, 54LS240LMQB, DM54LS240J, DM74LS240WM or DM74LS240N
See NS Package Number E20A, J20A, M20B, N20A or W20A

Dual-In-Line Package

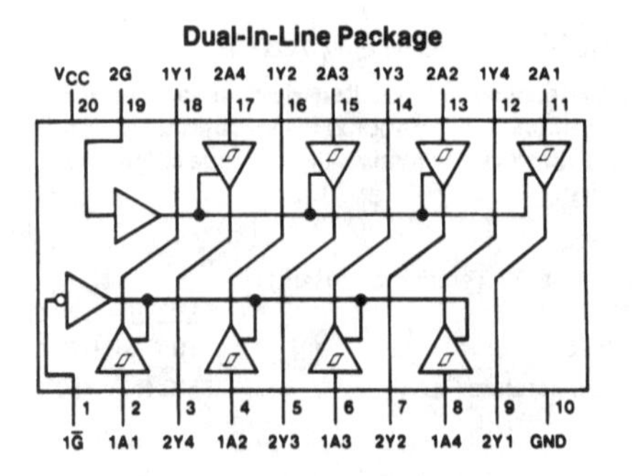

Order Number 54LS241DMQB, 54LS241FMQB, 54LS241LMQB, DM54LS241J, DM74LS241WM or DM74LS241N
See NS Package Number E20A, J20A, M20B, N20A or W20A

SWITCHING CHARACTERISTICS at V_{CC} = 5V and T_A = 25°C (See Section 1 for Test Waveforms and Output Load)

Symbol	Parameter	Conditions		DM54LS Max	DM74LS Max	Units
t_{PLH}	Propagation Delay Time Low to High Level Output	C_L = 45 pF, R_L = 667Ω	LS240	18	14	ns
			LS241	18	18	
t_{PHL}	Propagation Delay Time High to Low Level Output	C_L = 45 pF, R_L = 667Ω	LS240	18	18	ns
			LS241	18	18	
t_{PZL}	Output Enable Time to Low Level	C_L = 45 pF, R_L = 667Ω	LS240	30	30	ns
			LS241	30	30	
t_{PZH}	Output Enable Time to High Level	C_L = 45 pF, R_L = 667Ω	LS240	23	23	ns
			LS241	23	23	
t_{PLZ}	Output Disable Time from Low Level	C_L = 5 pF, R_L = 667Ω	LS240	25	25	ns
			LS241	25	25	
t_{PHZ}	Output Disable Time from High Level	C_L = 5 pF, R_L = 667Ω	LS240	18	18	ns
			LS241	18	18	
t_{PLH}	Propagation Delay Time Low to High Level Output	C_L = 150 pF, R_L = 667Ω	LS240		18	ns
			LS241		21	
t_{PHL}	Propagation Delay Time High to Low Level Output	C_L = 150 pF, R_L = 667Ω	LS240		22	ns
			LS241		22	
t_{PZL}	Output Enable Time to Low Level	C_L = 150 pF, R_L = 667Ω	LS240		33	ns
			LS241		33	
t_{PZH}	Output Enable Time to High Level	C_L = 150 pF, R_L = 667Ω	LS240		26	ns
			LS241		26	

Note: 54LS Output load is C_L = 50 pF for t_{PLH}, t_{PHL}, t_{PZL} and t_{PZH}.

FUNCTION TABLES

LS240

Inputs		Output
$\overline{G}$	A	Y
L	L	H
L	H	L
H	X	Z

LS241

Inputs				Outputs	
G	$\overline{G}$	1A	2A	1Y	2Y
X	L	L	X	L	
X	L	H	X	H	
X	H	X	X	Z	
H	X	X	L		L
H	X	X	H		H
L	X	X	X		Z

L = Low Logic Level
H = High Logic Level
X = Either Low or High Logic Level
Z = High Impedance

National DM74LS243 Quadruple Bus Transceiver

- Two-way asynchronous communication between data buses
- Pnp inputs reduce DC loading on bus line
- Hysteresis at data inputs improves noise margin

ABSOLUTE MAXIMUM RATINGS (Note)

If Military/Aerospace specified devices are required, please contact the National Semiconductor Sales Office/Distributors for availability and specifications.

Supply Voltage	7V
Input Voltage	
Any G	7V
A or B	5.5V
Operating Free Air Temperature Range	
DM74LS	0°C to +70°C
Storage Temperature Range	−65°C to +150°C

Note: *The "Absolute Maximum Ratings" are those values beyond which the safety of the device cannot be guaranteed. The device should not be operated at these limits. The parametric values defined in the "Electrical Characteristics" table are not guaranteed at the absolute maximum ratings. The "Recommended Operating Conditions" table will define the conditions for actual device operation.*

RECOMMENDED OPERATING CONDITIONS

Symbol	Parameter	DM74LS243			Units
		Min	Nom	Max	
V_{CC}	Supply Voltage	4.75	5	5.25	V
V_{IH}	High Level Input Voltage	2			V
V_{IL}	Low Level Input Voltage			0.8	V
I_{OH}	High Level Output Current			−15	mA
I_{OL}	Low Level Output Current			24	mA
T_A	Free Air Operating Temperature	0		70	°C

ELECTRICAL CHARACTERISTICS over recommended operating free air temperature range (unless otherwise noted)

Symbol	Parameter	Conditions	Min	Typ (Note 1)	Max	Units
V_I	Input Clamp Voltage	V_{CC} = Min, I_I = −18 mA			−1.5	V
HYS	Hysteresis ($V_{T+} - V_{T-}$) (Data Inputs Only)	V_{CC} = Min	0.2	0.4		V
V_{OH}	High Level Output Voltage	V_{CC} = Min, V_{IH} = Min, V_{IL} = Max, I_{OH} = −1 mA	2.7			V
		V_{CC} = Min, V_{IH} = Min, V_{IL} = Max, I_{OH} = −3 mA	2.4	3.4		V
		V_{CC} = Min, V_{IH} = Min, V_{IL} = 0.5V, I_{OH} = Max	2			V
V_{OL}	Low Level Output Voltage	V_{CC} = Min, V_{IL} = Max, V_{IH} = Min; I_{OL} = 12 mA			0.4	V
		V_{CC} = Min, V_{IL} = Max, V_{IH} = Min; I_{OL} = Max			0.5	V
I_{OZH}	Off-State Output Current, High Level Voltage Applied	V_{CC} = Max, V_{IL} = Max, V_{IH} = Min; V_O = 2.7V			40	μA
I_{OZL}	Off-State Output Current, Low Level Voltage Applied	V_{CC} = Max, V_{IL} = Max, V_{IH} = Min; V_O = 0.4V			−200	μA
I_I	Input Current at Maximum Input Voltage	V_{CC} = Max; V_I = 5.5V; A or B			0.1	mA
		V_{CC} = Max; V_I = 7V; Any G			0.1	mA
I_{IH}	High Level Input Current	V_{CC} = Max, V_I = 2.7V			20	μA
I_{IL}	Low Level Input Current	V_{CC} = Max, V_I = 0.4V			−0.2	mA
I_{OS}	Short Circuit Output Current	V_{CC} = Max (Note 2)	−40		−225	mA
I_{CC}	Supply Current	V_{CC} = Max, Outputs Open; Outputs High		22	38	mA
		V_{CC} = Max, Outputs Open; Outputs Low		29	50	mA
		V_{CC} = Max, Outputs Open; Outputs Disabled		32	54	mA

Note 1: All typicals are at V_{CC} = 5V, T_A = 25°C.

Note 2: Not more than one output should be shorted at a time, and the duration should not exceed one second.

CONNECTION DIAGRAM

Dual-In-Line Package

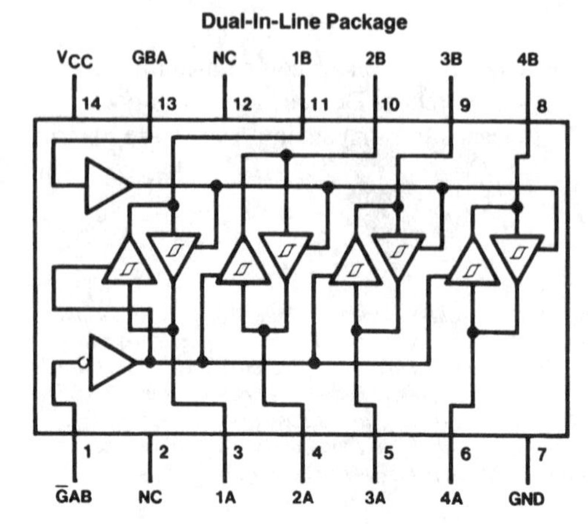

Order Number DM74LS243WM or DM74LS243N
See NS Package Number M14B or N14A

FUNCTION TABLE

Control Inputs		Data Port Status	
$\overline{GAB}$	GBA	A	B
H	H	O	I
L	H	*	*
H	L	ISOLATED	
L	L	I	O

*Possibly destructive oscillation may occur if the transceivers are enabled in both directions at once.

I = Input, O = Output.

H = High Logic Level, L = Low Logic Level.

SWITCHING CHARACTERISTICS at V_{CC} = 5V, T_A = 25°C (See Section 1 for Test Waveforms and Output Load)

Symbol	Parameter	Conditions	Min	Max	Units
t_{PLH}	Propagation Delay Time Low to High Level Output	C_L = 45 pF R_L = 667Ω		18	ns
t_{PHL}	Propagation Delay Time High to Low Level Output	C_L = 45 pF R_L = 667Ω		18	ns
t_{PZL}	Output Enable Time to Low Level	C_L = 45 pF R_L = 667Ω		30	ns
t_{PZH}	Output Enable Time to High Level	C_L = 45 pF R_L = 667Ω		23	ns
t_{PLZ}	Output Disable Time from Low Level	C_L = 5 pF R_L = 667Ω		25	ns
t_{PHZ}	Output Disable Time from High Level	C_L = 5 pF R_L = 667Ω		18	ns
t_{PLH}	Propagation Delay Time Low to High Level Output	C_L = 150 pF R_L = 667Ω		21	ns
t_{PHL}	Propagation Delay Time High to Low Level Output	C_L = 150 pF R_L = 667Ω		22	ns
t_{PZL}	Output Enable Time to Low Level	C_L = 150 pF R_L = 667Ω		33	ns
t_{PZH}	Output Enable Time to High Level	C_L = 150 pF R_L = 667Ω		26	ns

National 54LS244/DM74LS244

Octal Tri-State® Buffers/Line Drivers/Line Receivers

- Tri-State outputs drive bus lines directly
- Pnp inputs reduce DC loading on bus lines
- Hysteresis at data inputs improves noise margins
- Typical I_{OL} (sink current)
  54LS 12 mA
  74LS 24 mA
- Typical I_{OH} (source current)
  54LS −12 mA
  74LS −15 mA
- Typical propagation delay times
  Inverting 10.5 ns
  Noninverting 12 ns
- Typical enable/disable time: 18 ns
- Typical power dissipation (enabled)
  Inverting 130 mW
  Noninverting 135 mW

ABSOLUTE MAXIMUM RATINGS (Note)

If Military/Aerospace specified devices are required, please contact the National Semiconductor Sales Office/Distributors for availability and specifications.

Supply Voltage	7V
Input Voltage	7V
Operating Free Air Temperature Range	
54LS	−55°C to +125°C
DM74LS	0°C to +70°C
Storage Temperature Range	−65°C to +150°C

Note: *The "Absolute Maximum Ratings" are those values beyond which the safety of the device cannot be guaranteed. The device should not be operated at these limits. The parametric values defined in the "Electrical Characteristics" table are not guaranteed at the absolute maximum ratings. The "Recommended Operating Conditions" table will define the conditions for actual device operation.*

RECOMMENDED OPERATING CONDITIONS

Symbol	Parameter	54LS244			DM74LS244			Units
		Min	Nom	Max	Min	Nom	Max	
V_{CC}	Supply Voltage	4.5	5	5.5	4.75	5	5.25	V
V_{IH}	High Level Input Voltage	2			2			V
V_{IL}	Low Level Input Voltage			0.7			0.8	V
I_{OH}	High Level Output Current			−12			−15	mA
I_{OL}	Low Level Output Current			12			24	mA
T_A	Free Air Operating Temperature	−55		125	0		70	°C

ELECTRICAL CHARACTERISTICS over recommended operating free air temperature range (unless otherwise noted)

Symbol	Parameter	Conditions			Min	Typ (Note 1)	Max	Units
V_I	Input Clamp Voltage	V_{CC} = Min, I_I = −18 mA					−1.5	V
HYS	Hysteresis (V_{T+} − V_{T-}) Data Inputs Only	V_{CC} = Min			0.2	0.4		V
V_{OH}	High Level Output Voltage	V_{CC} = Min, V_{IH} = Min V_{IL} = Max, I_{OH} = −1 mA		DM74	2.7			V
		V_{CC} = Min, V_{IH} = Min V_{IL} = Max, I_{OH} = −3 mA		54LS/DM74	2.4	3.4		
		V_{CC} = Min, V_{IH} = Min V_{IL} = 0.5V, I_{OH} = Max		54LS/DM74	2			
V_{OL}	Low Level Output Voltage	V_{CC} = Min V_{IL} = Max V_{IH} = Min	I_{OL} = 12 mA	54LS/DM74			0.4	V
			I_{OL} = Max	DM74			0.5	
I_{OZH}	Off-State Output Current, High Level Voltage Applied	V_{CC} = Max V_{IL} = Max V_{IH} = Min	V_O = 2.7V				20	μA
I_{OZL}	Off-State Output Current, Low Level Voltage Applied		V_O = 0.4V				−20	μA
I_I	Input Current at Maximum Input Voltage	V_{CC} = Max	V_I = 7V (DM74) V_I = 10V (54LS)				0.1	mA
I_{IH}	High Level Input Current	V_{CC} = Max	V_I = 2.7V				20	μA
I_{IL}	Low Level Input Current	V_{CC} = Max	V_I = 0.4V		−0.5		−200	μA
I_{OS}	Short Circuit Output Current	V_{CC} = Max (Note 2)		54LS	−50		−225	mA
				DM74	−40			
I_{CC}	Supply Current	V_{CC} = Max, Outputs Open	Outputs High			13	23	mA
			Outputs Low			27	46	
			Outputs Disabled			32	54	

Note 1: All typicals are at V_{CC} = 5V, T_A = 25°C.

Note 2: Not more than one output should be shorted at a time, and the duration should not exceed one second.

SWITCHING CHARACTERISTICS at V_{CC} = 5V, T_A = 25°C (see Section 1 for Test Waveforms and Output Load)

Symbol	Parameter	Conditions	54LS Max	DM74LS Max	Units
t_{PLH}	Propagation Delay Time Low to High Level Output	C_L = 45 pF R_L = 667Ω	18	18	ns
t_{PHL}	Propagation Delay Time High to Low Level Output	C_L = 45 pF R_L = 667Ω	18	18	ns
t_{PZL}	Output Enable Time to Low Level	C_L = 45 pF R_L = 667Ω	30	30	ns
t_{PZH}	Output Enable Time to High Level	C_L = 45 pF R_L = 667Ω	23	23	ns
t_{PLZ}	Output Disable Time from Low Level	C_L = 5 pF R_L = 667Ω	25	25	ns
t_{PHZ}	Output Disable Time from High Level	C_L = 5 pF R_L = 667Ω	18	18	ns
t_{PLH}	Propagation Delay Time Low to High Level Output	C_L = 150 pF R_L = 667Ω		21	ns
t_{PHL}	Propagation Delay Time High to Low Level Output	C_L = 150 pF R_L = 667Ω		22	ns
t_{PZL}	Output Enable Time to Low Level	C_L = 150 pF R_L = 667Ω		33	ns
t_{PZH}	Output Enable Time to High Level	C_L = 150 pF R_L = 667Ω		26	ns

Note: 54LS Output Load is C_L = 50 pF for t_{PLH}, t_{PHL}, t_{PZL} and t_{PZH}.

CONNECTION DIAGRAM

Dual-In-Line Package

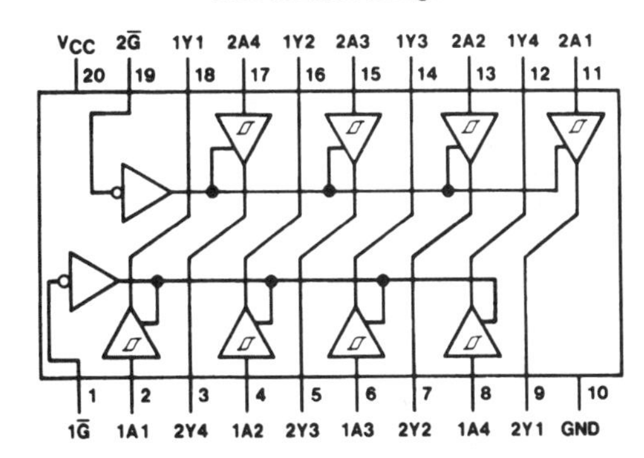

Order Number 54LS244DMQB, 54LS244FMQB, 54LS244LMQB, DM74LS244WM or DM74LS244N
See NS Package Number E20A, J20A, M20B, N20A or W20A

FUNCTION TABLE

Inputs		Output
$\overline{G}$	A	Y
L	L	L
L	H	H
H	X	Z

L = Low Logic Level
H = High Logic Level
X = Either Low or High Logic Level
Z = High Impedance

National 54LS245/DM54LS245/DM74LS245

Tri-State® Octal Bus Transceiver

- Bi-directional bus transceiver in a high-density 20-pin package
- Tri-state outputs drive bus lines directly
- Pnp inputs reduce dc loading on bus lines
- Hysteresis at bus inputs improve noise margins
- Typical propagation delay times, port-to-port: 8 ns
- Typical enable/disable times: 17 ns
- I_{OL} (sink current)
  - 54LS 12 mA
  - 74LS 24 mA
- I_{OH} (source current)
  - 54LS −12 mA
  - 74LS −15 mA

ABSOLUTE MAXIMUM RATINGS (Note)

If Military/Aerospace specified devices are required, please contact the National Semiconductor Sales Office/Distributors for availability and specifications.

Supply Voltage	7V
Input Voltage	
DIR or $\overline{G}$	7V
A or B	5.5V
Operating Free Air Temperature Range	
DM54LS and 54LS	−55°C to +125°C
DM74LS	0°C to +70°C
Storage Temperature Range	−65°C to +150°C

Note: *The "Absolute Maximum Ratings" are those values beyond which the safety of the device cannot be guaranteed. The device should not be operated at these limits. The parametric values defined in the "Electrical Characteristics" table are not guaranteed at the absolute maximum ratings. The "Recommended Operating Conditions" table will define the conditions for actual device operation.*

RECOMMENDED OPERATING CONDITIONS

Symbol	Parameter	DM54LS245			DM74LS245			Units
		Min	Nom	Max	Min	Nom	Max	
V_{CC}	Supply Voltage	4.5	5	5.5	4.75	5	5.25	V
V_{IH}	High Level Input Voltage	2			2			V
V_{IL}	Low Level Input Voltage			0.7			0.8	V
I_{OH}	High Level Output Current			−12			−15	mA
I_{OL}	Low Level Output Current			12			24	mA
T_A	Free Air Operating Temperature	−55		125	0		70	°C

ELECTRICAL CHARACTERISTICS over recommended operating free air temperature range (unless otherwise noted)

Symbol	Parameter	Conditions			Min	Typ (Note 1)	Max	Units
V_I	Input Clamp Voltage	V_{CC} = Min, I_I = −18 mA					−1.5	V
HYS	Hysteresis (V_{T+} − V_{T-})	V_{CC} = Min			0.2	0.4		V
V_{OH}	High Level Output Voltage	V_{CC} = Min, V_{IH} = Min V_{IL} = Max, I_{OH} = −1 mA		DM74	2.7			V
		V_{CC} = Min, V_{IL} = Min V_{IL} = Max, I_{OH} = −3 mA		DM54/DM74	2.4	3.4		
		V_{CC} = Min, V_{IH} = Min V_{IL} = 0.5V, I_{OH} = Max		DM54/DM74	2			
V_{OL}	Low Level Output Voltage	V_{CC} = Min V_{IL} = Max V_{IH} = Min	I_{OL} = 12 mA	DM74			0.4	V
			I_{OL} = Max	DM54			0.4	
				DM74			0.5	
I_{OZH}	Off-State Output Current, High Level Voltage Applied	V_{CC} = Max V_{IL} = Max V_{IH} = Min	V_O = 2.7V				20	μA
I_{OZL}	Off-State Output Current, Low Level Voltage Applied		V_O = 0.4V				−200	μA
I_I	Input Current at Maximum Input Voltage	V_{CC} = Max	A or B	V_I = 5.5V			0.1	mA
			DIR or $\overline{G}$	V_I = 7V			0.1	
I_{IH}	High Level Input Current	V_{CC} = Max, V_I = 2.7V					20	μA
I_{IL}	Low Level Input Current	V_{CC} = Max, V_I = 0.4V					−0.2	mA
I_{OS}	Short Circuit Output Current	V_{CC} = Max (Note 2)			−40		−225	mA
I_{CC}	Supply Current	Outputs High		V_{CC} = Max		48	70	mA
		Outputs Low				62	90	
		Outputs at Hi-Z				64	95	

Note 1: All typicals are at V_{CC} = 5V, T_A = 25°C.

Note 2: Not more than one output should be shorted at a time, not to exceed one second duration

SWITCHING CHARACTERISTICS V_{CC} = 5V, T_A = 25°C (See Section 1 for Test Waveforms and Output Load)

Symbol	Parameter	Conditions	DM54/74 LS245 Min	DM54/74 LS245 Max	Units
t_{PLH}	Propagation Delay Time, Low-to-High-Level Output	C_L = 45 pF R_L = 667Ω		12	ns
t_{PHL}	Propagation Delay Time, High-to-Low-Level Output			12	ns
t_{PZL}	Output Enable Time to Low Level			40	ns
t_{PZH}	Output Enable Time to High Level			40	ns
t_{PLZ}	Output Disable Time from Low Level	C_L = 5 pF R_L = 667Ω		25	ns
t_{PHZ}	Output Disable Time from High Level			25	ns
t_{PLH}	Propagation Delay Time, Low-to-High-Level Output	C_L = 150 pF R_L = 667Ω		16	ns
t_{PHL}	Propagation Delay Time, High-to-Low-Level Output			17	ns
t_{PZL}	Output Enable Time to Low Level			45	ns
t_{PZH}	Output Enable Time to High Level			45	ns

FUNCTION TABLE

Enable $\overline{G}$	Direction Control DIR	Operation
L	L	B data to A bus
L	H	A data to B bus
H	X	Isolation

H = High Level, L = Low Level, X = Irrelevant

CONNECTION DIAGRAM

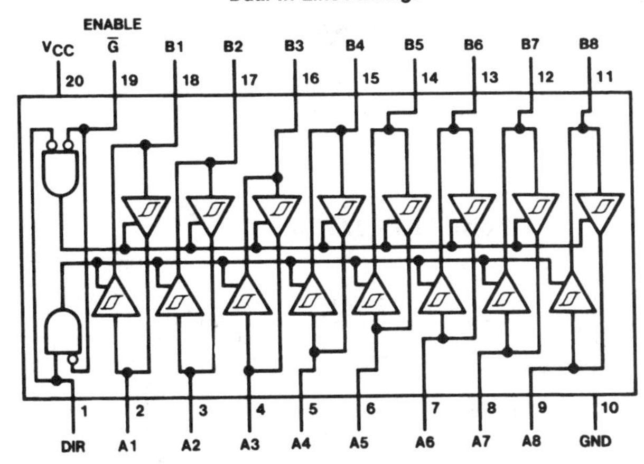

Order Number 54LS245DMQB, 54LS245FMQB, 54LS245LMQB, DM54LS245J, DM54LS245W, DM74LS245WM or DM74LS245N
See NS Package Number E20A, J20A, M20B, N20A or W20A

National DM54LS450/DM74LS450
16:1 Multiplexer
Preliminary Information

- 24-pin skinny DIP saves space
- Similar to 74150 (fat DIP)
- Low-current pnp inputs reduce loading

ABSOLUTE MAXIMUM RATINGS

If Military/Aerospace specified devices are required, please contact the National Semiconductor Sales Office/Distributors for availability and specifications.

Supply Voltage V_{CC}	7V
Input Voltage	5.5V
Off-State Output Voltage	5.5V
Storage Temperature	−65° to +150°C

OPERATING CONDITIONS

Symbol	Parameter	Military			Commercial			Units
		Min	Nom	Max	Min	Nom	Max	
V_{CC}	Supply Voltage	4.5	5	5.5	4.75	5	5.25	V
T_A	Operating Free-Air Temperature	−55		125*	0		75	°C

*Case temperature

ELECTRICAL CHARACTERISTICS Over Operating Conditions

Symbol	Parameter	Test Conditions			Min	Typ†	Max	Units
V_{IL}	Low-Level Input Voltage						0.8	V
V_{IH}	High-Level Input Voltage				2			V
V_{IC}	Input Clamp Voltage	V_{CC}=MIN	I_I=−18 mA				−1.5	V
I_{IL}	Low-Level Input Current	V_{CC}=MAX	V_I=0.4V				−0.25	mA
I_{IH}	High-Level Input Current	V_{CC}=MAX	V_I=2.4V				25	μA
I_I	Maximum Input Current	V_{CC}=MAX	V_I=5.5V				1	mA
V_{OL}	Low-Level Output Voltage	V_{CC}=MIN V_{IL}=0.8V V_{IH}=2V		I_{OL}=8 mA			0.5	V
V_{OH}	High-Level Output Voltage	V_{CC}=MIN V_{IL}=0.8V V_{IH}=2V	MIL	I_{OH}=−2 mA	2.4			V
			COM	I_{OH}=−3.2 mA				
I_{OS}	Output Short-Circuit Current*	V_{CC}=5.0V		V_O=0V	−30		−130	mA
I_{CC}	Supply Current	V_{CC}=MAX				60	100	mA

*No more than one output should be shorted at a time and duration of the short-circuit should not exceed one second.

†All typical values are at V_{CC}=5V, T_A=25°C.

SWITCHING CHARACTERISTICS Over Operating Conditions

Symbol	Parameter	Test Conditions (See Test Load)	Military			Commercial			Units
			Min	Typ	Max	Min	Typ	Max	
t_{PD}	Any Input to Y or W	$C_L = 50$ pF $R_1 = 560\Omega$ $R_2 = 1.1k\Omega$		25	45		25	40	ns

CONNECTION DIAGRAM

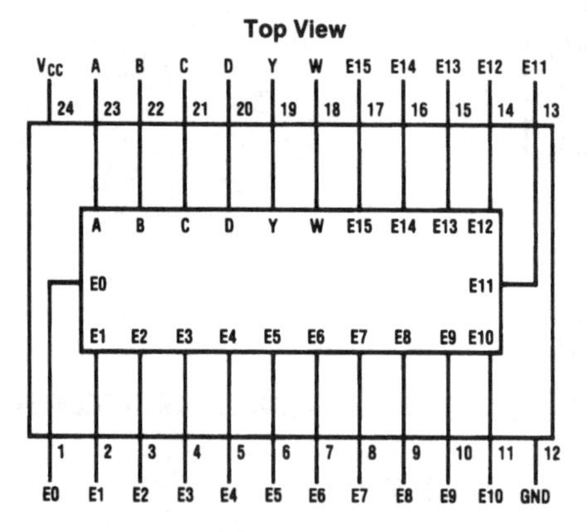

Order Number DM54LS450J, DM74LS450J, DM74LS450N or DM74LS450V
See NS Package Number J24F, N24C or V28A

FUNCTION TABLE

Input Select				Output	
D	C	B	A	W	Y
L	L	L	L	$\overline{E0}$	E0
L	L	L	H	$\overline{E1}$	E1
L	L	H	L	$\overline{E2}$	E2
L	L	H	H	$\overline{E3}$	E3
L	H	L	L	$\overline{E4}$	E4
L	H	L	H	$\overline{E5}$	E5
L	H	H	L	$\overline{E6}$	E6
L	H	H	H	$\overline{E7}$	E7
H	L	L	L	$\overline{E8}$	E8
H	L	L	H	$\overline{E9}$	E9
H	L	H	L	$\overline{E10}$	E10
H	L	H	H	$\overline{E11}$	E11
H	H	L	L	$\overline{E12}$	E12
H	H	L	H	$\overline{E13}$	E13
H	H	H	L	$\overline{E14}$	E14
H	H	H	H	$\overline{E15}$	E15

STANDARD TEST LOAD

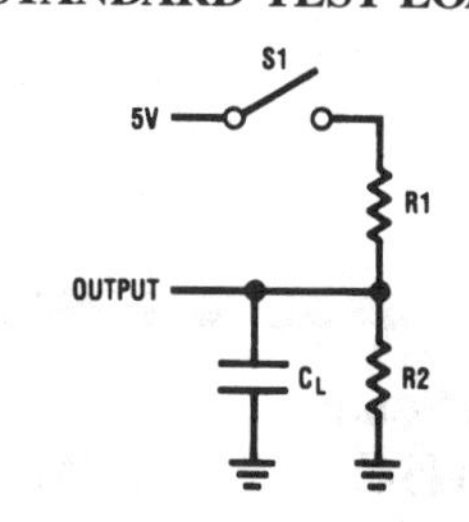

National DM54LS450A/DM74LS450
16:1 Multiplexer

- 24-pin skinny DIP saves space
- Similar to 74150 (fat DIP)
- Low-current pnp inputs reduce loading
- 15 ns typical propagation delay

ABSOLUTE MAXIMUM RATINGS (Note 1)

If Military/Aerospace specified devices are required, please contact the National Semiconductor Sales Office/Distributors for availability and specifications.

Supply Voltage V_{CC}	−0.5V to +7V (Note 2)
Input Voltage	−1.5V to +5.5V (Note 2)
Off-State Output Voltage	−1.5V to +5.5V (Note 2)
Input Current	−30.0 mA to +5.0 mA (Note 2)
Output Current (I_{OL})	+100 mA
Storage Temperature	−65°C to +150°C
Ambient Temperature with Power Applied	−65°C to +125°C
Junction Temperature with Power Applied	−65°C to +150°C
ESD Tolerance	2000V

C_{ZAP} = 100 pF
R_{ZAP} = 1500Ω
Test Method: Human Body Model
Test Specification: NSC SOP-5-028

RECOMMENDED OPERATING CONDITIONS

Symbol	Parameter	Military			Commercial			Units
		Min	Nom	Max	Min	Nom	Max	
V_{CC}	Supply Voltage	4.5	5	5.5	4.75	5	5.25	V
T_A	Operating Free-Air Temperature	−55		125	0		75	°C

SWITCHING CHARACTERISTICS Over Operating Conditions

Symbol	Parameter	Test Conditions	Military			Commercial			Units
			Min	Typ	Max	Min	Typ	Max	
T_{pd}	Input to Output	C_L = 50 pF			35			30	ns

ELECTRICAL CHARACTERISTICS Over Recommended Operating Conditions

Symbol	Parameter	Test Conditions			Min	Typ	Max	Units
V_{IL}	Low Level Input Voltage (Note 3)						0.8	V
V_{IH}	High Level Input Voltage (Note 3)				2			V
V_{IC}	Input Clamp Voltage	V_{CC} = Min, I = −18 mA					−1.5	V
I_{IL}	Low Level Input Current	V_{CC} = Max, V_I = 0.4V					−0.25	mA
I_{IH}	High Level Input Current	V_{CC} = Max, V_I = 2.4V					25	μA
I_I	Maximum Input Current	V_{CC} = Max, V_I = 5.5V					1	mA
V_{OL}	Low Level Output Voltage	V_{CC} = Min	I_{OL} = 8 mA				0.5	V
V_{OH}	High Level Output Voltage	V_{CC} = Min	I_{OH} = −2 mA	MIL	2.4			V
			I_{OH} = −3.2 mA	COM				
I_{OS}	Output Short-Circuit Current (Note 4)	V_{CC} = 5V, V_O = 0V			−30		−130	mA
I_{CC}	Supply Current	V_{CC} = Max, Outputs Open				60	100	mA

Note 1: Absolute maximum ratings are those values beyond which the device may be permanently damaged. Proper operation is not guaranteed outside the specified recommended operating conditions.

Note 2: Some device pins may be raised above these limits during programming operations according to the applicable specification.

Note 3: These are absolute voltages with respect to the ground pin on the device and include all overshoots due to system and/or tester noise. Do not attempt to test these values without suitable equipment.

Note 4: To avoid invalid readings in other parameter tests, it is preferable to conduct the I_{OS} test last. To minimize internal heating, only one output should be shorted at a time with maximum duration of 1.0 second each. Prolonged shorting of a high output may raise the chip temperature above normal and permanent damage may result.

CONNECTION DIAGRAM

Top View

V_{CC} A B C D Y W E15 E14 E13 E12 E11
24 23 22 21 20 19 18 17 16 15 14 13
A B C D Y W E15 E14 E13 E12
E0 E11
E1 E2 E3 E4 E5 E6 E7 E8 E9 E10
1 2 3 4 5 6 7 8 9 10 11 12
E0 E1 E2 E3 E4 E5 E6 E7 E8 E9 E10 GND

Order Number DM54LS450AJ, DM74LS450AJ, DM74LS450AN or DM74LS450AV
See NS Package Number J24F, N24C or V28A

FUNCTION TABLE

Input Select				Output	
D	C	B	A	W	Y
L	L	L	L	$\overline{E0}$	E0
L	L	L	H	$\overline{E1}$	E1
L	L	H	L	$\overline{E2}$	E2
L	L	H	H	$\overline{E3}$	E3
L	H	L	L	$\overline{E4}$	E4
L	H	L	H	$\overline{E5}$	E5
L	H	H	L	$\overline{E6}$	E6
L	H	H	H	$\overline{E7}$	E7
H	L	L	L	$\overline{E8}$	E8
H	L	L	H	$\overline{E9}$	E9
H	L	H	L	$\overline{E10}$	E10
H	L	H	H	$\overline{E11}$	E11
H	H	L	L	$\overline{E12}$	E12
H	H	L	H	$\overline{E13}$	E13
H	H	H	L	$\overline{E14}$	E14
H	H	H	H	$\overline{E15}$	E15

SCHEMATIC OF INPUTS AND OUTPUTS

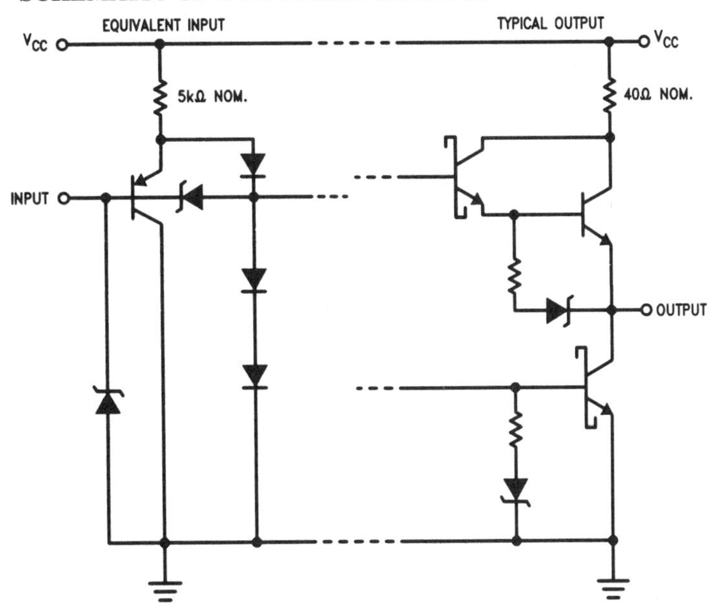

National DM54LS451/DM74LS451
Dual 8:1 Multiplexer

- 24-pin skinny DIP saves space
- Twice the density of 74LS151
- Low-current pnp inputs reduce loading

ABSOLUTE MAXIMUM RATINGS

If Military/Aerospace specified devices are required, please contact the National Semiconductor Sales Office/Distributors for availability and specifications.

Supply Voltage V_{CC}	7V
Input Voltage	5.5V
Off-State Output Voltage	5.5V
Storage Temperature	−65°C to +150°C

OPERATING CONDITIONS

Symbol	Parameter	Military			Commercial			Units
		Min	Nom	Max	Min	Nom	Max	
V_{CC}	Supply Voltage	4.5	5	5.5	4.75	5	5.25	V
T_A	Operating Free-Air Temperature	−55		125*	0		75	°C

*Case Temperature

ELECTRICAL CHARACTERISTICS Over Operating Conditions

Symbol	Parameter	Test Conditions		Min	Typ†	Max	Units
V_{IL}	Low-Level Input Voltage					0.8	V
V_{IH}	High-Level Input Voltage			2			V
V_{IC}	Input Clamp Voltage	V_{CC}=MIN	I_I=−18 mA			−1.5	V
I_{IL}	Low-Level Input Current	V_{CC}=MAX	V_I=0.4V			−0.25	mA
I_{IH}	High-Level Input Current	V_{CC}=MAX	V_I=2.4V			25	μA
I_I	Maximum Input Current	V_{CC}=MAX	V_I=5.5V			1	mA
V_{OL}	Low-Level Output Voltage	V_{CC}=MIN, V_{IL}=0.8V, V_{IH}=2V	I_{OL}=8 mA			0.5	V
V_{OH}	High-Level Output Voltage	V_{CC}=MIN, V_{IL}=0.8V, V_{IH}=2V	MIL I_{OH}=2 mA; COM I_{OH}=−3.2 mA	2.4			V
I_{OS}	Output Short-Circuit Current*	V_{CC}=5.0V	V_O=0V	−30		−130	mA
I_{CC}	Supply Current	V_{CC}=MAX			60	100	mA

*No more than one output should be shorted at a time and duration of the short-circuit should not exceed one second.

† All typical values are V_{CC}=5V, T_A=25°C.

SWITCHING CHARACTERISTICS Over Operating Conditions

Symbol	Parameter	Test Conditions (See Test Load)	Military			Commercial			Units
			Min	Typ	Max	Min	Typ	Max	
t_{PD}	Any Input to Y	C_L=50 pF, R_1=560Ω, R_2=1.1Ω		25	45		25	40	ns

CONNECTION DIAGRAM

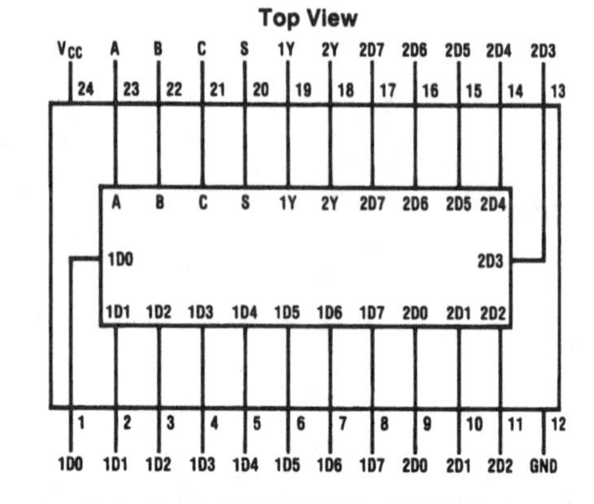

Order Number DM54LS451J, DM74LS451J, DM74LS451N or DM74LS451V
See NS Package Number J24F, N24C or V28A

FUNCTION TABLE

Inputs				Outputs
Select			Strobe	Y
C	B	A	S	
X	X	X	H	H
L	L	L	L	D0
L	L	H	L	D1
L	H	L	L	D2
L	H	H	L	D3
H	L	L	L	D4
H	L	H	L	D5
H	H	L	L	D6
H	H	H	L	D7

STANDARD TEST LOAD

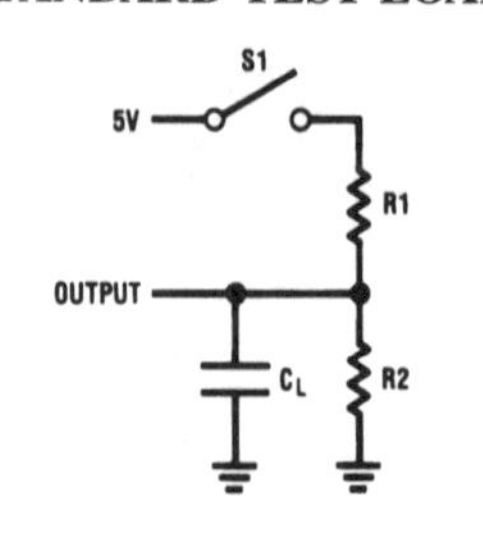

National DM54LS451A/DM74LS451A
Dual 8:1 Multiplexer

- 24-pin skinny DIP saves space
- Twice the density of 74LS151
- Low-current pnp inputs reduce loading
- 15 ns typical propagation delay

ABSOLUTE MAXIMUM RATINGS (Note 1)

If Military/Aerospace specified devices are required, please contact the National Semiconductor Sales Office/Distributors for availability and specifications.

Supply Voltage V_{CC}	−0.5V to +7V (Note 2)
Input Voltage	−1.5V to +5.5V (Note 2)
Off-State Output Voltage	−1.5V to +5.5V (Note 2)
Input Current	−30.0 mA to +5.0 mA (Note 2)
Output Current (I_{OL})	+100 mA
Storage Temperature	−65°C to +150°C
Ambient Temperature with Power Applied	−65°C to +125°C
Junction Temperature with Power Applied	−65°C to +150°C
ESD Tolerance	2000V

C_{ZAP} = 100 pF
R_{ZAP} = 1500Ω
Test Method: Human Body Model
Test Specification: NSC SOP-5-028

RECOMMENDED OPERATING CONDITIONS

Symbol	Parameter	Military			Commercial			Units
		Min	Nom	Max	Min	Nom	Max	
V_{CC}	Supply Voltage	4.5	5	5.5	4.75	5	5.25	V
T_A	Operating Free-Air Temperature	−55		125	0		75	°C

ELECTRICAL CHARACTERISTICS Over Recommended Operating Conditions

Symbol	Parameter	Test Conditions			Min	Typ	Max	Units
V_{IL}	Low Level Input Voltage (Note 3)						0.8	V
V_{IH}	High Level Input Voltage (Note 3)				2			V
V_{IC}	Input Clamp Voltage	V_{CC} = Min, I = −18 mA					−1.5	V
I_{IL}	Low Level Input Current	V_{CC} = Max, V_I = 0.4V					−0.25	mA
I_{IH}	High Level Input Current	V_{CC} = Max, V_I = 2.4V					25	μA
I_I	Maximum Input Current	V_{CC} = Max, V_I = 5.5V					1	mA
V_{OL}	Low Level Output Voltage	V_{CC} = Min	I_{OL} = 8 mA				0.5	V
V_{OH}	High Level Output Voltage	V_{CC} = Min	I_{OH} = −2 mA	MIL	2.4			V
			I_{OH} = −3.2 mA	COM				
I_{OS}	Output Short-Circuit Current (Note 4)	V_{CC} = 5V, V_O = 0V			−30		−130	mA
I_{CC}	Supply Current	V_{CC} = Max, Outputs Open				60	100	mA

Note 1: Absolute maximum ratings are those values beyond which the device may be permanently damaged. Proper operation is not guaranteed outside the specified recommended operating conditions.

Note 2: Some device pins may be raised above these limits during programming operations according to the applicable specification.

Note 3: These are absolute voltages with respect to the ground pin on the device and include all overshoots due to system and/or tester noise. Do not attempt to test these values without suitable equipment.

Note 4: To avoid invalid readings in other parameter tests, it is preferable to conduct the I_{OS} test last. To minimize internal heating, only one output should be shorted at a time with maximum duration of 1.0 second each. Prolonged shorting of a high output may raise the chip temperature above normal and permanent damage may result.

SWITCHING CHARACTERISTICS Over Recommended Operating Conditions

Symbol	Parameter	Test Conditions	Military			Commercial			Units
			Min	Typ	Max	Min	Typ	Max	
T_{pd}	Input to Output	C_L = 50 pF		15	30		15	25	ns

CONNECTION DIAGRAM

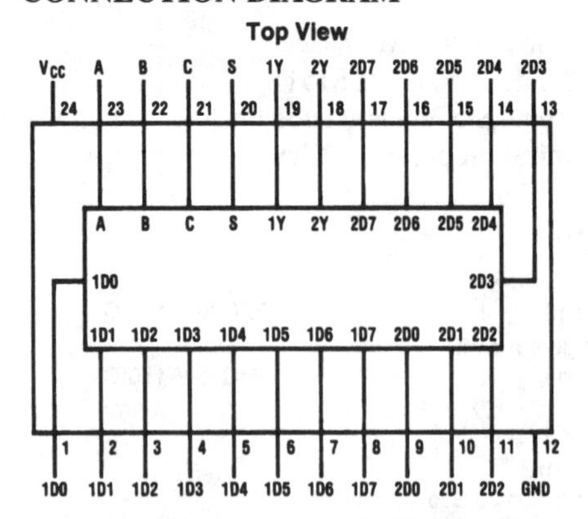

Order Number DM54LS451AJ, DM74LS451AJ, DM74LS451AN or DM74LS451AV
See NS Package Number J24F, N24C or V28A

FUNCTION TABLE

Inputs				Outputs
Select			Strobe	Y
C	B	A	S	
X	X	X	H	H
L	L	L	L	D0
L	L	H	L	D1
L	H	L	L	D2
L	H	H	L	D3
H	L	L	L	D4
H	L	H	L	D5
H	H	L	L	D6
H	H	H	L	D7

SCHEMATIC OF INPUTS AND OUTPUTS

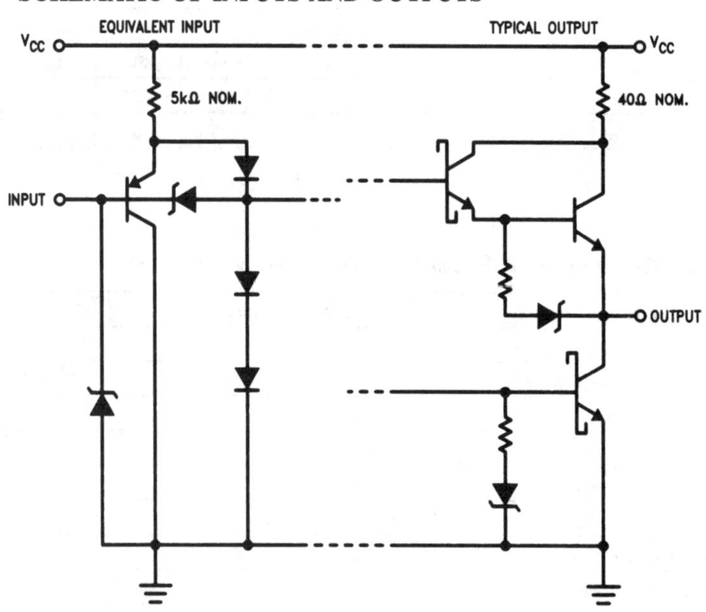

National DM54LS453/DM74LS453 Quad 4:1 Multiplexer

- 24-pin skinny DIP saves space
- Twice the density of 74LS153
- Low-current pnp inputs reduce loading

ABSOLUTE MAXIMUM RATINGS

If Military/Aerospace specified devices are required, please contact the National Semiconductor Sales Office/Distributors for availability and specifications.

Supply Voltage V_{CC}	7V
Input Voltage	5.5V
Off-State Output Voltage	5.5V
Storage Temperature	−66° to +150°C

OPERATING CONDITIONS

Symbol	Parameter	Military			Commercial			Units
		Min	Typ	Max	Min	Typ	Max	
V_{CC}	Supply Voltage	4.5	5	5.5	4.75	5	5.25	V
T_A	Operating Free-Air Temperature	−55		125*	0		75	°C

*Case temperature

ELECTRICAL CHARACTERISTICS Over Operating Conditions

Symbol	Parameter	Test Conditions			Min	Typ†	Max	Units
V_{IL}	Low-Level Input Voltage						0.8	V
V_{IH}	High-Level Input Voltage				2			V
V_{IC}	Input Clamp Voltage	V_{CC}=MIN	I_I=−18 mA				−1.5	V
I_{IL}	Low-Level Input Current	V_{CC}=MAX	V_I=0.4V				−0.25	mA
I_{IH}	High-Level Input Current	V_{CC}=MAX	V_I=2.4V				25	μA
I_I	Maximum Input Current	V_{CC}=MAX	V_I=5.5V				1	mA
V_{OL}	Low-Level Output Voltage	V_{CC}=MIN, V_{IL}=0.8V, V_{IH}=2V		I_{OL}=8 mA			0.5	V
V_{OH}	High-Level Output Voltage	V_{CC}=MIN, V_{IL}=0.8V, V_{IH}=2V	MIL	I_{OH}=−2 mA	2.4			V
			COM	I_{OH}=−3.2 mA				
I_{OS}	Output Short-Circuit Current*	V_{CC}=5.0V		V_O=0V	−30		−130	mA
I_{CC}	Supply Current	V_{CC}=MAX				60	100	mA

*No more than one output should be shorted at a time and duration of the short-circuit should not exceed one second.

†All typical values are at V_{CC}=5V, T_A=25°C

SWITCHING CHARACTERISTICS Over Operating Conditions

Symbol	Parameter	Test Conditions (See Test Load)	Military			Commercial			Units
			Min	Typ	Max	Min	Typ	Max	
t_{PD}	Any Input to Y	C_L=50 pF, R_1=560Ω, R_2=1.1 kΩ		25	45		25	40	ns

CONNECTION DIAGRAM

Top View

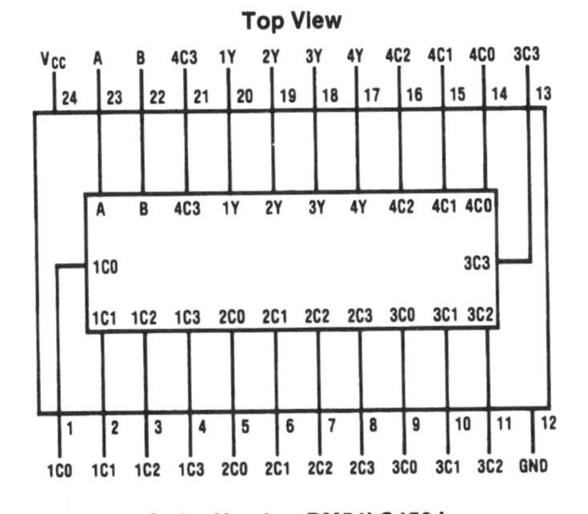

Order Number DM54LS453J, DM74LS453J or DM74LS453N
See NS Package Number J24F or N24C

STANDARD TEST LOAD

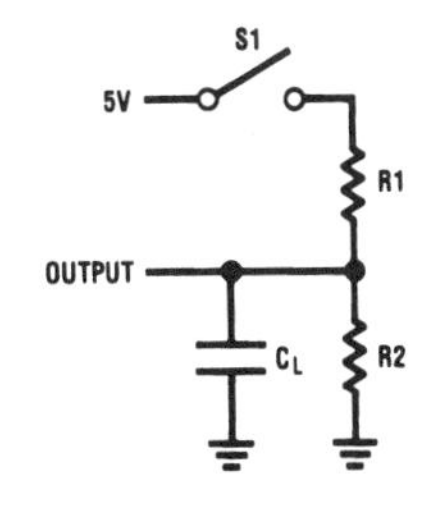

FUNCTION TABLE

INPUT SELECT		OUTPUTS
B	A	Y
L	L	C0
L	H	C1
H	L	C2
H	H	C3

National DM54LS453A/DM74LS453A Quad 4:1 Multiplexer

- 24-pin skinny DIP saves space
- Twice the density of 74LS153
- Low-current pnp inputs reduce loading
- 15 ns typical propagation delay

ABSOLUTE MAXIMUM RATINGS (Note 1)

If Military/Aerospace specified devices are required, please contact the National Semiconductor Sales Office/Distributors for availability and specifications.

Supply Voltage V_{CC}	−0.5V to +7V (Note 2)
Input Voltage	−1.5V to +5.5V (Note 2)
Off-State Output Voltage	−1.5V to +5.5V (Note 2)
Input Current	−30.0 mA to +5.0 mA (Note 2)
Output Current (I_{OL})	+100 mA
Storage Temperature	−65°C to +150°C
Ambient Temperature with Power Applied	−65°C to +125°C
Junction Temperature with Power Applied	−65°C to +150°C
ESD Tolerance	2000V

C_{ZAP} = 100 pF
R_{ZAP} = 1500Ω
Test Method: Human Body Model
Test Specification: NSC SOP5-028

RECOMMENDED OPERATING CONDITIONS

Symbol	Parameter	Military			Commercial			Units
		Min	Nom	Max	Min	Nom	Max	
V_{CC}	Supply Voltage	4.5	5	5.5	4.75	5	5.25	V
T_A	Operating Free-Air Temperature	−55		125	0		75	°C

ELECTRICAL CHARACTERISTICS

Symbol	Parameter	Test Conditions			Min	Typ	Max	Units
V_{IL}	Low Level Input Voltage (Note 3)						0.8	V
V_{IH}	High Level Input Voltage (Note 3)				2			V
V_{IC}	Input Clamp Voltage	V_{CC} = Min, I = −18 mA					−1.5	V
I_{IL}	Low Level Input Current	V_{CC} = Max, V_I = 0.4V					−0.25	mA
I_{IH}	High Level Input Current	V_{CC} = Max, V_I = 2.4V					25	μA
I_I	Maximum Input Current	V_{CC} = Max, V_I = 5.5V					1	mA
V_{OL}	Low Level Output Voltage	V_{CC} = Min	I_{OL} = 8 mA				0.5	V
V_{OH}	High Level Output Voltage	V_{CC} = Min	I_{OH} = −2 mA	MIL	2.4			V
			I_{OH} = −3.2 mA	COM				
I_{OS}	Output Short-Circuit Current (Note 4)	V_{CC} = 5V, V_O = 0V			−30		−130	mA
I_{CC}	Supply Current	V_{CC} = Max, Outputs Open				60	100	mA

Note 1: Absolute maximum ratings are those values beyond which the device may be permanently damaged. Proper operation is not guaranteed outside the specified recommended operating conditions.

Note 2: Some device pins may be raised above these limits during programming operations according to the applicable specification.

Note 3: These are absolute voltages with respect to the ground pin on the device and include all overshoots due to system and/or tester noise. Do not attempt to test these values without suitable equipment.

Note 4: To avoid invalid readings in other parameter tests, it is preferable to conduct the I_{OS} test last. To minimize internal heating, only one output should be shorted at a time with maximum duration of 1.0 second each. Prolonged shorting of a high output may raise the chip temperature above normal and permanent damage may result.

SWITCHING CHARACTERISTICS Over Operating Conditions

Symbol	Parameter	Test Conditions	Military			Commercial			Units
			Min	Typ	Max	Min	Typ	Max	
T_{pd}	Input to Output	C_L = 50 pF		15	30		15	25	ns

CONNECTION DIAGRAM

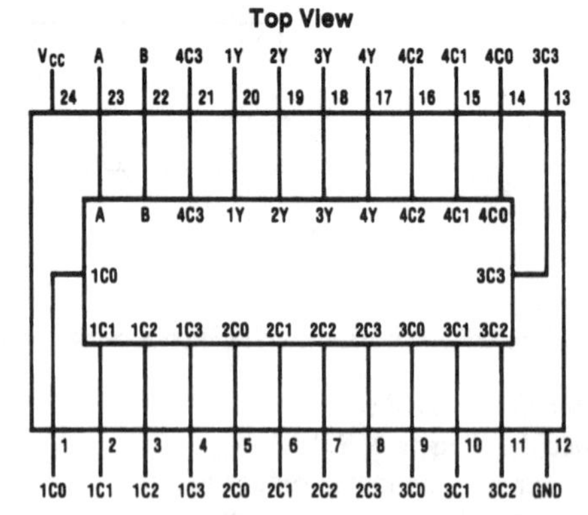

Order Number DM54LS453AJ, DM74LS453AJ, DM74LS453AN or DM74LS453AV See NS Package Number J24F, N24C or V28A

SCHEMATIC OF INPUTS AND OUTPUTS

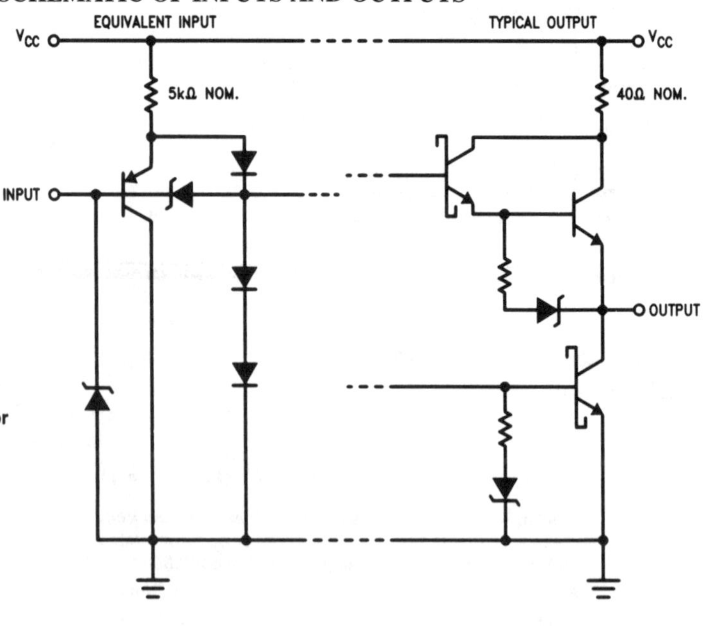

FUNCTION TABLE

Input Select		Outputs Y
B	A	
L	L	C0
L	H	C1
H	L	C2
H	H	C3

Raytheon RC4077

Precision Operational Amplifiers

- Ultra-low V_{OS}: 10 μV max
- Ultra-low V_{OS} drift: 0.3 μV/°C max
- Outstanding gain linearity
- High gain: 2500 V/mV min
- High CMRR: 120 dB min
- High PSRR: 110 dB min
- Low noise: 0.3 $\mu V_{p\text{-}p}$ (0.1 to 10 Hz)
- Low input bias current: 2.0 nA max
- Low power consumption: 50 mW max
- Replaces OP-07, 725, 108, 741 types

ABSOLUTE MAXIMUM RATINGS

PARAMETER	VALUE
Supply voltage	±22V
Input voltage	±22V
Differential input voltage	30V
Internal power dissipation	500 mW

ABSOLUTE MAXIMUM RATINGS (Continued)

PARAMETER	VALUE
Output short-circuit duration	Indefinite
Storage temperature range	−65 °C to +150 °C
Operating temperature range	
RM4077A	−55 °C to +125 °C
RC4077A,E,F	−25 °C to + 85 °C
Lead soldering temperature (Micro-Pak; 10 sec)	+260 °C
Lead soldering temperature (DIP, LCC, TO-99; 60 sec)	+300 °C

For supply voltages less than ±22V, the absolute maximum input voltage is equal to the supply voltage.

Observe package thermal characteristics.

The information contained herein has been carefully compiled; however, it shall not by implication or otherwise become part of the terms and conditions of any subsequent sale. Raytheon's liability shall be determined solely by its standard terms and conditions of sale. No representation as to application or use or that the circuits are either licensed or free from patent infringement is intended or implied. Raytheon reserves the right to change the circuitry and other data at any time without notice and assumes no liability for inadvertent errors.

8-Lead TO-99 Metal Can (Top View)

8-Lead Dual In-Line Package (Top View)

8-Lead Plastic Dual In-Line Micro-Pak (Top View)

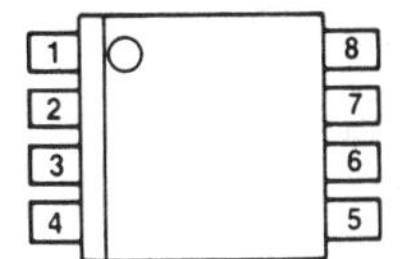

Pin	Function
1	V_{OS} Trim
2	-Input
3	+Input
4	$-V_S$ (Case)
5	NC
6	Output
7	$+V_S$
8	V_{OS} Trim

20-Pad LCC

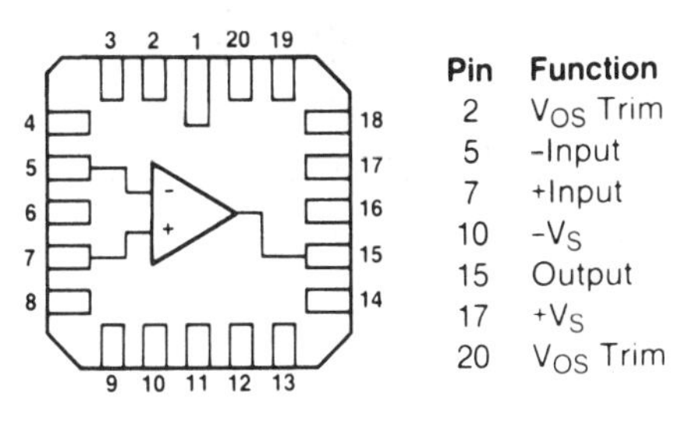

Pin	Function
2	V_{OS} Trim
5	-Input
7	+Input
10	$-V_S$
15	Output
17	$+V_S$
20	V_{OS} Trim

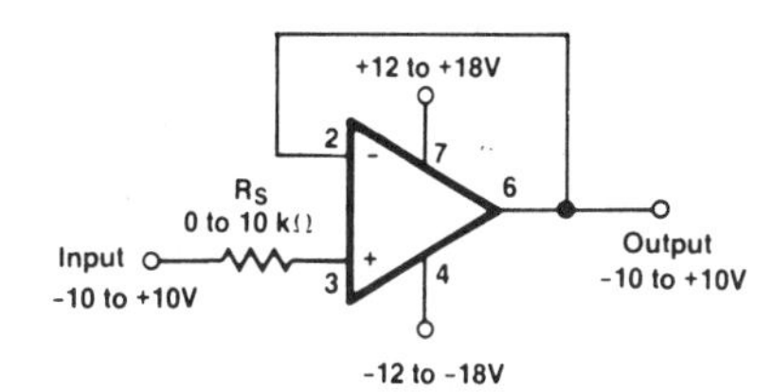

Large Signal Voltage Follower With 0.00063% Worst-Case Accuracy Error

OUTPUT ACCURACY

Error	RM4077A 25°C Max (μV)	RC4077F 25°C Max (μV)	RM4077A -55 to +125°C Max (μV)	RC4077F -25 to +85°C Max (μV)
Offset voltage	10	60	40	100
Bias current	15	28	40	60
CMRR	20	32	20	60
PSRR	18	18	18	30
Voltage gain	7	8	8	20
Worst case sum	70	146	126	270
Percent of full scale (= 20V)	.00035%	.00073%	.00063%	.0013%

PACKAGING INFORMATION

8-Lead TO-99 Metal Can

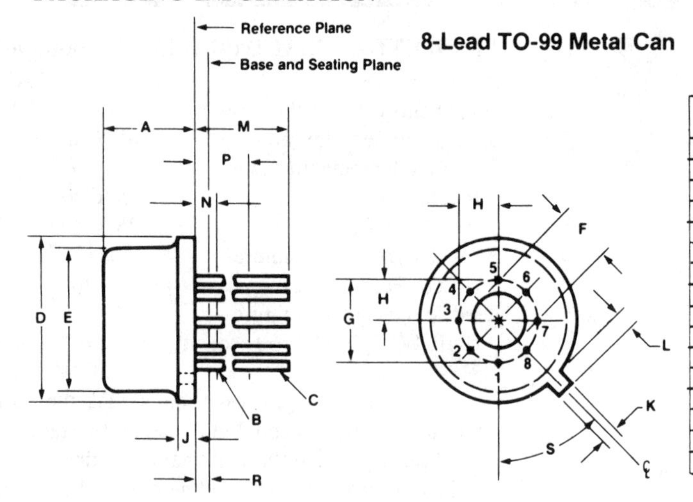

Dimension	Inches		Millimeters	
	Min.	Max.	Min.	Max.
A	.165	.185	4.19	4.70
B	.016	.019	.41	.48
C	.016	.021	.41	.53
D	.335	.370	8.51	9.40
E	.305	.335	7.75	8.51
F	.120	.160	3.05	4.06
G	.200 BSC		5.08 BSC	
H	.100 BSC		2.54 BSC	
J		.040		1.02
K	.027	.034	.69	.86
L	.027	.045	.69	1.14
M	.500	.750	12.70	19.05
N		.050		1.27
P	.250		6.35	
R	.010	.045	.25	1.14
S	45° BSC		45° BSC	

8-Lead Plastic Dual In-Line Package

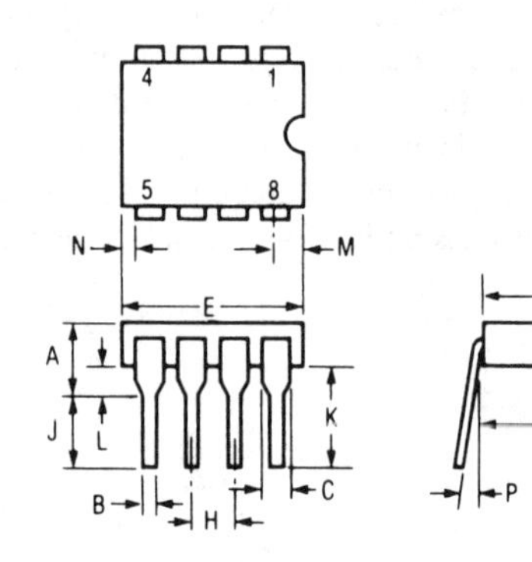

Dimension	Inches		Millimeters	
	Min.	Max.	Min.	Max.
A	.115	.125	2.92	3.17
B	.015	.021	0.38	0.53
C	.030	.070	0.76	1.78
D	.010	.015	0.25	0.38
E	.360	.400	9.14	10.16
F	.240	.260	6.09	6.60
G	.290	.310	7.37	7.87
H	.090	.110	2.29	2.79
J	.120	.135	3.05	3.43
K	.140	.165	3.56	4.18
L	.020	.030	0.51	0.75
M	.025	.050	0.64	1.27
N	.005		0.13	
P	0°	15°	0°	15°

8-Lead Plastic Dual In-Line Micro-Pak

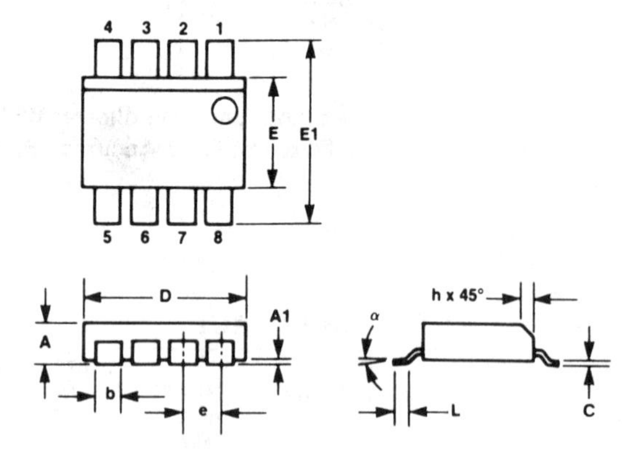

Dimension	Inches		Millimeters	
	Min.	Max.	Min.	Max.
A	.053	.069	1.35	175
A1	.004	.008	.10	.20
b	.014	.018	.350	A50
C	.007	.009	.19	22
D	.188	.197	4.80	5.00
E	.150	.158	3.80	400
e	.050 BSC		1.27 BSC	
E1	.228	.244	5.80	620
L	.021	.045	.508	1.143
X	0°	8°	0	8
h	.01	.02	.25	.50

Note: C Dimension does not include Hot Solder Dip thickness.

8-Lead Ceramic Dual In-Line Package

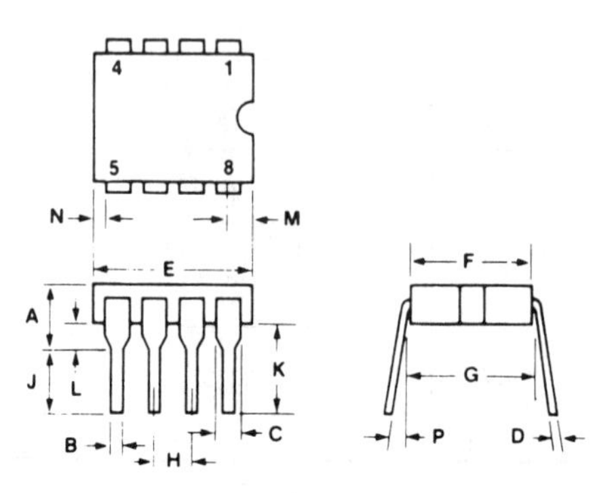

Dimension	Inches		Millimeters	
	Min.	Max.	Min.	Max.
A		.200		5.08
B	.014	.023	0.36	0.58
C	.030	.070	0.76	1.78
D	.008	.015	0.20	0.38
E		.390		9.91
F	.220	.310	5.59	7.87
G	.290	.320	7.37	8.13
H	.100 BSC		2.54 BSC	
J	.125	.200	3.18	5.08
K	.150		3.81	
L	.015	.060	0.38	1.52
M		.045		1.14
N	.005		0.13	
P	0°	15°	0°	15°

PACKAGING INFORMATION (Continued)

20-Pad LCC

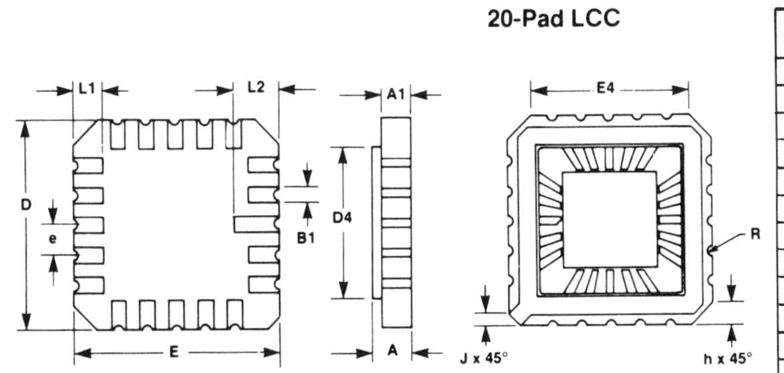

Dimension	Inches		Millimeters	
	Min.	Max.	Min.	Max.
A	0.064	0.086	1.63	2.18
A1	0.054	0.066	1.37	1.68
B1	0.022	0.028	0.56	0.71
D	0.342	0.358	8.69	9.09
D4		0.319		8.10
E	0.342	0.358	8.69	9.09
E4		0.319		8.10
e	0.050	BSC	1.27	BSC
h	0.040	REF	1.02	REF
J	0.020	REF	0.51	REF
L1	0.050	REF	1.27	REF
L2	0.077	0.093	1.96	2.36
R	0.009	REF	0.23	REF

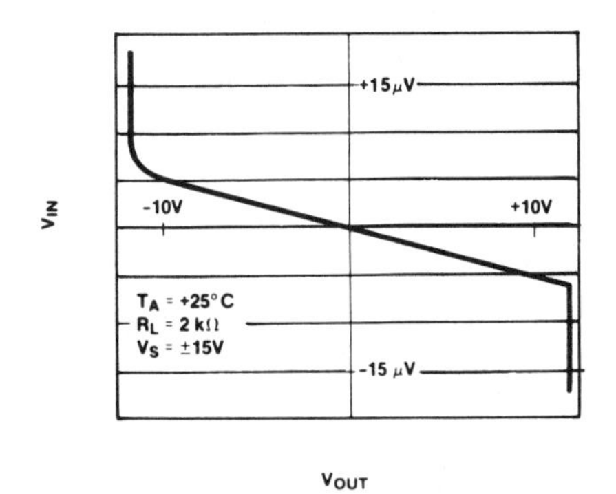

RM-4077 Open-Loop Gain Linearity

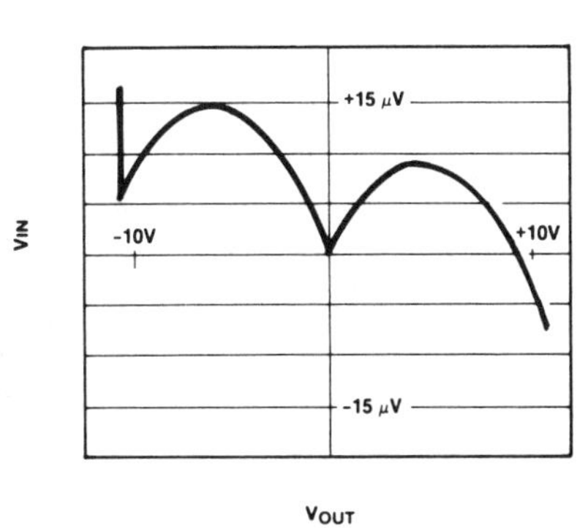

Typical Precision Op Amp Gain Linearity

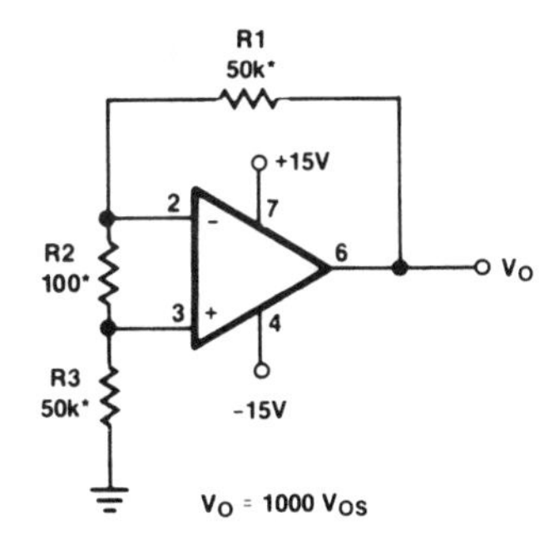

*Resistors must have low thermoelectric potential

Test Circuit for Offset Voltage and Its Drift With Temperature

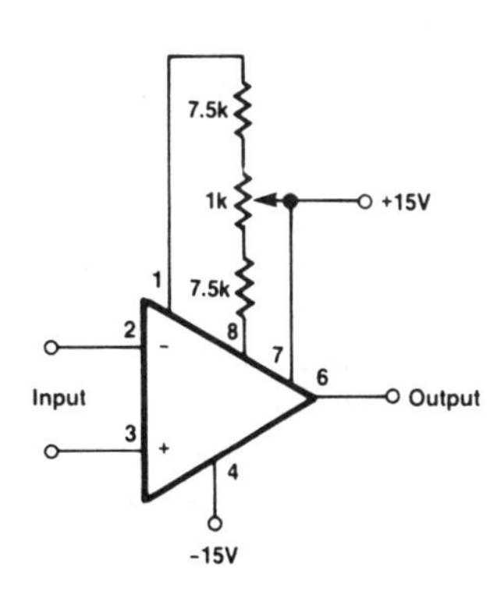

Improved Sensitivity Adjustment

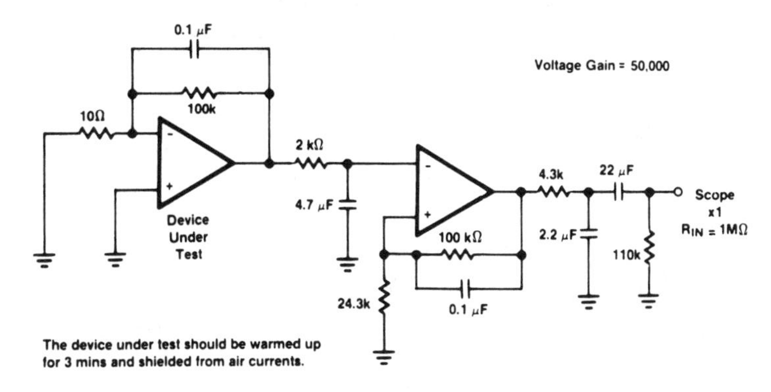

0.1 Hz to 10 Hz Noise Test Circuit
(peak to peak noise measured in 10 sec interval)

SIMPLIFIED SCHEMATIC DIAGRAM

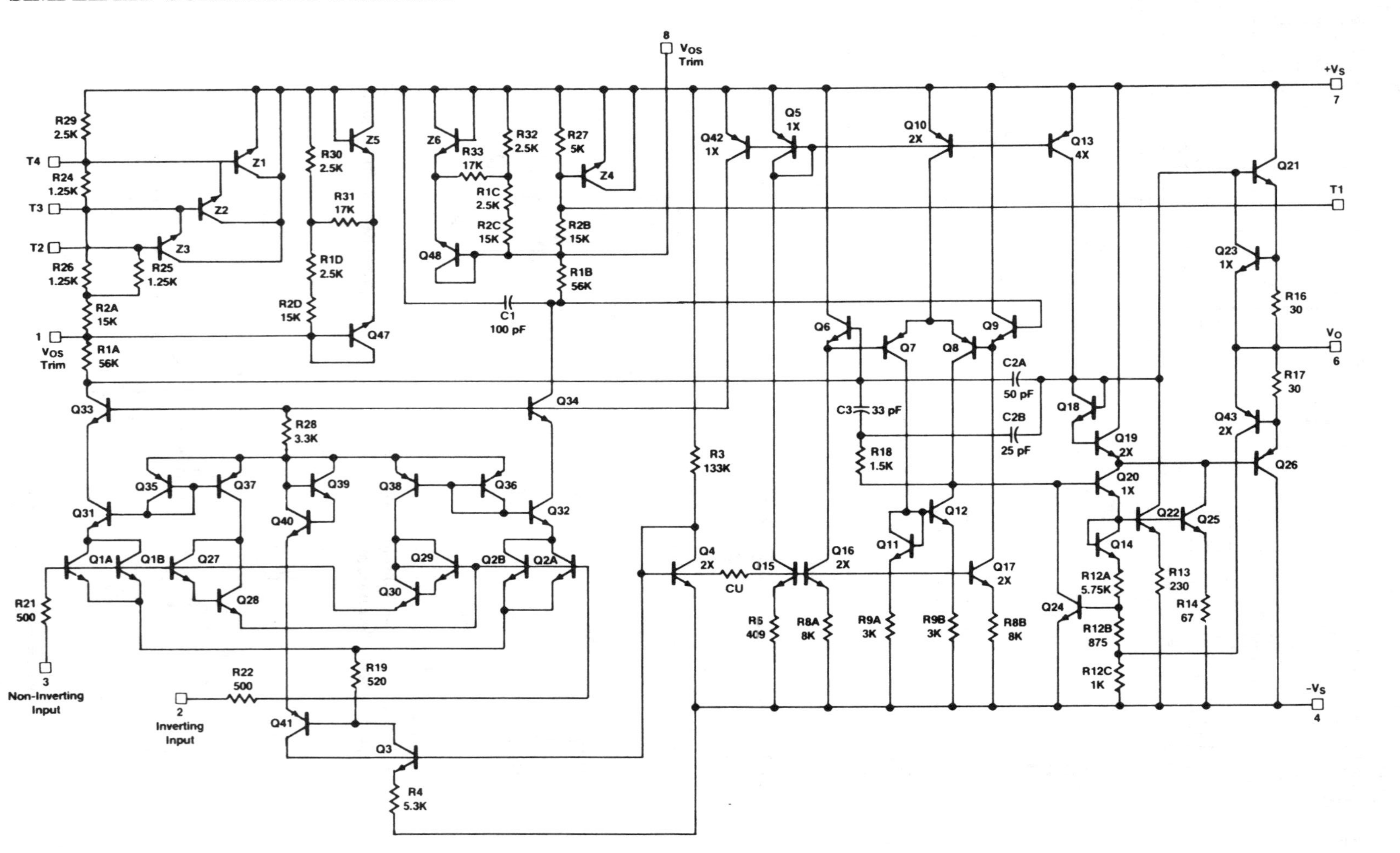

Raytheon
LP165/365
Micropower Programmable Quad Comparator

- Single programming resistor tailors power, input currents, speed, and other current characteristics
- Uncommitted emitters allow logic interface flexibility
- Wide supply voltage range or dual supplies (4V to 36V or ±2V to ±18V)
- Input common-mode range includes ground in single supply applications
- Low power consumption (10μW per comparator at V_S = 5V, I_{SET} = 0.5μA)

The LP165/365 consists of four independent voltage comparators constructed monolithically using a bipolar transistor fabrication process. Programmability gives the user ability to adjust supply current drain and so control power dissipation. At higher values of programming (I_{SET}), the supply current will increase, response times and output drive capability will improve, and input bias current will increase. At lower values of I_{SET}, supply current and power dissipation will decrease, the response time slows, and input bias currents improve. The uncommitted output emitter connection allows flexibility to interface with various logic families, such as TTL, DTL, CMOS, NMOS, and PMOS.

These comparators can be operated from a single or split power supply; the inputs have a common-mode range that includes the negative supply voltage (ground in single supply applications).

Applications include battery-powered circuits, threshold detectors, zero crossing detectors, multivibrators, VCOs, and digital interface circuits.

ABSOLUTE MAXIMUM RATINGS

PARAMETER	VALUE
Supply Voltage	36V or ±18V
Differential input voltage	36V
Input voltage	−0.3V to +36V (single supply)[1]
Output short circuit	
Duration to V_E	Indefinite[2]
V_{OUT} to V_E	$V_E - 7V \leq V_{OUT} \leq V_E + 36V$
Operating temperature range	
LP165	−55°C to +125°C
LP365/LP365A	0°C to +70°C
Storage temperature range	−65°C to +150°C
Lead soldering temperature (10 seconds)	+300°C

Notes:
1. The input voltage is not allowed to go 0.3V above V+ or −0.3V below V− as this will turn on a parasitic transistor causing large currents to flow through the device.
2. Short circuits from the output to V+ may cause excessive heating and eventual destruction. The current in the output leads and the V_E lead should not be allowed to exceed 30mA. The output should not be shortened to V− if $V_E \geq (V-) + 7V$.

PACKAGING INFORMATION

16-Lead
Ceramic Dual-in-Line

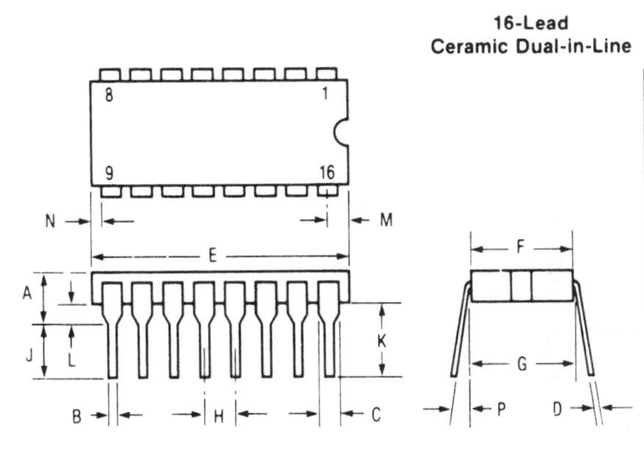

Dimension	Inches Min.	Inches Max.	Millimeters Min.	Millimeters Max.
A		.200		5.08
B	.014	.023	.36	.58
C	.030	.070	.76	1.78
D	.008	.015	.20	.38
E		.840		21.34
F	.220	.310	5.59	7.87
G	.290	.320	7.37	8.13
H	.100BSC		2.54BSC	
J	.125	.200	3.18	5.08
K	.150		3.81	
L	.015	.060	.38	1.52
M		.080		2.03
N	.005		.13	
P	0°	15°	0°	15°

16-Lead
Plastic Dual-in-Line

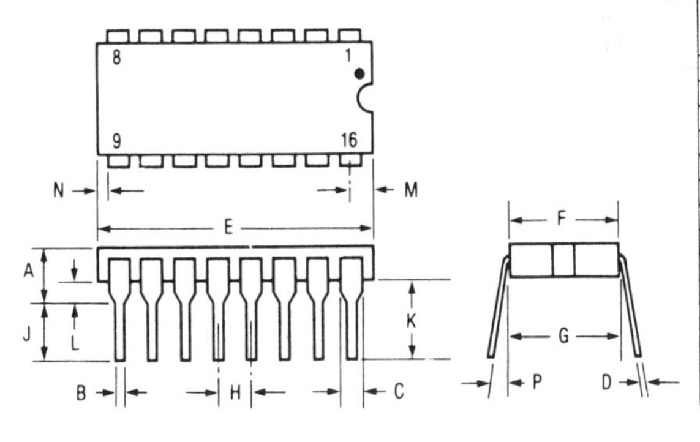

Dimension	Inches Min.	Inches Max.	Millimeters Min.	Millimeters Max.
A		.200		5.08
B	.014	.023	0.36	0.58
C	.030	.070	0.76	1.78
D	.008	.015	0.20	0.38
E	.740	.760	18.80	19.30
F	.240	.260	6.10	6.60
G	.290	.320	7.37	8.13
H	.100BSC		2.54BSC	
J	.125	.200	3.18	5.08
K	.135		3.43	
L	.015	.060	0.38	1.52
M	.020		0.51	
N	.005		0.13	
P	0°	15°	0°	15°

SIMPLIFIED SCHEMATIC DIAGRAM

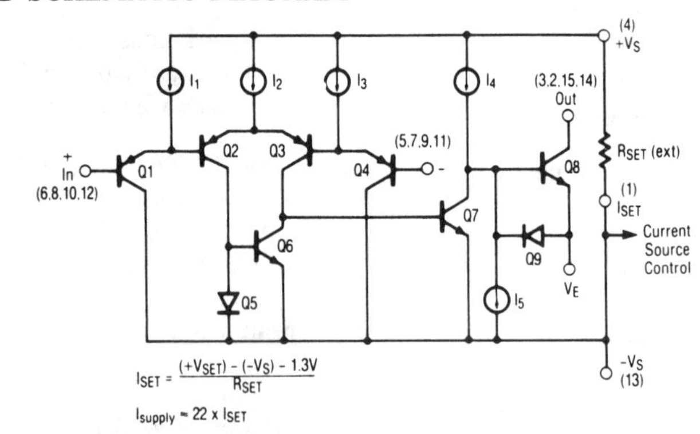

$I_{SET} = \frac{(+V_{SET}) - (-V_S) - 1.3V}{R_{SET}}$

$I_{supply} \approx 22 \times I_{SET}$

Current sources are programmed by I_{SET}
V_E is common to all 4 comparators

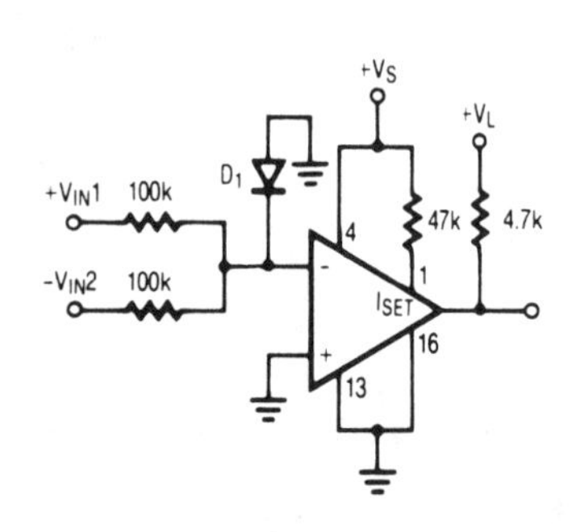

D_1 = Small signal Schottky or low V_D equivalent

Opposite Polarity Magnitude Comparator (Single Supply)

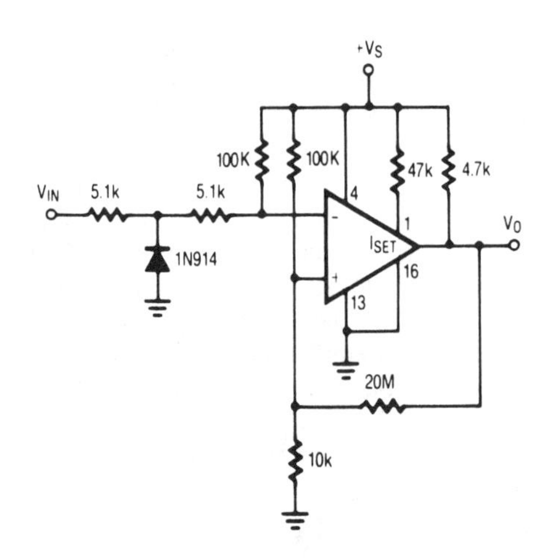

Zero Crossing Detector (Single Supply)

16-Lead
Dual In-Line Package
(Top View)

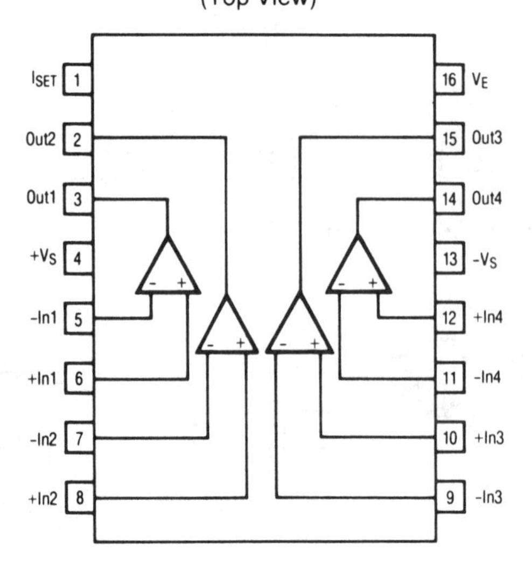

Pin	Function
1	Set Current
2	Output 2
3	Output 1
4	+ Supply Voltage
5	Inverting Input 1
6	Non-inverting Input 1
7	Inverting Input 2
8	Non-inverting Input 2
9	Inverting Input 3
10	Non-inverting Input 3
11	Inverting Input 4
12	Non-inverting Input 4
13	- Supply Voltage
14	Output 4
15	Output 3
16	Emitter Common

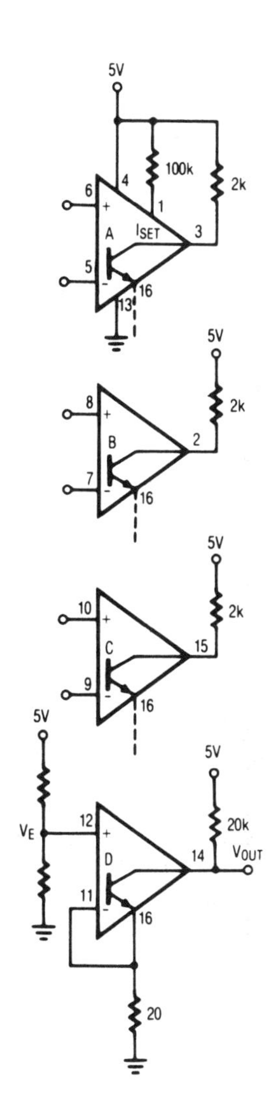

If you choose V_E = 25mV, 75mV, or 125mV, then V_{OUT} will fall if 1/3, 2/3 or all of the other three outputs are low.

Voting Comparator

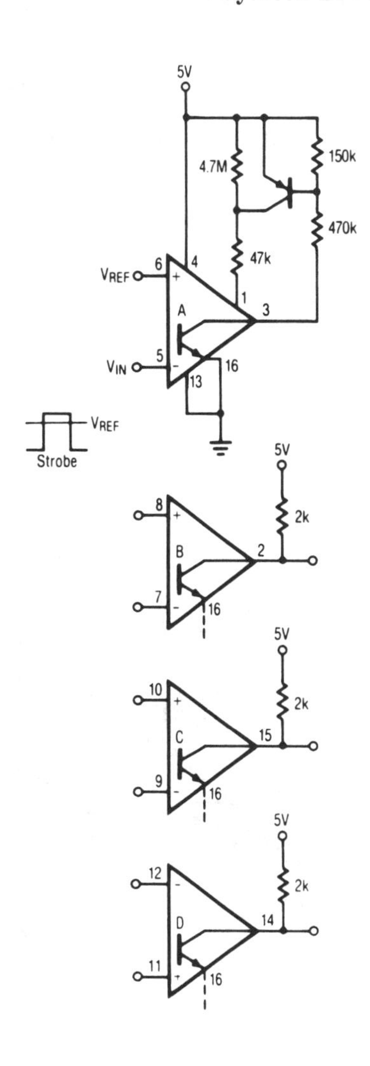

Comparators B, C, and D do not respond until activated by the signal applied to comparator A

Level Sensitive Strobe

TYPICAL PERFORMANCE CHARACTERISTICS

Supply Current vs. I_{SET}

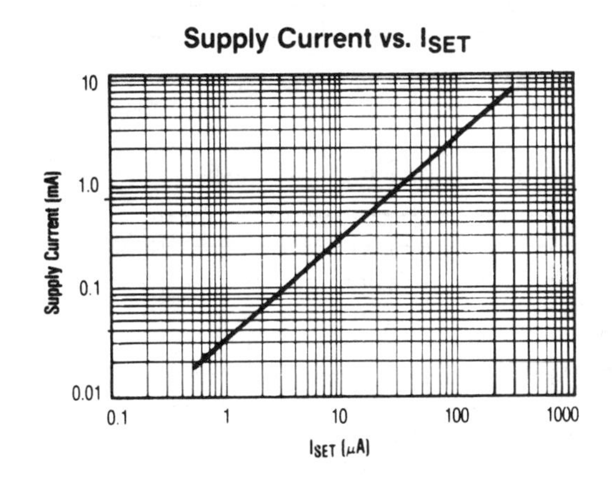

Supply Current vs. Supply Voltage

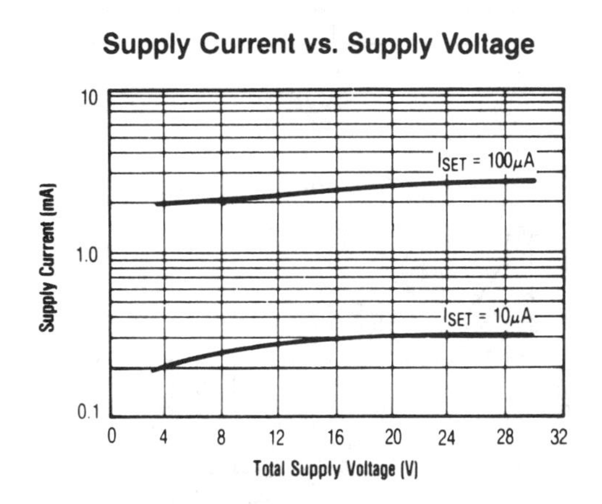

TYPICAL PERFORMANCE CHARACTERISTICS

Supply Current vs. Temperature

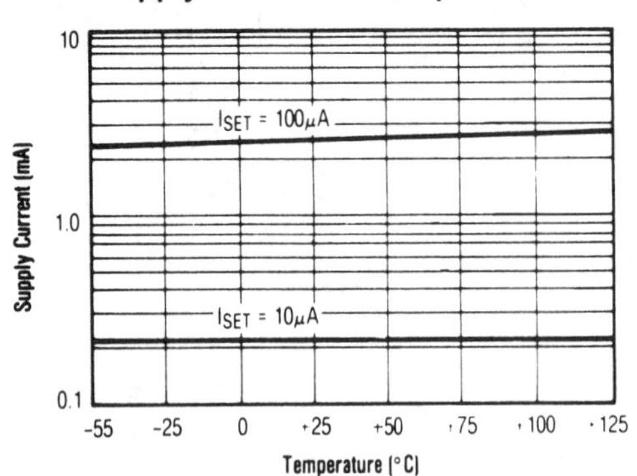

Input Bias Current vs. I_{SET}

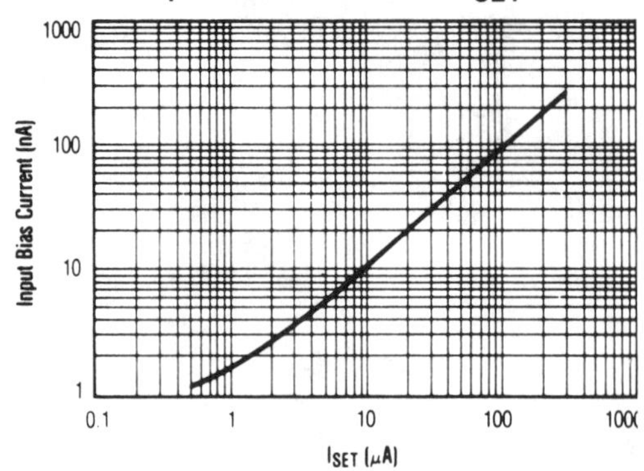

Output Saturation Voltage

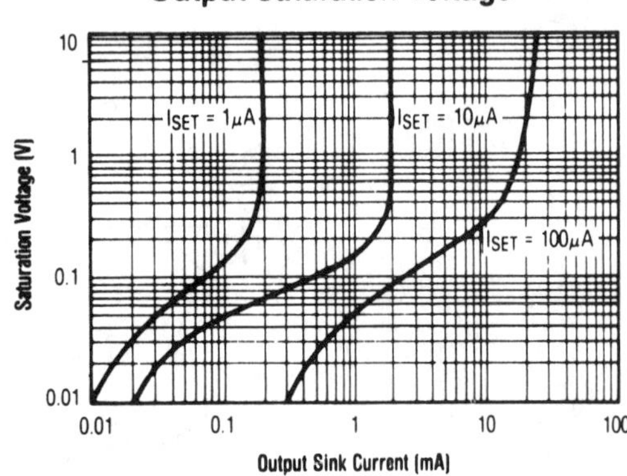

Voltage Gain vs. I_{SET}

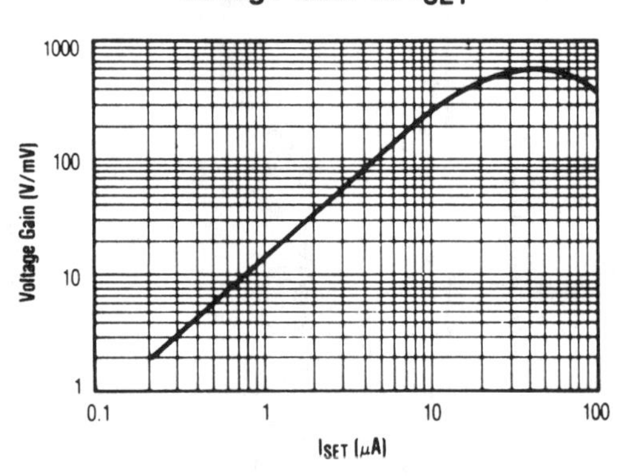

LP165/365 Response Time Negative Transition $T_A = +25°C$, ±5V, 5mV Overdrive

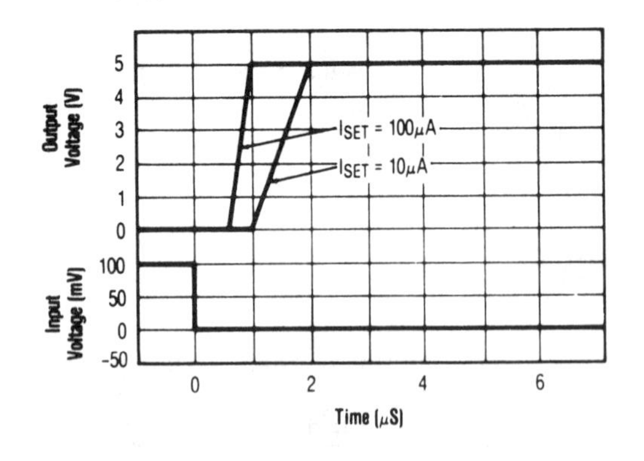

LP165/365 Response Time Positive Transition $T_A = +25°C$, ±5V, 5mV Overdrive

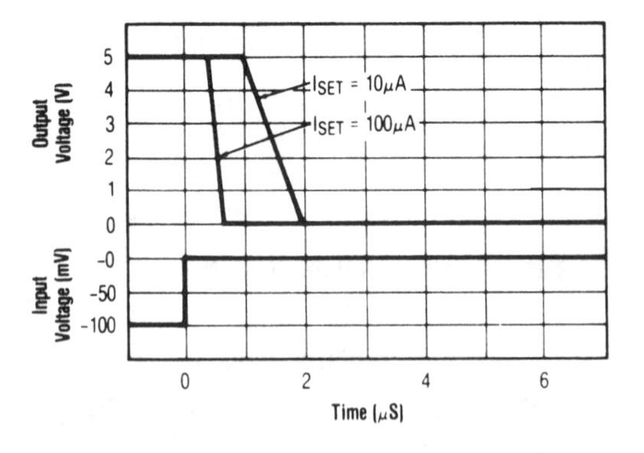

Response Time Negative Transition

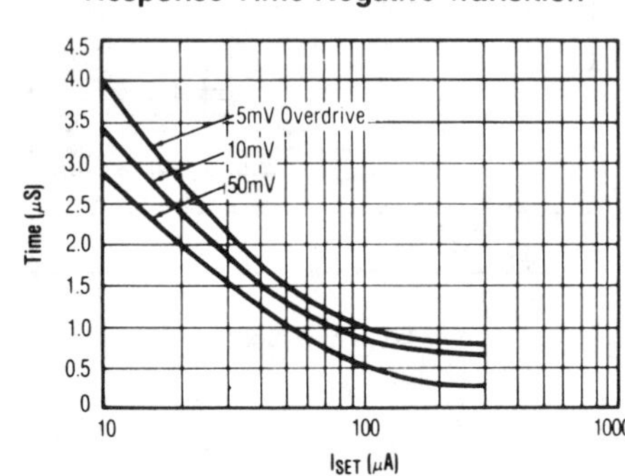

Response Time Positive Transition

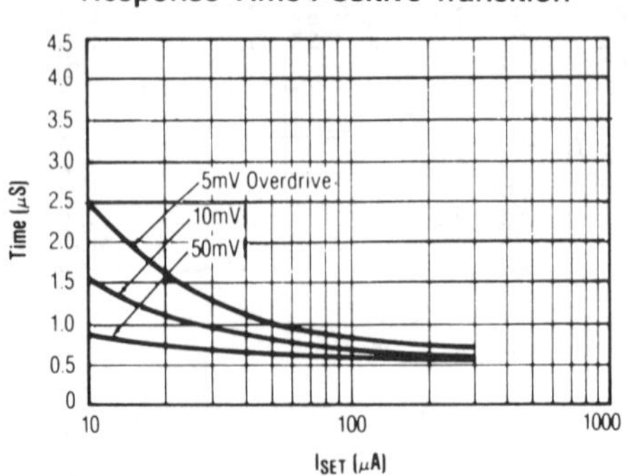

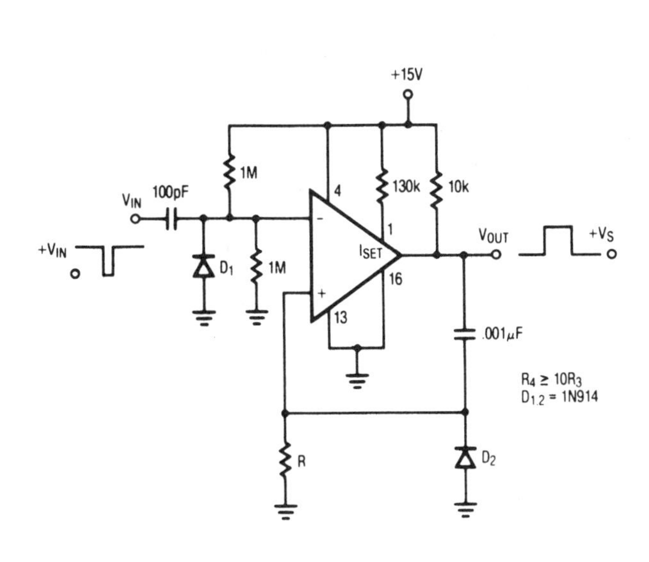

One Shot Multivibrator

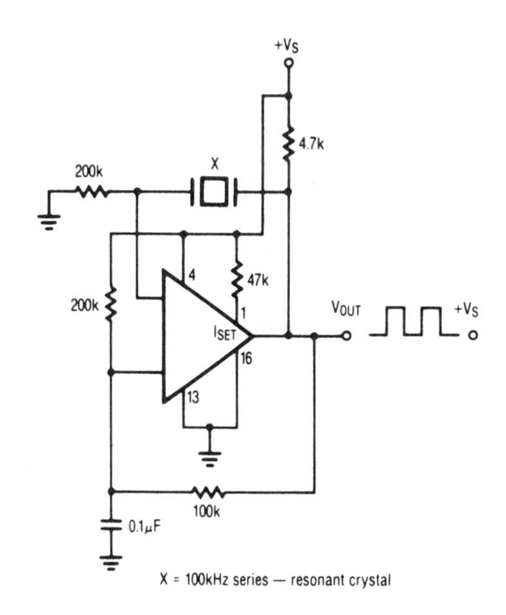

Crystal Controlled Oscillator (Single Supply)

Raytheon RC4444

4 × 4 × 2 Balanced Switching Crosspoint Array

- Low bidirectional R_{ON}
- High R_{OFF}
- Excellent matching of gates
- Low capacitance
- High rate firing
- Predictable holding current

The RC4444 is a monolithic dielectrically isolated crosspoint array arranged into a 4 × 4 × 2 matrix. The primary application is for balanced switching of 600Ω transmission lines. The ring and tip are selected by selective biasing of the P+ and P− gate.

Designed to replace reed relays in telephone switchboards, it does not require a constant gate drive to keep the SCR in the "ON" condition. It is several orders faster, with no bouncing, and has a much longer operating life than its mechanical counterpart.

The 16 SCR pairs with the gating system are packaged in a 24-pin dual in-line package.

The RC4444 is a monolithic pin-for-pin replacement for the MC3416 and MCBH7601.

ABSOLUTE MAXIMUM RATINGS

Operating Voltage[1] .... +25V
Operating Current per Crosspoint 100mA
Storage Temperature
Range .... -65°C to +150°C
Operating Temperature Range
RC4444 .... 0°C to +70°C
Lead Soldering Temperature
(60 Sec) .... +300°C

Notes: 1. Maximum voltage from anode to cathode.

THERMAL CHARACTERISTICS

	24-Lead Plastic DIP	24-Lead Ceramic DIP
Max. Junction Temp.	125°C	175°C
Max. P_D T_A < 50°C	555mW	1042mW
Therm. Res. θ_{JC}	—	60°C/W
Therm. Res. θ_{JA}	135°C/W	120°C/W
For T_A > 50°C Derate at	7.41mW per °C	8.33mW per °C

ELECTRICAL CHARACTERISTICS ($0°C \leq T_A \leq +70°C$ unless otherwise noted)

Parameters	Test Conditions	Min	Typ	Max	Units
Anode-Cathode Breakdown Voltage	I_{AK} = 25μA	25			V
Cathode-Anode Breakdown Voltage	I_{KA} = 25μA	25			V
Base-Cathode Breakdown Voltage	I_{BK} = 25μA	25			V
Cathode-Base Breakdown Voltage	I_{KB} = 25μA	25			V
Base-Emitter Breakdown Voltage	I_{BE} = 25μA	25			V
Emitter-Cathode Breakdown Voltage	I_{EK} = 25μA	25			V
OFF State Resistance	V_{AK} = 10V	100			MΩ
Dynamic ON Resistance	Center Current = 10mA	4.0		12	Ω
	Center Current = 20mA	2.0		10	
Holding Current		0.9		3.8	mA
Enable Current	V_{BE} = 1.5V	4.0			mA
Anode-Cathode ON Voltage	I_{AK} = 10mA			1.0	V
	I_{AK} = 20mA			1.1	
Gate Sharing Current Ratio at Cathodes	Under Select Conditions with Anodes Open	0.8		1.25	mA/mA
Inhibit Voltage	V_B = 3.0V			0.3	V
Inhibit Current	V_B = 3.0V			0.1	mA
OFF State Capacitance	V_{AK} = 0V			2.0	pF
Turn-ON Time				1.0	μS
Minimum Voltage Ramp	Which Could Fire the SCR Under Transient Conditions	800			V/μS

TYPICAL PERFORMANCE CHARACTERISTICS

Holding Current vs. Ambient Temperature

Anode-Cathode on Voltage vs. Current and Temperature

Difference in Anode-Cathode on Voltage (Between Associate Pairs of SCRs) vs. Anode-Cathode Current

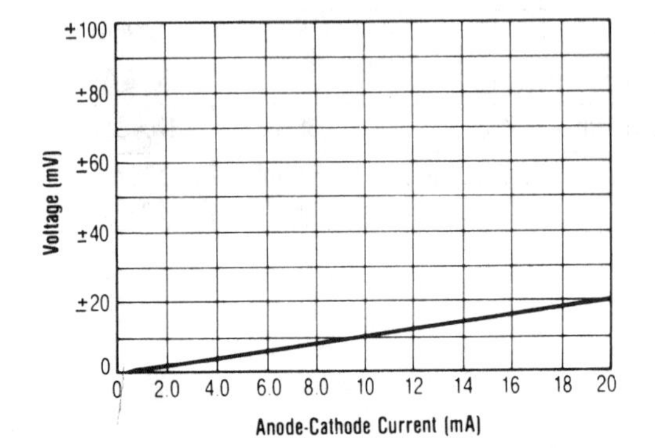

Off-Site Capacitance vs. Anode-Cathode Voltage

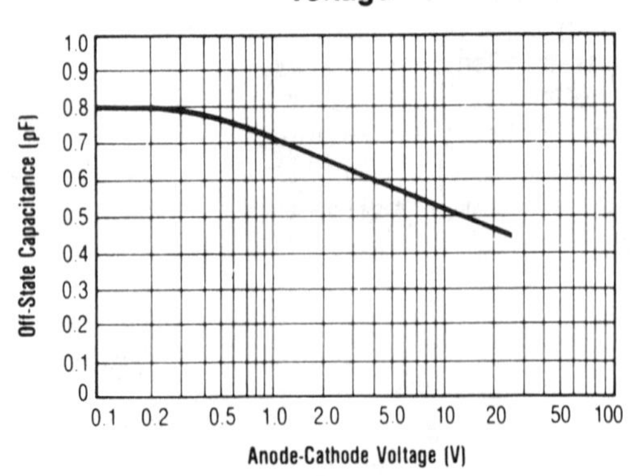

Dynamic on Resistance vs. Anode-Cathode Current

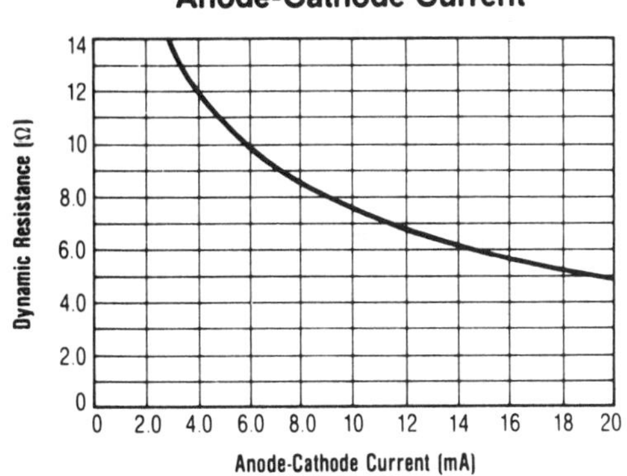

Dynamic on Resistance vs. Ambient Temperature

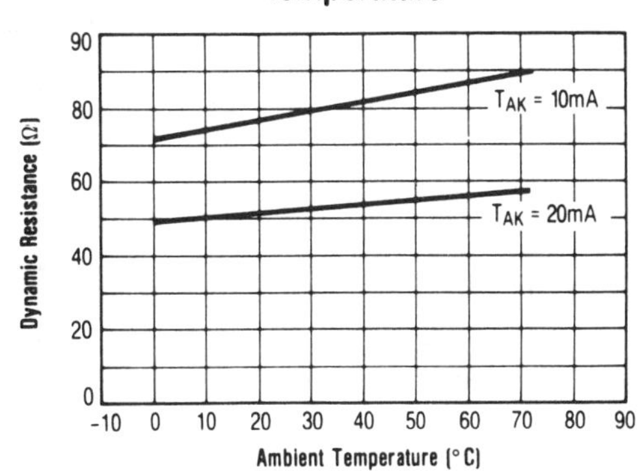

Crosstalk vs. Signal Frequency

-60
-70
-80
-90
-100
-110
-120
-130
Crosstalk (dB)
0.1 0.5 1.0 5.0 10 50 100
Signal Frequency (kHz)

Test Circuit for Crosstalk vs. Frequency

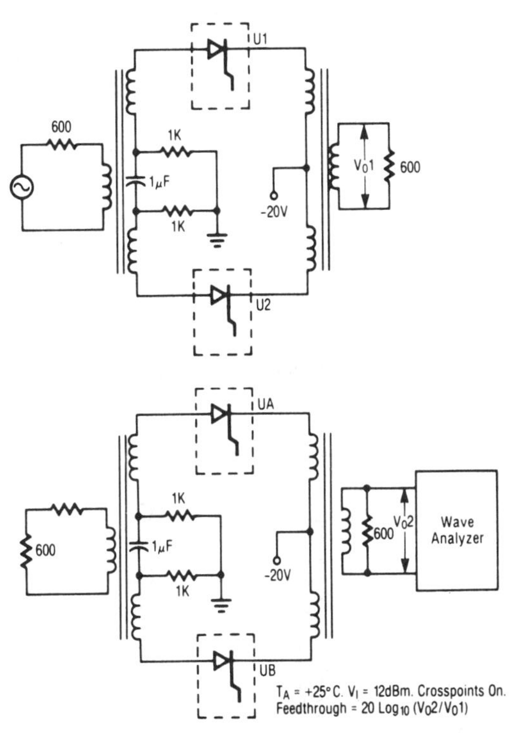

T_A = +25°C. V_I = 12dBm. Crosspoints On.
Feedthrough = 20 Log_{10} (V_O2/V_O1)

Feedthrough vs. Signal Frequency

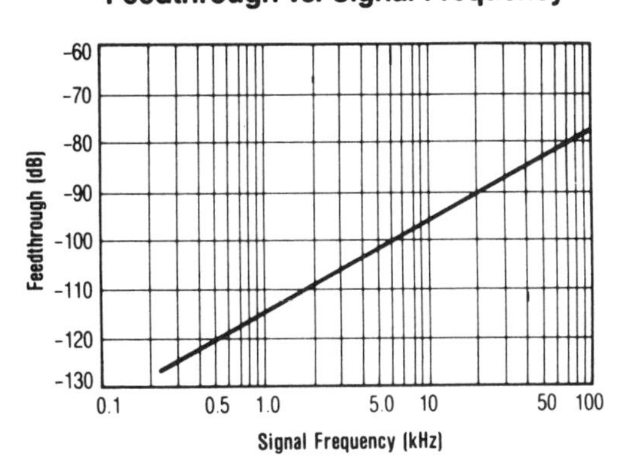

Test Circuit for Feedthrough vs. Frequency

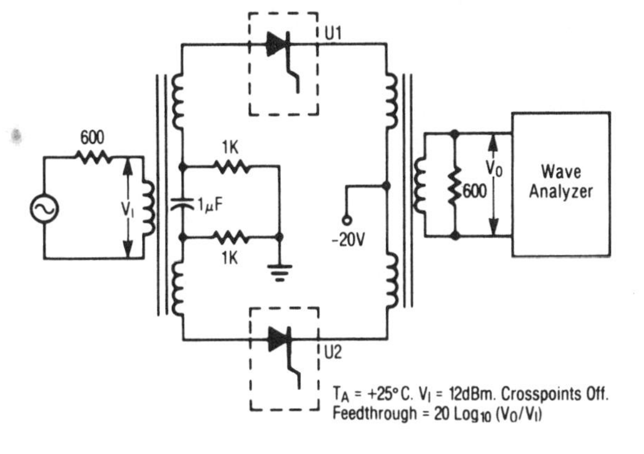

T_A = +25°C. V_I = 12dBm. Crosspoints Off.
Feedthrough = 20 Log_{10} (V_O/V_I)

SCHEMATIC DIAGRAM (1/16 Shown)

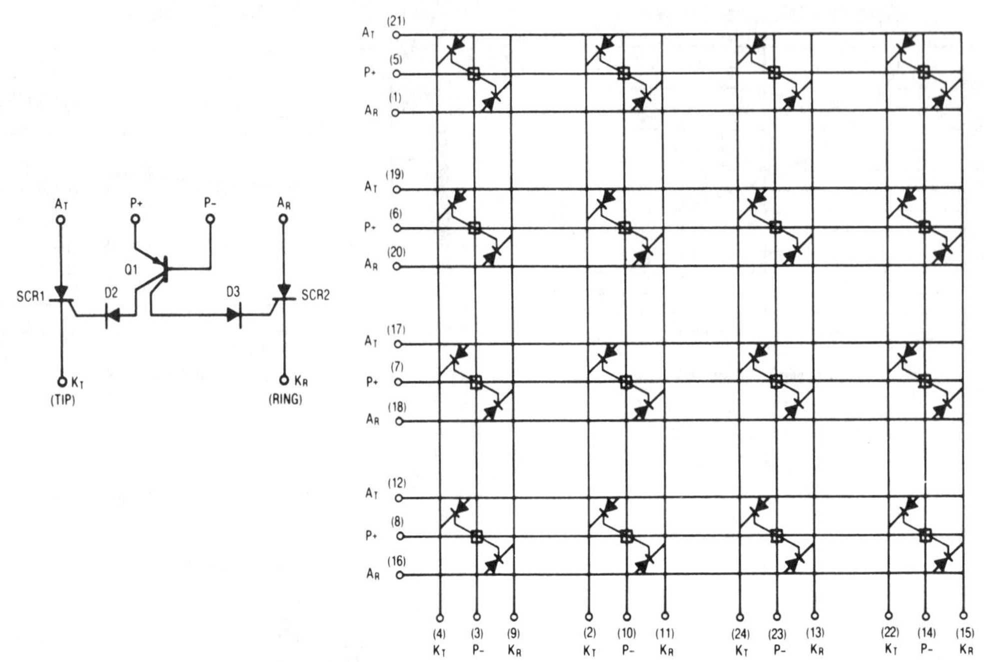

CONNECTION INFORMATION

24-Lead
Dual In-Line Package
(Top View)

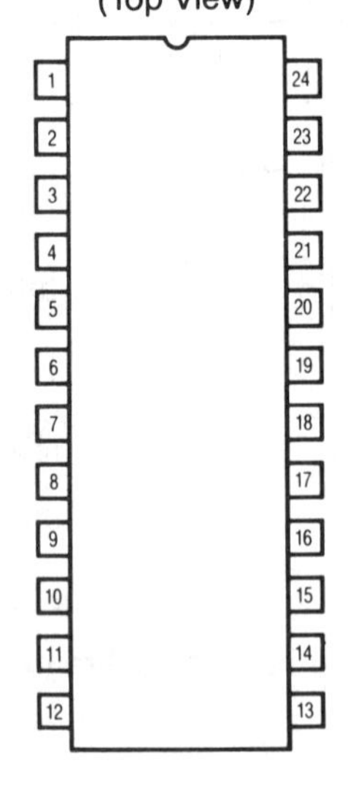

Pin	Function	Pin	Function
1	Anode A1	13	Cathode X1
2	Cathode Y2	14	Row Select W
3	Row Select Z	15	Cathode W1
4	Cathode Z2	16	Anode D1
5	Column Select A	17	Anode C2
6	Column Select B	18	Anode C1
7	Column Select C	19	Anode B2
8	Column Select D	20	Anode B1
9	Cathode Z1	21	Anode A2
10	Row Select Y	22	Cathode W2
11	Cathode Y1	23	Row Select X
12	Anode D2	24	Cathode X2

Raytheon XR-2207 Voltage-Controlled Oscillator

- Excellent temperature stability: 20ppm/°C
- Linear frequency sweep
- Adjustable duty cycle: 0.1% to 99.9%
- Two or four level FSK capability
- Wide sweep range: 1000: 1 min
- Logic compatible input and output levels
- Wide supply voltage range: ±4V to ±13V
- Low supply sensitivity: 0.15% V
- Wide frequency range: 0.01 Hz to 1 MHz
- Simultaneous triangle and squarewave outputs

APPLICATIONS

- FSK generation
- Voltage and current-to-frequency conversion
- Stable phase-locked loop
- Waveform generation triangle, sawtooth, pulse, squarewave
- FM and sweep generation

The XR-2207 is a monolithic voltage-controlled oscillator (VCO) integrated circuit featuring excellent frequency stability and a wide tuning range. The circuit provides simultaneous triangle and squarewave outputs over a frequency range of 0.01 Hz to 1 MHz. It is ideally suited for FM, FSK, and sweep or tone generation, as well as for phase-locked loop applications.

As shown in the schematic diagram, the circuit is comprised of four functional blocks: a variable-frequency oscillator which generates the basic periodic waveforms; four current switches actuated by binary keying inputs; and buffer amplifiers for both the triangle and squarewave outputs. The internal switches transfer the oscillator current to any of four external timing resistors to produce four discrete frequencies, which are selected according to the binary logic levels at the keying terminals (pins 8 and 9).

The XR-2207 has a typical drift specification of 20ppm/°C. The oscillator frequency can be linearly swept over a 1000:1 range with an external control voltage. The duty cycle of both the triangle and the squarewave outputs can be varied from 0.1% to 99.9% to generate stable pulse and sawtooth waveforms.

14-Lead Dual In-Line Package (Top View)

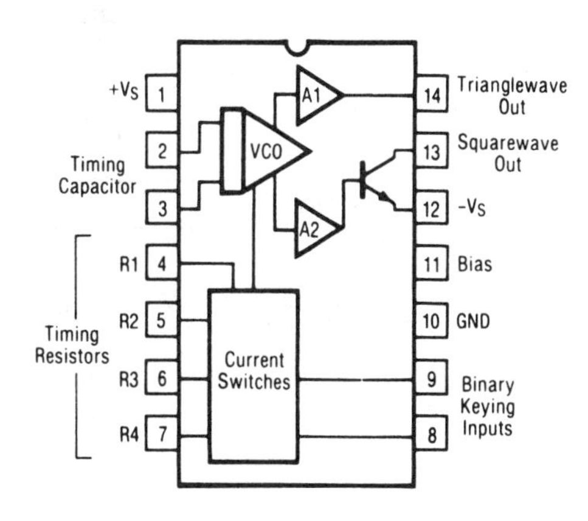

ABSOLUTE MAXIMUM RATINGS

Supply Voltage .... +26V
Storage Temperature Range .... -65°C to +150°C

THERMAL CHARACTERISTICS

	14-Lead Plastic DIP	14-Lead Ceramic DIP
Max. Junction Temp.	125°C	175°C
Max. P_D T_A < 50°C	468mW	1042mW
Therm. Res. θ_{JC}	—	60°C/W
Therm. Res. θ_{JA}	160°C/W	120°C/W
For T_A > 50°C Derate at	6.25mW per °C	8.33mW per °C

Ordering Information

Part Number	Package	Operating Temperature Range
XR2207CN	N	0°C to +70°C
XR2207N	N	-25°C to +85°C
XR2207MD XR2207MD/883B	D D	-55°C to +125°C -55°C to +125°C

Notes:
/883B suffix denotes Mil-Std-883, Level B processing
N = 14-lead plastic DIP
D = 14-lead ceramic DIP
Contact a Raytheon sales office or representative for ordering information on special package/temperature range combinations.

Parameters	Test Conditions	XR-2207			XR-2207C			Units
		Min	Typ	Max	Min	Typ	Max	
General Characteristics								
Supply Voltage Single Supply Split Supplies	See Typical Performance Characteristics	+8.0 ±4	+12 ±6	+26 ±13	+8.0 ±4	+12 ±6	+26 ±13	V
Supply Current Single Supply	Measured at pin 1, S1 open		5.0	7.0		5.0	8.0	mA
Split Supplies Positive	Measured at pin 1, S1 open		5.0	7.0		5.0	8.0	mA
Negative	Measured at pin 12, S1, S2 open		4.0	6.0		4.0	7.0	mA
Binary Keying Inputs								
Switching Threshold	Measured at pins 8 and 9. Refer to pin 10	1.4	2.2	2.8	1.4	2.2	2.8	V
Input Resistance			5.0			5.0		kΩ
Oscillator Section — Frequency Characteristics								
Upper Frequency Limit	C = 500pF, R3 = 2kΩ	0.5	1.0		0.5	1.0		MHz
Lower Practical Frequency	C = 50μF, R3 = 2Ω		0.01			0.01		Hz
Frequency Accuracy			±1.0	±3.0		±1.0	±5.0	% of f_0
Frequency Matching			0.5			0.5		% of f_0
Frequency Stability Vs. Temperature	$0°C < T_A < +75°C$		20	50		30		ppm/°C
Vs. Supply Voltage	$0°C < T_A < +75°C$		0.15			0.15		%/V
Sweep Range	R3 = 1.5kΩ for f_H R3 = 2MΩ for f_L	1000:1	3000:1			1000:1		f_H/f_L
Sweep Linearity 10:1 Sweep	C = 5000pF f_H = 10kHz, f_L = 1kHz		1.0	2.0		1.5		%
1000:1 Sweep	f_H = 100kHz, f_L = 100Hz		5.0			5.0		%
FM Distortion	±10% FM Deviation		0.1			0.1		%
Recommended Range of Timing Resistors	See Characteristic Curves	1.5		2000	1.5		2000	kΩ
Impedance at Timing Pins	Measured at pins 4, 5, 6 or 7		75			75		Ω
DC Level at Timing Terminals			10			10		mV
Output Characteristics								
Triangle Output Amplitude	Measured at pin 14	4	6		4	6		V_{p-p}
Impedance			10			10		Ω
DC Level	Referenced to pin 10		+100			+100		mV
Linearity	from 10% to 90% of swing		0.1			0.1		%
Squarewave Output Amplitude	Measured at pin 13, S2 Closed	11	12		11	12		V_{p-p}
Saturation Voltage	Referenced to pin 12		0.2	0.4		0.2	0.4	V
Rise Time	$C_L \leq 10pF$		200			200		nS
Fall Time	$C_L \leq 10pF$		20			20		nS

SCHEMATIC DIAGRAM

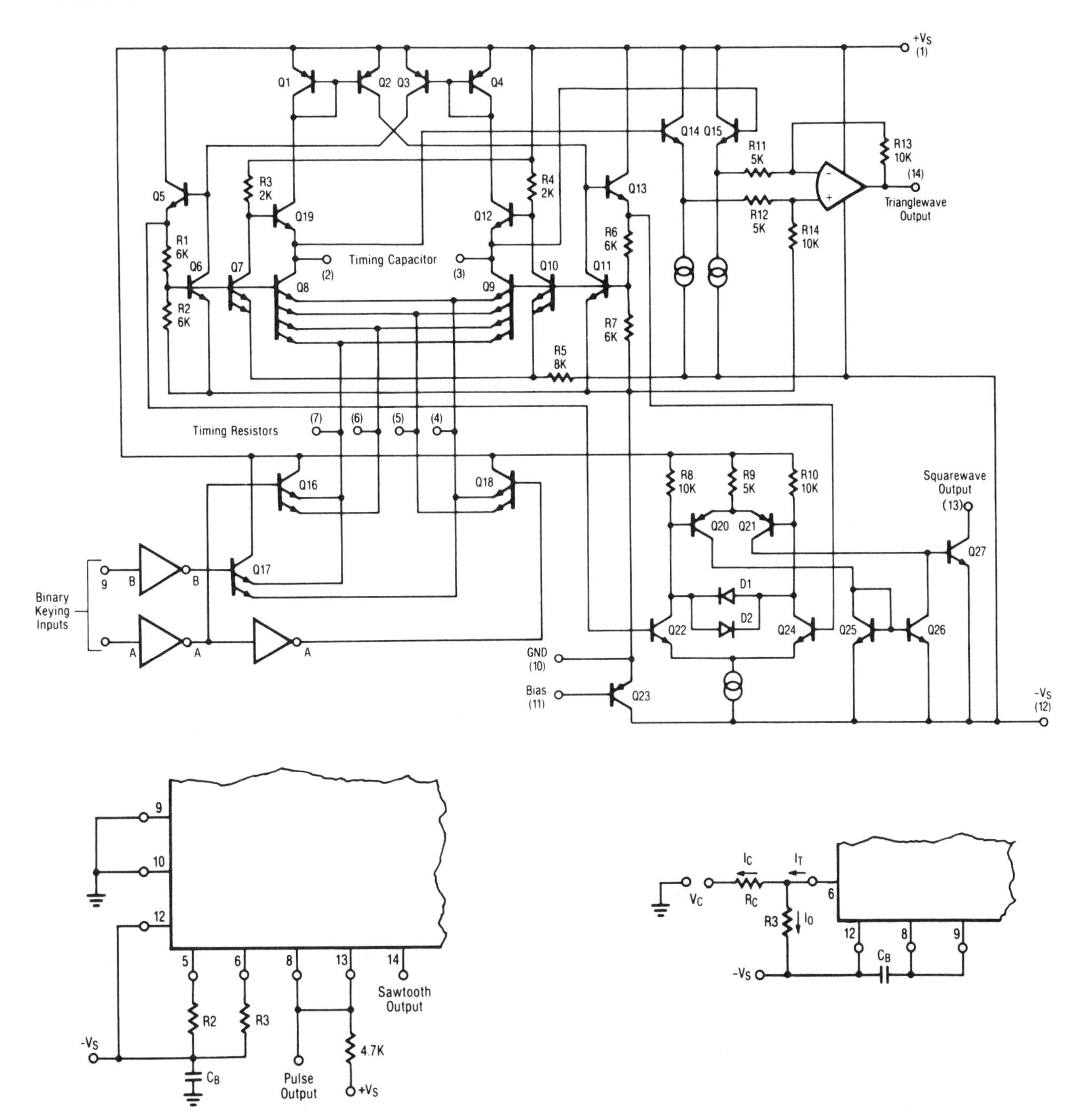

Pulse and Sawtooth Generation

Frequency Sweep Operation

Raytheon OP-27

Very Low Noise Operational Amplifier

- Very low noise
  Spectral noise density: $3.0\text{nV}/\sqrt{\text{Hz}}$
  1/f Noise corner frequency: 2.7Hz
- Very low V_{OS} drift
  0.2μV/Mo
  0.2μV/°C
- High gain: 1.8×10^6V/V
- High output drive capability: ±12V into 600Ω load
- High slew rate: $2.8\text{V}/\mu\text{S}$
- Wide gain bandwidth product: 8 MHz
- Good common-mode rejection ratio: 126dB
- Low input offset voltage: 10μV
- Minimum low-frequency noise: $0.08\mu V_{p\text{-}p}$ 0.1 Hz to 10 Hz
- Low input bias and offset currents: 10nA

The OP-27 is designed for instrumentation grade signal conditioning where low noise (both spectral density and burst), wide bandwidth, and high slew rate are required along with low input offset voltage, low input offset temperature coefficient, and low input bias currents. These features are all available in a device that is internally compensated for excellent phase margin (70°) in a unity-gain configuration. Digital nulling techniques performed at wafer sort make it feasible to guarantee temperature stable input offset voltages as low as 25μV. Input bias current cancellation techniques are used to obtain 10nA input bias currents.

The OP-27 is especially useful for instrumentation and professional-quality audio systems. Applying the slew rate vs. power bandwidth equation ($fp = SR/2\pi V_p$), the OP-27 will have an undistorted output up to its power bandwidth frequency of 34kHz, and an undistorted output of 8.0V_{p-p} at 100kHz. This device provides adequate performance for the most demanding high-fidelity applications.

In addition to providing superior performance for the professional audio market, the OP-27 design uniquely addresses the needs of the instrumentation designer. Power-supply rejection and common-mode rejection are both in excess of 120dB. A phase margin of 70° at unity gain guards against peaking (and ringing) in low-gain feedback circuits. Stable operation can be obtained with capacitive loads up to 2000pF[1]. Input offset voltage can be externally trimmed without affecting input offset voltage drift with temperature or time. The drift performance is, in fact, so good that the system designer must be cautioned that stray thermoelectric voltages generated by dissimilar metals at the contacts to the input terminals are enough to degrade its performance. For this reason, it is also important to keep both input terminals at the same relative temperature. The well-behaved temperature performance of the OP-27 has made it unnecessary to specify a commercial grade (0°C to +70°C). All grades of the OP-27 are specified to, at least, the industrial-grade (−25°C to +85°C) temperature range.

Note: 1. By decoupling the load capacitance with a series resistor of 50Ω or more load capacitances larger than 2000pF can be accommodated.

The information contained herein has been carefully compiled, however, it shall not by implication or otherwise become part of the terms and conditions of any subsequent sale. Raytheon's liability shall be determined solely by its standard terms and conditions of sale. No representation as to application or use or that the circuits are either licensed or free from patent infringement is intended or implied. Raytheon reserves the right to change the circuitry and other data at any time without notice and assumes no liability for inadvertent errors.

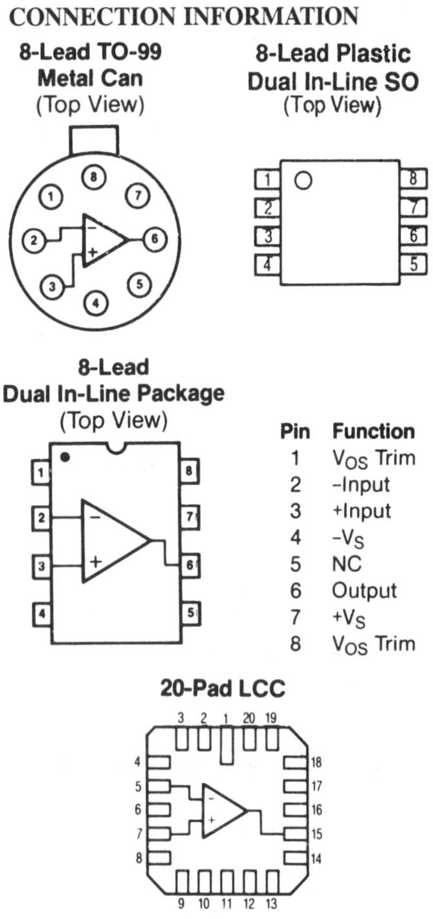

Pin	Function
2	V_{OS} Trim
5	-Input
7	+Input
10	$-V_S$
15	Output
17	$+V_S$
20	V_{OS} Trim

ORDERING INFORMATION

Part Number	Package	Operating Temperature Range
OP-27EN	N	0°C to +70°C
OP-27FN	N	0°C to +70°C
OP-27GN	N	0°C to +70°C
OP-27EM	M	0°C to +70°C
OP-27FM	M	0°C to +70°C
OP-27GM	M	0°C to +70°C
OP-27ED	D	-25°C to +85°C
OP-27FD	D	-25°C to +85°C
OP-27GD	D	-25°C to +85°C
OP-27ET	T	-25°C to +85°C
OP-27FT	T	-25°C to +85°C
OP-27GT	T	-25°C to +85°C
OP-27AD	D	-55°C to +125°C
OP-27AD/883	D	-55°C to +125°C
OP-27BD	D	-55°C to +125°C
OP-27BD/883	D	-55°C to +125°C
OP-27CD	D	-55°C to +125°C
OP-27CD/883	D	-55°C to +125°C
OP-27AT	T	-55°C to +125°C
OP-27AT/883	T	-55°C to +125°C
OP-27BT	T	-55°C to +125°C
OP-27BT/883	T	-55°C to +125°C
OP-27CT	T	-55°C to +125°C
OP-27CT/883	T	-55°C to +125°C
OP-27AL/883	L	-55°C to +125°C
OP-27BL/883	L	-55°C to +125°C

Notes:
/883B suffix denotes Mil-Std-883, Level B processing
N = 8-lead plastic DIP
D = 8 lead ceramic DIP
T = 8-lead metal can (TO-99)
L = 20-pad leadless chip carrier
M = 8-lead plastic SOIC
Contact a Raytheon sales office or representative for ordering information on special package/temperature range combinations.

ABSOLUTE MAXIMUM RATINGS

Supply Voltage.. ±22V
Input Voltage* .... ±22V
Differential Input Voltage 0.7V
Internal Power Dissipation** .... 658 mW
Output Short Circuit Duration .... Indefinite
Storage Temperature
Range .... -65°C to +150°C
Operating Temperature Range
OP-27A/B/C .... -55°C to +125°C
OP-27E/F/G (Hermetic) .... -25°C to +85°C
OP-27E/F/G (Plastic) .... 0°C to +70°C
Lead Soldering Temperature
(SO-8, 10 sec) .... +260°C
(DIP, LCC, TO-99; 60 sec) .... +300°C

*For supply voltages less than ±22V, the absolute maximum input voltage is equal to the supply voltage.
**Observe package thermal characteristics.

MASK PATTERN

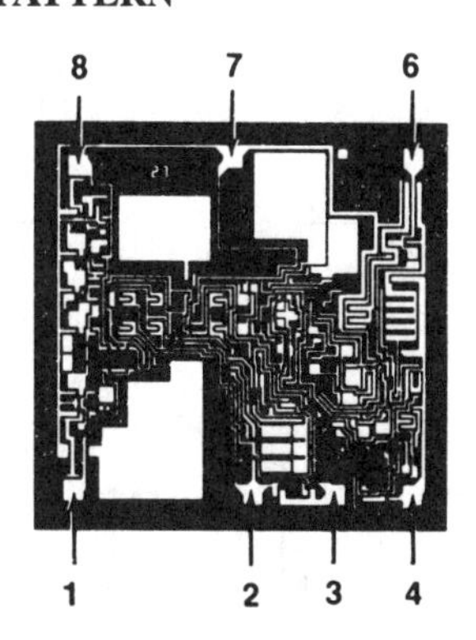

Die Size: 75 × 80 mils
Min. Pad Dimensions: 4 × 4 mils

THERMAL CHARACTERISTICS

	8-Lead Small Outline	8-Lead Ceramic DIP	8-Lead TO-99 Metal Can	20-Pad LCC	8-Lead Plastic DIP
Max. Junction Temp.	125°C	175°C	175°C	175°C	125°C
Max. P_D T_A <50°C	300 mW	833 mW	658 mW	925 mW	468 mW
Therm. Res θ_{JC}	—	45°C/W	50°C/W	37°C/W	—
Therm. Res. θ_{JA}	240°C/W	150°C/W	190°C/W	105°C/W	160°C/W
For T_A >50°C Derate at	4.17 mW/°C	8.33 mW/°C	5.26 mW/°C	7.0 mW/°C	6.25 mW/°C

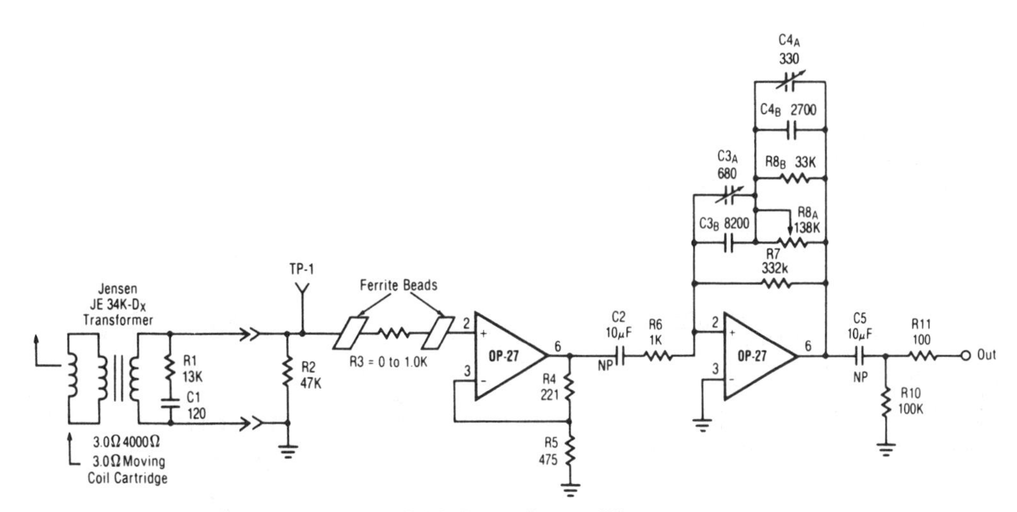

To test, disconnect transformer and inject signal at TP-1 **RIAA Phono Preamplifier**

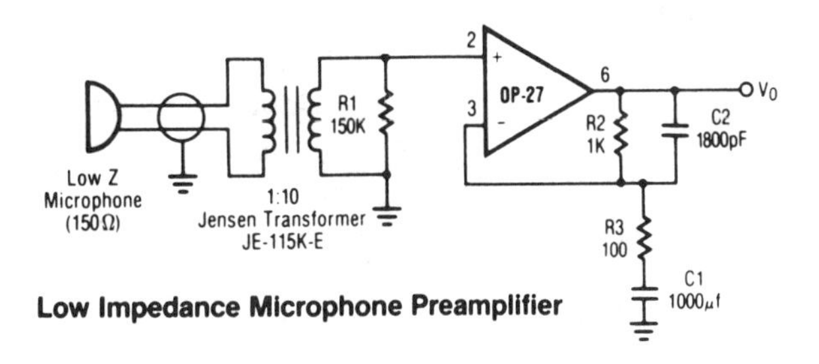

Low Impedance Microphone Preamplifier

SCHEMATIC DIAGRAM

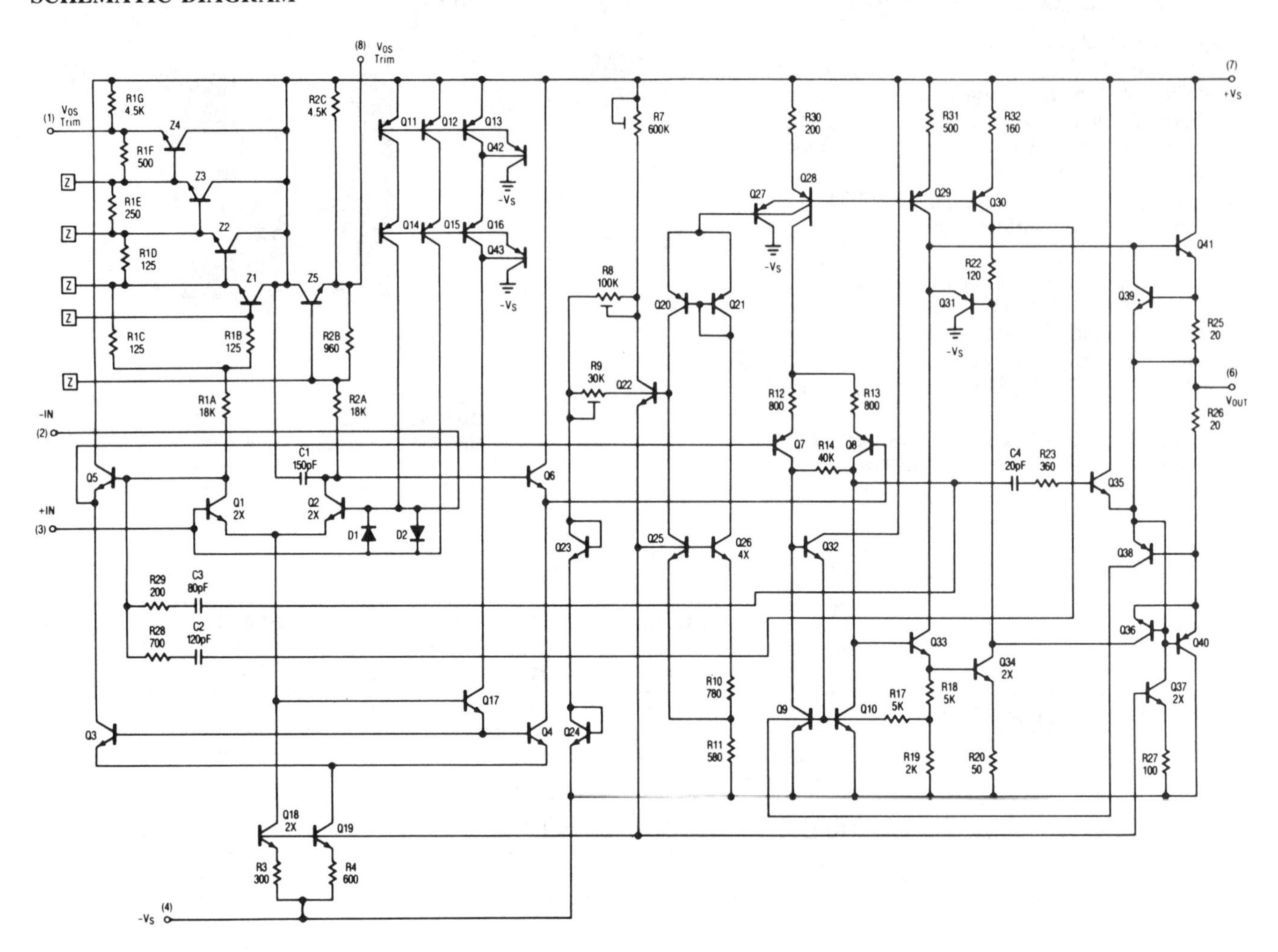

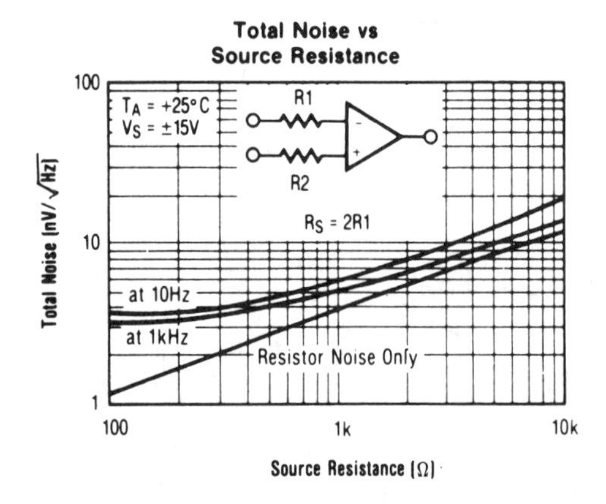

Voltage noise vs. source resistance for the difference amplifier. Noise performance shown is for V_S = ±15V, T_A = +25°C, and R_S = R1 + R2.

Total Noise vs. Source Resistance

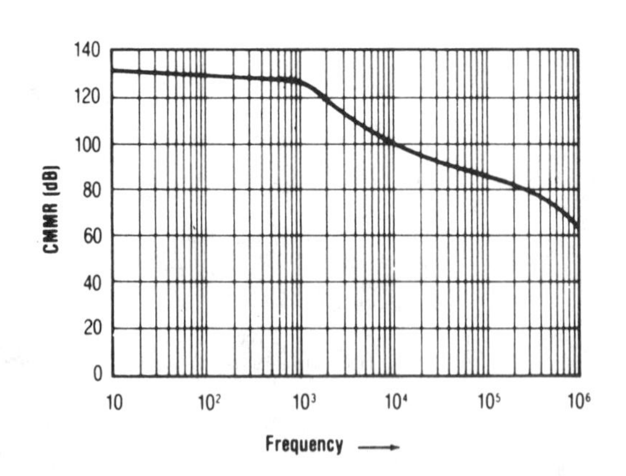

Common Mode Rejection Ratio vs. Frequency

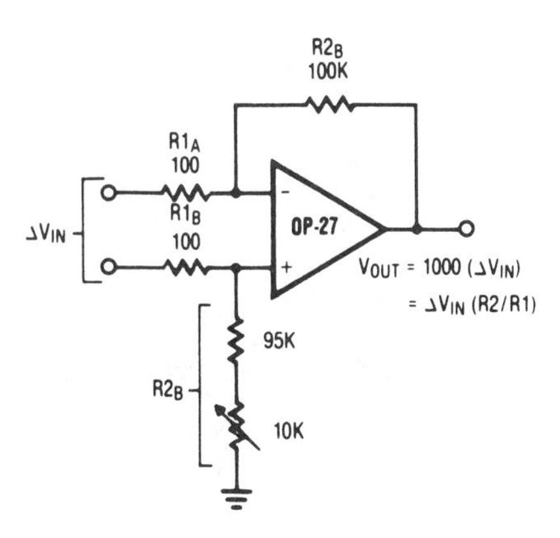

A Single Op Amp IC Difference Amplifier Using an OP-27. The Difference Amplifier is Connected for a Gain of 1000.

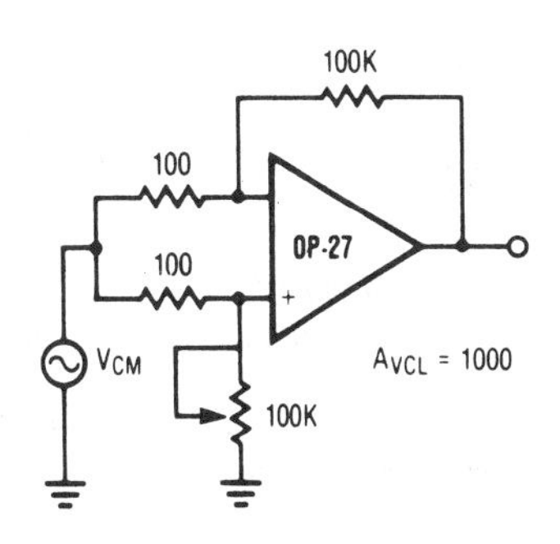

Common Mode Rejection Ratio Test Circuit

Raytheon OP-37

Very Low Noise Operational Amplifier

- Very low noise
  Spectral noise density: $3nV/\sqrt{Hz}$ 1/f noise corner frequency: 2.7Hz
- Very low V_{OS} drift
  $0.2\mu V/Mo$
  $0.2\mu V/°C$
- High gain: $1.8 \times 10^6 V/V$
- High output drive capability: $\pm 12V$ into 600Ω load
- High slew rate: $17V/\mu S$
- Wide gain bandwidth product: 63 MHz
- Good common-mode rejection ratio: 126dB
- Low input offset voltage: $10\mu V$
- Minimum low-frequency noise: $0.8\mu V_{p-p}$ 0.1Hz to 10Hz
- Low input bias and offset currents: 10nA

The OP-37 is designed for instrumentation-grade signal conditioning where low noise (both spectral density and burst), wide bandwidth, and high slew rate are required along with low input offset voltage, low input offset temperature coefficient, and low input bias currents in gains greater than or equal to ten. Digital nulling techniques performed at wafer sort make it feasible to guarantee temperature stable input offset voltages as low as $25\mu V$. Input bias current cancellation techniques are used to obtain 10nA input bias currents.

The OP-37 is especially useful for instrumentation and professional-quality audio applications in gains greater than or equal to ten. Applying the slew rate vs. power-bandwidth equation ($fp = SR/2\pi V_p$), the OP-37 will have an undistorted output up to its power-bandwidth frequency of 208kHz, with an undistorted output of $8V_{p-p}$ at 338kHz. This device provides performance adequate for the most demanding high-fidelity applications.

In addition to providing superior performance for the professional audio market, the OP-37 design uniquely addresses the needs of the instrumentation designer. Power-supply rejection and common-mode rejection are both typically 120dB. Input offset voltage can be externally trimmed without affecting input offset voltage drift with temperature or time. The drift performance is, in fact, so good that the system designer must be cautioned that stray thermoelectric voltages generated by dissimilar metals at the contacts to the input terminals are enough to degrade its performance. For this reason, it is also important to keep both input terminals at the same relative temperature. The well-behaved temperature performance of the OP-37 has made it unnecessary to specify a commercial grade (0 °C to +70 °C). All grades of the OP-37 are specified to, at least, the industrial-grade (−25 °C to + 85 °C) temperature range.

THERMAL CHARACTERISTICS

	8-Lead Small Outline	8-Lead Ceramic DIP	TO-99 8-Lead Metal Can	20-Pad LCC	8-Lead Plastic DIP
Max. Junction Temp.	125°C	175°C	175°C	175°C	125°C
Max. P_D T_A <50°C	300 mW	833 mW	658 mW	925 mW	468 mW
Therm. Res θ_{JC}	—	45°C/W	50°C/W	37°C/W	—
Therm. Res. θ_{JA}	240°C/W	150°C/W	190°C/W	105°C/W	160°C/W
For T_A >50°C Derate at	4.17 mW/°C	8.33 mW/°C	5.26 mW/°C	7.0 mW/°C	6.25 mW/°C

ABSOLUTE MAXIMUM RATINGS

Supply Voltage .... ±22V
Input Voltage* .... ±22V
Differential Input Voltage 0.7V
Internal Power Dissipation** .... 658 mW
Output Short Circuit Duration .... Indefinite
Storage Temperature
  Range .... -65°C to +150°C
Operating Temperature Range
  OP-37A/B/C .... -55°C to +125°C
  OP-37E/F/G (Hermetic) .... -25°C to +85°C
  OP-37E/F/G (Plastic) .... 0°C to +70°C
Lead Soldering Temperature
  (SO-8, 10 sec) .... +260°C
  (DIP, LCC, TO-99; 60 sec) .... +300°C

*For supply voltages less than ±22V, the absolute maximum input voltage is equal to the supply voltage.
**Observe package thermal characteristics.

MASK PATTERN

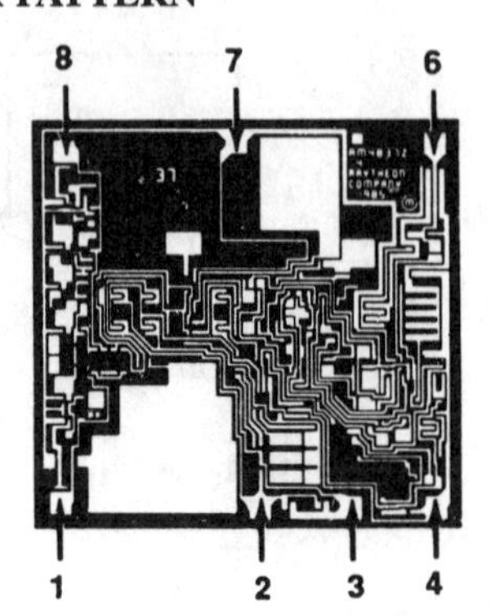

Die Size: 75 x 80 mils
Min. Pad Dimensions: 4 x 4 mils

CONNECTION INFORMATION

8-Lead TO-99 Metal Can
(Top View)

8-Lead Plastic Dual In-Line SO-8
(Top View)

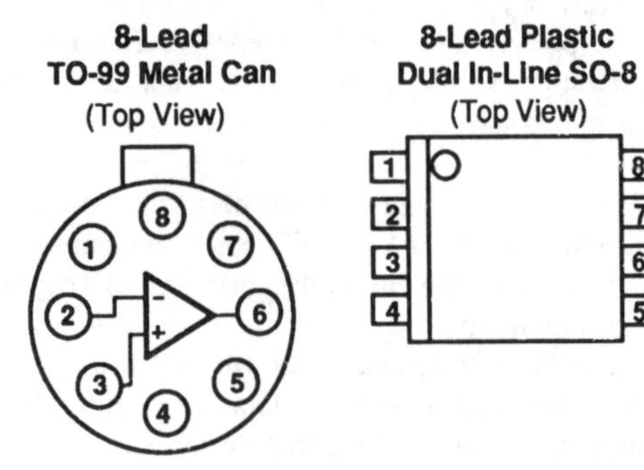

8-Lead Dual In-Line Package
(Top View)

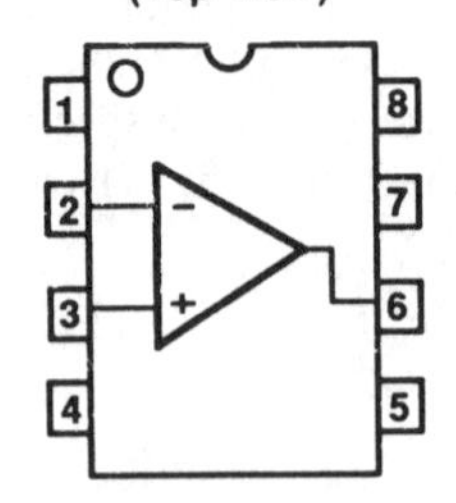

Pin	Function
1	V_{OS} Trim
2	-Input
3	+Input
4	$-V_S$
5	NC
6	Output
7	$+V_S$
8	V_{OS} Trim

20-Pad LCC
(Top View)

Pin	Function
2	V_{OS} Trim
5	-Input
7	+Input
10	$-V_S$
15	Output
17	$+V_S$
20	V_{OS} Trim

ORDERING INFORMATION

Part Number	Package	Operating Temperature Range
OP-37EN	N	0°C to +70°C
OP-37FN	N	0°C to +70°C
OP-37GN	N	0°C to +70°C
OP-37EM	M	0°C to +70°C
OP-37FM	M	0°C to +70°C
OP-37GM	M	0°C to +70°C
OP-37ED	D	-25°C to +85°C
OP-37FD	D	-25°C to +85°C
OP-37GD	D	-25°C to +85°C
OP-37ET	T	-25°C to +85°C
OP-37FT	T	-25°C to +85°C
OP-37GT	T	-25°C to +85°C
OP-37AD	D	-55°C to +125°C
OP-37AD/883B	D	-55°C to +125°C
OP-37BD	D	-55°C to +125°C
OP-37BD/883B	D	-55°C to +125°C
OP-37CD	D	-55°C to +125°C
OP-37CD/883B	D	-55°C to +125°C
OP-37AT	T	-55°C to +125°C
OP-37AT/883B	T	-55°C to +125°C
OP-37BT	T	-55°C to +125°C
OP-37BT/883B	T	-55°C to +125°C
OP-37CT	T	-55°C to +125°C
OP-37CT/883B	T	-55°C to +125°C
OP-37AL/883B	L	-55°C to +125°C
OP-37BL/883B	L	-55°C to +125°C

Notes:
/883B suffix denotes Mil-Std-883, Level B processing
N = 8-lead plastic DIP
D = 8 lead ceramic DIP
T = 8-lead metal can (TO-99)
L = 20-pad leadless chip carrier
M = 8-lead plastic SOIC
Contact a Raytheon sales office or representative for ordering information on special package/temperature range combinations.

SCHEMATIC DIAGRAM

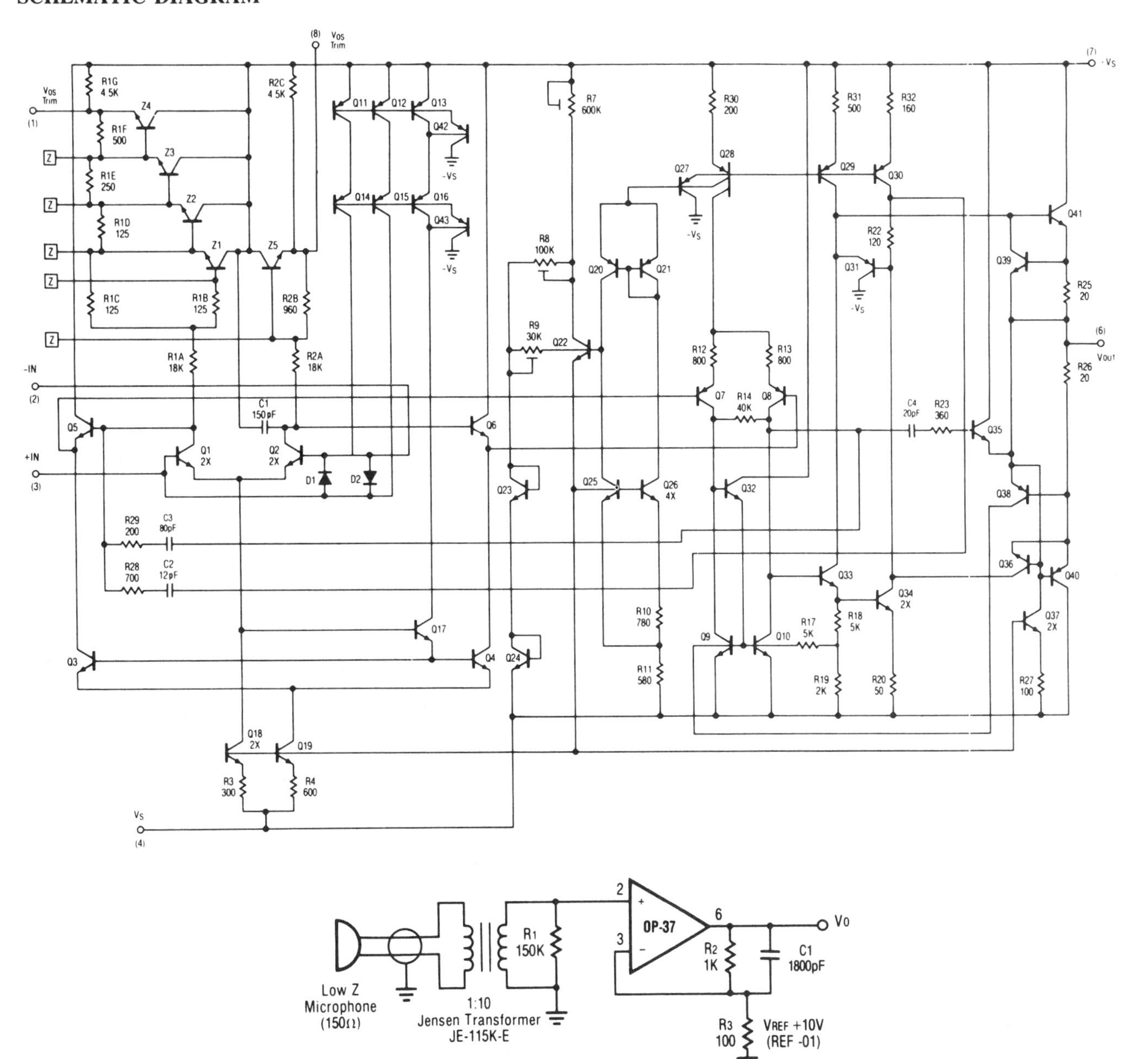

Low Impedance Microphone Preamplifier

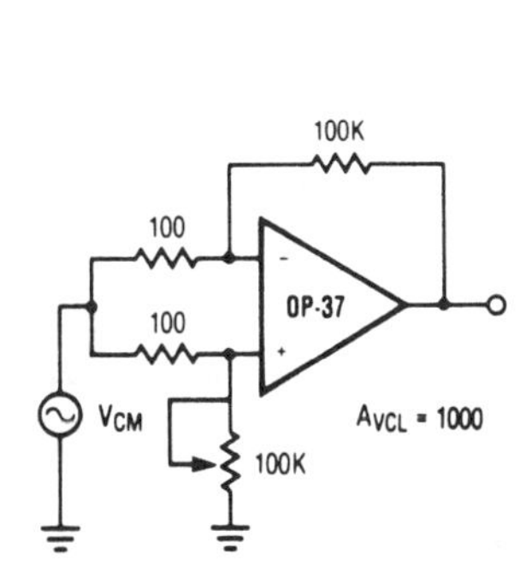

Common Mode Rejection Ratio Test Circuit

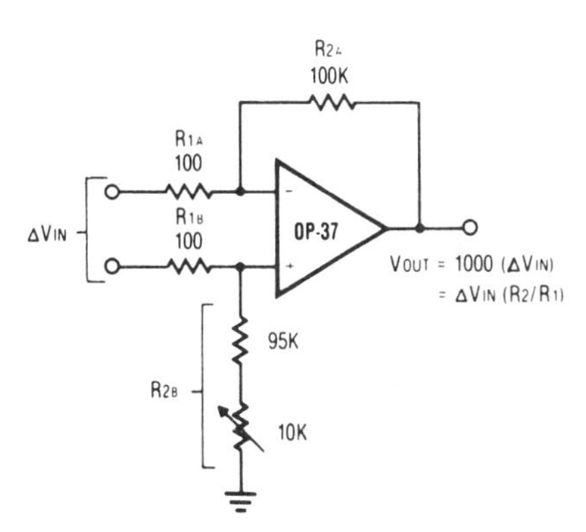

A Single Op Amp IC Difference Amplifier Using an OP-37. The Difference Amplifier is Connected for a Gain of 1000.

Raytheon OP-47

Very Low Noise Operational Amplifier

- Very low noise
  Spectral noise density: $3nV/\sqrt{Hz}$ 1/f noise corner frequency: 2.7 Hz
- Very low V_{OS} drift
  $0.2\mu V/Mo$
  $0.2\mu V/°C$
- High gain: $1.8 \times 10^6 V/V$
- High output drive capability: $\pm 12V$ into 600Ω load
- High slew rate: $50V/\mu S$ ($A_{VCL} > 400$)
- Wide gain bandwidth product: 70 MHz
- Good common mode rejection ratio: 126dB
- Low input offset voltage: $20\mu V$
- Minimum low frequency noise: $0.8\mu V_{p-p}$ 0.1 Hz to 10 Hz
- Low input bias and offset currents: 10nA

The OP-47 is designed for instrumentation-grade signal conditioning where low noise (both spectral density and burst), wide bandwidth, and high slew rate are required along with low input offset voltage, low input offset temperature coefficient, and low input bias currents in gains greater than or equal to 400. Digital nulling techniques performed at wafer sort make it feasible to guarantee temperature stable input offset voltages as low as $60\mu V$. Input bias current cancellation techniques are used to obtain 10nA input bias currents.

The OP-47 is especially useful for instrumentation and professional-quality audio applications in gains greater than or equal to 400. Applying the slew rate vs. power-bandwidth equation ($fp = SR/2\pi V_p$) the OP-47 will have an undistorted output of $8V_{p-p}$ of 900 kHz. This device provides performance adequate for the most demanding high-fidelity application.

In addition to providing superior performance for the professional audio market, the OP-47 design uniquely addresses the needs of the instrumentation designer. Power-supply rejection and common-mode rejection are both typically 120dB. Input offset voltage can be externally trimmed without affecting input offset voltage drift with temperature or time. The drift performance is, in fact, so good that the system designer must be cautioned that stray thermoelectric voltages generated by dissimilar metals at the contacts to the input terminals are enough to degrade its performance. For this reason, it is also important to keep both input terminals at the same relative temperature.

CONNECTION INFORMATION

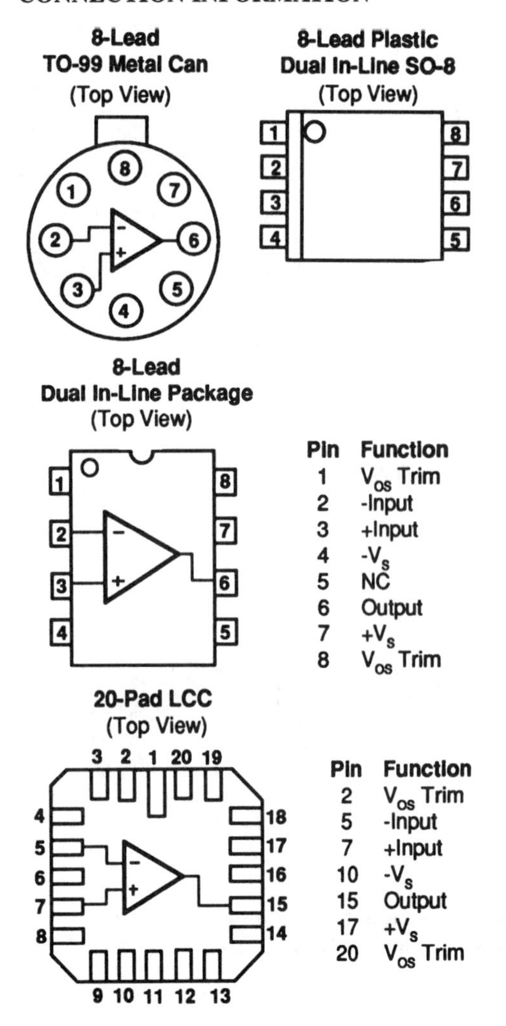

ORDERING INFORMATION

Part Number	Package	Operating Temperature Range
OP-47EN	N	0°C to +70°C
OP-47FN	N	0°C to +70°C
OP-47GN	N	0°C to +70°C
OP-47EM	M	0°C to +70°C
OP-47FM	M	0°C to +70°C
OP-47GM	M	0°C to +70°C
OP-47ED	D	-25°C to +85°C
OP-47FD	D	-25°C to +85°C
OP-47GD	D	-25°C to +85°C
OP-47ET	T	-25°C to +85°C
OP-47FT	T	-25°C to +85°C
OP-47GT	T	-25°C to +85°C
OP-47AD	D	-55°C to +125°C
OP-47AD/883B	D	-55°C to +125°C
OP-47BD	D	-55°C to +125°C
OP-47BD/883B	D	-55°C to +125°C
OP-47CD	D	-55°C to +125°C
OP-47CD/883B	D	-55°C to +125°C
OP-47AT	T	-55°C to +125°C
OP-47AT/883B	T	-55°C to +125°C
OP-47BT	T	-55°C to +125°C
OP-47BT/883B	T	-55°C to +125°C
OP-47CT	T	-55°C to +125°C
OP-47CT/883B	T	-55°C to +125°C
OP-47AL/883B	L	-55°C to +125°C
OP-47BL/883B	L	-55°C to +125°C

Notes:
/883B suffix denotes Mil-Std-883, Level B processing
N = 8-lead plastic DIP
D = 8 lead ceramic DIP
T = 8-lead metal can (TO-99)
L = 20-pad leadless chip carrier
M = 8-lead plastic SOIC
Contact a Raytheon sales office or representative for ordering information on special package/temperature range combinations.

ABSOLUTE MAXIMUM RATINGS

Supply Voltage ....±22V
Input Voltage* ....±22V
Differential Input Voltage 0.7V
Internal Power Dissipation** ....658 mW
Output Short Circuit Duration ....Indefinite
Storage Temperature
  Range ....-65°C to +150°C
Operating Temperature Range
  OP-47A/B/C ....-55°C to +125°C
  OP-47E/F/G (Hermetic) ....-25°C to +85°C
  OP-47E/F/G (Plastic) ....0°C to +70°C
Lead Soldering Temperature
  (SO-8, 10 sec) ....+260°C
  (DIP, LCC, TO-99; 60 sec) ....+300°C

*For supply voltages less than ±22V, the absolute maximum input voltage is equal to the supply voltage.
**Observe package thermal characteristics.

MASK PATTERN

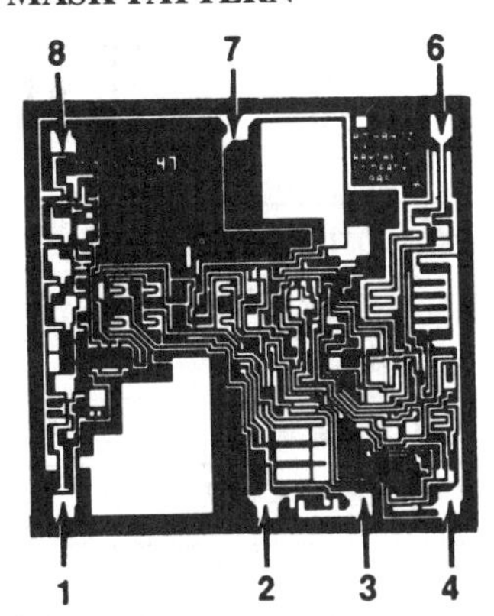

Die Size: 75 x 80 mils
Min Pad Dimensions: 4 x 4 mils

SCHEMATIC DIAGRAM

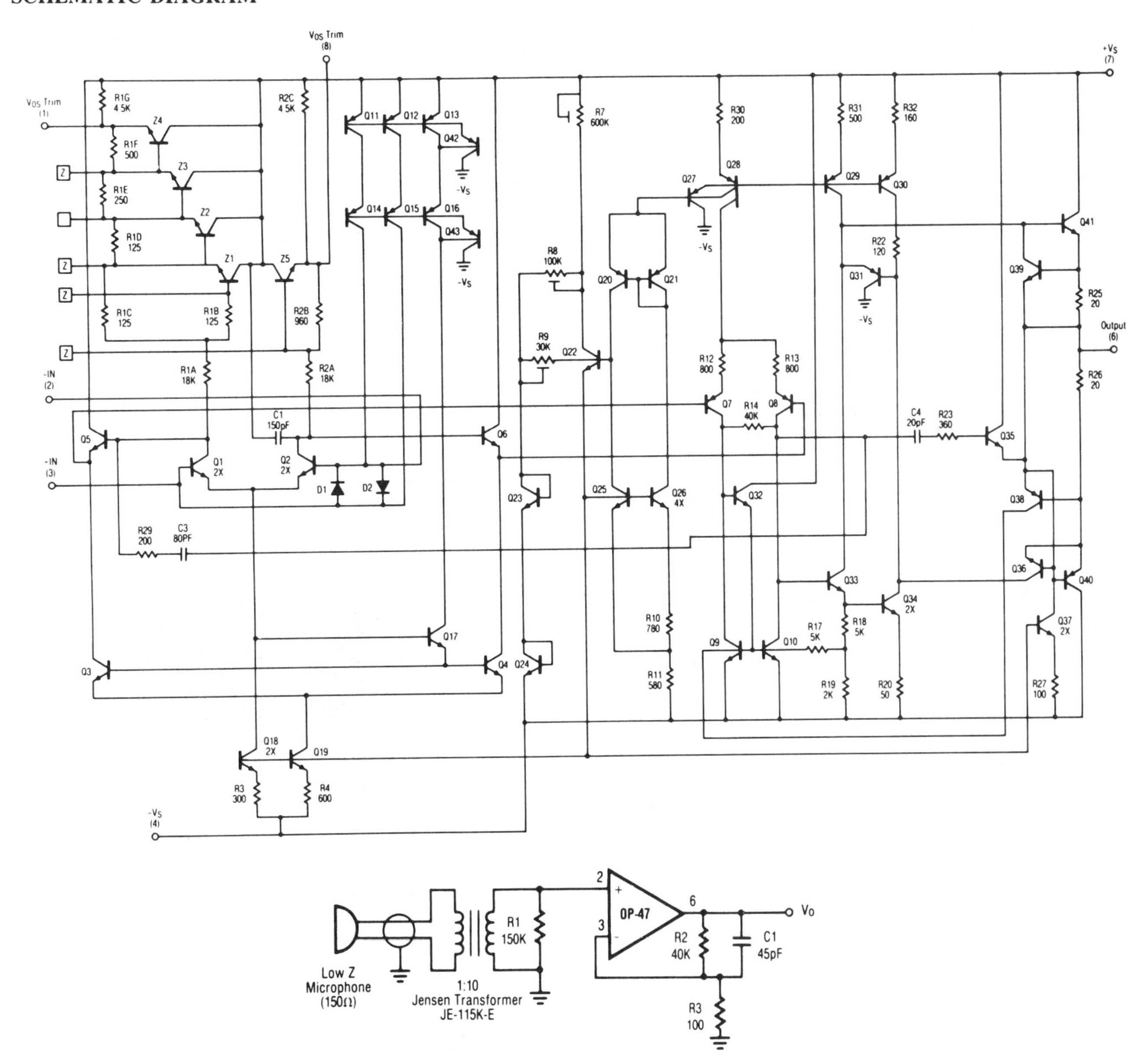

Low Impedance Microphone Preamplifier

THERMAL CHARACTERISTICS

	8-Lead Small Outline	8-Lead Ceramic DIP	TO-99 8-Lead Metal Can	20-Pad LCC	8-Lead Plastic DIP
Max. Junction Temp.	125°C	175°C	175°C	175°C	125°C
Max. P_D T_A <50°C	300 mW	833 mW	658 mW	925 mW	468 mW
Therm. Res θ_{JC}	—	45°C/W	50°C/W	37°C/W	—
Therm. Res. θ_{JA}	240°C/W	150°C/W	190°C/W	105°C/W	160°C/W
For T_A >50°C Derate at	4.17 mW/°C	8.33 mW/°C	5.26 mW/°C	7.0 mW/°C	6.25 mW/°C

Raytheon XR-2211

FSK Demodulator/Tone Decoder

- Wide frequency range: 0.01 Hz to 300 kHz
- Wide supply voltage range: 4.5V to 20V
- DTL/TTL/ECL logic compatibility
- FSK demodulation with carrier-detector
- Wide dynamic range: 2mV to $3V_{RMS}$
- Adjustable tracking range: ±1% to ±80%
- Excellent temperature stability: 20ppm/°C typical

APPLICATIONS

- FSK demodulation
- Data synchronization
- Tone decoding
- FM detection
- Carrier detection

The XR-2211 is a monolithic phase-locked loop (PLL) system especially designed for data communications. It is particularly well suited for FSK modem applications, and operates over a wide frequency range of 0.01 Hz to 300 kHz. It can accommodate analog signals between 2mV and 3V, and can interface with conventional DTL, TTL, and ECL logic families. The circuit consists of a basic PLL for tracking an input signal frequency within the passband, a quadrature phase detector, which provides carrier detection, and an FSK voltage comparator, which provides FSK demodulation. External components are used to independently set carrier frequency, bandwidth, and output delay.

ABSOLUTE MAXIMUM RATINGS

Supply Voltage .... +20V
Input Signal Level .... $3\ V_{RMS}$
Storage Temperature
 Range .... -65°C to +150°C
Operating Temperature Range
 XR2211MD .... -55°C to +125°C
 XR2211N .... -25°C to +85°C
 XR2211CN .... 0°C to +70°C
Lead Soldering Temperature
 (60 sec) .... +300°C

THERMAL CHARACTERISTICS

	14-Lead Plastic DIP	14-Lead Ceramic DIP
Max. Junction Temp.	125°C	175°C
Max. P_D T_A <50°C	468 mW	1042 mW
Therm. Res θ_{JC}	—	50°C/W
Therm. Res. θ_{JA}	160°C/W	120°C/W
For T_A >50°C Derate at	6.25 mW per °C	8.33 mw per °C

MASK PATTERN

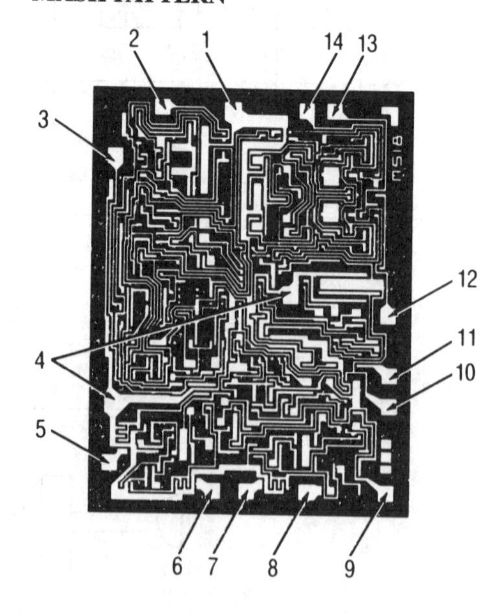

Die Size: 103 x 80 mils
Min. Pad Dimensions: 4 x 4 mils

ORDERING INFORMATION

Part Number	Package	Operating Temperature Range
XR2211CN	N	0°C to +70°C
XR2211N	N	-25°C to +85°C
XR2211MD XR2211MD/883B	D D	-55°C to +125°C -55°C to +125°C

Notes:
/883B suffix denotes Mil-Std-883, Level B processing
N = 14-lead plastic DIP
D = 14-lead ceramic DIP
Contact a Raytheon sales office or representative for ordering information on special package/temperature range combinations.

Recommended Component Values for Commonly Used FSK Bands

FSK Band	Component Values
300 Baud	$C0 = 0.039\mu F$ $C_F = 0.005\mu F$
$f_1 = 1070Hz$	$C1 = 0.01\mu F$ $R0 = 18k\Omega$
$f_2 = 1270Hz$	$R1 = 100k\Omega$
300 Baud	$C0 = 0.022\mu F$ $C_F = 0.005\mu F$
$f_1 = 2025Hz$	$C1 = 0.0047\mu F$ $R1 = 18k\Omega$
$f_2 = 2225Hz$	$R1 = 200k\Omega$
1200 Baud	$C0 = 0.027\mu F$ $C_F = 0.0022\mu F$
$f_1 = 1200Hz$	$C1 = 0.01\mu F$ $R0 = 18k\Omega$
$f_2 = 2200Hz$	$R1 = 30k\Omega$

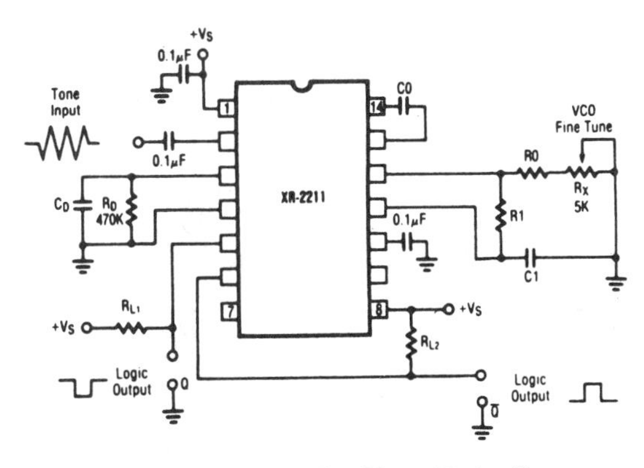

Circuit Connection for Tone Detection

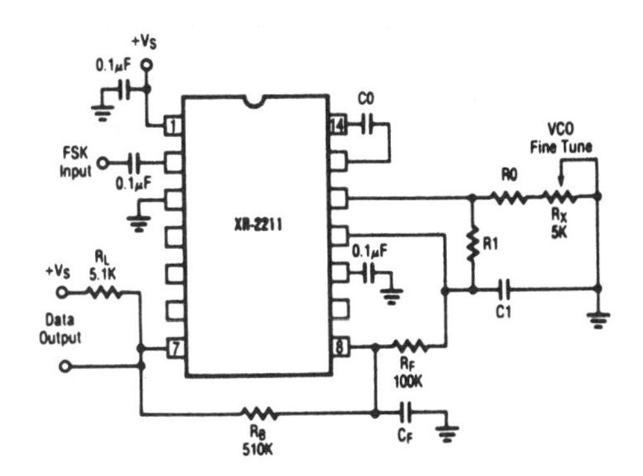

Circuit Connection for FSK Decoding

CONNECTION INFORMATION

14-Lead Dual In-Line Package (Top View)

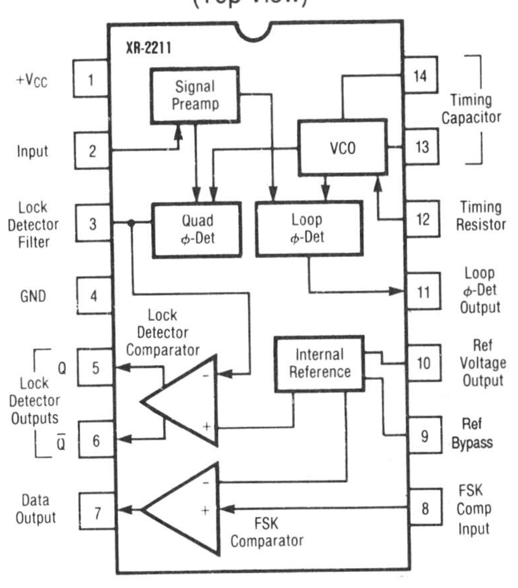

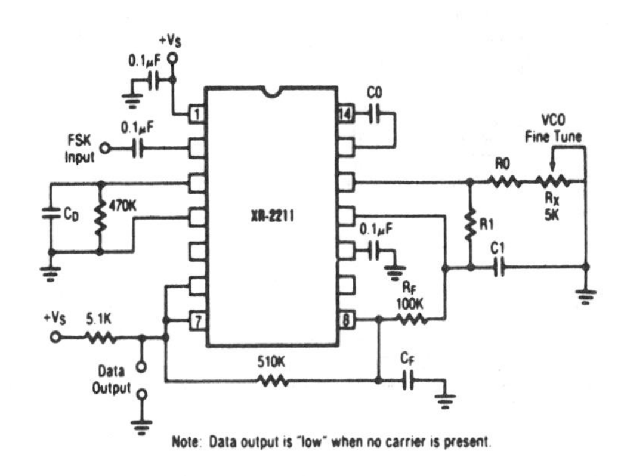

External Connectors for FSK Demodulation With Carrier Detect Capability

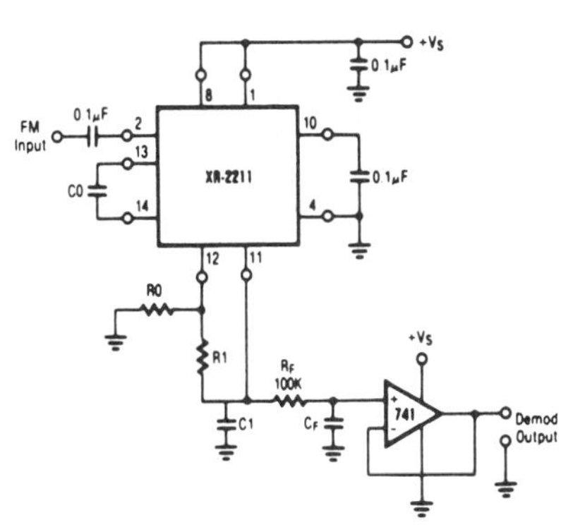

Linear FM Detector Using XR-2211 and an External Op Amp

ELECTRICAL CHARACTERISTICS (Test Conditions $+V_S$ = +12V, T_A = +25°C, R0 = 30kΩ C0 = 0.033μF.)

Parameters	Test Conditions	XR-2211/M			XR-2211C			Units
		Min	Typ	Max	Min	Typ	Max	
General								
Supply Voltage		4.5		20	4.5		20	V
Supply Current	R0 ≥ 10kΩ		4.0	9.0		5.0	11	mA
Oscillator								
Frequency Accuracy	Deviation from f_0 = 1/R0C0		±1.0	±3.0		±1.0		%
Frequency Stability Temperature Coefficient	R1 = ∞		±20	±50		±20		ppm/°C
Power Supply Rejection	$+V_S$ = 12 ±1V		0.05	0.5		0.05		%/V
	$+V_S$ = 5 ±0.5V		0.2			0.2		%/V
Upper Frequency Limit	R0 = 8.2kΩ, C0 = 400pF	100	300			300		kHz
Lowest Practical Operating Frequency	R0 = 2MΩ C0 = 50μF			0.01		0.01		Hz
Timing Resistor, R0 Operating Range		5.0		2000	5.0		2000	kΩ
Recommended Range		15		100	15		100	kΩ
Loop Phase Detector								
Peak Output Current	Meas. at Pin 11	±150	±200	±300	±100	±200	±300	μA
Output Offset Current			±1.0			±2.0		μA
Output Impedance			1.0			1.0		MΩ
Maximum Swing	Ref. to Pin 10	±4.0	±5.0		±4.0	±5.0		V
Quadrature Phase Detector								
Peak Output Current	Meas. at Pin 3	100	150			150		μA
Output Impedance			1.0			1.0		MΩ
Maximum Swing			11			11		V_{p-p}
Input Preamp								
Input Impedance	Meas. at Pin 2		20			20		kΩ
Input Signal Voltage Required to Cause Limiting			2.0	10		2.0		mV_{RMS}
Voltage Comparator								
Input Impedance	Meas. at Pins 3 & 8		2.0			2.0		MΩ
Input Bias Current			100			100		nA
Voltage Gain	R_L = 5.1kΩ	55	70		55	70		dB
Output Voltage Low	I_C = 3mA		300			300		mV
Output Leakage Current	V_0 = 12V		0.01			0.01		μA
Internal Reference								
Voltage Level	Meas. at Pin 10	4.9	5.3	5.7	4.75	5.3	5.85	V
Output Impedance			100			100		Ω

TYPICAL PERFORMANCE CHARACTERISTICS

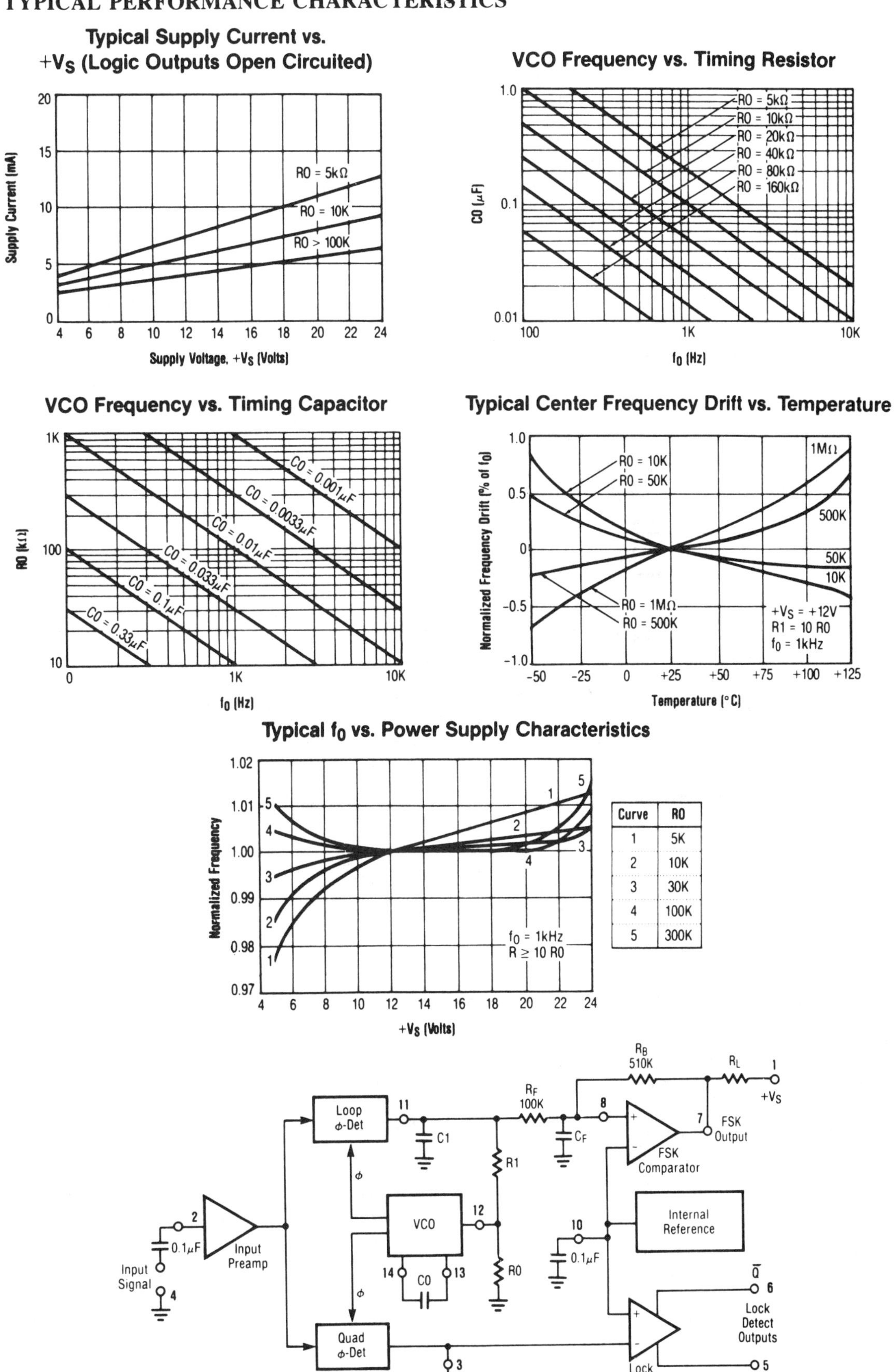

Curve	R0
1	5K
2	10K
3	30K
4	100K
5	300K

Generalized Circuit Connection for FSK and Tone Detection

SCHEMATIC DIAGRAM

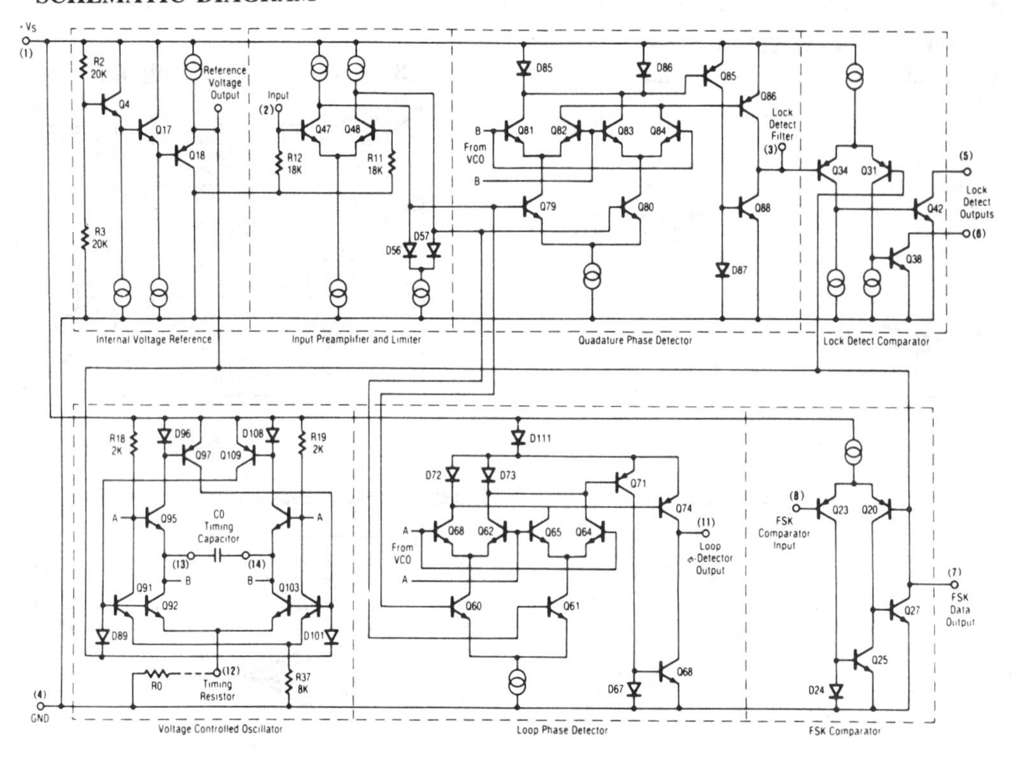

FUNCTIONAL BLOCK DIAGRAM

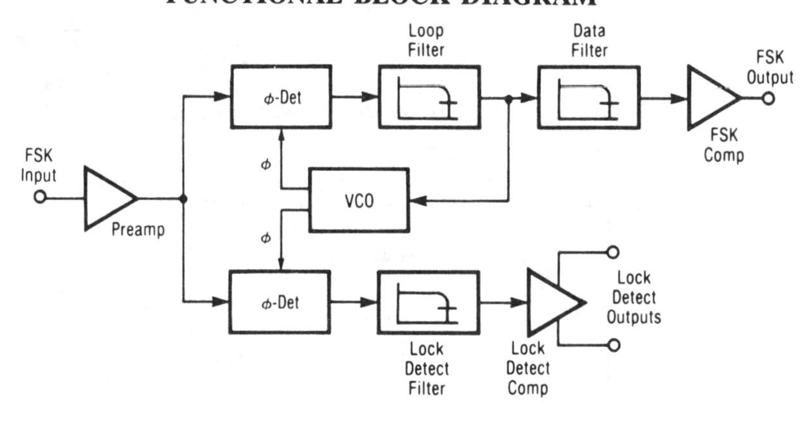

Rohm Corp. reserves the right to make changes to any product herein to improve reliability, function, or design. Rohm Corp. does not assume any liability arising out of the application or use of any product described herein, neither does it convey any license under its patent right nor the right of others.

This information is reproduced with permission of Rohm Corporation, Irvine, CA.

Rohm
BA718, BA728, BA728F
Dual Operational Amplifiers

- Can operate on a single power supply
- Low power consumption
- Pin configuration is identical to that of the 4558-type general-purpose op amps
- Supply voltage range for a single power supply is 3 to 18V
- Supply voltage range for dual power supply is ±1.5 to ±9V
- Output is short-circuit protected
- Output stage operates in class AB to minimize crossover distortion
- Small input bias current of 10 nA (typ.)
- Dual amplifiers in each package
- Internal phase compensation

APPLICATIONS

- Ground-sensing small-signal amplifiers
- Control amplifiers requiring high-phase margin, such as motor drivers
- Low-power, low-voltage operational amplifiers
- Capacitive-load driving amplifiers

The BA718, BA728, and BA728F are monolithic dual operational amplifiers; each chip contains two independent op amps with internal phase compensation. The devices feature a wide supply voltage range of 3 to 18V (1.5 to ±9V). They can operate on a single power supply and can include a negative voltage in the common-mode input voltage range.

The current consumption is small, 1.5 mA at V_{CC} = 6V and V_{EE} = −6 V, which is about a half that of the BA4558.

ABSOLUTE MAXIMUM RATINGS

Parameter	Symbol	Limits	Unit
Supply voltage	V_{CC}	18	V
Differential input voltage	V_{ID}	18	V
Common-mode input voltage range	V_{ICM}	−0.3 ~ 18	V
Power dissipation	Pd	450*1	mW
Operating temperature range	Topr	−20 ~ 75*2	°C
Storage temperature range	Tstg	−55 ~ 125	°C

*1 Derating is done at 4.5 mW/°C for operation above Ta=25°C.
*2 For an extended operating temperature range, consult your local ROHM representative.

ELECTRICAL CHARACTERISTICS (Ta=25°C, V_{CC}=6V, V_{EE}= −6V)

Parameter	Symbol	Min.	Typ.	Max.	Unit	Conditions
Input offset voltage	V_{IO}	—	2	10	mV	—
Input offset current	I_{IO}	—	1	50	nA	—
Input bias current	I_B	—	10	250	nA	—
Common-mode input voltage range	V_{ICM}	V_{EE}	—	$V_{CC}-1.5$	V	—
Quiescent current	I_Q	—	1.5	3.1	mA	—
Large signal voltage gain	A_V	86	100	—	dB	R_L=2kΩ
Output voltage amplitude	V_O	±3.0	±4.5	—	V	R_L=2kΩ
Common-mode rejection	CMR	70	90	—	dB	—
Supply voltage regulation	SVR	—	30	150	μV/V	—
Channel separation	S_{EP}	—	120	—	dB	—
Output current (SOURCE)	I_O source	—	20	—	mA	V_{IN}^+=1V, V_{IN}^-=0V
Output current (SINK)	I_O sink	—	20	—	mA	$V_{IN}-$=1V, V_{IN}^+=0V

*The input bias current flows out from the IC since a PNP transistor is used at the input.

CIRCUIT DIAGRAM

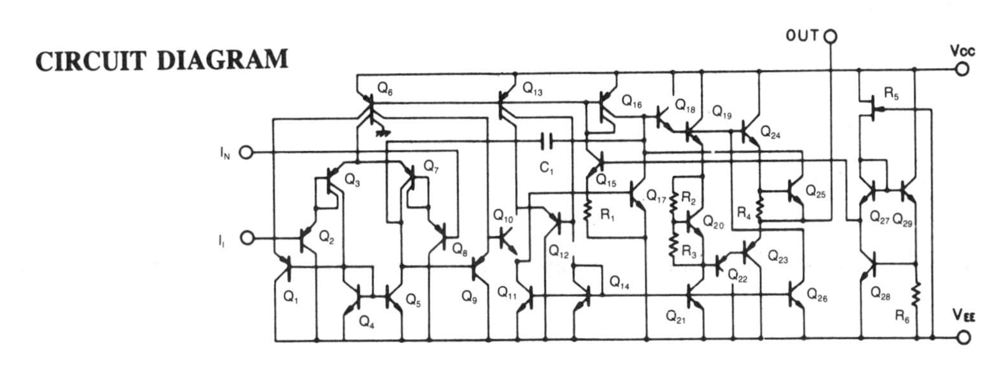

APPLICATION EXAMPLES

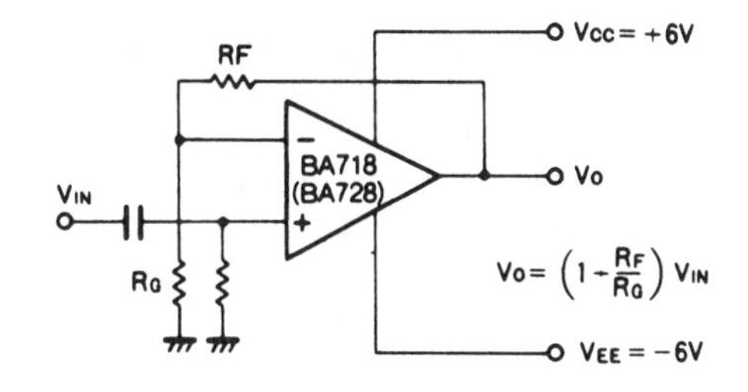

Noninverting amplifier

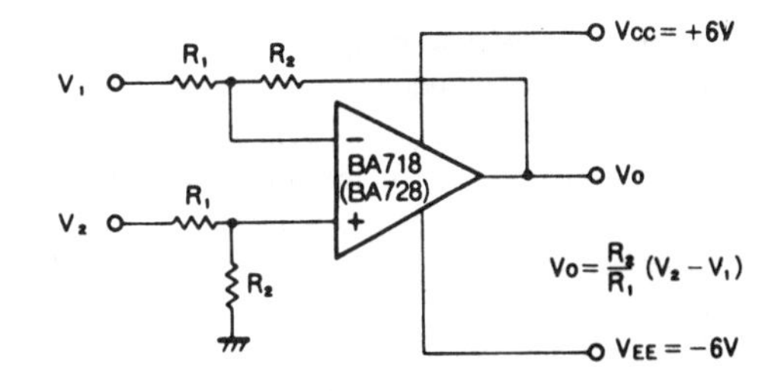

Differential amplifier

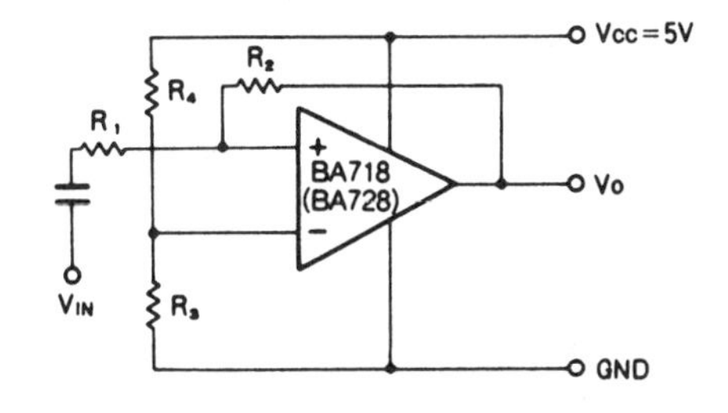

AC amplifier using a single supply

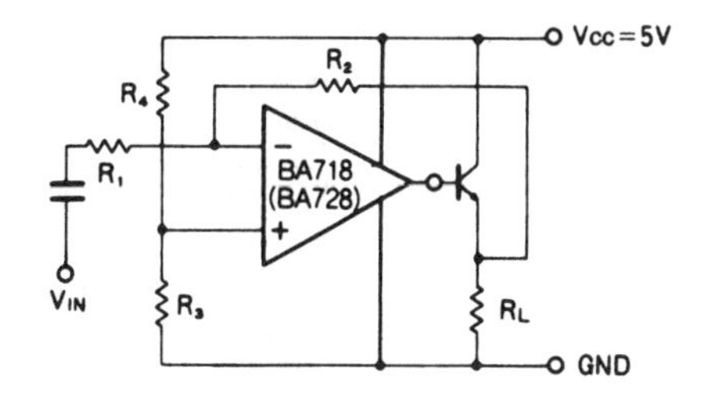

Booster circuit

BLOCK DIAGRAMS

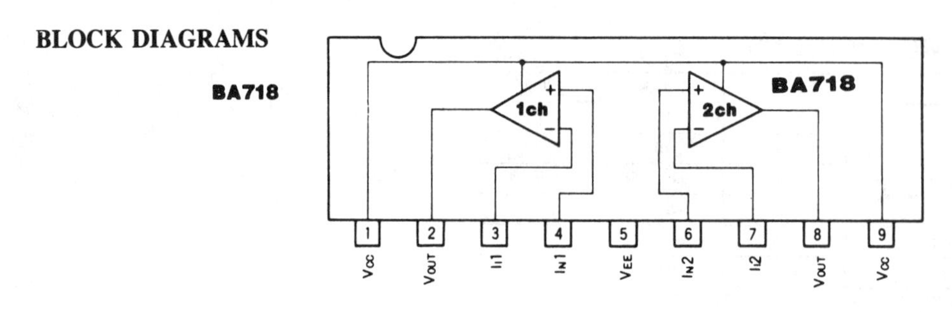

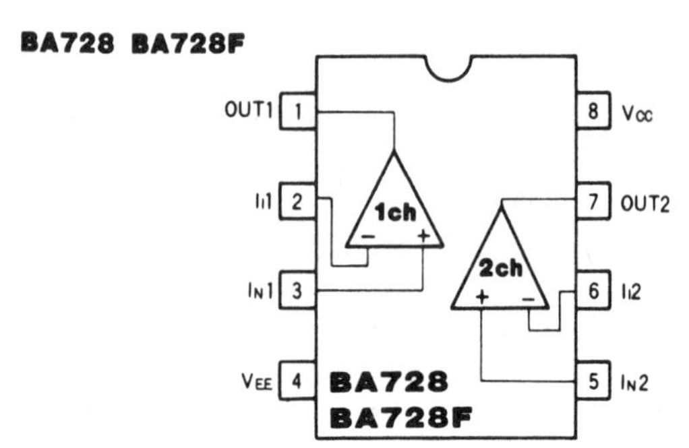

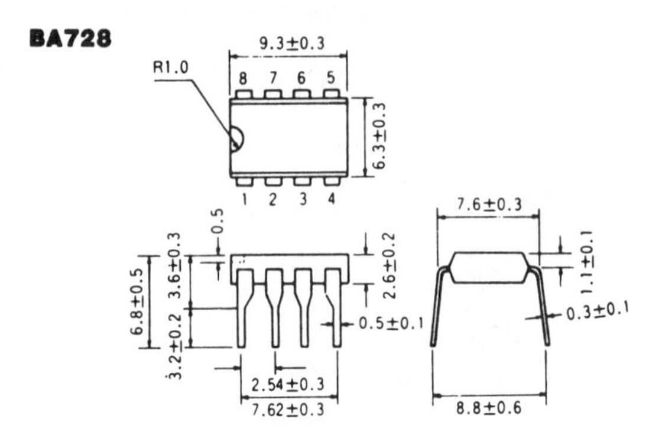

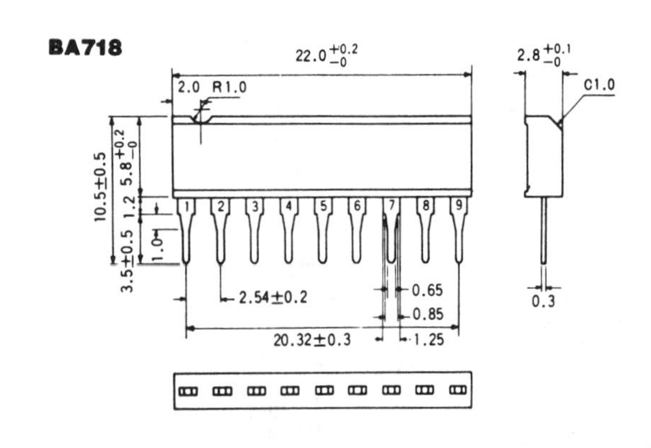

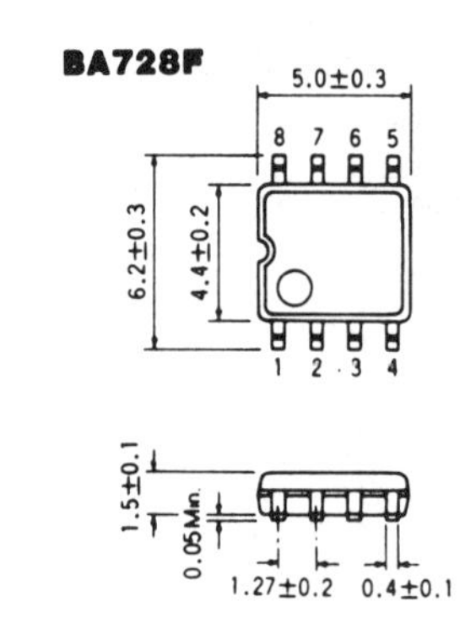

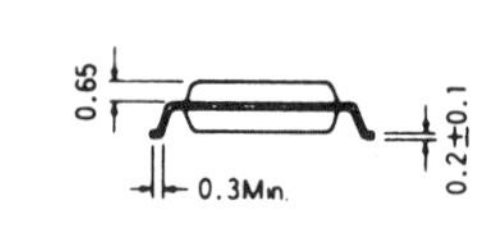

Rohm
BA10358, BA10358F, BA10358N
Low-Power Dual Operational Amplifiers

- Can operate on a single power supply
- Very small current consumption
- Level compatible with any type of logic device
- Supply voltage range is 3 to 30V for single power supply, and ±1.5 to ±15V for dual power supply
- Large dc voltage gain
- Wide band
- Pin configuration is identical to that of the 4558-type general-purpose op amps
- Pin compatible with the 358-type op amps

The BA10358, BA10358F, and BA10358N are dual monolithic operational amplifiers; each chip contains two independent, high-gain op amps with internal frequency compensation. The devices feature a wide supply voltage range of 3 to 30V (for single power supply). The small current consumption is constant over the entire supply voltage range. They come in three types of packages: 8-pin DIP (BA10358), 8-pin MF package (BA10358F), and 8-pin SIP (BA10358N).

ABSOLUTE MAXIMUM RATINGS (Ta=25°C)

Parameter		Symbol	Limits	Unit
Supply voltage		V_{CC}	32(±16)	V
Power dissipation	BA10358 BA10358F	Pd	500*1	mW
	BA10358N	Pd	900*2	
Differential input voltage		V_{ID}	32	V
Input voltage		V_{IN}	−0.3 ~ 32	V
Operating temperature range		Topr	−20 ~ 75*3	°C
Storage temperature range		Tstg	−55 ~ 125	°C

*1 Derating is done at 5mW (9mW*2)/°C for operation above Ta=25°C.
*3 For an extended operating temperature range, consult your local ROHM representative.

ELECTRICAL CHARACTERISTICS (Ta=25°C, V_{CC}=5V, V_{EE}=GND)

Parameter	Symbol	Min.	Typ.	Max.	Unit
Input offset voltage	V_{IO}	—	2	7	mV
Input offset current	I_{IO}	—	5	50	nA
Input bias current	I_B	—	45	250	nA
Large signal voltage gain	A_V	25	100	—	V/mV
Common-mode rejection	CMR	65	85	—	dB
Supply voltage regulation	SVR	65	100	—	dB
Maximum output voltage	V_{OH}	$V_{CC}-1.5$	—	—	V
Common-mode input voltage range	V_{ICM}	0	—	$V_{CC}-1.5$	V
Quiescent current	I_Q	—	0.7	1.2	mA

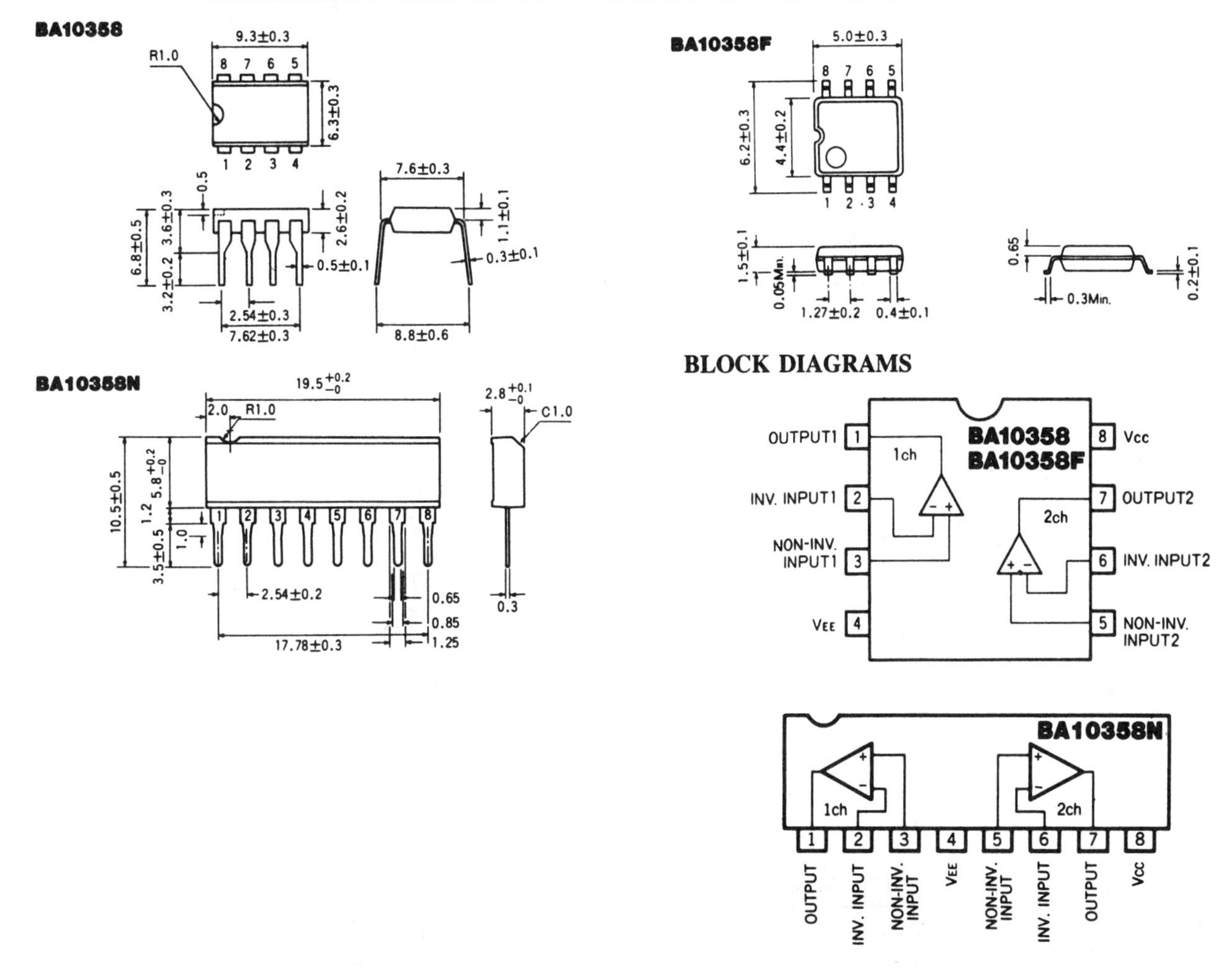

Rohm BA715, BA4558, BA4558F
Dual Operational Amplifiers

- Low power consumptions: Approx. 50 mW (typ.)
- Output is short-circuit protected
- Internal phase compensation
- No latchup
- Wide common-mode and differential input voltage ranges
- High gain and low noise
- BA715 has symmetrical pin arrangement allowing installation in opposite orientation

APPLICATIONS
- Active filters
- Audio amplifiers
- VCOs
- Other electronic circuits

The BA715, BA4558, and BA4558F are monolithic dual operational amplifiers; each chip contains two independent low-power op amps with internal phase compensation. They feature high speed, wide band, and low noise.

The devices have superior temperature characteristics, uniform gain-bandwidth product, and a wide range of applications. The BA4558 comes in an 8-pin DIP package and is pin-compatible with the 4558-type op amps. The BA715 comes in a 9-pin SIP package, and the BA4558F is available in an 8-pin MF package.

ELECTRICAL CHARACTERISTICS (Ta=25°C, V_{CC}=+15V, V_{EE}= −15V)

Parameter	Symbol	Min.	Typ.	Max.	Unit	Conditions
Input offset voltage	V_{IO}	—	0.5	6.0	mV	$R_S \leqq 10k\Omega$
Input offset current	I_{IO}	—	5	200	nA	—
Input bias current	I_B	—	60	500	nA	—
Large signal voltage gain	A_V	86	100	—	dB	$R_L \geqq 2k\Omega$, V_{OUT} = ±10V
Common-mode rejection	CMR	70	90	—	dB	$R_S \leqq 10k\Omega$
Supply voltage regulation	SVR	—	30	150	μV/V	$R_S \leqq 10k\Omega$
Maximum output voltage	V_{OM}	±12	±14	—	V	$R_L \geqq 10k\Omega$
Maximum output voltage	V_{OM}	±10	±13	—	V	$R_L \geqq 2k\Omega$
Common-mode input voltage range	V_{ICM}	±12	±14	—	V	—
Slew rate	SR	—	1.0	—	V/μs	A_V=1, $R_L \geqq 2k\Omega$
Channel separation	S_{EP}	—	105	—	dB	f=1kHz
Quiescent current	I_O	—	3.2	6	mA	$R_L = \infty$

CIRCUIT DIAGRAM

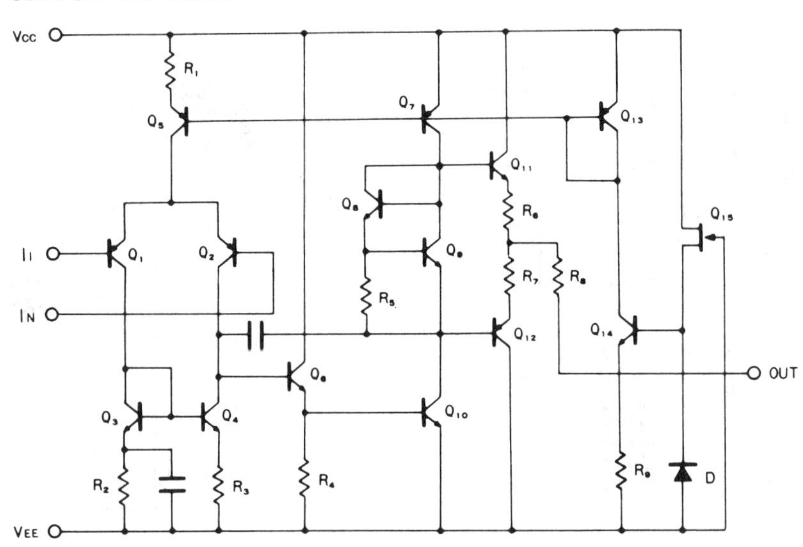

ABSOLUTE MAXIMUM RATINGS (Ta=25°C)

Parameter	Symbol	Limits	Unit
Supply voltage	V_{CC} V_{EE}	+18 −18	V
Power dissipation	Pd	500*1	mW
Operating temperature range	Topr	−20~40*2	°C
Storage temperature range	Tstg	−55~125	°C
Differential input voltage	V_{ID}	±30	V
Common-mode input voltage	V_{ICM}	±15*3	V

*1 Derating is done at 5mW/°C for operation above Ta=25°C.

*2 For an extended operating temperature range, consult your local ROHM representative.

*3 Allowable up to the supply voltage values. This is the rating when V_{CC}=+15V, V_{EE} = −15V

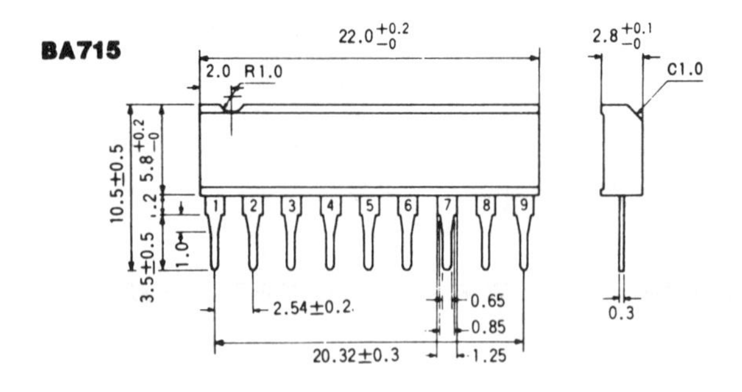

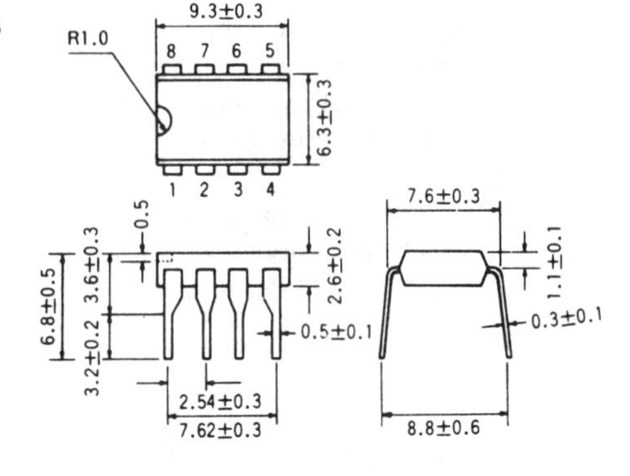

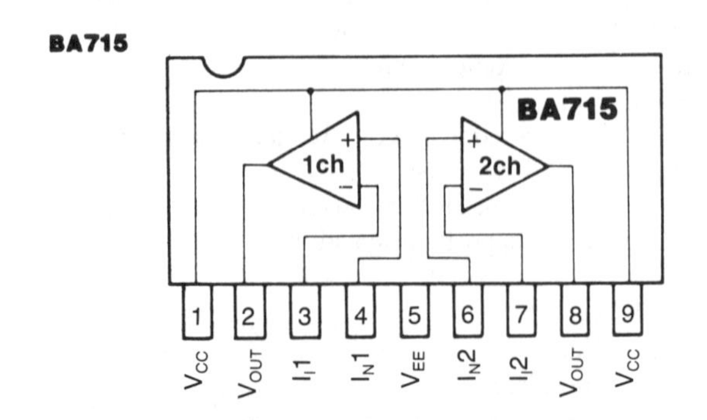

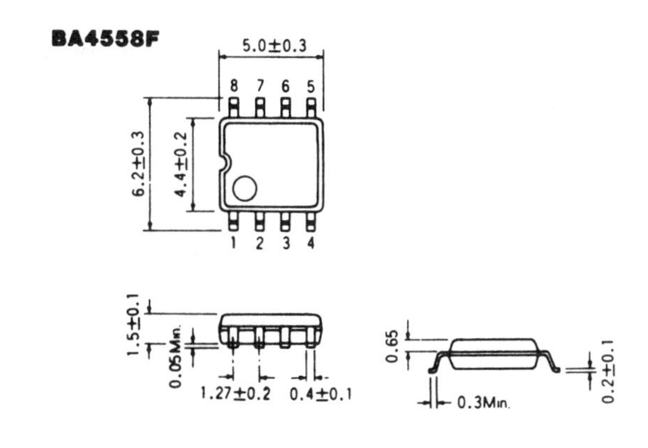

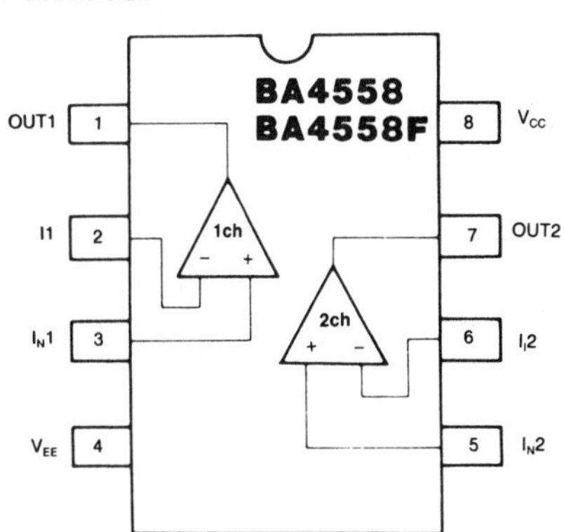

APPLICATION EXAMPLES

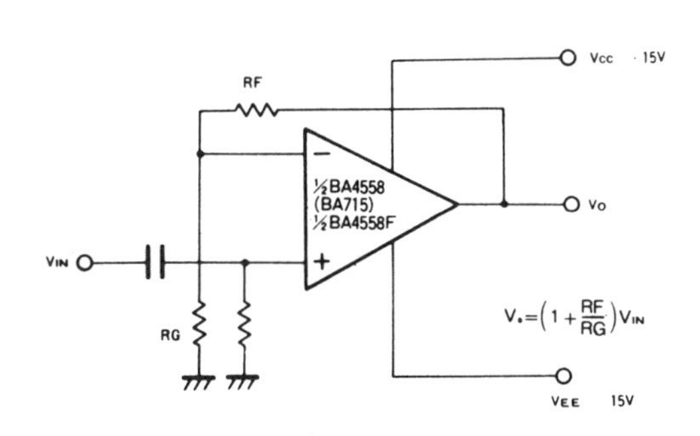

Noninverting amplifier

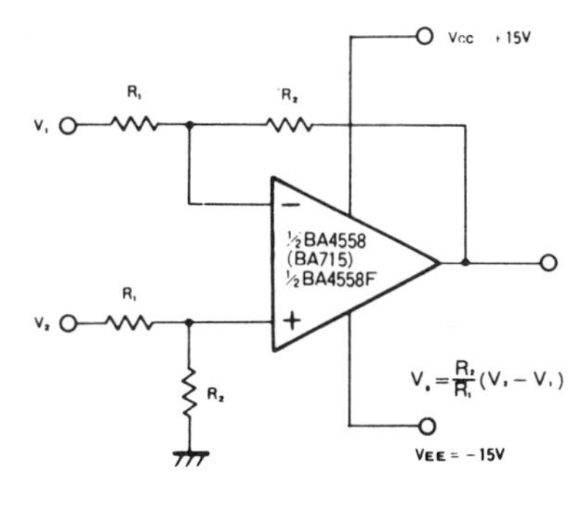

Differential amplifier

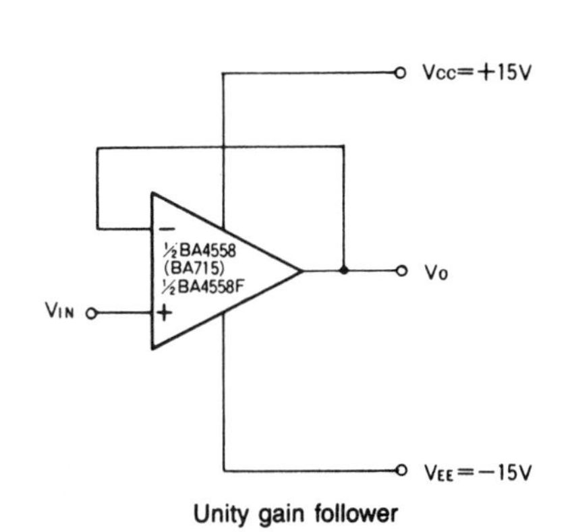

Unity gain follower

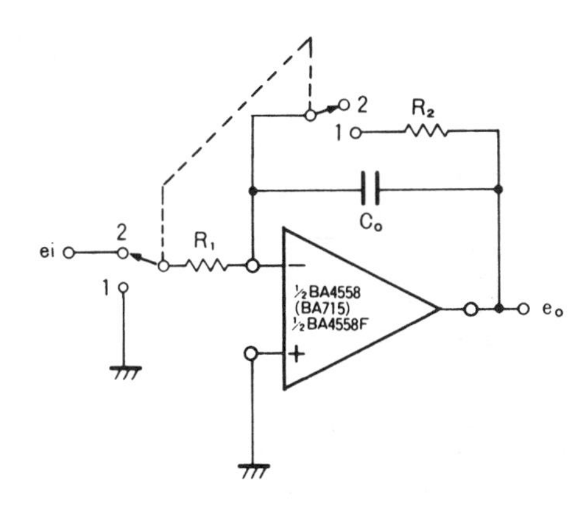

Inverting integrater

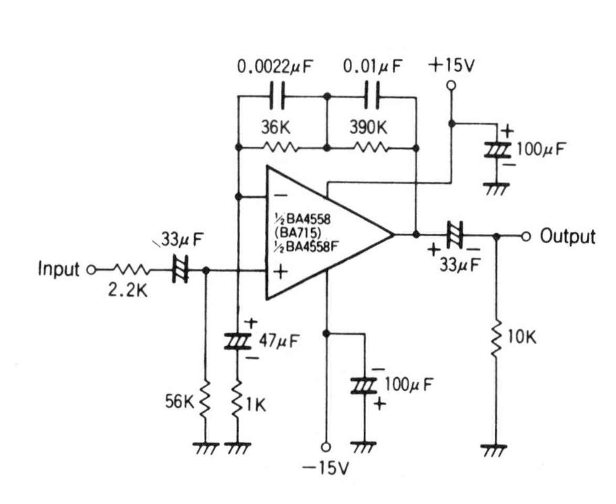

RIAA amplifier Av=32.5dB

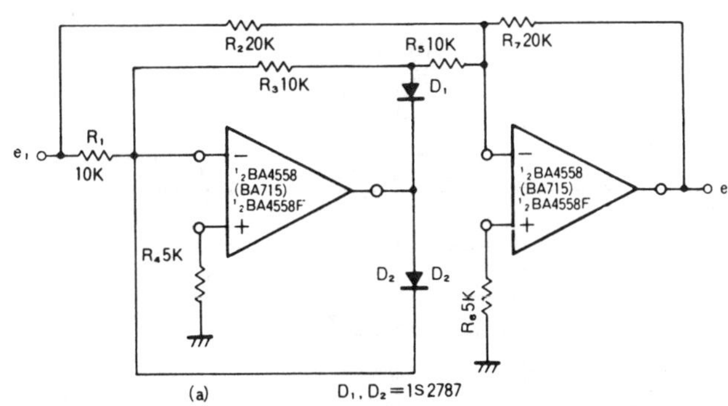

Absolute-value amplifier

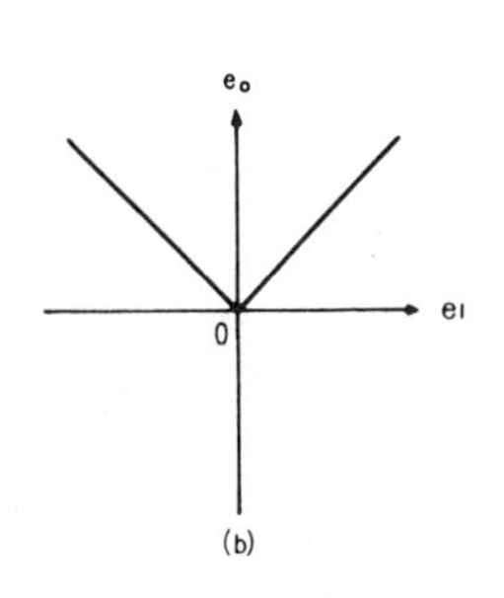

Output characteristics

Rohm
BA4560, BA4560F, BA4561
High Slew Rate Operational Amplifiers

The BA4560, BA4560F, and BA4561 are monolithic dual operational amplifiers. The devices are improved versions of the BA4558 and feature a higher output current capacity (approximately twice as large), higher slew rate (4V μs), larger gain-bandwidth product (10 MHz), and improved frequency response.

BA4560

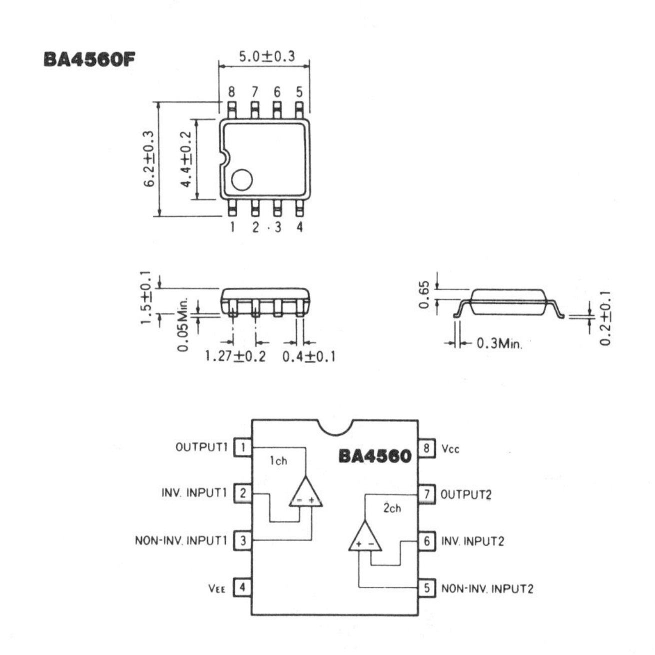

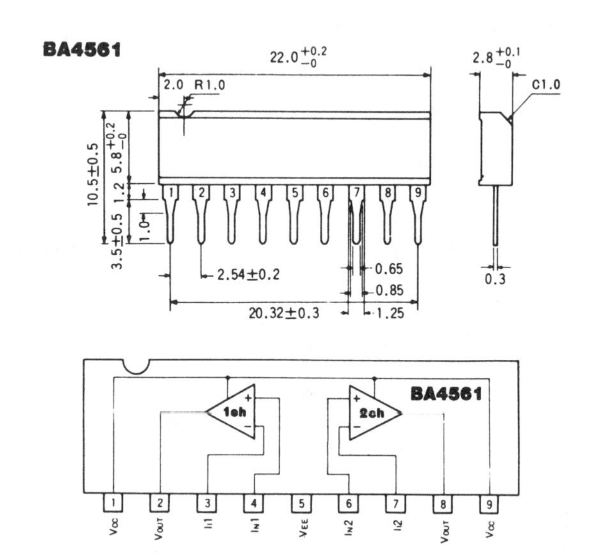

ABSOLUTE MAXIMUM RATINGS (Ta=25°C)

Parameter	Symbol	Limits	Unit
Supply voltage	V_{CC} V_{EE}	+18 −18	V
Power dissipation	Pd	500*1	mW
Differential input voltage	V_{ID}	±30	V
Common-mode input voltage	V_{ICM}	±15*2	V
Operating temperature rane	Topr	0~70*3	°C
Storage temperature range	Tstg	−55~125	°C

*1 Derating is done at 5mW/°C for operation above Ta=25°C.
*2 Allowable up to the supply voltage values. This is the rating when V_{CC}=15V, V_{EE} = −15V.
*3 For an extended operating temperature range, consult your local ROHM representative.

ELECTRICAL CHARACTERISTICS (Ta=25°C, V_{CC}=15V, V_{EE}= −15V)

Parameter	Symbol	Min.	Typ.	Max.	Unit
Input offset voltage	V_{IO}	—	±0.5	±6.0	mV
Input offset current	I_{IO}	—	±5	±200	nA
Input bias current	I_B	—	±50	±500	nA
Large signal voltage gain	A_V	86	100	—	dB
Common-mode rejection	CMR	70	90	—	dB
Supply voltage regulation	SVR	—	30	150	μV/V
Maximum output voltage	V_{OM}	±12	±14	—	V
Common-mode input voltage range	V_{ICM}	±12	±14	—	V
Slew rate	SR	—	4.0	—	V/μs
Equivalent input noise voltage	V_N	—	—	1.8	μV
Gain-bandwidth product	f_T	—	10	—	MHz

Rohm BA6110
Voltage-Controlled Operational Amplifier

- Low distortion (internal biasing diode for distortion reduction)
- Low noise
- Low offset voltage (V_{IO} = 3 mV max.)
- Internal output buffer
- gm programmable over 3 decades with high linearity

APPLICATIONS
- Electronic volume control
- Voltage-controlled impedance
- Voltage-controlled amplifiers (VCA)
- Voltage-controlled filters (VCF)
- Voltage-controlled oscillators (VCO)
- Multipliers
- Sample and hold networks
- Schmitt trigger circuit

The BA6110 is a monolithic, programmable operational amplifier featuring low noise and low offset.

Since its forward transconductance (gm) is programmable over a wide range with high linearity, the device is best suited for applications in voltage-controlled amplifiers (VCA), voltage-controlled filters (VCF), and voltage-controlled oscillators (VCO).

In conjunction with a distortion reduction circuit, the S/N (signal-to-noise) ratio of the device can be improved by 10 dB at 0.5% distortion. When used as a voltage-controlled amplifier (VCA), the device achieves an S/N ratio as high as 86 dB at 0.5% distortion.

The open-loop gain is programmable over a wide range. It is set by the control current and external gain-control resistor, R_L.

An internal low-impedance output buffer minimizes external component requirements.

ABSOLUTE MAXIMUM RATINGS (Ta=25°C)

Parameter	Symbol	Limits	Unit
Supply voltage	V_{CC}	34	V
Power dissipation	Pd	500*	mW
Operating temperature range	Topr	−20~70	°C
Storage temperature range	Tstg	−55~125	°C
Maximum control current	I_C	500	μA

*Derating is done at 5mW/°C for operation above Ta=25°C.

ELECTRICAL CHARACTERISTICS (Ta=25°C, V_{CC}=15V, V_{EE}=−15V)

Parameter	Symbol	Min.	Typ.	Max.	Unit	Conditions
Quiescent current	I_Q	0.9	3.0	6.0	mA	Icontrol=0μA
Bias current, pin 7	I_{7PIN}	—	0.8	5	μA	—
Total harmonic distortion	THD	—	0.2	1	%	Icontrol=200μA, Vi=5mVrms
Forward transconductance	gm	4800	8000	12000	μΩ	Icontrol=500μA
Maximum output voltage, pin 6	$\|V_{OM6}\|$	12	14	—	V	Icontrol=500μA
Maximum output voltage, pin 8	$\|V_{OM8}\|$	9	11	—	V	R_L=47KΩ
Maximum output current, pin 6	$\|I_{OM6}\|$	300	500	650	μA	Icontrol=500μA
Residual noise 1	VN1	—	−94	−90	dBm	Icontrol=0μA, BPF(30~20kHz, −3dB, −6dB/OCT)
Residual noise 2	VN2	—	−74	−66	dBm	Icontrol=200μA, BPF(30~20kHz, −3dB, −6dB/OCT)
Intermittent noise	VNP_2	—	10.5	11.5	dB	Icontrol=200μA, BPF(30~20kHz, −3dB, −6dB/OCT)
Leakage current	L(Leak)	—	−94	−75	dBm	Icontrol=0μA, V_{IN}=−30dBm f_{IN}=20kHz

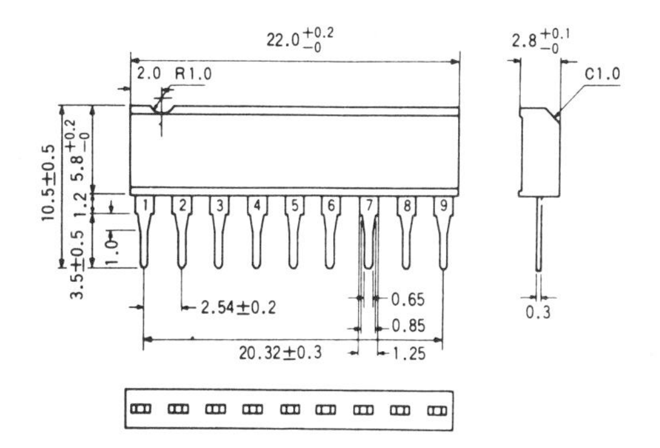

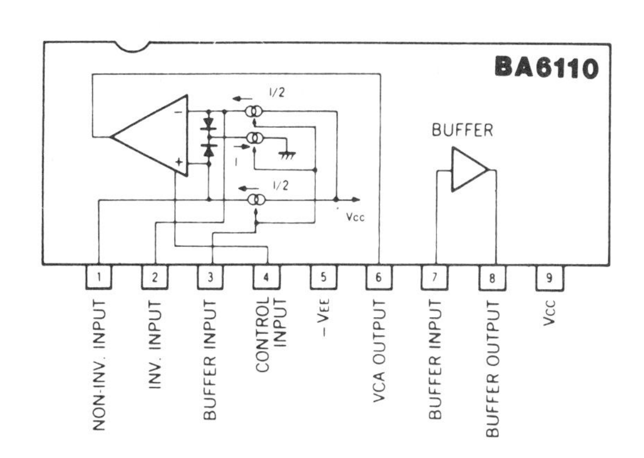

CIRCUIT DIAGRAM

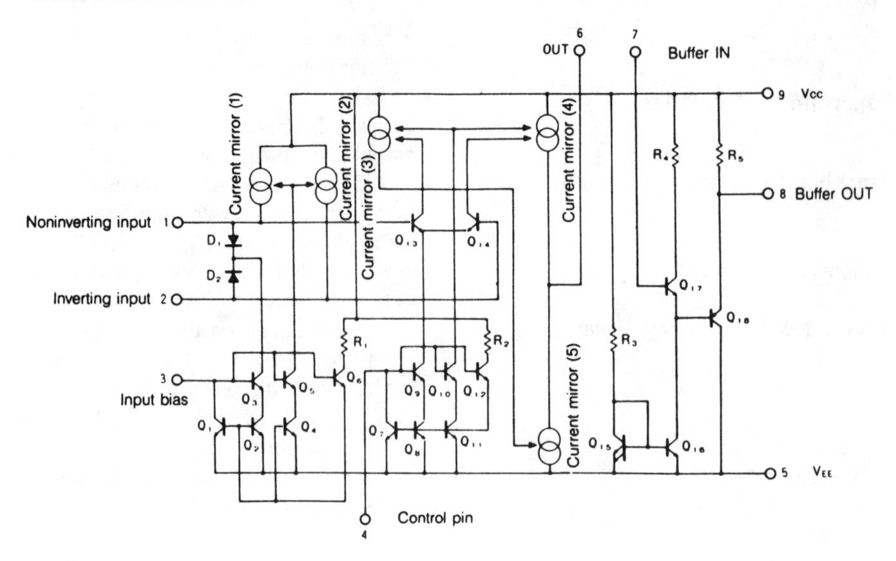

APPLICATION EXAMPLES

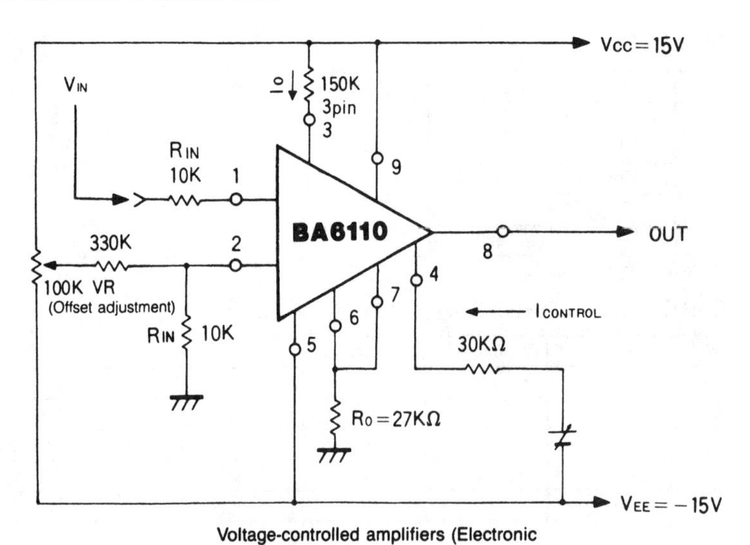

Voltage-controlled amplifiers (Electronic volume controls)

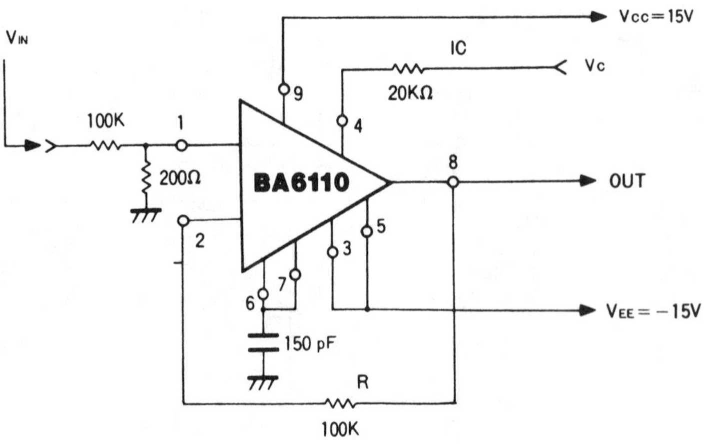

Voltage-controlled low-pass filter

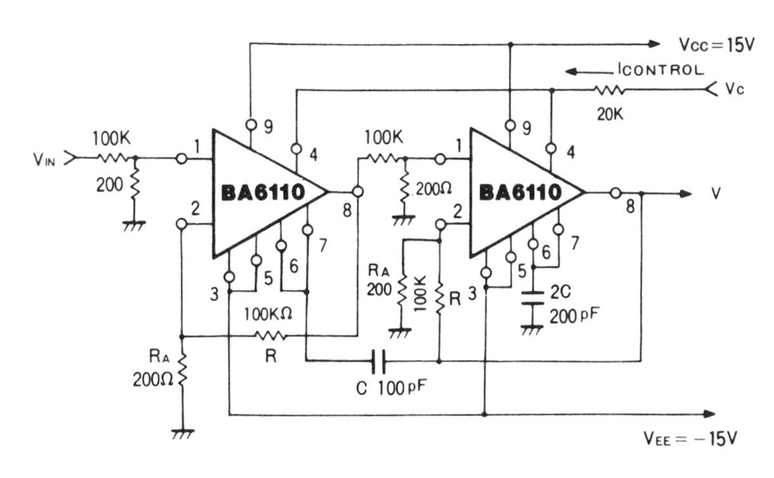

Voltage-controlled low-pass filter

Rohm BA10324, BA10324F
Quad Operational Amplifiers

- Wide supply voltage range and an ability to operate on a single power supply
  Supply voltage range is 3 to 32V for single power supply, ±1.5 to 16V for dual power supply
- Common-mode inputs are sensitive down to ground level
- Differential inputs can withstand voltages up to the supply voltage
- Low current consumption (I_Q = 0.8mA)
- Low offset voltage and current (V_{10} = 2mV, I_{10} = 5nA)
- Quad op amps with internal phase compensation implemented in a 14-pin DIP/MF package
- Pin-compatible with other manufacturers' 324-type op amps

APPLICATIONS

- Ground-sensing preamplifiers
- Active filters
- DC amplifiers
- Pulse generators, etc.

The BA10324 and BA10324F are monolithic quad operational amplifiers; each chip contains four independent op amps with internal phase compensation. The devices can operate on either a single or dual (±) power supply, and can be powered from a single +5V supply from a digital system. The applications include transducer amplifiers, dc amplifiers, and many other consumer and industrial purposes.

ABSOLUTE MAXIMUM RATINGS (Ta=25°C)

Parameter	Symbol	Limits	Unit
Supply voltage	V_{CC}	±16, 32	V
Power dissipation	Pd	600*1	mW
Operating temperature range	Topr	−40 ~ 85*2	°C
Storage temperature range	Tstg	−55 ~ +125	°C
Differential input voltage	V_{ID}	±V_{CC}	V
Common-mode input voltage range	V_{ICM}	−0.3 ~ V_{CC}	V

*1 Derating is done at 6mW/°C for operation above Ta=25°C.
*2 For an extended operating temperature range, consult your local ROHM representative.

RECOMMENDED OPERATING CONDITIONS (Ta=25°C)

Parameter	Symbol	Min.	Typ.	Max.	Unit	Conditions
Supply voltage (single power supply)	V_{CC}	3	—	32	V	—
Supply voltage (dual power supply)	V_{CC}	±1.5	—	±16	V	—

ELECTRICAL CHARACTERISTICS (Ta=25°C, V_{CC}=5V, V_{EE}=GND)

Parameter	Symbol	Min.	Typ.	Max.	Unit	Conditions
Quiescent current	I_Q	—	0.8	—	mA	$R_L=\infty$
Voltage gain	G_V	—	100	—	dB	$R_L \geq 2k\Omega$, V_{CC}=15V
Input offset voltage	V_{IO}	—	2	—	mV	Rs=50Ω
Input offset current	I_{IO}	—	5	—	nA	—
Input bias current	I_B	—	45	—	nA	—
Common-mode input voltage	V_{ICM}	0	—	$V_{CC}-1.5$	V	—
Common-mode rejection	CMRR	—	80	—	dB	—
Supply voltage regulation	SVRR	—	100	—	dB	—
Source current	I_{SOURCE}	—	40	—	mA	INPUT(+)=1V, INPUT(−)=0
Sink current	I_{SINK}	—	20	—	mA	INPUT(+)=0, INPUT(−)=1V
Channel separation	S_{EP}	—	120	—	dB	f=1kHz

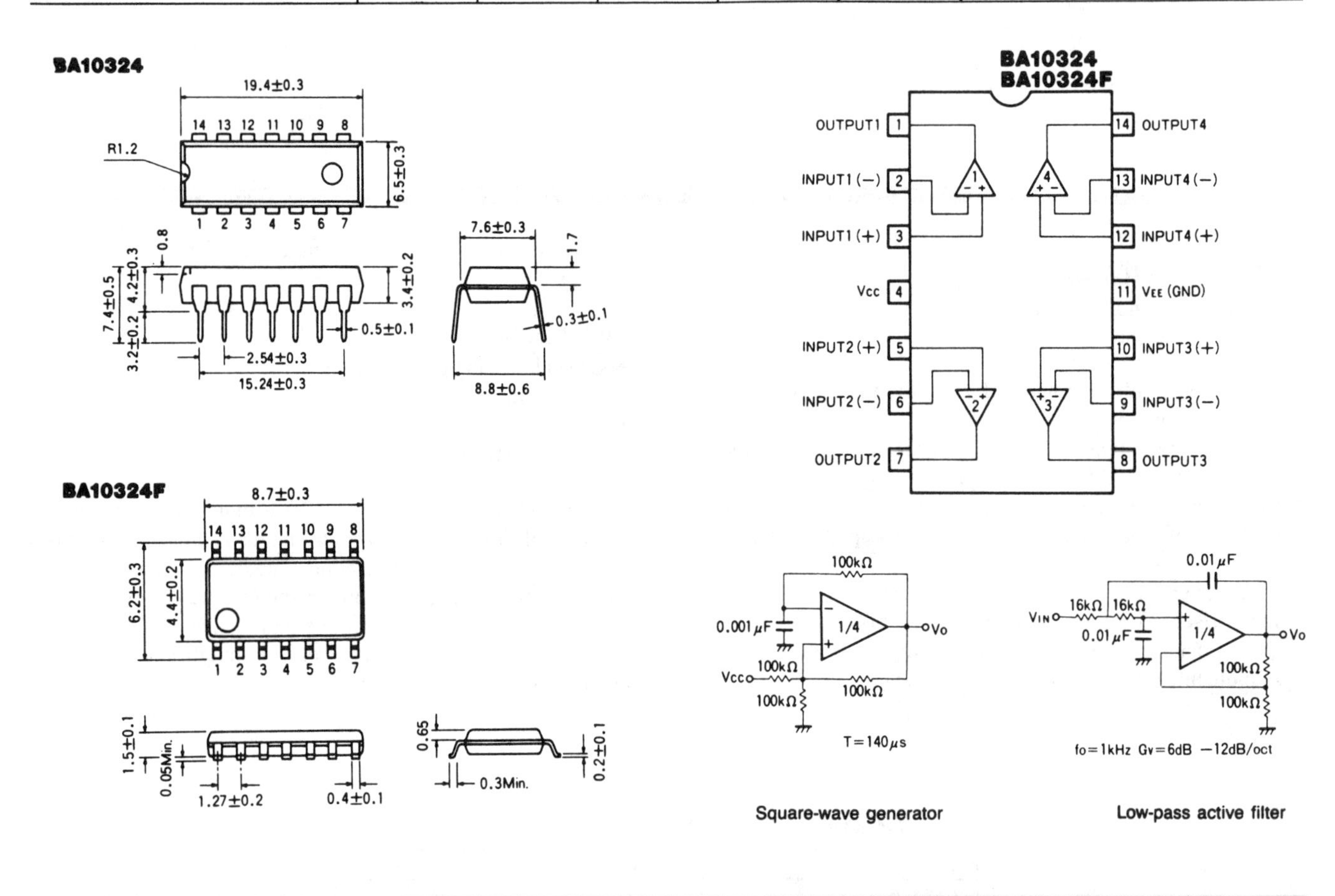

Square-wave generator

Low-pass active filter

Rohm BA1404, BA1404F
Stereo Transmitter

- Low-voltage, low-power design
- Stereo modulator, FM modulator, and transmitter implemented on a single chip
- Few external components required
- High separation (45dB type)

APPLICATIONS

- FM stereo transmitters
- Wireless microphones

The BA1404 and BA1404F are monolithic FM stereo transmitters. The devices contain a stereo modulator, an FM modulator, and an RF amplifier.

The stereo modulator creates a stereo composite signal (which consists of a main (L + R), sub (L − R), and pilot signals) from a 38-kHz quartz-controlled frequency.

The FM modulator oscillates a carrier in the FM broadcast band (76 to 108 MHz) and modulates it with the composite signal. The RF amplifier creates energy to emit the modulated FM signal. It also functions as a buffer for the FM modulator.

ABSOLUTE MAXIMUM RATINGS (Ta=25°C)

Parameter	Symbol	Limits	Unit
Supply voltage	V_{CC}	3.6	V
Power dissipation	Pd	500*	mW
Operating temperature range	Topr	−25~75	°C
Storage temperature range	Tstg	−50~125	°C

*Derating is done at 5mW/°C for operation above Ta=25°C.

RECOMMENDED OPERATING CONDITIONS

Parameter	Symbol	Min.	Typ.	Max.	Unit	Conditions
Supply voltage	V_{CC}	1	1.25	3	V	—

ELECTRICAL CHARACTERISTICS (Ta=25°C, V_{CC}=1.25V)

Parameter	Symbol	Min.	Typ.	Max.	Unit	Conditions
Quiescent current	I_Q	0.5	3	5	mA	—
Input impedance	Z_{IN}	360	540	720	Ω	f_{IN}=1kHz
Input gain	G_V	30	37	—	dB	V_{IN}=0.5mV
Channel balance	CB	—	—	2	dB	V_{IN}=0.5mV
MPX maximum output voltage	V_{OM}	200	—	—	mV p-p	THD≦3%
MPX 38kHz leakage	V_{OO}	—	1	—	mV	Quiescent condition
Pilot output voltage	V_{OP}	460	580	—	mV p-p	No-load
Channel separation	Sep	25	45	—	dB	with standard demodulator
Equivalent input noise voltage	V_{NIN}	—	1	—	μV rms	IHF-A at 38kHz stop
RF maximum output voltage	V_{OSC}	350	600	—	mV	—

BA1404

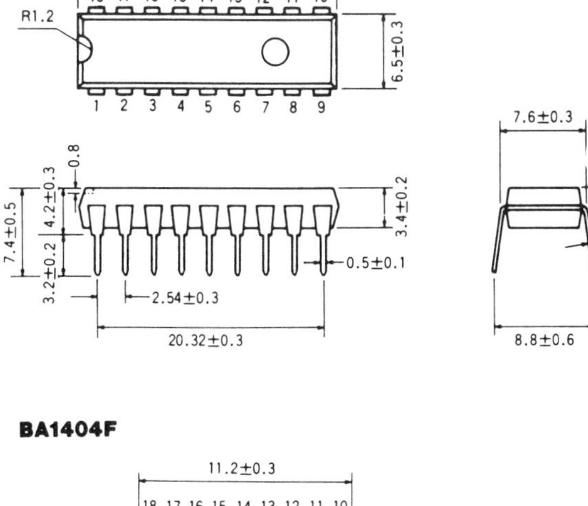

BA1404F

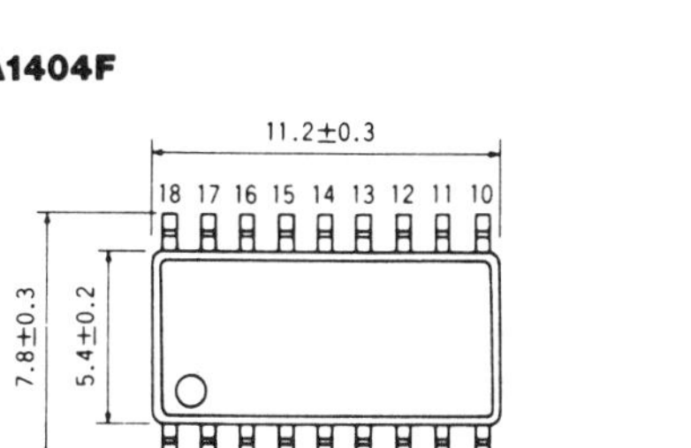

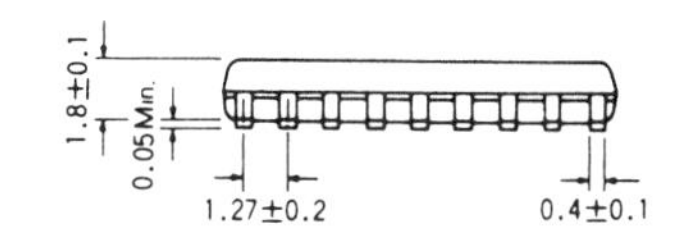

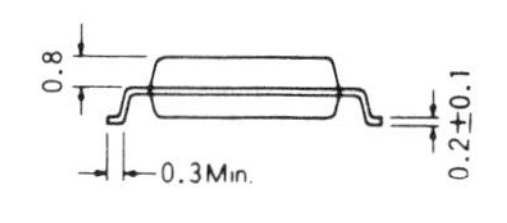

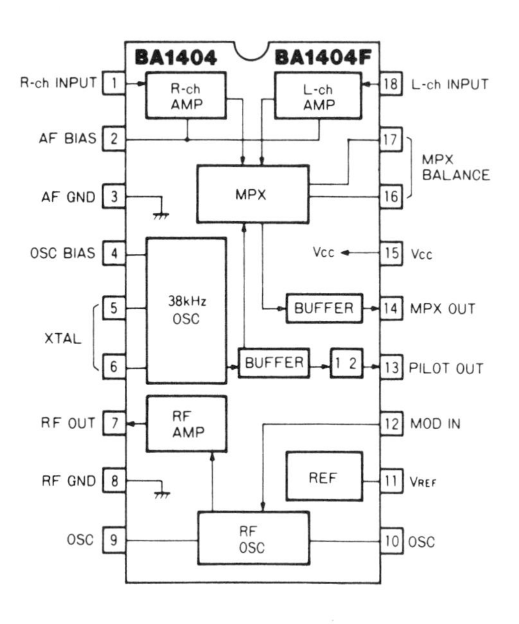

PRECAUTIONS

1. To match the frequency response of the transmitter with the FM broadcast receiver, use a pre-emphasis network with a time constant of 50 μs at the input of the AF amplifier. Use the following circuit and components:

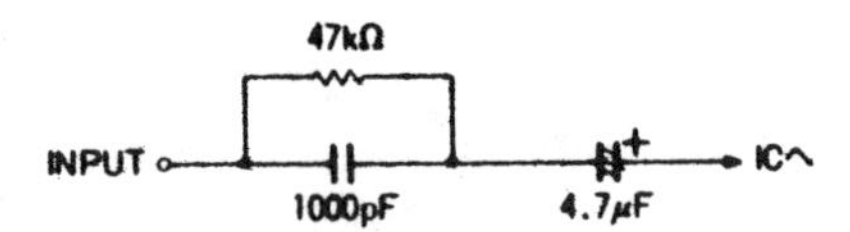

2. When synthesizing a composite signal from the stereo modulator output with pilot signal, channel separation might deteriorate unless the two signals are in-phase. Note this point if you change the constants of the external components connected to pins 12, 13, and/or 14.

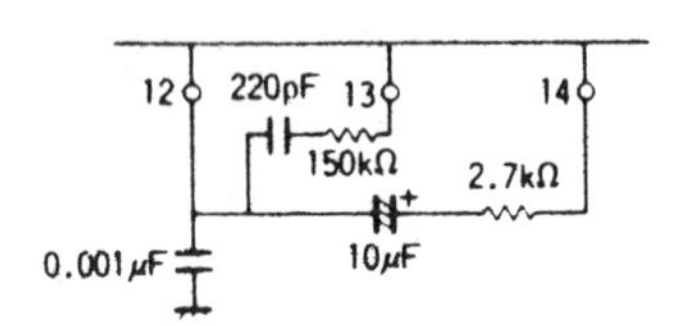

3. The carrier for an FM stereo signal can be modulated with an AF signal of up to 19 kHz. If impulse audio input, such as from an electronic musical instrument, is expected, use a low-pass filter at the input of the device to prevent beat interference or deterioration of separation.
4. Although the device ensures good separation, even if the balance control pins (16 and 17) are left open, it provides an even better separation if you connect around 50 kΩ across these pins to optimize the dc balance in the multiplex circuit.
5. The output voltage at pin 11 is internally set to (V_{CC} − 0.7) V.

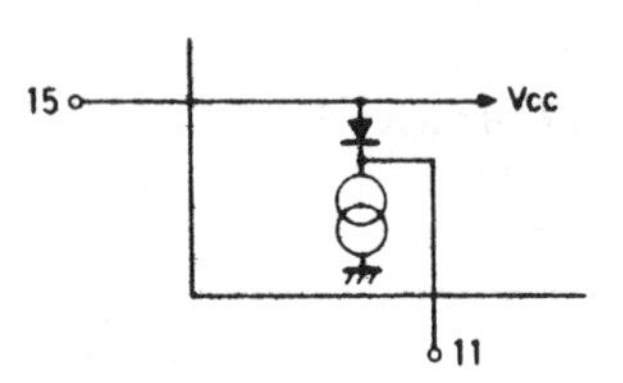

TEST CIRCUIT AND APPLICATION EXAMPLE

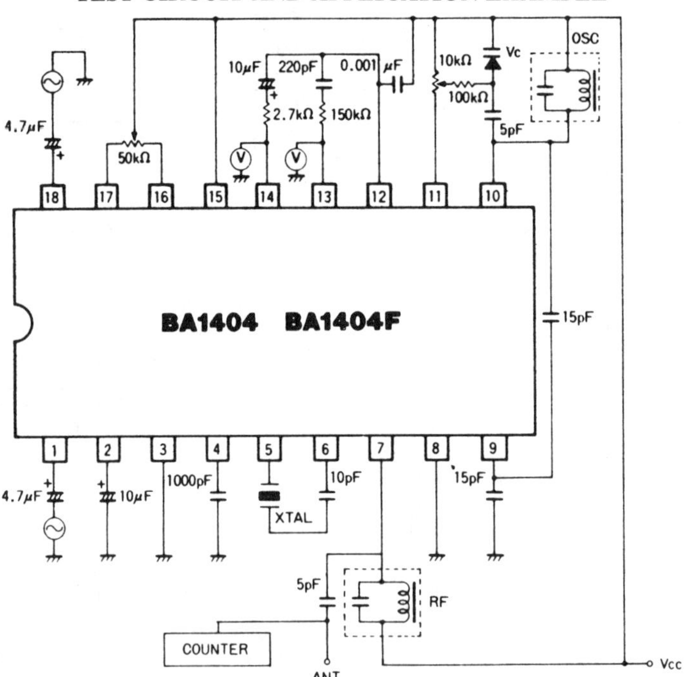

Coil data (OSC,RF common)

Bobbin	ϕ5mm with ferrite core
Coil	ϕ0.5mm enamel wire
Numbers of turns	2.25 turns
Capacity	47pF

Rohm
BA14741A, BA14741AF
Quad Operational Amplifiers

- Internal phase compensation
- Broad supply voltage range (±2 to ±18V)
- Standard quad op-amp pin configuration
- High gain with low noise

APPLICATIONS

- Active filters
- Audio amplifiers
- VCOs
- Other electronic applications

The BA14741A and BA14741AF are monolithic quad operational amplifiers each integrating four independent op amps with internal phase compensation. The devices can operate on either a single or dual (±) power supply.

Absolute Maximum Ratings

Parameter		Symbol	Limits	Unit	Conditions
Power dissipation	BA14741A	Pd	600*1	mW	—
	BA14741AF	Pd	450*2	mW	—

*1 Derating is done at 6mW (4.5mW*2)/°C for operation above Ta=25°C.

Electrical Characteristics (Unless otherwise specified, Ta=25°C, V_{CC}=15V, V_{EE}= −15V)

Parameter		Symbol	Min.	Typ.	Max.	Unit	Conditions
Input offset voltage		V_{IO}	—	1.0	—	mV	R_S=50Ω
Input offset current		I_{IO}	—	10	—	nA	—
Input bias current		I_B	—	60	—	nA	* —
Common-mode input voltage range		V_{ICM}	—	±13.5	—	V	—
Common-mode rejection		CMRR	—	100	—	dB	
Large signal voltage gain		A_V	—	100	—	V/mV	R_L≧2kΩ, V_O= ±10V
Supply voltage regulation		SVRR	—	100	—	dB	R_S=50Ω
Quiescent circuit current		I_Q	—	3.0	—	mA	R_L=∞, on all op-amps
Maximum output voltage		V_{OH}	—	±12.5	—	V	R_L=2kΩ
Maximum output current	(Source)	I_{OH}	—	20	—	mA	V_O= 0
	(Sink)	I_{OL}	—	10	—	mA	V_O= V_{CC}
Equivalent input noise voltage		V_N	—	2.0	—	μVrms	DIN/AUDIO FILTER, R_S=100Ω
Slew rate		SR	—	1.0	—	V/μs	R_L≧2kΩ, A_V=1
Channel separation		S_{EP}	—	100	—	dB	f=10kHz, R_S=1kΩ Equivalent input

*The input bias current flows out from the IC since a PNP transistor is used at the input.

DIMENSIONS

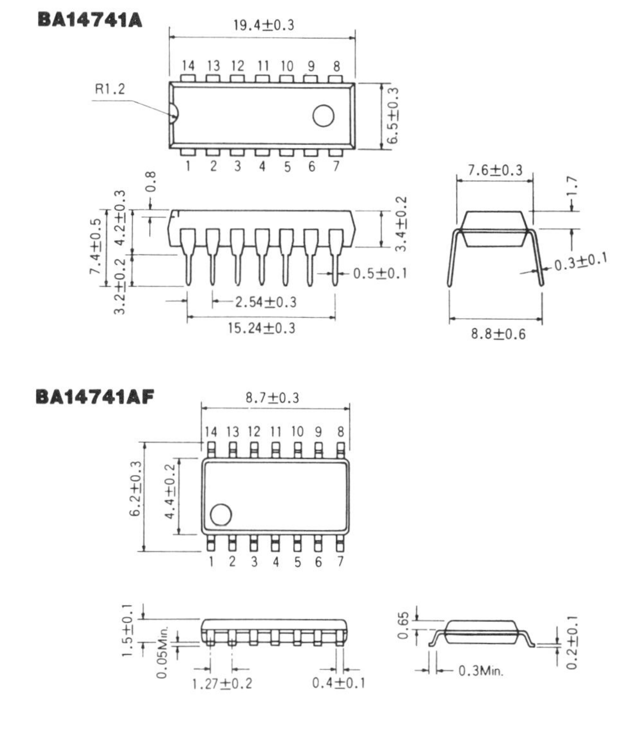

BLOCK DIAGRAM

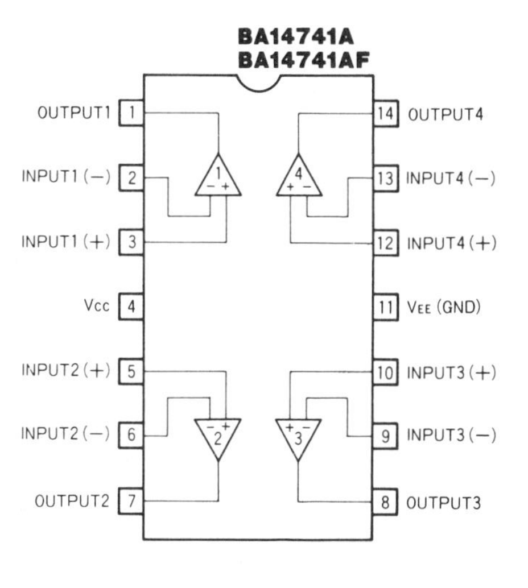

Rohm
BA15218, BA15218F, BA15218N
Dual Operational Amplifiers

- Can operate on a single low-voltage power supply
  Supply voltage range is 4 to 32V for single power supply, ±2 to ±16V for dual power supply
- Low noise (V_N=0.9μV typ.)
- High slew rate (3V/μs, GBW=7 MHz typ.)
- Low offset voltage (V_{IO}=0.5 mV typ.)
- High gain, low distortion (G_{VO}=110dB, THD=0.0015% typ.)
- Pin compatible with the 4588 and 4560-type general-purpose dual op amps
- BA15218F is available in an MF package for compact system design

APPLICATIONS
- Low-noise amplifiers
- Active filters
- Headphone amplifiers
- Pulse generators, etc.

The BA15218, BA1521F, and BA15218N are monolithic dual operational amplifiers featuring internal phase compensation, low noise, and low distortion. The devices can operate on either a single or dual power supply. They can be powered by a single +5V power source from a digital system.

Available in three types of packages, the devices allows flexible choice of mounting.

DIMENSIONS (Unit: mm)

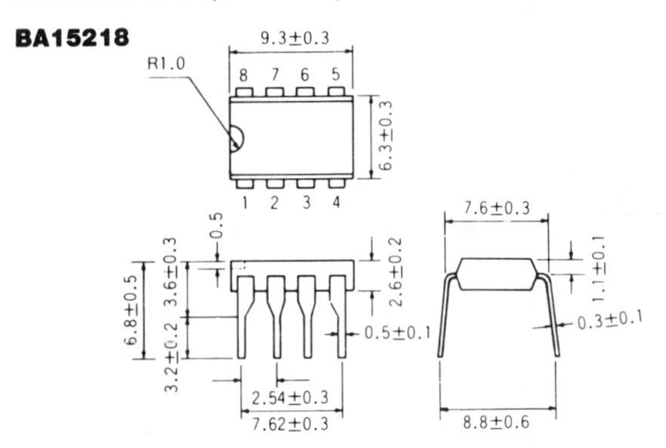

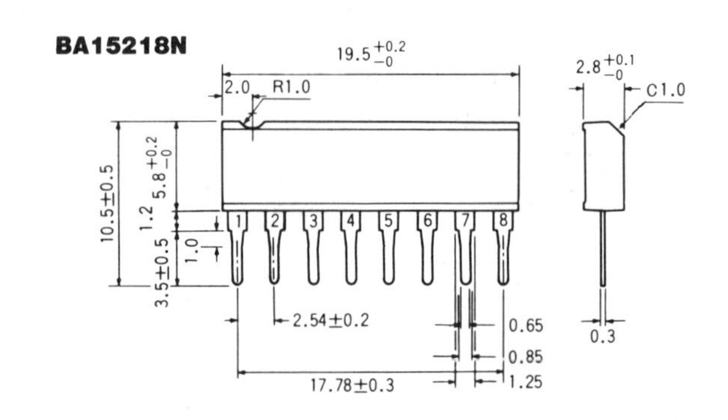

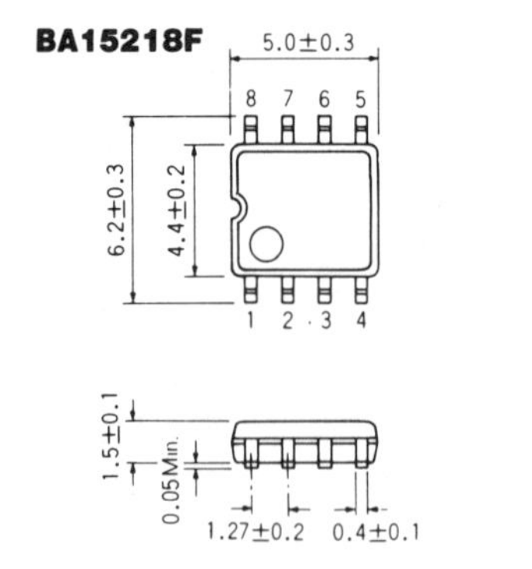

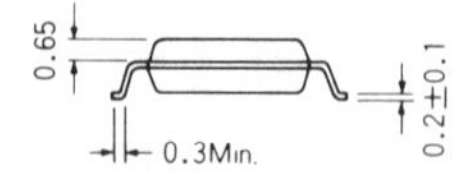

ABSOLUTE MAXIMUM RATINGS (Ta=25°C)

Parameter	Symbol	Limits	Unit
Supply voltage	V_{CC}	±18	V
Power dissipation	Pd	650*1	mW
Operating temperature range	Topr	−40~85*2	°C
Storage temperature range	Tstg	−55~125	°C
Differential input voltage	V_{ID}	±V_{CC}	V
Common-mode input voltage range	V_{ICM}	−V_{CC}~V_{CC}	V
Load current	$I_{O\,MAX}$	±50	mA

*1 Derating is done at 6.5mW/°C for operation above Ta=25°C.
*2 For an extended operating temperature range, consult your local ROHM representative.

Rohm BA15532

Dual Low-Noise Operational Amplifier

- High output current capacity
- High slew rate
- Low noise
- Pin compatible with the Signetics NE5532

The BA15532 is a low-noise dual operational amplifier designed specifically for use in high-quality audio equipment. The device features low noise, wide band, and high-output current capacity, and it can also be applied to instrumentation and control equipment.

ABSOLUTE MAXIMUM RATINGS (Ta=25°C)

Parameter	Symbol	Limits	Unit
Supply voltage	V_{CC} V_{EE}	+22 −22	V
Power dissipation	Pd	500*	mW
Differential input voltage	V_{ID}	±0.5	V
Common-mode input voltage	V_{ICM}	$-V_{EE} \sim V_{CC}$	V
Operating temperature range	Topr	−20 ~ 75	°C
Storage temperature range	Tstg	−55 ~ 125	°C

*Derating is done at 5mW/°C for operation above Ta=25°C.

ELECTRICAL CHARACTERISTICS (Ta=25°C, V_{CC}=15V, V_{EE}= −15V)

Parameter	Symbol	Min.	Typ.	Max.	Unit
Input offset voltage	V_{IO}	—	0.5	4.0	mA
Input offset current	I_{IO}	—	20	300	nA
Input bias current	I_B	—	500	1500	nA
Large signal voltage gain	A_V	25	100	—	V/mV
Common-mode rejection	CMR	70	100	—	dB
Supply voltage regulation	SVR	—	10	100	μV/V
Maximum output voltage	V_{OM}	±12	±13	—	V
Common-mode input voltage range	V_{ICM}	±12	±14	—	V
Slew rate	SR	—	13	—	V/μs
Channel separation	S_{EP}	—	120	—	dB
Quiescent current	I_Q	—	4	8	mA

RECOMMENDED OPERATING CONDITIONS

Supply voltage: single power supply 4.0 − 32.0V
dual power supply ±2 − ±16V

ELECTRICAL CHARACTERISTICS (Unless otherwise specified, Ta=25°C, V_{CC}=15V, V_{EE}= −15V)

Parameter	Symbol	Ratings			Unit	Conditions
		Min.	Typ.	Max.		
Quiescent current	I_Q	—	5.0	8.0	mA	V_{IN}=0, RL=∞
Input offset voltage	V_{IO}	—	0.5	5.0	mV	Rs≦10kΩ
Input offset current	I_{IO}	—	5	200	nA	
Input bias current	I_B	—	50	500	nA	*
Large signal voltage gain	A_V	86	110	—	dB	RL≧2kΩ, Vo= ±10V
Common-mode input voltage range	V_{ICM}	±12	±14	—	V	
Common-mode rejection	CMRR	70	90	—	dB	Rs≦10kΩ
Supply voltage regulation	SVRR	76	90	—	dB	Rs≦10kΩ
Maximum output voltage	V_{OH}	±12	±14	—	V	RL≧10kΩ
	V_{OL}	±10	±13	—	V	RL≧2kΩ
Slew rate	SR	—	3	—	V/μS	GV=0dB, RL=2kΩ
Gain-bandwidth product	GBW	—	10	—	MHz	f=10kHz
Equivalent input noise voltage	V_N	—	1.0	—	μVrms	Rs=1kΩ, BW=10Hz ~ 30kHz, RIAA
Channel separation	S_{EP}	—	120	—	dB	f=1kHz, equivalent input

*The input bias current flows out from the IC since a PNP transistor is used at the input.

DIMENSIONS (Unit: mm)

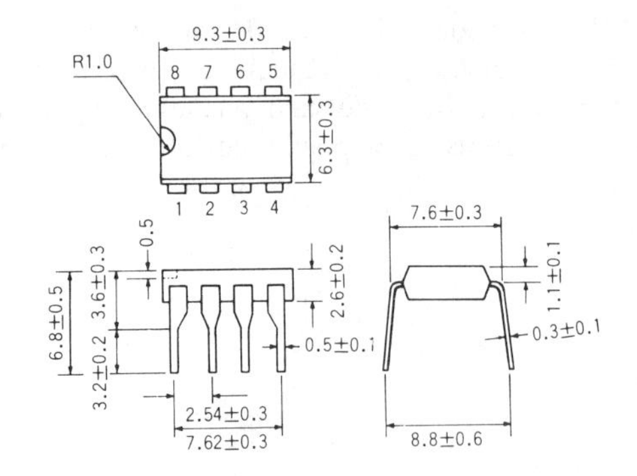

BLOCK DIAGRAM

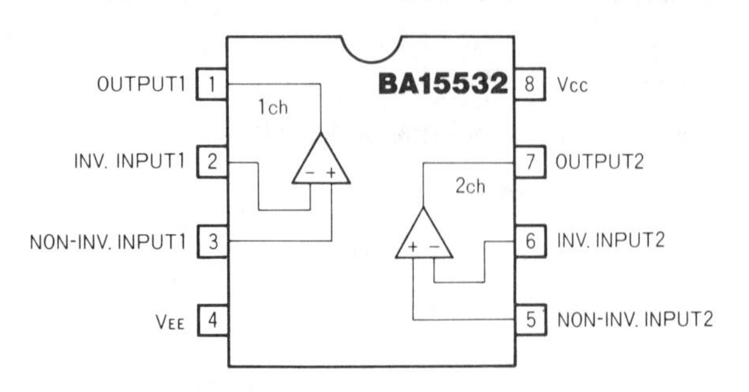

BLOCK DIAGRAMS

BA15218N

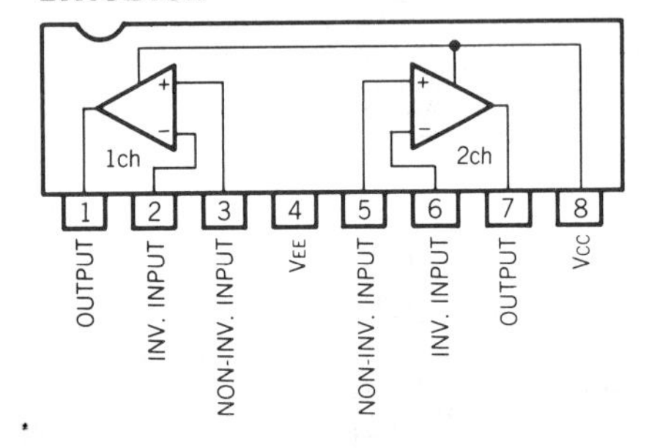

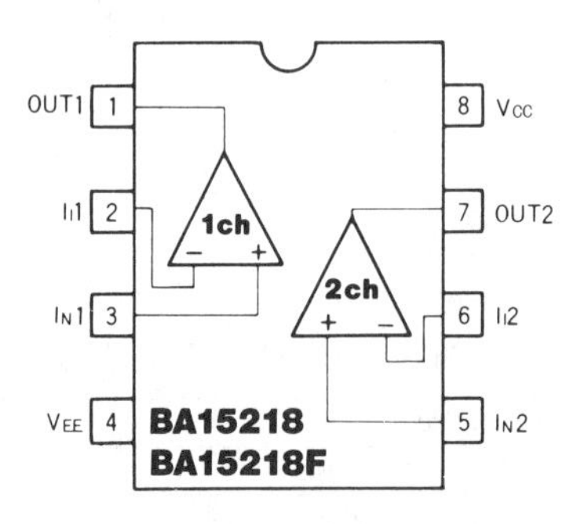

Rohm BU8304, BU8304F
Tone/Pulse Dialers

- Complies with the domestic standards; minimum pause = 965ms (10 pps), 483ms (20 pps)
- Tone and pulse dialer functions integrated on a single chip
- Requires no external power supply, and directly connects to the telephone line
- Low standby current
- Last number redial capability for up to 17 digits. The redial memory can be backed up from the telephone line.
- Internal 3.75 sec. digital pause timer allows cumulative pauses
- DTMF output amplitude is internally regulated and is not affected by supply voltage fluctuation
- Pulse break ratio is selectable from 63% or 66%; pulse rate is selectable from 10 pps and 20 pps
- During pulse dial, the device outputs a dial recognition tone for each dial key operation
- Internal power-on reset logic and voltage regulator
- The internal master oscillator uses an external 3.58-MHz crystal or ceramic resonator
- Allows the use of the standard 2-of-7 keypad

APPLICATIONS

- Telephone handsets
- Cordless telephone handsets, etc. in compliance with domestic standards.

The BU8304 and BU8304F are monolithic telephone dialers that combine tone and pulse dialer functions on a single chip.

RECOMMENDED OPERATING CONDITIONS (Ta=25°C)

Parameter	Symbol	Min.	Typ.	Max.	Unit	Conditions
Operating voltage	V_{DD}	2.5	—	5.5	V	—

ELECTRICAL CHARACTERISTICS (Ta=25°C, V_{DD}=3.5V)

Parameter	Symbol	Min.	Typ.	Max.	Unit	Conditions
Operating current	I_{DD}	—	1.0	2.0	mA	
Memory retention current	I_{MR}	—	0.1	0.75	μA	(V_{DD}=2.5V)
Keyboard debounce time		—	11.8	—	ms	
Key input time		40	—	—	ms	

PIN CONNECTIONS (TOP VIEW)

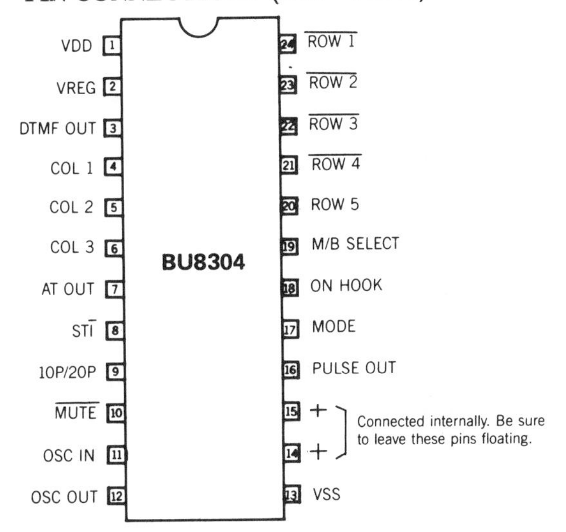

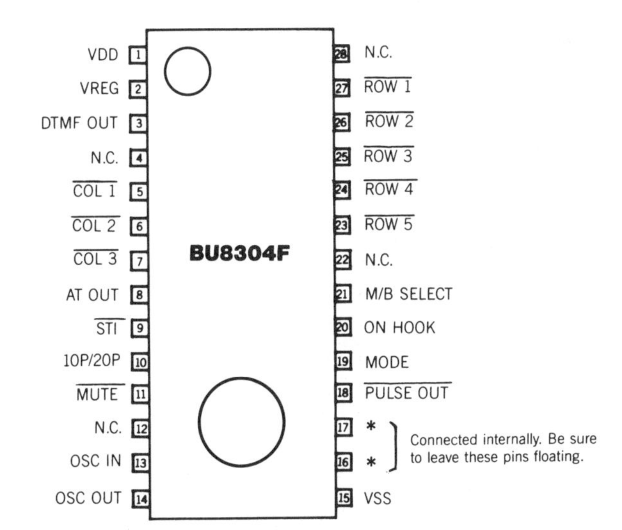

DIMENSIONS (Unit: mm)

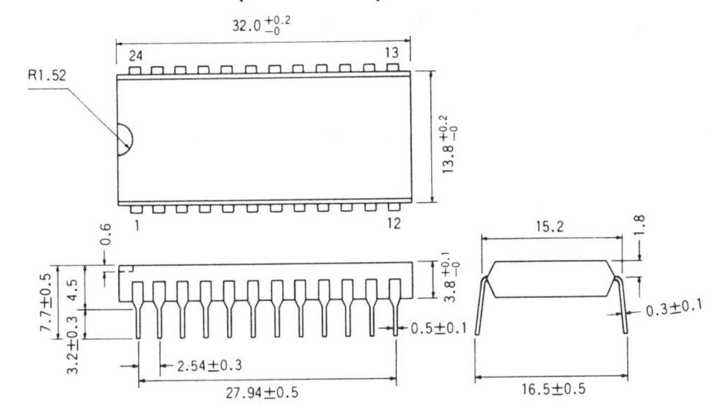

KEYPAD INPUT MATRIX

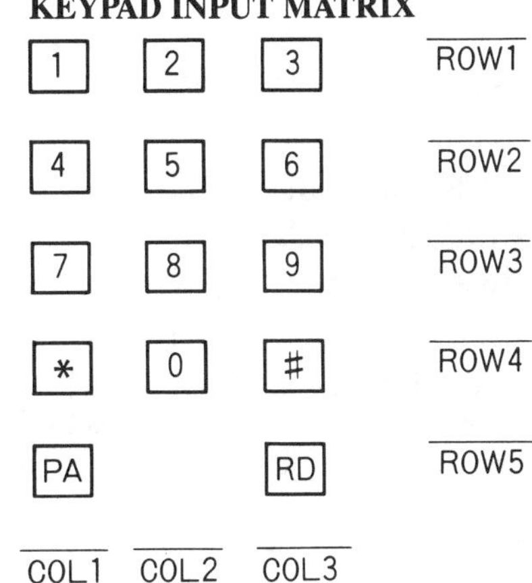

APPLICATION EXAMPLE

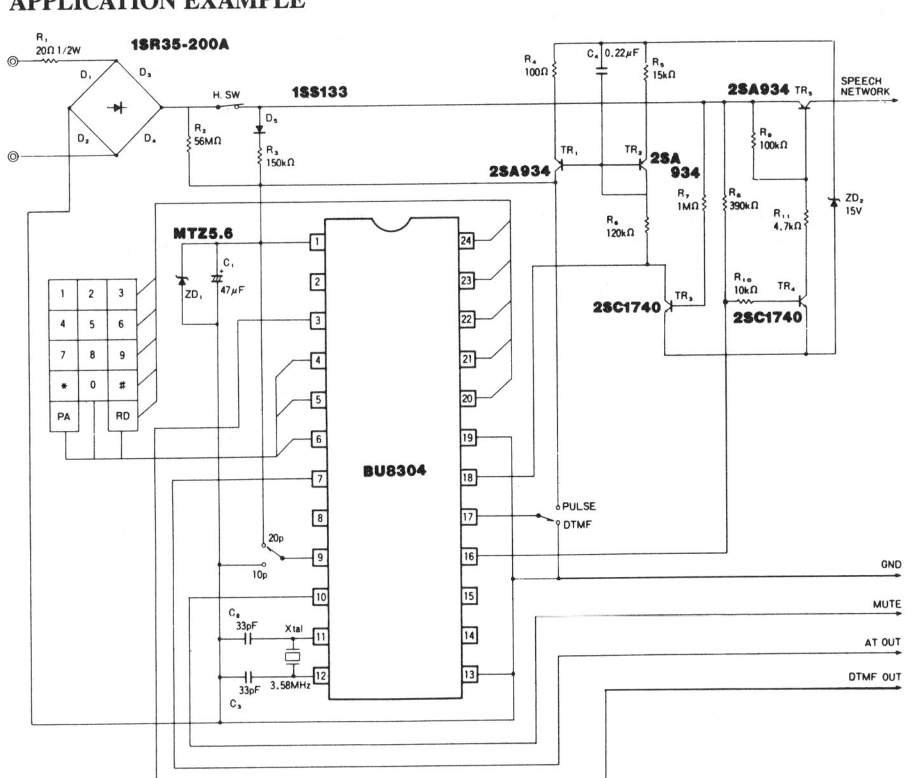

BLOCK DIAGRAMS

BU8304

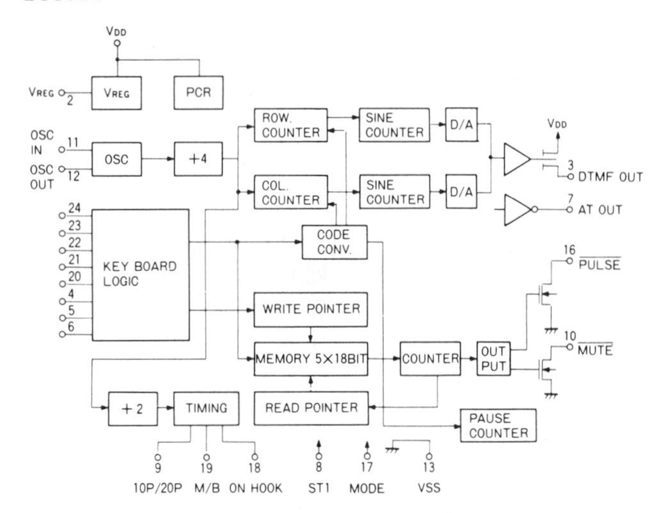

BU8304F

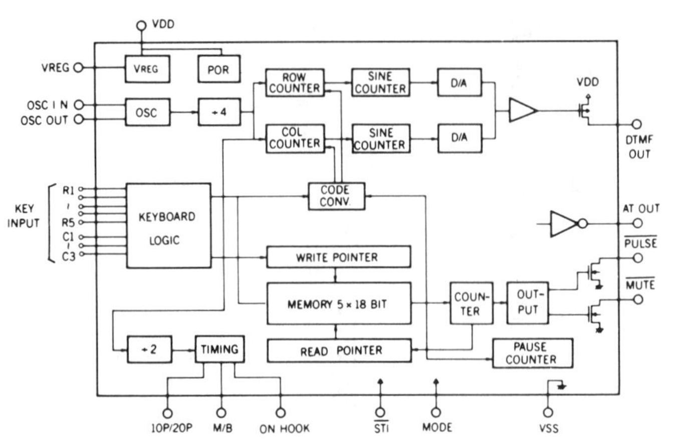

Rohm BU8320A, BU8320AF, BU8321, BU8321F, BU8322, BU8322F

20 Number Tone/Pulse Repertory Dialer

- With both pulse dialer and tone (DTMF) dialer functions. Pulse and tone switching operation is possible.
- Wide operating voltage range. Operate directly from telephone line current (1.5 to 5.5V for both pulse and tone).
- 50-kHz ceramic resonator used (CSB500E28)
- Up to 20 16-digit numbers can be stored in the repertory dial. In addition, there is a 32-digit redial memory. In tone mode, the memory is reduced one digit by privacy memory (i.e. 15-digit repertory memory and 31-digit redial memory).
- One-touch dialing and three-touch dialing are possible
- Redialing (last number dial) privacy
- Low memory retention current (I_{MR}=0.1μA typ.)
- Memory retention voltage: V_{MR}=1.0V min
- A digital pause timer is incorporated, allowing pauses to be stored and sent during dial. The t_{PA} is 3.7 seconds.
- Key confirmation tone (f_{AT}=926 Hz, t_{AT}=30 ms)
- Hookswitch function available
- Pressing keys is also allowed during hookswitch operation
- After hookswitch, a hookswitch pause is automatically inserted (t_{HXPA}=1.2 sec.)
- The output pulse rate can be 10 pps (t_{SDPP}=875 ms) or 20 pps (t_{IDPP}=486 ms)
- The pulse break ratios are as follows:

Type	BU8320A/ BU8320AF	BU8321/ BU8321F	BU8322/ BU8322F
Pulse break ratio	67%	60% and 67% can be switched	60%

- Two mute outputs are available
- The standard 2-of-7 (3 × 4) keyboard as well as 4 × 5 and 8 × 5 keyboards can be used
- 20-number repertory dialing is possible
- For BU8320A, BU8320AF, BU8322, and BU8322F, DTMF output distortion factor is as low as DIS=2% typ
- BU8321 and BU8321F have a terminal for switching write and read from the repertory memory
- BU8320AF, BU8321F, and BU8322F are housed in an MF 28-pin package to achieve thin, compact design.

ABSOLUTE MAXIMUM RATINGS (Ta=25°C)

Parameter	Symbol	Rating	Unit	Conditions
Supply voltage	V_{DD}	7.0	V	
Input voltage	V_{IN}	V_{SS}−0.3~V_{DD}+0.3	V	*1
Output voltage 1	V_{OUT1}	V_{SS}−0.3~V_{DD}+0.3	V	*2
Output voltage 2	V_{OUT2}	V_{SS}−0.3~7.0	V	*3
Power dissipation	Pd	600	mW	BU8320AF/BU8321F/ BU8322F
Power dissipation	Pd	700	mW	BU8320A/BU8321/ BU8322
Operating temperature range	Topr	−25~+60	°C	
Storage temperature range	Tstg	−55~+125	°C	

*1 Applies to ROW1~ROW5, COL1~COL8, ON HOOK, 10P/20P, MODE IN, OSC IN, F1, F2 RD/WR, 60%/67% pins.

*2 Applies to OSC OUT, AT OUT, DTMF OUT pins.

*3 Applies to MODE OUT, MUTE, DP MUTE, PULSE OUT pins.

ELECTRICAL CHARACTERISTICS (Unless otherwise specified, Ta=25°C, V_{DD}=3V)

Parameter	Symbol	Limits			Unit	Conditions
		Min.	Typ.	Max.		
Operating voltage	V_{DD}	1.5	3.0	5.5	V	
Operating current 1	I_{DDP}	—	0.15	0.5	mA	Output at no load PULSE
Operating current 2	I_{DDT}	—	0.4	1.0	mA	Output at no load TONE
Memory retention current	I_{MR}	—	0.1	0.75	μA	
Memory retention voltage	V_{MR}	1.0	—	—	V	
Input high voltage	V_{IH}	0.8V_{DD}	—	V_{DD}	V	V_{DD}=1.5~5.5V*1
Input low voltage	V_{IL}	V_{SS}	—	0.2V_{DD}	V	V_{DD}=1.5~5.5V*1
Input high current	I_{IH}	—	—	0.1	μA	V_{DD}=5.5V*2
Input low current	I_{IL}	—	—	0.1	μA	V_{DD}=5.5V*2
Key pull-up resistance	RKU	—	350	—	kΩ	ROW1~ROW5, COL1~COL8 pins
Key pull-down resistance	RKD	—	30	—	kΩ	ROW1~ROW5, COL1~COL8 pins
AT OUT sink current	I_{ATL}	250	—	—	μA	V_{DD}=1.5V, V_O=0.5V
AT OUT source current	I_{ATH}	−250	—	—	μA	V_{DD}=1.5V, V_O=1.0V
Output sink current	I_{OS}	250	—	—	μA	V_{DD}=1.5V, V_O=0.5V*3
Output leakage current	I_{OLKG}	—	—	1.0	μA	V_{DD}=5.5V*3

*1 Applies to $\overline{ROW1}$~$\overline{ROW5}$, $\overline{COL1}$~$\overline{COL8}$, ON HOOK 10P/20P, MODE IN, OSC IN RD/WR and 60%/67% pins.
*2 Applies to ON HOOK, 10P/20P, MODE IN RD/WR and 60%/67% pins.
*3 Applies to MODE OUT, $\overline{MUTE}$, $\overline{DP\ MUTE}$, $\overline{PULSE\ OUT}$ pins.

ELECTRICAL CHARACTERISTICS (Unless otherwise specified, Ta=25°C, V_{DD}=3V)

Parameter	Symbol	Limits			Unit	Conditions
		Min.	Typ.	Max.		
Oscillation frequency	f_{OSC}	—	500	—	kHz	
Oscillation start time	t_{OS}	—	0.4	1.0	ms	
Key debounce time	t_{DB}	—	14.6	—	ms	
Key input time	t_{KD}	40	—	—	ms	
Key release time	t_{KU}	5	—	—	ms	
Output pulse rate 1	PR_1	—	10.3	—	pps	10P/20P=L
Output pulse rate 2	PR_2	—	20.6	—	pps	10P/20P=H
Pulse break ratio 1	BR_1	—	66 2/3	—	%	60%/67%=H
Pulse break ratio 2	BR_2	—	60	—	%	60%/67%=L
Inter-digital pause 1	t_{IDPP1}	—	875	—	ms	10P/20P=L
Inter-digital pause 2	t_{IDPP2}	—	486	—	ms	10P/20P=H
Tone output time	t_{MF}	—	97	—	ms	*1
Tone inter-digital pause	t_{IDPM}	—	97	—	ms	*1
Tone output dividing error	Δf	—	—	0.21	%	
ROW tone output voltage	V_{OR}	73	103	133	mV_{P-P}	
COL tone output voltage	V_{OC}	91	130	169	mV_{P-P}	
High-band frequency pre-emphasis	PEHB	—	2	—	dB	
Tone output distortion	DIS	—	5	10	%	
Pause time	t_{PA}	—	3.7	—	sec	
Hooking time	t_{HK}	—	778	—	ms	
Hooking pause time	t_{HKPA}	—	1.2	—	sec	
Mute overlap time	t_{MO}	5	—	—	ms	
On-hook time	t_{OH}	10	—	—	ms	*2
Key confirmation tone frequency	f_{AT}	—	926	—	Hz	
Key confirmation tone time	t_{AT}	—	30	—	ms	

*1 When the key is being pressed during normal dialing, the tone is output continuously. The tone output time and inter-digital pause time continue for at least 97ms.
When repertory dialing and redialing, both the tone output time and interdigital pause time are set to 97ms.
*2 For operation of the handset by the switchhook, hold the ON HOOK pin at high for at least 10ms.

DIMENSIONS (Unit: mm)

BU8320A/BU8321/BU8322

37.1 +0.2/−0
28 15
R1.5
13.8 +0.2/−0
1 14
0.6
7.7±0.5
4.5
3.2±0.3
3.8 +0.1/−0
2.54±0.3
0.5±0.1
33.02±0.5
15.2
1.8
0.3±0.1
16.5±0.5

BU8320AF/BU8321F/BU8322F

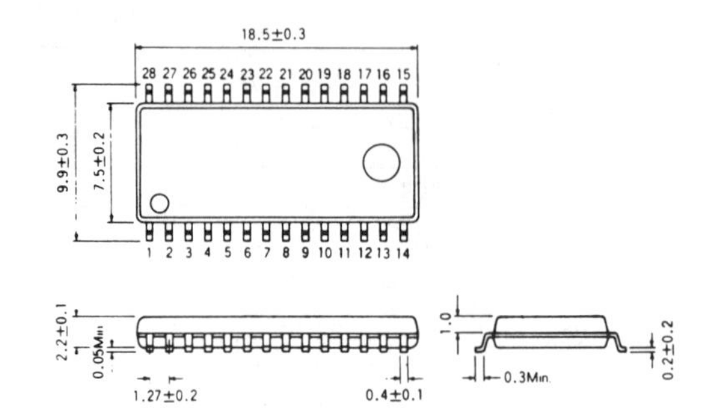

Differences Between BU8320A, BU8321, and BU8322

BU8320A and BU8321 have the same characteristics as BU8322 except for the functions of pin 17 and pin 18.

1. Comparing pins

pin 17

BU8320A BU8322	F2	The terminal for compensating DTMF operation amplifier phase
BU8321	60%/67%	The terminal for selecting pulse output break ratio

pin 18

BU8320A BU8322	F1	The terminal for compensating DTMF operation amplifier phase
BU8321	RD/WR	The input terminal for switching read and write from the repertory memory

2. Comparing electrical characteristics

Item	Symbol	BU8320A	BU8321	BU8322	Unit
		TYP	TYP	TYP	
Tone output distortion factor	DIS	2	5	2	%
Pulse break ratio 1	BR1	67	67	60	%
Pulse break ratio 2	BR2	—	60	—	%

Note: The pulse break ratio is fixed in BU8320A and BU8322.

3. Comparing circuits

BU8320A·BU8322

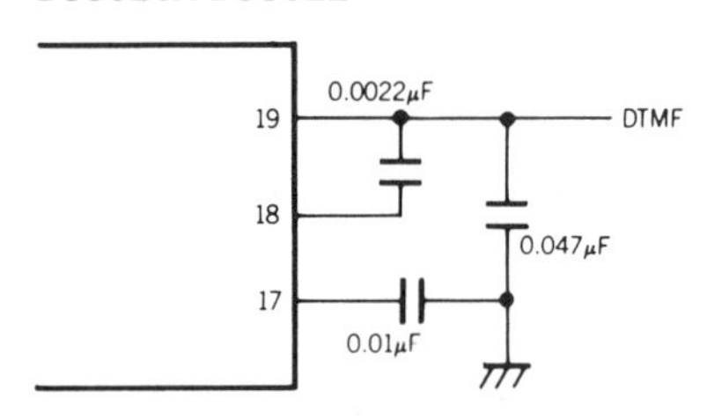

BU8321

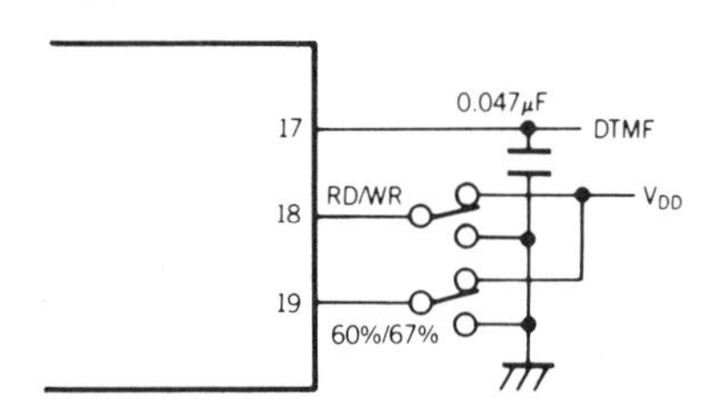

PIN CONNECTIONS

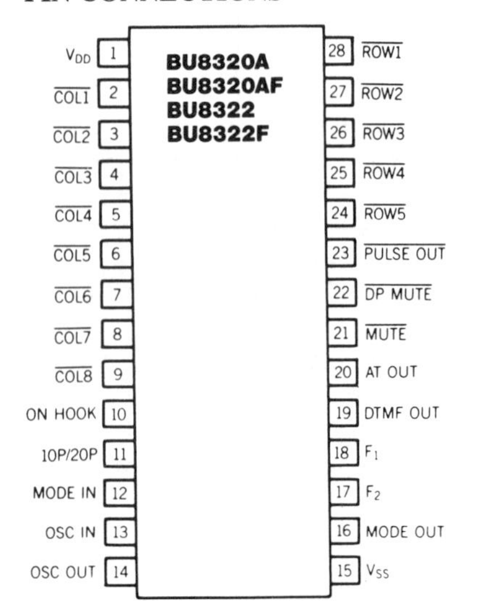

BU8321
BU8321F
VDD 1
COL1 2
COL2 3
COL3 4
COL4 5
COL5 6
COL6 7
COL7 8
COL8 9
ON HOOK 10
10P/20P 11
MODE IN 12
OSC IN 13
OSC OUT 14
28 ROW1
27 ROW2
26 ROW3
25 ROW4
24 ROW5
23 PULSE OUT
22 DP MUTE
21 MUTE
20 AT OUT
19 DTMF OUT
18 RD/WR
17 60%/67%
16 MODE OUT
15 VSS

EXAMPLE OF FULL KEYBOARD OPERATIONS

↑ OFF HOOK ↓ ON HOOK

Item	RD/WR in BU8321 BU8321F	Key operations	Dial signal outputs	Memory
Normal dialing MODE=high	L	↑ [1] [2] [3] [4]	1 2 3 4 PULSE	RD [1] [2] [3] [4]
Normal dialing MODE=low	L	↑ [1] [2] [3] [4]	1 2 3 4 DTMF	RD [T] [1] [2] [3] [4]
Normal dialing MODE=high	L	↑ [T] [1] [2] [3] [4]	1 2 3 4 DTMF	RD [T] [1] [2] [3] [4]
Access PAUSE	L	↑ [0] [PA] [1] [2] [3]	0␣1 2 3 3.7S	RD [0] [PA] [1] [2] [3]
Access PAUSE release halfway	L	↑ [0] [PA] [1] [2] [3] ↓ ↑ [R] [PA]	0␣1 2 3 3.7S 0␣1 2 3 ≦3.7S ↑ [PA] release halfway	RD [0] [PA] [1] [2] [3] RD [0] [PA] [1] [2] [3]
Redialing (1)	L	↑ [1] [2] [3] [4] ↓ ↑ [R]	1 2 3 4 1 2 3 4	RD [1] [2] [3] [4] RD [1] [2] [3] [4]
Redialing (2)	L	↑ [1] [2] ······ [2] [3] ↓ (1, 2 ······ 32, 33) ↑ [R]	1 2 ··············· 2 3 no outputs	RD [ ] [ ] [······] [ ] [ ] (1, 2 ······ 31, 32) RD [ ] [ ] [······] [ ] [ ] (1, 2 ······ 31, 32)
Redial inhibit (1)	L	↑ [1] [2] [3] [4] [R] [R] ↓ ↑ [R]	1 2 3 4 ↑ [R] [R] no outputs	RD [ ] [ ] [ ] RD [ ] [ ] [ ] [ ]
Redial inhibit (2)	L	↑ [1] [2] [3] [4] ↓ ↑ [R] [R] [R] ↓ ↑ [R]	1 2 3 4 1 2 3 4 ↑ [R] [R] no putputs	RD [1] [2] [3] [4] RD [ ] [ ] [ ] [ ] RD [ ] [ ] [ ] [ ]
Mode change by [T] key	L	↑ [1] [2] [T] [3] [4] ↓ (MODE IN="high") ↑ [R] (MODE IN="low")	1␣2 ␣ 3␣4 PULSE 3.7S DTMF 1␣2 ␣ 3␣4 PULSE 3.7S DTMF	RD [1] [2] [T] [3] [4] RD [1] [2] [T] [3] [4]
Writing repertory memory (1)	H	↓ [M] [1] [2] [3] [4] [M] [0] [0]	no outputs	RD [1] [2] [3] [4] M00 [1] [2] [3] [4]
Writing repertory memory (2)	H	↓ [M] [1] [2] [3] [4] ······ [6] [7] [M] [0] [1] (1, 2, 3, 4 ······ 16, 17)	no outputs	RD [1] [2] [3] [T] [······] [6] (1, 2, 3, 4 ······ 16) M01 [1] [2] [3] [4] [······] [6] (1, 2, 3, 4 ······ 16)
Writing repertory memory (3)	H	↓ [M] [1] [2] [3] [5] [M] [0] [0] ↓ [M] [5] [6] [7] [8] [M] [0] [1]	no outputs no outputs	RD [1] [2] [3] [4] M00 [1] [2] [3] [4] RD [5] [6] [7] [8] M01 [5] [6] [7] [8]
Reading repertory memory (1)	L	↑ [M] [0] [0] or ↑ [OT 00]	1 2 3 4 5	RD [1] [2] [3] [4] [5] M00 [1] [2] [3] [4] [5]
Reading repertory memory (2)	L	↑ [M] [0] [0] [M] [0] [1] or ↑ [OT 00] [OT 01]	1 2 3 4 5 6 7 8 9 0	RD [1] [2] [3] [4] [5] [6] [7] [8] [9] [0] M00 [1] [2] [3] [4] [5] M01 [6] [7] [8] [9] [0]
Reading repertory memory (3)	L	↑ [M] [0] [2] [1] [2] [3]	1 2 3 4 5 ↑ 1 2 3 [1] [2] [3]	RD [1] [2] [3] [4] [5] [1] [2] [3] M02 [1] [2] [3] [4] [5]
Reading repertory memory (4)	L	↑ [R] [0] [0] [R] [0] [1] [6] [7] or ↓ [OT 00] [OT 01] [6] [7]	1 2 3 4 5 6 ↑ 6 7 [6] [7]	RD [ ] [ ] [ ] [ ] [ ] (1, 2 ······ 31, 32) M00 [1] [2] [3] M01 [4] [5] [6]
Hookswitch	L	↑ [H]	Hookswitch operation	

EXAMPLE OF 2-OF-7 KEYBOARD OPERATIONS

Item	RD/WR in BU8321 BU8321F	Key operations	Dial signal outputs	Memory
Normal dialing MODE IN=high	L	↑ [7] [8] [9] [0]	7 8 9 0 PULSE	RD [7] [8] [9] [0]
Normal dialing MODE IN=low	L	↑ [0] [*M] [#R]	0 * # DTMF	RD [T] [0] [*] [#]
Redialing MODE IN=high	L	↑ [1] [2] [3] [4] ↑ [#R] [#R]	RD [1] [2] [3] [4] 1 2 3 4	RD [1] [2] [3] [4]
Redialing inhibit MODE IN=high	L	↑ [1] [2] [3] [4] [*M] [*M] ↓ ↑ [#R] [#R]	1 2 3 4 no outputs [*M] [*M]	RD [ ] [ ] [ ] [ ] RD [ ] [ ] [ ] [ ]
Writing repertory memory MODE IN=high	H	↓ [*M] [1] [2] [3] [4] [*M] [0] [0]	no outputs	M00 [1] [2] [3] [4]
Reading repertory memory MODE IN=high	L	↑ [#R] [0] [0] ↓	1 2 3 4	M00 [1] [2] [3] [4] RD [1] [2] [3] [4]
Access PAUSE MODE IN=high	L	↑ [0] [#R] [1] [2] [3]	0␣1 2 3 3.7S	RD [0] [PA] [1] [2] [3]
Access PAUSE release halfway MODE IN=high	L	↑ [0] [#R] [1] [2] [3] ↑ [#R] [#R] [#R]	0␣1 2 3 3.7S 0␣1 2 3 ≦3.7S release halfway	RD [0] [PA] [1] [2] [3] RD [0] [PA] [1] [2] [3]

TYPICAL APPLICATION CIRCUIT

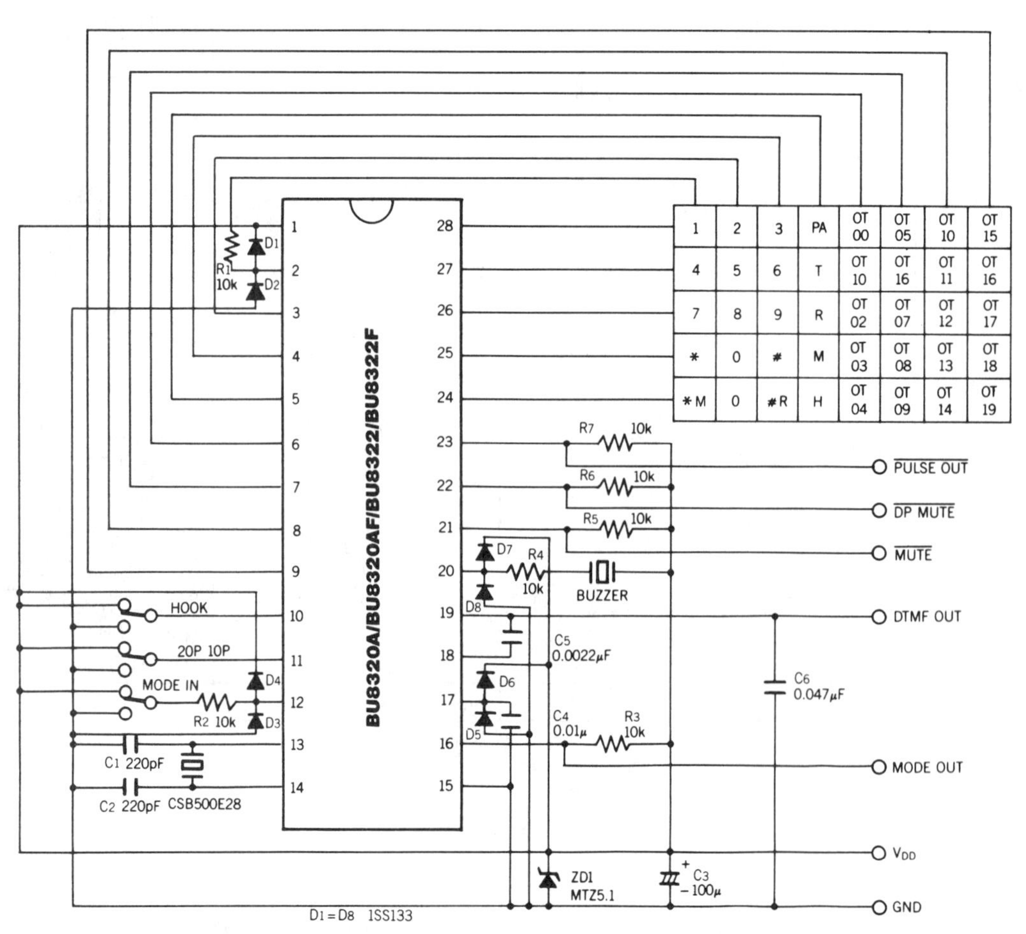

OFF-HOOK MEMORY-STORE APPLICATION

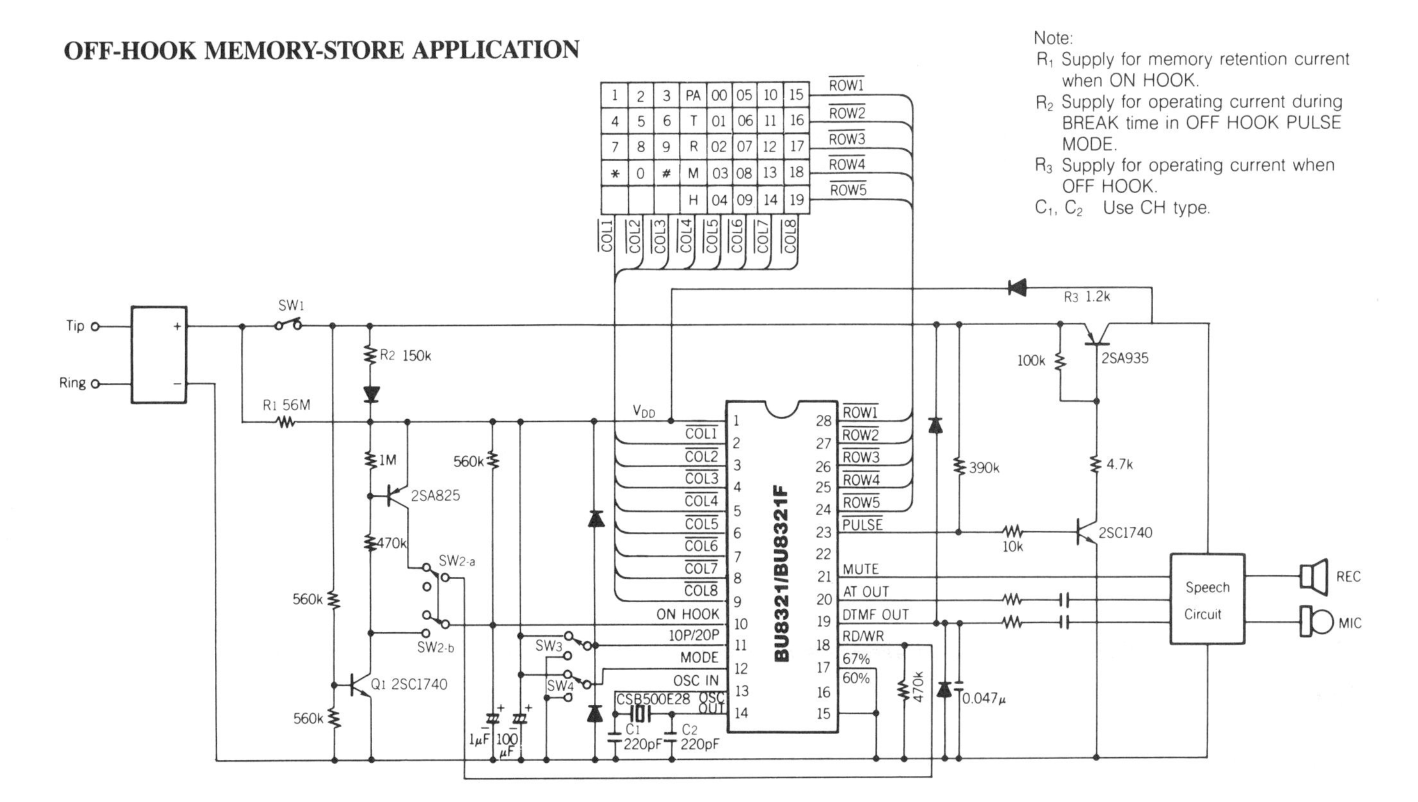

Note:
- R_1 Supply for memory retention current when ON HOOK.
- R_2 Supply for operating current during BREAK time in OFF HOOK PULSE MODE.
- R_3 Supply for operating current when OFF HOOK.
- C_1, C_2 Use CH type.

SW_1: HOOKSWITCH
SW_2: MEMORY READ/WRITE SELECT SWITCH

Q1 can replace by mechanical HOOKSWITCH.

Conditions		IC's state
SW_1	SW_2	
CLOSE	UP	MEMORY WRITE
CLOSE	DOWN	DIALING MEMORY READ
OPEN	UP	MEMORY RETENTION
OPEN	DOWN	MEMORY RETENTION

Operation example
1. MEMORY WRITE (OFF HOOK)
   SW_1→WIRTE [M][0][7][5][8][1][1][2][1][2][1][M][0][0]
2. MEMORY READ (OFF HOOK)
   SW_2→READ [M][0][0]

If key input terminals ($\overline{R1} \sim \overline{R5}$, $\overline{C1} \sim \overline{C8}$) are troubled by external noise, RF1 or EM1, connect 470pF capacitors between each key input terminal and GND.

BU8321 off hook memory store application

BLOCK DIAGRAMS

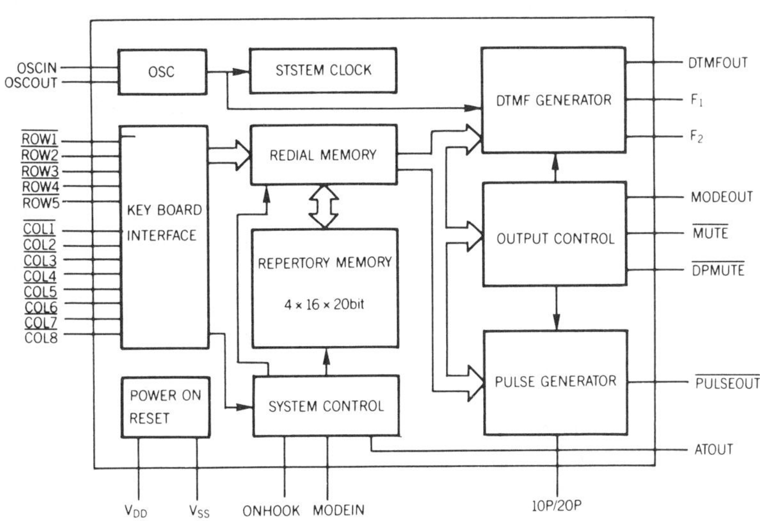

BU8320A/BU8320AF/BU8322/BU8322F

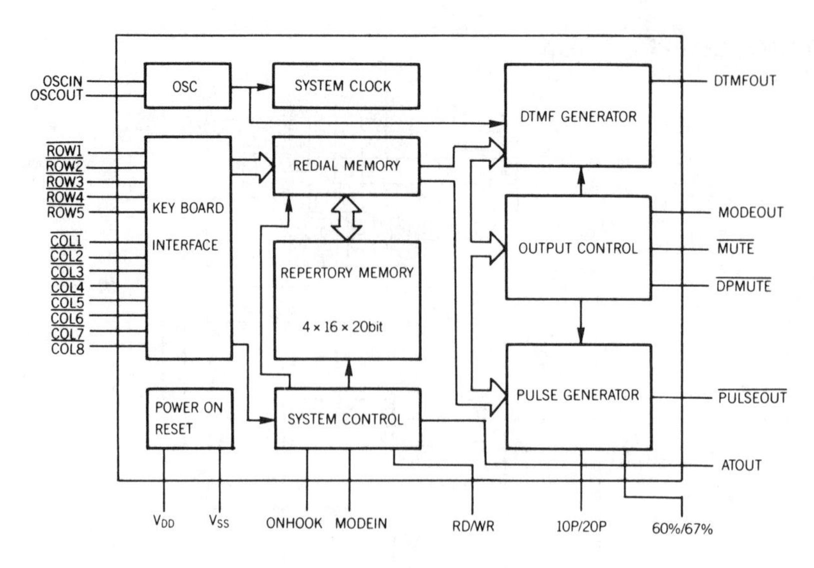

BU8321/BU8321F

Rohm BU8992
Pulse Dialer

- Connects directly to a telephone line
- Allows the use of the standard 2-of-7 keypad
- CMOS technology for low-power design
- Supply voltage range: 2.5 to 6.0V
- Output pulse make/break ratio is selectable form 60% or 66%
- Output pulse rate is selectable from 10 pps and 20 pps
- Dialing capability with the # or * key
- Internal RC oscillator

APPLICATIONS

- Telephone handsets
- Telephone-related equipment

The BU8992 is designed for use in telephone handsets to replace the conventional rotary dial with a pushbutton dial. The device is adaptable to different rotary dial specifications with simple modifications.

RECOMMENDED OPERATING CONDITIONS (Ta=25°C)

Operating voltage range V_{DD} (between pins 1 and 6)=2.5 ~ 6.0V
Recommended Operating Voltage V_{DD} (between pins 1 and 6)=3.5V

Parameter	Symbol	Min.	Typ.	Max.	Unit	Conditions
Operating voltage	V_{DD}	2.5	3.5	6.0	V	1-6pin

ELECTRICAL CHARACTERISTICS (Ta=25°C, V_{DD}=3.5V)

Parameter	Symbol	Min.	Typ.	Max.	Unit	Conditions
Operating current	I_{DD}	—	1.0	—	mA	—
Memory retention current	I_{MR}	—	0.5	—	μA	—

BLOCK DIAGRAM

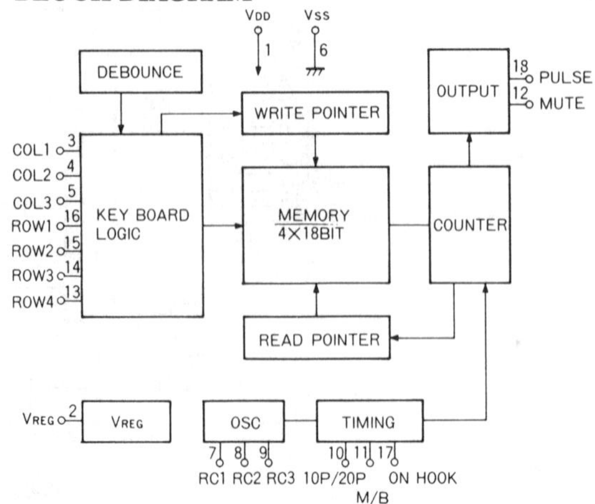

DIMENSIONS (Unit: mm)

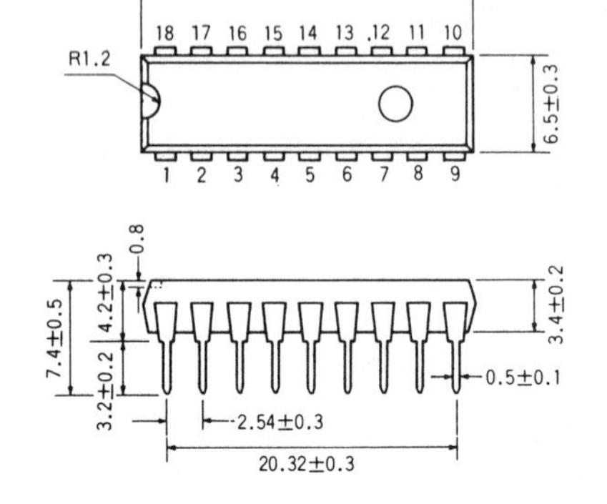

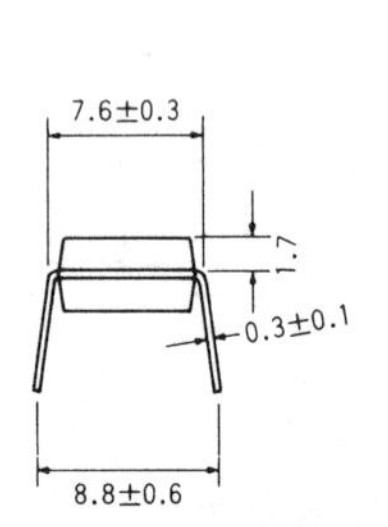

APPLICATION EXAMPLE

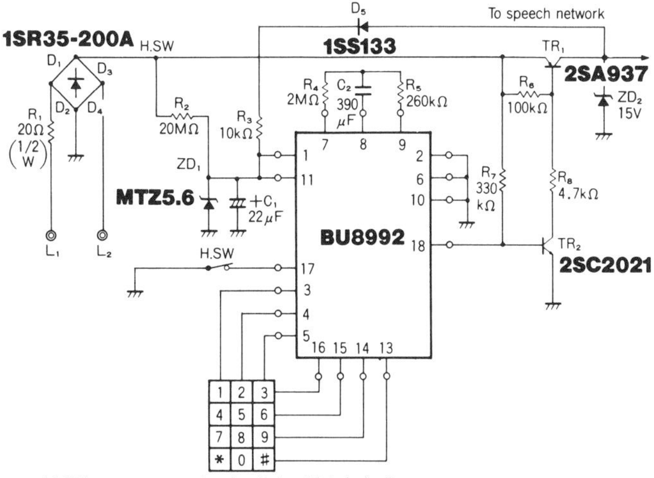

H.SW represents a hookswitch which is in the off-hook state. ZD2 to be used with large watt ratings.

Rohm BU8302A

Tone/Pulse Dialer

- Tone and pulse dialer functions integrated on a single chip
- Requires no external power supply, and directly connects to a telephone line
- Low standby current
- Last number redialing capability for up to 17 digits. The redial memory can be backed up from the telephone line.
- Internal 3.75 sec. digital pause timer allows cumulative pauses
- DTMF output amplitude is internally regulated and is not affected by supply voltage fluctuation
- Pulse break ratio is selectable from 63% or 66%; pulse rate is selectable from 10 pps and 20 pps
- During pulse dial, the device outputs a dial tone for each dial key operation
- Internal power-on reset logic and voltage regulator
- The internal master oscillator uses an external 3.58-MHz crystal or a ceramic resonator
- Allows the use of the standard 2-of-7 keypad

APPLICATIONS

- Telephone handsets
- Cordless telephones

The BU8302A is a monolithic telephone dialer combining tone and pulse dialer functions on a single chip.

The mode pin enables switching between tone (DTMF) and pulse modes. In the tone mode, all keys on the keypad are operative. During normal dialing, the number of key entry digits is not limited. Key entry is possible even during dialing; entry key data is sent following the dial number.

Key entry is also possible in pause mode; entry data is sent after the terminal exits the pause mode. The tone signal amplitude is internally regulated and is not affected by supply voltage fluctuation.

In the pulse mode, the * and # keys are made inoperative. Similar to the tone mode, the number of key-entry digits is not limited during normal dialing, and key entry is enabled during the dialing or pause mode. In this mode, the device outputs a dial tone for every key operation. The device provides the redial feature for up to 17 digits. Key entry is possible even during redialing; entry-key data is sent following the redial number.

In the pause mode, the device generates a pause interval of 3.75 seconds. The pause mode is available in both the tone and pulse modes. Key entry is possible in the pause mode. Because the pause mode is written as data into the redial memory, it is also available during redial operation.

RECOMMENDED OPERATING CONDITIONS (Ta=25°C)

Parameter	Symbol	Min.	Typ.	Max.	Unit	Conditions
Operating voltage	V_{DD}	2.5	—	6.0	V	—
Recommeded operating voltage	V_{DD}	—	3.5	—	V	—

ELECTRICAL CHARACTERISTICS (Ta=25°C, V_{DD}=3.5V)

Parameter	Symbol	Min.	Typ.	Max.	Unit	Conditions
Operating current	I_{DD}	—	1.0	2.0	mA	
Memory retention current	I_{MR}	—	0.1	0.75	μA	V_{DD}=2.5V
Keyboard debounce time	T_{DB}	—	11.8	—	ms	
Key input time	T_{KD}	40	—	—	ms	

DIMENSIONS (Unit: mm)

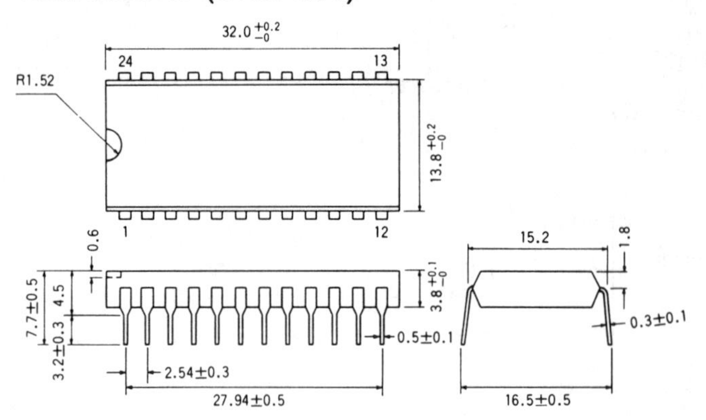

BLOCK DIAGRAM

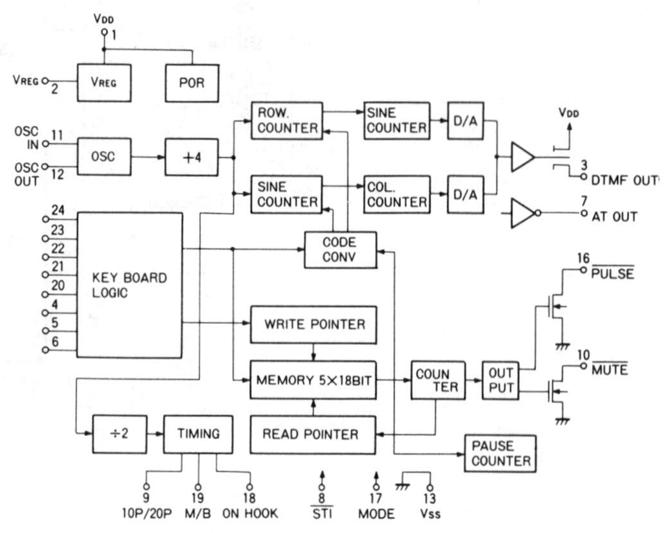

KEYBOARD FUNCTIONS

Key	Pulse mode	DTMF(tone) mode
0~9	0~9 figures data	0~9 figures data
*, #	Refuse key input	*, # data
PA	Pause key	Pause key
RD	Redial key	Redial key

KEY OPERATING EXAMPLE

Key operation	Dial output
1. OFF 0 7 5	0 7 5
ON	
OFF RD	0 7 5
ON	
OFF RD 3 1 1	0 7 5 3 1 1
2. OFF 0[1] 7[2] 5[3] .... 3[16] 1[17]	0[1] 7[2] 5[3] .... 3[16] 1[17]
ON	
OFF RD	0 7 5 .... 3 1
3. OFF 0 PA 0 7 5	0 3.75 sec. 0 7 5
ON	
OFF RD	0 3.75 sec. 0 7 5

OFF; OFF HOOK, ON; ON HOOK

PIN CONNECTIONS (TOP VIEW)

BU8302A

Pin	Name	Pin	Name
1	VDD	24	$\overline{ROW1}$
2	VREG	23	$\overline{ROW2}$
3	DTMF OUT	22	$\overline{ROW3}$
4	$\overline{COL1}$	21	$\overline{ROW4}$
5	$\overline{COL2}$	20	$\overline{ROW5}$
6	$\overline{COL3}$	19	M/B SELECT
7	AT OUT	18	ON HOOK
8	STI	17	MODE
9	10P/20P	16	$\overline{PULSE\ OUT}$
10	$\overline{MUTE}$	15	*
11	OSC IN	14	*
12	OSC OUT	13	Vss

* Leave these pins floating as they are connected internally.

KEYPAD INPUT MATRIX

1	2	3	$\overline{ROW1}$
4	5	6	$\overline{ROW2}$
7	8	9	$\overline{ROW3}$
*	0	#	$\overline{ROW4}$
PA		RD	$\overline{ROW5}$
$\overline{COL1}$	$\overline{COL2}$	$\overline{COL3}$	

APPLICATION EXAMPLE

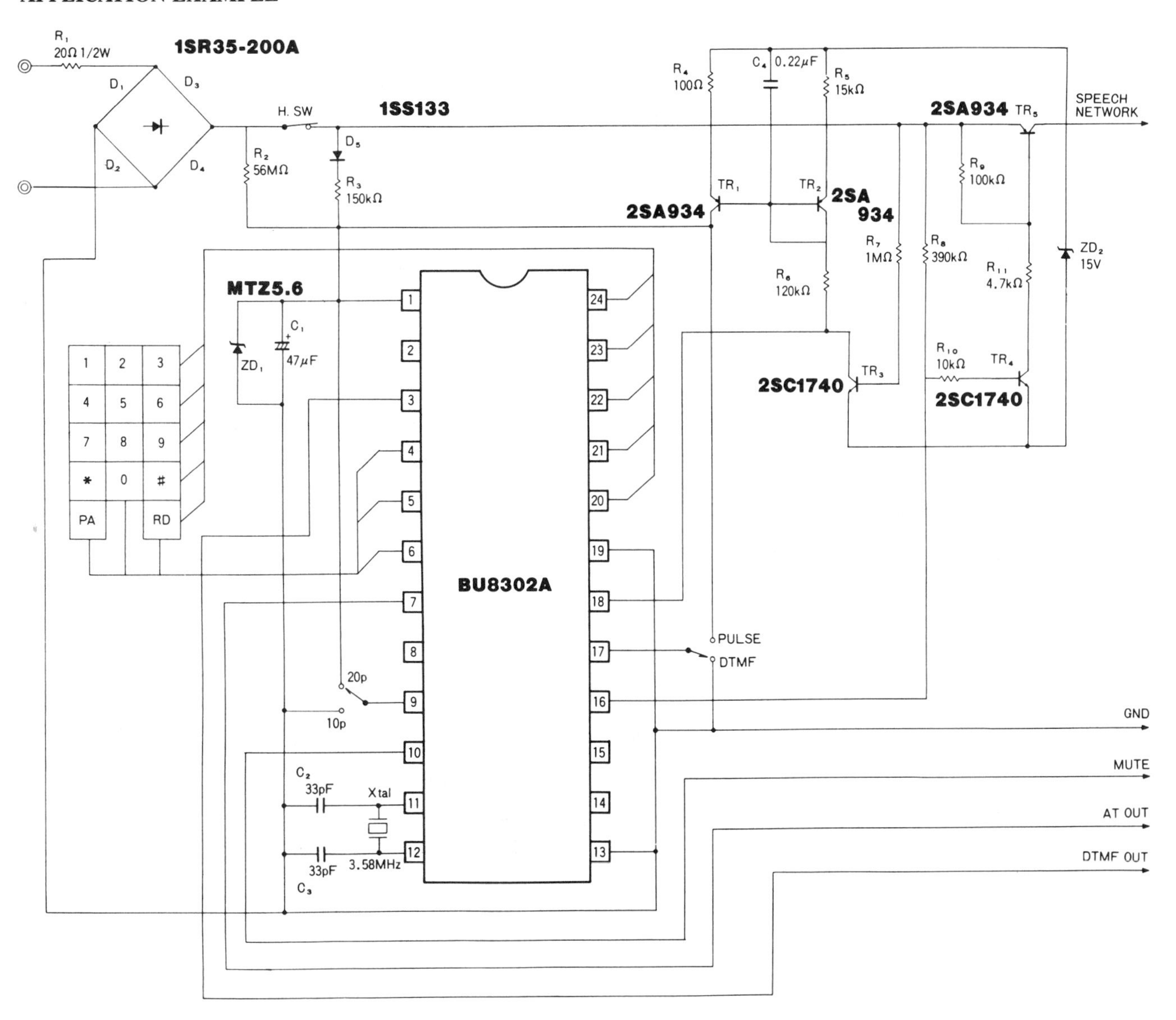

This information is reproduced with permission of Siliconix, Inc.

Siliconix DG540/541/542

Wideband/Video "T" Switches

- Wide bandwidth (500 MHz)
- Very low crosstalk (−85dB) and high off-isolation (−80dB) at 5 MHz
- "T" switch configuration
- TTL compatible
- Fast switching (t_{ON} <70 ns)
- ESD protection > ±4000V
- Low $r_{DS(ON)}$ <75 Ω
- Improved data throughput
- Low insertion loss
- Improved system performance
- Reduced board space
- Reduced power consumption

APPLICATIONS

- RF and video switching
- RGB switching
- Local- and wide-area networks
- Video routing
- Fast data acquisition
- ATE
- Radar/FLIR systems
- Video multiplexing

The DG540/541/542 are very high performance monolithic wideband/video switches designed for switching wide bandwidth analog and digital signals. By utilizing "T" switch configuration techniques on each channel, these devices achieve exceptionally low crosstalk and high off-isolation. The crosstalk and off-isolation of the DG540 are further improved by the introduction of extra GND pins between signal pins.

To achieve TTL compatibility, low channel capacitances and fast switching times, the DG540 family is built on the Siliconix proprietary D/CMOS process. Each switch conducts equally well in both directions when on.

The DG540 is available in 20-pin side braze, plastic, and PLCC packages. Packaging for the DG541 and DG542 includes 16-pin side braze, plastic, and small outline options. Performance grades include military. A suffix (−55 to 125°C) and industrial, and D suffix (−40 to 85°C) temperature ranges.

ABSOLUTE MAXIMUM RATINGS

Parameter		Rating
V+ to V-		-0.3 V to 21 V
V+ to GND		-0.3 V to 21 V
V- to GND		-19 V to +0.3 V
Digital Inputs		(V-) -0.3 V to (V+) +0.3 V or 20 mA, whichever occurs first
V_S, V_D		(V-) -0.3 V to (V-) +14 V or 20 mA, whichever occurs first
Continuous Current (Any Terminal)		20 mA
Current, S or D (Pulsed 1 ms, 10% duty cycle max)		40 mA
Storage Temperature	(A Suffix)	-65 to 150°C
	(D Suffix)	-65 to 125°C
Operating Temperature	(A Suffix)	-55 to 125°C
	(D Suffix)	-40 to 85°C

Power Dissipation (Package)*	
16-, 20-Pin Size Braze DIP**	900 mW
20-Pin Plastic DIP***	800 mW
20-Pin PLCC****	800 mW
16-Pin Plastic DIP*****	470 mW
16-Pin SO******	640 mW

* All leads welded or soldered to PC board.
** Derate 12 mW/°C above 75°C.
*** Derate 7 mW/°C above 25°C.
**** Derate 10 mW/°C above 75°C.
***** Derate 6.5 mW/°C above 25°C.
****** Derate 10 mW/°C above 75°C.

PIN CONFIGURATIONS, FUNCTIONAL BLOCK DIAGRAMS, AND TRUTH TABLES

Dual-In-Line Package

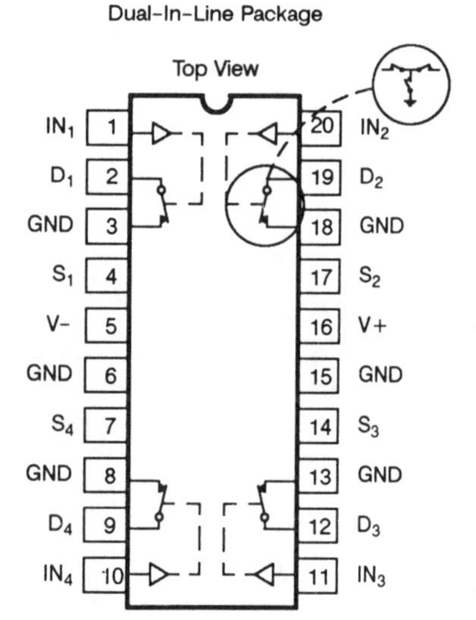

Order Numbers:
Side Braze: DG540AP, DG540AP/883
Plastic: DG540DJ

PLCC Package

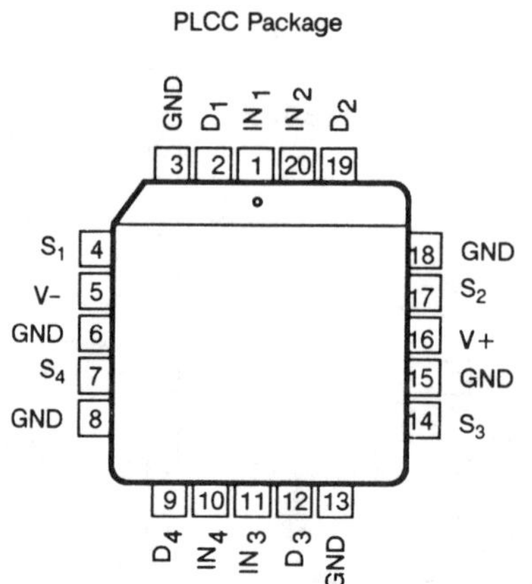

Top View

Order Number:
DG540DN

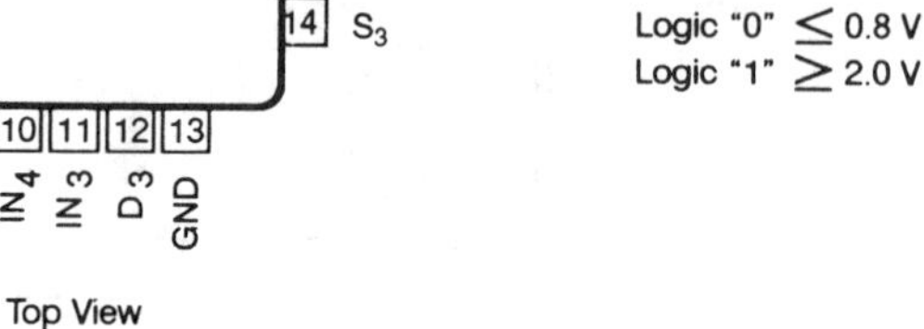

Logic	Switch
0	OFF
1	ON

Logic "0" ≤ 0.8 V
Logic "1" ≥ 2.0 V

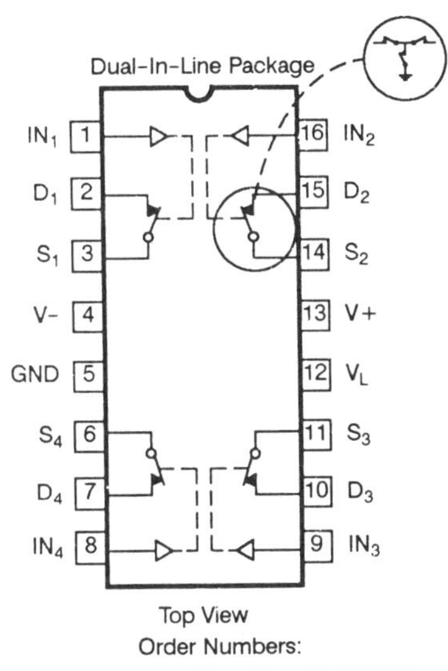

Top View
Order Numbers:
Side Braze: DG541AP, DG541AP/883
Plastic: DG541DJ

Narrow Body
SO Package

16 15 14 13 12 11 10 9
(Same pinout as DIP)
1 2 3 4 5 6 7 8

Top View

Order Numbers:
DG541DY

Logic	Switch
0	OFF
1	ON

Logic "0" ≤ 0.8 V
Logic "1" ≥ 2.0 V

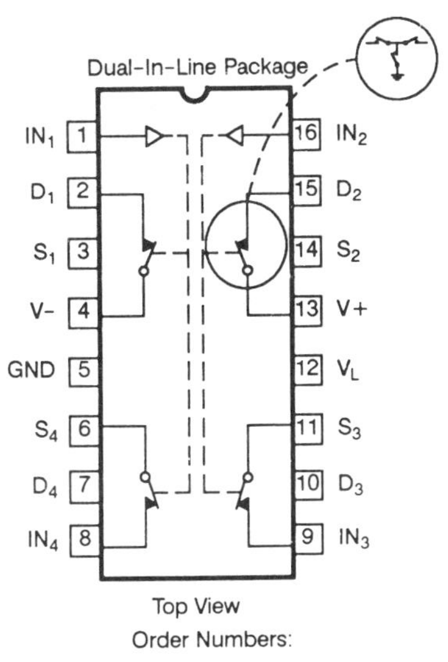

Top View
Order Numbers:
Side Braze: DG542AP, DG542AP/883
Plastic: DG542DJ

Narrow Body
SO Package

16 15 14 13 12 11 10 9
(Same pinout as DIP)
1 2 3 4 5 6 7 8

Top View

Order Numbers:
DG542DY

Logic	SW1 SW2	SW3 SW4
0	OFF	ON
1	ON	OFF

Logic "0" ≤ 0.8 V
Logic "1" ≥ 2.0 V

* Switches shown for logic "1" input.

TYPICAL CHARACTERISTICS

Supply Current vs. Temperature

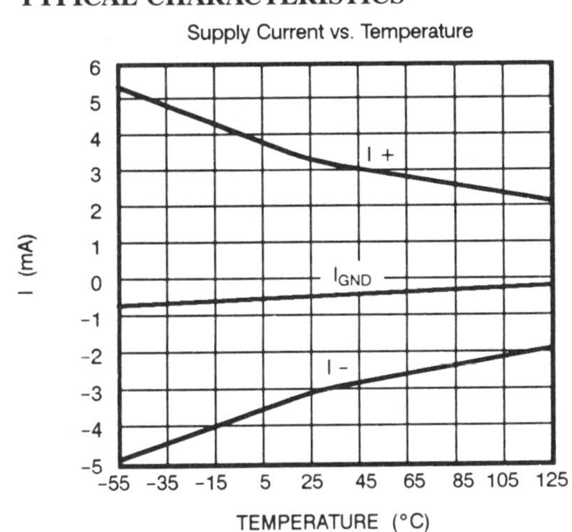

$I_{D(OFF)}$, $I_{S(OFF)}$ vs. Temperature

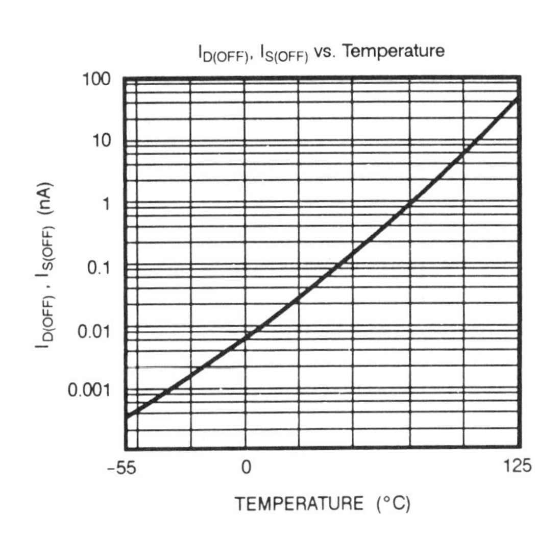

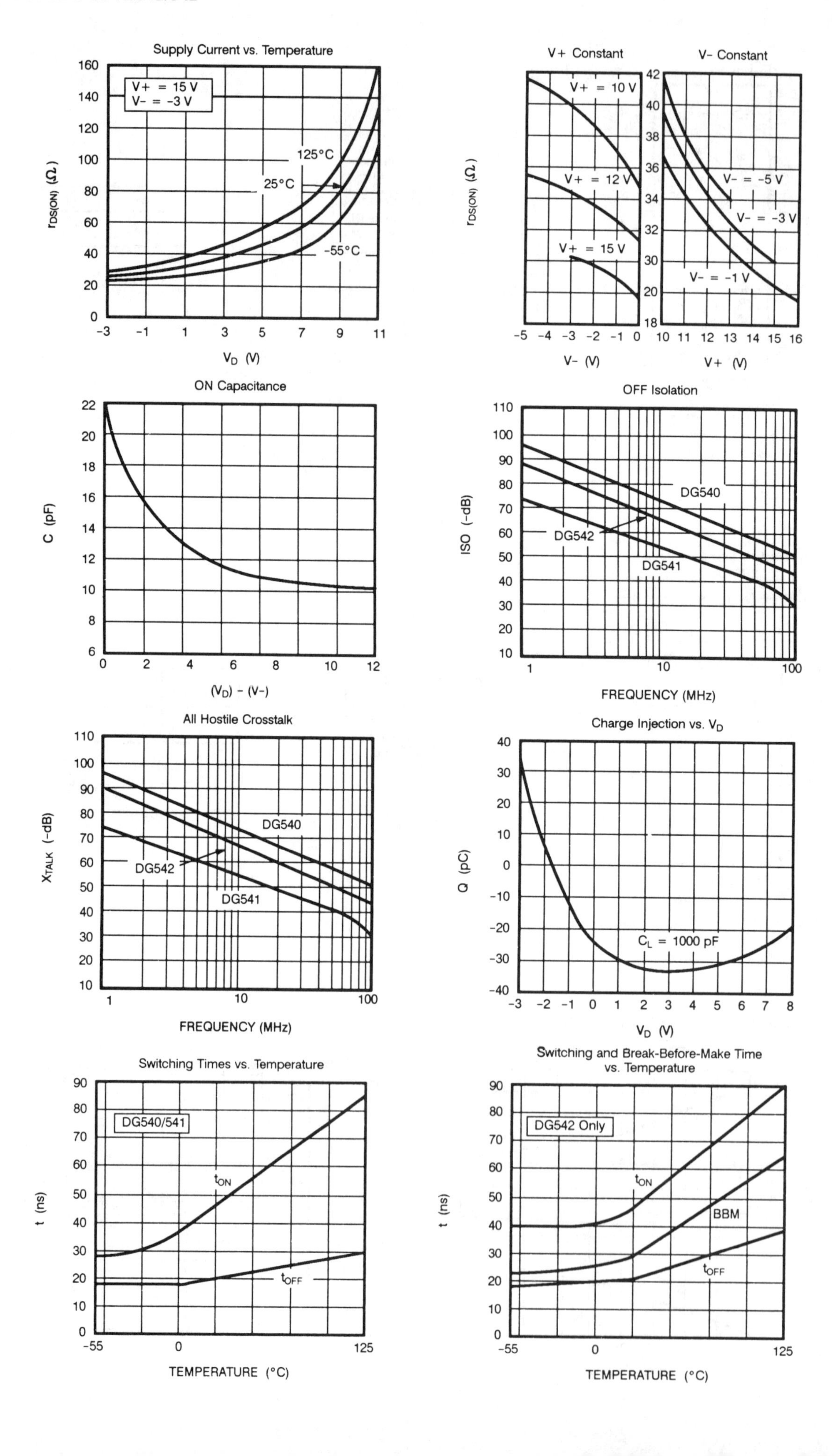
Supply Current vs. Temperature
V+ = 15 V
V– = -3 V
125°C
25°C
-55°C
$r_{DS(ON)}$ (Ω)
V_D (V)
V+ Constant
V– Constant
V+ = 10 V
V+ = 12 V
V+ = 15 V
V– = -5 V
V– = -3 V
V– = -1 V
V– (V)
V+ (V)
ON Capacitance
C (pF)
(V_D) – (V–)
OFF Isolation
ISO (–dB)
DG540
DG542
DG541
FREQUENCY (MHz)
All Hostile Crosstalk
X_{TALK} (–dB)
DG540
DG542
DG541
FREQUENCY (MHz)
Charge Injection vs. V_D
Q (pC)
C_L = 1000 pF
V_D (V)
Switching Times vs. Temperature
DG540/541
t_{ON}
t_{OFF}
t (ns)
TEMPERATURE (°C)
Switching and Break-Before-Make Time vs. Temperature
DG542 Only
t_{ON}
BBM
t_{OFF}
t (ns)
TEMPERATURE (°C)

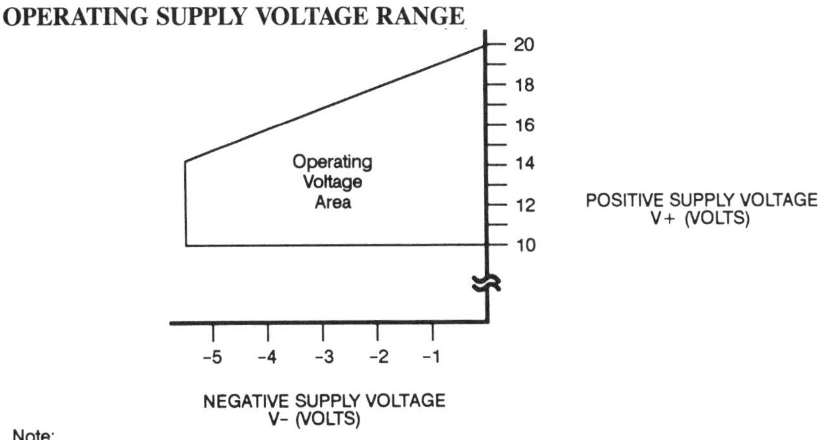

Note:

1. Both V+ and V- must have decoupling capacitors mounted as close as possible to the device pins. Typical decoupling capacitors would be 10 µF tantalum bead in parallel with 100 nF ceramic disc.

2. Production tested with V+ = 15 V and V- = -3 V.

DEVICE DESCRIPTION

The DG540/541/542 family of wideband switch offers true bidirectional switching of high-frequency analog or digital signals with minimum signal crosstalk, low insertion loss, and negligible nonlinearity distortion and group delay.

Built on the Siliconix D/CMOS process, these "T" switches provide excellent off-isolation with a bandwidth of around 500 MHz (350 MHz for DG541). Silicon-gate D/CMOS processing also yields fast switching speeds. An on-chip regulator circuit maintains TTL input compatibility over the whole operating supply-voltage range shown, easing control logic interfacing. Circuit layout is facilitated by the interchangeability of source and drain terminals.

A single switch on-channel exhibits both resistance ($r_{DS(ON)}$) and capacitance ($C_{S(ON)}$). This RC combination has an attenuation effect on the analog signal, which is frequency dependent (like an RC low-pass filter). The −3dB bandwidth of the DG540 is typically 500 MHz (into 50Ω). This measured figure of 500 MHz illustrates that the switch channel cannot be represented by a two-stage RC combination. The on capacitance of the channel is distributed along the on-resistance, and hence becomes a more complex multistage network of R's and C's making up the total $r_{DS(ON)}$ and $C_{S(ON)}$.

POWER SUPPLIES

A useful feature of the DG54X family is its power supply flexibility. It can be operated from a single positive supply (V+) if required (V− connected to ground). Note that the analog signal must not exceed V− by more than −0.3V to prevent forward biasing the substrate p-n junction. The use of the V− supply has a number of advantages:

- It allows flexibility in analog signal handling, i.e., with V− = −5V and V+ = 12V; up to ±5V ac signals can be controlled.
- The value of on capacitance ($C_{S(ON)}$) can be reduced. A property known as "the body-effect" on the DMOS switch devices causes various parametric effects to occur. One of these effects is the reduction in $C_{S(ON)}$ for an increasing V body-source. Note however that to increase V− normally requires V+ to be reduced (V+ to V− = 21V max). Reduction in V+ causes an increase in $r_{DS(DN)}$, hence a compromise must be achieved. It is also useful to note that tests indicate that optimum video linearity performance (e.g., differential phase and gain) occurs when V− is around −3V.
- V− eliminates the need to bias the analog signal using potential dividers and large coupling capacitors.

DECOUPLING

It is an established RF design practice to incorporate sufficient bypass capacitors in the circuit to decouple the power supplies to all active devices in the circuit. The dynamic performance of the DG54X is adversely affected by poor decoupling of power supply pins. Also, of even more significance, because the substrate of the device is connected to the negative supply, adequate decoupling of this pin is essential.

- Decoupling capacitors should be incorporated on all power supply pins (V+, V−).
- They should be mounted as close as possible to the device pins.
- Capacitors should be of a suitable type with good high-frequency characteristics—tantalum-bead and/or ceramic-disc types are adequate.
- Suitable decoupling capacitors are 1-to-10µF tantalum bead, plus 10-to-100nF ceramic or polyester.

BOARD LAYOUT

PCB layout rules for good high-frequency performance must also be observed to achieve the performance boasted by the DG540. Some tips for minimizing stray effects are:

- Use extensive ground planes on double-sided pcb, separating adjacent signal paths. Multilayer pcb is even better.
- Keep signal paths as short as practically possible, with all channel paths of near equal length.
- Careful arrangement of ground connections is also very important. Star-connected system grounds eliminate signal current, flowing through ground path parasitic resistance, from coupling between channels.

C_1 = 10 µF Tantalum
C_2 = 0.1 µF Polyester

Supply Decoupling

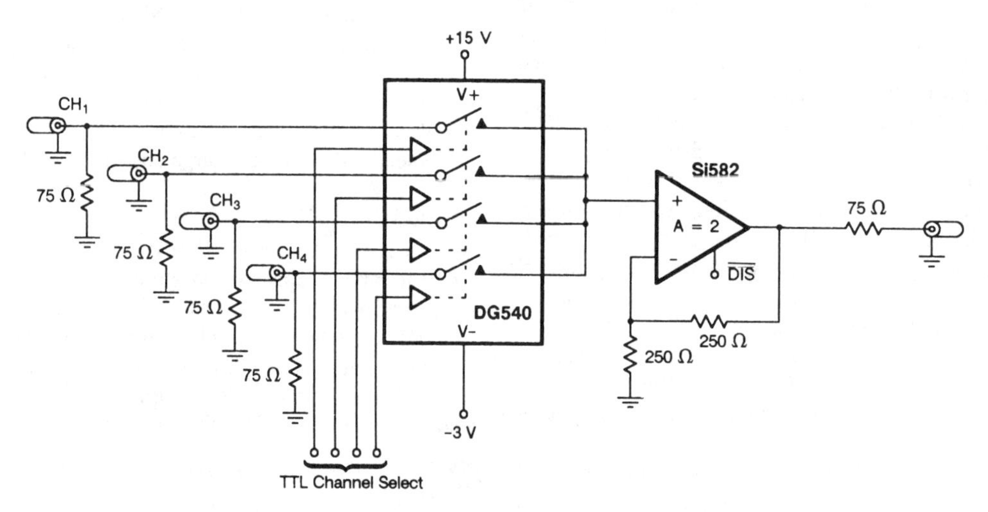

4 by 1 Video Multiplexing Using the DG540

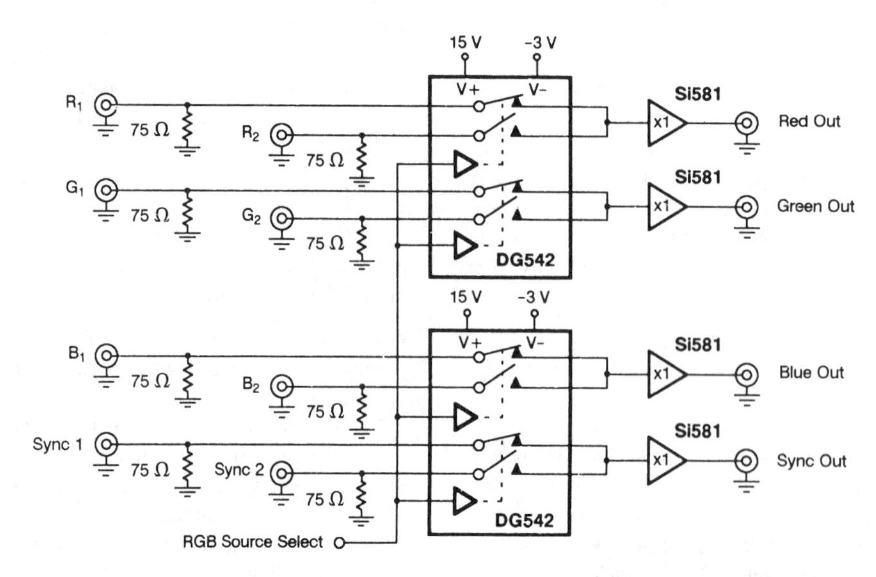

RGB Selector Switch Using Two DG542's

Siliconix DG201A/202

Quad Monolithic SPST CMOS Analog Switches

- ±15V input range
- Low off leakage ($I_{S(OFF)}$ 1nA)
- Low on-resistance
- 44V maximum supply ratings
- TTL and CMOS compatible
- Low power consumption
- Logic inputs accept negative voltages
- Wide input range
- Low distortion switching
- Can be driven from comparators or op amps without limiting resistors
- Multiple sourced

APPLICATIONS

- Disk drives
- Radar systems
- Communications systems
- Low-transient sample/holds

The DG201A and DG202 are quad SPST analog switches designed to provide accurate switching over a wide range of input signals. By combining a low on resistance and a wide signal range (±15V) with low charge-transfer, these devices are well-suited for industrial and military applications.

Built on Siliconix' high-voltage metal-gate process to achieve optimum switch performance, each switch conducts equally well in both directions when on. When off, these switches will block up to 30V peak-to-peak and have a 44V absolute maximum power supply rating. On resistance is very flat over the full ±15V analog range.

These two devices are differentiated by the type of switch actions, as shown in the functional block diagram. Package options for this series includes both the 16-pin plastic and CerDIP. Performance grades include the military A-suffix (−55 to 125 °C), industrial B-suffix (−25 to 85 °C), commercial C-suffix (0 to 70 °C) and extended industrial D-suffix (−40 to 85 °C) temperature ranges. Additionally, the DG201A is available in the 16-pin SO package. The DG441/DG442 upgrades are recommended for new designs.

ABSOLUTE MAXIMUM RATINGS

Voltages Referenced to V-

V+ .... 44 V

GND .... 25 V

Digital Inputs[e] V_S, V_D .... (V-) -2 V to (V+) +2 V .... or 20 mA, whichever occurs first.

Current, Any Terminal Except S or D .... 30 mA

Continuous Current, S or D .... 20 mA

Peak Current, S or D (Pulsed at 1 ms, 10% duty cycle max) .... 70 mA

Storage Temperature (K, Z Suffix) .... -65 to 150°C
(J, Y Suffix) .... -65 to 125°C

Operating Temperature (A Suffix) .... -55 to 125°C
(B Suffix) .... -25 to 85°C
(C Suffix) .... 0 to 70°C
(D Suffix) .... -40 to 85°C

Power Dissipation (Package)*

16-Pin CerDIP** .... 900 mW

16-Pin Plastic Dip*** .... 470 mW

20-Pin LCC**** .... 750 mW

16-Pin SO***** .... 640 mW

*Device mounted with all leads soldered or welded to PC board.
**Derate 12 mW/°C above 75°C.
***Derate 6.5 mW/°C above 25°C.
****Derate 10 mW/°C above 75°C.
*****Derate 7.6 mW/°C above 75°C

FUNCTIONAL BLOCK DIAGRAM, PIN CONFIGURATION, AND TRUTH TABLE

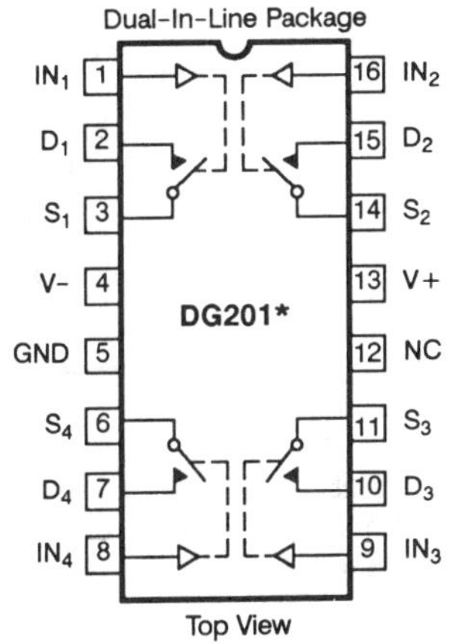

CerDIP: DG201AAK, DG201AAK/883
DG201ABK, DG201ACK
DG202AK/883

Plastic: DG201ACJ
DG202CJ

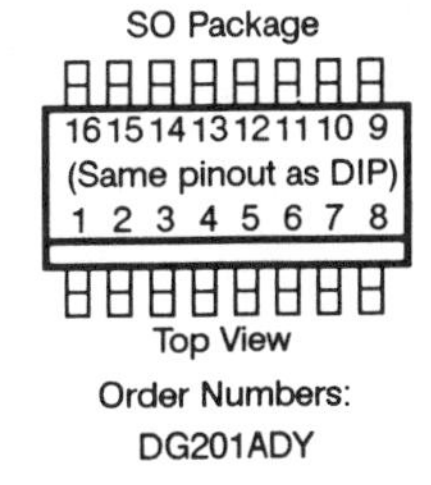

Logic	DG201A	DG202
0	ON	OFF
1	OFF	ON

Logic "0" ≤ 0.8 V
Logic "1" ≥ 2.4 V

*Switches shown for Logic "1" input

TYPICAL CHARACTERISTICS

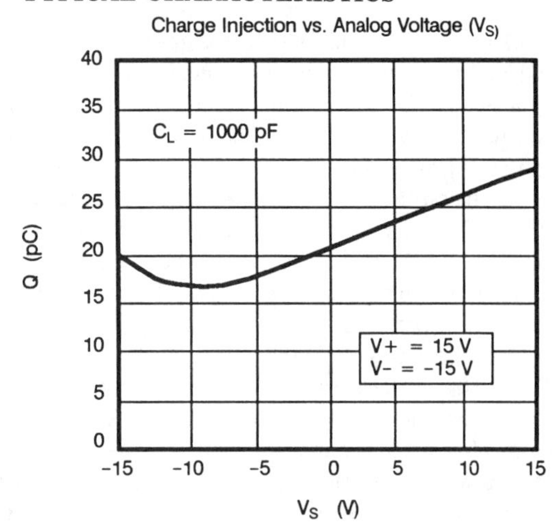

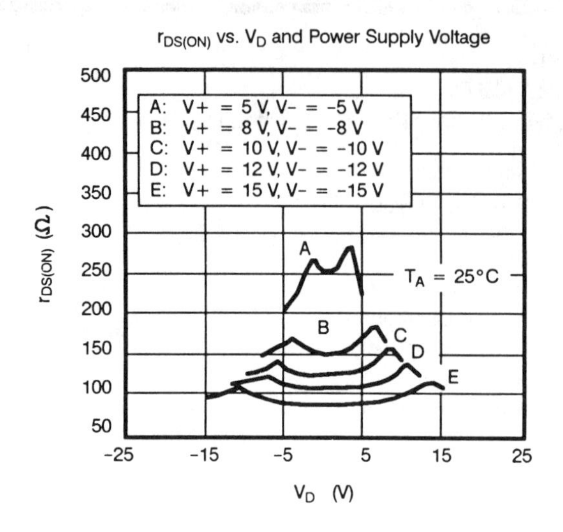

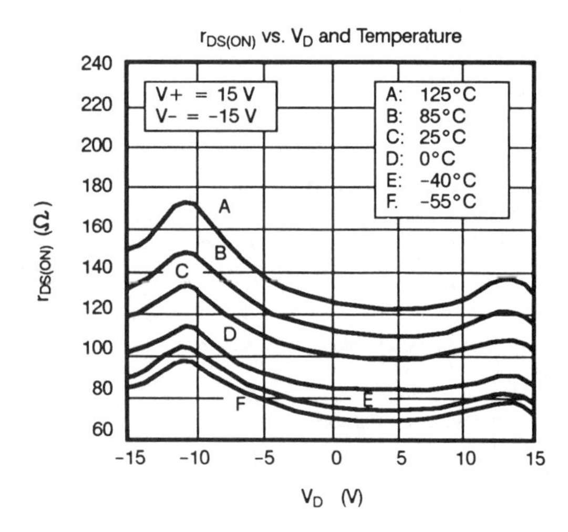

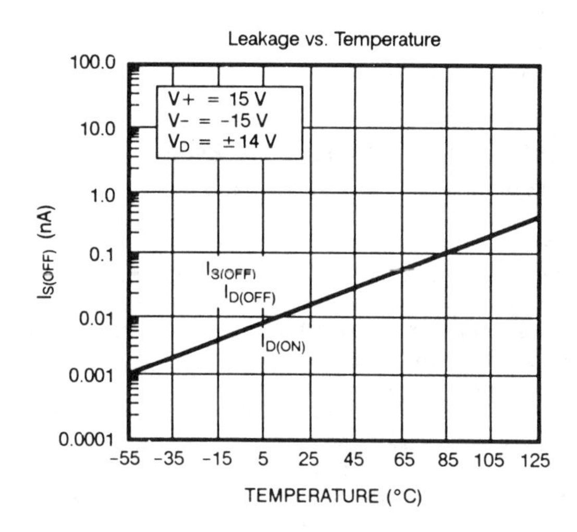

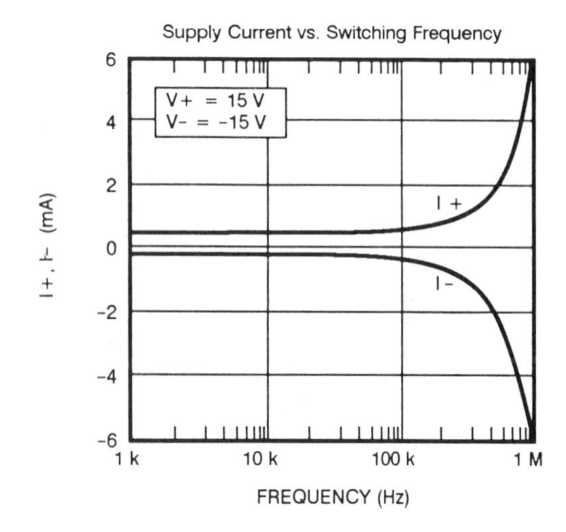

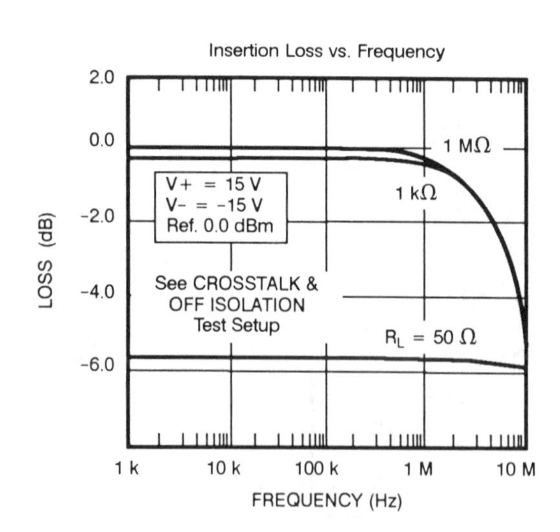

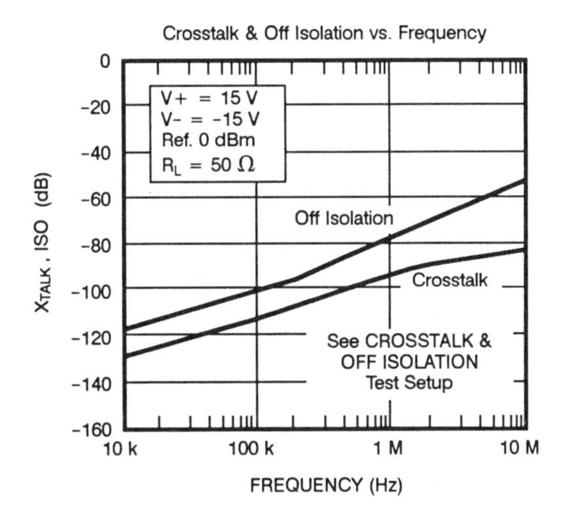

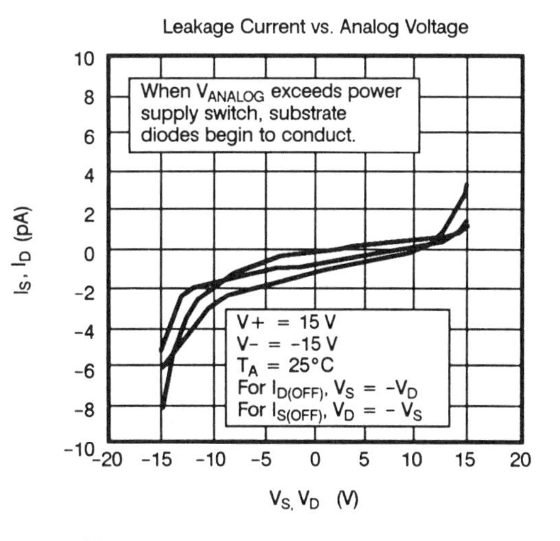

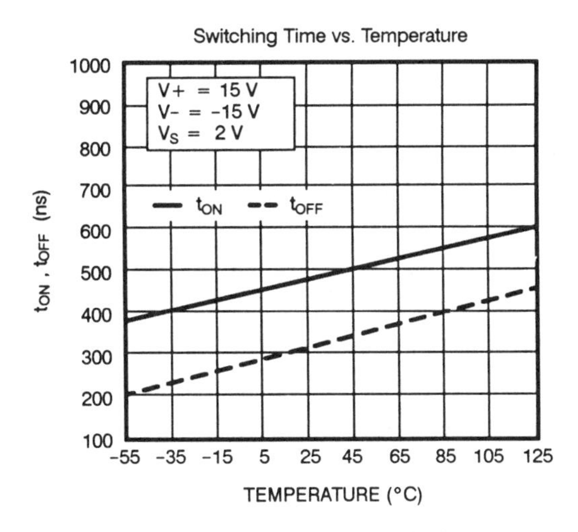

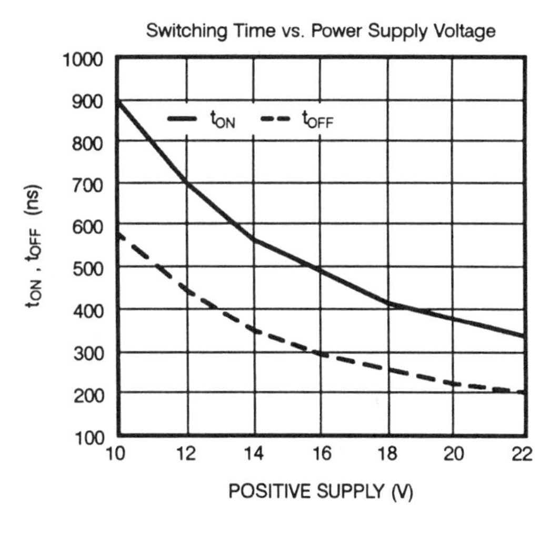

APPLICATIONS

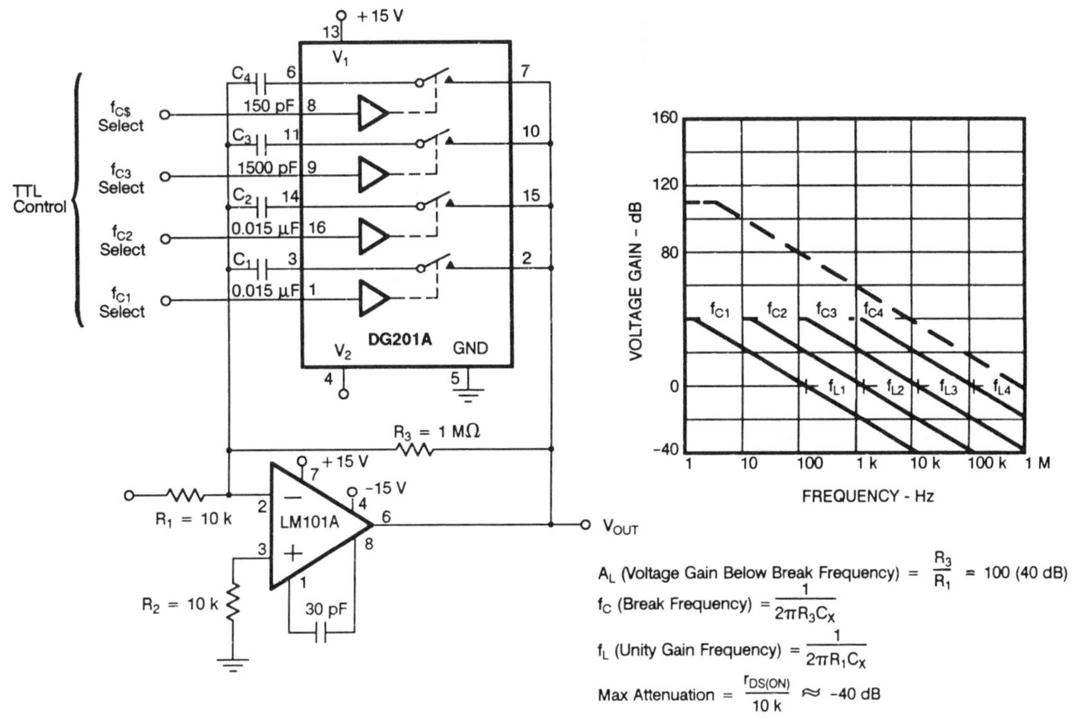

A_L (Voltage Gain Below Break Frequency) $= \frac{R_3}{R_1} = 100$ (40 dB)

f_C (Break Frequency) $= \frac{1}{2\pi R_3 C_X}$

f_L (Unity Gain Frequency) $= \frac{1}{2\pi R_1 C_X}$

Max Attenuation $= \frac{r_{DS(ON)}}{10\text{ k}} \approx -40$ dB

Active Low Pass Filter with Digitally Selected Break Frequency

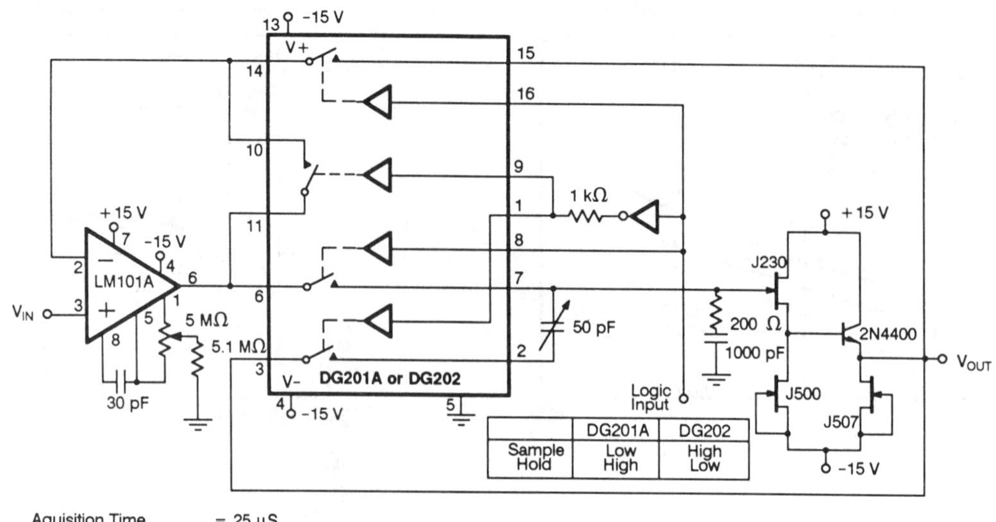

Sample-and-Hold

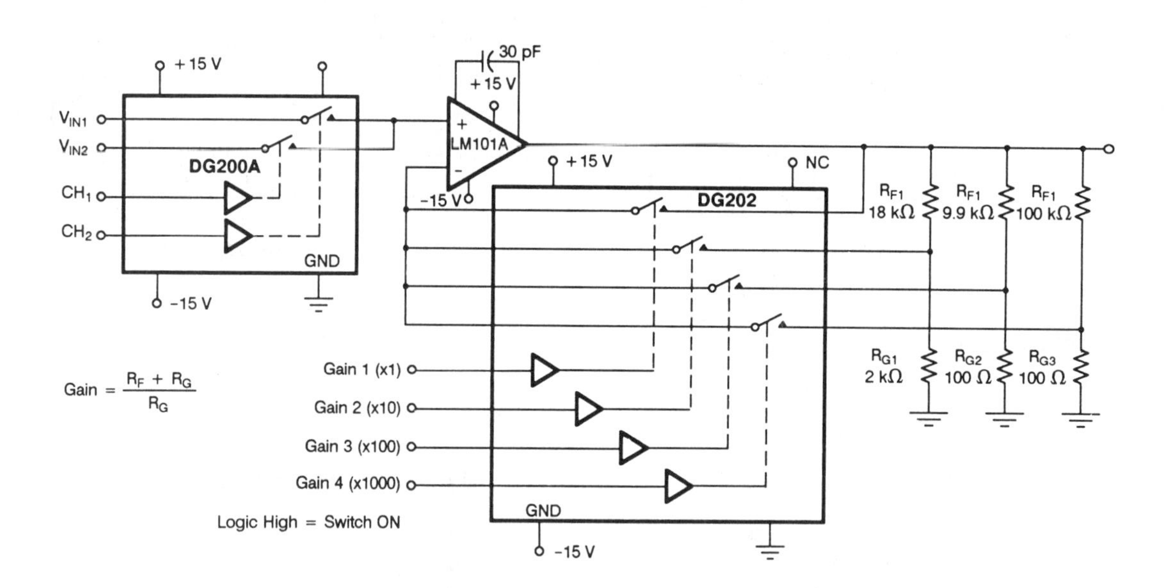

A Precision Amplifier with Digitally Programable Input and Gains

SCHEMATIC DIAGRAM (Typical channel)

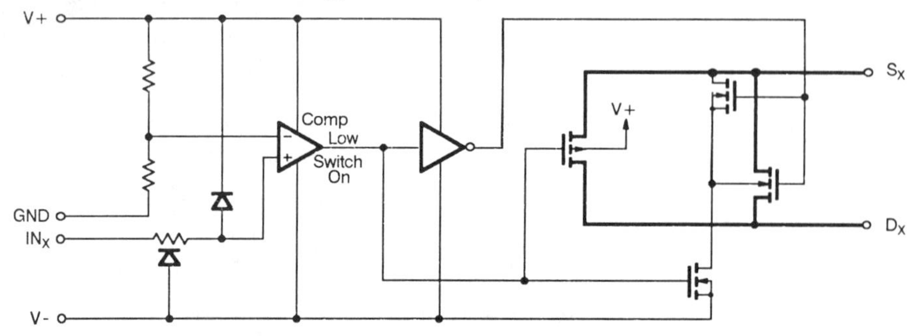

Siliconix DG211/212

Low-Cost 4-Channel Monolithic SPST CMOS Analog Switches

- ±15V analog signal range
- TTL compatibility
- Logic inputs accept negative voltages
- On-resistance <175Ω
- Wide signal range
- Simple logic interface
- Reduced power consumption

APPLICATIONS

- Disk drives
- Test equipment
- Communication systems

The DG211 and DG212 are low-cost quad single-pole single-throw analog switches for use in general-purpose switching applications in communication, instrumentation and process control. These devices differ only in that the digital control logic is inverted, as shown in the truth table. The use of both p- and n-channel devices minimizes on-resistance variations over the analog signal range.

Designed with Siliconix PLUS-40 CMOS process to combine low power dissipation with a high breakdown voltage rating of 40V, both switches will handle ±15V input signals with ease and have a continuous current rating of 20mA. An epitaxial layer prevents latchup.

Both devices feature true bidirectional performance (with no offset voltage) in the on condition, and will block signals to 30V peak-to-peak in the off condition. For new designs, silicon-gate DG444/445 upgrades are recommended.

Packaging for this series includes 16-pin plastic DIP and small outline options. Performance grades include both commercial C-suffix (0 to 70°C) and industrial D-suffix (−40 to 85°C) temperature ranges.

ABSOLUTE MAXIMUM RATINGS

Parameter	Rating
V+ to V-	44 V
V_{IN} to GND	V-, V+
V_L to GND	-0.3 V, 25 V
V_S or V_D to V+	0, -40 V
V_S or V_D to V-	0, 40 V
V+ to GND	25 V
V- to GND	-25 V
Current, Any Terminal Except S or D	30 mA
Continuous Current, S or D	20 mA
Peak Current, S or D (Pulsed at 1 ms, 10% duty cycle max)	70 mA
Storage Temperature	-65 to 125°C
Operating Temperature (C Suffix)	0 to 70°C
Operating Temperature (D Suffix)	40 to 85°C
Power Dissipation (Package)*	
16-Pin Plastic DIP**	470 mW
16-Pin Small Outline***	600 mW

*Device mounted with all leads soldered or welded to PC board.

**Derate 6.5 mW/°C above 25°C.

***Derate 7.6 mW/°C above 75°C.

FUNCTIONAL BLOCK DIAGRAM, PIN CONFIGURATION, AND TRUTH TABLES

Dual-In-Line Package

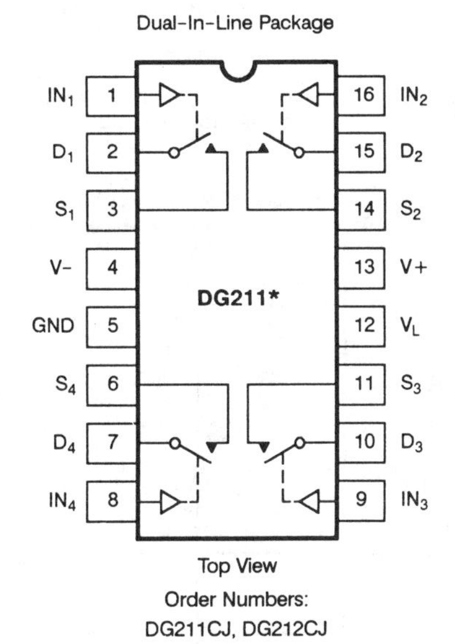

Top View

Order Numbers:
DG211CJ, DG212CJ

*Switches shown for Logic "1" input

SO Package

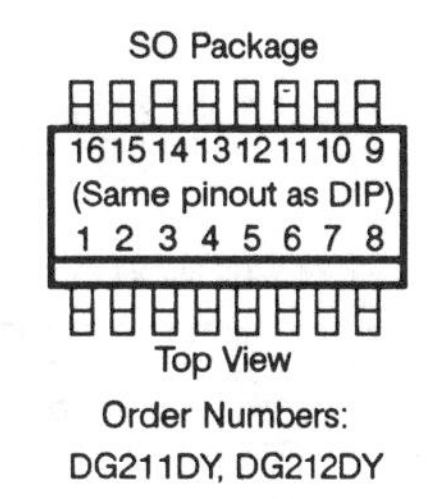

Top View

Order Numbers:
DG211DY, DG212DY

Truth Table

Logic	DG211	DG212
0	ON	OFF
1	OFF	ON

Logic "0" ≤ 0.8 V
Logic "1" ≥ 2.4 V

TYPICAL CHARACTERISTICS

The electrical characteristic table guarantees the DG211 and DG212 for operation at ±15 V, ±10%; however, functional operation occurs over the designed range of ±5 V to ±20 V power supplies. These characteristic graphs show the effect of device parameters over several parameter permutations including power supply variations. These graphs are for design aid only and are not subject to production testing.

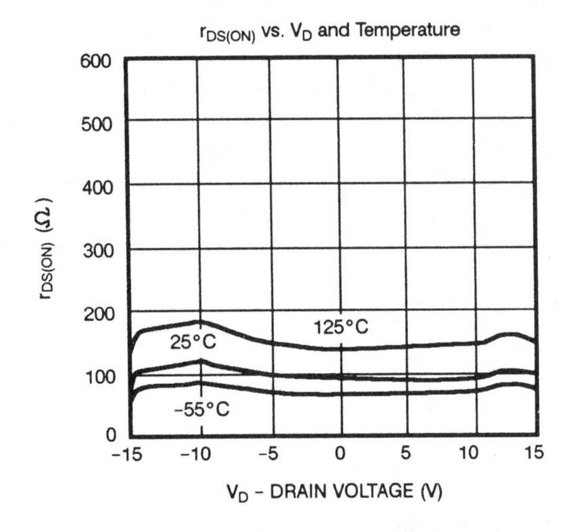

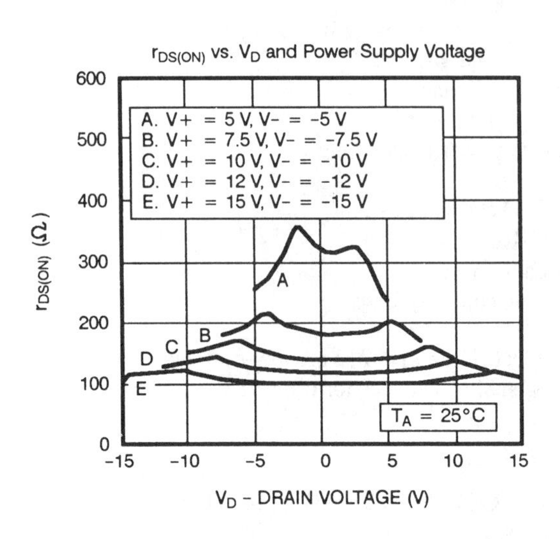

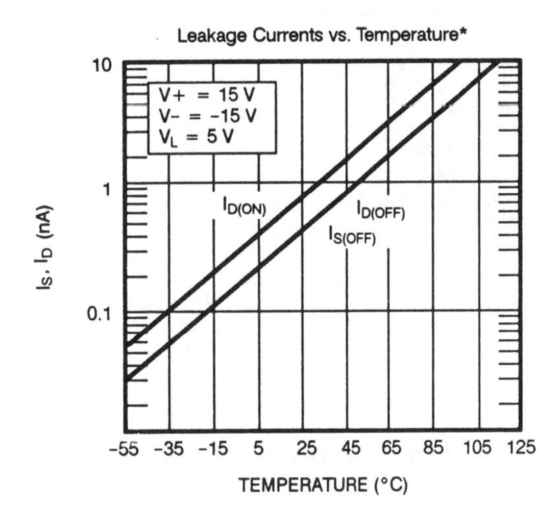

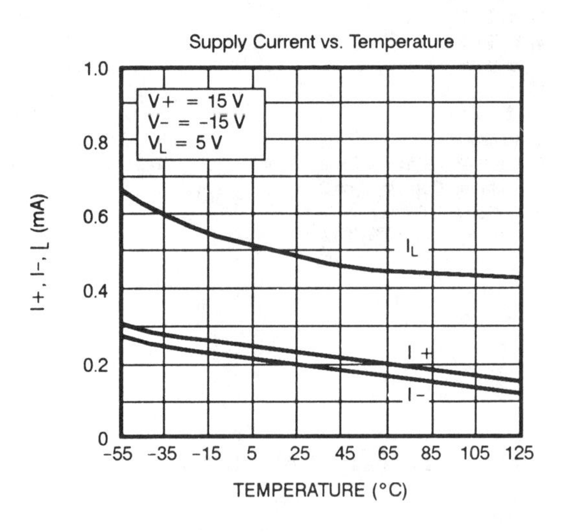

*The net leakage into the source or drain is the n-channel leakage minus the p-channel leakage. This difference can be positive, negative, or zero depending on the analog voltage and temperature, and will vary greatly from unit to unit.

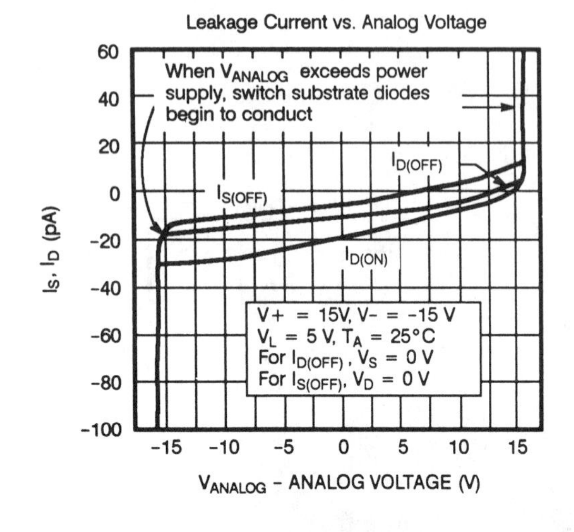

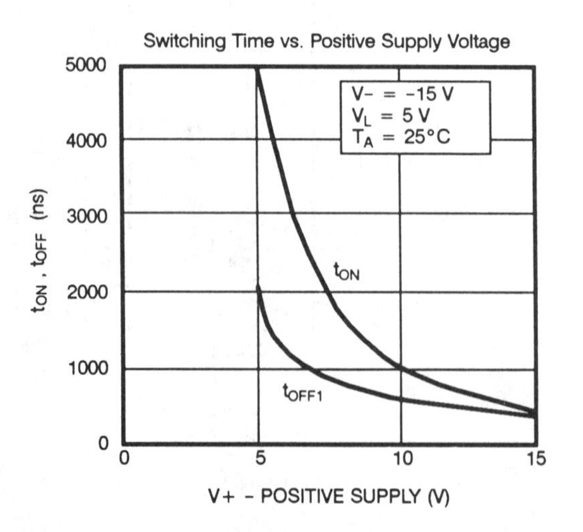

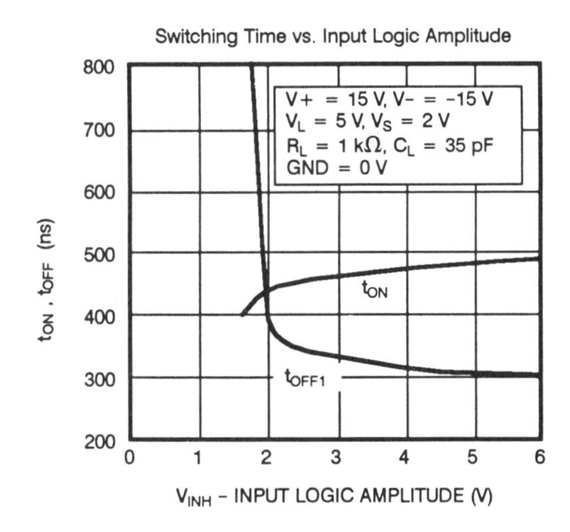

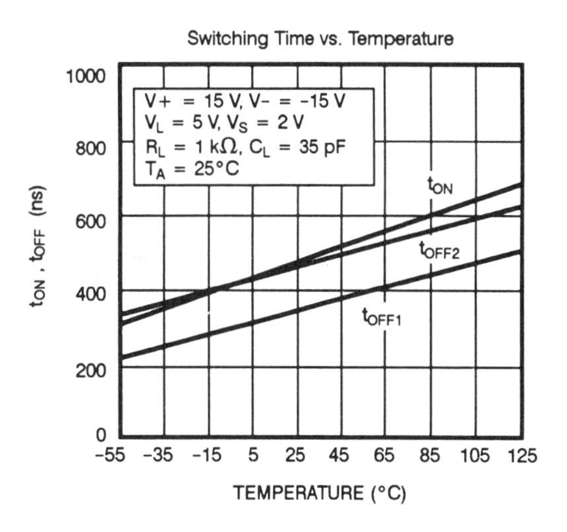

Typical delay, rise, fall settling times, and switching transients in this circuit.

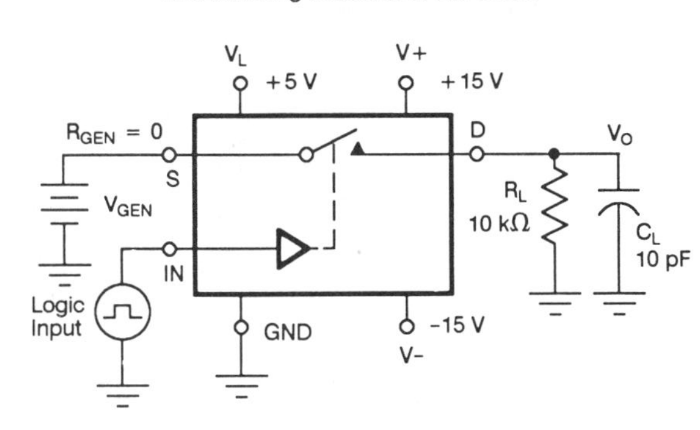

If V_{GEN}, R_L, or C_L is increased, there will be a proportional increase in rise and/or fall RC times.

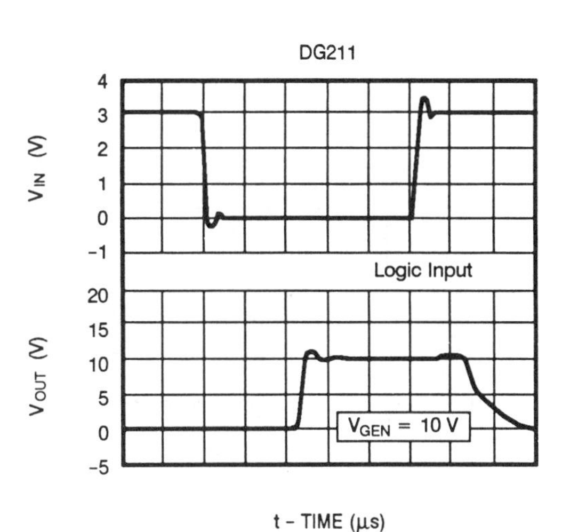

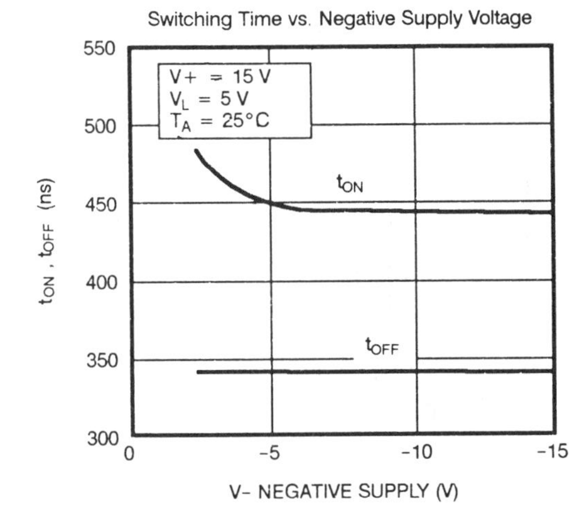

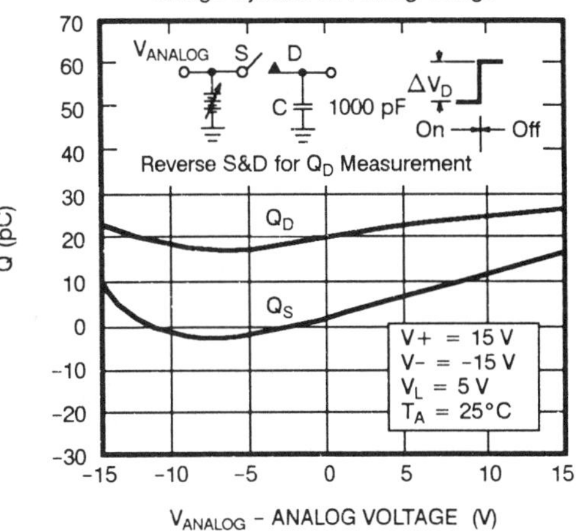

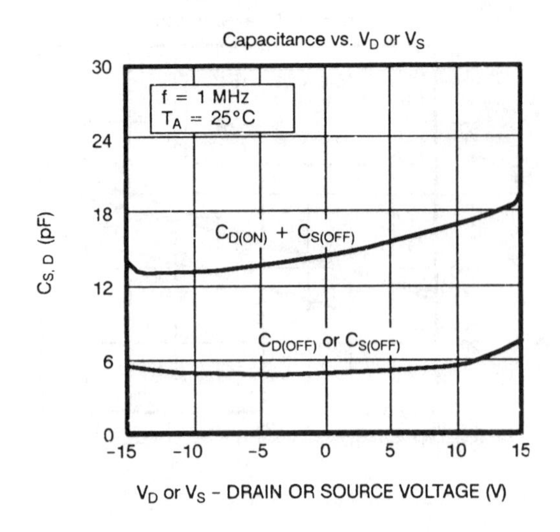

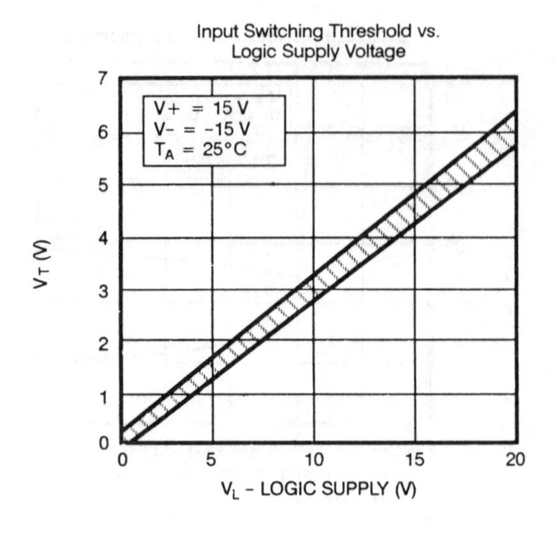

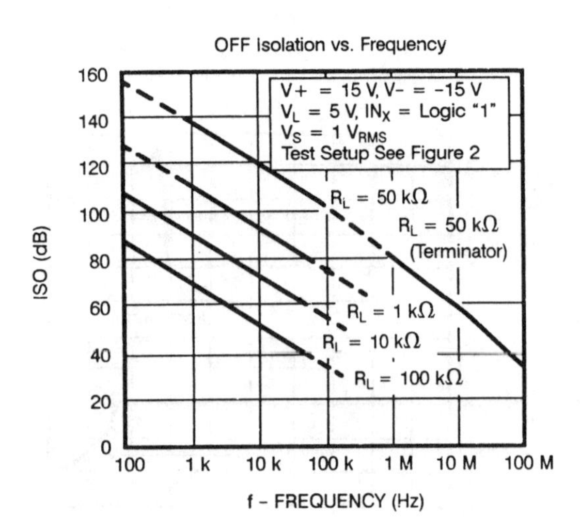

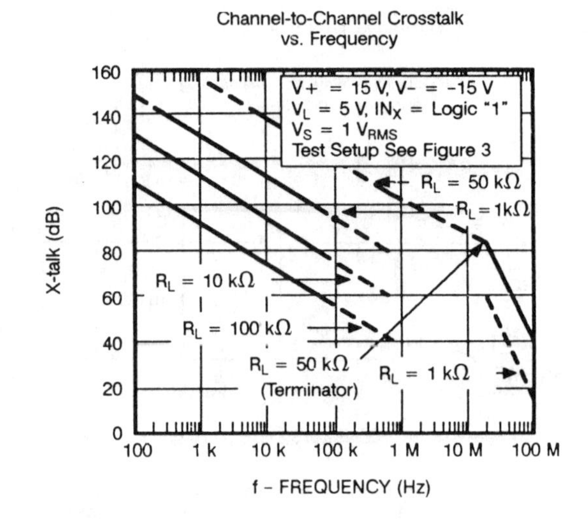

APPLICATIONS

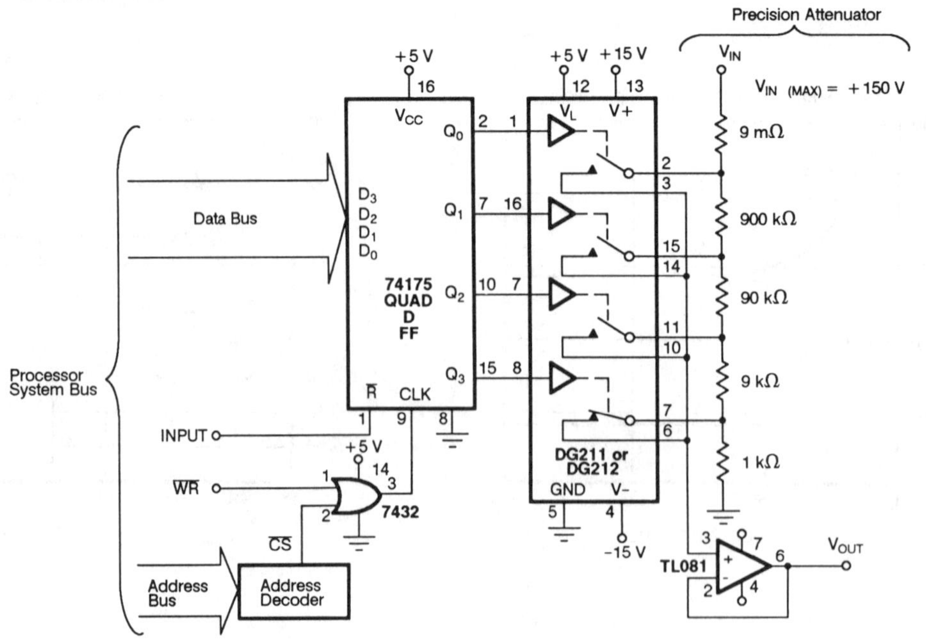

Microprocessor Controlled Analog Signal Attenuator

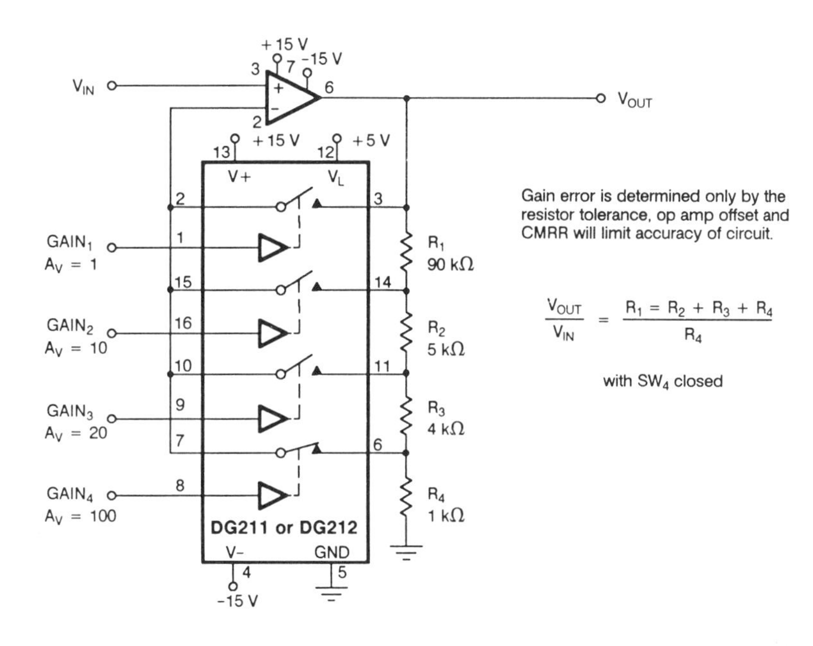

Precision-weighted Resistor Programmable-gain Amplifier

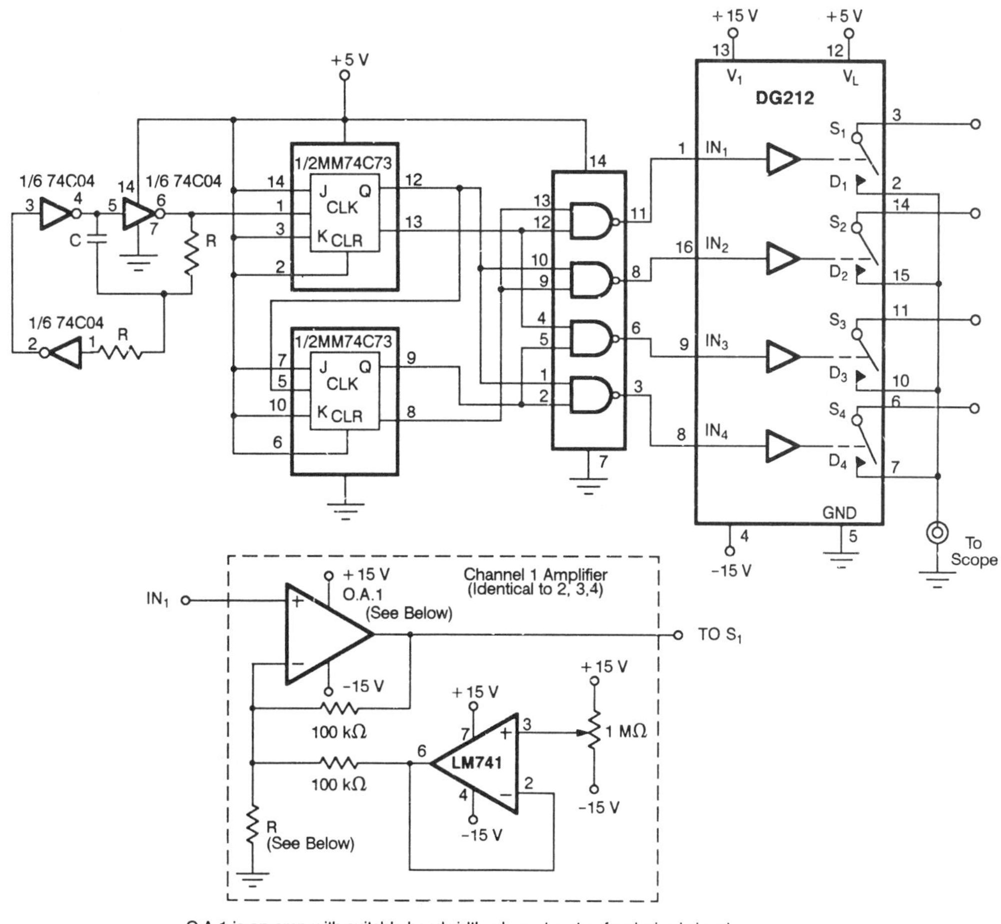

O.A.1 is op amp with suitable bandwidth, slew rate, etc., for desired signals.
R is added for extra gain according to formula: Voltage Gain $= 2 + \frac{100\ k}{R}$

The "Scope Extender" Which Displays 4-Channels Simultaneously on a Single Trace Scope

SCHEMATIC DIAGRAM (Typical Channel)

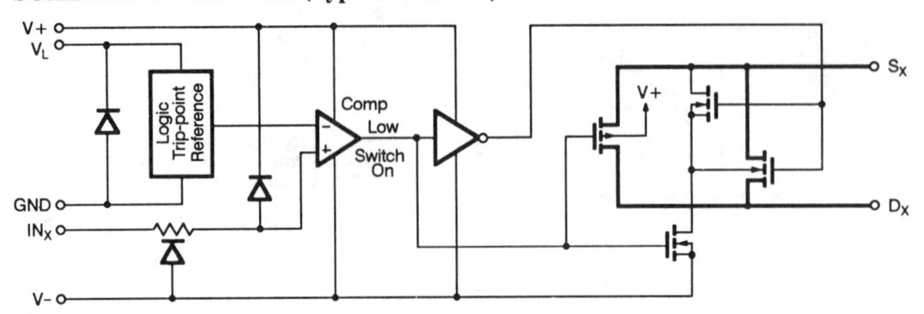

Siliconix DG406/407

16-Channel/Dual 8-Channel High-Performance CMOS Analog Multiplexers Preliminary Information

- Low $r_{DS(ON)}$ (100Ω max)
- Low charge injection (Q <20 pC typ.)
- Fast transition time (250 ns max)
- Low power (I_{SUPPLY} <75μA)
- Single-supply capability
- ESD protection > ±4000V
- Reduced switching errors
- Reduced glitching
- Improved data throughput
- Reduced power consumption
- Increased ruggedness

APPLICATIONS

- Data acquisition systems
- Audio signal routing and multiplexing/demultiplexing
- ATE systems
- Battery-operated systems
- High-rel systems
- Single-supply systems

The DG406 is a 16-channel single-ended analog multiplexer designed to connect 1 of 16 inputs to a common output, as determined by a 4-bit binary address. The DG407 is an 8-channel differential analog multiplexer designed to connect 1 of 8 differential inputs to a common dual output, as determined by its 3-bit binary address. Break-before-make switching action protects against momentary shorting of inputs.

An on channel conducts current equally well in both directions. In the off state, each channel blocks voltages up to the power supply rails. An enable (EN) function allows the user to reset the multiplexer/demultiplexer to all switches off for stacking several devices. All control inputs, address (A_X) and enable (EN) are TTL compatible over the full specified operating temperature range.

Applications for the DG406/407 include high-speed data acquisition, audio signal switching and routing. ATE systems, and avionics. High performance and low power dissipation make them ideal for battery-operated and remote-instrumentation applications.

Designed in the 44V silicon-gate CMOS process, the absolute maximum voltage rating is extended to 44V, allowing operation with ±20V supplies. Additionally single (12V) supply operation is allowed. An epitaxial layer prevents latchup. Both DG406 and DG407 are available in dual-in-line ceramic and plastic packages and are specified for operation over the military A-suffix (−55 to 125°C) and industrial D-suffix (−40 to 85°C) temperature ranges.

ABSOLUTE MAXIMUM RATINGS

Voltage Referenced to V−

Parameter	Rating
V+	44 V
GND	25 V
Digital Inputs [h], V_S, V_D	(V-) -2 V to (V+) +2 V or 20 mA, whichever occurs first.
Current (Any Terminal, Except S or D)	30 mA
Continuous Current, S or D	20 mA
Peak Current, S or D (Pulsed at 1 ms, 10% Duty Cycle Max)	40 mA

Operating Temperature	(A Suffix)	-55 to 125°C
	(D Suffix)	-40 to 85°C
Storage Temperature	(A Suffix)	-65 to 150°C
	(D Suffix)	-65 to 125°C

Power Dissipation (Package)*

28-Pin Ceramic DIP**	1200 mW
28-Pin Plastic DIP***	625 mW

*All leads soldered or welded to PC board.
**Derate 12 mW/°C above 75°C.
***Derate 6 mW/°C above 75°C.

FUNCTIONAL BLOCK DIAGRAM

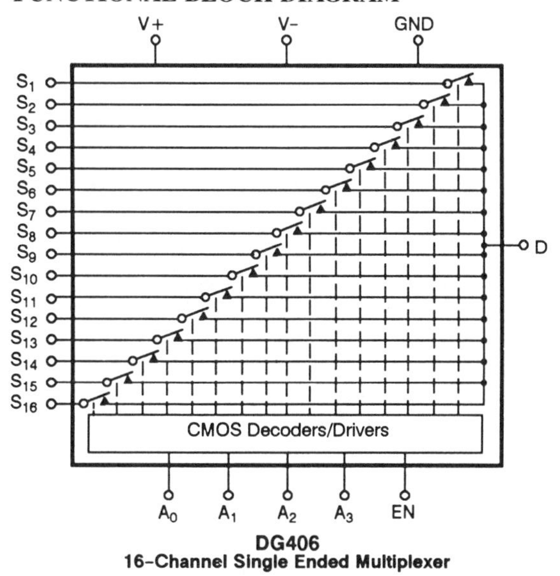

DG406
16-Channel Single Ended Multiplexer

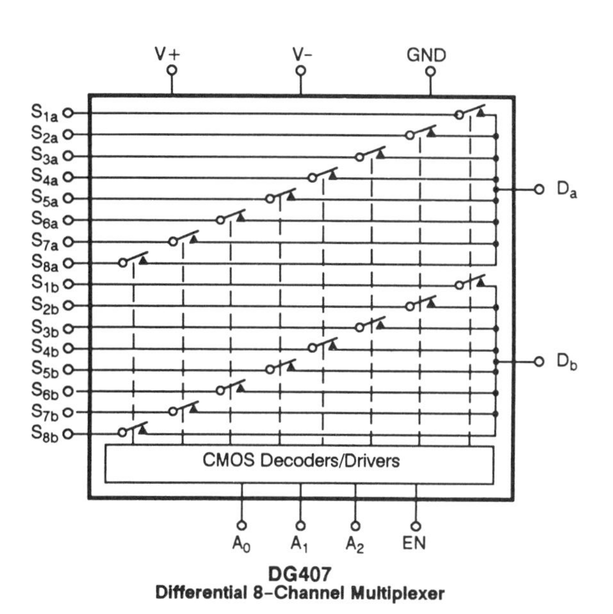

DG407
Differential 8-Channel Multiplexer

A_3	A_2	A_1	A_0	EN	ON Switch
X	X	X	X	0	None
0	0	0	0	1	1
0	0	0	1	1	2
0	0	1	0	1	3
0	0	1	1	1	4
0	1	0	0	1	5
0	1	0	1	1	6
0	1	1	0	1	7
0	1	1	1	1	8
1	0	0	0	1	9
1	0	0	1	1	10
1	0	1	0	1	11
1	0	1	1	1	12
1	1	0	0	1	13
1	1	0	1	1	14
1	1	1	0	1	15
1	1	1	1	1	16

A_2	A_1	A_0	EN	ON Switch
X	X	X	0	None
0	0	0	1	1
0	0	1	1	2
0	1	0	1	3
0	1	1	1	4
1	0	0	1	5
1	0	1	1	6
1	1	0	1	7
1	1	1	1	8

Logic "0" = $V_{AL} \leq 0.8$ V, Logic "1" = $V_{AH} \geq 2.4$ V

PIN CONFIGURATION

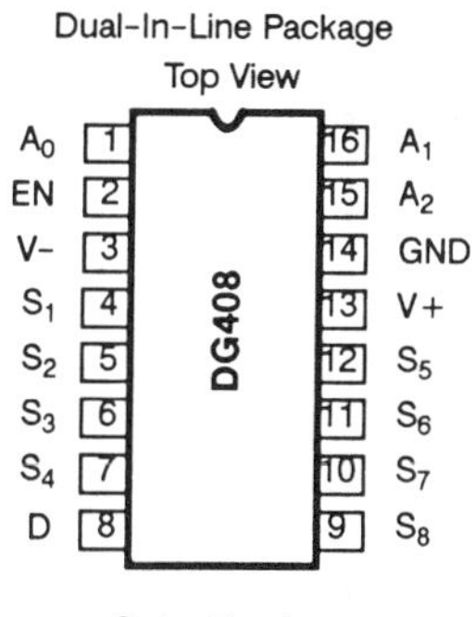

Order Numbers:
CerDIP: DG408AK
DG408AK/883
Plastic: DG408DJ

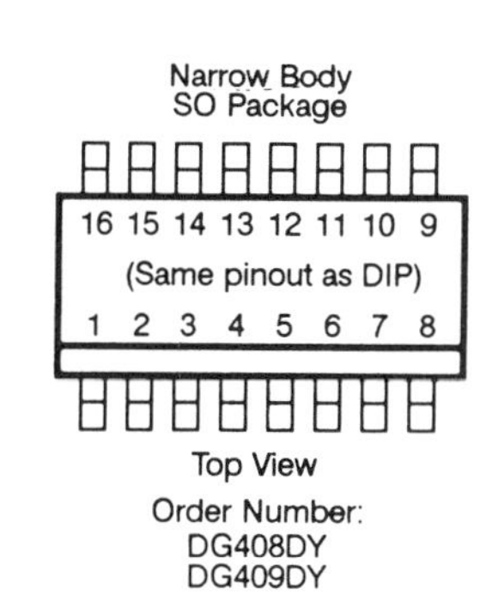

Order Number:
DG408DY
DG409DY

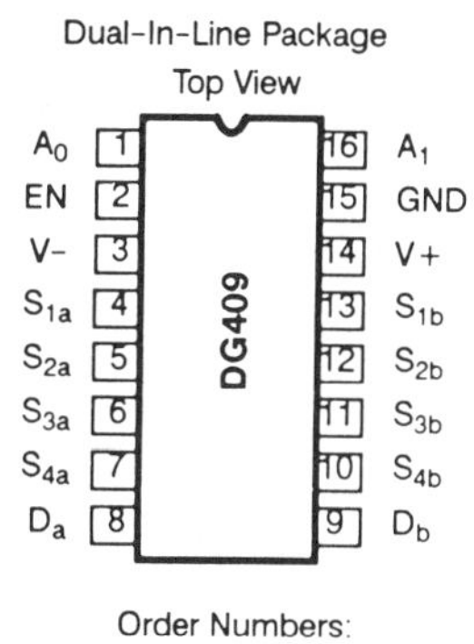

Order Numbers:
CerDIP: DG409AK
DG409AK/883
Plastic: DG409DJ

Siliconix DG458/459

8-Channel and Dual 4-Channel Fault Protected CMOS Analog Multiplexers

Preliminary Information

- Fault and overvoltage protection
- Fail safe with power loss (no latchup)
- Break-before-make switching
- TTL and CMOS compatible inputs
- All channels off when power off for signals up to ±35V
- Improved ruggedness
- Power-loss protected
- Prevents adjacent channel crosstalk
- Standard logic interface

APPLICATIONS

- Data acquisition systems
- Industrial process control
- Avionics test equipment
- High-rel control systems

The DG458 and DG458 are 8- and 4-channel multiplexers, respectively, incorporating fault protection. A series n-channel/p-channel/n-channel MOSFET structure provides device and signal-source protection in the event of power loss or overvoltages. Under fault conditions, the multiplexer input (or output) appears as an open circuit and only a few nanoamperes of leakage current will flow. This protects not only the multiplexer and the circuitry following it, but also protects the sensors or signal sources that drive the multiplexer.

The DG458 and DG459 multiplexers can withstand continuous overvoltage inputs up to ±35V. All digital inputs have TTL-compatible logic thresholds. Break-before-make operation prevents channel-to-channel interference.

The DG458 and DG459 are pin compatible with the industry-standard DG508A and DG509A multiplexers. The DG458/459 are offered in 16-pin plastic and CerDIP packages for operation over the extended industrial D-suffix (−40 to 85 °C) and military A-suffix (−55 to 125 °C) temperature ranges.

The Siliconix DG458/459 multiplexers are fully fault- and overvoltage-protected for continuous input voltages up to ±35V, whether or not voltage is applied to the power supply pins (V+, V−). These multiplexers are built on a high-voltage junction-isolated silicon-gate CMOS process. Two n-channel and one p-channel MOSFETs are connected in series to form each channel.

Within the normal analog signal range (±10V), the $r_{DS(ON)}$ variation as a function of analog signal voltage is comparable to that of the classic parallel n-mos and p-mos switches.

When the analog signal approaches or exceeds either supply rail, even for an on-channel, one of the three series MOSFETs gets cut-off, providing inherent protection against overvoltages, even if the multiplexer power supply voltages are lost. This protection is good up to the breakdown voltage of the respective series MOSFETs. Under fault conditions, only submicroamp leakage currents can flow in or out of the multiplexer. This not only provides protection for the multiplexer and succeeding circuitry, but it allows normal, undisturbed operation for all other channels. Additionally, in case of power loss to the multiplexer, the loading caused on the transducers and signal sources is insignificant, therefore redundant multiplexers can be used on critical applications, such as telemetry and avionics.

ABSOLUTE MAXIMUM RATINGS

Parameter	Rating
V+ to V-	44 V
V+ to GND	22 V
V- to GND	-25 V
V_{EN}, V_A Digital Input	(V-) -4 V to (V+) +4 V
V_S, Analog Input Overvoltage with Power ON	(V-) -20 V to (V+) +20 V
V_S, Analog Input Overvoltage with Power OFF	-35 V to +35 V
Continuous Current, S or D	20 mA
Peak Current, S or D (Pulsed at 1 ms, 10% Duty Cycle Max)	40 mA
Operating Temperature (A Suffix)	-55 to 125°C
Operating Temperature (D Suffix)	-40 to 85°C
Storage Temperature (A Suffix)	-65 to 150°C
Storage Temperature (D Suffix)	-65 to 125°C
Power Dissipation (Package)* 16-Pin Ceramic DIP**	1000 mW
16-Pin Plastic DIP***	600 mW

*All leads soldered or welded to PC board.
**Derate 12 mW/°C above 75°C.
***Derate 6.3 mW/°C above 25°C.

PIN CONFIGURATION

Dual-In-Line Package

Top View

Pin	DG458	Pin	DG458
1	A_0	16	A_1
2	EN	15	A_2
3	V-	14	GND
4	S_1	13	V+
5	S_2	12	S_5
6	S_3	11	S_6
7	S_4	10	S_7
8	D	9	S_8

Order Numbers:
CerDIP: DG458AK
Plastic: DG458DJ

Dual-In-Line Package

Top View

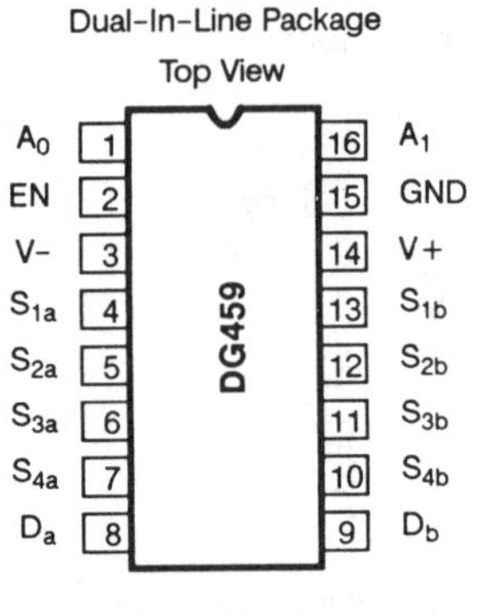

Order Numbers:
CerDIP: DG459AK
Plastic: DG459DJ

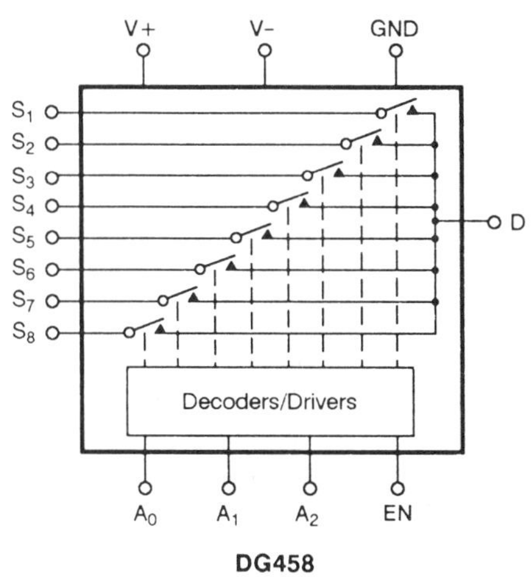

DG458
8-Channel Single Ended Multiplexer

DG459
Differential 4-Channel Multiplexer

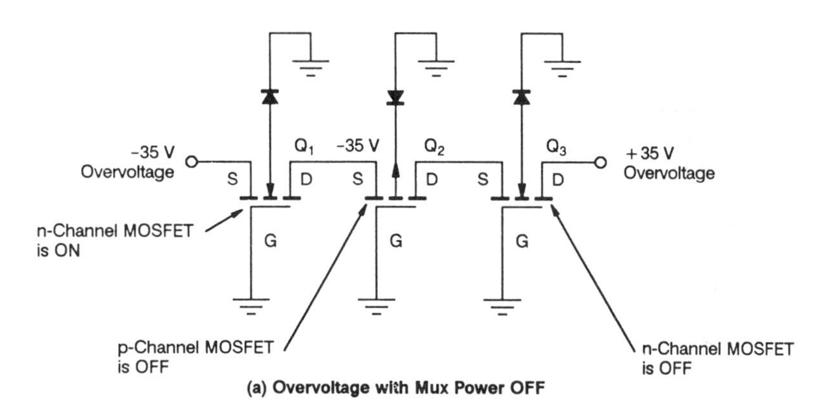

(a) Overvoltage with Mux Power OFF

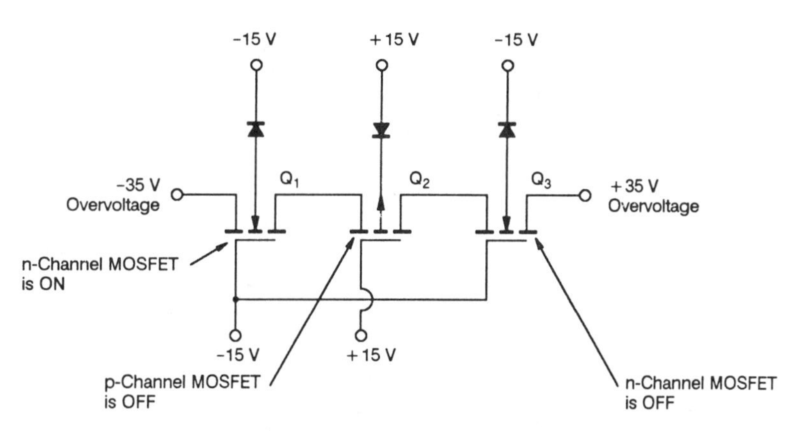

(b) Overvoltage with Mux Power ON

Overvoltage Protection

DG458

A_2	A_1	A_0	EN	On Switch
X	X	X	0	None
0	0	0	1	1
0	0	1	1	2
0	1	0	1	3
0	1	1	1	4
1	0	0	1	5
1	0	1	1	6
1	1	0	1	7
1	1	1	1	8

DG459

A_1	A_0	EN	On Switch
X	X	0	None
0	0	1	1
0	1	1	2
1	0	1	3
1	1	1	4

Logic "0" $V_{AL} \leq 0.8$ V, Logic "1" $V_{AH} \geq 2.4$ V

Sprague Series ULN-2000A High-Voltage, High-Current Darlington Array

DEVICE NUMBER DESIGNATION

$V_{CE(MAX)}$	50 V	50 V	95 V
$I_{C(MAX)}$	500 mA	600 mA	500 mA
Logic	**Type Number**		
General Purpose PMOS, CMOS	ULN-2001A	ULN-2011A	ULN-2021A
14-25 V PMOS	ULN-2002A	ULN-2012A	ULN-2022A
5 V TTL, CMOS	ULN-2003A	ULN-2013A	ULN-2023A
6-15 V CMOS, PMOS	ULN-2004A	ULN-2014A	ULN-2024A
High-Output TTL	ULN-2005A	ULN-2015A	ULN-2025A

ABSOLUTE MAXIMUM RATINGS

at +25°C Free-Air Temperature for any one Darlington pair (unless otherwise noted)

Output Voltage, V_{CE} (Series ULN-2000, 2010A) .... 50 V
(Series ULN-2020A) .... 95 V
Input Voltage, V_{IN} (Series ULN-2002, 2003, 2004A) .... 30 V
(Series ULN-2005A) .... 15 V
Continuous Collector Current, I_C (Series ULN-2000, 2020A) .... 500 mA
(Series ULN-2010A) .... 600 mA
Continuous Input Current, I_{IN} .... 25 mA
Power Dissipation, P_D (one Darlington pair) .... 1.0 W
(total package) .... 2.0 W*
Operating Ambient Temperature Range, T_A .... −20°C to +85°C
Storage Temperature Range, T_S .... −55°C to +150°C

*Derate at the rate of 16.67 mW/°C above +25°C.
Under normal operating conditions, these devices will sustain 350 mA per output with $V_{CE(SAT)}$ = 1.6 V at +70°C with a pulse width of 20 ms and a duty cycle of 34%.

SERIES ULN-2000A

ELECTRICAL CHARACTERISTICS AT +25°C (unless otherwise noted)

Characteristic	Symbol	Test Fig.	Applicable Devices	Test Conditions	Limits			
					Min.	Typ.	Max.	Units
Output Leakage Current	I_{CEX}	1A	All	V_{CE} = 50 V, T_A = 25°C	—	—	50	μA
				V_{CE} = 50 V, T_A = 70°C	—	—	100	μA
		1B	ULN-2002A	V_{CE} = 50 V, T_A = 70°C, V_{IN} = 6.0 V	—	—	500	μA
			ULN-2004A	V_{CE} = 50 V, T_A = 70°C, V_{IN} = 1.0 V	—	—	500	μA
Collector-Emitter Saturation Voltage	$V_{CE(SAT)}$	2	All	I_C = 100 mA, I_B = 250 μA	—	0.9	1.1	V
				I_C = 200 mA, I_B = 350 μA	—	1.1	1.3	V
				I_C = 350 mA, I_B = 500 μA	—	1.3	1.6	V
Input Current	$I_{IN(ON)}$	3	ULN-2002A	V_{IN} = 17 V	—	0.82	1.25	mA
			ULN-2003A	V_{IN} = 3.85 V	—	0.93	1.35	mA
			ULN-2004A	V_{IN} = 5.0 V	—	0.35	0.5	mA
				V_{IN} = 12 V	—	1.0	1.45	mA
			ULN-2005A	V_{IN} = 3.0 V	—	1.5	2.4	mA
	$I_{IN(OFF)}$	4	All	I_C = 500 μA, T_A = 70°C	50	65	—	μA
Input Voltage	$V_{IN(ON)}$	5	ULN-2002A	V_{CE} = 2.0 V, I_C = 300 mA	—	—	13	V
			ULN-2003A	V_{CE} = 2.0 V, I_C = 200 mA	—	—	2.4	V
				V_{CE} = 2.0 V, I_C = 250 mA	—	—	2.7	V
				V_{CE} = 2.0 V, I_C = 300 mA	—	—	3.0	V
			ULN-2004A	V_{CE} = 2.0 V, I_C = 125 mA	—	—	5.0	V
				V_{CE} = 2.0 V, I_C = 200 mA	—	—	6.0	V
				V_{CE} = 2.0 V, I_C = 275 mA	—	—	7.0	V
				V_{CE} = 2.0 V, I_C = 350 mA	—	—	8.0	V
			ULN-2005A	V_{CE} = 2.0 V, I_C = 350 mA	—	—	2.4	V
D-C Forward Current Transfer Ratio	h_{FE}	2	ULN-2001A	V_{CE} = 2.0 V, I_C = 350 mA	1000	—	—	
Input Capacitance	C_{IN}	—	All		—	15	25	pF
Turn-On Delay	t_{PLH}	—	All	0.5 E_{in} to 0.5 E_{out}	—	0.25	1.0	μs
Turn-Off Delay	t_{PHL}	—	All	0.5 E_{in} to 0.5 E_{out}	—	0.25	1.0	μs
Clamp Diode Leakage Current	I_R	6	All	V_R = 50 V, T_A = 25°C	—	—	50	μA
				V_R = 50 V, T_A = 70°C	—	—	100	μA
Clamp Diode Forward Voltage	V_F	7	All	I_F = 350 mA	—	1.7	2.0	V

SERIES ULN-2010A

ELECTRICAL CHARACTERISTICS AT +25°C (unless otherwise noted)

Characteristic	Symbol	Test Fig.	Applicable Devices	Test Conditions	Limits Min.	Limits Typ.	Limits Max.	Units
Output Leakage Current	I_{CEX}	1A	All	V_{CE} = 50 V, T_A = 25°C	—	—	50	μA
				V_{CE} = 50 V, T_A = 70°C	—	—	100	μA
		1B	ULN-2012A	V_{CE} = 50 V, T_A = 70°C, V_{IN} = 6.0 V	—	—	500	μA
			ULN-2014A	V_{CE} = 50 V, T_A = 70°C, V_{IN} = 1.0 V	—	—	500	μA
Collector-Emitter Saturation Voltage	$V_{CE(SAT)}$	2	All	I_C = 200 mA, I_B = 350 μA	—	1.1	1.3	V
				I_C = 350 mA, I_B = 500 μA	—	1.3	1.6	V
				I_C = 500 mA, I_B = 600 μA	—	1.7	1.9	V
Input Current	$I_{IN(ON)}$	3	ULN-2012A	V_{IN} = 17 V	—	0.82	1.25	mA
			ULN-2013A	V_{IN} = 3.85 V	—	0.93	1.35	mA
			ULN-2014A	V_{IN} = 5.0 V	—	0.35	0.5	mA
				V_{IN} = 12 V	—	1.0	1.45	mA
			ULN-2015A	V_{IN} = 3.0 V	—	1.5	2.4	mA
	$I_{IN(OFF)}$	4	All	I_C = 500 μA, T_A = 70°C	50	65	—	μA
Input Voltage	$V_{IN(ON)}$	5	ULN-2012A	V_{CE} = 2.0 V, I_C = 500 mA	—	—	17	V
			ULN-2013A	V_{CE} = 2.0 V, I_C = 250 mA	—	—	2.7	V
				V_{CE} = 2.0 V, I_C = 300 mA	—	—	3.0	V
				V_{CE} = 2.0 V, I_C = 500 mA	—	—	3.5	V
			ULN-2014A	V_{CE} = 2.0 V, I_C = 275 mA	—	—	7.0	V
				V_{CE} = 2.0 V, I_C = 350 mA	—	—	8.0	V
				V_{CE} = 2.0 V, I_C = 500 mA	—	—	9.5	V
			ULN-2015A	V_{CE} = 2.0 V, I_C = 500 mA	—	—	2.6	V
D-C Forward Current Transfer Ratio	h_{FE}	2	ULN-2011A	V_{CE} = 2.0 V, I_C = 350 mA	1000	—	—	
				V_{CE} = 2.0 V, I_C = 500 mA	900	—	—	
Input Capacitance	C_{IN}	—	All		—	15	25	pF
Turn-On Delay	t_{PLH}	—	All	0.5 E_{in} to 0.5 E_{out}	—	0.25	1.0	μs
Turn-Off Delay	t_{PHL}	—	All	0.5 E_{in} to 0.5 E_{out}	—	0.25	1.0	μs
Clamp Diode Leakage Current	I_R	6	All	V_R = 50 V, T_A = 25°C	—	—	50	μA
				V_R = 50 V, T_A = 70°C	—	—	100	μA
Clamp Diode Forward Voltage	V_F	7	All	I_F = 350 mA	—	1.7	2.0	V
				I_F = 500 mA	—	2.1	2.5	V

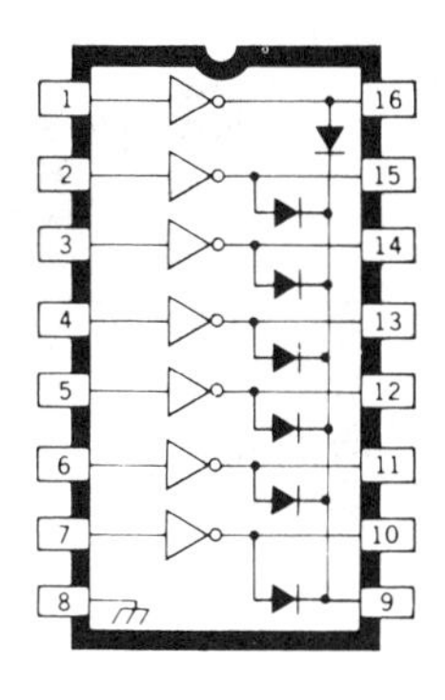

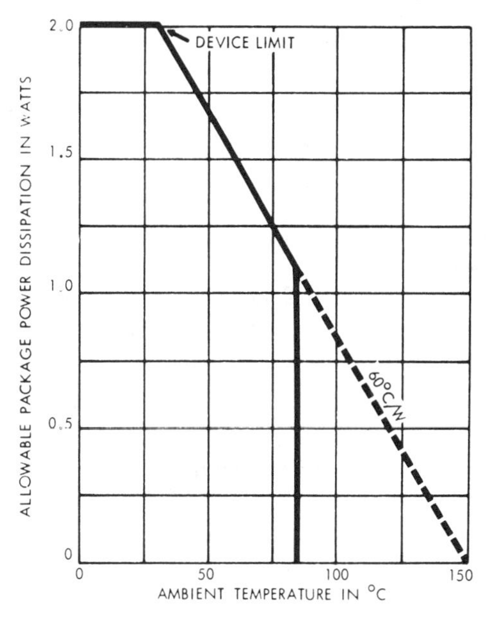

SERIES ULN-2020A

ELECTRICAL CHARACTERISTICS AT +25°C (unless otherwise noted)

Characteristic	Symbol	Test Fig.	Applicable Devices	Test Conditions	Min.	Typ.	Max.	Units
Output Leakage Current	I_{CEX}	1A	All	V_{CE} = 95 V, T_A = 25°C	—	—	50	μA
				V_{CE} = 95 V, T_A = 70°C	—	—	100	μA
		1B	ULN-2022A	V_{CE} = 95 V, T_A = 70°C, V_{IN} = 6.0 V	—	—	500	μA
			ULN-2024A	V_{CE} = 95 V, T_A = 70°C, V_{IN} = 1.0 V	—	—	500	μA
Collector-Emitter Saturation Voltage	$V_{CE(SAT)}$	2	All	I_C = 100 mA, I_B = 250 μA	—	0.9	1.1	V
				I_C = 200 mA, I_B = 350 μA	—	1.1	1.3	V
				I_C = 350 mA, I_B = 500 μA	—	1.3	1.6	V
Input Current	$I_{IN(ON)}$	3	ULN-2022A	V_{IN} = 17 V	—	0.82	1.25	mA
			ULN-2023A	V_{IN} = 3.85 V	—	0.93	1.35	mA
			ULN-2024A	V_{IN} = 5.0 V	—	0.35	0.5	mA
				V_{IN} = 12 V	—	1.0	1.45	mA
			ULN-2025A	V_{IN} = 3.0 V	—	1.5	2.4	mA
	$I_{IN(OFF)}$	4	All	I_C = 500 μA, T_A = 70°C	50	65	—	μA
Input Voltage	$V_{IN(ON)}$	5	ULN-2022A	V_{CE} = 2.0 V, I_C = 300 mA	—	—	13	V
			ULN-2023A	V_{CE} = 2.0 V, I_C = 200 mA	—	—	2.4	V
				V_{CE} = 2.0 V, I_C = 250 mA	—	—	2.7	V
				V_{CE} = 2.0 V, I_C = 300 mA	—	—	3.0	V
			ULN-2024A	V_{CE} = 2.0 V, I_C = 125 mA	—	—	5.0	V
				V_{CE} = 2.0 V, I_C = 200 mA	—	—	6.0	V
				V_{CE} = 2.0 V, I_C = 275 mA	—	—	7.0	V
				V_{CE} = 2.0 V, I_C = 350 mA	—	—	8.0	V
			ULN-2025A	V_{CE} = 2.0 V, I_C = 350 mA	—	—	2.4	V
D-C Forward Current Transfer Ratio	h_{FE}	2	ULN-2021A	V_{CE} = 2.0 V, I_C = 350 mA	1000	—	—	
Input Capacitance	C_{IN}	—	All		—	15	25	pF
Turn-On Delay	t_{PLH}	—	All	0.5 E_{in} to 0.5 E_{out}	—	0.25	1.0	μs
Turn-Off Delay	t_{PHL}	—	All	0.5 E_{in} to 0.5 E_{out}	—	0.25	1.0	μs
Clamp Diode Leakage Current	I_R	6	All	V_R = 95 V, T_A = 25°C	—	—	50	μA
				V_R = 95 V, T_A = 70°C	—	—	100	μA
Clamp Diode Forward Voltage	V_F	7	All	I_F = 350 mA	—	1.7	2.0	V

PARTIAL SCHEMATICS

Series ULN-2001A (each driver)

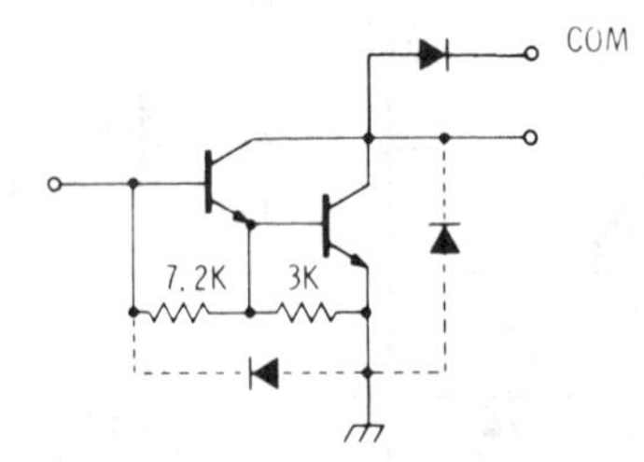

Series ULN-2002A (each driver)

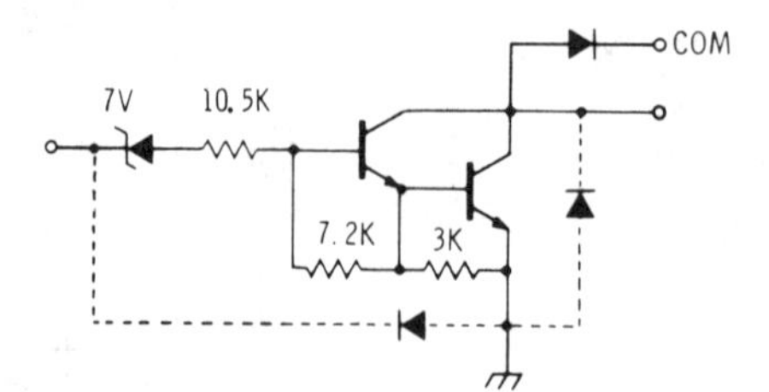

Series ULN-2003A (each driver)

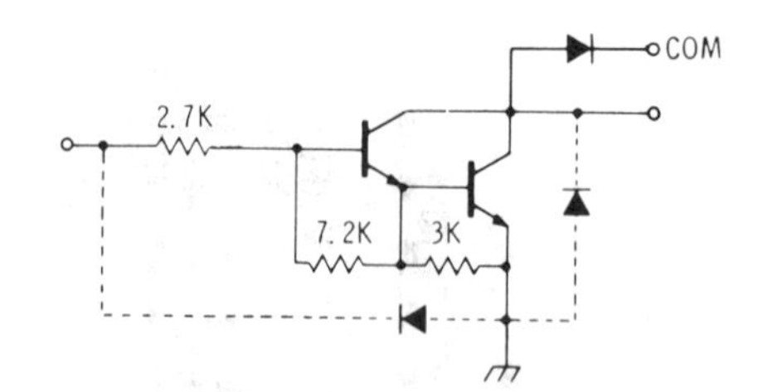

Series ULN-2004A (each driver)

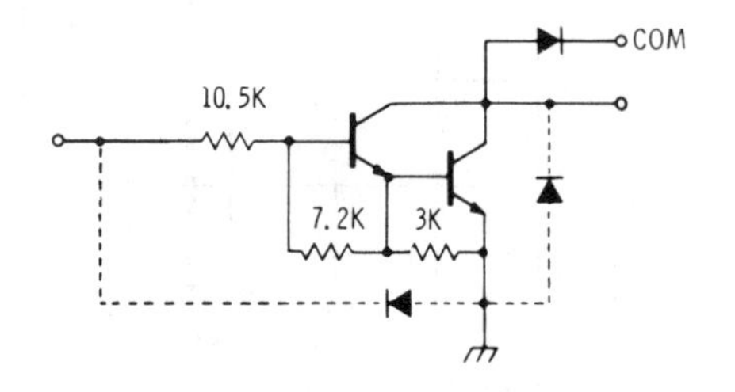

Series ULN-2005A (each driver)

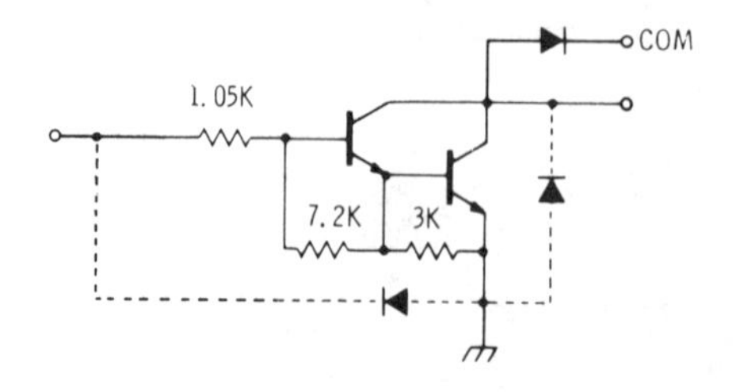

TEST FIGURES

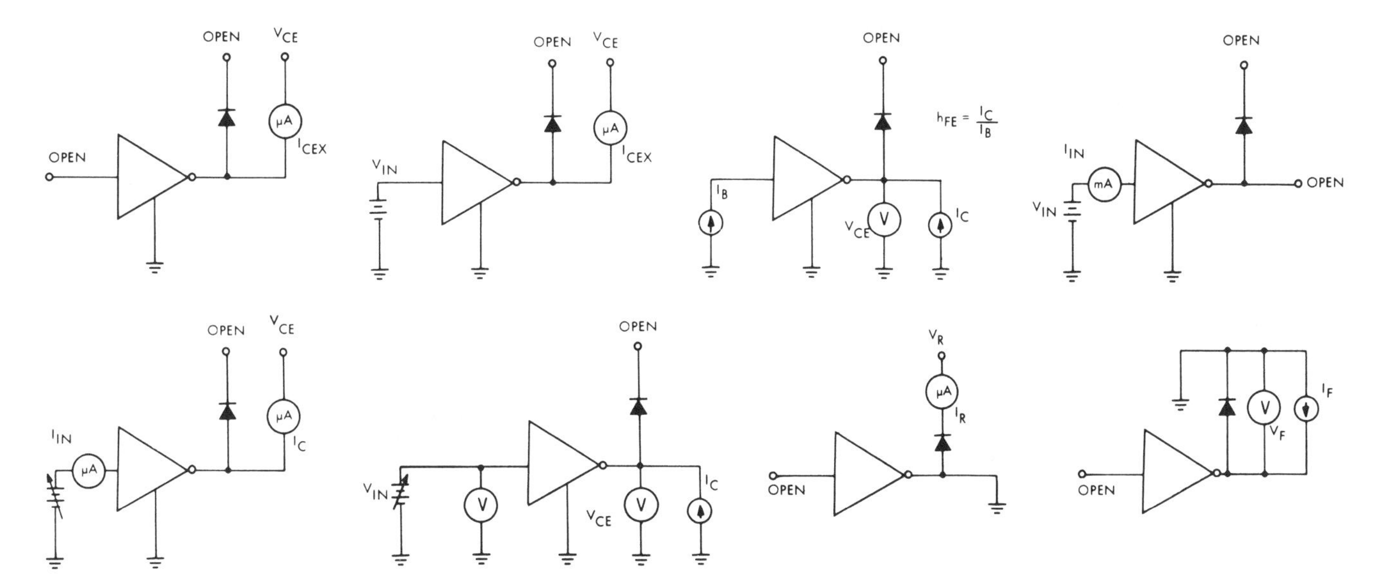

PEAK COLLECTOR CURRENT AS A FUNCTION OF DUTY CYCLE

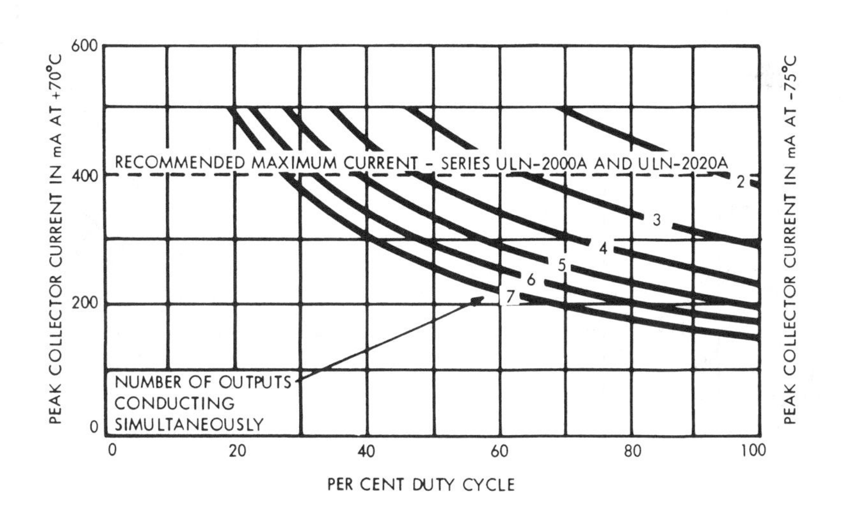

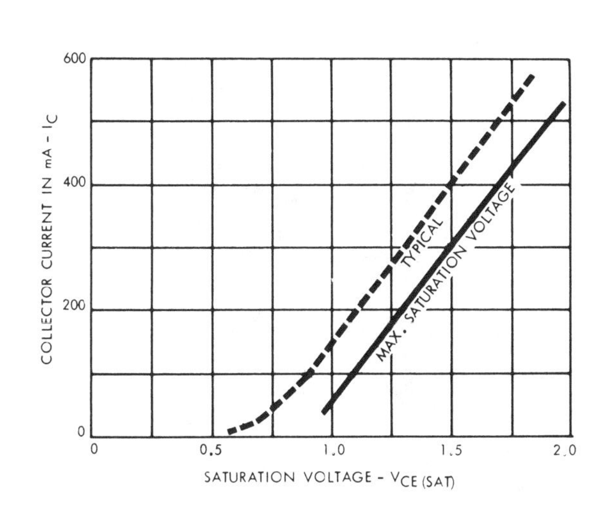

COLLECTOR CURRENT AS A FUNCTION OF INPUT CURRENT

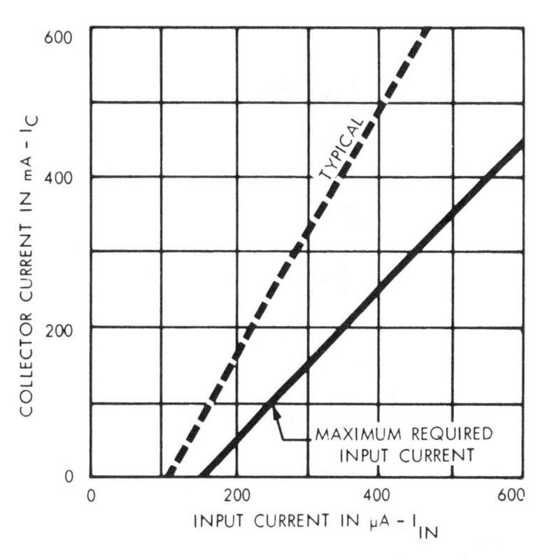

INPUT CURRENT AS A FUNCTION OF INPUT VOLTAGE

SERIES ULN-2002A

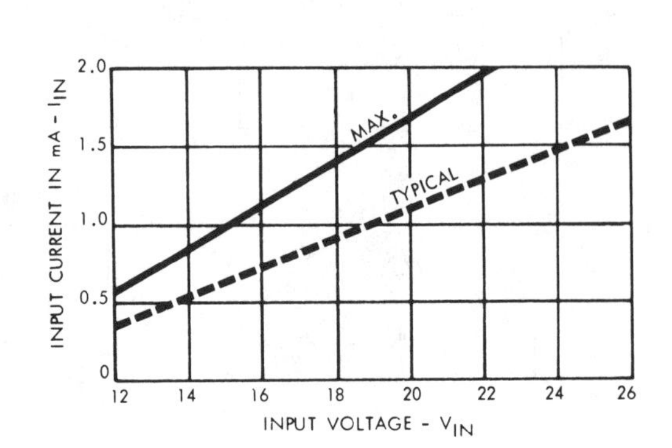

SERIES ULN-2003A

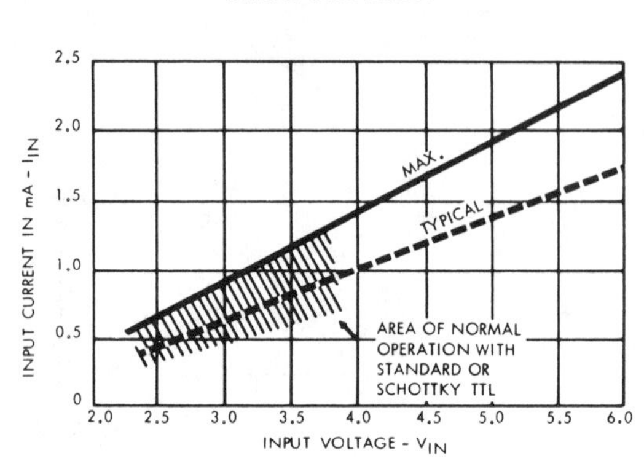

SERIES ULN-2004A

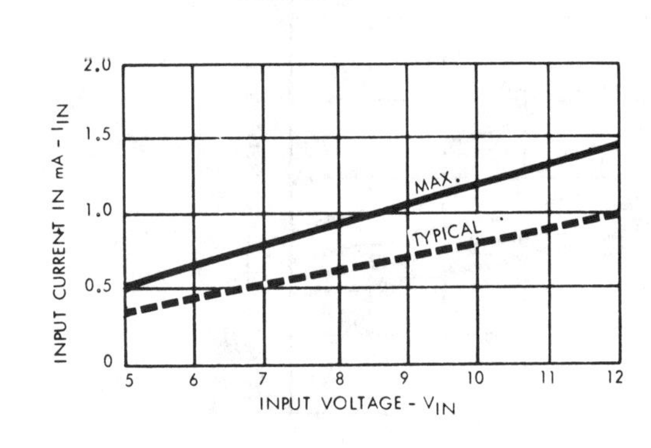

SERIES ULN-2005A

TYPICAL APPLICATIONS

PMOS TO LOAD

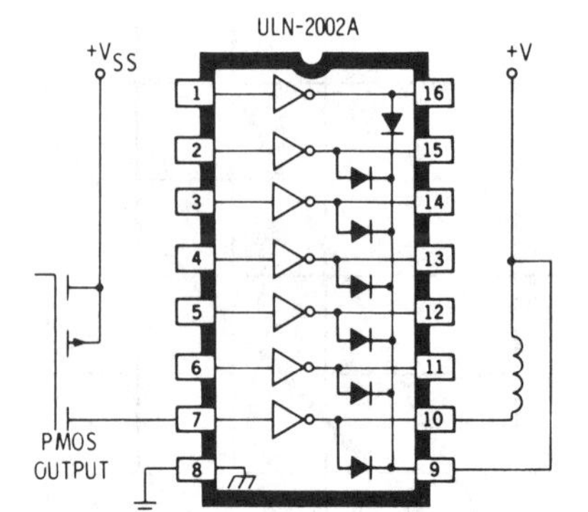

TTL TO LOAD

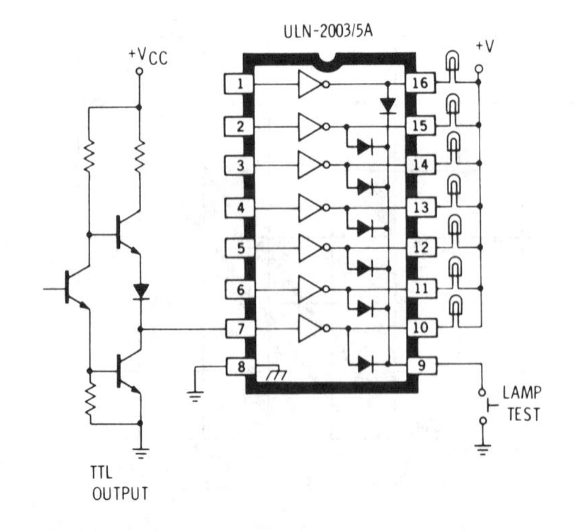

BUFFER FOR HIGH-CURRENT LOAD

USE OF PULL-UP RESISTORS TO INCREASE DRIVE CURRENT

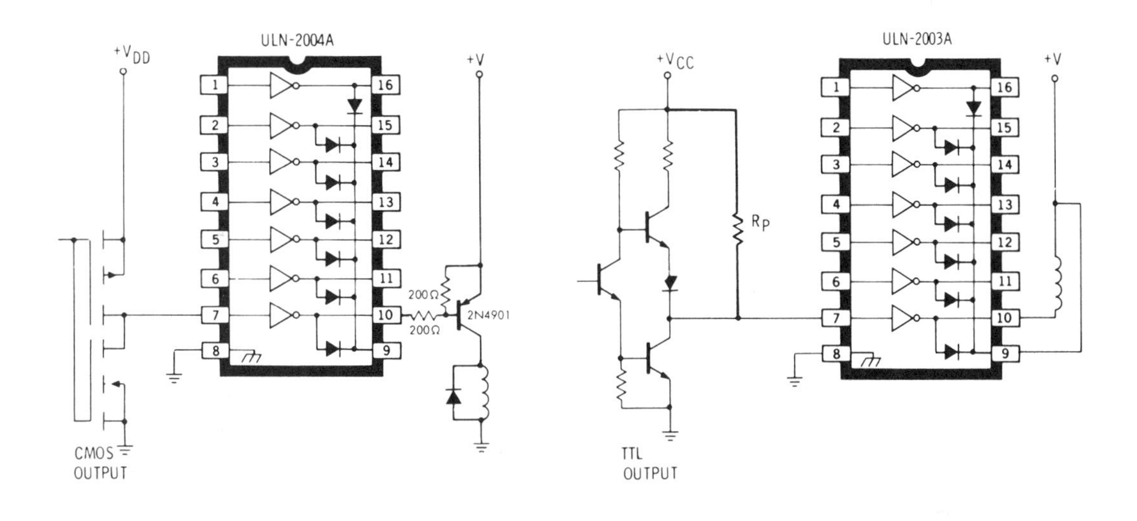

Sprague
Series ULN-2000L
High-Voltage, High-Current Darlington Arrays

ELECTRICAL CHARACTERISTICS at +25°C (unless otherwise noted)

Characteristic	Symbol	Applicable Devices	Test Conditions	Limits Min.	Typ.	Max.	Units
Output Leakage Current	I_{CEX}	All	V_{CE} = 50 V, T_A = 25°C	—	—	50	μA
			V_{CE} = 50 V, T_A = 70°C	—	—	100	μA
		ULN-2002L	V_{CE} = 50 V, T_A = 70°C, V_{IN} = 6.0 V	—	—	500	μA
		ULN-2004L	V_{CE} = 50 V, T_A = 70°C, V_{IN} = 1.0 V	—	—	500	μA
Collector-Emitter Saturation Voltage	$V_{CE(SAT)}$	All	I_C = 100 mA, I_B = 250 μA	—	0.9	1.1	V
			I_C = 200 mA, I_B = 350μA	—	1.1	1.3	V
			I_C = 350 mA, I_B = 5 μA	—	1.3	1.6	V
Input Current	$I_{IN(ON)}$	ULN-2002L	V_{IN} = 17 V	—	0.82	1.25	mA
		ULN-2003L	V_{IN} = 3.85 V	—	0.93	1.35	mA
		ULN-2004L	V_{IN} = 5.0 V	—	0.35	0.5	mA
			V_{IN} = 12 V	—	1.0	1.45	mA
		ULN-2005L	V_{IN} = 3.0 V	—	1.5	2.4	mA
	$I_{IN(OFF)}$	All	I_C = 500 μA, T_A = 70°C	50	65	—	μA
Input Voltage	$V_{IN(ON)}$	ULN-2002L	V_{CE} = 2.0 V, I_C = 300 mA	—	—	13	V
		ULN-2003L	V_{CE} = 2.0 V, I_C = 200 mA	—	—	2.4	V
			V_{CE} = 2.0 V, I_C = 250 mA	—	—	2.7	V
			V_{CE} = 2.0 V, I_C = 300 mA	—	—	3.0	V
		ULN-2004L	V_{CE} = 2.0 V, I_C = 125 mA	—	—	5.0	V
		ULN-2004L	V_{CE} = 2.0 V, I_C = 200 mA	—	—	6.0	V
			V_{CE} = 2.0 V, I_C = 275 mA	—	—	7.0	V
			V_{CE} = 2.0 V, I_C = 350 mA	—	—	8.0	V
		ULN-2005L	V_{CE} = 2.0 V, I_C = 350 mA	—	—	2.4	V
DC Forward Current Transfer Ratio	h_{FE}	ULN-2001L	V_{CE} = 2.0 V, I_C = 350 mA	1000	—	—	—
Input Capacitance	C_{IN}	All		—	15	25	pF
Turn-On Delay	t_{PLH}	All	0.5 E_{IN} to 0.5 E_{OUT}	—	0.25	1.0	μs
Turn-Off Delay	t_{PHL}	All	0.5 E_{IN} to 0.5 E_{OUT}	—	0.25	1.0	μs
Clamp Diode Leakage Current	I_R	All	V_R = 50 V, T_A = 25°C	—	—	50	μA
			V_R = 50 V, T_A = 70°C	—	—	100	μA
Clamp Diode Forward Voltage	V_F	All	I_F = 350 mA	—	1.7	2.0	V

ABSOLUTE MAXIMUM RATINGS

at +25°C Free-Air Temperature for any one Darlington pair (unless otherwise noted)

Output Voltage, V_{CE} .... 50 V
Input Voltage,
V_{IN} (ULN-2002, 2003, 2004L) .... 30 V
(ULN-2005L) .... 15 V
Continuous Collector Current, I_C .... 500 mA
Continuous Input Current, I_{IN} .... 25 mA
Power Dissipation, P_D (total package) .... 0.96 W*
Operating Ambient Temperature Range, T_A .. −20°C to +85°C
Storage Temperature Range, T_S .... −55°C to +150°C

*Derate at rate of 7.7 mW/°C above = 25°C.

Device Number Designation

$V_{CE(MAX)}$ $I_{C(MAX)}$	50 V 500 mA
Logic	Type Number
General Purpose PMOS, CMOS	ULN-2001L
14-25 V PMOS	ULN-2002L
5 V TTL, CMOS	ULN-2003L
6-15 V CMOS, PMOS	ULN-2004L
High-Output TTL	ULN-2005L

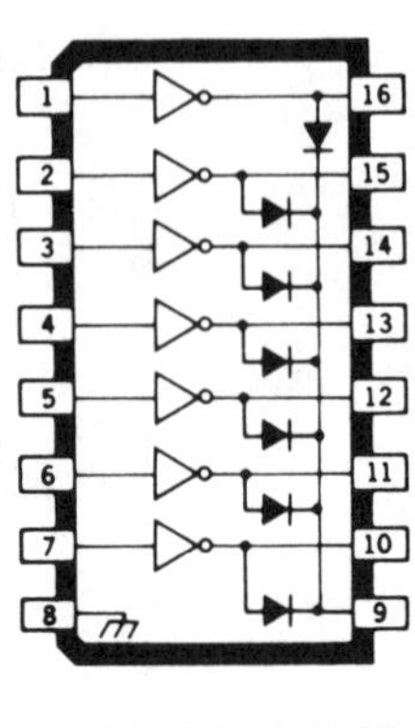

SOIC PACKAGE

PARTIAL SCHEMATICS

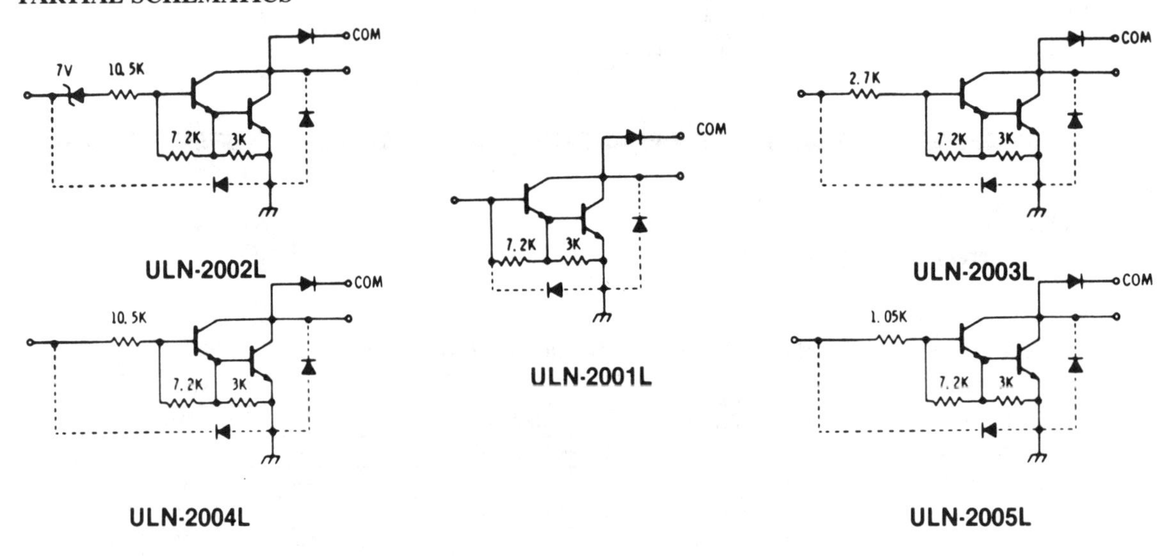

Sprague UDN-2522A Quad Bus Transceiver

- Driver output current to 300mA
- Driver output sustaining voltage of 50V
- Pulse-width discriminating receivers
- Internal receiver hysteresis
- Compatible with TTL and MOS logic
- Driver output clamp diodes

ABSOLUTE MAXIMUM RATINGS

AT T_A = +25°C

Driver Output Voltage, V_{CE} 70V
Driver Output Sustaining Voltage, $V_{CE(sus)}$ 50V
Driver Continuous Output Current, I_{OUT} 300mA
Driver Input Voltage, V_{IN} 5.5V
Receiver Output Current, I_{OUT} 50mA
Receiver Input Voltage, V_{IN} 70V
Supply Voltage, V_{CC} 7.0V
Package Power Dissipation, P_D .... See Graph
Operating Temperature Range, T_A .... −20°C to +85°C
Storage Temperature Range, T_S .... −55°C to +150°C

TYPICAL APPLICATION

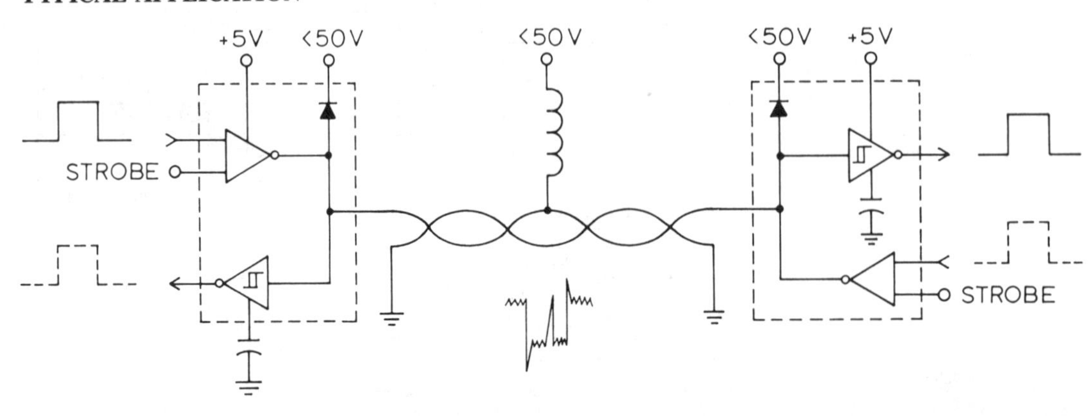

ELECTRICAL CHARACTERISTICS $T_A = +25°C$, Figure 1 and 2, 3, 4, or 5 as specified.

Characteristic	Test Conditions						Limits			
	V_{CC}	Input	Strobe	Out/In	C	Output	Min.	Typ.	Max.	Units
DRIVERS										
Output Leakage Current	4.5V	Open	0.8V	70V†	Open	Open	—	—	70	μA
Output Saturation Voltage	4.5V	2.0V	2.0V	210mA	Open	Open	—	0.2	0.4	V
	4.5V	2.0V	2.0V	300mA	Open	Open	—	0.4	0.6	V
Output Sat. Voltage Matching	4.5V	2.0V	2.0V	210mA	Open	Open	—	±20	±50	mV
Output Sustaining Voltage*	5.0V	3ℸ0V	2.0V	Fig. 2	Open	Open	50	—	—	V
Output Voltage	4.5V	0.8V	2.0V	−160μA	Open	Open	2.25	—	—	V
Logic Input Voltage	4.5V	—	—	Open	Open	Open	2.0	—	—	V
	4.5V	—	—	Open	Open	Open	—	—	0.8	V
Logic Input Current	5.5V	5.5V	2.0V	Open	Open	Open	—	—	20	μA
	5.5V	0.1V	2.0V	Open	Open	Open	−1.0	—	−20	μA
Strobe Input Current	5.5V	2.0V	5.5V	Open	Open	Open	—	—	50	μA
	5.5V	2.0V	0.1V	Open	Open	Open	—	—	−50	μA
Input Clamp Voltage	Open	−12mA	Open	Open	Open	Open	0	—	−1.6	V
Propagation Delay Time	5.5V	3ℸ0V	2.0V	Fig. 3	Open	Open	—	—	750	ns
	5.5V	2.0V	3ℸ0V	Fig. 3	Open	Open	—	—	750	ns
	5.5V	3ℸ0V	2.0V	Fig. 4	Open	Open	—	—	1.4	μs
	5.5V	2.0V	3ℸ0V	Fig. 4	Open	Open	—	—	1.4	μs
	4.5V	0ʃ3V	2.0V	Fig. 3	Open	Open	—	—	600	ns
	4.5V	2.0V	0ʃ3V	Fig. 3	Open	Open	—	—	600	ns
	4.5V	0ʃ3V	2.0V	Fig. 4	Open	Open	—	—	600	ns
	4.5V	2.0V	0ʃ3V	Fig. 4	Open	Open	—	—	600	ns
Output Rise Time	4.5V	3ℸ0V	2.0V	Fig. 3	Open	Open	—	—	175	ns
Ouput Fall Time	5.5V	0ʃ3V	2.0V	Fig. 3	Open	Open	—	—	175	ns
Clamp Diode Leakage Current	V_R = 70V			0.0V	Open	Open	—	—	70	μA
Clamp Diode Forward Voltage	I_F = 300mA				Open	Open	—	1.6	1.8	V
Supply Current (All Drivers)	5.5V	0.0V	0.8V	—	Open	Open	—	—	20	mA
	5.5V	2.0V	2.0V	—	Open	Open	—	—	50	mA
RECEIVERS										
Output Voltage	4.5V	0.8V	2.0V	Open	2.0V	4.0mA	—	—	0.5	V
	4.5V	2.0V	2.0V	Open	0.0V	−400μA	2.4	—	—	V
Short-Circuit Output Current	5.5V	2.0V	2.0V	Open	Open	0.0V	−2.0	—	−50	mA
Input Current	5.5V	Open	0.0V	4.0V	Open	Open	−250	—	—	μA
	4.5V	Open	0.0V	0.1V	Open	Open	—	—	−500	μA
Input Voltage	4.5V	Open	0.0V	—	Open	Low	2.0	—	—	V
	4.5V	Open	0.0V	—	Open	High	—	—	0.8	V
Input Voltage Hysteresis	4.5V	Open	0.0V	0ʃ3ℸ0V	Open		250	—	775	mV
Propagation Delay Time	4.5V	Open	0.0V	0ʃ3V	Open	Fig. 5	—	—	375	ns
	5.5V	Open	0.0V	3ℸ0V	Open	Fig. 5	—	—	375	ns
Output Fall Time	5.5V	Open	0.0V	0ʃ3V	Open	Fig. 5	—	—	75	ns
Output Rise Time	5.5V	Open	0.0V	3ℸ0V	Open	Fig. 5	—	—	75	ns
Noise Immunity	5.5V	Open	0.0V	0ʃ2V	0.1μF	Open	400	—	—	μs
	4.5V	Open	0.0V	0ʃ2V	0.1μF	Open	—	—	1400	μs

Note: Negative current is defined as coming out of (sourcing) the specified device pin.
$*V_{OUT(sus)}$ is measured with a 5 ms ON pulse, 12 ms after turn-OFF.
†Output clamp diode reverse-biased with V_K = 71 V.

TEST FIGURES

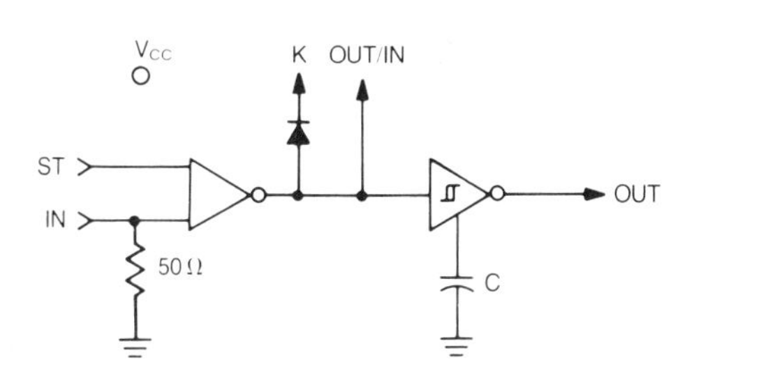

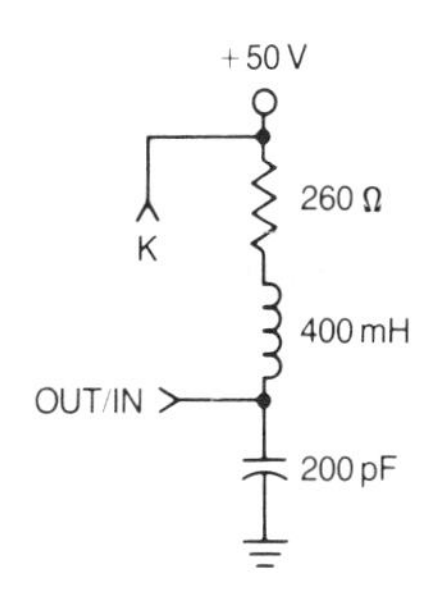

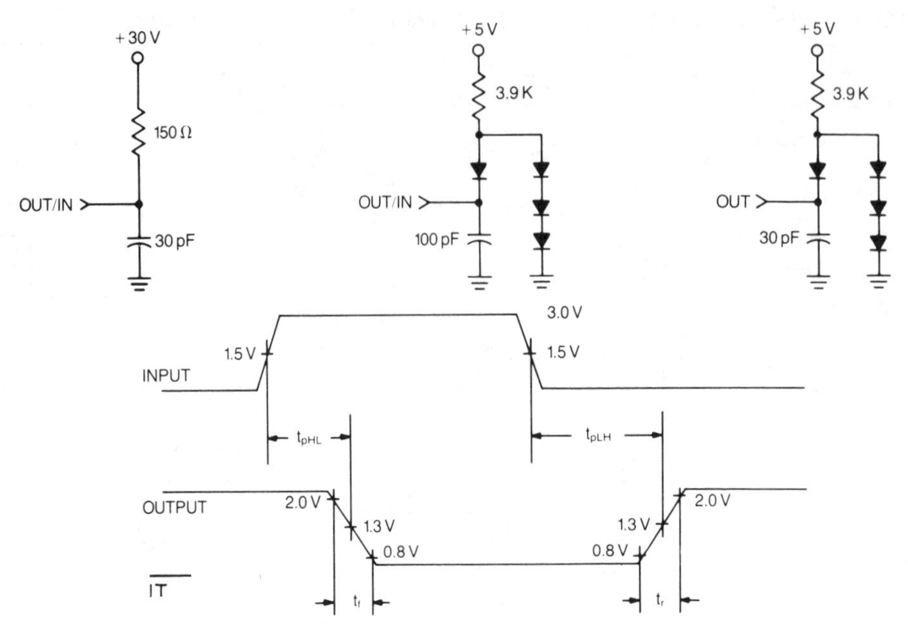

PARTIAL SCHEMATIC

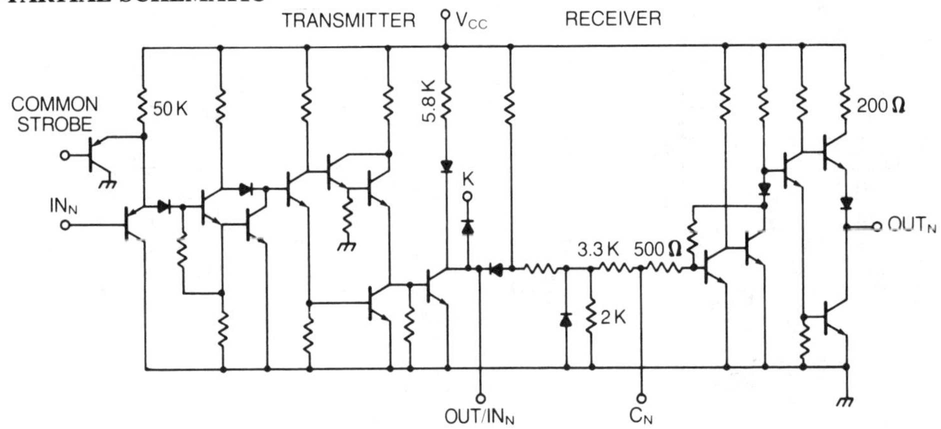

STROBE 1 ST
V_{CC} 20 V_{CC}
OUT/IN_1 2
19 OUT/IN_4
OUT_1 3
18 OUT_4
C_1 4
17 C_4
IN_1 5
16 IN_4
IN_2 6
15 IN_3
C_2 7
14 C_3
OUT_2 8
13 OUT_3
OUT/IN_2 9
12 OUT/IN_3
GROUND 10
K 11 K

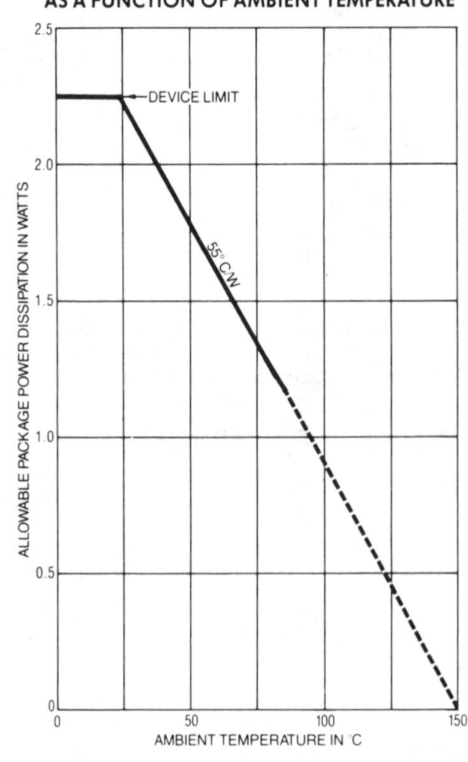

SECTION 3

CONTROL CIRCUITS

This information is reproduced with permission of Analog Devices, Inc.

Analog Devices
AD365
Programmable Gain and T/H DAS Amplifier

- Software programmable gain (1, 10, 100, 500)
- Low input noise (0.2μV p-p)
- Low gain error (0.05% max)
- Low nonlinearity (0.005% max)
- Low gain drift (10ppm/°C max)
- Low offset drift (2μV/°C RTI max)
- Fast setting (15μs @ gain 100)
- Small 16-pin metal DIP

APPLICATIONS

- Digitally controlled gain amplifier
- Auto-gain ranging amplifier
- Wide dynamic range measurement system
- Gain selection/channel amplifier
- Transducer/bridge amplifier
- Test equipment

The AD365 is a two-stage data acquisition system (DAS) front end consisting of a digitally selectable gain amplifier followed by an independent track/hold amplifier. The programmable-gain amplifier features differential inputs for excellent common-mode rejection, high open-loop gain for superior linearity, and fast settling for use in multiplexed high-speed systems. The track/hold amplifier features high open-loop gain for 12-bit compatible linearity, an internal hold capacitor for high reliability, and fast acquisition time for use with multichannel systems. Both amplifiers are capable of being used separately and are specified as independent function blocks.

The AD365 is comprised of the AD625 monolithic precision instrumentation amplifier to provide a precision differential input, the AD7502 monolithic CMOS multiplexer to handle gain switching, a precision thin-film resistor network, and the AD585 monolithic track-and-hold amplifier with internal-hold capacitor.

The input stage provides high common-mode rejection, low noise, fast settling at all gains, and low drift over temperature. The gains of 1, 10, 100, and 500 are digitally selected with the two gain control lines, which are 5V CMOS compatible.

The track-and-hold amplifier section is ideally suited for high-speed 12-bit applications where fast settling, low noise, and low sample-to-hold offset are critical. The T/H mode is controlled with a single input line that can be tied to the status output line of the accompanying A/D converter.

AD365 FUNCTIONAL BLOCK DIAGRAM

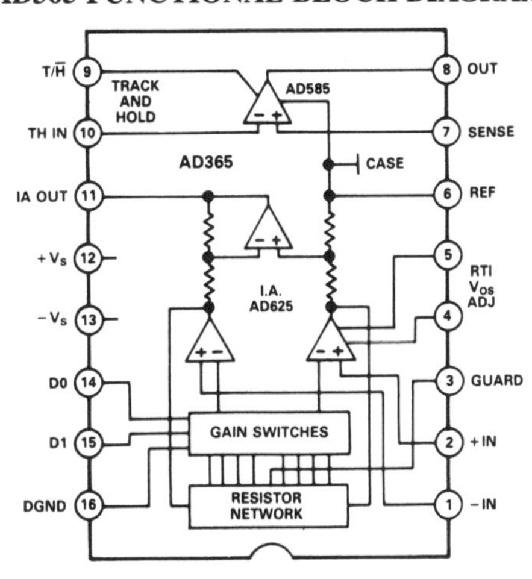

SPECIFICATIONS (typical @ V_S = ±15V, R_L = 2kΩ and T_A = +25°C unless otherwise specified)

AD365AM	Min	Typ	Max	Units
PGA GAIN				
Inaccuracy[1]				
@ G = 1, 10, 100		0.02	0.05	%
@ G = 500		0.04	0.1	%
Nonlinearity				
@ G = 1, 10, 100			0.005	%
@ G = 500			0.01	%
Drift				
@ G = 1		1	5	ppm/°C
@ G = 10, 100, 500		3	10	ppm/°C
PGA OFFSET (May be Nulled at Input and Output)				
Input Offset Voltage (RTI)		25	200	μV
vs. Temperature		0.1	2	μV/°C
vs. Common-Mode Voltage		0.5	3.2	μV/V
vs. Supply Voltage		1	10	μV/V
Output Offset Voltage (RTO)		1	5	mV
vs. Temperature		30	150	μV/°C
vs. Common-Mode Voltage		60	316	μV/V
vs. Supply Voltage		60	316	μV/V
PGA INPUT				
Common-Mode and Differential Impedance		$10^9 \| 5$		Ω‖pF
Differential Input Voltage, Linear	10	12		V
Common-Mode Voltage, Linear		$12 - V_{DIFF} \times G/2$		V
Input Stage Noise 0.1 to 10Hz		0.2		μV p-p
Input Stage Noise Density @ 1kHz		4		$nV/\sqrt{Hz}$
Bias Current		5	50	nA
vs. Temperature		50		pA/°C
Offset Current		2	20	nA
vs. Temperature		20		pA/°C
Noise Current (0.1 to 10Hz)		60		pA p-p
PGA OUTPUT				
Voltage 2kΩ Load	10	12		V
Output Impedance		0.2		Ω
Short Circuit Current		25		mA
Capacitive Load		500		pF
Output Stage Noise 0.1 to 10Hz		10		μV p-p
Output Stage Noise Density @ 1kHz		75		$nV/\sqrt{Hz}$
Guard Voltage		$(V_{+IN} + V_{-IN})/2$		V
Guard Offset		−550		mV
PGA DYNAMIC RESPONSE				
Small Signal −3dB				
G = 1		800		kHz
G = 10		400		kHz
G = 100		150		kHz
G = 500		40		kHz
Full Power Bandwidth G = 1 @ V_O = 20V p-p		60		kHz
Slew Rate		4		V/μs
Settling Time to 0.01% @ V_O = 20V p-p				
G = 1, 10		8	10	μs
G = 100		12	15	μs
G = 500		40	50	μs
Gain Switching Time		1.5		μs
Overdrive Recovery Time V_{IN} = 15V @ G = 1		7		μs
PGA DIGITAL INPUTS				
Logic Low	0		0.8	V
Logic High	3.0		$+V_S$	V
Current, I_{INH} or I_{INL}		0.01	1	μA

AD365AM	Min	Typ	Max	Units
TRACK AND HOLD AMPLIFIER SECTION				
TRANSFER CHARACTERISTICS				
Open Loop Gain V_O = 10V, R_L = 2k	100k	200k		V/V
Nonlinearity @ G = +1			0.005	% FSR
Output Voltage R_L = 2kΩ	10	12		V
Capacitive Load		100		pF
Short Circuit Current		25		mA
TRACK MODE DYNAMICS				
Acquisition Time to 0.01% 10V Step		2	3	μs
20V Step		4	5	μs
Small Signal Bandwidth −3dB		2		MHz
Full Power Bandwidth (20V p-p)		120		kHz
Slew Rate		10		V/μs
TRACK/HOLD SWITCHING				
Aperture Time		35		ns
Aperture Uncertainty		0.5		ns
Switching Transient		40		mV
Settling Time to 2mV		0.5		μs
HOLD MODE				
Droop Rate @ +25°C		0.3	1	V/sec
from $T_{AMBIENT}$ to T_{MAX}		Doubles/10°C		V/sec
Feedthrough		25		μV/V
Pedestal, Offset @ +25°C		2	3	mV
Over Temperature		3		mV
T/H ANALOG INPUT				
Bias Current		0.1	2	nA
Over Temperature		0.2	5	nA
Offset Voltage			2	mV
Over Temperature			3	mV
vs. Common Mode		25	100	μV/V
vs. Supplies		100	316	μV/V
Input Impedance		10^{12}‖10		Ω‖pF
Noise Density @ 1kHz		50		$nV/\sqrt{Hz}$
Noise 0.1Hz to 10Hz		10		μV p-p
T/H DIGITAL INPUT CHARACTERISTICS				
Logic Low (Hold Mode)	0		0.8	V
Logic High (Track Mode)	2.0		$+V_S$	V
Input Current		10	50	μA
AD365 POWER REQUIREMENTS				
Positive Supply Range	+11		+17	V
Negative Supply Range	−11		−17	V
Quiescent Current		12	16	mA
Power Dissipation		360	550	mW
Warm-Up Time to Specification		5		Minutes
Ambient Operating Temperature	−25		+85	°C
Package Thermal Resistance (θ_{ja})		60		°C/W
AD365 ABSOLUTE MAXIMUM RATINGS				
Positive Supply $+V_S$	−0.3		+17	V dc
Negative Supply $-V_S$	+0.3		−17	V dc
Analog Input Voltage	$-V_S$		$+V_S$	V
Analog Input Current	−10		+10	mA
Digital Input Voltage	−0.3		$+V_S$	V
T/H Differential V_{IN}			±30	V
Storage Temperature	−65		+150	°C
Lead Soldering, 10 Sec			300	°C
Short Circuit Duration		Indefinite		

TYPICAL CHARACTERISTICS (@ +25°unless otherwise noted)

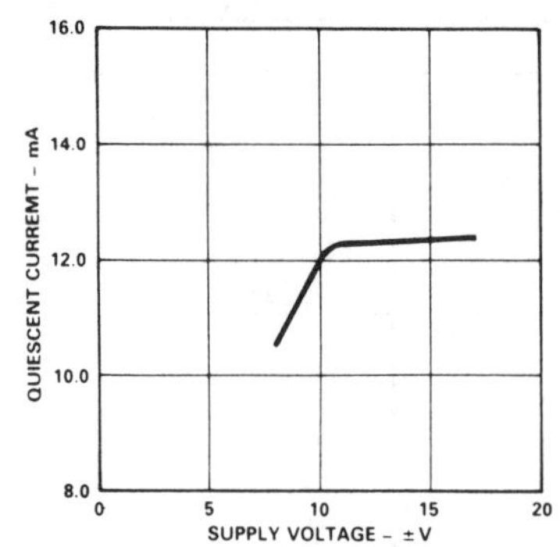

Figure 1. AD365 Quiescent Current vs. Supply Voltage

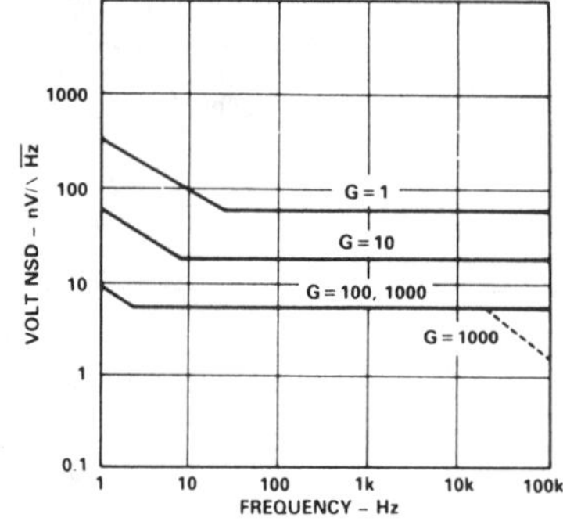

Figure 2. PGA RTI Noise Spectral Density vs. Gain

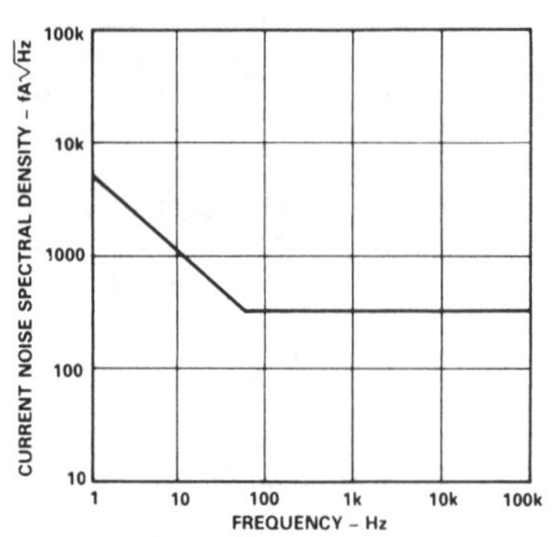

Figure 3. PGA Input Current Noise

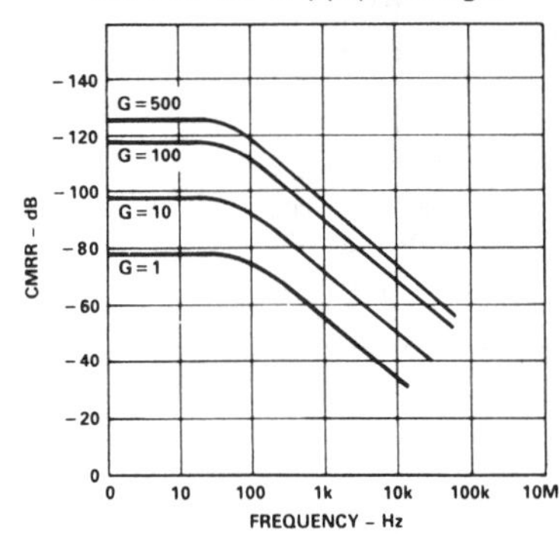

Figure 4. PGA CMRR vs. Frequency RTI, Zero to 1kΩ Source Imbalance

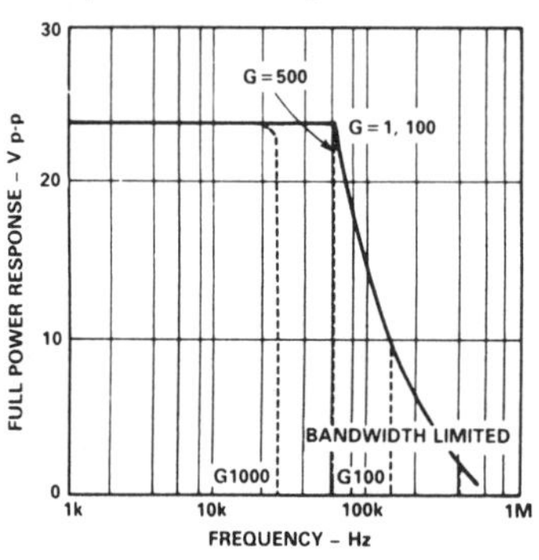

Figure 5. PGA Large Signal Frequency Response

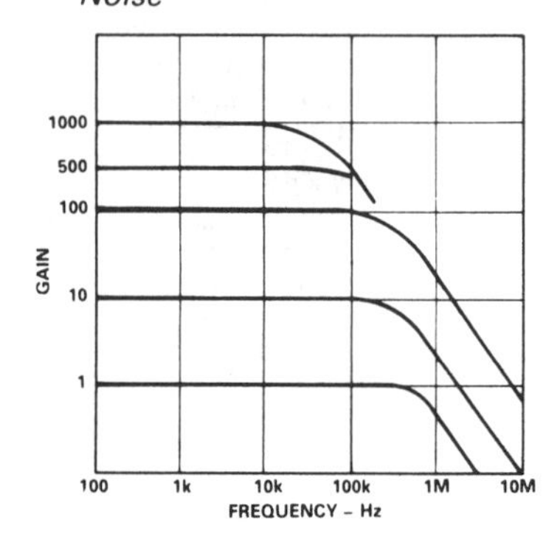

Figure 6. PGA Gain vs. Frequency

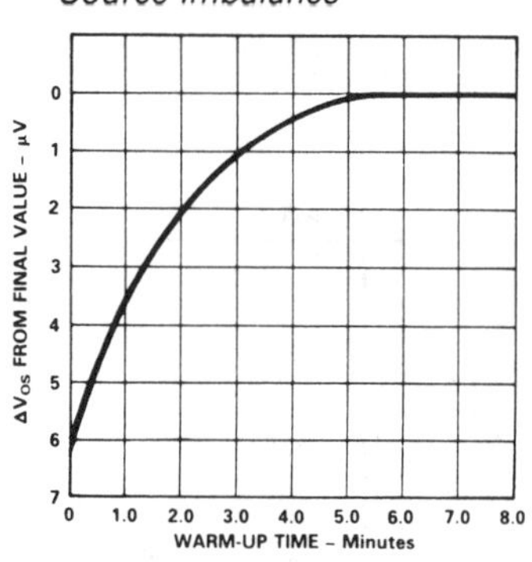

Figure 7. PGA Offset Voltage, RTI, Turn On Drift

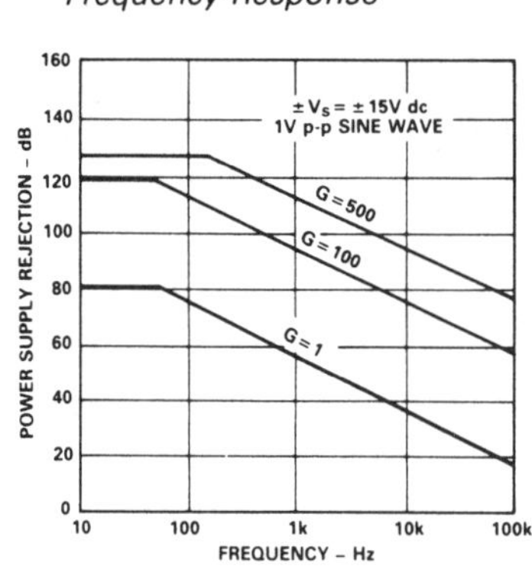

Figure 8. PGA PSRR vs. Frequency

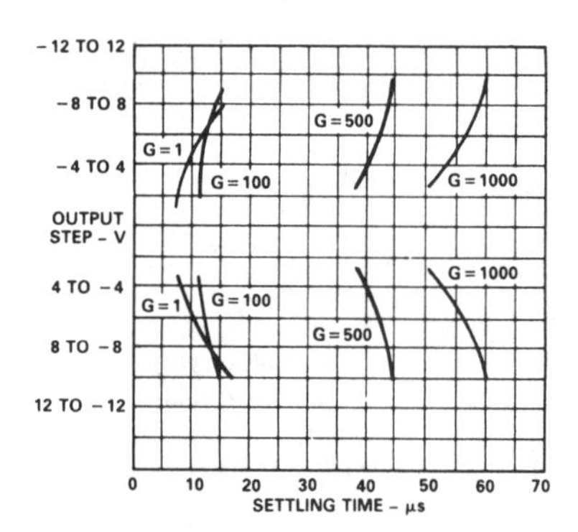

Figure 9. PGA Settling Time to 0.01%

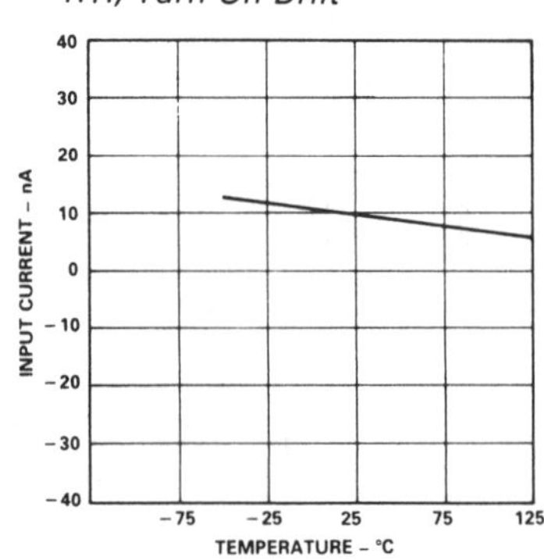

Figure 10. PGA Input Bias Current vs. Temperature

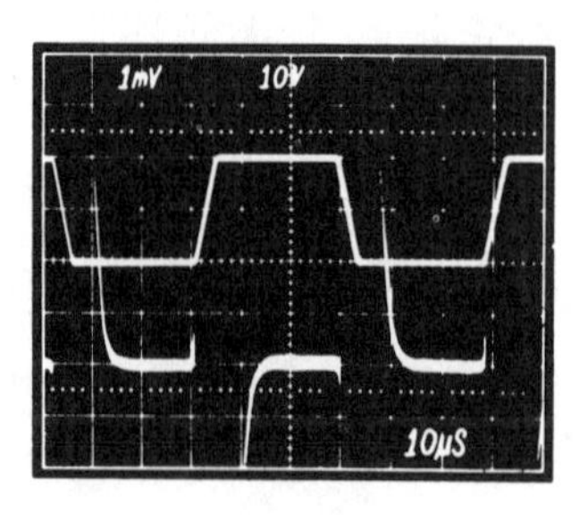

Figure 11. PGA Large Signal Pulse Response and Settling Time, G = 100

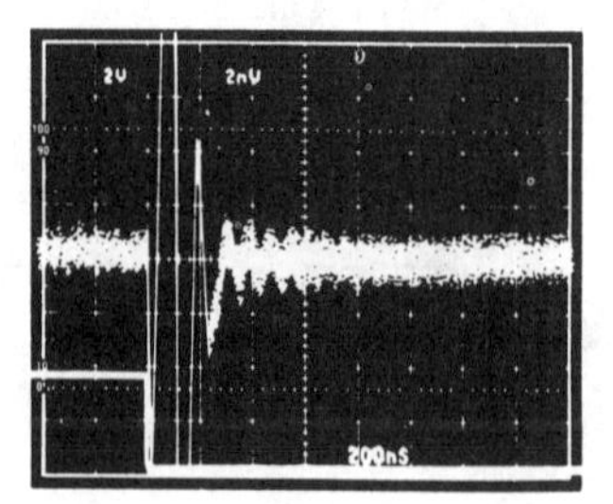

Figure 12. Sample-to-Hold Settling Time

Analog Devices AD521

Precision Instrumentation Amplifier

- Programmable gain from 0.1 to 1000
- Differential inputs
- High CMRR: 110dB min
- Low drift: 2μV/°C max (L)
- complete input protection, power ON and power OFF
- Functionally complete with the addition of two resistors
- Internally compensated
- Gain bandwidth product: 40 MHz
- Output current limited: 25mA
- Very low noise: 0.5μV p-p, 0.1Hz to 10Hz, RI @ G = 1000
- Chips are available

The AD521 is a second-generation low-cost monolithic IC instrumentation amplifier. As a true instrumentation amplifier, the AD521 is a gain block with differential inputs and an accurately programmable input/output gain relationship.

The AD521 IC instrumentation amplifier should not be confused with an operational amplifier, although several manufacturers (including Analog Devices) offer op amps that can be used as building blocks in variable-gain instrumentation amplifier circuits. Op amps are general-purpose components which, when used with precision-matched external resistors, can perform the instrumentation amplifier function.

An instrumentation amplifier is a precision differential voltage-gain device optimized for operation in a real world-environment, and is intended to be used wherever acquisition of a useful signal is difficult. It is characterized by high input impedance, balanced differential inputs, low bias currents, and high CMR.

As a complete instrumentation amplifier, the AD521 requires only two resistors to set its gain to any value between 0.1 and 1000. The ratio matching of these resistors does not affect the high CMRR (up to 120dB) or the high input impedance ($3 \times 10^9\Omega$) of the AD521. Furthermore, unlike most operational amplifier-based instrumentation amplifiers, the inputs are protected against overvoltages up to ±15 volts beyond the supplies.

The AD521 IC instrumentation amplifier is available in four different versions of accuracy and operating temperature range. The economical "J" grade, the low-drift "K" grade, and the lower drift, higher linearity "L" grade are specified from 0 to + 70°C. The "S" grade guarantees performance to specification over the extended temperature range: −55°C to +125°C.

- The AD521 is a true instrumentation amplifier in integrated circuit form, offering the user performance comparable to many modular instrumentation amplifiers at a fraction of the cost.
- The AD521 has low guaranteed-input offset voltage drift (2μV/°C for L grade) and low noise for precision high-gain applications.
- The AD521 is functionally complete with the addition of two resistors. Gain can be preset from 0.1 to more than 1000.
- The AD521 is fully protected for input levels up to 15V beyond the supply voltages and 30V differential at the inputs.
- Internally compensated for all gains, the AD521 also offers the user the provision for limiting bandwidth.
- Offset nulling can be achieved with an optional trim pot.
- The AD521 offers superior dynamic performance with a gain-bandwidth product of 40 MHz, full peak response of 100 kHz (independent of gain) and a settling time of 5μs to 0.1% of a 10V step.

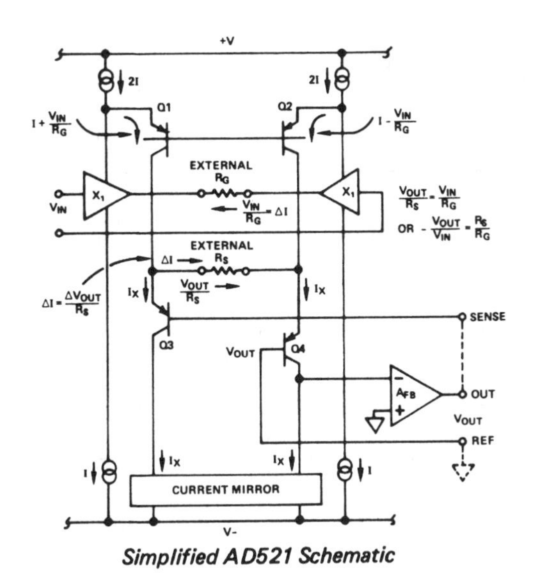

Simplified AD521 Schematic

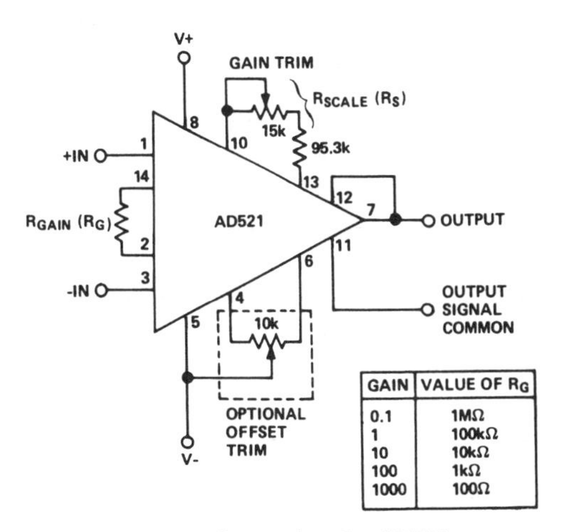

GAIN	VALUE OF R_G
0.1	1MΩ
1	100kΩ
10	10kΩ
100	1kΩ
1000	100Ω

Operating Connections for AD521

SPECIFICATIONS (typical @ $+V_S = \pm 15V$, $R_L = 2k\Omega$ & $T_A = +25^\circ C$ unless otherwise specified)

MODEL	AD521JD	AD521KD	AD521LD	AD521SD (AD521SD/883B)
GAIN				
Range (For Specified Operation, Note 1)	1 to 1000	*	*	*
Equation	$G = R_S/R_G$ V/V	*	*	*
Error from Equation	(±0.25−0.004G)%	*	*	*
Nonlinearity (Note 2)				
1≤G≤1000	0.2% max	*	0.1% max	*
Gain Temperature Coefficient	±(3 ±0.05G)ppm/°C	*	*	±(15 ±0.4G)ppm/°C
OUTPUT CHARACTERISTICS				
Rated Output	±10V, ±10mA min	*	*	*
Output at Maximum Operating Temperature	±10V @ 5mA min	*	*	*
Impedance	0.1Ω	*	*	*
DYNAMIC RESPONSE				
Small Signal Bandwidth (±3dB)				
G = 1	>2MHz	*	*	*
G = 10	300kHz	*	*	*
G = 100	200kHz	*	*	*
G = 1000	40kHz	*	*	*
Small Signal, ±1.0% Flatness				
G = 1	75kHz	*	*	*
G = 10	26kHz	*	*	*
G = 100	24kHz	*	*	*
G = 1000	6kHz	*	*	*
Full Peak Response (Note 3)	100kHz	*	*	*
Slew Rate, 1≤G≤1000	10V/µs	*	*	*
Settling Time (any 10V step to within 10mV of Final Value)				
G = 1	7µs	*	*	*
G = 10	5µs	*	*	*
G = 100	10µs	*	*	*
G = 1000	35µs	*	*	*
Differential Overload Recovery (±30V Input to within 10mV of Final Value) (Note 4)				
G = 1000	50µs	*	*	*
Common Mode Step Recovery (30V Input to within 10mV of Final Value) (Note 5)				
G = 1000	10µs	*	*	*
VOLTAGE OFFSET (may be nulled)				
Input Offset Voltage (V_{OS_I})	3mV max (2mV typ)	1.5mV max (0.5mV typ)	1.0mV max (0.5mV typ)	**
vs. Temperature	15µV/°C max (7µV/°C typ)	5µV/°C max (1.5µV/°C typ)	2µV/°C max	**
vs. Supply	3µV/%	*	*	*
Output Offset Voltage (V_{OS_O})	400mV max (200mV typ)	200mV max (30mV typ)	100mV max	**
vs. Temperature	400µV/°C max (150µV/°C typ)	150µV/°C max (50µV/°C typ)	75µV/°C max	**
vs. Supply (Note 6)	$0.005V_{OS_O}$/%	*	*	*
INPUT CURRENTS				
Input Bias Current (either input)	80nA max	40nA max	**	**
vs. Temperature	1nA/°C max	500pA/°C max	**	**
vs. Supply	2%/V	*	*	*
Input Offset Current	20nA max	10nA max	**	**
vs. Temperature	250pA/°C max	125pA/°C max	**	**
INPUT				
Differential Input Impedance (Note 7)	$3 \times 10^9 \Omega \| 1.8pF$	*	*	*
Common Mode Input Impedance (Note 8)	$6 \times 10^{10} \Omega \| 3.0pF$	*	*	*
Input Voltage Range for Specified Performance (with respect to ground)	±10V	*	*	*
Maximum Voltage without Damage to Unit, Power ON or OFF Differential Mode (Note 9)	30V	*	*	*
Voltage at either input (Note 9)	V_S ±15V	*	*	*
Common Mode Rejection Ratio, DC to 60Hz with 1kΩ source unbalance				
G = 1	70dB min (74dB typ)	74dB min (80dB typ)	**	**
G = 10	90dB min (94dB typ)	94dB min (100dB typ)	**	**
G = 100	100dB min (104dB typ)	104dB min (114dB typ)	**	**
G = 1000	100dB min (110dB typ)	110dB min (120dB typ)	**	**
NOISE				
Voltage RTO (p-p) @ 0.1Hz to 10Hz (Note 10)	$\sqrt{(0.5G)^2 + (225)^2}\mu V$	*	*	*
RMS RTO, 10Hz to 10kHz	$\sqrt{(1.2G)^2 + (50)^2}\mu V$	*	*	*
Input Current, rms, 10Hz to 10kHz	15pA (rms)	*	*	*
REFERENCE TERMINAL				
Bias Current	3µA	*	*	*
Input Resistance	10MΩ	*	*	*
Voltage Range	±10V	*	*	*
Gain to Output	1	*	*	*
POWER SUPPLY				
Operating Voltage Range	±5V to ±18V	*	*	*
Quiescent Supply Current	5mA max	*	*	*
TEMPERATURE RANGE				
Specified Performance	0 to +70°C	*	*	−55°C to +125°C
Operating	−25°C to +85°C	*	*	−55°C to +125°C
Storage	−65°C to +150°C	*	*	*
PACKAGE OPTION				
Ceramic (D-14)	AD521JD	AD521KD	AD521LD	AD521SD

NOTES

*Specifications same as AD521JD.
**Specifications same as AD521KD.
Specifications subject to change without notice.

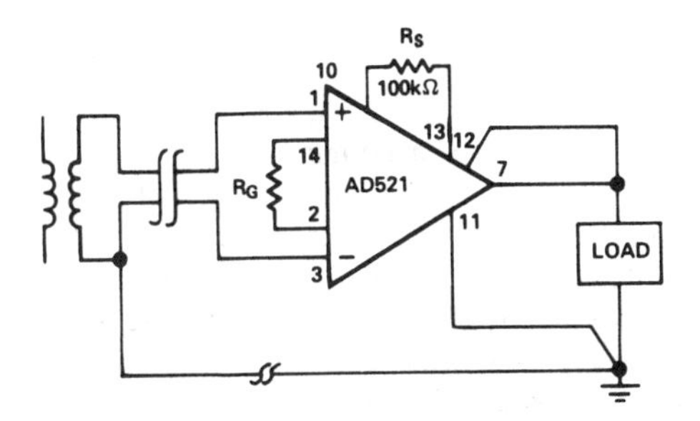

a). Transformer Coupled, Direct Return

AD521 PIN CONFIGURATION

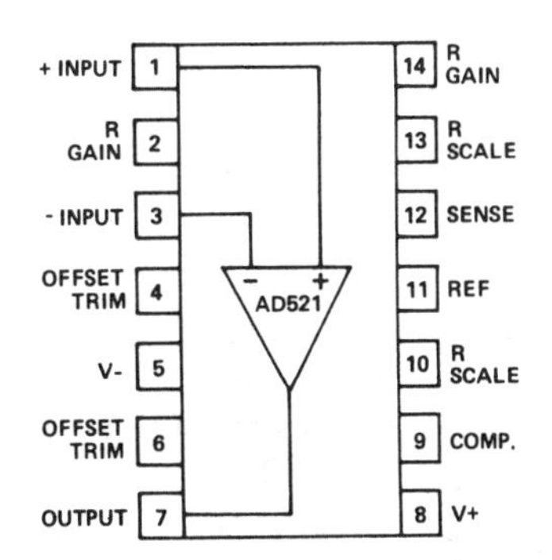

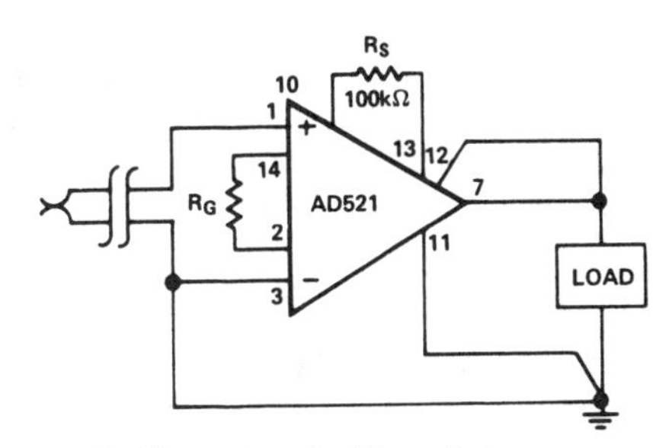

b). Thermocouple, Direct Return

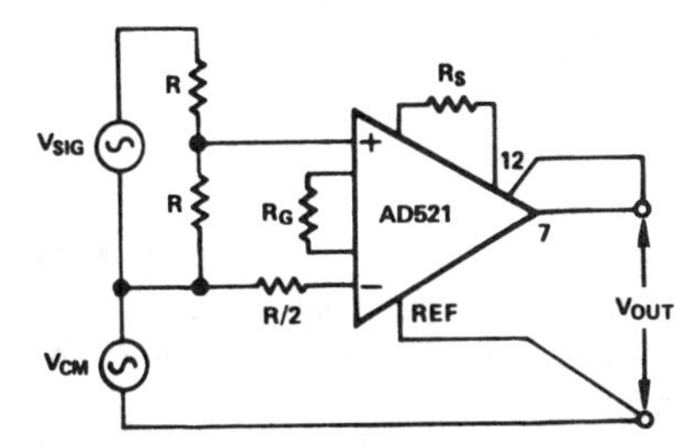

1. INCREASE R_G TO PICK UP GAIN LOST BY R DIVIDER NETWORK
2. INPUT SIGNAL MUST BE REDUCED IN PROPORTION TO POWER SUPPLY VOLTAGE LEVEL

Operating Conditions for $V_{IN} \approx V_S = 10V$

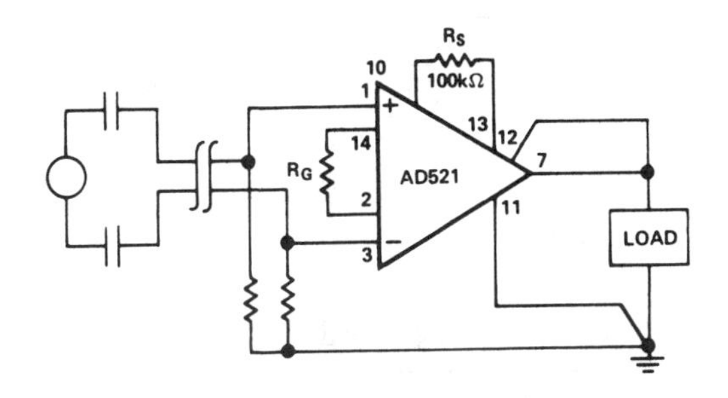

c). AC Coupled, Indirect Return

Ground Returns for "Floating" Transducers

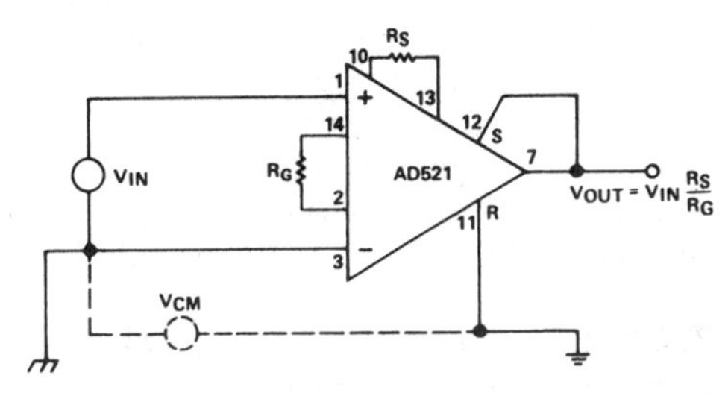

Ground loop elimination. The reference input, Pin 11, allows remote referencing of ground potential. Differences in ground potentials are attenuated by the high CMRR of the AD521.

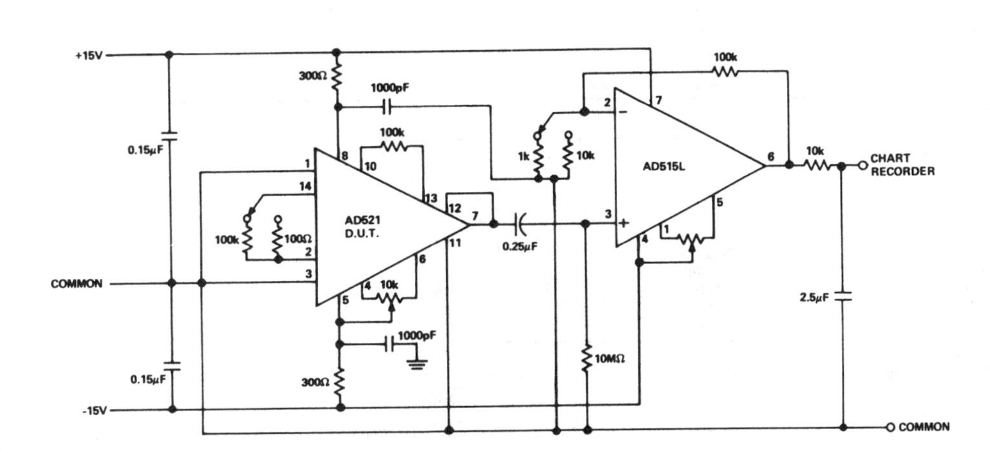

Test circuit for measuring peak to peak noise in the bandwidth 0.1Hz to 10Hz. Typical measurements are found by reading the maximum peak to peak voltage noise of the device under test (D.U.T.) for 3 observation periods of 10 seconds each.

Analog Devices AD522

High-Accuracy Data Acquisition Instrumentation Amplifier

- Low drift: 2.0μV/°C (AD522B)
- Low nonlinearity: 0.005% (G = 100)
- High CMRR: > 110dB (G = 1000)
- Low noise: 1.5μV p-p (0.1 to 100Hz)
- Low initial V_{OS}: 100μV (AD522B)
- Single-resistor gain programmable: 1 ≤ G ≤ 1000
- Output reference and sense terminals
- Data guard for improving ac CMR
- Internally compensated
- No external components except gain resistor
- Active trimmed offset, gain, and CMR

The AD522 is a precision IC instrumentation amplifier designed for data acquisition applications requiring high accuracy under worst-case operating conditions. An outstanding combination of high linearity, high common-mode rejection, low voltage drift, and low noise makes the AD522 suitable for use in many 12-bit data acquisition systems.

An instrumentation amplifier is usually employed as a bridge amplifier for resistance transducers (thermistors, strain gauges, etc.) found in process control, instrumentation, data processing, and medical testing. The operating environment is frequently characterized by low signal-to-noise levels, fluctuating temperatures, unbalanced input impedances, and remote locations that hinders recalibration

The AD522 was designed to provide highly accurate signal conditioning under these severe conditions. It provides output offset voltage drift of less than 10μV/°C, input offset voltage drift of less than 2.0 μV/°C, CMR above 80dB at unity gain (110dB at G = 1000), maximum gain nonlinearity of 0.001% at G = 1, and typical input impedance of $10^9\Omega$.

This excellent performance is achieved by combining a proven circuit configuration with state-of-the-art manufacturing technology that utilizes active laser trimming of tight-tolerance thin-film resistors to achieve low cost, small size, and high reliability. This combination of high value with no-compromise perlithic and modular instrumentation amplifiers, thus providing extremely cost-effective precision low-level amplification.

The AD522 is available in three versions with differing accuracies and operating temperature ranges: the "A", and "B" are specified from −25°C to +85°C, and the "S" is guaranteed over the extended aerospace temperature range of −55°C to +125°C. All versions are packaged in a 14-pin DIP and are supplied in a pin configuration similar to that of the popular AD521 instrumentation amplifier.

AD522 FUNCTIONAL BLOCK DIAGRAM

14-PIN DIP

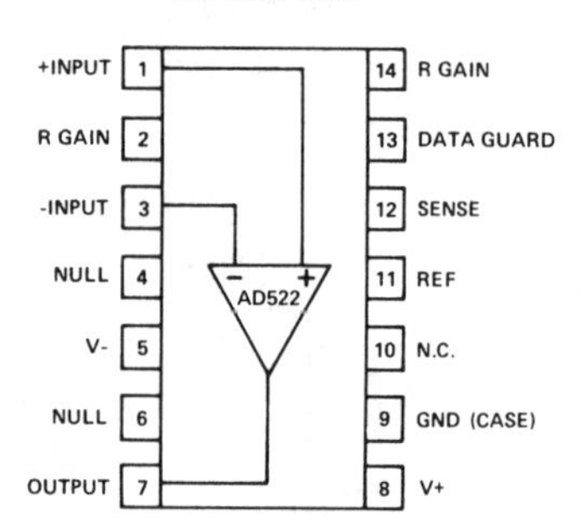

SPECIFICATIONS[1] (typical @ $+V_S$ = ±15V, R_L = 2kΩ & T_A = +25°C unless otherwise specified)

MODEL	AD522AD	AD522BD	AD522SD
GAIN			
Gain Equation	$1 + \frac{2(10^5)}{R_g}$	•	•
Gain Range	1 to 1000	•	•
Equation Error			
G = 1	0.2% max	0.05% max	••
G = 1000	1.0% max	0.2% max	••
Nonlinearity, max (see Fig. 4)			
G = 1	0.005%	0.001%	••
G = 1000	0.01%	0.005%	••
vs. Temp, max			
G = 1	2ppm/°C (1ppm/°C typ)	•	•
G = 1000	50ppm/°C (25ppm/°C typ)	•	•
OUTPUT CHARACTERISTICS			
Output Rating	±10V @ 5mA	•	•
DYNAMIC RESPONSE (see Fig. 6)			
Small Signal (−3dB)			
G = 1	300kHz	•	•
G = 100	3kHz	•	•
Full Power GBW	1.5kHz	•	•
Slew Rate	0.1V/μs	•	•
Settling Time to 0.1%, G = 100	0.5ms	•	•
to 0.01%, G = 100	5ms	•	•
to 0.01%, G = 10	2ms	•	•
to 0.01%, G = 1	0.5ms	•	•
VOLTAGE OFFSET			
Offsets Referred to Input			
Initial Offset Voltage (adjustable to zero)			
G = 1	±400μV max (±200μV typ)	±200μV max (±100μV typ)	±200μV max (±100μV typ)
vs. Temperature, max (see Fig. 3)			
G = 1	±50μV/°C (±10μV/°C typ)	±25μV/°C (±5μV/°C typ)	±100μV/°C (±10μV/°C typ)
G = 1000	±6μV/°C	±2μV/°C	±6μV/°C
1 < G < 1000	$\pm(\frac{50}{G} + 6)\mu V/°C$	$\pm(\frac{25}{G} + 2)\mu V/°C$	$\pm(\frac{100}{G} + 6)\mu V/°C$
vs. Supply, max			
G = 1	±20μV/%	•	•
G = 1000	±0.2μV/%	•	•

MODEL	AD522AD	AD522BD	AD522SD
INPUT CURRENTS			
Input Bias Current			
Initial max, +25°C	±25nA	•	•
vs. Temperature	±100pA/°C	•	•
Input Offset Current			
Initial max, +25°C	±20nA	•	•
vs. Temperature	±100pA/°C	•	•
INPUT			
Input Impedance			
Differential	$10^9\Omega$	•	•
Common Mode	$10^9\Omega$	•	•
Input Voltage Range			
Maximum Differential Input, Linear	±10V	•	•
Maximum Differential Input, Safe	±20V	•	•
Maximum Common Mode, Linear	±10V	•	•
Maximum Common Mode Input, Safe	±15V	•	•
Common Mode Rjection Ratio,			
Min @ ±10V, 1kΩ Source			
Imbalance (see Fig. 5)			
G = 1 (dc to 30Hz)	75dB (90dB typ)	80dB (100dB typ)	75dB (90dB typ)
G = 10 (dc to 10Hz)	90dB (100dB typ)	95dB (110dB typ)	90dB (110dB typ)
G = 100 (dc to 3Hz)	100dB (110dB typ)	100dB (120dB typ)	100dB (120dB typ)
G = 1000 (dc to 1Hz)	100dB (120dB typ)	110dB (>120dB typ)	100dB (>120dB typ)
G = 1 to 1000 (dc to 60Hz)	75dB (88dB typ)	80dB (88dB typ)	•
NOISE			
Voltage Noise, RTI (see Fig. 4)			
0.1Hz to 100Hz (p-p)			
G = 1	15μV	•	•
G = 1000	1.5μV	•	•
10Hz to 10kHz (rms)			
G = 1	15μV	•	•
TEMPERATURE RANGE			
Specified Performance	-25°C to +85°C	•	-55°C to +125°C
Operating	-55°C to +125°C	•	•
Storage	-65°C to +150°C	•	•
POWER SUPPLY			
Power Supply Range	±(5 to 18)V	•	•
Quiescent Current, max @ ±15V	±10mA	±8mA	••
PACKAGE OPTIONS[2]			
Ceramic[3] (DH-14A)	AD522AD	AD522BD	
Metal (DH-14B)			AD522SD

NOTES
[1] Specifications guaranteed after 10 minute warm-up.
[2] See Databooks for package outline information.
[3] Analog Devices reserves the right to ship metal package in lieu of the standard ceramic packages for A and B grades.

•Specifications same as AD522A.
••Specifications same as AD522B.
Specifications subject to change without notice.

Analog Devices AD524

Precision Instrumentation Amplifier

- Low noise: 0.3μV p-p 01.Hz to 10Hz
- Low nonlinearity: 0.003% (G = 1)
- High CMRR: 120dB (G = 1000)
- Low offset voltage: 50μV
- Low offset voltage drift: 0.5μV/°C
- Gain bandwidth product: 25 MHz
- Pin programmable gains of 1, 10, 100, and 1000
- Input protection, power on — power off
- No external components required
- Internally compensated
- MIL-STD-883B, chips, and plus parts available
- 16-pin ceramic DIP package and 20-terminal leadless chip carriers available

The AD524 is a precision monolithic instrumentation amplifier designed for data acquisition applications requiring high accuracy under worst-case operating conditions. An outstanding combination of high linearity, high common-mode rejection, low offset voltage drift, and low noise makes the AD524 suitable for use in many data acquisition systems.

The AD524 has an output offset voltage drift of less than 25μV/°C, input offset voltage drift of less than 0.5μV/°C, CMR above 90dB at unity gain (120dB at G = 1000) and maximum nonlinearity of 0.003% at G = 1. In addition to the outstanding dc specifications, the AD524 also has a 25 MHz gain bandwidth product (G = 100). To make it suitable for high-speed data acquisition systems, the AD524 has an output slew rate of 5V/μs and settles in 15μs to 0.01% for gains of 1 to 100.

As a complete amplifier, the AD524 does not require any external components for fixed gains of 1, 10, 100, and 1000. For other gain settings between 1 and 1000, only a single resistor is required. The AD524 input is fully protected for both power-on and power-off fault conditions.

The AD524 IC instrumentation amplifier is available in four different versions of accuracy and operating temperature range. The economical "A" grade, the low-drift "B" grade and lower drift, higher linearity "C" grade are specified from −25°C to +85°C. The "S" grade guarantees performance to specification over the extended temperature range −55°C to +125°C. Devices are available in a 16-pin ceramic DIP package and a 20-terminal leadless chip carrier.

1. The AD524 has guaranteed low-offset voltage, offset voltage drift and low noise for precision high-gain applications.
2. The AD524 is functionally complete with pin programmable gains of 1, 10, 100, and 1000, and single resistor programmable for any gain.
3. Input and output offset nulling terminals are provided for very high precision applications and to minimize offset voltage changes in gain ranging applications.
4. The AD524 is input protected for both power on and power off fault conditions.
5. The AD524 offers superior dynamic performance with a gain bandwidth product of 25 MHz, full-power response of 75 kHz and a settling time of 15μs to 0.0% of a 20V step (G = 100).

SPECIFICATIONS (@ $V_S = \pm 15V$, $R_L = 2k\Omega$ and $T_A = +25°C$ unless otherwise specified)

Model	AD524A Min	AD524A Typ	AD524A Max	AD524B Min	AD524B Typ	AD524B Max	AD524C Min	AD524C Typ	AD524C Max	AD524S Min	AD524S Typ	AD524S Max	Units
GAIN													
Gain Equation													
(External Resistor Gain Programming)		$\left[\frac{40{,}000}{R_G} + 1\right] \pm 20\%$			$\left[\frac{40{,}000}{R_G} + 1\right] \pm 20\%$			$\left[\frac{40{,}000}{R_G} + 1\right] \pm 20\%$			$\left[\frac{40{,}000}{R_G} + 1\right] \pm 20\%$		
Gain Range (Pin Programmable)	1 to 1000			1 to 1000			1 to 1000			1 to 1000			
Gain Error													
G = 1			±0.05			±0.03			±0.02			±0.05	%
G = 10			±0.25			±0.15			±0.1%			±0.25	%
G = 100			±0.5			±0.35			±0.25			±0.5	%
G = 1000			±2.0			±1.0			±0.5			±2.0	%
Nonlinearity													
G = 1			±0.01			±0.005			±0.003			±0.01	%
G = 10, 100			±0.01			±0.005			±0.003			±0.01	%
G = 1000			±0.01			±0.01			±0.01			±0.01	%
Gain vs. Temperature													
G = 1			5			5			5			5	ppm/°C
G = 10			15			10			10			10	ppm/°C
G = 100			35			25			25			25	ppm/°C
G = 1000			100			50			50			50	ppm/°C
VOLTAGE OFFSET (May be Nulled)													
Input Offset Voltage			250			100			50			100	μV
vs. Temperature			2			0.75			0.5			2.0	μV/°C
Output Offset Voltage			5			3			2.0			3.0	mV
vs. Temperature			100			50			25			50	μV/°C
Offset Referred to the Input vs. Supply													
G = 1	70			75			80			75			dB
G = 10	85			95			100			95			dB
G = 100	95			105			110			105			dB
G = 1000	100			110			115			110			dB
INPUT CURRENT													
Input Bias Current			±50			±25			±15			±50	nA
vs. Temperature		±100			±100			±100			±100		pA/°C
Input Offset Current			±35			+15			±10			±35	nA
vs. Temperature		±100			±100			±100			±100		pA/°C
INPUT													
Input Impedance													
Differential Resistance		10^9			10^9			10^9			10^9		Ω
Differential Capacitance		10			10			10			10		pF
Common Mode Resistance		10^9			10^9			10^9			10^9		Ω
Common Mode Capacitance		10			10			10			10		pF
Input Voltage Range													
Max Differ. Input Linear (V_D)		±10			±10			±10			±10		V
Max Common Mode Linear (V_{CM})		$12V - \left(\frac{G}{2} \times V_D\right)$			$12V - \left(\frac{G}{2} \times V_D\right)$			$12V - \left(\frac{G}{2} \times V_D\right)$			$12V - \left(\frac{G}{2} \times V_D\right)$		V
Common Mode Rejection dc to 60Hz with 1kΩ Source Imbalance													
G = 1	70			75			80			70			dB
G = 10	90			95			100			90			dB
G = 100	100			105			110			100			dB
G = 1000	110			115			120			110			dB
OUTPUT RATING													
V_{OUT}, $R_L = 2k\Omega$		±10			±10			±10			±10		V
DYNAMIC RESPONSE													
Small Signal − 3dB													
G = 1		1			1			1			1		MHz
G = 10		400			400			400			400		kHz
G = 100		150			150			150			150		kHz
G = 1000		25			25			25			25		kHz
Slew Rate		5.0			5.0			5.0			5.0		V/μs
Settling Time to 0.01%, 20V Step													
G = 1 to 100		15			15			15			15		μs
G = 1000		75			75			75			75		μs
NOISE													
Voltage Noise, 1kHz													
R.T.I.		7			7			7			7		$nV/\sqrt{Hz}$
R.T.O.		90			90			90			90		$nV/\sqrt{Hz}$
R.T.I., 0.1 to 10Hz													
G = 1		15			15			15			15		μV p-p
G = 10		2			2			2			2		μV p-p
G = 100, 1000		0.3			0.3			0.3			0.3		μV p-p
Current Noise													
0.1Hz to 10Hz		60			60			60			60		pA p-p

Model	AD524A			AD524B			AD524C			AD524S			Units
	Min	Typ	Max	Min	Typ	Max	Min	Typ	Max	Min	Typ	Max	
SENSE INPUT													
R_{IN}		20			20			20			20		kΩ ±20%
I_{IN}		15			15			15			15		μA
Voltage Range	±10			±10			±10			±10			V
Gain to Output		1			1			1			1		%
REFERENCE INPUT													
R_{IN}		40			40			40			40		kΩ ±20%
I_{IN}		15			15			15			15		μA
Voltage Range	±10			±10			10			10			V
Gain to Output		1			1			1			1		%
TEMPERATURE RANGE													
Specified Performance	25		+85	25		+85	25		+85	55		+125	°C
Storage	65		+150	65		+150	65		+150	65		+150	°C
POWER SUPPLY													
Power Supply Range	**±6**	±15	**±18**	**±6**	±15	**±18**	**±6**	±15	**±18**	**±6**	±15	**±18**	V
Quiescent Current		3.5	**5.0**		3.5	**5.0**		3.5	**5.0**		3.5	**5.0**	mA
PACKAGE OPTIONS													
16-Pin Ceramic (D-16)		AD524AD			AD524BD			AD524CD			AD524SD		
LCC (E-28A)		AD524AE			AD524BE			AD524CE			AD524SE		

NOTES

Specifications subject to change without notice.

Specifications shown in boldface are tested on all production units at final electrical test. Results from those tests are used to calculate outgoing quality levels. All min and max specifications are guaranteed, although only those shown in boldface are tested on all production units.

CONNECTION DIAGRAMS

Ceramic (D) Package

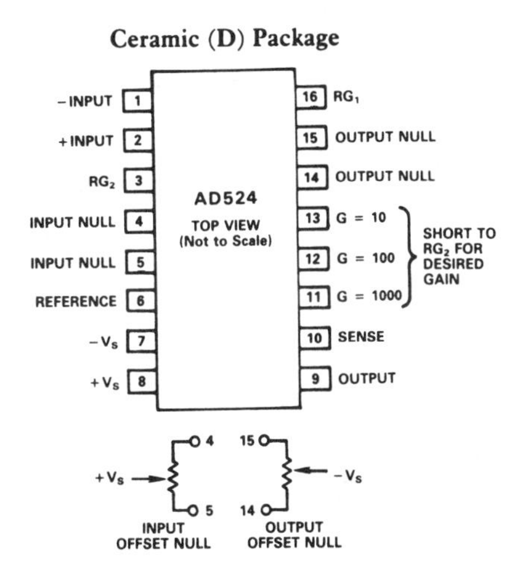

Leadless Chip Carrier (E) Package

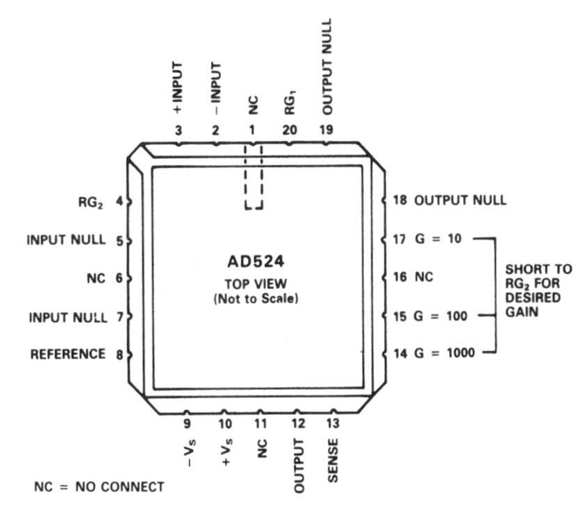

AD524 FUNCTIONAL BLOCK DIAGRAM

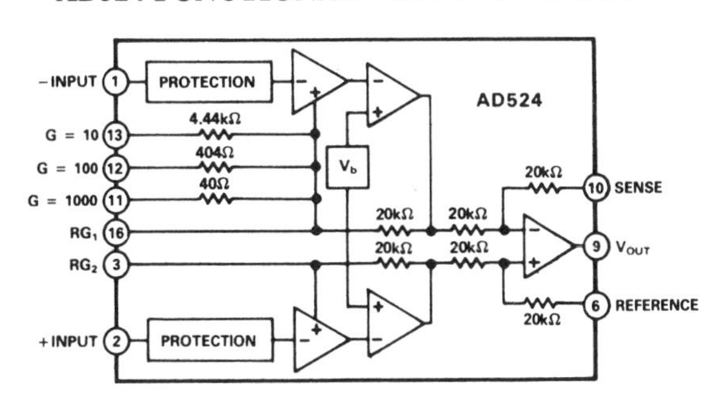

The first schematic shows a single application, in which the variation of the cold-junction voltage of a type J thermocouple-iron (+) — constantan — is compensated for by a voltage developed in series by the temperature-sensitive output current of an AD590 semiconductor temperature sensor.

The circuit is calibrated by adjusting R_T for proper output voltage with the measuring junction at a known reference temperature and the circuit near 25 °C. If resistors with low temperatures are used, compensation accuracy will be to within ±0.5 °C, for temperatures between +15 °C and +35 °C. Other thermocouple types might be accommodated with the standard resistance values shown in the table. For other ranges of ambient temperature, the equation in the figure can be solved for the optimum values of R_T and R_A.

The microprocessor-controlled data acquisition system shown in the second schematic includes both auto-zero and auto-gain capability. By dedicating two of the differential inputs, one to ground and one to the A/D reference, the proper program calibration cycles can eliminate both initial accuracy errors and accuracy errors over temperature. The auto-zero cycle, in this application, converts a number that appears to be ground and then writes that same number (8 bit) to the AD7524, which eliminates the zero error because its output has an inverted scale. The auto-gain cycle converts the A/D reference and compares it with full scale. A multiplicative correction factor is then computed and applied to subsequent readings.

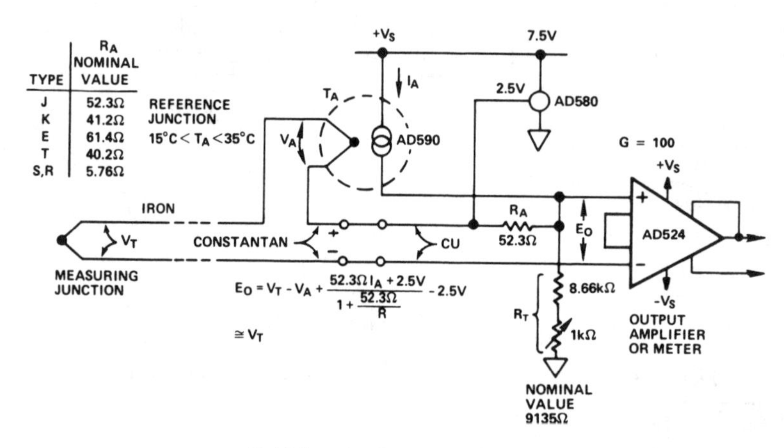

Cold-Junction Compensation

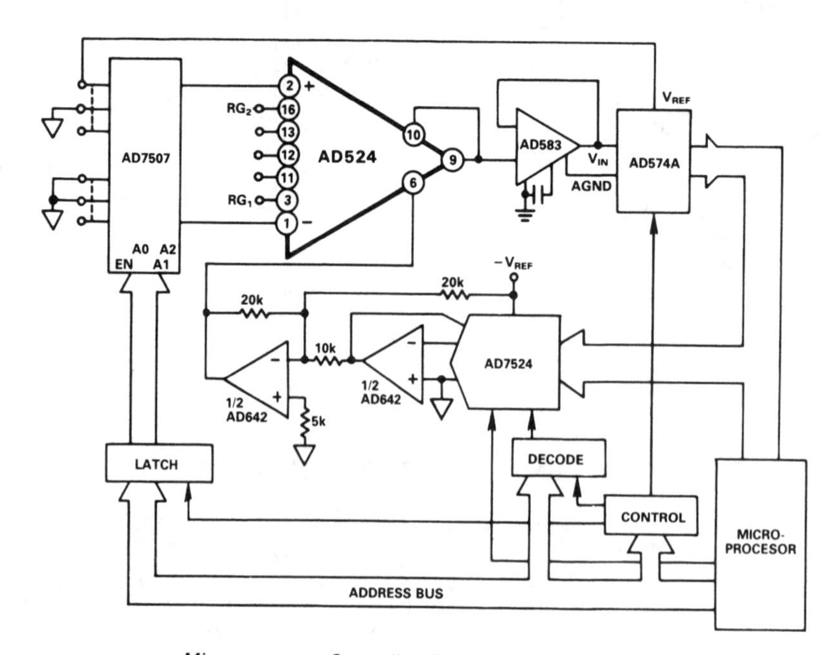

Microprocessor Controlled Data Acquisition System

Analog Devices AD526

Software-Programmable Gain Amplifier

- Digitally programmable binary gains from 1 to 16
- Two-chip cascade mode achieves binary gain from 1 to 256
- Gain error:
  - 0.01% max, gain = 1, 2, 4 (C grade)
  - 0.02% max, gain = 8, 16, (C grade)
  - 0.5ppm/°C drift over temperature
- Fast settling time
  - 10V signal change: 0.01% in 4.5μs (gain = 16)
  - gain change: 0.01% in 5.6μs (gain = 16)
- Low nonlinearity: ±0.005% FSR max (J grade)
- Excellent dc accuracy:
  - offset voltage: 0.5mV max (C grade)
  - offset voltage drift: 3μV/°C (C grade)
- TTL compatible digital inputs

The AD526 is a single-ended, monolithic software programmable gain amplifier (SPGA) that provides gains of 1, 2, 4, 8, and 16. It is complete, including amplifier, resistor network and TTL-compatible latched inputs, and requires no external components.

Low gain error and low nonlinearity make the AD526 ideal for precision instrumentation applications requiring programmable gain. The small-signal bandwidth is 350 kHz at a gain of 16. In addition, the AD526 provides excellent dc precision. The FET input stage results in a low bias current of 50pA. A guaranteed maximum input offset voltage of 0.5mV max (C grade) and low gain error (0.01%, G = 1, 2, 4, C grade) are accomplished using laser trimming technology.

To provide flexibility to the system designer, the AD526 can be operated in either latched or transparent mode. The force/sense configuration preserves accuracy when the output is connected to remote or low impedance loads.

The AD526 is offered in one commercial (0 to +70°C) grade (J) and three industrial grades (A, B, and C), which are specified from −40°C to +85°C. The S grade is specified from −55°C to +125°C. The military version is available processed to MIL-STD 883B, Rev C. The J grade is supplied in a 16-pin plastic DIP, and the other grades are offered in a 16-pin hermetic side-brazed ceramic DIP.

- **Dynamic range extension for ADC systems:** A single AD526 in conjunction with a 12-bit ADC can provide 96dB of dynamic range for ADC systems.
- **Gain ranging pre-amps:** The AD526 offers complete digital gain control with precise gains in binary steps from 1 to 16. Additional gains of 32, 64, 128, and 256 are possible by cascading two AD526s.

SPECIFICATIONS (@ V_S = ±15V, R_L = 2kΩ and T_A = +25°C unless otherwise specified)

	AD526J			AD526A			AD526B/S			AD526C			
Model	Min	Typ	Max	Min	Typ	Max	Min	Typ	Max	Min	Typ	Max	Units
GAIN													
Gain Range													
(Digitally Programmable)		1, 2, 4, 8, 16			1, 2, 4, 8, 16			1, 2, 4, 8, 16			1, 2, 4, 8, 16		
Gain Error													
G = 1			**0.05**			**0.02**			**0.01**			**0.01**	%
G = 2			**0.05**			**0.03**			**0.02**			**0.01**	%
G = 4			**0.10**			**0.03**			**0.02**			**0.01**	%
G = 8			**0.15**			**0.07**			**0.04**			**0.02**	%
G = 16			**0.15**			**0.07**			**0.04**			**0.02**	%
Gain Error Drift													
Over Temperature													
G = 1		0.5	2.0		0.5	2.0		0.5	2.0		0.5	2.0	ppm/°C
G = 2		0.5	2.0		0.5	2.0		0.5	2.0		0.5	2.0	ppm/°C
G = 4		0.5	3.0		0.5	3.0		0.5	3.0		0.5	3.0	ppm/°C
G = 8		0.5	5.0		0.5	5.0		0.5	5.0		0.5	5.0	ppm/°C
G = 16		1.0	5.0		1.0	5.0		1.0	5.0		1.0	5.0	ppm/°C
Gain Error (T_{min} to T_{max})													
G = 1			0.06			**0.03**			**0.02**			**0.015**	%
G = 2			0.06			**0.04**			**0.03**			**0.015**	%
G = 4			0.12			**0.04**			**0.03**			**0.015**	%
G = 8			0.17			**0.08**			**0.05**			**0.03**	%
G = 16			0.17			**0.08**			**0.05**			**0.03**	%
Nonlinearity													
G = 1			**0.005**			**0.005**			**0.005**			**0.0035**	% FSR
G = 2			0.001			0.001			0.001			0.001	% FSR
G = 4			0.001			0.001			0.001			0.001	% FSR
G = 8			0.001			0.001			0.001			0.001	% FSR
G = 16			0.001			0.001			0.001			0.001	% FSR
Nonlinearity (T_{min} to T_{max})													
G = 1			0.01			**0.01**			**0.01**			**0.007**	% FSR
G = 2			0.001			0.001			0.001			0.001	% FSR
G = 4			0.001			0.001			0.001			0.001	% FSR
G = 8			0.001			0.001			0.001			0.001	% FSR
G = 16			0.001			0.001			0.001			0.001	% FSR
VOLTAGE OFFSET, ALL GAINS													
Input Offset Voltage		0.4	**1.5**		0.25	**0.7**		0.25	**0.5**		0.25	**0.5**	mV
Input Offset Voltage Drift Over Temperature		5	20		3	10		3	10		3	10	μV/°C

Model	AD526J Min	AD526J Typ	AD526J Max	AD526A Min	AD526A Typ	AD526A Max	AD526B/S Min	AD526B/S Typ	AD526B/S Max	AD526C Min	AD526C Typ	AD526C Max	Units
Input Offset Voltage T_{min} to T_{max}			2.0			**1.0**			**0.8**			**0.8**	mV
Input Offset Voltage vs. Supply (V_S ± 10%)	80			**80**			**84**			**90**			dB
INPUT BIAS CURRENT													
Over Input Voltage Range ± 10V		50	**150**		50	**150**		50	**150**		50	**150**	pA
ANALOG INPUT CHARACTERISTICS													
Voltage Range (Linear Operation)	**±10**	±12		**±10**	±12		**±10**	±12		**±10**	±12		V
Capacitance		5			5			5			5		pF
RATED OUTPUT													
Voltage	**±10**	±12		**±10**	±12		**±10**	±12		**±10**	±12		V
Current (V_{OUT} = ± 10V)	**±5**	±10		**±5**	±10		**±5**	±10		**±5**	±10		mA
Short-Circuit Current	**15**	30		**15**	30		**15**	30		**15**	30		mA
DC Output Resistance		0.002			0.002			0.002			0.002		Ω
Load Capacitance (For Stable Operation)		700			700			700			700		pF
NOISE, ALL GAINS													
Voltage Noise, RTI 0.1Hz to 10Hz		3			3			3			3		μV p-p
Voltage Noise Density, RTI f = 10Hz		70			70			70			70		nV√Hz
f = 100Hz		60			60			60			60		nV√Hz
f = 1kHz		30			30			30			30		nV√Hz
f = 10kHz		25			25			25			25		nV√Hz
DYNAMIC RESPONSE													
– 3dB Bandwidth (Small Signal) G = 1		4.0			4.0			4.0			4.0		MHz
G = 2		2.0			2.0			2.0			2.0		MHz
G = 4		1.5			1.5			1.5			1.5		MHz
G = 8		0.65			0.65			0.65			0.65		MHz
G = 16		0.35			0.35			0.35			0.35		MHz
Signal Settling Time to 0.01% (ΔV_{OUT} = ± 10V) G = 1		2.1	4		2.1	4		2.1	4		2.1	4	μs
G = 2		2.5	5		2.5	5		2.5	5		2.5	5	μs
G = 4		2.7	5		2.7	5		2.7	5		2.7	5	μs
G = 8		3.6	7		3.6	7		3.6	7		3.6	7	μs
G = 16		4.1	7		4.1	7		4.1	7		4.1	7	μs
Full Power Bandwidth G = 1, 2, 4		0.10			0.10			0.10			0.10		MHz
G = 8, 16		0.35			0.35			0.35			0.35		MHz
Slew Rate G = 1, 2, 4	**4**	6		**4**	6		**4**	6		**4**	6		V/μs
G = 8, 16	**18**	24		**18**	24		**18**	24		**18**	24		V/μs
DIGITAL INPUTS (T_{min} to T_{max})													
Input Current (V_H = 5V)	60	100	140	60	100	140	60	100	140	60	100	140	μA
Logic "1"	2		6	2		6	2		6	2		6	V
Logic "0"	0		0.8	0		0.8	0		0.8	0		0.8	V
TIMING (V_L = 0.2V, V_H = 3.7V)													
A0, A1, A2 T_C	50			50			50			50			ns
T_S	30			30			30			30			ns
T_H	30			30			30			30			ns
B T_C	50			50			50			50			ns
T_S	40			40			40			40			ns
T_H	10			10			10			10			ns
TEMPERATURE RANGE													
Specified Performance	0		+ 70	– 40		+ 85	– 40/ – 55		+ 85/ + 125	– 40		+ 85	°C
Storage	– 65		+ 125	– 65		+ 150	– 65		+ 150	– 65		+ 150	°C
POWER SUPPLY													
Operating Range	**±4.5**		**±16.5**	**±4.5**		**±16.5**	**±4.5**		**±16.5**	**±4.5**		**±16.5**	V
Positive Supply Current		10	**14**		10	**14**		10	**14**		10	**14**	mA
Negative Supply Current		10	**13**		10	**13**		10	**13**		10	**13**	mA
PACKAGE OPTIONS													
Plastic (N-16)		AD526JN											
Ceramic DIP (D-16)					AD526AD			AD526BD AD526SD AD526SD/883B			AD526CD		

NOTE

FSR = Full-Scale Range = 20V.
RTI = Referred to Input.

Specifications subject to change without notice.

Specifications shown in boldface are tested on all production units at final electrical test. All min and max specifications are guaranteed, although only those shown in boldface are tested on all production units.

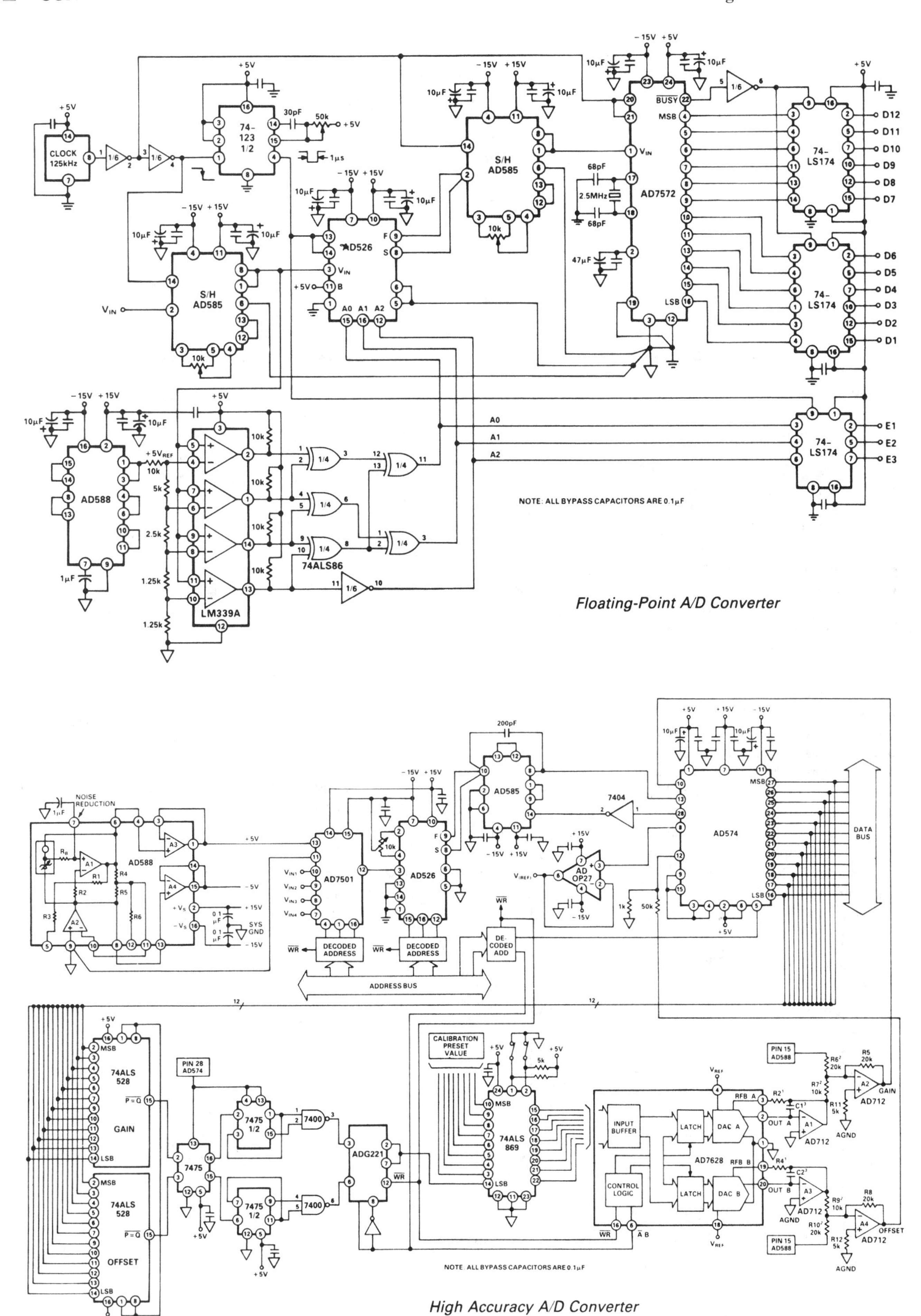

Floating-Point A/D Converter

High Accuracy A/D Converter

AD526 PIN CONFIGURATION

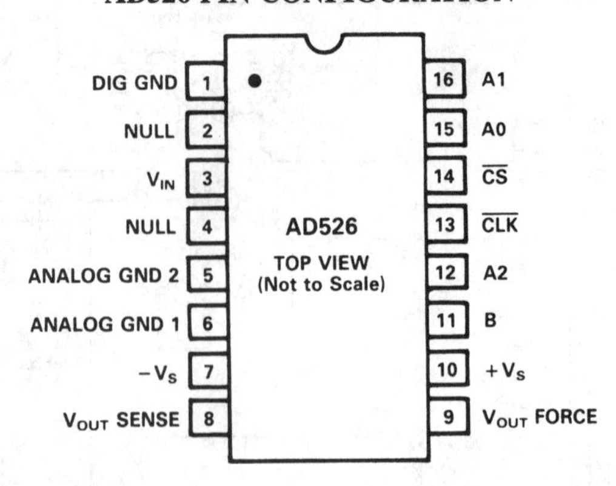

Analog Devices AD624

Precision Instrumentation Amplifier

- Low noise: 0.2μV p-p 0.1 Hz to 10 Hz
- Low gain TC: 5ppm max (G = 1)
- Low nonlinearity: 0.001% max (G = 1 to 200)
- High CMRR: 130dB min (G = 500 to 1000)
- Low input offset voltage: 25μV, max
- Low input offset voltage drift: 0.25μV/°C max
- Gain bandwidth product: 25 MHz
- Pin programmable gains of 1, 100, 200, 500, 1000
- No external components required
- Internally compensated

The AD624 is a high-precision low-noise instrumentation amplifier designed primarily for use with low-level transducers, including load cells, strain gauges, and pressure transducers. An outstanding combination of low noise, high-gain accuracy, low-gain temperature coefficient and high linearity make the AD624 ideal for use in high-resolution data acquisition systems.

The AD624C has an input offset voltage drift of less than 0.25μV/°C, output offset voltage drift of less than 10μV/°C, CMRR above 80dB at unity gain (130dB at G = 500) and a maximum nonlinearity of 0.001% at G = 1. In addition to these outstanding dc specifications, the AD624 exhibits superior ac performance as well. A 25-MHz gain bandwidth product 5V/μs slew rate and 15μs settling time permit the use of the AD624 in high-speed data acquisition applications.

The AD624 does not need any external components for pretrimmed gains of 1, 100, 200, 500, and 1000. Additional gains, such as 250 and 333, can be programmed within one percent accuracy with external jumpers. A single external resistor can also be used to set the 624's gain to any value in the range of 1 to 10,000.

- The AD624 offers outstanding noise performance. Input noise is typically less than $6nV/\sqrt{Hz}$ at 1 kHz.
- The AD624 is a functionally complete instrumentation amplifier. Pin programmable gains of 1, 100, 200, 500, and 1000 are provided on the chip. Other gains are achieved through the use of a single external resistor.
- The offset voltage, offset voltage drift, gain accuracy, and gain temperature coefficients are guaranteed for all pretrimmed gains.
- The AD624 provides totally independent input and output offset nulling terminals for high-precision applications. This minimizes the effect of offset voltage in gain ranging applications.
- A sense terminal is provided to enable the user to minimize the errors induced through long leads. A reference terminal is also provided to permit level shifting at the output.

AD624 FUNCTIONAL BLOCK DIAGRAM

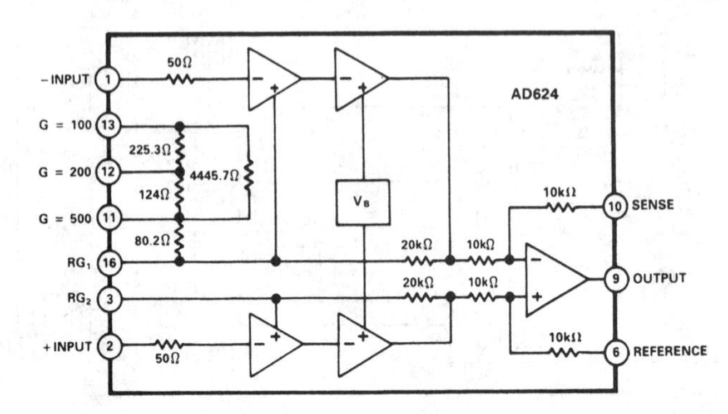

PIN CONFIGURATION

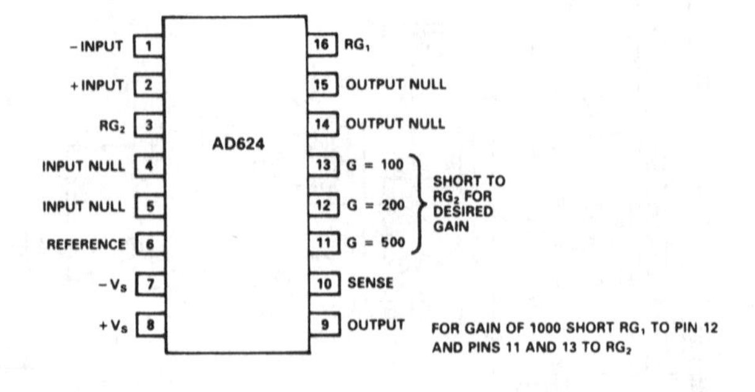

SPECIFICATIONS (@ V_S = ±15V, R_L = 2kΩ and T_A = +25°C unless otherwise specified)

Model	AD624A Min	AD624A Typ	AD624A Max	AD624B Min	AD624B Typ	AD624B Max	AD624C Min	AD624C Typ	AD624C Max	AD624S Min	AD624S Typ	AD624S Max	Units
GAIN													
Gain Equation													
(External Resistor Gain Programming)	$\left[\frac{40,000}{R_G}+1\right] \pm 20\%$			$\left[\frac{40,000}{R_G}+1\right] \pm 20\%$			$\left[\frac{40,000}{R_G}+1\right] \pm 20\%$			$\left[\frac{40,000}{R_G}+1\right] \pm 20\%$			
Gain Range (Pin Programmable)	1 to 1000			1 to 1000			1 to 1000			1 to 1000			
Gain Error													
G = 1			±0.05			±0.03			±0.02			±0.05	%
G = 100			±0.25			±0.15			±0.1			±0.25	%
G = 200, 500			±0.5			±0.35			±0.25			±0.5	%
G = 1000			±1.0			±1.0			±1.0			±1.0	%
Nonlinearity													
G = 1			±0.005			±0.003			±0.001			±0.005	%
G = 100, 200			±0.005			±0.003			±0.001			±0.005	%
G = 500, 1000			±0.005			±0.005			±0.005			±0.005	%
Gain vs. Temperature													
G = 1			5			5			5			5	ppm/°C
G = 100, 200			10			10			10			10	ppm/°C
G = 500, 1000			25			15			15			15	ppm/°C
VOLTAGE OFFSET (May be Nulled)													
Input Offset Voltage			200			75			25			75	μV
vs. Temperature			2			0.5			0.25			2.0	μV/°C
Output Offset Voltage			5			3			2			3	mV
vs. Temperature			50			25			10			50	μV/°C
Offset Referred to the Input vs. Supply													
G = 1	70			75			80			75			dB
G = 100, 200	95			105			110			105			dB
G = 500, 1000	100			110			115			110			dB
INPUT CURRENT													
Input Bias Current			±50			±25			±15			±50	nA
vs. Temperature		±50			±50			±50			±50		pA/°C
Input Offset Current			±35			±15			±10			±35	nA
vs. Temperature		±20			±20			±20			±20		pA/°C
INPUT													
Input Impedance													
Differential Resistance		10^9			10^9			10^9			10^9		Ω
Differential Capacitance		10			10			10			10		pF
Common-Mode Resistance		10^9			10^9			10^9			10^9		Ω
Common-Mode Capacitance		10			10			10			10		pF
Input Voltage Range													
Max Differ. Input Linear (V_D)		±10			±10			±10			±10		V
Max Common-Mode Linear (V_{CM})		$12V - \left(\frac{G}{2} \times V_D\right)$			$12V - \left(\frac{G}{2} \times V_D\right)$			$12V - \left(\frac{G}{2} \times V_D\right)$			$12V - \left(\frac{G}{2} \times V_D\right)$		V
Common-Mode Rejection dc to 60Hz with 1kΩ Source Imbalance													
G = 1	70			75			80			70			dB
G = 100, 200	100			105			110			100			dB
G = 500, 1000	110			120			130			110			dB
OUTPUT RATING													
V_{OUT}, R_L = 2kΩ		±10			±10			±10			±10		V
DYNAMIC RESPONSE													
Small Signal − 3dB													
G = 1		1			1			1			1		MHz
G = 100		150			150			150			150		kHz
G = 200		100			100			100			100		kHz
G = 500		50			50			50			50		kHz
G = 1000		25			25			25			25		kHz
Slew Rate		5.0			5.0			5.0			5.0		V/μs
Settling Time to 0.01%, 20V Step													
G = 1 to 200		15			15			15			15		μs
G = 500		35			35			35			35		μs
G = 1000		75			75			75			75		μs
NOISE													
Voltage Noise, 1kHz													
R.T.I.		4			4			4			4		nV/√Hz
R.T.O.		75			75			75			75		nV/√Hz
R.T.I., 0.1 to 10Hz													
G = 1		10			10			10			10		μV p-p
G = 100		0.3			0.3			0.3			0.3		μV p-p
G = 200, 500, 1000		0.2			0.2			0.2			0.2		μV p-p
Current Noise													
0.1Hz to 10Hz		60			60			60			60		pA p-p
SENSE INPUT													
R_{IN}	8	10	12	8	10	12	8	10	12	8	10	12	kΩ
I_{IN}		30			30			30			30		μA
Voltage Range	±10			±10			±10			±10			V
Gain to Output		1			1			1			1		%

Model	AD624A Min	AD624A Typ	AD624A Max	AD624B Min	AD624B Typ	AD624B Max	AD624C Min	AD624C Typ	AD624C Max	AD624S Min	AD624S Typ	AD624S Max	Units
REFERENCE INPUT													
R_{IN}	16	20	24	16	20	24	16	20	24	16	20	24	kΩ
I_{IN}		30			30			30			30		μA
Voltage Range	±10			±10			±10			±10			V
Gain to Output		1			1			1			1		%
TEMPERATURE RANGE													
Specified Performance	−25		+85	−25		+85	−25		+85	−55		+125	°C
Storage	−65		+150	−65		+150	−65		+150	−65		+150	°C
POWER SUPPLY													
Power Supply Range	**±6**	±15	**±18**	**±6**	±15	**±18**	**±6**	±15	**±18**	**±6**	±15	**±18**	V
Quiescent Current		3.5	5		3.5	5		3.5	5		3.5	5	mA
PACKAGE													
Ceramic (D-16)		AD624A			AD624B			AD624C			AD624S		

NOTES

Specifications subject to change without notice.

Specifications shown in boldface are tested on all production units at final electrical test. Results from those tests are used to calculate outgoing quality levels. All min and max specifications are guaranteed, although only those shown in boldface are tested on all production units.

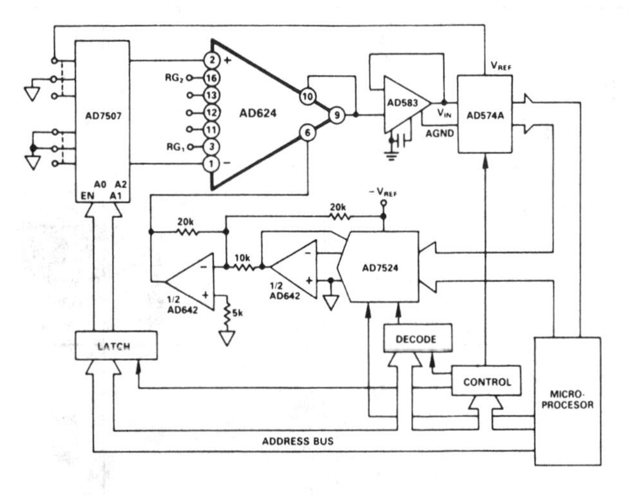

Microprocessor Controlled Data Acquisition System

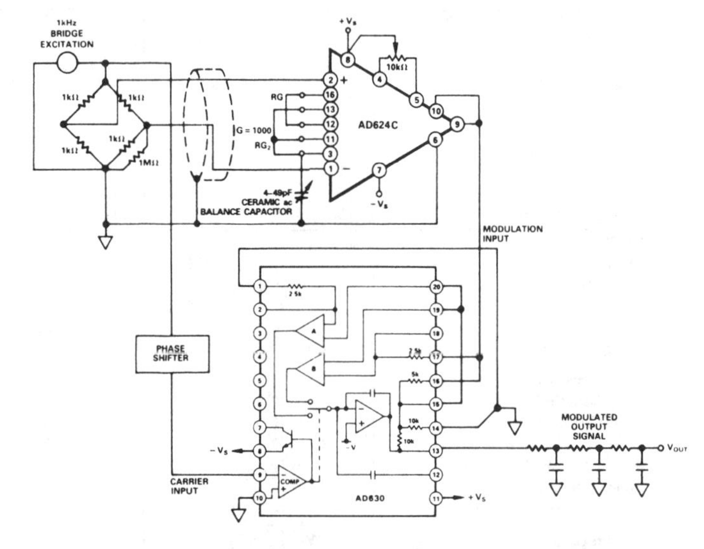

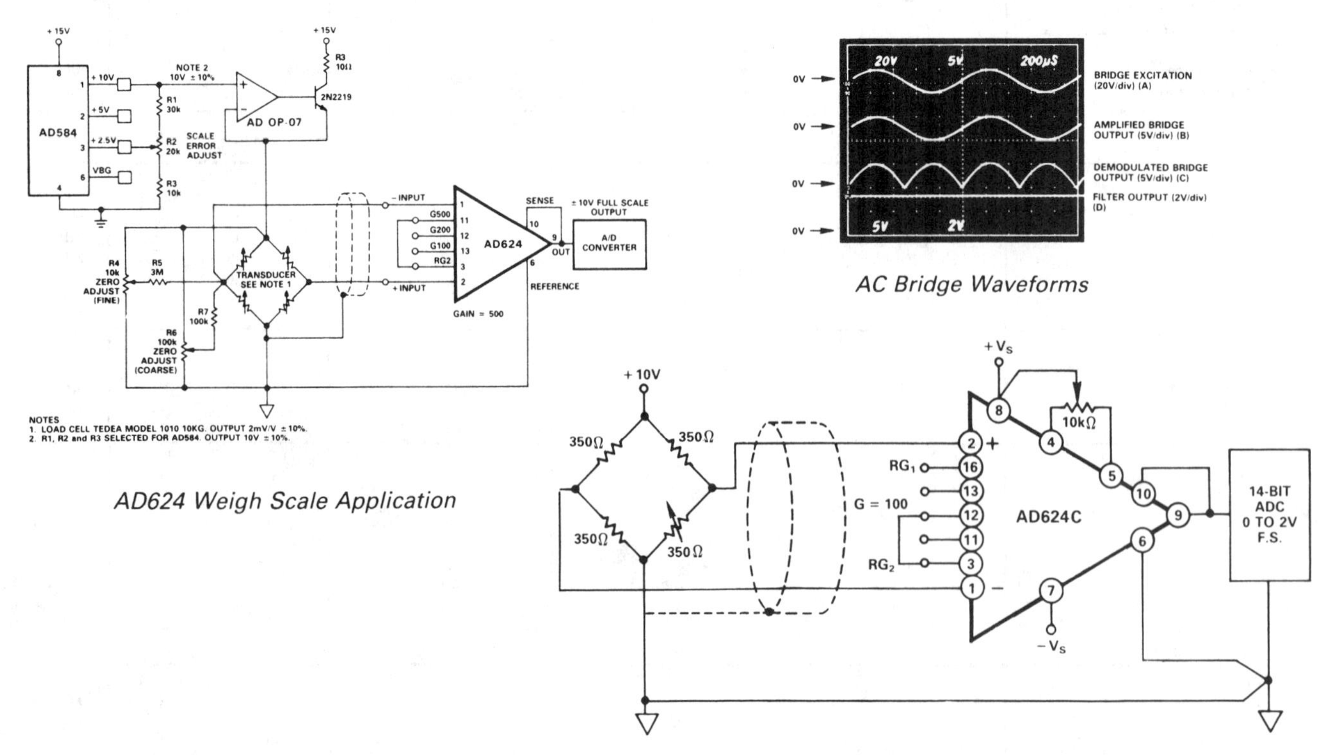

AD624 Weigh Scale Application

AC Bridge Waveforms

Typical Bridge Application

TYPICAL CHARACTERISTICS

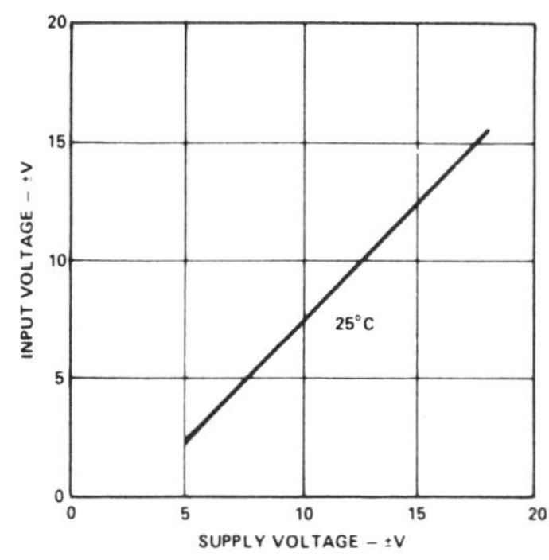

Figure 1. Input Voltage Range vs. Supply Voltage, G = 1

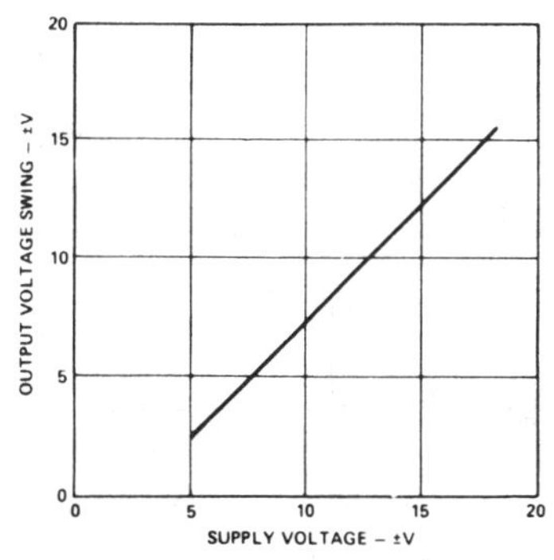

Figure 2. Output Voltage Swing vs. Supply Voltage

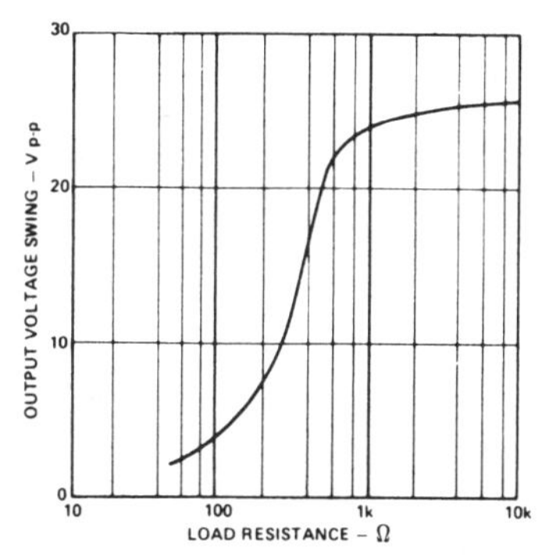

Figure 3. Output Voltage Swing vs. Resistive Load

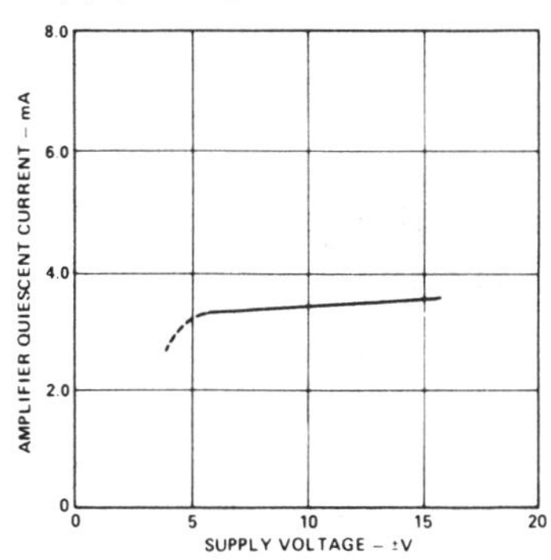

Figure 4. Quiescent Current vs. Supply Voltage

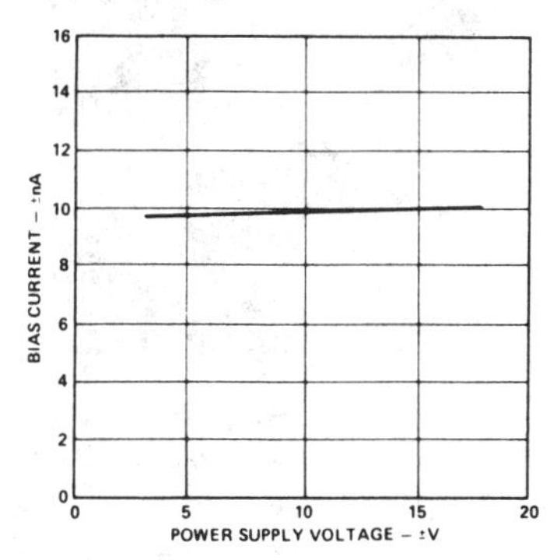

Figure 5. Input Bias Current vs. Supply Voltage

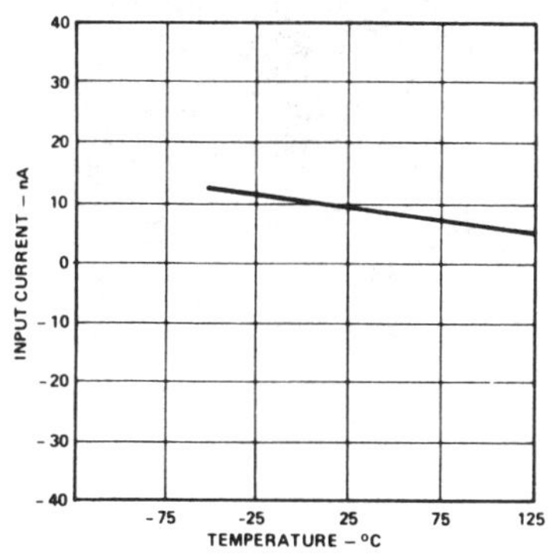

Figure 6. Input Bias Current vs. Temperature

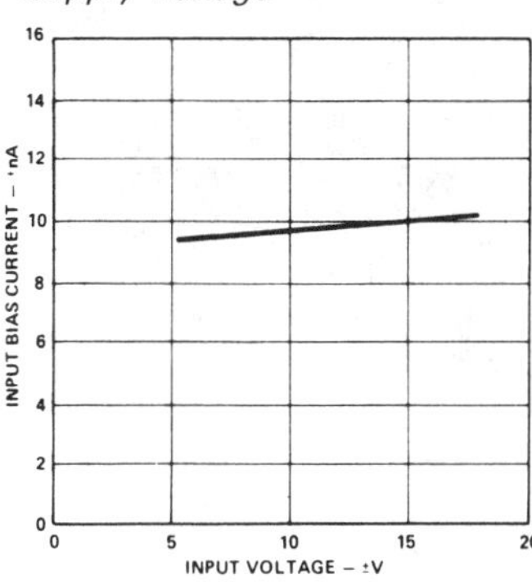

Figure 7. Input Bias Current vs. CMV

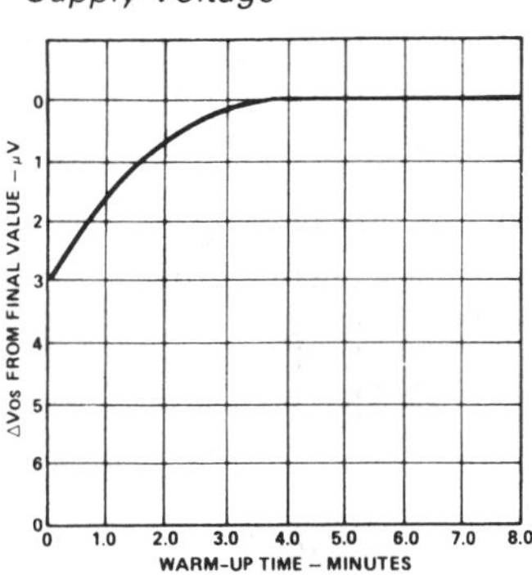

Figure 8. Offset Voltage, RTI, Turn On Drift

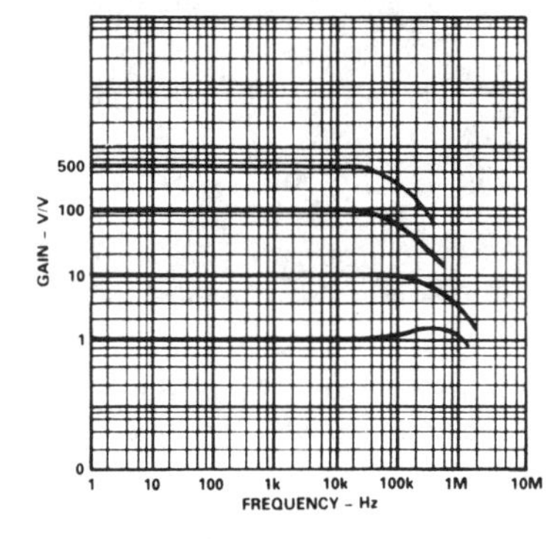

Figure 9. Gain vs. Frequency

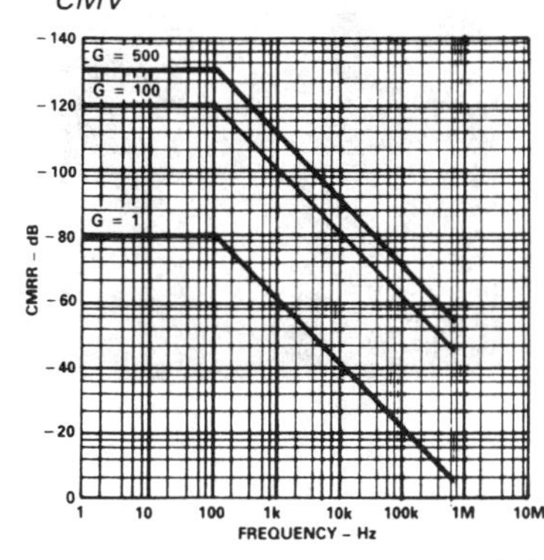

Figure 10. CMRR vs. Frequency RTI, Zero to 1k Source Imbalance

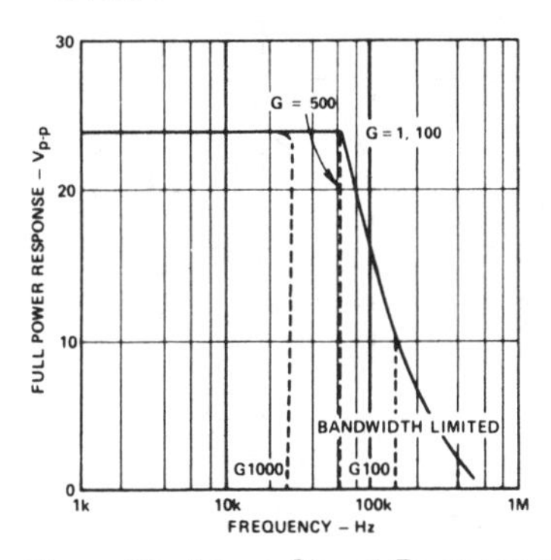

Figure 11. Large Signal Frequency Response

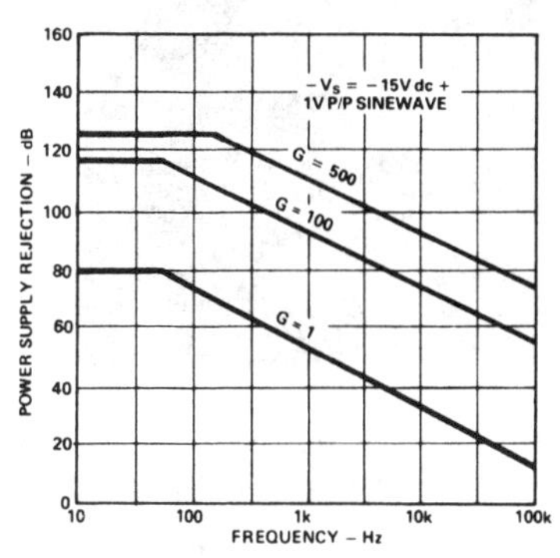

Figure 12. Positive PSRR vs. Frequency

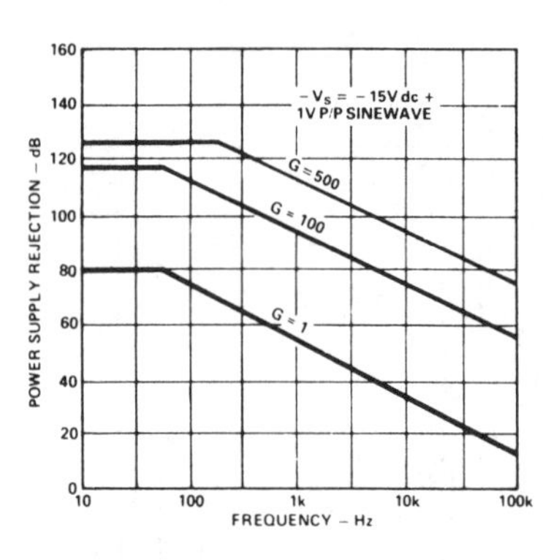

Figure 13. Negative PSRR vs. Frequency

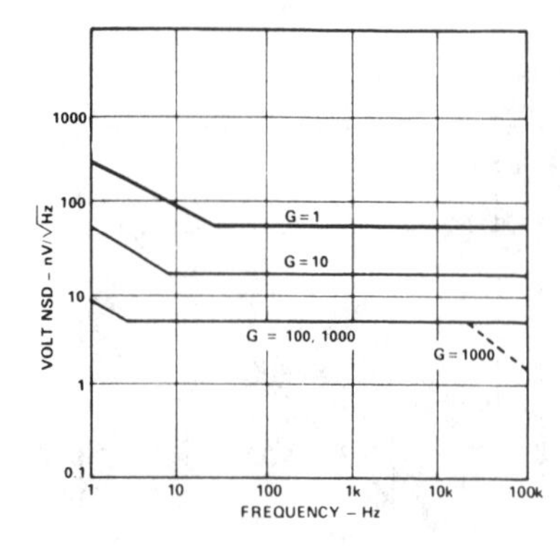

Figure 14. RTI Noise Spectral Density vs. Gain

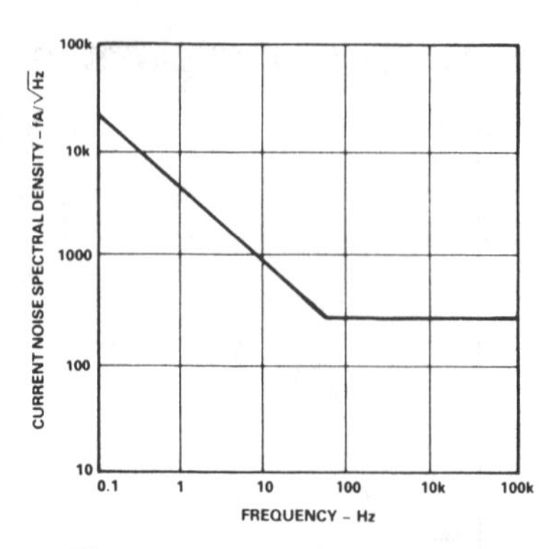

Figure 15. Input Current Noise

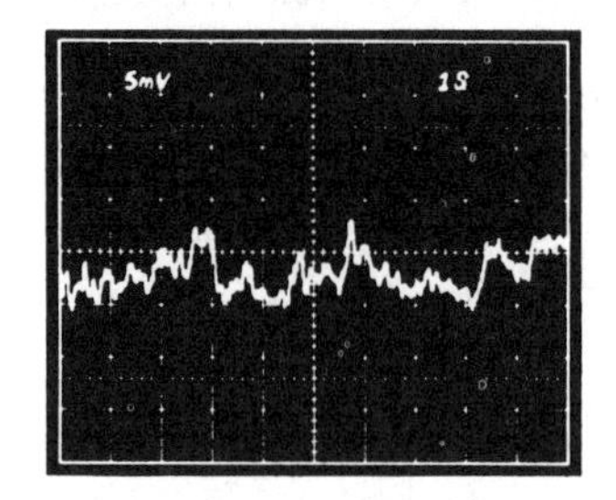

Figure 16. Low Frequency Voltage Noise – G = 1 (System Gain = 1000)

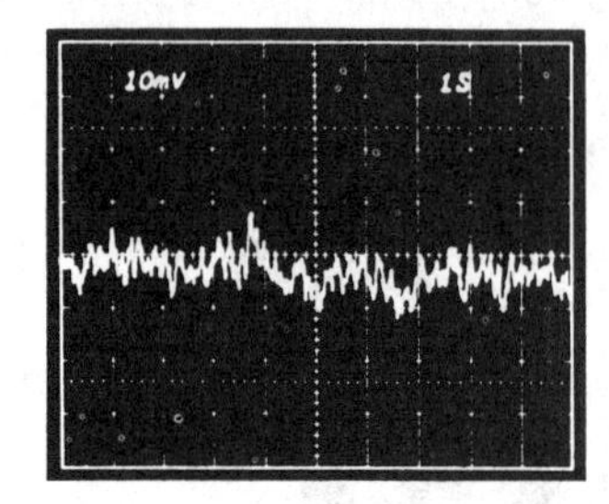

Figure 17. Low Frequency Voltage Noise – G = 1000 (System Gain = 100,000)

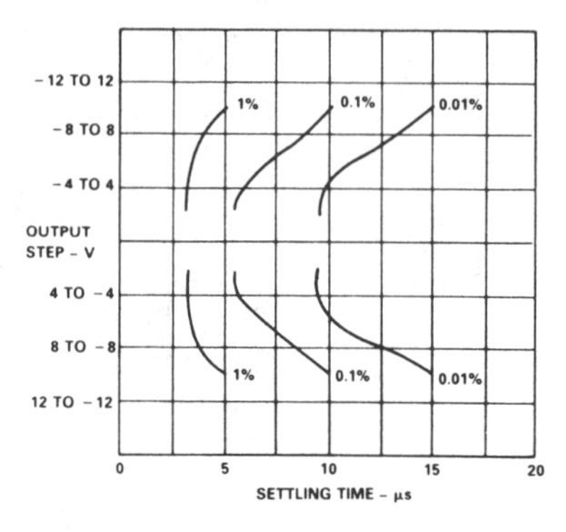

Figure 18. Settling Time Gain = 1

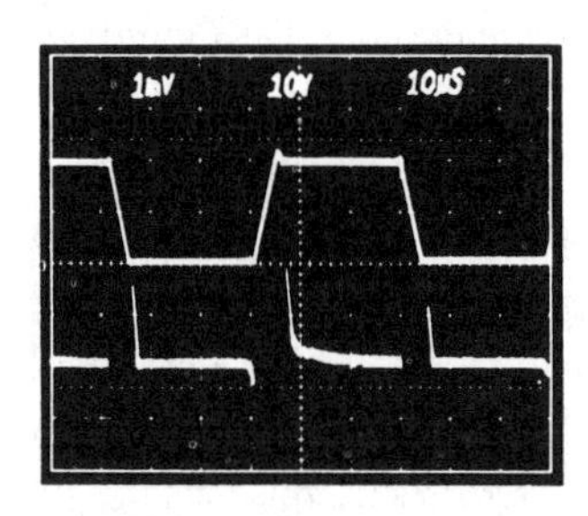

Figure 19. Large Signal Pulse Response and Settling Time – G = 1

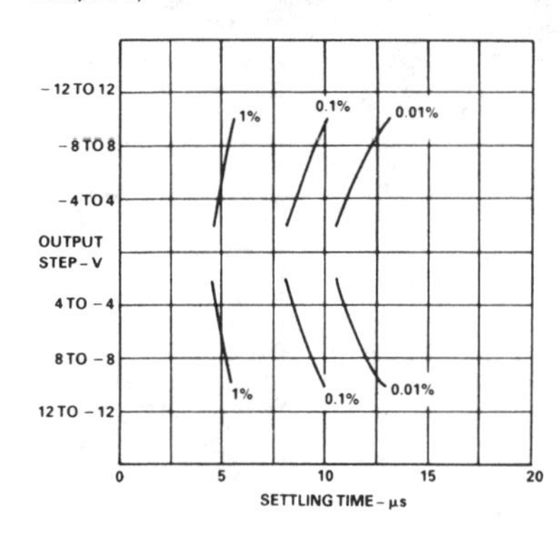

Figure 20. Settling Time Gain = 100

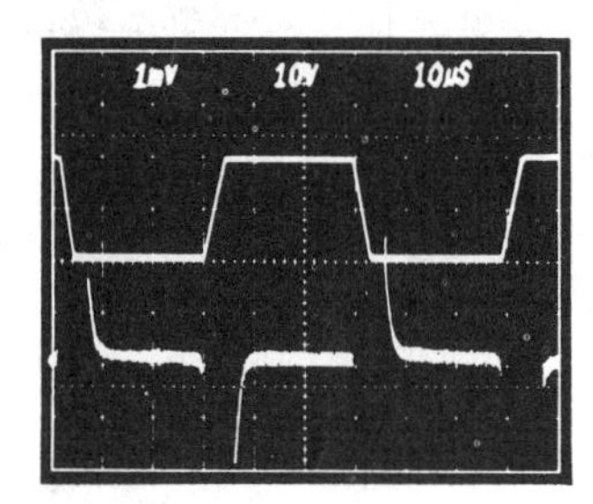

Figure 21. Large Signal Pulse Response and Settling Time G = 100

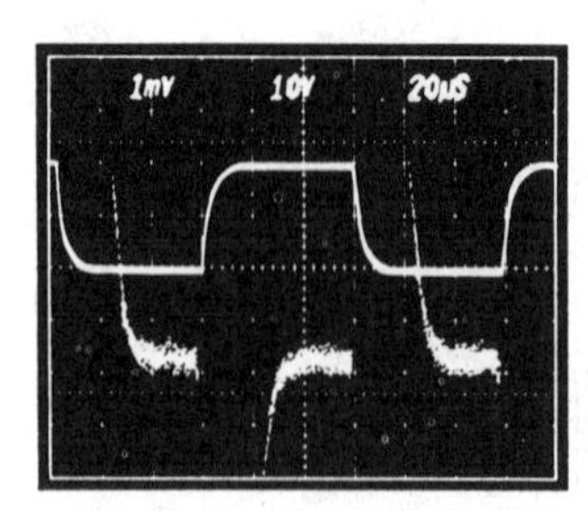

Figure 22. Range Signal Pulse Response and Settling Time G = 500

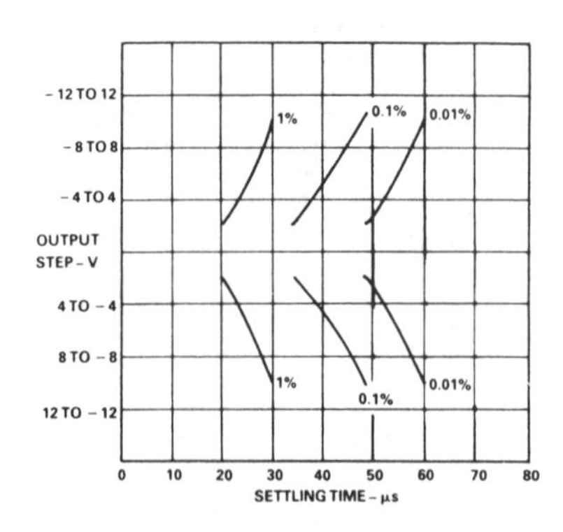

Figure 23. Settling Time Gain = 1000

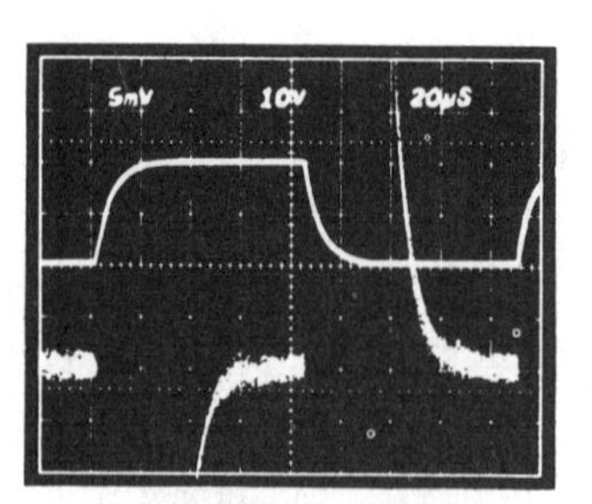

Figure 24. Large Signal Pulse Response and Settling Time G = 1000

Analog Devices AD625

Programmable-Gain Instrumentation Amplifier

- User programmed gains of 1 to 10,000
- Low gain error: 0.02% max
- Low gain TC: 5ppm/°C max
- Low nonlinearity: 0.001% max
- Low offset voltage: 25μV
- Low noise 4nV/$\sqrt{\text{Hz}}$ (at 1 kHz) RTI
- Gain bandwidth product: 25 MHz
- 16-pin ceramic or plastic DIP package
- MIL-standard parts available
- Low cost

The AD625 is a precision instrumentation amplifier specifically designed to fulfill circuits requiring nonstandard gains (i.e., gains not easily achievable with devices such as the AD524 and AD624) and circuits requiring a low-cost, precision software programmable-gain amplifier.

For low noise, high CMRR, and low drift, the AD625JN is the most cost-effective instrumentation amplifier solution available. An additional three resistors allow the user to set any gain from 1 to 10,000. The error contribution of the AD625JN is less than 0.05% gain error and under 5ppm/°C gain TC; performance limitations are primarily determined by the external resistors. Common-mode rejection is independent of the feedback resistor matching.

A software programmable-gain amplifier (SPGA) can be configured with the addition of a CMOS multiplexer (or other switch network), and a suitable resistor network. Because the on resistance of the switches is removed from the signal path, an AD625-based SPGA will deliver 12-bit precision, and can be programmed for any set of gains between 1 and 10,000, with completely user-selected gain steps.

For the highest precision, the AD625C offers an input offset voltage drift of less than 0.25μV/°C, output offset drift below 15μV/°C, and a maximum nonlinearity of 0.001% at G = 1. All grades exhibit excellent ac performance; a 25 MHz gain bandwidth product, 5Vμs slew rate, and 15 μs settling time.

The AD625 is available in three accuracy grades (A, B, C) for the industrial (−25°C to + 85°C) temperature range, two grades (J, K) for the commercial (0 to +70°C) temperature range, and one (S) grade rated over the extended (−55°C to +125°C) temperature range.

- The AD625 affords up to 16-bit precision for user-selected fixed gains from 1 to 10,000. Any gain in this range can be programmed by 3 external resistors.
- A 12-bit software programmable gain amplifier can be configured using the AD625, a CMOS multiplexer and a resistor network. Unlike previous instrumentation amplifier designs, the on resistance of a CMOS switch does not affect the gain accuracy.
- The gain accuracy and gain temperature coefficient of the amplifier circuit are primarily dependent on the user-selected external resistors.
- The AD625 provides totally independent input and output offset nulling terminals for high-precision applications. This minimizes the effects of offset voltage in gain-ranging applications.
- The proprietary design of the AD625 provides input voltage noise of 4nV/$\sqrt{\text{Hz}}$ at 1 kHz.
- External resistor matching is not required to maintain high common-mode rejection.

AD625 FUNCTIONAL BLOCK DIAGRAM

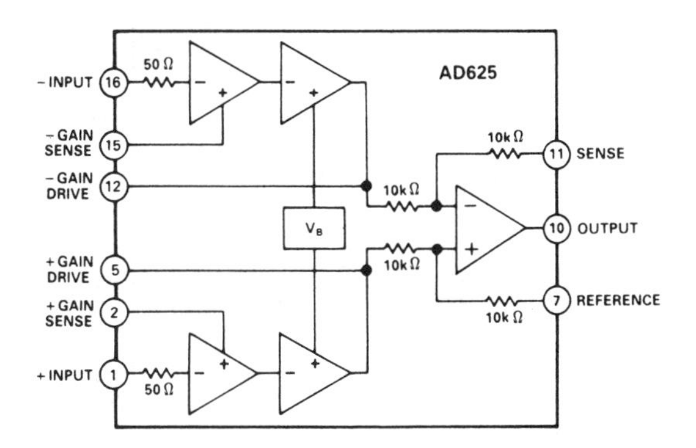

PIN CONFIGURATION

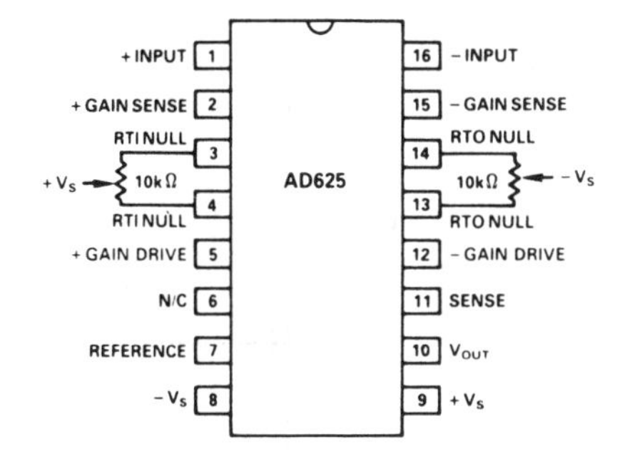

SPECIFICATIONS (typical @ $V_S = \pm 15V$, $R_L = 2k\Omega$ and $T_A = +25°C$ unless otherwise specified)

Model	AD625A/J/S Min	AD625A/J/S Typ	AD625A/J/S Max	AD625B/K Min	AD625B/K Typ	AD625B/K Max	AD625C Min	AD625C Typ	AD625C Max	Units
GAIN										
Gain Equation		$\frac{2R_F}{R_G}+1$			$\frac{2R_F}{R_G}+1$			$\frac{2R_F}{R_G}+1$		
Gain Range	1		10,000	1		10,000	1		10,000	
Gain Error[1]		±.035	**±0.05**		±0.02	**±0.03**		±0.01	**±0.02**	%
Nonlinearity, Gain = 1-256			±0.005			±0.002			±0.001	%
Gain>256			±0.01			±0.008			±0.005	%
Gain vs. Temp. Gain<1000[1]			5			5			5	ppm/°C
GAIN SENSE INPUT										
Gain Sense Current		300	**500**		150	**250**		50	**100**	nA
vs. Temperature		5	20		2	15		2	10	nA/°C
Gain Sense Offset Current		150	**500**		75	**250**		50	**100**	nA
vs. Temperature		2	15		1	10		1	5	nA/°C
VOLTAGE OFFSET (May be Nulled)										
Input Offset Voltage		50	**200**		25	**50**		10	**25**	μV
vs. Temperature		1	2/2		0.25	**0.50/1**		0.1	**0.25**	μV/°C
Output Offset Voltage		4	5		2	**3**		1	**2**	mV
vs. Temperature		20	50/50		10	**25/40**		10	**15**	μV/°C
Offset Referred to the Input vs. Supply										
G = 1	**70**	75		**75**	85		**80**	90		dB
G = 10	85	95		90	100		95	105		dB
G = 100	95	100		105	110		110	120		dB
G = 1000	**100**	110		**110**	120		**115**	140		dB
INPUT CURRENT										
Input Bias Current		±30	**±50**		±20	**±25**		±10	**±15**	nA
vs. Temperature		±50			±50			±50		pA/°C
Input Offset Current		±2	**±35**		±1	**±15**		±1	**±5**	nA
vs. Temperature		±20			±20			±20		pA/°C
INPUT										
Input Impedance										
Differential Resistance		1			1			1		GΩ
Differential Capacitance		4			4			4		pF
Common-Mode Resistance		1			1			1		GΩ
Common-Mode Capacitance		4			4			4		pF
Input Voltage Range										
Differ. Input Linear (V_D)			±10			±10			±10	V
Common-Mode Linear (V_{CM})		$12V - \left(\frac{G}{2} \times V_D\right)$			$12V - \left(\frac{G}{2} \times V_D\right)$			$12V - \left(\frac{G}{2} \times V_D\right)$		
Common-Mode Rejection Ratio dc to 60Hz with 1kΩ Source Imbalance										
G = 1	**70**	75		**75**	85		**80**	90		dB
G = 10	90	95		95	105		100	115		dB
G = 100	100	105		105	115		110	125		dB
G = 1000	**110**	115		**115**	125		**120**	140		dB
OUTPUT RATING		±10V @ 5mA			±10V @ 5mA			±10V @ 5mA		
DYNAMIC RESPONSE										
Small Signal −3dB										
G = 1 (R_F = 20kΩ)		650			650			650		kHz
G = 10		400			400			400		kHz
G = 100		150			150			150		kHz
G = 1000		25			25			25		kHz
Slew Rate		5.0			5.0			5.0		V/μs
Settling Time to 0.01%, 20V Step										
G = 1 to 200		15			15			15		μs
G = 500		35			35			35		μs
G = 1000		75			75			75		μs
NOISE										
Voltage Noise, 1kHz										
R.T.I.		4			4			4		nV/√Hz
R.T.O.		75			75			75		nV/√Hz
R.T.I., 0.1 to 10Hz										
G = 1		10			10			10		μV p-p
G = 10		1.0			1.0			1.0		μV p-p
G = 100		0.3			0.3			0.3		μV p-p
G = 1000		0.2			0.2			0.2		μV p-p
Current Noise										
0.1Hz to 10Hz		60			60			60		pA p-p

Model	AD625A/J/S Min	Typ	Max	AD625B/K Min	Typ	Max	AD625C Min	Typ	Max	Units
SENSE INPUT										
R_{IN}		10			10			10		kΩ
I_{IN}		30			30			30		μA
Voltage Range	± 10			± 10			± 10			V
Gain to Output		1 ± 0.01			1 ± 0.01			1 ± 0.01		%
REFERENCE INPUT										
R_{IN}		20			20			20		kΩ
I_{IN}		30			30			30		μA
Voltage Range	± 10			± 10			± 10			V
Gain to Output		1 ± 0.01			1 ± 0.01			1 ± 0.01		%
TEMPERATURE RANGE										
Specified Performance										
J/K Grades	0		+ 70	0		+ 70				°C
A/B/C Grades	− 25		+ 85	− 25		+ 85	− 25		+ 85	°C
S Grade	− 55		+ 125							
Storage	− 65		+ 150	− 65		+ 150	− 65		+ 150	°C
POWER SUPPLY										
Power Supply Range		± 6 to ± 18			± 6 to ± 18			± 6 to ± 18		V
Quiescent Current		3.5	5		3.5	5		3.5	5	mA
PACKAGE OPTIONS[2]										
Ceramic (D-16)		AD625AD/SD			AD625BD			AD625CD		
Plastic DIP (N-16)		AD625JN			AD625KN					

NOTES
[1]Gain error and gain TC are for the AD625 only. Resistor network errors will add to the specified errors.
[2]See Section 16 for package outline information.

Specifications subject to change without notice.

Specifications shown in boldface are tested on all production units at final electrical test. Results from those tests are used to calculate outgoing quality levels. All mim and max specifications are guaranteed, although only those shown in boldface are tested on all production units.

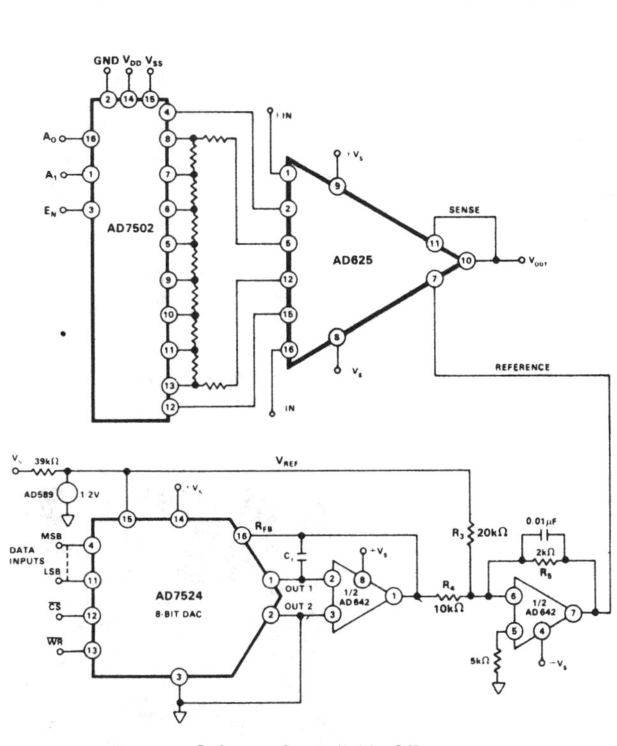

Software Controllable Offset

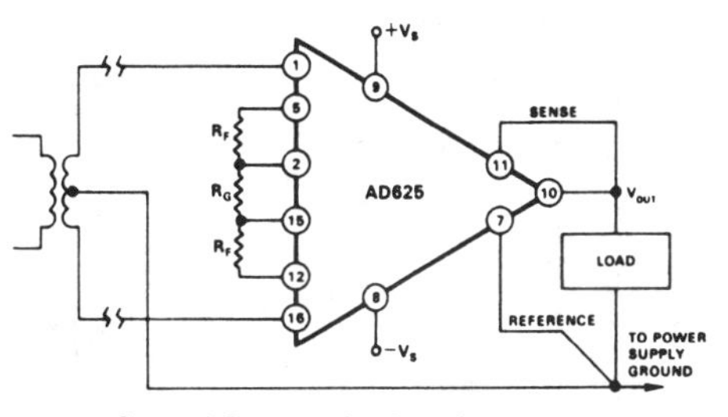

Ground Returns for Bias Currents with Transformer Coupled Inputs

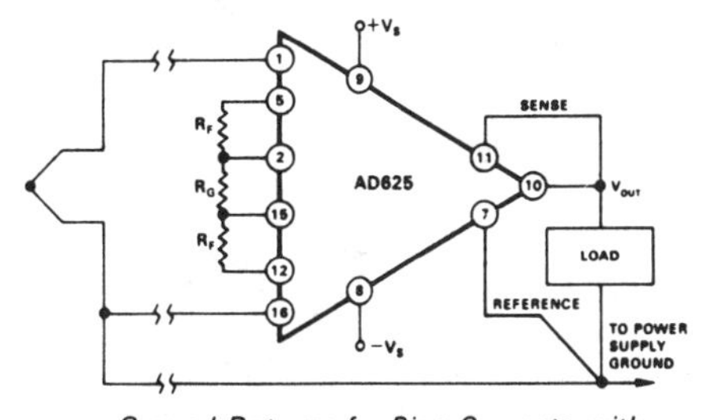

Ground Returns for Bias Currents with Thermocouple Input

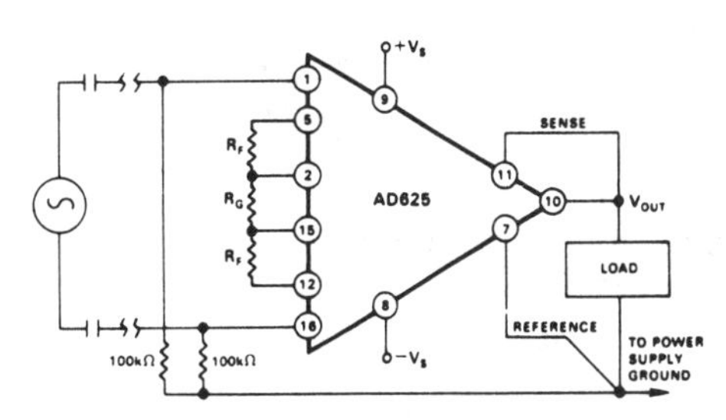

Ground Returns for Bias Currents with AC Coupled Inputs

TYPICAL CHARACTERISTICS

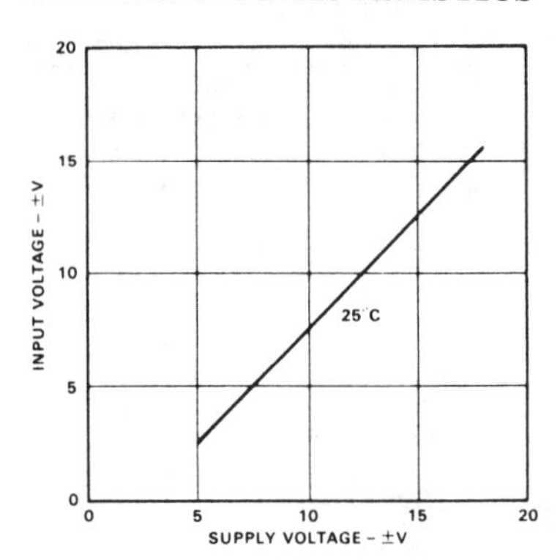

Figure 1. Input Voltage Range vs. Supply Voltage, G = 1

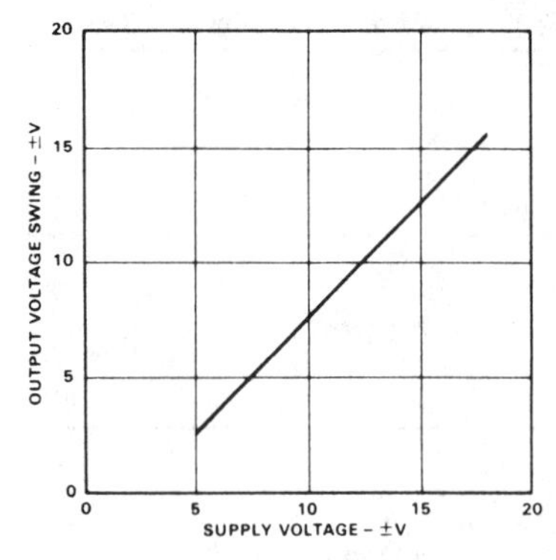

Figure 2. Output Voltage Swing vs. Supply Voltage

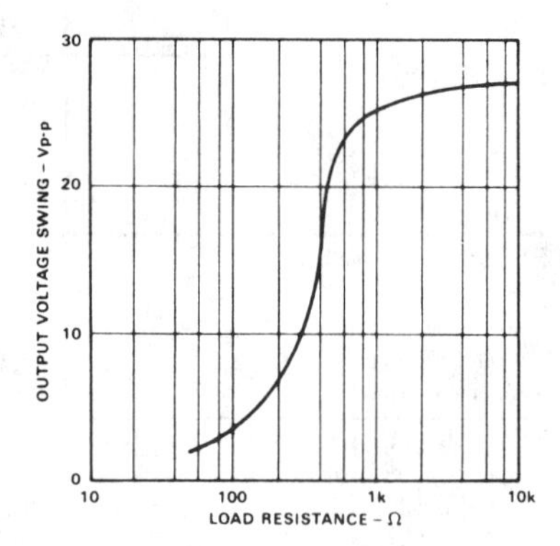

Figure 3. Output Voltage Swing vs. Resistive Load

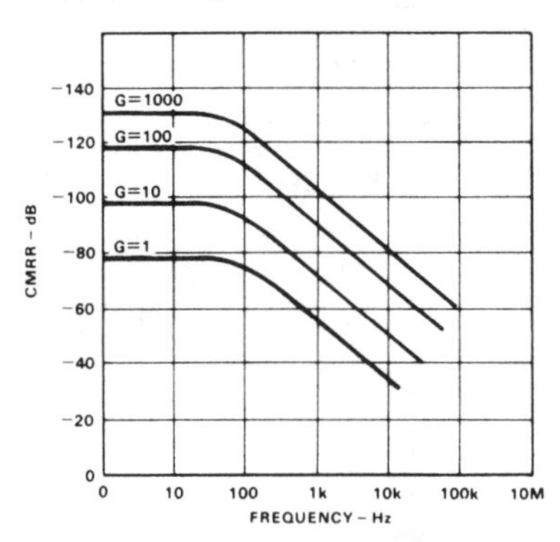

Figure 4. CMRR vs. Frequency RTI, Zero to 1kΩ Source Imbalance

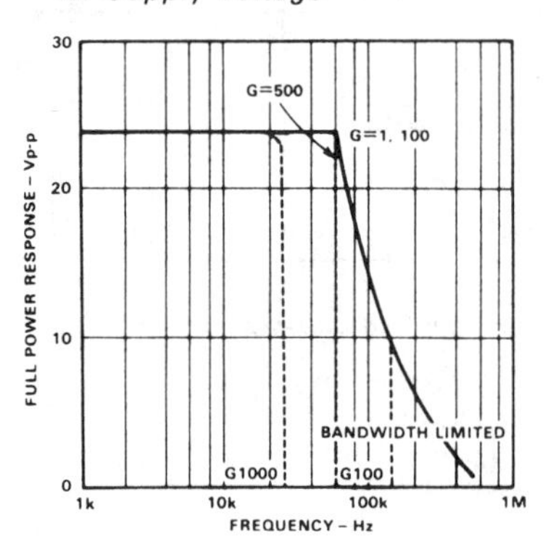

Figure 5. Large Signal Frequency Response

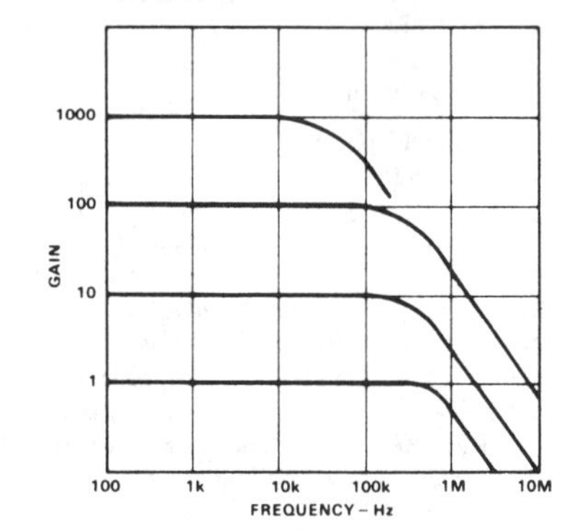

Figure 6. Gain vs. Frequency

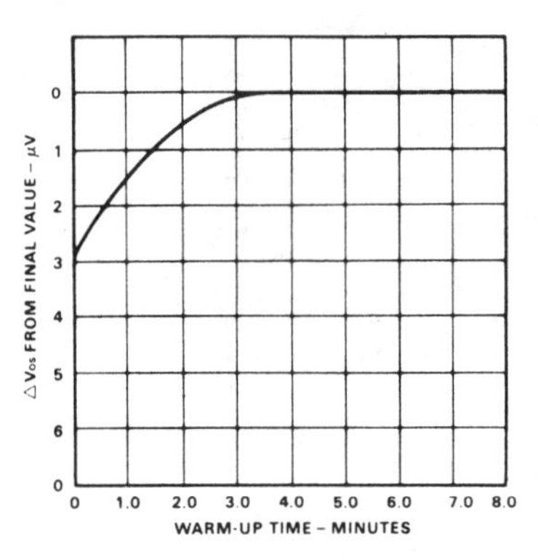

Figure 7. Offset Voltage, RTI, Turn On Drift

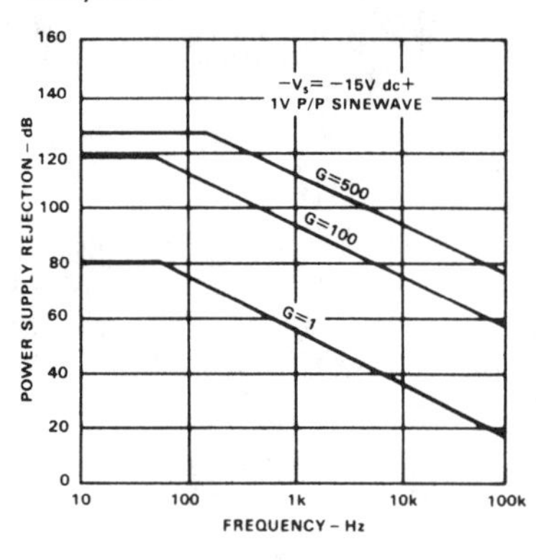

Figure 8. Negative PSRR vs. Frequency

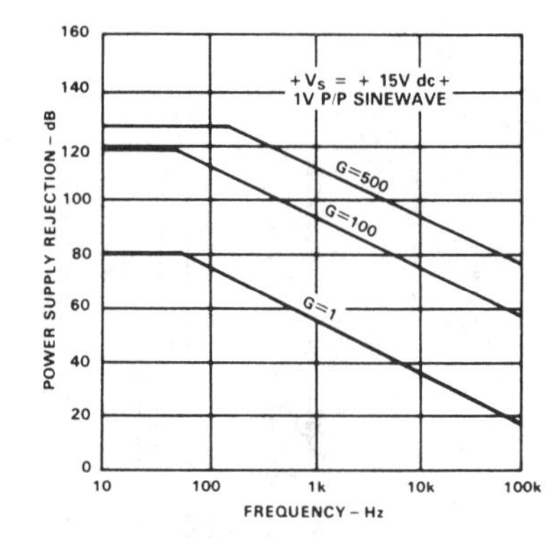

Figure 9. Positive PSRR vs. Frequency

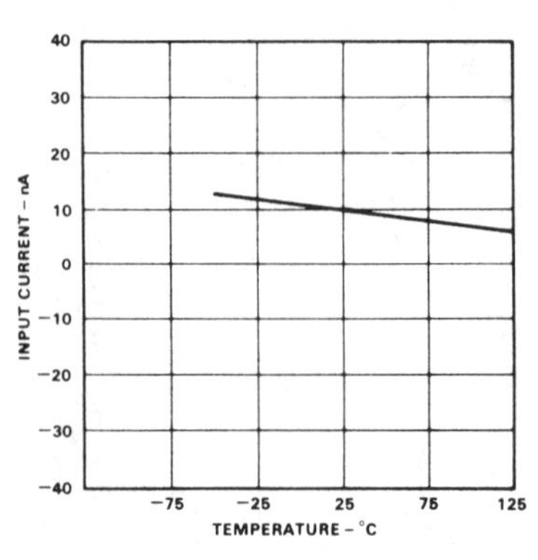

Figure 10. Input Bias Current vs. Temperature

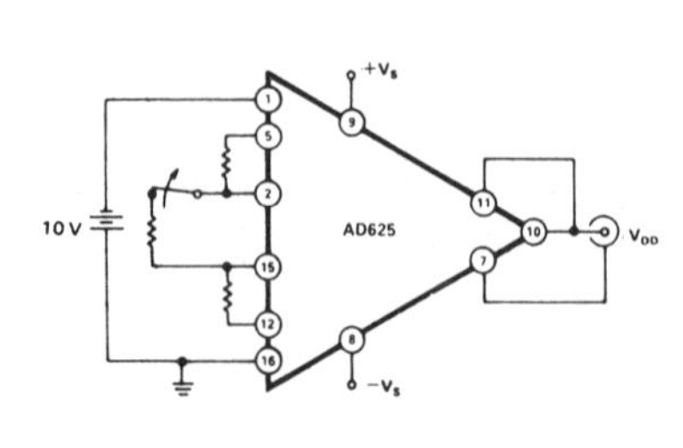

Figure 11. Overange and Gain Switching Test Circuit (G=8, G=1)

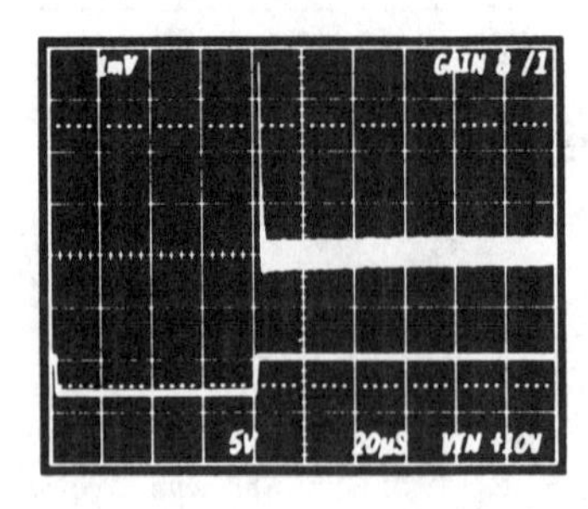

Figure 12. Gain Overange Recovery

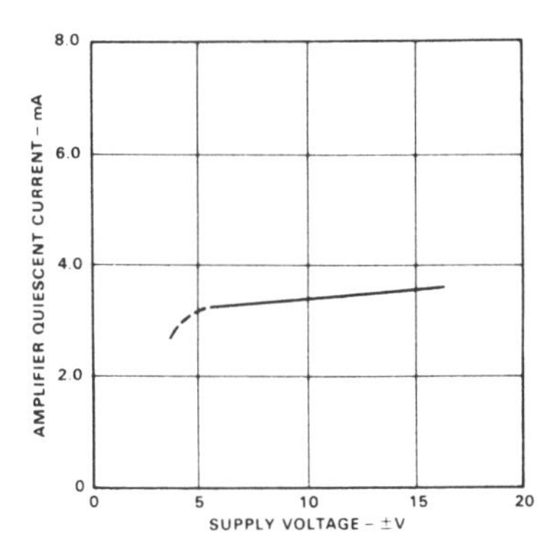

Figure 13. Quiescent Current vs. Supply Voltage

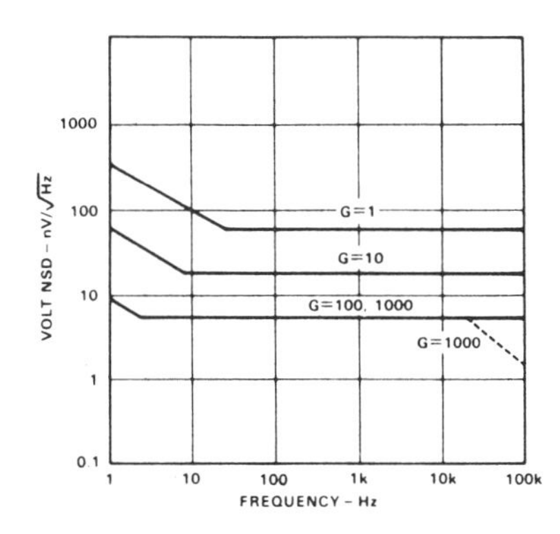

Figure 14. RTI Noise Spectral Density vs. Gain

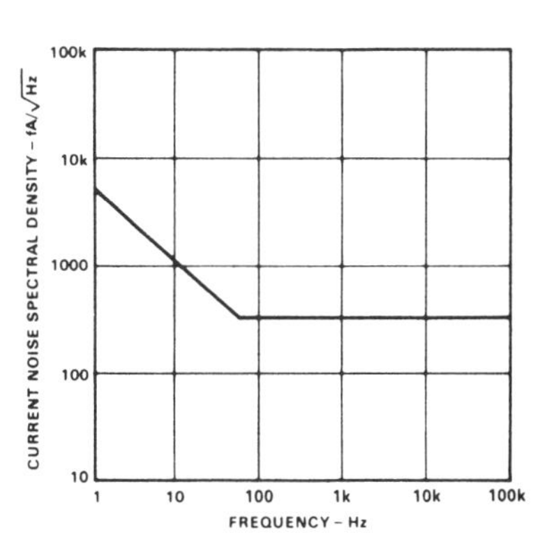

Figure 15. Input Current Noise

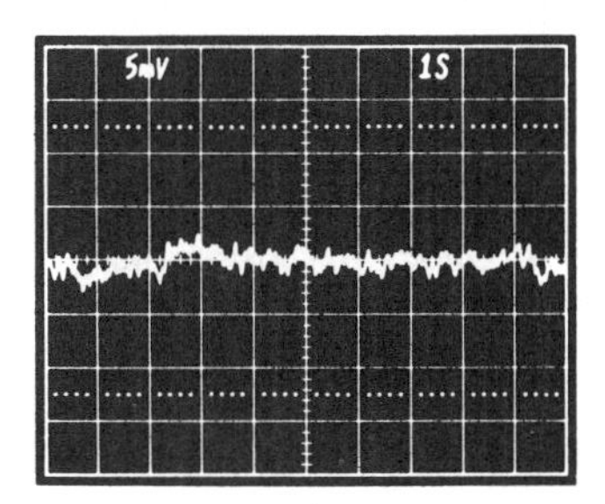

Figure 16. Low Frequency Voltage Noise, G=1 (System Gain=1000)

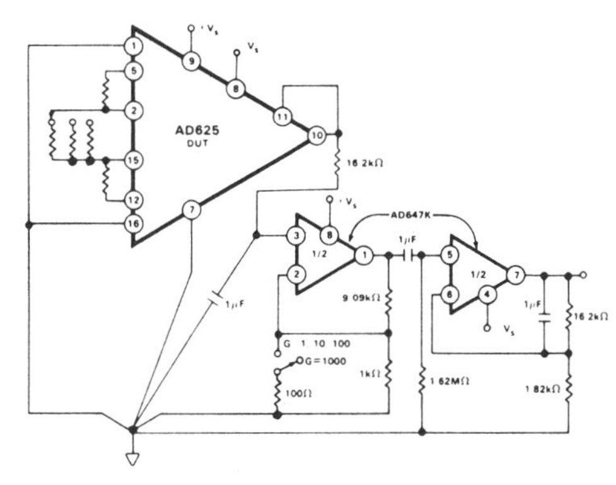

Figure 17. Noise Test Circuit

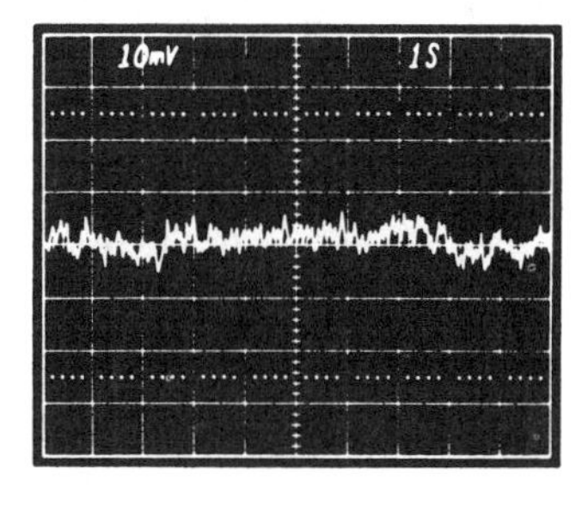

Figure 18. Low Frequency Voltage Noise, G=1000 (System Gain=100,000)

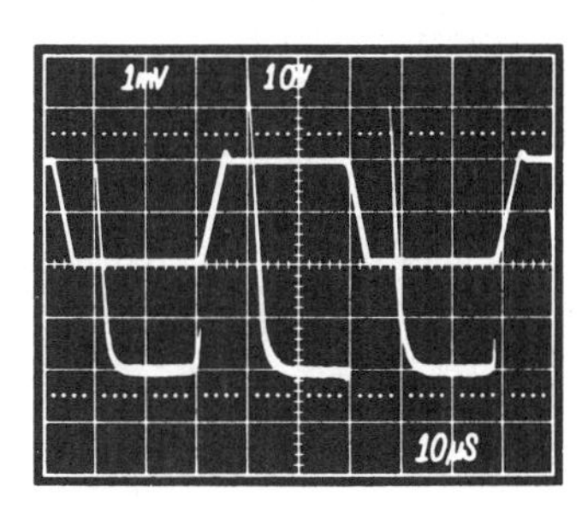

Figure 19. Large Signal Pulse Response and Settling Time, G=1

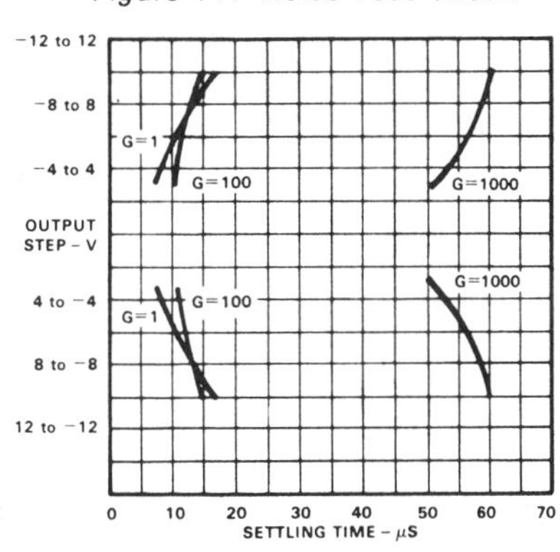

Figure 20. Settling Time to 0.01%

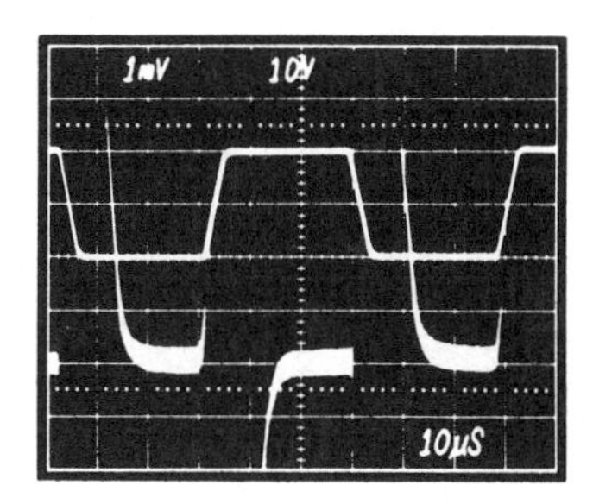

Figure 21. Large Signal Pulse Response and Settling Time, G = 100

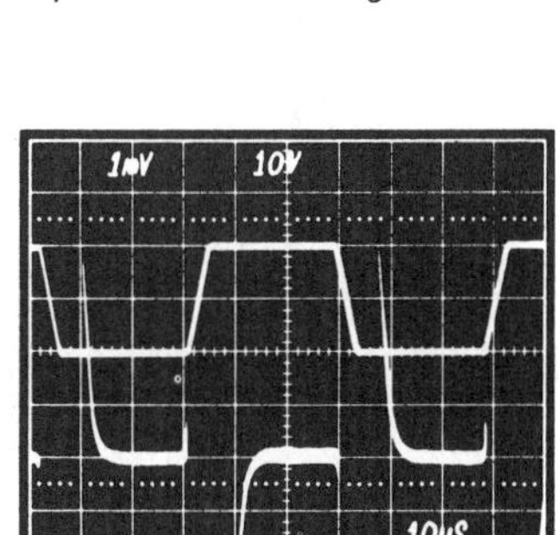

Figure 22. Large Signal Pulse Response and Settling Time, G = 10

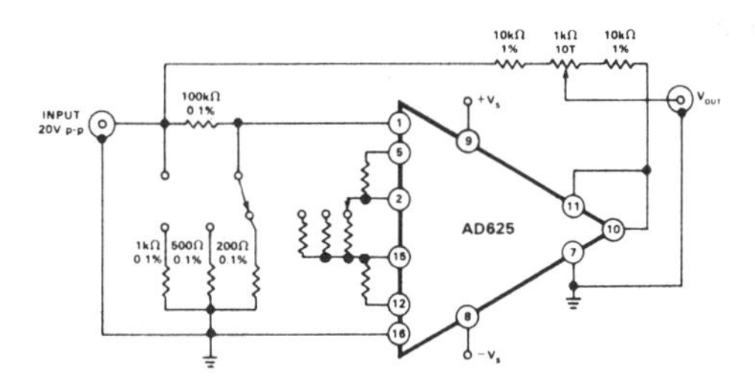

Figure 23. Settling Time Test Circuit

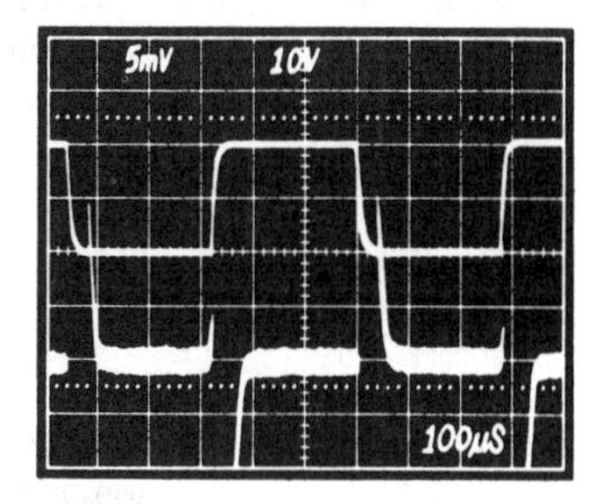

Figure 24. Large Signal Pulse Response and Settling Time, G=1000

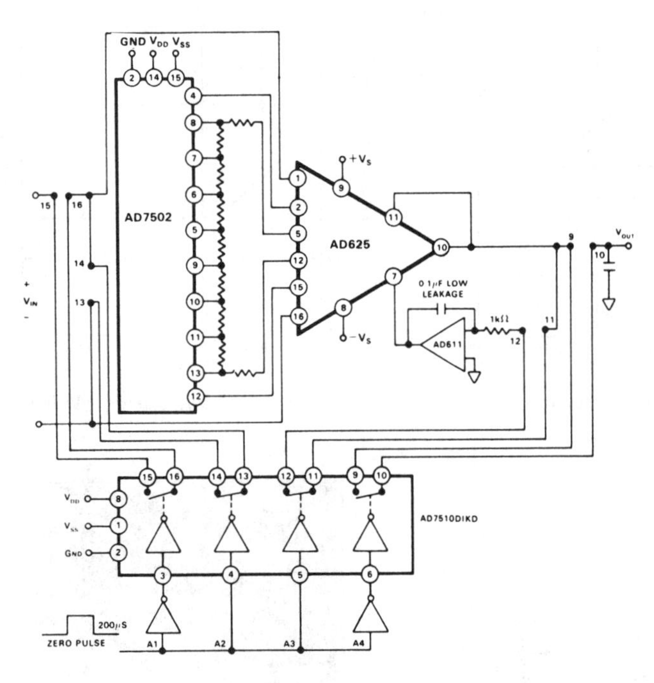

Auto-Zero Circuit

Circuit to Attenuate RF Interference

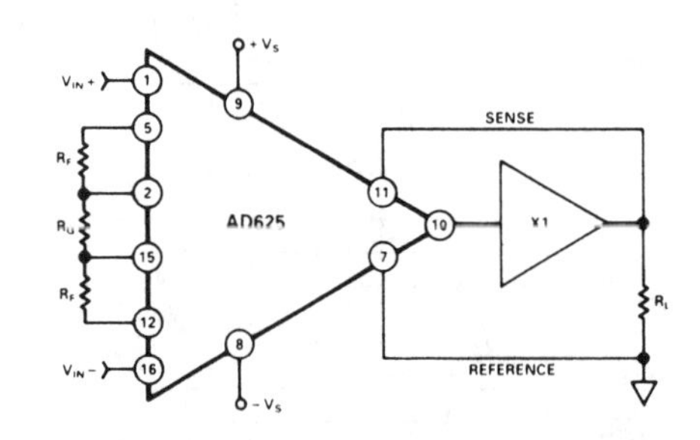

AD625 Instrumentation Amplifier with Output Current Booster

Analog Devices 1S74

Tachogenerator Output, Variable Resolution, Hybrid, Resolver-to-Digital Converter

- 40-pin hybrid
- Tachogenerator velocity output
- User-selectable resolution
- DC error output
- Sub LSB output
- Angle offset input
- Reference frequency of 2 kHz to 10 kHz
- Logic outputs for extension pitch counter

APPLICATIONS

- Numerical control of machine tools
- Feed forward velocity stabilizing loops
- Robotics
- Closed-loop motor drives
- Brushless tachometry
- Single board controllers

1S74 FUNCTIONAL BLOCK DIAGRAM

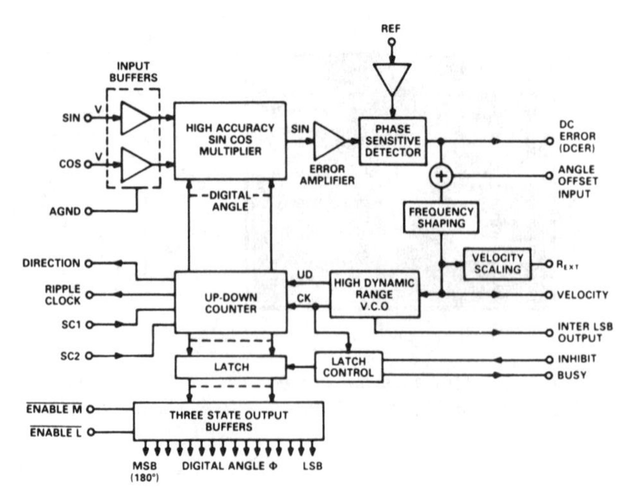

SPECIFICATIONS (typical for both commercial (5YO) and extended (4YO) temperature range options @ 25°C and ±15V or ±12V power supplies, unless otherwise noted)

Resolution	10 Bits	12 Bits	14 Bits	16 Bits	Units
RESOLVER INPUTS					
Signal Voltage	2.0 (±5%)	*	*	*	V rms
Reference Voltage	2.0 (+50% −20%)	*	*	*	V rms
Signal & Reference Frequency	2k – 10k	*	*	*	Hz
Signal Input Impedance	10 (min)	*	*	*	MΩ
Reference Input Impedance	125	*	*	*	kΩ
Allowable Phase Shift (Signal to Reference)	±10	*	*	*	Degrees
POSITION OUTPUT					
Resolution	10	12	14	16	Bits
1LSB	0.35	0.088	0.022	0.0055	Degrees
Accuracy (maximum error over temperature range)					
5YO	±25.0 (0.42)	±8.5 (0.14)	±5.3 (0.09)	±4.0 (0.07)	arc-mins (degrees)
	±0.12	±0.04	±0.025	±0.019	% F.S.
4YO	±25.0 (0.42)	±8.5 (0.14)	±5.3 (0.09)	±2.6 (0.04)	arc-mins (degrees)
	±0.12	±0.04	±0.025	±0.012	% F.S.
Digital Position Output Format	Parallel natural binary	*	*	*	
Load	6 (max)	*	*	*	LSTTL
Monotonicity	Guaranteed	*	*	*	
Repeatability	1	*	*	*	LSB
DATA TRANSFER					
Busy Output	Logic "Hi" when busy	*	*	*	
Load	6 (max)	*	*	*	LSTTL
Busy Width	380 (min) 530 (max)	*	*	*	ns
ENABLE INPUTS	Logic "Lo" to enable	*	*	*	
Load	1	*	*	*	LSTTL
Enable & Disable Times	250 (max)	*	*	*	ns
INHIBIT INPUT	Logic "Lo" to inhibit	*	*	*	
Load	1	*	*	*	LSTTL
Direction Output (DIR)	Logic "Hi" when counting up, logic "Lo" when counting down.				
Load	6 (max)	*	*	*	LSTTL
Ripple Clock (RC)	Negative pulse indicating when internal counters change from all "1's" to all "0's" or vice versa.				
Load	6 (max)	*	*	*	LSTTL
Width	1μ (max) 850n (min)	*	*	*	secs
DYNAMIC CHARACTERISTICS					
Tracking Rate					
with ±15V Supplies	40,800 (min)	10,200 (min)	2,550 (min)	630 (min)	rpm
with ±12V Supplies	34,680 (min)	8,670 (min)	2,168 (min)	536 (min)	rpm
Acceleration Constant K_a	220,000	*	*	*	sec^{-2}
Settling Time (179° step input)	25 (max)	35 (max)	60 (max)	120 (max)	ms
Bandwidth	230	*	*	*	Hz
VELOCITY OUTPUT					
Polarity	Positive for increasing angle	*	*	*	
Tachogenerator Voltage Scaling	0.25	1.00	4	16	V/K rpm
Scale Factor Accuracy	±1 (max)	*	*	*	% of output
Scale Factor Tempco	200 (max)	*	*	*	ppm/°C
Reversion Error	±0.2 (max)	*	*	*	%
Reversion Error Tempco	50 (max)	*	*	*	ppm/°C
Linearity	0.1	*	*	*	% of output
Over Full Temperature Range	0.25 (max)	*	*	*	% of output
Ripple and Noise					
Steady State (200Hz B/W)	100	150	300	1300	μV rms
Dynamic Ripple (av-pk)	0.5 (max)	*	*	*	% of output
Zero Offset	±500	*	*	*	μV
Zero Offset Tempco	50 (max)	*	*	*	μV/°C
Output Load	5 (min)	*	*	*	KΩ

Resolution	10 Bits	12 Bits	14 Bits	16 Bits	Units
SPECIAL FUNCTIONS					
dc Error Output Voltage	450	*	*	*	mV/deg
Inter LSB Output	±1 (±20%)	*	*	*	V/LSB
Load	1k (min)	*	*	*	Ω
Angle Offset Input (over operating temperature range)	320 (±10%)	*	*	*	nA/LSB
Maximum Input	32	*	*	*	LSB
POWER REQUIREMENTS					
Power Supplies					
$\pm V_S$	±15 (±5%) or ±12 (±5%)	*	*	*	V dc
+5V	+4.75 to +5.25	*	*	*	V dc
Power Supply Consumption					
$+V_S$	30 (max)	*	*	*	mA
$-V_S$	30 (max)	*	*	*	mA
+5V	125 (max)	*	*	*	mA
Power Dissipation	1.5 (max)	*	*	*	W
TEMPERATURE RANGE					
Operating 5YO Option	0 to +70	*	*	*	°C
4YO Option	−55 to +125	*	*	*	°C
Storage 5YO Option	−55 to +125	*	*	*	°C
4YO Option	−60 to +150	*	*	*	°C
DIMENSIONS					
5YO Option	2.1 × 1.1 × 0.195	*	*	*	Inches
	(53.5 × 28 × 4.95)	*	*	*	(mm)
4YO Option	2.14 × 1.14 × 0.18	*	*	*	Inches
	(54.5 × 29 × 4.6)	*	*	*	(mm)
WEIGHT	1 (28)	*	*	*	oz. (grams)

Specifications subject to change without notice.

ABSOLUTE MAXIMUM INPUTS (with respect to GND)

$+V_S$[1] . . . . 0V to +17V dc
$-V_S$[1] . . . . 0V to −17V dc
+5V[2] . . . . 0V to +6.0V dc
Reference . . . . ±17V dc
Sine . . . . ±17V dc
Cosine . . . . ±17V dc
Any Logical Input . . . . −0.4V to +5.5V dc

CAUTION:

1. Correct polarity voltages must be maintained on the $+V_S$ and $-V_S$ pins.
2. The +5 volt power supply must *never* go below GND potential.

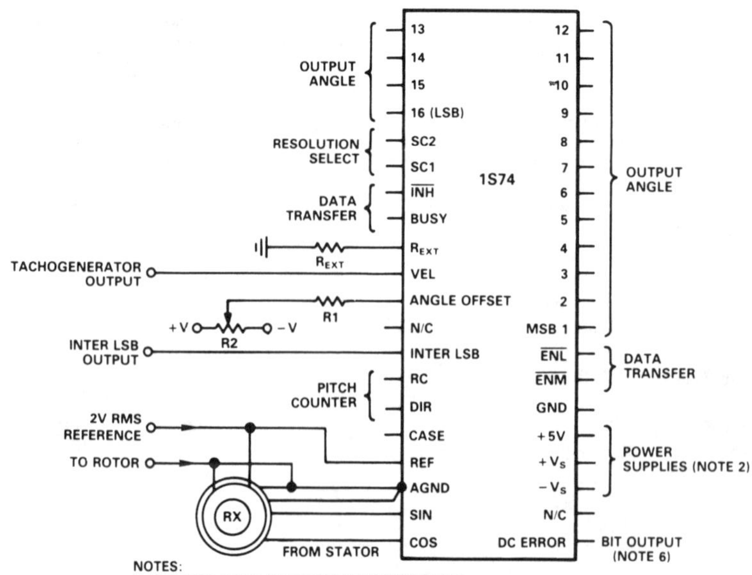

NOTES:
1. GND AND AGND ARE INTERNALLY CONNECTED.
2. EACH SUPPLY SHOULD BE DECOUPLED WITH 100nF CERAMIC CAPACITOR IN PARALLEL WITH A 6μF TANTALLUM CAPACITOR.
3. REXT IS EXTERNAL TACHOGENERATOR SENSITIVITY SCALING RESISTOR (IF REQUIRED) – SEE TEXT UNDER HEADING "VELOCITY OUTPUT"
4. R1 AND R2 ARE ANGLE OFFSET INPUT SCALING RESISTORS (IF REQUIRED) – SEE TEXT.
5. CASE PIN CONNECTED ON 460 OPTION ONLY.
6. POSSIBLE USE AS BUILT-IN TEST EQUIPMENT. (SEE HEADING "SPECIAL FUNCTIONS".)

Electrical Connections

Analog Devices
2S50
LVDT-to-Digital Converter

- Internal signal conditioning
- Direct conversion to digits
- Reference frequency 400 Hz or 1 kHz to 10 kHz
- High MTBF
- No external trims
- Absolute encoding

2S50 FUNCTIONAL BLOCK DIAGRAM

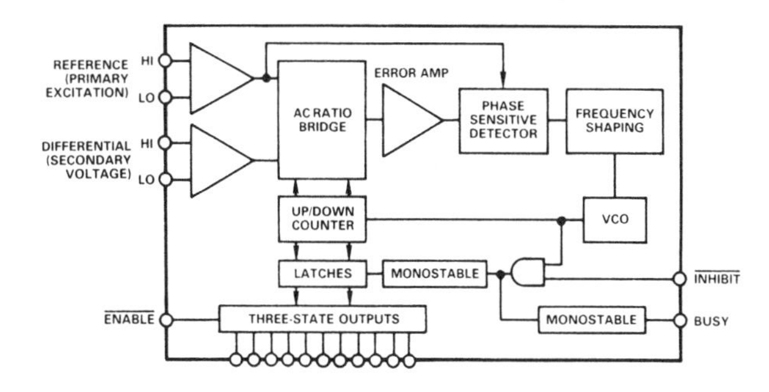

APPLICATIONS

- Industrial measurement and gauging
- Numerical control
- Avionic control systems
- Valves and actuators
- Limit sensing

SPECIFICATIONS (typical @ +25°C, unless otherwise noted)

Models	2S50/510	2S50/560	2S50/410	2S50/460
RESOLUTION	11 Bits	*	*	*
ACCURACY[1]	0.1% (Full Scale)	0.1%	0.2%	0.2%
LINEARITY	±1/2LSB	*	*	*
REFERENCE FREQUENCY	400Hz	1kHz–10kHz	400Hz	1kHz–10kHz
SIGNAL INPUTS[2]	2.5V rms	*	*	*
INPUT IMPEDANCE	5MΩ (min)	*	*	*
SLEW RATE (Min)	200LSB/ms	400LSB/ms	200LSB/ms	400LSB/ms
SETTLING TIME (99% FS Step)	50ms	25ms	50ms	25ms
ACCELERATION CONSTANT (k_a)	70,000	650,000	70,000	650,000
BUSY PULSE	1μs (max) 1 LS TTL Load	* *	* *	* *
INHIBIT INPUT	Logic "Lo" to Inhibit 1 LS TTL Load	* *	* *	* *
POWER DISSIPATION	550mW	*	*	*
POWER SUPPLIES[3]	−15V @ 18mA (typ) 25mA (max) +15V @ 18mA (typ) 25mA (max) +5V @ 3mA (max)	* * *	* * *	* * *
TEMPERATURE RANGE Operating Storage	 0 to +70°C −60°C to +150°C	 * *	 −55°C to +125°C *	 ** *
DIMENSIONS	1.72" × 1.1" × 0.205" (43.5 × 28.0 × 5.2mm)	* *	1.74" × 1.14" × 0.28" (44.2 × 28.9 × 7.1mm)	** **
WEIGHT	1 oz. (28g)	*	*	*
PACKAGE OPTIONS	DH-32E	DH-32E	M-32	M-32

NOTES
[1]Accuracy applies over ±20% signal voltage, ±20% excitation frequency and full temperature range, and for not greater than 3° phase error between reference and difference inputs.
[2]This is a nominal value.
[3]±12 volts to ±17 volts.

*Specifications same as 2S50/510.
**Specifications same as 2S50/410.
Specifications subject to change without notice.

ABSOLUTE MAXIMUM INPUTS (with respect to GND)

$+V_S$	0V to +17V dc
$-V_S$	0V to −17V dc
+5V	0V to +5.5V dc
Ref, Hi to Lo	±20V dc
Diff, Hi to Lo	±20V dc
Case to GND	±20V dc
Any Logical Input	−0.4V to +5.5V dc

PIN CONFIGURATION

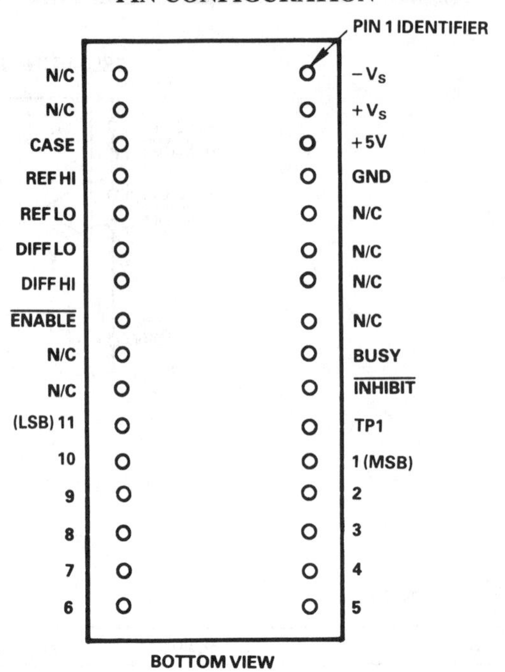

Analog Devices 2S54/2S56/2S58

High-Resolution LVDT-to-Digital Converters

- Direct conversion of LVDT and RVDT outputs into digital format
- Ratiometric conversion for extremely high stability
- High-resolution (14-16 bit) parallel digital output
- User-definable input gain
- Quadrature rejection
- Operation over 360 Hz to 11 kHz
- Linearity better than ±0.01%
- Internal bridge completion resistors
- 1 LSB repeatability
- 75% overrange capability
- Extended temperature-range versions

APPLICATIONS

- Direct LVDT/RVDT-to-digital conversion
- Industrial measurement and gauging
- Valve and actuator control
- Limit sensing
- Aircraft control systems
- Semiconductor wafer profiling
- AC-to-digital conversion

2S54, 2S56, AND 2S58 FUNCTIONAL BLOCK DIAGRAM

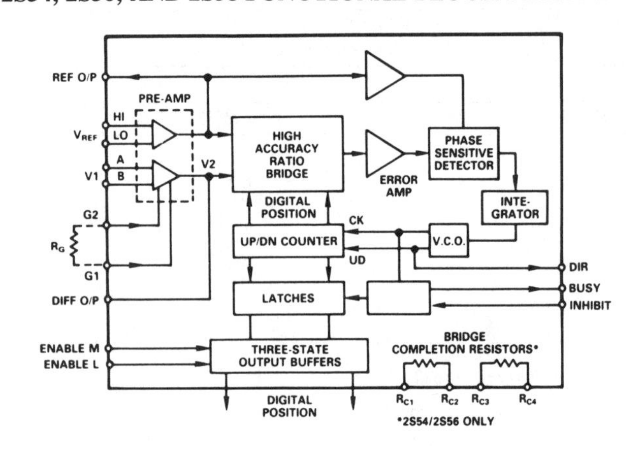

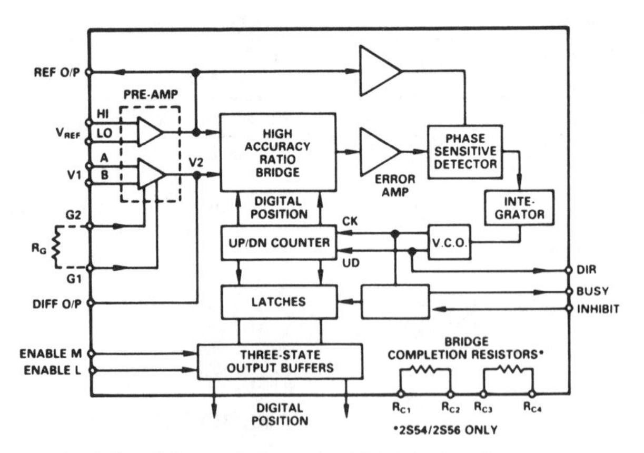

Principle of Operation of the 2S56 Series Converters

SPECIFICATIONS

Models	2S54	2S56	2S58	Comments	Units
DIGITAL OUTPUT				Output Coding Parallel Natural Binary	
Format	14-Bit Binary	16-Bit Binary	16-Bit Binary		
Overrange[2]	75% of FS	*	*		
INPUTS (DIFFERENTIAL)					
V_{REF}	2	*	*		V rms
V_2	2	*	*		V rms
V_1[3]	0.2 (min) 2.0 (max)	*	0.04 (min) 0.2 (max)	See "INPUT GAIN" and "SCALING INPUTS"	V rms
Input Gain	×1 to ×10	*	×10 to ×50		
Input Impedance (V_{REF}, V_1)[2]	6 (max)	*	*		MΩ
CMRR[2]					
@ ×1 Gain	100 (min)	*	NA		dB
@ ×10 Gain	100 (min)	*	120 (min)		dB
@ ×50 Gain	NA	NA	120 (min)		dB
BRIDGE COMPLETION RESISTORS[2]				(Only in 2S54/2S56)	
Value (XYO Options)	9990 (min) 10010 (max)	*	NA		Ω
Ratio Match	0.025	*	NA		%
Tracking Temperature Coefficient	2	*	NA		ppm/°C
REFERENCE FREQUENCY[2]					
50 Hz Bandwidth Option (2S54, 2S56)	360 (min) 5000 (max)	*	NA		Hz
140 Hz Bandwidth Option (2S54, 2S56)	1000 (min) 5000 (max)	*	NA		Hz
300 Hz Bandwidth Option (2S58 Only)	NA	NA	7000 (min) 11000 (max)		Hz
DIGITAL OUTPUT (BIT 1–BIT 16)					
Output Voltage				V_L = +5 V dc	
(Logic Low I_{OL}=8.0 mA)	0.4 (max)	*	*	Logic Low I_{OL} = 8.0 mA	V dc
(Logic High I_{OH}=−0.4 mA)	2.4 (min)	*	*	Logic High I_{OH} = −0.4 mA	V dc
Tristate Leakage Current				V_L = +5 V dc	
(V_{OZL}=0.4 V dc)	±20 (max)	*	*	Logic Low V_{OZL} = 0.4 V dc	μA
(V_{OZH}=2.4 V dc)	±20 (max)	*	*	Logic High V_{OZH} = 2.4V dc	μA
DIGITAL INPUT ($\overline{INHIBIT}$, $\overline{ENABLE\ M}$, $\overline{ENABLE\ L}$)					
Low Input Voltage	0.7 (max)	*	*	V_L = +5 V dc	V dc
High Input Voltage	2.0 (min)	*	*	V_L = +5 V dc	V dc
Low Input Current	−400 (max)	*	*	V_{IL} = 0.4 V dc	μA
High Input Current	20 (max)	*	*	V_{IH} = +2.4 V dc	μA
DATA TRANSFER[2]				See Figure 12	
BUSY Pulse Width	380 (min) 530 (max)	*	*		ns
BUSY Pulse Load[4]	6	*	*	BUSY Is "Hi" When Output Is Changing	LSTTL Loads
Enable/Disable Time	120 (typ) 220 (max)	*	*		ns
Data Setup Time	600	*	*		ns
ACCURACY[5]					
Conversion Accuracy	±0.7	±2.5	±1		LSB
Gain Accuracy[6,7]					
@ ×1 Gain					
0 to +70°C (5Y0)	±0.03 (max)	*	NA	2S54/2S56 Only	% FSR
−55°C to +125°C (4Y0)	±0.03 (max)	*	NA		% FSR
@ ×10 Gain					
0 to +70°C (5Y0)	±0.07 (max)	*	*	2S54/2S56 and 2S58	% FSR
−55°C to +125°C (4Y0)	±0.10 (max)	*	*		% FSR
@ ×50 Gain					
0 to +70°C (5Y0)	NA	NA	±0.09 (max)	2S58 Only	% FSR
−55°C to +125°C (4Y0)	NA	NA	±0.12 (max)		% FSR
Integral Linearity[6,7]					
0° Phase Shift, V_{REF} to V_1	±0.006 (max)	*	±0.00312 (max)	See "PHASE SHIFT AND QUADRATURE EFFECTS"	% FSR
1° Phase Shift, V_{REF} to V_1	±0.008 (max)	*	±0.00437 (max)		% FSR
5° Phase Shift, V_{REF} to V_1	±0.01 (max)	*	±0.00625 (max)		% FSR
Differential Linearity[6]	±0.5 (max)	*	*		LSB
Temperature Dependent Position Offset[2]	±0.04 (max)	*	*		% FSR
REPEATABILITY[5]					
Over 0 to +70°C[2]	±1	*	*		LSB
Hysterisis	0.5 (min) 1 (max)	*	*		LSB
DYNAMIC CHARACTERISTICS[5]					
Slew Rate[2]					
50 Hz Bandwidth Option (2S54,2S56)	150	*	NA		LSB/ms
140 Hz Bandwidth Option (2S54, 2S56)	360	*	NA		LSB/ms
300 Hz Bandwidth Option (2S58 Only)	NA	NA	688		LSB/ms

Models	2S54	2S56	2S58	Comments	Units
Settling Time (Half FS Step)					
50 Hz Bandwidth Option (2S54, 2S56)	160	300	NA	Half FS Step	ms
140 Hz Bandwidth Option (2S54, 2S56)	70	160	NA	Half FS Step	ms
300 Hz Bandwidth Option (2S58 Only)	NA	NA	65	Step From +FSR to −FSR	ms
ANALOG OUTPUTS					
DIFF O/P (Max Allowable Swing)	10	*	*		V p-p
REF O/P (Max Allowable Swing)	10	*	*		V p-p
POWER REQUIREMENTS					
$+V_S$	+15 ±5%	*	*		V dc
$-V_S$	−15 ±5%	*	*		V dc
+5 V	+5 ±5%	*	*		V dc
Supply Currents					
$\pm V_S$	25 (typ) 40 (max)	*	*	Quiescent Condition	mA
+5 V	105 (typ) 125 (max)	*	*		mA
Power Dissipation	1.3 (typ) 1.8 (max)	*	*		Watts
TEMPERATURE RANGE					
Operating	0 to +70 (5Y0 Option)	*	*		°C
	−55 to +125 (4Y0 Option)	*	*		°C
Storage	−55 to +125	*	*		°C
DIMENSIONS	2.14 × 1.14 × 0.18	*	*	See Packaging	Inches
	54.4 × 29.0 × 4.6	*	*	Specifications	mm
WEIGHT	1	*	*		Ounces
	28	*	*		Grams
PACKAGE OPTIONS	M-40	M-40	M-40		

NOTES
[1]Tested with nominal supply (±15 V dc, +5 V dc), reference/signal voltages and frequency.
[2]Guaranteed by design, test not required.
[3]V_1 is the signal input to the converter directly from the transducer. V_2 is the output of the internal gain stage. Because V_2 needs to be maintained at 2 V ±10% in order to meet the converter accuracy (see Note 5), the gain and the maximum value of V_1 should be carefully chosen. Furthermore, because the converter operates on the ratio of V_2 and V_{REF}, care should be taken to see these voltages are matched in order to achieve the full dynamic range of the converter.
[4]Maximum output current is 2.4 mA.
[5]Specified over the operating temperature range of the option and for:
a. ±10% difference in both V_{REF} and V_2 amplitudes
b. 10% harmonic distortion in V_{REF} and V_1.
c. The accuracy is specified for the preset gains of ×1, ×10, and ×50. For accuracy in the intermediate range, see Section "PHASE SHIFT AND QUADRATURE EFFECTS."
[6]Tested with input gains 1, 10 and 50 with V_1 attenuated by 1, 10 and 50, respectively.
[7]Full-Scale Range (FSR) is defined as $V_2 = +V_{REF}$ to $V_2 = -V_{REF}$. This would usually correspond to the utilized LVDT stroke.

*Specifications the same as the 2S54.
Specifications subject to change without notice.

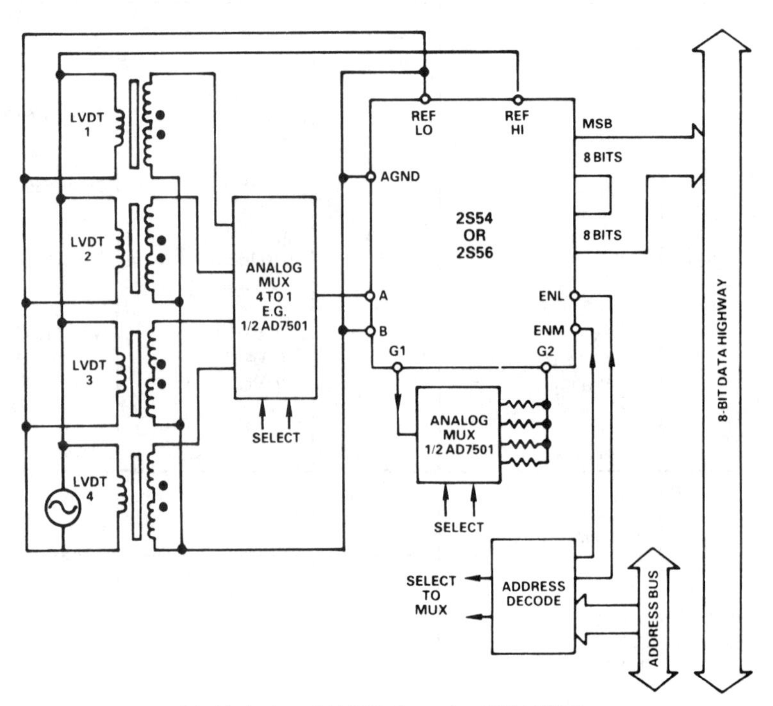

Multiplexing 4 LVDTs into the 2S54/2S56

Analog Devices 2S80
Variable Resolution, Monolithic Resolver-to-Digital Converter

- Monolithic (BiMOS II) tracking R/D converter
- 40-pin DIL package
- 10-, 12-, 14-, and 16-bit resolution set by user
- Ratiometric conversion
- Low power consumption: 300mW typ
- Dynamic performance set by user
- High max tracking rate 1040 rps (10 bits)
- Velocity output
- Military temperature range version

APPLICATIONS
- Brushless motor control
- Process control
- Numerical control of machine tools
- Robotics
- Axis control
- Military servo control

ABSOLUTE MAXIMUM INPUTS (with respect to GND)

$+V_S$[1]	0V to +14V dc
$-V_S$	0V to −14V dc
$+V_L$	0V to $+V_S$
Reference	+14V to $-V_S$
Sin	+14V to $-V_S$
Cos	+14V to $-V_S$
Any Logical Input	−0.4V to $+V_L$ dc
Demodulator Input	+14V to $-V_S$
Integrator Input	+14V to $-V_S$
VCO Input	+14V to $-V_S$

CAUTION:
1. Correct polarity voltages must be maintained on the $+V_S$ and $-V_S$ pins.

CONVERTER CHARACTERISTICS

Model	2S80	Units	Notes
RATIO MULTIPLIER			
Function	AC ERROR output represents the difference between the angle at the SIN and COS inputs compared to the position output angle.		
AC ERROR Output Scaling			
10-Bit Resolution	177.6	mV/bit	
12-Bit Resolution	44.4	mV/bit	
14-Bit Resolution	11.1	mV/bit	
16-Bit Resolution	2.775	mV/bit	
Accuracy			Maximum over temp. range.
JD and SD Options	±8	arc mins	
KD and TD Options	±4	arc mins	
LD and UD Options	±2	arc mins	
Differential Nonlinearity			
JD, KD, SD, TD Options	<1	Bits in 14	Guaranteed monotonic to 14 bits when connected in tracking mode.
LD, UD Options	<1	Bits in 16	Guaranteed monotonic to 16 bits when connected in tracking mode.
PHASE SENSITIVE DETECTOR			Specified over operating frequency range. Tested at 1kHz.
Output Offset Voltage	12 (max)	mV	
Gain of Signal (dc Out, rms In)			
In Phase w.r.t. Reference	−0.9 ± 2%		
In Quadrature w.r.t. Reference	± 0.02 (max)		
Input Bias Current	60 (typ), 150 (max)	nA	
Input Impedance	>1	MΩ	
Input Voltage Range	+8 to −8	V	
INTEGRATOR			
Open Loop Gain at 10kHz	60 ± 3	dB	
Dead Zone Current	100	nA/LSB	
Input Offset Voltage	1 (typ), 5 (max)	mV	
Input Bias Current	60 (typ), 150 (max)	nA	
Output Voltage Range (min)	+8 to −8	V	
Input Impedance	>1	MΩ	
Input Voltage Range	+8 to −8	V	
VCO			
Maximum Rate	1.1	MHz	
VCO Rate	7.4 ± 10%	kHz/μA	With ± 12V supplies.
VCO Rate Tempco	−0.05	%/°C	
Input Offset Voltage	1 (typ), 5 (max)	mV	
Input Bias Current	120 (typ), 300 (max)	nA	
Input Bias Current Tempco	−0.55	nA/°C	
Input Voltage Range	−8 to +8	V	
Linearity of Absolute Rate			
Over Full Range	± 1 (typ), ± 3 (max)	%	
Over 0 to 50% of Max Range	+1 (max)	%	
Reversion Error	<3 (max)	%	Symmetrical power supplies.
Sensitivity of Reversion Error to Symmetry of Power Supplies	8	%/V of Asymmetry	

Specifications subject to change without notice.

SPECIFICATIONS (typical at 25°C unless otherwise specified)

Model	2S80	Units	Notes
TYPICAL CONVERTER PERFORMANCE			
Resolution	10, 12, 14 or 16	bits	
Accuracy JD, SD Options	±8 + 1LSB	arc mins	Accuracy will be affected by the offset at the INTEGRATOR I/P.
KD, TD Options	±4 + 1LSB	arc mins	
LD, UD Options	±2 + 1LSB	arc mins	
Tracking Rate Range			
10-Bit Resolution	0 to 1040	rps	User Selected, max rate limited to 1/16 of the reference frequency.
12-Bit Resolution	0 to 260	rps	
14-Bit Resolution	0 to 65	rps	
16-Bit Resolution	0 to 16.25	rps	
Operating Frequency Range	50 to 20,000	Hz	
Repeatability of Position Output	1	LSB	
Bandwidth	User Selectable		
Velocity Signal			
Linearity			
Over Full Range	±1	% of output	See VCO spec.
Reversion Error	±1	%	With power supplies adjusted for best performance.
Zero Offset	+6	mV	For max tracking rate range. Depends on VCO I/P resistor (R6).
Zero Offset Tempco	−22	μV/°C	For max tracking rate range. Depends on VCO I/P resistor (R6).
Gain Scaling Accuracy	±10	% FSD	
Output Voltage	±8	V dc	
Noise and Ripple			
at LSB Rate	2	mV	
Dynamic Ripple (Peak)	1.5	% of mean output	
ANALOG INPUTS			
Protection	All analog inputs are diode protected against overvoltage at ±8V.		
REFERENCE INPUT			
Frequency	50–20,000	Hz	
Voltage Level Nominal	2	V rms	
Max	11	V peak	
Input Bias Current	60 (typ), 150 (max)	nA	
Input Impedance	>1	MΩ	
SIGNAL INPUTS (SIN, COS)			
Frequency	50–20,000	Hz	
Allowable Phase Shift (Signal to Reference)	10	Degrees	
Voltage Level	2, ±10%	V rms	
Input Bias Current	60 (typ), 150 (max)	nA	
Input Impedance	>1	MΩ	
Maximum Voltage Nominal	±8	V	
DIGITAL INPUTS	TTL Compatible		Except DATA LOAD and SHORT CYCLE INPUTS.
INHIBIT			
Sense	Logic LO to inhibit		
Time to Data Stable (After Negative Going Edge of INHIBIT)	600	ns	
DATA LOAD			
Sense	Internally pulled up to +12V. Unconnected for normal operation. Logic LO allows data to be loaded into the counters from the data lines.		Connect when multiplexing the 2S80 or when using as a control transformer. Ensure data lines are in high impedance state when loading data.
SHORT CYCLE INPUTS (SC1, SC2)			Internally pulled up to $+V_S$. Used to select the resolution of the converter. 0 = Digital Ground. Drive low with open collector TTL. 1 = Open Circuit (internally pulled up through 100kΩ).
	SC1	SC2	
For 10-Bit Resolution	0	0	
For 12-Bit Resolution	0	1	
For 14-Bit Resolution	1	0	
For 16-Bit Resolution	1	1	
BYTE SELECT			
Sense Logic HI	8 MSBs selected on data lines 1 to 8. LS Byte selected on data lines 9 to 16.		The size of the LS Byte will be between 2 and 8 bits depending on the resolution selected.
Logic LO	LS Byte selected on data lines 1 to 8 and 9 to 16.		
Time to Data Available (After Change in State)	150 (typ), 450 (max)	ns	

Model	2S80	Units	Notes
ENABLE			
Sense	Logic LO to enable position outputs. Logic HI position outputs in high impedance state.		
Enable and Disable Times	200 (typ), 550 (max)	ns	
ANALOG OUTPUTS			
Protection	Short-circuit output current limited to ± 8mA, ± 30%.		
Output Voltage Range, typ	+ 9 to − 9	V	With 1mA load.
max	+ 10.5 to − 10.5	V	
min	+ 8 to − 8	V	
DIGITAL OUTPUTS			
Format	V_L = + 5V	TTL Compatible	Voltage on V_L sets the voltage level of the digital outputs.
	V_L = + 12V	CMOS Compatible	
POSITION OUTPUTS			
Format	Three-state natural binary		
Resolution	10, 12, 14 or 16	bits	
Number of Data Lines	16		
Max Load	3	LSTTL	
Monotonicity			
JD, KD, SD, TD Options	Guaranteed to 14 bits		
LD and UD Options	Guaranteed to 16 bits		
DIRECTION			
Sense	Logic HI when counting up. Logic LO when counting down.		
Timing	Only changes, if required, at start of output position data cycle.		
Max Load	3	LSTTL	
RIPPLE CLOCK			
Sense	Positive going edge when counting up from all "1s" and when counting down from all "0s" as data changes.		
Timing	Edge occurs at least 300ns before change in DIR can occur.		
Width	300 (min)	ns	
Reset	By start of next data update.		
Max Load	3	LSTTL	
BUSY			
Sense	Logic HI when converter position output changing.		
Timing	Positive going edge 50ns before change in position output.		
Width typ	300	ns	
min	200	ns	
max	600	ns	
Max Load	3	LSTTL	
POWER SUPPLIES			The 2S80 may latch up if $+V_S$ is applied without $-V_S$.
Voltage Levels			
$+V_S$	+ 12 ± 10%	V	
$-V_S$	− 12 ± 10%	V	
$+V_L$	+ 5 to + 14	V	
Current			Over operating temperature range.
$+V_S$, $-V_S$ at 12V	12 (typ), 23 (max)	mA	
$+V_S$, $-V_S$ at 13.2V	19 (typ), 30 (max)	mA	
$+V_L$	0.5 (typ), 1.5 (max)	mA	
GENERAL			
Operating Temperature Range			
JD, KD, LD Options	0 to + 70	°C	
SD, TD, UD Options	− 55 to + 125	°C	
Storage Temperature Range (All Options)	− 60 to + 150	°C	
Weight	0.2 (5)	oz (grams)	

2S80 Connection Diagram

2S80 PIN CONFIGURATION

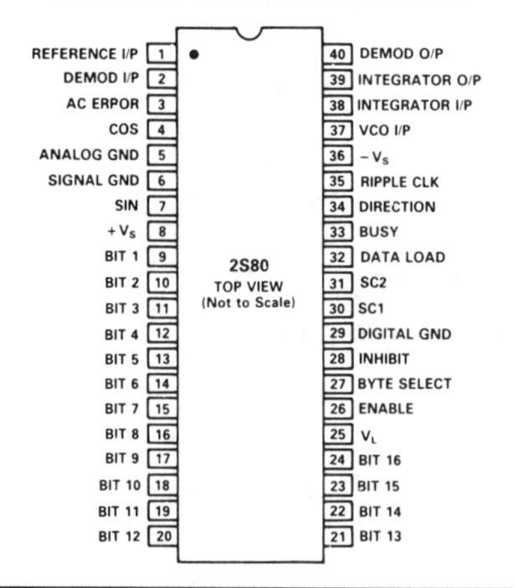

Analog Devices
2S81
Low-Cost, Monolithic 12-Bit Resolver-to-Digital Converter

- Low cost
- Monolithic construction
- 28-pin DIP package
- Ratiometric conversion
- Low power consumption: 300mW typical
- Dynamic performance set by user
- High tracking rate: 260 rps max
- Velocity output

APPLICATIONS

- Brushless motor control
- Programmable limit switches
- Process control
- Numerical control of machine tools
- Robotics
- Axis control

PIN CONFIGURATION

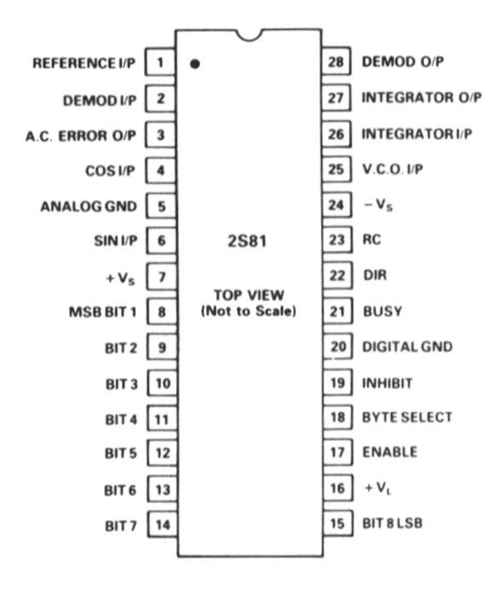

SPECIFICATIONS (typical at 25°C unless otherwise specified)

Model	2S81JD	Units	Notes
OVERALL CONVERTER SPECIFICATIONS (CONNECTED AS SHOWN IN FIGURE 1)			
Resolution	12	Bits	
Accuracy	±30 + 1LSB	Arc Minutes	Accuracy will be Affected by the Offset at the INTEGRATOR I/P.
Tracking Rate Range	0 to 260 (max)	rps	User Selected, Max Rate Limited at Lower Operating Frequencies.
Operating Frequency Range	400 to 20,000	Hz	
Repeatability of Position Output	1	LSB	
Bandwidth	User Selectable		
Velocity Signal			
Linearity Over Full Range	±1 (typ), ±3 (max)	%	
Over 0 to 6000 rpm	±1 (max)	%	
Reversion Error	±5 (max)	%	Symmetry of $-V_S$ and $+V_S$ Power Supplies to be within ±5%.
	±2 (max)	%	With $-V_S$ Adjusted for Best Performance.
Zero Offset (for 260 rps Max Tracking Rate)	±6 (typ) + 16 (max)	mV	Depends on VCO I/P Resistor (R6).
Zero Offset Tempco (for 260 rps Max Tracking Rate)	−22	μV/°C	Depends on VCO I/PResistor (R6).
Gain Scaling Accuracy	±10	% FSD	
Output Voltage	±8	V dc	
Noise and Ripple (av-pk)	1.5	%	
ANALOG INPUTS			
Protection	All Analog Inputs Are Diode Protected Against Overvoltage at ±8V		
REFERENCE INPUT			
Frequency	400 – 20,000	Hz	
Voltage Level Nominal	2	V rms	
max	11	V peak	
Input Bias Current	60 (typ), 150 (max)	nA	
Input Impedance	>1	MΩ	
SIGNAL INPUTS (SIN, COS)			
Frequency	400 – 20,000	Hz	
Allowable Phase Shift (Signal to Reference)	10	Degrees	
Voltage Level	2, ±10%	V rms	
Input Bias Current	60 (typ), 150 (max)	nA	
Input Impedance	>1	MΩ	
Maximum Voltage Nominal	±8	V	
DIGITAL INPUTS	TTL-Compatible		
INHIBIT			
Sense	Logic LO to Inhibit		
Time to Data Stable (After Negative Going Edge of INHIBIT)	1	μs	
BYTE SELECT			
Sense	Logic HI Selects 8MSBs on Pins 8-15 Logic LO Selects 4LSBs on Pins 8-11; Pins 12-15 Are Logic LO		
Data Available (After Change in State)	150 (typ), 450 (max)	ns	
ENABLE			
Sense	Logic LO to Enable Position Outputs Logic HI Position Outputs in High Impedance State		
Enable and Disable Times	200 (typ), 550 (max)	ns	
ANALOG OUTPUTS			
Protection	Short Circuit Output Current Limited to ±8mA, ±30% Output Voltage Range Will Be Degraded for Currents >3mA		
Output Voltage Range (typ)	+9 to −9	V	
(max)	+10.5 to −10.5	V	
(min)	+8 to −8	V	
DIGITAL OUTPUTS			
Format	V_L = +5V TTL Compatible V_L = +12V CMOS Compatible		Voltage on V_L Sets the Voltage Level of Digital Outputs.

Model	2S81JD	Units	Notes
POSITION OUTPUTS			
Format	Three-State Natural Binary		
Resolution	12	Bits	
Number of Data Lines	8		Pins 8 to 15
Max Load	3	LSTTL	
Monotonicity	Guaranteed		
DIRECTION (DIR)			
Sense	Logic "HI" When Counting Up Logic "LO" When Counting Down		
Timing	Only Changes, if Required, at Start of Output Position Data Update Cycle		
Max Load	3	LSTTL	
RIPPLE CLOCK (RC)			
Sense	Positive Going Edge When Counting Up from All "1s" and When Counting Down from All "0s" as Data Changes		
Timing	Edge Occurs at Least 300ns Before Change in DIR Can Occur		
Width (min)	300	ns	
Reset	By Start of Next Data Update		
Max Load	3	LSTTL	
BUSY			
Sense	Logic "HI" When Converter Position Output Changing		
Timing	Positive Going Edge 50ns Before Change in Position Output		
Width (typ)	300	ns	
(min)	200	ns	
(max)	600	ns	
Load, (max)	3	LSTTL	
POWER SUPPLIES			The Device May Latch Up If $+V_S$ is Applied without $-V_S$.
Voltage Levels			
$+V_S$	$+12 \pm 10\%$	V	
$-V_S$	$-12 \pm 10\%$	V	
$+V_L$[5]	$+5$ to $+14$	V	
Current			
$+V_S$	12 (typ), 23 (max)	mA	
$-V_S$	12 (typ), 23 (max)	mA	
$+V_L$	0.5 (typ), 1.5 (max)	mA	
Power Dissipation	300 (typ), 600 (max)	mW	
GENERAL			
Operating Temperature Range	0 to +70	°C	
Storage Temperature Range	−65 to +150	°C	
Weight	0.2 (5)	Oz. (Grams)	
CONVERTER CHARACTERISTICS			
RATIO MULTIPLIER			
Function	AC ERROR Output Represents the Difference between the Angle at the SIN and COS Inputs Compared to the Position Output Angle		
AC Error Output Scaling	44.4	mV/Bit	
Accuracy	30	Arc Minutes	
Differential Nonlinearity	±0.25 (max)	LSB	
PHASE SENSITIVE DETECTOR			Specified Over the Operating Frequency Range. Tested at 1kHz.
Output Offset Voltage (max)	15	mV	
Gain of Signal (dc out, rms in)			
In Phase w.r.t. Reference	$-0.9 \pm 2\%$		
In Quadrature w.r.t. Reference	±0.02 (max)		
Input Bias Current	60 (typ), 150 (max)	nA	
Input Impedance	>1	MΩ	
Input Voltage Range	+8 to −8	V	
INTEGRATOR			
Open Loop Gain at 10kHz	60 ± 3	dB	
Output Impedance at 10kHz (max)	0.5	Ω	
Dead Zone Current	100	nA/LSB	
Input Offset Voltage	1 (typ), 5 (max)	mV	
Input Bias Current	60 (typ), 150 (max)	nA	
Output Voltage Range (min)	+8 to −8	V	
Input Impedance	>1	MΩ	
Input Voltage Range	+8 to −8	V	

Model	2S81JD	Units	Notes
VCO			
Maximum Rate	1.1	MHz	
VCO Rate	7.4 ± 10%	kHz/μA	
Input Offset Voltage	1 (typ), 5 (max)	mV	
Input Bias Current	120 (typ), 300 (max)	nA	
Input Bias Current Tempco	−0.55	nA/°C	
Input Voltage Range	+8 to −8	V	
Reversion Error	±5	%	
Linearity of Absolute Rate	+3	%	
Sensitivity of VCO Rate in "Up Direction" to $-V_S$	−7	%/V	
Sensitivity of VCO Rate in "Down Direction" to $-V_S$	+2	%/V	

NOTE
Specifications subject to change without notice.

ABSOLUTE MAXIMUM INPUTS (with respect to GND)

$+V_S$[1] . . . . 0V to +14V dc
$-V_S$ . . . . 0V to −14V dc
$+V_L$ . . . . 0V to $+V_S$
Reference . . . . +14V to $-V_S$
Sin . . . . +14V to $-V_S$
Cos . . . . +14V to $-V_S$
Any Logic Input . . . . −0.4V to $+V_L$ dc
Demodulator Input . . . . +14V to $-V_S$
Integrator Input . . . . +14V to $-V_S$
VCO Input . . . . +14V to $-V_S$

CAUTION:
1. Correct polarity voltages must be maintained on the $+V_S$ and $-V_S$ pins.

ORDERING INFORMATION

Model	Package Option*	Temperature Range	Operating Frequency Range
2S81JD	D-28	0 to +70°C	400 to 20,000Hz

*See Section 14 for package outline information.

2S81 PIN CONFIGURATION

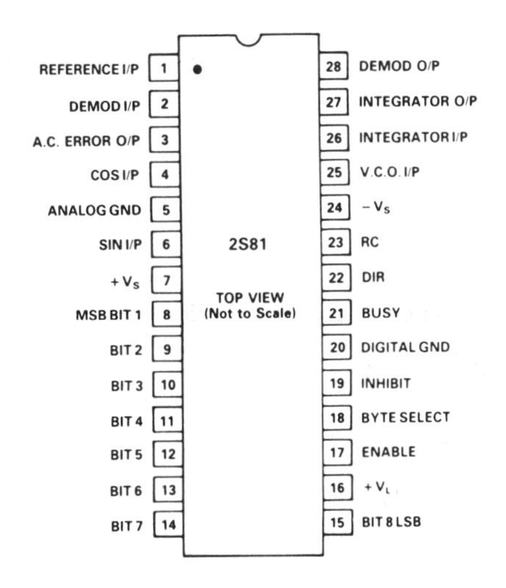

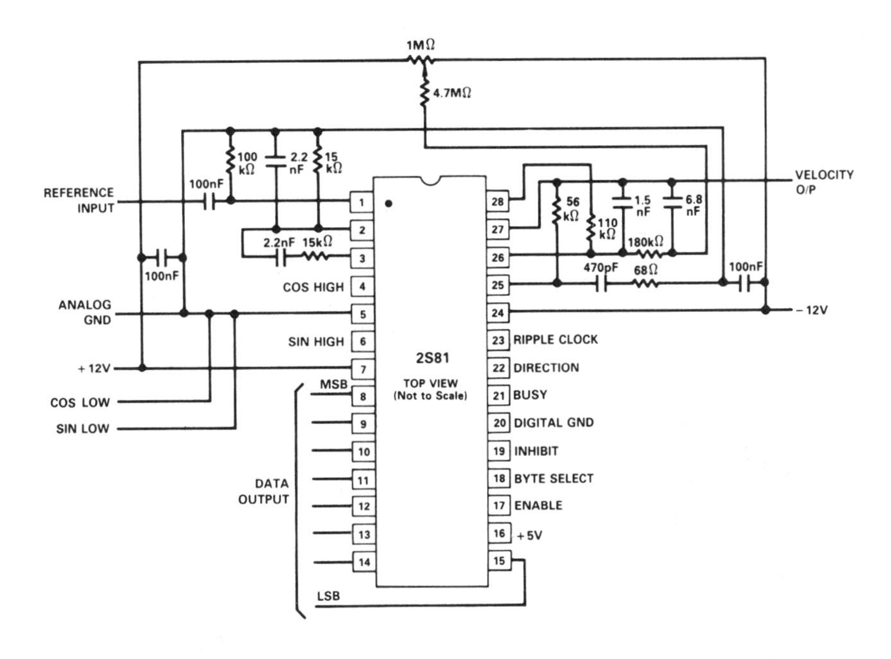

Typical Circuit for the 2S81

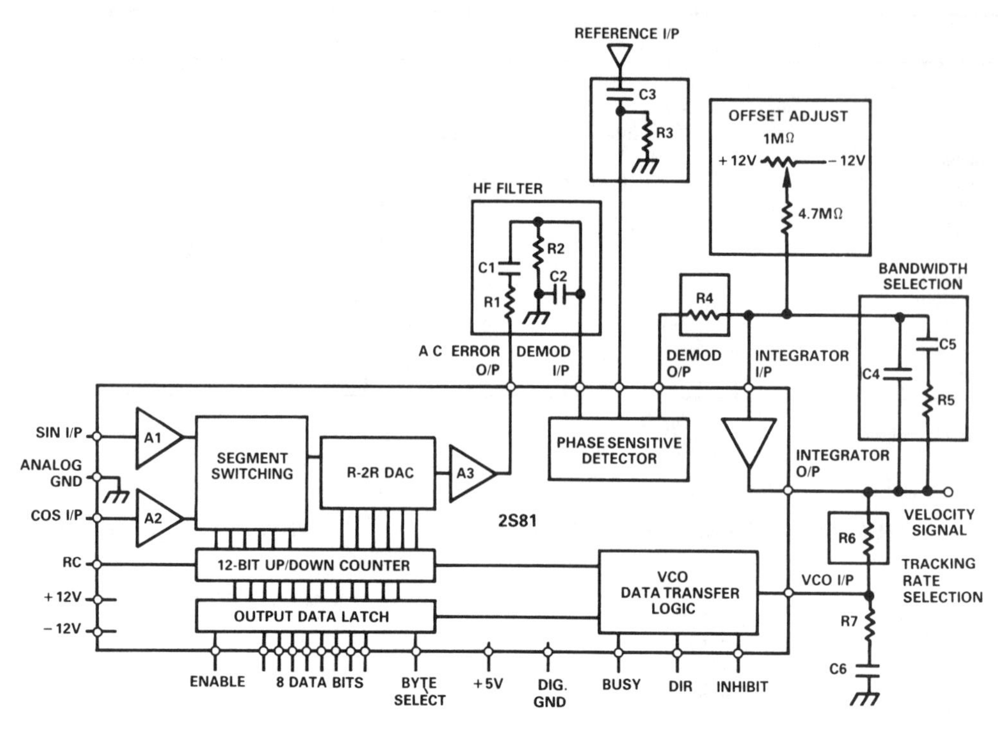

2S81 Connection Diagram

Analog Devices AD537

Integrated Circuit Voltage-to-Frequency Converter

- Low-cost A-D conversion
- Versatile input amplifier
  - Positive or negative voltage modes
  - Negative current mode
  - High input impedance, low drift
- Single supply, 5 to 36V
- Linearity: ±0.05% FS
- Low power: 1.2mA quiescent current
- Full-scale frequency up to 100 kHz
- 1.00V reference
- Thermometer output (1mV/K)
- F-V applications

AD537 PIN CONFIGURATIONS

"D" Package — TO-116 "H" Package — TO-100

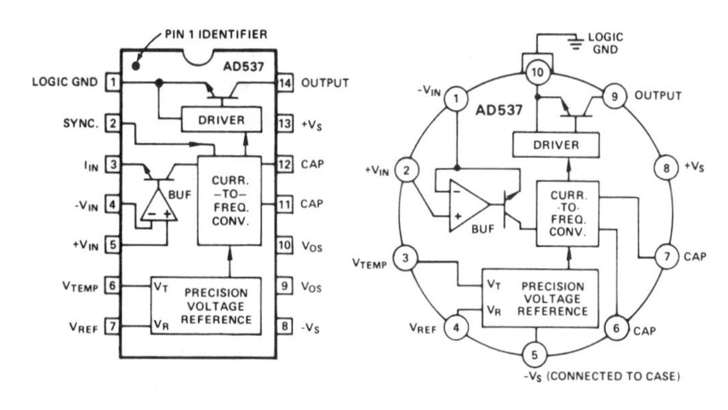

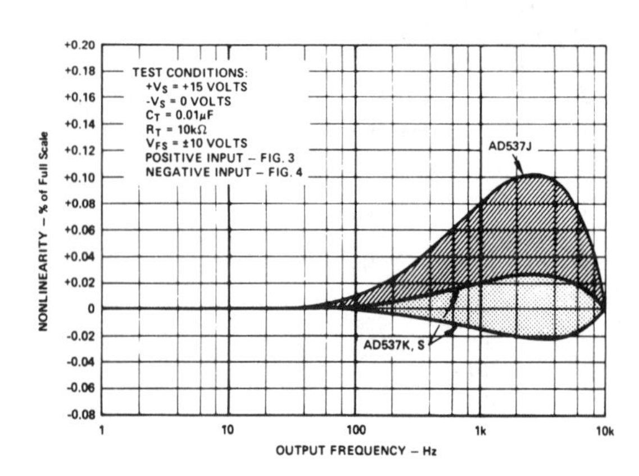

Typical Nonlinearity Error Envelopes with 10kHz F.S. Output

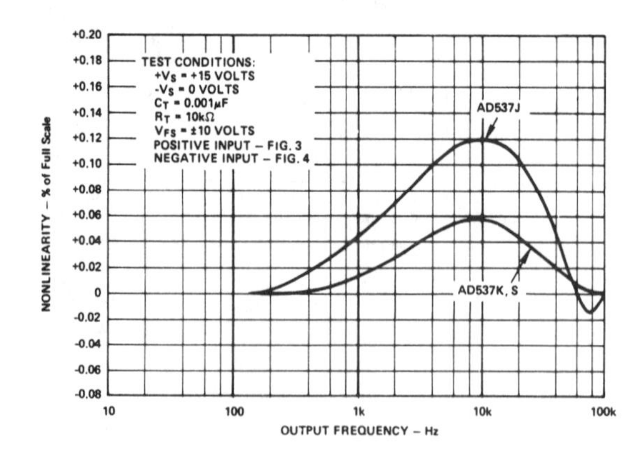

Typical Nonlinearity Error with 100kHz F.S. Output

SPECIFICATIONS (typical @ +25°C with V_S (total) = 5 to 36V, unless otherwise noted)

MODEL	AD537JH	AD537JD	AD537KD AD537KH	AD537SD[1] AD537SH[1]
CURRENT-TO-FREQUENCY CONVERTER				
Frequency Range	0 to 150kHz	*	*	*
Nonlinearity[1]				
f_{max} = 10kHz	0.15% max (0.1% typ)	*	0.07% max	**
f_{max} = 100kHz	0.25% max (0.15% typ)	*	0.1% max	**
Full Scale Calibration Error				
C = 0.01µF, I_{IN} = 1.000mA	±10% max	±7% max	±5% max	**
vs. Supply (f_{max} < 100kHz)	±0.1%/V max (0.01% typ)	*	*	*
vs. Temp. (T_{min} to T_{max})	±150ppm/°C max (50ppm typ)	*	50ppm/°C max (30ppm typ)[2]	150ppm/°C max
ANALOG INPUT AMPLIFIER				
(Voltage-to-Current Converter)				
Voltage Input Range				
Single Supply	0 to (+V_S - 4) Volts (min)	*	*	*
Dual Supply	-V_S to (+V_S - 4) Volts (min)	*	*	*
Input Bias Current				
(Either Input)	100nA	*	*	*
Input Resistance (Non-Inverting)	250MΩ	*	*	*
Input Offset Voltage				
(Trimmable in "D" Package Only)	5mV max	*	2mV max	**
vs. Supply	200µV/V max	100µV/V max	100µV/V max	**
vs. Temp. (T_{min} to T_{max})	5µV/°C	*	1µV/°C	10µV/°C max
Safe Input Voltage[3]	±V_S	*	*	*
REFERENCE OUTPUTS				
Voltage Reference				
Absolute Value	1.00 Volt ±5% max	*	*	*
vs. Temp. (T_{min} to T_{max})	50ppm/°C	*	100ppm/°C max[3]	**
vs. Supply	±0.03%/V max	*	*	*
Output Resistance[4]	380Ω	*	*	*
Absolute Temperature Reference[5]				
Nominal Output Level	1.00mV/K	*	*	*
Initial Calibration @ +25°C	298mV (±5mV)	*	298mV (±5mV max)	**
Slope Error from 1.00mV/K	±0.02mV/K	*	*	*
Slope Nonlinearity	±0.1K	*	*	*
Output Resistance[5]	900Ω	*	*	*
OUTPUT INTERFACE (Open Collector Output)				
(Symmetrical Square Wave)				
Output Sink Current in Logic "0"				
V_{OUT} = 0.4V max, T_{min} to T_{max})	20mA min	20mA min	20mA min	10mA min
Output Leakage Current in Logic "1"				
(T_{min} to T_{max})	200nA max	*	*	2µA max
Logic Common Level Range	-V_S to (+V_S - 4) Volts	*	*	*
Rise/Fall Times (C_T = 0.01µF)				
I_{IN} = 1mA	0.2µs	*	*	*
I_{IN} = 1µA	1µs	*	*	*
POWER SUPPLY				
Voltage, Rated Performance				
Single Supply	4.5V to 36V	*	*	*
Dual Supply	±5 to ±18V	*	*	*
Quiescent Current	1.2mA (2.5mA max)	*	*	*
TEMPERATURE RANGE				
Rated Performance	0 to +70°C	*	*	-55°C to +125°C
Storage	-65°C to +150°C	*	*	*
PACKAGE OPTIONS				
TO-116 Ceramic DIP (D-14)		AD537JD	AD537KD	AD537SD
TO-100 Header (H-10A)	AD537JH		AD537KH	AD537SH

NOTES
*Specifications same as AD537JH.
**Specifications same as AD537K.

Specifications subject to change without notice.

[1] Nonlinearity is specified for a current input level (I_{IN}) to the converter from 0.1 to 1000µA. Converter has 100% overrange capability up to I_{IN} = 2000µA with slightly reduced linearity. Nonlinearity is defined as deviation from a straight line from zero to full scale, expressed as a percentage of full scale.

[2] Guaranteed not tested.

[3] Maximum voltage input level is equal to the supply on either input terminal. However, large negative voltage levels can be applied to the negative terminal if the input is scaled to a nominal 1mA full scale through an appropriate value resistor (see Figure 2).

[4] Loading the 1.0 volt or 1mV/K outputs can cause a significant change in overall circuit performance, as indicated in the applications section. To maintain normal operation, these outputs should be operated into the external buffer or an external amplifier.

[5] Temperature reference output performance is specified from 0 to +70°C for "J" and "K" devices, -55°C to +125°C for "S" model.

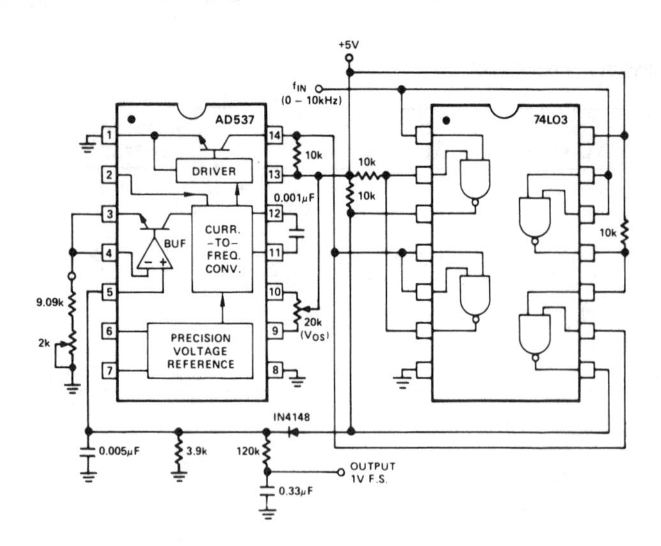

10kHz F-V Converter

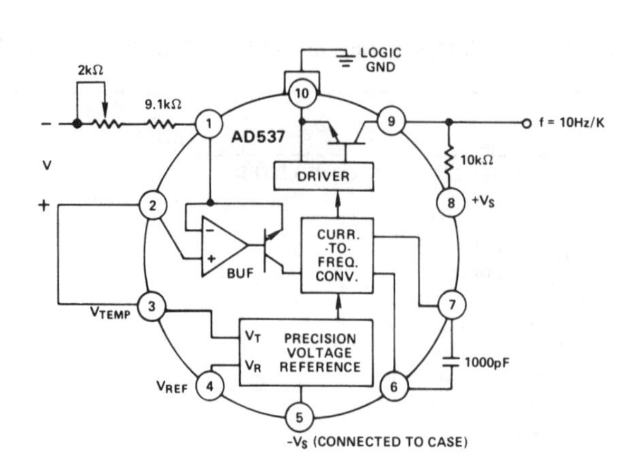

Absolute Temperature to Frequency Converter

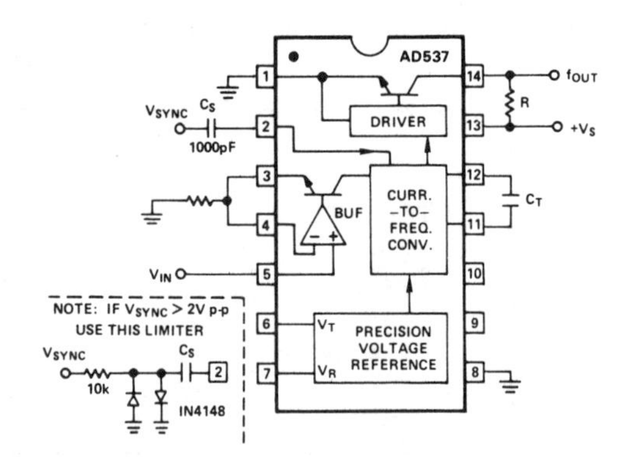

Connection for Synchronous Operation

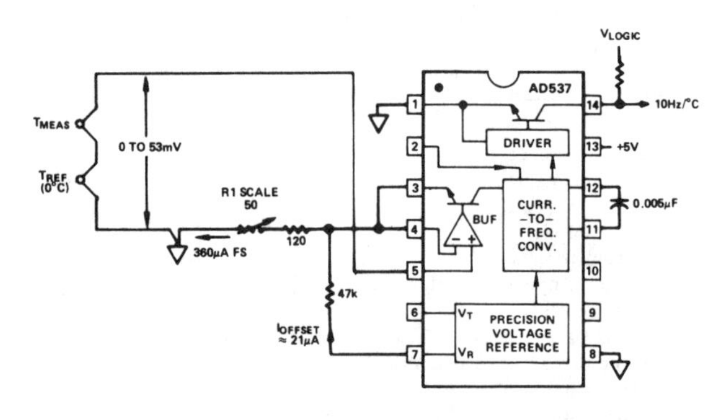

Thermocouple Interface with First-Order Linearization

Analog Devices AD650

Voltage-to-Frequency and Frequency-to-Voltage Converter

- V/F conversion to 1 MHz
- Reliable monolithic construction
- Very low nonlinearity
  - ○ 0.002% typ at 10 kHz
  - ○ 0.005% typ at 100 kHz
  - ○ 0.07% typ at 1 MHz
- Input offset trimmable to zero
- CMOS or TTL compatible
- Unipolar, Bipolar, or differential V/F
- V/F or F/V conversion
- Available in surface mount

ORDERING GUIDE

Part Number	Gain Tempco ppm/°C 100kHz	1MHz Linearity	Specified Temperature Range °C	Package
AD650JN	150 typ	0.1% typ	0 to +70	Plastic DIP
AD650KN	150 typ	0.1% max	0 to +70	Plastic DIP
AD650JP	150 typ	0.1% typ	0 to +70	PLCC
AD650KP	150 typ	0.1% max	0 to +70	PLCC
AD650AD	150 max	0.1% typ	−25 to +85	Ceramic
AD650BD	150 max	0.1% max	−25 to +85	Ceramic
AD650SD	150 max	0.1% max	−55 to +125	Ceramic

ABSOLUTE MAXIMUM RATINGS

Total Supply Voltage $+V_S$ to $-V_S$ 36V
Storage Temperature Ceramic . . . . −55°C to +165°C
Plastic . . . . −25°C to +125°C
Differential Input Voltage (Pins 2 & 3) . . . . ±10V
Maximum Input Voltage . . . . $\pm V_S$
Open Collector Output Voltage Above Digital GND . . 36V
Current 50mA
Amplifier Short Ckt to Ground . . . . Indefinite
Comparator Input Voltage (Pin 9) . . . . $\pm V_S$

SPECIFICATIONS (@ +25°C with V_S = ±15V unless otherwise noted)

Model	AD650J/AD650A			AD650K/AD650B			AD650S			Units
	Min	Typ	Max	Min	Typ	Max	Min	Typ	Max	
DYNAMIC PERFORMANCE										
Full Scale Frequency Range			1			1			1	MHz
Nonlinearity[1] f_{max} = 10kHz		0.002	0.005		0.002	0.005		0.002	0.005	%
100kHz		0.005	**0.02**		0.005	**0.02**		0.005	**0.02**	%
500kHz		0.02	0.05		0.02	0.05		0.02	0.05	%
1MHz		0.1			0.05	**0.1**		0.05	**0.1**	%
Full Scale Calibration Error[2], 100kHz		±5			±5			±5		%
1MHz		±10			±10			±5		%
vs. Supply[3]	−0.002		**+0.002**	−0.002		**+0.002**	−0.002		**+0.002**	% of FSR/V
vs. Temperature										
A, B, and S Grades										
at 10kHz			±75			±75			**±75**	ppm/°C
at 100kHz			±150			±150			**±150**	ppm/°C
J and K Grades										
at 10kHz		±75			±75					ppm/°C
at 100kHz		±150			±150					ppm/°C
BIPOLAR OFFSET CURRENT										
Activated by 1.24kΩ between pins 4 and 5	0.45	0.5	0.55	0.45	0.5	0.55	0.45	0.5	0.55	mA
DYNAMIC RESPONSE										
Maximum Settling Time for Full Scale Step Input	1 Pulse of New Frequency Plus 1μs			1 Pulse of New Frequency Plus 1μs			1 Pulse of New Frequency Plus 1μs			
Overload Recovery Time Step Input	1 Pulse of New Frequency Plus 1μs			1 Pulse of New Frequency Plus 1μs			1 Pulse of New Frequency Plus 1μs			
ANALOG INPUT AMPLIFIER (V/F Conversion)										
Current Input Range (Figure 1)	0		+0.6	0		+0.6	0		+0.6	mA
Voltage Input Range (Figure 5)	−10		0	−10		0	−10		0	V
Differential Impedance		2MΩ‖10pF			2MΩ‖10pF			2MΩ‖10pF		
Common Mode Impedance		1000MΩ‖10pF			1000MΩ‖10pF			1000MΩ‖10pF		
Input Bias Current										
Noninverting Input		40	**100**		40	**100**		40	**100**	nA
Inverting Input		±8	**±20**		±8	**±20**		±8	**±20**	nA
Input Offset Voltage (Trimmable to Zero)			**±4**			**±4**			**±4**	mV
vs. Temperature (T_{min} to T_{max})		±30				±30			±30	μV/°C
Safe Input Voltage		$\pm V_S$			$\pm V_S$			$\pm V_S$		C
COMPARATOR (F/V Conversion)										
Logic "0" Level	$-V_S$		−1	$-V_S$		−1	$-V_S$		+1	V
Logic "1" Level	0		$+V_S$	0		$+V_S$	0		$+V_S$	V
Pulse Width Range[4]	0.1		(0.3 × t_{OS})	0.1		(0.3 × t_{OS})	0.1		(0.3 × t_{OS})	μs
Input Impedance		250			250			250		kΩ
OPEN COLLECTOR OUTPUT (V/F Conversion)										
Output Voltage in Logic "0" I_{SINK} ≤ 8mA, T_{min} to T_{max}			0.4			0.4			0.4	V
Output Leakage Current in Logic "1"			**100**			**100**			**100**	nA
Voltage Range[5]	0		+36	0		+36	0		+36	V
AMPLIFIER OUTPUT (F/V Conversion)										
Voltage Range (1500Ω min load resistance)	0		+10	0		+10	0		+10	V
Source Current (750Ω max load resistance)	10			10			10			mA
Capacitive Load (Without Oscillation)			100			100			100	pF
POWER SUPPLY										
Voltage, Rated Performance	±9		**±18**	±9		**±18**	±9		**±18**	V
Quiescent Current			**8**			**8**			**8**	mA
TEMPERATURE RANGE										
Rated Performance – N Package	0		+70	0		+70				°C
D Package	−25		+85	−25		+85	−55		+125	°C
Storage – N Package	−25		+85	−25		+85				°C
D Package	−65		+150	−65		+150	−65		+150	°C
PACKAGE OPTIONS										
PLCC (P-20A)		AD650JP			AD650KP					
Plastic DIP (N-14)		AD650JN			AD650KN					
Ceramic DIP (D-14)		AD650AD			AD650BD			AD650SD		

NOTES

[1]Nonlinearity is defined as deviation from a straight line from zero to full scale, expressed as a fraction of full scale.

[2]Full scale calibration error adjustable to zero.

[3]Measured at full scale output frequency of 10kHz.

[4]Refer to F/V conversion section of the text.

[5]Referred to digital ground.

Specifications subject to change without notice.

Specifications shown in boldface are tested on all production units at final electrical test. Results from those tests are used to calculate outgoing quality levels. All min and max specifications are guaranteed, although only those shown in boldface are tested on all production units.

AD650 Pin Configuration

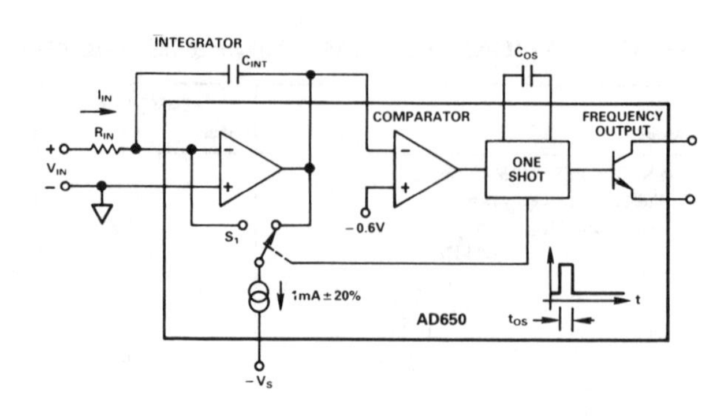

Block Diagram

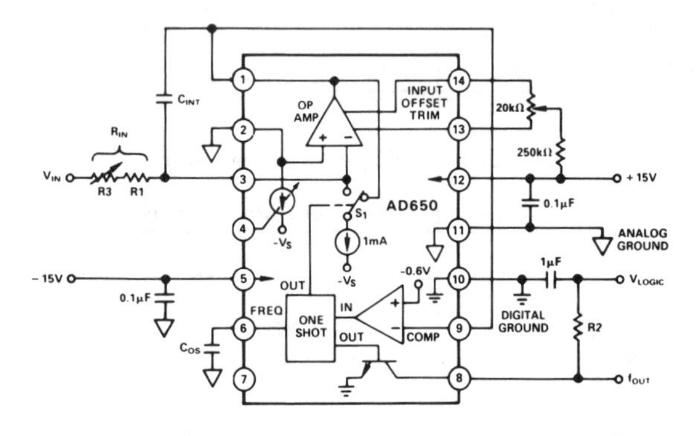

Connection Diagram for V/F Conversion, Positive Input Voltage

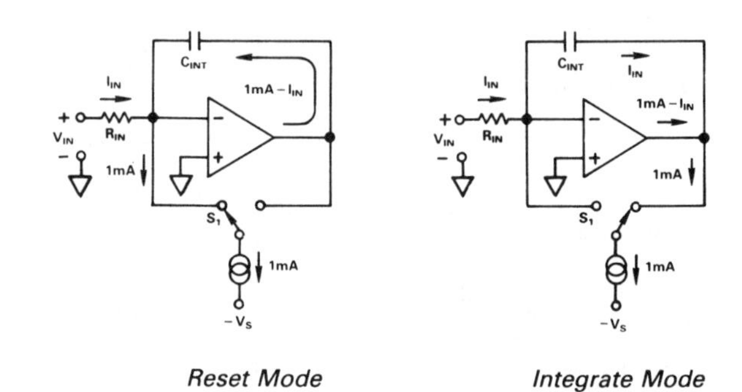

Reset Mode *Integrate Mode*

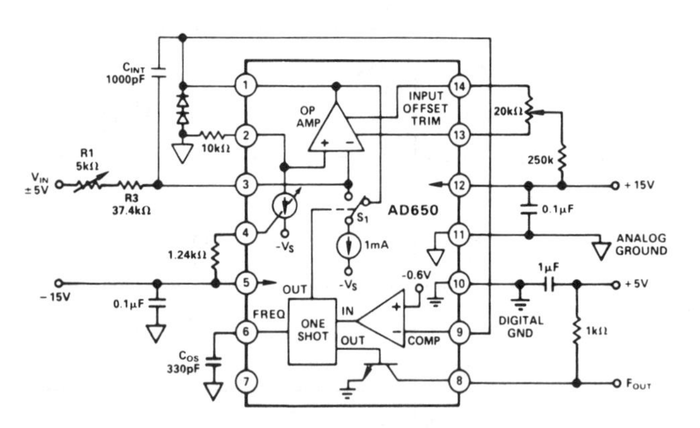

Connections for ± 5V Bipolar V/F with 0 to 100kHz TTL Output

Connection Diagram for V/F Conversion, Negative Input Voltage

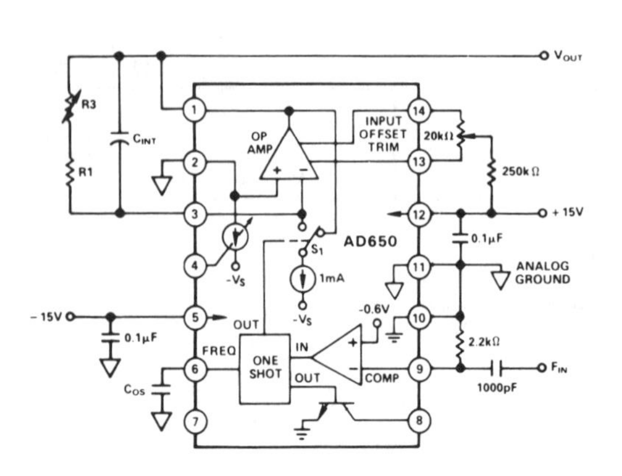

Connection Diagram for F/V Conversion

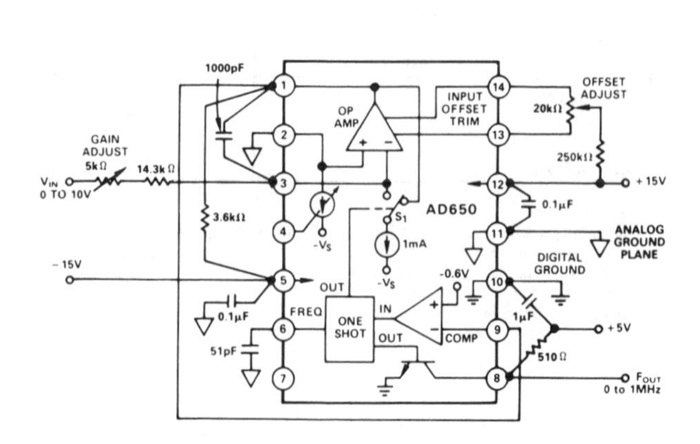

1MHz V/F Connection Diagram

Analog Devices AD652

Monolithic Synchronous Voltage-to-Frequency Converter

- Full-scale frequency (up to 2 MHz) set by external system clock
- Extremely low linearity error (0.005% max at 1 MHz FS, 0.02% max at 2 MHz FS)
- No critical external components required
- Accurate 5V reference voltage
- Low drift (25ppm/°C max)
- Dual- or single-supply operation
- Voltage or current input

AD652 PIN CONFIGURATIONS

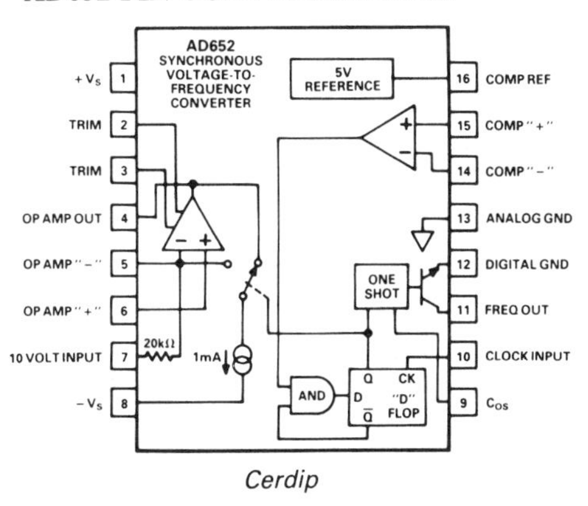

Cerdip

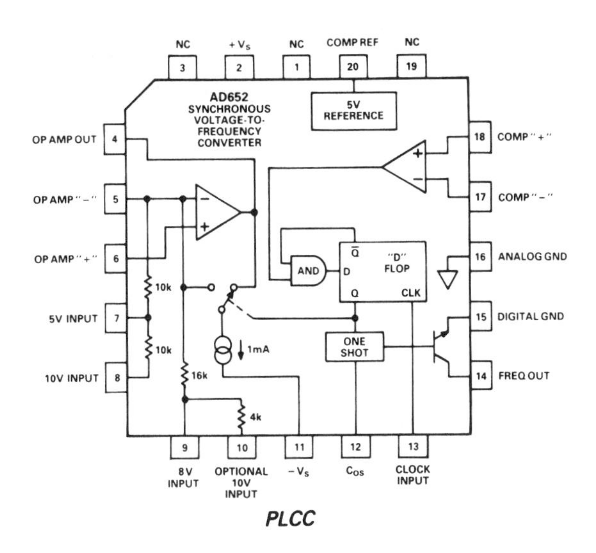

PLCC

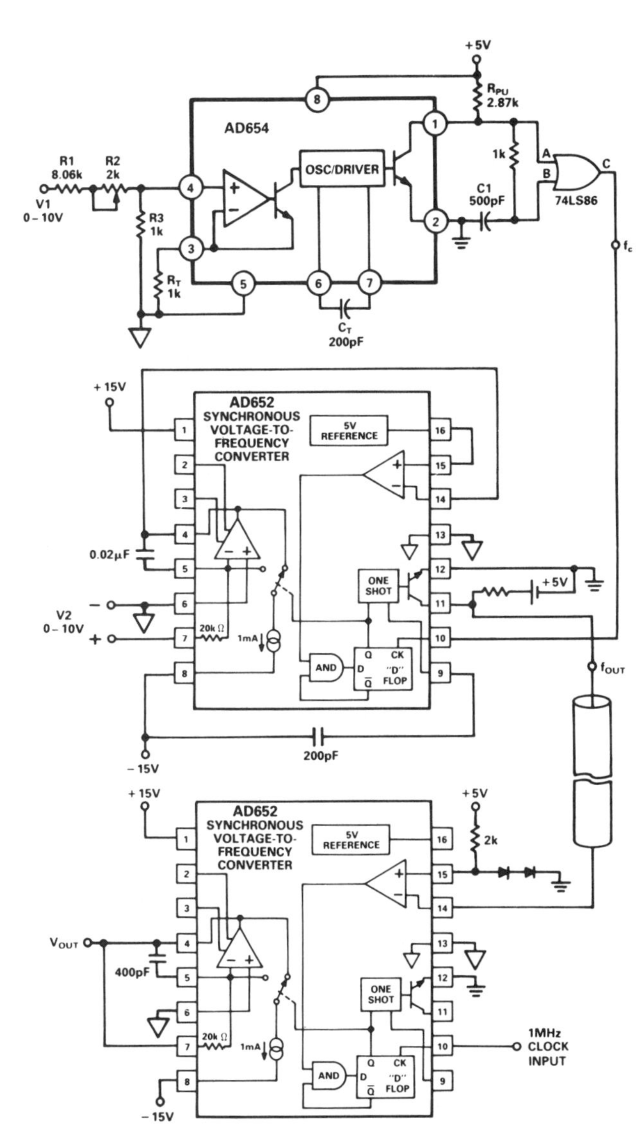

Frequency Output Multiplier

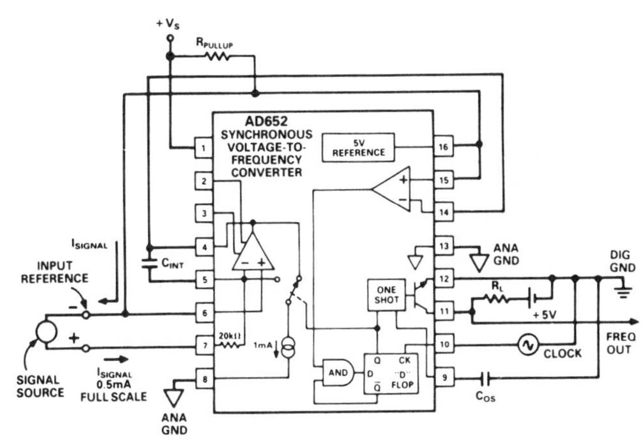

Single Supply Positive Voltage Input

SPECIFICATIONS (typical @ T_A = +25°C, V_S = ±15V, unless otherwise noted)

Parameter	AD652JP/AQ/SQ			AD652KP/BQ			Units
	Min	Typ	Max	Min	Typ	Max	
VOLTAGE-TO-FREQUENCY MODE							
Gain Error							
f_{CLOCK} = 200kHz		±0.5	±1		±0.25	±0.5	%
f_{CLOCK} = 1MHz		±0.5	±1		±0.25	±0.5	%
f_{CLOCK} = 4MHz		±0.5	±1.5		±0.25	±0.75	%
Gain Temperature Coefficient							
f_{CLOCK} = 200kHz		±25	±50		±15	±25	ppm/°C
f_{CLOCK} = 1MHz		±25	±50		±15	±25	ppm/°C
		±10	±25		±10	±15	ppm/°C[1]
f_{CLOCK} = 4MHz		±25	±75		±15	±50	ppm/°C
Power Supply Rejection Ratio		0.001	0.01		0.001	0.01	%/V
Linearity Error							
f_{CLOCK} = 200kHz		±0.002	±0.02		±0.002	±0.005	%
f_{CLOCK} = 1MHz		±0.002	±0.02		±0.002	±0.005	%
f_{CLOCK} = 2MHz		±0.01	±0.02		±0.002	±0.005	%
f_{CLOCK} = 4MHz		±0.02	±0.05		±0.01	±0.02	%
Offset (Transfer Function, RTI)		±1	±3		±1	±2	mV
Offset Temperature Coefficient		±10	±50		±10	±25	μV/°C
Response Time	One Period of New Output Frequency Plus One Clock Period.						
FREQUENCY-TO-VOLTAGE MODE							
Gain Error							
f_{IN} = 100kHz FS		±0.5	±1		±0.25	±0.5	%
Linearity Error							
f_{IN} = 100kHz FS		±0.002	±0.02		±0.002	±0.01	%
INPUT RESISTORS							
Cerdip (Figure 1a.) (0 to +10V FS Range)	19.8	20	20.2	19.8	20	20.2	kΩ
PLCC (Figure 1b.)							
Pin 8 to Pin 7	9.9	10	10.1	9.9	10	10.1	kΩ
Pin 7 to Pin 5 (0 to +5V FS Range)	9.9	10	10.1	9.9	10	10.1	kΩ
Pin 8 to Pin 5 (0 to +10V FS Range)	19.8	20	20.2	19.8	20	20.2	kΩ
Pin 9 to Pin 5 (0 to +8V FS Range)	15.8	16	16.2	15.8	16	16.2	kΩ
Pin 10 to Pin 5 (Auxiliary Input)	19.8	20	20.2	19.8	20	20.2	kΩ
Temperature Coefficient (All)		±50	±100		±50	±100	ppm/°C
INTEGRATOR OP AMP							
Input Bias Current							
Inverting Input (Pin 5)		±5	±20		±5	±20	nA
Noninverting Input (Pin 6)		20	50		20	50	nA
Input Offset Current		20	70		20	70	nA
Input Offset Current Drift		1	3		1	2	nA/°C
Input Offset Voltage		±1	±3		±1	±2	mV
Input Offset Voltage Drift		±10	±25		±10	±15	μV/°C
Open Loop Gain		86			86		dB
Common-Mode Input Range	$-V_S+5$		$+V_S-5$	$-V_S+5$		$+V_S-5$	V
CMRR	80			80			dB
Bandwidth	14	95		14	95		MHz
Output Voltage Range (Referred to Pin 6, R_l >= 5k)	−1		$(+V_S-4)$	−1		$(+V_S-4)$	V
COMPARATOR							
Input Bias Current		0.5	5		0.5	5	μA
Common-Mode Voltage	$-V_S+4$		$+V_S-4$	$-V_S+4$		$+V_S-4$	V
CLOCK INPUT							
Maximum Frequency	4	5		4	5		MHz
Threshold Voltage (Referred to Pin 12)		1.2			1.2		V
T_{min}-T_{max}	0.8		2.0	0.8		2.0	V
Input Current							
$(-V_S < V_{CLK} < +V_S)$		5	20		5	20	μA
Voltage Range	$-V_S$		$+V_S$	$-V_S$		$+V_S$	V
Rise Time			2			2	μs

Parameter	AD652JP/AQ/SQ Min	Typ	Max	AD652KP/BQ Min	Typ	Max	Units
OUTPUT STAGE							
V_{OL} (I_{OUT} = 10mA)			**0.4**			**0.4**	V
I_{OL}							
V_{OL}<0.8V			15			15	mA
V_{OL}<0.4V, T_{min}-T_{max}			8			8	mA
I_{OH} (Off Leakage)		0.01	**10**		0.01	**10**	μA
Delay Time, Positive Clock Edge to Output Pulse	**150**	200	**250**	**150**	200	**250**	ns
Fall Time (Load = 500pF and I_{SINK} = 5mA)		100			100		ns
Output Capacitance		5			5		pF
OUTPUT ONE-SHOT							
Pulse Width							
C_{OS} = 300pF	**1**	1.5	**2**	**1**	1.5	**2**	μs
C_{OS} = 1000pF	4	5	6	4	5	6	μs
REFERENCE OUTPUT							
Voltage	**4.950**	5.0	**5.050**	**4.975**	5.0	**5.025**	V
Drift			**100**			**50**	ppm/°C
Output Current							
Source T_{min} to T_{max}	**10**			**10**			mA
Sink	**100**	500		**100**	500		μA
Power Supply Rejection							
(Supply Range = ±12.5V to ±17.5V)			0.015			0.015	%/V
Output Impedance (Sourcing Current)		0.3	2		0.3	2	Ω
POWER SUPPLY							
Rated Voltage		±15			±15		V
Operating Range							
Dual Supplies	±6	±15	±18	±6	±15	±18	V
Single Supply ($-V_S$ = 0)	+12		+36	+12		+36	V
Quiescent Current		±11	**±15**		±11	**±15**	mA
Digital Common	$-V_S$		$+V_S-4$	$-V_S$		$+V_S-4$	V
Analog Common	$-V_S$		$+V_S$	$-V_S$		$+V_S$	V
TEMPERATURE RANGE							
Specified Performance							
JP, KP Grade	0		+70	0		+70	°C
AQ, BQ Grade	−40		+85	−40		+85	°C
SQ Grade	−55		+125				°C

NOTES
[1]Referred to internal V_{REF}. In PLCC package, tested on 10V input range only.
Specifications in **boldface** are 100% tested at final test and are used to measure outgoing quality levels.
Specifications subject to change without notice.

ABSOLUTE MAXIMUM RATINGS

Total Supply Voltage $+V_S$ to $-V_S$ 36V
Maximum Input Voltage (Figure 6) 36V
Maximum Output Current (Open Collector Output) . . 50mA
Amplifier Short Circuit to Ground . . . . Indefinite
Storage Temperature Range: Cerdip . . . . −65°C to +150°C
PLCC . . . . −65°C to +150°C

ORDERING GUIDE

Part Number	Gain Drift ppm/°C 100kHz	1MHz Linearity %	Specified Temperature Range °C	Package Options*
AD652JP	50 max	0.02 max	0 to +70	PLCC (P-20A)
AD652KP	25 max	0.005 max	0 to +70	PLCC (P-20A)
AD652AQ	50 max	0.02 max	−40 to +85	Cerdip (Q-16)
AD652BQ	25 max	0.005 max	−40 to +85	Cerdip (Q-16)
AD652SQ	50 max	0.02 max	−55 to +125	Cerdip (Q-16)

DEFINITIONS OF SPECIFICATIONS

GAIN ERROR – The gain of a voltage-to-frequency converter is that scale factor setting that provides the nominal conversion relationship, e.g. 1MHz full scale. The "gain error" is the difference in slope between the actual and ideal transfer functions for the V-F converter.

LINEARITY ERROR – The "linearity error" of a V-F is the deviation of the actual transfer function from a straight line passing through the endpoints of the transfer function.

GAIN TEMPERATURE COEFFICIENT – The gain temperature coefficient is the rate of change in full-scale frequency as a function of the temperature from +25°C to T_{min} or T_{max}.

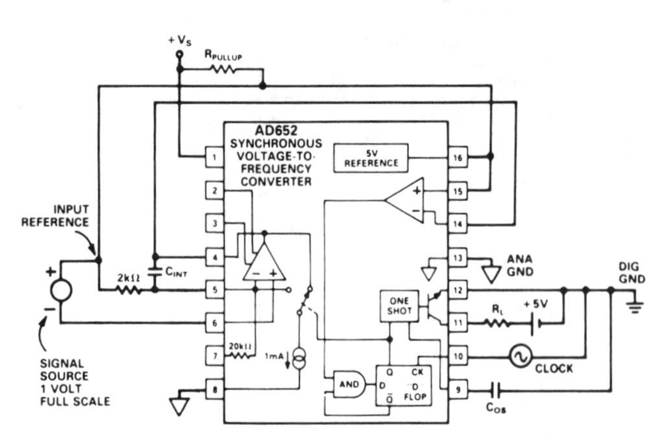

Single Supply Negative Voltage Input

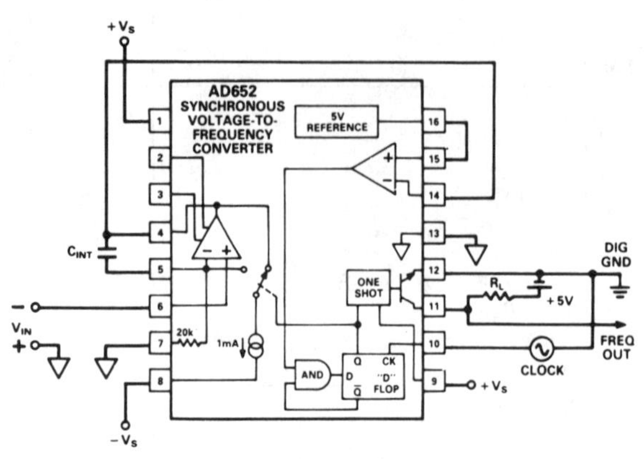

Negative Voltage Input

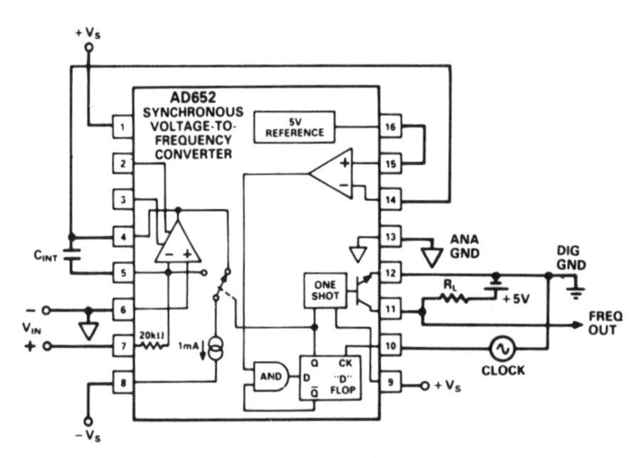

Standard V/F Connection for Positive Input Voltage with Dual Supply

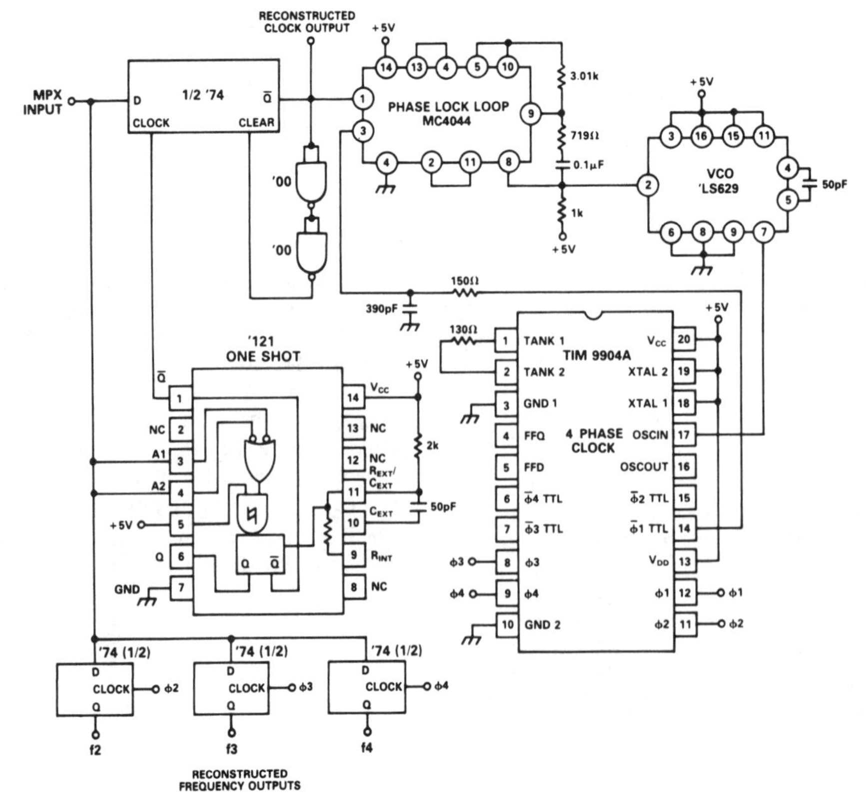

SVFC Demultiplexers

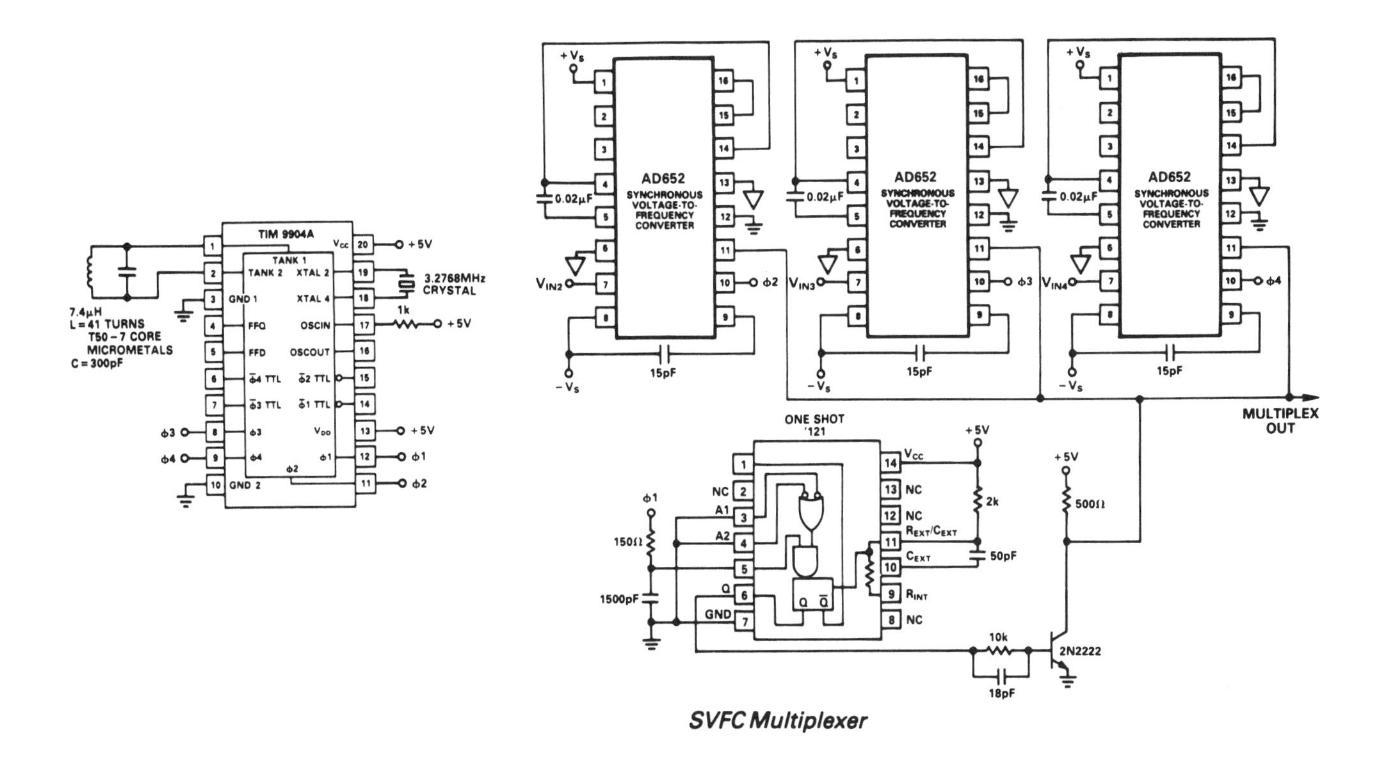

SVFC Multiplexer

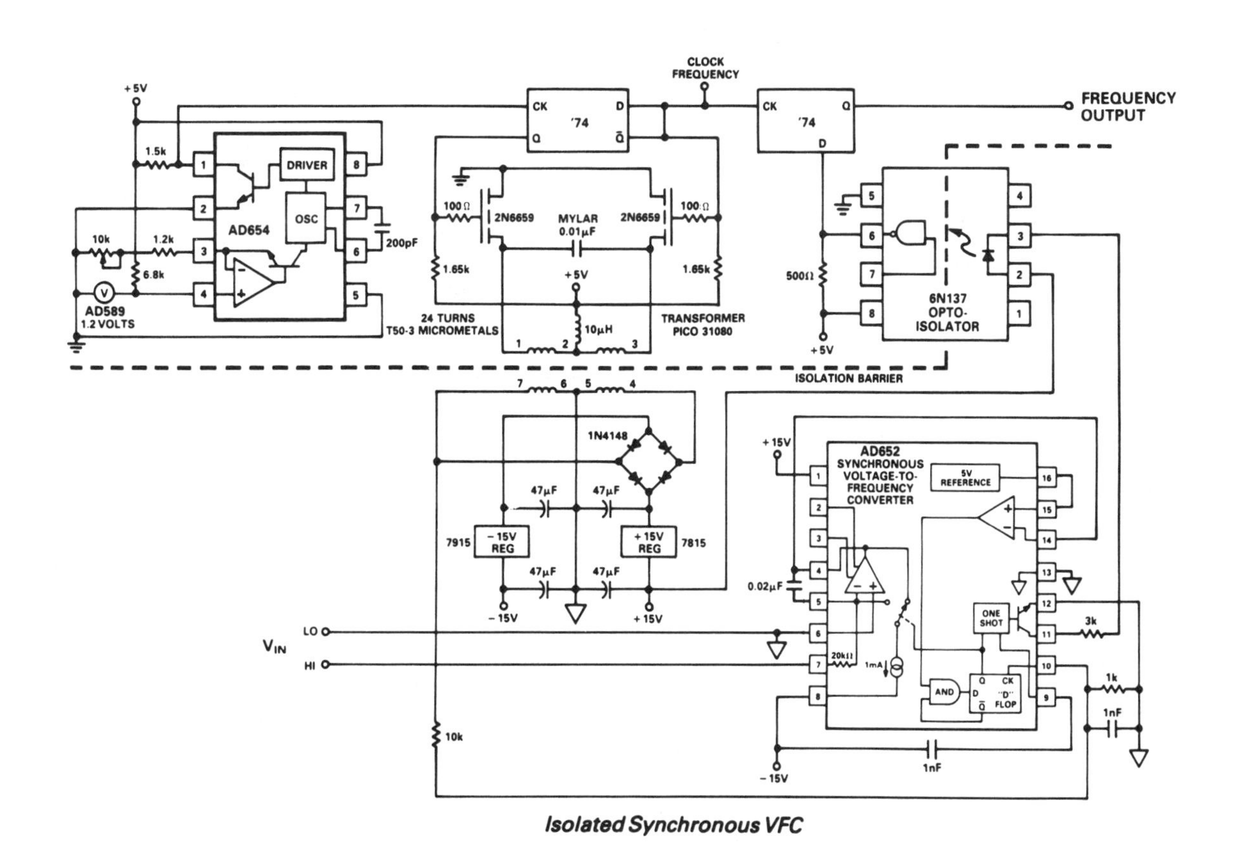

Isolated Synchronous VFC

Single Line Multiplexed Data Transmission Block Diagram

Analog Devices
AD654
Low-Cost Monolithic Voltage-to-Frequency Converter

- Low cost
- Single- or dual-supply, 5 to 36V, ±5 to ±18V
- Full-scale frequency up to 500 kHz
- Minimum number of external components needed
- Versatile input amplifier
  - Positive or negative voltage modes
  - Negative current mode
  - High input impedance, low drift
- Low power: 2.0mA quiescent current
- Low offset: 1mV

AD654 FUNCTIONAL BLOCK DIAGRAM

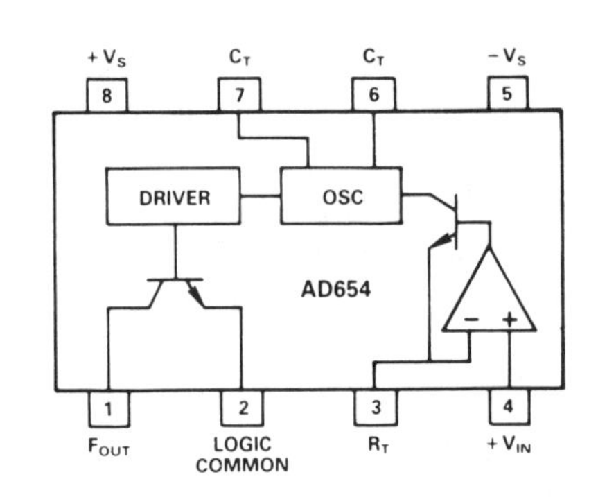

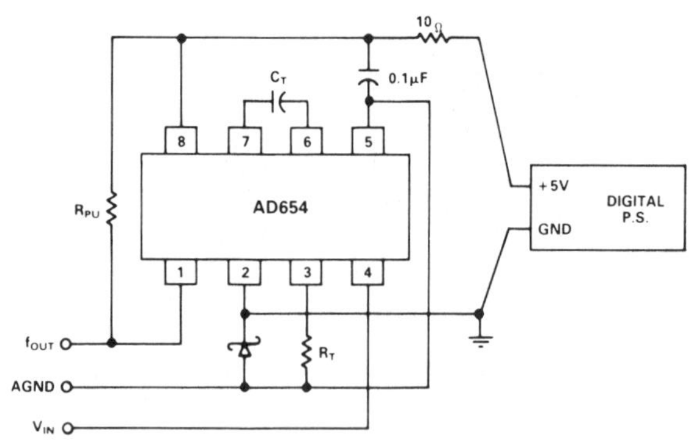

Proper Ground Scheme

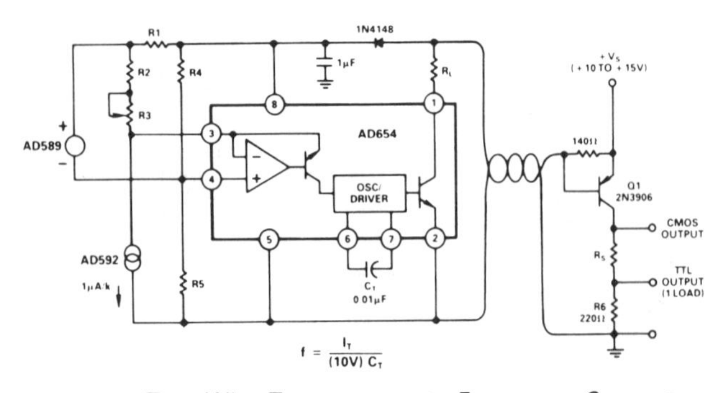

Two-Wire Temperature-to-Frequency Converter

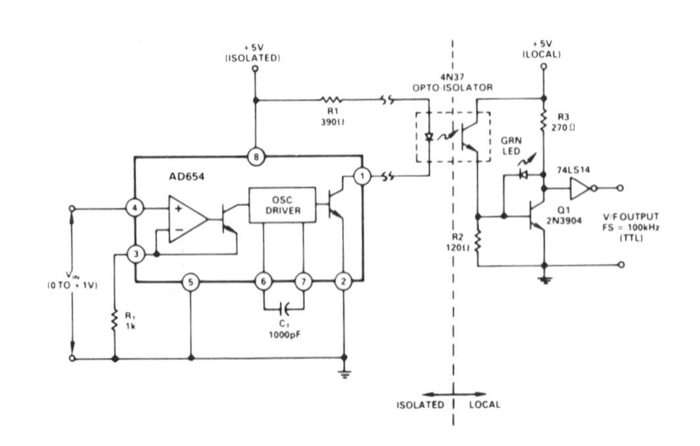

Optoisolator Interface

SPECIFICATIONS (t_{AMB} = +25°C and V_S (total) = 5 to 16.5V, unless otherwise specified. All testing done @ V_S = +5V).

Model	AD654JN/JR Min	Typ	Max	Units
CURRENT-TO-FREQUENCY CONVERTER				
Frequency Range	0		500	kHz
Nonlinearity[1]				
f_{max} = 250kHz		0.06	**0.1**	%
f_{max} = 500kHz		0.20	0.4	%
Full Scale Calibration Error				
C = 390pF, I_{IN} = 1.000mA	**−10**		**10**	%
vs. Supply (f_{max} ≤ 250kHz)				
V_S = +4.75 to +5.25V		0.20	**0.40**	%/V
V_S = +5.25 to +16.5V		0.05	**0.10**	%/V
vs. Temp (0 to 70°C)		50		ppm/°C
ANALOG INPUT AMPLIFIER				
(Voltage-to-Current Converter)				
Voltage Input Range				
Single Supply	0		(+V_S −4)	V
Dual Supply	−V_S		(+V_S −4)	V
Input Bias Current				
(Either Input)		30	**50**	nA
Input Offset Current		5		nA
Input Resistance (Non-Inverting)		250		MΩ
Input Offset Voltage		0.5	**1.0**	mV
vs. Supply				
V_S = +4.75 to +5.25V		0.1	**0.25**	mV/V
V_S = +5.25 to +16.5V		0.03	**0.1**	mV/V
vs. Temp (0 to 70°C)		4		μV/°C
OUTPUT INTERFACE (Open Collector Output)				
(Symmetrical Square Wave)				
Output Sink Current in Logic "0"[2]				
V_{OUT} = 0.4V max, 25°C	**10**	20		mA
V_{OUT} = 0.4V max, 0 to 70°C	5	10		mA
Output Leakage Current in Logic "1"		10	**100**	nA
0 to 70°C		50	500	nA
Logic Common Level Range	−V_S		(+V_S −4)	V
Rise/Fall Times (C_T = 0.01μF)				
I_{IN} = 1mA		0.2		μs
I_{IN} = 1μA		1		μs
POWER SUPPLY				
Voltage, Rated Performance	4.5		16.5	V
Voltage, Operating Range				
Single Supply	4.5		36	V
Dual Supply	±5		±18	V
Quiescent Current				
V_S (Total) = 5V		1.5	**2.5**	mA
V_S (Total) = 30V		2.0	**3.0**	mA
TEMPERATURE RANGE				
Operating Range	−40		85	°C
PACKAGE OPTION				
SOIC (R-8)		AD654JR		
Plastic DIP (N-8)		AD654JN		

NOTES
[1]At f_{max} = 250kHz; R_T = 1kΩ, C_T = 390pF, I_{IN} = 0-1mA.
f_{max} = 500kHz; R_T = 1kΩ, C_T = 200pF, I_{IN} = 0-1mA.
[2]The sink current is the amount of current that can flow into Pin 1 of the AD654 while maintaining a maximum voltage of 0.4V between Pin 1 and Logic Common.

Specifications shown in boldface are testeu on all production units at final electrical test. Results from those tests are used to calculate outgoing quality levels. All min and max specifications are guaranteed, although only those shown in boldface are tested on all production units.

Specifications subject to change without notice.

ABSOLUTE MAXIMUM RATINGS

Total Supply Voltage +V_S to −V_S 36V
Maximum Input Voltage
(Pins 3, 4) to −V_S . . . . −300mV to +V_S
Maximum Output Current
Instantaneous 50mA
Sustained 25mA
Logic Common to −V_S . . . . −500mV to (+V_S −4)
Storage Temperature Range . . . . −65°C to +150°C

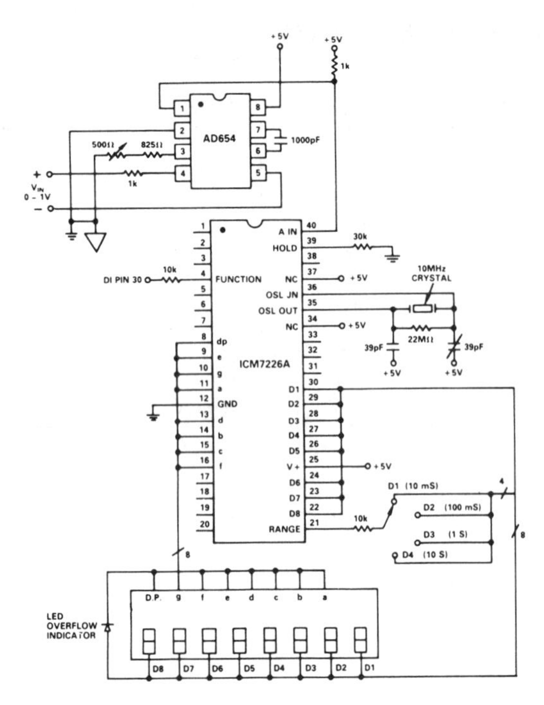

AD654 With Stand-Alone Frequency Counter/LED Display Driver

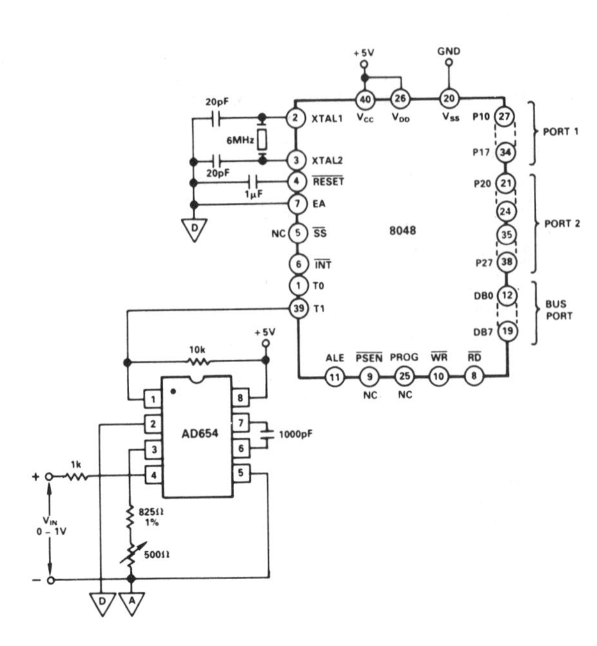

AD654 VFC as an ADC

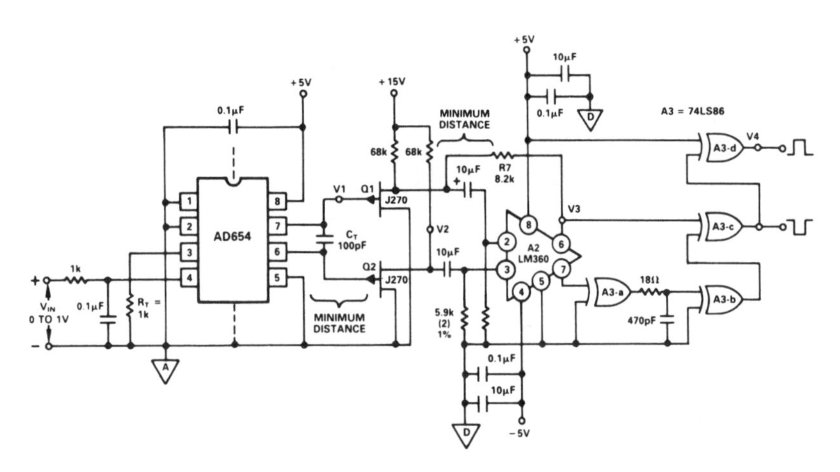

2MHz, Frequency Doubling V/F

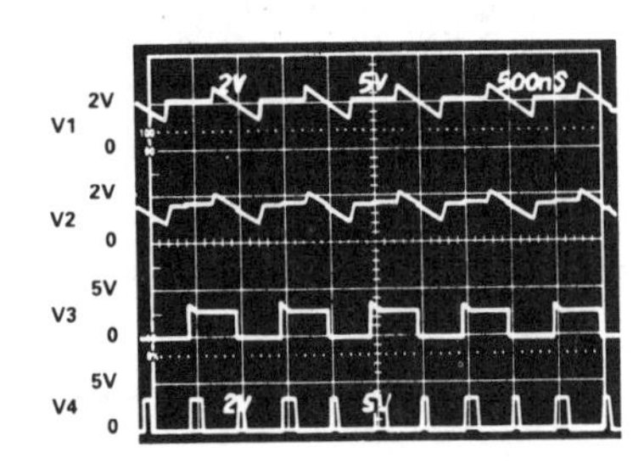

Waveforms of 2MHz Frequency Doubler

This information is reproduced with the permission of GEC Plessey Semiconductors.

Plessey TDA1085C Phase Control Integrated Circuit

- Powered direct from ac mains or dc line
- Flexible ramp generator to provide controlled acceleration and distribution period
- Actual speed derived from tachogenerator frequency or magnitude
- Control amplifier allowing loop gain control
- Symmetrical positive and negative wave firing of the triac
- Motor current limiting
- Fall safe in case of open-circuit tachogenerator
- Repeated triac pulses provided if triac unlatches.

ELECTRICAL CHARACTERISTICS

Test conditions (unless otherwise stated):
$T_{amb} = +25°C$
All potentials measured with respect to common (Pin 8)

Characteristic	Value Min.	Value Typ.	Value Max.	Units	Conditions
CURRENT CONSUMPTION					
Pin 9					
IC operating current		7.4	8.9	mA	Total current required is dependent on external circuitry.
VOLTAGE REGULATOR					
Pin 9					
Shunt regulating voltage		15.5	16	V	$I_9 + I_{10} = 10mA$
Monitor enable level		15.1		V	
Monitor disable level		14.5		V	
RAMP GENERATOR (See Fig.3)					
Pin 7					
Fast ramp current		1.2		mA	
Residual charging current		5		μA	During slow ramp period
Pin 5					
Speed program voltage range	0.08		13.5	V	
Bias current			-20	μA	
Pin 6					
Program distribute level	0		4	V	
Bias current			-20	μA	
Internal					
Low distribute level V_{RA}		V_6	1.2	V	Distribute levels referred to ramp generator.
High distribute level V_{RB}	$1.9V_6$	$2V_6$	$2.1V_6$	V	
FREQUENCY-ANALOG CONVERTER					
Pin 12					
Positive tacho input voltage			6	V	
Negative tacho input voltage			-3	V	
Minimum tacho input voltage	200			mV	Peak-Peak
Internal bias current		25		μA	
Pin 12 to Pin 11					
Conversion factor (typical)		7.5		mV/Hz	C pin 6 = 390pF,R pin 4 = 150kΩ
		15		mV/Hz	C pin 6 = 820pF,R pin 4 = 150kΩ
Conversion gain		10			
Linearity		±4		%	
CONTROL AMPLIFIER					
Pin 4					
Actual speed voltage limits	0		13.5	V	
Analogue input bias current			-350	nA	
Pin 4, 5 & 16					
Differential offset voltage	-60		+20	mV	$V_5 - V_4$ to give $I_{16} = 0$
Transconductance		300		μA/V	
Pin 16					
Output current drive		±100		μA	
FIRING PULSE TIMING					
Pin 2					
Voltage sync trip level		±50		μA	
Pin 1					
Current sync trip level		±50		μA	
Pin 16					
Phase control voltage swing		11.7		V	
Pin 13					
Firing pulse width		55		μs	R pin 15 = 300kΩ
Pulse repetition time		200		μs	C pin 14 = 47nF
Pin 14					
Ramp recharge current (I_R)		150		μA	
FIRING PULSE OUTPUT DRIVE					
Pin 13					
High output level		$V_{CC} - 4$		V	At 150mA drive current
Leakage current			30	μA	
LOAD CURRENT LIMITER					
Pin 3 & Pin 7					
Current gain		170	—		Reset of ramp generator
Pin 7					
Discharge current		35		mA	

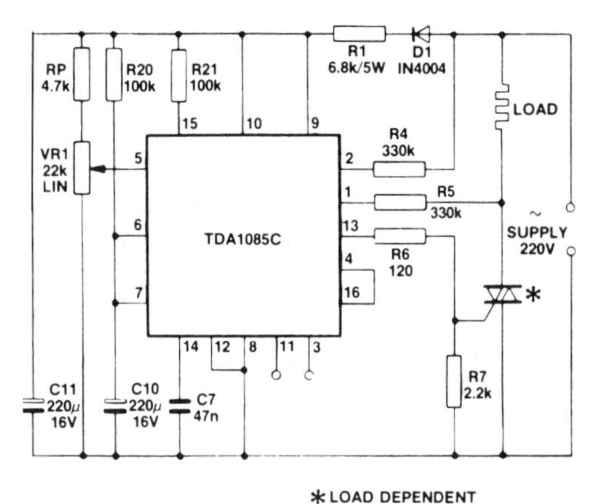

✱LOAD DEPENDENT

Manual phase control circuit

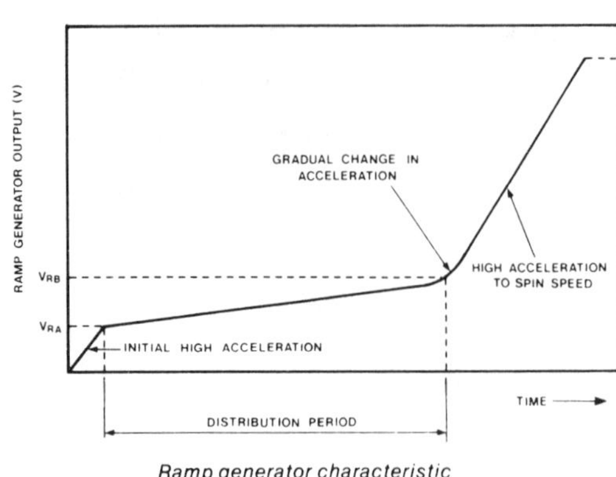

Ramp generator characteristic

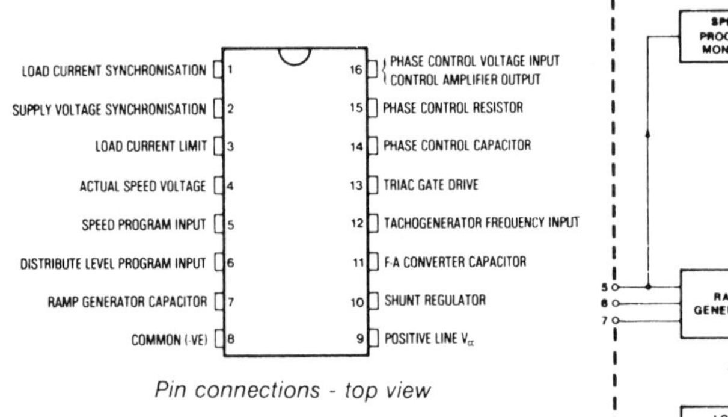

Pin connections - top view

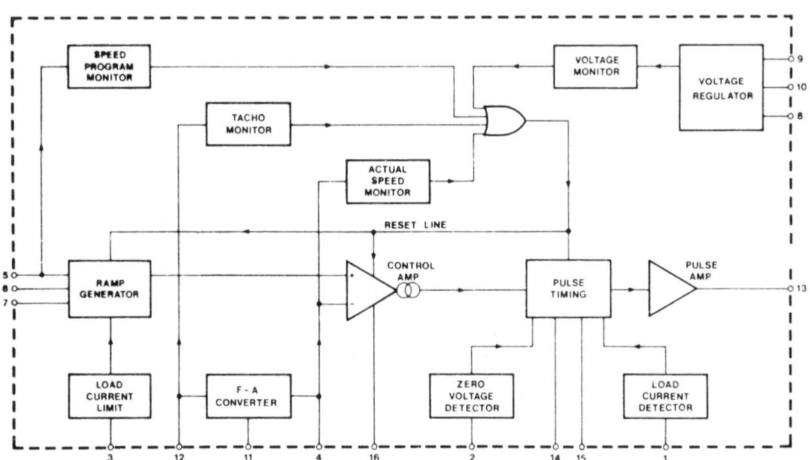

Block diagram of TDA1085C

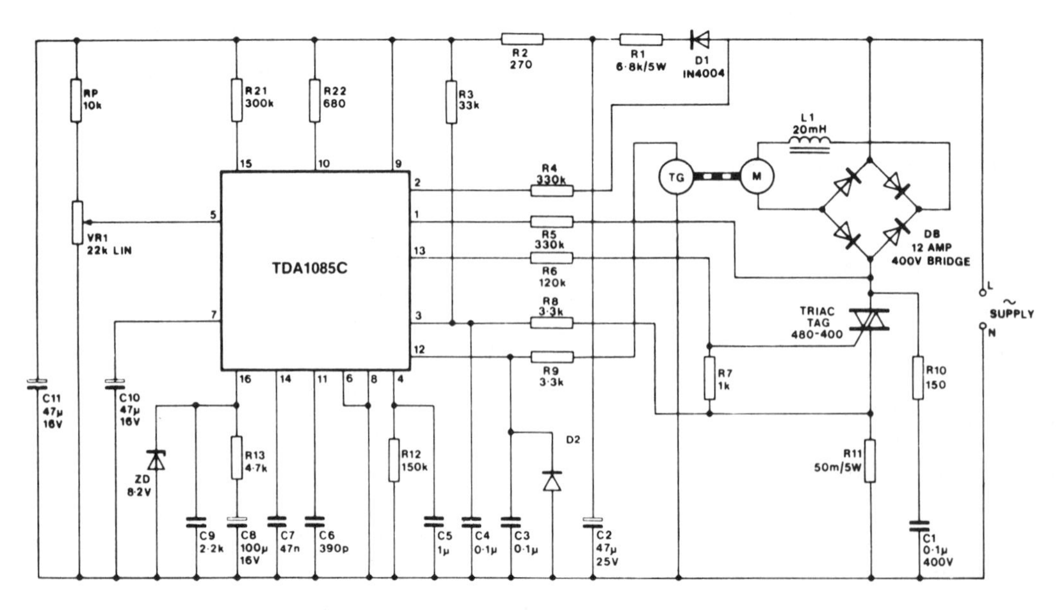

✱DIODE NOT REQUIRED IF TACHO VOLTAGE AT PIN 12 IS LESS THAN 6V P-P

Fixed field motor application circuit

ABSOLUTE MAXIMUM RATINGS

Electrical

Peak input current (I sync), pin 1:	±2mA
Peak input current (V sync), pin 2:	±2mA
Current drain, pin 3:	-5mA
Positive input voltage, pin 3:	6V
Analog voltage drive, pin 4:	V_{CC}
Speed reference voltage, pin 5:	V_{CC}
Distribute level, pin 6:	V_{CC}
IC Circuit current (pin 10 disconnected), pin 9:	10mA
Supply shunt regulating current, pin 10:	30mA
Tachogenerator (digital) drive input, pin 12:	-3 +0.1mA
Triac gate current, pin 13:	200mA
Phase timing current, pin 15:	1mA

Thermal

Operating ambient temperature:	0°C to +70°C
Storage temperature:	-55°C to +125°C

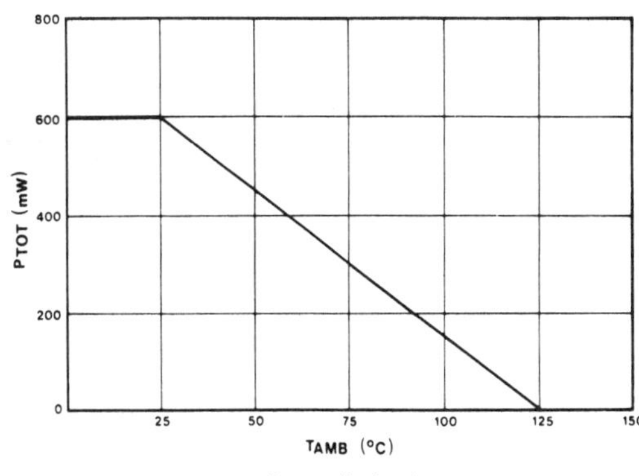

Power dissipation

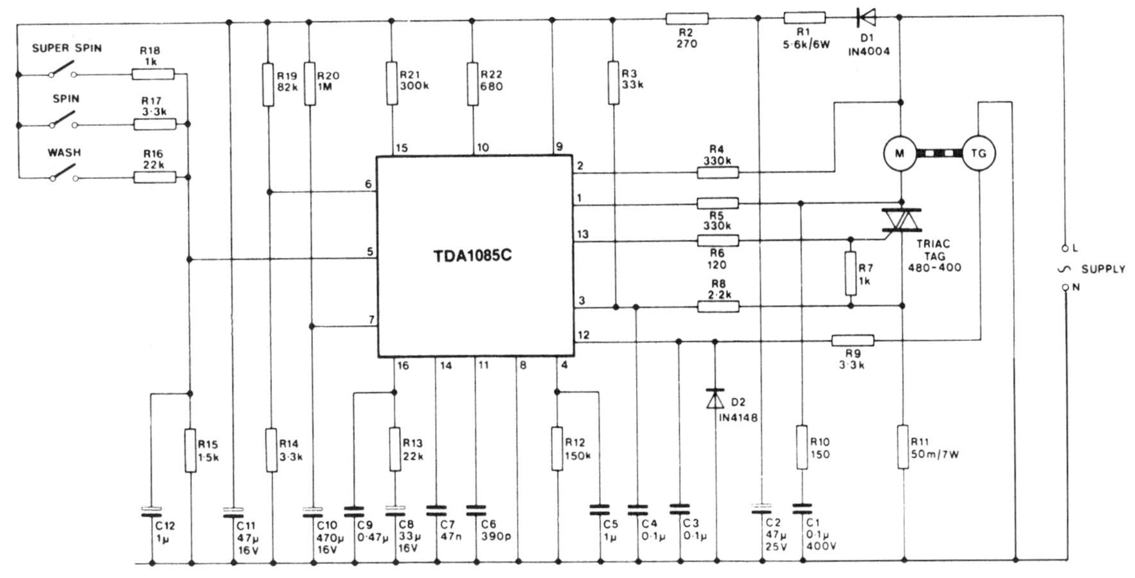

*DIODE NOT REQUIRED IF TACHO VOLTAGE AT PIN 12 IS LESS THAN 6V P-P

Universal motor application circuit

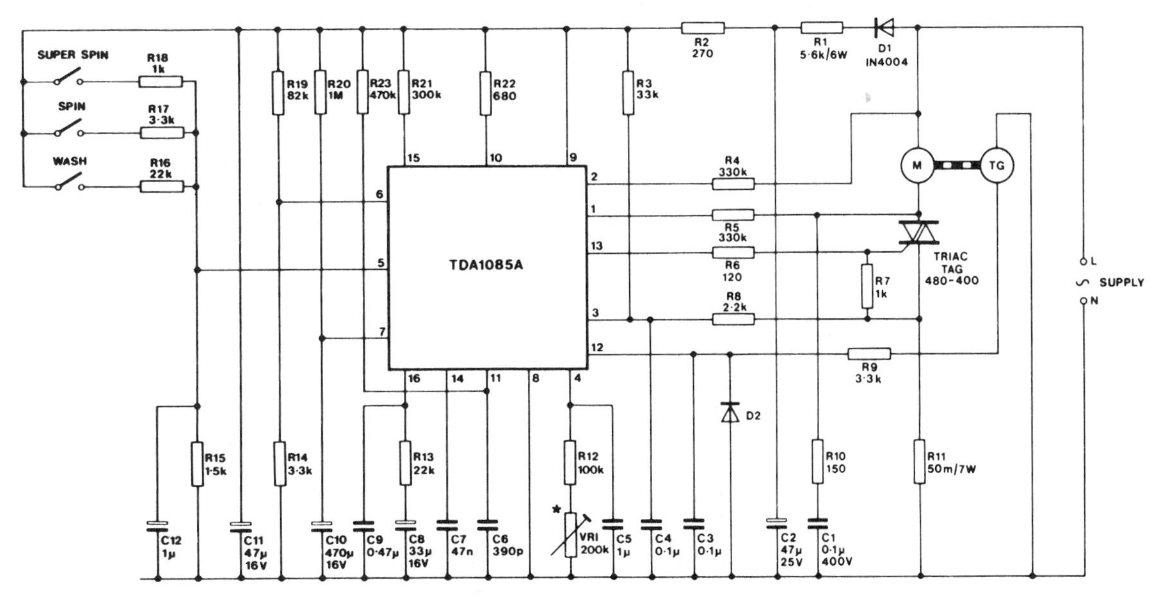

*DIODE NOT REQUIRED IF TACHO VOLTAGE AT PIN 12 IS LESS THAN 6V P-P

† Adjust for the highest speed.

Calibrated universal motor application circuit

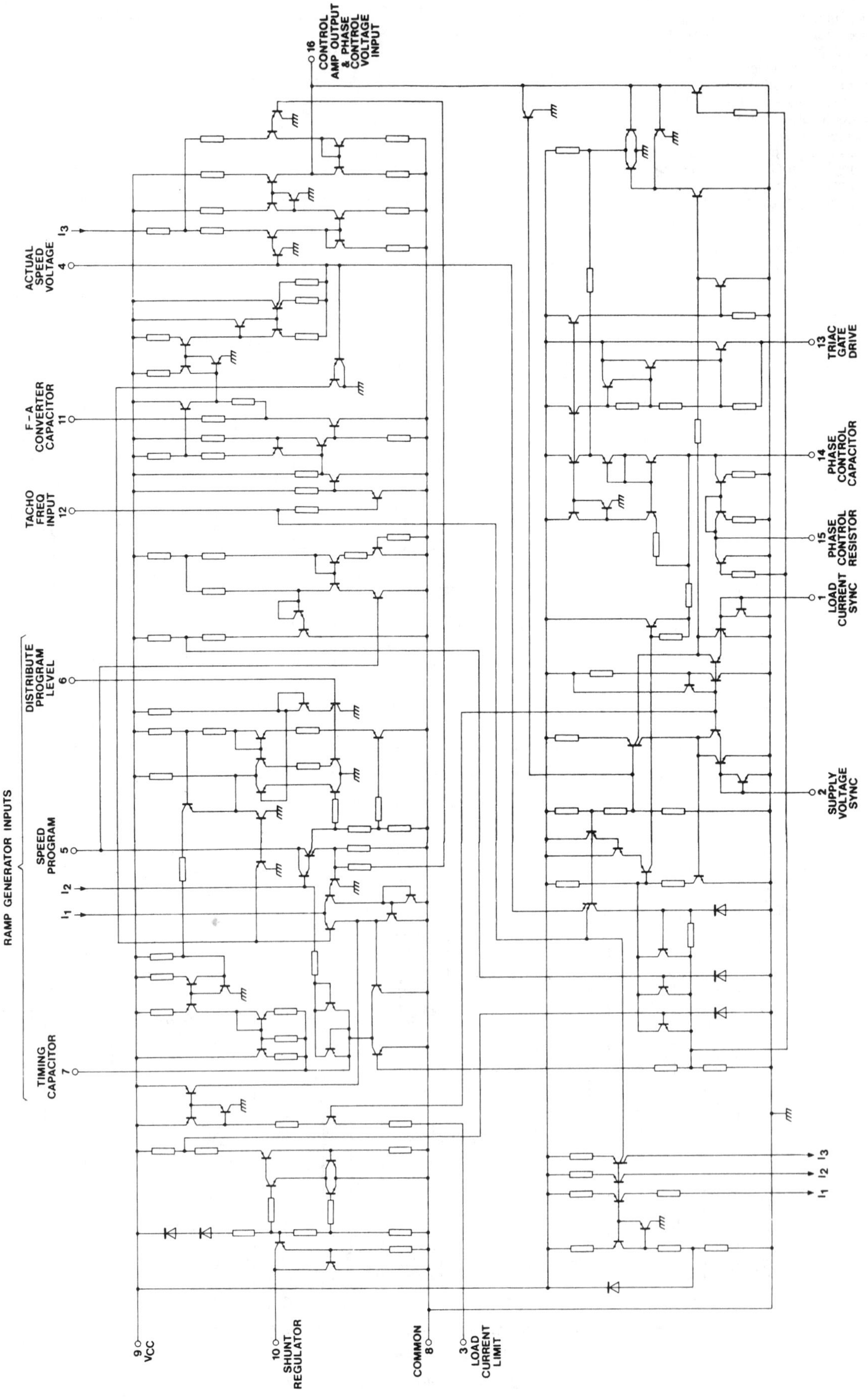

TDA1085C circuit diagram

Plessey TDA2086
Phase Control Integrated Circuit

- Power direct from ac mains or dc line
- 5V supply available for ancillary circuitry
- Low supply current consumption
- Average or peak load current limiting
- Ramp generator to provide controlled acceleration
- Negative triac firing pulses 100mA guaranteed minimum
- Warning LED drive circuit
- Actual speed derived from tachogenerator frequency or analog feedback
- Well-defined control voltage/phase angle relationship
- Inhibit input for use with thermistor temperature sensors

ELECTRICAL CHARACTERISTICS

Test conditions (unless otherwise stated):

T_{amb} = +25°C
All potentials measured with respect to common (Pin 3) (unless otherwise stated)
Pin numbers refer to DP16 package

Characteristic	Value			Units	Conditions
	Min.	Typ.	Max.		
CURRENT CONSUMPTION					
Pin 4					
IC Operating current		3.1	4.1	mA	Pin 4 voltage = 13.5V including triac gate drive current
SHUNT VOLTAGE REGULATOR					
Pin 4					
Regulating voltage	-16	-14.75	-13.5	V	Full temperature range
Voltage monitor enable level	-11		-9	V	
SERIES REGULATOR					
Pin 11					
Regulating voltage (Vreg)	-5.35	-5	-4.65	V	1mA external load
Temperature coefficient			±1	mV/°C	
External load			10	mA	
Regulation	-75		+75	mV	For 0-5mA external load change
RAMP GENERATOR					
Pin 9					
Capacitor charging current	25	30	35	μA	
Capacitor discharge current		25		μA	Load current limit in operation
Capacitor discharge current		10		mA	Load current inhibit in operation 5V on ramp C
Capacitor to actual speed voltage clamp	-0.8		+0.8	V	
SPEED PROGRAM CIRCUIT					
Pin 10					
Input voltage range	Vreg -0.5		0	V	
Input bias current			1	μA	
Zero power demand voltage	-100	-75	-50	mV	
FREQUENCY TO ANALOG CONVERTER					
Pin 15					
Tacho input voltage	100			mV	Peak value
Hysteresis	30	40	60	mV	
Bias current			10	μA	
Pin 15 to Pin 14					
Conversion factor (typical application)		0.5		mV/rpm	C pin 14 = 10nF, R pin 13 = 150k, 8 pole tacho 10000 rpm max.
Pin 4 to Pin 13					
Conversion gain		1			
ERROR AMPLIFIER					
Pin 9 and 13					
Input voltage range	Vreg		0	V	
Input bias current			0.5	μA	
Pin 10, 13 and 12					
Input offset voltage	-5		+15	mV	V10-V13 to give I_{12} = 0
Transconductance	80	100	120	μA/V	
Pin 12					
Output current drive	±20		±35	μA	

ELECTRICAL CHARACTERISTICS

Characteristic	Value			Units	Conditions
	Min.	Typ.	Max.		
FIRING PULSE TIMING					
Pin 7					
Voltage SYNC trip level	±35	±50	±65	μA	
Pin 6					
Current SYNC trip level	±35	±50	±65	μA	
Pin 12					
Phase control voltage swing	Vreg		0	V	
Pin 13					
Firing pulse width		50		μs	C pin 16 = 47nF
Pulse repetition time		100		μs	C pin 16 = 47nF, R pin 1 = 200k
FIRING PULSE OUTPUT					
Pin 2					
Drive current	100	125	150	mA	Pin 2 V = -3V
Leakage current			10μA		Pin 2 V = 0V
LOAD CURRENT LIMITING					
Pin 5					
Offset voltage			±20	mV	
Pin 5 and 8					
Current gain	0.475	0.5	0.525		Pin 5 current = 100μA
Pin 8					
Voltage for load current limit		-1V			(0.2 Vreg)
Voltage for load current inhibit		-1.5V			(0.3 Vreg)

ABSOLUTE MAXIMUM RATINGS

ELECTRICAL	Value	Units
Triac gate voltage pin 2	4	V
Repetitive peak input current pin 4	80	mA
Non repetitive peak input current pin 4 (tp 250μs)	200	mA
Peak input current pin 5 positive half cycle	2	mA
Repetitive peak input current pin 5 negative half cycle	80	mA
Non repetitive peak input current pin 5 negative half cycle (tp 250μs)	200	mA
Peak input current (I_{SYNC}) pin 6	±1	mA
Peak input current (V_{SYNC}) pin 7	±1	mA
Inhibit input voltage pin 8	Vreg	V
-5V regulator current pin 11	10	mA
Control amp input voltage pin 13	Vreg	V
Tacho input current pin 15	±20	mA
THERMAL		
Operating ambient temperature	0 to +85	°C
Storage temperature	-55 to +125	°C

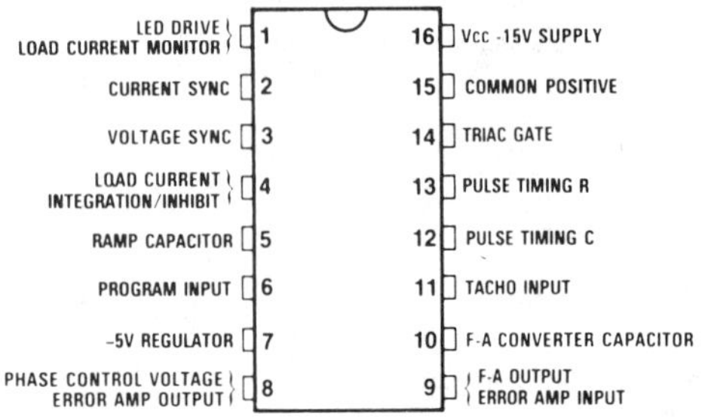

Pin	MP16	Pin	MP16
1	LED DRIVE / LOAD CURRENT MONITOR	16	Vcc -15V SUPPLY
2	CURRENT SYNC	15	COMMON POSITIVE
3	VOLTAGE SYNC	14	TRIAC GATE
4	LOAD CURRENT INTEGRATION/INHIBIT	13	PULSE TIMING R
5	RAMP CAPACITOR	12	PULSE TIMING C
6	PROGRAM INPUT	11	TACHO INPUT
7	-5V REGULATOR	10	F-A CONVERTER CAPACITOR
8	PHASE CONTROL VOLTAGE / ERROR AMP OUTPUT	9	F-A OUTPUT / ERROR AMP INPUT

MP16

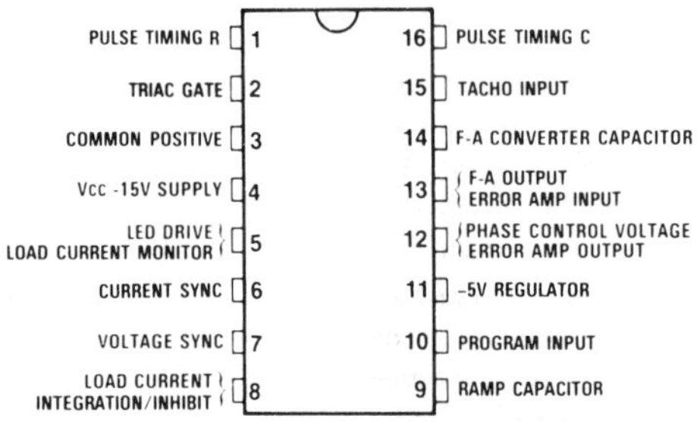

Pin	DP16	Pin	DP16
1	PULSE TIMING R	16	PULSE TIMING C
2	TRIAC GATE	15	TACHO INPUT
3	COMMON POSITIVE	14	F-A CONVERTER CAPACITOR
4	Vcc -15V SUPPLY	13	F-A OUTPUT / ERROR AMP INPUT
5	LED DRIVE / LOAD CURRENT MONITOR	12	PHASE CONTROL VOLTAGE / ERROR AMP OUTPUT
6	CURRENT SYNC	11	-5V REGULATOR
7	VOLTAGE SYNC	10	PROGRAM INPUT
8	LOAD CURRENT INTEGRATION/INHIBIT	9	RAMP CAPACITOR

DP16

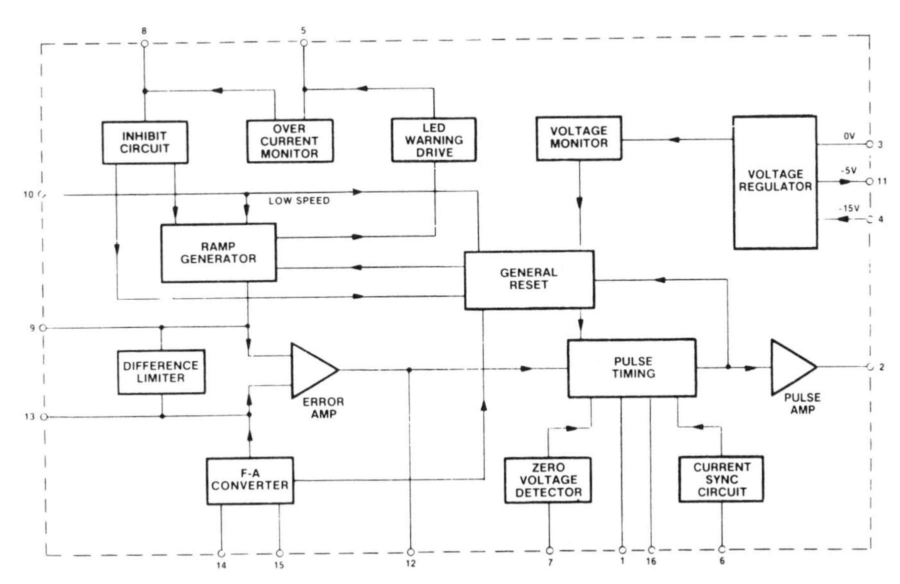

Block diagram of TDA2086

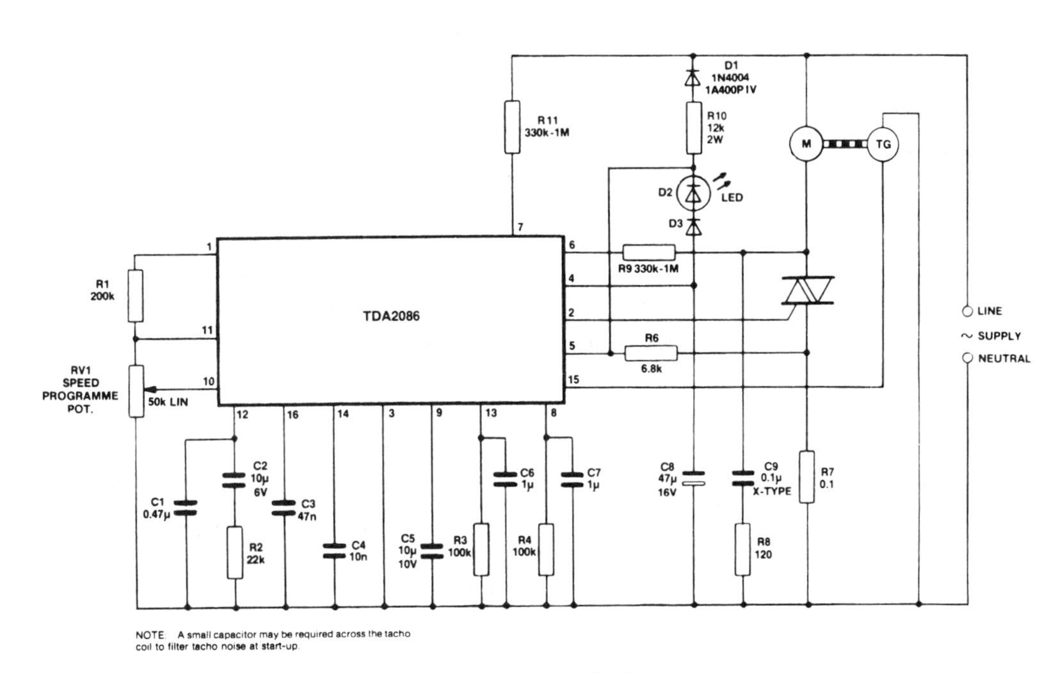

NOTE: A small capacitor may be required across the tacho coil to filter tacho noise at start-up.

Universal motor application

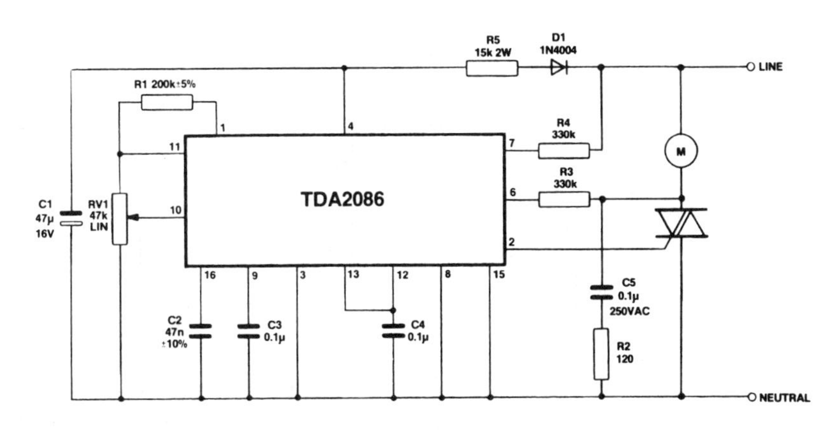

Open loop application, 240V

LINE
IN4004
15k
2W
SLOTTED
SWITCH
100μ
6V
470
TDA2086
7
5
4
15
11
47k
0.1μ
47μ
22k
NEUTRAL

Optical feedback application

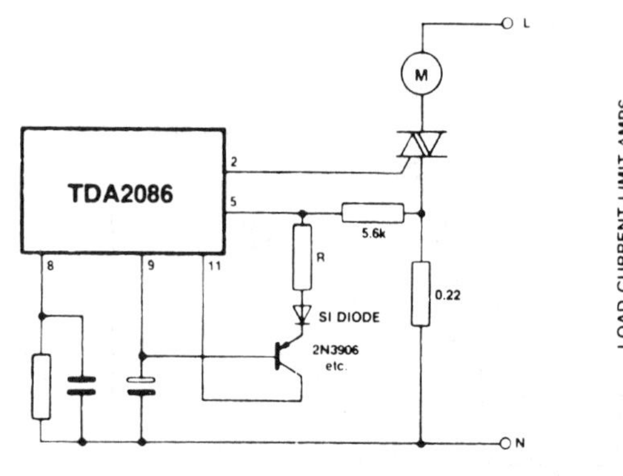

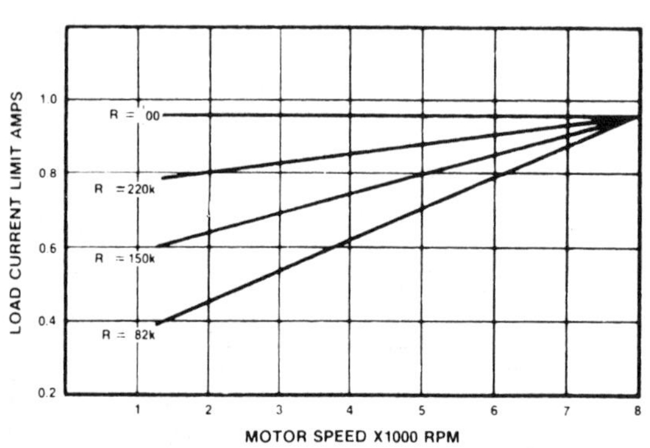

Current limit foldback, method 1

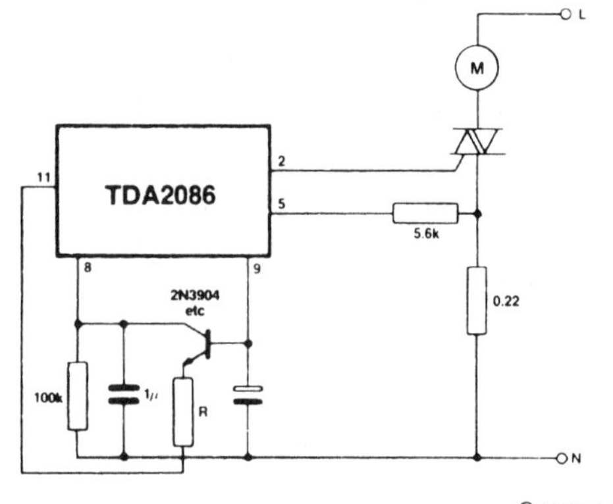

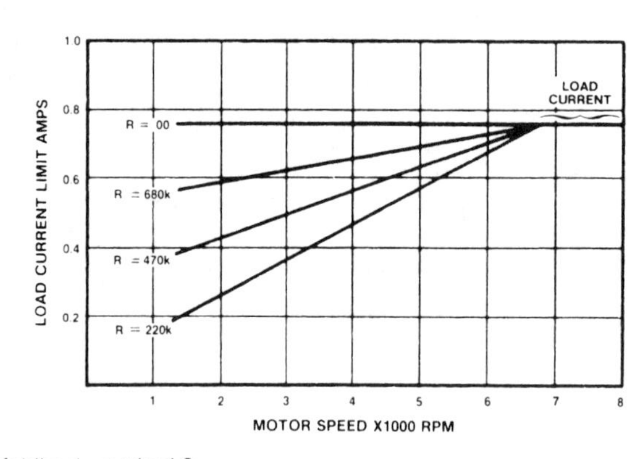

Current limit foldback, method 2

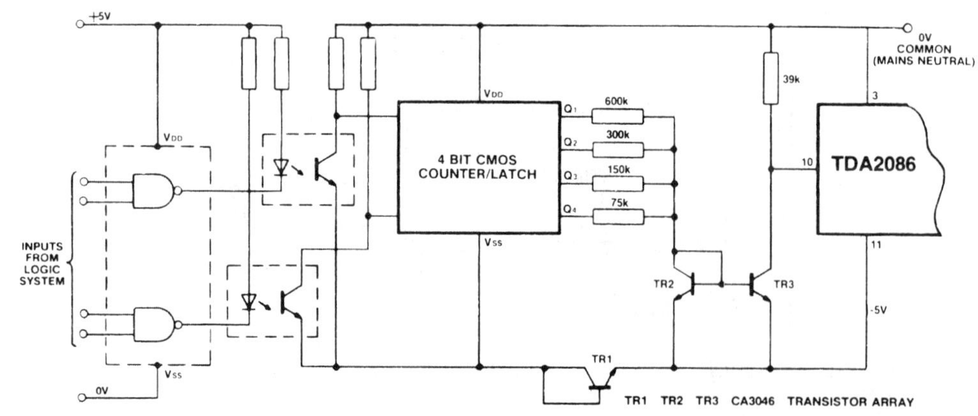

Interface to digital system

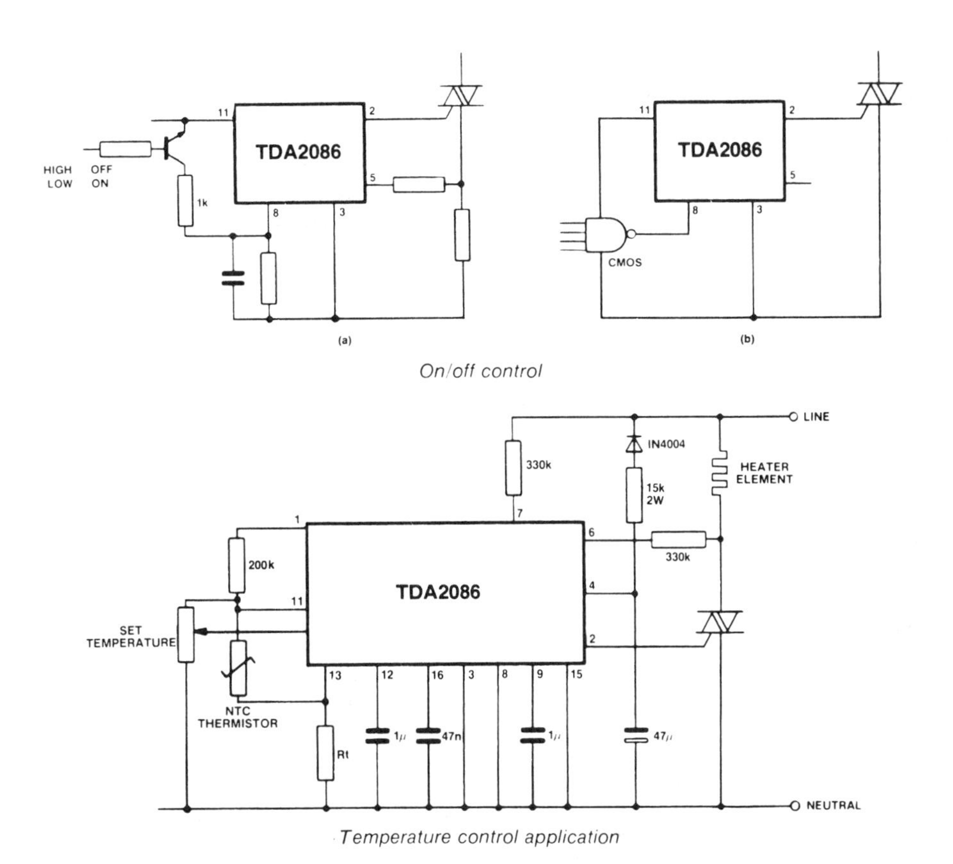

On/off control

Temperature control application

Plessey TDA2088

Phase Control Integrated Circuit for Current Feedback Applications

- Powered direct from ac mains or dc line
- −5V supply available for ancillary circuitry
- Low supply current consumption
- Negative triac firing pulses
- Guaranteed minimum 100mA triac drive current
- Well-defined control voltage/phase angle relationship
- Speed compensated by sensing motor current
- Simple optimization of control loop parameters

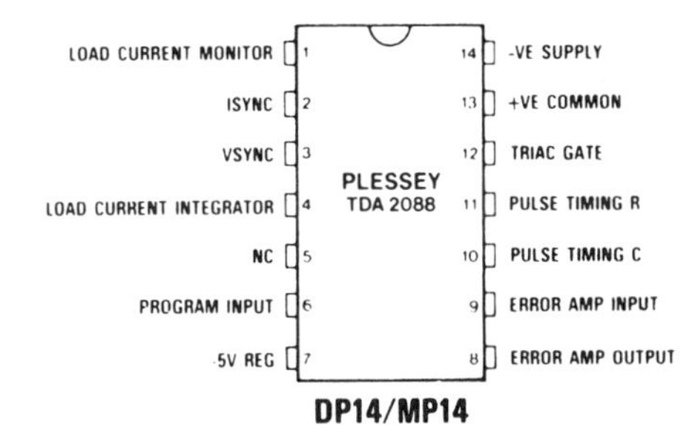

Pin connections - top view

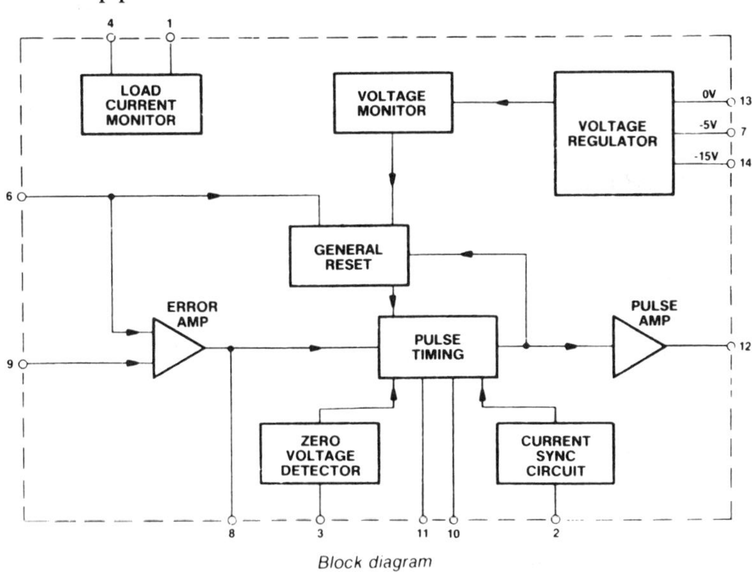

Block diagram

ELECTRICAL CHARACTERISTICS

Test conditions (unless otherwise stated):
T_{amb} = +25° C
All potentials measured with respect to common (Pin 13) (unless otherwise stated)

Characteristic	Value			Units	Conditions
	Min.	Typ.	Max.		
CURRENT CONSUMPTION					
Pin 14					
IC operating current		2.8	3.8	mA	Includes triac gate current for 50μs pulse
SHUNT VOLTAGE REGULATOR					
Pin 14					
Regulating voltage	-16	-14.75	-13.5	V	Full temperature range
Voltage monitor enable level	-11		-9	V	
SERIES REGULATOR					
Pin 7					
Regulating voltage (Vreg)	-5.35	-5	-4.65	V	1mA external load
Temperature coefficient			±1	mV/° C	
External load			10	mA	
Regulation	-75		+75	mV	For 0-5mA external load change
SPEED PROGRAM INPUT					
Pin 6					
Input voltage range	Vreg -0.5		0	V	
Input bias current			1	μA	
Zero power demand voltage	-100	-75	-50	mV	
ERROR AMPLIFIER					
Pin 6, 8 and 9					
Input offset voltage	-5		+15	mV	V_6 - V_9 to give I_8 = 0
Transconductance	80	100	120	μA/V	
Pin 8					
Output current drive	±20		±35	μA	
FIRING PULSE TIMING					
Pin 3					
Voltage SYNC trip level	±35	±50	±65	μA	
Pin 2					
Current SYNC trip level	±35	±50	±65	μA	
Pin 8					
Phase control voltage swing	Vreg		0	V	
Pin 10					
Firing pulse width		50		μs	C pin 10 = 47nF
Pulse repetition time		100		μs	C pin 10 = 47nF, R pin 11 = 200k
FIRING PULSE OUTPUT					
Pin 12					
Drive current	100	125	150	mA	Pin 12 V = -3V
Leakage current			10	μA	Pin 12 V = 0V
LOAD CURRENT SENSING					
Pin 1					
Offset voltage			±20	mV	
Pin 1 and 4					
Current gain	0.475	0.5	0.525		Pin 1 current = 100μA

ABSOLUTE MAXIMUM RATINGS

ELECTRICAL	Value	Units
Triac gate voltage pin 12	4	V
Repetitive peak input current pin 14	80	mA
Non repetitive peak input current pin 14 (tp = 250μs)	200	mA
Non repetitive peak input current pin 1 negative half cycle (tp 250μs)	200	mA
Peak input current (I_{SYNC}) pin 2	±1	mA
Peak input current (V_{SYNC}) pin 3	±1	mA
-5V regulator current pin 7	10	mA
Control amp input voltage pin 9	Vreg	V
THERMAL		
Operating ambient temperature	0 to +85	°C
Storage temperature	-55 to +125	°C

SYSTEM DESIGN WITH THE TDA2088

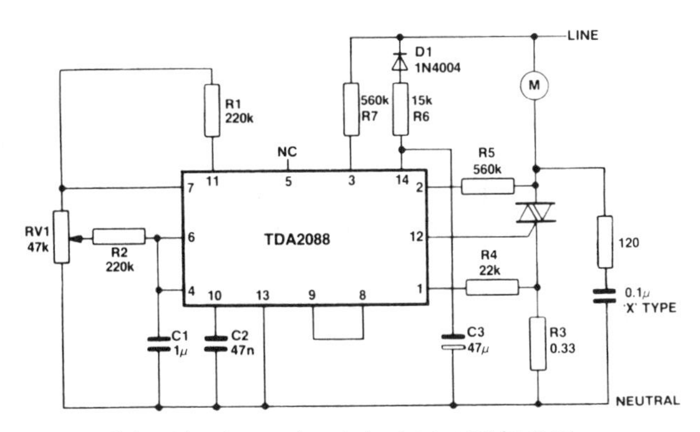

Universal motor speed control using current feedback

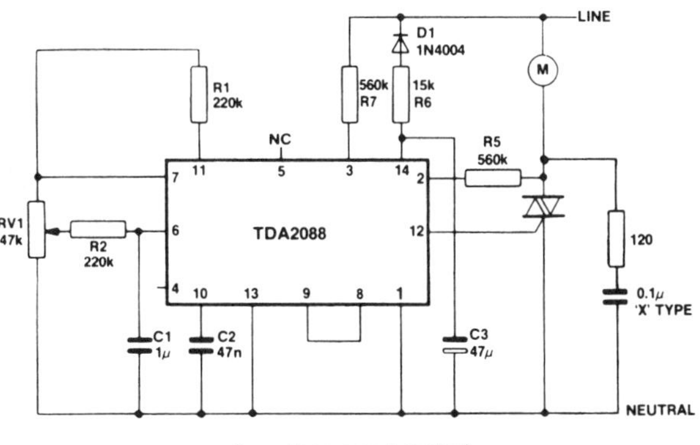

Open-loop speed control

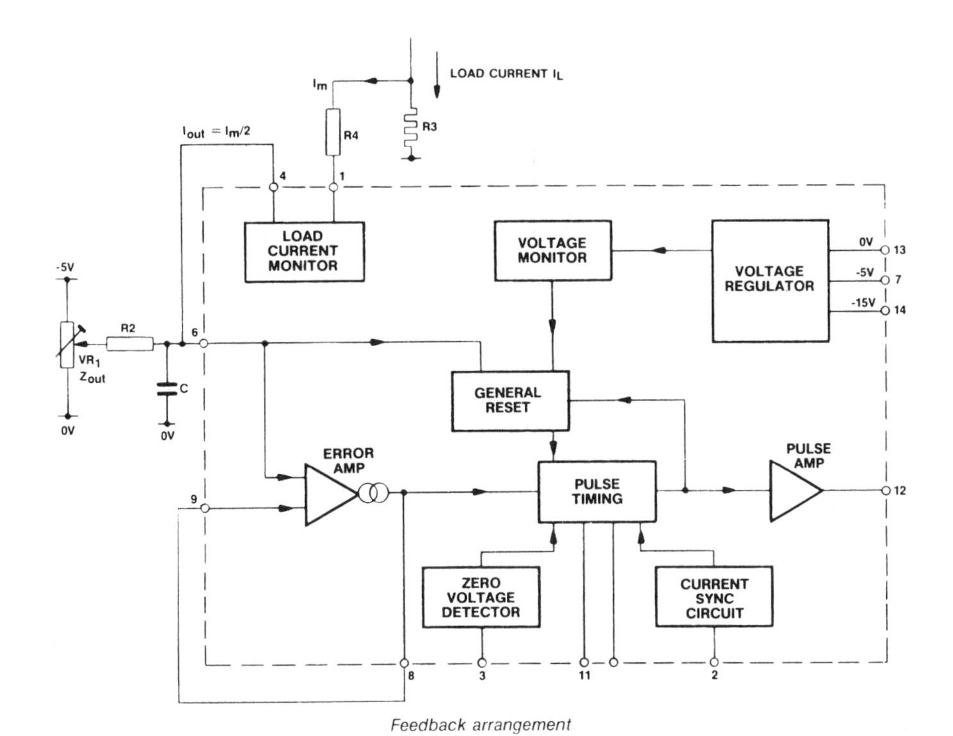

Feedback arrangement

Plessey TDA2090A

Zero-Voltage Switch

- 3-LED drive circuit indicates high, low, or in-band for controlled temperature
- Symmetrical negative triac firing pulses about the mains zero voltage points to minimize RFI
- Programmable switching rate, proportional band and LED indicator window
- −5V supply for sensing, thermistor bridge and ancillary control circuits
- Open-circuit sensor thermistor detector demands zero-power and lights over-temperature LED
- Powered direct from mains via current-limiting components or from dc line

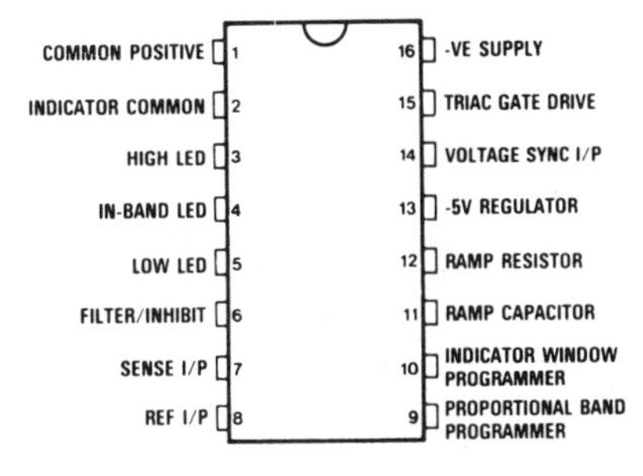

Pin connections - top view

ELECTRICAL CHARACTERISTICS

Characteristic	Value			Units	Conditions
	Min.	Typ.	Max.		
CURRENT CONSUMPTION					
Pin 16					
IC operating current		3.5	5.5	mA	Not including triac gate or bridge supply current
VOLTAGE MONITOR					
Pin 16					
Voltage monitor enable level	-11		-9	V	
SHUNT VOLTAGE REGULATOR (V_{EE})					
Pin 2					
Regulating voltage	-13.5	-13.5	-15.5	V	
SERIES REGULATOR					
Pin 13					
Regulating voltage (V_{reg})	-5.35	-5	-4.65	V	1mA external load
External current			5	mA	
Regulation	120		120	mV	For 0-5mA load change
CONTROL COMPARATOR					
Pins 6,7,8					
Proportional control band	±20	±50	±80	mV	Pin 9 = -0.5
Proportional control band	±140	±200	±260	mV	Pin 9 = -2V
Pins 7,8					
Input bias current			2	μA	
Hysteresis		10		mV	
Pin 7					
OPEN SENSOR inhibit level	20		40	mV	With respect to V_{reg}
INDICATOR WINDOW COMPARATORS					
Pins 7,8					
Indicator window	±50	±100	±150	mV	Pin 10 = -0.5V
Indicator window	±300	±400	±500	mV	Pin 10 = -2V
Indicator window hysteresis	10		30	mV	
FILTER/INHIBIT INPUT					
Pin 6					
Output drive current	±10		±50	μA	
Inhibit trip level	-3.5		-2.6	V	
LED DRIVE CIRCUIT					
Pins 3,4,5					
LED drive current			40	mA	
High output voltage		6.4		V	Output current = 20mA Pin 2 connected to common
Output leakage current			10	μA	Output voltage = V_{EE}
TRIAC PULSE AMPLIFIER					
Pin 15					
Drive current	50	75	95	mA	Pin 15 = -3V
Leakage current			10	μA	Pin 15 = 0V
WINDOW PROGRAMMER					
Pin 10					
Input bias current			2	μA	Pin 10 = 0V
PROPORTIONAL BAND PROGRAMMER					
Pin 9					
Input bias current			2	μA	Pin 9 = 0V
RAMP GENERATOR					
Pin 11					
Ramp Capacitor charge current	-12		-6	μA	With 470k resistor from
Ramp capacitor discharge current	6		12	μA	Pin 12 to 0V
Upper ramp trip voltage	-1.0		-2.5	V	
Lower ramp trip voltage	-6.5		-5.5	V	
Pin 12					
Ramp programming current	5		50	μA	
VOLTAGE SYNCHRONISATION					
Pin 14					
Voltage synchronisation trip level (I_{sync})	±20	±25	±30	μA	
Period pulse trip level	35	50	75	μA	

ABSOLUTE MAXIMUM RATINGS

	Value	Units
ELECTRICAL		
-14V shunt regulator repetitive peak input current pin 2	100	mA
Non repetitive peak input current pin 2 ($t_p < 250\mu s$)	250	mA
Repetitive peak input current pins 3,4,5	100	mA
Non repetitive peak input current pins 3,4,5 ($t_p < 250\mu s$)	250	mA
Peak input current pin 14	±5	mA
-5V regulator current pin 13	10	mA
Supply voltage pin 16	-18	V
Voltage on pins 6,7,8,9,10	V_{reg}	V
Triac gate voltage pin 15	4	V
Ramp current pin 12	0.5	mA
THERMAL		
Operating ambient temperature	0 to 60	°C
Storage temperature	-55 to +125	°C

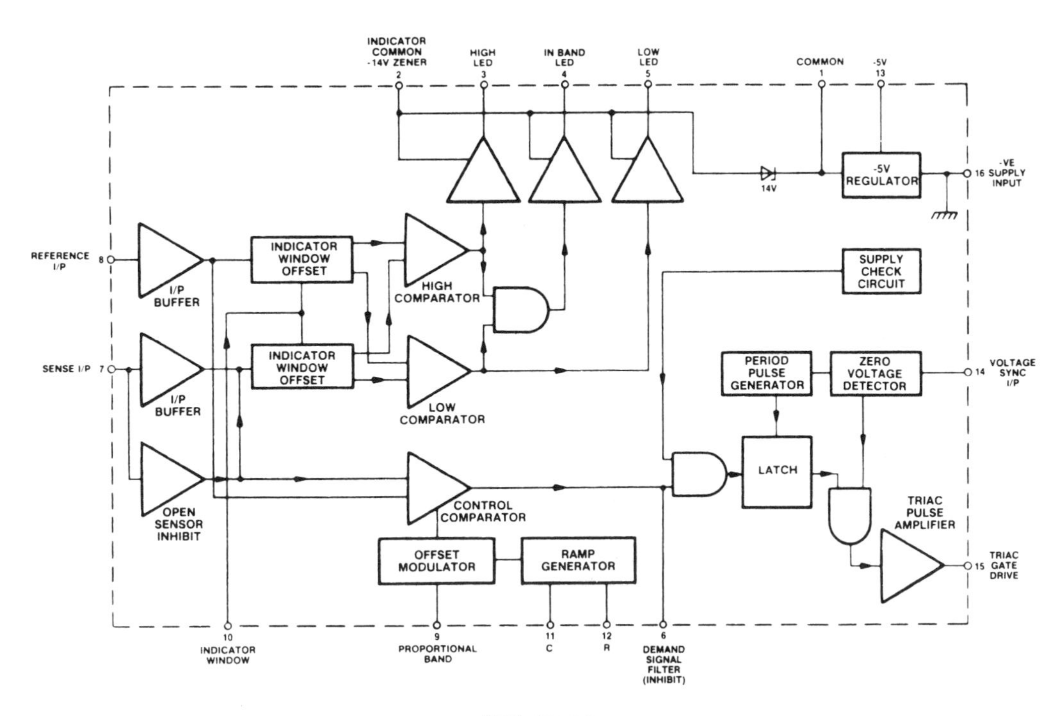

Block diagram

Plessey
ZN409CE
Precision Servo Integrated Circuit

- Low external component count
- Low quiescent current (7mA typical at 4.8V)
- Excellent voltage and temperature stability
- High output drive capability
- Consistent and repeatable performance
- Precision internal voltage stabilization
- Time-shared error pulse expansion
- Balanced deadband control
- Schmitt trigger input shaping
- Reversing relay output (dc motor speed control)

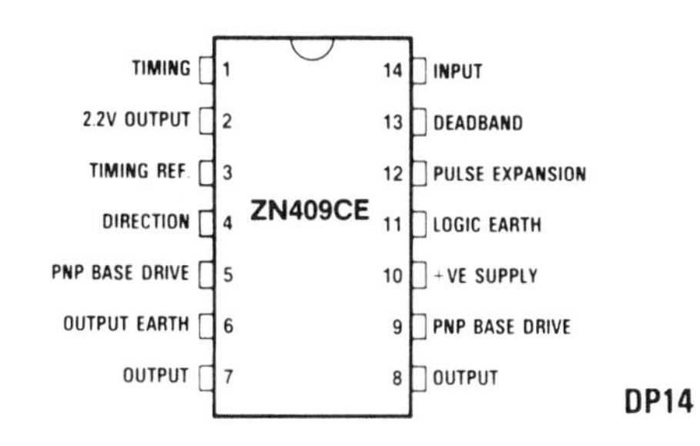

Pin connections - top view

ELECTRICAL CHARACTERISTICS

Test conditions (unless otherwise stated):
T_{amb} = 25°C, V_S = 5V

Characteristic	Value			Units	Test conditions
	Min.	Typ.	Max.		
Input threshold (lower)	1.15	1.25	1.35	V	Pin 14
Input threshold (upper)	1.4	1.5	1.6	V	Pin 14
Ratio upper/lower threshold	1:1	1:2	1:3		-10°C to +65°C
Input resistance	20	27	35	kΩ	
Input current	350	500	650	μA	
Regulator voltage	2.1	2.2	2.3	V	-10°C to +65°C, 1.3mA load current
Regulator supply rejection ratio	200	300	-		V_S = 3.5V to 6.5V $RSRR = \frac{dV_{IN}}{dV_{OUT}}$
Monostable linearity	-	3.5	4.0	%	±45°, R_P = 1.5kΩ, R_1 = 12kΩ
Monostable period temperature coefficient	-	+0.01	-	%/°C	Excluding R_T,C_T. R_P = 1.5kΩ, R_1 = 12kΩ (potentiometer slider set mid-way)
Output Schmitt deadband	±1	±1.5	±3	μs	C_E = 0.47μF
Minimum output pulse	2.5	3.5	4.5	ms	C_E = 0.47μF, R_E = 180kΩ
Error pulse for full drive	70	100	130	μs	15ms repetition rate C_E = 0.47μF, R_E = 180kΩ
Total deadband	±3.5	±5	±6.5	μs	C_D = 1000pF
PNP drive	40	55	70	mA	T = 25°C
	35	50	65	mA	T = -10°C
Output saturation voltage	-	300	400	mV	I_L = 400mA
Direction bistable output	2	2.8	3.6	mA	
Supply voltage range	3.5	5	6.5	V	
Supply current	4.6	6.7	10	mA	Quiescent
Total external current from regulator	1.3	-	-	mA	V_S = 3.5V
Peak voltage $V_{C\ EXT}$ (with respect	-	0.7	-	V	T = 25°C
to 2V regulated voltage)	-	0.5	-	V	T = -10°C

ABSOLUTE MAXIMUM RATINGS

Supply voltage	6.5V
Package dissipation	300mW
Operating temperature range	-20°C to +65°C
Storage temperature range	-65°C to +150°C

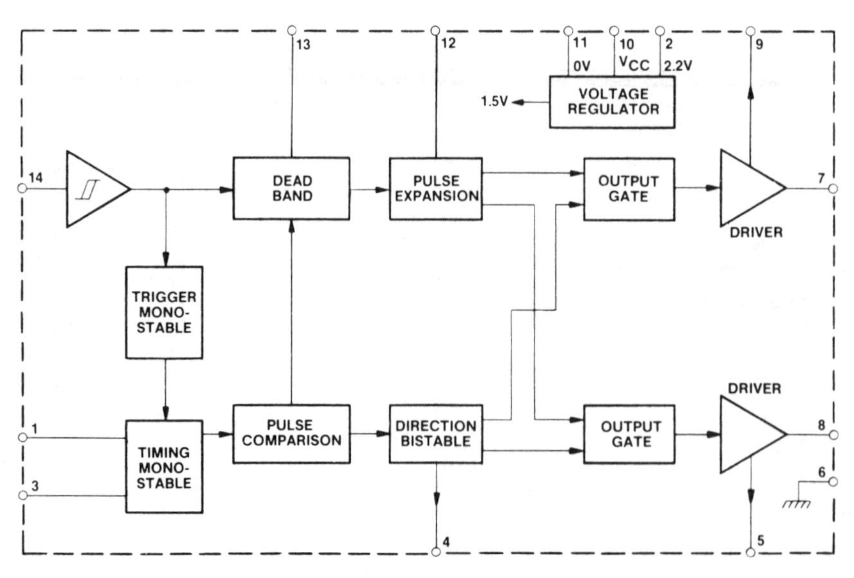

Block diagram for ZN409CE

Plessey ZN410E
Motor Speed Controller

- Direct supply from ac mains or dc power source
- Low external component count
- On-chip shunt regulator
- Soft-start ramp circuit
- Optional current limit or trip
- Magnetic pickup tacho input
- Circuit reset on power down
- Guaranteed full-cycle conduction with inductive load
- Negative firing triac pulses
- Low cost

ELECTRICAL CHARACTERISTICS

Test conditions (unless otherwise stated):
T_{amb} = 25°C

Characteristic	Value			Units	Conditions
	Min.	Typ.	Max.		
Voltage regulator					
Shunt regulator voltage (V_{REG})	4.7	-	5.6	V	I_{CC} = 5.5mA
Shunt regulator slope resistance	-	-	20	Ω	I_{CC} = 5.5mA
Supply current	3.5	-	25	mA	
F/V converter					
Tacho input threshold voltage	-	±100	-	mV	
Tacho input current	-	500	-	nA	$V_{pin\ 4}$ = V_{CC}
Tacho maximum input frequency	20	100	-	kHz	
Speed control					
Speed reference voltage (w.r.t. pin 6)	-	-1.28	-	V	I_{IN} = 12.8μA
Speed reference input current	-	-	25	μA	
Speed input voltage range (w.r.t. pin 6)	$-V_{pin\ 3}$	-	0	V	
Speed input current	-	0.05	1.0	μA	$V_{pin\ 1}$ = 5.0V
Soft start capacitor charge current	5.3	7.5	12.0	μA	
Current limit					
Current limit input threshold	(330)	425	465	mV	½ cycle mean AC
Limit capacitor time constant	75C	150C	300C	ms	$C_{pin\ 7}$ in μF
Current sync					
Current sync threshold current	-	±120	±165	μA	
Current sync asymmetry	-	-	±10	μA	
Current sync clamp voltage (w.r.t. pin 6)	-	±1450	-	mV	I_{IN} = ±1mA
Ramp generator					
Ramp input charge current	35	50	70	μA	V_{IN} = 0V w.r.t. pin 6
Ramp input optimum max. negative level	-	-1.45	-	V	w.r.t. pin 6
Ramp input discharge voltage		-0.75	-	V	w.r.t. pin 6
Ramp input discharge current	-	10	-	mA	$V_{pin\ 13}$ = -1.45V (w.r.t. pin 6)
Triac gate drive					
Output current (on state)	65	100	130	mA	$V_{pin\ 10}$ = 2.0V
Output current (off state)	-	-	20	μA	$V_{pin\ 10}$ = 5.0V
Output pulse width	50	100	150	μs	R_L = 100Ω. 50% level
Error amplifier					
Phase angle/error voltage relationship ($V_{pin\ 1}$ - $V_{pin\ 14}$)	-	2	-	°/mV	Ramp $V_{pin\ 13}$ = 500mV
Tacho filter input voltage range (w.r.t. pin 6)	-	-	-1.5	V	
Amplifier filter time constant	35C	50C	65C	ms	$C_{pin\ 13}$ in μF

ABSOLUTE MAXIMUM RATINGS

Shunt regulator current	25mA
Input voltage (pin 9 w.r.t. pin 6)	+5V max. -5V min.
Maximum input current (pins 4, 12)	±2mA
Output voltage (pin 10)	+7.5V max. -0.5V min.
Operating temperature range	0°C to 70°C
Storage temperature range	-55°C to +125°C

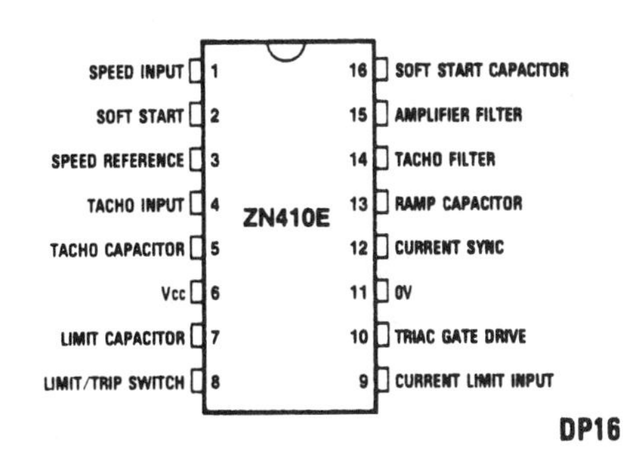

Pin connections - top view

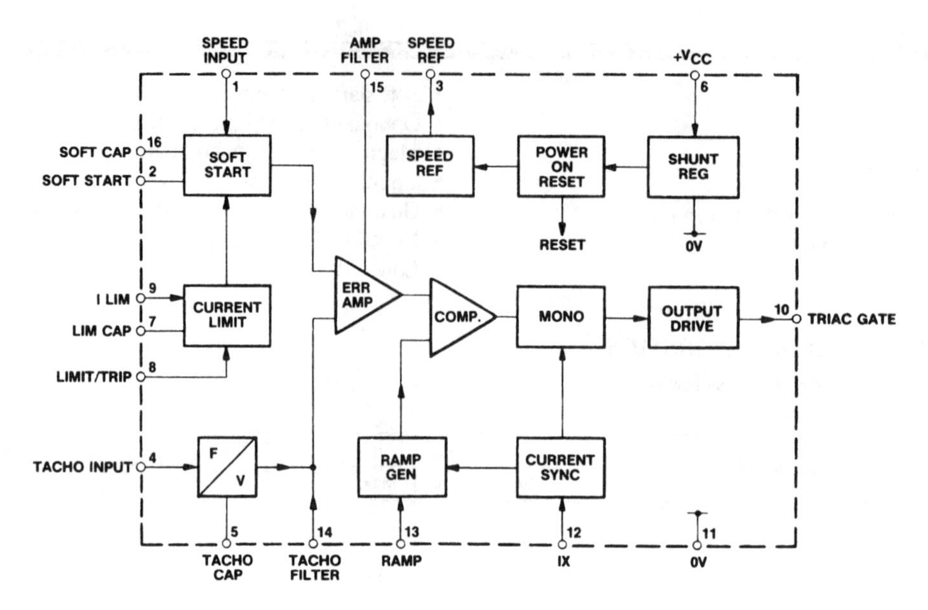

Block diagram of ZN410E

Plessey ZN411

Phase Control Integrated Circuit

- Direct supply from ac mains or dc power source
- Soft-start ramp circuit
- Negative triac firing pulses
- Triac retrigger facility
- Current limit
- Tacho input compatible with half-effect switch devices
- Electronic interlock and speed limit reverse mode

ABSOLUTE MAXIMUM RATINGS

All voltages measured with respect to -Vcc Pin 2.

Maximum shunt regulator current	25mA
Input voltage on pins 15 and 18	Maximum 7.5V
	Minimum -0.5V
Maximum input current on pins 13 and 14	±2mA
Output voltage on pin 3	Maximum 7.5V
	Minimum -0.5V
Operating temperature range	0°C to 70°C
Storage temperature range	-55°C to +125°C

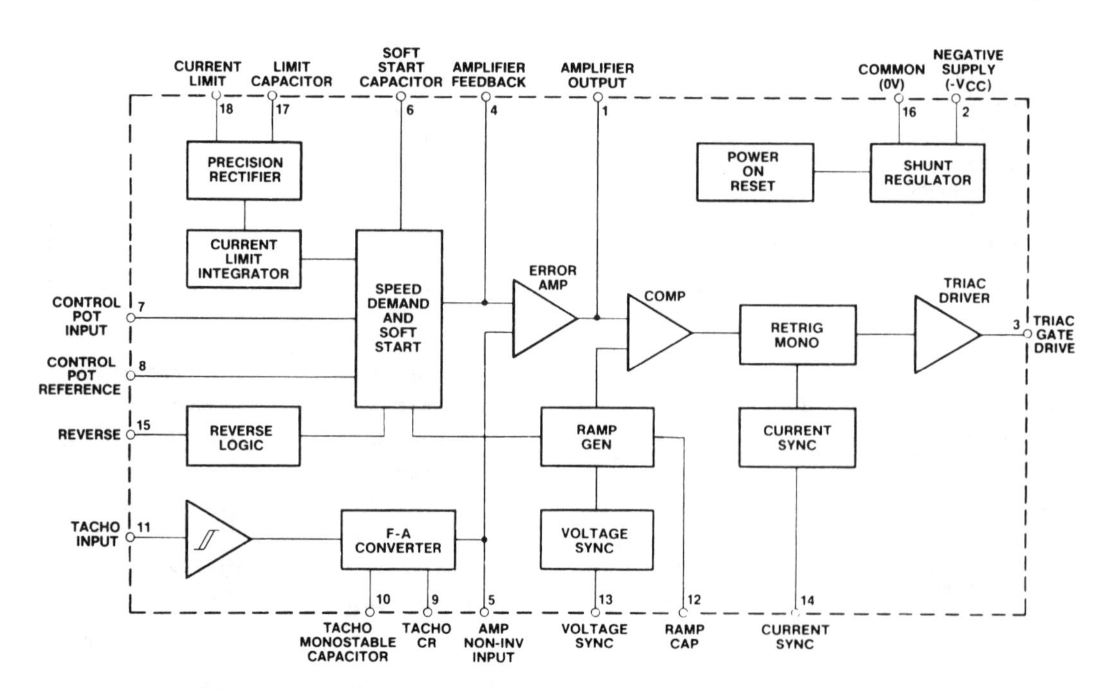

Block diagram of ZN411

ELECTRICAL CHARACTERISTICS

Test conditions (unless otherwise stated):
T_{amb} = 25°C. All voltages measured w.r.t. 0V Pin 16.

Characteristic	Value			Unit	Conditions
	Min.	Typ.	Max.		
Minimum operating voltage		-4.5		V	
Supply current I_{CC}		-4.0	-8.0	mA	$-V_{CC}$ = 4.5V All other pins O/C
Shunt regulator voltage V_S	-4.75	-5.1	-5.4	V	I_{CC} = 10mA
Shunt regulator slope resistance		4.0	10.0	Ω	I_{CC} = 10mA
Rev input threshold voltage Pin 15 V_{TH} High State V_{TL} Low state (w.r.t. $-V_{CC}$ Pin 2)	 1.1	 2.9 1.3	 3.4 1.7	 V V	
Rev input current Pin 15 I_{IH} High state I_{IL} Low state		 150	 -1	 μA μA	 V_{IN} = 0V V_{IN} = $-V_{CC}$
Tacho input threshold voltage Pin 11 V_{TH} High state V_{TL} Low state (w.r.t. $-V_{CC}$ Pin 2)		 2.9 1.3		 V V	
Tacho input current Pin 11 I_{IH} High state I_{IL} Low state		 20 -125		 μA μA	 V_{IN} = 0V V_{IN} = $-V_{CC}$
Control input voltage range Pin 7	V_{p8} (Note 1)		0	V	RV1 = 100k Pin 15 = 0V
Control input voltage range Pin 7 - Reverse mode	0.23V (Note 1)		0	V	RV1 = 100k Pin 15 = $-V_{CC}$
Control input current Pin 7		100		nA	V_{IN} = 0V
Control potentiometer input bias current Pin 8		20		μA	
Soft start capacitor charge current Pin 6		10		μA	
Current limiting input threshold voltage Pin 18		540		mV mean AC	C7 = 470nF
Error amplifier open loop DC voltage gain	29	40	51		
Error amplifier closed loop AC voltage gain	2.9	4.4	5.0		C3 = 220nF R1 = ∞ f = 1kHz
Minimum error amplifier output voltage Maximum error amplifier output voltage		-3.4 0		V V	RV1 = 100k
Voltage sync input, Pin 13 Positive input threshold current Negative input threshold current Clamp voltage (w.r.t. $-V_{CC}$)		 42 14 +1.4 -100		 μA μA V mV	 I_{IN} = +1mA I_{IN} = -1mA
Current synch. input Pin 14 Threshold current Clamp voltage (w.r.t. $-V_{CC}$)		 ±110 +1.4 -100		 μA V mV	 I_{IN} = +1mA I_{IN} = -1mA
Timing ramp amplitude, Pin 12	1.5	2.0	2.8	V pk-pk	Supply frequency at Pin 13 = 50Hz C5 = 1μF
Timing ramp capacitor charge current Pin 12	15	20	28	μA	
Triac gate pulse output Pin 3 output voltage (V_G)	-3.5			V	R_{LOAD} Pins 3-16 = 68Ω
Output current (On state) (I_G)	80	110	140	mA	V_{OUT} = -3V
Output current (Off state) (I_G)			20	μA	V_{OUT} = 0V
Output pulse width (t_G)	40	80	150	μs	
Minimum tacho input pulse width Pin 11		10		μs	

NOTE
1. V_{p8} = Voltage measured at pin 8 with RV1 = 100kohms. Typically = -2V.

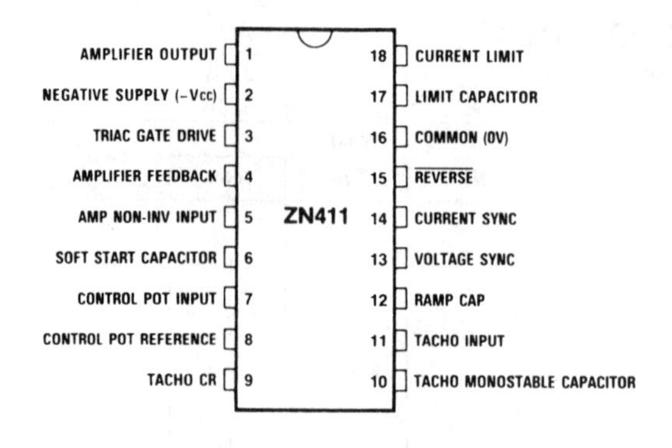

Pin connections

DP18

Plessey ZN1060E Switch-Mode Regulator Control

- Stabilized power supply
- Low supply voltage protection
- Linear pulse-width modulator
- Programmable duty cycle
- Programmable soft-start
- Double pulse suppression
- High-speed current limiting
- Loop fault protection
- Uncommitted error amplifier
- Overvoltage protection
- Remote on/off switching
- Secondary current monitoring
- Multiple device synchronization
- Core saturation protection

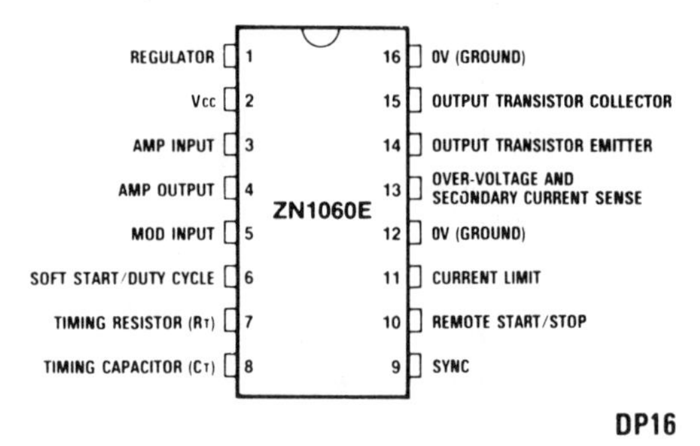

DP16

Pin connections - top view

ABSOLUTE MAXIMUM RATINGS

Dissipation	350mW
Output current sink	40mA
Collector supply voltage	6V
Operating temperature range	-20°C to +85°C
Storage temperature range	-55°C to +150°C

THERMAL CHARACTERISTICS

Chip to case θ_{JC}	65°C/W
Chip to ambient θ_{JA}	110°C/W

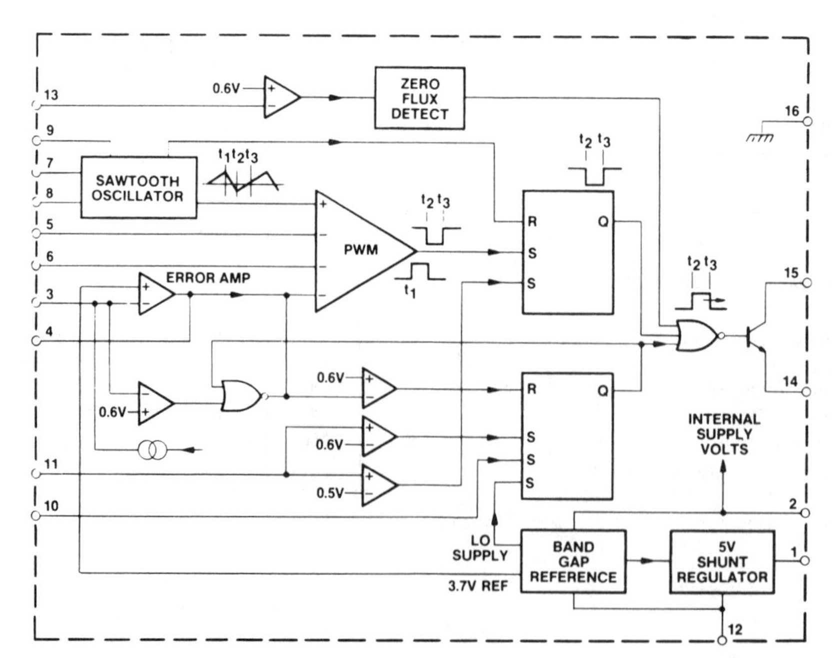

Block diagram of ZN1060E

ELECTRICAL CHARACTERISTICS

Test conditions (unless otherwise stated):
T_{amb} = 25°C, V_{CC} = 5V

Characteristic	Symbol	Value			Units	Test conditions
		Min.	Typ.	Max.		
Reference section						
Output voltage	V_{REF}	4.75	5	5.30	V	I_C = 20mA
Slope resistance	R_S	-	2	4	Ω	
Temperature coefficient	TC	-	100	-	ppm/°C	
Amplifier section						
Open loop gain	A_O	-	60	-	dB	
Input bias current	I_3	-	4	40	μA	
External feedback resistor	R_{3-12}	100	-	-	kΩ	
Reference voltage	V_{REF}	3.42	3.72	4.03	V	
Reference temperature coefficient	$\Delta V_{REF}/\Delta T$	-	100	-	ppm/°C	
Output voltage swing positive	V_{oh}	3	-	-	V	
Output voltage swing negative	V_{ol}	-	-	0.4	V	
Oscillator section						
Frequence range	f	50	-	100k	Hz	
External capacitor	C_T	1	-	-	nF	
External resistor (pin 7)	R_T	10	-	40	k	
Duty cycle range		0	-	98	%	
Sawtooth Upper level	V_{RH}	-	3	-		
Lower level	V_{RL}	-	1.1	-	V	
Modulator						
Modulator input current	I_{5-12}	-	4	40	μA	Voltage on pin 5 = 1V
Protection functions						
Pin 6 duty cycle limit control	V_6	39	41	43	% of V_Z	For 50% max. 15 to 50kHz
Pin 6 input current	I_{6-12}	-	0.5	20	μA	V_{IN} = 2V
Pin 1 low supply voltage protection threshold	V_{LT}	3.5	4	4.5	V	
Pin 3 feedback loop trip on threshold	V_{3-12}	472	600	720	mV	
Pin 3 pull up current	I_3	-	15	35	μA	
Pin 13 demagnetisation/overvoltage trip on threshold	V_{13-12}	-	600	-	mV	
Pin 13 input current	I_{x13}	-	1.3	5	mA	V_{IN} = 2V
External synchronisation						
Pin 9 Off	V_{9-12}	0	-	0.8	V	
On		2	-	5.25	V	
Sink current	I_{9-12}	-	-	100	μA	
Remote On/Off						
Pin 10 On	V_{10-12}	2	-	5	V	V_{CC} = 5V
Off	V_{10-12}	0	-	0.8	V	
Sink current	I_{10-12}	-	-	100	μA	V_{11-12} = 250mV
Current limit						
Pin 11						
Current limit	V_{11-12}	0.40	0.48	0.58	V	
Shutdown/slow start	V_{11-12}	0.47	0.60	0.72	V	
Sink current	I_{11-12}	-	-	450	μA	
Output stage						
Output current pin 15	I_{15}	40	-	-	mA	
Maximum emitter voltage pin 14	V_{14-12}	-	-	5	V	
Collector saturation voltage pin 15	V_{15-12}	-	0.4	-	V	
Supply current		-	12	20	mA	V_{1-12} = 5V

Plessey ZN1066E/J

Switching Regulator Control and Drive Unit

- Complete PWM power control circuitry
- Single-ended or push-pull totem pole-type outputs with ±120mA capability
- 0-100% duty-cycle control
- Feedback control guarantees nonoverlap of output pulses
- No dead-time setting required
- Output-frequency adjustable up to 500 kHz
- Independent control of output voltage and output current
- 2.6V stable reference ±50ppm/°C
- Inhibit and synchronizing input

ABSOLUTE MAXIMUM RATINGS

Electrical (at -55°C to +70°C)	
Supply current (Icc)	200mA
Main output drive currents	160mA total
Clock output current (sink)	25mA
Reference current (sink)	10mA
Ramp control current	1mA
Bais sourcing current	1mA
Thermal	
Operating temperature range:	
ZN1066J	-55°C to +125°C
ZN1066E	-40°C to +85°C
Storage temperature range	-65°C to +150°C

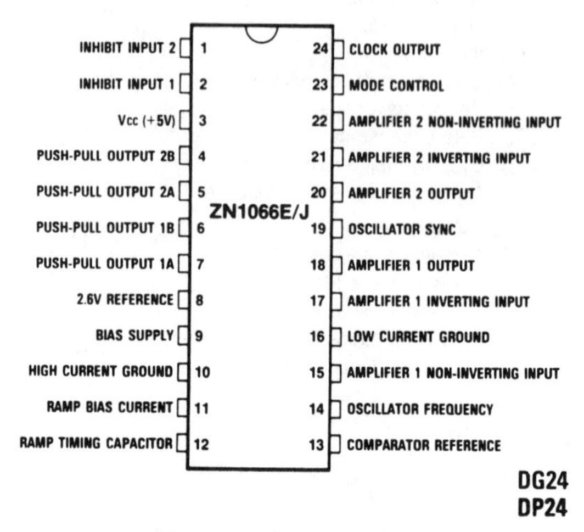

DG24
DP24

Pin connections - top view

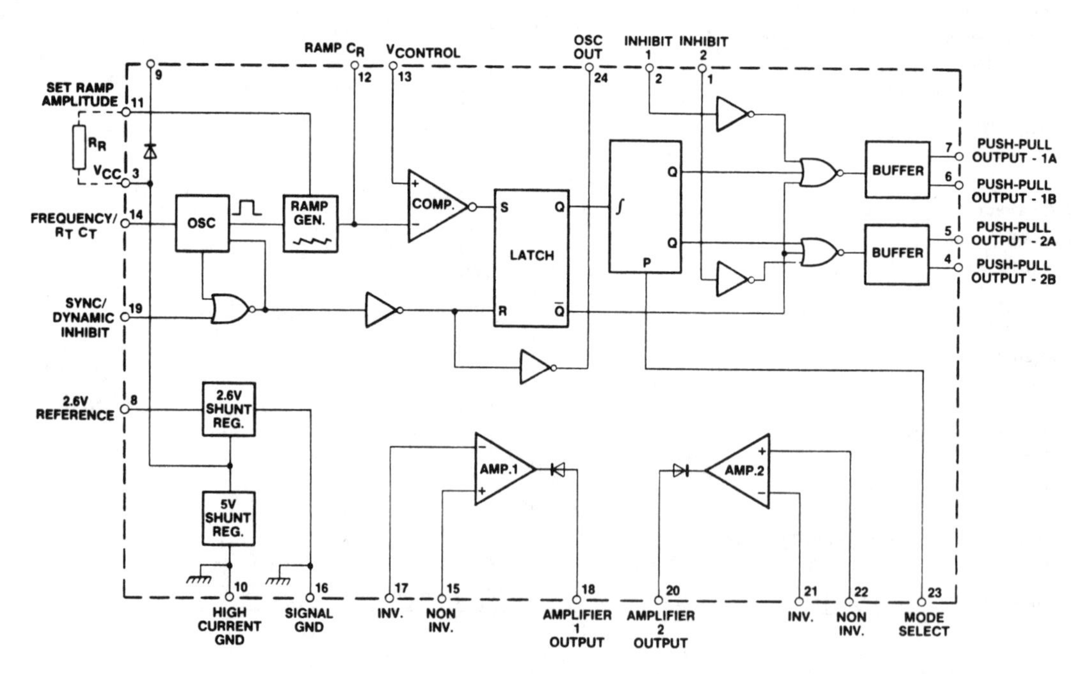

Block diagram of ZN1066E/J

ELECTRICAL CHARACTERISTICS

Test conditions (unless otherwise stated):
T_{amb} = 25°C

Characteristic	Value			Units	Conditions
	Min.	Typ.	Max.		
Shunt regulator section					
Output voltage (I_{CC} = 60mA)	4.75	5.0	5.25	V	See Notes 1 and 2
Voltage temperature coefficient (I_{CC} = 60mA)	-	100	-	ppm/°C	
Output impedance	-	1.5	3	Ω	
Supply current	-	40	-	mA	Shunt regulator just on
Amplifier section					
Open loop voltage gain	800	1200	-		
Input bias current	-	1	4	μA	
Input offset current	-	0.2	2	μA	
Input offset voltage	-	2	5	mV	
Offset voltage temperature coefficient	-	10	-	μV/°C	
Output low (sinking 1mA)	-	0.85	-	V	
Output high (sourcing 0.1mA)	4.7	-	-	V	With 1k pull up
Output impedance	-	5	-	kΩ	
Common mode range	1	-	2.8	V	
Comparator section					
Common mode range	1	-	4.3	V	
Delay to output drive (±50mV input)	-	0.17	0.3	μs	
Delay to output drive (±10mV input)	-	0.2	-	μs	
Input bias current	-	1	4	μA	
Input offset current	-	0.2	2	μA	
Reference section					
Reference voltage (at 1mA source)	2.4	2.55	2.7	V	See Notes 3 and 4
Temperature coefficient	-	50	-	ppm/°C	
Output impedance	-	1.5	-	Ω	
Mode control section					
Single ended operation control input logic '1' (outputs 1A and 1B)	2.4	-	-	V	May be connected direct to V_{CC}
Push pull operation control input logic '0' (all outputs)	-	-	0.4	V	0V or left open circuit
Cross couple inhibits section					
Input logic '1' enables outputs	-	0.07	0.2	mA	See Note 5
Input logic '0' inhibits outputs	-	-	0.4	V	
Oscillator section					
Maximum frequency range	5×10^{-4}	-	500	kHz	Minimum value of C_T = 1500pF
Initial accuracy	-	2	-	%	$R_T C_T$ constant
Temperature stability	-	1	-	%	Over temperature range -55°C to +125°C
Output pulse width	-	0.3	-	μs	C_T = 1500pF
Output logic '0' (sinking 10mA)	-	-	0.4	V	Buffered output pin 24
Output logic '1' (sourcing 1mA)	2.4	-	-	V	Buffered output in 24
Output section					
Output current	-	±60	-	mA	
Output logic '0' (sinking 60mA)	-	0.4	0.45	V	Each output
Output logic '1' (sourcing 60mA)	1.0	1.45	-	V	100mA max. under short circuit conditions
Total standby current					
V_{CC} at 2.5V, output current 4mA	-	17	-	mA	Operation from V_{REF} to V_{CC} is permissible
V_{CC} at 5V, with outputs open	-	40	-	mA	

NOTES
1. Decouple pin 3 to GND with 0.22microfarads as close to pins 3 and 10 as possible.
2. Pin 10 GND for 5V regulator and output buffers.
3. Decouple pin 8 to GND with 0.22microfarads minimum as close to pins 8 and 16 as possible. V_{REF} will supply 1mA maximum without additional bias. Maximum sink current is 10mA.
4. Pin 16 GND for oscillator, ramp generator, comparator, amplifiers and 2.5V reference.
5. The inhibit logic 1 current is the source current required to ensure digitally high operation. The base ground resistor is nominally 10kOhms. Catching diodes to the 5V rail are included on-chip.

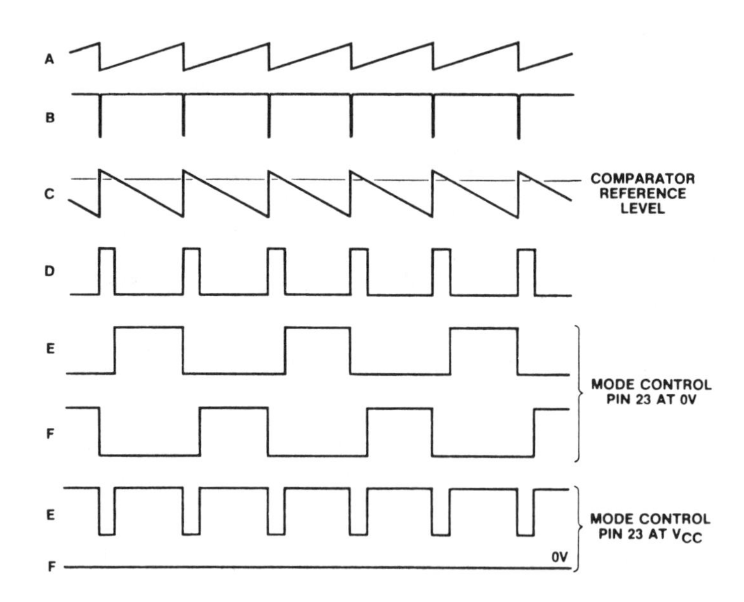

This information is reproduced with permission of LSI Computer Systems, Inc., 1235 Walt Whitman Road, Melville, NY 11747.

The information included herein is believed to be accurate and reliable. However, LSI Computer Systems, Inc. assumes no responsibilities for inaccuracies, nor for any infringements of patent rights of others which may result from its use.

LSI
LS7220
Digital Lock Circuit

- Stand-alone lock logic
- 5040, 4-digit combinations
- Out-of-sequence detection
- Direct LED and lock relay drive
- Chip enable (for automotive applications)
- Externally controlled convenience delay
- Save memory (for valet parking, etc.)
- Internal pull-down resistors on all inputs
- High noise immunity
- Low current consumption (40μA max @ 12Vdc)
- Single power supply operation (+5V to +18V)
- Momentary or static lock-control output
- All inputs protected

The LS7220 is a monolithic, ion-implanted MOS 4-key keyless lock. The circuit includes sequential logic for interpretation of correct key closure; out-of-sequence detection circuitry, and save memory.

MAXIMUM RATINGS: (Voltages Referenced to V_{DD})

Rating	Symbol	Value	Units
DC Supply Voltage	V_{ss}	+5 to +18	Vdc
Operating Temperature Range	TA	−25 to +70	°C
Storage Temperature Range	TSTG	−65 to +150	°C

QUIESCENT SUPPLY CURRENT:

(All inputs and outputs open)

Symbol	Vss	MAX	UNITS
I_{DD}	5Vdc	20	μA
	9Vdc	30	
	12Vdc	40	
	15Vdc	50	
	18Vdc	70	

INPUT CAPACITANCE: 10Pf

Parameter	Symbol	MAX	Units
Convenience Delay:			
Set-Up Time } See Note	TS	4	μsec
Hold Time } Below	TH	8	μsec
Input Lock Control:			
Output Delay	TLC	8	μsec
Input Pulse Width	TIW	25	μsec

NOTE: TS and TH were measured without any external capacitance at convenience delay (Pin 12) input.

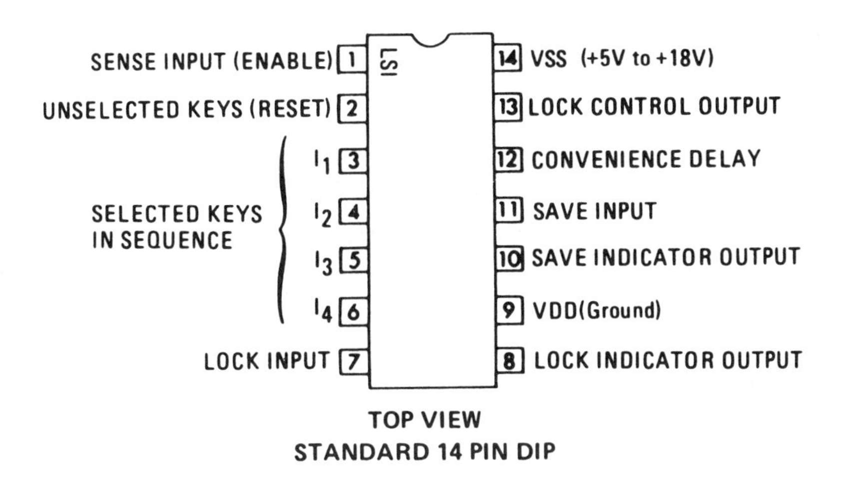

TOP VIEW
STANDARD 14 PIN DIP

Typical Hold Time Vs. VSS
(with 1μF convenience delay capacitor)

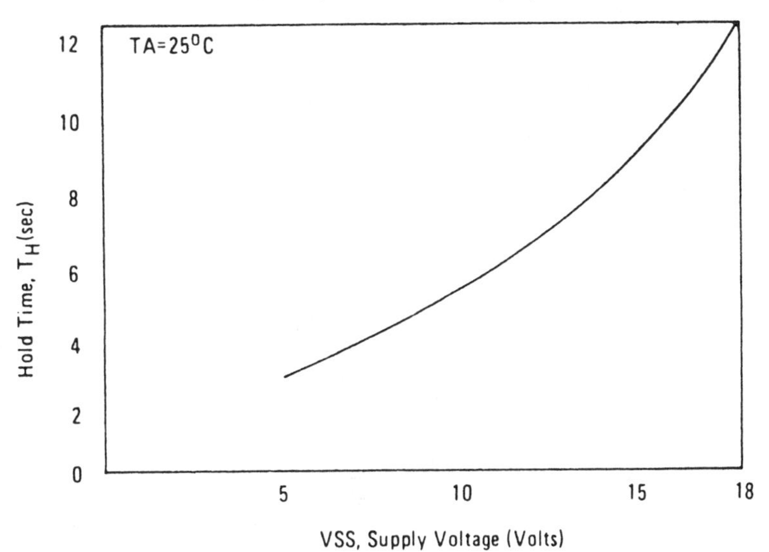

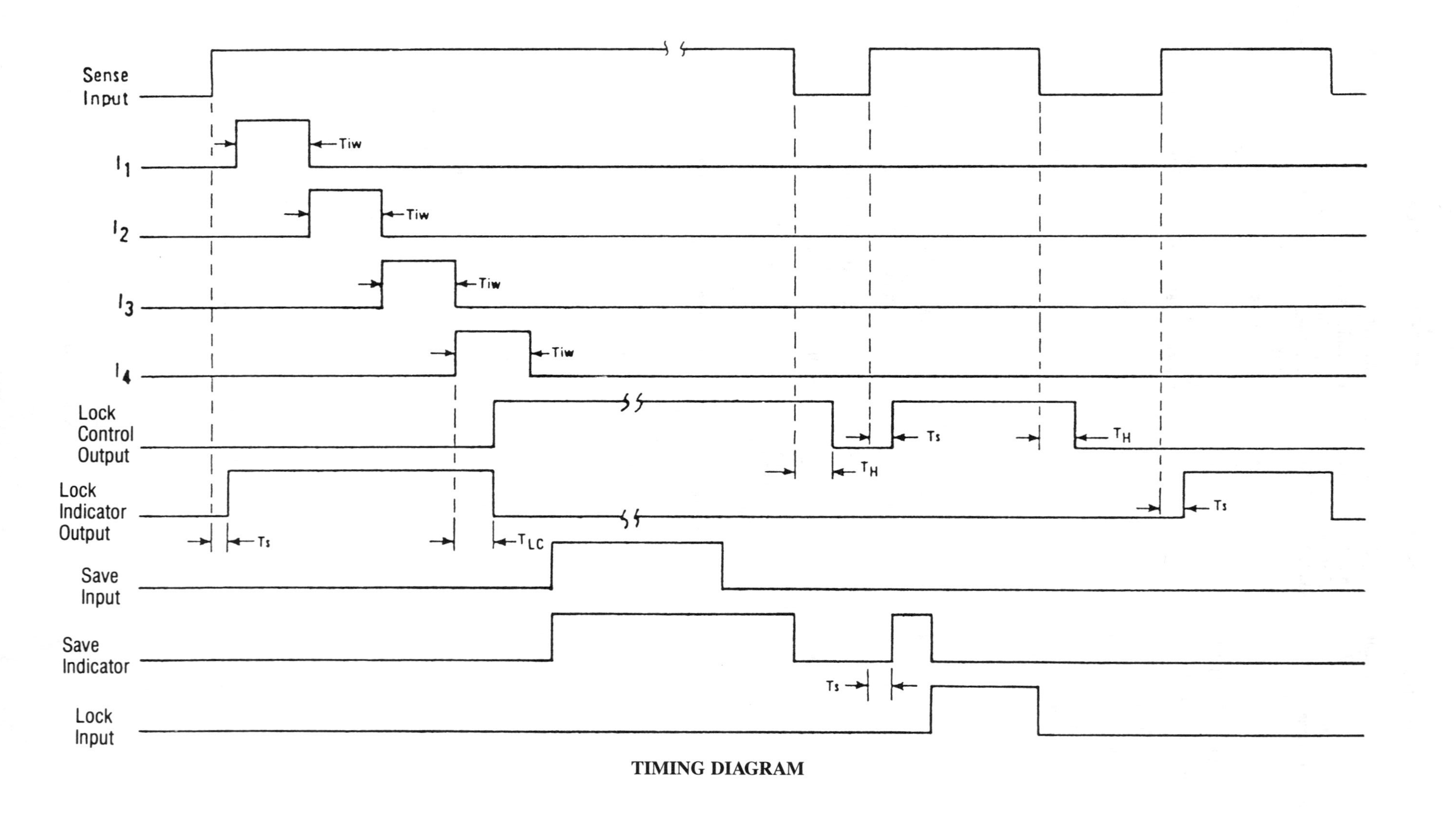

TIMING DIAGRAM

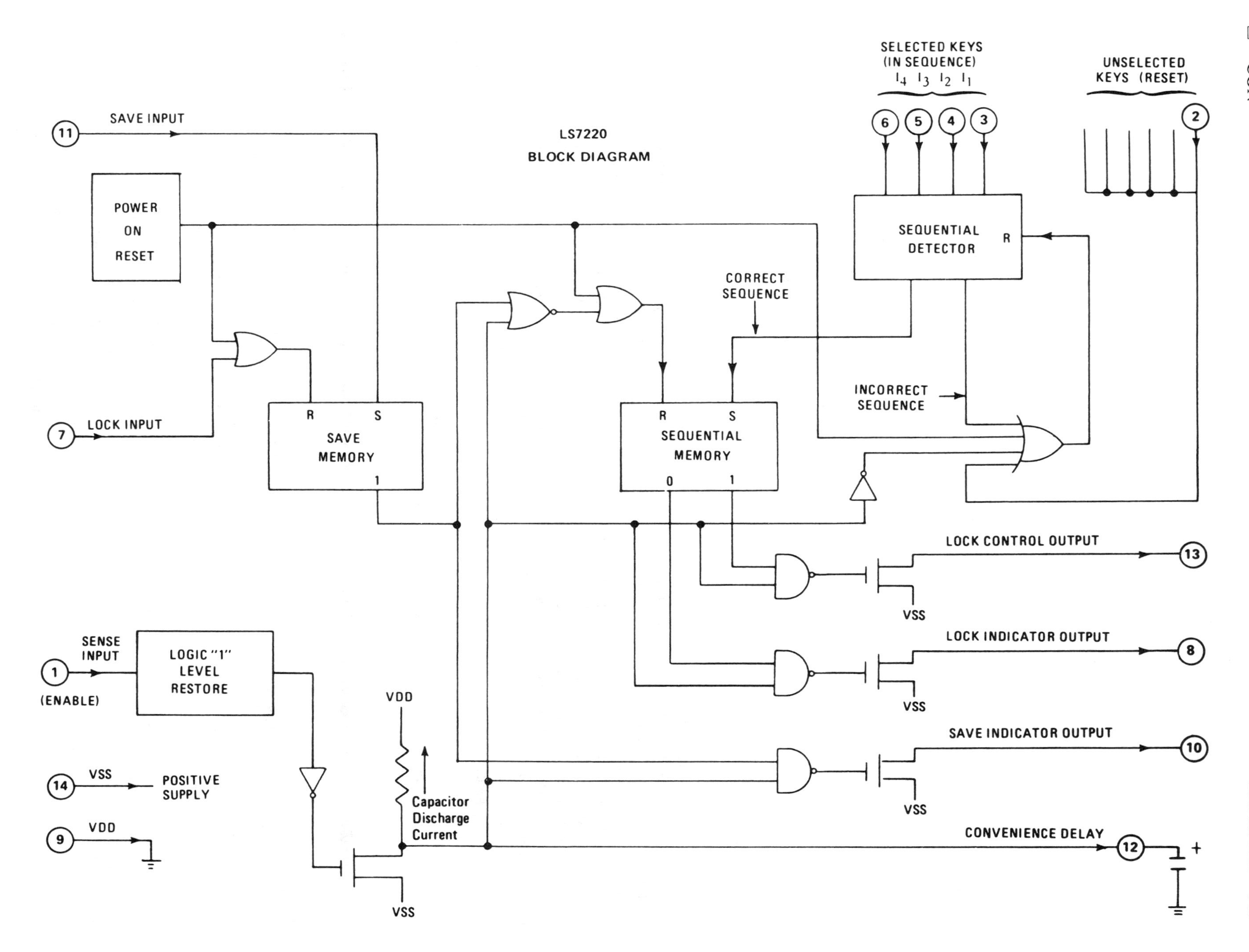
LS7220
BLOCK DIAGRAM
SAVE INPUT
11
POWER ON RESET
LOCK INPUT
7
R
S
SAVE MEMORY
1
SELECTED KEYS
(IN SEQUENCE)
I4 I3 I2 I1
6
5
4
3
UNSELECTED KEYS (RESET)
2
SEQUENTIAL DETECTOR
R
CORRECT SEQUENCE
INCORRECT SEQUENCE
R
S
SEQUENTIAL MEMORY
0
1
LOCK CONTROL OUTPUT
13
VSS
LOCK INDICATOR OUTPUT
8
VSS
SAVE INDICATOR OUTPUT
10
VSS
CONVENIENCE DELAY
12
+
SENSE INPUT
1
(ENABLE)
LOGIC "1" LEVEL RESTORE
VDD
Capacitor Discharge Current
VSS
14
VSS
POSITIVE SUPPLY
9
VDD

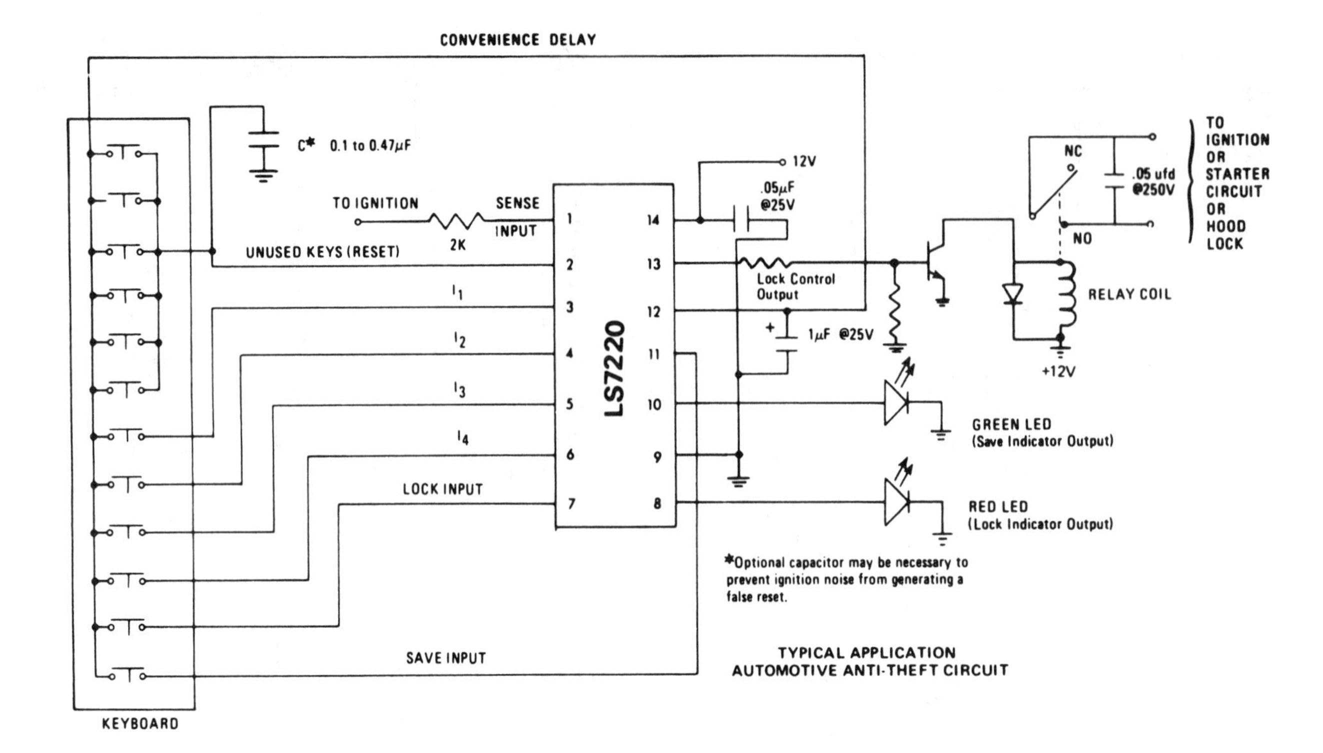

A typical automotive antitheft circuit is shown in the schematic diagram. When the ignition switch is turned on, the sense input (pin 1) goes high and the circuit is ready to accept the unlocking input sequence at I_1, I_2, I_3, and I_4 (pins 3, 4, 5, and 6, respectively). If the keys associated with these inputs are depressed exactly in sequence of I_1, I_2, I_3, and I_4, the lock control output (pin 13) will become on and the lock relay will be energized. This state will be indicated by the off condition of the lock indicator output (pin 8), which will render the red LED off (an indication of unlock condition). If the keys are depressed in any sequence other than as described above, the internal sequential detector will be reset and the entire sequence must be repeated.

In order to save the on condition of the lock control output when the ignition switch is turned off (i.e., when the sense input becomes low), the key associated with the save input (pin 11) will have to be depressed. The save status will be indicated by a high at the save indicator output (pin 10), which in turn will turn the green LED on. If the ignition switch is turned off while the green LED is on, all the output status will be preserved in the internal memory so that when the ignition switch is turned on again it will not need to go through the input sequence again. This feature could be used for valet parking and garage service.

Status saving can be cancelled by depressing the lock key followed by turning the ignition switch off for a time greater than the convenience delay. This will also turn off the lock control output.

LSI
LS7222
Keyboard-Programmable Digital Lock

- Stand-alone lock logic
- 38416, 4-digit combinations
- 3 different user-programmable codes
- Momentary static-lock control outputs
- Internal keyboard debounce circuit
- Tamper detection output
- Circuit status outputs
- Low current consumption (30μA max @ 12Vdc)
- Single power supply operation (+4 to +15Vdc)
- All inputs protected
- High noise immunity

The LS7222 is a programmable electronic lock implemented in a monolithic CMOS integrated circuit, packaged in a 20-pin DIP. The circuit contains all the necessary memory, decoder, and control logic to make a programmable "keyless" lock system to control electromechanical-type locks. Input is provided by a matrix keypad that has a maximum allowable size of 4 × 4.

The LS7222 can be programmed to recognize 3 different codes: one to lock (arm), one to unlock (disarm), and one to unlock and trigger an alarm (duress). Programming is done via the keypad inputs. Any entry from the keypad (when not in the program mode) that does not match one of the 3 programmed codes will cause the tamper output to become active.

The monolithic, low-power CMOS design of the LS7222 enables it to be designed into typical battery backed-up and automotive-type security systems.

MAXIMUM RATINGS: (Voltages references to VSS)

RATING	SYMBOL	VALUE	UNIT
DC supply voltage	VDD	+4 to +18	Vdc
Operating temperature range	TA	-25 to +70	°C
Storage temperature range	TSTG	-65 to +150	°C

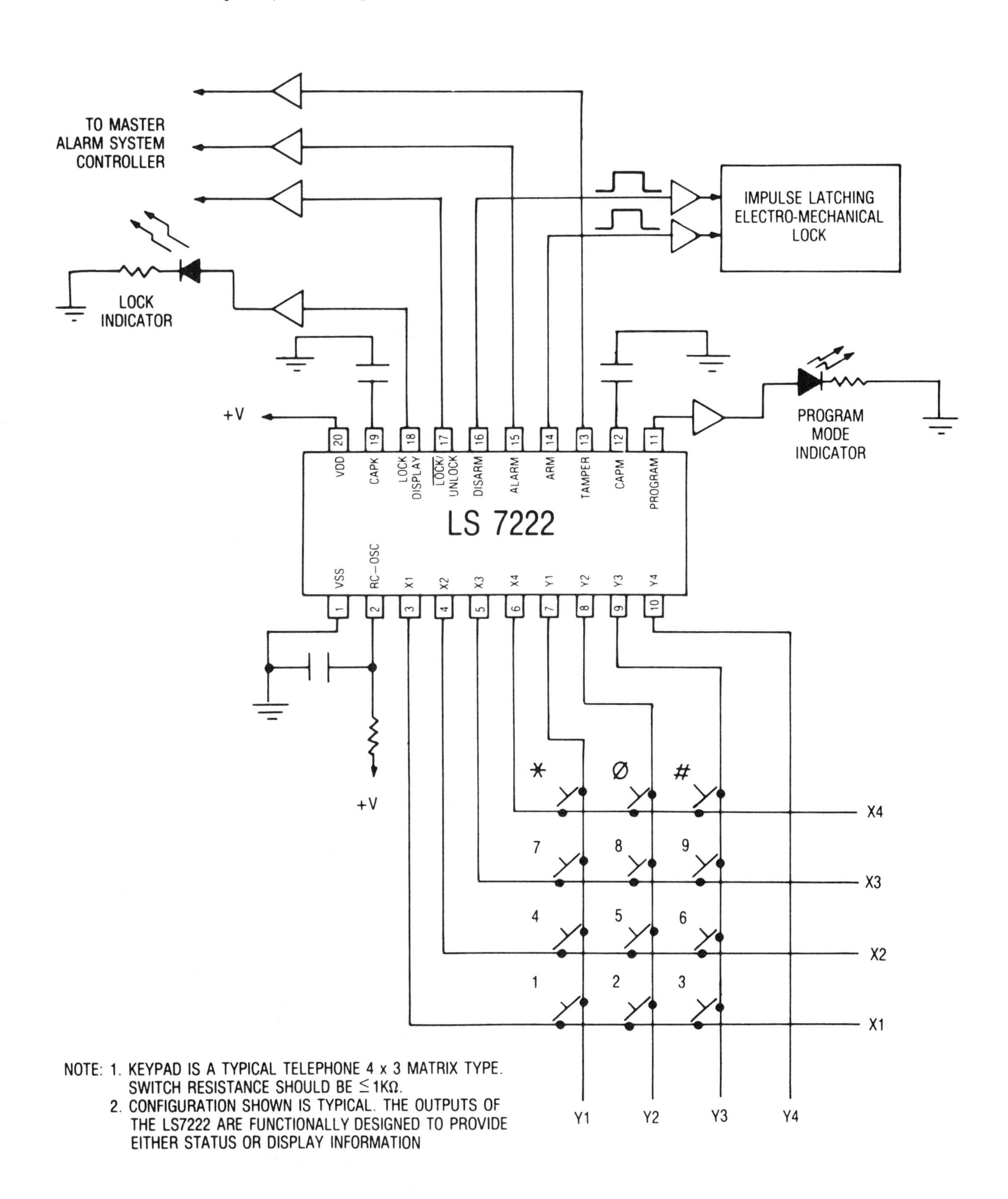

NOTE: 1. KEYPAD IS A TYPICAL TELEPHONE 4 x 3 MATRIX TYPE. SWITCH RESISTANCE SHOULD BE ≤1KΩ.
2. CONFIGURATION SHOWN IS TYPICAL. THE OUTPUTS OF THE LS7222 ARE FUNCTIONALLY DESIGNED TO PROVIDE EITHER STATUS OR DISPLAY INFORMATION

CONNECTION DIAGRAM — TOP VIEW
STANDARD 20 PIN PLASTIC DIP

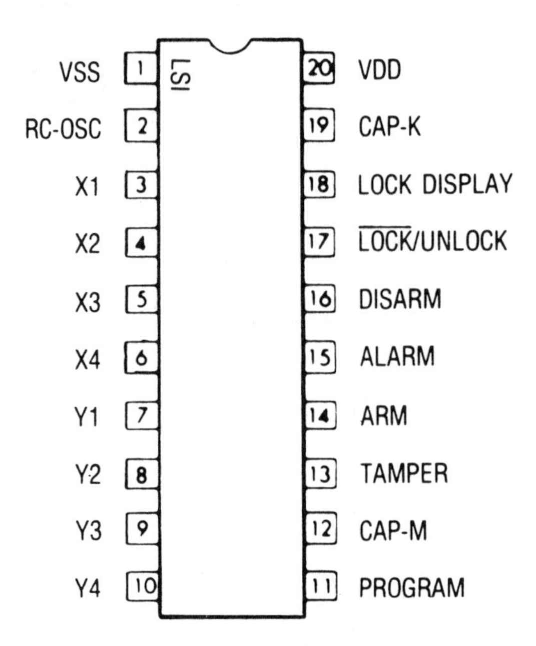

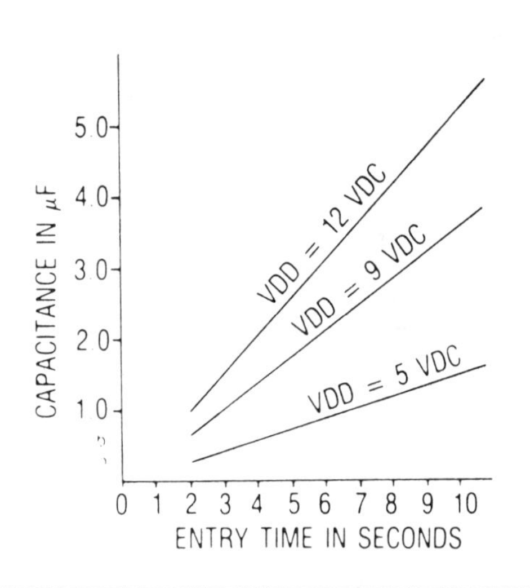

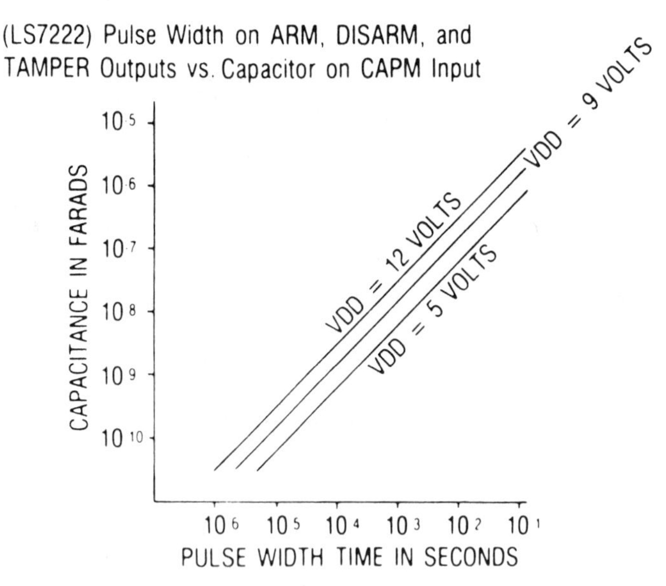

LSI
LS7223
Keyboard-Programmable Digital Lock

- Stand-alone lock logic
- 38416, 4-digit combinations
- 3 different user-programmable codes
- Momentary and static lock control outputs
- Internal keyboard debounce circuit
- Tamper detection output
- Circuit status outputs
- Low current consumption (30μA max @ 12Vdc)
- Single power supply operation (+4 to +15Vdc)
- All inputs protected
- High noise immunity

The LS7223 is a programmable electronic lock implemented in a monolithic CMOS integrated circuit, packaged in a 20-pin DIP. The circuit contains all the necessary memory, decoder, and control logic to make a programmable "keyless" lock system to control electromechanical-type locks. Input is provided by a matrix keypad that has a maximum allowable size of 4 × 4.

The LS7223 can be programmed to recognize 3 different codes: one to toggle an output, one to toggle an output and generate a pulse, and one to toggle an output and trigger an alarm. Programming is done via the keypad inputs. Any entry from the keypad (when not in the program mode) that does not match one of the 3 programmed codes, will cause the tamper output to become active.

The monolithic, low-power CMOS design of the LS7223 enables it to be designed into typical battery backed-up and automotive-type security systems.

MAXIMUM RATINGS: (Voltages references to VSS)

RATING	SYMBOL	VALUE	UNIT
DC supply voltage	VDD	+4 to +18	Vdc
Operating temperature range	TA	-25 to +70	°C
Storage temperature range	TSTG	-65 to +150	°C

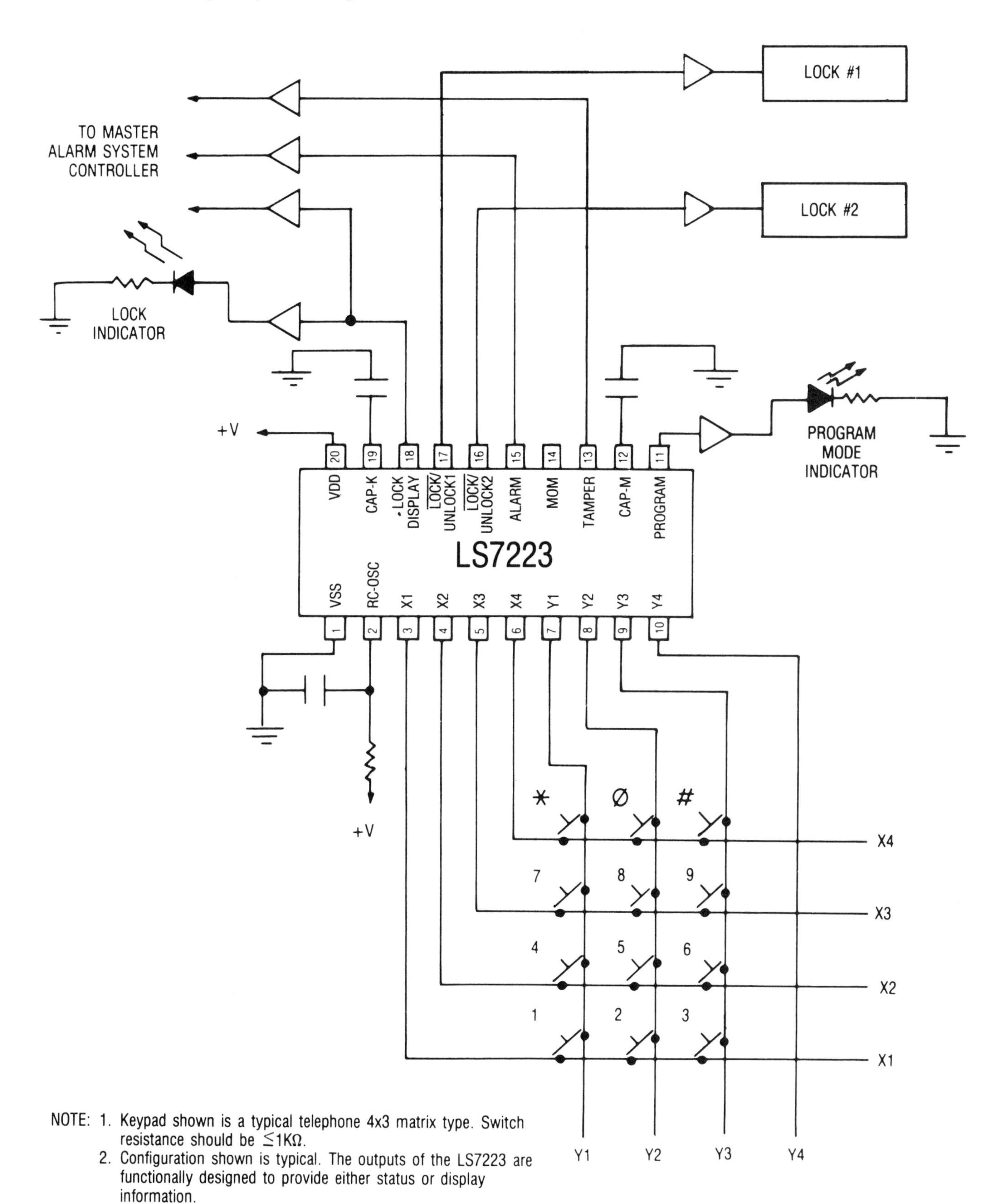

NOTE: 1. Keypad shown is a typical telephone 4x3 matrix type. Switch resistance should be ≤1KΩ.
2. Configuration shown is typical. The outputs of the LS7223 are functionally designed to provide either status or display information.

CONNECTION DIAGRAM — TOP VIEW
STANDARD 20 PIN PLASTIC DIP

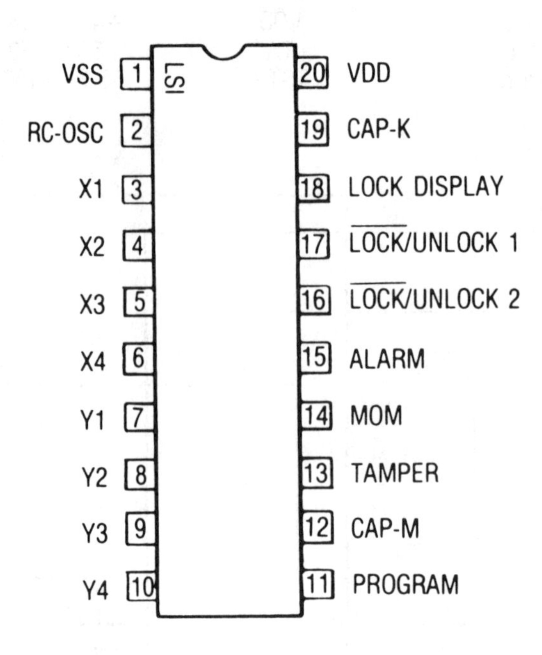

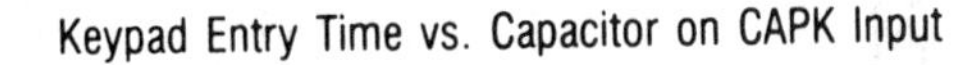

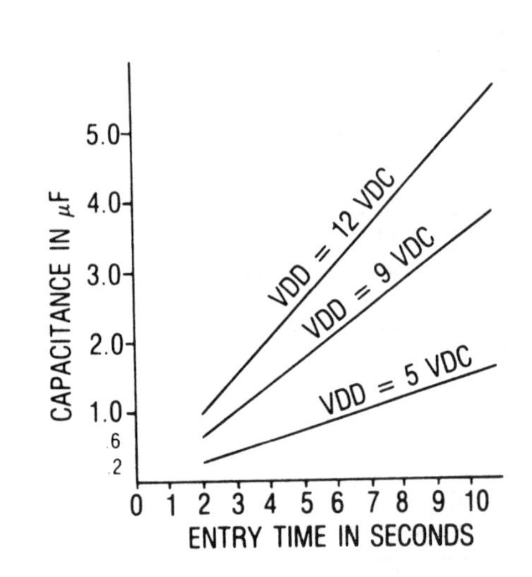

(LS7223) Pulse Width on MOMENTARY and TAMPER Outputs vs. Capacitor on CAPM Input

CAPACITANCE IN FARADS: 10^{-5}, 10^{-6}, 10^{-7}, 10^{-8}, 10^{-9}, 10^{-10}

PULSE WIDTH TIME IN SECONDS: 10^{-6}, 10^{-5}, 10^{-4}, 10^{-3}, 10^{-2}, 10^{-1}

VDD = 12 VOLTS; VDD = 9 VOLTS; VDD = 5 VOLTS

LSI
LS7225, LS7226
Digital Lock With Tamper Output

- Stand-alone lock logic
- 5040, 4-digit combination with a 10-number keyboard
- Out-of-sequence detection
- Tamper output, sequence enable input
- Direct LED and lock relay drive
- Externally controlled combination delay
- Internal pull-down resistors on all inputs
- High noise immunity
- Low current consumption (40μA max @ 12Vdc)
- Single power supply operation (+4V to +18V)
- Momentary or static lock control output
- Auxiliary delay circuitry included

The LS7225 is a monolithic, ion-implanted MOS 4-key keyless lock. The circuit includes sequential logic for interpretation of correct key closure, a momentary and static lock control output, out-of-sequence detection circuitry, and a tamper output.

MAXIMUM RATINGS: (Voltages Referenced to V_{DD})

Rating	Symbol	Value	Units
DC Supply Voltage	V_{SS}	+4 to +18	Vdc
Operating Temperature Range	TA	−25 to +70	°C
Storage Temperature Range	TSTG	−65 to +150	°C

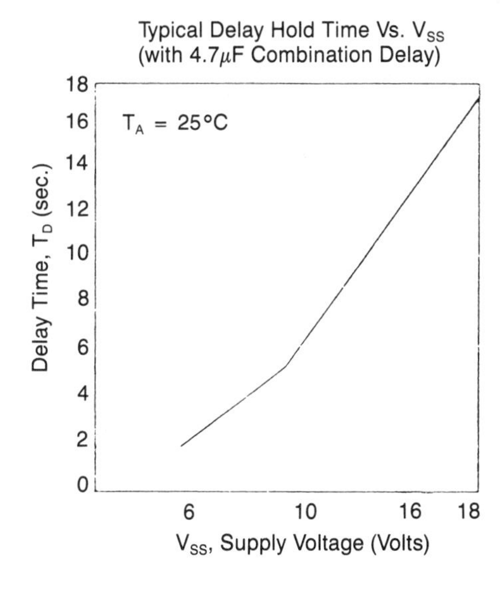

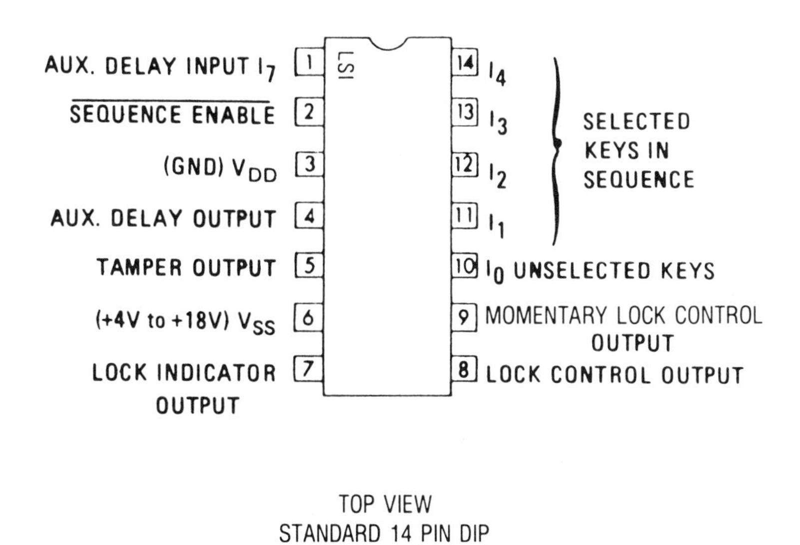

TOP VIEW
STANDARD 14 PIN DIP

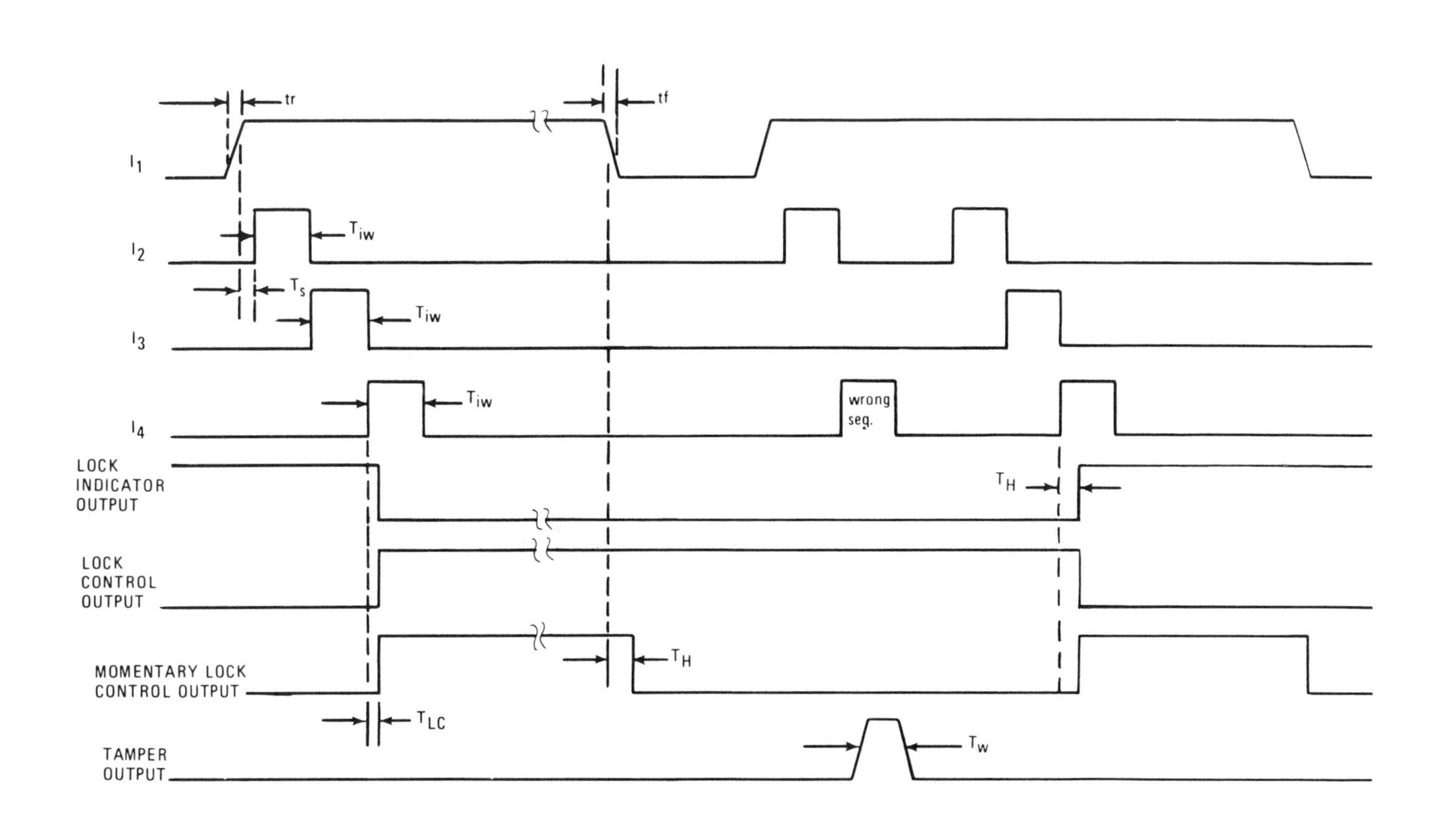

Timing Diagram

BLOCK DIAGRAM

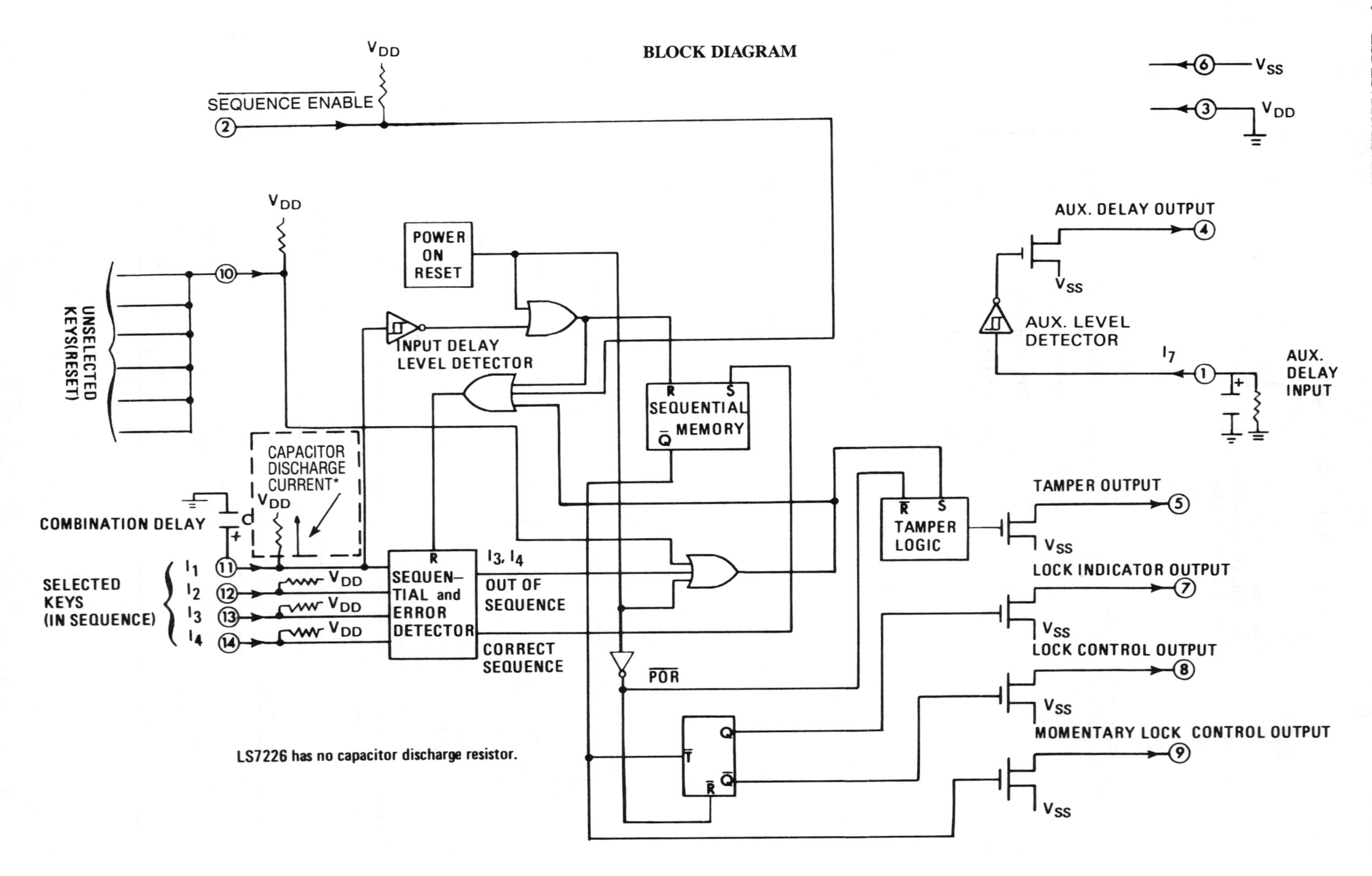

LS7226 has no capacitor discharge resistor.

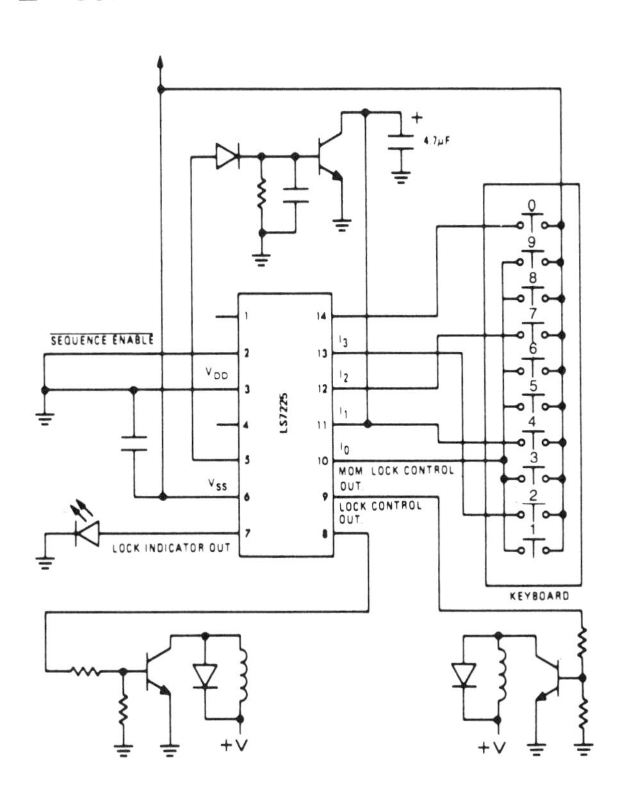

A typical circuit is shown in the schematic diagram.
When input I_1 (pin 11) goes high, the circuit is ready to accept the unlocking input sequence at I_2, I_3 and I_4 (pins 12, 13, and 14 respectively). If the Keys associated with these inputs are depressed exactly in sequence of I_1, I_2, I_3 and I_4, the lock control output (pin 8) will become ON, the momentary lock control output (pin 9) will be ON until input I_1 (pin 11) becomes low. The state ON of the lock control will be indicated by the OFF Condition of the lock indicator output (pin 7) which will render the LED OFF (an indication of unlock condition). If the Keys are depressed in any sequence other than as described above, the internal "SEQUENTIAL DETECTOR" will be reset and the entire sequence must be repeated. The lock control output is turned OFF by repeating the input sequence. The Momentary Lock Control output goes high each time the correct sequence is entered. The specific code shown is 4720.

TYPICAL APPLICATION OF LS7225 IN MACHINE OR AREA ACCESS

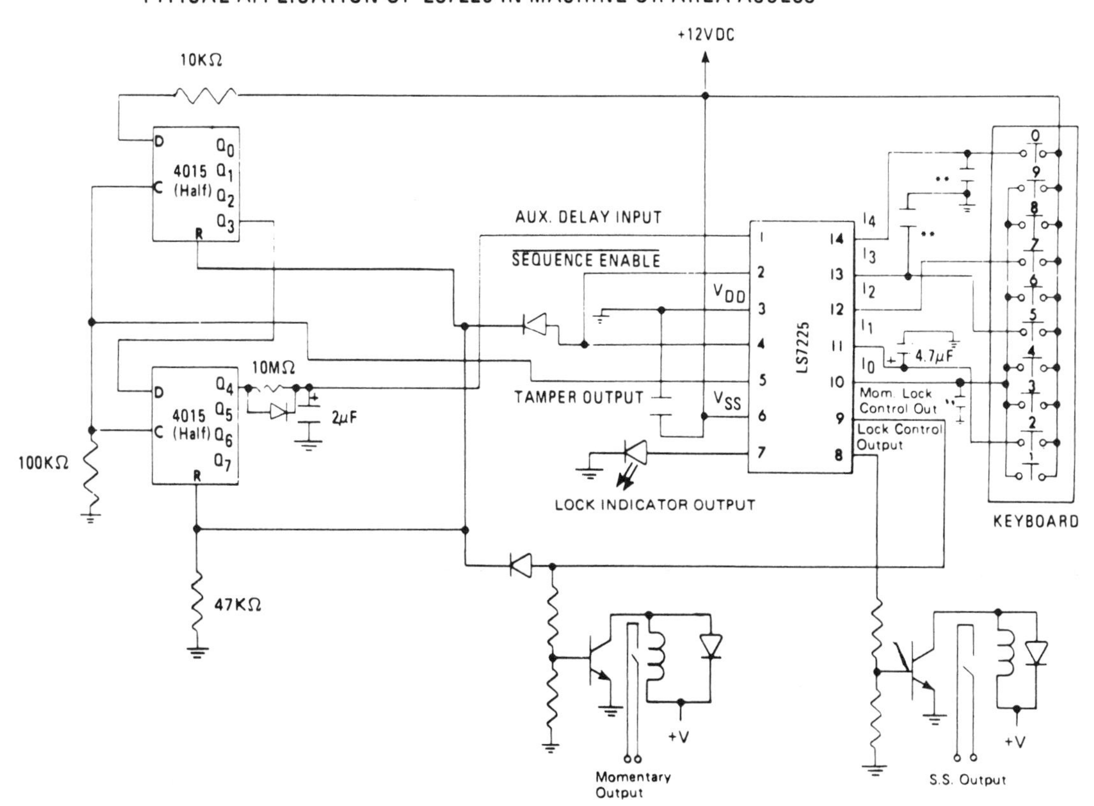

TYPICAL TAMPER LOCK DISABLE APPLICATION

A System with 12 seconds reset after 5 TAMPER outputs and support circuitry is shown. The specific code is 2750. If the Keys associated with the given code are not depressed exactly in sequence or if any Key associated with the unselected Keys (Reset) is depressed, the pulse from the Tamper output will clock the 4015 Dual 4-bit shift register which transmits a logical "1" from input D to output Q4 after five clock pulses. A logical "1" in Q4 will charge up the 2μF capacitor, turn on the AUX. DELAY OUTPUT for 12 seconds and keep the LS7225 in the RESET Mode vis SEQUENCE ENABLE (pin 2). After the 2μF capacitor discharges, pin 2 of LS7225 becomes a logical "0" and the keyless lock integrated circuit is ready to accept key inputs.

**NOTE: Due to mechanical keyboard bounce it may be necessary to include these (750pf) capacitors.

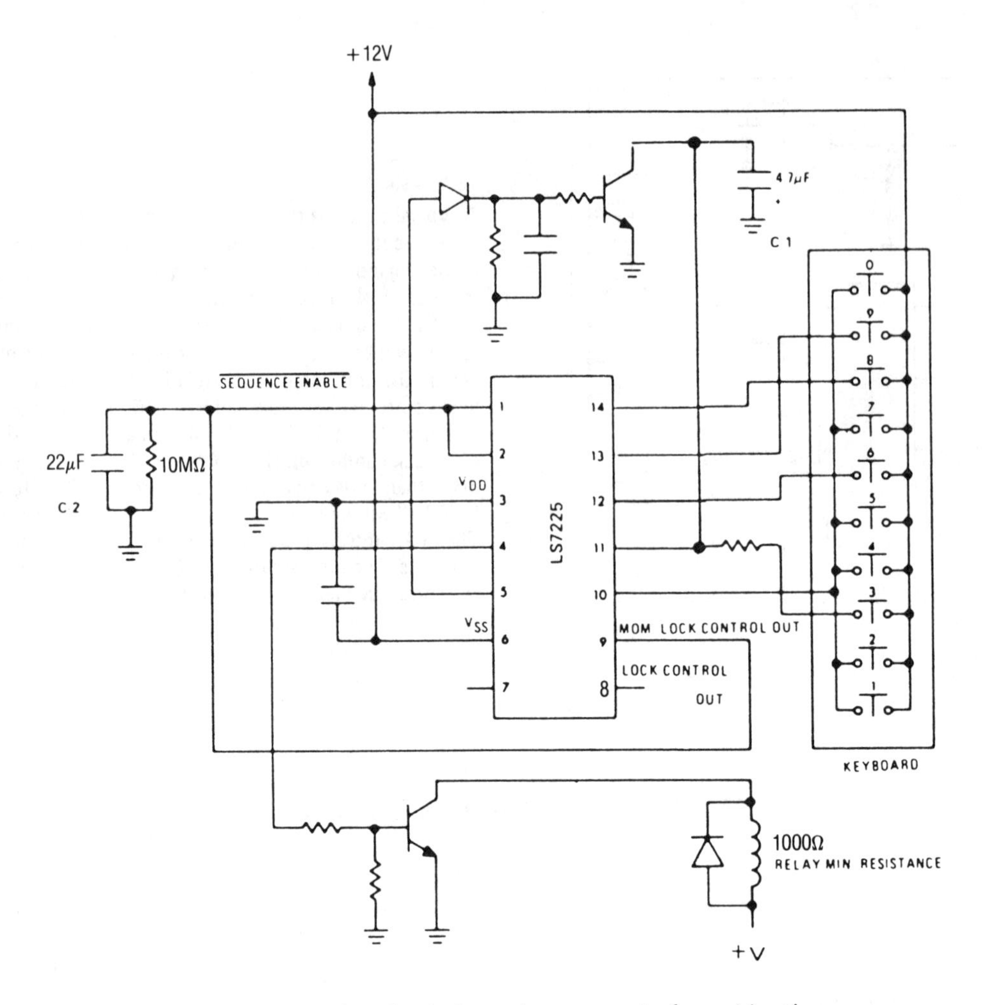

Typical application for independent control of combination (input) time and "UNLOCK" time.

C-1 determines input time.
C-2 determines "UNLOCK" time.

Note: With this configuration one tamper pulse is transmitted at the start of "UNLOCK" time.

LSI
LS7228, LS7229
Address Decoder/Two-Pushbutton Digital Lock

- Stand-alone lock logic
- 9-bit code determined by 9 parallel inputs
- Two options of code input available:
  - LS7228—Dual train pulsed input
  - LS7229—Two momentary switches
- Out-of-sequence disabling circuit
- Current-source lock-control output
- External controlled delay to set maximum interpulse time
- Single power supply operation (2.5V to 15.0V)
- Low standby current (15μA maximum)
- 16-pin dual-in-line plastic package
- Cascadable

LS7228/LS7229 are monolithic, ion-implanted MOS encoder circuits. Each circuit includes logic for interpretation of correct sequential key closure or pulse input and a momentary lock control output. An out-of-sequence detector will disable any further insertions, and a new sequence can be reapplied after a delay time, determined by an external R/C time constant.

The LS7228 utilizes a dual train input format where the input "one's" data is applied to pin 13 and the input "zero's" is applied to pin 14. The common input (pin 15) is not used. The LS7229 utilizes two momentary switches and pins 13, 14, and 15 in a manual operating mode.

MAXIMUM RATINGS:

PARAMETER	SYMBOL	VALUE	UNITS
Storage Temperature	T_{stg}	−65 to +150	°C
Operating Temperature	T_a	−25 to +70	°C
Voltage (any pin to V_{SS})	V_{max}	−30 to +0.5	VOLTS

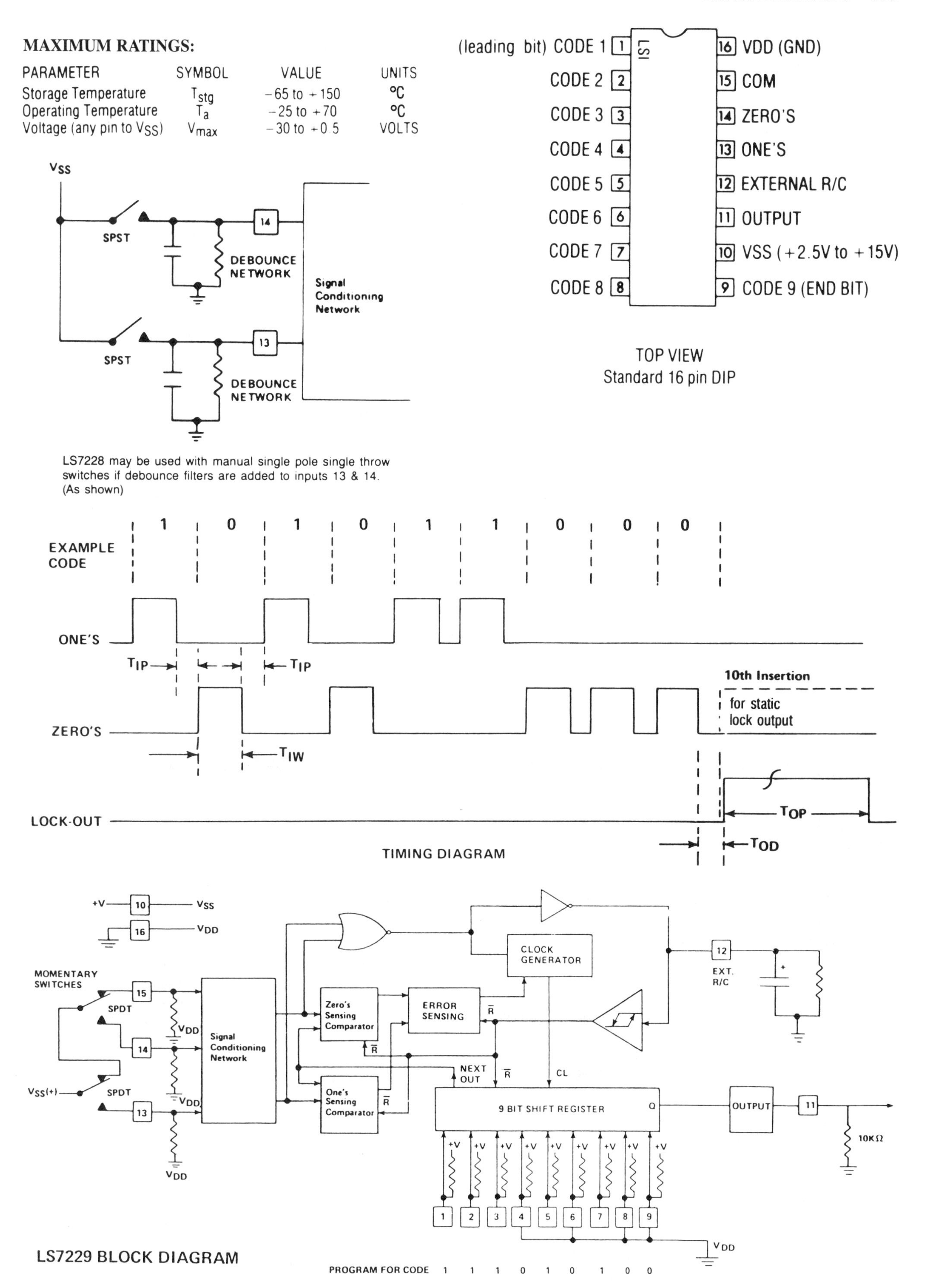

DUAL SWITCH CASCADING SERIES (LS7229)

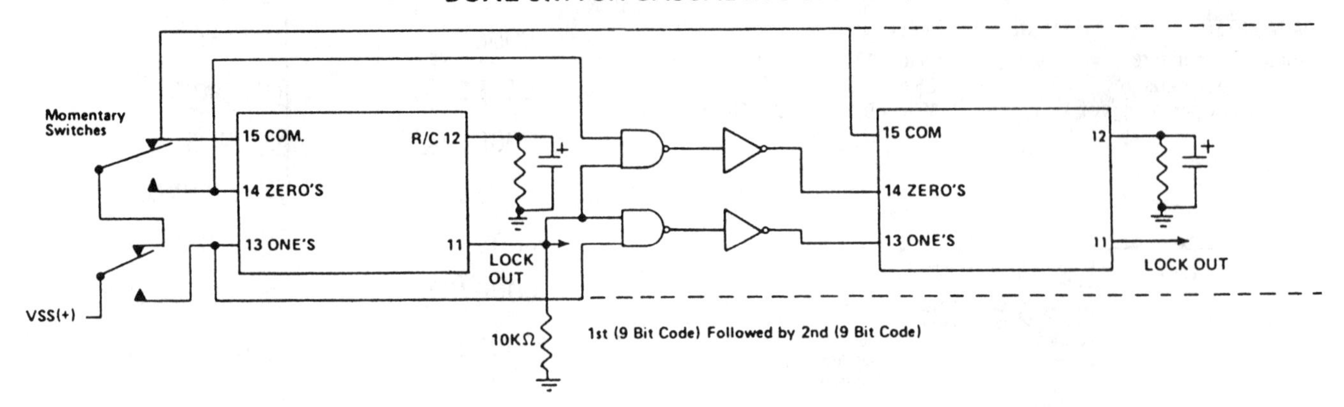

DUAL TRAIN CASCADING SERIES PARALLEL (LS7228)

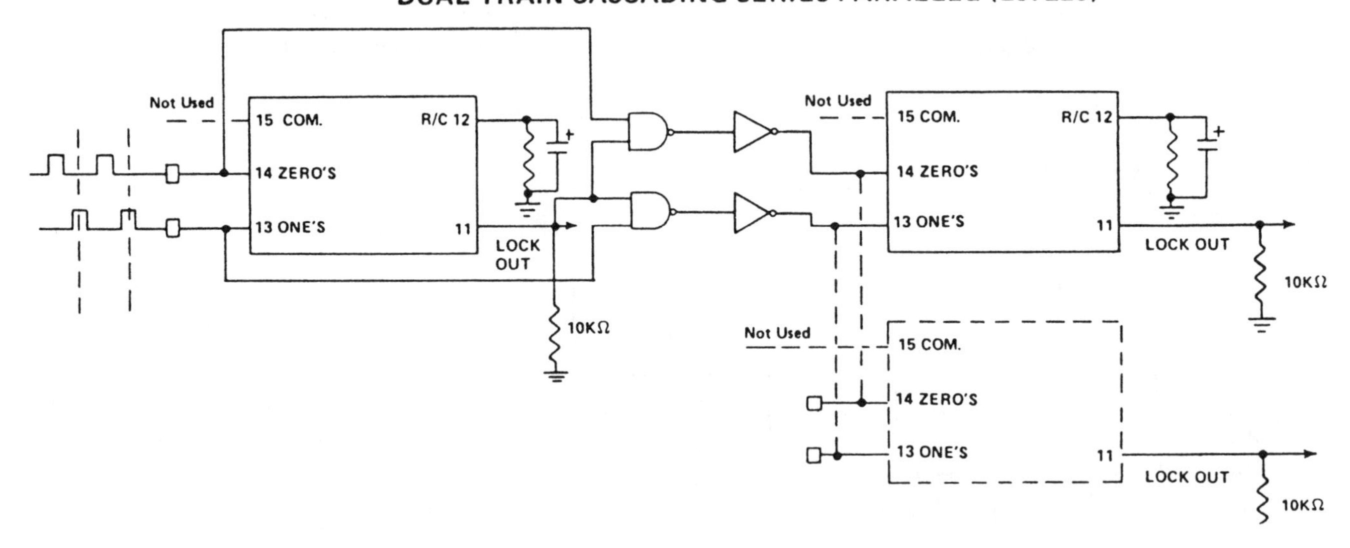

LSI LS7231-5

Touch Control Continuous Dimmer Light Switch and AC Motor-Speed Controller

- Phase-locked loop synchronization produces pure ac waveform across output load with no dc offset
- Provides on/off or brightness control of incandescent lamps and on/off or speed control of ac motors without the use of mechanical switches
- Controls brightness by controlling the ac duty cycle
- Provides speed control of ac motors, such as shaded pole and universal series motors
- Controls the "duty cycle" from 25% to 88% (on time angles for ac half cycles between 41° and 159° respectively)
- Operates at 50Hz/60Hz line frequency
- Provides control through transformers for low-voltage lighting applications
- Input for extensions or remote sensors
- Input for slow dimming
- 12 to 18Vdc supply voltage

LS7231 through LS7235 is a series of monolithic, ion-implanted MOS circuits that are specifically designed for brightness or on/off control of incandescent lamps or speed of ac motors used on the ac line. The outputs of these chips control the brightness of a lamp or speed of an ac motor by controlling the firing angle of a triac connected in series with the lamp or ac motor. All internal timings are synchronized with the line frequency by means of a built-in phase-locked loop circuit. The output occurs once every half-cycle of the line frequency. Within the half-cycle, the output can be positioned anywhere between 159° phase angle for maximum brightness/speed and 41° phase angle for minimum brightness/speed, in relation to the line frequency. The positioning of the output is controlled by applying a low level at the sensor input or a high level at the slave input.

These functions can be implemented with very few interface components, described in the application examples. When implemented in this manner, a touching of the sensor plate causes the lamp brightness or ac motor speed to change as follows:

- If the sensor is touched momentarily (32ms to 332ms), the lamp or ac motor is:
  - ○ turned off if it was on
  - ○ turned on if it was off. The brightness/speed resulting in either full brightness/speed or depending on the circuit type, a previous brightness/speed stored in the memory.

- If the sensor is touched for a prolonged time (more than 332ms) the light intensity changes slowly. As long as the touch is maintained, the change continues; the direction of change reverses whenever the maximum or minimum brightness is reached. The circuit also provides an input for slow dimming. By applying a slow clock to this input, the lamp can be dimmed slowly until total turn off occurs. This feature can be useful in children's bedroom lights.

ABSOLUTE MAXIMUM RATINGS

PARAMETER	SYMBOL	VALUE	UNITS
DC supply voltage	VSS	+20	Volt
Any input voltage	V_{IN}	VSS + .5	Volt
Operating temperature	T_A	0 to +80	°C
Storage temperature	Tstg	−65 to +150	°C

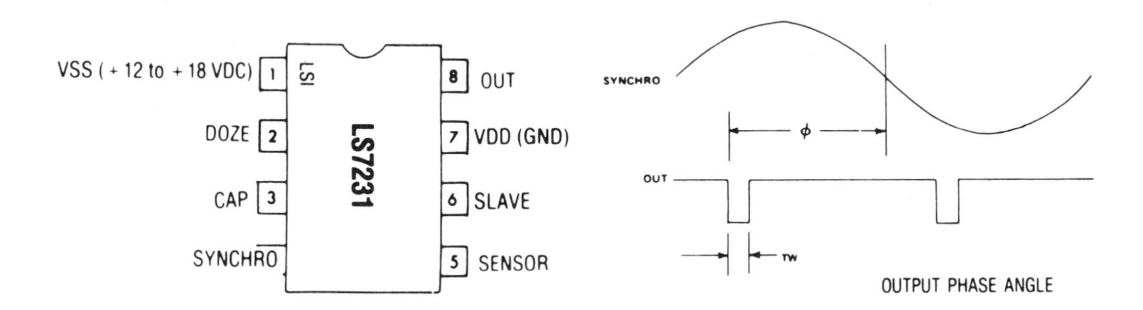

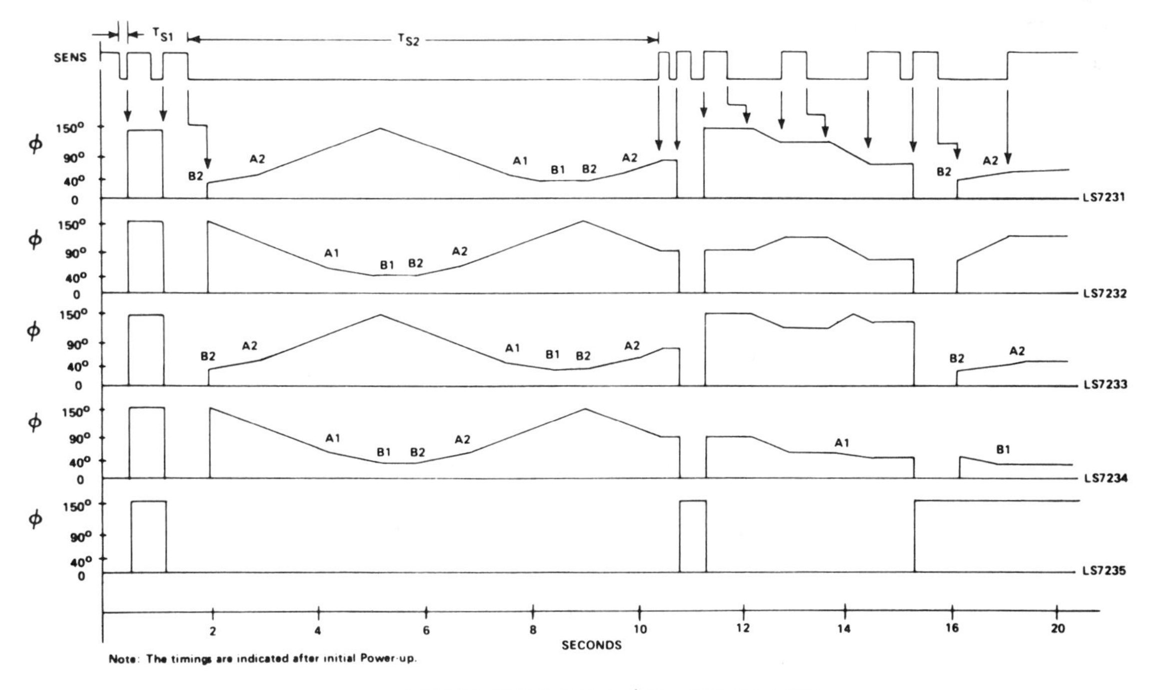

OUTPUT PHASE-ANGLE, φ Vs, SENSOR INPUT

TYPE	SENSOR (TOUCH) DURATION				DIMMING DIRECTION REVERSAL
	MOMENTARY (32ms to 332ms) (note 3)		PROLONGED (More than 332ms) (note 3)		
	PRE-TOUCH BRIGHTNESS	POST-TOUCH BRIGHTNESS	PRE-TOUCH BRIGHTNESS	POST-TOUCH BRIGHTNESS	
LS7231	Off	Max.	Off	Starts varying at Min.	N/A
	Max.	Off	Max.	Starts varying at Max.	N/A
	Intermediate	Off	Intermediate	Starts varying at Pre-Touch brightness	NO
LS7232	Off	Memory (See note 1)	Off	Starts varying at Memory (Note 2)	YES
	Max.	Off	Max.	Starts varying at Max.	N/A
	Intermediate	Off	Intermediate	Starts varying at Pre-Touch brightness	YES
LS7233	Off	Max.	Off	Starts varying at Min.	N/A
	Max.	Off	Max.	Starts varying at Max.	N/A
	Intermediate	Off	Intermediate	Starts varying at Pre-Touch brightness	YES
LS7234	Off	Memory (See note 1)	Off	Starts varying at Memory (Note 2)	NO
	Max.	Off	Max.	Starts varying at Max.	N/A
	Intermediate	Off	Intermediate	Starts varying at Pre-Touch brightness	NO
LS7235	Off	Max.	N/A	N/A	N/A
	Max.	Off	N/A	N/A	N/A

NOTE 1. "Memory" refers to the brightness stored in the memory. The brightness is stored in the memory when the light is turned off by momentary sensor touch. First time after power-up, momentary touch produces max. brightness.

NOTE 2. First time after power-up, prolonged touch causes intensity to vary starting at min.

NOTE 3. The time figure is based on 60Hz synchro frequency. For 50Hz the figures are 39ms and 399ms.

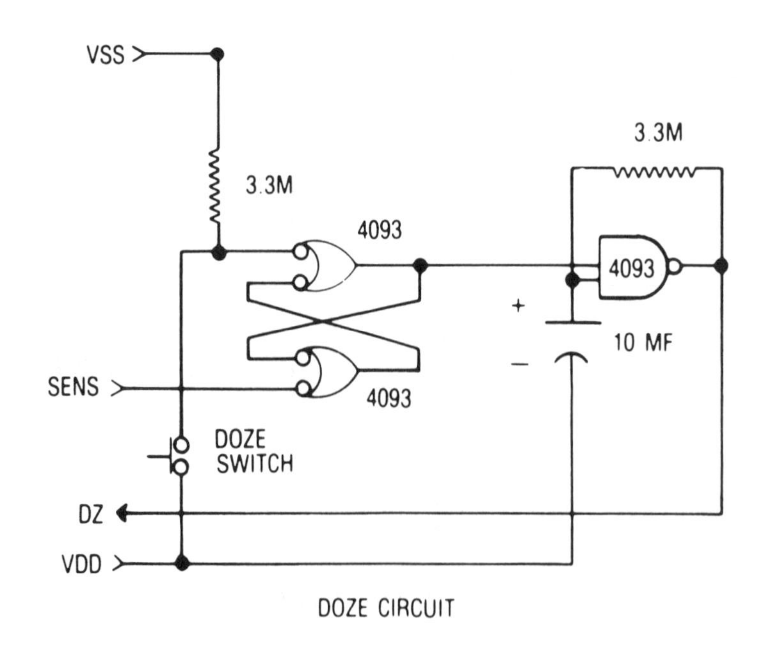

DOZE CIRCUIT

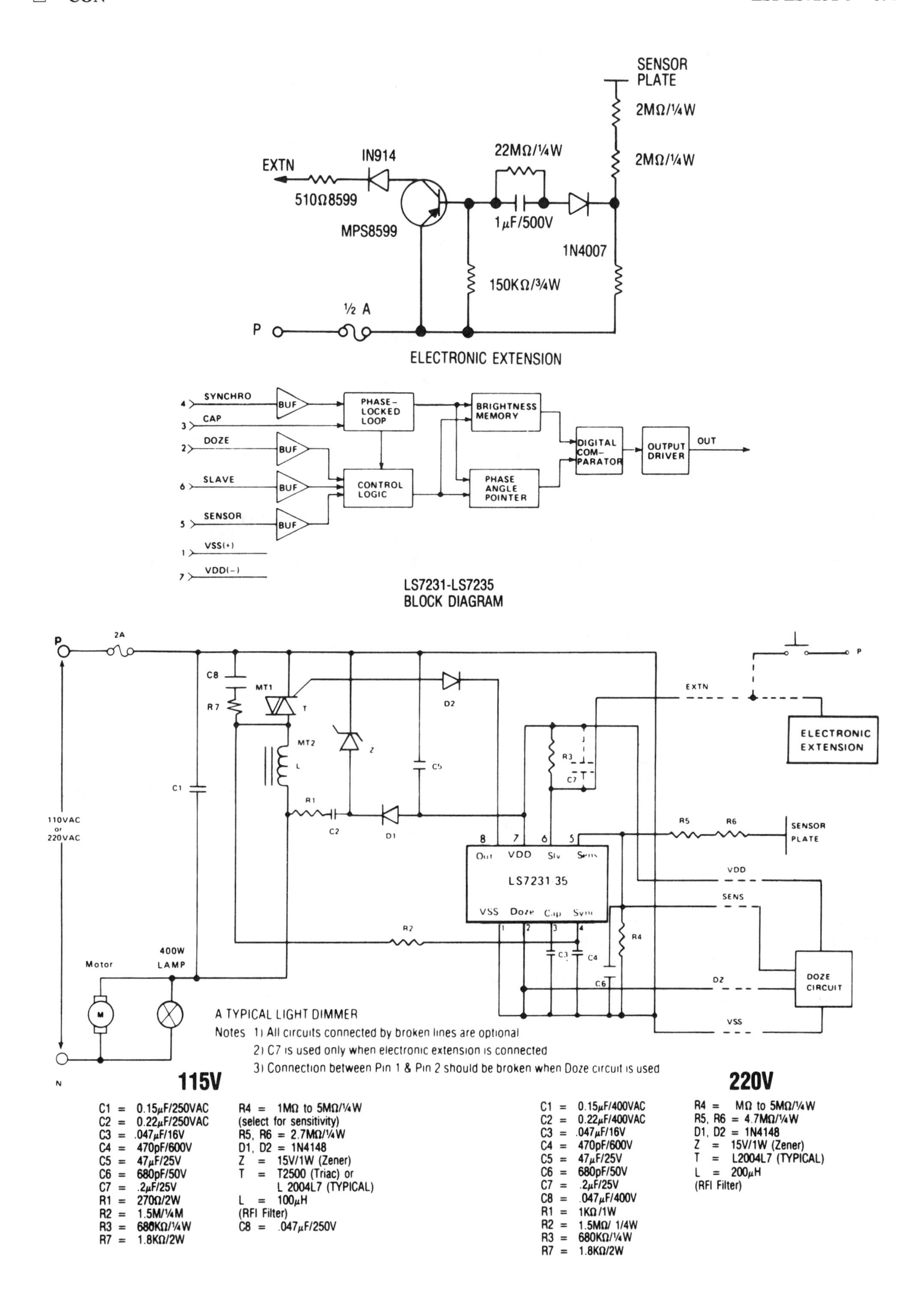
SENSOR PLATE
2MΩ/¼W
2MΩ/¼W
22MΩ/¼W
EXTN
IN914
510Ω8599
MPS8599
1μF/500V
1N4007
150KΩ/¾W
½ A
P
ELECTRONIC EXTENSION
SYNCHRO
CAP
DOZE
SLAVE
SENSOR
VSS(+)
VDD(−)
BUF
PHASE-LOCKED LOOP
CONTROL LOGIC
BRIGHTNESS MEMORY
PHASE ANGLE POINTER
DIGITAL COM-PARATOR
OUTPUT DRIVER
OUT
LS7231-LS7235
BLOCK DIAGRAM
2A
P
C8
R7
MT1
T
MT2
L
Z
C5
D2
C1
R1
C2
D1
110VAC or 220VAC
EXTN
ELECTRONIC EXTENSION
R3
C7
R5
R6
SENSOR PLATE
Out VDD Slv Sens
LS7231-35
VSS Doze Cap Sync
VDD
SENS
R2
R4
C3
C4
C6
DZ
DOZE CIRCUIT
VSS
Motor
400W LAMP
M
N
A TYPICAL LIGHT DIMMER
Notes 1) All circuits connected by broken lines are optional
2) C7 is used only when electronic extension is connected
3) Connection between Pin 1 & Pin 2 should be broken when Doze circuit is used
115V
C1 = 0.15μF/250VAC
C2 = 0.22μF/250VAC
C3 = .047μF/16V
C4 = 470pF/600V
C5 = 47μF/25V
C6 = 680pF/50V
C7 = .2μF/25V
R1 = 270Ω/2W
R2 = 1.5M/¼M
R3 = 680KΩ/¼W
R7 = 1.8KΩ/2W
R4 = 1MΩ to 5MΩ/¼W
(select for sensitivity)
R5, R6 = 2.7MΩ/¼W
D1, D2 = 1N4148
Z = 15V/1W (Zener)
T = T2500 (Triac) or
L 2004L7 (TYPICAL)
L = 100μH
(RFI Filter)
C8 = .047μF/250V
220V
C1 = 0.15μF/400VAC
C2 = 0.22μF/400VAC
C3 = .047μF/16V
C4 = 470pF/600V
C5 = 47μF/25V
C6 = 680pF/50V
C7 = .2μF/25V
C8 = .047μF/400V
R1 = 1KΩ/1W
R2 = 1.5MΩ/ 1/4W
R3 = 680KΩ/¼W
R7 = 1.8KΩ/2W
R4 = MΩ to 5MΩ/¼W
R5, R6 = 4.7MΩ/¼W
D1, D2 = 1N4148
Z = 15V/1W (Zener)
T = L2004L7 (TYPICAL)
L = 200μH
(RFI Filter)

LSI
LS7263
Brushless Dc Motor-Speed Controller

- Highly accurate speed regulation (±.1% derived from XTL-controlled time base)
- Rapid acceleration to speed with little overshoot
- Static braking
- 10V to 28V supply range
- Low-speed detection output
- Over-current detection logic
- Power-on reset
- Six outputs drive power switching bridge directly
- 187-pin dual-in-line package

The LS7263 is a monolithic, ion-implanted MOS circuit designed to control the speed of a 3-phase, brushless dc motor. This specific circuit is programmed for use in 2600-RPM applications. The circuit utilizes a 3.58-MHz crystal to provide its accurate speed regulation time base. Overcurrent circuitry is provided to protect the windings, associated drivers, and power supply. A positive braking feature is provided to effect rapid deceleration.

Speed corrections are made by measuring the time between tachometer inputs and varying the on time of the drive signal applied to each winding. A sampling window is generated using tachometer input time intervals during which crystal-derived clock pulses are accumulated. The contents of the accumulator provide the address of a look-up table that has been derived from the physical characteristics of the motor and the load. The look-up table output determines the amount of on time for each coil. Positive and negative signals are applied sequentially to each winding driver through the output decoder/driver section.

A static-type positive braking system shorts all winding together upon receipt of the brake input. This system creates an electrical load on the motor, thus causing rapid deceleration. An overcurrent condition, when sensed at the overcurrent detection input, disables all six winding outputs. Outputs will be reenabled upon removal of the overcurrent condition.

MAXIMUM RATINGS:

PARAMETER	SYMBOL	VALUE	UNITS
Storage Temperature	T_{stg}	−65 to +150	°C
Operating Temperature			
1. Plastic	T_{ap}	−25 to +70	°C
2. Ceramic	T_{ac}	−55 to +125	°C
Voltage (any pin to V_{SS})	V_{max}	−30 to + 0.5	VOLTS

TABLE 1A −01,

INPUTS A	B	C	OUTPUTS ENABLED	DRIVER A	DRIVER B	DRIVER C
0	0	0	$0_1, 0_5$	+	−	OFF
1	0	0	$0_3, 0_5$	OFF	−	+
1	1	0	$0_3, 0_4$	−	OFF	+
1	1	1	$0_2, 0_4$	−	+	OFF
0	1	1	$0_2, 0_6$	OFF	+	−
0	0	1	$0_1, 0_6$	+	OFF	−

DESCRIPTION OF AVAILABLE TYPES

TYPE	POLES	SENSOR SEPARATION	GAIN*
7263-01	4	60°	Medium
7263-02	8	120°	High
7263-03	4	120°	Medium
7263-07	8	120°	Medium

*Gain describes the change of output duty cycle as a function of change motor speed. For the high gain type, the duty cycle is caused to change from 0% to 100% over a 6 RPM motor speed change. For the medium gain type, the duty cycle changes from 0% to 100% when the motor speed changes by 40 RPM.

TABLE 1B −02, −03, −07

INPUTS A	B	C	OUTPUTS ENABLED	DRIVER A	DRIVER B	DRIVER C
0	0	1	$0_2, 0_6$	OFF	+	−
1	0	1	$0_2, 0_4$	−	+	OFF
1	0	0	$0_3, 0_4$	−	OFF	+
1	1	0	$0_3, 0_5$	OFF	−	+
0	1	0	$0_1, 0_5$	+	−	OFF
0	1	1	$0_1, 0_6$	+	OFF	−

Push pull drivers are made up of pairs of Outputs: 0_1 and 0_4 (Driver A), 0_2 and 0_5 (Driver B), 0_3 and 0_6 (Driver C).

Function	Pin	Pin	Function
LS DETECTS OUTPUT	1	18	TACHOMETER INPUT
BRAKE INPUT	2	17	$\overline{\text{OVERCURRENT DETECT}}$
OUTPUT 0_6	3	16	VDD (GND)
OUTPUT 0_5	4	15	FREQUENCY TEST POINT
OUTPUT 0_4	5	14	OSC OUT
OUTPUT 0_3	6	13	OSC IN
OUTPUT 0_2	7	12	INPUT C
OUTPUT 0_1	8	11	INPUT B
VSS (+10 to +28 VDC)	9	10	INPUT A

TOP VIEW

OUTPUT CURRENT
(DRIVING DARLINGTON PAIRS)

POWER SUPPLY VOLTS	20	15	10	7.5	5	2.5	mA
12	.1	.25	.56	.86	1.5	3.3	
15	.33	.51	.92	1.3	2.1	4.6	
18	*	.76	1.3	1.7	2.8	5.8	
21	*	*	1.6	2.2	3.3	7.0	
24	*	*	1.9	2.6	4.0	8.3	
28	*	*	*	3.2	4.9	9.9	

*causes excessive power dissipation.

RESISTANCE IN KILOHMS

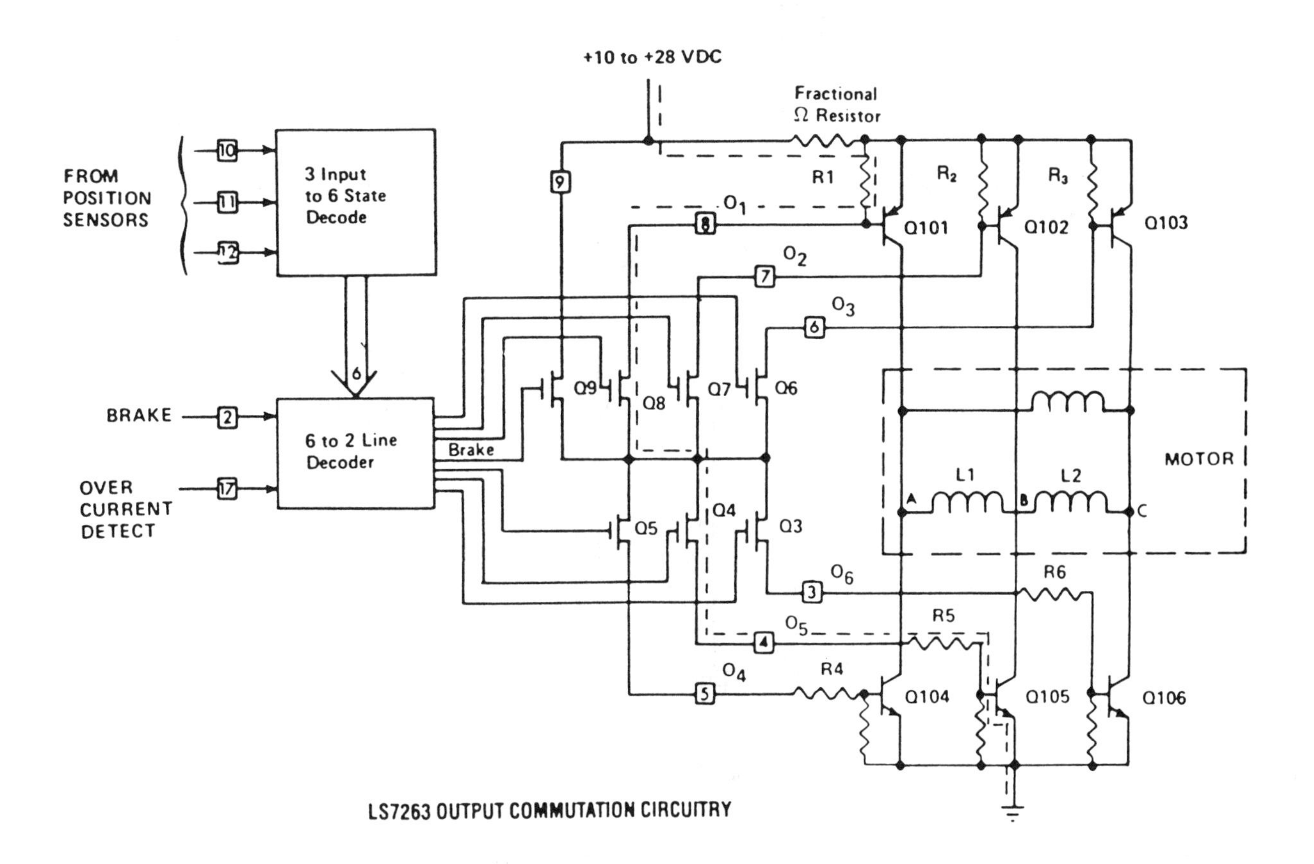

LS7263 OUTPUT COMMUTATION CIRCUITRY

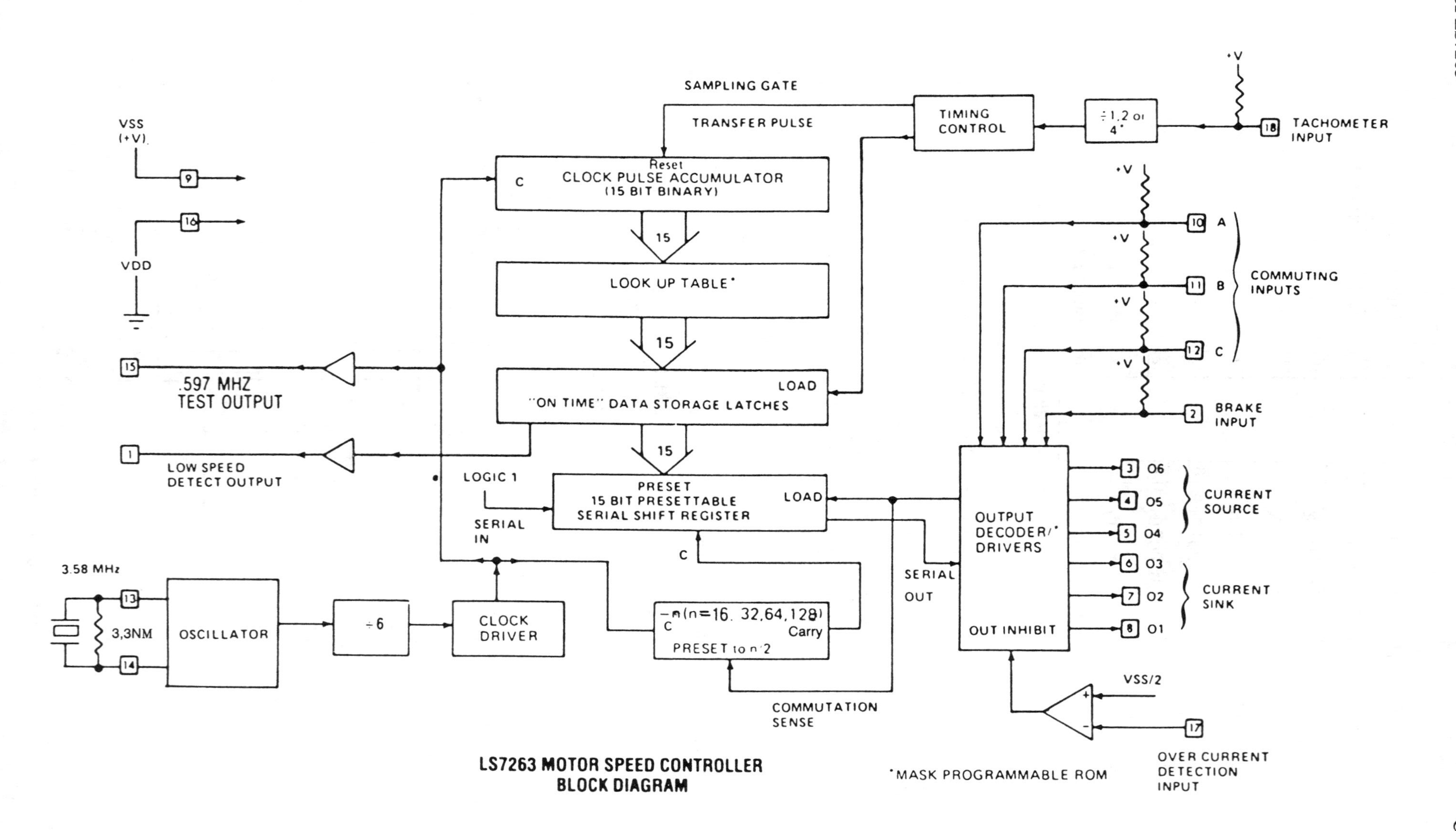

LS7263 MOTOR SPEED CONTROLLER BLOCK DIAGRAM

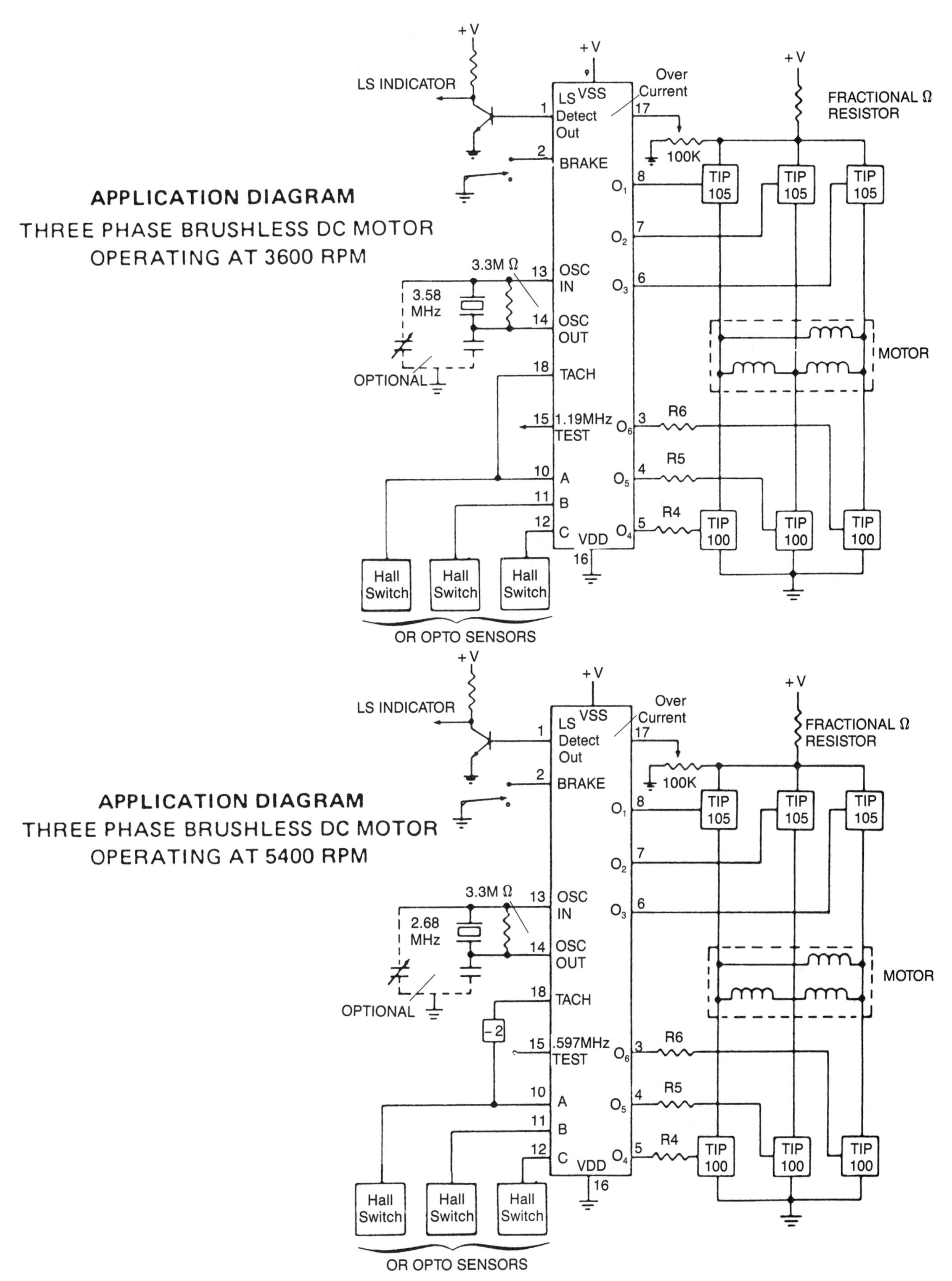

Tach input using ÷2 with output data rate doubled to achieve 5400 RPM operation.

LSI
LS7264
Four-Phase Brushless dc Motor-Speed Controller

- Highly accurate speed regulation (±.1%) derived from XTL controlled time base
- Rapid acceleration to speed with little overshoot
- Static braking
- 10V to 28V supply range
- Low-speed detection output
- Internal over current logic
- Power-on reset
- Four outputs drive power switching transistors directly
- 16-pin dual-in-line package

The LS7264 is a monolithic, ion-implanted MOS circuit designed to control the speed of a 4-phase, brushless, dc motor. This specific circuit is programmed for use in 3600-RPM applications. The circuit utilizes a 2.4576-MHz crystal to provide its accurate speed regulation time base. Overcurrent circuitry is provided to protect the windings, associated drivers, and the power supply. A static-braking feature is provided to effect rapid deceleration.

Speed corrections are made by measuring the time between tachometer inputs and varying the on-time of the drive signal applied to each winding. A sampling window is generated during which crystal-derived clock pulses are accumulated. The contents of the accumulator provide the address of a look-up table that has been derived from the physical characteristics of the motor and the load. The look-up table output determines the amount of on-time for each coil. Positive signals are applied sequentially to each winding driver through the output decoder/drive section.

A static-type, braking system shorts all winding together upon receipt of the brake input. This system creates an electrical load on the motor, thus causing rapid deceleration. An overcurrent condition, when sensed at the overcurrent detection input, disables all four winding outputs. Outputs will be reenabled upon removal of the overcurrent condition.

MAXIMUM RATINGS

PARAMETER	SYMBOL	VALUE	UNITS
Storage Temperature	T_{stg}	−65 to +150	°C
Operating Temperature			
1. Plastic	T_{ap}	−25 to +70	°C
2. Ceramic	T_{ac}	−55 to +125	°C
Voltage (any pin to V_{SS})	V_{max}	−30 to +0.5	VOLTS

LSI LS7264

Pin	Function	Pin	Function
1	LS DETECT OUTPUT	16	TACHOMETER INPUT
2	BRAKE INPUT	15	$\overline{\text{OVERCURRENT DETECT}}$
3	OUTPUT 0_1	14	VDD (GND)
4	OUTPUT 0_2	13	FREQUENCY TEST POINT
5	OUTPUT 0_3	12	OSC OUT
6	OUTPUT 0_4	11	OSC IN
7	VSS (+10 to +28 VDC)	10	$\overline{\text{FORWARD}}$/REVERSE
8	S1	9	INPUT S2

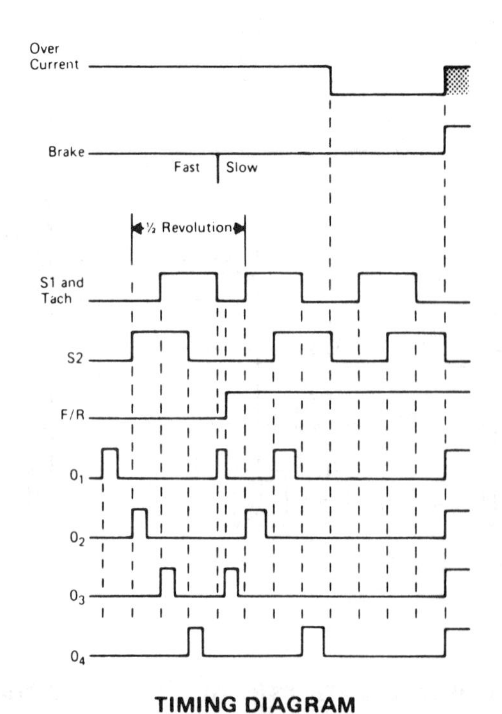

TIMING DIAGRAM
FOR LS7264-01

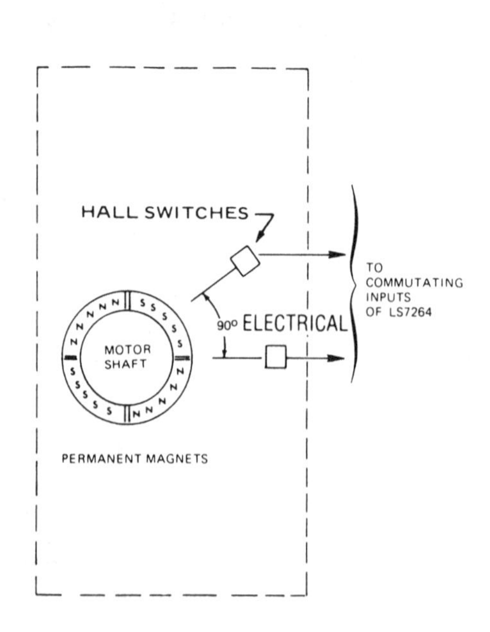

HALL SWITCH POSITIONING DIAGRAM
FOR LS7264-01

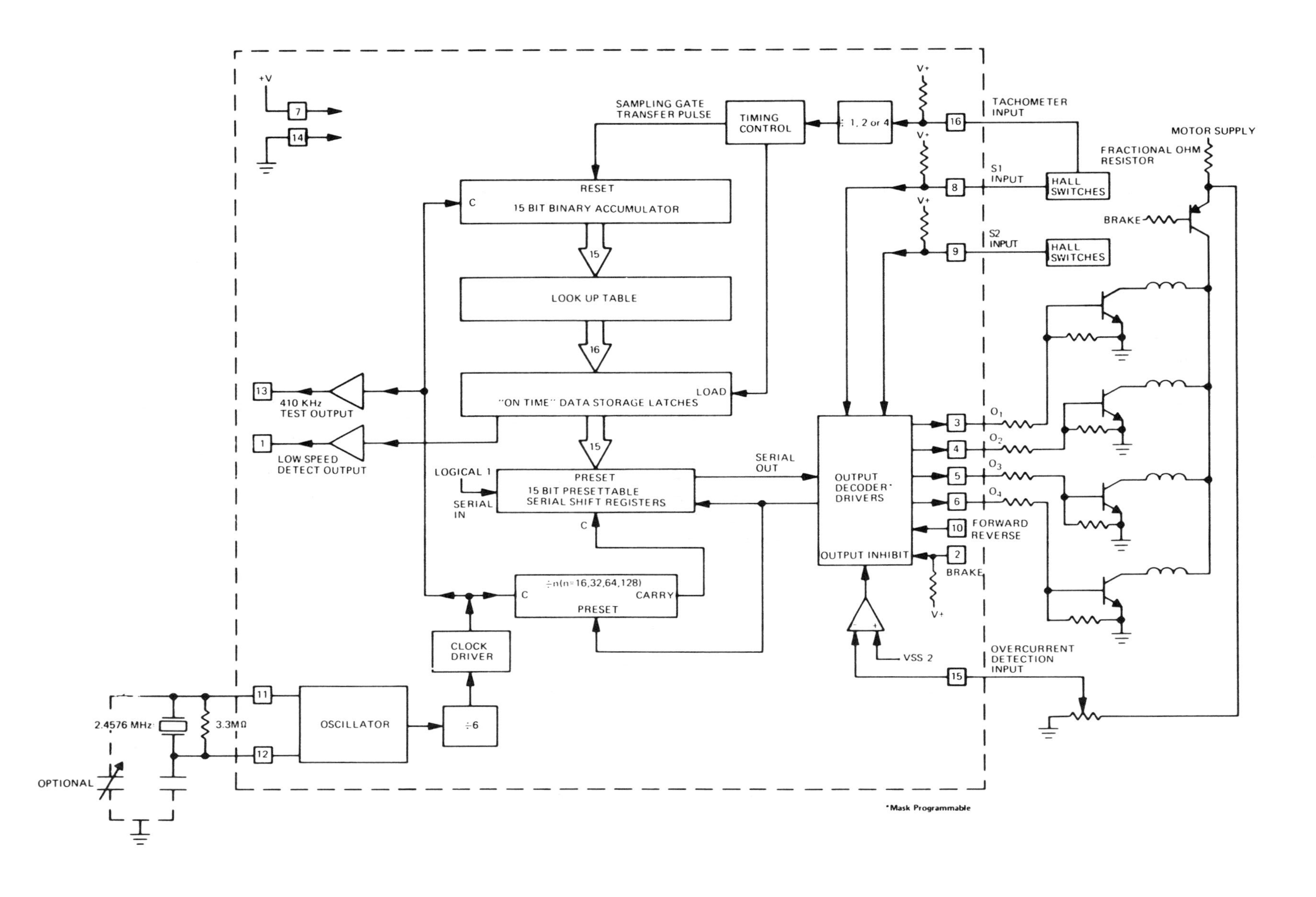

LS7264 MOTOR SPEED CONTROLLER

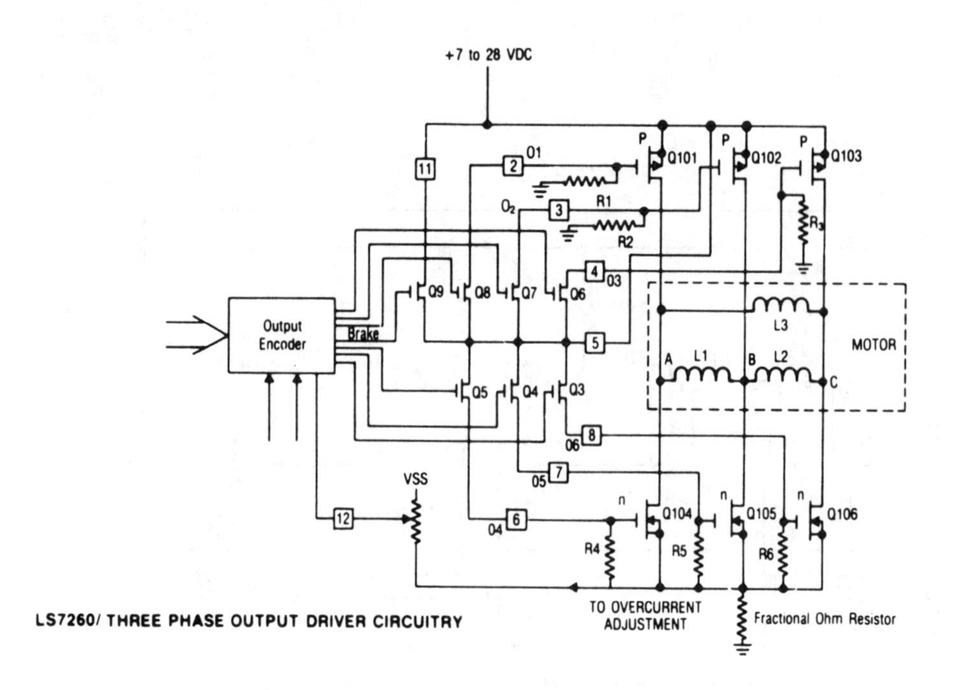

LS7260/ THREE PHASE OUTPUT DRIVER CIRCUITRY

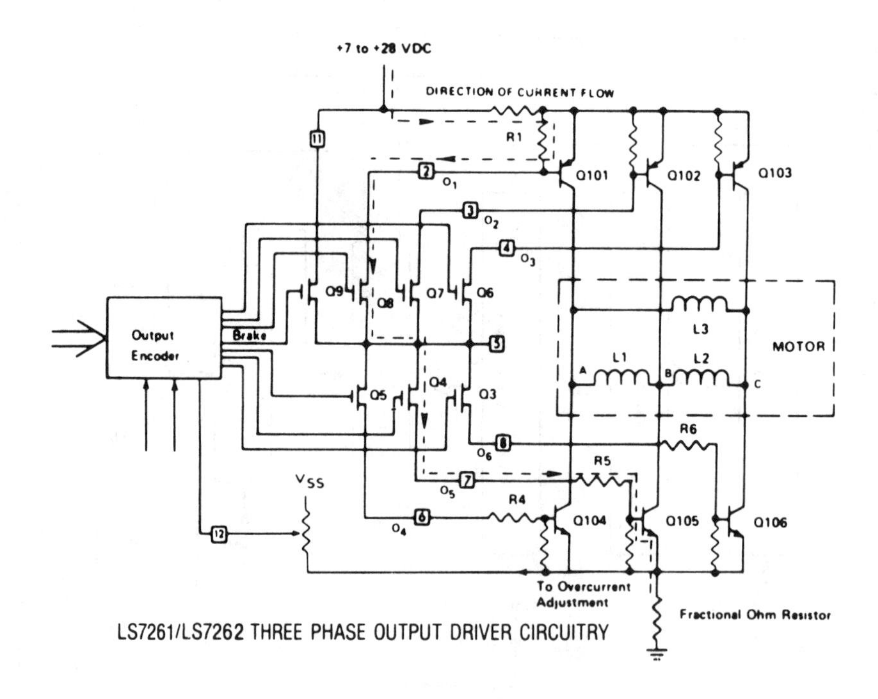

LS7261/LS7262 THREE PHASE OUTPUT DRIVER CIRCUITRY

OUTPUT COMMUTATION SEQUENCE
FOUR PHASE OPERATION
CS1=CS2=0
OUTPUTS ENABLED

S1	S2, S3	FORWARD/$\overline{\text{REVERSE}}$ = 1	FORWARD/$\overline{\text{REVERSE}}$ = 0
0	0	O_1	O_4
1	0	O_3	O_6
1	1	O_4	O_1
0	1	O_6	O_3

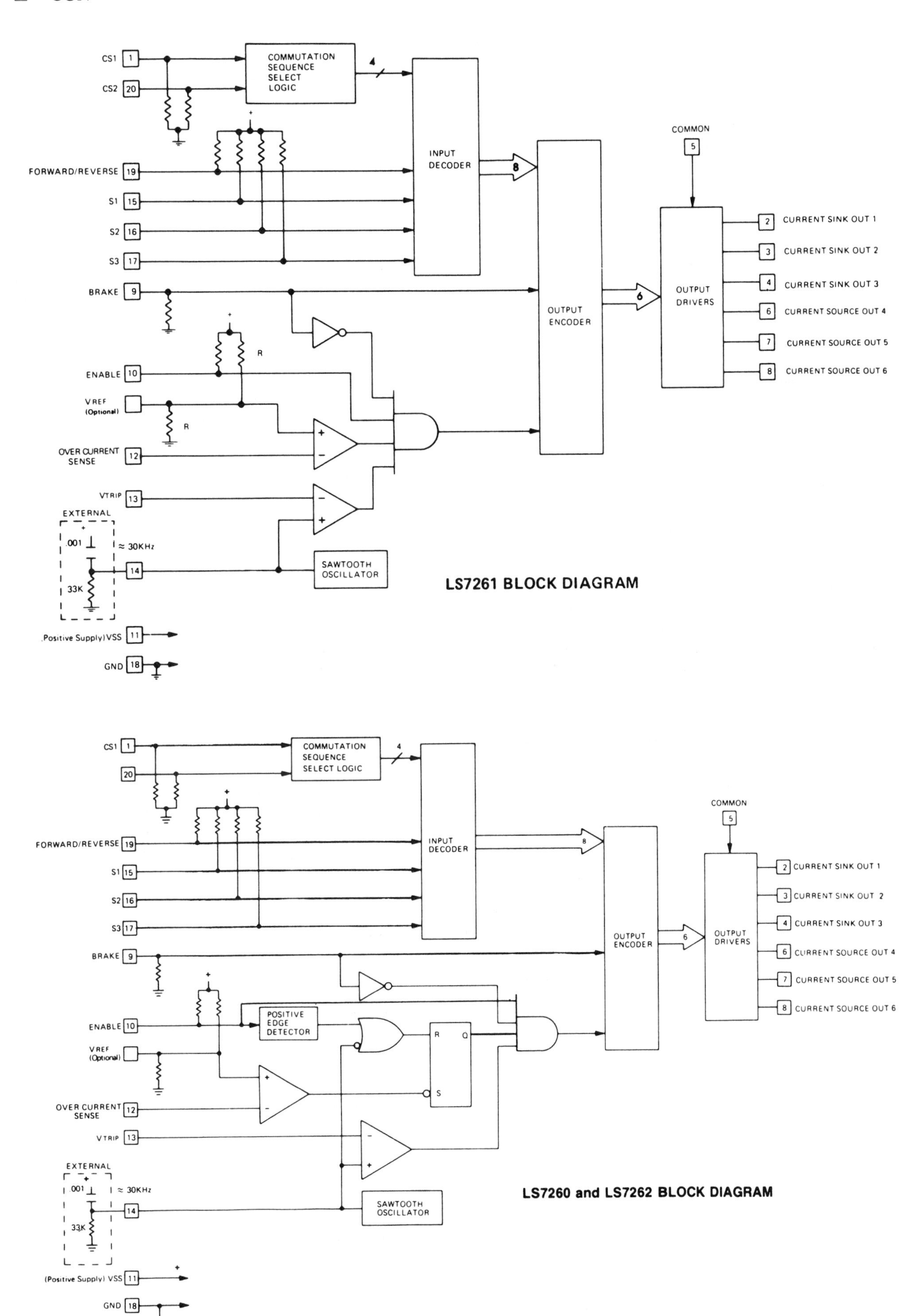

LS7261 BLOCK DIAGRAM

LS7260 and LS7262 BLOCK DIAGRAM

LSI
LS7331
Touch-Sensitive Light Dimmer and ac Motor-Speed Controller with Computer Control and Monitoring

- Provides speed control of ac motors and brightness control of incandescent lamps without the use of mechanical switches
- Control brightness/motor speed by controlling the ac "duty cycle," hence reducing the power dissipation
- Controls the "duty cycle" from 23% to 88% (on time angles for ac half cycles between 41° and 159° respectively
- Allows computer control of lamp or motor operation
- Provides outputs to computer indicating when lamp is at full brightness and when it is varying in brightness
- Has an output that indicates when loss of power has occurred
- Operates on 50/60-Hz line frequency
- Input for extensions or remote sensors
- Input for slow dimming
- 12V to 17Vdc supply voltage

LS7331 is a monolithic, ion-implanted MOS circuit that is specifically designed for the control of brightness of incandescent lamps or speed of ac motors used on the ac line. The outputs of these chips control the brightness of a lamp or speed of an ac motor by controlling the firing angle of a triac connected in series with the lamp or ac motor. All internal timings are synchronized with the line frequency by means of a built-in phase-locked loop circuit. The output occurs once every half-cycle of the line frequency. Within the half-cycle, the output can be positioned anywhere between 159° phase angle for maximum brightness/speed and 41° phase angle for minimum brightness/speed, in relation to the line frequency. The positioning of the output is controlled by applying a low level at the sensor input or a high level at the slave input. Alternately, the sensor input can be applied via a microprocessor or computer. The dim and full outputs are used to indicate the present state of the lamp or motor to the computer.

These functions can be implemented with very few interface components, which are described in the application examples. When implemented in this manner, a touching of the sensor plate or a control signal from the computer causes the lamp brightness or motor speed to change as follows:

- If the sensor is touched or a control signal is applied momentarily (32ms to 322ms), the lamp or motor is:
  - turned off if it was on
  - turned on if it was off. The brightness/speed to which the light/motor is turned on is either full brightness/speed, or depending on the circuit type, a previous brightness/speed stored in the memory.
- If the sensor is touched or the control signal is applied for a prolonged time (more than 322ms), the light intensity/speed changes slowly. As long as the touch is maintained, the change continues; the direction of change reverses whenever the maximum or minimum brightness/speed is reached.

The circuit also provides an input for slow dimming. By applying a slow clock to this input, the lamp can be dimmed slowly until total turn-off occurs. This feature can be useful in children's bedroom lights.

ABSOLUTE MAXIMUM RATINGS:

PARAMETER	SYMBOL	VALUE	UNITS
DC supply voltage	VSS	+20	Volt
Any pinput voltage	V_{IN}	VSS + .5	Volt
Operating Temperature	T_A	0 to + 80	°C
Storage Temperature	Tstg	-65 to +150	°C

CONNECTION DIAGRAM — TOP VIEW
STANDARD 14 PIN PLASTIC DIP

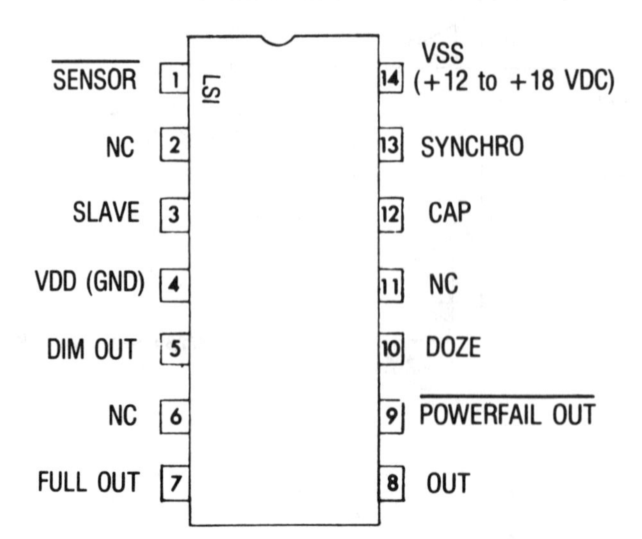

TYPE	SENSOR (TOUCH) DURATION			
	MOMENTARY (32ms to 332ms) (note 1)		PROLONGED (More than 332ms) (note 1)	
	PRE-TOUCH BRIGHTNESS	POST-TOUCH BRIGHTNESS	PRE-TOUCH BRIGHTNESS	POST-TOUCH BRIGHTNESS
LS 7331	Off	Max	Off	Starts varying at Min
	Max	Off	Max	Starts varying at Max
	Intermediate	Off	Intermediate	Starts varying at Pre-Touch brightness in same direction as previous prolonged touch

NOTE 1. The time figure is based on 60Hz synchro frequency. For 50Hz the figures are 39ms and 399ms.

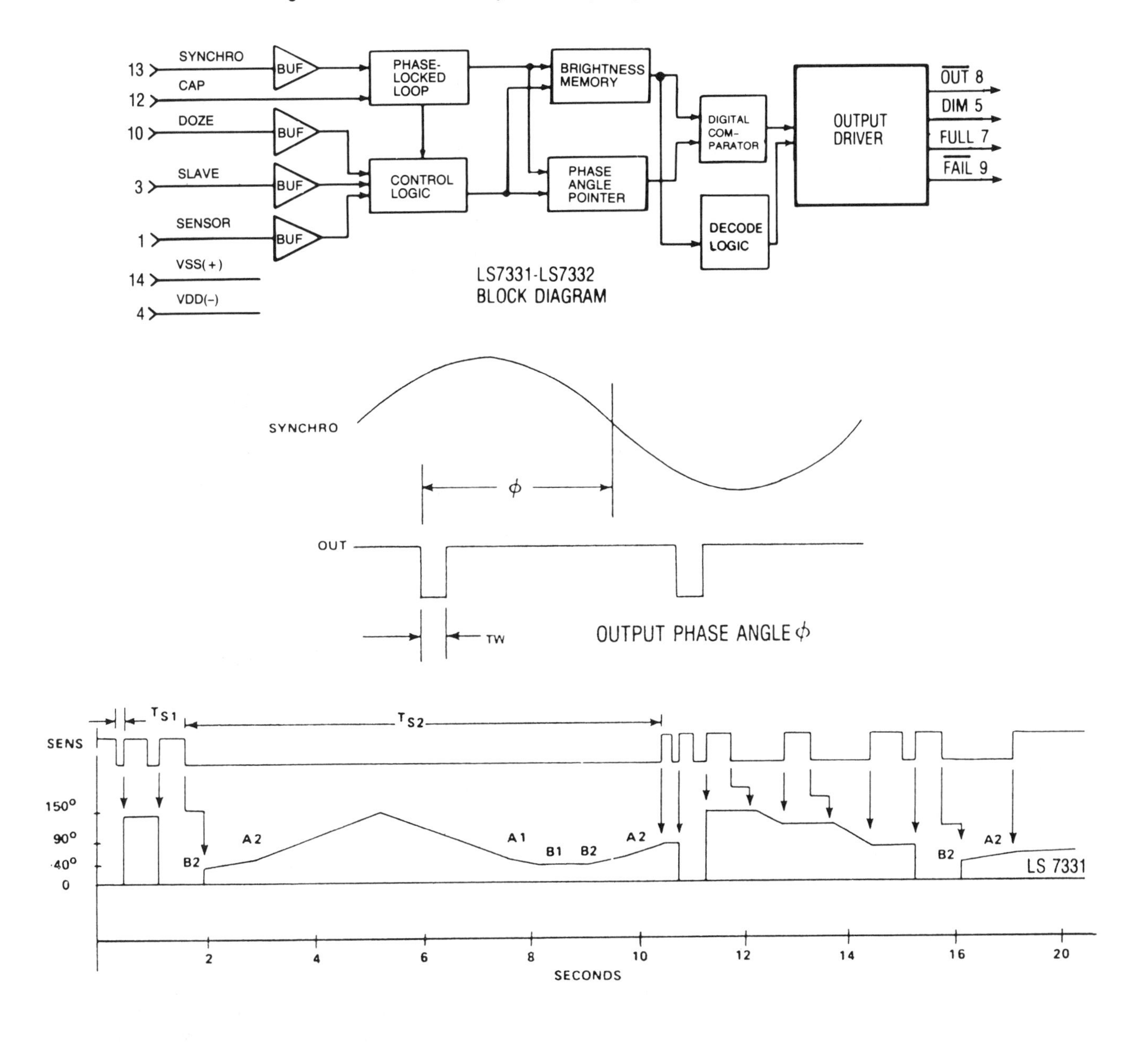

OUTPUT PHASE ANGLE, ø Vs, SENSOR OUTPUT

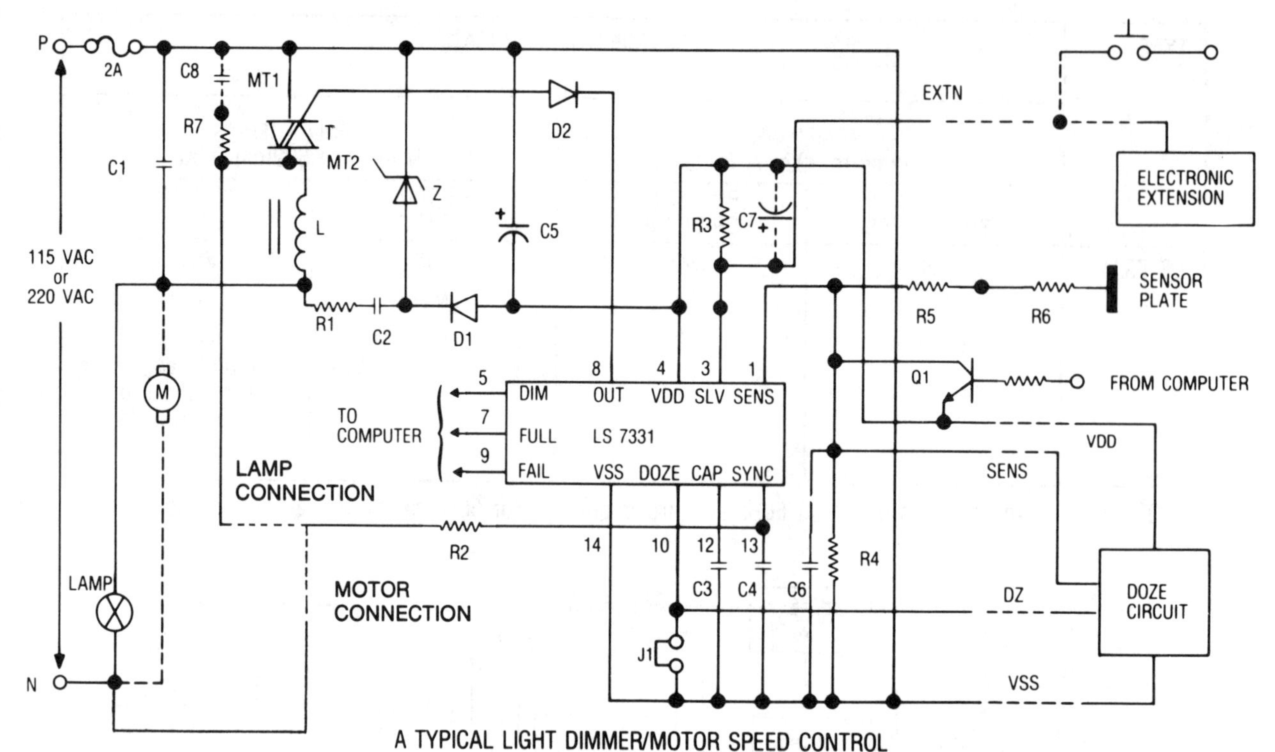

A TYPICAL LIGHT DIMMER/MOTOR SPEED CONTROL

Notes:
1) All circuits connected by broken lines are optional
2) C7 is used only when electronic extension is connected
3) Jumper between Pin 10 & VSS should be broken when Doze circuit is used.
4) Network C8-R7 is needed for inductive loads (such as motors) only.

115 VAC

C1 = 0.15μF/250VAC
C2 = 0.22μF/250VAC
C3 = 0.47μF/16V
C4 = 470pF/600V
C5 = 47μF/25V
C6 = 680pF/50V
C7 = 0.2μF/25V
C8 = 0.047μF/250V (see note #4)
R1 = 270Ω/2W
R2 = 1.5MΩ/¼W
R3 = 680KΩ/¼W
R4 = 1MΩ to 5MΩ/¼W (select for sensitivity)
R5, R6 = 2.7MΩ/¼W
R7 = 1.8KΩ/2W (see note -4)
Q1 = 2N2222 or equivalent
D1, D2 = 1N4148
Z = 15V/1W (Zener)
T = T2500D or Q4004L4 TRIAC (Typical)
L = 100μH (RFI Filter)

220 VAC

C1 = 0.15μF/250VAC
C2 = 0.22μF/250VAC
C3 = 0.47μF/16V
C4 = 470pF/600V
C5 = 47μF/25V
C6 = 680pF/50V
C7 = 0.2μF/25V
C8 = 0.047μF/250V (see note #4)
R1 = 270Ω/2W
R2 = 1.5MΩ/¼W
R3 = 680KΩ/¼W
R4 = 10KΩ¼W
R4 = 1MΩ to 5MΩ/¼W (select for sensitivity)
R5, R6 = 4.7MΩ/¼W
R7 = 1.8KΩ/2W (see note #4)
Q1 = 2N2222 or equivalent
D1, D2 = 1N4148
Z = 15V/1W (Zener)
T = Q5004L4 TRIAC (Typical)
L = 200μH (RFI Filter)

Rohm BA6208
Reversible Motor Driver

- Internal power transistors for motor driving (drive current 100mA typ.)
- Internal braking logic (which activates if both inputs A and B are set high)
- Internal surge suppressor
- Minimal standby current required when both control inputs A and B are set low
- Wide supply voltage range: 4.5 to 15.0V
- TTL-compatible inputs

The BA6208 is a monolithic motor driver with forward/reverse control capability. The device can provide (at the TTL logic level) auto-reverse control for cassette decks and radio cassette recorders, auto-return control for record player's tonearms, and many other reversible motor control applications. The BA6208 contains a forward/reverse control logic, forced-brake control logic, and forward/reverse drive output stage that has an output current capacity of I_0 = 200mA.

ABSOLUTE MAXIMUM RATINGS (Ta=25°C)

Parameter	Symbol	Limits	Unit
Supply voltage	V_{CC}	18	V
Power dissipation	Pd	700*	mW
Operating temperature range	Topr	−20 ~ 60	°C
Storage temperature range	Tstg	−55 ~ 125	°C
Maximum output current	I_{OUT}	500	mA

*Derating is done at 7mW/°C for operation above Ta=25°C.

ELECTRICAL CHARACTERISTICS (Ta=25°C, V_{CC}=9V)

Parameter	Symbol	Min.	Typ.	Max.	Unit	Conditions
Supply voltage range	V_{CC}	4.5	—	15	V	
Output current	I_O	200	—	—	mA	
Output saturation voltage	V_{CE}	—	—	1.6	V	I_O=100mA
Input high voltage	V_{IH}	2.0	—	—	V	
Input low voltage	V_{IL}	—	—	0.8	V	
Standby current	I_{ST}	—	—	0.4	mA	When both A and B inputs are low
Input high current	I_{IH}	—	—	400	μA	V_{IH}=4.5V

*Contains a diode which absorbs a minimum of 500 mA in response to external surge currents to the output terminal. A surge current is considered to be less than 10 ms and 1/10 duty.

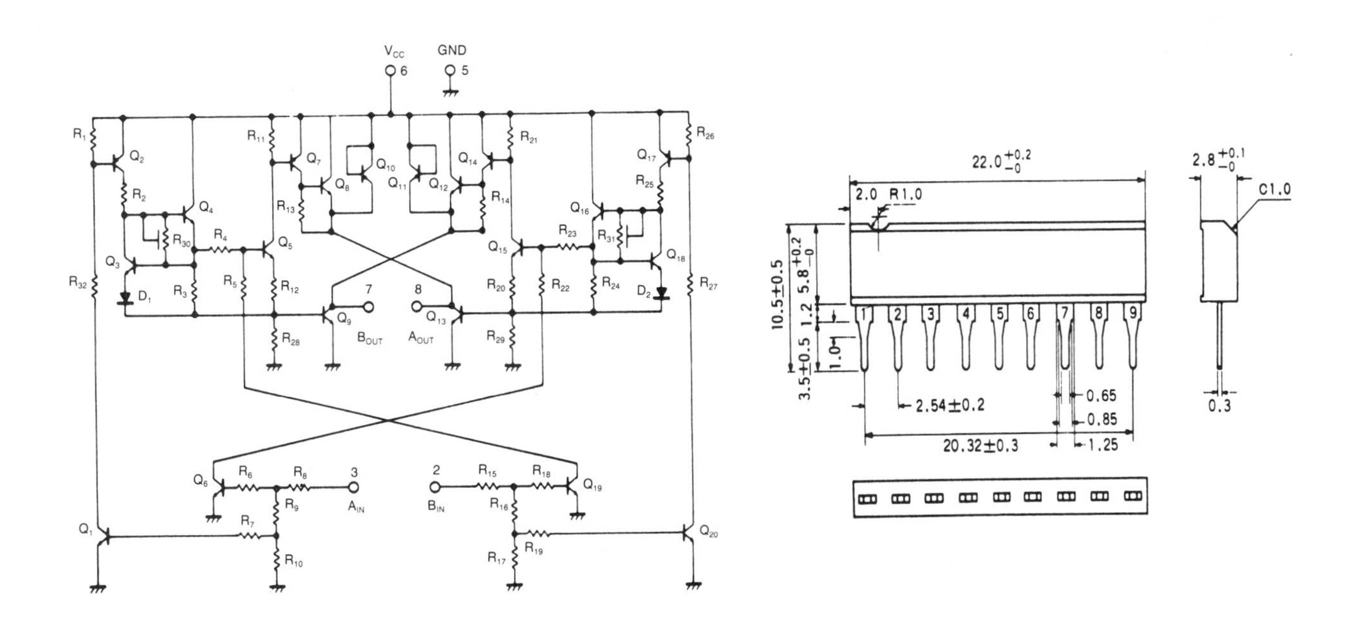

TEST CIRCUIT

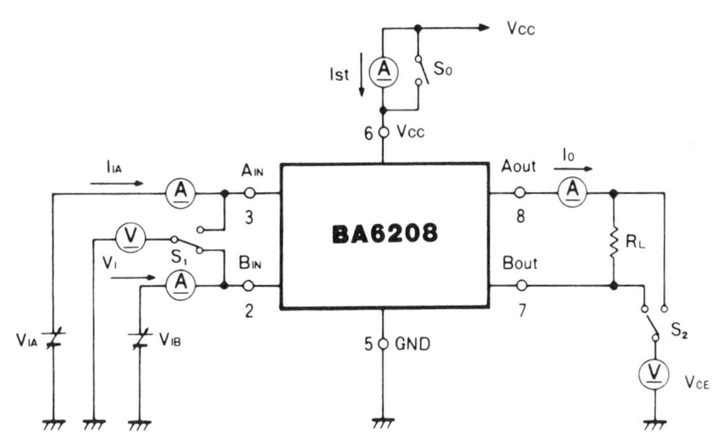

APPLICATION EXAMPLE

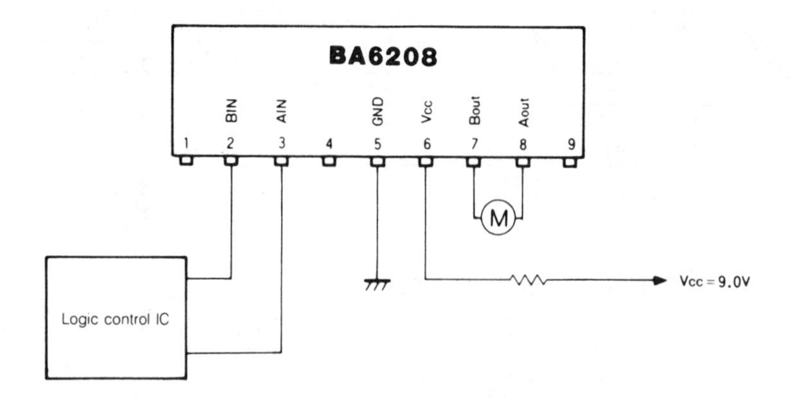

Rohm BA6218
Reversible Motor Driver

- Internal surge suppressor
- Small standby circuit current
- Wide supply voltage range (4.5 to 15.0V)
- TTL-compatible inputs

The BA6218 is a monolithic reversible motor driver with an output load current of 0.7A (max). The device has four output modes (forward, reverse, stop (idle), braking) selectable from two control inputs.

With separate ground lines for logic and power sections, the BA6218 allows the user to realize a reversible, variable-speed motor drive circuit just by adding an electronic governor to the output.

ABSOLUTE MAXIMUM RATINGS (Ta=25°C)

Parameter	Symbol	Limits	Unit
Supply voltage	V_{CC}	18	V
Power dissipation	Pd	800*	mW
Operating temperature range	Topr	−20 ~ 60	°C
Storage temperature range	Tstg	−55 ~ 125	°C
Maximum output current	I_O	0.7	A

*Derating is done at 8mW°/C for operation above Ta=25°C.

INPUT TRUTH TABLE

Pin 3(IN)	Pin 1(IN)	Pin 7(OUT)	Pin 9(OUT)
H	L	H	L
L	H	L	H
H	H	L	L
L	L	OPEN	OPEN

Note: High input is 2.0V (min.)
Low input is 0.8V (max.)

ELECTRICAL CHARACTERISTICS (Unless otherwise specified, Ta=25°C, V_{CC}=9V)

Parameter	Symbol	Typ.	Unit	Conditions
Output current	I_O	0.2	A	Minimum
Output saturation voltage	V_{CE}	1.6	V	I_O=0.2A (maximum)
Input high voltage	V_{IH}	2.0	V	Minimum
Input low vortage	V_{IL}	0.8	V	Maximum
Standby current	Ist	1	μA	When both A and B inputs are low
Input high current	I_{IH}	93	μA	V_{IN}=2.0V

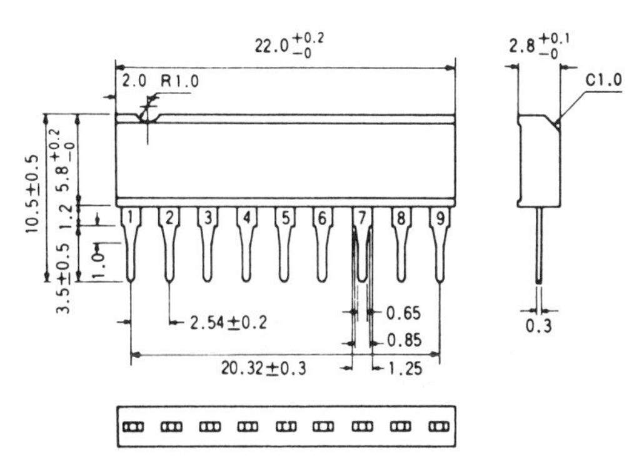

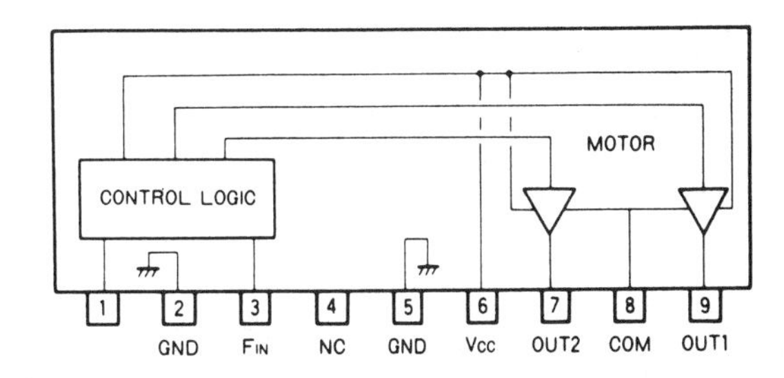

Rohm
BA6109
Reversible Motor Driver

- Internal motor-driving transistors with maximum rush current capacity of 800mA
- Internal braking feature
- Internal surge suppressor
- Inputs can directly interface with MOS LSI chips
- Minimum external component requirement
- Wide supply voltage range: 6 to 18V
- Available in 10-pin SIP for easy assembly

The BA6109 is a monolithic reversible motor driver with three output modes (forward, reverse, stop), selectable from two control inputs.

During the transition from forward or reverse mode into stop mode, the device absorbs counter electromotive force from the motor and brakes the motor. It also has an internal surge suppressor, which absorbs motor rush current that occurs during switching from one mode to another.

The output voltage is set by an external zener diode (connected across pin 4 and GND). The output transistors can tolerate a rush current of up to 800mA. This means that the device can directly drive various motors from 6V, 9V, and 12V driving voltages.

With an input sensitivity current of 50μA or less, the BA6109 can directly interface with CMOS and other logic devices.

ABSOLUTE MAXIMUM RATINGS (Ta=25°C)

Parameter	Symbol	Limits	Unit
Supply voltage	V_{CC}	18	V
Power dissipation	Pd	2200*1	mW
Operating temperature range	Topr	−25 ~ 75	°C
Storage temperature range	Tstg	−55 ~ 125	°C
Output current	I_{OUT}	800*2	mA
Input voltage range	V_{IN}	−0.3 ~ V_{CC}	V

*1 For details, refer to the derating curves.
*2 500μs pulse with a duty cycle of 1/100.

ELECTRICAL CHARACTERISTICS (Ta=25°C, V_{CC1}=12V)

Parameter	Symbol	Min.	Typ.	Max.	Unit	Conditions
Supply voltage range 1	$V_{CC}1$	6.0	—	18.0	V	—
Supply voltage range 2	$V_{CC}2$	—	—	18.0	V	—
Quiescent current	I_O	—	15.0	30.0	mA	Pins 5, 6:GND, $R_L=\infty$
Minimum input on-state current	I_{IN}	—	10.0	50.0	μA	$R_L=\infty$
Input threshold voltage	V_{INT}	0.7	—	2.0	V	$R_L=\infty$
Output leakage current	I_{OL}	—	—	1.0	mA	Pins 5, 6:GND, $R_L=\infty$
Output voltage	V_O	5.2	5.8	—	V	R_L=60Ω, ZD=7.4V

FIN	RIN	Vout1	Vout2
1	1	L	L
0	1	L	H
1	0	H	L
0	0	L	L

Input level 1 is 2.0V (min.)
Input level 0 is 0.7V (max.)

TEST CIRCUIT

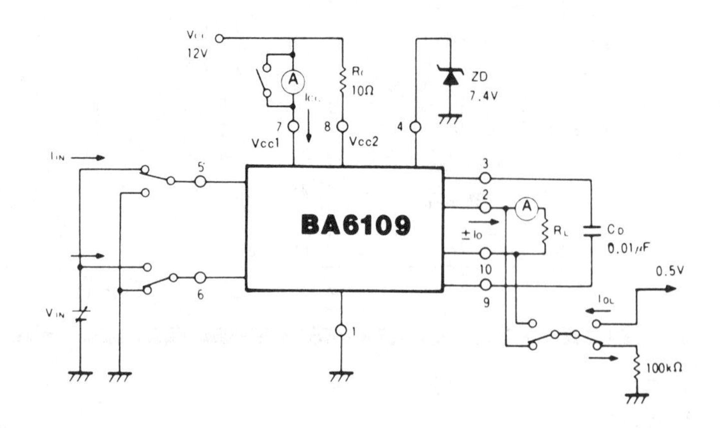

BLOCK DIAGRAM

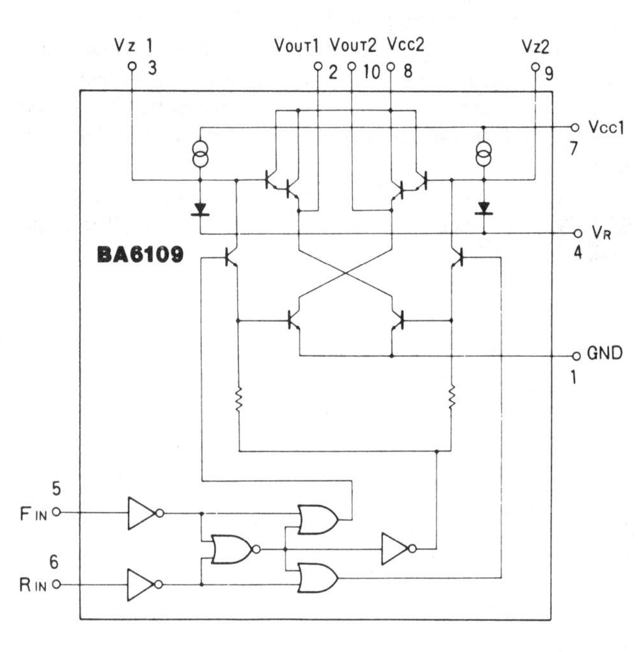

APPLICATION EXAMPLE

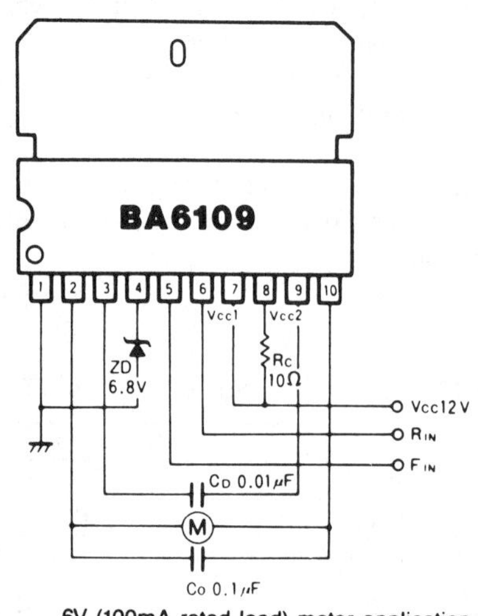

6V (100mA rated load) motor application

Rohm BA6209

Reversible Motor Driver

- Internal power transistors with large rush current capacity (1.6A max)
- Internal braking circuit for forced stop assures stable braking characteristic
- Internal surge suppressor that absorbs motor rush current
- Motor speed control input
- Small standby current (I_Q = 5.5mA typ. at V_{CC} = 12V)
- Stable reversing operation in either forward or reverse mode
- CMOS-compatible inputs

APPLICATIONS

- VCR tape loading motor drivers
- VCR cassette loading motor drivers
- VCR reel motor drivers
- VCR capstan motor drivers
- Other reversible motor driving purposes

The BA6209 is a reversible motor driver capable of directly driving a reversible motor, such as a VCR loading motor, capstan motor, and reel motor. With an internal surge suppressor, the device can tolerate a rush current of up to 1.6A, which can occur when the motor is changing operation direction or stopping.

The BA6209 also has an internal forced brake feature, which brakes the motor when the control inputs are both set high or low. Because the motor voltage can be controlled with the control pin (V_R), the BA6209 is suitable for driving reversible motors with a wide speed variation range.

ABSOLUTE MAXIMUM RATINGS (Ta=25°C)

Parameter	Symbol	Limits	Unit
Supply voltage	V_{CC}	18	V
Power dissipation	Pd	2200*	mW
Operating temperature range	Topr	−25 ~ 75	°C
Storage temperature range	Tstg	−55 ~ 125	°C
Output current	I_{OUT}	1600*	mA
Input voltage range	V_{IN}	−0.3 ~ V_{CC}	V

* 500μs pulse with a duty cycle of 1/100.

ELECTRICAL CHARACTERISTICS (Unless otherwise specified, Ta=25°C, V_{CC}=12V)

Parameter	Symbol	Min.	Typ.	Max.	Unit	Conditions
Supply voltage range 1	V_{CC} 1	6.0	—	18	V	—
Supply voltage range 2	V_{CC} 2	—	—	18	V	—
Quiescent current	I_Q	—	5.5	10	mA	Pins 5, 6:GND, $R_L = \infty$
Minimum input on-state current	I_{IN}	—	10	50	μA	$R_L = \infty$
Input threshold voltage	V_{INTH}	0.7	1.2	2.0	V	$R_L = \infty$
Output leakage current	I_{OL}	—	—	1.0	mA	Pins 5, 6:GND, $R_L = \infty$
Output voltage	V_{OUT}	6.6	7.2	—	V	R_L =60Ω, V_{ZD}=7.4V

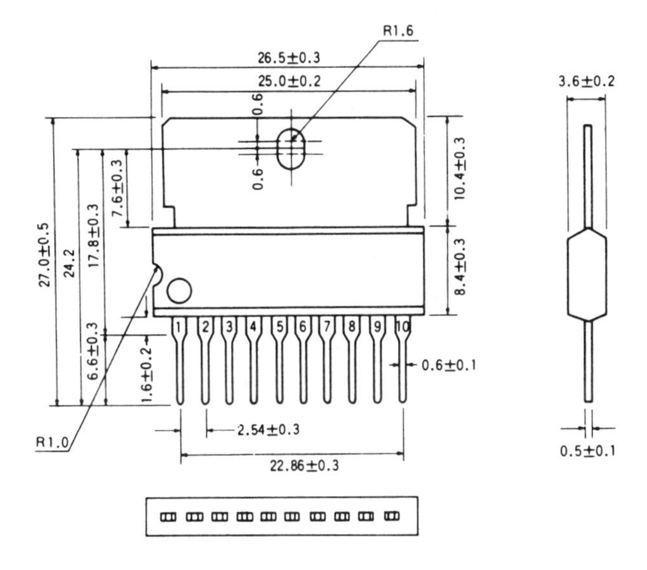

TEST CIRCUIT

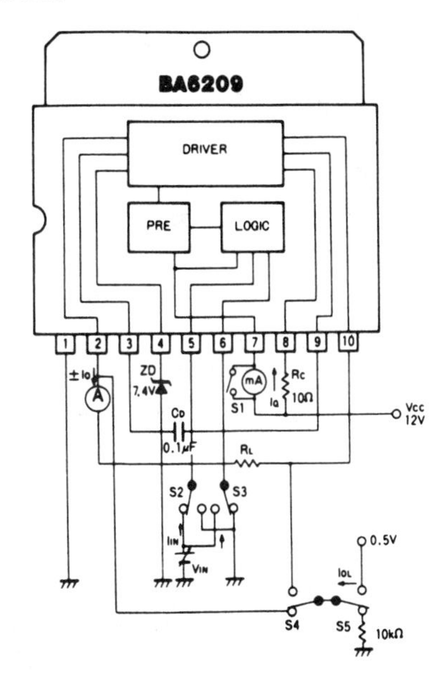

APPLICATION EXAMPLE

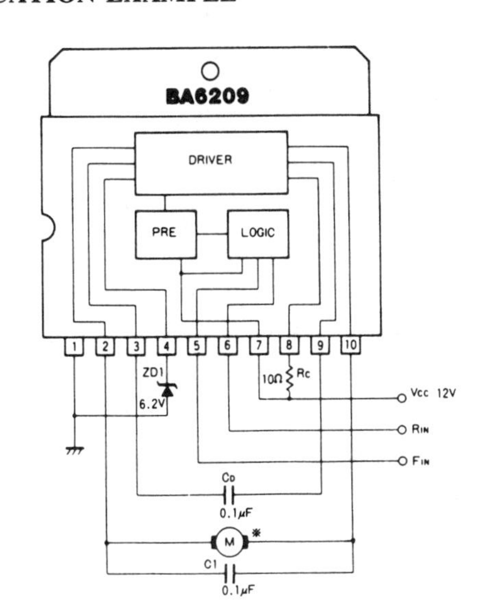

6V (100mA rated load) motor application

BLOCK DIAGRAM

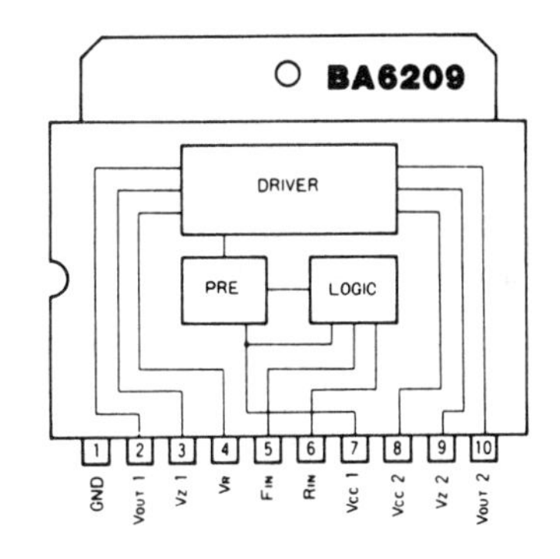

ELECTRICAL CHARACTERISTIC CURVES

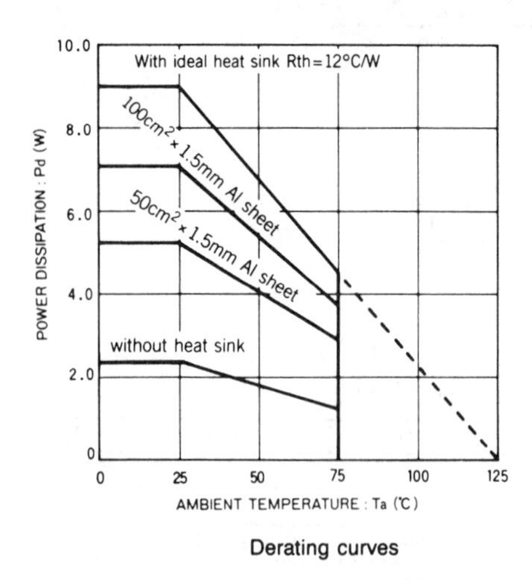

Derating curves

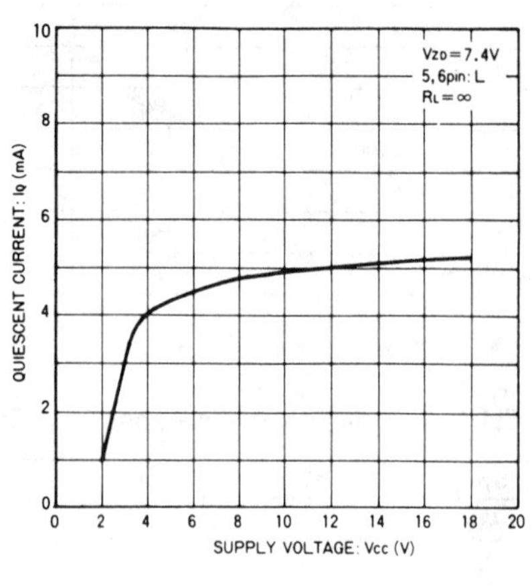

Quiescent current vs. supply voltage

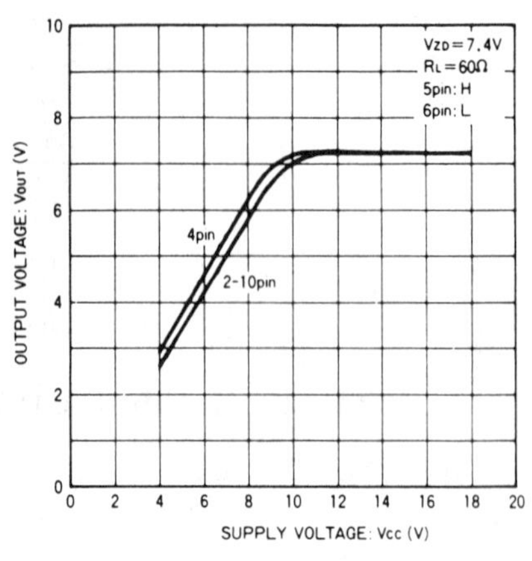

Maximum output voltage vs. supply voltage

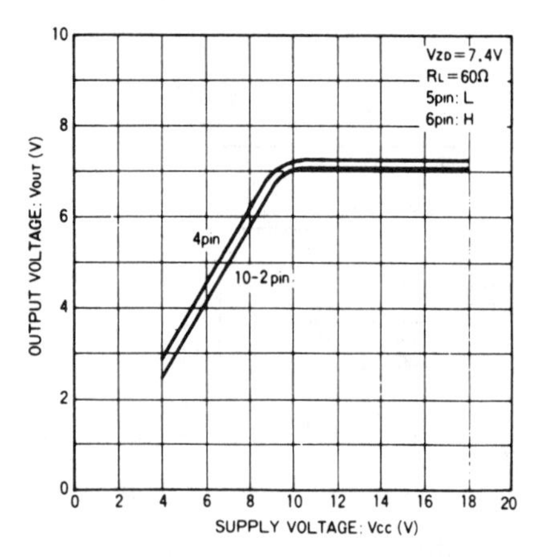

Maximum output voltage vs. supply voltage

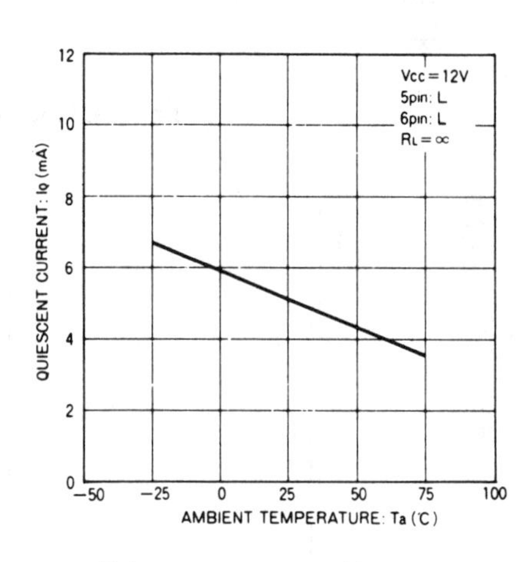

Quiescent current vs. ambient temperature

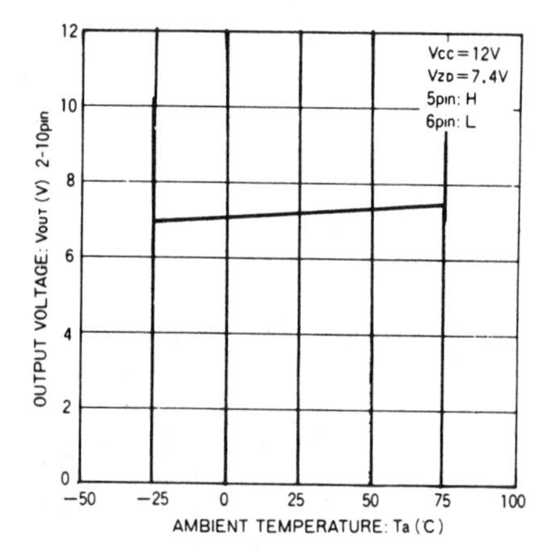

Output voltage vs. ambient temperature

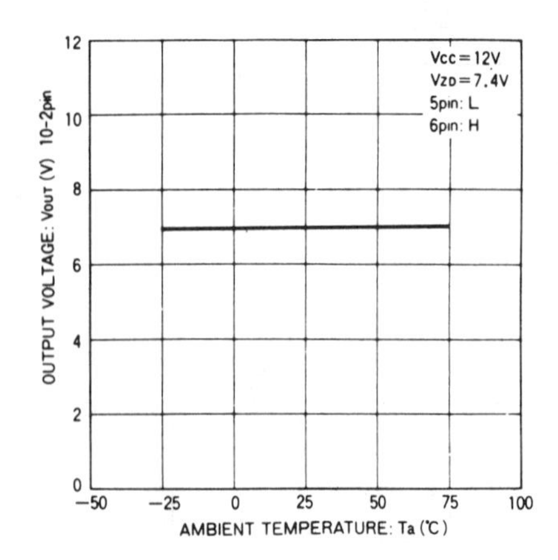

Output voltage vs. ambient temperature

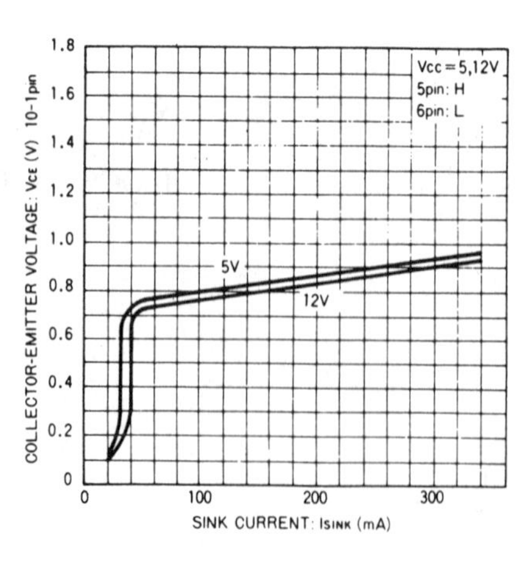

Output saturation voltage vs. sink current

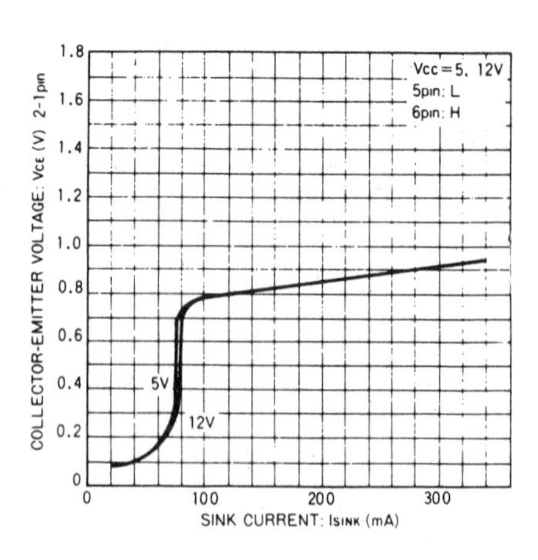

Output saturation voltage vs. sink current

ELECTRICAL CHARACTERISTIC CURVES

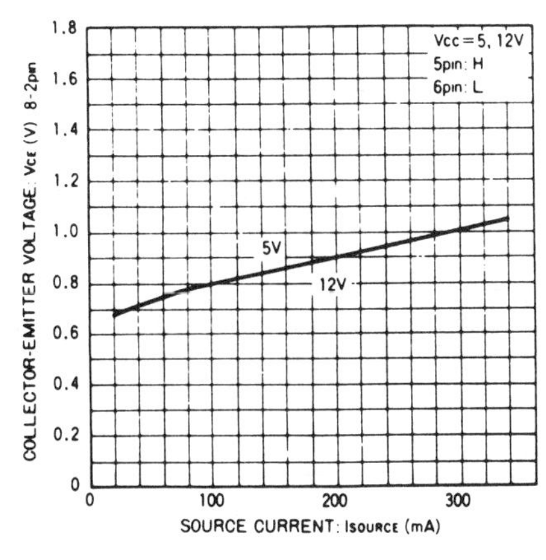

Output saturation voltage vs. source current

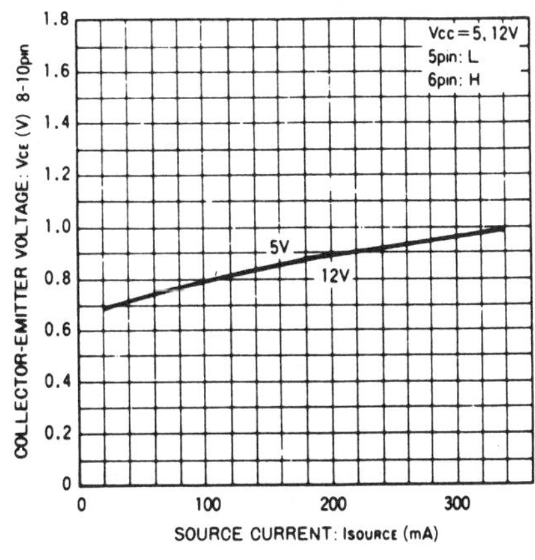

Output saturation voltage vs. source current

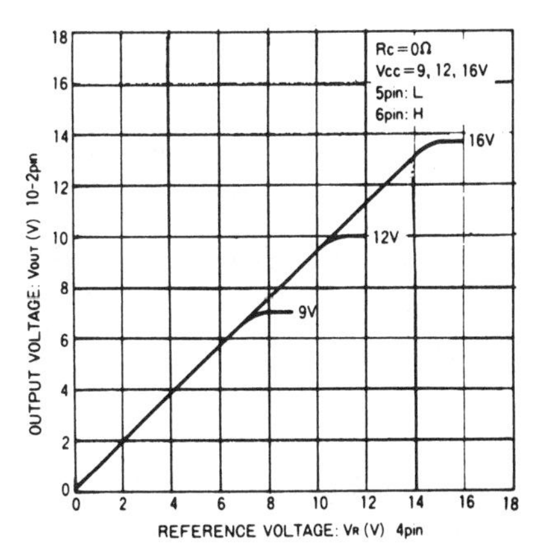

Output voltage vs. reference voltage

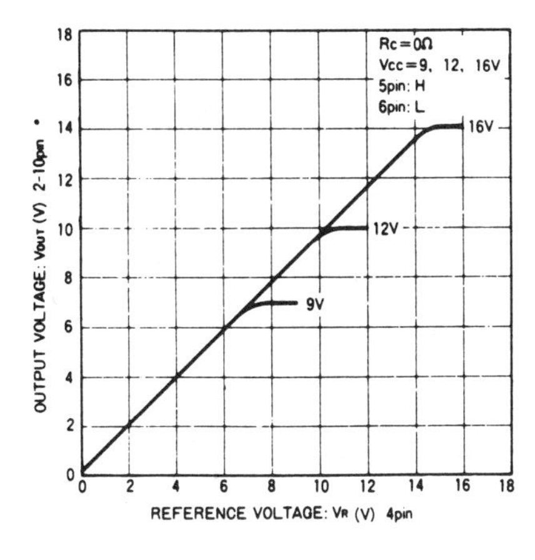

Output voltage vs. reference voltage

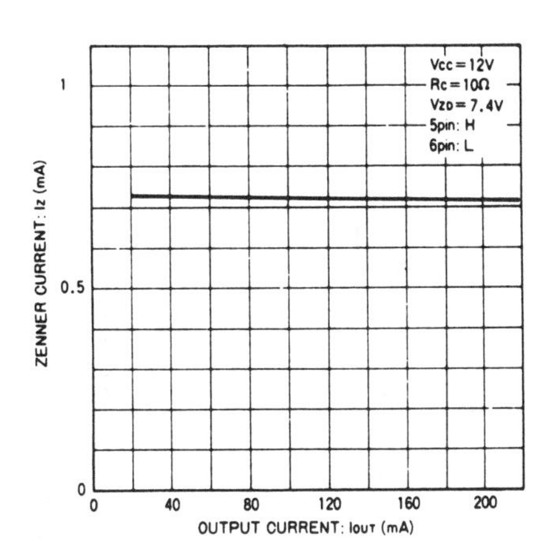

Zener current vs. output current

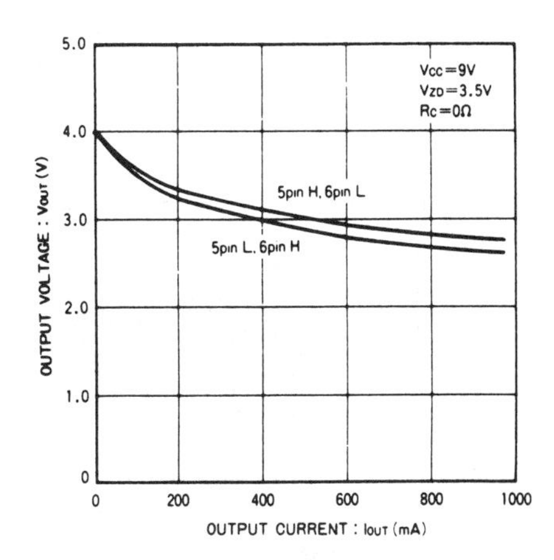

Output voltage vs. output current

Rohm BA6219
Reversible Motor Driver

- Low power design (I_{CC} = 1.2mA typ. at V_{CC} = 12V)
- Minimum external component requirement
- Four control modes available: forward, reverse, brake, and stop (idle)
- Internal motor driving power transistors (rush current: 2.2mA max)
- Internal surge suppressors
- Internal thermal shut-down circuit (which opens the output circuit if the chip temperature is raised to an abnormally high level because of motor lock-up, etc.)

APPLICATION
- Brush-type dc motor driver

The BA6219 is a reversible motor driver specifically designed for driving brush-type dc motors. The device has control pins that enable control of the rotation speed and the direction of rotation of the motor that it drives.

Continuous speed control is accomplished with an external operational amplifier, and step control of rotation speed can be achieved with zener diodes. The small power requirement of the BA6219 makes it best-suited for portable applications.

ABSOLUTE MAXIMUM RATINGS (Ta=25°C)

Parameter	Symbol	Min.	Typ.	Max.	Unit
Supply voltage	V_{CC}	—	24	—	V
Power dissipation	Pd	—	2200	—	mW
Maximum output current	I_O max	—	2.2	—	A

ELECTRICAL CHARACTERISTICS (Ta=25°C, V_{CC}=12V)

Parameter	Symbol	Min.	Typ.	Max.	Unit	Conditions
Circuit current	I_{CC}	—	1.2	—	mA	Pins 5, 6, H
Output high voltage	V_{OH}	6.5	—	—	V	Insert 6.8V zener diode between pin 4 and GND
Output low voltage	V_{OL}	—	—	1.2	V	
Input threshold voltage	Vth	—	2.0	—	V	"H"≧3V "L"≧1V

(Unit: mm)

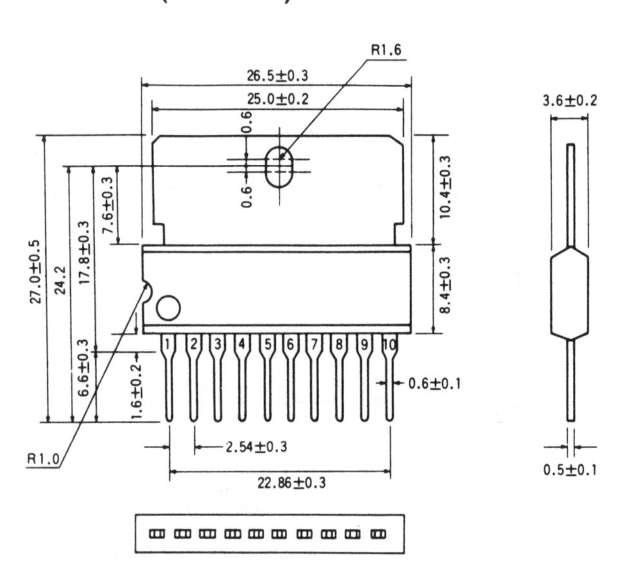

BLOCK DIAGRAM

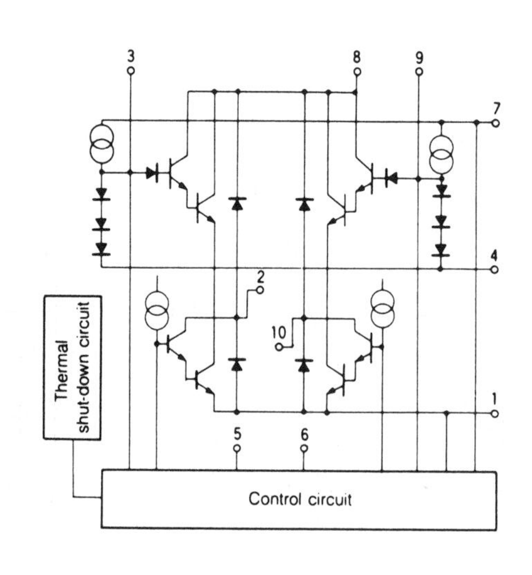

APPLICATION EXAMPLE

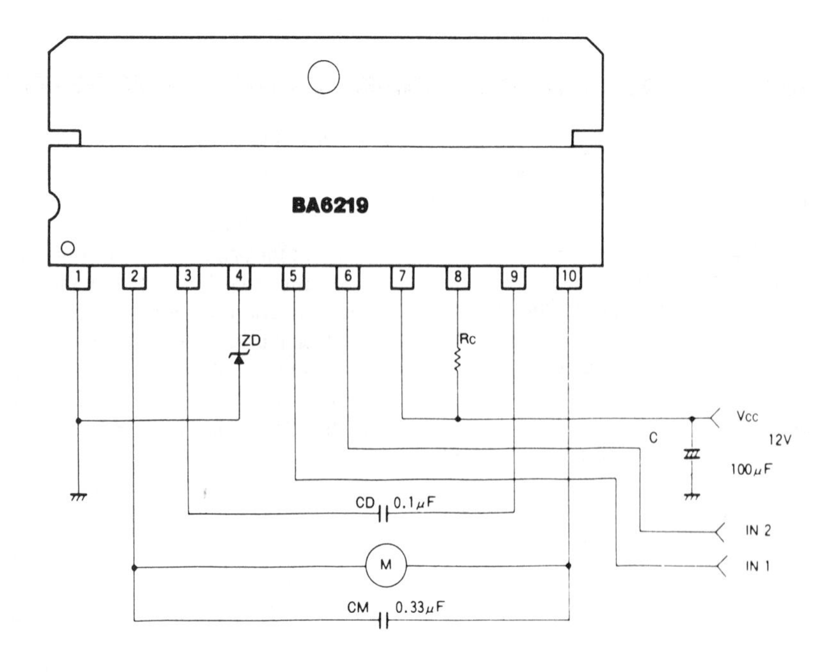

Rohm
BA2266A
Radio Servo Controller

- Internal channel divider
- Internal steering servo motor driver and driving motor servo
- Turning lights are linked to the steering
- Reverse lights are linked to the driving motor
- Turning light flashing interval can be varied by an external capacitor

APPLICATIONS

- Radio-controlled model cars
- Radio-controlled model motorcycles
- Radio-controlled model boats

The BA2266A is a monolithic radio servo controller designed for use in radio-controlled model cars.

With an internal channel divider, the device allows simultaneous control over the steering and motor speed including forward/reverse control. Because the BA2266A directly supplies control signals, energy loss is minimized and battery life is extended. The steering servo is linked to the turning lights. Also, when the driving motor is switched to the reverse position, the reverse lights are automatically lit.

Combined with a receiver chip, the BA2266A provides all the functions required for radio-controlled model cars.

ABSOLUTE MAXIMUM RATINGS (Ta=25°C)

Parameter	Symbol	Limits	Unit
Supply voltage	V_{CC}	10	V
Power dissipation	Pd	1100*	mW
Operating temperature range	Topr	−25~75	°C
Storage temperature range	Tstg	−55~125	°C

*Derating is done at 11mW/°C for operation above Ta=25°C.

ELECTRICAL CHARACTERISTICS (Ta=25°C, V_{CC}=6V)

Parameter	Symbol	Min.	Typ.	Max.	Unit	Conditions
Operating supply voltage	V_{CC}	3.8	6.0	9.5	V	—
Input threshold voltage	V_{IN}	—	0.7	1.0	V	—
Quiescent current	I_Q	—	7.0	20	mA	
Flasher drive current	I_{W1}	—	—	100	mA	—
Flasher drive current	I_{W2}	—	—	100	mA	—
Steering drive current 1	$I_{S1\cdot 2}$	—	—	370	mA	$t_1 > to_1$
Steering drive current 2	$I_{S2\cdot 1}$	—	—	370	mA	$t_1 < to_1$
Reverse lamp drive current	I_B	—	—	100	mA	$t_2 > to_2$

APPLICATION EXAMPLE

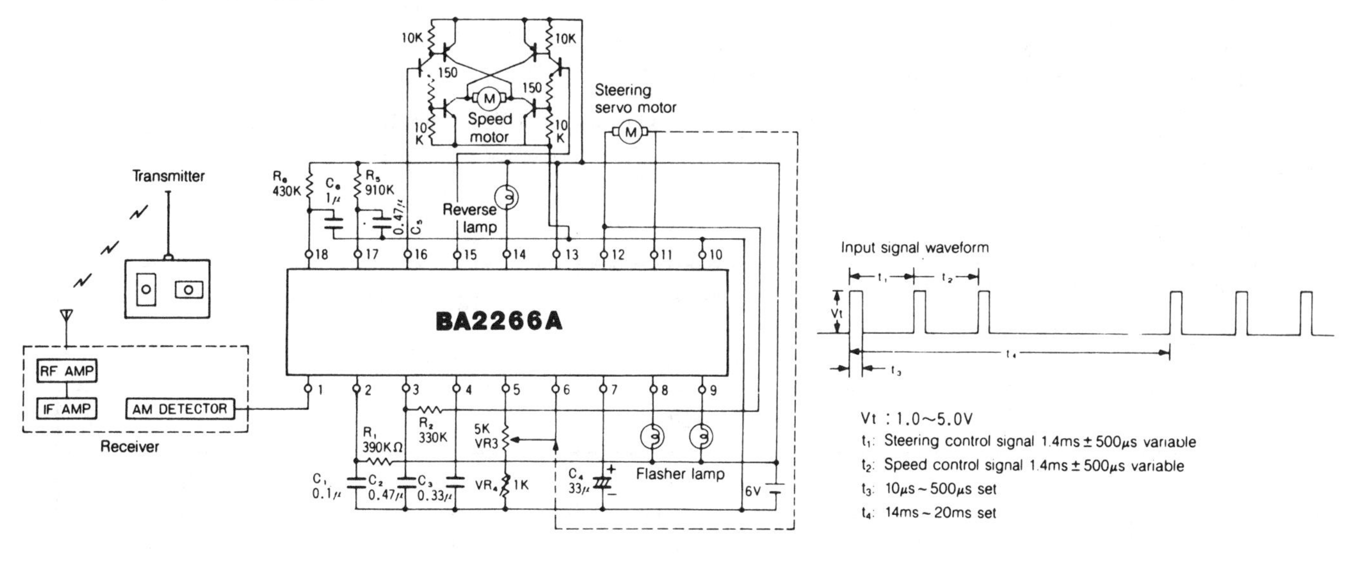

(Unit: mm)

22.9±0.3
18 17 16 15 14 13 12 11 10
R1.2
6.5±0.3
1 2 3 4 5 6 7 8 9
7.6±0.3
7.4±0.5
3.2±0.2
4.2±0.3
0.8
3.4±0.2
0.5±0.1
2.54±0.3
20.32±0.3
1.7
0.3±0.1
8.8±0.6

BLOCK DIAGRAM

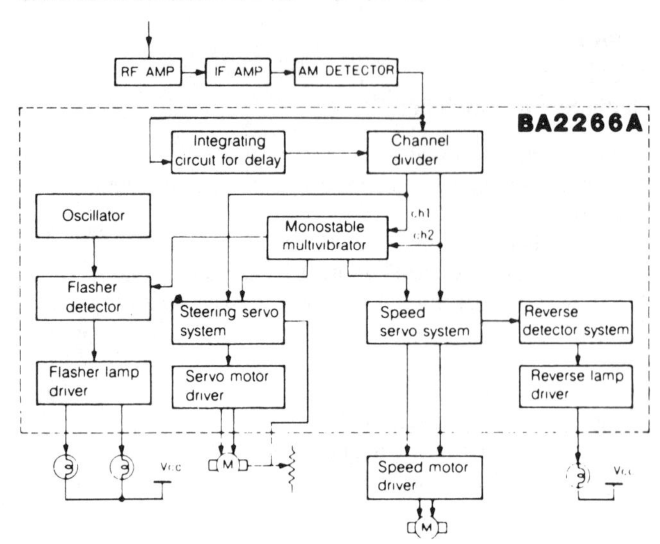

Rohm BR6116/BR6116F
2K × 8 High-Speed CMOS SRAM

- Single 5V supply and high density 24-pin package
- High speed: fast access time 70 ns/90ns/120ns (max.)
- Low-power standby standby : 5μW (typ.)
- Low-power operation operation : 250mW (typ.)
- Completely static RAM: no clock or timing strobe required
- Directly TTL compatible: all input and output
- Pin compatible with standard 16K EPROM/Mask ROM
- Equal access and cycle time

The BR6116 is a 16,384-bit static random access memory organized as 2048 words by 8 bits and operates from a single 5V supply. It is built with a high-performance CMOS process. A six-transistor full-CMOS memory cell provides low standby current and high-reliability. Inputs and three-state outputs are TTL compatible and allow for direct interfacing with common system bus structures. The BR6116 is available in both a standard 24-pin 600mil-DIP and a 24-pin SO package.

Rohm Corp. reserves the right to make changes to any product herein to improve reliability, function or design. Rohm Corp. does not assume any liability arising out of the application or use of any product described herein, neither does it convey any license under its patent right nor the rights of others.

ABSOLUTE MAXIMUM RATINGS*

Voltage on any pin relative to GND, V_T	−0.3 to +7.0V
Operating temperature, T_{opr}	0 to +70°C
Storage temperature (plastic), T_{stg}	−55 to +125°C
Temperature under bias, T_{bias}	−10 to +85°C
Power dissipation, P_T DIP 24	1.0W
SO 24	650mW

*COMMENTS

Stresses above those listed under "Absolute Maximum Ratings" could cause permanent damage to the device. These are stress ratings only. Functional operation of this device at these or any other conditions above those indicated in the operational sections of this specification is not implied and exposure to absolute maximum rating conditions for extended periods could affect device reliability.

AC CHARACTERISTICS

(V_{CC} = 5V ±5%, T_A = 0 to +70°C)

AC TEST CONDITIONS

Input pulse levels: 0 to 3.0V
Input rise and fall times: 5ns
Input and outputs timing reference levels: 1.5V
Output load: 1 TTL Gate and C_L = 100pF (including scope and jig)

TRUTH TABLE

$\overline{CS}$	$\overline{OE}$	$\overline{WE}$	Mode	V_{CC} Current	I/O Pin
H	x	x	standby	I_{SB}, I_{SBT}	High Z
L	L	H	Read	I_{CC}	Dout
L	H	H	Read	I_{CC}	High Z
L	x	L	Write	I_{CC}	Din

DC AND OPERATING CHARACTERISTICS

(V_{CC} = 5V ± 5%, GND = 0V, T_A = 0 to +70°C)

Item	Symbol	Test Conditions	6116-07			6116-09			6116-12			Units
			Min.	Typ.*	Max.	Min.	Typ.*	Max.	Min.	Typ.*	Max.	
Input Leakage Current	I_{L1}	V_{CC}=5.5V, V_{IN}=GND to V_{CC}	–	–	10	–	–	10	–	–	10	μA
Output Leakage Current	I_{LO}	$\overline{CS}$=V_{IH} or $\overline{OE}$=V_{IH}, $V_{I/O}$=GND to V_{CC}	–	–	10	–	–	10	–	–	10	μA
Operating Power Supply Current	I_{CC}	$\overline{CS}$=V_{IL}, $I_{I/O}$=0mA	–	50	100	–	50	100	–	50	100	mA
	I_{CC1}	V_{IH}=3.5V, V_{IL}=0.6V, $I_{I/O}$=0mA	–	45	–	–	45	–	–	45	–	mA
Dynamic Operating Current	I_{CC2}	Min. cycle, duty=100%	–	–	100	–	–	100	–	–	100	mA
Standby Power Supply Current	I_{SB}	$\overline{CS}$=V_{IH}	–	5	10	–	5	15	–	5	10	mA
	I_{SB1}	$\overline{CS}$≥V_{CC}-0.2V, V_{IN}≥V_{CC} -0.2V or V_{IN}≤0.2V	–	1	50	–	1	500	–	1	50	μA
Output Voltage	V_{OL}	I_{OL}=4mA	–	–	0.4	–	–	0.4	–	–	0.4	V
	V_{OH}	I_{OH}=-1.0mA	2.4	–	–	2.4	–	–	2.4	–	–	V
Input Voltage	V_{IH}		2.2	3.5	5.8	2.2	3.5	5.8	2.2	3.5	5.8	V
	V_{IL}		-0.3	–	+0.8	-0.3	–	+0.8	-0.3	–	+0.8	V

*V_{CC}=5V, T_A=25°C

PIN CONFIGURATION

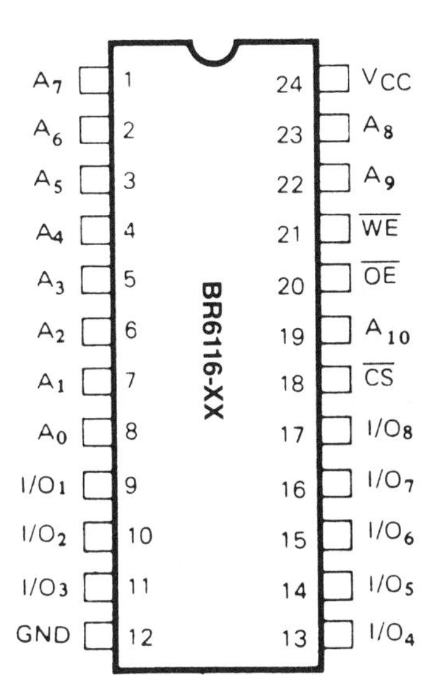

BLOCK DIAGRAM

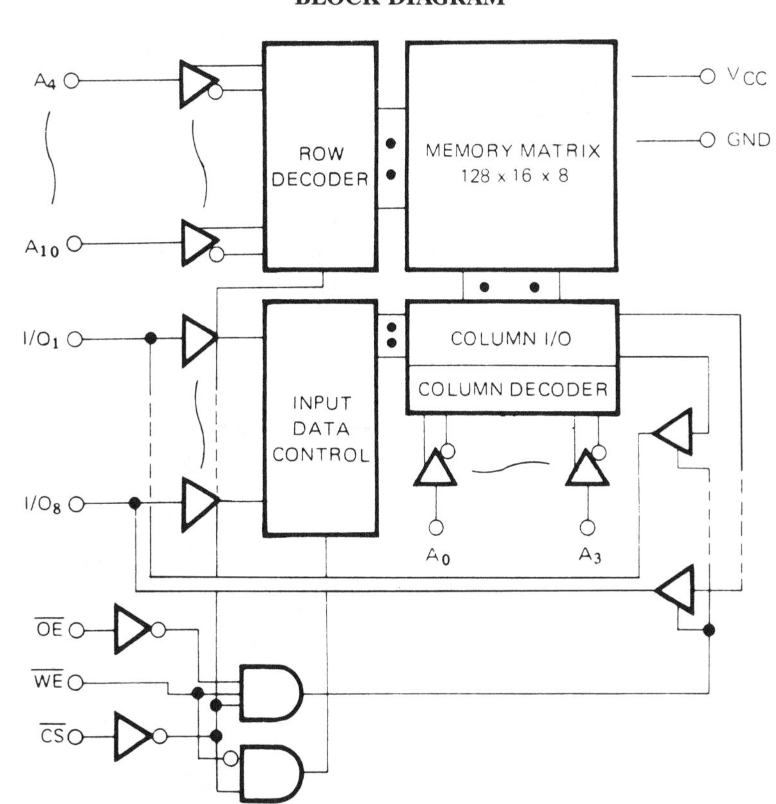

TIMING WAVEFORM

READ CYCLE (1)[(1)(5)]

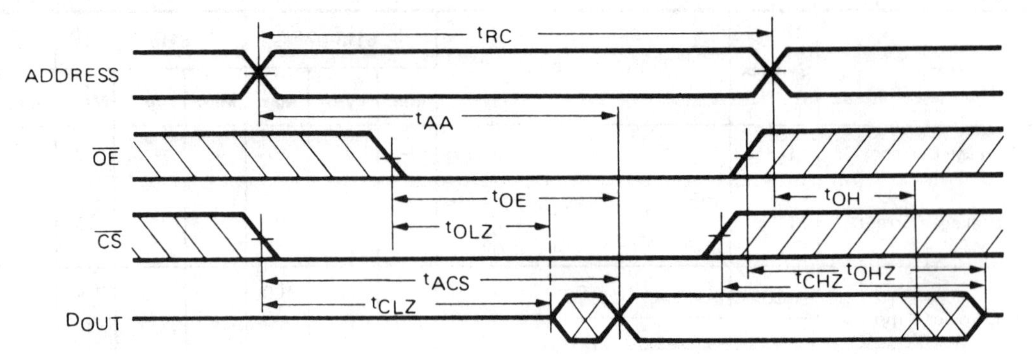

READ CYCLE (2)[(1)(2)(4)]

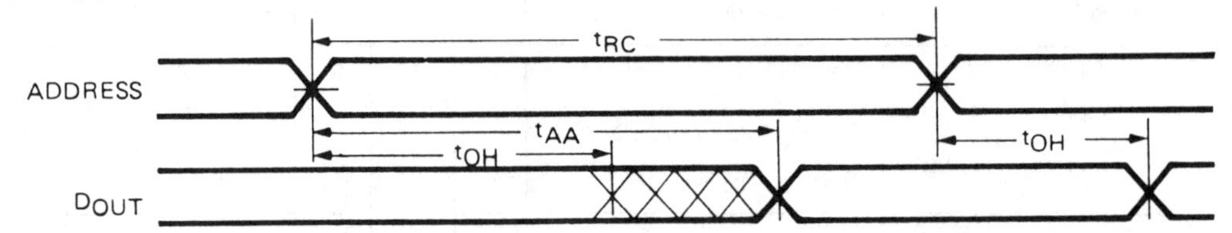

READ CYCLE (3)[(1)(3)(4)(5)]

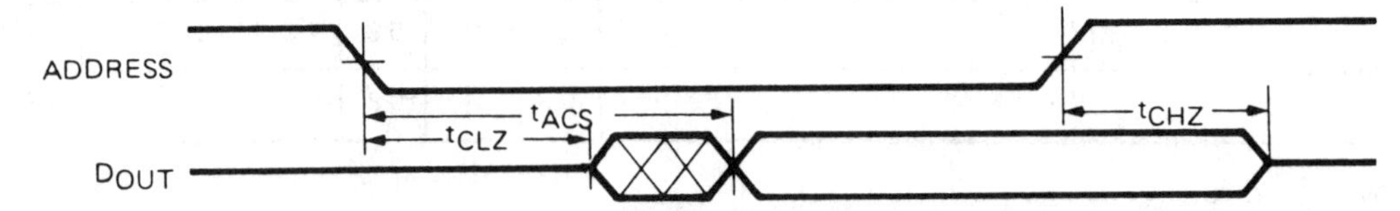

Notes:

1. $\overline{WE}$ is High for Read Cycle.
2. Device is continuously selected, $\overline{CS} = V_{IL}$.
3. Address Valid prior to or coincident with $\overline{CS}$ transition Low.
4. $\overline{OE} = V_{IL}$.
5. Transition is measured ±500mV from steady state. This parameter is sampled and not 100% tested.

WRITE CYCLE (1)

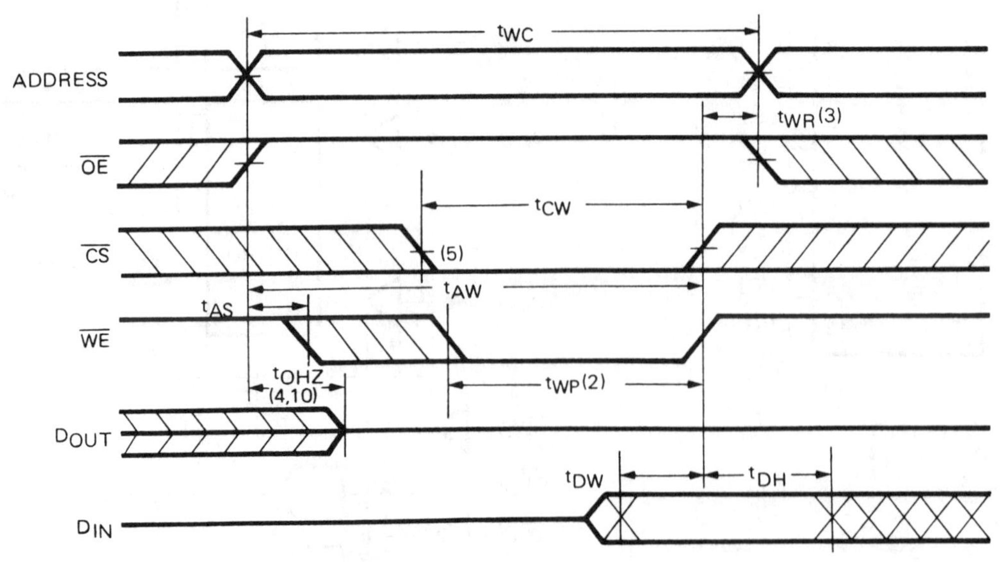

WRITE CYCLE (2) (1) (6)

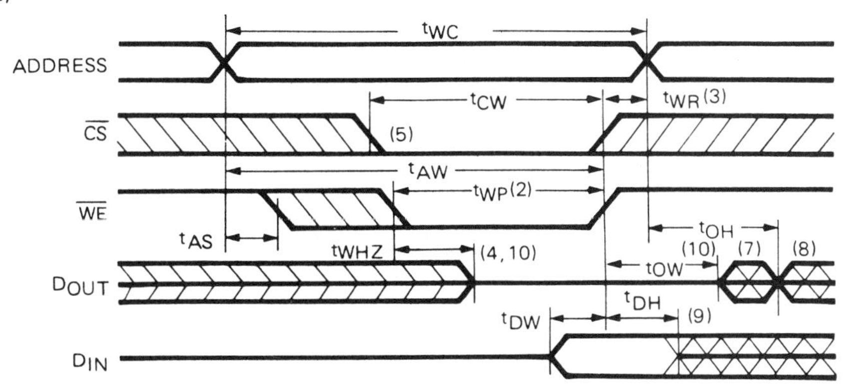

Notes:
1. $\overline{WE}$ must be high during all address transitions.
2. A write occurs during the overlap (tWP) of a low $\overline{CS}$ and a low $\overline{WE}$.
3. tWR is measured from the earlier of $\overline{CS}$ or $\overline{WE}$ going high to the end of write cycle.
4. During this period, I/O pins are in the output state so the input signals of opposite phase to the outputs must not be applied.
5. If the $\overline{CS}$ low transition occurs simultaneously with the $\overline{WE}$ low transitions or after the $\overline{WE}$ transition, outputs remain in a high impedance state.
6. $\overline{OE}$ is continuously low ($\overline{OE} = V_{IL}$).
7. D_{OUT} is the same phase of write data of this write cycle.
8. D_{OUT} is the read data of next address.
9. If $\overline{CS}$ is low during this period, I/O pins are in the output state. Then the data input signals of opposite phase to the outputs must not be applied to them.
10. Transition is measured ± 500mV from steady state. This parameter is sampled and not 100% tested.

Data Retention Characteristics over the operating temperature range

Symbol	Parameter	Test Conditions	Min.	Typ.(1)	Max.	Units
V_{DR}	V_{CC} for Retention Data	$\overline{CS} = V_{CC}$	2.0	–	–	V
I_{CCDR}	Data Retention Current	$V_{IN} = 0V$ or V_{CC}	–	2	20	μA
t_{CDR}	Chip Deselect to Data Retention Time	$V_{CC} = 2.0V$, $CS = V_{CC}$	0	–	–	ns
t_R	Operation Recovery Time	$V_{IN} = 0V$ or V_{CC}	t_{RC}(2)	–	–	ns

1. $V_{CC} = 2V$, $T_A = +25°C$
2. I_{RC} = Read Cycle Time

Timing Waveform Low V_{CC} Data Retention Waveform

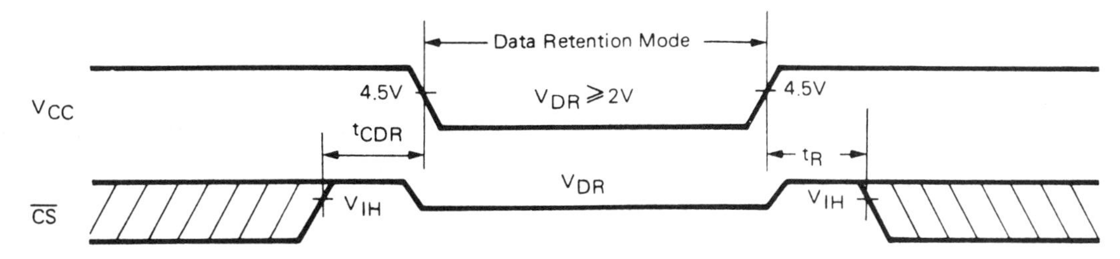

Ordering Information

Part Number	Access Time (Max.)	Package
BR6116-12	120 ns	Plastic
BR6116-09	90 ns	Plastic
BR6116-07	70 ns	Plastic

PACKAGE INFORMATION — mm (inch)

24 — LEAD DUAL IN-LINE; PLASTIC

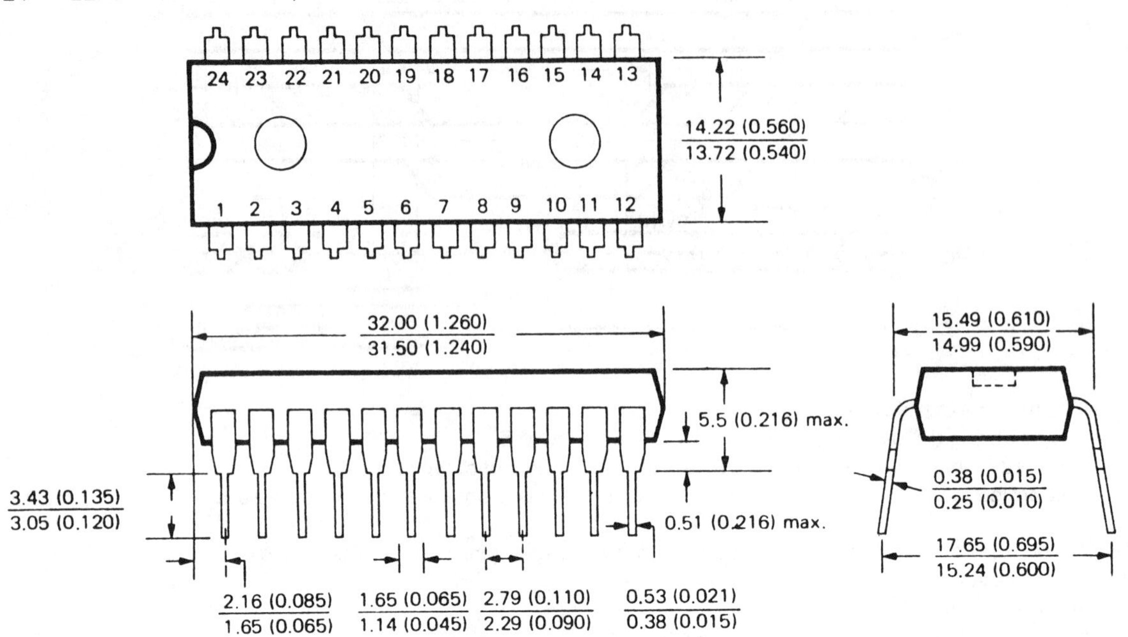

24 Pin Small Outline

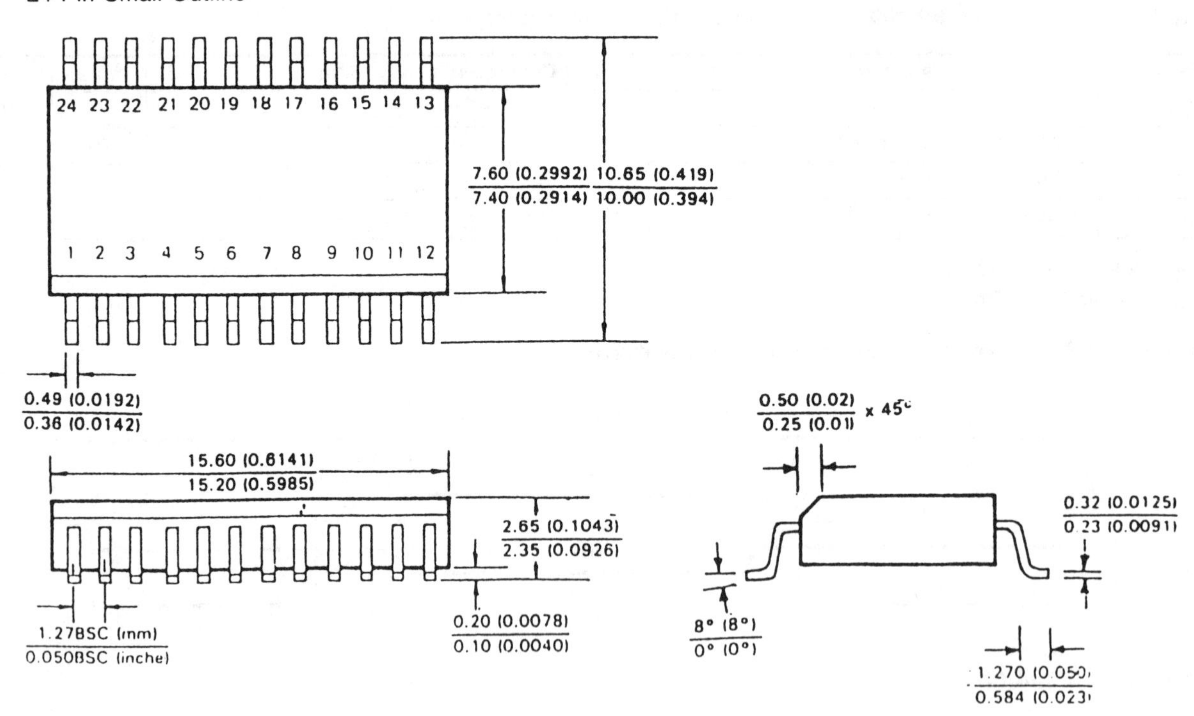

Rohm BR6264

8K × 8 CMOS SRAM

- High-speed 100/120ns (max.)
- Low-power dissipation
  Standard version: operating: 90mA max.
  standby:2mA max.
  Low-power version: operating: 85mA max.
  standby: 100μA max.
- Single 5V power supply
- Fully static operation: no clock or refreshing required
- TTL compatible: all inputs and outputs
- Common I/O using three-state outputs
- Output enable and two chip enable inputs for easy application
- Data retention supply voltage: 2V min. (BR6264-10L/12L)
- Standard 28-pin plastic DIP and SOP packages

The BR6264 is a high-speed, low-power 65,536-bit static random-access memory organized as 8,192 words by 8 bits and operates from a single 5V supply. It is built with Rohm's high-performance twin-tub CMOS process.

Inputs and three-state outputs are TTL compatible and allow for direct interfacing with common system bus structures. Two chip select inputs are provided for battery back-up application, and an output enable input is included for easy interface.

Data retention is guaranteed at a power supply voltage as low as 2V (BR6264-10L/12L). The BR6264 is packaged in a standard 28-pin plastic dual-in-line package. The BR6264F is packaged in a standard 28-pin plastic small-outline IC package.

ABSOLUTE MAXIMUM RATINGS*

Terminal voltage with respect to GND	−0.5 to +7.0V
Temperature under bias	−10 to +125°C
Storage temperature	−40 to +150°C
Power dissipation	1.0W/SOP 0.7W
Dc output current	20mA

*COMMENTS

Stresses greater than those listed under Absolute Maximum Ratings could cause permanent damage to the device. This is a stress rating only and functional operation of the device at these or any other conditions above those indicated in the operational sections of this specification is not implied. Exposure to absolute maximum rating conditions for extended periods could affect reliability.

DC ELECTRICAL CHARACTERISTICS (V_{CC} = 5V ± 10%, GND = 0V, T_A = 0 to +70°C)

Symbol	Parameter	Test Conditions	BR6264-10 Min.	Typ.*	Max.
$\lvert I_{LI} \rvert$	Input Leakage Current	V_{IN} = GND to V_{CC}	–	–	2
$\lvert I_{LO} \rvert$	Output Leakage Current	$\overline{CS_1} = V_{IH}$ or $CS_2 = V_{IL}$ or $\overline{OE} = V_{IH}$, $V_{I/O}$ = GND to V_{CC}	–	–	2
I_{CC}	Operating Power Supply Current	$\overline{CS_1} = V_{IL}$, $CS_2 = V_{IH}$, $I_{I/O}$ = 0mA	–	50	90
I_{CC1}	Average Operating Current	Min. Duty Cycle = 100%, $\overline{CS_1} = V_{IL}$, $CS_2 = V_{IH}$	–	50	90
I_{SB}	Standby Power Supply Current	$\overline{CS_1} = V_{IH}$ or $CS_2 = V_{IL}$, $I_{I/O}$ = 0mA	–	–	15
I_{SB1}**		$\overline{CS_1} \geqslant V_{CC} - 0.2V$, $V_{IN} \geqslant V_{CC} - 0.2V$ or $V_{IN} \leqslant 0.2V$	–	–	2
I_{SB2}**		$CS_2 \leqslant 0.2V$, $V_{IN} \geqslant V_{CC} - 0.2V$ or $V_{IN} \leqslant 0.2V$	–	–	2
V_{OL}	Output Voltage	I_{OL} = 4mA	–	–	0.4
V_{OH}		I_{OH} = −1.0mA	2.4	–	–

Symbol	BR6264-10L Min.	Typ.*	Max.	BR6264-12 Min.	Typ.*	Max.	BR6264-12L Min.	Typ.*	Max.	Unit
$\lvert I_{LI} \rvert$	–	–	2	–	–	2	–	–	2	μA
$\lvert I_{LO} \rvert$	–	–	2	–	–	2	–	–	2	μA
I_{CC}	–	45	85	–	50	90	–	45	85	mA
I_{CC1}	–	45	85	–	50	90	–	45	85	mA
I_{SB}	–	–	15	–	–	15	–	–	15	mA
I_{SB1}**	–	0.01	0.1	–	–	2	–	0.01	0.1	mA
I_{SB2}**	–	0.01	0.1	–	–	2	–	0.01	0.1	mA
V_{OL}	–	–	0.4	–	–	0.4	–	–	0.4	V
V_{OH}	2.4	–	–	2.4	–	–	2.4	–	–	V

* Typical limits are at V_{CC} = 5.0V, T_A = 25°C and specified loading.
** V_{IL} min = −0.3V

PIN NAME

No.	Symbol	Function
1	NC	No connection
2-10, 21, 23-25	A_0-A_{12}	Address input
11-13, 15-19	I/O_1-I/O_8	Data input/output
14	GND	Ground
20	$\overline{CS_1}$	Chip select input, active low
22	$\overline{OE}$	Output enable input
26	CS_2	Chip select input, active high
27	$\overline{WE}$	Write enable input
28	V_{CC}	+5V Power supply

RECOMMENDED DC OPERATING CONDITIONS

(T_A=0 to +70°C)

Symbol	Parameter	Min.	Typ.	Max.	Unit
V_{CC}	Supply Voltage	4.5	5.0	5.5	V
GND	Supply Voltage	0	0	0	V
V_{IH}	Input High Voltage	2.2	3.5	6.0	V
V_{IL}	Input Low Voltage	−0.5	0	0.8	V

CAPACITANCE (T_A = 25°C, f = 1.0MHZ)

Symbol	Parameter	Conditions	Max.	Unit
C_{IN}	Input Capacitance	V_{IN} = 0V	6	pF
$C_{I/O}$	Input/Output Capacitance	$V_{I/O}$ = 0V	8	pF

* This parameter is sampled and not 100% tested.

AC TEST CONDITIONS

Input Pulse Levels	0V to 3.0V
Input Rise and Fall Times	5ns
Input and Ouput Timing Reference Level	1.5V
Output Load	1 TTL Gate and C_L = 30pF (including scope and jig)

AC ELECTRICAL CHARACTERISTICS

Symbol	Parameter		BR6264-10/10L Min.	BR6264-10/10L Max.	BR6264-12/12L Min.	BR6264-12/12L Max.	Unit
Read Cycle							
t_{RC}	Read Cycle Time		100	–	120	–	ns
t_{AA}	Address Access Time		–	100	–	120	ns
t_{ACS1}	Chip Select Access Time	$\overline{CS_1}$	–	100	–	120	ns
t_{ACS2}	Chip Select Access Time	CS_2	–	100	–	120	ns
t_{OE}	Output Enable to Output Valid		–	50	–	60	ns
t_{CLZ1}	Chip Selection to Output in Low Z	$\overline{CS_1}$	5	–	5	–	ns
t_{CLZ2}	Chip Selection to Output in Low Z	CS_2	5	–	5	–	ns
t_{OLZ}	Output Enable to Output in Low Z		5	–	5	–	ns
t_{CHZ1}	Chip Deselection to Output in High Z	$\overline{CS_1}$	0	35	0	40	ns
t_{CHZ2}	Chip Deselection to Output in High Z	CS_2	0	35	0	40	ns
t_{OHZ}	Output Disable to Output in High Z		0	35	0	35	ns
t_{OH}	Output Hold from Address Change		5	–	5	–	ns
Write Cycle							
t_{WC}	Write Cycle Time		100	–	120	–	ns
t_{CW}	Chip Selection to End of Write		80	–	85	–	ns
t_{AS}	Address Setup Time		0	–	0	–	ns
t_{AW}	Address Valid to End of Write		80	–	85	–	ns
t_{WP}	Write Pulse Width		60	–	70	–	ns
t_{WR1}	Write Recovery Time	$\overline{CS_1}$, $\overline{WE}$	5	–	5	–	ns
t_{WR2}	Write Recovery Time	CS_2	5	–	5	–	ns
t_{WHZ}	Write to Output in High Z		0	35	0	40	ns
t_{DW}	Data to Write Time Overlap		40	–	45	–	ns
t_{DH}	Data Hold from Write Time		5	–	5	–	ns
t_{OHZ}	Output Disable to Output in High Z		0	35	0	35	ns
t_{OW}	Output Active from End of Write		5	–	5	–	ns

NOTES: t_{CHZ}, t_{OHZ} and t_{WHZ} are defined as the time at which the outputs achieve the open circuit condition and are not referred to output voltage levels.

TRUTH TABLE

Mode	$\overline{WE}$	$\overline{CS}_1$	CS_2	$\overline{OE}$	I/O Operation	V_{CC} Current
Not Selected	X	H	X	X	High Z	I_{SB}, I_{SB1}
(Power Down)	X	X	L	X	High Z	I_{SB}, I_{SB2}
Output Disabled	H	L	H	H	High Z	I_{CC}, I_{CC1}
Read	H	L	H	L	D_{OUT}	I_{CC}, I_{CC1}
Write	L	L	H	X	D_{IN}	I_{CC}, I_{CC1}

PIN CONFIGURATIONS

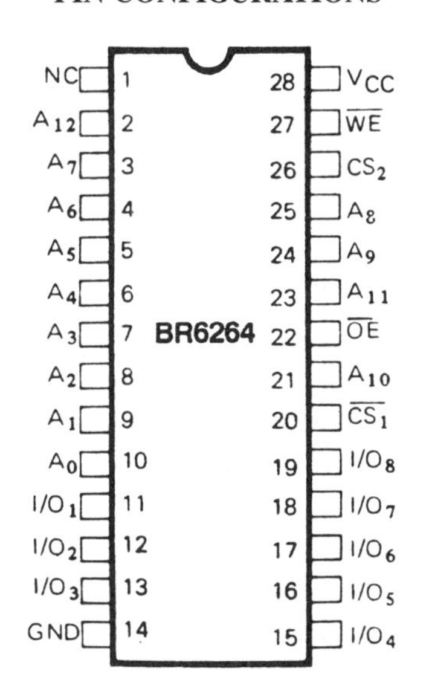

BLOCK DIAGRAM

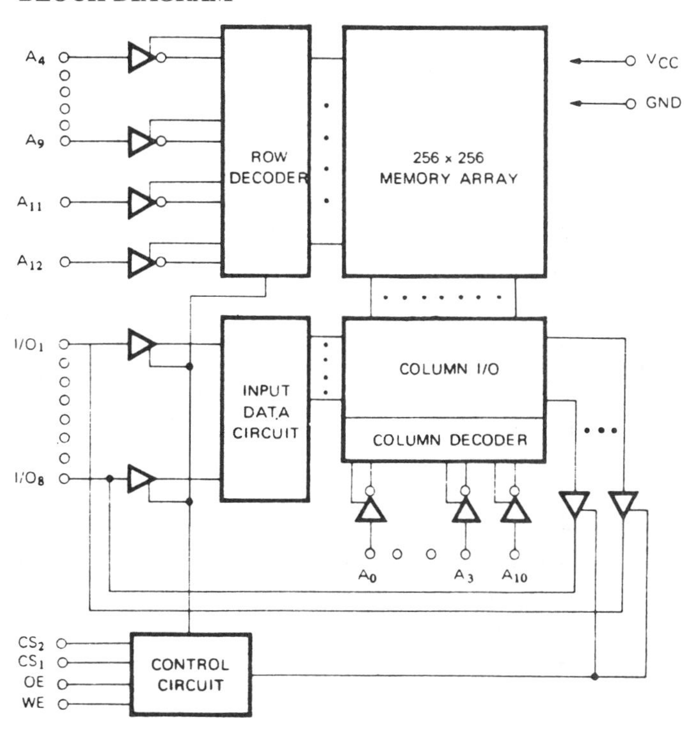

PACKAGE INFORMATION – mm (inch)

28 LEAD DUAL IN-LINE; PLASTIC

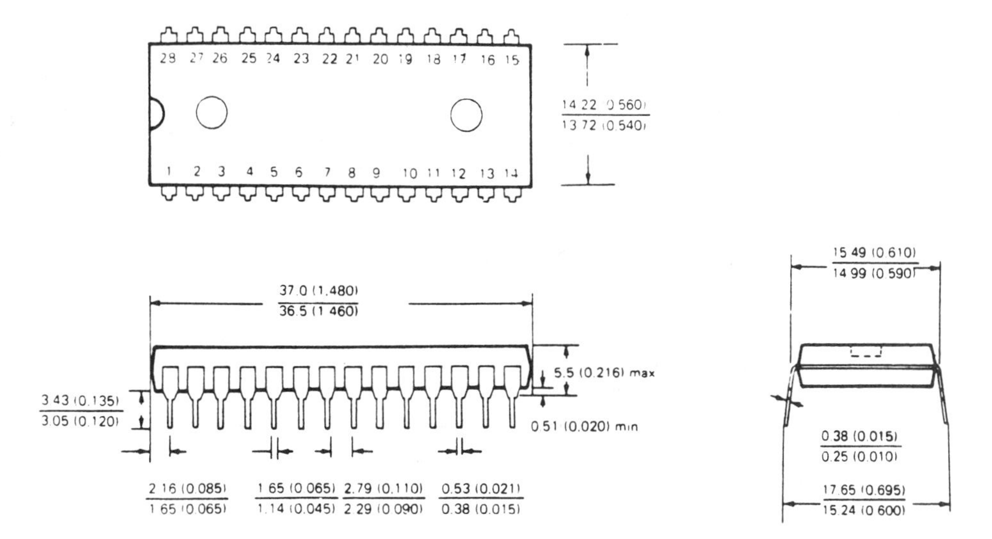

28 Pin Small Outline Unit: mm (inch)

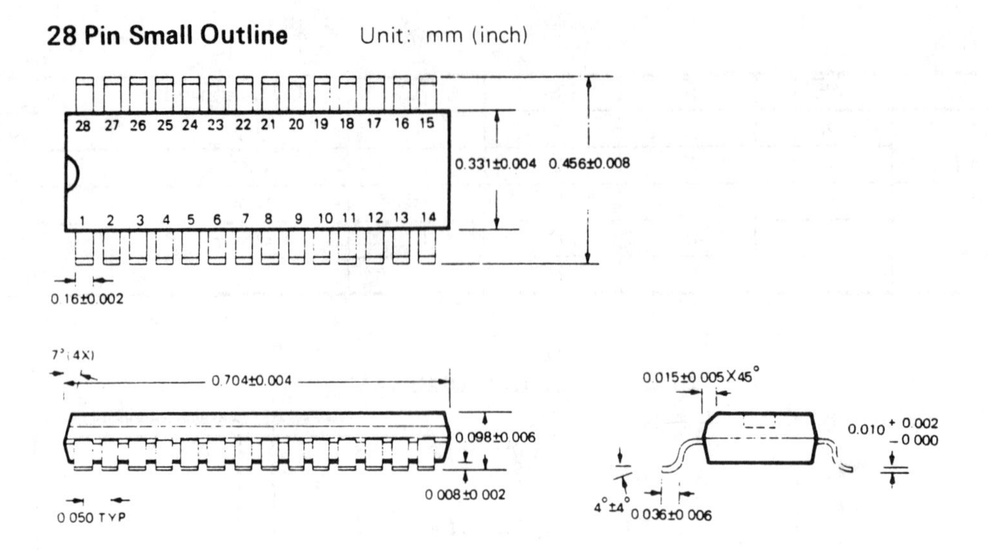

DATA RETENTION CHARACTERISTICS (T_A = 0 to +70°C; L version only)

Symbol	Parameter	Test Conditions	Min.	Typ.*	Max.	Unit
V_{DR_1}	V_{CC} for Data Retention	$\overline{CS_1} \geqslant V_{CC} - 0.2V$, $V_{IN} \geqslant V_{CC} - 0.2V$ or $V_{IN} \leqslant 0.2V$	2.0	–	–	V
V_{DR_2}		$CS_2 \leqslant 0.2V$, $V_{IN} \geqslant V_{CC} - 0.2V$ or $V_{IN} \leqslant 0.2V$	2.0	–	–	V
I_{CCDR_1}	Data Retention Current	$\overline{CS_1} \geqslant V_{CC} - 0.2V$, $V_{IN} \geqslant V_{CC} - 0.2V$ or $V_{IN} \leqslant 0.2V$	–	2	50	μA
I_{CCDR_2}		$CS_2 \leqslant 0.2V$, $V_{IN} \geqslant V_{CC} - 0.2V$ or $V_{IN} \leqslant 0.2V$	–	2	50	μA
t_{CDR}	Chip Deselect to Data Retention Time	See Retention Waveform	0	–	–	ns
t_R	Operation Recovery Time		t_{RC}**	–	–	ns

* V_{CC} = 2V, T_A = +25°C
** t_{RC} = Read Cycle Time

Low V_{CC} Data Retention Waveform (1) ($\overline{CS_1}$ Controlled)

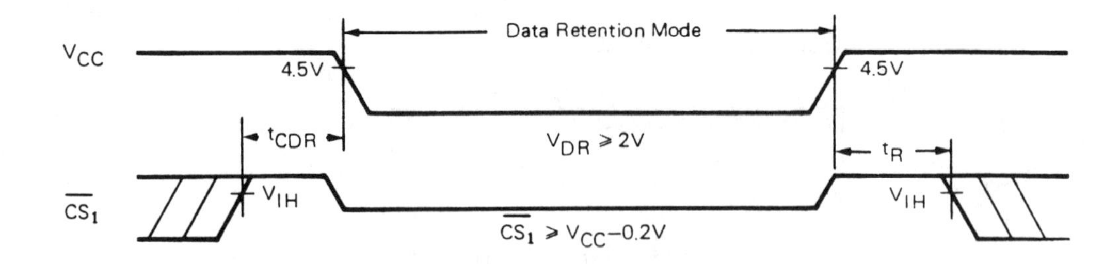

Low V_{CC} Data Retention Waveform (2) (CS_2 Controlled)

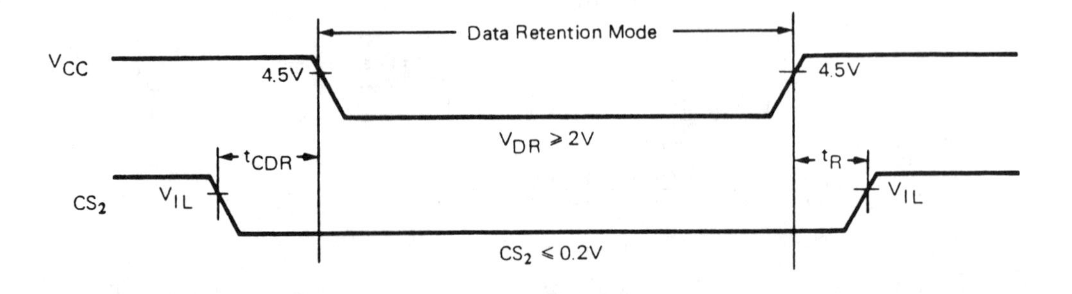

Timing Waveform of Read Cycle No. 1[1,2,4]

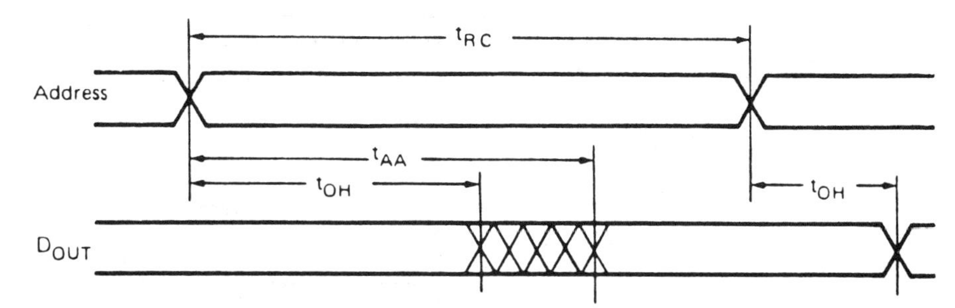

Read Cycle 2[1,3,4,6]

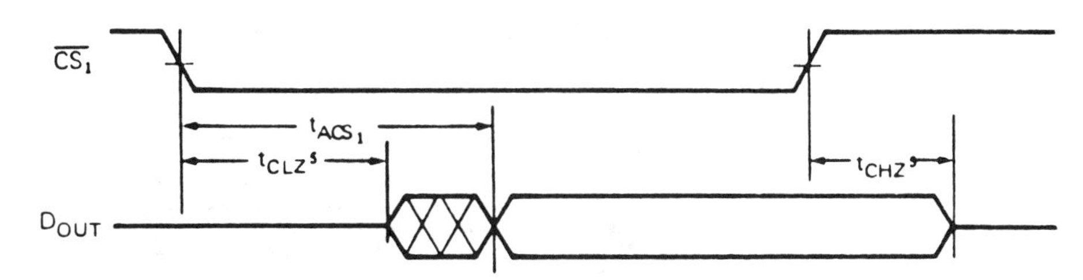

Read Cycle 3[1,4,7]

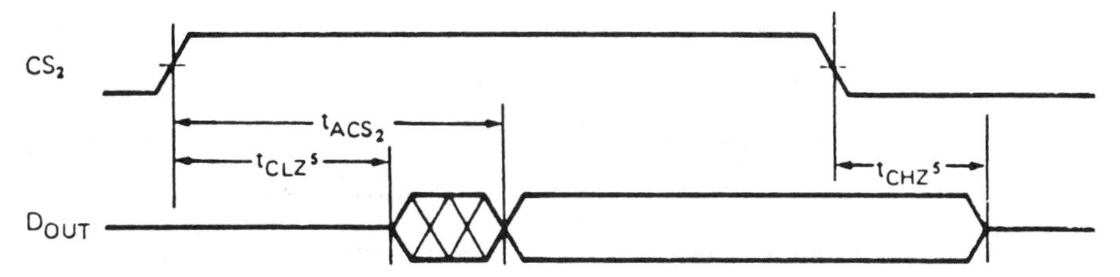

Read Cycle 4[1]

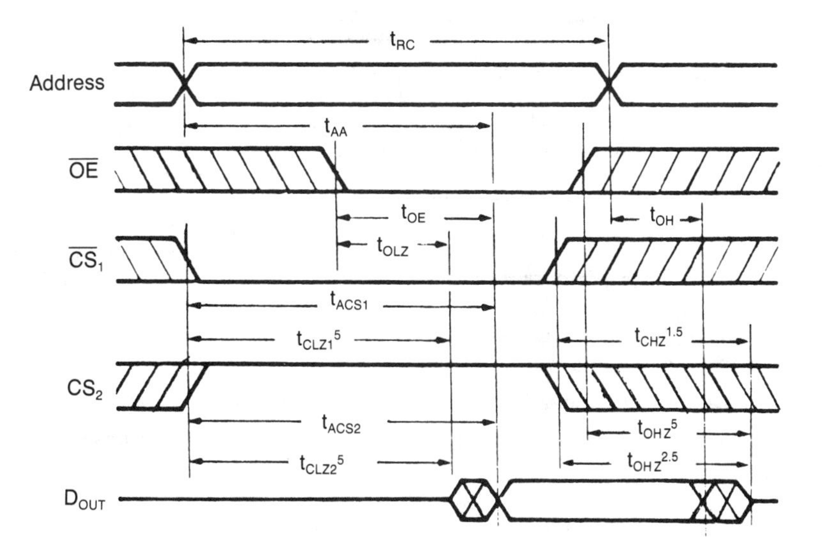

Notes: 1. $\overline{WE}$ is high for READ cycle.
2. Device is continuously selected $\overline{CS_1} = V_{IL}$ and $CS_2 = V_{IH}$.
3. Address valid prior to or coincident with $\overline{CS_1}$ transition low.
4. $\overline{OE} = V_{IL}$.
5. Transition is measured ± 500mV from steady state. This parameter is sampled and not 100% tested.
6. CS_2 is high.
7. $\overline{CS_1}$ is low.

Timing Waveforms of Write Cycle 1[1]

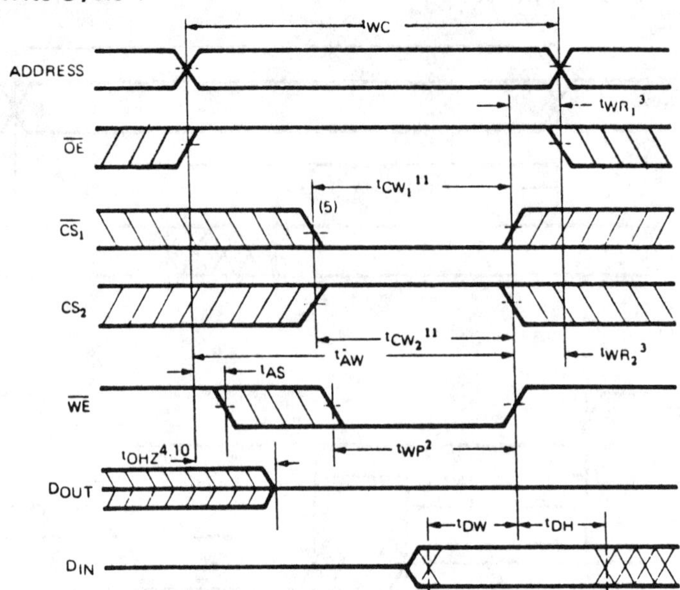

Write Cycle 2[1,6]

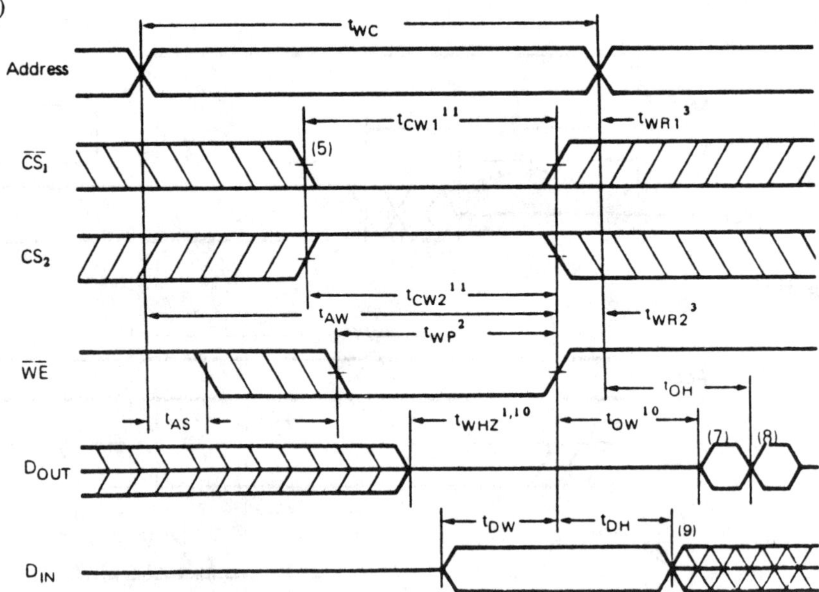

Notes:
1. $\overline{WE}$ must be high during address transitions.
2. A write occurs during the overlap (t_{WP}) of a low $\overline{CS_1}$, a high CS_2 and a low $\overline{WE}$.
3. t_{WR} is measured from the earlier of $\overline{CS_1}$ or $\overline{WE}$ going high or CS_2 going low to the end of write cycle.
4. During this period, I/O pins are in the output state so that the input signals of opposite phase to the outputs must not be applied.
5. If the $\overline{CS_1}$ low transition or the CS_2 high transition occurs simultaneously with the $\overline{WE}$ low transitions or after the $\overline{WE}$ transition, outputs remain in a high impedance state.
6. $\overline{OE}$ is continuously low ($\overline{OE} = V_{IL}$).
7. D_{OUT} is the same phase of write data of this write cycle.
8. D_{OUT} is the read data of next address.
9. if $\overline{CS_1}$ is low and CS_2 is high during this period, I/O pins are in the output state. Then the data input signals of opposite phase to the outputs must not be applied to them.
10. Transition is measured ± 500mV from steady state. This parameter is sampled and not 100% tested.
11. t_{cw} is measured from the later of $\overline{CS_1}$ going low or CS_2 going high to the end of write.

ORDERING INFORMATION

Access Time (ns)	Ordering Code	Operating Current Max. (mA)	Standby Current Max. (mA)	Package Type
100	BR6264-10	90	2	DIP-28
	BR6264-10L	85	0.1	DIP-28
	BR6264 F-10L	85	0.1	SO-28
120	BR6264-12	90	2	DIP-28
	BR6264-12L	85	0.1	DIP-28
	BR6264 F-12L	85	0.1	SO-28

Rohm BR93C46

1024-Bit Serial Electrically Erasable PROM with 2V Read Capability

- State of the art architecture
  - Nonvolatile data storage
  - Single supply −5V operation
  - Full TTL-compatible inputs and outputs
  - Auto increment for continuous read
- Hardware and software write protection
  - Defaults to write-disabled state at power up
  - Software instructions for write-enable/disable
- Low power consumption:
  - 1mA active (typ.),
  - 1μA standby (typ.)
- Low-voltage read operations
  - Reliable read operations down to 2.0 volts
- Advanced low-voltage CMOS E^2PROM technology
- Versatile, easy-to-use interface
  - Self-timed programming cycle
  - Automatic erase-before-write
  - Programming status indicator
  - Word and bulk erasable
  - Stop SK anytime for power savings
- Durable and reliable
  - 10-year data retention after 100K write cycles
  - Minimum of 100,000 write cycles per word
  - Unlimited read cycles
  - ESD protection

The BR93C46 is a low-cost 1024-bit, nonvolatile, serial E^2ROM. It is fabricated using Rohm's advanced CMOS E^2PROM technology. The BR93C46 provides efficient nonvolatile read/write memory arranged as 64 registers of 16 bits each. Seven 9-bit instructions control the operation of the device, which includes read, write, and mode-enable functions. The data-out pin (DO) indicates the status of the device during the self-timed nonvolatile programming cycle.

The self-timed write cycle includes an automatic erase-before-write capability. To protect against inadvertent writes, the write instruction is accepted only while the chip is in the write-enabled state. Data is written in 16 bits per write instruction into the selected register. If chip select (CS) is brought high after initiation of the write cycle, the data output (DO) pin will indicate the ready/busy status of the chip.

ABSOLUTE MAXIMUM RATINGS

Temperature under bias:	93C46	0 to +70 °C
	93C46-E	−40 to +85 °C
Storage temperature		−65 to +125 °C
Voltage with respect to ground		−0.3 to +6.5V

Note: These are stress ratings only. Appropriate conditions for operating these devices are given elsewhere in this specification. Stresses beyond those listed here could permanently damage the part. Prolonged exposure to maximum ratings could affect device reliability.

The BR93C46 is ideal for high-volume applications requiring low power and low-density storage. This device uses a low-cost, space-saving 8-pin package. Candidate applications include robotics, alarm devices, electronic locks, meters, and instrumentation settings.

The BR93C46 is designed for applications requiring up to 100,000 write cycles. It provides 10 years of secure data retention, without power after the execution of 100,000 write cycles.

DC ELECTRICAL CHARACTERISTICS

T_A = 0°C to +70°C or -40°C to +85°C for 93C46-E, V_{CC} = 5V ±10%

Symbol	Parameter	Conditions	93C46		93C46-E		Units
			Min	Max	Min	Max	
I_{CC1}	Operating Current CMOS Input Levels	CS = Vcc, SK = 1MHz		2		2	mA
I_{CC2}	Operating Current TTL Input Levels	CS = V_{IH}, SK = 1MHz		5		5	mA
I_{SB}	Standby Current	CS = DI = SK =0V		2		2	μA
I_{IL}	Input Leakage	V_{IN} = 0V to Vcc, CS, SK, DI	-1	1	-1	1	μA
I_{OL}	Output Leakage	V_{OUT} = 0V to Vcc, CS = 0V	-1	1	-1	1	μA
V_{IL}	Input Low Voltage		-0.1	0.8	-0.1	0.8	V
V_{IH}	Input High Voltage		2	Vcc	2	Vcc	V
V_{OL1}	Output Low Voltage	I_{OL} = 2.1mA TTL		0.4		0.4	V
V_{OH1}	Output High Voltage	I_{OH} = -400μA TTL	2.4		2.4		V
V_{OL2}	Output Low Voltage	I_{OL} = 10μA CMOS		0.2		0.2	V
V_{OH2}	Output High Voltage	I_{OH} = -10μA CMOS	Vcc-0.2		Vcc-0.2		V

AC ELECTRICAL CHARACTERISTICS

T_A = 0°C to +70°C or -40°C to +85°C for 93C46-E, V_{CC} = 5V ±10%

Symbol	Parameter	Conditions	93C46		93C46-E		Units
			Min	Max	Min	Max	
f_{SK}	SK Clock Frequency		0	1	0	1	MHz
t_{SKH}	SK High Time		400		400		ns
t_{SKL}	SK Low Time		250		250		ns
t_{CS}	Minimum CS Low Time		250		250		ns
t_{CSS}	CS Setup Time	Relative to SK	50		50		ns
t_{DIS}	DI Setup Time	Relative to SK	100		100		ns
t_{CSH}	CS Hold Time	Relative to SK	0		0		ns
t_{DIH}	DI Hold Time	Relative to SK	100		100		ns
t_{PD1}	Output Delay to "1"	AC Test		500		500	ns
t_{PD0}	Output Delay to "0"	AC Test		500		500	ns
t_{SV}	CS to Status Valid	AC Test C_L = 100pF		500		500	ns
t_{DF}	CS to DO in 3-state	CS = V_{IL}		100		100	ns
t_{WP}	Write Cycle Time			10		10	ms

CAPACITANCE

T_A = 25°C, f = 250KHz

Symbol	Parameter	Max	Units
C_{OUT}	Output Capacitance	5	pF
C_{IN}	Input Capacitance	5	pF

INSTRUCTION SET

Instruction	Start Bit	OP Code	Address	Input Data
READ	1	10	(A5-A0)	
WEN (Write Enable)	1	00	11XXXX	
WRITE	1	01	(A5-A0)	D15-D0
WRALL (Write All Registers)	1	00	01XXXX	D15-D0
WDS (Write Disable)	1	00	00XXXX	
ERASE	1	11	(A5-A0)	
ERAL (Erase All Registers)	1	00	10XXXX	

PIN CONFIGURATIONS

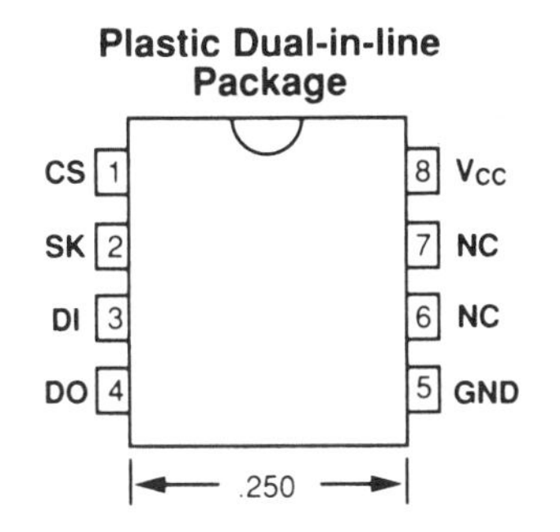

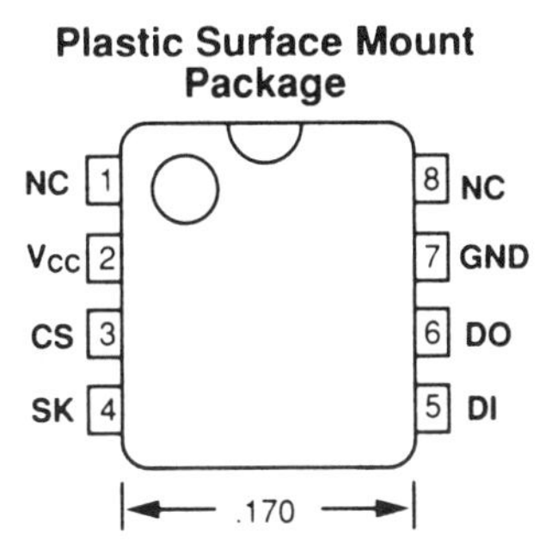

PIN NAMES

CS	Chip Select
SK	Serial Data Clock
DI	Serial Data Input
DO	Serial Data Output
GND	Ground
Vcc	Power Supply
NC	Not Connected

BLOCK DIAGRAM

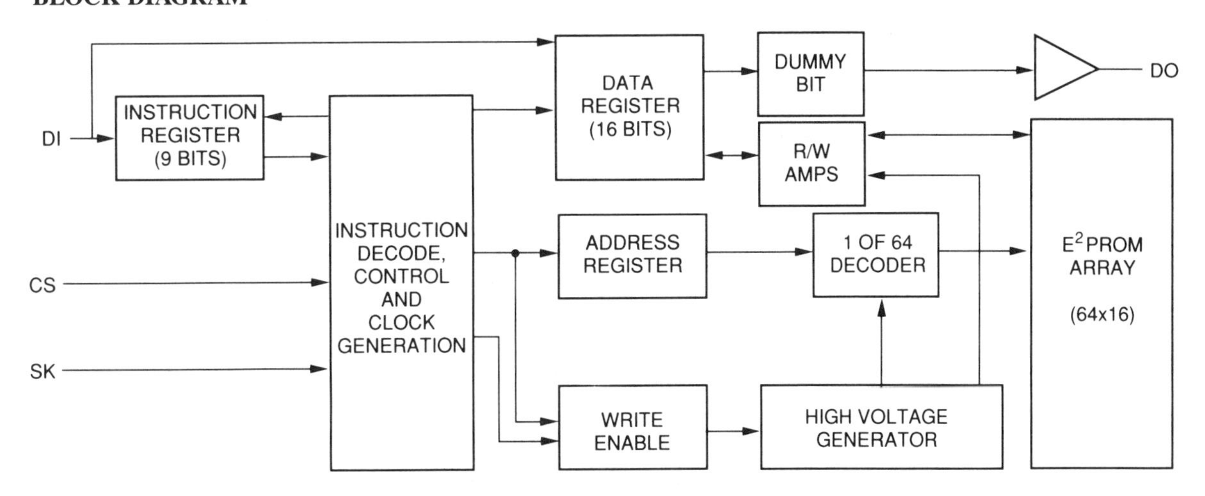

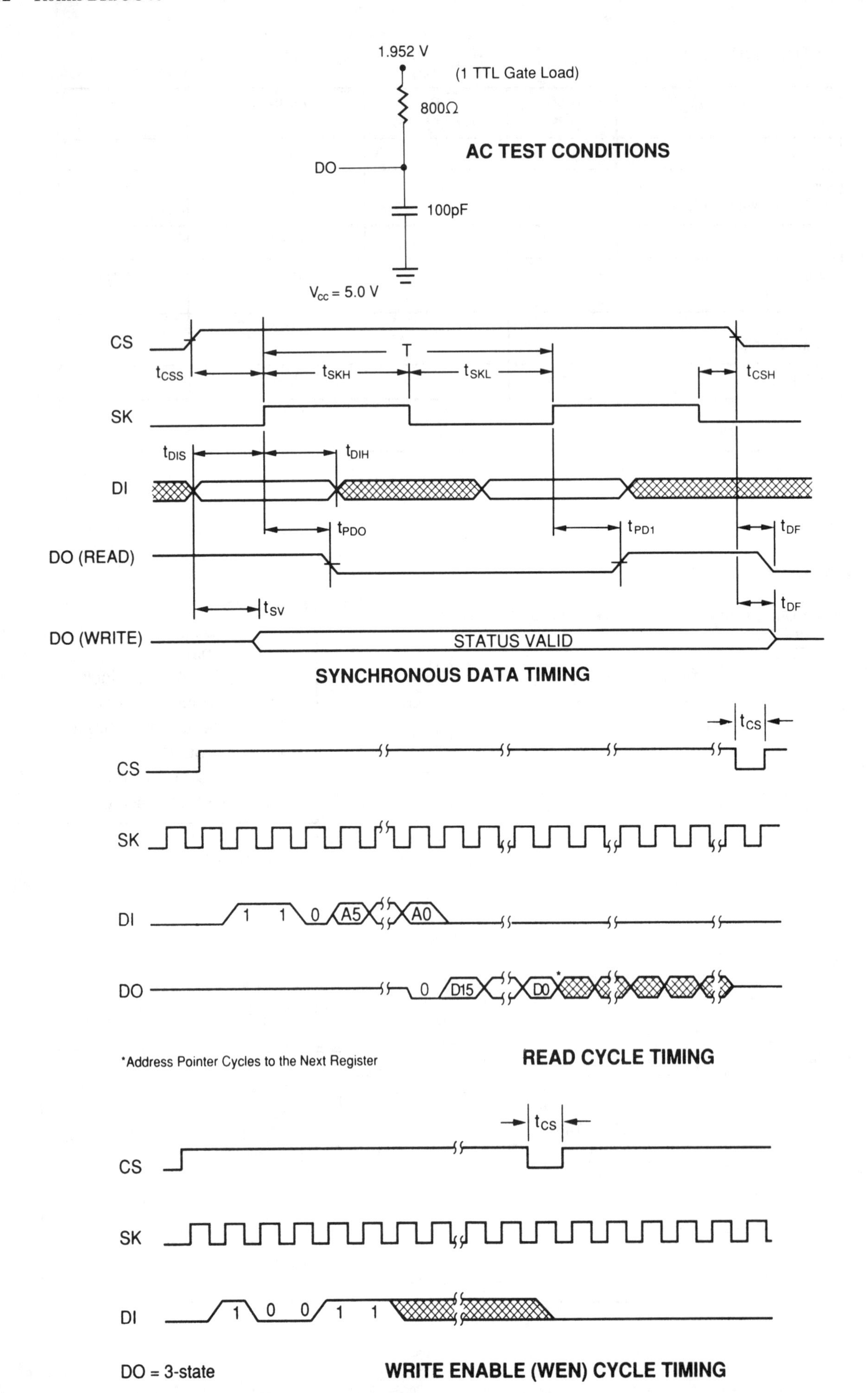
1.952 V
(1 TTL Gate Load)
800Ω
DO
100pF
Vcc = 5.0 V
AC TEST CONDITIONS
CS
SK
DI
DO (READ)
DO (WRITE)
T
tCSS
tSKH
tSKL
tCSH
tDIS
tDIH
tPDO
tPD1
tDF
tSV
STATUS VALID
SYNCHRONOUS DATA TIMING
tCS
1 1 0 A5 A0
0 D15 D0
*Address Pointer Cycles to the Next Register
READ CYCLE TIMING
1 0 0 1 1
DO = 3-state
WRITE ENABLE (WEN) CYCLE TIMING

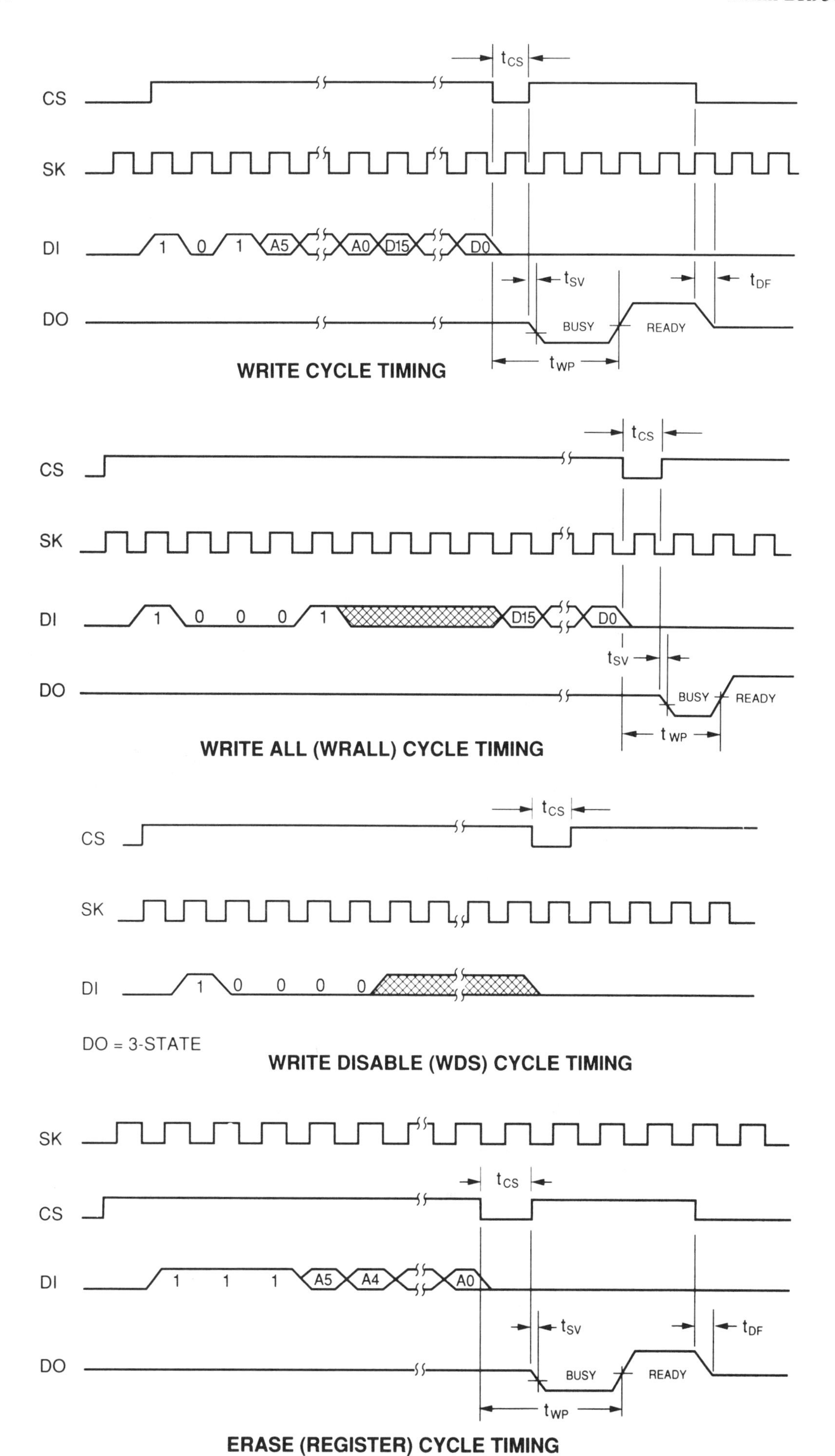
CS
SK
DI
1
0
1
A5
A0
D15
D0
DO
t_{CS}
t_{SV}
t_{DF}
BUSY
READY
t_{WP}
WRITE CYCLE TIMING
CS
SK
DI
1
0
0
0
1
D15
D0
DO
t_{CS}
t_{SV}
BUSY
READY
t_{WP}
WRITE ALL (WRALL) CYCLE TIMING
CS
SK
DI
1
0
0
0
0
t_{CS}
DO = 3-STATE
WRITE DISABLE (WDS) CYCLE TIMING
SK
CS
DI
1
1
1
A5
A4
A0
DO
t_{CS}
t_{SV}
t_{DF}
BUSY
READY
t_{WP}
ERASE (REGISTER) CYCLE TIMING

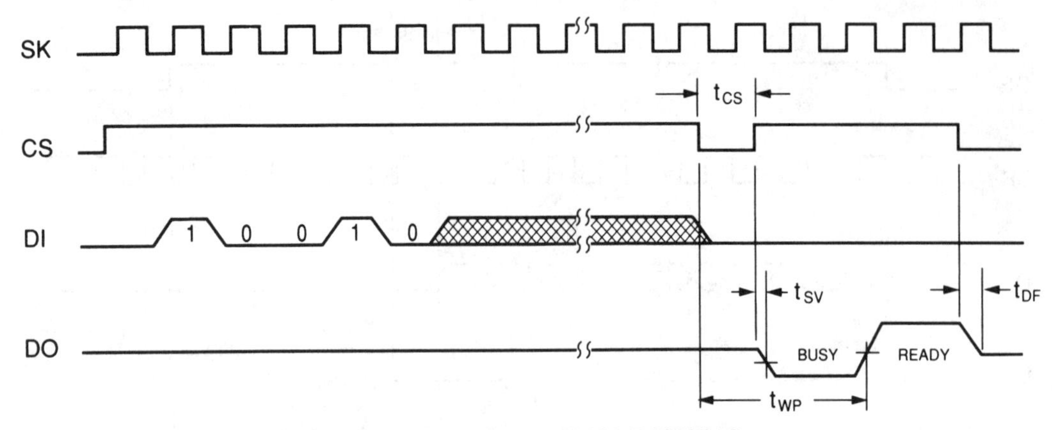

ERASE ALL (ERAL) CYCLE TIMING

PACKAGE DIAGRAMS

Plastic Dual-in-line Package (PDIP)

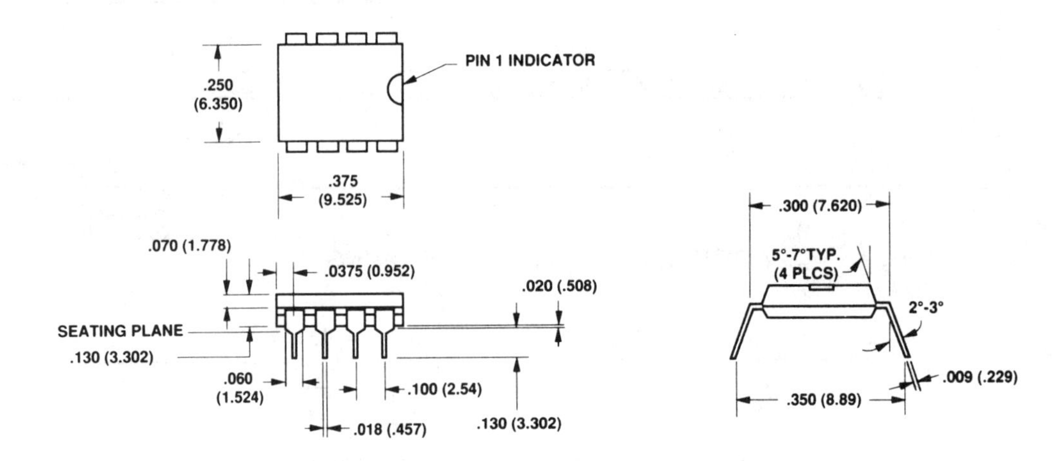

Plastic Surface Mount Package

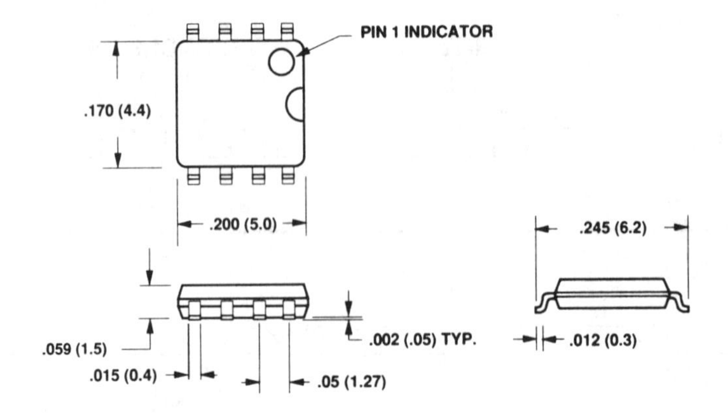

ORDERING INFORMATION

Standard Configurations

Prefix	Part Type	Package Type	Operating Voltage	Operating Teperature
BR	93C46	PDIP, SOIC	5V	Commercial Extended

*CONTACT ROHM FOR YOUR SPECIAL TEMPERATURE AND PACKAGING REQUIREMENTS

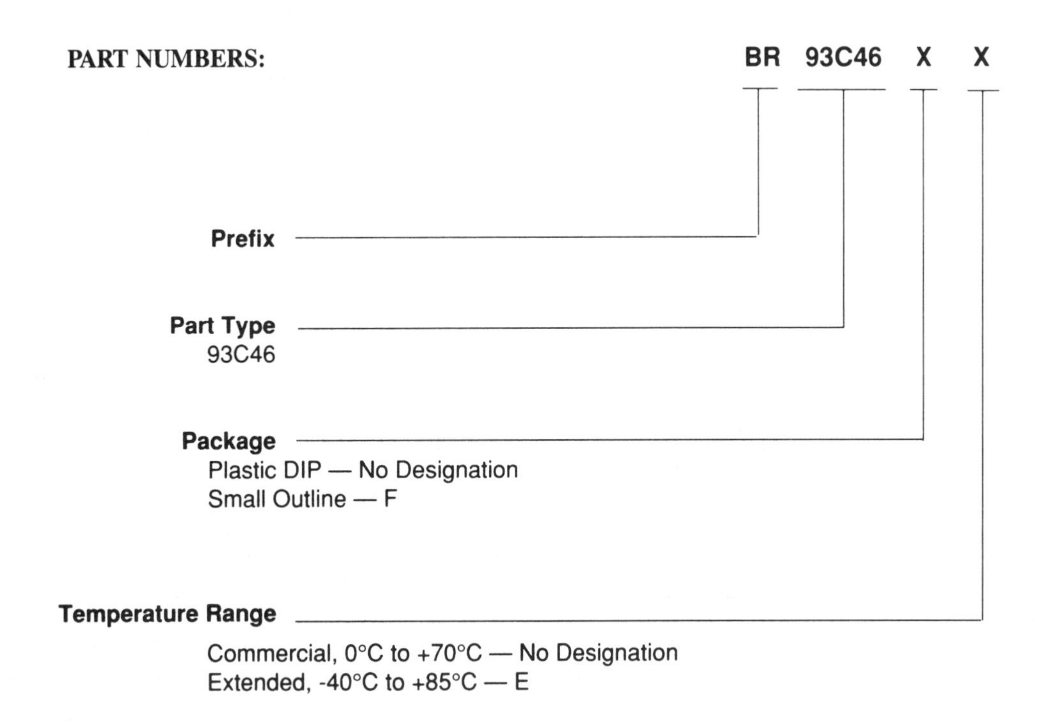

NOTE: SOIC extended temperature package marking is a white dot.

Rohm BR93CS46
1024-Bit Serial (3V and 5V) Electrically Erasable PROM Preliminary Information

- State-of-the-art architecture
  - ○ Nonvolatile data storage
  - ○ 3V to 5V operation
  - ○ 1024, accessed in 16-bit registers
  - ○ Auto-increment of register address, for continuous read
- Hardware and software write protection
  - ○ Defaults to write-disabled state at power up
  - ○ Pin-enabled writes to memory and protect register
  - ○ Software instructions for write-enable/disable
  - ○ Temporary or permanent protection of selected registers
- Low power consumption
  - ○ 1mA active (typ.)
  - ○ μA standby (typ.)
- Advanced low-voltage CMOS EEPROM technology
- Versatile easy-to-use interface
  - ○ Self-timed programming cycle
  - ○ Automatic erase-before-write
  - ○ Programming status indicator
- Durable and reliable
  - ○ 10-year data retention
  - ○ Up to 100,000 write cycles

Rohm Corp. reserves the right to make changes to any product herein to improve reliability, function, or design. Rohm Corp. does not assume any liability arising out of the application or use of any product described herein, neither does it convey any license under its patent right nor the rights or others.

OVERVIEW

The BR93CS46 is a 16-bit serial EEPROM. The device provides 64, 16-bit registers. Any number of the registers can be protected against data modification by programming the on-chip protect register. This register holds the address of the lowest memory register to be protected. The value in the protect register can be frozen, ensuring that the selected range of registers can never be altered.

The read instruction uses an address pointer that automatically increments when all 16 bits in one register have been clocked out. If clocking continues, successive registers are accessed, allowing the BR93CS46 memory to function as nonvolatile shift registers. Data can be read as a continuous data stream or as individual registers, providing from 16 to 1024 bits.

The self-timed write cycle (one register per write) includes an automatic erase-before-write capability. To protect against inadvertent writes, the write instruction is accepted only while program enable (PE) is held high, and only functions if the selected address is less than the address in the protect register. Data is written 16 bits per instruction, into the selected register. If chip select (CS) is brought high after initiation of the write cycle, the data output (DO) pin will indicate the ready/busy status of the chip.

APPLICATIONS

The BR93CS46 is ideal for high-volume applications requiring low-power and low-density storage. This device uses a low-cost space-saving 8-pin package. Candidate applications include robotics, alarm devices, electronic locks, meters, and instrumentation settings.

ENDURANCE AND DATA RETENTION

The BR93CS46 is designed for applications requiring up to 100,000 write cycles. It provides 10 years of secure data retention, with or without power applied.

ABSOLUTE MAXIMUM RATINGS

Operating temperature range	−40 to +85°C
Storage temperature	−65 to +125°C
Voltage with respect to ground	−0.3 to +6.5V

Note: These are stress ratings only. Appropriate conditions for operating these devices are given elsewhere in this specification. Stresses beyond those listed here could permanently damage the part. Prolonged exposure to maximum ratings could affect device reliability.

DC ELECTRICAL CHARACTERISTICS

V_{CC} = 3 V ±10%

Symbol	Parameter	Conditions	93CS46-3		93CS46-3E		Units
			Min	Max	Min	Max	
I_{CC1}	Operating Current	CS = V_{IH}, SK = 1KHz CMOS Input Levels		2		2	mA
I_{CC2}	Operating Current	CS = V_{IH}, SK = 1KHz TTL Input Levels		3		3	mA
I_{CC3}	Standby Current	PRE = CS = 0V DI = SK = PE = Vcc		2		2	µA
I_{IL}	Input Leakage	V_{IN} = 0V to Vcc	-10.0	10.0	-20.0	20.0	µA
I_{OL}	Output Leakage	V_{OUT} = 0V to Vcc	-1	1	-1	1	µA
V_{IL}	Input Low Voltage		-0.1	0.15Vcc		0.15 Vcc	V
V_{IH}	Input High Voltage		0.8 Vcc	Vcc	0.8 Vcc	Vcc	V
V_{OL2}	Output Low Voltage	I_{OL} = 1.0mA CMOS		0.3		0.3	V
V_{OH2}	Output High Voltage	I_{OH} = -10µA CMOS	Vcc-0.2		Vcc-0.2		V

DC ELECTRICAL CHARACTERISTICS

V_{CC} = 5V ±10%

Symbol	Parameter	Conditions	93CS46		93CS46-E		Units
			Min	Max	Min	Max	
I_{CC1}	Operating Current	CS = V_{IH}, SK = 1MHz CMOS Input Levels		2		2	mA
I_{CC2}	Operating Current	CS = V_{IH}, SK = 1MHz TTL Input Levels		5		5	mA
I_{CC3}	Standby Current	PRE = CS = 0V DI = SK = PE = Vcc		2		2	µA
I_{IL}	Input Leakage	V_{IN} = 0V to Vcc	-10.0	10.0	-10.0	1.0	µA
I_{OL}	Output Leakage	V_{OUT} = 0V to Vcc	-10.0	10.0	-10.0	10.0	µA
V_{IL}	Input Low Voltage		-0.1	0.8	-0.1	0.8	V
V_{IH}	Input High Voltage		2	Vcc	2	Vcc	V
V_{OL1}	Output Low Voltage	I_{OL} = 2.1mA TTL		0.4		0.4	V
V_{OH1}	Output High Voltage	I_{OH} = -400µA TTL	2.4		2.4		V
V_{OL2}	Output Low Voltage	I_{OL} = 10µA CMOS		0.2		0.2	V
V_{OH2}	Output High Voltage	I_{OH} = -10µA CMOS	Vcc-0.2		Vcc-0.2		V

AC ELECTRICAL CHARACTERISTICS

Vcc = 3 V ±10%

Symbol	Parameter	Conditions	93CS46-3		93CS46-3E		Units
			Min	Max	Min	Max	
f_{SK}	SK Clock Frequency		0	250	0	250	KHz
t_{SKH}	SK High Time		1		1		µs
t_{SKL}	SK Low Time		1		1		µs
t_{CS}	Minimum CS Low Time		1		1		µs
t_{CSS}	CS Setup Time	Relative to SK	200		200		ns
t_{PRES}	PRE Setup Time	Relative to SK	200		200		ns
t_{PES}	PE setup Time	Relative to SK	200		200		ns
t_{DIS}	DI Setup Time	Relative to SK	400		400		ns
t_{CSH}	CS Hold Time	Relative to SK	0		0		ns
t_{PEH}	PE Hold Time	Relative to CS	200		200		ns
t_{PREH}	PRE Hold Time	Relative to CS	200		200		ns
t_{DIH}	DI Hold Time	Relative to SK	400		400		ns
t_{PD1}	Output Delay to "1"	AC Test		2		2	µs
t_{PD0}	Output Delay to "0"	AC Test		2		2	µs
t_{SV}	CS to Status Valid	AC Test		2		2	µs
t_{DF}	CS to DO in 3-state	CS = V_{IL}		400		400	ns
t_{WP}	Write Cycle Time			20		20	ms

AC ELECTRICAL CHARACTERISTICS

Vcc = 5V ±10%

Symbol	Parameter	Conditions	93CS46		93CS46-E		Units
			Min	Max	Min	Max	
f_{SK}	SK Clock Frequency		0	1	0	1	MHz
t_{SKH}	SK High Time		400		400		ns
t_{SKL}	SK Low Time		250		250		ns
t_{CS}	Minimum CS Low Time		250		250		ns
t_{CSS}	CS Setup Time	Relative to SK	50		50		ns
t_{PRES}	PRE Setup Time	Relative to SK	50		50		ns
t_{PES}	PE setup Time	Relative to SK	50		50		ns
t_{DIS}	DI Setup Time	Relative to SK	100		100		ns
t_{CSH}	CS Hold Time	Relative to SK	0		0		ns
t_{PEH}	PE Hold Time	Relative to CS	50		50		ns
t_{PREH}	PRE Hold Time	Relative to CS	50		50		ns
t_{DIH}	DI Hold Time	Relative to SK	100		100		ns
t_{PD1}	Output Delay to "1"	AC Test		500		500	ns
t_{PD0}	Output Delay to "0"	AC Test		500		500	ns
t_{SV}	CS to Status Valid	AC Test		500		500	ns
t_{DF}	CS to DO in 3-state	CS = V_{IL}		100		100	ns
t_{WP}	Write Cycle Time			10		10	ms

CAPACITANCE

TA = 25°C, f = 1MHz

Symbol	Parameter	Max	Units
C_{OUT}	Output Capacitance	5	pF
C_{IN}	Input Capacitance	5	pF

INSTRUCTION SET

Instruction	Start Bit	OP Code	Address	Input Data	PRE Pin	PE Pin
READ	1	10	(A5-A0)		0	X
WEN (Write Enable)	1	00	11XXXX		0	1
WRITE	1	01	(A5-A0)	D15-D0	0	1
WRALL (Write All Registers)	1	00	01XXXX	D15-D0	0	1
WDS (Write Disable)	1	00	00XXXX		0	X
PRREAD (Protect Register Read)	1	10	XXXXXX		1	X
PREN (Protect Register Enable)	1	00	11XXXX		1	1
PRCLEAR (Protect Register Clear)	1	11	111111		1	1
PRWRITE (Protect Register Write)	1	01	(A5-A0)		1	1
PRDS (Protect Register Disable)	1	00	000000		1	1
ERASE	1	11	(A5-A0)		0	1
ERAL (Erase All Registers)	1	00	10XXXX		0	1

PIN CONFIGURATIONS

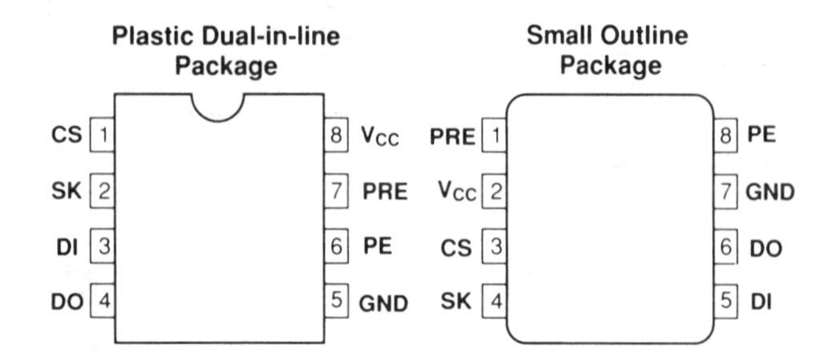

PIN NAMES

CS	Chip Select
SK	Serial Data Clock
DI	Serial Data Input
DO	Serial Data Output
GND	Ground
PE	Program Enable
PRE	Protect Register Enable
Vcc	Power Supply

BLOCK DIAGRAM

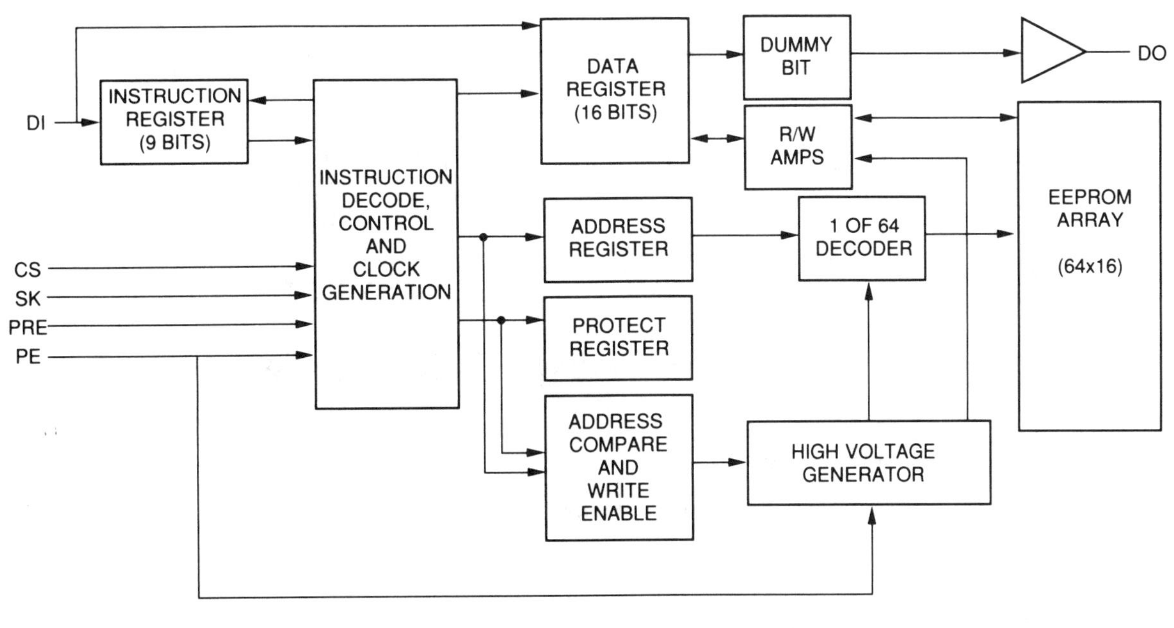

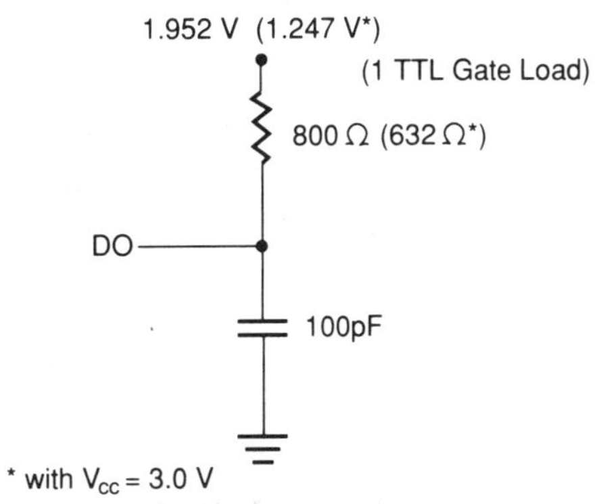

AC TEST CONDITIONS

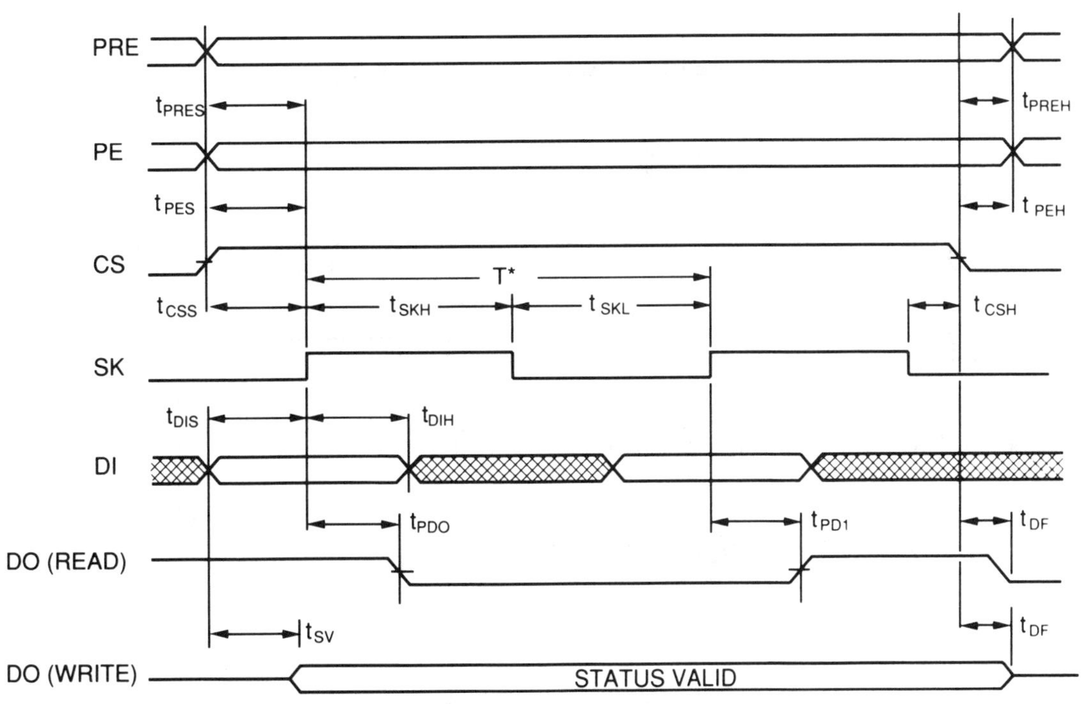

SYNCHRONOUS DATA TIMING

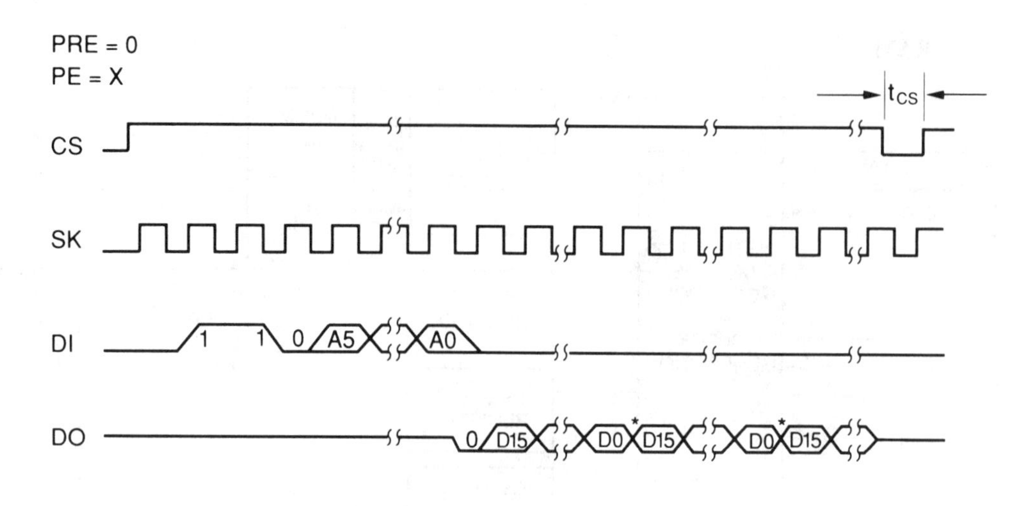

*Address Pointer Cycles to Next Register

READ CYCLE TIMING

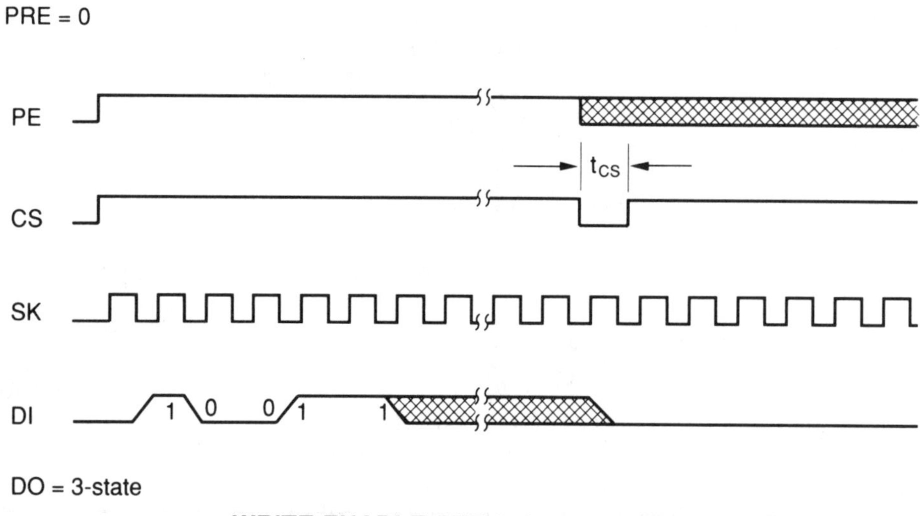

WRITE ENABLE (WEN) CYCLE TIMING

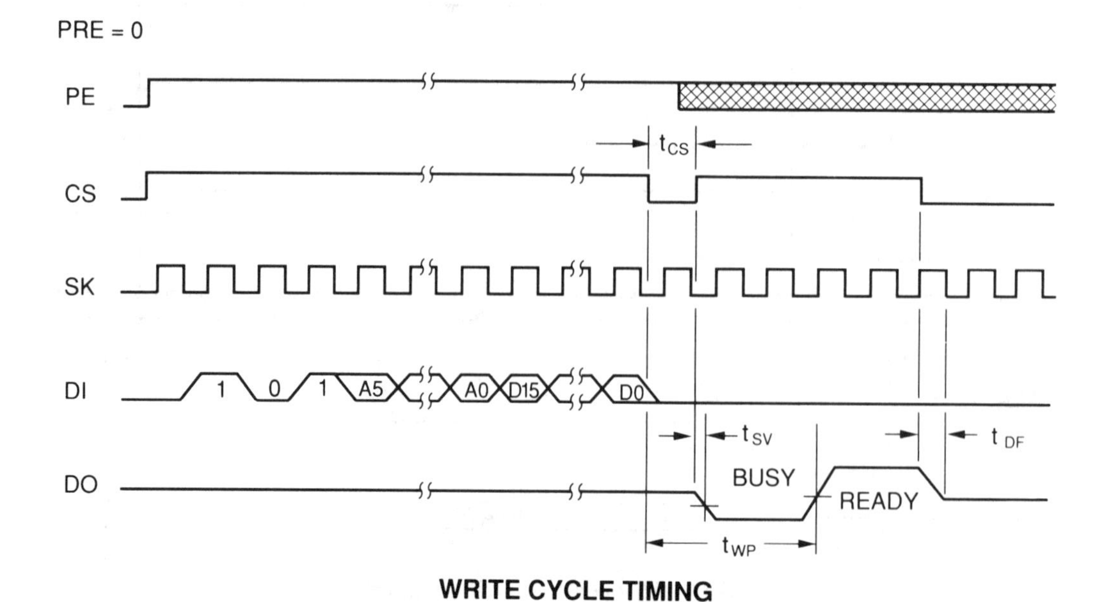

WRITE CYCLE TIMING

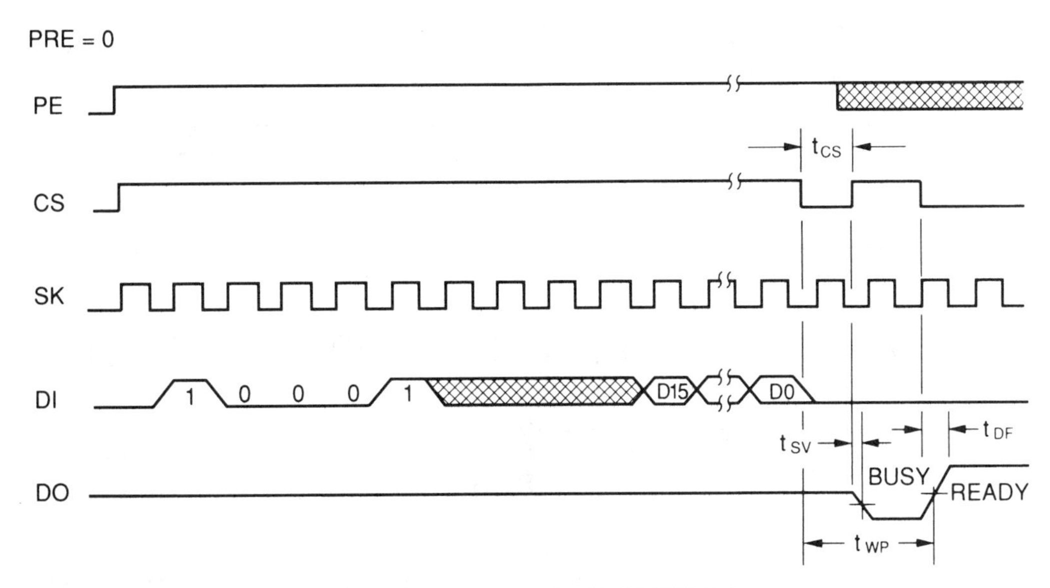

WRITE ALL (WRALL) CYCLE TIMING

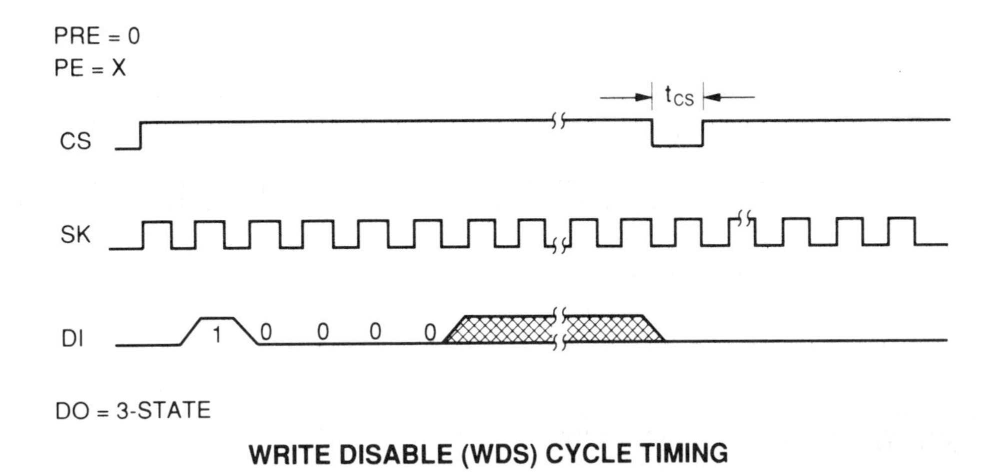

WRITE DISABLE (WDS) CYCLE TIMING

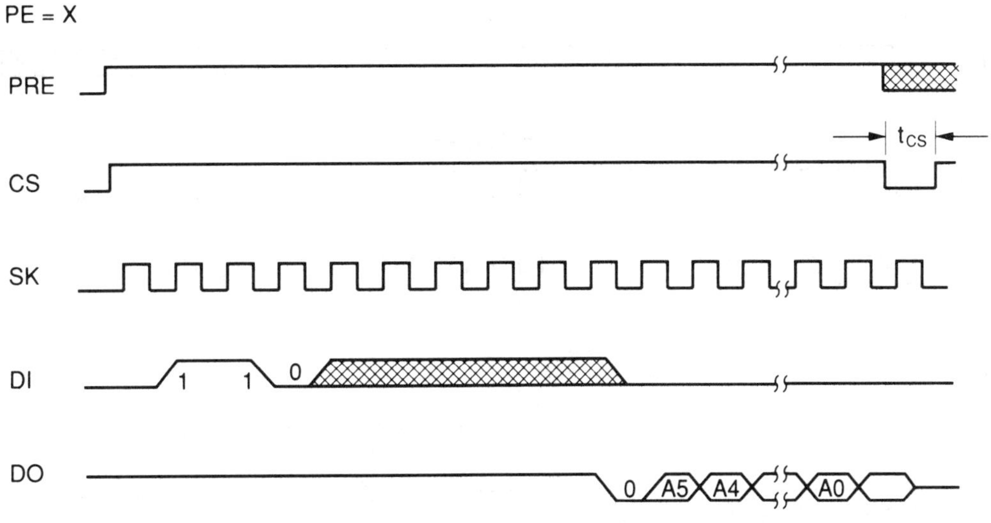

PE = X

PRE

t_{CS}

CS

SK

DI 1 1 0

DO 0 A5 A4 A0

PROTECT REGISTER READ (PRREAD) CYCLE TIMING

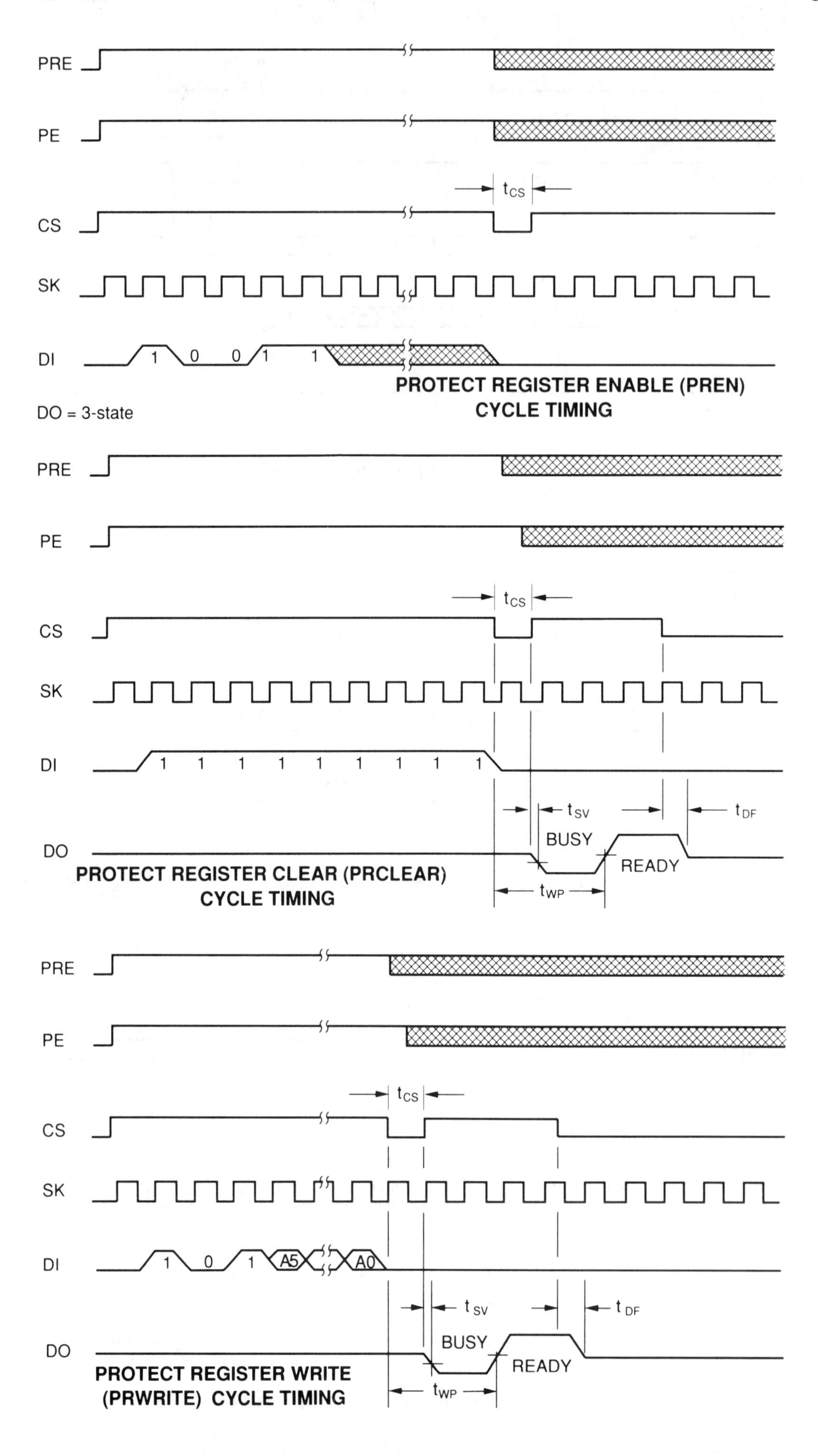
PRE
PE
t_{CS}
CS
SK
DI
1 0 0 1 1
PROTECT REGISTER ENABLE (PREN)
CYCLE TIMING
DO = 3-state
PRE
PE
t_{CS}
CS
SK
DI
1 1 1 1 1 1 1 1 1
t_{SV}
t_{DF}
DO
BUSY
READY
PROTECT REGISTER CLEAR (PRCLEAR)
CYCLE TIMING
t_{WP}
PRE
PE
t_{CS}
CS
SK
DI
1 0 1 A5 A0
t_{SV}
t_{DF}
DO
BUSY
READY
PROTECT REGISTER WRITE
(PRWRITE) CYCLE TIMING
t_{WP}

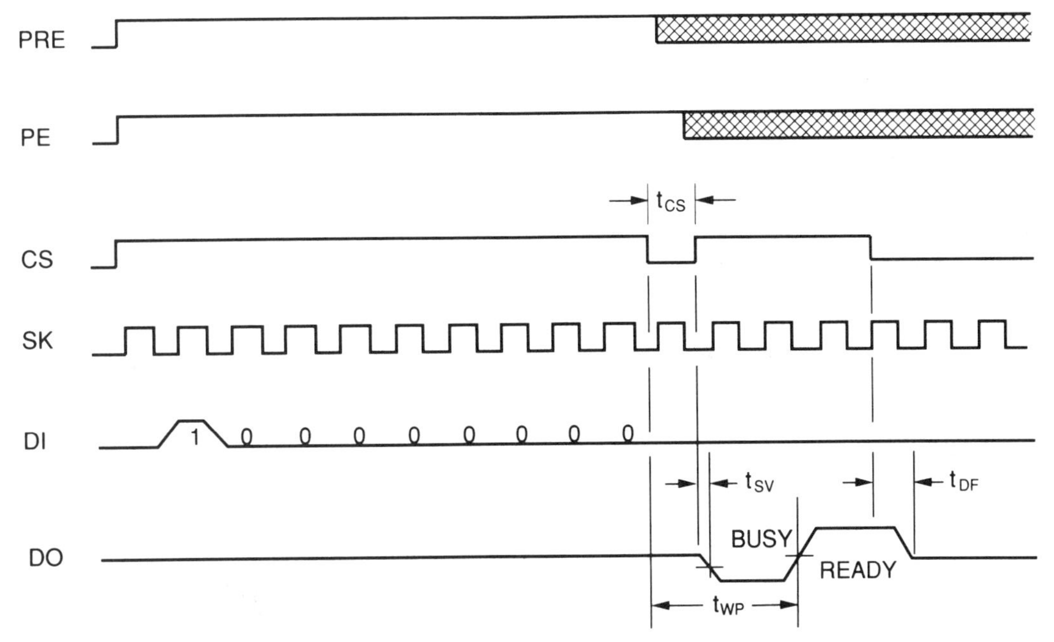

PROTECT REGISTER DISABLE (PRDS) CYCLE TIMING

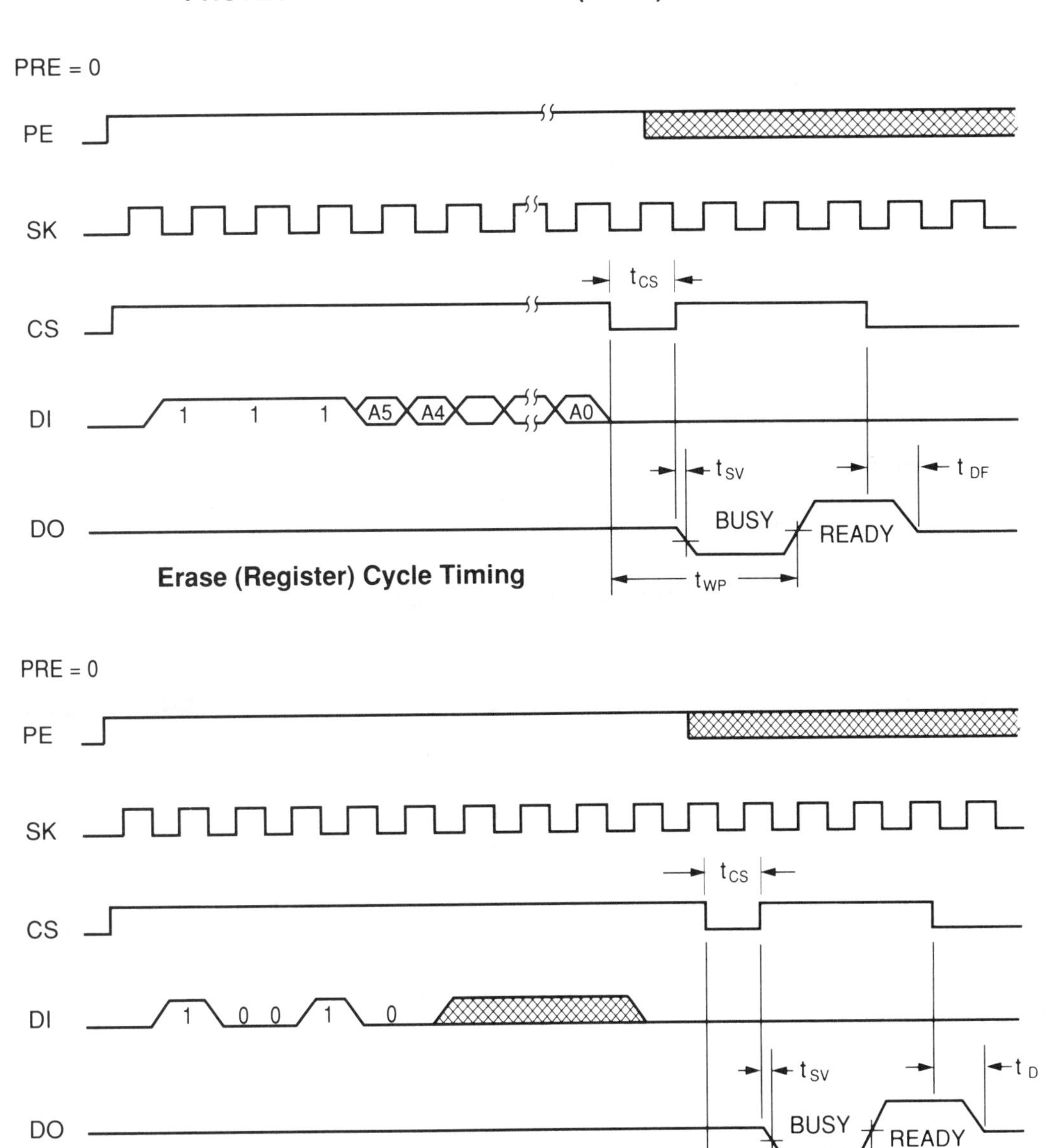

Erase (Register) Cycle Timing

Erase All Cycle Timing

ORDERING INFORMATION

Standard Configurations

Prefix	Part Type	Package Type	Operating Voltage	Operating Teperature
BR	93CS46	PDIP, SOIC	5V, 3V	Commercial Industrial

*CONTACT ROHM FOR YOUR SPECIAL TEMPERATURE AND PACKAGING REQUIREMENTS

PART NUMBERS:

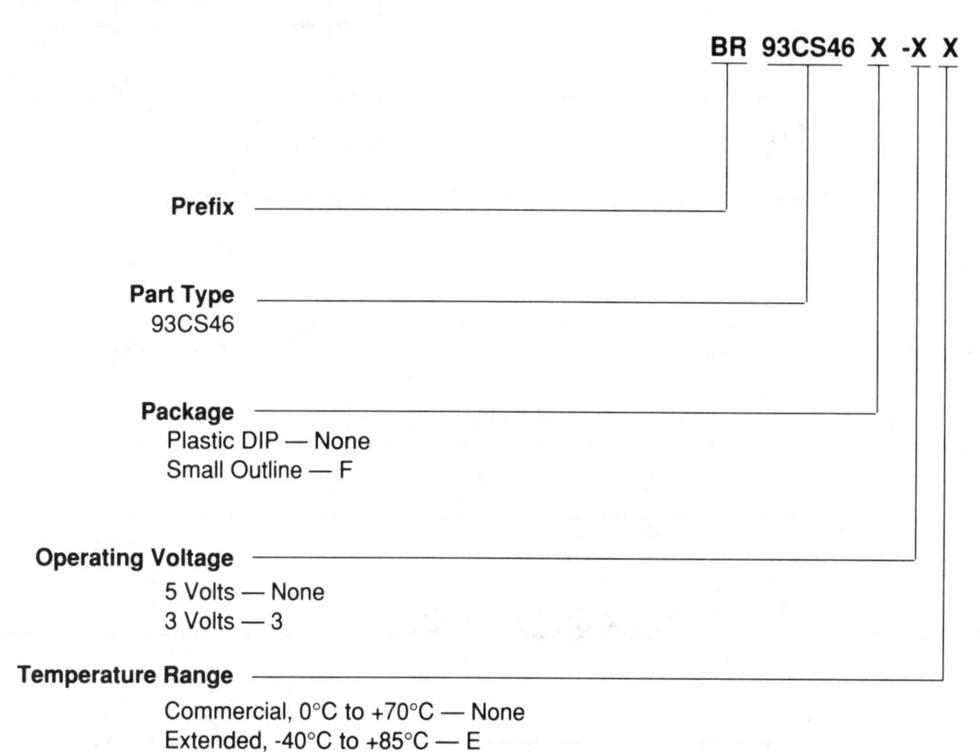

NOTE: SOIC extended temperature package marking is a white dot.

Rohm BA612

Large Current Driver

- Contains five Darlington transistor arrays
- Large current-driving capability (400mA)
- Inputs/outputs arranged on adjacent pins to allow easy board layout
- Large current amplification ratio
- Directly compatible with MOS devices

APPLICATIONS

- Solenoid hammer drivers
- Relay drivers
- LED drivers

The BA612 is primarily designed as a large-current electric hammer driver and contains five Darlington transistor arrays with input resistors. Available in a 14-pin DIP with single-input orientation, the device allows easy board layout.

ABSOLUTE MAXIMUM RATINGS (Ta=25°C)

Parameter	Symbol	Limits	Unit
Supply voltage	V_{CC}	30	V
Power dissipation	Pd	550*	mW
Operating temperature range	Topr	−25 ~ 75	°C
Storage temperature range	Tstg	−55 ~ 125	°C
Collector current	I_C	400	mA
Input voltage (+)	V_{IN+}	30	V
Input voltage (−)	V_{IN-}	−0.5	V

*Derating is done at 5.5mW/°C for operation above Ta=25°C.

ELECTRICAL CHARACTERISTICS (Ta=25°C)

Parameter	Symbol	Min.	Typ.	Max.	Unit	Conditions
Supply voltage range	V_{CC}	—	—	26	V	—
Output leakage current	I_L	—	—	100	μA	V_{CC}=26V, V_{IN}=OV
Output current (1 circuit)	I_{OUT}	—	—	350	mA	Only 1 circuit is ON
Output current (5 circuits)	I_{OUT}		Fig. 5		—	Current per circuit, when 5 circuits are ON simultaneously
Collector current	V_{CE}(sat)	—	—	2.2	V	Iout=350mA, V_{IN}=17V
DC forward current transfer ratio	h_{FE}	2000	—	—	—	—
Input current	I_{IN}	—	0.63	—	mA	V_{IN}=17V, I_{OUT}=OmA

CIRCUIT DIAGRAM

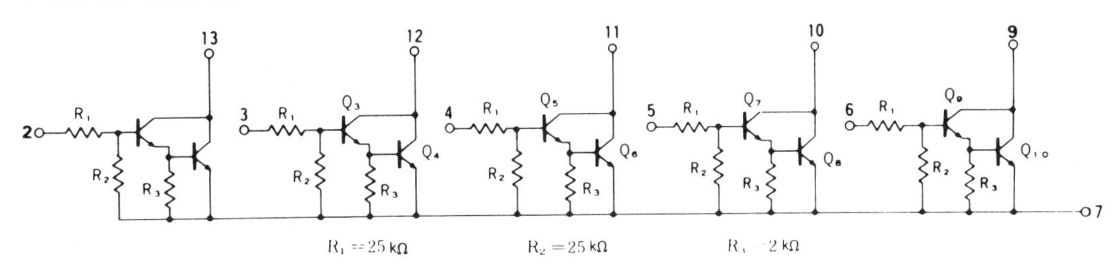

DIMENSIONS (Unit: mm)

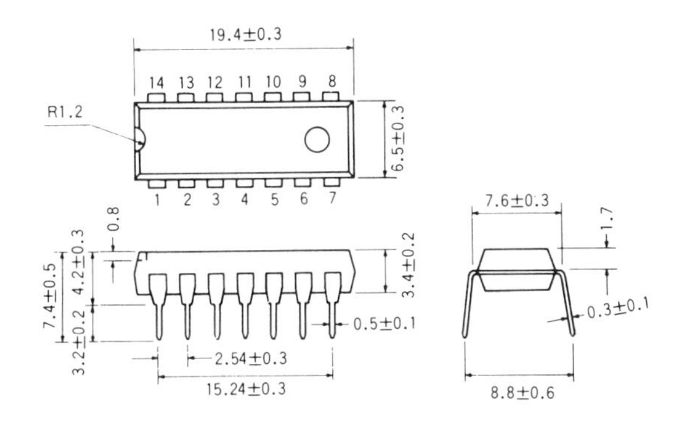

BLOCK DIAGRAM

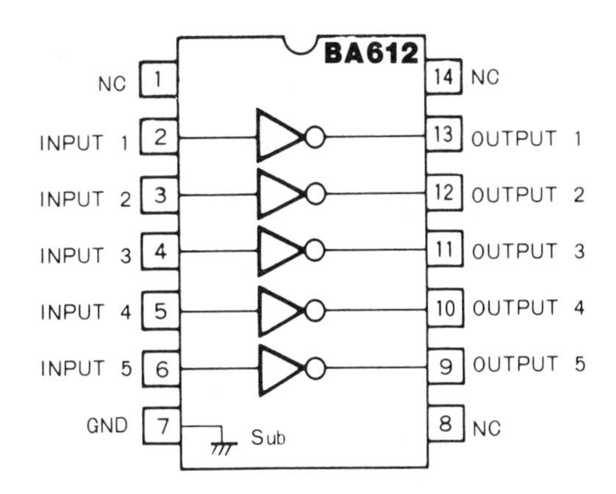

ELECTRICAL CHARACTERISTIC CURVES

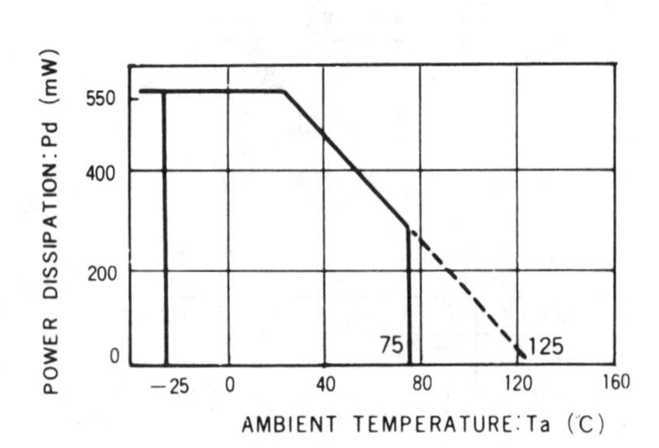

Power dissipation vs. ambient temperature

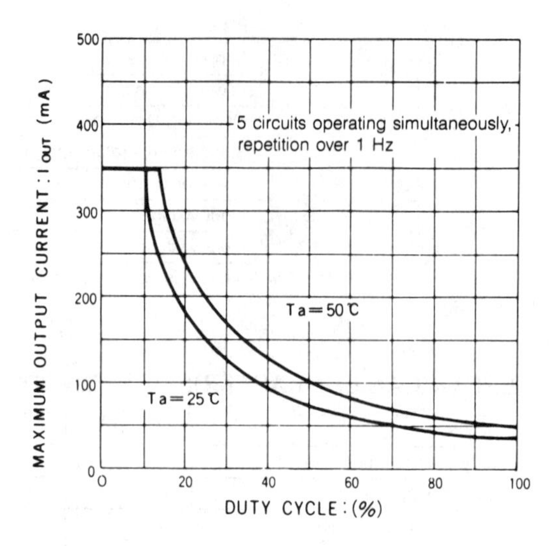

Maximum output current vs. duty cycle

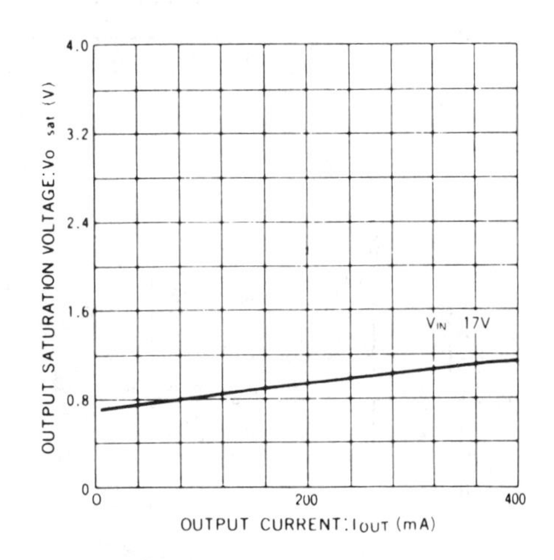

Output saturation voltage vs. output current

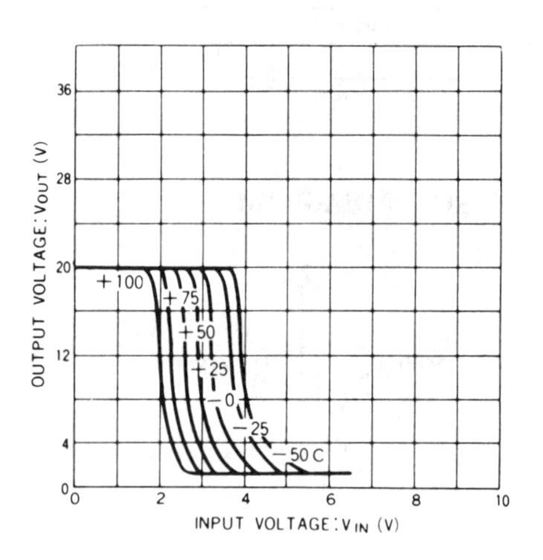

Output voltage vs. input voltage

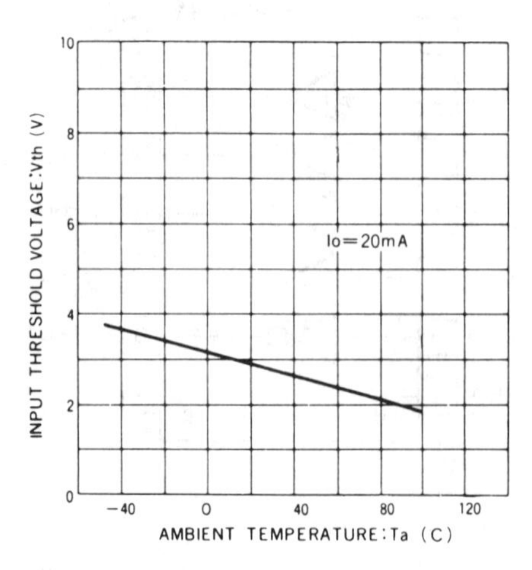

Input threshold voltage vs. ambient temperature

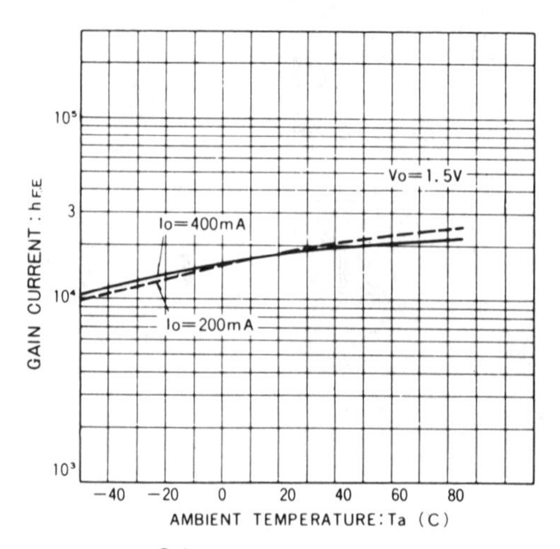

Gain current vs. ambient temperature (when $V_O = 1.5$ V)

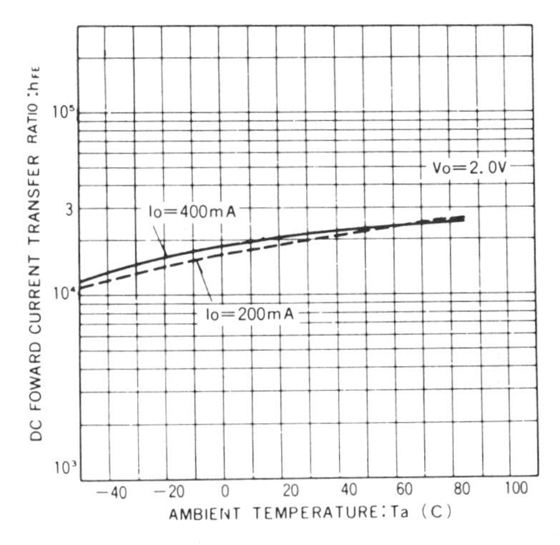

DC forward current transfer ratio vs. ambient temperature (when $V_O = 2$ V)

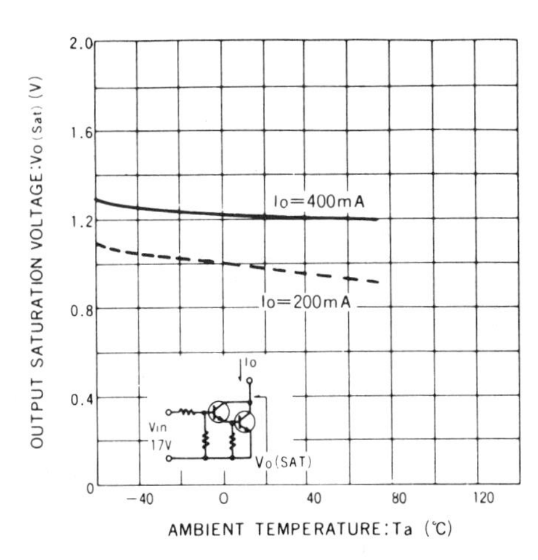

Output saturation voltage vs. ambient temperature

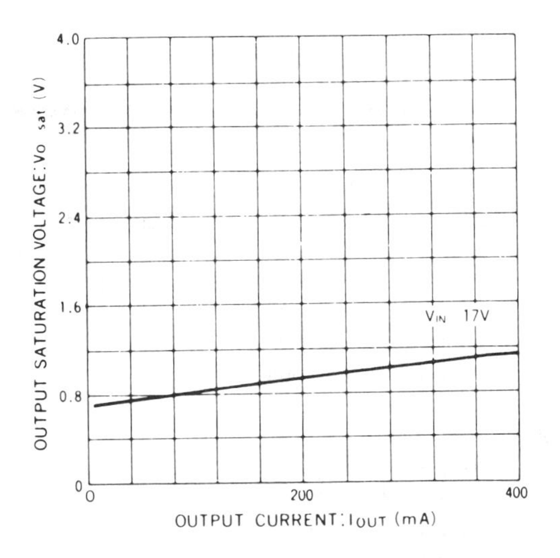

Output saturation voltage vs. output current

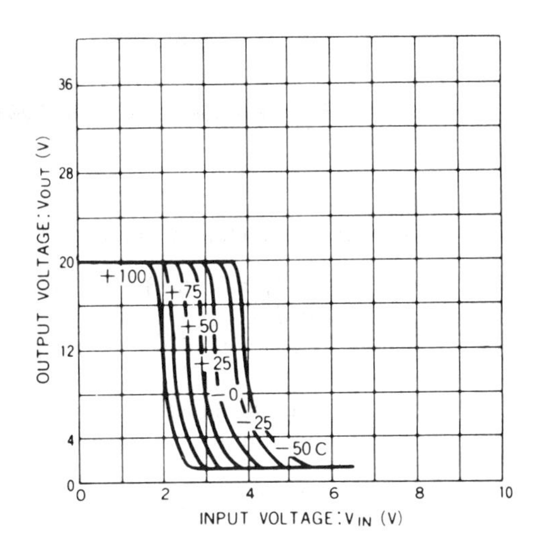

Output voltage vs. input voltage

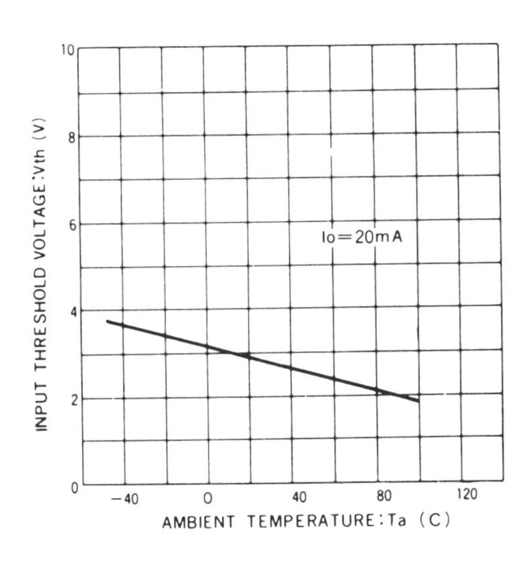

Input threshold voltage vs. ambient temperature

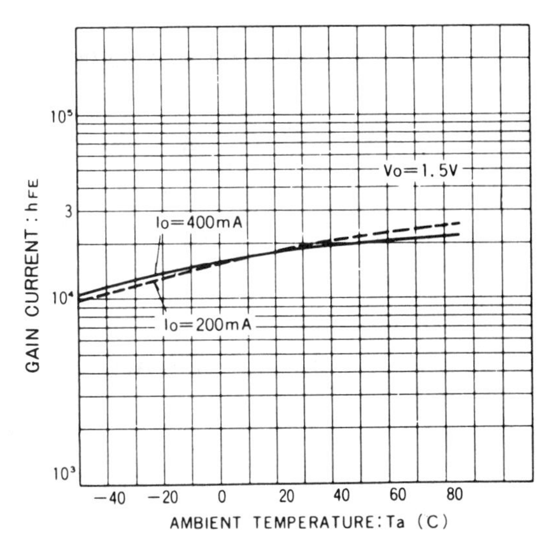

Gain current vs. ambient temperature (when $V_O = 1.5$ V)

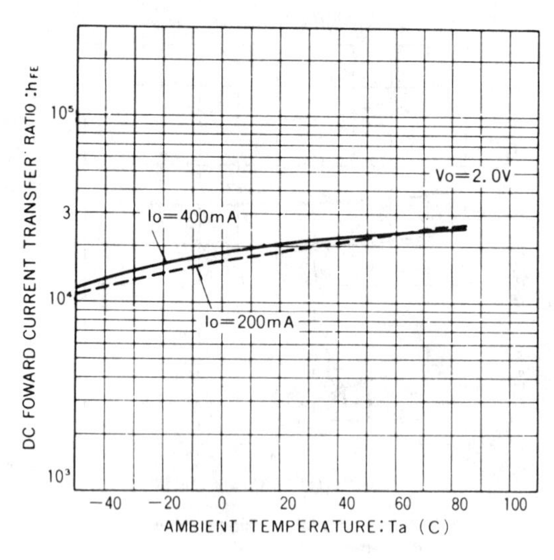

DC forward current transfer ratio vs. ambient temperature (when V_O=2 V)

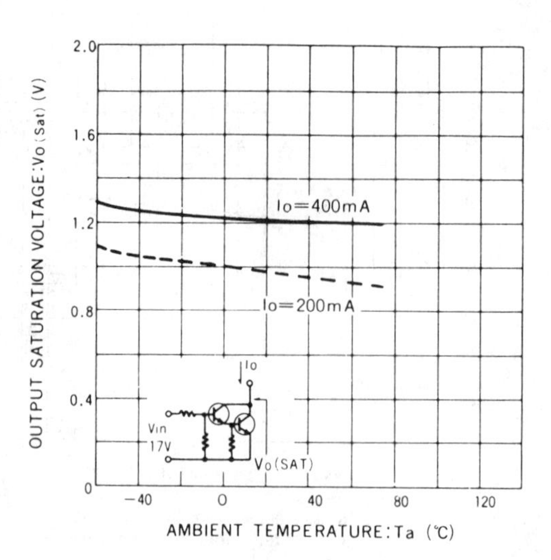

Output saturation voltage vs. ambient temperature

Rohm BA12001

High-Voltage High-Current Darlington Transistor Array

- Large outputs sink current (I_o = 500mA)
- High output voltage (V_o = 50V)
- Seven Darlington transistor arrays implemented on a single chip
- Internal surge suppressor in output stage
- External base current-limiting resistors enables the user to choose the base currents
- Pin-compatible with the XR2201 by EXAR and the μPA2001C by NEC.

APPLICATIONS

- LED, lamp, relay and solenoid drivers
- Device interface

The BA12001 is a high-voltage high-current transistor array containing seven Darlington transistors. With internal surge suppressors for inductive load driving (such as relay coils), the device requires few external components.

Also featuring high output voltage (50V) and high output sink current (500mA), the BA12001 is suitable for interfacing with electric screwdrivers and other types of power tools.

PRECAUTIONS

The BA12001 is a general-purpose transistor array directly interfaceable with PMOS, CMOS, TTL, and other general-logic devices. To limit the base currents to 25mA or below, the device requires series current-limiting resistors at its inputs.

The load should be connected across the driver output and power source. To protect the device from transient spikes, pull up the COM pin (pin 9) to the power-supply level.

ABSOLUTE MAXIMUM RATINGS (Ta=25°C)

Parameter	Symbol	Limits	Unit
Supply voltage	V_{CC}	50	V
Output current	I_O	500	mA/unit
Output voltage	V_O	50	V
Input current	I_I	25	mA/unit

RECOMMENDED OPERATING CONDITIONS (Ta=25°C)

Parameter	Symbol	Typ.	Unit	Conditions
Output saturation current	$V_{CE(sat)}$	0.9	V	I_O=100mA

Note: Input voltage and input current of BA12001 depend on externally connected resistance.

DIMENSIONS Unit: mm)

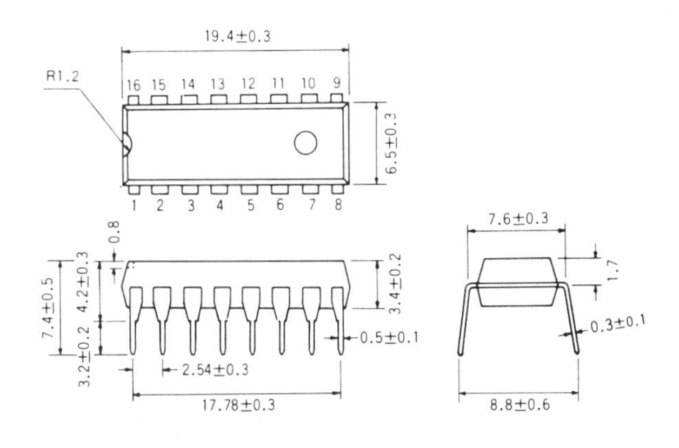

BLOCK DIAGRAM

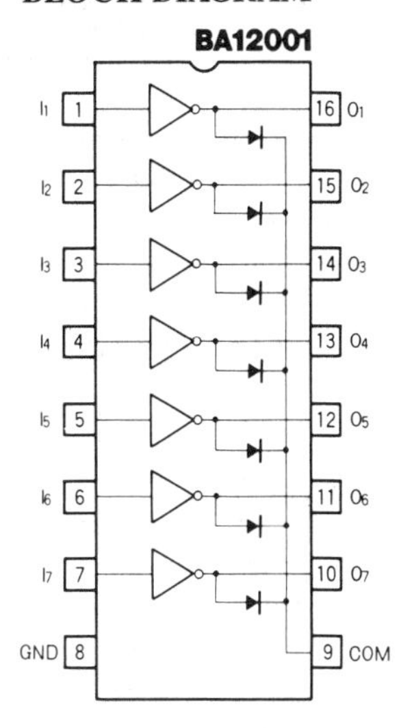

CIRCUIT DIAGRAM

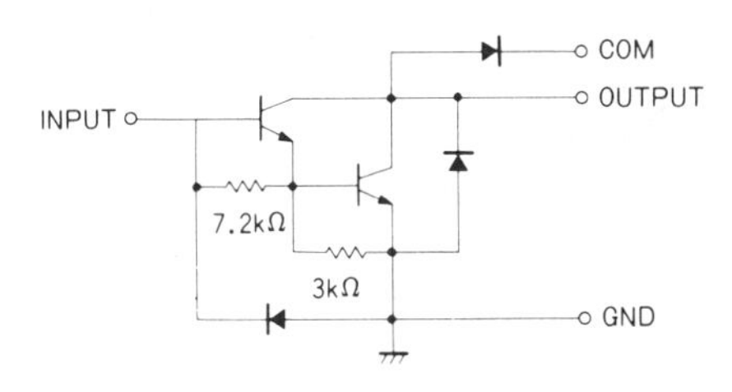

1/7 BA12001 circuit

APPLICATION EXAMPLES

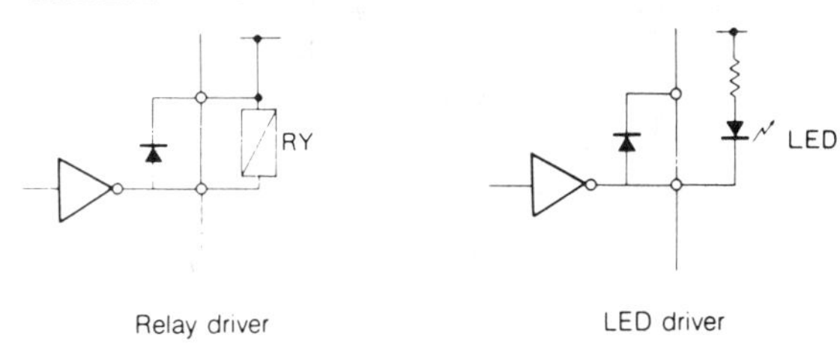

Relay driver

LED driver

Rohm
BA12003, BA12004
7-Channel Drivers

- High output sink current (I_{OUT} = 500mA max.)
- High output voltage (V_{OUT} = 50V max.)
- Seven Darlington transistor arrays implemented on a single chip
- Internal surge suppressor in output stage
- Pin-compatible with the XR-2203/XR-2204 by EXAR and the μPA2003C/μPA2004C by NEC.

APPLICATIONS

- LED, lamp, relay, and solenoid drivers
- Device interface

The BA12003 and BA12004 are high-voltage high-current transistor arrays, each containing seven Darlington transistors. With internal surge suppressors and base current limiting resistor for inductive load driving (such as relay coils), the devices require few external components. Also featuring high-output voltage (50V) and high-output sink current (500mA), the BA12003 and BA12004 are suited for interfacing with other drivers and power devices.

ABSOLUTE MAXIMUM RATINGS (Ta=25°C)

Parameter	Symbol	Limits	Unit
Supply voltage	V_{CE}	50	V
Power dissipation	Pd	1100*1	mW
Operating temperature range	Topr	−25 ~ 75	°C
Storage temperature range	Tstg	−55 ~ 125	°C
Input voltage	V_{IN}	−0.5 ~ 30	V
Output current	I_{OUT}	500	mA/unit
Ground pin current	I_{GND}	2.3*2	A
Diode reverse voltage	V_R	60	V
Diode forward current	I_F	500	mA

*1 Derating is done at 11mW/°C for operation above Ta=25°C.
*2 Pulse width ≦20ms, duty cycle ≦10%, same current through 7 circuits.

RECOMMENDED OPERATING CONDITIONS (Ta=25°C)

Parameter	Symbol	Min.	Typ.	Max.	Unit
Output current	I_{OUT}	—	—	350	mA
Supply voltage	V_{CE}	—	—	50	V
Input voltage	V_{IN}	−0.5	—	30	V
Power dissipation	Pd	—	—	1100	mW

ELECTRICAL CHARACTERISTICS (Unless otherwise specified, Ta=25°C)

Parameter		Symbol	Min.	Typ.	Max.	Unit	Conditions
Output leakage current		I_L	—	—	10	μA	V_{CE}=50V
DC forward current transfer ratio		h_{FE}	1000	—	—	—	V_{CE}=2.0V, I_{OUT}=350mA
Output saturation voltage		$V_{CE(sat)}$	—	—	1.1	V	I_{OUT}=100mA, I_{IN}=250μA
Output saturation voltage		$V_{CE(sat)}$	—	—	1.3	V	I_{OUT}=200mA, I_{IN}=350μA
Output saturation voltage		$V_{CE(sat)}$	—	—	1.6	V	I_{OUT}=350mA, I_{IN}=500μA
Input voltage		V_{IN}	—	—	2.0(5.0)	V	V_{CE}=2.0V, I_{OUT}=100mA
Input voltage		V_{IN}	—	—	2.4(6.0)	V	V_{CE}=2.0V, I_{OUT}=200mA
Input voltage		V_{IN}	—	—	3.4(8.0)	V	V_{CE}=2.0V, I_{OUT}=350mA
Input current	BA12003	I_{IN}	—	—	1.35	mA	V_{IN}=3.85V
	BA12004		—	—	0.5		V_{IN}=5.0V
Diode reverse current		I_R	—	—	50	μA	V_R=50V
Diode forward voltage		V_F	—	—	2.0	V	I_F=350mA
Input capacity		C_{IN}	—	70	—	pF	V_{IN}=0, f=1MHz

Values within ( ) are for BA12003

DIMENSIONS (Unit: mm)

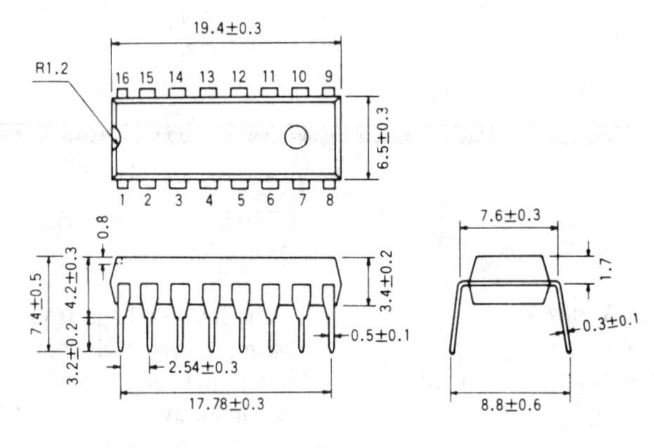

BLOCK DIAGRAM

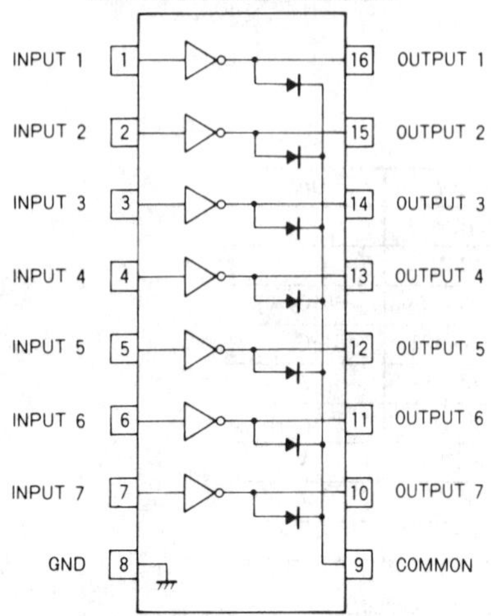

CIRCUIT DIAGRAMS

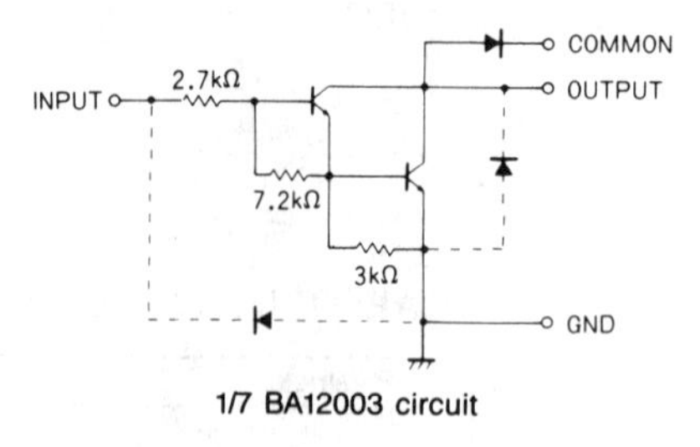

1/7 BA12003 circuit

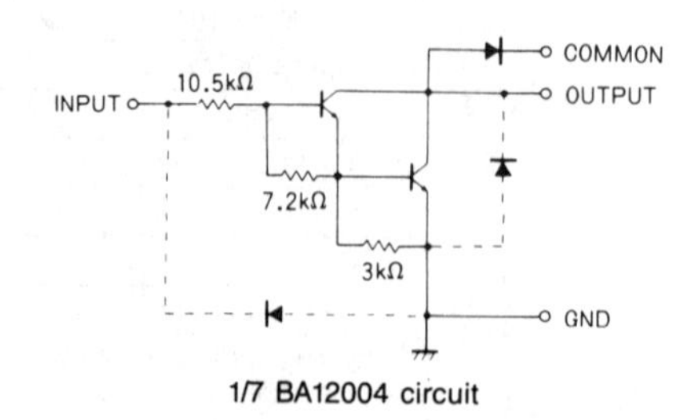

1/7 BA12004 circuit

ELECTRICAL CHARACTERISTIC CURVES (BA12003)

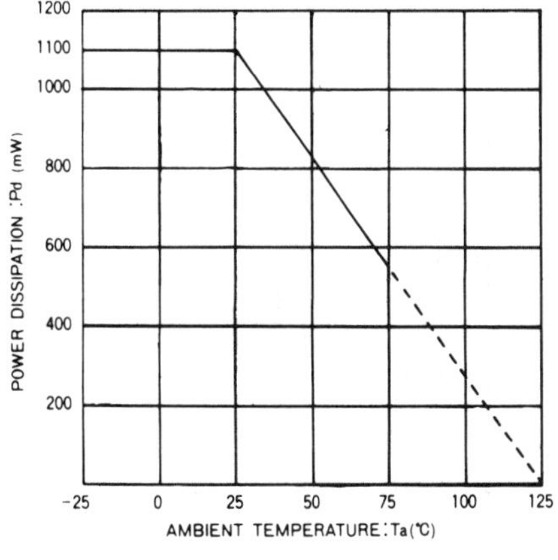

Power dissipation vs. ambient temperature

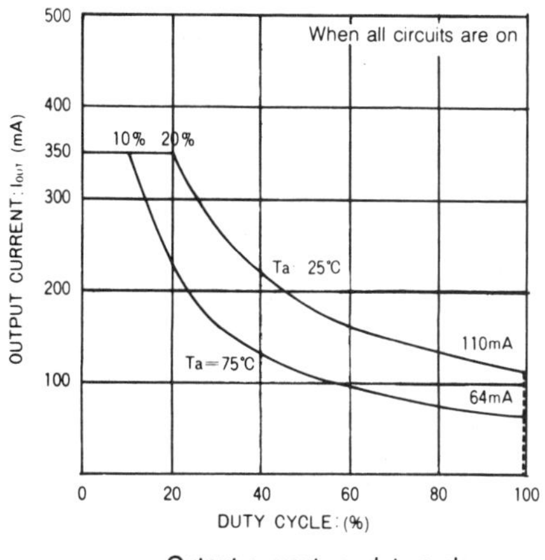

Output current vs. duty cycle

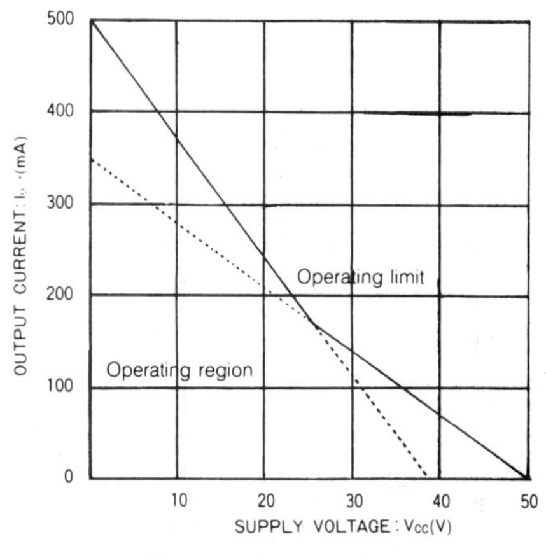

Output current vs. supply voltage

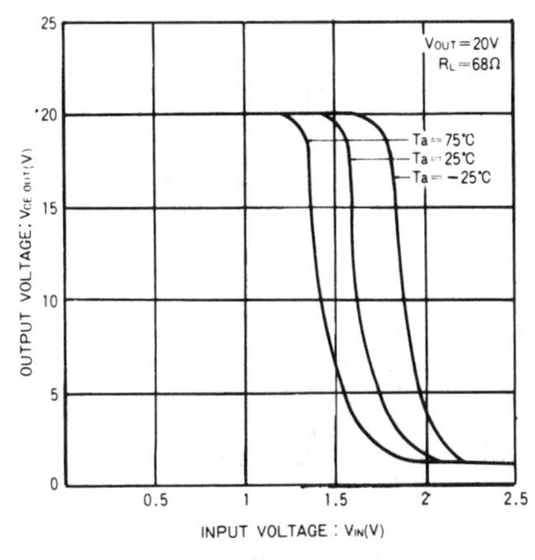

Output voltage vs. input voltage

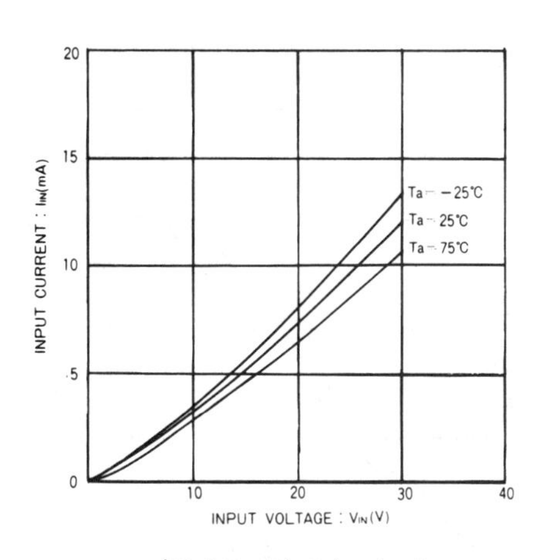

Input current vs. input voltage

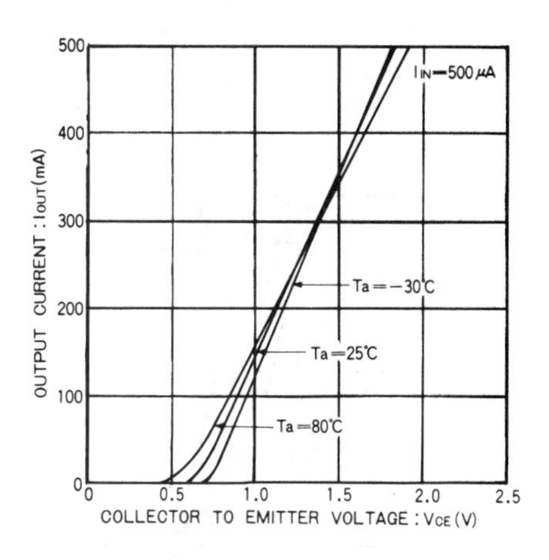

Output current vs. collector to emitter voltage

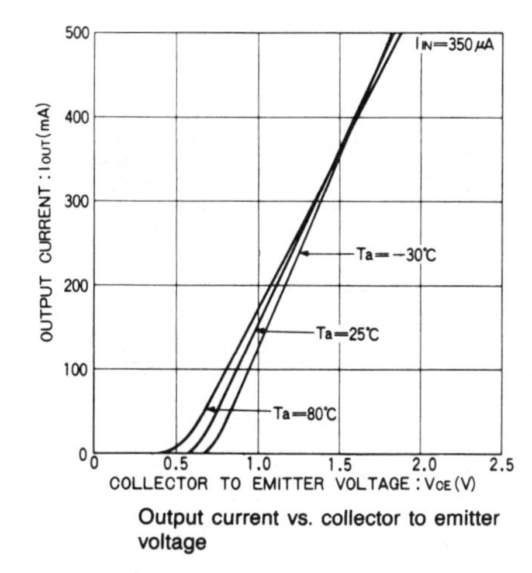

Output current vs. collector to emitter voltage

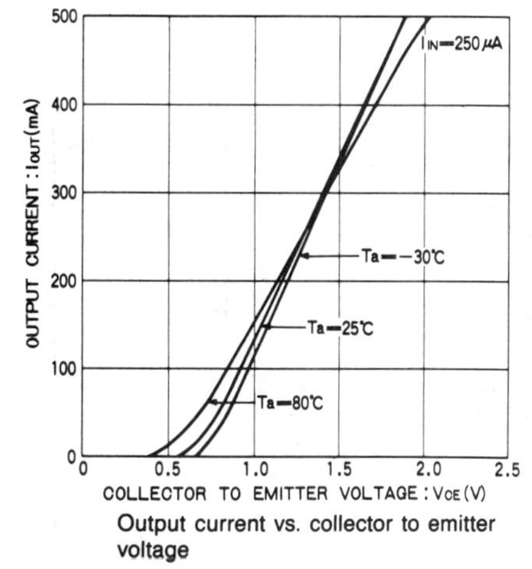

Output current vs. collector to emitter voltage

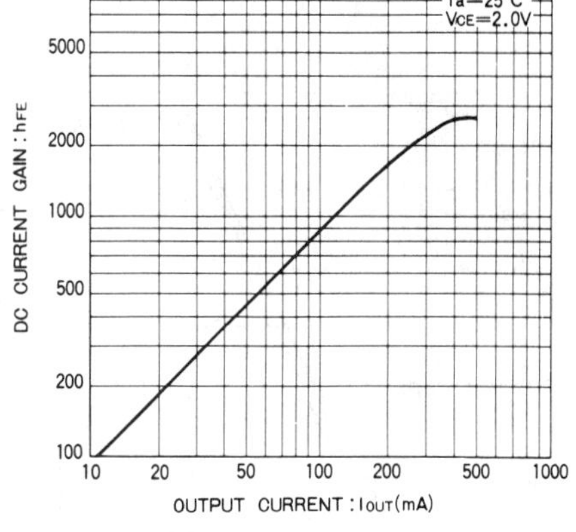

DC current gain vs. output current

ELECTRICAL CHARACTERISTIC CURVES (BA12004)

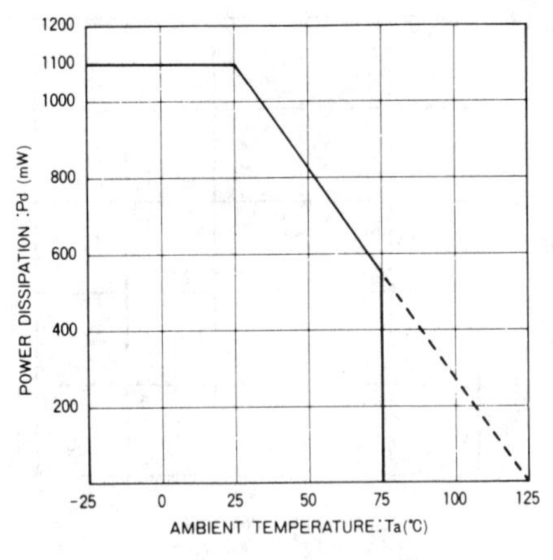

Power dissipation vs. ambient temperature

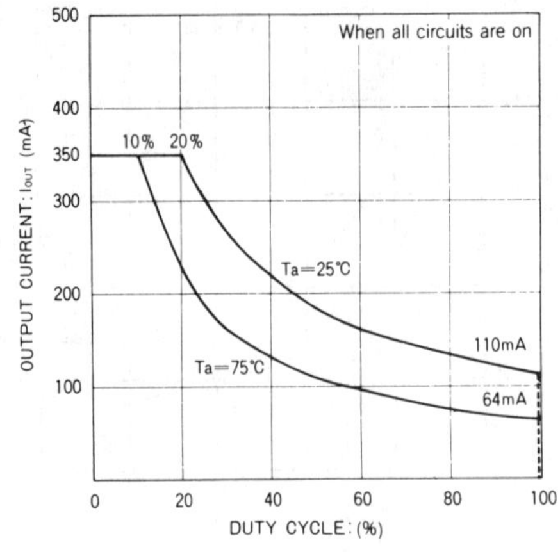

Output current vs. duty cycle

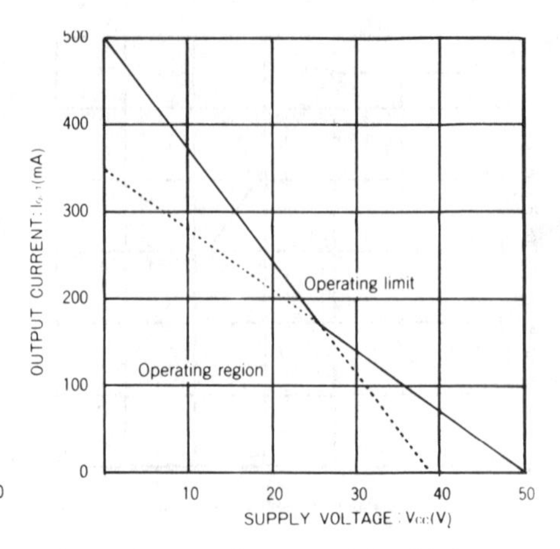

Output current vs. supply voltage

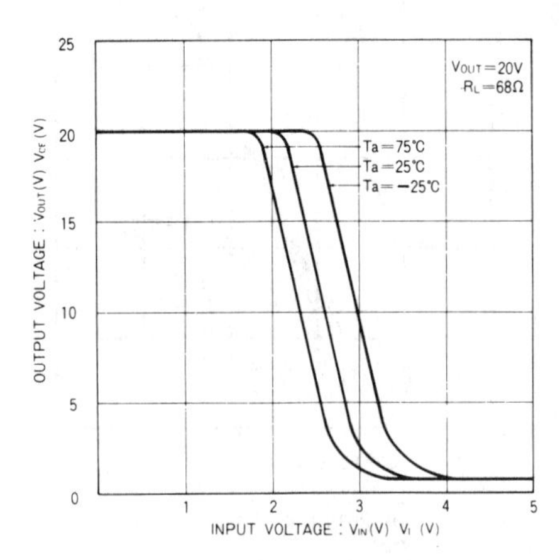

Output voltage vs. input voltage

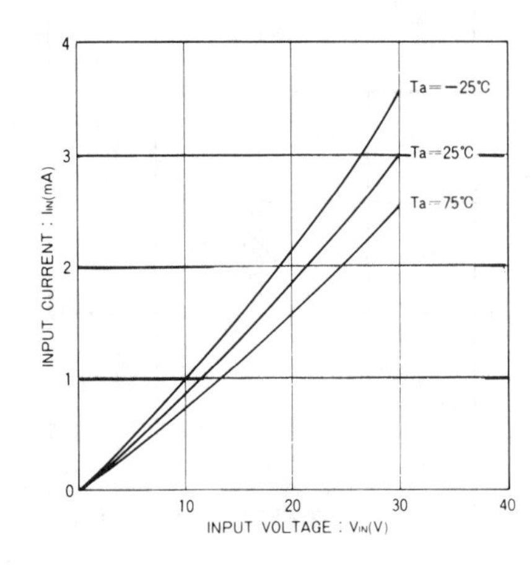

Input current vs. input voltage

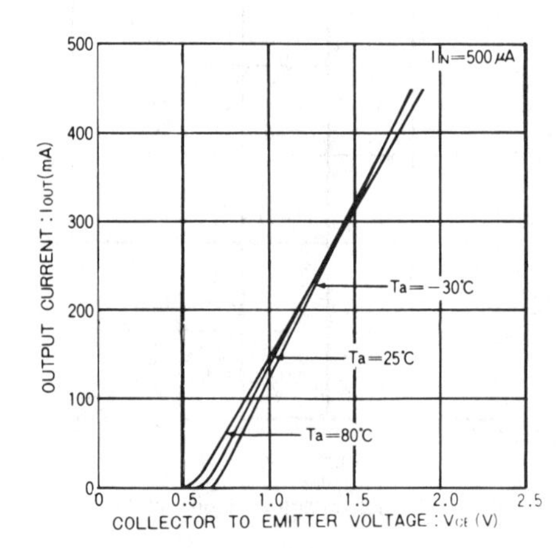

Output current vs. collector to emitter voltage

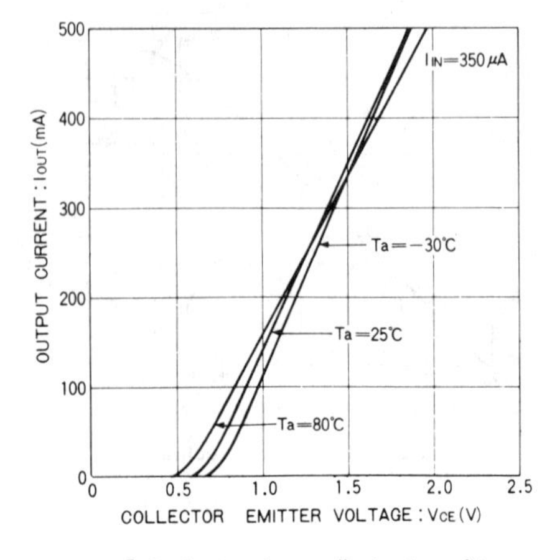

Output current vs. collector to emitter voltage

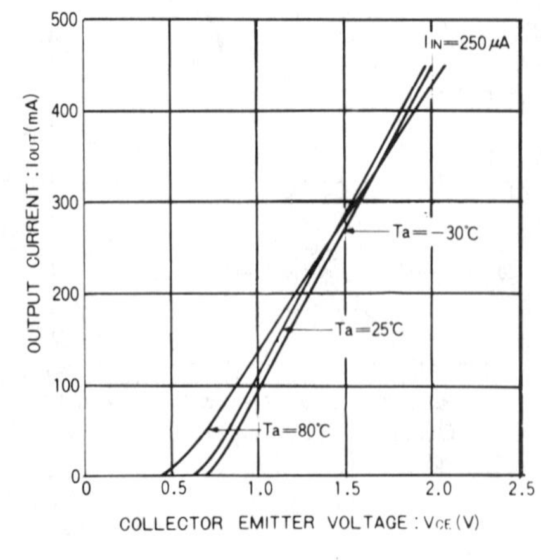

Output current vs. collector to emitter voltage

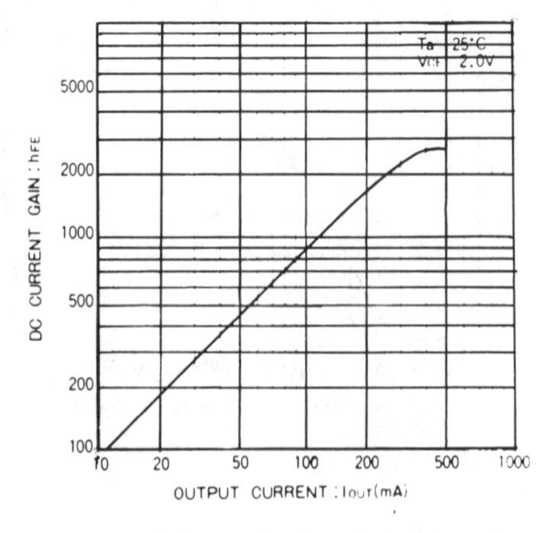

DC current gain vs. output current

Rohm
BA618
LED Driver

- Contains seven drivers
- Up to 100mA of current driving capacity
- Input/output pins arranged on adjacent pin arrays to allow easy board layout
- Directly compatible with TTL devices

APPLICATIONS
- LED drivers
- Relay drivers

The BA618 is a seven-segment LED display driver containing seven independent drivers with active high logic. Available in a 16-pin DIP with all inputs configured on one pin array, the device allows easy board layout design.

ABSOLUTE MAXIMUM RATINGS (Ta=25°C)

Parameter	Symbol	Limits	Unit
Supply voltage	V_{CC}	16	V
Power dissipation	Pd	500*	mW
Operating temperature range	Topr	−30~75	°C
Storage temperature range	Tstg	−55~125	°C
Maximum drive current	I_{OUT}	100	mA
Allowable input voltage range	V_{IN}	−0.5~16	V

*Derating is done at 5mW/°C for operation above Ta=25°C.

ELECTRICAL CHARACTERISTICS (Ta=25°C, V_{CC}=10V, R_L=100Ω)

Parameter	Symbol	Min.	Typ.	Max.	Unit	Conditions
Circuit current (output off)	$I_{CC(OFF)}$	—	—	500	μA	—
Input current (output on)	$I_{IN(ON)}$	—	0.4	0.8	mA	V_{IN}=5V, $V_{OUT}\geqq$8.5V
Input voltage (output on)	$V_{IN(ON)}$	—	1.9	2.5	V	$V_{OUT}\geqq$8.5V(R_L=200Ω)
Input voltage (output on)	$V_{IN(OFF)}$	0.8	1.5	—	V	$V_{OUT}\leqq$3.0mV, 100Ω
Output voltage (input on)	$V_{OUT(ON)}$	8.5	8.9	—	V	V_{IN}=3.0V
Output current (input off)	$I_{OUT(OFF)}$	—	—	30	μA	V_{IN}=0.8V

BLOCK DIAGRAM

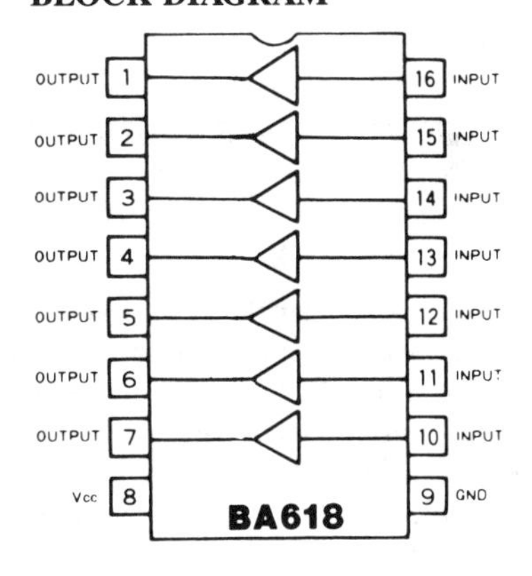

DIMENSIONS (Unit: mm)

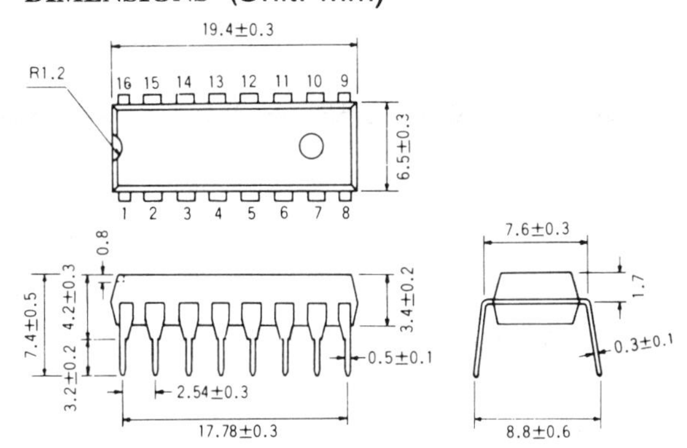

CIRCUIT DIAGRAM

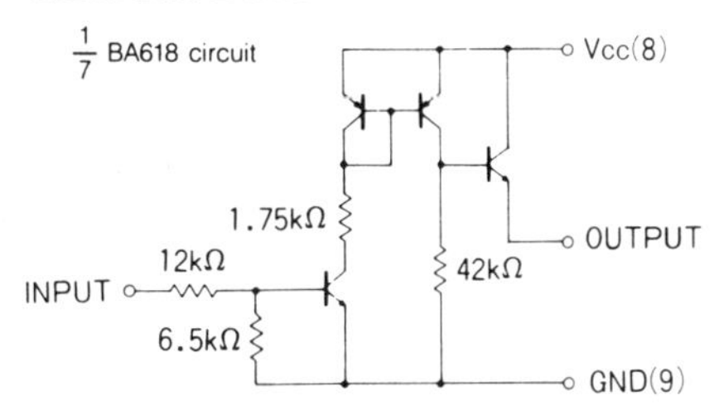

ELECTRICAL CHARACTERISTIC CURVES

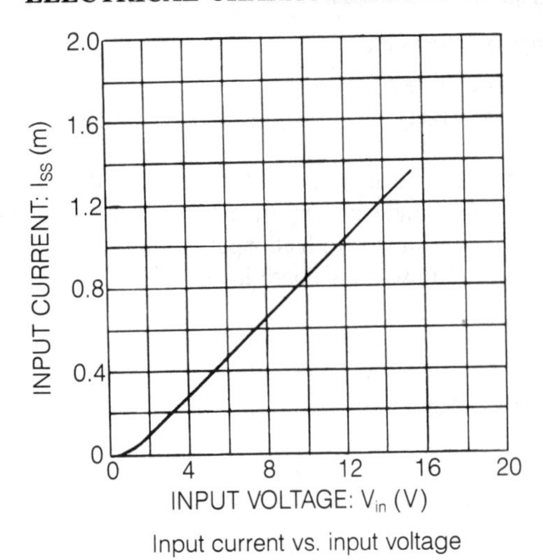

Input current vs. input voltage

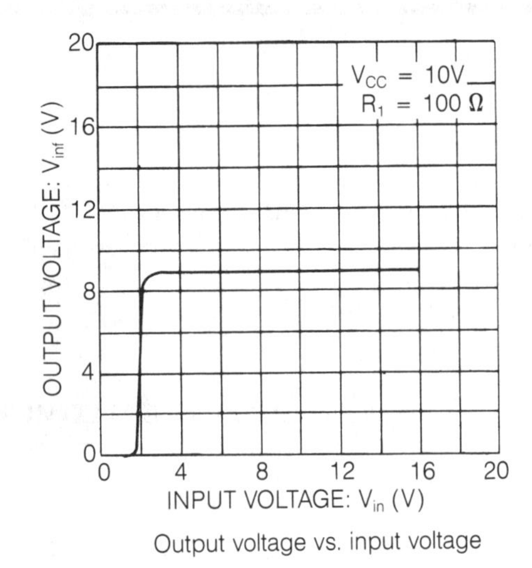

Output voltage vs. input voltage

APPLICATION EXAMPLES

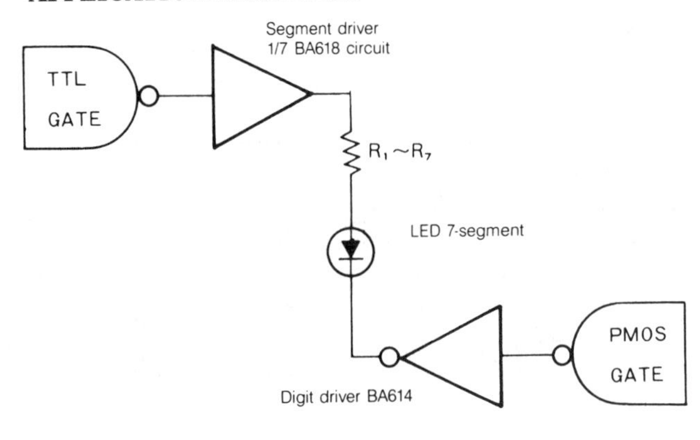

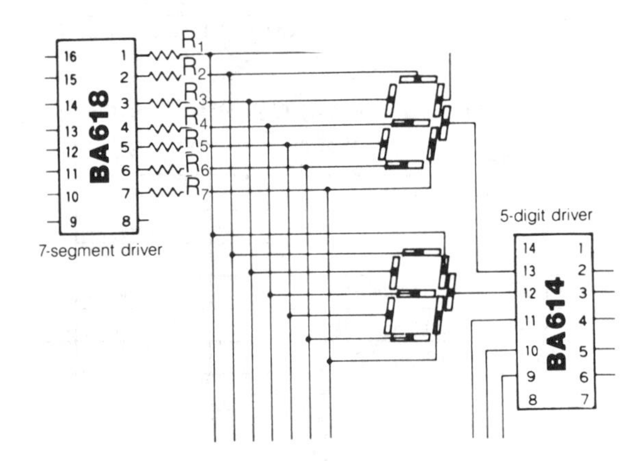

7-segment, 5-digit LED driver circuit

Siliconix DG200A
Dual Monolithic SPST CMOS Analog Switch

- ±15V input signal range
- 44V maximum supply ratings
- On-resistance <70Ω
- TTL and CMOS compatibility
- 2500V ESD protection
- Wide dynamic range
- Simple interfacing
- Reduced external component count

APPLICATIONS
- Servo control switching
- Programmable-gain amplifiers
- Audio switching

The DG200A is a dual single-pole, single-throw analog switch designed to provide general-purpose switching of analog signals. This device is ideally suited for designs requiring a wide analog voltage range coupled with low on-resistance.

The DG200A is designed on Siliconix' improved PLUS-40 CMOS process which includes sandwich passivation and 2500V ESD protection to MIL-M-3015.7 for ruggedness. An epitaxial layer prevents latchup.

Each switch conducts equally well in both directions when on, and blocks up to 30 volts peak-to-peak when off. In the on condition, this bidirectional switch introduces no offset voltage of its own.

Packaging for the DG200A include a 14-pin CerDIP, metal can, and plastic DIP options. Performance grades include military A-suffix (−55 to 125 °C), industrial B-suffix (−25 to 85 °C), and commercial C-suffix (0 to 70 °C) temperature ranges.

ABSOLUTE MAXIMUM RATINGS

V+ to V- . . . . 44 V

GND to V- . . . . 25 V

Digital Inputs[1], V_S, V_D . . . . (V-) -2 V to (V+) +2 V or . . . . 30 mA, whichever occurs first.

Current (Any Terminal) Continuous . . . . 30 mA

Current S or D (Pulsed at 1 ms, 10% Duty Cycle Max) . . . . 100 mA

Operating Temperature (A Suffix) . . . . -55 to 125°C
(B Suffix) . . . . -25 to 85°C
(C Suffix) . . . . 0 to 70°C

Storage Temperature (A & B Suffix) . . . . . . -65 to 150°C
(C Suffix) . . . . . . -65 to 125°C

Power Dissipation (Package)*

Metal Can** . . . . 450 mW

14-Pin Ceramic DIP*** . . . . 825 mW

14-Pin Plastic DIP**** . . . . 470 mW

*All leads soldered or welded to PC board.
**Derate 6 mW/°C above 75°C.
***Derate 11 mW/°C above 75°C.
****Derate 6.5 mW/°C above 25°C.

PIN CONFIGURATION

Dual-In-Line Package

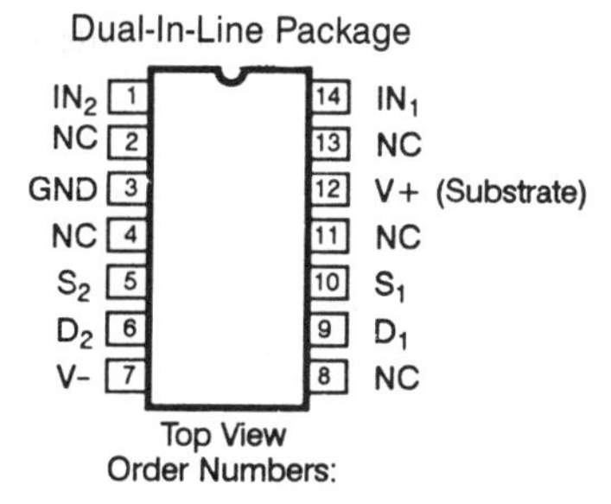

Top View
Order Numbers:

CerDIP: DG200AAK, DG200AAK/883
DG200ABK
Plastic: DG200ACJ

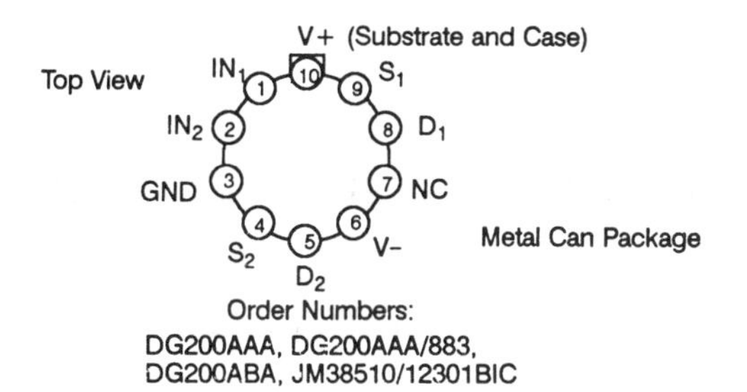

Metal Can Package

Order Numbers:
DG200AAA, DG200AAA/883,
DG200ABA, JM38510/12301BIC

FUNCTIONAL BLOCK DIAGRAM

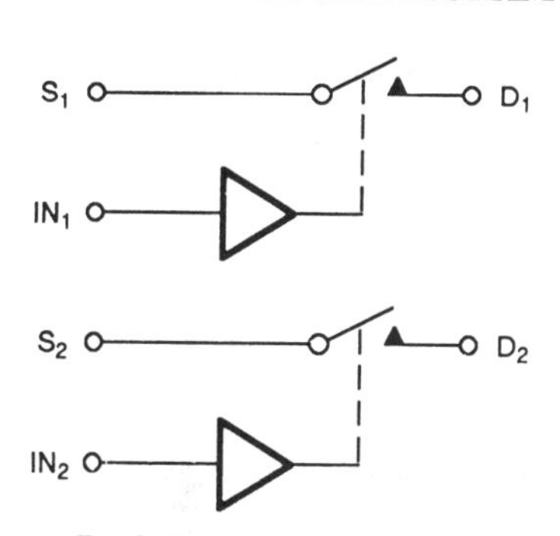

Two SPST Switches per Package*

Truth Table

Logic	Switch
0	ON
1	OFF

Logic "0" ≤ 0.8 V
Logic "1" ≥ 2.4 V

* Switches Shown for Logic "1" Input

TYPICAL CHARACTERISTICS

$r_{DS(ON)}$ vs. V_D and Power Supply Voltage

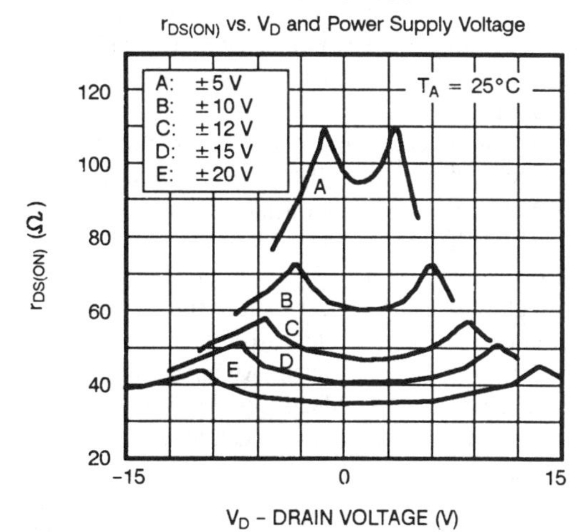

Leakage Currents vs. Analog Voltage

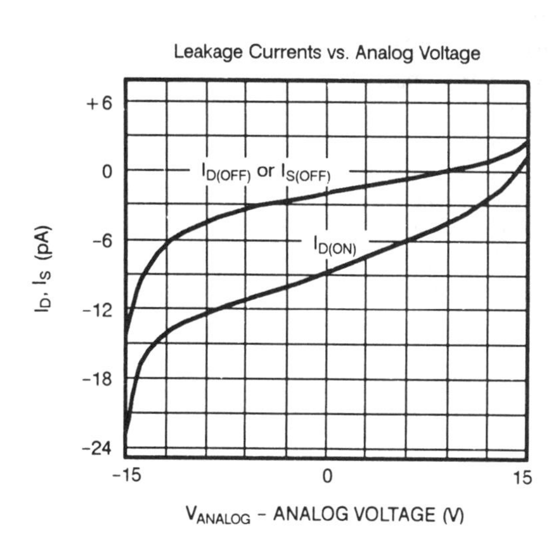

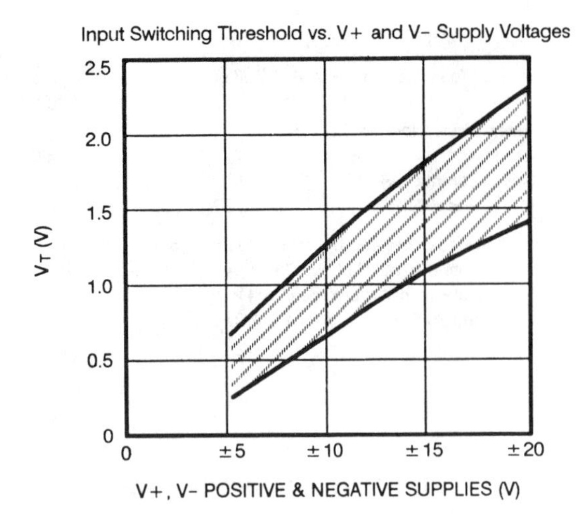

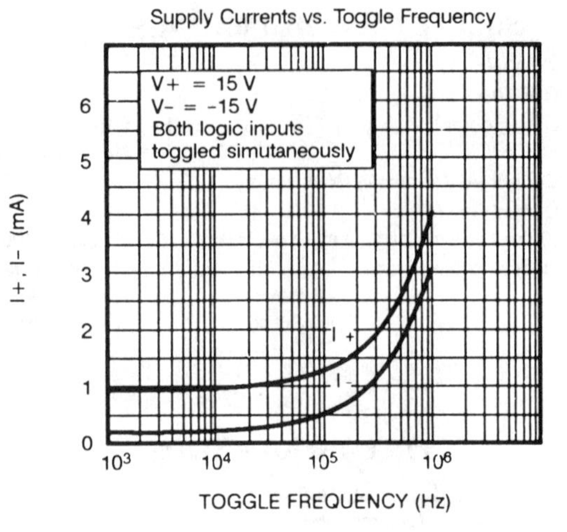

SCHEMATIC DIAGRAM

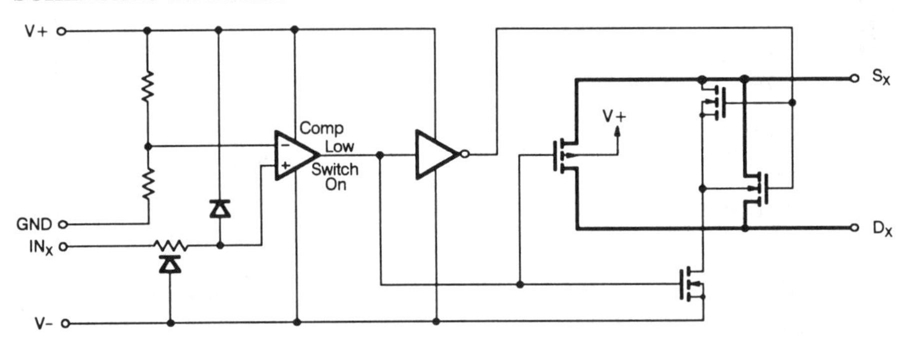

Siliconix DG485

Low-Power CMOS Octal Analog Switch Array

- ±15V input range
- On-resistance <85Ω
- Serial date input/output
- Low-power (P_D < 105μW)
- TTL and CMOS compatible
- Any combination of 8 SPST to the output
- ESD protection > ±4000V
- Devices can be chained for system expansion
- Reduced control wires
- Reduced board space
- Low signal distortion
- Reduced switch errors
- Reduced power supply
- Simple interfacing
- Improved reliability

APPLICATIONS

- Audio switching and routing
- Audio teleconferencing
- Serial data acquisition and process control
- Battery and remote systems
- Automotive, avionics, and ATE systems
- Summing node amplifiers

The DG485 is an analog switch array that can be used as a low-power 8-channel multiplexer for use in serial control applications. Any, all or none of the 8 switches can be closed at any given time. Combining low on-resistance ($r_{DS(ON)}$ <85Ω) and fast switching t_{ON}<200ns), the DG485 is ideally suited for data acquisition, process control, communication, and avionic applications.

The control data is input serially into the shift register with each clock pulse. The shift register contents can be latched-in via LD at any point into an octal latch, which in turn controls all switches. $\overline{RS}$ resets the shift register, forcing all latch inputs to a low condition. The serial input (D_{IN}) and serial output (D_{OUT}) allow chaining of arrays for large systems.

Built on the Siliconix high-voltage silicon-gate process, the DG485 has a wide 44V range. An epitaxial layer prevents latchup. Each channel conducts equally well in either direction when on and blocks up to 30V peak-to-peak when off.

Packaging for the DG485 consists of the 18-pin CerDIP, plastic DIP and 20-pin PLCC for surface mount. Temperature ranges available are military A-suffix (−55 to 125 °C) and industrial D-suffix (−40 to 85 °C).

ABSOLUTE MAXIMUM RATINGS

Voltages Referenced to V-

V+ .... 44 V

GND .... 25 V

Digital Inputs[1] V_S, V_D .... (V-) -2 V to (V+) + 2 V or 30 mA, whichever occurs first

Continuous Current (Any Terminal) .... 30 mA

Current, S or D (Pulsed 1 ms, 10% duty cycle) .... 100 mA

Storage Temperature (A Suffix) .... -65 to 150°C
(D Suffix) .... -65 to 125°C

Operating Temperature (A Suffix) .... -55 to 125°C
(D Suffix) .... -40 to 85°C

Power Dissipation (Package)*

18-Pin CerDIP** .... 600 mW
18-Pin Plastic DIP*** .... 470 mW
20-Pin PLCC**** .... 450 mW

*All leads welded or soldered to PC Board.
**Derate 9.2 mW/°C above 75°C.
***Derate 16.5 mW/°C above 25°C.
****Derate 6 mW/°C above 75°C.

1 Signals on S_x, D_x, or IN_x exceeding V+ or V- will be clamped by internal diodes. Limit forward diode current to maximum current ratings.

FUNCTIONAL BLOCK DIAGRAM AND TRUTH TABLES

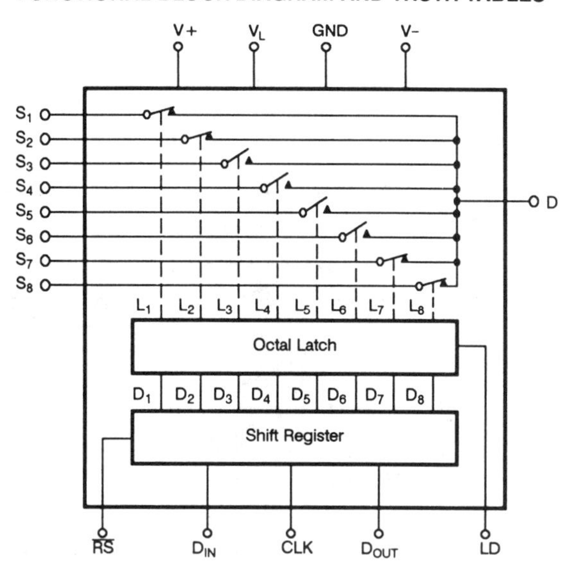

$\overline{RS}$	CLK	D_{IN}	D_1	D_N
1	rising edge	0	0	D_{N-1}
1	rising edge	1	1	D_{N-1}
1	falling edge	X	D_1	D_N (No Change)
0	X	X	0	0

The CLK input is edge triggered

LD	D_N	L_N	SW_N
rising edge	0	0	OFF
rising edge	1	1	ON
falling edge	D_n	L_n	(No Change)

The LD input is level triggered

PIN CONFIGURATIONS

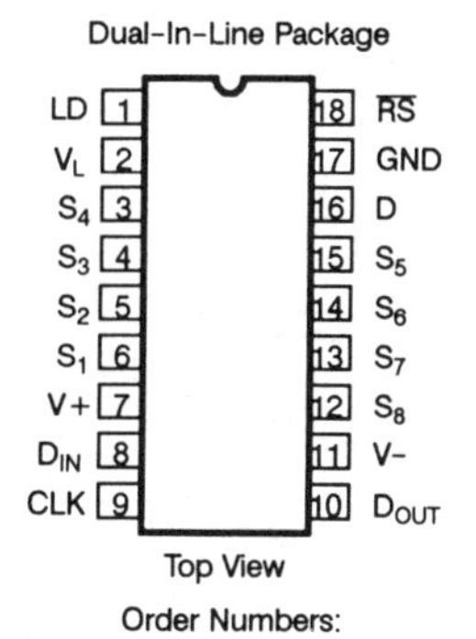

Order Numbers:
CerDIP: DG485AK, DG485AK/883
Plastic: DG485DJ

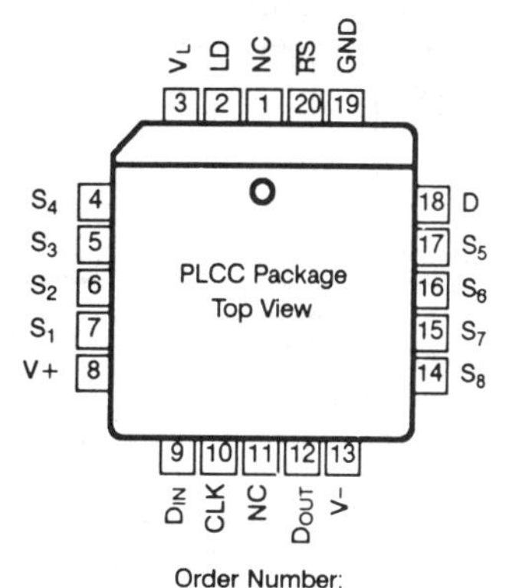

Order Number:
DG485DN

TYPICAL CHARACTERISTICS

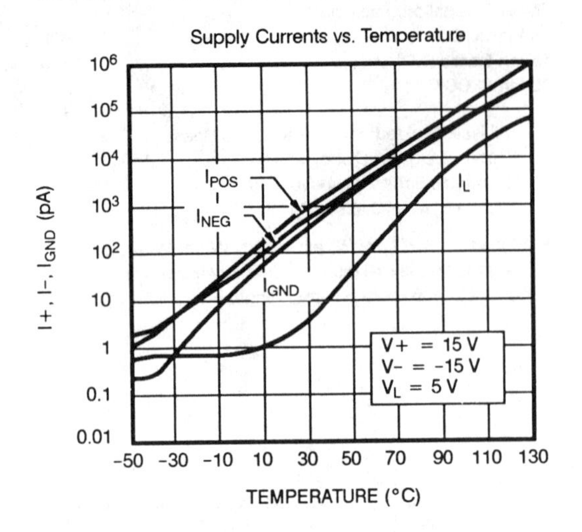

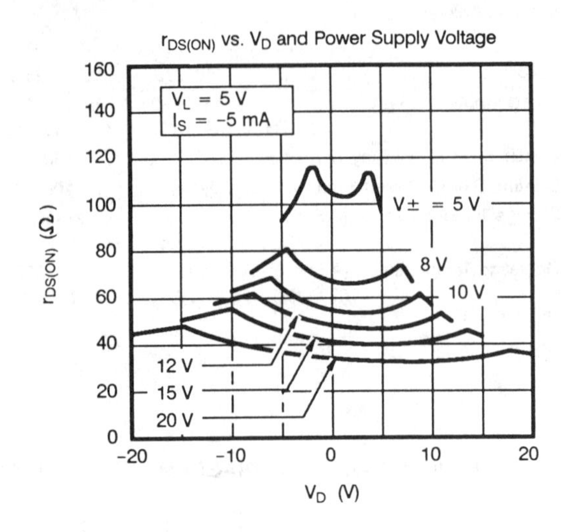

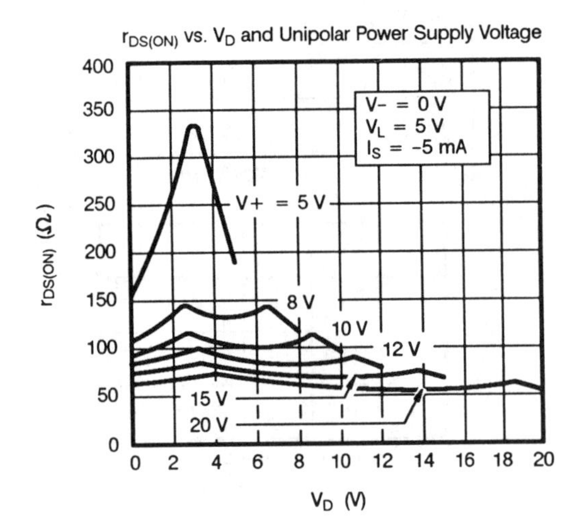

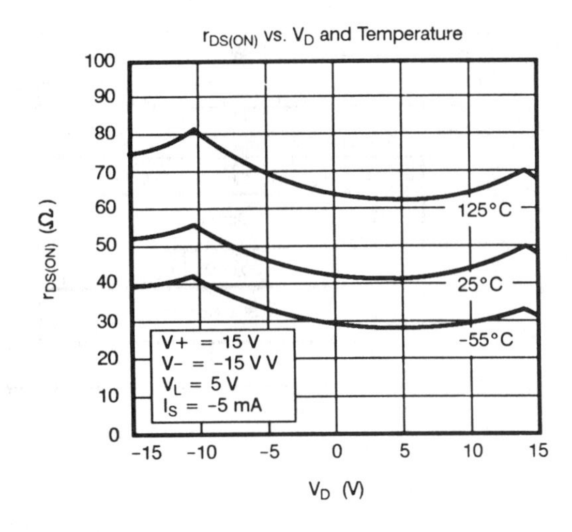

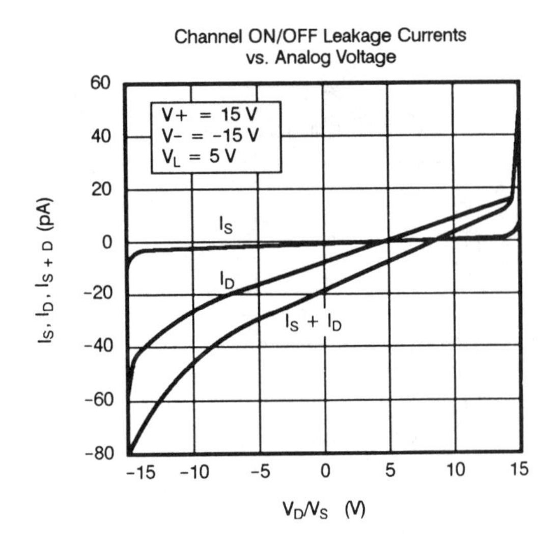

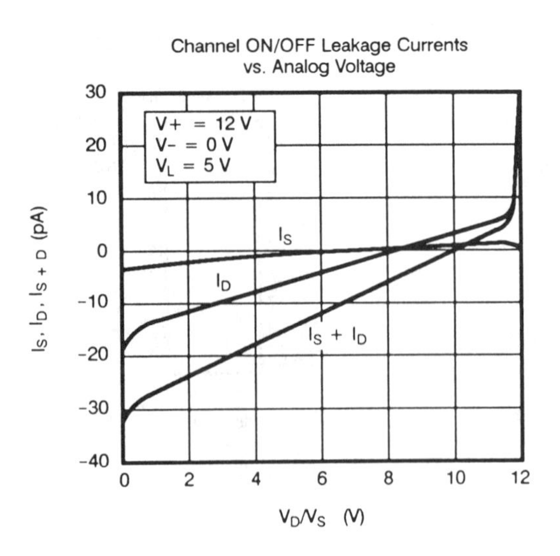

TYPICAL CHARACTERISTICS (cont'd)

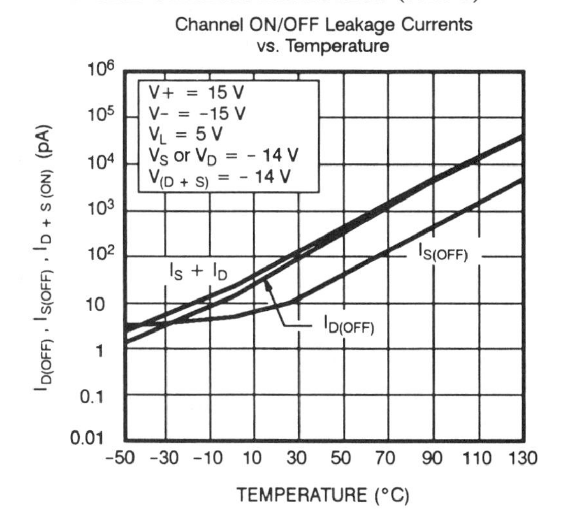

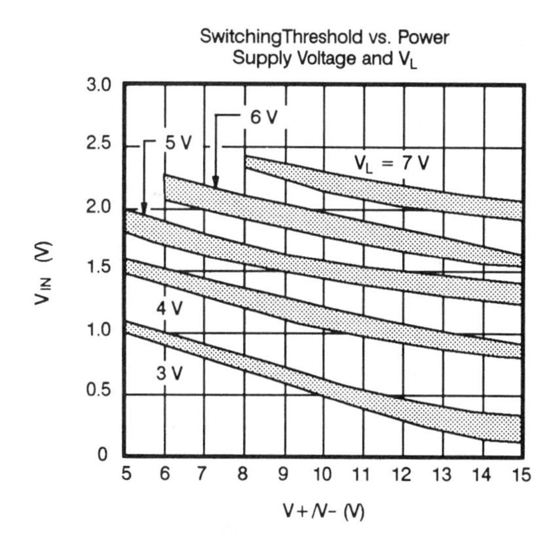

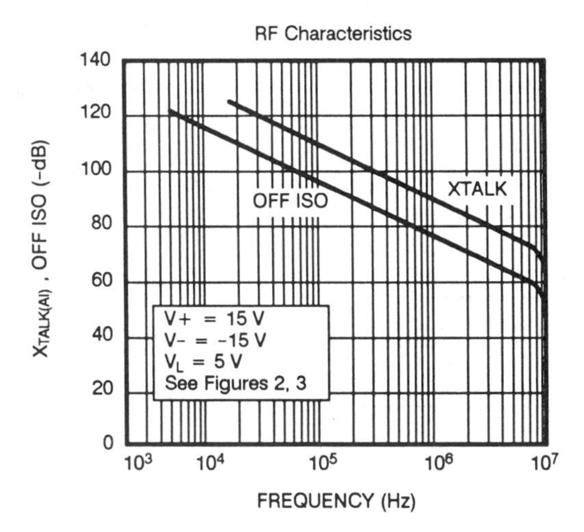

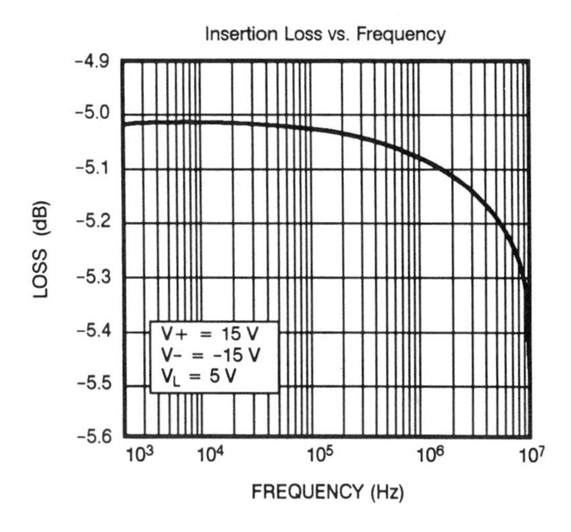

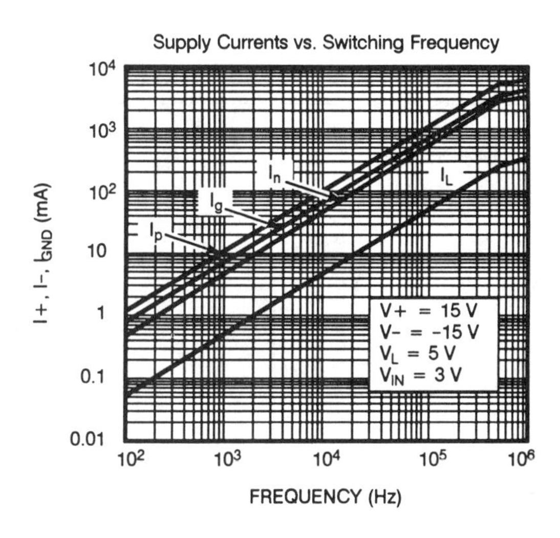

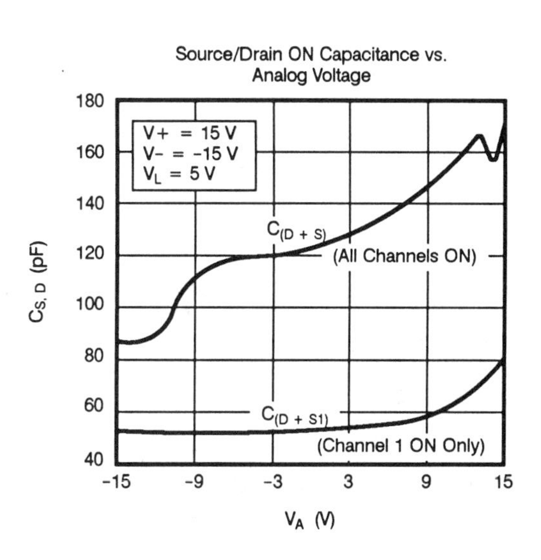

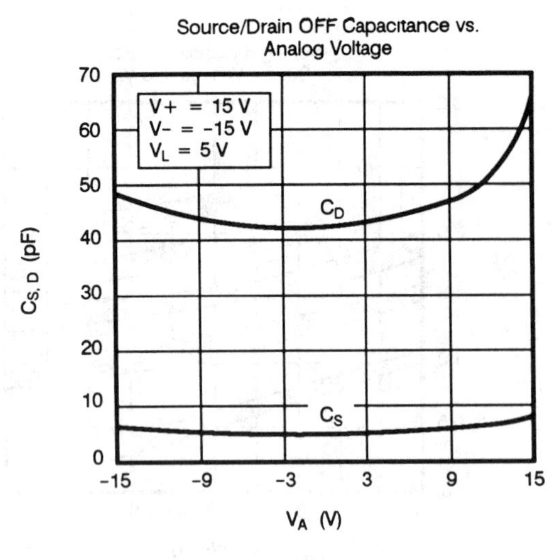

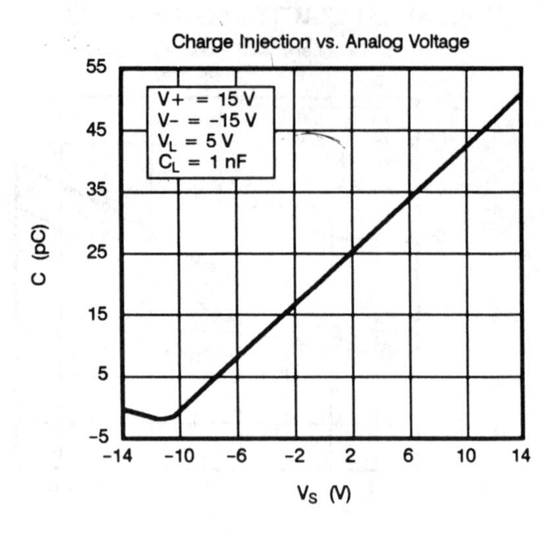

TYPICAL INPUT TIMING REQUIREMENTS

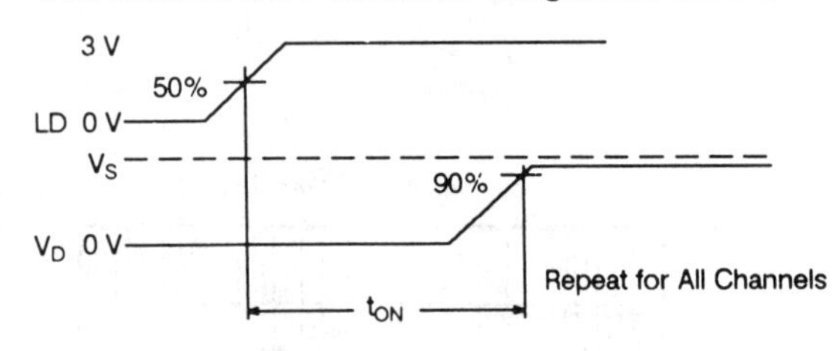

t_{ON} from LD

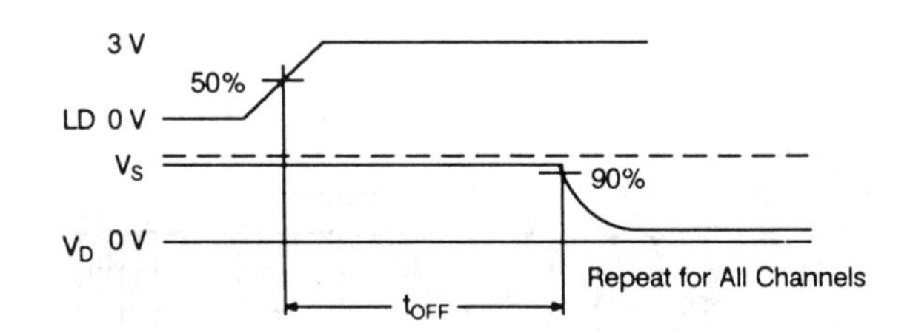

t_{OFF} from LD

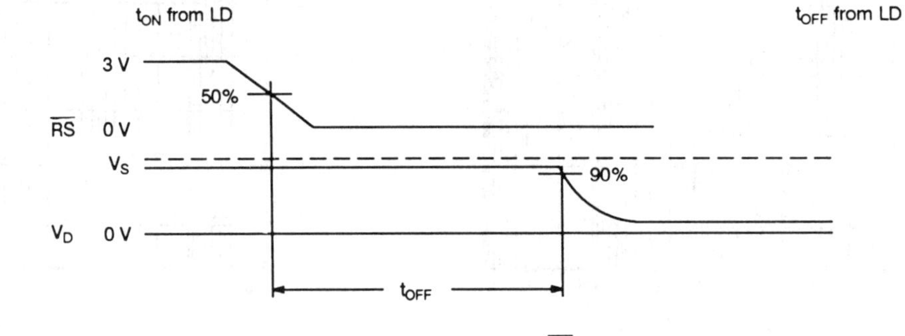

t_{OFF} from $\overline{RS}$

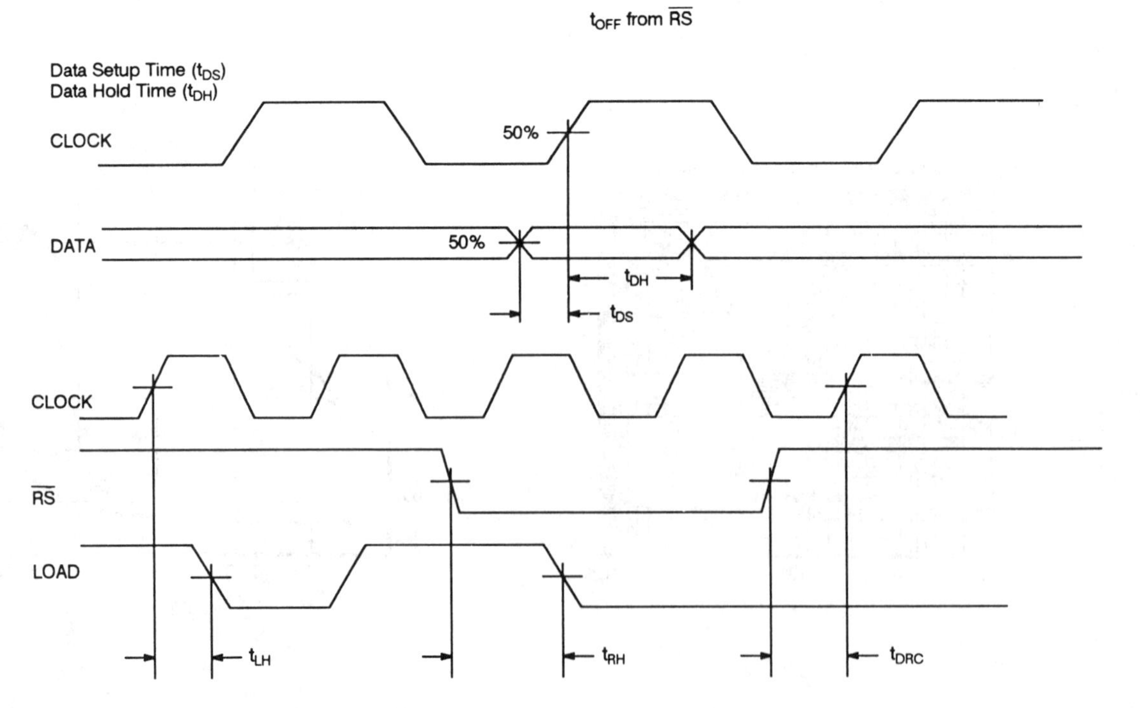

Timing Relationships

TIMING DIAGRAM

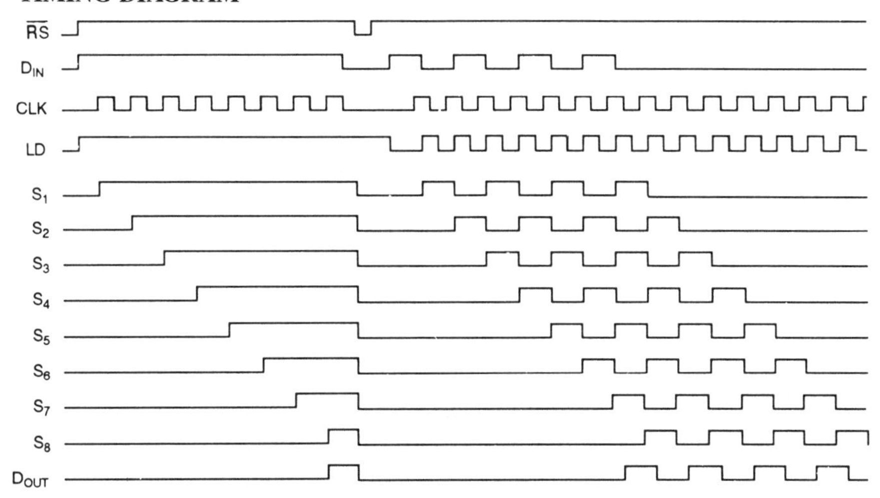

S_1 - S_8 and D_{OUT} are expected output with the drain connected high. The sources require pull-down of 1 kΩ.

APPLICATIONS

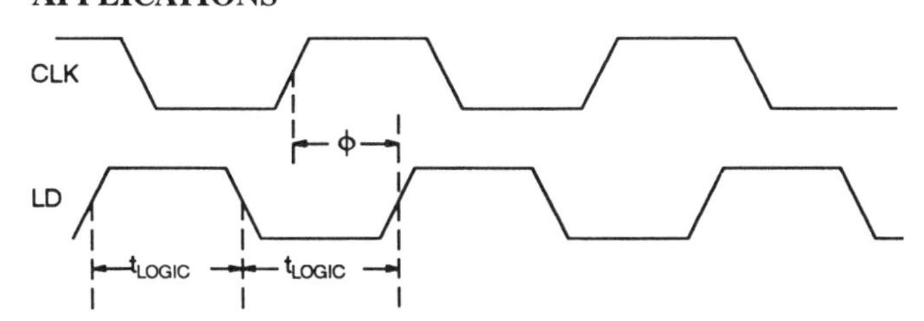

φ = for CLK and LD inputs of the same frequency. The recommended phase delay of LD from CLK is ½ t_{LOGIC} to t_{LOGIC}:

$t_{LOGIC(MIN)}$:	80 ns at 25°C	V+ = 15 V
	150 ns at 125°C	V- = -15 V
		GND = 0 V

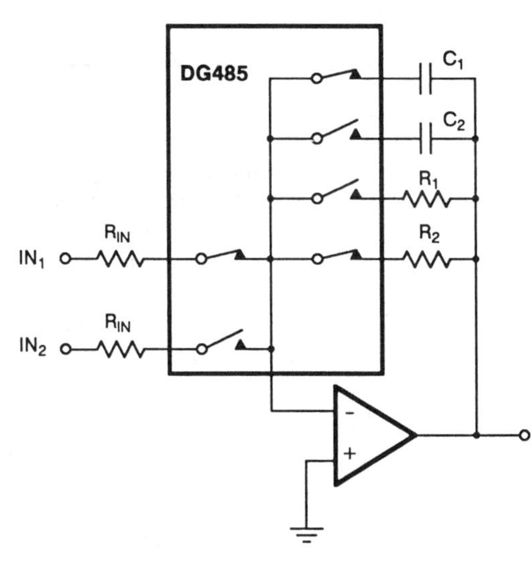

Multi-Function circuit Provides Input Selection, Gain Ranging and Filtering with One DG485

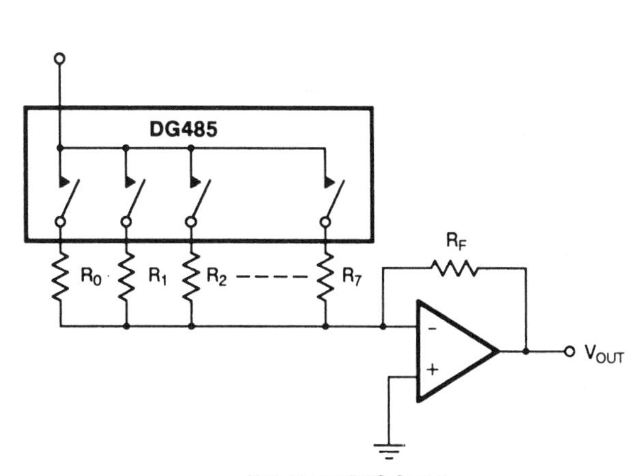

Non-Linear DAC Circuit

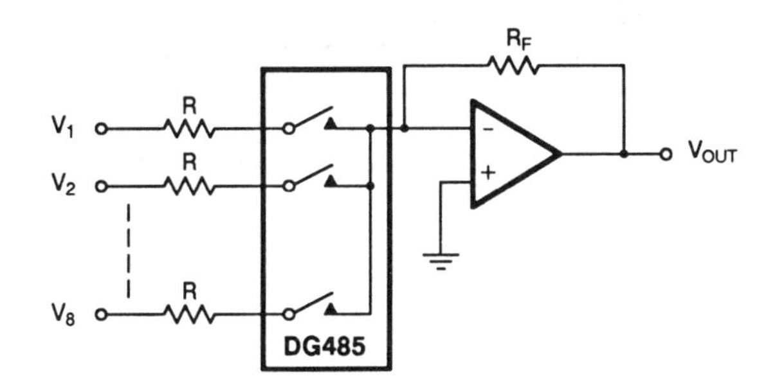

Summing Node Mixer

Multi-Channel Sampling and TDM Application

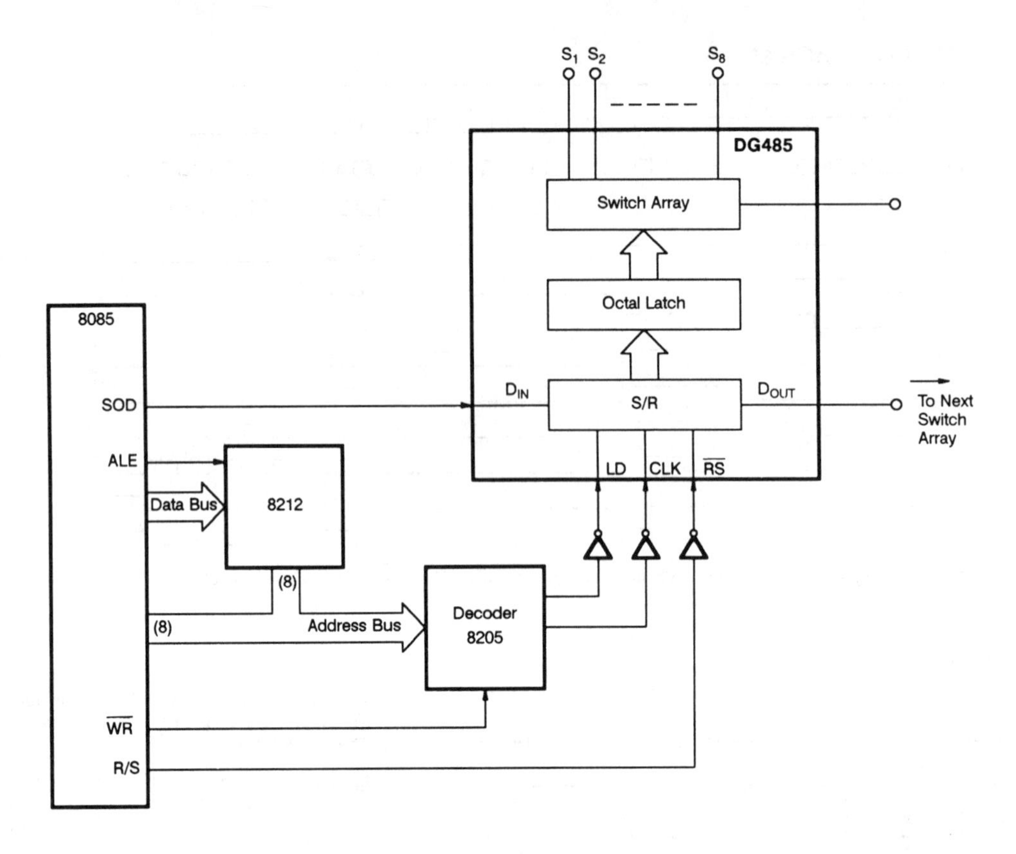

Direct Serial Interface (8085)

Siliconix DG528/529

8-Channel and Dual 4-Channel Latchable Multiplexers

- TTL compatible
- 44V power supply
- On-board address latches
- Low $r_{DS(ON)}$ (270 Ω typ.)
- Break-before-make
- Improved ESD protection >2500V
- Easily interfaced
- Increased analog signal range
- Microprocessor bus compatible
- Improved system accuracy
- Reduced crosstalk

APPLICATIONS

- Data acquisition systems
- Automatic test equipment
- Avionics and military systems
- Communication systems
- Microprocessor-controlled analog systems
- Audio signal multiplexing

DG528/529 analog multiplexers have on-chip address and control latches to simplify design in microprocessor based applications. Break-before-make switching action protects against momentary shorting of the input signals. The DG528/529 are built on the improved PLUS-40 CMOS process, which includes sandwich passivation and improved ESD protection (up to 2500V) for ruggedness.

DG528 is an 8-channel single-ended analog multiplexer designed to connect 1 of 8 inputs to a common output, as determined by a 3-bit binary address (A_0, A_1, A_2). DG529, a 4-channel dual analog multiplexer, is designed to connect 1 of 4 differential inputs to a common differential output, as determined by its 2-bit binary address (A_0, A_1) logic. An epitaxial layer prevents latchup.

The on-board TTL-compatible address latches simplify the digital interface design and reduce board space in bus-controlled systems, such as data acquisition systems, process controls, avionics, and ATE. The DG528 is available in 18-pin CerDIP in the military A-suffix (−55 to 125 °C), industrial B-suffix (−25 to 85 °C) and commercial C-suffix (0 to 70 °C) temperature ranges, and in the plastic DIP for commercial temperature operation. The DG528 is also available in surface-mount PLCC-20 for the extended industrial range (−40 to 85 °C).

ABSOLUTE MAXIMUM RATINGS

Voltage Referenced to V−

V+ .... 44 V

GND .... 25 V

Digital Inputs[h], V_S, V_D .... (V−) −2 V to (V+) +2 V or 20 mA, whichever occurs first.

Current (Any Terminal Except S or D) .... 30 mA

Continuous Current, S or D .... 20 mA

Peak Current, S or D (Pulsed at 1 ms, 10% Duty Cycle Max) .... 40 mA

Operating Temperature
- (A Suffix) .... −55 to 125°C
- (B Suffix) .... −25 to 85°C
- (C Suffix) .... 0 to 70°C
- (D Suffix) .... −40 to 85°C

Storage Temperature
- (A & B Suffix) .... −65 to 150°C
- (C & D Suffix) .... −65 to 125°C

Power Dissipation (Package)*

18-Pin Ceramic DIP** .... 900 mW

18-Pin Plastic DIP*** .... 470 mW

20-Pin PLCC**** .... 800 mW

*All leads soldered or welded to PC board.
**Derate 12 mW/°C above 75°C.
***Derate 6.3 mW/°C above 75°C.
****Derate 10 mW/°C above 75°C.

FUNCTIONAL BLOCK DIAGRAMS AND TRUTH TABLES

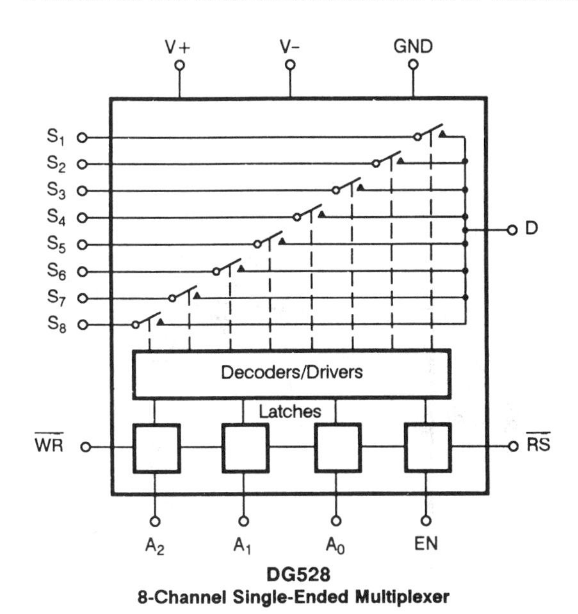

DG528
8-Channel Single-Ended Multiplexer

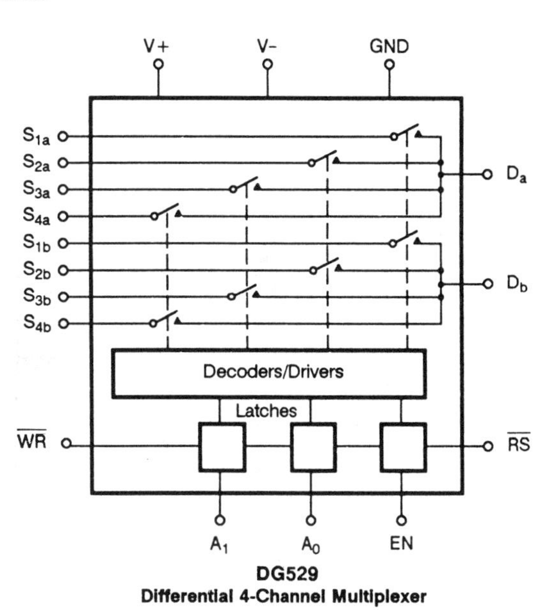

DG529
Differential 4-Channel Multiplexer

A_2	A_1	A_0	EN	$\overline{WR}$	$\overline{RS}$	On Switch
Latching						
X	X	X	X	↑	1	Maintains previous switch condition
Reset						
X	X	X	X	X	0	NONE (latches cleared)
Transparent Operation						
X	X	X	0	0	1	NONE
0	0	0	1	0	1	1
0	0	1	1	0	1	2
0	1	0	1	0	1	3
0	1	1	1	0	1	4
1	0	0	1	0	1	5
1	0	1	1	0	1	6
1	1	0	1	0	1	7
1	1	1	1	0	1	8

A_1	A_0	EN	$\overline{WR}$	$\overline{RS}$	On Switch
Latching					
X	X	X	↑	1	Maintains previous switch condition
Reset					
X	X	X	X	0	NONE (latches cleared)
Transparent Operation					
X	X	0	0	1	NONE
0	0	1	0	1	1
0	1	1	0	1	2
1	0	1	0	1	3
1	1	1	0	1	4

Logic "0" = $V_{AL} \leq 0.8$ V, Logic "1" = $V_{AH} \geq 2.4$ V

PIN CONFIGURATION

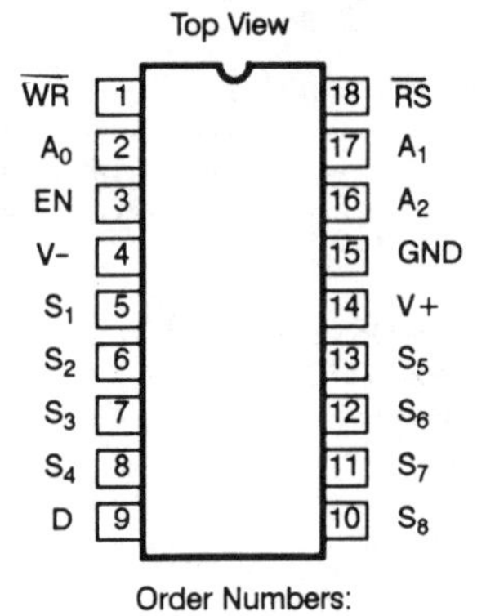

Order Numbers:

CerDIP: DG528AK, DG528AK/883
DG528BK, DG528CK

Plastic: DG528CJ

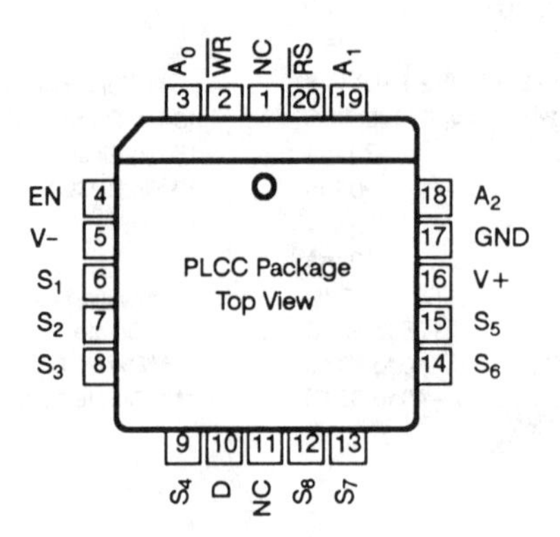

Order Number:

DG528DN

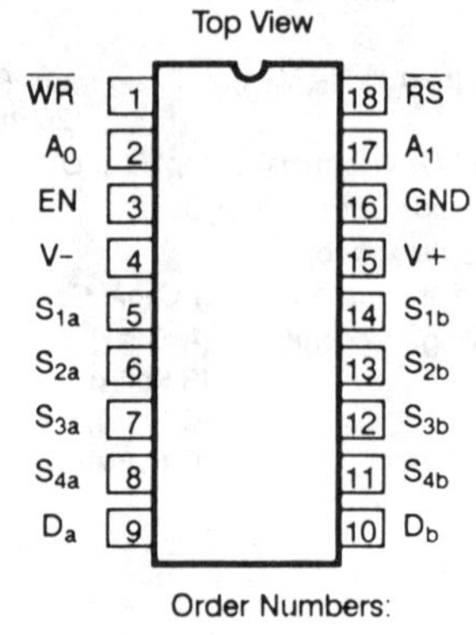

Order Numbers:

CerDIP: DG529AK/883
DG529BK

Plastic: DG529CJ

TYPICAL CHARACTERISTICS

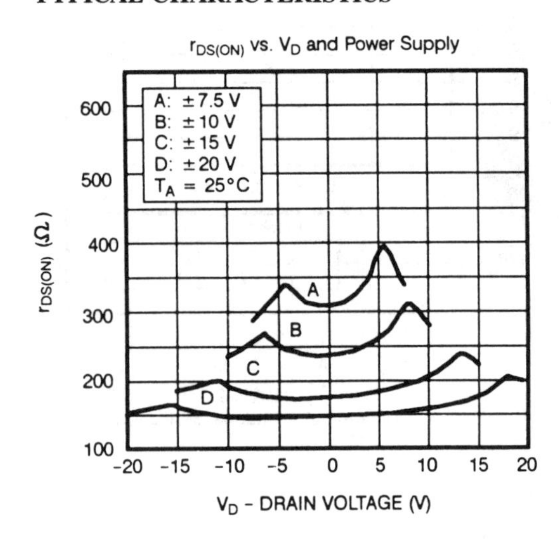

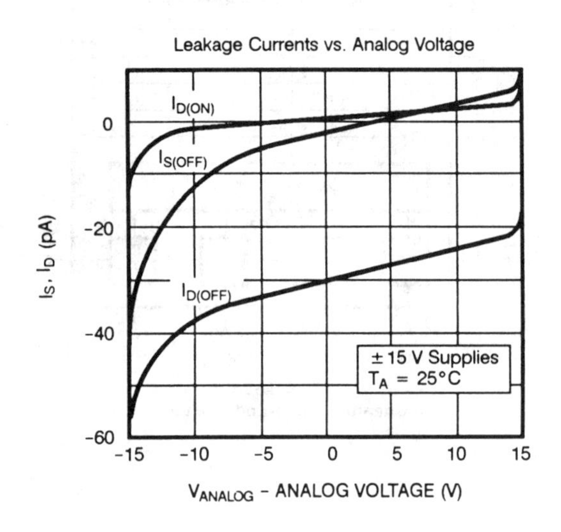

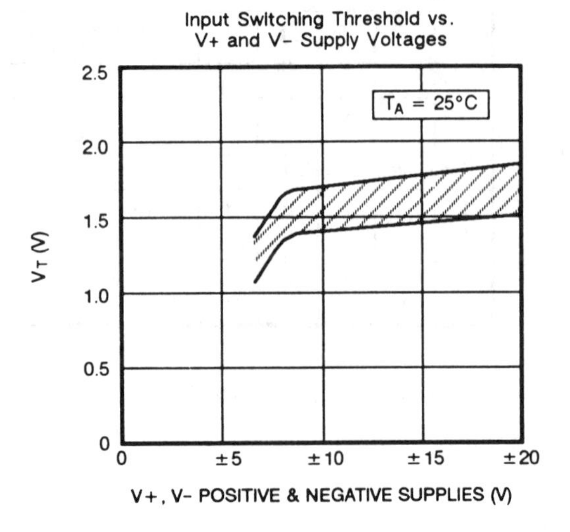

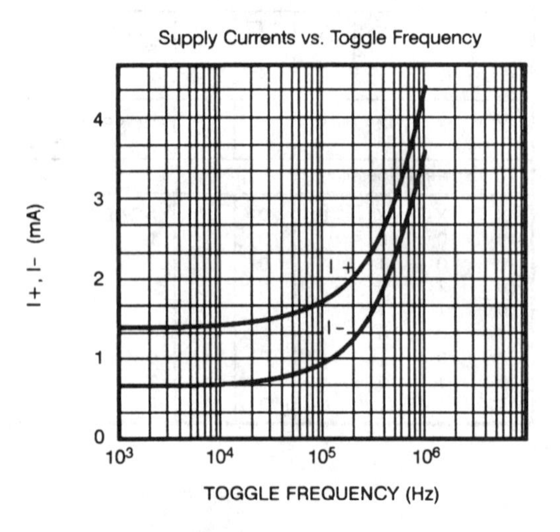

TIMING DIAGRAMS

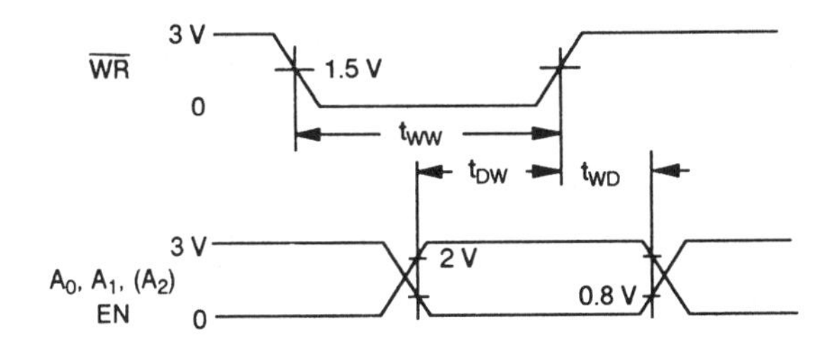

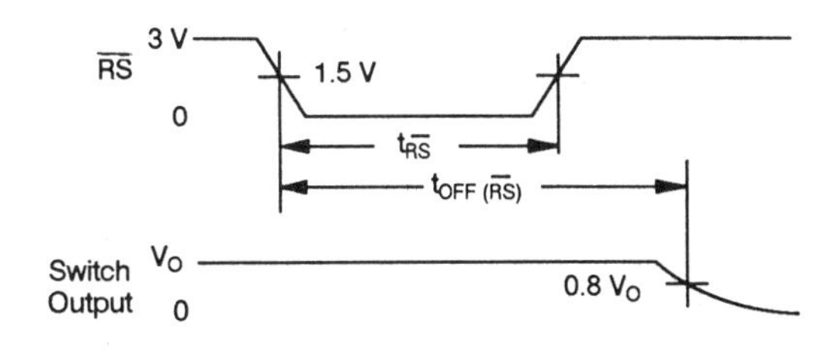

This information is reproduced with permission of Sprague Electric Company, Sensor Division, Concord, NH.

Sprague UHP-400, UHP-400-1, and UHP-500

Power and Relay Drivers

- Inputs compatible with DTL/TTL
- 500mA output current-sink capability
- Pinning compatible with 54/74 logic series
- Transient-protected outputs on relay drivers
- High-voltage output:
  100V Series UHP-500
  70V Series UHP-400-1
  40V Series UHP-400

Series UHP-400, UHP-400-1, and UHP-500 power and relay drivers are bipolar integrated circuits with logic and high-current switching transistors on the same chip. Each output transistor is capable of sinking 500mA in the on state.

UHP Part Numbers			Function
400	400-1	500	Quad 2-Input AND
402	402-1	502	Quad 2-Input OR
403	403-1	503	Quad OR for Inductive Loads
406	406-1	506	Quad AND for Inductive Loads
407	407-1	507	Quad NAND for Inductive Loads
408	408-1	508	Quad 2-Input NAND
432	432-1	532	Quad 2-Input NOR
433	433-1	533	Quad NOR for Inductive Loads

ABSOLUTE MAXIMUM RATINGS

Supply Voltage, V_{CC}	7 V
Input Voltage, V_{IN}	5.5 V
Output Off-State Voltage, V_{OFF}	
Series UHP-400	40 V
Series UHP-400-1	70 V
Series UHP-500	100 V
Output On-State Sink Current, I_{ON} (one driver)	500 mA
(total package)	1 A
Suppression Diode Off-State Voltage, V_R	
Series UHP-400	40 V
Series UHP-400-1	70 V
Series UHP-500	100 V
Suppression Diode On-State Current, I_F	500 mA
Operating Free-Air Temperature Range, T_A	−20°C to +85°C
Storage Temperature Range, T_S	−65°C to +150°C

RECOMMENDED OPERATING CONDITIONS

	Min.	Nom.	Max.	Units
Supply Voltage (V_{CC})	4.75	5.0	5.25	V
Operating Temperature Range	0	+25	+85	°C
Current into Any Output (ON State)	—	—	250	mA

SWITCHING CHARACTERISTICS at $T_A = +25°C, V_{CC} = 5.0\,V$

Characteristic	Series	Test Conditions (Note 3)	Limits Min.	Typ.	Max.	Units
Turn-On Delay Time (t_{pd0})	UHP-400	$V_S = 40\,V, R_L = 265\Omega$ (6 W)	—	200	500	ns
	UHP-400-1	$V_S = 70\,V, R_L = 465\Omega$ (10 W)	—	200	500	ns
	UHP-500	$V_S = 100\,V, R_L = 670\Omega$ (15 W)	—	200	500	ns
Turn-Off Delay Time (t_{pd1})	UHP-400	$V_S = 40\,V, R_L = 265\Omega$ (6 W)	—	300	750	ns
	UHP-400-1	$V_S = 70\,V, R_L = 465\Omega$ (10 W)	—	300	750	ns
	UHP-500	$V_S = 100\,V, R_L = 670\Omega$ (15 W)	—	300	750	ns

NOTES:
1. Each input tested separately.
2. Voltage values shown in the test-circuit waveforms are with respect to network ground terminal.
3. C_L = 15 pF. Capacitance value specified includes probe and test fixture capacitance.

INPUT PULSE CHARACTERISTICS

$V_{in(0)} = 0\,V$	$t_f = 7.0$ ns	$t_p = 1.0\,\mu s$
$V_{in(1)} = 3.5\,V$	$t_r = 14$ ns	PRR = 500 kHz

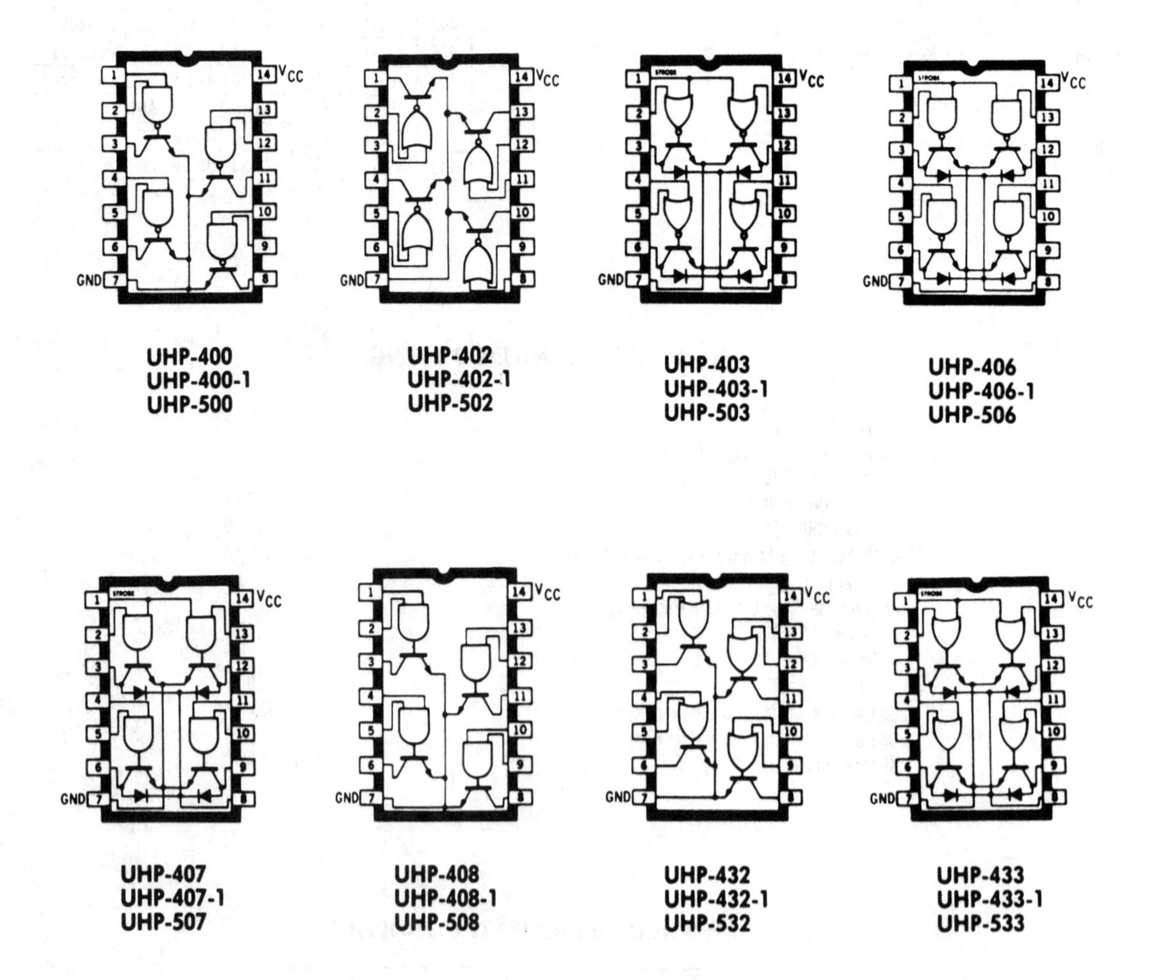

UHP-400, UHP-400-1, and UHP-500
Quad 2-Input AND Power Drivers

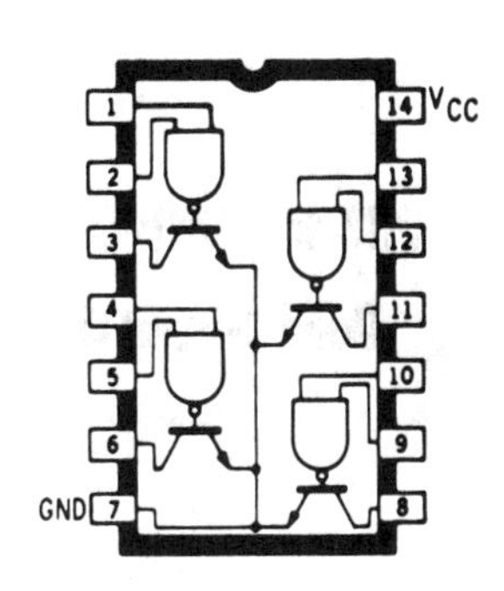

ELECTRICAL CHARACTERISTICS over operating temperature range (unless otherwise noted)

Characteristic	Symbol	Applicable Devices	Test Conditions V_{CC}	Driven Input	Other Input	Output	Limits Min.	Typ.	Max.	Units
Output Reverse Current	I_{CEX}	UHP-400	Min.	2.0 V	2.0 V	40 V	—	—	50	µA
		UHP-400-1	Min.	2.0 V	2.0 V	70 V	—	—	50	µA
		UHP-500	Min.	2.0 V	2.0 V	100 V	—	—	50	µA
Output Voltage	$V_{CE(SAT)}$	All	Min.	0.8 V	4.75 V	150 mA	—	—	0.5	V
			Min.	0.8 V	4.75 V	250 mA	—	—	0.7	V
Supply Current (Notes 1, 2 and 4)	$I_{CC(1)}$	All	Max.	5.0 V	5.0 V	—	—	4.0	6.0	mA
	$I_{CC(0)}$	All	Max.	0 V	0 V	—	—	17.5	24.5	mA
Input Voltage	$V_{IN(1)}$	All	Min.	—	—	—	2.0	—	—	V
	$V_{IN(0)}$	All	Min.	—	—	—	—	—	0.8	V
Input Current (Note 3)	$I_{IN(0)}$	All	Max.	0.4 V	4.5 V	—	—	−0.55	−0.8	mA
	$I_{IN(1)}$	All	Max.	2.4 V	0 V	—	—	—	40	µA
			Max.	5.5 V	0 V	—	—	—	1.0	mA

1. Typical values at V_{CC} = 5.0 V.
2. Each gate.
3. Each input tested separately.
4. T_A = +25°C.

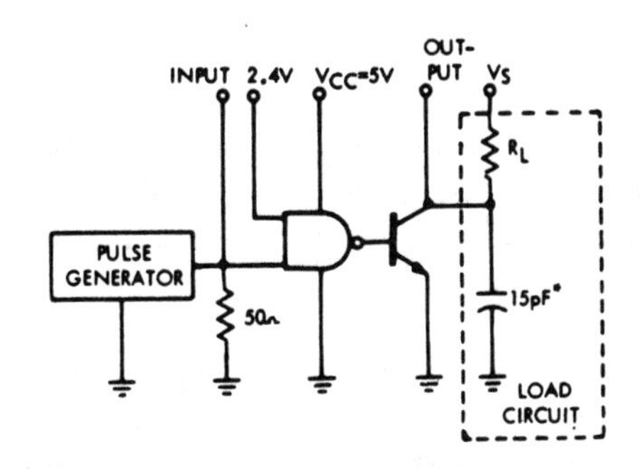

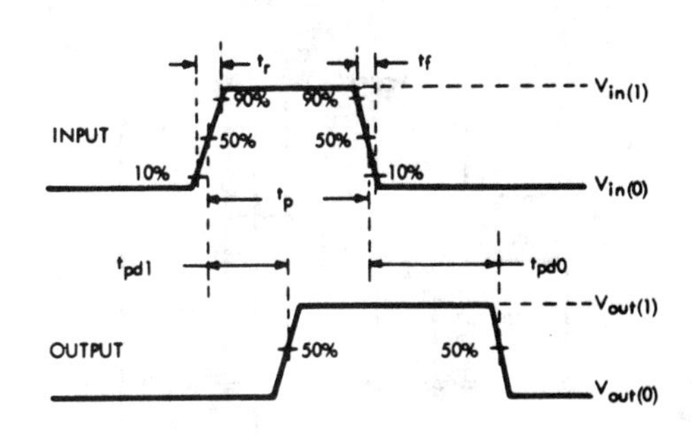

*Includes probe and test fixture capacitance.

UHP-402, UHP-402-1, and UHP-502
Quad 2-Input OR Power Drivers

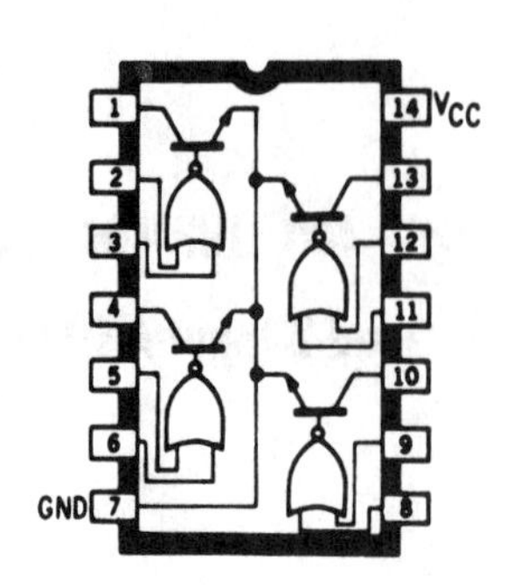

ELECTRICAL CHARACTERISTICS over operating temperature range (unless otherwise noted)

Characteristic	Symbol	Applicable Devices	Test Conditions				Limits			
			V_{CC}	Driven Input	Other Input	Output	Min.	Typ.	Max.	Units
Output Reverse Current	I_{CEX}	UHP-402	Min.	2.0 V	0 V	40 V	—	—	50	μA
		UHP-402-1	Min.	2.0 V	0 V	70 V	—	—	50	μA
		UHP-502	Min.	2.0 V	0 V	100 V	—	—	50	μA
Output Voltage	$V_{CE(SAT)}$	All	Min.	0.8 V	0.8 V	150 mA	—	—	0.5	V
			Min.	0.8 V	0.8 V	250 mA	—	—	0.7	V
Supply Current (Notes 1, 2 and 4)	$I_{CC(1)}$	All	Max.	5.0 V	5.0 V	—	—	4.1	6.3	mA
	$I_{CC(0)}$	All	Max.	0 V	0 V	—	—	18	25	mA
Input Voltage	$V_{IN(1)}$	All	Min.	—	—	—	2.0	—	—	V
	$V_{IN(0)}$	All	Min.	—	—	—	—	—	0.8	V
Input Current (Note 3)	$I_{IN(0)}$	All	Max.	0.4 V	4.5 V	—	—	−0.55	−0.8	mA
	$I_{IN(1)}$	All	Max.	2.4 V	0 V	—	—	—	40	μA
			Max.	5.5 V	0 V	—	—	—	1.0	mA

1. Typical values at V_{CC} = 5.0 V.
2. Each gate.
3. Each input tested separately.
4. T_A = +25°C.

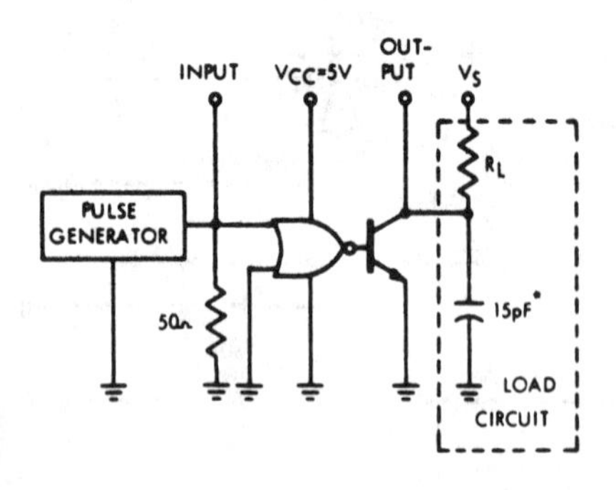

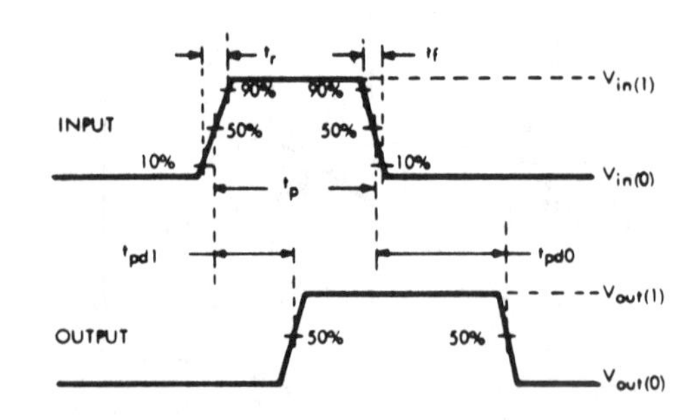

*Includes probe and test fixture capacitance.

UHP-403, UHP-403-1, and UHP-503
Quad OR Relay Drivers

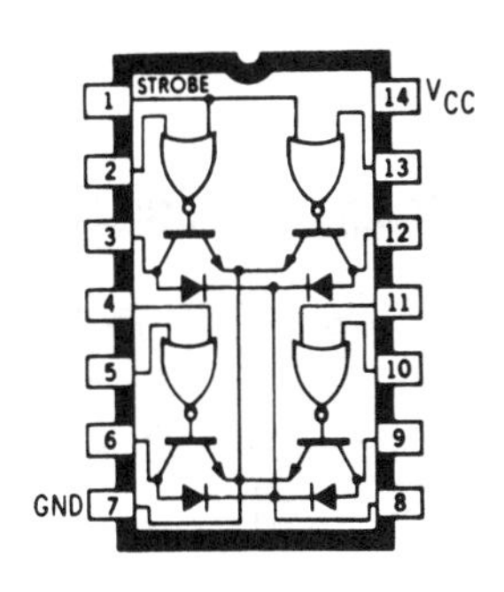

ELECTRICAL CHARACTERISTICS **over operating temperature range (unless otherwise noted)**

Characteristic	Symbol	Applicable Devices	Test Conditions				Limits			
			V_{CC}	Driven Input	Other Input	Output	Min.	Typ.	Max.	Units
Output Reverse Current	I_{CEX}	UHP-403	Min.	2.0 V	0 V	40 V	—	—	100	µA
		UHP-403-1	Min.	2.0 V	0 V	70 V	—	—	100	µA
		UHP-503	Min.	2.0 V	0 V	100 V	—	—	100	µA
Diode Leakage Current (Note 5)	I_R	All	Nom.	0 V	0 V	Open	—	—	200	µA
Diode Forward Voltage Drop (Note 6)	V_F	All	Nom.	5.0 V	5.0 V	—	—	1.5	1.75	V
Output Voltage	$V_{CE(SAT)}$	All	Min.	0.8 V	0.8 V	150 mA	—	—	0.5	V
			Min.	0.8 V	0.8 V	250 mA	—	—	0.7	V
Supply Current (Notes 1, 2 and 4)	$I_{CC(1)}$	All	Max.	5.0 V	5.0 V	—	—	4.1	6.3	mA
	$I_{CC(0)}$	All	Max.	0 V	0 V	—	—	18	25	mA
Input Voltage	$V_{IN(1)}$	All	Min.	—	—	—	2.0	—	—	V
	$V_{IN(0)}$	All	Max.	—	—	—	—	—	0.8	V
Input Current at All Inputs Except Strobe (Note 3)	$I_{IN(0)}$	All	Max.	0.4 V	4.5 V	—	—	−0.55	−0.8	mA
	$I_{IN(1)}$	All	Max.	2.4 V	0 V	—	—	—	40	µA
			Max.	5.5 V	0 V	—	—	—	1.0	mA
Input Current at Strobe (Note 3)	$I_{IN(0)}$	All	Max.	0.4 V	4.5 V	—	—	−1.1	−1.6	mA
	$I_{IN(1)}$	All	Max.	2.4 V	0 V	—	—	—	100	µA
			Max.	5.5 V	0 V	—	—	—	1.0	mA

1. Typical values at V_{CC} = 5.0 V.
2. Each gate.
3. Each input tested separately.
4. T_A = +25°C.
5. Diode leakage current measured at $V_R = V_{OFF(MIN)}$.
6. Diode forward voltage drop measured at I_F = 200 mA.

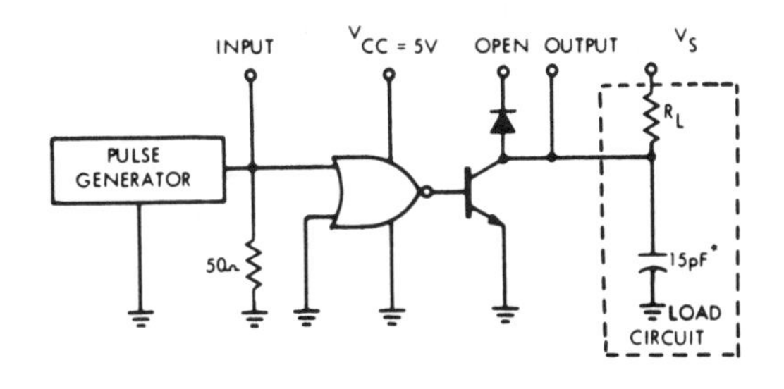

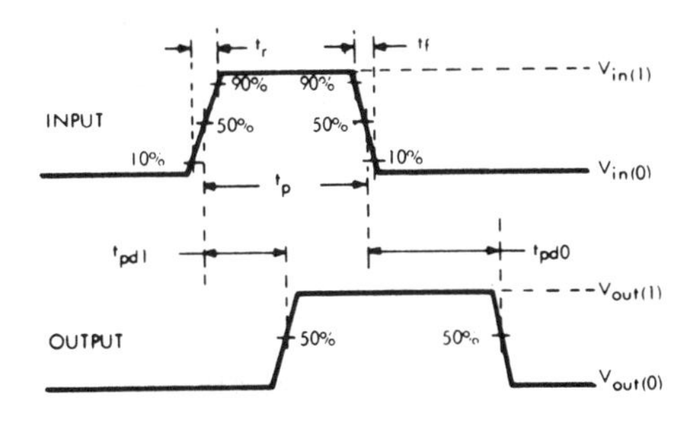

*Includes probe and test fixture capacitance.

UHP-406, UHP-406-1, and UHP-506
Quad AND Relay Drivers

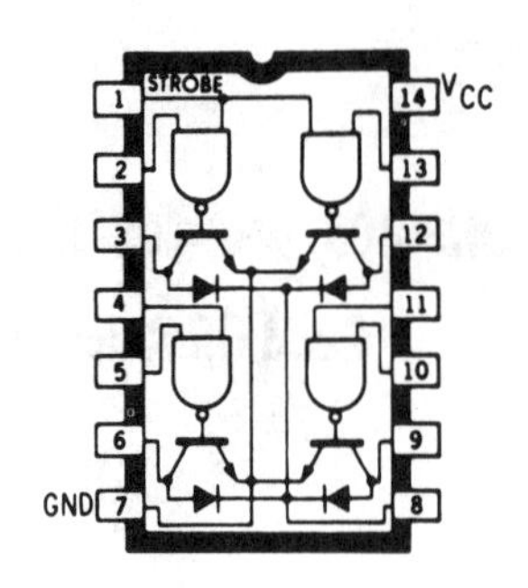

ELECTRICAL CHARACTERISTICS over operating temperature range (unless otherwise noted)

Characteristic	Symbol	Applicable Devices	Test Conditions				Limits			
			V_{CC}	Driven Input	Other Input	Output	Min.	Typ.	Max.	Units
Output Reverse Current	I_{CEX}	UHP-406	Min.	2.0 V	2.0 V	40 V	—	—	100	μA
		UHP-406-1	Min.	2.0 V	2.0 V	70 V	—	—	100	μA
		UHP-506	Min.	2.0 V	2.0 V	100 V	—	—	100	μA
Diode Leakage Current (Note 5)	I_R	All	Nom.	0 V	0 V	Open	—	—	200	μA
Diode Forward Voltage Drop (Note 6)	V_F	All	Nom.	5.0 V	5.0 V	—	—	1.5	1.75	V
Output Voltage	$V_{CE(SAT)}$	All	Min.	0.8 V	4.75 V	150 mA	—	—	0.5	V
			Min.	0.8 V	4.75 V	250 mA	—	—	0.7	V
Supply Current (Notes 1, 2 and 4)	$I_{CC(1)}$	All	Max.	5.0 V	5.0 V	—	—	4.0	6.0	mA
	$I_{CC(0)}$	All	Max.	0 V	0 V	—	—	17.5	24.5	mA
Input Voltage	$V_{IN(1)}$	All	Min.	—	—	—	2.0	—	—	V
	$V_{IN(0)}$	All	Min.	—	—	—	—	—	0.8	V
Input Current at All Inputs Except Strobe (Note 3)	$I_{IN(0)}$	All	Max.	0.4 V	4.5 V	—	—	−0.55	−0.8	mA
	$I_{IN(1)}$	All	Max.	2.4 V	0 V	—	—	—	40	μA
			Max.	5.5 V	0 V	—	—	—	1.0	mA
Input Current at Strobe (Note 3)	$I_{IN(0)}$	All	Max.	0.4 V	4.5 V	—	—	−1.1	−1.6	mA
	$I_{IN(1)}$	All	Max.	2.4 V	0 V	—	—	—	100	μA
			Max.	5.5 V	0 V	—	—	—	1.0	mA

1. Typical values at V_{CC} = 5.0 V.
2. Each gate.
3. Each input tested separately.
4. T_A = +25°C.
5. Diode leakage current measured at $V_R = V_{OFF(MIN)}$.
6. Diode forward voltage drop measured at I_F = 200 mA.

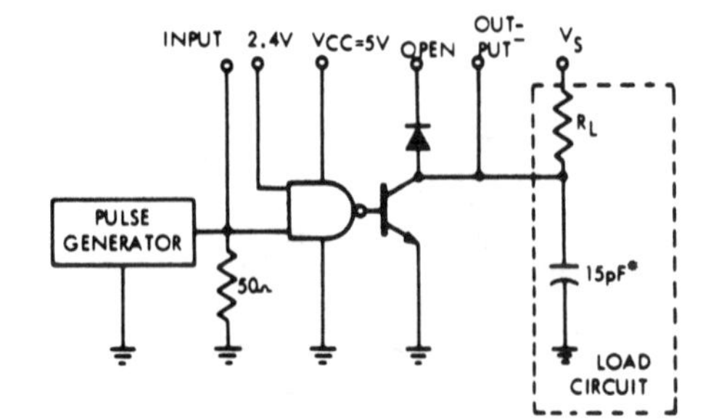

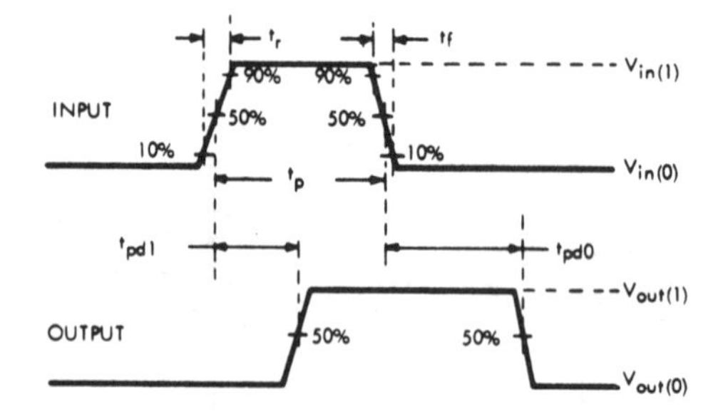

*Includes probe and test fixture capacitance.

UHP-407, UHP-407-1, and UHP-507
Quad NAND Relay Drivers

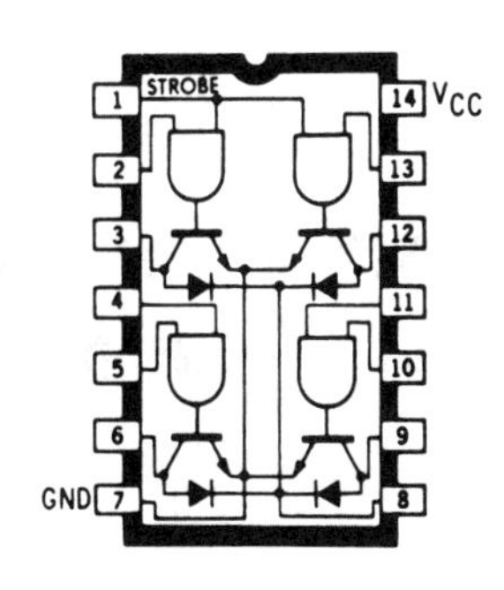

ELECTRICAL CHARACTERISTICS over operating temperature range (unless otherwise noted)

Characteristic	Symbol	Applicable Devices	Test Conditions V_{CC}	Driven Input	Other Input	Output	Limits Min.	Typ.	Max.	Units
Output Reverse Current	I_{CEX}	UHP-407	Min.	0.8 V	4.75 V	40 V	—	—	100	μA
		UHP-407-1	Min.	0.8 V	4.75 V	70 V	—	—	100	μA
		UHP-507	Min.	0.8 V	4.75 V	100 V	—	—	100	μA
Diode Leakage Current (Note 5)	I_R	All	Nom.	5.0 V	5.0 V	Open	—	—	200	μA
Diode Forward Voltage Drop (Note 6)	V_F	All	Nom.	0 V	0 V	—	—	1.5	1.75	V
Output Voltage	$V_{CE(SAT)}$	All	Min.	2.0 V	2.0 V	150 mA	—	—	0.5	V
			Min.	2.0 V	2.0 V	250 mA	—	—	0.7	V
Supply Current (Notes 1, 2 and 4)	$I_{CC(1)}$	All	Max.	0 V	0 V	—	—	6.0	7.5	mA
	$I_{CC(0)}$	All	Max.	5.0 V	5.0 V	—	—	20	26.5	mA
Input Voltage	$V_{IN(1)}$	All	Min.	—	—	—	2.0	—	—	V
	$V_{IN(0)}$	All	Min.	—	—	—	—	—	0.8	V
Input Current at All Inputs Except Strobe (Note 3)	$I_{IN(0)}$	All	Max.	0.4 V	4.5 V	—	—	−0.55	−0.8	mA
	$I_{IN(1)}$	All	Max.	2.4 V	0 V	—	—	—	40	μA
			Max.	5.5 V	0 V	—	—	—	1.0	mA
Input Current at Strobe (Note 3)	$I_{IN(0)}$	All	Max.	0.4 V	4.5 V	—	—	−1.1	−1.6	mA
	$I_{IN(1)}$	All	Max.	2.4 V	0 V	—	—	—	100	μA
			Max.	5.5 V	0 V	—	—	—	1.0	mA

1. Typical values at V_{CC} = 5.0 V.
2. Each gate.
3. Each input tested separately.
4. T_A = +25°C.
5. Diode leakage current measured at $V_R = V_{OFF(MIN)}$.
6. Diode forward voltage drop measured at I_F = 200 mA.

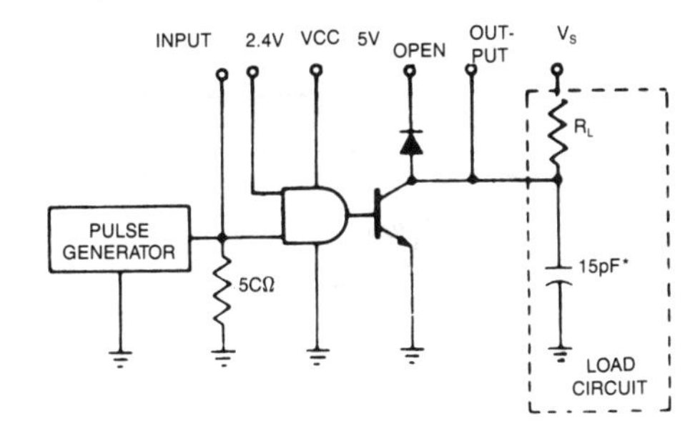

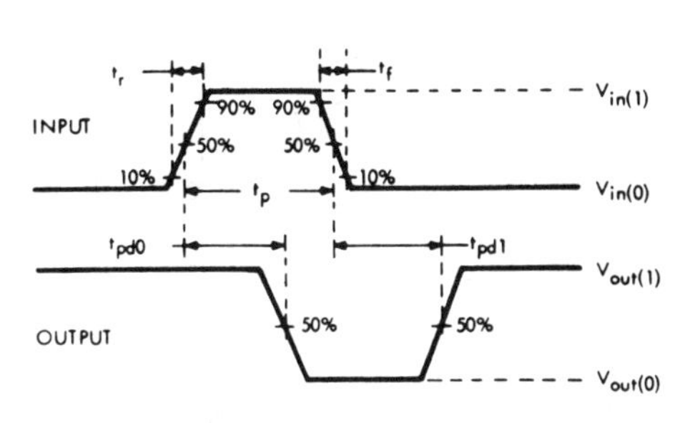

*Includes probe and test fixture capacitance.

UHP-408, UHP-408-1, and UHP-508
Quad 2-Input NAND Power Drivers

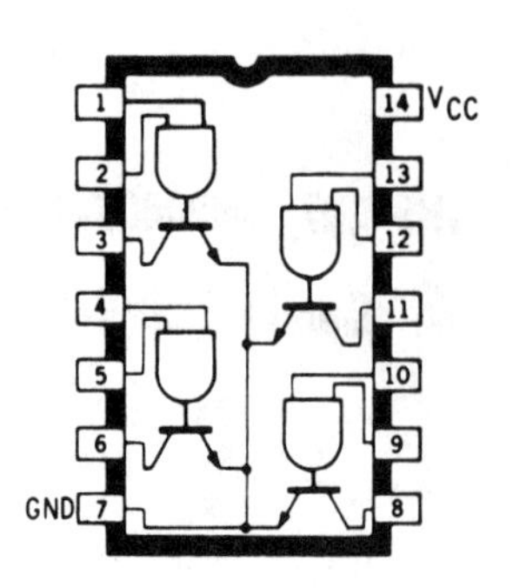

ELECTRICAL CHARACTERISTICS over operating temperature range (unless otherwise noted)

Characteristic	Symbol	Applicable Devices	Test Conditions: V_{CC}	Driven Input	Other Input	Output	Limits: Min.	Typ.	Max.	Units
Output Reverse Current	I_{CEX}	UHP-408	Min.	0.8 V	4.75 V	40 V	—	—	50	μA
		UHP-408-1	Min.	0.8 V	4.75 V	70 V	—	—	50	μA
		UHP-508	Min.	0.8 V	4.75 V	100 V	—	—	50	μA
Output Voltage	$V_{CE(SAT)}$	All	Min.	2.0 V	2.0 V	150 mA	—	—	0.5	V
			Min.	2.0 V	2.0 V	250 mA	—	—	0.7	V
Supply Current (Notes 1, 2 and 4)	$I_{CC(1)}$	All	Max.	0 V	0 V	—	—	6.0	7.5	mA
	$I_{CC(0)}$	All	Max.	5.0 V	5.0 V	—	—	20	26.5	mA
Input Voltage	$V_{IN(1)}$	All	Min.	—	—	—	2.0	—	—	V
	$V_{IN(0)}$	All	Min.	—	—	—	—	—	0.8	V
Input Current (Note 3)	$I_{IN(0)}$	All	Max.	0.4 V	4.5 V	—	—	−0.55	−0.8	mA
	$I_{IN(1)}$	All	Max.	2.4 V	0 V	—	—	—	40	μA
			Max.	5.5 V	0 V	—	—	—	1.0	mA

1. Typical values at V_{CC} = 5.0 V.
2. Each gate.
3. Each input tested separately.
4. T_A = +25°C.

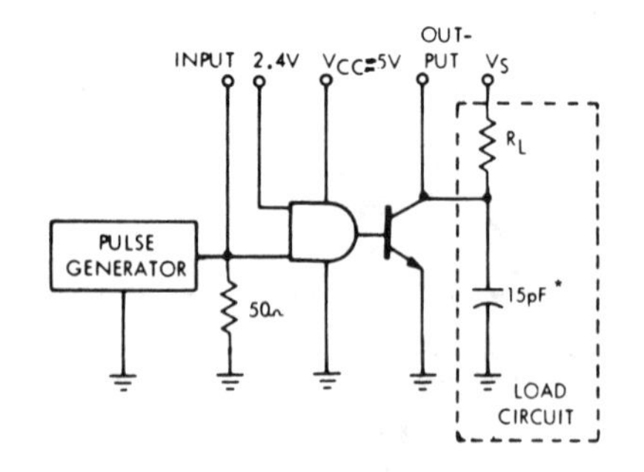

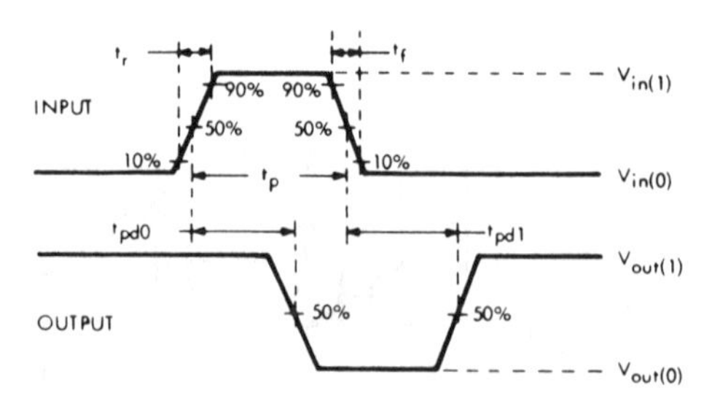

*Includes probe and test fixture capacitance.

UHP-432, UHP-432-1, and UHP-532
Quad 2-Input NOR Power Drivers

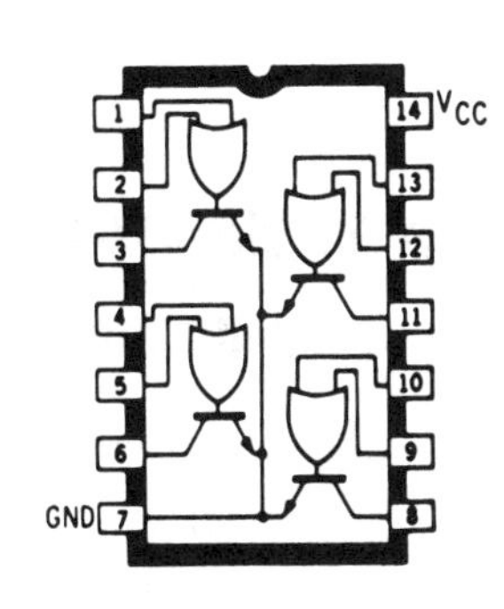

ELECTRICAL CHARACTERISTICS over operating temperature range (unless otherwise noted)

Characteristic	Symbol	Applicable Devices	Test Conditions				Limits			
			V_{CC}	Driven Input	Other Input	Output	Min.	Typ.	Max.	Units
Output Reverse Current	I_{CEX}	UHP-432	Min.	0.8 V	0.8 V	40 V	—	—	50	μA
		UHP-432-1	Min.	0.8 V	0.8 V	70 V	—	—	50	μA
		UHP-532	Min.	0.8 V	0.8 V	100 V	—	—	50	μA
Output Voltage	$V_{CE(SAT)}$	All	Min.	2.0 V	0 V	150 mA	—	—	0.5	V
			Min.	2.0 V	0 V	250 mA	—	—	0.7	V
Supply Current (Notes 1, 2 and 4)	$I_{CC(1)}$	All	Max.	0 V	0 V	—	—	6.0	7.5	mA
	$I_{CC(0)}$	All	Max.	5.0 V	5.0 V	—	—	20	25	mA
Input Voltage	$V_{IN(1)}$	All	Min.	—	—	—	2.0	—	—	V
	$V_{IN(0)}$	All	Min.	—	—	—	—	—	0.8	V
Input Current (Note 3)	$I_{IN(0)}$	All	Max.	0.4 V	4.5 V	—	—	−0.55	−0.8	mA
	$I_{IN(1)}$	All	Max.	2.4 V	0 V	—	—	—	40	μA
			Max.	5.5 V	0 V	—	—	—	1.0	mA

1. Typical values at V_{CC} = 5.0 V.
2. Each gate.
3. Each input tested separately.
4. T_A = +25°C.

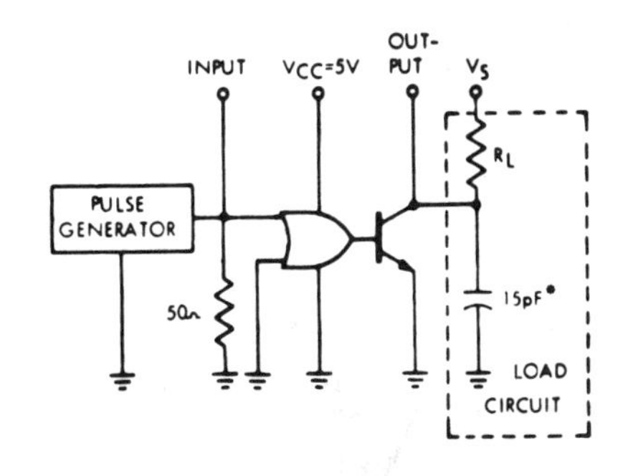

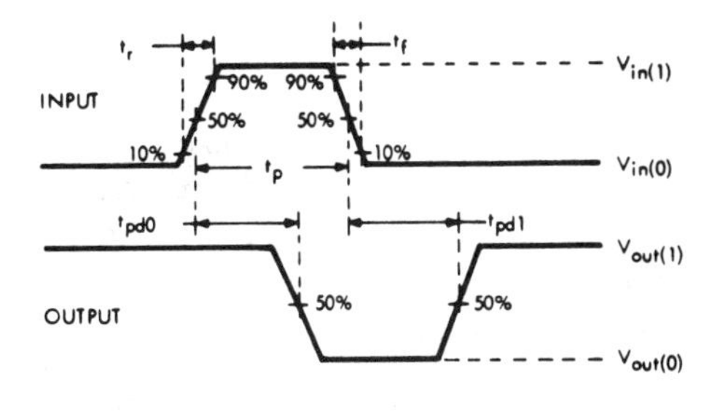

*Includes probe and test fixture capacitance.

UHP-433, UHP-433-1, and UHP-533 Quad NOR Relay Drivers

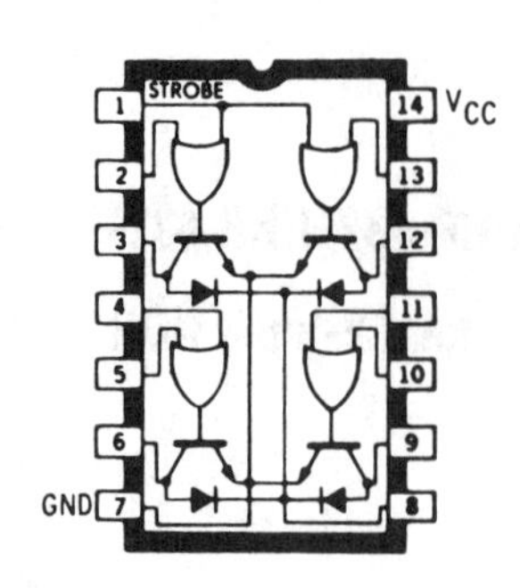

ELECTRICAL CHARACTERISTICS over operating temperature range (unless otherwise noted)

Characteristic	Symbol	Applicable Devices	Test Conditions				Limits			
			V_{CC}	Driven Input	Other Input	Output	Min.	Typ.	Max.	Units
Output Reverse Current	I_{CEX}	UHP-433	Min.	0.8 V	0.8 V	40 V	—	—	100	μA
		UHP-433-1	Min.	0.8 V	0.8 V	70 V	—	—	100	μA
		UHP-533	Min.	0.8 V	0.8 V	100 V	—	—	100	μA
Diode Leakage Current (Note 5)	I_R	All	Nom.	5.0 V	5.0 V	Open	—	—	200	μA
Diode Forward Voltage Drop (Note 6)	V_F	All	Nom.	0 V	0 V	—	—	1.5	1.75	V
Output Voltage	$V_{CE(SAT)}$	All	Min.	2.0 V	0 V	150 mA	—	—	0.5	V
			Min.	2.0 V	0 V	250 mA	—	—	0.7	V
Supply Current (Notes 1, 2 and 4)	$I_{CC(1)}$	All	Max.	0 V	0 V	—	—	6.0	7.5	mA
	$I_{CC(0)}$	All	Max.	5.0 V	5.0 V	—	—	20	25	mA
Input Voltage	$V_{IN(1)}$	All	Min.	—	—	—	2.0	—	—	V
	$V_{IN(0)}$	All	Min.	—	—	—	—	—	0.8	V
Input Current at All Inputs Except Strobe (Note 3)	$I_{IN(0)}$	All	Max.	0.4 V	4.5 V	—	—	−0.55	−0.8	mA
	$I_{IN(1)}$	All	Max.	2.4 V	0 V	—	—	—	40	μA
			Max.	5.5 V	0 V	—	—	—	1.0	mA
Input Current at Strobe (Note 3)	$I_{IN(0)}$	All	Max.	0.4 V	4.5 V	—	—	−1.1	−1.6	mA
	$I_{IN(1)}$	All	Max.	2.4 V	0 V	—	—	—	100	μA
			Max.	5.5 V	0 V	—	—	—	1.0	mA

1. Typical values at V_{CC} = 5.0 V.
2. Each gate.
3. Each input tested separately.
4. T_A = +25°C.
5. Diode leakage current measured at $V_R = V_{OFF(MIN)}$.
6. Diode forward voltage drop measured at I_F = 200 mA.

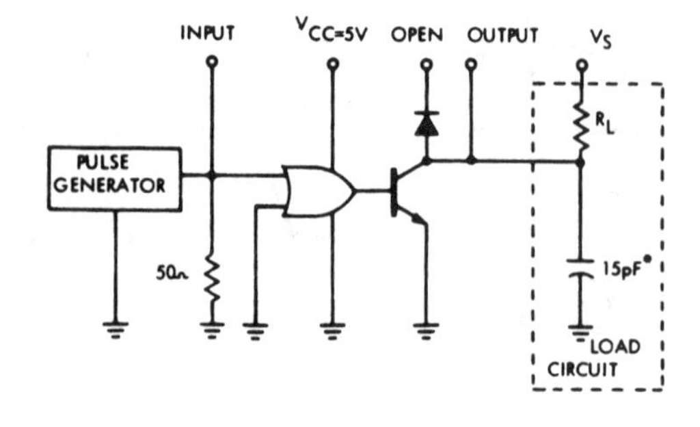

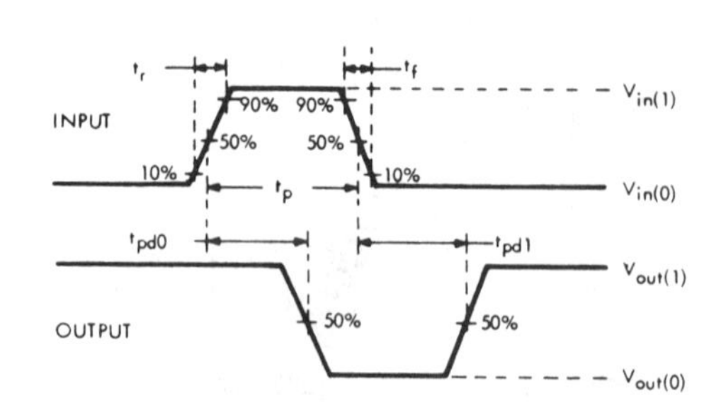

*Includes probe and test fixture capacitance.

Sprague
ULS-2000H and ULS-2000R
High-Voltage, High-Current Darlington Arrays

- TTL-, DTL-, PMOS-, or CMOS-compatible inputs
- Peak output current to 600mA
- Transient-protected outputs
- Side-brazed hermetic package
- CerDIP hermetic package
- High-reliability screening to MIL-STD-883, Class B
- −55 °C to +125 °C temperature range

Comprised of seven silicon npn Darlington power drivers on a common monolithic substrate, Series ULS-2000H and ULS-2000R arrays drive relays, solenoids, lamps, and other devices in high-reliability military or aerospace applications with up to 3A of output current per package.

These devices are screened to MIL-STD-883, Class B and are supplied in either the popular ceramic/metal side-brazed 16-pin hermetic package (suffix H) or ceramic/glass cerDIP hermetic package (suffix R). Both package styles conform to the dimensional requirements of MIL-M-38510 and are rates for operation over the full military temperature range of −55 °C to +125 °C. Reverse-bias burn-in and 100% high-reliability screening are standard.

The integrated circuits described here permit the circuit designer to select the optimal device for any application. In addition to the two package styles, there are five input characteristics, two output-voltage ratings, and two output-current ratings. The appropriate part for specific applications can be determined from the Device Part Number Designation chart. All units have open-collector outputs and on-chip diodes for inductive-load transient suppression.

DEVICE PART NUMBER DESIGNATION

$V_{CE(MAX)}$	50 V	50 V	95 V
$I_{C(MAX)}$	500 mA	600 mA	500 mA
Logic	Part Number		
General Purpose PMOS, CMOS	ULS-2001*	ULS-2011*	ULS-2021*
14-25 V PMOS	ULS-2002*	ULS-2012*	ULS-2022*
5 V TTL, CMOS	ULS-2003*	ULS-2013*	ULS-2023*
6-15 V CMOS, PMOS	ULS-2004*	ULS-2014*	ULS-2024*
High-Output TTL	ULS-2005*	ULS-2015*	ULS-2025*

*Complete part number includes a final letter to indicate package (H = ceramic/metal size-brazed, R = ceramic/glass cer-DIP).

ABSOLUTE MAXIMUM RATINGS

Output Voltage, V_{CE}	
(ULS-200X*, ULS-201X*)	50 V
(ULS-202X*)	95 V
Input Voltage, V_{IN}	
(ULS-20X2, X3, X4*)	30 V
(ULS-20X5*)	15 V
Peak Output Current, I_{OUT}	
(ULS-200X*, ULS-202X*)	500 mA
(ULS-201X*)	600 mA
Ground Terminal Current, I_{GND}	3.0 A
Continuous Input Current, I_{IN}	25 mA
Power Dissipation, P_D	
(one Darlington pair)	1.0 W
(total package)	See Graph
Operating Temperature Range, T_A	−55°C to +125°C
Storage Temperature Range, T_S	−65°C to +150°C

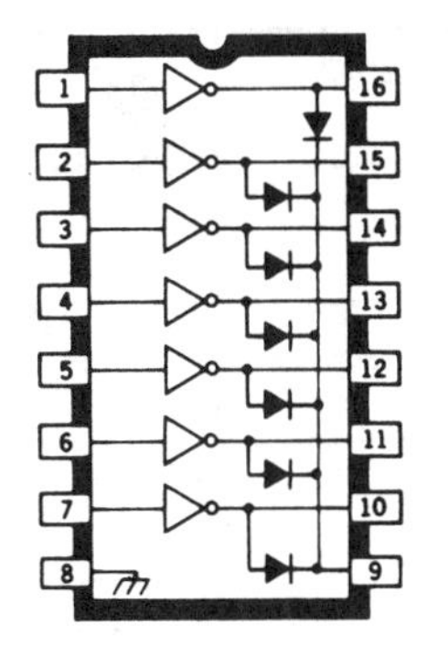

ALLOWABLE PACKAGE POWER DISSIPATION

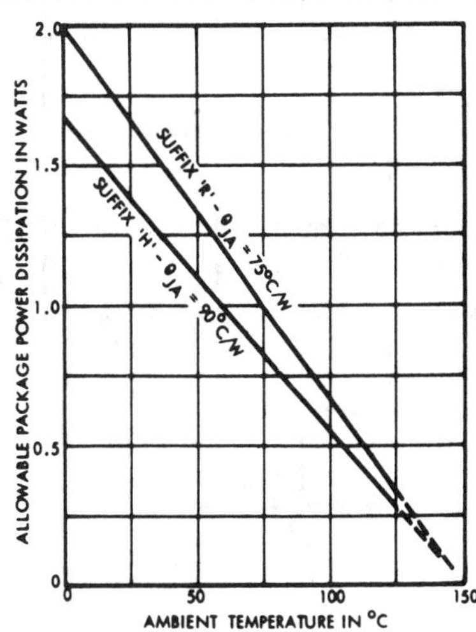

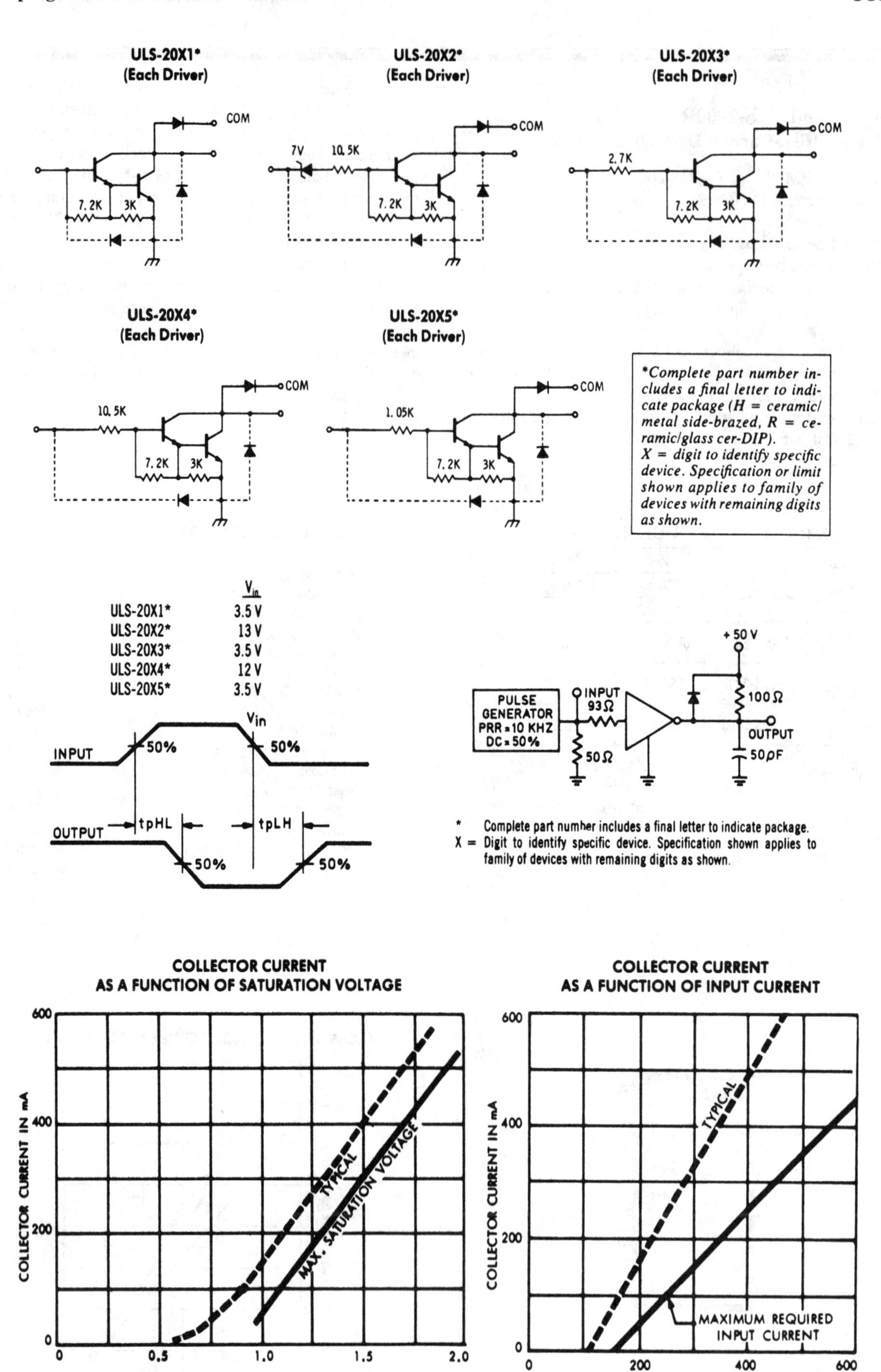
ULS-20X1*
(Each Driver)
COM
7.2K
3K
ULS-20X2*
(Each Driver)
7V
10.5K
COM
7.2K
3K
ULS-20X3*
(Each Driver)
2.7K
COM
7.2K
3K
ULS-20X4*
(Each Driver)
10.5K
COM
7.2K
3K
ULS-20X5*
(Each Driver)
1.05K
COM
7.2K
3K
*Complete part number includes a final letter to indicate package (H = ceramic/metal side-brazed, R = ceramic/glass cer-DIP).
X = digit to identify specific device. Specification or limit shown applies to family of devices with remaining digits as shown.
Vin
ULS-20X1* 3.5 V
ULS-20X2* 13 V
ULS-20X3* 3.5 V
ULS-20X4* 12 V
ULS-20X5* 3.5 V
Vin
INPUT
50%
50%
OUTPUT
tpHL
tpLH
50%
50%
+50 V
PULSE GENERATOR PRR = 10 KHZ DC = 50%
INPUT
93Ω
50Ω
100Ω
OUTPUT
50pF
* Complete part number includes a final letter to indicate package.
X = Digit to identify specific device. Specification shown applies to family of devices with remaining digits as shown.
COLLECTOR CURRENT AS A FUNCTION OF SATURATION VOLTAGE
COLLECTOR CURRENT IN mA
600
400
200
0
0
0.5
1.0
1.5
2.0
TYPICAL
MAX. SATURATION VOLTAGE
COLLECTOR-EMITTER SATURATION VOLTAGE IN VOLTS
COLLECTOR CURRENT AS A FUNCTION OF INPUT CURRENT
COLLECTOR CURRENT IN mA
600
400
200
0
0
200
400
600
TYPICAL
MAXIMUM REQUIRED INPUT CURRENT
INPUT CURRENT IN μA

SERIES ULS-2000H

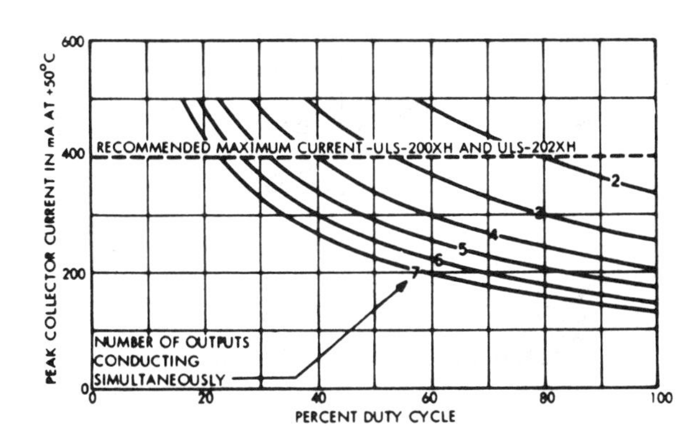

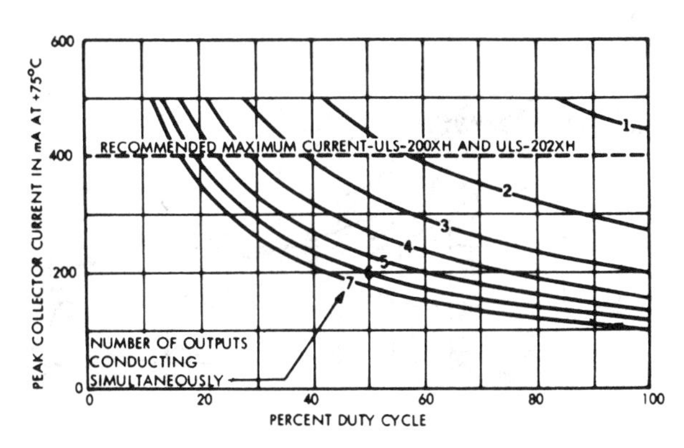

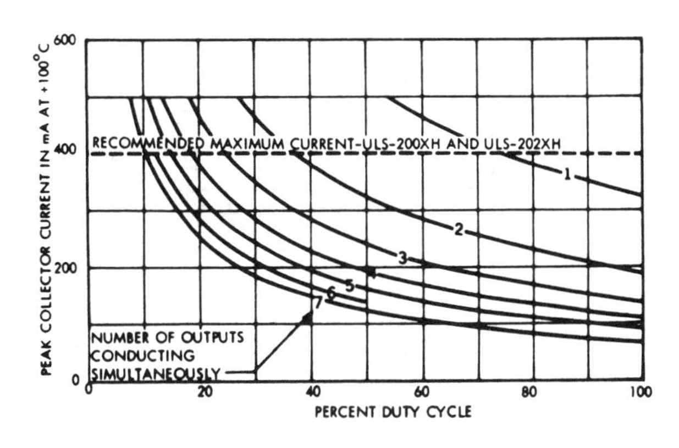

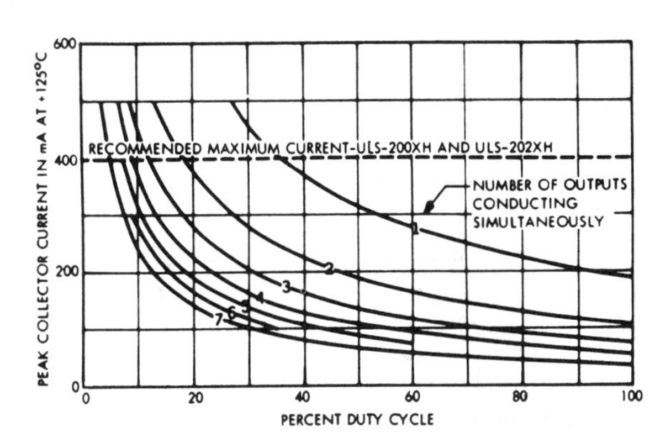

X = digit to identify specific device. Specification or limit shown applies to family of devices with remaining digits as shown.

SERIES ULS-2000R

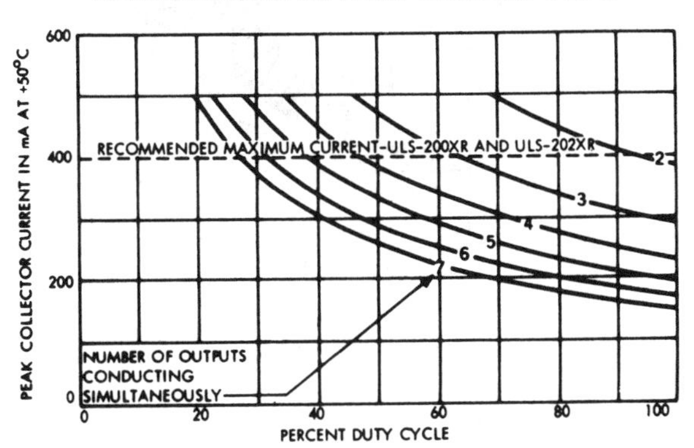

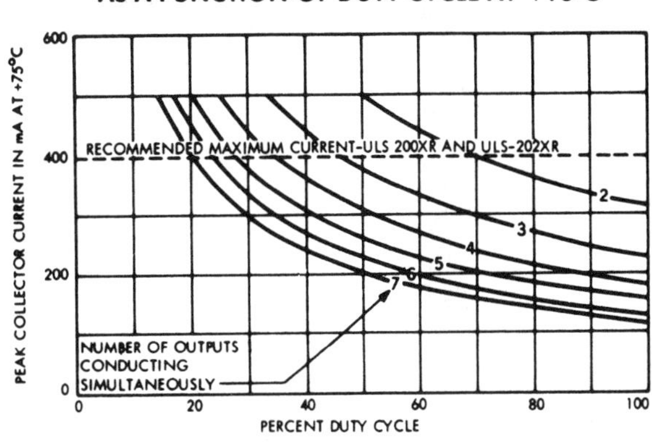

Series ULS-2000R (Continued)

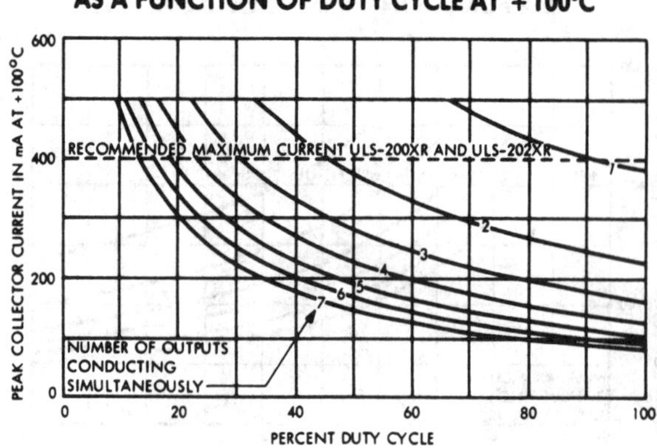

PEAK COLLECTOR CURRENT AS A FUNCTION OF DUTY CYCLE AT +125°C

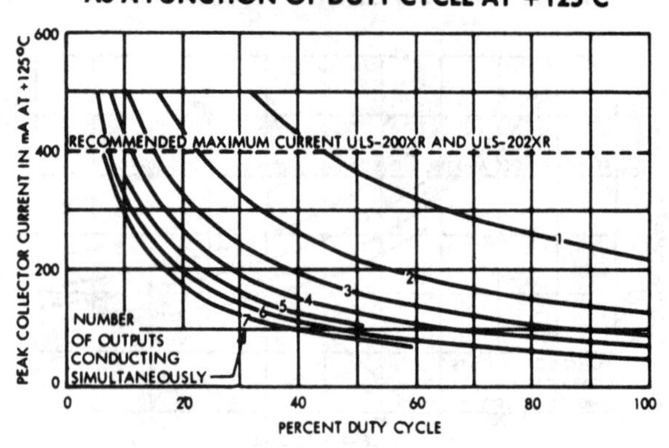

X = digit to identify specific device. Specification or limit shown applies to family of devices with remaining digits as shown.

INPUT CURRENT AS A FUNCTION OF INPUT VOLTAGE

ULS-20X2

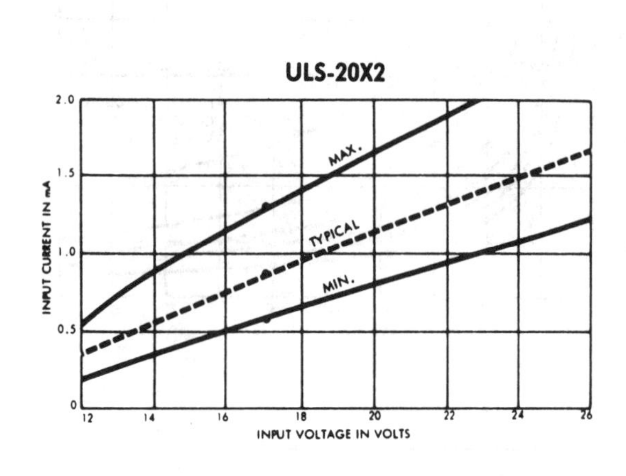

ULS-20X3

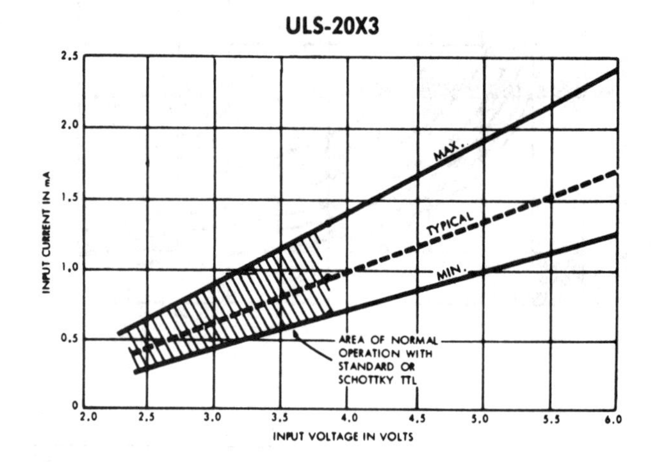

ULS-20X4

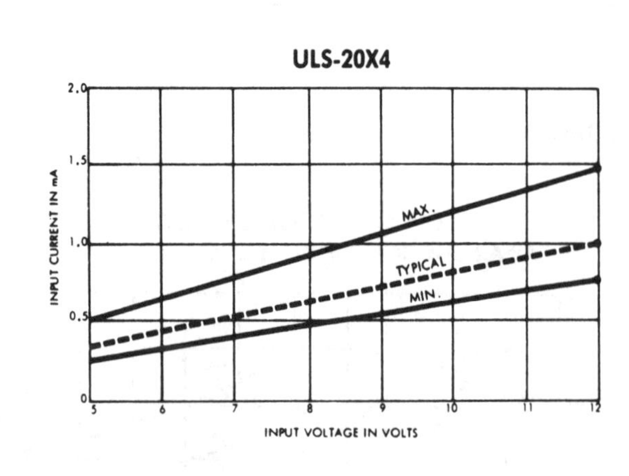

ULS-20X5

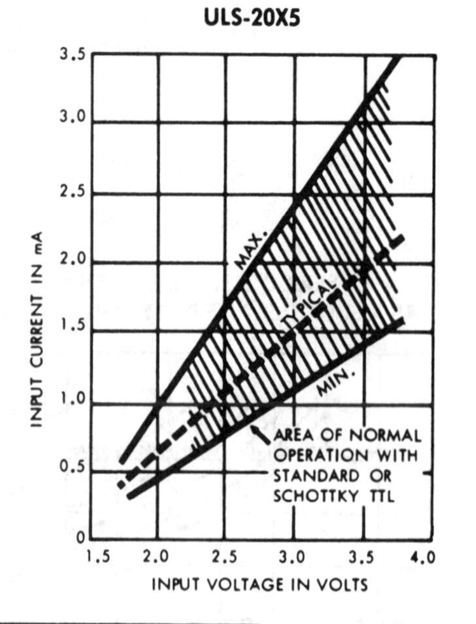

X = digit to identify specific device. Specification or limit shown applies to family of devices with remaining digits as shown.

Sprague
ULS-2064H through ULS2077H
1.25A Quad Darlington Switches

- TTL-, DTL-, PMOS-, or CMOS-compatible units
- Transient-protected outputs
- Hermetically sealed packages
- High-reliability screening to MIL-STD-883, Class B

Intended for military, aerospace, and related applications, ULS-206H through ULS-2077H quad Darlington switches interface between low-level logic and a variety of peripheral power loads, such as relays, solenoids, dc and stepping motors, multiplexed LED and incandescent displays, heaters, and similar loads of up to 400 watts (1.25A per output, 80V, 12.5% duty cycle, +50 °C). The devices are specified with a minimum output breakdown of 50 volts (35 volts sustaining at 100mA) or 80 volts (50 volts sustaining), and a saturated output current specifications of 1.25A.

The ULS-2064/65/68/69H switches are designed for use with TTL, DTL, Schottky TTL, and 5V CMOS logic. The ULS-2066/67/70/77H are intended for use with 6V to 15V CMOS and PMOS logic. These devices include integral transient-suppression diodes for use with inductive loads.

Types ULS-2068H and ULS-2069H incorporate a pre-driver stage operating from a low current, 5V supply. The pre-driver for the ULS-2070H and ULS-2071H operates from a low-current, 12V supply. The input drive requirements for these devices are reduced, while still allowing the outputs to switch currents up to 1.5A.

The ULS-2074H through ULS-2077H switches are intended for use in emitter-follower applications. These circuits are identical with the ULS-2064H through ULS-2067H, except for the uncommitted emitters and the omission of the suppression diodes.

Reverse-bias burn-in and 100% high-reliability screening are standard for all side-brazed hermetic integrated circuits from Sprague. Those devices previously manufactured as the ULS-2064H through ULS-2077H are now screened to the additional requirements of MIL-STD-883, Class B, and are so marked.

These quad Darlington switches are supplied in 16-pin ceramic/metal side-brazed hermetic packages. On special order, economical ceramic/glass cerDIP hermetic packages can be specified by changing the part number suffix—from "H" to "R". Both package styles conform to the dimensional requirements of MIL-M-38510 and are rated for operation over the military temperature range of −55 °C to +125 °C.

ABSOLUTE MAXIMUM RATINGS
at 25°C Free-Air Temperature for any one driver (unless otherwise noted)

Output Voltage, V_{CEX}	See Below
Output Sustaining Voltage, $V_{CE(SUS)}$	See Below
Output Current, I_{OUT}	1.5 A
Input Voltage, V_{IN}	See Below
Input Current, I_B	25 mA
Supply Voltage, V_S (ULS-2068/69H)	10 V
(ULS-2070/71H)	20 V
Total Package Power Dissipation	See Graph
Power Dissipation, P_D/Output	2.2 W
Operating Ambient Temperature Range, T_A	−55°C to +125°C
Storage Temperature Range, T_S	−65°C to +150°C

ALLOWABLE PACKAGE POWER DISSIPATION AS A FUNCTION OF TEMPERATURE

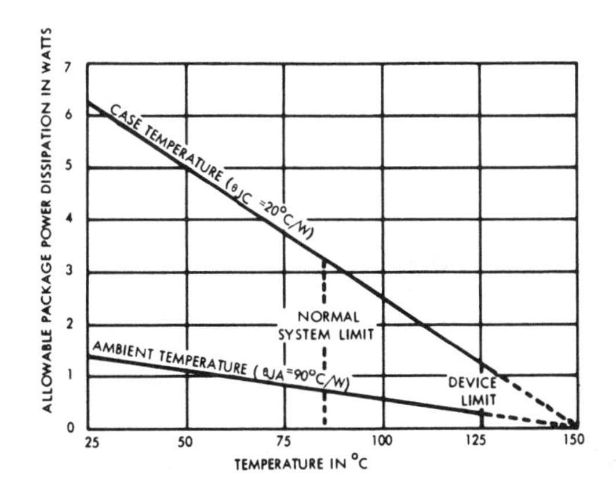

Type Number	V_{CEX} (Max.)	$V_{CE(SUS)}$ (Min.)	V_{IN} (Max.)	Application
ULS-2064H	50 V	35 V	15 V	TTL, DTL, Schottky TTL,
ULS-2065H	80 V	50 V	15 V	and 5 V CMOS
ULS-2066H	50 V	35 V	30 V	6 to 15 V CMOS
ULS-2067H	80 V	50 V	30 V	and PMOS
ULS-2068H	50 V	35 V	15 V	TTL, DTL, Schottky TTL,
ULS-2069H	80 V	50 V	15 V	and 5 V CMOS
ULS-2070H	50 V	35 V	30 V	6 to 15 V CMOS
ULS-2071H	80 V	50 V	30 V	and PMOS
ULS-2074H	50 V	35 V	30 V	General-Purpose
ULS-2075H	80 V	50 V	60 V	
ULS-2076H	50 V	35 V	30 V	6 to 15 V CMOS
ULS-2077H	80 V	50 V	60 V	and PMOS

ULS-2064H THROUGH ULS-2067H

PARTIAL SCHEMATIC

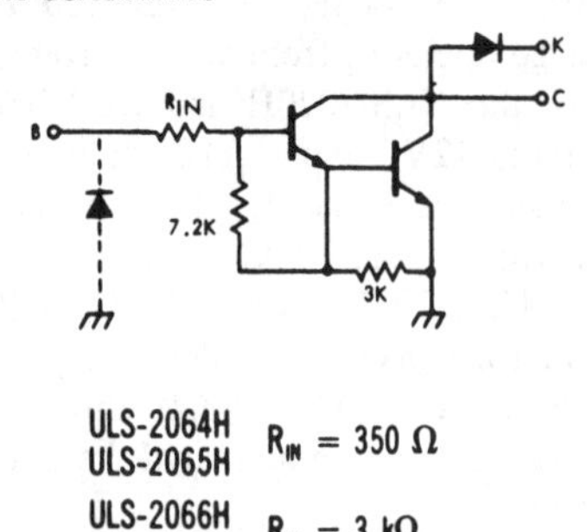

ULS-2064H
ULS-2065H R_{IN} = 350 Ω

ULS-2066H
ULS-2067H R_{IN} = 3 kΩ

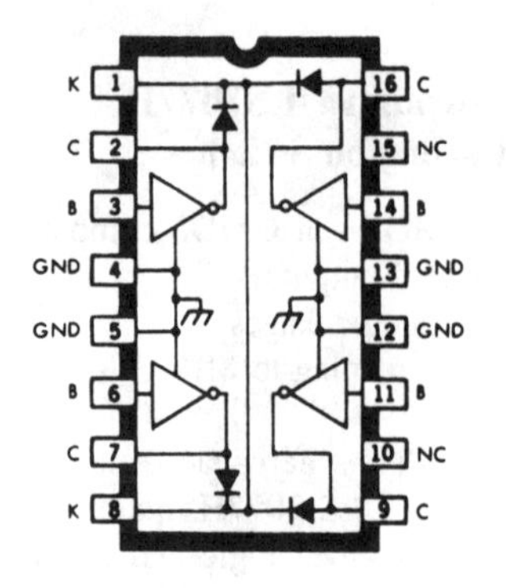

ULS-2068H THROUGH ULS-2071H

PARTIAL SCHEMATIC

ULS-2068H
ULS-2069H R_{IN} = 2.5 kΩ, R_S = 900 Ω

ULS-2070H
ULS-2071H R_{IN} = 11.6 kΩ, R_S = 3.4 kΩ

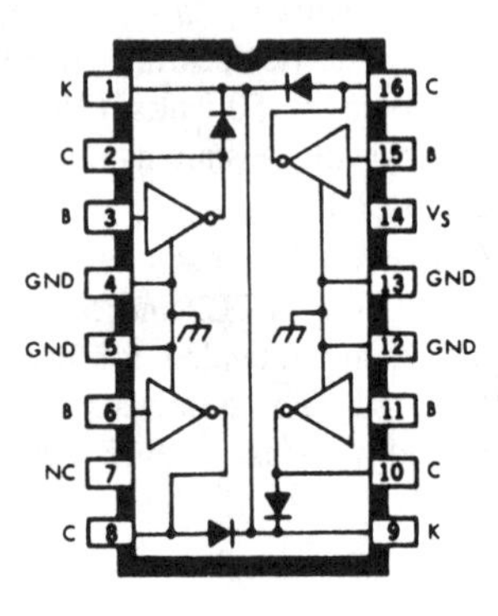

ULS-2074H THROUGH ULS-2077H

PARTIAL SCHEMATIC

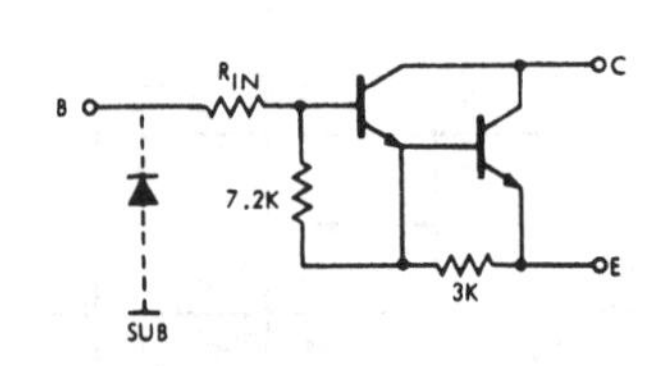

ULS-2074H
ULS-2075H R_{IN} = 350 Ω

ULS-2076H
ULS-2077H R_{IN} = 3 kΩ

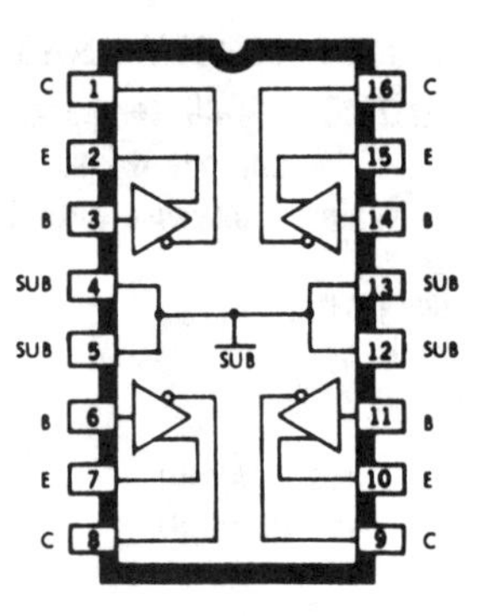

Sprague UDN-2543B
Quad NAND-Gate Power Driver

- 1.0A output current
- Output voltage to 60V
- Low output-saturation voltage
- Integral output-suppression diodes
- Efficient input/output pin structure
- TTL-, CMOS-, PMOS-, NMOS-compatible
- Over-current protected

Providing interface between low-level signal processing circuits and power loads to 240W, the UDN-2543B quad power driver combines NAND logic gates and high-current bipolar outputs. Each of the four independent outputs can sink up to 1A in the on state. The outputs have a minimum breakdown voltage of 60V and a sustaining voltage of 35V. Inputs are compatible with most TTL, DTL, LSTTL, and 5V CMOS and PMOS logic systems.

Over-current protection has been designed into the UDN-2543B and typically occurs at 1A. It protects the device from output short-circuits with supply voltages of up to 25V. When the maximum driver output current is reached, that output stage is driven linearly. If the overcurrent condition continues, that output driver's thermal limiting will operate, limiting the driver's power dissipation and junction temperature. The outputs also include transient-suppression diodes for use with inductive loads, such as relays, solenoids, and dc stepping motors. In display applications, the diodes can be used for the lamp-test function.

The UDN-2543B is supplied in a 16-pin dual in-line plastic package with heatsink contact tabs. The lead configuration allows easy attachment of an inexpensive heatsink and fits a standard integrated circuit socket or a printed wiring board layout.

ABSOLUTE MAXIMUM RATINGS
at $T_A = +25°C$

Output Voltage, V_{CE}	60 V
Over-Current Protected Output Voltage, V_{CE}	25 V
Min. Output Sustaining Voltage, $V_{CE(SUS)}$	35 V
Output Current, I_{OUT}	1.0 A*
Supply Voltage, V_{CC}	7.0 V
Input Voltage, V_{IN}	18 V
Package Power Dissipation, P_D	See Graph
Operating Temperature Range, T_A	−20°C to +85°C
Storage Temperature Range, T_S	−55°C to +150°C

*Outputs are current limited at approximately 1.0 A per driver and junction temperature limited if current in excess of 1.0 A is attempted.

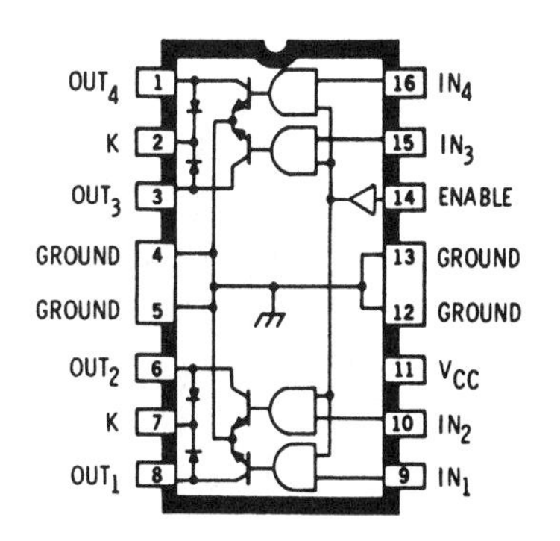

ELECTRICAL CHARACTERISTICS AT $T_A = +25°C$, $V_{CC} = 4.75$ V to 5.25 V (unless otherwise noted)

Characteristic	Symbol	Test Conditions	Limits Min.	Limits Max.	Units
Output Leakage Current	I_{CEX}	$V_{OUT} = 60$ V, $V_{IN} = 0.8$ V, $V_{ENABLE} = 2.0$ V	–	100	µA
		$V_{OUT} = 60$ V, $V_{IN} = 2.0$ V, $V_{ENABLE} = 0.8$ V	–	100	µA
Output Sustaining Voltage	$V_{CE(SUS)}$	$I_{OUT} = 100$ mA, $V_{IN} = V_{ENABLE} = 0.8$ V	35	–	V
Output Saturation Voltage	$V_{CE(SAT)}$	$I_{OUT} = 100$ mA, $V_{IN} = V_{ENABLE} = 2.0$ V	–	200	mV
		$I_{OUT} = 400$ mA, $V_{IN} = V_{ENABLE} = 2.0$ V	–	400	mV
		$I_{OUT} = 700$ mA, $V_{IN} = V_{ENABLE} = 2.0$ V	–	600	mV
Input Voltage	Logic 1	$V_{IN(1)}$ or $V_{ENABLE(1)}$	2.0	–	V
	Logic 0	$V_{IN(0)}$ or $V_{ENABLE(0)}$	–	0.8	V
Input Current	Logic 1	$V_{IN(1)}$ or $V_{ENABLE(1)} = 2.0$ V	–	10	µA
	Logic 0	$V_{IN(0)}$ or $V_{ENABLE(0)} = 0.8$ V	–	-10	µA
Total Supply Current	I_{CC}	$I_{OUT} = 700$ mA, $V_{IN}^* = V_{ENABLE} = 2.0$ V	–	65	mA
		Outputs Open, $V_{IN}^* = 0.8$ V, $V_{ENABLE} = 2.0$ V	–	15	mA
Clamp Diode Forward Voltage	V_F	$I_F = 1.0$ A	–	1.6	V
		$I_F = 1.5$ A	–	2.0	V
Clamp Diode Leakage Current	I_R	$V_R = 60$ V, $V_{IN} = V_{ENABLE} = 2.0$ V, $D_1 + D_2$ or $D_3 + D_4$	–	50	µA

All inputs simultaneously, all other tests are performed with each input tested separately.

ALLOWABLE AVERAGE PACKAGE POWER DISSIPATION AS A FUNCTION OF TEMPERATURE

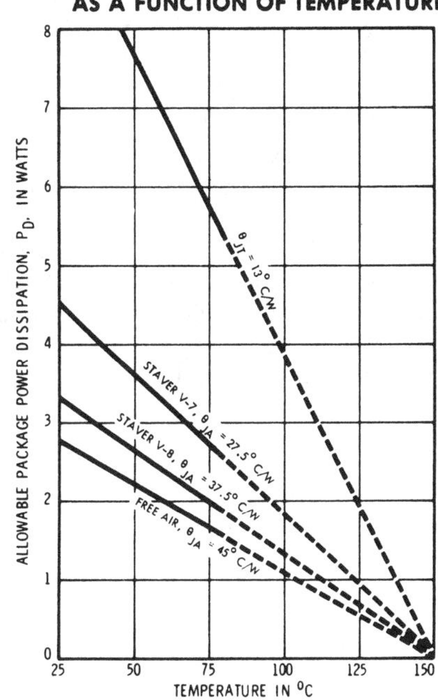

FUNCTIONAL BLOCK DIAGRAM
(1 of 4 Channels)

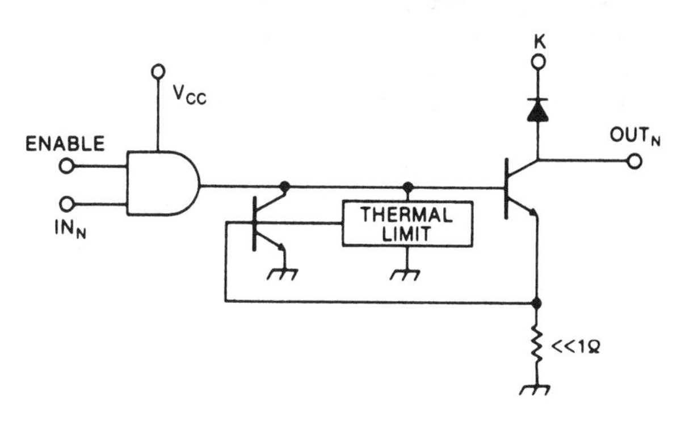

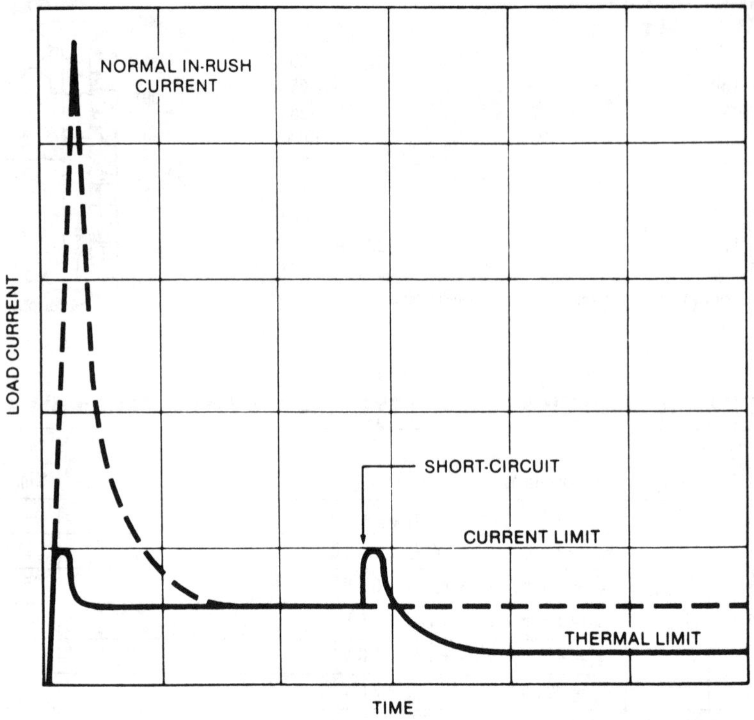

NORMAL IN-RUSH CURRENT
LOAD CURRENT
SHORT-CIRCUIT
CURRENT LIMIT
THERMAL LIMIT
TIME

Sprague UDN-2580A
8-Channel Source Drivers

- TTL-, CMOS-, PMOS-, NMOS-compatible
- High-output current ratings
- Internal transient suppression
- Efficient input/output pin structure

This versatile family of integrated circuits, originally designed to link NMOS logic with high-current inductive loads, will work with many combinations of logic- and load-voltage levels, meeting interface requirements beyond the capabilities of standard logic buffers.

Series UDN-2580A source drivers can drive incandescent, LED, or vacuum fluorescent displays. Internal transient-suppression diodes permit the drivers to be used with inductive loads.

Type UDN-2580A is a high-current source driver used to switch the ground ends of loads that are directly connected to a negative supply. Typical loads are telephone relays, PIN diodes, and LEDs.

Type UDN-2585A is a driver designed for applications requiring low-output saturation voltages. Typical loads are low-voltage LEDs and incandescent displays. The eight non-Darlington outputs will simultaneously sustain continuous load currents of −120mA at ambient temperatures to +70 °C.

Type UDN-2588A, a high-current source driver similar to Type UDN-2580A, has separate logic and driver supply lines. Its eight drivers can serve as an interface between positive logic (TTL, CMOS, PMOS) or negative logic (NMOS) and either negative or split-load supplies.

Types UDN-2580A and UDN-2588A are rated for operation with output voltages of up to 50V. Selected devices, carrying the suffix "-1" on the Sprague part number, have maximum ratings of 80V.

Types UDN-2580A and UDN-2585A are furnished in 18-pin dual-in-line plastic packages; type UDN-2588A is supplied in a 20-pin dual-in-line plastic package. All input connections are on one side of the packages, output pins on the other, to simplify printed wiring board layout.

ABSOLUTE MAXIMUM RATINGS

at 25°C Free-Air Temperature
for Any One Driver
(unless otherwise noted)

	UDN-2580A	UDN-2580A-1	UDN-2585A	UDN-2588A	UDN-2588A-1
Output Voltage, V_{CE}	50 V	80 V	25 V	50 V	80 V
Supply Voltage, V_S (ref. sub.)	50 V	80 V	25 V	50 V	80 V
Supply Voltage, V_{CC} (ref. sub.)	—	—	—	50 V	80 V
Input Voltage, V_{IN} (ref. V_S)	−30 V	−30 V	−20 V	−30 V	−30 V
Total Current, $I_{CC} + I_S$	−500 mA	−500 mA	−250 mA	−500 mA	−500 mA
Substrate Current, I_{SUB}	3.0 A	3.0 A	2.0 A	3.0 A	3.0 A

Allowable Power Dissipation, P_D (single output)	1.0 W
(total package)	2.2 W*
Operating Temperature Range, T_A	−20°C to +85°C
Storage Temperature Range, T_S	−55°C to +150°C

*Derate at the rate of 18 mW/°C above 25°C

TYPICAL OPERATING VOLTAGES

V_S	$V_{IN(ON)}$	$V_{IN(OFF)}$	V_{CC}	$V_{EE(MAX)}$	Device Type
0 V	−15 V to −3.6 V	−0.5 V to 0 V	NA	−25 V	UDN-2585A
				−50 V	UDN-2580A
				−80 V	UDN-2580A-1
+5 V	0 V to +1.4 V	+4.5 V to +5 V	NA	−20 V	UDN-2585A
				−45 V	UDN-2580A
				−75 V	UDN-2580A-1
			≤5 V	−45 V	UDN-2588A
				−75 V	UDN-2588A-1
+12 V	0 V to +8.4 V	+11.5 V to +12 V	NA	−13 V	UDN-2585A
				−38 V	UDN-2580A
				−68 V	UDN-2580A-1
			≤12 V	−38 V	UDN-2588A
				−68 V	UDN-2588A-1
+15 V	0 V to +11.4 V	+14.5 V to +15 V	NA	−10 V	UDN-2585A
				−35 V	UDN-2580A
				−65 V	UDN-2580A-1
			≤15 V	−35 V	UDN2588A
				−65 V	UDN-2588A-1

NOTE: The substrate must be tied to the most negative point in the external circuit to maintain isolation between drivers and to provide for normal circuit operation.

TYPICAL APPLICATIONS

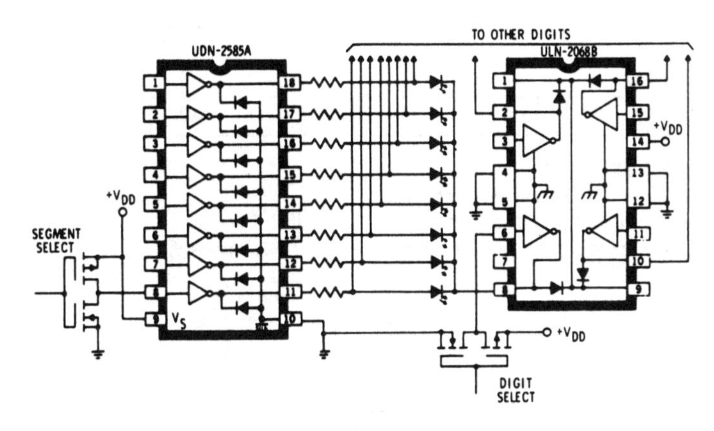

COMMON-CATHODE LED DRIVER

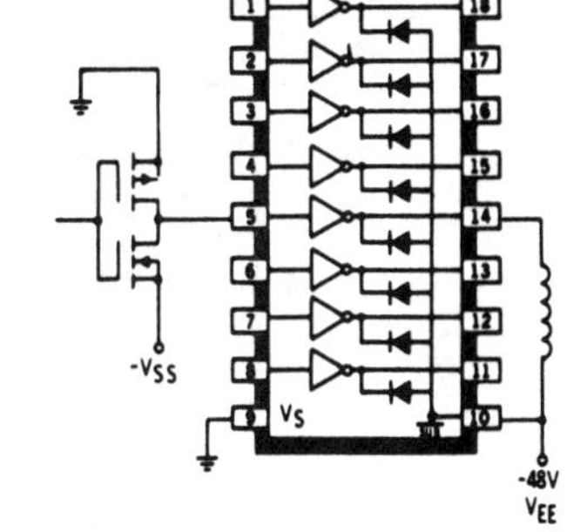

TELECOMMUNICATIONS RELAY DRIVER (Negative Logic)

TYPICAL APPLICATIONS (Continued)

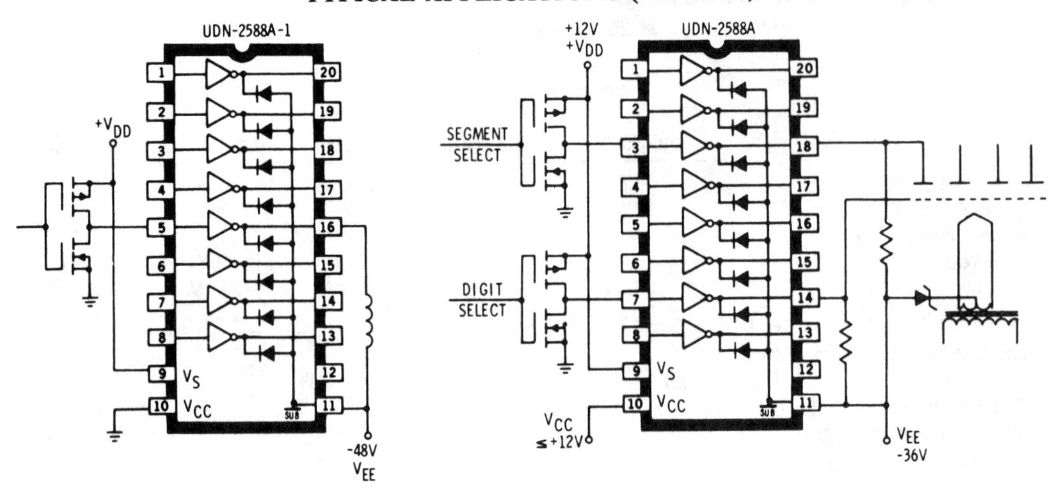

TELECOMMUNICATIONS RELAY DRIVER (Positive Logic)

VACUUM FLUORESCENT DISPLAY DRIVER (Split Supply)

UDN-2585A

ELECTRICAL CHARACTERISTICS at T_A = +25°C, V_S = 0 V, V_{EE} = −20 V (unless otherwise noted)

Characteristic	Symbol	Test Conditions	Limits Min.	Limits Max.	Units
Output Leakage Current	I_{CEX}	V_{IN} = −0.5 V, V_{OUT} = V_{EE} = −25 V	—	50	μA
		V_{IN} = −0.4 V, V_{OUT} = V_{EE} = −25 V, T_A = 70°C	—	100	μA
Output Sustaining Voltage	$V_{CE(SUS)}$	V_{IN} = −0.4 V, I_{OUT} = −25 mA, Note 1	15	—	V
Output Saturation Voltage	$V_{CE(SAT)}$	V_{IN} = −4.6 V, I_{OUT} = −60 mA	—	1.1	V
		V_{IN} = −4.6 V, I_{OUT} = −120 mA	—	1.2	V
Input Current	$I_{IN(ON)}$	V_{IN} = −4.6 V, I_{OUT} = −120 mA	—	−1.6	mA
		V_{IN} = −14.6 V, I_{OUT} = −120 mA	—	−5.0	mA
Input Voltage	$V_{IN(ON)}$	I_{OUT} = −120 mA, V_{CE} ≤1.2 V, Note 3	—	−4.6	V
	$V_{IN(OFF)}$	I_{OUT} = −100 μA, T_A = 70°C	−0.4	—	V
Clamp Diode Leakage Current	I_R	V_R = 25 V, T_A = 70°C	—	50	μA
Clamp Diode Forward Voltage	V_F	I_F = 120 mA	—	2.0	V
Input Capacitance	C_{IN}		—	25	pF
Turn-On Delay	t_{PHL}	0.5 E_{IN} to 0.5 E_{OUT}	—	5.0	μs
Turn-Off Delay	t_{PLH}	0.5 E_{IN} to 0.5 E_{OUT}	—	5.0	μs

NOTES: 1. Pulsed test, t_p ≤300 μs, duty cycle ≤2%.
2. Negative current is defined as coming out of the specified device pin.
3. The $V_{IN(ON)}$ voltage limit guarantees a minimum output source current per the specified conditions.
4. The substrate must always be tied to the most negative point and must be at least 4.0 V below V_S.

PARTIAL SCHEMATIC

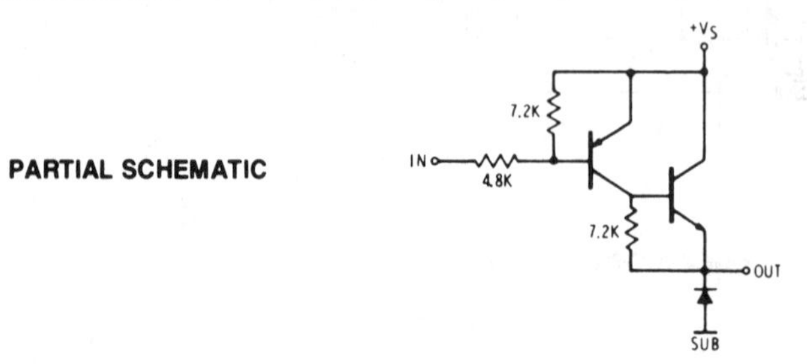

UDN-2580A
UDN-2580A-1

PARTIAL SCHEMATIC

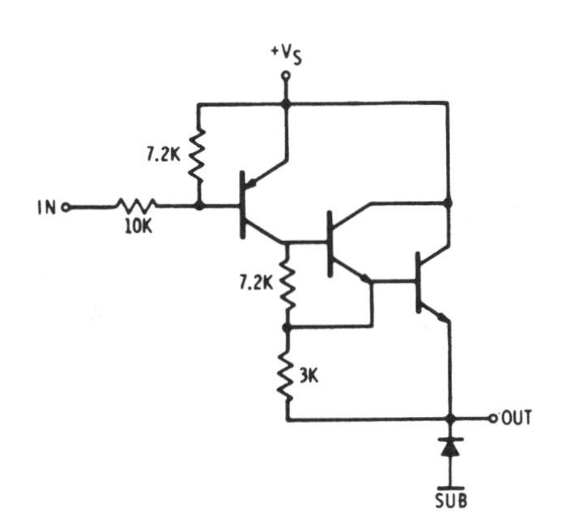

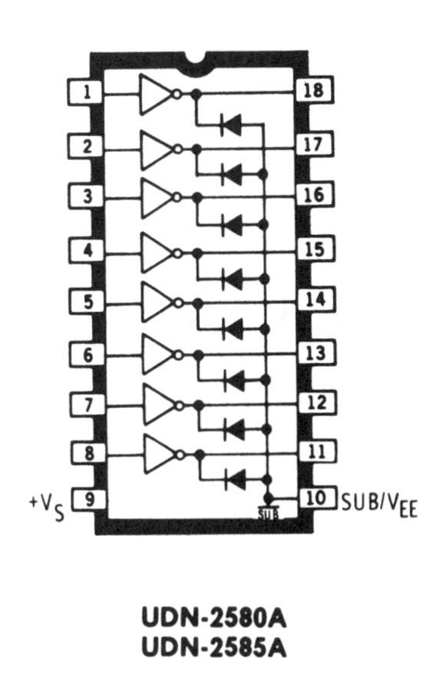

UDN-2580A
UDN-2585A

ELECTRICAL CHARACTERISTICS at T_A = +25°C, V_S = 0 V, V_{EE} = −45 V (unless otherwise noted)

Characteristic	Symbol	Applicable Devices	Test Conditions	Limits Min.	Limits Max.	Units
Output Leakage Current	I_{CEX}	UDN-2580A	V_{IN} = −0.5 V, V_{OUT} = V_{EE} = −50 V	—	50	μA
			V_{IN} = −0.4 V, V_{OUT} = V_{EE} = −50 V, T_A = 70°C	—	100	μA
		UDN-2580A-1	V_{IN} = −0.5 V, V_{OUT} = V_{EE} = −80 V	—	50	μA
			V_{IN} = −0.4 V, V_{OUT} = V_{EE} = −80 V, T_A = 70°C	—	100	μA
Output Sustaining Voltage	$V_{CE(SUS)}$	UDN-2580A	V_{IN} = −0.4 V, I_{OUT} = −25 mA, Note 1	35	—	V
		UDN-2580A-1	V_{IN} = −0.4 V, V_{EE} = −75 V, I_{OUT} = −25 mA, Note 1	50	—	V
Output Saturation Voltage	$V_{CE(SAT)}$	Both	V_{IN} = −2.4 V, I_{OUT} = −100 mA	—	1.8	V
			V_{IN} = −3.0 V, I_{OUT} = −225 mA	—	1.9	V
			V_{IN} = −3.6 V, I_{OUT} = −350 mA	—	2.0	V
Input Current	$I_{IN(ON)}$	Both	V_{IN} = −3.6 V, I_{OUT} = −350 mA	—	−500	μA
			V_{IN} = −15 V, I_{OUT} = −350 mA	—	−2.1	mA
	$I_{IN(OFF)}$	Both	I_{OUT} = −500 μA, T_A = 70°C, Note 3	−50	—	μA
Input Voltage	$V_{IN(ON)}$	Both	I_{OUT} = −100 mA, V_{CE} ≤1.8 V, Note 4	—	−2.4	V
			I_{OUT} = −225 mA, V_{CE} ≤1.9 V, Note 4	—	−3.0	V
			I_{OUT} = −350 mA, V_{CE} ≤2.0 V, Note 4	—	−3.6	V
	$V_{IN(OFF)}$	Both	I_{OUT} = −500 μA, T_A = 70°C	−0.2	—	V
Clamp Diode Leakage Current	I_R	UDN-2580A	V_R = 50 V, T_A = 70°C	—	50	μA
		UDN-2580A-1	V_R = 80 V, T_A = 70°C	—	50	μA
Clamp Diode Forward Voltage	V_F	Both	I_F = 350 mA	—	2.0	V
Input Capacitance	C_{IN}	Both		—	25	pF
Turn-On Delay	t_{PHL}	Both	0.5 E_{IN} to 0.5 E_{OUT}	—	5.0	μs
Turn-Off Delay	t_{PLH}	Both	0.5 E_{IN} to 0.5 E_{OUT}	—	5.0	μs

NOTES: 1. Pulsed test, t_p ≤300 μs, duty cycle ≤2%.
2. Negative current is defined as coming out of the specified device pin.
3. The $I_{IN(OFF)}$ current limit guarantees against partial turn-on of the output.
4. The $V_{IN(ON)}$ voltage limit guarantees a minimum output source current per the specified conditions.
5. The substrate must always be tied to the most negative point and must be at least 4.0 V below V_S.

UDN-2588A
UDN-2588A-1

PARTIAL SCHEMATIC

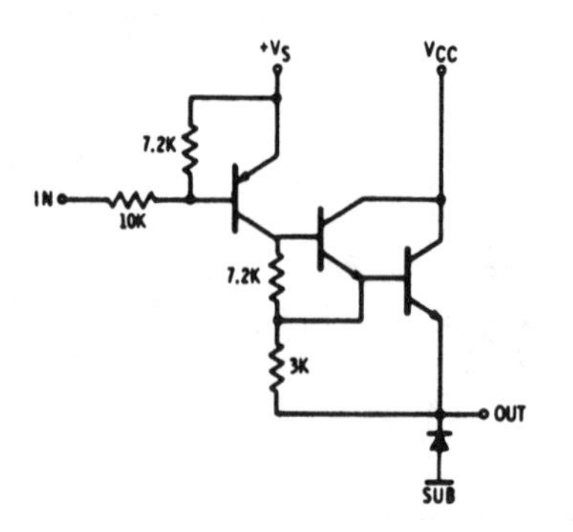

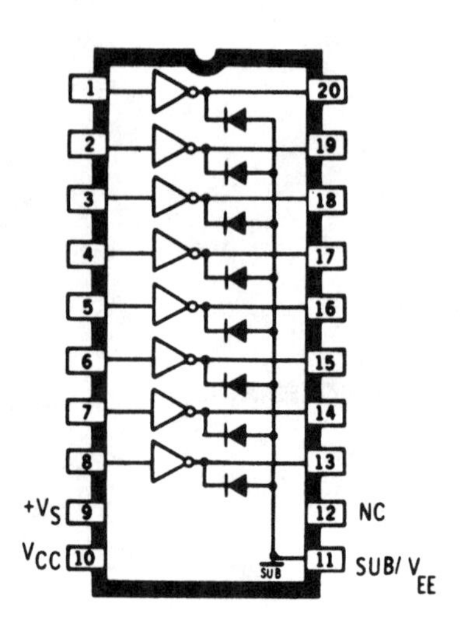

UDN-2588A

ELECTRICAL CHARACTERISTICS at $T_A = +25°C$, $V_S = 5.0$ V, $V_{CC} = 5.0$ V, $V_{EE} = -40$ V (unless otherwise noted)

Characteristic	Symbol	Applicable Devices	Test Conditions	Limits Min.	Limits Max.	Units
Output Leakage Current	I_{CEX}	UDN-2588A	$V_{IN} \geq 4.5$ V, $V_{OUT} = V_{EE} = -45$ V	—	50	μA
			$V_{IN} \geq 4.6$ V, $V_{OUT} = V_{EE} = -45$ V, $T_A = 70°C$	—	100	μA
		UDN-2588A-1	$V_{IN} \geq 4.5$ V, $V_{OUT} = V_{EE} = -75$ V	—	50	μA
			$V_{IN} \geq 4.6$ V, $V_{OUT} = V_{EE} = -75$ V, $T_A = 70°C$	—	100	μA
Output Sustaining Voltage	$V_{CE(SUS)}$	UDN-2588A	$V_{IN} \geq 4.6$ V, $I_{OUT} = -25$ mA, Note 1	35	—	V
		UDN-2588A-1	$V_{IN} \geq 4.6$ V, $V_{EE} = -70$ V, $I_{OUT} = -25$ mA, Note 1	50	—	V
Output Saturation Voltage	$V_{CE(SAT)}$	Both	$V_{IN} = 2.6$ V, $I_{OUT} = -100$ mA, Ref. V_{CC}	—	1.8	V
			$V_{IN} = 2.0$ V, $I_{OUT} = -225$ mA, Ref. V_{CC}	—	1.9	V
			$V_{IN} = 1.4$ V, $I_{OUT} = -350$ mA, Ref. V_{CC}	—	2.0	V
Input Current	$I_{IN(ON)}$	Both	$V_{IN} = 1.4$ V, $I_{OUT} = -350$ mA	—	−500	μA
			$V_S = 15$ V, $V_{EE} = -30$ V, $V_{IN} = 0$ V, $I_{OUT} = -350$ mA	—	−2.1	mA
	$I_{IN(OFF)}$	Both	$I_{OUT} = -500$ A, $T_A = 70°C$, Note 3	−50	—	μA
Input Voltage	$V_{IN(ON)}$	Both	$I_{OUT} = -100$ mA, $V_{CE} \leq 1.8$ V, Note 4	—	2.6	V
			$I_{OUT} = -225$ mA, $V_{CE} \leq 1.9$ V, Note 4	—	2.0	V
			$I_{OUT} = -350$ mA, $V_{CE} \leq 2.0$ V, Note 4	—	1.4	V
	$V_{IN(OFF)}$	Both	$I_{OUT} = -500$ μA, $T_A = 70°C$	4.8	—	V
Clamp Diode Leakage Current	I_R	UDN-2588A	$V_R = 50$ V, $T_A = 70°C$	—	50	μA
		UDN-2588A-1	$V_R = 80$ V, $T_A = 70°C$	—	50	μA
Clamp Diode Forward Voltage	V_F	Both	$I_F = 350$ mA	—	2.0	V
Input Capacitance	C_{IN}	Both		—	25	pF
Turn-On Delay	t_{PHL}	Both	0.5 E_{IN} to 0.5 E_{OUT}	—	5.0	μs
Turn-Off Delay	t_{PLH}	Both	0.5 E_{IN} to 0.5 E_{OUT}	—	5.0	μs

NOTES:
1. Pulsed test, $t_p \leq 300$ μs, duty cycle $\leq 2\%$.
2. Negative current is defined as coming out of the specified device pin.
3. The $I_{IN(OFF)}$ current limit guarantees against partial turn-on of the output.
4. The $V_{IN(ON)}$ voltage limit guarantees a minimum output source current per the specified conditions.
5. The substrate must always be tied to the most negative point and must be at least 4.0 V below V_S.
6. V_{CC} must never be more positive than V_S.

TYPICAL APPLICATIONS

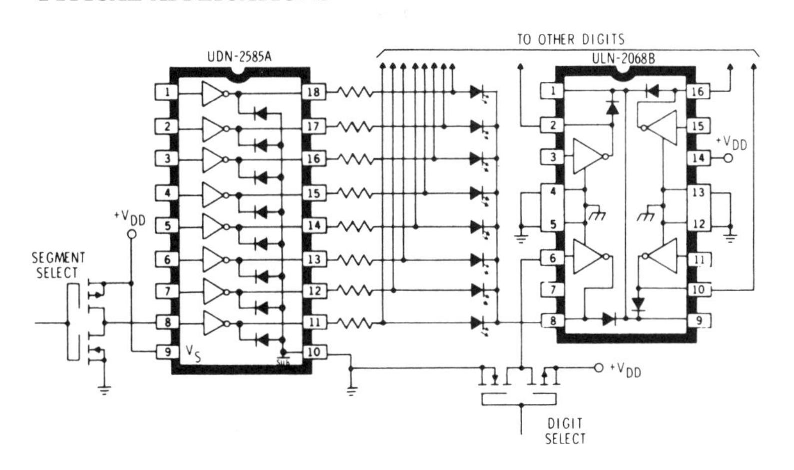

COMMON-CATHODE LED DRIVER

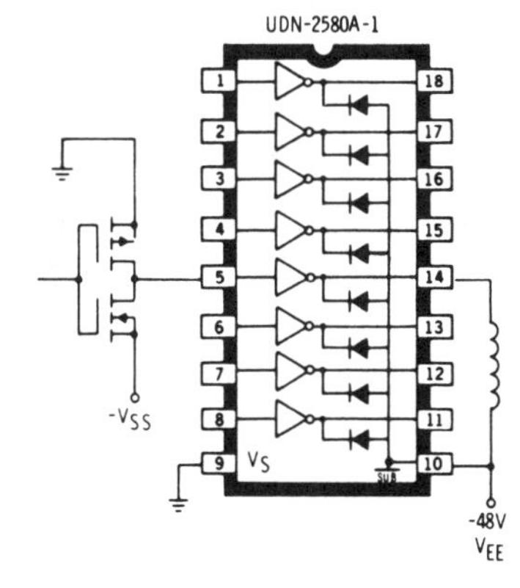

TELECOMMUNICATIONS RELAY DRIVER (Negative Logic)

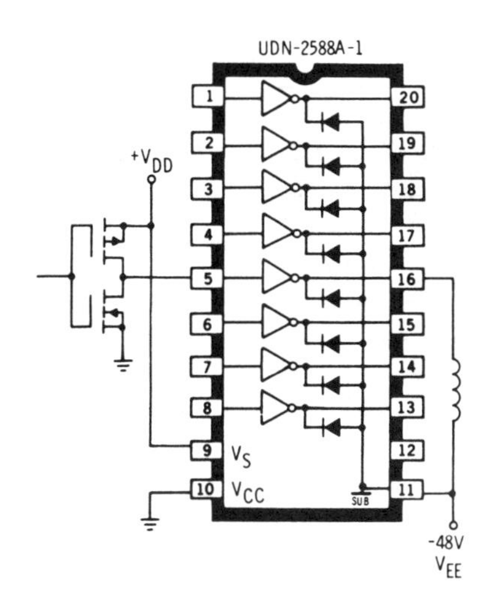

TELECOMMUNICATIONS RELAY DRIVER (Positive Logic)

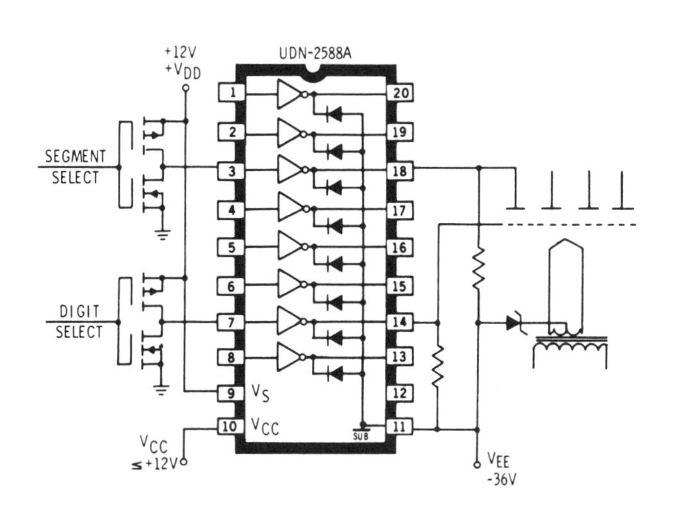

VACUUM FLUORESCENT DISPLAY DRIVER (Split Supply)

Sprague UDN-2595A
8-Channel Current-Sink Driver

- 200mA current rating
- Low saturation voltage
- TTL-, CMOS-, NMOS-compatible
- Efficient input/output pin format
- 18-pin dual-in-line plastic package

ABSOLUTE MAXIMUM RATINGS

at 25°C Free-Air Temperature
for any one driver
(unless otherwise noted)

Parameter	Rating
Output Voltage, V_{CE}	20 V
Supply Voltage, V_S	20 V
Input Voltage, V_{IN}	20 V
Output Collector Current, I_C	200 mA
Ground Terminal Current, I_{GND}	1.6 A
Allowable Power Dissipation, P_D	
(single output)	1.0 W
(total package)	2.2 W*
Operating Temperature Range, T_A	−20°C to +85°C
Storage Temperature Range, T_S	−55°C to +150°C

*Derate at the rate of 18 mW/°C above +25°C.

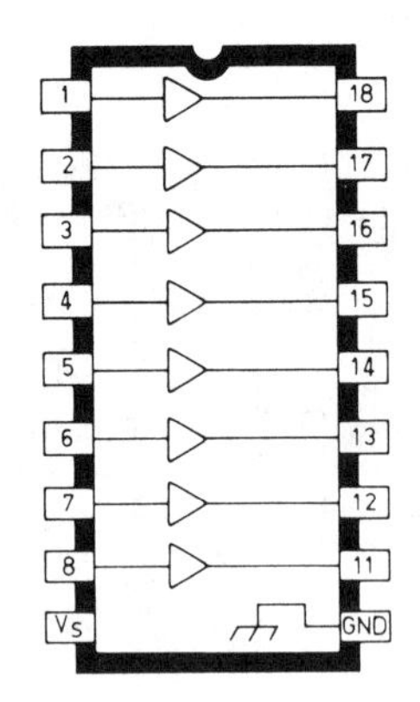

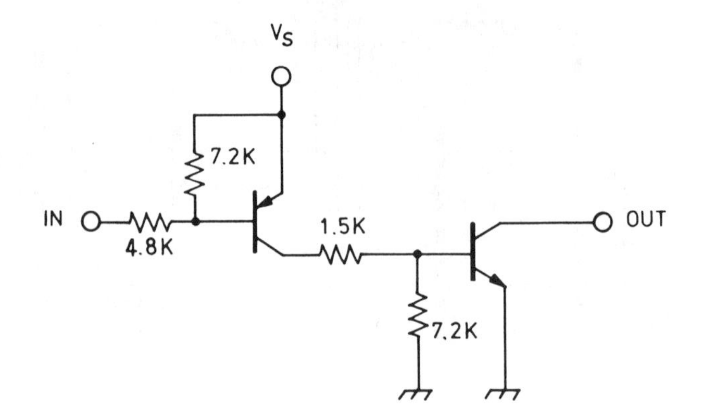

ELECTRICAL CHARACTERISTICS at T_A = +25°C, V_S = 5.0 V (unless otherwise noted).

Characteristic	Symbol	Test Conditions	Limits Min.	Limits Max.	Units
Output Leakage Current	I_{CEX}	$V_{IN} \geq 4.5$ V, $V_{OUT} = 20$ V, $T_A = 25°C$	—	50	µA
		$V_{IN} \geq 4.6$ V, $V_{OUT} = 20$ V, $T_A = 70°C$	—	100	µA
Output Saturation Voltage	$V_{CE(SAT)}$	$V_{IN} = 0.4$ V, $I_{OUT} = 50$ mA	—	0.5	V
		$V_{IN} = 0.4$ V, $I_{OUT} = 100$ mA	—	0.6	V
Input Current	$I_{IN(ON)}$	$V_{IN} = 0.4$ V, $I_{OUT} = 100$ mA	—	−1.6	mA
		$V_{IN} = 0.4$ V, $I_{OUT} = 100$ mA, $V_S = 15$ V	—	−5.0	mA
Input Voltage	$V_{IN(ON)}$	$I_{OUT} = 100$ mA, $V_{OUT} \leq 0.6$ V, $V_S = 5$ V	—	0.4	V
	$V_{IN(OFF)}$	$I_{OUT} = 100$ µA, $T_A = 70°C$	4.6	—	V
Input Capacitance	C_{IN}		—	25	pF
Supply Current	I_{SS}	$V_{IN} = 0.4$ V, $I_{OUT} = 100$ mA	—	6.0	mA
		$V_{IN} = 0.4$ V, $I_{OUT} = 100$ mA, $V_S = 15$ V	—	20	mA

NOTES:
1. Negative current is defined as coming out of the specified device pin.
2. The $V_{IN(ON)}$ voltage limit guarantees a minimum output sink current per the specified conditions.
3. I_{SS} is measured with any one of eight drivers turned ON.

Sprague UDN-2596A through UDN-2599A
8-Channel Saturated Sink Drivers

- Low output on voltages
- Up to 1.0A sink capability
- 50V min. output breakdown
- Output transient-suppression diodes
- Output pull-down for fast turn off
- TTL-, CMOS-compatible inputs

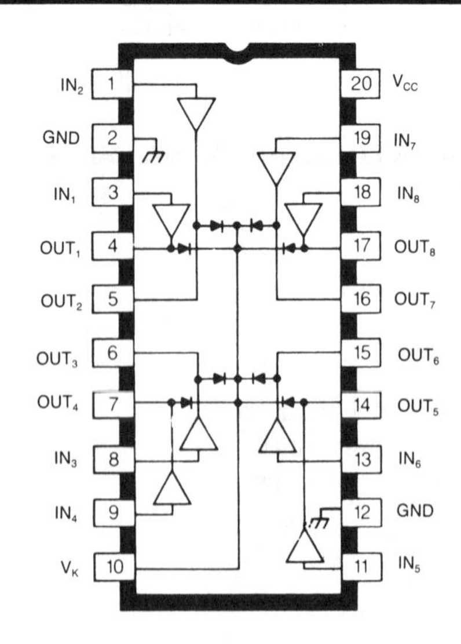

ONE OF EIGHT DRIVERS

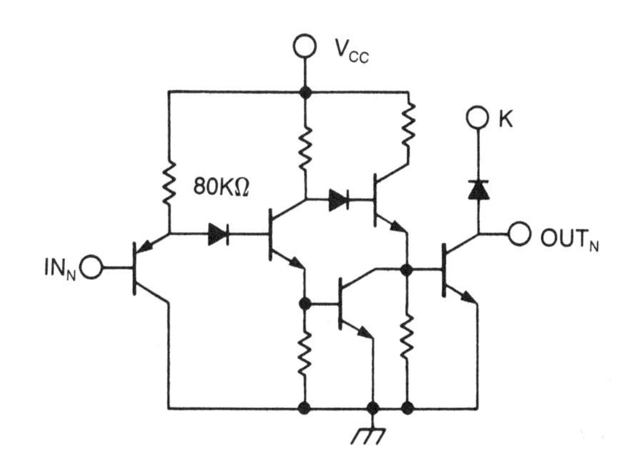

ABSOLUTE MAXIMUM RATINGS

at $T_A = +25°C$

Output Voltage, V_{CE}	50V
Output Current, I_{OUT}	
(UDN-2596/98A)	500mA
(UDN-2597/99A)	1.0A
Supply Voltage, V_{CC}	
(UDN-2596/97A)	7.0V
(UDN-2598/99A)	15V
Input Voltage, V_{IN}	
(UDN-2596/97A)	7.0V
(UDN-2598/99A)	15V
Package Power Dissipation, P_D	2.27W*
Operating Temperature Range, T_A	−20°C to +85°C
Storage Temperature Range, T_S	−65°C to +150°C

*Derate at the rate of 18.2mW/°C above T_A = 25°C

ELECTRICAL CHARACTERISTICS at $T_A = +25°C$, V_{CC} = 5.0V (UDN-2596/97A) or 12V (UDN-2598/99A)

Characteristics	Symbol	Applicable Devices*	Test Conditions	Limits Min.	Limits Max.	Units
Output Leakage Current	I_{CEX}	All	V_{OUT} = 50V, V_{IN} = 2.4V	—	10	μA
Output Sustaining Voltage	$V_{CE(sus)}$	2596/98	I_{OUT} = 300mA, L = 2mH	35	—	V
		2597/99	I_{OUT} = 750mA, L = 2mH	35	—	V
Output Saturation Voltage	$V_{CE(SAT)}$	2596/98	I_{OUT} = 300mA	—	0.5	V
		2597/99	I_{OUT} = 750mA	—	1.0	V
Clamp Diode Leakage Current	I_R	All	V_R = 50V	—	10	μA
Clamp Diode Forward Voltage	V_F	2596/98	I_F = 300mA	—	1.8	V
		2597/99	I_F = 750mA	—	1.8	V
Logic Input Current	$I_{IN(0)}$	2596/97	V_{IN} = 0.8V	—	−15	μA
		2598/99	V_{IN} = 0.8V	—	−50	μA
	$I_{IN(1)}$	2596/97	V_{IN} = 2.4V	—	10	μA
		2598/99	V_{IN} = 12V	—	10	μA
Supply Current (per driver)	$I_{CC(ON)}$	2596/98	V_{IN} = 0.8V	—	6.0	mA
		2597/99	V_{IN} = 0.8V	—	22	mA
	$I_{CC(OFF)}$	2596/97	V_{IN} = 2.4V	—	1.3	mA
		2598/99	V_{IN} = 2.4V	—	2.0	mA
Turn-On Delay	t_{pd0}	All	$0.5E_{IN}$ to $0.5E_{OUT}$	—	3.0	μs
Turn-Off Delay	t_{pd1}	All	$0.5E_{IN}$ to $0.5E_{OUT}$	—	2.0	μs

*Complete part number includes prefix UDN- and suffix A, e.g. UDN-2596A.

TYPICAL APPLICATION

DUAL STEPPER MOTOR DRIVE

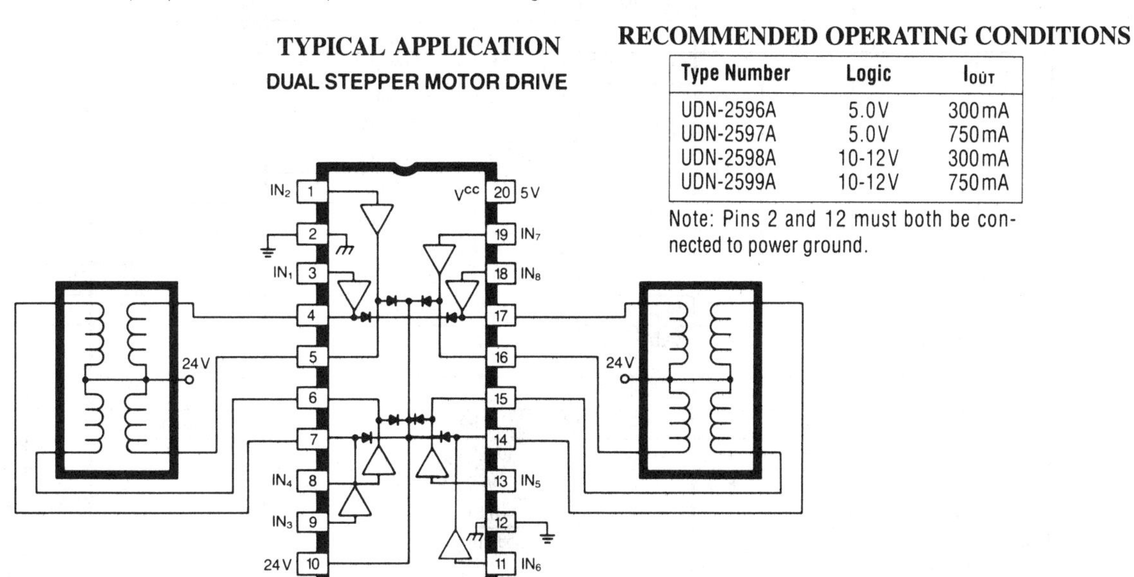

RECOMMENDED OPERATING CONDITIONS

Type Number	Logic	I_{OUT}
UDN-2596A	5.0V	300mA
UDN-2597A	5.0V	750mA
UDN-2598A	10-12V	300mA
UDN-2599A	10-12V	750mA

Note: Pins 2 and 12 must both be connected to power ground.

Sprague
Series ULN-2800A
High-Voltage, High-Current Darlington Transistor Arrays

DEVICE TYPE NUMBER DESIGNATION

$V_{CE(MAX)}$ = $I_{C(MAX)}$ =	50 V 500 mA	50 V 600 mA	95 V 500 mA
	Type Number		
General Purpose PMOS, CMOS	ULN-2801A	ULN-2811A	ULN-2821A
14 - 25 V PMOS	ULN-2802A	ULN-2812A	ULN-2822A
5 V TTL, CMOS	ULN-2803A	ULN-2813A	ULN-2823A
6 - 15 V CMOS, PMOS	ULN-2804A	ULN-2814A	ULN-2824A
High Output TTL	ULN-2805A	ULN-2815A	ULN-2825A

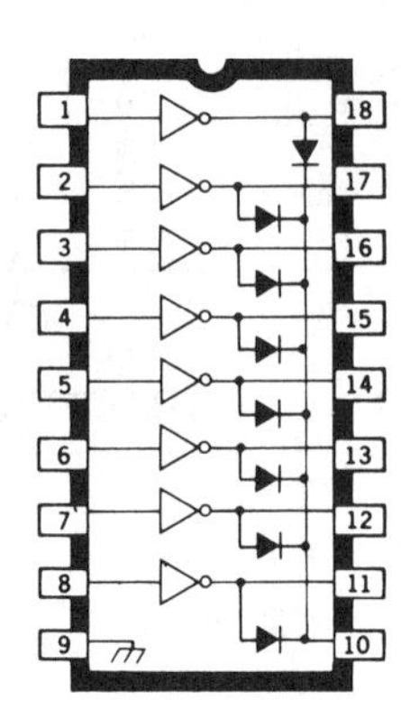

SERIES ULN-2800A

ELECTRICAL CHARACTERISTICS at 25°C (unless otherwise noted)

Characteristic	Symbol	Test Fig.	Applicable Devices	Test Conditions	Limits Min.	Typ.	Max.	Units
Output Leakage Current	I_{CEX}	1A	All	V_{CE} = 50 V, T_A = 25°C	—	—	50	μA
				V_{CE} = 50 V, T_A = 70°C	—	—	100	μA
		1B	ULN-2802A	V_{CE} = 50 V, T_A = 70°C, V_{IN} = 6.0 V	—	—	500	μA
			ULN-2804A	V_{CE} = 50 V, T_A = 70°C, V_{IN} = 1.0 V	—	—	500	μA
Collector-Emitter Saturation Voltage	$V_{CE(SAT)}$	2	All	I_C = 100 mA, I_B = 250 μA	—	0.9	1.1	V
				I_C = 200 mA, I_B = 350 μA	—	1.1	1.3	V
				I_C = 350 mA, I_B = 500 μA	—	1.3	1.6	V
Input Current	$I_{IN(ON)}$	3	ULN-2802A	V_{IN} = 17 V	—	0.82	1.25	mA
			ULN-2803A	V_{IN} = 3.85 V	—	0.93	1.35	mA
			ULN-2804A	V_{IN} = 5.0 V	—	0.35	0.5	mA
				V_{IN} = 12 V	—	1.0	1.45	mA
			ULN-2805A	V_{IN} = 3.0 V	—	1.5	2.4	mA
	$I_{IN(OFF)}$	4	All	I_C = 500 μA, T_A = 70°C	50	65	—	μA
Input Voltage	$V_{IN(ON)}$	5	ULN-2802A	V_{CE} = 2.0 V, I_C = 300 mA	—	—	13	V
			ULN-2803A	V_{CE} = 2.0 V, I_C = 200 mA	—	—	2.4	V
				V_{CE} = 2.0 V, I_C = 250 mA	—	—	2.7	V
				V_{CE} = 2.0 V, I_C = 300 mA	—	—	3.0	V
			ULN-2804A	V_{CE} = 2.0 V, I_C = 125 mA	—	—	5.0	V
				V_{CE} = 2.0 V, I_C = 200 mA	—	—	6.0	V
				V_{CE} = 2.0 V, I_C = 275 mA	—	—	7.0	V
				V_{CE} = 2.0 V, I_C = 350 mA	—	—	8.0	V
			ULN-2805A	V_{CE} = 2.0 V, I_C = 350 mA	—	—	2.4	V
D-C Forward Current Transfer Ratio	h_{FE}	2	ULN-2801A	V_{CE} = 2.0 V, I_C = 350 mA	1000	—	—	
Input Capacitance	C_{IN}	–	All		—	15	25	pF
Turn-On Delay	t_{ON}	–	All	0.5 E_{in} to 0.5 E_{out}	—	0.25	1.0	μs
Turn-Off Delay	t_{OFF}	–	All	0.5 E_{in} to 0.5 E_{out}	—	0.25	1.0	μs
Clamp Diode Leakage Current	I_R	6	All	V_R = 50 V, T_A = 25°C	—	—	50	μA
				V_R = 50 V, T_A = 70°C	—	—	100	μA
Clamp Diode Forward Voltage	V_F	7	All	I_F = 350 mA	—	1.7	2.0	V

ABSOLUTE MAXIMUM RATINGS at 25°C Free-Air Temperature

for any one Darlington pair (unless otherwise noted)

Output Voltage, V_{CE} (Series ULN-2800, 2810A) .... 50 V
(Series ULN-2820A) .... 95 V
Input Voltage, V_{IN} (Series ULN-2802, 2803, 2804A) .... 30 V
(Series ULN-2805A) .... 15 V
Continuous Collector Current, I_C (Series ULN-2800, 2820A) .... 500 mA
(Series ULN-2810A) .... 600 mA
Continuous Base Current, I_B .... 25 mA
Power Dissipation, P_D (one Darlington pair) .... 1.0 W
(total package) .... 2.25 W*
Operating Ambient Temperature Range, T_A .... −20°C to +85°C
Storage Temperature Range, T_S .... −55°C to +150°C

*Derate at the rate of 18.18mW/°C above 25°C.
Under normal operating conditions, these devices will sustain 350 mA per output with $V_{CE(SAT)}$ = 1.6 V at 50°C with a pulse width of 20 ms and a duty cycle of 40%.

SERIES ULN-2810A

ELECTRICAL CHARACTERISTICS at 25°C (unless otherwise noted)

Characteristic	Symbol	Test Fig.	Applicable Devices	Test Conditions	Limits Min.	Typ.	Max.	Units
Output Leakage Current	I_{CEX}	1A	All	V_{CE} = 50 V, T_A = 25°C	—	—	50	μA
				V_{CE} = 50 V, T_A = 70°C	—	—	100	μA
		1B	ULN-2812A	V_{CE} = 50 V, T_A = 70°C, V_{IN} = 6.0 V	—	—	500	μA
			ULN-2814A	V_{CE} = 50 V, T_A = 70°C, V_{IN} = 1.0 V	—	—	500	μA
Collector-Emitter Saturation Voltage	$V_{CE(SAT)}$	2	All	I_C = 200 mA, I_B = 350 μA	—	1.1	1.3	V
				I_C = 350 mA, I_B = 500 μA	—	1.3	1.6	V
				I_C = 500 mA, I_B = 600 μA	—	1.7	1.9	V
Input Current	$I_{IN(ON)}$	3	ULN-2812A	V_{IN} = 17 V	—	0.82	1.25	mA
			ULN-2813A	V_{IN} = 3.85 V	—	0.93	1.35	mA
			ULN-2814A	V_{IN} = 5.0 V	—	0.35	0.5	mA
				V_{IN} = 12 V	—	1.0	1.45	mA
			ULN-2815A	V_{IN} = 3.0 V	—	1.5	2.4	mA
	$I_{IN(OFF)}$	4	All	I_C = 500 μA, T_A = 70°C	50	65	—	μA
Input Voltage	$V_{IN(ON)}$	5	ULN-2812A	V_{CE} = 2.0 V, I_C = 500 mA	—	—	17	V
			ULN-2813A	V_{CE} = 2.0 V, I_C = 250 mA	—	—	2.7	V
				V_{CE} = 2.0 V, I_C = 300 mA	—	—	3.0	V
				V_{CE} = 2.0 V, I_C = 500 mA	—	—	3.5	V
			ULN-2814A	V_{CE} = 2.0 V, I_C = 275 mA	—	—	7.0	V
				V_{CE} = 2.0 V, I_C = 350 mA	—	—	8.0	V
				V_{CE} = 2.0 V, I_C = 500 mA	—	—	9.5	V
			ULN-2815A	V_{CE} = 2.0 V, I_C = 500 mA	—	—	2.6	V
D-C Forward Current Transfer Ratio	h_{FE}	2	ULN-2811A	V_{CE} = 2.0 V, I_C = 350 mA	1000	—	—	
				V_{CE} = 2.0 V, I_C = 500 mA	900	—	—	
Input Capacitance	C_{IN}	—	All		—	15	25	pF
Turn-On Delay	t_{ON}	—	All	0.5 E_{in} to 0.5 E_{out}	—	0.25	1.0	μs
Turn-Off Delay	t_{OFF}	—	All	0.5 E_{in} to 0.5 E_{out}	—	0.25	1.0	μs
Clamp Diode Leakage Current	I_R	6	All	V_R = 50 V, T_A = 25°C	—	—	50	μA
				V_R = 50 V, T_A = 70°C	—	—	100	μA
Clamp Diode Forward Voltage	V_F	7	All	I_F = 350 mA	—	1.7	2.0	V
				I_F = 500 mA	—	2.1	2.5	V

SERIES ULN-2820A

ELECTRICAL CHARACTERISTICS at 25°C (unless otherwise noted)

Characteristic	Symbol	Test Fig.	Applicable Devices	Test Conditions	Limits Min.	Limits Typ.	Limits Max.	Units
Output Leakage Current	I_{CEX}	1A	All	V_{CE} = 95 V, T_A = 25°C	—	—	50	μA
				V_{CE} = 95 V, T_A = 70°C	—	—	100	μA
		1B	ULN-2822A	V_{CE} = 95 V, T_A = 70°C, V_{IN} = 6.0 V		—	500	μA
			ULN-2824A	V_{CE} = 95 V, T_A = 70°C, V_{IN} = 1.0 V		—	500	μA
Collector-Emitter Saturation Voltage	$V_{CE(SAT)}$	2	All	I_C = 100 mA, I_B = 250 μA		0.9	1.1	V
				I_C = 200 mA, I_B = 350 μA		1.1	1.3	V
				I_C = 350 mA, I_B = 500 μA		1.3	1.6	V
Input Current	$I_{IN(ON)}$	3	ULN-2822A	V_{IN} = 17 V	—	0.82	1.25	mA
			ULN-2823A	V_{IN} = 3.85 V	—	0.93	1.35	mA
			ULN-2824A	V_{IN} = 5.0 V	—	0.35	0.5	mA
				V_{IN} = 12 V	—	1.0	1.45	mA
			ULN-2825A	V_{IN} = 3.0 V	—	1.5	2.4	mA
	$I_{IN(OFF)}$	4	All	I_C = 500 μA, T_A = 70°C	50	65	—	μA
Input Voltage	$V_{IN(ON)}$	5	ULN-2822A	V_{CE} = 2.0 V, I_C = 300 mA	—	—	13	V
			ULN-2823A	V_{CE} = 2.0 V, I_C = 200 mA	—	—	2.4	V
				V_{CE} = 2.0 V, I_C = 250 mA	—	—	2.7	V
				V_{CE} = 2.0 V, I_C = 300 mA	—	—	3.0	V
			ULN-2824A	V_{CE} = 2.0 V, I_C = 125 mA	—	—	5.0	V
				V_{CE} = 2.0 V, I_C = 200 mA	—	—	6.0	V
				V_{CE} = 2.0 V, I_C = 275 mA	—	—	7.0	V
				V_{CE} = 2.0 V, I_C = 350 mA	—	—	8.0	V
			ULN-2825A	V_{CE} = 2.0 V, I_C = 350 mA	—	—	2.4	V
D-C Forward Current Transfer Ratio	h_{FE}	2	ULN-2821A	V_{CE} = 2.0 V, I_C = 350 mA	1000	—	—	
Input Capacitance	C_{IN}		All		—	15	25	pF
Turn-On Delay	t_{ON}		All	0.5 E_{in} to 0.5 E_{out}	—	0.25	1.0	μs
Turn-Off Delay	t_{OFF}		All	0.5 E_{in} to 0.5 E_{out}	—	0.25	1.0	μs
Clamp Diode Leakage Current	I_R	6	All	V_R = 95 V, T_A = 25°C	—	—	50	μA
				V_R = 95 V, T_A = 70°C	—	—	100	μA
Clamp Diode Forward Voltage	V_F	7	All	I_F = 350 mA	—	1.7	2.0	V

TEST FIGURES

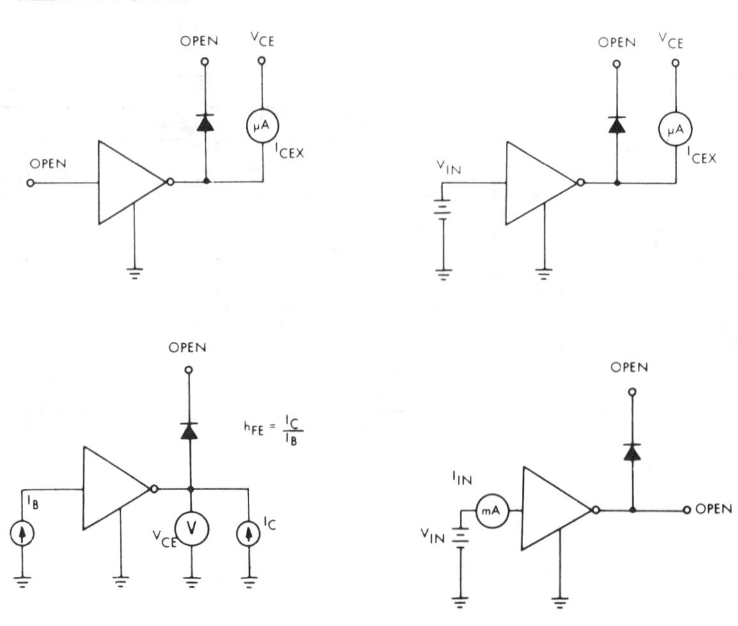

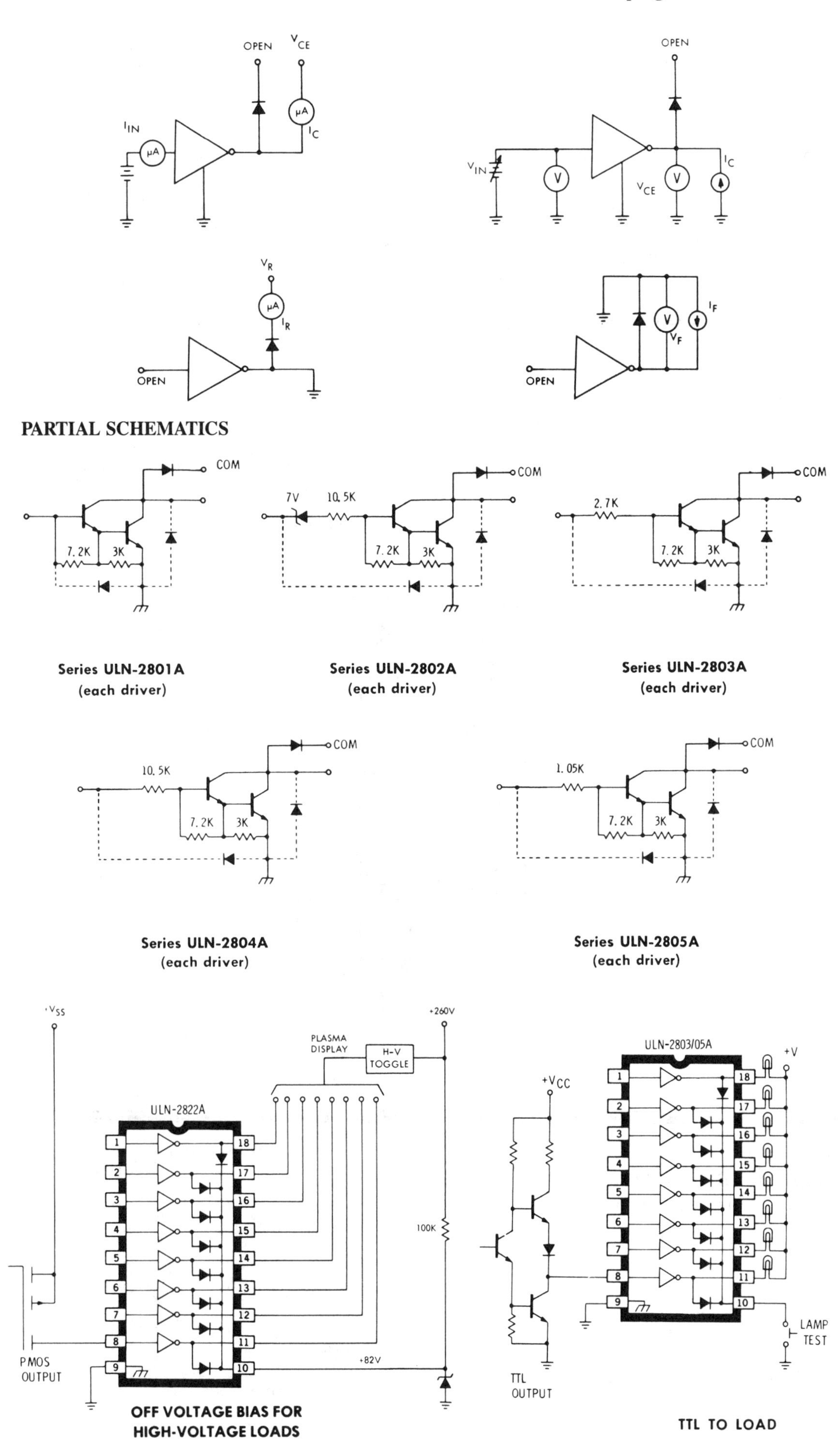
OPEN
V_{CE}
I_{IN}
μA
I_C
OPEN
V_{IN}
V
V_{CE}
I_C
V_R
μA
I_R
OPEN
I_F
V
V_F
OPEN
PARTIAL SCHEMATICS
COM
7.2K
3K
Series ULN-2801A
(each driver)
7V
10.5K
COM
7.2K
3K
Series ULN-2802A
(each driver)
2.7K
COM
7.2K
3K
Series ULN-2803A
(each driver)
10.5K
COM
7.2K
3K
Series ULN-2804A
(each driver)
1.05K
COM
7.2K
3K
Series ULN-2805A
(each driver)
+V_{SS}
+260V
PLASMA
DISPLAY
H-V
TOGGLE
ULN-2822A
100K
+82V
PMOS
OUTPUT
OFF VOLTAGE BIAS FOR
HIGH-VOLTAGE LOADS
ULN-2803/05A
+V
+V_{CC}
TTL
OUTPUT
LAMP
TEST
TTL TO LOAD

BUFFER FOR HIGHER CURRENT LOADS

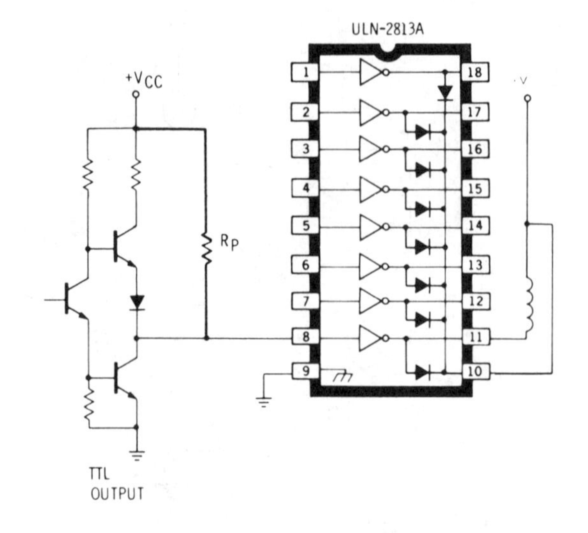

USE OF PULL-UP RESISTORS TO INCREASE DRIVE CURRENT

TYPICAL DISPLAY INTERFACE

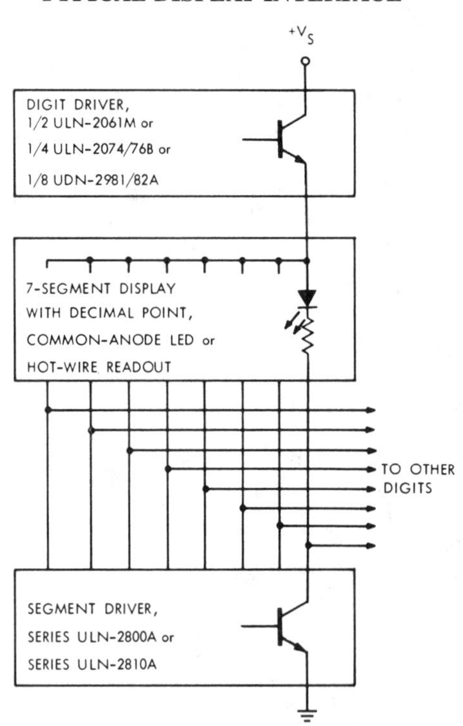

Sprague
UDN-2933B and UDN-2934B
3-Channel Half-Bridge Motor Drivers

- Output currents to 1A
- Output voltages to 30V
- Low output-saturation voltage
- Transient-protected outputs
- Tri-state outputs
- TTL- or CMOS-compatible inputs
- Reliable monolithic construction

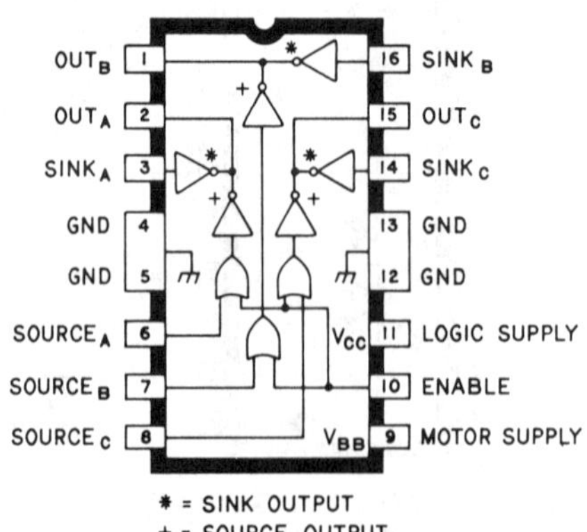

ABSOLUTE MAXIMUM RATINGS

at +25°C Free-Air Temperature

Motor Supply Voltage, V_{BB} .... 30 V
Logic Supply Voltage Range, V_{CC}
(UDN-2933B) .... 4.5 V to 7.0 V
(UDN-2934B) .... 10 V to 15 V
Logic Input Voltage, V_{IN} .... V_{CC}
Output Current, I_{OUT} .... ±1.0 A
Package Power Dissipation, P_D .... See Graph
Operating Temperature Range, T_A .... −20°C to +85°C
Storage Temperature Range, T_S .... −55°C to +150°C

ELECTRICAL CHARACTERISTICS at T_A = +25°C, V_{BB} = 30 V, V_{CC} = 5 V (UDN-2933B) or V_{CC} = 12 V (UDN-2934B), $T_{TAB} \leq$ +70°C

Characteristic	Symbol	Applicable Devices	Test Condtions	Limits			Units
				Min.	Typ.	Max.	
Output Leakage Current	I_{CEX}	Both	All Drivers OFF, V_{OUT} = 0 V	—	−5.0	−100	μA
			All Drivers OFF, V_{OUT} = 30 V	—	5.0	100	μA
Output Saturation Voltage	$V_{CE(SAT)}$	Both	I_{OUT} = −100 mA	—	0.80	1.1	V
			I_{OUT} = 100 mA	—	0.08	0.2	V
			I_{OUT} = −250 mA	—	0.90	1.2	V
			I_{OUT} = 250 mA	—	0.13	0.3	V
			I_{OUT} = −500 mA	—	1.1	1.5	V
			I_{OUT} = 500 mA	—	0.25	0.6	V
			I_{OUT} = −800 mA	—	1.3	1.8	V
			I_{OUT} = 800 mA	—	0.45	0.8	V
Output Sustaining Voltage	$V_{CE(SUS)}$	Both	I_{OUT} = ±800 mA, L = 3 mH	30	—	—	V
Motor Supply Current	I_{BB}	Both	All Drivers OFF	—	50	200	μA
			1 Source + 1 Sink ON, No Loads	—	1.0	1.3	mA
Clamp Diode Forward Voltage	V_F	Both	I_F = 500 mA	—	1.3	2.0	V
			I_F = 800 mA	—	1.3	2.0	V
Logic Input Voltage	$V_{IN(1)}$	UDN-2933B		2.4	—	—	V
		UDN-2934B		8.0	—	—	V
	$V_{IN(0)}$	UDN-2933B		—	—	0.8	V
		UDN-2934B		—	—	4.0	V
Logic Input Current	$I_{IN(1)}$	UDN-2933B	V_{IN} = 2.4 V	—	<1.0	10	μA
		UDN-2934B	V_{IN} = 8.0 V	—	<1.0	10	μA
	$I_{IN(0)}$	Both	V_{IN} = 0.8 V	—	−50	−300	μA
Logic Supply Current	I_{CC}	Both	All Drivers OFF	—	1.7	3.0	mA
			1 Source + 1 Sink ON	—	30	40	mA
Output Rise Time	t_r	Both	I_{OUT} = −500 mA, V_{BB} = 20 V	—	250	—	ns
			I_{OUT} = 500 mA, V_{BB} = 20 V	—	30	—	ns
Output Fall Time	t_f	Both	I_{OUT} = −500 mA, V_{BB} = 20 V	—	500	—	ns
			I_{OUT} = 500 mA, V_{BB} = 20 V	—	50	—	ns

NOTES: 1. Each driver is tested separately.
2. Positive (negative) current is defined as going into (coming out of) the specified device pin.

FUNCTIONAL BLOCK DIAGRAM

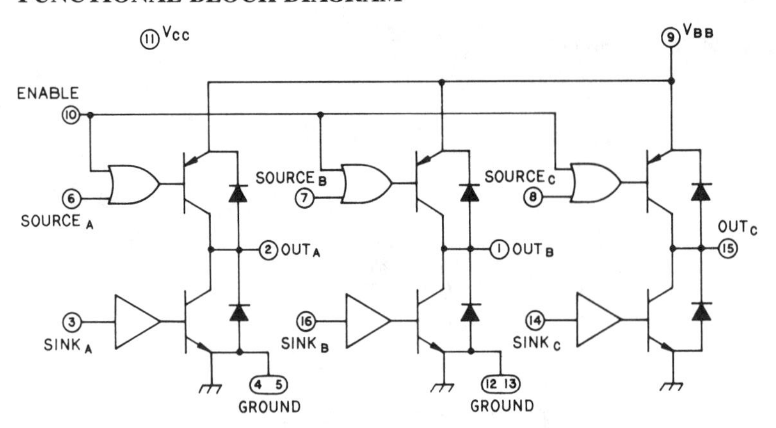

TRUTH TABLE

Sink Driver Input	Source Driver Input	Enable Input	Output
Low	Low	Low	High
Low	High	Low	Open
High	Low	Low	Disallowed
High	High	Low	Low
High	Any	High	Low
Low	Any	High	Open

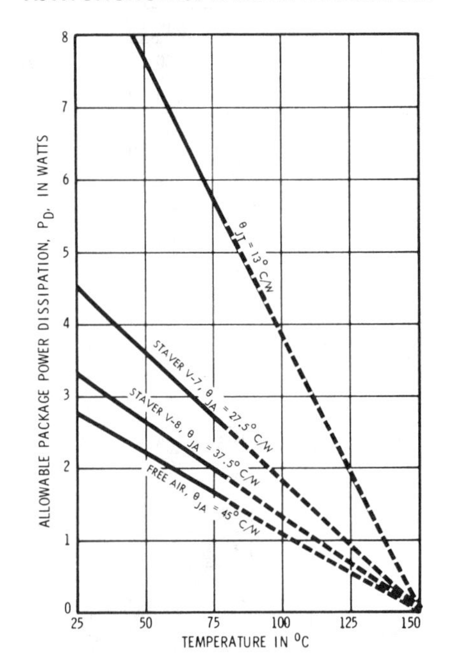

TYPICAL COMMUTATION SEQUENCE

Drivers ON*	Motor Current	Elec. Degrees
1+4	AB	0
1+6	−CA	60
3+6	BC	120
3+2	−AB	180
5+2	CA	240
5+4	−BC	300

*Enable input must be low; Source drivers are turned ON with a logic low, sink drivers are turned ON with a logic high.

Sprague UDN-2956A and UDN-2957A

High-Voltage, High-Current Source Drivers

- 500mA output source current
- 50V output sustaining voltage
- Output transient protection
- 6-16V PMOS, CMOS input: UDN-2956A
- TTL, DT, 5V CMOS input: UDN-2957A
- Plastic or CerDIP package

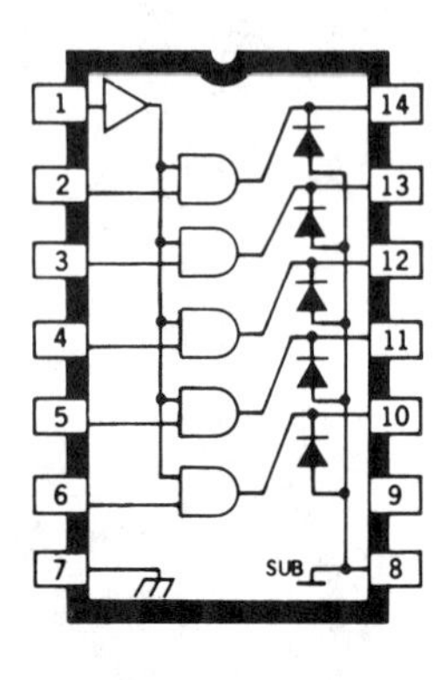

ABSOLUTE MAXIMUM RATINGS

at +25°C Free-Air Temperature (Reference Pin 7)

Supply Voltage, V_{EE} . . . . −80 V
Input Voltage, V_{IN} (UDN-2956A) . . . . . . +20 V
(UDN-2957A) . . . . . . +10 V
Output Current, I_{OUT} . . . . −500 mA
Power Dissipation, P_D (any one driver) . . . . . . 1.0 W
(total package) . . . . . . 2.0 W*
Operating Temperature Range, T_A . . . . −20°C to +85°C
Storage Temperature Range, T_S . . . . −55°C to +150°C

*Derate at the rate of 16.67 mW/°C above 25°C.

ELECTRICAL CHARACTERISTICS

at T_A = + 25°C, V_{ENABLE} = V_{IN} (unless otherwise specified)

Characteristic	Symbol	Applicable Devices	Test Conditions	Limit
Output Leakage Current	I_{CEX}	UDN-2956A	V_{IN} = V_{ENABLE} = 0.4 V, V_{OUT} = −80 V, T_A = +70°C	−200 μA Max.
			V_{IN} = 0.4 V, V_{ENABLE} = 15 V, V_{OUT} = −80 V, T_A = +70°C	−200 μA Max.
			V_{IN} = 15 V, V_{ENABLE} = 0.4 V, V_{OUT} = −80 V, T_A = +70°C	−200 μA Max.
		UDN-2957A	V_{IN} = V_{ENABLE} = 0.4 V, V_{OUT} = −80 V, T_A = +70°C	−200 μA Max.
			V_{IN} = 0.4 V, V_{ENABLE} = 3.85 V, V_{OUT} = −80 V, T_A = +70°C	−200 μA Max.
			V_{IN} = 3.85 V, V_{ENABLE} = 0.4 V, V_{OUT} = −80 V, T_A = 70°C	−200 μA Max.
Collector-Emitter Saturation Voltage	$V_{CE(SAT)}$	UDN-2956A	V_{IN} = 6.0 V, I_{OUT} = − 100 mA	− 1.20 V Max.
			V_{IN} = 7.0 V, I_{OUT} = − 175 mA	− 1.35 V Max.
			V_{IN} = 10 V, I_{OUT} = −350 mA	− 1.70 V Max.
		UDN-2957A	V_{IN} = 2.4 V, I_{OUT} = −100 mA	− 1.20 V Max.
			V_{IN} = 2.7 V, I_{OUT} = −175 mA	− 1.35 V Max.
			V_{IN} = 3.9 V, I_{OUT} = −350 mA	− 1.70 V Max.
Input Current	$I_{IN(ON)}$	UDN-2956A	V_{IN} = 6.0 V, V_{OUT} = −2.0 V	650 μA Max.
			V_{IN} = 15 V, V_{OUT} = −2.0 V	1.85 mA Max.
		UDN-2957A	V_{IN} = 2.4 V, V_{OUT} = −2.0 V	675 μA Max.
			V_{IN} = 3.85 V, V_{OUT} = −2.0 V	1.40 mA Max.
	$I_{IN(OFF)}$	ALL	I_{OUT} = −500 μA, T_A = +70°C	50 μA Min.
Output Source Current	I_{OUT}	UDN-2956A	V_{IN} = 5.0 V, V_{OUT} = −2.0 V	−125 mA Min.
			V_{IN} = 6.0 V, V_{OUT} = −2.0 V	−200 mA Min.
			V_{IN} = 7.0 V, V_{OUT} = −2.0 V	−250 mA Min.
			V_{IN} = 8.0 V, V_{OUT} = −2.0 V	−300 mA Min.
			V_{IN} = 9.0 V, V_{OUT} = −2.0 V	−350 mA Min.
		UDN-2957A	V_{IN} = 2.4 V, V_{OUT} = −2.0 V	−125 mA Min.
			V_{IN} = 2.7 V, V_{OUT} = −2.0 V	−200 mA Min.
			V_{IN} = 3.0 V, V_{OUT} = −2.0 V	−250 mA Min.
			V_{IN} = 3.3 V, V_{OUT} = −2.0 V	−300 mA Min.
			V_{IN} = 3.6 V, V_{OUT} = −2.0 V	−350 mA Min.
Output Sustaining Voltage	$V_{CE(SUS)}$	UDN-2956A	V_{IN} = 0.4 V, I_{OUT} = −25 mA	50 V Min.
		UDN-2957A	V_{IN} = 0.4 V, I_{OUT} = −25 mA	50 V Min.
Clamp Diode Leakage Current	I_R	ALL	V_R = 80 V	50 μA Max.
Clamp Diode Forward Voltage	V_F	ALL	I_F = 350 mA	2.0 V Max.
Turn-On Delay	t_{ON}	ALL	0.5 E_{in} to 0.5 E_{out}, R_L = 400 Ω C_T = 25 pF	4.0 μs Max.
Turn-Off Delay	t_{OFF}	ALL	0.5 E_{in} to 0.5 E_{out}, R_L = 400 Ω C_T = 25 pF	10 μs Max.

Input current as a function of input voltage

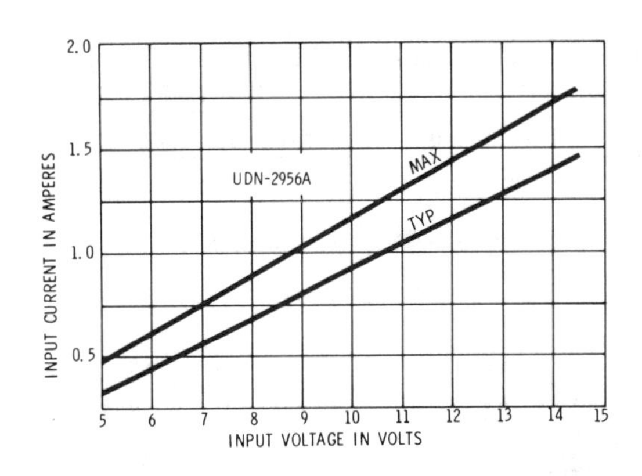

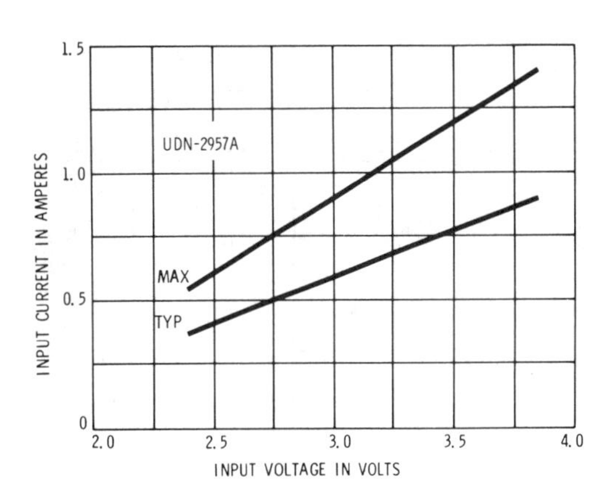

ALLOWABLE PEAK OUTPUT CURRENT AS A FUNCTION OF DUTY CYCLE

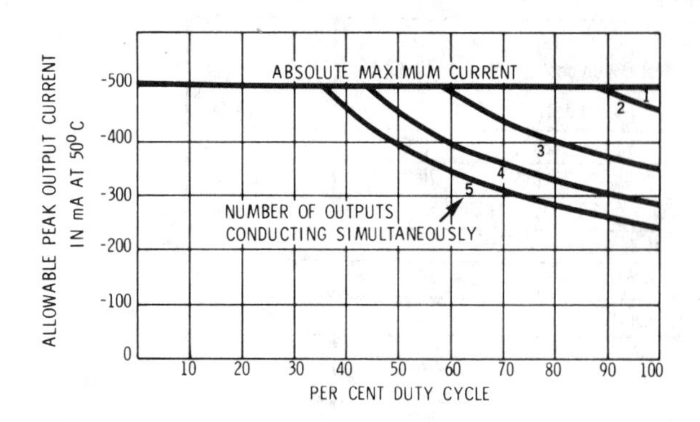

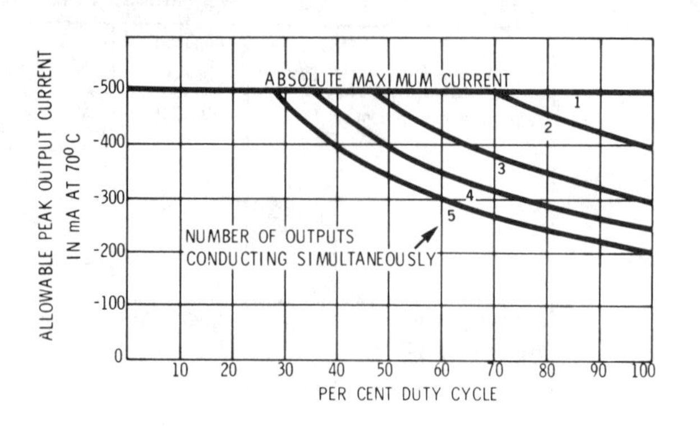

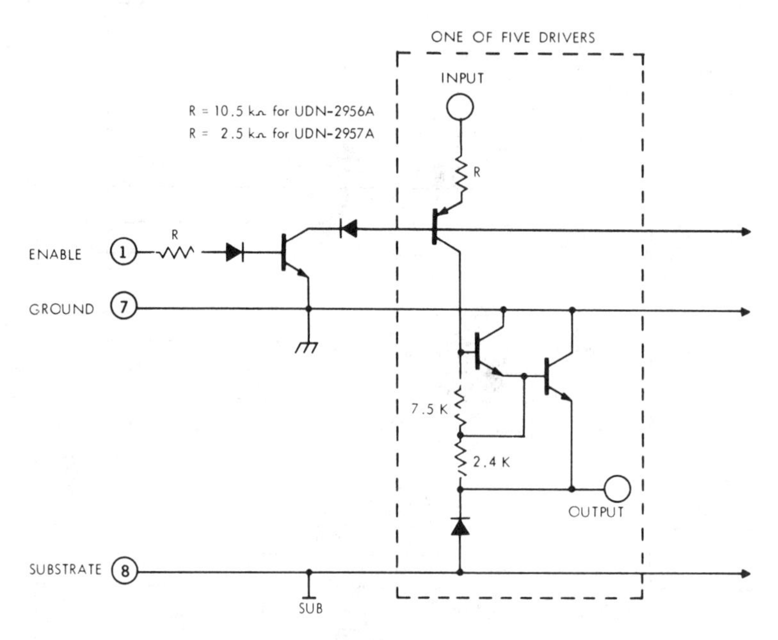

Sprague ULN-2435A, ULN-2445A, ULN-2455A
Automotive Lamp Monitors

- No standby power
- Integral to wiring assembly
- Fail-safe
- Reverse voltage protected
- Internal transient protection
- Dual-in-line plastic packages

ELECTRICAL CHARACTERISTICS at T_A = +25°C, V_{CC} = V_{IN} = 10 to 16 V (unless otherwise shown)

Characteristic	Test Pins ULN-2435/45A	Test Pins ULN-2455A	Test Conditions	Min.	Typ.	Max.	Units
Output Leakage Current	1, 7, 10, 13, 15, 16	1, 4, 8, 11	V_{OUT} = 80 V, $\Delta V_{IN} < 7$ mV	—	—	100	μA
Output Saturation Voltage	1, 7, 10, 13, 15, 16	1, 4, 8, 11	I_{OUT} = 5 mA, $\Delta V_{IN} > 20$ mV	—	0.8	1.0	V
			I_{OUT} = 30 mA, $\Delta V_{IN} > 20$ mV	—	1.4	2.0	V
Differential Switch Voltage	2-3, 8-9, 11-12, 17-18	2-3, 5-6, 9-10, 12-13	Absolute Value $V_{(2)} - V_{(3)}$	7.0	13	20	mV
Input Current	4	NA	$V_{IN} = V_{CC}$ = 16 V	—	—	500	μA
	5	NA	$V_{IN} = V_{CC}$ = 16 V	—	—	15	mA
	6	NA	V_{IN} = 0 V, V_{CC} = 16 V	—	—	−1.0	mA
	2, 8, 11, 17	2, 5, 9, 12	$\Delta V_{IN} = V_{(2)} - V_{(3)}$ = +30 mV	150	300	800	μA
	3, 9, 12, 18	3, 6, 10,1 3	$\Delta V_{IN} = V_{(2)} - V_{(3)}$ = −30 mV	0.5	1.7	3.5	mA

ABSOLUTE MAXIMUM RATINGS

at +25°C Free-Air Temperature

Supply Voltage, V_{CC} . . . . 30 V
Peak Supply Voltage, V_{CC} (0.1 s) . . . . 80 V
Peak Reverse Voltage, V_R . . . . 30 V
Output Current, I_{OUT} . . . . 35 mA
Package Power Dissipation, P_D (ULN-2435/45A) . . . . . . 2.3 W*
(ULN-2455A) . . . . 2.0 W**
Operating Temperature Range, T_A . . . . −40°C to +85°C
Storage Temperature Range, T_S . . . . −65°C to +150°C

*Derate at the rate of 18.18 mW/°C above T_A = +25°C.
**Derate at the rate of 16.67 mW/°C above T_A = +25°C.

FUNCTIONAL BLOCK DIAGRAM

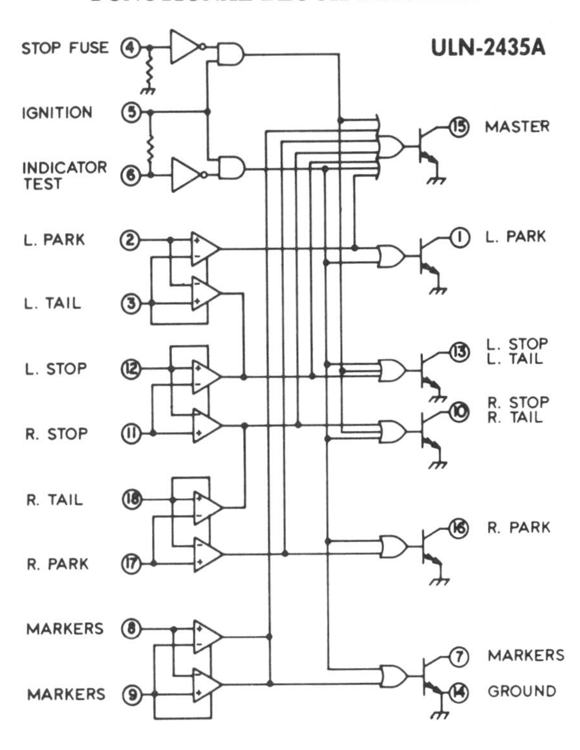

ULN-2455A

FUNCTIONAL BLOCK DIAGRAM

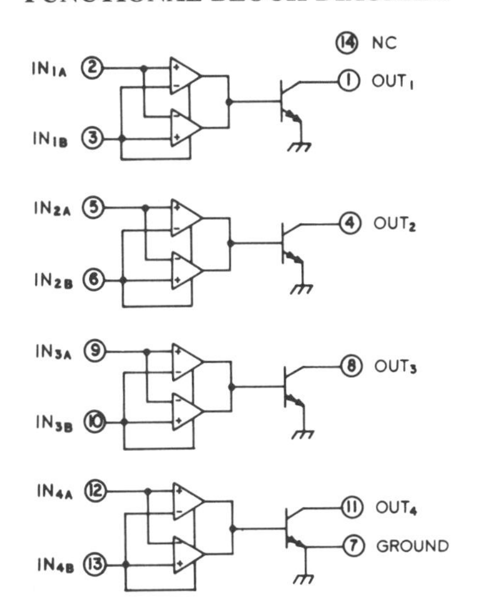

BASIC BRIDGE MONITORING SYSTEM

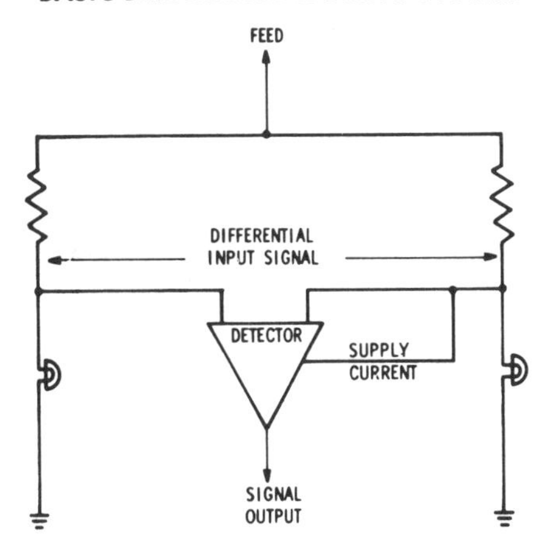

TYPICAL SWITCH CHARACTERISTICS

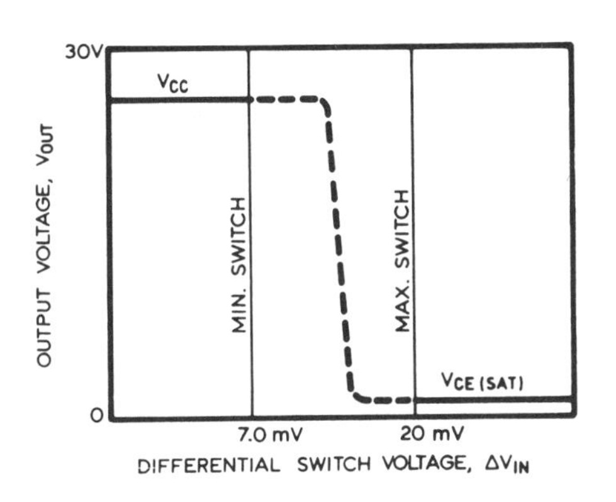

ULN-2445A

FUNCTIONAL BLOCK DIAGRAM

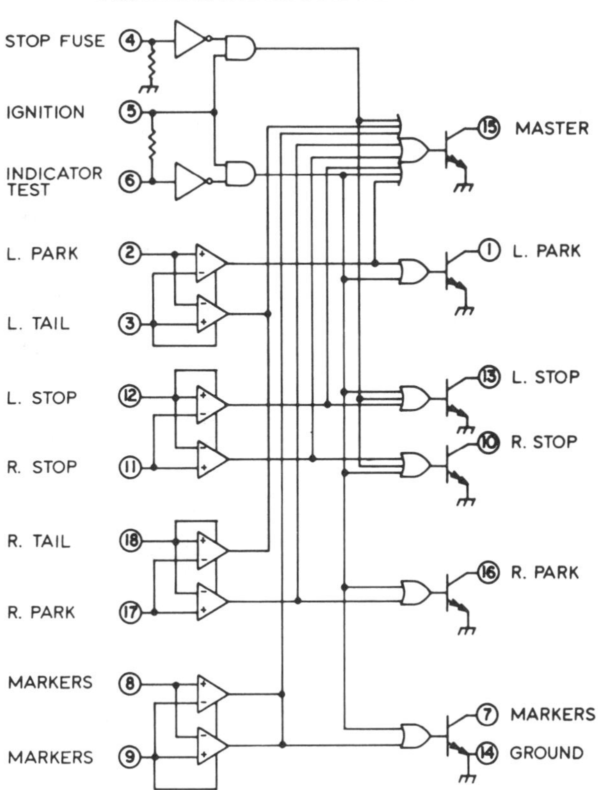

TYPICAL APPLICATIONS

AUTOMOTIVE LAMP MONITOR

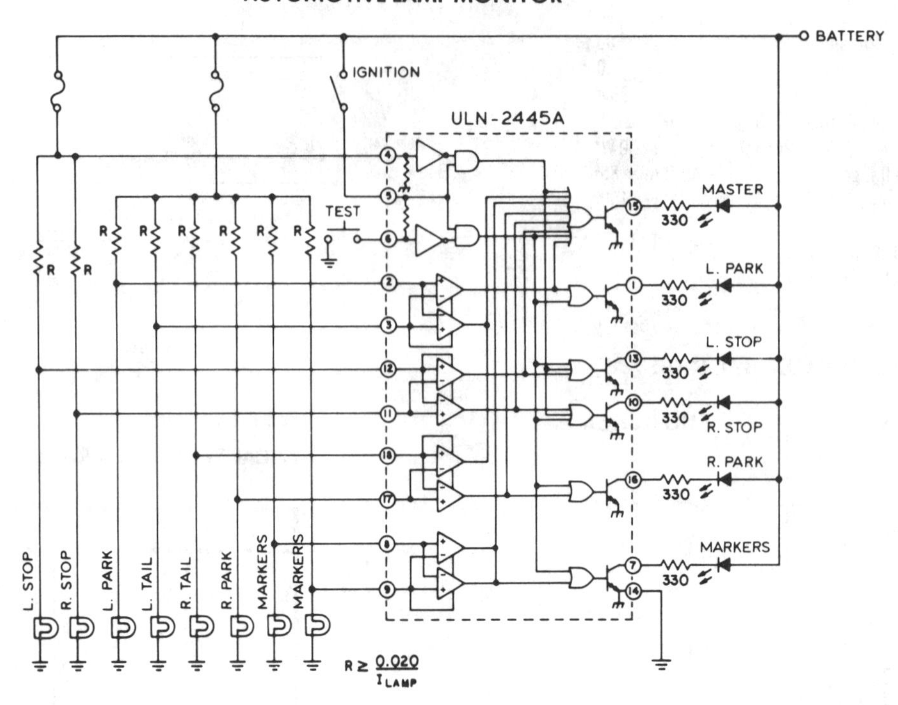

QUAD LAMP MONITOR

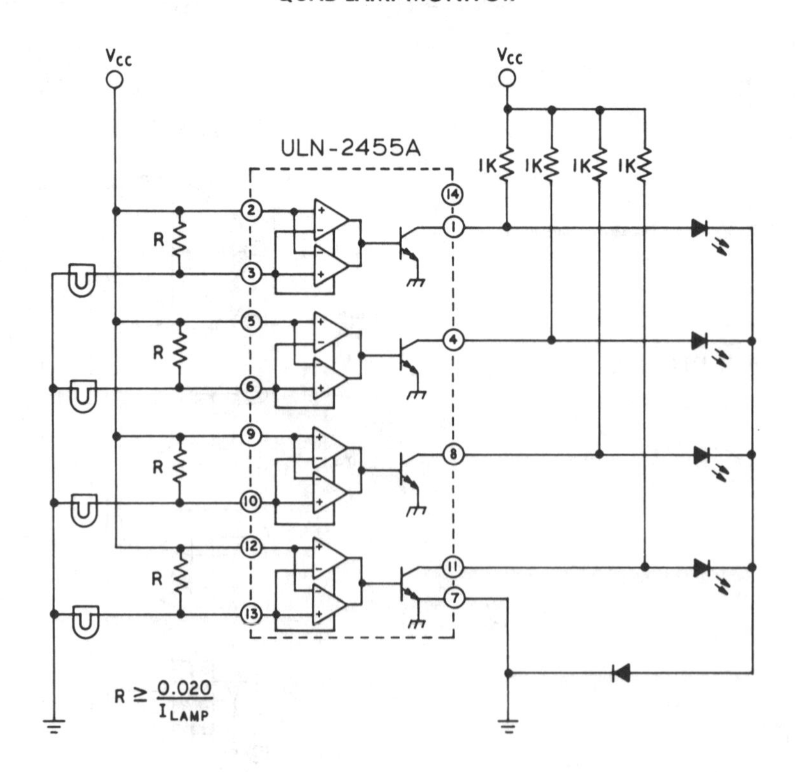

POWER SUPPLY SUPERVISORY CIRCUIT

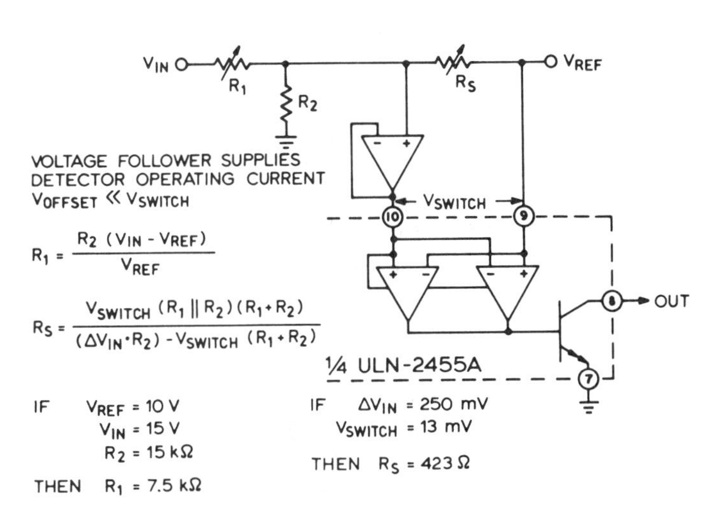

SIMPLIFIED SCHEMATIC
(One of 4 differential sense amplifiers)

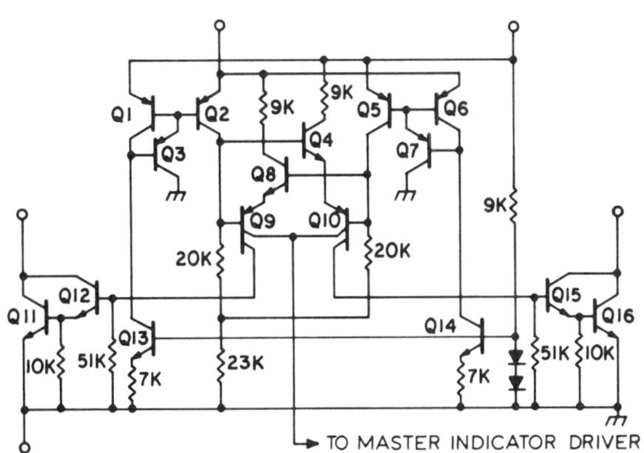

Sprague ULN-2457A and ULN-2457L Quad Lamp Monitors for 24V Systems

- 18 to32V operation
- No standby power
- Integral to wiring assembly
- Fail-safe
- Reverse-voltage protected
- Internal transient protection

ABSOLUTE MAXIMUM RATINGS
at +25°C Free-Air Temperature

Supply Voltage, V_{CC} . . . . 34 V
Peak Supply Voltage, V_{CC} (0.1 s) . . . . 80 V
Peak Reverse Voltage, V_R . . . . 30 V
Output Current, I_{OUT} . . . . 35 mA
Package Power Dissipation, P_D . . . . See Graph
Operating Temperature Range, T_A . . . . −40°C to +85°C
Storage Temperature Range, T_S . . . . −65°C to +150°C

ELECTRICAL CHARACTERISTICS at T_A = +25°C, V_{CC} = V_{IN} = 18 to 32 V (unless otherwise shown)

Characteristic	Test Pins	Test Conditions	Limits Min.	Limits Typ.	Limits Max.	Units
Output Leakage Current	1, 4, 8, 11	V_{OUT} = 80 V, $\Delta V_{IN} < 7$ mV	—	—	100	µA
Output Saturation Voltage	1, 4, 8, 11	I_{OUT} = 5 mA, $\Delta V_{IN} > 20$ mV	—	0.8	1.0	V
		I_{OUT} = 30 mA, $\Delta V_{IN} > 20$ mV	—	1.4	2.0	V
Differential Switch Voltage	2-3, 5-6, 9-10, 12-13	Absolute Value $V_{(2)} - V_{(3)}$	7.0	13	20	mV
Input Current	2, 5, 9, 12	$\Delta V_{IN} = V_{(2)} - V_{(3)} = +30$ mV	150	300	800	µA
	3, 6, 10, 13	$\Delta V_{IN} = V_{(2)} - V_{(3)} = -30$ mV	0.5	1.7	3.5	mA

BASIC BRIDGE MONITORING SYSTEM

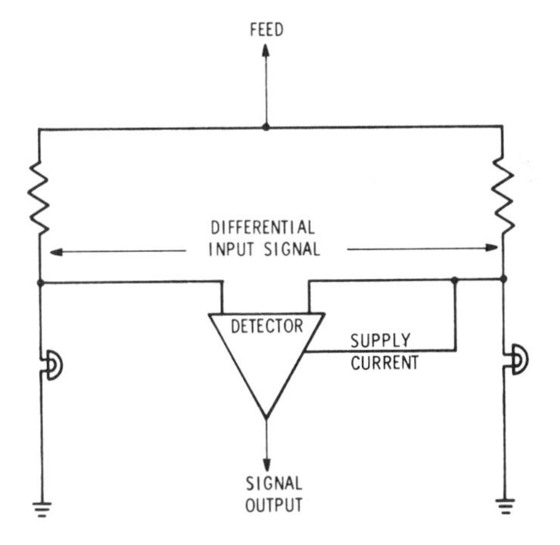

TYPICAL SWITCH CHARACTERISTICS

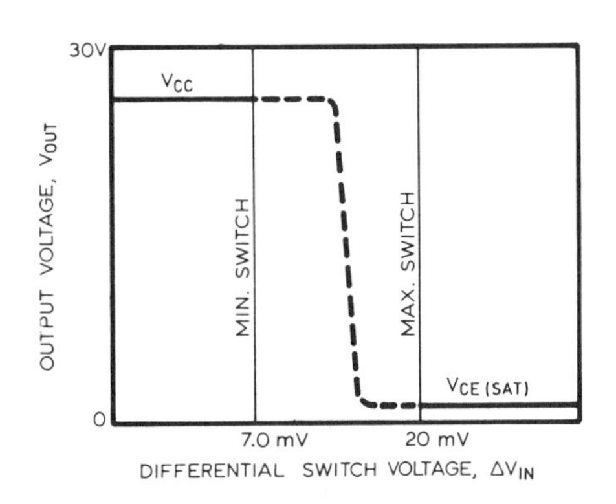

ALLOWABLE PACKAGE POWER DISSIPATION AS A FUNCTION OF TEMPERATURE

ALLOWABLE PACKAGE POWER DISSIPATION, P_D, IN WATTS

2.0
1.5
1.0
0.5
0

14-LEAD 'A' θ_{JA} = 60°C/W
14-LEAD 'L' θ_{JA} = 125°C/W

25 50 75 100 125 150
TEMPERATURE IN °C

SIMPLIFIED SCHEMATIC

(One of 4 differential sense amplifiers)

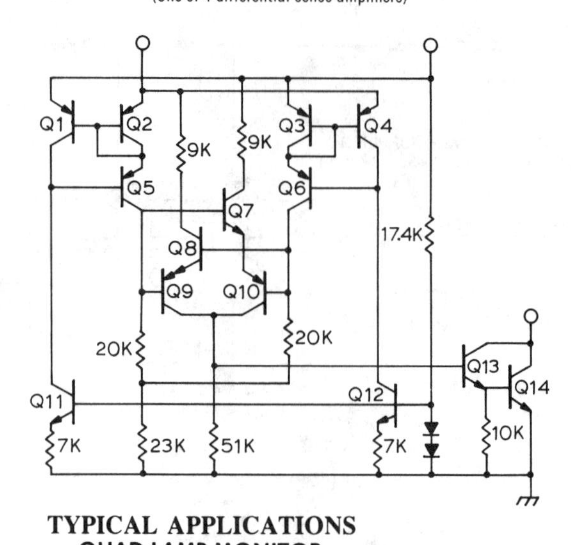

FUNCTIONAL BLOCK DIAGRAM

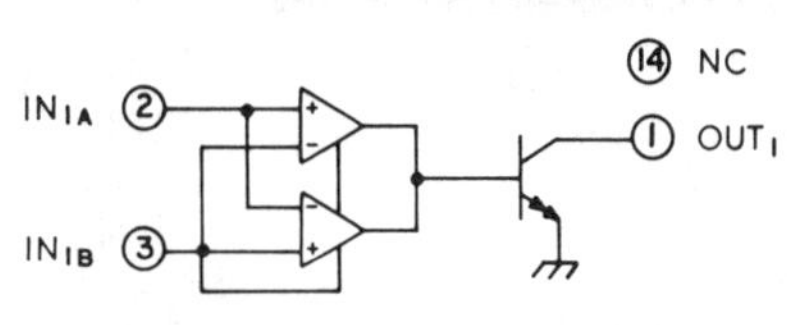

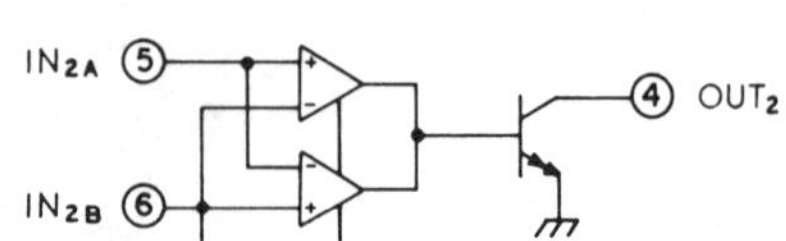

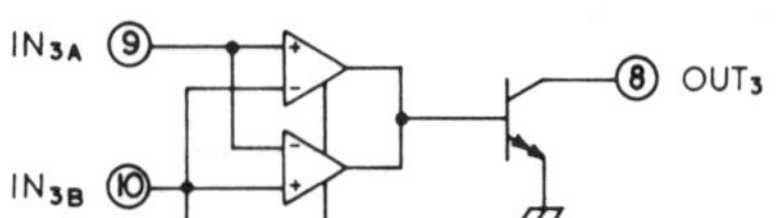

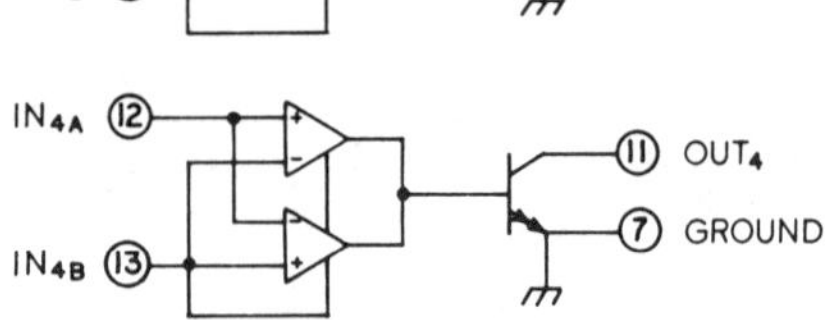

TYPICAL APPLICATIONS

QUAD LAMP MONITOR

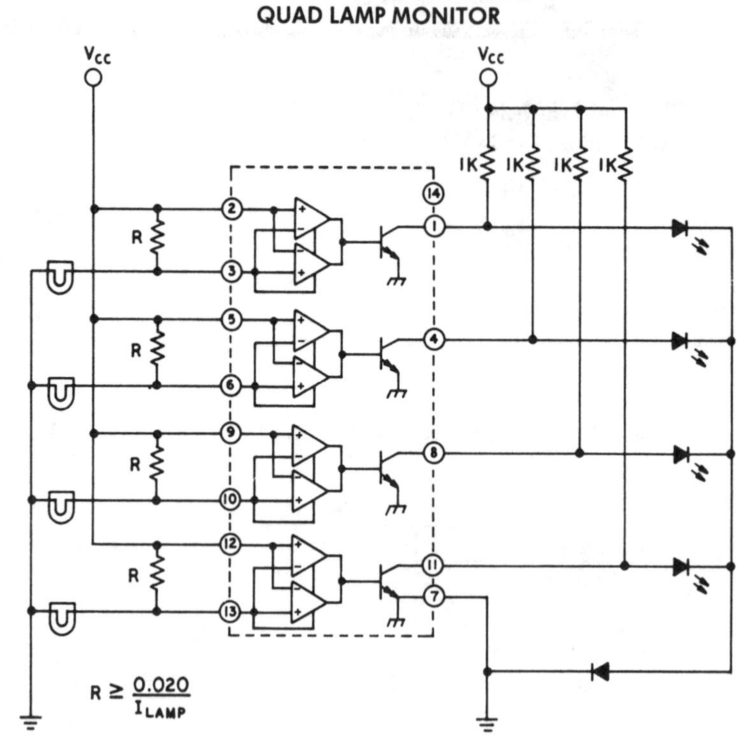

POWER SUPPLY SUPERVISORY CIRCUIT

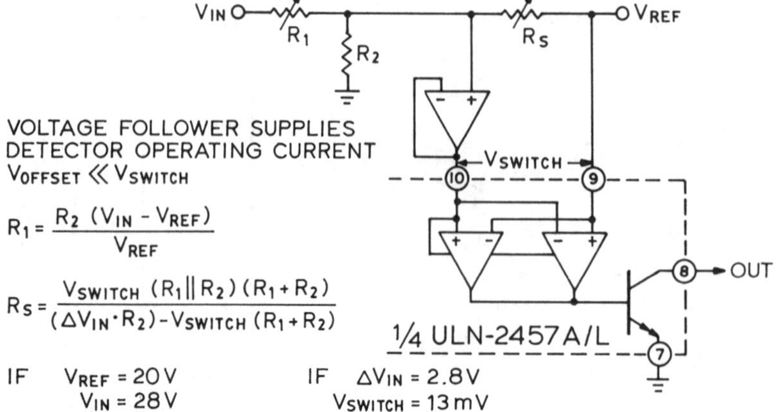

VOLTAGE FOLLOWER SUPPLIES DETECTOR OPERATING CURRENT
$V_{OFFSET} \ll V_{SWITCH}$

$$R_1 = \frac{R_2 (V_{IN} - V_{REF})}{V_{REF}}$$

$$R_S = \frac{V_{SWITCH} (R_1 \| R_2)(R_1 + R_2)}{(\Delta V_{IN} \cdot R_2) - V_{SWITCH} (R_1 + R_2)}$$

IF $V_{REF} = 20V$
$V_{IN} = 28V$
$R_2 = 15k\Omega$

THEN $R_1 = 6.0k\Omega$

IF $\Delta V_{IN} = 2.8V$
$V_{SWITCH} = 13mV$

THEN $R_S = 28\Omega$

Sprague UDN-2935Z and UDN-2950Z
Bipolar Half-Bridge Motor Drivers

- 3.5A peak output
- 37V min. output breakdown
- Output transient protection
- Tri-state outputs
- TTL-, CMOS-, PMOS-, NMOS-compatible inputs
- Internal thermal shutdown
- High-speed chopper (to 100 kHz)
- UDN-2935Z replaces SG3635P
- UDN-2950Z replaces UDN-2949Z, SN75605
- TO-220-style packages

ABSOLUTE MAXIMUM RATINGS

Supply Voltage Range, V_S .... 8.0 V to 35 V
Output Voltage Range, V_{OUT} .... −2.0 V to V_S + 2.0 V
Input Voltage Range, V_{IN} .... −0.3 V to +7.0 V
Peak Output Current (100 ms, 10% d-c), I_{OP} .... ±3.5 A
Continuous Output Current, I_{OUT} .... ±2.0 A
Package Power Dissipation, P_D .... See Graph
Operating Temperature Range, T_A .... −20°C to +85°C
Storage Temperature Range, T_S .... −55°C to +85°C

ELECTRICAL CHARACTERISTICS at T_A = +25°C, T_{TAB} = +70°C, V_S = 35 V (unless otherwise noted)

Characteristic	Source Driver Input, Pin 2		Sink Driver Input, Pin 5	Output, Pin 4	Other	Limits		
	UDN-2935Z	UDN-2950Z				Min.	Max.	Units
Output Leakage Current	2.4 V	0.8 V	2.4 V	0 V	—	—	−500	μA
	2.4 V	0.8 V	2.4 V	35 V	—	—	500	μA
Output Sustaining Voltage	2.4 V	0.8 V	0.8 to 2.4 V	2.0 A	Fig. 1	35	—	V
Output Saturation Voltage	0.8 V	2.4 V	2.4 V	−2.0 A	—	33	—	V
	2.4 V	0.8 V	0.8 V	2.0 A	—	—	2.0	V
Output Source Current	0.8 V	2.4 V	2.4 V	—	—	−2.0	—	A
Output Sink Current	2.4 V	0.8 V	0.8 V	—	—	2.0	—	A
Input Open-Circuit Voltage	−250 μA	−250 μA	−250 μA	—	—	—	7.5	V
Input Current	—	2.4 V	2.4 V	NC	—	—	−700	μA
	2.4 V	—	2.4 V	NC	—	—	10	μA
	0.8 V	0.8 V	0.8 V	NC	—	—	−1.6	mA
Propagation Delay	2.4 V	0.8 V	0.8 to 2.4 V	2.0 A	—	—	750	ns
	0.8 to 2.4 V	2.4 to 0.8 V	2.4 V	2.0 A	—	—	2.0	μs
Clamp Diode Forward Voltage	NC	NC	NC	2.0 A	—	—	2.2	V
Supply Current	0.8 V	2.4 V	NC	NC	—	—	35	mA

NOTE: Positive (negative) current is defined as going into (coming out of) the specified device pin.

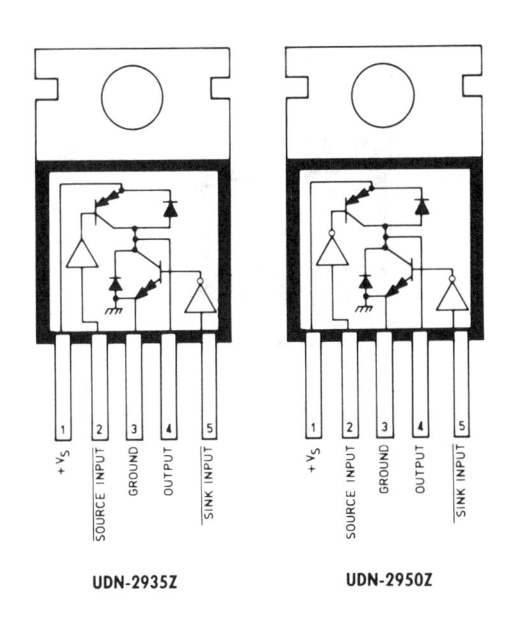

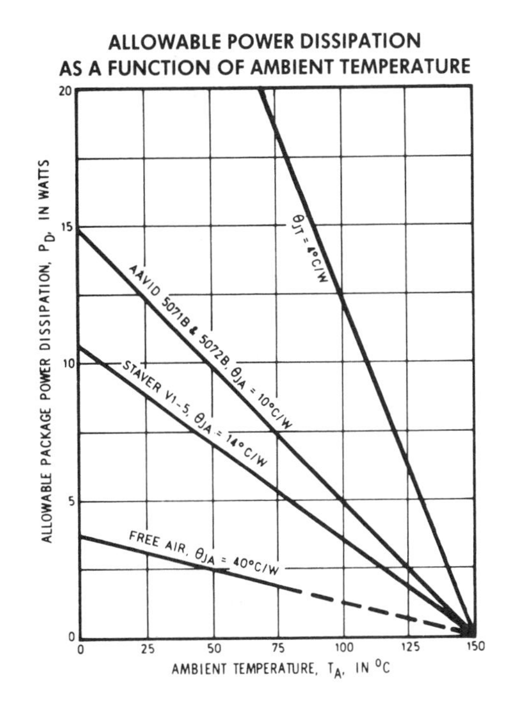

TRUTH TABLE

Source Driver, Pin 2	Sink Driver, Pin 5	Output, Pin 4 UDN-2935Z	Output, Pin 4 UDN-2950Z
Low	Low	High	Low
Low	High	High	High Z
High	Low	Low	High
High	High	High Z	High

FUNCTIONAL BLOCK DIAGRAMS

UDN-2935Z

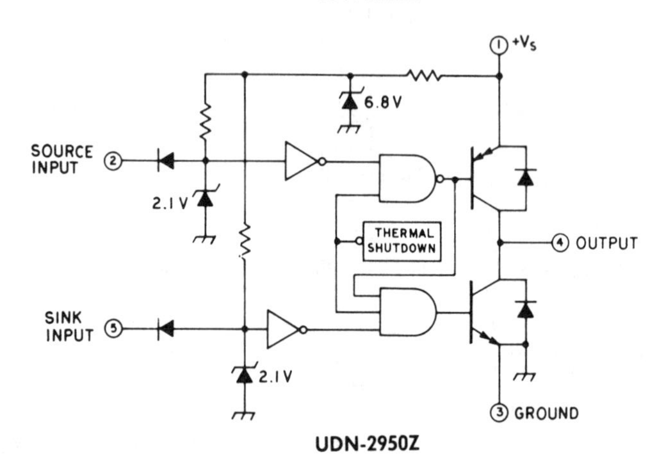

UDN-2950Z

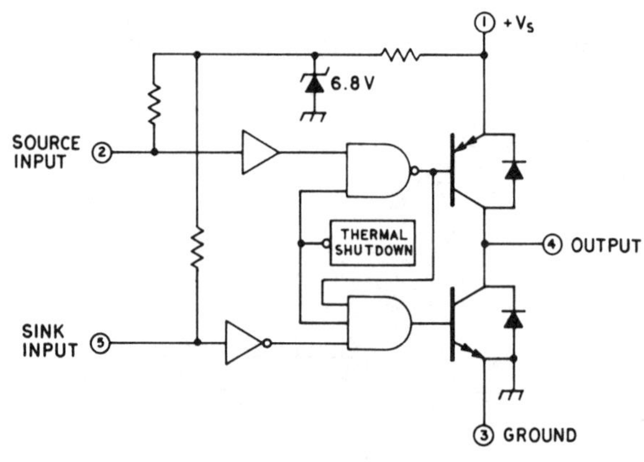

TEST FIGURE 1

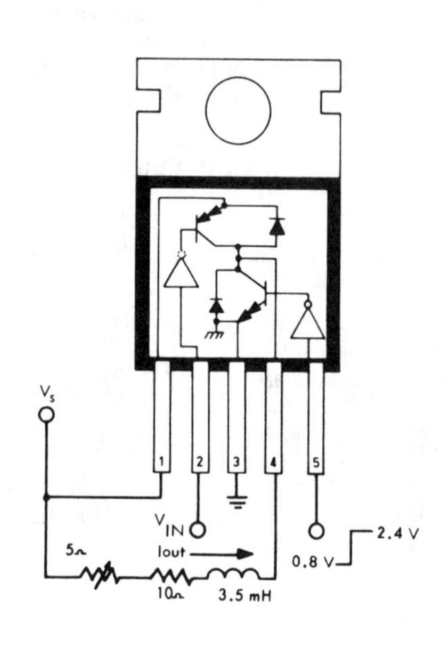

RECOMMENDED TIMING CONDITIONS

(UDN-2950Z shown)

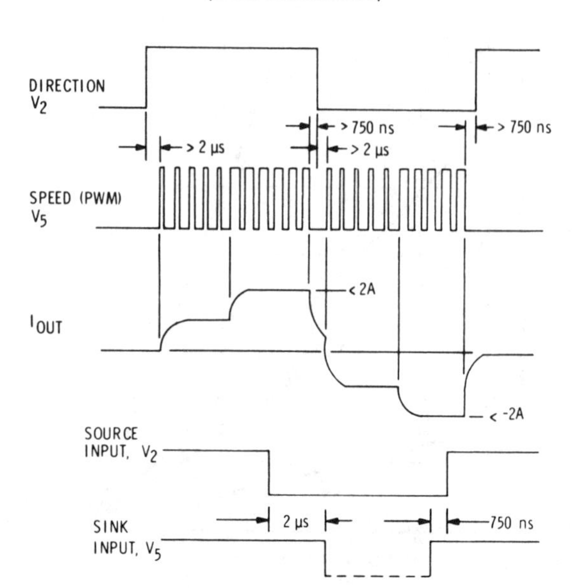

TEST FIGURE 2

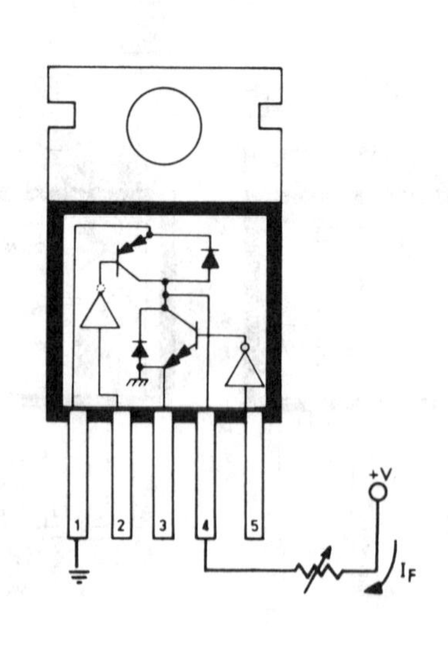

TYPICAL APPLICATIONS

3-PHASE BRUSHLESS DC MOTOR DRIVE

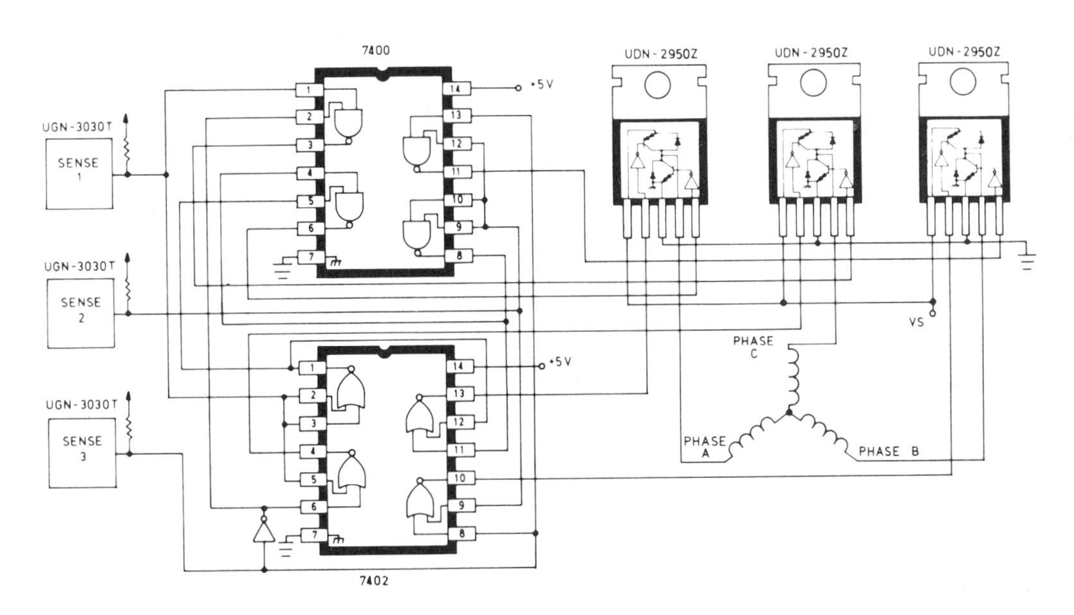

SINGLE-WINDING DC OR STEPPER MOTOR DRIVE

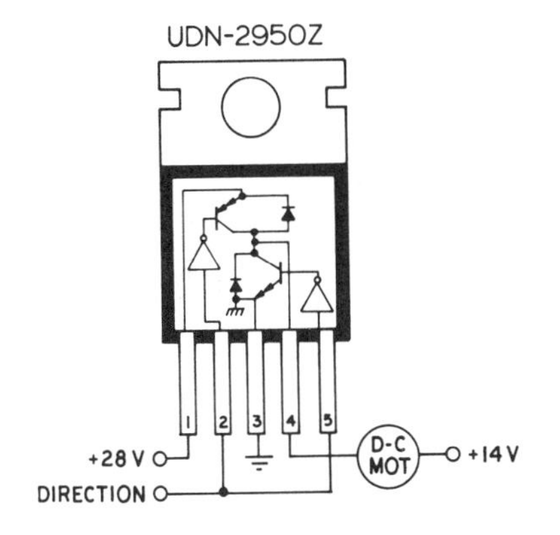

FULL-BRIDGE DC SERVO MOTOR DRIVE

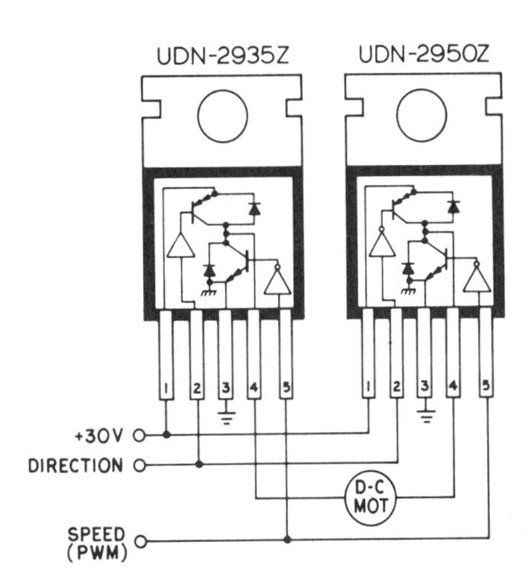

Sprague UDN-2936W and UDN-2937W

3-Phase Brushless DC Motor Controllers

- 10 to 45V operation
- ±4A peak output current
- Internal clamp diodes
- Internal PWM current control
- 60° commutation decoding logic
- Thermal shutdown protection
- Compatible with single-ended or differential hall-effect sensors
- Braking and direction control (UDN-2936W only)

ABSOLUTE MAXIMUM RATINGS

at $T_{TAB} \leq +70°C$

Supply Voltage, V_{BB}	45V
Output Current, I_{OUT} (continuous)	±3A
(peak)	±4A
Input Voltage Range, V_{IN}	−0.3V to 15V
Threshold Voltage, V_{THS}	15V
Package Power Dissipation, P_D	See Graph
Operating Temperature Range, T_A	−20°C to +85°C
Storage Temperature Range, T_S	−55°C to +150°C

NOTE: Output current rating may be limited by duty cycle, ambient temperature, and heat sinking. Under any set of conditions, do not exceed the specified peak current and a junction temperature of +150°C.

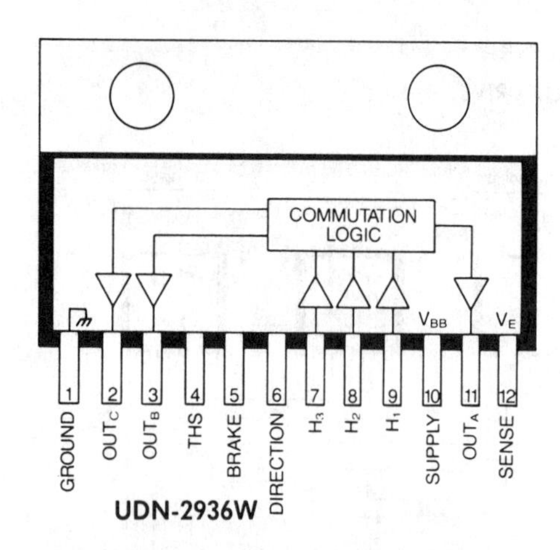

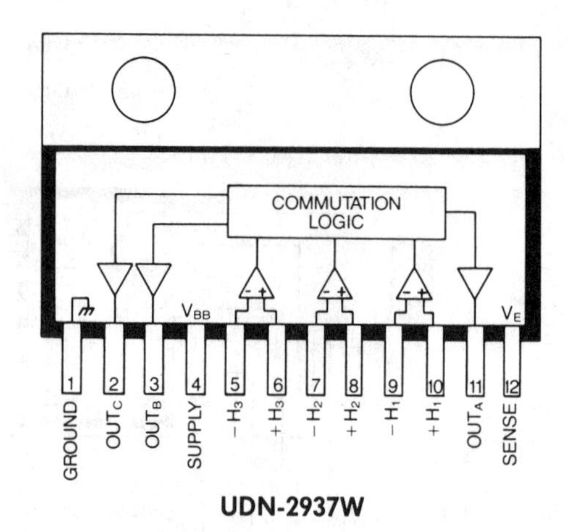

ELECTRICAL CHARACTERISTICS $T_A = +25°C, T_{TAB} \leq 70°C, V_{BB} = 45V$

Characteristic	Symbol	Test Conditions	Limits Min.	Typ.	Max.	Units
Supply Voltage Range	V_{BB}	Operating	10	—	45	V
Supply Current	I_{BB}	Outputs Open	—	52	60	mA
		$V_{BRAKE} = 0.8V$, UDN-2936W Only	—	54	60	mA
Thermal Shutdown Temperature	T_J		—	165	—	°C
Thermal Shutdown Hysteresis	ΔT_J		—	25	—	°C
Output Drivers						
Output Leakage Current	I_{CEX}	$V_{OUT} = V_{BB}$	—	—	50	μA
		$V_{OUT} = 0V$	—	—	−50	μA
Output Saturation Voltage	$V_{CE(SAT)}$	$I_{OUT} = -1A$	—	1.7	1.9	V
		$I_{OUT} = +1A$	—	1.1	1.3	V
		$I_{OUT} = -2A$	—	1.9	2.1	V
		$I_{OUT} = +2A$	—	1.4	1.6	V
		$I_{OUT} = -3A$	—	2.35	2.50	V
		$I_{OUT} = +3A$	—	1.85	2.00	V
Output Sustaining Voltage	$V_{CE(sus)}$	$I_{OUT} = \pm 3A$, L = 2mH	45	—	—	V
Clamp Diode Forward Voltage	V_F	$I_F = 2A$	—	1.8	2.0	V
Clamp Diode Leakage Current	I_R	$V_R = 45V$	—	—	50	μA
Output Switching Time	t_r	$I_{OUT} = \pm 2A$, Resistive Load	—	2.0	—	μs
	t_f	$I_{OUT} = \pm 2A$, Resistive Load	—	2.0	—	μs
Turn-ON Delay (Resistive Load)	t_{on}	Source Drivers, 0 to −2A	—	1.25	—	μs
		Sink Drivers, 0 to +2A	—	1.9	—	μs
Turn-OFF Delay (Resistive Load)	t_{off}	Source Drivers, −2A to 0	—	1.7	—	μs
		Sink Drivers, +2A to 0	—	0.9	—	μs
UDN-2936W Control Logic						
Logic Input Voltage	$V_{IN(1)}$	V_{DIR} or V_{BRAKE}	2.0	—	—	V
	$V_{IN(0)}$	V_{DIR} or V_{BRAKE}	—	—	0.8	V
Sensor Input Voltage Threshold	V_{IN}	H_1, H_2, or H_3	—	2.5	—	V
Input Current	$I_{IN(1)}$	$V_{DIR} = 2V$	—	150	200	μA
		$V_{BRAKE} = 2V$	—	<1.0	5.0	μA
		$V_H = 5V$	—	−190	−220	μA
	$I_{IN(0)}$	$V_{DIR} = 0.8V$	—	35	50	μA
		$V_{BRAKE} = 0.8V$	—	−5.0	−20	μA
		$V_H = 0.8V$	—	−0.64	−1.0	mA
	I_{THS}	$V_{THS} \geq 3.0V$	—	−8.0	−15	μA
		$V_{THS} < 3.0V$, $V_{SENSE} < V_{THS}/10.5$	—	−15	−30	μA
		$V_{THS} < 3.0V$, $V_{SENSE} > V_{THS}/9.5$	140	200	260	μA
Current Limit Threshold		V_{THS}/V_{SENSE} at trip point, $V_{THS} < 3.0V$	9.5	10	10.5	
Default Sense Trip Voltage	V_{SENSE}	$V_{THS} \geq 3.0V$	270	300	330	mV
Default Hysteresis		$V_{THS} \geq 3.0V$	—	7.5	—	%
Deadtime	t_d	BRAKE or DIRECTION	—	2.0	—	μs
UDN-2937W Control Logic						
Input Common-Mode Voltage Range	V_{CM}		1.5	2.0	4.0	V
Input Voltage Hysteresis	$V_{IN(HYS)}$		—	10	—	mV
Input Current	I_{IN}	$V_{IN} = 5V$	—	12	20	μA
Sense Trip Voltage	V_{SENSE}		270	300	330	mV
Hysteresis			—	7.5	—	%

FUNCTIONAL BLOCK DIAGRAM
UDN-2936W

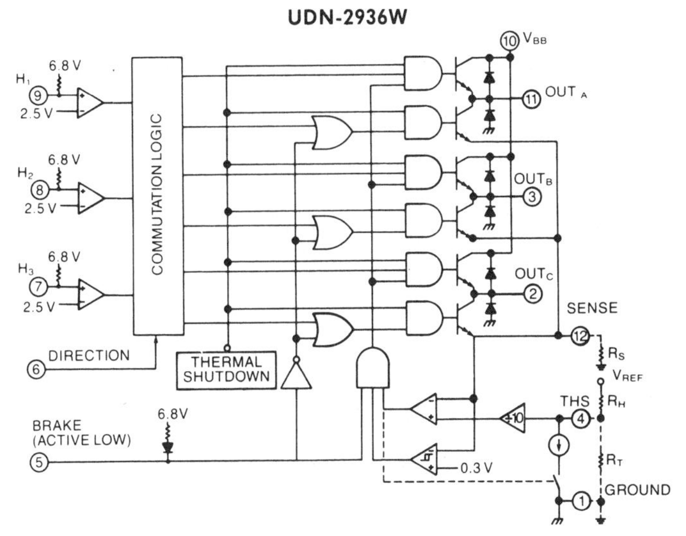

FUNCTIONAL BLOCK DIAGRAM
UDN-2937W

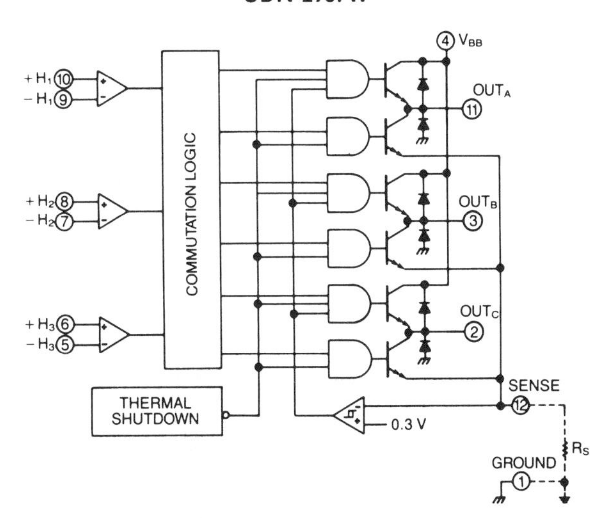

COMMUTATION TRUTH TABLE
UDN-2936W

Hall Sensor Inputs					Outputs		
H_1	H_2	H_3	Direction	Brake	OUT_A	OUT_B	OUT_C
High	High	High	Low	High	Z	Low	High
High	High	Low	Low	High	High	Low	Z
High	Low	Low	Low	High	High	Z	Low
Low	Low	Low	Low	High	Z	High	Low
Low	Low	High	Low	High	Low	High	Z
Low	High	High	Low	High	Low	Z	High
High	High	High	High	High	Z	High	Low
High	High	Low	High	High	Low	High	Z
High	Low	Low	High	High	Low	Z	High
Low	Low	Low	High	High	Z	Low	High
Low	Low	High	High	High	High	Low	Z
Low	High	High	High	High	High	Z	Low
X	X	X	X	Low	Low	Low	Low

X = Irrelevant
Z = High Impedance

TYPICAL HALL EFFECT SENSOR LOCATIONS

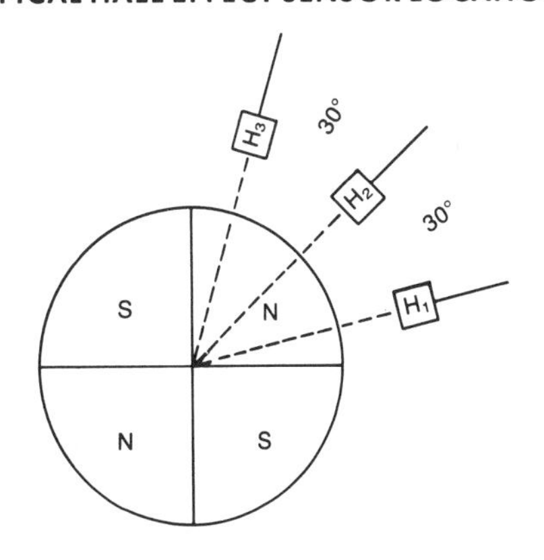

COMMUTATION TRUTH TABLE
UDN-2937W

Hall Sensor Inputs*			Outputs		
$+H_1$	$+H_2$	$+H_3$	OUT_A	OUT_B	OUT_C
High	High	High	Z	Low	High
High	High	Low	High	Low	Z
High	Low	Low	High	Z	Low
Low	Low	Low	Z	High	Low
Low	Low	High	Low	High	Z
Low	High	High	Low	Z	High

* Inputs are with respect to $-H_N$ inputs.
Z = High Impedance

TYPICAL APPLICATION

UDN-2936W

COMMUTATION LOGIC

V_{BB} V_E

DIR.

+40V

3 HALL EFFECT SENSORS

BRAKE

0.1 Ω SENSE

+5 V

ALLOWABLE AVERAGE PACKAGE POWER DISSIPATION AS A FUNCTION OF TEMPERATURE

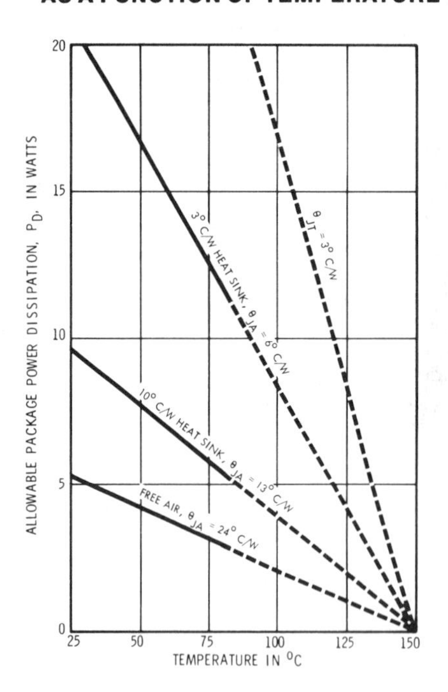

SECTION 4

DATA-CONVERSION AND PROCESSING

This information is reproduced with permission of Analog Devices, Inc.

Analog Devices AD380 Wideband, Fast-Settling, FET-Input Op Amp

- High output current: 50mA @ ±10V
- Fast settling to 0.1%: 130ns
- High slew rate: 330V/μs
- High gain-bandwidth product: 300 MHz
- High unity gain bandwidth: 40 MHz
- Low offset voltage (1mV for AD380K, L, S)

The AD380 is a hybrid operational amplifier that combines the low input bias current advantages of a FET input stage with the high slew rate and line-driving capability of a fast, high-power output amplifier.

The AD380 has a slew rate of 330V/μs and will output ±10V at ±50mA. A single external compensation capacitor allows the user to optimize the bandwidth, slew rate, or settling time for the given application.

A true differential input ensures equally superior performance in all system designs, whether they are inverting, noninverting, or differential.

The AD380 is especially designed for use in applications, such as fast A/D, D/A and sampling circuits, that require fast and smooth settling and FET input parameters.

The AD380 is offered in three commercial versions, J, K, and L specified from 0 to +70°C and one extended-temperature version, the S, specified from −55°C to +125°C. All grades are packaged in hermetically sealed TO-8-style cans.

AD380 FUNCTIONAL BLOCK DIAGRAM

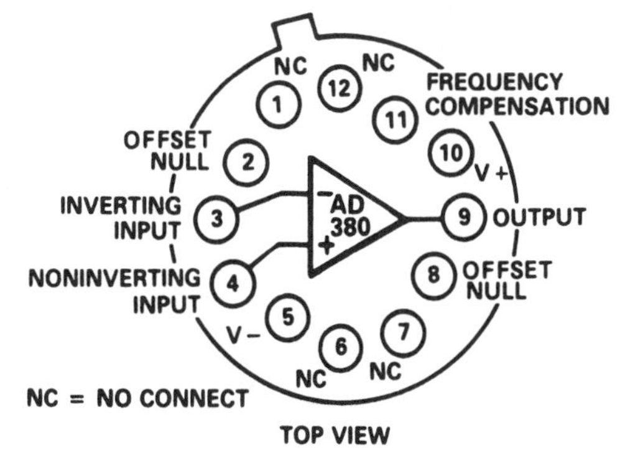

TYPICAL CHARACTERISTICS

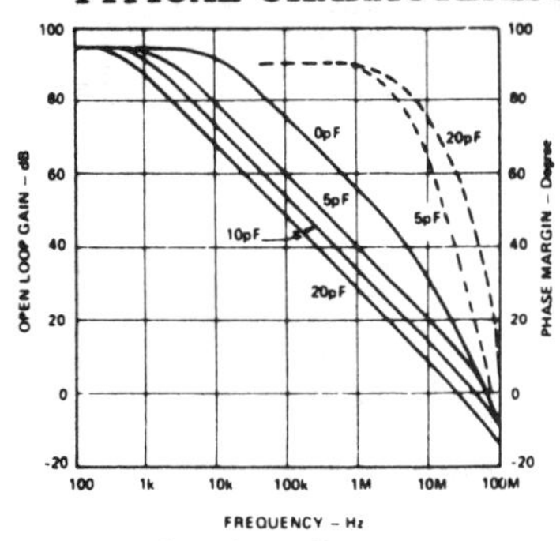

Open Loop Frequency Response

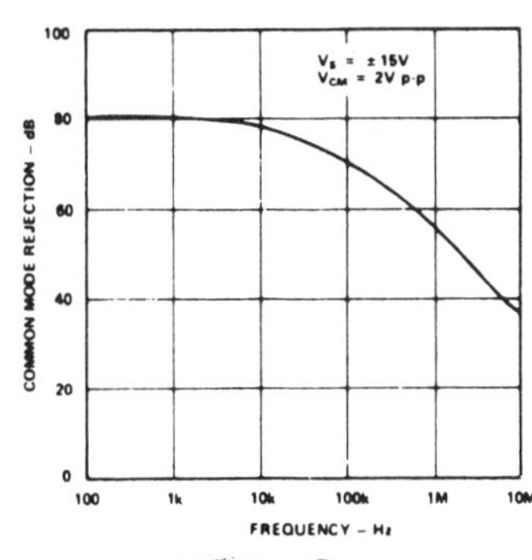

CMRR vs. Frequency

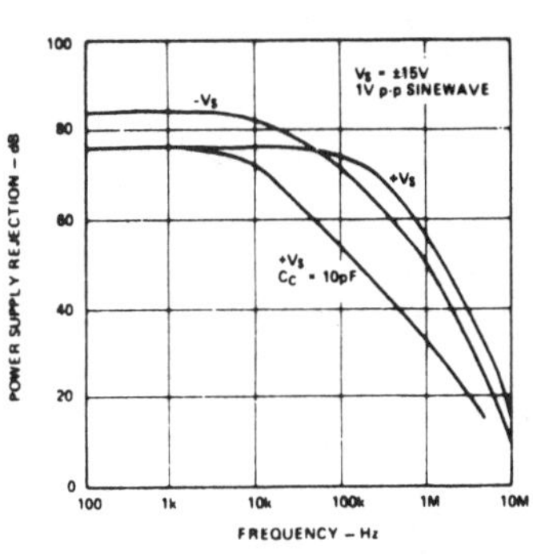

PSRR vs. Frequency

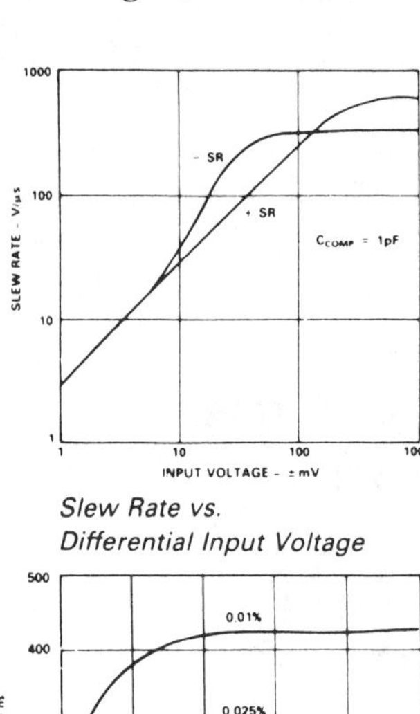

Slew Rate vs.
Differential Input Voltage

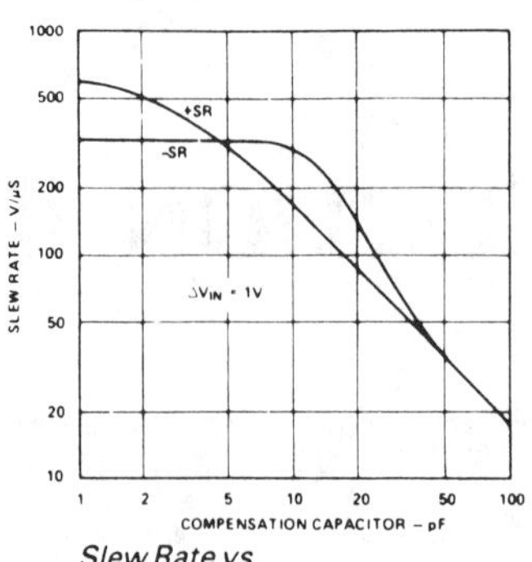

Slew Rate vs.
Compensation Capacitor

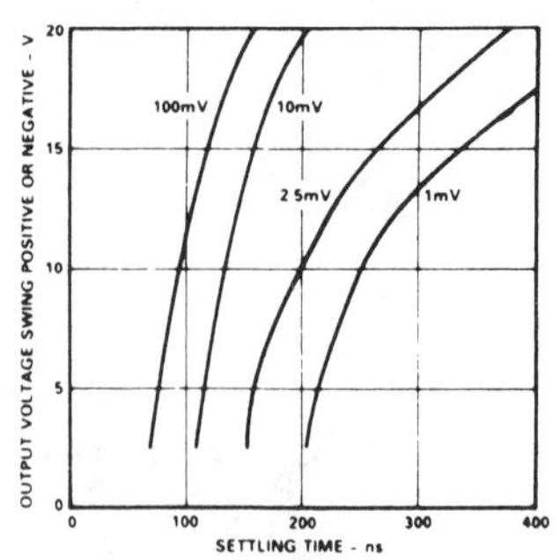

Output Settling Time vs.
Output Voltage Swing and Error

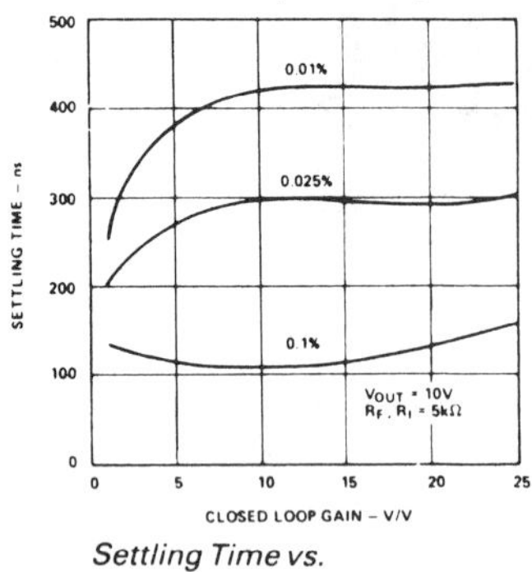

Settling Time vs.
Closed Loop Gain

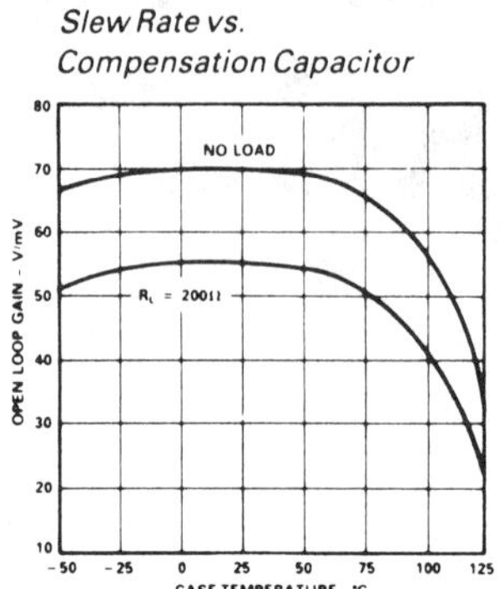

Gain vs. Temperature

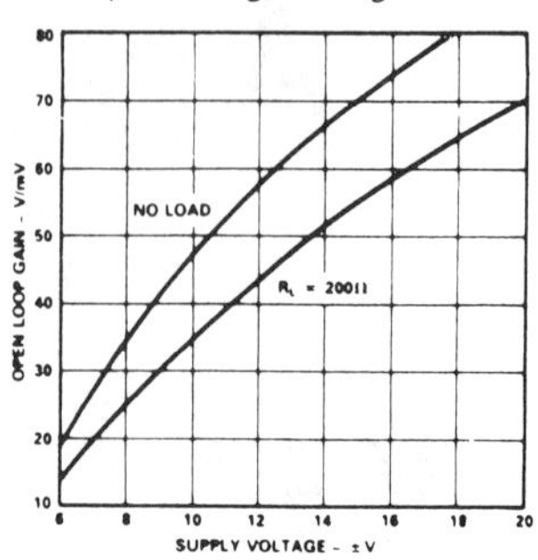

Gain vs. Supply Voltage

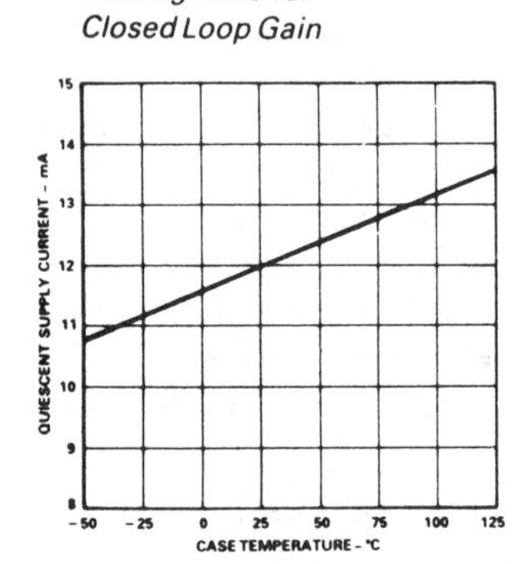

Supply Current vs.
Temperature

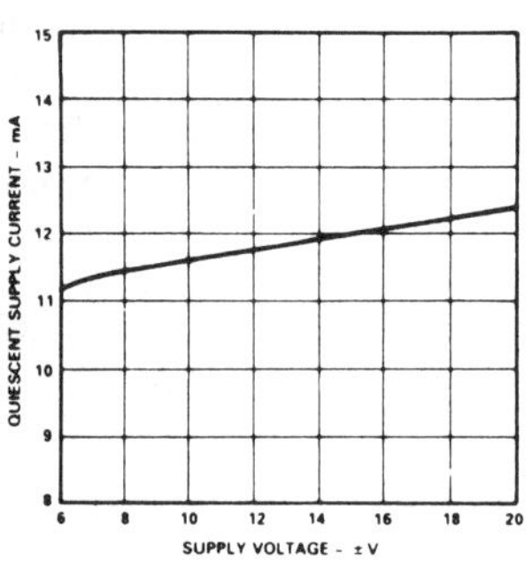

Supply Current vs.
Supply Voltage

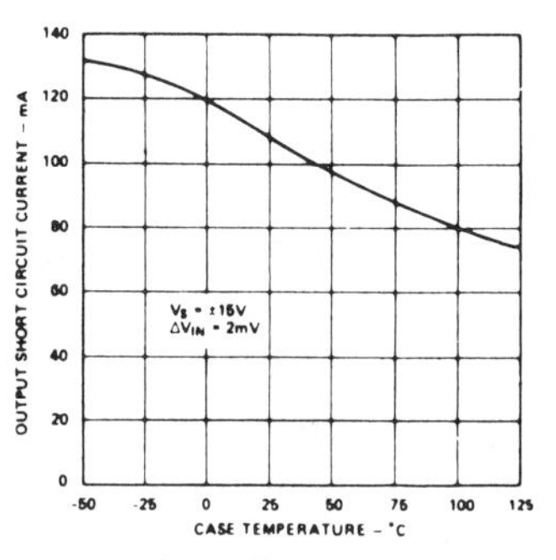

I_{SC} vs. Temperature

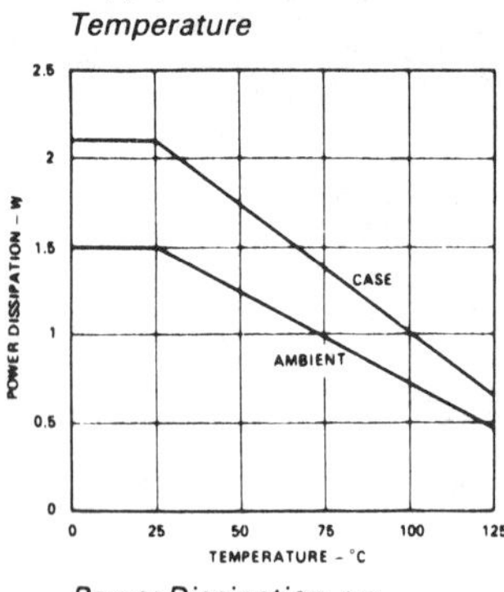

Power Dissipation vs.
Temperature

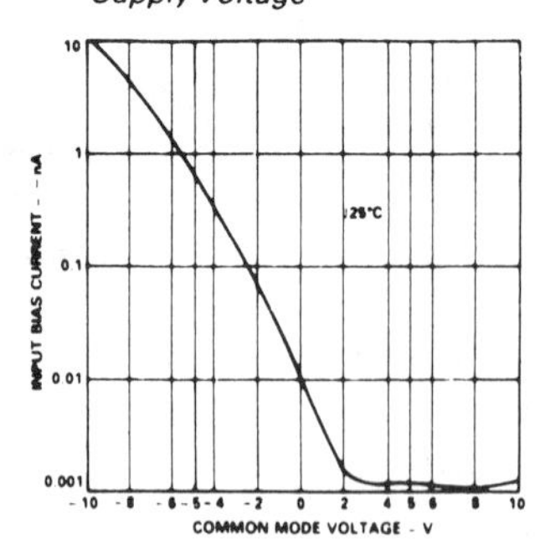

Input Bias Current vs.
Common Mode Voltage

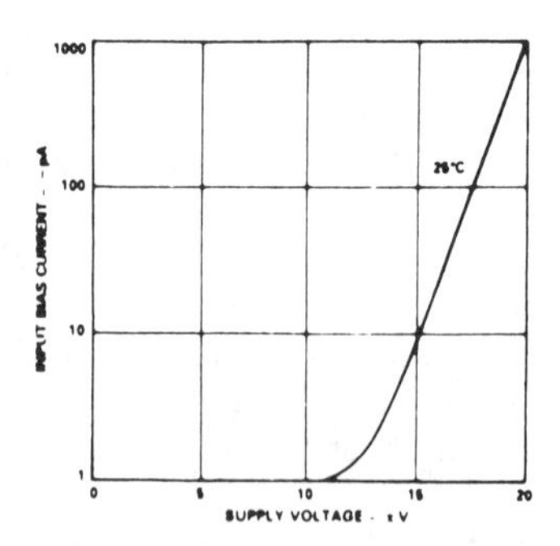

Input Bias Current vs.
Supply Voltage

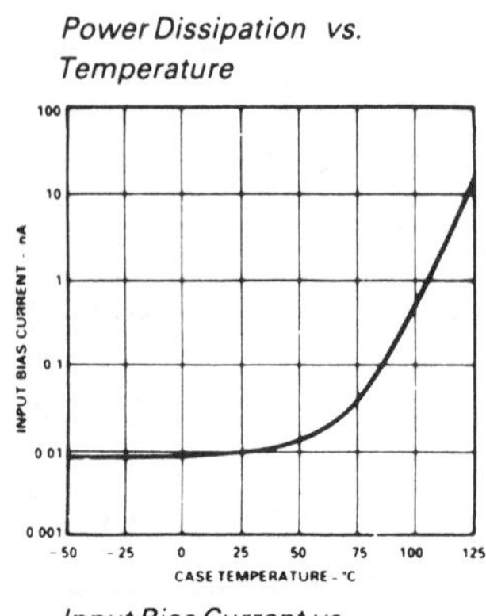

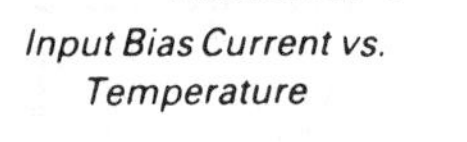
Input Bias Current vs.
Temperature

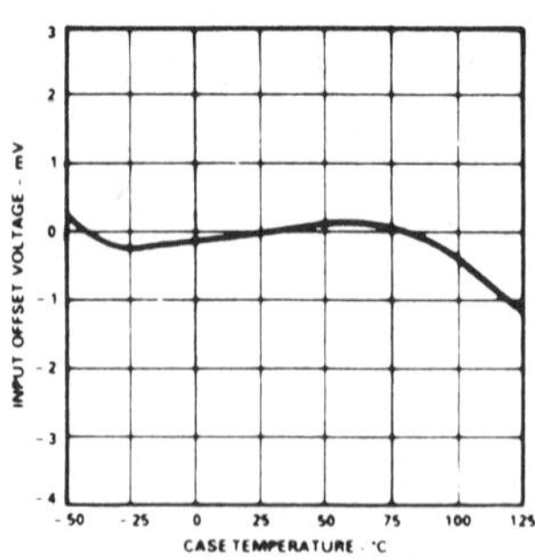

Offset Voltage vs.
Temperature

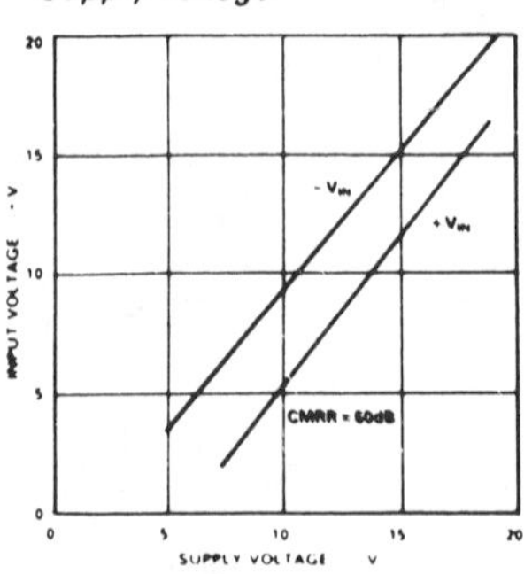

Input Voltage Range vs.
Supply Voltage

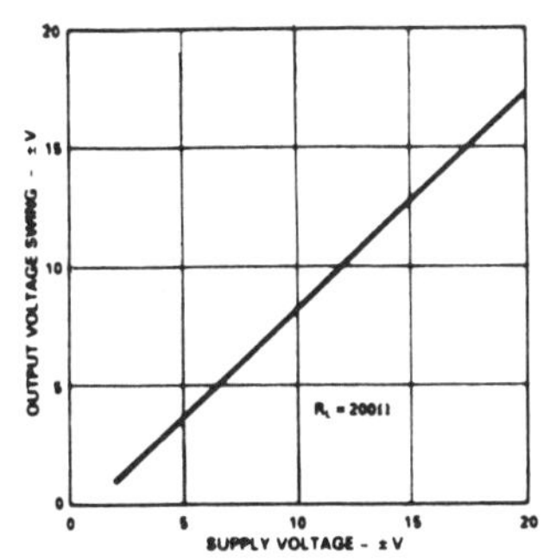

Output Voltage Swing vs.
Supply Voltage

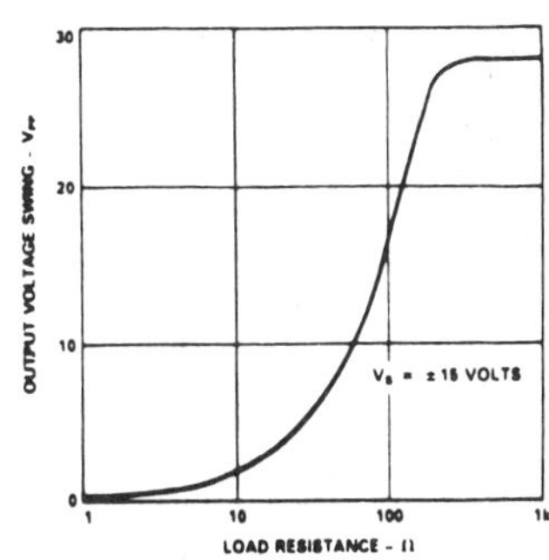

Output Voltage Swing vs.
Load Resistance

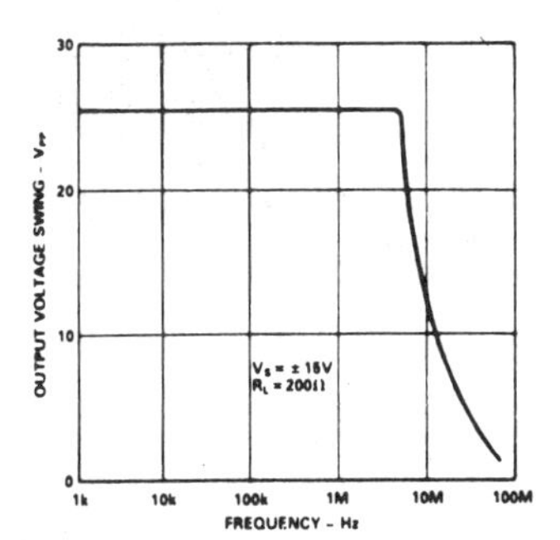

Large Signal Frequency
Response

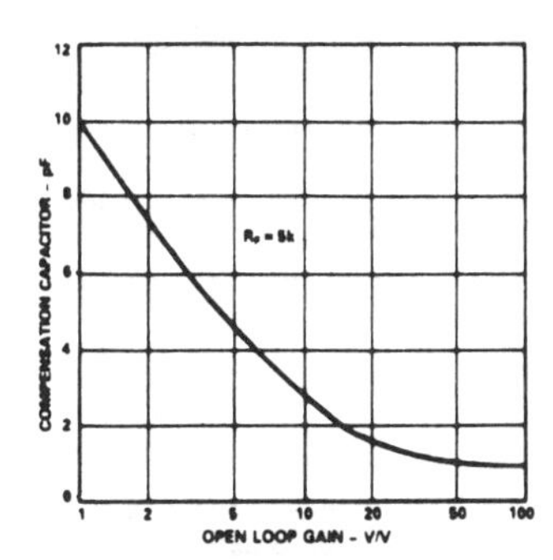

Recommended Compensation
Capacitor vs. Closed Loop Gain

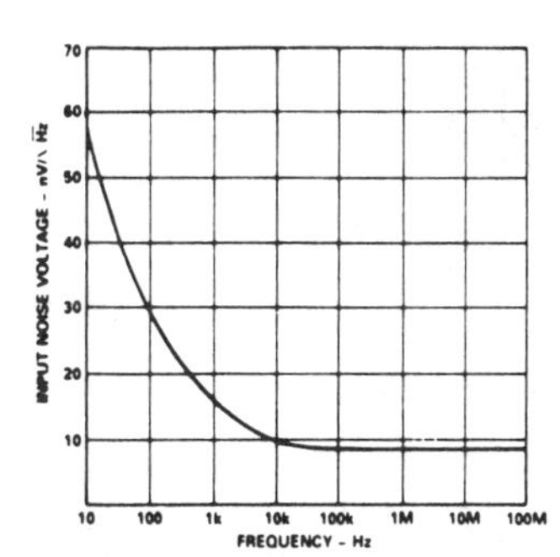

Input Noise Voltage
Spectral Density

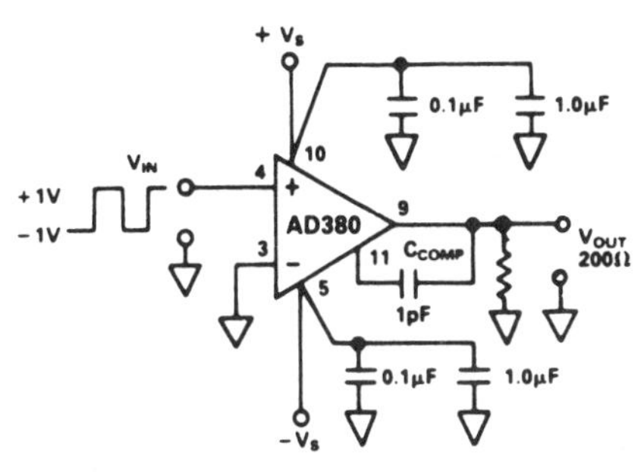

Overdrive Recovery
Test Circuit

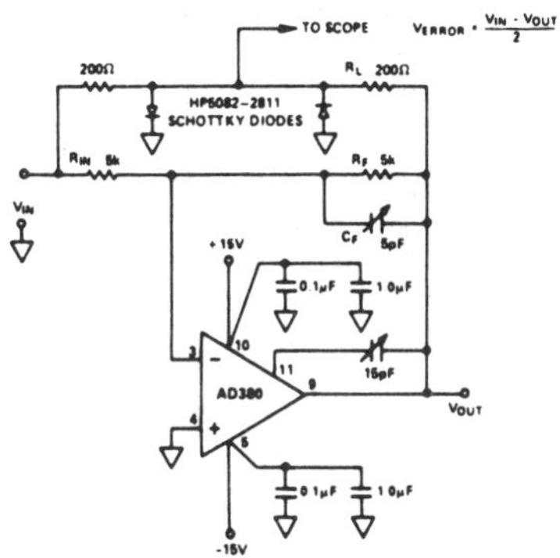

Unity Gain Inverter
Settling Time Test Circuit

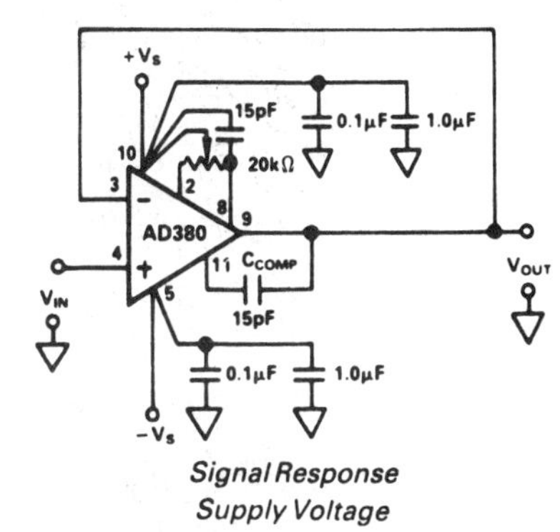

Signal Response
Supply Voltage

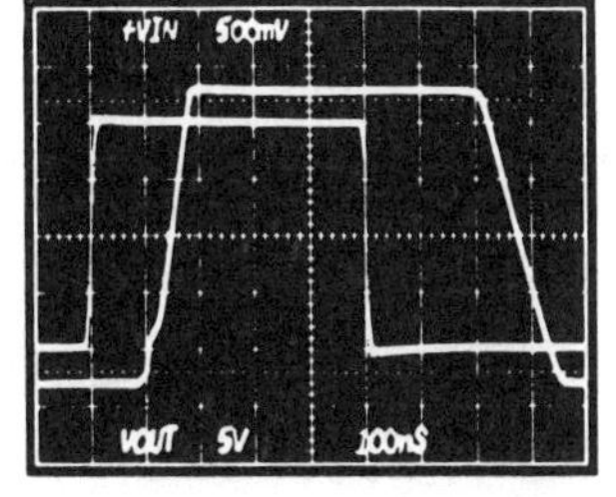

Overdrive Recovery
Response (Symmetrial 20ns
Version Available)

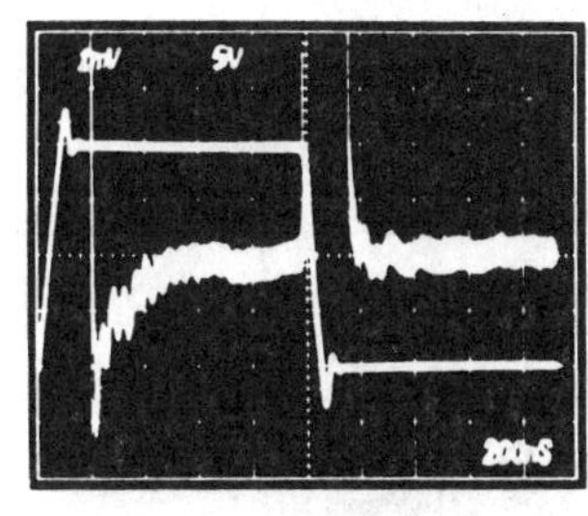

Unity Gain Inverter
Large Signal Response

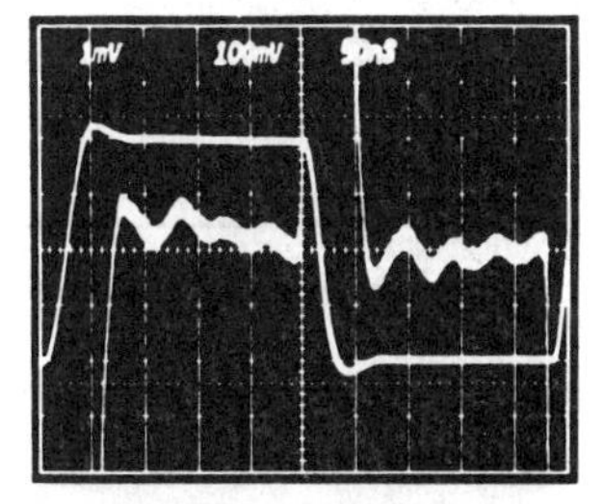

Unity Gain Inverter
Small Signal Response

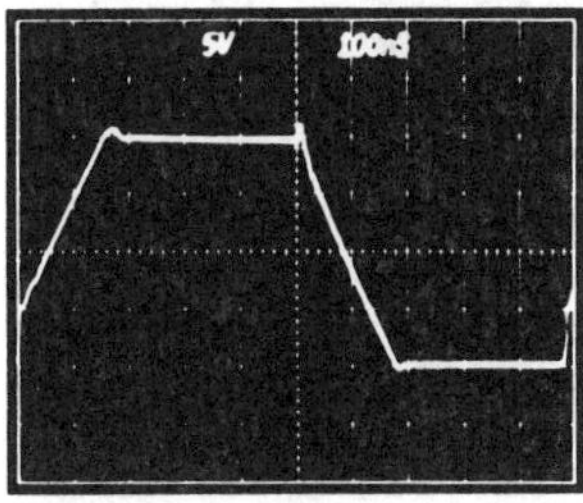

Unity Gain Buffer
Large Signal Response

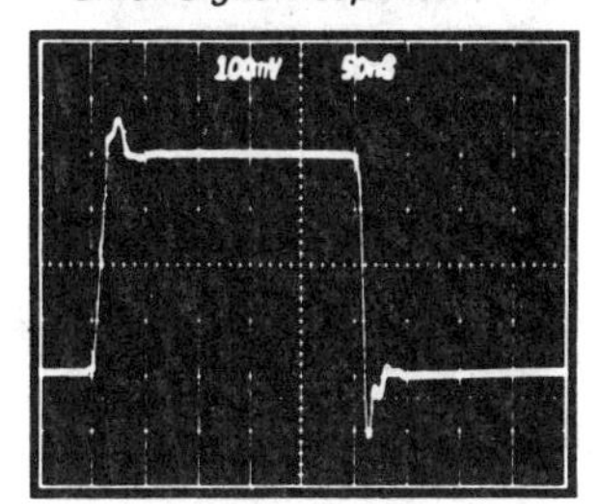

Unity Gain Small

*Optional Differential Input Components Used to Reject Noise Between Input Ground and the A/D Analog Ground.

Fast-Settling Buffer

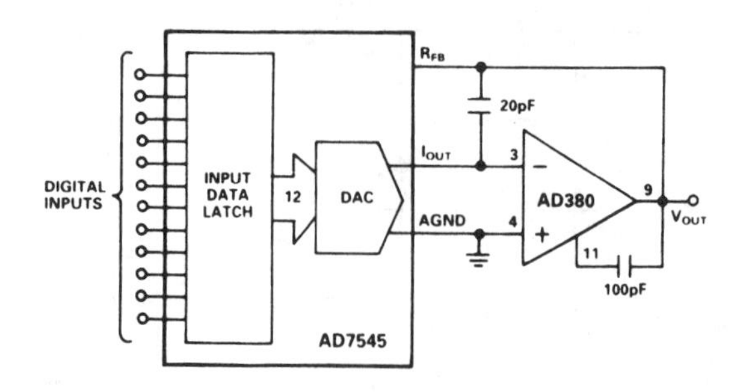

CMOS DAC Output Amplifier

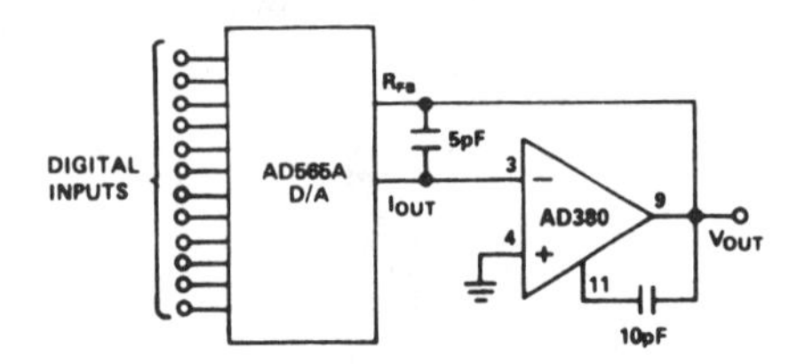

12-Bit Voltage Output DAC Circuit Settles to 1/2LSB in 300ns

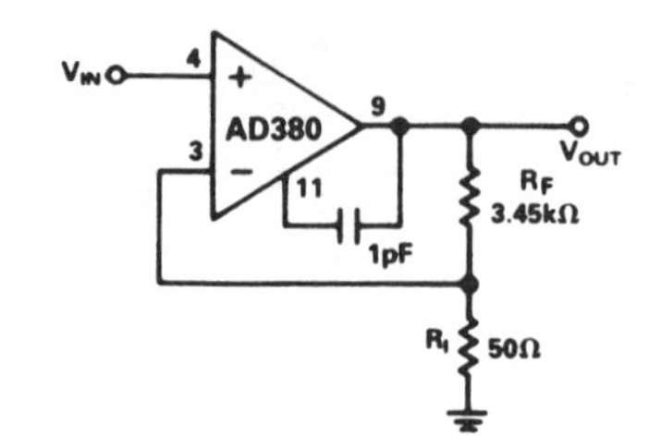

Video Amplifier

SPECIFICATIONS (typical @ +25°C and V_S = ±15V dc unless otherwise specified)

MODEL	AD380JH	AD380KH	AD380LH	AD380SH
OPEN LOOP GAIN				
V_{OUT} = ±10V, no load	40,000 min	*	*	*
V_{OUT} = ±10V, $R_L \geq 200\Omega$	25,000 min	*	*	*
OUTPUT CHARACTERISTICS				
Voltage @ R_L = 200Ω, T_A = min to max	±12V (±10V min)	*	*	*
Output Impedance (Open Loop)	100Ω	*	*	*
Short Circuit Current	100mA	*	*	*
DYNAMIC RESPONSE				
Unity Gain, Small Signal	40MHz	*	*	*
Gain-Bandwidth Product, f = 100kHz, C_C = 1pF	300MHz (200MHz min)	*	*	*
Full Power Response	6MHz	*	*	*
Slew Rate, C_C = 1pF, 20V Swing	330V/μs (200V/μs min)	*	*	*
Settling Time: 10V Step to 1%	90ns	*	*	*
10V Step to 0.1%	130ns	*	*	*
10V Step to 0.01%	250ns	250ns (400ns max)	**	**
INPUT OFFSET VOLTAGE	2.0mV max	1.0mV max	**	**
vs. Temperature[1], T_A = min to max	50μV/°C max	20μV/°C max	10μV/°C max	50μV/°C max
vs. Supply	1mV/V max	*	*	*
INPUT BIAS CURRENT				
Either Input, Initial[2]	10pA (100pA max)	*	*	*
Input Offset Current	5pA	*	*	*
INPUT IMPEDANCE				
Differential	$10^{11}\Omega\|6pF$	*	*	*
Common Mode	$10^{11}\Omega\|6pF$	*	*	*
INPUT VOLTAGE RANGE				
Differential[3]	±20V	*	*	*
Common Mode	±12V (±10V min)	*	*	*
Common Mode Rejection, V_{IN} = ±10V	60dB min	*	*	*
POWER SUPPLY				
Rated Performance	±15V	*	*	*
Operating	±(6 to 20)V	*	*	*
Quiescent Current	12mA (15mA max)	*	*	*
VOLTAGE NOISE				
0.1Hz to 100Hz	3.3μV p-p (0.5μV rms)	*	*	*
100Hz to 10kHz	6.6μV p-p (1μV rms)	*	*	*
10kHz to 1MHz	40μV p-p (6μV rms)	*	*	*
TEMPERATURE RANGE				
Operating, Rated Performance	0 to +70°C	*	*	−55°C to +125°C
Storage	−65°C to +150°C	*	*	*
Thermal Resistance θ_{JA}	100°C/W	*	*	*
θ_{JC}	70°C/W	*	*	*
PACKAGE OPTION[4]				
TO-8 Style	H-12A	*	*	*

NOTES
[1]Input Offset Voltage Drift is specified with the offset voltage unnulled. Nulling will induce an additional 3μV/°C/mV of offset nulled.
[2]Bias Current specifications are guaranteed maximum at either input at T_{CASE} = +25°C.
[3]Defined as the maximum safe voltage between inputs such that neither exceeds ±10V from ground.

Analog Devices AD509
High-Speed Fast-Settling IC Op Amp

- Fast settling time
  - 0.1% in 500ns max
  - 0.01% in 2.5μs max
- High slew rate: 100V/μs min
- Low I_{OS}: 25nA max
- Guaranteed V_{OS} drift: 30μV/°C max
- High CMRR: 80dB min
- Drives 500pF
- Low price

APPLICATIONS
- D/A and A/D conversion
- Wideband amplifiers
- Multiplexers
- Pulse amplifiers

The AD509J, AD509K, and AD509S are monolithic operational amplifiers specifically designed for applications requiring fast settling times to high accuracy. Other comparable dynamic parameters include a small-signal bandwidth of 20 MHz, slew rate of 100V/μs min and a full power response of 150 kHz min. The devices are internally compensated for all closed-loop gains greater than 3, and are compensated with a single capacitor for lower gains.

The input characteristics of the AD509 are consistent with 0.01% accuracy over limited-temperature ranges; offset current is 25nA max, offset voltage is 8mV max, nullable to zero, and offset voltage drift is limited to 30μV/°C max. PSRR and CMRR are typically 90dB.

The AD509 is designed for use with high-speed D/A or A/D converters, where the minimum conversion time is limited by the amplifier settling time. If 0.01% accuracy of conversion is required, a conversion cannot be made in a shorter period than the time required for the amplifier to settle to within 0.01% of its final value.

All devices are supplied in the TO-99 package. The AD509J and AD509K are specified for 0 to +70°C temperature range; the AD509S is specified for operation from −55°C to +125°C.

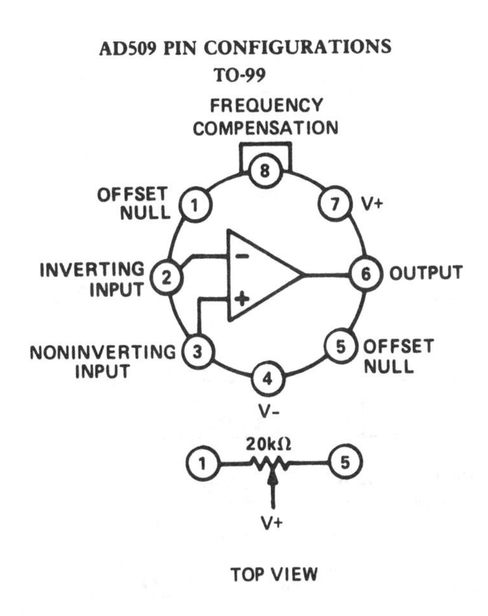

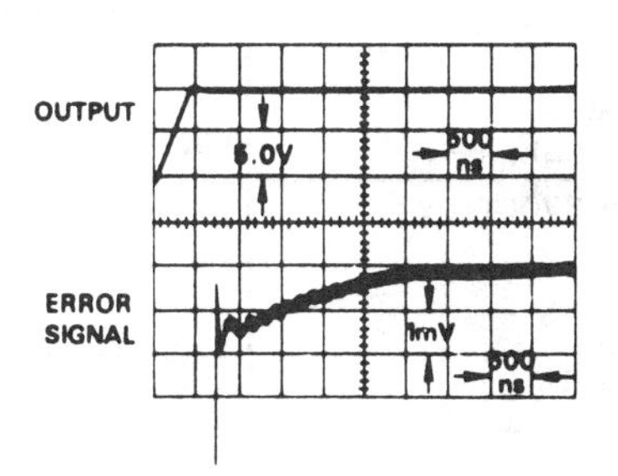

Settling Time of AD509

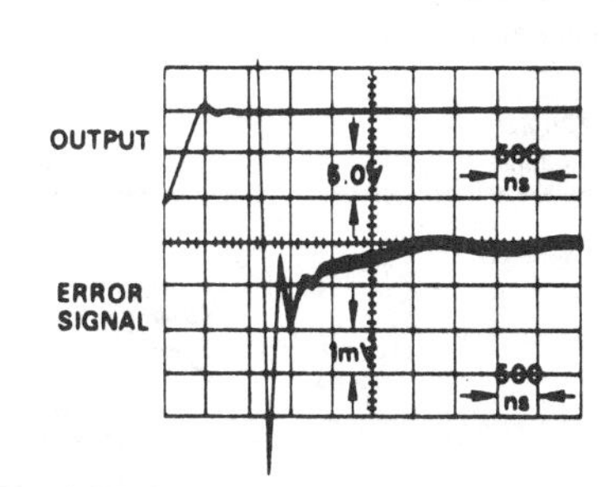

AD509 with 500pF Capacitive Load

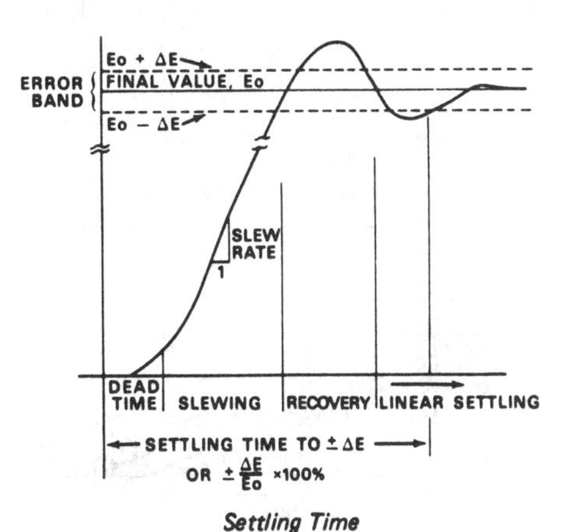

Settling Time

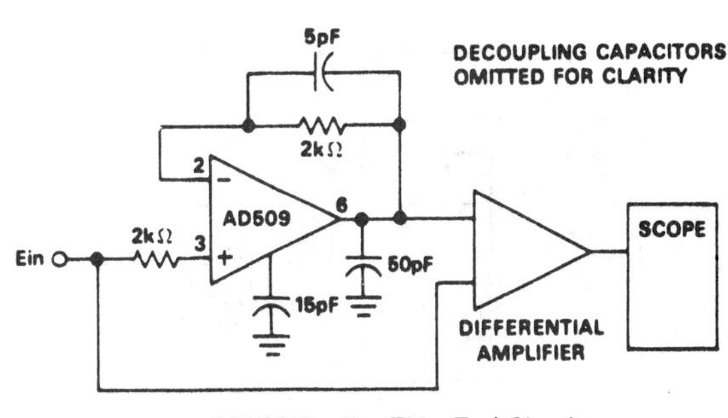

AD509 Settling Time Test Circuit

SPECIFICATIONS (@ +25°C and V_S = ±15V dc unless otherwise specified)

Model	AD509J Min	AD509J Typ	AD509J Max	AD509K Min	AD509K Typ	AD509K Max	AD509S Min	AD509S Typ	AD509S Max	Units
OPEN LOOP GAIN										
V_O = ±10V, R_L > 2kΩ	**7,500**	15,000		**10,000**	15,000		**10,000**	15,000		V/V
T_{min} to T_{max}, R_L 2kΩ	5,000			7,500			**7,500**			V/V
OUTPUT CHARACTERISTICS										
Voltage @ R_L = 2kΩ, T_{min} to T_{max}	**±10**	±12		**±10**	±12		**±10**	±12		V
FREQUENCY RESPONSE										
Unity Gain Small Signal		20			20			20		MHz
Full Power Response	1.2	1.6		1.5	2.0		1.5	2.0		MHz
Slew Rate, Unity Gain	80	120		80	120		100	120		V/μs
Settling Time										
to 0.1%		200			200			200	500	ms
to 0.01%		1.0			1.0			1.0	2.5	μs
INPUT OFFSET VOLTAGE										
Initial Offset		5	**10**		4	**8**		4	**8**	mV
Input Offset Voltage T_{min} to T_{max}			14			11			**11**	mV
Input Offset Voltage vs. Supply, T_{min} to T_{max}			**200**			**100**			**100**	μV/V
INPUT BIAS CURRENT										
Initial		125	**250**		100	**200**		100	**200**	nA
T_{min} to T_{max}			500			400			**400**	nA
INPUT OFFSET CURRENT										
Initial		20	50		10	25		10	25	nA
T_A = min to max			100			50			50	nA
INPUT IMPEDANCE										
Differential	40	100		50	100		50	100		MΩ
INPUT VOLTAGE RANGE										
Differential		±15			±15			±15		V
Common Mode		±10			±10			±10		V
Common Mode Rejection	**74**	90		**80**	90		**80**	90		dB
INPUT NOISE VOLTAGE										
f = 10Hz		100			100			100		nV/√Hz
f = 100Hz		30			30			30		nV/√Hz
f = 100kHz		19			19			19		nV/√Hz
POWER SUPPLY										
Rated Performance		±15			±15			±15		V
Operating	±5		±20	±5		±20	±5		±20	V
Quiescent Current		4	**6**		4	**6**		4	**6**	mA
TEMPERATURE RANGE										
Operating, Rated Performance	0		+70	0		+70	−55		+70	°C
Storage	−65		+150	−65		+150	−65		+150	°C
PACKAGE OPTION[1]										
TO-99 Style (H-08A)		AD509JH			AD509KH			AD509SH		

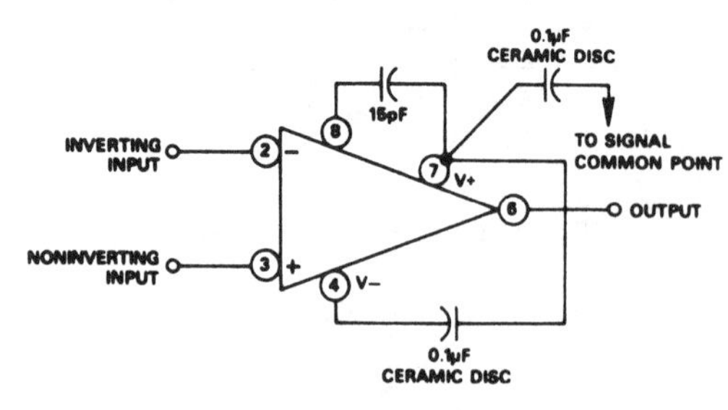

Configuration for Unity Gain Applications

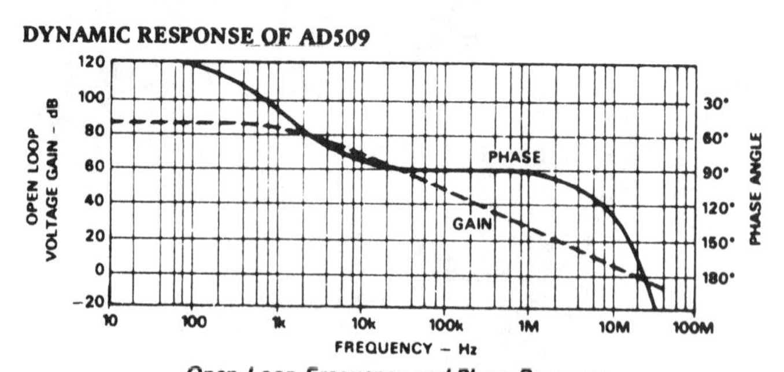

Open Loop Frequency and Phase Response

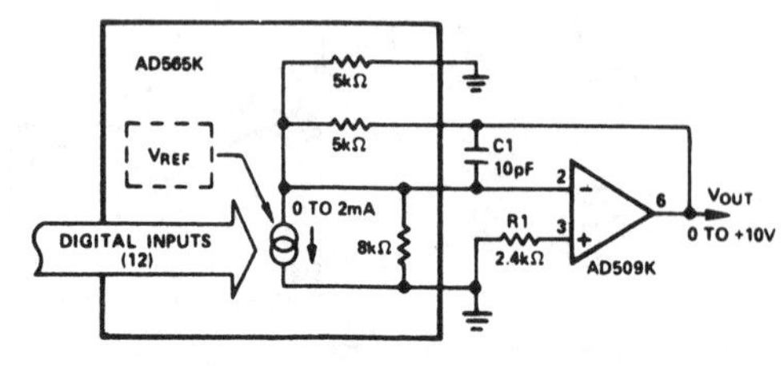

AD509 as an Output Amplifier for a Fast Current-Output D-to-A Converter

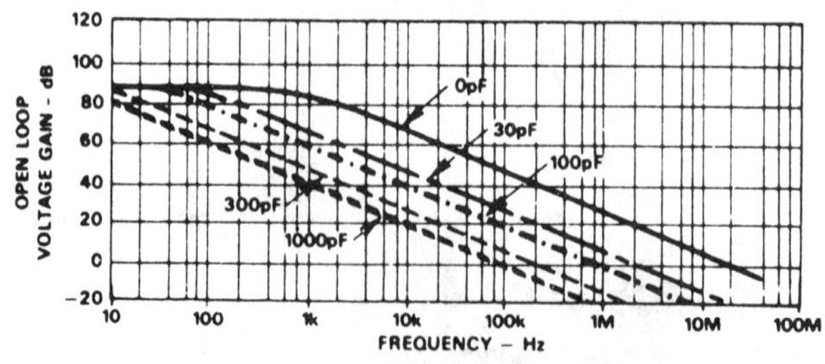

Open Loop Frequency Response for Various C_c's

Analog Devices AD542/AD544/AD547

High-Performance BiFET Operational Amplifiers

- Ultralow drift: 1μV/°C—AD547L
- Low offset voltage: 0.25mV—AD547L
- Low input bias currents: 25pA max, warmed-up
- Low quiescent current: 1.5mA
- Low noise: 2μV p-p
- High open-loop gain: 110dB
- High slew rate: 13V/μs
- Fast settling to ±0.01%: 3μs
- Low total harmonic distortion: 0.0025%

The BiFET series are precision monolithic FET-input operational amplifiers fabricated with the most advanced BiFET and laser trimming technologies. The series offers bias currents significantly lower than currently available BiFET devices, 25pA max, warmed-up.

In addition, the offset voltage is laser trimmed to less than 0.25mV on the AD547L which is achieved by utilizing Analog's exclusive laser-wafer trimming (LWT) process. When combined with the AD547's low offset voltage drift (1μV/°C), these features offer the user IC performance truly superior to existing BiFET op amps—at low, BiFET pricing.

The AD542 or AD547 is recommended for any operational amplifier application requiring excellent dc performance at low to moderate costs. Precision instrument front ends requiring accurate amplification of millivolt level signals from megohm source impedances will benefit from the device's excellent combination of low offset voltage and drift, low bias current, and low 1/f noise. High common-mode rejection (80dB, min on the "K" and "L" versions) and high open-loop gain—even under heavy loading—ensures better than "12-bit" linearity in high-impedance buffer applications.

The AD544 is recommended for any operational amplifier application requiring excellent ac and dc performance at low cost. The 2-MHz bandwidth and low offset of the AD544 make it the first choice as an output amplifier for current output D/A converters, such as the AD7541 12-bit CMOS DAC.

Devices in this series are available in four versions: the "J," "K," and "L" are specified over the 0 to +70°C temperature range and the "S" over the −55°C to +125°C operating temperature range. All devices are packaged in the hermetically-sealed TO-99 metal can.

APPLICATION NOTES

The BiFET series was designed for high-performance op-amp applications that require true dc precision. To capitalize on all of the performance available from the BiFETs, some practical error sources should be considered.

The bias currents of JFET input amplifiers double with every 10°C increase in chip temperature. Therefore, minimizing the junction temperature of the chip will result in extending the performance limits of the device.

- Heat dissipation as a result of power consumption is the main contributor to self-heating and can be minimized by reducing the power supplies to the lowest level allowed by the application.
- The effects of output loading should be carefully considered. Greater power dissipation increases bias currents and decreases open-loop gain.

AD542, AD544, AD547 FUNCTIONAL BLOCK DIAGRAM

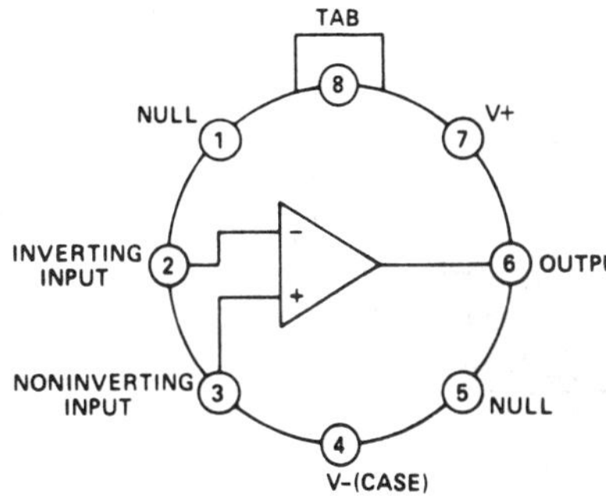

TAB
NULL 1
8
V+ 7
INVERTING INPUT 2
6 OUTPUT
NONINVERTING INPUT 3
5 NULL
4
V-(CASE)
TOP VIEW

GENERAL PURPOSE D/A CONVERTERS

Model	Resolution Bits	Settling Time μs	MDAC	Current Out	Voltage Out	On-Chip Reference	No. of DACs	μP Interface	Double Buffered Inputs	Notes
AD DAC71/2–I	16	1		X		X				
AD569	16	3	X		X			X	X	8/16-bit bus compatible
AD DAC71/2–V	16	5			X					
AD1145	16	6			X			X	X	w/ readback function
DAC1136	16	8			X	X				
AD1147/1148	16	20			X	X		X		μP correction capability
DAC1138	18	10			X	X				
AD1139	18	40			X	X		X		
AD7111	0.375dB	4.5	X	X						8-bit LOGDAC

SPECIFICATIONS (@ +25°C and V_S = ±15V dc)

Model	AD542 Min	AD542 Typ	AD542 Max	AD544 Min	AD544 Typ	AD544 Max	AD547 Min	AD547 Typ	AD547 Max	Units
OPEN LOOP GAIN[1]										
V_{OUT} = ±10V $R_L \geq 2k\Omega$										
J	100,000			30,000			100,000			V/V
K, L, S	250,000			50,000			250,000			V/V
$T_A = T_{min}$ to T_{max}										
J	100,000			20,000			100,000			V/V
S	100,000			20,000			100,000			V/V
K, L	250,000			40,000			250,000			V/V
OUTPUT CHARACTERISTICS										
V_{OUT} = R_L = 2kΩ										
$T_A = T_{min}$ to T_{max}	±10	±12		±10	±12		±10	±12		Volts
V_{OUT} = R_L = 10kΩ										
$T_A = T_{min}$ to T_{max}	±12	±13		±12	±13		±12	±13		Volts
Short Circuit Current		25			25			25		mA
FREQUENCY RESPONSE										
Unity Gain, Small Signal		1.0			2.0			1.0		MHz
Full Power Response		50			200			50		kHz
Slew Rate, Unity Gain	2.0	3.0		8.0	13.0		2.0	3.0		V/μs
Total Harmonic Distortion					0.0025					%
INPUT OFFSET VOLTAGE[2]										
J			2.0			2.0			1.0	mV
K			1.0			1.0			0.5	mV
L			0.5			0.5			0.25	mV
S			1.0			1.0			0.5	mV
vs. Temperature[3]										
J			20			20			5	μV/°C
K			10			10			2	μV/°C
L			5			5			1	μV/°C
S			15			15			5	μV/°C
vs. Supply, $T_A = T_{min}$ to T_{max}										
J			200			200			200	μV/V
K, L, S			100			100			100	μV/V
INPUT BIAS CURRENT[4]										
Either Input										
J			50			50			50	pA
K, L, S		10	25		10	25		10	25	pA
Input Offset Current										
J		5	15		5	15		5	15	pA
K, L, S		2	15		2	15		2	15	pA
INPUT IMPEDANCE										
Differential		10^{12} Ω‖6pF			10^{12} Ω‖6pF			10^{12} Ω‖6pF		
Common Mode		10^{12} Ω‖3pF			10^{12} Ω‖3pF			10^{12} Ω‖3pF		
INPUT VOLTAGE[5]										
Differential		±20			±20			±20		Volts
Common Mode	±10	±12		±10	±12		±10	±12		Volts
Common-Mode Rejection										
V_{IN} = ±10V										
J	76			76			76			dB
K, L, S	80			80			80			dB
POWER SUPPLY										
Rated Performance		±15			±15			±15		Volts
Operating	±5		±18	±5		±18	±5		±18	Volts
Quiescent Current		1.1	1.5		1.8	2.5		1.1	1.5	mA
VOLTAGE NOISE										
0.1–10Hz										
J		2.0			2.0			2.0		μV p-p
K, L, S		2.0			2.0				4.0	μV p-p
10Hz		70			35			70		nV/√Hz
100Hz		45			22			45		nV/√Hz
1kHz		30			18			30		nV/√Hz
10kHz		25			16			25		nV/√Hz
TEMPERATURE RANGE										
Operating, Rated Performance										
J, K, L		0 to +70			0 to +70			0 to +70		°C
S		55 to +125			55 to +125			55 to +125		°C
Storage		65 to +150			65 to +165			65 to +165		°C
PACKAGE OPTIONS[6]										
TO-99 (H-08A)		AD542JH, AD542KH AD542LH, AD542SH			AD544JH, AD544KH AD544LH, AD544SH			AD547JH, AD547KH AD547LH, AD547SH		

TYPICAL CHARACTERISTICS

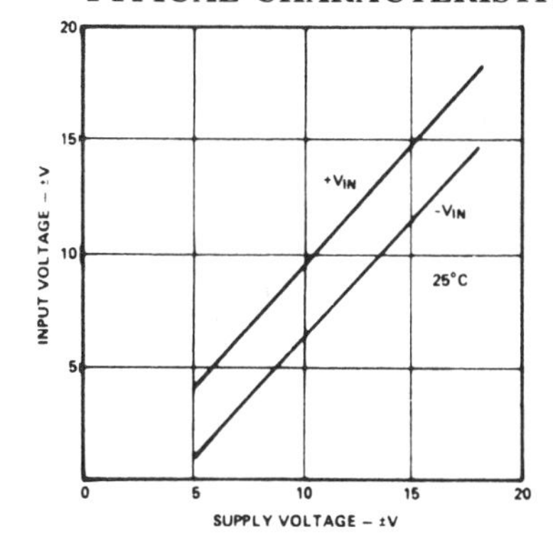

Input Voltage Range vs. Supply Voltage

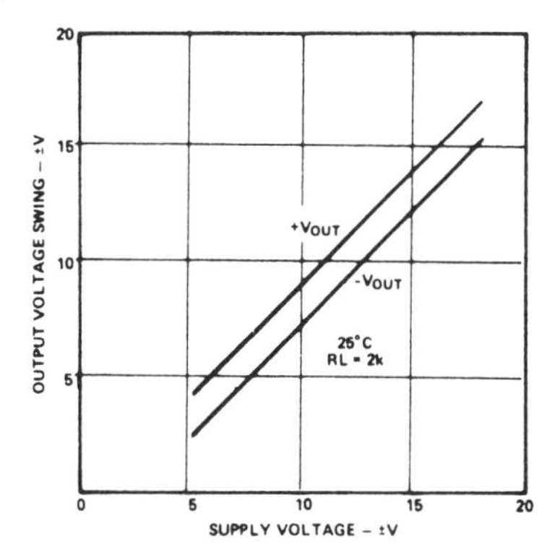

Output Voltage Swing vs. Supply Voltage

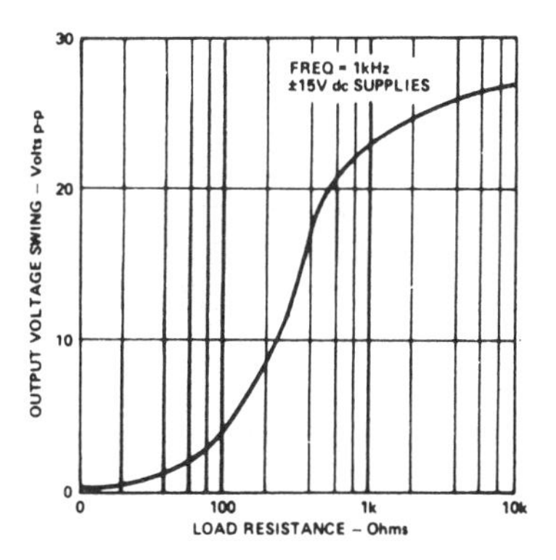

Output Voltage Swing vs. Resistive Load

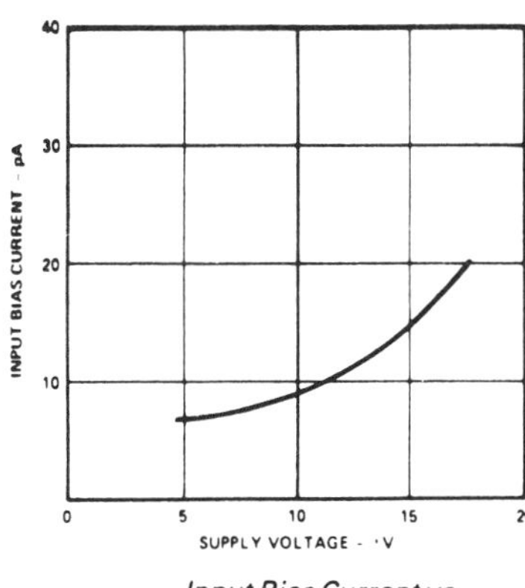

Input Bias Current vs. Supply Voltage

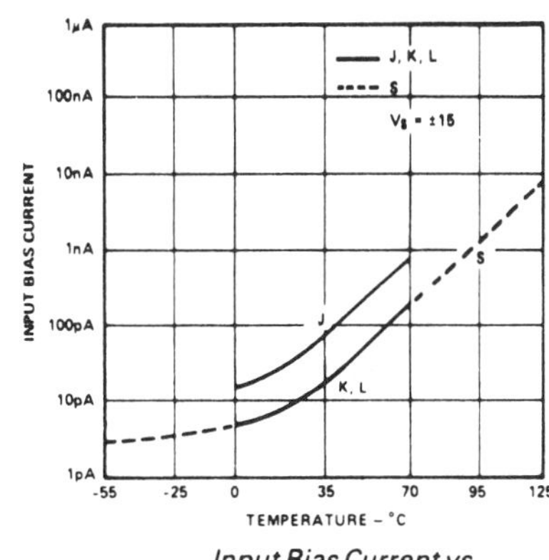

Input Bias Current vs. Temperature

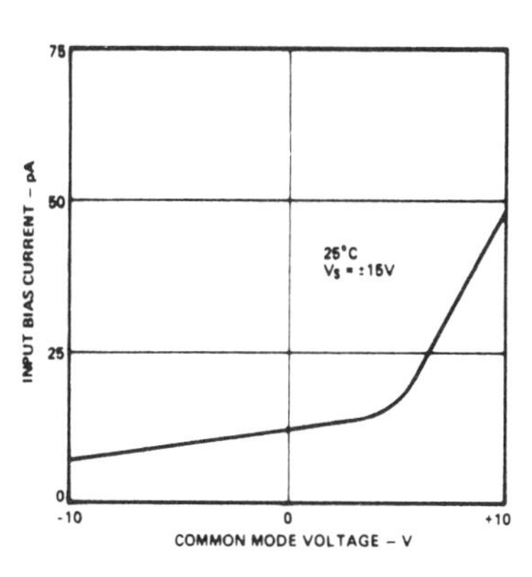

Input Bias Current vs. CMV

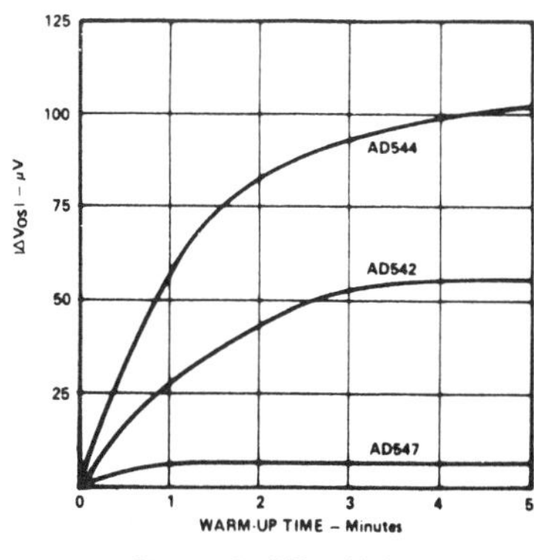

Change in Offset Voltage vs. Warm-Up Time

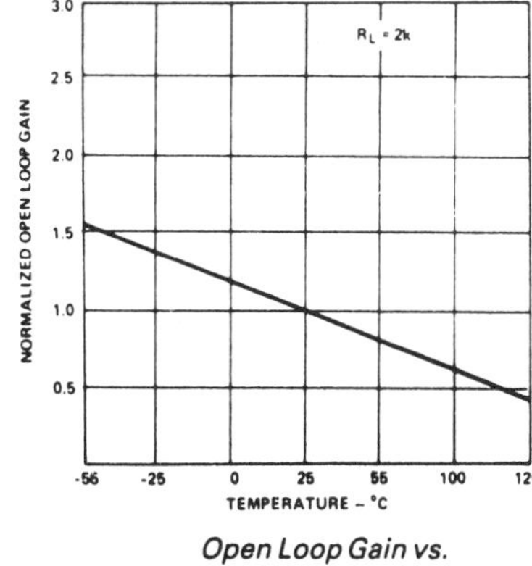

Open Loop Gain vs. Temperature

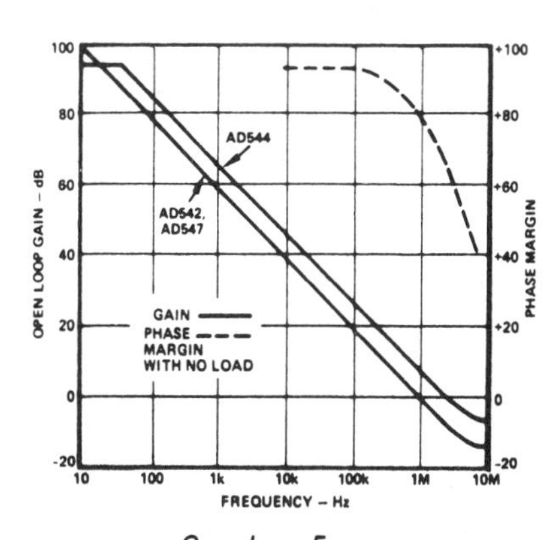

Open Loop Frequency Response

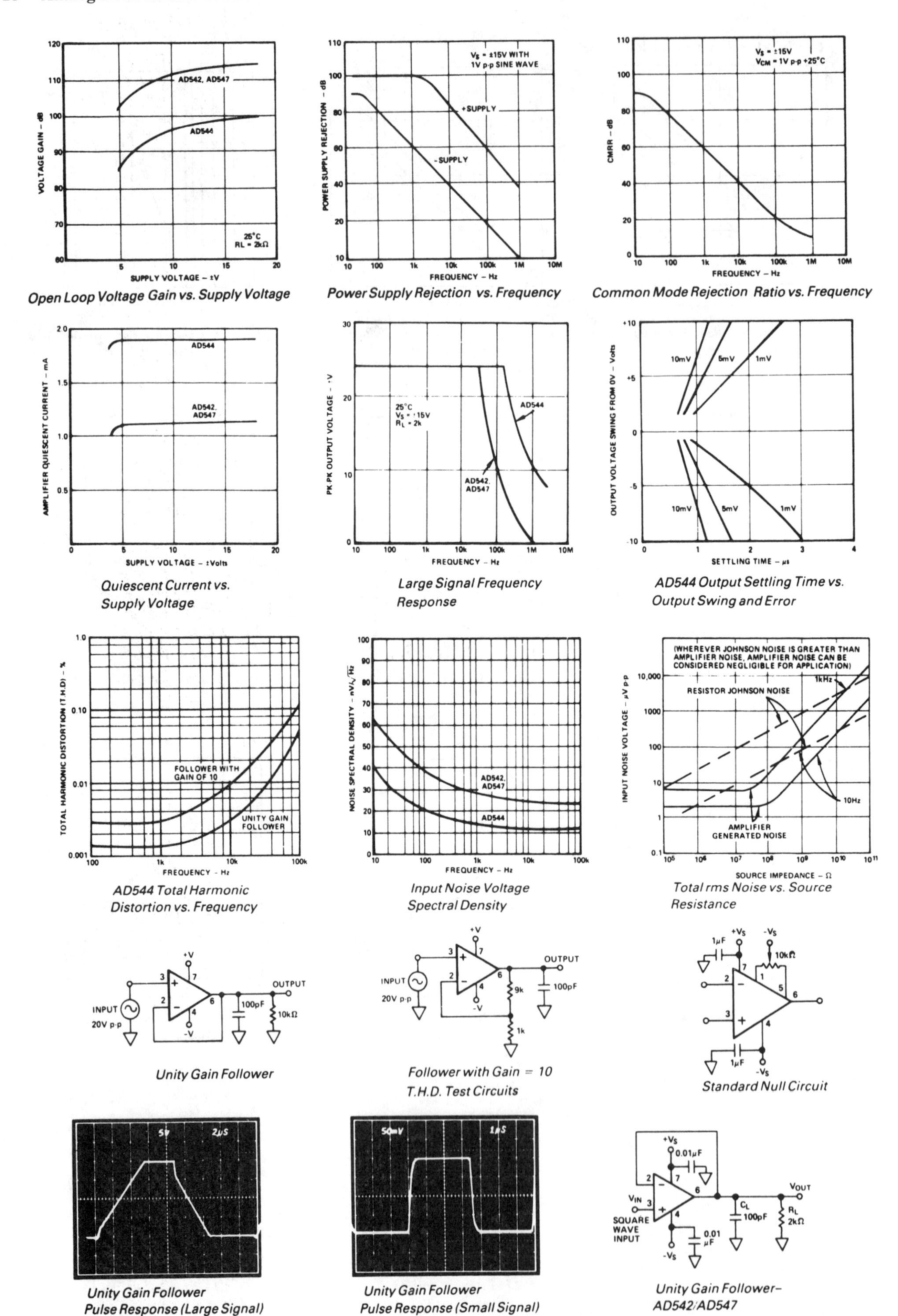

Open Loop Voltage Gain vs. Supply Voltage

Power Supply Rejection vs. Frequency

Common Mode Rejection Ratio vs. Frequency

Quiescent Current vs. Supply Voltage

Large Signal Frequency Response

AD544 Output Settling Time vs. Output Swing and Error

AD544 Total Harmonic Distortion vs. Frequency

Input Noise Voltage Spectral Density

Total rms Noise vs. Source Resistance

Unity Gain Follower

Follower with Gain = 10

T.H.D. Test Circuits

Standard Null Circuit

Unity Gain Follower Pulse Response (Large Signal)

Unity Gain Follower Pulse Response (Small Signal)

Unity Gain Follower–AD542/AD547

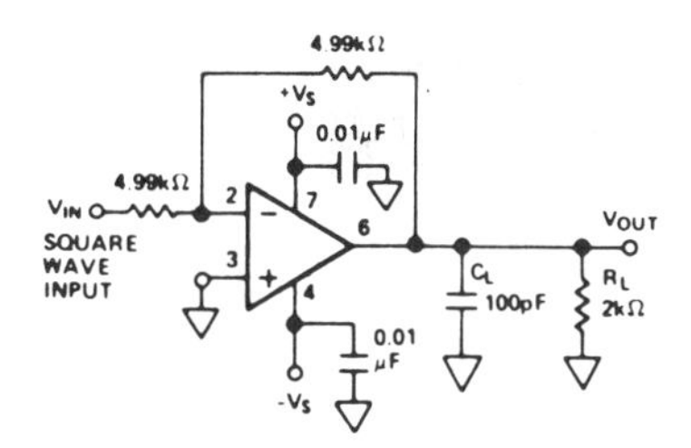

Unity Gain Inverter– AD542/AD547

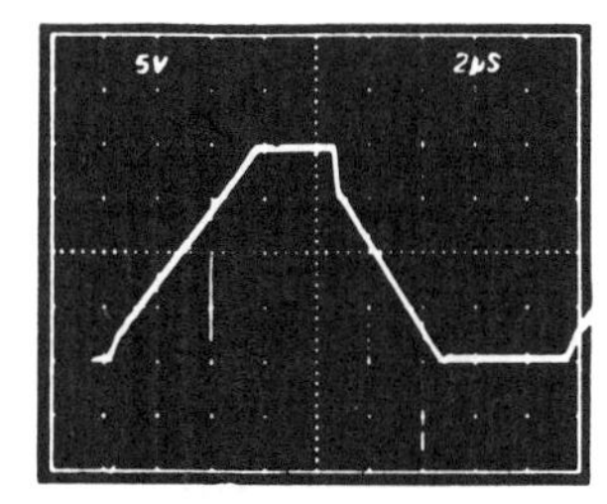

Unity Gain Inverter Pulse Response (Large Signal)

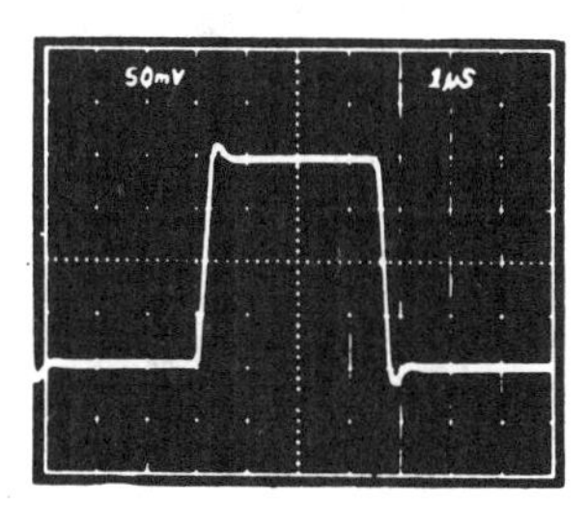

Unity Gain Inverter Pulse Response (Small Signal)

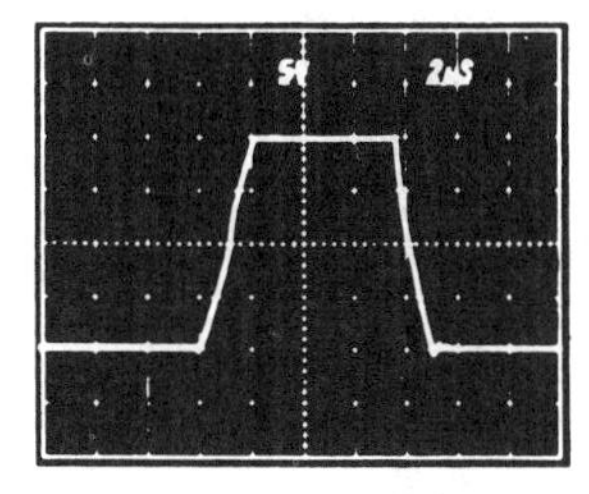

Unity Gain Follower Pulse Response (Large Signal)

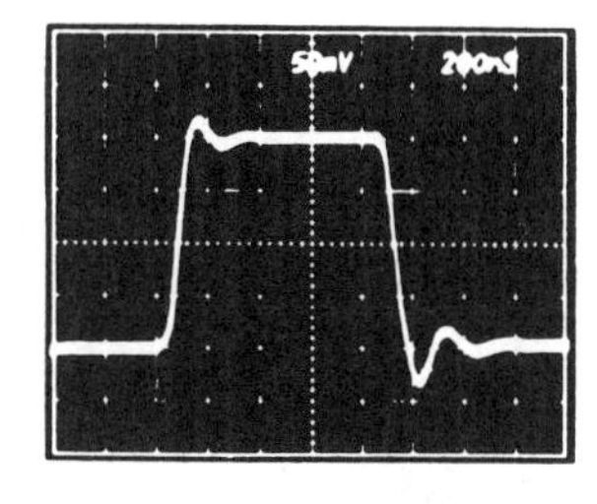

Unity Gain Follower Pulse Response (Small Signal)

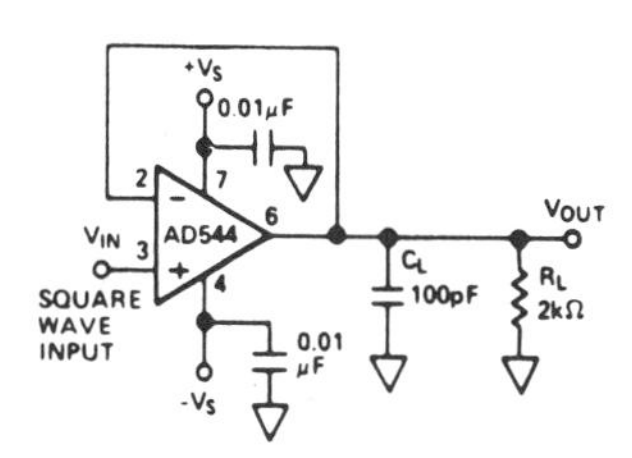

Unity Gain Follower

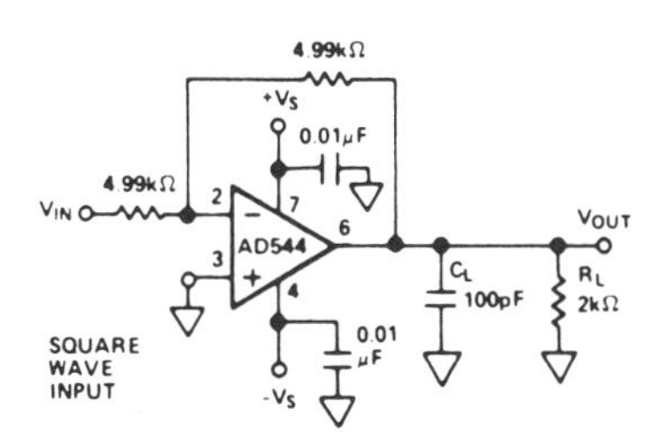

Unity Gain Inverter

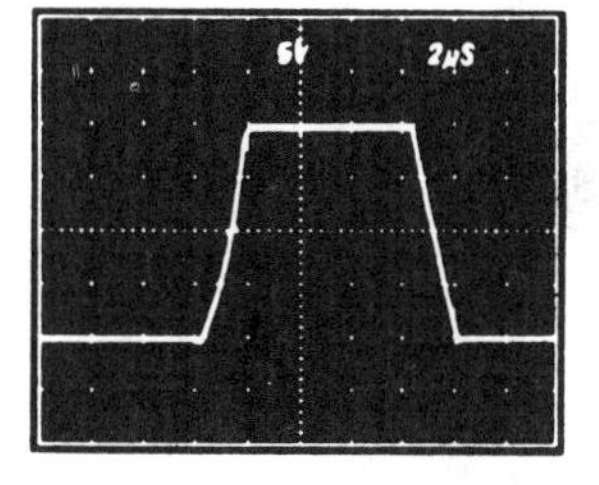

Unity Gain Inverter Pulse Response (Large Signal)

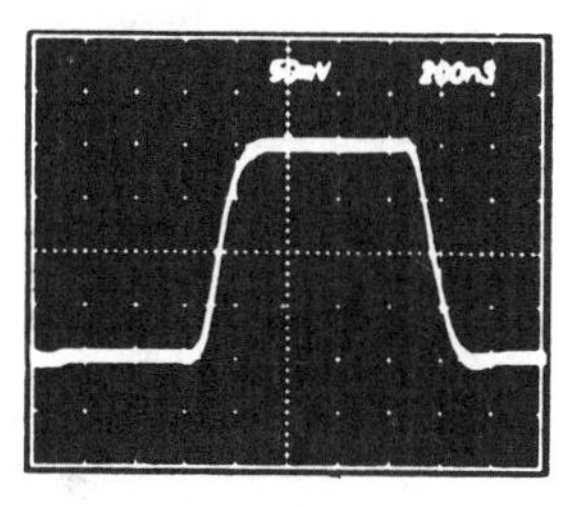

Unity Gain Inverter Pulse Response (Small Signal)

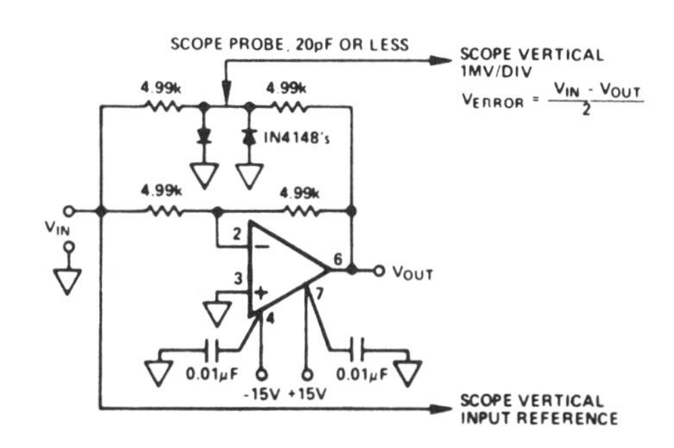

Settling Time Test Circuit

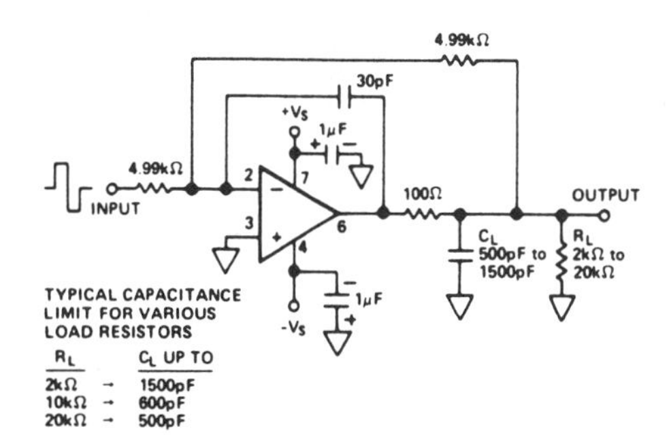

Circuit for Driving a Large Capacitance Load

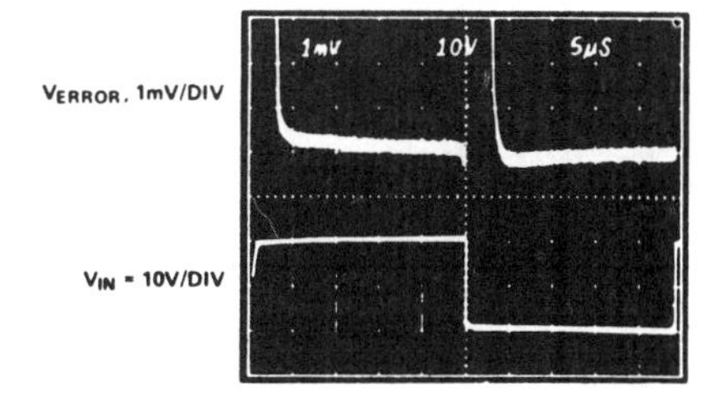

Settling Characteristic Detail – AD544

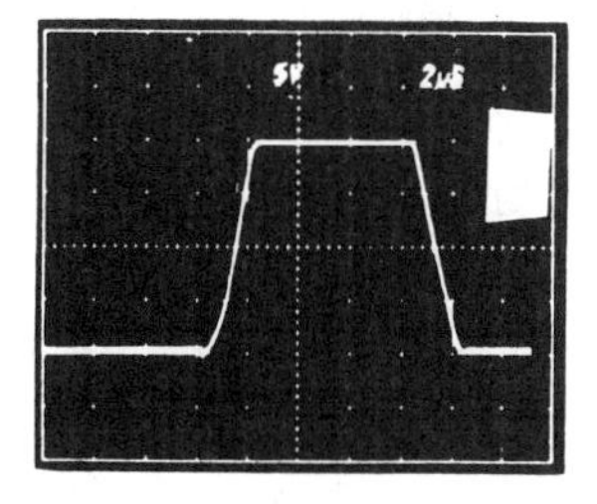

Transient Response $R_L = 2k\Omega$ $C_L = 500pF$–AD544

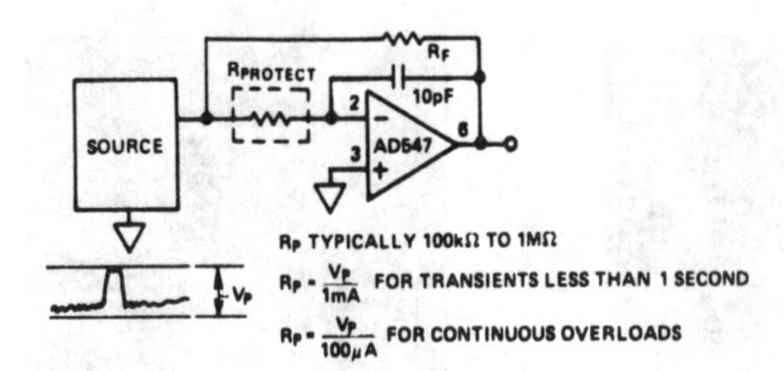

Input Protection

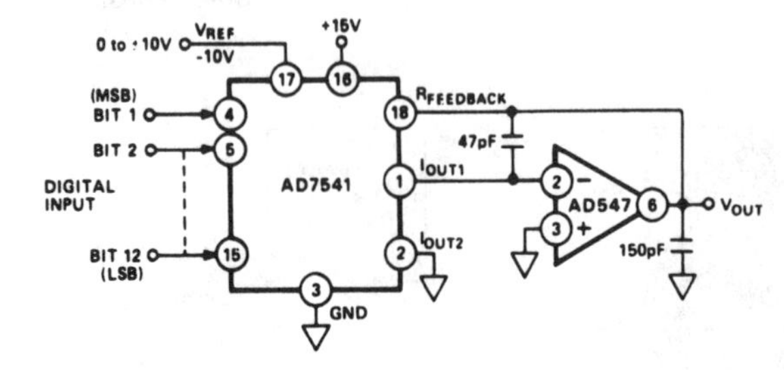

AD547 Used as DAC Output Amplifier

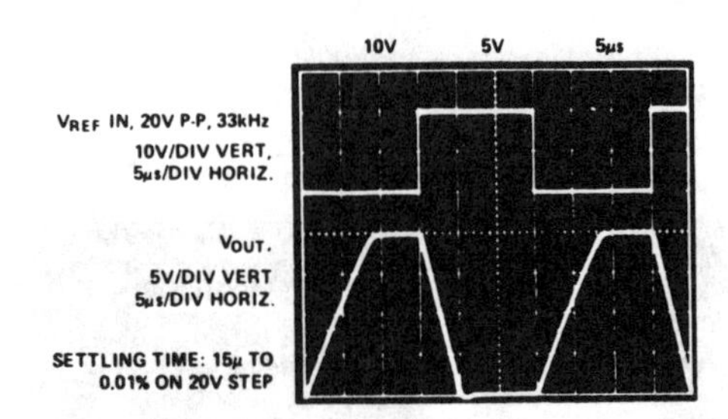

Voltage Output DAC Settling Characteristic

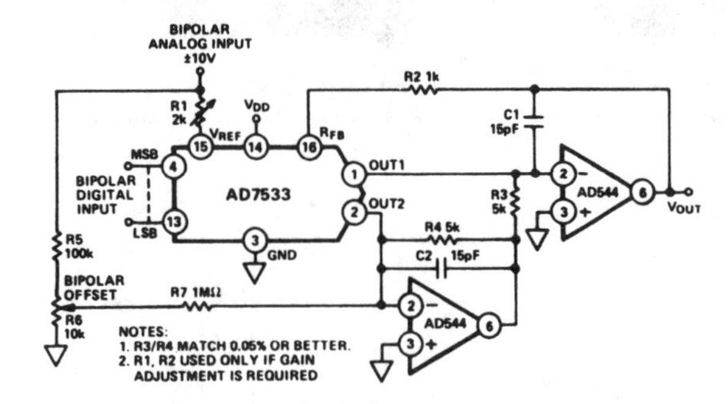

AD544 Used as DAC Output Amplifiers

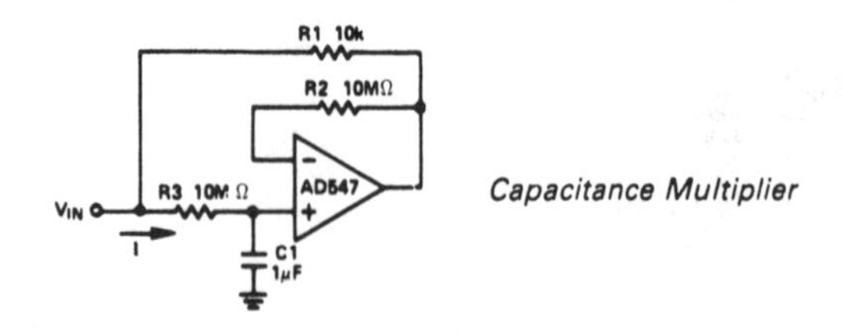

Capacitance Multiplier

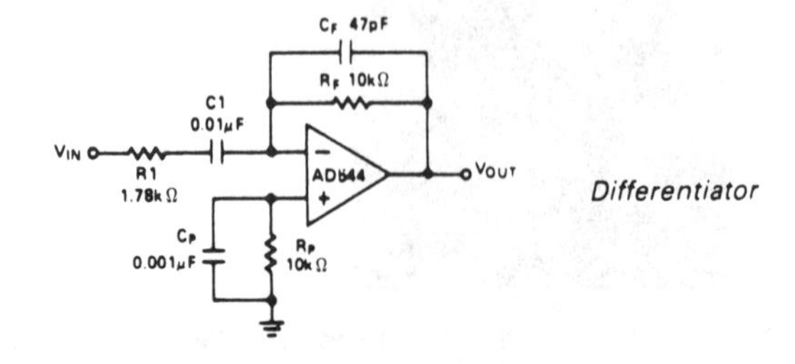

Differentiator

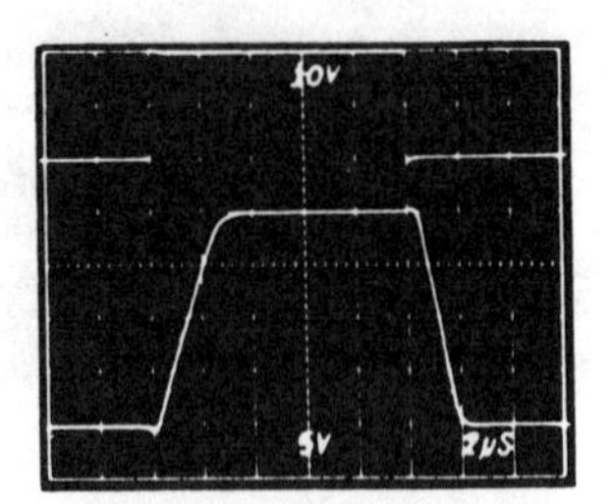

Large Signal Response

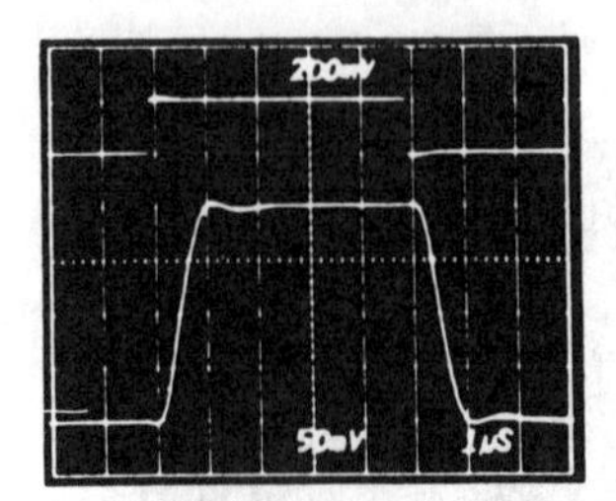

Small Signal Response

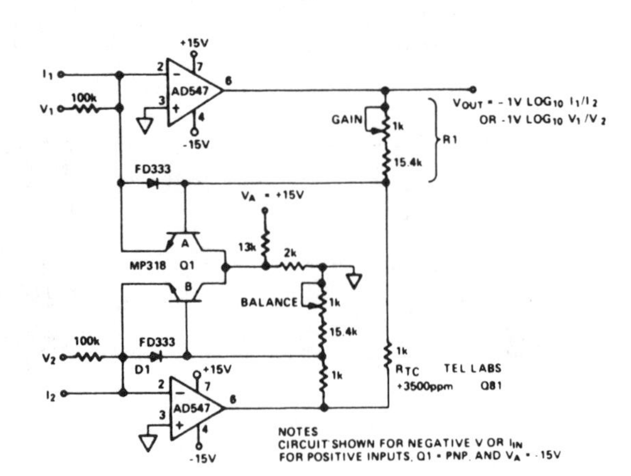

Log-Ratio Amplifier

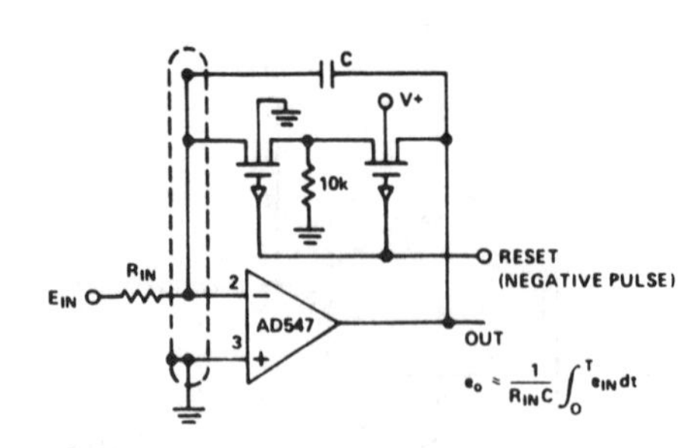

$$e_o = \frac{1}{R_{IN}C}\int_0^T e_{IN}\,dt$$

Low Drift Integrator and Low-Leakage Guarded Reset

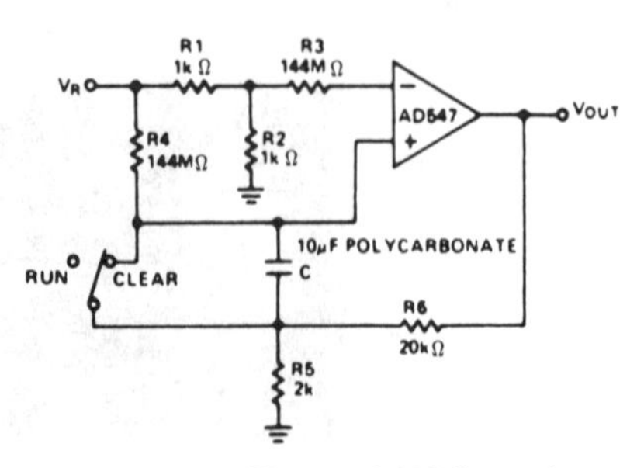

Long Interval Timer – 1,000 Seconds

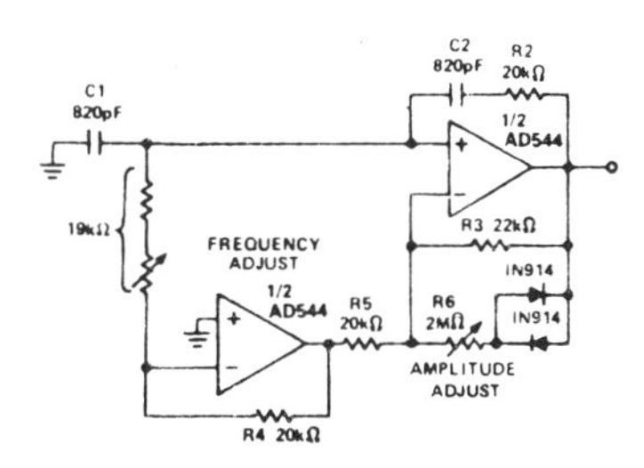

Wien-Bridge Oscillator – $f_o = 10kHz$

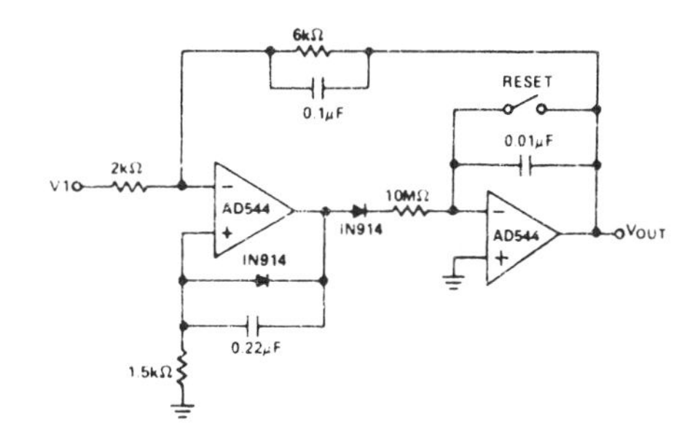

Positive Peak Detector

SELECTION GUIDE: D/A CONVERTERS

GENERAL PURPOSE D/A CONVERTERS

Model	Resolution Bits	Settling Time µs	MDAC	Current Out	Voltage Out	On-Chip Reference	No. of DACs	µP Interface	Double Buffered Inputs	Notes
AD9768	8	0.005		X		X				100MHz update rate
AD7524	8	0.1	X	X				X		
AD7528	8	0.2	X	X			2	X		
AD557	8	0.8			X	X		X		Low cost
AD7569	8	1			X	X		X		w/8-bit 2µs ADC
AD558	8	3			X	X		X		
AD7225	8	5			X		4	X	X	
AD7228	8	5			X		8	X		Common reference input
AD7224	8	7			X			X	X	
AD7226	8	7			X		4	X		Common reference input
AD7628	8	–	X	X			2	X		
AD561	10	0.25		X		X				
AD7533	10	0.6	X	X						
AD568	12	0.035		X		X				
HDS-1250	12	0.035		X		X				
AD668	12	0.05	X	X						
HDM-1210	12	0.085		X		X				
AD565A	12	0.25		X		X				
AD DAC80-I	12	0.30		X		X				
AD566A	12	0.35		X						
AD7541A	12	0.6	X	X						
AD7548	12	1	X	X				X	X	8-bit data bus
AD562	12	1.5		X						
AD563	12	1.5		X		X				
AD7537	12	1.5	X	X			2	X	X	8-bit data bus
AD7547	12	1.5	X	X			2	X		
AD7549	12	1.5	X	X			2	X	X	4-bit data bus
HDD-1206	12	2			X	X				w/deglitcher
AD DAC80-V	12	3			X	X				
AD392	12	4			X	X	4	X	X	w/readback function
AD667	12	4			X	X		X	X	4/8/16-bit bus compatible
AD767	12	4			X	X		X		
AD7845	12	5	X		X			X		
AD390	12	8			X	X	4	X	X	
AD664	12	10	X		X		4	X	X	w/reset and readback functions
AD7245	12	10			X	X		X	X	
AD7248	12	10			X	X		X	X	8-bit data bus
AD394/395	12	15	X		X		4	X		
AD7542	12	–	X	X				X	X	4-bit data bus
AD7543	12	–	X	X				X		Serial load
AD7545	12	–	X	X				X		
AD7545A	12	–	X	X				X		
AD7534	14	1.5	X	X				X	X	8-bit data bus
AD7535	14	1.5	X	X				X	X	8/16-bit bus compatible
AD7536	14	1.5	X	X				X	X	8/16-bit bus compatible
AD7538	14	1.5	X	X				X	X	
AD396	14	15.0	X		X		4	X	X	

SELECTION GUIDE
DIGITAL-TO-ANALOG CONVERTERS

VOLTAGE OUTPUT DACs

Model	Res Bits	Settling Time µs typ	Bus Interface Bits[1]	Reference Voltage Int/Ext (M)[2]	Package Options[3]	Temp Range[4]	Page	Comments
AD557	8	0.8	8, µP	Int	N, P	C	2–43	Lowest Cost 8-Bit DACPORT™. Single +5 V Supply
AD7569	8	1	8, µP	Int	E, N, P, Q	C, I, M	3–237	CMOS, Complete 8-Bit DAC/ADC/SHA/ Reference
AD558	8	3	8, µP	Int	D, E, N, P	C, M	2–47	10 V Out DACPORT. Single or Dual Supply
AD7224	8	7	8, µP	2–12.5 V, Ext	E, N, P, Q	C, I, M	2–189	CMOS, Low Cost 8-Bit DAC
HDD-1206	12	2	12	Int	M, W	C, M	2–459	Deglitched Voltage Output
*AD662	12	3	12, µp	2.56 V, Int	N, Q	C, I, M	2–95	Complete 12-Bit DACPORT™. Single +5 V Supply
AD DAC80-V	12	3	12	6.3 V, Int	D	C	2–411	Improved Industry Standard
AD DAC85-V	12	3	12	6.3 V, Int	D	I, M	2–411	Improved Industry Standard
AD DAC87-V	12	3	12	6.3 V, Int	D	I, M	2–411	Improved Industry Standard
AD667	12	3	4/8/12, µP	10 V, Int	D, E, N, P	C, I, M	2–123	Highest Accuracy Complete 12-Bit DAC
AD767	12	3	12, µP	10 V, Int	D, N	C, I, M	2–135	Fastest Interface Complete 12-Bit DAC
*AD7848	12	4	12, µP	3 V, Int	E, N, P, Q	C, I, M	2–371	CMOS, Complete 12-Bit DAC with DSP Interface
AD7845	12	5	12, µP	Ext (M)	E, N, P, Q	C, I, M	2–345	CMOS, 12-Bit Multiplying DAC with Output Amplifier
AD7245	12	10	12, µP	5 V, Int	E, N, P, Q	C, I, M	2–221	CMOS, 12-Bit Complete DAC, Parallel Load
AD7248	12	10	8, µP	5 V, Int	E, N, P, Q	C, I, M	2–221	CMOS, 12-Bit Complete DAC, Byte Load
*AD7840	14	4	14/Serial, µP	3 V, Int	E, N, P, Q	C, I, M	2–329	CMOS, 14-Bit Complete DAC, Parallel or Serial Load
*AD1856	16	1.5	Serial, µP	Int	N	C	2–161	16-Bit PCM Audio DAC
AD569	16	3	8/16, µP	±5 V, Ext (M)	D, N	I, M	2–83	Monolithic, 16-Bit Monotonic DAC
AD DAC71-V	16	5	16	6.3 V, Int	D, H	C	2–407	High Resolution 16-Bit DAC
AD DAC72-V	16	5	16	6.3 V, Int	D, H	C, I	2–407	High Resolution 16-Bit DAC
*AD7846	16	6	16, µP	Ext (M)	D, E, N, P	C, I, M	2–357	CMOS, 16-Bit Multiplying DAC with Readback Capability
AD1145	16	6	8/16/Serial, µP	3–6 V, Ext	G, PLLCC[4]	C	2–149	
DAC1136	16	8	16	6 V, Int	Module	I	2–453	High Resolution and Accuracy
AD1147	16	20	16, µP	10 V, Int	D	I	2–155	8-Bit Latched Input DAC for Offset and Gain Adjust
AD1148	16	20	16, µP	10 V, Int	D	I	2–155	Separate 8-Bit Bus for Offset and Gain Adjust DACs
*AD1860	18	1.5	Serial, µP	Int	N	C	2–171	18-Bit PCM Audio DAC
DAC1138	18	10	18	6 V, Int	Module	C	2–453	High Resolution and Accuracy
AD1139	18	40	8, µP	–10 V, Int	D	C	2–143	True 18-Bit Accuracy

DACPORT is a trademark of Analog Devices, Inc.

CURRENT OUTPUT DACS

Model	Res Bits	Settling Time µs typ	Bus Interface Bits[1]	Reference Volt Int/Ext (M)[2]	Package Options[3]	Temp Range[4]	Page	Comments
AD9768	8	0.005	8, µP	−1.26 V, Int	D, E	C, M	2–403	Ultrahigh Speed, ECL Compatible, 20 mA Output Current
AD7524	8	0.1	8, µP	Ext (M)	E, N, P, Q	C, I, M	2–235	CMOS, Low Cost, 8-Bit Multiplying DAC with Latch
AD561	10	0.25	10	Int	D, N	C, M	2–55	Industry Standard 10-Bit DAC, JAN Part Available
AD7533	10	0.6	10	Ext (M)	E, N, P, Q	C, I, M	2–245	CMOS, Low Cost 10-Bit Multiplying DAC
AD568	12	0.035	12	Int	Q	C, M	2–71	Highest Accuracy 12-Bit Ultrahigh Speed DAC
HDS-1250	12	0.035	12	Int	D, M	C, M	2–477	Ultrahigh Speed 12-Bit DAC
AD668	12	0.05	12	Ext (M)	Q	C, M	2–131	Multiplying 12-Bit Ultrahigh Speed DAC
HDM-1210	12	0.085	12	Int	D	I, M	2–471	Ultrahigh Speed 12-Bit Multiplying DAC
AD565A	12	0.25	12	10 V, Int	D	C, I, M	2–63	Industry Workhorse High Speed DAC. JAN Part Available
AD DAC80-I	12	0.3	12	6.3 V, Int	D	C	2–411	Industry Standard, High Speed DAC
AD DAC85-I	12	0.3	12	6.3 V, Int	D	I, M	2–411	Improved Industry Standard
AD DAC87-I	12	0.3	12	6.3 V, Int	D	I, M	2–411	Improved Industry Standard
AD566A	12	0.35	12	10 V, Ext	D	C, M	2–63	High Speed DAC
AD7541A	12	0.6	12	Ext (M)	E, N, P, Q	C, I, M	2–275	CMOS, 12-Bit Multiplying DAC
AD7548	12	1	8, µP	Ext (M)	E, N, P, Q	C, I, M	2–305	CMOS, Byte Load 12-Bit DAC, Specified with Single and Dual Supplies
AD562	12	1.5	12	Ext	D	C, I, M	2–59	Industry Standard, JAN Part Available
AD563	12	1.5	12	2.5 V, Int	D	C, M	2–59	Industry Standard
AD7542	12	2.0	4, µP	Ext (M)	D, E, N, P	C, I, M	2–281	CMOS, Nibble Load 12-Bit Multiplying DAC

CURRENT OUTPUT DACS

Model	Res Bits	Settling Time µs typ	Bus Interface Bits[1]	Reference Volt Int/Ext (M)[2]	Package Options[3]	Temp Range[4]	Page	Comments
AD7543	**12**	**2.0**	**Serial, µP**	**Ext (M)**	**D, E, N, P, Q**	**C, I, M**	**2–289**	**CMOS, Serial Load 12-Bit Multiplying DAC**
AD7545	12	2.0	12, µP	Ext (M)	E, N, P, Q	C, I, M	2–293	CMOS, Parallel Load 12-Bit Multiplying DAC
AD7545A	12	1.0	12, µP	Ext (M)	E, N, P, Q	C, I, M	2–297	CMOS, Improved AD7545
AD7534	**14**	**1.5**	**8, µP**	**Ext (M)**	**D, N, P**	**C, I, M**	**2–251**	**CMOS, Byte Load**
AD7535	**14**	**1.5**	**8/14, µP**	**Ext (M)**	**D, E, N, P**	**C, I, M**	**2–255**	**CMOS, Parallel or Byte Load**
AD7536	**14**	**1.5**	**8/14, µP**	**Ext (M)**	**D, E, N, P**	**C, I, M**	**2–259**	**CMOS, Parallel or Byte Load, Bipolar Output**
AD7538	**14**	**1.5**	**14, µP**	**Ext (M)**	**N, Q**	**C, I, M**	**2–267**	**CMOS, Parallel Load**
***AD1856**	**16**	**0.35**	**Serial, µP**	**Int**	**N**	**C**	**2–161**	**16-Bit PCM Audio DAC**
AD DAC71-I	16	1	16	6.3 V, Int	D, H	C	2–407	High Resolution 16-Bit DAC
AD DAC72-I	16	1	16	6.3 V, Int	D, H	C, I	2–407	High Resolution 16-Bit DAC
***AD1860**	**18**	**0.35**	**Serial, µP**	**Int**	**N**	**C**	**2–171**	**18-Bit PCM Audio DAC**

VIDEO DACS

Model	Res Bits	Update Rate MHz min	Settling Time ns max	Reference Voltage Int/Ext[1]	Package Options[2]	Temp Range[3]	Page	Comments
AD9702	4	125	5	Ext	D, W	I	2–391	RGB Output, TTL or ECL Interface
*ADV476	6	66, 50, 35			N	C	2–431	CMOS, Triple 6-Bit Color Palette RAM-DAC
*ADV471	6	80, 50, 35			P	C	2–441	CMOS, Triple 6-Bit Color Palette RAM-DAC
AD9703	8	300	6	Int	D, W	I, M	2–395	Synchronous Composite Functions, Designed for High Resolution Screens, 300 MHz Update Rate
AD9701	8	225	8	Int	E, Q	I, M	2–385	Low Power, Low Glitch Impulse, Synchronous Composite Functions, 250 MHz Update Rate
HDG-0805	8	150	9	Int	D, W	I, M	2–463	−5.2 V Power Supply
AD9700	8	100	12	Ext	D, W	I, M	2–379	Single −5.2 V Power Supply, On-Chip Reference
*ADV478	8	80, 50, 35			P	C	2–441	CMOS, Triple 8-Bit Color Palette RAM-DAC
*ADV453	8	66, 40			N, P	C	2–421	CMOS, Triple 8-Bit Color Palette RAM-DAC
HDG-0807	8	50	14	Int	D, W	I	2–467	TTL-Compatible Inputs
*AD9713	12	80	25	−1.2 V, Int	N, P	C	2–399	TTL Compatible Inputs, Low Glitch Energy
*AD9712	12	100	25	−1.2 V, Int	N, P	C	2–399	ECL Compatible Inputs, Low Glitch Energy

LOGDACS™

Model	Res dB	Full Scale Range dB	Accuracy dB	Package Options[2]	Temp Range[3]	Page	Comments
AD7111	0.375	88.5	0.17	E, N, Q	C, I, M	C2–183	Low Distortion

LOGDAC is a trademark of Analog Devices, Inc.

MULTIPLE DACS

Model	Res Bits	Out Mode V/I	Settling Time µs typ	Bus Interface Bits[4]	Reference Volt Int/Ext[1]	# DACs	Package Options[2]	Temp Range[3]	Page	Comments
*AD7669	8	V	1	8, µP	Int	2	N, P	C, I, M	3–237	CMOS, Complete 8-Bit Dual DAC/ADC/SHA/Reference
*AD7769	8	V	2.5	8, µP	Ext	2	N, P	C, I	3–329	CMOS, Complete 8-Bit Dual DAC/2-Channel ADC
AD7225	8	V	5	8, µP	2–12.5 V, Ext	4	E, N, P, Q	C, I, M	2–193	CMOS, Separate Reference for Each DAC
AD7228	8	V	5	8, µP	2–10 V, Ext	8	E, N, P, Q	C, I, M	2–205	CMOS, Specified with Single and Dual Supplies, Skinny 20-Pin Package
AD7226	8	V	7	8, µP	2–12.5 V, Ext	4	E, N, P, Q, R	C, I, M	2–199	CMOS, No User Trims, Specified with Single and Dual Supplies
AD392	12	V	4	12, µP	Int	4	M	C	2–21	Fast Bus Access Time (<40ns), Data Readback Capability
AD390	12	V	8	12, µP	+10 V, Int	4	D	C, M	2–13	Factory Trimmed Gain and Offset
*AD7237	12	V	10	8, µP	+5 V, Int	2	N, P, Q	C, I, M	2–213	CMOS, Complete 12-Bit Dual, Byte Load
*AD7247	12	V	10	12, µP	+5 V, Int	2	N, P, Q	C, I, M	2–213	CMOS, Complete 12-Bit Dual, Parallel Load
AD664	12	V	10	12, µP	±14.5 V, Ext (M)	4	D, E, N, P	C, I, M	2–103	Readback, Reset, Low Power Quad DAC

Model	Res Bits	Out Mode V/I	Settling Time µs typ	Bus Interface Bits[4]	Reference Volt Int/Ext[1]	# DACs	Package Options[2]	Temp Range[3]	Page	Comments
AD394	**12**	**V**	**15**	**12, µP**	**±11 V, Ext (M)**	**4**	**D**	**C, M**	**2–27**	**Four Independent Reference Inputs, Precision Amps for Bipolar Output**
AD395	**12**	**V**	**15**	**12, µP**	**±11 V, Ext (M)**	**4**	**D**	**C, M**	**2–27**	**Four Independent Reference Inputs, Precision Amps for Unipolar Output**
AD396	**14**	**V**	**15**	**8, µP**	**±11 V, Ext (M)**	**4**	**D**	**C, M**	**2–35**	**Four Independent Reference Inputs, Bipolar Output, Simultaneous Update**
AD7528	**8**	**I**	**0.2**	**8, µP**	**Ext (M)**	**2**	**E, N, P, Q, R**	**C, I, M**	**2–241**	**CMOS, +5 V to +15 V Operation, TTL Compatible at V_{DD} = 5 V**
AD7628	8	I	0.35	8, µP	Ext (M)	2	E, N, P, Q	C, I, M	2–325	CMOS, +12 V to +15 V Operation, TTL Compatible at V_{DD} = 12 V to 15 V
AD7537	12	I	1.5	8, µP	Ext (M)	2	E, N, P, Q	C, I, M	2–263	CMOS, Byte Load, Double Buffered
AD7547	12	I	1.5	12, µP	Ext (M)	2	E, N, P, Q	C, I, M	2–301	CMOS, Parallel Load
AD7549	12	I	1.5	4, µP	Ext (M)	2	D, E, N, P	C, I, M	2–317	CMOS, Nibble Load, Double Buffered

[1]Ext indicates external reference with the range of voltages listed where applicable. Ext (M) indicates external reference with multiplying capability. Int indicates reference is internal. A voltage value is given if the reference is pinned out.
[2]Package Options: D–Side-Brazed Dual-In-Line Ceramic; E–Leadless Chip Carrier; M–Metal Hermetic Dual-In-Line; N–Plastic Molded Dual-In-Line; P–Plastic Leaded Chip Carrier (PLCC); Q–Cerdip; R–Small Outline Plastic (SOIC); W–Ceramic/Glass Dual-In-Line, Non-Hermetic.
[3]Temperature Ranges: C–Commercial, 0 to +70°C; I–Industrial, −40°C to +85°C (Some older products −25°C to +85°C); M–Military, −55°C to +125°C.
[4]This column lists the data format for the bus with "µP" indicating microprocessor capability—i.e., for a 12-bit converter 8/12, µP indicates that the data can be formatted for an 8-bit bus or can be in parallel (12 bits) and is microprocessor compatible.
Boldface Type: Product recommended for new design.
*New product since the publication of the 1987/1988 Databooks.

Analog Devices AD390*

Quad 12-Bit Microprocessor-Compatible D/A Converter

- Four complete 12-bit DACs in one IC package
- Linearity error ±1/2LSB T_{min}—T_{max} (AD390K, T)
- Factory-trimmed gain and offset
- Buffered voltage output
- Monotonicity guaranteed over full temperature range
- Double-buffered data latches
- Includes reference and buffer
- Fast settling: 8µs max to ±1/2LSB

The AD390 contains four 12-bit high-speed voltage-output digital-to-analog converters in a compact 28-pin hybrid package. The design is based on a proprietary latched 12-bit DAC chip, which reduces chip count and provides high reliability. The AD390 is ideal for systems requiring digital control of many analog voltages, where board space is at a premium. Such applications include automatic test equipment, process controllers, and vectorscan displays.

The AD390 is laser-trimmed to ±1/2LSB max nonlinearity (AD390KD, TD) and absolute accuracy of ±0.05 percent of full scale. The high initial accuracy is made possible by the use of thin-film scaling resistors on the monolithic DAC chips. The internal-buried zener voltage reference provides excellent temperature-drift characteristics (20ppm/°C) and an initial tolerance of ±0.03% maximum. The internal reference buffer allows a single common reference to be used for multiple AD390 devices in large systems.

The individual DACs are accessed by the $\overline{CS1}$ through $\overline{CS4}$ control inputs and the $\overline{A0}$ and $\overline{A1}$ lines. These control signals permit the registers of the four DACs to be loaded sequentially and the outputs to be simultaneously updated.

The AD390 outputs are calibrated for a ±10V output range with positive-true offset binary input coding. A 0-to-+10V version is available on special order.

The AD390 is packaged in a 28-lead ceramic package and is specified for operation over the 0 to +70°C and −55°C to +125°C temperature range.

*Covered by patent numbers 3,803,590; 3,890,611; 3,932,863; 3,978,473, 4,020,486; other patents pending.

AD390 FUNCTIONAL BLOCK DIAGRAM

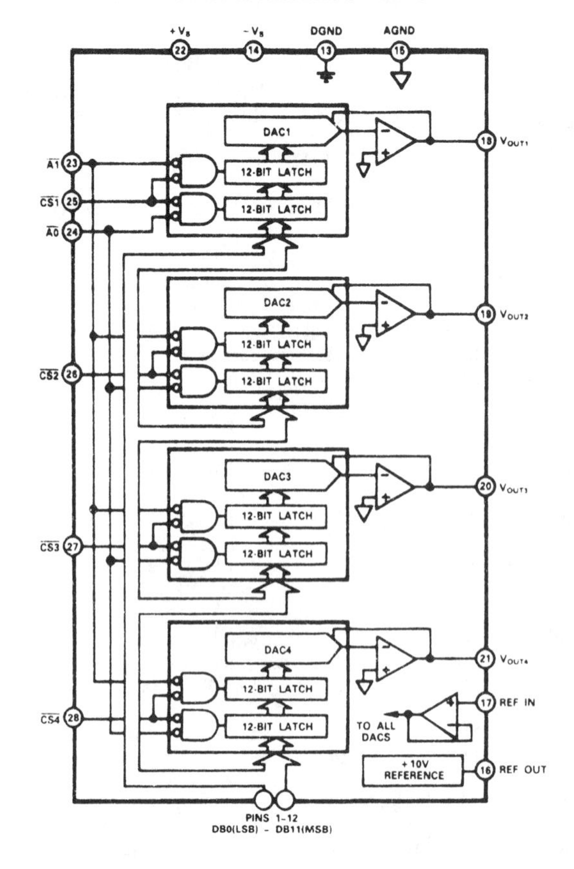

PIN CONFIGURATION TOP VIEW

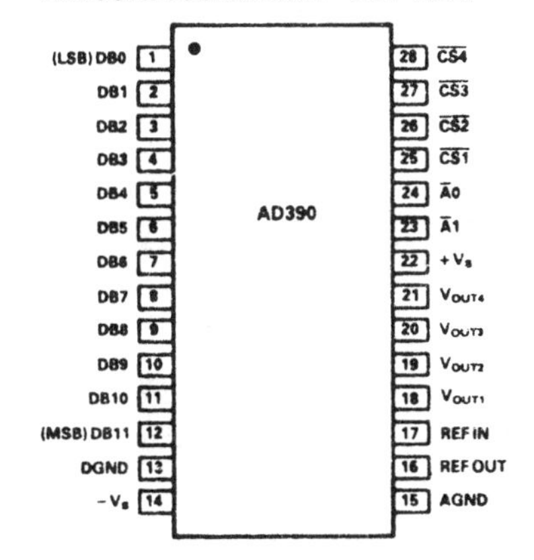

ABSOLUTE MAXIMUM RATINGS

$+V_S$ to DGND	0 to +18V
$-V_S$ to DGND	0 to −18V
Digital Inputs (Pins 1-12, 23-28) to DGND	−10 to +7V
Ref In to DGND	$\pm V_S$
AGND to DGND	±0.6V
Analog Outputs (Pins 16, 18-21)	
	Indefinite Short to AGND or DGND
	Momentary Short to $\pm V_S$

SPECIFICATIONS (T_A = +25°C, V_S = ±15V unless otherwise specified)

Model	AD390JD/SD Min	AD390JD/SD Typ	AD390JD/SD Max	AD390KD/TD Min	AD390KD/TD Typ	AD390KD/TD Max	Units
DATA INPUTS (Pins 1-12 and 23-28)[1]							
Except Pin 24 TTL or 5 Volt CMOS							
Input Voltage							
Bit ON (Logic "1")	+2.0		+5.5	+2.0		+5.5	V
Bit OFF (Logic "0")			+0.8			+0.8	V
Input Current (Pin 24 is 3 × Larger)							
Bit ON (Logic "1")		500	1200		500	1200	μA
Bit OFF (Logic "0")		150	400		150	400	μA
RESOLUTION			12			12	Bits
OUTPUT[2]							
Voltage Range[3]			±10			±10	V
Current			5			5	mA
Settling Time (to ±½LSB)		4	8		4	8	μs
ACCURACY							
Gain Error (w/ext. 10.000V reference)		±0.05	±0.1		±0.025	±0.05	% of FSR[4]
Offset		±0.025	±0.05		±0.012	±0.025	% of FSR
Linearity Error		±1/4	±3/4		±1/8	±1/2	LSB
Differential Linearity Error		±1/2	±3/4		±1/4	±1/2	LSB
TEMPERATURE DRIFT							
Gain (internal reference)			±40			±20	ppm/°C
(external reference)			±10			±5	ppm/°C
Zero			±10			±5	ppm/°C
Linearity Error $T_{min} - T_{max}$		±1/2	±3/4		±1/4	±1/2	LSB
Differential Linearity	MONOTONICITY GUARANTEED OVER FULL TEMPERATURE RANGE						
CROSS TALK[5]		0.1			0.1		LSB
REFERENCE OUTPUT							
Voltage (without load)	9.997	10.000	10.003	9.997	10.000	10.003	V
Current (available for external use)	2.5	3.5		2.5	3.5		mA
REFERENCE INPUT							
Input Resistance		10^{10}			10^{10}		Ω
Voltage Range	5		11	5		11	V
POWER REQUIREMENTS							
Voltage[6]	±13.5	±15	±16.5	±13.5	±15	±16.5	V
Current							
$+V_S$		12	20		12	20	mA
$-V_S$		−75	−90		−75	−90	mA
POWER SUPPLY GAIN SENSITIVITY							
$+V_S$		0.002	0.006		0.002	0.006	%FS/%
$-V_S$		0.0025	0.006		0.0025	0.006	%FS/%
TEMPERATURE RANGE							
Operating (Full Specifications) J, K	0		+70	0		+70	°C
S, T	−55		+125	−55		+125	°C
Storage	−65		+150	−65		+150	°C

NOTES
[1]Timing specifications appear in data sheets.
[2]The AD390 outputs are guaranteed stable for load capacitances up to 300pF.
[3]±10V range is standard. A 0 to 10V version is available on special order. Consult the factory.
[4]FSR means Full Scale Range and is equal to 20V for a ±10V range.
[5]Crosstalk is defined as the change in any one output as a result of any other output being driven from −10V to +10V into a 2kΩ load.
[6]The AD390 can be used with supply voltage as low as ±11.4V.

Specifications subject to change without notice.

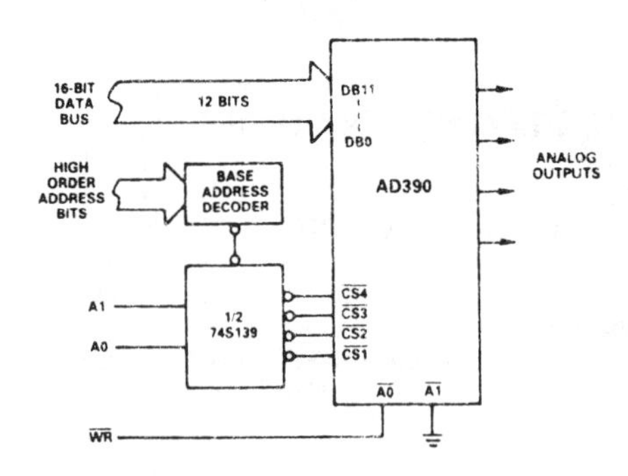

AD390–16-Bit Bus Interface

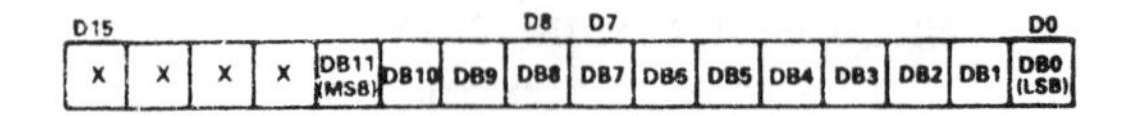

a. *Right-Justified Data* $(0 \leq D \leq 4095)$;

$$V_{OUT} = -10V + (4.883mV \times D)$$

D15							D8	D7							D0
DB11 (MSB)	DB10	DB9	DB8	DB7	DB6	DB5	DB4	DB3	DB2	DB1	DB0 (LSB)	X	X	X	X

b. *Left-Justified Data* $\left(0 \leq D \leq \frac{65520}{65536}\right)$;

$$V_{OUT} = -10V + (20V \times D)$$

12-Bit Data Formats for 16-Bit Bus

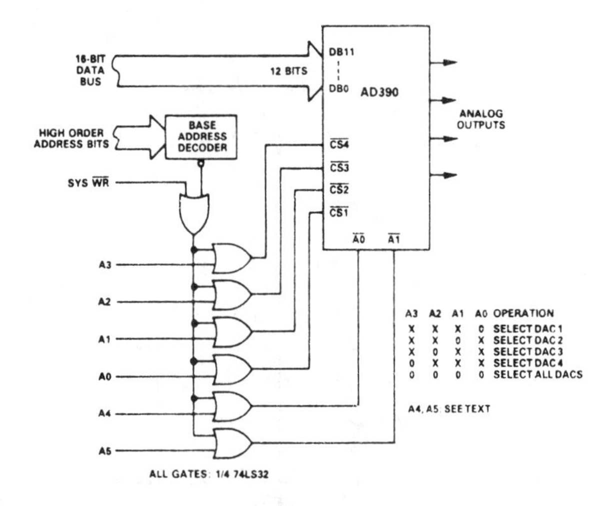

Alternate 16-Bit Bus Interface

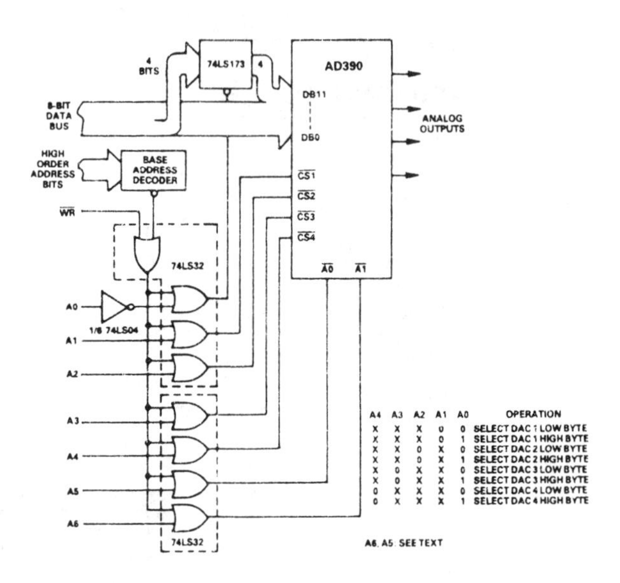

AD390–8-Bit Bus Interface Connections

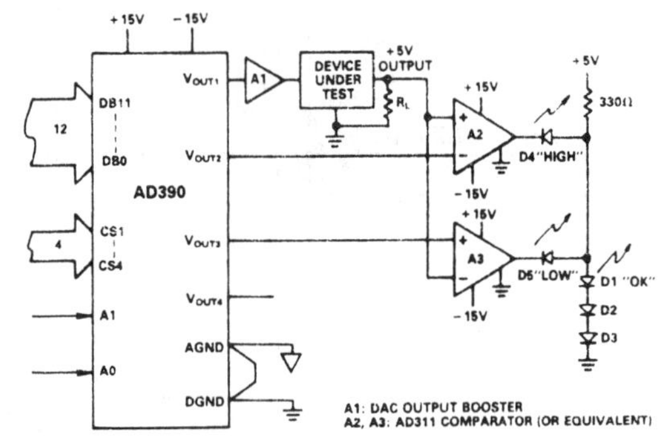

Programmable Window Comparator Used In Power Supply Testing

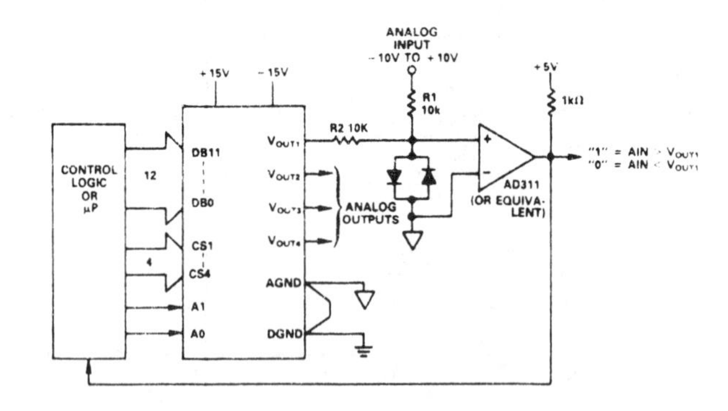

Using One AD390 Output for A/D Conversion

Analog Devices AD392

Complete Quad 12-Bit D/A Converter with Readback

- Data readback capability
- Four complete, voltage output, 12-bit DACs in one 32-pin hermetic package
- Fast bus access: 40ns max, T_{min}-T_{max}
- Asynchronous reset to zero volts
- Minimum of two TTL load drive (readback mode)
- Double-buffered data latches
- Monotonicity guaranteed T_{min}-T_{max}
- Linearity error $\pm 1/2$LSB
- Low digital-to-analog feedthrough, 2nV sec typ
- Factory-trimmed gain and offset
- Low cost

The AD392 is a quad 12-bit high-speed voltage output digital-to-analog converter with readback in a 32-pin hermetically sealed package. The design is based on a custom IC interface to complete 12-bit DAC chips, which reduces chip count and provides high reliability. The AD392 is ideal for systems requiring digital control of many analog voltages and for the monitoring of these analog voltages, especially where board space is a premium. Such applications include ATE, robotics, process controllers, and precision filters.

Featuring maximum access time of 40ns, the AD392 is capable of interfacing to the fastest of microprocessors. The readback capability provides a diagnostic check between the data sent from the microprocessor and the actual data received and transferred to the DAC. When $\overline{RESET}$ is low, all four DACs are simultaneously set to (bipolar) zero providing a known starting point.

The AD392 is laser-trimmed to $\pm 1/2$LSB integral linearity and ± 1LSB max differential linearity at ± 25 °C. Monotonicity is guaranteed over the full operating temperature range. The high initial accuracy and stability over temperature are made possible by the use of precision thin-film resistors.

The individual DAC registers are accessed by the address lines A0 and A1 and control lines $\overline{CS}$ and 2ND UP. These control signals permit the registers of the four DACs to be loaded sequentially and the outputs to be simultaneously updated.

The AD392 outputs are calibrated for a ± 10V output range with positive true offset binary input coding.

The AD392 is packaged in a 32-lead ceramic package and is hermetically sealed. The AD392 is specified for operation over the 0 to +70 °C temperature range.

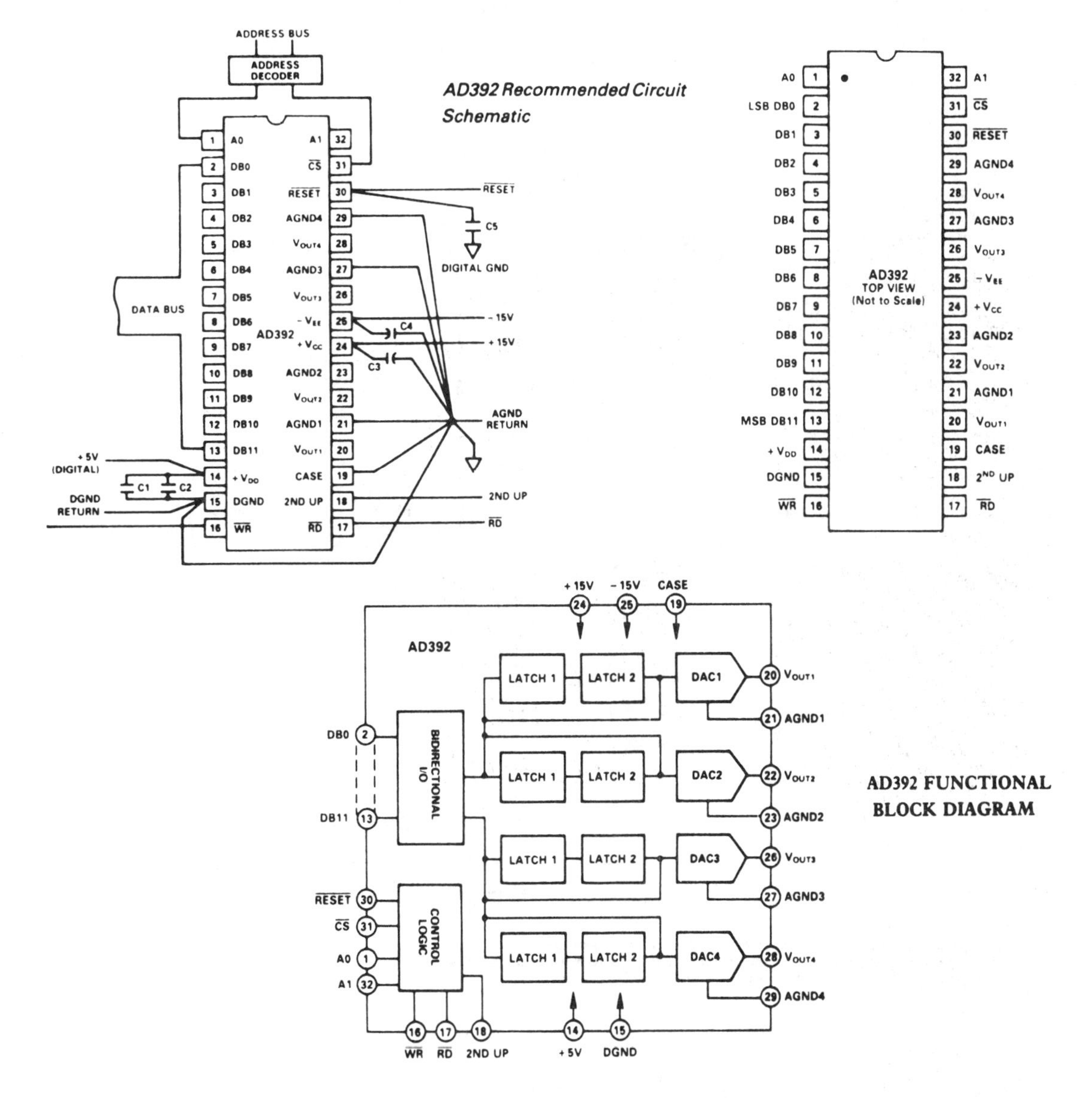

AD392 Recommended Circuit Schematic

AD392 FUNCTIONAL BLOCK DIAGRAM

ABSOLUTE MAXIMUM RATINGS*

$+V_{CC}$ to AGND (Any DAC)	0 to +18V
$-V_{EE}$ to AGND (Any DAC)	0 to −18V
$+V_{DD}$ to DGND	−0.3V to +7V
Digital Inputs to DGND (Pins 1-13, 16-18, 30-32)	−0.3V to +7V
Analog Outputs (Pins 20, 22, 26, 28) Short Circuit Duration ($+V_{CC}$, $-V_{EE}$ or AGND)	Indefinite
Storage Temperature	−65°C to +150°C

*Stresses above those listed under "Absolute Maximum Ratings" may cause permanent damage to the device. This is a stess rating only and functional operation at or above this specification is not implied. Exposure to above maximum rating conditions for extended periods may affect device reliability.

CAUTION:

ESD (Electro-Static-Discharge) sensitive device. The digital control inputs are diode protected; however, permanent damage may occur on unconnected devices subject to high energy electrostatic fields. Unused devices must be stored in conductive foam or shunts. The protective foam should be discharged to the destination socket before devices are removed.

SPECIFICATIONS (V_{CC} = +15V, V_{EE} = −15V, V_{DD} = +5V, T_A +25°C, unless otherwise specified)

Parameter	AD392 Min	Typ	Max	Units	Comments
DATA INPUTS (Pins 1-13, 16-18, 30-32)					
TTL Compatible					
Input Voltage					
Bit ON (Logic "1")	+2.0		$+V_{DD}$	V	V_{DD} = 5.25V
Bit OFF (Logic "0")	DGND		+0.8	V	V_{DD} = 4.75V
Input Current					
+25°C	−2		+2	μA	V_{IN} = V_{DD} or GND
T_{min} to T_{max}	−20		+20	μA	V_{IN} = V_{DD} or GND
RESOLUTION			12	Bits	
OUTPUT					
Bidirectional Outputs (Pins 2-13)					
Voltage Output Low (I_{OL} = +4.0mA)	0		+0.4	V	
Voltage Output High (I_{OH} = −4.0mA)	+2.4		V_{DD}	V	
Tristate Output Leakage					
T_{min} to T_{max}	−20		+20	μA	See Note 1
DAC Output Voltage Range		±10		V	
Current Range	−5		+5	mA	
Short Circuit Current			+40	mA	
STATIC ACCURACY					
Gain Error	−0.1	±0.05	+0.1	% of FSR	
Offset	−0.05	±0.025	+0.05	% of FSR	
Bipolar Zero		±0.025		% of FSR	
Integral Linearity Error	−0.5	±0.25	+0.5	LSB	
Differential Linearity Error	−1	±0.5	+1	LSB	
TEMPERATURE PERFORMANCE					
Gain Drift	−25	±20	+25	ppm FSR/°C	
Offset Drift	−25	±20	+25	ppm FSR/°C	
Integral Linearity Error					
T_{min} to T_{max}	−1		+1	LSB	
Differential Linearity Error	– Monotonicity Guaranteed Over Full Temperature Range –				
AC ANALOG PERFORMANCE					
Settling Time (to ± 1/2LSB)					
Change All Register Inputs					
From +5V to 0V/0V to +5V			4	μs	See Note 2
For LSB Change		1	2	μs	
Slew		10		V/μs	
Digital-to-Analog Glitch Impulse		2		nV sec	See Note 3
Crosstalk		0.1		LSB	See Note 4
POWER REQUIREMENTS					
$+V_{CC}$, $-V_{EE}$	±13.5		±16.5	V	
$+V_{DD}$	+4.5		+5.5	V	
Current (All Digital Inputs DGND or $+V_{DD}$ ONLY, No Load)					
I_{CC}		26	44	mA	
I_{EE}		62	82	mA	
I_{DD}		7.2	13	mA	
Power Dissipation		1356	1955	mW	See Note 5
POWER SUPPLY GAIN SENSITIVITY					
$+V_{CC}$, V_{DD}, $-V_{EE}$			0.002	%FS/%V_S	See Note 6
TEMPERATURE RANGE					
Operating (Full Specifications)	0		+70	°C	
Storage	−65		+150	°C	

NOTES
[1] V_{OUT} = V_{DD} or DGND.
[2] Referenced to trailing rising edge of $\overline{WR}$.
[3] Digital-to-Analog Glitch Impulse: This is a measure of the amount of charge injected from the digital inputs to the analog outputs when the inputs change state. Specified as the area of the glitch in nV secs.
[4] Crosstalk is defined as the change in any one output as a result of any other output being driven from −10V to +10V into a 2kΩ load.
[5] θ_{JC} approximately 10°C/W.
[6] $+V_{CC}$, $+V_{DD}$, $-V_{EE}$ are ± 10%.

Specifications subject to change without notice.

Analog Devices AD394/AD395

μP-Compatible Multiplying Quad 12-Bit D/A Converters

- Four complete 12-bit CMOS DACs with buffer registers
- Linearity error ±1/2LSB T_{min}-T_{max} (AD394, AD395K, T)
- Factory-trimmed gain and offset
- Precision output amplifiers for V_{OUT}
- Full four quadrant multiplication per DAC
- Monotonicity guaranteed over full temperature range
- Fast settling: 15μs max to ±1/2LSB
- Available to MIL-STD-883 (see ADI Military Catalog)

The AD394 and AD395 contain four 12-bit high-speed low-power voltage-output multiplying digital-to-analog converters in a compact 28-pin hybrid package. The design is based on a proprietary latched 12-bit CMOS DAC chip, which reduces chip count and provides high reliability. The AD394 and AD395 both are ideal for systems requiring digital control of many analog voltages where board space is at a premium and low power consumption a necessity. Such applications include automatic test equipment, process controllers, and vector stroke displays.

Both the AD394 and AD395 are laser-trimmed to ±1/2LSB max differential and integral linearity (AD394, AD395K, T) and full-scale accuracy of ±0.05 percent at 25°C. The high initial accuracy is made possible by the use of precision laser trimmed thin-film scaling resistors.

The individual DAC registers are accessed by the $\overline{CS1}$ through $\overline{CS4}$ control pins. These control signals allow any combination of the DAC select matrix to occur. Once selected, the DAC is loaded with a single 12-bit wide word. The 12-bit parallel digital input interfaces to most 12- and 16-bit bus systems.

The AD394 outputs (V_{REFIN} = +10V) provide a ±10V bipolar output range with positive-true offset binary input coding. The AD395 outputs (V_{REFIN} = −10V) provide a 0V to +10V unipolar output range with straight binary input coding.

Both the AD394 and the AD395 are packaged in a 28-lead metal package and are available for operation over the 0 to +70°C and −55°C to +125°C temperature ranges.

ABSOLUTE MAXIMUM RATINGS*

$+V_S$ to DGND	−0.3V to +17V
$-V_S$ to DGND	+0.3V to −17V
Digital Inputs (Pins 1-16) to DGND	−0.3V to +7V
V_{REFIN} to DGND	±25V
AGND to DGND	±0.6V
Analog Outputs (Pins 18, 21, 24, 27)	Indefinite Short to AGND or DGND Momentary Short to $\pm V_S$

*Stresses above those listed under "Absolute Maximum Ratings" may cause permanent damage to the device. This is a stress rating only and functional operation of the device at these or any other conditions above those indicated in the operational sections of this specification is not implied. Exposure to absolute maximum rating conditions for extended periods may affect device reliability.

CAUTION:

ESD (Electro-Static Discharge) sensitive device. The digital control inputs are diode protected; however, permanent damage may occur on unconnected devices subject to high energy electrostatic fields. Unused devices must be stored in conductive foam or shunts. The protective foam should be discharged to the destination socket before devices are removed.

SPECIFICATIONS (T_A = +25°C, V_{REFIN} = 10V, V_S = ±15V unless otherwise specified)

Model	AD394JD/SD[1] AD395JD/SD			AD394KD/TD[1] AD395KD/TD			
	MIN	TYP	MAX	MIN	TYP	MAX	UNITS
DATA INPUTS (Pins 1-16)[2]							
TTL or 5 Volt CMOS Compatible							
Input Voltage							
Bit ON (Logic "1")	+2.4		+5.5	+2.4		+5.5	V
Bit OFF (Logic "0")	0		+0.8	0		+0.8	V
Input Current		±4	±40		±4	±40	μA
RESOLUTION			12			12	Bits
OUTPUT							
Voltage Range[3]							
AD394		$\pm V_{REFIN}$			$\pm V_{REFIN}$		V
AD395		0V to $-(V_{REFIN})$			0V to $-(V_{REFIN})$		V
Current	5			5			mA
STATIC ACCURACY							
Gain Error		±0.05	±0.1		±0.025	±0.05	% of FSR[4]
Offset		±0.025	±0.05		±0.012	±0.025	% of FSR
Bipolar Zero (AD394)		±0.025			±0.012		% of FSR
Integral Linearity Error[5]		±1/4	±3/4		±1/8	±1/2	LSB
Differential Linearity Error		±1/2	±3/4		±1/4	±1/2	LSB
TEMPERATURE PERFORMANCE							
Gain Drift			±10			±5	ppm FSR/°C
Offset Drift			±10			±5	ppm FSR/°C
Integral Linearity Error[5]							
T_{min} to T_{max}		±1/2	±3/4		±1/4	±1/2	LSB
Differential Linearity Error	MONOTONICITY GUARANTEED OVER FULL TEMPERATURE RANGE						
REFERENCE INPUTS							
Input Resistance	5		25	5		25	kΩ
Voltage Range	−11		+11	−11		+11	V
DYNAMIC PERFORMANCE							
Settling Time (to ±1/2LSB)							
V_{REFIN} = +10V, Change All Digital Inputs from +5.0V to 0V		10	15		10	15	μs

SPECIFICATIONS (T_A = +25°C, V_{REFIN} = 10V, V_S = ±15V unless otherwise specified)

Model	AD394JD/SD[1] AD395JD/SD MIN	TYP	MAX	AD394KD/TD[1] AD395KD/TD MIN	TYP	MAX	UNITS
V_{REFIN} = 0 to 5V Step, All Digital Inputs = 0V		10	15		10	15	µs
Reference Feedthrough Error[6]							
AD395		5			5		mV p-p
AD394							
Digital-to-Analog Glitch Impulse[7]		250			250		nV sec
Crosstalk							
Digital Input (Static)[8]		0.1			0.1		LSB
Reference[9]		2.0			2.0		mV p-p
POWER REQUIREMENTS							
Supply Voltage[10]	±13.5		±16.5	±13.5		±16.5	V
Current (All Digital Inputs 0V or +5V)							
$+V_S$		20	22		20	22	mA
$-V_S$		18	28		18	28	mA
Power Dissipation		570	750		570	750	mW
POWER SUPPLY GAIN SENSITIVITY							
$+V_S$		0.002	0.006		0.002	0.006	%FS/%
$-V_S$		0.0025	0.006		0.0025	0.006	%FS/%
TEMPERATURE RANGE							
Operating (Full Specifications) J, K	0		+70	0		+70	°C
S, T	−55		+125	−55		+125	°C
Storage	−65		+150	−65		+150	°C

NOTES
[1]AD394 and AD395 S and T grades are available to MIL-STD-883, Method 5008, Class B. See Analog Devices Military Catalog (1987) for proper part number and detail specification.
[2]Timing specifications appear in data sheets.
[3]Code tables and graphs appear on Theory of Operation page.
[4]FSR means Full Scale Range and is equal to 20V for a ± 10V bipolar range and 10V for 0 to 10V unipolar range.
[5]Integral nonlinearity is a measure of the maximum deviation from a straight line passing though the endpoints of the DAC transfer function.
[6]For AD395 (unipolar), DAC register loaded with 0000 0000 0000, V_{REFIN} = 20V p-p, 10kHz sinewave. For AD394 (bipolar), V_{REFIN} = 20V p-p, 60 and 400Hz.
[7]This is a measure of the amount of charge injected from the digital inputs to the analog outputs when the inputs change state. It is usually specified as the area of the glitch in nVs and is measured with V_{REFIN} = AGND.
[8]Digital crosstalk is defined as the change in any one output's steady state value as a result of any other output being driven from V_{OUTMIN} to V_{OUTMAX} into a 2kΩ load by means of varying the digital input code.
[9]Reference crosstalk is defined as the change in any one output as a result of any other output being driven from V_{OUTMIN} to V_{OUTMAX} @10kHz into a 2kΩ load by means of varying the amplitude of the reference signal.
[10]The AD394 and the AD395 can be used with supply voltages as low as ± 11.4V.

Specifications subject to change without notice.

AD394/AD395 FUNCTIONAL BLOCK DIAGRAMS

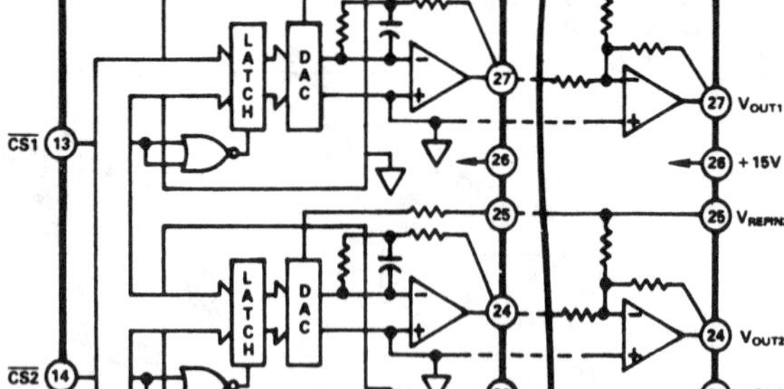
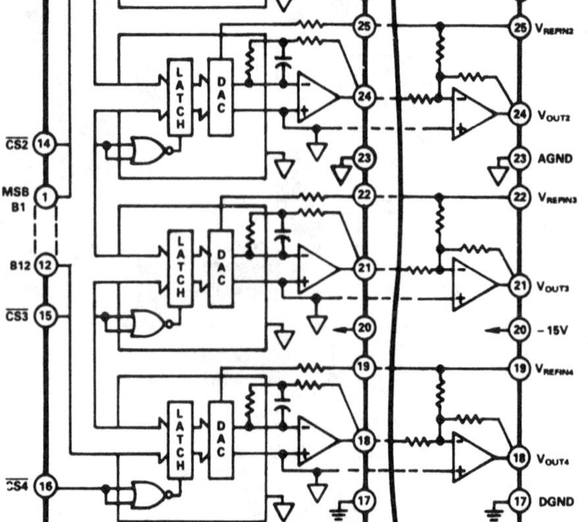

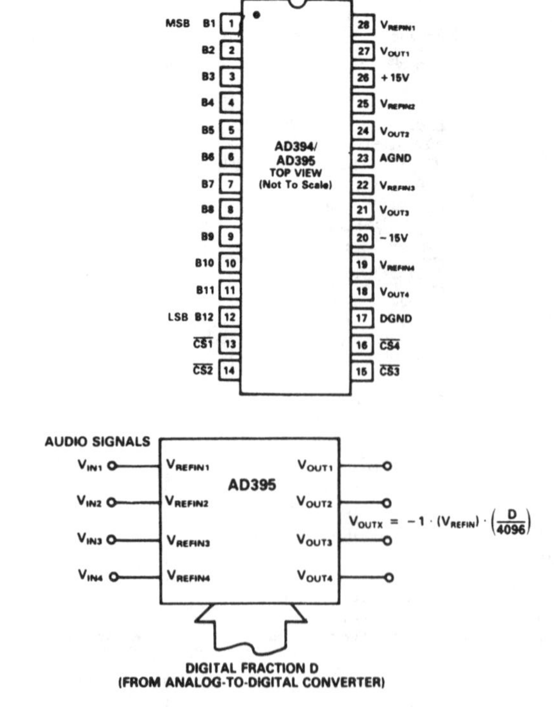

AD395 as a Multiplier or Attenuator

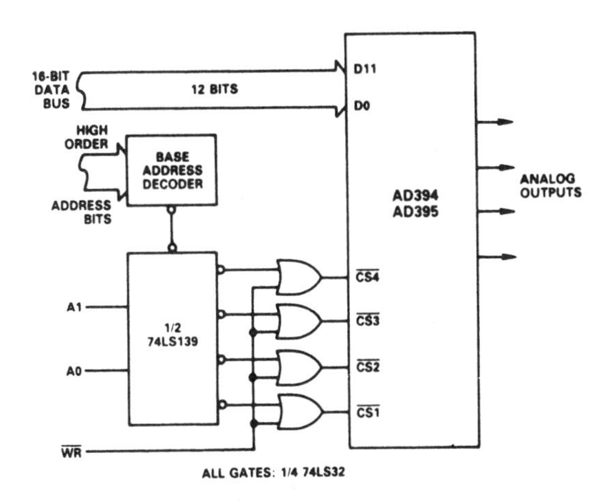

AD394, AD395 16-Bit Bus Interface

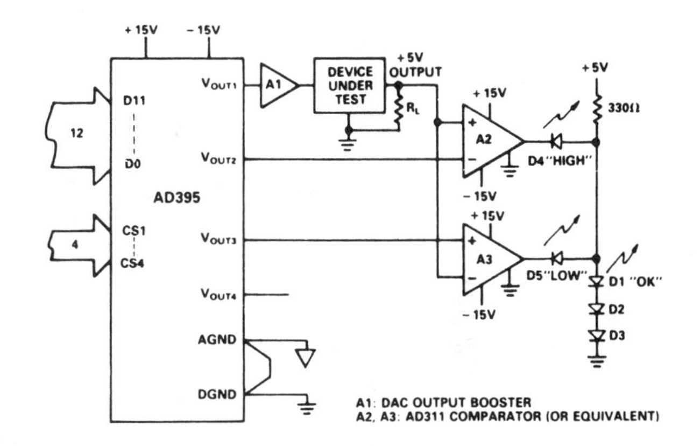

Programmable Window Comparator Used in Power-Supply Testing

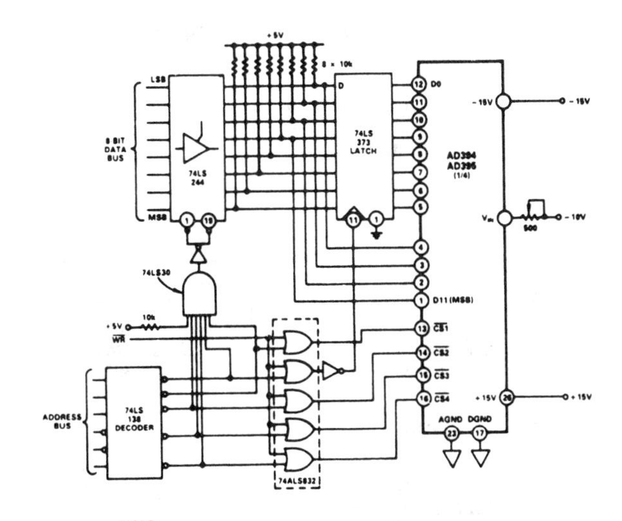

AD394, AD395 8-Bit Data Bus Interface

Analog Devices AD396

μP-Compatible Multiplying Quad 14-Bit D/A Converter

- Four pretrimmed 14-bit CMOS DACs
- Double buffered for simultaneous update
- Precision output amplifiers for voltage out
- Full four quadrant multiplication-independently pinned-out DAC reference
- Monotonicity guaranteed over full MIL temp. range
- Low power: 780mW max
- Small 28-lead hermetic double DIP package
- MIL-STD-883 processing available

The AD396 is a high-speed microprocessor compatible quad 14-bit digital-to-analog converter. The AD396 contains four 14-bit low-power multiplying digital-to-analog converters followed by precision voltage-output amplifiers all in a compact 28-pin hybrid package. The design is based on a proprietary latched 14-bit CMOS DAC chip, which reduces chip count and provides high reliability .

The AD396 (K, T) is laser-trimmed to ±1LSB max differential and integral linearity, and to full-scale accuracy of ±0.05 percent at 25 °C. The high initial accuracy is made possible by the use of precision laser trimmed thin-film scaling resistors.

The individual DAC registers are accessed by the $\overline{CS1}$ through $\overline{CS4}$ control pins. These control signals allow any combination of the DAC select matrix to occur. Once selected, the DAC is loaded with right-justified data in two bytes from an 8-bit data bus. Standard chip select and memory write logic is used to access the DACs. Address lines A0 and A1 control internal register loading and transfer.

The AD396 outputs (V_{REF} = +10V) provide a ±10V bipolar output range with positive-true offset binary input coding.

The AD396 is packaged in a 28-lead double DIP package and is available for operation over the 0 to +70 °C and −55 °C to +125 °C temperature ranges.

The AD396 is for systems requiring digital control of many analog voltages where board space is at a premium and low power consumption is a necessity. Such applications include automatic test equipment, process controllers, and vector stroke displays.

ABSOLUTE MAXIMUM RATINGS*

$+V_S$ to DGND	−0.3V to +17V
$-V_S$ to DGND	+0.3V to −17V
Digital Inputs (Pins 1-16) to DGND	−0.3V to +7V
V_{REFIN} to DGND	±25V
AGND to DGND	+0.3V to $+V_S$

Analog Outputs (Pins 18, 21, 24, 27)
Indefinite Short to AGND or DGND
Momentary Short to $\pm V_S$

*Stresses above those listed under "Absolute Maximum Ratings" may cause permanent damage to the device. This is a stress rating only and functional operation of the device at these or any other conditions above those indicated in the operational sections of this specification is not implied. Exposure to absolute maximum rating conditions for extended periods may affect device reliability.

CAUTION:

ESD (Electro-Static Discharge) sensitive device. The digital control inputs are diode protected; however, permanent damange may occur on unconnected devices subject to high energy electrostatic fields. Unused devices must be stored in conductive foam or shunts. The protective foam should be discharged to the destination socket before devices are removed.

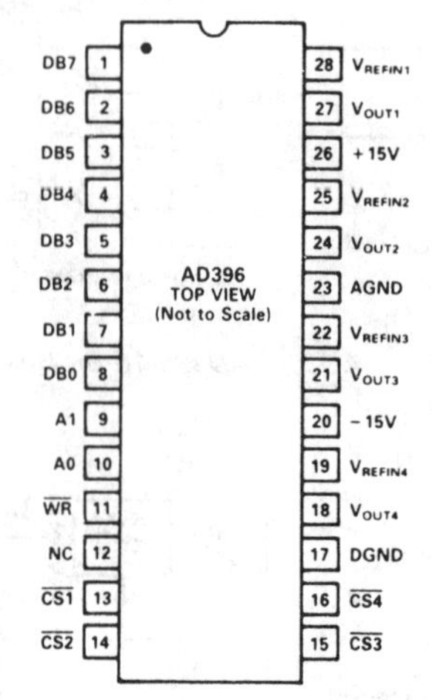

AD396 FUNCTIONAL BLOCK DIAGRAM

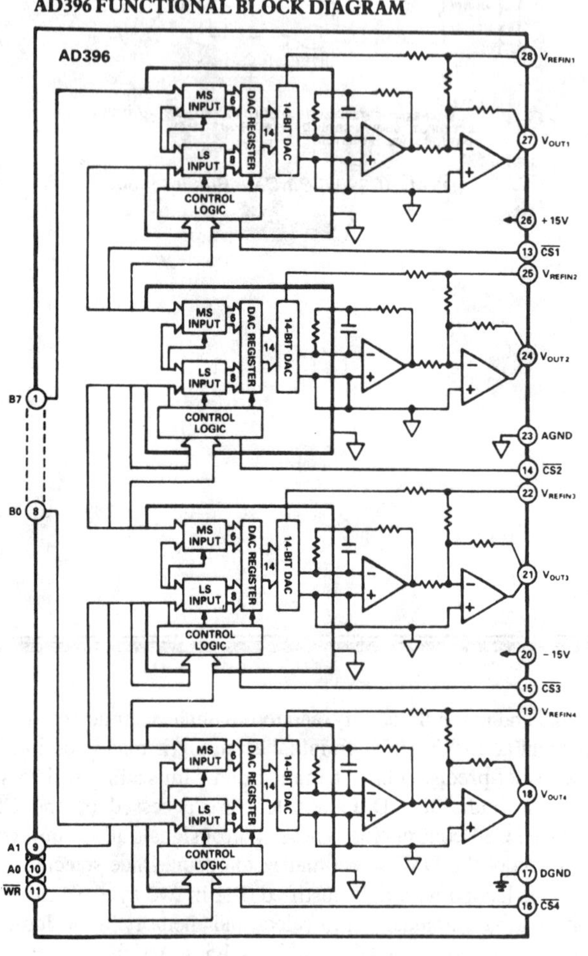

PIN	FUNCTION	DESCRIPTION
1	DB7	DATA BIT 7
2	DB6	DATA BIT 6
3	DB5	DATA BIT 5/DATA BIT 13 (DAC MSB)
4	DB4	DATA BIT 5/DATA BIT 12
5	DB3	DATA BIT 3/DATA BIT 11
6	DB2	DATA BIT 2/DATA BIT 10
7	DB1	DATA BIT 1/DATA BIT 9
8	DB0	DATA BIT 0/DATA BIT 8
9	A1	ADDRESS LINE 0
10	A0	ADDRESS LINE 1
11	$\overline{WR}$	WRITE INPUT. ACTIVE LOW
12	NC	NO CONNECTION
13	$\overline{CS1}$	CHIP SELECT DAC 1. ACTIVE LOW
14	$\overline{CS2}$	CHIP SELECT DAC 2. ACTIVE LOW
15	$\overline{CS3}$	CHIP SELECT DAC 3. ACTIVE LOW
16	$\overline{CS4}$	CHIP SELECT DAC 4. ACTIVE LOW
17	DGND	DIGITAL GROUND
18	V_{OUT4}	DAC 4 VOLTAGE OUTPUT
19	V_{REFIN4}	DAC 4 REFERENCE INPUT
20	−15V	−15V SUPPLY INPUT
21	V_{OUT3}	DAC 3 VOLTAGE OUTPUT
22	V_{REFIN3}	DAC 3 REFERENCE INPUT
23	AGND	ANALOG GROUND
24	V_{OUT2}	DAC 2 VOLTAGE OUTPUT
25	V_{REFIN2}	DAC 2 REFERENCE INPUT
26	+15V	+15V SUPPLY INPUT
27	V_{OUT1}	DAC 1 VOLTAGE OUTPUT
28	V_{REFIN1}	DAC 1 REFERENCE INPUT

SPECIFICATIONS ($T_A = +25°C$, $V_{REFIN} = 10V$, $V_S = \pm 15V$ unless otherwise specified)

Model	AD396JD/SD[1]			AD396KD/TD[1]			
	MIN	TYP	MAX	MIN	TYP	MAX	UNITS
DATA INPUTS (Pins 1-16)[2]							
TTL or 5 Volt CMOS Compatible							
Input Voltage							
Bit ON (Logic "1")	+2.4		+5.5	+2.4		+5.5	V
Bit OFF (Logic "0")	0		+0.8	0		+0.8	V
Input Current		±4	±40		±4	±40	μA
RESOLUTION			14			14	Bits
OUTPUT							
Voltage Range[3]		$\pm V_{REFIN}$			$\pm V_{REFIN}$		V
Current	5			5			mA
STATIC ACCURACY							
Gain Error		±0.05	±0.1		±0.025	±0.05	% of FSR[4]
Offset		±0.025	±0.05		±0.012	±0.025	% of FSR
Bipolar Zero		±0.025			±0.012		% of FSR
Integral Linearity Error[5]		±1	±2		±1/2	±1	LSB
Differential Linearity Error		±1/2	±1		±1/2	±1	LSB

SPECIFICATIONS (T_A = +25°C, V_{REFIN} = 10V, V_S = ±15V unless otherwise specified)

Model	AD396JD/SD[1] MIN	TYP	MAX	AD396KD/TD[1] MIN	TYP	MAX	UNITS
TEMPERATURE PERFORMANCE							
Gain Drift			±10			±5	ppm FSR/°C
Offset Drift			±10			±5	ppm FSR/°C
Integral Linearity Error[5]							
0 to +70°C		±1	±2		±1/2	±1	LSB
−55°C to +125°C		±2	±4		±1	±2	LSB
Differential Linearity Error	MONOTONICITY GUARANTEED OVER FULL TEMPERATURE RANGE						
REFERENCE INPUTS							
Input Resistance	5		25	5		25	kΩ
Voltage Range	−11		+11	−11		+11	V
DYNAMIC PERFORMANCE							
Settling Time (to ±1/2LSB)							
V_{REFIN} = +10V, Change All Digital Inputs from +5.0V to 0V		10	15		10	15	μs
V_{REFIN} = 0 to 5V Step, All Digital Inputs = 0V		10	15		10	15	μs
Reference Feedthrough Error[6]		5			5		mV p-p
Digital-to-Analog Glitch Impulse[7]		250			250		nV sec
Crosstalk							
Digital Input (Static)[8]		1/2			1/2		LSB
Reference[9]		4.0			4.0		mV p-p
POWER REQUIREMENTS							
Supply Voltage[10]	±13.5		±16.5	±13.5		±16.5	V
Current (All Digital Inputs 0V or +5V)							
$+V_S$		20	22		20	22	mA
$-V_S$		18	28		18	28	mA
Power Dissipation		570	780		570	780	mW
POWER SUPPLY GAIN SENSITIVITY							
$+V_S$		0.002	0.006		0.002	0.006	%FS/%
$-V_S$		0.0025	0.006		0.0025	0.006	%FS/%
TEMPERATURE RANGE							
Operating (Full Specifications) J, K	0		+70	0		+70	°C
S, T	−55		+125	−55		+125	°C
Storage	−65		+150	−65		+150	°C

NOTES

[1]AD396S and T grades are available to MIL-STD-883, Method 5008, Class B.

[2]Timing specifications appear in data sheets.

[3]Code tables and graphs appear in data sheets.

[4]FSR means Full Scale Range and is equal to 20V for a ±10V bipolar range.

[5]Integral nonlinearity is a measure of the maximum deviation from a straight line passing through the endpoints of the DAC transfer function.

[6]For AD396 (bipolar), DAC register loaded with 00 0000 0000 0000, V_{REFIN} = 20V p-p, 60 and 400Hz.

[7]This is a measure of the amount of charge injected from the digital inputs to the analog outputs when the inputs change state. It is usually specified as the area of the glitch in nVs and is measured with V_{RFIN} = AGND.

[8]Digital crosstalk is defined as the change in any one output's steady state value as a result of any other output being driven from V_{OUTMIN} to V_{OUTMAX} in a 2kΩ load by means of varying the digital input code.

[9]Reference crosstalk is defined as the change in any one output as a result of any other output being driven from V_{OUTMIN} to V_{OUTMAX} @ 10kHz into a 2kΩ load by means of varying the amplitude of the reference signal.

[10]The AD396 can be used with supply voltages as low as ±11.4V.

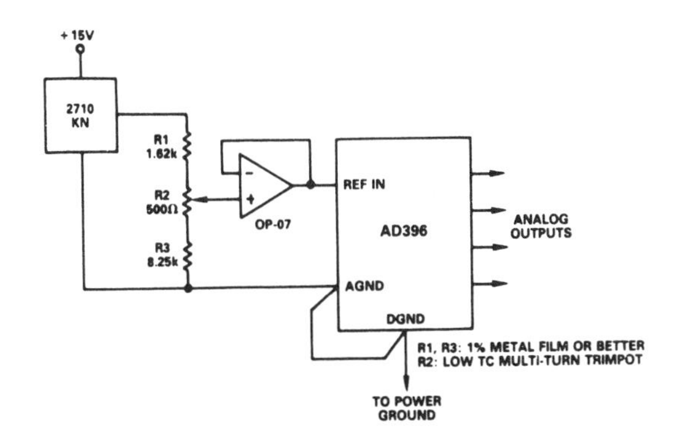

Connections for ±8.192V Full Scale (Recommended for ±12V Power Supplies)

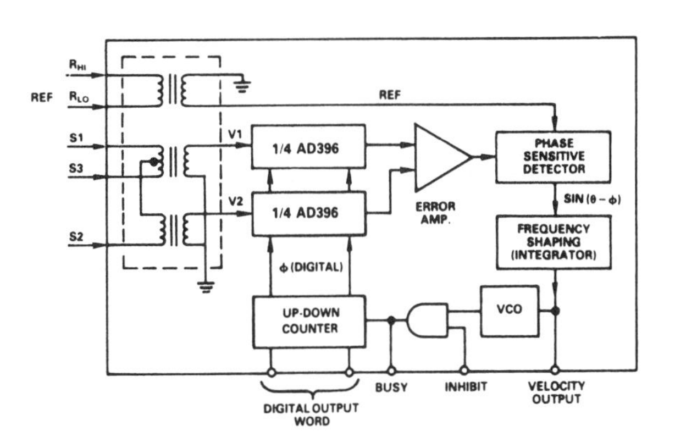

A Tracking Synchro-to-Digital Converter

Analog Devices AD557

DACPORT* Low-Cost Complete μP-Compatible 8-Bit DAC

- Complete 8-bit DAC
- Voltage output—0 to 2.56V
- Internal precision band-gap reference
- Single-supply operation: +5V (±10%)
- Full microprocessor interface
- Fast: 1μs voltage settling to ±1/2LSB
- Low power: 75mW
- No user trims required
- Guaranteed monotonic over temperature
- All errors specified T_{min} to T_{max}
- Small 16-pin DIP or 20-pin PLCC package
- Low cost

The AD557 DACPORT™ is a complete voltage-output 8-bit digital-to-analog converter, including output amplifier, full microprocessor interface and precision voltage reference on a single monolithic chip. No external components or trims are required to interface, with full accuracy, an 8-bit data bus to an analog system.

The low cost and versatility of the AD557 DACPORT are the result of continued development in monolithic bipolar technologies.

The complete microprocessor interface and control logic is implemented with integrated injection logic (I²L), an extremely dense and low-power logic structure that is process-compatible with linear bipolar fabrication. The internal precision voltage reference is the patented low-voltage band-gap circuit, which permits full-accuracy performance on a single +5V power supply. Thin-film silicon-chromium resistors provide the stability required for guaranteed monotonic operation over the entire operating temperature range, while laser-wafer trimming of these thin-film resistors permits absolute calibration at the factory to within ±2.5LSB; thus, no user-trims for gain or offset are required. A new circuit design provides voltage settling to ±1/2LSB for a full-scale step in 800ns.

The AD557 is available in two package configurations. The AD557JN is packaged in a 16-pin plastic, 0.3″-wide DIP. For surface-mount applications, the AD557JP is packaged in a 20-pin JEDEC standard PLCC. Both versions are specified over the operating temperature range of 0 to +70°C.

*DACPORT is a trademark of Analog Devices, Inc. Covered by U.S. Patent Nos. 3,887,863; 3,685,045; 4,323,795; other patents pending.

AD557 FUNCTIONAL BLOCK DIAGRAM

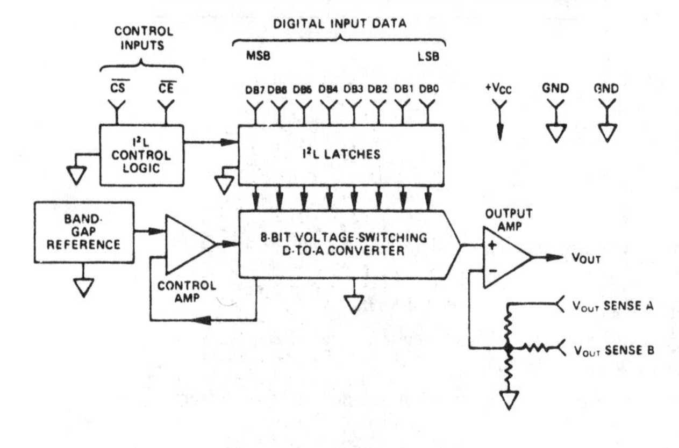

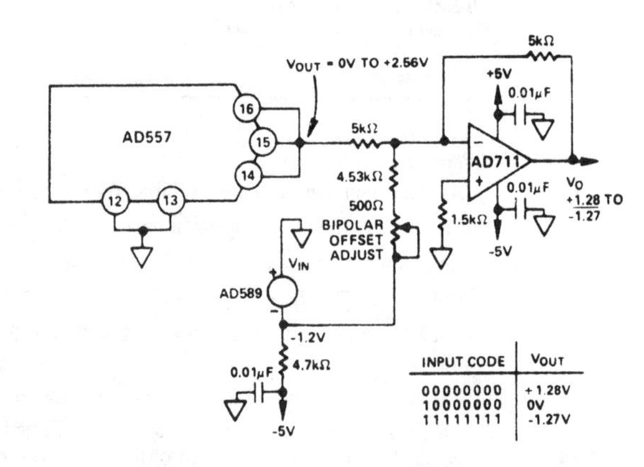

Bipolar Operation of AD557 from ±5V Supplies

ABSOLUTE MAXIMUM RATINGS*

V_{CC} to Ground	0V to +18V
Digital Inputs (Pins 1-10)	0 to +7.0V
V_{OUT}	Indefinite Short to Ground Momentary Short to V_{CC}
Power Dissipation	450mW
Storage Temperature Range	
N/P (Plastic) Packages	−25°C to +100°C
Lead Temperature (soldering, 10 sec)	300°C
Thermal Resistance	
Junction to Ambient/Junction to Case	
N/P (Plastic) Packages	140/55°C/W

*Stresses above those listed under "Absolute Maximum Ratings" may cause permanent damage to the device. This is a stress rating only and functional operation of the device at these or any other conditions above those indicated in the operational sections of this specification is not implied. Exposure to absolute maximum rating conditions for extended periods may affect device reliability.

SPECIFICATIONS (@ T_A = +25°C, V_{CC} = +5V unless otherwise specified)

Model	AD557J Min	Typ	Max	Units
RESOLUTION			8	Bits
RELATIVE ACCURACY[1]				
0 to +70°C		±1/2	1	LSB
OUTPUT				
Ranges		0 to +2.56		V
Current Source	+5			mA
Sink		Internal Passive Pull-Down to Ground[2]		
OUTPUT SETTLING TIME[3]		0.8	1.5	μs
FULL SCALE ACCURACY[4]				
@25°C		±1.5	±2.5	LSB
T_{min} to T_{max}		±2.5	±4.0	LSB
ZERO ERROR				
@25°C			±1	LSB
T_{min} to T_{max}			±3	LSB
MONOTONICITY[5]				
T_{min} to T_{max}		**Guaranteed**		
DIGITAL INPUTS				
T_{min} to T_{max}				
Input Current			**±100**	μA
Data Inputs, Voltage				
Bit On – Logic "1"	2.0			V
Bit On – Logic "0"	0		0.8	V
Control Inputs, Voltage				
On – Logic "1"	2.0			V
On – Logic "0"	0		0.8	V
Input Capacitance		4		pF
TIMING				
t_W Strobe Pulse Width	225			ns
T_{min} to T_{max}	**300**			ns
t_{DH} Data Hold Time	10			ns
T_{min} to T_{max}	**10**			ns
t_{DS} Data Setup Time	225			ns
T_{min} to T_{max}	**300**			ns
POWER SUPPLY				
Operating Voltage Range (V_{CC})				
2.56 Volt Range	**+4.5**		**+5.5**	V
Current (I_{CC})		15	**25**	mA
Rejection Ratio			**0.03**	%/%
POWER DISSIPATION, V_{CC} = 5V		75	**125**	mW
OPERATING TEMPERATURE RANGE	0		+70	°C

NOTES
[1]Relative Accuracy is defined as the deviation of the code transition points from the ideal transfer point on a straight line from the zero to the full scale of the device.
[2]Passive pull-down resistance is 2kΩ.
[3]Settling time is specified for a positive-going full-scale step to ±1/2LSB. Negative-going steps to zero are slower, but can be improved with an external pull-down.
[4]The full-scale output voltage is 2.55V and is guaranteed with a +5V supply.
[5]A monotonic converter has a maximum differential linearity error of ±1LSB.
Specifications shown in **boldface** are tested on all production units at final electrical test.
Specifications subject to change without notice.

DIP

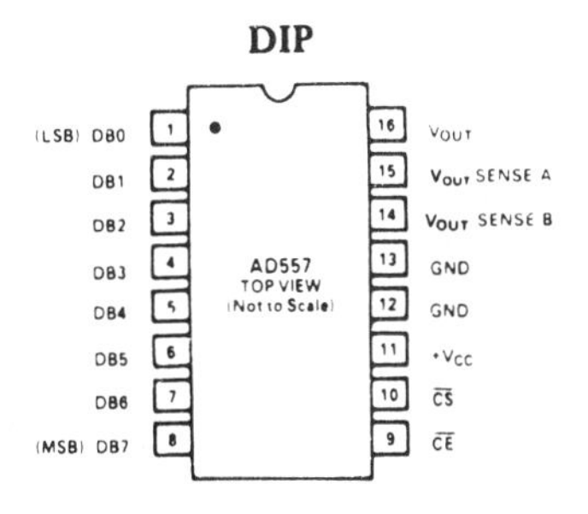

PLCC

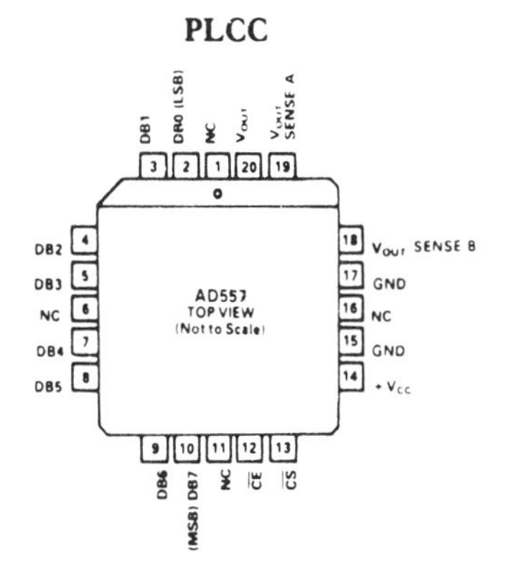

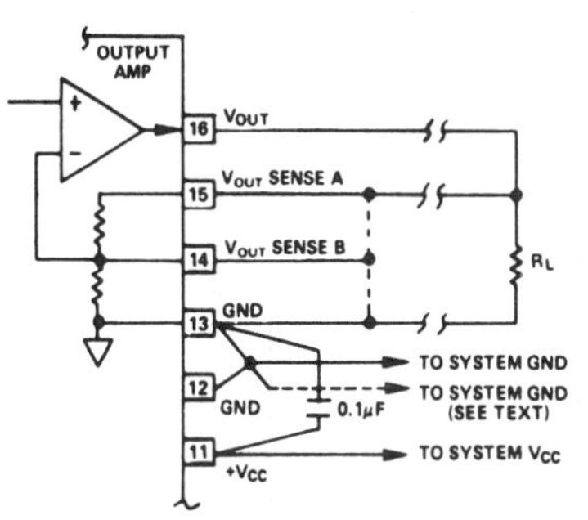

Recommended Grounding and Bypassing

Analog Devices AD558*
DACPORT Low-Cost Complete μP-Compatible 8-Bit DAC

- Complete 8-bit DAC
- Voltage output: 2 calibrated ranges
- Internal precision band-gap reference
- Single-supply operation: +5V to +15V
- Full microprocessor interface
- Fast: 1μs voltage settling to ±1/2LSB
- Low power: 75mW
- No user trims
- Guaranteed monotonic over temperature
- All errors specified T_{min} to T_{max}
- Small 16-pin DIP or PLCC package
- Single laser-wafer-trimmed chip for hybrids
- Low cost

The AD558 DACPORT™ is a complete voltage-output 8-bit digital-to-analog converter, including output amplifier, full microprocessor interface, and precision voltage reference on a single monolithic chip. No external components or trims are required to interface, with full accuracy, an 8-bit data bus to an analog system.

The performance and versatility of the DACPORT is a result of several recently-developed monolithic bipolar technologies. The complete microprocessor interface and control logic is implemented with integrated injection logic (I^2L), an extremely dense and low-power logic structure that is process-compatible with linear bipolar fabrication. The internal precision voltage reference is the patented low-voltage band-gap circuit, which permits full-accuracy performance on a single +5V to +15V power supply. Thin-film silicon-chromium resistors provide the stability required for guaranteed monotonic operation over the entire operating temperature range (all grades), while recent advances in laser-wafer-trimming of these thin-film resistors permit absolute calibration at the factory to within ±1LSB; thus no user-trims for gain or offset are required. A new circuit design provides voltage settling to ±1/2LSB for a full-scale step in 800ms.

The AD558 is available in four performance grades. The

AD558J and K are specified for use over the 0 to +70 °C temperature range, while the AD558S and T grades are specified for −55 °C to +125 °C operation. The "J" and "K" grades are available either in 16-pin plastic (N) or hermetic ceramic (D) DIPs. They are also available in 20-pin JEDEC standard PLCC packages. The "S" and "T" grades are available in 16-pin hermetic ceramic DIP packages.

*Covered by U.S. Patent Nos. 3,887,863; 3,685,045; 4,323,795; Patents Pending. DACPORT is a trademark of Analog Devices, Inc.

SPECIFICATIONS (T_A = +25°C, V_{REFIN} = 10V, V_S = ±15V unless otherwise specified)

Model	AD558J			AD558K			AD558S[1]			AD558T[1]			
	Min	Typ	Max	Min	Typ	Max	Min	Typ	Max	Min	Typ	Max	Units
RESOLUTION			8			8			8			8	Bits
RELATIVE ACCURACY[2]													
0 to +70°C			±1/2			±1/4			±1/2			±1/4	LSB
−55°C to +125°C									±3/4			±3/8	LSB
OUTPUT													
Ranges[3]		0 to +2.56			0 to +2.56			0 to +2.56			0 to +2.56		V
		0 to +10			0 to +10			0 to +10			0 to +10		V
Current Source	+5			+5			+5			+5			mA
Sink		Internal Passive Pull-Down to Ground[4]			Internal Passive Pull-Down to Ground			Internal Passive Pull-Down to Ground			Internal Passive Pull-Down to Ground		
OUTPUT SETTLING TIME[5]													
0 to 2.56 Volt Range		0.8	1.5		0.8	1.5		0.8	1.5		0.8	1.5	μs
0 to 10 Volt Range[4]		2.0	3.0		2.0	3.0		2.0	3.0		2.0	3.0	μs
FULL SCALE ACCURACY[6]													
@ 25°C			±1.5			±0.5			±1.5			±0.5	LSB
T_{min} to T_{max}			±2.5			±1			±2.5			±1	LSB
ZERO ERROR													
@ 25°C			±1			±1/2			±1			±1/2	LSB
T_{min} to T_{max}			±2			±1			±2			±1	LSB
MONOTONICITY[7]													
T_{min} to T_{max}		Guaranteed			Guaranteed			Guaranteed			Guaranteed		
DIGITAL INPUTS													
T_{min} to T_{max}													
Input Current			±100			±100			±100			100	μA
Data Inputs, Voltage													
Bit On - Logic "1"	2.0			2.0			2.0			2.0			V
Bit On - Logic "0"	0		0.8	0			0			0			V
Control Inputs, Voltage													
On - Logic "1"	2.0			2.0			2.0			2.0			V
On - Logic "0"	0		0.8	0		0.8	0		0.8	0		0.8	V
Input Capacitance		4			4			4			4		pF
TIMING													
t_W Strobe Pulse Width	200			200			200			200			ns
T_{min} to T_{max}	270			270			270			270			ns
t_{DH} Data Hold Time	10			10			10			10			ns
T_{min} to T_{max}	10			10			10			10			ns
t_{DS} Data Set-Up Time	200			200			200			200			ns
T_{min} to T_{max}	270			270			270			270			ns
POWER SUPPLY													
Operating Voltage Range (V_{CC})													
2.56 Volt Range	+4.5		+16.5	+4.5		+16.5	+4.5		+16.5	+4.5		+16.5	V
10 Volt Range	+11.4		+16.5	+11.4		+16.5	+11.4		+16.5	+11.4		+16.5	V
Current (I_{CC})		15	25		15	25		15	25		15	25	mA
Rejection Ratio			0.03			0.03			0.03			0.03	%/%
POWER DISSIPATION, V_{CC} = 5V		75	125		75	125		75	125		75	125	mW
V_{CC} = 15V		225	375		225	375		225	375		225	375	mW
OPERATING TEMPERATURE RANGE	0		+70	0		+70	−55		+125	−55		+125	°C

NOTES
[1]The AD558 S & T grades are available processed and screened to MIL-STD-883 Class B. Consult Analog Devices' Military Databook for details.
[2]Relative Accuracy is defined as the deviation of the code transition points from the ideal transfer point on a straight line from the zero to the full scale of the device.
[3]Operation of the 0 to 10 volt output range requires a minimum supply voltage of +11.4 volts.
[4]Passive pull-down resistance is 2kΩ for 2.56 volt range, 10kΩ for 10 volt range.
[5]Settling time is specified for a positive-going full-scale step to ±1/2LSB. Negative-going steps to zero are slower, but can be improved with an external pull-down.
[6]The full range output voltage for the 2.56 range is 2.55V and is guaranteed with a +5V supply, for the 10V range, it is 9.960V guaranteed with a +15V supply.
[7]A monotonic converter has a maximum differential linearity error of ±1LSB.
Specifications shown in **boldface** are tested on all production units at final electrical test.
Specifications subject to change without notice.

ABSOLUTE MAXIMUM RATINGS

V_{CC} to Ground	0V to +18V
Digital Inputs (Pins 1-10)	0 to +7.0V
V_{OUT}	Indefinite Short to Ground Momentary Short to V_{CC}
Power Dissipation	450mW
Storage Temperature Range	
N/P (Plastic) Packages	-25°C to +100°C
D (Ceramic) Package	-55°C to +150°C
Lead Temperature (soldering, 10 second)	300°C
Thermal Resistance	
Junction to Ambient/Junction to Case	
D (Ceramic) Package	100/30°C/W
N/P (Plastic) Packages	140/55°C/W

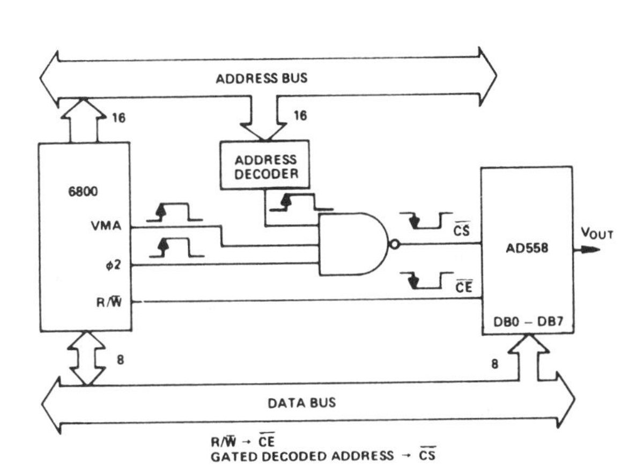

a. 6800/AD558 Interface

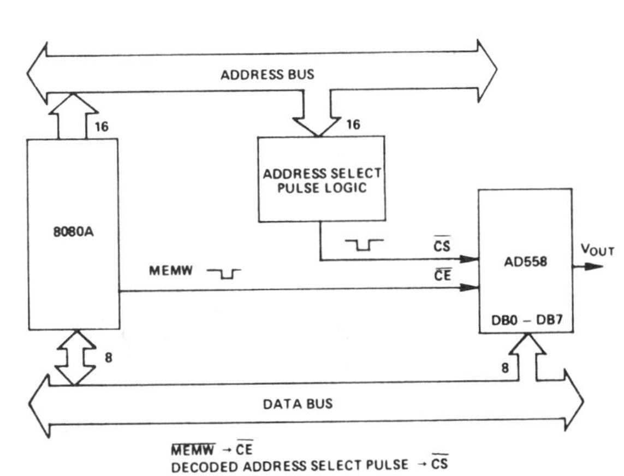

b. 8080A/AD558 Interface

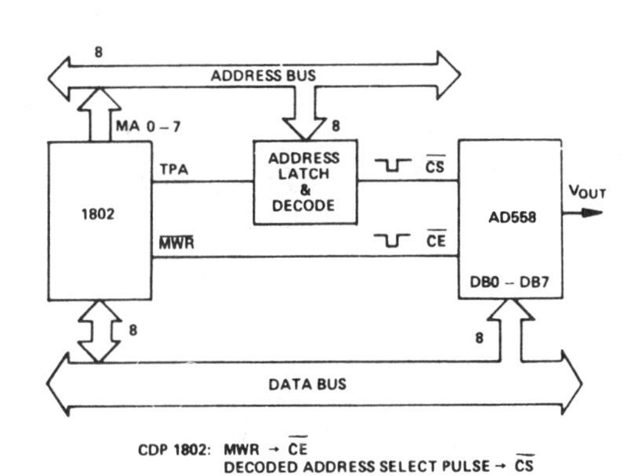

c. 1802/AD558 Interface

Interfacing the AD558 to Microprocessors

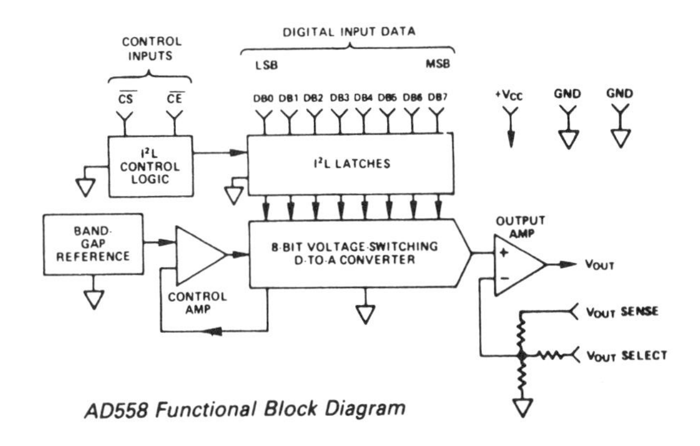

AD558 Functional Block Diagram

AD558 PIN CONFIGURATION (DIP)

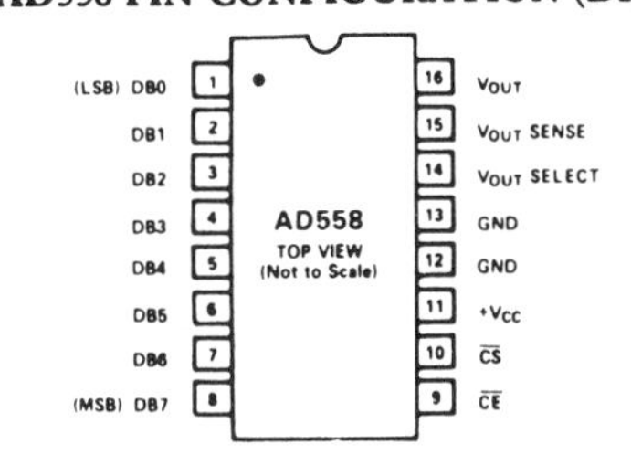

AD558 PIN CONFIGURATION (PLCC)

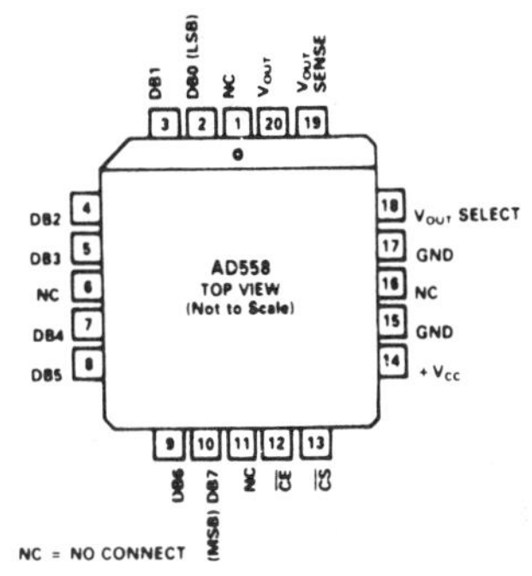

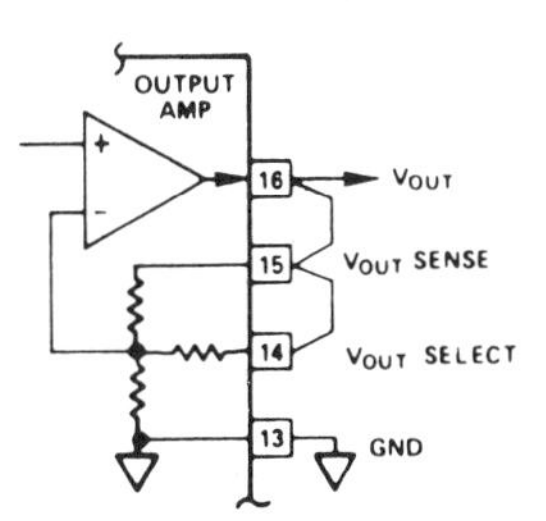

a. 0V to 2.56V Output Range

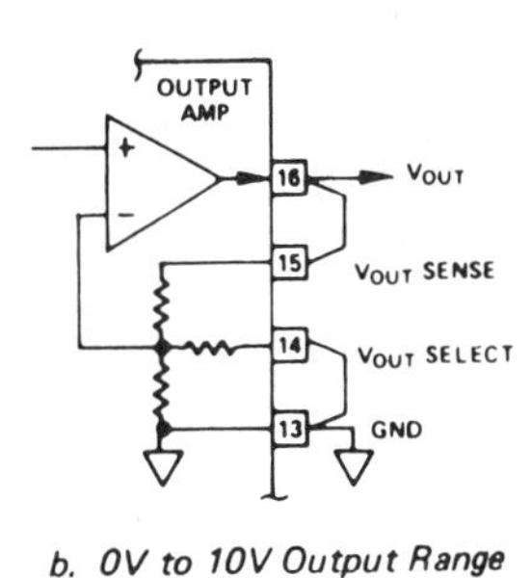

b. 0V to 10V Output Range

Connection Diagrams

Analog Devices AD561*

Low-Cost 10-Bit Monolithic D/A Converter

- Complete current output converter
- High stability buried zener reference
- Laser trimmed to high accuracy (1/4LSB max error, AD561K, T)
- Trimmed output application resistors for 0 to +10, ±5V ranges
- Fast settling: 250ns to 1/2LSB
- Guaranteed monotonicity over full operating temperature range
- TTL/DTL and CMOS compatible (positive true logic)
- Single-chip monolithic construction
- Available in chip form

The AD561 is an integrated circuit 10-bit digital-to-analog converter combined with a high-stability voltage reference fabricated on a single monolithic chip. Using 10 precision high-speed current-steering switches, a control amplifier, voltage reference, and laser-trimmed thin-film SiCr resistor network, the device produces a fast, accurate analog output current. Laser-trimmed output application resistors are also included to facilitate accurate, stable current-to-voltage conversion; they are trimmed to 0.1% accuracy, thus eliminating external trimmers in many situations.

Several important technologies combine to make the AD561 the most accurate and most stable 10-bit DAC available. The low temperature coefficient high-stability thin-film network is trimmed at the wafer level by a fine resolution laser system to 0.01% typical linearity. This results in an accuracy specification of ±1/4LSB max for the K and T versions, and 1/2LSB max for the J and S versions.

The AD561 also incorporates a low-noise high-stability subsurface zener diode to produce a reference voltage with excellent long term stability and temperature cycle characteristics which challenge the best discrete zener references. A temperature-compensation circuit is laser-trimmed to allow custom correction of the temperature coefficient of each device. This results in a typical full-scale temperature coefficient of 15 ppm/°C; the T.C. is tested and guaranteed to 30ppm/°C max for the K and T versions, 60ppm/°C max for the S, and 80ppm/°C for the J.

The AD561 is available in four performance grades. The AD561J and K are specified for use over the 0 to +70°C temperature range and are available in either a 16-pin hermetically-sealed ceramic DIP or a 16-pin molded plastic DIP. The AD561S and T grades are specified for the −55°C to +125°C range and are available in the ceramic package.

*Covered by Patent Nos.: 3,940,760; 3,747,088; RE 28,633; 3,803,590; RE 29,619; 3,961,326; 4,141,004; 4,213,806; 4,136,349.

AD561 FUNCTIONAL BLOCK DIAGRAM

TO-116

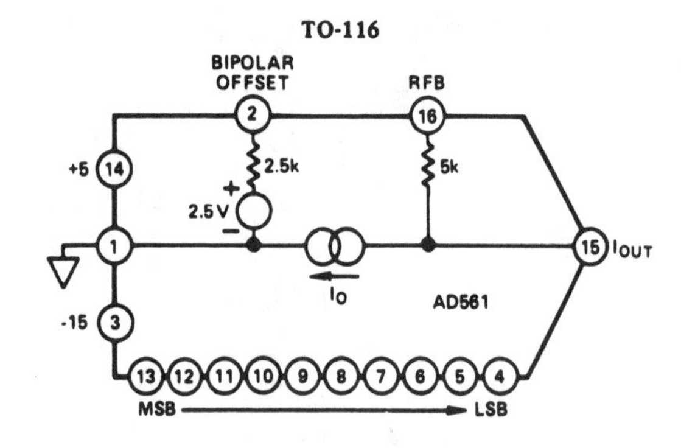

PIN CONFIGURATION TOP VIEW

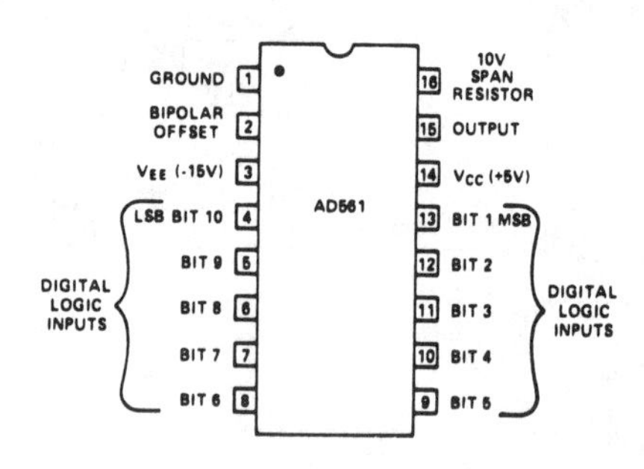

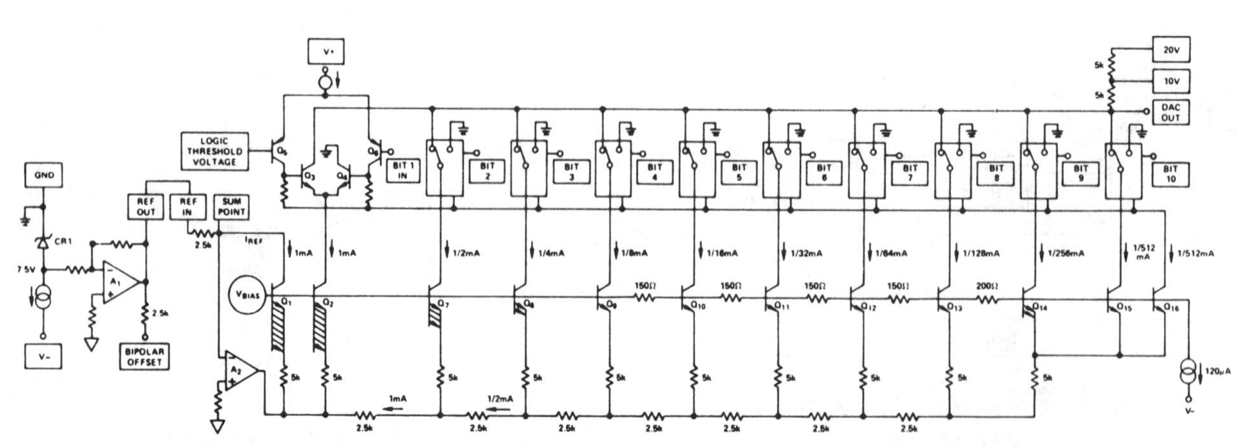

Circuit Diagram Showing Reference, Control Amplifier, Switching Cell, R-2R Ladder, and Bit Arrangement of AD561

SPECIFICATIONS (T_A = +25°C, V_{CC} ± +5V, V_{EE} = −15V, unless otherwise specified)

MODEL	AD561J MIN	AD561J TYP	AD561J MAX	AD561K MIN	AD561K TYP	AD561K MAX	UNITS
RESOLUTION		10 Bits			10 Bits		
ACCURACY (Error Relative to Full Scale)		±1/4 (0.025)	±1/2 (0.05)		±1/8 (0.012)	±1/4 (0.025)	LSB % of F.S.
DIFFERENTIAL NONLINEARITY		±1/2			±1/4	±1/2	LSB
DATA INPUTS							
TTL, V_{CC} = +5V							
Bit ON Logic "1"	+2.0			*			V
Bit OFF Logic "0"			+0.8			*	V
CMOS, 10V ≤ V_{CC} ≤ 16.5V (See Figure 1)							
Bit ON Logic "1"	70% V_{CC}			*			V
Bit OFF Logic "0"			30% V_{CC}			*	V
Logic Current (Each Bit) (T_{min} to T_{max})							
Bit ON Logic "1"		+5	+100		*	*	nA
Bit OFF Logic "0"		-5	-25		*	*	μA
OUTPUT							
Current							
Unipolar	1.5	2.0	2.4	*	*	*	mA
Bipolar	±0.75	±1.0	±1.2	*	*	*	mA
Resistance (Exclusive of Application Resistors)		40M			*		Ω
Unipolar Zero (All Bits OFF)		0.01	0.05		*	*	% of F.S.
Capacitance		25			*		pF
Compliance Voltage	-2	-3	+10	*	*	*	V
SETTLING TIME TO 1/2LSB							
All Bits ON-to-OFF or OFF-to-ON		250			*		ns
POWER REQUIREMENTS							
V_{CC}, +4.5V dc to +16.5V dc		8	10		*	*	mA
V_{EE}, −10.8V dc to −16.5V dc		12	16		*	*	mA
POWER SUPPLY GAIN SENSITIVITY							
V_{CC}, +4.5V dc to +16.5V dc		2	10		*	*	ppm of F.S./%
V_{EE}, −10.8V dc to −16.5V dc		4	25		*	*	ppm of F.S./%
TEMPERATURE RANGE							
Operating		0 to +70			*	*	°C
Storage ("D" Package)		−65 to +150			*	*	°C
("N" Package)		−25 to +85			*	*	°C
TEMPERATURE COEFFICIENTS							
With Internal Reference							
Unipolar Zero		1	10		1	5	ppm of F.S./°C
Bipolar Zero		2	20		2	10	ppm of F.S./°C
Full Scale		15	80		15	30	ppm of F.S./°C
Differential Nonlinearity		2.5			2.5		ppm of F.S./°C
MONOTONICITY		Guaranteed over full operating temp. range			Guaranteed over full operating temp. range		
PROGRAMMABLE OUTPUT RANGES		0 to +10 -5 to +5			* *		V V
CALIBRATION ACCURACY							
Full Scale Error with Fixed 25Ω Resistor		±0.1			*		% of F.S.
Bipolar Zero Error with Fixed 10Ω Resistor		±0.1			*		% of F.S.
CALIBRATION ADJUSTMENT RANGE							
Full Scale (With 50Ω Trimmer)		±0.5			*		% of F.S.
Bipolar Zero (With 50Ω Trimmer)		±0.5			*		% of F.S.

NOTES
*Specifications same as AD561J specs.
Specifications subject to change without notice.

MODEL	AD561S MIN	AD561S TYP	AD561S MAX	AD561T MIN	AD561T TYP	AD561T MAX	UNITS
RESOLUTION		10 Bits			10 Bits		
ACCURACY (Error Relative to Full Scale)		±1/4 (0.025)	±1/2 (0.05)		±1/8 (0.012)	±1/4 (0.025)	LSB % of F.S.
DIFFERENTIAL NONLINEARITY		±1/2			±1/4	±1/2	LSB
DATA INPUTS							
TTL, V_{CC} = +5V							
Bit ON Logic "1"	+2.0			••			V
Bit OFF Logic "0"			+0.8			••	V
CMOS, 10V ≤ V_{CC} ≤ 16.5V (See Figure 1)							
Bit ON Logic "1"	70% V_{CC}			••			V
Bit OFF Logic "0"			30% V_{CC}			••	V
Logic Current (Each Bit) (T_{min} to T_{max})							
Bit ON Logic "1"		+20	+100		••	••	nA
Bit OFF Logic "0"		-25	-100		••	••	μA
OUTPUT							
Current							
Unipolar	1.5	2.0	2.4	••	••	••	mA
Bipolar	±0.75	±1.0	±1.2	••	••	••	mA
Resistance (Exclusive of Application Resistors)		40M			••		Ω
Unipolar Zero (All Bits OFF)		0.01	0.05		••	••	% of F.S.
Capacitance		25			••		pF
Compliance Voltage	-2	-3	+10	••	••	••	V
SETTLING TIME TO 1/2LSB							
All Bits ON-to-OFF or OFF-to-ON		250			••		ns
POWER REQUIREMENTS							
V_{CC}, +4.5V dc to +16.5V dc		6	10		••	••	mA
V_{EE}, -10.8V dc to -16.5V dc		11	16		••	••	mA
POWER SUPPLY GAIN SENSITIVITY							
V_{CC}, +4.5V dc to +16.5V dc		2	10		••	••	ppm of F.S./%
V_{EE}, -10.8V dc to -16.5V dc		4	25		••	••	ppm of F.S./%
TEMPERATURE RANGE							
Operating		-55 to +125			••	••	°C
Storage		-65 to +150			••	••	°C
TEMPERATURE COEFFICIENTS							
With Internal Reference							
Unipolar Zero		1	10		1	5	ppm of F.S./°C
Bipolar Zero		2	20		2	10	ppm of F.S./°C
Full Scale		15	60		15	30	ppm of F.S./°C
Differential Nonlinearity		2.5			2.5		ppm of F.S./°C
MONOTONICITY		Guaranteed over full operating temp. range			Guaranteed over full operating temp. range		
PROGRAMMABLE OUTPUT RANGES		0 to +10 -5 to +5			•• ••		V V
CALIBRATION ACCURACY							
Full Scale Error with Fixed 25Ω Resistor		±0.1			••		% of F.S.
Bipolar Zero Error with Fixed 10Ω Resistor		±0.1			••		% of F.S.
CALIBRATION ADJUSTMENT RANGE							
Full Scale (With 50Ω Trimmer)		±0.5			••		% of F.S.
Bipolar Zero (With 50Ω Trimmer)		±0.5			••		% of F.S.

NOTES

****Specifications same as AD561S specs.**

Specifications subject to change without notice.

Analog Devices AD562/AD563
IC 12-Bit D/A Converters

- True 12-bit accuracy
- Guaranteed monotonicity over full temperature range
- Hermetic 14-pin DIP
- TTL/DTL and CMOS compatibility
- Positive true logic

The AD562/AD563 are monolithic 12-bit digital-to-analog converters consisting of especially designed precision bipolar switches and control amplifiers and compatible high-stability silicon chromium thin-film resistors. The AD563 also includes its own internal voltage reference.

A unique combination of advanced circuit design, high-stability SiCr thin-film resistor processing and laser-trimming technology provide the AD562/AD563 with true 12-bit accuracy. The maximum error at +25 °C is limited to ± 1/2LSB on all versions and monotonicity is guaranteed over the full operating temperature range.

The AD562 and AD563 are recommended for high-accuracy 12-bit D/A converter applications where true 12-bit performance is required, but low cost and small size are considerations. Both devices are also ideal for use in constructing A/D conversion systems and as building blocks for higher resolution D/A systems. J and K versions are specified for operation over the 0 to +70 °C temperature range, the S and T for operation over the extended temperature range, −55 °C to +125 °C.

AD562, AD563 PIN CONFIGURATIONS

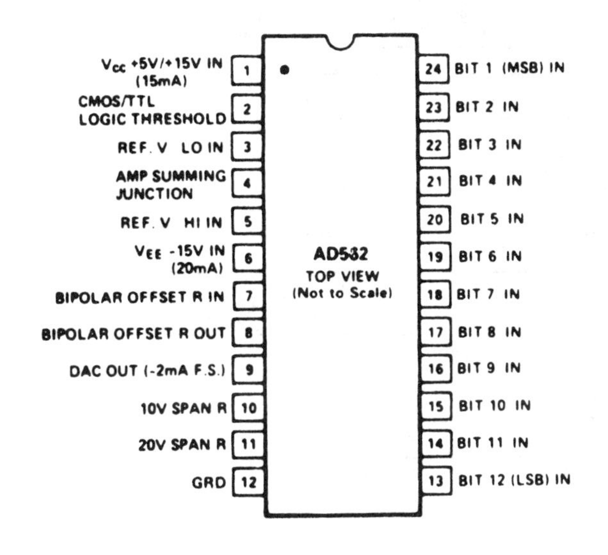

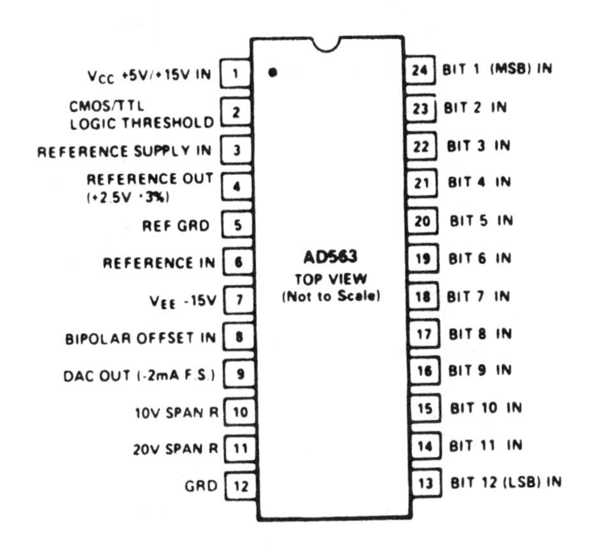

SPECIFICATIONS (T_A = +25°C, unless otherwise specified)

MODEL	AD562KD/BIN AD562KD/BCD	AD562AD/BIN AD562AD/BCD	AD562SD/BIN AD562SD/BCD
DATA INPUTS (positive True, Binary (BCD) and Offset Binary (BCD))			
TTL, V_{CC} = +5V, Pin 2			
Open Circuit			
Bit ON Logic "1"	+2.0V	•	•
Bit OFF Logic "0"	+0.8V max	•	•
CMOS, 4.75 ≤ V_{CC} ≤ 15.8, Pin 2 Tied to Pin 1			
Bit ON Logic "1"	70% V_{CC} min	•	•
Bit OFF Logic "0"	30% V_{CC} max	•	•
Logic Current (Each Bit)			
Bit ON Logic "1"	+20nA typ, +100nA max	•	•
Bit OFF Logic "0"	-50μA typ, -100μA max	•	•
OUTPUT			
Current			
Unipolar	-1.6mA min, -2.0mA typ, -2.4mA max	•	•
Bipolar	±0.8mA min, ±1.0mA typ, ±1.2mA max	•	•
Resistance (Exclusive of Span Resistors)	5.3kΩ min, 6.6kΩ typ, 7.9kΩ max	•	•
Unipolar Zero (All Bits OFF)	0.01% of F.S. typ, 0.05% of F.S. max	•	•
Capacitance	33pF typ	•	•
Compliance Voltage	-1.5V to +10V typ	•	•
RESOLUTION			
Binary	12 Bits	•	•
BCD	3 Digits	•	•
ACCURACY (Error Relative to Full Scale)			
Binary	±1/2LSB max	•	±1/4LSB max
BCD	±1/2LSB max	•	±1/10LSB max
DIFFERENTIAL NONLINEARITY	±1/2LSB max	•	•
SETTLING TIME TO 1/2LSB			
All Bits ON-to-OFF or OFF-to-ON	1.5μs typ	•	•
POWER REQUIREMENTS			
V_{CC}, +4.75 to +15.8V dc	15mA typ, 18mA max	•	•
V_{EE}, -15V dc ±5%	20mA typ, 25mA max	•	•
POWER SUPPLY GAIN SENSITIVITY			
V_{CC} @ +5V dc	2ppm of F.S./% max	•	•
V_{CC} @ +15V dc	2ppm of F.S./% max	•	•
V_{EE} @ -15V dc	6ppm of F.S./% max	•	•
TEMPERATURE RANGE			
Operating	0 to +70°C typ	-25°C to +85°C	-55°C to +125°C
Storage	-65°C to +150°C typ	•	•
TEMPERATURE COEFFICIENT			
Unipolar Zero	2ppm of F.S./°C max	•	•
Bipolar Zero	4ppm of F.S./°C max	•	•
Gain	5ppm of F.S./°C max	•	•
Differential Nonlinearity	2ppm of F.S./°C	•	1ppm of F.S./°C
MONOTONICITY	Guaranteed Over Full Operating Temperature Range	•	•
EXTERNAL ADJUSTMENTS[1]			
Gain Error with Fixed 50Ω Resistor	±0.2% of F.S. typ	•	•
Bipolar Zero Error with Fixed 50Ω Resistor	±0.1% of F.S. typ	•	•
Gain Adjustment Range	±0.25% of F.S. typ	•	•
Binary Bipolar Zero Adjustments Range	±0.25% of F.S. typ	•	•
BCD Bipolar Offset Adjustment Range	±0.17% of F.S. typ	•	•
PROGRAMMABLE OUTPUT RANGES	0 to +5V typ	•	•
	-2.5V to +2.5V typ	•	•
	0V to +10V typ	•	•
	-5V to +5V typ	•	•
	-10V to +10V typ	•	•
REFERENCE INPUT			
Input Impedance	20kΩ typ	•	•

*Specifications same as AD562KD. **Specifications same as AD563KD. ***Specifications same as AD563JD. [1] Device calibrated with internal reference.
Specifications subject to change without notice.

AD563JD/BIN AD563JD/BCD	AD563KD/BIN AD563KD/BCD	AD563SD/BIN AD563SD/BCD	AD563TD/BIN AD563TD/BCD
•	•	•	•
•	•	•	•
•	•	•	•
•	•	•	•
•	•	•	•
•	•	•	•
•	•	•	•
•	•	•	•
•	•	•	•
•	•	•	•
•	•	•	•
•	•	•	•
•	•	•	•
•	•	•	•
•	±1/4LSB	••	••
•	±1/4LSB	••	••
•	•	•	•
•	•	•	•
15mA typ, 20mA max	•••	•••	•••
•	•	•	•
3ppm of F.S./% typ, 10ppm of F.S./% max	•••	•••	•••
3ppm of F.S./% typ, 10ppm of F.S./% max	•••	•••	•••
14ppm of F.S./% typ, 25ppm of F.S./% max	•••	•••	•••
•	•	-55°C to +125°C	-55°C to +125°C
•	•	•	•
With Internal Reference			
1ppm of F.S./°C typ, 2ppm of F.S./°C max	•••	•••	•••
10ppm of F.S./°C max	•••	•••	•••
50ppm of F.S./°C max	20ppm of F.S./°C max	30ppm of F.S./°C max	10ppm of F.S./°C max
•	•	•	•
•	•	•	•
With Fixed 10Ω Resistor			
±0.2% of F.S. typ	•••	•••	•••
•	•	•	•
•	•	•	•
•	•	•	•
•	•	•	•
•	•	•	•
•	•	•	•
•	•	•	•
•	•	•	•
•	•	•	•
5kΩ typ	•••	•••	•••

SELECTION GUIDE
ANALOG-TO-DIGITAL CONVERTERS

GENERAL PURPOSE ADCs

Model	Res Bits	Conv Time μs	Int SHA BW kHz[1]	Reference Voltage Int/Ext[2]	Bus Interface Bits[3]	Package Options[4]	Temp Range[5]	Page	Comments
AD7821	8	0.66	100	0–5 V, Ext	8, μP	N, P, Q	C, I	3–367	CMOS, Bipolar or Unipolar Operation
AD7569	8	2	200	Int	8, μP	E, N, P, Q	C, I, M	3–233	CMOS, Complete I/O Port with DAC, ADC, SHA, Amps and Reference
*AD7669	8	2	200	Int	8, μP	N, P	C, I, M	3–233	CMOS, Complete I/O Port with 2 DACS, ADC, SHA, Amps and Reference
AD7820	8	2	7	0–5 V, Ext	8, μP	E, N, P, Q, R	C, I, M	3–357	CMOS, 8-Bit Sampling ADC
*AD7769	8	2.5	200	Ext	8, μP	N, P	C, I	3–325	CMOS, Two-Channel ADC/DAC with Output Amplifiers
AD7824	8	2.5	10	0–5 V, Ext	8, μP	N, Q	C, I, M	3–379	CMOS, 4 Channel, 8-Bit Sampling ADC
AD7828	8	2.5	10	0–5 V, Ext	8, μP	E, N, P, Q	C, I, M	3–379	CMOS, 8 Channel, 8-Bit Sampling ADC
AD7575	8	5	50	1.23 V, Ext	8, μP	E, N, P, Q	C, I, M	3–265	CMOS, Low Cost
AD670	8	10		Int	8, μP	D, E, N, P	C, I, M	3–69	Single Supply, Including In-Amp and Reference
AD7576	8	10		1.23 V, Ext	8, μP	E, N, P, Q	C, I, M	3–269	CMOS, Low Cost
AD570	8	25		Int	8	D	C, M	3–15	
AD673	8	30		Int	8, μP	D, N, P	C, M	3–81	
AD7581	8	66.7		−5 V–(−15 V), Ext	8, μP	D, N	C, I	3–295	CMOS 8-Bit ADC
AD579	10	1.8		10 V, Int	10/Serial	D, N	C, I	3–63	
AD7579	10	18.5	25	2.5 V, Ext	8, μP	E, N, P, Q	C, I, M	3–279	CMOS, Low Cost 10-Bit Sampling ADC
AD7580	10	18.5	25	2.5 V, Ext	10, μP	E, N, P, Q	C, I, M	3–279	CMOS, Low Cost 10-Bit Sampling ADC
AD571	10	25		Int	10	D	C, M	3–15	
AD573	10	30		Int	8/10, μP	D, N, P	C, M	3–29	
AD575	10	30		Int	Serial	D, N	C, M	3–49	
*AD9005	12	0.1	38000	Int	12	M	C, M	3–459	Complete 12-Bit ADC with T/H, Reference and Timing Circuitry
CAV-1205	12	0.2	15000	Int	12	Card	C	3–567	12-Bit, 5MSPS Eurocard
MOD-1205	12	0.5	15000	Int	12	Card	C	3–593	12-Bit, 5MSPS Video ADC
AD9003	12	1	10000	Int	12	M	C	3–451	12-Bit, 1MSPS ADC. Single 40-Pin DIP
HAS-1201	12	1	2000	Int	12	M	C, M	3–573	12-Bit, 1MSPS ADC
HAS-1202A	12	1.6		Int	12	D	C, I, M	3–579	12-Bit, 641kHz ADC
HAS-1204	12	2	7000	Int	12	M	C, M	3–583	12-Bit, 500kHz ADC. Single 40-Pin DIP
HAS-1202	12	2.9		Int	12	D	C, I, M	3–579	12-Bit, 349kHz ADC
AD578	12	3		10 V, Int	12	D, N	C, M	3–57	Complete, 3μs, 12-Bit ADC
AD7672	12	3		−5 V, Ext	12, μP	E, N, P, Q	C, I, M	3–309	CMOS, Unipolar or Bipolar, −12 V, +5 V Supply
AD678	12	4	X	5 V, Int	12, μP	D, N, P	C	3–99	BiMOS, High-Impedance, High-Bandwidth Sampling Input, 10 V Range
*AD1678	12	4	500	Int	8/12, μP	D, N, P	C	3–191	BiMOS, 12-Bit Sampling ADC, ac Characterized
AD5240	12	5		6.3 V, Int	12	D	C, I	3–547	Industry Standard
AD7572	12	5		−5.25 V, Int	12, μP	E, N, P, Q	C, I, M	3–253	CMOS 12-Bit ADC
AD1332	12	8	125	−5 V, Int	12, μP	D	I	9–31	Complete 12-Bit 125kHz Sampling ADC for Digital Signal Processing
AD7870	12	10	X	3 V, Int	8/12/Serial, μP	E, N, P, Q	C, I, M	3–391	CMOS, 100kHz Throughput—
AD7878	12	10	X	3 V, Int	12, μP	E, N, P, Q	C, I, M	3–419	CMOS, 100kHz Throughput, On-Chip FIFO. Serial, Parallel or Byte Output
*AD7772	12	10	X	5.25, Int	Serial, μP	E, N, P, Q	C, I, M	3–341	CMOS, Serial Output 12-Bit ADC
*AD1334	12	15	235	−5 V, Int	12, μP	D	I	9–49	Four Channel 65kHz 12-Bit Sampling ADC for Digital Signal Processing
AD ADC84	12	10		6.3 V, Int	12	D	C	3–547	Industry Standard
AD ADC85	12	10		6.3 V, Int	12	D	C, I	3–547	Industry Standard
AD5210	12	13		−10 V, Int/Ext	12	D	I, M	3–227	Industry Standard
AD674A	12	15		10 V, Int	12, μP	D	C, M	3–89	Complete 12-Bit ADC
AD368	12	15	40–1000	6.3 V, Int	12	D	I, M	9–19	Complete 12-Bit ADC with Programmable Gains of 1, 8, 64, 512
AD369	12	15	40–1000	6.3 V, Int	12	M	I	9–19	Complete 12-Bit ADC with Programmable Gains of 1, 10, 100, 500
AD572	12	25		10 V, Int	12	D, M	I, M	3–21	12-Bit Successive Approximation ADC
AD ADC80	12	30		6.3 V, Int	12	D	I	3–539	Industry Standard
AD574A	12	35		10 V, Int	8/12, μP	D, E, N, P	C, M	3–37	Complete ADC with Reference and Clock
AD363	12	40	X	10 V, Int	12, μP	D	C, M	9–5	High Speed 16-Channel, 12-Bit DAS
AD364	12	50	X	10 V, Int	12, μP	D	C, M	9–5	16-Channel, 12-Bit DAS with Three-State Buffered Output
AD5200	12	50		−10 V, Int/Ext	12	D	I, M	3–227	Industry Standard
AD7578	12	100		5 V, Ext	12, μP	D, N	C, I, M	3–273	CMOS, 1LSB Total Unadjusted Error

GENERAL PURPOSE ADCs

Model	Res Bits	Conv Time μs	Int SHA BW kHz[1]	Reference Voltage Int/Ext[2]	Bus Interface Bits[3]	Package Options[4]	Temp Range[5]	Page	Comments
AD7582	12	100		5 V, Ext	12, μP	D, E, N, P	C, I, M	3–303	CMOS, 1LSB Total Unadjusted Error
HAS-1409	14	9	200–800	Int	14	M	C	3–587	14-Bit, 125kHz ADC. Single 40-Pin DIP
AD679	14	10	X	5 V, Int	14, μP	D, N, P	C	3–111	BiMOS, High Impedance, High Bandwidth Sampling Input, 10 V Input Range
*AD1679	14	10	500	5 V, Int	8/14, μP	D, N, P	C	3–203	14-Bit BiMOS Sampling ADC, ac Characterized
*AD7871	14	10	X	3 V, Int	8/14/Serial, μP	E, N, P, Q	C, I, M	3–407	CMOS, 14-Bit, 100kHz Sampling ADC
ADC1131	14	12		Int	14	Module	C	3–555	14-Bit, High Speed ADC
ADC1130	14	25		Int	14	Module	C	3–555	14-Bit, High Speed ADC
DAS1152	14	40	X	10 V, Int	16	Module	I	9–73	14-Bit High Accuracy Sampling ADC
DAS1157	14	55	X	10 V, Int	16	Module	I	9–77	Low Power, 14-Bit Sampling ADC
DAS1153	15	50	X	10 V, Int	16	Module	I	9–73	15-Bit High Accuracy Sampling ADC
DAS1158	15	55	X	10 V, Int	16	Module	I	9–77	Low Power, 15-Bit Sampling ADC
*AD1377	16	10	X	Int	16, Serial	D	C	3–175	Complete 16-Bit Converter. Industry Standard Pin Out
AD1376	16	15		Int	16, Serial	V	C	3–167	Complete, High Speed 16-Bit ADC Operation over −25°C to +85°C
AD1380	16	20	900	Int	16, Serial	D	C	3–183	Low Cost, 16-Bit Sampling ADC Operation over −55°C to +85°C Temperature Range
ADC1140	16	35		10 V, Int	16	Module	C	3–559	16-Bit ADC, Operates over −25°C to +85°C Temperature Range
AD ADC71	16	50		6.3 V, Int	16	D, M	C	3–531	Industry Standard
AD ADC72	16	50		6.3 V, Int	16	D, M	C, I	3–531	Industry Standard
DAS1159	16	55	X	10 V, Int	16	D	I	9–77	Low Power, 16-Bit Sampling ADC
AD1170	18	1000		5 V, Int	24	D	C	3–147	7 to 22-Bit Programmable Integrating ADC
AD1175K	22	50ms		6.95 V, Int/Ext	24	Module	C	3–159	High Accuracy, 22-Bit Integrating ADC
AD7821	8	0.66	100	0–5 V, Ext	8, μP	N, P, Q	C, I	3–367	CMOS, Bipolar or Unipolar Operation
AD7569	8	2	200	Int	8, μP	E, N, P, Q	C, I, M	3–233	CMOS, Complete I/O Port with DAC, ADC, SHA, AMPs, & Reference
*AD7669	8	2	200	Int	8, μP	N, P	C, I, M	3–233	CMOS, Complete I/O Port with 2 DACs, ADC, SHA, AMPs, & Reference
AD7820	8	2	7	0–5 V, Ext	8, μP	E, N, P, Q, R	C, I, M	3–357	CMOS, 8-Bit Sampling ADC
AD7824	8	2.5	10	0–5 V, Ext	8, μP	N, Q	C, I, M	3–379	CMOS, 4 Channel, 8-Bit Sampling ADC
AD7828	8	2.5	10	0–5 V, Ext	8, μP	E, N, P, Q	C, I, M	3–379	CMOS, 8 Channel, 8-Bit Sampling ADC
AD7575	8	5	50	1.23 V, Ext	8, μP	E, N, P, Q	C, I, M	3–265	CMOS, Low Cost
AD7579	10	18.5	25	2.5 V, Ext	8, μP	E, N, P, Q	C, I, M	3–279	CMOS, Low Cost 10-Bit Sampling ADC
AD7580	10	18.5	25	2.5 V, Ext	10, μP	E, N, P, Q	C, I, M	3–279	CMOS, Low Cost 10-Bit Sampling ADC
*AD9005	12	0.1	38000	Int	12	M	C, M	3–459	Complete 12-Bit ADC with T/H, Reference and Timing Circuitry
CAV-1205	12	0.2	15000	Int	12	Card	C	3–567	12-Bit, 5MSPS Eurocard
MOD-1205	12	0.5	15000	Int	12	Card	C	3–593	12-Bit, 5MSPS Video ADC
AD9003	12	1	10000	Int	12	M	C	3–451	12-Bit, 1MSPS ADC, Single 40-Pin DIP
HAS-1201	12	1	2000	Int	12	M	C, M	3–573	12-Bit, 1MSPS ADC
HAS-1204	12	2	7000	Int	12	M	C, M	3–583	12-Bit 500kHz. ADC Single 40-Pin DIP
AD678	12	4	500	5 V, Int	8/12, μP	D, N, P	C, M	3–99	BiMOS, High Impedance High Bandwidth Sampling Input, 10 V Range
*AD1678	12	4	500	Int	8/12, μP	D, N, P	C	3–195	BiMOS, 12-Bit Sampling ADC, ac Characterized
AD1332	12	8	125	−5 V, Int	12, μP	D	I	9–31	Complete 12-Bit 125kHz Sampling ADC for Digital Signal Processing
AD7870	12	8	500	3 V, Int	8/12/Serial, μP	N, P, Q	C, I, M	3–391	CMOS, 100kHz Throughput Rate
AD7878	12	8	500	3 V, Int	12, μP	E, N, P, Q	C, I, M	3–419	CMOS, 100kHz Throughput, On-Chip FIFO
*AD1334	12	15	235	−5 V, Int	12, μP	D	I	9–49	Four Channel 65kHz 12-Bit Sampling ADC for Digital Signal Processing
AD368	12	15	40–1000	6.3 V, Int	12	D	I, M	9–19	Complete 12-Bit ADC with Programmable Gains of 1, 8, 64, 512
AD369	12	15	40–1000	6.3 V, Int	12	M	I	9–19	Complete 12-Bit ADC with Programmable Gains of 1, 10, 100, 500
AD363	12	40	X	10 V, Int	12, μP	D	C, M	9–5	16-Channel, 12-Bit DAS
AD364	12	50	X	10 V, Int	12, μP	D	C, M	9–5	High Speed, 16-Channel, 12-Bit DAS with Three-State Buffered Output

SAMPLING ADCs

Model	Res Bits	Conv Time μs max	SHA BW kHz typ[1]	Reference Volt Int/Ext[2]	Bus Interface Bits[3]	Package Options[4]	Temp Range[5]	Page	Comments
HAS-1409	14	9	200–800	Int	14	M	C	3–587	125kHz Word Rates; Includes T/H
AD679	14	10	500	5 V, Int	8, μP	D, N, P	C, M	3–111	BiMOS, High-Impedance High-Bandwidth Sampling Input, 10 V Input Range
*AD1679	14	10	500	5 V, Int	8, μP	D, N, P	C	3–203	14-Bit BiMOS Sampling ADC, ac Characterized
*AD779	14	10	500	5 V, Int	14, μP	D, N	C, M	3–135	BiMOS, High-Impedance High-Bandwidth Sampling Input, 10 V Input Range
*AD1779	14	10	500	5 V, Int	14, μP	D, N	C, M	3–215	14-Bit BiMOS Sampling ADC, ac Characterized
*AD7871	14	10	X	3 V, Int	8/14/Serial, μP	N, P, Q	C, I, M	3–407	CMOS, 14-Bit, 100kHz Sampling ADC
*AD7872	14	12	X	3 V, Int	Serial, μP	N, Q	C, I, M	3–407	CMOS, 14-Bit, Sampling ADC with Serial Output
DAS1152	14	40	X	10 V, Int	14	D	I	9–73	14-Bit High Accuracy Sampling ADC
DAS1157	14	55	X	10 V, Int	14	D	I	9–77	Low Power, 14-Bit Sampling ADC
DAS1153	15	50	X	10 V, Int	15	D	I	9–73	15-Bit High Accuracy Sampling ADC
DAS1158	15	55	X	10 V, Int	15	D	I	9–77	Low Power, 15-Bit Sampling ADC
AD1380	16	20	900	Int	16/Serial	D	C	3–183	Low Cost, 16-Bit Sampling ADC. Operation Over −55°C to +85°C Temperature Range
DAS1159	16	55	X	10 V, Int	16	D	I	9–77	Low Power, 16-Bit Sampling ADC

MULTIPLEXED ADCs

Model	Res Bits	# Chan	Conv Time μs	SHA BW kHz	Reference Volt Int/Ext[2]	Bus Interface Bits[3]	Package Options[4]	Temp Range[5]	Page	Comments
AD7824	8	4	2.5	10	0–5 V, Ext	8, μP	N, Q	C, I, M	3–379	CMOS, On-Chip Track-Hold
AD7828	8	8	2.5	10	0–5 V, Ext	8, μP	E, N, P, Q	C, I, M	3–379	CMOS, On-Chip Track-Hold
AD7581	8	8	66.7		−5 V−(−15 V), Ext	8, μP	D, N	C, I	3–295	CMOS
AD363	12	16	40		10 V, Int	12, μP	D	C, M	9–5	High Speed, 16-Channel, 12-Bit DAS
AD364	12	16	50		10 V, Int	12, μP	D	C, M	9–5	16-Channel, 12-Bit DAS with Three-State Buffers
AD7582	12	4	100		4 V–6 V, Ext	12, μP	D, E, N, P	C, I, M	3–303	CMOS, 1LSB Total Unadjusted Error

VIDEO ADCs

Model	Res Bits	Throughput Rate MSPS min	Full Power BW MHz typ	Reference Voltage Int/Ext[2]	Bus Interface Bits[3]	Package Options[4]	Temp Range[5]	Page	Comments
AD9688	4	175	100	0.16–6 V, Ext	4	E, Q	I, M	3–525	Second Source to AM688, Overrange Bits, Stackable to 8 Bits
*AD9006	6	470	250 (min)	±1 V, Ext	6, μP	E, Z	C, M	3–467	470MSPS, 6-Bit ADC. 8.5pF Input Capacitance
*AD9016	6	470	250 (min)	±1 V, Ext	6, μP	E, Z	C, M	3–467	470MSPS, 6-Bit ADC with On-Board Demultiplexing Circuitry
AD9000	6	50	20	0.5–2 V, Ext	6	D, E	C, M	3–435	MIL-STD-883, Rev. C, Devices Available. Low Error Rate
*AD9028	8	300	250	-2 V, Ext	8	E	C, M	3–497	300 MSPS, 8-Bit ADC, Guaranteed Dynamic Performance
*AD9038	8	300	250	-2 V, Ext	Dual 8	E	C, M	3–497	300 MSPS, 5-Bit ADC with On-Board 1:2 Demultiplexed Data Outputs
AD770	8	200	250	±2 V, Ext	8	D	C, M	3–123	High Bandwidth, Error Correction
AD9002	8	125	115 (Sm. Sig.)	0.1−(−2.1) Ext	8	D, E	I, M	3–443	Single Supply, Low Power, Low Input Capacitance, MIL-STD-883, Rev. C Device Available
*AD9011	8	100	80	Int	8	M	C, M	3–483	8-Bit, 100MSPS ADC with On-Board Amp and Reference, Multiple Gain Selection
AD9012	8	75	180	−2 V, Ext	8	Q, E	I, M	3–489	TTL Compatible Outputs
*AD9048	8	35	15	−2 V, Ext	8, μP	N, P, Q, Z	C, M	3–509	35MSPS, 8-Bit Video ADC, 16pF Input Capacitance
AD9502	8	13	7.5	Int	8	M	I	3–517	RS-170 Video Frame Grabber. Digitizes RS-170, NTSC, PAL Signals

Model	Res Bits	Throughput Rate MSPS min	Full Power BW MHz typ	Reference Voltage Int/Ext[2]	Bus Interface Bits[3]	Package Options[4]	Temp Range[5]	Page	Comments
CAV-1040	10	40	20	Int	10	Card	C	3–563	Excellent Dynamic Performance over Frequency
CAV-1040A	**10**	**40**	**40**	**Int**	**10**	**Card**	**C**	**3–563**	**Higher Input Bandwidth Version of CAV-1040**
CAV-1220	**12**	**20**	**35**	**Int**	**12**	**Card**	**C**	**3–569**	**Fastest 12-Bit A/D Converter Available**

[1]X indicates that the internal SHA bandwidth is not specified in kHz.

[2]Ext indicates external reference with the range of voltages listed where applicable. Ext (M) indicates external reference with multiplying capability. Int indicates reference is internal. A voltage value is given if the reference is pinned out.

[3]This column lists the data format for the bus with "μP" indicating microprocessor capability—i.e., for a 12-bit converter 8/12, μP indicates that the data can be formatted for an 8-bit bus or can be in parallel (12 bits) and is microprocessor compatible.

[4]Package Options: D—Side-Brazed Dual-In-Line Ceramic; E—Leadless Chip Carrier; M—Metal Hermetic Dual-In-Line; N—Plastic Molded Dual-In-Line; P— Plastic Leaded Chip Carrier (PLCC); Q—Cerdip; R—Small Outline Plastic (SOIC).

[5]Temperature Ranges: C—Commercial, 0 to +70 °C; I—Industrial, −40 °C to +85 °C (Some older products −25 °C to +85 °C); M—Military, −55 °C to +125 °C.

Boldface Type: Product recommended for new design.

*New product since the publication of the 1987/1988 Databooks.

Analog Devices AD368/AD369

Complete 12-Bit A/D Converters with Programmable Gain

- Low-cost data acquisition systems including:
  - Programmable gain instrumentation amplifier
  - Track-and-hold amplifier
  - 12-bit A/D converter
- Digitally controlled gains:
  - AD368 gains = 1, 8, 64, 512
  - AD3698 gains = 1, 10, 100, 500
- 50 kHz throughput rate
- Small size: 28-pin hermetic double DIP
- Guaranteed no missing codes over specified temperature
- True 12-bit linear; error ≤ 1/2LSB (B-grade)
- Unipolar or bipolar operation
- MIL-STD-883B screening available

APPLICATIONS

- Microprocessor-based data acquisition
- Wide dynamic range measurement systems
- Analytical and medical instruments
- Multichannel systems with high-/low-level signals

The AD368/AD369 are low-cost wide dynamic range data-acquisition systems which condition and subsequently convert an analog signal into a 12-bit digital word. They include a programmable gain amplifier, a track-and-hold amplifier, and a 12-bit analog-to-digital converter—all in a 28-pin dual-in-line package.

The digitally programmable-gain amplifier (PGA) of the AD368 enables the user to select binary-based gains of 1, 8, 64, and 512. These gain steps are especially useful in extending system dynamic range in DSP applications. The PGA of the AD369, with gains of 1, 10, 100, and 500, allows the user to choose full-scale input voltage ranges of 10V, 1V, 100mV, and 20mV, respectively. In addition, the precision differential input of the PGA provides the AD368/AD369 with excellent common-mode rejection.

The track-and-hold amplifier (T/H) features excellent linearity, low noise, and an internal hold capacitor.

The successive approximation analog-to-digital converter (ADC) features true 12-bit operation, with 0.012% max nonlinearity (B-grade). The user can select bipolar or unipolar operation to digitize both ac and dc input signals.

The AD368/AD369 provide a completely specified (industrial and military temperature ranges) and tested function in a space-saving 28-pin hermetic package for system designers with cost, space, and time constraints.

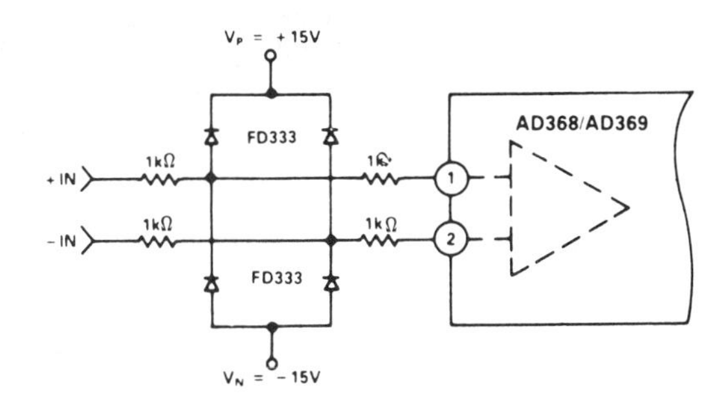

Input Protection Circuit for AD368/AD369

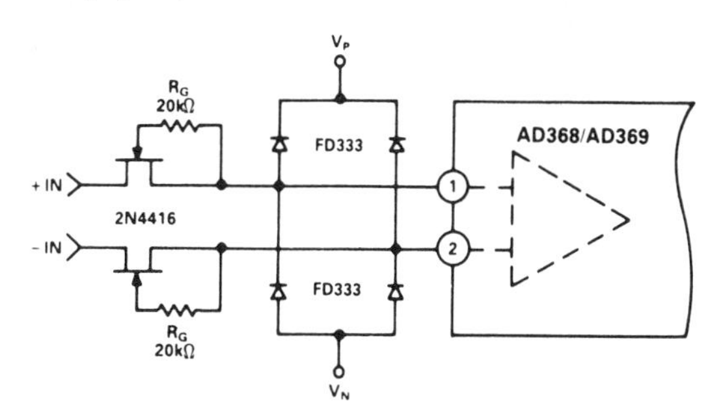

Low Noise Input Protection Circuit for AD368/ AD369

AD368/AD369 FUNCTIONAL DIAGRAM AND PIN DESIGNATION

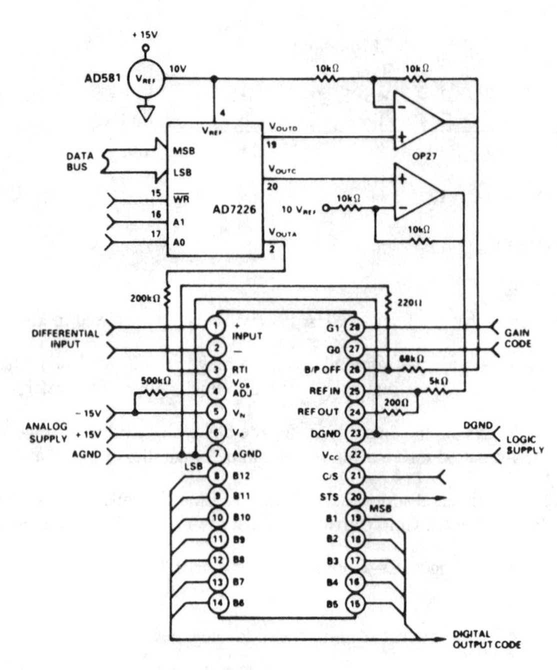

AD368/AD369 in the Unipolar Mode with D/A Circuit Replacing Trimpots

ABSOLUTE MAXIMUM RATINGS

Parameter	Min	Max	Units
Positive Supply, V_P	−0.3	+17	V
Negative Supply, V_N	+0.3	−17	V
Digital-to-Analog Ground	−1	+1	V
Logic Supply	−0.3	+7	V
Analog Input (Either)	V_N	V_P	V
Analog Input Current	−20	+20	mA
Lead Soldering, 10 sec		+300	°C
Operating Temperature Range			
to Specifications: A, B	−25	+85	°C
S	−55	+125	°C
Storage Temperature	−65	+150	°C

SPECIFICATIONS (typical @ +25°C, V_S = ±15V, +5V, R_{SPAN} = 63Ω and R(B/P) = 31Ω unless otherwise noted)

	AD368AD/SD AD369AD/SD			AD368BD AD369BD			
Parameter	**Min**	**Typ**	**Max**	**Min**	**Typ**	**Max**	**Units**
ANALOG INPUT							
Voltage Range, Unipolar (G = 1)	0		+10	*		*	V
Voltage Range, Bipolar (G = 1)	−5		+5	*		*	V
Common-Mode Voltage		12 − (V_{DIFF} × G/2)			*		V
Resistance		10^9			*		Ω
Capacitance		5			*		pF
Bias Current (I_B)		10	50		*	25	nA
I_B vs. Temperature		50			*		pA/°C
Input Offset Current (I_{OS})		2	20		*	10	nA
I_{OS} vs. Temperature		20			*		pA/°C
Noise Current (0.1 to 10Hz)		60			*		pA p-p
Output Offset Voltage (V_{OS})[1]		5 + 0.02 × G	25 + 0.2 × G		*	10 + 0.1 × G	mV
V_{OS} vs. Temperature		70 + 0.2 × G	300 + 2.0 × G		*	*	μV/°C
V_{OS} vs. Common-Mode Voltage[2]		60 + 0.5 × G	320 + 3.2 × G		*	150 + 1.5 × G	μV/V
V_{OS} vs. Supply Voltage[3]		100 + 1.0 × G	2300 + 10 × G		*	1000 + 4 × G	μV/V
Output Noise Voltage (rms)							
G = 1		250			*		μV
G = 8, 10		260			*		μV
G = 64, 100		340			*		μV
G = 512, 500		600			*		μV
DIGITAL INPUTS[4]							
V_{IH}	3.0		V_{CC}	*		*	V
V_{IL}	0.0		0.8	*		*	V
I_{IH}, I_{IL}		0.01	1.0		*	*	μA
C/S Pulse Width	50			*			ns
DIGITAL OUTPUTS, 12-BIT PARALLEL							
V_{OH} @ I_{OH} = −40μA	3.6	5.0		*	*		V
V_{OL} @ I_{OL} = 1.6mA		0.2	0.4		*	*	V

Parameter	AD368AD/SD AD369AD/SD			AD368BD AD369BD			
	Min	Typ	Max	Min	Typ	Max	Units
SIGNAL DYNAMICS							
Conversion Time (t_C)		12	15		*	*	µs
t_C vs. Temperature		−10			*		ns/°C
System Throughput Rate[5]							
G = 1, 8, 10			50			*	kHz
G = 64, 100			50			*	kHz
G = 512, 500			20			*	kHz
Gain Switching Time		1.5	2.0		*	*	µs
PGA Settling Time (to 1/2LSB)							
G = 1, 8, 10		8	10		*	*	µs
G = 64, 100		12	15		*	*	µs
G = 512, 500		40	50		*	*	µs
Amplifier −3dB Bandwidth							
G = 1		1000			*		kHz
G = 8, 10		400			*		kHz
G = 64, 100		150			*		kHz
G = 512, 500		40			*		kHz
T/H̄ Acquisition Time (t_{ACQ} to 1/2LSB)			3			*	µs
T/H̄ Aperture Delay Time (t_{AP})		140	250		*	*	ns
t_{AP} vs. Temperature		−0.3			*		ns/°C
Aperture Jitter		1			*		ns
ACCURACY							
Integral Nonlinearity		0.30	0.75		*	0.5	LSB
Differential Nonlinearity (DNL)[6]		0.30	0.90		*	0.5	LSB
Gain Error @ G = 1		0.05	0.5		*	0.2	%
@ Other Gains Referred to G = 1[7]		0.01	0.1		*	0.05	%
Gain vs. Temperature @ G = 1		3	30		*	*	ppm/°C
@ Other Gains Referred to G = 1		3	10		*	*	ppm/°C
Gain vs. Supply Voltage							
V_P ± 10%		10	30		*	*	ppm/%
V_N ± 10%		5	30		*	*	ppm/%
V_{CC} ± 10%		5	15		*	*	ppm/%
MONOTONIC TEMPERATURE RANGE							
12 Bits	−25		+85	−25		+85	°C
	−55 (S Grade)		+85 (S Grade)				°C
10 Bits	−55 (S Grade)		+125 (S Grade)				°C
REFERENCE							
Voltage (V_{REF})	6.28	6.30	6.32	*	*	*	V
V_{REF} vs. Temperature			20			*	ppm/°C
Internal Resistance		2			*		Ω
External Load			0.5			*	mA
POWER REQUIREMENTS							
Positive Supply Range	+13.5	15	16.5	*	*	*	V
Negative Supply Range	−13.5	−15	−16.5	*	*	*	V
Logic Supply Range	4.5	5.0	5.5	*	*	*	V
Supply Current, V_{IN} = 10V, f_C = 50kHz							
+15V		15	20		*	*	mA
−15V		30	35	*	*		mA
+5V		20	30		*	*	mA
Power Consumption		775			*		mW
THERMAL RESISTANCE (J-A)		25			*		°C/W
PACKAGE OPTION							
DH-28A		AD368AD/SD AD369AD/SD			AD368BD AD369BD		

NOTES

*Same specifications as A Grade.

[1]Offset voltage applies to both bipolar and unipolar operating modes.

[2]V_{CM} = ±10V.

[3]V_S = ±10%.

[4]For digital inputs, pull-up resistors needed (typ 5kΩ) when interfacing with TTL/DTL logic.

[5]Assumes pipelining, i.e., signal is inputted to I.A. when T/H goes into hold mode, allowing voltage to settle concurrently with A/D conversion (see timing diagram).

[6]Includes T/H droop rate.

[7]This is gain error (% FS) after error at G = 1 is cancelled by adjustment. Without adjustment, total error becomes:
E (Total) = E (G = 1) + E (G = 8/10, 64/100, or 512/500).

Specifications subject to change without notice.

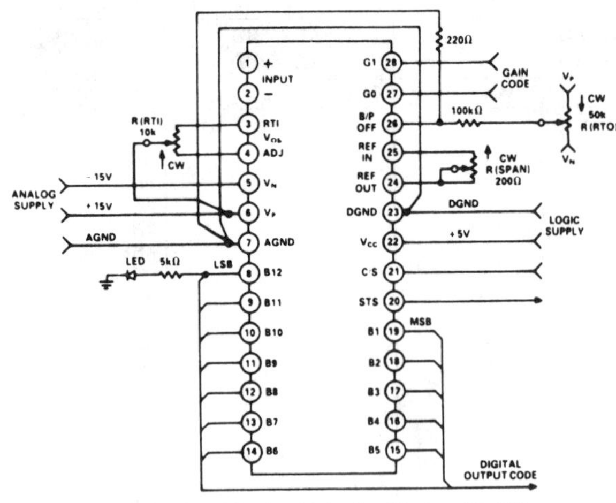

AD368/AD369 in the Unipolar Mode with RTI V_{OS}, RTO V_{OS} and Span Trimpots

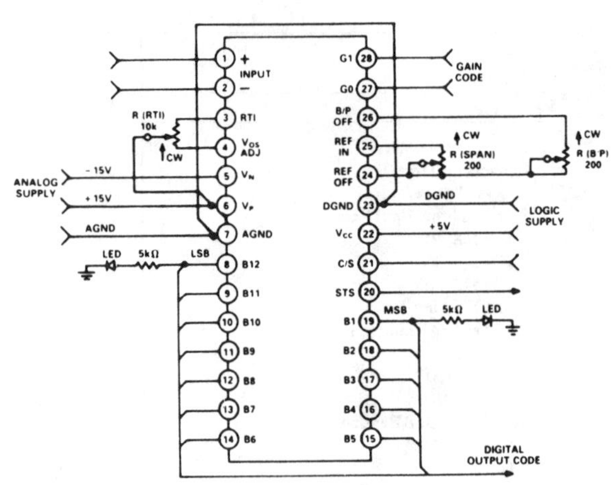

AD368/AD369 in the Bipolar Mode with Offset and Gain Trimpots

Analog Devices AD570/AD571*
8- and 10-Bit Analog-to-Digital Converters

- Complete A/D converters with reference and clock
  - AD570: 8 bit
  - AD571: 10 bit
- Fast successive approximation conversion: 25μs
- No missing codes over temperature
- Digital multiplexing: 3 state outputs
- 18-pin ceramic DIP
- Low-cost monolithic construction

The AD570/AD571 are successive approximation A/D converters consisting of a DAC, voltage reference, clock, comparator, successive approximation register and output buffers—all fabricated on a single chip. No external components are required to perform full-accuracy conversion of 25μs.

The AD570/AD571 incorporate advanced integrated circuit design and processing technologies. They employ I^2L (integrated logic) processing in the fabrication of the SAR function. Laser trimming of the high-stability SiCr thin-film resistor ladder network ensures high accuracy, which is maintained with a temperature-compensated subsurface Zener reference.

Operating on supplies of +5V to +15V and −15V, the AD570/AD571 will accept analog inputs of 0 to +10V, unipolar or ±5V bipolar, externally selectable. As the BLANK and $\overline{\text{CONVERT}}$ input is driven low, the three-state outputs will be open and a conversion will commence. Upon completion of the conversion, the $\overline{\text{DATA READY}}$ line will go low and the data will appear at the output. Pulling the BLANK and $\overline{\text{CONVERT}}$ high blanks the outputs and readies the device for the next conversion.

The devices are available in two versions; the "J" and "K" specified for the 0 to +70 °C temperature range. The "S" guarantees the specified accuracy and no missing codes from −55 °C to +125 °C.

*Covered by Patent Nos. 3,940,760; 4,213,806; 4,136,349.

SPECIFICATIONS (T_A = +25°C, V+ = +5V, V− = −12V or −15V, all voltages measured with respect to digital common, unless otherwise indicated)

Model	AD570J Min	AD570J Typ	AD570J Max	AD570S Min	AD570S Typ	AD570S Max	Units
RESOLUTION[1]			8			8	Bits
RELATIVE ACCURACY T_{min} to T_{max}			±1/2			±1/2	LSB
FULL-SCALE CALIBRATION		±2			±2		LSB
UNIPOLAR OFFSET			±1/2			±1/2	LSB
BIPOLAR ZERO			±1/2			±1/2	LSB
DIFFERENTIAL NONLINEARITY T_{min} to T_{max}	8			8			Bits
TEMPERATURE RANGE	0		+70	−55		+125	°C
TEMPERATURE COEFFICIENTS							
Unipolar Offset			±1			±1	LSB
Bipolar Offset			±1			±1	LSB
Full-Scale Calibration			±2			±2	LSB
POWER SUPPLY REJECTION							
CMOS Positive Supply +13.5V≤V+≤+16.5V	–	–	–	–	–	–	LSB

SPECIFICATIONS

($T_A = 25°C$, V+ = +5V, V− = −12V or −15V, all voltages measured with respect to digital common, unless otherwise indicated)

Model	AD570J Min	AD570J Typ	AD570J Max	AD570S Min	AD570S Typ	AD570S Max	Units
TTL Positive Supply							
+4.5V≤V+≤+5.5V			±2			±2	LSB
Negative Supply							
−16.0V≤V−≤−13.5V			±2			±2	LSB
ANALOG INPUT IMPEDANCE	**3.0**	5.0	**7.0**	**3.0**	5.0	**7.0**	kΩ
ANALOG INPUT RANGES							
Unipolar	0		+10	0		+10	V
Bipolar	−5		+5	−5		+5	V
OUTPUT CODING							
Unipolar	Positive True Binary			Positive True Binary			
Bipolar	Positive True Offset Binary			Positive True Offset Binary			
LOGIC OUTPUT							
Output Sink Current							
(V_{OUT} = 0.4V max, T_{min} to T_{max})	**3.2**			**3.2**			mA
Output Source Current							
(V_{OUT} = 2.4V max, T_{min} to T_{max})	**0.5**			**0.5**			mA
Output Leakage			**±40**			**±40**	μA
LOGIC INPUTS							
Input Current			**±100**			**±100**	μA
Logic "1"	**2.0**			**2.0**			V
Logic "0"			**0.8**			**0.8**	V
CONVERSION TIME							
T_{min} to T_{max}	**15**	25	**40**	**15**	25	**40**	μs
POWER SUPPLY							
V+	**+4.5**	+5.0	**+7.0**	**+4.5**	+5.0	**+7.0**	V
V−	**−12.0**	−15	**−16.5**	**−12.0**	−15	**−16.5**	V
OPERATING CURRENT							
V+		7	**10**		7	**10**	mA
V−		9	**15**		9	**15**	mA
PACKAGE OPTION							
Ceramic (D-18)		AD570JD			AD579SD		

NOTES

[1]The AD570 is a selected version of the AD571 10-bit A-to-D converter. Only TTL logic inputs should be connected to Pins 1 and 18 (or no connection made) or damage may result.

Specifications subject to change without notice.

Specifications shown in boldface are tested on all production units at final electrical test. Results from those tests are used to calculate outgoing quality levels. All min and max specifications are guaranteed, although only those shown in boldface are tested on all production units.

SPECIFICATIONS

(T_A = +25°C, V+ = +5V, V− = −12V or −15V, all voltages measured with respect to digital common, unless otherwise indicated)

Model	AD571J Min	AD571J Typ	AD571J Max	AD571K Min	AD571K Typ	AD571K Max	AD571S Min	AD571S Typ	AD571S Max	Units
RESOLUTION			10			10			10	Bits
RELATIVE ACCURACY, T_A			**±1**			**±1/2**			**±1**	LSB
T_{min} to T_{max}			**±1**			**±1/2**			**±1**	LSB
FULL-SCALE CALIBRATION		±2			±2			±2		LSB
UNIPOLAR OFFSET			**±1**			**±1/2**			**±1**	LSB
BIPOLAR ZERO			**±1**			**±1/2**			**±1**	LSB
DIFFERENTIAL NONLINEARITY, T_A	**10**			**10**			**10**			Bits
T_{min} to T_{max}	**9**			**10**			**10**			Bits
TEMPERATURE RANGE	0		+70	0		+70	55		+125	°C
TEMPERATURE COEFFICIENTS										
Unipolar Offset			±2			±1			±2	LSB
Bipolar Offset			±2			±1			±2	LSB
Full-Scale Calibration			±4			±2			±5	LSB
POWER SUPPLY REJECTION										
CMOS Positive Supply										
+13.5V≤V+≤+16.5V	–	–	–			±1	–	–	–	LSB
TTL Positive Supply										
+4.5V≤V+≤+5.5V			±2			±1			±2	LSB
Negative Supply										
−16.0V≤V−≤−13.5V			±2			±1			±2	LSB
ANALOG INPUT IMPEDANCE	**3.0**	5.0	**7.0**	**3.0**	5.0	**7.0**	**3.0**	5.0	7.0	kΩ

Model	AD571J Min	AD571J Typ	AD571J Max	AD571K Min	AD571K Typ	AD571K Max	AD571S Min	AD571S Typ	AD571S Max	Units
ANALOG INPUT RANGES										
Unipolar	0		+10	0		+10	0		+10	V
Bipolar	-5		+5	-5		+5	5		+5	V
OUTPUT CODING										
Unipolar	Positive True Binary			Positive True Binary			Positive True Binary			
Bipolar	Positive True Offset Binary			Positive True Offset Binary			Positive True Offset Binary			
LOGIC OUTPUT										
Output Sink Current (V_{OUT} = 0.4V max, T_{min} to T_{max})	**3.2**			**3.2**			**3.2**			mA
Output Source Current[1] (V_{OUT} = 2.4V max, T_{min} to T_{max})	**0.5**			**0.5**			**0.5**			mA
Output Leakage			**±40**			**±40**			**±40**	μA
LOGIC INPUTS										
Input Current			**±100**			**±100**			**±100**	μA
Logic "1"	**2.0**			**2.0**			**2.0**			V
Logic "0"			**0.8**			**0.8**			**0.8**	V
CONVERSION TIME										
T_{min} to T_{max}	**15**	25	**40**	**15**	25	**40**	**15**	25	**40**	μs
POWER SUPPLY										
V+	**+4.5**	+5.0	**+7.0**	**+4.5**	+5.0	**+16.5**	**+4.5**	+5.0	**+7.0**	V
V−	**−12.0**	−15	**−16.5**	**−12.0**	−15	**−16.5**	**−12.0**	15	**−16.5**	V
OPERATING CURRENT										
V+		7	**10**		7	**10**		7	**10**	mA
V−		9	**15**		9	**15**		9	**15**	mA
PACKAGE OPTION										
Ceramic (D-18)	AD571JD			AD571KD			AD571SD			

NOTES
[1]The data output lines have active pull-ups to source 0.5mA. The $\overline{\text{DATA READY}}$ line is open collector with a nominal 6kΩ internal pull-up resistor.
Specifications subject to change without notice.

Specifications shown in boldface are tested on all production units at final electrical test. Results from those tests are used to calculate outgoing quality levels. All min and max specifications are guaranteed, although only those shown in boldface are tested on all production units.

ABSOLUTE MAXIMUM RATINGS

V+ to Digital Common AD570J, S/AD571J, S	0 to +7V
AD571K	0 to +16.5V
V− to Digital Common	0 to −16.0V
Analog Common to Digital Common	±1V
Analog Input to Analog Common	±15V
Control Inputs	0 to V+
Digital Outputs (Blank Mode)	0 to V+
Power Dissipation	800mW

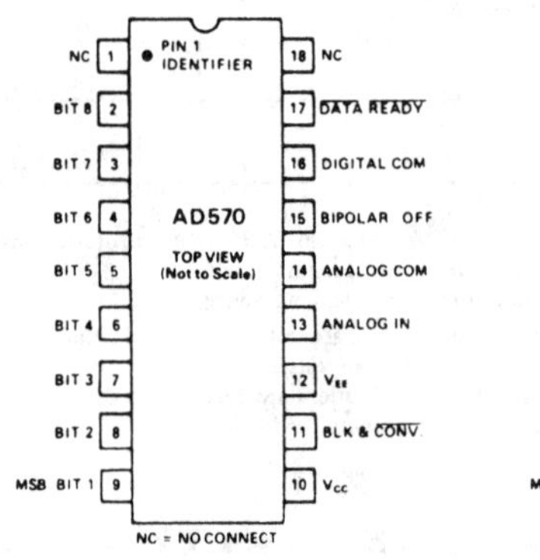

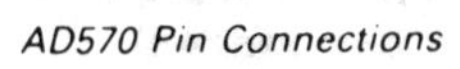
AD570 Pin Connections

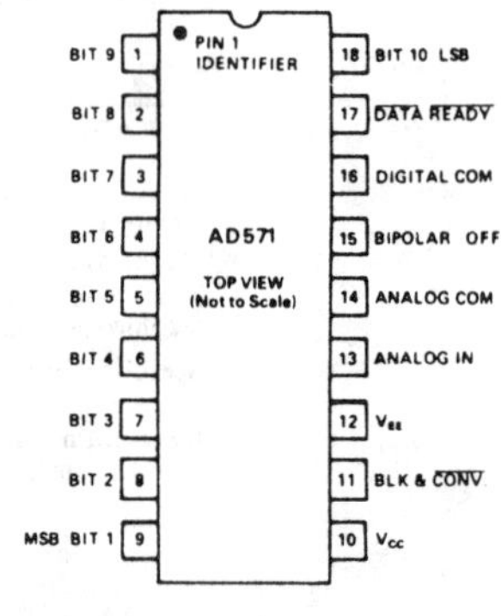

AD571 Pin Connections

AD570/AD571 FUNCTIONAL BLOCK DIAGRAM

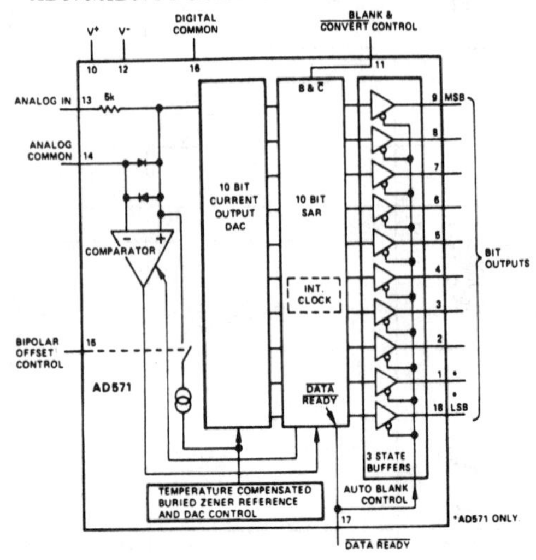

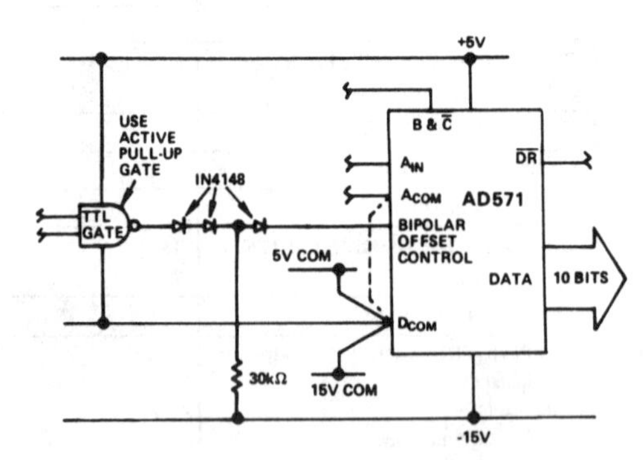

Bipolar Offset Controlled by Logic Gate
Gate Output = 1 Unipolar 0 − 10V Input Range
Gate Output = 0 Bipolar ±5V Input Range

Analog Devices AD572

12-Bit Successive Approximation Integrated Circuit A/D Converter

- Performance
  - True 12-bit operation: max nonlinearity < ±0.012%
  - Low gain T.C.: < ±15ppm/°C (AD572B)
  - Low power: 900mW
  - Fast conversion time: <25µs
  - Monotonic feedback DAC guarantees no missing codes
- Versatility
  - Aerospace temperature range: −55°C to +125°C (AD527S)
  - Positive-true serial or parallel logic outputs
  - Short-cycle capability
- Value
  - Precision +10V reference for external application
  - Internal buffer amplifier
  - High-reliability package

The AD572 is a complete 12-bit successive approximation analog-to-digital converter that includes an internal clock, reference, comparator, and buffer amplifier. Its hybrid IC design utilizes MSI digital and linear monolithic chips and active laser trimming of high-stability thin-film resistors to provide superior performance, flexibility, and ease of use, combined with IC size, price, and reliability.

Important performance characteristics of the AD572 include a maximum linearity error at 25°C of ±0.12%, gain T.C. below 15ppm/°C, typical power dissipation of 900mW, and conversion time of less than 25µs. Of considerable significance in aerospace applications is the guaranteed performance from −55°C to +125°C of the AD572S. Monotonic operation of the feedback D/A converter guarantees no missing output codes over temperature ranges of 0 to +70°C, −25°C to +85°C, and −55°C to +125°C.

The design of the AD572 includes scaling resistors that provide analog input signal ranges of ±2.5V, ±5.0V, ±10V, 0 to +5V, or 0 to +10V. Adding flexibility and value are the +10V precision reference, which also can be used for external applications, and the input buffer amplifier. All digital signals are fully DTL and TTL compatible, and the data output is positive-true and available in either serial or parallel form.

The new ceramic AD572 package reduces the predicted failure rate by a factor of two. The new package integrates the device substrate and package in a single ceramic element to eliminate a number of bond wires and interconnections.

The AD572 is available in three versions with differing guaranteed performance characteristics and operating temperature ranges; "A" and "B" are specified from −25°C to +85°C, and "S" from −55°C to +125°C.

An alternate offset adjust circuit, which contributes negligible offset tempco if metal film resistors (tempco < 100 ppm/°C) are used.

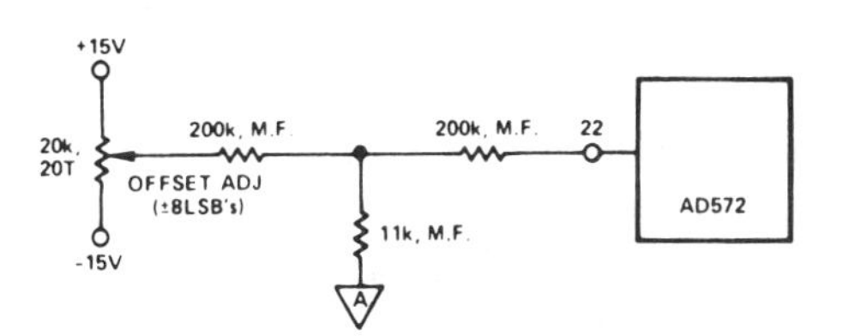

Low Tempco Zero Adj Circuit

SPECIFICATIONS (typical @ +25°C, ±15V and +5V unless otherwise noted)

MODEL	AD572AD	AD572BD	AD572SD
RESOLUTION	12 Bits	•	•
ANALOG INPUTS			
Voltage Ranges			
Bipolar	±2.5, ±5.0, ±10.0V	•	•
Unipolar	0 to +5, 0 to +10V	•	•
Impedance (Direct Input)			
0 to +5V, ±2.5V	2.5kΩ	•	•
0 to +10V, ±5V	5.0kΩ	•	•
±10V	10kΩ	•	•
Buffer Amplifier			
Impedance (min)	100MΩ	•	•
Bias Current	50nA	•	•
Settling Time			
to 0.01% of FSR for 20V step	2µs	•	•
DIGITAL INPUTS			
Convert Command	Note 1	•	•
Logic Loading	1 TTL Load	•	•
TRANSFER CHARACTERISTICS			
Gain Error (Note 2)	±0.05% FSR (Adj to Zero)	•	•
Unipolar Offset Error	±0.05% FSR (Adj to Zero)	•	•
Bipolar Offset Error	±0.1% FSR (Adj to Zero)	•	•
Linearity Error (max)	0.012% FSR	•	•

MODEL	AD572AD	AD572BD	AD572SD
Inherent Quantization Error	±½ LSB	*	*
Differential Linearity Error	±½ LSB	*	*
No Missing Codes	Guaranteed: 0 to +70°C	Guaranteed: -25°C to +85°C	Guaranteed: -55°C to +125°C
Power Supply Sensitivity			
±15V	±0.002% FSR/%ΔV_S	*	*
±5V	±0.001% FSR/%ΔV_S	*	*
TEMPERATURE COEFFICIENTS			
Gain (max)	±30ppm/°C (-25°C to +85°C)	±15ppm/°C (-25°C to +85°C)	±15ppm/°C (-25°C to +85°C) ±25ppm/°C (-55°C to +125°C)
Unipolar Offset	±3ppm FSR/°C	±5ppm FSR/°C (max)	**
Bipolar Offset (max)	±15ppm FSR/°C	±7ppm FSR/°C	**
Linearity	±3ppm FSR/°C	±2ppm FSR/°C	**
CONVERSION TIME (max)	25µs	*	*
DIGITAL OUTPUTS (All Codes Positive-True)			
Parallel Data			
Unipolar Code	Binary	*	*
Bipolar Code	Offset Binary/Two's Complement	*	*
Output Drive	2 TTL Loads	*	*
Serial Data (NRZ format)			
Unipolar Code	Binary	*	*
Bipolar Code	Offset Binary	*	*
Output Drive	2 TTL Loads	*	*
Status	Logic "1" during Conversion	*	*
$\overline{\text{Status}}$	Logic "0" during Conversion	*	*
Output Drive	2 TTL Loads	*	*
Internal Clock			
Output Drive	2 TTL Loads	*	*
Frequency	500kHz	*	*
INTERNAL REFERENCE VOLTAGE	+10.00V, ±10mV typ	*	*
Max External Current	±1mA	*	*
Voltage Temperature Coefficient (max)	±20ppm/°C	±10ppm/°C	*
POWER REQUIREMENTS			
Supply Voltages/Currents	+15V, ±5% @ +25mA (40 max)	*	*
	-15V, ±5% @ -20mA (35 max)	*	*
	+5V, ±5% @ +80mA (150 max)	*	*
Total Power Dissipation	925mW	*	*
TEMPERATURE RANGE			
Specification	-25°C to +85°C	*	-55°C to +125°C
Operating	-55°C to +125°C	*	*
Storage	-55°C to +150°C	*	*

NOTES
***Same specification as AD572AD.**
****Same specification as AD572BD.**

Specifications subject to change without notice.

Note 1 Positive pulse 200ns wide (min). Leading edge ("0" to "1") resets registers. Trailing edge ("1" to "0") initiates conversion.

Note 2 With 50Ω, 1% fixed resistor in place of Gain Adjust pot.

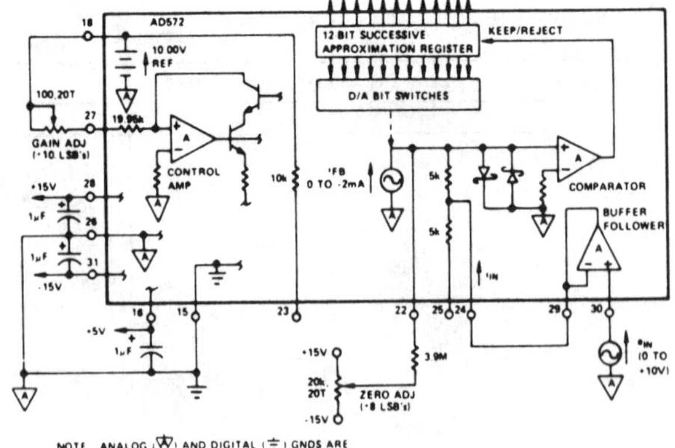

Analog and Power Connections for Unipolar 0 to +10V Input Range with Buffer Follower

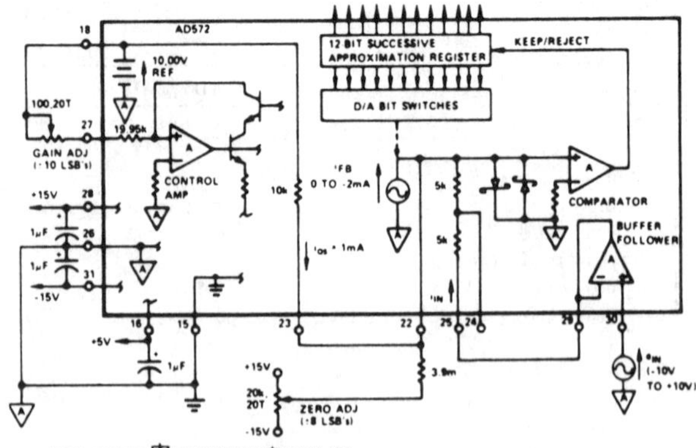

Analog and Power Connections for Bipolar -10V to +10V Input Range with Buffer Follower

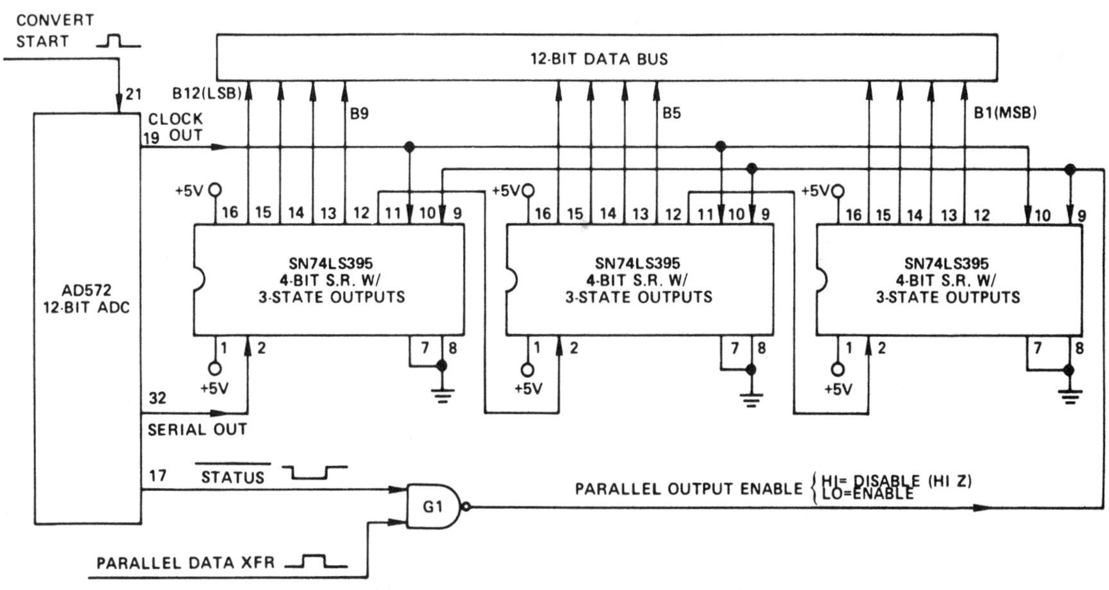

Serial Data Transfer Into Shift Register With Parallel Output to Data Bus

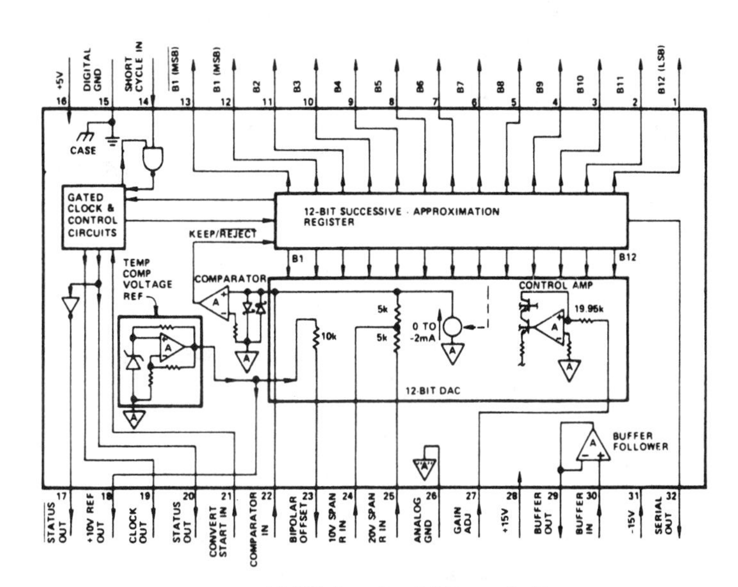

AD572 Functional Diagram & Pinout

Analog Devices AD573*

10-Bit A/D Converter

- Complete 10-bit A/D converter with reference, clock, and comparator
- Full 8- or 16-bit microprocessor bus interface
- Fast successive approximation conversion: 20μs typ
- No missing codes over temperature
- Operates on +5V and −12V to −15V supplies
- Low-cost monolithic construction

The AD573 is a complete 10-bit successive approximation analog to digital converter consisting of a DAC, voltage reference, clock, comparator, successive approximation register (SAR) and three-state output buffers—all fabricated on a single chip. No external components are required to perform a full-accuracy 10-bit conversion in 20μs.

The AD573 incorporates advanced integrated circuit design and processing technologies. The successive approximation function is implemented with I^2L (integrated injection logic). Laser trimming of the high-stability SiCr thin-film resistor ladder network ensures high accuracy, which is maintained with a temperature-compensated subsurface Zener reference.

Operating on supplies of +5V and −12V to −15V, the AD573 will accept analog inputs of 0 to +10V or −5V to +5V. The trailing edge of a positive pulse on the CONVERT line initiates the 20μs conversion cycle. $\overline{\text{DATA READY}}$ indicates completion of the conversion. $\overline{\text{HIGH BYTE ENABLE (HBE)}}$ and $\overline{\text{LOW BYTE ENABLE (LBE)}}$ control the 8-bit and 2-bit three state output buffers.

The AD573 is available in two versions for the 0 to +70 °C temperature range, the AD573J and AD573K. The AD573S guarantees ±1LSB relative accuracy and no missing codes from −55 °C to +125 °C.

Two package configurations are offered. All versions are also offered in a 20-pin hermetically sealed ceramic DIP. The AD573J and AD573K are also available in a 20-pin plastic DIP.

*Protected by U.S. Patent Nos.: 3,940,760; 4,213,806; 4,136,349; 4,400,689; and 4,400,690.

ABSOLUTE MAXIMUM RATINGS

Parameter	Rating
V+ to Digital Common	0 to +7V
V− to Digital Common	0 to −16.5V
Analog Common to Digital Common	±1V
Analog Input to Analog Common	±15V
Control Inputs	0 to V+
Digital Outputs (High Impedance State)	0 to V+
Power Dissipation	800mW

AD573 FUNCTIONAL BLOCK DIAGRAM

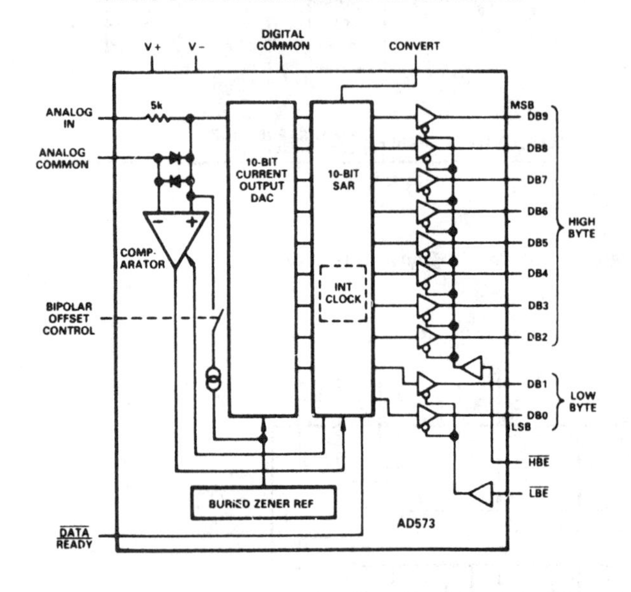

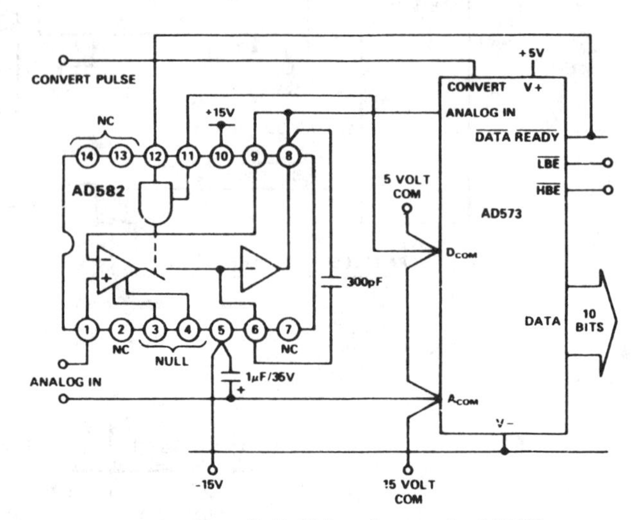

Sample-Hold Interface to the AD573

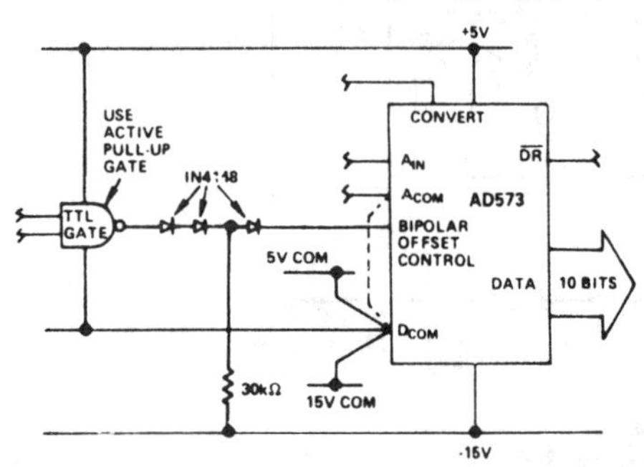

Bipolar Offset Controlled by Logic Gate
Gate Output = 1 Unipolar 0 – 10V Input Range
Gate Output = 0 Bipolar ±5V Input Range

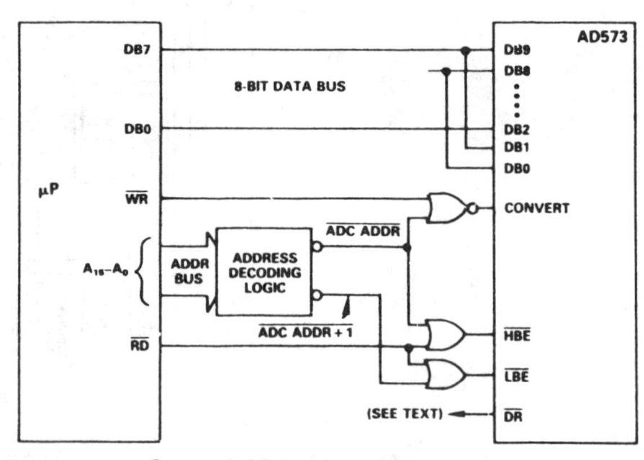

General AD573 Interface to 8-Bit Microprocessor

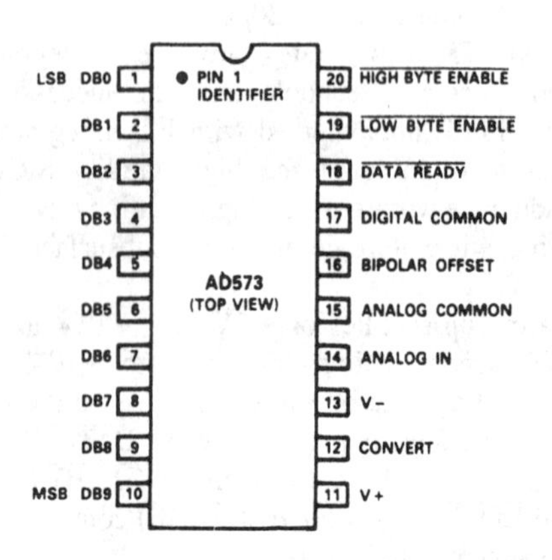

AD573 Pin Connections

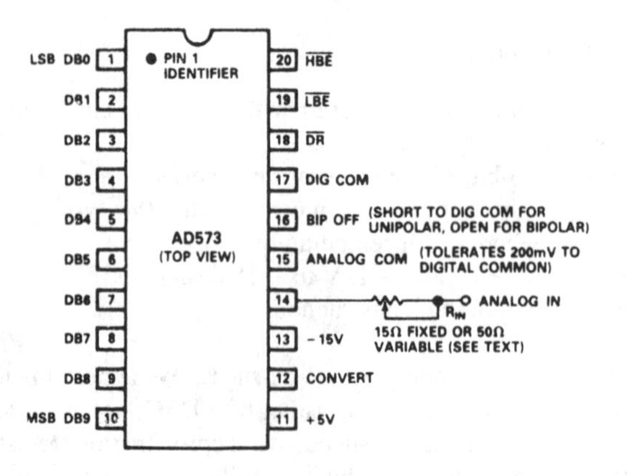

Standard AD573 Connections

SPECIFICATIONS (T_A = 25°C, V+ = +5V, V− = −12V or −15V, all voltages measured with respect to digital common, unless otherwise indicated)

Model	AD573J Min	AD573J Typ	AD573J Max	AD573K Min	AD573K Typ	AD573K Max	AD573S Min	AD573S Typ	AD573S Max	Units
RESOLUTION		10			10			10		Bits
RELATIVE ACCURACY[1]			±1			±1/2			±1	LSB
$T_A = T_{min}$ to T_{max}			±1			±1/2			±1	LSB
FULL SCALE CALIBRATION[2]		±2			±2				±2	LSB
UNIPOLAR OFFSET			±1			±1/2			±1	LSB
BIPOLAR OFFSET			±1			±1/2			±1	LSB
DIFFERENTIAL NONLINEARITY[3]	**10**			**10**			**10**			Bits
$T_A = T_{min}$ to T_{max}	**9**			**10**			**10**			Bits
TEMPERATURE RANGE	0		+70	0		+70	−55		+125	°C
TEMPERATURE COEFFICIENTS[4]										
Unipolar Offset			±2			±1			±2	LSB
Bipolar Offset			±2			±1			±2	LSB
Full Scale Calibration[2]			±4			±2			±5	LSB
POWER SUPPLY REJECTION										
Positive Supply										
+4.5≤V+≤+5.5V			±2			±1			±2	LSB
Negative Supply										
−15.75V≤V−≤−14.25V			±2			±1			±2	LSB
−12.6V≤V−≤−11.4V			±2			±1			±2	LSB
ANALOG INPUT IMPEDANCE	3.0	5.0	7.0	3.0	5.0	7.0	3.0	5.0	7.0	kΩ
ANALOG INPUT RANGES										
Unipolar	0		+10	0		+10	0		+10	V
Bipolar	−5		+5	−5		+5	−5		+5	V
OUTPUT CODING										
Unipolar	Positive True Binary			Positive True Binary			Positive True Binary			
Bipolar	Positive True Offset Binary			Positive True Offset Binary			Positive True Offset Binary			
LOGIC OUTPUT										
Output Sink Current										
(V_{OUT} = 0.4V max, T_{min} to T_{max})	**3.2**			**3.2**			**3.2**			mA
Output Source Current[5]										
(V_{OUT} = 2.4V max, T_{min} to T_{max})	**0.5**			**0.5**			**0.5**			mA
Output Leakage			**±40**			**±40**			**±40**	μA
LOGIC INPUTS										
Input Current			**±100**			**±100**			**±100**	μA
Logic "1"	**2.0**			**2.0**			**2.0**			V
Logic "0"			**0.8**			**0.8**			**0.8**	V
CONVERSION TIME										
$T_A = T_{min}$ to T_{max}	**10**	20	**30**	**10**	20	**30**	**10**	20	**30**	μs
POWER SUPPLY										
V+	**+4.5**	+5.0	**+7.0**	**+4.5**	+5.0	**+7.0**	**+4.5**	+5.0	**+7.0**	V
V−	**−11.4**	−15	**−16.5**	**+11.4**	−15	**−16.5**	**−11.4**	−15	**−16.5**	V
OPERATING CURRENT										
V+		15	**20**		15	**20**		15	**20**	mA
V−		9	**15**		9	**15**		9	**15**	mA

NOTES

[1]Relative accuracy is defined as the deviation of the code transition points from the ideal transfer point on a straight line from the zero to the full scale of the device.

[2]Full-scale calibration is guaranteed trimmable to zero with an external 50Ω potentiometer in place of the 15Ω fixed resistor. Full scale is defined as 10 volts minus 1LSB, or 9.990 volts.

[3]Defined as the resolution for which no missing codes will occur.

[4]Change from +25°C value from +25°C to T_{min} or T_{max}.

[5]The data output lines have active pull-ups to source 0.5mA. The DATA READY line is open collector with a nominal 6kΩ internal pull-up resistor.

Specifications subject to change without notice.

Specifications shown in boldface are tested on all production units at final electrical test. Results from those tests are used to calculate outgoing quality levels. All min and max specifications are guaranteed, although only those shown in boldface are tested on all production units.

Analog Devices AD574A*

Complete 12-Bit A/D Converter

- Complete 12-bit A/D converter with reference and clock
- 8- and 16-bit microprocessor bus interface
- Guaranteed linearity over temperature
  0 to +70°C—AD574J, K, L
  −55°C to +125°C—AD574AS, T, U
- No missing codes over temperature
- 35μs maximum conversion time
- Buried zener reference for long-term stability and low-gain T.C. 10ppm/°C max AD574AL, 12.5ppm/°C max AD574AU
- Ceramic DIP, plastic DIP, or PLCC package

The AD574A is a complete 12-bit successive-approximation analog-to-digital converter with 3-state output-buffer circuitry for direct interface to an 8- or 16-bit microprocessor bus.

A high-precision voltage reference and clock are included on-chip, and the circuit guarantees full-rated performance without external circuitry or clock signals.

The AD574A design is implemented using Analog Devices' Bipolar/I²L process, and all analog and digital functions are integrated on one chip. Offset, linearity, and scaling errors are minimized by active laser-trimming of thin-film resistors at the wafer stage. The voltage reference uses an implanted buried Zener for low noise and low drift. On the digital side, I²L logic is used for the successive-approximation register, control circuitry, and three-state output buffers.

The AD574A is available in six different grades. The AD574AJ, K, and L grades are specified for operation over the 0 to +70°C temperature range. The AD574AS, T, and U are specified for the −55°C to +125°C range. All grades are available in a 28-pin hermetically-sealed ceramic DIP. The J, K, and L grades are also available in a 28-pin plastic DIP and PLCC.

The S, T, and U grades are available with optional processing to MIL-STD-883C Class B. The Analog Devices' Military Products Databook should be consulted for details on /883B testing of the AD574A.

*Protected by U.S. Patent Nos. 3,803,590; 4,213,806; 4,511,413; RE 28,633.

- The AD574A interfaces to most 8- or 16-bit microprocessors. Multiple-mode three-state output buffers connect directly to the data bus while the read and convert commands are taken from the control bus. The 12 bits of output data can be read either as one 12-bit word or as two 8-bit bytes (one with 8 data bits, the other with 4 data bits and 4 trailing zeros).
- The precision laser-trimmed scaling and bipolar offset resistors provide four calibrated ranges: 0 to +10 and 0 to +20V unipolar, −5 to +5 and −10 to +10V bipolar. Typical bipolar offset and full-scale calibration errors of ±0.1% can be trimmed to zero with one external component each.
- The internal buried Zener reference is trimmed to 10.00V with 0.2% maximum error and 15ppm/°C typical T.C. The reference is available externally and can drive up to 1.5mA beyond the requirements of the reference and bipolar offset resistors.

SPECIFICATIONS (@ +25°C with V_{CC} = +15V or +12V, V_{LOGIC} = +5V, V_{EE} = −15V or −12V unless otherwise indicated)

Model	AD574AJ Min	AD574AJ Typ	AD574AJ Max	AD574AK Min	AD574AK Typ	AD574AK Max	AD574AL Min	AD574AL Typ	AD574AL Max	Units
RESOLUTION			12			12			12	Bits
LINEARITY ERROR @ +25°C			±1			±1/2			±1/2	LSB
T_{min} to T_{max}			±1			±1/2			±1/2	LSB
DIFFERENTIAL LINEARITY ERROR (Minimum resolution for which no missing codes are guaranteed)										
T_{min} to T_{max}	11			12			12			Bits
UNIPOLAR OFFSET (Adjustable to zero)			±2			±1			±1	LSB
BIPOLAR OFFSET (Adjustable to zero)			±4			±4			±2	LSB
FULL-SCALE CALIBRATION ERROR (with fixed 50Ω resistor from REF OUT to REF IN) (Adjustable to zero)			0.25			0.25			0.125	% of F.S.
TEMPERATURE RANGE	0		+70	0		+70	0		+70	°C
TEMPERATURE COEFFICIENTS (Using internal reference)										
T_{min} to T_{max}										
Unipolar Offset			±2 (10)			±1 (5)			±1 (5)	LSB (ppm/°C)
Bipolar Offset			±2 (10)			±1 (5)			±1 (5)	LSB (ppm/°C)
Full-Scale Calibration			±9 (50)			±5 (27)			±2 (10)	LSB (ppm/°C)
POWER SUPPLY REJECTION										
Max change in Full Scale Calibration										
V_{CC} = 15V ± 1.5V or 12V ± 0.6V			±2			±1			±1	LSB
V_{LOGIC} = 5V ± 0.5V			±1/2			±1/2			±1/2	LSB
V_{EE} = −15V ± 1.5V or −12V ± 0.6V			±2			±1			±1	LSB
ANALOG INPUT										
Input Ranges										
Bipolar	−5		+5	−5		+5	−5		+5	Volts
	−10		+10	−10		+10	−10		+10	Volts
Unipolar	0		+10	0		+10	0		+10	Volts
	0		+20	0		+20	0		+20	Volts
Input Impedance										
10 Volt Span	3	5	7	3	5	7	3	5	7	kΩ
20 Volt Span	6	10	14	6	10	14	6	10	14	kΩ
DIGITAL CHARACTERISTICS[1] (T_{min}–T_{max})										
Inputs[2] (CE, $\overline{CS}$, R/$\overline{C}$, A_0)										
Logic "1" Voltage	+2.0		+5.5	+2.0		+5.5	+2.0		+5.5	Volts
Logic "0" Voltage	−0.5		+0.8	−0.5		+0.8	−0.5		+0.8	Volts

NOTES
[1]Detailed Timing Specifications appear in the Timing Section.
[2]12/$\overline{8}$ Input is not TTL-compatible and must be hard wired to V_{LOGIC} or Digital Common.
[3]The reference should be buffered for operation on ± 12V supplies.

Specifications shown in boldface are tested on all production units at final electrical test. Results from those tests are used to calculate outgoing quality levels. All min and max specifications are guaranteed, although only those shown in boldface are tested on all production units.

Specifications subject to change without notice.

Model	AD574AS Min	AD574AS Typ	AD574AS Max	AD574AT Min	AD574AT Typ	AD574AT Max	AD574AU Min	AD574AU Typ	AD574AU Max	Units
Current	−20		+20	−20		+20	−20		+20	μA
Capacitance		5			5			5		pF
Outputs (DB11–DB0, STS)										
Logic "1" Voltage ($I_{SOURCE} \leq 500\mu A$)	+2.4			+2.4			+2.4			Volts
Logic "0" Voltage ($I_{SINK} \leq 1.6mA$)			+0.4			+0.4			+0.4	Volts
Leakage (DB11–DB0, High-Z State)	−20		+20	−20		+20	−20		+20	μA
Capacitance		5			5			5		pF
POWER SUPPLIES										
Operating Range										
V_{LOGIC}	+4.5		+5.5	+4.5		+5.5	+4.5		+5.5	Volts
V_{CC}	+11.4		+16.5	+11.4		+16.5	+11.4		+16.5	Volts
V_{EE}	−11.4		−16.5	−11.4		−16.5	−11.4		−16.5	Volts
Operating Current										
I_{LOGIC}		30	40		30	40		30	40	mA
I_{CC}		2	5		2	5		2	5	mA
I_{EE}		18	30		18	30		18	30	mA
POWER DISSIPATION		390	725		390	725		390	725	mW
INTERNAL REFERENCE VOLTAGE	9.98	10.0	10.02	9.98	10.0	10.02	9.99	10.0	10.01	Volts
Output current (available for external loads)[3]			1.5			1.5			1.5	mA
(External load should not change during conversion)										
PACKAGE OPTIONS										
Ceramic (D-28)		AD574ASD			AD574AKD			AD574ALD		
Plastic (N-28)		AD574AJN			AD574AKN			AD574ALN		
PLCC (P-28A)		AD574AJP			AD574AKP					
RESOLUTION			12			12			12	Bits
LINEARITY ERROR @ +25°C			±1			±1/2			±1/2	LSB
T_{min} to T_{max}			±1			±1			±1	LSB
DIFFERENTIAL LINEARITY ERROR										
(Minimum resolution for which no missing codes are guaranteed)										
T_{min} to T_{max}	11			12			12			Bits
UNIPOLAR OFFSET (Adjustable to zero)			±2			±1			±1	LSB
BIPOLAR OFFSET (Adjustable to zero)			±4			±4			±2	LSB
FULL-SCALE CALIBRATION ERROR										
(with fixed 50Ω resistor from REF OUT to REF IN)										
(Adjustable to zero)			0.25			0.25			0.125	% of F.S.
TEMPERATURE RANGE	−55		+125	−55		+125	−55		+125	°C
TEMPERATURE COEFFICIENTS										
(Using internal reference)										
T_{min} to T_{max}										
Unipolar Offset			±2 (5)			±1 (2.5)			±1 (2.5)	LSB (ppm/°C)
Bipolar Offset			±4 (10)			±2 (5)			±1 (2.5)	LSB (ppm/°C)
Full-Scale Calibration			±20 (50)			±10 (25)			±5 (12.5)	LSB (ppm/°C)
POWER SUPPLY REJECTION										
Max change in Full Scale Calibration										
V_{CC} = 15V ± 1.5V or 12V ± 0.6V			±2			±1			±1	LSB
V_{LOGIC} = 5V ± 0.5V			±1/2			±1/2			±1/2	LSB
V_{EE} = −15V ± 1.5V or −12V ± 0.6V			±2			±1			±1	LSB
ANALOG INPUT										
Input Ranges										
Bipolar	−5		+5	−5		+5	−5		+5	Volts
	−10		+10	−10		+10	−10		+10	Volts
Unipolar	0		+10	0		+10	0		+10	Volts
	0		+20	0		+20	0		+20	Volts
Input Impedance										
10 Volt Span	3	5	7	3	5	7	3	5	7	kΩ
20 Volt Span	6	10	14	6	10	14	6	10	14	kΩ
DIGITAL CHARACTERISTICS[1] (T_{min}–T_{max})										
Inputs[2] (CE, $\overline{CS}$, R/$\overline{C}$, A_0)										
Logic "1" Voltage	+2.0		+5.5	+2.0		+5.5	+2.0		+5.5	Volts
Logic "0" Voltage	−0.5		+0.8	−0.5		+0.8	−0.5		+0.8	Volts
Current	−20		+20	−20		+20	−20		+20	μA
Capacitance		5			5			5		pF
Outputs (DB11–DB0, STS)										
Logic "1" Voltage ($I_{SOURCE} \leq 500\mu A$)	+2.4			+2.4			+2.4			Volts
Logic "0" Voltage ($I_{SINK} \leq 1.6mA$)			+0.4			+0.4			+0.4	Volts
Leakage (DB11–DB0, High-Z State)	−20		+20	−20		+20	−20		+20	μA
Capacitance		5			5			5		pF

NOTES

[1]Detailed Timing Specifications appear in the Timing Section.
[2]12/$\overline{8}$ Input is not TTL-compatible and must be hard wired to V_{LOGIC} or Digital Common.
[3]The reference should be buffered for operation on ± 12V supplies.

Specifications subject to change without notice.

Specifications shown in boldface are tested on all production units at final electrical test. Results from those tests are used to calculate outgoing quality levels. All min and max specifications are guaranteed, although only those shown in boldface are tested on all production units.

Model	AD574AJ Min	Typ	Max	AD574AK Min	Typ	Max	AD574AL Min	Typ	Max	Units
POWER SUPPLIES										
Operating Range										
V_{LOGIC}	+4.5		+5.5	+4.5		+5.5	+4.5		+5.5	Volts
V_{CC}	+11.4		+16.5	+11.4		+16.5	+11.4		+16.5	Volts
V_{EE}	−11.4		−16.5	−11.4		−16.5	−11.4		−16.5	Volts
Operating Current										
I_{LOGIC}		30	**40**		30	**40**		30	**40**	mA
I_{CC}		2	5		2	5		2	5	mA
I_{EE}		18	**30**		18	**30**		18	**30**	mA
POWER DISSIPATION		390	725		390	725		390	725	mW
INTERNAL REFERENCE VOLTAGE	**9.98**	10.0	**10.02**	**9.98**	10.0	**10.02**	**9.99**	10.0	**10.01**	Volts
Output current (available for external loads)[3]			1.5			1.5			1.5	mA
(External load should not change during conversion)										
PACKAGE OPTIONS										
Ceramic (D-28)		AD574ASD			AD574ATD			AD574AUD		

NOTES
[1]Detailed Timing Specifications appear in the Timing Section.
[2]12/$\overline{8}$ Input is not TTL-compatible and must be hard wired to V_{LOGIC} or Digital Common.
[3]The reference should be buffered for operation on ± 12V supplies.

Specifications shown in boldface are tested on all production units at final electrical test. Results from those tests are used to calculate outgoing quality levels. All min and max specifications are guaranteed, although only those shown in boldface are tested on all production units.

Specifications subject to change without notice.

ABSOLUTE MAXIMUM RATINGS*

(Specifications apply to all grades, except where noted)

V_{CC} to Digital Common . . . . 0 to +16.5V
V_{EE} to Digital Common . . . . 0 to −16.5V
V_{LOGIC} to Digital Common . . . . 0 to +7V
Analog Common to Digital Common . . . . ±1V
Control Inputs (CE, $\overline{CS}$, A_O, 12/$\overline{8}$, R/$\overline{C}$) to Digital Common . . −0.5V to V_{LOGIC} +0.5V
Analog Inputs (REF IN, BIP OFF, $10V_{IN}$) to Analog Common . . . . V_{EE} to V_{CC}
$20V_{IN}$ to Analog Common . . . . ±24V
REF OUT . . . . Indefinite short to common
Momentary short to V_{CC}
Chip Temperature . . . . 175°C
Power Dissipation 825mW
Lead Temperature, Soldering . . . . +300°C, 10 sec.
Storage Temperature (Ceramic) . . . . −65°C to +150°C
(Plastic) . . . . −25°C to +100°C

*Stresses above those listed under "Absolute Maximum Ratings" may cause permanent damage to the device. This is a stress rating only and functional operation of the device at these or any other conditions above those indicated in the operational sections of this specification is not implied. Exposure to absolute maximum rating conditions for extended periods may affect device reliability.

AD574A ORDERING GUIDE

Model*	Temp. Range	Linearity Error Max (T_{min} to T_{max})	Resolution No Missing Codes (T_{min} to T_{max})	Max Full Scale T.C. (ppm/°C)
AD574AJ(X)	0 to +70°C	±1LSB	11 Bits	50.0
AD574AK(X)	0 to +70°C	±1/2LSB	12 Bits	27.0
AD574AL(X)	0 to +70°C	±1/2LSB	12 Bits	10.0
AD574AS(X)	−55°C to +125°C	±1LSB	11 Bits	50.0
AD574AT(X)	−55°C to +125°C	±1LSB	12 Bits	25.0
AD574AU(X)	−55°C to +125°C	±1LSB	12 Bits	12.5

NOTES
*X = Package designator. Available packages are:
D (D-28) for all grades.
E (E-28) for J, K, S, T, U grades.
N (N-28) for J, K, and L grades.
P for PLCC in J, K grades.
Example: AD574AKN is K grade in plastic DIP.

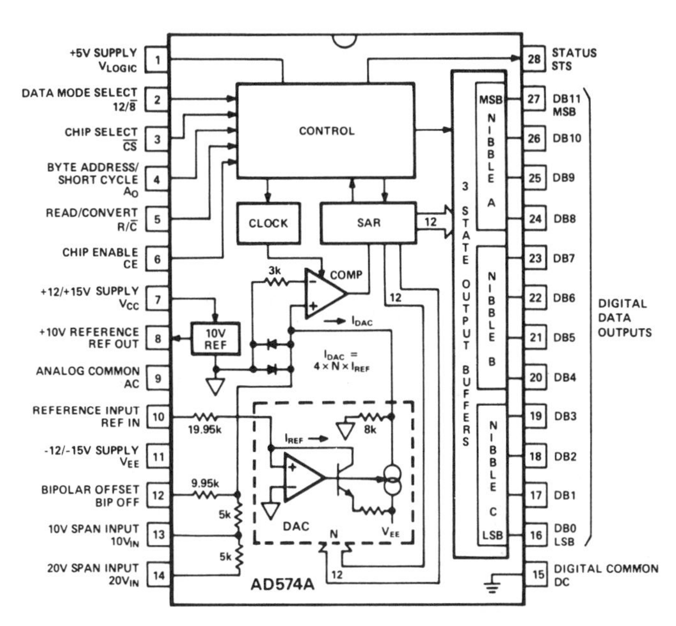
AD574A Block Diagram and Pin Configuration

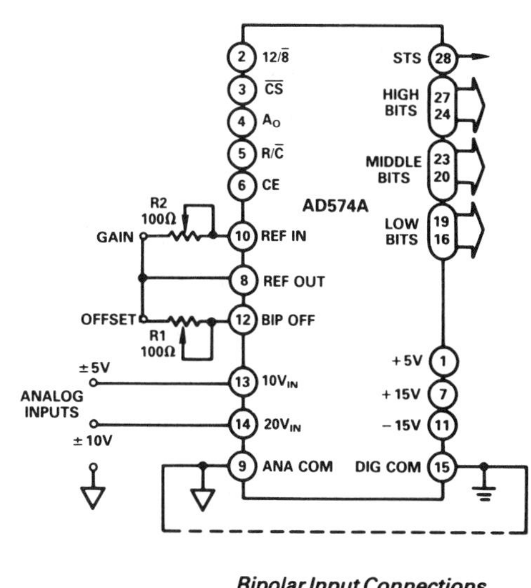
Bipolar Input Connections

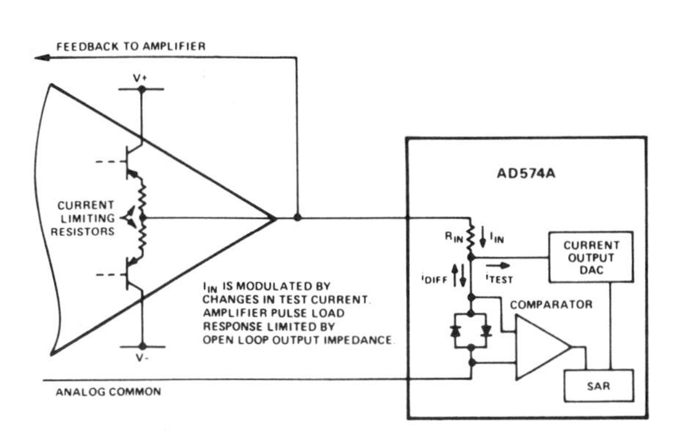
Op Amp – AD574A Interface

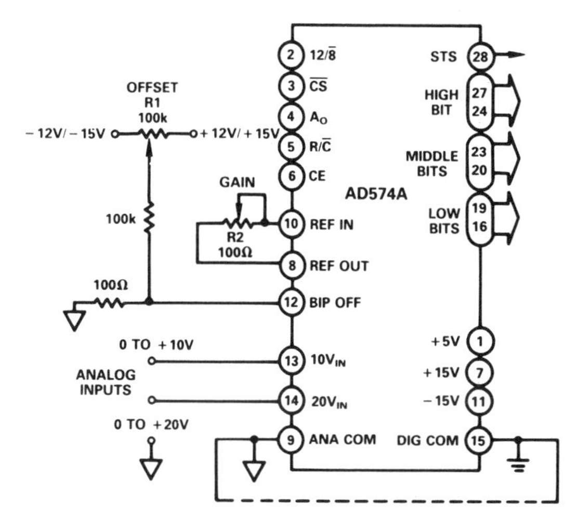
Unipolar Input Connections

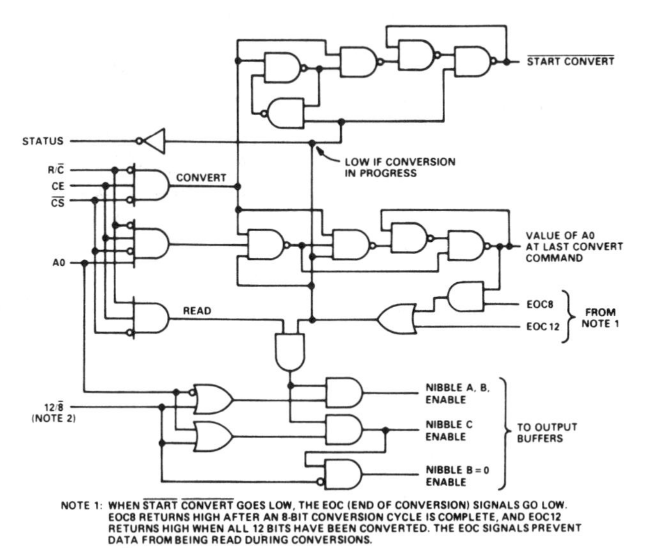
NOTE 1: WHEN START CONVERT GOES LOW, THE EOC (END OF CONVERSION) SIGNALS GO LOW. EOC8 RETURNS HIGH AFTER AN 8-BIT CONVERSION CYCLE IS COMPLETE, AND EOC12 RETURNS HIGH WHEN ALL 12 BITS HAVE BEEN CONVERTED. THE EOC SIGNALS PREVENT DATA FROM BEING READ DURING CONVERSIONS.

NOTE 2: 12/8 IS NOT A TTL-COMPATIBLE INPUT AND SHOULD ALWAYS BE WIRED DIRECTLY TO V_{LOGIC} OR DIGITAL COMMON.

AD574A Control Logic

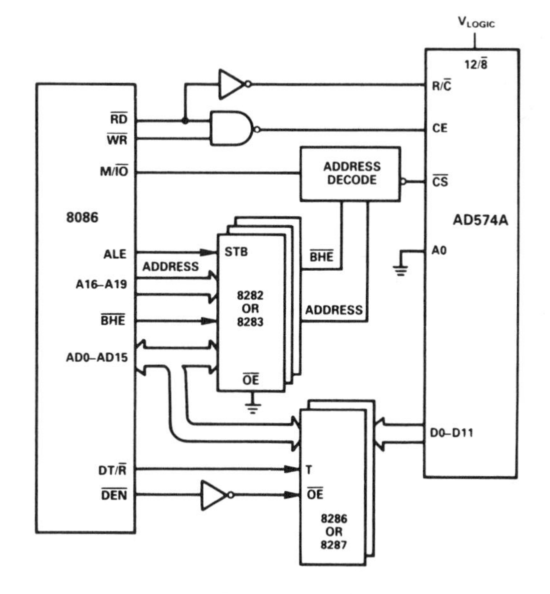
8086 – AD574A with Buffered Bus Interface

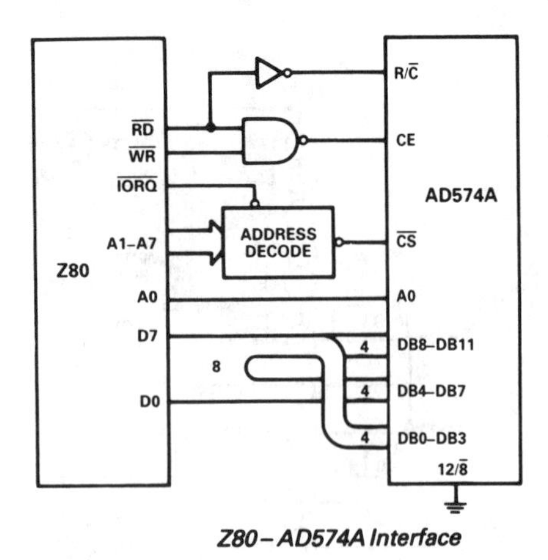

Z80-AD574A Interface

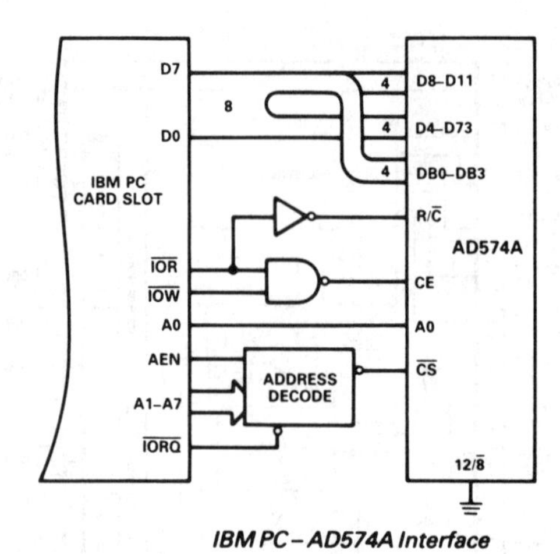

IBM PC-AD574A Interface

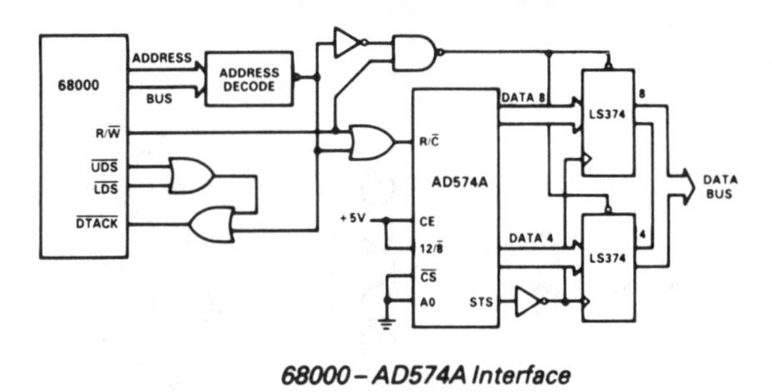

68000-AD574A Interface

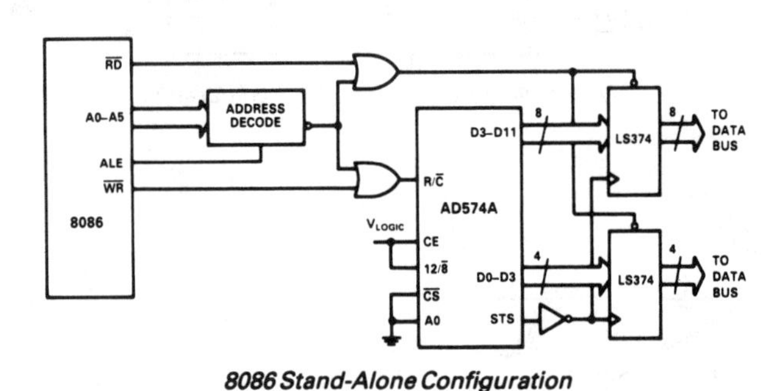

8086 Stand-Alone Configuration

Analog Devices AD673*
8-Bit A/D Converter

- Complete 8-bit A/D converter with reference, clock, and comparator
- 30μs maximum conversion time
- Full 8- or 16-bit microprocessor bus interface
- Unipolar and bipolar inputs
- No missing codes over temperature
- Operates on +5V and −12V to −15V supplies

The AD673 is a complete 8-bit successive-approximation analog-to-digital converter consisting of a DAC, voltage reference, clock, comparator, successive approximation register (SAR), and three-state output buffers—all fabricated on a single chip. No external components are required to perform a full-accuracy 8-bit conversion in 20μs.

The AD673 incorporates advanced integrated circuit design and processing technologies. The successive approximation function is implemented with I^2L (integrated injection logic). Laser trimming of the high-stability SiCr thin-film resistor ladder network insures high accuracy, which is maintained with a temperature-compensated subsurface Zener reference.

Operating on supplies of +5V and −12C to −15V, the AD673 will accept analog inputs of 0 to +10V or −5V to +5V. The trailing edge of a positive pulse on the CONVERT line initiates the 20μs conversion cycle. $\overline{\text{DATA READY}}$ indicates completion of the conversion.

The AD673 is available in two versions. The AD673J as specified over the 0 to +70°C temperature range and the AD673S guarantees ±1/2LSB relative accuracy and no missing codes from −55°C to +125°C.

Two package configurations are offered. All versions are also offered in a 20-pin hermetically sealed ceramic DIP. The AD673J is also available in a 20-pin plastic DIP.

*Protected by U.S. Patent Nos.; 3,940,760; 4,213,806; 4,136,349; 4,400,689; and 4,400,690

SPECIFICATIONS (T_A = 25°C, V+ = +5V, V− = −12V or −15V, all voltages measured with respect to digital common, unless otherwise indicated)

Model	AD673J Min	AD673J Typ	AD673J Max	AD673S Min	AD673S Typ	AD673S Max	Units
RESOLUTION		8			8		Bits
RELATIVE ACCURACY,[1]			±1/2			±1/2	LSB
$T_A = T_{min}$ to T_{max}			±1/2			±1/2	LSB
FULL SCALE CALIBRATION[2]		±2			±2		LSB
UNIPOLAR OFFSET			±1/2			±1/2	LSB
BIPOLAR OFFSET			±1/2			±1/2	LSB
DIFFERENTIAL NONLINEARITY,[3]	**8**			**8**			Bits
$T_A = T_{min}$ to T_{max}	**8**			**8**			Bits
TEMPERATURE RANGE	0		+70	−55		+125	°C
TEMPERATURE COEFFICIENTS							
Unipolar Offset			**±1**			**±1**	LSB
Bipolar Offset			**±1**			**±1**	LSB
Full Scale Calibration[2]			**±2**			**±2**	LSB
POWER SUPPLY REJECTION							
Positive Supply							
+4.5≤V+≤+5.5V			**±2**			**±2**	LSB
Negative Supply							
−15.75V≤V−≤−14.25V			**±2**			**±2**	LSB
−12.6V≤V−≤−11.4V			**±2**			**±2**	LSB
ANALOG INPUT IMPEDANCE	3.0	5.0	7.0	3.0	5.0	7.0	kΩ
ANALOG INPUT RANGES							
Unipolar	0		+10	0		+10	V
Bipolar	−5		+5	−5		+5	V
OUTPUT CODING							
Unipolar	Positive True Binary			Positive True Binary			
Bipolar	Positive True Offset Binary			Positive True Offset Binary			
LOGIC OUTPUT							
Output Sink Current							
(V_{OUT} = 0.4V max, T_{min} to T_{max})	**3.2**			**3.2**			mA
Output Source Current[4]							
(V_{OUT} = 2.4V min, T_{min} to T_{max})	**0.5**			**0.5**			mA
Output Leakage			**±40**			**±40**	μA
LOGIC INPUTS							
Input Current			**±100**			**±100**	μA
Logic "1"	**2.0**			**2.0**			V
Logic "0"			**0.8**			**0.8**	V
CONVERSION TIME, T_A and T_{min} to T_{max}	10	20	**30**	10	20	**30**	μs
POWER SUPPLY							
V+	**+4.5**	+5.0	**+7.0**	**+4.5**	+5.0	**+7.0**	V
V−	−11.4	−15	−16.5	−11.4	−15	**−16.5**	V
OPERATING CURRENT							
V+		15	**20**		15	**20**	mA
V−		9	**15**		9	**15**	mA

NOTES

[1]Relative accuracy is defined as the deviation of the code transition points from the ideal transfer point on a straight line from the zero to the full scale of the device.

[2]Full scale calibration is guaranteed trimmable to zero with an external 200Ω potentiometer in place of the 15Ω fixed resistor.
Full scale is defined as 10 volts minus 1LSB, or 9.961 volts.

[3]Defined as the resolution for which no missing codes will occur.

[4]The data output lines have active pull-ups to source 0.5mA. The DATA READY line is open collector with a nominal 6kΩ internal pull-up resistor.

Specifications subject to change without notice.

Specifications shown in boldface are tested on all production units at final electrical test. Results from those tests are used to calculate outgoing quality levels. All min and max specifications are guaranteed, although only those shown in boldface are tested on all production units.

AD673 Functional Block Diagram

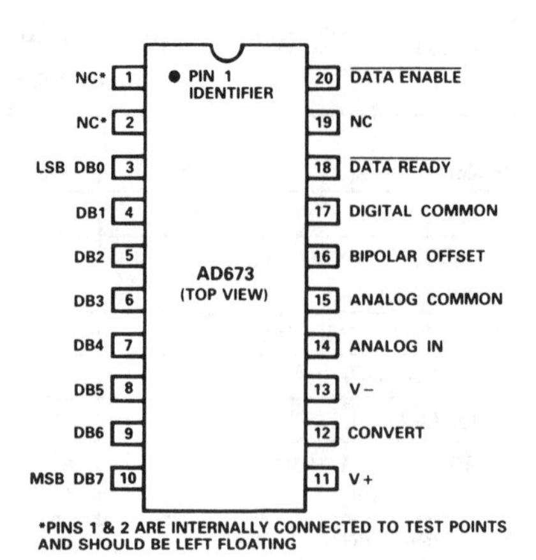

AD673 Pin Connections

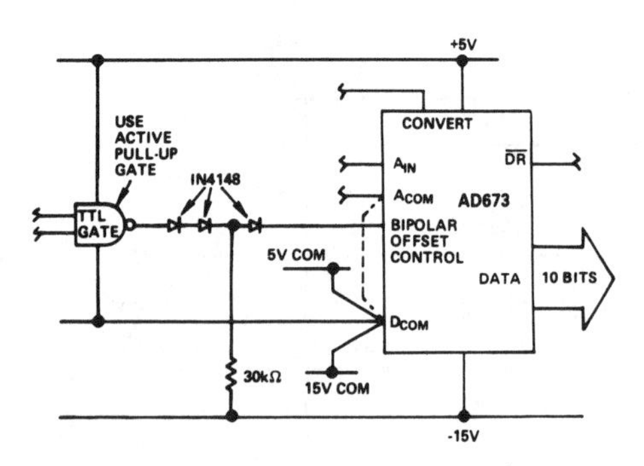

Bipolar Offset Controlled by Logic Gate
Gate Output = 1 Unipolar 0–10V Input Range
Gate Output = 0 Bipolar ±5V Input Range

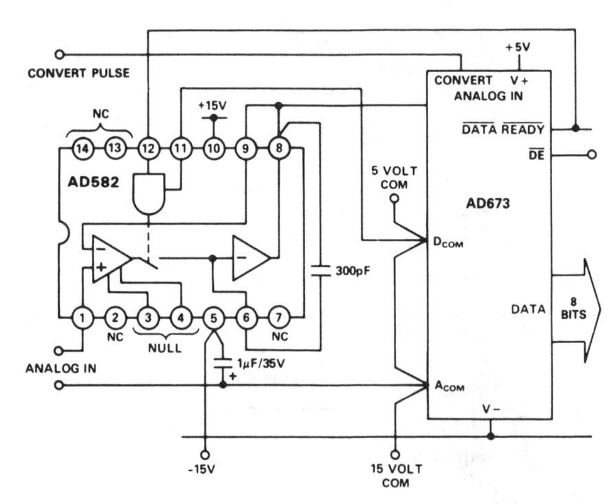

Sample-Hold Interface to the AD673

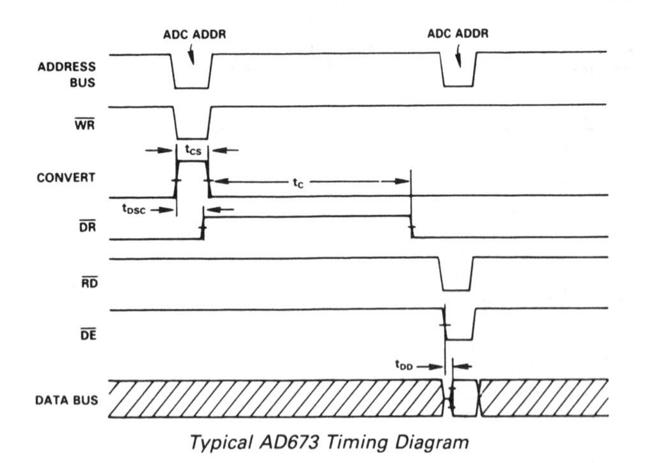

Typical AD673 Timing Diagram

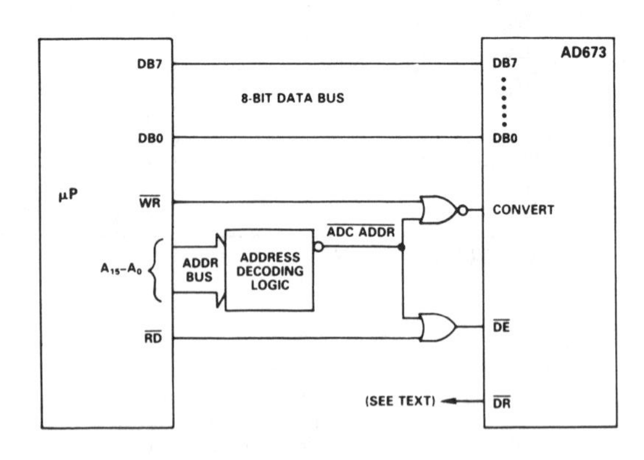

General AD673 Interface to 8-Bit Microprocessor

Analog Devices AD674A*

Complete 12-Bit A/D Converter

- Complete 12-bit A/D converter with reference and clock
- Faster version of AD574A
- 8- and 16-bit bus interface
- No missing codes over temperature
- 15μs max conversion time
- ±12V and ±15V operation
- Unipolar and bipolar inputs
- DIP package

The AD674A is a complete 12-bit successive-approximation analog-to-digital converter with three-state output buffer circuitry for direct interface to an 8- and 16-bit microprocessor bus. A high-precision voltage reference and clock are included on-chip, and the circuit requires only power supplies and control signals for operation.

The AD674A is pin compatible with the industry-standard AD574A, but it offers faster conversion time and bus-access speed. The AD674A design is implemented with two LSI chips, each containing both analog and digital circuitry, resulting in the maximum performance and flexibility at the lowest cost. The chips are laser trimmed at the wafer stage to obtain full-rated performance without external trims.

The AD674A is available in six different grades. The AD674AJ, K, and L grades are specified for operation over the 0 to +70 °C temperature range. The AD674AS, T, and U are specified for the −55 °C to +125 °C range. All grades are available in a 28-pin hermetically sealed ceramic DIP.

The S, T, and U grades are also available with optional processing to MIL-STD-883C Class B in 28-pin DIP. The Analog Devices Military Products Databook should be consulted for details on /883B testing of the AD674A.

*Protected by U.S. Patent Nos.: 3,803,590; 4,213,806; 4,511,413; RE 28,633.

SPECIFICATIONS (@ =25°C with V_{CC} = +15V or +12V, V_{LOGIC} = +5V, V_{EE} = −15V or −12V unless otherwise indicated)

Model	AD674AJ			AD674AK			AD674AL			Units
	Min	Typ	Max	Min	Typ	Max	Min	Typ	Max	
RESOLUTION			12			12			12	Bits
LINEARITY ERROR			±1			±1/2			±1/2	LSB
T_{min} to T_{max}			±1			±1/2			±1/2	LSB
DIFFERENTIAL LINEARITY ERROR (Minimum resolution for which no missing codes are guaranteed) T_{min} to T_{max}	11			12			12			Bits
UNIPOLAR OFFSET (Adjustable to zero)			±2			±2			±2	LSB
BIPOLAR OFFSET (Adjustable to zero)			±10			±4			±4	LSB
FULL-SCALE CALIBRATION ERROR (with fixed 50Ω resistor from REF OUT to REF IN) (Adjustable to zero)		0.1	0.25		0.1	0.25		0.1	0.25	% of F.S.
TEMPERATURE RANGE	0		+70	0		+70	0		+70	°C
TEMPERATURE COEFFICIENTS (Using internal reference) T_{min} to T_{max}										
Unipolar Offset			±2(10)			±1(5)			±1(5)	LSB (ppm/°C)
Bipolar Offset			±2(10)			±1(5)			±1(5)	LSB (ppm/°C)
Full-Scale Calibration			±9(50)			±5(27)			±2(10)	LSB (ppm/°C)
POWER SUPPLY REJECTION Max change in Full Scale Calibration										
V_{CC} = 15V ±1.5V or 12V ±0.6V			±2			±1			±1	LSB
V_{LOGIC} = 5V ±0.5V			±1/2			±1/2			±1/2	LSB
V_{EE} = −15V ±1.5V or −12V ±0.6V			±2			±1			±1	LSB
ANALOG INPUT										
Input Ranges										
Bipolar	−5		+5	−5		+5	−5		+5	Volts
	−10		+10	−10		+10	−10		+10	Volts
Unipolar	0		+10	0		+10	0		+10	Volts
	0		+20	0		+20	0		+20	Volts
Input Impedance										
10 Volt Span	3	5	7	3	5	7	3	5	7	kΩ
20 Volt Span	6	10	14	6	10	14	6	10	14	kΩ
DIGITAL CHARACTERISTICS[1] (T_{min} to T_{max})										
Inputs										
Logic "1" Voltage	+2.0		+5.5	+2.0		+5.5	+2.0		+5.5	Volts
Logic "0" Voltage	−0.5		+0.8	−0.5		+0.8	−0.5		+0.8	Volts
Current	−100		+100	−100		+100	−100		+100	μA
Capacitance		5			5			5		pF
Outputs (DB11–DB0, STS)										
Logic "1" Voltage ($I_{SOURCE} \leq 500\mu A$)	+2.4			+2.4			+2.4			Volts
Logic "0" Voltage ($I_{SINK} \leq 1.6mA$)			+0.4			+0.4			+0.4	Volts
Leakage (DB11–DB0, High-Z State)	−20		+20	−20		+20	−20		+20	μA
Capacitance		5			5			5		pF

NOTES

[1]Detailed Timing Specifications appear in data sheets.

[2]The reference should be buffered for operation on ±12V supplies.

Specifications subject to change without notice.

Specifications shown in boldface are tested on all production units at final electrical test. Results from those tests are used to calculate outgoing quality levels. All min and max specifications are guaranteed, although only those shown in boldface are tested on all production units.

Model	AD674AS Min	AD674AS Typ	AD674AS Max	AD674AT Min	AD674AT Typ	AD674AT Max	AD674AU Min	AD674AU Typ	AD674AU Max	Units
POWER SUPPLIES										
Operating Range										
V_{LOGIC}	+4.5		+5.5	+4.5		+5.5	+4.5		+5.5	Volts
V_{CC}	+11.4		+16.5	+11.4		+16.5	+11.4		+16.5	Volts
V_{EE}	−11.4		−16.5	−11.4		−16.5	−11.4		−16.5	Volts
Operating Current										
I_{LOGIC}		30	40		30	40		30	40	mA
I_{CC}		2	5		2	5		2	5	mA
I_{EE}		18	29		18	29		18	29	mA
POWER DISSIPATION		390	720		390	720		390	720	mW
INTERNAL REFERENCE VOLTAGE	9.9	10.0	10.1	9.9	10.0	10.1	9.9	10.0	10.1	Volts
Output current (available for external loads)[2] (External load should not change during conversion)			2.0			2.0			2.0	mA
RESOLUTION			12			12			12	Bits
LINEARITY ERROR			±1			±1/2			±1/2	LSB
T_{min} to T_{max}			±1			±1			±1	LSB
DIFFERENTIAL LINEARITY ERROR (Minimum resolution for which no missing codes are guaranteed)										
T_{min} to T_{max}	11			12			12			Bits
UNIPOLAR OFFSET (Adjustable to zero)			±2			±2			±2	LSB
BIPOLAR OFFSET (Adjustable to zero)			±10			±4			±4	LSB
FULL-SCALE CALIBRATION ERROR (with fixed 50Ω resistor from REF OUT to REF IN) (Adjustable to zero)		0.1	0.25		0.1	0.25		0.1	0.25	% of F.S.
TEMPERATURE RANGE	−55		+125	−55		+125	−55		+125	°C
TEMPERATURE COEFFICIENTS (Using internal reference)										
T_{min} to T_{max}										
Unipolar Offset			±2 (5)			±1 (2.5)			±1 (2.5)	LSB (ppm/°C)
Bipolar Offset			±4 (10)			±2 (5)			±1 (2.5)	LSB (ppm/°C)
Full-Scale Calibration			±20 (50)			±10 (25)			±5 (12.5)	LSB (ppm/°C)
POWER SUPPLY REJECTION										
Max change in Full Scale Calibration										
V_{CC} = 15V ±1.5V or 12V ±0.6V			±2			±1			±1	LSB
V_{LOGIC} = 5V ±0.5V			±1/2			±1/2			±1/2	LSB
V_{EE} = −15V ±1.5V or −12V ±0.6V			±2			±1			±1	LSB
ANALOG INPUT										
Input Ranges										
Bipolar	−5		+5	−5		+5	−5		+5	Volts
	−10		+10	−10		+10	−10		+10	Volts
Unipolar	0		+10	0		+10	0		+10	Volts
	0		+20	0		+20	0		+20	Volts
Input Impedance										
10 Volt Span	3	5	7	3	5	7	3	5	7	kΩ
20 Volt Span	6	10	14	6	10	14	6	10	14	kΩ
DIGITAL CHARACTERISTICS[1] (T_{min} to T_{max})										
Inputs										
Logic "1" Voltage	+2.0		+5.5	+2.0		+5.5	+2.0		+5.5	Volts
Logic "0" Voltage	−0.5		+0.8	−0.5		+0.8	−0.5		+0.8	Volts
Current	−100		+100	−100		+100	−100		+100	μA
Capacitance		5			5			5		pF
Outputs (DB11–DB0, STS)										
Logic "1" Voltage ($I_{SOURCE} \leq 500\mu A$)	+2.4			+2.4			+2.4			Volts
Logic "0" Voltage ($I_{SINK} \leq 1.6mA$)			+0.4			+0.4			+0.4	Volts
Leakage (DB11–DB0, High-Z State)	−20		+20	−20		+20	−20		+20	μA
Capacitance		5			5			5		pF
POWER SUPPLIES										
Operating Range										
V_{LOGIC}	+4.5		+5.5	+4.5		+5.5	+4.5		+5.5	Volts
V_{CC}	+11.4		+16.5	+11.4		+16.5	+11.4		+16.5	Volts
V_{EE}	−11.4		−16.5	−11.4		−16.5	−11.4		−16.5	Volts

NOTES

[1]Detailed Timing Specifications appear in data sheets.

[2]The reference should be buffered for operation on ±12V supplies.

Specifications subject to change without notice.

Specifications shown in boldface are tested on all production units at final electrical test. Results from those tests are used to calculate outgoing quality levels. All min and max specifications are guaranteed, although only those shown in boldface are tested on all production units.

Model	AD674AS Min	AD674AS Typ	AD674AS Max	AD674AT Min	AD674AT Typ	AD674AT Max	AD674AU Min	AD674AU Typ	AD674AU Max	Units
Operating Current										
I_{LOGIC}		30	**40**		30	**40**		30	**40**	mA
I_{CC}		2	**5**		2	**5**		2	**5**	mA
I_{EE}		18	**29**		18	**29**		18	**29**	mA
POWER DISSIPATION		390	**720**		390	**720**		390	**720**	mW
INTERNAL REFERENCE VOLTAGE	**9.9**	10.0	**10.1**	**9.9**	10.0	**10.1**	**9.9**	10.0	**10.1**	Volts
Output current (available for external loads)[2] (External load should not change during conversion)			2.0			2.0			2.0	mA

NOTES
[1]Detailed Timing Specifications appear in data sheets.

[2]The reference should be buffered for operation on ±12V supplies.

Specifications subject to change without notice.

Specifications shown in boldface are tested on all production units at final electrical test. Results from those tests are used to calculate outgoing quality levels. All min and max specifications are guaranteed, although only those shown in boldface are tested on all production units.

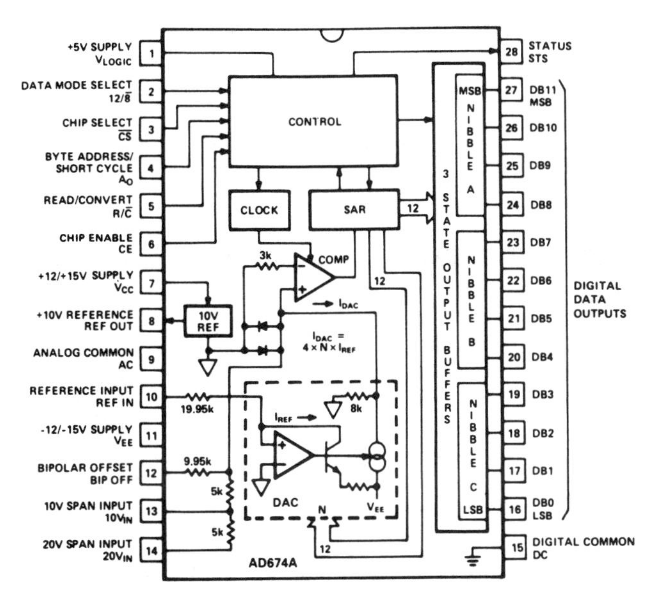

Block Diagram of AD674A 12-Bit A-to-D Converter

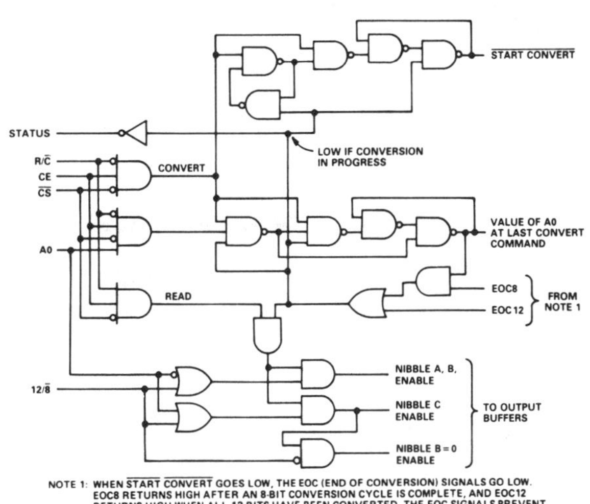

AD674A Control Logic

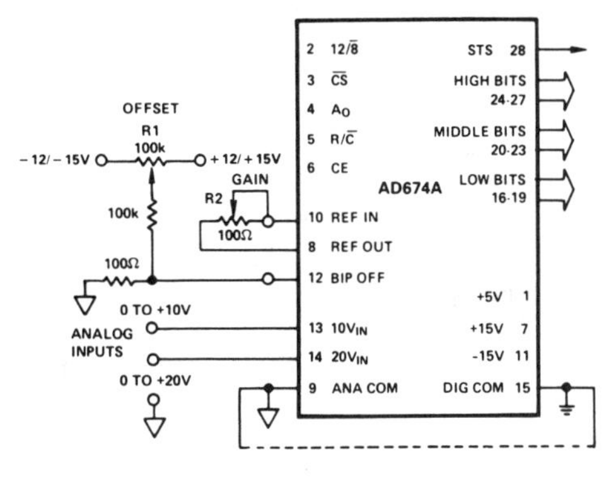

Unipolar Input Connections

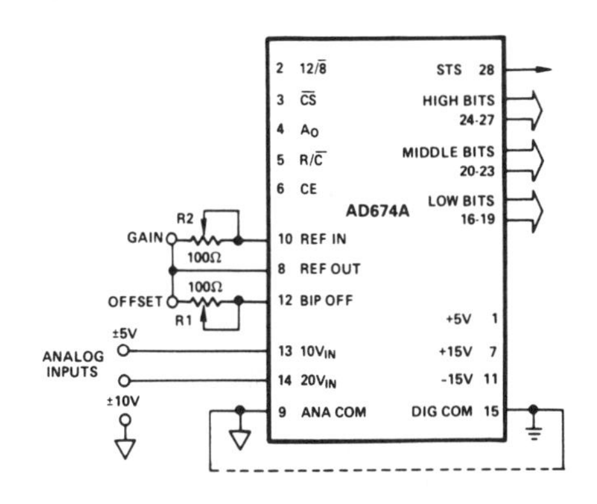

Bipolar Input Connections

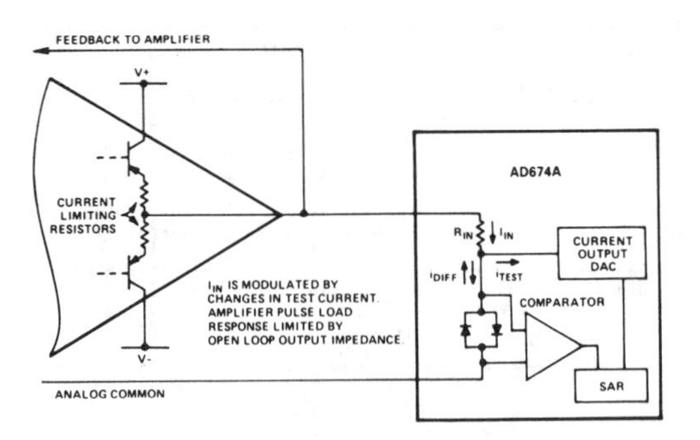

Op Amp – AD674A Interface

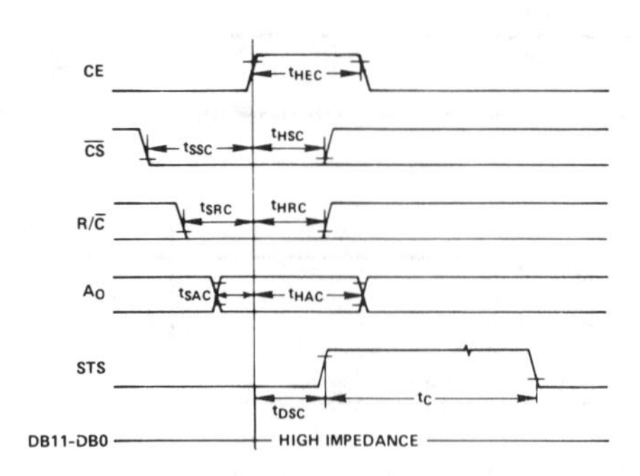

Convert Start Timing

CONVERT START TIMING—FULL CONTROL MODE

Symbol	Parameter	Min	Typ	Max	Units
t_{DSC}	STS Delay from CE			200	ns
t_{HEC}	CE Pulse Width	50			ns
t_{SSC}	$\overline{CS}$ to CE Setup	50			ns
t_{HSC}	$\overline{CS}$ Low During CE High	50			ns
t_{SRC}	R/$\overline{C}$ to CE Setup	50			ns
t_{HRC}	R/$\overline{C}$ Low During CE High	50			ns
t_{SAC}	A_O to CE Setup	0			ns
t_{HAC}	A_O Valid During CE High	50			ns
t_C	Conversion Time				
	8-Bit Cycle	6	8	10	µs
	12-Bit Cycle	9	12	15	µs

CE	$\overline{CS}$	R/$\overline{C}$	12/$\overline{8}$	A_O	Operation
0	X	X	X	X	None
X	1	X	X	X	None
1	0	0	X	0	Initiate 12-Bit Conversion
1	0	0	X	1	Initiate 8-Bit Conversion
1	0	1	1	X	Enable 12-Bit Parallel Output
1	0	1	0	0	Enable 8 Most Significant Bits
1	0	1	0	1	Enable 4LSBs + 4 Trailing Zeroes

AD674A Truth Table

This information is reproduced with the permission of GEC Plessey Semiconductors.

Plessey MJ1410 8 Bit Format Converter

FUNCTIONAL DESCRIPTION

Pin No.	Title	Function
1	H (Data inputs)	Data i/p H
2	G (Data inputs)	Data i/p G
3	F (Data inputs)	Data i/p F
4	E (Data inputs)	Data i/p E
5	D (Data inputs)	Data i/p D
6	C (Data inputs)	Data i/p C
7	B (Data inputs)	Data i/p B
8	A (Data inputs)	Data i/p A
9	Vcc	Positive supply, 5V ± 5%
10	2^2	Counter preset i/p bit 2 — The counter is preset to the data on these i/ps on the 3rd positive clock edge following a negative edge on the 'sync' input.
11	2^1	Counter preset i/p bit 1
12	2^0	Counter preset i/p bit 0
13	SYNC	A negative edge on this i/p initiates the counter preset sequence which causes the conversion cycle to start in the register which corresponds to the binary value of the counter preset i/ps.
14	CLOCK	System clock
15	GND	Zero volts
16	0 (Data outputs)	Three state data o/p '0'
17	1 (Data outputs)	Three state data o/p '1'
18	2 (Data outputs)	Three state data o/p '2'
19	3 (Data outputs)	Three state data o/p '3'
20	4 (Data outputs)	Three state data o/p '4'
21	5 (Data outputs)	Three state data o/p '5'
22	6 (Data outputs)	Three state data o/p '6'
23	7 (Data outputs)	Three state data o/p '7'
24	O/P $\overline{EN}$	A logic '1' on this i/p forces all the data outputs to a high impedance state.

ELECTRICAL CHARACTERISTICS

Test conditions (unless otherwise stated): V_{CC} = 5V, T_{amb} = 22°C ± 2°C, Test circuit: Fig.6.
Supply voltage V_{CC} = 5V ± 10%. Ambient operating temperature T_{amb} = 10°C to +70°C

STATIC CHARACTERISTICS

Characteristic	Symbol	Pins	Value			Units	Conditions
			Min.	Typ.	Max.		
Low level I/P voltage	V_{IL}	1,2,3,4, 5,6,7,8, 10,11,12, 13,14,24	-0.3		0.8	Volts	
High level I/P voltage	V_{IH}	1,2,3,4, 5,6,7,8, 10,11,12, 13,14,24	2.5		V_{CC}	Volts	
Low level I/P current/high level I/P current	I_{IN}	1,2,3,4, 5,6,7,8, 10,11,12, 13,14,24		1	50	μA	
Low level O/P voltage	V_{OL}	16,17,18, 19,20,21, 22,23			0.5	Volts	I_{SYNC} = 1.6mA
High level O/P voltage	V_{OH}	16,17,18, 19,20,21, 22,23	2.5			Volts	I_{SOURCE} = 100uA
Low level O/P current sink capability	I_{OL}	16,17,18, 19,20,21, 22,23	-1.6			mA	
High level O/P current source capability	I_{OH}	16,17,18, 19,20,21, 22,23	100			μA	
OFF state O/P current	$I_{OFF\ L}$	16,17,18, 19,20,21, 22,23			40	μA	V_{OUT} = GND
	$I_{OFF\ H}$	16,17,18, 19,20,21, 22,23			-40	μA	V_{OUT} = V_{CC}
Power dissipation	P_{DISS}		90		500	mW	V_{CC} = 5.5V

DYNAMIC CHARACTERISTICS

Characteristic	Symbol	Value			Units	Conditions
		Min.	Typ.	Max.		
Max.clock frequency	F_{max}	2.4		10	MHz	
Min. clock frequency	F_{min}	0			MHz	
Sync. pulse width (positive)	t_{SPP}	60			ns	
Sync. pulse width (negative)	t_{SPN}	100			ns	
Lead of sync. clocking edge on positive clock edge	t_{SL}	130			ns	
Set up time of counter inputs (2^0,2^1,2^2)	t_{SC}	70			ns	
Hold time of counter inputs	t_{HC}	60			ns	
Set up time of data inputs (A-H)	t_{SD}	80			ns	
Hold time of data inputs	t_{HD}	85			ns	
Propagation delay, data out valid from output $\overline{\text{ENABLE}}$ low	t_{PDE}			100	ns	
Propagation delay, data out disabled from output $\overline{\text{ENABLE}}$ high	t_{PDD}			100	ns	
Propagation delay, clock to data out valid	t_{PCD}			200	ns	

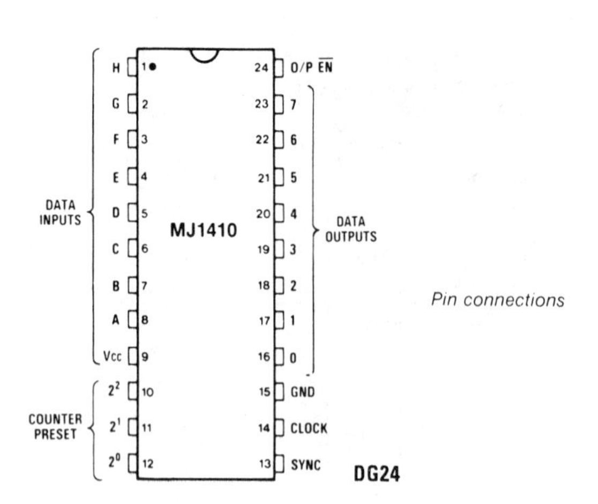

Pin connections

ABSOLUTE MAXIMUM RATINGS

Voltage on any pin w.r.t. ground = 7V max.
Storage temperature = -55°C to +125°C

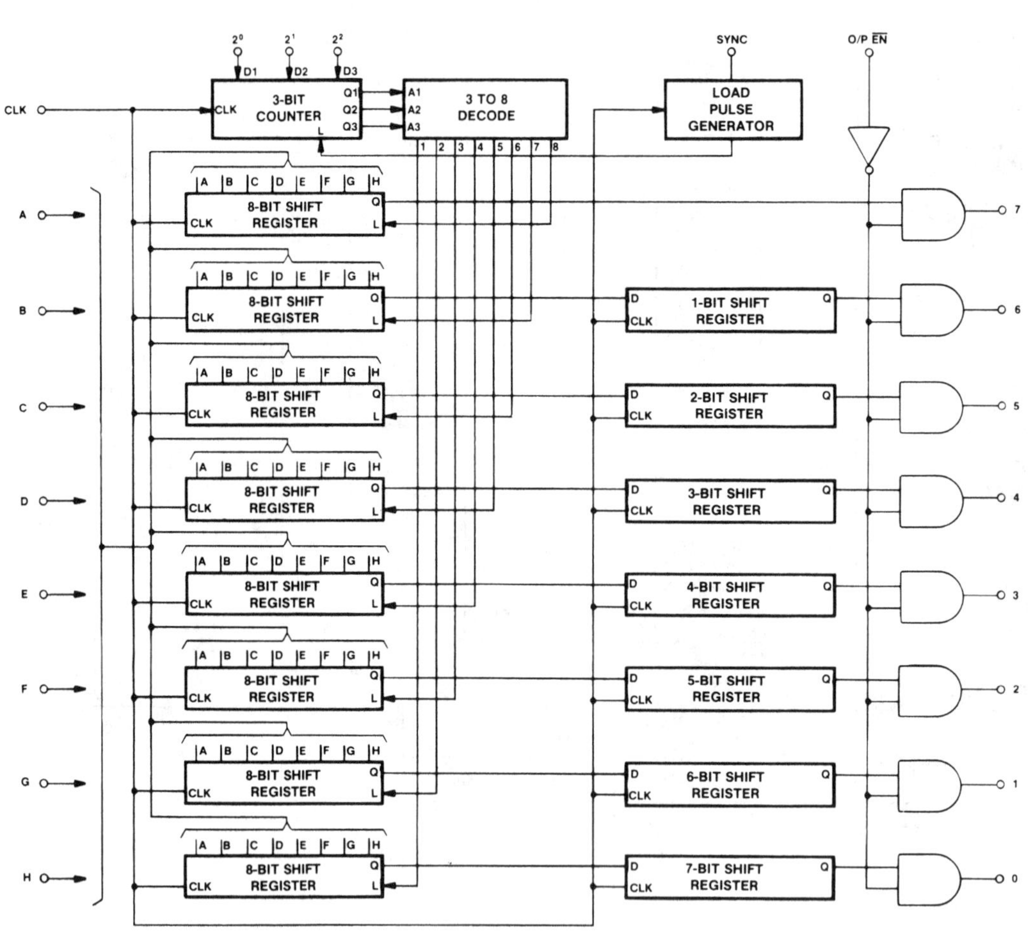

Block diagram

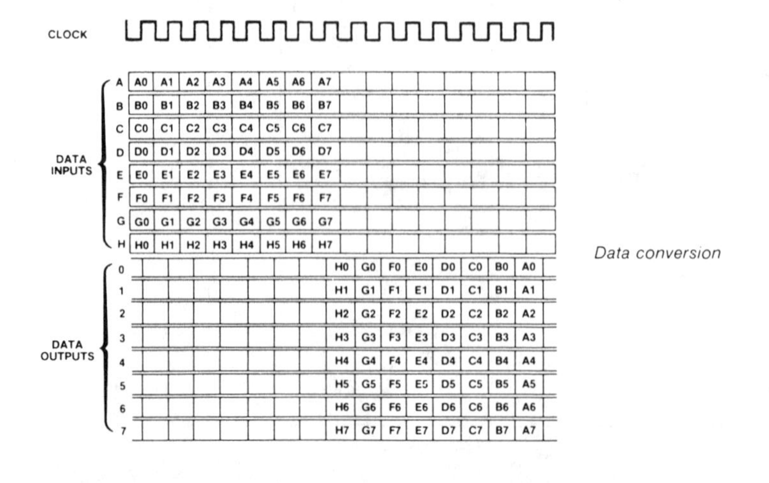

Data conversion

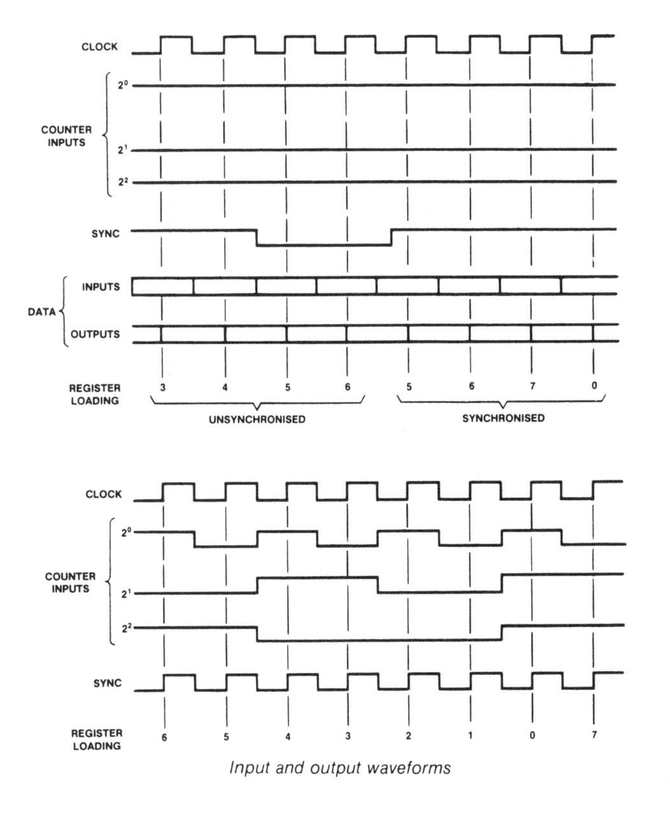

Input and output waveforms

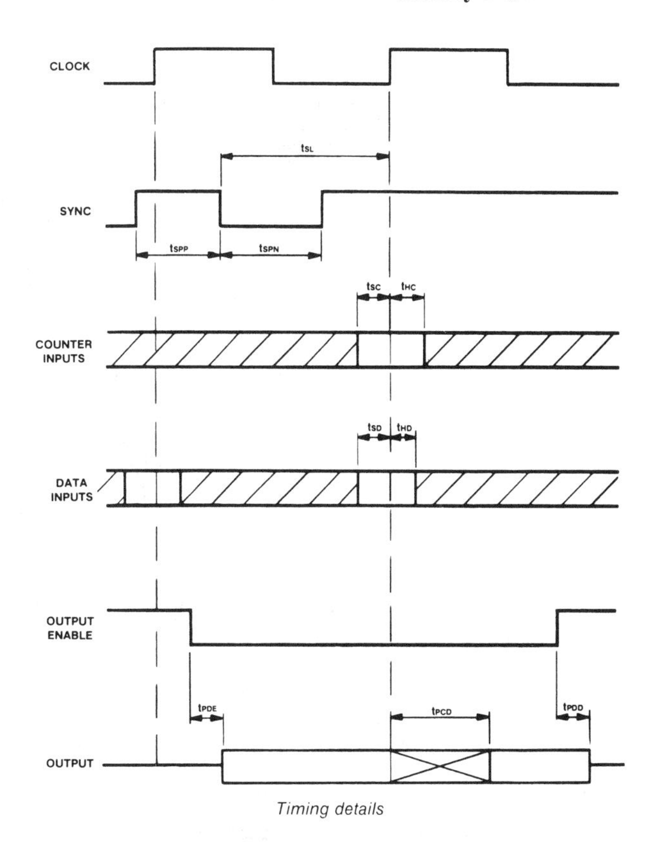

Timing details

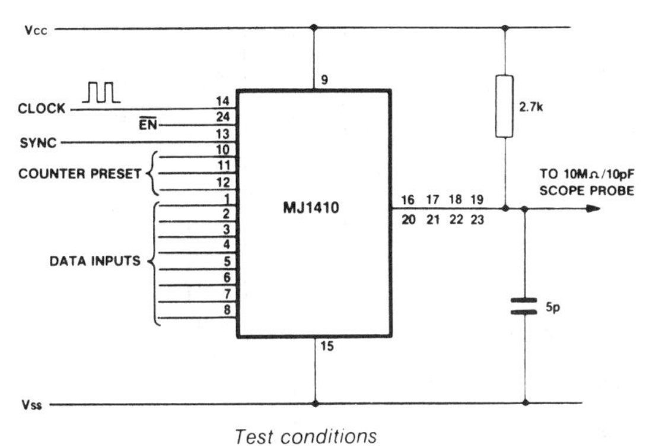

Test conditions

Plessey MJ1440
HDB3 Encoder/Decoder

- 5V ±5% supply −50mA max
- HDB3 encoding and decoding to CCITT rec. G703
- Asynchronous operation
- Simultaneous encoding and decoding
- Clock recovery signal generated from incoming HDB3 data
- Loop-back control
- HDB3 error monitor
- "All ones" error monitor
- Loss of input alarm (all zeros detector)
- Decode Data in NRZ form

ABSOLUTE MAXIMUM RATINGS

The absolute maximum ratings are limiting values above which operating life could be shortened or specified parameters could be degraded.

Electrical Ratings

+Vcc	7V
Inputs	Vcc + 0.5V Gnd — 0.3V
Outputs	Vcc, Gnd — 0.3V

Thermal Ratings

Max junction temperature		175 °C
Thermal resistance:	chip to case	chip to amb.
	40 °C/Watt	120 °C/Watt

ELECTRICAL CHARACTERISTICS

Test conditions (unless otherwise stated):
Supply voltage, V_{CC} = 5V ±0.25V
Ambient temperature, T_{amb} = 0°C to +70°C

Static characteristics

Characteristic	Symbol	Pins	Value			Units	Conditions
			Min	Typ	Max		
Low level input voltage	V_{IL}	1,2,5,6 10,11,12,13	-0.3		0.8	V	
Low level input current	I_{IL}	1,2,5,6 10,11,12,13			50	μA	V_{IL} = 0V
High level input voltage	V_{IH}	1,2,5,6 10,11,12,13	2.5		V_{cc}	V	
High level input current	I_{IH}	1,2,5,6 10,11,12,13			50	μA	V_{IH} = 5V
Low level output voltage	V_{OL}	10,14,15			0.5	V	Isink = 80μA
		3,4,7,9			0.4	V	Isink = 1.6mA
High level output voltage	V_{OH}	3,4,7,9	2.7			V	Isource = 60μA
		14,15	2.8			V	Isource = 2mA
		10	2.8			V	Isource = 1mA
Supply current	I_{cc}			20	50	mA	All inputs to 0V All outputs open circuit

Dynamic Characteristics

Characteristic	Symbol	Value			Units
		Min.	Typ.	Max.	
Max. Clock (Encoder) frequency	$fmax_{enc}$	4.0			MHz
Max. Clock (Decoder) frequency	$fmax_{dec}$	2.2			MHz
Propagation delay Clock (Encoder) to O_1, O_2	tpd1A/B			100	ns
Rise and Fall times O_1, O_2				20	ns
tpd1A-tpd1B				20	ns
Propagation delay Clock (Encoder) to Clock	tpd3			150	ns
Setup time of NRZ data in to Clock (Encoder)	ts3	30			ns
Hold time of NRZ data in	th3	55			ns
Propagation delay A_{in}, B_{in} to Clock	tpd2			150	ns
Propagation delay Clock (Decoder) to loss of input				150	ns
Propagation delay Clock (Decoder) to error	tpd4			200	ns
Propagation delay $\overline{\text{Reset AIS}}$ to AIS	tpd5			200	ns
Propagation delay Clock (Decoder) to NRZ data out	tpd6			150	ns
Setup time of A_{in}, B_{in} to Clock (Decoder)	ts1	75			ns
Hold time of A_{in}, B_{in} to Clock (Decoder)	th1	5			ns
Hold time of $\overline{\text{Reset AIS}}$ = '0'	th2	100			ns
Setup time Clock (Decoder) to $\overline{\text{Reset AIS}}$	ts2	200			ns
Setup time $\overline{\text{Reset AIS}}$ = 1 to Clock (Decoder)	ts2′	0			ns

NOTES
1. Encoded HDB3 outputs (O_1, O_2) are delayed by 3½ clock periods from NRZ data in (Fig.3).
2. The decoded NRZ output is delayed by 3 clock periods from the HDB3 inputs (A_{IN}, B_{IN}) (Fig.4).

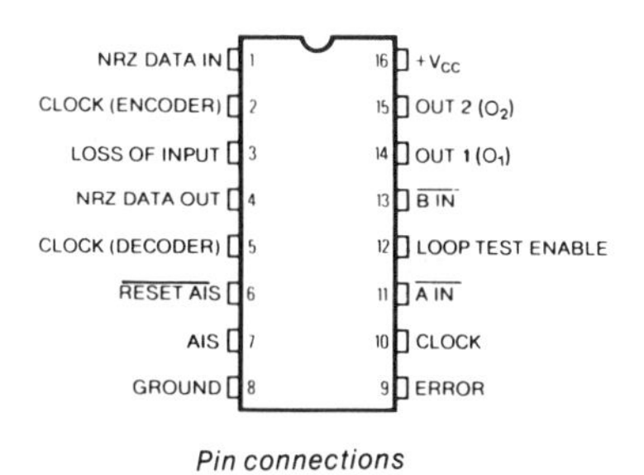

Pin connections

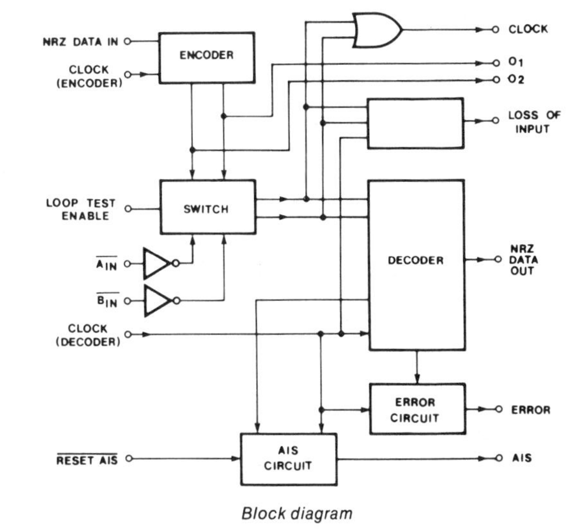

Block diagram

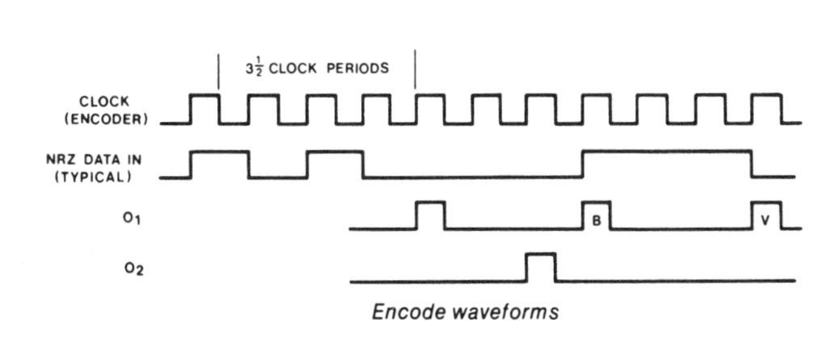

Encode waveforms

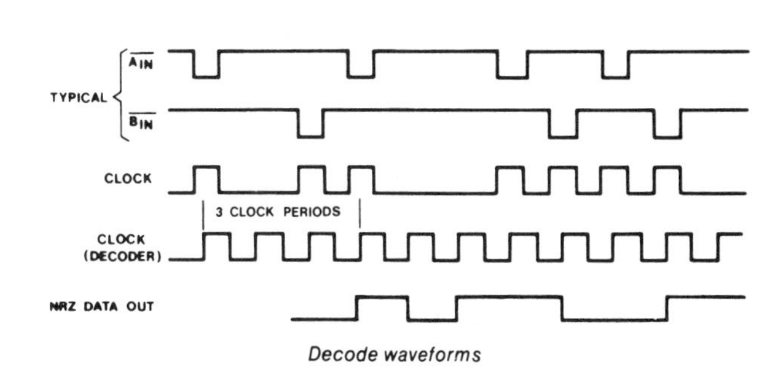

Decode waveforms

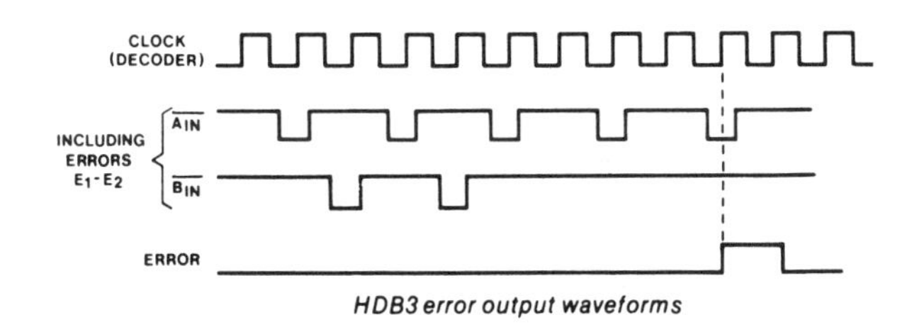

HDB3 error output waveforms

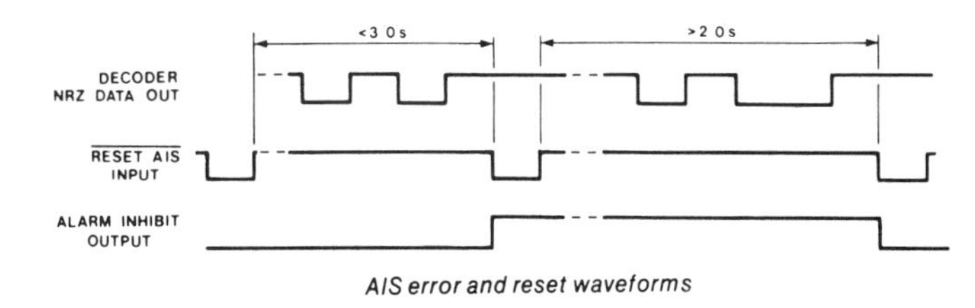

AIS error and reset waveforms

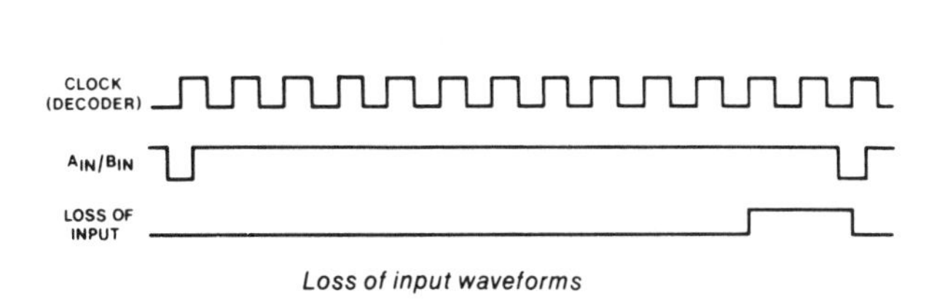

Loss of input waveforms

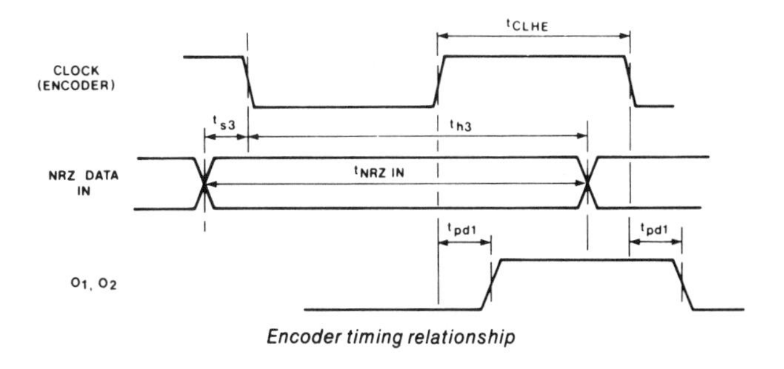

Encoder timing relationship

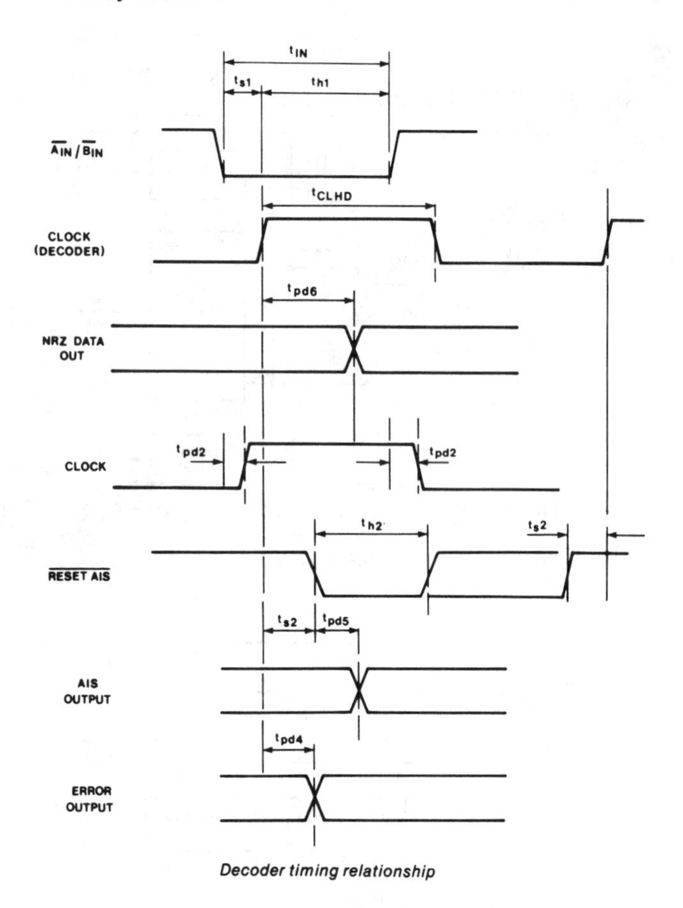

Decoder timing relationship

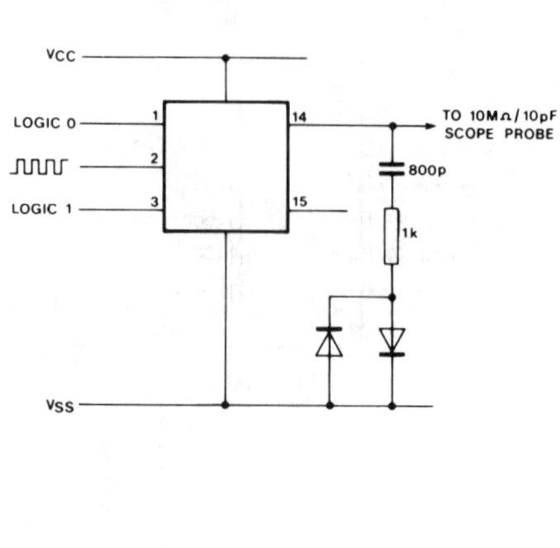

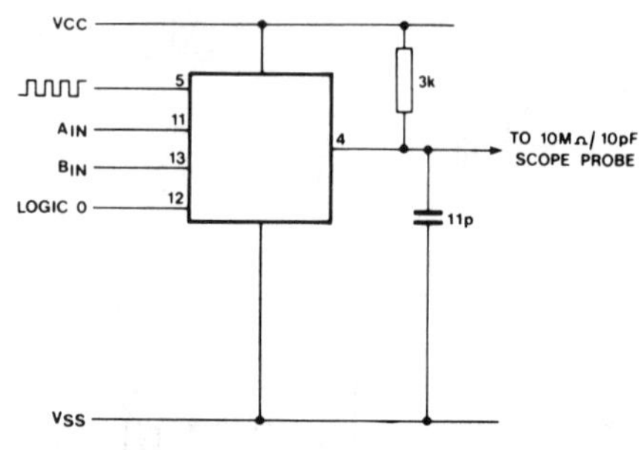

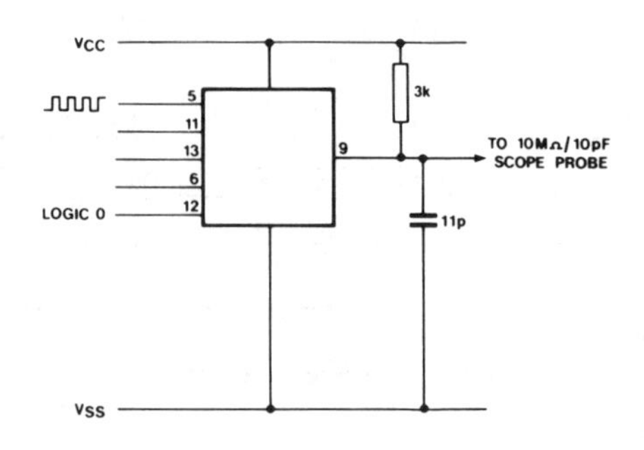

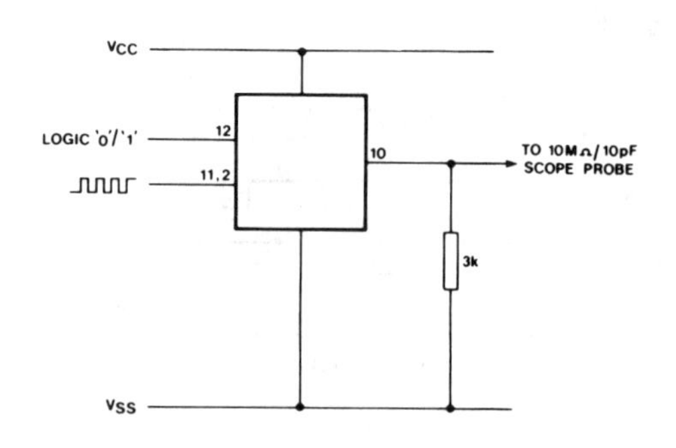

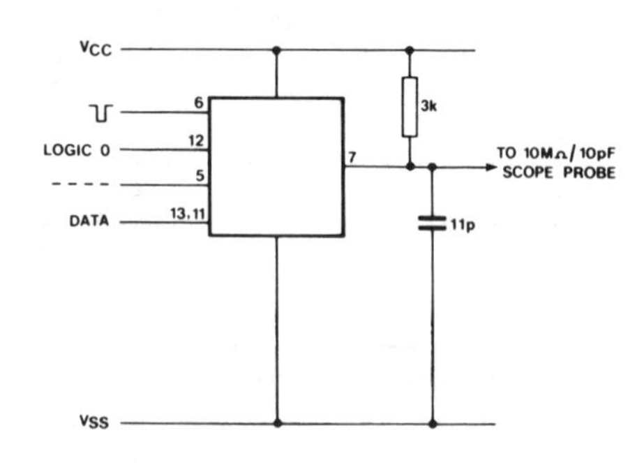

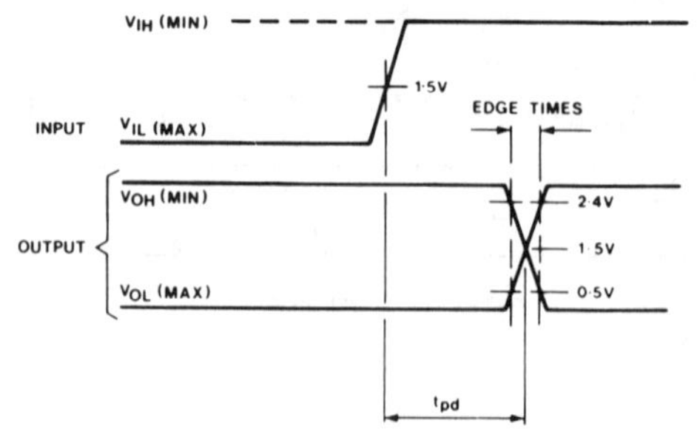

Test timing definitions

Plessey MV3506

A-Law Filter/Codec

MV3507

μ-Law Filter/Codec

MV3507A

μ-Law Filter/Codec with A/B Signalling

MV3508

A-Law Filter/Codec with Optional Squelch

MV3509

μ-Law Filter/Codec with Optional Squelch

- Low-power CMOS 80mW (operating) 10mW (standby)
- Meets or exceeds AT & T3, and CCITT G.711, G.712 and G.733 specifications
- Input analog filter eliminates need for external antialiasing prefilter
- Uncommitted input and output op amps for programming gain
- Output op amp provides ±3.1V into a 1200-ohm load or can be switched off for reduced power (70mW)
- Encoder has dual-speed auto-zero loop for fast acquisition on power-up
- Low absolute group delay = 410 microseconds at 1 kHz

ABSOLUTE MAXIMUM RATINGS

Exceeding these ratings may cause permanent damage.
Functional operation under these conditions is not implied.

Positive supply voltage V_{DD}	-0.5V to +6.0V
Analog ground V_{AGND}	-0.1V to +0.1V
Negative supply voltage V_{SS}	-6.0V to +0.5V
Storage temperature T_S	-65°C to +150°C
Voltage at digital or analog pins V_P	V_{SS} -0.3V to V_{DD} +0.3V
Package power dissipation P	1000mW

ELECTRICAL CHARACTERISTICS

Test conditions - Voltages are with respect to digital ground (V_{DGND})

Characteristic	Symbol	Value			Units
		Min.	Typ.(1)	Max.	
Digital supply voltage	V_{DD}	4.75	5	5.25	V
Negative supply voltage	V_{SS}	-5.25	-5	-4.75	V
Analog ground voltage	V_{AGND}	-0.1	0	0.1	V
Ambient temperature	V_{AMB}	0		70	°C
Input low voltage - digital inputs	V_{IL}	0	0.4	0.8	V
Input high voltage - digital inputs	V_{IH}	2.0	2.4	V_{DD}	V
System clock frequency					
CLK SEL tied to V_{DD}	f_s	2047.90	2048	2048.10	kHz
CLK SEL tied to D GND		255.99	256	256.01	
CLK SEL tied to V_{SS}		1549.92	1544	1544.08	
Capacitive loading - digital outputs	C_{LD}	0		100	pF
Pull-up resistance for PCM OUT pin	R_{PU}	510			Ω
Analog input voltage	V_{IA}	V_{AGND} -3.1		V_{AGND} +3.1	V
Capacitive loading - analog outputs	C_{LA}			50	pF
Resistive loading - V_{OUT} pin	R_{VOUT}	1200			Ω
Resistive loading - V_{IN} pin	R_{VIN}	10			kΩ
Resistive loading - FLT OUT pin	$R_{RLT\ OUT}$	20			kΩ

POWER SUPPLY REQUIREMENTS - V_{DD} = 5V, V_{SS} = -5V

Characteristic	Symbol	Value			Units	Conditions
		Min.	Typ.(1)	Max.		
Power dissipation - normal	P_N		80	110	mW	Unloaded
Power dissipation - without output amp.	P_{WA}		70		mW	Unloaded
Power dissipation - standby	P_S		10	20	mW	Unloaded

STATIC CHARACTERISTICS Voltages are with respect to digital ground (V_{DGND})

Characteristic	Symbol	Value			Units	Conditions
		Min.	Typ.(1)	Max.		
Pin capacitance	C_{PIN}		7	15	pF	
Input leakage current	I_{IL}			1	μA	$0 < V < V_{DD}$
Input source current - inputs with pull-ups	I_{IS}			600	μA	$0 < V < V_{DD}$
Output high voltage	V_{OH}	2.4		V_{DD}	V	I_{OH} (Source) = 40μA
Output low voltage	V_{OL}	0		0.4	V	I_{OL} (Sink) = 1.6mA
Output leakage current	I_{OL}			10	μA	$0 < V < V_{DD}$
Analog input resistance	R_{IA}	100			kΩ	
Analog output voltage	V_{OA}	V_{AGND} -3.1		V_{AGND} +3.1	V	

DIGITAL SWITCHING CHARACTERISTICS - System Clock

Characteristic	Symbol	Value			Units	Conditions
		Min.	Typ.(1)	Max.		
System clock rise time	t_{SR}		50		ns	
System clock high period	t_{SH}	0.4/fs		0.6/fs	s	
System clock fall time	t_{SF}		50		ns	
System clock low period	t_{SL}	0.4/fs		0.6/fs	s	

DIGITAL SWITCHING CHARACTERISTICS - Receive Strobe and Clock

Characteristic	Symbol	Value			Units	Conditions
		Min.	Typ.(1)	Max.		
Receive strobe frequency	f_{RS}	7.99996	8	8.00004	kHz	Phase-locked with system clock
Receive strobe falling set-up time	t_{RSFS}	120			ns	
Receive strobe early jitter	t_{RSEJ}			200	ns	
Receive strobe late jitter	t_{RSLJ}			100	ns	
Receive strobe falling hold time	t_{RSFH}	220			ns	
Receive clock frequency	f_{RC}	63.9997		2048.01	kHz	Phase-locked with receive strobe
Receive clock rise time	t_{RCR}			100	ns	
Receive clock high period	t_{RCH}	0.4/f_{RC}		0.6/f_{RC}	s	
Receive clock fall time	t_{RCF}			100	ns	
Receive clock low period	t_{RCL}	0.4/f_{RC}		0.6/f_{RC}	s	
Receive clock early jitter	t_{RCEJ}			200	ns	
Receive clock late jitter	t_{RSLJ}			100	ns	

ANALOG CHANNEL CHARACTERISTICS - μ-Law

Characteristic	Symbol	Value			Units	Conditions
		Min.	Typ.(1)	Max.		
0dBm0 level (see Note 2)	0dBm0	5.3	5.8	6.3	dBm	±5V, 25°C
Variation in 0dBm0 level	Δ_{dBm0}	-0.3	0	0.3	dB	Over test conditions
Weighted idle channel noise	ICN_W		5	17	dBrnc0	AT&T D3 (see Note 3)
Single frequency idle channel noise	ICN_{SF}			-60	dBm0	AT&T D3
Weighted receive idle channel noise	ICN_{WR}			15	dBrnc0	AT&T D3
Spurious out-band noise	N_{SOB}			-28	dBm0	AT&T D3
Spurious in-band noise	N_{SIB}			-40	dBm0	AT&T D3
Two tone interdemodulation	IMD_{2T}			-35	dBm0	AT&T D3
Tone + power inter-demodulation	IMD_{TP}			-49	dBm0	AT&T D3
Crosstalk attenuation between V_{IN} and V_{OUT}	A_X	75	80		dB	AT&T D3

NOTES
1. Typical figures are for design aid only. They are not guaranteed and not subject to production testing.
2. The typical 0dBm0 level of 5.8dBm corresponds to an RMS voltage of 1.51V and a maximum coding level of 3.1V.
3. The maximum value reduces to -68dBm0p without squelch (MV3508 with TST/SE pin unconnected).
4. The maximum value reduces to 22dBrnc0 without squelch (MV3509 with TST/SE pin unconnected).

DIGITAL SWITCHING CHARACTERISTICS - Transmit Strobe and Clock

Characteristic	Symbol	Value			Units	Conditions
		Min.	Typ.(1)	Max.		
Transmit strobe frequency	f_{TS}	7.99996	8	8.00004	kHz	Phase-locked with system clock
Transmit strobe falling set-up time	t_{TSFS}	120			ns	
Transmit strobe early jitter	t_{TSEJ}			200	ns	
Transmit strobe late jitter	t_{TSLJ}			100	ns	
Transmit strobe falling hold time	t_{TSFH}	220			ns	
Transmit clock frequency	f_{TC}	63.9997		2048.01	kHz	Phase-locked with transmit strobe
Transmit clock rise time	t_{TCR}			100	ns	
Transmit clock high period	t_{TCH}	$0.4/f_{TC}$		$0.6/f_{TC}$	s	
Transmit clock fall time	t_{TCF}			100	ns	
Transmit clock low period	t_{TCL}	$0.4/f_{TC}$		$0.6/f_{TC}$	s	
Transmit clock early jitter	t_{TCEJ}			200	ns	
Transmit clock late jitter	t_{TCLJ}			100	ns	

ANALOG CHANNEL CHARACTERISTICS - Filter Delays

Characteristic	Symbol	Value			Units	Conditions
		Min.	Typ.(1)	Max.		
Transmit filter delay	t_{TFD}			182	μs	1kHz
Receive filter delay	t_{RFD}			110	μs	1kHz

ANALOG CHANNEL CHARACTERISTICS - A-Law

Characteristic	Symbol	Value			Units	Conditions
		Min.	Typ.(1)	Max.		
0dBm0 level (see Note 2)	0dBm0	5.3	5.8	6.3	dBm	±5V, 25°C
Variation in 0dBm0 level	Δ_{0dBm0}	-0.3	0	0.3	dB	Over test conditions
Weighted idle channel noise	ICN_W		-85	-73	dBm0p	CCITT G.712, §5.1 (see Note 3)
Single frequency idle channel noise	ICN_{SF}			-60	dBm0	CCITT G.712, §5.2
Weighted receive idle channel noise	ICN_{WR}			-78	dBm0p	CCITT G.712, §5.3
Spurious out-band noise	N_{SOB}			-30	dBm0	CCITT G.712, §7.1
Spurious in-band noise	N_{SIB}			-40	dBm0	CCITT G.712, §10
Two tone interdemodulation	IMD_{2T}			-35	dBm0	CCITT G.712, §8.1
Tone + power inter-demodulation	IMD_{TP}			-49	dBm0	CCITT G.712, §8.2
Crosstalk attenuation between V_{IN} and V_{OUT}	A_X	75	80		dB	CCITT G.712, §12

DIGITAL SWITCHING CHARACTERISTICS - Transmit Data

Characteristic	Symbol	Value			Units	Conditions
		Min.	Typ.(1)	Max.		
PCM output holt time	t_{POH}	0	50		ns	
PCM output delay	t_{POD}		100	150	ns	

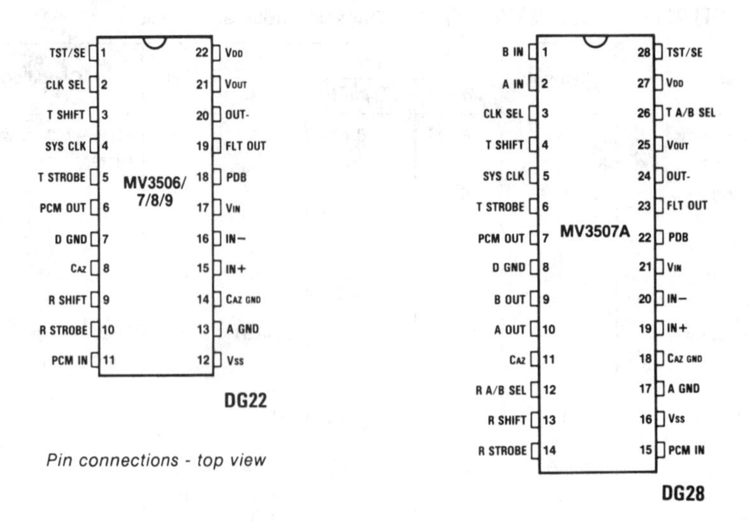

Pin connections - top view

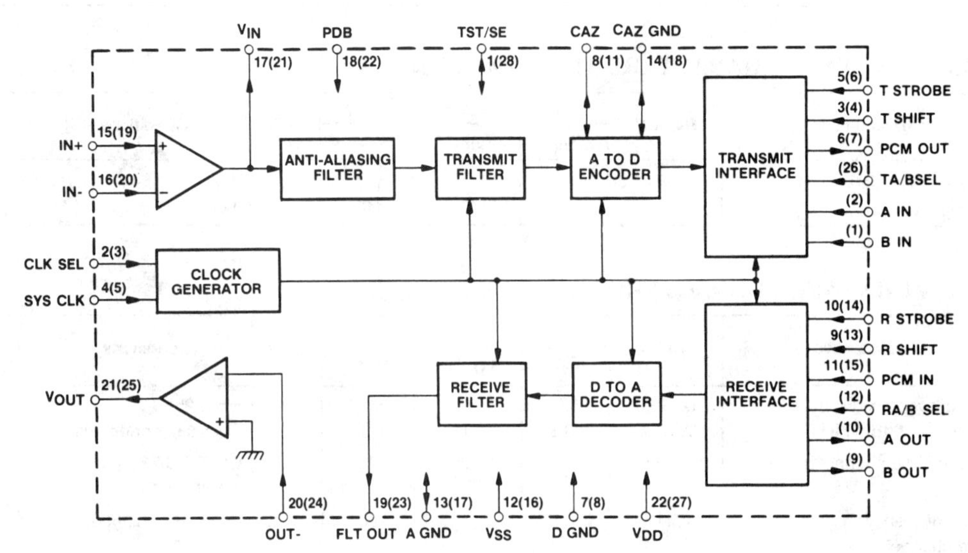

Functional block diagram (pin numbers for the MV3507A are in brackets)

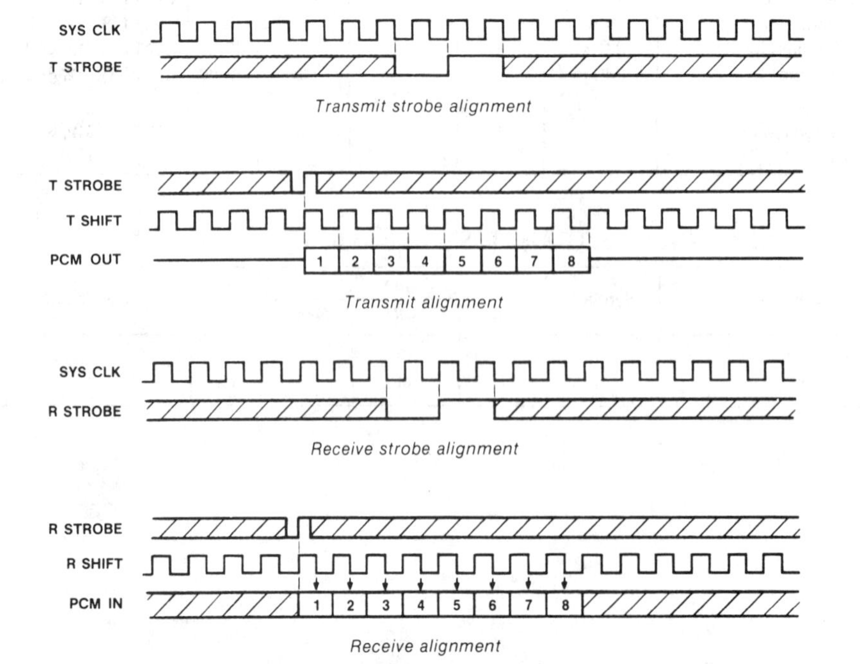

Transmit strobe alignment

Transmit alignment

Receive strobe alignment

Receive alignment

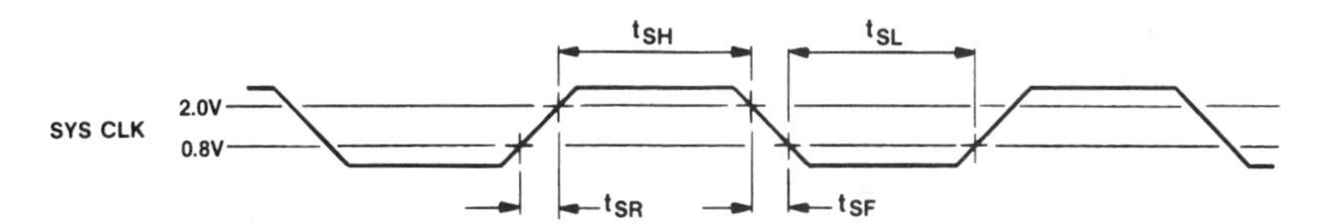

Timing - system clock

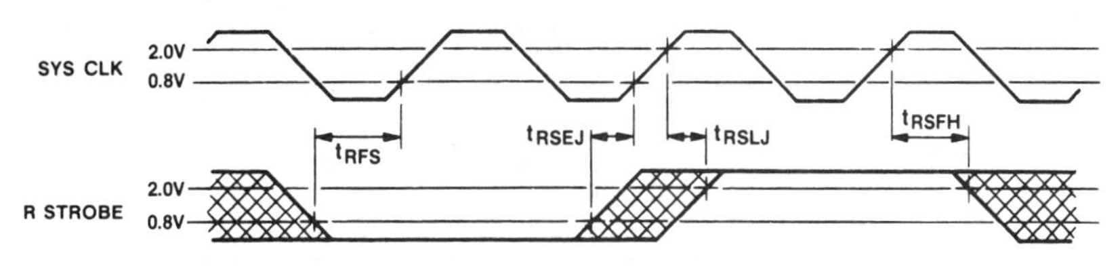

Timing - receive strobe

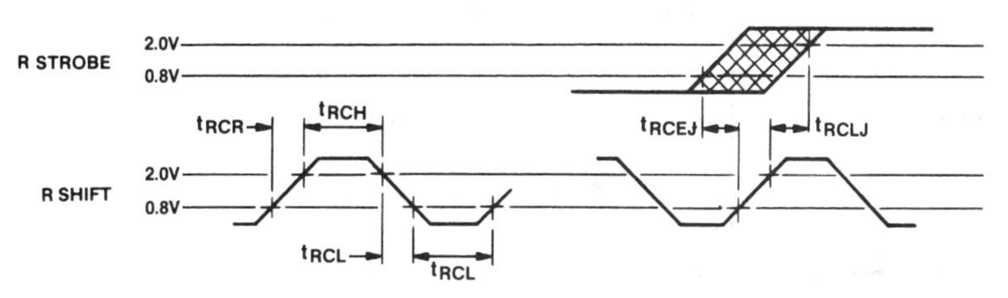

Timing - receive clock

DIGITAL SWITCHING CHARACTERISTICS - Receive Data

Characteristic	Symbol	Value			Units	Conditions
		Min.	Typ.(1)	Max.		
PCM input set-up time	t_{PIS}	60			ns	
PCM input hold time	t_{PIH}	60			ns	

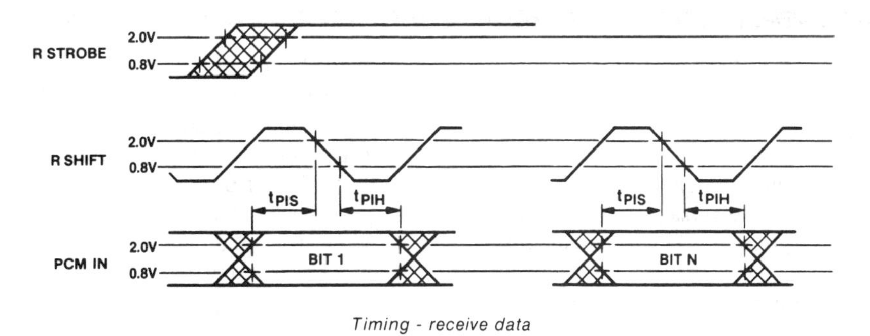

Timing - receive data

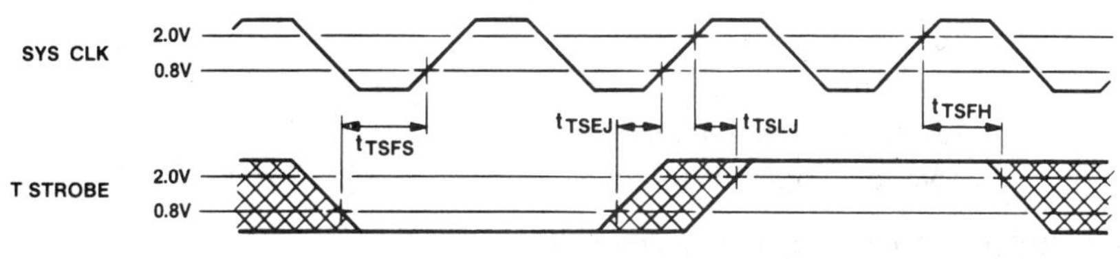

Timing - receive strobe

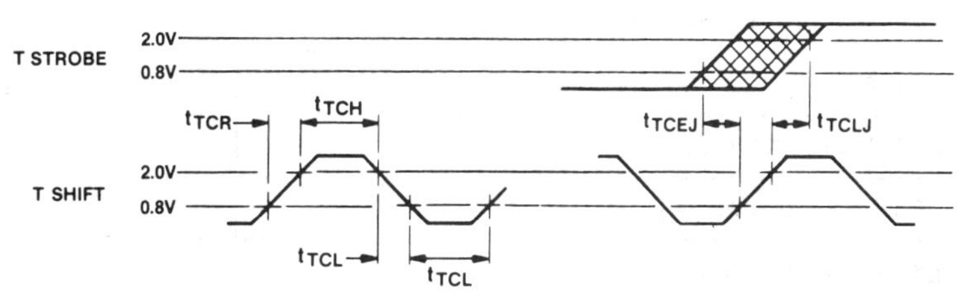

Timing - receive clock

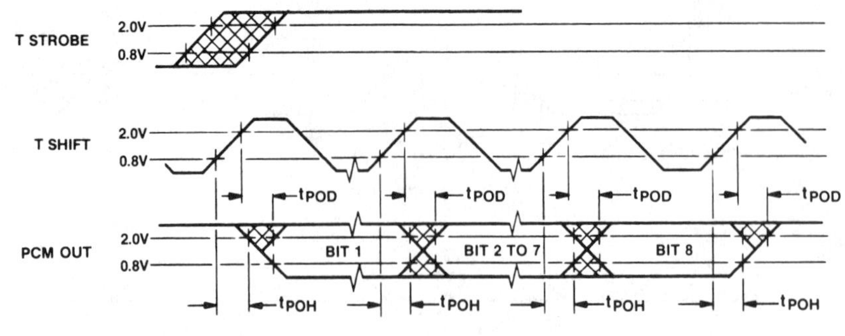

Timing - transmit data

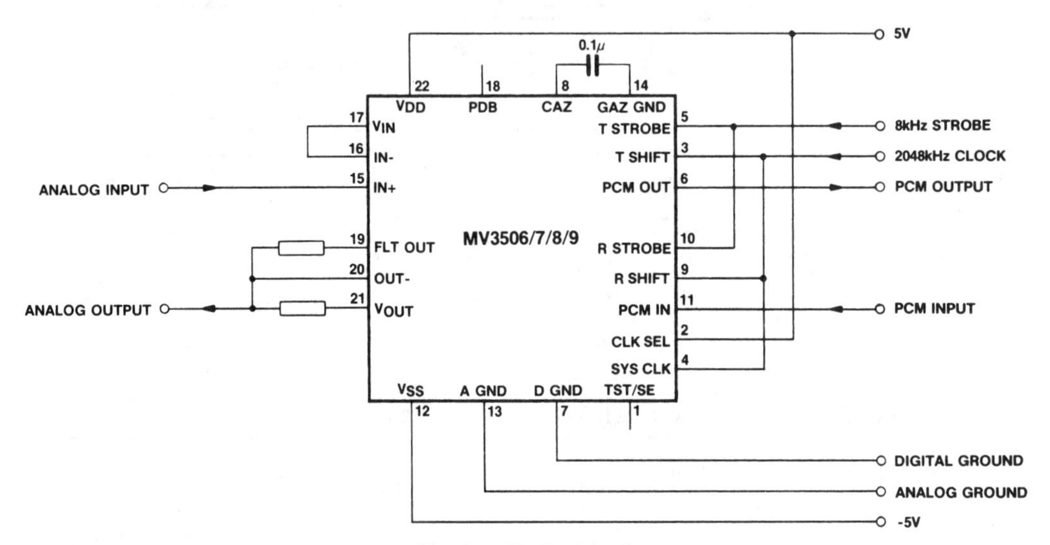

Simple application circuit

Plessey ZNPCM1
Single-Channel Codec

- Converts a delta-sigma modulated digital pulse stream into compressed 'A' law pcm and vice-versa
- Enables realization of a single-channel codec circuit with minimum component usage
- Pin-selectable input/output interface providing either single-channel operation at 64K bit/s (2,048 kHz external clock) or up to 2,048K bit/s (2,048 kHz external clock) for multichannel burst format
- Encoder and decoder can be clocked asynchronously (useful for prom multiplex applications).
- Optional alternate digit inversion
- Electrically and pin compatible with AY-3-9900
- Fully TTL compatible
- Single 5V supply

ABSOLUTE MAXIMUM RATINGS

Supply Voltage, V_{CC} .. .. +7 Volts
Input Voltage, V_{IN} .. .. +5.5 Volts
Operating Temperature Range .. 0°C to +70°C
Storage Temperature Range .. .. –65°C to +150°C

RECOMMENDED OPERATING CONDITIONS

Parameter	Min.	Nom.	Max.	Unit
Supply Voltage, V_{CC}	4.75	5.0	5.25	V
High-level Output Current, I_{OH}	—	—	–400	μA
Low-level Output Current, I_{OL}	—	—	4	mA
Operating Temperature Range, T_{amb}	0	—	70	°C

ELECTRICAL CHARACTERISTICS (over recommended operating temperature range).

	Parameter	Test Conditions	Min.	Typ.	Max.	Unit
V_{IH}	High level input voltage		2.5	—	—	V
V_{IL}	Low level input voltage		—	—	0.8	V
V_{OH}	High level output voltage	V_{CC} = Min., I_{OH} = Max.	2.4	3.5	—	V
V_{OL}	Low level output voltage	V_{CC} = Min., I_{OL} = Max.	—	—	0.4	V
I_{IH}	High level input current	V_{CC} = Max., V_{IH} = Min.	—	0.2	0.4	mA
I_{IL}	Low level input current	V_{CC} = Max., V_{IL} = Max.	—	–1	–10	μA
I_{CC}	Supply current	V_{CC} = Max.	—	80	110	mA
t_{vw}	Encoder timing vector pulse width		—	488	—	ns
t_{vv}	Encoding timing vector pulse width with edge variation		—	—	100	ns
t_{ww}	Decoder timing waveform pulse width		10	15.6	—	μs
f_{max}	Operating frequency		2.048	4	—	MHz
t_r & t_f	Rise and fall times	0.4V to 3V Transition	5	--	40	ns
t_{pw}	Pulse width	Between 1.5V levels	200	—	—	ns
C_I	Input capacitance		—	—	10	pF

PIN CONFIGURATIONS

Pin	Notation	Comments
1	0V	
2	MS	MODE SELECT (Note 1) Logic 0 = External pcm I/O interface timing Logic 1 = Internal pcm I/O interface timing
3	DS1	DECODER SELECT 1 and 2 (Note 2)
4	DS2	A two bit binary word selects required digit delay between encoder and decoder. DS1 DS2 Digit Delay 0 0 0 0 1 1 1 0 2 1 1 3
5	ADI	ALTERNATE DIGIT INVERSION Logic 0 = No. ADI Logic 1 = ADI
6	N.C.	NO CONNECTION
7	0V	
8	V_{CC}	
9	DSMO	DELTA-SIGMA MODULATED OUTPUT SIGNAL
10	SGN	SIGN BIT OUTPUT Sign bit from the encoder, used to operate on the delta-sigma modulator to reduce d.c. offset effects.

PIN CONFIGURATIONS *(continued)*

Pin	Notation	Comments
11	DSMI	DELTA-SIGMA MODULATED INPUT
12	SRF	SPECTRAL REDISTRIBUTION FUNCTION Output signal used to operate on the delta-sigma modulator to reduce low frequency quantisation noise.
13	PCMO	PCM OUTPUT
14	SGBI	SIGNALLING BIT INPUT Facility for adding signalling bit(s) to the output pcm stream.
15	ETV	ENCODER TIMING VECTOR A pulse defining the beginning of each frame used to maintain encoder timing.
16	PCMI	PCM INPUT
17	SGBO	SIGNALLING BIT OUTPUT Serial output for extracting signalling bit(s) from the incoming pcm stream.
18	DTW	DECODER TIMING WAVEFORM A pulse used to indicate to the decoder when the input pcm stream is in the input register (required only when external shift clocks are used).
19	SCE	ENCODER SHIFT CLOCK Used to control the output of serial pcm data from the encoder (when MS is low).
20	CLKD	DECODER MAIN CLOCK
21	CLKE	ENCODER MAIN CLOCK
22	SCD	DECODER SHIFT CLOCK Used to control the input of the serial pcm data to the decoder (when MS is low).
23	N.C.	NO CONNECTION
24	I.C.	INTERNAL CONNECTION Make no external connection to this pin.

Notes:

1. With MS low (logic 0) serial PCM transmission is under the control of an externally generated shift clock SCE which can vary from 64 kHz to 2,048 kHz. The timing of this input function allows the insertion of a number of signalling bits into the PCM stream via the SGB1 input. In the high (logic 1) state the 8 bit PCM codeword will be transmitted at a rate of 64K bit/sec and each codeword will occupy the full 125 μs frame period with the leading edge of the first bit occurring at a time defined by the ETV pulse.

2. Delays through the transmission network, normally under the control of transmission switches, may cause the decoder input pulse stream to be delayed in time by a number of digits from the original transmitted pulse. To compensate for this delay two control inputs, DS1 and DS2, are provided. Consequently when MS is in the high state discrete digit delays of 0 to 3 periods may be selected resulting in a controlled shift of decoder timing in order to re-align Bit 1 in its correct position in the input register.

   When using an externally generated clock (i.e. MS in low state) an input shift clock (SCD) and timing waveform (DTW) are required to ensure that Bit 1 of the input codeword occupies its correct position in the input shift register.

FUNCTIONAL DIAGRAM

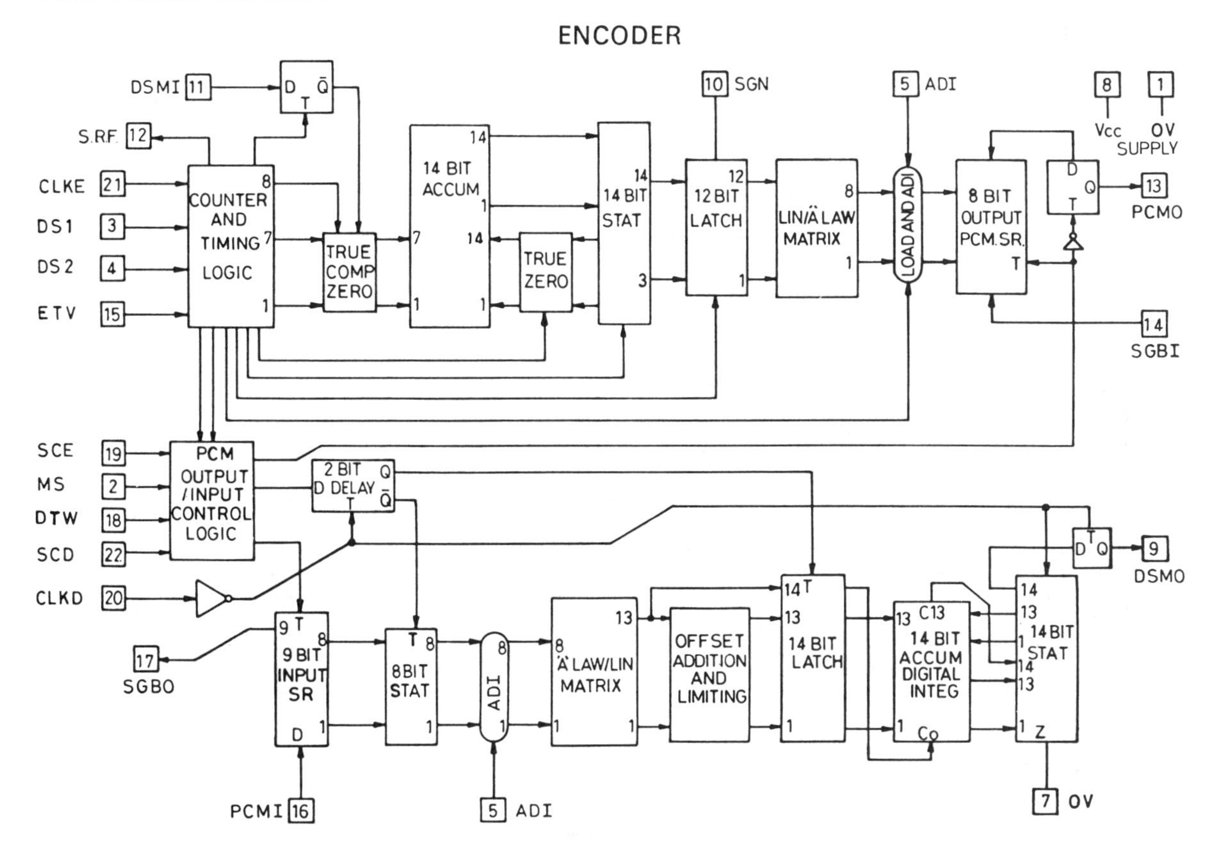

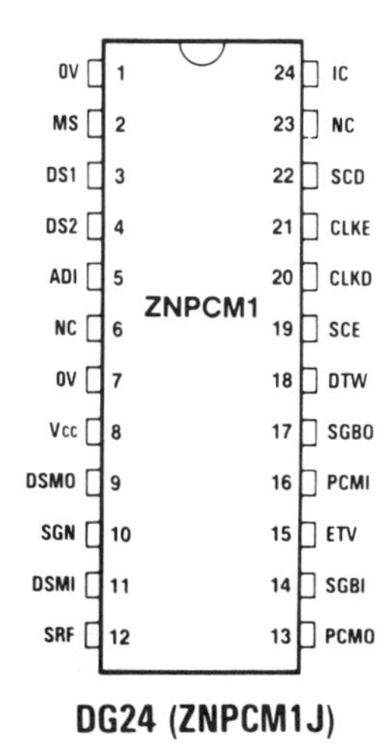

DG24 (ZNPCM1J)

Pin connections - top view

Plessey ZNPCM2
Delta Sigma Modulator/Demodulator

- Converts analog input signal into delta-sigma modulated (DSM) signal to be used as input for ZNPCM1 codec IC
- Converts DSM signal produced by ZNPCM1 into level-defined digital pulse stream for conversion to the analog equivalent signal using low-pass filter techniques
- High signal-to-noise ratio
- Monolithic integrated circuit combining digital and analog circuitry
- Single 5V supply
- 18-lead ceramic or modulated DIL package

Designed using the same technology as the ZNPCM1 Codec I.C., the ZNPCM2 combines both linear and digital circuits on the same monolithic I.C. Packaged in an 18-lead ceramic (ZNPCM2J) DIL, the device is designed to operate over the temperature range 0°C to +70°C.

ABSOLUTE MAXIMUM RATINGS

Supply Voltage, V_{CC} . . . . +7 volts
Digital Input Voltage, $V_{IN(D)}$ . . . . +5.5 volts
Analogue Input Voltage, $V_{IN(A)}$ . . . . 4 volts pk-to-pk
Operating Temperature Range . . . . 0°C to +70°C
Storage Temperature Range . . . . −65°C to +150°C

RECOMMENDED OPERATING CONDITIONS

Parameter	Min.	Nom.	Max.	Unit
Supply Voltage	4.75	5.0	5.25	V
High-level Output Current, I_{OH} (Digital Outputs)	—	—	−400	μA
Low-level Output Current, I_{OL} (Digital Outputs)	—	—	4	mA
Analogue Output Impedance, Z_{AO}	—	100	—	Ω
Operating Temperature Range, T_{amb}	0	—	70	°C

ELECTRICAL CHARACTERISTICS (over the recommended operating temperature range).

(a) Digital Inputs and Outputs.

	Parameter	Test Conditions	Min.	Typ.	Max.	Unit
V_{IH}	High-level input voltage		2.3	—	—	V
V_{IL}	Low-level input voltage		—	—	0.8	V
V_{OH}	High-level output voltage	V_{CC} = Min, I_{OH} = Max.	2.4	4	—	V
V_{OL}	Low-level output voltage	V_{CC} = Min, I_{OL} = Max.	—	—	0.4	V
I_{IH}	High-level input current	V_{CC} = Max, V_{IH} = Min.	—	0.2	0.4	mA
I_{IL}	Low-level input current	V_{CC} = Max, V_{IL} = Min.	—	—	10	μA
I_{CC}	Supply current	V_{CC} = Max.	—	24	—	mA
f	Operating frequency		—	2,048	—	kHz
t_r & t_f	Rise and fall time	0.4V – 3.0V transition	5	—	40	ns
t_{pd}	Propagation delay	Clock $Ø_E$ or $Ø_D$ to DSM output 2.5V level	—	40	60	ns
t_{pw}	Pulse width	Between 1.5V levels	200	—	—	ns

(b) Analogue Input and Output.

	Parameter	Test Conditions	Min.	Typ.	Max.	Unit
V_{IN}	Analogue Input Voltage for 0dBm0	Peak-to-peak	—	1.4	—	V
Z_{IN}	Analogue Input Impedance	Measured at 1kHz	80	100	—	kΩ
V_C	D.C. Voltage across C_{11}	V_{CC} = Max.	—	±3.0	±5.0	mV

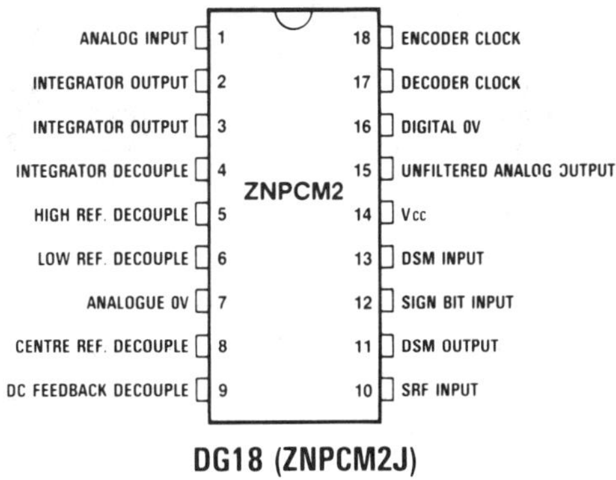

DG18 (ZNPCM2J)
Pin connections - top view

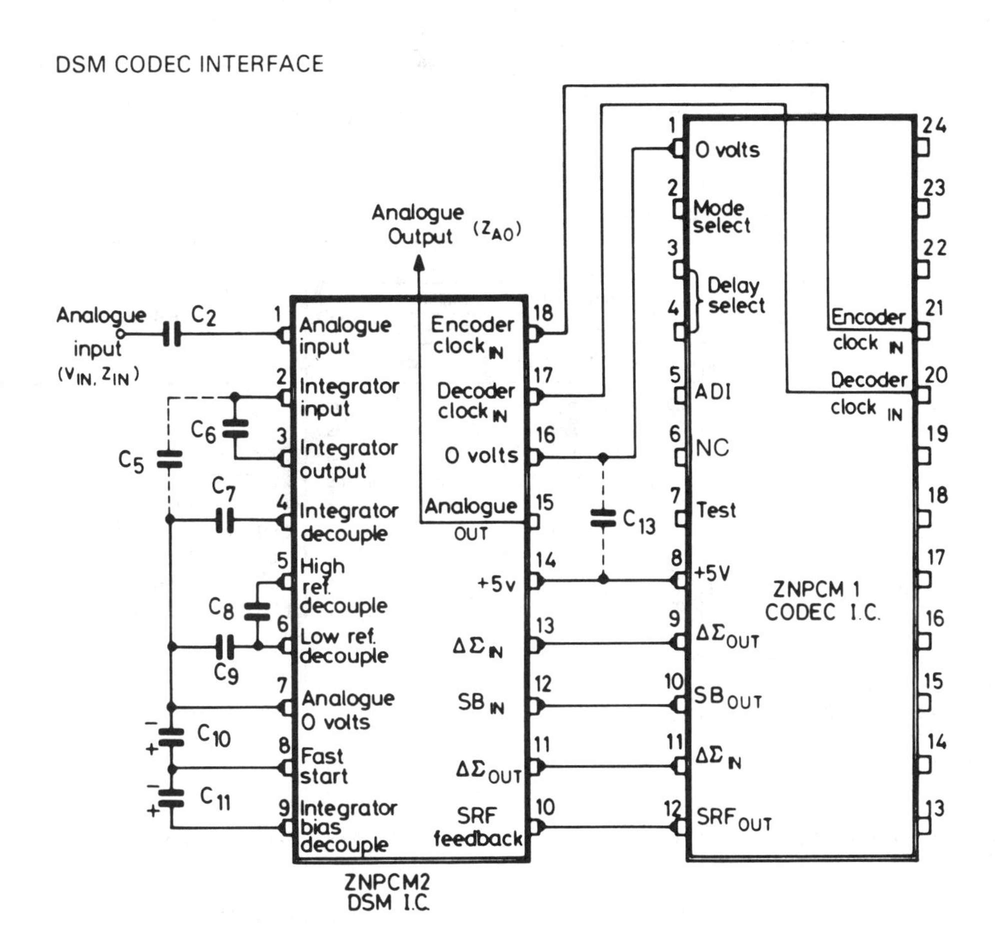

Plessey PDSP1640/PDSP1640A
40 MHz Address Generator

- 40-MHz 8-bit address generator (PDSP1640A)
- Fast cascade logic gives 20-MHz operation at 24 bits (PDSP1640A)
- Five on-chip, user-programmable registers
- Output mask logic
- 2 micron CMOS
- 300mW maximum power dissipation
- 28-pin DIL, LCC, or HC package

APPLICATIONS

- DSP address generation
- Database addressing
- DMA controllers
- Modulo counting

ELECTRICAL CHARACTERISTICS

Test conditions (unless otherwise stated):

T_{amb} (Industrial) = -40°C to +85°C, V_{CC} = 5.0V ± 10%, GND = 0V
T_{amb} (Commercial) = 0°C to +70°C, V_{CC} = 5.0V ± 5%, GND = 0V
T_{amb} (Military) = -55°C to +125°C, V_{CC} = 5.0V ± 10%, GND = 0V

STATIC CHARACTERISTICS

Characteristic	Symbol	Value			Units	Conditions
		Min.	Typ.	Max.		
Output high voltage	V_{OH}	2.4			V	I_{OH} = 8mA
Output low voltage	V_{OL}			0.6	V	I_{OL} = -8mA
Input high voltage	V_{IH}	2.2			V	
Input low voltage	V_{IL}			0.8	V	
Input leakage current	V_{IL}	-10		10	μA	GND ⩽Vin⩽V_{CC}
Output leakage current	I_{OZ}	-50		50	μA	GND ⩽V_{OUT}⩽V_{CC} = V_{CC} max.
Output short cct current (Note 2)	I_{OS}	40		250	mA	V_{CC} = max.
Input capacitance	C_{IN}		9		pF	LC and HC packages
			12		pF	DG and DP packages

SWITCHING CHARACTERISTICS

Characteristic	Value								Units	Conditions
	Industrial			Commercial			Military			
	PDSP1640 B0			PDSP1640A C0			PDSP1640 A0			
	Min.	Typ.	Max.	Min.	Typ.	Max.	Min.	Max.		
CKL frequency			20		45			20	MHz	
CLK high period	20			12	6		20		ns	
CLK low period	15			10	8		15		ns	
CLK to CO			44		25	34		44	ns	1 LSTTL +5pF load
CLK to DO			34		21	27		34	ns	2 LSTTL +20pF load Opcode 3
CLK to DO			28		19	22		28	ns	2 LSTTL +20pF load Remaining Opcodes
CLK to COMP			35		19	30		35	ns	50pF load (Opcode 0,1,2)
CI to CO			20		12	16		20	ns	1 LSTTL +5pF load
Setup DI to CLK	10			10	5		10		ns	
Hold DI to CLK	3			3	0		3		ns	
Setup CI to CLK	20			15	8		20		ns	
Hold CI to CLK	3			3	0		3		ns	
Setup I to CLK	15			10	5		15		ns	
Hold I to CLK	3			3	0		3		ns	
Setup CCEN to CLK	30			18	15		30		ns	
Hold CCEN to CLK	0			0	0		0		ns	
$\overline{OE}$ high to DO High Z			30			20		30	ns	See $\overline{OE}$ test diagrams
$\overline{OE}$ low to DO/COMP valid			22			16		22	ns	
V_{CC} current			20			40		20	mA	V_{CC} = Max. outputs unloaded CLK Freq = Max.

ABSOLUTE MAXIMUM RATINGS (Note 1)

Supply voltage V_{CC}	-0.5 to 7.0V
Input voltage V_{IN}	-0.9 to V_{CC} +0.9V
Output voltage V_{OUT}	-0.9 to V_{CC} +0.9V
Clamp diode current per pin I_K (see Note 2)	±18mA
Static discharge voltage (HMB)	500V
Storage temperature range T_S	-65°C to +150°C
Ambient temperature with power applied T_{amb}	
Military	-55°C to +125°C
Industrial	-40°C to +85°C
Commercial	0°C to +70°C
Junction temperature	150°C
Package power dissipation	1000mW

NOTES
1. Exceeding these ratings may cause permanent damage. Functional operation under these conditions is not implied.
2. Maximum dissipation or 1 second should not be exceeded, only one output to be tested at any one time.
3. Exposure to absolute maximum ratings for extended periods may affect device reliability.

THERMAL CHARACTERISTICS

Package Type	θ_{JC} °C/W	θ_{JA} °C/W
DG	12	40
LC	13	56
HC	13	56

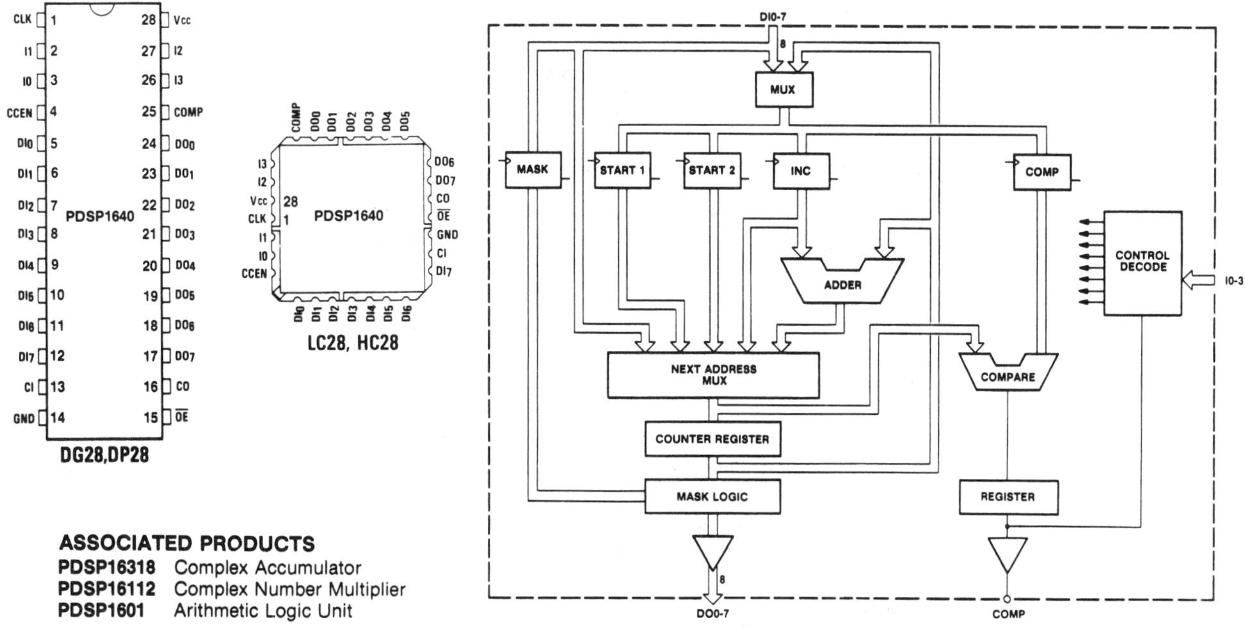

ASSOCIATED PRODUCTS

PDSP16318 Complex Accumulator
PDSP16112 Complex Number Multiplier
PDSP1601 Arithmetic Logic Unit

Pin connections - top view

PDSP1640 simplified block diagram

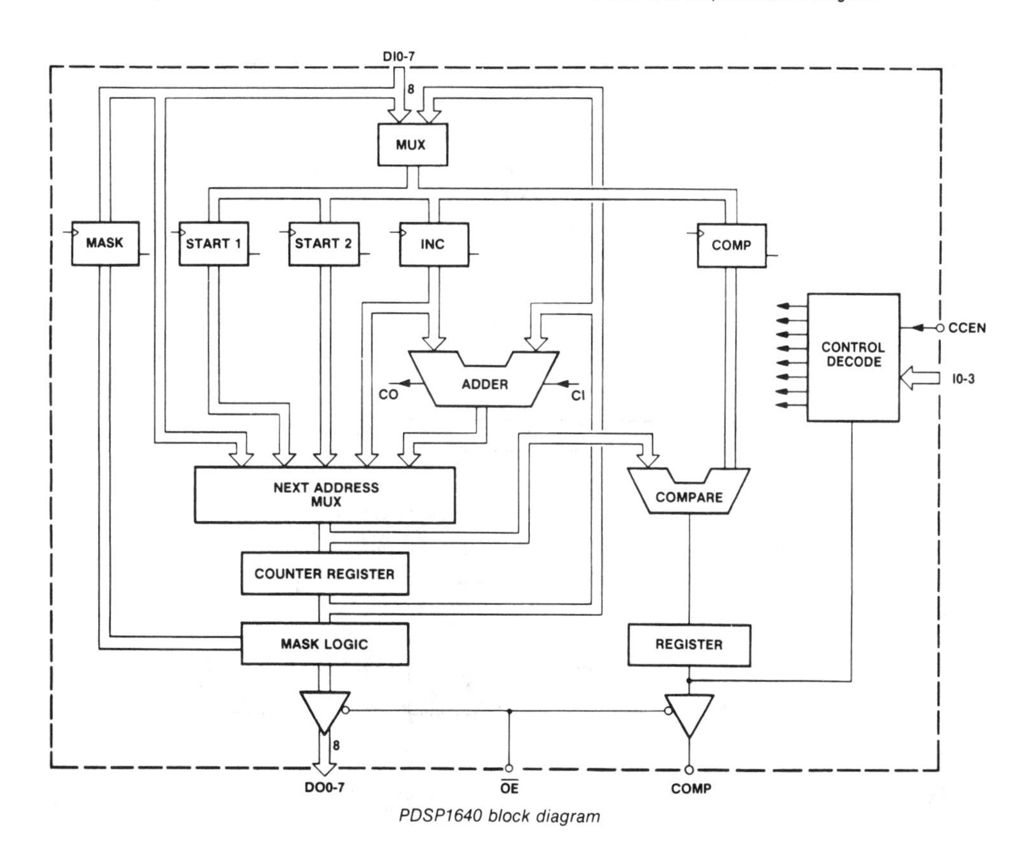

PDSP1640 block diagram

CCEN COMP PDSP1640 CI
CCEN COMP CO PDSP1640 CI
CCEN COMP CO PDSP1640

Chaining to 24 bits

PIN DESCRIPTIONS

Symbol	Pin No.	Pin name and description
CLK	1	**Common clock** to all registered internal elements. All registers are loaded, and outputs change on the rising edge of CLK.
I0-3	3,2, 27,26	**Instruction inputs.** The 16 instructions executable by the PDSP1640 are encoded onto these four lines. The instruction for any cycle must be valid at the inputs prior to the rising edge of CLK defining the start of the cycle in which the instruction is to be executed. The I0-3 inputs are internally latched by the rising edge of CLK.
CCEN	4	**Conditional Instruction Enable.** The Conditional Instructions during the current cycle are enabled if CCEN goes high before the end of the cycle. CCEN may be controlled directly by microcode or, where multiple 1640's are used, this input is used for expansion. See Figs.6 and 7.
COMP	25	**Comparator Flag Output.** This indicates that the comparator has detected an 'equal to' condition, COMP changes when CLK goes HIGH.
CI	13	**Carry In.** Carry in to least significant bit of the 8-bit adder.
CO	16	**Carry Out.** Carry out from the MSB of the adder.
DI0-7	5-12	**Data Inputs.** 8-bit data input to PDSP1640. The data on this port is loaded into the on-board registers on the rising edge of CLK.
DO0-7	24-17	**Data Outputs.** The 8-bit output from the counter. The output changes on the rising edge of CLK.
$\overline{OE}$	15	**3-State Output Control.** When high, this signal forces the DO0-7 and COMP outputs into a high-impedance state.
GND	14	0V supply.
Vcc	28	+5V supply.

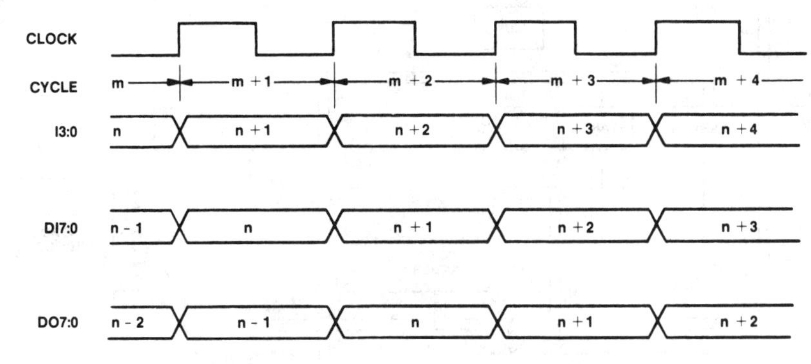

EXAMPLE
The result on DO7:0 on cycle m + 3 is generated from data (DI7:0) applied on cycle m + 2 and the instruction (I3:0) applied on cycle m + 1.

Register delays

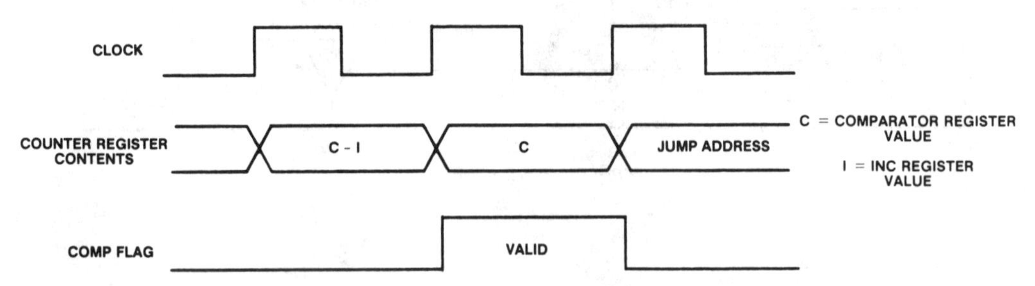

NOTE
A jump can also be forced by following LCPCR with CCJXX, provided the LCPCR instruction is held for two cycles prior to CCJXX.

Comparison flag timing

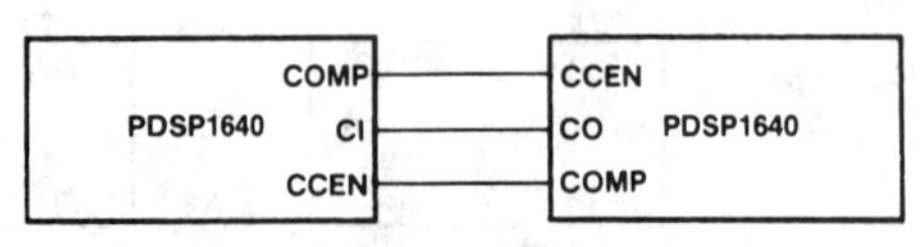

Chaining to 16 bits

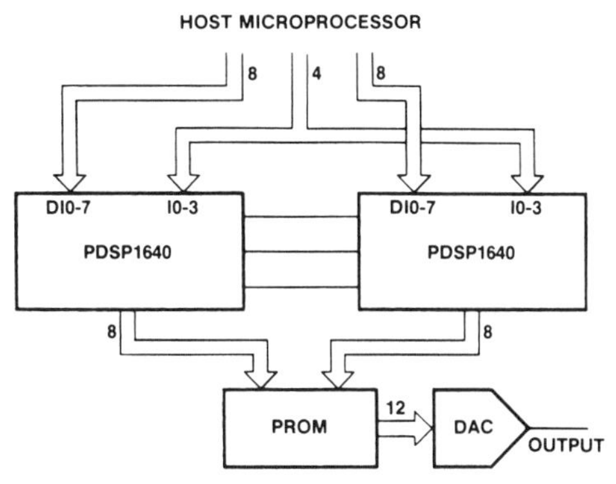

Arbitrary waveform generation

Test	Waveform - measurement level
Delay from output high to output high impedance	V_H 0.5V
Delay from output low to output high impedance	V_L 0.5V
Delay from output high impedance to output low	1.5V 0.5V
Delay from output high impedance to output high	1.5V 0.5V

NOTES
1. V_H - Voltage reached when output driven high.
2. V_L - Voltage reached when output driven low.

I_{OL}
1.5V
DUT
100p
I_{OH}

Three state delay measurement load

Plessey PDSP16112/PDSP16112A
16 × 12 Bit Complex Multiplier

- 20-MHz complex number (16 + 16) × (12 + 12) multiplication
- Pipelined architecture
- Power dissipation only 500mW
- TTL-compatible inputs
- Advanced 2-micron CMOS process

APPLICATIONS

- Digital filtering
- Fast fourier transforms
- Radar and sonar processing
- Instrumentation
- Automation
- Image processing

ASSOCIATED PRODUCTS

- PDSP1601 Arithmetic logic unit
- PDSP1640 40-MHz address generator
- PDSP16318 Complex accumulator
- PDSP16330 Pythagoras processor

ELECTRICAL CHARACTERISTICS

Test conditions (unless otherwise stated):
T_{amb} (Industrial) = -40°C to +85°C, V_{CC} = 5.0V ± 10%, GND = 0V
T_{amb} (Military) = -55°C to +125°C, V_{CC} = 5.0V ± 10%, GND = 0V

STATIC CHARACTERISTICS

Characteristic	Symbol	Value						Units	Conditions
		PDSP16112			PDSP16112A				
		Min.	Typ.	Max.	Min.	Typ.	Max.		
Output high voltage	V_{OH}	2.4			2.4			V	I_{OH} = 4mA
Output low voltage	V_{OL}			0.6			0.6	V	I_{OL} = 4mA
Input high voltage	V_{IH}	2.2			2.2			V	
Input low voltage	V_{IL}			0.8			0.8	V	
Input leakage current *	I_L	-1.0		+0.01	-1.0		+0.01	mA	GND ≤ V_{IN} ≤ V_{CC}
Output short circuit current	I_{OS}	15			20			mA	V_{CC} = max.
Input capacitance	C_I		10			10		pF	

* All inputs have a nominal 10K pull-up resistor to V_{CC}.

AC CHARACTERISTICS

Characteristic	Symbol	Value Industrial						Value Military		Units	Conditions
		PDSP16112			PDSP16112A						
		Min.	Typ.	Max.	Min.	Typ.	Max.	Min.	Max.		
V_{CC} current	I_{CC}			90			170		90	mA	V_{CC} = max Outputs unloaded f_{CLK} = max
Max. CLK frequency	f_{CLK}	10			20			10		MHz	
Min. CLK frequency				DC			DC		DC		
Input setup time	t_{su}			30			20		30	ns	
Input hold time	t_{ih}			5			5		5	ns	
CLK to output delay	t_d			50			30		50	ns	
CLK Mark/Space ratio		40		60	40		60	40	60	%	
Drive capability		2 x LSTTL +20pF									

ABSOLUTE MAXIMUM RATINGS (Note)

Supply voltage V_{CC}	-0.5V to 7.0V
Input voltage V_{IN}	-0.9V to V_{CC} +0.9V
Output voltage V_{OUT}	-0.9V to V_{CC} +0.9V
Clamp diode current per pin I_K (see Note 2)	±18mA
Static discharge voltage	500V
Storage temperature range T_S	-65°C to +150°C
Junction temperature	150°C
Ambient temperature with power applied T_{amb}	
Industrial	-40°C to +85°C
Military	-55°C to +125°C
Package power dissipation P_{TOT}	1000mW

NOTES
1. Exceeding these ratings may cause permanent damage. Functional operation under these conditions is not implied.
2. Maximum dissipation or 1 second should not be exceeded, only one output to be tested at any one time.
3. Exposure to absolute maximum ratings for extended periods may affect device reliability.

THERMAL CHARACTERISTICS

Package Type	θ_{JC} °C/W	θ_{JA} °C/W
AC	11	35

PIN OUT—FUNCTION TO PIN

PDSP16112/A

Symbol	Pin No.	Symbol	Pin No.	Symbol	Pin No.	Symbol	Pin No.
PR00	D13	PR09	A11	PI00	D1	PI09	B4
PR01	D12	PR10	C10	PI01	D2	PI10	A4
PR02	C13	PR11	B10	PI02	C1	PI11	C5
PR03	B13	PR12	A10	PI03	B1	PI12	B5
PR04	D11	PR13	C9	PI04	D3	PI13	A5
PR05	C12	PR14	B9	PI05	C2	PI14	C6
PR06	A12	PR15	A9	PI06	B3	PI15	B6
PR07	C11	PR16	C8	PI07	A3	PI16	A6
PR08	B11	CLK	L7	PI08	C4	CLK	B7
XR00	F12	XI00	E1	YR00	M8	YI00	M6
XR01	F13	XI01	F3	YR01	L8	YI01	L6
XR02	G13	XI02	F2	YR02	N9	YI02	N5
XR03	G11	XI03	F1	YR03	M9	YI03	M5
XR04	G12	XI04	G2	YR04	L9	YI04	L5
XR05	H13	XI05	G1	YR05	N10	YI05	N4
XR06	H12	XI06	H1	YR06	M10	YI06	M4
XR07	H11	XI07	H2	YR07	L10	YI07	L4
XR08	J13	XI08	H3	YR08	N11	YI08	N3
XR09	J12	XI09	J1	YR09	M11	YI09	M3
XR10	J11	XI10	J2	YR10	L11	YI10	L3
XR11	K13	XI11	J3	YR11	L12	YI11	K3
XR12	K12	XI12	K1	NC	B2	NC	M12
XR13	L13	XI13	K2	NC	L2	NC	M2
XR14	M13	XI14	L1	VCC	A1	NC	E11
XR15	K11	XI15	M1	VCC	G3	NC	C3
GND	N12	GND	C7	VCC	E2	GND	N8
GND	N7	GND	A2	VCC	A13	GND	N6
GND	M7	GND	E12	VCC	E13	GND	F11
GND	N2	GND	E3	VCC	N1	IC	B8
GND	A8	GND	B12	VCC	N13	IC	A7

NOTE
IC = Internally connected - do not connect to these pins.
All inputs are internally connected to Vcc by 10k (nominal) resistors.

PIN OUT—PIN TO FUNCTION

	1	2	3	4	5	6	7	8	9	10	11	12	13
A	VCC	GND	PI07	PI10	PI13	PI16	IC	GND	PR15	PR12	PR09	PR06	VCC
B	PI03	NC	PI06	PI09	PI12	PI15	CLK	IC	PR14	PR11	PR08	GND	PR03
C	PI02	PI05	NC	PI08	PI11	PI14	GND	PR16	PR13	PR10	PR07	PR05	PR02
D	PI00	PI01	PI04								PR04	PR01	PR00
E	XI00	VCC	GND								NC	GND	VCC
F	XI03	XI02	XI01								GND	XR00	XR01
G	XI05	XI04	VCC								XR03	XR04	XR02
H	XI06	XI07	XI08								XR07	XR06	XR05
J	XI09	XI10	XI11								XR10	XR09	XR08
K	XI12	XI13	YI11								XR15	XR12	XR11
L	XI14	NC	YI10	YI07	YI04	YI01	CLK	YR01	YR04	YR07	YR10	YR11	XR13
M	XI15	NC	YI09	YI06	YI03	YI00	GND	YR00	YR03	YR06	YR09	NC	XR14
N	VCC	GND	YI08	YI05	YI02	GND	GND	GND	YR02	YR05	YR08	GND	VCC

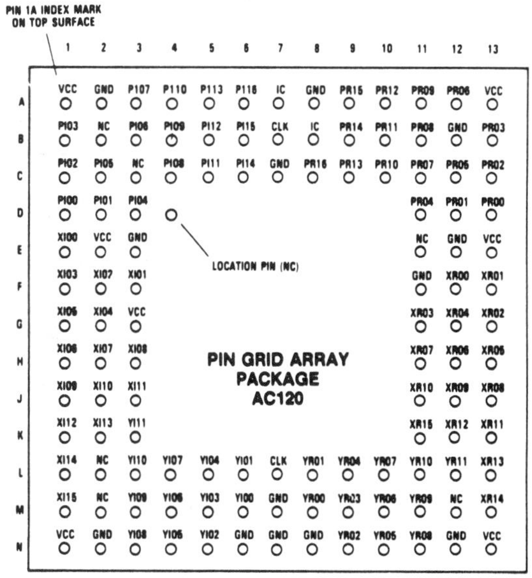

PIN 1A INDEX MARK ON TOP SURFACE

LOCATION PIN (NC)

PIN GRID ARRAY PACKAGE AC120

XRxx : X REAL INPUTS
XIxx : X IMAGINARY INPUTS
YRxx : Y REAL INPUTS
YIxx : Y IMAGINARY INPUTS
PRxx : PRODUCT REAL OUTPUTS
PIxx : PRODUCT IMAGINARY OUTPUTS

Pin connections - top view

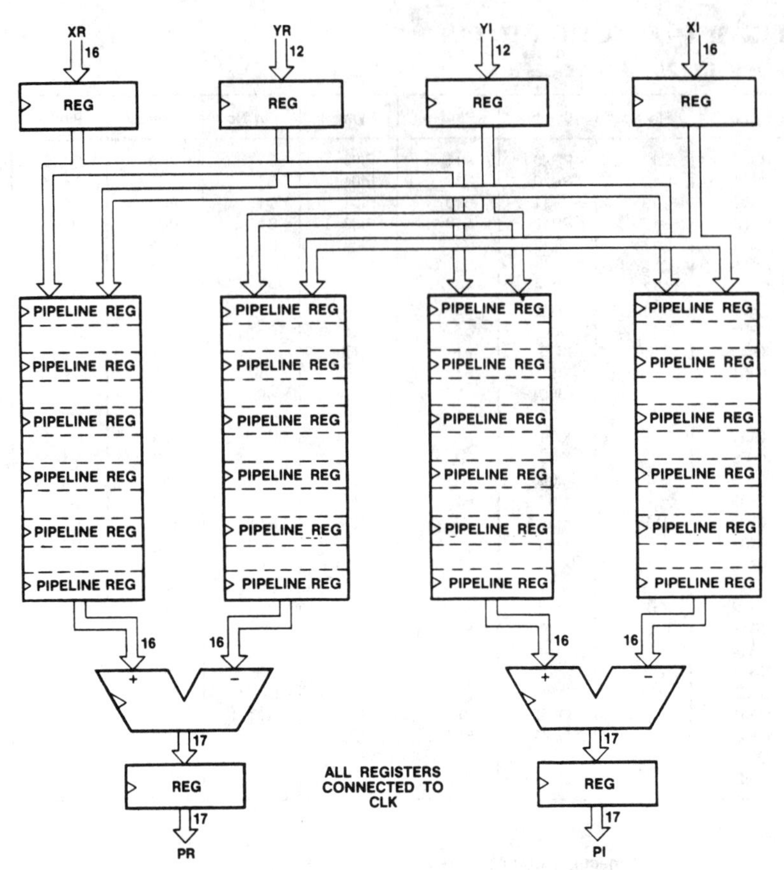

Pipelined multiplier structure

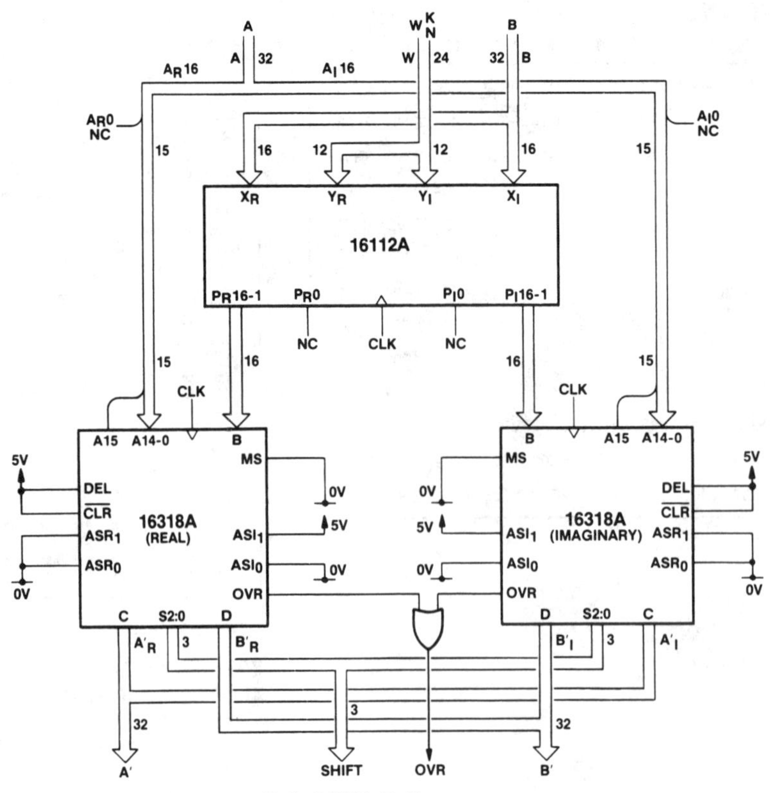

Radix 2 DIT butterfly processor

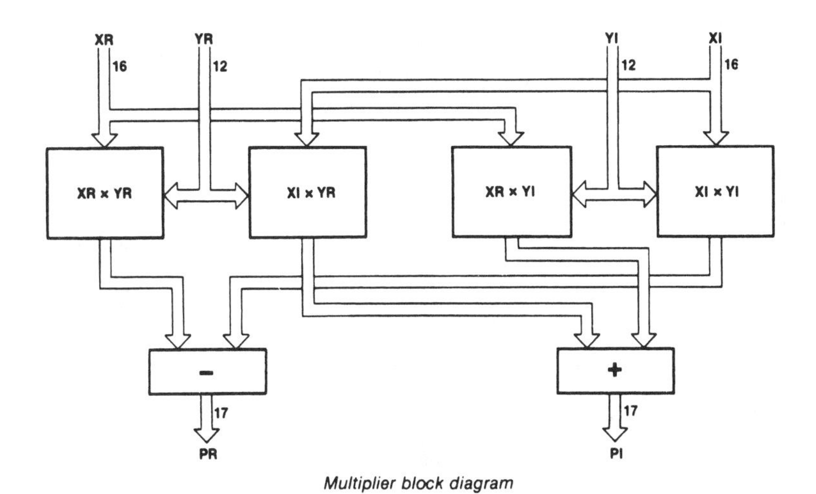

Multiplier block diagram

Plessey PDSP16316, PDSP16316A
Complex Accumulator

- Full 20-MHz throughput in FFT applications
- Four independent 16-bit I/O ports
- 20-bit addition or accumulation
- Two's complement fractional arithmetic
- Full compatible with PDSP16112 complex multiplier
- On-chip shift structures for result scaling
- Overflow detection
- Independent three-state outputs and clock enables for 2-port 20-MHz operation
- 2 micron CMOS
- 500mW maximum power dissipation
- 84-pin PGA package

APPLICATIONS

- High-speed complex FFT or DFT's
- Complex finite impulse response (FIR) filtering
- Complex conjugation
- Complex correlation/convolution

ASSOCIATED PRODUCTS

- PDSP16112 16 × 12 complex multiplier
- PDSP1601 Arithmetic logic unit
- PDSP1640 40-MHz address generator
- PDSP16330 Pythagoras processor
- PDSP16116 16 × 16 complex multiplier

ABSOLUTE MAXIMUM RATINGS (Note 1)

Parameter	Rating
Supply voltage V_{CC}	-0.5V to 7.0V
Input voltage V_{IN}	-0.9V to V_{CC} +0.9V
Output voltage V_{OUT}	-0.9V to V_{CC} +0.9V
Clamp diode current per pin I_K (see Note 2)	18mA
Static discharge voltage (HMB) V_{STAT}	500V
Storage temperature range T_S	-65°C to +150°C
Ambient temperature with power applied T_{amb}	
Industrial	-40°C to +85°C
Military	-55°C to +125°C
Junction temperature	150°C
Package power dissipation P_{TOT}	1000mW

NOTES
1. Exceeding these ratings may cause permanent damage. Functional operation under these conditions is not implied.
2. Maximum dissipation or 1 second should not be exceeded, only one output to be tested at any one time.
3. Exposure to absolute maximum ratings for extended periods may affect device reliability.

THERMAL CHARACTERISTICS

Package Type	θ_{JC} °C/W	θ_{JA} °C/W
LC	12	35
AC	12	36

ELECTRICAL CHARACTERISTICS

Test conditions (unless otherwise stated):

T_{amb} (Industrial) = -40°C to +85°C, V_{CC} = 5.0V ± 10%, GND = 0V
T_{amb} (Military) = -55°C to +125°C, V_{CC} = 5.0V ± 10%, GND = 0V

STATIC CHARACTERISTICS

Characteristic	Symbol	Value			Units	Conditions
		Min.	Typ.	Max.		
Output high voltage	V_{OH}	2.4		-	V	I_{OH} = 3.2mA
Output low voltage	V_{OL}	-		0.4	V	I_{OL} = -3.2mA
Input high voltage	V_{IH}	2.0		-	V	
Input low voltage	V_{IL}	-		0.8	V	
Input leakage current	I_{IL}	-5		+5	μA	GND ≤ V_{IN} ≤ V_{CC}
Input capacitance	C_{IN}	-	-	5	pF	
Output leakage current	I_{OZ}	-10	-	+10	μA	GND ≤ V_{OUT} ≤ V_{CC}
Output S/C current	I_{OS}	10	-	100	mA	V_{CC} = Max
Input capacitance	C_I	-	9		pF	

SWITCHING CHARACTERISTICS

Characteristic	Value Industrial						Value Military		Units	Conditions
	PDSP16316			PDSP16316A						
	Min.	Typ.	Max.	Min.	Typ.	Max.	Min.	Max.		
Clock period	100		-	50		-	100		ns	
Clock high time	20		-	15		-	20		ns	
Clock low time	20		-	15		-	20		ns	
A15:0, B15:0 setup to clock rising edge	8		-	5		-	8		ns	
A15:0, B15:0 hold after clock rising edge	8		-	5		-	8		ns	
DEL, ASR, ASI, MSR, MSI, S2:0 setup to clock rising edge	10		-	5		-	10		ns	
DEL, ASR, ASI, MSR, MSI, S2:0 hold after clock rising edge	10		-	6		-	10		ns	
$\overline{CEA}$, $\overline{CEB}$ setup to clock falling edge	0		-	0		-	0		ns	
$\overline{CEA}$, $\overline{CEB}$ hold after clock rising edge	0		-	0		-	0		ns	
Clock rising edge to OVR, C15:0, D15:0	-		40	-		30		40	ns	2 x LSTTL +20pF
$\overline{OEC}$ low to C15:0 high data valid	-		40	-		30		40	ns	
$\overline{OEC}$ low to C15:0 low data valid	-		40	-		30		40	ns	
$\overline{OEC}$ high to C15:0 high impedance	-		40	-		30		40	ns	
$\overline{OED}$ low to D15:0 high data valid	-		40	-		30		40	ns	
$\overline{OED}$ low to D15:0 low data valid	-		40	-		30		40	ns	
$\overline{OED}$ high to D15:0 high impedance	-		40	-		30		40	ns	
V_{CC} current	-		50	-		100		50	mA	V_{CC} = max Outputs unloaded f_{CLK} = max

ASR or ASI ASX1	ASX0	ALU Function
0	0	A + B
0	1	A
1	0	A - B
1	1	B - A

DEL	Delay Mux Control
0	A port input
1	Delayed A port input

MSR	Real Mux Control
0	B port input
1	C accumulator

MSI	Imag' Mux Control
0	Del mux output
1	D accumulator

S2:0			Adder result																			
S2	S1	S0	19	18	17	16	15	14	13	12	11	10	9	8	7	6	5	4	3	2	1	0
0	0	0	15	14	13	12	11	10	9	8	7	6	5	4	3	2	1	0				
0	0	1		15	14	13	12	11	10	9	8	7	6	5	4	3	2	1	0			
0	1	0			15	14	13	12	11	10	9	8	7	6	5	4	3	2	1	0		
0	1	1				15	14	13	12	11	10	9	8	7	6	5	4	3	2	1	0	
1	0	0					15	14	13	12	11	10	9	8	7	6	5	4	3	2	1	0
1	0	1						15	14	13	12	11	10	9	8	7	6	5	4	3	2	1
1	1	0							15	14	13	12	11	10	9	8	7	6	5	4	3	2
1	1	1								15	14	13	12	11	10	9	8	7	6	5	4	3

NOTE
This table shows the portion of the adder result passed to the D15:0 and C15:0 outputs. Where fewer than 16 adder bits are selected, the output data is padded with zeros.

PIN DESCRIPTIONS

Symbol	Type	Description
A15:0	Input	**Data** presented to this input is loaded into the input register on the rising edge of CLK. A15 is the MSB.
B15:0	Input	**Data** presented to this input is loaded into the input register on the rising edge of CLK. B15 is the MSB and has the same weighting as A15.
C15:0	Output	New **data** appears on this output after the rising edge of CLK. C15 is the MSB.
D15:0	Output	New **data** appears on this output after the rising edge of CLK. D15 is the MSB.
CLK	Input	**Common Clock** to all internal registers
$\overline{CEA}$	Input	**Clock enable:** when low the clock to the A input register is enabled.
$\overline{CEB}$	Input	**Clock enable:** when low the clock to the B input register is enabled.
$\overline{OEC}$	Input	**Output enable:** Asynchronous 3-state output control: The C outputs are in a high impedance state when this input is high.
$\overline{OED}$	Input	**Output enable:** Asynchronous 3-state output control: The D outputs are in a high impedance state when this input is high.
OVR	Output	**Overflow flag:** This flag will go high in any cycle during which either the output data overflows the number range selected or either of the adder results overflow. A new OVR appears after the rising edge of the CLK.
ASR1:0	Input	**Add/subtract Real:** Control input for the 'Real' adder. This input is latched by the rising edge of clock.
ASI1:0	Input	**Add/subtract Imag:** Control input for the 'Imag' adder. This input is latched by the rising edge of clock.
MSR	Input	**Mux select Real:** Control input for the 'Real' adder mux. This input is latched by the rising edge of CLK. When high the feedback path is selected.
MSI	Input	**Mux select Imag:** Control input for the 'Imag' adder mux. This input is latched by the rising edge of CLK. When high the feedback path is selected.
S2:0	Input	**Scaling control:** This input selects the 16-bit field from the 20-bit adder result that is routed to the outputs. This input is latched by the rising edge of CLK.
DEL	Input	**Delay Control:** This input selects the delayed input to the real adder for operations involving the PDSP16112. This input is latched by the rising edge of CLK.
VCC	Power	**+5V supply:** Both Vcc pins must be connected.
GND	Ground	**0V supply:** Both GND pins must be connected.

LC Pin	AC Pin	Function	LC Pin	AC Pin	Function	LC Pin	AC Pin	Function	LC Pin	AC Pin	Function
75	B2	D7	12	K2	C7	33	K10	A1	54	B10	B10
76	C2	D8	13	K3	C6	34	J10	A2	55	B9	B9
77	B1	D9	14	L2	C5	35	K11	A3	56	A10	B8
78	C1	D10	15	L3	C4	36	J11	A4	57	A9	B7
79	D2	GND	16	K4	C3	37	H10	A5	58	B8	B6
80	D1	Vcc	17	L4	C2	38	H11	A6	59	A8	B5
81	E3	D11	18	J5	C1	39	F10	A7	60	B6	B4
82	E2	D12	19	K5	C0	40	G10	A8	61	B7	B3
83	E1	D13	20	L5	$\overline{OED}$	41	G11	A9	62	A7	B2
84	F2	D14	21	K6	$\overline{OEC}$	42	G9	A10	63	C7	B1
1	F3	D15	22	J6	S2	43	F9	A11	64	C6	B0
2	G3	C15	23	J7	S1	44	F11	A12	65	A6	CLK
3	G1	C14	24	L7	S0	45	E11	A13	66	A5	$\overline{CEB}$
4	G2	C13	25	K7	MSI	46	E10	A14	67	B5	OVR
5	F1	C12	26	L6	ASI1	47	E9	A15	68	C5	D0
6	H1	Vcc	27	L8	ASI0	48	D11	$\overline{CEA}$	69	A4	D1
7	H2	GND	28	K8	DEL	49	D10	B15	70	B4	D2
8	J1	C11	29	L9	MSR	50	C11	B14	71	A3	D3
9	K1	C10	30	L10	ASR1	51	B11	B13	72	A2	D4
10	J2	C9	31	K9	ASR0	52	C10	B12	73	B3	D5
11	L1	C8	32	L11	A0	53	A11	B11	74	A1	D6

PIN 1A INDEX MARK ON TOP SURFACE

AC84

LC84

Pin connections - bottom view

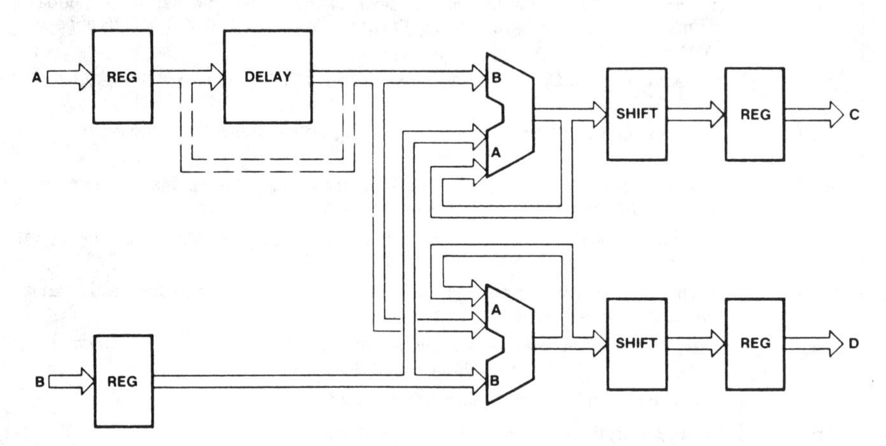

PDSP16316 simplified block diagram

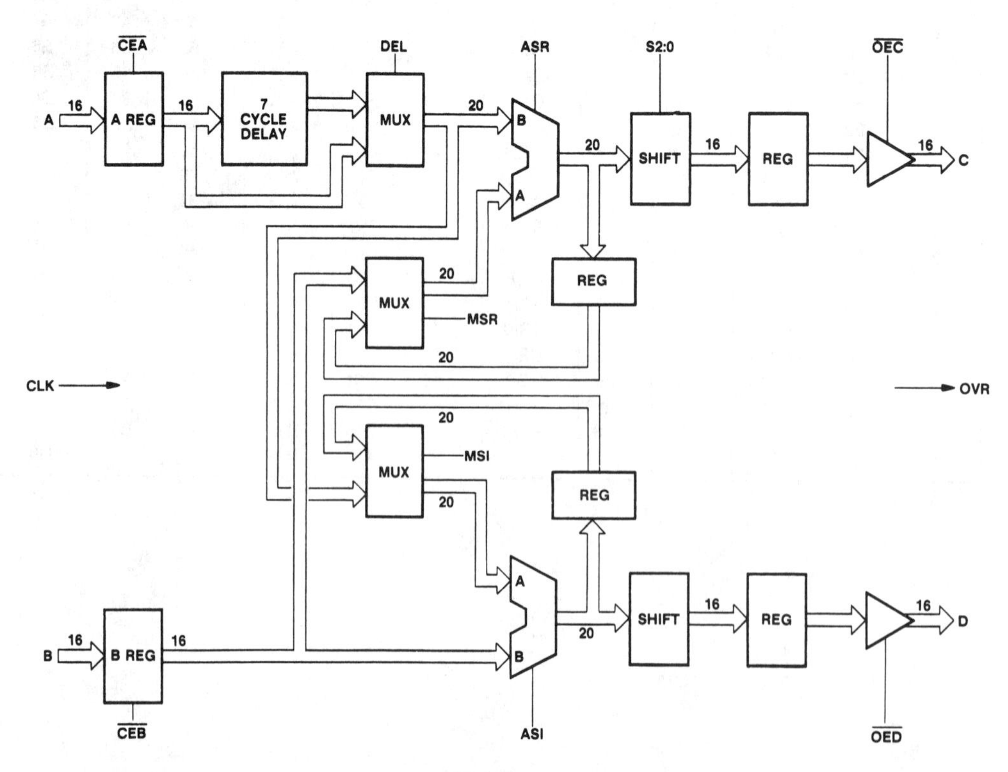

Block diagram

Plessey PDSP16318/PDSP16318A
Complex Accumulator
Preliminary Information

- Full 20 MHz throughput in FFT applications
- Four independent 16-bit I/O ports
- 20-bit addition or accumulation
- Fully compatible with PDSP16112 complex multiplier
- On-chip shift structures for result scaling
- Overflow detection
- Independent three-state outputs and clock enables for 2-port 2 MHz operation
- 1.5 micron CMOS
- 500mW maximum power dissipation
- 84-pin PGA package

APPLICATIONS

- High-speed complex FFT or DFT's
- Complex finite impulse response (FIR) filtering
- Complex conjugation
- Complex correlation/convolution

ASSOCIATED PRODUCTS

- PDSP16112 16 × 12 complex multiplier
- PDSP16116 16 × 16 complex multiplier
- PDSP1601 Arithmetic logic unit/barrel shifter
- PDSP1640 40 MHz address generator
- PDSP16330 Pythagoras processor

ABSOLUTE MAXIMUM RATINGS (Note 1)

Supply voltage V_{CC}	-0.5V to 7.0V
Input voltage V_{IN}	-0.9V to V_{CC} +0.9V
Output voltage V_{OUT}	-0.9V to V_{CC} +0.9V
Clamp diode current per pin I_K (see Note 2)	18mA
Static discharge voltage (HMB) V_{STAT}	500V
Storage temperature range T_S	-65°C to +150°C
Ambient temperature with power applied T_{amb}	
Industrial	-40°C to +85°C
Military	-55°C to +125°C
Junction temperature	150°C
Package power dissipation P_{TOT}	1000mW

NOTES
1. Exceeding these ratings may cause permanent damage. Functional operation under these conditions is not implied.
2. Maximum dissipation or 1 second should not be exceeded, only one output to be tested at any one time.
3. Exposure to absolute maximum ratings for extended periods may affect device reliability.

THERMAL CHARACTERISTICS

Package Type	θ_{JC} °C/W	θ_{JA} °C/W
LC	12	35
AC	12	36

ELECTRICAL CHARACTERISTICS

Test conditions (unless otherwise stated):
T_{amb} (Industrial) = -40°C to +85°C, V_{CC} = 5.0V ± 10%, GND = 0V
T_{amb} (Military) = -55°C to +125°C, V_{CC} = 5.0V ± 10%, GND = 0V

SWITCHING CHARACTERISTICS

Characteristic	Value Industrial						Value Military		Units	Conditions
	PDSP16318			PDSP16318A						
	Min.	Typ.	Max.	Min.	Typ.	Max.	Min.	Max.		
Clock period	100		-	50		-	100		ns	
Clock high time	20		-	15		-	20		ns	
Clock low time	20		-	15		-	20		ns	
A15:0, B15:0 setup to clock rising edge	8		-	5		-	8		ns	
A15:0, B15:0 hold after clock rising edge	8		-	5		-	8		ns	
DEL, ASR, ASI, MS, $\overline{CLR}$, S2:0 setup to clock rising edge	10		-	5		-	10		ns	
DEL, ASR, ASI, MS, $\overline{CLR}$, S2:0 hold after clock rising edge	10		-	6		-	10		ns	
$\overline{CEA}$, $\overline{CEB}$ setup to clock falling edge	0		-	0		-	0		ns	
$\overline{CEA}$, $\overline{CEB}$ hold after clock rising edge	0		-	0		-	0		ns	
Clock rising edge to OVR, C15:0, D15:0	-		40	-		30		40	ns	2 x LSTTL +20pF
$\overline{OEC}$ low to C15:0 high data valid	-		40	-		30		40	ns	
$\overline{OEC}$ low to C15:0 low data valid	-		40	-		30		40	ns	
$\overline{OEC}$ high to C15:0 high impedance	-		40	-		30		40	ns	
$\overline{OED}$ low to D15:0 high data valid	-		40	-		30		40	ns	
$\overline{OED}$ low to D15:0 low data valid	-		40	-		30		40	ns	
$\overline{OED}$ high to D15:0 high impedance	-		40	-		30		40	ns	
V_{CC} current	-		50	-		100		50	mA	V_{CC} = max Outputs unloaded f_{CLK} = max

STATIC CHARACTERISTICS

Characteristic	Symbol	Value			Units	Conditions
		Min.	Typ.	Max.		
Output high voltage	V_{OH}	2.4		-	V	I_{OH} = 3.2mA
Output low voltage	V_{OL}	-		0.4	V	I_{OL} = -3.2mA
Input high voltage	V_{IH}	2.0		-	V	
Input low voltage	V_{IL}	-		0.8	V	
Input leakage current	I_{IL}	-5		+5	μA	GND ⩽ V_{IN} ⩽ V_{CC}
Input capacitance	C_{IN}	-	-	5	pF	
Output leakage current	I_{OZ}	-10	-	+10	μA	GND ⩽ V_{OUT} ⩽ V_{CC}
Output S/C current	I_{OS}	10	-	100	mA	V_{CC} = Max
Input capacitance	C_I	-	9		pF	

PIN DESCRIPTIONS

Symbol	Type	Description
A15:0	Input	**Data** presented to this input is loaded into the input register on the rising edge of CLK. A15 is the MSB.
B15:0	Input	**Data** presented to this input is loaded into the input register on the rising edge of CLK. B15 is the MSB and has the same weighting as A15.
C15:0	Output	New **data** appears on this output after the rising edge of CLK. C15 is the MSB.
D15:0	Output	New **data** appears on this output after the rising edge of CLK. D15 is the MSB.
CLK	Input	**Common Clock** to all internal registers
$\overline{CEA}$	Input	**Clock enable:** when low the clock to the A input register is enabled.
$\overline{CEB}$	Input	**Clock enable:** when low the clock to the B input register is enabled.
$\overline{OEC}$	Input	**Output enable:** Asynchronous 3-state output control: The C outputs are in a high impedance state when this input is high.
$\overline{OED}$	Input	**Output enable:** Asynchronous 3-state output control: The D outputs are in a high impedance state when this input is high.
OVR	Output	**Overflow flag:** This flag will go high in any cycle during which either the output data overflows the number range selected or either of the adder results overflow. A new OVR appears after the rising edge of the CLK.
ASR1:0	Input	**Add/subtract Real:** Control input for the 'Real' adder. This input is latched by the rising edge of clock.
ASI1:0	Input	**Add/subtract Imag:** Control input for the 'Imag' adder. This input is latched by the rising edge of clock.
$\overline{CLR}$	Input	**Accumulator Clear:** Common accumulator clear for both Adder/Subtractor units. This input is latched by the rising edge of CLK.
MS	Input	**Mux select:** Control input for both adder multiplexers. This input is latched by the rising edge of CLK. When high the feedback path is selected.
S2:0	Input	**Scaling control:** This input selects the 16-bit field from the 20-bit adder result that is routed to the outputs. This input is latched by the rising edge of CLK.
DEL	Input	**Delay Control:** This input selects the delayed input to the real adder for operations involving the PDSP16112. This input is latched by the rising edge of CLK.
VCC	Power	**+5V supply:** Both Vcc pins must be connected.
GND	Ground	**0V supply:** Both GND pins must be connected.

Pin	Function	Pin	Function	Pin	Function	Pin	Function	Pin	Function
B2	D7	J1	C11	K7	MS	G9	A10	A8	B5
C2	D8	K1	C10	L6	ASI1	F9	A11	B6	B4
B1	D9	J2	C9	L8	ASI0	F11	A12	B7	B3
C1	D10	L1	C8	K8	DEL	E11	A13	A7	B2
D2	GND	K2	C7	L9	$\overline{CLR}$	E10	A14	C7	B1
D1	Vcc	K3	C6	L10	ASR1	E9	A15	C6	B0
E3	D11	L2	C5	K9	ASR0	D11	$\overline{CEA}$	A6	CLK
E2	D12	L3	C4	L11	A0	D10	B15	A5	$\overline{CEB}$
E1	D13	K4	C3	K10	A1	C11	B14	B5	OVR
F2	D14	L4	C2	J10	A2	B11	B13	C5	D0
F3	D15	J5	C1	K11	A3	C10	B12	A4	D1
G3	C15	K5	C0	J11	A4	A11	B11	B4	D2
G1	C14	L5	$\overline{OED}$	H10	A5	B10	B10	A3	D3
G2	C13	K6	$\overline{OEC}$	H11	A6	B9	B9	A2	D4
F1	C12	J6	S2	F10	A7	A10	B8	B3	D5
H1	Vcc	J7	S1	G10	A8	A9	B7	A1	D6
H2	GND	L7	S0	G11	A9	B8	B6		

ASR or ASI ASX1	ASX0	ALU Function
0	0	A + B
0	1	A
1	0	A - B
1	1	B - A

MS	Real and Imag' Mux Control
0	B port input/Del mux output
1	C accumulator/D accumulator

DEL	Delay Mux Control
0	A port input
1	Delayed A port input

S2:0			Adder result																			
S2	S1	S0	19	18	17	16	15	14	13	12	11	10	9	8	7	6	5	4	3	2	1	0
0	0	0	15	14	13	12	11	10	9	8	7	6	5	4	3	2	1	0				
0	0	1		15	14	13	12	11	10	9	8	7	6	5	4	3	2	1	0			
0	1	0			15	14	13	12	11	10	9	8	7	6	5	4	3	2	1	0		
0	1	1				15	14	13	12	11	10	9	8	7	6	5	4	3	2	1	0	
1	0	0					15	14	13	12	11	10	9	8	7	6	5	4	3	2	1	0
1	0	1						15	14	13	12	11	10	9	8	7	6	5	4	3	2	1
1	1	0							15	14	13	12	11	10	9	8	7	6	5	4	3	2
1	1	1								15	14	13	12	11	10	9	8	7	6	5	4	3

NOTE
This table shows the portion of the adder result passed to the D15:0 and C15:0 outputs. Where fewer than 16 adder bits are selected, the output data is padded with zeros.

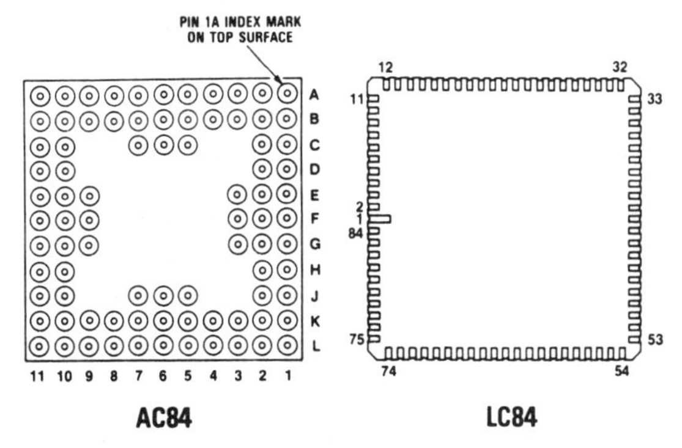

Pin connections - bottom view

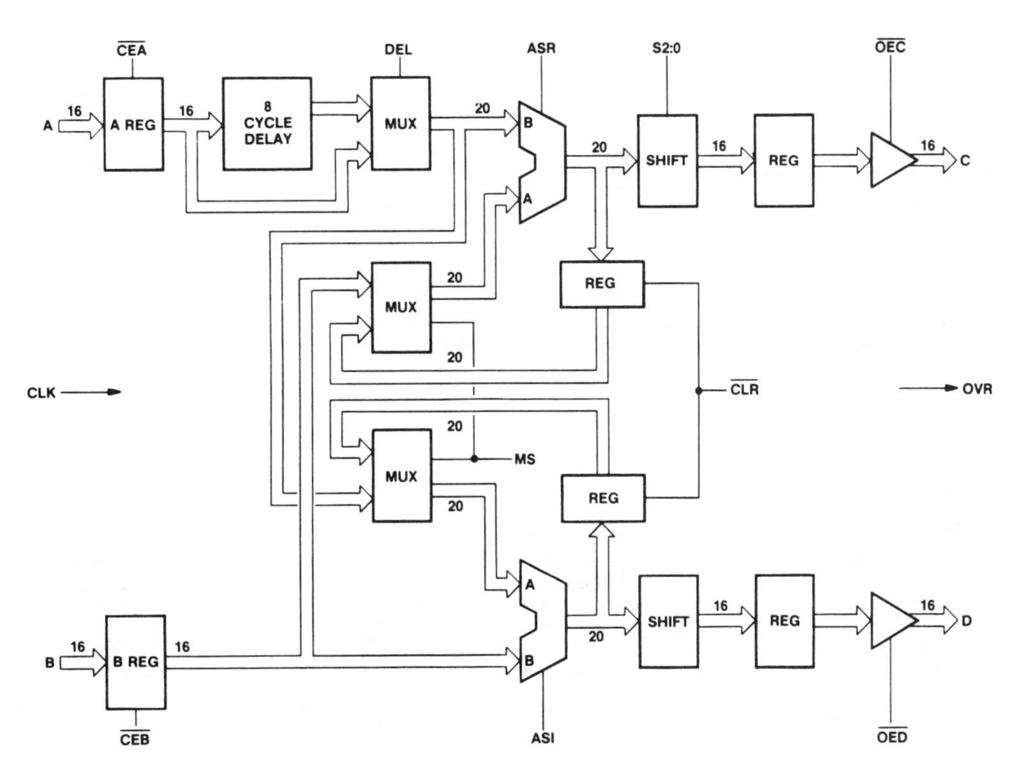

Block diagram

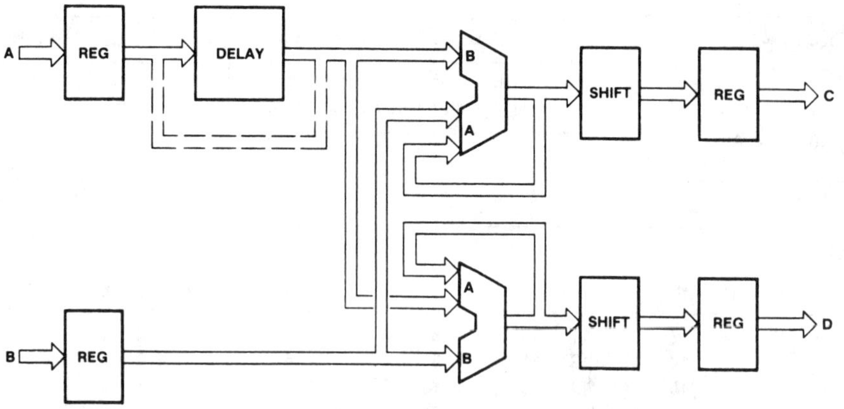

PDSP16318 simplified block diagram

Reprinted with permission from Microelectronics Communications Handbook Issue 7, copyright 1990 by Mitel Corporation.

Mitel
ISO-CMOS ST-BUS™ FAMILY MT8950
Data Codec

- Transparent coding and decoding of 0 to 8, 9.6 and 19.2 kbps data
- Coding compatible to PCM voice channels at 56/64 kbps in ST-BUS format
- Automatic line polarity detection and correction
- Loopback facility for test purposes
- Selectable data formats: RZ or NRZ
- Eight user-selectable modes of operation
- Low-power ISO-CMOS technology

APPLICATIONS

- Transparent coder/decoder for synchronous and asynchronous data
- Data terminal (RS-232, etc.) to ST-BUS interface
- Data switching on digital PBXs
- Channel banks/TDM multiplexers

ABSOLUTE MAXIMUM RATINGS* - Voltages are with respect to ground (V_{SS}) unless otherwise stated

	Parameter	Symbol	Min	Max	Units
1	Supply Voltage	V_{DD}	-0.5	7.0	V
2	DC Input Voltage	V_{IN}	V_{SS}-0.3	V_{DD}+0.3	V
3	DC Output Voltage	V_{OUT}	V_{SS}-0.3	V_{DD}+0.3	V
4	Input Diode Current ($V_I<0$ or $V_I>V_{DD}$)	I_{IK}		± 10	mA
5	Output Diode Current ($V_O<0$ or $V_O>V_{DD}$)	I_{OK}		± 20	mA
6	DC Output Current, per pin	I_O		± 25	mA
7	DC Supply or Ground Current	I_{DD}/I_{SS}		± 50	mA
8	StorageTemperature	T_{ST}	-65	150	°C
9	Package Power Dissipation (CERDIP) T_A= 25° C	P_D		1.0	W

*Exceeding these values may cause permanent damage. Functional operation under these conditions is not implied.

RECOMMENDED OPERATING CONDITIONS - Voltages are with respect to ground (V_{SS}) unless otherwise stated

	Characteristics	Sym	Min	Typ‡	Max	Units	Test Conditions
1	Supply voltage	V_{DD}	4.5	5.0	5.5	V	
2	Operating Frequency	f_{CK}		2.048		MHz	
3	Operating Temperature	T_A	0		70	°C	

‡ Typical figures are at 25°C and are for design aid only: not guaranteed and not subject to production testing.

DC ELECTRICAL CHARACTERISTICS

V_{DD}=5.0V±10%; V_{SS}=0V; T_A=0°C to 70°C - Voltages are with respect to ground (V_{SS}) unless otherwise stated

		Characteristics	Sym	Min	Typ‡	Max	Units	Test Conditions
1	SUP	Quiescent supply current	I_{QS}			150	μA	All outputs unloaded All inputs @ V_{SS}
2	SUP	Operating supply current	I_{DD}			1.0	mA	All outputs unloaded. Input pins 2 and 3 clocked at 2.048 MHz. Pins 1,8,15,20,21 @ V_{SS} Pins 5,6,7,9,10,13 and 22 @V_{DD}
3	INPUTS	TTL inputs [1] HIGH voltage LOW voltage	 V_{IH} V_{IL}	 2.0 V_{SS}		 V_{DD} 0.8	 V V	
4	INPUTS	CMOS inputs [2] HIGH voltage LOW voltage	 V_{IH} V_{IL}	 3.5 V_{SS}		 V_{DD} 1.5	 V V	
5	INPUTS	CMOS Schmitt inputs [3] HIGH voltage LOW voltage	 V_{IH} V_{IL}	 3.0 V_{SS}		 V_{DD} 1.0	 V V	
6	INPUTS	SPi Comparator ON Voltage	V_{T+}	2.25	2.5	2.75	V	V_{DD} = 5V
7	INPUTS	SPi Comparator OFF Voltage	V_{T-}	2.0			V	V_{DD} = 5V
8	INPUTS	Input Leakage Current	I_{IN}		±1	±10	μA	V_{DD} = 5V

‡ Typical figures are at 25°C and are for design aid only: not guaranteed and not subject to production testing.

1. Include DSTi, CSTi, C2i, $\overline{F1i}$ and SCLK
2. Include DF
3. Include $\overline{RxE}$,D_X1, D_X2 and $\overline{PRST}$

AC ELECTRICAL CHARACTERISTICS† Voltages are with respect to ground (V_{SS}) unless otherwise stated

	Characteristics	Sym	Min	Typ‡	Max	Units	Test Conditions
1	C2i Clock Frequency	f_{CK}	2.028	2.048	2.068	MHz	
2	C2i Clock Rise Time	t_{CR}			50	ns	
3	C2i Clock Fall Time	t_{CF}			50	ns	
4	Clock Duty Cycle (C2i & SCLK)			50		%	
5	SCLK Clock Frequency	f_{SCLK}	0	0.6	128	kHz	
6	SCLK Clock Rise Time	t_{SCLKR}			50	ns	
7	SCLK Clock Fall Time	t_{SCLKF}			50	ns	
8	$\overline{F1i}$ and $\overline{CA}$ Rise Time	t_{ER}			100	ns	
9	$\overline{F1i}$ and $\overline{CA}$ Fall Time	t_{EF}			100	ns	
10	$\overline{F1i}$ and $\overline{CA}$ Setup Time	t_{ES}	25			ns	
11	$\overline{F1i}$ and $\overline{CA}$ Hold Time	t_{EH}	-25		25	ns	
12	DSTo Rise Time	t_{OR}			100	ns	Note 1
13	DSTo Fall Time	t_{OF}			100	ns	Note 1
14	Propagation Delay From Clock (C2i) To Output (DSTo) enable.	t_{PZH} t_{PZL}			125	ns	Note 1
15	Propagation Delay From Clock (C2i) To Output (DSTo) .	t_{PLH} t_{PHL}			125	ns	Note 1
16	Input Rise Time (DSTi, CSTi)	t_{IR}			100	ns	
17	Input Fall Time (DSTi, CSTi)	t_{IF}			100	ns	
18	DSTi, CSTi Setup Time	t_{ISH} t_{ISL}	0			ns	
19	DSTi, CSTi Hold Time	t_{IH}	90			ns	
20	$\overline{PRST}$ Low Time		488			ns	

† Timing is over recommended temperature & Power Supply voltages

‡ Typical figures are at 25°C and are for design aid only: not guaranteed and not subject to production testing.

Note 1: R_L=10KΩ to V_{DD}, C_L=150 pF to V_{SS}

DC ELECTRICAL CHARACTERISTICS

V_{DD} = 5.0V ± 10%; V_{SS} = 0V; T_A = 0°C to 70°C - Voltages are with respect to ground (V_{SS}) unless otherwise stated

		Characteristics	Sym	Min	Typ‡	Max	Units	Test Conditions
1	OUTPUTS	Output LOW Voltage	V_{OL}			0.05	V	$\lvert I_O \rvert < 1.0\ \mu A$ $V_{DD} = 5V$
2		Output HIGH Voltage	V_{OH}	4.95			V	$\lvert I_O \rvert < 1.0\ \mu A$ $V_{DD} = 5V$
3		Output LOW Current (On all outputs except DSTo)	I_{OL}	2.2	2.8		mA	$V_{OL} = 0.4V$
4		Output HIGH Current (On all outputs except DSTo)	I_{OH}	-3.5	-4.2		mA	$V_{OH} = 2.4V$
5		Output LOW Current (On DSTo output)	I_{OL}	8.9	11.1		mA	$V_{OL} = 0.4V$
6		Output HIGH Current (On DSTo output)	I_{OH}	-14.0	-16.8		mA	$V_{OH} = 2.4V$
7		Output Leakage Current	I_{OZ}		±1	±10	μA	

‡ Typical figures are at 25°C and are for design aid only: not guaranteed and not subject to production testing.

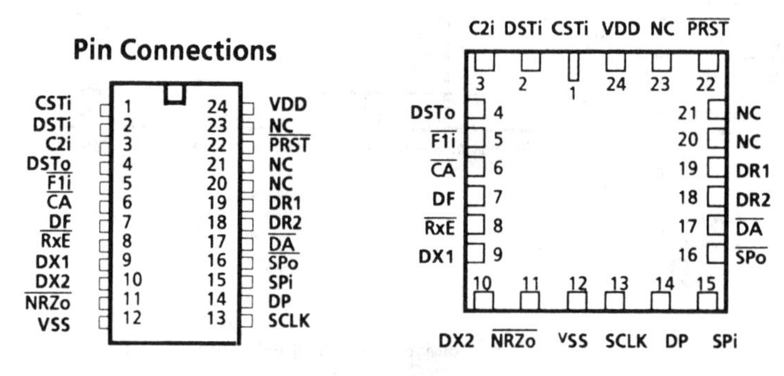

Ordering Information 0°C to 70°C

MT8950AC 24 Pin Ceramic
MT8950AY 24 Pin LCC

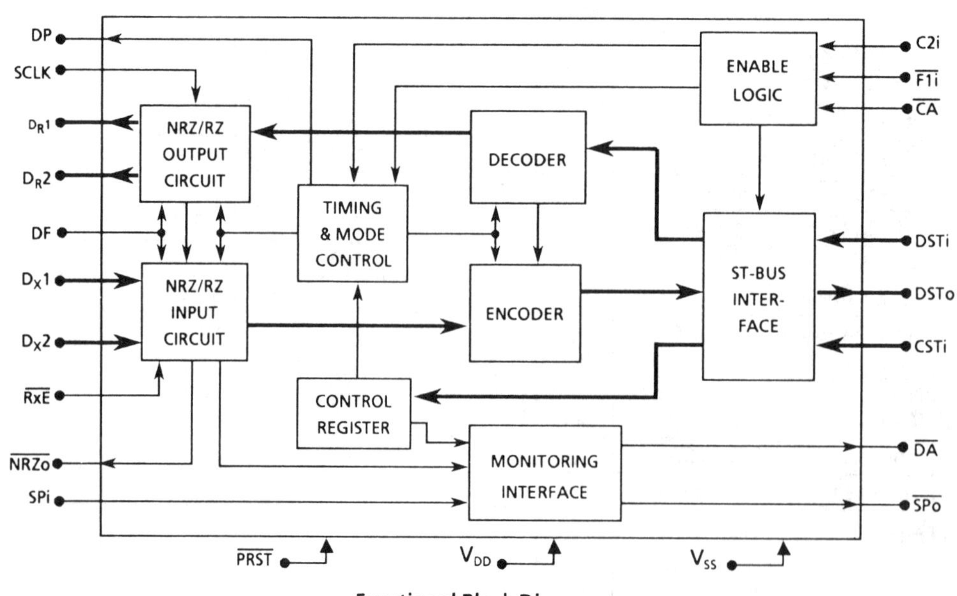

Functional Block Diagram

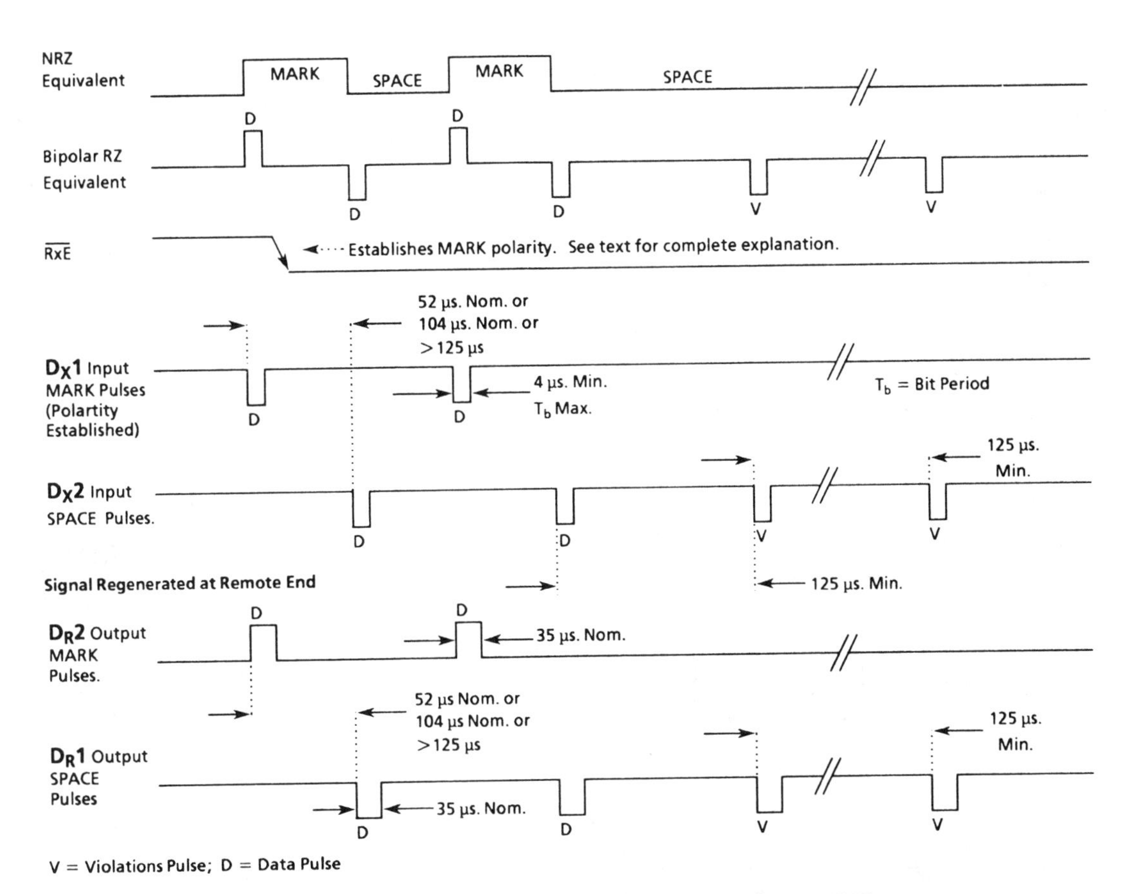

Example Input/output Waveform in the RZ format (DF= HIGH)

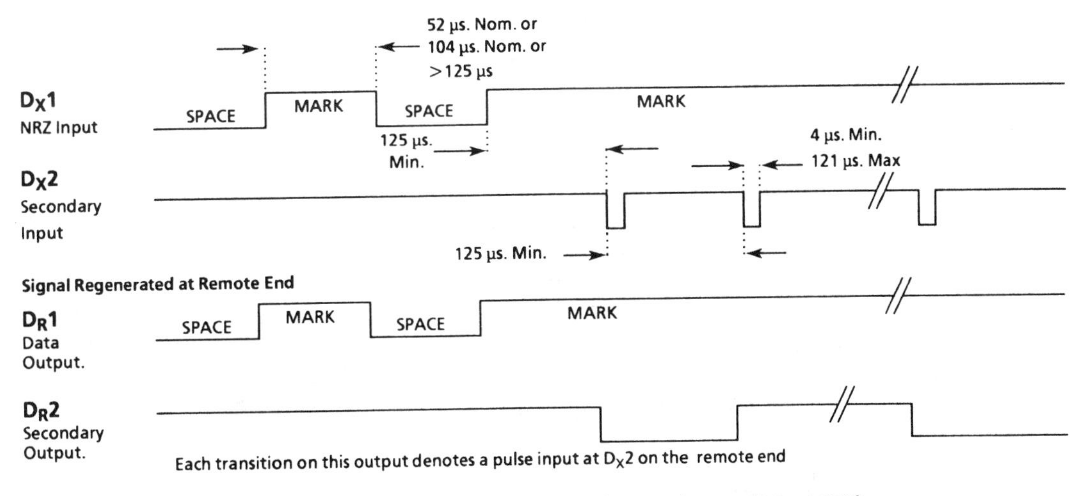

Example Input/Output Waveform in the NRZ format (DF =LOW)

Voice-Data integration using the Data Codec

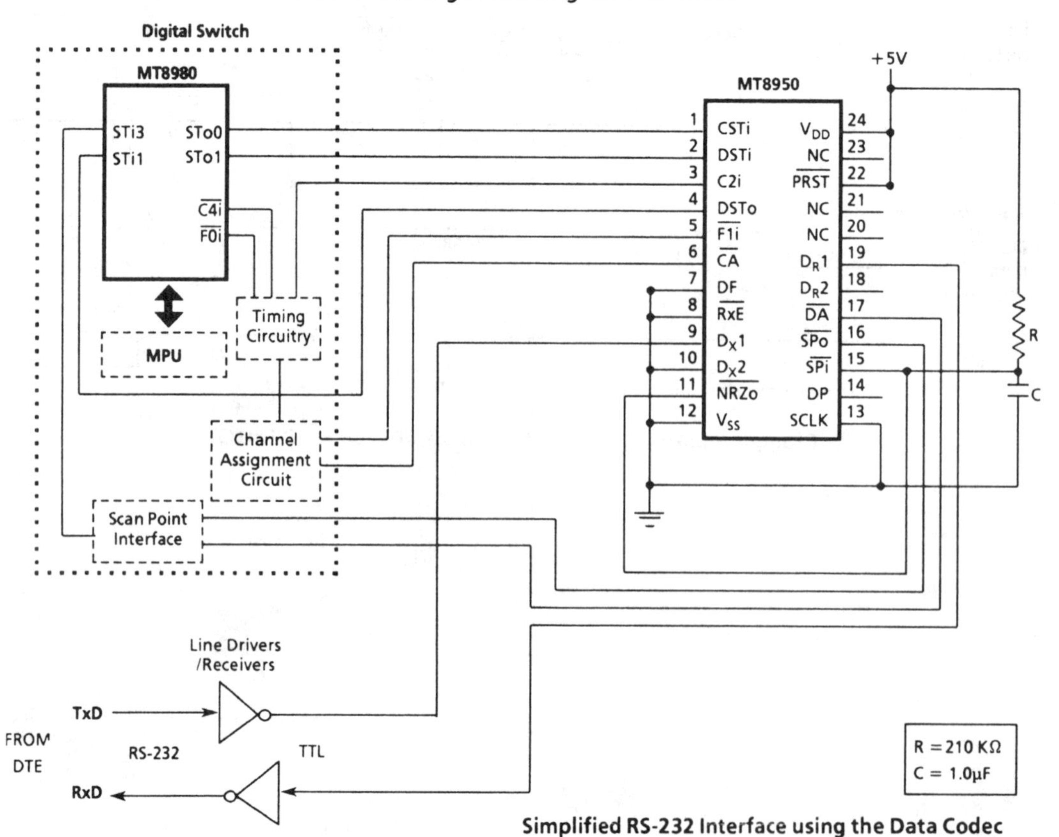

Simplified RS-232 Interface using the Data Codec

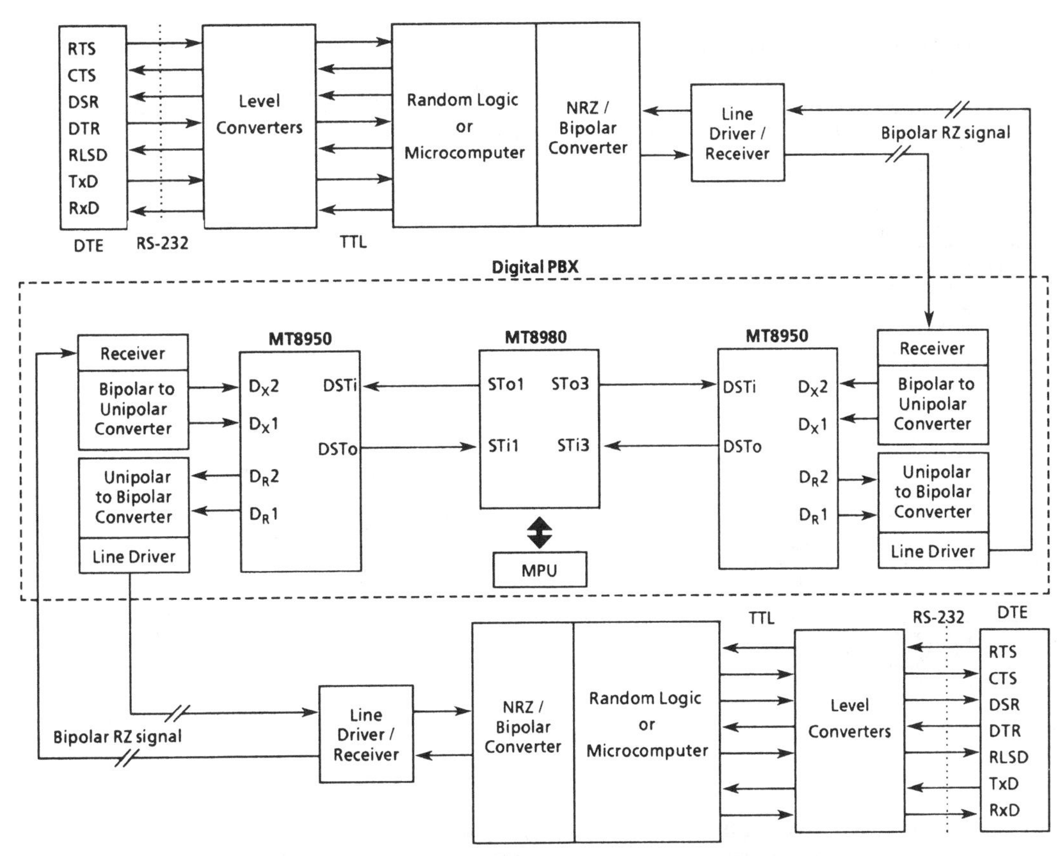

Block Diagram Illustration of a Scheme to Submultiplex RS-232 Control Signals

Mitel
ST-BUS™ FAMILY MH89500
R-Interface Module (RIM)
Preliminary Information

- Compatible with ECMA.102, CCITT 1.461 (X.30), and 1.463 (V.110)
- V.24, X.20, and X.21 terminal interfaces
- Synchronous operation at 600, 1200, 2400, 4800, 9600, 19200, and 64000 bit/s
- Asynchronous operation at 50 - 600, 1200, 2400, 4800, 9600, and 19200 bit/s
- User-programmable network port interface
- Flexible-format subrate multiplexing
- Automatic entry to and exit from data-transfer state
- Frame sync monitoring and auto resync
- Local and remote loopback modes
- User independent clock control outputs
- User access to E-bits
- Optional automatic user-rate selection
- Standalone or MPU peripheral mode operation
- Single +5V supply low-power CMOS

ABSOLUTE MAXIMUM RATINGS*

	Parameters	Symbol	Min	Max	Units
1	Supply voltage	V_{DD}	−0.3	7.0	V
2	All Input voltages	V_I	−0.3	V_{DD} +0.3	V
3	All output voltages	V_O	−0.3	V_{DD} +0.3	V
4	Storage temperate range	T_S	−55	125	°C
5	Input current (any pin)	I_{SS}		±10	mA
6	Continuous power dissipation	P_D		800	mW

*Exceeding these values may cause permanent damage. Functional operation under these conditions is not implied.

RECOMMENDED OPERATING CONDITIONS‡ Voltages are with respect to ground (V_{SS}) unless otherwise stated.

		Characteristics	Sym	Min	Typ‡	Max	Units	Test Conditions
1		Supply Voltage	V_{DD}	4.5	5.0	5.5	V	
2		Operating Temperature	T_{OP}	0		+70	°C	

‡ Typical figures are at 25°C and are for design aid only: not guaranteed and not subject to production testing.

DC ELECTRICAL CHARACTERISTICS- Voltages are with respect to ground (V_{SS}) unless otherwise stated.

		Characteristics		Sym	Min	Typ‡	Max	Units	Test Conditions
1		Operating Supply Voltage		V_{DD}	4.5	5.0	5.5	V	
2		Operating Supply Current		I_{DD}		10.0		mA	Outputs Unloaded NCLK/$\overline{\text{C4i}}$ = 4.096 MHz
3		Powerdown Supply Current		I_Q		1.5		mA	
4	I N P U T S	Input High Voltage (pins 21, 23)	$\overline{\text{RST}}$ XTAL2	V_{IH}	$0.7V_{DD}$ $0.8V_{DD}$			V	
5		Input High Voltage (Except pins 21, 23)		V_{IH}	2.0 $0.7V_{DD}$			V	TTL Inputs * CMOS Inputs **
6		Input Low Voltage (pins 21, 23)	$\overline{\text{RST}}$ XTAL2	V_{IL}			$0.3V_{DD}$ $0.2V_{DD}$	V	
7		Input Low Voltage (Except pins 21, 23)		V_{IL}			0.8 $0.2V_{DD}$	V	TTL Inputs * CMOS Inputs **
8		Input Leakage Current		I_{IZ}		10.0		μA	$V_{DD} = 5.0V = V_{IN}$
9		Input Capacitance		C_I		15.0		pF	
10	O / P	Output High Voltage		V_{OH}	V_{DD}-0.1			V	$I_{OH} = -20\mu A$
11		Output Low Voltage		V_{OL}			0.1	V	$I_{OL} = 20\mu A$
12		Output Capacitance		C_O		15.0		pF	

‡ Typical figures are at 25°C and are for design aid only: not guaranteed and not subject to production testing.

* TTL inputs: A0, A1, $\overline{\text{RD}}$, $\overline{\text{WR}}$, D0-D7.

** CMOS inputs: PWD/$\overline{\text{CS}}$, $\overline{\text{DTR}}$, $\overline{\text{RTS}}$, TxD, NRxD, NCLK, NFRM, $\overline{\text{RST}}$

AC ELECTRICAL CHARACTERISTICS - Voltages are with respect to ground (V_{SS}) unless otherwise stated.

		Characteristics	Sym	Min	Typ‡	Max	Units	Test Conditions
1		Output Rise Time	t_{or}	0.8 1.1		5 11	ns ns	$C_L = 15pF, V_{DD} = 5V, R_L = \infty$ $C_L = 50pF, V_{DD} = 5V, R_L = \infty$
2		Output Fall Time	t_{of}	1.6 2.7		12 25	ns ns	$C_L = 15pF, V_{DD} = 5V, R_L = \infty$ $C_L = 50pF, V_{DD} = 5V, R_L = \infty$
3		XTAL2 Input Rise Time	t_{xir}			20	ns	
4		XTAL2 Input Fall Time	t_{xif}			10	ns	
5		XTAL2 Input Duty Cycle	TCC	45	50	55	%	
6		XTAL Frequency	f		6.16± 100ppm		MHz	Parallel Resonance

‡ Typical figures are at 25°C and are for design aid only: not guaranteed and not subject to production testing.

Pin Connections

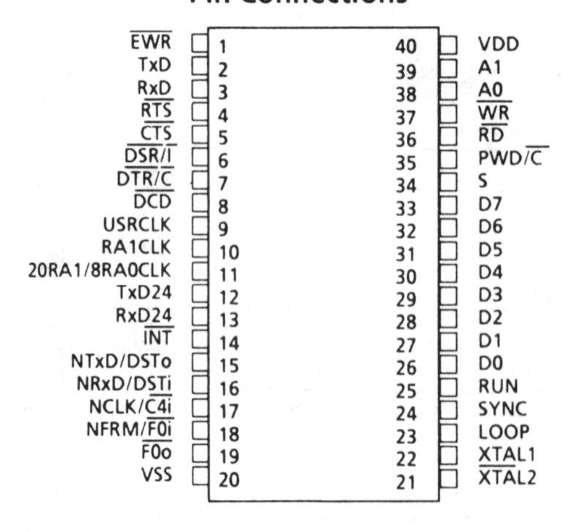

Ordering Information

MH89500 40 pin DIL Hybrid

0 °C to 70 °C

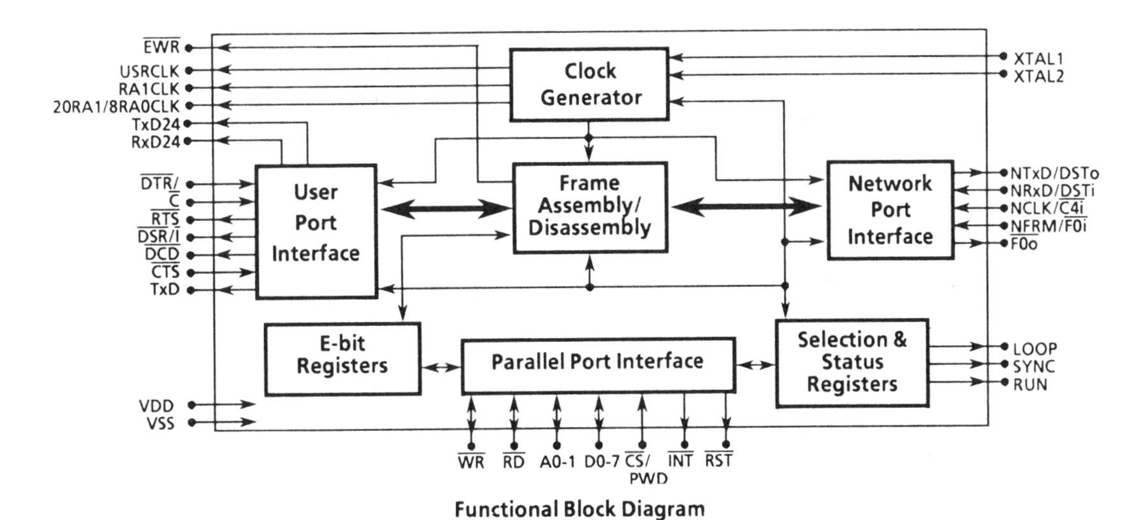

Functional Block Diagram

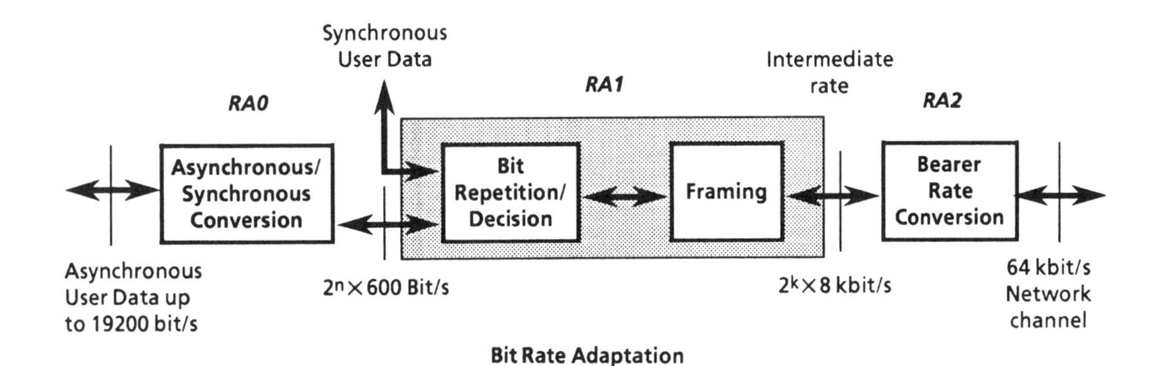

Bit Rate Adaptation

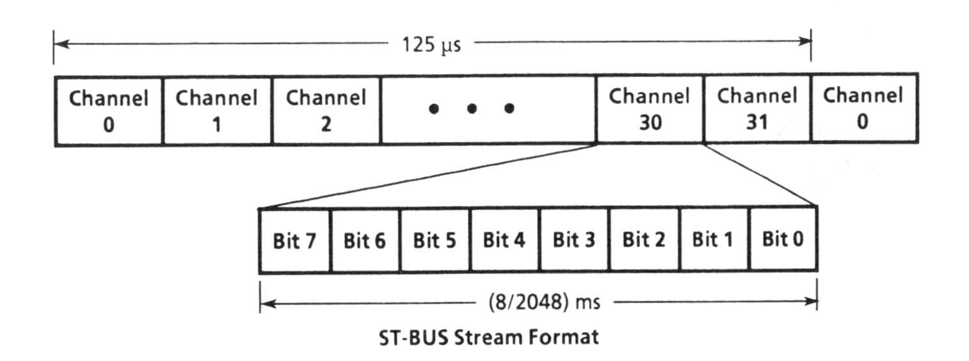

ST-BUS Stream Format

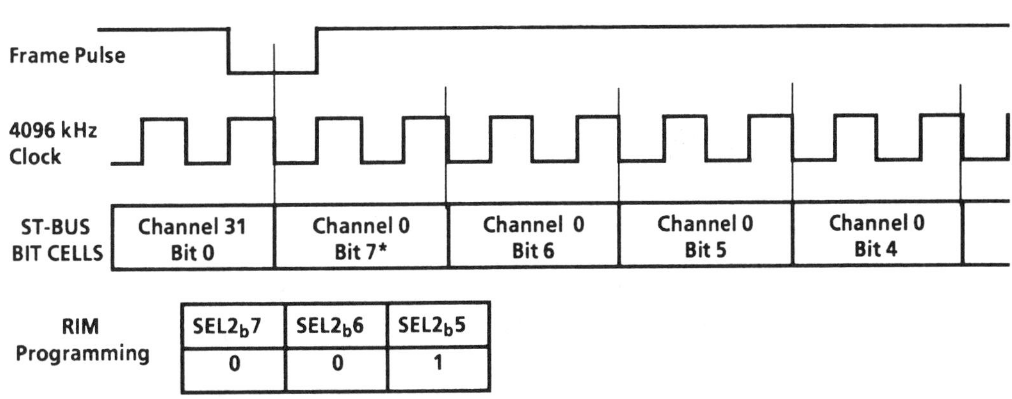

** The order of bit transmission in the standard ST-BUS format is opposite in the RIM*

RIM Programming for Clock & Frame Alignment for ST-BUS Streams

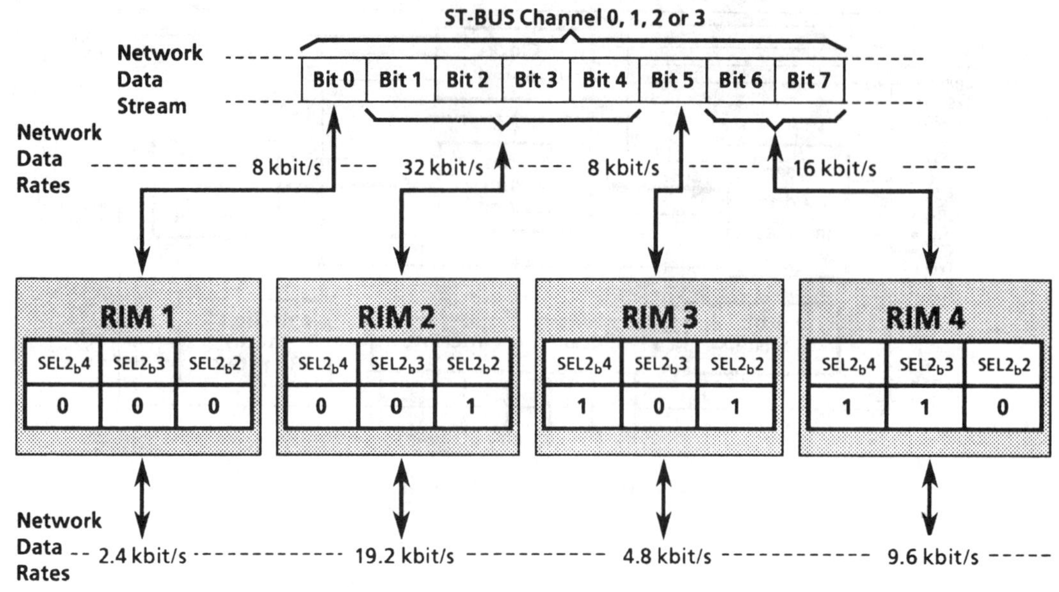

Example of Subrate Stream Multiplexing

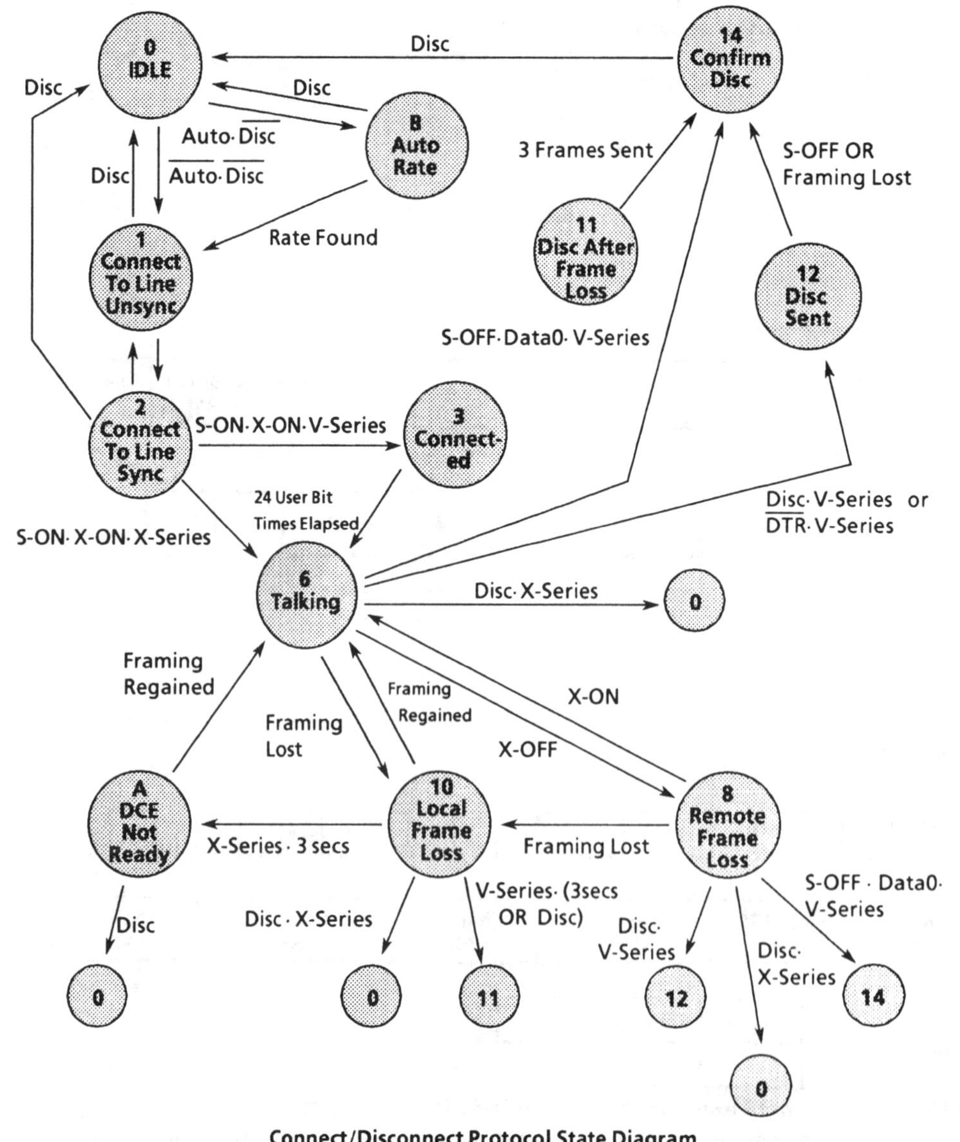

Connect/Disconnect Protocol State Diagram

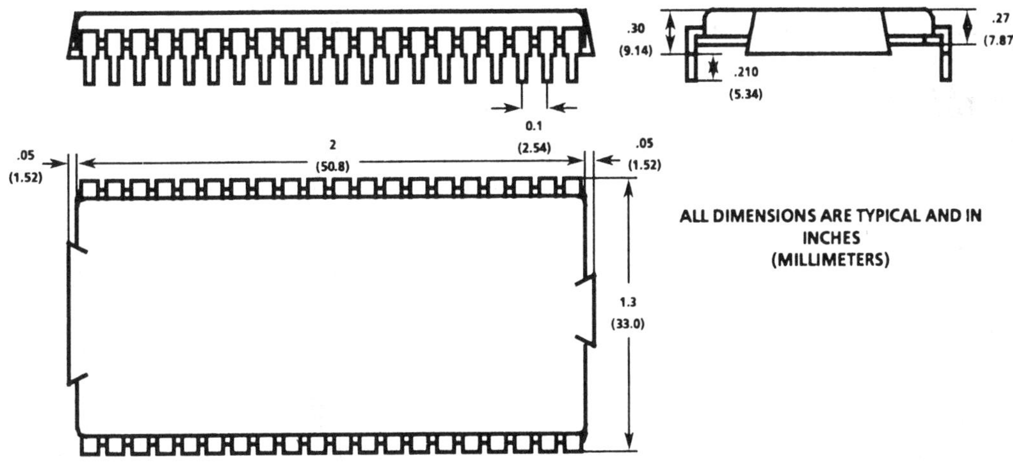

Physical Dimensions of 40 Pin Dual in Line Hybrid Package

Octet #	BIT POSITION NUMBER							
	1	2	3	4	5	6	7	8
0	0	0	0	0	0	0	0	0
1	1	D1	D2	D3	D4	D5	D6	S1
2	1	D7	D8	D9	D10	D11	D12	X
3	1	D13	D14	D15	D16	D17	D18	S3
4	1	D19	D20	D21	D22	D23	D24	S4
5	1	E1	E2	E3	E4	E5	E6	E7
6	1	D25	D26	D27	D28	D29	D30	S6
7	1	D31	D32	D33	D34	D35	D36	X
8	1	D37	D38	D39	D40	D41	D42	S8
9	1	D43	D43	D45	D46	D47	D48	S9

Frame Structure

0	0	0	0	0	0	0	0
1	D1	D1	D2	D2	D3	D3	S1
1	D4	D4	D5	D5	D6	D6	X
1	D7	D7	D8	D8	D9	D9	S3
1	D10	D10	D11	D11	D12	D12	S4
1	1	1	0	E4	E5	E6	E7
1	D13	D13	D14	D14	D15	D15	S6
1	D16	D16	D17	D17	D18	D18	X
1	D19	D19	D20	D20	D21	D21	S8
1	D22	D22	D23	D23	D24	D24	S9

Adaptation of 2400 bit/s User Rate to 8 kbit/s Intermediate Rate

0	0	0	0	0	0	0	0
1	D1	D1	D1	D1	D1	D1	S1
1	D1	D1	D2	D2	D2	D2	X
1	D2	D2	D2	D2	D3	D3	S3
1	D3	D3	D3	D3	D3	D3	S4
1	1	0	0	E4	E5	E6	M
1	D4	D4	D4	D4	D4	D4	S6
1	D4	D4	D5	D5	D5	D5	X
1	D5	D5	D5	D5	D6	D6	S8
1	D6	D6	D6	D6	D6	D6	S9

Adaptation of 600 bit/s User Rate to 8 kbit/s Intermediate Rate

0	0	0	0	0	0	0	0
1	D1	D2	D3	D4	D5	D6	S1
1	D7	D8	D9	D10	D11	D12	X
1	D13	D14	D15	D16	D17	D18	S3
1	D19	D20	D21	D22	D23	D24	S4
1	0	1	1	E4	E5	E6	E7
1	D25	D26	D27	D28	D29	D30	S6
1	D31	D32	D33	D34	D35	D36	X
1	D37	D38	D39	D40	D41	D42	S8
1	D43	D44	D45	D46	D47	D48	S9

Adaptation of 4800 bit/s to 8 kbit/s, 9600 bit/s to 16 kbit/s, 19200 bit/s to 32 kbit/s

0	0	0	0	0	0	0	0
1	D1	D1	D1	D1	D2	D2	S1
1	D2	D2	D3	D3	D3	D3	X
1	D4	D4	D4	D4	D5	D5	S3
1	D5	D5	D6	D6	D6	D6	S4
1	0	1	0	E4	E5	E6	E7
1	D7	D7	D7	D7	D8	D8	S6
1	D8	D8	D9	D9	D9	D9	X
1	D10	D10	D10	D10	D11	D11	S8
1	D11	D11	D12	D12	D12	D12	S9

Adaptation of 1200 bit/s User Rate to 8 kbit/s Intermediate Rate

Mitel
ST-BUS™ FAMILY MH89750
T1 Framer and Interface

- MITEL ST-BUS compatible
- Interface between a 2048 kbit/s, (ST-BUS), serial data stream and a bidirectional DS1 link
- Robbed-bit signalling or clear-channel capabilities
- Insertion and detection of S bit
- Selectable B8ZS/Jammed-bit zero-code replacement
- AMI encoding and decoding
- Per channel control
- Programmable digital attenuation
- 1.544-MHz clock extraction from received line
- Three-line equalization circuits for direct drive of 0-150 ft, 150-450 ft, 450-750 ft, plus a 6dB pad
- Two-frame elastic buffering
- One uncommitted scan point and drive point
- Compatible with AT&T T.A. #34

Pin Connections

Name	Pin	Pin	Name
		40	TL
VCC	2	39	TI
LA	3	38	RL
LB	4	37	RI
VDD	5	36	TXT
RxR	6	35	EIT
RxT	7	34	EA
RxA (overbar)	8	33	EB
RxB (overbar)	9	32	EC
DSTi	10	31	RCHT
CSTi	11	30	RCLT
E1.5o	12	29	TXR
XCtl	13	28	EIR
E8Ko	14	27	SW
CSTo	15	26	RCHR
C2i	16	25	RCLR
XS1	17	24	RxD
F0i (overbar)	18	23	DSTo
C1.5i	19	22	OUTA
OUTB	20	21	VSS

Ordering Information 0°C to 70°C

MH89750 40 Pin DIL Hybrid Module

APPLICATIONS

- High-speed data link using DS1 transmission link
- PBX or computer to DS1 interface

ABSOLUTE MAXIMUM RATINGS*

	Parameter	Symbol	Min	Max	Units
1	Power Supplies With respect to V_{SS}	V_{CC}	-.3	15	V
		V_{DD}	-.3	7	V
2	Voltage on any pin other than supplies and OUTA or OUTB		V_{SS}-0.3	V_{DD}+0.3	V
3	Voltage on OUTA or OUTB			15	V
4	Current at any pin other than OUTA OUTB and supplies			20	mA
5	Current at OUTA and OUTB			200	mA
6	Storage Temperature	T_{ST}	-20	85	°C

*Exceeding these values may cause permanent damage. Functional operation under these conditions is not implied.

DC ELECTRICAL CHARACTERISTICS - Clocked operation over recommended temperature ranges

		Parameters	Sym	Min	Typ‡	Max	Units	Test Conditions
1	Inputs	Supply Current	I_{CC}		10	20	mA	Outputs Unloaded
			I_{DD}		15	30	mA	Outputs Unloaded
2		Input High Voltage	V_{IH}	2.0			V	Digital Inputs
3		Input Low Voltage	V_{IL}			0.8	V	Digital Inputs
4		Input Leakage Current	I_{IL}		±1	±10	μA	Digital Inputs V_{IN}=0to V_{DD}
5	Outputs	Output High Voltage Digital	V_{OH}	2.4			V	I_{OL}=10mA
6		Output High Leakage	I_{OL}			500	nA	
7		Output High Current Digital Except E1544	I_{OH}	10	20		mA	Source Current V_{OH}=2.4V
			I_{OH}	8	16		mA	Source Current V_{OH}=3.0V
8		Output Low Voltage Digital	V_{OL}			0.4	V	I_{OL}=2mA
		OUTA or OUTB	V_{OL}			.25	V	I_{OL}=10mA
9		Output Low Current Digital Except E1544	I_{OL}	2	10		mA	Sink Current V_{OL}=0.4V
			I_{OL}	6	30		mA	Sink Current V_{OL}=2.0V
10		Input Impedance RxT to RxR	Z_{IN}		400		Ω	
		RxT or RxR to Gnd			1K		Ω	

‡ Typical figures are at 25°C and are for design aid only: not guaranteed and not subject to production testing

RECOMMENDED OPERATING CONDITIONS - Voltages are with respect to ground (V_{SS}) unless otherwise stated.

		Parameters	Sym	Min	Typ‡	Max	Units	Test Conditions
1	Inputs	Operating Temperature	T_{OP}	0		70	°C	
2		Power Supplies	V_{CC}	11.4	12	12.6	V	
			V_{DD}	4.5	5.0	5.5	V	
3		Input High Voltage	V_{IH}	2.4		V_{DD}	V	Digital Inputs
			V_{IH}		3.0		V	Line Inputs
4		Input Low Voltage	V_{IL}	V_{SS}		0.4	V	Digital Inputs
			V_{IL}		0.3		V	Line Inputs

‡ Typical figures are at 25°C and are for design aid only: not guaranteed and not subject to production testing.

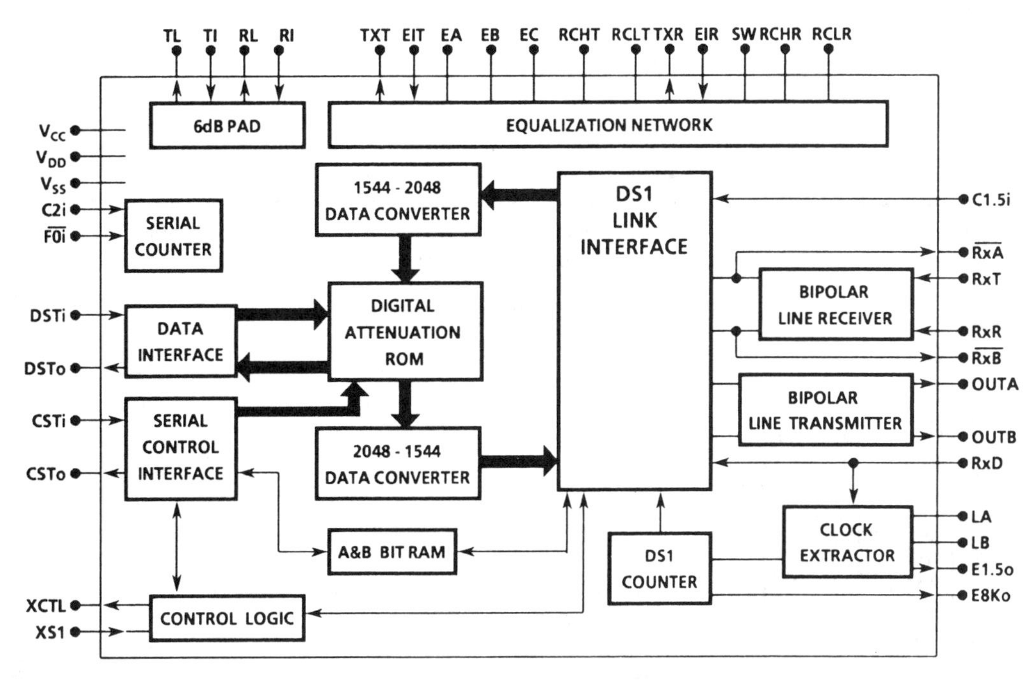

Block Diagram

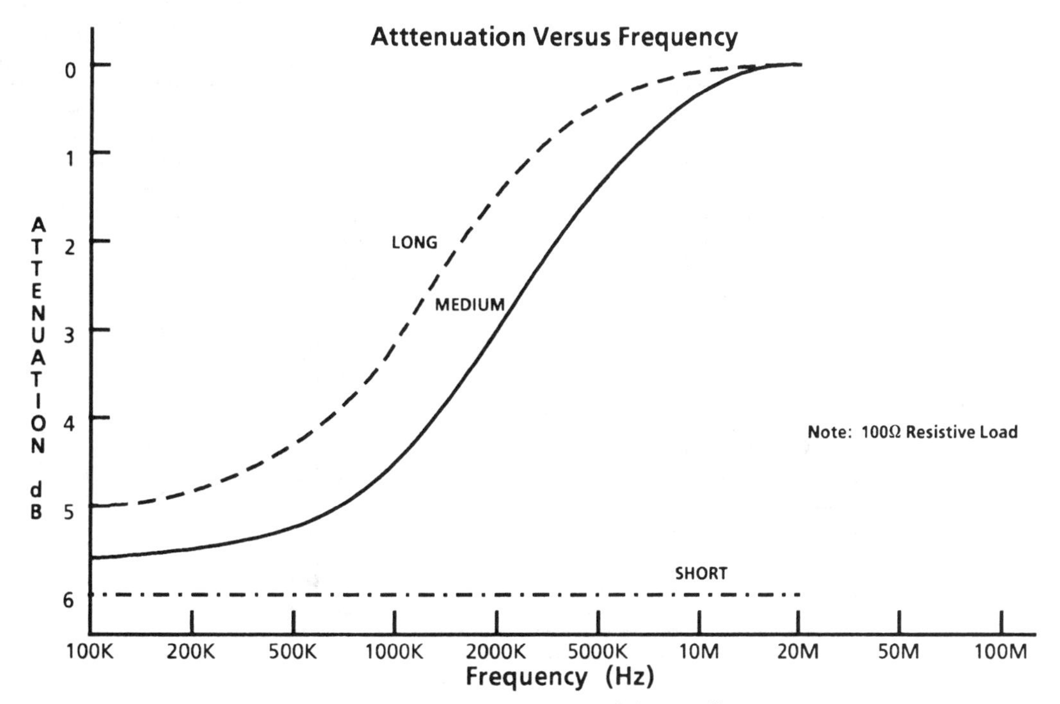

Typical Frequency Response of the Equalizers

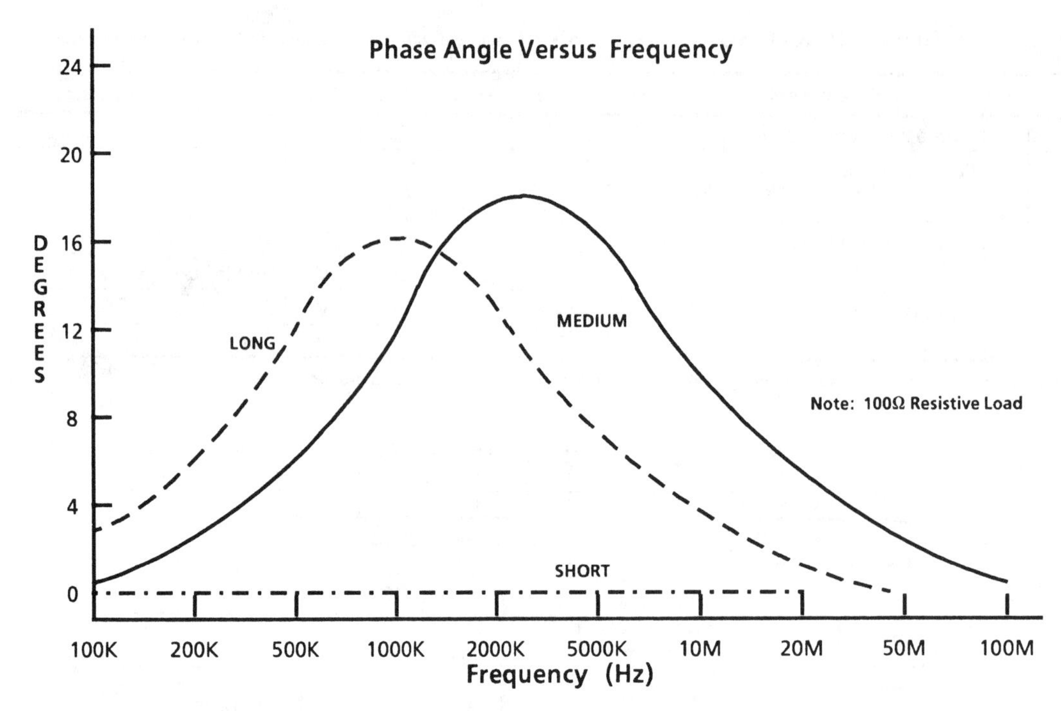

Typical Phase Response of the Equalizers

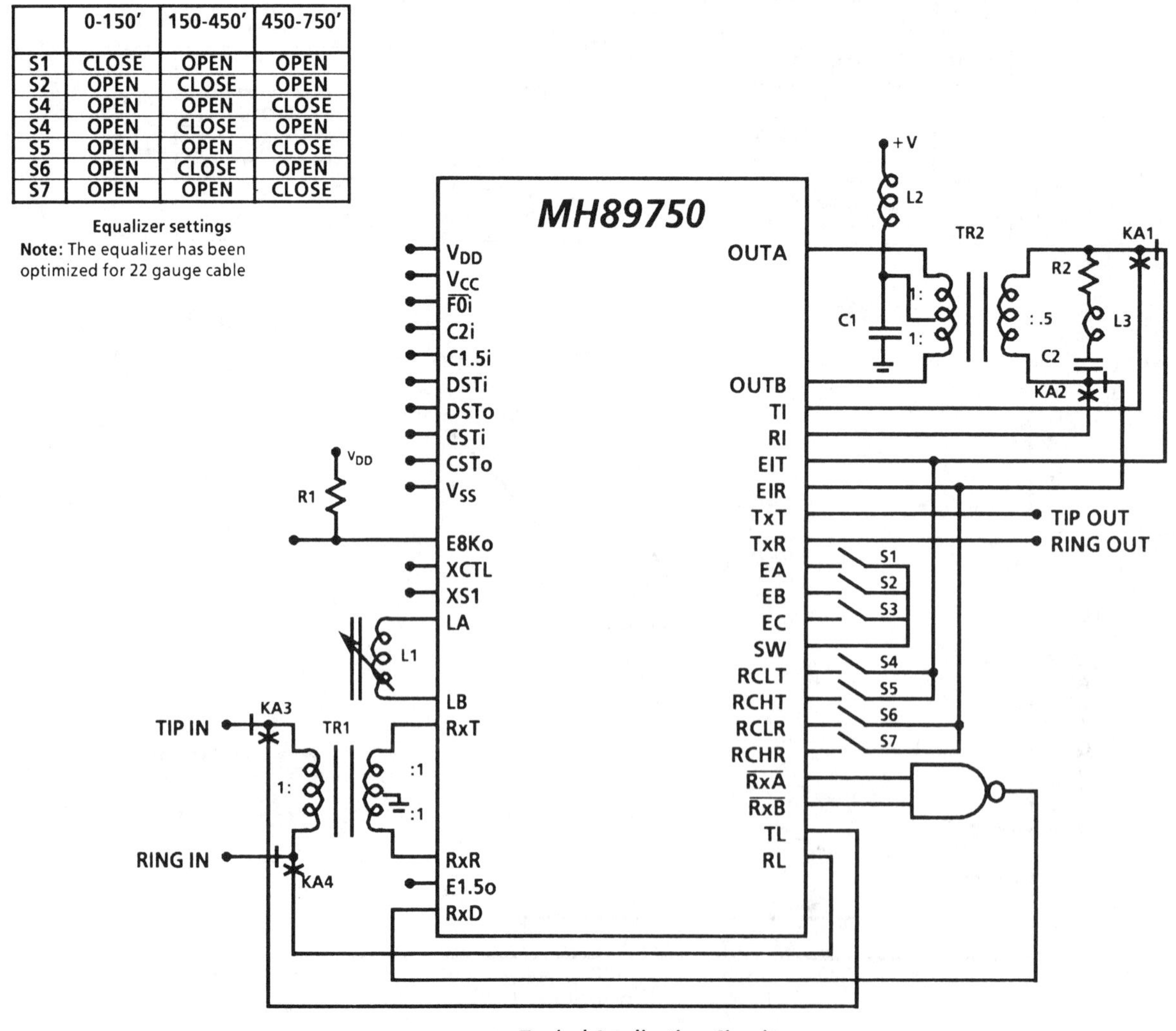

	0-150'	150-450'	450-750'
S1	CLOSE	OPEN	OPEN
S2	OPEN	CLOSE	OPEN
S4	OPEN	OPEN	CLOSE
S4	OPEN	CLOSE	OPEN
S5	OPEN	OPEN	CLOSE
S6	OPEN	CLOSE	OPEN
S7	OPEN	OPEN	CLOSE

Equalizer settings
Note: The equalizer has been optimized for 22 gauge cable

Typical Application Circuit

DSTi	X 0	1	2	3	X 4	5	6	7	X 8	9	10	11	X 12	13	14	15	X 16	17	18	19	X 20	21	22	23	X 24	25	26	27	X 28	29	30	31
DS1		1	2	3		4	5	6		7	8	9		10	11	12		13	14	15		16	17	18		19	20	21		22	23	24

Relationship Between Input DSTi Channels and Transmitted DS1 Channels
- X Denotes Unused DSTi Channels
Note: Functional representation only

DS1		1	2	3		4	5	6		7	8	9		10	11	12		13	14	15		16	17	18		19	20	21		22	23	24
DSTo	0 X	1	2	3	4 X	5	6	7	8 X	9	10	11	12 X	13	14	15	16 X	17	18	19	20 X	21	22	23	24 X	25	26	27	28 X	29	30	31

Relationship Between Received DS1 Channels and Output DSTo Channels
- X Denotes Unused DSTo Channels
Note: Functional representation only

CSTi	0	1	2	X 3	4	5	6	X 7	8	9	10	X 11	12	13	14	C 15	16	17	18	X 19	20	21	22	X 23	24	25	26	X 27	28	29	30	X 31
DSTi	1	2	3	4	5	6	7	8	9	10	11	12	13	14	15	16	17	18	19	20	21	22	23	24	25	26	27	28	29	30	31	0

Relationship Between Input CSTi Channels and Controlled DSTi Channels
- X Denotes Unused CSTi & DSTi Channels. Unused CSTi input channels must be set to $00
- C Denotes Master Control Channel
Note: Functional representation only

DS1	1	2	3		4	5	6		7	8	9		10	11	12		13	14	15		16	17	18		19	20	21		22	23	24	
CSTo	0	1	2	3 X	4	5	6	7 X	8	9	10	11 X	12	13	14	15 S	16	17	18	19 X	20	21	22	23 X	24	25	26	27 X	28	29	30	31 X

Relationship Between Received DS1 Channels and Output CSTo Channels
- X Denotes Unused CSTo Channels
- S Denotes Master Status Channel
Note: Functional representation only

Mitel MT8976 T1/ESF Framer

- D3/D4 or ESF framing and SLC-96 compatible
- 2-frame elastic buffer with 32 μsec jitter buffer
- Insertion and detection of A, B, C, D bits. Signalling freeze, optional debounce
- Selectable B8ZS, jammed-bit (ZCS) or no-zero code suppression
- Yellow-alarm and blue-alarm signal capabilities
- Bipolar violation count, F_T error count, CRC error count
- Selectable robbed-bit signalling
- Frame and superframe sync signals, Tx and Rx
- AMI encoding and decoding
- Per channel, overall, and remote loop around
- Digital phase detector between T1 line and ST-BUS
- One uncommitted scan point and drive point
- Pin compatible with MT8979
- ST-BUS compatible

APPLICATIONS

- DS1/ESF digital trunk interfaces
- Computer to PBX interfaces (DMI and CPI)
- High-speed computer to computer data links

ABSOLUTE MAXIMUM RATINGS*

	Parameter	Symbol	Min	Max	Units
1	Power Supplies with respect to V_{SS}	V_{DD}	-0.3	7	V
2	Voltage on any pin other than supplies		V_{SS}-0.3	V_{DD}+0.3	V
3	Current at any pin other than supplies			40	mA
4	Storage Temperature	T_{ST}	-55	125	°C
5	Package Power Dissipation	P		800	mW

*Exceeding these values may cause permanent damage. Functional operation under these conditions is not implied.

RECOMMENDED OPERATING CONDITIONS- Voltages are with respect to ground (V_{SS}) unless otherwise stated

		Characteristics	Sym	Min	Typ‡	Max	Units	Test Conditions
1	Inputs	Operating Temperature	T_{OP}	-40		85	°C	
2		Power Supplies	V_{DD}	4.5	5.0	5.5	V	
3		Input High Voltage	V_{IH}	2.4		V_{DD}	V	For 400 mV noise margin
4		Input Low Voltage	V_{IL}	V_{SS}		0.4	V	For 400 mV noise margin

‡ Typical figures are at 25°C and are for design aid only: not guaranteed and not subject to production testing.

DC ELECTRICAL CHARACTERISTICS

- Clocked operation over recommended temperature ranges and power supply voltages.

		Parameters	Sym	Min	Typ‡	Max	Units	Test Conditions
1		Supply Current	I_{DD}		6	10	mA	Outputs Unloaded
2	Inputs	Input High Voltage	V_{IH}	2.0			V	Digital Inputs
3		Input Low Voltage	V_{IL}			0.8	V	Digital Inputs
4		Input Leakage Current	I_{IL}		±1	±10	μA	Digital Inputs V_{IN}=0 to V_{DD}
5		Schmitt Trigger Input (XSt)	V_{T+}			4.0	V	
			V_{T-}	1.5			V	
6	Outputs	Output High Current	I_{OH}	7	20		mA	Source Current, V_{OH}=2.4V
7		Output Low Current	I_{OL}	2	10		mA	Sink Current, V_{OL}=0.4V

‡ Typical figures are at 25°C and are for design aid only: not guaranteed and not subject to production testing.

AC ELECTRICAL CHARACTERISTICS†-Capacitance

	Characteristics	Sym	Min	Typ‡	Max	Units	Test Conditions
1	Input Pin Capacitance	C_I		10		pF	
2	Output Pin Capacitance	C_O		10		pF	

† Timing is over recommended temperature & power supply voltages

‡ Typical figures are at 25°C and are for design aid only: not guaranteed and not subject to production testing.

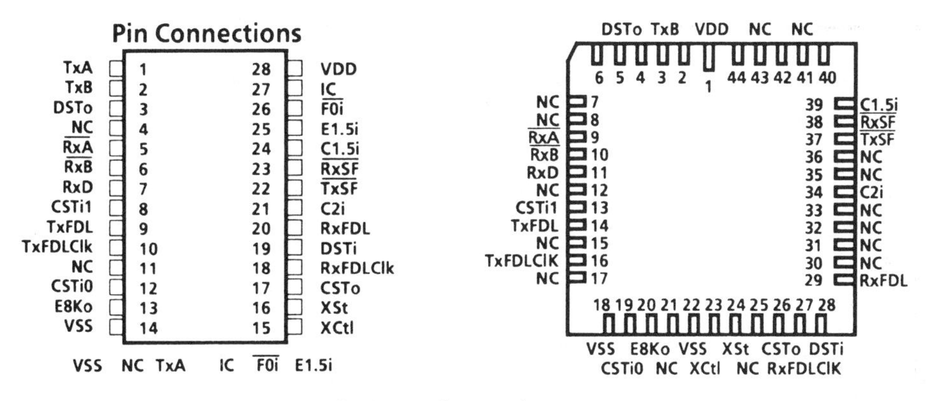

Ordering Information

MT8976AC/AE	28 Pin CERDIP/PDIP
MT8976AY/AP	44 Pin LCC/PLCC

-40 °C to 85°C

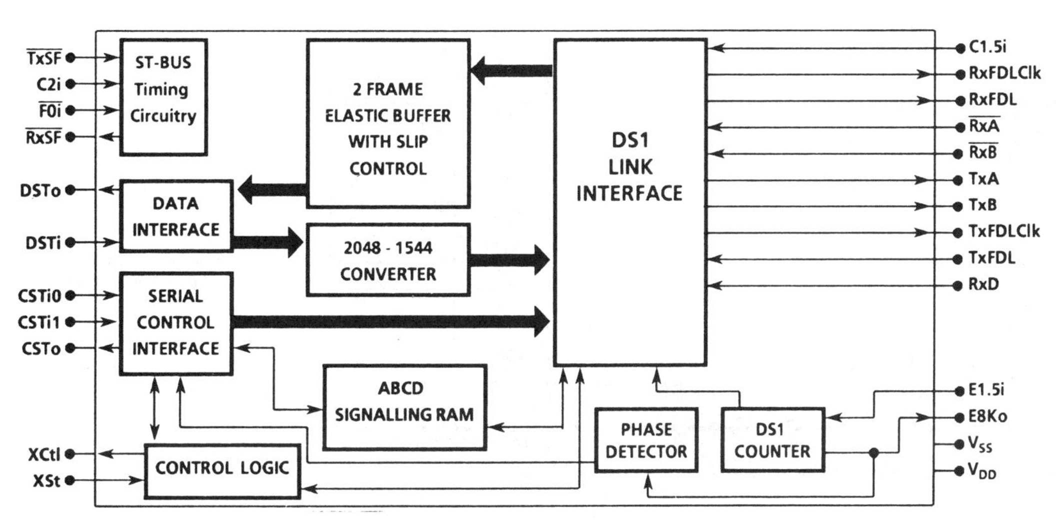

Functional Block Diagram

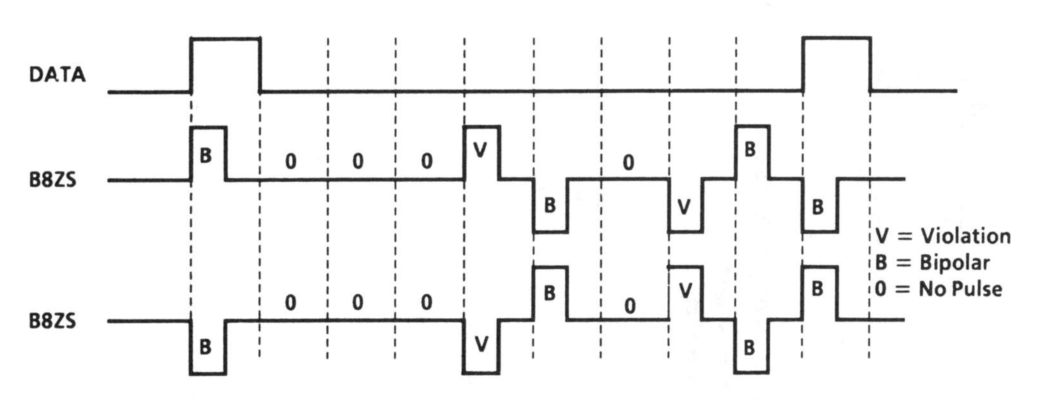

B8ZS Output Coding

Frame #	F_T	F_S†	Notes	Frame #	F_T	F_S†	Notes
1	1		Resynchronization Data Bits	37	1		X = Concentrator Field Bits
2		0		38		X	
3	0			39	0		
4		0		40		X	
5	1			41	1		
6		0		42		X	
7	0			43	0		
8		1		44		X	
9	1			45	1		
10		1		46		X	
11	0			47	0		S = Spoiler Bits
12		1		48		S	
13	1			49	1		
14		0		50		S	
15	0			51	0		
16		0		52		S	
17	1			53	1		
18		0		54		C	C = Maintenance Field Bits
19	0			55	0		
20		1		56		C	
21	1			57	1		
22		1		58		C	
23	0			59	0		A = Alarm Field Bits
24		1		60		A	
25	1		X = Concentrator Field Bits	61	1		
26		X		62		A	
27	0			63	0		L = Line Switch Field Bits
28		X		64		L	
29	1			65	1		
30		X		66		L	
31	0			67	0		
32		X		68		L	
33	1			69	1		
34		X		70		L	
35	0			71	0		S = Spoiler Bits
36		X		72		S	

SLC-96 Framing Pattern

†Note: The F_S pattern has to be supplied by the user

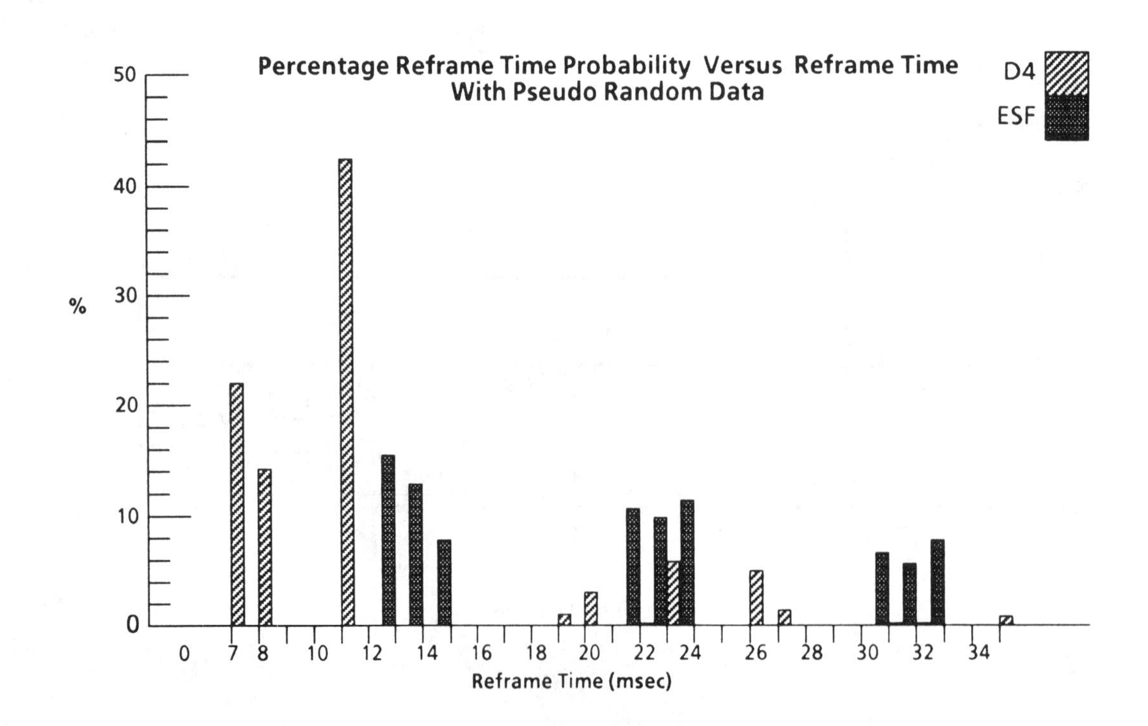

DSTi	0	1	2	3	4	5	6	7	8	9	10	11	12	13	14	15	16	17	18	19	20	21	22	23	24	25	26	27	28	29	30	31
	X				X				X				X				X				X				X				X			
DS1		1	2	3		4	5	6		7	8	9		10	11	12		13	14	15		16	17	18		19	20	21		22	23	24

ST-BUS CHANNEL VERSUS DS1 CHANNEL TRANSMITTED

DSTo	0	1	2	3	4	5	6	7	8	9	10	11	12	13	14	15	16	17	18	19	20	21	22	23	24	25	26	27	28	29	30	31
	X				X				X				X				X				X				X				X			
DS1		1	2	3		4	5	6		7	8	9		10	11	12		13	14	15		16	17	18		19	20	21		22	23	24

ST-BUS CHANNEL VERSUS DS1 CHANNEL RECEIVED

CSTi0	0	1	2	3	4	5	6	7	8	9	10	11	12	13	14	15	16	17	18	19	20	21	22	23	24	25	26	27	28	29	30	31
	PCCW 1	PCCW 1	PCCW 1	X	PCCW 1	PCCW 1	PCCW 1	X	PCCW 1	PCCW 1	PCCW 1	X	PCCW 1	PCCW 1	PCCW 1	MCW1	PCCW 1	PCCW 1	PCCW 1	X	PCCW 1	PCCW 1	PCCW 1	X	PCCW 1	PCCW 1	PCCW 1	X	PCCW 1	PCCW 1	PCCW 1	MCW2
DS1	1	2	3		4	5	6		7	8	9		10	11	12		13	14	15		16	17	18		19	20	21		22	23	24	

PCCW = PER CHANNEL CONTROL WORD
MCW1/2 = MASTER CONTROL WORD 1/2

ST-BUS CHANNEL VERSUS DS1 CHANNEL CONTROLLED

CSTi1	0	1	2	3	4	5	6	7	8	9	10	11	12	13	14	15	16	17	18	19	20	21	22	23	24	25	26	27	28	29	30	31
	PCCW 2	PCCW 2	PCCW 2	X	PCCW 2	PCCW 2	PCCW 2	X	PCCW 2	PCCW 2	PCCW 2	X	PCCW 2	PCCW 2	PCCW 2	X	PCCW 2	PCCW 2	PCCW 2	X	PCCW 2	PCCW 2	PCCW 2	X	PCCW 2	PCCW 2	PCCW 2	X	PCCW 2	PCCW 2	PCCW 2	X
DS1	1	2	3		4	5	6		7	8	9		10	11	12		13	14	15		16	17	18		19	20	21		22	23	24	

PCCW = PER CHANNEL CONTROL WORD

ST-BUS CHANNEL VERSUS DS1 CHANNEL CONTROLLED

CSTo	0	1	2	3	4	5	6	7	8	9	10	11	12	13	14	15	16	17	18	19	20	21	22	23	24	25	26	27	28	29	30	31
	PCSW	PCSW	PCSW	PSW	PCSW	PCSW	PCSW	X	PCSW	PCSW	PCSW	X	PCSW	PCSW	PCSW	MSW1	PCSW	PCSW	PCSW	X	PCSW	PCSW	PCSW	X	PCSW	PCSW	PCSW	X	PCSW	PCSW	PCSW	MSW2
DS1	1	2	3		4	5	6		7	8	9		10	11	12		13	14	15		16	17	18		19	20	21		22	23	24	

PCSW = PER CHANNEL STATUS WORD
PSW = PHASE STATUS WORD
MSW = MASTER STATUS WORD

ST-BUS VERSUS DS1 CHANNEL STATUS

ST-BUS Channel Allocations

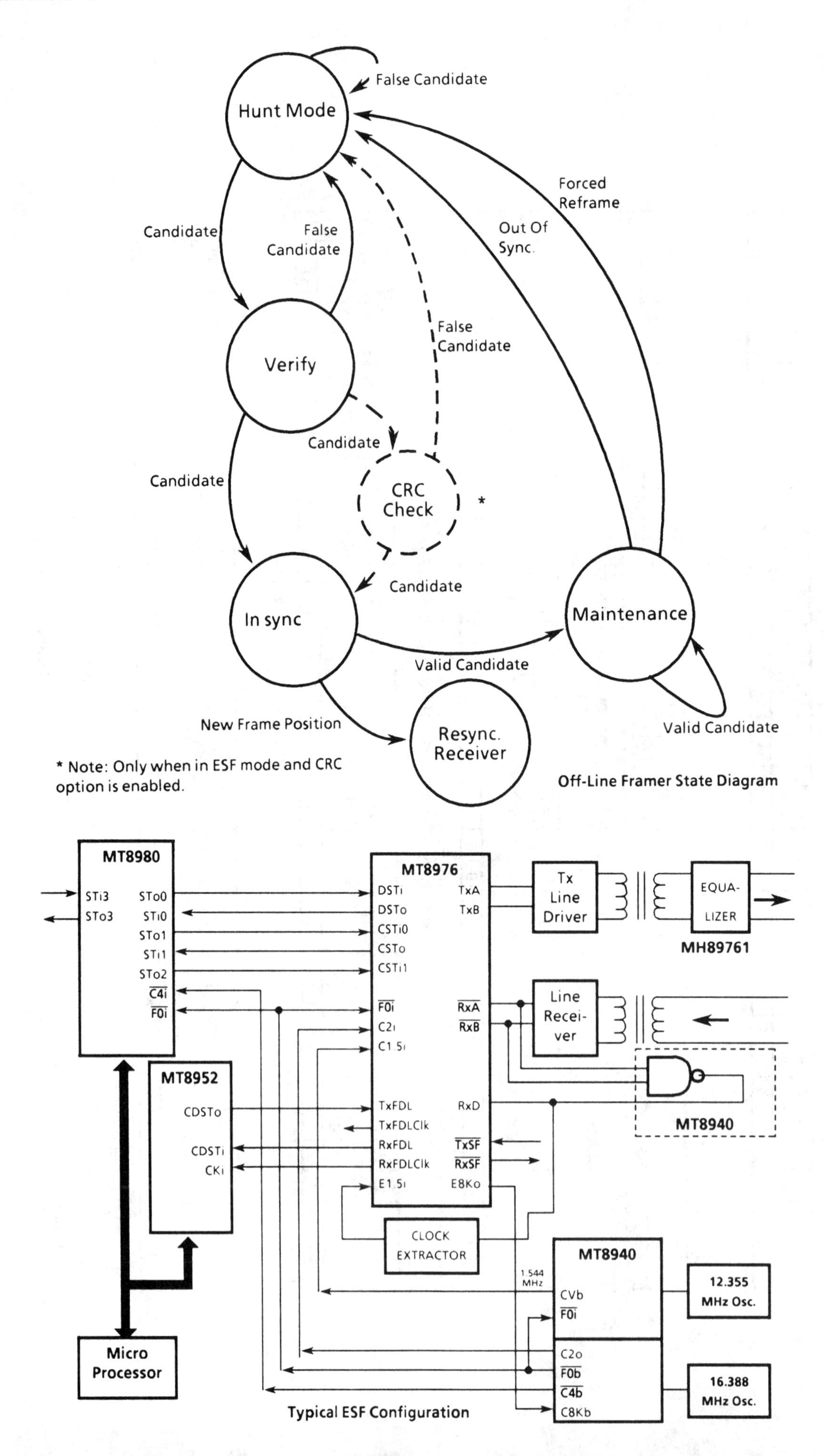

Off-Line Framer State Diagram

Typical ESF Configuration

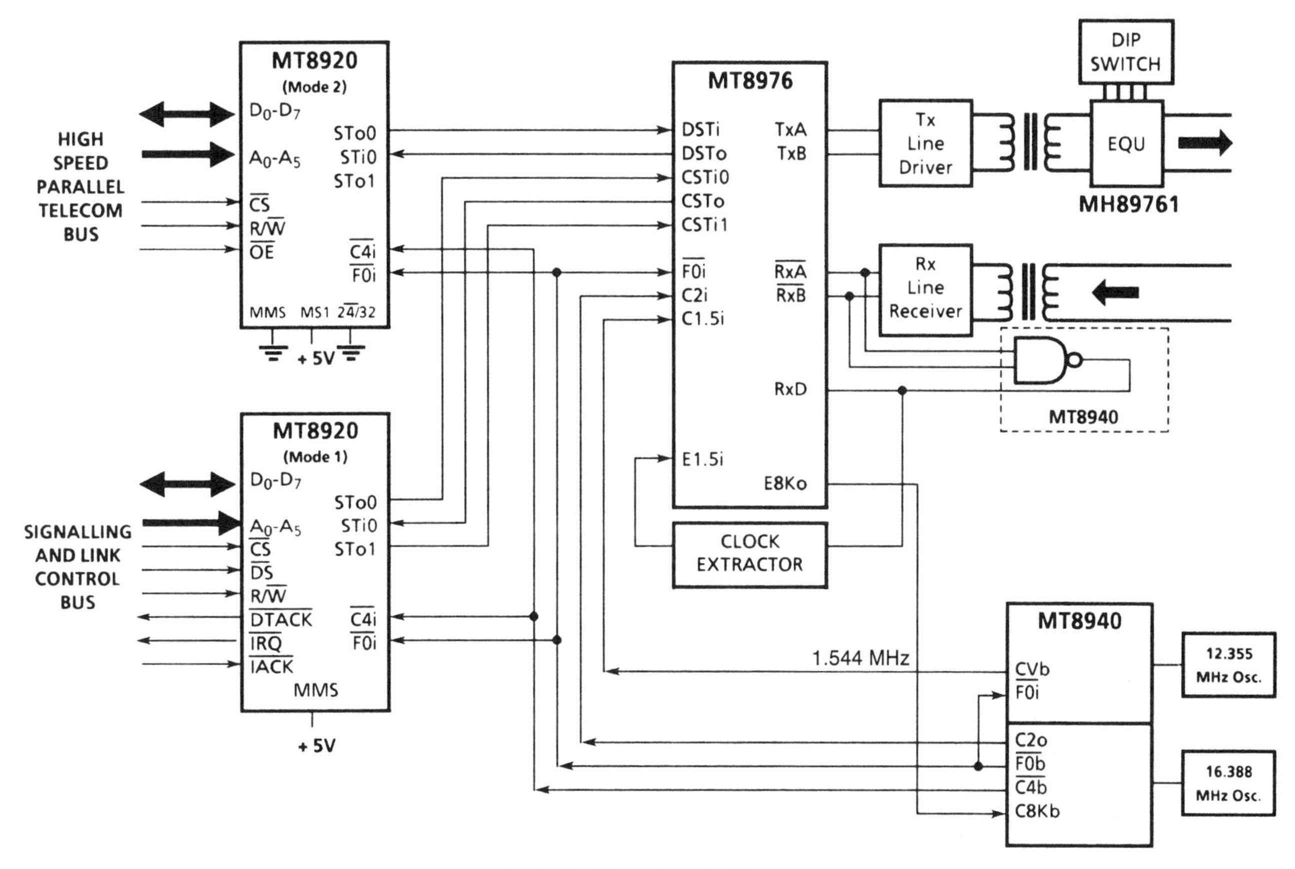

Using the MT8976 in a Parallel Bus Environment

LIFE SUPPORT POLICY

National's products are not authorized for use as critical components in life support devices or systems without the express written approval of the president of National Semiconductor Corporation. As used herein:

1. Life support devices or systems are devices or systems which, (a) are intended for surgical implant into the body, or (b) support or sustain life, and whose failure to perform, when properly used in accordance with instructions for use provided in the labeling, can be reasonably expected to result in a significant injury to the user.

2. A critical component is any component of a life support device or system whose failure to perform can be reasonably expected to cause the failure of the life support device or system, or to affect its safety or effectiveness.

National 54LS138/DM54LS138/DM74LS138, 54LS139/DM54LS139/DM74LS139 Decoders/Demultiplexers

- Designed specifically for high speed:
  - Memory decoders
  - Data transmission systems
- LS138 3-to-8-line decoders incorporates 3 enable inputs to simplify cascading and/or data reception
- LS139 contains two fully independent 2-to-4-line decoders/demultiplexers
- Schottky clamped for high performance
- Typical propagation delay (3 levels of logic)
  - LS138 21ns
  - LS193 21ns
- Typical power dissipation
  - LS138 32mW
  - LS139 34mW
- Alternate military/aerospace devices (54LS138, 54LS139) are available.

ABSOLUTE MAXIMUM RATINGS (Note)

If Military/Aerospace specified devices are required, please contact the National Semiconductor Sales Office/Distributors for availability and specifications.

Supply Voltage	7V
Input Voltage	7V
Operating Free Air Temperature Range	
DM54LS and 54LS	−55°C to +125°C
DM74LS	0°C to +70°C
Storage Temperature Range	−65°C to +150°C

Note: *The "Absolute Maximum Ratings" are those values beyond which the safety of the device cannot be guaranteed. The device should not be operated at these limits. The parametric values defined in the "Electrical Characteristics" table are not guaranteed at the absolute maximum ratings. The "Recommended Operating Conditions" table will define the conditions for actual device operation.*

RECOMMENDED OPERATING CONDITIONS

Symbol	Parameter	DM54LS138			DM74LS138			Units
		Min	Nom	Max	Min	Nom	Max	
V_{CC}	Supply Voltage	4.5	5	5.5	4.75	5	5.25	V
V_{IH}	High Level Input Voltage	2			2			V
V_{IL}	Low Level Input Voltage			0.7			0.8	V
I_{OH}	High Level Output Current			−0.4			−0.4	mA
I_{OL}	Low Level Output Current			4			8	mA
T_A	Free Air Operating Temperature	−55		125	0		70	°C

'LS138 ELECTRICAL CHARACTERISTICS

over recommended operating free air temperature range (unless otherwise noted)

Symbol	Parameter	Conditions		Min	Typ (Note 1)	Max	Units
V_I	Input Clamp Voltage	V_{CC} = Min, I_I = −18 mA				−1.5	V
V_{OH}	High Level Output Voltage	V_{CC} = Min, I_{OH} = Max, V_{IL} = Max, V_{IH} = Min	DM54	2.5	3.4		V
			DM74	2.7	3.4		
V_{OL}	Low Level Output Voltage	V_{CC} = Min, I_{OL} = Max, V_{IL} = Max, V_{IH} = Min	DM54		0.25	0.4	V
			DM74		0.35	0.5	
		I_{OL} = 4 mA, V_{CC} = Min	DM74		0.25	0.4	
I_I	Input Current @ Max Input Voltage	V_{CC} = Max, V_I = 7V				0.1	mA
I_{IH}	High Level Input Current	V_{CC} = Max, V_I = 2.7V				20	μA
I_{IL}	Low Level Input Current	V_{CC} = Max, V_I = 0.4V				−0.36	mA
I_{OS}	Short Circuit Output Current	V_{CC} = Max (Note 2)	DM54	−20		−100	mA
			DM74	−20		−100	
I_{CC}	Supply Current	V_{CC} = Max (Note 3)			6.3	10	mA

Note 1: All typicals are at V_{CC} = 5V, T_A = 25°C.

Note 2: Not more than one output should be shorted at a time, and the duration should not exceed one second.

Note 3: I_{CC} is measured with all outputs enabled and open.

'LS138 SWITCHING CHARACTERISTICS

at V_{CC} = 5V and T_A = 25°C (See Section 1 for Test Waveforms and Output Load)

Symbol	Parameter	From (Input) To (Output)	Levels of Delay	R_L = 2 kΩ				Units
				C_L = 15 pF		C_L = 50 pF		
				Min	Max	Min	Max	
t_{PLH}	Propagation Delay Time Low to High Level Output	Select to Output	2		18		27	ns
t_{PHL}	Propagation Delay Time High to Low Level Output	Select to Output	2		27		40	ns
t_{PLH}	Propagation Delay Time Low to High Level Output	Select to Output	3		18		27	ns
t_{PHL}	Propagation Delay Time High to Low Level Output	Select to Output	3		27		40	ns
t_{PLH}	Propagation Delay Time Low to High Level Output	Enable to Output	2		18		27	ns
t_{PHL}	Propagation Delay Time High to Low Level Output	Enable to Output	2		24		40	ns
t_{PLH}	Propagation Delay Time Low to High Level Output	Enable to Output	3		18		27	ns
t_{PHL}	Propagation Delay Time High to Low Level Output	Enable to Output	3		28		40	ns

RECOMMENDED OPERATING CONDITIONS

Symbol	Parameter	DM54LS139			DM74LS139			Units
		Min	Nom	Max	Min	Nom	Max	
V_{CC}	Supply Voltage	4.5	5	5.5	4.75	5	5.25	V
V_{IH}	High Level Input Voltage	2			2			V
V_{IL}	Low Level Input Voltage			0.7			0.8	V
I_{OH}	High Level Output Current			−0.4			−0.4	mA
I_{OL}	Low Level Output Current			4			8	mA
T_A	Free Air Operating Temperature	−55		125	0		70	°C

CONNECTION DIAGRAMS

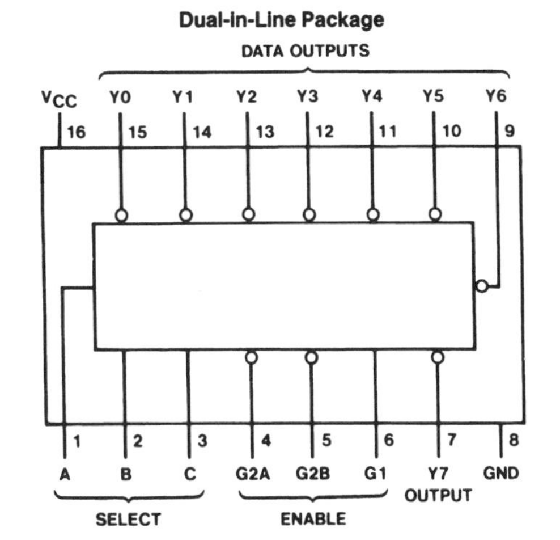

Order Number 54LS138DMQB, 54LS138FMQB, 54LS138LMQB, DM54LS138J, DM54LS138W, DM74LS138M or DM74LS138N See NS Package Number E20A, J16A, M16A, N16E or W16A

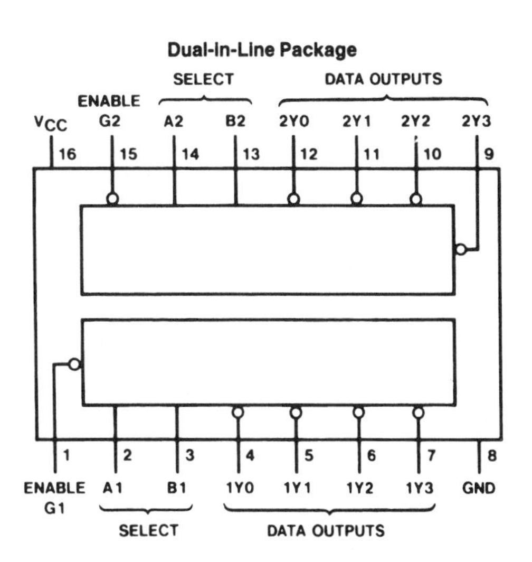

Order Number 54LS139DMQB, 54LS139FMQB, 54LS139LMQB, DM54LS139J, DM54LS139W, DM74LS139M or DM74LS139N See NS Package Number E20A, J16A, M16A, N16E or W16A

'LS139 ELECTRICAL CHARACTERISTICS

over recommended operating free air temperature range (unless otherwise noted)

Symbol	Parameter	Conditions		Min	Typ (Note 1)	Max	Units
V_I	Input Clamp Voltage	V_{CC} = Min, I_I = −18 mA				−1.5	V
V_{OH}	High Level Output Voltage	V_{CC} = Min, I_{OH} = Max, V_{IL} = Max, V_{IH} = Min	DM54	2.5	3.4		V
			DM74	2.7	3.4		
V_{OL}	Low Level Output Voltage	V_{CC} = Min, I_{OL} = Max V_{IL} = Max, V_{IH} = Min	DM54		0.25	0.4	V
			DM74		0.35	0.5	
		I_{OL} = 4 mA, V_{CC} = Min	DM74		0.25	0.4	
I_I	Input Current @ Max Input Voltage	V_{CC} = Max, V_I = 7V				0.1	mA
I_{IH}	High Level Input Current	V_{CC} = Max, V_I = 2.7V				20	μA
I_{IL}	Low Level Input Current	V_{CC} = Max, V_I = 0.4V				−0.36	mA
I_{OS}	Short Circuit Output Current	V_{CC} = Max (Note 2)	DM54	−20		−100	mA
			DM74	−20		−100	
I_{CC}	Supply Current	V_{CC} = Max (Note 3)			6.8	11	mA

Note 1: All typicals are at V_{CC} = 5V, T_A = 25°C.

Note 2: Not more than one output should be shorted at a time, and the duration should not exceed one second.

Note 3: I_{CC} is measured with all outputs enabled and open.

'LS139 SWITCHING CHARACTERISTICS

at V_{CC} = 5V and T_A = 25°C (See Section 1 for Test Waveforms and Output Load)

Symbol	Parameter	From (Input) To (Output)	R_L = 2 kΩ				Units
			C_L = 15 pF		C_L = 50 pF		
			Min	Max	Min	Max	
t_{PLH}	Propagation Delay Time Low to High Level Output	Select to Output		18		27	ns
t_{PHL}	Propagation Delay Time High to Low Level Output	Select to Output		27		40	ns
t_{PLH}	Propagation Delay Time Low to High Level Output	Enable to Output		18		27	ns
t_{PHL}	Propagation Delay Time High to Low Level Output	Enable to Output		24		40	ns

FUNCTION TABLES

LS138

Inputs					Outputs							
Enable		Select										
G1	G2*	C	B	A	Y0	Y1	Y2	Y3	Y4	Y5	Y6	Y7
X	H	X	X	X	H	H	H	H	H	H	H	H
L	X	X	X	X	H	H	H	H	H	H	H	H
H	L	L	L	L	L	H	H	H	H	H	H	H
H	L	L	L	H	H	L	H	H	H	H	H	H
H	L	L	H	L	H	H	L	H	H	H	H	H
H	L	L	H	H	H	H	H	L	H	H	H	H
H	L	H	L	L	H	H	H	H	L	H	H	H
H	L	H	L	H	H	H	H	H	H	L	H	H
H	L	H	H	L	H	H	H	H	H	H	L	H
H	L	H	H	H	H	H	H	H	H	H	H	L

* G2 = G2A + G2B

H = High Level, L = Low Level, X = Don't Care

LS139

Inputs			Outputs			
Enable	Select					
G	B	A	Y0	Y1	Y2	Y3
H	X	X	H	H	H	H
L	L	L	L	H	H	H
L	L	H	H	L	H	H
L	H	L	H	H	L	H
L	H	H	H	H	H	L

H = High Level, L = Low Level, X = Don't Care

LOGIC DIAGRAMS

LS138

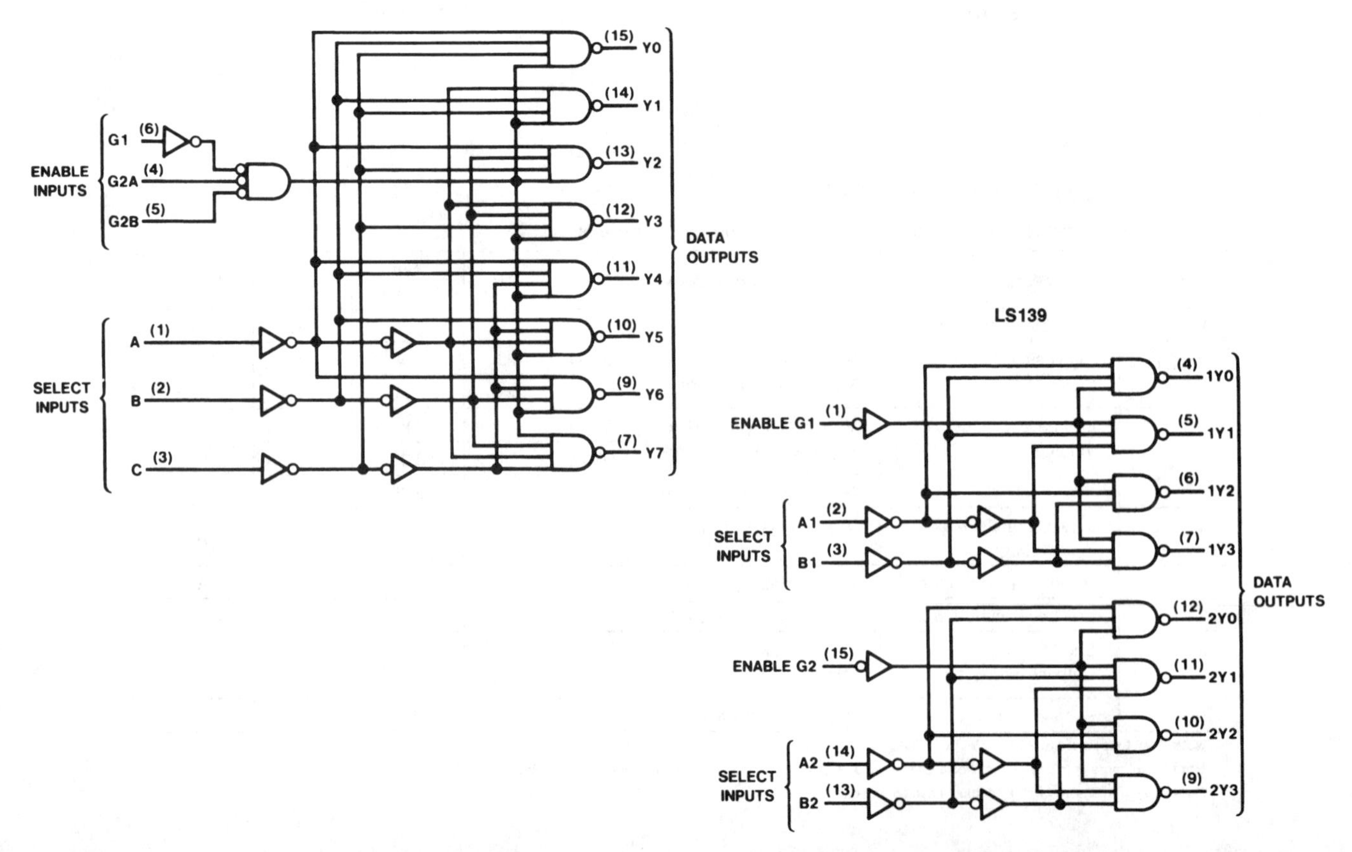

National 54LS151/DM54LS151/DM74LS151 Data Selector/Multiplexer

- Select one-of-eight data lines
- Performs parallel-to-serial conversion
- Permits multiplexing from N lines to one line
- Also for use as Boolean function generator
- Typical average propagation delay time data input to W output 12.5ns
- Typical power dissipation 30mW
- Alternate military/aerospace device (54LS151) is available.

ABSOLUTE MAXIMUM RATINGS (Note)

If Military/Aerospace specified devices are required, please contact the National Semiconductor Sales Office/Distributors for availability and specifications.

Supply Voltage	7V
Input Voltage	7V
Operating Free Air Temperature Range	
DM54LS and 54LS	−55°C to +125°C
DM74LS	0°C to +70°C
Storage Temperature Range	−65°C to +150°C

Note: *The "Absolute Maximum Ratings" are those values beyond which the safety of the device cannot be guaranteed. The device should not be operated at these limits. The parametric values defined in the "Electrical Characteristics" table are not guaranteed at the absolute maximum ratings. The "Recommended Operating Conditions" table will define the conditions for actual device operation.*

RECOMMENDED OPERATING CONDITIONS

Symbol	Parameter	DM54LS151			DM74LS151			Units
		Min	Nom	Max	Min	Nom	Max	
V_{CC}	Supply Voltage	4.5	5	5.5	4.75	5	5.25	V
V_{IH}	High Level Input Voltage	2			2			V
V_{IL}	Low Level Input Voltage			0.7			0.8	V
I_{OH}	High Level Output Current			−0.4			−0.4	mA
I_{OL}	Low Level Output Current			4			8	mA
T_A	Free Air Operating Temperature	−55		125	0		70	°C

ELECTRICAL CHARACTERISTICS

over recommended operating free air temperature range (unless otherwise noted)

Symbol	Parameter	Conditions		Min	Typ (Note 1)	Max	Units
V_I	Input Clamp Voltage	V_{CC} = Min, I_I = −18 mA				−1.5	V
V_{OH}	High Level Output Voltage	V_{CC} = Min, I_{OH} = Max V_{IL} = Max, V_{IH} = Min	DM54	2.5	3.4		V
			DM74	2.7	3.4		
V_{OL}	Low Level Output Voltage	V_{CC} = Min, I_{OL} = Max V_{IL} = Max, V_{IH} = Min	DM54		0.25	0.4	V
			DM74		0.35	0.5	
		I_{OL} = 4 mA, V_{CC} = Min	DM74		0.25	0.4	
I_I	Input Current @ Max Input Voltage	V_{CC} = Max, V_I = 7V				0.1	mA
I_{IH}	High Level Input Current	V_{CC} = Max, V_I = 2.7V				20	μA
I_{IL}	Low Level Input Current	V_{CC} = Max, V_I = 0.4V				−0.4	mA
I_{OS}	Short Circuit Output Current	V_{CC} = Max (Note 2)	DM54	−20		−100	mA
			DM74	−20		−100	
I_{CC}	Supply Current	V_{CC} = Max (Note 3)			6	10	mA

Note 1: All typicals are at V_{CC} = 5V, T_A = 25°C.

Note 2: Not more than one output should be shorted at a time, and the duration should not exceed one second.

Note 3: I_{CC} is measured with all outputs open, strobe and data select inputs at 4.5V, and all other inputs open.

SWITCHING CHARACTERISTICS at V_{CC} = 5V and T_A = 25°C (See Section 1 for Test Waveforms and Output Load)

Symbol	Parameter	From (Input) To (output)	R_L = 2 kΩ				Units
			C_L = 15 pF		C_L = 50 pF		
			Min	Max	Min	Max	
t_{PLH}	Propagation Delay Time Low to High Level Output	Select (4 Levels) to Y		43		46	ns
t_{PHL}	Propagation Delay Time High to Low Level Output	Select (4 Levels) to Y		30		36	ns
t_{PLH}	Propagation Delay Time Low to High Level Output	Select (3 Levels) to W		23		25	ns
t_{PHL}	Propagation Delay Time High to Low Level Output	Select (3 Levels) to W		32		40	ns
t_{PLH}	Propagation Delay Time Low to High Level Output	Strobe to Y		42		44	ns
t_{PHL}	Propagation Delay Time High to Low Level Output	Strobe to Y		32		40	ns
t_{PLH}	Propagation Delay Time Low to High Level Output	Strobe to W		24		27	ns
t_{PHL}	Propagation Delay Time High to Low Level Output	Strobe to W		30		36	ns
t_{PLH}	Propagation Delay Time Low to High Level Output	D0 thru D7 to Y		32		35	ns
t_{PHL}	Propagation Delay Time High to Low Level Output	D0 thru D7 to Y		26		33	ns
t_{PLH}	Propagation Delay Time Low to High Level Output	D0 thru D7 to W		21		25	ns
t_{PHL}	Propagation Delay Time High to Low Level Output	D0 thru D7 to W		20		27	ns

LOGIC DIAGRAM

LS151

STROBE (ENABLE) (7)
DATA INPUTS: D0 (4), D1 (3), D2 (2), D3 (1), D4 (15), D5 (14), D6 (13), D7 (12)
(5) OUTPUT Y
(6) OUTPUT W
$\bar{A}$ A $\bar{B}$ B $\bar{C}$ C

CONNECTION DIAGRAM

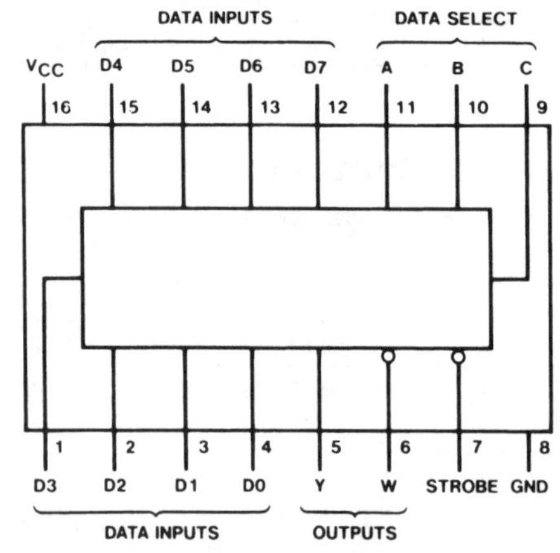

Order Number 54LS151DMQB, 54LS151FMQB, 54LS151LMQB, DM54LS151J, DM54LS151W, DM74LS151M or DM74LS151N
See NS Package Number E20A, J16A, M16A, N16E or W16A

Address Buffers for 54LS151/74LS151

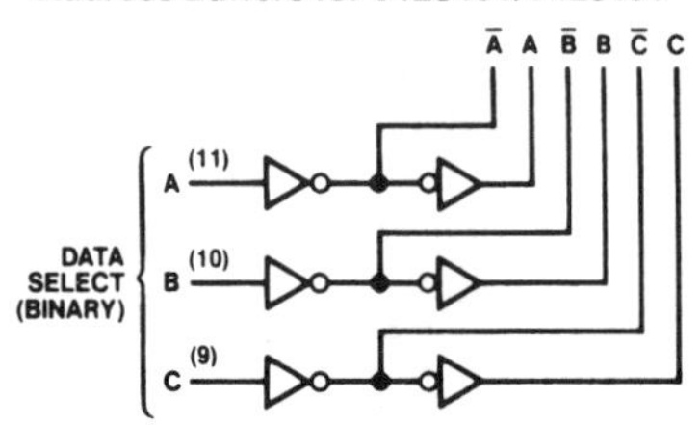

TRUTH TABLE

Inputs				Outputs	
Select			Strobe S	Y	W
C	B	A			
X	X	X	H	L	H
L	L	L	L	D0	$\overline{D0}$
L	L	H	L	D1	$\overline{D1}$
L	H	L	L	D2	$\overline{D2}$
L	H	H	L	D3	$\overline{D3}$
H	L	L	L	D4	$\overline{D4}$
H	L	H	L	D5	$\overline{D5}$
H	H	L	L	D6	$\overline{D6}$
H	H	H	L	D7	$\overline{D7}$

H = High Level, L = Low Level, X = Don't Care
D0, D1...D7 = the level of the respective D input

National 54LS152
8-Input Multiplexer

ABSOLUTE MAXIMUM RATINGS (Note)

If Military/Aerospace specified devices are required, please contact the National Semiconductor Sales Office/Distributors for availability and specifications.

Supply Voltage	7V
Input Voltage	5.5V
Operating Free Air Temperature Range 54LS	−55°C to +25°C
Storage Temperature Range	−65°C to +150°C

Note: *The "Absolute Maximum Ratings" are those values beyond which the safety of the device cannot be guaranteed. The device should not be operated at these limits. The parametric values defined in the "Electrical Characteristics" table are not guaranteed at the absolute maximum ratings. The "Recommended Operating Conditions" table will define the conditions for actual device operation.*

RECOMMENDED OPERATING CONDITIONS

Symbol	Parameter	54LS152			Units
		Min	Nom	Max	
V_{CC}	Supply Voltage	4.5	5	5.5	V
V_{IH}	High Level Input Voltage	2			V
V_{IL}	Low Level Input Voltage			0.7	V
I_{OH}	High Level Output Current			−0.4	mA
I_{OL}	Low Level Output Current			4.0	mA
T_A	Free Air Operating Temperature	−55		125	°C

ELECTRICAL CHARACTERISTICS over recommended operating free air temperature (unless otherwise noted)

Symbol	Parameter	Conditions	Min	Typ (Note 1)	Max	Units
V_I	Input Clamp Voltage	V_{CC} = Min, I_I = −18 mA			−1.5	V
V_{OH}	High Level Output Voltage	V_{CC} = Min, I_{OH} = Max, V_{IL} = Max, V_{IH} = Min	2.5	3.4		V
V_{OL}	Low Level Output Voltage	V_{CC} = Min, I_{OL} = Max, V_{IH} = Min, V_{IL} = Max			0.4	V
I_I	Input Current @ Max Input Voltage	V_{CC} = Max, V_I = 10.0V			0.1	mA
I_{IH}	High Level Input Current	V_{CC} = Max, V_I = 2.7V			20	μA
I_{IL}	Low Level Input Current	V_{CC} = Max, V_I = 0.5V	−30		−400	μA
I_{OS}	Short Circuit Output Current	V_{CC} = Max (Note 2)	−20		−100	mA
I_{CC}	Supply Current	V_{CC} = Max (Note 3)			9	mA

Note 1: All typicals are at V_{CC} = 5V, T_A = 25°C.

Note 2: Not more than one output should be shorted at a time, and the duration should not exceed one second.

SWITCHING CHARACTERISTICS V_{CC} = +5.0V, T_A = +25°C (See Section 1 for test waveforms and output load)

Symbol	Parameter	C_L = 15 pF		Units
		Min	Max	
t_{PLH} t_{PHL}	Propagation Delay, Sn to $\overline{Z}$		23 32	ns
t_{PLH} t_{PHL}	Propagation Delay, In to $\overline{Z}$		21 20	ns

CONNECTION DIAGRAM

Dual-In-Line Package

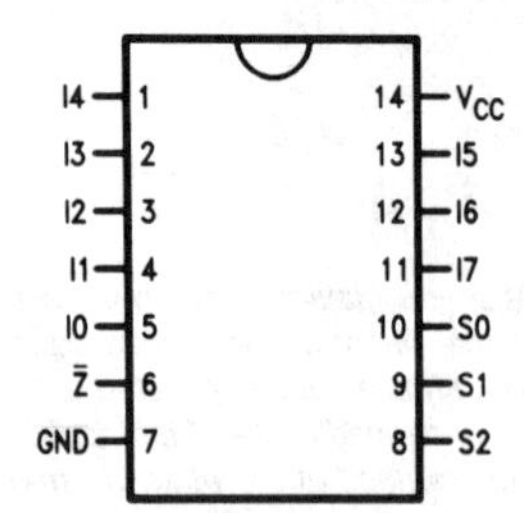

Order Number 54LS152FMQB
See NS Package Number W14B

LOGIC SYMBOL

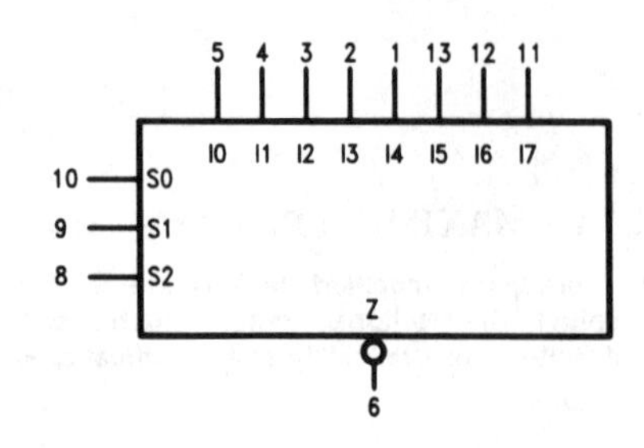

V_{CC} = Pin 14
GND = Pin 7

Pin Names	Description
I0–I7	Data Inputs
S0–S2	Select Inputs
$\overline{Z}$	Inverted Data Outputs

TRUTH TABLE

Inputs			Output
S2	S1	S0	$\overline{Z}$
L	L	L	$\overline{I0}$
L	L	H	$\overline{I1}$
L	H	L	$\overline{I2}$
L	H	H	$\overline{I3}$
H	L	L	$\overline{I4}$
H	L	H	$\overline{I5}$
H	H	L	$\overline{I6}$
H	H	H	$\overline{I7}$

H = HIGH Voltage Level
L = LOW Voltage Level

LOGIC DIAGRAM

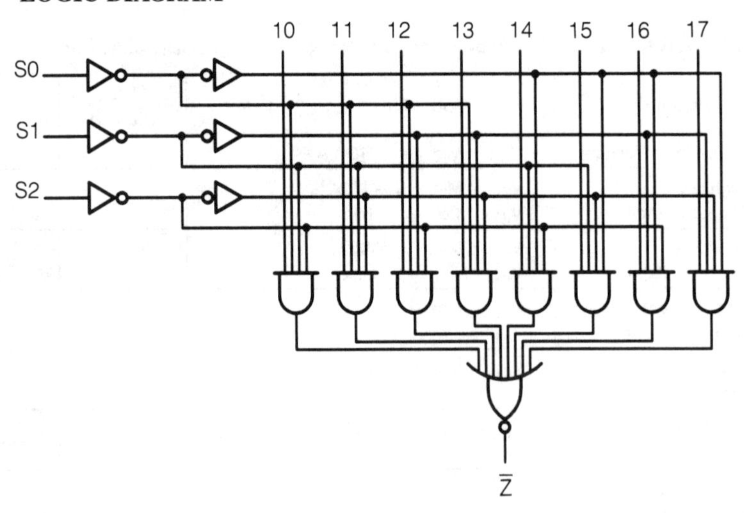

National 54LS153/DM54LS153/DM74LS153

Dual 4-Line to 1-Line Data Selectors/Multiplexers

- Permits multiplexing from N lines to 1 line
- Performs at parallel-to-serial conversion
- Strobe (enable) line provided for cascading (N lines to n lines)
- High fan-out, low-impedance, totem-pole outputs
- Typical average propagation delay times
  - From data: 14ns
  - From strobe: 19ns
  - From select: 22ns
- Typical power dissipation: 31mW
- Alternate military/aerospace device (54LS153) is available.

ABSOLUTE MAXIMUM RATINGS (Note)

If Military/Aerospace specified devices are required, please contact the National Semiconductor Sales Office/Distributors for availability and specifications.

Supply Voltage	7V
Input Voltage	7V
Operating Free Air Temperature Range	
DM54LS and 54LS	−55°C to +125°C
DM74LS	0°C to +70°C
Storage Temperature Range	−65°C to +150° C

Note: *The "Absolute Maximum Ratings" are those values beyond which the safety of the device cannot be guaranteed. The device should not be operated at these limits. The parametric values defined in the "Electrical Characteristics" table are not guaranteed at the absolute maximum ratings. The "Recommended Operating Conditions" table will define the conditions for actual device operation.*

RECOMMENDED OPERATING CONDITIONS

Symbol	Parameter	DM54LS153			DM74LS153			Units
		Min	Nom	Max	Min	Nom	Max	
V_{CC}	Supply Voltage	4.5	5	5.5	4.75	5	5.25	V
V_{IH}	High Level Input Voltage	2			2			V
V_{IL}	Low Level Input Voltage			0.7			0.8	V
I_{OH}	High Level Output Current			−0.4			−0.4	mA
I_{OL}	Low Level Output Current			4			8	mA
T_A	Free Air Operating Temperature	−55		125	0		70	°C

ELECTRICAL CHARACTERISTICS over recommended operating free air temperature range (unless otherwise noted)

Symbol	Parameter	Conditions		Min	Typ (Note 1)	Max	Units
V_I	Input Clamp Voltage	V_{CC} = Min, I_I = −18 mA				−1.5	V
V_{OH}	High Level Output Voltage	V_{CC} = Min, I_{OH} = Max V_{IL} = Max, V_{IH} = Min	DM54	2.5	3.4		V
			DM74	2.7	3.4		
V_{OL}	Low Level Output Voltage	V_{CC} = Min, I_{OL} = Max V_{IL} = Max, V_{IH} = Min	DM54		0.25	0.4	V
			DM74		0.35	0.5	
		I_{OL} = 4 mA, V_{CC} = Min	DM74		0.25	0.4	
I_I	Input Current @ Max Input Voltage	V_{CC} = Max, V_I = 7V				0.1	mA
I_{IH}	High Level Input Current	V_{CC} = Max, V_I = 2.7V				20	μA
I_{IL}	Low Level Input Current	V_{CC} = Max, V_I = 0.4V				−0.36	mA
I_{OS}	Short Circuit Output Current	V_{CC} = Max (Note 2)	DM54	−20		−100	mA
			DM74	−20		−100	
I_{CC}	Supply Current	V_{CC} = Max (Note 3)			6.2	10	mA

Note 1: All typicals are at V_{CC} = 5V, T_A = 25° C.

Note 2: Not more than one output should be shorted at a time, and the duration should not exceed one second.

Note 3: I_{CC} is measured with all outputs open and all other inputs grounded.

SWITCHING CHARACTERISTICS-at V_{CC} = 5V and T_A = 25°C (See Section 1 for Test Waveforms and Output Load)

Symbol	Parameter	From (Input) to (Output)	R_L = 2 kΩ				Units
			C_L = 15 pF		C_L = 50 pF		
			Min	Max	Min	Max	
t_{PLH}	Propagation Delay Time Low to High Level Output	Data to Y		15		20	ns
t_{PHL}	Propagation Delay Time High to Low Level Output	Data to Y		26		35	ns
t_{PLH}	Propagation Delay Time Low to High Level Output	Select to Y		29		35	ns
t_{PHL}	Propagation Delay Time High to Low Level Output	Select to Y		38		45	ns
t_{PLH}	Propagation Delay Time Low to High Level Output	Strobe to Y		24		30	ns
t_{PHL}	Propagation Delay Time High to Low Level Output	Strobe to Y		32		40	ns

CONNECTION DIAGRAM

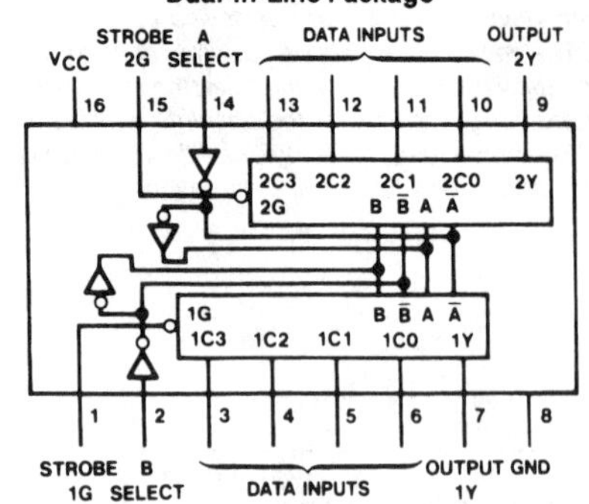

Order Number 54LS153DMQB, 54LS153FMQB, 54LS153LMQB, DM54LS153J, DM54LS153W, DM74LS153M or DM74LS153N
See NS Package Number E20A, J16A, M16A, N16E or W16A

LOGIC DIAGRAM

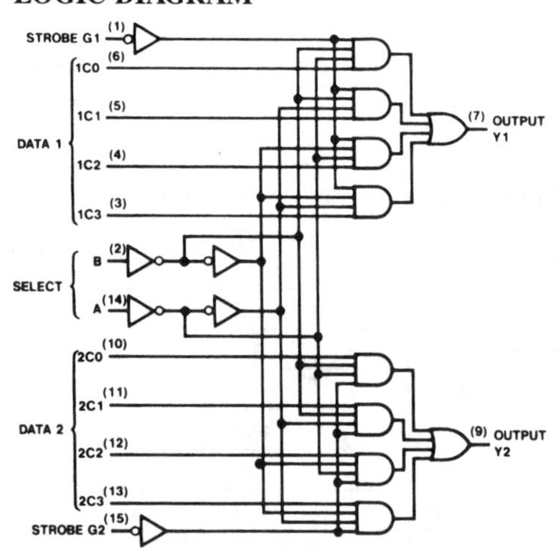

FUNCTION TABLE

Select Inputs		Data Inputs				Strobe	Output
B	A	C0	C1	C2	C3	G	Y
X	X	X	X	X	X	H	L
L	L	L	X	X	X	L	L
L	L	H	X	X	X	L	H
L	H	X	L	X	X	L	L
L	H	X	H	X	X	L	H
H	L	X	X	L	X	L	L
H	L	X	X	H	X	L	H
H	H	X	X	X	L	L	L
H	H	X	X	X	H	L	H

Select inputs A and B are common to both sections.
H = High Level, L = Low Level, X = Don't Care

National DM54LS154/DM74LS154

4-Line to 16-Line Decoders/Demultiplexers

- Decodes 4 binary-coded inputs into one of 16 mutually exclusive outputs
- Performs the demultiplexing function by distributing data from one input line to any one of 16 outputs
- Input clamping diodes simplify system design
- High fan-out low-impedance totem-pole outputs
- Typical propagation delay
  - 3 levels of logic 23ns
  - Strobe 19ns
- Typical power dissipation: 45mW

ABSOLUTE MAXIMUM RATINGS (Note)

If Military/Aerospace specified devices are required, please contact the National Semiconductor Sales Office/Distributors for availability and specifications.

Supply Voltage	7V
Input Voltage	7V
Operating Free Air Temperature Range	
DM54LS	−55°C to +125°C
DM74LS	0°C to +70°C
Storage Temperature Range	−65°C to +150°C

Note: *The "Absolute Maximum Ratings" are those values beyond which the safety of the device cannot be guaranteed. The device should not be operated at these limits. The parametric values defined in the "Electrical Characteristics" table are not guaranteed at the absolute maximum ratings. The "Recommended Operating Conditions" table will define the conditions for actual device operation.*

RECOMMENDED OPERATING CONDITIONS

Symbol	Parameter	DM54LS154			DM74LS154			Units
		Min	Nom	Max	Min	Nom	Max	
V_{CC}	Supply Voltage	4.5	5	5.5	4.75	5	5.25	V
V_{IH}	High Level Input Voltage	2			2			V
V_{IL}	Low Level Input Voltage			0.7			0.8	V
I_{OH}	High Level Output Current			−0.4			−0.4	mA
I_{OL}	Low Level Output Current			4			8	mA
T_A	Free Air Operating Temperature	−55		125	0		70	°C

ELECTRICAL CHARACTERISTICS over recommended operating free air temperature range (unless otherwise noted)

Symbol	Parameter	Conditions		Min	Typ (Note 1)	Max	Units
V_I	Input Clamp Voltage	V_{CC} = Min, I_I = −18 mA				−1.5	V
V_{OH}	High Level Output Voltage	V_{CC} = Min, I_{OH} = Max, V_{IL} = Max, V_{IH} = Min	DM54	2.5	3.4		V
			DM74	2.7	3.4		
V_{OL}	Low Level Output Voltage	V_{CC} = Min, I_{OL} = Max, V_{IL} = Max, V_{IH} = Min	DM54		0.25	0.4	V
			DM74		0.35	0.5	
		I_{OL} = 4 mA, V_{CC} = Min	DM74		0.25	0.4	
I_I	Input Current @ Max Input Voltage	V_{CC} = Max, V_I = 7V				0.1	mA
I_{IH}	High Level Input Current	V_{CC} = Max, V_I = 2.7V				20	μA
I_{IL}	Low Level Input Current	V_{CC} = Max, V_I = 0.4V				−0.4	mA
I_{OS}	Short Circuit Output Current	V_{CC} = Max (Note 2)	DM54	−20		−100	mA
			DM74	−20		−100	
I_{CC}	Supply Current	V_{CC} = Max (Note 3)			9	14	mA

Note 1: All typicals are at V_{CC} = 5V, T_A = 25°C.

Note 2: Not more than one output should be shorted at a time, and the duration should not exceed one second.

Note 3: I_{CC} is measured with all outputs open and all inputs grounded.

SWITCHING CHARACTERISTICS at V_{CC} = 5V and T_A = 25°C (See Section 1 for Test Waveforms and Output Load)

Symbol	Parameter	From (Input) To (Output)	R_L = 2 kΩ				Units
			C_L = 15 pF		C_L = 50 pF		
			Min	Max	Min	Max	
t_{PLH}	Propagation Delay Time Low to High Level Output	Data to Output		30		35	ns
t_{PHL}	Propagation Delay Time High to Low Level Output	Data to Output		30		35	ns
t_{PLH}	Propagation Delay Time Low to High Level Output	Strobe to Output		20		25	ns
t_{PHL}	Propagation Delay Time High to Low Level Output	Strobe to Output		25		35	ns

FUNCTION TABLE

Inputs						Outputs															
G1	G2	D	C	B	A	0	1	2	3	4	5	6	7	8	9	10	11	12	13	14	15
L	L	L	L	L	L	L	H	H	H	H	H	H	H	H	H	H	H	H	H	H	H
L	L	L	L	L	H	H	L	H	H	H	H	H	H	H	H	H	H	H	H	H	H
L	L	L	L	H	L	H	H	L	H	H	H	H	H	H	H	H	H	H	H	H	H
L	L	L	L	H	H	H	H	H	L	H	H	H	H	H	H	H	H	H	H	H	H
L	L	L	H	L	L	H	H	H	H	L	H	H	H	H	H	H	H	H	H	H	H
L	L	L	H	L	H	H	H	H	H	H	L	H	H	H	H	H	H	H	H	H	H
L	L	L	H	H	L	H	H	H	H	H	H	L	H	H	H	H	H	H	H	H	H
L	L	L	H	H	H	H	H	H	H	H	H	H	L	H	H	H	H	H	H	H	H
L	L	H	L	L	L	H	H	H	H	H	H	H	H	L	H	H	H	H	H	H	H
L	L	H	L	L	H	H	H	H	H	H	H	H	H	H	L	H	H	H	H	H	H
L	L	H	L	H	L	H	H	H	H	H	H	H	H	H	H	L	H	H	H	H	H
L	L	H	L	H	H	H	H	H	H	H	H	H	H	H	H	H	L	H	H	H	H
L	L	H	H	L	L	H	H	H	H	H	H	H	H	H	H	H	H	L	H	H	H
L	L	H	H	L	H	H	H	H	H	H	H	H	H	H	H	H	H	H	L	H	H
L	L	H	H	H	L	H	H	H	H	H	H	H	H	H	H	H	H	H	H	L	H
L	L	H	H	H	H	H	H	H	H	H	H	H	H	H	H	H	H	H	H	H	L
L	H	X	X	X	X	H	H	H	H	H	H	H	H	H	H	H	H	H	H	H	H
H	L	X	X	X	X	H	H	H	H	H	H	H	H	H	H	H	H	H	H	H	H
H	H	X	X	X	X	H	H	H	H	H	H	H	H	H	H	H	H	H	H	H	H

H = High Level, L = Low Level, X = Don't Care

CONNECTION AND LOGIC DIAGRAMS

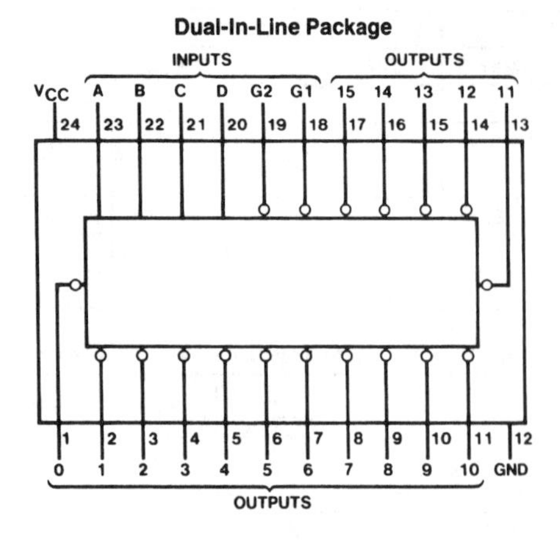

Order Number DM54LS154J, DM74LS154WM or DM74LS154N
See NS Package Number J24A, M24B or N24A

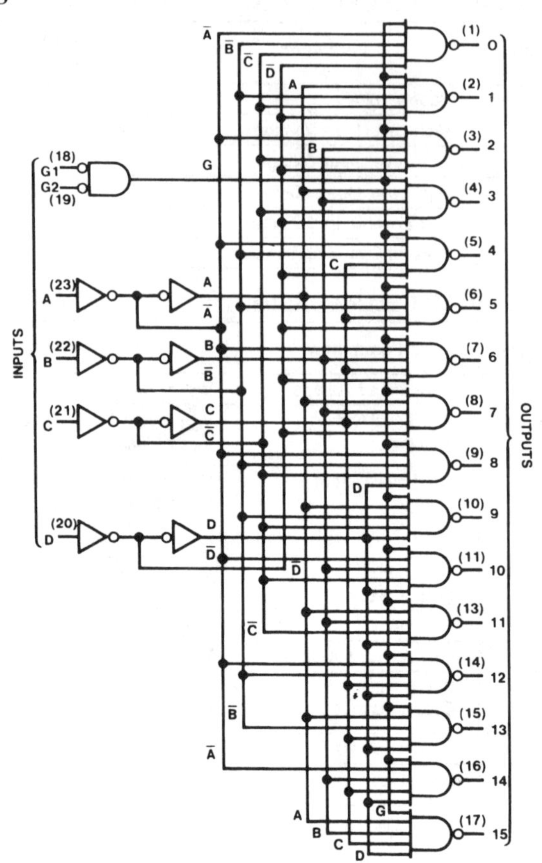

National 54LS155/DM54LS155/DM74LS155, 54LS156/DM54LS156/DM74LS156

Dual 2-Line to 4-Line Decoders/Demultiplexers

- Individual strobes simplify cascading for decoding or demultiplexing larger words
- Input clamping diodes simplify system design
- Choice of outputs:
  - Totem-pole (LS155)
  - Open-collector (LS156)
- Alternate military/aerospace device (54LS155/156) is available.

APPLICATIONS

- Dual 2-to-4-line decoder
- Dual 1-to-4-line demultiplexer
- 3-to-8-line decoder
- 1-to-8-line demultiplexer

ABSOLUTE MAXIMUM RATINGS (Note)

If Military/Aerospace specified devices are required, please contact the National Semiconductor Sales Office/Distributors for availability and specifications.

Supply Voltage	7V
Input Voltage	7V
Operating Free Air Temperature Range	
DM54LS and 54LS	−55°C to +125°C
DM74LS	0°C to +70°C
Storage Temperature Range	−65°C to +150°C

Note: *The "Absolute Maximum Ratings" are those values beyond which the safety of the device cannot be guaranteed. The device should not be operated at these limits. The parametric values defined in the "Electrical Characteristics" table are not guaranteed at the absolute maximum ratings. The "Recommended Operating Conditions" table will define the conditions for actual device operation.*

RECOMMENDED OPERATING CONDITIONS

Symbol	Parameter	DM54LS155			DM74LS155			Units
		Min	Nom	Max	Min	Nom	Max	
V_{CC}	Supply Voltage	4.5	5	5.5	4.75	5	5.25	V
V_{IH}	High Level Input Voltage	2			2			V
V_{IL}	Low Level Input Voltage			0.7			0.8	V
V_{OH}	High Level Output Current			−0.4			−0.4	mA
I_{OL}	Low Level Output Current			4			8	mA
T_A	Free Air Operating Temperature	−55		125	0		70	°C

'LS155 ELECTRICAL CHARACTERISTICS

over recommended operating free air temperature range (unless otherwise noted)

Symbol	Parameter	Conditions		Min	Typ (Note 1)	Max	Units
V_I	Input Clamp Voltage	V_{CC} = Min, I_I = −18 mA				−1.5	V
V_{OH}	High Level Output Voltage	V_{CC} = Min, I_{OH} = Max V_{IL} = Max, V_{IH} = Min	DM54	2.5	3.4		V
			DM74	2.7	3.4		
V_{OL}	Low Level Output Voltage	V_{CC} = Min, I_{OL} = Max V_{IL} = Max, V_{IH} = Min	DM54		0.25	0.4	V
			DM74		0.35	0.5	
		I_{OL} = 4 mA, V_{CC} = Min	DM74		0.25	0.4	
I_I	Input Current @ Max Input Voltage	V_{CC} = Max, V_I = 7V				0.1	mA
I_{IH}	High Level Input Current	V_{CC} = Max, V_I = 2.7V				20	μA
I_{IL}	Low Level Input Current	V_{CC} = Max, V_I = 0.4V				−0.36	mA
I_{OS}	Short Circuit Output Current	V_{CC} = Max (Note 2)	DM54	−20		−100	mA
			DM74	−20		−100	
I_{CC}	Supply Current	V_{CC} = Max (Note 3)			6.1	10	mA

Note 1: All typicals are at V_{CC} = 5V, T_A = 25° C.

Note 2: Not more than one output should be shorted at a time, and the duration should not exceed one second.

Note 3: I_{CC} is measured with all outputs open, A,B, and C1 inputs at 4.5V, and C2, G1, and G2 inputs grounded.

'LS155 SWITCHING CHARACTERISTICS

at V_{CC} = 5V and T_A = 25°C (See Section 1 for Test Waveforms and Output Load)

Symbol	Parameter	From (Input) To (Output)	R_L = 2 kΩ				Units
			C_L = 15 pF		C_L = 50 pF		
			Min	Max	Min	Max	
t_{PLH}	Propagation Delay Time Low to High Level Output	A, B, C2, G1 or G2 to Y		18		22	ns
t_{PHL}	Propagation Delay Time High to Low Level Output	A, B, C2, G1 or G2 to Y		27		35	ns
t_{PLH}	Propagation Delay Time Low to High Level Output	A or B to Y		18		24	ns
t_{PHL}	Propagation Delay Time High to Low Level Output	A or B to Y		27		35	ns
t_{PLH}	Propagation Delay Time Low to High Level Output	C1 to Y		20		24	ns
t_{PHL}	Propagation Delay Time High to Low Level Output	C1 to Y		27		35	ns

RECOMMENDED OPERATING CONDITIONS

Symbol	Parameter	DM54LS156			DM74LS156			Units
		Min	Nom	Max	Min	Nom	Max	
V_{CC}	Supply Voltage	4.5	5	5.5	4.75	5	5.25	V
V_{IH}	High Level Input Voltage	2			2			V
V_{IL}	Low Level Input Voltage			0.7			0.8	V
V_{OH}	High Level Output Voltage			5.5			5.5	V
I_{OL}	Low Level Output Current			4			8	mA
T_A	Free Air Operating Temperature	−55		125	0		70	°C

'LS156 ELECTRICAL CHARACTERISTICS

over recommended operating free air temperature range (unless otherwise noted)

Symbol	Parameter	Conditions		Min	Typ (Note 1)	Max	Units
V_I	Input Clamp Voltage	V_{CC} = Min, I_I = −18 mA				−1.5	V
I_{CEX}	High Level Output Current	V_{CC} = Min, V_O = 5.5V V_{IL} = Max, V_{IH} = Min				100	μA
V_{OL}	Low Level Output Voltage	V_{CC} = Min, I_{OL} = Max V_{IL} = Max, V_{IH} = Min	DM54		0.25	0.4	V
			DM74		0.35	0.5	
		I_{OL} = 4 mA, V_{CC} = Min	DM74		0.25	0.4	
I_I	Input Current @ Max Input Voltage	V_{CC} = Max, V_I = 7V				0.1	mA
I_{IH}	High Level Input Current	V_{CC} = Max, V_I = 2.7V				20	μA
I_{IL}	Low Level Input Current	V_{CC} = Max, V_I = 0.4V				−0.36	mA
I_{CC}	Supply Current	V_{CC} = Max (Note 2)			6.1	10	mA

Note 1: All typicals are at V_{CC} = 5V, T_A = 25° C.

Note 2: I_{CC} is measured with all outputs open, A, B, and C1 inputs at 4.5V, and C2, G1, and G2 grounded.

'LS156 SWITCHING CHARACTERISTICS

at V_{CC} = 5V and T_A = 25°C (See Section 1 for Test Waveforms and Output Load)

Symbol	Parameter	From (Input) To (Output)	R_L = 2 kΩ				Units
			C_L = 15 pF		C_L = 50 pF		
			Min	Max	Min	Max	
t_{PLH}	Propagation Delay Time Low to High Level Output	A, B, C2, G1 or G2 to Y		28		53	ns
t_{PHL}	Propagation Delay Time High to Low Level Output	A, B, C2, G1 or G2 to Y		33		43	ns
t_{PLH}	Propagation Delay Time Low to High Level Output	A or B to Y		28		53	ns
t_{PHL}	Propagation Delay Time High to Low Level Output	A or B to Y		33		43	ns
t_{PLH}	Propagation Delay Time Low to High Level Output	C1 to Y		28		53	ns
t_{PHL}	Propagation Delay Time High to Low Level Output	C1 to Y		34		43	ns

LOGIC DIAGRAM

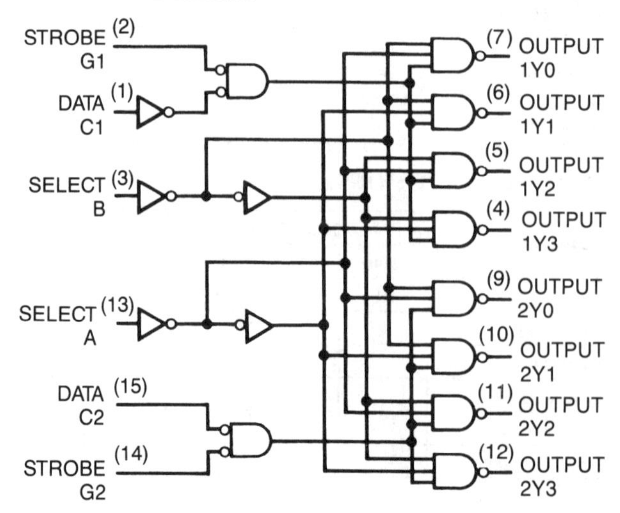

CONNECTION DIAGRAM AND FUNCTION TABLES

Dual-In-Line Package

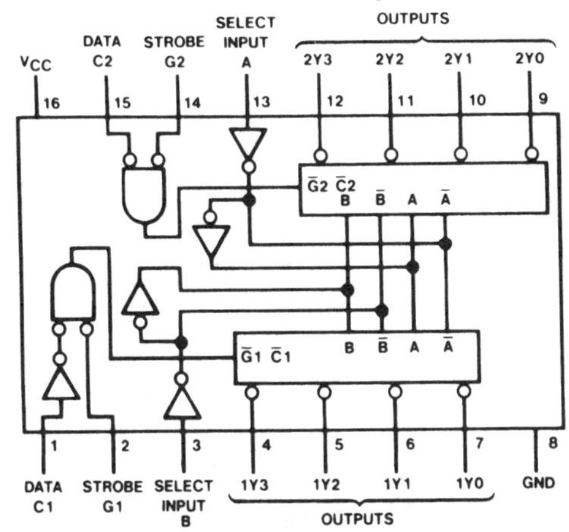

Order Number 54LS155DMQB, 54LS155FMQB, 54LS155LMQB, DM54LS155J, DM54LS155W, DM74LS155M, DM74LS155N, 54LS156DMQB, 54LS156FMQB, DM54LS156J, DM54LS156W, DM74LS156M or DM74LS156N
See NS Package Number E20A, J16A, M16A, N16E or W16A

3-Line-to-8-Line Decoder or 1-Line-to-8-Line Demultiplexer

Inputs				Outputs							
Select			Strobe Or Data	(0)	(1)	(2)	(3)	(4)	(5)	(6)	(7)
C†	B	A	G‡	2Y0	2Y1	2Y2	2Y3	1Y0	1Y1	1Y2	1Y3
X	X	X	H	H	H	H	H	H	H	H	H
L	L	L	L	L	H	H	H	H	H	H	H
L	L	H	L	H	L	H	H	H	H	H	H
L	H	L	L	H	H	L	H	H	H	H	H
L	H	H	L	H	H	H	L	H	H	H	H
H	L	L	L	H	H	H	H	L	H	H	H
H	L	H	L	H	H	H	H	H	L	H	H
H	H	L	L	H	H	H	H	H	H	L	H
H	H	H	L	H	H	H	H	H	H	H	L

2-Line-to-4-Line Decoder or 1-Line-to-4-Line Demultiplexer

Inputs				Outputs			
Select		Strobe	Data				
B	A	G1	C1	1Y0	1Y1	1Y2	1Y3
X	X	H	X	H	H	H	H
L	L	L	H	L	H	H	H
L	H	L	H	H	L	H	H
H	L	L	H	H	H	L	H
H	H	L	H	H	H	H	L
X	X	X	L	H	H	H	H

Inputs				Outputs			
Select		Strobe	Data				
B	A	G2	C2	2Y0	2Y1	2Y2	2Y3
X	X	H	X	H	H	H	H
L	L	L	L	L	H	H	H
L	H	L	L	H	L	H	H
H	L	L	L	H	H	L	H
H	H	L	L	H	H	H	L
X	X	X	H	H	H	H	H

†C = inputs C1 and C2 connected together
‡G = inputs G1 and G2 connected together
H = high level, L = low level, X = don't care

National 54LS157/DM54LS157/DM74LS157, 54LS158/DM54LS158/DM74LS158
Quad 2-Line to 1-Line Data Selectors/Multiplexers

- Buffered inputs and outputs
- Typical propagation time
  - ○ LS157 9ns
  - ○ LS158 7ns
- Typical power dissipation
  - ○ LS157 49mW
  - ○ LS158 24mW
- Alternate military/aerospace device (54LS157, 54LS158) is available.

ABSOLUTE MAXIMUM RATINGS

If Military/Aerospace specified devices are required, please contact the National Semiconductor Sales Office/Distributors for availability and specifications.

Supply Voltage	7V
Input Voltage	7V
Operating Free Air Temperature Range	
DM54LS and 54LS	−55°C to +125°C
DM74LS	0°C to +70°C
Storage Temperature Range	−65°C to +150°C

Note: *The "Absolute Maximum Ratings" are those values beyond which the safety of the device cannot be guaranteed. The device should not be operated at these limits. The parametric values defined in the "Electrical Characteristics" table are not guaranteed at the absolute maximum ratings. The "Recommended Operating Conditions" table will define the conditions for actual device operation.*

RECOMMENDED OPERATING CONDITIONS

Symbol	Parameter	DM54LS157			DM74LS157			Units
		Min	Nom	Max	Min	Nom	Max	
V_{CC}	Supply Voltage	4.5	5	5.5	4.75	5	5.25	V
V_{IH}	High Level Input Voltage	2			2			V
V_{IL}	Low Level Input Voltage			0.7			0.8	V
I_{OH}	High Level Output Current			−0.4			−0.4	mA
I_{OL}	Low Level Output Current			4			8	mA
T_A	Free Air Operating Temperature	−55		125	0		70	°C

'LS157 ELECTRICAL CHARACTERISTICS

over recommended operating free air temperature range (unless otherwise noted)

Symbol	Parameter	Conditions		Min	Typ (Note 1)	Max	Units
V_I	Input Clamp Voltage	V_{CC} = Min, I_I = −18 mA				−1.5	V
V_{OH}	High Level Output Voltage	V_{CC} = Min, I_{OH} = Max, V_{IL} = Max, V_{IH} = Min	DM54	2.5	3.4		V
			DM74	2.7	3.4		
V_{OL}	Low Level Output Voltage	V_{CC} = Min, I_{OL} = Max, V_{IL} = Max, V_{IH} = Min	DM54		0.25	0.4	V
			DM74		0.35	0.5	
		I_{OL} = 4 mA, V_{CC} = Min	DM74		0.25	0.4	
I_I	Input Current @ Max Input Voltage	V_{CC} = Max, V_I = 7V	S or G			0.2	mA
			A or B			0.1	
I_{IH}	High Level Input Current	V_{CC} = Max, V_I = 2.7V	S or G			40	μA
			A or B			20	
I_{IL}	Low Level Input Current	V_{CC} = Max, V_I = 0.4V	S or G			−0.8	mA
			A or B			−0.4	
I_{OS}	Short Circuit Output Current	V_{CC} = Max (Note 2)	DM54	−20		−100	mA
			DM74	−20		−100	
I_{CC}	Supply Current	V_{CC} = Max (Note 3)			9.7	16	mA

Note 1: All typicals are at V_{CC} = 5V, T_A = 25°C.

Note 2: Not more than one output should be shorted at a time, and the duration should not exceed one second.

Note 3: I_{CC} is measured with 4.5V applied to all inputs and all outputs open.

'LS157 SWITCHING CHARACTERISTICS

at V_{CC} = 5V and T_A = 25°C (See Section 1 for Test Waveforms and Output Load)

Symbol	Parameter	From (Input) To (Output)	R_L = 2 kΩ				Units
			C_L = 15 pF		C_L = 50 pF		
			Min	Max	Min	Max	
t_{PLH}	Propagation Delay Time Low to High Level Output	Data to Y		14		18	ns
t_{PHL}	Propagation Delay Time High to Low Level Output	Data to Y		14		23	ns
t_{PLH}	Propagation Delay Time Low to High Level Output	Strobe to Y		20		24	ns
t_{PHL}	Propagation Delay Time High to Low Level Output	Strobe to Y		21		30	ns
t_{PLH}	Propagation Delay Time Low to High Level Output	Select to Y		23		28	ns
t_{PHL}	Propagation Delay Time High to Low Level Output	Select to Y		27		32	ns

RECOMMENDED OPERATING CONDITIONS

Symbol	Parameter	DM54LS158			DM74LS158			Units
		Min	Nom	Max	Min	Nom	Max	
V_{CC}	Supply Voltage	4.5	5	5.5	4.75	5	5.25	V
V_{IH}	High Level Input Voltage	2			2			V
V_{IL}	Low Level Input Voltage			0.7			0.8	V
I_{OH}	High Level Output Current			−0.4			−0.4	mA
I_{OL}	Low Level Output Current			4			8	mA
T_A	Free Air Operating Temperature	−55		125	0		70	°C

'LS158 ELECTRICAL CHARACTERISTICS

over recommended operating free air temperature range (unless otherwise noted)

Symbol	Parameter	Conditions		Min	Typ (Note 1)	Max	Units
V_I	Input Clamp Voltage	V_{CC} = Min, I_I = −18 mA				−1.5	V
V_{OH}	High Level Output Voltage	V_{CC} = Min, I_{OH} = Max V_{IL} = Max, V_{IH} = Min	DM54	2.5	3.4		V
			DM74	2.7	3.4		
V_{OL}	Low Level Output Voltage	V_{CC} = Min, I_{OL} = Max V_{IL} = Max, V_{IH} = Min	DM54		0.25	0.4	V
			DM74		0.35	0.5	
		I_{OL} = 4 mA, V_{CC} = Min	DM74		0.25	0.4	
I_I	Input Current @ Max Input Voltage	V_{CC} = Max V_I = 7V	S or G			0.2	mA
			A or B			0.1	
I_{IH}	High Level Input Current	V_{CC} = Max V_I = 2.7V	S or G			40	µA
			A or B			20	
I_{IL}	Low Level Input Current	V_{CC} = Max V_I = 0.4V	S or G			−0.8	mA
			A or B			−0.4	
I_{OS}	Short Circuit Output Current	V_{CC} = Max (Note 2)	DM54	−20		−100	mA
			DM74	−20		−100	
I_{CC}	Supply Current	V_{CC} = Max (Note 3)			4.8	8	mA

Note 1: All typicals are at V_{CC} = 5V, T_A = 25°C.

Note 2: Not more than one output should be shorted at a time, and the duration should not exceed one second.

Note 3: I_{CC} is measured with 4.5V applied to all inputs and all outputs open.

'LS158 SWITCHING CHARACTERISTICS

at V_{CC} = 5 v and T_A = 25°C (See Section 1 for Test Waveforms and Output Load)

Symbol	Parameter	From (Input) To (Output)	R_L = 2 kΩ				Units
			C_L = 15 pF		C_L = 50 pF		
			Min	Max	Min	Max	
t_{PLH}	Propagation Delay Time Low to High Level Output	Data to Y		12		18	ns
t_{PHL}	Propagation Delay Time High to Low Level Output	Data to Y		12		21	ns
t_{PLH}	Propagation Delay Time Low to High Level Output	Strobe to Y		17		23	ns
t_{PHL}	Propagation Delay Time High to Low Level Output	Strobe to Y		18		28	ns
t_{PLH}	Propagation Delay Time Low to High Level Output	Select to Y		20		24	ns
t_{PHL}	Propagation Delay Time High to Low Level Output	Select to Y		24		36	ns

LOGIC DIAGRAMS

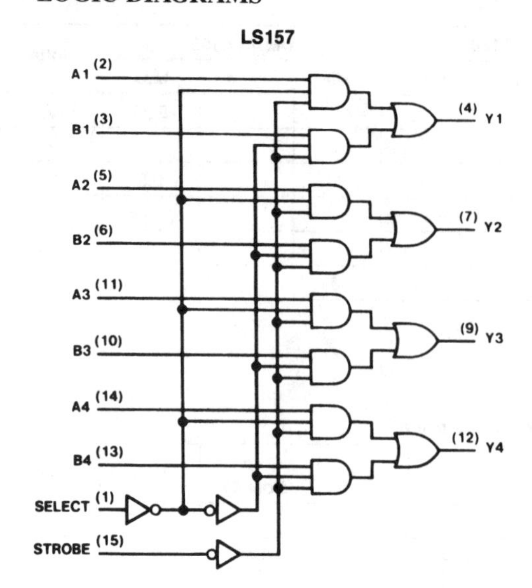

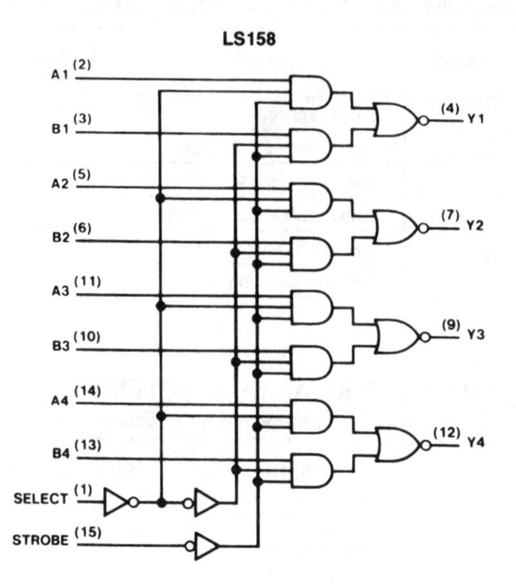

CONNECTION DIAGRAMS

Dual-In-Line Package

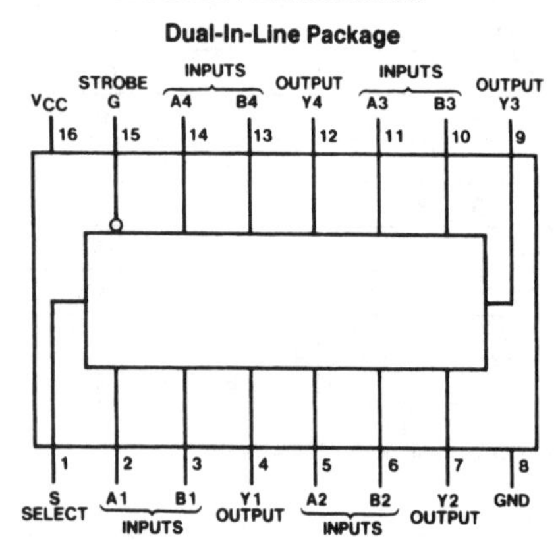

Order Number 54LS157DMQB, 54LS157FMQB, 54LS157LMQB, DM54LS157J, DM54LS157W, DM74LS157M or DM74LS157N
See NS Package Number E20A, J16A, M16A, N16E or W16A

Dual-In-Line Package

VCC, STROBE G, INPUTS A4 B4, OUTPUT Y4, INPUTS A3 B3, OUTPUT Y3
16 15 14 13 12 11 10 9
1 2 3 4 5 6 7 8
S SELECT, A1 B1 INPUTS, Y1 OUTPUT, A2 B2 INPUTS, Y2 OUTPUT, GND

Order Number 54LS158DMQB, 54LS158FMQB, 54LS158LMQB, DM54LS158J, DM54LS158W, DM74LS158M or DM74LS158N
See NS Package Number E20A, J16A, M16A, N16E or W16A

FUNCTION TABLE

Inputs				Output Y	
Strobe	Select	A	B	LS157	LS158
H	X	X	X	L	H
L	L	L	X	L	H
L	L	H	X	H	L
L	H	X	L	L	H
L	H	X	H	H	L

H = High Level, L = Low Level, X = Don't Care

National DM54LS251/DM74LS251
Tri-State® Data Selectors/Multiplexers

- Tri-state version of LS151
- Interface directly with system bus
- Perform parallel-to-serial conversion
- Permit multiplexing from N-lines to one line
- Complementary outputs provide true and inverted data
- Maximum number of common outputs
  - 54LS 49
  - 74LS 129
- Typical propagation delay time (D to Y)
  - 54LS 17ns
  - 74LS 17ns
- Typical power dissipation
  - 54LS 35mW
  - 74LS 35mW

ABSOLUTE MAXIMUM RATINGS

If Military/Aerospace specified devices are required, please contact the National Semiconductor Sales Office/Distributors for availability and specifications.

Supply Voltage	7V
Input Voltage	7V
Operating Free Air Temperature Range	
DM54LS	−55°C to +125°C
DM74LS	0°C to +70°C
Storage Temperature Range	−65°C to +150°C

Note: *The "Absolute Maximum Ratings" are those values beyond which the safety of the device cannot be guaranteed. The device should not be operated at these limits. The parametric values defined in the "Electrical Characteristics" table are not guaranteed at the absolute maximum ratings. The "Recommended Operating Conditions" table will define the conditions for actual device operation.*

RECOMMENDED OPERATING CONDITIONS

Symbol	Parameter	DM54LS251			DM74LS251			Units
		Min	Nom	Max	Min	Nom	Max	
V_{CC}	Supply Voltage	4.5	5	5.5	4.75	5	5.25	V
V_{IH}	High Level Input Voltage	2			2			V
V_{IL}	Low Level Input Voltage			0.7			0.8	V
I_{OH}	High Level Output Current			−1			−2.6	mA
I_{OL}	Low Level Output Current			12			24	mA
T_A	Free Air Operating Temperature	−55		125	0		70	°C

ELECTRICAL CHARACTERISTICS over recommended operating free air temperature range (unless otherwise noted)

Symbol	Parameter	Conditions		Min	Typ (Note 1)	Max	Units
V_I	Input Clamp Voltage	V_{CC} = Min, I_I = −18 mA				−1.5	V
V_{OH}	High Level Output Voltage	V_{CC} = Min, I_{OH} = Max, V_{IL} = Max, V_{IH} = Min	DM54	2.4	3.4		V
			DM74	2.4	3.1		
V_{OL}	Low Level Output Voltage	V_{CC} = Min, I_{OL} = Max, V_{IL} = Max, V_{IH} = Min	DM54		0.25	0.4	V
			DM74		0.35	0.5	
		I_{OL} = 12 mA, V_{CC} = Min	DM74		0.25	0.4	
I_I	Input Current @ Max Input Voltage	V_{CC} = Max, V_I = 7V				0.1	mA
I_{IH}	High Level Input Current	V_{CC} = Max, V_I = 2.7V				20	μA
I_{IL}	Low Level Input Current	V_{CC} = Max, V_I = 0.4V				−0.4	mA
I_{OZH}	Off-State Output Current with High Level Output Voltage Applied	V_{CC} = Max, V_O = 2.7V, V_{IH} = Min, V_{IL} = Max				20	μA
I_{OZL}	Off-State Output Current with Low Level Output Voltage Applied	V_{CC} = Max, V_O = 0.4V, V_{IH} = Min, V_{IL} = Max				−20	μA
I_{OS}	Short Circuit Output Current	V_{CC} = Max (Note 2)	DM54	−20		−100	mA
			DM74	−20		−100	
I_{CC1}	Supply Current	V_{CC} = Max (Note 3)			6.1	10	mA
I_{CC2}	Supply Current	V_{CC} = Max (Note 4)			7.1	12	mA

Note 1: All typicals are at V_{CC} = 5V, T_A = 25°C.

Note 2: Not more than one output should be shorted at a time, and the duration should not exceed one second.

Note 3: I_{CC1} is measured with the outputs open, STROBE grounded, and all other inputs at 4.5V.

Note 4: I_{CC2} is measured with the outputs open and all inputs at 4.5V.

SWITCHING CHARACTERISTICS at V_{CC} = 5V and T_A = 25°C (See Section 1 for Test Waveforms and Output Load)

Symbol	Parameter	From (Input) to (Output)	$R_L = 667\Omega$				Units
			C_L = 45 pF		C_L = 150 pF		
			Min	Max	Min	Max	
t_{PLH}	Propagation Delay Time Low to High Level Output	A, B, C (4 Levels) to Y		45		53	ns
t_{PHL}	Propagation Delay Time High to Low Level Output	A, B, C (4 Levels) to Y		45		53	ns
t_{PLH}	Propagation Delay Time Low to High Level Output	A, B, C (3 Levels) to W		33		38	ns
t_{PHL}	Propagation Delay Time High to Low Level Output	A, B, C (3 Levels) to W		33		42	ns
t_{PLH}	Propagation Delay Time Low to High Level Output	D to Y		28		35	ns
t_{PHL}	Propagation Delay Time High to Low Level Output	D to Y		28		38	ns
t_{PLH}	Propagation Delay Time Low to High Level Output	D to W		15		25	ns
t_{PHL}	Propagation Delay Time High to Low Level Output	D to W		15		25	ns
t_{PZH}	Output Enable Time to High Level Output	Strobe to Y		45		60	ns
t_{PZL}	Output Enable Time to Low Level Output	Strobe to Y		40		51	ns
t_{PHZ}	Output Disable Time from High Level Output (Note 1)	Strobe to Y		45			ns
t_{PLZ}	Output Disable Time from Low Level Output (Note 1)	Strobe to Y		25			ns
t_{PZH}	Output Enable Time to High Level Output	Strobe to W		27		40	ns
t_{PZL}	Output Enable Time to Low Level Output	Strobe to W		40		47	ns
t_{PHZ}	Output Disable Time from High Level Output (Note 1)	Strobe to W		55			ns
t_{PLZ}	Output Disable Time from Low Level Output (Note 1)	Strobe to W		25			ns

Note 1: C_L = 5 pF

CONNECTION DIAGRAM

Dual-In-Line Package

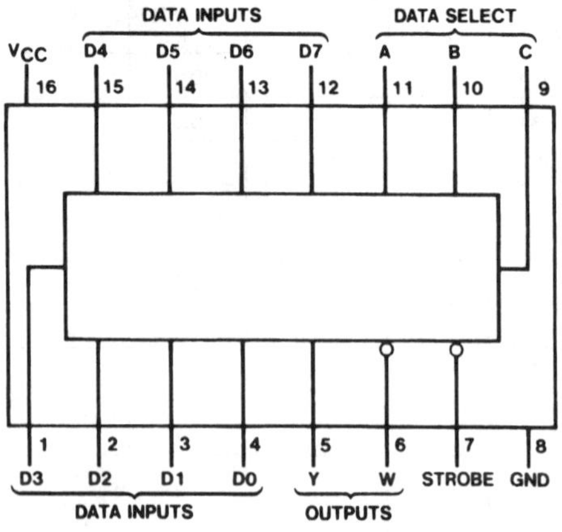

DATA INPUTS: D4 (15), D5 (14), D6 (13), D7 (12); DATA SELECT: A (11), B (10), C (9); V_{CC} (16)

DATA INPUTS: D3 (1), D2 (2), D1 (3), D0 (4); OUTPUTS: Y (5), W (6); STROBE (7); GND (8)

Order Number DM54LS251J, DM54LS251W, DM74LS251M or DM74LS251N
See NS Package Number J16A, M16A, N16E or W16A

FUNCTION TABLE

Inputs				Outputs	
Select			Strobe	Y	W
C	B	A	S		
X	X	X	H	Z	Z
L	L	L	L	D0	$\overline{D0}$
L	L	H	L	D1	$\overline{D1}$
L	H	L	L	D2	$\overline{D2}$
L	H	H	L	D3	$\overline{D3}$
H	L	L	L	D4	$\overline{D4}$
H	L	H	L	D5	$\overline{D5}$
H	H	L	L	D6	$\overline{D6}$
H	H	H	L	D7	$\overline{D7}$

H = High Logic Level, L = Low Logic Level,
X = Don't Care, Z = High Impedance (Off)
D0, D1 . . . D7 = The level of the respective D input

LOGIC DIAGRAM

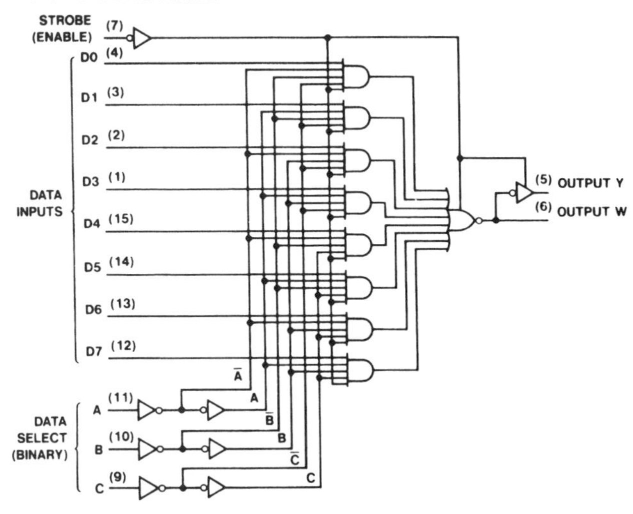

National 54LS253/DM54LS253/DM74LS253

Tri-state® Data Selectors/Multiplexers

- Tri-state version of LS153 with same pinout
- Schottky-diode-clamped transistors
- Permit multiplexing from N-lines to one line
- Performs parallel-to-serial conversion
- Strobe/output control
- High fanout totem-pole outputs
- Typical propagation delay
  - Data to output 12ns
  - Select to output 21ns
- Typical power dissipation 35mW
- Alternate military/aerospace device (54LS253) is available.

ABSOLUTE MAXIMUM RATINGS (Note)

If Military/Aerospace specified devices are required, please contact the National Semiconductor Sales Office/Distributors for availability and specifications.

Supply Voltage	7V
Input Voltage	7V
Operating Free Air Temperature Range	
DM54LS and 54LS	−55°C to +125°C
DM74LS	0°C to +70°C
Storage Temperature Range	−65°C to +150°C

Note: *The "Absolute Maximum Ratings" are those values beyond which the safety of the device cannot be guaranteed. The device should not be operated at these limits. The parametric values defined in the "Electrical Characteristics" table are not guaranteed at the absolute maximum ratings. The "Recommended Operating Conditions" table will define the conditions for actual device operation.*

RECOMMENDED OPERATING CONDITIONS

Symbol	Parameter	DM54LS253			DM74LS253			Units
		Min	Nom	Max	Min	Nom	Max	
V_{CC}	Supply Voltage	4.5	5	5.5	4.75	5	5.25	V
V_{IH}	High Level Input Voltage	2			2			V
V_{IL}	Low Level Input Voltage			0.7			0.8	V
I_{OH}	High Level Output Current			−1			−2.6	mA
I_{OL}	Low Level Output Current			12			24	mA
T_A	Free Air Operating Temperature	−55		125	0		70	°C

ELECTRICAL CHARACTERISTICS over recommended operating free air temperature range (unless otherwise noted)

Symbol	Parameter	Conditions		Min	Typ (Note 1)	Max	Units
V_I	Input Clamp Voltage	V_{CC} = Min, I_I = −18 mA				−1.5	V
V_{OH}	High Level Output Voltage	V_{CC} = Min, I_{OH} = Max, V_{IL} = Max, V_{IH} = Min	DM54	2.4	3.4		V
			DM74	2.4	3.1		
V_{OL}	Low Level Output Voltage	V_{CC} = Min, I_{OL} = Max, V_{IL} = Max, V_{IH} = Min	DM54			0.4	V
			DM74			0.5	
		I_{OL} = 12 mA, V_{CC} = Min	DM74			0.4	
I_I	Input Current @ Max Input Voltage	V_{CC} = Max, V_I = 7V				0.1	mA
I_{IH}	High Level Input Current	V_{CC} = Max, V_I = 2.7V				20	μA
I_{IL}	Low Level Input Current	V_{CC} = Max, V_I = 0.4V				−0.4	mA
I_{OZH}	Off-State Output Current with High Level Output Voltage Applied	V_{CC} = Max, V_O = 2.7V, V_{IH} = Min, V_{IL} = Max				20	μA
I_{OZL}	Off-State Output Current with Low Level Output Voltage Applied	V_{CC} = Max, V_O = 0.4, V_{IH} = Min, V_{IL} = Max				−20	μA
I_{OS}	Short Circuit Output Current	V_{CC} = Max (Note 2)	DM54	−20		−100	mA
			DM74	−20		−100	
I_{CC1}	Supply Current	V_{CC} = Max (Note 3)			7	12	mA
I_{CC2}	Supply Current	V_{CC} = Max (Note 4)			8.5	14	mA

Note 1: All typicals are at V_{CC} = 5V, T_A = 25°C.

Note 2: Not more than one output should be shorted at a time, and the duration should not exceed one second.

Note 3: I_{CC1} is measured with all outputs open, and all the inputs grounded.

Note 4: I_{CC2} is measured with the outputs open, OUTPUT CONTROL at 4.5V and all other inputs grounded.

SWITCHING CHARACTERISTICS at V_{CC} = 5V and T_A = 25°C (See Section 1 for Test Waveforms and Output Load)

Symbol	Parameter	From (Input) To (Output)	R_L = 667Ω				Units
			C_L = 45 pF		C_L = 150 pF		
			Min	Max	Min	Max	
t_{PLH}	Propagation Delay Time Low to High Level Output	Data to Y		25		35	ns
t_{PHL}	Propagation Delay Time High to Low Level Output	Data to Y		20		30	ns
t_{PLH}	Propagation Delay Time Low to High Level Output	Select to Y		45		54	ns
t_{PHL}	Propagation Delay Time High to Low Level Output	Select to Y		32		44	ns
t_{PZH}	Output Enable Time to High Level Output	Output Control to Y		18		32	ns
t_{PZL}	Output Enable Time to Low Level Output	Output Control to Y		23		35	ns
t_{PHZ}	Output Disable Time from High Level Output (Note 1)	Output Control to Y		41			ns
t_{PLZ}	Output Disable Time from Low Level Output (Note 1)	Output Control to Y		27			ns

Note 1: C_L = 5 pF.

CONNECTION DIAGRAM

Dual-In-Line Package

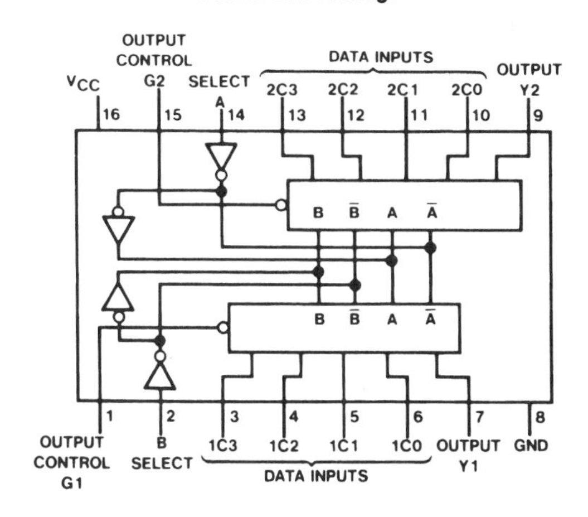

Order Number 54LS253DMQB, 54LS253FMQB, 54LS253LMQB, DM54LS253J, DM54LS253W, DM74LS253M or DM74LS253N
See NS Package Number E20A, J16A, M16A, N16E or W16A

LOGIC DIAGRAM

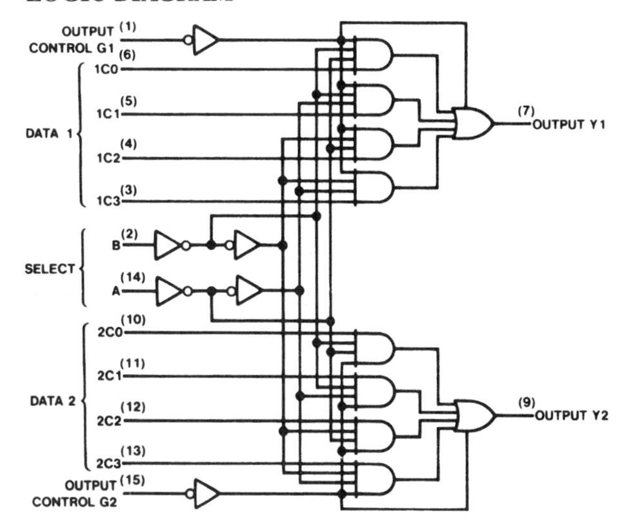

FUNCTION TABLE

Select Inputs		Data Inputs				Output Control	Output
B	A	C0	C1	C2	C3	G	Y
X	X	X	X	X	X	H	Z
L	L	L	X	X	X	L	L
L	L	H	X	X	X	L	H
L	H	X	L	X	X	L	L
L	H	X	H	X	X	L	H
H	L	X	X	L	X	L	L
H	L	X	X	H	X	L	H
H	H	X	X	X	L	L	L
H	H	X	X	X	H	L	H

Address Inputs A and B are common to both sections.
H = High Level, L = Low Level, X = Don't Care, Z = High Impedance (off).

National 54LS257A/DM54LS257B/DM74LS257B, 54LS258A/DM54LS258B/DM74LS258B

Tri-state® Quad 2-Data Selectors/Multiplexers

- Tri-state versions LS157 and LS158 with same pinouts
- Schottky-clamped for significant improvement in ac performance
- Provides bus interface from multiple sources in high-performance systems
- Average propagation delay from data input 12ns
- Typical power dissipation
  - LS257B 50mW
  - LS258B 35mW
- Alternate military/aerospace devices (54LS257A/54LS258A) are available.

ABSOLUTE MAXIMUM RATINGS (Note)

If Military/Aerospace specified devices are required, please contact the National Semiconductor Sales Office/Distributors for availability and specifications.

Supply Voltage	7V
Input Voltage	7V
Operating Free Air Temperature Range	
DM54LS and 54LS	−55°C to +125°C
DM74LS	0°C to +70°C
Storage Temperature Range	−65°C to +150°C

Note: *The "Absolute Maximum Ratings" are those values beyond which the safety of the device cannot be guaranteed. The device should not be operated at these limits. The parametric values defined in the "Electrical Characteristics" table are not guaranteed at the absolute maximum ratings. The "Recommended Operating Conditions" table will define the conditions for actual device operation.*

RECOMMENDED OPERATING CONDITIONS

Symbol	Parameter	DM54LS257B			DM74LS257B			Units
		Min	Nom	Max	Min	Nom	Max	
V_{CC}	Supply Voltage	4.5	5	5.5	4.75	5	5.25	V
V_{IH}	High Level Input Voltage	2			2			V
V_{IL}	Low Level Input Voltage			0.7			0.8	V
I_{OH}	High Level Output Current			−1			−2.6	mA
I_{OL}	Low Level Output Current			12			24	mA
T_A	Free Air Operating Temperature	−55		125	0		70	°C

'LS257B ELECTRICAL CHARACTERISTICS

over recommended operating free air temperature range (unless otherwise noted)

Symbol	Parameter	Conditions		Min	Typ (Note 1)	Max	Units
V_I	Input Clamp Voltage	V_{CC} = Min, I_I = −18 mA				−1.5	V
V_{OH}	High Level Output Voltage	V_{CC} = Min, I_{OH} = Max V_{IL} = Max, V_{IH} = Min	DM54	2.4	3.4		V
			DM74	2.4	3.1		
V_{OL}	Low Level Output Voltage	V_{CC} = Min, I_{OL} = Max V_{IL} = Max, V_{IH} = Min	DM54		0.25	0.4	V
			DM74		0.35	0.5	
		I_{OL} = 12 mA, V_{CC} = Min	DM74		0.25	0.4	
I_I	Input Current @ Max Input Voltage	V_{CC} = Max, V_I = 7V	Select			0.2	mA
			Other			0.1	
I_{IH}	High Level Input Current	V_{CC} = Max, V_I = 2.7V	Select			40	μA
			Other			20	
I_{IL}	Low Level Input Current	V_{CC} = Max, V_I = 0.4V	Select			−0.8	mA
			Other			−0.4	
I_{OZH}	Off-State Output Current with High Level Output Voltage Applied	V_{CC} = Max, V_O = 2.7V V_{IH} = Min, V_{IL} = Max				20	μA
I_{OZL}	Off-State Output Current with Low Level Output Voltage Applied	V_{CC} = Max, V_O = 0.4V V_{IH} = Min, V_{IL} = Max				−20	μA
I_{OS}	Short Circuit Output Current	V_{CC} = Max (Note 2)	DM54	−20		−100	mA
			DM74	−20		−100	
I_{CCH}	Supply Current with Outputs High	V_{CC} = Max (Note 3)			5.9	10	mA
I_{CCL}	Supply Current with Outputs Low	V_{CC} = Max (Note 3)			9.2	16	mA
I_{CCZ}	Supply Current with Outputs Disabled	V_{CC} = Max (Note 3)			12	19	mA

Note 1: All typicals are at V_{CC} = 5V, T_A = 25°C.

Note 2: Not more than one output should be shorted at a time, and the duration should not exceed one second.

Note 3: I_{CC} is measured with all outputs open and all possible inputs grounded, while achieving the stated output conditions.

'LS257B SWITCHING CHARACTERISTICS

at V_{CC} = 5V and T_A = 25°C (See Section 1 for Test Waveforms and Output Load)

Symbol	Parameter	From (Input) To (Output)	R_L = 667Ω				Units
			C_L = 45 pF		C_L = 150 pF		
			Min	Max	Min	Max	
t_{PLH}	Propagation Delay Time Low to High Level Output	Data to Output		18		27	ns
t_{PHL}	Propagation Delay Time High to Low Level Output	Data to Output		18		27	ns
t_{PLH}	Propagation Delay Time Low to High Level Output	Select to Output		28		35	ns
t_{PHL}	Propagation Delay Time High to Low Level Output	Select to Output		35		42	ns
t_{PZH}	Output Enable Time to High Level Output	Output Control to Y		15		27	ns
t_{PZL}	Output Enable Time to Low Level Output	Output Control to Y		28		38	ns
t_{PHZ}	Output Disable Time from High Level Output (Note 1)	Output Control to Y		26			ns
t_{PLZ}	Output Disable Time from Low Level Output (Note 1)	Output Control to Y		25			ns

Note 1: C_L = 5 pF.

RECOMMENDED OPERATING CONDITIONS

Symbol	Parameter	DM54LS258B			DM74LS258B			Units
		Min	Nom	Max	Min	Nom	Max	
V_{CC}	Supply Voltage	4.5	5	5.5	4.75	5	5.25	V
V_{IH}	High Level Input Voltage	2			2			V
V_{IL}	Low Level Input Voltage			0.7			0.8	V
I_{OH}	High Level Output Current			−1			−2.6	mA
I_{OL}	Low Level Output Current			12			24	mA
T_A	Free Air Operating Temperature	−55		125	0		70	°C

'LS258B ELECTRICAL CHARACTERISTICS

over recommended operating free air temperature range (unless otherwise noted)

Symbol	Parameter	Conditions		Min	Typ (Note 1)	Max	Units
V_I	Input Clamp Voltage	V_{CC} = Min, I_I = −18 mA				−1.5	V
V_{OH}	High Level Output Voltage	V_{CC} = Min, I_{OH} = Max V_{IL} = Max, V_{IH} = Min	DM54	2.4	3.4		V
			DM74	2.4	3.1		
V_{OL}	Low Level Output Voltage	V_{CC} = Min, I_{OL} = Max V_{IL} = Max, V_{IH} = Min	DM54		0.25	0.4	V
			DM74		0.35	0.5	
		I_{OL} = 12 mA, V_{CC} = Min	DM74		0.25	0.4	
I_I	Input Current @ Max Input Voltage	V_{CC} = Max, V_I = 7V	Select			0.2	mA
			Other			0.1	
I_{IH}	High Level Input Current	V_{CC} = Max, V_I = 2.7V	Select			40	μA
			Other			20	

'LS258B ELECTRICAL CHARACTERISTICS

over recommended operating free air temperature range (unless otherwise noted) (Continued)

Symbol	Parameter	Conditions		Min	Typ (Note 1)	Max	Units
I_{IL}	Low Level Input Current	V_{CC} = Max, V_I = 0.4V	Select			−0.8	mA
			Other			−0.4	
I_{OZH}	Off-State Output Current with High Level Output Voltage Applied	V_{CC} = Max, V_O = 2.7V V_{IH} = Min, V_{IL} = Max				20	μA
I_{OZL}	Off-State Output Current with Low Level Output Voltage Applied	V_{CC} = Max, V_O = 0.4V V_{IH} = Min, V_{IL} = Max				−20	μA
I_{OS}	Short Circuit Output Current	V_{CC} = Max (Note 2)	DM54	−20		−100	mA
			DM74	−20		−100	
I_{CCH}	Supply Current with Outputs High	V_{CC} = Max (Note 3)			4.1	7	mA
I_{CCL}	Supply Current with Outputs Low	V_{CC} = Max (Note 3)			9	14	mA
I_{CCZ}	Supply Current with Outputs Disabled	V_{CC} = Max (Note 3)			12	19	mA

Note 1: All typicals are at V_{CC} = 5V, T_A = 25°C.

Note 2: Not more than one output should be shorted at a time, and the duration should not exceed one second.

Note 3: I_{CC} is measured with all outputs open and all possible inputs grounded, while achieving the stated output conditions.

'LS258B SWITCHING CHARACTERISTICS

at V_{CC} = 5V and T_A = 25°C (See Section 1 for Test Waveforms and Output Load)

Symbol	Parameter	From (Input) To (Output)	R_L = 667Ω				Units
			C_L = 45 pF		C_L = 150 pF		
			Min	Max	Min	Max	
t_{PLH}	Propagation Delay Time Low to High Level Output	Data to Output		18		27	ns
t_{PHL}	Propagation Delay Time High to Low Level Output	Data to Output		18		27	ns
t_{PLH}	Propagation Delay Time Low to High Level Output	Select to Output		28		35	ns
t_{PHL}	Propagation Delay Time High to Low Level Output	Select to Output		35		42	ns
t_{PZH}	Output Enable Time to High Level Output	Output Control to Y		15		27	ns
t_{PZL}	Output Enable Time to Low Level Output	Output Control to Y		28		38	ns
t_{PHZ}	Output Disable Time from High Level Output (Note 4)	Output Control to Y		26			ns
t_{PLZ}	Output Disable Time from Low Level Output (Note 4)	Output Control to Y		25			ns

Note 4: C_L = 5 pF.

CONNECTION DIAGRAMS

Dual-In-Line Package

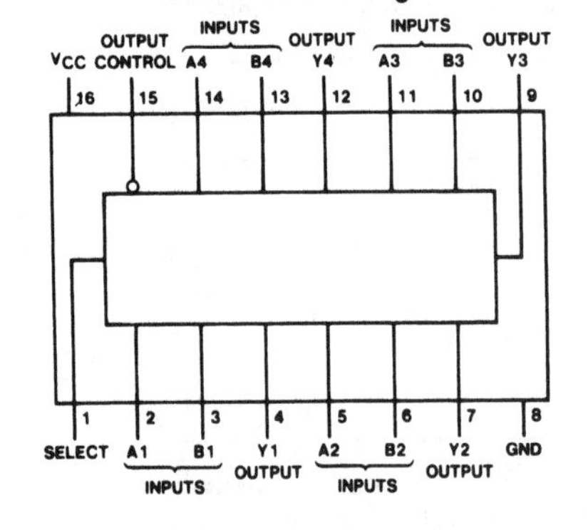

Order Number 54LS257ADMQB, 54LS257AFMQB, 54LS257ALMQB, DM54LS257BJ, DM54LS257BW, DM74LS257BM or DM74LS257BN
See NS Package Number E20A, J16A, M16A, N16E or W16A

Dual-In-Line Package

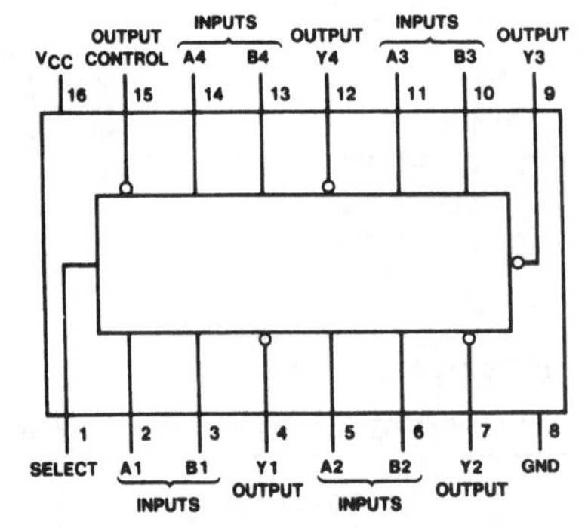

Order Number 54LS258ADMQB, 54LS258AFMQB, 54LS258ALMQB, DM54LS258BJ, DM54LS258BW, DM74LS258BM or DM74LS258BN
See NS Package Number E20A, J16A, M16A, N16E or W16A

FUNCTION TABLE

Inputs				Output Y	
Output Control	Select	A	B	LS257	LS258
H	X	X	X	Z	Z
L	L	L	X	L	H
L	L	H	X	H	L
L	H	X	L	L	H
L	H	X	H	H	L

H = High Level, L = Low Level, X = Don't Care,
Z = High Impedance (off)

LOGIC DIAGRAMS

LS257B

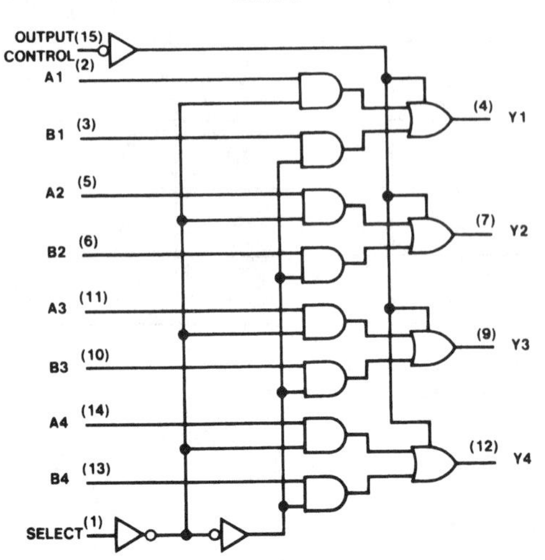

LS258B

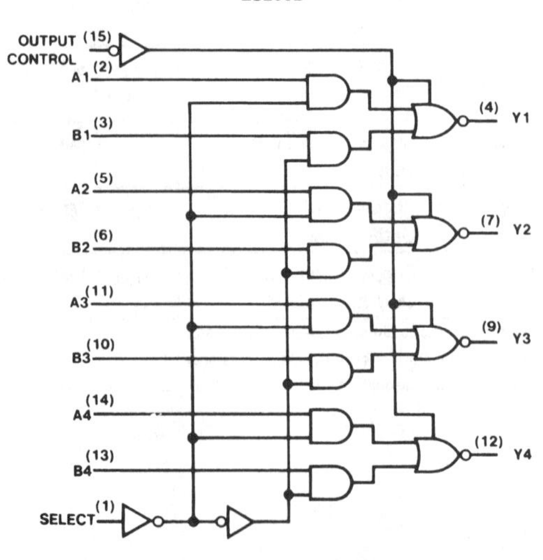

National 54LS322/DM74LS322

8-Bit Serial/Parallel Register with Sign Extend

ABSOLUTE MAXIMUM RATINGS (Note)

If Military/Aerospace specified devices are required, please contact the National Semiconductor Sales Office/Distributors for availability and specifications.

Supply Voltage	7V
Input Voltage	10V
Operating Free Air Temperature Range	
54LS	−55°C to +125°C
DM74LS	0°C to +70°C
Storage Temperature Range	−65°C to +150°C

Note: *The "Absolute Maximum Ratings" are those values beyond which the safety of the device cannot be guaranteed. The device should not be operated at these limits. The parametric values defined in the "Electrical Characteristics" table are not guaranteed at the absolute maximum ratings. The "Recommended Operating Conditions" table will define the conditions for actual device operation.*

RECOMMENDED OPERATING CONDITIONS

Symbol	Parameter	54LS322			DM74LS322			Units
		Min	Nom	Max	Min	Nom	Max	
V_{CC}	Supply Voltage	4.5	5	5.5	4.75	5	5.25	V
V_{IH}	High Level Input Voltage	2			2			V
V_{IL}	Low Level Input Voltage			0.7			0.8	V
I_{OH}	High Level Output Current			−0.4			−0.4	mA
I_{OL}	Low Level Output Current			4			8	mA
T_A	Free Air Operating Temperature	−55		125	0		70	°C
t_s (H) t_s (L)	Setup Time HIGH or LOW $\overline{RE}$ to CP	24 24			24 24			ns
t_h (H) t_h (L)	Hold Time HIGH or LOW $\overline{RE}$ to CP	5 5			0 0			ns
t_s (H) t_s (L)	Setup Time HIGH or LOW D0, D1 or I/O_n to CP	15 15			10 10			ns
t_h (H) t_h (L)	Hold Time HIGH or LOW D0, D1 or I/O_n to CP	5 5			0 0			ns
t_s (H) t_s (L)	Setup Time HIGH or LOW $\overline{SE}$ to CP	15 15			15 15			ns
t_h (H) t_h (L)	Hold Time HIGH or LOW $\overline{SE}$ to CP	0 0			0 0			ns
t_s (H) t_s (L)	Setup Time HIGH or LOW S$\overline{P}$ to CP	24 24			24 24			ns
t_s (H) t_s (L)	Setup Time HIGH or LOW S to CP	15 15			15 15			ns
t_h (H) t_h (L)	Hold Time HIGH or LOW S or S$\overline{P}$ to CP	0 0			0 0			ns
t_w (H)	CP Pulse Width HIGH	15			15			ns
t_w (L)	$\overline{MR}$ Pulse Width LOW	15			15			ns
t_{rec}	Recovery Time $\overline{MR}$ to CP	15			15			ns

ELECTRICAL CHARACTERISTICS

Over recommended operating free air temperature range (unless otherwise noted)

Symbol	Parameter	Conditions			Min	Typ (Note 1)	Max	Units
V_I	Input Clamp Voltage	V_{CC} = Min, I_I = −18 mA					−1.5	V
V_{OH}	High Level Output Voltage	V_{CC} = Min, I_{OH} = Max, V_{IL} = Max	54LS		2.5			V
			DM74		2.7	3.4		
V_{OL}	Low Level Output Voltage	V_{CC} = Min, I_{OL} = Max, V_{IH} = Min	54LS				0.4	V
			DM74			0.35	0.5	
		I_{OL} = 4 mA, V_{CC} = Min	DM74			0.25	0.4	
I_I	Input Current @ Max Input Voltage	V_{CC} = Max, V_I = 10V					0.1	mA
			S Input				0.2	
			SE Input				0.3	
I_{IH}	High Level Input Current	V_{CC} = Max, V_I = 2.7V					20	μA
			S Input				40	
			SE Input				60	
I_{IL}	Low Level Input Current	V_{CC} = Max, V_I = 0.4V					−0.4	mA
		V_{CC} = Max, V_I = 0.4V	S Input				−0.8	
		V_{CC} = Max, V_I = 0.4V	SE Input				−1.2	
I_{OS}	Short Circuit Output Current	V_{CC} = Max (Note 2)	54LS	I/On	−30		−130	mA
				Qn	−20		−100	
			DM74		−20		−100	
I_{CC}	Supply Current	V_{CC} = Max					60	mA
I_{OZH}	TRI-STATE Output Off Current HIGH	V_{CC} = V_{CCH}, V_{OZH} = 2.7V					40	μA
I_{OZL}	TRI-STATE Output Off Current LOW	V_{CC} = V_{CCH}, V_{OZL} = 0.4V					−0.4	mA

Note 1: All typicals are at V_{CC} = 5V, T_A = 25°C.

Note 2: Not more than one output should be shorted at a time, and the duration should not exceed one second.

SWITCHING CHARACTERISTICS

V_{CC} = +5.0V, T_A = +25°C (See Section 1 for waveforms and load configurations)

Symbol	Parameter	R_L = 2 kΩ, C_L = 15 pF				Units
		54LS		DM74LS		
		Min	Max	Min	Max	
f_{max}	Maximum Clock Frequency	35		35		MHz
t_{PLH} t_{PHL}	Propagation Delay CP to I/O_n**		25 35		25 34	ns
t_{PLH} t_{PHL}	Propagation Delay CP to Q0		26 28		26 29	ns
t_{PHL}	Propagation Delay $\overline{MR}$ to I/O_n**		35		34.1	ns
t_{PHL}	Propagation Delay $\overline{MR}$ to Q0		28		28	ns
t_{PZH} t_{PZL}	Output Enable Time $\overline{OE}$ to I/O_n**		18 25		21.5 23.9	ns

**C_L = 50 pF

SWITCHING CHARACTERISTICS

V_{CC} = +5.0V, T_A = +25°C (See Section 1 for waveforms and load configurations)

Symbol	Parameter	C_L = 15 pF				Units
		54LS		DM74LS		
		Min	Max	Min	Max	
t_{PHZ} t_{PLZ}	Output Disable Time $\overline{OE}$ to I/O_n*		15 20		15 15	ns
t_{PZH} t_{PZL}	Output Enable Time S/$\overline{P}$ to I/O_n**		22 30		25.2 25.8	ns
t_{PHZ} t_{PLZ}	Output Disable Time S$\overline{P}$ to I/O_n*		23 23		40.2 26.8	ns

*C_L = 5 pF

**C_L = 50 pF

Mode Table

Mode	Inputs							Outputs								
	$\overline{MR}$	$\overline{RE}$	S/$\overline{P}$	$\overline{SE}$	S	$\overline{OE}$*	CP	I/O7	I/O6	I/O5	I/O4	I/O3	I/O2	I/O1	I/O0	Q0
Clear	L	X	X	X	X	L	X	L	L	L	L	L	L	L	L	L
	L	X	X	X	X	H	X	Z	Z	Z	Z	Z	Z	Z	Z	L
Parallel Load	H	L	L	X	X	X	↗	I7	I6	I5	I4	I3	I2	I1	I0	I0
Shift Right	H	L	H	H	L	L	↗	D0	O7	O6	O5	O4	O3	O2	O1	O1
	H	L	H	H	H	L	↗	D1	O7	O6	O5	O4	O3	O2	O1	O1
Sign Extend	H	L	H	L	X	L	↗	O7	O7	O6	O5	O4	O3	O2	O1	O1
Hold	H	H	X	X	X	L	↗	NC	NC	NC	NC	NC	NC	NC	NC	NC

*When the $\overline{OE}$ input is HIGH, all I/O_n terminals are at the high-impedance state; sequential operation or clearing of the register is not affected.

Note 1: I7–I0 = The level of the steady-state input at the respective I/O terminal is loaded into the flip-flop while the flip-flop outputs (except Q0) are isolated from the I/O terminal.

Note 2: D0, D1 = The level of the steady-state inputs to the serial multiplexer input.

Note 3: O7–O0 = The level of the respective Q_n flip-flop prior to the last Clock LOW-to-HIGH transition.

NC = No Change Z = High-Impedance Output State H = HIGH Voltage Level L = LOW Voltage Level

CONNECTION DIAGRAM

Dual-In-Line Package

Pin	Name	Pin	Name
1	$\overline{RE}$	20	V_{CC}
2	S/$\overline{P}$	19	S
3	D0	18	$\overline{SE}$
4	I/O7	17	D1
5	I/O5	16	I/O6
6	I/O3	15	I/O4
7	I/O1	14	I/O2
8	$\overline{OE}$	13	I/O0
9	$\overline{MR}$	12	Q0
10	GND	11	CP

Order Number 54LS322DMQB, 54LS322FMQB, DM74LS322WM or DM74LS322N
See NS Package Number J20A, M20B, N20A or W20A

LOGIC SYMBOL

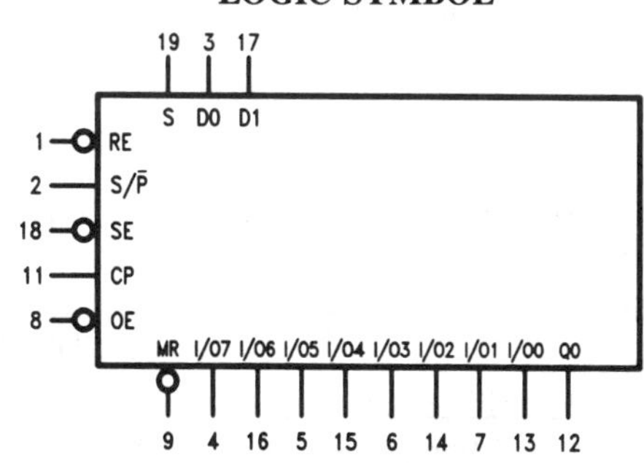

V_{CC} = Pin 20
GND = Pin 10

Pin Names	Description
$\overline{RE}$	Register Enable Input (Active LOW)
S/$\overline{P}$	Serial (HIGH) or Parallel (LOW) Mode Control Input
$\overline{SE}$	Sign Extend Input (Active LOW)
S	Serial Data Select Input
D0, D1	Serial Data Inputs
CP	Clock Pulse Input (Active Rising Edge)
$\overline{MR}$	Asynchronous Master Reset Input (Active LOW)
$\overline{OE}$	TRI-STATE Output Enable Input (Active LOW)
Q0	Bi-State Serial Output
I/O0–I/O7	Multiplexed Parallel Inputs or TRI-STATE Parallel Outputs

LOGIC DIAGRAM

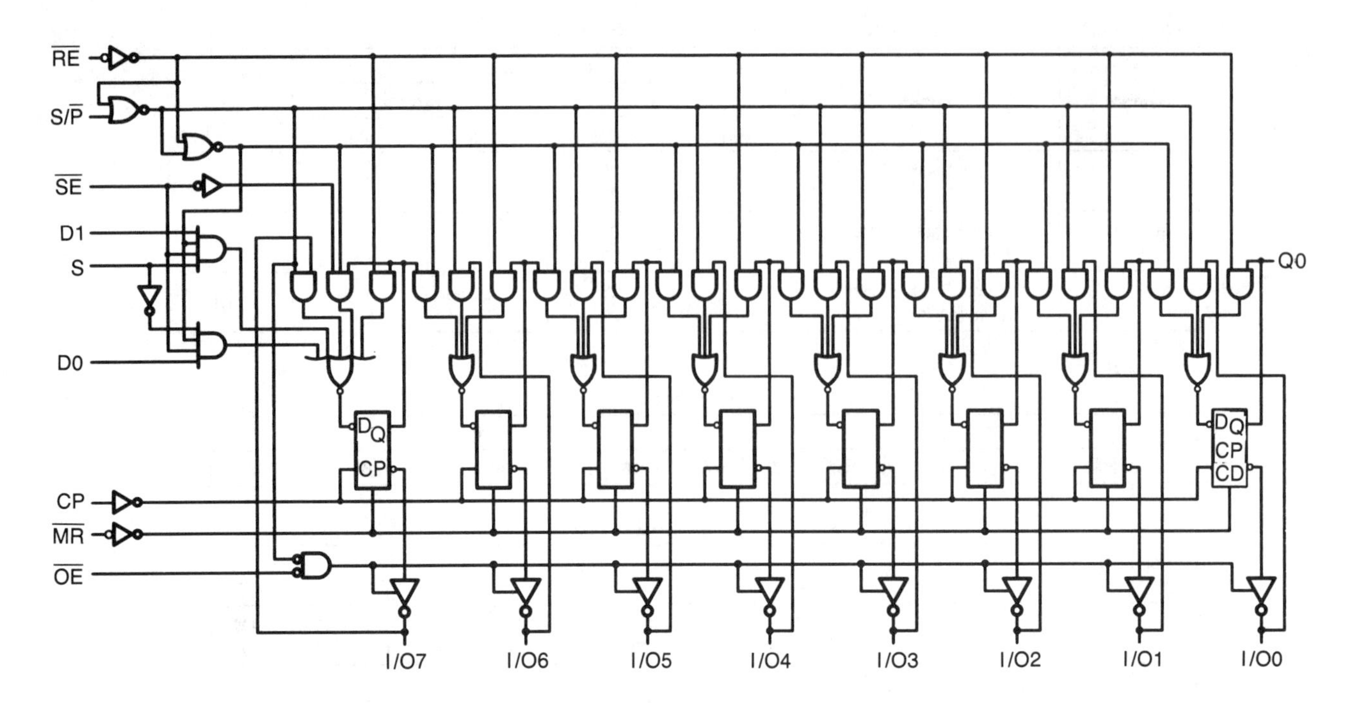

National
54LS323/DM74LS323
8-Bit Universal Shift/Storage Register with Synchronous Reset and Common I/O Pins

- Common I/O for reduced pin count
- Four operation modes: shift left, shift right, parallel load, and store
- Separate continuous inputs and outputs from Q0 and Q7 allow easy cascading
- Fully synchronous reset
- Tri-state outputs for bus-oriented applications

RECOMMENDED OPERATING CONDITIONS

Symbol	Parameter	54LS323			DM74LS323			Units
		Min	Nom	Max	Min	Nom	Max	
V_{CC}	Supply Voltage	4.5	5	5.5	4.75	5	5.25	V
V_{IH}	High Level Input Voltage	2			2			V
V_{IL}	Low Level Input Voltage			0.7			0.8	V
I_{OH}	High Level Output Current			−0.4			−0.4	mA
I_{OL}	Low Level Output Current			4			8	mA
T_A	Free Air Operating Temperature	−55		125	0		70	°C
t_s (H) t_s (L)	Setup Time HIGH or LOW S0 or S1 to CP	24 24			24 24			ns
t_h (H) t_h (L)	Hold Time HIGH or LOW S0 or S1 to CP	5 5			0 0			ns
t_s (H) t_s (L)	Setup Time HIGH or LOW I/O_n, D_S0, D_S7 to CP	15 15			10 10			ns
t_h (H) t_h (L)	Hold Time HIGH or LOW I/O_n, D_S0, D_S7 to CP	5 5			0 0			ns
t_s (H) t_s (L)	Setup Time HIGH or LOW $\overline{SR}$ to CP	30 20			15 15			ns
t_h (H) t_h (L)	Hold Time HIGH or LOW $\overline{SR}$ to CP	0 0			0 0			ns
t_w (H) t_w (L)	CP Pulse Width HIGH or LOW	15 15			15 15			ns

ABSOLUTE MAXIMUM RATINGS (Note)

If Military/Aerospace specified devices are required, please contact the National Semiconductor Sales Office/Distributors for availability and specifications.

Supply Voltage	7V
Input Voltage	10V
Operating Free Air Temperature Range	
54LS	−55°C to +125°C
DM74LS	0°C to +70°C
Storage Temperature Range	−65°C to +150°C

Note: *The "Absolute Maximum Ratings" are those values beyond which the safety of the device cannot be guaranteed. The device should not be operated at these limits. The parametric values defined in the "Electrical Characteristics" table are not guaranteed at the absolute maximum ratings. The "Recommended Operating Conditions" table will define the conditions for actual device operation.*

ELECTRICAL CHARACTERISTICS

Over recommended operating free air temperature range (unless otherwise noted)

Symbol	Parameter	Conditions		Min	Typ (Note 1)	Max	Units
V_I	Input Clamp Voltage	V_{CC} = Min, I_I = −18 mA				−1.5	V
V_{OH}	High Level Output Voltage	V_{CC} = Min, I_{OH} = Max, V_{IL} = Max	54LS	2.5			V
			DM74	2.7	3.4		
V_{OL}	Low Level Output Voltage	V_{CC} = Min, I_{OL} = Max, V_{IH} = Min	54LS			0.4	V
			DM74		0.35	0.5	
		I_{OL} = 4 mA, V_{CC} = Min	DM74		0.25	0.4	
I_I	Input Current @ Max Input Voltage	V_{CC} = Max, V_I = 10V				0.1	mA
			S_n Inputs			0.2	mA
I_{IH}	High Level Input Current	V_{CC} = Max, V_I = 2.7V				20	μA
			S_n Inputs			40	μA
I_{IL}	Low Level Input Current	V_{CC} = Max, V_I = 0.4V				−0.4	mA
			S_n Inputs			−0.8	mA
I_{OS}	Short Circuit Output Current	V_{CC} = Max (Note 2)	54LS	−20		−100	mA
			DM74	−20		−100	
I_{CC}	Supply Current	V_{CC} = Max				60	mA
I_{OZH}	TRI-STATE Output Off Current HIGH	V_{CC} = V_{CCH}, V_{OZH} = 2.7V				40	μA
I_{OZL}	TRI-STATE Output Off Current LOW	V_{CC} = V_{CCH}, V_{OZL} = 0.4V				−400	μA

Note 1: All typicals are at V_{CC} = 5V, T_A = 25°C.

Note 2: Not more than one output should be shorted at a time, and the duration should not exceed one second.

SWITCHING CHARACTERISTICS

V_{CC} = +5.0V, T_A = +25°C (See Section 1 for waveforms and load configurations)

Symbol	Parameter	54LS323 C_L = 15 pF		DM74LS323 R_L = 2 kΩ, C_L = 15 pF		Units
		Min	Max	Min	Max	
f_{max}	Maximum Input Frequency	35		35		MHz
t_{PLH}	Propagation Delay CP to Q0 or Q7		26		23	ns
t_{PHL}			28		25	
t_{PLH}	Propagation Delay CP to I/O_n		25		25	ns
t_{PHL}			35		29	
t_{PZH}	Output Enable Time C_L = 50 pF		18		18	ns
t_{PZL}			25		23	
t_{PHZ}	Output Disable Time C_L = 5 pF		15		15	ns
t_{PLZ}			20		15	

LOGIC DIAGRAM

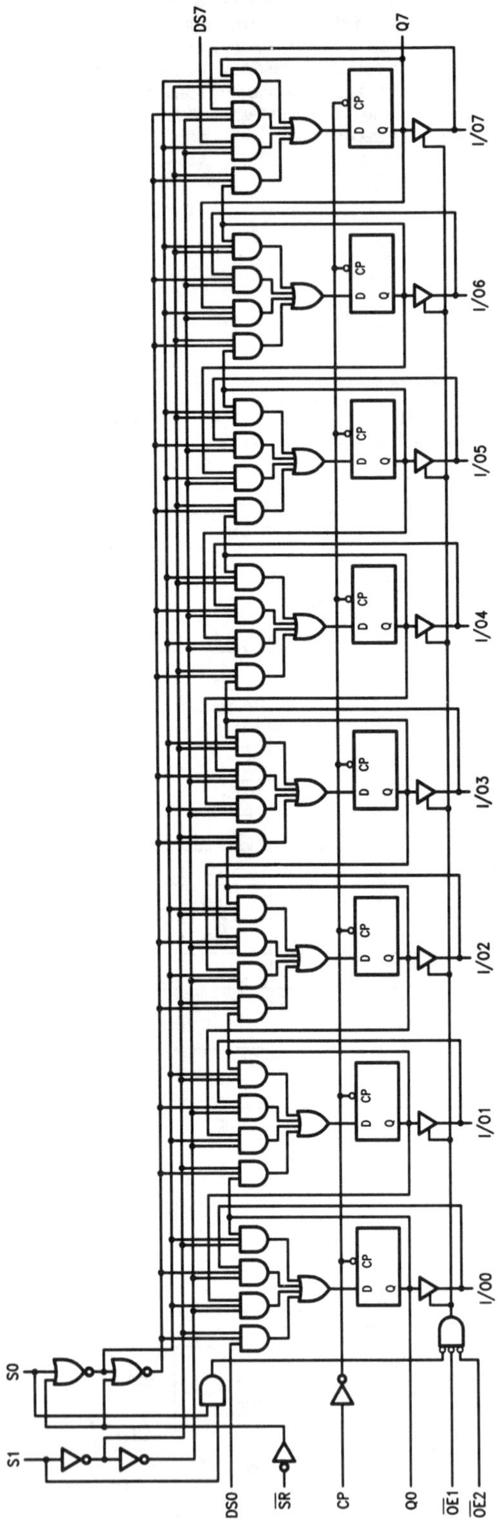

CONNECTION DIAGRAM

Dual-In-Line Package

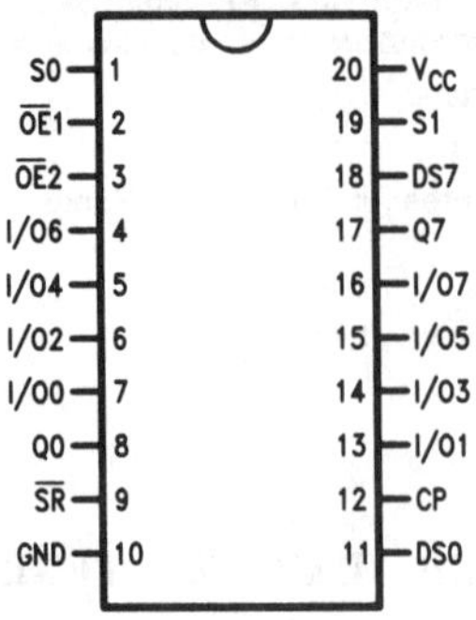

Pin	Name	Pin	Name
1	S0	20	V_{CC}
2	$\overline{OE}1$	19	S1
3	$\overline{OE}2$	18	DS7
4	I/O6	17	Q7
5	I/O4	16	I/O7
6	I/O2	15	I/O5
7	I/O0	14	I/O3
8	Q0	13	I/O1
9	$\overline{SR}$	12	CP
10	GND	11	DS0

Order Number 54LS323DMQB, 54LS323FMQB, DM74LS323WM or DM74LS323N
See NS Package Number J20A, M20B, N20A or W20A

Pin Names	Description
CP	Clock Pulse Input (Active Rising Edge)
D_S0	Serial Data Input for Right Shift
D_S7	Serial Data Input for Left Shift
S0, S1	Mode Select Inputs
$\overline{SR}$	Synchronous Reset Input (Active LOW)
$\overline{OE}1$, $\overline{OE}2$	TRI-STATE Output Enable Inputs (Active LOW)
I/O0–I/O7	Parallel Data Inputs or TRI-STATE Parallel Outputs
Q0, Q7	Serial Outputs

National 54LS347/DM74LS347 BCD to 7-Segment Decoder/Driver

ABSOLUTE MAXIMUM RATINGS (Note)

If Military/Aerospace specified devices are required, please contact the National Semiconductor Sales Office/Distributors for availability and specifications.

Supply Voltage	7V
Input Voltage	7V
Operating Free Air Temperature Range	
54LS	−55°C to +125°C
DM74LS	0°C to +70°C
Storage Temperature Range	−65°C to +150°C

Note: *The "Absolute Maximum Ratings" are those values beyond which the safety of the device cannot be guaranteed. The device should not be operated at these limits. The parametric values defined in the "Electrical Characteristics" table are not guaranteed at the absolute maximum ratings. The "Recommended Operating Conditions" table will define the conditions for actual device operation.*

RECOMMENDED OPERATING CONDITIONS

Symbol	Parameter	54LS347			DM74LS347			Units
		Min	Nom	Max	Min	Nom	Max	
V_{CC}	Supply Voltage	4.5	5	5.5	4.75	5	5.25	V
V_{IH}	High Level Input Voltage	2			2			V
V_{IL}	Low Level Input Voltage			0.7			0.8	V
I_{OH}	High Level Output Voltage			−50			−50	μA
I_{OL}	Low Level Output Current			12			24	mA
T_A	Free Air Operating Temperature	−55		125	0		70	°C

ELECTRICAL CHARACTERISTICS over recommended operating free air temperature range (unless otherwise noted)

Symbol	Parameter	Conditions		Min	Typ (Note 1)	Max	Units
V_I	Input Clamp Voltage	V_{CC} = Min, I_I = −18 mA				−1.5	V
V_{OH}	High Level Output Voltage	V_{CC} = Min, V_{OH} = Max, V_{IL} = Max	54LS	2.5			V
			DM74	2.7			
V_{OL}	Low Level Output Voltage	V_{CC} = Min, I_{OL} = Max, V_{IH} = Min	54LS			0.4	V
			DM74			0.5	
		I_{OL} = 4 mA, V_{CC} = Min	DM74			0.4	
I_I	Input Current @ Max Input Voltage	V_{CC} = Max, V_I = 10V				0.1	mA
I_{IH}	High Level Input Current	V_{CC} = Max, V_I = 2.7V				20	μA
I_{IL}	Low Level Input Current	V_{CC} = Max, V_I = 0.4V		−0.03		−0.4	mA
		BI/$\overline{RBO}$ Input		−0.09		−1.2	mA
I_{OS}	Short Circuit Output Current	V_{CC} = Max (Note 2)	54LS	−0.3		−2.0	mA
			DM74	−0.3		−2.0	
I_{CC}	Supply Current	V_{CC} = Max				13	mA
I_{OFF}		Segment Outputs, V_O = 7V				250	μA

SWITCHING CHARACTERISTICS

at V_{CC} = 5V and T_A = 25°C (See Section 1 for Test Waveforms and Output Loading)

Symbol	Parameter	C_L = 15 pF		Units
		Min	Max	
t_{PLH}	Propagation Delay A_n to $\overline{a}$–$\overline{g}$		100	ns
t_{PHL}			100	ns
t_{PLH}	Propagation Delay $\overline{RBI}$ to $\overline{a}$–$\overline{g}$		100	ns
t_{PHL}			100	ns

Note 1: All typicals are at V_{CC} = 5V, T_A = 25°C.

Note 2: Not more than one output should be shorted at a time, and the duration should not exceed one second.

CONNECTION DIAGRAM

Dual-In-Line Package

Pin	Name	Pin	Name
1	A0	16	V_{CC}
2	A1	15	$\overline{f}$
3	$\overline{LT}$	14	$\overline{g}$
4	$\overline{BI}/\overline{RBO}$	13	$\overline{a}$
5	$\overline{RBI}$	12	$\overline{b}$
6	A2	11	$\overline{c}$
7	A3	10	$\overline{d}$
8	GND	9	$\overline{e}$

Order Number 54LS347DMQB, 54LS347FMQB, DM74LS347M or DM74LS347N
See NS Package Number J16A, M16A, N16E or W16A

LOGIC SYMBOL

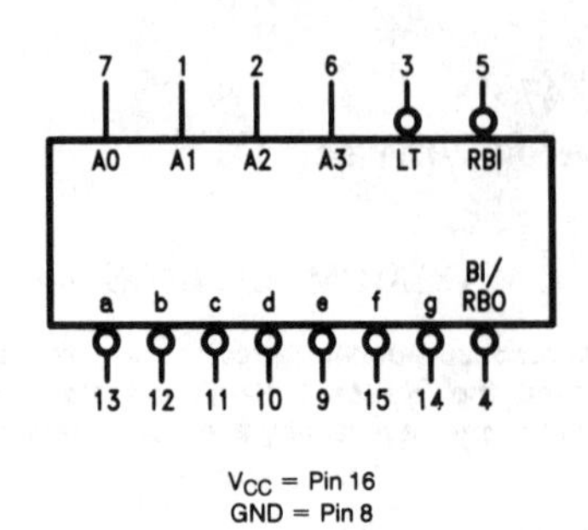

Pin Names	Description
A0–A3	BCD Inputs
$\overline{RBI}$	Ripple Blanking Input (Active LOW)
$\overline{LT}$	Lamp Test Input (Active LOW)
$\overline{BI}/\overline{RBO}$	Blanking Input (Active LOW) or Ripple Blanking Output (Active LOW)
$\overline{a}$–$\overline{g}$	*Segment Outputs (Active LOW)

*OC—Open Collector

National 54LS352/DM74LS352

Dual 4-Line to 1-Line Data Selectors/Multiplexers

- Inverting version of DM54/74LS153
- Permits multiplexing from N lines to 1 line
- Performs parallel-to-serial conversion
- Strobe (enable) line provided for cascading (N lines to n lines)
- High fan-out, low-impedance, totem-pole outputs
- Typical average propagation delay times
  - From data: 15ns
  - From strobe: 19ns
  - From select: 22ns
- Typical power dissipation: 31mW

ABSOLUTE MAXIMUM RATINGS (Note)

If Military/Aerospace specified devices are required, please contact the National Semiconductor Sales Office/Distributors for availability and specifications.

Supply Voltage	7V
Input Voltage	7V
Operating Free Air Temperature Range	
54LS	−55° C to +125°C
DM74LS	0° C to +70°C
Storage Temperature Range	−65° C to +150° C

Note: *The "Absolute Maximum Ratings" are those values beyond which the safety of the device cannot be guaranteed. The device should not be operated at these limits. The parametric values defined in the "Electrical Characteristics" table are not guaranteed at the absolute maximum ratings. The "Recommended Operating Conditions" table will define the conditions for actual device operation.*

RECOMMENDED OPERATING CONDITIONS

Symbol	Parameter	54LS352			DM74LS352			Units
		Min	Nom	Max	Min	Nom	Max	
V_{CC}	Supply Voltage	4.5	5	5.5	4.75	5	5.25	V
V_{IH}	High Level Input Voltage	2			2			V
V_{IL}	Low Level Input Voltage			0.7			0.8	V
I_{OH}	High Level Output Current			−0.4			−0.4	mA
I_{OL}	Low Level Output Current			12			8	mA
T_A	Free Air Operating Temperature	−55		125	0		70	°C

ELECTRICAL CHARACTERISTICS over recommended operating free air temperature range (unless otherwise noted)

Symbol	Parameter	Conditions		Min	Typ (Note 1)	Max	Units
V_I	Input Clamp Voltage	V_{CC} = Min, I_I = −18 mA				−1.5	V
V_{OH}	High Level Output Voltage	V_{CC} = Min, I_{OH} = Max, V_{IL} = Max, V_{IH} = Min	54LS	2.5			V
			DM74	2.7	3.4		
V_{OL}	Low Level Output Voltage	V_{CC} = Min, I_{OL} = Max, V_{IL} = Max, V_{IH} = Min	54LS			0.4	V
			DM74		0.35	0.5	
		I_{OL} = 4 mA, V_{CC} = Min	DM74		0.25	0.4	
I_I	Input Current @ Max Input Voltage	V_{CC} = Max, V_I = 10V	54LS			0.1	mA
		V_{CC} = Max, V_I = 7V	DM74				
I_{IH}	High Level Input Current	V_{CC} = Max, V_I = 2.7V				20	μA
I_{IL}	Low Level Input Current	V_{CC} = Max, V_I = 0.4V				−0.4	mA
I_{OS}	Short Circuit Output Current	V_{CC} = Max (Note 2)	54LS	−20		−100	mA
			DM74	−20		−100	
I_{CC}	Supply Current	V_{CC} = Max (Note 3)			6.2	10	mA

Note 1: All typicals are at V_{CC} = 5V, T_A = 25°C.

Note 2: Not more than one output should be shorted at a time, and the duration should not exceed one second.

Note 3: I_{CC} is measured with all outputs open and all other inputs at ground.

SWITCHING CHARACTERISTICS at V_{CC} = 5V and T_A = 25°C (See Section 1 for Test Waveforms and Output Load)

Symbol	Parameter	From (Input) To (Output)	54LS C_L = 15 pF		DM74LS C_L = 50 pF, R_L = 2 kΩ		Units
			Min	Max	Min	Max	
t_{PLH}	Propagation Delay Time Low to High Level Output	Data to Y		12		24	ns
t_{PHL}	Propagation Delay Time High to Low Level Output	Data to Y		12		35	ns
t_{PLH}	Propagation Delay Time Low to High Level Output	Select to Y		22		33	ns
t_{PHL}	Propagation Delay Time High to Low Level Output	Select to Y		38		47	ns
t_{PLH}	Propagation Delay Time Low to High Level Output	Strobe to Y		15		29	ns
t_{PHL}	Propagation Delay Time High to Low Level Output	Strobe to Y		20		41	ns

CONNECTION DIAGRAM

Dual-In-Line Package

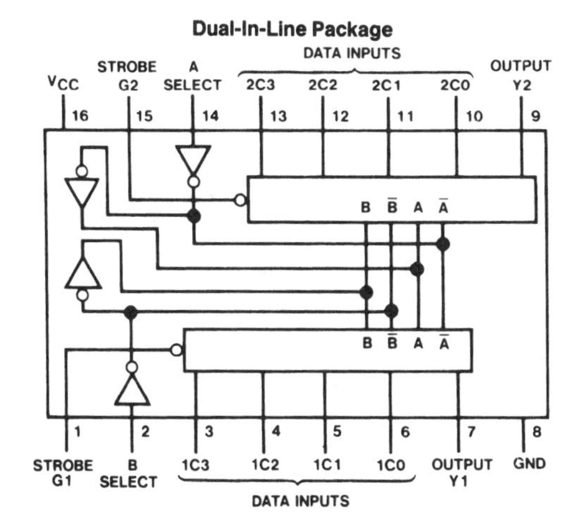

Order Number 54LS352DMQB, 54LS352FMQB, DM74LS352M or DM74LS352N
See NS Package Number J16A, M16A, N16E or W16A

FUNCTION TABLE

Select Inputs		Data Inputs				Strobe	Output
B	A	C0	C1	C2	C3	G	Y
X	X	X	X	X	X	H	H
L	L	L	X	X	X	L	H
L	L	H	X	X	X	L	L
L	H	X	L	X	X	L	H
L	H	X	H	X	X	L	L
H	L	X	X	L	X	L	H
H	L	X	X	H	X	L	L
H	H	X	X	X	L	L	H
H	H	X	X	X	H	L	L

Select inputs A and B are common to both sections.

H = High Level, L = Low Level, X = Don't Care

LOGIC DIAGRAM

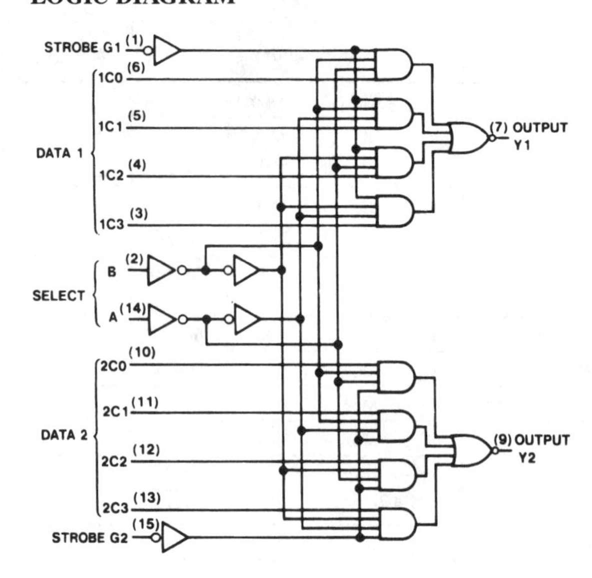

National 54LS353/DM74LS353

Dual 4-Input Multiplexer with Tri-state® Outputs

- Inverted version of 'LS253
- Schottky process for high speed
- Multifunction capability

ABSOLUTE MAXIMUM RATINGS (Note)

If Military/Aerospace specified devices are required, please contact the National Semiconductor Sales Office/Distributors for availability and specifications.

Supply Voltage	7V
Input Voltage	7V
Operating Free Air Temperature Range	
54LS	−55°C to +125°C
DM74LS	0°C to +70°C
Storage Temperature Range	−65°C to +150°C

Note: *The "Absolute Maximum Ratings" are those values beyond which the safety of the device cannot be guaranteed. The device should not be operated at these limits. The parametric values defined in the "Electrical Characteristics" table are not guaranteed at the absolute maximum ratings. The "Recommended Operating Conditions" table will define the conditions for actual device operation.*

RECOMMENDED OPERATING CONDITIONS

Symbol	Parameter	54LS353			DM74LS353			Units
		Min	Nom	Max	Min	Nom	Max	
V_{CC}	Supply Voltage	4.5	5	5.5	4.75	5	5.25	V
V_{IH}	High Level Input Voltage	2			2			V
V_{IL}	Low Level Input Voltage			0.7			0.8	V
I_{OH}	High Level Output Current			−1.0			−2.6	mA
I_{OL}	Low Level Output Current			12			24	mA
T_A	Free Air Operating Temperature	−55		125	0		70	°C

ELECTRICAL CHARACTERISTICS over recommended operating free air temperature range (unless otherwise noted)

Symbol	Parameter	Conditions		Min	Typ (Note 1)	Max	Units
V_I	Input Clamp Voltage	V_{CC} = Min, I_I = −18 mA				−1.5	V
V_{OH}	High Level Output Voltage	V_{CC} = Min, I_{OH} = Max, V_{IL} = Max	54LS	2.5			V
			DM74	2.7			
V_{OL}	Low Level Output Voltage	V_{CC} = Min, I_{OL} = Max, V_{IH} = Min	54LS			0.4	V
			DM74			0.5	
		I_{OL} = 4 mA, V_{CC} = Min	DM74			0.4	
I_I	Input Current @ Max Input Voltage	V_{CC} = Max, V_I = 10V				0.1	mA
I_{IH}	High Level Input Current	V_{CC} = Max, V_I = 2.7V				20	μA
I_{IL}	Low Level Input Current	V_{CC} = Max, V_I = 0.4V				−0.4	mA
I_{OS}	Short Circuit Output Current	V_{CC} = Max (Note 2)	54LS	−30		−130	mA
			DM74	−30		−130	
I_{CCL}	Supply Current Outputs HIGH	V_{CC} = Max, In, Sn, $\overline{OE}$n = GND				12	mA
I_{CCZ}	Supply Current Outputs OFF	V_{CC} = Max, $\overline{OE}$n = 4.5V In, Sn = GND				14	mA
I_{OZH}	TRI-STATE Output OFF Current HIGH	V_{CC} = V_{CCH} V_{OZH} = 2.7V				20	μA
I_{OZL}	TRI-STATE Output OFF Current LOW	V_{CC} = V_{CCH} V_{OZL} = 0.4V				−20	μA

Note 1: All typicals are at V_{CC} = 5V, T_A = 25°C.

Note 2: Not more than one output should be shorted at a time, and the duration should not exceed one second.

SWITCHING CHARACTERISTICS V_{CC} = +5.0V, T_A = +25°C (See Section 1 for test waveforms and output loads)

Symbol	Parameter	R_L = 2 kΩ, C_L = 50 pF		Units
		Min	Max	
t_{PLH} t_{PHL}	Propagation Delay Sn to $\overline{Z}$n		24 32	ns
t_{PLH} t_{PHL}	Propagation Delay In to $\overline{Z}$n		15 15	ns
t_{PZH} t_{PZL}	Output Enable Time $\overline{OE}$ to Zn		18 18	ns
t_{PHZ} t_{PLZ}	Output Disable Time $\overline{OE}$ to Zn		18 18	ns

TRUTH TABLE

Select Inputs		Data Inputs				Output Enable	Output
S0	S1	I0	I1	I2	I3	$\overline{OE}$	$\overline{Z}$
X	X	X	X	X	X	H	(Z)
L	L	L	X	X	X	L	H
L	L	H	X	X	X	L	L
H	L	X	L	X	X	L	H
H	L	X	H	X	X	L	L
L	H	X	X	L	X	L	H
L	H	X	X	H	X	L	L
H	H	X	X	X	L	L	H
H	H	X	X	X	H	L	L

Address inputs S0 and S1 are common to both sections.

H = HIGH Voltage Level
L = LOW Voltage Level
X = Immaterial
(Z) = High Impedance

CONNECTION DIAGRAM

Dual-In-Line Package

Pin	Name	Pin	Name
1	$\overline{OE}_a$	16	V_{CC}
2	S1	15	$\overline{OE}_b$
3	I_{3a}	14	S0
4	I_{2a}	13	I_{3b}
5	I_{1a}	12	I_{2b}
6	I_{0a}	11	I_{1b}
7	$\overline{Z}_a$	10	I_{0b}
8	GND	9	$\overline{Z}_b$

Order Number 54LS353DMQB, 54LS353FMQB, DM74LS353M or DM74LS353N
See NS Package Number J16A, M16A, N16E or W16A

LOGIC SYMBOL

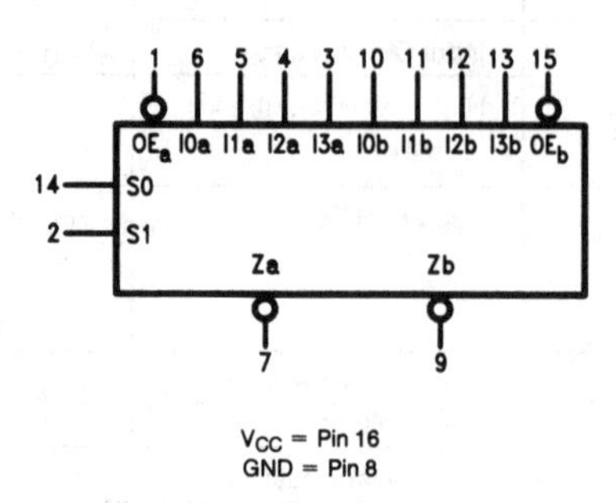

V_{CC} = Pin 16
GND = Pin 8

Pin Names	Description
I0a–I3a	Side A Data Inputs
I0b–I3b	Side B Data Inputs
S0, S1	Common Select Inputs
$\overline{OE}_a$	Side A Output Enable Input (Active Low)
$\overline{OE}_b$	Side B Output Enable Input (Active Low)
$\overline{Z}_a$, $\overline{Z}_b$	TRI-STATE Outputs (Inverted)

LOGIC DIAGRAM

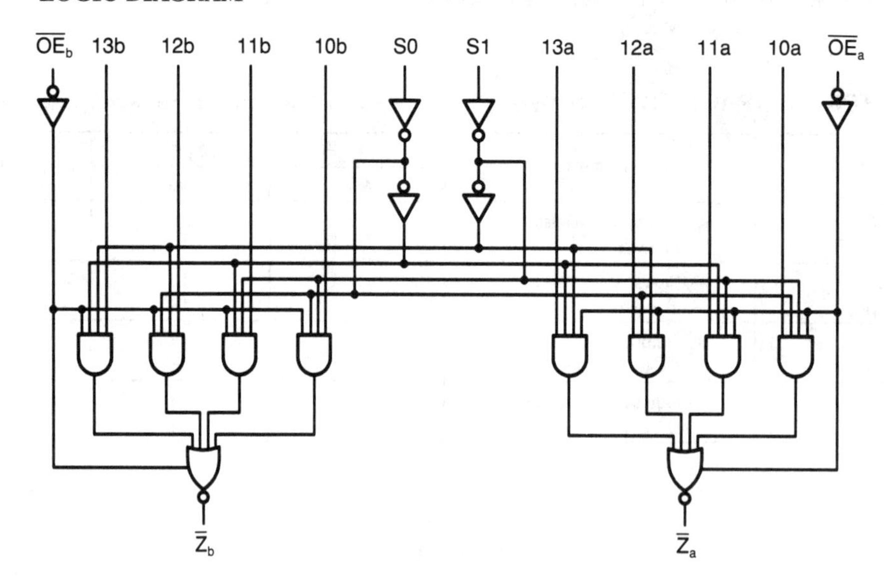

National DM54S138/DM74S138, DM54S139/DM74S139 Decoders/Demultiplexers

- Designed specifically for high speed:
  memory decoders
  data transmission systems
- S138 3-to-8-line decoders incorporates 3 enable inputs to simplify cascading and/or data reception
- S139 contains two fully independent 2-to-4-line decoders/demultiplexers
- Schottky clamped for high performance
- Typical propagation delay time (3 levels of logic)
  - S138 8ns
  - S139 7.5ns
- Typical power dissipation
  - S138 245mW
  - S139 300mW

ABSOLUTE MAXIMUM RATINGS (Note)

If Military/Aerospace specified devices are required, please contact the National Semiconductor Sales Office/Distributors for availability and specifications.

Supply Voltage	7V
Input Voltage	5.5V
Operating Free Air Temperature Range	
DM54S	−55° C to +125°C
DM74S	0° C to +70°C
Storage Temperature Range	−65° C to +150° C

Note: *The "Absolute Maximum Ratings" are those values beyond which the safety of the device cannot be guaranteed. The device should not be operated at these limits. The parametric values defined in the "Electrical Characteristics" table are not guaranteed at the absolute maximum ratings. The "Recommended Operating Conditions" table will define the conditions for actual device operation.*

RECOMMENDED OPERATING CONDITIONS

Symbol	Parameter	DM54S138,S139			DM74S138,S139			Units
		Min	Nom	Max	Min	Nom	Max	
V_{CC}	Supply Voltage	4.5	5	5.5	4.75	5	5.25	V
V_{IH}	High Level Input Voltage	2			2			V
V_{IL}	Low Level Input Voltage			0.8			0.8	V
I_{OH}	High Level Output Current			−1			−1	mA
I_{OL}	Low Level Output Current			20			20	mA
T_A	Free Air Operating Temperature	−55		125	0		70	°C

ELECTRICAL CHARACTERISTICS over recommended operating free air temperature (unless otherwise noted)

Symbol	Parameter	Conditions		Min	Typ (Note 1)	Max	Units
V_I	Input Clamp Voltage	V_{CC} = Min, I_I = −18 mA				−1.2	V
V_{OH}	High Level Output Voltage	V_{CC} = Min, I_{OH} = Max V_{IL} = Max, V_{IH} = Min	DM54	2.5	3.4		V
			DM74	2.7	3.4		
V_{OL}	Low Level Output Voltage	V_{CC} = Min, I_{OL} = Max V_{IH} = Min, V_{IL} = Max				0.5	V
I_I	Input Current @ Max Input Voltage	V_{CC} = Max, V_I = 5.5V				1	mA
I_{IH}	High Level Input Current	V_{CC} = Max, V_I = 2.7V				50	μA
I_{IL}	Low Level Input Current	V_{CC} = Max, V_I = 0.5V				−2	mA
I_{OS}	Short Circuit Output Current	V_{CC} = Max (Note 2)	DM54	−40		−100	mA
			DM74	−40		−100	
I_{CC}	Supply Current (S138)	V_{CC} = Max (Note 3)			49	74	mA
I_{CC}	Supply Current (S139)	V_{CC} = Max (Note 3)			60	90	mA

Note 1: All typicals are at V_{CC} = 5V, T_A = 25°C.

Note 2: Not more than one output should be shorted at a time, and the duration should not exceed one second.

Note 3: I_{CC} is measured with all outputs enabled and open.

FUNCTION TABLES

S138

Inputs					Outputs							
Enable		Select										
G1	G2*	C	B	A	Y0	Y1	Y2	Y3	Y4	Y5	Y6	Y7
X	H	X	X	X	H	H	H	H	H	H	H	H
L	X	X	X	X	H	H	H	H	H	H	H	H
H	L	L	L	L	L	H	H	H	H	H	H	H
H	L	L	L	H	H	L	H	H	H	H	H	H
H	L	L	H	L	H	H	L	H	H	H	H	H
H	L	L	H	H	H	H	H	L	H	H	H	H
H	L	H	L	L	H	H	H	H	L	H	H	H
H	L	H	L	H	H	H	H	H	H	L	H	H
H	L	H	H	L	H	H	H	H	H	H	L	H
H	L	H	H	H	H	H	H	H	H	H	H	L

* G2 = G2A + G2B

H = high level, L = low level, X = don't care (either low or high logic level)

S139

Inputs			Outputs			
Enable	Select					
G	B	A	Y0	Y1	Y2	Y3
H	X	X	H	H	H	H
L	L	L	L	H	H	H
L	L	H	H	L	H	H
L	H	L	H	H	L	H
L	H	H	H	H	H	L

H = high level, L = low level, X = don't care (either low or high logic level)

'S138 SWITCHING CHARACTERISTICS

at V_{CC} = 5V and T_A = 25°C (See Section 1 for Test Waveforms and Output Load)

Symbol	Parameter	From (Input) to (Output)	Levels of Delay	$R_L = 280\Omega$, C_L = 15 pF Min	$R_L = 280\Omega$, C_L = 15 pF Max	$R_L = 280\Omega$, C_L = 50 pF Min	$R_L = 280\Omega$, C_L = 50 pF Max	Units
t_{PLH}	Propagation Delay Time Low to High Level Output	Select to Output	2		7		9	ns
t_{PHL}	Propagation Delay Time High to Low Level Output	Select to Output	2		10.5		14	ns
t_{PLH}	Propagation Delay Time Low to High Level Output	Select to Output	3		12		14	ns
t_{PHL}	Propagation Delay Time High to Low Level Output	Select to Output	3		12		15	ns
t_{PLH}	Propagation Delay Time Low to High Level Output	Enable to Output	2		8		10	ns
t_{PHL}	Propagation Delay Time High to Low Level Output	Enable to Output	2		11		14	ns
t_{PLH}	Propagation Delay Time Low to High Level Output	Enable to Output	3		11		13	ns
t_{PHL}	Propagation Delay Time High to Low Level Output	Enable to Output	3		11		14	ns

'S139 SWITCHING CHARACTERISTICS

at V_{CC} = 5V and T_A = 25°C (See Section 1 for Test Waveforms and Output Load)

Symbol	Parameter	From (Input) to (Output)	Levels of Delay	$R_L = 280\Omega$, C_L = 15 pF Min	$R_L = 280\Omega$, C_L = 15 pF Max	$R_L = 280\Omega$, C_L = 50 pF Min	$R_L = 280\Omega$, C_L = 50 pF Max	Units
t_{PLH}	Propagation Delay Time Low to High Level Output	Select to Output	2		7.5		10	ns
t_{PHL}	Propagation Delay Time High to Low Level Output	Select to Output	2		10		13	ns
t_{PLH}	Propagation Delay Time Low to High Level Output	Select to Output	3		12		13	ns
t_{PHL}	Propagation Delay Time High to Low Level Output	Select to Output	3		12		15	ns
t_{PLH}	Propagation Delay Time Low to High Level Output	Enable to Output	2		8		10	ns
t_{PHL}	Propagation Delay Time High to Low Level Output	Enable to Output	2		10		13	ns

LOGIC DIAGRAMS

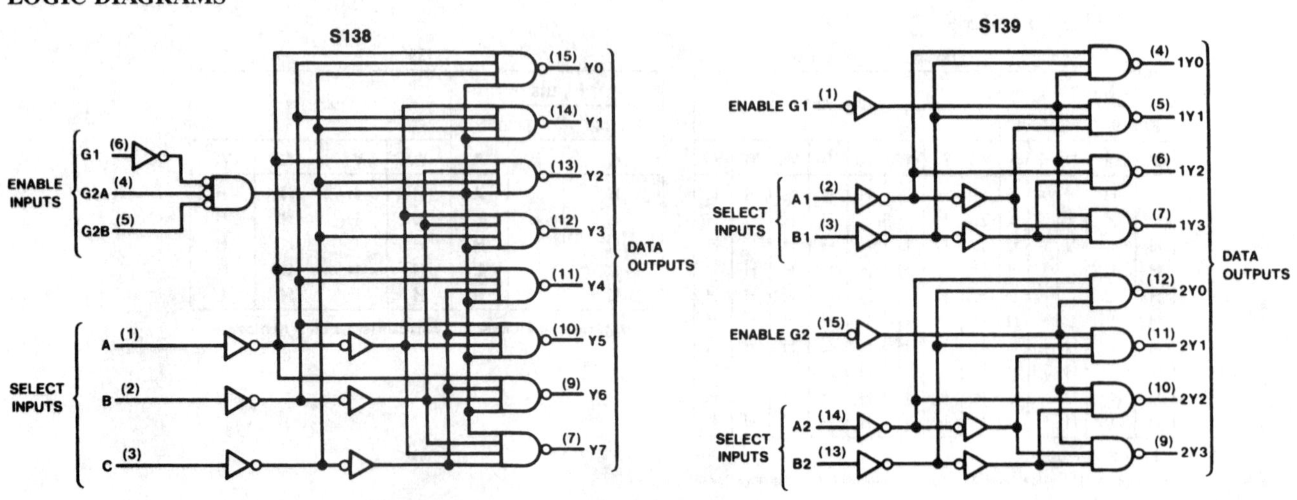

CONNECTION DIAGRAMS

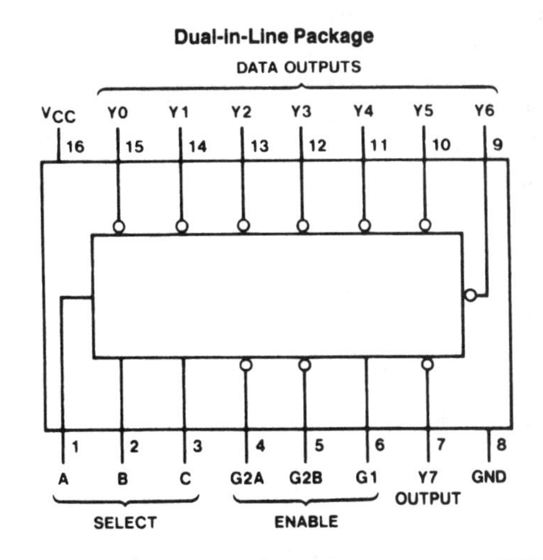

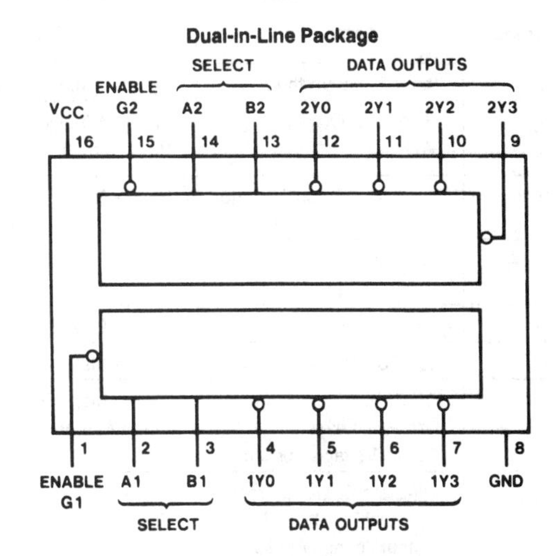

Order Number DM54S138J, DM54S139J, DM54S138W, DM54139W, DM74S138N or DM74S139N
See NS Package Number J16A, N16E or W16A

National DM54S251/DM74S251
Tri-state® 1 of 8-Line Data Selector/Multiplexer

- Tri-state version of S151
- Interface directly with system bus
- Perform parallel-to-serial conversion
- Permit multiplexing from N-lines to one line
- Complementary outputs provide true and inverted data
- Max no. of common outputs
  - 54S 39
  - 74S 129
- Typical propagation delay time (D to Y) 8ns
- Typical power dissipation: 275mW

ABSOLUTE MAXIMUM RATINGS (Note)

If Military/Aerospace specified devices are required, please contact the National Semiconductor Sales Office/Distributors for availability and specifications.

Supply Voltage	7V
Input Voltage	5.5V
Operating Free Air Temperature Range	
DM54S	−55°C to +125°C
DM74S	0°C to +70°C
Storage Temperature Range	−65°C to +150°C

Note: *The "Absolute Maximum Ratings" are those values beyond which the safety of the device cannot be guaranteed. The device should not be operated at these limits. The parametric values defined in the "Electrical Characteristics" table are not guaranteed at the absolute maximum ratings. The "Recommended Operating Conditions" table will define the conditions for actual device operation.*

RECOMMENDED OPERATING CONDITIONS

Symbol	Parameter	DM54S251			DM74S251			Units
		Min	Nom	Max	Min	Nom	Max	
V_{CC}	Supply Voltage	4.5	5	5.5	4.75	5	5.25	V
V_{IH}	High Level Input Voltage	2			2			V
V_{IL}	Low Level Input Voltage			0.8			0.8	V
I_{OH}	High Level Output Current			−2			−6.5	mA
I_{OL}	Low Level Output Current			20			20	mA
T_A	Free Air Operating Temperature	−55		125	0		70	°C

ELECTRICAL CHARACTERISTICS over recommended operating free air temperature (unless otherwise noted)

Symbol	Parameter	Conditions		Min	Typ (Note 1)	Max	Units
V_I	Input Clamp Voltage	V_{CC} = Min, I_I = −18 mA				−1.2	V
V_{OH}	High Level Output Voltage	V_{CC} = Min, I_{OH} = Max V_{IL} = Max, V_{IH} = Min	DM54	2.4	3.4		V
			DM74	2.4	3.2		
V_{OL}	Low Level Output Voltage	V_{CC} = Min, I_{OL} = Max V_{IH} = Min, V_{IL} = Max				0.5	V
I_I	Input Current @ Max Input Voltage	V_{CC} = Max, V_I = 5.5V				1	mA
I_{IH}	High Level Input	V_{CC} = Max, V_I = 2.7V				50	μA
I_{IL}	Low Level Input Current	V_{CC} = Max, V_I = 0.5V				−2	mA
I_{OZH}	Off-State Output Current with High Level Output Voltage Applied	V_{CC} = Max, V_O = 2.4 V_{IH} = Min, V_{IL} = Max				50	μA
I_{OZL}	Off-State Output Current with Low Level Output Voltage Applied	V_{CC} = Max, V_O = 0.5 V_{IH} = Min, V_{IL} = Max				−50	μA
I_{OS}	Short Circuit Output Current	V_{CC} = Max (Note 2)	DM54	−40		−100	mA
			DM74	−40		−100	
I_{CC}	Supply Current	V_{CC} = Max (Note 3)			55	85	mA

Note 1: All typicals are at V_{CC} = 5V, T_A = 25°C.
Note 2: Not more than one output should be shorted at a time, and the duration should not exceed one second.
Note 3: I_{CC} is measured with the outputs open and all inputs at 4.5V.

SWITCHING CHARACTERISTICS at V_{CC} = 5V T_A = 25°C (See Section 1 for Test Waveforms and Output Load)

Symbol	Parameter	From (Input) To (Output)	R_L = 280Ω				Units
			C_L = 15 pF		C_L = 50 pF		
			Min	Max	Min	Max	
t_{PLH}	Propagation Delay Time Low to High Level Output	A, B, or C (4 Levels) to Y		18		21	ns
t_{PHL}	Propagation Delay Time High to Low Level Output	A, B, or C (4 Levels) to Y		19.5		23	ns
t_{PLH}	Propagation Delay Time Low to High Level Output	A, B, or C (3 Levels) to W		15		18	ns
t_{PHL}	Propagation Delay Time High to Low Level Output	A, B, or C (3 Levels) to W		13.5		17	ns
t_{PLH}	Propagation Delay Time Low to High Level Output	D to Y		12		15	ns
t_{PHL}	Propagation Delay Time High to Low Level Output	D to Y		12		15	ns
t_{PLH}	Propagation Delay Time Low to High Level Output	D to W		7		10	ns
t_{PHL}	Propagation Delay Time High to Low Level Output	D to W		7		10	ns
t_{PZH}	Output Enable Time to High Level Output	Strobe to Y				19.5	ns
t_{PZL}	Output Enable Time to Low Level Output	Strobe to Y				21	ns
t_{PHZ}	Output Disable Time to High Level Output (Note 1)	Strobe to Y		8.5			ns
t_{PLZ}	Output Disable Time to Low Level Output (Note 1)	Strobe to Y		14			ns
t_{PZH}	Output Enable Time to High Level Output	Strobe to W				19.5	ns
t_{PZL}	Output Enable Time to Low Level Output	Strobe to W				21	ns
t_{PHZ}	Output Disable Time to High Level Output (Note 1)	Strobe to W		8.5			ns
t_{PLZ}	Output Disable Time to Low Level Output (Note 1)	Strobe to W		14			ns

Note 1: C_L = 5 pF.

CONNECTION DIAGRAM

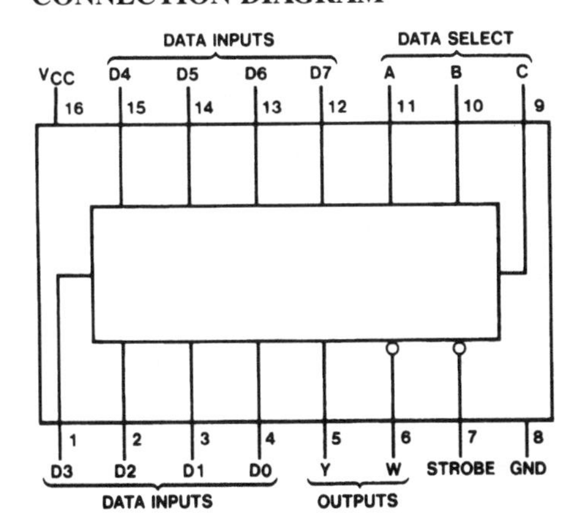

Order Number DM54S251J or DM74S251N
See NS Package Number J16A or N16E

FUNCTION TABLE

Inputs				Outputs	
Select			Strobe S	Y	W
C	B	A			
X	X	X	H	Z	Z
L	L	L	L	D0	$\overline{D0}$
L	L	H	L	D1	$\overline{D1}$
L	H	L	L	D2	$\overline{D2}$
L	H	H	L	D3	$\overline{D3}$
H	L	L	L	D4	$\overline{D4}$
H	L	H	L	D5	$\overline{D5}$
H	H	L	L	D6	$\overline{D6}$
H	H	H	L	D7	$\overline{D7}$

H = High Logic Level, L = Low Logic Level
X = Don't Care, Z = High Impedance (Off)
D0, D1 . . . D7 = The Level of the respective D input

LOGIC DIAGRAM

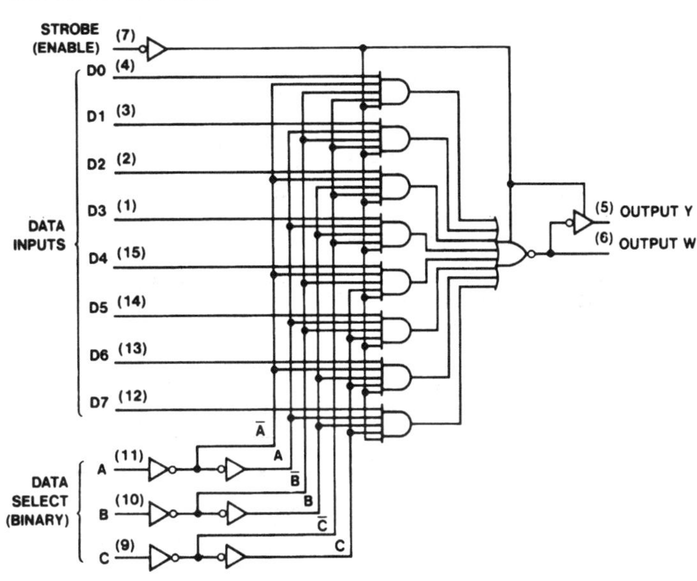

National DM54S253/DM74S253
Dual Tri-State® 1 of 4-Line Data Selectors/Multiplexers

- Tri-state version of S153 with same pinout
- Schottky-diode-clamped transistors
- Permits multiplexing from N lines to 1 line
- Performs parallel T-serial conversion
- Strobe/output control
- High fan-out totem-pole outputs
- Typical propagation delay
  - From data to output: 6ns
  - From select to output: 12ns
- Typical power dissipation: 275mW

ABSOLUTE MAXIMUM RATINGS (Note)

If Military/Aerospace specified devices are required, please contact the National Semiconductor Sales Office/Distributors for availability and specifications.

Supply Voltage	7V
Input Voltage	5.5V
Operating Free Air Temperature Range	
DM54S	−55°C to +125°C
DM74S	0°C to +70°C
Storage Temperature Range	−65°C to +150°C

Note: *The "Absolute Maximum Ratings" are those values beyond which the safety of the device cannot be guaranteed. The device should not be operated at these limits. The parametric values defined in the "Electrical Characteristics" table are not guaranteed at the absolute maximum ratings. The "Recommended Operating Conditions" table will define the conditions for actual device operation.*

RECOMMENDED OPERATING CONDITIONS

Symbol	Parameter	DM54S253			DM74S253			Units
		Min	Nom	Max	Min	Nom	Max	
V_{CC}	Supply Voltage	4.5	5	5.5	4.75	5	5.25	V
V_{IH}	High Level Input Voltage	2			2			V
V_{IL}	Low Level Input Voltage			0.8			0.8	V
I_{OH}	High Level Output Current			−2			−6.5	mA
I_{OL}	Low Level Output Current			20			20	mA
T_A	Free Air Operating Temperature	−55		125	0		70	°C

ELECTRICAL CHARACTERISTICS over recommended operating free air temperature (unless otherwise noted)

Symbol	Parameter	Conditions		Min	Typ (Note 1)	Max	Units
V_I	Input Clamp Voltage	V_{CC} = Min, I_I = −18 mA				−1.2	V
V_{OH}	High Level Output Voltage	V_{CC} = Min, I_{OH} = Max V_{IL} = Max, V_{IH} = Min	DM54	2.4	3.4		V
			DM74	2.4	3.2		
V_{OL}	Low Level Output Voltage	V_{CC} = Min, I_{OL} = Max V_{IH} = Min, V_{IL} = Max				0.5	V
I_I	Input Current @ Max Input Voltage	V_{CC} = Max, V_I = 5.5V				1	mA
I_{IH}	High Level Input Current	V_{CC} = Max, V_I = 2.7V				50	μA
I_{IL}	Low Level Input Current	V_{CC} = Max, V_I = 0.5V				−2	mA
I_{OZH}	Off-State Output Current with High Level Output Voltage Applied	V_{CC} = Max, V_O = 2.4V V_{IH} = Min, V_{IL} = Max				50	μA
I_{OZL}	Off-State Output Current with Low Level Output Voltage Applied	V_{CC} = Max, V_O = 0.5V V_{IH} = Min, V_{IL} = Max				−50	μA
I_{OS}	Short Circuit Output Current	V_{CC} = Max (Note 2)	DM54	−40		−100	mA
			DM74	−40		−100	
I_{CC}	Supply Current	V_{CC} = Max (Note 3)			55	70	mA

Note 1: All typicals are at V_{CC} = 5V, T_A = 25°C.

Note 2: Not more than one output should be shorted at a time, and the duration should not exceed one second.

Note 3: I_{CC} is measured with all outputs open.

SWITCHING CHARACTERISTICS at V_{CC} = 5V and T_A = 25°C (See Section 1 for Test Waveforms and Output Load)

Symbol	Parameter	From (Input) To (Output)	R_L = 280Ω				Units
			C_L = 15 pF		C_L = 50 pF		
			Min	Max	Min	Max	
t_{PLH}	Propagation Delay Time Low to High Level Output	Data to Y		9		12	ns
t_{PHL}	Propagation Delay Time High to Low Level Output	Data to Y		9		12	ns
t_{PLH}	Propagation Delay Time Low to High Level Output	Select to Y		18		21	ns
t_{PHL}	Propagation Delay Time High to Low Level Output	Select to Y		18		21	ns
t_{PZH}	Output Enable Time to High Level Output	Output Control to Y		16.5		19.5	ns
t_{PZL}	Output Enable Time to Low Level Output	Output Control to Y		18		21	ns
t_{PHZ}	Output Disable Time to High Level Output (Note 1)	Output Control to Y		9.5			ns
t_{PLZ}	Output Disable Time to Low Level Output (Note 1)	Output Control to Y		15			ns

Note 1: C_L = 5 pF.

CONNECTION DIAGRAM

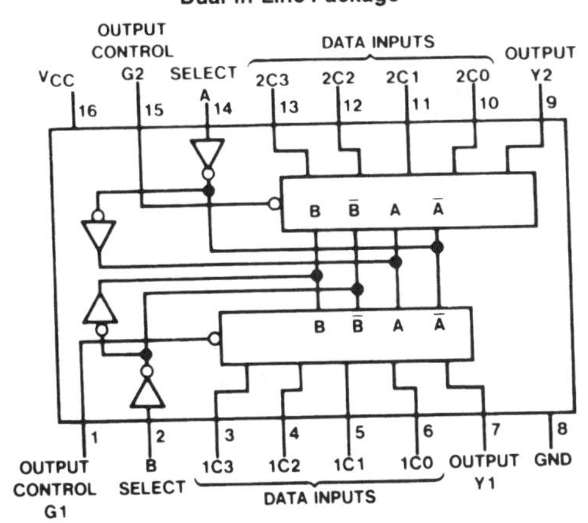

Order Number DM54S253J, DM54S253W or DM74S253N
NS Package Number J16A, N16E or W16A

FUNCTION TABLE

Select Inputs		Data Inputs				Output Control	Output
B	A	C0	C1	C2	C3	G	Y
X	X	X	X	X	X	H	Z
L	L	L	X	X	X	L	L
L	L	H	X	X	X	L	H
L	H	X	L	X	X	L	L
L	H	X	H	X	X	L	H
H	L	X	X	L	X	L	L
H	L	X	X	H	X	L	H
H	H	X	X	X	L	L	L
H	H	X	X	X	H	L	H

Address inputs A and B are common to both sections.

H = High Level, L = Low Level, X = Don't Care, Z = High Impedance

LOGIC DIAGRAM

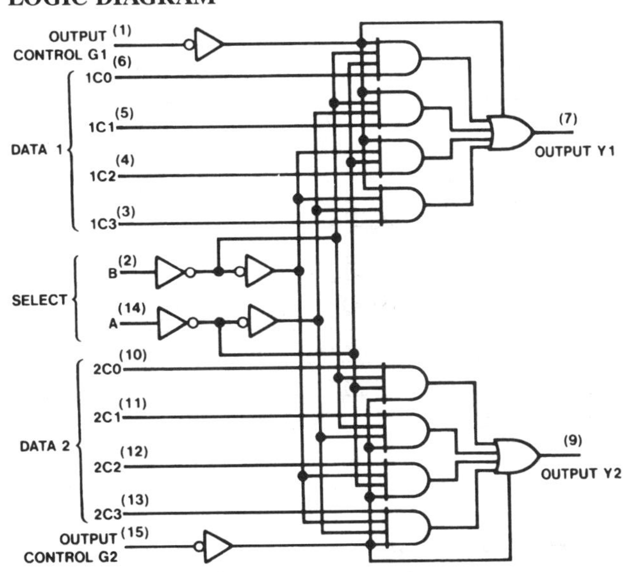

National DM54S257/DM74S257, DM54S258/DM74S258

Tri-State® Quad 1 of 2 Data Selectors/Multiplexers

- Tri-state versions S157, S158, with same pinouts
- Schottky-clamped for significant improvement in ac performance
- Provides bus interface from multiple sources in high-performance systems
- Average propagation delay from data input
  - S257 4.8ns
  - S258 4ns
- Typical power dissipation
  - S257 320mW
  - S258 280mW

ABSOLUTE MAXIMUM RATINGS (Note)

If Military/Aerospace specified devices are required, please contact the National Semiconductor Sales Office/Distributors for availability and specifications.

Supply Voltage	7V
Input Voltage	5.5V
Operating Free Air Temperature Range	
DM54S	−55°C to +125°C
DM74S	0°C to +70°C
Storage Temperature Range	−65°C to +150°C

Note: *The "Absolute Maximum Ratings" are those values beyond which the safety of the device cannot be guaranteed. The device should not be operated at these limits. The parametric values defined in the "Electrical Characteristics" table are not guaranteed at the absolute maximum ratings. The "Recommended Operating Conditions" table will define the conditions for actual device operation.*

RECOMMENDED OPERATING CONDITIONS

Symbol	Parameter	DM54S257			DM74S257			Units
		Min	Nom	Max	Min	Nom	Max	
V_{CC}	Supply Voltage	4.5	5	5.5	4.75	5	5.25	V
V_{IH}	High Level Input Voltage	2			2			V
V_{IL}	Low Level Input Voltage			0.8			0.8	V
I_{OH}	High Level Output Current			−2			−6.5	mA
I_{OL}	Low Level Output Current			20			20	mA
T_A	Free Air Operating Temperature	−55		125	0		70	°C

'S257 ELECTRICAL CHARACTERISTICS over recommended operating free air temperature (unless otherwise noted)

Symbol	Parameter	Conditions		Min	Typ (Note 1)	Max	Units
V_I	Input Clamp Voltage	V_{CC} = Min, I_I = −18 mA				−1.2	V
V_{OH}	High Level Output Voltage	V_{CC} = Min, I_{OH} = Max V_{IL} = Max, V_{IH} = Min	DM54	2.4	3.4		V
			DM74	2.4	3.2		
V_{OL}	Low Level Output Voltage	V_{CC} = Min, I_{OL} = Max V_{IH} = Min, V_{IL} = Max				0.5	V
I_I	Input Current @ Max Input Voltage	V_{CC} = Max, V_I = 5.5V				1	mA
I_{IH}	High Level Input Current	V_{CC} = Max V_I = 2.7V	Select			100	μA
			Other			50	
I_{IL}	Low Level Input Current	V_{CC} = Max, V_I = 0.5V	Select			−4	mA
			Other			−2	
I_{OZH}	Off-State Output Current with High Level Output Voltage Applied	V_{CC} = Max, V_O = 2.4V V_{IH} = Min, V_{IL} = Max				50	μA
I_{OZL}	Off-State Output Current with Low Level Output Voltage Applied	V_{CC} = Max, V_O = 0.5V V_{IH} = Min, V_{IL} = Max				−50	μA
I_{OS}	Short Circuit Output Current	V_{CC} = Max (Note 2)	DM54	−40		−100	mA
			DM74	−40		−100	
I_{CCH}	Supply Current with Outputs High	V_{CC} = Max (Note 3)			44	68	mA
I_{CCL}	Supply Current with Outputs Low	V_{CC} = Max (Note 3)			60	93	mA
I_{CCZ}	Supply Current with Outputs Disabled	V_{CC} = Max (Note 3)			64	99	mA

Note 1: All typicals are at V_{CC} = 5V, T_A = 25°C.

Note 2: Not more than one output should be shorted at a time, and the duration should not exceed one second.

Note 3: I_{CC} is measured with all outputs open and all possible inputs grounded, while achieving the stated output conditions.

'S257 SWITCHING CHARACTERISTICS

at V_{CC} = 5V and T_A = 25°C (See Section 1 for Test Waveforms and Output Load)

Symbol	Parameter	From (Input) To (Output)	R_L = 280Ω				Units
			C_L = 15 pF		C_L = 50 pF		
			Min	Max	Min	Max	
t_{PLH}	Propagation Delay Time Low to High Level Output	Data to Output		7.5		11	ns
t_{PHL}	Propagation Delay Time High to Low Level Output	Data to Output		6.5		10	ns
t_{PLH}	Propagation Delay Time Low to High Level Output	Select to Output		15		16	ns
t_{PHL}	Propagation Delay Time High to Low Level Output	Select to Output		15		16	ns
t_{PZH}	Output Enable Time to High Level Output	Output Control to Y		19.5		23	ns
t_{PZL}	Output Enable Time to Low Level Output	Output Control to Y		21		24	ns
t_{PHZ}	Output Disable Time to High Level Output (Note 1)	Output Control to Y		8.5			ns
t_{PLZ}	Output Disable Time to Low Level Output (Note 1)	Output Control to Y		14			ns

Note 1: C_L = 5 pF.

RECOMMENDED OPERATING CONDITIONS

Symbol	Parameter	DM54S258			DM74S258			Units
		Min	Nom	Max	Min	Nom	Max	
V_{CC}	Supply Voltage	4.5	5	5.5	4.75	5	5.25	V
V_{IH}	High Level Input Voltage	2			2			V
V_{IL}	Low Level Input Voltage			0.8			0.8	V
I_{OH}	High Level Output Current			−2			−6.5	mA
I_{OL}	Low Level Output Current			20			20	mA
T_A	Free Air Operating Temperature	−55		125	0		70	°C

'S258 ELECTRICAL CHARACTERISTICS

over recommended operating free air temperature (unless otherwise noted)

Symbol	Parameter	Conditions		Min	Typ (Note 1)	Max	Units
V_I	Input Clamp Voltage	V_{CC} = Min, I_I = −18 mA				−1.2	V
V_{OH}	High Level Output Voltage	V_{CC} = Min, I_{OH} = Max V_{IL} = Max, V_{IH} = Min	DM54	2.4	3.4		V
			DM74	2.4	3.2		
V_{OL}	Low Level Output Voltage	V_{CC} = Min, I_{OL} = Max V_{IH} = Min, V_{IL} = Max				0.5	V
I_I	Input Current @ Max Input Voltage	V_{CC} = Max, V_I = 5.5V				1	mA
I_{IH}	High Level Input Current	V_{CC} = Max, V_I = 2.7V	Select			100	μA
			Other			50	
I_{IL}	Low Level Input Current	V_{CC} = Max, V_I = 0.5V	Select			−4	mA
			Other			−2	
I_{OZH}	Off-State Output Current with High Level Output Voltage Applied	V_{CC} = Max, V_O = 2.4V V_{IH} = Min, V_{IL} = Max				50	μA
I_{OZL}	Off-State Output Current with Low Level Output Voltage Applied	V_{CC} = Max, V_O = 0.5V V_{IH} = Min, V_{IL} = Max				−50	μA
I_{OS}	Short Circuit Output Current	V_{CC} = Max (Note 2)	DM54	−40		−100	mA
			DM74	−40		−100	
I_{CCH}	Supply Current with Outputs High	V_{CC} = Max (Note 3)			36	56	mA
I_{CCL}	Supply Current with Outputs Low	V_{CC} = Max (Note 3)			52	81	mA
I_{CCZ}	Supply Current with Outputs Disabled	V_{CC} = Max (Note 3)			56	87	mA

Note 1: All typicals are at V_{CC} = 5V, T_A = 25°C.

Note 2: Not more than one output should be shorted at a time, and the duration should not exceed one second.

Note 3: I_{CC} is measured with all outputs open and all possible inputs grounded, while achieving the stated output conditions.

FUNCTION TABLE

Inputs				Output Y	
Output Control	Select	A	B	S257	S258
H	X	X	X	Z	Z
L	L	L	X	L	H
L	L	H	X	H	L
L	H	X	L	L	H
L	H	X	H	H	L

H = High Level, L = Low Level, X = Don't Care

Z = High Impedance (off)

'S258 SWITCHING CHARACTERISTICS

at V_{CC} = 5V and T_A = 25°C (See Section 1 for Test Waveforms and Output Load)

Symbol	Parameter	From (Input) To (Output)	R_L = 280Ω				Units
			C_L = 15 pF		C_L = 50 pF		
			Min	Max	Min	Max	
t_{PLH}	Propagation Delay Time Low to High Level Output	Data to Output		6		9	ns
t_{PHL}	Propagation Delay Time High to Low Level Output	Data to Output		6		9	ns
t_{PLH}	Propagation Delay Time Low to High Level Output	Select to Output		12		15	ns
t_{PHL}	Propagation Delay Time High to Low Level Output	Select to Output		12		15	ns
t_{PZH}	Output Enable Time to High Level Output	Output Control to Y		19.5		23	ns
t_{PZL}	Output Enable Time to Low Level Output	Output Control to Y		21		24	ns
t_{PHZ}	Output Disable Time to High Level Output (Note 1)	Output Control to Y		8.5			ns
t_{PLZ}	Output Disable Time to Low Level Output (Note 1)	Output Control to Y		14			ns

Note 1: C_L = 5 pF.

CONNECTION DIAGRAMS

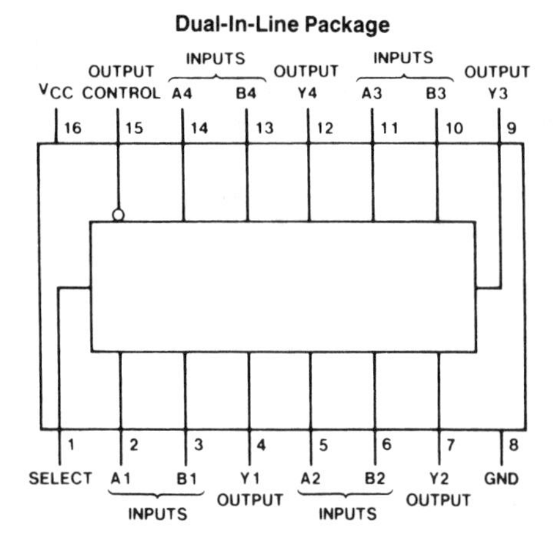

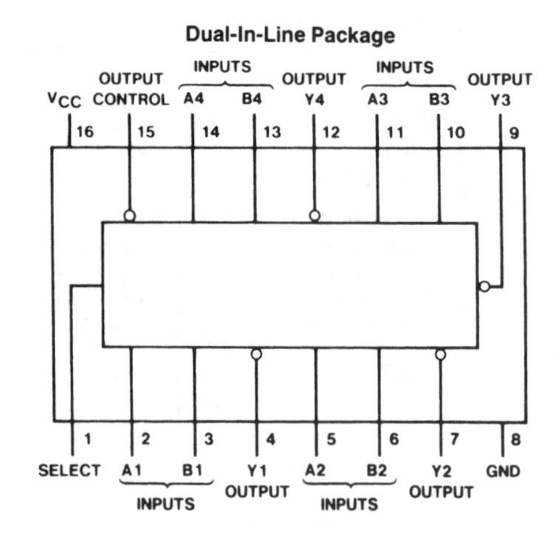

Order Number DM54S257J, DM54S258J, DM54S257W, DM74S257N or DM74S258N
See NS Package Number J16A, N16E or W16A

National
DM54S280/DM74S280
9-Bit Parity Generators/Checkers

- Generates either odd or even parity for nine data lines
- Cascadable for N-bits
- Can be used to upgrade existing systems using MSI parity circuits
- Typical data-to-output delay: 14ns

ABSOLUTE MAXIMUM RATINGS (Note)

If Military/Aerospace specified devices are required, please contact the National Semiconductor Sales Office/Distributors for availability and specifications.

Supply Voltage	7V
Input Voltage	5.5V
Operating Free Air Temperature Range	
DM54S	−55°C to +125°C
DM74S	0°C to +70°C
Storage Temperature Range	−65°C to +150°C

Note: *The "Absolute Maximum Ratings" are those values beyond which the safety of the device cannot be guaranteed. The device should not be operated at these limits. The parametric values defined in the "Electrical Characteristics" table are not guaranteed at the absolute maximum ratings. The "Recommended Operating Conditions" table will define the conditions for actual device operation.*

RECOMMENDED OPERATING CONDITIONS

Symbol	Parameter	DM54S280			DM74S280			Units
		Min	Nom	Max	Min	Nom	Max	
V_{CC}	Supply Voltage	4.5	5	5.5	4.75	5	5.25	V
V_{IH}	High Level Input Voltage	2			2			V
V_{IL}	Low Level Input Voltage			0.8			0.8	V
I_{OH}	High Level Output Current			−1			−1	mA
I_{OL}	Low Level Output Current			20			20	mA
T_A	Free Air Operating Temperature	−55		125	0		70	°C

ELECTRICAL CHARACTERISTICS over recommended operating free air temperature range (unless otherwise noted)

Symbol	Parameter	Conditions		Min	Typ (Note 1)	Max	Units
V_I	Input Clamp Voltage	V_{CC} = Min, I_I = −18 mA				−1.2	V
V_{OH}	High Level Output Voltage	V_{CC} = Min, I_{OH} = Max V_{IL} = Max, V_{IH} = Min		2.7	3.4		V
V_{OL}	Low Level Output Voltage	V_{CC} = Min, I_{OL} = Max V_{IH} = Min, V_{IL} = Max				0.5	V
I_I	Input Current @ Max Input Voltage	V_{CC} = Max, V_I = 5.5V				1	mA
I_{IH}	High Level Input Current	V_{CC} = Max, V_I = 2.7V				50	μA
I_{IL}	Low Level Input Current	V_{CC} = Max, V_I = 0.5V				−2	mA
I_{OS}	Short Circuit Output Current	V_{CC} = Max (Note 2)	DM54	−40		−100	mA
			DM74	−40		−100	
I_{CC}	Supply Current	V_{CC} Max (Note 3)			67	105	mA

Note 1: All typicals are at V_{CC} = 5V, T_A = 25°C.

Note 2: Not more than one output should be shorted at a time, and the duration should not exceed one second.

Note 3: I_{CC} is measured with all inputs grounded and all outputs open.

SWITCHING CHARACTERISTICS at V_{CC} = 5V and T_A = 25°C (See Section 1 for Test Waveforms and Output Load)

Symbol	Parameter	From (Input) To (Output)	R_L = 280Ω C_L = 15 pF		R_L = 280Ω C_L = 50 pF		Units
			Min	Max	Min	Max	
t_{PLH}	Propagation Delay Time Low to High Level Output	Data to Σ Even		21		24	ns
t_{PHL}	Propagation Delay Time High to Low Level Output	Data to Σ Even		18		21	ns
t_{PLH}	Propagation Delay Time Low to High Level Output	Data to Σ Odd		21		24	ns
t_{PHL}	Propagation Delay Time High to Low Level Output	Data to Σ Odd		18		21	ns

CONNECTION DIAGRAM

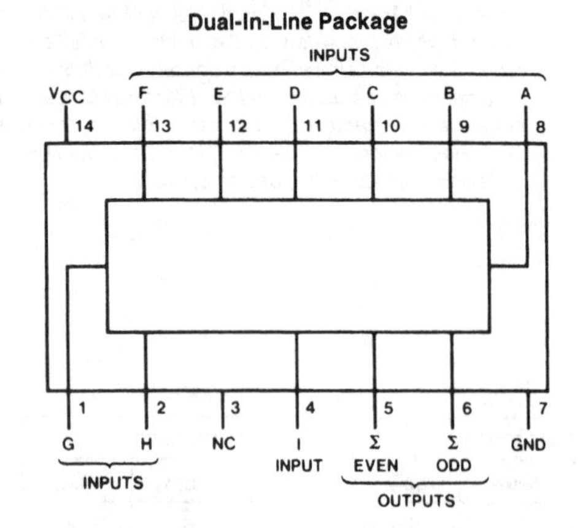

Order Number DM54S280J, DM54S280W, DM74S280M or DM74S280N
See NS Package Number J14A, M14A, N14A or W14B

FUNCTION TABLE

Number of Inputs (A Thru I) that are High	Outputs	
	Σ Even	Σ Odd
0, 2, 4, 6, 8	H	L
1, 3, 5, 7, 9	L	H

LOGIC DIAGRAM

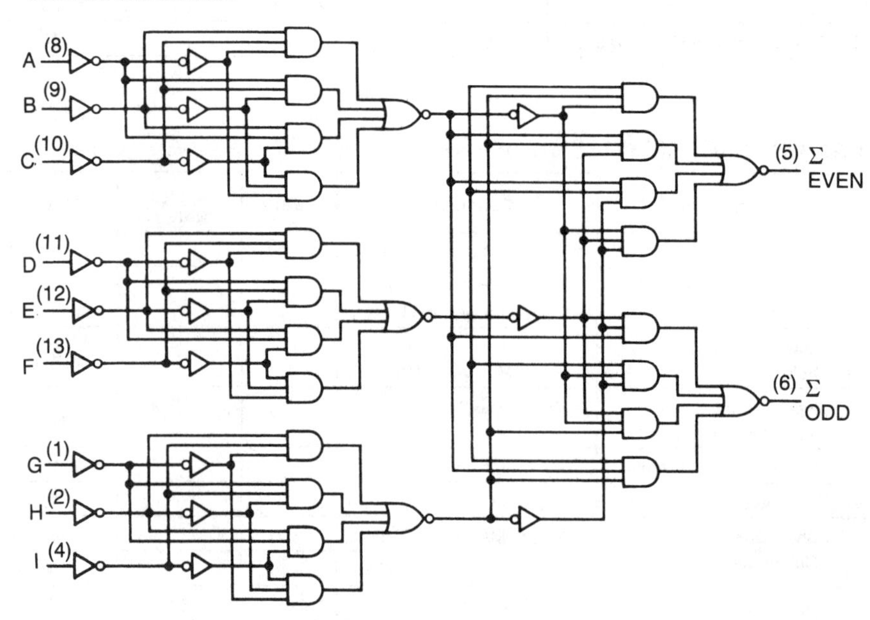

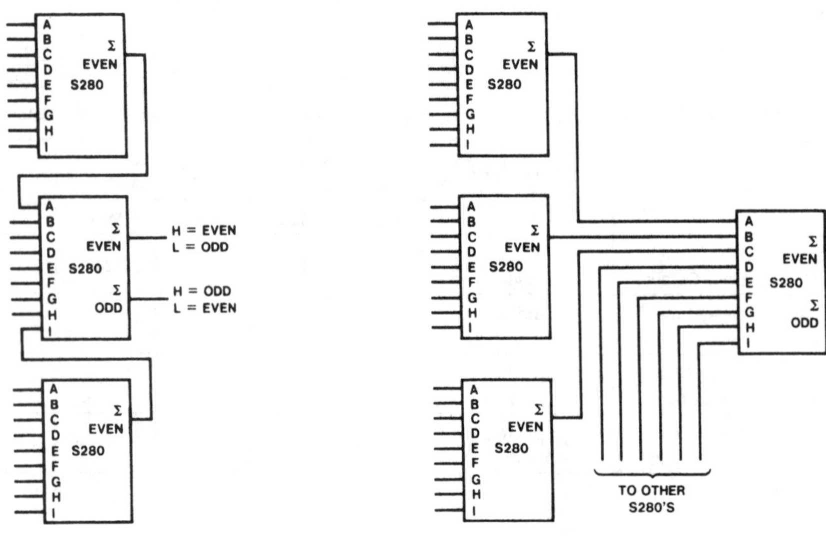

25-Line Parity/Generator Checker

81-Line Parity/Generator Checker

This information is reproduced with permission of Rohm Corporation, Irvine, CA.

Rohm BA1610
FSK Linear Modem

- Zero-crossing switching from reception into transmission mode
- Half-duplex transmission and reception functions implemented on a single chip
- Wide supply voltage range (V_{CC} = 5 to 14V) allows operation on a single power supply
- Wide frequency range (f max. = 200 kHz)
- High temperature stability: 100ppm/°C
- Directly interfaceable with DTL/TTL/ECL logic
- Recommended supply voltage: V_{CC} = 5V

APPLICATIONS
- Modem telephones
- Home automation equipment
- Data communications, POS, and PC communications

The BA1610 is a single-chip FSK linear modem with half-duplex scheme. The device is capable of both transmission and reception. With internal PLL, it outputs data without being affected by ambient noise, only if a signal within the target frequency band is received.

ABSOLUTE MAXIMUM RATINGS (Ta=25°C)

Parameter	Symbol	Limits	Unit
Supply voltage	V_{CC}	14	V
Power dissipation	Pd	600	mW
Operating temperature range	Topr	−10 ~ 60	°C
Storage temperature range	Tstg	−55 ~ 125	°C
Maximum input level	$V_{IN\ MAX}$	3	V p-p
Maximum load current, pin 12	$I_{SAT\ MAX}$	5	mA

ELECTRICAL CHARACTERISTICS (Ta=25°C)

Parameter	Symbol	Min.	Typ.	Max.	Unit	Conditions
Current consumption	I_{CC}	6.8	9.0	11.2	mA	when receiving
Frequency accuracy	f_C	—	±2.0	±3.2	%/fo	—
Frequency temperature stability	ft	—	100	—	ppm/°C	Ta=0 ~ 55°C
Frequency supply voltage stability	f_{VCC}	—	±0.1	—	%/V	V_{CC}=12V±1V
Maximum frequency	f_{MAX}	—	—	200	kHz	—
Internal stability voltage	Vref	2.6	3.0	3.3	V	—
Internal bias voltage	V_B	1.40	1.55	1.72	V	—
Tx/Rx low voltage	V_L	—	—	0.8	V	—
Tx/Rx high voltage	V_H	2.0	—	—	V	—
Data low input voltage	V_{DL}	—	—	0.8	V	—
Data high input voltage	V_{DH}	2.0	—	—	V	—
Sine-wave output voltage	V_{OUT}	0.85	1	—	V p-p	—
Receiving input sensitivity	VS	—	2.0	—	mV rms	—

Dimensions (Unit: mm)

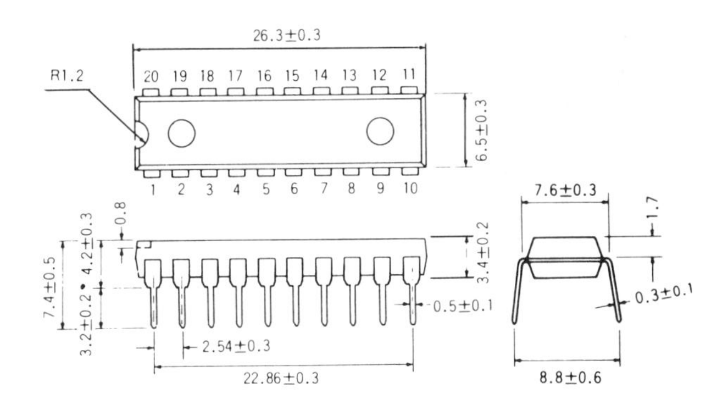

Block Diagram

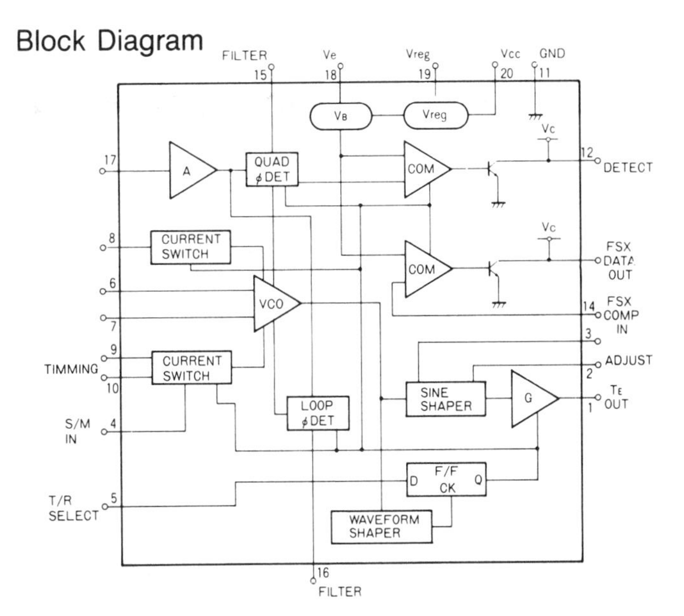

Application Example

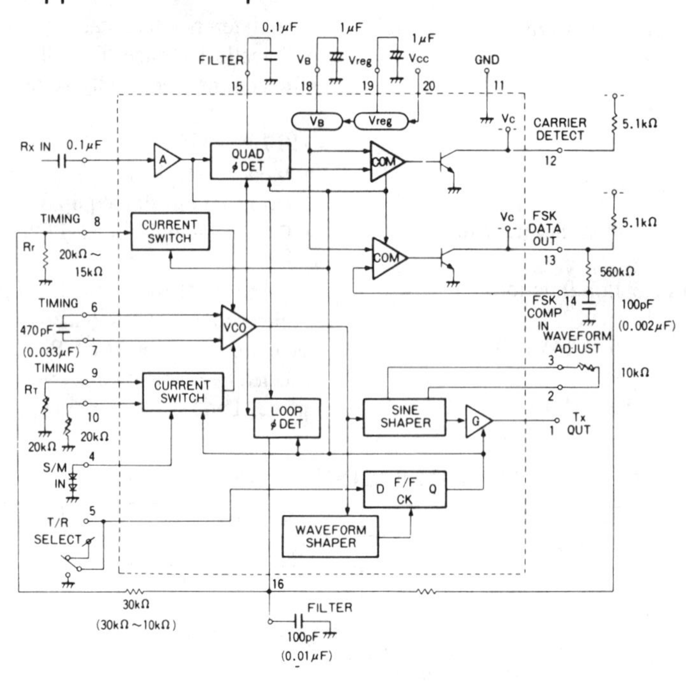

Rohm
BA9101, BA9101B, BA9101S
8-Bit Successive Comparison A/D Converters

- Complete monolithic A/D converter with internal reference and clock sources
- Tri-state outputs directly connect to bus lines
- Easily controlled by microprocessor
- Allows easy bipolar operation
- External clock can be used
- Settling time: 20μs
- Analog input: 0 to 5V, ±2.5V

APPLICATIONS
- Voice signal processing
- Instrumentation and control equipment
- Thermometers
- Medical equipment

The BA9101 is a single-chip, 8-bit, monolithic A/D converter with internal reference voltage and clock sources. The device requires no external components for basic operations.

The Tri-state outputs allow the device to directly connect to a data bus. Conversion start, data read, and clock timings can be externally controlled.

ABSOLUTE MAXIMUM RATINGS (Ta=25°C)

Parameter	Symbol	Limits	Unit
Supply voltage (pin 22)	V_{CC}	6	V
Supply voltage (pin 1)	V_{EE}	−8.5	V
Power dissipation	Pd	500*	mW
Operating temperature range	Topr	−25 ~ 75	°C
Storage temperature range	Tstg	−55 ~ 125	°C

*Derating is done at 5.0mW/°C for operation above Ta=25°C.

ELECTRICAL CHARACTERISTICS (Ta=25°C, V_{CC}=5V, V_{EE}= −7V)

Parameter		Symbol	Min.	Typ.	Max.	Unit	Conditions
Analog input voltage range		V_{IU}	—	5	—	V	—
Analog input voltage range (bipolar)		V_{IB}	—	±2.5	—	V	—
Analog input resistance		V_I	2	3	4	kΩ	—
Input voltage changing from 00000000 to 00000001		V_{OI}	10.5	19.5	28.5	mV	V_{REF}=2.0V
Input voltage changing from 11111110 to 11111111		V_{FF}	4.78	4.98	5.08	V	V_{REF}=2.0V
Linearity error	BA9101	E_L	—	—	±0.5	LSB	—
	BA9101B		—	—	±0.75		—
	BA9101S		—	—	±1.0		—
Supply voltage characteristics		PSS	—	1000	—	ppm $_{FSR}$/V	—
Temperature characteristics		ATS	—	100	—	ppm $_{FSR}$/°C	—
Settling time		ts	—	20	30	μs	—
Input high voltage		V_{IH}	2.3	—	—	V	—
Input low voltage		V_{IL}	—	—	0.8	V	—
Input high current		I_{IH}	—	—	400	μA	—
Output high voltage		V_{OH}	2.4	—	—	V	—
Output low voltage		V_{OL}	—	—	0.4	V	—
Maximum clock frequency		f_{CLOCK}	400	—	—	kHz	—
Circuit supply current		I_{CC}	—	15	22	mA	—
		I_{EE}	—	15	22	mA	—
Supply voltage range		V_{CC}	4.5	—	5.5	V	—
		V_{EE}	−6.3	—	−7.7	V	—
Reference output voltage		V_{REF}	2.005	—	—	V	V_{REF} adjust resistance R=50 kΩ
			—	—	1.990	V	V_{REF} adjust resistance R=0 kΩ

BLOCK DIAGRAM

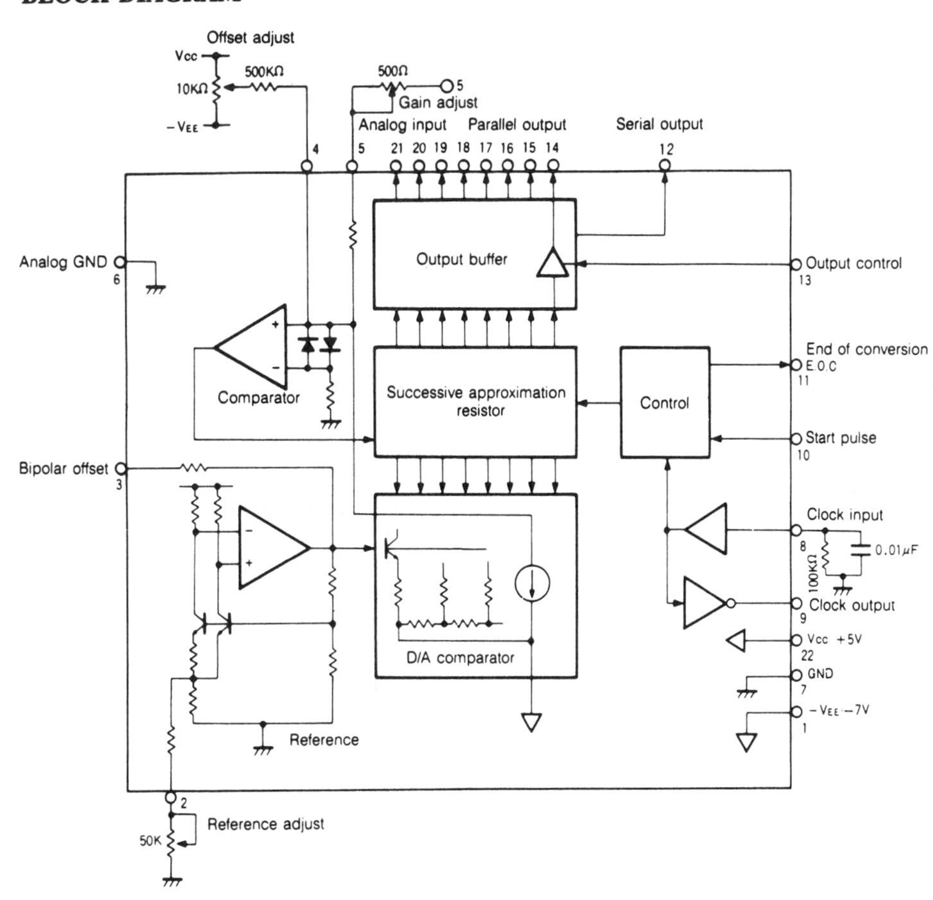

TIMING DIAGRAM

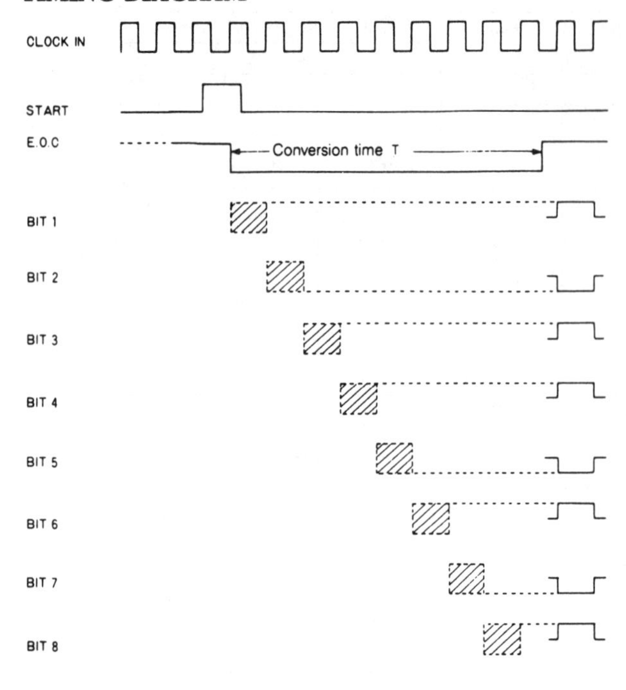

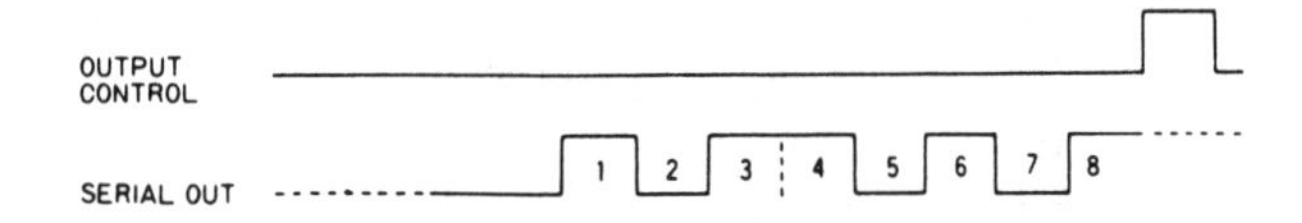

EXAMPLE CIRCUIT for CHECKING BASIC OPERATION

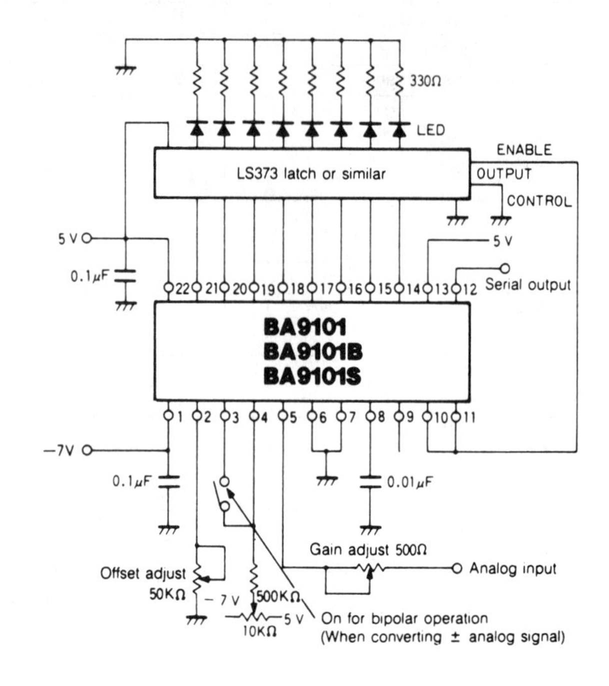

PIN CONNECTIONS

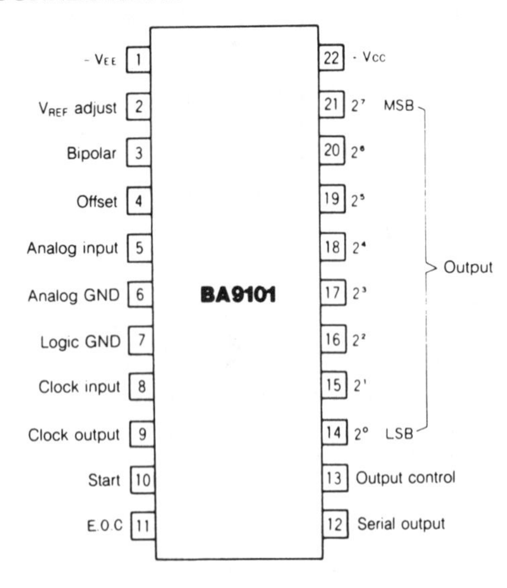

LOGIC CODE

1. Unipolar operation

Analog input	Digital output	
FS-LSB	D_7 1 1 1 1 1 1 1 1 D_0	
FS-2LSB	1 1 1 1 1 1 1 0	$1LSB = \frac{FS}{256}$
1/2FS	1 0 0 0 0 0 0 0	
1LSB	0 0 0 0 0 0 0 1	
0	0 0 0 0 0 0 0 0	

2. Bipolar operation

Analog input	Digital output	
+(FS-1LSB)	D_7 1 1 1 1 1 1 1 1 D_0	
+(FS-2LSB)	1 1 1 1 1 1 1 0	
0	1 0 0 0 0 0 0 0	$1LSB = \frac{2FS}{256}$
−1LSB	0 1 1 1 1 1 1 1	
−(FS-1LSB)	0 0 0 0 0 0 0 1	
−FS	0 0 0 0 0 0 0 0	

FS: full scale

BA9101/BA9101B/BA9101S

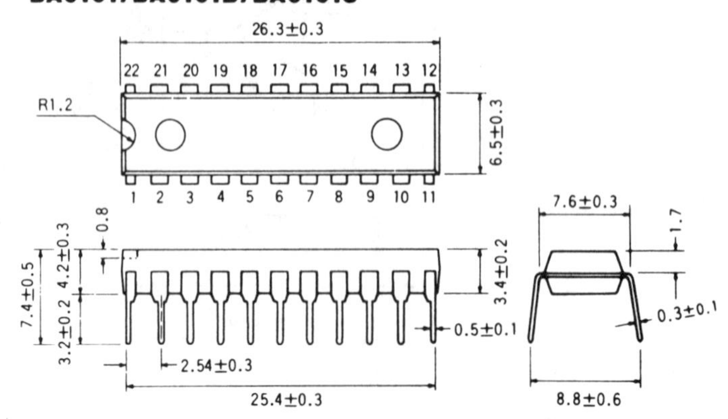

Rohm
BA9201, BA9201F
8-Bit D/A Converters with Latch

- Internal reference voltage source
- Internal input data latch to enable easy control by microprocessors
- Settling time: 500ns

APPLICATIONS
- Instrumentation and control equipment
- Digital audio systems
- Electronic musical instruments
- Signal generators
- Servo controllers

The BA9201 is a complete monolithic, 8-bit D/A converter with internal reference voltage source and input data latch. The reference voltage source is implemented in an independent block so that an external reference voltage source can also be used. The input data latch allows the user to build a system requiring multiple D/A converters.

PRECAUTION
The BA9201 and BA9201F come in different packages with different pin configurations.

ABSOLUTE MAXIMUM RATINGS (Ta=25°C)

Parameter	Symbol	Limits	Unit
Supply voltage (pin 18)	V_{CC}	6	V
Supply voltage (pin 1)	V_{EE}	−8.5	V
Power dissipation	Pd	500*	mW
Operating temperature range	Topr	−25 ~ 75	°C
Storage temperature range	Tstg	−55 ~ 125	°C

*Derating is done at 5mW/°C for operation above Ta=25°C.

CODE FORMATS

Digital inputs	Analog outputs	
D_7 D_0	I_0	$\overline{I_0}$
1 1 1 1 1 1 1 1	1.992mA	0.000mA
1 1 1 1 1 1 1 0	1.984	0.008
1 0 0 0 0 0 0 0	1.000	0.992
0 1 1 1 1 1 1 1	0.992	1.000
0 0 0 0 0 0 0 1	0.008	1.984
0 0 0 0 0 0 0 0	0.000	1.992

ELECTRICAL CHARACTERISTICS (Ta=25°C, V_{CC}=5, V_{EE}= −7V)

Parameter	Symbol	Min.	Typ.	Max.	Unit	Conditions
Resolution	RES	8	8	8	bits	—
Nonlinearity	N.L	—	—	±1/2	LSB	—
Full-scale current	I_{FS}	1.90	1.992	2.10	mA	—
Full-scale temperature coefficient	TCI_{FS}	—	±50	—	ppm/°C	When using an external reference voltage supply
Full-scale nonsymmetry current	I_{FSS}	—	—	±10	μA	I_0-$\overline{I_0}$
Settling time	ts	—	500	—	ns	—
Internal reference voltage supply	V_{REF}	2.005	—	—	V	Pins 3-9 R=50 kΩ
		—	—	1.99	V	Pins 3-9 R=0Ω
Internal reference voltage supply temperature characteristics	TCV_{REF}	—	±100	—	ppm/°C	After adjusting to V_{REF}=2.00V
Input high voltage	V_{IH}	2.3	—	—	V	—
Input low voltage	V_{IL}	—	—	0.8	V	—
Input high current	I_{IH}	—	—	400	μA	—
Circuit supply current (pin 18)	I_{CC}	—	7	—	mA	—
Circuit supply current (pin 1)	I_{EE}	—	+12	—	mA	—
Supply voltage range (pin 18)	V_{CC}	4.5	—	5.5	V	—
Supply voltage range (pin 1)	V_{EE}	−6.3	—	−7.7	V	—

BLOCK DIAGRAM

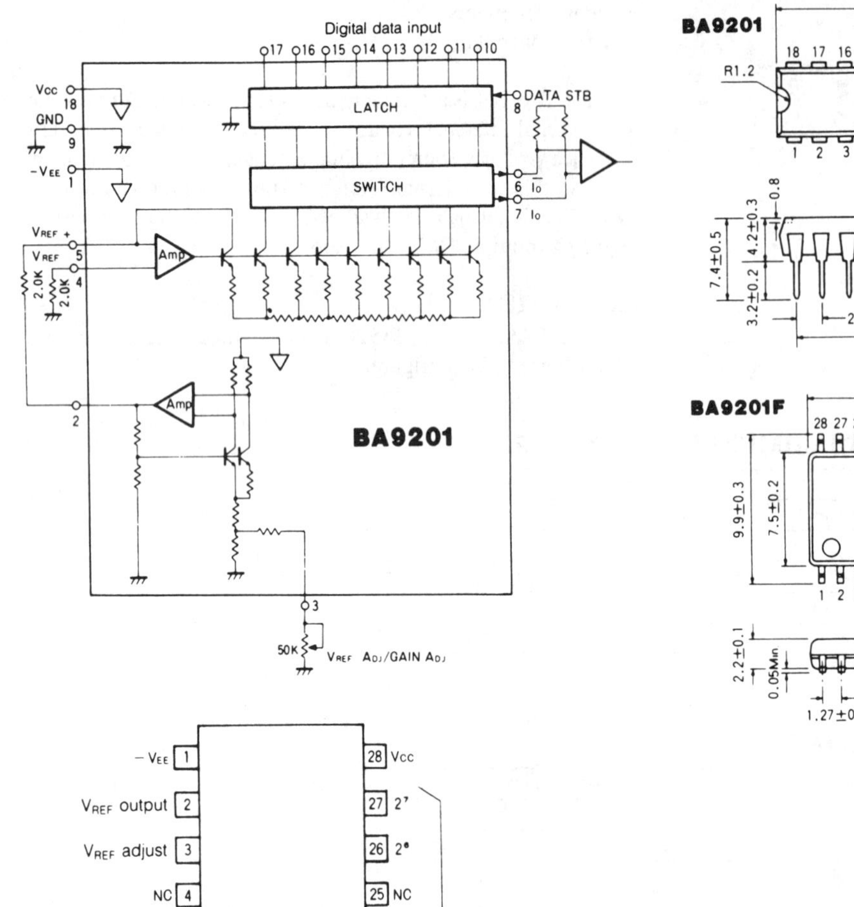

DIMENSIONS (UNIT: MM)

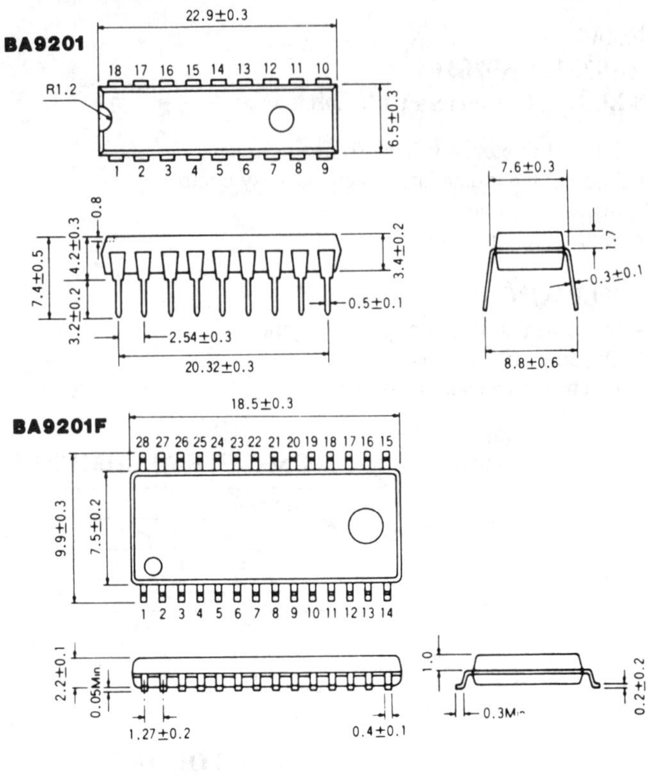

BA9201F pin connections:

Pin	Function	Pin	Function
1	$-V_{EE}$	28	Vcc
2	V_{REF} output	27	2^7
3	V_{REF} adjust	26	2^6
4	NC	25	NC
5	NC	24	NC
6	$V_{REF -}$ input	23	NC
7	$V_{REF +}$ input	22	2^5
8	NC	21	2^4
9	$\overline{\text{Analog output}}$	20	2^3
10	NC	19	NC
11	NC	18	NC
12	Analog output	17	2^2
13	Strobe input	16	2^1
14	GND	15	2^0

Pins 15–17, 20–22, 26, 27: Digital input

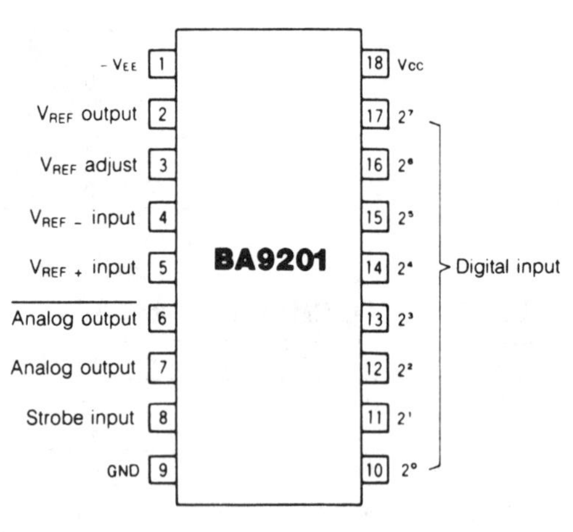

Rohm
BA9211, BA9211F
10-Bit D/A Converters with Internal Reference Voltage Supply

- 10-bit accuracy
- Short settling time: 250ns
- Internal reference voltage supply. This voltage supply is implemented in a separate block so that external reference can also be used.
- Multiplying operation capability
- Available in DIP and MF packages

APPLICATIONS

- Programmable-gain amplifiers
- Programmable attenuators
- Signal generators
- Servo controllers
- Digitally-programmable power supplies
- Electronic musical instruments

The BA9211 and BA9211F are monolithic 10-bit D/A converters featuring an internal reference voltage supply and capable of multiplying operation. With high conversion speed, the devices are primarily intended for use in digital-to-analog control equipment.

THERMAL DERATING CURVE

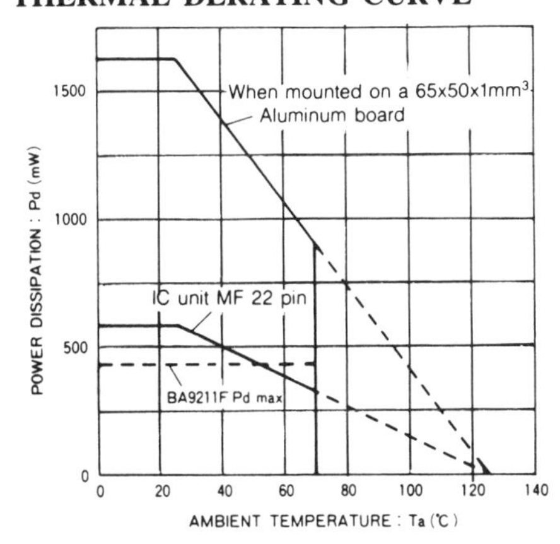

The BA9211F requires a maximum power of approx. 400 mW at an operating temperature of 0 to 70°C, so mount it on a board with good thermal conductivity.

ABSOLUTE MAXIMUM RATINGS (Ta=25°C)

Parameter	Symbol	Limits	Unit
Supply voltage	V_{CC} V_{EE}	+5.5 −15	V
Power dissipation	Pd	1650*	mW
Operating temperature range	Topr	0~70	°C
Storage temperature range	Tstg	−55~125	°C
Reference input pins V_{16}, V_{17}	V_{16} V_{17}	V_{EE}~V_{CC}	V
Logic input pin	V_{IN}	−5~V_{CC}	V

*Above figures with 65 × 50 × 1mm³ aluminum board loading unit of 550mW.

ELECTRICAL CHARACTERISTICS (Unless otherwise specified, Ta=25°C, V_{CC}=+5V, V_{EE}= −12V)

Parameter	Symbol	Min.	Typ.	Max.	Unit	Conditions
Resolution	—	10	10	10	bits	—
Monotonicity	—	10	10	10	bits	—
Differential nonlinearity	D.N.L.	10	—	—	bits	—
Nonlinearity	N.L	—	—	±0.05	%FS	—
Full-scale current	I_{FS}	—	3.996	—	mA	—
Full-scale current temperature coefficient	TCI_{FS}	—	±50	—	ppm/°C	—
Full-scale symmetry current	I_{FSS}	—	—	±2.0	μA	$I_{FSS}=I_{FS}-\overline{I_{FS}}$
Zero scale current	I_{ZS}	—	—	0.1	μA	—
Settling time	ts	—	250	—	ns	—
Input high voltage	V_{IH}	2.0	—	—	V	—
Input low voltage	V_{IL}	—	—	0.8	V	—
Logic input current	I_{IN}	—	—	70	μA	—
Reference voltage supply	V_{REF}	1.96	2.10	2.24	V	—
Reference voltage supply temperature coefficient	TCV_{REF}	—	±50	—	ppm/°C	—
Reference input current	I_{REF}	0.2	1.0	1.1	mA	—
Reference input bias current	I_{17}	—	−0.5	−2.0	μA	—
Reference input slew rate	dI/dt	4.0	8.0	—	mA/μs	C=0 pF
Supply voltage dependency	P_{SS}+	—	—	±0.001	%FS/%	V_{CC}=4.75~5.25V, V_{EE}= −12V
Supply voltage dependency	P_{SS}−	—	—	±0.001	%FS/%	V_{CC}=5.0V, V_{EE}= −11~−15V
Supply voltage range	V_{CC}	4.75	5.0	5.25	V	V_{OUT}=0V
Supply voltage range	V_{EE}	−10.8	−12.0	−13.2	V	V_{OUT}=0V
Circuit supply current	I_{CC}	—	13.0	17.0	mA	V_{CC}=5V, V_{EE}= −12V
Circuit supply current	I_{EE}	—	−16.0	−23.0	mA	V_{CC}=5V, V_{EE}= −12V
Power consumption	Pd	—	257	393	mW	V_{CC}=5V, V_{EE}= −12V

APPLICATION EXAMPLE

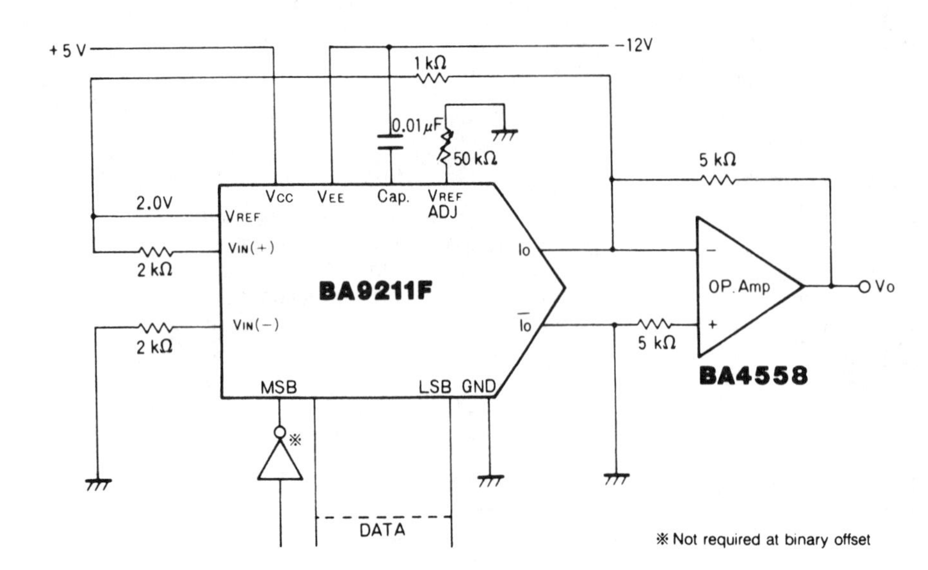

MSB DATA LSB	Vo
0 1 1 1 1 1 1 1 1 1	9.980V
0 1 1 1 1 1 1 1 1 0	9.960V
0 0 0 0 0 0 0 0 0 1	0.020V
0 0 0 0 0 0 0 0 0 0	0.000V
1 1 1 1 1 1 1 1 1 1	−0.020V
1 0 0 0 0 0 0 0 0 1	−9.980V
1 0 0 0 0 0 0 0 0 0	−10.000V

Simple D/A converter

DIMENSIONS (UNIT: MM)

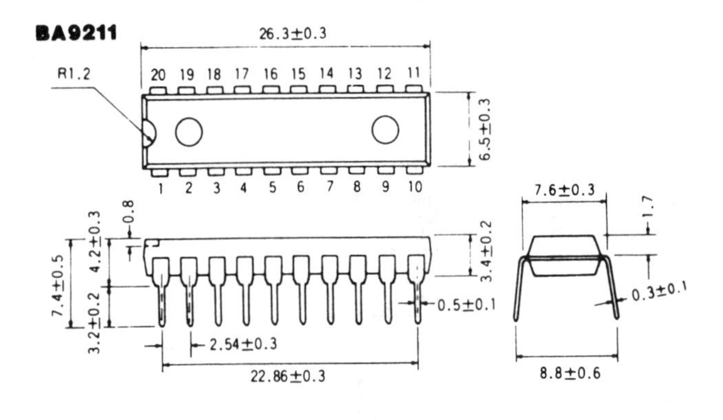

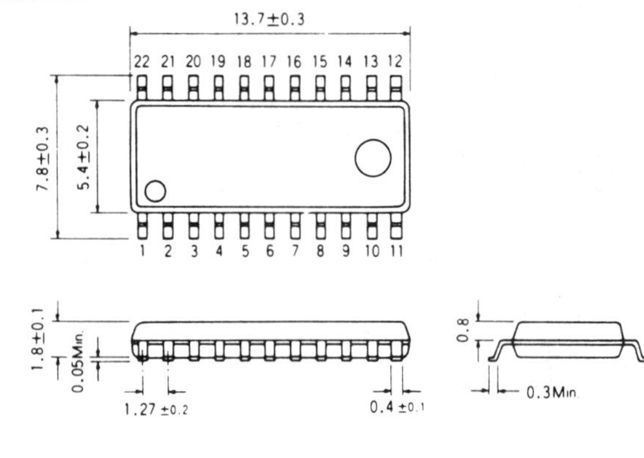

BLOCK DIAGRAM

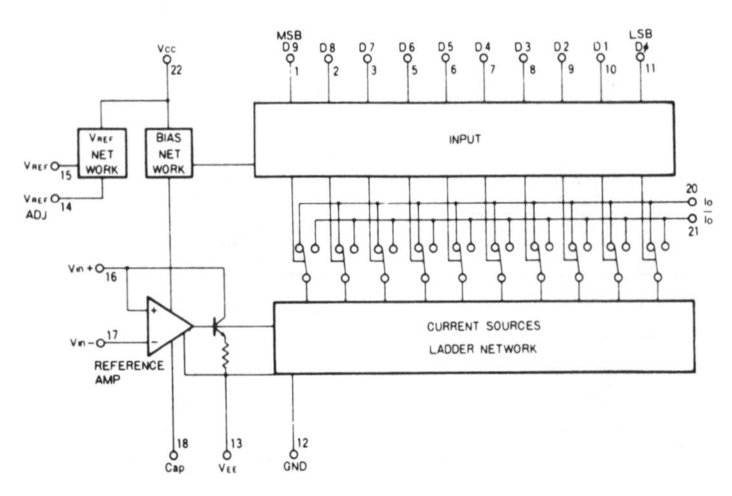

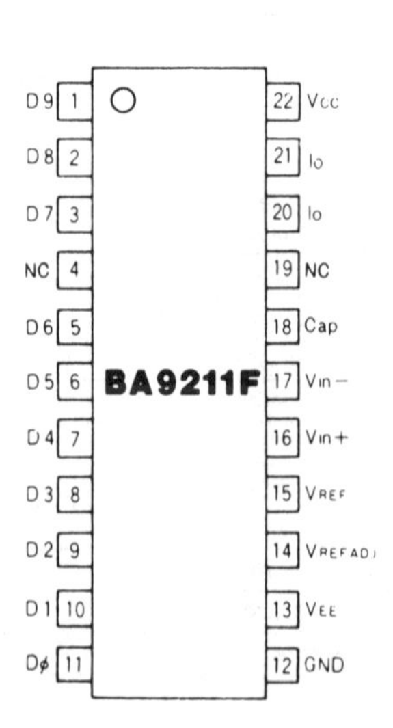

Rohm
BA9221, BA9221F
12-Bit D/A Converters

- 12-bit monotonicity
- Short settling time: 250 ns
- Full-scale current: I_{FS} = 4mA
- Multiplying operation capability
- Differential current output

APPLICATIONS

- Digital audio systems
- Digitally-programmable attenuators
- Servo controllers
- Programmable power supplies
- A/D converters

The BA9221 and BA9221F are monolithic 12-bit D/A converters that are capable of multiplying operation. The conversion time is a short 250 ns. The devices are primarily designed for use in the analog output stage of digital audio systems or high-speed digital-to-analog control systems.

PRECAUTION

The BA9221 and BA9221F come in different packages with different pin configurations.

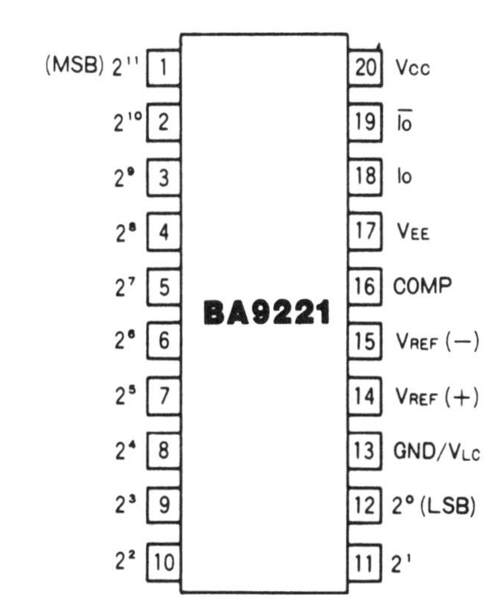

ABSOLUTE MAXIMUM RATINGS (Ta=25°C)

Parameter	Symbol	Limits	Unit
Supply voltage	V_{CC} V_{EE}	+7 −18	V
Power dissipation	Pd	600*	mV
Operating temperature range	Topr	−25 ~ 70	°C
Storage temperature range	Tstg	−55 ~ 125	°C
Reference input pins V_{14}, V_{15}	V_{14}, V_{15}	V_{EE} ~ V_{CC}	V
Logic input pin	V_{IN}	−5 ~ V_{CC}	V

*Derating is done at 6mW/°C for operation above Ta=25°C.

Electrical Characteristics (Unless otherwise specified, Ta=25°C, V_{CC} = +5V, V_{EE} = −15V)

Parameter	Symbol	Min.	Typ.	Max.	Unit	Conditions
Resolution	—	12	12	12	bits	—
Monotonicity	—	12	12	12	bits	—
Differential nonlinearity	D.N.L	—	—	±1	LSB	—
Nonlinearity	N.L	—	—	0.05	%FS	—
Full-scale current	I_{FS}	—	3.999	—	mA	V_{REF} = 10.000V
Full-scale current temperature coefficient	TCI_{FS}	—	±10	—	ppm/°C	—
Full-scale symmetry current	I_{FSS}	—	—	±2.0	μA	$I_{FS}-\overline{I_{FS}}$
Zero scale current	I_{ZS}	—	—	0.1	μA	—
Settling time	ts	—	250	—	ns	Ta=25°C
Input high voltage	V_{IH}	2.0	—	—	V	—
Input low voltage	V_{IL}	—	—	0.8	V	—
Logic input current	I_{IN}	—	—	60	μA	V_{IN} = −5V ~ +5V
Reference voltage supply input current	I_{REF}	0.2	1.0	1.1	mA	—
Reference voltage supply input bias current	I_{15}	—	−0.5	−2.0	μA	—
Reference voltage supply input slew rate	dI/dt	4.0	8.0	—	mA/μs	R_{14} = 800Ω, C_C = 0pF
Supply voltage dependency	$P_{SS}+$	—	—	±0.0001	%FS/%	V_{CC} = 4.5V ~ 5.5V, V_{EE} = −15V
Supply voltage dependency	$P_{SS}-$	—	—	±0.0001	%FS/%	V_{EE} = −13.5V ~ −16.5V, V_{CC} = 5V
Supply voltage range	V_{CC}	4.5	—	5.5	V	V_{OUT} = 0V
Supply voltage range	V_{EE}	−18	—	−10.8	V	V_{OUT} = 0V
Operating circuit current	I_{CC}	—	11.0	18.0	mA	V_{CC} = 5V, V_{EE} = −15V
Operating circuit current	I_{EE}	—	−16.0	−22.0	mA	V_{CC} = 5V, V_{EE} = −15V
Power consumption	Pd	—	295	420	mW	V_{CC} = 5V, V_{EE} = −15V

APPLICATION EXAMPLES

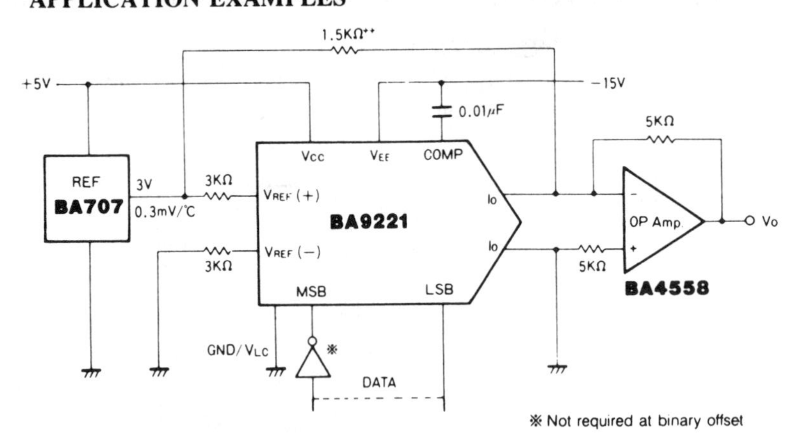

MSB DATA LSB	Vo
0 1 1 1 1 1 1 1 1 1 1 1	9.9951
0 1 1 1 1 1 1 1 1 1 1 0	9.9902
0 0 0 0 0 0 0 0 0 0 0 1	0.0049
0 0 0 0 0 0 0 0 0 0 0 0	0.0000
1 1 1 1 1 1 1 1 1 1 1 1	−0.0049
1 0 0 0 0 0 0 0 0 0 0 1	−9.9951
1 0 0 0 0 0 0 0 0 0 0 0	−10.0000

Simple D/A converter

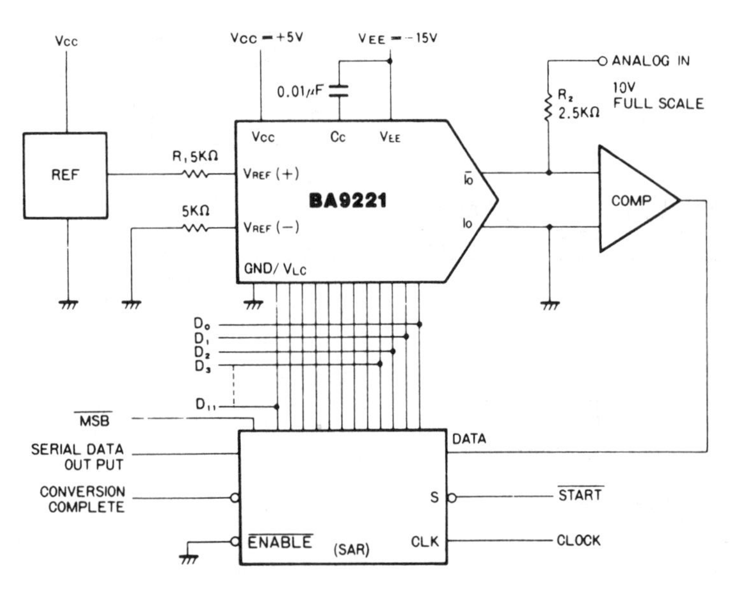

Note: Gain adjustment to be set by inserting VR into R_1 or R_2.

Simple A/D converter

DIMENSIONS (UNIT: MM)

BA9221

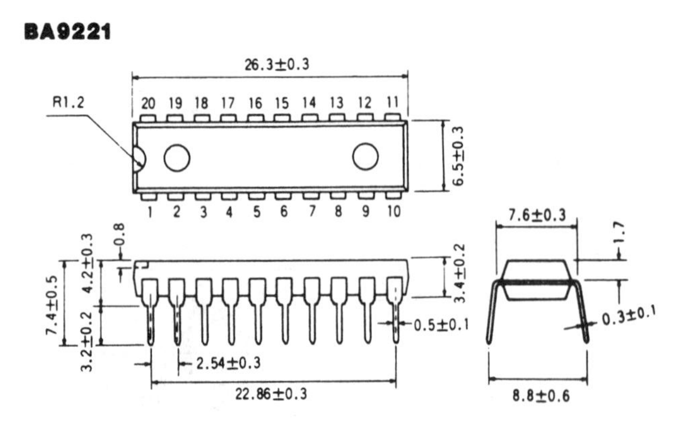

BA9221F

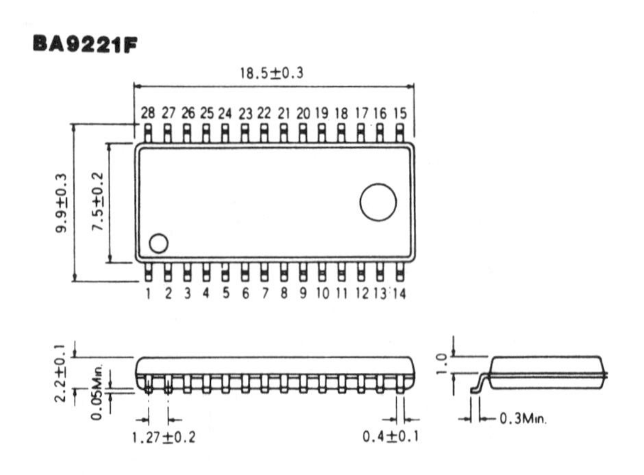

BLOCK DIAGRAM

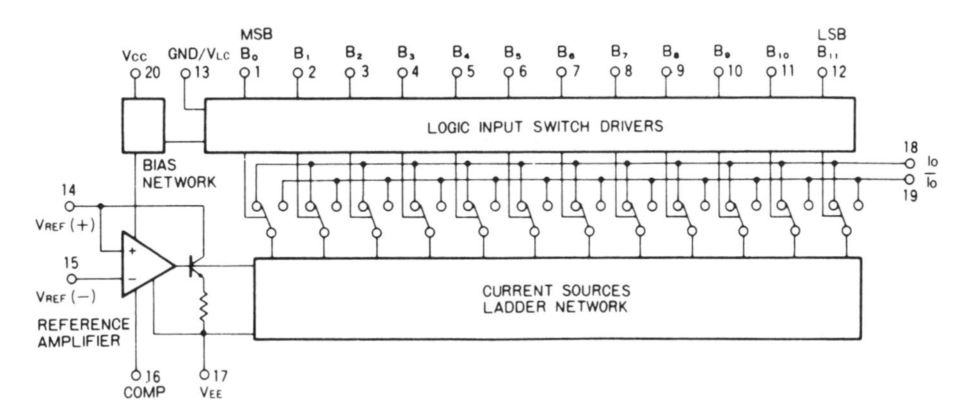

Rohm BA6590S

Centronics Interface

- All the functions required for the Centronics interface implemented on a single chip
- TTL-compatible inputs and outputs
- Operating voltage range is the same as that of TTL logic: 5 ±0.25V
- Available in a 42-pin shrink DIP to save board space
- Centronics: One of the interface standards that specifies the communication protocol between a printer and its host system. Centronics is a trademark of the Centronics company.

APPLICATIONS

- Printers (thermal, dot, ink-jet, etc.)

The BA6590S is a PC/printer interface that complies with the Centronics interface standards. The device consists of an 8-bit D-F/F (received data buffer), D-F/Fs for handshake signals, and output buffer (with open-collector outputs). The operating voltage range is 4.75 to 5.25V, which is compatible with TTL logics. The input and output levels are also TTL compatible.

ABSOLUTE MAXIMUM RATINGS (Ta=25°C)

Parameter	Symbol	Limits	Unit
Supply voltage	V_{CC}	7	V
Power dissipation	Pd	700	mW
Operating temperature range	Topr	0 ~ 70	°C
Storage temperature range	Tstg	− 55 ~ 125	°C
Input voltage range	V_{IN}	− 0.3 ~ 7.0	V

DC ELECTRICAL CHARACTERISTICS (Unless otherwise specified, Ta=25°C, V_{CC}=5V)

Parameter	Symbol	Min.	Typ.	Max.	Unit	Conditions
Supply voltage	V_{CC}	4.75	5	5.25	V	—
Current consumption	I_{CC1}	—	30	50	mA	$\overline{OC}$ = L, all inputs L, V_{CC} = 5V
Current consumption	I_{CC2}	—	30	50	mA	$\overline{OC}$ = L, all inputs H, V_{CC} = 5V
Current consumption	I_{CC3}	—	30	50	mA	$\overline{OC}$ = H, all inputs L, V_{CC} = 5V
Current consumption	I_{CC4}	—	30	50	mA	$\overline{OC}$ = H, all inputs H, V_{CC} = 5V
High input voltage	V_{IH}	2	—	—	V	All input pins except pin 10
Low input voltage	V_{IL}	—	—	0.8	V	All input pins except pin 10
Forward threshold voltage	V_T	1.4	1.6	1.9	V	Pin 10
Reverse threshold voltage	V_T	0.5	0.8	1.1	V	Pin 10
Hysteresis voltage	$V_{T+} - V_{T-}$	0.4	0.8	—	V	Pin 10
High input current	I_{IH1}	—	—	− 600	μA	V_{CC} = 5.25V, V_I = 2.7V *1
High input current	I_{IH2}	—	—	20	μA	V_{CC} = 5.25V, pin 10 *2
Low input current	I_{IL1}	—	—	− 1.3	mA	V_{CC} = 5.25V, V_I = 0.4V *1
Low input current	I_{IL2}	—	—	− 0.4	mA	V_{CC} = 5.25V, pin 10 *2
High output voltage	V_{OH1}	2.4	3.1	—	V	V_{CC} = 4.75V, I_{OH} = − 2.6mA *3
High output voltage	V_{OH2}	2.7	3.4	—	V	V_{CC} = 4.75V, I_{ON} = − 400μA *4
High output voltage	V_{OH3}	2.7	3.4	—	V	V_{CC} = 4.75V, I_{OH} = − 200μA *5, *6
Low output voltage	V_{OL1}	—	0.35	0.5	V	V_{CC} = 4.75V, I_{OL} = 12mA *3
Low output voltage	V_{OL2}	—	0.25	0.4	V	V_{CC} = 4.75V, I_{OL} = 8mA *4, *5, *6
"Z" output current	I_{OZU}	—	—	20	μA	V_{CC} = 5.25V, V_{IH} = 2V, V_O = 2.7V *3
"Z" output current	I_{OZL}	—	—	− 20	μA	V_{CC} = 5.25V, V_{IH} = 2V, V_O = 0.4V *3

*1 Pins 2 ~ 9, 18 ~ 20 *3 Pins 34 ~ 41 *5 Pins 12, 14 ~ 17
*2 Pins 11, 26 ~ 31, 33 *4 Pins 24, 25, 32 *6 Pin 23 (6.8kΩ pull-up resistor)

AC ELECTRICAL CHARACTERISTICS (Unless otherwise specified, Ta=25°C, V_{CC}=5V)

Parameter	Symbol	Min.	Typ.	Max.	Unit	Conditions
STROBE input time width	t_{ST}	0.5	—	—	μs	—
Data output delay time	t_{PLHST}	—	0.6	1.2	μs	$\overline{\text{STROBE}}$→DATA OUT
Data output delay time	t_{PHLST}	—	0.6	1.2	μs	$\overline{\text{STROBE}}$→DATA OUT
Data output delay time	t_{PZH}	—	50	85	ns	OC→DATA OUT
Data output delay time	t_{PZL}	—	50	85	ns	OC→DATA OUT
Data output deay time	t_{PHZ}	—	70	110	ns	OC→DATA OUT
Data output delay time	t_{PLZ}	—	70	110	ns	OC→DATA OUT
BUSY output delay time	t_{PHLBU}	—	0.7	1.4	μs	$\overline{\text{STROBE}}$→$\overline{\text{BUSY}}$
BUSY output delay time	t_{PHLP}	—	0.4	0.8	μs	$\overline{\text{PRE-SET}}$→$\overline{\text{BUSY}}$
BUSY output delay time	t_{PLHC}	—	0.6	1.2	μs	CLEAR→$\overline{\text{BUSY}}$
BUSY output delay time	t_{PLHB}	—	0.8	1.6	μs	$\overline{\text{PRE-SET}}$→BUSY
BUSY output delay time	t_{PHLB}	—	0.6	1.2	μs	CLEAR→BUSY
BUSY output delay time	t_{PLHCL}	—	1.0	2.0	μs	CLEAR→BUSY
BUSY output delay time	t_{PLHSB}	—	0.8	1.6	μs	SEL. BUSY→BUSY
BUSY output delay time	t_{PHLSB}	—	0.65	1.3	μs	SEL. BUSY→BUSY
BUSY output delay time	t_{PHLA}	—	1.15	2.3	μs	$\overline{\text{ACK}}$→BUSY
BUSY output delay time	t_{PLHTB}	—	1.1	2.2	μs	$\overline{\text{STROBE}}$→BUSY
Inverter delay time	t_{PLH}	—	0.4	0.8	μs	—
Inverter delay time	t_{PHL}	—	50	200	ns	—
Buffer delay time	t_{PLH}	—	0.45	0.8	μs	—
Buffer delay time	t_{PHL}	—	100	200	ns	—
CLEAR pulse width	t_{CL}	210	—	—	ns	—

BLOCK DIAGRAM

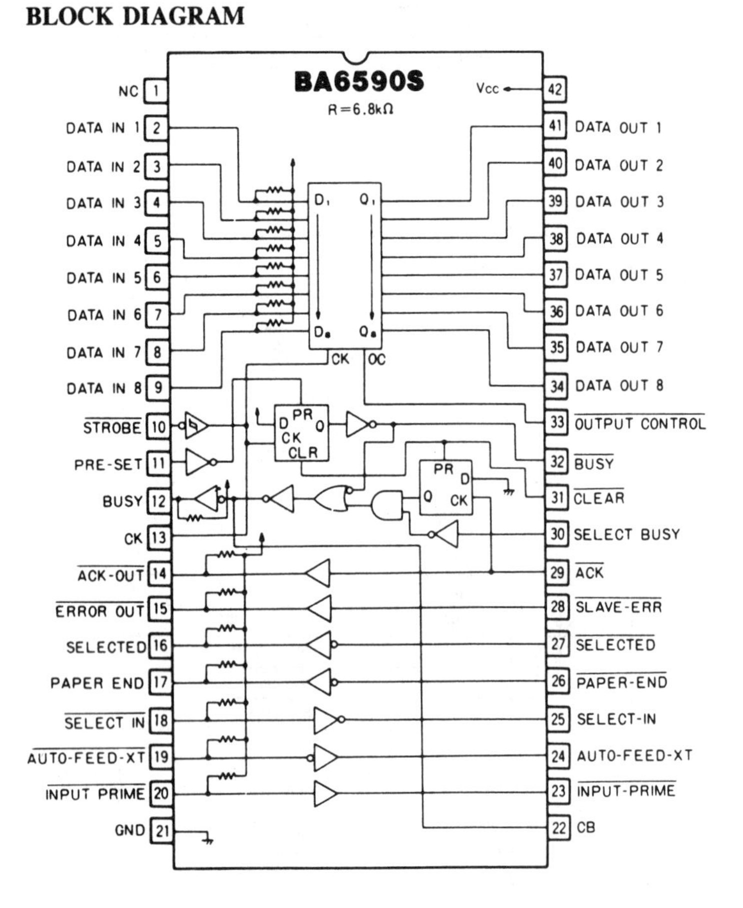

STANDARD CENTRONICS SPECIFICATION CONNECTOR PIN ARRAY

Pin No.	Signal Name	BA6590S pin No.
1	$\overline{\text{STROBE}}$	10
2	DATA 1	2
3	DATA 2	3
4	DATA 3	4
5	DATA 4	5
6	DATA 5	6
7	DATA 6	7
8	DATA 7	8
9	DATA 8	9
10	$\overline{\text{ACK}}$	14
11	BUSY	12
12	PAPER END	17
13	SELECTED	16
14	NC	—
15	NC	—
16	SIGNAL GND	21
17	FG	—
18	+5V dc	—
19	TWISTED PAIR GND	—
20	TWISTED PAIR GND	—
21	TWISTED PAIR GND	—
22	TWISTED PAIR GND	—
23	TWISTED PAIR GND	—
24	TWISTED PAIR GND	—
25	TWISTED PAIR GND	—
26	TWISTED PAIR GND	—
27	TWISTED PAIR GND	—
28	TWISTED PAIR GND	—
29	TWISTED PAIR GND	—
30	TWISTED PAIR GND	—
31	$\overline{\text{INPUT PRIME}}$	20
32	$\overline{\text{ERROR}}$	15
33	SIGNAL GND	—
34	NC	—
35	NC	—
36	NC	—

Each manufacturer has its own signal names.

DIMENSIONS (UNIT: MM)

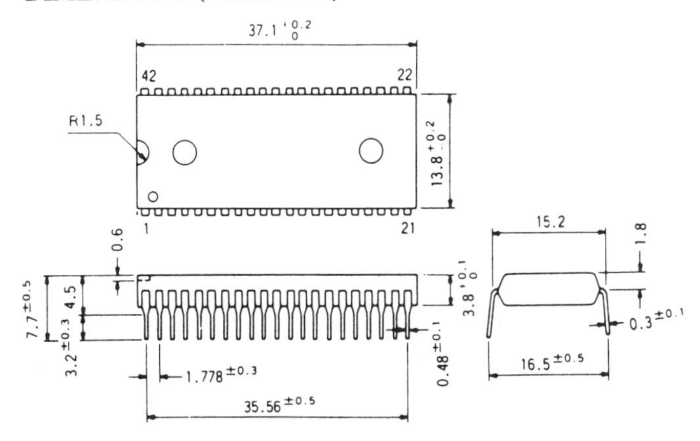

APPLICATION EXAMPLE

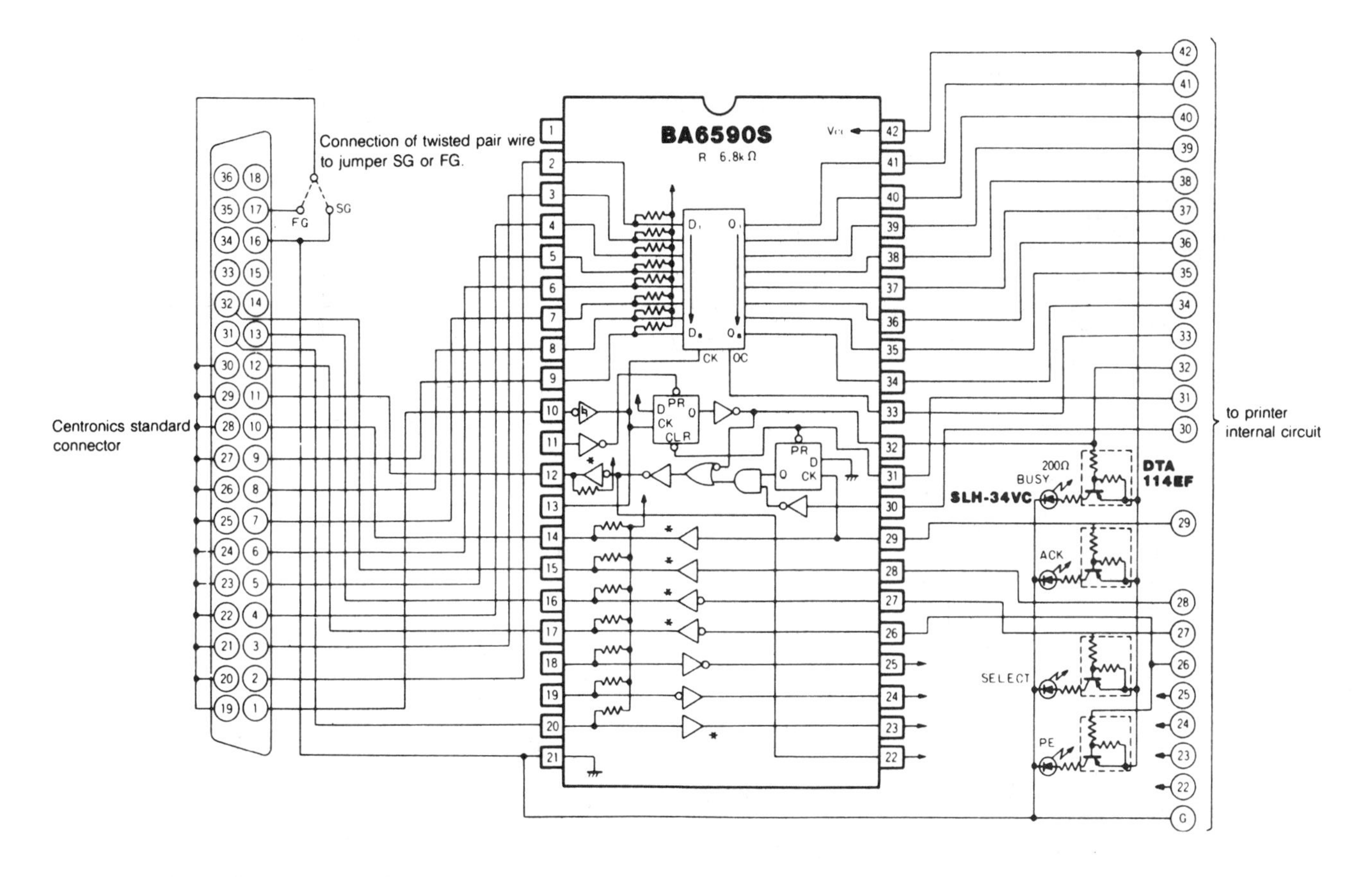

Rohm BA820
8-Bit Serial-In, Parallel-Out Driver

- Driving capacity of 200mA max
- Current required in non-driving periods can be reduced by controlling the strobe input with a drive timing pulse
- Tandem connection is possible, the serial data output is the input for the next driver
- Simplified wiring
- Inputs compatible with TTL and CMOS devices

APPLICATIONS

- Thermal print head drivers
- LED character display drivers

The BA820 is a monolithic driver with a single serial data input and both a single serial and eight parallel data outputs. It is designed for driving thermal heads or LED character displays.

The BA820 has clock (C), data (D_1), and strobe (S) inputs. Serial input data is shifted into the device at the leading edge of the clock. The 8-bit data, which has been set into the device, appears at the parallel O_0-O_7 outputs with the strobe timing. The data pulse width is identical to the strobe pulse width. The serial data output (D_0) is used for tandem connection of more than one device. It provides the output of the last stage of the shift register and is connected to the input of the next driver. The parallel data outputs can be increased in 8-bit increments by sharing the clock and strobe with additional drivers.

ABSOLUTE MAXIMUM RATINGS (Ta=25°C)

Parameter	Symbol	Limits	Unit
Supply voltage	V_{CC}	5.6*1	V
Power dissipation	Pd	550*2	mW
Operating temperature range	Topr	−20~75	°C
Storage temperature range	Tstg	−55~125	°C
Input voltage	V_{IN}	−0.3~6.0	V

*1 12V max. at output pins O_0-O_7.
*2 Derating is done at 5.5mW/°C for operation above Ta=25°C.

ELECTRICAL CHARACTERISTICS (Ta=25°C, V_{CC}=5V)

Parameter	Symbol	Min.	Typ.	Max.	Unit	Conditions
Supply voltage range	V_{CC}	4.5	5.0	5.5	V	V_{CC} pin
Quiescent current	I_Q	—	2	4	mA	When all the data values are "0"
Input high voltage	V_{IH}	2	—	—	V	—
Input low voltage	V_{IL}	—	—	0.8	V	—
Input high current	I_{IH}	—	—	0.4	mA	V_{IN}=4.5V
Maximum output pin impressed voltage	$V_{O\,OFF}$	—	—	16	V	Output O_0~O_7, I_O=10μA
Output saturation voltage	$V_{O\,ON}$	—	—	1.3	V	At I_O=100mA sink current
Output current	I_{OL}	—	—	200	mA	—
Data output voltage	V_{DOH}	2.4	—	—	V	R_L=10kΩ
Maimum data transfer speed	f_D	200	—	—	kHz	V_{IH}=2.0V, V_{IL}=0.8V
Minimum set-up time	t_1	—	—	1	μs	〃
Minimum shift pulse width	t_2	—	—	1	μs	〃
Minimum timing period	t_3	—	—	3	μs	〃

DIMENSIONS (Unit: mm)

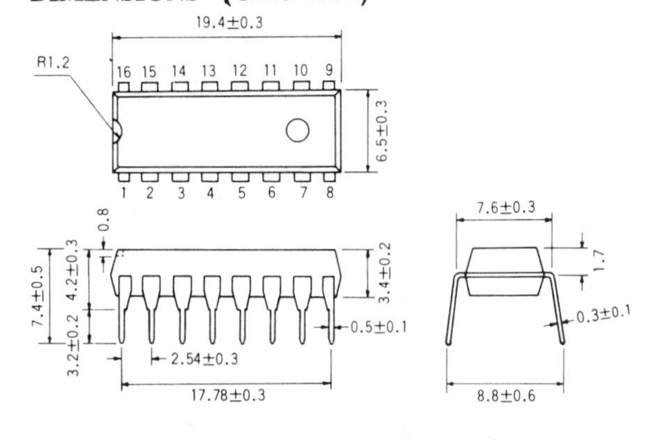

OUTPUT CONDITIONS

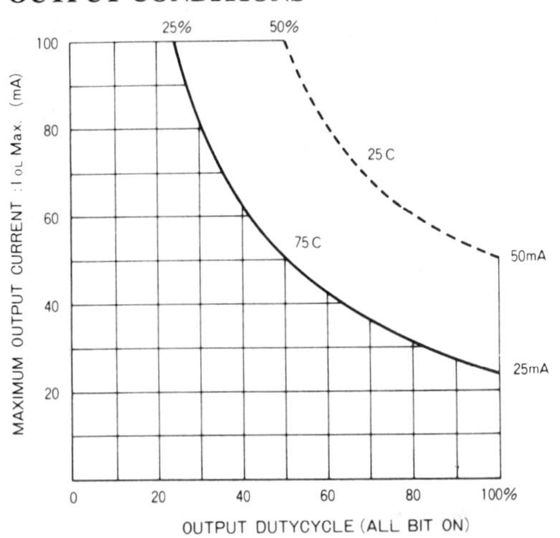

BLOCK DIAGRAM

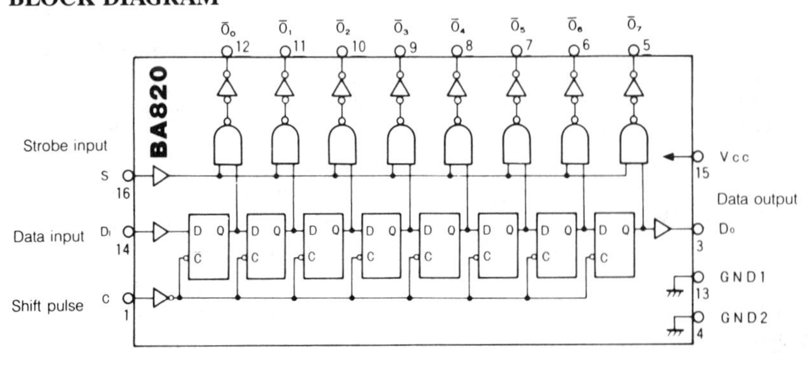

INPUT/OUTPUT CIRCUIT

Input circuit | Output circuit | Data output circuit

1, 14, 16pin | 5, 6, 7, 8, 9, 10, 11, 12pin | 3pin

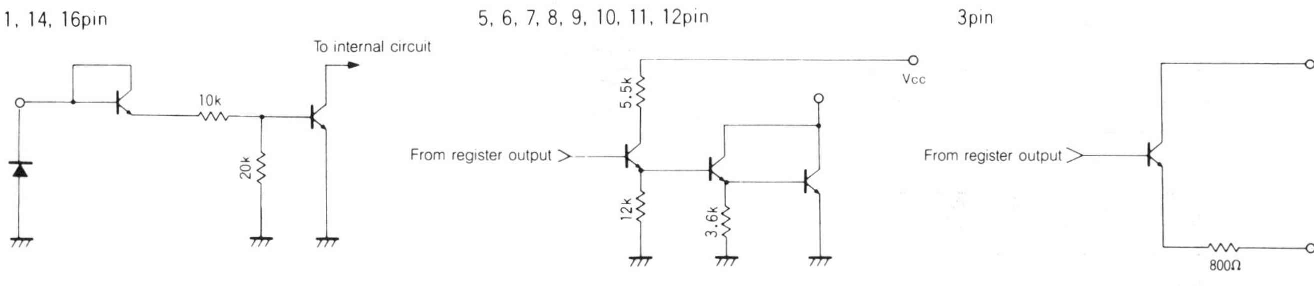

INPUT CONDITIONS

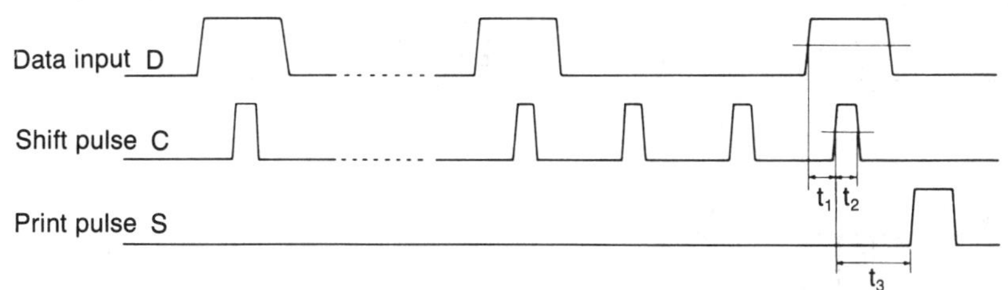

APPLICATIONS AND NUMBER OF CHIPS FOR THE 768 ELEMENT

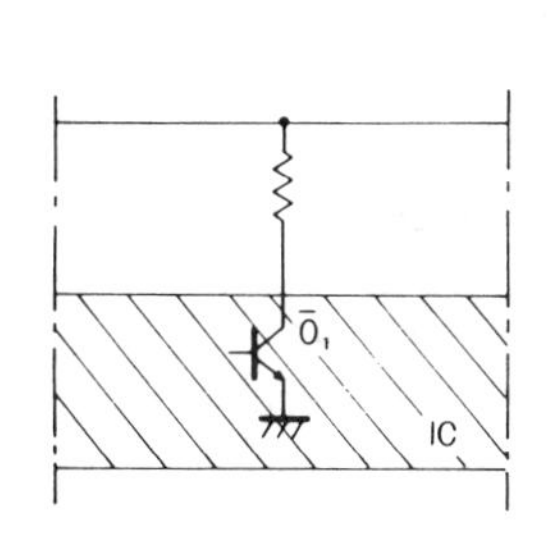

(1) For 1-phase using 96 chips

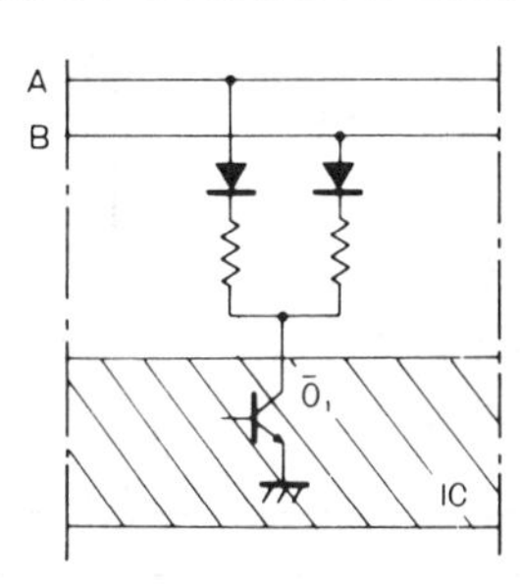

(2) For 2-phase using 48 chips

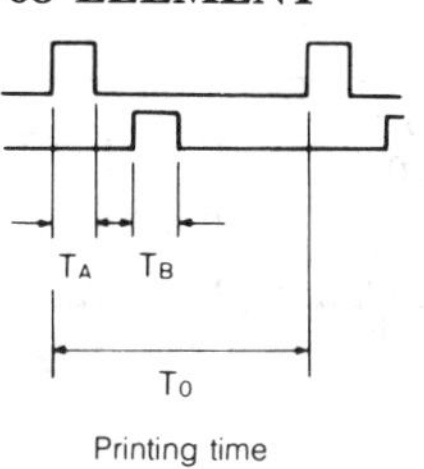

(3) For 3-phase using 32 chips

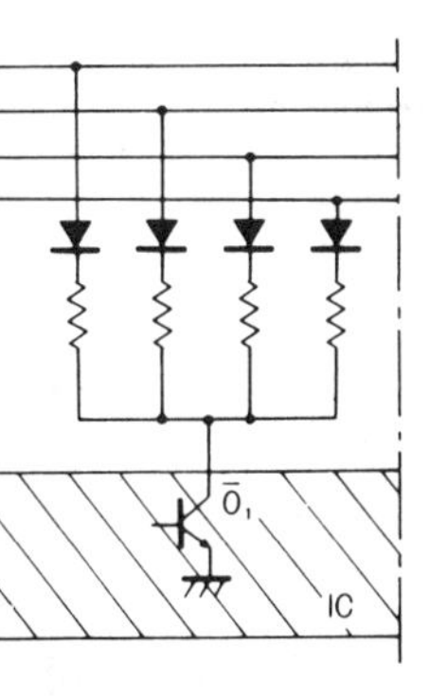

(4) For 4-phase using 24 chips

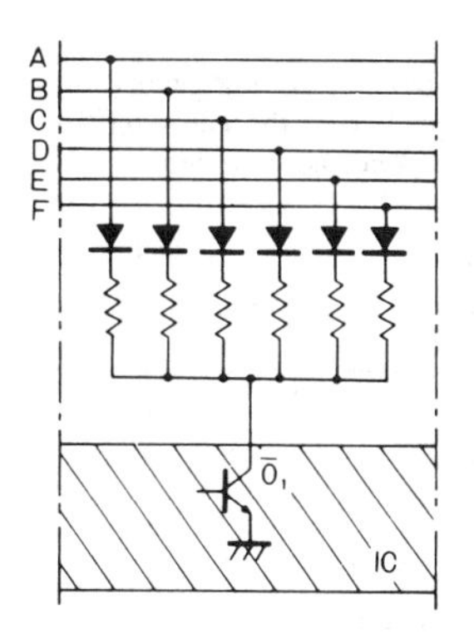

(5) For 6-phase using 16 chips

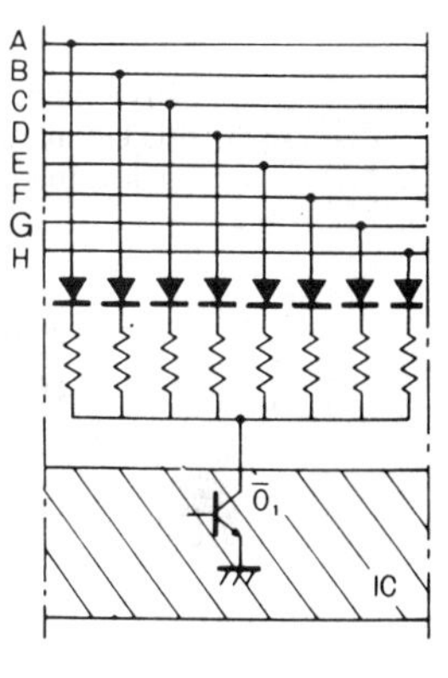

(6) For 8-phase using 12 chips

Note:
Observe the condition

$$\text{Pd Max} > \left(\frac{(T_A + T_B + \ldots)}{T_O} \times V_{CE(sat)} \times I_O \right)$$

where, I_O: Total current with 8 outputs on
$V_{CE(sat)}$: Voltage with 8 outputs on
T_O: Time for 1 cycle
T_A: Total time for 1 print pulse

PIN DESCRIPTION

Pin No.	Pin	Symbol	Features
1	SHIFT PULSE	C	Shift pulse of shift register
14	DATA INPUT	D_1	Data input of shift register, stored at leading edge of shift pulse
16	STROBE	S	Outputs content of shift register when "1"
12	OUTPUT	$\bar{O}_0$	Goes low when the content of shift register is "1" at the output of the first bit.
11	〃	$\bar{O}_1$	Goes low when the content of shift register is "1" at the output of the second bit.
10	〃	$\bar{O}_2$	Goes low when the content of shift register is "1" at the output of the third bit.
9	〃	$\bar{O}_3$	Goes low when the content of shift register is "1" at the output of the fourth bit.
8	〃	$\bar{O}_4$	Goes low when the content of shift register is "1" at the output of the fifth bit.
7	〃	$\bar{O}_5$	Goes low when the content of shift register is "1" at the output of the sixth bit.
6	〃	$\bar{O}_6$	Goes low when the content of shift register is "1" at the output of the seventh bit.
5	〃	$\bar{O}_7$	Goes low when the content of shift register is "1" at the output of the eighth bit.
3	DATA OUTPUT	D_O	Signal from the output circuit of O_7 and becomes the input for the next stage.
15	V_{CC}	V_{CC}	Typically 5.0V (±10%)
13	GND	GND_1	Ground for output circuits of O_0-O_3
4	GND	GND_2	Ground for output circuits of O_4-O_7
2	NO CONNECTION	NC	—

APPLICATION EXAMPLE

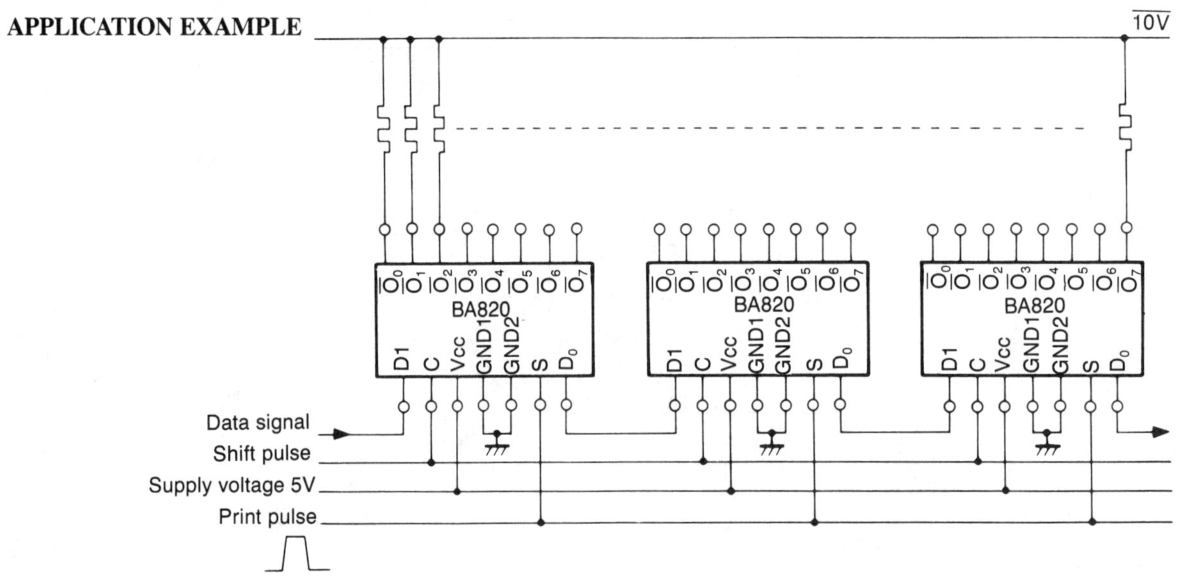

Note: Application circuit diagram for thermal printer
(timing is 1-phase)

This example uses the strobe input for print timing. The advantage of this circuit configuration is that the common line (large current) for the heating elements need not be switched.

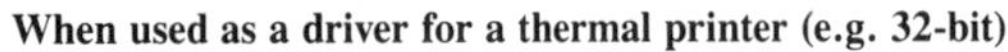

When used as a driver for a thermal printer (e.g. 32-bit)

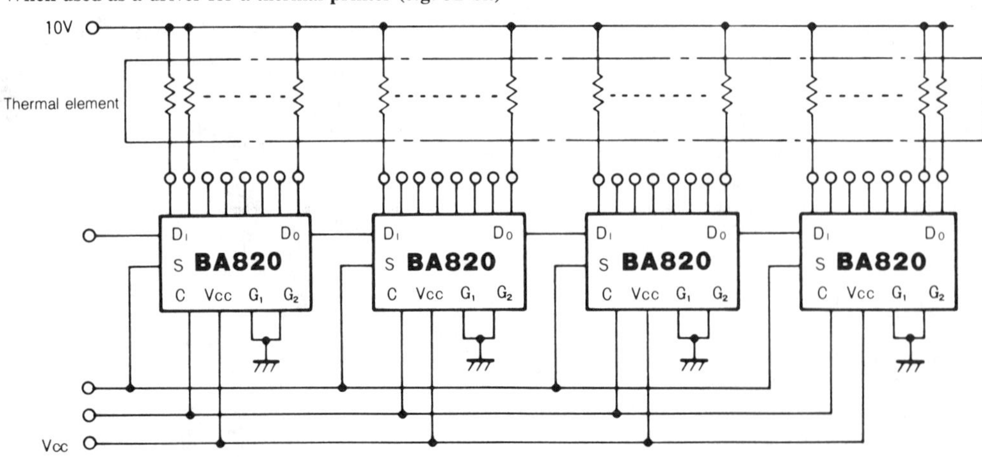

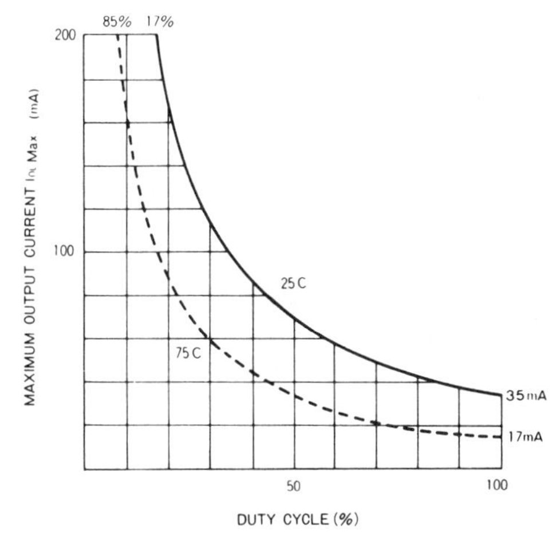

Maximum output current vs. duty cycle

TIMING DIAGRAM

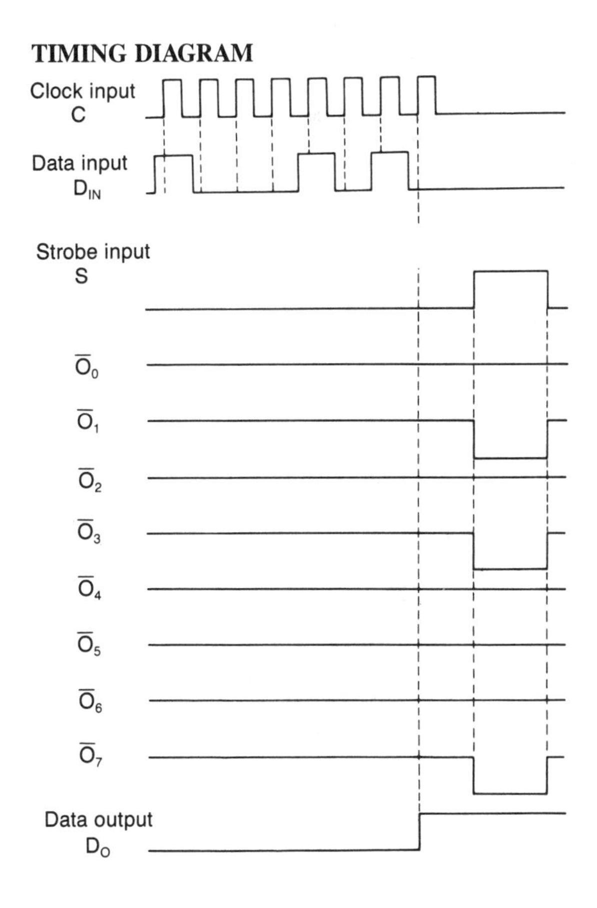

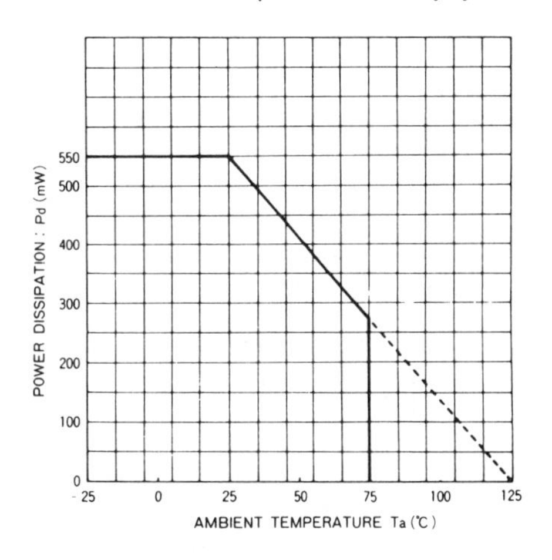

Power dissipation vs. ambient temperature

Both GND pins (pins 4 and 13) should be connected to the GND line. when driving inductive loads, use surge clamping diodes in parallel with the loads.

Siliconix DG411/412/413

Precision Monolithic Quad SPST CMOS Analog Switches

- $\pm$15V input range
- On-resistance <35Ω
- Fast switching action t_{ON} < 175ns
- Ultra low power (P_D <35μW)
- TTL, CMOS compatible
- ESD protection > $\pm$4000V
- Single supply capability
- Widest dynamic range
- Low signal errors and distortion
- Break-before-make switching action
- Simple interfacing

APPLICATIONS

- Precision automatic test equipment
- Precision data acquisition
- Communication systems
- Battery-operated systems

The DG411 series of monolithic quad analog switches was designed to provide high-speed low-error switching of precision analog signals. Combining low power (<35μW) with high speed (t_{ON} < 175ns), the DG411 family is ideally suited for portable and battery-powered industrial and military applications.

To achieve high-voltage ratings and superior switching performance, the DG411 series was built on Siliconix's high-voltage silicon-gate process. An epitaxial layer prevents latchup.

Each switch conducts equally well in both directions when on, and blocks up to the supplies when off. On-resistance is very flat over the full $\pm$15V analog range, rivaling JFET performance without the inherent dynamic range limitation.

The three devices in this series are differentiated by the type of switch action, as shown in the functional block diagrams. Package options include the 16-pin CerDIP, plastic and small outline (SO) packages. Performance grades include both the industrial D-suffix (−40 to 85°C), and the military A-suffix (−55 to 125°C) temperature ranges.

ABSOLUTE MAXIMUM RATINGS

V+ to V- . . . . 44 V
GND to V- . . . . 25 V
V_L . . . . (GND -0.3 V) (V+) +0.3 V
Digital Inputs, V_S, V_D[1] . . . . (V-) -2 V to (V+) +2 V
or 30 mA, whichever occurs first
Continuous Current (Any Terminal) . . . . 30 mA
Current, S or D (Pulsed 1 ms, 10% Duty Cycle) . . . . 100 mA
Storage Temperature (A Suffix) . . . . -65 to 150°C
(D Suffix) . . . . -65 to 125°C
Operating Temperature (A Suffix) . . . . -55 to 125°C
(D Suffix) . . . . -40 to 85°C

Power Dissipation (Package)*
16-Pin Plastic DIP** . . . . 470 mW
16-Pin CerDIP*** . . . . 900 mW
16-Pin SO**** . . . . 600 mW

*All leads welded or soldered to PC board.
**Derate 6 mW/°C above 25°C.
***Derate 12 mW/°C above 75°C.
****Derate 7.6 mW/°C above 75°C.

[1]Signals on S_X, D_X, or IN_X exceeding V+ or V- will be clamped by internal diodes. Limit forward diode current to maximum current ratings.

FUNCTIONAL BLOCK DIAGRAM AND PIN CONFIGURATION

Dual-In-Line Package

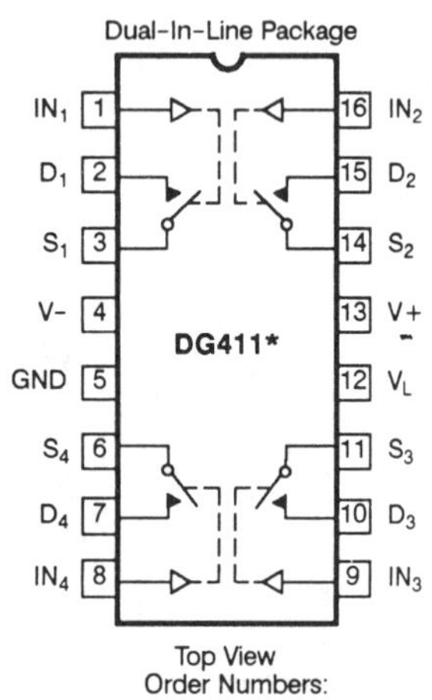

Top View
Order Numbers:
CerDIP: DG411AK, DG411AK/883
DG412AK, DG412AK/883
Plastic: DG411DJ, DG412DJ

Narrow Body
SO Package

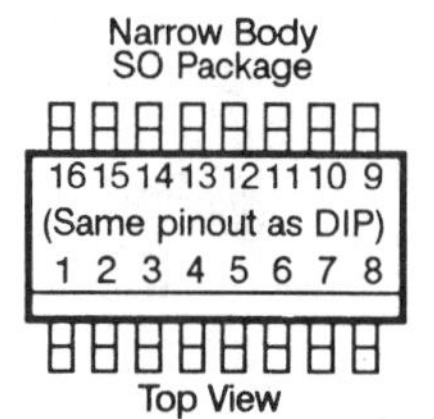

Top View

Order Numbers:
DG411DY, DG412DY

Four SPST Switches per Package
Truth Table

Logic	Switch DG411	Switch DG412
0	ON	OFF
1	OFF	ON

Logic "0" $\leq$ 0.8 V
Logic "1" $\geq$ 2.4 V

* Switches shown for logic "1" input.

Dual-In-Line Package

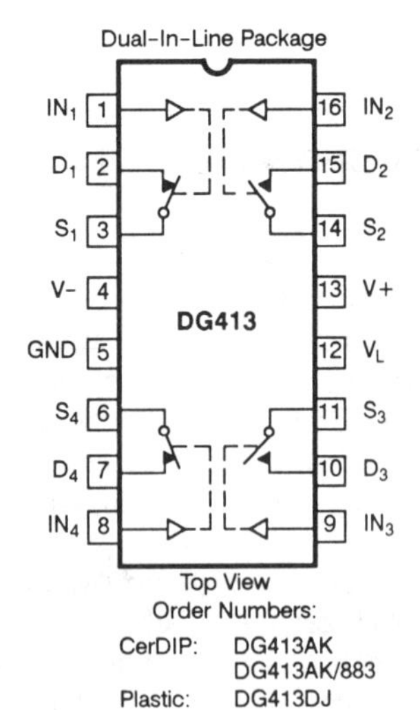

Top View
Order Numbers:
CerDIP: DG413AK
DG413AK/883
Plastic: DG413DJ

SO Package

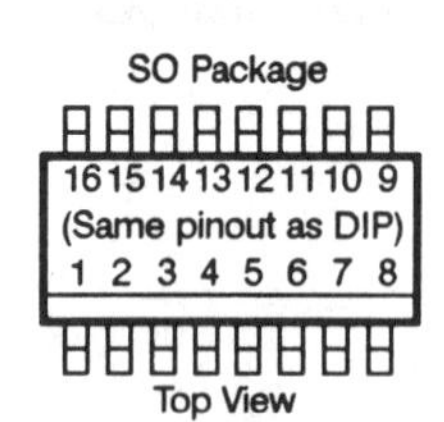

Top View

Order Number:
DG413DY

DG413

Four SPST Switches per Package

Truth Table*

Logic	Switch 1, 4	Switch 2, 3
0	OFF	ON
1	ON	OFF

Logic "0" $\leq$ 0.8 V
Logic "1" $\geq$ 2.4 V

* Switches shown for logic "1" input.

TYPICAL CHARACTERISTICS

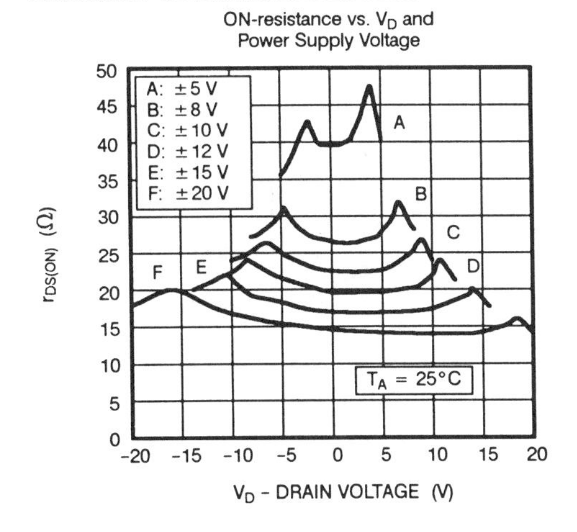

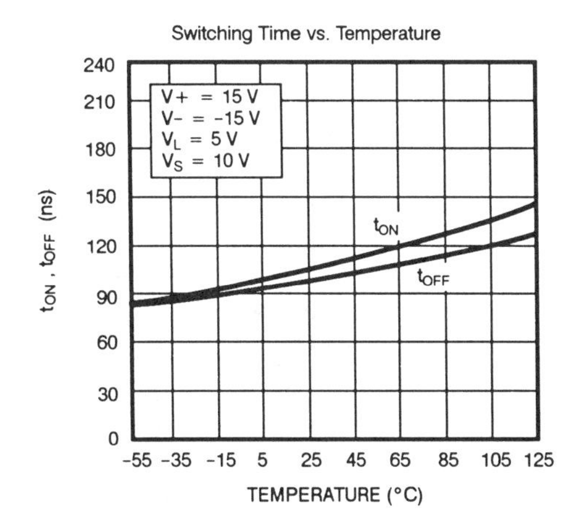

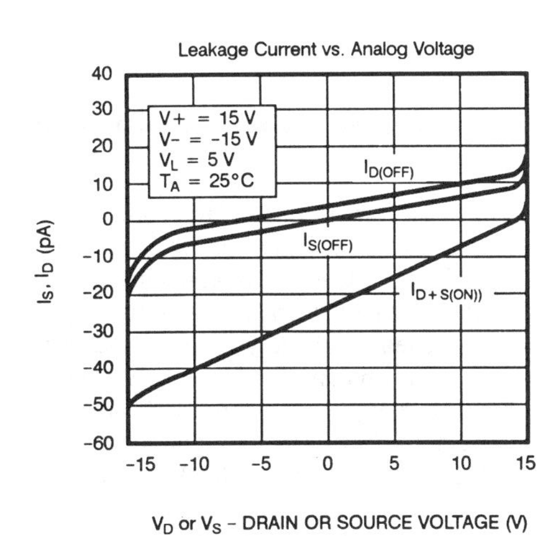

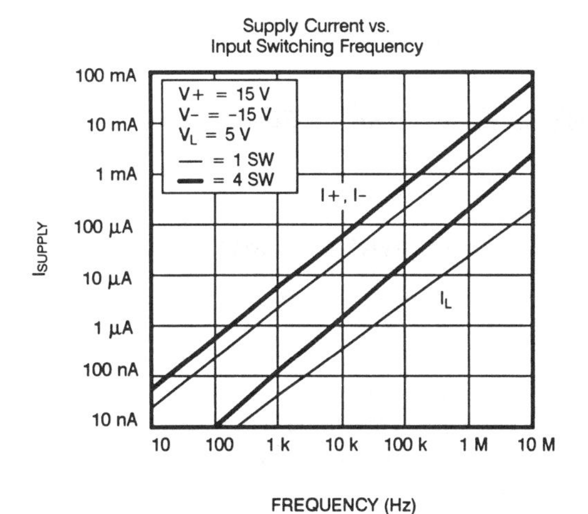

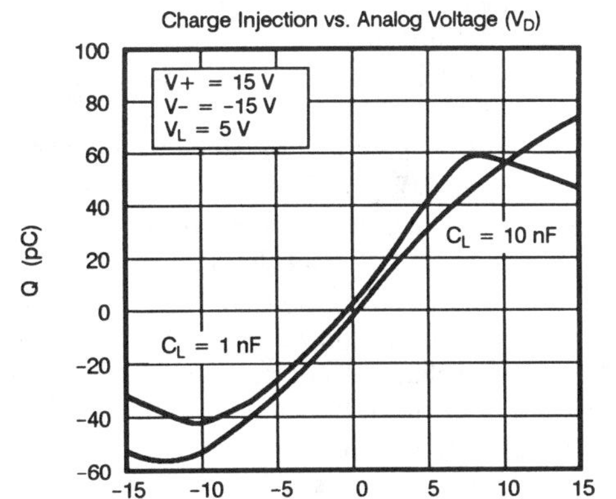

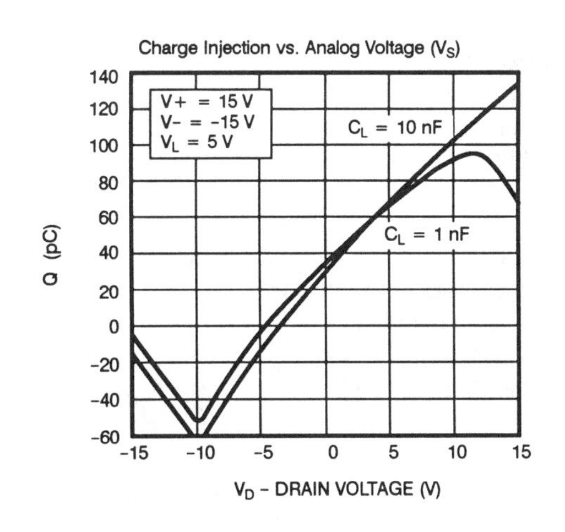

SCHEMATIC DIAGRAM (Typical Channel)

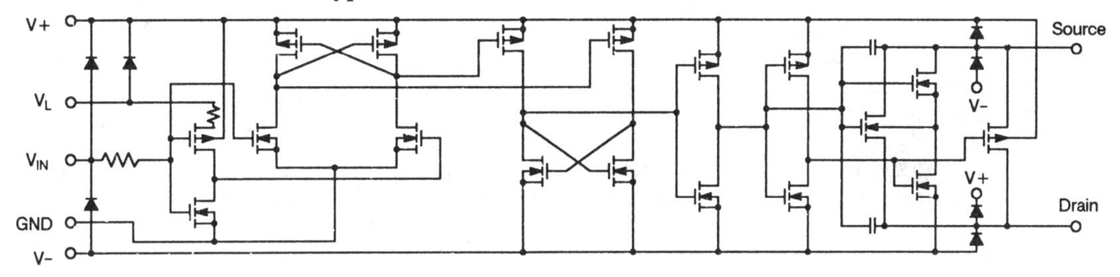

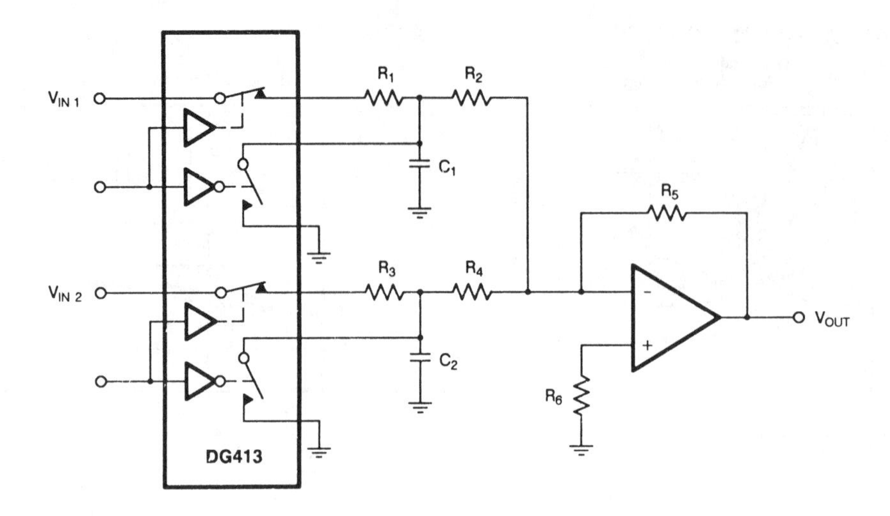

Summing Amplifier

Siliconix DG421/423/425

Low-Power, High-Speed, Latchable CMOS Analog Switches

- Latched inputs
- ± 15V input range
- On-resistance $<35\Omega$
- Fast switching action $t_{ON} < 250$ ns
- Micropower requirements ($P_D < 35\mu W$)
- TTL, CMOS compatible
- Wide dynamic range
- μP compatible
- Reduced component count
- Low signal errors and distortion
- Break-before-make switching action
- Battery operation

APPLICATIONS

- High performance data bus switching
- Precision sample-and-hold circuits
- Digital filters
- μP-controlled analog systems
- Portable instruments

The DG421 series of dual monolithic analog switches features latchable logic inputs which simplify interfacing with microprocessors. This series combines fast switching speed ($t_{ON} < 250$ns), and low on-resistance ($r_{DS(ON)} < 35\ \Omega$), making it ideally suited for battery-powered industrial and military applications that require μP-compatible analog switches.

To achieve high-voltage ratings and superior switching performance, the DG421 series is built on Siliconix's high-voltage silicon-gate CMOS process. Break-before-make is guaranteed for the DG423. An epitaxial layer prevents latchup.

Each switch conducts equally well in both directions when on and blocks up to 30V peak-to-peak when off. On-resistance is nearly flat over the full ± 15V analog range, rivaling JFET performance without the inherent dynamic range and supply-voltage limitations.

When $\overline{WR}$ is set low the input data latches become transparent. When $\overline{WR}$ goes high, the latches store the logic control data. The $\overline{RS}$ pin is used to reset all the switches in the circuit to the default value (all inputs low) when it is set low.

This family offers three devices, which are differentiated by switch action as shown in the functional block diagrams. Packaging includes 16-pin plastic DIP and CerDIP. Performance grades include both the industrial D-suffix (-40 to 85 °C) and the military A-suffix (-55 to 125 °C) temperature ranges. Additionally, a 20-pin PLCC is available for the DG423.

ABSOLUTE MAXIMUM RATINGS

Voltages Referenced to V−

Parameter		Rating
V+		44 V
GND		25 V
V_L		(GND −0.3 V) to (V+) +0.3 V
Digital Inputs[1] V_S, V_D		V− minus 2 V to (V+ plus 2 V) or 30 mA, whichever occurs first
Continuous Current (Any Terminal)		40 mA
Current, S or D (Pulsed 1 ms, 10% duty)		100 mA
Storage Temperature	(A Suffix)	−65 to 150°C
	(D Suffix)	−65 to 125°C
Operating Temperature	(A Suffix)	−55 to 125°C
	(D Suffix)	−40 to 85°C

Power Dissipation (Package)*

16-Pin Plastic DIP**	470 mW
16-Pin CerDIP***	900 mW
20-Pin PLCC****	800 mW

*All leads welded or soldered to PC Board.
**Derate 6 mW/°C above 75°C.
***Derate 12 mW/°C above 75°C.
****Derate 10 mW/°C above 75°C.

[1]Signals on S_X, D_X, or IN_X exceeding V+ or V− will be clamped by internal diodes. Limit forward diode current to maximum current ratings.

PIN CONFIGURATIONS, FUNCTIONAL BLOCK DIAGRAMS, AND TRUTH TABLES

Dual-In-Line Package

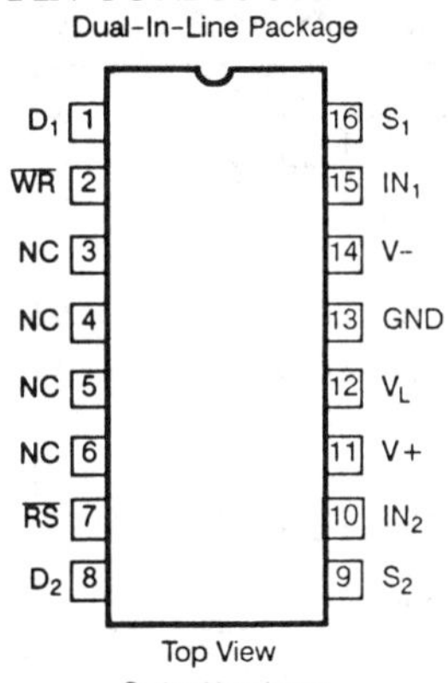

Top View

Order Numbers:

CerDIP: DG421AK, DG421AK/883
Plastic: DG421DJ

DG421

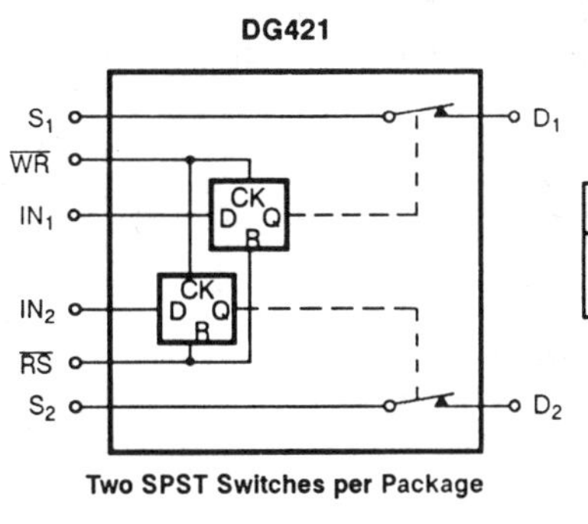

Two SPST Switches per Package

Truth Table*

$\overline{WR}$	$\overline{RS}$	IN_X	Switch
0	1	0 1	OFF ON

Logic "0" $\leq$ 0.8 V
Logic "1" $\geq$ 2.4 V

*Switches Shown for Logic "I" Input

Dual-In-Line Package

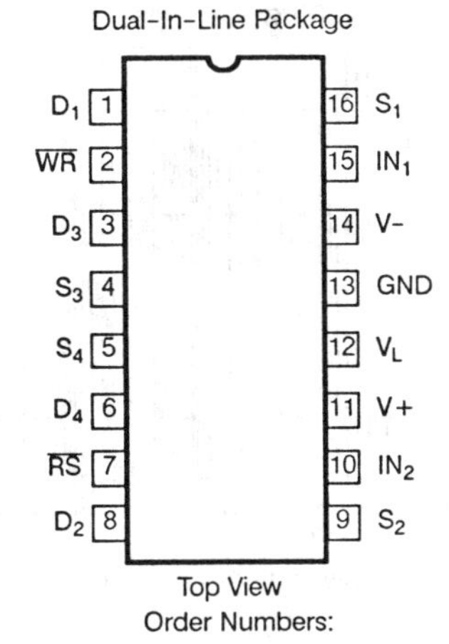

Top View

Order Numbers:

CerDIP: DG423AK, DG425AK
DG423AK/883, DG425AK/883
Plastic: DG423DJ, DG425DJ

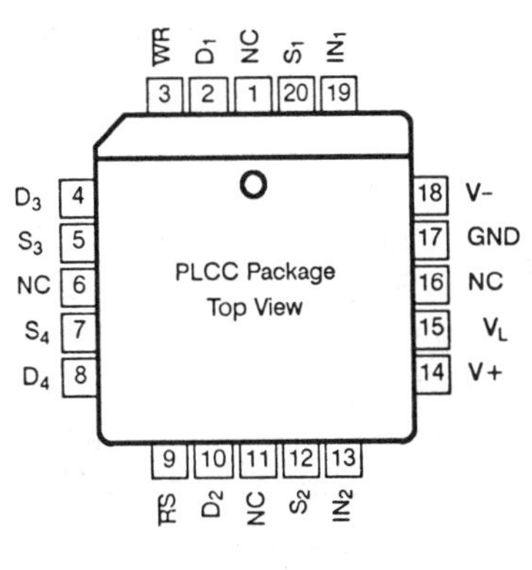

Order Number:

DG423DN

DG423

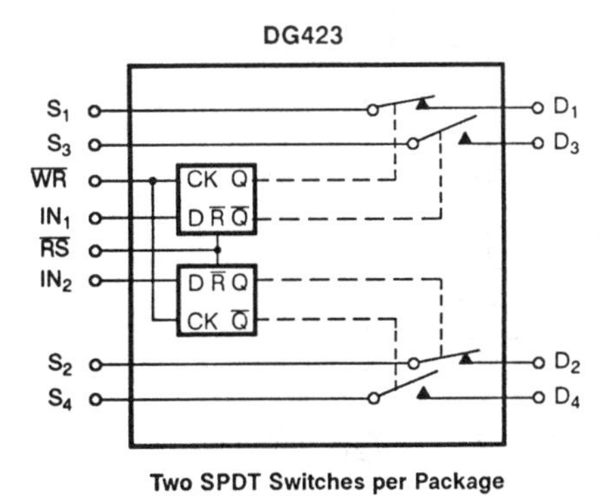

Two SPDT Switches per Package

DG425

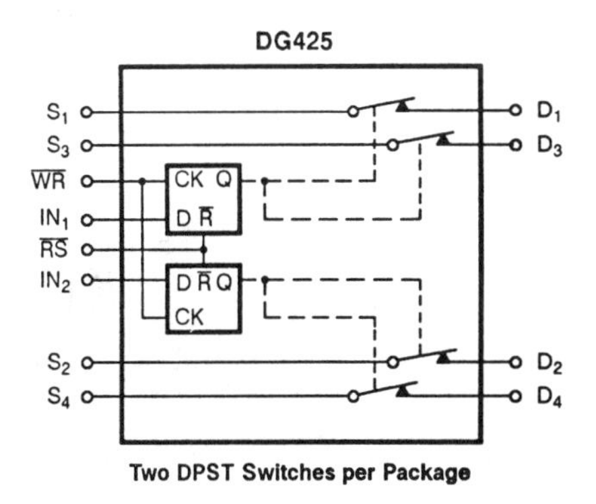

Two DPST Switches per Package

Truth Table

$\overline{WR}$	$\overline{RS}$	IN_X	SW 1, 2	SW 3, 4
0	1	0 1	OFF ON	ON OFF

Logic "0" $\leq$ 0.8 V
Logic "1" $\geq$ 2.4 V

Truth Table

$\overline{WR}$	$\overline{RS}$	IN_X	Switch
0	1	0 1	OFF ON

Logic "0" $\leq$ 0.8 V
Logic "1" $\geq$ 2.4 V

Latch Operation Truth Table

IN_X	$\overline{RS}$	$\overline{WR}$	Latch/Switch X
X	1	0	Transparent Latch Operation
X	1	⌐ (rising edge)	Control Data Latched-in, Switches On or Off as Selected by Last IN_X
X	0	X	All Latches Reset, Switches On or OFF as When IN_X = 0, $\overline{WR}$ = 0, $\overline{RS}$ = 1
X	⌐ (falling edge)	X	

TYPICAL CHARACTERISTICS

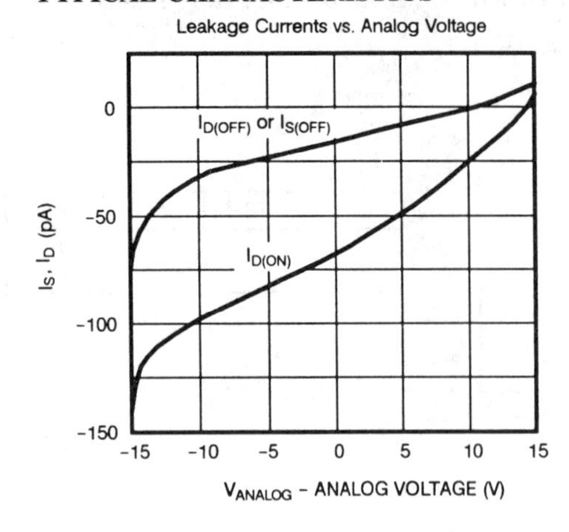

Leakage Currents vs. Analog Voltage

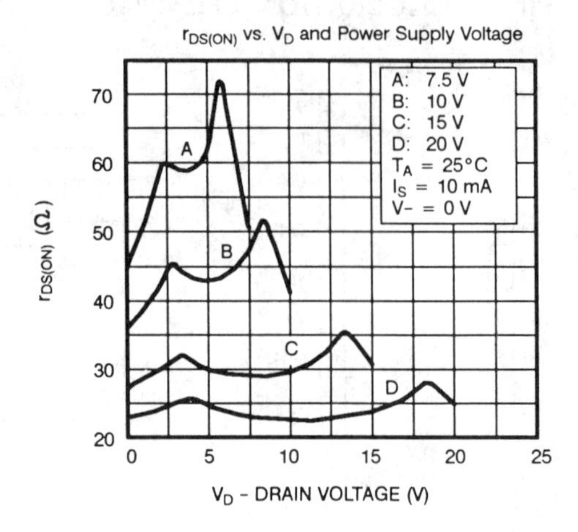

$r_{DS(ON)}$ vs. V_D and Power Supply Voltage

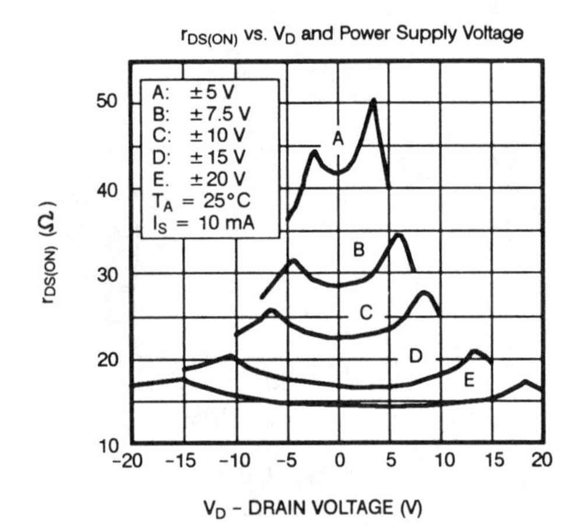

$r_{DS(ON)}$ vs. V_D and Power Supply Voltage

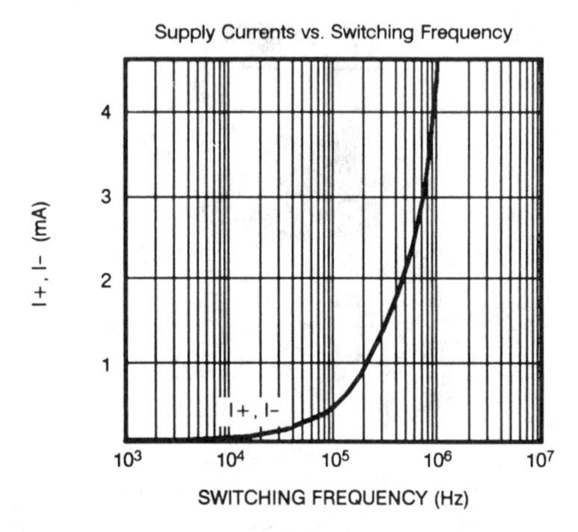

Supply Currents vs. Switching Frequency

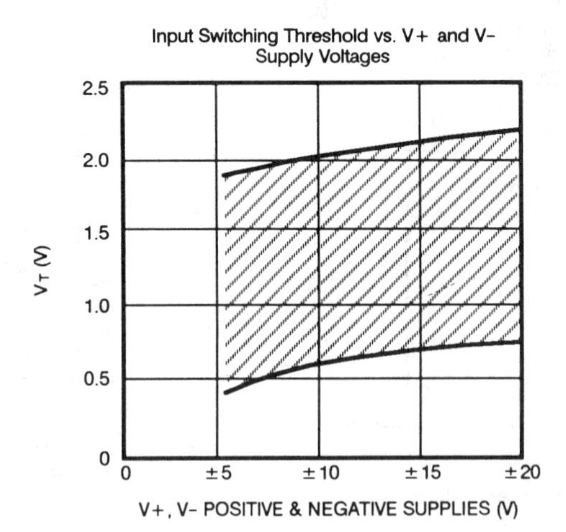

Input Switching Threshold vs. V+ and V– Supply Voltages

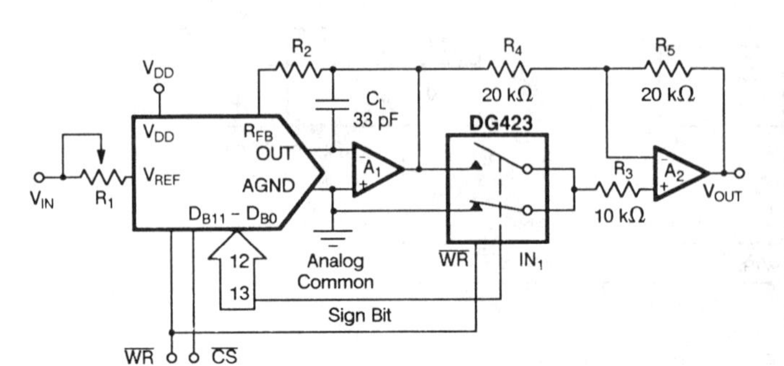

12-Bit Plus Sign Magnitude D/A Converter

12-Bit Plus Sign Magitude Code Table

Sign Bit	Digital Input MSB LSB	Analog Output (V_{OUT})
0	1111 1111 1111	+(4095/4096)V_{IN}
0	0000 0000 0000	0 Volts
1	0000 0000 0000	0 Volts
1	1111 1111 1111	+(4095/4096)V_{IN}

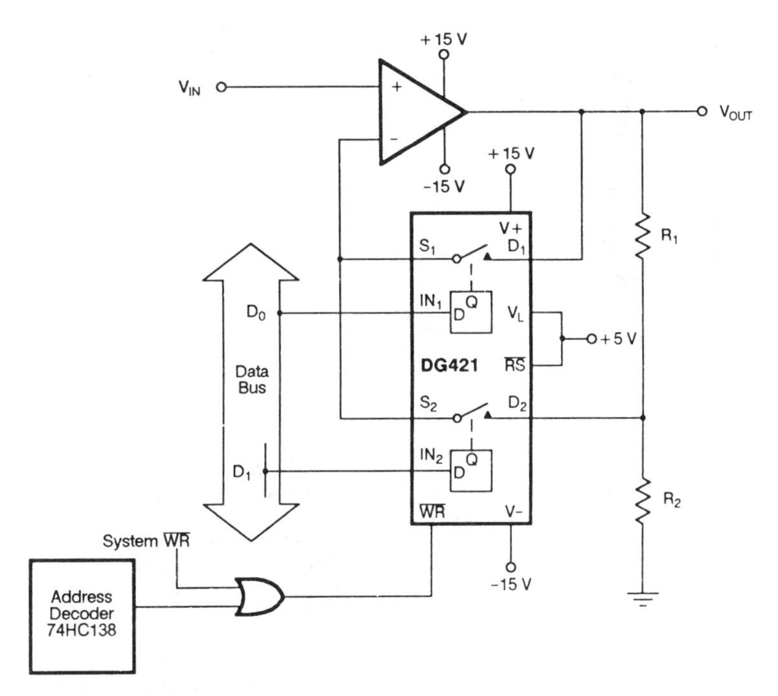

Bus-Controlled Precision Gain-Ranging Circuit

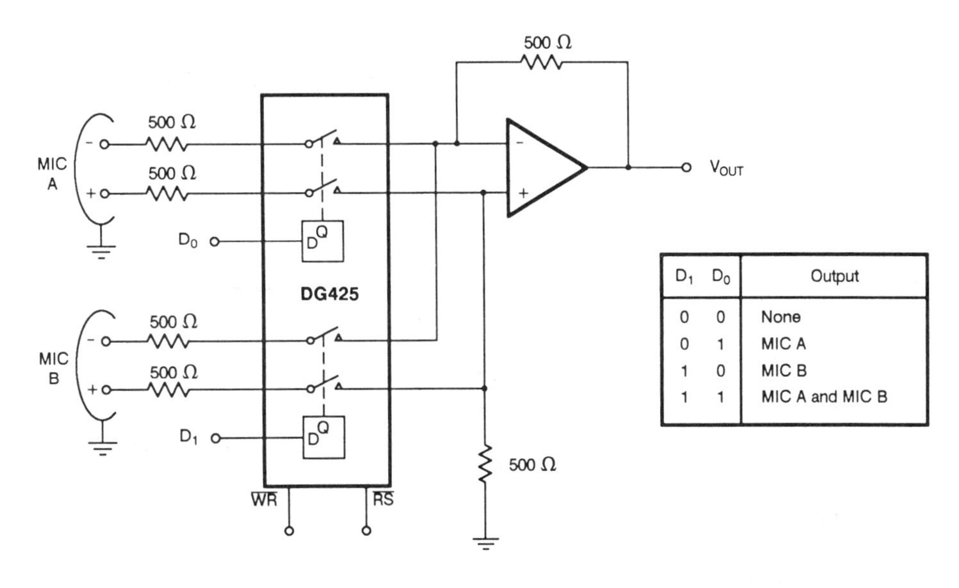

D_1	D_0	Output
0	0	None
0	1	MIC A
1	0	MIC B
1	1	MIC A and MIC B

Bus-Controlled Selector for Balanced-Line Microphones

Siliconix DG441/442
Monolithic Quad SPST CMOS Analog Switches

- On-resistance $<85\Omega$
- Low power consumption (P_D <1.6mW)
- Fast switching action
  - ○ t_{ON} <250ns
  - ○ t_{OFF} <120ns (DG441)
- ESD Protection > ±4000V
- Low charge injection
- DG201A/DG202 upgrades
- TTL, CMOS compatible
- Single-supply capability
- Low signal errors and distortion
- Reduced power supply
- Faster throughput
- Improved reliability
- Reduced pedestal error
- Simplifies retrofit
- Simple interfacing

APPLICATIONS
- Audio switching
- Battery-operated systems
- Data acquisition
- Hi-rel systems
- Sample-and-hold circuits
- Communication systems
- Automatic test equipment

The DG441 series of monolithic quad analog switches was designed to provide high speed, low error switching of analog and audio signals. Combining low on-resistance ($<85\Omega$) with high speed ($t_{ON} < 250$ns), the DG441 series is ideally suited for upgrading DG201A/DG202 sockets. Charge injection has been minimized on the drain for use in sample-and-hold circuits.

To achieve high voltage ratings and superior switching performance, the DG441 series is built on Siliconix's high-voltage silicon-gate process. An epitaxial layer prevents latchup.

Each switch conducts equally well in both directions when on, and blocks up to 30V peak-to-peak when off. On-resistance is very flat over the full ± 5V analog range.

The two devices in this series are differentiated by the type of switch action, as shown in the functional block diagrams for each. The DG441/442 are available in 16-pin plastic, CerDIP, and SO packages. Performance grades include both the military A-suffix (-55 to 125 °C) and industrial D-suffix (-40 to 85 °C) temperature range.

ABSOLUTE MAXIMUM RATINGS

V+ to V- .... 44 V

GND to V- .... 25 V

Digital Inputs[1] V_S, V_D .... (V-) -2 V to (V+) +2 V or 30 mA, whichever occurs first

Continuous Current (Any Terminal) .... 30 mA

Current, S or D (Pulsed 1 ms, 10% duty cycle) .... 100 mA

Storage Temperature (A Suffix) .... -65 to 150°C
(D Suffix) .... -65 to 125°C

Operating Temperature (A Suffix) .... -55 to 125°C
(D Suffix) .... -40 to 85°C

Power Dissipation (Package)*

16-Pin Plastic DIP** .... 450 mW

16-Pin CerDIP*** .... 900 mW

16-Pin Narrow Body SO**** .... 600 mW

*All leads welded or soldered to PC Board.
**Derate 6 mW/°C above 75°C.
***Derate 12 mW/°C above 75°C.
****Derate 7.6 mW/°C above 75°C.

[1]Signals on S_X, D_X, or IN_X exceeding V+ or V- will be clamped by internal diodes. Limit forward diode current to maximum current ratings.

FUNCTIONAL BLOCK DIAGRAM, PIN CONFIGURATION, AND TRUTH TABLE

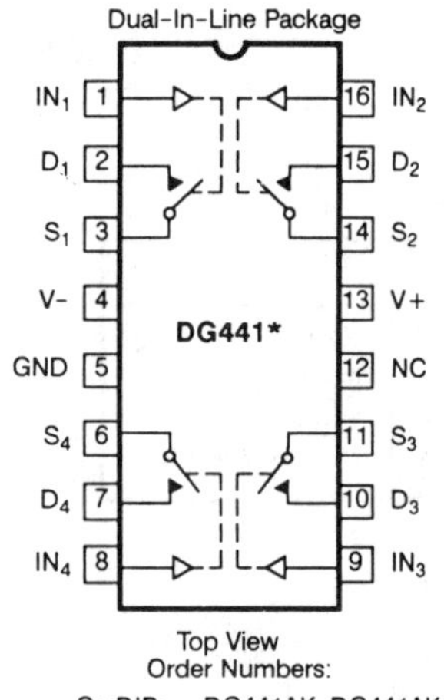

Top View
Order Numbers:
CerDIP: DG441AK, DG441AK/883
DG442AK, DG442AK/883
Plastic: DG441DJ, DG442DJ

Narrow Body SO Package

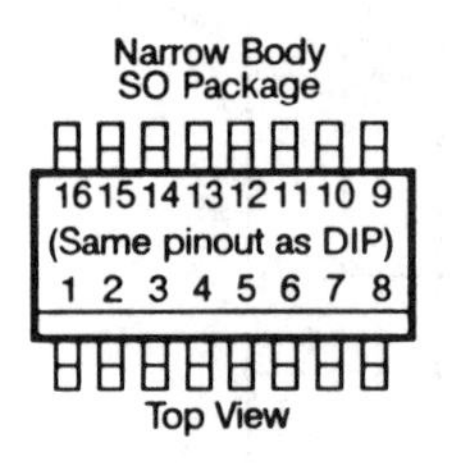

Top View

Order Numbers:
DG441DY, DG442DY

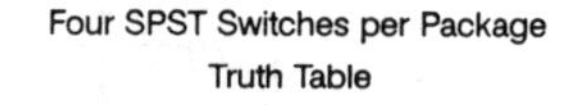

Logic	Switch DG441*	Switch DG442
0	ON	OFF
1	OFF	ON

Logic "0" ≤ 0.8 V
Logic "1" ≥ 2.4 V

* Switches shown for logic "1" input.

TYPICAL CHARACTERISTICS

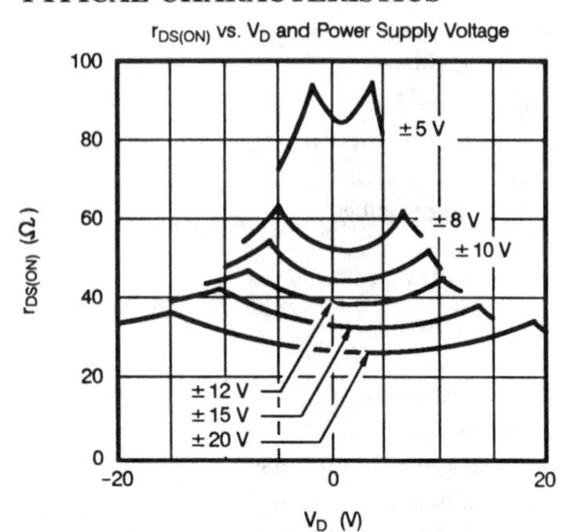

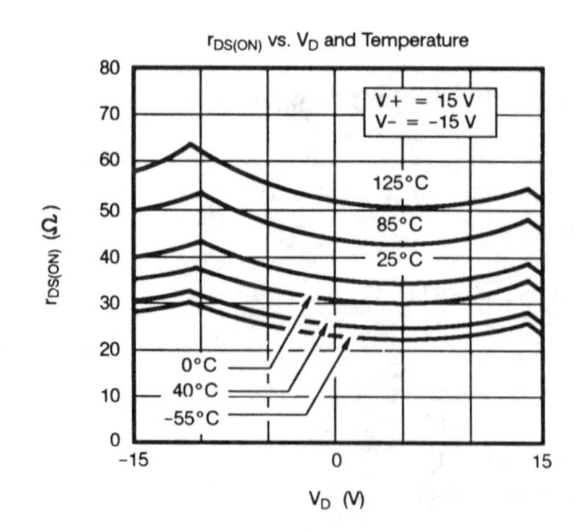

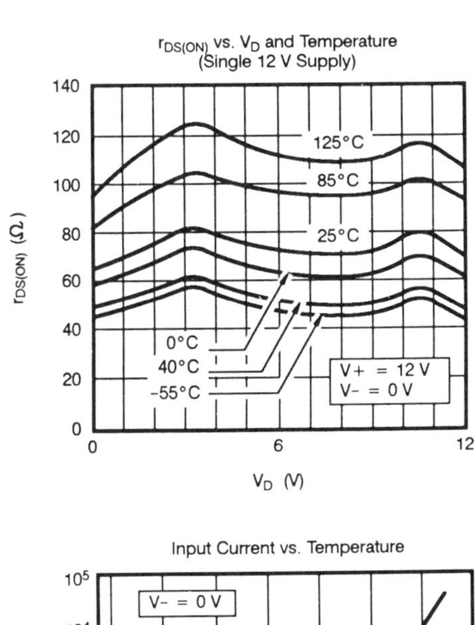
$r_{DS(ON)}$ vs. V_D and Temperature
(Single 12 V Supply)
$r_{DS(ON)}$ (Ω)
125°C
85°C
25°C
0°C
40°C
-55°C
V+ = 12 V
V- = 0 V
V_D (V)

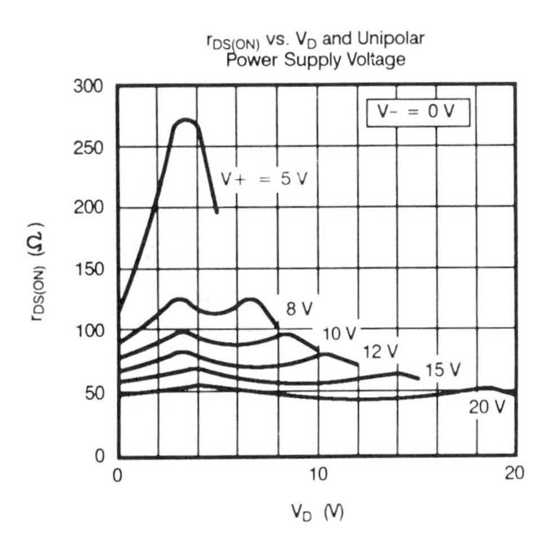
$r_{DS(ON)}$ vs. V_D and Unipolar
Power Supply Voltage
$r_{DS(ON)}$ (Ω)
V- = 0 V
V+ = 5 V
8 V
10 V
12 V
15 V
20 V
V_D (V)

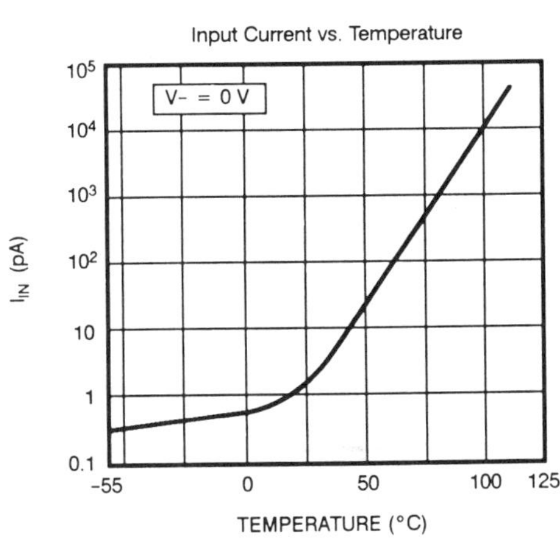
Input Current vs. Temperature
I_{IN} (pA)
V- = 0 V
TEMPERATURE (°C)

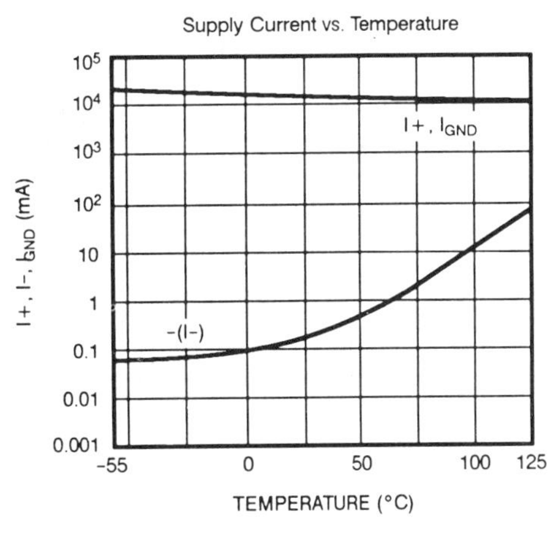
Supply Current vs. Temperature
I+, I-, I_{GND} (mA)
I+, I_{GND}
-(I-)
TEMPERATURE (°C)

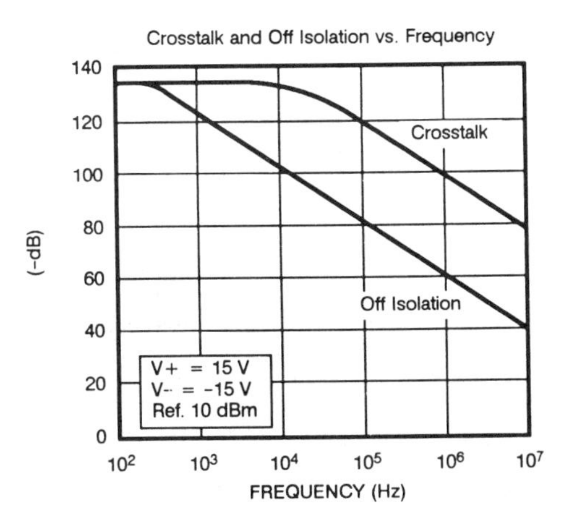
Crosstalk and Off Isolation vs. Frequency
(-dB)
Crosstalk
Off Isolation
V+ = 15 V
V- = -15 V
Ref. 10 dBm
FREQUENCY (Hz)

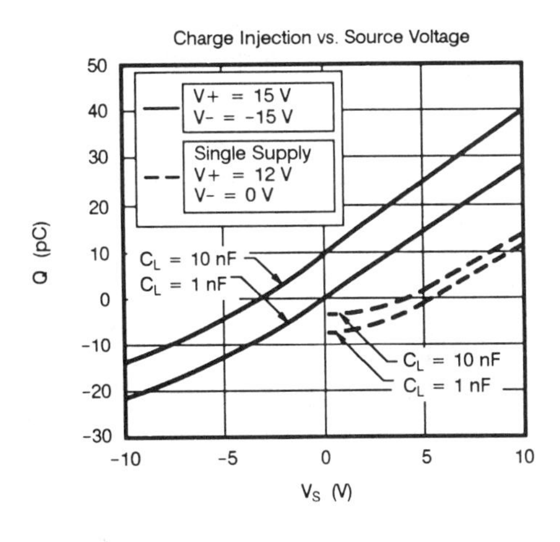
Charge Injection vs. Source Voltage
Q (pC)
V+ = 15 V
V- = -15 V
Single Supply
V+ = 12 V
V- = 0 V
C_L = 10 nF
C_L = 1 nF
C_L = 10 nF
C_L = 1 nF
V_S (V)

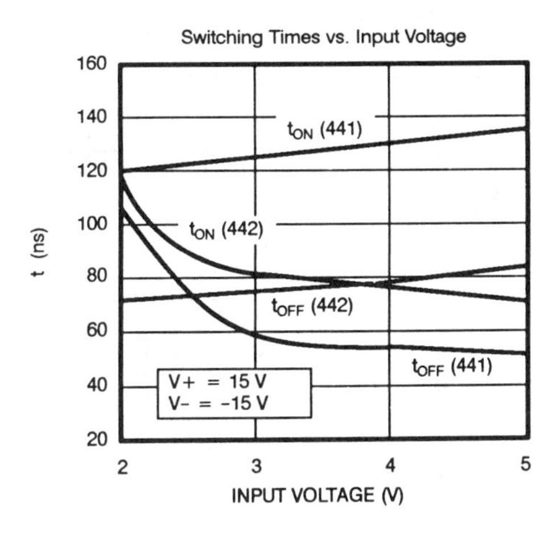
Switching Times vs. Input Voltage
t (ns)
t_{ON} (441)
t_{ON} (442)
t_{OFF} (442)
t_{OFF} (441)
V+ = 15 V
V- = -15 V
INPUT VOLTAGE (V)

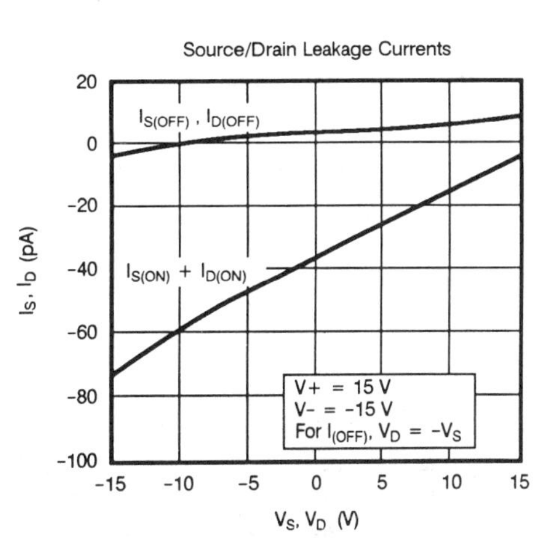
Source/Drain Leakage Currents
I_S, I_D (pA)
$I_{S(OFF)}$, $I_{D(OFF)}$
$I_{S(ON)}$ + $I_{D(ON)}$
V+ = 15 V
V- = -15 V
For $I_{(OFF)}$, V_D = $-V_S$
V_S, V_D (V)

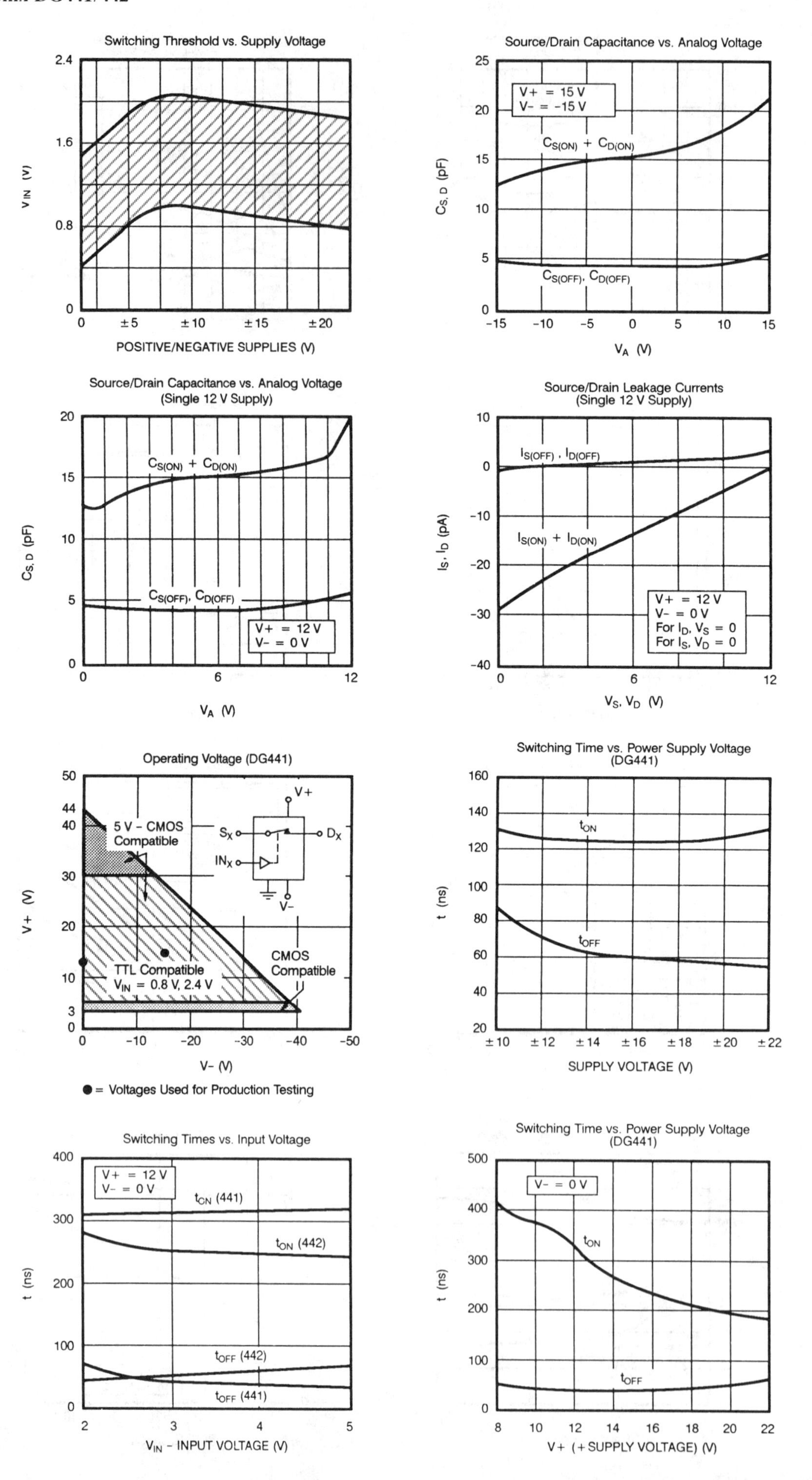
Switching Threshold vs. Supply Voltage
2.4
1.6
0.8
0
V IN (V)
0
±5
±10
±15
±20
POSITIVE/NEGATIVE SUPPLIES (V)
Source/Drain Capacitance vs. Analog Voltage
V+ = 15 V
V– = –15 V
CS(ON) + CD(ON)
CS(OFF), CD(OFF)
CS, D (pF)
VA (V)
Source/Drain Capacitance vs. Analog Voltage
(Single 12 V Supply)
CS(ON) + CD(ON)
CS(OFF), CD(OFF)
V+ = 12 V
V– = 0 V
VA (V)
Source/Drain Leakage Currents
(Single 12 V Supply)
IS(OFF), ID(OFF)
IS(ON) + ID(ON)
IS, ID (pA)
V+ = 12 V
V– = 0 V
For ID, VS = 0
For IS, VD = 0
VS, VD (V)
Operating Voltage (DG441)
5 V - CMOS
Compatible
SX
DX
INX
V+
V–
TTL Compatible
VIN = 0.8 V, 2.4 V
CMOS
Compatible
V+ (V)
V– (V)
● = Voltages Used for Production Testing
Switching Time vs. Power Supply Voltage
(DG441)
tON
tOFF
t (ns)
SUPPLY VOLTAGE (V)
Switching Times vs. Input Voltage
V+ = 12 V
V– = 0 V
tON (441)
tON (442)
tOFF (442)
tOFF (441)
t (ns)
VIN – INPUT VOLTAGE (V)
Switching Time vs. Power Supply Voltage
(DG441)
V– = 0 V
tON
tOFF
V+ (+SUPPLY VOLTAGE) (V)

SCHEMATIC DIAGRAM (Typical Channel)

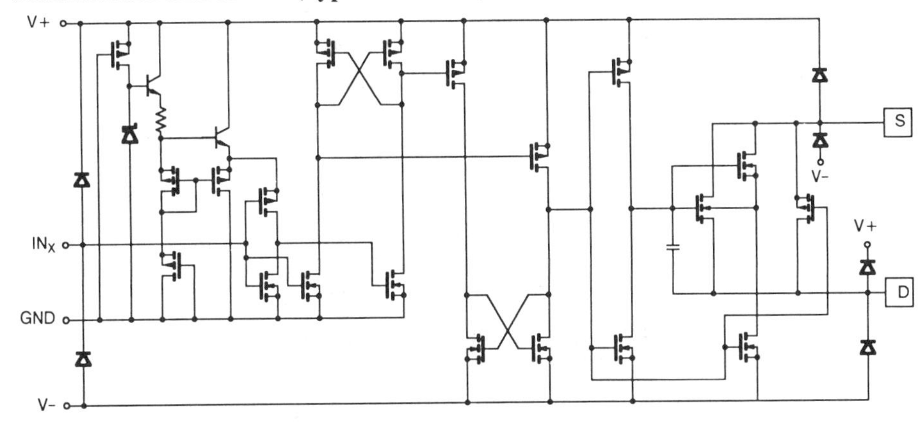

APPLICATIONS

V+ Positive Supply Voltage (V)	V- Negative Supply Voltage (V)	$V_{INH(MIN)}$ HIGH Logic Input Voltage (V)	$V_{INL(MAX)}$ LOW Logic Input Voltage (V)	V_S or V_D Analog Signal Range (V)
20	-20	2.4	0.8	-20 to 20
15	-15	2.4	0.8	-15 to 15
10	-10	2.4	0.8	-10 to 10
5	-5	2.0	0.5	-5 to 5
12	0	2.4	0.8	0 to 12
5	0	2.0	0.5	0 to 5

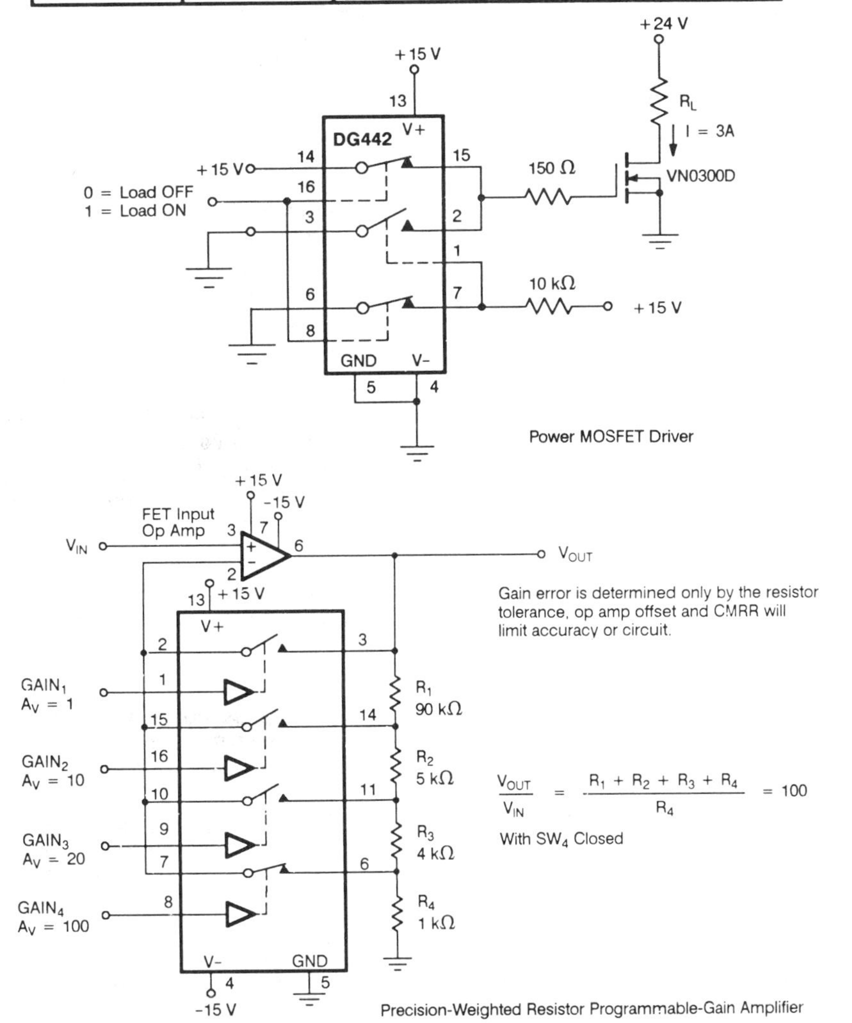

Power MOSFET Driver

Precision-Weighted Resistor Programmable-Gain Amplifier

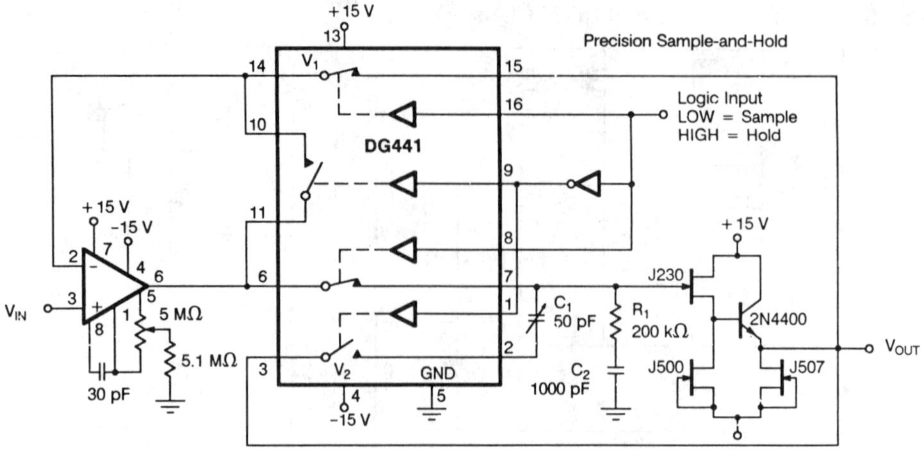

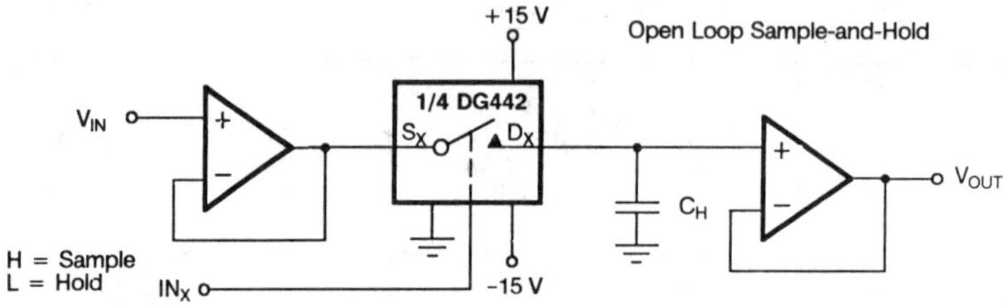

Precision Sample-and-Hold

Open Loop Sample-and-Hold

Siliconix DG444/445

Low-Cost Quad SPST CMOS Analog Switches

- On-resistance $<85\Omega$
- Low power consumption (P_D <35mW)
- Fast switching action
  t_{ON} <250ns
  t_{OFF} <120ns (DG444)
- Low cost
- ESD protection $> \pm 4000$V
- Low charge injection
- DG211/DG212 upgrades
- TTL, CMOS compatible
- Low signal errors and distortion
- Reduced power supply
- Faster throughput
- Improved reliability
- Reduced pedestal error
- Simple interfacing

APPLICATIONS

- Audio and video switching
- Battery-operated systems
- Data acquisition
- Hi-rel systems
- Sample-and-hold circuits
- Communication systems
- Automatic test equipment
- Single-supply circuits

The DG444 series of monolithic quad analog switches was designed to provide high-speed, low-error switching of analog signals. Combining low power ($<35\mu$W) with high speed (t_{ON} <250ns), the DG444 series is ideally suited for upgrading DG21/212 sockets. Charge injection has been minimized on the drain for use in sample-and-hold circuits.

To achieve high-voltage ratings and superior switching performance, the DG444 series is built on Siliconix's high-voltage silicon-gate process. An epitaxial layer prevents latchup.

Each switch conducts equally well in both directions when on, and blocks up to 30V peak-to-peak when off. On-resistance is very flat over the full ± 15V analog range.

The two devices in this series are differentiated by the type of switch action as shown in the functional block diagrams for each. Packaging options include the 16-pin plastic and small outline. The performance grade for this series is the industrial D-suffix (-40 to 85 °C) temperature range.

ABSOLUTE MAXIMUM RATINGS

Parameter	Rating
V+ to V-	44 V
GND to V-	25 V
V_L	(GND -0.3 V) to (V+) + 0.3 V
Digital Inputs[1], V_S, V_D	(V-) -2 V to (V+) +2 V or 30 mA, whichever occurs first
Continuous Current (Any Terminal)	30 mA
Current, S or D (Pulsed 1 ms, 10% duty cycle)	100 mA
Storage Temperature (D Suffix)	-65 to 125°C
Operating Temperature (D Suffix)	-40 to 85°C
Power Dissipation (Package)*	
16-Pin Plastic DIP**	450 mW
16-Pin Narrow Body SO****	600 mW

*All leads welded or soldered to PC Board.
**Derate 6 mW/°C above 75°C.
***Derate 12 mW/°C above 75°C.

[1]Signals on S_X, D_X, or IN_X exceeding V+ or V- will be clamped by internal diodes. Limit forward diode current to maximum current ratings.

FUNCTIONAL BLOCK DIAGRAM, PIN CONFIGURATION, AND TRUTH TABLE

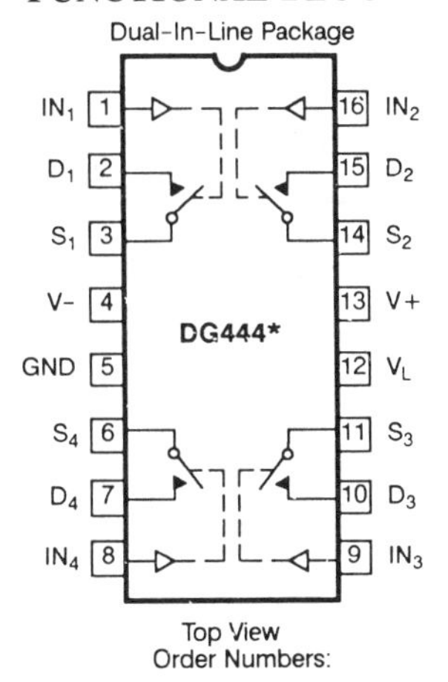

Order Numbers:

Plastic: DG444DJ, DG445DJ

Narrow Body SO Package

16 15 14 13 12 11 10 9
(Same pinout as DIP)
1 2 3 4 5 6 7 8

Top View

Order Numbers: DG444DY, DG445DY

Four SPST Switches per Package
Truth Table

Logic	Switch	
	DG444*	DG445
0	ON	OFF
1	OFF	ON

Logic "0" ≤ 0.8 V
Logic "1" ≥ 2.4 V

* Switches shown for logic "1" input.

TYPICAL CHARACTERISTICS

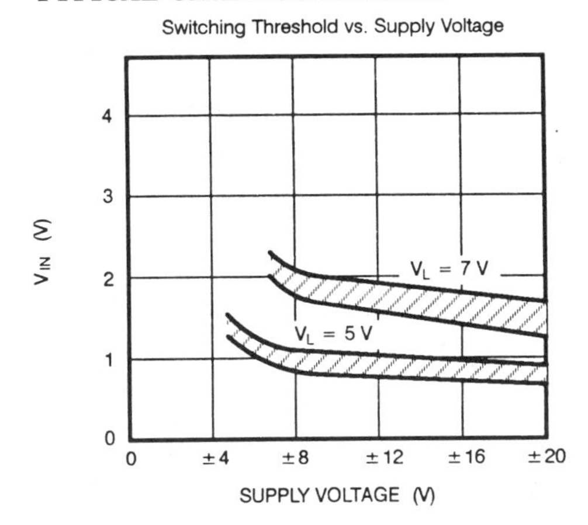

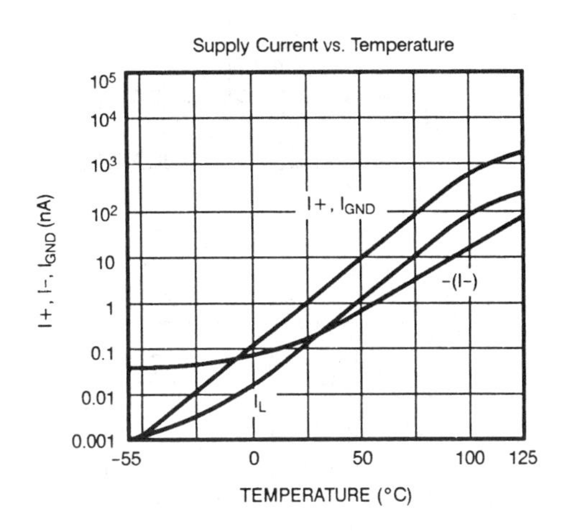

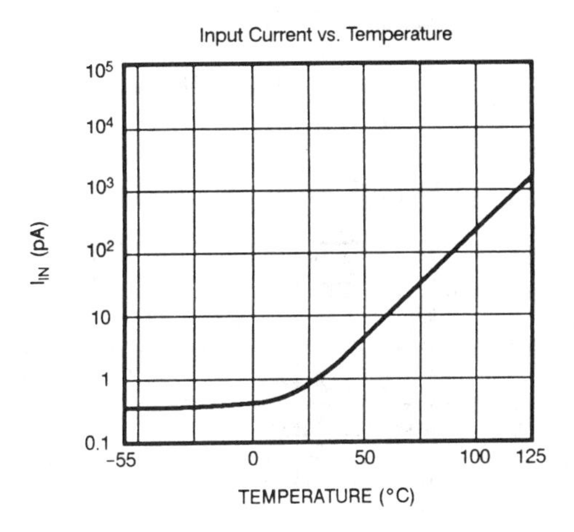

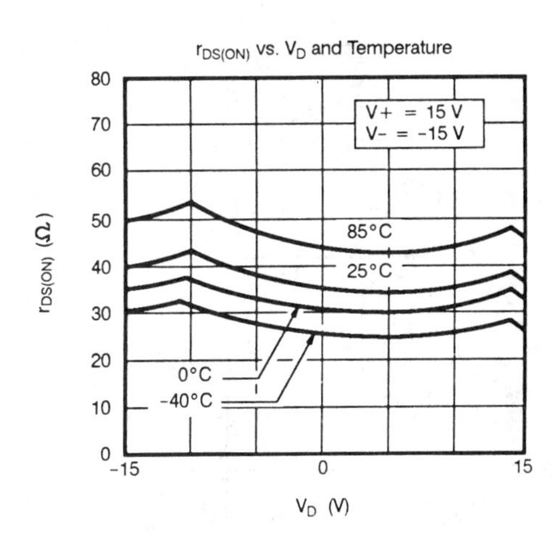

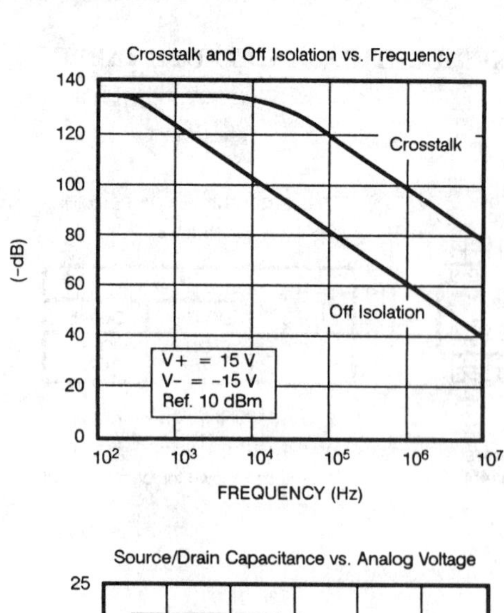
Crosstalk and Off Isolation vs. Frequency
Crosstalk
Off Isolation
V+ = 15 V
V– = –15 V
Ref. 10 dBm
(–dB)
FREQUENCY (Hz)

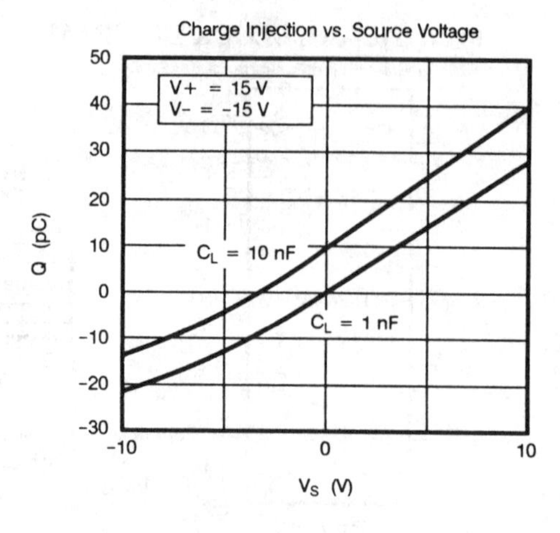
Charge Injection vs. Source Voltage
V+ = 15 V
V– = –15 V
CL = 10 nF
CL = 1 nF
Q (pC)
VS (V)

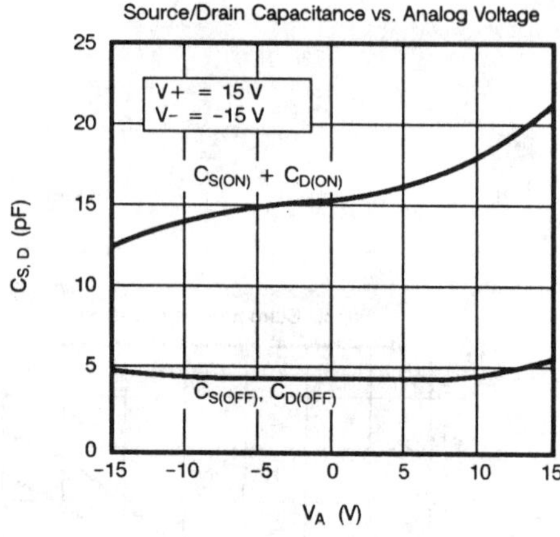
Source/Drain Capacitance vs. Analog Voltage
V+ = 15 V
V– = –15 V
CS(ON) + CD(ON)
CS(OFF), CD(OFF)
CS, D (pF)
VA (V)

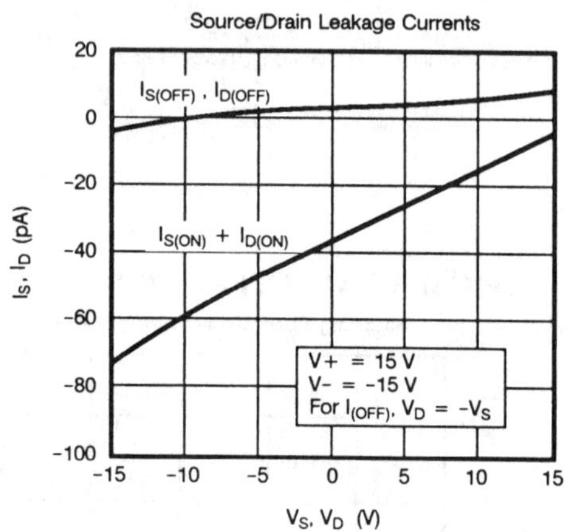
Source/Drain Leakage Currents
IS(OFF), ID(OFF)
IS(ON) + ID(ON)
V+ = 15 V
V– = –15 V
For I(OFF), VD = –VS
IS, ID (pA)
VS, VD (V)

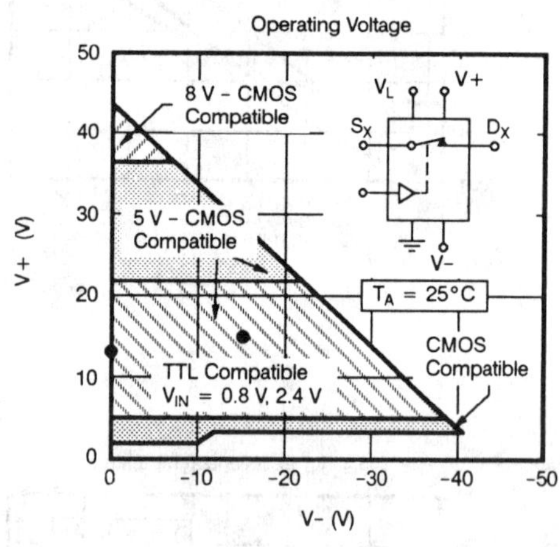
Operating Voltage
8 V - CMOS
Compatible
5 V - CMOS
Compatible
TTL Compatible
VIN = 0.8 V, 2.4 V
CMOS
Compatible
TA = 25°C
VL
V+
SX
DX
V–
V+ (V)
V– (V)

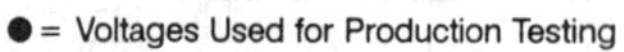
● = Voltages Used for Production Testing

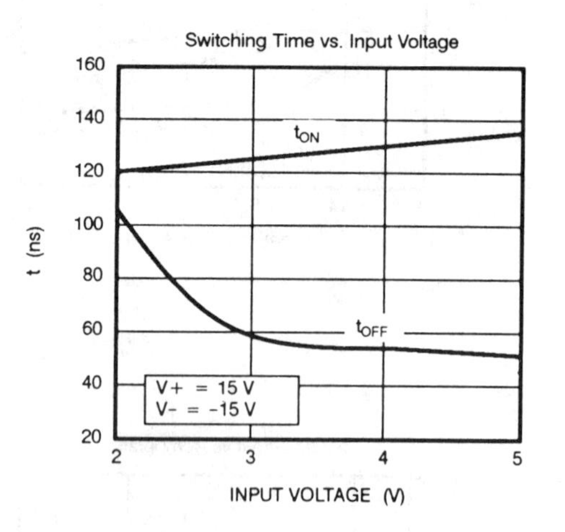
Switching Time vs. Input Voltage
tON
tOFF
V+ = 15 V
V– = –15 V
t (ns)
INPUT VOLTAGE (V)

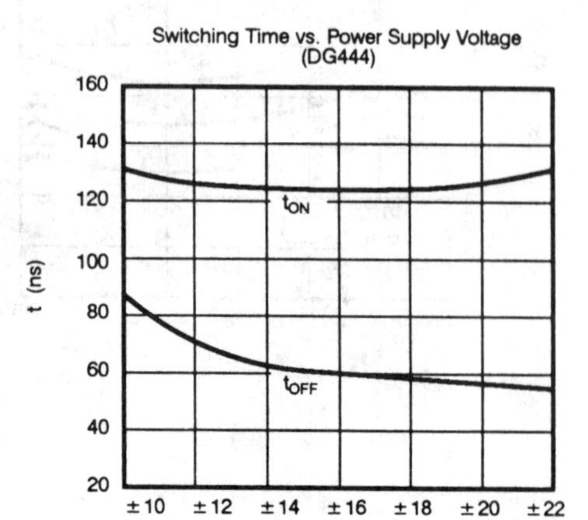
Switching Time vs. Power Supply Voltage
(DG444)
tON
tOFF
t (ns)
SUPPLY VOLTAGE (V)

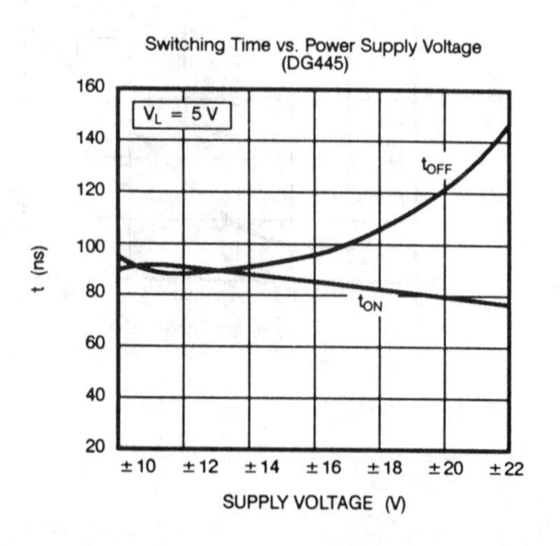
Switching Time vs. Power Supply Voltage
(DG445)
VL = 5 V
tOFF
tON
t (ns)
SUPPLY VOLTAGE (V)

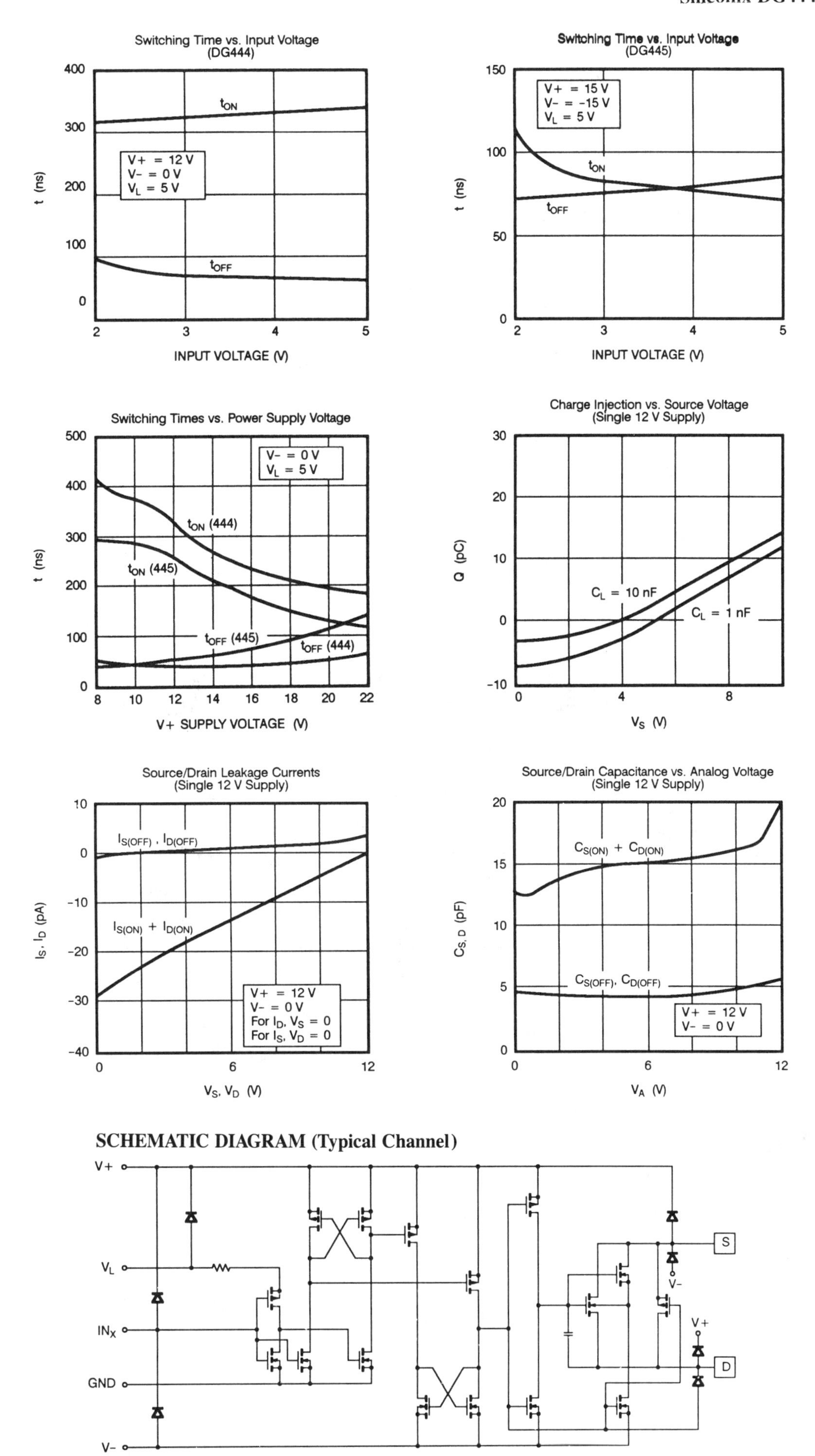

SCHEMATIC DIAGRAM (Typical Channel)

APPLICATIONS

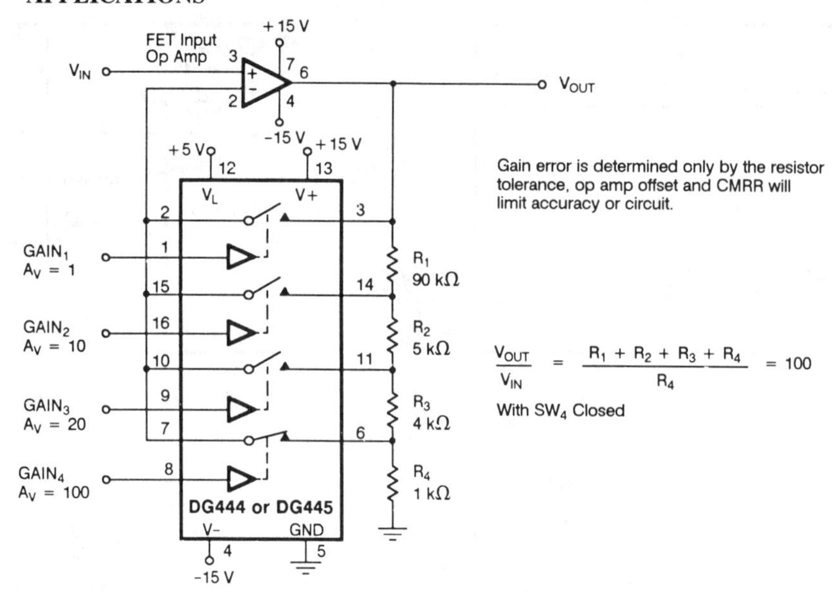

Precision-Weighted Resistor Programmable-Gain Amplifier

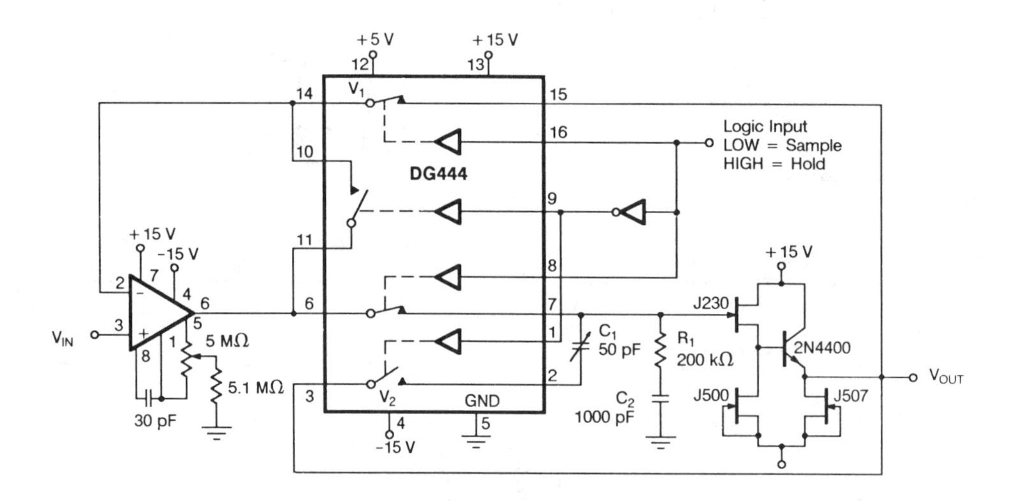

Precision Sample-and-Hold

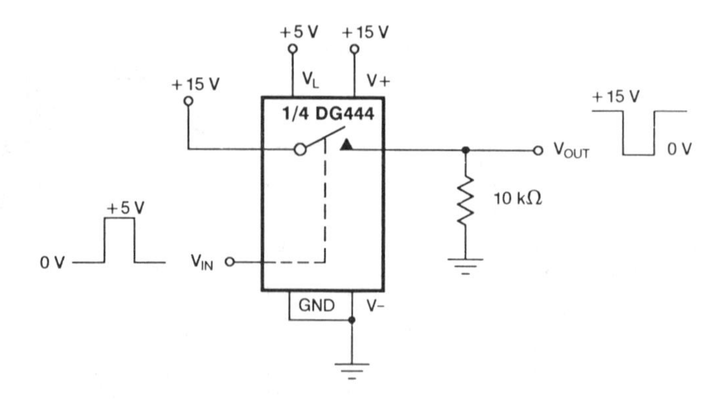

Level Shifter

SECTION 5

LOGIC CIRCUITS

This information is reproduced with permission of Analog Devices, Inc.

Analog Devices
AD532
Internally Trimmed Integrated Circuit Multiplier

- Pretrimmed too ±1.0% (AD532K)
- No external components required
- Guaranteed ±1.0% max 4-quadrant error (AD532K)
- Diff inputs for $(X_1-X_2)(Y_1-Y_2)/10V$ transfer function
- Monolithic construction, low cost

APPLICATIONS
- Multiplication, division, squaring
- Square rooting
- Algebraic computation
- Power measurements
- Instrumentation applications
- Available in chip form

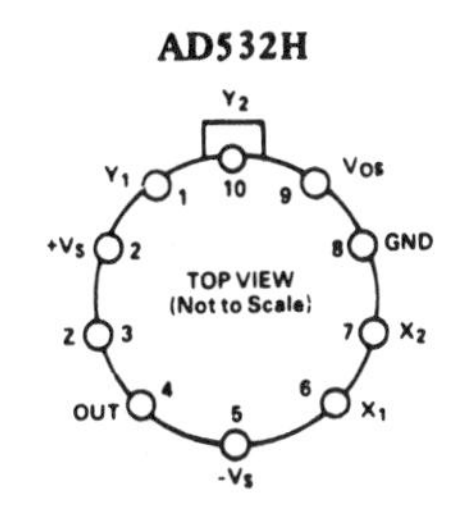

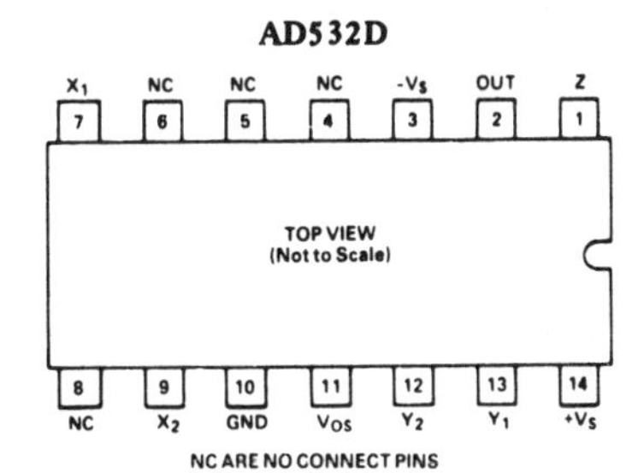

The AD532 is the first pretrimmed single-chip monolithic multiplier/divider. It guarantees a maximum multiplying error of ±1.0% and a ±10V output voltage without the need for any external trimming resistors or output op amp. Because the AD532 is internally trimmed, its simplicity of use provides design engineers with an attractive alternative to modular multipliers, and its monolithic construction provides significant advantages in size, reliability, and economy. Further, the AD532 can be used as a direct replacement for other IC multipliers that require external trim networks (such as the AD530).

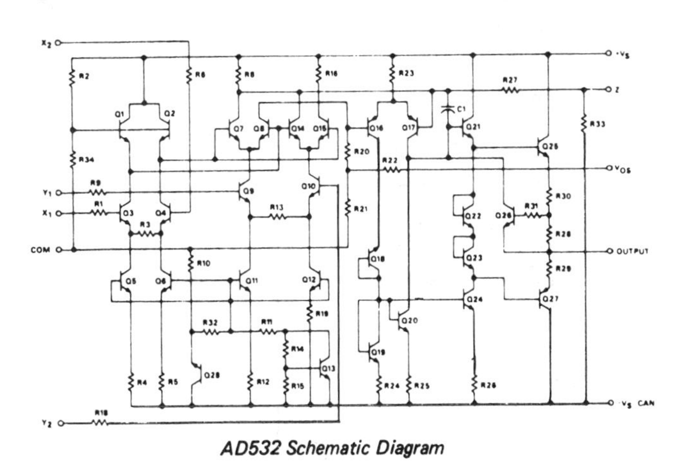

AD532 Schematic Diagram

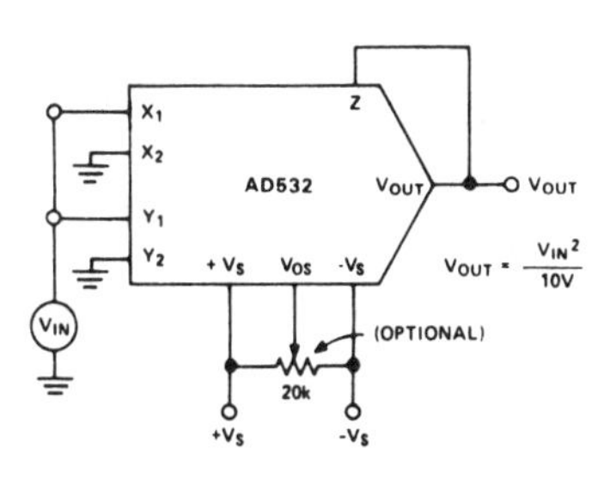

Squarer Connection

SPECIFICATIONS (@ +25°C, V_S = ±15V, R≥2kΩ V_{OS} grounded)

Model	AD532J Min	AD532J Typ	AD532J Max	AD532K Min	AD532K Typ	AD532K Max	AD532S Min	AD532S Typ	AD532S Max	Units
MULTIPLIER PERFORMANCE										
Transfer Function		$\frac{(X_1 - X_2)(Y_1 - Y_2)}{10V}$			$\frac{(X_1 - X_2)(Y_1 - Y_2)}{10V}$			$\frac{(X_1 - X_2)(Y_1 - Y_2)}{10V}$		
Total Error (−10V≤X, Y≤ +10V)		±1.5	**±2.0**		±0.7	**±1.0**		±0.5	**±1.0**	%
T_A = min to max		±2.5			±1.5				**±4.0**	%
Total Error vs Temperature		±0.04			±0.03			±0.01	**±0.04**	%/°C
Supply Rejection (±15V ±10%)		±0.05			±0.05			±0.05		%/%
Nonlinearity, X (X = 20V pk-pk, Y = 10V)		±0.8			±0.5			±0.5		%
Nonlinearity, Y (Y = 20V pk-pk, X = 10V)		±0.3			±0.2			±0.2		%
Feedthrough, X (Y Nulled, X = 20V pk-pk 50Hz)		50	**200**		30	**100**		30	**100**	mV
Feedthrough, Y (X Nulled, Y = 20V pk-pk 50Hz)		30	**150**		25	**80**		25	**80**	mV
Feedthrough vs. Temp.		2.0			1.0			1.0		mV p-p/°C
Feedthrough vs. Power Supply		±0.25			±0.25			±0.25		mV/%
DYNAMICS										
Small Signal BW (V_{OUT} = 0.1 rms)		1			1			1		MHz
1% Amplitude Error		75			75			75		kHz
Slew Rate (V_{OUT} 20 pk-pk)		45			45			45		V/µs
Settling Time (to 2%, ΔV_{OUT} = 20V)		1			1			1		µs
NOISE										
Wideband Noise f = 5Hz to 10kHz		0.6			0.6			0.6		mV (rms)
f = 5Hz to 5MHz		3.0			3.0			3.0		mV (rms)
OUTPUT										
Output Voltage Swing	±10	±13		±10	±13		±10	±13		V
Output Impedance (f≤1kHz)		1			1			1		Ω
Output Offset Voltage		±40				**±30**			**±30**	mV
Output Offset Voltage vs. Temp.		0.7			0.7				**2.0**	mV/°C
Output Offset Voltage vs. Supply		±2.5			±2.5			±2.5		mV/%
INPUT AMPLIFIERS (X, Y and Z)										
Signal Voltage Range (Diff. or CM Operating Diff)		±10			±10			±10		V
CMRR	**40**			**50**			**50**			dB
Input Bias Current										
X, Y Inputs		3			1.5	**4**		1.5	**4**	µA
X, Y Inputs T_{min} to T_{max}		10			8			8		µA
Z Input		±10			±5	**±15**		±5	**±15**	µA
Z Input T_{min} to T_{max}		±30			±25			±25		µA
Offset Current		±0.3			±0.1			±0.1		µA
Differential Resistance		10			10			10		MΩ
DIVIDER PERFORMANCE										
Transfer Function ($X_1 > X_2$)		10V Z/($X_1 - X_2$)			10V Z/($X_1 - X_2$)			10V Z/($X_1 - X_2$)		
Total Error										
(V_X = −10V, −10V≤V_Z≤ +10V)		±2			±1			±1		%
(V_X = −1V, −10V≤V_Z≤ +10V)		±4			±3			±3		%
SQUARE PERFORMANCE										
Transfer Function		$\frac{(X_1 - X_2)^2}{10V}$			$\frac{(X_1 - X_2)^2}{10V}$			$\frac{(X_1 - X_2)^2}{10V}$		
Total Error		±0.8			±0.4			±0.4		%
SQUARE-ROOTER PERFORMANCE										
Transfer Function		$-\sqrt{10VZ}$			$-\sqrt{10VZ}$			$-\sqrt{10VZ}$		
Total Error (0V≤V_Z≤10V)		±1.5			±1.0			±1.0		%
POWER SUPPLY SPECIFICATIONS										
Supply Voltage										
Rated Performance		±15			±15			±15		V
Operating	±10		**±18**	±10		**±18**	±10		±22	V
Supply Current										
Quiescent		4	**6**		4	**6**		4	**6**	mA
PACKAGE OPTIONS										
TO-116 (D-14)	AD532JD			AD532KD			AD532SD			
TO-100 (H-10A)	AD532JH			AD532KH			AD532SH			

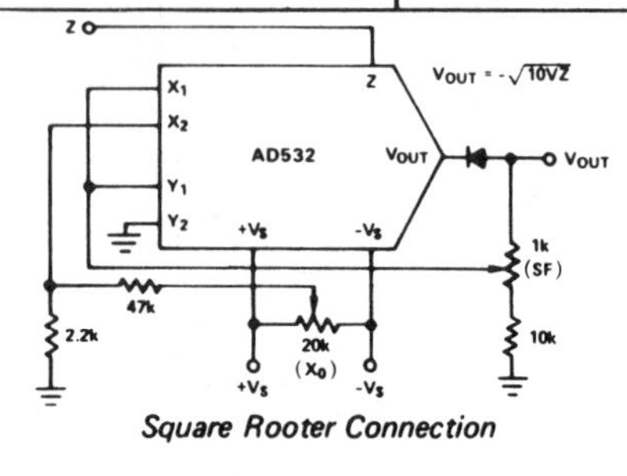

Square Rooter Connection

FUNCTIONAL BLOCK DIAGRAM

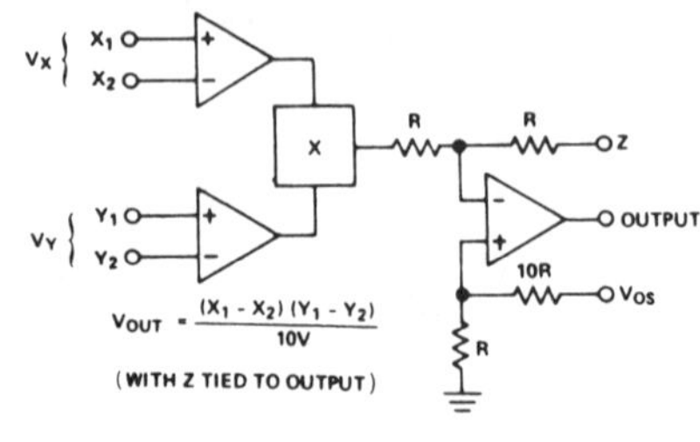

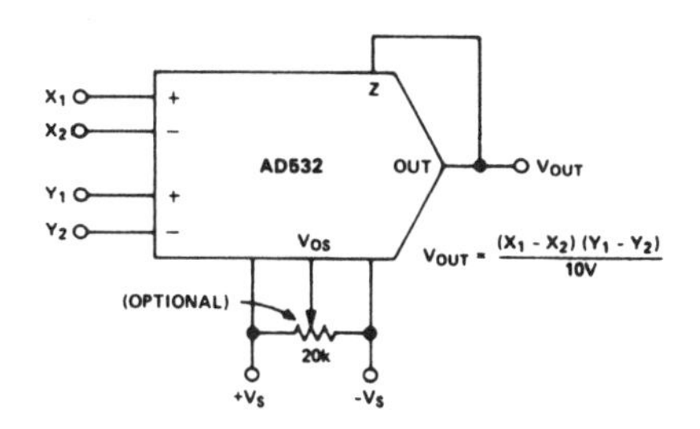

Multiplier Connection

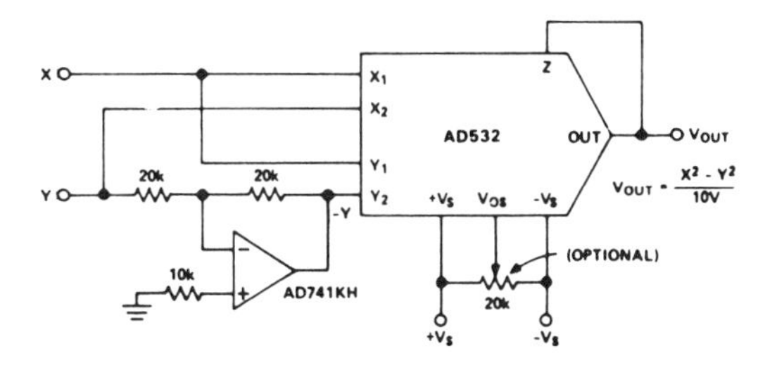

Differential of Squares Connection

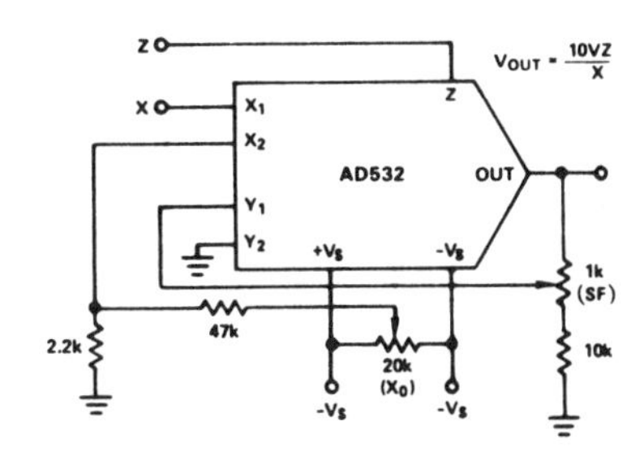

Divider Connection

Analog Devices AD534

Internally Trimmed Precision IC Multiplier

- Pretrimmed to ±0.25% max 4-quadrant error (AD534L)
- All inputs (X, Y and Z) differential, high impedance for $[(X_1-X_2)(Y_1-Y_2)/10V]+Z_2$ transfer function
- Scale-factor adjustable to provide up to ×100 gain
- Low noise design: 90μV rms, 100 Hz−10kHz
- Low-cost monolithic construction
- Excellent long-term stability

APPLICATIONS

- High-quality analog signal processing
- Differential ratio and percentage computations
- Algebraic and trigonometric function synthesis
- Wideband high-crest rms-to-dc conversion
- Accurate voltage-controlled oscillators and filters
- Available in chip form

The AD534 is a monolithic laser-trimmed four-quadrant multiplier divider having accuracy specifications previously found only in expensive hybrid or modular products. A maximum multiplication error of ±0.25% is guaranteed for the AD534L without any external trimming. Excellent supply rejection, low temperature coefficients, and long-term stability of the on-chip thin-film resistors and buried-zener reference preserve accuracy even under adverse conditions of use. It is the first multiplier to offer fully differential, high-impedance operation on all inputs, including the Z-input, a feature that greatly increases its flexibility and ease of use. The scale factor is pretrimmed to the standard value of 10.00V; by means of an external resistor that can be reduced to values as low as 3V.

The wide spectrum of applications and the availability of several grades commend this multiplier as the first choice for all new designs. The AD534J (±1%) max error), AD534K (±0.5% max) and AD534L (±0.25% max) are specified for operation over the 0 to +70°C temperature range. The AD534S (±1% max) and AD534T (±0.5% max) are specified over the extended temperature range, −55°C to +125°C. All grades are available in hermetically sealed TO-100 metal cans and TO-116 ceramic DIP packages.

ABSOLUTE MAXIMUM RATINGS

	AD534J, K, L	AD534S, T
Supply Voltage	±18V	±22V
Internal Power Dissipation	500mW	*
Output Short-Circuit to Ground	Indefinite	*
Input Voltages, X_1 X_2 Y_1 Y_2 Z_1 Z_2	$\pm V_S$	*
Rated Operating Temperature Range	0 to +70°C	-55°C to +125°C
Storage Temperature Range	-65°C to +150°C	*
Lead Temperature, 60s soldering	+300°C	*

*Same as AD534J specs.

AD534 FUNCTIONAL BLOCK DIAGRAM

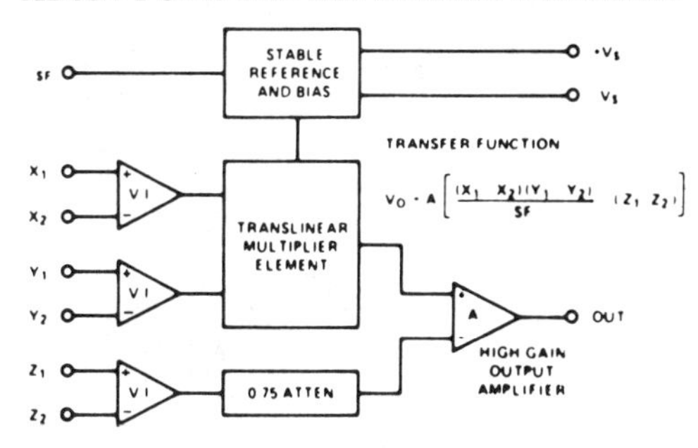

OPTIONAL TRIMMING CONFIGURATION

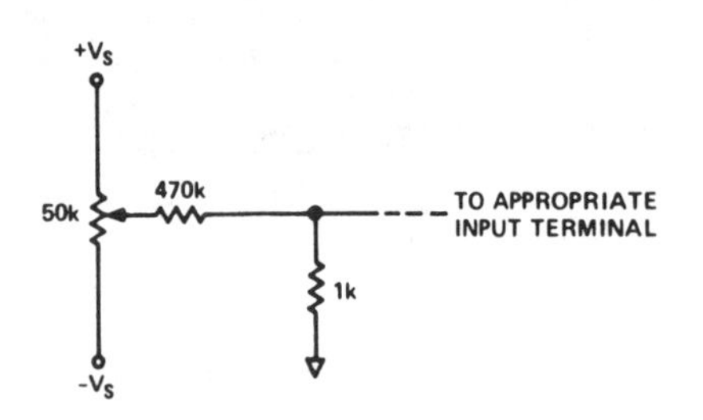

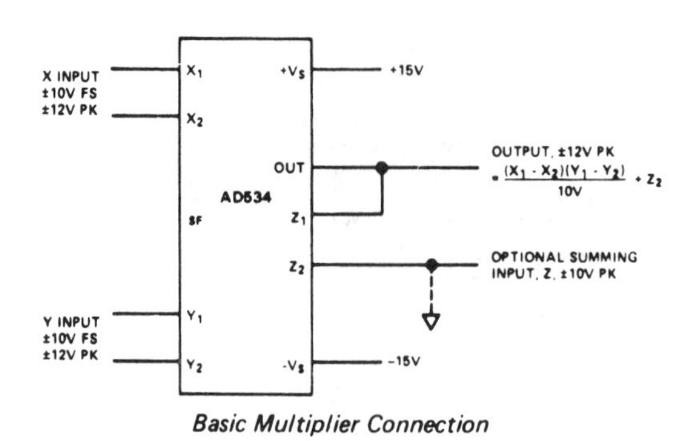

Basic Multiplier Connection

SPECIFICATIONS ($T_A = +25°C$, $\pm V_S = 15V$, $R \geq 2k\Omega$)

Model	AD534J Min	AD534J Typ	AD534J Max	AD534K Min	AD534K Typ	AD534K Max	AD534L Min	AD534L Typ	AD534L Max	Units
MULTIPLIER PERFORMANCE										
Transfer Function		$\frac{(X_1 - X_2)(Y_1 - Y_2)}{10V} + Z_2$			$\frac{(X_1 - X_2)(Y_1 - Y_2)}{10V} + Z_2$			$\frac{(X_1 - X_2)(Y_1 - Y_2)}{10V} + Z_2$		
Total Error[1] ($-10V \leq X, Y \leq +10V$)			±1.0			±0.5			±0.25	%
T_A = min to max		±1.5			±1.0			±0.5		%
Total Error vs Temperature		±0.022			±0.015			±0.008		%/°C
Scale Factor Error (SF = 10.000V Nominal)[2]		±0.25			±0.1			±0.1		%
Temperature-Coefficient of Scaling Voltage		±0.02			±0.01			±0.005		%/°C
Supply Rejection (±15V ±1V)		±0.01			±0.01			±0.01		%
Nonlinearity, X (X = 20V pk-pk, Y = 10V)		±0.4			±0.2	±0.3		±0.10	±0.12	%
Nonlinearity, Y (Y = 20V pk-pk, X = 10V)		±0.2			±0.1	±0.1		±0.005	±0.1	%
Feedthrough[3], X (Y Nulled, X = 20V pk-pk 50Hz)		±0.3			±0.15	±0.3		±0.05	±0.12	%
Feedthrough[3], Y (X Nulled, Y = 20V pk-pk 50Hz)		±0.01			±0.01	±0.1		±0.003	±0.1	%
Output Offset Voltage		±5	±30		±2	±15		±2	±10	mV
Output Offset Voltage Drift		200			100			100		μV/°C
DYNAMICS										
Small Signal BW, (V_{OUT} = 0.1 rms)		1			1			1		MHz
1% Amplitude Error (C_{LOAD} = 1000pF)		50			50			50		kHz
Slew Rate (V_{OUT} 20 pk-pk)		20			20			20		V/μs
Settling Time (to 1%, ΔV_{OUT} = 20V)		2			2			2		μs
NOISE										
Noise Spectral-Density SF = 10V		0.8			0.8			0.8		$\mu V/\sqrt{Hz}$
SF = 3V[4]		0.4			0.4			0.4		$\mu V/\sqrt{Hz}$
Wideband Noise f = 10Hz to 5MHz		1			1			1		mV/rms
f = 10Hz to 10kHz		90			90			90		μV/rms
OUTPUT										
Output Voltage Swing	±11			±11			±11			V
Output Impedance (f ≤ 1kHz)		0.1			0.1			0.1		Ω
Output Short Circuit Current (R_L = 0, T_A = min to max)		30			30			30		mA
Amplifier Open Loop Gain (f = 50Hz)		70			70			70		dB
INPUT AMPLIFIERS (X, Y and Z)										
Signal Voltage Range (Diff. or CM		±10			±10			±10		V
Operating Diff.)		±12			±12			±12		V
Offset Voltage X, Y		±5	±20		±2	±10		±2	±10	mV
Offset Voltage Drift X, Y		100			50			50		μV/°C
Offset Voltage Z		±5	±30		±2	±15		±2	±10	mV
Offset Voltage Drift Z		200			100			100		μV/°C
CMRR	60	80		70	90		70	90		dB
Bias Current		0.8	2.0		0.8	2.0		0.8	2.0	μA
Offset Current		0.1			0.1			0.05	0.2	μA
Differential Resistance		10			10			10		MΩ
DIVIDER PERFORMANCE										
Transfer Function ($X_1 > X_2$)		$10V\frac{(Z_2 - Z_1)}{(X_1 - X_2)} + Y_1$			$10V\frac{(Z_2 - Z_1)}{(X_1 - X_2)} + Y_1$			$10V\frac{(Z_2 - Z_1)}{(X_1 - X_2)} + Y_1$		
Total Error[1] ($X = 10V$, $-10V \leq Z \leq +10V$)		±0.75			±0.35			±0.2		%
($X = 1V$, $-1V \leq Z \leq +1V$)		±2.0			±1.0			±0.8		%
($0.1V \leq X \leq 10V$, $-10V \leq Z \leq 10V$)		±2.5			±1.0			±0.8		%
SQUARE PERFORMANCE										
Transfer Function		$\frac{(X_1 - X_2)^2}{10V} + Z_2$			$\frac{(X_1 - X_2)^2}{10V} + Z_2$			$\frac{(X_1 - X_2)^2}{10V} + Z_2$		
Total Error ($-10V \leq X \leq 10V$)		±0.6			±0.3			±0.2		%
SQUARE-ROOTER PERFORMANCE										
Transfer Function ($Z_1 \leq Z_2$)		$\sqrt{10V(Z_2 - Z_1)} + X_2$			$\sqrt{10V(Z_2 - Z_1)} + X_2$			$\sqrt{10V(Z_2 - Z_1)} + X_2$		
Total Error[1] ($1V \leq Z \leq 10V$)		±1.0			±0.5			±0.25		%
POWER SUPPLY SPECIFICATIONS										
Supply Voltage										
Rated Performance		±15			±15			±15		V
Operating	±8		±18	±8		±18	±8		±18	V
Supply Current										
Quiescent		4	6		4	6		4	6	mA
PACKAGE OPTIONS										
TO-100 (H-10A)		AD534JH			AD534KH			AD534LH		
TO-116 (D-14)		AD534JD			AD534KD			AD534LD		

Model	AD534S Min	AD534S Typ	AD534S Max	AD534T Min	AD534T Typ	AD534T Max	Units
MULTIPLIER PERFORMANCE							
Transfer Function		$\frac{(X_1 - X_2)(Y_1 - Y_2)}{10V} + Z_2$			$\frac{(X_1 - X_2)(Y_1 - Y_2)}{10V} + Z_2$		
Total Error[1] ($-10V \leq X, Y \leq +10V$)			±1.0			±0.5	%
T_A = min to max			±2.0			±1.0	%
Total Error vs Temperature			±0.02			±0.01	%/°C
Scale Factor Error		±0.25			±0.1		%
Temperature-Coefficient of Scaling Voltage		±0.02				±0.005	%/°C
Supply Rejection (±15V ±1V)		±0.01			±0.01		%
Nonlinearity, X (X = 20V pk-pk, Y = 10V)		±0.4			±0.2	±0.3	%
Nonlinearity, Y (Y = 20V pk-pk, X = 10V)		±0.2			±0.1	±0.1	%
Feedthrough[3], X (Y Nulled, X = 20V pk-pk 50Hz)		±0.3			±0.15	±0.3	%
Feedthrough[3], Y (X Nulled, Y = 20V pk-pk 50Hz)		±0.01			±0.01	±0.1	%
Output Offset Voltage		±5	±30		±2	±15	mV
Output Offset Voltage Drift			500			300	μV/°C
DYNAMICS							
Small Signal BW, (V_{OUT} = 0.1 rms)		1			1		MHz
1% Amplitude Error (C_{LOAD} = 1000pF)		50			50		kHz
Slew Rate (V_{OUT} 20 pk-pk)		20			20		V/μs
Settling Time (to 1%, ΔV_{OUT} = 20V)		2			2		μs
NOISE							
Noise Spectral-Density SF = 10V		0.8			0.8		$\mu V/\sqrt{Hz}$
		0.4			0.4		$\mu V/\sqrt{Hz}$
Wideband Noise f = 10Hz to 5MHz		1.0			1.0		mV/rms
f = 10Hz to 10kHz		90			90		μV/rms
OUTPUT							
Output Voltage Swing	±11			±11			V
Output Impedance (f ≤ 1kHz)		0.1			0.1		Ω
Output Short Circuit Current R_L = 0, T_A = min to max)		30			30		mA
Amplifier Open Loop Gain (f = 50Hz)		70			70		dB
Signal Voltage Range (Diff. or CM		±10			±10		V
Operating Diff.)		±12			±12		V
Offset Voltage X, Y		±5	±20		±2	±10	mV
Offset Voltage Drift X, Y		100			150		μV/°C
Offset Voltage Z		±5	±30		±2	±15	mV
Offset Voltage Drift Z			500			300	μV/°C
CMRR	60	80		70	90		dB
Bias Current		0.8	2.0		0.8	2.0	μA
Offset Current		0.1			0.1		μA
Differential Resistance		10			10		MΩ
DIVIDER PERFORMANCE							
Transfer Function ($X_1 > X_2$)		$10V\frac{(Z_2 - Z_1)}{(X_1 - X_2)} + Y_1$			$10V\frac{(Z_2 - Z_1)}{(X_1 - X_2)} + Y_1$		
Total Error[1] (X = 10V, $-10V \leq Z \leq +10V$)		±0.75			±0.35		%
(X = 1V, $-1V \leq Z \leq +1V$)		±2.0			±1.0		%
($0.1V \leq X \leq 10V$, $-10V \leq Z \leq 10V$)		±2.5			±1.0		%
SQUARE PERFORMANCE							
Transfer Function		$\frac{(X_1 - X_2)^2}{10V} + Z_2$			$\frac{(X_1 - X_2)^2}{10V} + Z_2$		
Total Error ($-10V \leq X \leq 10V$)		±0.6			±0.3		%
SQUARE-ROOTER PERFORMANCE							
Transfer Function ($Z_1 \leq Z_2$)		$\sqrt{10V(Z_2 - Z_1)} + X_2$			$\sqrt{10V(Z_2 - Z_1)} + X_2$		
Total Error ($1V \leq Z \leq 10V$)		±1.0			±0.5		%
POWER SUPPLY SPECIFICATIONS							
Supply Voltage							
Rated Performance		±15			±15		V
Operating	±8		±22	±8		±22	V
Supply Current							
Quiescent		4	6		4	6	mA
TO-100 Package (H-10A)		AD534SH			AD534TH		
TO-116 Package (D-14)		AD534SD			AD534TD		

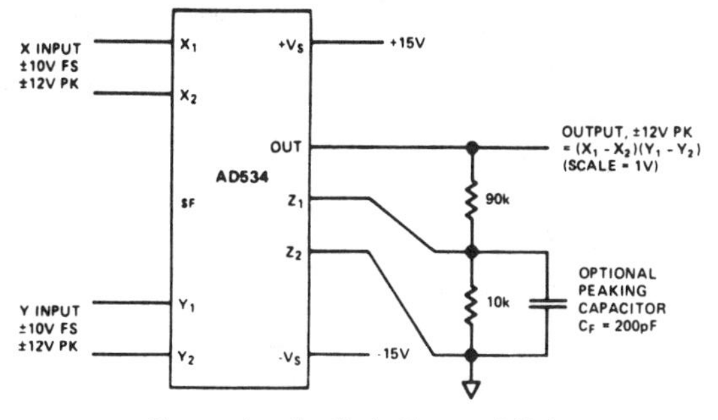

Connections for Scale-Factor of Unity

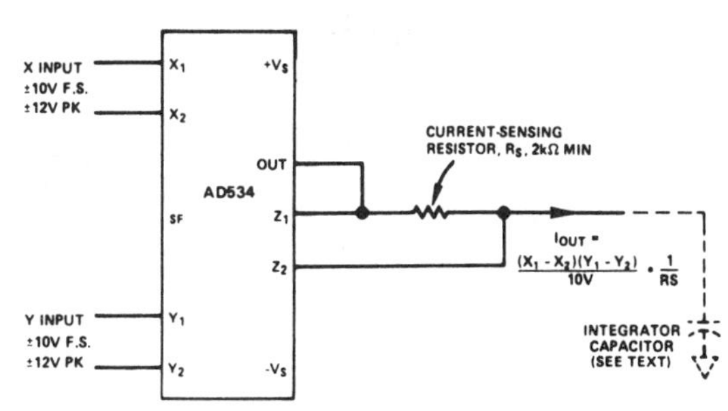

Conversion of Output to Current

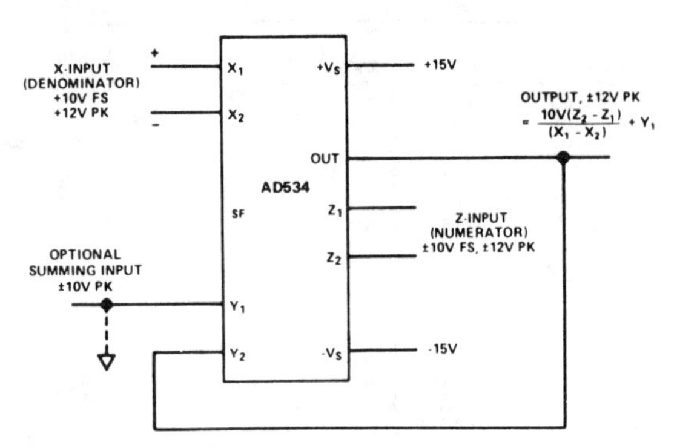

Basic Divider Connection

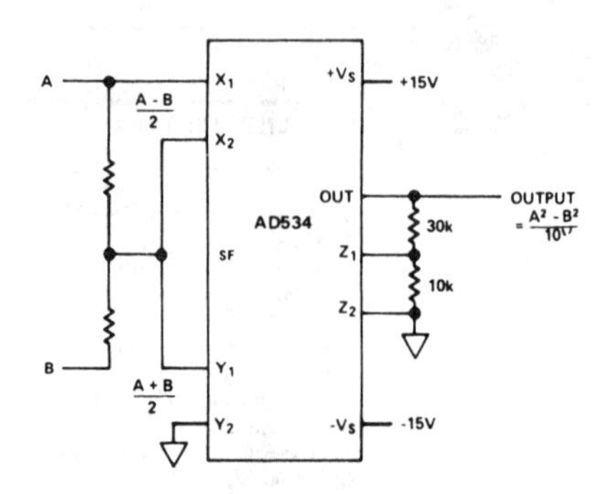

Difference-of-Squares

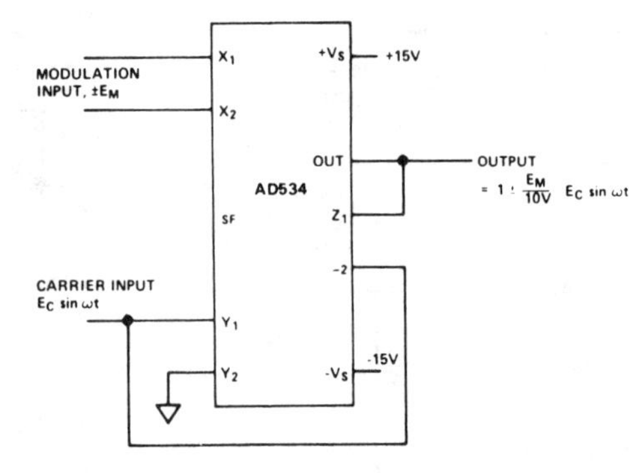

THE SF PIN OR A Z-ATTENUATOR CAN BE USED TO PROVIDE OVERALL SIGNAL AMPLIFICATION. OPERATION FROM A SINGLE SUPPLY IS POSSIBLE; BIAS Y_2 TO $V_S/2$.

Linear AM Modulator

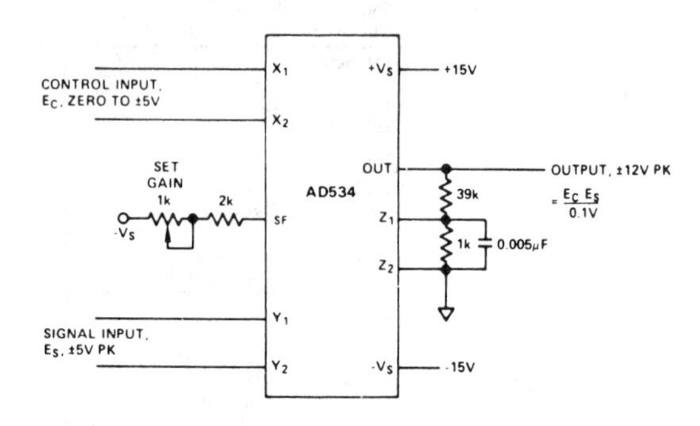

NOTES:
1) GAIN IS X10 PER VOLT OF E_C, ZERO TO X50
2) WIDEBAND (10Hz – 30kHz) OUTPUT NOISE IS 3mV RMS, TYP CORRESPONDING TO A F.S. S/N RATIO OF 70dB
3) NOISE REFERRED TO SIGNAL INPUT, WITH E_C = ±5V, IS 60µV RMS, TYP
4) BANDWIDTH IS DC TO 20kHz, -3dB, INDEPENDENT OF GAIN

Voltage-Controlled Amplifier

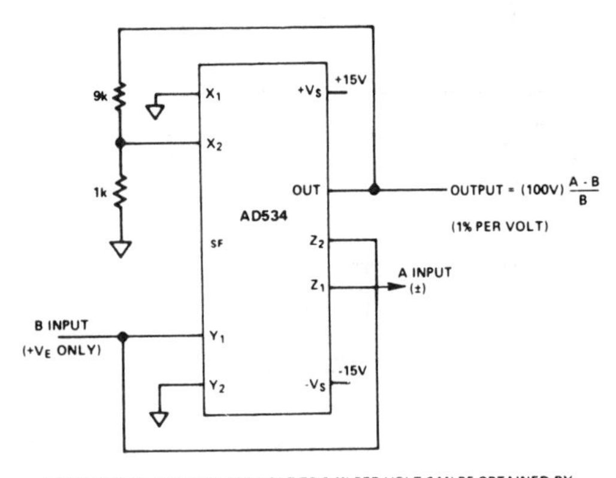

OTHER SCALES, FROM 10% PER VOLT TO 0.1% PER VOLT CAN BE OBTAINED BY ALTERING THE FEEDBACK RATIO.

Percentage Computer

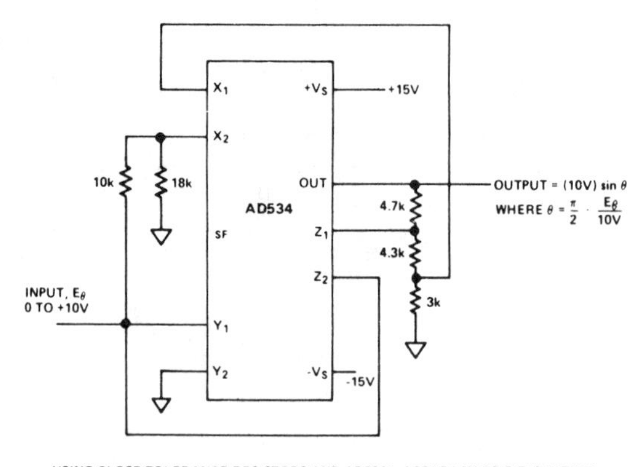

USING CLOSE TOLERANCE RESISTORS AND AD534L, ACCURACY OF FIT IS WITHIN ±0.5% AT ALL POINTS. θ IS IN RADIANS.

Sine-Function Generator

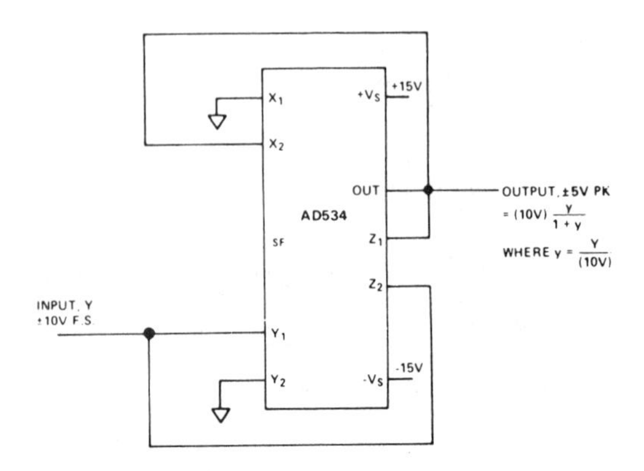

Bridge-Linearization Function

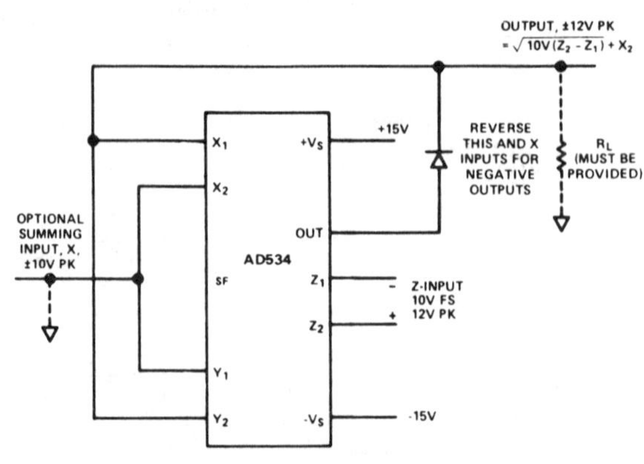

Square-Rooter Connection

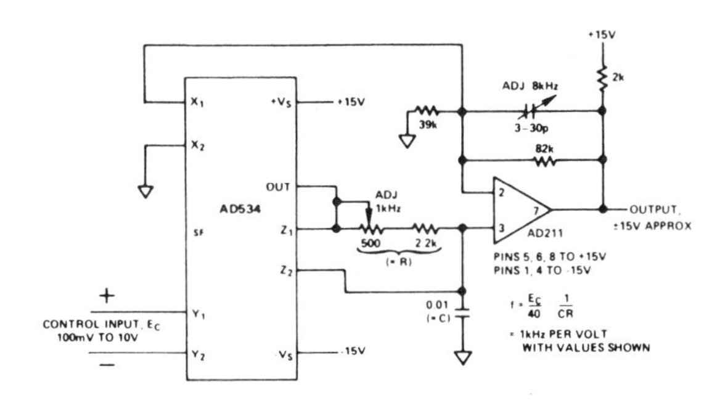

CALIBRATION PROCEDURE:

WITH E_C = 1.0V, ADJUST POT TO SET f = 1.000kHz. WITH E_C = 8.0V, ADJUST TRIMMER CAPACITOR TO SET f = 8.000kHz. LINEARITY WILL TYPICALLY BE WITHIN ±0.1% OF F.S. FOR ANY OTHER INPUT.

DUE TO DELAYS IN THE COMPARATOR, THIS TECHNIQUE IS NOT SUITABLE FOR MAXIMUM FREQUENCIES ABOVE 10kHz. FOR FREQUENCIES ABOVE 10kHz THE AD537 VOLTAGE TO FREQUENCY CONVERTER IS RECOMMENDED.

A TRIANGLE-WAVE OF ±5V PK APPEARS ACROSS THE 0.01μF CAPACITOR; IF USED AS AN OUTPUT, A VOLTAGE-FOLLOWER SHOULD BE INTERPOSED.

Differential-Input Voltage-to-Frequency Converter

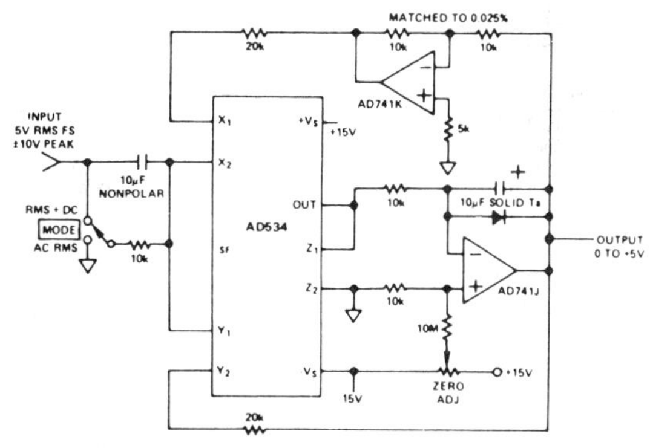

CALIBRATION PROCEDURE:

WITH 'MODE' SWITCH IN 'RMS + DC' POSITION, APPLY AN INPUT OF +1.00VDC. ADJUST ZERO UNTIL OUTPUT READS SAME AS INPUT. CHECK FOR INPUTS OF ±10V, OUTPUT SHOULD BE WITHIN ±0.05% (5mV).

ACCURACY IS MAINTAINED FROM 60Hz to 100kHz, AND IS TYPICALLY HIGH BY 0.5% AT 1MHz FOR V_{IN} = 4V RMS (SINE, SQUARE OR TRIANGULAR WAVE).

PROVIDED THAT THE PEAK INPUT IS NOT EXCEEDED, CREST FACTORS UP TO AT LEAST TEN HAVE NO APPRECIABLE EFFECT ON ACCURACY.

INPUT IMPEDANCE IS ABOUT 10kΩ; FOR HIGH (10MΩ) IMPEDANCE, REMOVE MODE SWITCH AND INPUT COUPLING COMPONENTS.

FOR GUARANTEED SPECIFICATIONS THE AD536A AND AD636 IS OFFERED AS A SINGLE PACKAGE RMS-TO-DC CONVERTER.

Wideband, High-Crest Factor, RMS-to-DC Converter

Analog Devices AD539
Wideband Dual-Channel Linear Multiplier/Divider

- Two quadrant multiplication/division
- Two independent signal channels
- Signal bandwidth of 60 MHz (I_{OUT})
- Linear control channel bandwidth of 5 MHz
- Low distortion (to 0.01%)
- Fully-calibrated monolithic circuit

APPLICATIONS

- Precise high-bandwidth AGC and VCA systems
- Voltage-controlled filters
- Video-signal processing
- High-speed analog division
- Automatic signal-leveling
- Square-law gain/loss control

The AD539 is a low-distortion analog multiplier having two identical signal channels (Y1 and Y2), with a common X-input providing linear control of gain. Excellent ac characteristics up to video frequencies and a 3dB bandwidth of over 60 MHz are provided. Although intended primarily for applications where speed is important, the circuit exhibits good static accuracy in "computational" applications. Scaling is accurately determined by a band-gap voltage reference and all critical parameters are laser-trimmed during manufacture.

The full bandwidth can be realized over most of the gain range using the AD539 with simple resistive loads of up to 100Ω. Output voltage is restricted to a few hundred millivolts under these conditions. Using external op amps, such as the AD539 in conjunction with the on-chip scaling resistors, accurate multiplication can be achieved, with bandwidths typically as high as 50 MHz.

The two channels provide flexibility. In single-channel applications, they can be used in parallel to double the output current in series to achieve a square-law gain function with a control range of over 100dB, or differentially to reduce distortion. Alternatively, they can be used independently, as in audio stereo applications, with low crosstalk between channels. Voltage-controlled filters and oscillators using the "state-variable" approach are easily designed, taking advantage of the dual channels and common control. The AD539 can also be configured as a divider with signal bandwidths up to 15 MHz.

Power consumption is only 135mW using the recommended ±5V supplies. The AD539 is available in three versions: the "J" and "K" grades are specified for 0 to +70°C operation and "S" grade is guaranteed over the extended range of −55°C to +125°C. The J and K grades are available in either a hermetic ceramic DIP (D) or a low-cost plastic DIP (N); the S grade is available only in ceramic.

AD539 FUNCTIONAL BLOCK DIAGRAM

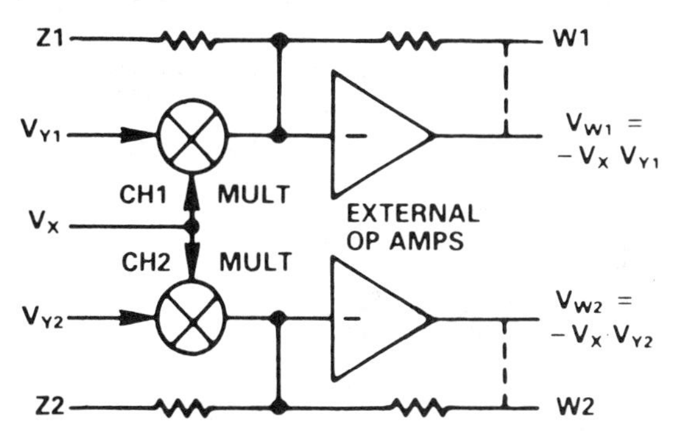

SPECIFICATIONS (@ $T_A = 25°C$, $V_S = \pm 5V$, unless otherwise specified)

Parameter	Conditions	AD539J Min	AD539J Typ	AD539J Max	AD539K Min	AD539K Typ	AD539K Max	AD539S Min	AD539S Typ	AD539S Max	Units
SIGNAL-CHANNEL DYNAMICS											
Minimal Configuration	Reference Figure 6a										
Bandwidth, −3dB	$R_L = 50\Omega$, $C_C = 0.01\mu F$	30	60		30	60		30	60		MHz
Maximum Output	$+0.1V < V_X < +3V$, $V_Yac = 1V$ rms		−10			−10			−10		dBm
Feedthrough, f<1MHz	$V_X = 0$, $V_Yac = 1.5V$ rms		−75			−75			−75		dBm
f = 20MHz			−55			−55			−55		dBm
Differential Phase Linearity											
$-1V < V_Ydc < +1V$	f = 3.58MHz, $V_X = +3V$,		±0.2			±0.2			±0.2		Degrees
$-2V < V_Ydc < +2V$	$V_Yac = 100mV$		±0.5			±0.5			±0.5		Degrees
Group Delay	$V_X = +3V$, $V_Yac = 1V$ rms, f = 1MHz		4			4			4		ns
Standard Dual-Channel Multiplier	Reference Figure 2										
Maximum Output	$V_X = +3V$, $V_Yac = 1.5V$ rms		4.5			4.5			4.5		V
Feedthrough, f<100kHz	$V_X = 0$, $V_Yac = 1.5V$ rms		1			1			1		mV rms
Crosstalk (CH1 to CH2)	$V_{Y1} = 1V$ rms, $V_{Y2} = 0$ $V_X = +3V$, f<100kHz		−40			−40			−40		dB
RTO Noise, 10Hz to 1MHz	$V_X = +1.5V$, $V_Y = 0$, Figure 2		200			200			200		$nV/\sqrt{Hz}$
THD + Noise, $V_X = +1V$,	f = 10kHz, $V_Yac = 1V$ rms		0.02			0.02			0.02		%
$V_Y = +3V$	f = 10kHz, $V_Yac = 1V$ rms		0.04			0.04			0.04		%
Wide Band Two-Channel Multiplier	Figure 2										
Bandwidth, −3dB (LH0032)	$+0.1V < V_X < +3V$, $V_Yac = 1V$ rms		25			25			25		MHz
Maximum Output $V_X = +3V$	$V_Yac = 1.5V$ rms, f = 3MHz		4.5			4.5			4.5		V rms
Feedthrough $V_X = 0V$	$V_Yac = 1.0V$ rms, f = 3MHz		14			14			14		mV rms
Wide Band Single Channel VCA (AD5539)	Reference Figure 8										
Bandwidth, −3dB	$+0.1V < V_X < +3V$, $V_Yac = 1V$ rms		50			50			50		MHz
Maximum Output	75Ω Load		±1			±1			±1		V
Feedthrough	$V_X = -0.01$, f = 5MHz		−54			−54			−54		dB
CONTROL CHANNEL DYNAMICS											
Bandwidth, −3dB	$C_C = 3000pF$, $V_Xdc = +1.5V$, $V_Xac = 100mV$ rms		5			5			5		MHz
SIGNAL INPUTS, V_{Y1} & V_{Y2}											
Nominal Full-Scale Input			±2			±2			±2		V
Operational Range, Degraded Performance	$-V_S > 7V$	±4.2			±4.2			±4.2			V
Input Resistance			400			400			400		kΩ
Bias Current			10	30		10	20		10	30	μA
Offset Voltage	$V_X = +3V$, $V_Y = 0$		5	20		5	10		5	20	mV
(T_{min} to T_{max})			10			5			15	35	mV
Power Supply Sensitivity	$V_X = +3V$, $V_Y = 0$		2			2			2		mV/V
CONTROL INPUT, V_X											
Nominal Full-Scale Input			+3.0			+3.0			+3.0		V
Operational Range, Degraded Performance		+3.2			+3.2			+3.2			V
			500			500			500		Ω
Offset Voltage			1	4		1	2		1	4	mV
(T_{min} to T_{max})			3			2			2	5	mV
Power Supply Sensitivity			30			30			30		μV/V
Gain	(Figure 2)										
Absolute Gain Error	$V_X = +0.1V$ to $+3.0V$ and		0.2	0.4		0.1	0.2		0.2	0.4	dB
(T_{min} to T_{max})	$V_y = \pm 2V$		0.3			0.15			0.25	0.5	dB
Full-Scale Output Current	$V_X = +3V$, $V_Y = \pm 2V$		±1			±1			±1		mA
Peak Output Current	$V_X = +3.3V$, $V_Y = \pm 5V$, $V_S = \pm 7.5V$	±2	±2.8		±2	±2.8		±2	±2.8		mA
Output Offset Current	$V_X = 0$, $V_Y = 0$		0.2	1.5		0.2	1.5		0.2	1.5	μA
	Figure 2, $V_X = 0$, $V_Y = 0$		3	10		3	10		3	10	mV
			1.2			1.2			1.2		kΩ
Scaling Resistors											
CH1	Z1, W1 to CH1		6			6			6		kΩ
CH2	Z2, W2 to CH2		6			6			6		kΩ
Multiplier Transfer Function, Either Channel	(Figure 2)		$V_W = -V_X \cdot V_Y/V_U$			$V_W = -V_X \cdot V_Y/V_U$			$V_W = -V_X \cdot V_Y/V_U$		
Multiplier Scaling Voltage, V_U		0.98	1.0	1.02	0.99	1.0	1.01	0.98	1.0	1.02	V
Accuracy			0.5	2		0.5	1		0.5	2	%
(T_{min} to T_{max})			1			0.5			1.0	3	%
Power Supply Sensitivity			0.04			0.04			0.04		%/V
	$V_X <= +3V$, $-2V < V_Y < 2V$		1	2.5		0.6	1.5		1	2.5	% FSR
T_{min} to T_{max}			2			1			2	4	%
Control Feedthrough	$V_X = 0$ to $+3V$, $V_Y = 0$		25	60		15	30		15	60	mV
T_{min} to T_{max}			30			15			60	120	mV
TEMPERATURE RANGE											
Rated Performance		0		+70	0		+70	−55		+125	°C
POWER SUPPLIES											
Operational Range		±4.5		±16.5	±4.5		±16.5	±4.5		±16.5	V
Current Consumption											
$+V_S$			8.5	10.2		8.5	10.2		8.5	10.2	mA
$-V_S$			18.5	22.2		18.5	22.2		18.5	22.2	mA
Plastic (N-16)			AD539JN			AD539KN					
TO-116 (D-16)			AD539JD			AD539KD			AD539SD		

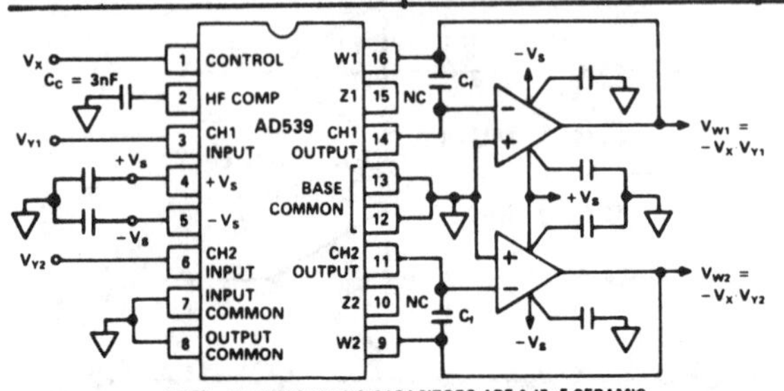

NOTE: ALL DECOUPLING CAPACITORS ARE 0.47μF CERAMIC.

Standard Dual-Channel Multiplier

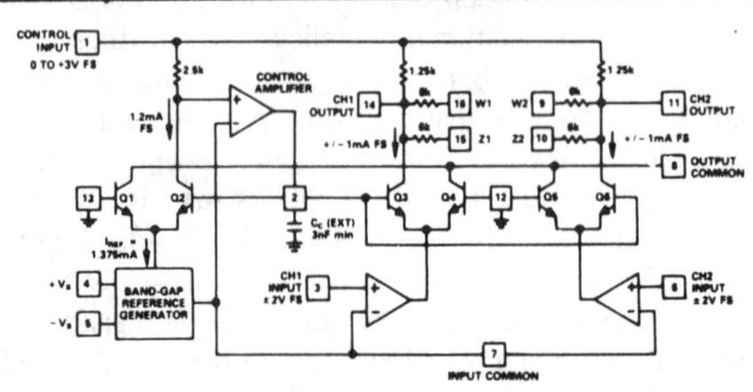

Simplified Schematic of AD539 Multiplier

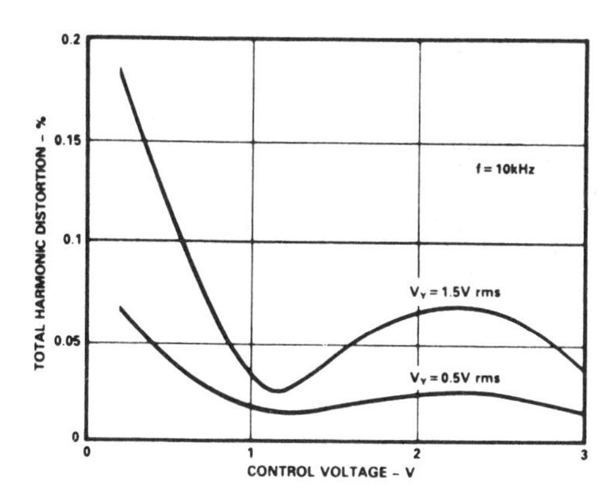

Total Harmonic Distortion vs. Control Voltage

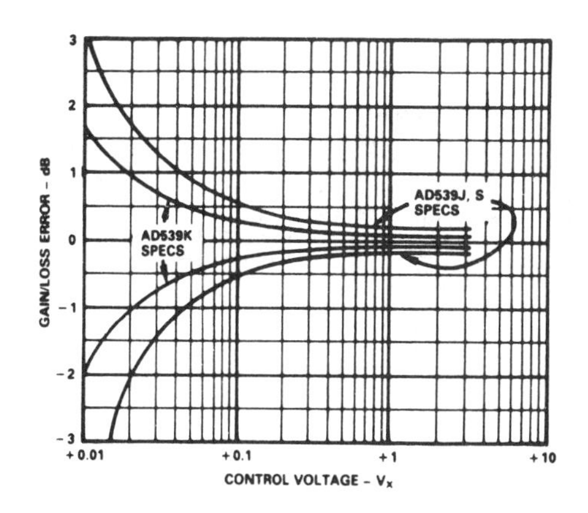

Maximum ac Gain Error Boundaries

Load Resistance	50Ω	75Ω	100Ω	150Ω	600Ω	O/C
FS Output Voltage	±92.6mV 65.5mV rms	±134mV 94.7mV rms	±172mV 122mV rms	±242mV 171mV rms	±612mV 433mV rms	±1V *
FS Output-Power in Load	0.086mW −10.5dBm	0.12mW −9.2dBm	0.15mW −8.3dBm	0.195mW −7.1dBm	0.312mW −5.05dBm	– –
Pk Output Voltage	±210mV 148mV rms	±300mV 212mV rms	±388mV 274mV rms	±544mV 385mV rms	±1V *	±1V *
Pk Output-Power in Load	0.44mW −7dBm	0.6mW −4.4dBm	0.75mW −2.5dBm	1mW 0dBm	±1V *	±1V *
Effective Scaling Voltage, V_U'	67.5V	46.7V	36.3V	25.8V	10.2V	5V

*Peak negative voltage swing limited by output compliance.

Summary of Performance for Minimal Configuration

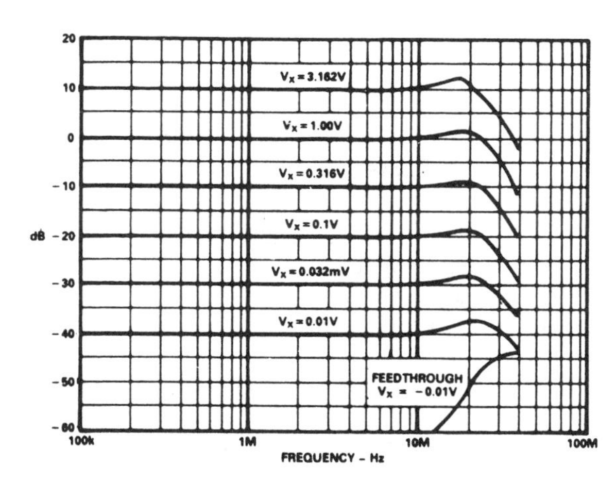

Multiplier HF Response
Op Amps Using ADLH0032

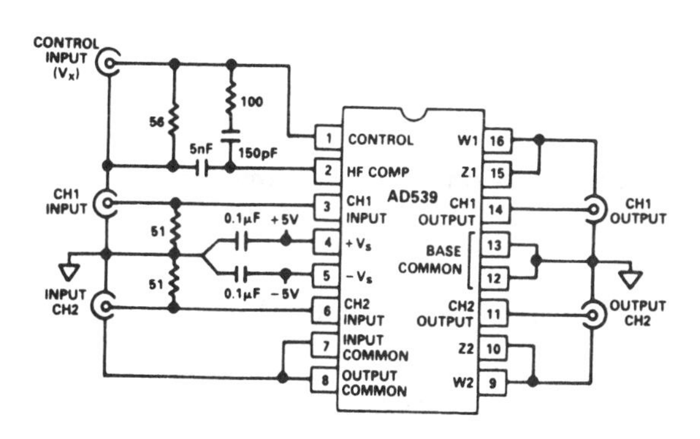

High-Speed Differential Configuration

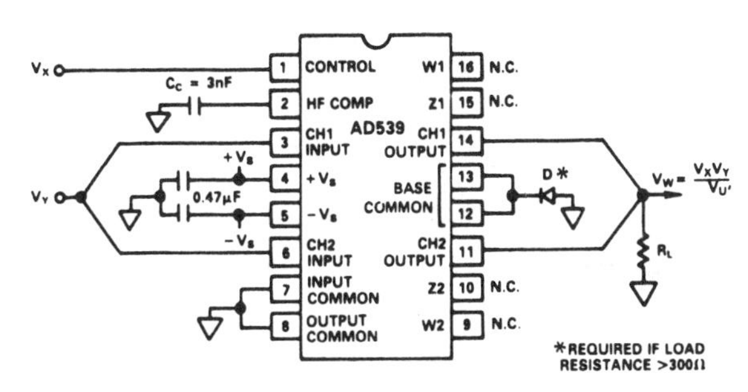

Minimal Single-Channel Multiplier

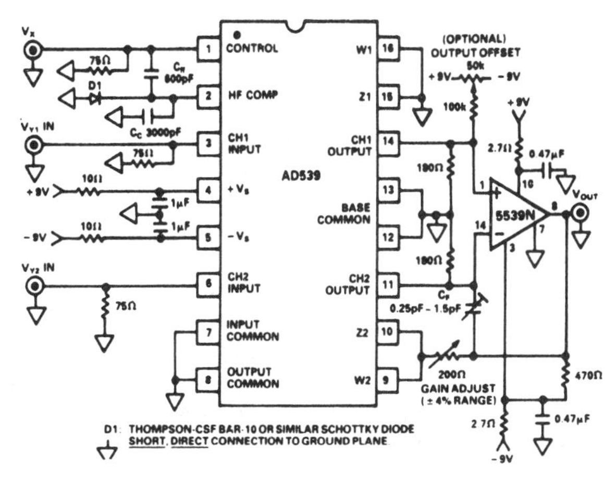

A Wide Bandwidth Voltage-Controlled Amplifier

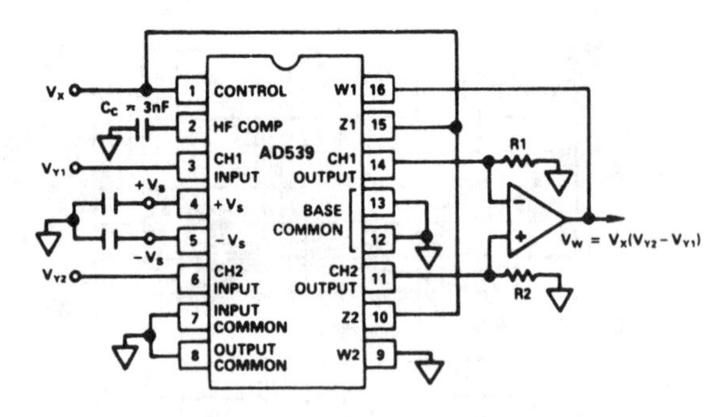

Low-Distortion Differential Configuration

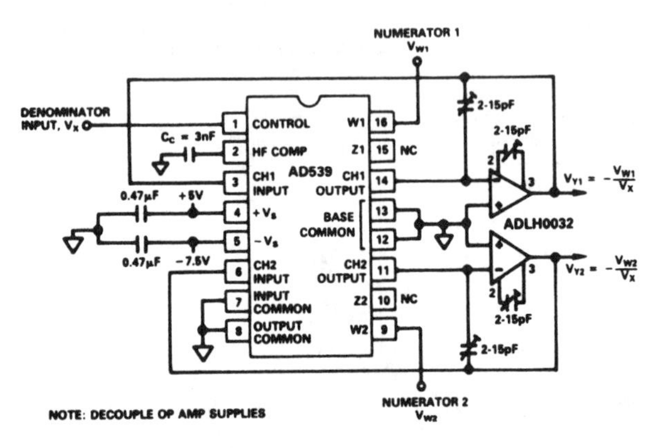

Two-Channel Divider with 1V Scaling

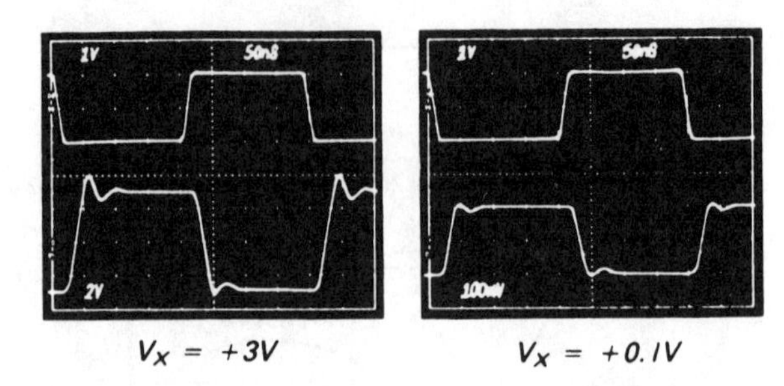

Multiplier Pulse Response Using ADLH0032 Op Amps

Op Amp Supply Voltages	±15V	±9V	±10V
Op Amp Compensation Capacitor	None	None	1-5pF
Feedback Capacitor, C_F	None	0.25-1.5pF	1-4pF
−3dB Bandwidth, V_X = +1V	900kHz	50MHz	25MHz
Load Capacitance	<1nF	<10pF	<100pF
HF Feedthrough,			
V_X = −0.01V, f = 5MHz	N/A	−54dB	−70dB
rms Output Noise,			
V_X = +1V, BW 10Hz-10kHz	50μV	40μV	30μV
V_X = +1V, BW 10Hz-5MHz	120μV	620μV	500μV

In all cases, 0.47μF ceramic supply-decoupling capacitors were used at each IC pin, the AD539 supplies were ±5V and the control-compensation capacitor C_C was 3nF.

Summary of Operating Conditions and Performance for the AD539 When Used with Various External Op-Amp Output Amplifiers

Analog Devices AD632
Internally Trimmed Precision IC Multiplier

- Pretrimmed to ±0.5% max 4-quadrant error
- All inputs (X, Y and Z) differential, high impedance for $[(X_1-X_2)(Y_1)-Y_2)/]+Z_2$ transfer function
- Scale-factor adjustable to provide up to ×10 gain
- Low-noise design: 90μV rms, 10 Hz−10 kHz
- Low-cost monolithic construction
- Excellent long-term stability

APPLICATIONS

- High-quality analog signal processing
- Differential ratio and percentage computations
- Algebraic and trigonometric function synthesis
- Accurate voltage-controlled oscillators and filters

The AD632 is an internally-trimmed monolithic four-quadrant multiplier/divider. The AD632B has a maximum multiplying error of ±0.5% without external trims.

Excellent supply rejection, low temperature coefficients, and long-term stability of the on-chip thin-film resistors and buried zener reference preserve accuracy, even under adverse conditions. The simplicity and flexibility of use provide an attractive alternative to the solution of complex control functions

The AD632 is pin-for-pin compatible with the industry-standard AD532 with improved specifications and a fully differential high-impedance Z-input. The AD632 is capable of providing gains of up to ×10, frequently eliminating the need for separate instrumentation amplifiers to precondition the inputs. The AD632 can be effectively employed as a variable-gain differential input amplifier with high common-mode rejection. The effectiveness of the variable-gain capability is enhanced by the inherent low noise of the AD632: 90μV rms.

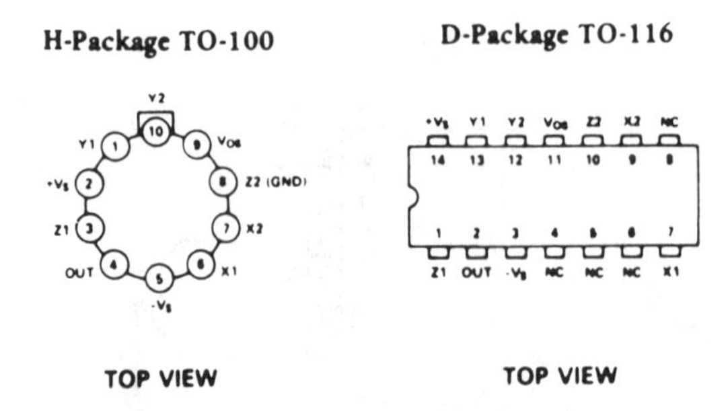

SPECIFICATIONS (@ +25°C, $V_S = \pm 15V$, $R \geq 2k\Omega$ unless otherwise noted)

Model	AD632A Min	AD632A Typ	AD632A Max	AD632B Min	AD632B Typ	AD632B Max	AD632S Min	AD632S Typ	AD632S Max	AD632T Min	AD632T Typ	AD632T Max	Units
MULTIPLIER PERFORMANCE													
Transfer Function		$\frac{(X_1-X_2)(Y_1-Y_2)}{10V}+Z_2$			$\frac{(X_1-X_2)(Y_1-Y_2)}{10V}+Z_2$			$\frac{(X_1-X_2)(Y_1-Y_2)}{10V}+Z_2$			$\frac{(X_1-X_2)(Y_1-Y_2)}{10V}+Z_2$		
Total Error (−10V<X, Y<+10V)			±1.0			±0.5			±1.0			±0.5	%
T_A = min to max		±1.5			±1.0				±2.0			±1.0	%
Total Error vs Temperature		±0.022			±0.015				±0.02			±0.01	%/°C
Scale Factor Error (SF = 10.000V Nominal)		±0.25			±0.1			±0.25			±0.1		%
Temperature-Coefficient of Scaling-Voltage		±0.02			±0.01			±0.2				±0.005	%/°C
Supply Rejection (±15V ±1V)		±0.01			±0.01			±0.01			±0.01		%
Nonlinearity, X (X = 20V pk-pk, Y = 10V)		±0.4			±0.2	±0.3		±0.4			±0.2	±0.3	%
Nonlinearity, Y (Y = 20V pk-pk, X = 10V)		±0.2			±0.1	±0.1		±0.2			±0.1	±0.1	%
Feedthrough[3], X (Y Nulled, X = 20V pk-pk 50Hz)		±0.3			±0.15	±0.3		±0.3			±0.15	±0.3	%
Feedthrough[3], Y (X Nulled, Y = 20V pk-pk 50Hz)		±0.01			±0.01	±0.1		±0.01			±0.01	±0.1	%
Output Offset Voltage		±5	±30		±2	±15		±5	±30		±2	±15	mV
Output Offset Voltage Drift		200			100				500			300	μV/°C
DYNAMICS													
Small Signal BW, (V_{OUT} = 0.1rms)		1			1			1			1		MHz
1% Amplitude Error (C_{LOAD} = 1000pF)		50			50			50			50		kHz
Slew Rate (V_{OUT} 20 pk-pk)		20			20			20			20		V/μs
Settling Time (to 1%, ΔV_{OUT} = 20V)		2			2			2			2		μs
NOISE													
Noise Spectral-Density SF = 10V		0.8			0.8			0.8			0.8		$\mu V/\sqrt{Hz}$
SF = 3V		0.4			0.4			0.4			0.4		$\mu V/\sqrt{Hz}$
Wideband Noise A = 10Hz to 5MHz		1.0			1.0			1.0			1.0		μV/rms
P = 10Hz to 10kHz		90			90			90			90		μV/rms
OUTPUT													
Output Voltage Swing	±11			±11			±11			±11			V
Output Impedance (f≤1kHz)		0.1			0.1			0.1			0.1		Ω
Output Short Circuit Current (R_L = 0, T_A = min to max)		30			30			30			30		mA
Amplifier Open Loop Gain (f = 50Hz)		70			70			70			70		dB
INPUT AMPLIFIERS (X, Y and Z)													
Signal Voltage Range (Diff. or CM		±10			±10			±10			±10		V
Operating Diff.)		±12			±12			±12			±12		V
Offset Voltage X, Y		±5	±20		±2	±10		±5	±20		±2	±10	mV
Offset Voltage Drift X, Y		100			50			100			150		μV/°C
Offset Voltage Z		±5	±30		±2	±15		±5	±30		±2	±15	mV
Offset Voltage Drift Z		200			100				500			300	μV/°C
CMRR	60	80		70	90		60	80		70	90		dB
Bias Current		0.8	2.0		0.8	2.0		0.8	2.0		0.8	2.0	μA
Offset Current		0.1			0.1			0.1			0.1		μA
Differential Resistance		10			10			10			10		MΩ
DIVIDER PERFORMANCE													
Transfer Function ($X_1 > X_2$)		$10V\frac{(Z_2-Z_1)}{(X_1-X_2)}+Y_1$			$10V\frac{(Z_2-Z_1)}{(X_1-X_2)}+Y_1$			$10V\frac{(Z_2-Z_1)}{(X_1-X_2)}+Y_1$			$10V\frac{(Z_2-Z_1)}{(X_1-X_2)}+Y_1$		
Total Error													
(X = 10V, −10V≤Z≤+10V)		±0.75			±0.35			±0.75			±0.35		%
(X = 1V, −1V≤Z≤+1V)		±2.0			±1.0			±2.0			±1.0		%
(0.1V≤X≤10V, −10V≤Z≤10V)		±2.5			±1.0			±2.5			±1.0		%
SQUARER PERFORMANCE													
Transfer Function		$\frac{(X_1-X_2)^2}{10V}+Z_2$			$\frac{(X_1-X_2)^2}{10V}+Z_2$			$\frac{(X_1-X_2)^2}{10V}+Z_2$			$\frac{(X_1-X_2)^2}{10V}+Z_2$		
Total Error (−10V≤X≤10V)		±0.6			±0.3			±0.6			±0.3		%
SQUARE-ROOTER PERFORMANCE													
Transfer Function, ($Z_1 \leq Z_2$)		$\sqrt{10V(Z_2-Z_1)}+X_2$			$\sqrt{10V(Z_2-Z_1)}+X_2$			$\sqrt{10V(Z_2-Z_1)}+X_2$			$\sqrt{10V(Z_2-Z_1)}+X_2$		
Total Error[1] (1V≤Z≤10V)		±1.0			±0.5			±1.0			±0.5		%
POWER SUPPLY SPECIFICATIONS													
Supply Voltage													
Rated Performance		±15			±15			±15			±15		V
Operating	±8		±18	±8		±18	±8		±22	±8		±22	V
Supply Current													
Quiescent		4	6		4	6		4	6		4	6	mA
PACKAGE OPTIONS													
TO-100 (H-10A)		AD632AH			AD632BH			AD632SH			AD632TH		
TO-116 (D-14)		AD632AD			AD632BD			AD632SH			AD632TD		

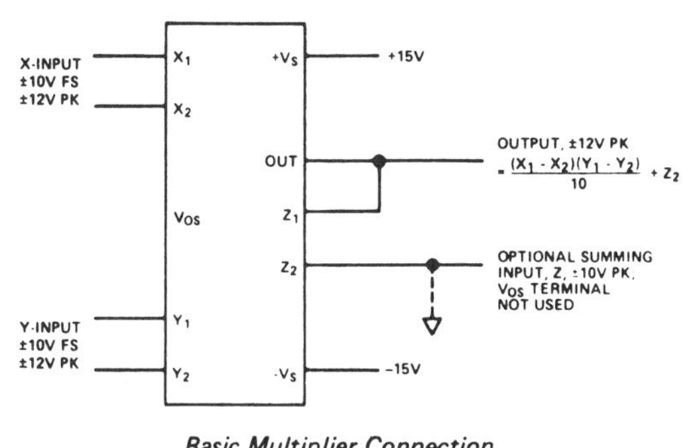

Basic Multiplier Connection

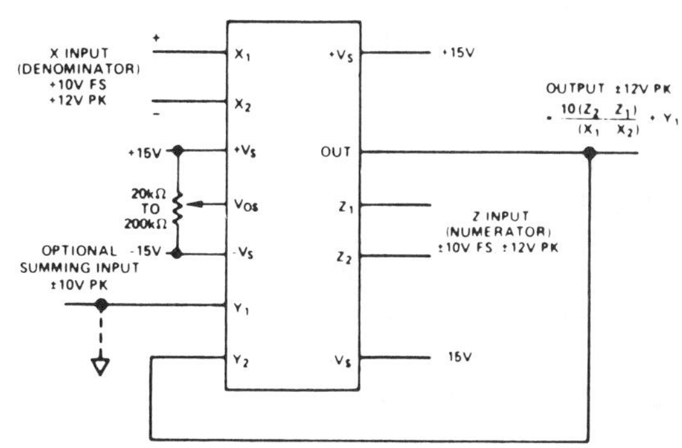

Basic Divider Connection

TYPICAL PERFORMANCE CURVES (typical at +25°C with $\pm V_S$ = 15V)

AC Feedthrough vs. Frequency

Frequency Response vs. Divider Denominator Input Voltage

Frequency Response as a Multiplier

CHIP DIMENSIONS & PAD LAYOUT

For further information, consult factory

Analog Devices ADG201A/ADG202A CMOS Quad SPST Switches

- 44V supply maximum rating
- ±15V analog signal range
- Low R_{ON} (60Ω)
- Low leakage (0.5nA)
- Extended plastic temperature range (−40°C to +85°C)
- Low power dissipation (33mW)
- Standard 16-pin DIPs and 20-terminal surface-mount packages
- Superior second source:
  - ADG201A replaces DG201A, HI-201
  - ADG202A replaces DG202

- **Extended signal range** These switches are fabricated on an enhanced LC²MOS process, resulting in high breakdown and an increased analog signal range of ±15V.
- **Single-supply operation** For applications where the analog signal is unipolar (0V to 15V), the switches can be operated from a single +15V supply.
- **Low leakage** Leakage currents in the range of 500pA make these switches suitable for high-precision circuits. The added feature of Break-before-make allows for multiple outputs to be tied together for multiplexer applications while keeping leakage errors to a minimum.

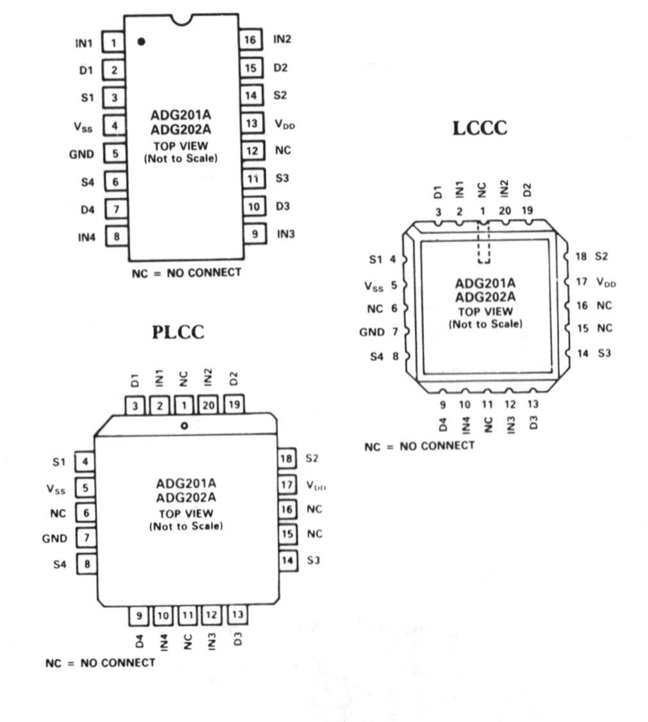

SPECIFICATIONS (V_{DD} = +15V, V_{SS} = −15V, unless otherwise noted)

Parameter	K Version 25°C	K Version −40°C to +85°C	B Version 25°C	B Version −40°C to +85°C	T Version 25°C	T Version −55°C to +125°C	Units	Test Conditions
ANALOG SWITCH								
Analog Signal Range	±15	±15	±15	±15	±15	±15	Volts	
R_{ON}	60		60		60		Ω typ	−10V ≤ V_S ≤ +10V
	90	145	90	145	90	145	Ω max	I_{DS} = 1.0mA Test Circuit 1
R_{ON} vs. V_D (V_S)	20		20		20		% typ	
R_{ON} Drift	0.5		0.5		0.5		%/°C typ	
R_{ON} Match	5		5		5		% typ	V_S = 0V, I_{DS} = 1mA
I_S (OFF)	0.5		0.5		0.5		nA typ	V_D = ±14V; V_S ∓ 14V; Test Circuit 2
OFF Input Leakage	2	100	2	100	1	100	nA max	
I_D (OFF)	0.5		0.5		0.5		nA typ	V_D = ±14V; V_S = ∓14V; Test Circuit 2
OFF Output Leakage	2	100	2	100	1	100	nA max	
I_D (ON)	0.5		0.5		0.5		nA typ	V_D = ±14V; Test Circuit 3
ON Channel Leakage	2	200	2	200	1	200	nA max	
DIGITAL CONTROL								
V_{INH}, Input High Voltage		2.4		2.4		2.4	V min	
V_{INL}, Input Low Voltage		0.8		0.8		0.8	V max	
I_{INL} or I_{INH}		1		1		1	μA max	
DYNAMIC CHARACTERISTICS								
t_{OPEN}	30		30		30		ns typ	
t_{ON}[1]	300		300		300		ns max	Test Circuit 4
t_{OFF}[1]	250		250		250		ns max	Test Circuit 4
OFF Isolation	80		80		80		dB typ	V_S = 10V(p-p); f = 100kHz R_L = 75Ω; Test Circuit 6
Channel-to-Channel Crosstalk	80		80		80		dB typ	Test Circuit 7
C_S (OFF)	5		5		5		pF typ	
C_D (OFF)	5		5		5		pF typ	
C_D, C_S (ON)	16		16		16		pF typ	
C_{IN} Digital Input Capacitance	5		5		5		pF typ	
Q_{INJ} Charge Injection	20		20		20		pC typ	R_S = 0Ω; C_L = 1000pF; V_S = 0V Test Circuit 5
POWER SUPPLY								
I_{DD}	0.6		0.6		0.6		mA typ	Digital Inputs = V_{INL} or V_{INH}
I_{DD}		2		2		2	mA max	
I_{SS}	0.1		0.1		0.1		mA typ	
I_{SS}		0.2		0.2		0.2	mA max	
Power Dissipation		33		33		33	mW max	

NOTES
[1]Sample tested at 25°C to ensure compliance.
Specifications subject to change without notice.

ABSOLUTE MAXIMUM RATINGS*

(T_A = +25°C unless otherwise stated)

V_{DD} to V_{SS} 44V
V_{DD} to GND 25V
V_{SS} to GND . . . . −25V
Analog Inputs[1]
  Voltage at S, D . . . . V_{SS} −0.3V to V_{DD} +0.3V
  Continuous Current, S or D 30mA
  Pulsed Current S or D
    1ms Duration, 10% Duty Cycle 70mA
Digital Inputs[1]
  Voltage at IN . . . . V_{SS} −2V to V_{DD} +2V or 20mA, Whichever Occurs First

Power Dissipation (Any Package)
  Up to +75°C 470mW
    Derates above +75°C by . . . . 6mW/°C
Operating Temperature
  Commercial (K Version) . . . . −40°C to +85°C
  Industrial (B Version) . . . . −40°C to +85°C
  Extended (T Version) . . . . −55°C to +125°C
Storage Temperature Range . . . . −65°C to +150°C
Lead Temperature (Soldering 10sec) . . . . +300°C

NOTE
[1]Overvoltage at IN, S or D will be clamped by diodes. Current should be limited to the Maximum Rating above.

*COMMENT: Stresses above those listed under "Absolute Maximum Ratings" may cause permanent damage to the device. This is a stress rating only and functional operation of the device at these or any other conditions above those indicated in the operational sections of this specification is not implied. Exposure to absolute maximum rating conditions for extended periods may affect device reliability. Only one Absolute Maximum Rating may be applied at any one time.

CAUTION

ESD (electrostatic discharge) sensitive device. The digital control inputs are diode protected; however, permanent damage may occur on unconnected devices subject to high energy electrostatic fields. Unused devices must be stored in conductive foam or shunts. The protective foam should be discharged to the destination socket before devices are removed.

ADG201A/ADG202A FUNCTIONAL BLOCK DIAGRAM

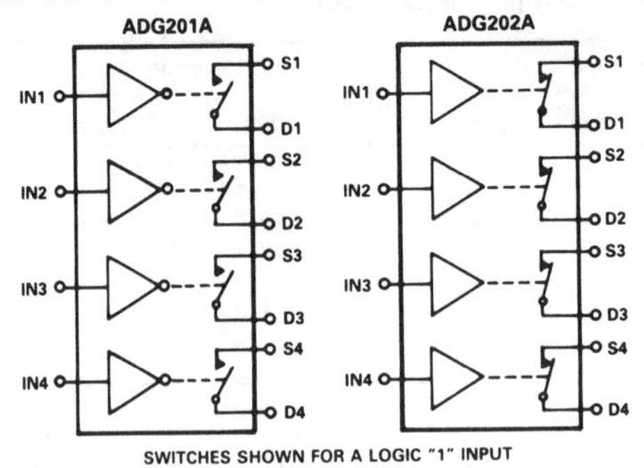

Analog Devices ADG201HS

LC²MOS High-Speed Quad SPST Switch

- 50 ns max switching time over full temperature range
- Low R_{ON} (30Ω typ)
- Single supply specifications for +10.8V to +16.5V operation
- Extended plastic temperature range (−40°C to +85°C)
- Break-before-make switching
- Low leakage (100pA typ)
- 44V supply max rating
- ADG201HS (K, B, T) replaces HI-201HS
- ADG201HS (J, A, S) replaces DG271

- **50 ns max t_{ON} and t_{OFF}** The ADG201HS top grades (K, B, T) have guaranteed 50 ns max turn-on and turn-off times over the full operating temperature range. The lower grades (J, A, S) have guaranteed 75 ns switching times over the full operating temperature range.
- **Single supply specifications** The ADG201HS is fully specified for applications that require a single positive power supply in the +10.8V to +16.5V range.
- **Low leakage** Leakage currents in the range of 100pA makes these switches suitable for high-precision circuits. The added feature of break-before-make allows for multiple outputs to be tied together for multiplexer applications while keeping leakage errors to a minimum.

ABSOLUTE MAXIMUM RATINGS*

(T_A = 25°C unless otherwise noted)

Parameter	Rating
V_{DD} to V_{SS}	44V
V_{DD} to GND	−0.3V, 25V
V_{SS} to GND[1]	+0.3V, −25V
Analog Inputs[2]	
Voltage at S, D	V_{SS} −2V to V_{DD} +2V or 20mA, Whichever Occurs First
Continuous Current, S or D	20mA
Pulsed Current S or D	
1ms Duration, 10% Duty Cycle	70mA
Digital Inputs[2]	
Voltage at IN	V_{SS} −4V to V_{DD} +4V or 20mA, Whichever Occurs First
Power Dissipation (Any Package)	
Up to +75°C	470mW
Derates above +75°C by	6mW/°C
Operating Temperature	
Commerical (J, K Version)	−40°C to +85°C
Industrial (A, B Version)	−40°C to +85°C
Extended (S, T Version)	−55°C to +125°C
Storage Temperature Range	−65°C to +150°C
Lead Temperature (Soldering 10sec)	+300°C

NOTES

[1]If V_{SS} is open circuited with V_{DD} and GND applied, the V_{SS} pin will be pulled positive, exceeding the Absolute Maximum Ratings. If this possibility exists, a Schottky diode from V_{SS} to GND (cathode end to GND) ensures that the Absolute Maximum Ratings will be observed.

[2]Overvoltage at IN, S or D, will be clamped by diodes. Current should be limited to the maximum rating above.

*COMMENT: Stresses above those listed under "Absolute Maximum Ratings" may cause permanent damage to the device. This is a stress rating only and functional operation of the device at these or any other conditions above those indicated in the operational sections of this specification is not implied. Exposure to absolute maximum rating conditions for extended periods may affect device reliability.

CAUTION

ESD (electrostatic discharge) sensitive device. The digital control inputs are diode protected; however, permanent damage may occur on unconnected devices subject to high energy electrostatic fields. Unused devices must be stored in conductive foam or shunts. The protective foam should be discharged to the destination socket before devices are removed.

PIN CONFIGURATIONS

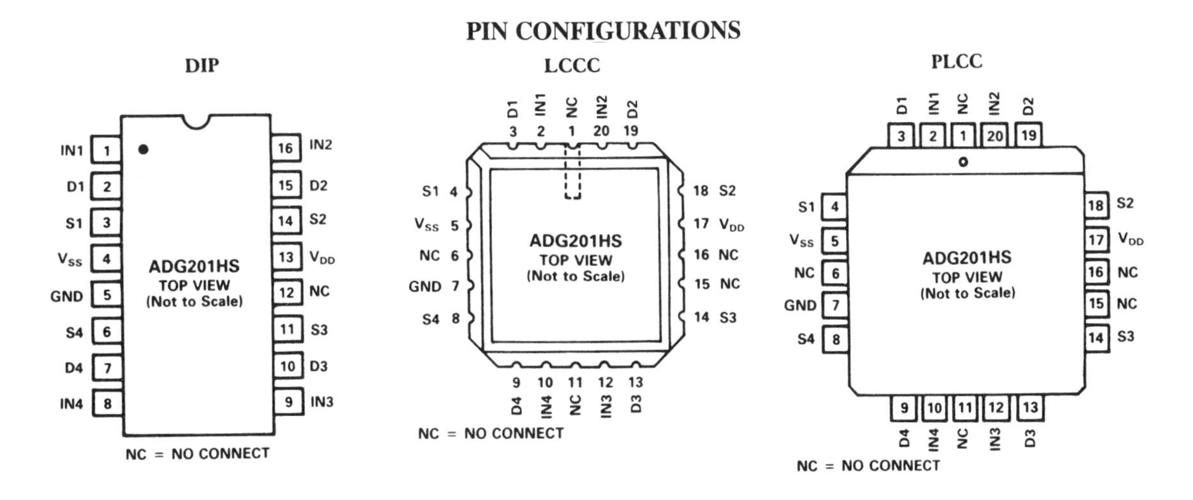

SPECIFICATIONS

Dual Supply (V_{DD} = +13.5V to +16.5V, V_{SS} = −13.5V to −16.5V, GND = 0V, V_{IN} = 3V (Logic High Level) or 0.8V (Logic Low Level) unless otherwise noted)

Parameter	Version	+25°C	$T_{min}-T_{max}$[1]	Units	Comments
ANALOG SWITCH					
Analog Signal Range	All	V_{SS}	V_{SS}	Vmin	
	All	V_{DD}	V_{DD}	V max	
R_{ON}	All	30	–	Ω typ	$-10V \le V_S \le +10V$, I_{DS} = 1mA; Test Circuit 1
	All	50	75	Ω max	
R_{ON} Drift	All	0.5	–	%/°C typ	$-10V \le V_S \le +10V$, I_{DS} = 1mA
R_{ON} Match	All	3	–	% typ	$-10V \le V_S \le +10V$, I_{DS} = 1mA
I_S (OFF), Off Input Leakage[2]	All	0.1		nA typ	$V_D = \pm 14V$; $V_S = \mp 14V$; Test Circuit 2
	J, K, A, B	1	20	nA max	
	S, T	1	60	nA max	
I_D (OFF), Off Output Leakage[2]	All	0.1		nA typ	$V_D = \pm 14V$; $V_S = \mp 14V$; Test Circuit 2
	J, K, A, B	1	20	nA max	
	S, T	1	60	nA max	
I_D (ON), On Channel Leakage[2]	All	0.1		nA typ	$V_D = \pm 14V$; Test Circuit 3
	J, K, A, B	1	20	nA max	
	S, T	1	60	nA max	
DIGITAL CONTROL					
V_{INH}, Input High Voltage	All	2.4	2.4	V min	
V_{INL}, Input Low Voltage	All	0.8	0.8	V max	
I_{INL} or I_{INH}	All	1	1	μA max	
C_{IN}	All	8	8	pF max	
DYNAMIC CHARACTERISTICS					
t_{ON}	K, B, T	50	50	ns max	Test Circuit 4
	J, A, S	75	75	ns max	
t_{OFF1}	K, B, T	50	50	ns max	Test Circuit 4
	J, A, S	75	75	ns max	
t_{OFF2}	All	150	–	ns typ	Test Circuit 4
t_{OPEN}	All	5	5	ns typ	$t_{ON}-t_{OFF1}$; Test Circuit 4
Output Settling Time to 0.1%	All	180	–	ns typ	V_{IN} = 3V to 0V; Test Circuit 4
OFF Isolation	All	72	–	dB typ	V_S = 3V rms, f = 100kHz, R_L = 1kΩ; C_L = 10pF; Test Circuit 5
Channel-to-Channel Crosstalk	All	86	–	dB typ	V_S = 3V rms, f = 100kHz, R_L = 1kΩ; C_L = 10pF; Test Circuit 6
Q_{INJ}, Charge Injection	All	10	–	pC typ	R_S = 0Ω, V_S = 0V; Test Circuit 7
C_S (OFF)	All	10	–	pF typ	
C_D (OFF)	All	10	–	pF typ	
C_D, C_S (ON)	All	30	–	pF typ	
C_{DS} (OFF)	All	0.5	–	pF typ	
POWER SUPPLY					
I_{DD}	All	10	10	mA max	
I_{SS}	All	6	6	mA max	
Power Dissipation	All	240	240	mW max	V_{DD} = +15V, V_{SS} = −15V

NOTES

[1]Temperature ranges are as follows: ADG201HSJ, K; −40°C to +85°C
ADG201HSA, B; −40°C to +85°C
ADG201HSS, T; −55°C to +125°C

[2]Leakage specifications apply with a V_D (V_S) of ±14V or with a V_D (V_S) of 0.5V within the supply voltages (V_{DD}, V_{SS}), whichever is the minimum.

Specifications subject to change without notice.

SINGLE SUPPLY (V_{DD} = +10.8V to +16.5V, V_{SS} = GND = 0V, V_{IN} = 3V [Logic High Level] or 0.8V [Logic Low Level] unless otherwise noted)

Parameter	Version	+25°C	$T_{min}-T_{max}$	Units	Comments
ANALOG SWITCH					
Analog Signal Range	All	V_{SS}	V_{SS}	V min	
	All	V_{DD}	V_{DD}	V max	
R_{ON}	All	65	–	Ω typ	0V≤V_S≤+10V, I_{DS} = 1mA; Test Circuit 1
	All	90	120	Ω max	
R_{ON} Drift	All	0.5	–	%/°C typ	0V≤V_S≤+10V, I_{DS} = 1mA
R_{ON} Match	All	3	–	% typ	0V≤V_S≤+10V, I_{DS} = 1mA
I_S (OFF), Off Input Leakage[1]	All	0.1		nA typ	V_D = +10V/+0.5V; V_S = +0.5V/+10V; Test Circuit 2
	J, K, A, B	1	20	nA max	
	S, T	1	60	nA max	
I_D (OFF), Off Output Leakage[1]	All	0.1		nA typ	V_D = +10V/+0.5V; V_S = +0.5V/+10V; Test Circuit 2
	J, K, A, B	1	20	nA max	
	S, T	1	60	nA max	
I_D (ON), On Channel Leakage[1]	All	0.1		nA typ	V_D = +10V/+0.5V; Test Circuit 3
	J, K, A, B	1	20	nA max	
	S, T	1	60	nA max	
DIGITAL CONTROL					
V_{INH}, Input High Voltage	All	2.4	2.4	V min	
V_{INL}, Input Low Voltage	All	0.8	0.8	V max	
I_{INL} or I_{INH}	All	1	1	μA max	
C_{IN}	All	8	8	pF max	
DYNAMIC CHARACTERISTICS					
t_{ON}	K, B, T	50	70	ns max	Test Circuit 4
	J, A, S	75	90	ns max	
t_{OFF1}	K, B, T	50	70	ns max	Test Circuit 4
	J, A, S	75	90	ns max	
t_{OFF2}	All	150	–	ns typ	Test Circuit 4
t_{OPEN}	All	5	5	ns typ	$t_{ON}-t_{OFF1}$; Test Circuit 4
Output Settling Time to 0.1%	All	180	–	ns typ	V_{IN} = 3V to 0V; Test Circuit 4
OFF Isolation	All	72	–	dB typ	V_S = 3V rms, f = 100kHz, R_L = 1kΩ; C_L = 10pF; Test Circuit 5
Channel-to-Channel Crosstalk	All	86	–	dB typ	V_S = 3V rms, f = 100kHz, R_L = 1kΩ; C_L = 10pF; Test Circuit 6
Q_{INJ}, Charge Injection	All	10	–	pC typ	R_S = 0Ω, V_S = 0V; Test Circuit 7
C_S (OFF)	All	10	–	pF typ	
C_D (OFF)	All	10	–	pF typ	
C_D, C_S (ON)	All	30	–	pF typ	
C_{DS} (OFF)	All	0.5	–	pF typ	
POWER SUPPLY					
I_{DD}	All	10	10	mA max	
Power Dissipation	All	150	150	mW max	V_{DD} = +15V

NOTE
[1]The leakage specifications degrade marginally (typically 1nA at 25°C) with V_D (V_S) = V_{SS}.
Specifications subject to change without notice.

Analog Devices ADG211A/ADG212A
LC²MOS Quad SPST Switches

- 44V supply maximum rating
- ±15V analog signal range
- Low R_{ON} (115Ω max)
- Low leakage (0.5nA typ)
- Single-supply operation possible
- Extended plastic temperature range (−40°C to +85°C)
- TTL/CMOS compatible
- Standard 16-pin DIPs and 20-terminal PLCC packages
- Superior second source:
  - ADG211A replaces DG211
  - ADG212A replaces DG212

- **Extended signal range** These switches are fabricated on an enhanced LC²MOS process, resulting in high breakdown and an increased analog signal range of ±15V.
- **Single supply operation** For applications where the analog signal is unipolar (0V to 15V), the switches can be operated from a single +15V supply.
- **Low leakage** Leakage currents in the range of 500pA make these switches suitable for high-precision circuits. The added feature of break-before-make allows for multiple outputs to be tied together for multiplexer applications while keeping leakage errors to a minimum.

SPECIFICATIONS (V_{DD} = +15V, V_{SS} = −15V, V_L = 5V, unless otherwise noted)

Parameter	K Version 25°C	K Version −40°C to +85°C	Units	Test Conditions
ANALOG SWITCH				
Analog Signal Range	±15	±15	Volts	
R_{ON}	115	175	Ωmax	$-10V \leq V_S \leq +10V$, I_{DS} = 1mA, Test Circuit 1
R_{ON} vs. $V_D(V_S)$	20		% typ	
R_{ON} Drift	0.5		%/°C typ	
R_{ON} Match	5		% typ	V_S = 0V, I_{DS} = 1mA
I_S (OFF)	0.5		nA typ	$V_D = \pm 14V$; $V_S \mp 14V$; Test Circuit 2
OFF Input Leakage	5	100	nA max	
I_D (OFF)	0.5		nA typ	$V_D = \pm 14V$; $V_S = \mp 14V$; Test Circuit 2
OFF Output Leakage	5	100	nA max	
I_D (ON)	0.5		nA typ	$V_D = \pm 14V$; Test Circuit 3
ON Channel Leakage	5	200	nA max	
DIGITAL CONTROL				
V_{INH}, Input High Voltage		2.4	V min	TTL Compatibility is Independent of V_L
V_{INL}, Input Low Voltage		0.8	V max	
I_{INL} or I_{INH}		1	μA max	
C_{IN}, Digital Input Capacitance	5		pF typ	
DYNAMIC CHARACTERISTICS				
t_{OPEN}[1]	30		ns typ	Test Circuit 4
t_{ON}[1]	600		ns max	Test Circuit 5
t_{OFF}[1]	450		ns max	Test Circuit 5
OFF Isolation	80		dB typ	V_S = 10V(p-p); f = 100kHz R_L = 75Ω; Test Circuit 6
Channel-to-Channel Crosstalk	80		dB typ	Test Circuit 7
C_S (OFF)	5		pF typ	
C_D (OFF)	5		pF typ	
C_S, C_D (ON)	16		pF typ	
Q_{INJ}, Charge Injection	20		pC typ	R_S = 0Ω; C_L = 1000pF; V_S = 0V Test Circuit 8
POWER SUPPLY				
I_{DD}	0.6		mA typ	Digital Inputs = V_{INL} or V_{INH}
I_{DD}	1		mA max	
I_{SS}	0.1		mA typ	
I_{SS}	0.2		mA max	
I_L	0.9		mA max	

NOTE
[1]Sample tested at 25°C to ensure compliance.
Specifications subject to change without notice.

PIN CONFIGURATIONS

DIP

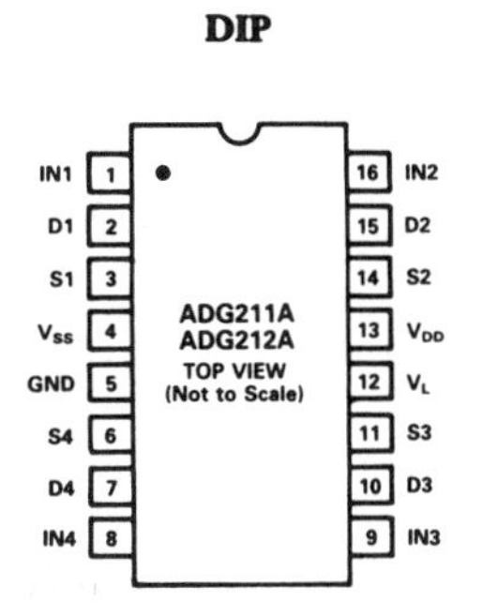

PLCC

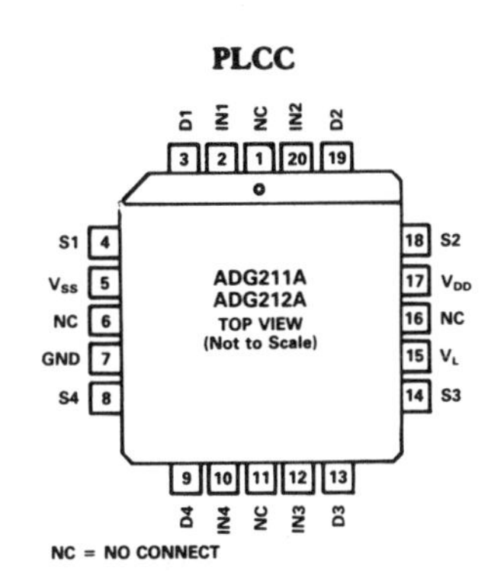

ADG211A/ADG212A FUNCTIONAL BLOCK DIAGRAMS

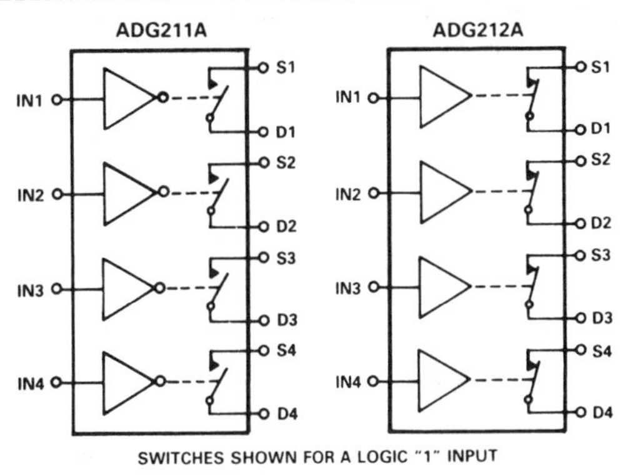

ABSOLUTE MAXIMUM RATINGS*

(T_A = 25°C unless otherwise stated)

V_{DD} to V_{SS}	44V
V_{DD} to GND	25V
V_{SS} to GND	−25V
V_L to GND	−0.3V, 25V
Analog Inputs[1]	
Voltage at S, D	V_{SS} −0.3V to V_{DD} +0.3V
Continuous Current, S or D	30mA
Pulsed Current S or D	
1ms Duration, 10% Duty Cycle	70mA
Digital Inputs[1]	
Voltage at IN	V_{SS} −2V to V_{DD} +2V or 20mA, Whichever Occurs First
Power Dissipation (Any Package)	
Up to +75°C	470mW
Derates above +75°C by	6mW/°C
Operating Temperature	−40°C to +85°C
Storage Temperature Range	−65°C to +150°C
Lead Temperature (Soldering 10sec)	+300°C

NOTE
[1]Overvoltage at IN, S or D will be clamped by diodes. Current should be limited to the Maximum Rating above.

*COMMENT: Stresses above those listed under "Absolute Maximum Ratings" may cause permanent damage to the device. This is a stress rating only and functional operation of the device at these or any other conditions above those indicated in the operational sections of this specification is not implied. Exposure to absolute maximum rating conditions for extended periods may affect device reliability. Only one Absolute Maximum Rating may be applied at any one time.

CAUTION

ESD (electrostatic discharge) sensitive device. The digital control inputs are diode protected; however, permanent damage may occur on unconnected devices subject to high energy electrostatic fields. Unused devices must be stored in conductive foam or shunts. The protective foam should be discharged to the destination socket before devices are removed.

Analog Devices ADG221/ADG222
CMOS Quad SPST Switches

- 44V Supply maximum rating
- ±15V analog signal range
- Low R_{ON} (60Ω)
- Low leakage (0.5nA)
- Extended plastic temperature range (−40°C to +85°C)
- Low power dissipation (25.5mW)
- μP, TTL, CMOS compatible
- Standard 16-pin DIPs and 20-terminal surface-mount packages
- Superior DG221 replacement
- **Easily Interfaced** Digital inputs are latched with a $\overline{WR}$ signal for microprocessor interfacing. A 5V regulated supply is internally generated permitting wider tolerances on the supplies without affecting the TTL digital input switching levels.
- **Single-supply operation** For applications where the analog signal is unipolar (0V to 15V), the switches can be operated from a single +15V supply.
- **Low leakage** Leakage currents in the range of 500pA make these switches suitable for high-precision circuits. The added feature of break-before-make allows for multiple outputs to be tied together for multiplexer applications while keeping leakage errors to a minimum.

ADG221/ADG222 FUNCTIONAL BLOCK DIAGRAM

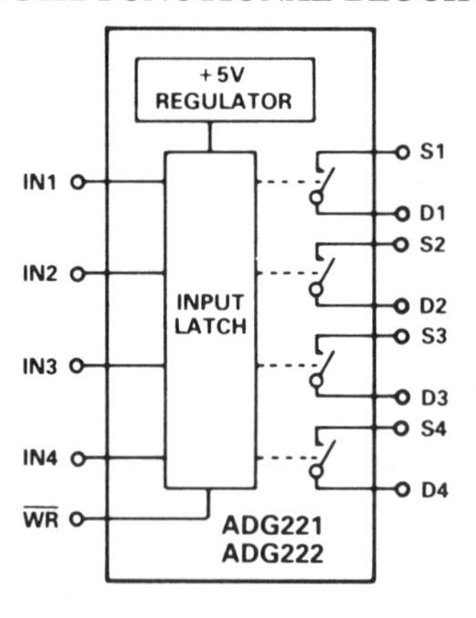

PIN CONFIGURATIONS

DIP

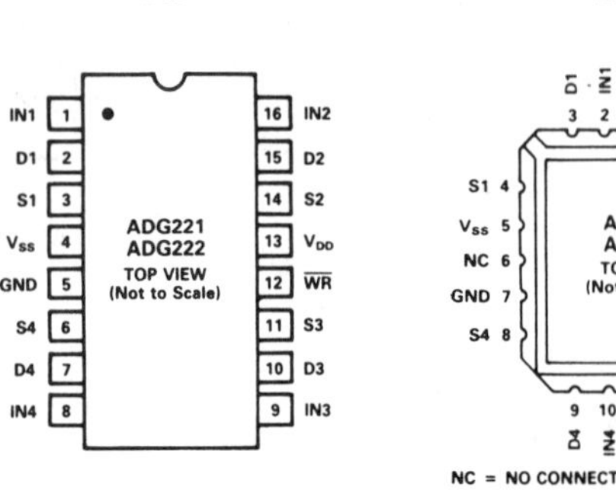

LCCC

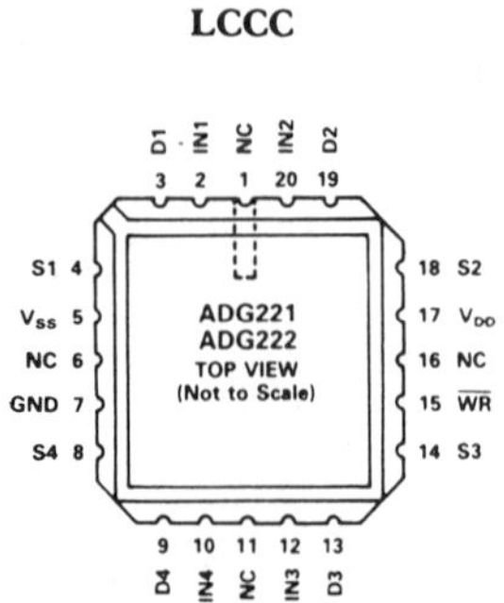

PLCC

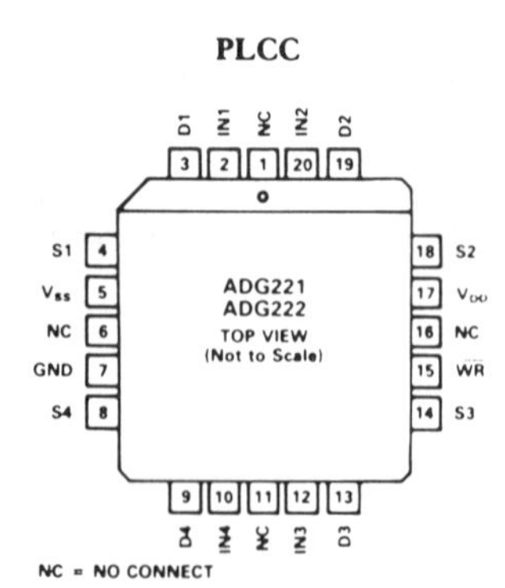

ABSOLUTE MAXIMUM RATINGS

(T_A = +25°C unless otherwise stated)

V_{DD} to V_{SS} 44V
V_{DD} to GND 25V
V_{SS} to GND . . . . −25V
Analog Inputs[1]
  Voltage at S, D . . . . V_{SS} −0.3V to V_{DD} +0.3V
  Continuous Current, S or D 30mA
  Pulsed Current S or D
    1ms Duration, 10% Duty Cycle 70mA
Digital Inputs[1]
  Voltage at IN, $\overline{WR}$ . . . . V_{SS} −2V to V_{DD} +2V or 20mA, Whichever Occurs First

Power Dissipation (Any Package)
  Up to +75°C 470mW
  Derates above +75°C by . . . . 6mW/°C
Operating Temperature
  Commercial (K Version) . . . . −40°C to +85°C
  Industrial (B Version) . . . . −40°C to +85°C
  Extended (T Version) . . . . −55°C to +125°C
Storage Temperature . . . . −65°C to +150°C
Lead Temperature (Soldering 10sec) . . . . +300°C

NOTE
[1]Overvoltage at IN, $\overline{WR}$, S or D will be clamped by diodes. Current should be limited to the Maximum Rating above.

*COMMENT: Stresses above those listed under "Absolute Maximum Ratings" may cause permanent damage to the device. This is a stress rating only and functional operation of the device at these or any other conditions above those indicated in the operational sections of this specification is not implied. Exposure to absolute maximum rating conditions for extended periods may affect device reliability. Only one Absolute Maximum Rating may be applied at any one time.

CAUTION

ESD (electrostatic discharge) sensitive device. The digital control inputs are diode protected; however, permanent damage may occur on unconnected devices subject to high energy electrostatic fields. Unused devices must be stored in conductive foam or shunts. The protective foam should be discharged to the destination socket before devices are removed.

SPECIFICATIONS (V_{DD} = +15V, V_{SS} = −15V, unless otherwise noted)

Parameter	K Version 25°C	K Version −40°C to +85°C	B Version 25°C	B Version −40°C to +85°C	T Version 25°C	T Version −55°C to +125°C	Units	Test Conditions
ANALOG SWITCH								
Analog Signal Range	±15	±15	±15	±15	±15	±15	Volts	
R_{ON}	60		60		60		Ω typ	$-10V \le V_S \le +10V$
	90	145	90	145	90	145	Ω max	I_{DS} = 1.0mA Test Circuit 1
R_{ON} vs. V_D (V_S)	20		20		20		% typ	
R_{ON} Drift	0.5		0.5		0.5		%/°C typ	
R_{ON} Match	5		5		5		% typ	V_S = 0V, I_{DS} = 1mA
I_S (OFF)	0.5		0.5		0.5		nA typ	V_D = ±14V; V_S ∓14V; Test Circuit 2
OFF Input Leakage	2	100	2	100	1	100	nA max	
I_D (OFF)	0.5		0.5		0.5		nA typ	V_D = ±14V; V_S = ∓14V; Test Circuit 2
OFF Output Leakage	2	100	2	100	1	100	nA max	
I_D (ON)	0.5		0.5		0.5		nA typ	V_D = ±14V; Test Circuit 3
ON Channel Leakage	2	200	2	200	1	200	nA max	
DIGITAL CONTROL								
V_{INH}, Input High Voltage		2.4		2.4		2.4	V min	
V_{INL}, Input Low Voltage		0.8		0.8		0.8	V max	
I_{INL} or I_{INH}		1		1		1	µA max	
DYNAMIC CHARACTERISTICS								
t_{OPEN}	30		30		30		ns typ	
t_{ON}[1]	300		300		300		ns max	Test Circuit 4
t_{OFF}[1]	250		250		250		ns max	Test Circuit 4
t_W[1] Write Pulse Width		100		100	100	120	ns min	
t_S[1] Digital Input Setup Time		100		100	100	120	ns min	
t_H[1] Digital Input Hold Time		20		20	20	20	ns min	
OFF Isolation	80		80		80		dB typ	V_S = 10V (p-p); f = 100kHz R_L = 75Ω; Test Circuit 6
Channel-to-Channel Crosstalk	80		80		80		dB typ	Test Circuit 7
C_S (OFF)	5		5		5		pF typ	
C_D (OFF)	5		5		5		pF typ	
C_D, C_S (ON)	16		16		16		pF typ	
C_{IN} Digital Input Capacitance	5		5		5		pF typ	
Q_{INJ} Charge Injection	20		20		20		pC typ	R_S = 0Ω; C_L = 1000pF; V_S = 0V Test Circuit 5
POWER SUPPLY								
I_{DD}	0.6		0.6		0.6		mA typ	Digital Inputs = V_{INL} or V_{INH}
I_{DD}		1.5		1.5		1.5	mA max	
I_{SS}	0.1		0.1		0.1		mA typ	
I_{SS}		0.2		0.2		0.2	mA max	
Power Dissipation		25.5		25.5		25.5	mW max	

NOTE
[1]Sample tested at 25°C to ensure compliance.
t_{ON}, t_{OFF} are the same for both IN and $\overline{WR}$ digital input changes.
Specifications subject to change without notice.

Plessey SP8600A and B
250 MHz ÷ 44

- Open collector output
- Ac-coupled input
- Supply voltage: −5.2V
- Power consumption: 85mW
- Max. input frequency: 250 MHz
- Temperature range:
  - −55 °C to +125 °C (A Grade)
  - −30 °C to +70 °C (B Grade)

ABSOLUTE MAXIMUM RATINGS

Supply voltage	-10V
Output voltage (Pins 1 and 3)	V_{EE} +14V
Storage temperature range	-55 °C to +175 °C
Max. junction temperature	+175 °C
Max. clock I/P voltage	2.5V p-p

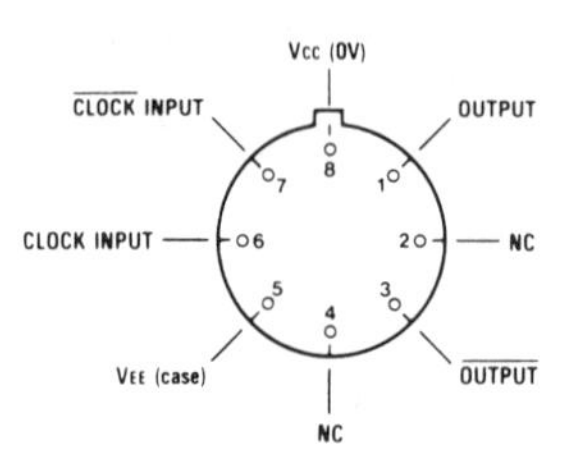

Pin connections - bottom view

ELECTRICAL CHARACTERISTICS

Supply voltage: V_{EE} = -5.2V ± 0.25V V_{CC} = 0V
Temperature: A Grade T_{amb} = -55°C to +125°C
B Grade T_{amb} = -30°C to +70°C

Characteristic	Symbol	Value		Units	Conditions
		Min.	Max.		
Maximum frequency (sinewave input)	f_{max}	250		MHz	Input = 400-800mV
Minimum frequency (sinewave input)	f_{min}		25	MHz	Input = 400-800mV
Power supply current	I_{EE}		25	mA	V_{EE} = -5.2V
Output current	I_{OUT}	1.65		mA	

NOTES
1. Unless otherwise stated the electrical characteristics are guaranteed over specified supply, frequency and temperature range.
2. The dynamic test circuit is shown in Fig. 5.

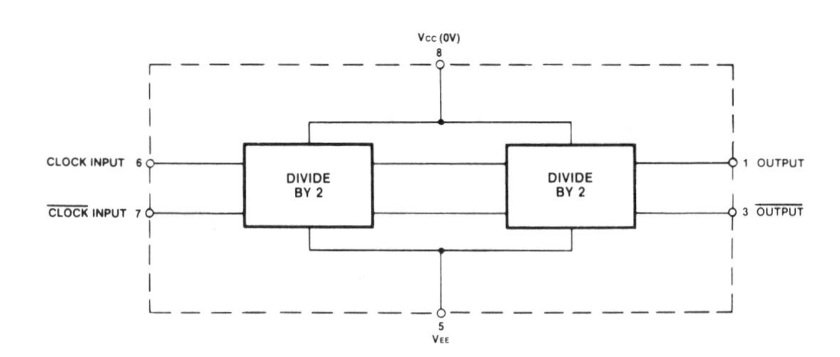

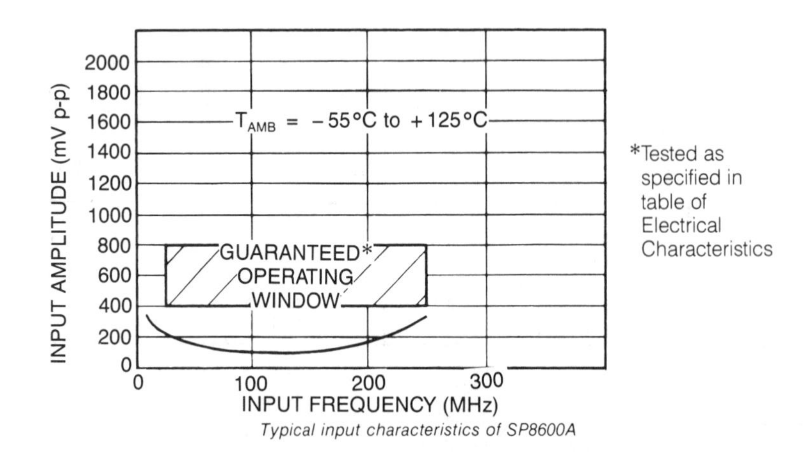

Typical input characteristics of SP8600A

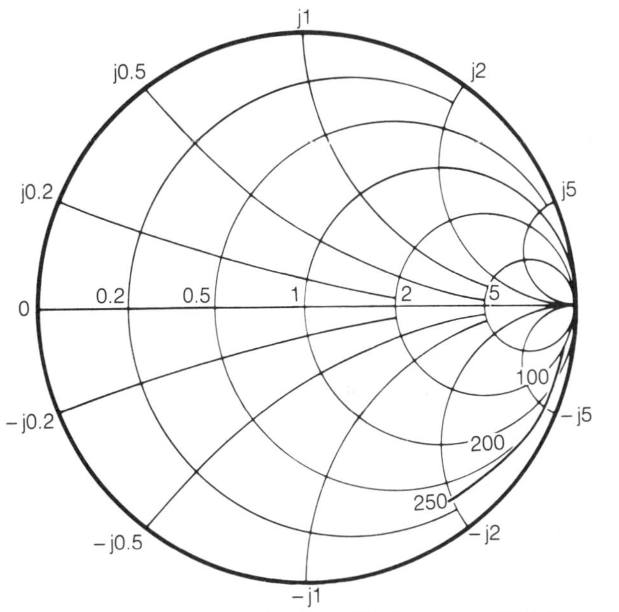

Typical input impedance: supply voltage -5.2V, temperature 25° C, frequencies in MHz, impedances normalised to 50 ohms

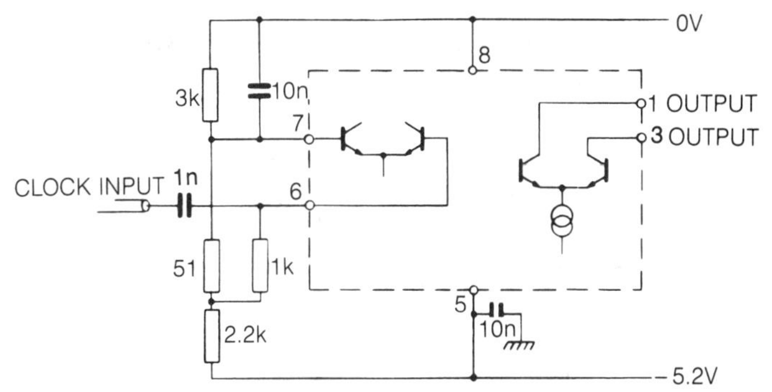

Biasing to prevent oscillation under no signal conditions

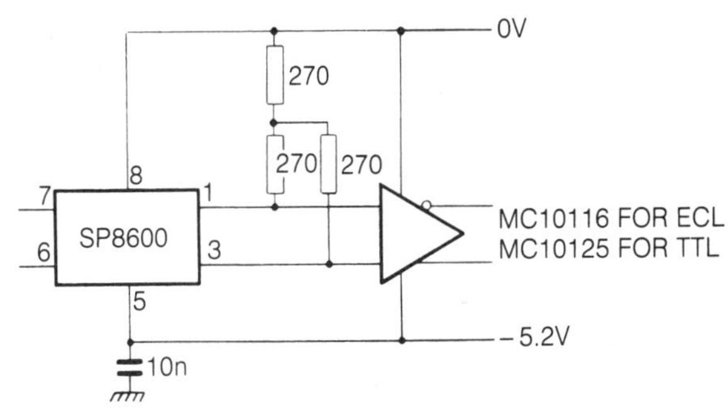

Interfacing to ECL and Schottky TTL

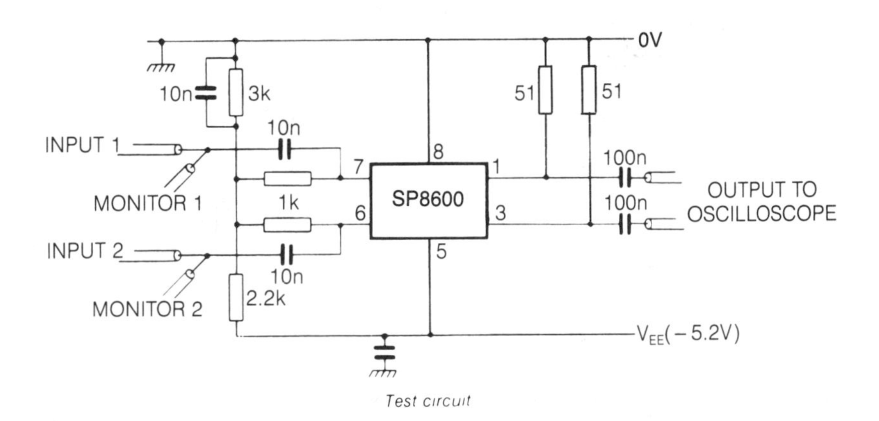

Test circuit

Plessey SP8602A and B 500 MHz ÷ 2 SP8604A and B 300 MHz ÷ 2

- ECL compatible outputs
- Ac-coupled inputs (internal bias)
- Supply voltage: –5.2V
- Power consumption: 85mW
- Temperature range:
  - –55 °C to +125 °C (A grade)
  - –30 °C to +70 °C (B grade)

ABSOLUTE MAXIMUM RATINGS

Supply voltage	-8V
Output current	10mA
Storage temperature range	-55°C to +175°C
Max. junction temperature	+175°C
Max. clock I P voltage	2.5V p-p

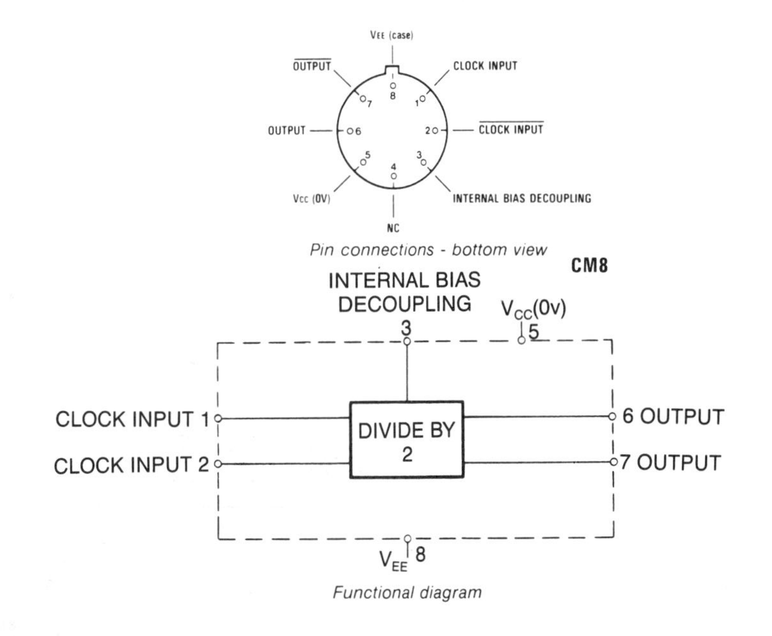

Pin connections - bottom view

Functional diagram

ELECTRICAL CHARACTERISTICS

Supply voltage: V_{CC} = 0V, V_{EE} = -5.2V ± 0.25V
Temperature: T_{amb} A Grade = -55°C to +125°C
B Grade = -30°C to +70°C

Characteristics	Symbol	Value		Units	Grade	Conditions	Notes
		Min.	Max.				
Maximum frequency (sinewave input)	f_{max}	500		MHz	SP8602	Input = 400-800mV p-p	
		300		MHz	SP8604		
Minimum frequency (sinewave input)	f_{min}		40	MHz	All	Input = 400-800mV p-p	
Power supply current	I_{EE}		18	mA	All	V_{EE} = 5.2V Outputs unloaded	
Output low voltage	V_{OL}	-1.8	-1.4	V	All	V_{EE} = -5.2V	Note 4
Output high voltage	V_{OH}	-0.85	-0.7	V	All	V_{EE} = -5.2V	Note 4
Minimum output swing	V_{OUT}	400		mV	All	V_{EE} = -5.2V	

NOTES
1. Unless otherwise stated the electrical characteristics shown above are guaranteed over specified supply, frequency and temperature range.
2. The temperature coefficients of V_{OH} = +1.63mV/°C and V_{OL} = +0.34mV/°C but these are not tested.
3. The test configuration for dynamic testing is shown in Fig.5.
4. Tested at 25°C only.

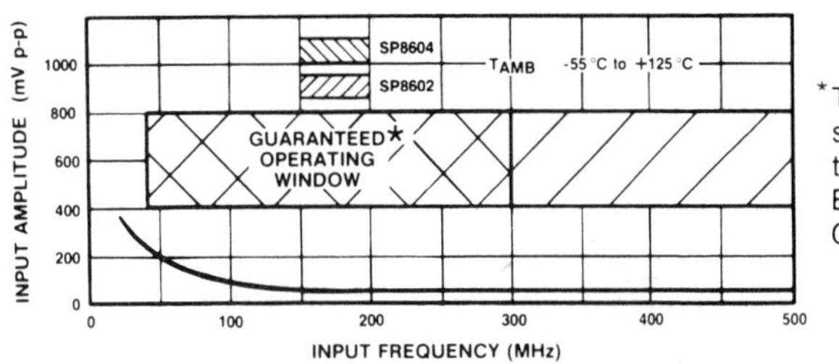

*Tested as specified in table of Electrical Characteristics

Typical characteristic of SP8602 and SP8604

Typical input impedance. Test conditions: supply voltage -5.2V, ambient temperature 25° C, frequencies in MHz, impedances normalised to 50 ohms.

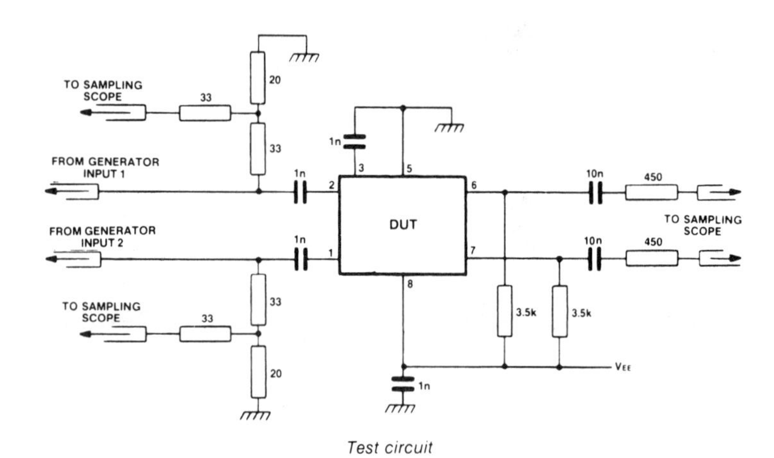

Test circuit

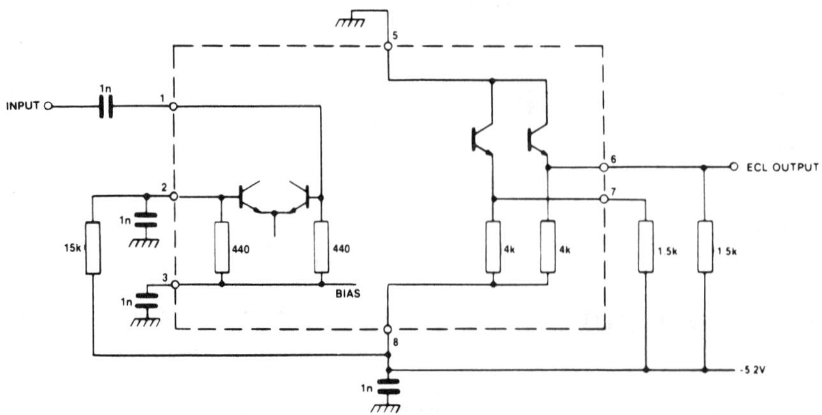

Typical application showing interfacing

Plessey
SP8605A and B
1000 MHz ÷ 2
SP8606A and B
1300 MHz ÷ 2

- ECL-compatible outputs
- Ac-coupled inputs (internal bias)
- Supply voltage: −5.2V
- Power consumption: 320 mW
- Max. input frequency: 1300 MHz (SP8606)
- Temperature range:
  - A grade: −55 °C to +110 °C (125 °C with suitable heatsink)
  - B grade: 0 °C to +70 °C

ABSOLUTE MAXIMUM RATINGS

Supply voltage	-8V
Output current	15mA
Storage temperature range	-55 °C to +150 °C
Max. junction temperature	+175 °C
Max. clock I/P voltage	2.5V p-p

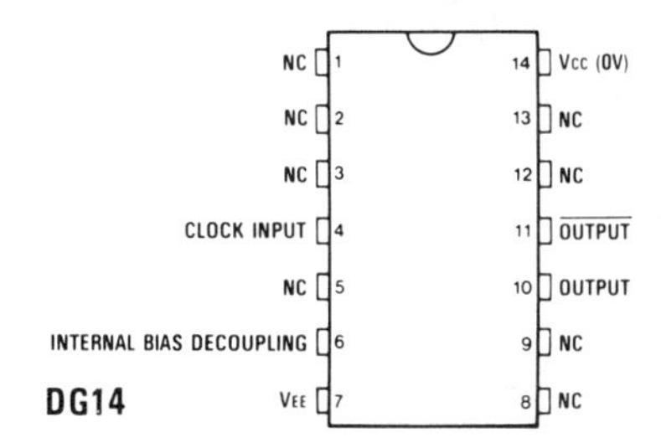

Pin connections - top view

ELECTRICAL CHARACTERISTICS

Supply voltage: V_{CC} = 0V, V_{EE} = -5.2V ± 0.25V
Temperature: A Grade T_{case} = -55°C to +125°C (Note 2)
B Grade T_{amb} = 0°C to +70°C

Characteristics	Symbol	Value Min.	Value Max.	Units	Grade	Conditions	Notes
Maximum frequency (sinewave input)	f_{max}	1.0		GHz	SP8605A,B	Input = 400-1200mV p-p	Note 7
		1.3		GHz	SP8606A	Input = 800-1200mV p-p	Note 7
		1.3		GHz	SP8606B	Input = 400-1200mV p-p	Note 7
Minimum frequency (sinewave input)	f_{min}		150	MHz	All	Input = 600-1200mV p-p	Note 5
Current consumption	I_{EE}		100	mA	All	V_{EE} = -5.45V Outputs unloaded	Note 6
Output low voltage	V_{OL}	-1.92	-1.62	V	All	V_{EE} = -5.2V Outputs loaded with 430Ω(25° C)	
Output high voltage	V_{OH}	-0.93	-0.75	V	All	V_{EE}= -5.2V Outputs loaded with 430Ω(25° C)	
Minimum output swing	V_{OUT}	500		mV	All	V_{EE} = -5.2V Outputs loaded with 430 ohms	Note 6

NOTES
1. Unless otherwise stated the electrical characteristics shown above are guaranteed over specified supply, frequency and temperature range.
2. The A grade devices must be used with a heat sink to maintain chip temperature below +175°C when operating in an ambient of +125°C.
3. The temperature coefficients of V_{OH} = +1.2mV/°C and V_{OL} = +0.24mV/°C but these are not tested.
4. The test configuration for dynamic testing is shown in Fig.5.
5. Tested at 25°C and +125°C only (+70°C for B grade).
6. Tested at 25°C only.
7. Tested at +125°C only (+70°C for B grade).

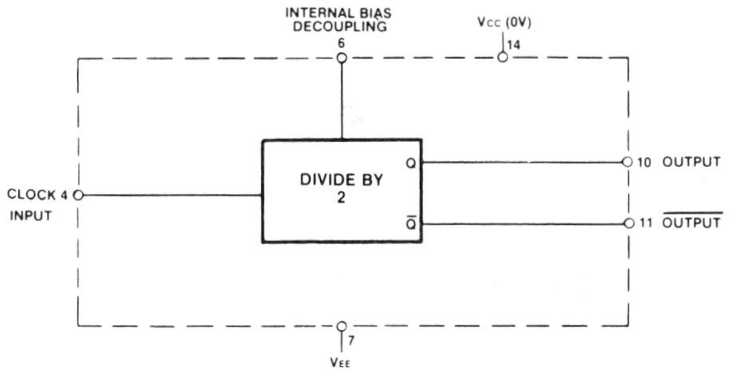

Functional diagram

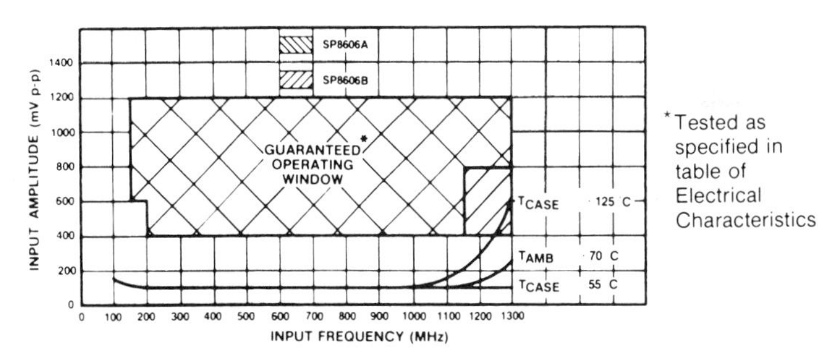

Typical characteristic of SP8606

Typical input impedance. Test conditions: supply voltage -5.2V, ambient temperature 25°C, frequencies in MHz, impedances normalised to 50 ohms.

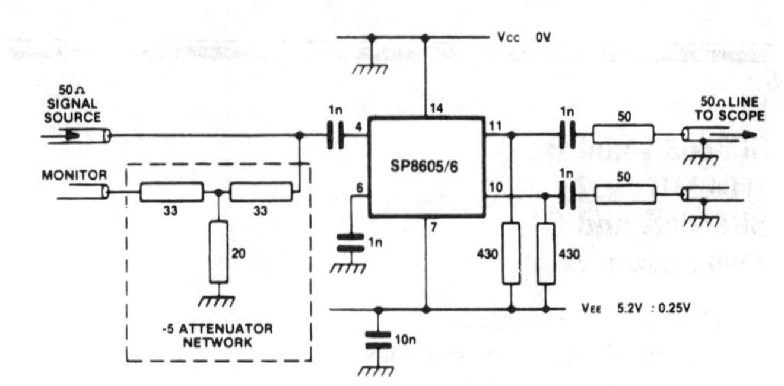

Toggle frequency test circuit

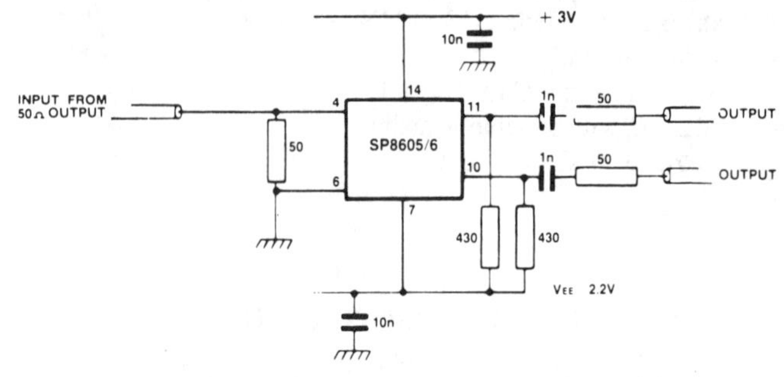

Circuit for using the input signal about ground potential

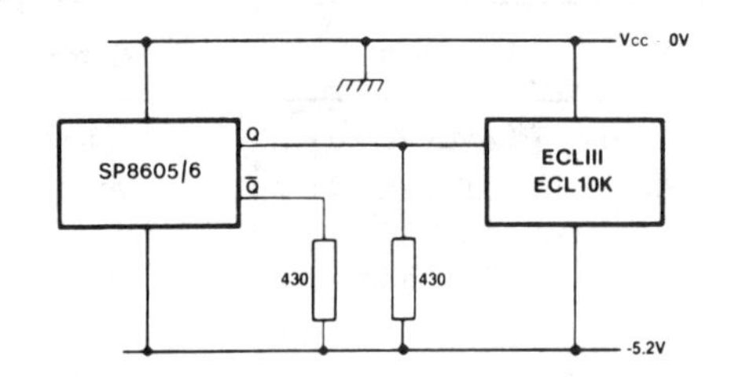

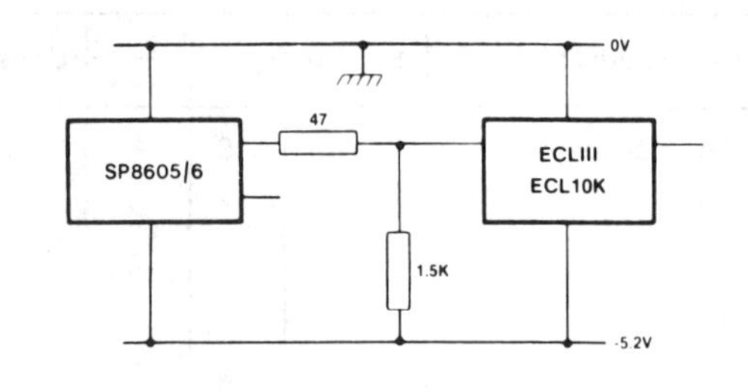

Interfacing SP8605/6 to ECL 10K and ECL III

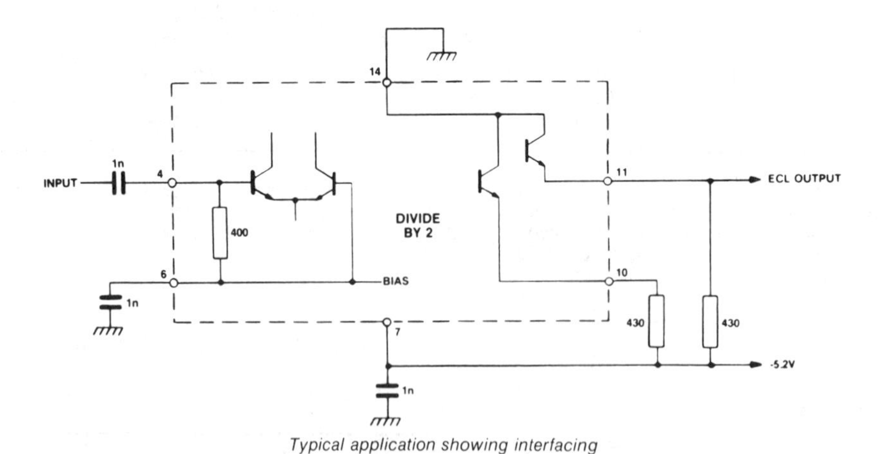

Typical application showing interfacing

Plessey SP8670A and B 600 MHz ÷ 8

- ECL-compatible outputs
- Ac-coupled inputs (internal bias)
- Supply voltage: −5.2V
- Power consumption: 300 mW
- Temperature range:
  - ○ −55 °C to +125 °C (A grade)
  - ○ −30 °C to +70 °C (B grade)

ABSOLUTE MAXIMUM RATINGS

Supply voltage	-8V
Output current	10mA
Storage temperature range	-55 °C to +150 °C
Max. junction temperature	+175 °C
Max. clock I/P voltage	2.5V p-p

ELECTRICAL CHARACTERISTICS

Supply voltage: V_{CC} = 0V, V_{EE} = -5.2V ± 0.25V
Temperature: A Grade T_{amb} = -55°C to +125°C
B Grade T_{amb} = -30°C to +70°C

Characteristics	Symbol	Value Min.	Value Max.	Units	Conditions	Notes
Maximum frequency (sinewave input)	f_{max}	600		MHz	Input = 400-800mV p-p	
Minimum frequency (sinewave input)	f_{min}		40	MHz	Input = 400-800mV p-p	Note 4
Power supply current	I_{EE}		60	mA	V_{EE} = -5.2V	Note 4
Output low voltage	V_{OL}	-1.8	-1.5	V	V_{EE} = -5.2V (25°C)	
Output high voltage	V_{OH}	-0.85	-0.7	V	V_{EE} = -5.2V (25°C)	
Minimum output swing	V_{OUT}	500		mV	V_{EE} = -5.2V	

NOTES
1. Unless otherwise stated the electrical characteristics shown above are guaranteed over specified supply, frequency and temperature range.
2. The temperature coefficients of V_{OH} = +1.63mV/°C and V_{OL} = +0.94mV/°C but these are not tested.
3. The test configuration for dynamic testing is shown in Fig.5.
4. Tested at 25°C only.

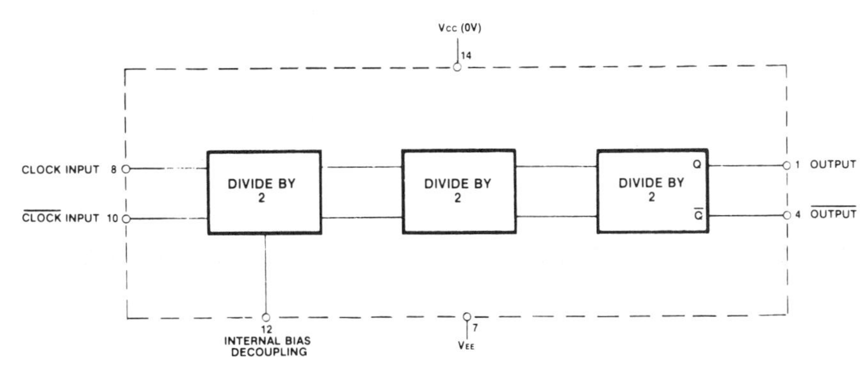

Functional diagram

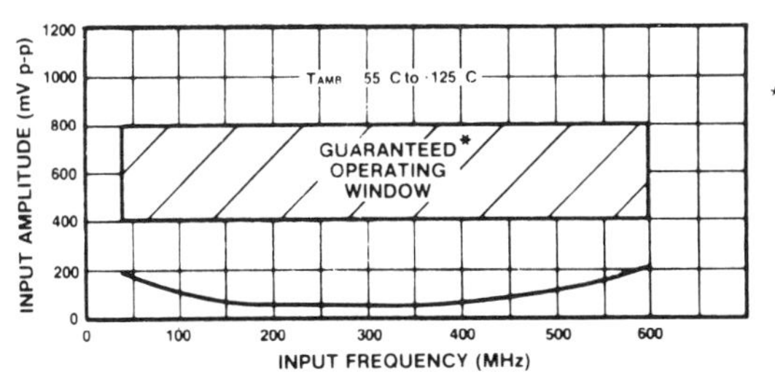

*Tested as specified in table of Electrical Characteristics

Typical input characteristic of SP8670A

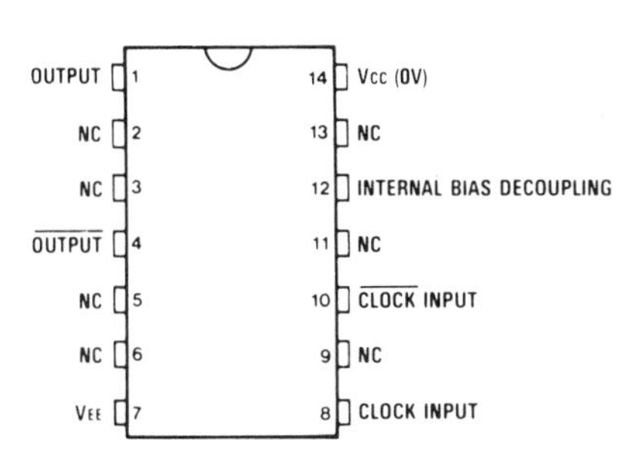

DG14

Pin connections - top view

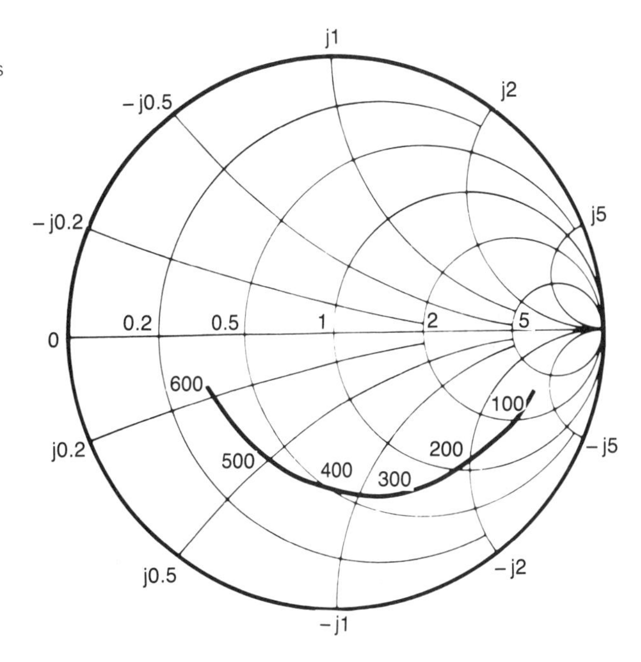

Test circuit

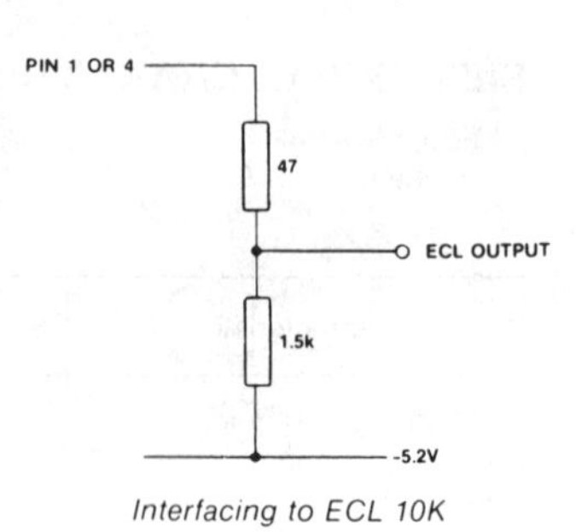

Interfacing to ECL 10K

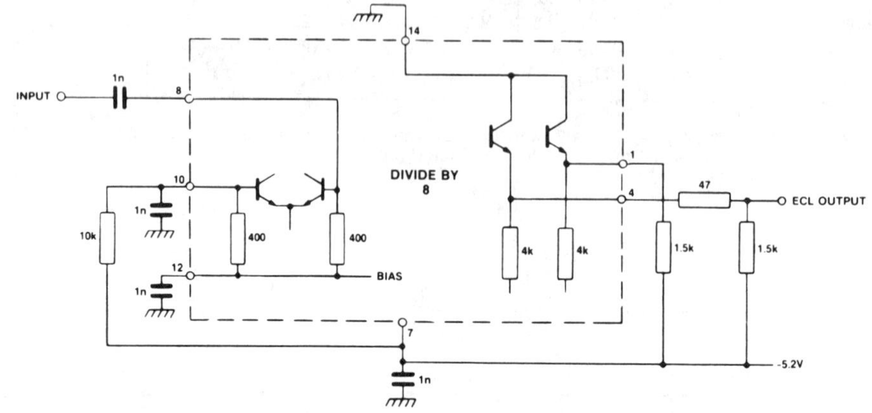

Typical application showing interfacing

Plessey SP8678B 1500 MHz ÷ 8

- ECL-compatible output
- Ac-coupled input
- Clock-inhibit input
- Supply voltage: −6.8V
- Power consumption: 475mW
- Temperature range: 0 °C to +70 °C

ABSOLUTE MAXIMUM RATINGS

Supply voltage	-8V
Output current	20mA
Storage temperature range	-55 °C to +150 °C
Max. junction temperature	+175 °C
Max. clock I/P voltage	2.5V p-p

ELECTRICAL CHARACTERISTICS

Supply voltage: V_{CC} = 0V V_{EE} = -6.8V ± 0.3V
T_{amb} (B grade) = 0°C to +70°C

Characteristic	Symbol	Value Min.	Value Max.	Units	Conditions	Notes
Maximum frequency (sinewave input)	f_{max}	1.5		GHz	Input = 600 - 1200mV p-p	Note 5
Minimum frequency (sinewave input)	f_{min}		150	MHz	Input = 600 - 1200mV p-p	Note 6
Current consumption	I_{EE}		95	mA	V_{EE} = -6.8V	Note 6
Output low voltage	V_{OL}	-1.87	-1.5	V	V_{EE} = -6.8V(25° C)	
Output high voltage	V_{OH}	-0.87	-0.7	V	V_{EE} = -6.8V(25° C)	
Minimum output swing	V_{OUT}	500		mV		Note 5
Clock inhibit high threshold voltage	V_{INBH}	-0.96		V	V_{EE} = -6.8V(25° C)	
Clock inhibit low threshold voltage	V_{INBL}		-1.62	V	V_{EE} = -6.8V(25° C)	

NOTES
1. Unless otherwise stated, the electrical characteristics are guaranteed over specified supply, frequency and temperature range
2. The test configuration for dynamic testing is shown in Fig. 6.
3. The temperature coefficient of V_{OH} = +1.3mV/°C and V_{OL} = +0.5mV/°C but these are not tested.
4. The temperature coefficient of V_{INB} = +0.8mV/°C but this is not tested.
5. Tested at 25°C and 70°C only.
6. Tested at 25°C only.

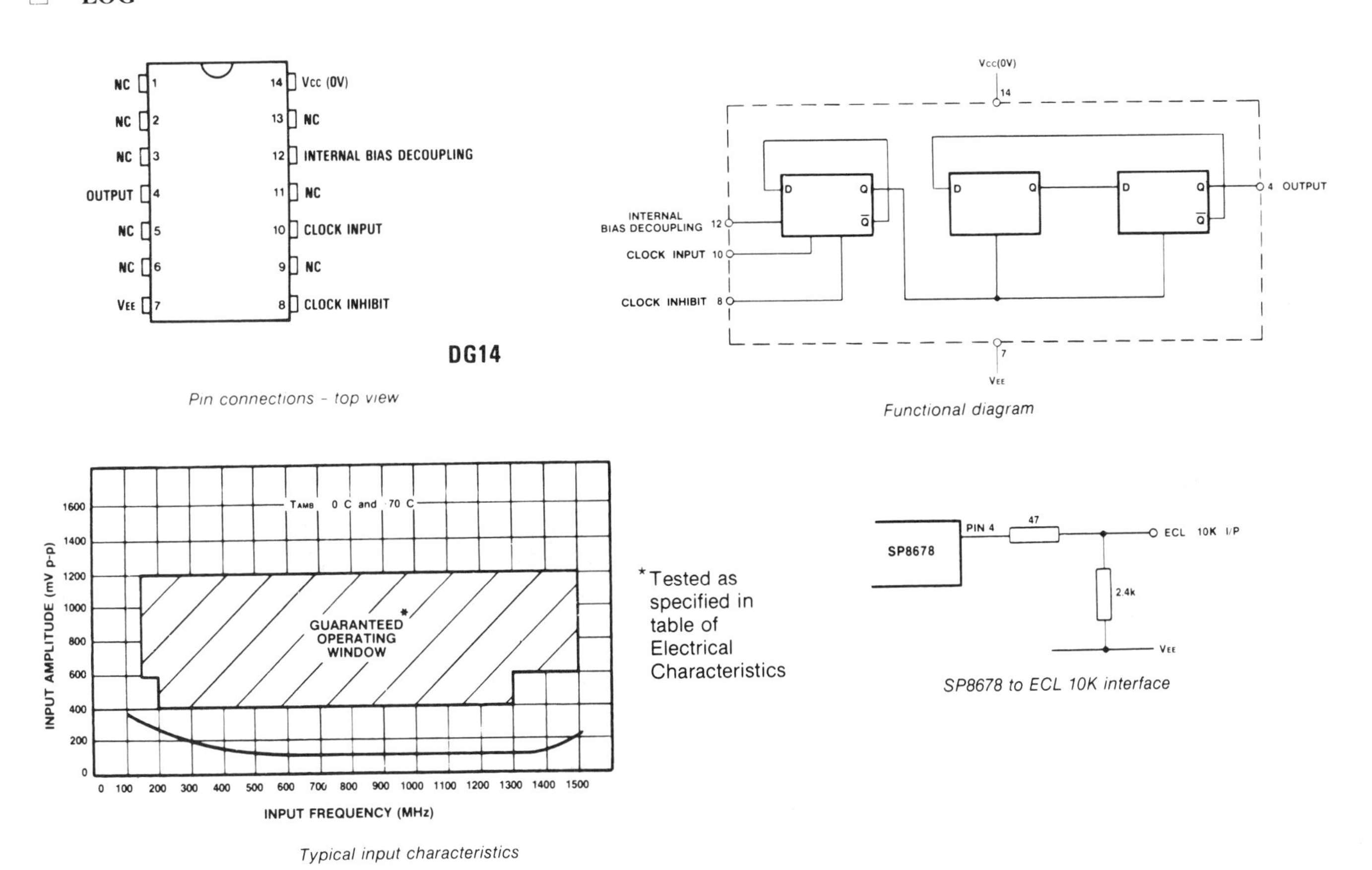

Pin connections - top view

Functional diagram

Typical input characteristics

SP8678 to ECL 10K interface

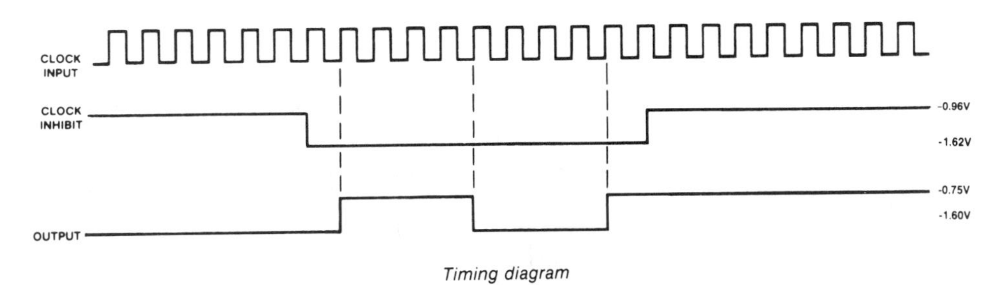

Timing diagram

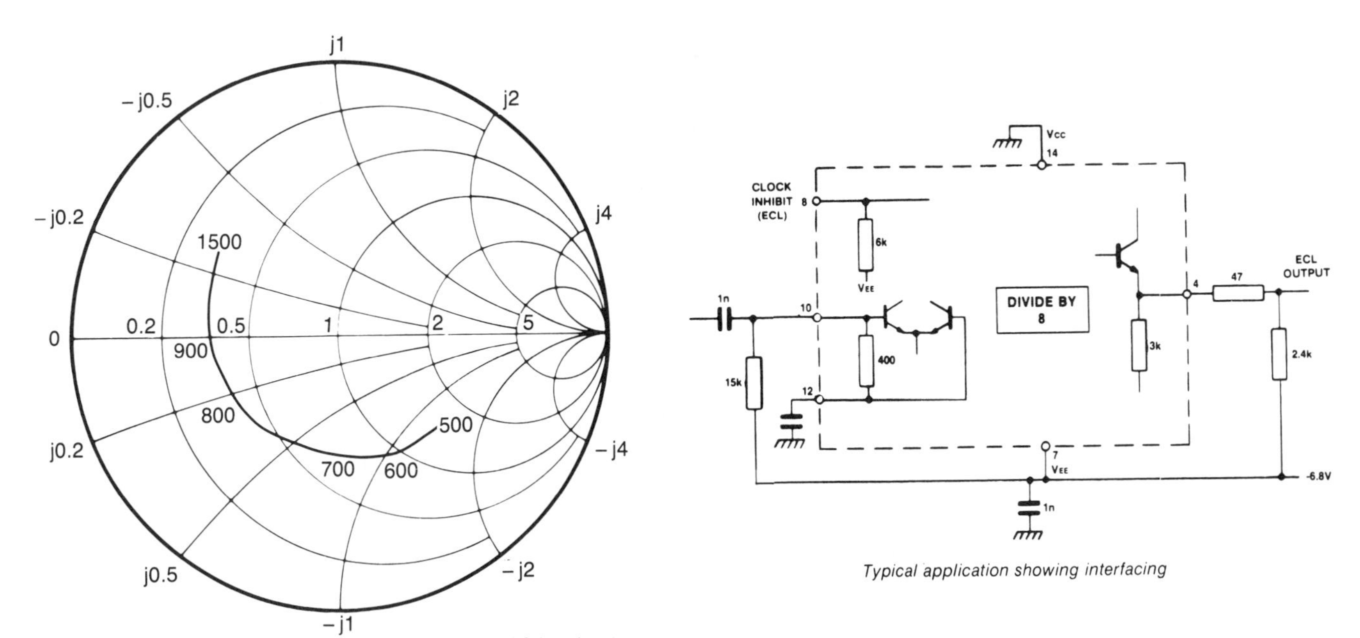

Typical input impedance. Test conditions: supply voltage -6.8V, ambient temperature 25°C, frequencies in MHz, impedances normalised to 50 ohms.

Typical application showing interfacing

Plessey SP8680A

600 MHz ÷ 10/11

- Very high speed: 650 MHz Typ.
- ECL- and TTL-compatible outputs
- Dc or ac clocking
- Clock enable
- Divide by 10 or 11
- Asynchronous master set
- Equivalent to Fairchild 11C90
- Supply voltage: 5V +0.5V −0.25V or −5V −0.5V +0.25V
- Power consumption: 420mW
- Temperature: −55 °C to +125 °C

ABSOLUTE MAXIMUM RATINGS

Supply voltage	8V
ECL output source current	50mA
Storage temperature range	-55° C to +150° C
Max. junction temperature	+175° C
TTL output sink current	30mA
Max. clock I/P voltage	2.5V p-p

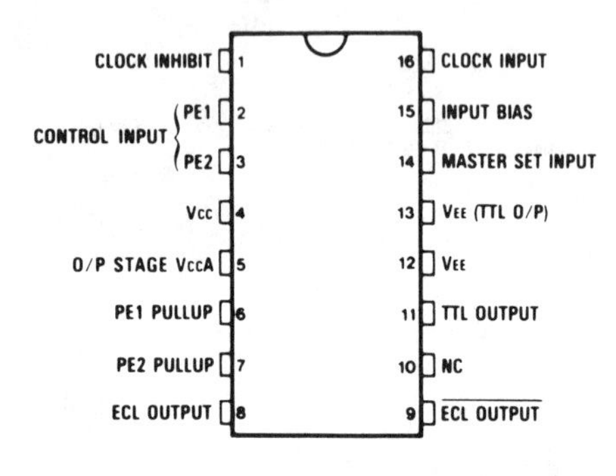

DG16

Pin connections - top view

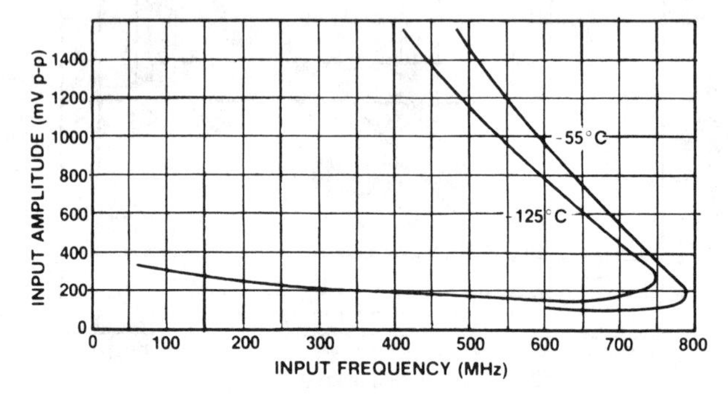

Typical input sensitivity SP8680

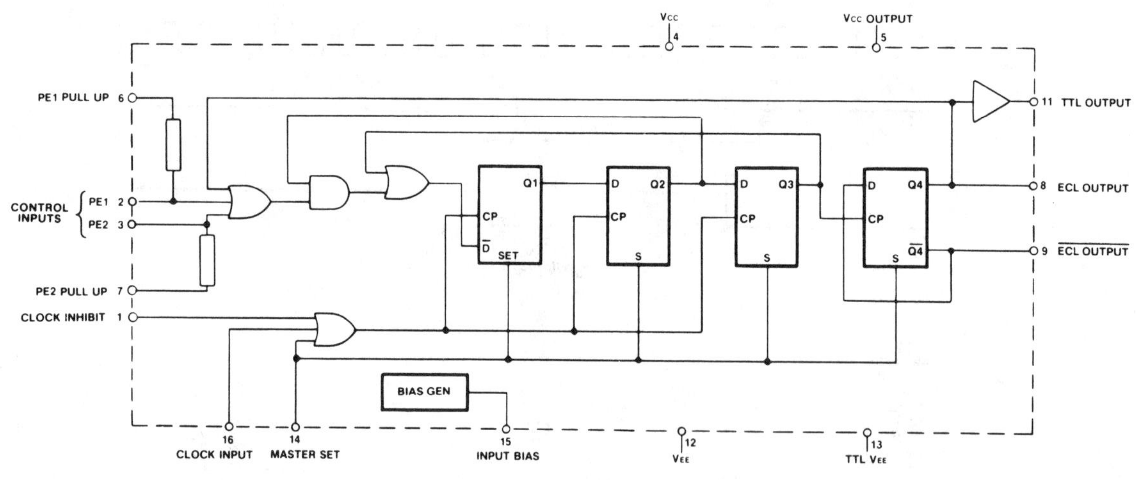

Functional diagram

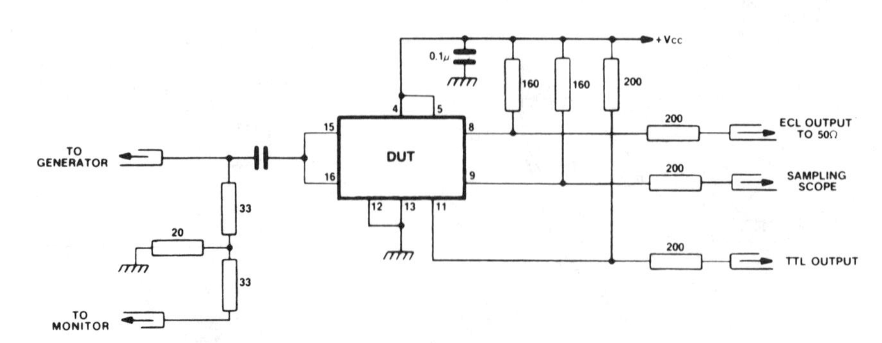

Test circuit

ELECTRICAL CHARACTERISTICS
TTL OPERATION

Supply voltage: V_{CC} = V_{CCA} = 4.75 to 5.5V V_{EE} = 0V
Temperature: T_{amb} = -55°C to +125°C

Characteristics	Symbol	Value		Units	Conditions	Notes
		Min.	Max.			
Maximum frequency sinewave input	f_{max}		550	MHz	Clock input AC coupled = 350mV p-p	Note 4
Minimum frequency sinewave input	f_{min}	10		MHz	Clock input AC coupled = 600mV p-p	Note 5
Power supply current	I_{EE}		105	mA	V_{CC} = V_{CC} max. Pins 6,7,13 open circuit	Note 4
Power supply current including TTL stage	I_{EE}		111	mA	V_{CC} = V_{CC} max. Pins 6,7 open circuit	Note 4
TTL output high voltage	V_{OH}	2.3		V	V_{CC} = V_{CC} min. I_{OH} = -640μA	Note 4
TTL output low voltage	V_{OL}		0.5	V	V_{CC} = V_{CC} max. I_{OL} = -20mA	Note 4
Input high voltage PE1 and PE2 inputs	V_{INH}	3.9		V	V_{CC} = 5.0V (25°C)	
Input low voltage PE1 and PE2 inputs	V_{INL}		3.5	V	V_{CC} = 5.0V (25°C)	
Input low current PE1 and PE2 inputs	I_{IL}	-4		mA	V_{CC} = V_{CC} max. (25°C) Pins 6,7 = V_{CC} V_{IN} = 0.4V	
Propagation delay CP to Q TTL	t_{pHL} t_{pLH}	6	14	ns	V_{CC} = 5.0V (25°C)	Note 5
Propagation delay MS to Q TTL	t_p		17	ns	V_{CC} = 5.0V (25°C)	Note 5
Mode control set-up time	t_s	4		ns	V_{CC} = 5.0V (25°C)	Note 5
Mode control release time	t_r	4		ns	V_{CC} = 5.0V (25°C)	Note 5
TTL output rise time (20% - 80%)	t_{TLH}		5	ns	V_{CC} = 5.0V (25°C)	Note 5
TTL output fall time (80% - 20%)	t_{THL}		5	ns	V_{CC} = 5.0V (25°C)	Note 5

ELECTRICAL CHARACTERISTICS
ECL OPERATION

Supply Voltage: V_{EE} = -4.75V to -5.5V V_{CC} = 0V
Temperature: T_{amb} = -55°C to +125°C

Characteristics	Symbol	Value		Units	Conditions	Notes
		Min.	Max.			
Maximum frequency sinewave input	f_{max}		550	MHz	Clock input AC coupled = 350mV p-p	Note 4
Minimum frequency sinewave input	f_{min}	10		MHz	Clock input AC coupled = 600mV p-p	Note 5
Power supply current	I_{EE}		105	mA	V_{CC} = V_{CC} max. Pins 6,7,13 open circuit	Note 4
ECL output high voltage	V_{OH}	-0.93	-0.78	V	V_{EE} = -5.2V (25°C) Load = 100Ω to -2V	
ECL output low voltage	V_{OL}	-1.85	-1.62	V	V_{EE} = -5.2V (25°C) Load = 100Ω to -2V	
Input high voltage	V_{INH}	-1.095	-0.81	V	V_{EE} = -5.2V (25°C)	
Input low voltage	V_{INL}	-1.85	-1.475	V	V_{EE} = -5.2V (25°C)	
Input low currents	I_{IL}	0.5		μA	25°C	
Input high current						
Clock and MS	I_H		400	μA	V_{IN} = -1.85V (25°C)	
PE1 and PE2	I_H		250	μA	V_{IN} = -0.8V (25°C)	
Propagation delay CP to Q4	t_{pHL}		4	ns	Load = 100Ω to -2V (25°C)	Note 5
	t_{pLH}		3	ns		
Propagation delay MS to Q4	t_{pLH}		6	ns	25°C	Note 5
Mode control set-up time	t_s	4		ns	25°C	Note 5
Mode control release time	t_r	4		ns	25°C	Note 5
ECL output rise time (20% - 80%)	t_{TLH}		2	ns	25°C	Note 5
ECL output fall time (80% - 20%)	t_{THL}		2	ns	25°C	Note 5

NOTES
1. Unless otherwise stated, the electrical characteristics are guaranteed over specified supply, frequency and temperature range.
2. The temperature coefficient of V_{OH} = +1.2mV/°C, V_{OL} = +0.24mV/°C and of V_{IN} = +0.8mV/°C but these are not tested.
3. The test configuration for dynamic testing is shown in Fig.6.
4. Tested at 25°C and +125°C only.
5. Guaranteed but not tested.

CLOCK INPUT

tr ts

$\overline{PE}$ INPUT

Q4 AND TTL 6 5 5

MS	CLOCK INH.	PE1	PE2	OUTPUT RESPONSE
H	X	X	X	ALL OUTPUTS SET HIGH
L	H	X	X	HOLD
L	L	L	L	÷11
L	L	H	L	÷10
L	L	L	H	÷10
L	L	H	H	÷10

X = DON'T CARE

Truth table and timing diagram SP8680

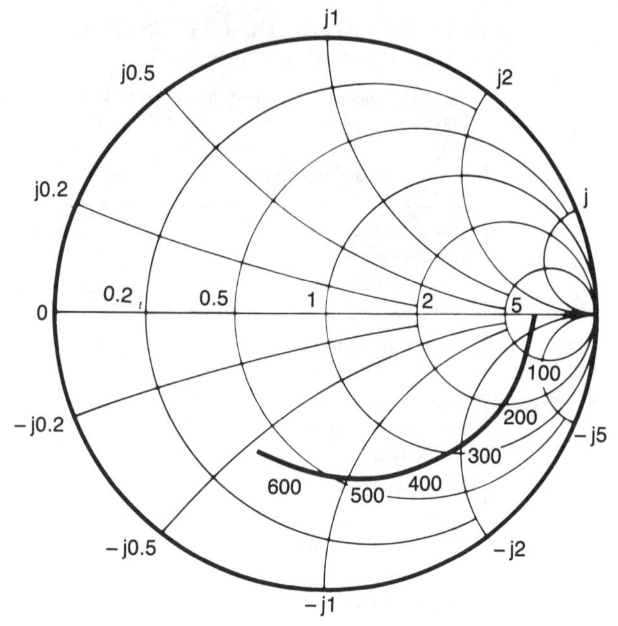

Typical input impedance. Test conditions: supply voltage 5V, ambient temperature 25 °C, frequencies in MHz, impedances normalised to 50 ohms.

NOTE:

The set-up time t_s is defined as minimum time that can elapse between L→H transition of control input and the next L→H clock pulse transition to ensure that the ÷10 mode is obtained.

The release time t_r is defined as the minimum time that can elapse between a H→L transition of control input and the next L→H clock pulse transition to ensure that the ÷11 mode is obtained.

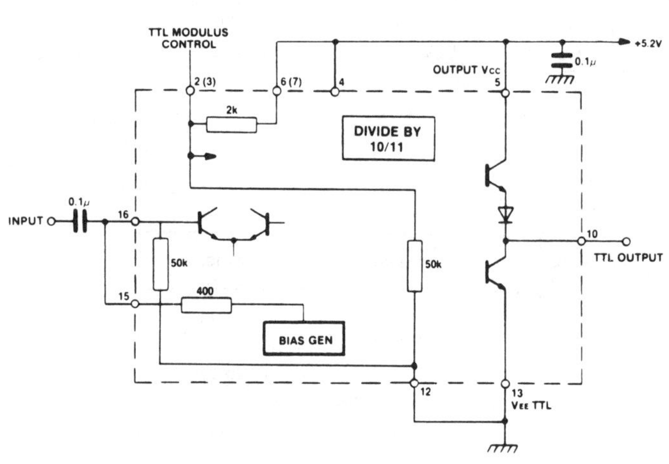

Typical application showing interfacing

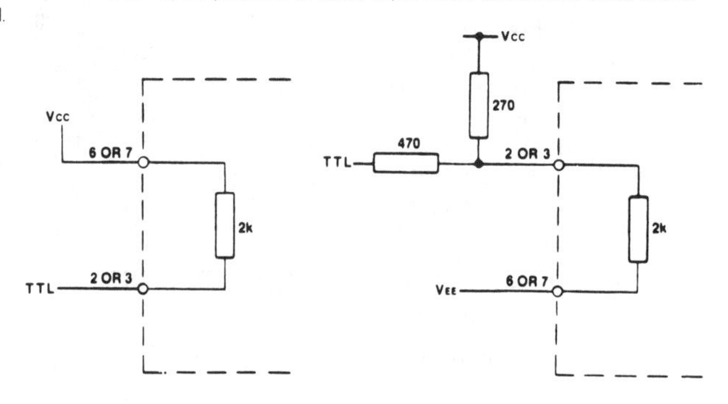

TTL interface to PE1 and PE2

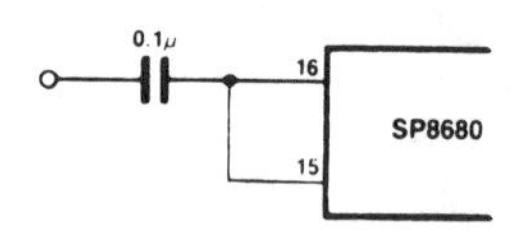

AC coupled input

Plessey SP8680B
600 MHz ÷ 10/11

- Very high speed: 650 MHz typ.
- ECL- and TTL-compatible outputs
- Dc or ac clocking
- Clock enable
- Divide by 10 or 11
- Asynchronous master set
- Equivalent to Fairchild 11C90
- Supply voltage: 5V +0.5V −0.25V or −5V −0.5V +0.25V
- Power consumption: 420 mW
- Temperature: −40 °C to +85 °C

ABSOLUTE MAXIMUM RATINGS

Supply voltage	8V
ECL output source current	50mA
Storage temperature range	-55 °C to +125 °C
Max. junction temperature	+175 °C
TTL output sink current	30mA
Max. clock I/P voltage	2.5V p-p

ELECTRICAL CHARACTERISTICS TTL OPERATION

Test conditions (unless otherwise stated):
T_{amb} = -40°C to +85°C Supply voltage: V_{CC} = V_{CCA} = 4.75 to 5.5V V_{EE} = 0V

Characteristics	Symbol	Value		Units	Conditions	Notes
		Min.	Max.			
Maximum frequency sinewave input	f_{max}		575	MHz	Clock input AC coupled = 350mV p-p	Note 4
Minimum frequency sinewave input	f_{min}	10		MHz	Clock input AC coupled = 600mV p-p	
Power supply current	I_{EE}		105	mA	V_{CC} = V_{CC} max. Pins 6,7,13 open circuit	
Power supply current including TTL stage	I_{EE}		111	mA	V_{CC} = V_{CC} max. Pins 6,7 open circuit	
TTL output high voltage	V_{OH}	2.3		V	V_{CC} = V_{CC} min. I_{OH} = -640µA	
TTL output low voltage	V_{OL}		0.5	V	V_{CC} = V_{CC} max. I_{OL} = -20mA	
Input high voltage PE1 and PE2 inputs	V_{INH}	3.9		V	V_{CC} = 5.0V (25°C)	
Input low voltage PE1 and PE2 inputs	V_{INL}		3.5	V	V_{CC} = 5.0V (25°C)	
Input low current PE1 and PE2 inputs	I_{IL}	-4		mA	V_{CC} = V_{CC} max. (25°C) Pins 6,7 = V_{CC} V_{IN} = 0.4V	
Propagation delay CP to Q TTL	t_{pHL} t_{pLH}	6	14	ns	V_{CC} = 5.0V (25°C)	Note 4
Propagation delay MS to Q TTL	t_p		17	ns	V_{CC} = 5.0V (25°C)	Note 4
Mode control set-up time	t_s	4		ns	V_{CC} = 5.0V (25°C)	Note 4
Mode control release time	t_r	4		ns	V_{CC} = 5.0V (25°C)	Note 4
TTL output rise time (20 % - 80 %)	t_{TLH}		5	ns	V_{CC} = 5.0V (25°C)	Note 4
TTL output fall time (80 % - 20 %)	t_{THL}		5	ns	V_{CC} = 5.0V (25°C)	Note 4

ELECTRICAL CHARACTERISTICS ECL OPERATION

Test conditions (unless otherwise stated):
T_{amb} = -40°C to +85°C Supply Voltage: V_{EE} = -4.75V to -5.5V V_{CC} = 0V

Characteristics	Symbol	Value		Units	Conditions	Notes
		Min.	Max.			
Maximum frequency sinewave input	f_{max}		575	MHz	Clock input AC coupled = 350mV p-p	Note 4
Minimum frequency sinewave input	f_{min}	10		MHz	Clock input AC coupled = 600mV p-p	
Power supply current	I_{EE}		105	mA	V_{CC} = V_{CC} max. Pins 6,7,13 open circuit	
ECL output high voltage	V_{OH}	-0.93	-0.78	V	V_{EE} = -5.2V (25°C) Load = 100Ω to -2V	
ECL output low voltage	V_{OL}	-1.85	-1.62	V	V_{EE} = -5.2V (25°C) Load = 100Ω to -2V	
Input high voltage	V_{INH}	-1.095	-0.81	V	V_{EE} = -5.2V (25°C)	
Input low voltage	V_{INL}	-1.85	-1.475	V	V_{EE} = -5.2V (25°C)	

Characteristic	Symbol	Value		Units	Conditions	Notes
		Min.	Max.			
Input low currents	I_{IL}	0.5		µA	25°C	
Input high current						
Clock and MS	I_H		400	µA	V_{IN} = -1.85V(25°C)	
PE1 and PE2	I_H		250	µA	V_{IN} = -0.8V(25°C)	
Propagation delay CP to Q4	t_{pLH}		3	ns	Load = 100Ω to -2V(25°C)	Note 4
Propagation delay MS to Q4	t_{pLH}		6	ns	25°C	Note 4
Mode control set-up time	t_s	4		ns	25°C	Note 4
Mode control release time	t_r	4		ns	25°C	Note 4
ECL output rise time (20 % - 80 %)	t_{TLH}		2	ns	25°C	Note 4
ECL output fall time (80 % - 20 %)	t_{THL}		2	ns	25°C	Note 4

NOTES
1. Unless otherwise stated, the electrical characteristics are guaranteed over specified supply, frequency and temperature range.
2. The temperature coefficient of V_{OH} = +1.2mV/°C, V_{OL} = +0.25mV/°C and of V_{IN} = +0.8mV/°C but these are not tested.
3. The test configuration for dynamic testing is shown in Fig.5
4. Guaranteed but not tested.

Functional diagram

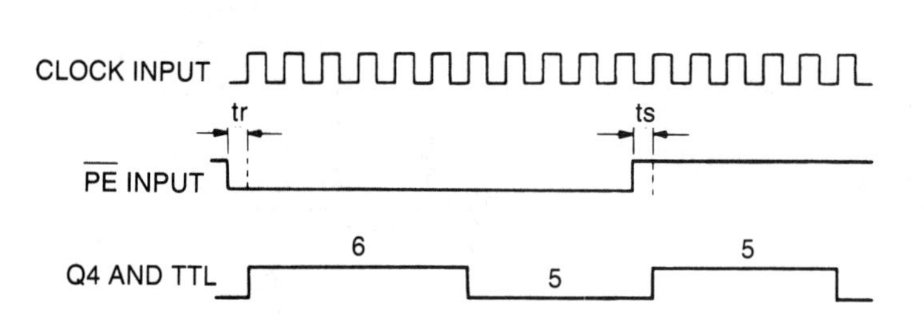

MS	CLOCK INH.	PE1	PE2	OUTPUT RESPONSE
H	X	X	X	ALL OUTPUTS SET HIGH
L	H	X	X	HOLD
L	L	L	L	÷11
L	L	H	L	÷10
L	L	L	H	÷10
L	L	H	H	÷10

X = DON'T CARE

NOTE:

The set-up time t_s is defined as minimum time that can elapse between L→H transition of control input and the next L→H clock pulse transition to ensure that the ÷10 mode is obtained.

The release time t_r is defined as the minimum time that can elapse between a H→L transition of control input and the next L→H clock pulse transition to ensure that the ÷11 mode is obtained.

Truth table and timing diagram SP8680

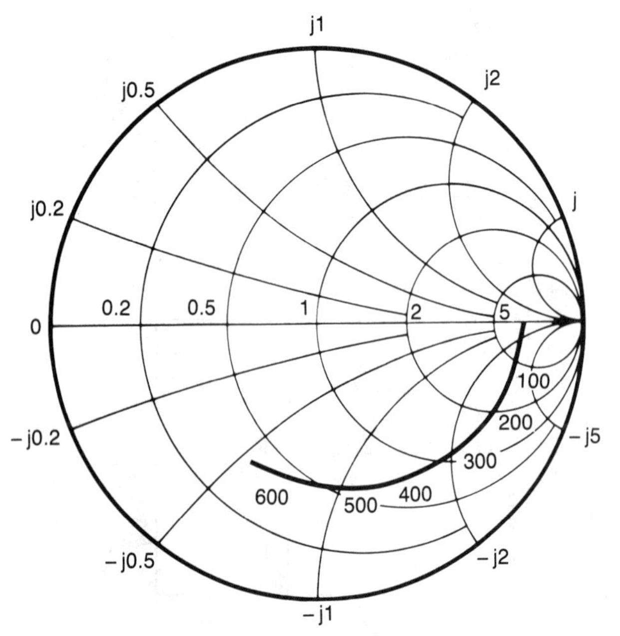

Typical input impedance. Test conditions: supply voltage 5V, ambient temperature 25° C, frequencies in MHz, impedances normalised to 50 ohms.

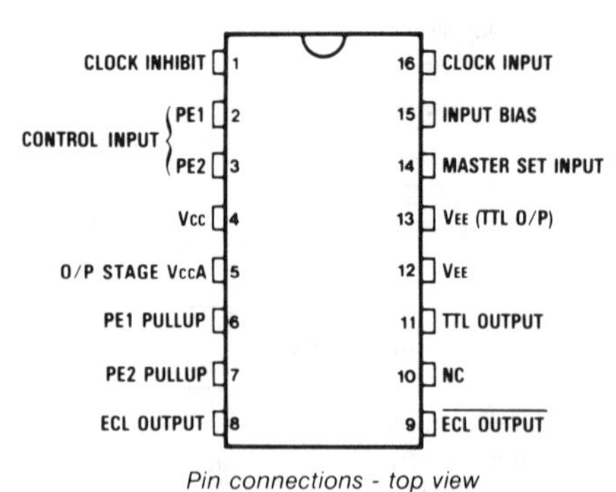

Pin connections - top view

DP16

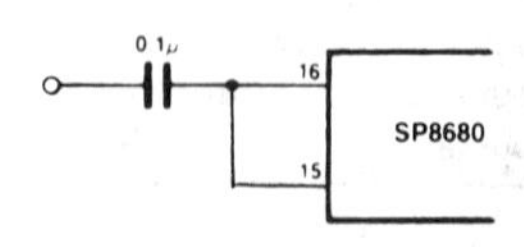

AC coupled input

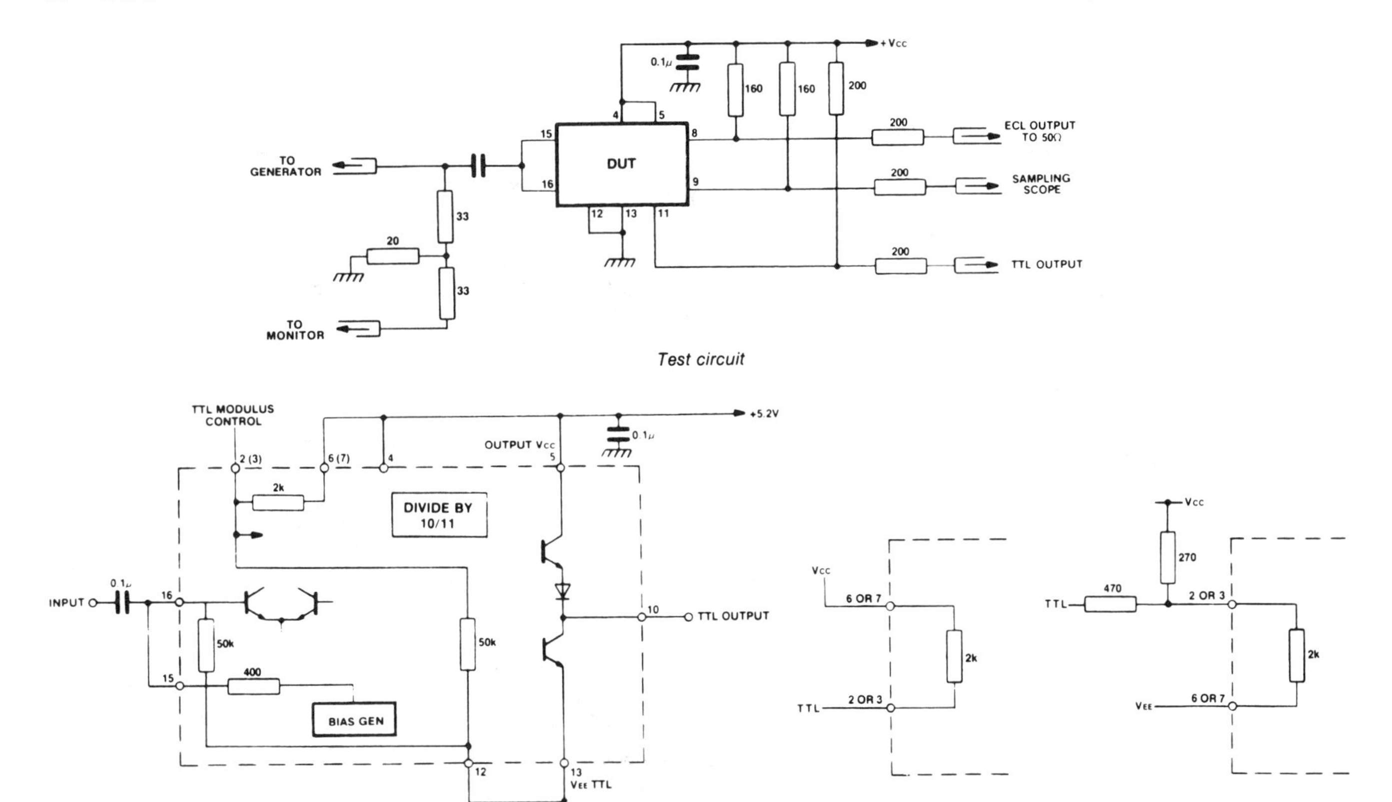

Test circuit

Typical application showing interfacing

TTL interface to PE1 and PE2

Plessey SP8685A and B 500 MHz ÷ 10/11

- Divides by 10 and 11
- Ac-coupled input (internal bias)
- ECL-compatible output
- Supply voltage: −5.2V
- Power consumption: 300mW
- Temperature range:
  - −55°C to +125°C (A grade)
  - −30°C to +70°C (B grade)

ABSOLUTE MAXIMUM RATINGS

Supply voltage	-8V
Output current	20mA
Storage temperature range	-55°C to +150°C
Max. junction temperature	+175°C
Max. clock I/P voltage	2.5V p-p

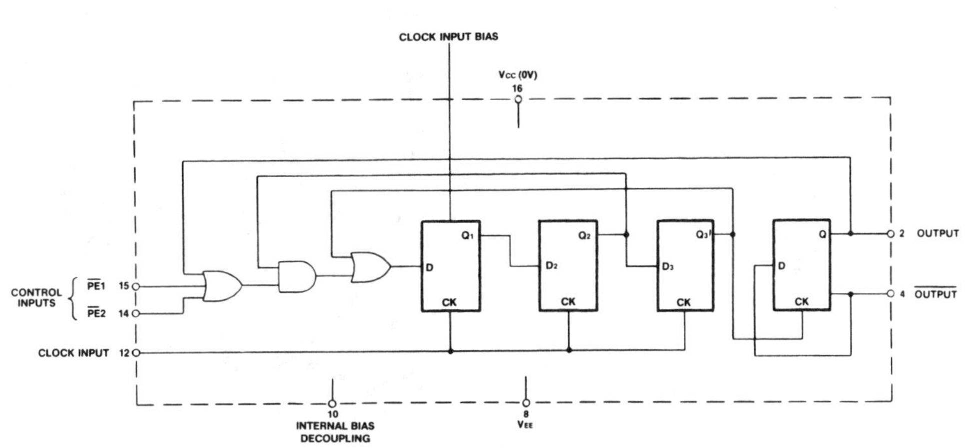

Functional diagram

ELECTRICAL CHARACTERISTICS

Supply Voltage: V_{CC} = 0V V_{EE} = -5.2V ± 0.25V
Temperature: A Grade T_{amb} = -55°C to +125°C
B Grade T_{amb} = -30°C to +70°C

Characteristic	Symbol	Value		Units	Conditions	Notes
		Min.	Max.			
Maximum frequency (sinewave input)	f_{max}	500		MHz	Input = 400-800mV p-p	
Minimum frequency (sinewave input)	f_{min}		50	MHz	Input = 400-800mV p-p	Note 6
Power supply current	I_{EE}		70	mA	V_{EE} = -5.2V	Note 6
Output high voltage	V_{OH}	-0.87	-0.7	V	V_{EE} = -5.2V (25°C)	
Output low voltage	V_{OL}	-1.8	-1.5	V	V_{EE} = -5.2V (25°C)	
$\overline{PE}$ input high voltage	V_{INH}	-0.93		V	V_{EE} = -5.2V (25°C)	
$\overline{PE}$ input low voltage	V_{INL}		-1.62	V	V_{EE} = -5.2V (25°C)	
Clock to output delay	t_p		6	ns		Note 7
Set-up time	t_s	2		ns		Note 7
Release time	t_r	2		ns		Note 7

NOTES
1. Unless otherwise stated, the electrical characteristics shown above are guaranteed over specified supply, frequency and temperature range
2. The temperature coefficient of V_{OH} = +1.63mV/°C, V_{OL} = +0.94mV/°C and of V_{IN} = +1.22mV/°C but these are not tested
3. The test configuration for dynamic testing is shown
4. The set up time t_s is defined as minimum time that can elapse between L → H transition of control input and the next L → H clock pulse transition to ensure that ÷10 is obtained.
5. The release time t_r is defined as the minimum time that can elapse between H → L transition of the control input and the next L → H clock pulse transition to ensure that the ÷11 mode is obtained.
6. Tested at 25°C only.
7. Guaranteed but not tested.

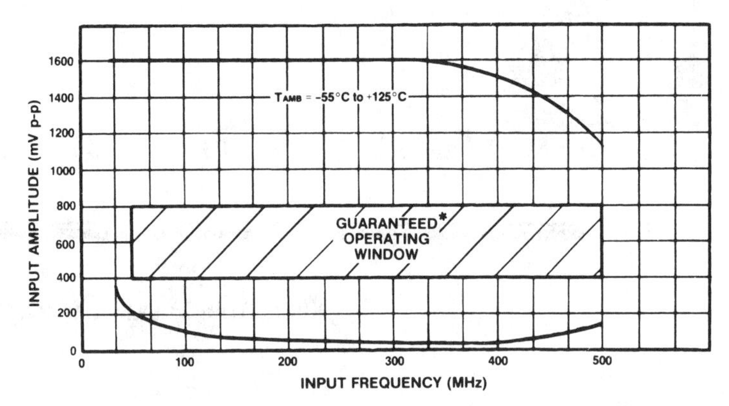

*Tested as specified in table of Electrical Characteristics

Typical input characteristic SP8685A

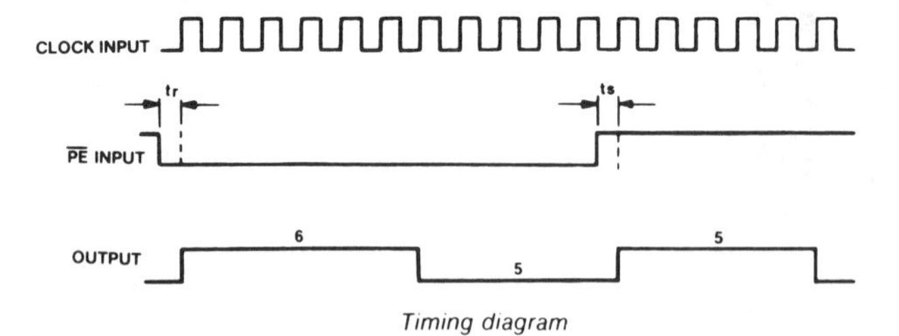

Timing diagram

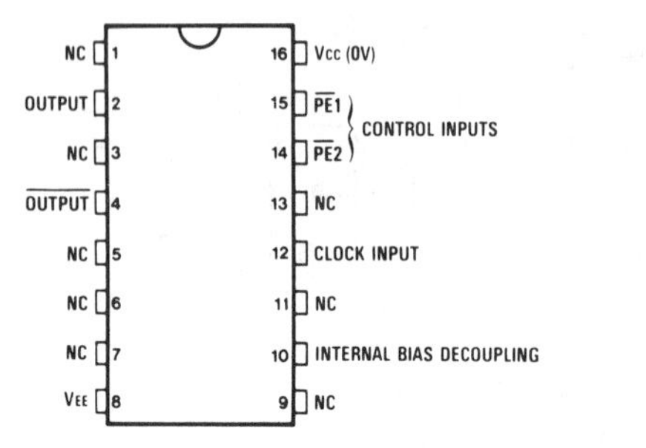

DG16

Pin connections - top view

Typical input impedance. Test conditions: supply voltage -5.2V, ambient temperature 25° C, frequencies in MHz, impedances normalised to 50 ohms.

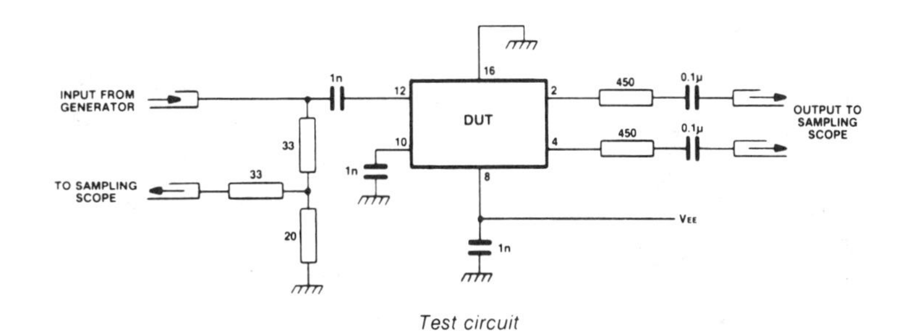

Test circuit

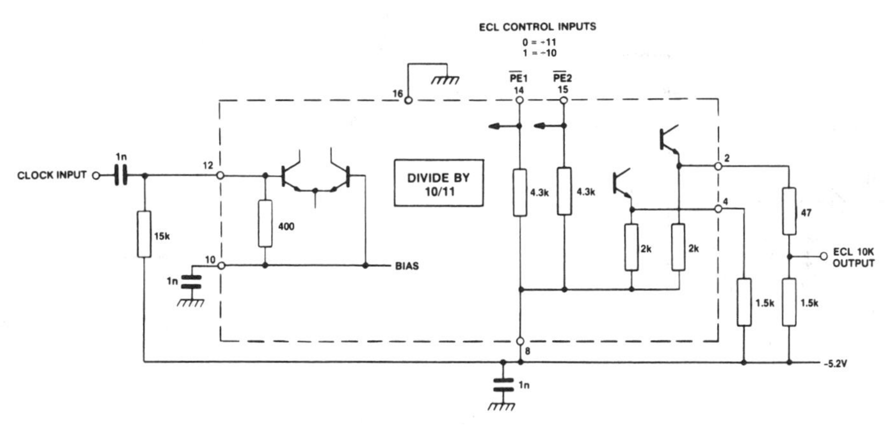

Typical application showing interfacing

Plessey SP8650A and B
600 MHz ÷ 16

- ECL-compatible outputs
- Ac-coupled inputs (internal bias)
- Supply voltage: −5.2V
- Power consumption: 300 mW
- Temperature range:
  - −55 °C to +125 °C (A grade)
  - −30 °C to +70 °C (B grade)

ABSOLUTE MAXIMUM RATINGS

Supply voltage	-8V
Output current	10mA
Storage temperature range	-55 °C to +150 °C
Max. junction temperature	+175 °C
Max. clock I/P voltage	2.5V p-p

ELECTRICAL CHARACTERISTICS

Supply voltage: V_{CC} = 0V, V_{EE} = -5.2V ± 0.25V
Temperature: A Grade T_{amb} = -55°C to +125°C
B Grade T_{amb} = -30°C to +70°C

Characteristics	Symbol	Value Min.	Value Max.	Units	Conditions	Notes
Maximum frequency (sinewave input)	f_{max}	600		MHz	Input = 400-800mV p-p	
Minimum frequency (sinewave input)	f_{min}		40	MHz	Input = 400-800mV p-p	Note 4
Power supply current	I_{EE}		60	mA		Note 4
Output low voltage	V_{OL}	-1.8	-1.5	V	V_{EE} = -5.2V (25°C)	
Output high voltage	V_{OH}	-0.85	-0.7	V	V_{EE} = -5.2V (25°C)	

NOTES
1. Unless otherwise stated the electrical characteristics shown above are guaranteed over specified supply, frequency and temperature range.
2. The temperature coefficients of V_{OH} = +1.63mV/°C and V_{OL} = +0.94mV/°C but these are not tested.
3. The test configuration for dynamic testing is shown in Fig.5.
4. Tested at 25° only.

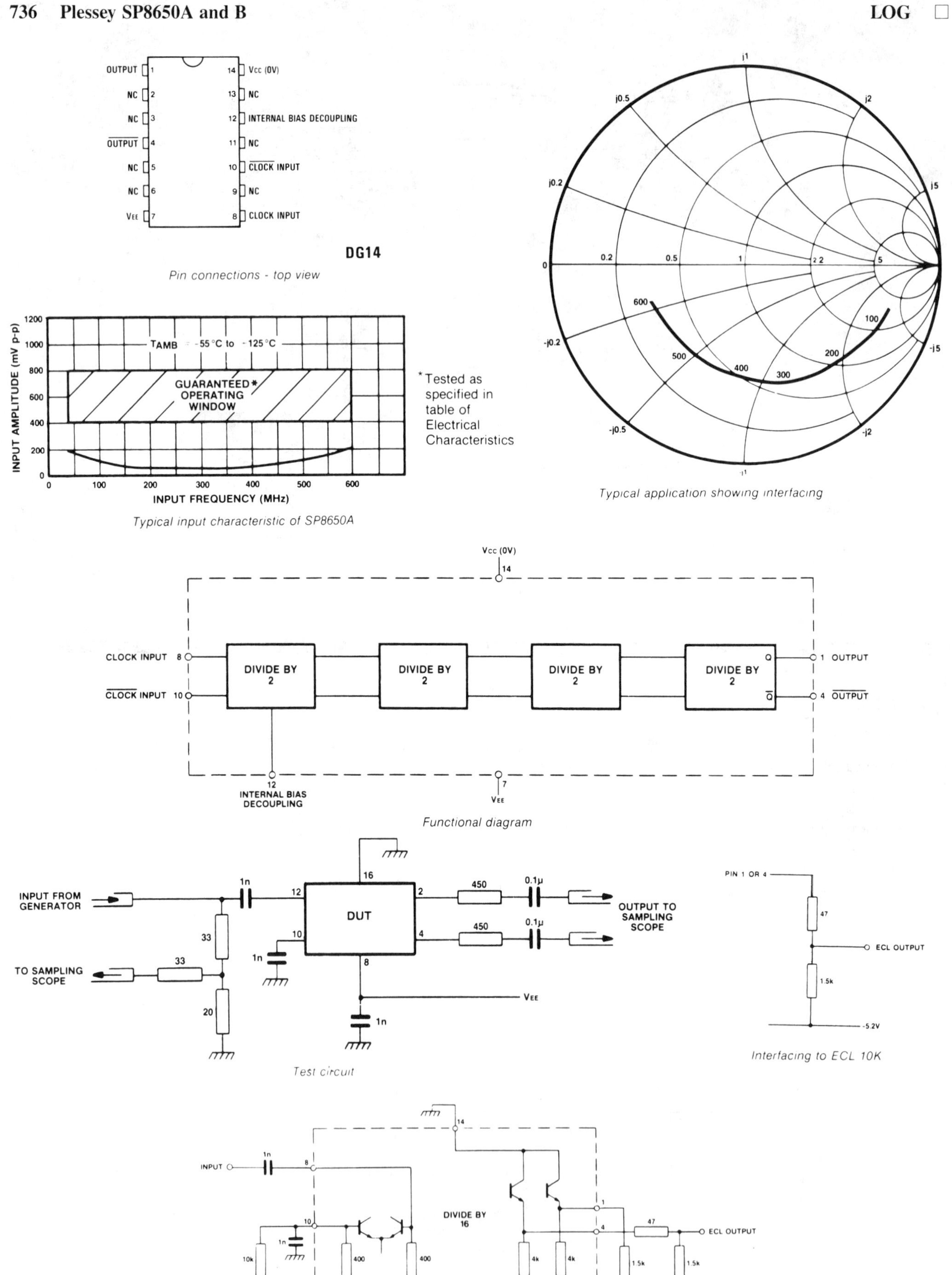

Pin connections - top view

Typical input characteristic of SP8650A

Typical application showing interfacing

Functional diagram

Test circuit

Interfacing to ECL 10K

Plessey
SP8655 A and B
200 MHz ÷ 32

SP8657A and B
200 MHz ÷ 20

SP8659A and B
200 MHz ÷ 16

- AC-coupled inputs
- Low-power consumption
- Open collector output CMOS and TTL compatible
- Supply voltage: 5.0V
- Power consumption: 50mW
- Temperature range:
  - −55 °C to +125 °C (A grade)
  - −30 °C to +70 °C (B grade)

ABSOLUTE MAXIMUM RATINGS

Supply voltage	8V
Open collector output voltage	12V
Storage temperature range	-55 °C to +150 °C
Max. junction temperature	+175 °C
Max. clock I/P voltage	2.5V p-p
Output sink current	10mA

ELECTRICAL CHARACTERISTICS

Supply voltage: V_{CC} = 5.0V ± 0.25V V_{EE} = 0V
Temperature: A grade T_{amb} = -55°C to +125°C
B grade T_{amb} = -30°C to +70°C

Characteristic	Symbol	Value		Units	Conditions
		Min.	Max.		
Maximum frequency (sinewave input)	f_{max}	200		MHz	Input = 400 - 800mV
Minimum frequency (sinewave input)	f_{min}		40	MHz	Input = 400 - 800mV
Power supply current	I_{EE}		13	mA	V_{CC} = 5.25V
Output high voltage	V_{OH}	7.5		V	V_{CC} = 5V Note 4 Pin 4 = 1.5kΩ to 10V
Output low voltage	V_{OL}		400	mV	V_{CC} = 5V Pin 4 = 1.5kΩ to 10V

NOTES
1. Unless otherwise stated the electrical characteristics are guaranteed over specified supply, frequency and temperature range
2. The dynamic test circuit is shown in Fig.5.
3. Above characteristics are not tested at 25°C (tested at low and high temperature only).
4. Open collector output not to be used above 15MHz. C_{load} ≤ 5pF.

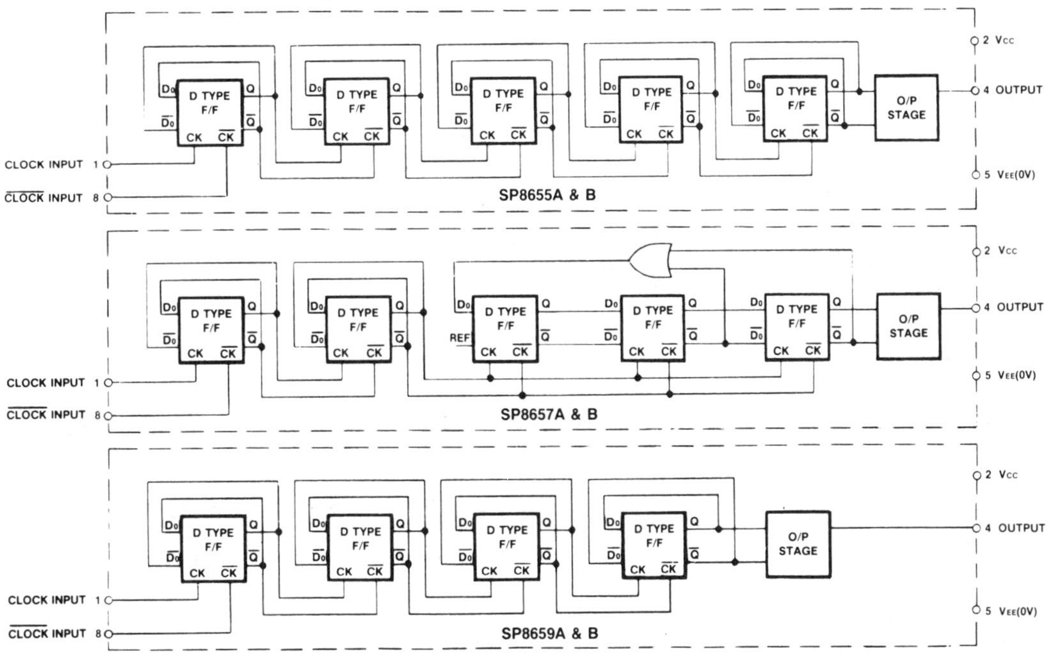

Functional diagram

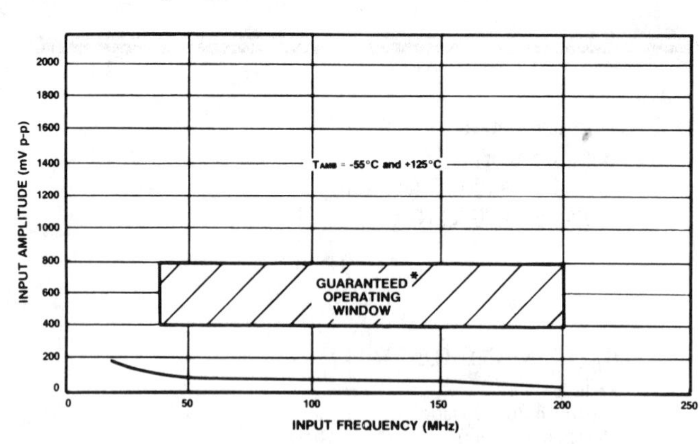

Typical input characteristics

*Tested as specified in table of Electrical Characteristics

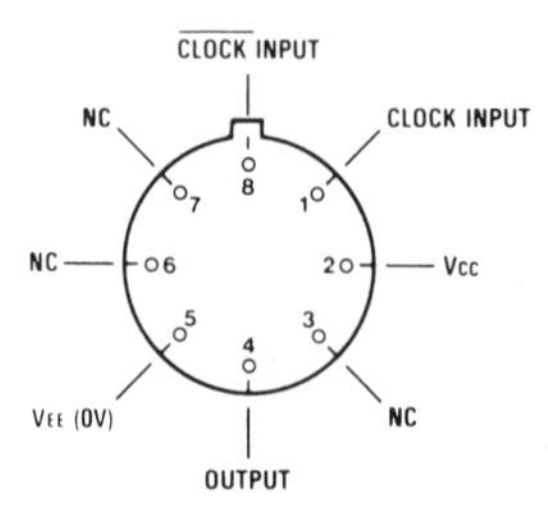

CM8

Pin connections - bottom view

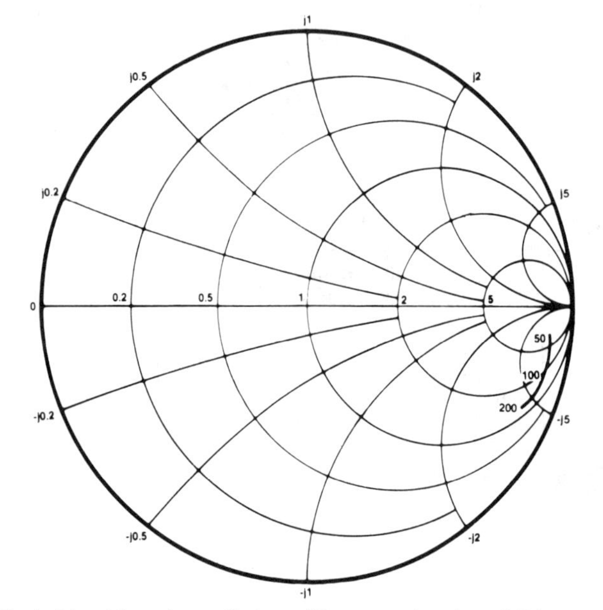

Typical input impedance. Test conditions: supply voltage 5.0V, ambient temperature 25°C, frequencies in MHz, impedances normalised to 50 ohms.

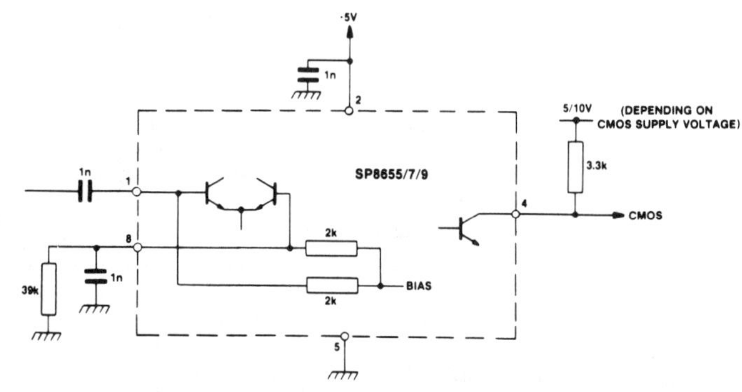

Typical application showing interfacing

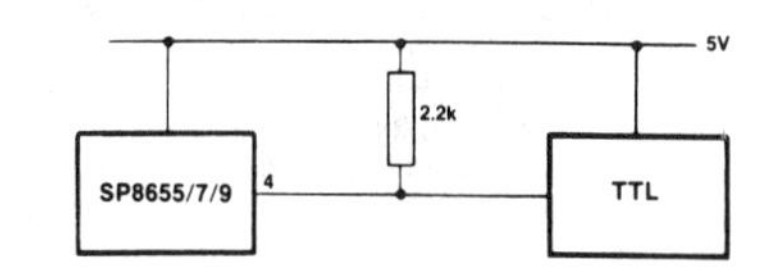

Interfacing to TTL. Load not to exceed 3 TTL unit loads.

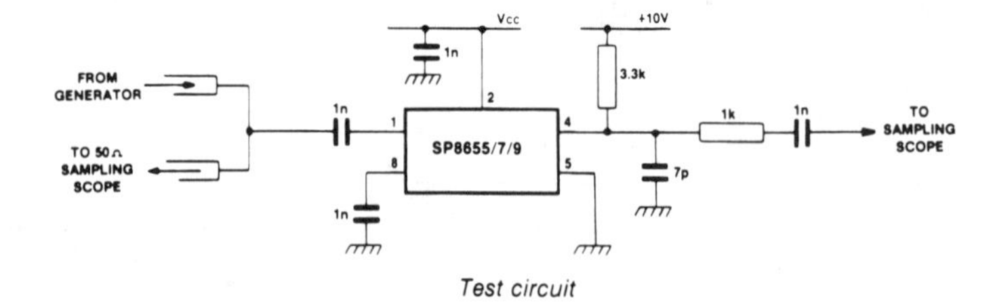

Test circuit

Plessey SP8799

225 MHz ÷ 10/11 Two Modulus Divider

- Very low power
- Control input and output CMOS/TTL compatible
- Ac-coupled input
- Operation up to 9.5V using internal regulator
- Supply voltage: +5.2V or 6.8 to 9.5V
- Power consumption: 26 mW typical
- Temperature range: −40°C to +85°C

ABSOLUTE MAXIMUM RATINGS

Supply voltage	6.0V Pins 7 & 8 tied
	13.5V Pin 8, Pin 7 decoupled
Storage temperature range	-55°C to +125°C
Max. junction temperature	+175°C
Max. clock input voltage	2.5V p-p
Vcc2	Max. 10V

ELECTRICAL CHARACTERISTICS

Test conditions (unless otherwise stated):

Supply voltage: V_{CC} 1 & 2 = 5.2V ± 0.25V or 6.8V to 9.5V

V_{EE} = 0V; Temperature T_{amb} = -40°C to +85°C

Characteristic	Symbol	Value		Units	Notes	Conditions
		Min.	Max.			
Maximum frequency (sinewave input)	f_{max}	225		MHz	Note 4	Input = 200-800mV p-p
Minimum frequency (sinewave input)	f_{min}		20	MHz	Note 4	Input = 400-800mV p-p
Power supply current	I_{EE}		7	mA	Note 4	
Control input high voltage	V_{INH}	4		V	Note 4	
Control input low voltage	V_{INL}		2	V	Note 4	
Output high voltage	V_{OH}	2.4		V	Note 4	Pins 2, 7 and 8 linked V_{CC} = 4.95V I_{OH} = 100μA
Output low voltage	V_{OL}		0.5	V	Note 4	Pin 2 linked to 8 and 7 I_{OL} = 1.6mA
Set up time	t_s	14		ns	Note 3	25°C
Release time	t_r	20		ns	Note 3	25°C
Clock to output propagation time	t_p		45	ns	Note 3	25°C

NOTES
1. Unless otherwise stated the electrical characteristics are guaranteed over full specified supply, frequency and temperature range.
2. The test configuration for dynamic testing is shown in Fig.6.
3. Guaranteed but not tested.
4. Tested only at 25°C.

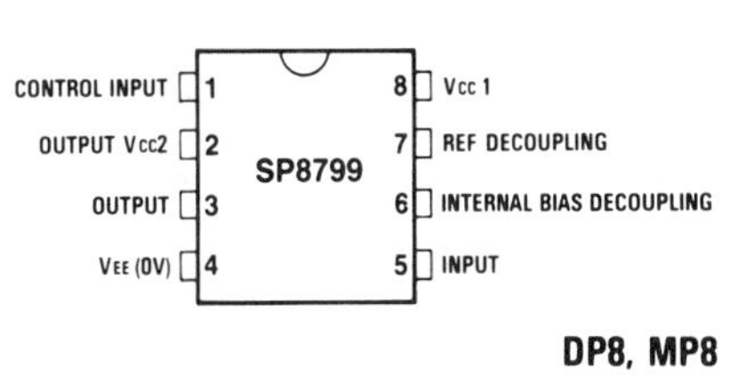

Pin connections - top view

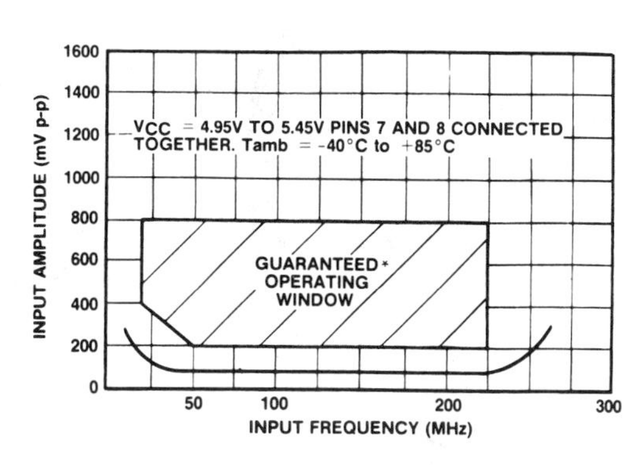

*Tested as specified in table of Electrical Characteristics

Input sensitivity SP8799

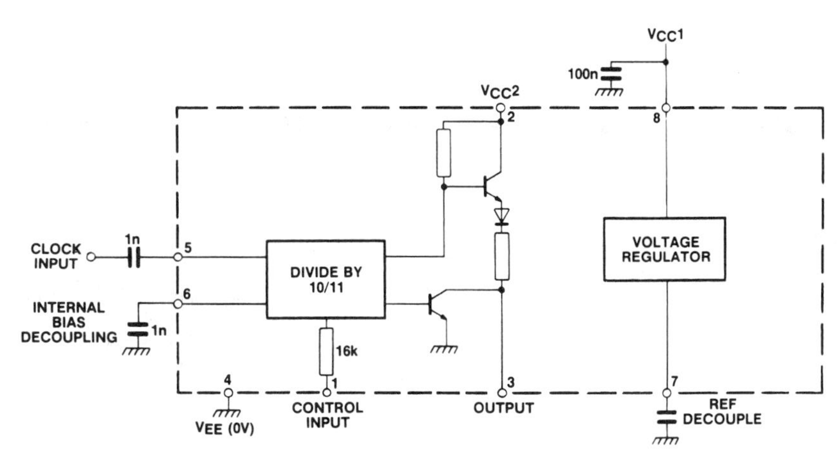

Functional diagram SP8799

CLOCK INPUT
CONTROL INPUT
OUTPUT
t_r t_s 5 6 5 5

TRUTH TABLE FOR CONTROL INPUTS

Control input	Division ratio
0	11
1	10

NOTES

The set-up time t_s is defined as minimum time that can elapse between L→H transition of control input and next L→H clock pulse transition to ensure ÷10 mode is selected.

The release time t_r is defined as minimum time that can elapse between H→L transition of the control input and the next L→H clock pulse transition to ensure the ÷11 mode is selected.

Timing diagram SP8799

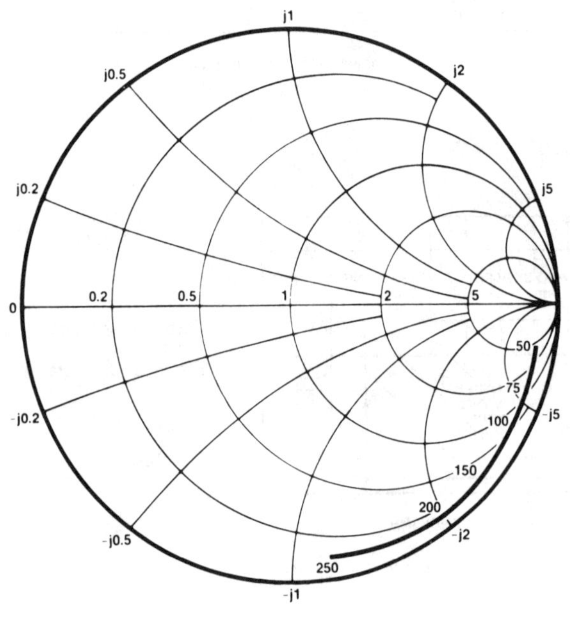

Typical input impedance. Test conditions: supply voltage 5.2V, ambient temperature 25°C, frequencies in MHz, impedances normalised to 50 ohms.

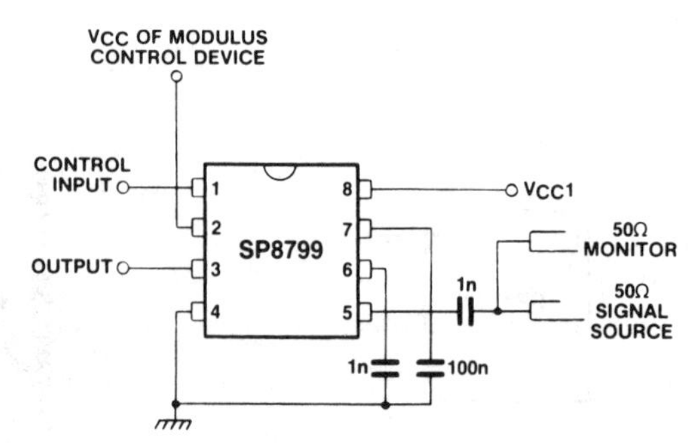

Toggle frequency test circuit

Plessey SP8799A

200 MHz ÷ 10/11 Two Modulus Divider

- Very low power
- Control input and output CMOS/TTL compatible
- Ac-coupled input
- Supply voltage: +5.2V
- Power consumption: 26 mW typical
- Temperature range: −55 °C to +125 °C

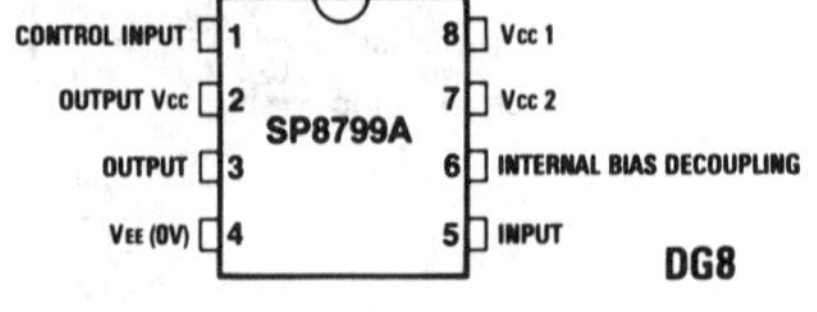

Pin connections - top view

ABSOLUTE MAXIMUM RATINGS

Supply voltage:	6.0V Pins 7 & 8 tied
Storage temperature range:	-55°C to +150°C
Max. junction temperature:	+175°C
Max. clock input voltage:	2.5V p-p

ELECTRICAL CHARACTERISTICS

Test conditions (unless otherwise stated):
Supply voltage: V_{CC} 1 & 2 = 5.2V ± 0.25V; V_{EE} = 0V; Temperature T_{amb} = -55°C to +125°C

Characteristic	Symbol	Value		Units	Notes	Conditions
		Min.	Max.			
Maximum frequency	f_{max}	200		MHz		Input = 200-400mV p-p
(sinewave input)		150		MHz	Note 3	Input = 200-800mV p-p
Minimum frequency	f_{min}		20	MHz	Note 3	Input = 400mV p-p
(sinewave input)			50	MHz		Input = 200mV p-p
Power supply current	I_{EE}		7	mA	Note 4	
Control input high voltage	V_{INH}	4	5.2	V	Note 4	
Control input low voltage	V_{INL}		2	V	Note 4	
Output high voltage	V_{OH}	2.4		V	Note 4	Pins 2, 7 and 8 linked V_{CC} = 4.95V I_{OH} = 100μA
Output low voltage	V_{OL}		0.5	V	Note 4	Pin 2 linked to 8 and 7 I_{OL} = 1.6mA
Set up time	t_s	14		ns	Note 3	25°C
Release time	t_r	20		ns	Note 3	25°C
Clock to output propagation time	t_p		45	ns	Note 3	25°C

NOTES
1. Unless otherwise stated the electrical characteristics are guaranteed over full specified supply, frequency and temperature range
2. The test configuration for dynamic testing is shown.
3. Guaranteed but not tested.
4. Tested at 25°C only.

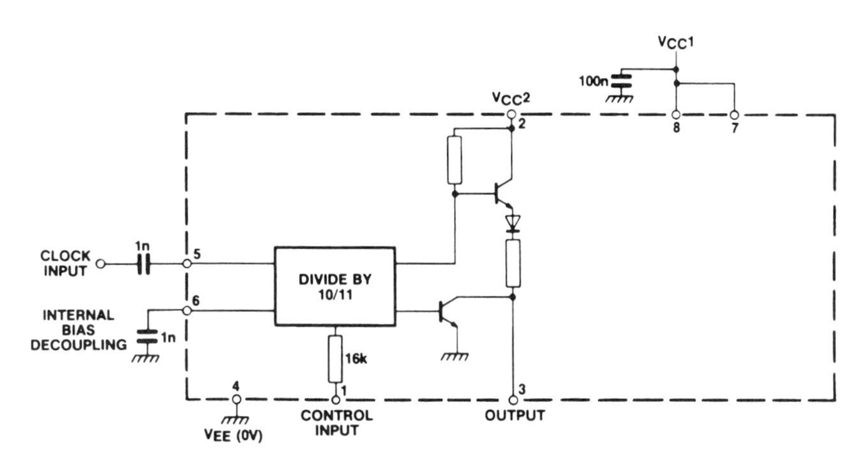

Function diagram SP8799A

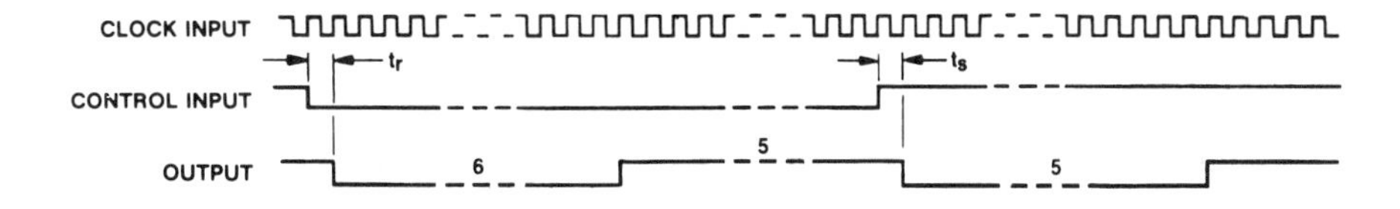

TRUTH TABLE FOR CONTROL INPUTS

Control inputs	Division ratio
0	11
1	10

NOTES
The set-up time t_s is defined as minimum time that can elapse between L→H transition of control input and next L→H clock pulse transition to ensure ÷10 mode is selected.
The release time t_r is defined as minimum time that can elapse between H→L transition of the control input and the next L→H clock pulse transition to ensure the ÷11 mode is selected.

Timing diagram SP8799A

*Tested as specified in table of Electrical Characteristics

Input sensitivity SP8799A

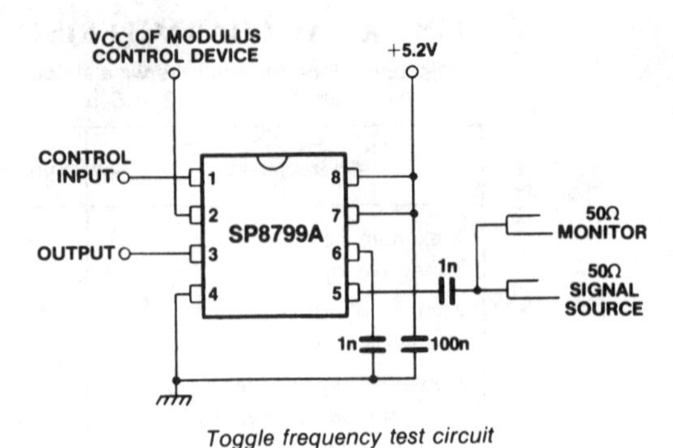

Toggle frequency test circuit

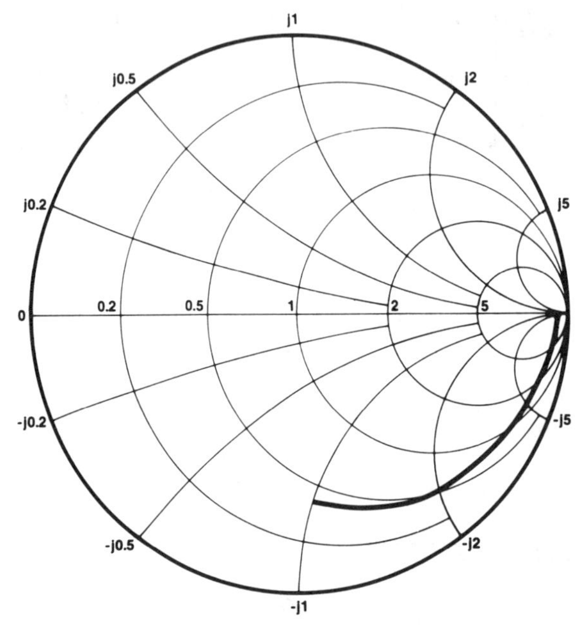

Typical input impedance. Test conditions: supply voltage 5.2V, ambient temperature 25° C, frequencies in MHz, impedances normalised to 50 ohms.

Plessey SP8802A

3.3 GHz ÷ 2 Fixed Modulus Divider

- Very high speed operation: 3.3 GHz
- Silicon technology for low-phase noise (typically better than −140dBc/Hz at 10 kHz)
- Specified over the full military temperature range
- Low power dissipation: 420mW (typ.)
- 5V single-supply operation
- High input sensitivity
- Very wide operating frequency range

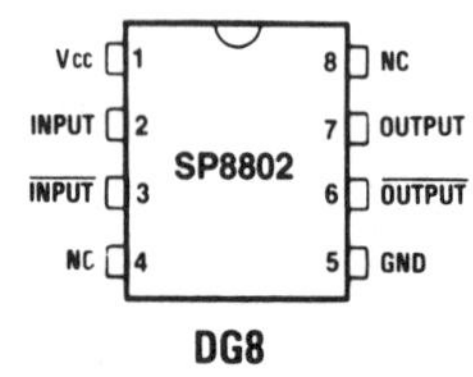

Pin connections - top view

ABSOLUTE MAXIMUM RATINGS

Supply voltage Vcc	6.5V
Clock input voltage	2.5V p-p
Storage temperature range	-55 °C to +150 °C
Junction temperature	+175 °C

THERMAL CHARACTERISTICS

θ_{JA} = 150 °C/W

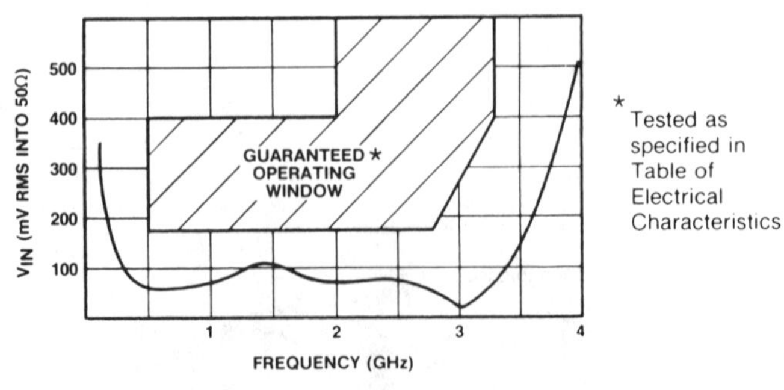

Typical input sensitivity

ELECTRICAL CHARACTERISTICS

Test conditions (unless otherwise stated):
T_{amb} = -55°C to +125°C, V_{CC} = 4.75V to 5.25V (See Note)

Characteristic	Pin	Value			Units	Conditions
		Min.	Typ.	Max.		
Supply current	1		84	100	mA	V_{CC} = 5V
Input sensitivity	2,3					RMS sinewave
0.5GHz to 2.8GHz				175	mV	Measured in 50Ω
3.3GHz				400	mV	system. See Figs. 3 & 4
Input impedance (series equivalent)	2,3		50		Ω	
			2		pF	
Output voltage with f_{in} = 1000MHz	6,7	0.8	1		V p-p	V_{CC} = 5V
Output voltage with f_{in} = 3GHz	6,7		0.35		V p-p	V_{CC} = 5V load as Fig.4

NOTE
Devices must be used with a suitable heatsink to maintain chip temperature below 175°C when operating at T_{amb} > 100°C.

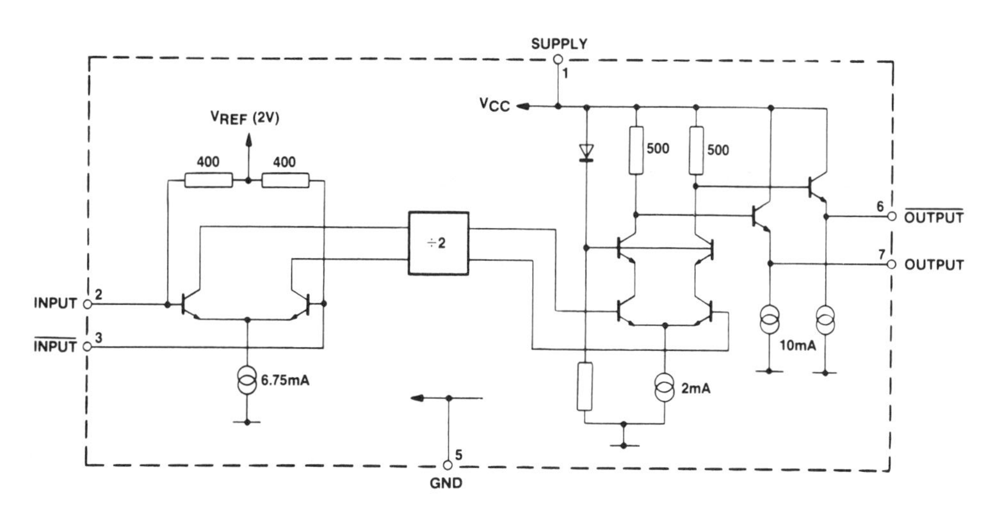

SP8802A block diagram

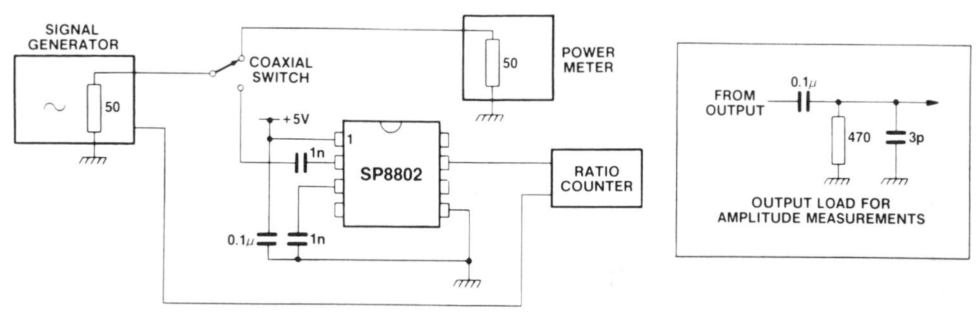

Test circuit

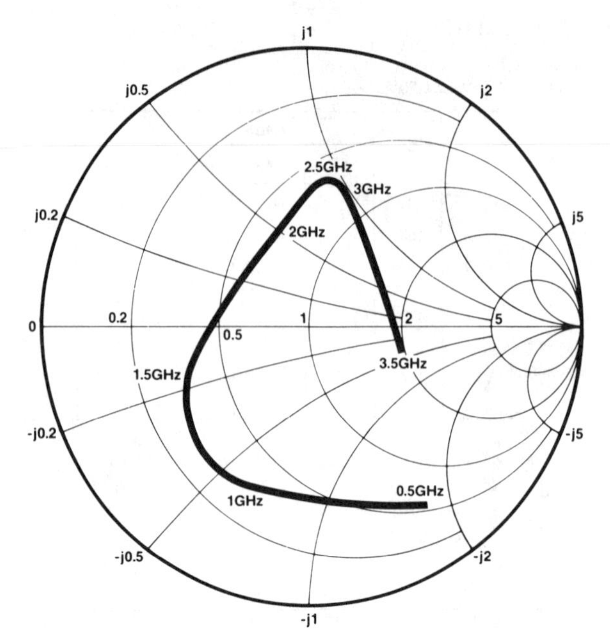

Typical input impedance

Plessey SP8804A
3.3 GHz ÷ 4 Fixed Modulus Divider

- Very high speed operation: 3.3 GHz
- Silicons technology for low-phase noise (typically better than −140dBc/Hz at 10 kHz)
- Specified over the full military temperature range
- Low power dissipation: 370 mW (typ.)
- 5V single-supply operation
- High input sensitivity
- Very wide operating frequency range

ABSOLUTE MAXIMUM RATINGS

Supply voltage V_{CC}	6.5V
Clock input voltage	2.5V p-p
Storage temperature range	-55°C to +150°C
Junction temperature	+175°C

THERMAL CHARACTERISTICS

θ_{JA} = 150°C/W

ELECTRICAL CHARACTERISTICS

Test conditions (unless otherwise stated):
T_{amb} = -55°C to +125°C, V_{CC} = 4.75V to 5.25V

Characteristic	Pin	Value			Units	Conditions
		Min.	Typ.	Max.		
Supply current	1		74	90	mA	V_{CC} = 5V
Input sensitivity	2,3					RMS sinewave
0.5GHz to 2.8GHz				175	mV	Measured in 50Ω
3.3GHz				400	mV	system.
Input impedance (series equivalent)	2,3		50		Ω	
			2		pF	
Output voltage with f_{in} = 1000MHz	6,7	0.8	1		V p-p	V_{CC} = 5V
Output voltage with f_{in} = 3GHz	6,7		0.25		V p-p	V_{CC} = 5V load as

NOTE
Devices must be used with a suitable heatsink to maintain chip temperature below 175°C when operating at T_{amb} >105°C.

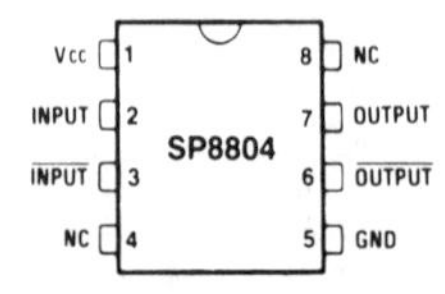

Pin connections - top view

SUPPLY
1
V_{CC}
V_{REF} (2V)
400
400
÷4
500
500
6 OUTPUT
7 OUTPUT
INPUT 2
INPUT 3
6.75mA
2mA
5mA
5
GND

SP8804A block diagram

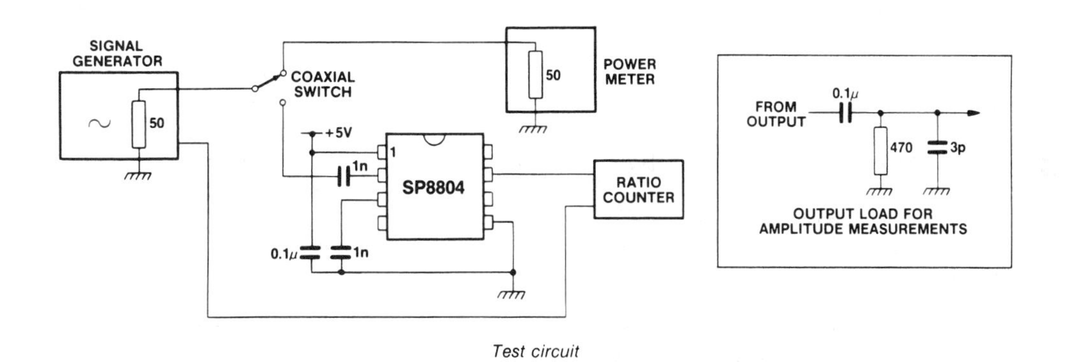

Test circuit

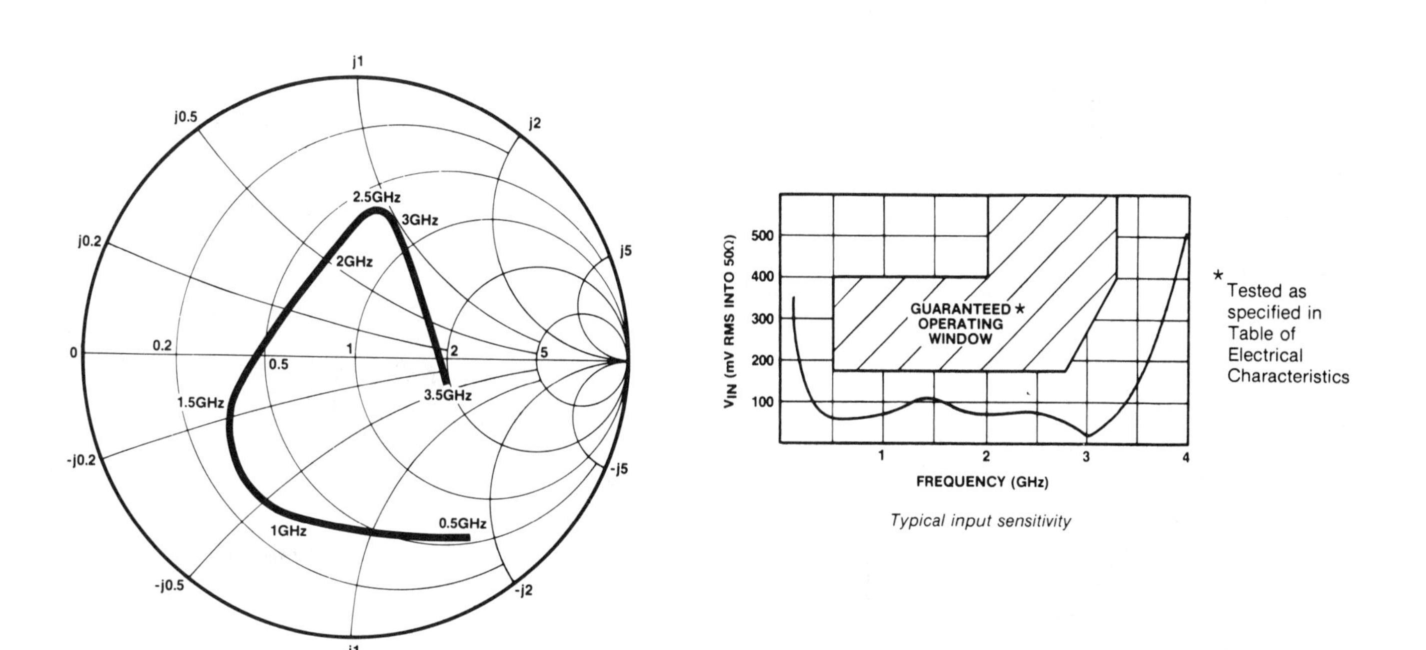

Typical input sensitivity

Typical input impedance

Plessey SP8808A
3.3 GHz ÷ 8 Fixed Modulus Divider

- Very high speed operation: 3.3 GHz
- Silicon technology for low-phase noise (typically better than −140dBc/Hz at 10 kHz
- Specified over the full military temperature range
- Low power dissipation: 345 mW (typ.)
- 5V single-supply operation
- High input sensitivity
- Very wide operating frequency range

ABSOLUTE MAXIMUM RATINGS

Supply voltage V_{CC}	6.5V
Clock input voltage	2.5V p-p
Storage temperature range	-55°C to +150°C
Junction temperature	+175°C

THERMAL CHARACTERISTICS

θ_{JA} = 150°C/W

ELECTRICAL CHARACTERISTICS

Test conditions (unless otherwise stated):
T_{amb} = -55°C to +125°C, V_{CC} = 4.75V to 5.25V (See Note)

Characteristic	Pin	Value			Units	Conditions
		Min.	Typ.	Max.		
Supply current	1		69	85	mA	V_{CC} = 5V
Input sensitivity	2,3					RMS sinewave
0.5GHz to 2.8GHz				175	mV	Measured in 50Ω
3.3GHz				400	mV	system.
Input impedance (series equivalent)	2,3		50		Ω	
			2		pF	
Output voltage with f_{in} = 1000MHz	6,7	0.8	1		V p-p	V_{CC} = 5V
Output voltage with f_{in} = 3GHz	6,7		0.4		V p-p	V_{CC} = 5V load as

NOTE
Devices must be used with a suitable heat sink to maintain chip temperature below 175°C when operating at T_{amb} >110°C.

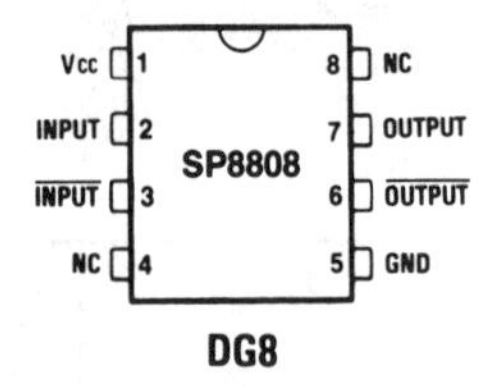

Pin connections - top view

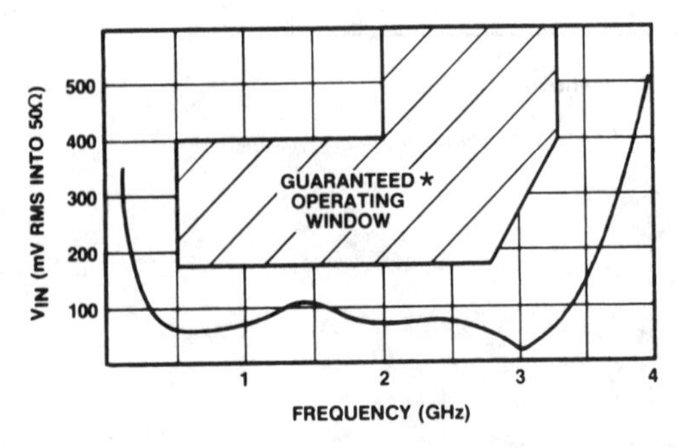

Typical input sensitivity

* Tested as specified in Table of Electrical Characteristics

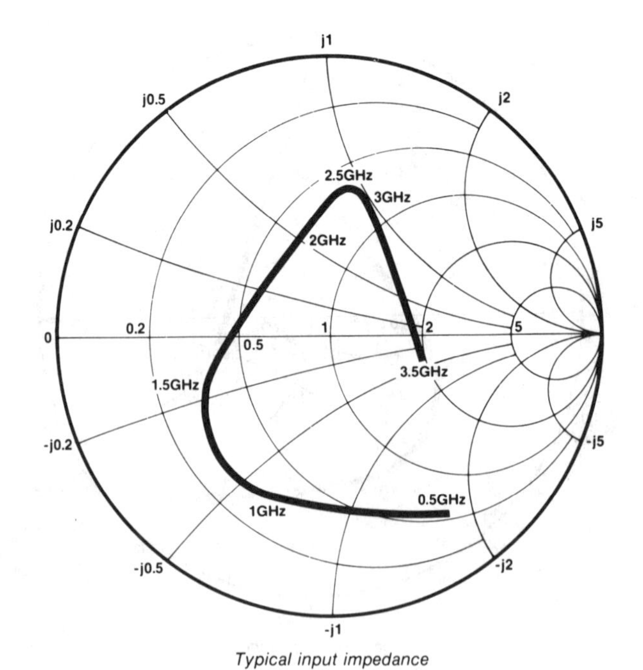

Typical input impedance

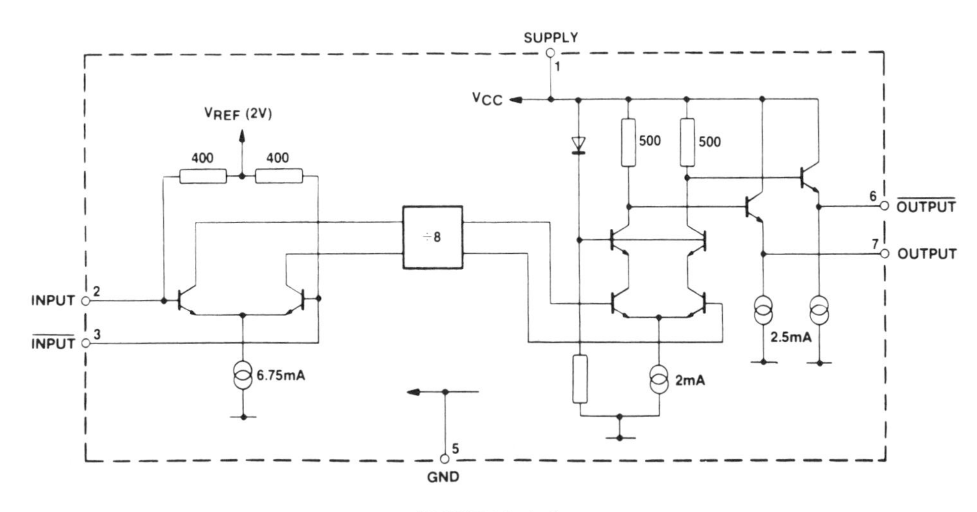

SP8808A block diagram

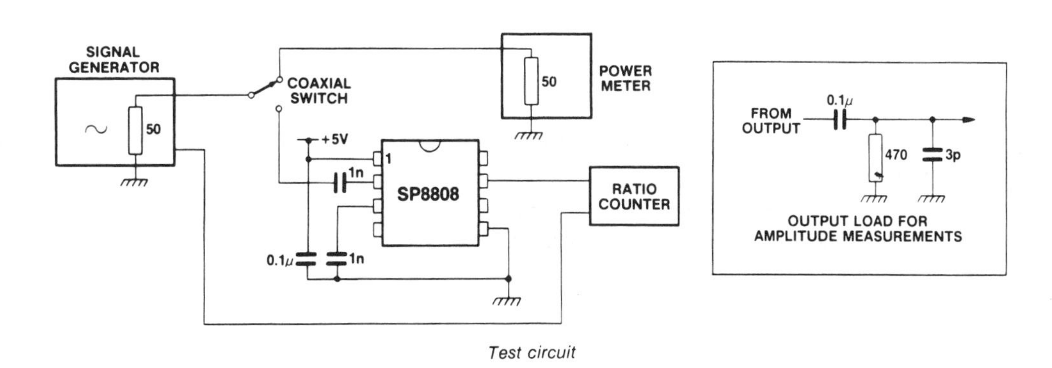

Test circuit

This information is reproduced with permission of LSI Computer Systems, Inc., 1235 Walt Whitman Road, Melville, NY 11747.

LSI
Red Series
Complementary MOS (CMOS) Divider Circuits

RED SERIES

RED 5/6	Divide by 5 or 6
RED 50/60	Divide by 50 or 60
RED 100/120	Divide by 100 or 120
RED 300/360	Divide by 300 or 360
RED 500/600	Divide by 500 or 600
RED 3000/3600	Divide by 3000 or 3600

- Clock input pulse shaper accepts 50-Hz/60-Hz sine wave directly
- Fully static counter operation
- +4.5V to +15V operation ($V_{DD}-V_{SS}$)
- Low power dissipation
- High noise immunity
- Reset
- Input enable
- 50-Hz/60-Hz division select input
- Output low-power TTL compatible at +4.5V operation
- All inputs protected
- Square-wave output (except for ÷5)

Time base generator from either 50-Hz or 60-Hz line frequency to produce:

10 pulses per second	(RED 5/6)
1 pulse per second	(RED 50/60)
1 pulse per 2 seconds	(RED 100/120)
1 pulse per 1 minute	(RED 300/360)
1 pulse per 10 seconds	(RED 500/600)
1 pulse per minute	(RED 3000/3600)

The counter advances by one on each negative transition of the input clock pulse as long as the enable signal is "high" and the reset signal is "low." When the enable signal is "low" the input clock pulses will be inhibited and the counter will be held at the state it was in, prior to bringing the enable "low." A "high" reset signal clears the counter to zero count.

Depending on the devices used, a "low" on the division select input will cause a divide by 6, 60, 120, 360, 600, or 3600. A "high" on the division select will cause a divide by 5, 50, 100, 300, 500, or 3000.

MAXIMUM RATINGS:

PARAMETER	SYMBOL	VALUE
Dc supply voltage:	V_{DD}	+18 to −0.5 V_{dc}
Input voltage:	V_{IN}	V_{DD} to V_{SS} V_{dc}
Oper. temp. range:	T_A	−40 to +85 °C
Storage temp. range:	T_{stg}	−65 to +150 °C

The information included herein is believed to be accurate and reliable. However, LSI Computer Systems, Inc. assumes no responsibilities for inaccuracies, nor any infringements of patent rights of others which may result from its use.

ELECTRICAL CHARACTERISTICS: (TA = 25° unless otherwise specified)

TEST CONDITIONS: Vss = 0V
Output Capacitance Load = 15 pF
Input Rise and Fall times = 20 ns, except clock Rise and Fall times

		VDD	Min	Max	Units
	Quiescent Device Current:	5V		10	uA
		10V		20	uA
	Output Voltage, Low Level:	5V		0.01	Volts
		10V		0.01	Volts
	High Level:	5V	4.99		Volts
		10V	9.99		Volts
	Clock Input Voltage, Low Level	5V		1	Volts
		10V		2	Volts
	High Level	5V	4		Volts
		10V	8		Volts
	Input Noise Immunity (except clock):	5V	1.5		Volts
	(Low and High)	10V	3.0		Volts
	Output Drive Current:				
Full Temp. Range	N Channel Sink Current:	4.5V	0.18		mA
	(Vout = Vss + .4V)	10V	0.45		mA
	P Channel Source Current:	4.5V	0.3		mA
	(Vout = VDD - 1V)	10V	0.75		mA

	VDD	Min	Max	Units
Input Capacitance:	(Any Input)		5	pF
Clock Rise and Fall Time:	5V	No Maximum Limit		
	10V	No Maximum Limit		
Clock Frequency:	5V	DC	600	KHz
	10V	DC	1200	KHz
Input Clock Pulse Width:	5V	800		ns
	10V	400		ns
Output Rise and Fall Time:	5V		225	ns
	10V		150	ns
Propagation Delay to Output:	5V		1500	ns
	10V		750	ns
Enable Set-up Time:	5V		300	ns
	10V		150	ns
Reset Pulse Width:	5V	800		ns
	10V	400		ns
Reset Removal Time:	5V		1200	ns
	10V		600	ns
Reset Propagation Delay to Output:	5V		1400	ns
	10V		700	ns

ENABLE SIGNAL TIMING CONSIDERATION

If the Enable signal switches Low during a positive clock phase and then switches High during a negative clock phase, a false count will be registered.

To prevent this from happening, the Enable signal should not switch Low during a positive clock phase unless the switch to High also occurs during a positive clock phase. The Enable signal should normally be switched during a negative clock phase.

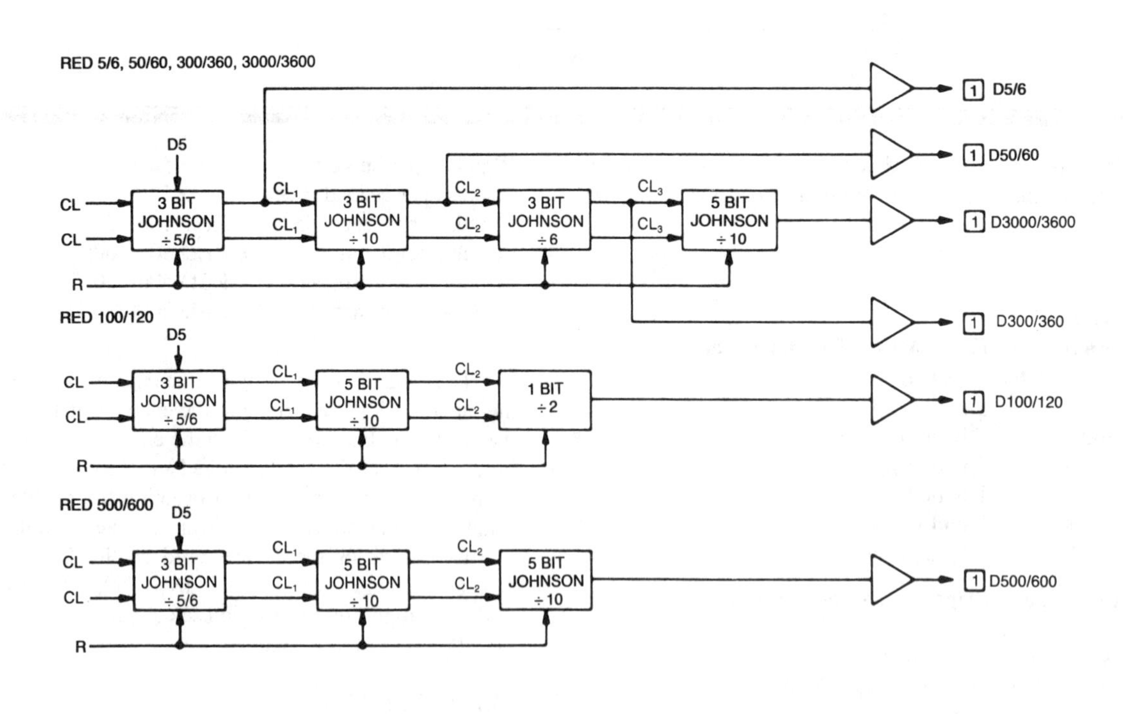

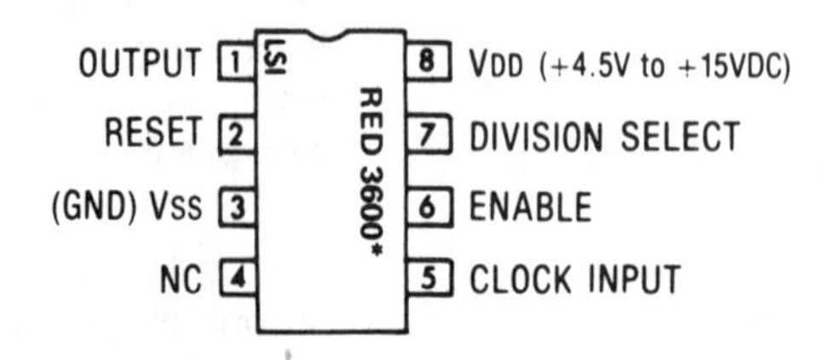

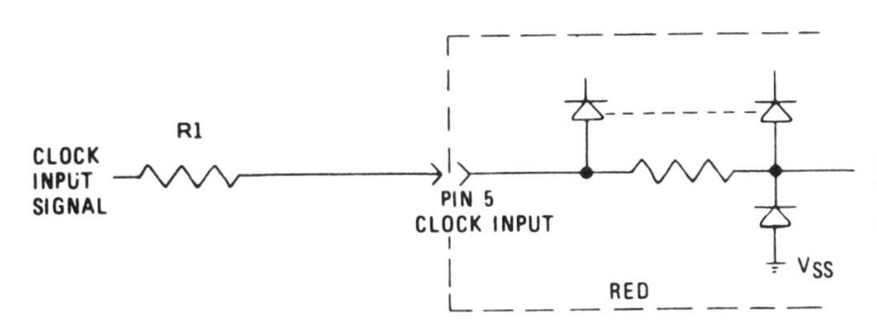

TYPICAL CLOCK INPUT

If input signals are less than V_{SS} or greater than V_{DD}, a series input resistor, R1, should be used to limit the maximum input current to 2 milliamperes.

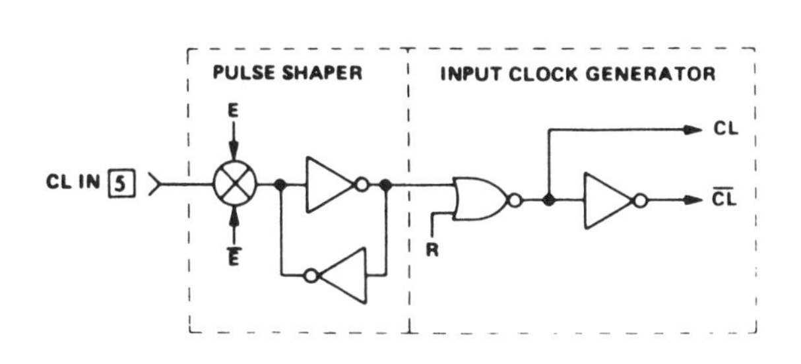

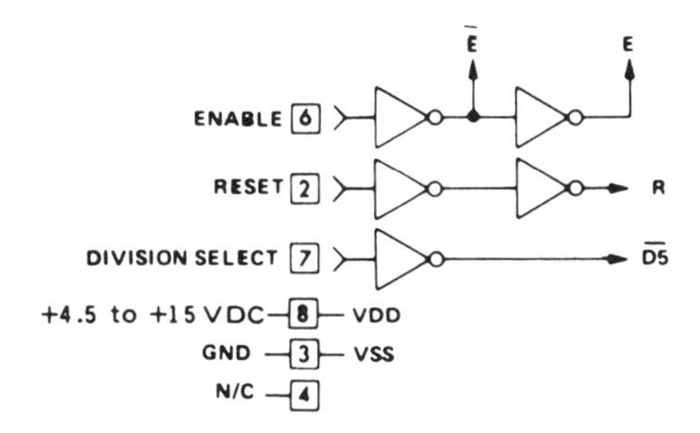

LSI RDD104

Selectable 4-Decade CMOS Divider

- Selectable divide by 10, 100, 1000, or 10,000
- Clock input shaping network accepts fast- or slow-edge inputs
- Active oscillator network for external crystal
- Square-wave output
- Output TTL compatible at +4.5V operation
- High noise immunity
- Reset
- All inputs protected
- 4.5 to 15V operation
- Low power dissipation

The RDD104 is a monolithic CMOS (Complementary MOS) four-decade divider circuit that advances on each negative transition of the input clock pulse. When the reset input is high the circuit is cleared to zero. The clock input is applied to a three stage amplifier network whose output is brought out so that an external crystal network can be used to form an oscillator circuit. If the clock output is not used, the amplifier acts as an input buffer. Two select inputs are provided, which enables the circuit to divide by 10, 100, 1000, or 10,000.

Minimum Part Oscillator Circuit

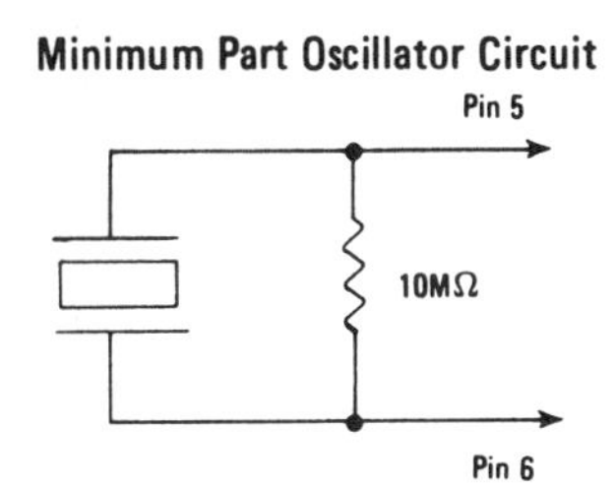

RDD104 BLOCK DIAGRAM

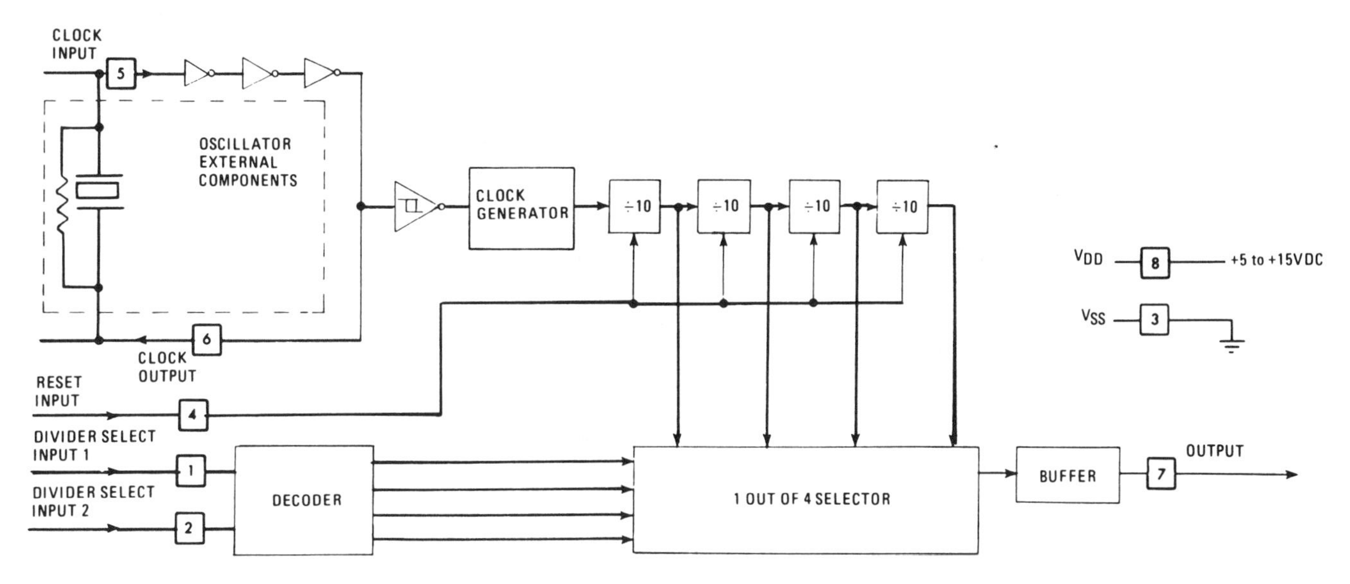

DIVIDER SELECT INPUTS		OUTPUT DIVISION
SELECT 2 (Pin 2)	SELECT 1 (Pin 1)	
0	0	10,000
0	1	1,000
1	0	100
1	1	10

Typical Oscillator Circuit 1 MHz and Below with Trim

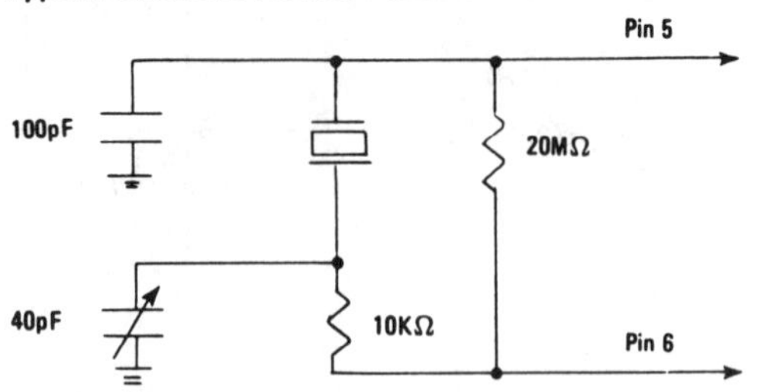

Pin Connections RDD104:

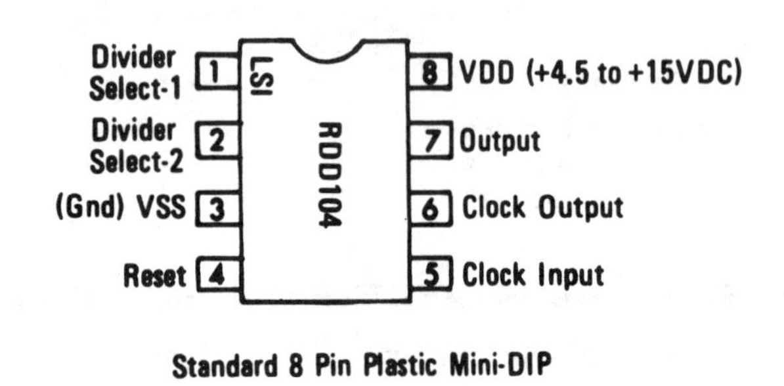

Standard 8 Pin Plastic Mini-DIP

Typical Oscillator Circuit Above 1 MHz with Trim

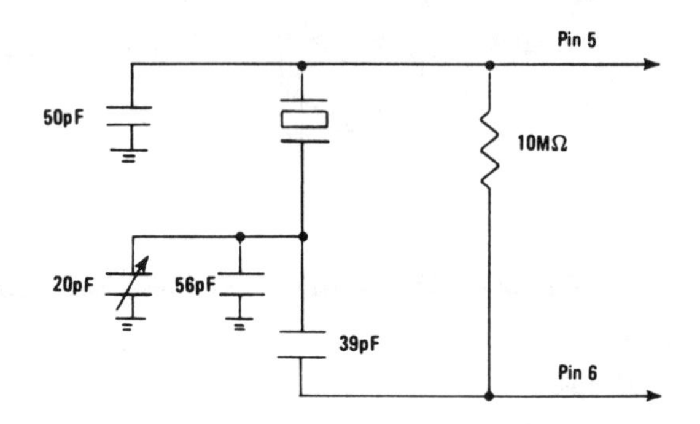

MAXIMUM RATINGS

PARAMETER	SYMBOL	VALUE	UNITS
Storage Temperature	T_{stg}	−65 to +150	°C
Operating Temperature	T_A	−40 to +85	°C
DC Supply Voltage	V_{DD}	+18 to −0.5	VDC
Input Voltage	V_{IN}	V_{DD} to V_{SS}	VDC

DC ELECTRICAL CHARACTERISTICS

(V_{SS} = 0 Volts, C Load = 50pF) Input rise and fall times = 20ns except for clock

	V_{DD}	TEMPERATURE (°C)			UNITS
		−40	+25	+85	
Quiescent Device Current	4.5V	10	10	300	μA Max
	10V	20	20	600	μA Max
Output Voltage, Low Level	4.5V	.01	.01	.05	V Min
	10V	.01	.01	.05	V Min
High Level	4.5V	4.49	4.49	4.45	V Max
	10V	9.99	9.99	9.95	V Max
Input Noise Immunity	4.5V	1.3	1.3	1.3	V Min
(Low and High)	10V	3.0	3.0	3.0	V Min
Output Drive Current					
N Channel Sink Current	4.5V	2.3	1.9	1.6	mA Min
(V_{OUT}=V_{SS}+.4V)	10V	5.0	4.0	3.5	mA Min
P Channel Source Current	4.5V	1.1	.95	.8	mA Min
(V_{OUT} = V_{DD} − 1V)	10V	2.5	2.1	1.8	mA Min
Input Capacitance (any input)			5.0		pF Max

DYNAMIC ELECTRICAL CHARACTERISTICS

(C Load = 50pF, Input Rise and Fall Times = 20ns Except for Clock)

	V_{DD}	MIN	TYP	MAX	Units
Clock Input Frequency	4.5V	DC		1.5	MHz
	10V	DC		4.0	MHz
	15V	DC		6.0	MHz
Clock Input Rise and Fall Times	4.5 to 15V		No Limit		
Clock Output Rise and Fall Time CL=15pF	4.5V			140	ns
	10V			70	ns
Clock Output Propagation Delay CL=15pF	4.5V			300	ns
	10V			150	ns
Output Rise & Fall Times	4.5V			400	ns
	10V			200	ns
Propagation Delay to Output	4.5V			1500	ns
	10V			750	ns
Reset Pulse Width	4.5V	800			ns
	10V	400			ns
Reset Removal Time	4.5V			500	ns
	10V			250	ns
Reset Propagation Delay to Output	4.5V			1400	ns
	10V			700	ns
Select Input Setup Time	4.5V			800	ns
	10V			400	ns

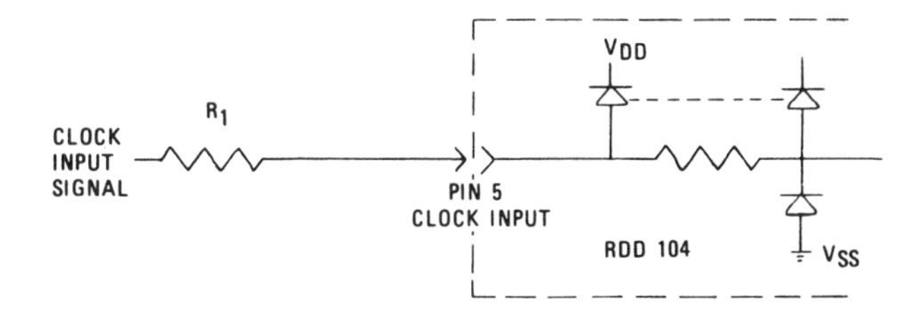

TYPICAL INPUT

If input signals are less than V_{SS} or greater than V_{DD}, a series input resistor, R1, should be used to limit the maximum input current to 2 milliamperes.

LSI LS7100

BCD to 7-Segment Latch/Decoder/Driver

- Up to −50V segment output
- All inputs are TTL or CMOS compatible
- Internal pull-down resistors on all inputs
- Operating voltage range from −5V to −60V

The LS7100 is a monolithic ion-implanted MOS BCD to 7-segment latched decoder/driver capable of driving displays over a wide voltage range.

This circuit is specifically intended to drive large light-scattering liquid crystal displays.

ABSOLUTE MAXIMUM RATINGS

(All voltages referenced to V_{SS}, Pin 16)

	SYMBOL	VALUE	UNIT
DC Supply Voltage	V_{DD}	+0.3 to −60	V
Common In	V_{CI}	+0.3 to −60	V
All other inputs	V_{IN}	+0.3 to −30	V
Operating Temperature	T_A	−40 to +70	°C
Storage Temperature	Tstg	−65 to +125	°C

CHANNEL ON RESISTANCE, R_{ON}

LS7100 FUNCTIONAL DIAGRAM

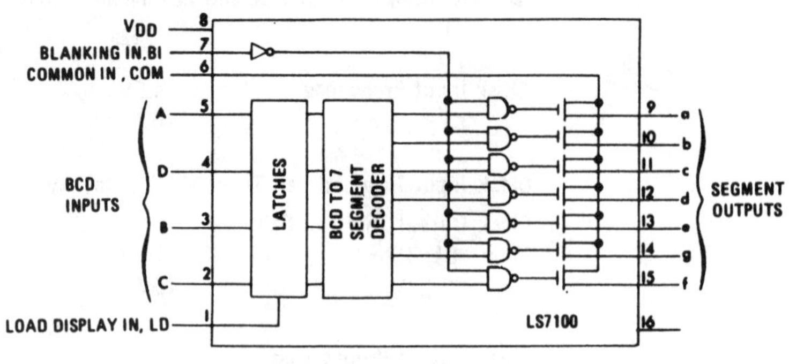

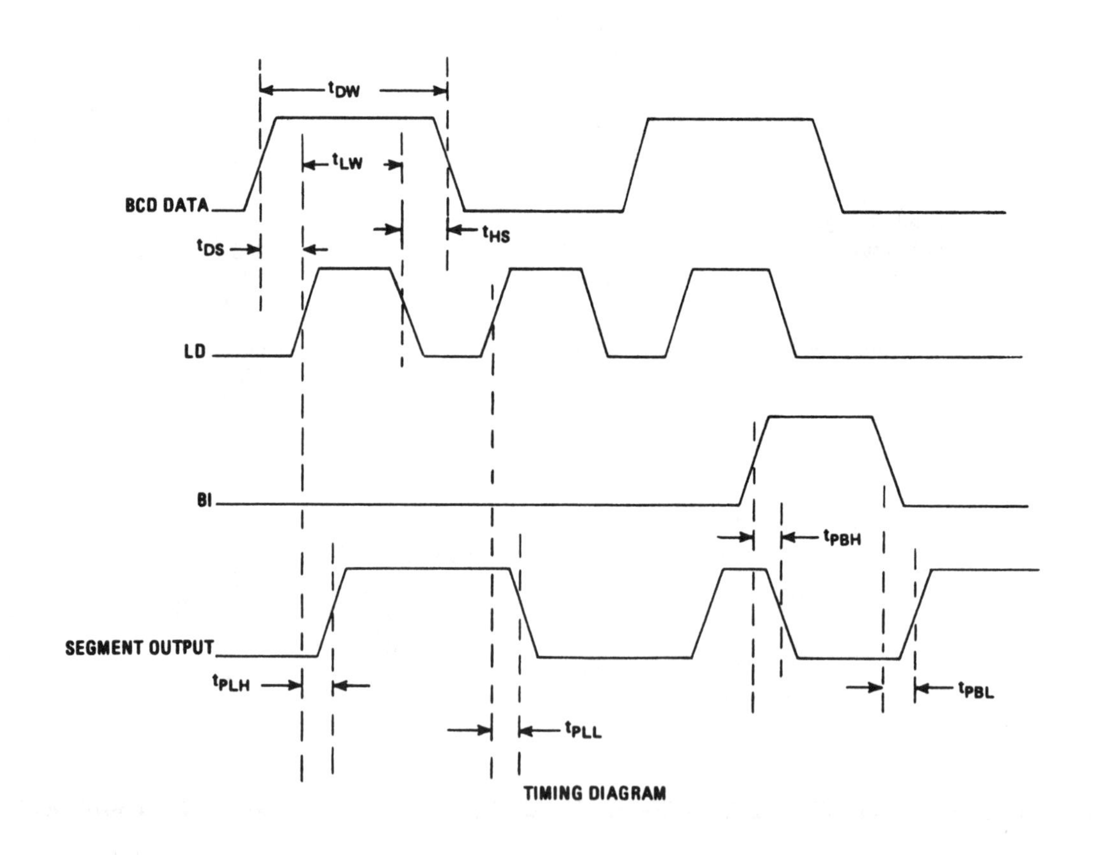

TIMING DIAGRAM

DISPLAY FORMAT

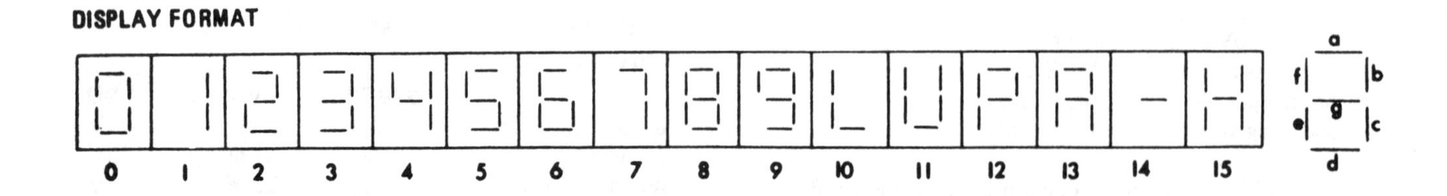

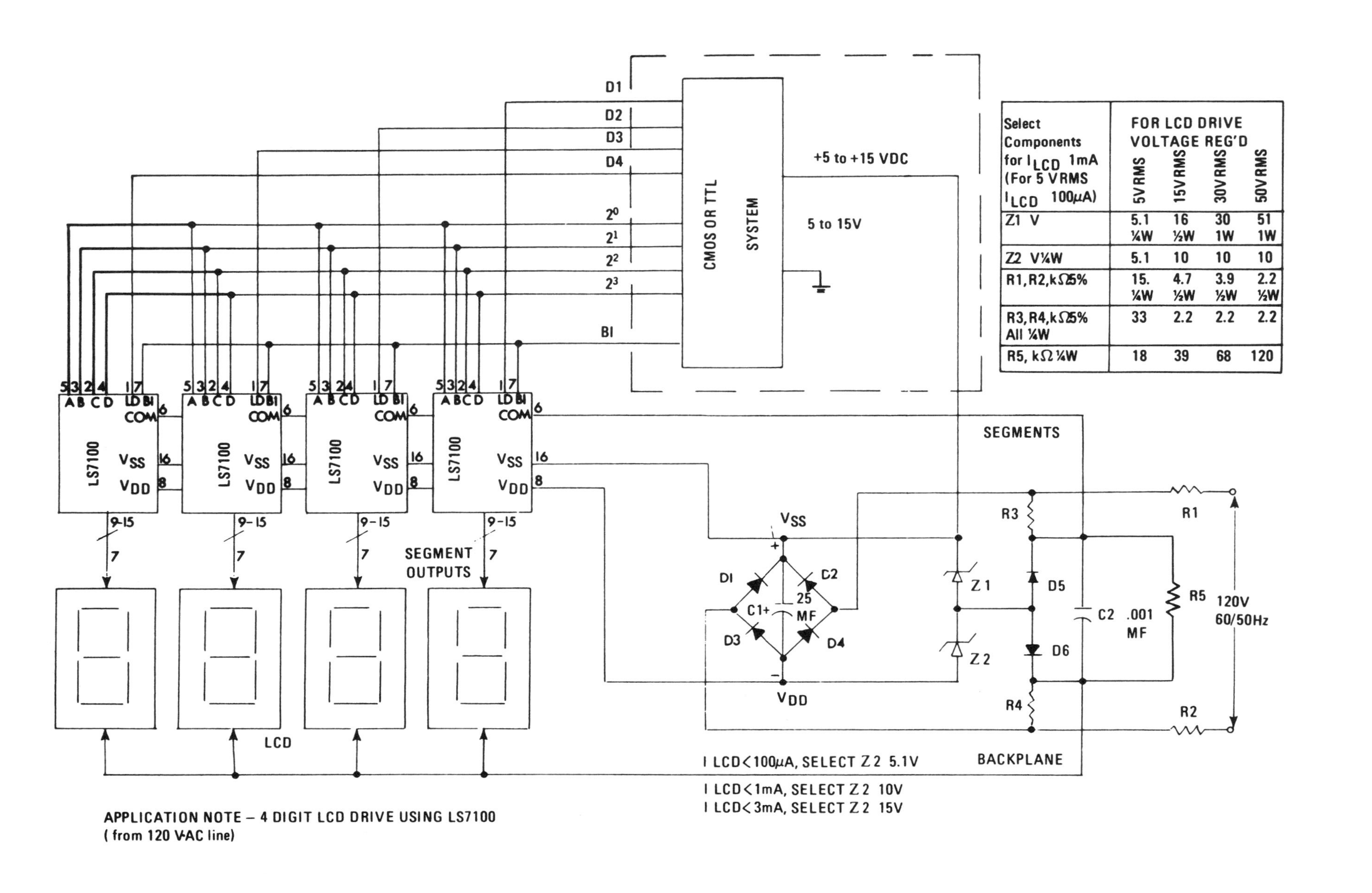

Select Components for I LCD 1mA (For 5 VRMS I LCD 100µA)	FOR LCD DRIVE VOLTAGE REG'D 5VRMS	15VRMS	30VRMS	50VRMS
Z1 V	5.1 ¼W	16 ½W	30 1W	51 1W
Z2 V¼W	5.1	10	10	10
R1, R2, kΩ5%	15. ¼W	4.7 ½W	3.9 ½W	2.2 ½W
R3, R4, kΩ5% All ¼W	33	2.2	2.2	2.2
R5, kΩ ¼W	18	39	68	120

APPLICATION NOTE – 4 DIGIT LCD DRIVE USING LS7100
(from 120 V-AC line)

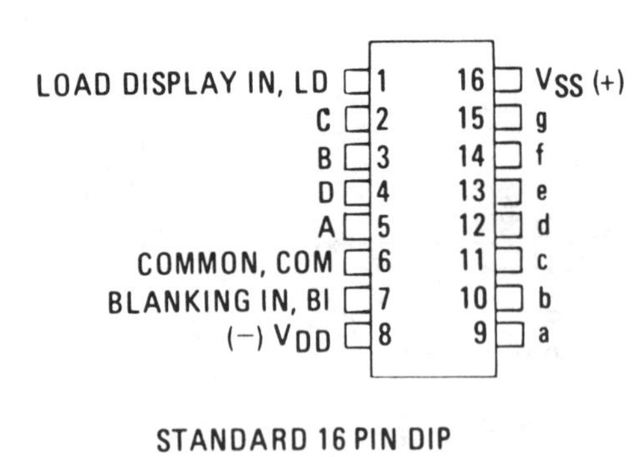

STANDARD 16 PIN DIP

BLANKING (BI) INPUT

Blanking of the display is provided by the BI. When BI is high, all FET switches are opened thereby turning off display segments. When BI is low, the selected FET switches are closed.

BI has an internal pull-down (to logic "0") resistor.

INPUT INTERFACE

LS7100 inputs can be interfaced with TTL, CMOS, NMOS or PMOS outputs by connecting VSS to the positive terminal (output logic "1", reference supply) of the TTL, CMOS, NMOS or PMOS supply.

LSI
LS7110
Binary Addressable Latched 8-Channel Demultiplexer/Driver

- Up to −50 volts outputs
- All inputs are TTL and CMOS compatible
- Internal pull-down resistors on all inputs
- Operating voltage range from −5 to −60V

LS7110 inputs can be interfaced with TTL, CMOS, NMOS, or PMOS outputs by connecting V_{SS} to the positive terminal of the TTL, CMOS, NMOS, or PMOS supply.

ABSOLUTE MAXIMUM RATINGS

(All voltages referenced to V_{SS}, Pin 16)

PARAMETER	SYMBOL	VALUE	UNITS
DC Supply Voltage	V_{DD}	+.3 to −60	V
Common In	V_{CI}	+.3 to −60	V
All other inputs	V_{IN}	+.3 to −30	V
Operating Temperature	T_A	−40 to +85	°C
Storage Temperature	T_{stg}	−65 to +125	°C

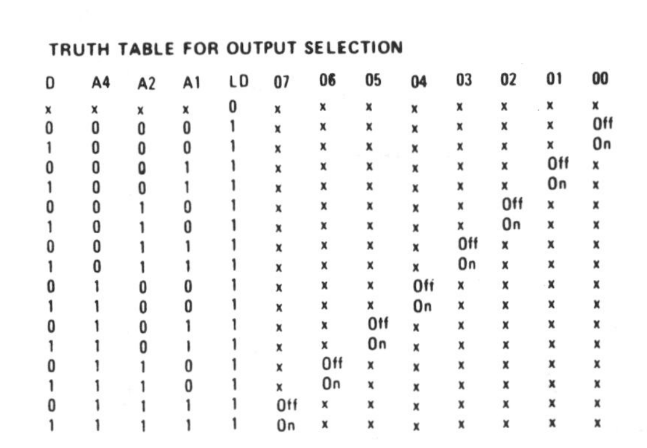

TRUTH TABLE FOR OUTPUT SELECTION

D	A4	A2	A1	LD	07	06	05	04	03	02	01	00
x	x	x	x	0	x	x	x	x	x	x	x	x
0	0	0	0	1	x	x	x	x	x	x	x	Off
1	0	0	0	1	x	x	x	x	x	x	x	On
0	0	0	1	1	x	x	x	x	x	x	Off	x
1	0	0	1	1	x	x	x	x	x	x	On	x
0	0	1	0	1	x	x	x	x	x	Off	x	x
1	0	1	0	1	x	x	x	x	x	On	x	x
0	0	1	1	1	x	x	x	x	Off	x	x	x
1	0	1	1	1	x	x	x	x	On	x	x	x
0	1	0	0	1	x	x	x	Off	x	x	x	x
1	1	0	0	1	x	x	x	On	x	x	x	x
0	1	0	1	1	x	x	Off	x	x	x	x	x
1	1	0	1	1	x	x	On	x	x	x	x	x
0	1	1	0	1	x	Off	x	x	x	x	x	x
1	1	1	0	1	x	On	x	x	x	x	x	x
0	1	1	1	1	Off	x	x	x	x	x	x	x
1	1	1	1	1	On	x	x	x	x	x	x	x

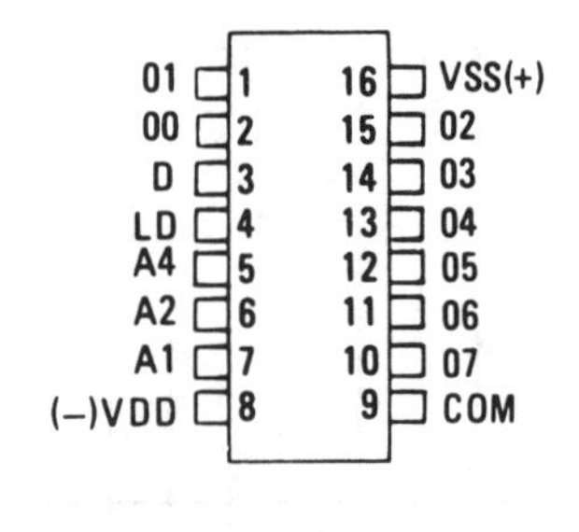

STANDARD 16 PIN DIP

FUNCTIONAL DIAGRAM

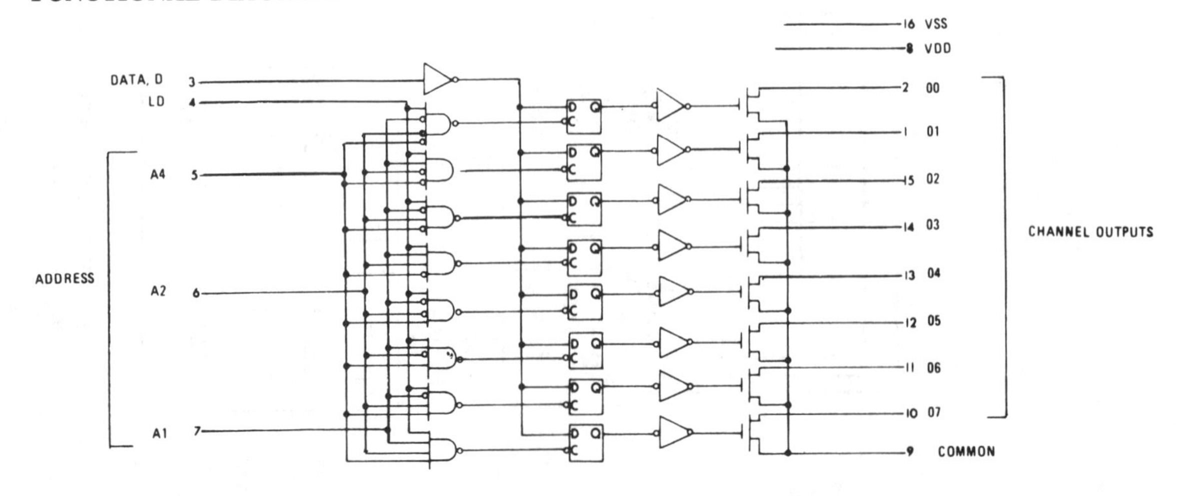

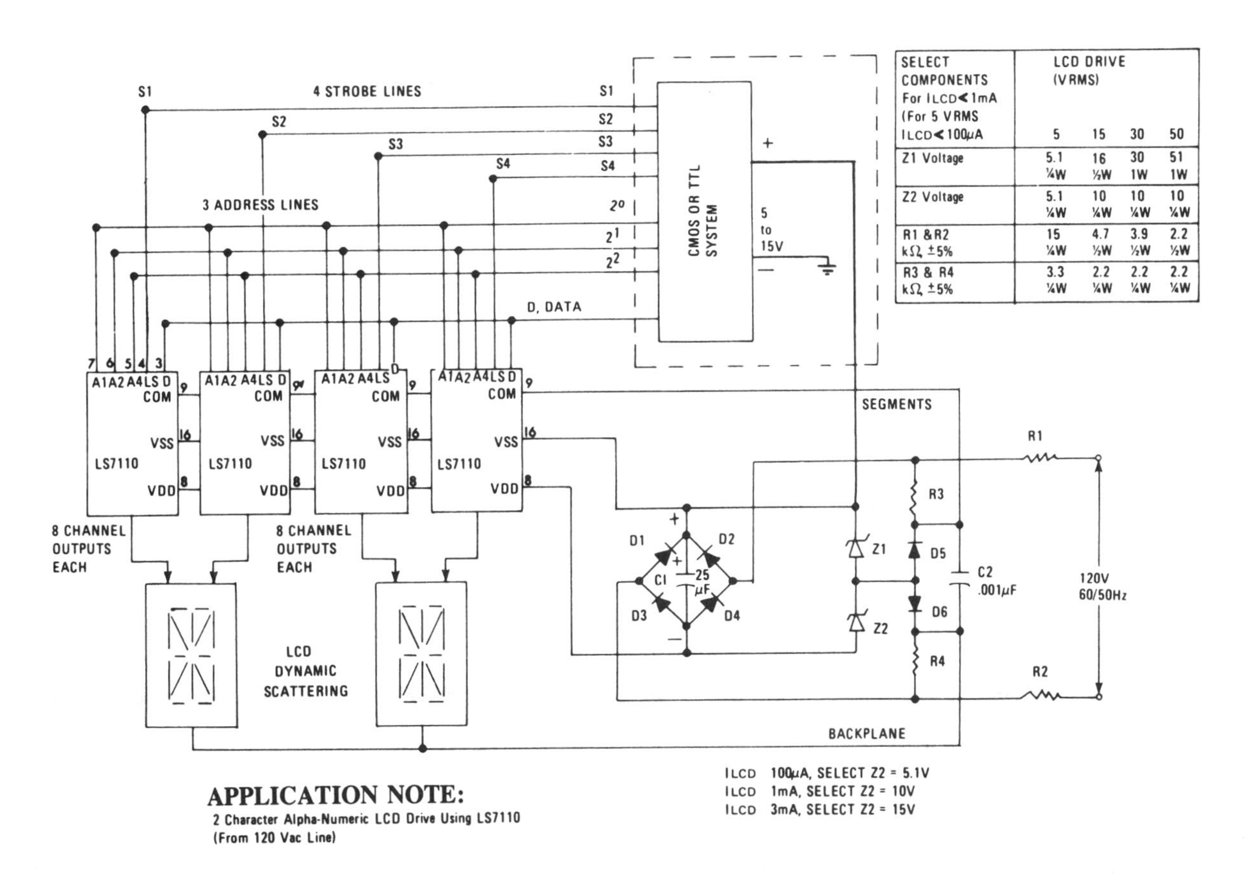

SELECT COMPONENTS For ILCD < 1mA (For 5 VRMS ILCD < 100µA	LCD DRIVE (VRMS) 5	15	30	50
Z1 Voltage	5.1 ¼W	16 ½W	30 1W	51 1W
Z2 Voltage	5.1 ¼W	10 ¼W	10 ¼W	10 ¼W
R1 & R2 kΩ ±5%	15 ¼W	4.7 ½W	3.9 ½W	2.2 ½W
R3 & R4 kΩ ±5%	3.3 ¼W	2.2 ¼W	2.2 ¼W	2.2 ¼W

APPLICATION NOTE:
2 Character Alpha-Numeric LCD Drive Using LS7110
(From 120 Vac Line)

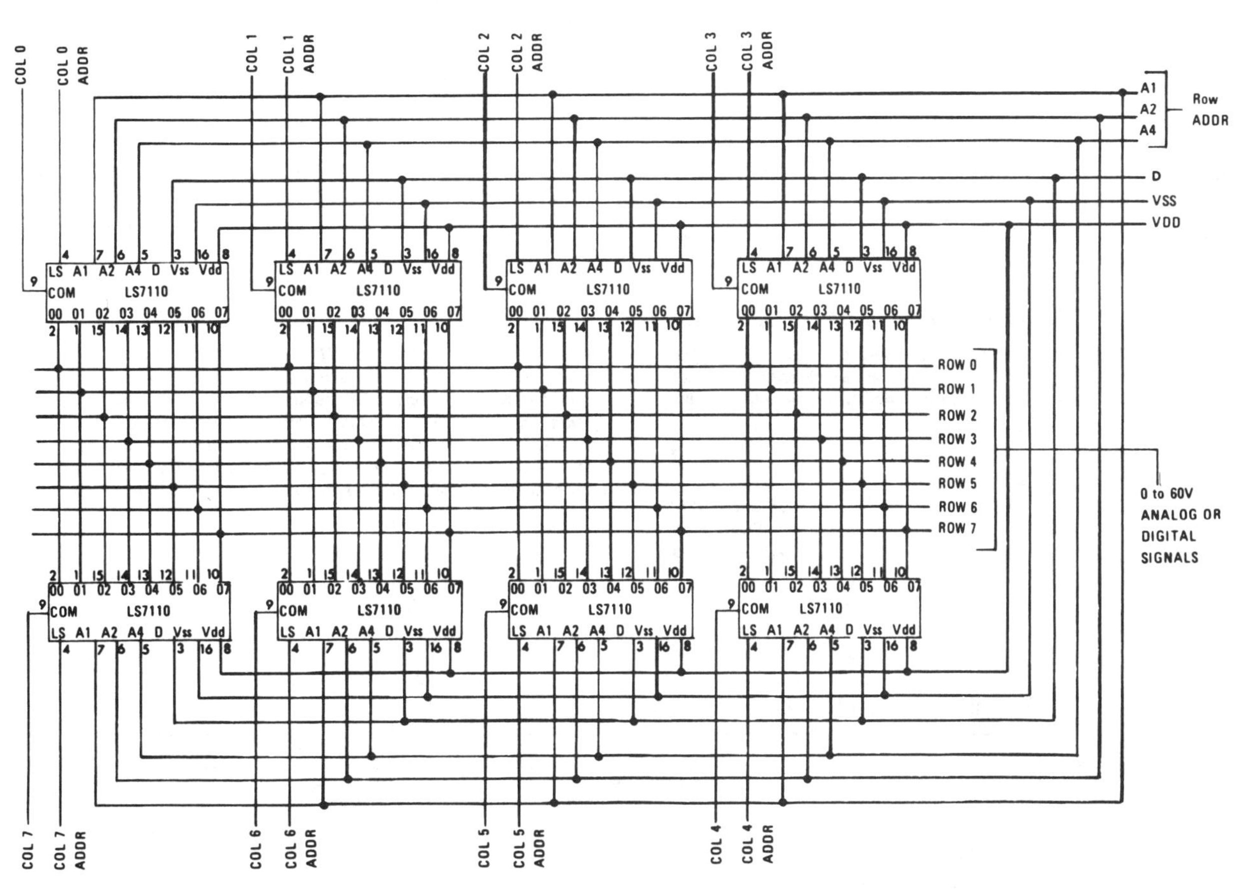

APPLICATION NOTE: 8 × SWITCHBOARD MATRIX USING LS7110

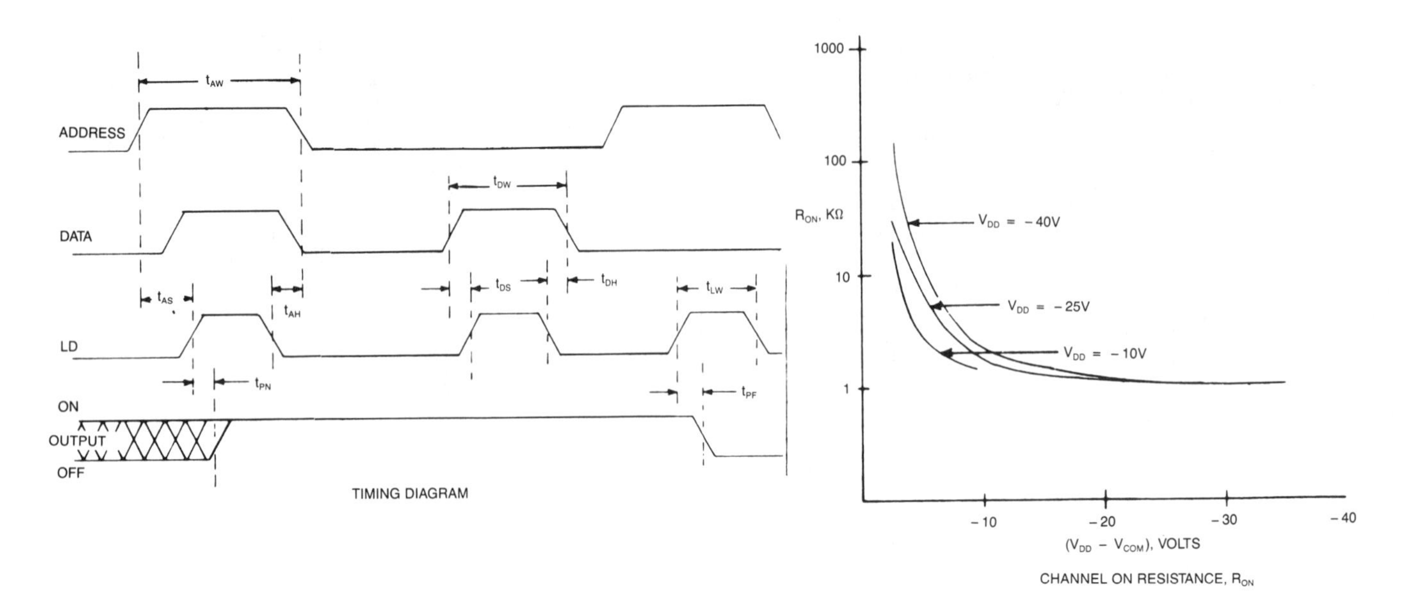

This information is reproduced with permission of National Semiconductor Corporation.

LIFE SUPPORT POLICY

National's products are not authorized for use as critical components in life support devices or systems without the express written approval of the president of National Semiconductor Corporation. As used herein:

1. Life support devices or systems are devices or systems which, (a) are intended for surgical implant into the body, or (b) support or sustain life, and whose failure to perform, when properly used in accordance with instructions for use provided in the labeling, can be reasonably expected to result in a significant injury to the user.

2. A critical component is any component of a life support device or system whose failure to perform can be reasonably expected to cause the failure of the life support device or system, or to affect its safety or effectiveness.

National 54LS00/DM54LS00/DM74LS00

Quad 2-Input NAND Gates

- This device contains four independent gates each of which performs the logic NAND function.
- Alternate military/aerospace device (54LS00) is available.

ABSOLUTE MAXIMUM RATINGS (Note)

If Military/Aerospace specified devices are required, please contact the National Semiconductor Sales Office/Distributors for availability and specifications.

Supply Voltage	7V
Input Voltage	7V
Operating Free Air Temperature Range	
DM54LS and 54LS	−55°C to +125°C
DM74LS	0°C to +70°C
Storage Temperature Range	−65°C to +150°C

Note: *The "Absolute Maximum Ratings" are those values beyond which the safety of the device cannot be guaranteed. The device should not be operated at these limits. The parametric values defined in the "Electrical Characteristics" table are not guaranteed at the absolute maximum ratings. The "Recommended Operating Conditions" table will define the conditions for actual device operation.*

RECOMMENDED OPERATING CONDITIONS

Symbol	Parameter	DM54LS00			DM74LS00			Units
		Min	Nom	Max	Min	Nom	Max	
V_{CC}	Supply Voltage	4.5	5	5.5	4.75	5	5.25	V
V_{IH}	High Level Input Voltage	2			2			V
V_{IL}	Low Level Input Voltage			0.7			0.8	V
I_{OH}	High Level Output Current			−0.4			−0.4	mA
I_{OL}	Low Level Output Current			4			8	mA
T_A	Free Air Operating Temperature	−55		125	0		70	°C

ELECTRICAL CHARACTERISTICS over recommended operating free air temperature range (unless otherwise noted)

Symbol	Parameter	Conditions		Min	Typ (Note 1)	Max	Units
V_I	Input Clamp Voltage	V_{CC} = Min, I_I = −18 mA				−1.5	V
V_{OH}	High Level Output Voltage	V_{CC} = Min, I_{OH} = Max, V_{IL} = Max	DM54	2.5	3.4		V
			DM74	2.7	3.4		
V_{OL}	Low Level Output Voltage	V_{CC} = Min, I_{OL} = Max, V_{IH} = Min	DM54		0.25	0.4	V
			DM74		0.35	0.5	
		I_{OL} = 4 mA, V_{CC} = Min	DM74		0.25	0.4	
I_I	Input Current @ Max Input Voltage	V_{CC} = Max, V_I = 7V				0.1	mA
I_{IH}	High Level Input Current	V_{CC} = Max, V_I = 2.7V				20	μA
I_{IL}	Low Level Input Current	V_{CC} = Max, V_I = 0.4V				−0.36	mA
I_{OS}	Short Circuit Output Current	V_{CC} = Max (Note 2)	DM54	−20		−100	mA
			DM74	−20		−100	
I_{CCH}	Supply Current with Outputs High	V_{CC} = Max			0.8	1.6	mA
I_{CCL}	Supply Current with Outputs Low	V_{CC} = Max			2.4	4.4	mA

SWITCHING CHARACTERISTICS at V_{CC} = 5V and T_A = 25°C (See Section 1 for Test Waveforms and Output Load)

Symbol	Parameter	R_L = 2 kΩ				Units
		C_L = 15 pF		C_L = 50 pF		
		Min	Max	Min	Max	
t_{PLH}	Propagation Delay Time Low to High Level Output	3	10	4	15	ns
t_{PHL}	Propagation Delay Time High to Low Level Output	3	10	4	15	ns

Note 1: All typicals are at V_{CC} = 5V, T_A = 25°C.

Note 2: Not more than one output should be shorted at a time, and the duration should not exceed one second.

CONNECTION DIAGRAM

Dual-In-Line Package

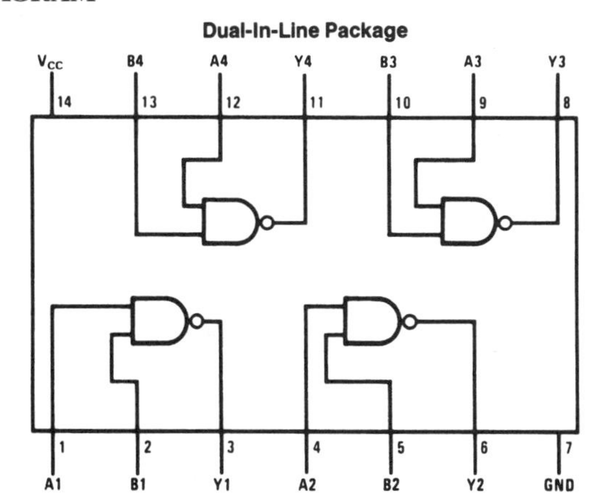

Order Number 54LS00DMQB, 54LS00FMQB, 54LS00LMQB, DM54LS00J, DM54LS00W, DM74LS00M or DM74LS00N
See NS Package Number E20A, J14A, M14A, N14A or W14B

FUNCTION TABLE

$Y = \overline{AB}$

Inputs		Output
A	B	Y
L	L	H
L	H	H
H	L	H
H	H	L

H = High Logic Level
L = Low Logic Level

National 54LS02/DM54LS02/DM74LS02
Quad 2-Input NOR Gates

- This device contains four independent gates, each of which performs the logic NOR function.
- Alternate military/aerospace device (54LS02) is available.

ABSOLUTE MAXIMUM RATINGS (Note)

If Military/Aerospace specified devices are required, please contact the National Semiconductor Sales Office/Distributors for availability and specifications.

Supply Voltage	7V
Input Voltage	7V
Operating Free Air Temperature Range	
DM54LS and 54LS	−55°C to +125°C
DM74LS	0°C to +70°C
Storage Temperature Range	−65°C to +150°C

Note: *The "Absolute Maximum Ratings" are those values beyond which the safety of the device cannot be guaranteed. The device should not be operated at these limits. The parametric values defined in the "Electrical Characteristics" table are not guaranteed at the absolute maximum ratings. The "Recommended Operating Conditions" table will define the conditions for actual device operation.*

RECOMMENDED OPERATING CONDITIONS

Symbol	Parameter	DM54LS02			DM74LS02			Units
		Min	Nom	Max	Min	Nom	Max	
V_{CC}	Supply Voltage	4.5	5	5.5	4.75	5	5.25	V
V_{IH}	High Level Input Voltage	2			2			V
V_{IL}	Low Level Input Voltage			0.7			0.8	V
I_{OH}	High Level Output Current			−0.4			−0.4	mA
I_{OL}	Low Level Output Current			4			8	mA
T_A	Free Air Operating Temperature	−55		125	0		70	°C

ELECTRICAL CHARACTERISTICS over recommended operating free air temperature range (unless otherwise noted)

Symbol	Parameter	Conditions		Min	Typ (Note 1)	Max	Units
V_I	Input Clamp Voltage	V_{CC} = Min, I_I = −18 mA				−1.5	V
V_{OH}	High Level Output Voltage	V_{CC} = Min, I_{OH} = Max, V_{IL} = Max	DM54	2.5	3.4		V
			DM74	2.7	3.4		
V_{OL}	Low Level Output Voltage	V_{CC} = Min, I_{OL} = Max, V_{IH} = Min	DM54		0.25	0.4	V
			DM74		0.35	0.5	
		I_{OL} = 4 mA, V_{CC} = Min	DM74		0.25	0.4	
I_I	Input Current @ Max Input Voltage	V_{CC} = Max, V_I = 7V				0.1	mA
I_{IH}	High Level Input Current	V_{CC} = Max, V_I = 2.7V				20	μA
I_{IL}	Low Level Input Current	V_{CC} = Max, V_I = 0.4V				−0.40	mA
I_{OS}	Short Circuit Output Current	V_{CC} = Max (Note 2)	DM54	−20		−100	mA
			DM74	−20		−100	
I_{CCH}	Supply Current with Outputs High	V_{CC} = Max			1.6	3.2	mA
I_{CCL}	Supply Current with Outputs Low	V_{CC} = Max			2.8	5.4	mA

SWITCHING CHARACTERISTICS at V_{CC} = 5V and T_A = 25°C (See Section 1 for Test Waveforms and Output Load)

Symbol	Parameter	R_L = 2 kΩ				Units
		C_L = 15 pF		C_L = 50 pF		
		Min	Max	Min	Max	
t_{PLH}	Propagation Delay Time Low to High Level Output		13		18	ns
t_{PHL}	Propagation Delay Time High to Low Level Output		10		15	ns

Note 1: All typicals are at V_{CC} = 5V, T_A = 25°C.

Note 2: Not more than one output should be shorted at a time, and the duration should not exceed one second.

CONNECTION DIAGRAM

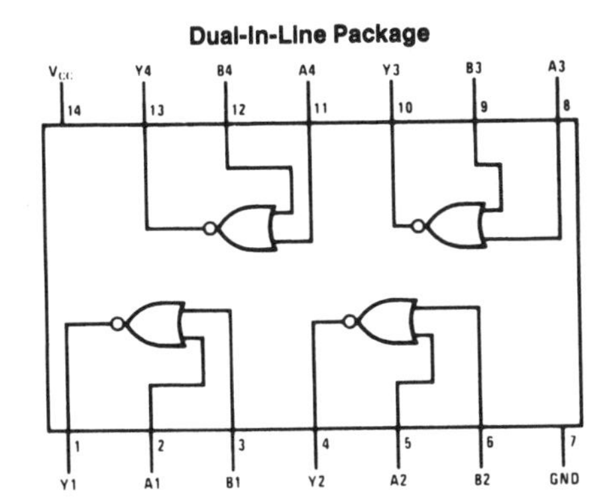

Order Number 54LS02DMQB, 54LS02FMQB, 54LS02LMQB, DM54LS02J, DM54LS02W, DM74LS02M or DM74LS02N See NS Package Number E20A, J14A, M14A, N14A or W14B

FUNCTION TABLE

$Y = \overline{A + B}$

Inputs		Output
A	B	Y
L	L	H
L	H	L
H	L	L
H	H	L

H = High Logic Level
L = Low Logic Level

National 54LS04/DM54LS04/DM74LS04
Hex Inverting Gates

- This device contains six independent gates each of which performs the logic invert function.
- Alternate military/aerospace device (54LS04) is available.

ABSOLUTE MAXIMUM RATINGS (Note)

If Military/Aerospace specified devices are required, please contact the National Semiconductor Sales Office/Distributors for availability and specifications.

Supply Voltage	7V
Input Voltage	7V
Operating Free Air Temperature Range	
DM54LS and 54LS	−55°C to +125°C
DM74LS	0°C to +70°C
Storage Temperature Range	−65°C to +150°C

Note: *The "Absolute Maximum Ratings" are those values beyond which the safety of the device cannot be guaranteed. The device should not be operated at these limits. The parametric values defined in the "Electrical Characteristics" table are not guaranteed at the absolute maximum ratings. The "Recommended Operating Conditions" table will define the conditions for actual device operation.*

RECOMMENDED OPERATING CONDITIONS (Note)

Symbol	Parameter	DM54LS04			DM74LS04			Units
		Min	Nom	Max	Min	Nom	Max	
V_{CC}	Supply Voltage	4.5	5	5.5	4.75	5	5.25	V
V_{IH}	High Level Input Voltage	2			2			V
V_{IL}	Low Level Input Voltage			0.7			0.8	V
I_{OH}	High Level Output Current			−0.4			−0.4	mA
I_{OL}	Low Level Output Current			4			8	mA
T_A	Free Air Operating Temperature	−55		125	0		70	°C

SWITCHING CHARACTERISTICS at V_{CC} = 5V and T_A = 25°C (See Section 1 for Test Waveforms and Output Load)

Symbol	Parameter	$R_L = 2\ k\Omega$				Units
		C_L = 15 pF		C_L = 50 pF		
		Min	Max	Min	Max	
t_{PLH}	Propagation Delay Time Low to High Level Output	3	10	4	15	ns
t_{PHL}	Propagation Delay Time High to Low Level Output	3	10	4	15	ns

Note 1: All typicals are at V_{CC} = 5V, T_A = 25°C.

Note 2: Not more than one output should be shorted at a time, and the duration should not exceed one second.

ELECTRICAL CHARACTERISTICS over recommended operating free air temperature range (unless otherwise noted)

Symbol	Parameter	Conditions		Min	Typ (Note 1)	Max	Units
V_I	Input Clamp Voltage	V_{CC} = Min, I_I = −18 mA				−1.5	V
V_{OH}	High Level Output Voltage	V_{CC} = Min, I_{OH} = Max, V_{IL} = Max	DM54	2.5	3.4		V
			DM74	2.7	3.4		
V_{OL}	Low Level Output Voltage	V_{CC} = Min, I_{OL} = Max, V_{IH} = Min	DM54		0.25	0.4	V
			DM74		0.35	0.5	
		I_{OL} = 4 mA, V_{CC} = Min	DM74		0.25	0.4	
I_I	Input Current @ Max Input Voltage	V_{CC} = Max, V_I = 7V				0.1	mA
I_{IH}	High Level Input Current	V_{CC} = Max, V_I = 2.7V				20	μA
I_{IL}	Low Level Input Current	V_{CC} = Max, V_I = 0.4V				−0.36	mA
I_{OS}	Short Circuit Output Current	V_{CC} = Max (Note 2)	DM54	−20		−100	mA
			DM74	−20		−100	
I_{CCH}	Supply Current with Outputs High	V_{CC} = Max			1.2	2.4	mA
I_{CCL}	Supply Current with Outputs Low	V_{CC} = Max			3.6	6.6	mA

CONNECTION DIAGRAM

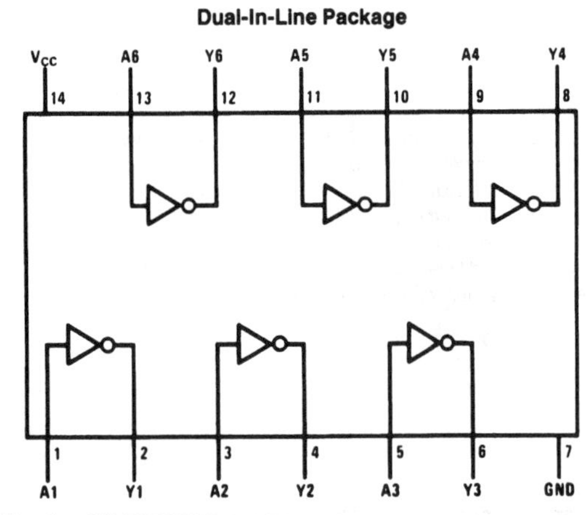

Order Number 54LS04DMQB, 54LS04FMQB, 54LS04LMQB, DM54LS04J, DM54LS04W, DM74LS04M or DM74LS04N See NS Package Number E20A, J14A, M14A, N14A or W14B

FUNCTION TABLE

$Y = \overline{A}$

Input	Output
A	Y
L	H
H	L

H = High Logic Level
L = Low Logic Level

National 54LS20/DM54LS20/DM74LS20
Dual 4-Input NAND Gates

- This device contains two independent gates each of which performs the logic NAND function.
- Alternate military/aerospace device (54LS20) is available.

ABSOLUTE MAXIMUM RATINGS (Note)

If Military/Aerospace specified devices are required, please contact the National Semiconductor Sales Office/Distributors for availability and specifications.

Supply Voltage	7V
Input Voltage	7V
Operating Free Air Temperature Range	
DM54LS and 54LS	−55°C to +125°C
DM74LS	0°C to +70°C
Storage Temperature Range	−65°C to +150°C

Note: *The "Absolute Maximum Ratings" are those values beyond which the safety of the device cannot be guaranteed. The device should not be operated at these limits. The parametric values defined in the "Electrical Characteristics" table are not guaranteed at the absolute maximum ratings. The "Recommended Operating Conditions" table will define the conditions for actual device operation.*

RECOMMENDED OPERATING CONDITIONS

Symbol	Parameter	DM54LS20			DM74LS20			Units
		Min	Nom	Max	Min	Nom	Max	
V_{CC}	Supply Voltage	4.5	5	5.5	4.75	5	5.25	V
V_{IH}	High Level Input Voltage	2			2			V
V_{IL}	Low Level Input Voltage			0.7			0.8	V
I_{OH}	High Level Output Current			−0.4			−0.4	mA
I_{OL}	Low Level Output Current			4			8	mA
T_A	Free Air Operating Temperature	−55		125	0		70	°C

ELECTRICAL CHARACTERISTICS over recommended operating free air temperature range (unless otherwise noted)

Symbol	Parameter	Conditions		Min	Typ (Note 1)	Max	Units
V_I	Input Clamp Voltage	V_{CC} = Min, I_I = −18 mA				−1.5	V
V_{OH}	High Level Output Voltage	V_{CC} = Min, I_{OH} = Max, V_{IL} = Max	DM54	2.5	3.4		V
			DM74	2.7	3.4		
V_{OL}	Low Level Output Voltage	V_{CC} = Min, I_{OL} = Max, V_{IH} = Min	DM54		0.25	0.4	V
			DM74		0.35	0.5	
		I_{OL} = 4 mA, V_{CC} = Min	DM74		0.25	0.4	
I_I	Input Current @ Max Input Voltage	V_{CC} = Max, V_I = 7V				0.1	mA
I_{IH}	High Level Input Current	V_{CC} = Max, V_I = 2.7V				20	μA
I_{IL}	Low Level Input Current	V_{CC} = Max, V_I = 0.4V				−0.36	mA
I_{OS}	Short Circuit Output Current	V_{CC} = Max (Note 2)	DM54	−20		−100	mA
			DM74	−20		−100	
I_{CCH}	Supply Current with Outputs High	V_{CC} = Max			0.4	0.8	mA
I_{CCL}	Supply Current with Outputs Low	V_{CC} = Max			1.2	2.2	mA

SWITCHING CHARACTERISTICS at V_{CC} = 5V and T_A = 25°C (See Section 1 for Test Waveforms and Output Load)

Symbol	Parameter	R_L = 2 kΩ				Units
		C_L = 15 pF		C_L = 50 pF		
		Min	Max	Min	Max	
t_{PLH}	Propagation Delay Time Low to High Level Output	3	10	4	15	ns
t_{PHL}	Propagation Delay Time High to Low Level Output	3	10	4	15	ns

Note 1: All typicals are at V_{CC} = 5V, T_A = 25°C.

Note 2: Not more than one output should be shorted at a time, and the duration should not exceed one second.

CONNECTION DIAGRAM

Dual-In-Line Package

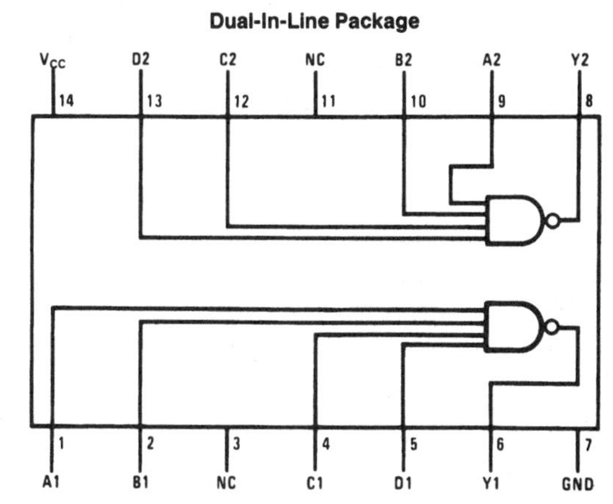

Order Number 54LS20DMQB, 54LS20FMQB, 54LS20LMQB, DM54LS20J, DM54LS20W, DM74LS20M or DM74LS20N
See NS Package Number E20A, J14A, M14A, N14A or W14B

FUNCTION TABLE

Y = $\overline{ABCD}$

Inputs				Output
A	B	C	D	Y
X	X	X	L	H
X	X	L	X	H
X	L	X	X	H
L	X	X	X	H
H	H	H	H	L

H = High Logic Level
L = Low Logic Level
X = Either Low or High Logic Level

National 54LS21/DM54LS21/DM74LS21

Dual 4-Input AND Gates

- This device contains two independent gates each of which performs the logic AND function.
- Alternate military/aerospace device (54LS21) is available.

ABSOLUTE MAXIMUM RATINGS (Note)

If Military/Aerospace specified devices are required, please contact the National Semiconductor Sales Office/Distributors for availability and specifications.

Supply Voltage	7V
Input Voltage	7V
Operating Free Air Temperature Range	
DM54LS and 54LS	−55°C to +125°C
DM74LS	0°C to +70°C
Storage Temperature Range	−65°C to +150°C

Note: *The "Absolute Maximum Ratings" are those values beyond which the safety of the device cannot be guaranteed. The device should not be operated at these limits. The parametric values defined in the "Electrical Characteristics" table are not guaranteed at the absolute maximum ratings. The "Recommended Operating Conditions" table will define the conditions for actual device operation.*

RECOMMENDED OPERATING CONDITIONS

Symbol	Parameter	DM54LS21			DM74LS21			Units
		Min	Nom	Max	Min	Nom	Max	
V_{CC}	Supply Voltage	4.5	5	5.5	4.75	5	5.25	V
V_{IH}	High Level Input Voltage	2			2			V
V_{IL}	Low Level Input Voltage			0.7			0.8	V
I_{OH}	High Level Output Current			−0.4			−0.4	mA
I_{OL}	Low Level Output Current			4			8	mA
T_A	Free Air Operating Temperature	−55		125	0		70	°C

ELECTRICAL CHARACTERISTICS over recommended operating free air temperature range (unless otherwise noted)

Symbol	Parameter	Conditions		Min	Typ (Note 1)	Max	Units
V_I	Input Clamp Voltage	V_{CC} = Min, I_I = −18 mA				−1.5	V
V_{OH}	High Level Output Voltage	V_{CC} = Min, I_{OH} = Max, V_{IH} = Min	DM54	2.5	3.4		V
			DM74	2.7	3.4		
V_{OL}	Low Level Output Voltage	V_{CC} = Min, I_{OL} = Max, V_{IL} = Max	DM54		0.25	0.4	V
			DM74		0.35	0.5	
		I_{OL} = 4 mA, V_{CC} = Min	DM74		0.25	0.4	
I_I	Input Current @ Max Input Voltage	V_{CC} = Max, V_I = 7V				0.1	mA
I_{IH}	High Level Input Current	V_{CC} = Max, V_I = 2.7V				20	μA
I_{IL}	Low Level Input Current	V_{CC} = Max, V_I = 0.4V				−0.36	mA
I_{OS}	Short Circuit Output Current	V_{CC} = Max (Note 2)	DM54	−20		−100	mA
			DM74	−20		−100	
I_{CCH}	Supply Current with Outputs High	V_{CC} = Max			1.2	2.4	mA
I_{CCL}	Supply Current with Outputs Low	V_{CC} = Max			2.2	4.4	mA

SWITCHING CHARACTERISTICS at V_{CC} = 5V and T_A = 25°C (See Section 1 for Test Waveforms and Output Load)

Symbol	Parameter	R_L = 2 kΩ				Units
		C_L = 15 pF		C_L = 50 pF		
		Min	Max	Min	Max	
t_{PLH}	Propagation Delay Time Low to High Level Output	4	13	6	18	ns
t_{PHL}	Propagation Delay Time High to Low Level Output	3	11	5	18	ns

Note 1: All typicals are at V_{CC} = 5V, T_A = 25°C.

Note 2: Not more than one output should be shorted at a time, and the duration should not exceed one second.

CONNECTION DIAGRAM

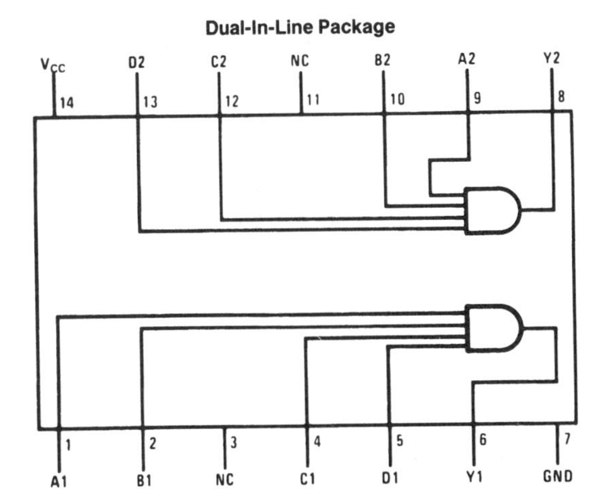

Order Number 54LS21DMQB, 54LS21FMQB, 54LS21LMQB, DM54LS21J, DM54LS21W, DM74LS21M or DM74LS21N
See NS Package Number E20A, J14A, M14A, N14A or W14B

FUNCTION TABLE

Y = ABCD

Inputs				Output
A	**B**	**C**	**D**	**Y**
X	X	X	L	L
X	X	L	X	L
X	L	X	X	L
L	X	X	X	L
H	H	H	H	H

H = High Logic Level
L = Low Logic Level
X = Either Low or High Logic Level

National 54LS22/DM74LS22
Dual 4-Input NAND Gate with Open-Collector Output

ABSOLUTE MAXIMUM RATINGS (Note)

If Military/Aerospace specified devices are required, please contact the National Semiconductor Sales Office/Distributors for availability and specifications.

Supply Voltage	7V
Input Voltage	7V
Operating Free Air Temperature Range	
54LS	−55°C to +125°C
DM74LS	0°C to +70°C
Storage Temperature Range	−65°C to +150°C

Note: *The "Absolute Maximum Ratings" are those values beyond which the safety of the device cannot be guaranteed. The device should not be operated at these limits. The parametric values defined in the "Electrical Characteristics" table are not guaranteed at the absolute maximum ratings. The "Recommended Operating Conditions" table will define the conditions for actual device operation.*

ELECTRICAL CHARACTERISTICS over recommended operating free air temperature range (unless otherwise noted)

Symbol	Parameter	Conditions		Min	Typ (Note 1)	Max	Units
V_I	Input Clamp Voltage	V_{CC} = Min, I_I = −18 mA				−1.5	V
I_{CEX}	High Level Output Current	V_{CC} = Min, V_O = 5.5V, V_{IL} = Max				100	μA
V_{OL}	Low Level Output Voltage	V_{CC} = Min, I_{OL} = Max, V_{IH} = Min	54LS			0.4	V
			DM74			0.5	
		I_{OL} = 4 mA, V_{CC} = Min	DM74			0.4	
I_I	Input Current @ Max Input Voltage	V_{CC} = Max, V_I = 5.5V				0.1	mA
I_{IH}	High Level Input Current	V_{CC} = Max, V_I = 2.7V				20	μA
I_{IL}	Low Level Input Current	V_{CC} = Max, V_I = 0.4V				−0.4	mA
I_{CCH}	Supply Current Outputs High	V_{CC} = Max, V_{IN} = GND				0.8	mA
I_{CCL}	Supply Current Outputs Low	V_{CC} = Max, V_{IN} = Open				2.2	mA

Note 1: All typicals are at V_{CC} = 5V, T_A = 25°C.

RECOMMENDED OPERATING CONDITIONS

Symbol	Parameter	54LS22			DM74LS22			Units
		Min	Nom	Max	Min	Nom	Max	
V_{CC}	Supply Voltage	4.5	5	5.5	4.75	5	5.25	V
V_{IH}	High Level Input Voltage	2			2			V
V_{IL}	Low Level Input Voltage			0.7			0.8	V
V_{OH}	High Level Output Voltage			5.5			5.5	mA
I_{OL}	Low Level Output Current			4			8	mA
T_A	Free Air Operating Temperature	−55		125	0		70	°C

SWITCHING CHARACTERISTICS

at V_{CC} = +5.0V, T_A = +25°C (See Section 1 for test waveforms and output load)

Symbol	Parameter	$R_L = 2\ k\Omega, C_L = 15\ pF$		Units
		Min	Max	
t_{PLH}	Propagation Delay Time Low to High Level Output		22	ns
t_{PHL}	Propagation Delay Time High to Low Level Output		24	ns

CONNECTION DIAGRAM

Dual-In-Line Package

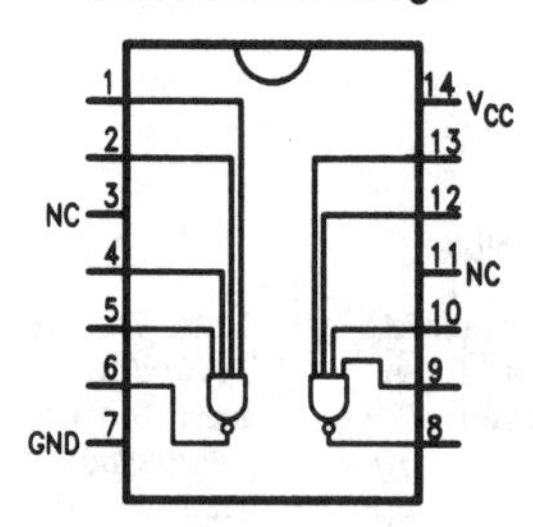

Order Number 54LS22DMQB, 54LS22FMQB, DM74LS22M or DM74LS22N
See NS Package Number J14A, M14A, N14A or W14B

National 54LS26/DM74LS26 Quad 2-Input NAND Gates with High-Voltage Open-Collector Outputs

ABSOLUTE MAXIMUM RATINGS (Note)

If Military/Aerospace specified devices are required, please contact the National Semiconductor Sales Office/Distributors for availability and specifications.

Supply Voltage	7V
Input Voltage	7V
Output Voltage	15V
Operating Free Air Temperature Range	
54LS	−55°C to +125°C
DM74LS	0°C to +70°C
Storage Temperature Range	−65°C to +150°C

PULL-UP RESISTOR EQUATIONS

$$R_{MAX} = \frac{V_O\,(Min) - V_{OH}}{N_1\,(I_{OH}) + N_2\,(I_{IH})}$$

$$R_{MIN} = \frac{V_O\,(Max) - V_{OL}}{I_{OL} - N_3\,(I_{IL})}$$

Where: $N_1\,(I_{OH})$ = total maximum output high current for all outputs tied to pull-up resistor

$N_2\,(I_{IH})$ = total maximum input high current for all inputs tied to pull-up resistor

$N_3\,(I_{IL})$ = total maximum input low current for all inputs tied to pull-up resistor

Note: *The "Absolute Maximum Ratings" are those values beyond which the safety of the device cannot be guaranteed. The device should not be operated at these limits. The parametric values defined in the "Electrical Characteristics" table are not guaranteed at the absolute maximum ratings. The "Recommended Operating Conditions" table will define the conditions for actual device operation.*

RECOMMENDED OPERATING CONDITIONS

Symbol	Parameter	54LS26			DM74LS26			Units
		Min	Nom	Max	Min	Nom	Max	
V_{CC}	Supply Voltage	4.5	5	5.5	4.75	5	5.25	V
V_{IH}	High Level Input Voltage	2			2			V
V_{IL}	Low Level Input Voltage			0.7			0.8	V
V_{OH}	High Level Output Voltage			15			15	V
I_{OL}	Low Level Output Current			4			8	mA
T_A	Free Air Operating Temperature	−55		125	0		70	°C

ELECTRICAL CHARACTERISTICS over recommended operating free air temperature range (unless otherwise noted)

Symbol	Parameter	Conditions		Min	Typ (Note 1)	Max	Units
V_I	Input Clamp Voltage	V_{CC} = Min, I_I = −18 mA				−1.5	V
I_{CEX}	High Level Output Current	V_{CC} = Min, V_{IL} = Max	V_O = 15V			1000	μA
			V_O = 12V			50	
V_{OL}	Low Level Output Voltage	V_{CC} = Min, I_{OL} = Max, V_{IH} = Min	54LS			0.4	V
			DM74		0.35	0.5	
		I_{OL} = 4 mA, V_{CC} = Min	DM74		0.25	0.4	
I_I	Input Current @ Max Input Voltage	V_{CC} = Max, V_I = 5.5V				0.1	mA
I_{IH}	High Level Input Current	V_{CC} = Max, V_I = 2.7V				20	μA
I_{IL}	Low Level Input Current	V_{CC} = Max, V_I = 0.4V	54LS			−0.40	mA
			DM74			−0.36	
I_{CCH}	Supply Current with Outputs High	V_{CC} = Max			0.8	1.6	mA
I_{CCL}	Supply Current with Outputs Low	V_{CC} = Max			2.4	4.4	mA

SWITCHING CHARACTERISTICS at V_{CC} = 5V and T_A = 25°C (See Section 1 for Test Waveforms and Output Load)

Symbol	Parameter	R_L = 2 kΩ, C_L = 15 pF Max 54LS	R_L = 2 kΩ, C_L = 15 pF Max DM74	Units
t_{PLH}	Propagation Delay Time Low to High Level Output	27	32	ns
t_{PHL}	Propagation Delay Time High to Low Level Output	18	28	ns

Note 1: All typicals are at V_{CC} = 5V, T_A = 25°C.

CONNECTION DIAGRAM

Dual-In-Line Package

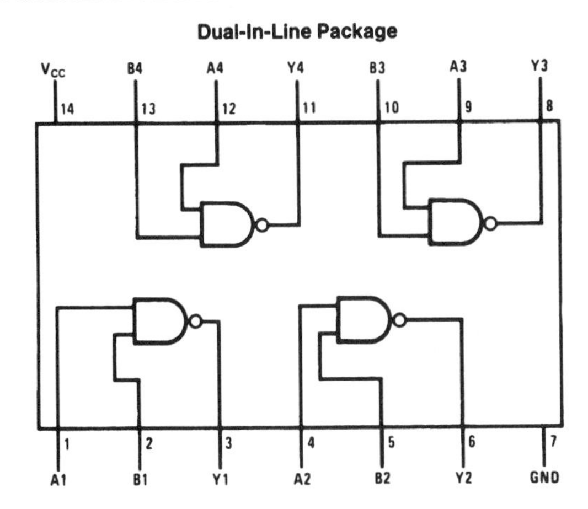

Order Number 54LS26DMQB, 54LS26FMQB, DM74LS26M or DM74LS26N
See NS Package Number J14A, M14A, N14A or W14B

FUNCTION TABLE

$Y = \overline{AB}$

Inputs A	Inputs B	Output Y
L	L	H
L	H	H
H	L	H
H	H	L

H = High Logic Level
L = Low Logic Level

National 54LS27/DM54LS27/DM74LS27
Triple 3-Input NOR Gates

- This device contains three independent gates each of which performs the logic NOR function.
- Alternate Military/aerospace device (54LS27) is available.

ABSOLUTE MAXIMUM RATINGS (Note)

If Military/Aerospace specified devices are required, please contact the National Semiconductor Sales Office/Distributors for availability and specifications.

Supply Voltage	7V
Input Voltage	7V
Operating Free Air Temperature Range	
DM54LS and 54LS	−55°C to +125°C
DM74LS	0°C to +70°C
Storage Temperature Range	−65°C to +150°C

Note: *The "Absolute Maximum Ratings" are those values beyond which the safety of the device cannot be guaranteed. The device should not be operated at these limits. The parametric values defined in the "Electrical Characteristics" table are not guaranteed at the absolute maximum ratings. The "Recommended Operating Conditions" table will define the conditions for actual device operation.*

RECOMMENDED OPERATING CONDITIONS

Symbol	Parameter	DM54LS27			DM74LS27			Units
		Min	Nom	Max	Min	Nom	Max	
V_{CC}	Supply Voltage	4.5	5	5.5	4.75	5	5.25	V
V_{IH}	High Level Input Voltage	2			2			V
V_{IL}	Low Level Input Voltage			0.7			0.8	V
I_{OH}	High Level Output Current			−0.4			−0.4	mA
I_{OL}	Low Level Output Current			4			8	mA
T_A	Free Air Operating Temperature	−55		125	0		70	°C

ELECTRICAL CHARACTERISTICS over recommended operating free air temperature range (unless otherwise noted)

Symbol	Parameter	Conditions		Min	Typ (Note 1)	Max	Units
V_I	Input Clamp Voltage	V_{CC} = Min, I_I = −18 mA				−1.5	V
V_{OH}	High Level Output Voltage	V_{CC} = Min, I_{OH} = Max, V_{IL} = Max	DM54	2.5	3.4		V
			DM74	2.7	3.4		
V_{OL}	Low Level Output Voltage	V_{CC} = Min, I_{OL} = Max, V_{IH} = Min	DM54		0.25	0.4	V
			DM74		0.35	0.5	
		I_{OL} = 4 mA, V_{CC} = Min	DM74		0.25	0.4	
I_I	Input Current @ Max Input Voltage	V_{CC} = Max, V_I = 7V				0.1	mA
I_{IH}	High Level Input Current	V_{CC} = Max, V_I = 2.7V				20	μA
I_{IL}	Low Level Input Current	V_{CC} = Max, V_I = 0.4V				−0.36	mA
I_{OS}	Short Circuit Output Current	V_{CC} = Max (Note 2)	DM54	−20		−100	mA
			DM74	−20		−100	
I_{CCH}	Supply Current with Outputs High	V_{CC} = Max			2	4	mA
I_{CCL}	Supply Current with Outputs Low	V_{CC} = Max			3.4	6.8	mA

SWITCHING CHARACTERISTICS at V_{CC} = 5V and T_A = 25°C (See Section 1 for Test Waveforms and Output Load)

Symbol	Parameter	R_L = 2 kΩ				Units
		C_L = 15 pF		C_L = 50 pF		
		Min	Max	Min	Max	
t_{PLH}	Propagation Delay Time Low to High Level Output	3	13	5	18	ns
t_{PHL}	Propagation Delay Time High to Low Level Output	3	10	4	15	ns

Note 1: All typicals are at V_{CC} = 5V, T_A = 25°C.

Note 2: Not more than one output should be shorted at a time, and the duration should not exceed one second.

CONNECTION DIAGRAM

Dual-In-Line Package

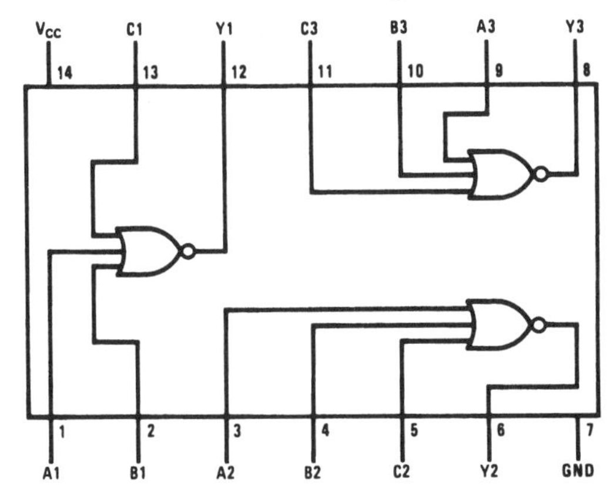

Order Number 54LS27DMQB, 54LS27FMQB, 54LS27LMQB, DM54LS27J, DM54LS27W, DM74LS27M or DM74LS27N
See NS Package Number E20A, J14A, M14A, N14A or W14B

FUNCTION TABLE

$Y = \overline{A + B}$

Inputs		Output
A	**B**	**Y**
L	L	H
L	H	L
H	L	L
H	H	L

H = High Logic Level
L = Low Logic Level

National 54LS28/DM74LS28 Quad 2-Input NOR Buffer

ABSOLUTE MAXIMUM RATINGS (Note)

If Military/Aerospace specified devices are required, please contact the National Semiconductor Sales Office/Distributors for availability and specifications.

Supply Voltage	7V
Input Voltage	7V
Operating Free Air Temperature Range	
54LS	−55°C to +125°C
DM74LS	0°C to +70°C
Storage Temperature Range	−65°C to +150°C

Note: *The "Absolute Maximum Ratings" are those values beyond which the safety of the device cannot be guaranteed. The device should not be operated at these limits. The parametric values defined in the "Electrical Characteristics" table are not guaranteed at the absolute maximum ratings. The "Recommended Operating Conditions" table will define the conditions for actual device operation.*

ELECTRICAL CHARACTERISTICS over recommended operating free air temperature range (unless otherwise noted)

Symbol	Parameter	Conditions		Min	Typ (Note 1)	Max	Units
V_I	Input Clamp Voltage	V_{CC} = Min, I_I = −18 mA				−1.5	V
V_{OH}	High Level Output Voltage	V_{CC} = Min, I_{OH} = Max, V_{IL} = Max	54LS	2.5			V
			DM74	2.7			
V_{OL}	Low Level Output Voltage	V_{CC} = Min, I_{OL} = Max, V_{IH} = Min	54LS			0.4	V
			DM74			0.5	
		I_{OL} = 12 mA, V_{CC} = Min	DM74			0.4	
I_I	Input Current @ Max Input Voltage	V_{CC} = Max, V_I = 10V				0.1	mA
I_{IH}	High Level Input Current	V_{CC} = Max, V_I = 2.7V				20	μA
I_{IL}	Low Level Input Current	V_{CC} = Max, V_I = 0.4V				−0.4	mA
I_{OS}	Short Circuit Output Current	V_{CC} = Max (Note 2)	54LS	−30		−130	mA
			DM74	−30		−130	
I_{CCH}	Supply Current with Outputs High	V_{CC} = Max				3.6	mA
I_{CCL}	Supply Current with Outputs Low	V_{CC} = Max				13.8	mA

Note 1: All typicals are at V_{CC} = 5V, T_A = 25°C.

Note 2: Not more than one output should be shorted at a time, and the duration should not exceed one second.

RECOMMENDED OPERATING CONDITIONS

Symbol	Parameter	54LS28			DM74LS28			Units
		Min	Nom	Max	Min	Nom	Max	
V_{CC}	Supply Voltage	4.5	5	5.5	4.75	5	5.25	V
V_{IH}	High Level Input Voltage	2			2			V
V_{IL}	Low Level Input Voltage			0.7			0.8	V
I_{OH}	High Level Output Current			−1.2			−1.2	mA
I_{OL}	Low Level Output Current			12			24	mA
T_A	Free Air Operating Temperature	−55		125	0		70	°C

SWITCHING CHARACTERISTICS

at V_{CC} = +5.0V, T_A = +25°C (See Section 1 for test waveforms and output load)

Symbol	Parameter	R_L = 2 kΩ C_L = 15 pF		Units
		Min	Max	
t_{PLH}	Propagation Delay Time Low to High Level Output		20	ns
t_{PHL}	Propagation Delay Time High to Low Level Output		20	ns

CONNECTION DIAGRAM

Dual-In-Line Package

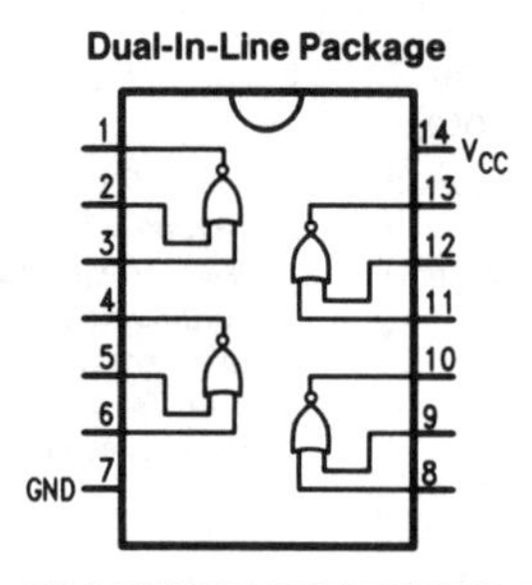

Order Number 54LS28DMQB, 54LS28FMQB, 54LS28LMQB, DM74LS28M or DM74LS28N
See NS Package Number E20A, J14A, M14A, N14A or W14B

National 54LS33/DM74LS33 Quad 2-Input NOR Buffer with Open-Collector Outputs

ABSOLUTE MAXIMUM RATINGS (Note)

If Military/Aerospace specified devices are required, please contact the National Semiconductor Sales Office/Distributors for availability and specifications.

Supply Voltage	7V
Input Voltage	7V
Output Voltage	7V
Operating Free Air Temperature Range	
54LS	−55°C to +125°C
DM74LS	0°C to +70°C
Storage Temperature Range	−65°C to +150°C

Note: *The "Absolute Maximum Ratings" are those values beyond which the safety of the device cannot be guaranteed. The device should not be operated at these limits. The parametric values defined in the "Electrical Characteristics" table are not guaranteed at the absolute maximum ratings. The "Recommended Operating Conditions" table will define the conditions for actual device operation.*

RECOMMENDED OPERATING CONDITIONS

Symbol	Parameter	54LS33			DM74LS33			Units
		Min	Nom	Max	Min	Nom	Max	
V_{CC}	Supply Voltage	4.5	5	5.5	4.75	5	5.25	V
V_{IH}	High Level Input Voltage	2			2			V
V_{IL}	Low Level Input Voltage			0.7			0.8	V
V_{OH}	High Level Output Voltage			5.5			5.5	V
I_{OL}	Low Level Output Current			12			24	mA
T_A	Free Air Operating Temperature	−55		125	0		70	°C

ELECTRICAL CHARACTERISTICS over recommended operating free air temperature range (unless otherwise noted)

Symbol	Parameter	Conditions		Min	Typ (Note 1)	Max	Units
V_I	Input Clamp Voltage	V_{CC} = Min, I_I = −18 mA				−1.5	V
I_{CEX}	High Level Output Current	V_{CC} = Min, V_O = 5.5V, V_{IL} = Max				100	μA
V_{OL}	Low Level Output Voltage	V_{CC} = Min, I_{OL} = Max, V_{IH} = Min	54LS			0.4	V
			DM74			0.5	
		I_{OL} = 12 mA, V_{CC} = Min	DM74			0.4	
I_I	Input Current @ Max Input Voltage	V_{CC} = Max, V_I = 7V				0.1	mA
I_{IH}	High Level Input Current	V_{CC} = Max, V_I = 2.7V				20	μA
I_{IL}	Low Level Input Current	V_{CC} = Max, V_I = 0.4V				−0.4	mA
I_{CCH}	Supply Current with Outputs High	V_{CC} = Max, V_{IN} = GND				3.6	mA
I_{CCL}	Supply Current with Outputs Low	V_{CC} = Max, V_{IN} = Open				13.8	mA

Note 1: All typicals are at V_{CC} = 5V, T_A = 25°C.

SWITCHING CHARACTERISTICS V_{CC} = 5V and T_A = 25°C (See Section 1 for Test Waveforms and Output Load)

Symbol	Parameter	R_L = 2 kΩ, C_L = 15 pF		Units
		Min	Max	
t_{PLH}	Propagation Delay Time Low to High Level Output		22	ns
t_{PHL}	Propagation Delay Time High to Low Level Output		22	ns

CONNECTION DIAGRAM

Dual-In-Line Package

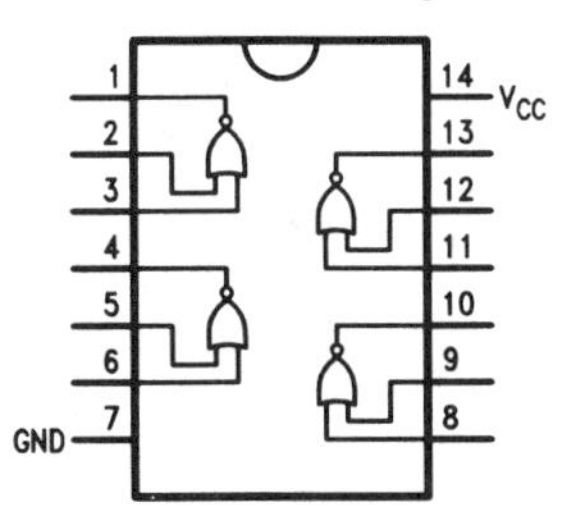

Order Number 54LS33DMQB, 54LS33FMQB, DM74LS33M or DM74LS33N
See NS Package Number J14A, M14A, N14A or W14B

National 54LS37/DM74LS37
Quad 2-Input NAND Buffers

ABSOLUTE MAXIMUM RATINGS (Note)

If Military/Aerospace specified devices are required, please contact the National Semiconductor Sales Office/Distributors for availability and specifications.

Supply Voltage	7V
Input Voltage	7V
Operating Free Air Temperature Range	
54LS	−55°C to +125°C
DM74LS	0°C to +70°C
Storage Temperature Range	−65°C to +150°C

Note: *The "Absolute Maximum Ratings" are those values beyond which the safety of the device cannot be guaranteed. The device should not be operated at these limits. The parametric values defined in the "Electrical Characteristics" table are not guaranteed at the absolute maximum ratings. The "Recommended Operating Conditions" table will define the conditions for actual device operation.*

RECOMMENDED OPERATING CONDITIONS

Symbol	Parameter	54LS37			DM74LS37			Units
		Min	Nom	Max	Min	Nom	Max	
V_{CC}	Supply Voltage	4.5	5	5.5	4.75	5	5.25	V
V_{IH}	High Level Input Voltage	2			2			V
V_{IL}	Low Level Input Voltage			0.7			0.8	V
I_{OH}	High Level Output Current			−1.2			−1.2	mA
I_{OL}	Low Level Output Current			12			24	mA
T_A	Free Air Operating Temperature	−55		125	0		70	°C

ELECTRICAL CHARACTERISTICS over recommended operating free air temperature range (unless otherwise noted)

Symbol	Parameter	Conditions		Min	Typ (Note 1)	Max	Units
V_I	Input Clamp Voltage	V_{CC} = Min, I_I = −18 mA				−1.5	V
V_{OH}	High Level Output Voltage	V_{CC} = Min, I_{OH} = Max, V_{IL} = Max	54LS	2.5			V
			DM74LS	2.7	3.4		
V_{OL}	Low Level Output Voltage	V_{CC} = Min, I_{OL} = Max, V_{IH} = Min	54LS			0.4	V
			DM74LS		0.35	0.5	
		I_{OL} = 12 mA, V_{CC} = Min			0.25	0.4	
I_I	Input Current @ Max Input Voltage	V_{CC} = Max, V_I = 10V (54LS), V_I = 7V (DM74LS)				0.1	mA
I_{IH}	High Level Input Current	V_{CC} = Max, V_I = 2.7V				20	μA
I_{IL}	Low Level Input Current	V_{CC} = Max, V_I = 0.4V	54LS			−0.40	mA
			DM74LS			−0.36	
I_{OS}	Short Circuit Output Current	V_{CC} = Max (Note 2)	54LS	−30		−130	mA
			DM74LS	−20		−100	
I_{CCH}	Supply Current with Outputs High	V_{CC} = Max			0.9	2	mA
I_{CCL}	Supply Current with Outputs Low	V_{CC} = Max			6	12	mA

SWITCHING CHARACTERISTICS at V_{CC} = 5V and T_A = 25°C (See Section 1 for Test Waveforms and Output Load)

Symbol	Parameter	54LS		DM74LS		Units
		C_L = 50 pF		C_L = 150 pF, R_L = 667Ω		
		Min	Max	Min	Max	
t_{PLH}	Propagation Delay Time Low to High Level Output		20	4	18	ns
t_{PHL}	Propagation Delay Time High to Low Level Output		20	4	21	ns

Note 1: All typicals are at V_{CC} = 5V, T_A = 25°C.

Note 2: Not more than one output should be shorted at a time, and the duration should not exceed one second.

CONNECTION DIAGRAM

Dual-In-Line Package

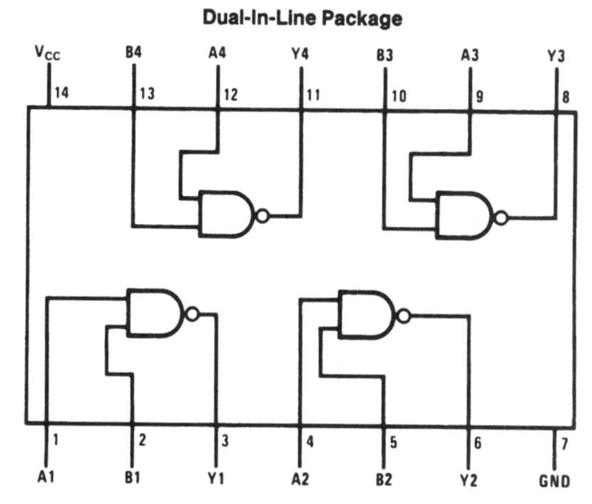

Order Number 54LS37DMQB, 54LS37FMQB, 54LS37LMQB, DM74LS37M or DM74LS37N
See NS Package Number E20A, J14A, M14A, N14A or W14B

FUNCTION TABLE

$Y = \overline{AB}$

Inputs		Output
A	B	Y
L	L	H
L	H	H
H	L	H
H	H	L

H = High Logic Level
L = Low Logic Level

National 54LS38/DM54LS38/DM74LS38
Quad 2-Input NAND Buffers with Open-Collector Outputs

- This device contains four independent gates, each of which performs the logic NAND function. The open-collector outputs requires external pull-up resistors for proper logical operation.
- Alternate Military/aerospace device (54LS388) is available.

PULL-UP RESISTOR EQUATIONS

$$R_{MAX} = \frac{V_{CC}\,(Min) - V_{OH}}{N_1\,(I_{OH}) + N_2\,(I_{IH})}$$

$$R_{MIN} = \frac{V_{CC}\,(Max) - V_{OL}}{I_{OL} - N_3\,(I_{IL})}$$

Where: N_1 (I_{OH}) = total maximum output high current for all outputs tied to pull-up resistor

N_2 (I_{IH}) = total maximum input high current for all inputs tied to pull-up resistor

N_3 (I_{IL}) = total maximum input low current for all inputs tied to pull-up resistor

ABSOLUTE MAXIMUM RATINGS (Note)

If Military/Aerospace specified devices are required, please contact the National Semiconductor Sales Office/Distributors for availability and specifications.

Supply Voltage	7V
Input Voltage	7V
Output Voltage	7V
Operating Free Air Temperature Range	
DM54LS and 54LS	−55°C to +125°C
DM74LS	0°C to +70°C
Storage Temperature Range	−65°C to +150°C

Note: *The "Absolute Maximum Ratings" are those values beyond which the safety of the device cannot be guaranteed. The device should not be operated at these limits. The parametric values defined in the "Electrical Characteristics" table are not guaranteed at the absolute maximum ratings. The "Recommended Operating Conditions" table will define the conditions for actual device operation.*

RECOMMENDED OPERATING CONDITIONS

Symbol	Parameter	DM54LS38			DM74LS38			Units
		Min	Nom	Max	Min	Nom	Max	
V_{CC}	Supply Voltage	4.5	5	5.5	4.75	5	5.25	V
V_{IH}	High Level Input Voltage	2			2			V
V_{IL}	Low Level Input Voltage			0.7			0.8	V
V_{OH}	High Level Output Voltage			5.5			5.5	V
I_{OL}	Low Level Output Current			12			24	mA
T_A	Free Air Operating Temperature	−55		125	0		70	°C

ELECTRICAL CHARACTERISTICS over recommended operating free air temperature range (unless otherwise noted)

Symbol	Parameter	Conditions		Min	Typ (Note 1)	Max	Units
V_I	Input Clamp Voltage	V_{CC} = Min, I_I = −18 mA				−1.5	V
I_{CEX}	High Level Output Current	V_{CC} = Min, V_O = 5.5V V_{IL} = Max				250	μA
V_{OL}	Low Level Output Voltage	V_{CC} = Min, I_{OL} = Max V_{IH} = Min	DM54		0.25	0.4	V
			DM74		0.35	0.5	
		I_{OL} = 12 mA, V_{CC} = Min	DM74		0.25	0.4	
I_I	Input Current @ Max Input Voltage	V_{CC} = Max, V_I = 7V				0.1	mA
I_{IH}	High Level Input Current	V_{CC} = Max, V_I = 2.7V				20	μA
I_{IL}	Low Level Input Current	V_{CC} = Max, V_I = 0.4V				−0.36	mA
I_{CCH}	Supply Current with Outputs High	V_{CC} = Max			0.9	2	mA
I_{CCL}	Supply Current with Outputs Low	V_{CC} = Max			6	12	mA

SWITCHING CHARACTERISTICS at V_{CC} = 5V and T_A = 25°C (See Section 1 for Test Waveforms and Output Load)

Symbol	Parameter	R_L = 667Ω				Units
		C_L = 45 pF		C_L = 150 pF		
		Min	Max	Min	Max	
t_{PLH}	Propagation Delay Time Low to High Level Output		22		48	ns
t_{PHL}	Propagation Delay Time High to Low Level Output		22		29	ns

Note 1: All typicals are at V_{CC} = 5V, T_A = 25°C.

CONNECTION DIAGRAM

Dual-In-Line Package

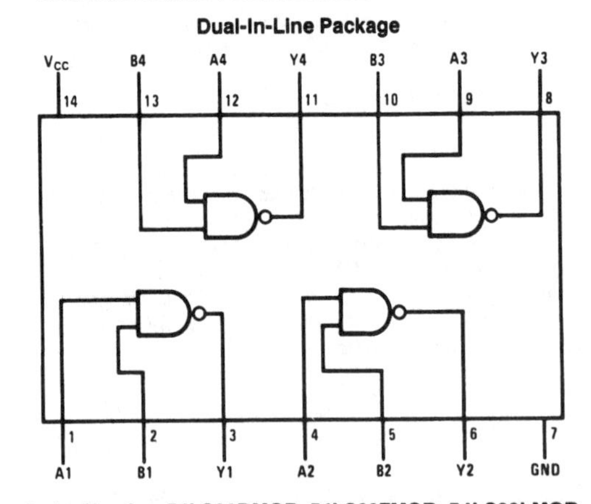

Order Number 54LS38DMQB, 54LS38FMQB, 54LS38LMQB, DM54LS38J, DM74LS38M or DM74LS38N
See NS Package Number E20A, J14A, M14A, N14A or W14B

FUNCTION TABLE

Y = $\overline{AB}$

Inputs		Output
A	B	Y
L	L	H
L	H	H
H	L	H
H	H	L

H = High Logic Level
L = Low Logic Level

National
54LS40/DM74LS40
Dual 4-Input NAND Buffer

ABSOLUTE MAXIMUM RATINGS (Note)

If Military/Aerospace specified devices are required, please contact the National Semiconductor Sales Office/Distributors for availability and specifications.

Supply Voltage	7V
Input Voltage	7V
Operating Free Air Temperature Range	
54LS	−55°C to +125°C
DM74LS	0°C to +70°C
Storage Temperature Range	−65°C to +150°C

Note: *The "Absolute Maximum Ratings" are those values beyond which the safety of the device cannot be guaranteed. The device should not be operated at these limits. The parametric values defined in the "Electrical Characteristics" table are not guaranteed at the absolute maximum ratings. The "Recommended Operating Conditions" table will define the conditions for actual device operation.*

RECOMMENDED OPERATING CONDITIONS

Symbol	Parameter	54LS40			DM74LS40			Units
		Min	Nom	Max	Min	Nom	Max	
V_{CC}	Supply Voltage	4.5	5	5.5	4.75	5	5.25	V
V_{IH}	High Level Input Voltage	2			2			V
V_{IL}	Low Level Input Voltage			0.7			0.8	V
I_{OH}	High Level Output Current			−1.2			−1.2	mA
I_{OL}	Low Level Output Current			12			24	mA
T_A	Free Air Operating Temperature	−55		125	0		70	°C

ELECTRICAL CHARACTERISTICS over recommended operating free air temperature range (unless otherwise noted)

Symbol	Parameter	Conditions		Min	Typ (Note 1)	Max	Units
V_I	Input Clamp Voltage	V_{CC} = Min, I_I = −18 mA				−1.5	V
V_{OH}	High Level Output Voltage	V_{CC} = Min, I_{OH} = Max, V_{IL} = Max	54LS	2.5			V
			DM74	2.7			
V_{OL}	Low Level Output Voltage	V_{CC} = Min, I_{OL} = Max, V_{IH} = Min	54LS			0.4	V
			DM74			0.5	
		I_{OL} = 4 mA, V_{CC} = Min	DM74			0.4	
I_I	Input Current @ Max Input Voltage	V_{CC} = Max, V_I = 10V				0.1	mA
I_{IH}	High Level Input Current	V_{CC} = Max, V_I = 2.7V				20	μA
I_{IL}	Low Level Input Current	V_{CC} = Max, V_I = 0.4V				−1.6	mA
I_{OS}	Short Circuit Output Current	V_{CC} = Max (Note 2)	54LS	−30		−130	mA
			DM74	−30		−130	
I_{CCH}	Supply Current with Outputs High	V_{CC} = Max, V_{IN} = GND				1.0	mA
I_{CCL}	Supply Current with Outputs Low	V_{CC} = Max, V_{IN} = OPEN				6.0	mA

Note 1: All typicals are at V_{CC} = 5V, T_A = 25°C.

Note 2: Note more than one output should be shorted at a time, and the duration should not exceed one second.

SWITCHING CHARACTERISTICS

V_{CC} = +5.0V, T_A = +25°C (See Section 1 for test waveforms and output load)

Symbol	Parameter	R_L = 2 kΩ, C_L = 15 pF		Units
		Min	Max	
t_{PLH}	Propagation Delay Time Low to High Level Output		24	ns
t_{PHL}	Propagation Delay Time High to Low Level Output		24	ns

CONNECTION DIAGRAMS

Dual-In-Line Package

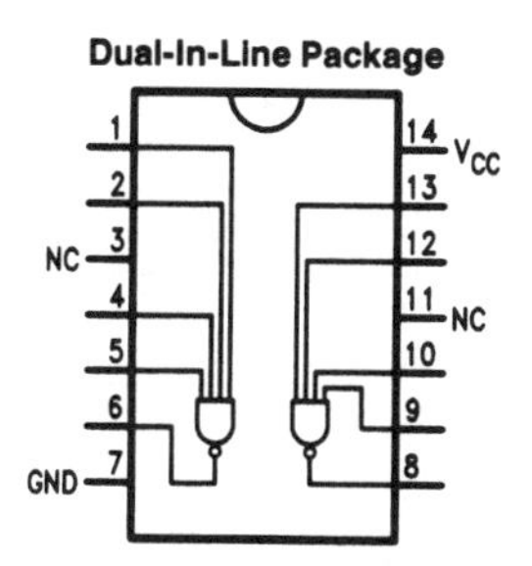

Order Number 54LS40DMQB, 54LS40FMQB, 54LS40LMQB, DM74LS40M or DM74LS40N
See NS Package Number E20A, J14A, M14A, N14A or W14B

National
54LS54/DM74LS54
4-Wide, 2-Input AND-OR-INVERT Gate

ABSOLUTE MAXIMUM RATINGS (Note)

If Military/Aerospace specified devices are required, please contact the National Semiconductor Sales Office/Distributors for availability and specifications.

Supply Voltage	7V
Input Voltage	7V
Operating Free Air Temperature Range	
54LS	−55°C to +125°C
DM74LS	0°C to +70°C
Storage Temperature Range	−65°C to +150°C

Note: *The "Absolute Maximum Ratings" are those values beyond which the safety of the device cannot be guaranteed. The device should not be operated at these limits. The parametric values defined in the "Electrical Characteristics" table are not guaranteed at the absolute maximum ratings. The "Recommended Operating Conditions" table will define the conditions for actual device operation.*

RECOMMENDED OPERATING CONDITIONS

Symbol	Parameter	54LS54			DM74LS54			Units
		Min	Nom	Max	Min	Nom	Max	
V_{CC}	Supply Voltage	4.5	5	5.5	4.75	5	5.25	V
V_{IH}	High Level Input Voltage	2			2			V
V_{IL}	Low Level Input Voltage			0.7			0.8	V
I_{OH}	High Level Output Voltage			−0.4			−0.4	mA
I_{OL}	Low Level Output Current			4			8	mA
T_A	Free Air Operating Temperature	−55		125	0		70	°C

ELECTRICAL CHARACTERISTICS over recommended operating free air temperature range (unless otherwise noted)

Symbol	Parameter	Conditions		Min	Typ (Note 1)	Max	Units
V_I	Input Clamp Voltage	V_{CC} = Min, I_I = −18 mA				−1.5	V
V_{OH}	High Level Output Voltage	V_{CC} = Min, I_{OH} = Max, V_{IL} = Max	54LS	2.5			V
			DM74LS	2.7			
V_{OL}	Low Level Output Voltage	V_{CC} = Min, I_{OL} = Max, V_{IH} = Min	54LS			0.4	V
			DM74LS			0.5	
		I_{OL} = 4 mA, V_{CC} = Min	DM74LS			0.4	
I_I	Input Current @ Max Input Voltage	V_{CC} = Max, V_I = 10V				0.1	mA
I_{IH}	High Level Input Current	V_{CC} = Max, V_I = 2.7V				20	μA
I_{IL}	Low Level Input Current	V_{CC} = Max, V_I = 0.4V				−0.4	mA
I_{OS}	Short Circuit Output Current	V_{CC} = Max (Note 2)	54LS	−20		−100	mA
			DM74LS	−20		−100	
I_{CCH}	Supply Current with Outputs High	V_{CC} = Max, V_{IN} = GND				1.6	mA
I_{CCL}	Supply Current with Outputs Low	V_{CC} = Max, V_{IN} = Open				2.0	mA

SWITCHING CHARACTERISTICS at V_{CC} = 5V and T_A = 25°C (See Section 1 for Test Waveforms and Output Load)

Symbol	Parameter	C_L = 15 pF, R_L = 2 kΩ		Units
		Min	Max	
t_{PLH}	Propagation Delay Time Low to High Level Output		15	ns
t_{PHL}	Propagation Delay Time High to Low Level Output		15	ns

Note 1: All typicals are at V_{CC} = 5V, T_A = 25°C.

Note 2: Not more than one output should be shorted at a time, and the duration should not exceed one second.

CONNECTION DIAGRAM

Dual-In-Line Package

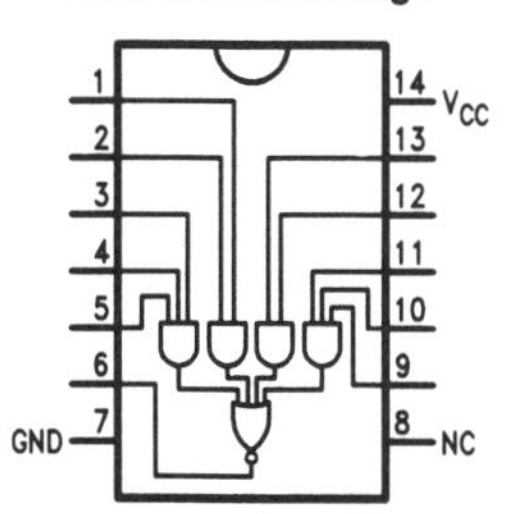

Order Number 54LS54DMQB, 54LS54FMQB, DM74LS54M or DM74LS54N
See NS Package Number J14A, M14A, N14A or W14B

National 54LS55/DM74LS55
2-Wide, 4-Input AOI Gate

ABSOLUTE MAXIMUM RATINGS (Note)

If Military/Aerospace specified devices are required, please contact the National Semiconductor Sales Office/Distributors for availability and specifications.

Supply Voltage	7V
Input Voltage	7V
Operating Free Air Temperature Range	
54LS	−55°C to +125°C
DM74LS	0°C to +70°C
Storage Temperature Range	−65°C to +150°C

Note: *The "Absolute Maximum Ratings" are those values beyond which the safety of the device cannot be guaranteed. The device should not be operated at these limits. The parametric values defined in the "Electrical Characteristics" table are not guaranteed at the absolute maximum ratings. The "Recommended Operating Conditions" table will define the conditions for actual device operation.*

RECOMMENDED OPERATING CONDITIONS

Symbol	Parameter	54LS55			DM74LS55			Units
		Min	Nom	Max	Min	Nom	Max	
V_{CC}	Supply Voltage	4.5	5	5.5	4.75	5	5.25	V
V_{IH}	High Level Input Voltage	2			2			V
V_{IL}	Low Level Input Voltage			0.7			0.8	V
I_{OH}	High Level Output Current			−0.4			−0.4	mA
I_{OL}	Low Level Output Current			4			8	mA
T_A	Free Air Operating Temperature	−55		125	0		70	°C

SWITCHING CHARACTERISTICS

V_{CC} = +5.0V, T_A = +25°C (See Section 1 for Test Waveforms and Output Load)

Symbol	Parameter	C_L = 15 pF, R_L = 2 kΩ		Units
		Min	Max	
t_{PLH}	Propagation Delay Time		15	ns
t_{PHL}			15	

ELECTRICAL CHARACTERISTICS over recommended operating free air temperature range (unless otherwise noted)

Symbol	Parameter	Conditions		Min	Typ (Note 1)	Max	Units
V_I	Input Clamp Voltage	V_{CC} = Min, I_I = −18 mA				−1.5	V
V_{OH}	High Level Output Voltage	V_{CC} = Min, I_{OH} = Max, V_{IL} = Max	54LS	2.5			V
			DM74	2.7			
V_{OL}	Low Level Output Voltage	V_{CC} = Min, I_{OL} = Max, V_{IH} = Min	54LS			0.4	V
			DM74			0.5	
		I_{OL} = 4 mA, V_{CC} = Min	DM74			0.4	
I_I	Input Current @ Max Input Voltage	V_{CC} = Max, V_I = 10V				0.1	mA
I_{IH}	High Level Input Current	V_{CC} = Max, V_I = 2.7V				20	μA
I_{IL}	Low Level Input Current	V_{CC} = Max, V_I = 0.4V				−0.4	mA
I_{OS}	Short Circuit Output Current	V_{CC} = Max (Note 2)	54LS	−20		−100	mA
			DM74	−20		−100	
I_{CCH}	Supply Current with Outputs High	V_{CC} = Max, V_{IN} = GND				0.8	mA
I_{CCL}	Supply Current with Outputs Low	V_{CC} = Max, V_{IN} = Open				1.3	mA

Note 1: All typicals are at V_{CC} = 5V, T_A = 25°C.

Note 2: Not more than one output should be shorted at a time, and the duration should not exceed one second.

CONNECTION DIAGRAM

Dual-In-Line Package

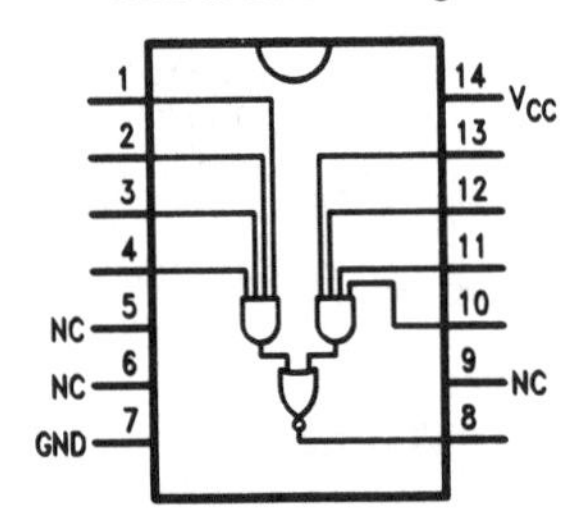

Order Number 54LS55DMQB, 54LS55FMQB, DM74LS55M or DM74LS55N
See NS Package Number J14A, M14A, N14A or W14B

National DM54LS107A/DM74LS107A
Dual Negative-Edge-Triggered Master-Slave J-K Flip-Flops with Clear and Complementary Outputs

ABSOLUTE MAXIMUM RATINGS (Note)

If Military/Aerospace specified devices are required, please contact the National Semiconductor Sales Office/Distributors for availability and specifications.

Supply Voltage	7V
Input Voltage	7V
Operating Free Air Temperature Range	
DM54LS	−55°C to +125°C
DM74LS	0°C to +70°C
Storage Temperature Range	−65°C to +150°C

Note: *The "Absolute Maximum Ratings" are those values beyond which the safety of the device cannot be guaranteed. The device should not be operated at these limits. The parametric values defined in the "Electrical Characteristics" table are not guaranteed at the absolute maximum ratings. The "Recommended Operating Conditions" table will define the conditions for actual device operation.*

RECOMMENDED OPERATING CONDITIONS

Symbol	Parameter		DM54LS107A Min	DM54LS107A Nom	DM54LS107A Max	DM74LS107A Min	DM74LS107A Nom	DM74LS107A Max	Units
V_{CC}	Supply Voltage		4.5	5	5.5	4.75	5	5.25	V
V_{IH}	High Level Input Voltage		2			2			V
V_{IL}	Low Level Input Voltage				0.7			0.8	V
I_{OH}	High Level Output Current				−0.4			−0.4	mA
I_{OL}	Low Level Output Current				4			8	mA
f_{CLK}	Clock Frequency (Note 2)		0		30	0		30	MHz
f_{CLK}	Clock Frequency (Note 3)		0		25	0		25	MHz
t_W	Pulse Width (Note 2)	Clock High	20			20			ns
		Clear Low	25			25			
t_W	Pulse Width (Note 3)	Clock High	25			25			ns
		Clear Low	30			30			
t_{SU}	Setup Time (Notes 1 & 2)		20↓			20↓			ns
t_{SU}	Setup Time (Notes 1 & 3)		25↓			25↓			ns
t_H	Hold Time (Notes 1 & 2)		0↓			0↓			ns
t_H	Hold Time (Notes 1 & 3)		5↓			5↓			ns
T_A	Free Air Operating Temperature		−55		125	0		70	°C

Note 1: The symbol (↓) indicates the falling edge of the clock pulse is used for reference.

Note 2: C_L = 15 pF, R_L = 2 kΩ, T_A = 25°C and V_{CC} = 5V.

Note 3: C_L = 50 pF, R_L = 2 kΩ, T_A = 25°C and V_{CC} = 5V.

ELECTRICAL CHARACTERISTICS over recommended operating free air temperature range (unless otherwise noted)

Symbol	Parameter	Conditions		Min	Typ (Note 1)	Max	Units
V_I	Input Clamp Voltage	V_{CC} = Min, I_I = −18 mA				−1.5	V
V_{OH}	High Level Output Voltage	V_{CC} = Min, I_{OH} = Max V_{IL} = Max, V_{IH} = Min	DM54	2.5	3.4		V
			DM74	2.7	3.4		
V_{OL}	Low Level Output Voltage	V_{CC} = Min, I_{OL} = Max V_{IL} = Max, V_{IH} = Min	DM54		0.25	0.4	V
			DM74		0.35	0.5	
		I_{OL} = 4mA, V_{CC} = Min	DM74		0.25	0.4	
I_I	Input Current @ Max Input Voltage	V_{CC} = Max, V_I = 7V	J, K			0.1	mA
			Clear			0.3	
			Clock			0.4	
I_{IH}	High Level Input Current	V_{CC} = Max V_I = 2.7V	J, K			20	μA
			Clear			60	
			Clock			80	
I_{IL}	Low Level Input Current	V_{CC} = Max V_I = 0.4V	J, K			−0.4	mA
			Clear			−0.8	
			Clock			−0.8	
I_{OS}	Short Circuit Output Current	V_{CC} = Max (Note 2)	DM54	−20		−100	mA
			DM74	−20		−100	
I_{CC}	Supply Current	V_{CC} = Max (Note 3)			4	6	mA

SWITCHING CHARACTERISTICS at V_{CC} = 5V and T_A = 25°C (See Section 1 for Test Waveforms and Output Load)

Symbol	Parameter	From (Input) To (Output)	R_L = 2 kΩ				Units
			C_L = 15 pF		C_L = 50 pF		
			Min	Max	Min	Max	
f_{MAX}	Maximum Clock Frequency		30		25		MHz
t_{PLH}	Propagation Delay Time Low to High Level Output	Preset to Q		20		24	ns
t_{PHL}	Propagation Delay Time High to Low Level Output	Preset to $\overline{Q}$		20		28	ns
t_{PLH}	Propagation Delay Time Low to High Level Output	Clear to $\overline{Q}$		20		24	ns
t_{PHL}	Propagation Delay Time High to Low Level Output	Clear to Q		20		28	ns
t_{PLH}	Propagation Delay Time Low to High Level Output	Clock to Q or $\overline{Q}$		20		24	ns
t_{PHL}	Propagation Delay Time High to Low Level Output	Clock to Q or $\overline{Q}$		20		28	ns

Note 1: All typicals are at V_{CC} = 5V, T_A = 25°C.

Note 2: Not more than one output should be shorted at a time, and the duration should not exceed one second. For devices, with feedback from the outputs, where shorting the outputs to ground may cause the outputs to change logic state an equivalent test may be performed where V_O = 2.25V and 2.125V for DM54 and DM74 series, respectively, with the minimum and maximum limits reduced by one half from their stated values. This is very useful when using automatic test equipment.

Note 3: With all inputs open, I_{CC} is measured with the Q and $\overline{Q}$ outputs high in turn. At the time of measurement the clock is grounded.

CONNECTION DIAGRAM

Dual-In-Line Package

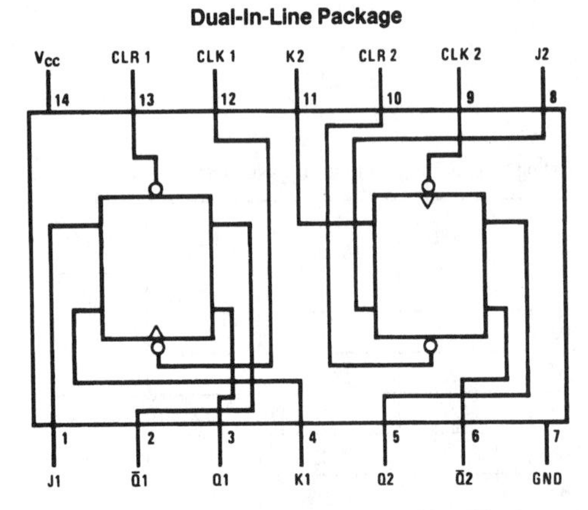

Order Number DM54LS107AJ, DM54LS107AW, DM74LS107AM or DM74LS107AN
See NS Package Number J14A, M14A, N14A or W14B

FUNCTION TABLE

Inputs				Outputs	
CLR	CLK	J	K	Q	$\overline{Q}$
L	X	X	X	L	H
H	↓	L	L	Q_0	$\overline{Q}_0$
H	↓	H	L	H	L
H	↓	L	H	L	H
H	↓	H	H	Toggle	
H	H	X	X	Q_0	$\overline{Q}_0$

H = High Logic Level
X = Either Low or High Logic Level
L = Low Logic Level
↓ = Negative going edge of pulse.
Q_0 = The output logic level before the indicated input conditions were established.
Toggle = Each output changes to the complement of its previous level on each falling edge of the clock pulse.

National 54LS109/DM54LS109A/DM74LS109A

Dual Positive-Edge-Triggered J-$\overline{K}$ Flip-Flops with Preset, Clear, and Complementary Outputs

ABSOLUTE MAXIMUM RATINGS (Note)

If Military/Aerospace specified devices are required, please contact the National Semiconductor Sales Office/Distributors for availability and specifications.

Supply Voltage	7V
Input Voltage	7V
Operating Free Air Temperature Range	
DM54LS and 54LS	−55°C to +125°C
DM74LS	0°C to +70°C
Storage Temperature Range	−65°C to +150°C

Note: *The "Absolute Maximum Ratings" are those values beyond which the safety of the device cannot be guaranteed. The device should not be operated at these limits. The parametric values defined in the "Electrical Characteristics" table are not guaranteed at the absolute maximum ratings. The "Recommended Operating Conditions" table will define the conditions for actual device operation.*

RECOMMENDED OPERATING CONDITIONS

Symbol	Parameter		DM54LS109A Min	DM54LS109A Nom	DM54LS109A Max	DM74LS109A Min	DM74LS109A Nom	DM74LS109A Max	Units
V_{CC}	Supply Voltage		4.5	5	5.5	4.75	5	5.25	V
V_{IH}	High Level Input Voltage		2			2			V
V_{IL}	Low Level Input Voltage				0.7			0.8	V
I_{OH}	High Level Output Current				−0.4			−0.4	mA
I_{OL}	Low Level Output Current				4			8	mA
f_{CLK}	Clock Frequency (Note 2)		0		25	0		25	MHz
f_{CLK}	Clock Frequency (Note 3)		0		20	0		20	MHz
t_W	Pulse Width (Note 2)	Clock High	18			18			ns
		Preset Low	15			15			
		Clear Low	15			15			
t_W	Pulse Width (Note 3)	Clock High	25			25			ns
		Preset Low	20			20			
		Clear Low	20			20			
t_{SU}	Setup Time (Notes 1 & 2)	Data High	30 ↑			30 ↑			ns
		Data Low	20 ↑			20 ↑			
t_{SU}	Setup Time (Notes 1 & 3)	Data High	35 ↑			35 ↑			ns
		Data Low	25 ↑			25 ↑			
t_H	Hold Time (Note 4)		0 ↑			0 ↑			ns
T_A	Free Air Operating Temperature		−55		125	0		70	°C

Note 1: The symbol (↑) indicates the rising edge of the clock pulse is used for reference.

Note 2: C_L = 15 pF, R_L = 2 kΩ, T_A = 25°C and V_{CC} = 5V.

Note 3: C_L = 50 pF, R_L = 2 kΩ, T_A = 25°C and V_{CC} = 5V.

Note 4: T_A = 25°C and V_{CC} = 5V.

ELECTRICAL CHARACTERISTICS over recommended operating free air temperature range (unless otherwise noted)

Symbol	Parameter	Conditions		Min	Typ (Note 1)	Max	Units
V_I	Input Clamp Voltage	V_{CC} = Min, I_I = −18 mA				−1.5	V
V_{OH}	High Level Output Voltage	V_{CC} = Min, I_{OH} = Max, V_{IL} = Max, V_{IH} = Min	DM54	2.5	3.4		V
			DM74	2.7	3.4		
V_{OL}	Low Level Output Voltage	V_{CC} = Min, I_{OL} = Max, V_{IL} = Max, V_{IH} = Min	DM54		0.25	0.4	V
			DM74		0.35	0.5	
		I_{OL} = 4 mA, V_{CC} = Min	DM74		0.25	0.4	
I_I	Input Current @ Max Input Voltage	V_{CC} = Max, V_I = 7V	J, $\overline{K}$			0.1	mA
			Clock			0.1	
			Preset			0.2	
			Clear			0.2	
I_{IH}	High Level Input Current	V_{CC} = Max, V_I = 2.7V	J, $\overline{K}$			20	μA
			Clock			20	
			Preset			40	
			Clear			40	
I_{IL}	Low Level Input Current	V_{CC} = Max, V_I = 0.4V	J, $\overline{K}$			−0.4	mA
			Clock			−0.4	
			Preset			−0.8	
			Clear			−0.8	
I_{OS}	Short Circuit Output Current	V_{CC} = Max (Note 2)	DM54	−20		−100	mA
			DM74	−20		−100	
I_{CC}	Supply Current	V_{CC} = Max (Note 3)			4	8	mA

SWITCHING CHARACTERISTICS at V_{CC} = 5V and T_A = 25°C (See Section 1 for Test Waveforms and Output Load)

Symbol	Parameter	From (Input) To (Output)	R_L = 2 kΩ				Units
			C_L = 15 pF		C_L = 50 pF		
			Min	Max	Min	Max	
f_{MAX}	Maximum Clock Frequency		25		20		MHz
t_{PLH}	Propagation Delay Time Low to High Level Output	Clock to Q or $\overline{Q}$		25		35	ns
t_{PHL}	Propagation Delay Time High to Low Level Output	Clock to Q or $\overline{Q}$		30		35	ns
t_{PLH}	Propagation Delay Time Low to High Level Output	Clear to $\overline{Q}$		25		35	ns
t_{PHL}	Propagation Delay Time High to Low Level Output	Clear to Q		30		35	ns
t_{PLH}	Propagation Delay Time Low to High Level Output	Preset to Q		25		35	ns
t_{PHL}	Propagation Delay Time High to Low Level Output	Preset to $\overline{Q}$		30		35	ns

Note 1: All typicals are at V_{CC} = 5V, T_A = 25°C.

Note 2: Not more than one output should be shorted at a time, and the duration should not exceed one second. For devices, with feedback from the outputs, where shorting the outputs to ground may cause the outputs to change logic state an equivalent test may be performed where V_O = 2.25V and 2.125V for DM54 and DM74 series, respectively, with the minimum and maximum limits reduced by one half from their stated values. This is very useful when using automatic test equipment.

Note 3: I_{CC} is measured with all outputs open, with CLOCK grounded after setting the Q and $\overline{Q}$ outputs high in turn.

CONNECTION DIAGRAM

Dual-In-Line Package

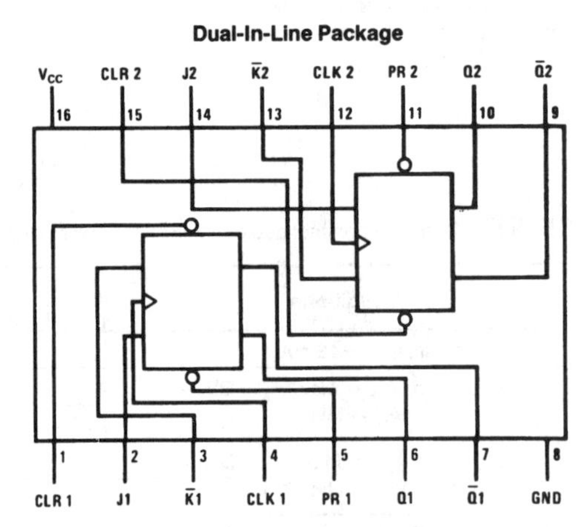

Order Number 54LS109DMQB, 54LS109FMQB, DM54LS109AJ, DM54LS109AW, DM74LS109AM or DM74LS109AN
See NS Package Number J16A, M16A, N16E or W16A

FUNCTION TABLE

Inputs					Outputs	
PR	CLR	CLK	J	$\overline{K}$	Q	$\overline{Q}$
L	H	X	X	X	H	L
H	L	X	X	X	L	H
L	L	X	X	X	H*	H*
H	H	↑	L	L	L	H
H	H	↑	H	L	Toggle	
H	H	↑	L	H	Q_0	$\overline{Q}_0$
H	H	↑	H	H	H	L
H	H	L	X	X	Q_0	$\overline{Q}_0$

H = High Logic Level
L = Low Logic Level
X = Either Low or High Logic Level
↑ = Rising Edge of Pulse
* = This configuration is nonstable; that is, it will not persist when preset and/or clear inputs return to their inactive (high) state.
Q_0 = The output logic level of Q before the indicated input conditions were established.
Toggle = Each output changes to the complement of its previous level on each active transition of the clock pulse.

National 54LS112/DM54LS112A/DM74LS112A

Dual Negative-Edge-Triggered Master-Slave J-K Flip-Flops with Preset, Clear, and Complementary Outputs

ABSOLUTE MAXIMUM RATINGS (Note)

If Military/Aerospace specified devices are required, please contact the National Semiconductor Sales Office/Distributors for availability and specifications.

Supply Voltage	7V
Input Voltage	7V
Operating Free Air Temperature Range	
DM54LS and 54LS	−55°C to +125°C
DM74LS	0°C to +70°C
Storage Temperature Range	−65°C to +150°C

Note: *The "Absolute Maximum Ratings" are those values beyond which the safety of the device cannot be guaranteed. The device should not be operated at these limits. The parametric values defined in the "Electrical Characteristics" table are not guaranteed at the absolute maximum ratings. The "Recommended Operating Conditions" table will define the conditions for actual device operation.*

RECOMMENDED OPERATING CONDITIONS

Symbol	Parameter		DM54LS112A			DM74LS112A			Units
			Min	Nom	Max	Min	Nom	Max	
V_{CC}	Supply Voltage		4.5	5	5.5	4.75	5	5.25	V
V_{IH}	High Level Input Voltage		2			2			V
V_{IL}	Low Level Input Voltage				0.7			0.8	V
I_{OH}	High Level Output Current				−0.4			−0.4	mA
I_{OL}	Low Level Output Current				4			8	mA
f_{CLK}	Clock Frequency (Note 2)		0		30	0		30	MHz
f_{CLK}	Clock Frequency (Note 3)		0		25	0		25	MHz
t_W	Pulse Width (Note 2)	Clock High	20			20			ns
		Preset Low	25			25			
		Clear Low	25			25			
t_W	Pulse Width (Note 3)	Clock High	25			25			ns
		Preset Low	30			30			
		Clear Low	30			30			
t_{SU}	Setup Time (Notes 1 and 2)		20↓			20↓			ns
t_{SU}	Setup Time (Notes 1 and 3)		25↓			25↓			ns
t_H	Hold Time (Notes 1 and 2)		0↓			0↓			ns
t_H	Hold Time (Notes 1 and 3)		5↓			5↓			ns
T_A	Free Air Operating Temperature		−55		125	0		70	°C

Note 1: The symbol (↓) indicates the falling edge of the clock pulse is used for reference.

Note 2: C_L = 15 pF, R_L = 2 kΩ, T_A = 25°C and V_{CC} = 5V.

Note 3: C_L = 50 pF, R_L = 2 kΩ, T_A = 25°C and V_{CC} = 5V.

ELECTRICAL CHARACTERISTICS over recommended operating free air temperature range (unless otherwise noted)

Symbol	Parameter	Conditions		Min	Typ (Note 1)	Max	Units
V_I	Input Clamp Voltage	V_{CC} = Min, I_I = −18 mA				−1.5	V
V_{OH}	High Level Output Voltage	V_{CC} = Min, I_{OH} = Max V_{IL} = Max, V_{IH} = Min	DM54	2.5	3.4		V
			DM74	2.7	3.4		
V_{OL}	Low Level Output Voltage	V_{CC} = Min, I_{OL} = Max V_{IL} = Max, V_{IH} = Min	DM54		0.25	0.4	V
			DM74		0.35	0.5	
		I_{OL} = 4 mA, V_{CC} = Min	DM74		0.25	0.4	
I_I	Input Current @ Max Input Voltage	V_{CC} = Max, V_I = 7V	J, K			0.1	mA
			Clear			0.3	
			Preset			0.3	
			Clock			0.4	

ELECTRICAL CHARACTERISTICS over recommended operating free air temperature range (unless otherwise noted) (Continued)

Symbol	Parameter	Conditions		Min	Typ (Note 1)	Max	Units
I_{IH}	High Level Input Current	V_{CC} = Max, V_I = 2.7V	J, K			20	μA
			Clear			60	
			Preset			60	
			Clock			80	
I_{IL}	Low Level Input Current	V_{CC} = Max, V_I = 0.4V	J, K			−0.4	mA
			Clear			−0.8	
			Preset			−0.8	
			Clock			−0.8	
I_{OS}	Short Circuit Output Current	V_{CC} = Max (Note 2)	DM54	−20		−100	mA
			DM74	−20		−100	
I_{CC}	Supply Current	V_{CC} = Max (Note 3)			4	6	mA

SWITCHING CHARACTERISTICS at V_{CC} = 5V and T_A = 25°C (See Section 1 for Test Waveforms and Output Load)

Symbol	Parameter	From (Input) To (Output)	R_L = 2 kΩ				Units
			C_L = 15 pF		C_L = 50 pF		
			Min	Max	Min	Max	
f_{MAX}	Maximum Clock Frequency		30		25		MHz
t_{PLH}	Propagation Delay Time Low to High Level Output	Preset to Q		20		24	ns
t_{PHL}	Propagation Delay Time High to Low Level Output	Preset to $\overline{Q}$		20		28	ns
t_{PLH}	Propagation Delay Time Low to High Level Output	Clear to $\overline{Q}$		20		24	ns
t_{PHL}	Propagation Delay Time High to Low Level Output	Clear to Q		20		28	ns
t_{PLH}	Propagation Delay Time Low to High Level Output	Clock to Q or $\overline{Q}$		20		24	ns
t_{PHL}	Propagation Delay Time High to Low Level Output	Clock to Q or $\overline{Q}$		20		28	ns

Note 1: All typicals are at V_{CC} = 5V, T_A = 25°C.

Note 2: Not more than one output should be shorted at a time, and the duration should not exceed one second. For devices, with feedback from the outputs, where shorting the outputs to ground may cause the outputs to change logic state an equivalent test may be performed where V_O = 2.25V and 2.125V for DM54 and DM74 series, respectively, with the minimum and maximum limits reduced by one half from their stated values. This is very useful when using automatic test equipment.

Note 3: With all outputs open, I_{CC} is measured with the Q and $\overline{Q}$ outputs high in turn. At the time of measurement the clock is grounded.

CONNECTION DIAGRAM

Dual-In-Line Package

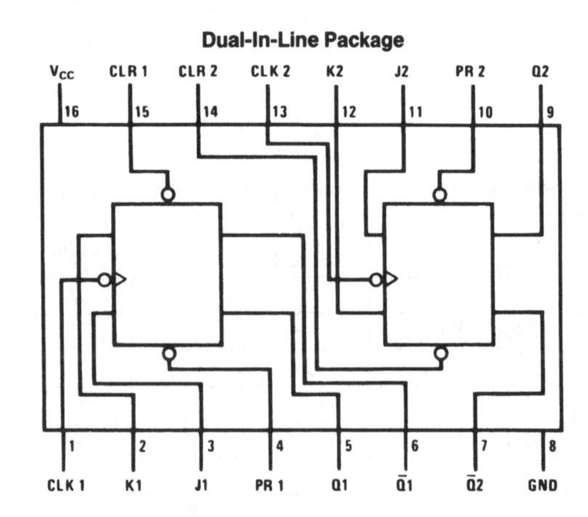

Order Number 54LS112DMQB, 54LS112FMQB, 54LS112LMQB, DM54LS112AJ, DM54LS112AW, DM74LS112AM or DM74LS112AN
See NS Package Number E20A, J16A, M16A, N16E or W16A

FUNCTION TABLE

Inputs					Outputs	
PR	CLR	CLK	J	K	Q	$\overline{Q}$
L	H	X	X	X	H	L
H	L	X	X	X	L	H
L	L	X	X	X	H*	H*
H	H	↓	L	L	Q_0	$\overline{Q}_0$
H	H	↓	H	L	H	L
H	H	↓	L	H	L	H
H	H	↓	H	H	Toggle	
H	H	H	X	X	Q_0	$\overline{Q}_0$

H = High Logic Level

L = Low Logic Level

X = Either Low or High Logic Level

↓ = Negative Going Edge of Pulse

* = This configuration is nonstable; that is, it will not persist when preset and/or clear inputs return to their inactive (high) level.

Q_0 = The output logic level before the indicated input conditions were established.

Toggle = Each output changes to the complement of its previous level on each falling edge of the clock pulse.

National 54LS113
Dual JK Edge-Triggered Flip-Flop

ABSOLUTE MAXIMUM RATINGS (Note)

If Military/Aerospace specified devices are required, please contact the National Semiconductor Sales Office/Distributors for availability and specifications.

Supply Voltage	7V
Input Voltage	5.5V
Operating Free Air Temperature Range	
54LS	−55°C to +125°C
Storage Temperature Range	−65°C to +150°C

Note: *The "Absolute Maximum Ratings" are those values beyond which the safety of the device cannot be guaranteed. The device should not be operated at these limits. The parametric values defined in the "Electrical Characteristics" table are not guaranteed at the absolute maximum ratings. The "Recommended Operating Conditions" table will define the conditions for actual operation.*

RECOMMENDED OPERATING CONDITIONS

Symbol	Parameter	54LS113			Units
		Min	Nom	Max	
V_{CC}	Supply Voltage	4.5	5	5.5	V
V_{IH}	High Level Input Voltage	2			V
V_{IL}	Low Level Input Voltage			0.7	V
I_{OH}	High Level Output Current			−0.4	mA
I_{OL}	Low Level Output Current			4	mA
T_A	Free Air Operating Temperature	−55		125	°C
t_s (H) t_s (L)	Setup Time J_n or K_n to $\overline{CP}_n$	20 20			ns
t_h (H) t_h (L)	Hold Time J_n or K_n to $\overline{CP}_n$	0 0			ns
t_w (H) t_w (L)	$\overline{CP}_n$ Pulse Width	20 15			ns
t_w (L)	$\overline{S}_{Dn}$ Pulse Width LOW	15			ns

ELECTRICAL CHARACTERISTICS over recommended operating free air temperature (unless otherwise noted)

Symbol	Parameter	Conditions		Min	Typ (Note 1)	Max	Units
V_I	Input Clamp Voltage	V_{CC} = Min, I_I = −18 mA				−1.5	V
V_{OH}	High Level Output Voltage	V_{CC} = Min, I_{OH} = Max, V_{IL} = Max, V_{IH} = Min		2.5			V
V_{OL}	Low Level Output Voltage	V_{CC} = Min, I_{OL} = Max, V_{IH} = Min, V_{IL} = Max				0.4	V
I_I	Input Current @ Max Input Voltage	V_{CC} = Max, V_I = 5.5V	J, K			0.1	mA
			SD			0.3	
			CP			0.4	
I_{IH}	High Level Input Current	V_{CC} = Max, V_I = 2.7V	J, K			20	μA
			SD			60	
			CP			80	
I_{IL}	Low Level Input Current	V_{CC} = Max, V_I = 0.5V	J, K	−30		−400	μA
			CP, SD	−60		−800	
I_{OS}	Short Circuit Output Current	V_{CC} = Max (Note 2)		−20		−100	mA
I_{CC}	Supply Current	V_{CC} = Max (Note 3)				8	mA

Note 1: All typicals are at V_{CC} = 5V, T_A = 25°C.

Note 2: Not more than one output should be shorted at a time, and the duration should not exceed one second.

Note 3: I_{CC} is measured with all outputs open and all inputs grounded.

SWITCHING CHARACTERISTICS V_{CC} = +5.0V, T_A = +25°C (See Section 1 for test waveforms and output load)

Symbol	Parameter	54LS113 C_L = 15 pF Min	Max	Units
f_{max}	Maximum Clock Frequency	30		MHz
t_{PLH} t_{PHL}	Propagation Delay $\overline{CP}_n$ to Q_n or $\overline{Q}_n$		16 24	ns
t_{PLH} t_{PHL}	Propagation Delay $\overline{S}_{Dn}$ to Q_n or $\overline{Q}_n$		16 24	ns

CONNECTION DIAGRAM

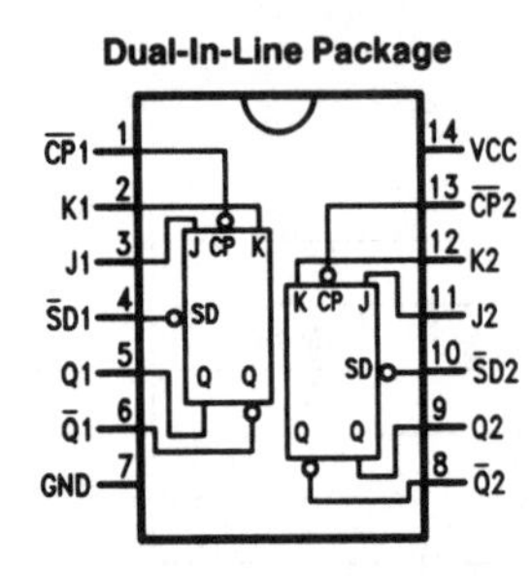

Order Number 54LS113DMQB, 54LS113FMQB or 54LS113LMQB
See NS Package Number E20A, J14A or W14B

LOGIC SYMBOL

V_{CC} = Pin 14
GND = Pin 7

TRUTH TABLE

Inputs		Output
@ t_n		@ t_n + 1
J	K	Q
L	L	Q_n
L	H	L
H	L	H
H	H	$\overline{Q}_n$

t_n = Bit Time before Clock Pulse
t_n + 1 = Bit Time after Clock Pulse
H = HIGH Voltage Level
L = LOW Voltage Level

Asynchronous Input:
Low input to $\overline{S}_D$ sets Q to HIGH level
Set is independent of clock

Pin Names	Description
J1, J2, K1, K2	Data Inputs
$\overline{CP}$1, $\overline{CP}$2	Clock Pulse Inputs (Active Falling Edge)
$\overline{SD}$1, $\overline{SD}$2	Direct Set Inputs (Active LOW)
Q1, Q2, $\overline{Q}$1, $\overline{Q}$2	Outputs

National 54LS114

Dual JK Negative Edge-Triggered Flip-Flop with Common Clocks and Clears

ABSOLUTE MAXIMUM RATINGS (Note)

If Military/Aerospace specified devices are required, please contact the National Semiconductor Sales Office/Distributors for availability and specifications.

Supply Voltage	7V
Input Voltage	7V
Operating Free Air Temperature Range	
54LS	−55°C to +125°C
Storage Temperature Range	−65°C to +150°C

Note: *The "Absolute Maximum Ratings" are those values beyond which the safety of the device cannot be guaranteed. The device should not be operated at these limits. The parametric values defined in the "Electrical Characteristics" table are not guaranteed at the absolute maximum ratings. The "Recommended Operating Conditions" table will define the conditions for actual device operation.*

RECOMMENDED OPERATING CONDITIONS

Symbol	Parameter	54LS114			Units
		Min	Nom	Max	
V_{CC}	Supply Voltage	4.5	5	5.5	V
V_{IH}	High Level Input Voltage	2			V
V_{IL}	Low Level Input Voltage			0.7	V
I_{OH}	High Level Output Current			−0.4	mA
I_{OL}	Low Level Output Current			4	mA
T_A	Free Air Operating Temperature	−55		125	°C
t_s (H) t_s (L)	Setup Time Jn or Kn to $\overline{CP}$	20 20			ns
t_h (H) t_h (L)	Hold Time Jn or Kn to $\overline{CP}$	0 0			ns
t_w (H) t_w (L)	$\overline{CP}$ Pulse Width	20 15			ns
t_w	$\overline{CD}$ or $\overline{SD}$n Pulse Width	15			ns

ELECTRICAL CHARACTERISTICS Over recommended operating free air temperature range (unless otherwise noted)

Symbol	Parameter	Conditions	Min	Typ (Note 1)	Max	Units
V_I	Input Clamp Voltage	V_{CC} = Min, I_I = −18 mA			−1.5	V
V_{OH}	High Level Output Voltage	V_{CC} = Min, I_{OH} = Max, V_{IL} = Max	2.5			V
V_{OL}	Low Level Output Voltage	V_{CC} = Min, I_{OL} = Max, V_{IH} = Min			0.4 0.5	V
I_I	Input Current @ Max Input Voltage	V_{CC} = Max, V_I = 10V; Jn, Kn Inputs SD1, SD2 Inputs CD Input CP Input			0.1 0.3 0.6 0.8	mA mA mA mA
I_{IH}	High Level Input Current	V_{CC} = Max, V_I = 2.7V; Jn, Kn Inputs SD1, SD2 Inputs CD Input CP Input			20 60 120 160	μA μA μA μA
I_{IL}	Low Level Input Current	V_{CC} = Max, V_I = 0.4V Jn, Kn Inputs SD1, SD2 Inputs CD Input CP Input			−0.4 −0.8 −1.6 −1.44	mA mA mA mA
I_{OS}	Short Circuit Output Current	V_{CC} = Max (Note 2)	−20		−100	mA
I_{CC}	Supply Current	V_{CC} = Max, V_{CP} = 0V			8.0	mA

Note 1: All typicals are at V_{CC} = 5V, T_A = 25°C.

Note 2: Not more than one output should be shorted at a time, and the duration should not exceed one second.

SWITCHING CHARACTERISTICS

V_{CC} = +5.0V, T_A = +25°C (See Section 1 for Test Waveforms and Output Load)

Symbol	Parameter	R_L = 2k, C_L = 15 pF		Units
		Min	Max	
f_{max}	Maximum Count Frequency	30		MHz
t_{PLH} t_{PHL}	Propagation Delay $\overline{CP}$ to Q or $\overline{Q}$		16 24	ns
t_{PLH} t_{PHL}	Propagation Delay $\overline{CD}$ or $\overline{SD}$n to Q or $\overline{Q}$		16 24	ns

CONNECTION DIAGRAM

Dual-In-Line Package

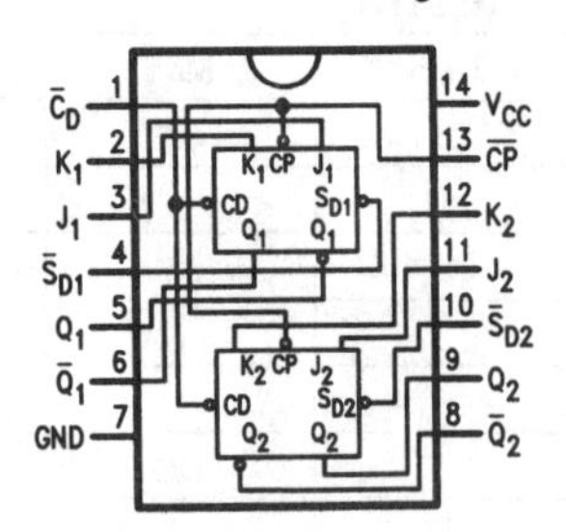

Order Number 54LS114DMQB, 54LS114FMQB or 54LS114LMQB
See NS Package Number E20A, J14A or W14B

LOGIC SYMBOL

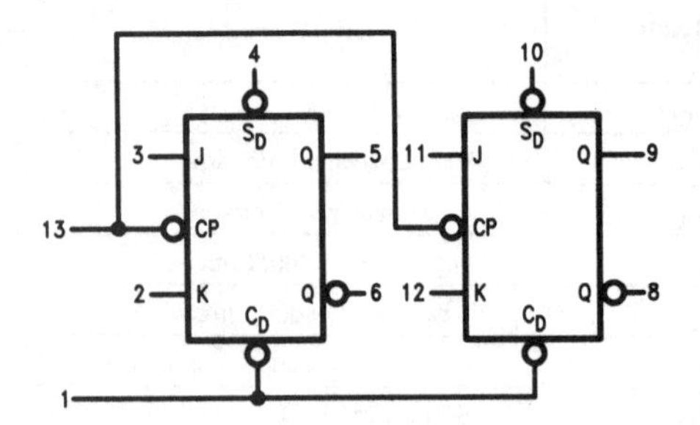

V_{CC} = Pin 14
GND = Pin 7

Pin Names	Description
J1, J2, K1, K2	Data Inputs
$\overline{CP}$	Clock Pulse Input (Active Falling Edge)
$\overline{CD}$	Direct Clear Input (Active LOW)
$\overline{SD}1$, $\overline{SD}2$	Direct Set Inputs (Active LOW)
Q1, Q2, $\overline{Q}1$, $\overline{Q}2$	Outputs

TRUTH TABLE

Asynchronous Inputs:
LOW input to $\overline{SD}$ sets Q to HIGH level
LOW input to $\overline{CD}$ sets Q to LOW level
Clear and Set are independent of clock
Simultaneous LOW on $\overline{CD}$ and $\overline{SD}$ makes both Q and $\overline{Q}$ HIGH

H = HIGH Voltage Level
L = LOW Voltage Level
t_n = Bit time before clock pulse.
t_{n+1} = Bit time after clock pulse.

Inputs		Output
@ t_n		@ t_{n+1}
J	K	Q
L	L	Qn
L	H	L
H	L	H
H	H	$\overline{Q}n$

National DM54ALS00A/DM74ALS00A Quad 2-Input NAND Gate

ABSOLUTE MAXIMUM RATINGS

If Military/Aerospace specified devices are required, please contact the National Semiconductor Sales Office/Distributors for availability and specifications.

Supply Voltage	7V
Input Voltage	7V
Operating Free Air Temperature Range	
DM54ALS	−55°C to +125°C
DM74ALS	0°C to +70°C
Storage Temperature Range	−65°C to +150°C
Typical θ_{JA}	
N Package	86.5°C/W
M Package	116.0°C/W

Note: *The "Absolute Maximum Ratings" are those values beyond which the safety of the device cannot be guaranteed. The device should not be operated at these limits. The parametric values defined in the "Electrical Characteristics" table are not guaranteed at the absolute maximum ratings. The "Recommended Operating Conditions" table will define the conditions for actual device operation.*

RECOMMENDED OPERATING CONDITIONS

Symbol	Parameter	DM54ALS00A			DM74ALS00A			Units
		Min	Nom	Max	Min	Nom	Max	
V_{CC}	Supply Voltage	4.5	5	5.5	4.5	5	5.5	V
V_{IH}	High Level Input Voltage	2			2			V
V_{IL}	Low Level Input Voltage			0.7			0.8	V
I_{OH}	High Level Output Current			−0.4			−0.4	mA
I_{OL}	Low Level Output Current			4			8	mA
T_A	Free Air Operating Temperature	−55		125	0		70	°C

ELECTRICAL CHARACTERISTICS

over recommended operating free air temperature range. All typical values are measured at V_{CC} = 5V, T_A = 25°C.

Symbol	Parameter	Conditions		Min	Typ	Max	Units
V_{IK}	Input Clamp Voltage	V_{CC} = 4.5V, I_I = −18 mA				−1.5	V
V_{OH}	High Level Output Voltage	I_{OH} = −0.4 mA V_{CC} = 4.5V to 5.5V		V_{CC} − 2			V
V_{OL}	Low Level Output Voltage	V_{CC} = 4.5V	54/74ALS I_{OL} = 4 mA		0.25	0.4	V
			74ALS I_{OL} = 8 mA		0.35	0.5	V
I_I	Input Current at Max Input Voltage	V_{CC} = 5.5V, V_{IH} = 7V				0.1	mA
I_{IH}	High Level Input Current	V_{CC} = 5.5V, V_{IH} = 2.7V				20	μA
I_{IL}	Low Level Input Current	V_{CC} = 5.5V, V_{IL} = 0.4V				−0.1	mA
I_O	Output Drive Current	V_{CC} = 5.5V	V_O = 2.25V	−30		−112	mA
I_{CC}	Supply Current	V_{CC} = 5.5V	Outputs High		0.43	0.85	mA
			Outputs Low		1.62	3	mA

SWITCHING CHARACTERISTICS over recommended operating free air temperature range (Note 1)

Symbol	Parameter	Conditions	DM54ALS00A		DM74ALS00A		Units
			Min	Max	Min	Max	
t_{PLH}	Propagation Delay Time Low to High Level Output	V_{CC} = 4.5V to 5.5V R_L = 500Ω C_L = 50 pF	3	15	3	11	ns
t_{PHL}	Propagation Delay Time High to Low Level Output		2	9	2	8	ns

Note 1: See Section 1 for test waveforms and output load.

CONNECTION DIAGRAM

Dual-In-Line Package

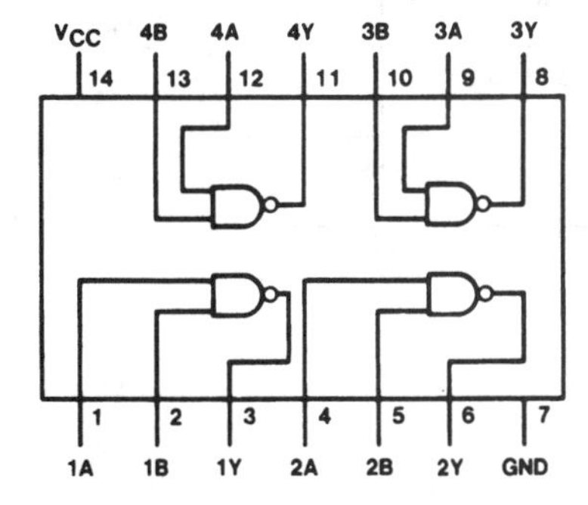

Order Number DM54ALS00AJ, DM74ALS00AM, DM74ALS00AN or DM74ALS00ASJ
See NS Package Number J14A, M14A, M14D or N14A

FUNCTION TABLE

$Y = \overline{AB}$

Inputs		Output
A	B	Y
L	L	H
L	H	H
H	L	H
H	H	L

H = High Logic Level
L = Low Logic Level

National DM74ALS01
Quad 2-Input NAND Gate with Open-Collector Outputs

ABSOLUTE MAXIMUM RATINGS

Supply Voltage	7V
Input Voltage	7V
High Level Output Voltage	7V
Operating Free Air Temperature Range	
DM74ALS	0°C to +70°C
Storage Temperature Range	−65°C to +150°C
Typical θ_{JA}	
N Package	86.5°C/W
M Package	116.0°C/W

Note: *The "Absolute Maximum Ratings" are those values beyond which the safety of the device cannot be guaranteed. The device should not be operated at these limits. The parametric values defined in the "Electrical Characteristics" table are not guaranteed at the absolute maximum ratings. The "Recommended Operating Conditions" table will define the conditions for actual device operation.*

RECOMMENDED OPERATING CONDITIONS

Symbol	Parameter	DM74ALS01			Units
		Min	Nom	Max	
V_{CC}	Supply Voltage	4.5	5	5.5	V
V_{IH}	High Level Input Voltage	2			V
V_{IL}	Low Level Input Voltage			0.8	V
V_{OH}	High Level Output Voltage			5.5	V
I_{OL}	Low Level Output Current			8	mA
T_A	Free Air Operating Temperature	0		70	°C

ELECTRICAL CHARACTERISTICS

over recommended operating free air temperature range. All typical values are measured at V_{CC} = 5V, T_A = 25°C.

Symbol	Parameter	Conditions		Min	Typ	Max	Units
V_{IK}	Input Clamp Voltage	V_{CC} = 4.5V, I_I = −18 mA				−1.5	V
I_{OH}	High Level Output Current	V_{CC} = 4.5V, V_{OH} = 5.5V				100	μA
V_{OL}	Low Level Output Voltage	V_{CC} = 4.5V	I_{OL} = 4 mA		0.25	0.4	V
			I_{OL} = 8 mA		0.35	0.5	V
I_I	Input Current @ Max. Input Voltage	V_{CC} = 5.5V, V_{IH} = 7V				0.1	mA
I_{IH}	High Level Input Current	V_{CC} = 5.5V, V_{IH} = 2.7V				20	μA
I_{IL}	Low Level Input Current	V_{CC} = 5.5V, V_{IL} = 0.4V				−0.1	mA
I_{CC}	Supply Current	V_{CC} = 5.5V	Outputs High		0.43	0.85	mA
			Outputs Low		1.62	3	mA

SWITCHING CHARACTERISTICS

over recommended operating free air temperature range (Note 1).

Symbol	Parameter	Conditions	DM74ALS01		Units
			Min	Max	
t_{PLH}	Propagation Delay Time Low to High Level Output	V_{CC} = 4.5V to 5.5V R_L = 2 kΩ C_L = 50 pF	23	54	ns
t_{PHL}	Propagation Delay Time High to Low Level Output		4	28	ns

Note 1: See Section 1 for test waveforms and output load.

CONNECTION DIAGRAM

Dual-In-Line Package

V_{CC} 4Y 4B 4A 3Y 3B 3A
14 13 12 11 10 9 8
1 2 3 4 5 6 7
1Y 1A 1B 2Y 2A 2B GND

Order Number DM74ALS01M or DM74ALS01N
See NS Package Number M14A or N14A

FUNCTION TABLE

$Y = \overline{AB}$

Inputs		Output
A	B	Y
L	L	H
L	H	H
H	L	H
H	H	L

H = High Logic Level
L = Low Logic Level

National DM54ALS02/DM74ALS02
Quad 2-Input NOR Gate

ABSOLUTE MAXIMUM RATINGS

If Military/Aerospace specified devices are required, please contact the National Semiconductor Sales Office/Distributors for availability and specifications.

Supply Voltage 7V

Input Voltage 7V

Operating Free Air Temperature Range
- DM54ALS −55°C to +125°C
- DM74ALS 0°C to +70°C

Storage Temperature Range −65°C to +150°C

Typical θ_{JA}
- N Package 86.5°C/W
- M Package 116.0°C/W

Note: *The "Absolute Maximum Ratings" are those values beyond which the safety of the device cannot be guaranteed. The device should not be operated at these limits. The parametric values defined in the "Electrical Characteristics" table are not guaranteed at the absolute maximum ratings. The "Recommended Operating Conditions" table will define the conditions for actual device operation.*

RECOMMENDED OPERATING CONDITIONS

Symbol	Parameter	DM54ALS02			DM74ALS02			Units
		Min	Nom	Max	Min	Nom	Max	
V_{CC}	Supply Voltage	4.5	5	5.5	4.5	5	5.5	V
V_{IH}	High Level Input Voltage	2			2			V
V_{IL}	Low Level Input Voltage			0.7			0.8	V
I_{OH}	High Level Output Current			−0.4			−0.4	mA
I_{OL}	Low Level Output Current			4			8	mA
T_A	Free Air Operating Temperature	−55		125	0		70	°C

ELECTRICAL CHARACTERISTICS

over recommended operating free air temperature range. All typical values are measured at V_{CC} = 5V, T_A = 25°C.

Symbol	Parameter	Conditions		Min	Typ	Max	Units
V_{IK}	Input Clamp Voltage	V_{CC} = 4.5V, I_I = −18 mA				−1.5	V
V_{OH}	High Level Output Voltage	I_{OH} = −0.4 mA V_{CC} = 4.5V to 5.5V		V_{CC} − 2			V
V_{OL}	Low Level Output Voltage	V_{CC} = 4.5V	54/74ALS I_{OL} = 4 mA		0.25	0.4	V
			74ALS I_{OL} = 8 mA		0.35	0.5	V
I_I	Input Current @ Max. Input Voltage	V_{CC} = 5.5V, V_{IH} = 7V				0.1	mA
I_{IH}	High Level Input Current	V_{CC} = 5.5V, V_{IH} = 2.7V				20	μA
I_{IL}	Low Level Input Current	V_{CC} = 5.5V, V_{IL} = 0.4V				−0.1	mA
I_O	Output Drive Current	V_{CC} = 5.5V	V_O = 2.25V	−30		−112	mA
I_{CC}	Supply Current	V_{CC} = 5.5V	Outputs High		0.85	2.2	mA
			Outputs Low		2.16	4	mA

SWITCHING CHARACTERISTICS

over recommended operating free air temperature range (Note 1).

Symbol	Parameter	Conditions	DM54ALS02		DM74ALS02		Units
			Min	Max	Min	Max	
t_{PLH}	Propagation Delay Time Low to High Level Output	V_{CC} = 4.5V to 5.5V R_L = 500Ω C_L = 50 pF	1	16	3	12	ns
t_{PHL}	Propagation Delay Time High to Low Level Output		1	7.5	3	10	ns

Note 1: See Section 1 for test waveforms and output load.

CONNECTION DIAGRAM

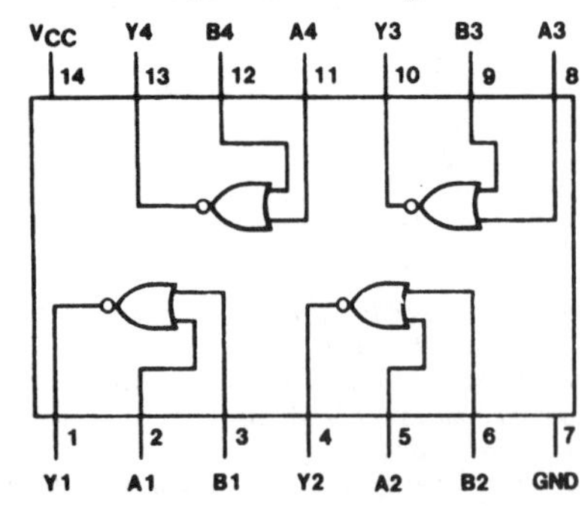

Order Number DM54ALS02J, DM74ALS02M, DM74ALS02N or DM74ALS02SJ
See NS Package Number J14A, M14A, M14D or N14A

FUNCTION TABLE

$Y = \overline{A + B}$

Inputs		Output
A	**B**	**Y**
L	L	H
L	H	L
H	L	L
H	H	L

H = High Logic Level
L = Low Logic Level

National DM54ALS04A/DM74ALS04B
Hex Inverter

ABSOLUTE MAXIMUM RATINGS

If Military/Aerospace specified devices are required, please contact the National Semiconductor Sales Office/Distributors for availability and specifications.

Supply Voltage	7V
Input Voltage	7V
Operating Free Air Temperature Range	
DM54ALS	−55°C to +125°C
DM74ALS	0°C to +70°C
Storage Temperature Range	−65°C to +150°C
Typical θ_{JA}	
N Package	88.0°C/W
M Package	118.5°C/W

Note: *The "Absolute Maximum Ratings" are those values beyond which the safety of the device cannot be guaranteed. The device should not be operated at these limits. The parametric values defined in the "Electrical Characteristics" table are not guaranteed at the absolute maximum ratings. The "Recommended Operating Conditions" table will define the conditions for actual device operation.*

RECOMMENDED OPERATING CONDITIONS

Symbol	Parameter	DM54ALS04A			DM74ALS04B			Units
		Min	Nom	Max	Min	Nom	Max	
V_{CC}	Supply Voltage	4.5	5	5.5	4.5	5	5.5	V
V_{IH}	High Level Input Voltage	2			2			V
V_{IL}	Low Level Input Voltage			0.7			0.8	V
I_{OH}	High Level Output Current			−0.4			−0.4	mA
I_{OL}	Low Level Output Current			4			8	mA
T_A	Free Air Operating Temperature	−55		125	0		70	°C

ELECTRICAL CHARACTERISTICS

over recommended operating free air temperature range. All typical values are measured at V_{CC} = 5V, T_A = 25°C.

Symbol	Parameter	Conditions		Min	Typ	Max	Units
V_{IK}	Input Clamp Voltage	V_{CC} = 4.5V, I_I = −18 mA				−1.2	V
V_{OH}	High Level Output Voltage	I_{OH} = −0.4 mA V_{CC} = 4.5V to 5.5V		V_{CC} − 2			V
V_{OL}	Low Level Output Voltage	V_{CC} = 4.5V	54/74ALS I_{OL} = 4 mA		0.25	0.4	V
			74ALS I_{OL} = 8 mA		0.35	0.5	V
I_I	Input Current @ Max. Input Voltage	V_{CC} = 5.5V, V_{IH} = 7V				0.1	mA
I_{IH}	High Level Input Current	V_{CC} = 5.5V, V_{IH} = 2.7V				20	μA
I_{IL}	Low Level Input Current	V_{CC} = 5.5V, V_{IL} = 0.4V				−0.1	mA
I_O	Output Drive Current	V_{CC} = 5.5V	V_O = 2.25V	−30		−112	mA
I_{CC}	Supply Current	V_{CC} = 5.5V	Outputs High		0.65	1.1	mA
			Outputs Low		2.4	4.2	mA

SWITCHING CHARACTERISTICS

over recommended operating free air temperature range (Note 1).

Symbol	Parameter	Conditions	DM54ALS04A		DM74ALS04B		Units
			Min	Max	Min	Max	
t_{PLH}	Propagation Delay Time Low to High Level Output	V_{CC} = 4.5V to 5.5V R_L = 500Ω C_L = 50 pF	3	13	3	11	ns
t_{PHL}	Propagation Delay Time High to Low Level Output		2	9	2	8	ns

Note 1: See Section 1 for test waveforms and output load.

CONNECTION DIAGRAM

Dual-In-Line Package

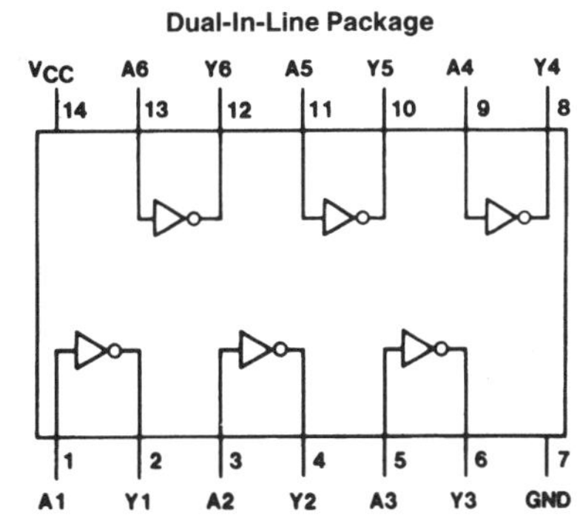

Order Number DM54ALS04AJ, DM74ALS04BM, DM74ALS04BN or DM74ALS04BSJ
See NS Package Number J14A, M14A, M14D or N14A

FUNCTION TABLE

$Y = \overline{A}$

Input	Output
A	Y
L	H
H	L

H = High Logic Level
L = Low Logic Level

National DM74ALS05A
Hex Inverter with Open Collector Outputs

ABSOLUTE MAXIMUM RATINGS

Supply Voltage	7V
Input Voltage	7V
High Level Output Voltage	7V
Operating Free Air Temperature Range	
DM74ALS	0°C to +70°C
Storage Temperature Range	−65°C to +150°C
Typical θ_{JA}	
N Package	88.0°C/W
M Package	118.5°C/W

Note: *The "Absolute Maximum Ratings" are those values beyond which the safety of the device cannot be guaranteed. The device should not be operated at these limits. The parametric values defined in the "Electrical Characteristics" table are not guaranteed at the absolute maximum ratings. The "Recommended Operating Conditions" table will define the conditions for actual device operation.*

RECOMMENDED OPERATING CONDITIONS

Symbol	Parameter	DM74ALS05A			Units
		Min	Nom	Max	
V_{CC}	Supply Voltage	4.5	5	5.5	V
V_{IH}	High Level Input Voltage	2			V
V_{IL}	Low Level Input Voltage			0.8	V
V_{OH}	High Level Output Voltage			5.5	V
I_{OL}	Low Level Output Current			8	mA
T_A	Free Air Operating Temperature	0		70	°C

ELECTRICAL CHARACTERISTICS

over recommended operating free air temperature range. All typical values are measured at V_{CC} = 5V, T_A = 25°C.

Symbol	Parameter	Conditions		Min	Typ	Max	Units
V_{IK}	Input Clamp Voltage	V_{CC} = 4.5V, I_I = −18 mA				−1.5	V
I_{OH}	High Level Output Current	V_{CC} = 4.5V, V_{OH} = 5.5V				100	μA
V_{OL}	Low Level Output Voltage	V_{CC} = 4.5V	I_{OL} = 4 mA		0.25	0.4	V
			I_{OL} = 8 mA		0.35	0.5	V
I_I	Input Current @ Max Input Voltage	V_{CC} = 5.5V, V_{IH} = 7V				0.1	mA
I_{IH}	High Level Input Current	V_{CC} = 5.5V, V_{IH} = 2.7V				20	μA
I_{IL}	Low Level Input Current	V_{CC} = 5.5V, V_{IL} = 0.4V				−0.1	mA
I_{CC}	Supply Current	V_{CC} = 5.5V	Outputs High		0.65	1.1	mA
			Outputs Low		2.4	4.2	mA

SWITCHING CHARACTERISTICS

over recommended operating free air temperature range (Note 1)

Symbol	Parameter	Conditions	DM74ALS05A		Units
			Min	Max	
t_{PLH}	Propagation Delay Time Low to High Level Output	V_{CC} = 4.5V to 5.5V, R_L = 2 kΩ, C_L = 50 pF	23	54	ns
t_{PHL}	Propagation Delay Time High to Low Level Output		4	14	ns

Note 1: See Section 1 for test waveforms and output load.

CONNECTION DIAGRAM

Dual-In-Line Package

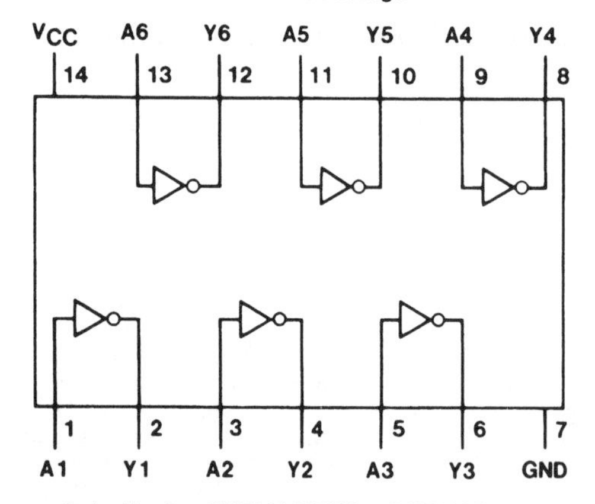

Order Number DM74ALS05AM or DM74ALS05AN
See NS Package Number M14A or N14A

FUNCTION TABLE

Y = $\overline{A}$

Input	Output
A	Y
L	H
H	L

H = High Logic Level
L = Low Logic Level

National DM54ALS08/DM74ALS08
Quad 2-Input AND Gate

ABSOLUTE MAXIMUM RATINGS

If Military/Aerospace specified devices are required, please contact the National Semiconductor Sales Office/Distributors for availability and specifications.

Supply Voltage	7V
Input Voltage	7V
Operating Free Air Temperature Range	
DM54ALS	−55°C to +125°C
DM74ALS	0°C to +70°C
Storage Temperature Range	−65°C to +150°C
Typical θ_{JA}	
N Package	86.5°C/W
M Package	116.0°C/W

Note: *The "Absolute Maximum Ratings" are those values beyond which the safety of the device cannot be guaranteed. The device should not be operated at these limits. The parametric values defined in the "Electrical Characteristics" table are not guaranteed at the absolute maximum ratings. The "Recommended Operating Conditions" table will define the conditions for actual device operation.*

RECOMMENDED OPERATING CONDITIONS

Symbol	Parameter	DM54ALS08			DM74ALS08			Units
		Min	Nom	Max	Min	Nom	Max	
V_{CC}	Supply Voltage	4.5	5	5.5	4.5	5	5.5	V
V_{IH}	High Level Input Voltage	2			2			V
V_{IL}	Low Level Input Voltage			0.7			0.8	V
I_{OH}	High Level Output Current			−0.4			−0.4	mA
I_{OL}	Low Level Output Current			4			8	mA
T_A	Free Air Operating Temperature	−55		125	0		70	°C

ELECTRICAL CHARACTERISTICS

over recommended operating free air temperature range. All typical values are measured at V_{CC} = 5V, T_A = 25°C.

Symbol	Parameter	Conditions		Min	Typ	Max	Units
V_{IK}	Input Clamp Voltage	V_{CC} = 4.5V, I_I = −18 mA				−1.5	V
V_{OH}	High Level Output Voltage	I_{OH} = −0.4 mA, V_{CC} = 4.5V to 5.5V		V_{CC} − 2			V
V_{OL}	Low Level Output Voltage	V_{CC} = 4.5V	54/74ALS I_{OL} = 4 mA		0.25	0.4	V
			74ALS I_{OL} = 8 mA		0.35	0.5	V
I_I	Input Current @ Max. Input Voltage	V_{CC} = 5.5V, V_{IH} = 7V				0.1	mA
I_{IH}	High Level Input Current	V_{CC} = 5.5V, V_{IH} = 2.7V				20	μA
I_{IL}	Low Level Input Current	V_{CC} = 5.5V, V_{IL} = 0.4V				−0.1	mA
I_O	Output Drive Current	V_{CC} = 5.5V	V_O = 2.25V	−30		−112	mA
I_{CC}	Supply Current	V_{CC} = 5.5V	Outputs High		1.3	2.4	mA
			Outputs Low		2.2	4	mA

SWITCHING CHARACTERISTICS

over recommended operating free air temperature range (Note 1).

Symbol	Parameter	Conditions	DM54ALS08		DM74ALS08		Units
			Min	Max	Min	Max	
t_{PLH}	Propagation Delay Time Low to High Level Output	V_{CC} = 4.5V to 5.5V, R_L = 500Ω, C_L = 50 pF	4	14	4	14	ns
t_{PHL}	Propagation Delay Time High to Low Level Output		3	12.5	3	10	ns

Note 1: See Section 1 for test waveforms and output load.

CONNECTION DIAGRAM

Dual-In-Line Package

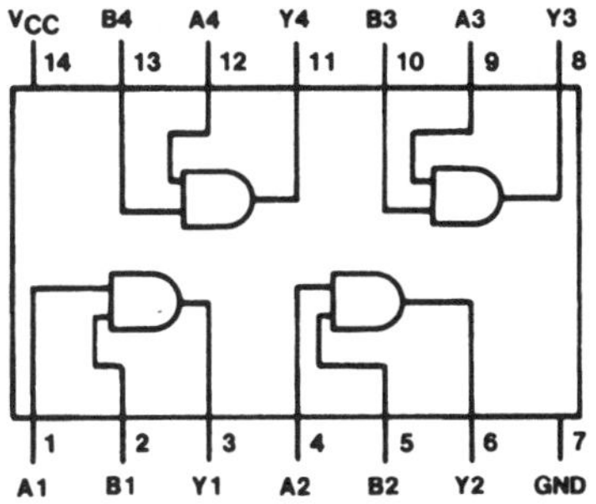

V_{CC} B4 A4 Y4 B3 A3 Y3
14 13 12 11 10 9 8

1 2 3 4 5 6 7
A1 B1 Y1 A2 B2 Y2 GND

Order Number DM54ALS08J, DM74ALS08M, DM74ALS08N or DM74ALS08SJ
See NS Package Number J14A, M14A, M14D or N14A

FUNCTION TABLE

Y = AB

Inputs		Output
A	B	Y
L	L	L
L	H	L
H	L	L
H	H	H

H = High Logic Level
L = Low Logic Level

National DM74ALS09
Quad 2-Input AND Gate with Open Collector Outputs

ABSOLUTE MAXIMUM RATINGS

Supply Voltage	7V
Input Voltage	7V
High Level Output Voltage	7V
Operating Free Air Temperature Range	
DM74ALS	0°C to +70°C
Storage Temperature Range	−65°C to +150°C
Typical θ_{JA}	
N Package	86.5°C/W
M Package	116.0°C/W

Note: *The "Absolute Maximum Ratings" are those values beyond which the safety of the device cannot be guaranteed. The device should not be operated at these limits. The parametric values defined in the "Electrical Characteristics" table are not guaranteed at the absolute maximum ratings. The "Recommended Operating Conditions" table will define the conditions for actual device operation.*

RECOMMENDED OPERATING CONDITIONS

Symbol	Parameter	DM74ALS09			Units
		Min	Nom	Max	
V_{CC}	Supply Voltage	4.5	5	5.5	V
V_{IH}	High Level Input Voltage	2			V
V_{IL}	Low Level Input Voltage			0.8	V
V_{OH}	High Level Output Voltage			5.5	V
I_{OL}	Low Level Output Current			8	mA
T_A	Free Air Operating Temperature	0		70	°C

ELECTRICAL CHARACTERISTICS

over recommended operating free air temperature range. All typical values are measured at V_{CC} = 5V, T_A = 25°C.

Symbol	Parameter	Conditions		Min	Typ	Max	Units
V_{IK}	Input Clamp Voltage	V_{CC} = 4.5V, I_I = −18 mA				−1.5	V
I_{OH}	High Level Output Current	V_{CC} = 4.5V, V_{OH} = 5.5V				100	μA
V_{OL}	Low Level Output Voltage	V_{CC} = 4.5V	I_{OL} = 4 mA		0.25	0.4	V
			I_{OL} = 8 mA		0.35	0.5	V
I_I	Input Current @ Max Input Voltage	V_{CC} = 5.5V, V_{IH} = 7V				0.1	mA
I_{IH}	High Level Input Current	V_{CC} = 5.5V, V_{IH} = 2.7V				20	μA
I_{IL}	Low Level Input Current	V_{CC} = 5.5V, V_{IL} = 0.4V				−0.1	mA
I_{CC}	Supply Current	V_{CC} = 5.5V	Outputs High		1.3	2.4	mA
			Outputs Low		2.2	4	mA

SWITCHING CHARACTERISTICS

over recommended operating free air temperature range (Note 1).

Symbol	Parameter	Conditions	DM74ALS09		Units
			Min	Max	
t_{PLH}	Propagation Delay Time Low to High Level Output	V_{CC} = 4.5V to 5.5V, R_L = 2 kΩ, C_L = 50 pF	23	54	ns
t_{PHL}	Propagation Delay Time High to Low Level Output		5	15	ns

Note 1: See Section 1 for test waveforms and output load.

CONNECTION DIAGRAM

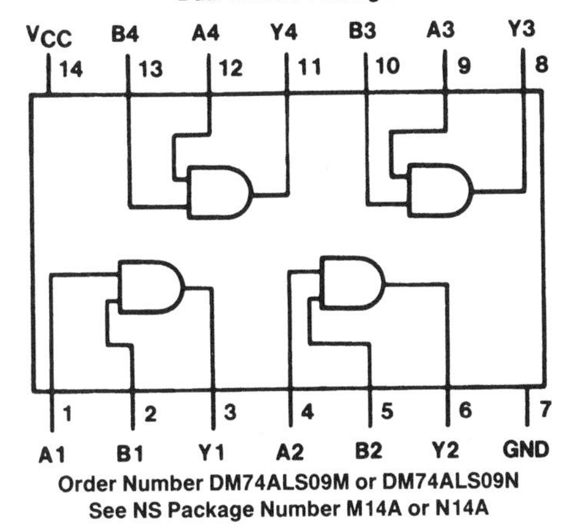

FUNCTION TABLE

Y = AB

Inputs		Output
A	B	Y
L	L	L
L	H	L
H	L	L
H	H	H

H = High Logic Level
L = Low Logic Level

This information is reproduced with permission of Rohm Corporation, Irvine, CA.

Rohm BU74HC Series CMOS Logic ICs

- Low power consumption
- Wide supply voltage range (2 ~ 6V)
- High input impedance
- High fan-out
- Capable of directly driving LS-TTL10
- High speed

The BU74HC is a series of CMOS ICs characterized by low voltage and low power consumption. In addition to a wide supply voltage range, the BU74HC is compatible with the general-purpose 74HC series. Another feature of the series is that it can drive LS-TTL ICs directly. The BU74HC is available in standard DIP and MF (mini-flat) packages.

BU74HC SERIES PRODUCT SUMMARY

Category	Type	Function	Package	
			Configuration	No. of pins
Gates	BU74HC00	Quad 2-input NAND gate	DIP/MF	14
	BU74HC02	Quad 2-input NOR gate	DIP/MF	14
	BU74HC08	Quad 2-input AND gate	DIP/MF	14
	☆BU74HC14	Hex Schmitt trigger	DIP/MF	14
	BU74HC86	Quad 2-input exclusive OR Gate	DIP/MF	14
	☆BU74HC132	Quad 2-input Schmitt trigger	DIP/MF	14
	BU74HC266	Quad 2-input exclusive NOR gate	DIP/MF	14
Latches	☆BU74HC373	Octal tristate noninverting D-type transparent latch	DIP/MF	20
	☆BU74HC533	Octal tristate noninverting D-type transparent latch	DIP/MF	20
Flip-flops	BU74HC73	Dual J-K flip-flop with reset	DIP/MF	14
	BU74HC74	Dual D-type flip-flop with set & reset	DIP/MF	14
	BU74HC76	Dual J-K flip-flop with set & reset	DIP/MF	16
	☆BU74HC174	Hex D-type flip-flop with common clock & reset	DIP/MF	16
	☆BU74HC374	Octal tristate noninverting D-type flip-flop	DIP/MF	20
	☆BU74HC534	Octal tristate noninverting D-type flip-flop	DIP/MF	20
Digital data selectors/multiplexers	☆BU74HC157	Quad 2-input data selector/multiplexer	DIP/MF	16
	☆BU74HC158	Quad 2-input data selector/multiplexer with inverting output	DIP/MF	16
Decoders	BU74HC138	1-OF-8 decoder/demultiplexer	DIP/MF	16
	BU74HC139	Dual 1-OF-4 decoder/demultiplexer	DIP/MF	16
Counters	BU74HC160	Presettable BCD counter	DIP/MF	16
	BU74HC161	Presettable binary counter	DIP/MF	16
	☆BU74HC162	Presettable BCD counter	DIP/MF	16
	☆BU74HC163	Presettable binary counter	DIP/MF	16
Buffers/inverters	BU74HCU04	Hex unbuffered inverter	DIP/MF	14
	☆BU74HC240	Octal tristate inverting buffer/line driver/line receiver	DIP/MF	20
	☆BU74HC241	Octal tristate noninverting buffer/line driver/line receiver	DIP/MF	20
	☆BU74HC244	Octal tristate noninverting buffer/line driver/line receiver	DIP/MF	20
	☆BU74HC367	Hex noninverting buffer	DIP/MF	16
	☆BU74HC368	Hex inverting buffer	DIP/MF	16

ABSOLUTE MAXIMUM RATINGS

Parameter	Symbol	Limits	Unit
Supply voltage	V_{CC}	−0.5 ~ 7.0	V
Input voltage	V_{IN}	0.5 ~ V_{CC} +0.5	V
Output voltage	V_{OUT}	−0.5 ~ V_{CC} +0.5	V
Input current	I_{IN}	±20	mA
Output current	I_{OUT}	±25	mA
Circuit current	I_{CC}	±50	mA
Power dissipation	Pd	500*	mW
Storage temperature	Tstg	−65 ~ 150	°C

*Derating is done at 5.0mW/°C for operation above Ta=25°C.

RECOMMENDED OPERATING CONDITIONS (T_A = 25°C)

Parameter	Symbol	Limits	Unit	Conditions
Supply voltage	V_{CC}	2.0 ~ 6.0	V	—
Input, output voltage	V_{IN}, V_{OUT}	0 ~ V_{CC}	V	—
Operating temperature	Topr	−40 ~ 85*	°C	—
Output rise time, fall time	tr, tf	~500	ns	—

*For an extended operating temperature range, consult your local ROHM representative.

ELECTRICAL CHARACTERISTICS/DC CHARACTERISTICS

Parameter	Symbol	Min.	Typ.	Max.	Unit	Conditions V_{CC}(V)		
Input high voltage	V_{IH}	1.5	1.2	—	V	2.0	—	V_{OUT} =0.1V or V_{CC} −0.1V $\|I_{OUT}\|$ =20μA
		3.15	2.4	—		4.5	—	
		4.2	3.2	—		6.0	—	
Input low voltage	V_{IL}	—	0.6	0.3	V	2.0	—	V_{OUT} =0.1V or V_{CC} −0.1V $\|I_{OUT}\|$ =20μA
		—	1.8	0.9		4.5	—	
		—	2.4	1.2		6.0	—	
Output high voltage	V_{OH}	1.9	1.998	—	V	2.0	V_{IH} or V_{IL}	I_{OUT} = −20μA
		4.4	4.499	—		4.5		
		5.9	5.999	—		6.0		
Output low voltage	V_{OL}	—	0.002	0.1	V	2.0	V_{IN} or V_{IL}	I_{OUT} =20μA
		—	0.001	0.1		4.5		
		—	0.001	0.1		6.0		
Input current	I_{IN}	−0.1	0.00001	0.1	μA	6.0	V_{CC} or GND	—
Power consumption	I_{CC}	—	—	4	μA	6.0	V_{CC} or GND	I_{OUT} =0μA

ELECTRICAL CHARACTERISTICS/SWITCHING CHARACTERISTICS ($T_{Ae-25^{\circ}C}$)

Parameter	Symbol	Min.	Typ.	Max.	Unit	Conditions V_{CC}(V)	V_{IN} (V)
Max. clock frequency	f_{MAX}	5	11	—	MHz	2.0	50% DUTY CYCLE C_L =50pF INPUT:tr=tf=6ns
		27	54	—		4.5	
		32	64	—		6.0	
Low-to-high propagation delay time Clock → Q,Q̄	t_{PLH}	—	88	175	ns	2.0	C_L =50pF INPUT:tr=tf=6ns
		—	18	35		4.5	
		—	15	30		6.0	
High-to-low propagation delay time Clock → Q,Q̄	t_{PHL}	—	88	175	ns	2.0	C_L =50pF INPUT:tr=tf=6ns
		—	18	35		4.5	
		—	15	30		6.0	
Low-to-high propagation delay time SET, RESET → Q,Q̄	t_{PLH}	—	115	230	ns	2.0	C_L =50pF INPUT:tr=tf=6ns
		—	23	46		4.5	
		—	20	39		6.0	
High-to-low propagation delay time SET, RESET → Q,Q̄	t_{PHL}	—	115	230	ns	2.0	C_L =50pF INPUT:tr=tf=6ns
		—	23	46		4.5	
		—	20	39		6.0	
Output rise time Output fall time	t_{TLH}, t_{THL}	—	38	75	ns	2.0	OUTPUT:C_L =50pF INPUT:tr=tf=6ns
		—	8	15		4.5	
		—	6	13		6.0	
Maximum input capacitance	C_{IN}	—	5	10	pF	—	—

16-pin MF

20-pin MF

(Unit: mm)

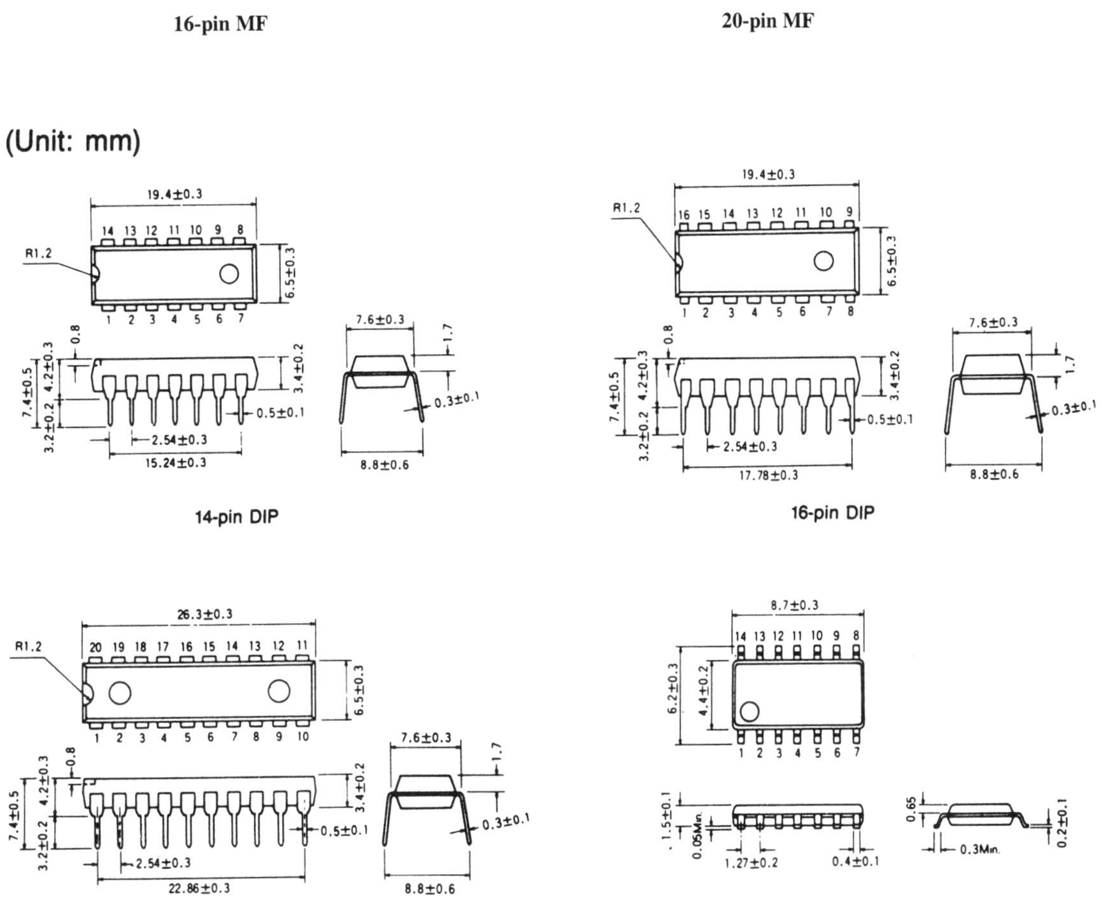

14-pin DIP

16-pin DIP

20-pin DIP

14-pin MF

ELECTRICAL CHARACTERISTIC CURVES

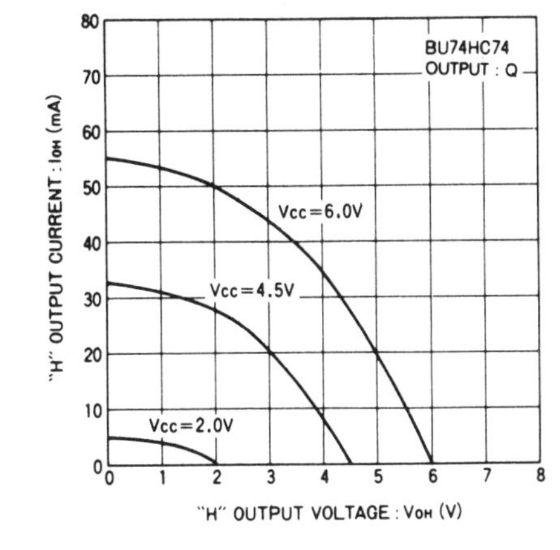

Output high current vs. output high voltage

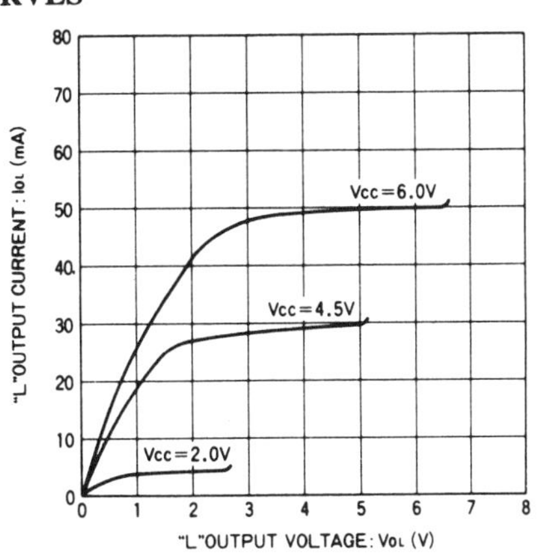

Output low current vs. output low voltage

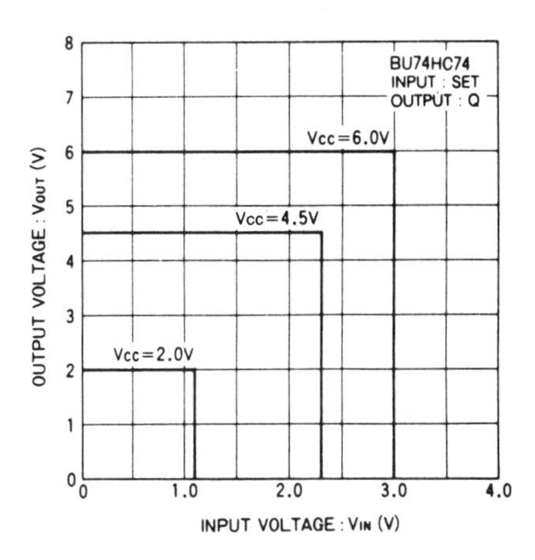

Output voltage vs. input voltage

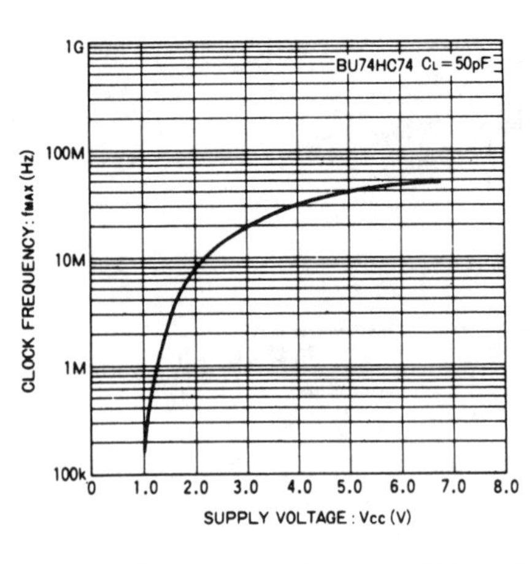

Maximum clock frequency supply voltage

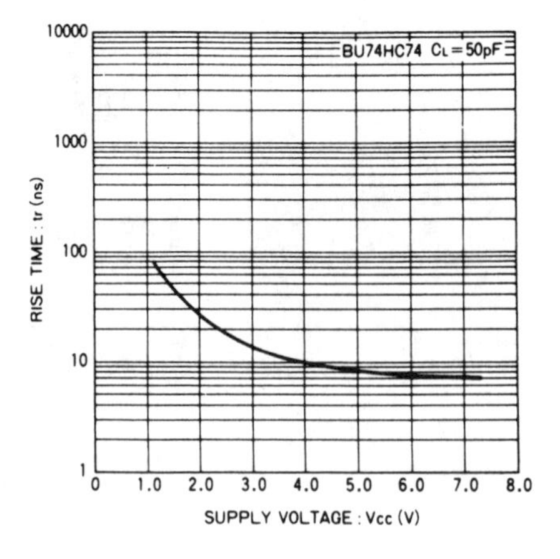

Rise time vs. supply voltage

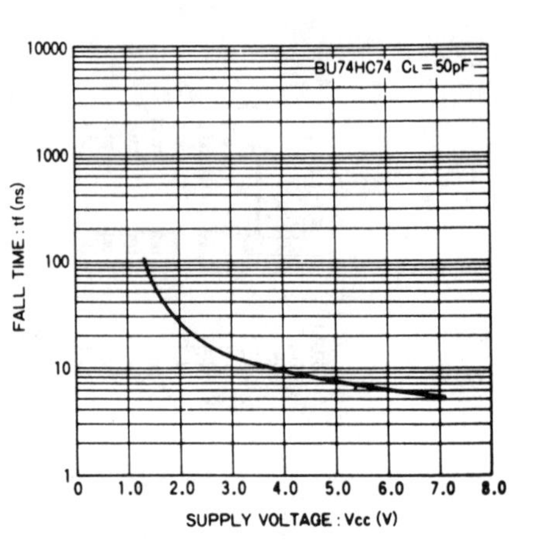

Fall time vs. supply voltage

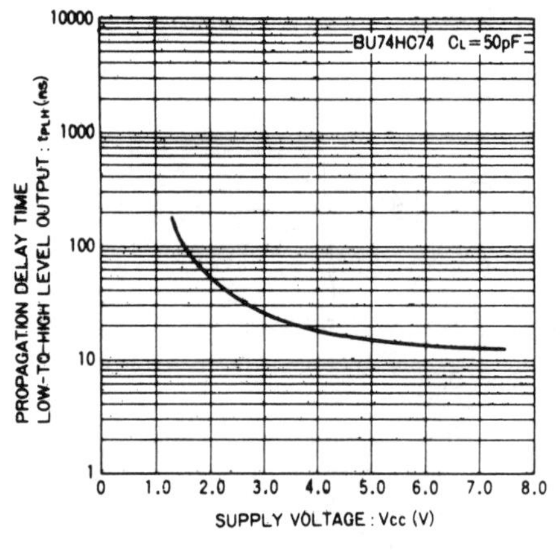

Low-to-high propagation delay time vs. supply voltage

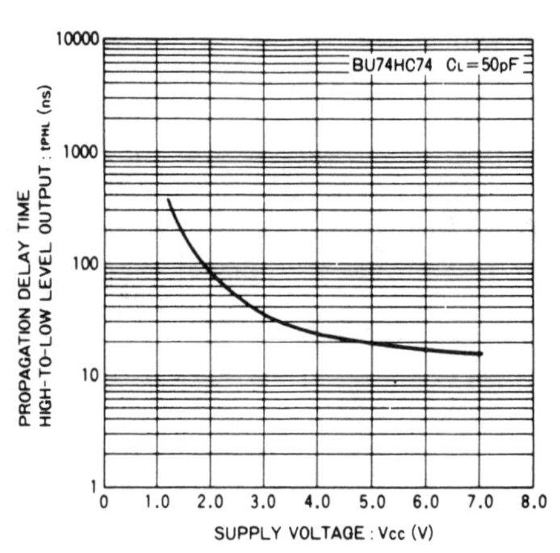

High-to-low propagation delay time vs. supply voltage

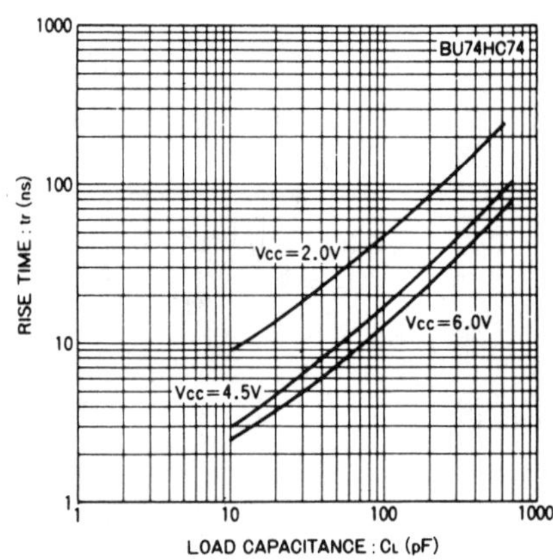

Rise time vs. load capacitance

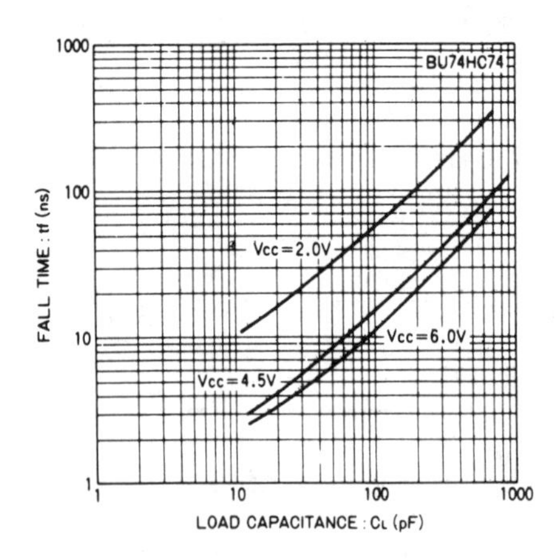

Fall time vs. load capacitance

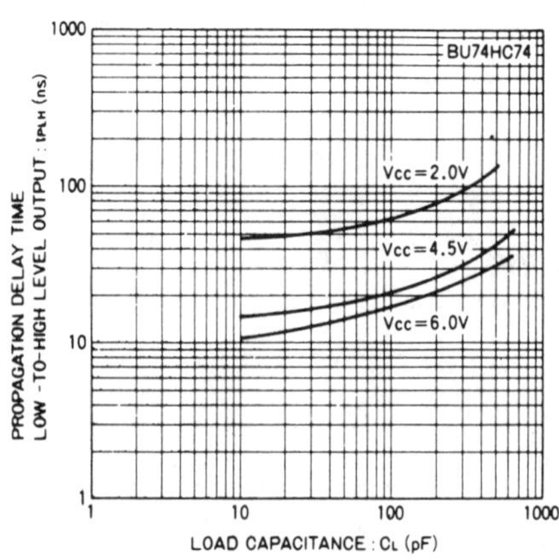

Low-to-high propagation delay vs. load capacitance

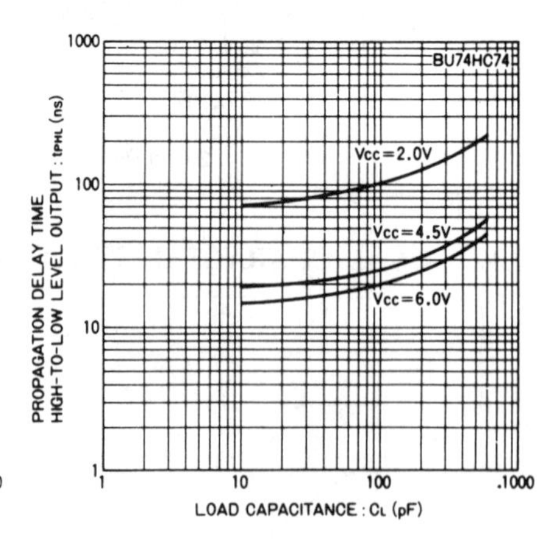

High-to-low propagation delay time vs. load capacitance

BLOCK DIAGRAMS

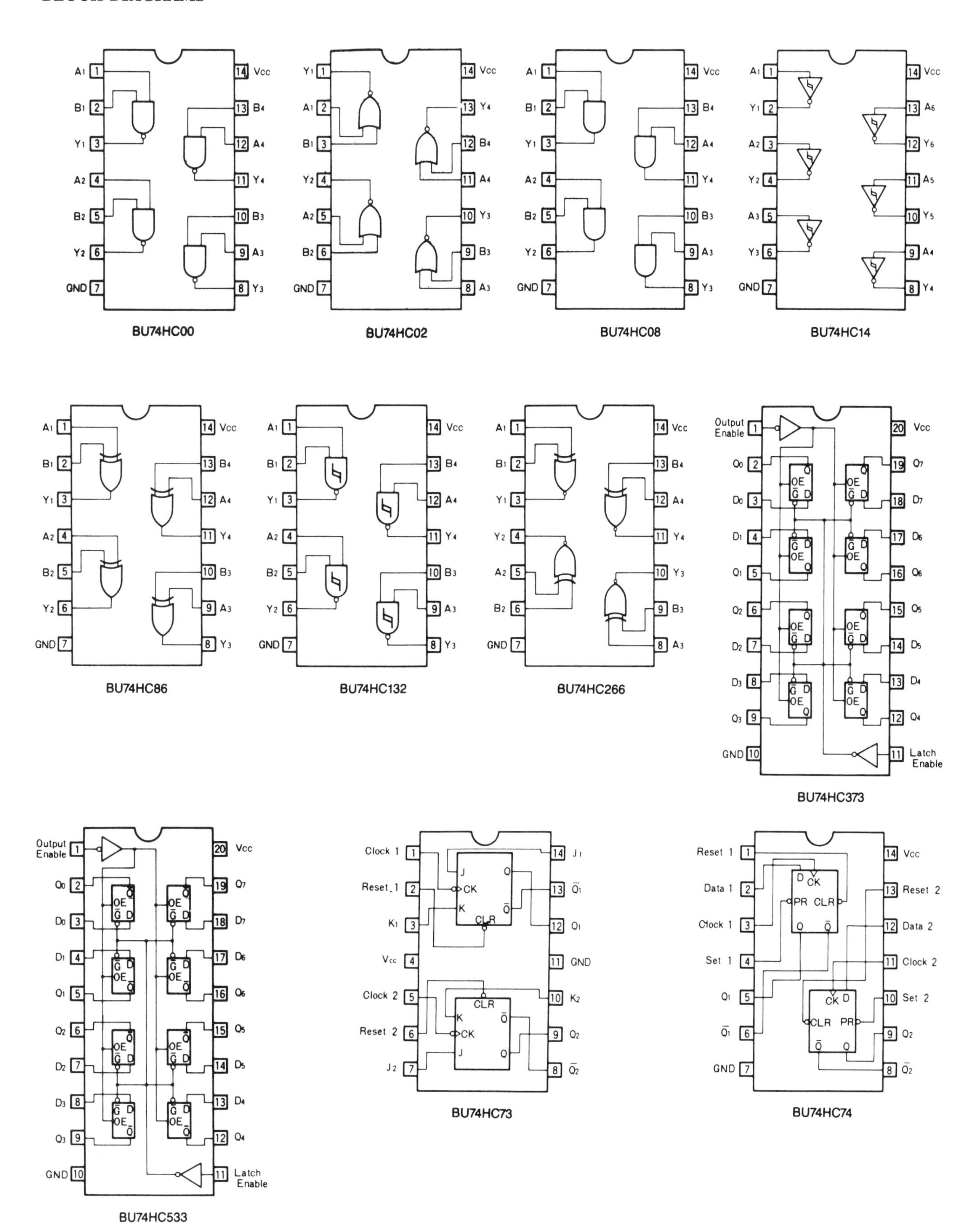

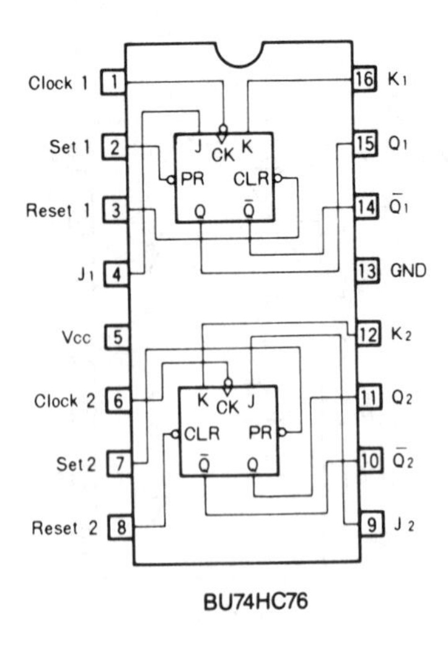

BU74HC76

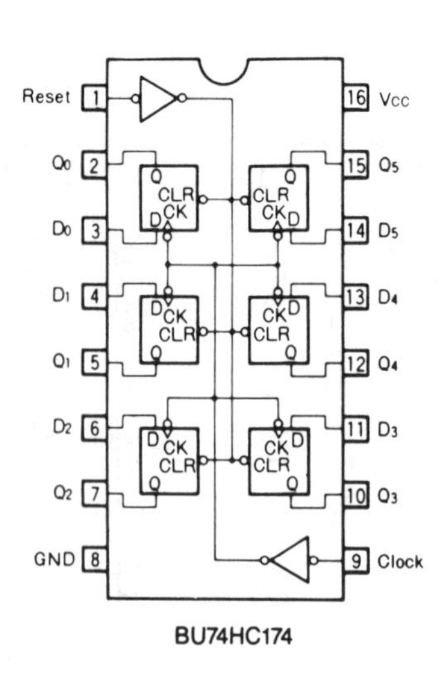

BU74HC174

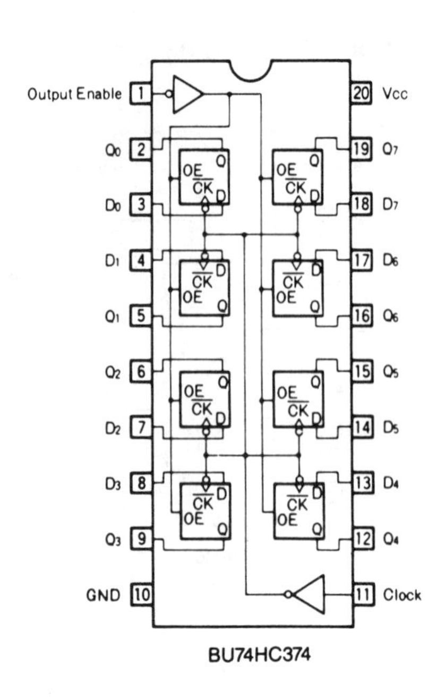

BU74HC374

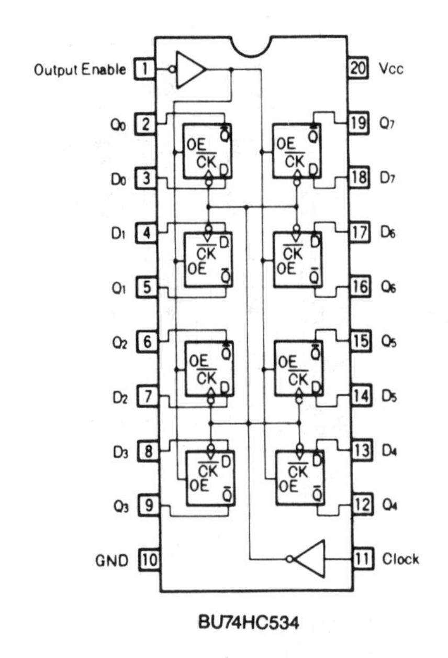

BU74HC534

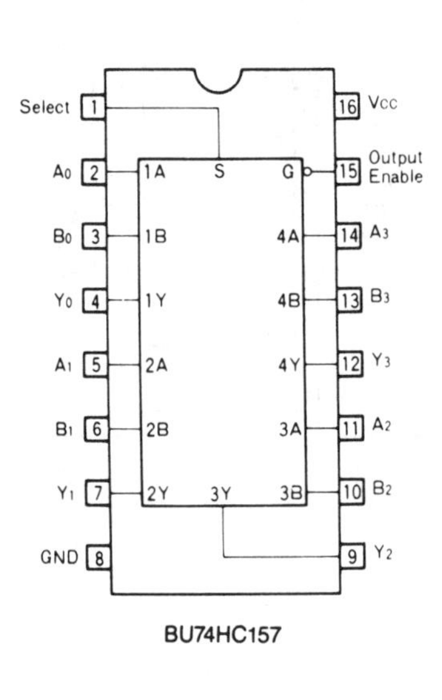

BU74HC157

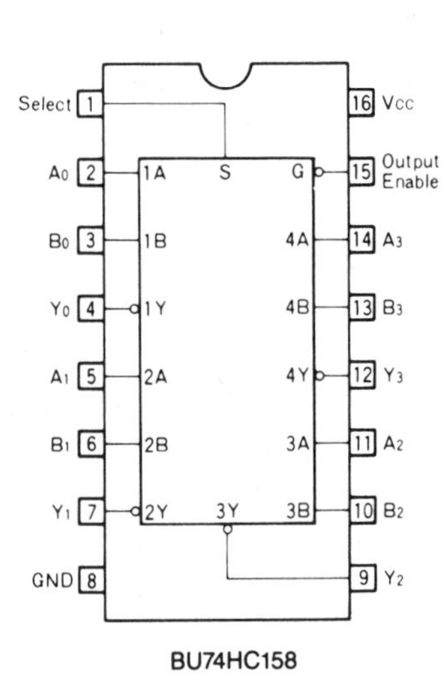

BU74HC158

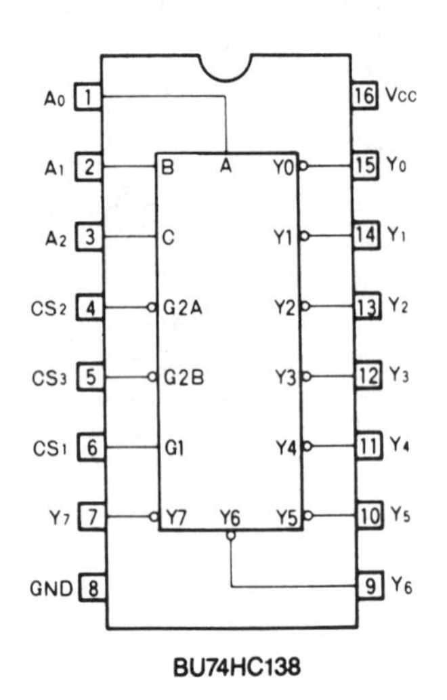

BU74HC138

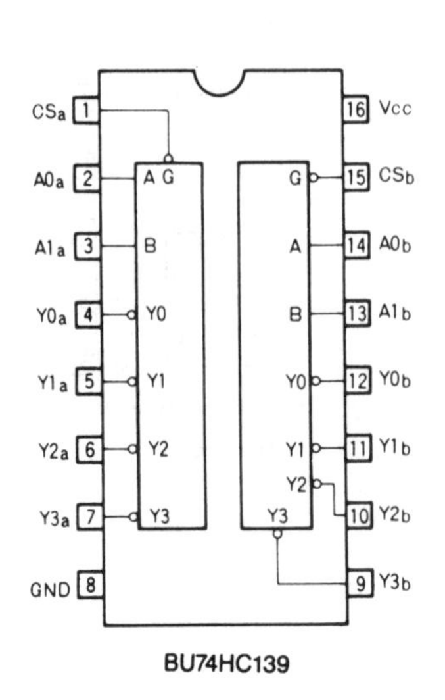

BU74HC139

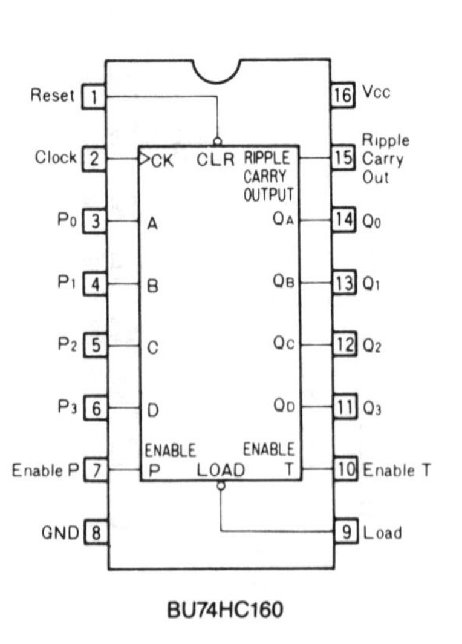

BU74HC160

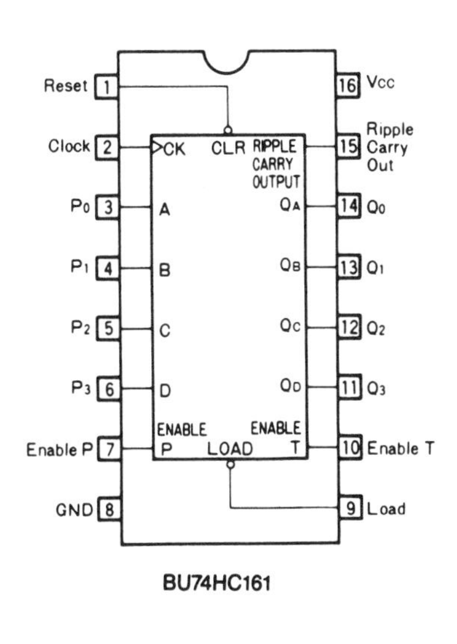

BU74HC161

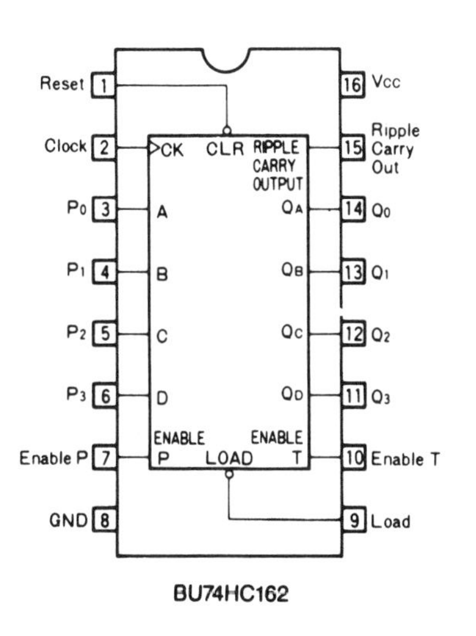

BU74HC162

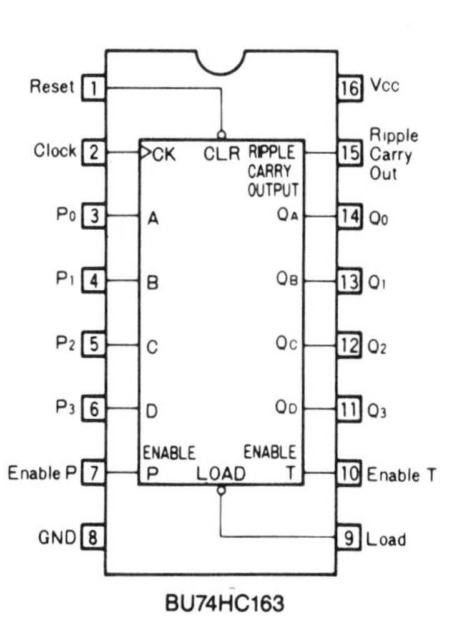

BU74HC163

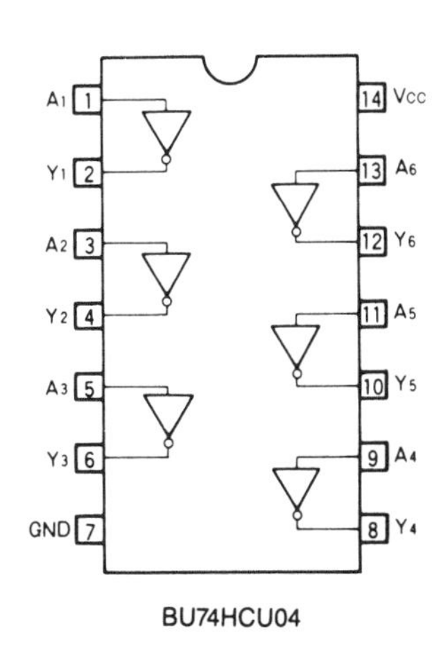

BU74HCU04

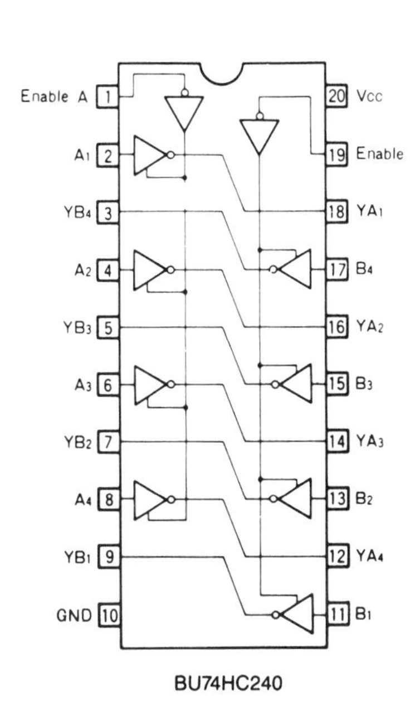

BU74HC240

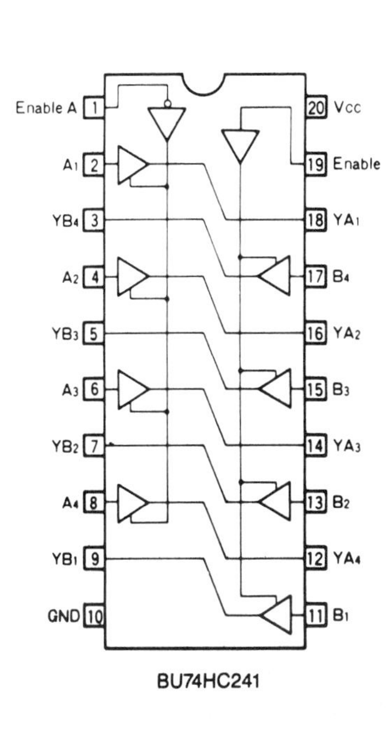

BU74HC241

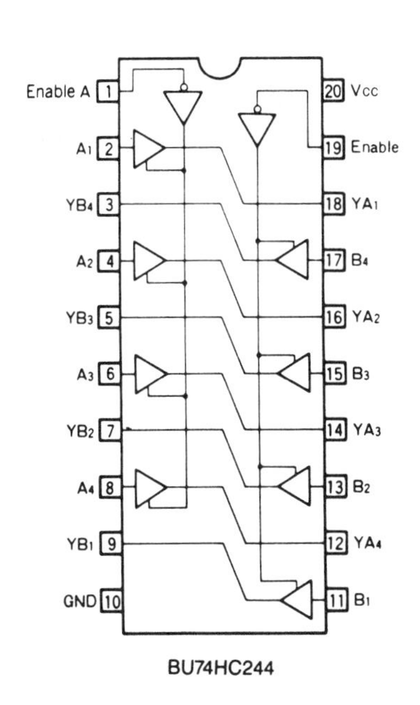

BU74HC244

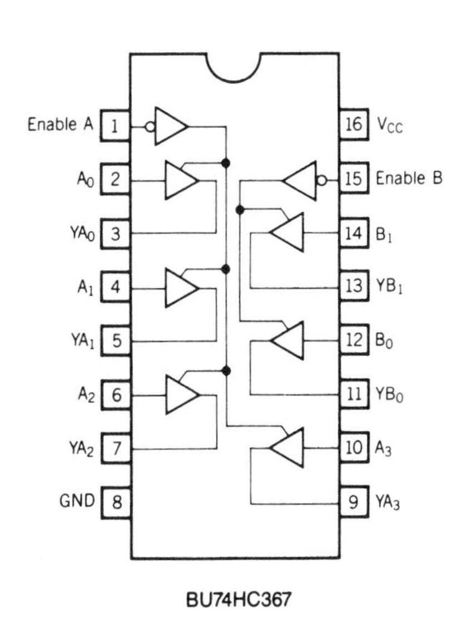

BU74HC367

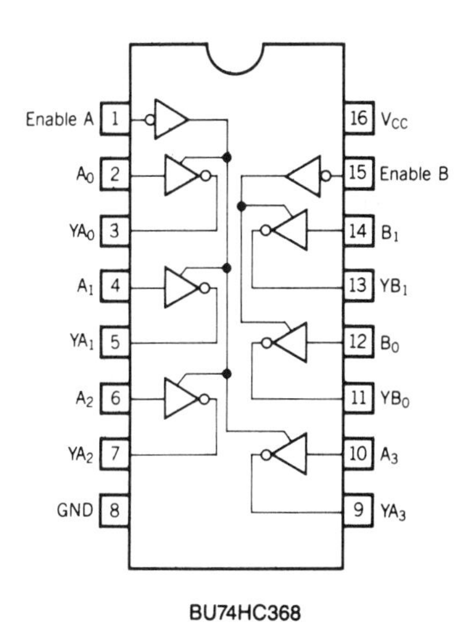

BU74HC368

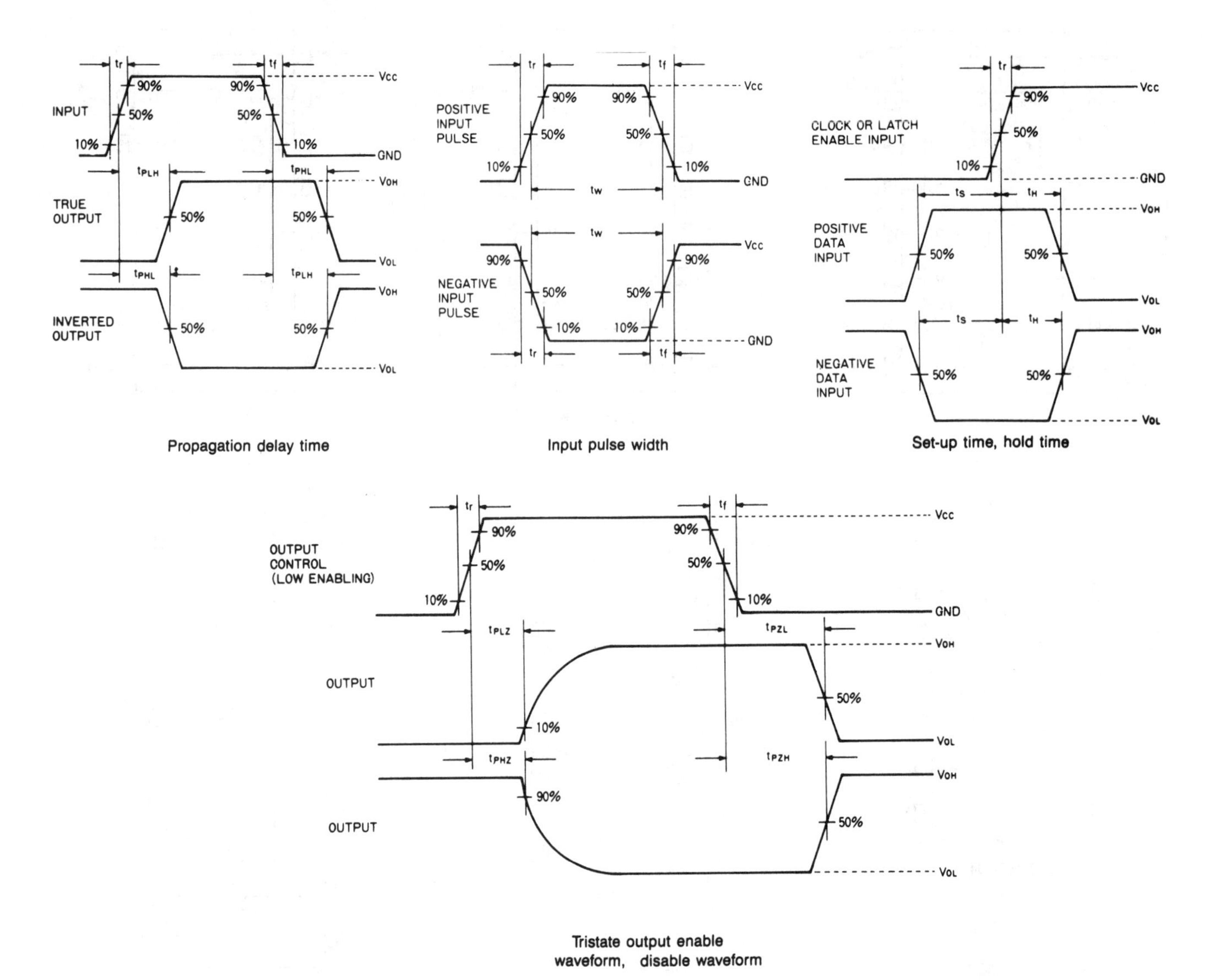

Propagation delay time

Input pulse width

Set-up time, hold time

Tristate output enable waveform, disable waveform

Rohm BU4000B Series CMOS Logic ICs

- Low power consumption
- Wide supply voltage range
- High input impedance
- High fan-out
- Capable of directly driving LS-TTL1 and LS-TTL2

The BU4000B is a series of CMOS ICs characterized by low voltage and low power consumption. In addition to a wide supply range, the BU4000B is compatible with the general-purpose 4000B series. Another feature of the series is that it can drive LS-TTL ICs directly. The BU4000B is available in standard DIP and mini-flat (MF) packages.

ABSOLUTE MAXIMUM RATINGS(T_A = 25°C)

Parameter	Symbol	Limits	Unit
Supply voltage	V_{DD}	−0.3 ~ 16	V
Input voltage	V_{IN}	−0.3 ~ V_{DD} +0.3	V
Power dissipation	Pd	550(for both DIP and MF)	mW
Storage temperature	Tstg	−55 ~ 150	°C

RECOMMENDED OPERATING CONDITIONS

Parameter	Symbol	Limits	Unit	Conditions
Supply voltage	V_{DD}	3 ~ 16	V	—
Input voltage	V_{IN}	0 ~ V_{DD}	V	—
Operating temperature	Topr	−40 ~ 85*	°C	—

*For an extended operating temperature range, consult your local ROHM representative.

ELECTRICAL CHARACTERISTICS/SWITCHING CHARACTERISTICS (T_A = 25°C, V_{SS} = 0V)

Parameter	Symbol	Min.	Typ.	Max.	Unit	Conditions V_{DD}	
Output rise time	Tr	—	180	400	ns	5	C_L = 50pF
		—	90	200		10	
		—	65	160		15	
Output fall time	Tf	—	100	200	ns	5	C_L = 50pF
		—	50	100		10	
		—	40	80		15	
Maximum clock frequency	fφmax.	—	2.0	—	MHz	5	C_L = 50pF
		—	6.0	—		10	
		—	7.5	—		15	

BU4000B SERIES PRODUCT SUMMARY

Category	Type	Function	Package Configuration	No. of pins
Analog switches/multiplexers	BU4016B	Quad bilateral switch	DIP/MF	14
	BU4066B	Quad bilateral switch	DIP/MF	14
	BU4051B	8-channel analog multiplexer/demultiplexer	DIP/MF	16
	BU4052B	Dual 4-channel analog multiplexer/demultiplexer	DIP/MF	16
	BU4053B	Triple 2-channel analog multiplexer/demultiplexer	DIP/MF	16
	BU4551B	Quad 2-channel analog multiplexer/demultiplexer	DIP/MF	16
Gates	BU4001B	Quad 2-input NOR gate	DIP/MF	14
	BU4011B	Quad 2-input NAND gate	DIP/MF	14
	BU4030B	Quad exclusive-OR gate	DIP/MF	14
	BU4070B	Quad exclusive-OR gate	DIP/MF	14
	BU4081B	Quad 2-input AND gate	DIP/MF	14
	BU4093B	Quad 2-input NAND Schmitt trigger	DIP/MF	14
	☆BU4049UB	Hex inverting buffer/converter	DIP/MF	16
	BU4069UB	Hex inverter	DIP/MF	14
	☆BU4503B	Hex tristate buffer	DIP/MF	16
	BU4584B	Hex Schmitt trigger	DIP/MF	14
Flip-flops	BU4013B	Dual D-type flip-flop	DIP/MF	14
	☆BU4027B	Dual J-K master-slave flip-flop	DIP/MF	16
Shift registers/counters	BU4015B	Dual 4-bit static shift register	DIP/MF	16
	☆BU4021B	8-stage static shift register	DIP/MF	16
	☆BU4516B	Binary up/down counter	DIP/MF	16
Monostable multivibrator	BU4538B	Dual precision monostable multivibrator	DIP/MF	16
Decoder	BU4028B	BCD to decimal decoder	DIP/MF	16
Latch	BU4042B	Quad D-latch	DIP/MF	16
Others	☆BU4007UB	Dual complementary pair plus inverter	DIP/MF	14

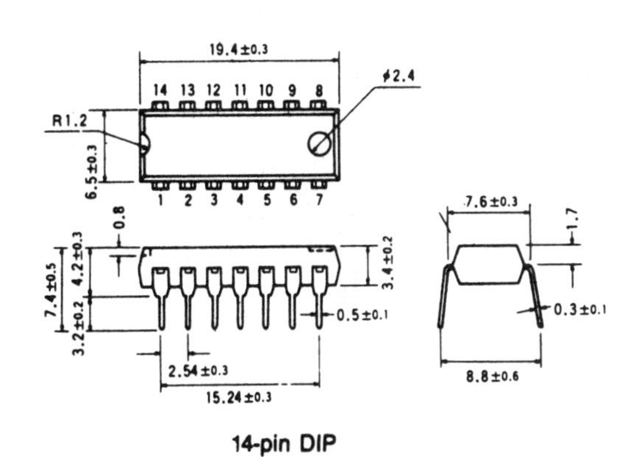

14-pin DIP

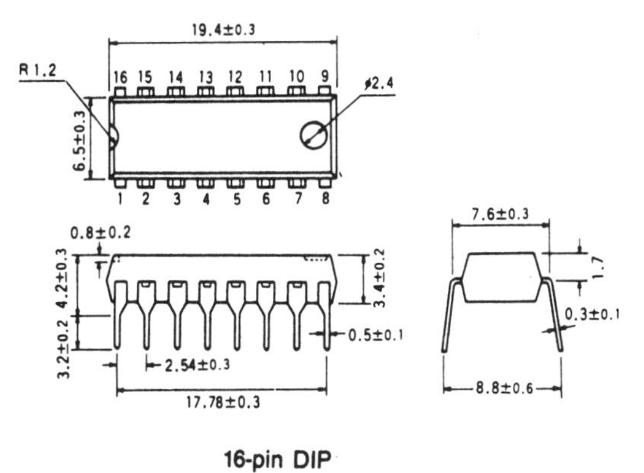

16-pin DIP

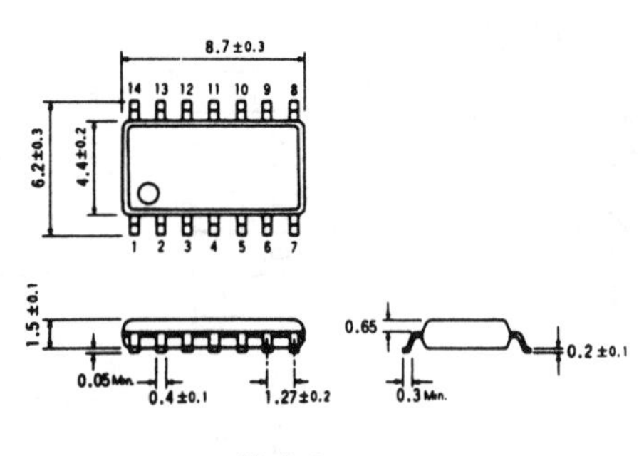

14-pin MF

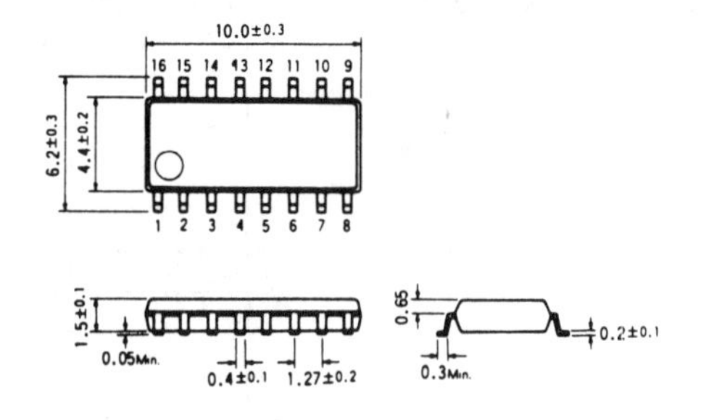

16-pin MF

ELECTRICAL CHARACTERISTICS/DC CHARACTERISTICS (Ta=25°C, V_{SS}=0V)

Parameter	Symbol	Min	Typ.	Max.	Unit	Conditions		
						V_{DD}(V)	V_{IN} (V)	
Input high voltage (Type B)	V_{IH}	3.5	2.75	—	V	5	—	—
		7.0	5.50	—		10		
		11.0	8.25	—		15		
Input high voltage (Type UB)		4.0	2.75	—		5	—	—
		8.0	5.50	—		10		
		12.5	8.25	—		15		
Input low voltage (Type B)	V_{IL}	—	2.25	1.5	V	5	—	—
		—	4.50	3.0		10		
		—	6.75	4.0		15		
Input low voltage (Type UB)		—	2.25	1.0		5	—	—
		—	4.50	2.0		10		
		—	6.75	2.5		15		
Output high voltage	V_{OH}	4.95	5.00	—	V	5	0, V_{DD}	I_O=0mA
		9.95	10.00	—		10		
		14.95	15.00	—		15		
Output low voltage	V_{OL}	—	0.00	0.05	V	5	0, V_{DD}	I_O=0mA
		—	0.00	0.05		10		
		—	0.00	0.05		15		
Output high current	I_{OH}	−0.16	—	—	mA	5	0, V_{DD}	V_{OH}=4.6V
		−0.4	—	—		10		V_{OH}=9.5V
		−1.2	—	—		15		V_{OH}=13.5V
Output low current	I_{OL}	0.44	—	—	mA	5	0, V_{DD}	V_{OL}=0.4V
		1.1	—	—		10		V_{OL}=0.5V
		3.0	—	—		15		V_{OL}=1.5V
Input current	I_{IN}	—	—	±0.3	μA	5	0, V_{DD}	—
Power consumption (Gates)	I_{DD}	—	—	1.0	μA	5	0, V_{DD}	—
		—	—	2.0		10		
		—	—	4.0		15		
Power consumption (Flip-fops, buffers)		—	—	4.0		5		
		—	—	8.0		10		
		—	—	16.0		15		
Power consumption (MSI)		—	—	20		5		
		—	—	40		10		
		—	—	80		15		

ELECTRICAL CHARACTERISTIC CURVES

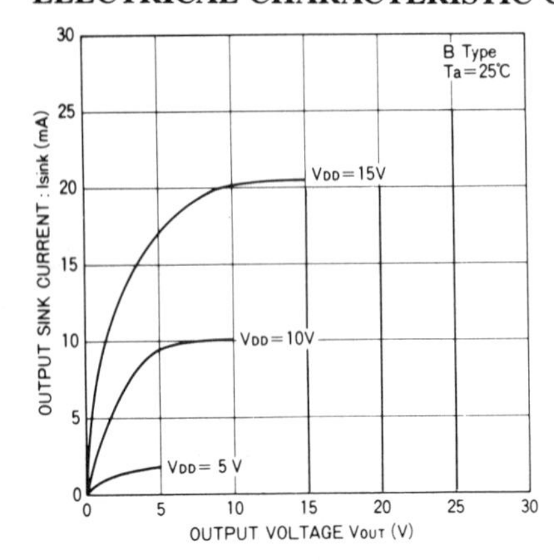

Output sink current vs. output voltage

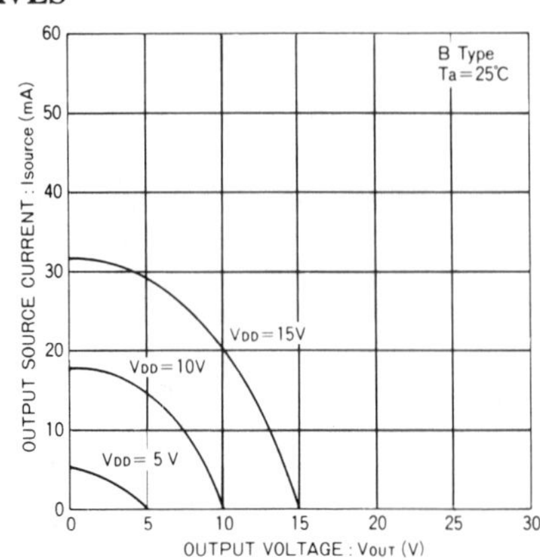

Output source current vs. output voltage

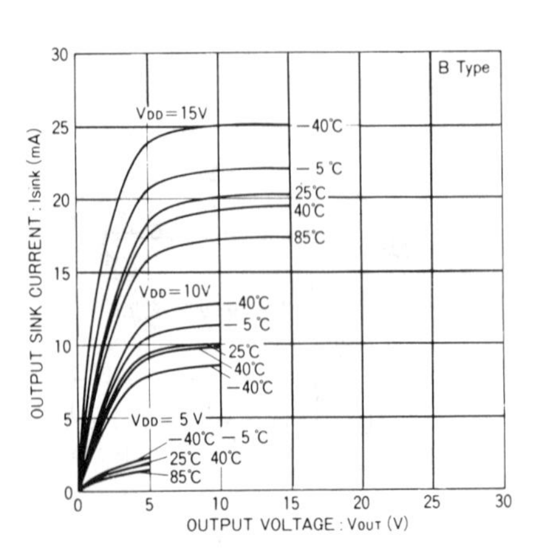

Output sink current vs. output voltage

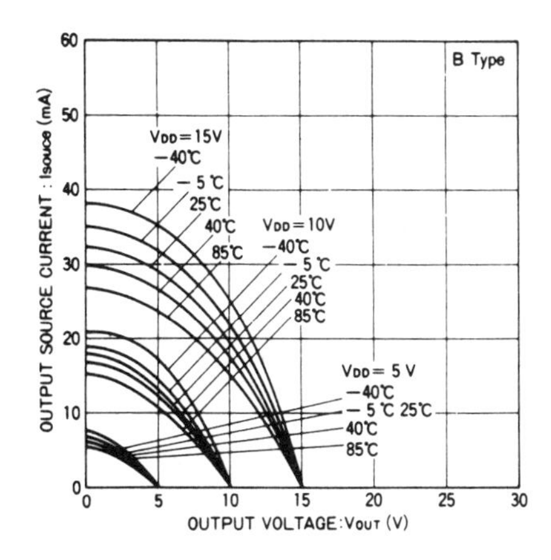

Output source current vs. output voltage

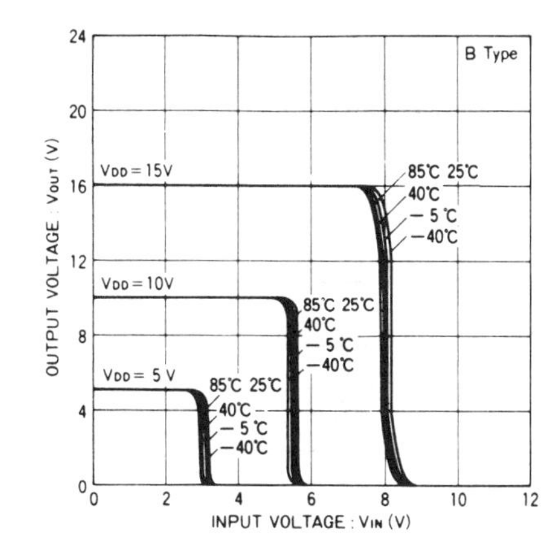

Output voltage vs. input voltage

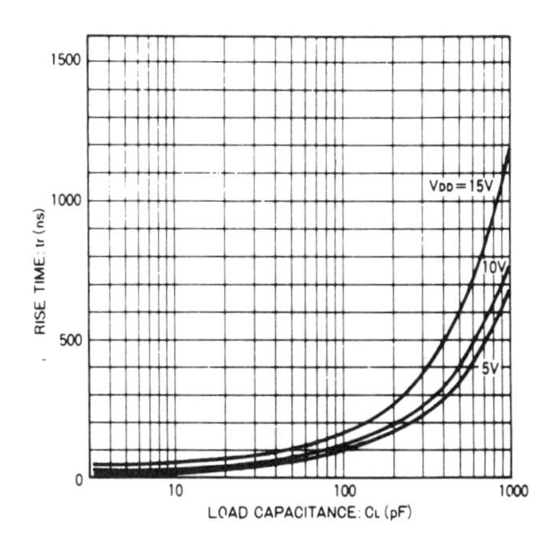

Rise time vs. load capacitance

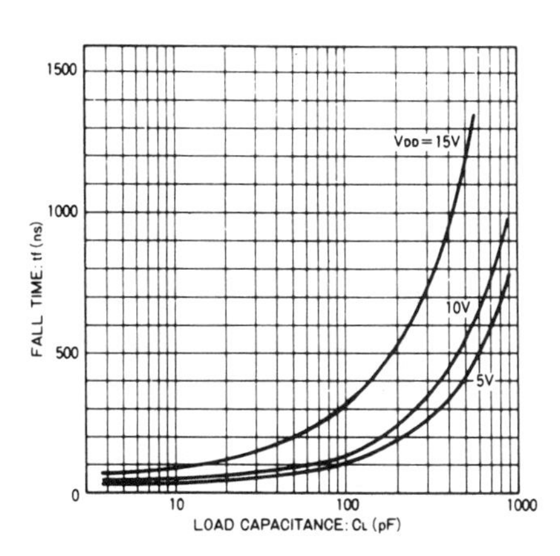

Fall time vs. load capacitance

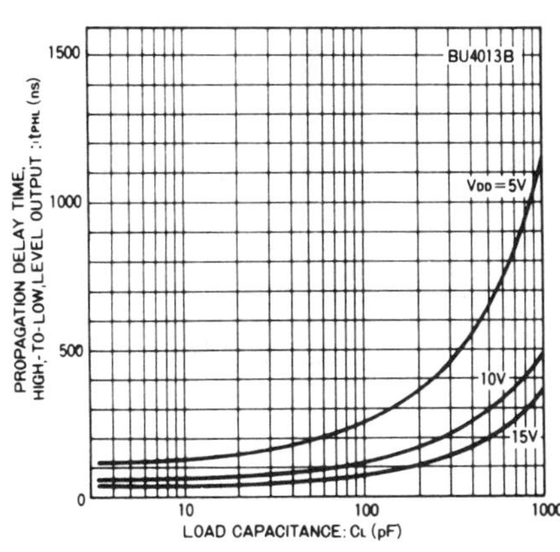

Propagation delay time, high-to-low output vs. load capacitance

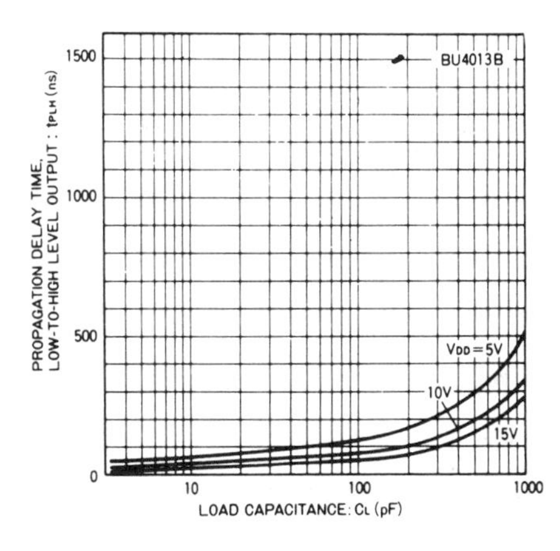

Propagation delay time, low-to-high output vs. load capacitance

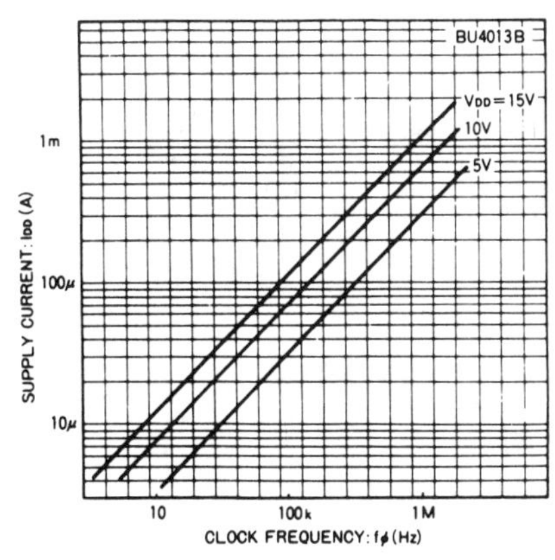

Supply current vs. clock frequency

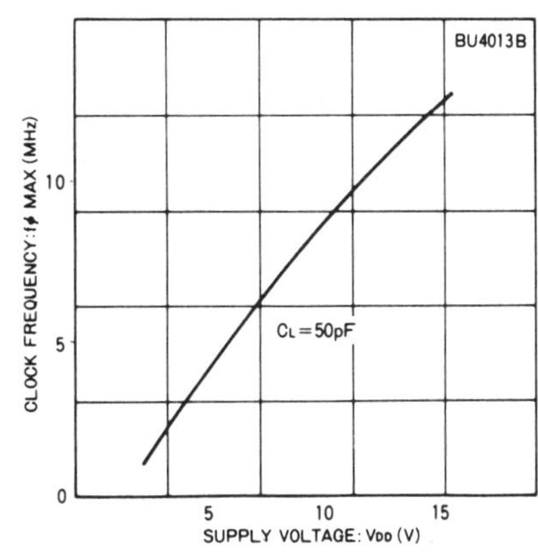

Clock frequency vs. supply voltage

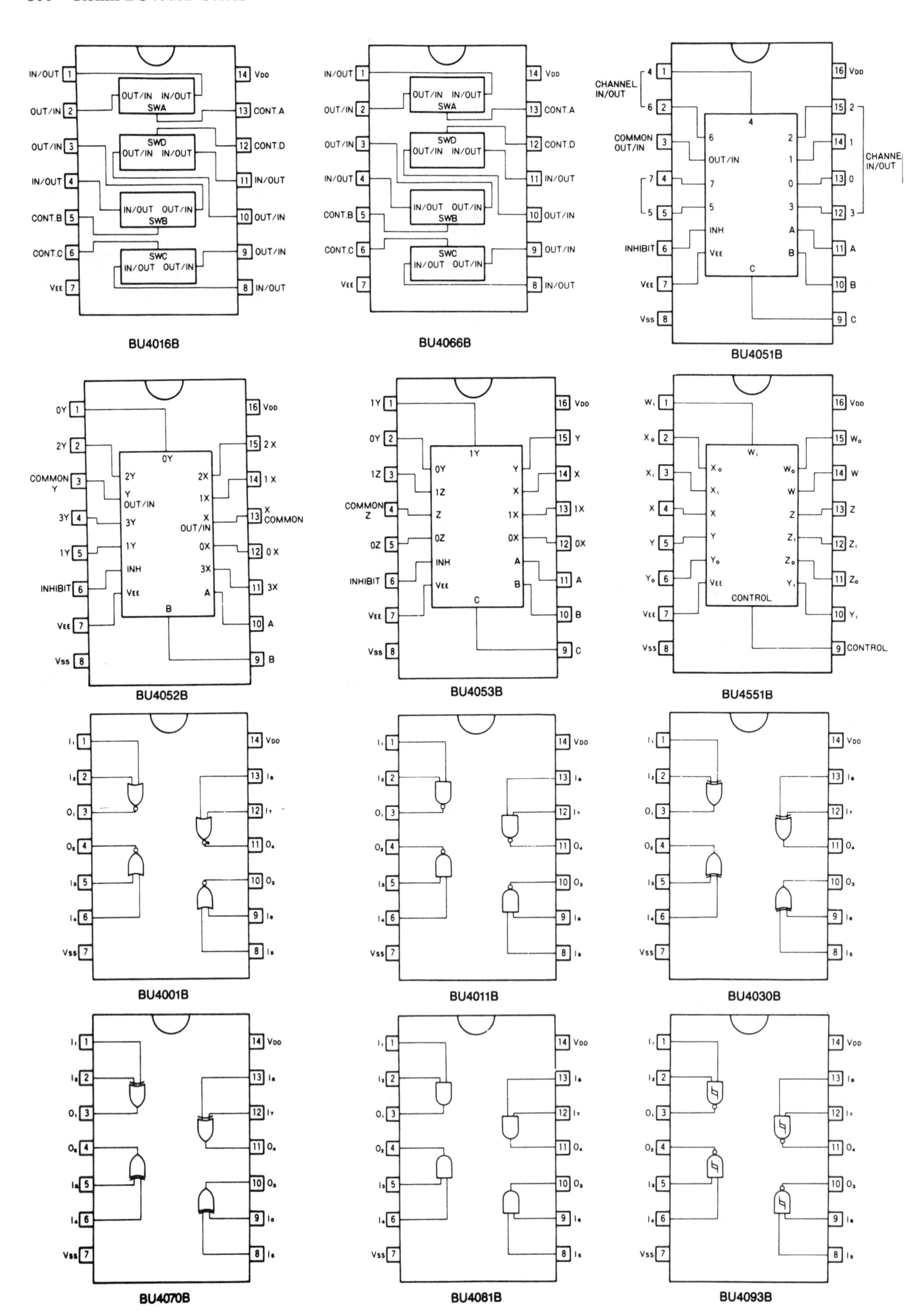
IN/OUT
OUT/IN
CONT.A
CONT.B
CONT.C
CONT.D
SWA
SWB
SWC
SWD
VDD
VEE
VSS
CHANNEL IN/OUT
COMMON OUT/IN
INHIBIT
COMMON Y
X COMMON
COMMON Z
CONTROL
BU4016B
BU4066B
BU4051B
BU4052B
BU4053B
BU4551B
BU4001B
BU4011B
BU4030B
BU4070B
BU4081B
BU4093B

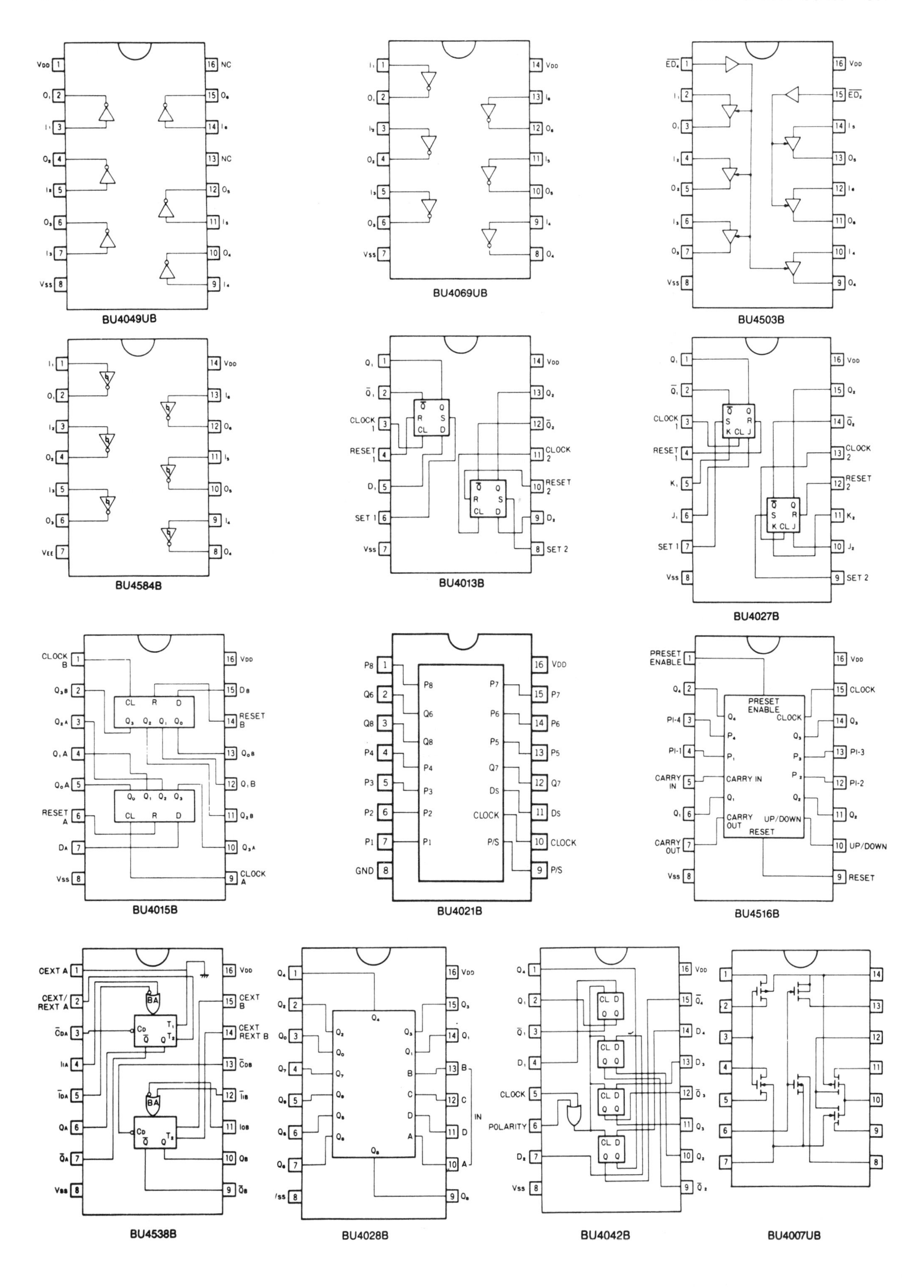
BU4049UB
BU4069UB
BU4503B
BU4584B
BU4013B
BU4027B
BU4015B
BU4021B
BU4516B
BU4538B
BU4028B
BU4042B
BU4007UB

Rohm
BA634 BA634F
T Flip-Flops with Reset

- Reset pin provided
- Low power consumption: 1 mA (typ.)
- Small difference in power consumptions at output high and low enables easy power supply design

APPLICATIONS
- Electronic organs
- Digital clocks
- Control equipment
- Digital equipment in general

The BA634 is a toggle-type flip-flop with high noise margin. It is available in a 5-pin SIP package. The output circuit has a totem-pole configuration.

ABSOLUTE MAXIMUM RATINGS

Parameter	Symbol	Limits	Unit
Supply voltage	V_{EE}	−16	V
Power dissipation	Pd	150*	mW
Operating temperature range	Topr	−10~60	°C
Storage temperature range	Tstg	−55~125	°C

*Derating is done at 1.5mW/°C for operation above Ta=25°C.

ELECTRICAL CHARACTERISTICS (Ta=25°C, V_{EE}= −12V)

Parameter	Symbol	Min.	Typ.	Max.	Unit	Conditions
Input voltage range	V_{IL}	−12	—	0	V	$R_L=\infty$
Set threshold voltage	V_{TH}	−4.4	−2.8	−2.1	V	$R_L=\infty$
Maximum frequency	F_{OP}	100	500	—	kHz	$R_L=\infty$
Quiescent current	I_{Q1}	—	1.0	2.2	mA	V_{IN}=H, V_O=H, RESET=H, $R_L=\infty$
Quiescent current	I_{Q2}	—	1.3	2.7	mA	V_{IN}=L, V_O=L, RESET=H
Quiescent current	I_{Q3}	—	2.4	4.3	mA	V_{IN}=H, RESET=L
Low output voltage	V_{OL}	—	−11.4	−11	V	R_L=39k
High output voltage	V_{OH1}	−1.9	−1.35	—	V	R_L=10k
High output voltage	V_{OH2}	−1.0	−0.7	—	V	R_L=100k
Output voltage rise time	Tr	—	3	8	μs	$R_L=\infty$
Output voltage fall time	Tf	—	1	3	μs	$R_L=\infty$
Reset threshold voltage	Vthr	−6.0	3.3	−2.0	V	$R_L=\infty$
Reset maximum frequency	Fopr	20	—	—	kHz	$R_L=\infty$

CIRCUIT DIAGRAM

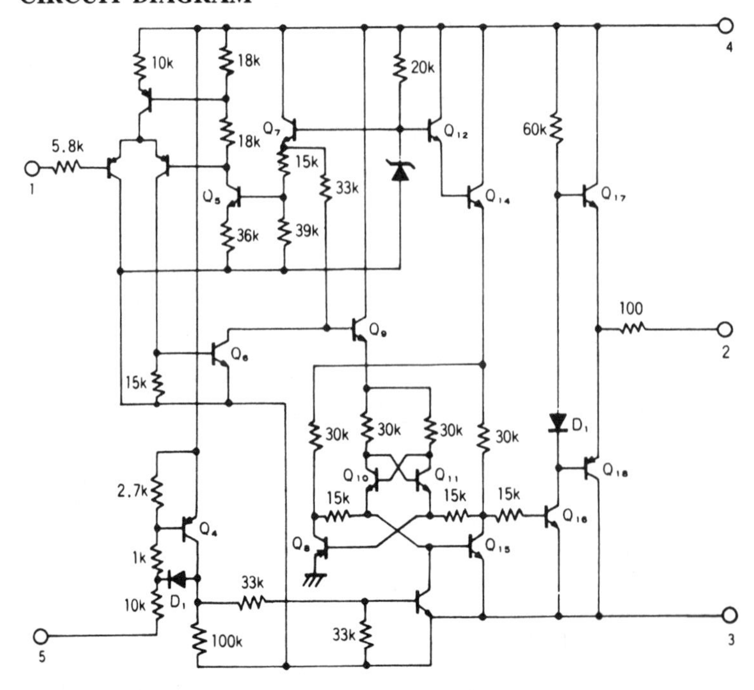

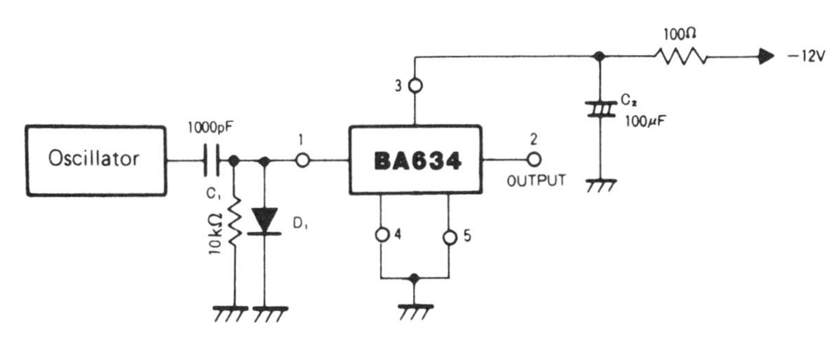

Application for clock

Capacitor C_1 is an input coupling capacitor.
Diode D_1 prevents the input pin from swinging on the positive side. Capacitor C_2 is a line filter.

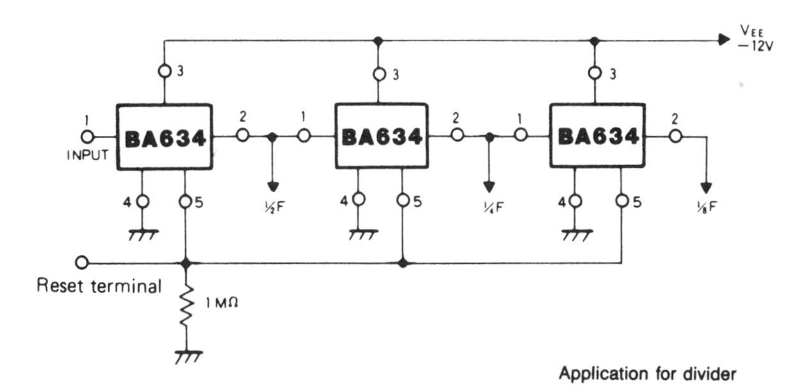

Application for divider

The device is reset by pulling down the reset pin to V_{EE}. Reset operation overrides all other operations. During reset, dividing operation is stopped and the output is set low.

BLOCK DIAGRAM

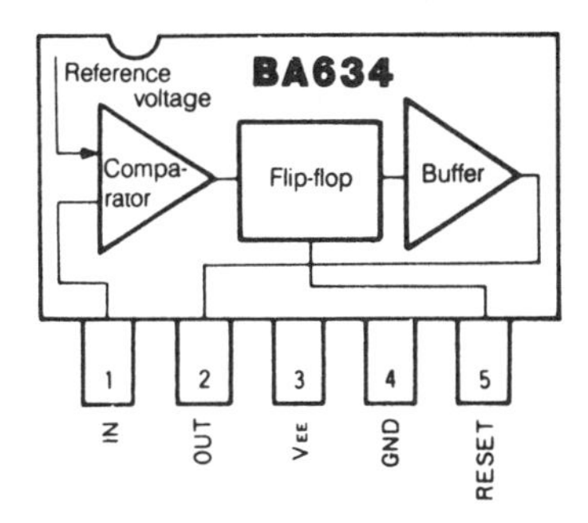

(Unit: mm)

Assembled in the mini-flat package at your request.

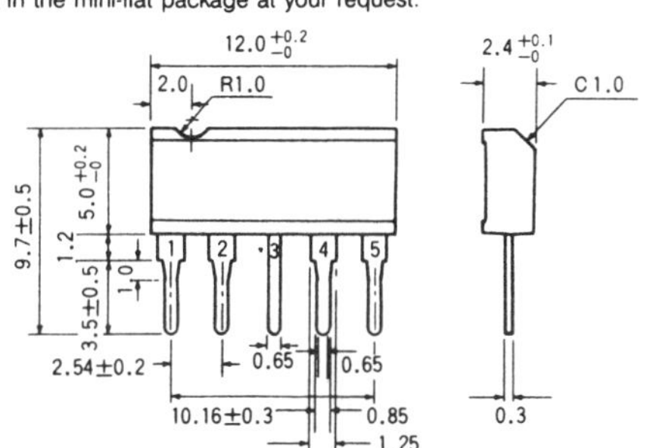

PRECAUTION

The BA634 is also available in a mini-flat package (BA634F) at customer's request. Note that the BA634 and BA634F come in different packages with different pin configurations.

SECTION 6

MICROCOMPUTERS

This information is produced with permission of Analog Devices, Inc.

Analog Devices
AD538
Real-Time Analog Computational Unit (ACU)

- $V_{OUT} = V_Y \left(\frac{V_Z}{V_X}\right)^m$ transfer function
- Wide dynamic range (denominator) −1000:1
- Simultaneous multiplication and division
- Resistor-programmable powers and roots
- No external trims required
- Low input offsets $<100\mu V$
- Low error ±0.25% of reading (100:1 range)
- +2V and +10V on-chip references
- Monolithic construction

APPLICATIONS

- One- or two-quadrant multi/div
- Log ratio computation
- Squaring/square rooting
- Trigonometric function approximations
- Linearization via curve fitting
- Precision AGC
- Power functions

The AD538 is a monolithic real-time computational circuit, which provides precision analog multiplication, division, and exponentiation. The combination of low input and output offset voltages and excellent linearity results in accurate computation over an unusually wide input dynamic range. Laser wafer trimming makes multiplication and division with errors as low as 0.25% of reading possible, while typical output offsets of $100\mu V$ or less add to the overall off-the-shelf performance level. Real-time analog signal processing is further enhanced by the device's 400-kHz bandwidth.

The AD538's overall transfer function is $V_O = V_Y (V_z/V_X)^m$. Programming a particular function is via pin strapping. No external components are required for one-quadrant (positive input) multiplication and division. Two-quadrant (bipolar numerator) division is possible with the use of external level shifting and scaling resistors. The desired scale factor for both multiplication and division can be set using the on-chip +2V or +10V references, or they can be controlled externally to provide simultaneous multiplication and division. Exponentiation with an m value from 0.2 to 5 can be implemented with the addition of one or two external resistors.

ABSOLUTE MAXIMUM RATINGS

Supply Voltage	±18V
Internal Power Dissipation	250mW
Output Short Circuit-to-Ground	Indefinite
Input Voltages V_X, V_Y, V_Z	$(+V_S - 1V)$, −1V
Input Currents I_X, I_Y, I_Z, I_O	1mA
Operating Temperature Range	−25°C to +85°C
Storage Temperature Range	−65°C to +150°
Lead Temperature, Storage	60 sec, +300°C

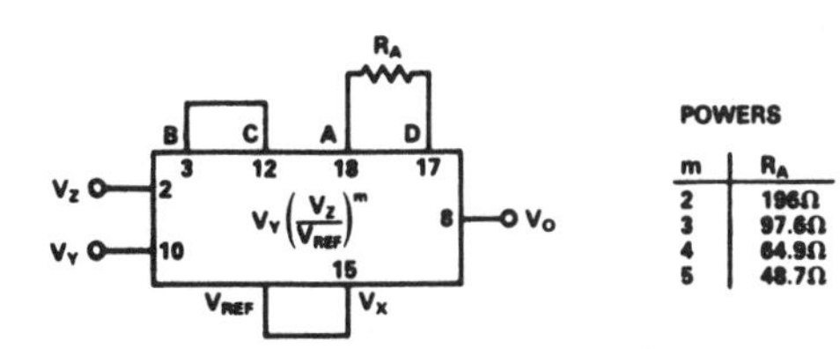

m	R_A
2	196Ω
3	97.6Ω
4	64.9Ω
5	48.7Ω

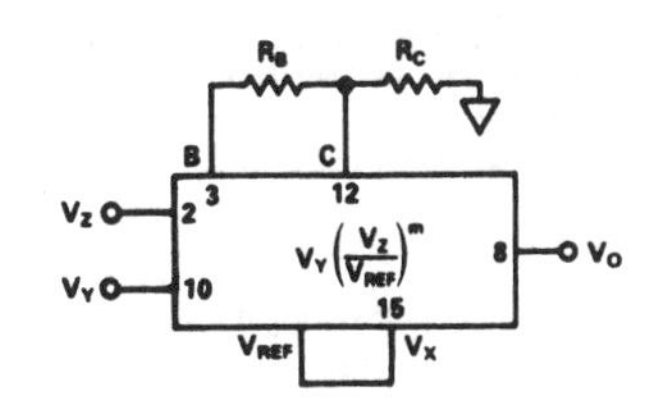

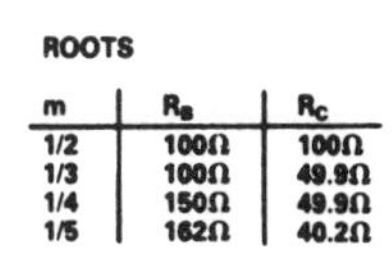

ROOTS

m	R_B	R_C
1/2	100Ω	100Ω
1/3	100Ω	49.9Ω
1/4	150Ω	49.9Ω
1/5	162Ω	40.2Ω

Basic Configurations and Transfer Functions for the AD538

SPECIFICATIONS ($V_S = \pm15V, T_A = 25°C$ unless otherwise specified)

Parameters	Conditions	AD538AD			AD538BD			AD538SD			Units
		Min	Typ	Max	Min	Typ	Max	Min	Typ	Max	
MULTIPLIER/DIVIDER PERFORMANCE											
Nominal Transfer Function	$10V \geq V_X, V_Y, V_Z \geq 0$	$V_O = V_Y\left(\frac{V_Z}{V_X}\right)^m$			$V_O = V_Y\left(\frac{V_Z}{V_X}\right)^m$			$V_O = V_Y\left(\frac{V_Z}{V_X}\right)^m$			
	$400\mu A \geq I_X, I_Y, I_Z \geq 0$	$V_O = 25k\Omega \times I_Y\left(\frac{I_Z}{I_X}\right)^m$			$V_O = 25k\Omega \times I_Y\left(\frac{I_Z}{I_X}\right)^m$			$V_O = 25k\Omega \times I_Y\left(\frac{I_Z}{I_X}\right)^m$			
Total Error Terms 100:1 Input Range[1]	$100mV \leq V_X \leq 10V$		±0.5	**±1**		±0.25	**±0.5**		±0.5	**±1**	% of Reading +
	$100mV \leq V_Y \leq 10V$		±200	**±500**		±100	**±250**		±200	**±500**	µV
	$100mV \leq V_Z \leq 10V$										
	$V_Z \leq 10\,V_X$, m = 1.0										
	$T_A = T_{min}$ to T_{max}		±1	**±2**		±0.5	**±1**		±1.25	**±2.5**	% of Reading +
			±450	**±750**		±350	**±500**		±750	**±1000**	µV
Wide Dynamic Range[2]	$10mV \leq V_X \leq 10V$		±1	**±2**		±0.5	**±1**		±1	**±2**	% of Reading +
	$1mV \leq V_Y \leq 10V$		±200	**±500**		±100	**±250**		±200	**±500**	µV +
	$0V \leq V_Z \leq 10V$		±100	**±250**		±750	**±150**		±200	**±250**	$\mu V \times (V_Y + V_Z)/V_X$
	$V_Z \leq 10V_X$, m = 1.0										
	$T_A = T_{min}$ to T_{max}		±1.5	**±3**		±1	**±2**		±2	**±4**	% of Reading +
			±450	**±750**		±350	**±500**		±750	**±1000**	µV +
			±450	**±750**		±350	**±500**		±750	**±1000**	$\mu V \times (V_Y + V_Z)/V_X$
Exponent (m) Range	$T_A = T_{min}$ to T_{max}	0.2		5	0.2		5	0.2		5	
OUTPUT CHARACTERISTICS											
Offset Voltage	$V_Y = 0, V_C = -600mV$		±200	**±500**		±100	**±250**		±200	**±500**	µV
	$T_A = T_{min}$ to T_{max}		±450	**±750**		±350	**±500**		±750	**±1000**	µV
Output Voltage Swing	$R_L = 2k\Omega$	−11		**±11**	−11		**+11**	−11		**+11**	V
Output Current		5	10		5	10		5	10		mA
FREQUENCY RESPONSE											
Slew Rate			1.4			1.4			1.4		V/µs
Small Signal Bandwidth	$100mV \leq V_Y, V_Z, V_X \leq 10V$		400			400			400		kHz
VOLTAGE REFERENCE											
Accuracy	V_{REF} = 10V or 2V		±25	**±50**		±15	**±25**		±25	**±50**	mV
Additional Error	$T_A = T_{min}$ or T_{max}		±20	**±30**		±20	**±30**		±30	**±50**	mV
Output Current	V_{REF} = 10V to 2V	**1**	2.5		**1**	2.5		**1**	2.5		mA
Power Supply Rejection											
$+2V = V_{REF}$	$\pm4.5V \leq V_S \leq \pm18V$		300	**600**		300	**600**		300	**600**	µV/V
$+10V = V_{REF}$	$\pm13V \leq V_S \leq \pm18V$		200	**500**		200	**500**		200	**500**	µV/V
POWER SUPPLY											
Rated	$R_L = 2k\Omega$		±15			±15			±15		V
Operating Range[3]		**±4.5**		**±18**	**±4.5**		**±18**	**±4.5**		**±18**	V
PSRR	$\pm4.5V < V_S < \pm18V$		0.05	**0.1**		0.05	**0.1**		0.05	**0.1**	%/V
	$V_X = V_Y = V_Z = 1V$										
	$V_{OUT} = 1V$										
Quiescent Current			4.5	**7**		4.5	**7**		4.5	**7**	mA
TEMPERATURE RANGE											
Rated		−25		+85	−25		+85	−55		+125	°C
Storage		−65		+150	−65		+150	−65		+150	°C

NOTES

[1]Over the 100mV to 10V operating range total error is the sum of a percent of reading term and an output offset. With this input dynamic range the input offset contribution to total error is negligible compared to the percent of reading error. Thus, it is specified indirectly as a part of the percent of reading error.

[2]The most accurate representation of total error with low level inputs is the summation of a percent of reading term, an output offset and an input offset multiplied by the incremental gain $(V_Y + V_Z)/V_X$.

[3]When using supplies below ±13V the 10V reference pin *must* be connected to the 2V pin in order for the AD538 to operate correctly.

Specifications subject to change without notice.

Specifications shown in boldface are tested on all production units at final electrical test. Results from those tests are used to calculate outgoing quality levels. All min and max specifications are guaranteed, although only those shown in boldface are tested on all production units.

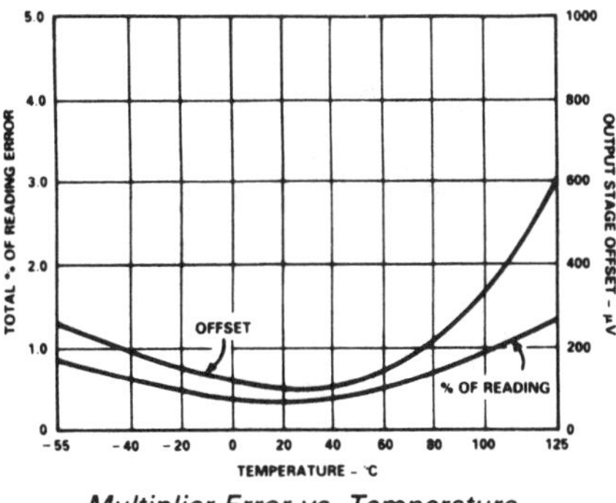

Multiplier Error vs. Temperature
(100mV < V_X, V_Y, V_Z ≤ 10V)

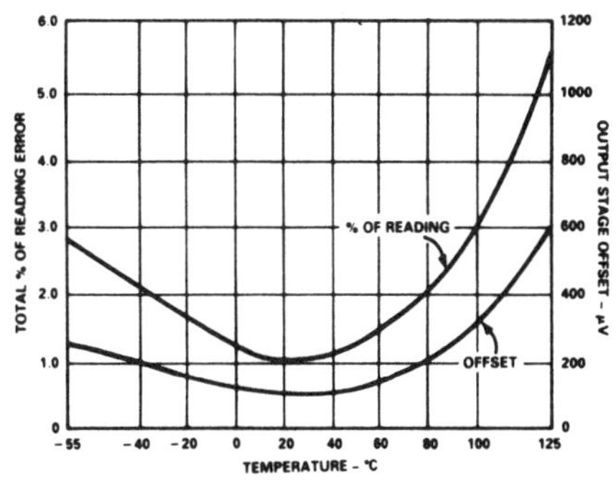

Multiplier Error vs. Temperature
(10mV < V_X, V_Y, V_Z ≤ 100mV)

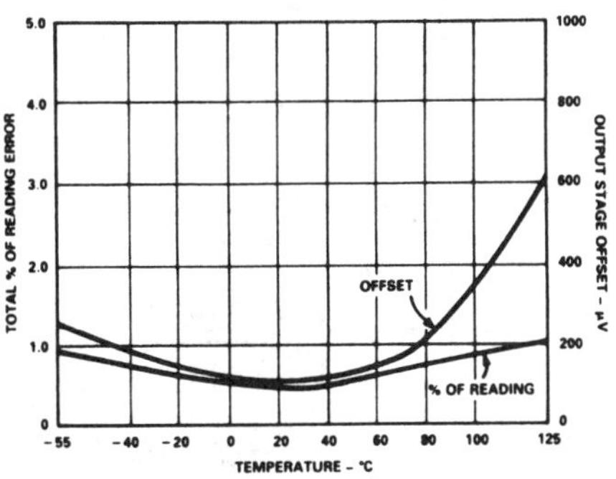

Divider Error vs. Temperature
(100mV < V_X, V_Y, V_Z ≤ 10V)

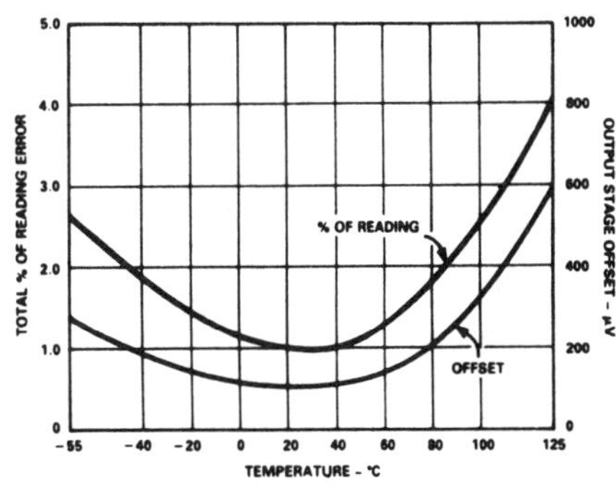

Divider Error vs. Temperature
(10mV ≤ V_X, V_Y, V_Z ≤ 100mV)

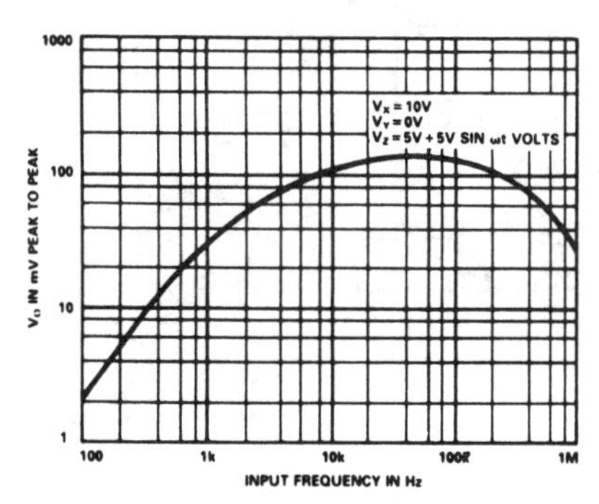

V_Z Feedthrough vs. Frequency

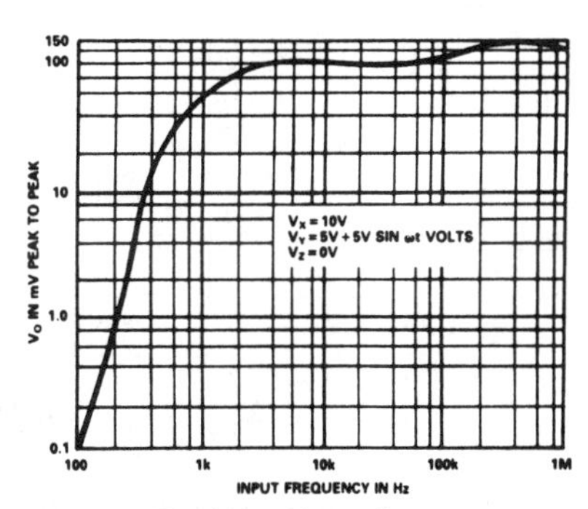

V_Y Feedthrough vs. Frequency

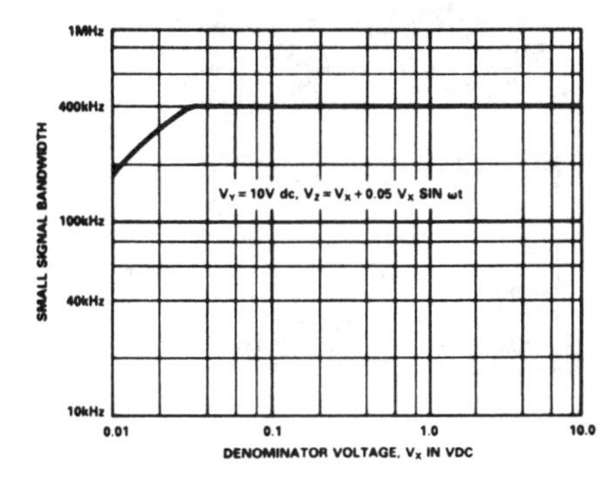

Small Signal Bandwidth vs Denominator Voltage
(One-Quadrant Mult/Div)

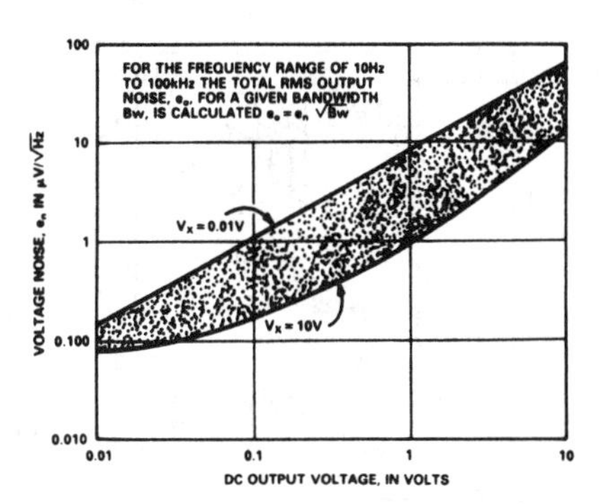

1kHz Output Noise Spectral Density vs. dc Output Voltage

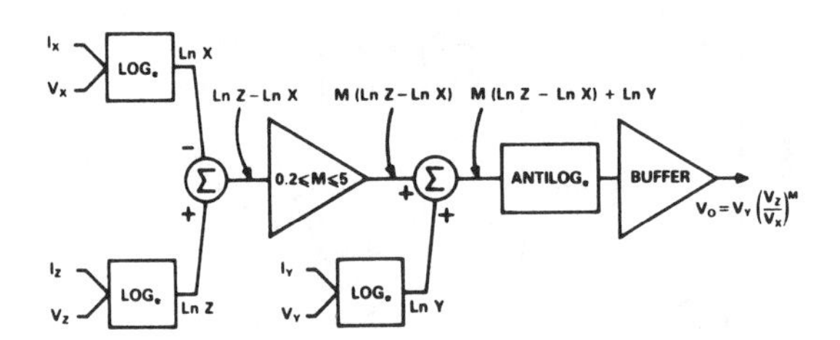

Model Circuit

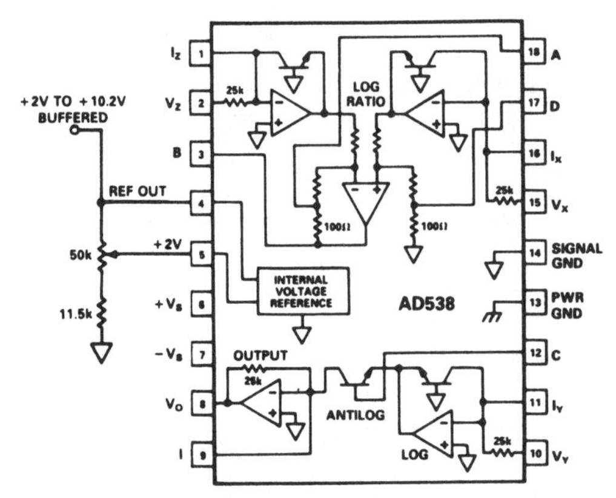

+2V to +10.2V Adjustable Reference

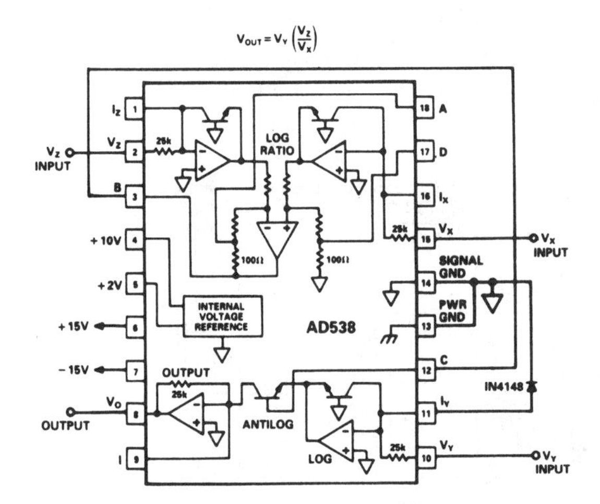

One-Quadrant Combination Multiplier/Divider

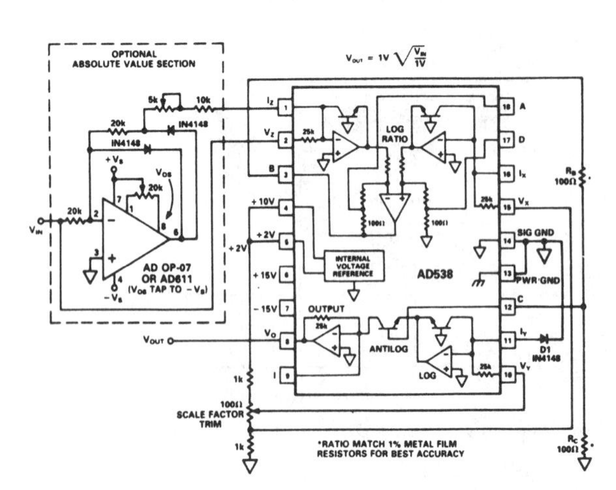

Square Root Circuit

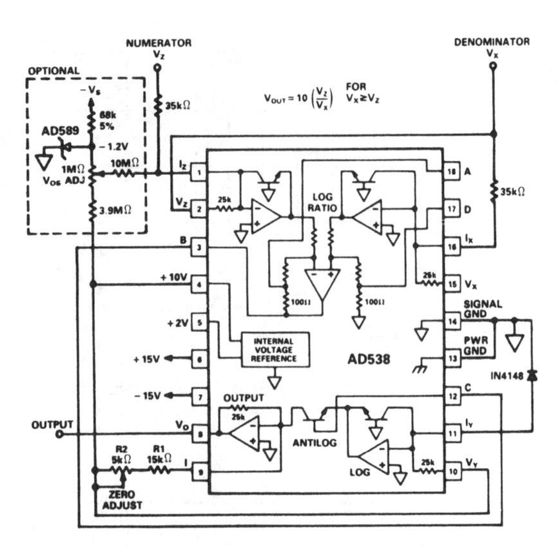

Two-Quadrant Division with 10V Scaling

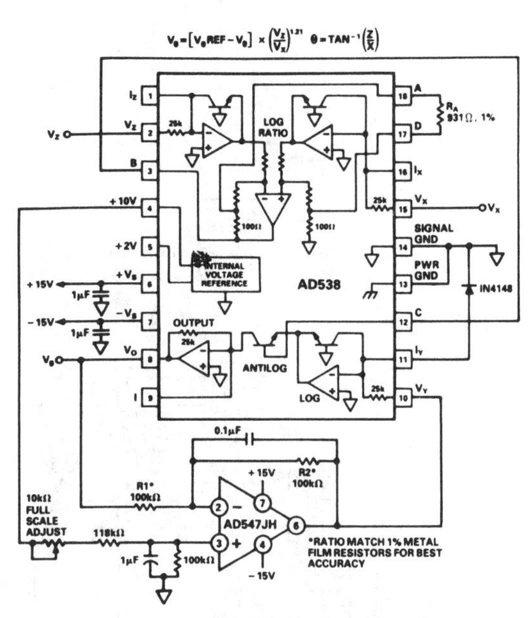

The Arc-Tangent Function

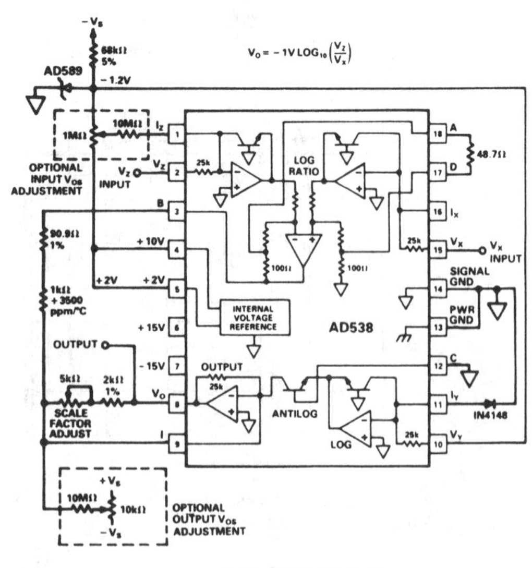

Log Ratio Circuit

Plessey PDSP1601/PDSP1601A

Augmented Arithmetic Logic Unit

- 16-bit, 32-Instruction 20-MHz ALU
- 16-bit, 20-MHz logical, arithmetic or barrel shifter
- Independent ALU and shifter operation
- 4 × 16-bit on-chip Scratched Registers
- Multiprecision operation; e.g., 200ns 64-bit accumulate
- Three-port structure with three internal feedback paths eliminates I/O bottlenecks
- Block floating point support
- 2-micron CMOS
- 300m/W maximum power dissipation
- 84-pin grid array or 84 contact LCC packages

APPLICATIONS

- Digital signal processing
- Array processing
- Graphics
- Database addressing
- High-speed arithmetic processors

ASSOCIATED PRODUCTS

PDSP16112	Complex multiplier
PDSP1640	40-MHz address generator
PDSP16116	16 × 16 complex multiplier
PDSP16318	Complex accumulator
PDSP16330	Pythagoras processor

ABSOLUTE MAXIMUM RATINGS (Note 1)

Supply Voltage V_{CC}	-0.5 to 7.0V
Input Voltage V_{in}	-0.9 to V_{CC} +0.9V
Output Voltage V_{out}	-0.9 to V_{CC} +0.9V
Clamp diode current per pin Ik (See Note 2)	±18mA
Static discharge voltage (HMB)	500V
Storage temperature T_S	-65° C to +150° C
Ambient temperature with power applied T_{amb}	
Military	-40 °C to +125 °C
Industrial	-40 °C to +85 °C
Package power dissipation P_{TOT}	
AC	1000mW
LC	1000mW

NOTES
1. Exceeding these ratings may cause permanent damage. Functional operation under these conditions is not implied.
2. Maximum dissipation or 1 second should not be exceeded, only one output to be tested at any one time.

SWITCHING CHARACTERISTICS

Characteristic	Value				Units	Conditions
	PDSP1601		PDSP1601A			
	Min.	Max.	Min.	Max.		
CLK rising edge to C-PORT		40		25	ns	2 x LSTTL +20pF
CLK rising edge to CO		100		50	ns	1 x LSTTL +5pF
CLK rising edge to BFP		100		50	ns	1 x LSTTL +5pF
Setup $\overline{CEA}$ or $\overline{CEB}$ to CLK rising edge	30		15		ns	
Hold $\overline{CEA}$ or $\overline{CEB}$ to CLK rising edge		0		0	ns	
Setup A or B port inputs to CLK rising edge	40		20		ns	
Hold A or B port inputs to CLK rising edge		0		0	ns	
Setup MSA0-1, MSB, MSS, MSC, RA0-2	40		20		ns	
RS0-2, IA0-4, IS0-3 to CLK rising edge					ns	
Hold RS0-2, IA0-4, IS0-3 to CLK rising edge		0		0	ns	
Setup SV to CLK rising edge	40		20		ns	Input mode
Hold SV to CLK rising edge		0		0	ns	Input mode
CLK rising edge to SV		100		50	ns	20pF load
						SV o/p mode
$\overline{OE}$ ↑ C-PORT ↓ Z					ns	2 x LSTTL +20pF
$\overline{OE}$ ↑ C-PORT ↑ Z					ns	2 x LSTTL +20pF
$\overline{OE}$ ↓ C-PORT Z ↑					ns	2 x LSTTL +20pF
$\overline{OE}$ ↓ C-PORT Z ↓					ns	2 x LSTTL +20pF
Clock period (ALU & Barrel Shifter, serial mode)	200		100		ns	
Clock period (ALU & Barrel Shifter, parallel mode)	100		50		ns	
Clock high time	40		20		ns	
Clock low time	40		20		ns	

ELECTRICAL CHARACTERISTICS

Test conditions (unless otherwise stated):
T_{amb} = -40°C to +85°C V_{CC} = 5.0V ± 10% Ground = 0V

Static Characteristics

Characteristic	Symbol	Value			Units	Conditions
		Min.	Typ.	Max.		
Output high voltage	V_{OH}	2.4			V	I_{OH} = 8mA
Output low voltage	V_{OL}			0.4	V	I_{OL} = -8mA
Input high voltage	V_{IH}	2.0			V	
Input low voltage	V_{IL}			0.8	V	
Input leakage current	I_{IL}	-10		+10	μA	GND < V_{IN} < V_{CC}
V_{CC} current	I_{CC}			60	mA	T_{amb} = -40°C to +85°C
Output leakage current	I_{OZ}	-50		+50	μA	GND < V_{OUT} < V_{CC}
Output S/C current	I_{OS}	15		80	mA	V_{CC} = Max
Input capacitance	C_{IN}		5		pF	

PIN DESCRIPTIONS

LC Pin	AC Pin	Function	LC Pin	AC Pin	Function	LC Pin	AC Pin	Function	LC Pin	AC Pin	Function
1	C6	IA4	22	F3	GND	43	J6	IS0	64	F9	GND
2	A6	MSB	23	G3	MSA0	44	J7	IS1	65	F11	C8
3	A5	MSS	24	G1	MSA1	45	L7	IS2	66	E11	C9
4	B5	B15	25	G2	A15	46	K7	IS3	67	E10	C10
5	C5	B14	26	F1	A14	47	L6	SV0	68	E9	C11
6	A4	B13	27	H1	A13	48	L8	SV1	69	D11	C12
7	B4	B12	28	H2	A12	49	K8	SV2	70	D10	C13
8	A3	B11	29	J1	A11	50	L9	SV3	71	C11	C14
9	A2	B10	30	K1	A10	51	L10	$\overline{SVOE}$	72	B11	C15
10	B3	B9	31	J2	A9	52	K9	RS0	73	C10	$\overline{OE}$
11	A1	B8	32	L1	A8	53	L11	RS1	74	A11	BFP
12	B2	B7	33	K2	A7	54	K10	VCC	75	B10	VCC
13	C2	B6	34	K3	A6	55	J10	RS2	76	B9	CO
14	B1	B5	35	L2	A5	56	K11	C0	77	A10	RA0
15	C1	B4	36	L3	A4	57	J11	C1	78	A9	RA1
16	D2	B3	37	K4	A3	58	H10	C2	79	B8	RA2
17	D1	B2	38	L4	A2	59	H11	C3	80	A8	CI
18	E3	B1	39	J5	A1	60	F10	C4	81	B6	IA0
19	E2	B0	40	K5	A0	61	G10	C5	82	B7	IA1
20	E1	$\overline{CEB}$	41	L5	$\overline{CEA}$	62	G11	C6	83	A7	IA2
21	F2	CLK	42	K6	MSC	63	G9	C7	84	C7	IA3

ALU instructions

1a. ARITHMETIC INSTRUCTIONS

Inst	IA4-AI0	Mnemonic	Operation	Function	Mode
00	00000	CLRXX	RESET	CLEAR ALL REGISTERS	------
01	00001	MIAX1	MINUS A	NA Plus 1	LSBYTE
02	00010	MIACI	MINUS A	NA Plus CI	CASCADE
03	00011	MIACO	MINUS A	NA Plus CO	MULTICYCLE
04	00100	A2SGN	A/2	A/2 Sign Extend	MSBYTE
05	00101	A2RAL	A/2	A/2 with RAL LSB	MULTICYCLE
06	00110	A2RAR	A/2	A/2 with RAR LSB	MULTICYCLE
07	00111	A2RSX	A/2	A/2 with RSX LSB	MULTICYCLE
08	01000	APBCI	A PLUS B	A Plus B Plus CI	CASCADE
09	01001	APBCO	A PLUS B	A Plus B Plus CO	MULTICYCLE
0A	01010	AMBX1	A MINUS B	A Plus NB Plus 1	LSBYTE
0B	01011	AMBCI	A MINUS B	A Plus NB Plus CI	CASCADE
0C	01100	AMBCO	A MINUS B	A Plus NB Plus CO	MULTICYCLE
0D	01101	BMAX1	B MINUS A	NA Plus B Plus 1	LSBYTE
0E	01110	BMACI	B MINUS A	NA Plus B Plus CI	CASCADE
0F	01111	BMACO	B MINUS A	NA Plus B Plus CO	MULTICYCLE

1b. LOGICAL INSTRUCTIONS

Inst	IA4-IA0	Mnemonic	Operation	Function
10	10000	ANXAB	A AND B	A.B
11	10001	ANANB	A AND NB	A.NB
12	10010	ANNAB	NA AND B	NA.B
13	10011	ORXAB	A OR B	A + B
14	10100	ORNAB	NA OR B	NA + B
15	10101	XORAB	A XOR B	A XOR B
16	10110	PASXA	PASS A	A
17	10111	PASNA	INVERT A	NA

1c. CONTROL INSTRUCTIONS

Inst	IA4-IA0	Mnemonic	Operation
18	11000	SBFOV	Set BFP Flag to OVR, Force ALU output to zero
19	11001	SBFU1	Set BFP Flag to UND 1 Force ALU output to zero
1A	11010	SBFU2	Set BFP Flag to UND 2 Force ALU output to zero
1B	11011	SBFZE	Set BFP Flag to ZERO Force ALU output to zero
1C	11100	OPONE	Output 0001 Hex
1D	11101	OPBYT	Output 00FF Hex
1E	11110	OPNIB	Output 000F Hex
1F	11111	OPALT	Output 5555 Hex

KEY

A = A Input to ALU
B = B Input to ALU
CI = External Carry in to ALU
CO = Internally Registered Carry out from ALU
RAL = ALU Register (Left)
RAR = ALU Register (Right)
RSX = Shifter Register (Left or Right)

MNEMONICS

CLRXX	Clear All Registers to zero		
MIAXX	Minus A,	XX	= Carry in to LSB
A2XXX	A Divided by 2,	XXX	= Source of MSB
APBXX	A Plus B,	XX	= Carry in to LSB
AMBXX	A Minus B,	XX	= Carry in to LSB
BMAXX	B Minus A,	XX	= Carry in to LSB
ANX-Y	AND	X	= Operand 1, Y = Operand 2
ORX-Y	OR	X	= Operand 1, Y = Operand 2
XORXY	Exclusive OR	X	= Operand 1, Y = Operand 2
PASXX	Pass	XX	= Operand
SBFXX	Set BFP Flag	XX	= Function
OPXXX	Output Constant	XXX	= Value

Barrel shifter instructions

Inst	IS3-IS0	Mnemonic	Function	I/O
0	0000	LSRSV	Logical Shift Right by SV	I
1	0001	LSLSV	Logical Shift Left by SV	I
2	0010	BSRSV	Barrel Shift Right by SV	I
3	0011	BSLSV	Barrel Shift Left by SV	I
4	0100	LSRR1	Logical Shift Right by R1	X
5	0101	LSLR1	Logical Shift Left by R1	X
6	0110	LSRR2	Logical Shift Right by R2	X
7	0111	LSLR2	Logical Shift Left by R2	X
8	1000	LR1SV	Load Register 1 From SV	I
9	1001	LR2SV	Load Register 2 From SV	I
A	1010	ASRSV	Arithmetic Shift Right by SV	I
B	1011	ASRR1	Arithmetic Shift Right by R1	X
C	1100	ASRR2	Arithmetic Shift Right by R2	X
D	1101	NRMXX	Normalise Output PE	O
E	1110	NRMR1	Normalise Output PE, Load R1	O
F	1111	NRMR2	Normalise Output PE, Load R2	O

Table 2

KEY

SV = Shift Value
R1 = Register 1
R2 = Register 2
PE = Priority Encoder Output
I => SV Port operates as an Input
O => SV Port operates as an Output
X => SV Port in a High Impedance State

MNEMONICS

LSXYY	Logical Shift,	X	= Direction	YY = Source of Shift Value
BSXYY	Barrel Shift,	X	= Direction	YY = Source of Shift Value
ASXYY	Arithmetic Shift,	X	= Direction	YY = Source of Shift Value
LXXYY	Load	XX	= Target	YY = Source
NRMYY	Normalise by PE, Output PE value on SV Port, Load YY Reg			

PIN DESCRIPTIONS

Symbol	Pin No. (LC84 Package)	Description
MSB	2	**ALU B-input multiplexer select control.**[1] This input is latched internally on the rising edge of CLK.
MSS	3	**Shifter Input multiplexer select control.**[1] This input is latched internally on the rising edge of CLK.
B15 - B0	4 - 19	**B Port data input**. Data presented to this port is latched into the input register on the rising edge of CLK. B15 is the MSB.
$\overline{\text{CEB}}$	20	**Clock enable, B Port input register**. When low the clock to this register is enabled.
CLK	21	**Common clock to all internal registered elements**. All registers are loaded, and outputs change on the rising edge of CLK.
MSA0 - MSA1	23 - 24	**ALU A-input multiplexer select control.**[1] These inputs are latched internally on the rising edge of CLK.
A15 - A0	25 - 40	**A Port data input**. Data presented to this port is latched into the input register on the rising edge of CLK. A15 is the MSB.
$\overline{\text{CEA}}$	41	**Clock enable, A Port input register**. When low the clock to this register is enabled.
MSC	42	**C-Port multiplexer select control.**[1] This input is latched internally on the rising edge of CLK.
IS0 - IS3	43 - 46	**Instruction inputs to Barrel Shifter, IS3 = MSB.**[1] These inputs are latched internally on the rising edge of CLK.
SV0 - SV3	47 - 50	**Shift Value I/O Port**. This port is used as an input when shift values are supplied from external sources, and as an output when Normalise operations are invoked. The I/O functions are determined by the IS0 - IS3 instruction inputs, and by the $\overline{\text{SVOE}}$ control. The shift value is latched internally on the rising edge of CLK.
$\overline{\text{SVOE}}$	51	**SV Output enable**. When high the SV port can only operate as an input. When low the SV port can act as an input or as an output, according to the IS0 - IS3 instruction. This pin should be tied high or low, depending upon the application.
RS0, RS1, RS2	52 - 53 55	**Instruction inputs to Barrel Shifter registers.**[1] These inputs are latched internally on the rising edge of CLK.
C0 - C15	56 - 63 65 - 72	**C Port data output**. Data output on this port is selected by the C output multiplexer. C15 is the MSB.
$\overline{\text{OE}}$	73	**Output enable**. The C Port outputs are in a high impedance condition when this control is high.
BFP	74	**Block Floating Point Flag** from ALU, active high.
CO	76	**Carry out** from MSB of ALU.
RA0 - RA2	77 - 79	**Instruction inputs to ALU registers.**[1] These inputs are latched internally on the rising edge of CLK.
CI	80	**Carry in** to LSB of ALU.
IA0 - IA3 IA4	81 - 84 1	**Instruction inputs to ALU,**[1] IA4 = MSB. These inputs are latched internally on the rising edge of CLK.
Vcc	54 & 75	**+5V supply**. Both Vcc pins must be connected
GND	22 & 64	**0V supply**. Both GND pins must be connected.

NOTES
1. All instructions are executed in the cycle commencing with the rising edge of the CLK which latches the inputs.

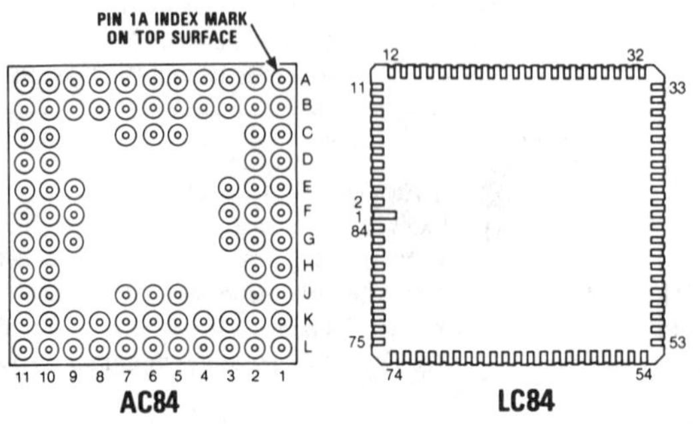

Pin connections - bottom view

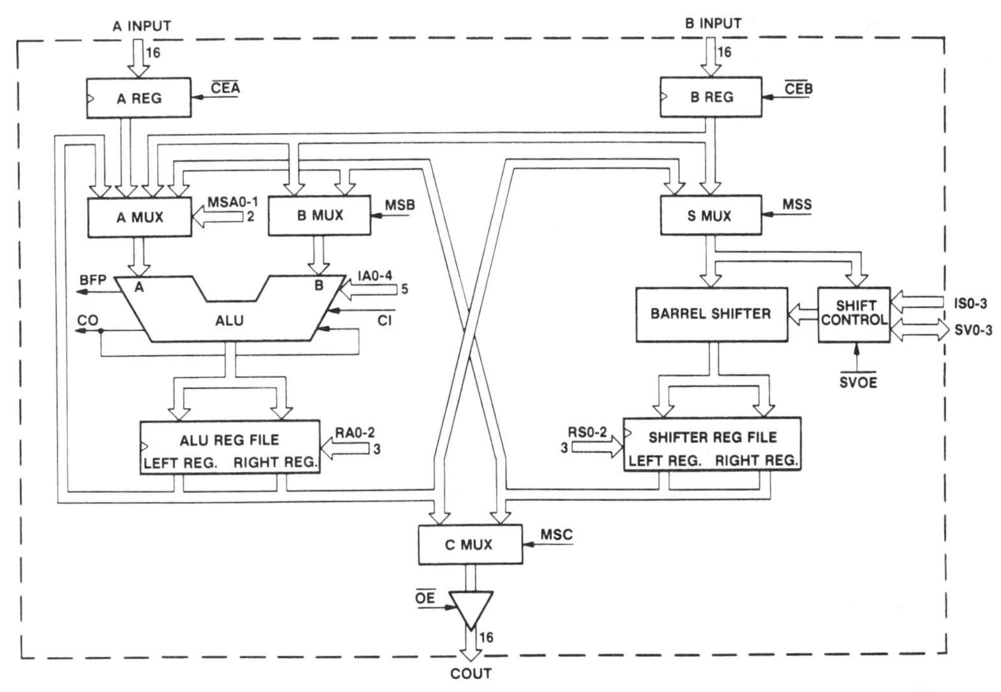

PDSP1601 block diagram

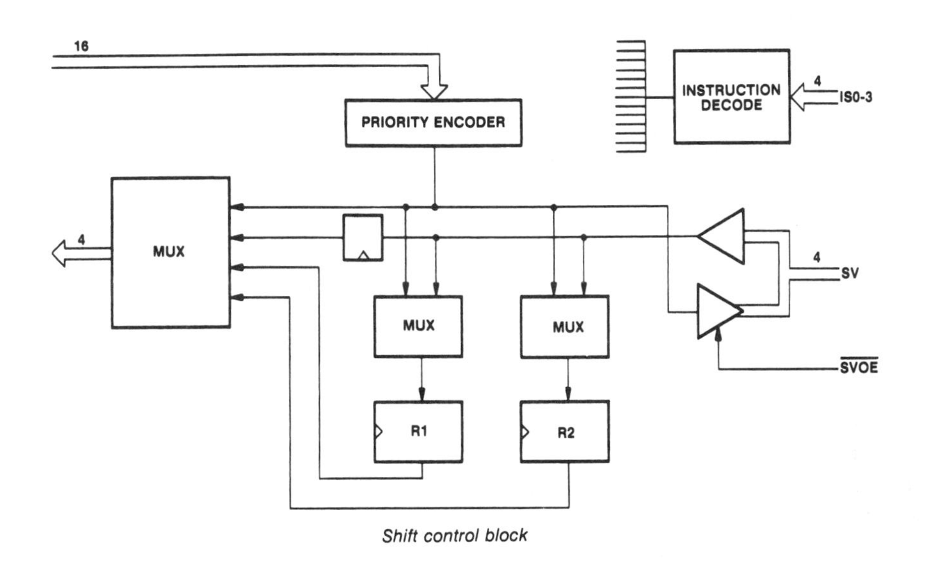

Shift control block

Plessey NJ8821/NJ8821B

Frequency synthesizer (Microprocessor interface) with Resettable counters

- Low power consumption
- High-performance sample-and-hold phase detector
- Microprocessor compatible
- >10-MHz input frequency

ABSOLUTE MAXIMUM RATINGS

Supply voltage (V_{DD} - V_{SS})	-0.5V to 7V
Input voltage	
Open drain O/P (pin 3)	7V
All other pins	V_{SS} -0.3V to V_{DD} +0.3V
Storage temperature	-65°C to +150°C (DG Package, NJ8821B)
Storage temperature	-55°C to +125°C (DP and MP packages, NJ8821)

ELECTRICAL CONDITIONS

Test conditions (unless otherwise stated):
V_{DD}-V_{SS} 5V ± 0.5V
Temperature range NJ8821: -30°C to +70°C, NJ8821B: -40°C to +85°C

DC Characteristics at V_{DD} = 5V

Characteristics	Value			Units	Conditions
	Min.	Typ.	Max.		
Supply current		3.5	5.5	mA	FOSC, FIN = 10MHz } 0 to 5V square wave
		0.7	1.5	mA	FOSC, FIN = 1.0MHz } 0 to 5V square wave
MODULUS CONTROL OUT					
High level	4.6			V	I_{source} 1mA
Low level			0.4	V	I_{sink} 1mA
LOCK DETECT OUT					
Low level			0.4	V	I_{sink} 4mA
Open drain pull-up voltage			8	V	
PDB Output					
High level	4.6			V	I_{source} 5mA
Low level			0.4	V	I_{sink} 5mA
3-state leakage			±0.1	μA	
INPUT LEVELS					
Data Inputs					
High level	4.25			V	TTL compatible
Low level			0.75	V	See note 1
Program Enable Input					
High level	4.25			V	
Low level			0.75	V	
DS INPUTS					
High level	4.25			V	
Low level			0.75	V	

AC Characteristics

Characteristics	Value			Units	Conditions
	Min.	Typ.	Max.		
FIN/OSC inputs	200			mV RMS	10MHz AC coupled sinewave
Max. operating freq. OSC/FIN inputs	10.6			MHz	V_{DD} = 5V, Input squarewave V_{DD}-V_{SS}. Note 4
Propagation delay, clock to modulus control		30	50	ns	Note 2
Strobe pulse width external mode, $t_{w(ST)}$	2			μs	
Data set-up time, $t_{s(DATA)}$	1			μs	
Data hold time, $t_{H(DATA)}$	1			μs	
Address set-up time, t_{SE}	1			μs	
Address hold time, t_{HE}	1			μs	
Digital phase detector propagation delay		500		ns	
Gain programming resistor, RB	5			kΩ	
Hold capacitor, CH			1	nF	Note 3
Output resistance PDA			5	kΩ	
Digital phase detector gain		1		V/Rad	

NOTES
1. Data inputs have internal 'pull-up' resistors to enable them to be driven from TTL outputs.
2. All counters have outputs directly synchronous with their respective clock rising edges.
3. Finite output resistance of internal voltage follower and 'on' resistance of sample switch driving this pin will add a finite time-constant to the loop. A 1nF hold capacitor will give a maximum time-constant of 5 microseconds.
4. Operation at up to 15MHz is possible with a full logic swing but is not guaranteed.

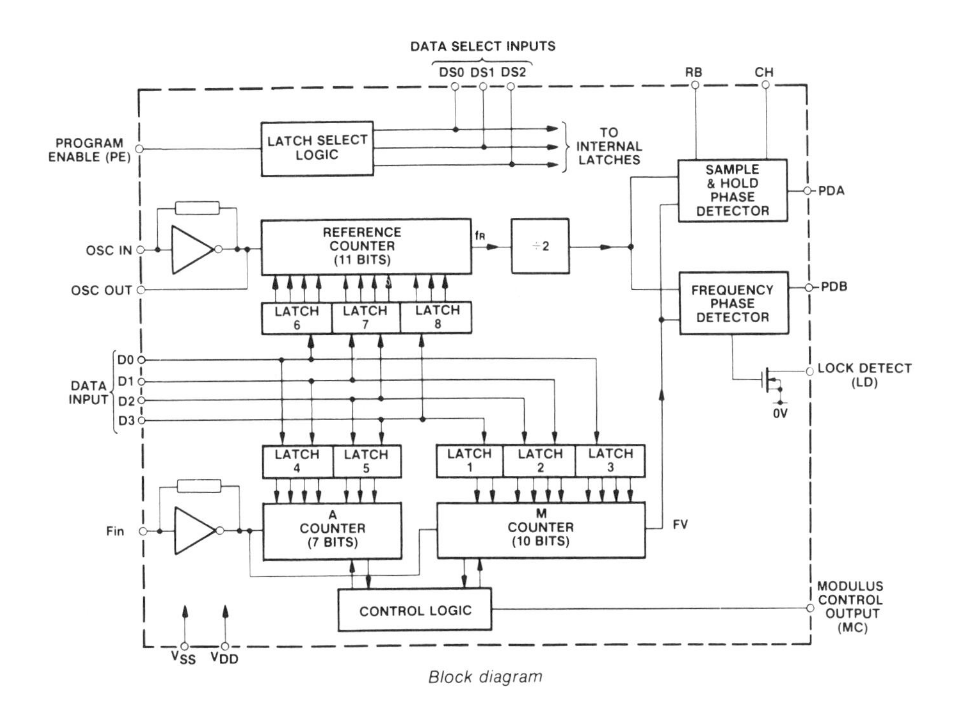

Block diagram

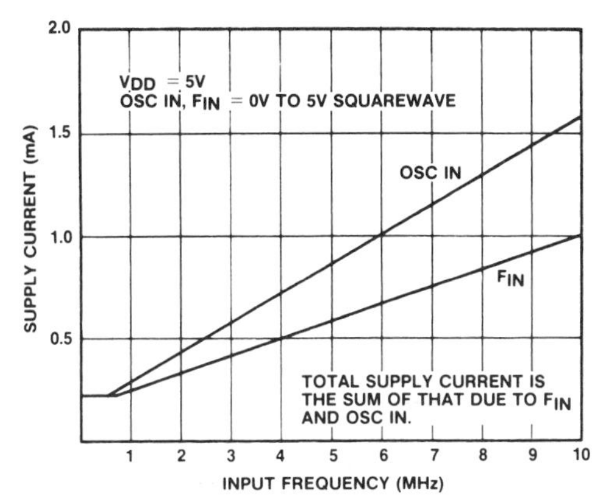

Typical supply current versus input frequency

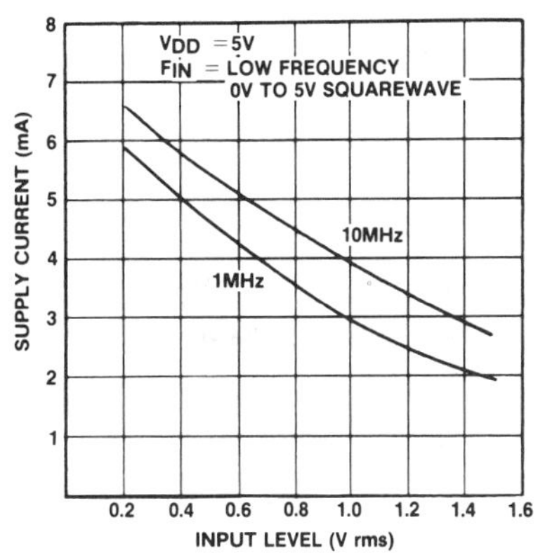

Typical supply current versus input level, Osc In

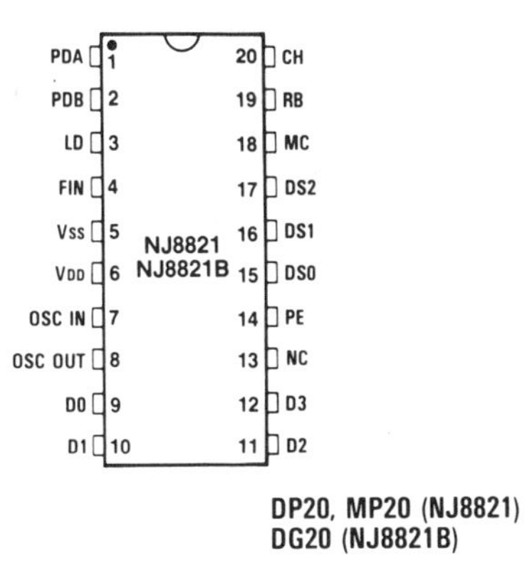

DP20, MP20 (NJ8821)
DG20 (NJ8821B)

Pin connections

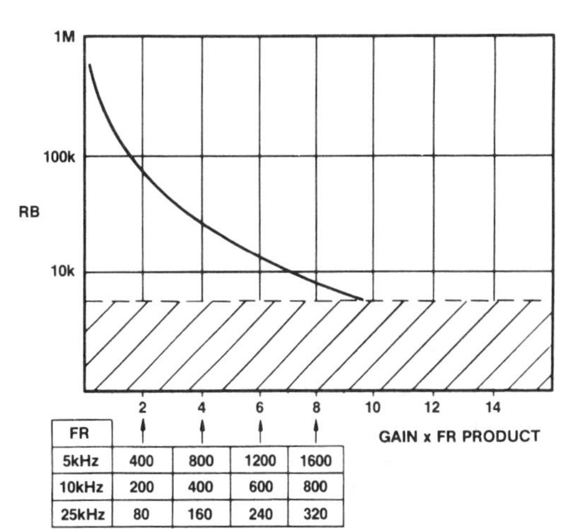

FR	2	4	6	8
5kHz	400	800	1200	1600
10kHz	200	400	600	800
25kHz	80	160	240	320

RB versus gain and reference frequency

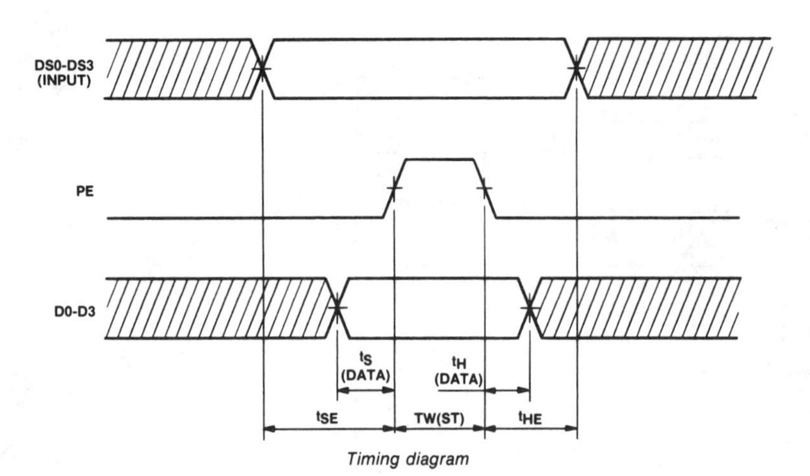

Timing diagram

WORD	DS2	DS1	DS0	D3	D2	D1	D0
1	0	0	0	M1	M0	-	-
2	0	0	1	M5	M4	M3	M2
3	0	1	0	M9	M8	M7	M6
4	0	1	1	A3	A2	A1	A0
5	1	0	0	-	A6	A5	A4
6	1	0	1	R3	R2	R1	R0
7	1	1	0	R7	R6	R5	R4
8	1	1	1	-	R10	R9	R8

Data map

Plessey NJ8821A
Frequency Synthesizer (Microprocessor Interface) With Resettable Counters

- Low power consumption
- High-performance sample-and-hold phase detector
- Microprocessor compatible
- >10 MHz input frequency
- Military temperature range (−55 °C to +125 °C)

ABSOLUTE MAXIMUM RATINGS

Supply voltage (V_{DD} - V_{SS})	-0.5V to 7V
Input voltage	
Open drain O/P (pin 3)	7V
All other pins	V_{SS} -0.3V to V_{DD} +0.3V
Storage temperature	-65°C to +150°C

ELECTRICAL CHARACTERISTICS

Test conditions (unless otherwise stated):
V_{DD} - V_{SS} 5V ± 0.5V
Temperature range -55°C to +125°C

DC Characteristics at V_{DD} = 5V

Characteristics	Value Min.	Value Typ.	Value Max.	Units	Conditions
Supply current		3.5	7.0	mA	FOSC, FIN = 10MHz } 0 to 5V square wave
		0.7	2.0	mA	FOSC, FIN = 1.0MHz }
MODULUS CONTROL OUT					
High level	4.6			V	I_{source} 1mA
Low level			0.4	V	I_{sink} 1mA
LOCK DETECT OUT					
Low level			0.4	V	I_{sink} 4mA
Open drain pull-up voltage			8	V	
PDB Output					
High level	4.6			V	I_{source} 4mA
Low level			0.4	V	I_{sink} 4mA
3-state leakage			±0.1	μA	
INPUT LEVELS					
Data Inputs					
High level	4.25			V	TTL compatible
Low level			0.75	V	See note 1
Program Enable Input					
High level	4.25			V	
Low level			0.75	V	
DS INPUTS					
High level	4.25			V	
Low level			0.75	V	

AC Characteristics

Characteristics	Value			Units	Conditions
	Min.	Typ.	Max.		
FIN/OSC inputs	200			mV RMS	10MHz AC coupled sinewave
Max. operating freq. OSC/FIN inputs	10.6			MHz	V_{DD} = 5V, Input squarewave V_{DD}-V_{SS}.Note 4
Propagation delay, clock to modulus control		30	50	ns	Note 2
Strobe pulse width external mode, $t_{w(ST)}$	2			μs	
Data set-up time, $t_{s(DATA)}$	1			μs	
Data hold time, $t_{H(DATA)}$	1			μs	
Address set-up time, t_{SE}	1			μs	
Address hold time, t_{HE}	1			μs	
Digital phase detector propagation delay		500		ns	
Gain programming resistor, RB	5			kΩ	
Hold capacitor, CH			1	nF	Note 3
Output resistance PDA			5	kΩ	
Digital phase detector gain		1		V/Rad	

NOTES
1. Data inputs have internal 'pull-up' resistors to enable them to be driven from TTL outputs.
2. All counters have outputs directly synchronous with their respective clock rising edges.
3. Finite output resistance of internal voltage follower and 'on' resistance of sample switch driving this pin will add a finite time-constant to the loop. A 1nF hold capacitor will give a maximum time-constant of 5 microseconds.
4. Operation at up to 15MHz is possible with a full logic swing but is not guaranteed.

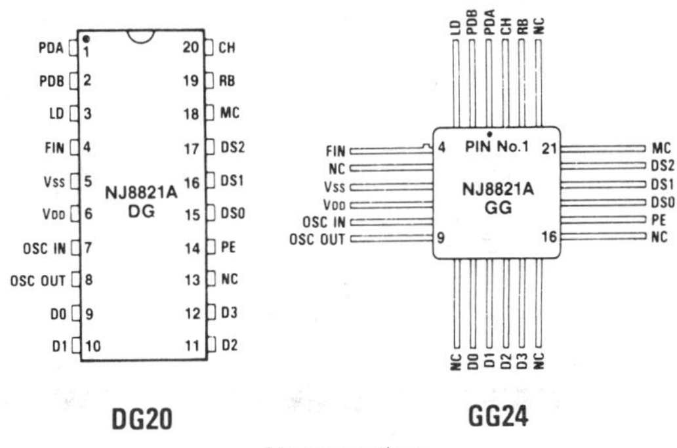

Pin connections

WORD	DS2	DS1	DS0	D3	D2	D1	D0
1	0	0	0	M1	M0	-	-
2	0	0	1	M5	M4	M3	M2
3	0	1	0	M9	M8	M7	M6
4	0	1	1	A3	A2	A1	A0
5	1	0	0	-	A6	A5	A4
6	1	0	1	R3	R2	R1	R0
7	1	1	0	R7	R6	R5	R4
8	1	1	1	-	R10	R9	R8

Data map

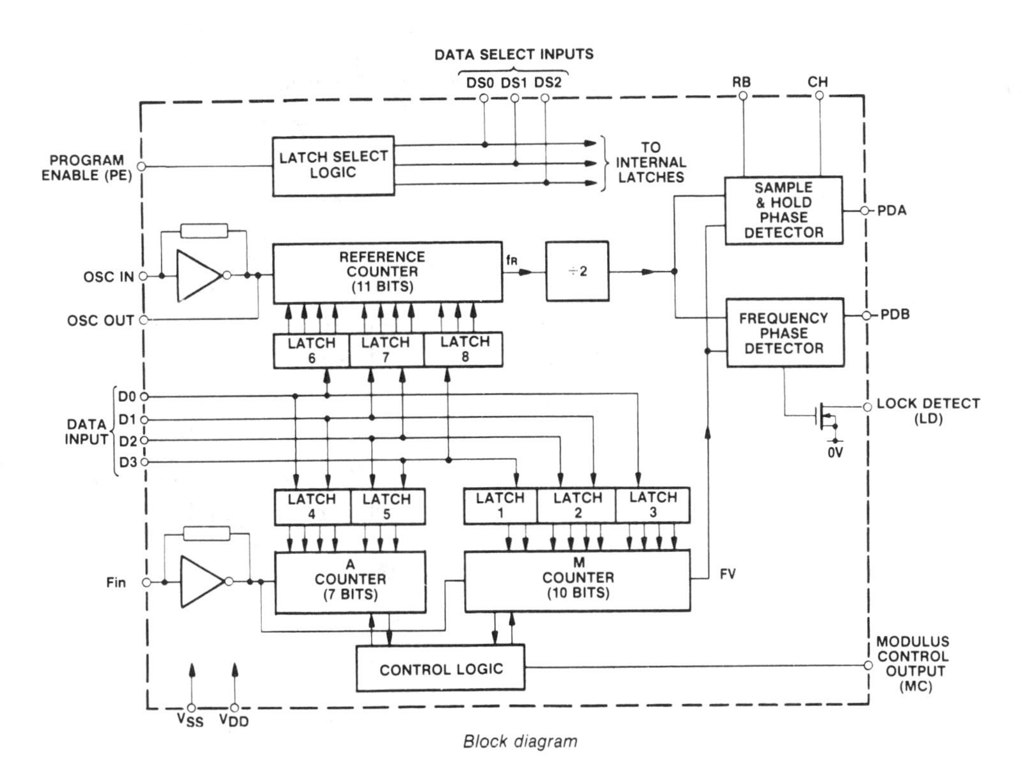

Block diagram

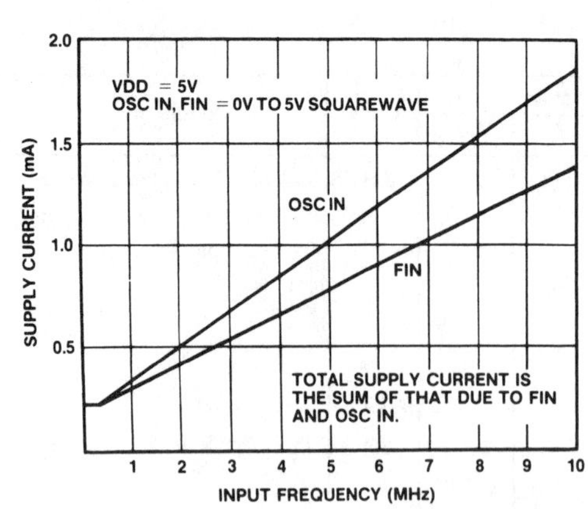

Typical supply current versus input frequency

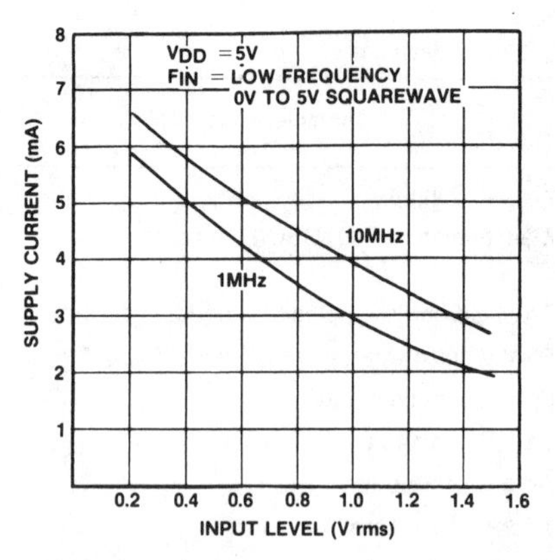

Typical supply current versus input level, Osc In

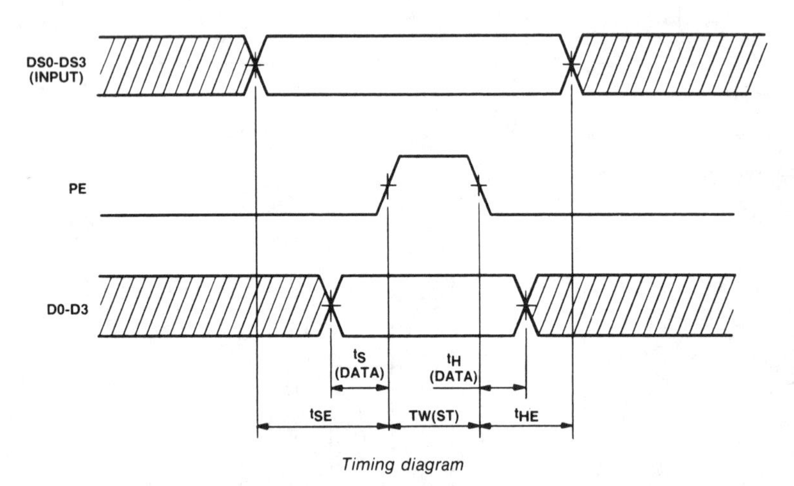

Timing diagram

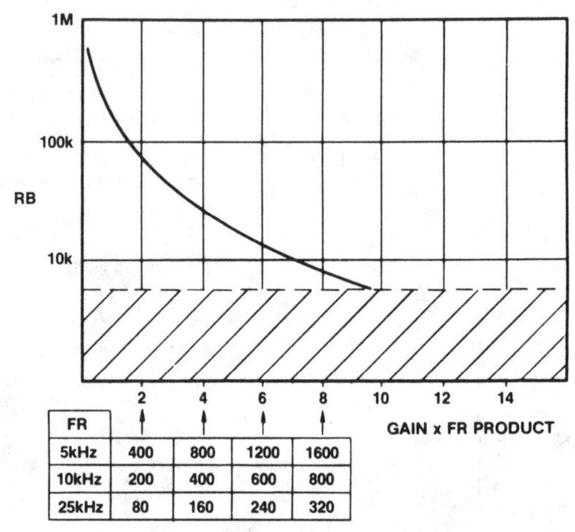

FR				
5kHz	400	800	1200	1600
10kHz	200	400	600	800
25kHz	80	160	240	320

RB versus gain and reference frequency

Plessey NJ8822/NJ8822B
Frequency Synthesizer (Microprocessor Serial Interface) With Resettable Counters

- Low power consumption
- High-performance sample-and-hold phase detector
- Serial input with fast-update feature
- >10-MHz input frequency

ELECTRICAL CHARACTERISTICS

Test conditions (unless otherwise stated):

V_{DD}-V_{SS} 5V ± 0.5V

Temperature range: NJ8822 -30°C to +70°C, NJ8822B -40°C to +85°C

DC Characteristics at V_{DD} = 5V

Characteristics	Value Min.	Value Typ.	Value Max.	Units	Conditions
Supply current			5.5	mA	FOSC, FIN = 10MHz } 0 to 5V square wave
			1.5	mA	FOSC, FIN = 1MHz
MODULUS CONTROL OUT					
High level	4.6			V	I_{source} 1mA
Low level			0.4	V	I_{sink} 1mA
LOCK DETECT OUT					
Low level			0.4	V	I_{sink} 4mA
Open drain pull-up voltage			8	V	
PDB OUTPUT					
High level	4.6			V	I_{source} 5mA
Low level			0.4	V	I_{sink} 5mA
3-state leakage			±0.1	μA	

AC Characteristics

Characteristics	Value			Units	Conditions
	Min.	Typ.	Max.		
FIN/OSC inputs	200			mV RMS	10MHz AC coupled sinewave
Max. operating freq. OSC/FIN inputs	10			MHz	V_{DD} = 5V, Input squarewave V_{DD}-V_{SS}, 25°C
Propagation delay, clock to modulus control		30	50	ns	Note 2
Programming inputs					
Clock high time, t_{CH}	0.5			μs	All timing periods are referenced to the negative transition of the clock waveform
Clock low time, t_{CL}	0.5			μs	
Enable set-up time, t_{ES}	0.2		t_{CH}	μs	
Enable hold time, t_{EH}	0.2			μs	
Data set-up time, t_{DS}	0.2			μs	
Data hold time, t_{DH}	0.2			μs	
Clock rise and fall times	0.2			μs	
Positive going threshold, V_{T+}	3			V	Note 1
Negative going threshold, V_{T-}			2	V	
Phase Detector					
Digital phase detector propagation delay		500		ns	
Gain programming resistor, RB	5			kΩ	
Hold capacitor, CH			1	nF	Note 3
Programming capacitor, CAP			1	nF	
Output resistance, PDA			5	kΩ	

NOTES
1. Data, Clock and Enable inputs are high impedance Schmitt buffers without ull up resistors. They are therefore not TTL compatible.
2. All counters have outputs directly synchronous with their respective clock rising edges.
3. The finite output resistance of the internal voltage follower and 'on' resistance of sample switch driving this pin will add a finite time-constant to the loop. A 1nF hold capacitor will give a maximum time-constant of 5 microseconds.
4. The inputs to the device should be at logic '0' when power is applied if latch up conditions are to be avoided. This includes the signal/osc. frequency inputs.

ABSOLUTE MAXIMUM RATING

Supply voltage (V_{DD}-V_{SS}) -0.5V to 7V
Input voltage
Open drain O/P (pin 3 (DG) pin 4 (MP)) 7V
All other pins V_{SS} -0.3V to V_{DD} +0.3V

Storage temperature -55°C to +125°C (DP and MP packages, NJ8822)
Storage temperature -65°C to +150°C (DG package, NJ8822B)

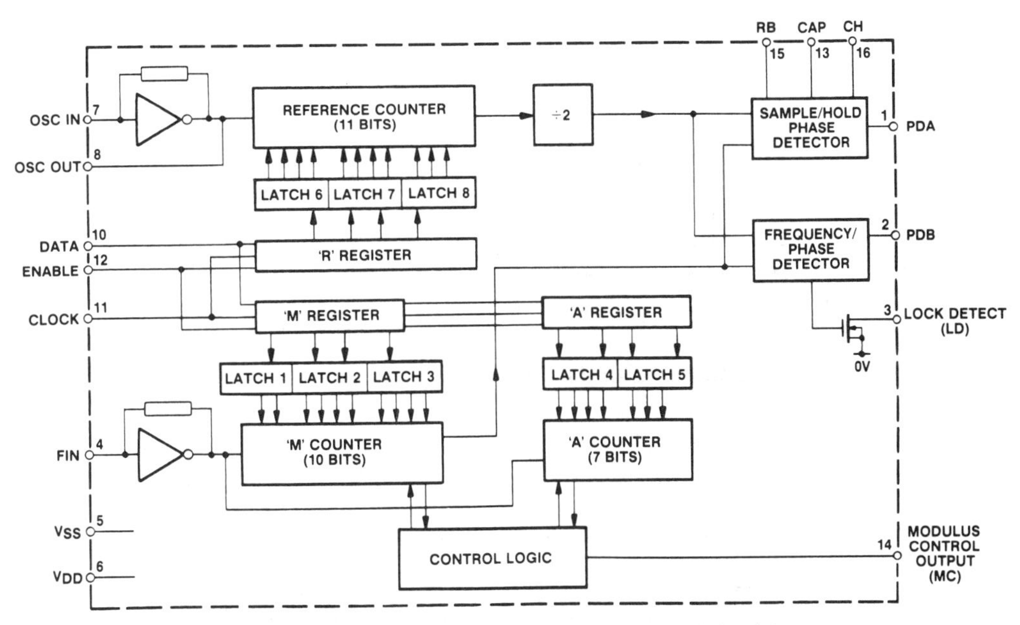

Block diagram. Pin numbers for MP package are shown in brackets.

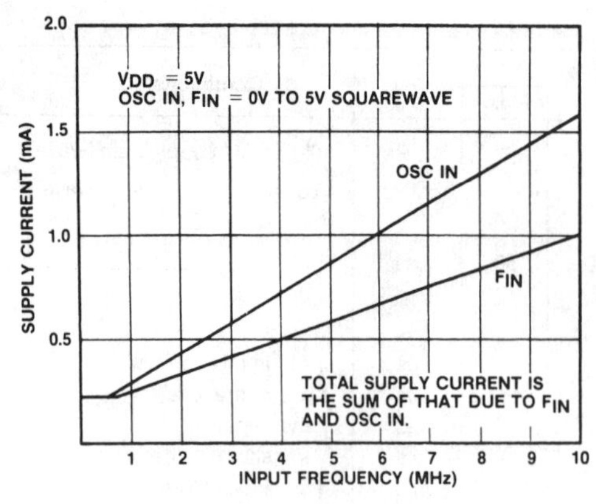

Typical supply current v. input frequency

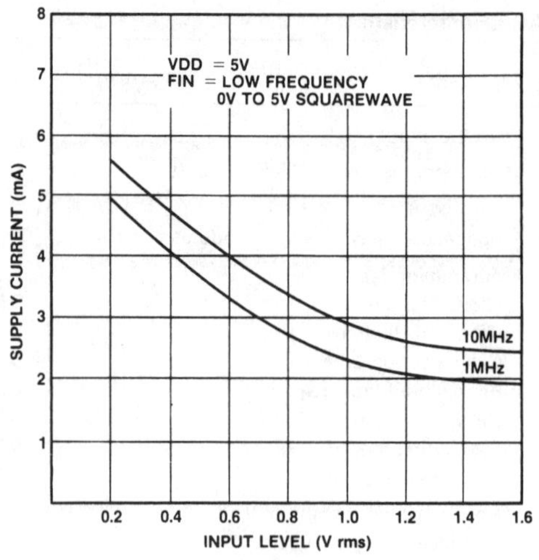

Typical supply current v. input level, Osc In

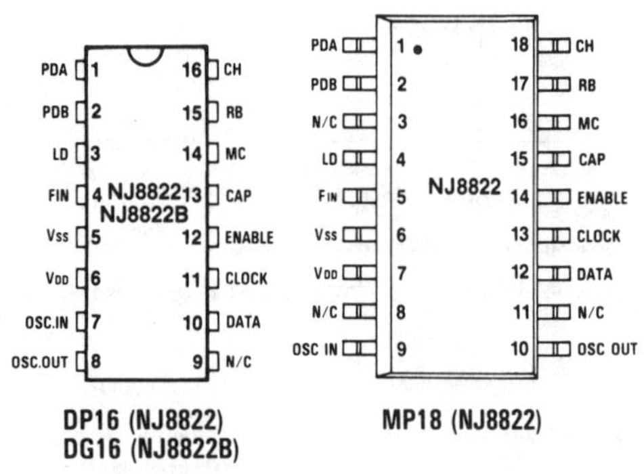

Pin connections - top view, not to scale

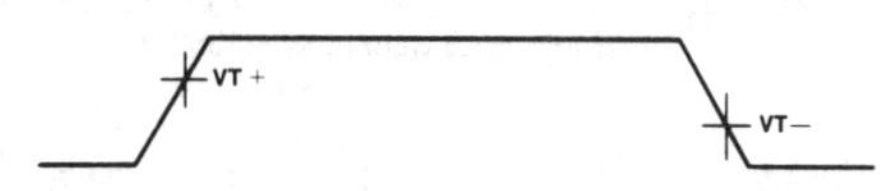

Timing diagram showing voltage thresholds

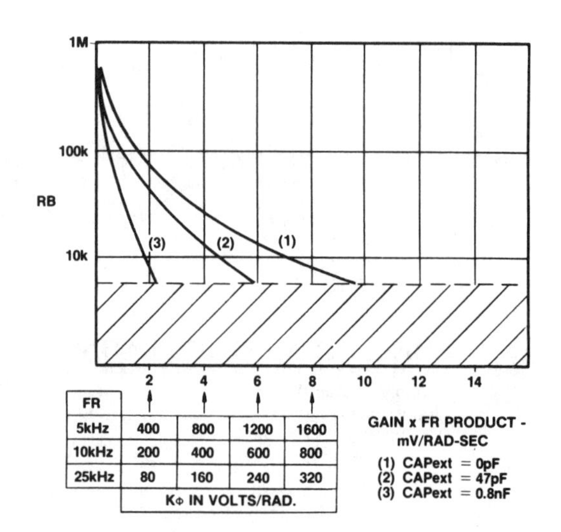

FR				
5kHz	400	800	1200	1600
10kHz	200	400	600	800
25kHz	80	160	240	320
	KΦ IN VOLTS/RAD.			

RB v. gain and reference frequency

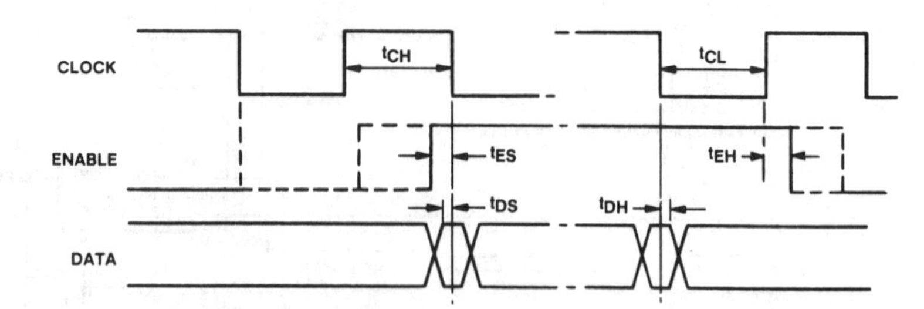

Timing diagram showing timing periods required for correct operation

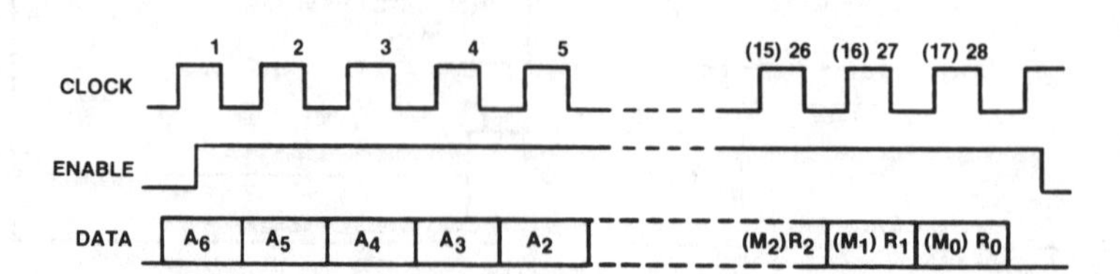

Timing diagram showing programming details

Plessey NJ8822A

Frequency Synthesizer (Microprocessor Serial Interface) With Resettable Counters

- Low power consumption
- High-performance sample-and-hold phase detector
- Serial input with fast-update feature
- >10-MHz input frequency
- Military temperature range (−55 °C to +125 °C)

ABSOLUTE MAXIMUM RATINGS

Supply voltage (V_{DD}-V_{SS})	-0.5V to 7V
Input voltage	
Open drain O/P (pin 3)	7V
All other pins	V_{SS} -0.3V to V_{DD} +0.3V
Storage temperature	-65°C to +150°C

ELECTRICAL CHARACTERISTICS

Test conditions (unless otherwise stated):
V_{DD}-V_{SS} 5V ± 0.5V
Temperature range -55°C to +125°C

DC Characteristics at V_{DD} = 5V

Characteristics	Value			Units	Conditions
	Min.	Typ.	Max.		
Supply current		6.3	7.0	mA	FOSC, FIN = 10MHz } 0 to 5V square wave
		0.7	2.0	mA	FOSC, FIN = 1MHz
MODULUS CONTROL OUT					
High level	4.6			V	I_{source} 1mA
Low level			0.4	V	I_{sink} 1mA
LOCK DETECT OUT					
Low level			0.4	V	I_{sink} 4mA
Open drain pull-up voltage			8	V	
PDB OUTPUT					
High level	4.6			V	I_{source} 4mA
Low level			0.4	V	I_{sink} 4mA
3-state leakage			±0.1	μA	

AC Characteristics

Characteristics	Value			Units	Conditions
	Min.	Typ.	Max.		
FIN/OSC inputs	200			mV RMS	10MHz AC coupled sinewave
Max. operating freq. OSC/FIN inputs	10			MHz	V_{DD} = 5V, Input squarewave V_{DD}-V_{SS}, 25°C
Propagation delay, clock to modulus control		30	50	ns	Note 2
Programming inputs					
Clock high time, t_{CH}	0.5			μs	All timing periods are referenced to the negative transition of the clock waveform
Clock low time, t_{CL}	0.5			μs	
Enable set-up time, t_{ES}	0.2		t_{CH}	μs	
Enable hold time, t_{EH}	0.2			μs	
Data set-up time, t_{DS}	0.2			μs	
Data hold time, t_{DH}	0.2			μs	
Clock rise and fall times	0.2			μs	
Positive going threshold, V_T+	3			V	Note 1
Negative going threshold, V_T-			2	V	
Phase Detector					
Digital phase detector propagation delay		500		ns	
Gain programming resistor, RB	5			kΩ	
Hold capacitor, CH			1	nF	Note 3
Programming capacitor, CAP			1	nF	
Output resistance, PDA			5	kΩ	

NOTES
1. Data, Clock and Enable inputs are high impedance Schmitt buffers without pull up resistors. They are therefore not TTL compatible.
2. All counters have outputs directly synchronous with their respective clock rising edges.
3. The finite output resistance of the internal voltage follower and 'on' resistance of sample switch driving this pin will add a finite time-constant to the loop. A 1nF hold capacitor will give a maximum time-constant of 5 microseconds.
4. The inputs to the device should be at logic '0' when power is applied if latch up conditions are to be avoided. This includes the signal/osc. frequency inputs.

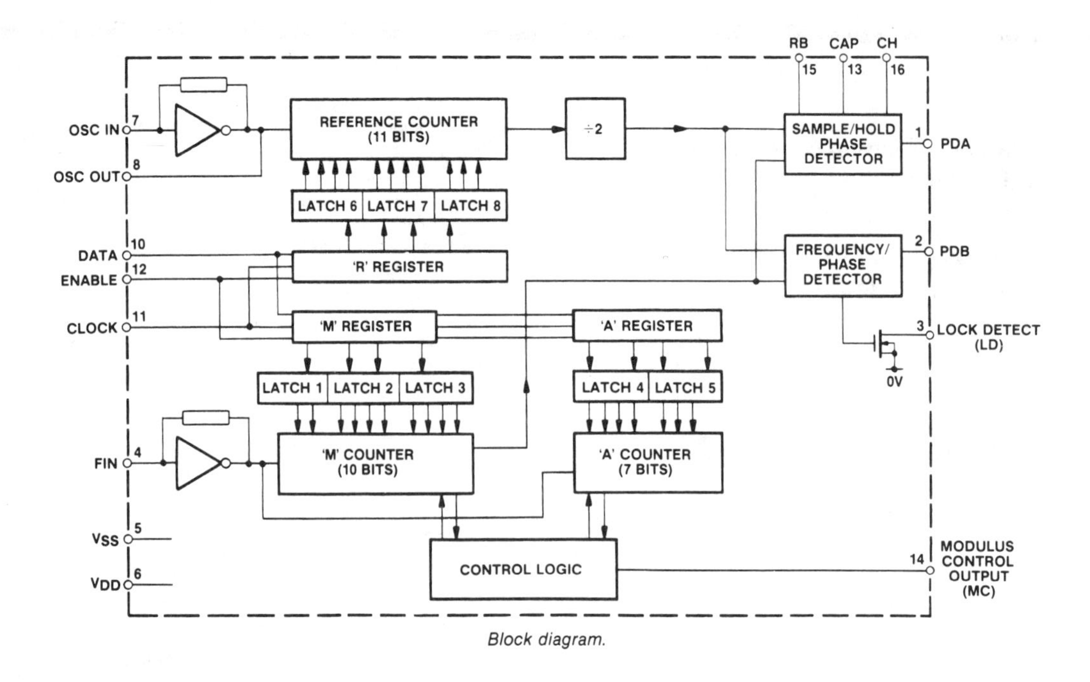

Block diagram.

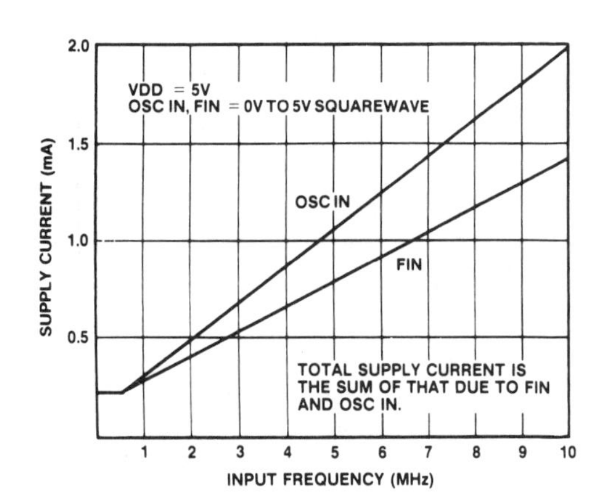

Typical supply current v. input frequency

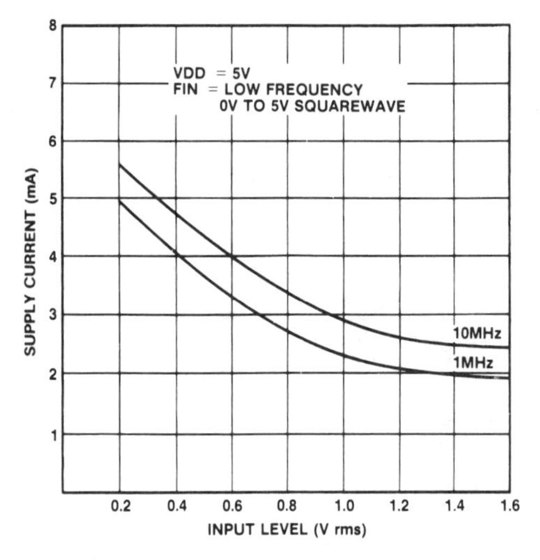

Typical supply current v. input level, Osc In

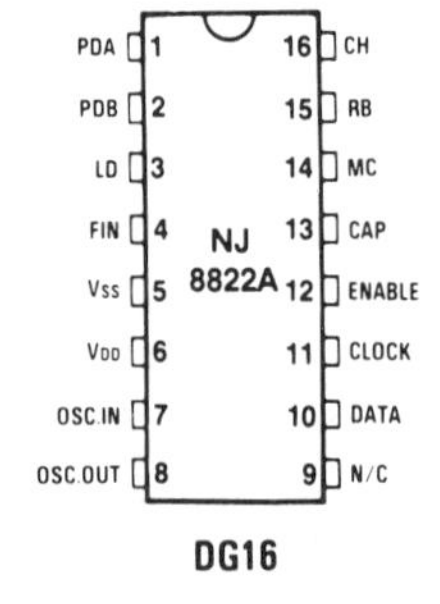

Pin connections - top view, not to scale

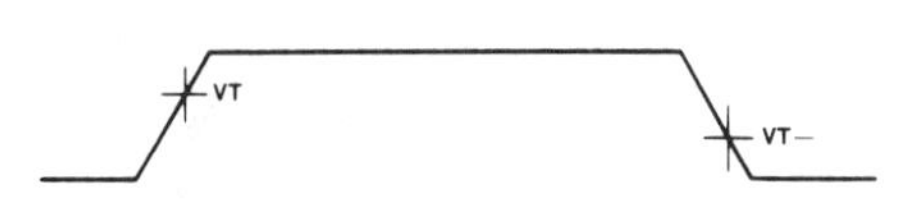

Timing diagram showing voltage thresholds

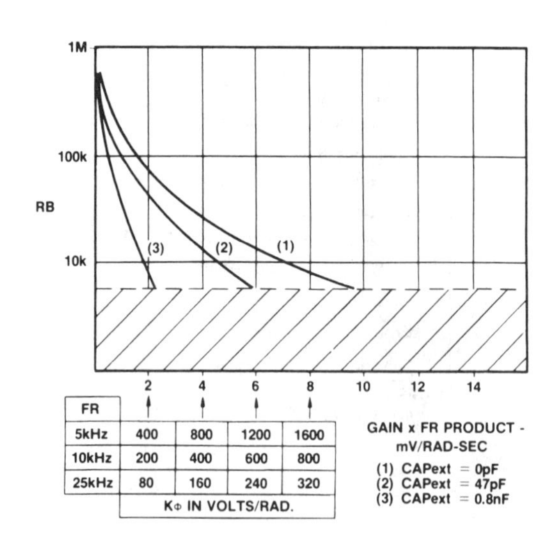

FR				
5kHz	400	800	1200	1600
10kHz	200	400	600	800
25kHz	80	160	240	320
	KΦ IN VOLTS/RAD.			

RB v. gain and reference frequency

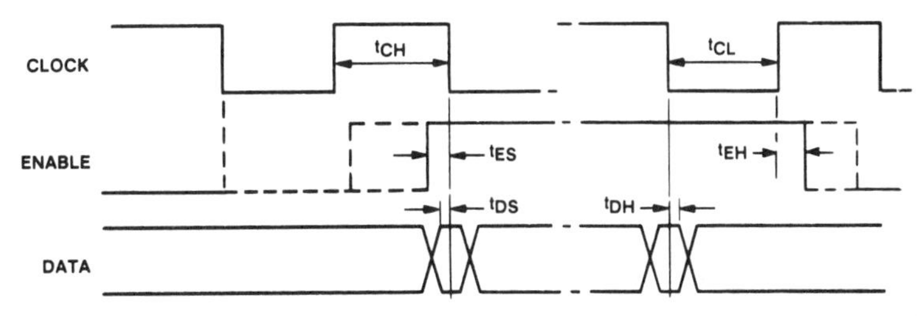

Timing diagram showing timing periods required for correct operation

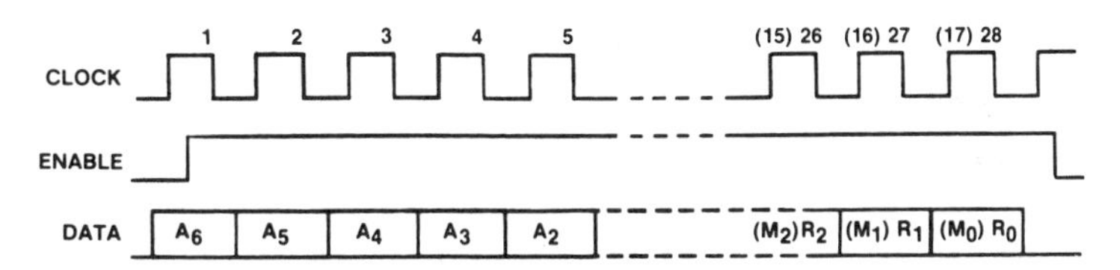

Timing diagram showing programming details

Plessey NJ8823, NJ8823B

Frequency Synthesizer (Microprocessor Interface) With Nonresettable Counters

- Low power consumption
- High-performance sample-and-hold phase detector
- Microprocessor Compatible
- >10-MHz input frequency
- Fast lock-up time

ABSOLUTE MAXIMUM RATINGS

Supply voltage (V_{DD} - V_{SS})	-0.5V to 7V
Input voltage	
Open drain O/P (pin 3)	7V
All other pins	V_{SS} -0.3V to V_{DD} +0.3V
Storage temperature	-65°C to +150°C (DG Package, NJ8823B)
Storage temperature	-55°C to +125°C (DP and MP Packages, NJ8823)

ELECTRICAL CHARACTERISTICS

Test conditions (unless otherwise stated):
V_{DD}-V_{SS} 5V ± 0.5V
Temperature range: NJ8823 -30°C to +70°C, NJ8823B -40°C to +85°C

DC Characteristics at V_{DD} = 5V

Characteristics	Value			Units	Conditions
	Min.	Typ.	Max.		
Supply current		3.5	5.5	mA	FOSC, FIN = 10MHz } 0 to 5V square wave
		0.7	1.5	mA	FOSC, FIN = 1.0MHz }
MODULUS CONTROL OUT					
High level	4.6			V	I_{source} 1mA
Low level			0.4	V	I_{sink} 1mA
LOCK DETECT OUT					
Low level			0.4	V	I_{sink} 4mA
Open drain pull-up voltage			8	V	
PDB Output					
High level	4.6			V	I_{source} 5mA
Low level			0.4	V	I_{sink} 5mA
3-state leakage			±0.1	μA	
INPUT LEVELS					
Data Inputs					
High level	4.25			V	TTL compatible
Low level			0.75	V	See note 1
Program Enable Input					
High level	4.25			V	
Low level			0.75	V	
DS INPUTS					
High level	4.25			V	
Low level			0.75	V	

AC Characteristics

Characteristics	Value			Units	Conditions
	Min.	Typ.	Max.		
FIN/OSC inputs	200			mV RMS	10MHz AC coupled sinewave
Max. operating freq. OSC/FIN inputs	10.6			MHz	V_{DD} = 5V, Input squarewave V_{DD}-V_{SS}.Note 4
Propagation delay, clock to modulus control		30	50	ns	Note 2
Strobe pulse width external mode, $t_{W(ST)}$	2			μs	
Data set-up time, $t_{S(DATA)}$	1			μs	
Data hold time, $t_{H(DATA)}$	1			μs	
Address set-up time, t_{SE}	1			μs	
Address hold time, t_{HE}	1			μs	
Digital phase detector propagation delay		500		ns	
Gain programming resistor, RB	5			kΩ	
Hold capacitor, CH			1	nF	Note 3
Output resistance PDA			5	kΩ	
Digital phase detector gain		1		V/Rad	

NOTES
1. Data inputs have internal 'pull-up' resistors to enable them to be driven from TTL outputs.
2. All counters have outputs directly synchronous with their respective clock rising edges.
3. Finite output resistance of internal voltage follower and 'on' resistance of sample switch driving this pin will add a finite time-constant to the loop. A 1nF hold capacitor will give a maximum time-constant of 5 microseconds.
4. Operation at up to 15MHz is possible with a full logic swing but is not guaranteed.

Pin	Name	Pin	Name
1	PDA	20	CH
2	PDB	19	RB
3	LD	18	MC
4	FIN	17	DS2
5	Vss	16	DS1
6	VDD	15	DS0
7	OSC IN	14	PE
8	OSC OUT	13	NC
9	D0	12	D3
10	D1	11	D2

NJ8823 NJ8823B

DP20, MP20 (NJ8823)
DG20 (NJ8823B)

Pin connections

WORD	DS2	DS1	DS0	D3	D2	D1	D0
1	0	0	0	M1	M0	-	-
2	0	0	1	M5	M4	M3	M2
3	0	1	0	M9	M8	M7	M6
4	0	1	1	A3	A2	A1	A0
5	1	0	0	-	A6	A5	A4
6	1	0	1	R3	R2	R1	R0
7	1	1	0	R7	R6	R5	R4
8	1	1	1	-	R10	R9	R8

Data map

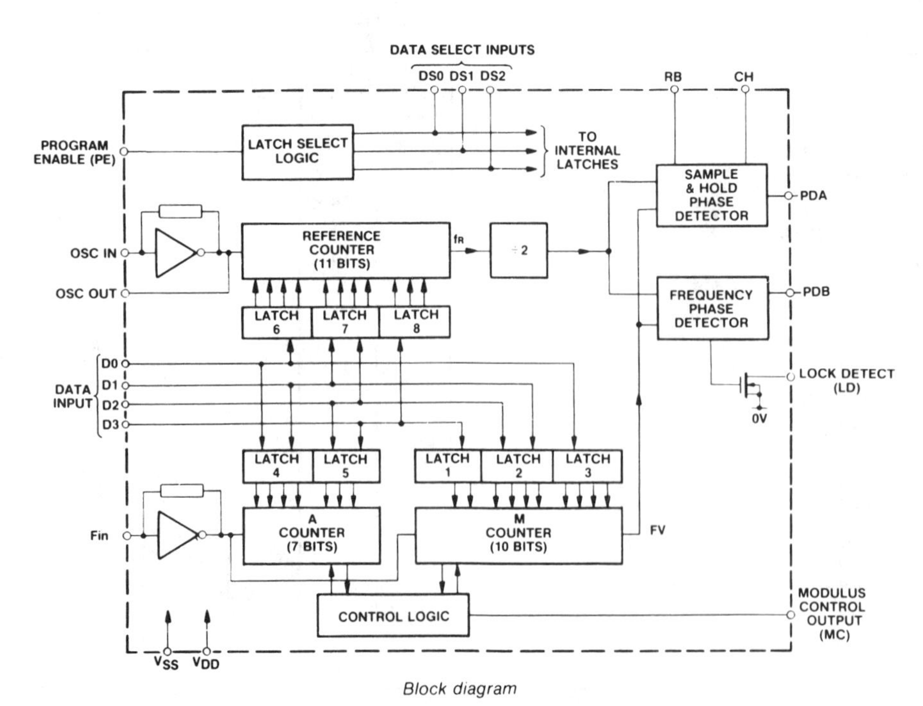

Block diagram

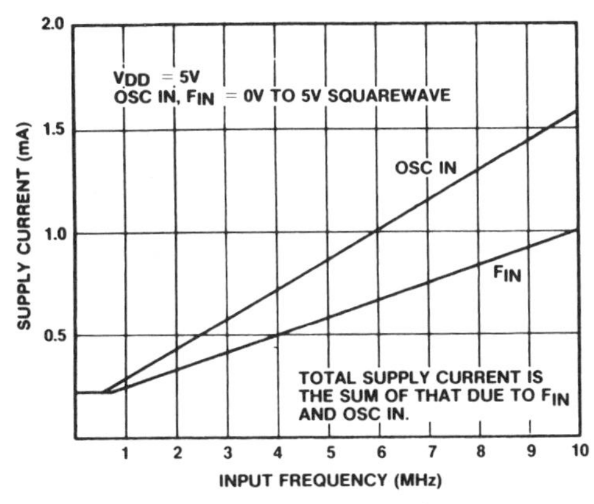

Typical supply current versus input frequency

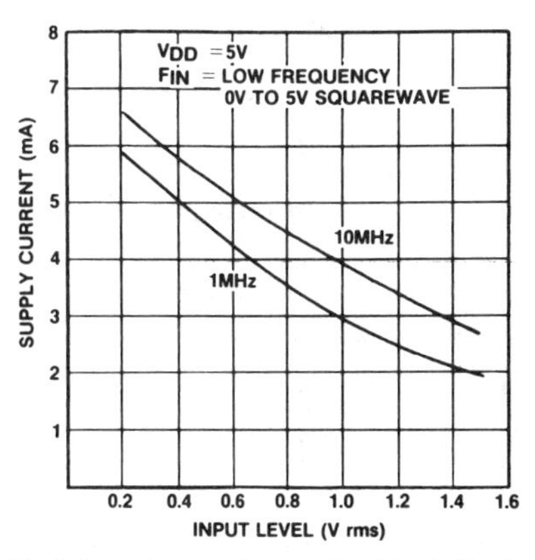

Typical supply current versus input level, Osc In

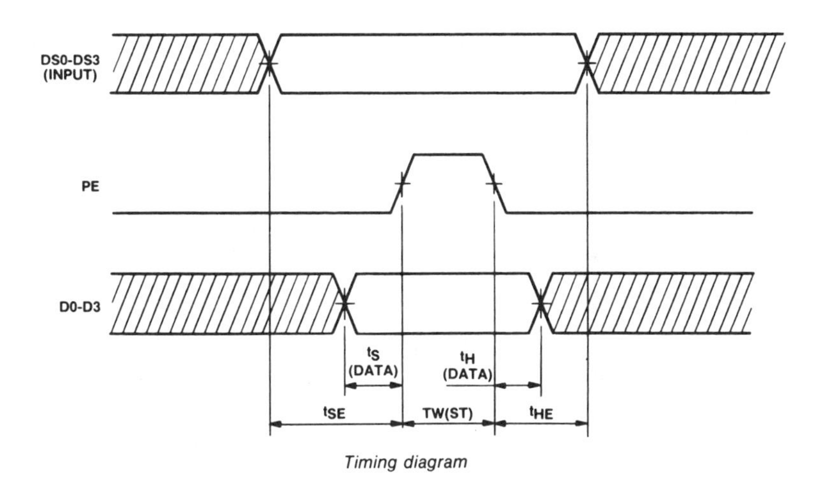

Timing diagram

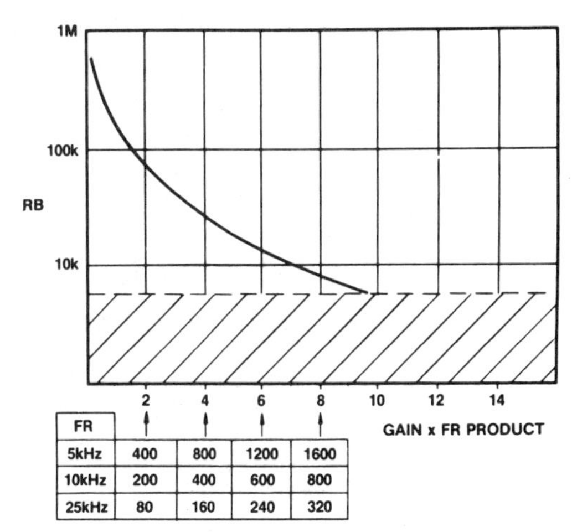

RB versus gain and reference frequency

Plessey NJ8824/NJ8824B Frequency Synthesizer (Microprocessor Serial Interface) With Nonresettable Counters

- Low power consumption
- High-performance sample-and-hold phase detector
- Serial input with fast-update feature
- >10-MHz input frequency
- Fast lock-up time

ELECTRICAL CHARACTERISTICS

Test conditions (unless otherwise stated):
V_{DD}-V_{SS} 5V ± 0.5V
Temperature range: NJ8824 -30°C to +70°C, NJ8824B -40°C to +85°C

DC Characteristics at V_{DD} = 5V

Characteristics	Value			Units	Conditions
	Min.	Typ.	Max.		
Supply current			5.5	mA	FOSC, FIN = 10MHz } 0 to 5V square wave
			1.5	mA	FOSC, FIN = 1MHz
MODULUS CONTROL OUT					
High level	4.6			V	I_{source} 1mA
Low level			0.4	V	I_{sink} 1mA
LOCK DETECT OUT					
Low level			0.4	V	I_{sink} 4mA
Open drain pull-up voltage			8	V	
PDB OUTPUT					
High level	4.6			V	I_{source} 5mA
Low level			0.4	V	I_{sink} 5mA
3-state leakage			±0.1	μA	

AC Characteristics

Characteristics	Value Min.	Value Typ.	Value Max.	Units	Conditions
FIN/OSC inputs	200			mV RMS	10MHz AC coupled sinewave
Max. operating freq. OSC/FIN inputs	10			MHz	V_{DD} = 5V, Input squarewave V_{DD}-V_{SS}, 25°C
Propagation delay, clock to modulus control		30	50	ns	Note 2
Programming inputs					
Clock high time, t_{CH}	0.5			μs	All timing periods are referenced to the negative transition of the clock waveform
Clock low time, t_{CL}	0.5			μs	
Enable set-up time, t_{ES}	0.2		t_{CH}	μs	
Enable hold time, t_{EH}	0.2			μs	
Data set-up time, t_{DS}	0.2			μs	
Data hold time, t_{DH}	0.2			μs	
Clock rise and fall times	0.2			μs	
Positive going threshold, V_T +	3			V	Note 1
Negative going threshold, V_T-			2	V	
Phase Detector					
Digital phase detector propagation delay		500		ns	
Gain programming resistor, RB	5			kΩ	
Hold capacitor, CH			1	nF	Note 3
Programming capacitor, CAP			1	nF	
Output resistance, PDA			5	kΩ	

NOTES
1. Data, Clock and Enable inputs are high impedance Schmitt buffers without pull up resistors. They are therefore not TTL compatible.
2. All counters have outputs directly synchronous with their respective clock rising edges.
3. The finite output resistance of the internal voltage follower and 'on' resistance of sample switch driving this pin will add a finite time-constant to the loop. A 1nF hold capacitor will give a maximum time-constant of 5 microseconds.
4. The inputs to the device should be at logic '0' when power is applied if latch up conditions are to be avoided. This includes the signal/osc. frequency inputs.

ABSOLUTE MAXIMUM RATINGS

Supply voltage (V_{DD}-V_{SS}) -0.5V to 7V
Input voltage
Open drain O/P (pin 3 (DG) pin 4 (MP)) 7V
All other pins V_{SS} -0.3V to V_{DD} +0.3V

Storage temperature -55°C to +125°C (DP and MP packages, NJ8824)
Storage temperature -65°C to +150°C (DG packages, NJ8824B)

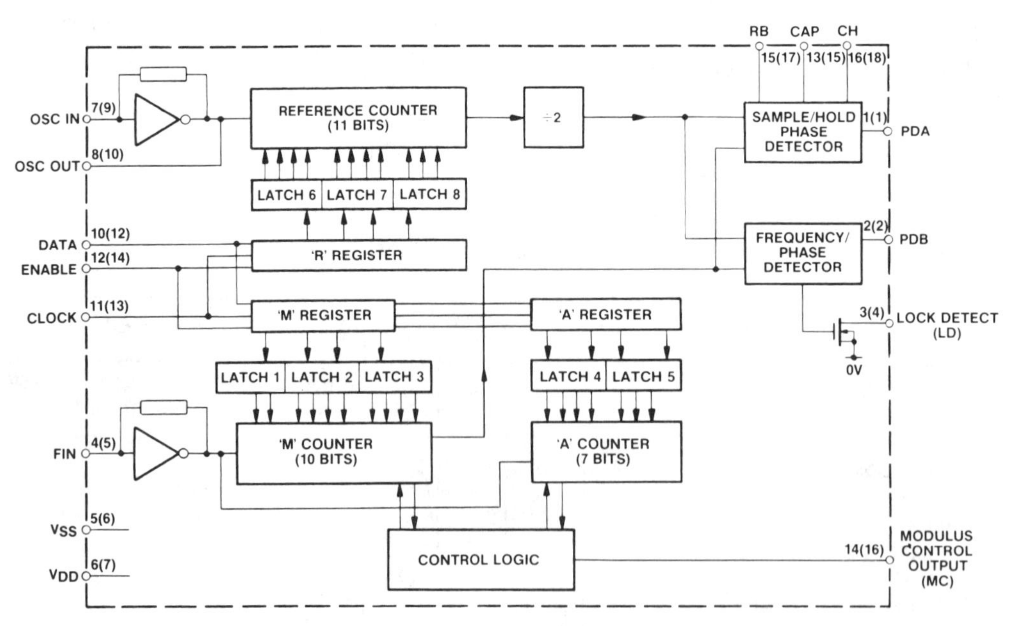

Block diagram. Pin numbers for MP package are shown in brackets.

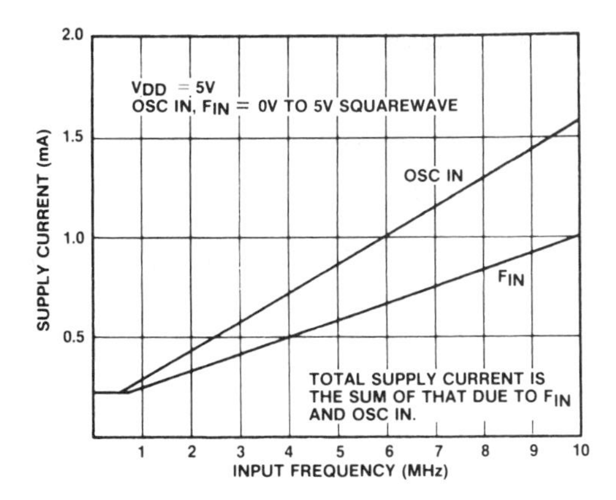

Typical supply current v. input frequency

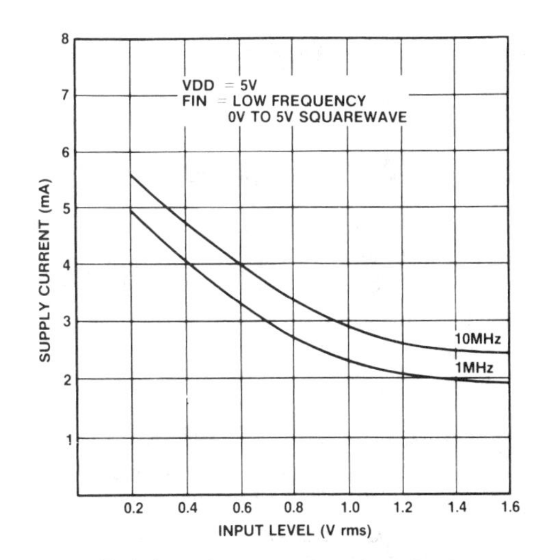

Typical supply current v. input level, Osc In

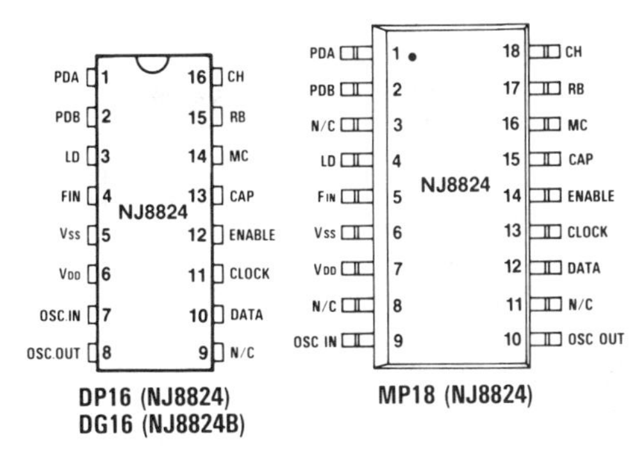

Pin connections - top view, not to scale

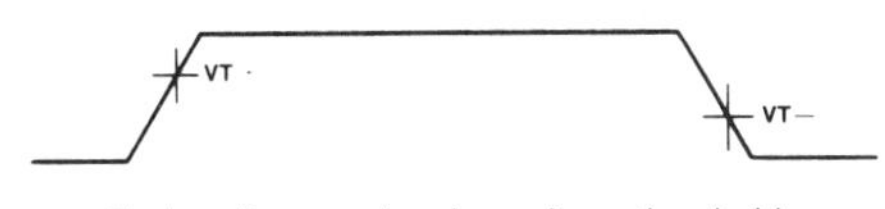

Timing diagram showing voltage thresholds

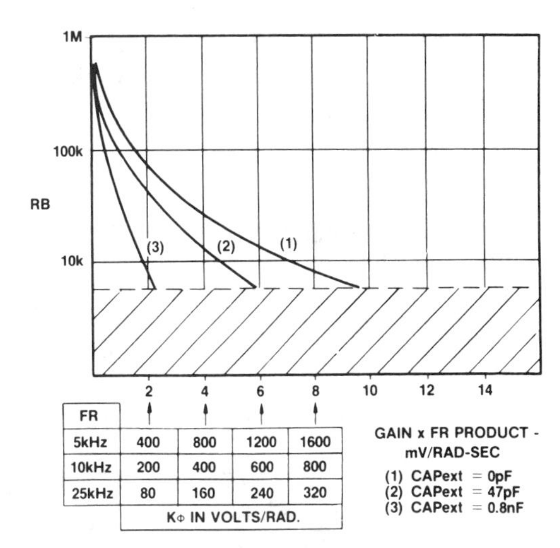

FR	2	4	6	8
5kHz	400	800	1200	1600
10kHz	200	400	600	800
25kHz	80	160	240	320
	KΦ IN VOLTS/RAD.			

RB v. gain and reference frequency

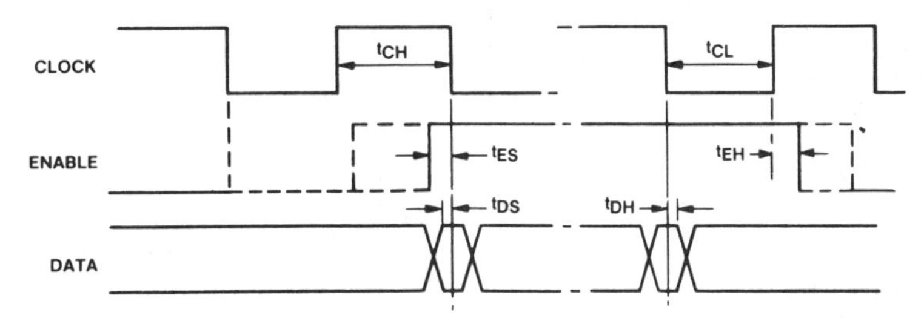

Timing diagram showing timing periods required for correct operation

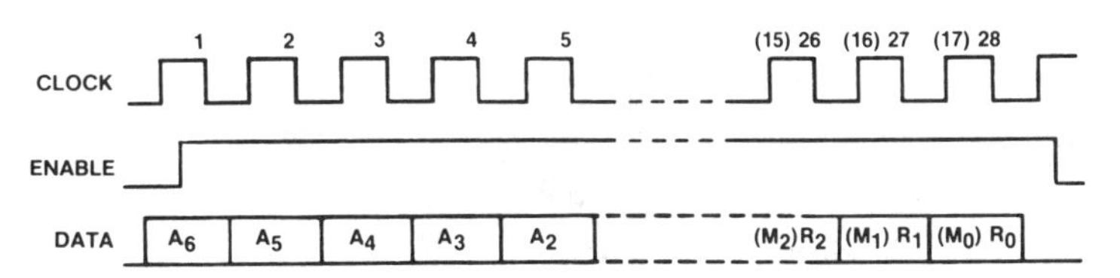

Timing diagram showing programming details

Plessey NJ88C25

Frequency Synthesizer (Microprocessor Serial Interface)

- 3.0 to 5.0V supply range
- Low power consumption
- High-performance sample-and-hold phase detector
- Serial input with fast-update feature
- >20-MHz input frequency

ABSOLUTE MAXIMUM RATINGS

Supply voltage (V_{DD} - V_{SS})	-0.5V to 7V
Input voltage	
Open drain O/Ps (pins 3 & 4)	7V
All other pins	V_{SS} -0.3V to V_{DD} +0.3V
Storage temperature	-65°C to +150°C

ELECTRICAL CHARACTERISTICS

Test conditions (unless otherwise stated):
V_{DD} - V_{SS} 2.7V to 5.5V, Temperature Range -30°C to +70°C

DC Characteristics at V_{DD} = 5.0V

Characteristic	Value			Units	Conditions
	Min.	Typ.	Max.		
Supply current		5.5	TBA	mA	FOSC, FIN = 20MHz } 0 to 5V square wave
		0.7	TBA	mA	FOSC, FIN = 1MHz
		3.7	TBA	mA	FOSC, FIN = 10MHz
Modulus Control out, BAND 0, BAND 1					
High level	V_{DD}-0.4			V	I_{source} 1mA
Low level			0.4	V	I_{sink} 1mA
Lock Detect Out, FV					
Low level			0.4	V	I_{sink} 4mA
Open drain pull-up voltage			7	V	
PDB output					
High level	4.6			V	I_{source} 4mA
Low level			0.4	V	I_{sink} 4mA
3-state leakage			±0.1	μA	

AC Characteristics

Characteristic	Value			Units	Conditions
	Min.	Typ.	Max.		
FIN/OSC outputs	200			mV RMS	20MHz AC coupled sinewave
Max. operating freq. OSC/FIN inputs	20			MHz	V_{DD} = 5V, 0 to 5V square wave
Propagation delay, clock to modulus control		30	50	ns	Note 2
Programming inputs					
Clock high time, t_{CH}	0.5			μs	
Clock low time, t_{CL}	0.5			μs	
Enable set-up time, t_{ES}	0.2		t_{CH}	μs	Note 5
Enable hold time, t_{EH}	0.2			μs	
Data set-up time, t_{DS}	0.2			μs	
Data hold time, t_{DH}	0.2			μs	
Clock rise and fall times	0.2			μs	
Positive going threshold, V_{T+}	3			V	TTL compatible
Negative going threshold, V_{T-}			2	V	
Digital phase detector propagation delay		500		ns	
Gain programming resistor, RB	5			kΩ	
Hold capacitor, CH			1	nF	Note 3
Programming capacitor, CAP			1	nF	
Output resistance, PDA			5	kΩ	

NOTES
1. Data inputs have internal 'pull-up' resistors to enable them to be driven from TTL outputs.
2. All counters have outputs directly synchronous with their respective clock rising edges.
3. The finite output resistance of the internal voltage follower and 'on' resistance of sample switch driving this pin will add a finite time-constant to the loop. A 1nF hold capacitor will give a maximum time constant of 5 microseconds.
4. The inputs to the device should be at logic '0' when power is applied if latch up conditions are to be avoided. This includes the signal/osc. frequency inputs.
5. Clock to enable set up time is variable, dependent on frequency of OSC. IN, it needs to be specified in terms of OSC. IN frequency, clock high time (t_{CH}) and clock low time (t_{CL}). Enable set-up time, t_{ES} must meet following conditions: $4 \times 1/\text{OSC. IN} \leq t_{ES} < (t_{CH} + t_{CL})$.

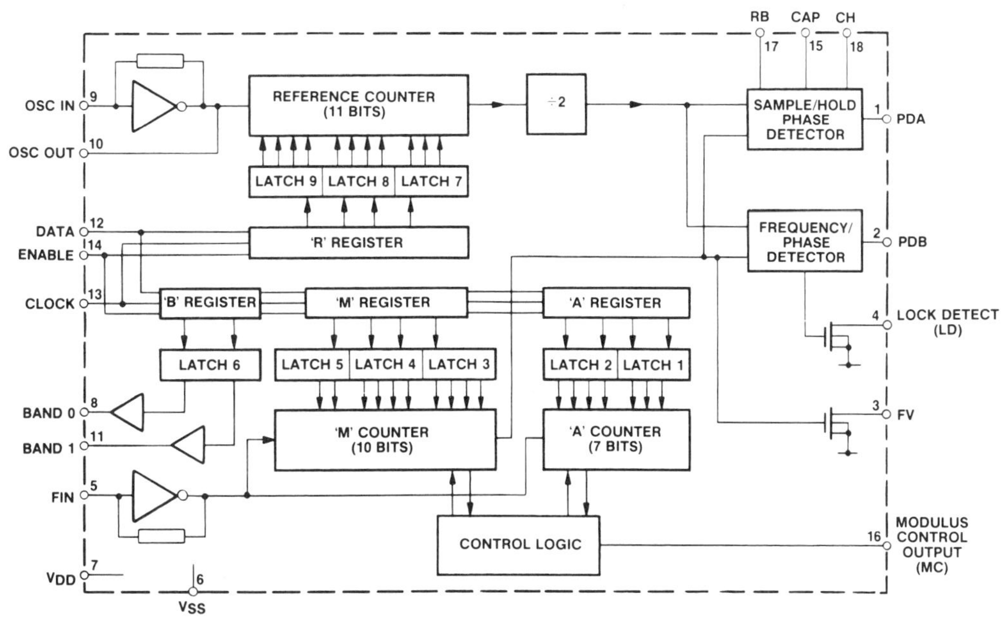

Block Diagram

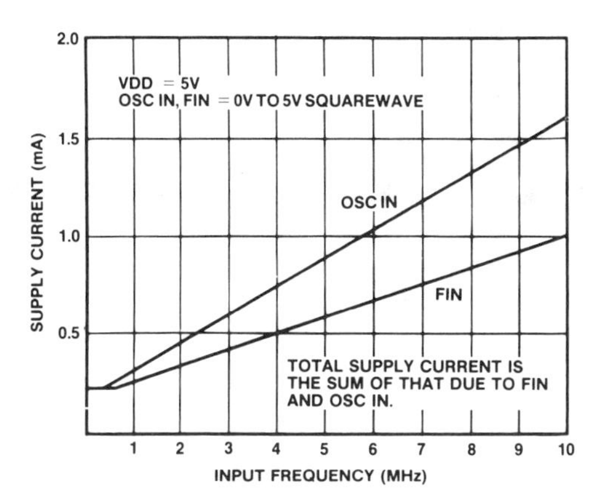

Typical supply current versus input frequency

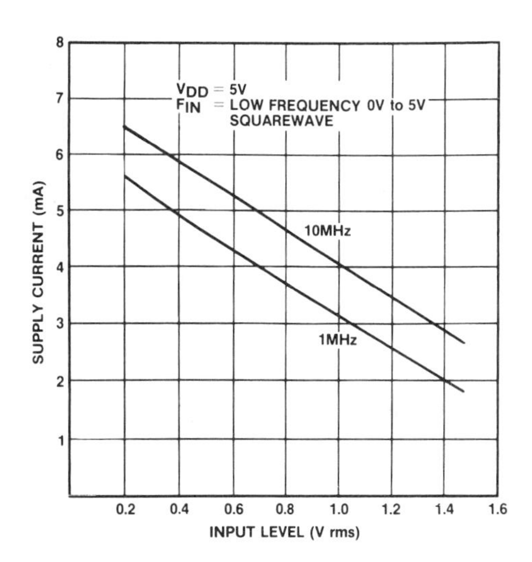

Typical supply current versus input level, Osc In

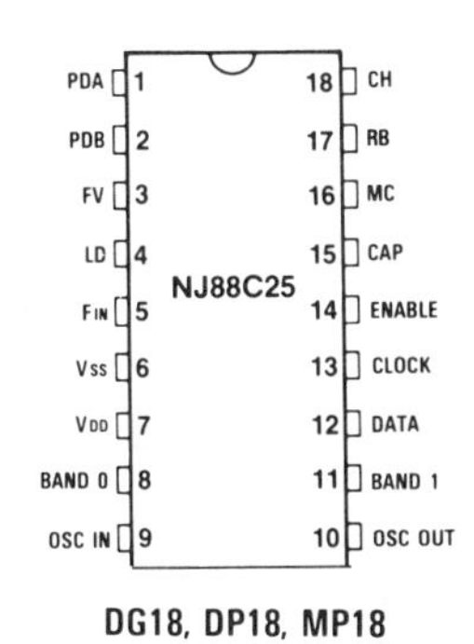

DG18, DP18, MP18

Pin connections - top view

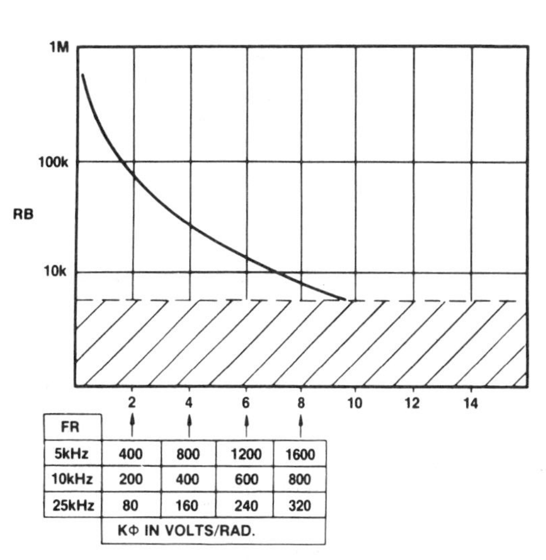

FR	2	4	6	8
5kHz	400	800	1200	1600
10kHz	200	400	600	800
25kHz	80	160	240	320
	KΦ IN VOLTS/RAD.			

RB versus gain and reference frequency

Timing diagram showing voltage thresholds

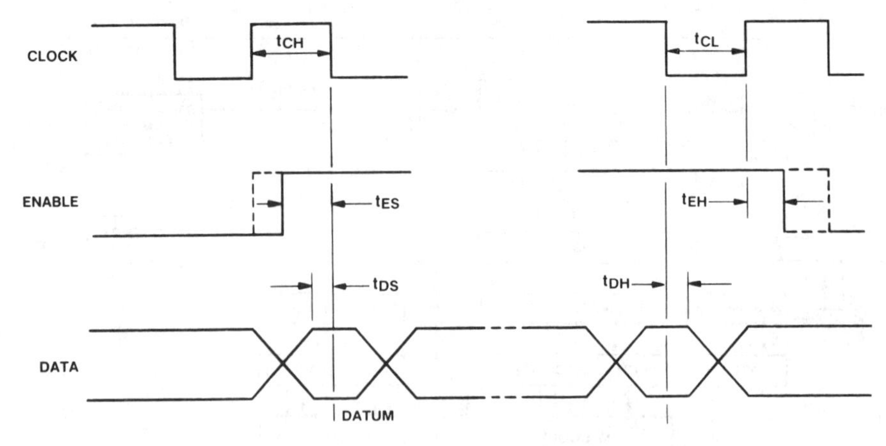

Timing diagram showing timing periods required for correct operation

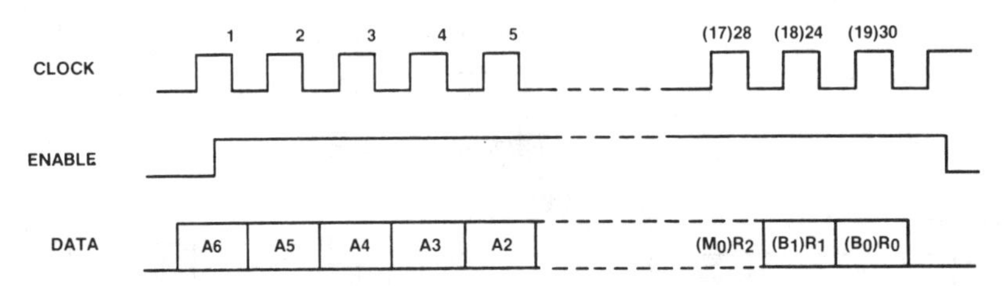

Timing diagram showing programming details

Mitel MT8880/MT8880-1/MT8880-2
Integrated DTMF Transceiver

- Complete DTMF transmitter/receiver
- Central office quality
- Low power consumption
- Microprocessor port
- Adjustable guard time
- Automatic tone burst mode
- Call progress mode

APPLICATIONS
- Credit card systems
- Paging systems
- Repeater systems/mobile radio
- Interconnect dialers
- Personal computers

ABSOLUTE MAXIMUM RATINGS*

	Parameter	Symbol	Min	Max	Units
1	Power supply voltage V_{DD}-V_{SS}	V_{DD}		6	V
2	Voltage on any pin	V_I	$V_{SS}-0.3$	$V_{DD}+0.3$	V
3	Current at any pin (Except V_{DD} and V_{SS})			10	mA
4	Storage temperature	T_{ST}	−65	+150	°C
5	Package power dissipation	P_D		1000	mW

*Exceeding these values may cause permanent damage. Functional operation under these conditions is not implied.

RECOMMENDED OPERATING CONDITIONS - Voltages are with respect to ground (V_{SS}) unless otherwise stated.

	Parameter	Sym	Min	Typ‡	Max	Units	Test Conditions
1	Positive power supply	V_{DD}	4.75	5.00	5.25	V	
2	Operating temperature	T_O	−40		+85	°C	
3	Φ2 clock frequency	f_C	0.001		1.0	MHz	
4	Crystal clock frequency	f_{CLK}	3.575965	3.579545	3.583124	MHz	

‡ Typical figures are at 25 °C and for design aid only: not guaranteed and not subject to production testing.

DC ELECTRICAL CHARACTERISTICS† – f_c = 3.579545 MHz, $\phi 2$ = 1 MHz, V_{SS} = 0 V.

		Characteristics	Sym	Min	Typ‡	Max	Units	Test Conditions
1	SUP	Operating supply voltage	V_{DD}	4.75	5.0	5.25	V	
2		Operating supply current	I_{DD}			10	mA	
3		Power consumption	P_C			52.5	mW	
4	INPUTS	High level input voltage (OSC1)	V_{IHO}	3.5			V	
5		Low level input voltage (OSC1)	V_{ILO}			1.5	V	
6		Steering threshold voltage	V_{TSt}	2.2	2.3	2.5	V	V_{DD} = 5 V
7	OUTPUTS	Low level output voltage (OSC2)	V_{OLO}			0.1	V	No load
8		High level output voltage (OSC2)	V_{OHO}	4.9			V	No load
9		Output leakage current (IRQ)	I_{OZ}		1	10	μA	V_{OH} = 2.4 V
10		V_{Ref} output voltage	V_{Ref}	2.4	2.5	2.7	V	No load
11		V_{Ref} output resistance	R_{OR}			1	kΩ	
12	DATA BUS	Low level input voltage	V_{IL}			0.8	V	
13		High level input voltage	V_{IH}	2.0			V	
14		Source current	I_{OH}	-1.4	-6.6		mA	V_{OH} = 2.4V
15		Sink current	I_{OL}	2.0	4.0		mA	V_{OL} = 0.4V
16		Input leakage current	I_{IZ}			10	μA	V_{IN} = 0.4 to 2.4 V
17	ESt and StGt	Source current	I_{OH}	-0.02	-0.9		mA	V_{OH} = 4.6V
18		Sink current	I_{OL}	1.0	2.7		mA	V_{OL} = 0.4V
19	$\overline{IRQ}/\overline{CP}$	Sink current	I_{OL}	4	16		mA	V_{OL} = 0.4V

† Characteristics are over recommended operating conditions unless otherwise stated.
‡ Typical figures are at 25 °C, V_{DD} = 5V and for design aid only: not guaranteed and not subject to production testing.

ELECTRICAL CHARACTERISTICS
GAIN SETTING AMPLIFIER - Voltages are with respect to ground (V_{SS}) unless otherwise stated, V_{SS} = 0 V, V_{DD} = 5V, T_O = 25°C.

	Characteristics	Sym	Min	Typ‡	Max	Units	Test Conditions
1	Input leakage current	I_{IN}		±100		nA	$V_{SS} \leq V_{IN} \leq V_{DD}$.
2	Input resistance	R_{IN}		10		MΩ	
3	Input offset voltage	V_{OS}		25		mV	
4	Power supply rejection	PSRR		60		dB	1 kHz
5	Common mode rejection	CMRR		60		dB	-3.0 V ≤ V_{IN} ≤ 3.0 V
6	DC open loop voltage gain	A_{VOL}		65		dB	
7	Unity gain bandwidth	BW		1.5		MHz	
8	Output voltage swing	V_O		4.5		V_{pp}	R_L ≥ 100 kΩ to V_{SS}
9	Allowable capacitive load (GS)	C_L		100		pF	
10	Allowable resistive load (GS)	R_L		50		kΩ	
11	Common mode range	V_{CM}		3.0		V_{pp}	No Load

‡ Typical figures are at 25°C and for design aid only: not guaranteed and not subject to production testing.

MT8880-1 AC ELECTRICAL CHARACTERISTICS † - Voltages are with respect to ground (V_{SS}) unless otherwise stated.

		Characteristics	Sym	Min	Typ	Max	Units	Notes*
1	RX	Valid input signal levels (each tone of composite signal)		-31			dBm	1,2,3,5,6,9
				21.8			mV_{RMS}	1,2,3,5,6,9
						+1	dBm	1,2,3,5,6,9
						869	mV_{RMS}	1,2,3,5,6,9
2		Input Signal Level Reject		-37			dBm	1,2,3,5,6,9
				10.9			mV_{RMS}	1,2,3,5,6,9

† V_{DD} = 5 V, V_{SS} = 0, T_A = 25°C, Φ2 = 1MHz and f_C = 3.579545 MHz using test circuit shown in Figure 15.

MT8880/8880-2 AC ELECTRICAL CHARACTERISTICS †Voltages are with respect to ground (V_{SS}) unless otherwise stated.

		Characteristics	Sym	Min	Typ‡	Max	Units	Notes*
1	RX	Valid Input signal levels (each tone of composite signal)		−29			dBm	1,2,3,5,6,9
				27.5			mV_{RMS}	1,2,3,5,6,9
						+1	dBm	1,2,3,5,6,9
						869	mV_{RMS}	1,2,3,5,6,9

† V_{DD}=5 V, V_{SS}= 0, T_A=25°C, Φ2=1MHz and f_C=3.579545 MHz using test circuit shown in Figure 15.

AC ELECTRICAL CHARACTERISTICS † Voltages are with respect to ground (V_{SS}) unless otherwise stated.

V_{SS}= 0 V, Φ2=1 MHz f_C=3.579545 MHz.

		Characteristics	Sym	Min	Typ‡	Max	Units	Notes*
1	RX	Positive twist accept				6	dB	2,3,6,9
2		Negative twist accept				6	dB	2,3,6,9
3		Freq. deviation accept		±1.5% ± 2Hz				2,3,5,9
4		Freq. deviation reject		±3.5%				2,3,5
5		Third tone tolerance			-16		dB	2,3,4,5,9,10
6		Noise tolerance			-12		dB	2,3,4,5,7,9,10
7		Dial tone tolerance			22		dB	2,3,4,5,8,9,11

†Characteristics are over recommended operating conditions unless otherwise stated
‡ Typical figures are at 25°C, V_{DD} = 5V, and for design aid only: not guaranteed and not subject to production testing
*See "Notes" following AC Electrical Characteristics Tables

AC ELECTRICAL CHARACTERISTICS†- Voltages are with respect to ground (V_{SS}) unless otherwise stated.

		Characteristics	Sym	Min	Typ‡	Max	Units	Conditions
1	RX	Tone present detect time	t_{DP}	3	11	14	ms	Note 12
2		Tone absent detect time	t_{DA}	0.5	4	8.5	ms	Note 12
3		Tone duration accept	t_{REC}			40	ms	User adjustable
4		Tone duration reject	$t_{\overline{REC}}$	20			ms	User adjustable
5		Interdigit pause accept	t_{ID}			40	ms	User adjustable
6		Interdigit pause reject	t_{DO}	20			ms	User adjustable
7		Delay St to b3	t_{PStb3}		13		μs	
8		Delay St to RX_0-RX_3	t_{PStRX}		8		μs	
9	TX	Tone burst duration	t_{BST}	50		52	ms	DTMF mode
10		Tone pause duration	t_{PS}	50		52	ms	DTMF mode
11		Tone burst duration (extended)	t_{BSTE}	100		104	ms	Call Progress mode
12		Tone pause duration (extended)	t_{PSE}	100		104	ms	Call Progress mode
13	TONE OUT	High group output level	V_{HOUT}	-6.1		-2.1	dBm	R_L=10kΩ
14		Low group output level	V_{LOUT}	-8.1		-4.1	dBm	R_L=10kΩ
15		Pre-emphasis	dB_P		2	3	dB	R_L=10kΩ
16		Output distortion (Single Tone)	THD		-25		dB	25 kHz Bandwidth R_L=10 kΩ
17		Frequency deviation	f_D		±0.7	±1.5	%	f_C=3.579545 MHz
18		Output load resistance	R_{LT}	10		50	kΩ	
19	MPU INTERFACE	Φ2 cycle period	t_{CYC}	1		1000	μs	
20		Φ2 high pulse width	t_{CH}	450			ns	
21		Φ2 low pulse width	t_{CL}	430			ns	
22		Φ2 rise and fall time	t_R, t_F			25	ns	
23		Address, R/$\overline{W}$ hold time	t_{AH}, t_{RWH}	26			ns	
24		Address, R/$\overline{W}$ setup time (before Φ2)	t_{AS}, t_{RWS}	23			ns	
25		Data hold time (read)	t_{DHR}	22			ns	*
26		Φ2 to valid data delay (read)	t_{DDR}			290	ns	200 pF load
27		Data setup time (write)	t_{DSW}	45			ns	
28		Data hold time (write)	t_{DHW}	10			ns	
29		Input Capacitance (data bus)	C_{IN}		5		pF	
30		Output Capacitance ($\overline{IRQ}$/CP)	C_{OUT}		5		pF	

AC ELECTRICAL CHARACTERISTICS† (Cont'd)-Voltages are with respect to ground (V_{SS}) unless otherwise stated

		Characteristics	Sym	Min	Typ‡	Max	Units	Notes*
31	DTMF CLK	Crystal/clock frequency	f_C	3.5759	3.5795	3.5831	MHz	
32		Clock input rise time	t_{LHCL}			110	ns	Ext. clock
33		Clock input duty cycle	t_{HLCL}			110	ns	Ext. clock
34		Clock input duty cycle	DC_{CL}	40	50	60	%	Ext. clock
35		Capacitive load (OSC2)	C_{LO}			30	pF	

†Timing is over recommended temperature & power supply voltages. $f_C = 3.579545$ MHz
‡ Typical figures are at 25°C and for design aid only: not guaranteed and not subject to production testing.
* The data bus output buffers are no longer sourcing or sinking current by t_{DHR}.

NOTES: 1) dBm = decibels above or below a reference power of 1 mW into a 600 ohm load.
2) Digit sequence consists of all 16 DTMF tones.
3) Tone duration = 40 ms. Tone pause = 40 ms.
4) Nominal DTMF frequencies are used.
5) Both tones in the composite signal have an equal amplitude.
6) The tone pair is deviated by ± 1.5 % ± 2 Hz.
7) Bandwidth limited (3 kHz) Gaussian noise.
8) The precise dial tone frequencies are 350 and 440 Hz (± 2 %).
9) For an error rate of less than 1 in 10,000.
10) Referenced to the lowest amplitude tone in the DTMF signal.
11) Referenced to the minimum valid accept level.
12) For guard time calculation purposes.

AC ELECTRICAL CHARACTERISTICS†-Call Progress

- Voltages are with respect to ground (V_{SS}) unless otherwise stated

$V_{SS} = 0$ V, $\Phi 2 = 1$ MHz $f_C = 3.579545$ MHz

		Characteristics	Sym	Min	Typ‡	Max	Units	Notes*
1		Lower freq. (ACCEPT)	f_{LA}		320		Hz	@ −25 dBm
2		Upper freq. (ACCEPT)	f_{HA}		510		Hz	@ −25 dBm
3		Lower freq. (REJECT)	f_{LR}		290		Hz	@ −25 dBm
4		Upper freq. (REJECT)	f_{HR}		540		Hz	@ −25 dBm

†Characteristics are over recommended operating conditions unless otherwise stated
‡ Typical figures are at 25°C, V_{DD} = 5V, and for design aid only: not guaranteed and not subject to production testing
*See "Notes" AC Electrical Characteristics Tables

PIN CONNECTIONS

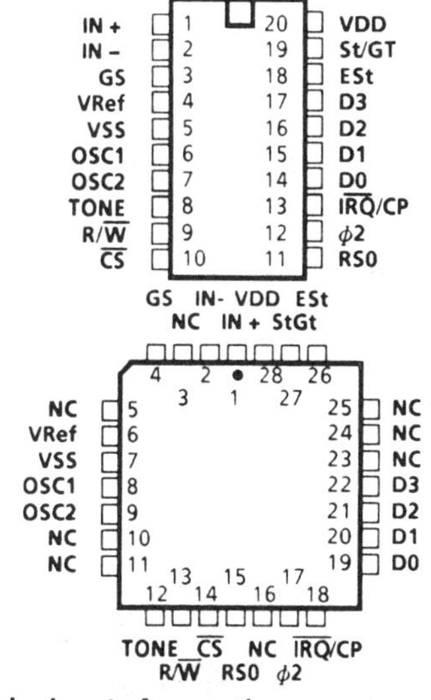

Ordering Information -40°C to +85°C

MT8880/-1/-2AE 20 Pin Plastic DIP
MT8880/-1/-2AC 20 Pin Ceramic DIP
MT8880/-1/-2AP 28 Pin Plastic LCC

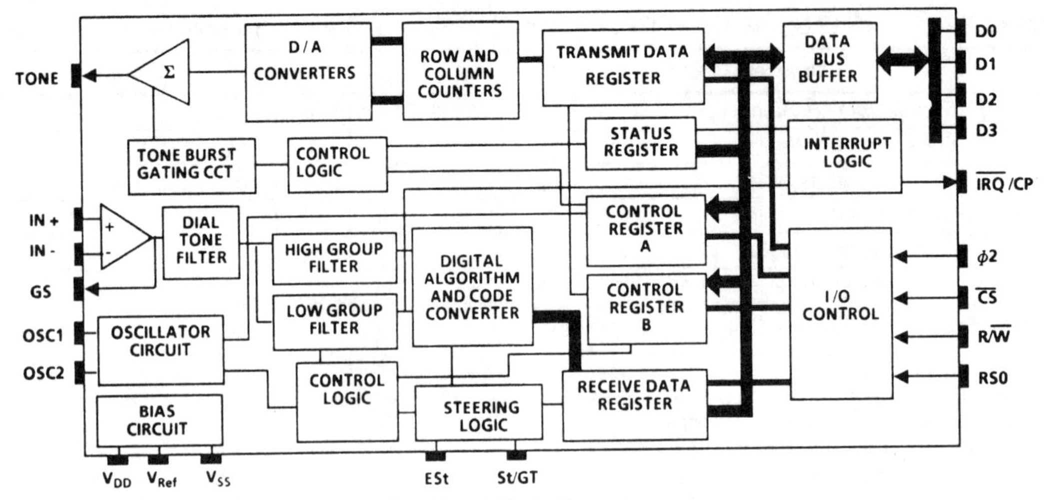

Functional Block Diagram

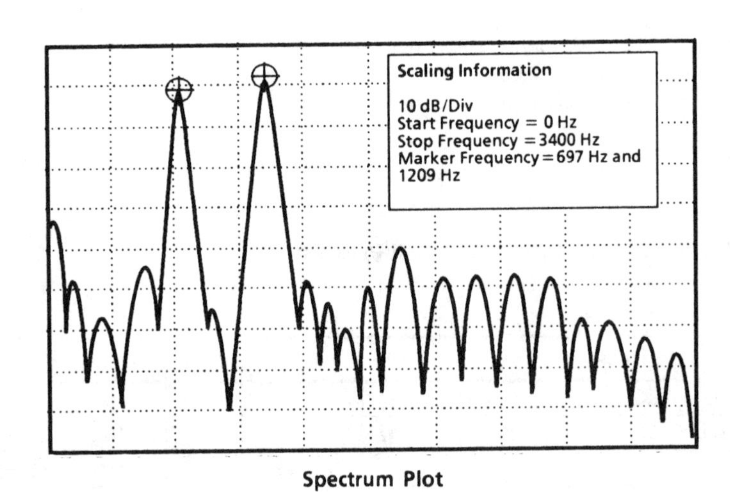

Spectrum Plot

MT8880

DTMF/CP INPUT

DTMF OUTPUT

To μP or μC

Notes:
R1, R2 = 100 kΩ 1%
R3 = 374k Ω 1 %
R4 = 3 kΩ 10 %
R_L = 10 k Ω (min)
C1 = 100 nF 5 %
C2 = 100 nF 5 %
C3 = 100 nF 10 % *
C4 = 10 nF 10 %
X-tal = 3.579545 MHz

* Microprocessor based systems can inject undesirable noise into the supply rails. The performance of the MT8880 can be optimized by keeping noise on the supply rails to a minimum. The decoupling capacitor (C3) should be connected close to the device and ground loops should be avoided.

Application Circuit (Single-Ended Input)

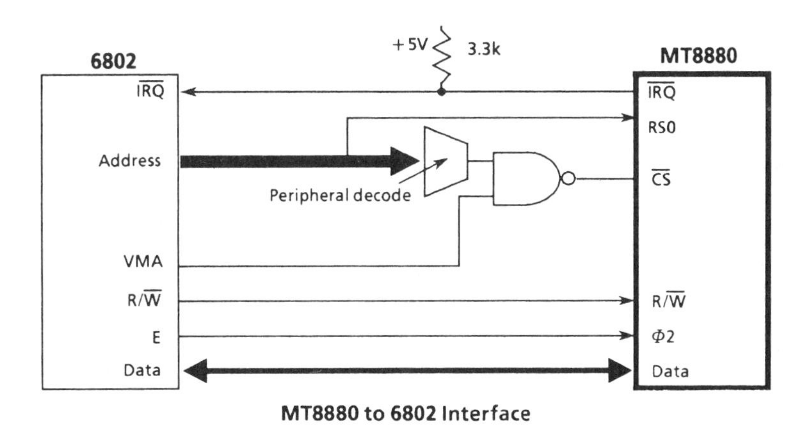

MT8880 to 6802 Interface

Mitel MT8952B HDLC Protocol Controller

- Formats data as per X.25(CCITT) level-2 standards.
- Go-ahead sequence generation and detection.
- Single-byte address recognition.
- Microprocessor port and directly accessible registers for flexible operation and control
- 19-byte FIFO in both send and receive paths.
- Handshake signals for multiplexing data links.
- High-speed serial clocked output (2.5 Mbps).
- ST-BUS compatibility with programmable channel selection for data and separate timeslot for control information.
- Independent watchdog timer.
- Facility to disable protocol functions.
- Low-power ISO-CMOS technology.

APPLICATIONS

- Data-link controllers and protocol generators.
- Digital sets, PBXs, and private packet networks.
- D-channel controller for ISDN basic access.
- C-channel controller to digital network interface circuits (typically MT8972).
- Interprocessor communication.

ABSOLUTE MAXIMUM RATINGS*

	Parameter		Symbol	Min	Max	Units
1	Supply voltage		V_{DD}	−0.3	7.0	V
2	Voltage on any pin (other than supply pins)		V_I	$V_{SS}-0.3$	$V_{DD}+0.3$	V
3	Current on any pin (other than supply pins)		I_I / I_O		±25	mA
4	DC Supply or ground current		I_{DD} / I_{SS}		±50	mA
5	Storage temperature		T_{ST}	−65	150	°C
6	Package power dissipation	Cerdip	P_D		1.0	W
		Plastic	P_D		0.6	W

*Exceeding these values may cause permanent damage. Functional operation under these conditions is not implied.

RECOMMENDED OPERATING CONDITIONS -Voltages are with respect to ground (V_{SS}) unless otherwise stated

		Characteristics	Sym	Min	Typ‡	Max	Units	Test Conditions
1		Supply voltage	V_{DD}	4.75	5.0	5.25	V	
2		Input HIGH voltage	V_{IH}	2.4		V_{DD}	V	For a Noise Margin of 400 mV
3		Input LOW voltage	V_{IL}	V_{SS}		0.4	V	For a Noise Margin of 400 mV
4		Frequency of operation	f_{CL}			5.0	MHz	When clock input is at twice the bit rate.
5		Operating temperature	T_A	−40	25	85	°C	

‡ Typical figures are at 25°C and are for design aid only: not guaranteed and not subject to production testing.

DC ELECTRICAL CHARACTERISTICS -Voltages are with respect to ground (VSS) unless otherwise stated.

V_{DD}=5V±5%, V_{SS}=0V, T_A=−40 to 85°C

		Characteristics	Sym	Min	Typ‡	Max	Units	Test Conditions
1		Supply current (Quiescent)	I_{DD}		1	10	μA	Outputs unloaded and clock input (CKi) grounded
2		Supply current (Operational)	I_{DD}		0.4	1.0	mA	*See below
3	INPUT	Input HIGH voltage	V_{IH}	2.0			V	
4	INPUT	Input LOW voltage	V_{IL}			0.8	V	
5	INPUT	Input leakage current	I_{IZ}			10	μA	
6	INPUT	Input capacitance	C_{in}		10		pF	
7	INPUT	HIGH switching point for Schmitt Trigger ($\overline{RST}$) input	V_{T+}		4.0		V	
8	INPUT	LOW switching point for Schmitt Trigger ($\overline{RST}$) input	V_{T-}		1.0		V	
9	INPUT	Hysteresis on Schmitt Trigger ($\overline{RST}$) input	V_H		0.5		V	
10	OUTPUT	Output HIGH current (on all the outputs except $\overline{IRQ}$)	I_{OH}	−5	−16		mA	V_{OH}=2.4 V
11	OUTPUT	Output LOW current (on all the outputs including $\overline{IRQ}$)	I_{OL}	5	10		mA	V_{OL}=0.4 V
12	OUTPUT	Output capacitance	C_o		15		pF	

‡ Typical figures are at 25°C and are for design aid only: not guaranteed and not subject to production testing.

* Outputs unloaded. Input pins 12 and 25 clocked at 2048 kHz. All other inputs at V_{SS}.

AC ELECTRICAL CHARACTERISTICS† -Microprocessor Interface

Voltages are with respect to ground (V_{SS}) unless otherwise stated

		Characteristics	Sym	Min	Typ‡	Max	Units	Test Conditions
1		Delay between $\overline{CS}$ and E clock	t_{CSE}	0			ns	
2		Cycle time	t_{CYC}	205			ns	
3		E Clock pulse width HIGH	t_{EWH}	145			ns	
4		E Clock pulse width LOW	t_{EWL}	60			ns	
5		Read/Write setup time	t_{RWS}	20			ns	
6		Read/Write hold time	t_{RWH}	10			ns	
7		Address setup time	t_{AS}	20			ns	
8		Address hold time	t_{AH}	60			ns	
9		Data setup time (write)	t_{DSW}	35			ns	
10		Data hold time (write)	t_{DHW}	10			ns	
11		E clock to valid data delay	t_{DZL} t_{DZH}			145	ns	Test load circuit 1 (Fig. 11) C_L=200pF
12		Data hold time (read)	t_{DLZ} t_{DHZ}	10		60	ns	Test load circuit 3 (Fig. 11)

† Timing is over recommended temperature & power supply voltages (V_{DD}=5V±5%, V_{SS}=0V, T_A=−40 to 85°C)

‡ Typical figures are at 25°C and are for design aid only: not guaranteed and not subject to production testing.

PIN CONNECTIONS

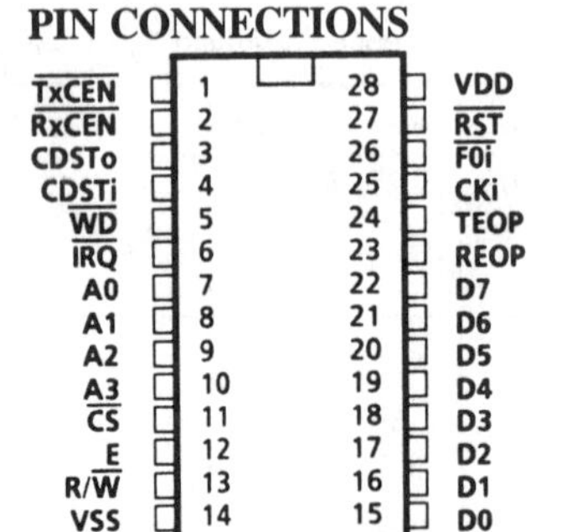

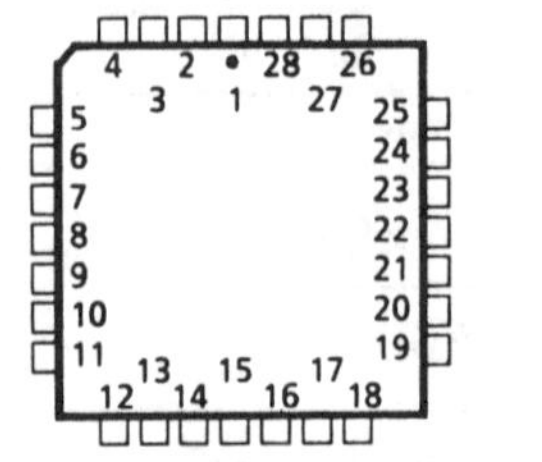

Note: Plastic J-Lead pin naming same as DIP

Ordering Information −40°C to 85°C

MT8952BC	28 Pin Cerdip
MT8952BE	28 Pin Plastic
MT8952BP	28 Pin PLCC

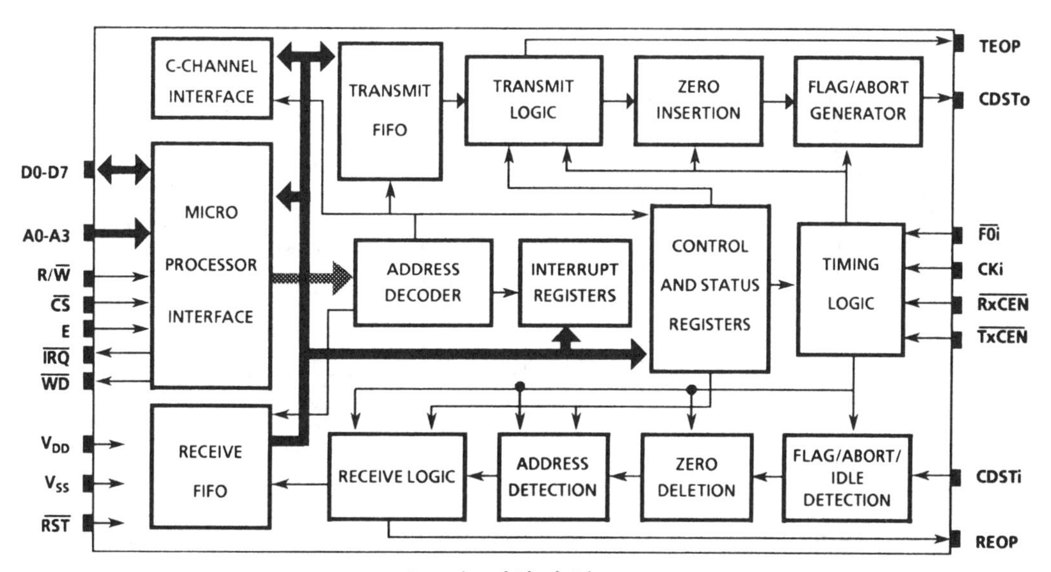

Functional Block Diagram

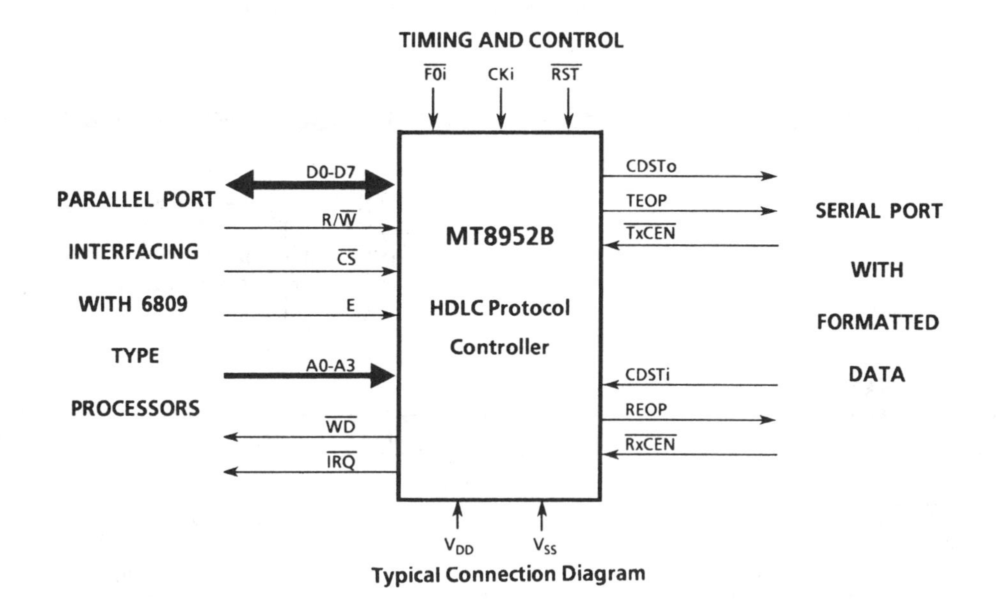

Typical Connection Diagram

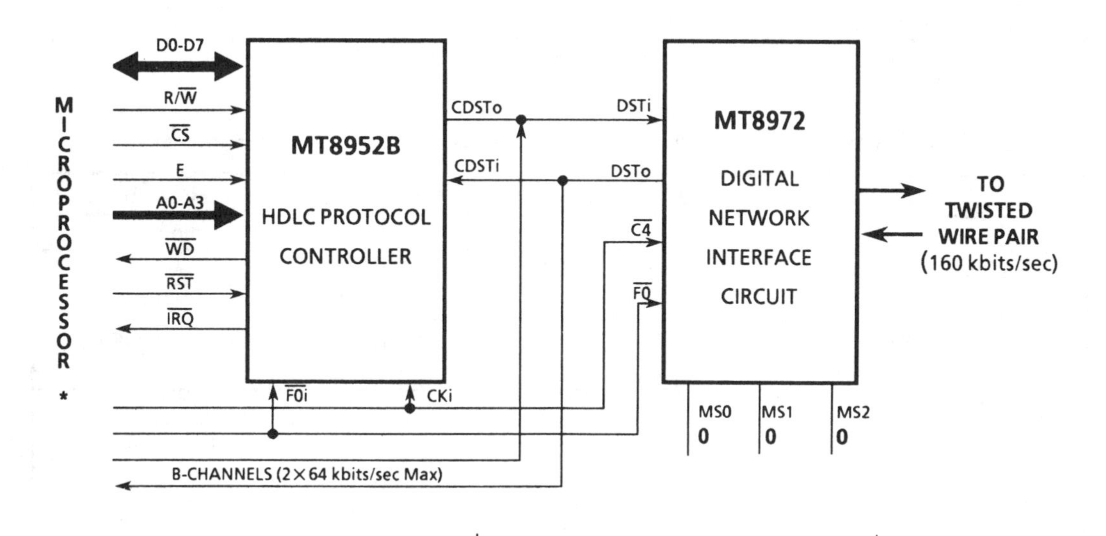

*Note: See MSAN 119 *"How to Interface Mitel Components to Microprocessors"* for microprocessor interfacing.

HDLC Protocol Controller at the Primary End of the Link

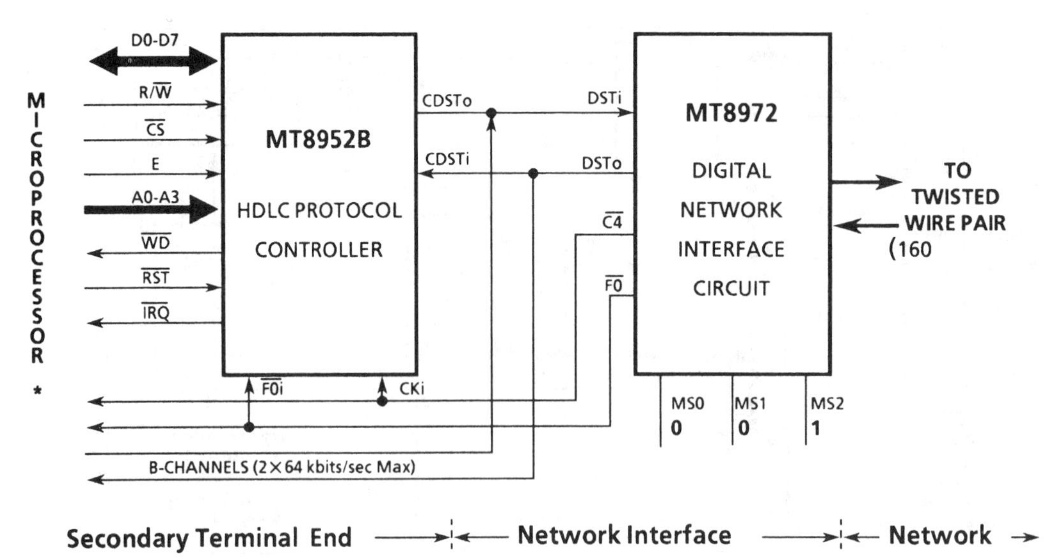

*Note: See MSAN 119 *"How to Interface Mitel Components to Microprocessors"* for microprocessor interfacing.

HDLC Protocol Controller at the Secondary End of the Link

Mitel MT8980D Digital Switch

- Mitel ST-BUS compatible
- 8-Line×32-channel inputs
- 8-Line×32-channel outputs
- 256 ports nonblocking switch
- Single power supply (+5 V)
- Low power consumption: 30 mW Typ.
- Microprocessor-control interface
- Three-state serial outputs

ABSOLUTE MAXIMUM RATINGS*

	Parameter	Symbol	Min	Max	Units
1	V_{DD} - V_{SS}		−0.3	7	V
2	Voltage on Digital Inputs	V_I	$V_{SS}-0.3$	$V_{DD}+0.3$	V
3	Voltage on Digital Outputs	V_O	$V_{SS}-0.3$	$V_{DD}+0.3$	V
4	Current at Digital Outputs	I_O		40	mA
5	Storage Temperature	T_S	−65	+150	°C
6	Package Power Dissipation	P_D		2	W

*Exceeding these values may cause permanent damage. Functional operation under these conditions is not implied.

DC ELECTRICAL CHARACTERISTICS - Voltages are with respect to ground (V_{SS}) unless otherwise stated

		Characteristics	Sym	Min	Typ‡	Max	Units	Test Conditions
1	INPUTS	Supply Current	I_{DD}		6	10	mA	Outputs unloaded
2		Input High Voltage	V_{IH}	2.0			V	
3		Input Low Voltage	V_{IL}			0.8	V	
4		Input Leakage	I_{IL}			5	μA	V_I between V_{SS} and V_{DD}
5		Input Pin Capacitance	C_I		8		pF	
6	OUTPUTS	Output High Voltage	V_{OH}	2.4			V	I_{OH} = 10 mA
7		Output High Current	I_{OH}	10	15		mA	Sourcing. V_{OH}=2.4V
8		Output Low Voltage	V_{OL}			0.4	V	I_{OL} = 5 mA
9		Output Low Current	I_{OL}	5	10		mA	Sinking. V_{OL} = 0.4V
10		High Impedance Leakage	I_{OZ}			5	μA	V_O between V_{SS} and V_{DD}
11		Output Pin Capacitance	C_O		8		pF	

‡ Typical figures are at 25°C and are for design aid only: not guaranteed and not subject to production testing.

RECOMMENDED OPERATING CONDITIONS - Voltages are with respect to ground (V_{SS}) unless otherwise stated

		Characteristics	Sym	Min	Typ‡	Max	Units	Test Conditions
1		Operating Temperature	T_{OP}	-40		+85	°C	
2		Positive Supply	V_{DD}	4.75		5.25	V	
3		Input Voltage	V_I	0		V_{DD}	V	

‡ Typical figures are at 25°C and are for design aid only: not guaranteed and not subject to production testing.

PIN CONNECTIONS

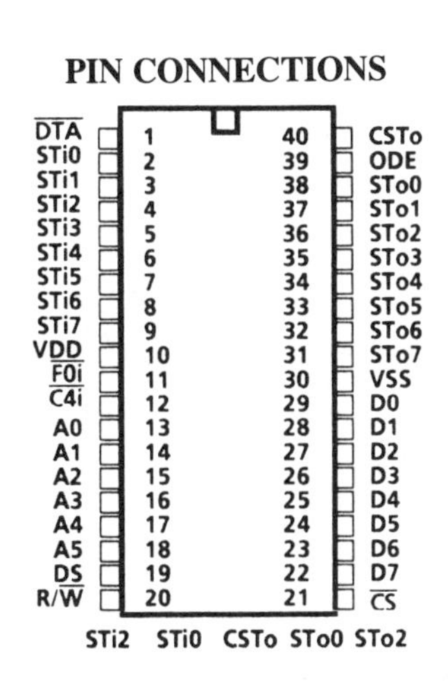

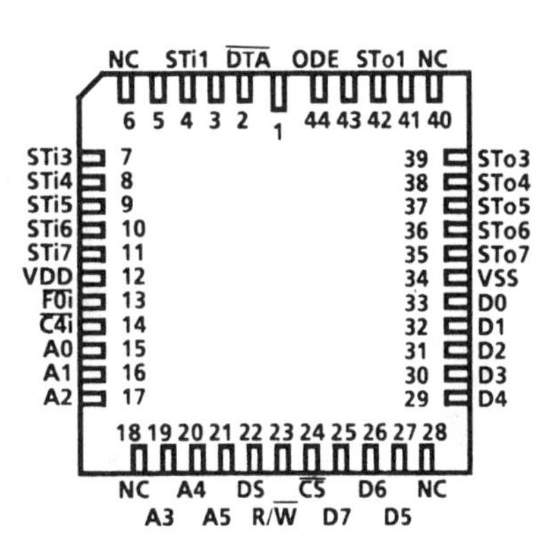

Ordering Information -40°C to +85°C

MT8980DC 40 Pin Ceramic DIL (Cerdip)
MT8980DE 40 Pin Plastic DIL
MT8980DP 44 Pin Plastic PLCC

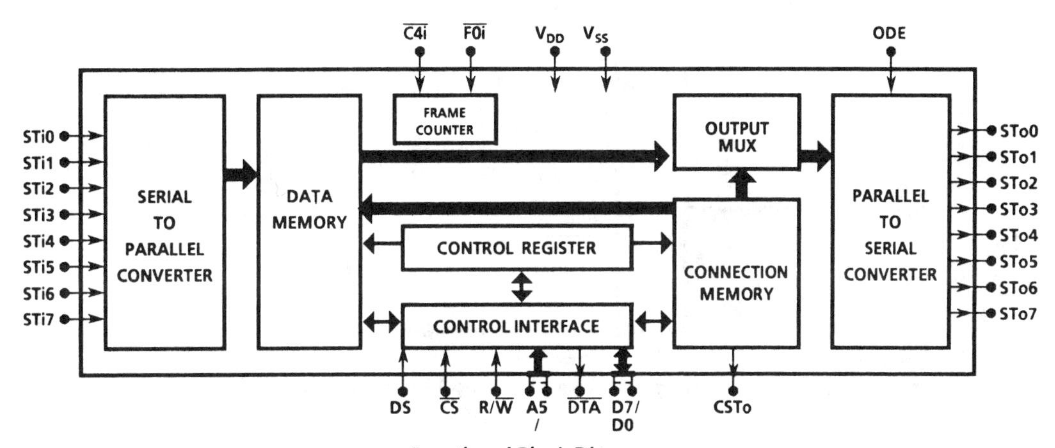

Functional Block Diagram

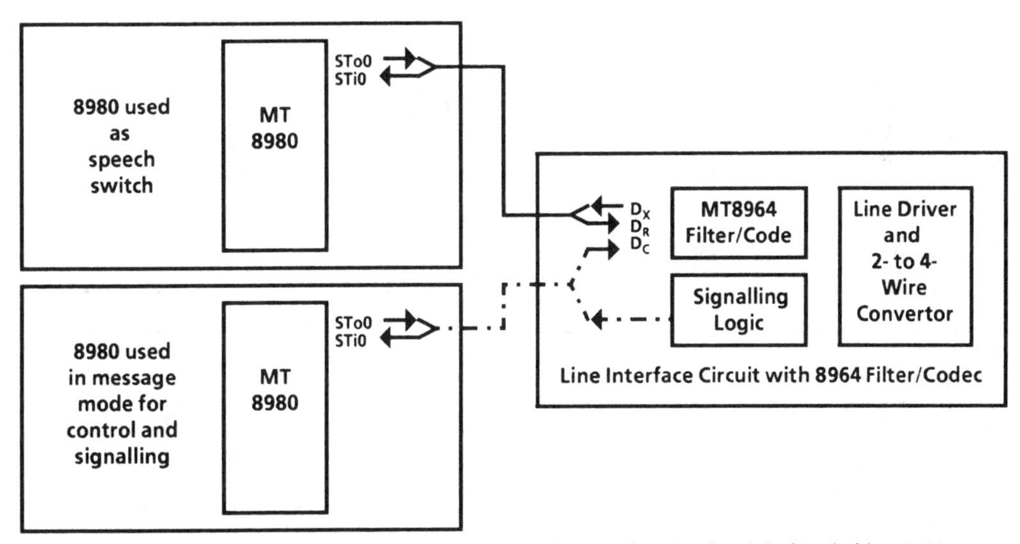

Example of Typical Interface between 8980s and 8964s for Simple Digital Switching System

Example Architecture of a Simple Digital Switching System

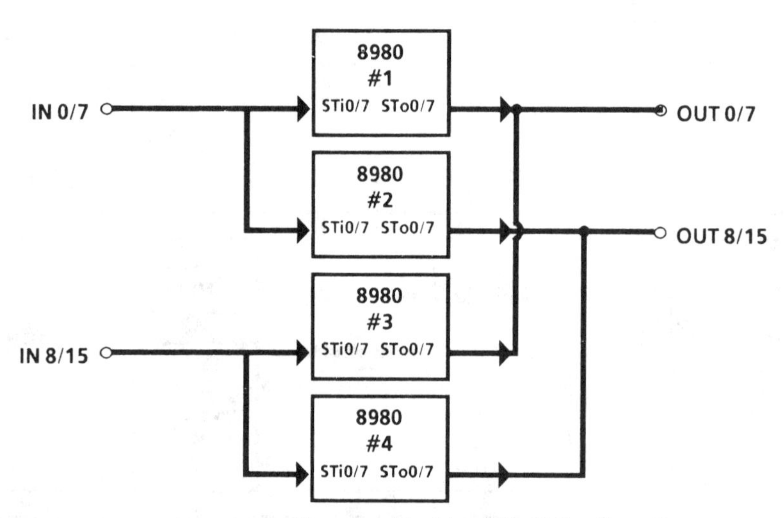

Four 8980s Arranged in a Non-Blocking 16 x 16 Configuration

Mitel MT8981D Digital Switch

- Mitel ST-BUS compatible
- 4-Line × 32-channel inputs
- 4-Line × 32-channel outputs
- 128 ports nonblocking switch
- Single power supply (+5 V)
- Low power consumption: 30 mW Typ.
- Microprocessor-control interface
- Three-state serial outputs

ABSOLUTE MAXIMUM RATINGS*

	Parameter	Symbol	Min	Max	Units
1	V_{DD} - V_{SS}		−0.3	7	V
2	Voltage on Digital Inputs	V_I	$V_{SS}-0.3$	$V_{DD}+0.3$	V
3	Voltage on Digital Outputs	V_O	$V_{SS}-0.3$	$V_{DD}+0.3$	V
4	Current at Digital Outputs	I_O		40	mA
5	Storage Temperature	T_S	−65	+150	°C
6	Package Power Dissipation	P_D		2	W

*Exceeding these values may cause permanent damage. Functional operation under these conditions is not implied.

RECOMMENDED OPERATING CONDITIONS - Voltages are with respect to ground (V_{SS}) unless otherwise stated

		Characteristics	Sym	Min	Typ‡	Max	Units	Test Conditions
1		Operating Temperature	T_{OP}	-40		+85	°C	
2		Positive Supply	V_{DD}	4.75		5.25	V	
3		Input Voltage	V_I	0		V_{DD}	V	

‡ Typical figures are at 25°C and are for design aid only: not guaranteed and not subject to production testing.

DC ELECTRICAL CHARACTERISTICS - Voltages are with respect to ground (V_{SS}) unless otherwise stated

		Characteristics	Sym	Min	Typ‡	Max	Units	Test Conditions
1	INPUTS	Supply Current	I_{DD}		6	10	mA	Outputs unloaded
2		Input High Voltage	V_{IH}	2.0			V	
3		Input Low Voltage	V_{IL}			0.8	V	
4		Input Leakage	I_{IL}			5	µA	V_I between V_{SS} and V_{DD}
5		Input Pin Capacitance	C_I		8		pF	
6	OUTPUTS	Output High Voltage	V_{OH}	2.4			V	I_{OH} = 10 mA
7		Output High Current	I_{OH}	10	15		mA	Sourcing. V_{OH}=2.4V
8		Output Low Voltage	V_{OL}			0.4	V	I_{OL} = 5 mA
9		Output Low Current	I_{OL}	5	10		mA	Sinking. V_{OL} = 0.4V
10		High Impedance Leakage	I_{OZ}			5	µA	V_O between V_{SS} and V_{DD}
11		Output Pin Capacitance	C_O		8		pF	

‡ Typical figures are at 25°C and are for design aid only: not guaranteed and not subject to production testing.

PIN CONNECTIONS

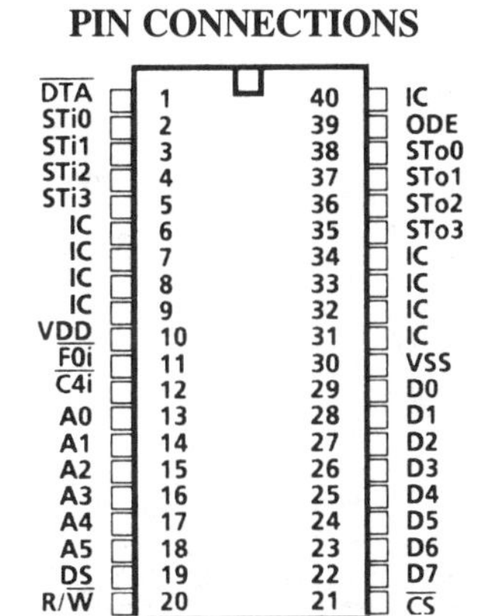

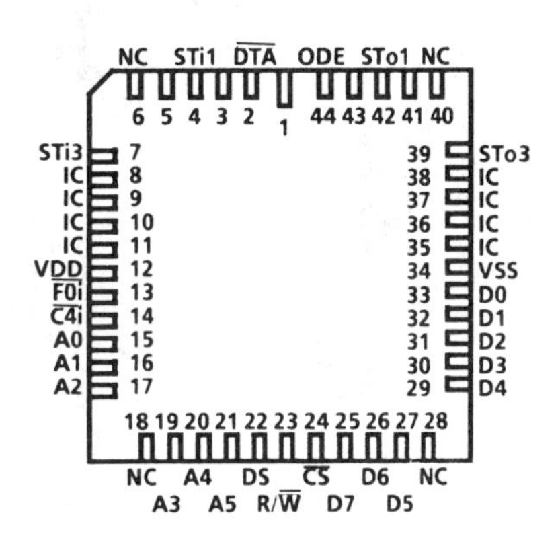

Ordering Information -40°C to +85°C

MT8981DC	40 Pin Ceramic DIL (Cerdip)
MT8981DE	40 Pin Plastic DIL
MT8981DP	44 Pin Plastic PLCC

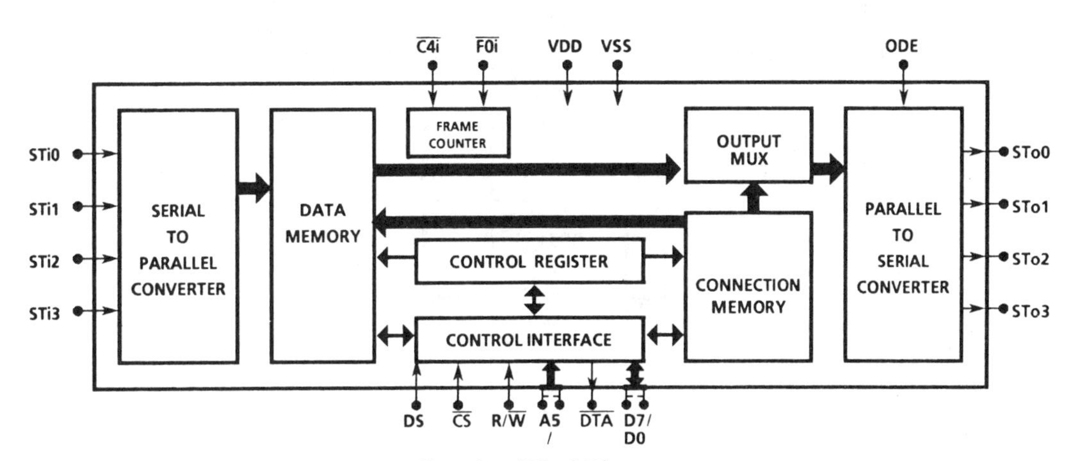

Functional Block Diagram

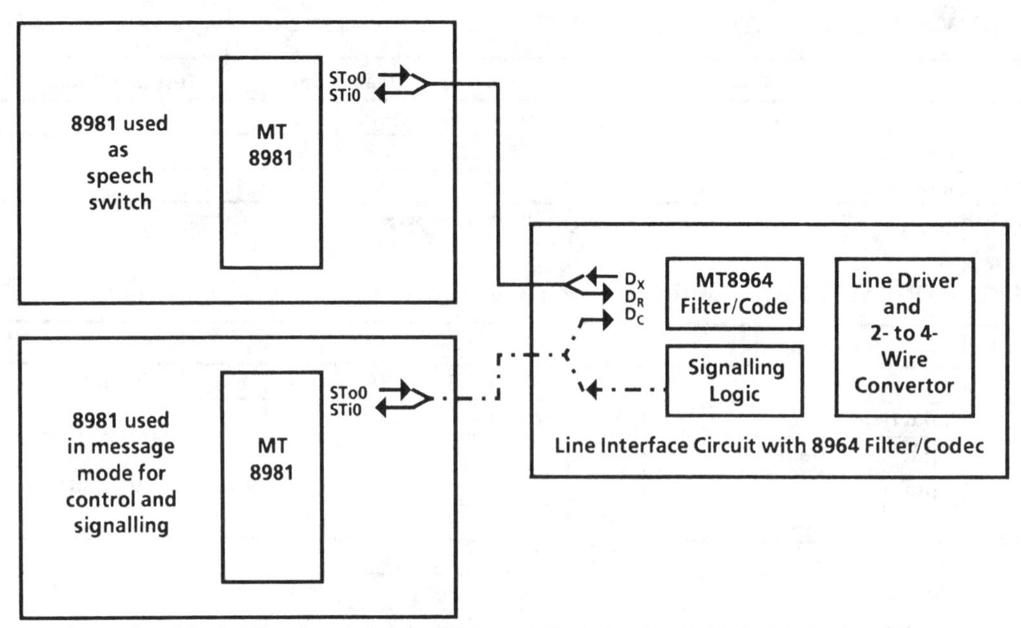

Example of Typical Interface between 8981s and 8964s for Simple Digital Switching System

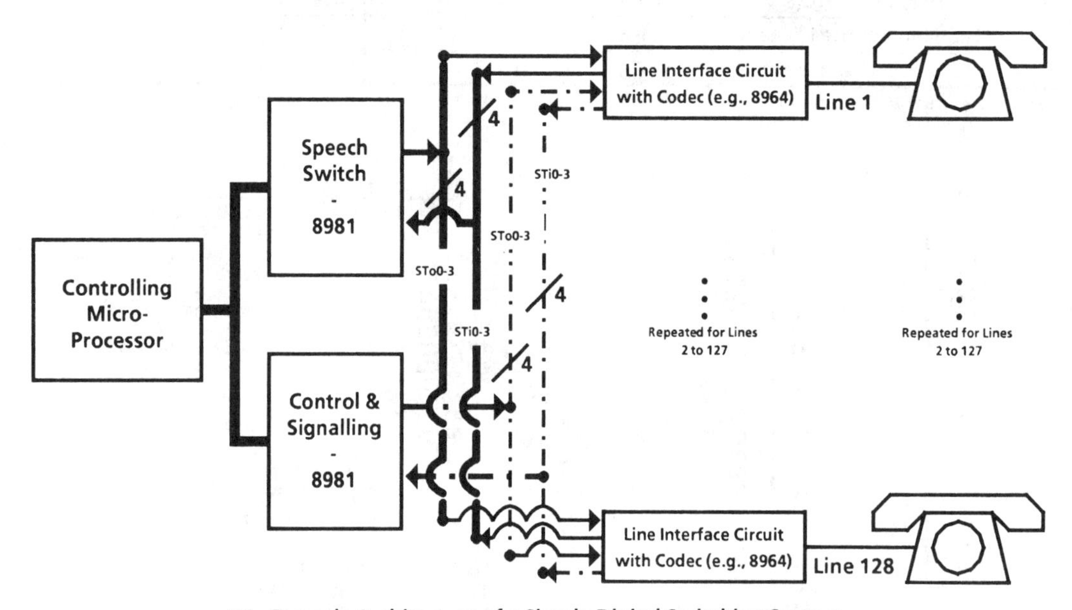

14 - Example Architecture of a Simple Digital Switching System

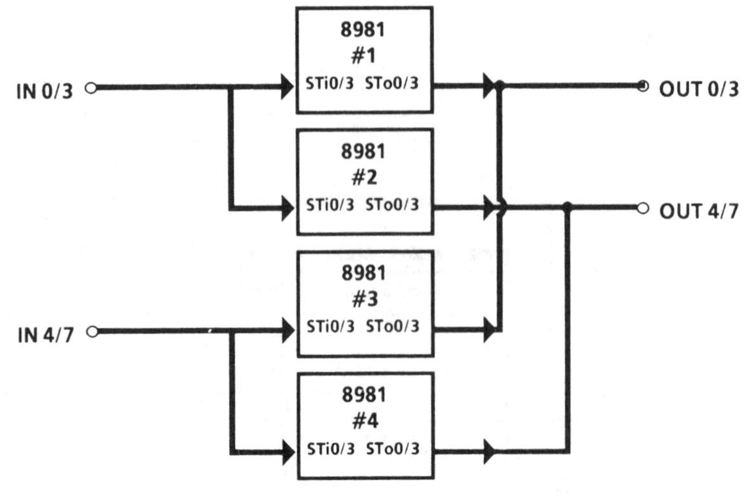

Four 8981s Arranged in a Non-Blocking 8 x 8 Configuration

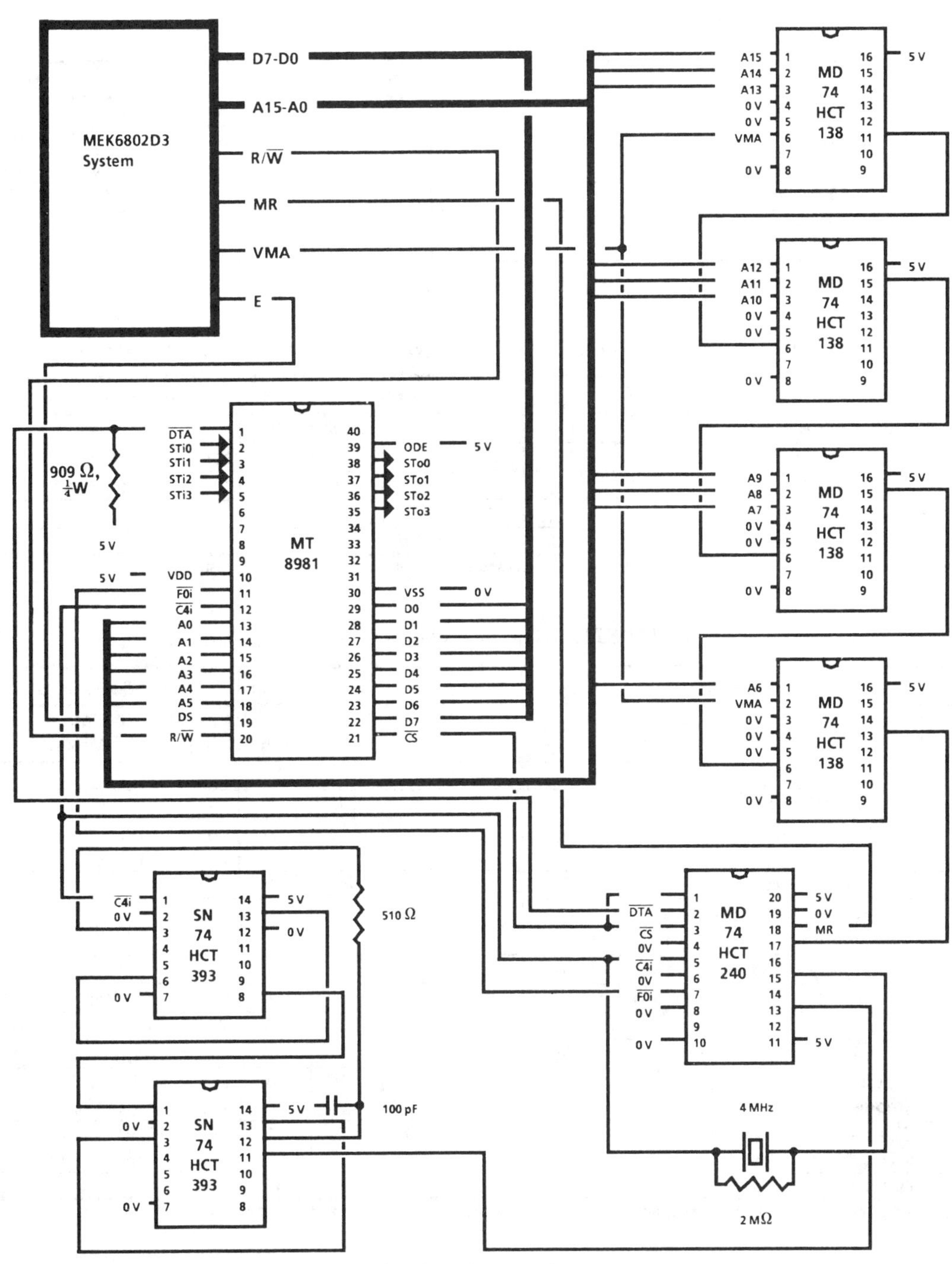

Application Circuit with 6802

Mitel MT8920/MT8920-1
ST-BUS Parallel Access Circuit

- High-speed parallel access to the serial ST-BUS
- Parallel bus to serial bus interface optimized for 68000 μP (mode 1)
- Fast dual-port RAM access (mode 2)
  Access time: 120 nsec - MT8920-1
  180 nsec - MT8920
- Parallel bus controller (mode-3) no external controller required
- Flexible interrupt capabilities—two independent/programmable interrupt sources with auto-vectoring
- Selectable 24- and 32-channel operation
- Programmable loop-around modes
- Low-power ISO-CMOS technology

APPLICATIONS

- Parallel control/data access to T1/CEPT digital trunk interfaces.
- Digital signal processor interface to ST-BUS.
- Computer to DPABX link.
- Voice store and forward systems.
- Interprocessor communications.

ABSOLUTE MAXIMUM RATINGS* - Voltages are with respect to ground (V_{SS}) unless otherwise stated

	Parameter		Symbol	Min	Max	Units
1	Supply Voltage		V_{DD}	−0.3	7.0	V
2	Voltage on any I/O pin			−0.3	V_{DD} + 0.3	V
3	Current on any I/O pin		$I_{I/O}$		±25	mA
4	Storage Temperature		T_{ST}	−55	125	°C
5	Package Power Dissipation	Cerdip Plastic	P_D P_D		1000 600	mW mW

*Exceeding these values may cause permanent damage. Functional operation under these conditions is not implied.

RECOMMENDED OPERATING CONDITIONS - Voltages are with respect to ground (V_{SS}) unless otherwise stated

	Characteristics	Sym	Min	Typ‡	Max	Units	Test Conditions
1	Supply Voltage	V_{DD}	4.75	5.0	5.25	V	
2	Input High Voltage	V_{IH}	2.4		V_{DD}	V	for 400mV noise margin
3	Input Low Voltage	V_{IL}	0		0.4	V	for 400mV noise margin
4	Operating Temperature	T_A	-40	25	85	°C	
5	Operating Clock Frequency	f_{CK}		4.096		MHz	

‡ Typical figures are at 25°C and are for design aid only: not guaranteed and not subject to production testing.

DC ELECTRICAL CHARACTERISTICS - Voltages are with respect to ground (V_{SS}) unless otherwise stated

	Characteristics		Sym	Min	Typ‡	Max	Units	Test Conditions
1	Supply Current	Static Dynamic	I_{CCS} I_{CCD}		10 5	 10	µA mA	outputs unloaded @f_{CK} = 4.096 MHz
2	Input High Voltage		V_{IH}	2.0			V	
3	Input Low Voltage		V_{IL}			0.8	V	
4	Input Leakage Current		I_Z			±10	µA	V_{DD}=5.25V, V_{IN}=V_{SS} to V_{DD}
5	Input capacitance		C_{IN}			10	pF	
6	Schmitt trigger input high (MMS)		V_{T+}	3.8	3.0		V	
7	Schmitt trigger input low (MMS)		V_{T-}		2.0	1.0	V	
8	Schmitt trigger hysteresis (MMS)		V_H		1.0	0.8	V	
9	Output high current (except $\overline{IRQ}$)		I_{OH}	10	15		mA	V_{OH} = 2.4V, V_{DD} = 4.75V
10	Output low current (except $\overline{IRQ}$)		I_{OL}	5	10		mA	V_{OL} = 0.4V, V_{DD} = 4.75V
11	$\overline{IRQ}$, $\overline{DTACK}$, $\overline{BUSY}$ Sink Current		I_{OL}	10	15		mA	V_{OL} = 0.4V, V_{DD} = 4.75V
12	Tristate Leakage A_4-A_0, $\overline{OE}$,$\overline{WE}$ (mode 3)		I_{OZ}		±1	±10	µA	V_{DD} = 5.25V V_{OUT} = V_{SS} to V_{DD}
13	Open drain off-state current $\overline{IRQ}$, $\overline{DTACK}$, $\overline{BUSY}$		I_{OFF}		±1	±20	µA	V_{DD} = 5.25V V_{OUT} = V_{DD}
14	Output capacitance		C_O			15	pF	

‡ Typical figures are at 25°C and are for design aid only: not guaranteed and not subject to production testing.

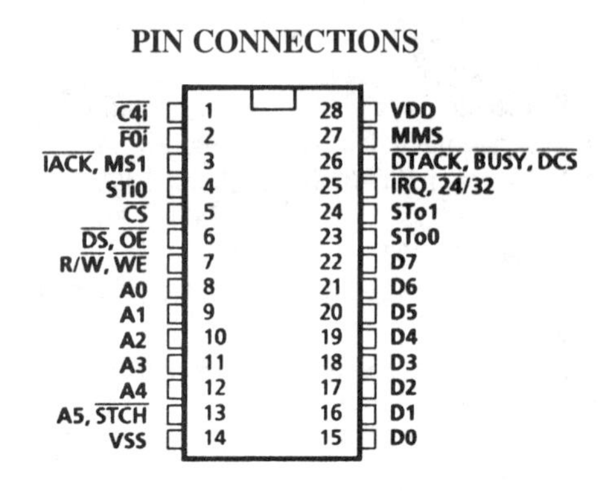

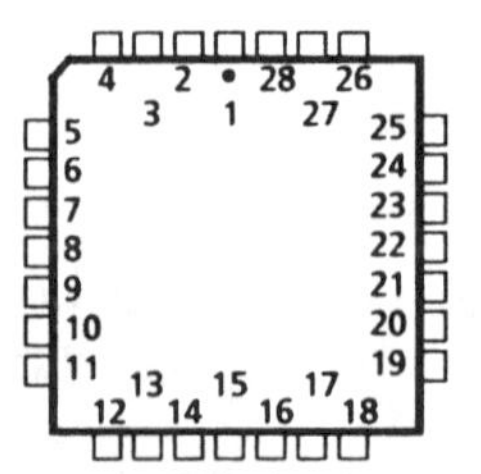

Note: Plastic J-Lead pin naming same as DIP

Ordering Information -40°C to 85°C

MT8920/MT8920-1AE 28 Pin Plastic DIP
MT8920/MT8920-1AC 28 Pin Ceramic DIP
MT8920/MT8920-1AP 28 Pin Plastic J-Lead

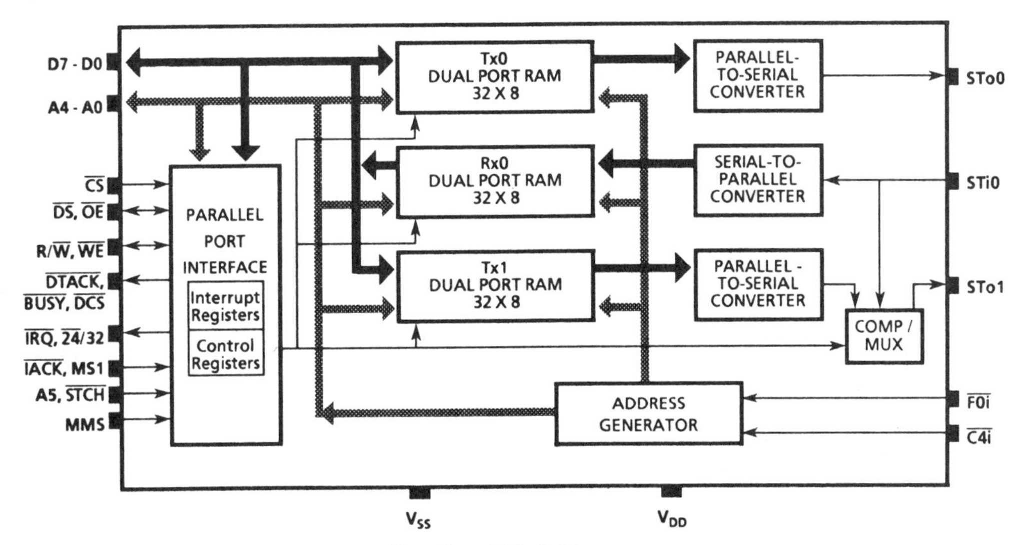

Functional Block Diagram

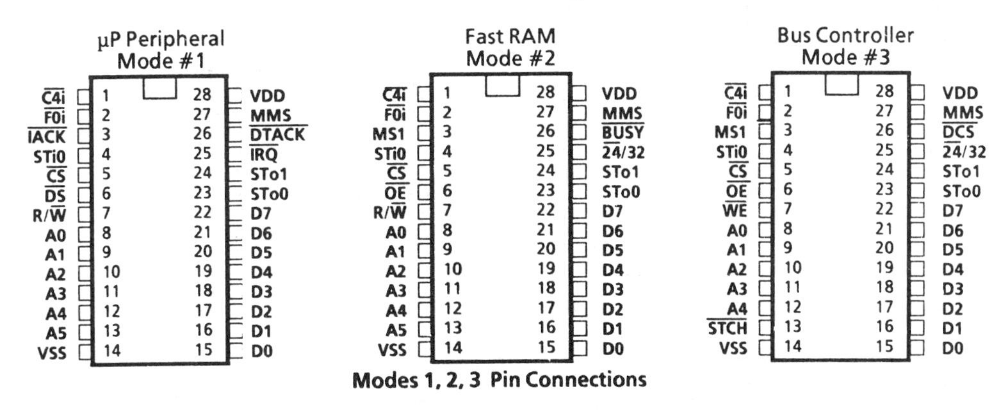

Modes 1, 2, 3 Pin Connections

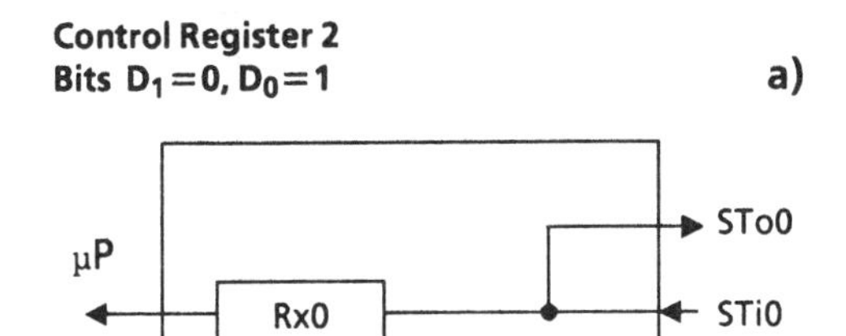

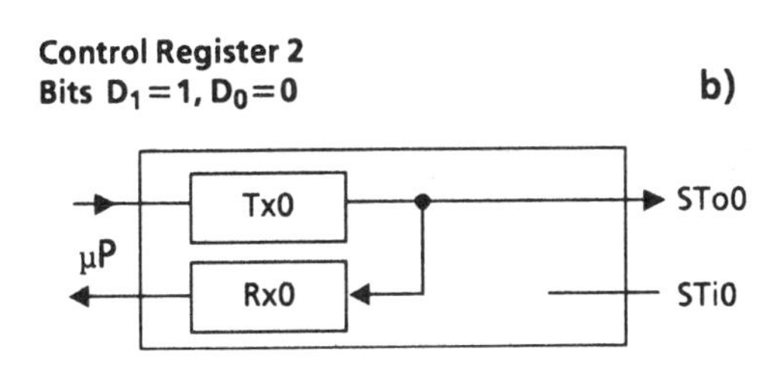

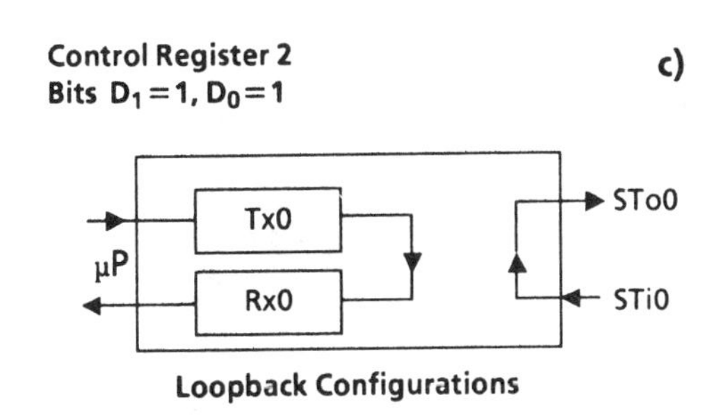

Loopback Configurations

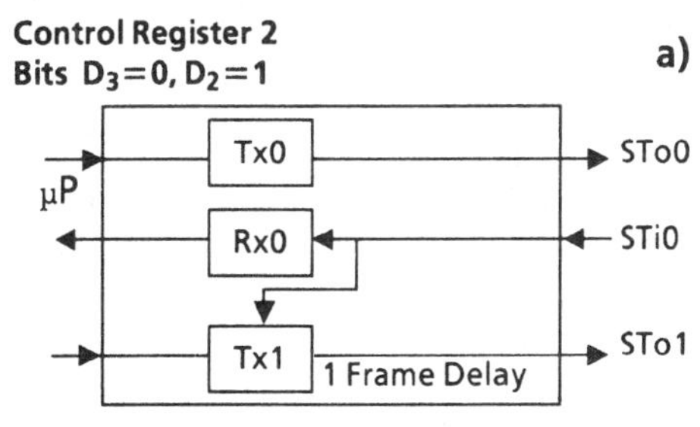

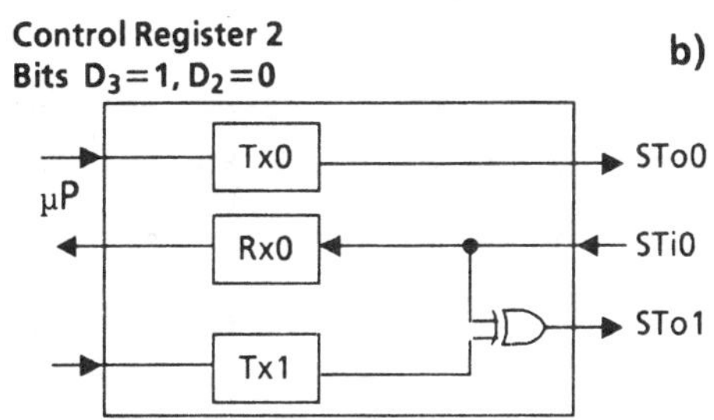

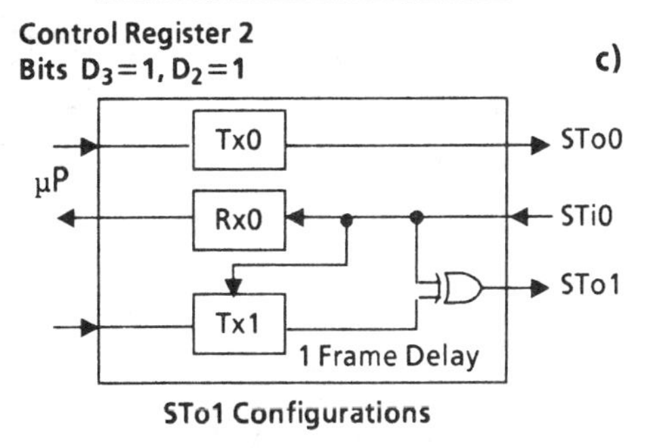

STo1 Configurations

STi0	0 X	1	2	3	4 X	5	6	7	8 X	9	10	11	12 X	13	14	15	16 X	17	18	19	20 X	21	22	23	24 X	25	26	27	28 X	29	30	31
RELATIVE RAM LOCATION		0	1	2		3	4	5		6	7	8		9	10	11		12	13	14		15	16	17		18	19	20		21	22	23

RELATIVE Rx RAM ADDRESS vs. ST-BUS CHANNEL - 24 CHANNEL MODE

STo0 STo1	0 X	1	2	3	4 X	5	6	7	8 X	9	10	11	12 X	13	14	15	16 X	17	18	19	20 X	21	22	23	24 X	25	26	27	28 X	29	30	31
RELATIVE RAM LOCATION		0	1	2		3	4	5		6	7	8		9	10	11		12	13	14		15	16	17		18	19	20		21	22	23

RELATIVE Tx RAM ADDRESS vs. ST-BUS CHANNEL - 24 CHANNEL MODE

STi0	0	1	2	3	4	5	6	7	8	9	10	11	12	13	14	15	16	17	18	19	20	21	22	23	24	25	26	27	28	29	30	31
RELATIVE RAM LOCATION	0	1	2	3	4	5	6	7	8	9	10	11	12	13	14	15	16	17	18	19	20	21	22	23	24	25	26	27	28	29	30	31

RELATIVE Rx RAM ADDRESS vs. ST-BUS CHANNEL - 32 CHANNEL MODE

STo0 STo1	0	1	2	3	4	5	6	7	8	9	10	11	12	13	14	15	16	17	18	19	20	21	22	23	24	25	26	27	28	29	30	31
RELATIVE RAM LOCATION	0	1	2	3	4	5	6	7	8	9	10	11	12	13	14	15	16	17	18	19	20	21	22	23	24	25	26	27	28	29	30	31

RELATIVE Tx RAM ADDRESS vs. ST-BUS CHANNEL - 32 CHANNEL MODE

X- unused channels marked X transmit FF_{16}

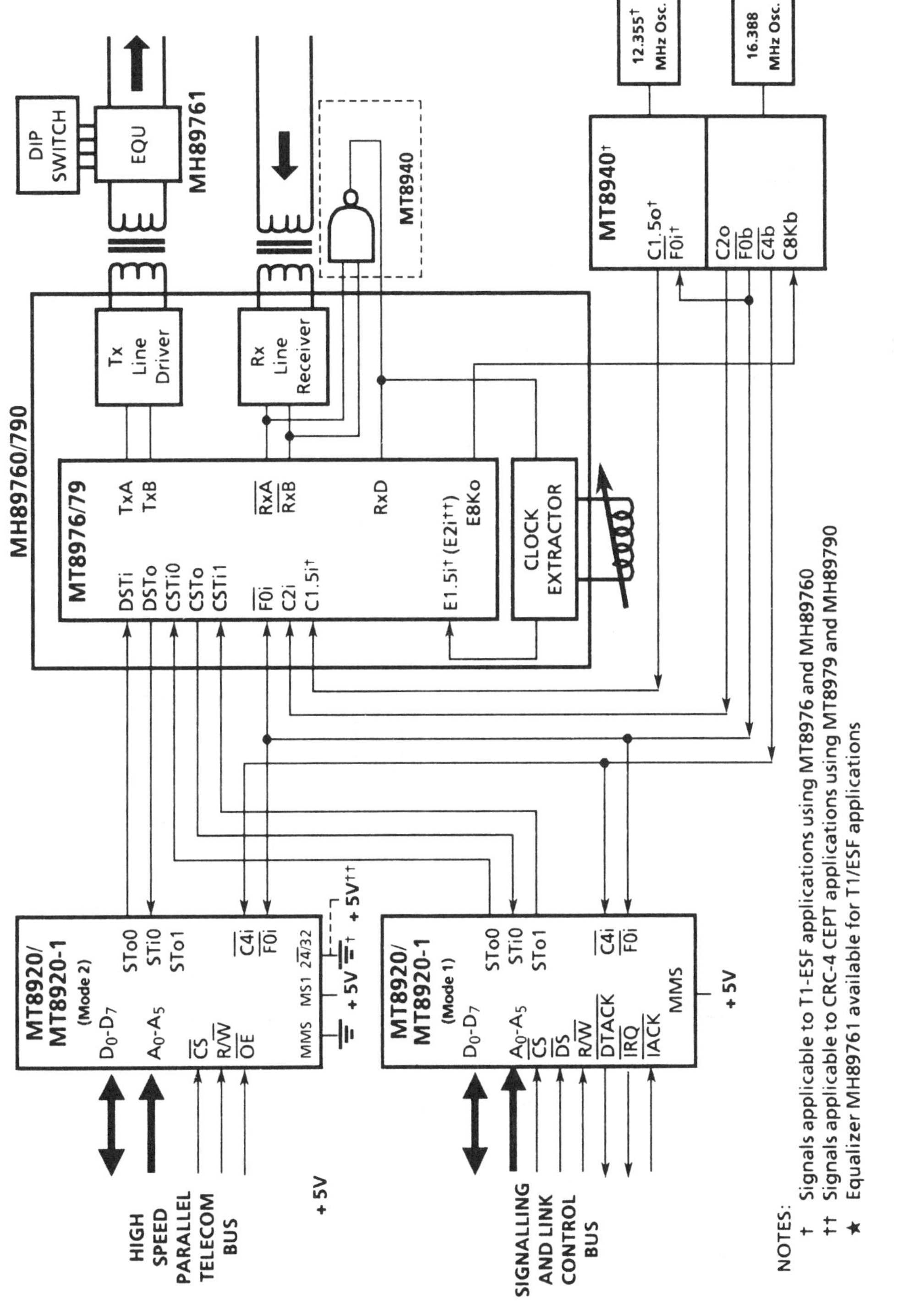

Typical T1-ESF / CRC-4 CEPT Digital Trunk Configuration

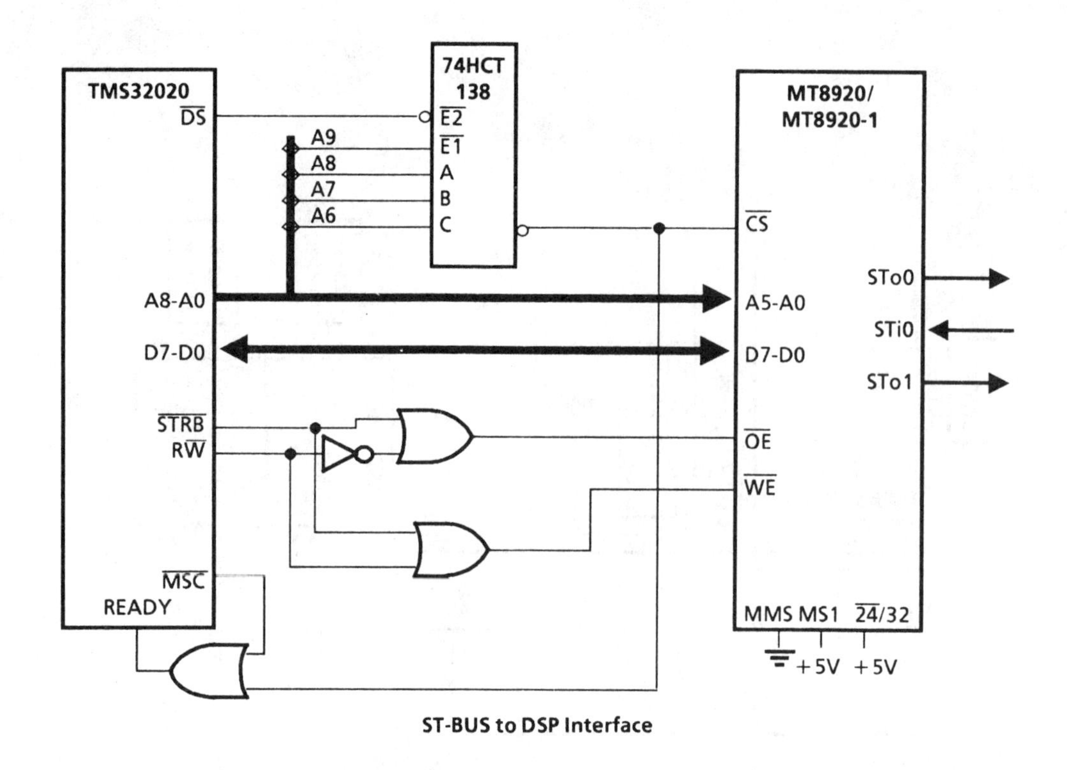

ST-BUS to DSP Interface

Mitel
Application Note MSAN-119
How to Interface Mitel Components to Microprocessors

Processor / Component	6800	6802	6809	68000	68010	68008	8085	8086	8088	Z80	Z8002
MD65SC51	4	2	3	5	5	5	6	10	10	8	12
MT8880	4	2	3	5	5	5	6,7	9,10	9,10	8	11,12
MT8952	4	2	3	5	5	5	6	10	10	8	12
MT8980	15	14	13	16	16	16	17	19	19	18	20
MT8981	15	14	13	16	16	16	17	19	19	18	20
MT8920	23	22	21	24	24	24	25	27	27	26	28

Table of Circuits Cross Referenced by Microprocessor and Mitel Part Number

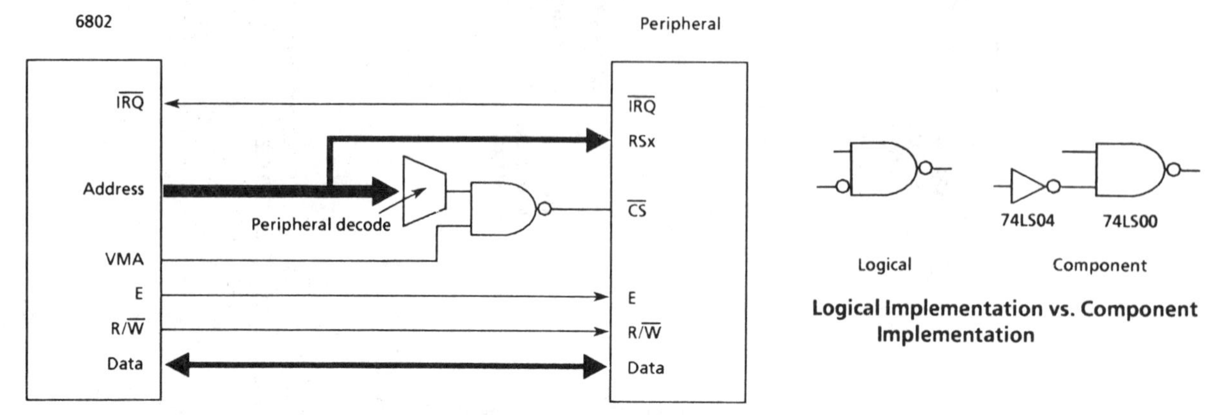

Interfacing the MD65SC51, MT8952, and the MT8880 to the 6802

Logical Implementation vs. Component Implementation

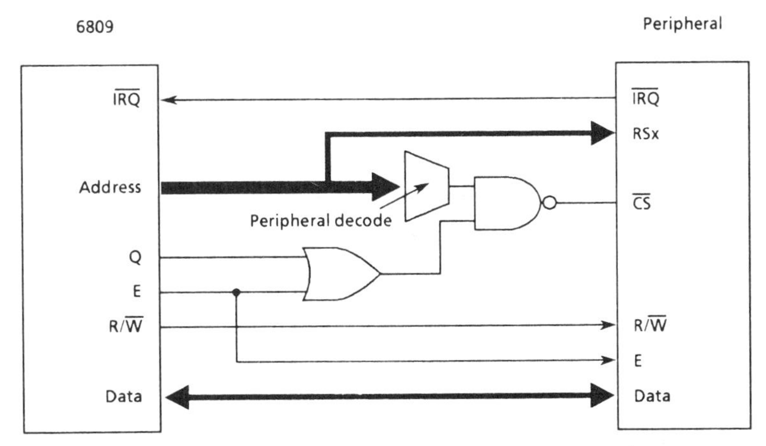

Interfacing the MD65SC51, MT8952, and the MT8880 to the 6809

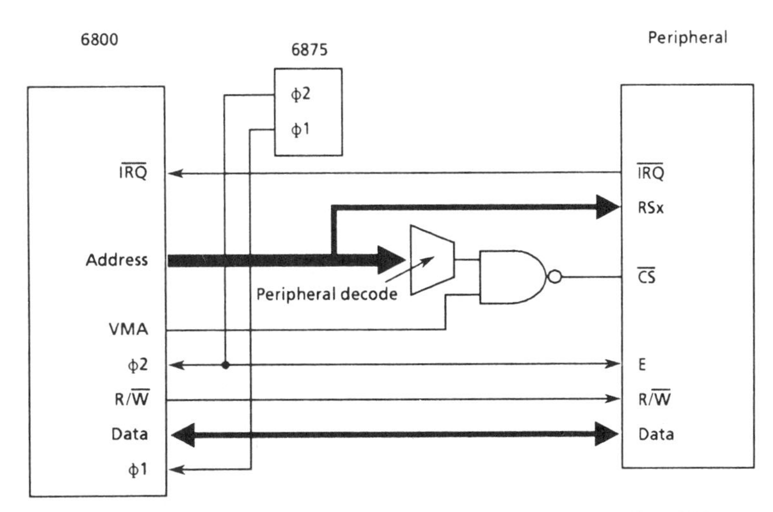

Interfacing the MD65SC51, MT8952, and the MT8880 to the 6800

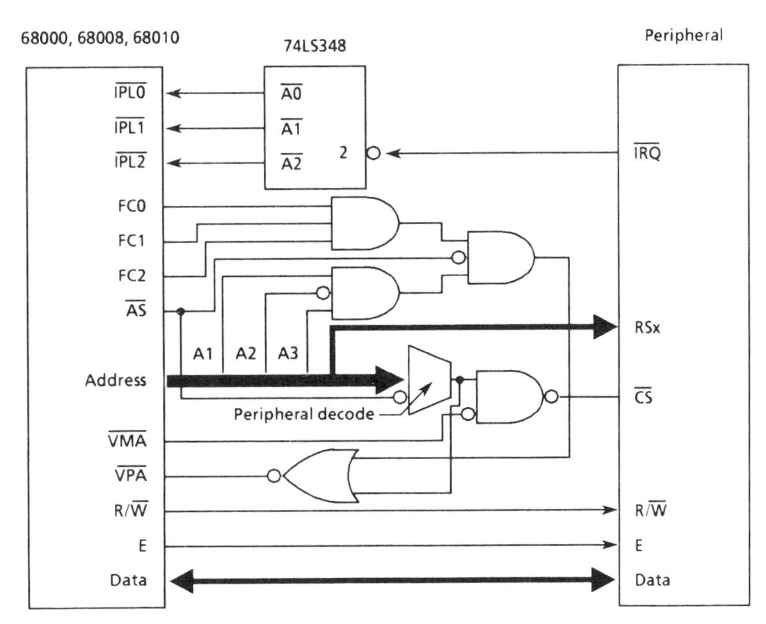

Interfacing the MD65SC51, MT8952, and the MT8880 to the 68000, 68010 and the 68008

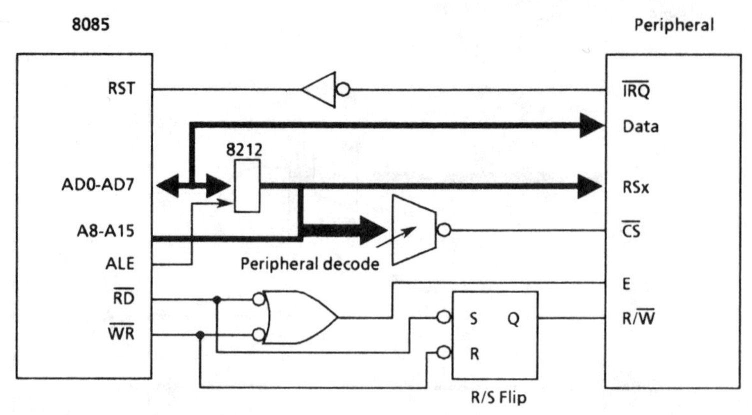

Interfacing the MD65SC51, MT8952, and the MT8880 to the 8085

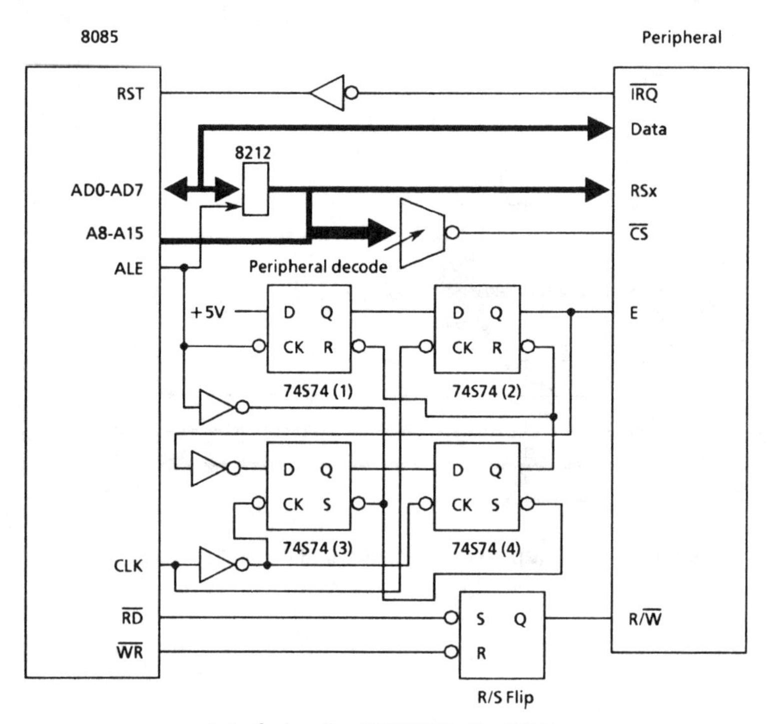

Interfacing the MT8880 to the 8085

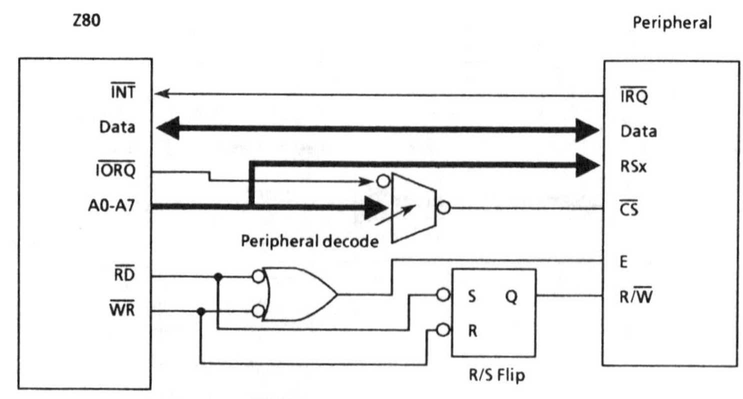

Interfacing the MD65SC51, MT8952, and the MT8880 to the Z80

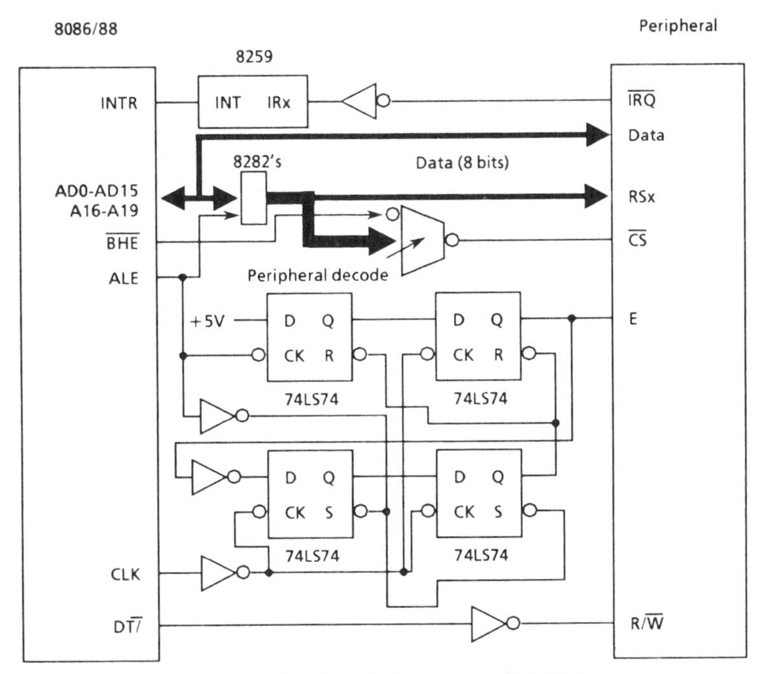

Interfacing the MT8880 to the 8086/88

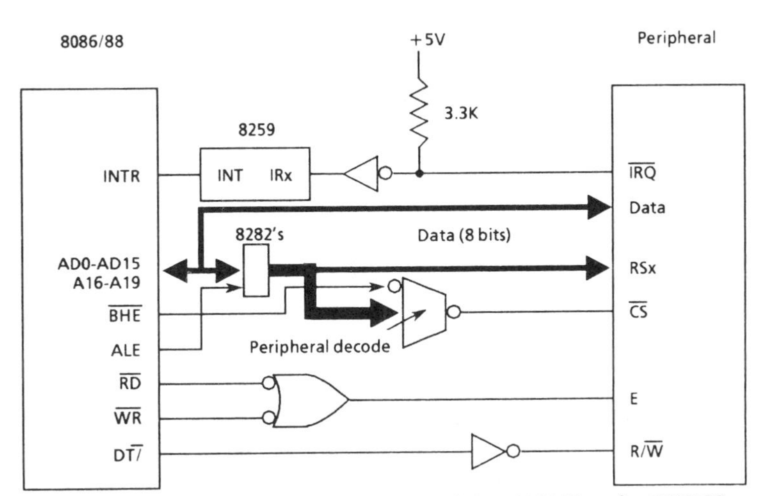

Interfacing the MD65SC51, MT8952, and the MT8980 to the 8086/88

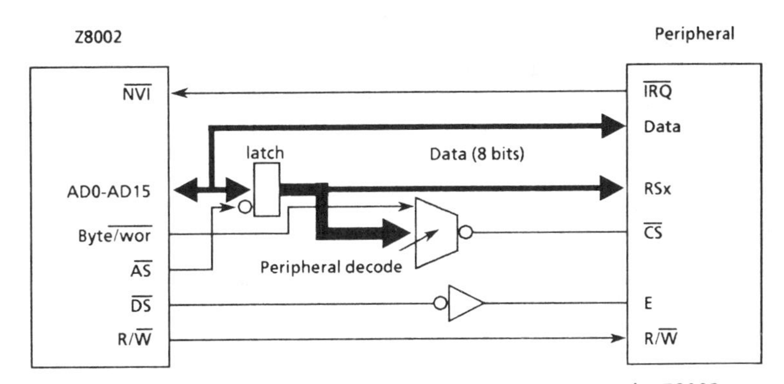

Interfacing the MD65SC51, MT8952 and the MT8880 to the Z8002

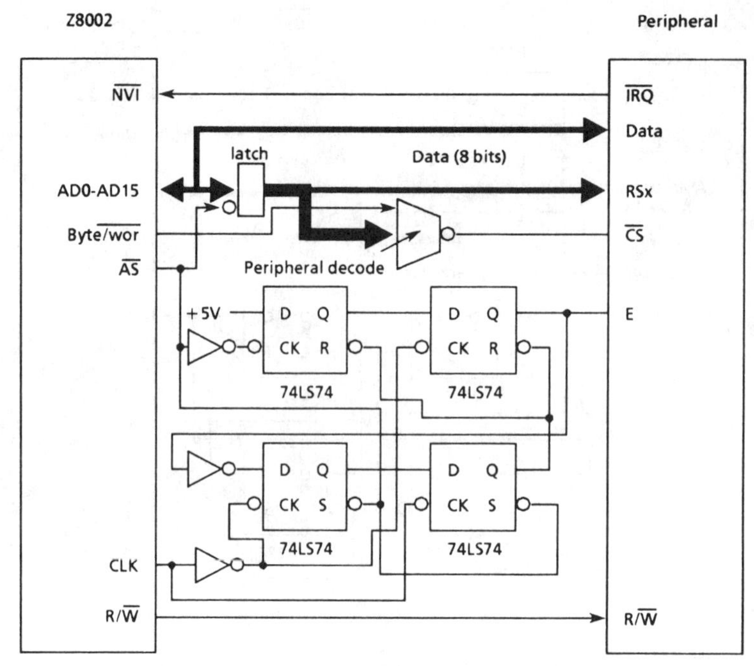

Interfacing the MT8880 to the Z8002

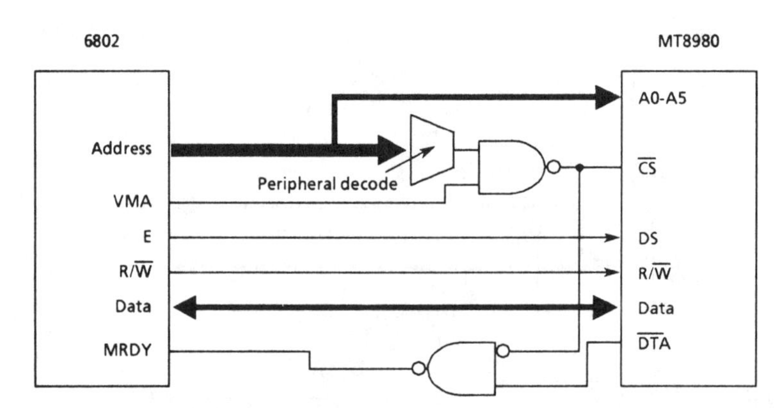

Interfacing the MT8980 and the MT8981 to the 6802

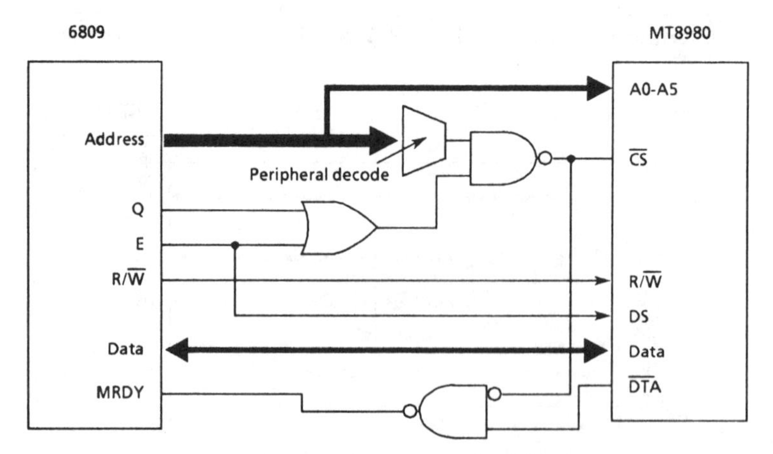

Interfacing the MT8980 and the MT8981 to the 6809

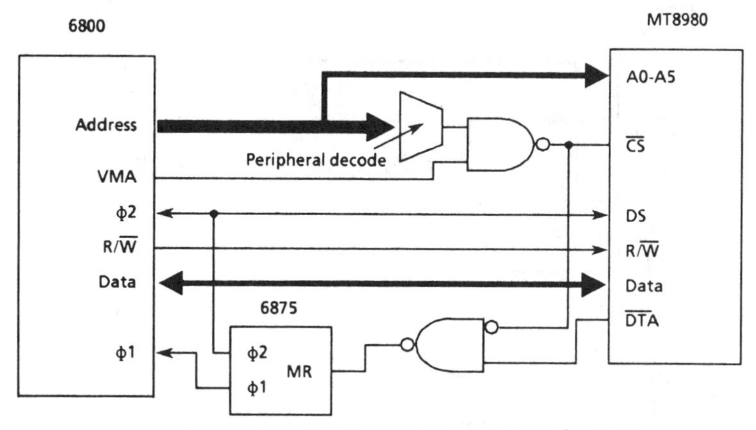

Interfacing the MT8980 and the MT8981 to the 6800

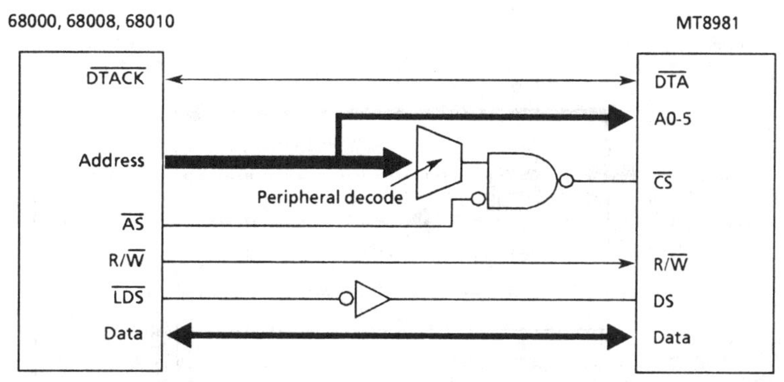

Interfacing the MT8980 and the MT8981 to the 68000, 68010 and the 68008

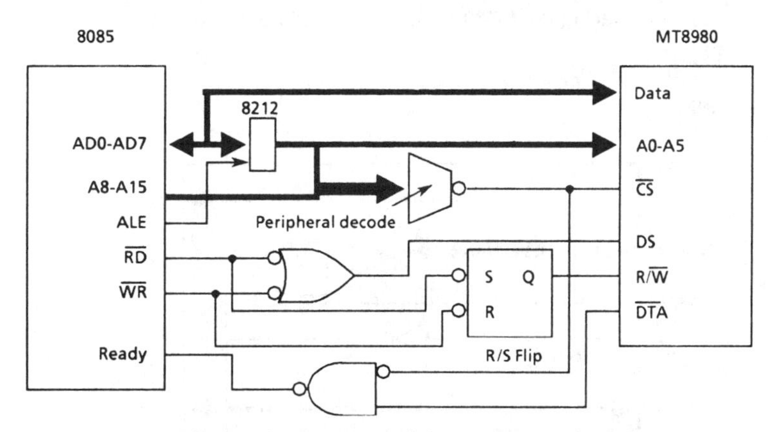

Interfacing the MT8980 and the MT8981 to the 8085

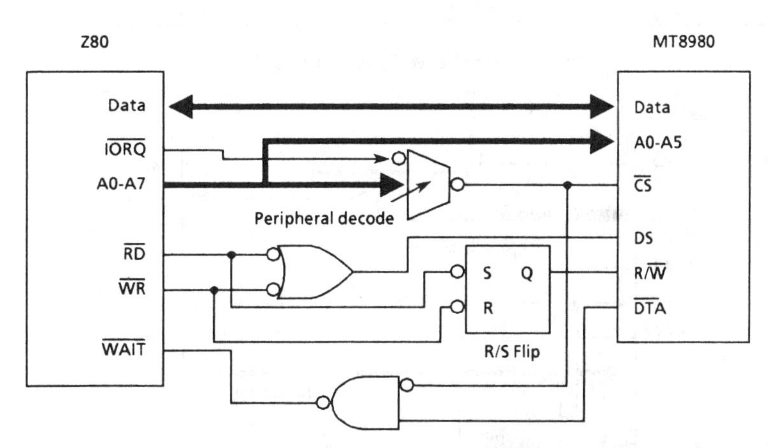

Interfacing the MT8980 and the MT8981 to the Z80

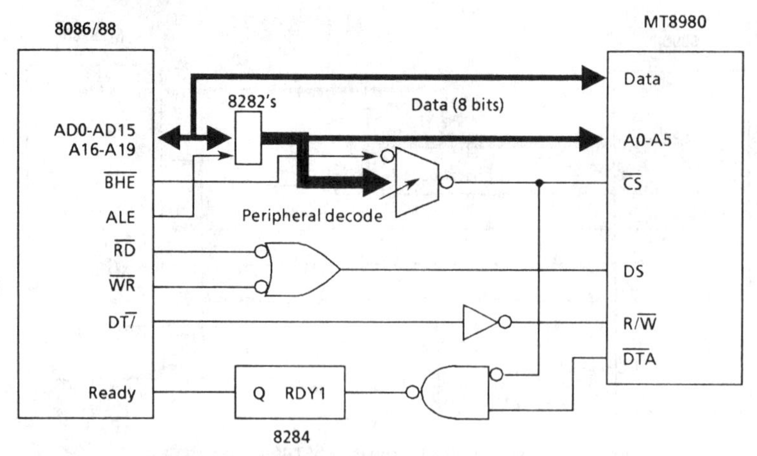

Interfacing the MT8980 and the MT8981 to the 8086/88

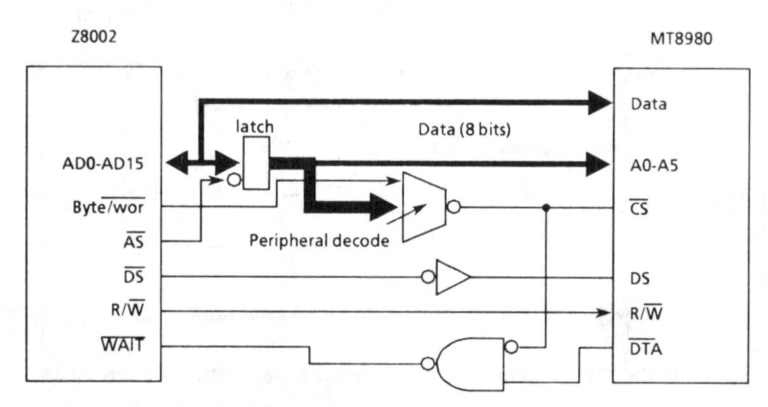

Interfacing the MT8980 and the MT8981 to the Z8002

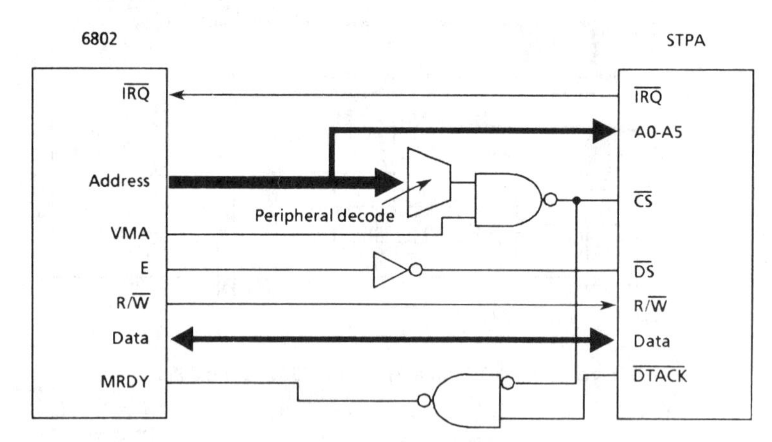

Interfacing the MT8920 to the 6802

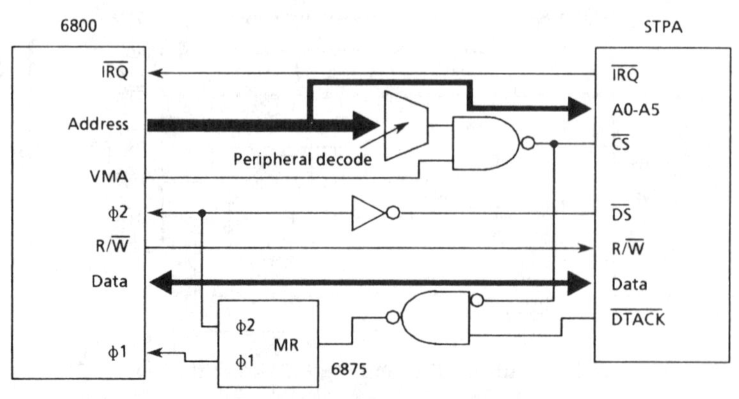

Interfacing the MT8920 to the 6800

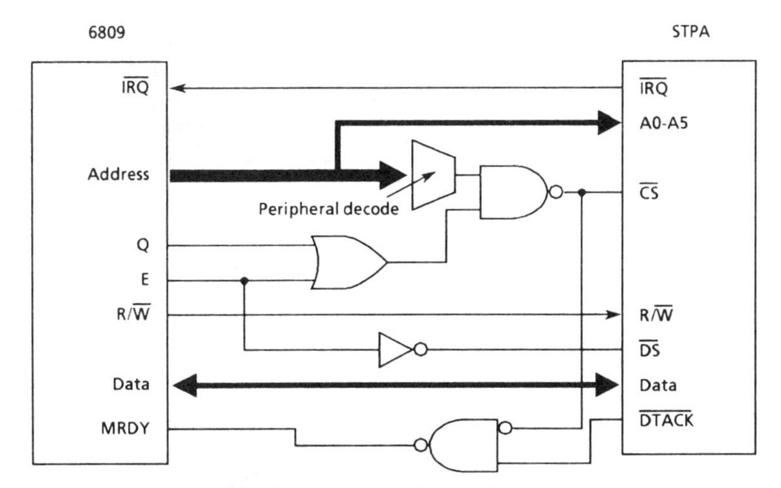

Interfacing the MT8920 to the 6809

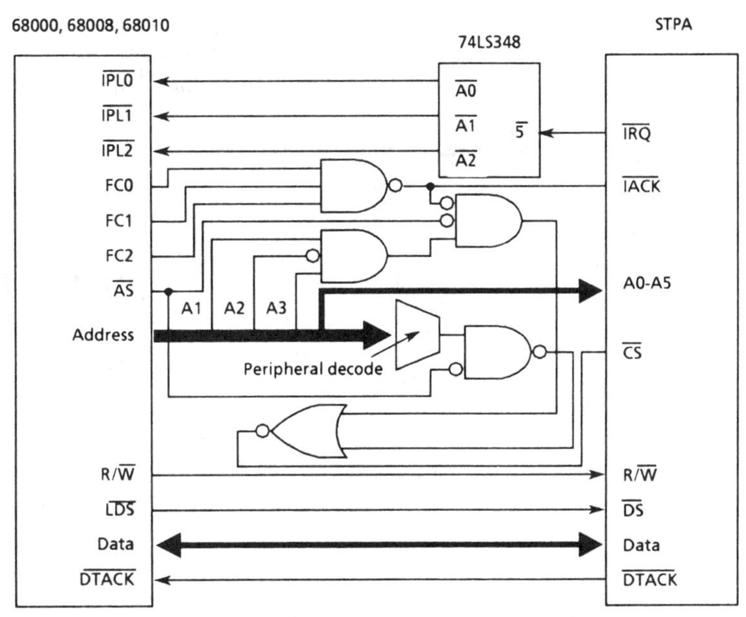

Interfacing the MT8920 to the 68000, 68010 and the 68008

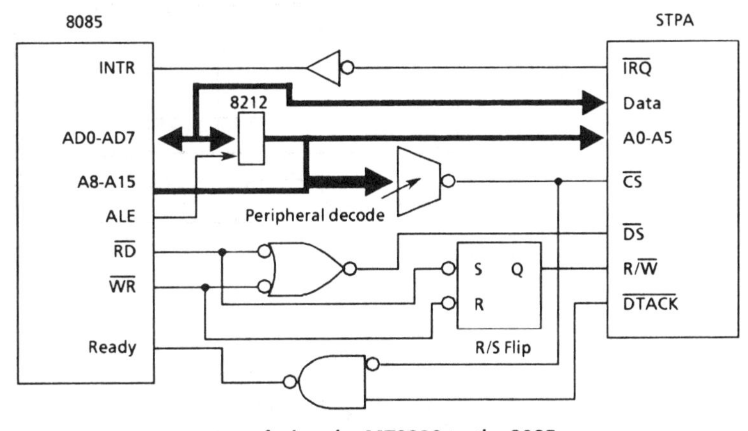

Interfacing the MT8920 to the 8085

Z80
STPA
INT
Data
IORQ
A0-A7
RD
WR
WAIT
IRQ
Data
A0-A5
CS
DS
R/W
DTACK
Peripheral decode
S Q
R
R/S Flip

Interfacing the MT8920 to the Z80

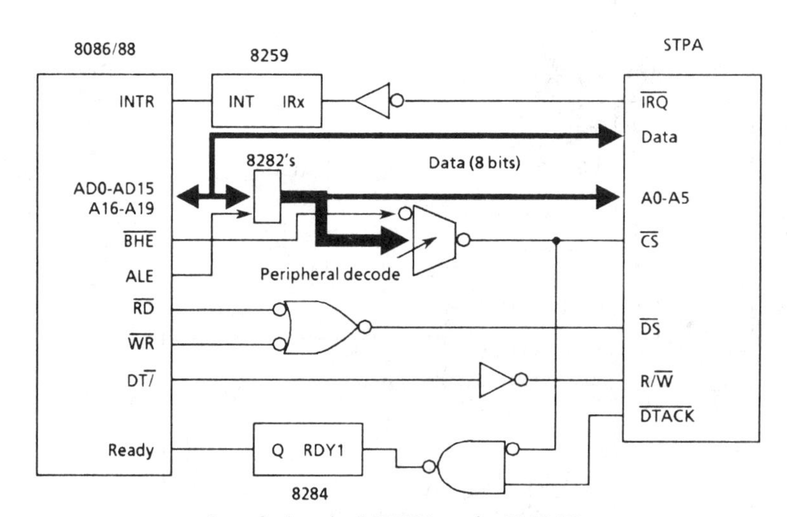

Interfacing the MT8920 to the 8086/88

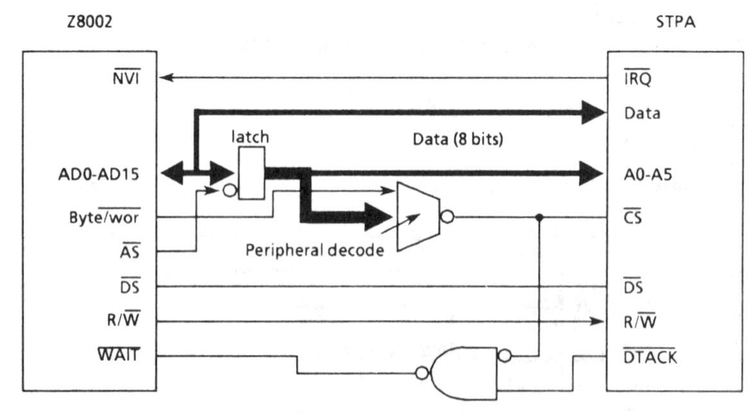

Interfacing the STPA to the Z8002

LIFE SUPPORT POLICY

National's products are not authorized for use as critical components in life support devices or systems without the express written approval of the president of National Semiconductor Corporation. As used herein:

1. Life support devices or systems are devices or systems which, (a) are intended for surgical implant into the body, or (b) support or sustain life, and whose failure to perform, when properly used in accordance with instructions for use provided in the labeling, can be reasonably expected to result in a significant injury to the user.

2. A critical component is any component of a life support device or system whose failure to perform can be reasonably expected to cause the failure of the life support device or system, or to affect its safety or effectiveness.

NATIONAL DM74ALS242C/DM74ALS243A

Quad Tri-state® Bidirectional Bus Driver

- Advanced low-power oxide-isolated ion-implanted Schottky TTL process
- Functional and pin compatible with the 74LS counterpart
- Improved switching performance with less power dissipation compared with the 74LS counterpart
- Switching response specified into 500Ω and 50pF load
- Switching response specifications guaranteed over full temperature and V_{CC} supply range
- Pnp input design reduces input loading
- Low level drive current: 74ALS = 24 mA

ABSOLUTE MAXIMUM RATINGS

Supply Voltage, V_{CC}	7V
Input Voltage	
Dedicated Inputs	7V
I/O Ports	5.5V
Operating Free Air Temperature Range	
DM74ALS	0 to +70°C
Storage Temperature Range	−65°C to +150°C
Typical θ_{JA}	
N Package	78.0°C/W
M Package	111.5°C/W

Note: *The "Absolute Maximum Ratings" are those values beyond which the safety of the device cannot be guaranteed. The device should not be operated at these limits. The parametric values defined in the "Electrical Characteristics" table are not guaranteed at the absolute maximum ratings. The "Recommended Operating Conditions" table will define the conditions for actual device operation.*

ELECTRICAL CHARACTERISTICS over recommended operating free-air temperature (unless otherwise specified)

Symbol	Parameter	Conditions		DM74ALS242C, 243A			Units
				Min	Typ	Max	
V_{IK}	Input Clamp Voltage	V_{CC} = 4.5V, I_I = −18 mA				−1.2	V
V_{OH}	High Level Output	V_{CC} = 4.5V to 5.5V	I_{OH} = −0.4 mA	V_{CC} − 2			V
		V_{CC} = 4.5V	I_{OH} = −3 mA	2.4			V
			I_{OH} = Max	2			V
V_{OL}	Low Level Output Voltage	V_{CC} = 4.5V I_{OL} = 54ALS (Max)			0.25	0.4	V
		I_{OL} = 74ALS (Max)			0.35	0.5	V
I_I	Input Current at Max Input Voltage	V_{CC} = 5.5V, V_I = 7V (5.5V for I/O Ports)				0.1	mA
I_{IH}	High Level Input Current	V_{CC} = 5.5V, V_I = 2.7V (Note 1)				20	μA
I_{IL}	Low Level Input Current	V_{CC} = 5.5V, V_{IL} = 0.4V (Note 1)				−0.1	mA
I_O	Output Drive Current	V_{CC} = 5.5V, V_O = 2.25V		−30		−112	mA
I_{CC}	Supply Current	V_{CC} = 5.5V, ALS242C Active Outputs High			10	16	mA
		Active Outputs Low			14	21	mA
		Outputs TRI-STATE			12	19	mA
		V_{CC} = 5.5V, ALS243A Active Outputs High			15	25	mA
		Active Outputs Low			20	30	mA
		Outputs TRI-STATE			21	32	mA

Note 1: For the I/O ports, the parameters I_{IH} and I_{IL} include the TRI-STATE output currents (I_{OZH} and I_{OZL}).

RECOMMENDED OPERATING CONDITIONS

Symbol	Parameter	DM74ALS242C, 243A			Units
		Min	Typ	Max	
V_{CC}	Supply Voltage	4.5	5	5.5	V
V_{IH}	High Level Input Voltage	2			V
V_{IL}	Low Level Input Voltage			0.8	V
I_{OH}	High Level Output Current			−15	mA
I_{OL}	Low Level Output Current			24	mA
T_A	Operating Free-Air Temperature	0		70	°C

'ALS242C SWITCHING CHARACTERISTIC over recommended operating free-air temperature range (Note 1)

Symbol	Parameter	Conditions	From (Input)	To (Output)	74ALS242C		Units
					Min	Max	
t_{PLH}	Propagation Delay Time Low to High Level Output	V_{CC} = 4.5V to 5.5V, C_L = 50 pF, R1 = 500Ω, R2 = 500Ω, T_A = Min to Max	A or B	B or A	2	11	ns
t_{PHL}	Propagation Delay Time High to Low Level Output				2	10	ns
t_{PZH}	Output Enable Time to High Level Output		$\overline{G}$AB	B	4	18	ns
t_{PZL}	Output Enable Time to Low Level Output				7	21	ns
t_{PHZ}	Output Disable Time to High Level Output		$\overline{G}$AB	B	2	14	ns
t_{PLZ}	Output Disable Time to Low Level Output				2	15	ns
t_{PZH}	Output Enable Time to High Level Output		GBA	A	4	18	ns
t_{PZL}	Output Enable Time to Low Level Output				7	21	ns
t_{PHZ}	Output Disable Time from High Level Output		GBA	A	2	14	ns
t_{PLZ}	Output Disable Time from Low Level Output				2	15	ns

'ALS243A SWITCHING CHARACTERISTICS over recommended operating free-air temperature range (Note 1)

Symbol	Parameter	Conditions	From (Input)	To (Output)	74ALS243A		Units
					Min	Max	
t_{PLH}	Propagation Delay Time Low to High Level Output	V_{CC} = 4.5V to 5.5V, C_L = 50 pF, R1 = 500Ω, R2 = 500Ω, T_A = Min to Max	A or B	B or A	4	11	ns
t_{PHL}	Propagation Delay Time High to Low Level Output				4	11	ns
t_{PZH}	Output Enable Time to High Level Output		$\overline{G}$AB	B	7	20	ns
t_{PZL}	Output Enable Time to Low Level Output				7	20	ns
t_{PHZ}	Output Disable Time to High Level Output		$\overline{G}$AB	B	2	14	ns
t_{PLZ}	Output Disable Time to Low Level Output				3	22	ns
t_{PZH}	Output Enable Time to High Level Output		GBA	A	7	20	ns
t_{PZL}	Output Enable Time to Low Level Output				7	20	ns
t_{PHZ}	Output Disable Time from High Level Output		GBA	A	2	14	ns
t_{PLZ}	Output Disable Time from Low Level Output				3	22	ns

Note 1: See Section 1 for test waveforms and output loads.

CONNECTION DIAGRAM

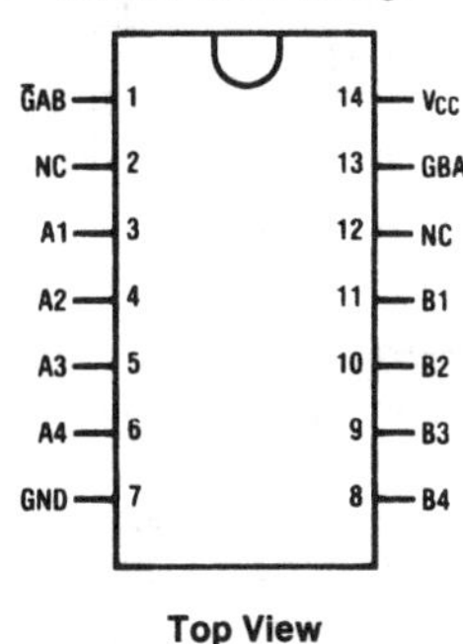

Top View

Order Number DM74ALS242CM, DM74ALS242CN, DM74ALS243AM or DM74ALS243AN
See NS Package Number M14A or N14A

FUNCTION TABLE

Inputs		'ALS242C	'ALS243A
$\overline{G}$AB	GBA		
L	L	$\overline{A}$ to B	A to B
H	H	$\overline{B}$ to A	B to A
H	L	Isolation	Isolation
L	H	Latch A and B (A = $\overline{B}$)	Latch A and B (A = B)

National DM54ALS244A/DM74ALS244A
Octal Tri-state® Bus Driver

- Advanced low-power oxide-isolated ion-implanted Schottky TTL process
- Functional and pin compatible with the DM54/74LS counterpart
- Improved switching performance with less power dissipation compared with the DM54/74LS counterpart
- Switching response specified into 500Ω and 50pF load
- Switching response specifications guaranteed over full temperature and V_{CC} supply range
- Pnp input design reduces input loading
- Low-level drive current: 54ALS = 12 mA, 74ALS = 24 mA

ABSOLUTE MAXIMUM RATINGS

If Military/Aerospace specified devices are required, please contact the National Semiconductor Sales Office/Distributors for availability and specifications.

Supply Voltage, V_{CC}	7V
Input Voltage	7V
Voltage Applied to Disabled Output	5.5V
Operating Free Air Temperature Range	
DM54ALS	−55°C to +125°C
DM74ALS	0 to +70°C
Storage Temperature Range	−65°C to +150°C
Typical θ_{JA}	
N Package	60.5°C/W
M Package	79.8°C/W

Note: *The "Absolute Maximum Ratings" are those values beyond which the safety of the device cannot be guaranteed. The device should not be operated at these limits. The parametric values defined in the "Electrical Characteristics" table are not guaranteed at the absolute maximum ratings. The "Recommended Operating Conditions" table will define the conditions for actual device operation.*

RECOMMENDED OPERATING CONDITIONS

Symbol	Parameter	DM54ALS244A			DM74ALS244A			Units
		Min	Typ	Max	Min	Typ	Max	
V_{CC}	Supply Voltage	4.5	5	5.5	4.5	5	5.5	V
V_{IH}	High Level Input Voltage	2			2			V
V_{IL}	Low Level Input Voltage			0.7			0.8	V
I_{OH}	High Level Output Current			−12			−15	mA
I_{OL}	Low Level Output Current			12			24	mA
T_A	Operating Free-Air Temperature	−55		125	0		70	°C

ELECTRICAL CHARACTERISTICS over recommended operating free air temperature (unless otherwise specified)

Symbol	Parameter	Conditions		DM54ALS244A			DM74ALS244A			Units
				Min	Typ	Max	Min	Typ	Max	
V_{IK}	Input Clamp Voltage	$V_{CC} = 4.5V$, $I_I = -18$ mA				−1.5			−1.5	V
V_{OH}	High Level Output Voltage	$V_{CC} = 4.5V$ to 5.5V	$I_{OH} = -0.4$ mA	$V_{CC} - 2$			$V_{CC} - 2$			V
		$V_{CC} = 4.5V$	$I_{OH} = -3$ mA	2.4			2.4			V
			I_{OH} = Max	2			2			V
V_{OL}	Low Level Output Voltage	$V_{CC} = 4.5V$ I_{OL} = 54ALS (Max)			0.25	0.4				V
		I_{OL} = 74ALS (Max)			—	—		0.35	0.5	V
I_I	Input Current at Max Input Voltage	$V_{CC} = 5.5V$, $V_I = 7V$				0.1			0.1	mA
I_{IH}	High Level Input Current	$V_{CC} = 5.5V$, $V_I = 2.7V$				20			20	μA
I_{IL}	Low Level Input Current	$V_{CC} = 5.5V$, $V_{IL} = 0.4V$				−0.1			−0.1	mA
I_O	Output Drive Current	$V_{CC} = 5.5V$, $V_O = 2.25V$		−30		−112	−30		−112	mA
I_{OZH}	High Level TRI-STATE Output Current	$V_{CC} = 5.5V$, $V_O = 2.7V$				20			20	μA
I_{OZL}	Low Level TRI-STATE Output Current	$V_{CC} = 5.5V$, $V_O = 0.4V$				−20			−20	μA
I_{CC}	Supply Current	$V_{CC} = 5.5V$ Outputs High			9	15		9	15	mA
		Outputs Low			15	24		15	24	mA
		Outputs TRI-STATE			17	27		17	27	mA

SWITCHING CHARACTERISTICS over recommended operating free-air temperature range

Symbol	Parameter	From (Input)	To (Output)	Conditions	54ALS244A		74ALS244A		Units
					Min	Max	Min	Max	
t_{PLH}	Propagation Delay Time Low to High Level Output	A	Y	$V_{CC} = 4.5V$ to 5.5V, $C_L = 50$ pF, R1 = 500Ω, R2 = 500Ω, T_A = Min to Max	1	16	3	10	ns
t_{PHL}	Propagation Delay Time High to Low Level Output	A	Y		3	12	3	10	ns
t_{PZH}	Output Enable Time to High Level Output	$\overline{G}$	Y		1	26	3	20	ns
t_{PZL}	Output Enable Time to Low Level Output	$\overline{G}$	Y		1	24	3	20	ns
t_{PHZ}	Output Disable Time from High Level Output	$\overline{G}$	Y		2	10	2	10	ns
t_{PLZ}	Output Disable Time from Low Level Output	$\overline{G}$	Y		1	21	1	13	ns

CONNECTION DIAGRAM

Dual-In-Line Package

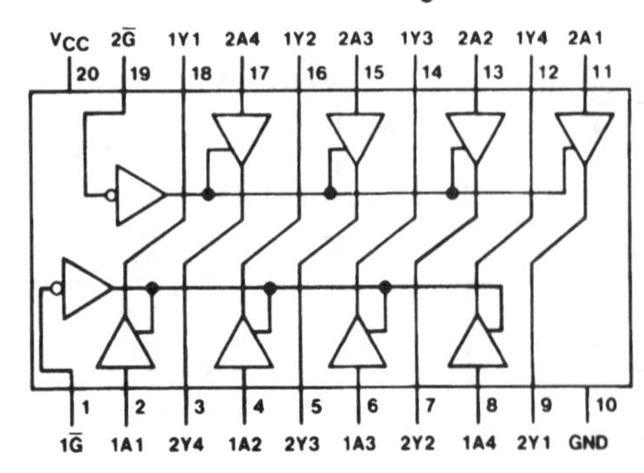

Top View

Order Number DM54ALS244AJ, DM74ALS244AWM, DM74ALS244AN or DM74ALS244ASJ
See NS Package Number J20A, M20B, M20D or N20A

FUNCTION TABLE

Input		Output
$\overline{G}$	A	Y
L	L	L
L	H	H
H	X	Z

H = High Level Logic State
L = Low Level Logic State
X = Don't Care (Either Low or High Level Logic State)
Z = High Impedance (Off) State

National DM54ALS245A/DM74ALS245A
Octal Tri-state® Bus Transceiver

- Advanced oxide-isolated, ion-implanted Schottky TTL process
- Noninverting logic output
- Glitch-free bus during power up and down
- Tri-state outputs independently controlled on A and B buses
- Low output impedance to drive terminated transmission lines to 133Ω
- Switching response specified into 500Ω/50pF
- Specified to interface with CMOS at $V_{OH} = V_{CC} - 2V$
- Pnp inputs to reduce input loading
- Switching specifications guaranteed over full temperature and V_{CC} range

ABSOLUTE MAXIMUM RATINGS

If Military/Aerospace specified devices are required, please contact the National Semiconductor Sales Office/Distributors for availability and specifications.

Supply Voltage	7V
Input Voltage	
Control Inputs	7V
I/O Ports	5.5V
Operating Free Air Temperature Range	
DM54ALS	−55°C to +125°C
DM74ALS	0°C to +70°C
Storage Temperature Range	−65°C to +150°C
Typical θ_{JA}:	
N Package	53.0°C/W
M Package	72.0°C/W

Note: *The "Absolute Maximum Ratings" are those values beyond which the safety of the device cannot be guaranteed. The device should not be operated at these limits. The parametric values defined in the "Electrical Characteristics" table are not guaranteed at the absolute maximum ratings. The "Recommended Operating Conditions" table will define the conditions for actual device operation.*

RECOMMENDED OPERATING CONDITIONS

Symbol	Parameter	DM54ALS245A			DM74ALS245A			Units
		Min	Typ	Max	Min	Typ	Max	
V_{CC}	Supply Voltage	4.5	5	5.5	4.5	5	5.5	V
V_{IH}	High Level Input Voltage	2			2			V
V_{IL}	Low Level Input Voltage			0.7			0.8	V
I_{OH}	High Level Output Current			−12			−15	mA
I_{OL}	Low Level Output Current			12			24	mA
T_A	Operating Free Air Temperature	−55		125	0		70	°C

ELECTRICAL CHARACTERISTICS

over recommended operating free air temperature range. All typical values are measured at V_{CC} = 5V, T_A = 25°C.

Symbol	Parameter	Conditions	Min	Typ	Max	Units
V_{IK}	Input Clamp Voltage	V_{CC} = 4.5V, I_{IN} = −18 mA			−1.5	V
V_{OH}	High Level Output Voltage	V_{CC} = 4.5V, I_{OH} = −3 mA	2.4	3.2		V
		V_{CC} = 4.5V, I_{OH} = Max	2	2.3		V
		I_{OH} = −0.4 mA, V_{CC} = 4.5V to 5.5V	V_{CC} − 2			V
V_{OL}	Low Level Output Voltage	V_{CC} = 4.5V, 54/74ALS I_{OL} = 12 mA		0.25	0.4	V
		V_{CC} = 4.5V, 74ALS I_{OL} = 24 mA		0.35	0.5	V
I_I	Input Current at Max Input Voltage	V_{CC} = 5.5V, V_{IN} = 7V, Control Inputs			0.1	mA
		V_{CC} = 5.5V, V_{IN} = 5.5V, A or B Ports			0.1	mA
I_{IH}	High Level Input Current	V_{CC} = 5.5V, V_{IN} = 2.7V			20	μA
I_{IL}	Low Level Input Current	V_{CC} = 5.5V, V_{IN} = 0.4V			−0.1	mA
I_O	Output Drive Current	V_{CC} = 5.5V, V_{OUT} = 2.25V	−30		−112	mA
I_{CC}	54ALS245A Supply Current	V_{CC} = 5.5V, Outputs High		30	48	mA
		V_{CC} = 5.5V, Outputs Low		38	60	mA
		V_{CC} = 5.5V, TRI-STATE		38	63	mA
I_{CC}	74ALS245A Supply Current	V_{CC} = 5.5V, Outputs High		30	45	mA
		V_{CC} = 5.5V, Outputs Low		36	55	mA
		V_{CC} = 5.5V, TRI-STATE		38	58	mA

SWITCHING CHARACTERISTICS over recommended operating free air temperature range

Symbol	Parameter	Circuit Configuration	DM54ALS245A		DM74ALS245A		Units
			Min	Max	Min	Max	
t_{PLH}	Propagation Delay Time High-to-Low Level Output	IN A OR B → B OR A OUT	1	19	3	10	ns
t_{PHL}	Propagation Delay Time High-to-Low Level Output		1	14	3	10	ns
t_{PZL}	Output Enable Time to Low Level	IN $\bar{G}$; A OR B OUT	2	29	5	20	ns
t_{PZH}	Output Enable Time to High Level		2	30	5	20	ns
t_{PLZ}	Output Disable Time from Low Level		2	30	4	15	ns
t_{PHZ}	Output Disable Time from High Level		2	14	2	10	ns

CONNECTION DIAGRAM

Dual-In-Line Package

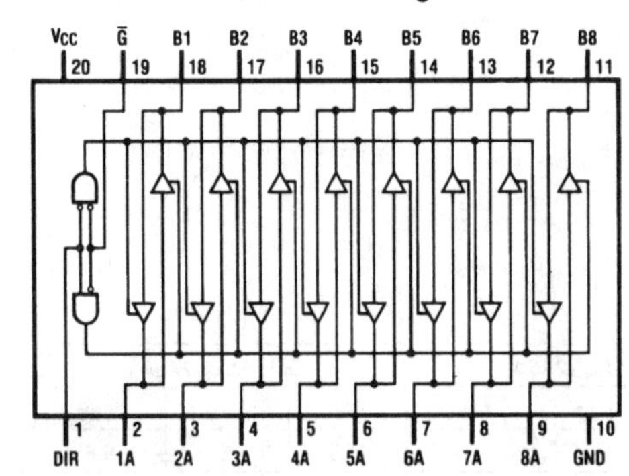

Order Number DM54ALS245AJ, DM74ALS245AWM, DM74ALS245AWN or DM74ALS245ASJ
See NS Package Number J20A, M20B, M20D or N20A

FUNCTION TABLE

Control Inputs		Operation
$\bar{G}$	DIR	
L	L	B Data to A Bus
L	H	A Data to B Bus
H	X	Hi-Z

H = High Logic Level
L = Low Logic Level
X = Either High or Low Logic Level

National DM74ALS465A/466A/467A/468A
Octal Tri-state® Bidirectional Bus Driver

- Advanced low-power ioxide-isolated ion-implanted Schottky TTL process
- Functional and pin compatible with the DM54/74LS counterpart and the DM71/81LS95, 96, 97, 98
- Improved switching performance with less power dissipation compared with the DM54/74LS counterpart
- Switching response specified into 500Ω/50pF load
- Switching response specifications guaranteed over full temperature and V_{CC} supply range
- Pnp input design reduces input loading

ABSOLUTE MAXIMUM RATINGS

Supply Voltage, V_{CC}	7V
Input Voltage	7V
Output Voltage (Disabled)	5.5V
Operating Free Air Temperature Range	
DM74ALS	0°C to +70°C
Storage Temperature Range	−65°C to +150°C
Typical θ_{JA}	
N Package	60.5°C/W
M Package	79.8°C/W

Note: *The "Absolute Maximum Ratings" are those values beyond which the safety of the device cannot be guaranteed. The device should not be operated at these limits. The parametric values defined in the "Electrical Characteristics" table are not guaranteed at the absolute maximum ratings. The "Recommended Operating Conditions" table will define the conditions for actual device operation.*

RECOMMENDED OPERATING CONDITIONS

Symbol	Parameter	Min	Typ	Max	Units
V_{CC}	Supply Voltage	4.5	5	5.5	V
V_{IH}	High Level Input Voltage	2			V
V_{IL}	Low Level Input Voltage			0.8	V
I_{OH}	High Level Output Current			−15	mA
I_{OL}	Low Level Output Current			24	mA
T_A	Operating Free Air Temperature	0		70	°C

ELECTRICAL CHARACTERISTICS

over recommended operating free air temperature range (unless otherwise specified)

Symbol	Parameter	Conditions		Min	Typ	Max	Units
V_{IK}	Input Clamp Voltage	$V_{CC} = 4.5V$, $I_I = -18$ mA				−1.5	V
V_{OH}	High Level Output Voltage	$V_{CC} = 4.5V$ to $5.5V$	$I_{OH} = -0.4$ mA	$V_{CC} - 2$			V
		$V_{CC} = 4.5V$	$I_{OH} = -3$ mA	2.4			V
			I_{OH} = Max	2			V
V_{OL}	Low Level Output	I_{OL} = Max			0.35	0.5	V
I_I	Input Current at Max Input Voltage	$V_{CC} = 5.5V$, $V_I = 7V$				0.1	mA
I_{IH}	High Level Input Current	$V_{CC} = 5.5V$, $V_I = 2.7V$				20	μA
I_{IL}	Low Level Input Current	$V_{CC} = 5.5V$, $V_{IL} = 0.4V$				−0.1	mA
I_O	Output Drive Current	$V_{CC} = 5.5V$, $V_O = 2.25V$		−30		−112	mA
I_{OZH}	High Level TRI-STATE Output Current	$V_{CC} = 5.5V$, $V_O = 2.7V$				20	μA
I_{OZL}	Low Level TRI-STATE Output Current	$V_{CC} = 5.5V$, $V_O = 0.4V$				−20	μA
I_{CC}	Supply Current	$V_{CC} = 5.5V$, ALS465A Outputs High Outputs Low Outputs TRI-STATE		 11 19 23		 16 28 33	mA
		$V_{CC} = 5.5V$, ALS466A Outputs High Outputs Low Outputs TRI-STATE		 7 16 19		 10 24 27	mA
		$V_{CC} = 5.5V$, ALS467A Outputs High Outputs Low Outputs TRI-STATE		 11 19 23		 16 28 33	mA
		$V_{CC} = 5.5V$, ALS468A Outputs High Outputs Low Outputs TRI-STATE		 7 16 19		 10 24 27	mA

'ALS465A AND 'ALS467A SWITCHING CHARACTERISTICS

over recommended operating free air temperature range

Parameter	Conditions	From (Input)	To (Output)	Min	Max	Units
t_{PLH}	$V_{CC} = 4.5V$ to $5.5V$, $C_L = 50$ pF, $R1 = 500\Omega$, $R2 = 500\Omega$, T_A = Min to Max	A	Y	2	13	ns
t_{PHL}				4	12	ns
t_{PZH}		$\overline{G}$	Any Y	4	23	ns
t_{PZL}				5	25	ns
t_{PHZ}		$\overline{G}$	Any Y	2	10	ns
t_{PLZ}				3	18	ns

'ALS466A AND 'ALS468A SWITCHING CHARACTERISTICS

over recommended operating free air temperature range

Parameter	Conditions	From (Input)	To (Output)	Min	Max	Units
t_{PLH}	V_{CC} = 4.5V to 5.5V, C_L = 50 pF, R1 = 500Ω, R2 = 500Ω, T_A = Min to Max	A	Y	3	12	ns
t_{PHL}				2	9	ns
t_{PZH}		$\overline{G}$	Any Y	4	16	ns
t_{PZL}				7	23	ns
t_{PHZ}		$\overline{G}$	Any Y	2	10	ns
t_{PLZ}				2	17	ns

CONNECTION DIAGRAMS

Dual-In-Line Package

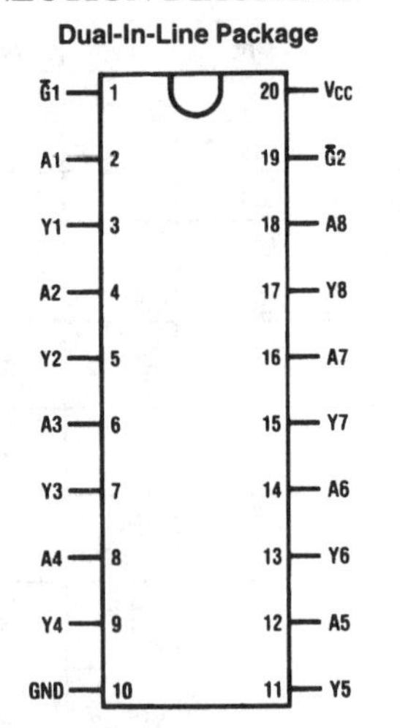

Top View

Order DM74ALS465AWM, DM74ALS466AWM, DM74ALS465AN or DM74ALS466AN
See NS Package Number M20B or N20A

Dual-In-Line Package

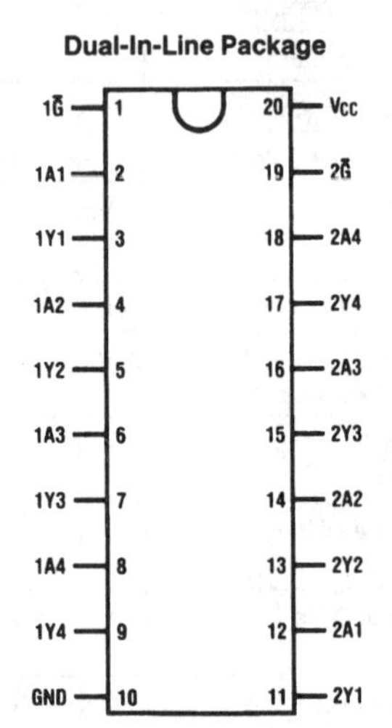

Top View

Order Number DM74ALS467AWM, DM74ALS468AWM, DM74ALS467AN or DM74ALS468AN
See NS Package Number M20B or N20A

National DM7AS640

Tri-state® Octal Bus Transceiver

- Switching specifications at 50pF
- Switching specifications guaranteed over full temperature and V_{CC} range
- Advanced oxide-isolated ion-implanted Schottky TTL process
- Functionally and pin-for-pin compatible with Schottky, low-power Schottky, and advanced low-power Schottky counterparts
- Improved ac performance over Schottky, low-power Schottky, and advanced low-power Schottky counterparts
- Tri-state outputs independently controlled on A and B buses
- Low output impedance drive-to-drive terminated transmission lines to 133Ω
- Specified to interface with CMOS at $V_{OH} = V_{CC} - 2V$

ABSOLUTE MAXIMUM RATINGS

Supply Voltage	7V
Input Voltage	
Control Inputs	7V
I/O Ports	5.5V
Operating Free Air Temperature Range	0°C to +70°C
Storage Temperature Range	−65°C to +150°C
Typical θ_{JA}	
N Package	51.5°C
M Package	69.0°C

Note: *The "Absolute Maximum Ratings" are those values beyond which the safety of the device cannot be guaranteed. The device should not be operated at these limits. The parametric values defined in the "Electrical Characteristics" table are not guaranteed at the absolute maximum ratings. The "Recommended Operating Conditions" table will define the conditions for actual device operation.*

RECOMMENDED OPERATING CONDITIONS

Symbol	Parameter	Min	Typ	Max	Units
V_{CC}	Supply Voltage	4.5	5	5.5	V
V_{IH}	High Level Input Voltage	2			V
V_{IL}	Low Level Input Voltage			0.8	V
I_{OH}	High Level Output Current			−15	mA
I_{OL}	Low Level Output Current			64	mA
T_A	Free Air Operating Temperature	0		70	°C

ELECTRICAL CHARACTERISTICS

over recommended operating free air temperature range (unless otherwise noted)

Symbol	Parameter	Conditions		Min	Typ (Note 1)	Max	Units
V_I	Input Clamp Voltage	V_{CC} = Min, I_I = −18 mA				−1.2	V
V_{OH}	High Level Output Voltage	V_{CC} = 4.5V to 5.5V, I_{OH} = −2 mA		V_{CC} − 2			V
		V_{CC} = 4.5V, I_{OH} = −3 mA		2.4			V
		V_{CC} = 4.5V, I_{OH} = Max		2.4			V
V_{OL}	Low Level Output Voltage	V_{CC} = Min, I_{OL} = Max			0.35	0.55	V
I_I	Input Current at Max Input Voltage	V_{CC} = Max, V_I = 7V, (V_I = 5.5V for A or B Ports)				0.1	mA
I_{IH}	High Level Input Current	V_{CC} = Max V_I = 2.7V (Note 2)	Control Inputs			20	μA
			A or B Ports			70	
I_{IL}	Low Level Input Current	V_{CC} = Max, V_I = 0.4V (Note 2)	Control Inputs			−0.5	mA
			A or B Ports			−0.75	
I_O	Output Drive Current	V_{CC} = Max, V_O = 2.25V		−50		−150	mA
I_{CCH}	Supply Current with Outputs High	V_{CC} = Max			37	58	mA
I_{CCL}	Supply Current with Outputs Low				78	123	mA
I_{CCZ}	Supply Current with Outputs in TRI-STATE				51	80	mA

Note 1: All typicals are at V_{CC} = 5.0V, T_A = 25°C.

Note 2: For I/O ports, the parameters I_{IH} and I_{IL} include the off-state output current, I_{OZH} and I_{OZL}.

SWITCHING CHARACTERISTICS

over recommended operating free air temperature range (unless otherwise noted)

Symbol	Parameter	From (Input)	To (Output)	V_{CC} = Min to Max, C_L = 50 pF, R_1 = R_2 = 500Ω		Units
				Min	Max	
t_{PLH}	Propagation Delay Time Low to High Level Output	A or B	B or A	2	7	ns
t_{PHL}	Propagation Delay Time High to Low Level Output	A or B	B or A	2	6	ns
t_{PZH}	Output Enable Time to High Level Output	$\overline{G}$	A or B	2	8	ns
t_{PZL}	Output Enable Time to Low Level Output	$\overline{G}$	A or B	2	10	ns
t_{PHZ}	Output Disable Time from High Level Output	$\overline{G}$	A or B	2	8	ns
t_{PLZ}	Output Disable Time from Low Level Output	$\overline{G}$	A or B	2	13	ns

CONNECTION DIAGRAM

Dual-In-Line Package

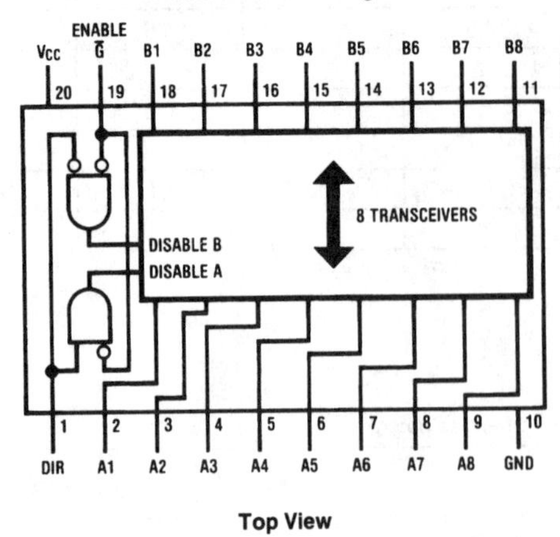

Top View

Order Number DM74AS640N or DM74AS640WM
See NS Package Number M20B or N20A

FUNCTION TABLE

Control Inputs		Operation
G̅	DIR	
L	L	B̅ Data to A Bus
L	H	A̅ Data to B Bus
H	X	Isolation

LOGIC DIAGRAM

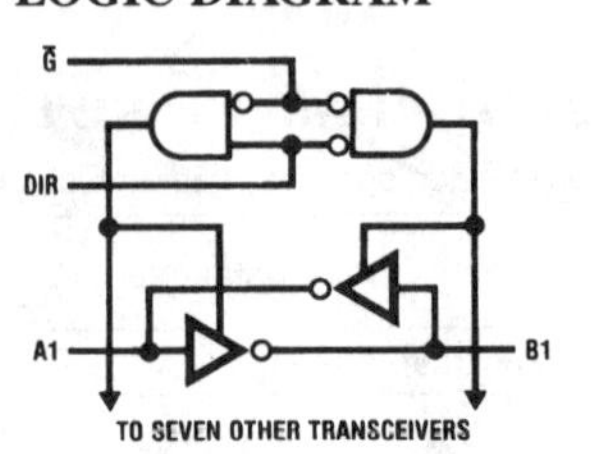

National DM7AS645
Tri-state® Octal Bus Transceiver

- Switching specifications at 50pF
- Switching specifications guaranteed over full temperature and V_{CC} range
- Advanced oxide-isolated ion-implanted Schottky TTL process
- Functionally and pin-for-pin compatible with Schottky, low-power Schottky, and advanced low-power Schottky TTL counterpart
- Improved ac performance over Schottky, low-power Schottky, and advanced low-power Schottky counterparts
- Tri-state outputs independently controlled on A and B buses
- Low output impedance drive-to-drive terminated transmission lines to 133Ω
- Specified to interface with CMOS at $V_{OH} = V_{CC} - 2\,V$

ABSOLUTE MAXIMUM RATINGS

Supply Voltage	7V
Input Voltage	
Control Inputs	7V
I/O Ports	5.5V
Operating Free Air Temperature Range	0°C to +70°C
Storage Temperature Range	−65°C to +150°C
Typical θ_{JA}	
N Package	51.5°C/W
M Package	69.0°C/W

Note: *The "Absolute Maximum Ratings" are those values beyond which the safety of the device cannot be guaranteed. The device should not be operated at these limits. The parametric values defined in the "Electrical Characteristics" table are not guaranteed at the absolute maximum ratings. The "Recommended Operating Conditions" table will define the conditions for actual device operation.*

RECOMMENDED OPERATING CONDITIONS

Symbol	Parameter	Min	Typ	Max	Units
V_{CC}	Supply Voltage	4.5	5	5.5	V
V_{IH}	High Level Input Voltage	2			V
V_{IL}	Low Level Input Voltage			0.8	V
I_{OH}	High Level Output Current			−15	mA
I_{OL}	Low Level Output Current			64	mA
T_A	Free Air Operating Temperature	0		70	°C

ELECTRICAL CHARACTERISTICS

over recommended operating free air temperature range (unless otherwise noted)

Symbol	Parameter	Conditions		Min	Typ (Note 1)	Max	Units
V_I	Input Clamp Voltage	V_{CC} = Min, I_I = −18 mA				−1.2	V
V_{OH}	High Level Output Voltage	V_{CC} = 4.5V to 5.5V, I_{OH} = −2 mA		$V_{CC}-2$			V
		V_{CC} = 4.5V, I_{OH} = −3 mA		2.4			V
		V_{CC} = 4.5V, I_{OH} = Max		2.4			V
V_{OL}	Low Level Output Voltage	V_{CC} = Min, I_{OL} = Max			0.35	0.55	V
I_I	Input Current at Max Input Voltage	V_{CC} = Max, V_I = 7V, (V_I = 5.5V for A or B Ports)				0.1	mA
I_{IH}	High Level Input Current	V_{CC} = Max V_I = 2.7V (Note 2)	Control Inputs			20	μA
			A or B Ports			70	
I_{IL}	Low Level Input Current	V_{CC} = Max, V_I = 0.4V (Note 2)	Control Inputs			−0.5	mA
			A or B Ports			−0.75	
I_O	Output Drive Current	V_{CC} = Max, V_O = 2.25V		−50		−150	mA
I_{CCH}	Supply Current with Outputs High	V_{CC} = Max			62	97	mA
I_{CCL}	Supply Current with Outputs Low				95	149	mA
I_{CC}	Supply Current with Outputs in TRI-STATE				79	123	mA

Note 1: All typicals are at V_{CC} = 5.0V, T_A = 25°C.

Note 2: For I/O ports, the parameters I_{IH} and I_{IL} include the off-state output current, I_{OZH} and I_{OZL}.

SWITCHING CHARACTERISTICS over recommended operating free air temperature range (unless otherwise noted)

Symbol	Parameter	From (Input)	To (Output)	V_{CC} = Min to Max, C_L = 50 pF, R_1 = R_2 = 500Ω		Units
				Min	Max	
t_{PLH}	Propagation Delay Time Low to High Level Output	A or B	B or A	2	9.5	ns
t_{PHL}	Propagation Delay Time High to Low Level Output	A or B	B or A	2	9	ns
t_{PZH}	Output Enable Time to High Level Output	$\overline{G}$	A or B	2	11	ns
t_{PZL}	Output Enable Time to Low Level Output	$\overline{G}$	A or B	2	10	ns
t_{PHZ}	Output Disable Time from High Level Output	$\overline{G}$	A or B	2	7	ns
t_{PLZ}	Output Disable Time from Low Level Output	$\overline{G}$	A or B	2	12	ns

CONNECTION DIAGRAM

Dual-In-Line Package

Top View

Order Number DM74AS645WM or DM74AS645N
See NS Package Number M20B or N20A

FUNCTION TABLE

Control Inputs		Operation
$\overline{G}$	DIR	
L	L	B Data to A Bus
L	H	A Data to B Bus
H	X	Isolation

LOGIC DIAGRAM

DM74AS645

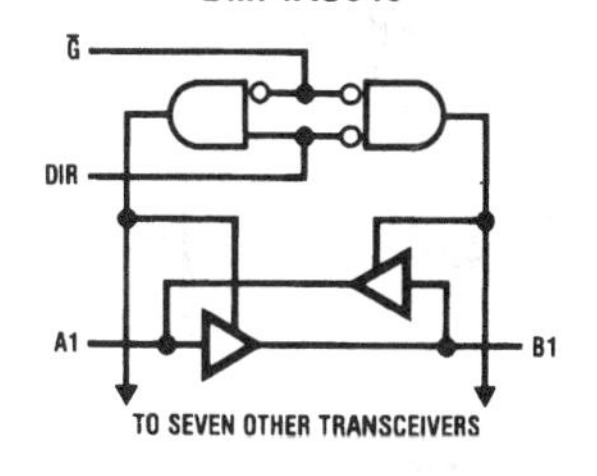

National DM74AS646/DM74AS648

Octal Bus Transceiver and Register

- Switching specifications at 50pF
- Switching specifications guaranteed over full temperature and V_{CC} range
- Advanced oxide-isolated ion-implanted Schottky TTL process
- Functionally and pin-for-pin compatible with LS TTL counterpart
- Tri-state® buffer-type outputs drive bus lines directly

ABSOLUTE MAXIMUM RATINGS

Supply Voltage	7V
Input Voltage	
Control Inputs	7V
I/O Ports	5.5V
Operating Free Air Temperature Range	0°C to +70°C
Storage Temperature Range	−65°C to +150°C
Typical θ_{JA}	
N Package	41.1°C/W
M Package	81.5°C/W

Note: *The "Absolute Maximum Ratings" are those values beyond which the safety of the device cannot be guaranteed. The device should not be operated at these limits. The parametric values defined in the "Electrical Characteristics" table are not guaranteed at the absolute maximum ratings. The "Recommended Operating Conditions" table will define the conditions for actual device operation.*

RECOMMENDED OPERATING CONDITIONS

Symbol	Parameter		DM74AS646, 648			Units
			Min	Nom	Max	
V_{CC}	Supply Voltage		4.5	5	5.5	V
V_{IH}	High Level Input Voltage		2			V
V_{IL}	Low Level Input Voltage				0.8	V
I_{OH}	High Level Output Current				−15	mA
I_{OL}	Low Level Output Current				48	mA
f_{CLK}	Clock Frequency		0		90	MHz
t_W	Width of Clock Pulse	High	5			ns
		Low	6			ns
t_{SU}	Data Setup Time		6↑			ns
t_H	Data Hold Time		0↑			ns
T_A	Free Air Operating Temperature		0		70	°C

The (↑) arrow indicates the positive edge of the Clock is used for reference.

ELECTRICAL CHARACTERISTICS

over recommended operating free air temperature range. All typical values are measured at V_{CC} = 5V, T_A = 25°C.

Symbol	Parameter	Conditions			Min	Typ	Max	Units
V_{IK}	Input Clamp Voltage	V_{CC} = 4.5V, I_I = −18 mA					−1.2	V
V_{OH}	High Level Output Voltage	V_{CC} = 4.5V, V_{IL} = Max V_{IH} = Min		I_{OH} = Max	2			V
				I_{OH} = −3 mA	2.4	3.2		
		V_{CC} = 4.5V to 5.5V, I_{OH} = −2 mA			V_{CC} − 2			
V_{OL}	Low Level Output Voltage	V_{CC} = 4.5V, V_{IL} = Min V_{IH} = 2V, I_{OL} = Max				0.35	0.5	V
I_I	Input Current @ Max Input Voltage	V_{CC} = 5.5V	V_I = 7V	Control Inputs			0.1	mA
			V_I = 5.5V	A or B Ports			0.1	
I_{IH}	High Level Input Current	V_{CC} = 5.5V, V_{IH} = 2.7V (Note 1)		Control Inputs			20	μA
				A or B Ports			70	
I_{IL}	Low Level Input Current	V_{CC} = 5.5V, V_{IL} = 0.4V (Note 1)		Control Inputs			−0.5	mA
				A or B Ports			−0.75	
I_O	Output Drive Current	V_{CC} = 5.5V, V_O = 2.25V			−30		−112	mA
I_{CC}	Supply Current	V_{CC} = 5.5V	'AS646	Outputs High		120	195	mA
				Outputs Low		130	211	
				Outputs Disabled		130	211	
			'AS648	Outputs High		110	185	
				Outputs Low		120	195	
				Outputs Disabled		120	195	

Note 1: For I/O ports, the parameters I_{IH} and I_{IL} include the off-state current, I_{OZH} and I_{OZL}.

'AS646 SWITCHING CHARACTERISTICS over recommended operating free air temperature range

Symbol	Parameter	Conditions	From (Input)	To (Output)	DM74AS646		Units
					Min	Max	
f_{MAX}	Maximum Clock Frequency	V_{CC} = 4.5V to 5.5V, $R_1 = R_2 = 500\Omega$, C_L = 50 pF (Note 1)			90		MHz
t_{PLH}	Propagation Delay Time Low to High Level Output		CBA or CAB	A or B	2	8.5	ns
t_{PHL}	Propagation Delay Time High to Low Level Output				2	9	ns
t_{PLH}	Propagation Delay Time Low to High Level Output		A or B	B or A	2	9	ns
t_{PHL}	Propagation Delay Time High to Low Level Output				1	7	ns
t_{PLH}	Propagation Delay Time Low to High Level Output		SBA or SAB	A or B	2	11	ns
t_{PHL}	Propagation Delay Time High to Low Level Output				2	9	ns
t_{PZH}	Output Enable Time to High Level Output		Enable $\overline{G}$	A or B	2	9	ns
t_{PZL}	Output Enable Time to Low Level Output				3	14	ns
t_{PHZ}	Output Disable Time from High Level Output				2	9	ns
t_{PLZ}	Output Disable Time from Low Level Output				2	9	ns
t_{PZH}	Output Enable Time to High Level Output		DIR	A or B	3	16	ns
t_{PZL}	Output Enable Time to Low Level Output				3	18	ns
t_{PHZ}	Output Disable Time from High Level Output				2	10	ns
t_{PLZ}	Output Disable Time from Low Level Output				2	10	ns

CONNECTION DIAGRAM

Dual-In-Line Package

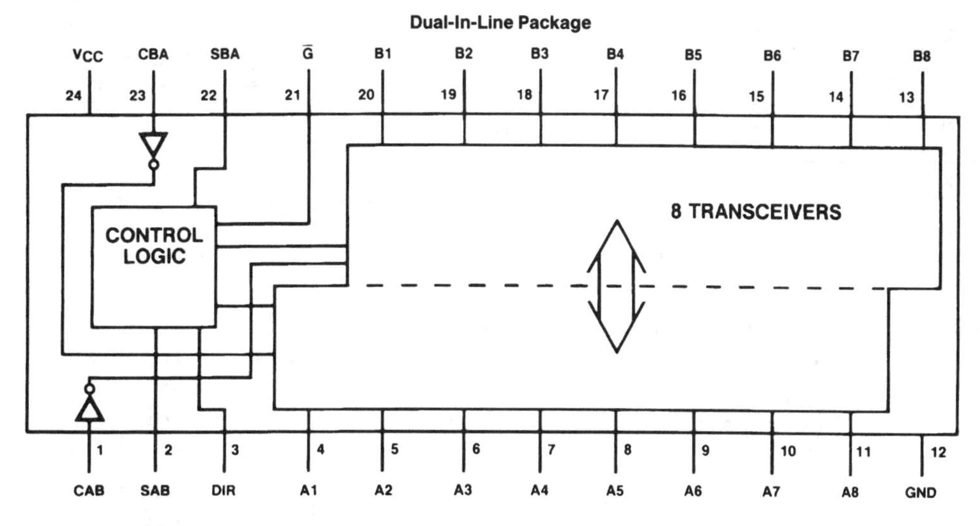

Order Number DM74AS646NT, DM74AS646WM, DM74AS648NT or DM74AS648WM
See NS Package Number M24B or N24C

'AS648 SWITCHING CHARACTERISTICS over recommended operating free air temperature range

Symbol	Parameter	Conditions	From (Input)	To (Output)	DM74AS648 Min	DM74AS648 Max	Units
f_{MAX}	Maximum Clock Frequency	V_{CC} = 4.5V to 5.5V, $R_1 = R_2 = 500\Omega$ C_L = 50 pF (Note 1)			90		MHz
t_{PLH}	Propagation Delay Time Low to High Level Output		CAB or CBA	A or B	2	8.5	ns
t_{PHL}	Propagation Delay Time High to Low Level Output				2	9	ns
t_{PLH}	Propagation Delay Time Low to High Level Output		A or B	B or A	2	8	ns
t_{PHL}	Propagation Delay Time High to Low Level Output				1	7	ns
t_{PLH}	Propagation Delay Time Low to High Level Output		SBA or SAB	A or B	2	11	ns
t_{PHL}	Propagation Delay Time High to Low Level Output				2	9	ns
t_{PZH}	Output Enable Timo to High Level Output		Enable $\overline{G}$	A or B	2	9	ns
t_{PZL}	Output Enable Time to Low Level Output				3	15	ns
t_{PHZ}	Output Disable Time from High Level Output				2	9	ns
t_{PLZ}	Output Disable Time from Low Level Output				2	9	ns
t_{PZH}	Output Enable Time to High Level Output		DIR	A or B	3	16	ns
t_{PZL}	Output Enable Time to Low Level Output				3	18	ns
t_{PHZ}	Output Disable Time from High Level Output				2	10	ns
t_{PLZ}	Output Disable Time from Low Level Output				2	10	ns

FUNCTION TABLE

Inputs						Data I/O*		Operation or Function	
$\overline{G}$	DIR	CAB	CBA	SAB	SBA	A1 thru A8	B1 thru B8	'AS646	'AS648
H	X X	H or L ↑	H or L ↑	X X	X X	Input	Input	Isolation, Hold Storage Store A and B Data	Isolation, Hold Storage Store A and B Data
L	L L	X X	X H or L	X X	L H	Output	Input	Real Time B Data to A Bus Stored B Data to A Bus	Real Time $\overline{B}$ Data to A Bus Stored $\overline{B}$ Data to A Bus
L	H H	X H or L	X X	L H	X X	Input	Output	Real Time A Data to B Bus Stored A Data to B Bus	Real Time $\overline{A}$ Data to B Bus Stored $\overline{A}$ Data to B Bus
X X	X X	↑ X	X ↑	X X	X X	Input Unspecified*	Unspecified* Input	Store A, B Unspecified* Store B, A Unspecified*	Store A, B Unspecified* Store B, A Unspecified*

H—high level; L—low level; X—irrelevant; ↑—low-to-high level transition

*The data output functions may be enabled or disabled by various signals at the $\overline{G}$ and DIR inputs. Data input functions are always enabled, i.e., data at the bus pins will be stored on every low-to-high transition on the clock inputs.

DIFFERENT MODES OF CONTROL FOR AS646, AS648

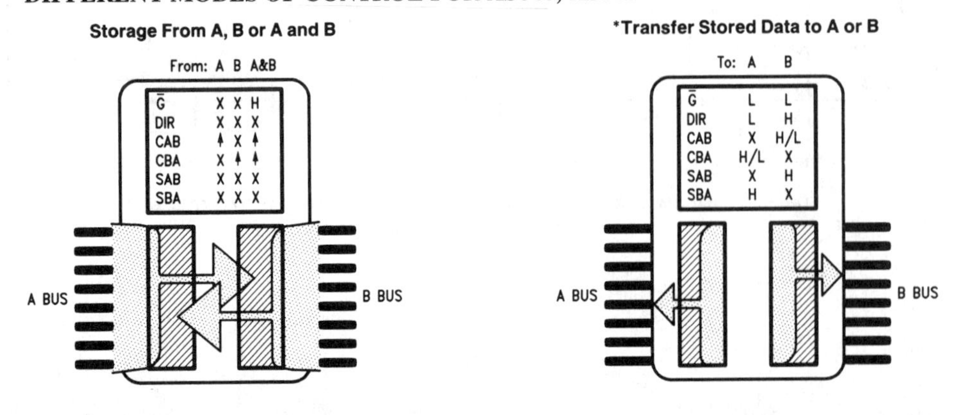

*Real-Time Transfer Bus A to Bus B

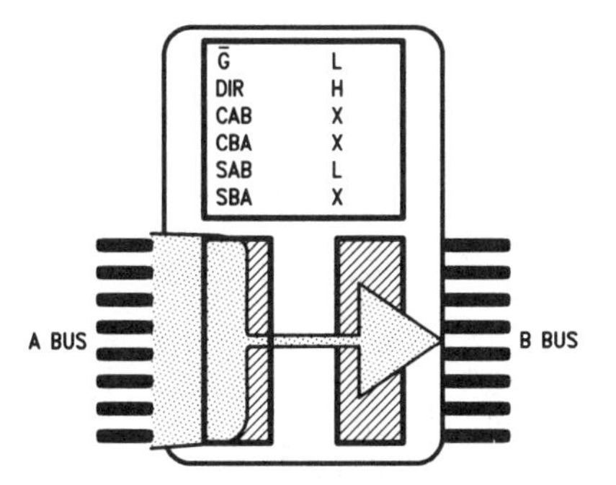

*Real-Time Transfer Bus B to Bus A

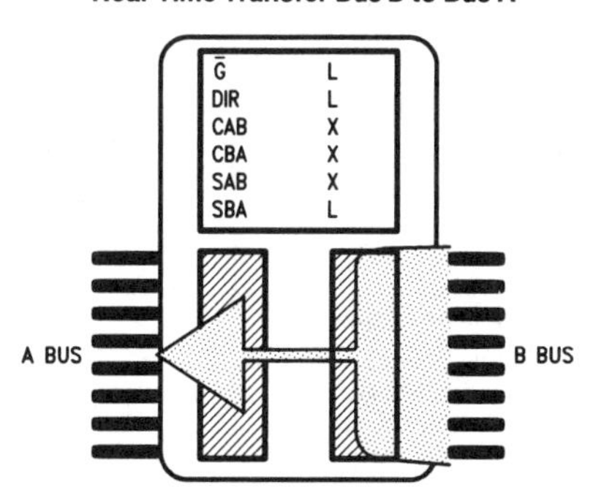

*The complement of A and B data are stored and transferred for AS648

National DM7AS651/DM74AS652

Octal Bus Transceiver and Register

- Switching specifications at 50pF
- Switching specifications guaranteed over full temperature and V_{CC} range
- Advanced oxide-isolated ion-implanted Schottky TTL process
- Tri-state® buffer-type outputs drive bus lines directly

ABSOLUTE MAXIMUM RATINGS

Supply Voltage	7V
Input Voltage	
Control Inputs	7V
I/O Ports	5.5V
Operating Free Air Temperature Range	0°C to +70°C
Storage Temperature Range	−65°C to +150°C
Typical θ_{JA}	
N Package	41.1°C/W
M Package	81.5°C/W

Note: *The "Absolute Maximum Ratings" are those values beyond which the safety of the device cannot be guaranteed. The device should not be operated at these limits. The parametric values defined in the "Electrical Characteristics" table are not guaranteed at the absolute maximum ratings. The "Recommended Operating Conditions" table will define the conditions for actual device operation.*

RECOMMENDED OPERATING CONDITIONS

Symbol	Parameter		Min	Nom	Max	Units
V_{CC}	Supply Voltage		4.5	5	5.5	V
V_{IH}	High Level Input Voltage		2			V
V_{IL}	Low Level Input Voltage				0.8	V
I_{OH}	High Level Output Current				−15	mA
I_{OL}	Low Level Output Current				48	mA
f_{CLK}	Clock Frequency		0		90	MHz
t_{WCLK}	Width of Enable Pulse	High	5			ns
		Low	6			
t_{SU}	Data Setup Time		6			ns
t_H	Data Hold Time		0			ns
T_A	Operating free Air Temperature		0		70	°C

The (↑) arrow indicates the positive edge of the Clock is used for reference.

ELECTRICAL CHARACTERISTICS

over recommended operating free air temperature range. All typical values are measured at V_{CC} = 5V, T_A = 25°C.

Symbol	Parameter	Conditions			Min	Typ	Max	Units
V_{IK}	Input Clamp Voltage	V_{CC} = 4.5V, I_I = −18 mA					−1.2	V
V_{OH}	High Level Output Voltage	V_{CC} = 4.5V		I_{OH} = Max	2			V
				I_{OH} = −3 mA	2.4	3.2		
		V_{CC} = 4.5V to 5.5V		I_{OH} = −2 mA	V_{CC} − 2			
V_{OL}	Low Level Output Voltage	V_{CC} = 4.5V, I_{OL} = Max				0.35	0.5	V
I_I	Input Current at Max Input Voltage	V_{CC} = 5.5V	V_I = 7V	Control Inputs			0.1	mA
			V_I = 5.5V	A or B Ports			0.1	
I_{IH}	High Level Input Current	V_{CC} = 5.5V, V_{IH} = 2.7V		Control Inputs			20	μA
				A or B Ports			70	
I_{IL}	Low Level Input Current	V_{CC} = 5.5V, V_{IL} = 0.4V		Control Inputs			−0.5	mA
				A or B Ports			−0.75	
I_O	Output Drive Current	V_{CC} = 5.5V, V_O = 2.25V			−30		−112	mA
I_{CC}	Supply Current	V_{CC} = 5.5V	'AS651	Outputs High		110	185	mA
				Outputs Low		120	195	
				Outputs Disabled		130	195	
			'AS652	Outputs High		120	195	
				Outputs Low		130	211	
				Outputs Disabled		130	211	

'AS651 SWITCHING CHARACTERISTICS over recommended operating free air temperature range

Symbol	Parameter	Conditions	From	To	Min	Max	Units
f_{MAX}	Maximum Clock Frequency	V_{CC} = 4.5V to 5.5V R_1 = R_2 = 500Ω C_L = 50 pF			90		MHz
t_{PLH}	Propagation Delay Time Low to High Level Output		CBA or CAB	A or B	2	8.5	ns
t_{PHL}	Propagation Delay Time High to Low Level Output				2	9	ns
t_{PLH}	Propagation Delay Time Low to High Level Output		A or B	B or A	2	8	ns
t_{PHL}	Propagation Delay Time High to Low Level Output				1	7	ns
t_{PLH}	Propagation Delay Time Low to High Level Output		SBA or SAB	A or B	2	11	ns
t_{PHL}	Propagation Delay Time High to Low Level Output				2	9	ns
t_{PZH}	Output Enable Time to High Level Output		Enable $\overline{G}$BA	A	2	10	ns
t_{PZL}	Output Enable Time to Low Level Output				3	16	ns
t_{PHZ}	Output Disable Time from High Level Output				2	9	ns
t_{PLZ}	Output Disable Time from Low Level Output				2	9	ns
t_{PZH}	Output Disable Time to High Level Output		Enable GAB	B	3	11	ns
t_{PZL}	Output Disable Time to Low Level Output				3	16	ns
t_{PHZ}	Output Disable Time from High Level Output				2	10	ns
t_{PLZ}	Output Disable Time from Low Level Output				2	11	ns

CONNECTION DIAGRAM

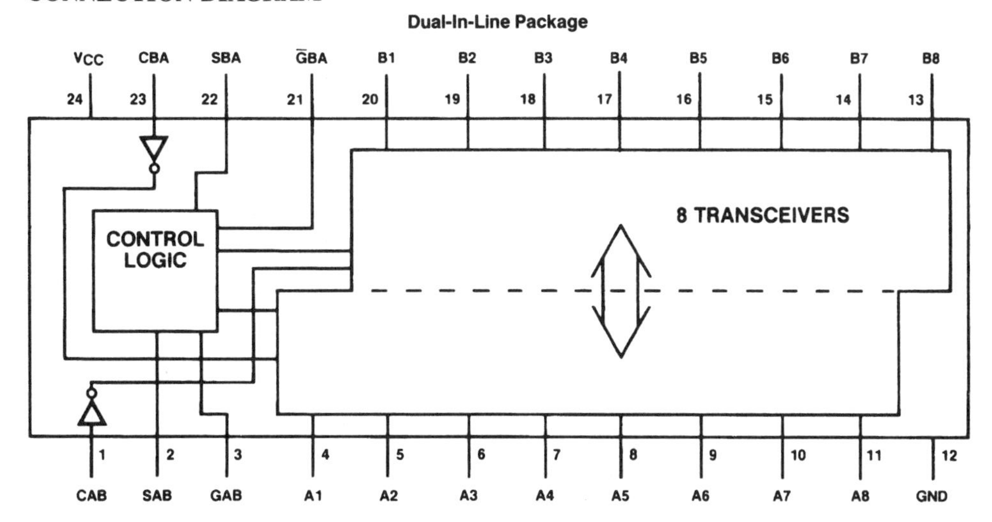

Order Number DM74AS651NT, DM74AS651WM, DM74AS652NT or DM74AS652WM
See NS Package Number N24C or M24B

'AS652 SWITCHING CHARACTERISTICS over recommended operating free air temperature range

Symbol	Parameter	Conditions	From	To	Min	Max	Units
f_{MAX}	Maximum Clock Frequency	V_{CC} = 4.5V to 5.5V $R_1 = R_2 = 500\Omega$ C_L = 50 pF			90		MHz
t_{PLH}	Propagation Delay Time Low to High Level Output		CBA or CAB	A or B	2	8.5	ns
t_{PHL}	Propagation Delay Time High to Low Level Output				2	9	ns
t_{PLH}	Propagation Delay Time Low to High Level Output		A or B	B or A	2	9	ns
t_{PHL}	Propagation Delay Time High to Low Level Output				1	7	ns
t_{PLH}	Propagation Delay Time Low to High Level Output		SBA or SAB	A or B	2	11	ns
t_{PHL}	Propagation Delay Time High to Low Level Output				2	9	ns
t_{PZH}	Output Enable Time to High Level Output		Enable $\overline{G}$BA	A	2	10	ns
t_{PZL}	Output Enable Time to Low Level Output				3	16	ns
t_{PHZ}	Output Disable Time from High Level Output				2	9	ns
t_{PLZ}	Output Disable Time from Low Level Output				2	9	ns
t_{PZH}	Output Disable Time to High Level Output		Enable GAB	B	3	11	ns
t_{PZL}	Output Disable Time to Low Level Output				3	16	ns
t_{PHZ}	Output Disable Time from High Level Output				2	10	ns
t_{PLZ}	Output Disable Time from Low Level Output				2	11	ns

FUNCTION TABLE

INPUTS						DATA I/O*		OPERATION OR FUNCTION	
GAB	$\overline{GBA}$	CAB	CBA	SAB	SBA	A1 THRU A8	B1 THRU B8	'AS651	'AS652
L L	H H	H or L ↑	H or L ↑	X X	X X	Input	Input	Isolation Store A and B Data	Isolation Store A and B Data
L L	L L	X X	X H or L	X X	L H	Output	Input	Real Time $\overline{B}$ Data to A Bus Stored $\overline{B}$ Data to A Bus	Real Time B Data to A Bus Stored B Data to A Bus
H H	H H	X H or L	X X	L H	X X	Input	Output	Real Time $\overline{A}$ Data to B Bus Stored $\overline{A}$ Data to B Bus	Real Time A Data to B Bus Stored A Data to B Bus
H	L	H or L	H or L	H	H	Output	Output	Stored $\overline{A}$ Data to B Bus & Stored $\overline{B}$ Data to A Bus	Stored A Data to B Bus & Stored B Data to A Bus
X H	H H	↑ ↑	H or L ↑	X X(1)	X X	Input Input	Unspecified* Output	Store A, Hold B Store A in both registers	Store A, Hold B Store A in both registers
L L	X L	H or L ↑	↑ ↑	X X	X X(1)	Unspecified* Output	Input Input	Hold A, Store B Store B in both registers	Hold A, Store B Store B in both registers

Note 1: If the select control is low, the clocks can occur simultaneously. If the select control is high, the clocks must be staggered in order to load both registers.

H—high level L—low level X—irrelevant ↑—low-to-high transition

*The data output functions may be enabled or disabled by various signals at the GAB and $\overline{G}$BA inputs. Data input functions are always enabled, i.e., data at the bus pins will be stored on every low-to-high transition on the clock inputs.

National 54LS366A/DM74LS366A
Hex Tri-state® Inverting Buffers

ABSOLUTE MAXIMUM RATINGS (Note)

If Military/Aerospace specified devices are required, please contact the National Semiconductor Sales Office/Distributors for availability and specifications.

Supply Voltage	7V
Input Voltage	7V
Operating Free Air Temperature Range	
DM54LS	−55°C to +125°C
DM74LS	0°C to +70°C
Storage Temperature Range	−65°C to +150°C

Note: *The "Absolute Maximum Ratings" are those values beyond which the safety of the device cannot be guaranteed. The device should not be operated at these limits. The parametric values defined in the "Electrical Characteristics" table are not guaranteed at the absolute maximum ratings. The "Recommended Operating Conditions" table will define the conditions for actual device operation.*

RECOMMENDED OPERATING CONDITIONS

Symbol	Parameter	54LS366A			DM74LS366A			Units
		Min	Nom	Max	Min	Nom	Max	
V_{CC}	Supply Voltage	4.5	5	5.5	4.75	5	5.25	V
V_{IH}	High Level Input Voltage	2			2			V
V_{IL}	Low Level Input Voltage			0.7			0.8	V
I_{OH}	High Level Output Current			−1			−2.6	mA
I_{OL}	Low Level Output Current			12			24	mA
T_A	Free Air Operating Temperature	−55		125	0		70	°C

ELECTRICAL CHARACTERISTICS recommended operating free air temperature range (unless otherwise noted)

Symbol	Parameter	Conditions		Min	Typ (Note 1)	Max	Units
V_I	Input Clamp Voltage	V_{CC} = Min, I_I = −18 mA				−1.5	V
V_{OH}	High Level Output Voltage	V_{CC} = Min, I_{OH} = Max V_{IL} = Max, V_{IH} = Min		2.4	3.4		V
V_{OL}	Low Level Output Voltage	V_{CC} = Min, I_{OL} = Max V_{IL} = Max, V_{IH} = Min	54LS		0.25	0.4	V
			DM74		0.35	0.5	
		I_{OL} = 12 mA, V_{CC} = Min	DM74		0.25	0.4	
I_I	Input Current @ Max Input Voltage	V_{CC} = Max, V_I = 7V	DM74			0.1	mA
		V_{CC} = Max, V_I = 10.0V	54LS				
I_{IH}	High Level Input Current	V_{CC} = Max, V_I = 2.7V				20	μA
I_{IL}	Low Level Input Current	V_{CC} = Max, V_I = 0.5V (Note 4)	A Input			−20	μA
		V_{CC} = Max, V_I = 0.4V (Note 5)	A Input			−0.4	mA
		V_{CC} = Max, V_I = 0.4V	$\overline{G}$ Input			−0.4	
I_{OZH}	Off-State Output Current with High Level Output Voltage Applied	V_{CC} = Max, V_O = 2.4V V_{IH} = Min, V_{IL} = Max				20	μA
I_{OZL}	Off-State Output Current with Low Level Output Voltage Applied	V_{CC} = Max, V_O = 0.4V V_{IH} = Min, V_{IL} = Max				−20	μA
I_{OS}	Short Circuit Output Current	V_{CC} = Max (Note 2)	54LS	−30		−130	mA
			DM74	−20		−100	
I_{CC}	Supply Current	V_{CC} = Max (Note 3)			12	21	mA

Note 1: All typicals are at V_{CC} = 5V, T_A = 25°C.

Note 2: Not more than one output should be shorted at a time, and the duration should not exceed one second.

Note 3: I_{CC} is measured with the DATA inputs grounded and the OUTPUT CONTROLS at 4.5V.

Note 4: Both $\overline{G}$ inputs are at 2V.

Note 5: Both $\overline{G}$ inputs at 0.4V.

SWITCHING CHARACTERISTICS at V_{CC} = 5V and T_A = 25°C (See Section 1 for Test Waveforms and Output Load)

Symbol	Parameter	54LS		DM74LS		Units
		C_L = 50 pF		C_L = 150 pF R_L = 667Ω		
		Min	Max	Min	Max	
t_{PLH}	Propagation Delay Time Low to High Level Output		12		25	ns
t_{PHL}	Propagation Delay Time High to Low Level Output		22		25	ns
t_{PZH}	Output Enable Time to High Level Output		24		35	ns
t_{PZL}	Output Enable Time to Low Level Output		30		40	ns
t_{PHZ}	Output Disable Time from High Level Output (Note 6)		25			ns
t_{PLZ}	Output Disable Time from Low Level Output (Note 6)		20			ns

Note 6: C_L = 5 pF.

CONNECTION DIAGRAM

Dual-In-Line Package

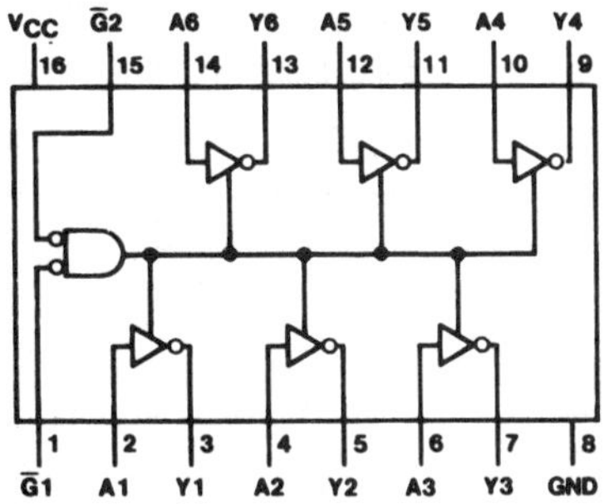

V_{CC} 16, $\overline{G2}$ 15, A6 14, Y6 13, A5 12, Y5 11, A4 10, Y4 9

$\overline{G1}$ 1, A1 2, Y1 3, A2 4, Y2 5, A3 6, Y3 7, GND 8

Order Number 54LS366ADMQB, 54LS366AFMQB, 54LS366ALMQB, DM74LS366AM or DM74LS366AN See NS Package Number E20A, J16A, M16A, N16E or W16A

FUNCTION TABLE

$Y = \overline{A}$

Inputs			Output
$\overline{G1}$	$\overline{G2}$	A	Y
H	X	X	Hi-Z
X	H	X	Hi-Z
L	L	L	H
L	L	H	L

H = High Logic Level

L = Low Logic Level

X = Either Low or High Logic Level

Hi-Z = TRI-STATE (Outputs are disabled)

National 54LS367A/DM54LS3676A/DM74LS367A
Hex Tri-state® Buffers

ABSOLUTE MAXIMUM RATINGS (Note)

If Military/Aerospace specified devices are required, please contact the National Semiconductor Sales Office/Distributors for availability and specifications.

Supply Voltage 7V
Input Voltage 7V
Operating Free Air Temperature Range
DM54LS −55°C to +125°C
DM74LS 0°C to +70°C
Storage Temperature Range −65°C to +150°C

Note: *The "Absolute Maximum Ratings" are those values beyond which the safety of the device cannot be guaranteed. The device should not be operated at these limits. The parametric values defined in the "Electrical Characteristics" table are not guaranteed at the absolute maximum ratings. The "Recommended Operating Conditions" table will define the conditions for actual device operation.*

RECOMMENDED OPERATING CONDITIONS

Symbol	Parameter	DM54LS367A			DM74LS367A			Units
		Min	Nom	Max	Min	Nom	Max	
V_{CC}	Supply Voltage	4.5	5	5.5	4.75	5	5.25	V
V_{IH}	High Level Input Voltage	2			2			V
V_{IL}	Low Level Input Voltage			0.7			0.8	V
I_{OH}	High Level Output Current			−1			−2.6	mA
I_{OL}	Low Level Output Current			12			24	mA
T_A	Free Air Operating Temperature	−55		125	0		70	°C

ELECTRICAL CHARACTERISTICS over recommended operating free air temperature range (unless otherwise noted)

Symbol	Parameter	Conditions		Min	Typ (Note 1)	Max	Units
V_I	Input Clamp Voltage	V_{CC} = Min, I_I = −18 mA				−1.5	V
V_{OH}	High Level Output Voltage	V_{CC} = Min, I_{OH} = Max V_{IL} = Max, V_{IH} = Min		2.4	3.4		V
V_{OL}	Low Level Output Voltage	V_{CC} = Min, I_{OL} = Max V_{IL} = Max, V_{IH} = Min	DM54		0.25	0.4	V
			DM74		0.35	0.5	
		I_{OL} = 12 mA, V_{CC} = Min	DM74		0.25	0.4	
I_I	Input Current @ Max Input Voltage	V_{CC} = Max, V_I = 7V				0.1	mA
I_{IH}	High Level Input Current	V_{CC} = Max, V_I = 2.7V				20	μA
I_{IL}	Low Level Input Current	V_{CC} = Max, V_I = 0.5V (Note 4)	A Input			−20	μA
		V_{CC} = Max, V_I = 0.4V (Note 5)	A Input			−0.4	mA
		V_{CC} = Max, V_I = 0.4V	$\overline{G}$ Input			−0.4	
I_{OZH}	Off-State Output Current with High Level Output Voltage Applied	V_{CC} = Max, V_O = 2.4V V_{IH} = Min, V_{IL} = Max				20	μA
I_{OZL}	Off-State Output Current with Low Level Output Voltage Applied	V_{CC} = Max, V_O = 0.4V V_{IH} = Min, V_{IL} = Max				−20	μA
I_{OS}	Short Circuit Output Current	V_{CC} = Max (Note 2)	DM54	−20		−100	mA
			DM74	−20		−100	
I_{CC}	Supply Current	V_{CC} = Max (Note 3)			14	24	mA

Note 1: All typicals are at V_{CC} = 5V, T_A = 25°C.

Note 2: Not more than one output should be shorted at a time, and the duration should not exceed one second.

Note 3: I_{CC} is measured with the DATA inputs grounded and the OUTPUT CONTROLS at 4.5V.

Note 4: Both $\overline{G}$ inputs are at 2V.

Note 5: Both $\overline{G}$ inputs at 0.4V.

SWITCHING CHARACTERISTICS at V_{CC} = 5V and T_A = 25°C (See Section 1 for Test Waveforms and Output Load)

Symbol	Parameter	R_L = 667Ω				Units
		C_L = 50 pF		C_L = 150 pF		
		Min	Max	Min	Max	
t_{PLH}	Propagation Delay Time Low to High Level Output		16		25	ns
t_{PHL}	Propagation Delay Time High to Low Level Output		16		25	ns
t_{PZH}	Output Enable Time to High Level Output		30		40	ns
t_{PZL}	Output Enable Time to Low Level Output		30		40	ns
t_{PHZ}	Output Disable Time from High Level Output (Note 6)		20			ns
t_{PLZ}	Output Disable Time from Low Level Output (Note 6)		20			ns

Note 6: C_L = 5 pF.

CONNECTION DIAGRAM

Dual-In-Line Package

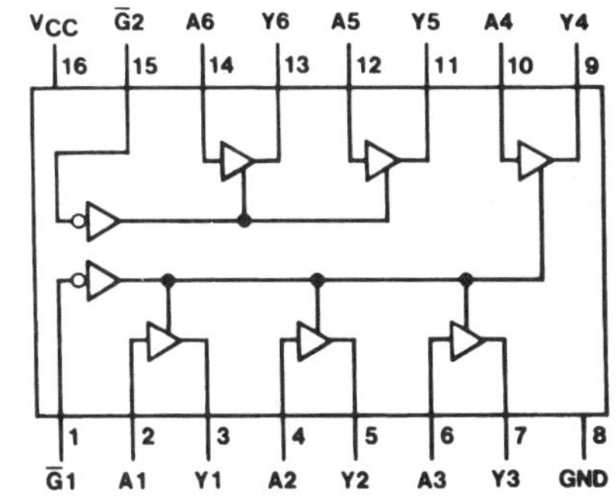

Order Number 54LS367ADMQB, 54LS367AFMQB, 54LS367ALMQB, DM54LS367AJ, DM54LS367AW, DM74LS367AM or DM74LS367AN
See NS Package Number E20A, J16A, M16A, N16E or W16A

FUNCTION TABLE

Y = A

Inputs		Output
A	$\overline{G}$	Y
L	L	L
H	L	H
X	H	Hi-Z

H = High Logic Level
L = Low Logic Level
X = Either Low or High Logic Level
Hi-Z = TRI-STATE (Outputs are disabled)

National DM54S181/DM74S181 Arithmetic Logic Unit/Function Generators

- Arithmetic operating modes:
  - Addition
  - Subtraction
  - Shift operand A one position
  - Magnitude comparison
  - Plus twelve other arithmetic operations
- Logic function modes:
  - EXCLUSIVE-OR
  - Comparator
  - AND, NAND, OR, NOR
  - Plus ten other logic operations
- Full look-ahead for high-speed operations on long words

Number of Bits	Typical Addition Times	Package Count		Carry Method Between ALU's
		Arithmetic/ Logic Units	Look Ahead Carry Generators	
1 to 4	20 ns	1	0	None
5 to 8	30 ns	2	0	Ripple
9 to 16	30 ns	3 or 4	1	Full Look-Ahead
17 to 64	50 ns	5 to 16	2 to 5	Full Look-Ahead

Pin Number	2	1	23	22	21	20	19	18	9	10	11	13	7	16	15	17
Active-High Data (Table I)	A0	B0	A1	B1	A2	B2	A3	B3	F0	F1	F2	F3	$\overline{C}_n$	$\overline{C}_n+4$	X	Y
Active-Low Data (Table II)	$\overline{A}0$	$\overline{B}0$	$\overline{A}1$	$\overline{B}1$	$\overline{A}2$	$\overline{B}2$	$\overline{A}3$	$\overline{B}3$	$\overline{F}0$	$\overline{F}1$	$\overline{F}2$	$\overline{F}3$	C_n	C_n+4	$\overline{P}$	$\overline{G}$

Input C_n	Output C_n+4	Active-High Data (Figure 1)	Active-Low Data (Figure 2)
H	H	A ≤ B	A ≤ B
H	L	A ≤ B	A ≤ B
L	H	A ≤ B	A ≤ B
L	L	A ≤ B	A ≤ B

CONNECTION DIAGRAM

Dual-In-Line Package

Order Number DM54S181J or DM74S181N
See NS Package Number J24A or N24A

PIN DESIGNATIONS

Designation	Pin Nos.	Function
A3, A2, A1, A0	19, 21, 23, 2	Word A Inputs
B3, B2, B1, B0	18, 20, 22, 1	Word B Inputs
S3, S2, S1, S0	3, 4, 5, 6	Function-Select Inputs
C_n	7	Inv. Carry Input
M	8	Mode Control Input
F3, F2, F1, F0	13, 11, 10, 9	Function Outputs
A = B	14	Comparator Output
P	15	Carry Propagate Output
C_n+4	16	Inv. Carry Output
G	17	Carry Generate Output
V_{CC}	24	Supply Voltage
GND	12	Ground

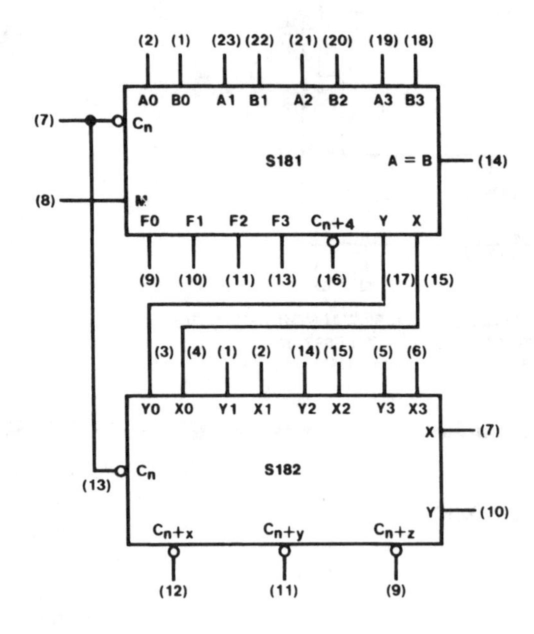

TABLE I

Selection				Active High Data		
				M = H Logic Functions	M = L; Arithmetic Operations	
S3	S2	S1	S0		C_n = H (no carry)	C_n = L (with carry)
L	L	L	L	F = $\overline{A}$	F = A	F = A Plus 1
L	L	L	H	F = $\overline{A + B}$	F = A + B	F = (A + B) Plus 1
L	L	H	L	F = $\overline{A}B$	F = A + $\overline{B}$	F = (A + $\overline{B}$) Plus 1
L	L	H	H	F = 0	F = Minus 1 (2's Compl)	F = Zero
L	H	L	L	F = $\overline{AB}$	F = A Plus A$\overline{B}$	F = A Plus A$\overline{B}$ Plus 1
L	H	L	H	F = $\overline{B}$	F = (A + B) Plus A$\overline{B}$	F = (A + B) Plus A$\overline{B}$ Plus 1
L	H	H	L	F = A ⊕ B	F = A Minus B Minus 1	F = A Minus B
L	H	H	H	F = A$\overline{B}$	F = A$\overline{B}$ Minus 1	F = A$\overline{B}$
H	L	L	L	F = $\overline{A}$ + B	F = A Plus AB	F = A Plus AB Plus 1
H	L	L	H	F = $\overline{A \oplus B}$	F = A Plus B	F = A Plus B Plus 1
H	L	H	L	F = B	F = (A + $\overline{B}$) Plus AB	F = (A + $\overline{B}$) Plus AB Plus 1
H	L	H	H	F = AB	F = AB Minus 1	F = AB
H	H	L	L	F = 1	F = A Plus A*	F = A Plus A Plus 1
H	H	L	H	F = A + $\overline{B}$	F = (A + B) Plus A	F = (A + B) Plus A Plus 1
H	H	H	L	F = A + B	F = (A + $\overline{B}$) Plus A	F = (A + $\overline{B}$) Plus A Plus 1
H	H	H	H	F = A	F = A Minus 1	F = A

*Each bit is shifted to the next more significant position.

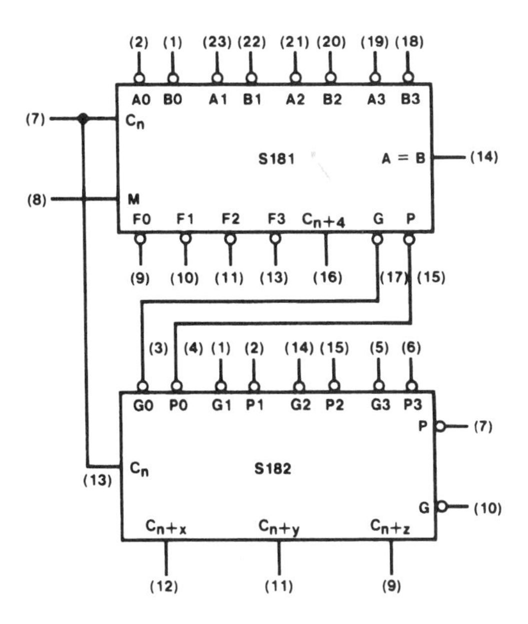

TABLE II

Selection				Active Low Data		
				M = H	M = L; Arithmetic Operations	
S3	S2	S1	S0	Logic Functions	C_n = L (no carry)	C_n = H (with carry)
L	L	L	L	$F = \overline{A}$	F = A Minus 1	F = A
L	L	L	H	$F = \overline{AB}$	F = AB Minus 1	F = AB
L	L	H	L	$F = \overline{A} + B$	F = $A\overline{B}$ Minus 1	$F = A\overline{B}$
L	L	H	H	F = 1	F = Minus 1 (2's Compl)	F = Zero
L	H	L	L	$F = \overline{A + B}$	F = A Plus $(A + \overline{B})$	F = A Plus $(A + \overline{B})$ Plus 1
L	H	L	H	$F = \overline{B}$	F = AB Plus (A + B)	F = AB Plus $(A + \overline{B})$ Plus 1
L	H	H	L	$F = \overline{A \oplus B}$	F = A Minus B Minus 1	F = A Minus B
L	H	H	H	$F = A + \overline{B}$	$F = A + \overline{B}$	F = $(A + \overline{B})$ Plus 1
H	L	L	L	$F = \overline{A}B$	F = A Plus (A + B)	F = A Plus (A + B) Plus 1
H	L	L	H	$F = A \oplus B$	F = A Plus B	F = A Plus B Plus 1
H	L	H	L	F = B	F = $A\overline{B}$ Plus (A + B)	F = $A\overline{B}$ Plus (A + B) Plus 1
H	L	H	H	F = A + B	F = A + B	F = (A + B) Plus 1
H	H	L	L	F = 0	F = A Plus A*	F = A Plus A Plus 1
H	H	L	H	$F = A + A\overline{B}$	F = AB Plus A	F = AB Plus A Plus 1
H	H	H	L	F = AB	F = $A\overline{B}$ Plus A	F = $A\overline{B}$ Plus A Plus 1
H	H	H	H	F = A	F = A	F = A Plus 1

*Each bit is shifted to the next more significant position.

ABSOLUTE MAXIMUM RATINGS (Note)

If Military/Aerospace specified devices are required, please contact the National Semiconductor Sales Office/Distributors for availability and specifications.

Supply Voltage	7V
Input Voltage	5.5V
Output Voltage (A = B Output)	5.5V
Operating Free Air Temperature Range	
DM54S	−55°C to +125°C
DM74S	0°C to +70°C
Storage Temperature Range	−65°C to +150°C

Note: *The "Absolute Maximum Ratings" are those values beyond which the safety of the device cannot be guaranteed. The device should not be operated at these limits. The parametric values defined in the "Electrical Characteristics" table are not guaranteed at the absolute maximum ratings. The "Recommended Operating Conditions" table will define the conditions for actual device operation.*

SWITCHING CHARACTERISTICS V_{CC} = 5V, T_A = 25°C

Symbol	Parameter	Conditions	From (Input)	To (Output)	DM54/74 S181 R_L = 280Ω, C_L = 15 pF Min	Max	R_L = 280Ω, C_L = 50 pF Min	Max	Units
t_{PLH}	Propagation Delay Time, Low-to-High Level Output		C_n	C_n+4		10.5		14	ns
t_{PHL}	Propagation Delay Time, High-to-Low Level Output					10.5		14	
t_{PLH}	Propagation Delay Time, Low-to-High Level Output	M = 0V, S0 = S3 = 4.5V S1 = S2 = 0V ($\overline{SUM}$ mode)	Any A or B	C_n+4		18.5		22	ns
t_{PHL}	Propagation Delay Time, High-to-Low Level Output					18.5		22	
t_{PLH}	Propagation Delay Time, Low-to-High Level Output	M = 0V, S0 = S3 = 0V S1 = S2 = 4.5V ($\overline{DIFF}$ mode)	Any A or B	C_n+4		23		27	ns
t_{PHL}	Propagation Delay Time, High-to-Low Level Output					23		27	
t_{PLH}	Propagation Delay Time, Low-to-High Level Output	M = 0V ($\overline{SUM}$ or $\overline{DIFF}$ mode)	C_n	Any F		12		14	ns
t_{PHL}	Propagation Delay Time, High-to-Low Level Output					12		14	
t_{PLH}	Propagation Delay Time, Low-to-High Level Output	M = 0V, S0 = S3 = 4.5V S1 = S2 = 0V ($\overline{SUM}$ mode)	Any A or B	G		12		15	ns
t_{PHL}	Propagation Delay Time, High-to-Low Level Output					12		15	
t_{PLH}	Propagation Delay Time, Low-to-High Level Output	M = 0V, S0 = S3 = 0V S1 = S2 = 4.5V ($\overline{DIFF}$ mode)	Any A or B	G		15		19	ns
t_{PHL}	Propagation Delay Time, High-to-Low Level Output					15		20	
t_{PLH}	Propagation Delay Time, Low-to-High Level Output	M = 0V, S0 = S3 = 4.5V S1 = S2 = 0V ($\overline{SUM}$ mode)	Any A or B	P		12		15	ns
t_{PHL}	Propagation Delay Time, High-to-Low Level Output					12		15	
t_{PLH}	Propagation Delay Time, Low-to-High Level Output	M = 0V, S0 = S3 = 0V S1 = S2 = 4.5V ($\overline{DIFF}$ mode)	Any A or B	P		15		19	ns
t_{PHL}	Propagation Delay Time, High-to-Low Level Output					15		20	
t_{PLH}	Propagation Delay Time, Low-to-High Level Output	M = 0V, S0 = S3 = 4.5V S1 = S2 = 0V ($\overline{SUM}$ mode)	A_i or B_i	F_i		16.5		20	ns
t_{PHL}	Propagation Delay Time, High-to-Low Level Output					16.5		20	
t_{PLH}	Propagation Delay Time, Low-to-High Level Output	M = 0V, S0 = S3 = 0V S1 = S2 = 4.5V ($\overline{DIFF}$ mode)	A_i or B_i	F_i		20		24	ns
t_{PHL}	Propagation Delay Time, High-to-Low Level Output					22		24	
t_{PLH}	Propagation Delay Time, Low-to-High Level Output	M = 4.5V (logic mode)	A_i or B_i	F_i		20		24	ns
t_{PHL}	Propagation Delay Time, High-to-Low Level Output					22		24	
t_{PLH}	Propagation Delay Time, Low-to-High Level Output	M = 0V, S0 = S3 = 0V S1 = S2 = 4.5V ($\overline{DIFF}$ mode)	Any A or B	A = B		23		26	ns
t_{PHL}	Propagation Delay Time, High-to-Low Level Output					30		33	

RECOMMENDED OPERATING CONDITIONS

Symbol	Parameter	DM54S181 Min	Nom	Max	DM74S181 Min	Nom	Max	Units
V_{CC}	Supply Voltage	4.5	5	5.5	4.75	5	5.25	V
V_{IH}	High Level Input Voltage	2			2			V
V_{IL}	Low Level Input Voltage			0.8			0.8	V
V_{OH}	High Level Output Voltage (A = B Output)			5.5			5.5	V
I_{OH}	High Level Output Current (All Except A = B)			−1			−1	mA
I_{OL}	Low Level Output Current			20			20	mA
T_A	Free Air Operating Temperature	−55		125	0		70	°C

ELECTRICAL CHARACTERISTICS over recommended operating free air temperature (unless otherwise noted)

Symbol	Parameter	Conditions		Min	Typ (Note 1)	Max	Units
V_I	Input Clamp Voltage	V_{CC} = Min, I_I = −18 mA				−1.2	V
I_{CEX}	High Level Output Current (A = B Output)	V_{CC} = Min, V_O = 5.5V V_{IL} = Max, V_{IH} = Min				250	μA
V_{OH}	High Level Output Voltage (All Except A = B)	V_{CC} = Min, I_{OH} = Max V_{IL} = Max, V_{IH} = Min	DM54	2.5	3.4		V
			DM74	2.7	3.4		
V_{OL}	Low Level Output Voltage	V_{CC} = Min, I_{OL} = Max V_{IH} = Min, V_{IL} = Max				0.5	V
I_I	Input Current @ Max Input Voltage	V_{CC} = Max, V_I = 5.5V				1	mA
I_{IH}	High Level Input Current	V_{CC} = Max V_I = 2.7V	Mode			50	μA
			A or B			150	
			S			200	
			Carry			250	
I_{IL}	Low Level Input Current	V_{CC} = Max V_I = 0.5V	Mode			−2	mA
			A or B			−6	
			S			−8	
			Carry			−10	
I_{OS}	Short Circuit Output Current (Any Output Except A = B)	V_{CC} = Max (Note 2)		−40		−100	mA
I_{CC}	Supply Current	V_{CC} = Max (Note 3)			120	220	mA

Note 1: All typicals are at V_{CC} = 5V, T_A = 25°C.

Note 2: Not more than one output should be shorted at a time, and the duration should not exceed one second.

Note 3: I_{CC} is measured for the following conditions: A. S0 through S3, M, and A inputs at 4.5V, all other inputs grounded and all outputs open. B. S0 through S3 and M inputs at 4.5V, all other inputs grounded and all outputs open.

National DM54S182/DM74S182 Look-Ahead Carry Generators

ABSOLUTE MAXIMUM RATINGS (Note)

If Military/Aerospace specified devices are required, please contact the National Semiconductor Sales Office/Distributors for availability and specifications.

Supply Voltage	7V
Input Voltage	5.5V
Operating Free Air Temperature Range	
DM54S	−55°C to +125°C
DM74S	0°C to +70°C
Storage Temperature Range	−65°C to +150°C

Note: *The "Absolute Maximum Ratings" are those values beyond which the safety of the device cannot be guaranteed. The device should not be operated at these limits. The parametric values defined in the "Electrical Characteristics" table are not guaranteed at the absolute maximum ratings. The "Recommended Operating Conditions" table will define the conditions for actual device operation.*

RECOMMENDED OPERATING CONDITIONS

Symbol	Parameter	DM54S182			DM74S182			Units
		Min	Nom	Max	Min	Nom	Max	
V_{CC}	Supply Voltage	4.5	5	5.5	4.75	5	5.25	V
V_{IH}	High Level Input Voltage	2			2			V
V_{IL}	Low Level Input Voltage			0.8			0.8	V
I_{OH}	High Level Output Current			−1			−1	mA
I_{OL}	Low Level Output Current			20			20	mA
T_A	Free Air Operating Temperature	−55		125	0		70	°C

ELECTRICAL CHARACTERISTICS over recommended operating free air temperature (unless otherwise noted)

Symbol	Parameter	Conditions		Min	Typ (Note 1)	Max	Units
V_I	Input Clamp Voltage	V_{CC} = Min, I_I = −18 mA				−1.2	V
V_{OH}	High Level Output Voltage	V_{CC} = Min, I_{OH} = Max, V_{IL} = Max, V_{IH} = Min	DM54	2.5	3.4		V
			DM74	2.7	3.4		
V_{OL}	Low Level Output Voltage	V_{CC} = Min, I_{OL} = Max, V_{IH} = Min, V_{IL} = Max				0.5	V
I_I	Input Current @ Max Input Voltage	V_{CC} = Max, V_I = 5.5V				1	mA
I_{IH}	High Level Input Current	V_{CC} = Max, V_I = 2.7V	P0, P1 or G3			200	μA
			P3			100	
			P2			150	
			C_n			50	
			G0, G2			350	
			G1			400	
I_{IL}	Low Level Input Current	V_{CC} = Max, V_I = 0.5V	P0, P1 or G3			−8	mA
			P3			−4	
			P2			−6	
			C_n			−2	
			G0, G2			−14	
			G1			−16	
I_{OS}	Short Circuit Output Current	V_{CC} = Max (Note 2)	DM54	−40		−100	mA
			DM74	−40		−100	
I_{CCH}	Supply Current with Outputs High	V_{CC} = Max (Note 3)	DM54		39	55	mA
			DM74		39	55	
I_{CCL}	Supply Currents with Outputs Low	V_{CC} = Max (Note 4)	DM54		69	99	mA
			DM74		69	109	

Note 1: All typicals are at V_{CC} = 5V, T_A = 25°C.

Note 2: Not more than one output should be shorted at a time, and the duration should not exceed one second.

Note 3: I_{CCH} is measured with all outputs open, inputs P3 and G3 at 4.5V, and all other inputs grounded.

Note 4: I_{CCL} is measured with all outputs open, inputs G0, G1, and G2 at 4.5V, and all other inputs grounded.

SWITCHING CHARACTERISTICS at V_{CC} = 5V and T_A = 25°C

Symbol	Parameter	From (Input) To (Output)	R_L = 280Ω				Units
			C_L = 15 pF		C_L = 50 pF		
			Min	Max	Min	Min	
t_{PLH}	Propagation Delay Time Low to High Level Output	GN or PN to C_n + x, y, z		7		10	ns
t_{PHL}	Propagation Delay Time High to Low Level Output	GN or PN to C_n + x, y, z		7		11	ns
t_{PLH}	Propagation Delay Time Low to High Level Output	GN or PN to G		7.5		11	ns
t_{PHL}	Propagation Delay Time High to Low Level Output	GN or PN to G		10.5		14	ns
t_{PLH}	Propagation Delay Time Low to High Level Output	PN to P		6.5		10	ns
t_{PHL}	Propagation Delay Time High to Low Level Output	PN to P		10		14	ns
t_{PLH}	Propagation Delay Time Low to High Level Output	C_n to C_n + x, y, z		10		13	ns
t_{PHL}	Propagation Delay Time High to Low Level Output	C_n to C_n + x, y, z		10.5		14	ns

CONNECTION DIAGRAM

Dual-In-Line Package

INPUTS: P2 (15), G2 (14), C_n (13)

OUTPUTS: C_{n+x} (12), C_{n+y} (11), G (10), C_{n+z} (9)

V_{CC} (16)

INPUTS: G1 (1), P1 (2), G0 (3), P0 (4), G3 (5), P3 (6)

OUTPUT: P (7)

GND (8)

Order Number DM54S182J or DM74S182N
See NS Package Number J16A or N16E

PIN DESIGNATIONS

Designation	Pin Nos.	Function
G0, G1, G2, G3	3, 1, 14, 5	Active Low Carry Generate Inputs
P0, P1, P2, P3	4, 2, 15, 6	Active Low Carry Propagate Inputs
C_n	13	Carry Input
C_{n+x}, C_{n+y}, C_{n+z}	12, 11, 9	Carry Outputs
G	10	Active Low Carry Generate Output
P	7	Active Low Carry Propagate Output
V_{CC}	16	Supply Voltage
GND	8	Ground

TYPICAL APPLICATION

64-Bit ALU, Full-Carry Look Ahead in Three Levels

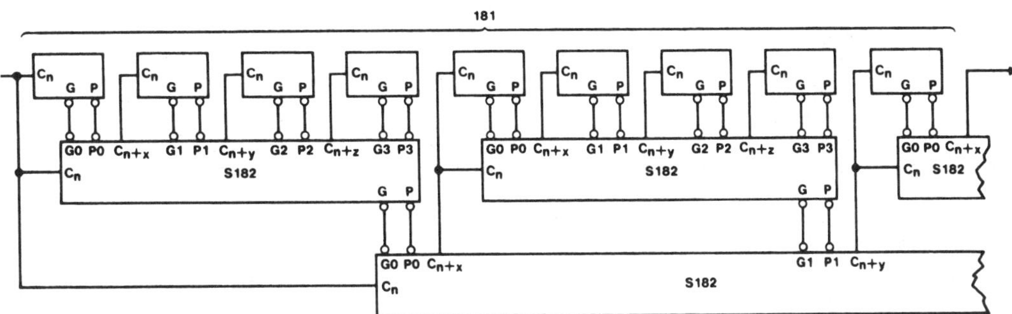

A and B inputs, and F outputs of 181 are not shown.

National DM54S194/DM74S194
4-Bit Bidirectional Universal Shift Registers

- Parallel inputs and outputs
- Four operating modes:
  - ○ Synchronous parallel load
  - ○ Right shift
  - ○ Left shift
  - ○ Do nothing
- Positive edge-triggered clocking
- Direct overriding clear
- Typical clock frequency: 105 MHz
- Typical power dissipation: 425 mW

ABSOLUTE MAXIMUM RATINGS (Note)

If Military/Aerospace specified devices are required, please contact the National Semiconductor Sales Office/Distributors for availability and specifications.

Supply Voltage	7V
Input Voltage	5.5V
Operating Free Air Temperature Range	
DM54S	−55°C to +125°C
DM74S	0°C to +70°C
Storage Temperature Range	−65°C to +150°C

Note: *The "Absolute Maximum Ratings" are those values beyond which the safety of the device cannot be guaranteed. The device should not be operated at these limits. The parametric values defined in the "Electrical Characteristics" table are not guaranteed at the absolute maximum ratings. The "Recommended Operating Conditions" table will define the conditions for actual device operation.*

RECOMMENDED OPERATING CONDITIONS

Symbol	Parameter		DM54S194 Min	DM54S194 Nom	DM54S194 Max	DM74S194 Min	DM74S194 Nom	DM74S194 Max	Units
V_{CC}	Supply Voltage		4.5	5	5.5	4.75	5	5.25	V
V_{IH}	High Level Input Voltage		2			2			V
V_{IL}	Low Level Input Voltage				0.8			0.8	V
I_{OH}	High Level Output Current				−1			−1	mA
I_{OL}	Low Level Output Current				20			20	mA
f_{CLK}	Clock Frequency (Note 1)		0	105	70	0	105	70	MHz
f_{CLK}	Clock Frequency (Note 2)		0	90	60	0	90	60	MHz
t_W	Pulse Width (Note 3)	Clock	7			7			ns
		Clear	12			12			
t_{SU}	Setup Time (Note 3)	Mode	11			11			ns
		Data	5			5			
t_H	Hold Time (Note 3)		3			3			ns
t_{REL}	Clear Release Time (Note 3)		9			9			ns
T_A	Free Air Operating Temperature		−55		125	0		70	°C

Note 1: C_L = 15 pF, R_L = 280Ω, T_A = 25°C and V_{CC} = 5V.

Note 2: C_L = 50 pF, R_L = 280Ω, T_A = 25°C and V_{CC} = 5V.

Note 3: T_A = 25°C and V_{CC} = 5V.

ELECTRICAL CHARACTERISTICS over recommended operating free air temperature (unless otherwise noted)

Symbol	Parameter	Conditions		Min	Typ (Note 4)	Max	Units
V_I	Input Clamp Voltage	V_{CC} = Min, I_I = −18 mA				−1.2	V
V_{OH}	High Level Output Voltage	V_{CC} = Min, I_{OH} = Max V_{IL} = Max, V_{IH} = Min	DM54	2.5	3.4		V
			DM74	2.7	3.4		
V_{OL}	Low Level Output Voltage	V_{CC} = Min, I_{OL} = Max V_{IH} = Min, V_{IL} = Max				0.5	V
I_I	Input Current @ Max Input Voltage	V_{CC} = Max, V_I = 5.5V				1	mA
I_{IH}	High Level Input Current	V_{CC} = Max, V_I = 2.7V				50	μA
I_{IL}	Low Level Input Current	V_{CC} = Max, V_I = 0.5V				−2	mA
I_{OS}	Short Circuit Output Current	V_{CC} = Max (Note 5)	DM54	−40		−100	mA
			DM74	−40		−100	
I_{CC}	Supply Current	V_{CC} = Max (Note 6)			85	135	mA

Note 4: All typicals are at V_{CC} = 5V, T_A = 25°C.

Note 5: Not more than one output should be shorted at a time, and the duration should not exceed one second.

Note 6: With all outputs open, inputs A through D grounded, and 4.5V applied to S0, S1, CLEAR, and the SERIAL inputs, I_{CC} is tested with a momentary ground, then 4.5V applied to CLOCK.

SWITCHING CHARACTERISTICS at V_{CC} = 5V and T_A = 25°C

Symbol	Parameter	From (Input) To (Output)	R_L = 280Ω, C_L = 15 pF Min	R_L = 280Ω, C_L = 15 pF Max	R_L = 280Ω, C_L = 50 pF Min	R_L = 280Ω, C_L = 50 pF Max	Units
f_{MAX}	Maximum Clock Frequency		70		60		MHz
t_{PLH}	Propagation Delay Time Low to High Level Output	Clock to Q		12		15	ns
t_{PHL}	Propagation Delay Time High to Low Level Output	Clock to Q		16.5		20	ns
t_{PHL}	Propagation Delay Time High to Low Level Output	Clear to Q		18.5		23	ns

FUNCTION TABLE

Inputs										Outputs			
Clear	Mode		Clock	Serial		Parallel				Q_A	Q_B	Q_C	Q_D
	S1	S0		Left	Right	A	B	C	D				
L	X	X	X	X	X	X	X	X	X	L	L	L	L
H	X	X	L	X	X	X	X	X	X	Q_{A0}	Q_{B0}	Q_{C0}	Q_{D0}
H	H	H	↑	X	X	a	b	c	d	a	b	c	d
H	L	H	↑	X	H	X	X	X	X	H	Q_{An}	Q_{Bn}	Q_{Cn}
H	L	H	↑	X	L	X	X	X	X	L	Q_{An}	Q_{Bn}	Q_{Cn}
H	H	L	↑	H	X	X	X	X	X	Q_{Bn}	Q_{Cn}	Q_{Dn}	H
H	H	L	↑	L	X	X	X	X	X	Q_{Bn}	Q_{Cn}	Q_{Dn}	L
H	L	L	X	X	X	X	X	X	X	Q_{A0}	Q_{B0}	Q_{C0}	Q_{D0}

H = High Level (steady state). L = Low Level (steady state). X = Don't Care (any input, including transitions).

↑ = Transition from low to high level.

a, b, c, d = The level of steady state input at inputs A, B, C, or D, respectively.

Q_{A0}, Q_{B0}, Q_{C0}, Q_{D0} = The level of Q_A, Q_B, Q_C, or Q_D, respectively, before the indicated steady state input conditions were established.

Q_{An}, Q_{Bn}, Q_{Cn}, Q_{Dn} = The level of Q_A, Q_B, Q_C respectively, before the most recent ↑ transition of the clock.

CONNECTION DIAGRAM

Dual-In-Line Package

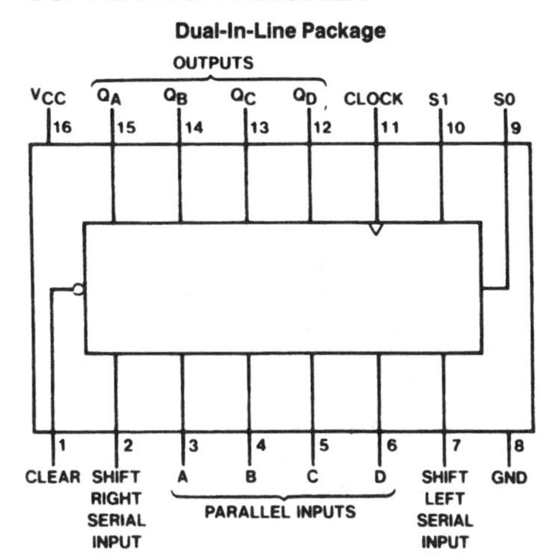

Order Number DM54S194J or DM74S194N
See NS Package Number J16A or N16E

National DM54S195/DM74S195

4-Bit Parallel-Access Shift Registers

- Synchronous parallel load
- Positive edge-triggered clocking
- Parallel inputs and outputs from each flip-flop
- Direct overriding clear
- J and $\overline{K}$ inputs to first stage
- Complementary outputs from last stage
- For use in high-performance:
  - accumulators/processors
  - serial-to-parallel, parallel-to-serial converters
- Typical clock frequency: 105 MHz
- Typical power dissipation: 350 mW

ABSOLUTE MAXIMUM RATINGS (Note)

If Military/Aerospace specified devices are required, please contact the National Semiconductor Sales Office/Distributors for availability and specifications.

Supply Voltage	7V
Input Voltage	5.5V
Operating Free Air Temperature Range	
DM54S	−55°C to +125°C
DM74S	0°C to +70°C
Storage Temperature Range	−65°C to +150°C

Note: *The "Absolute Maximum Ratings" are those values beyond which the safety of the device cannot be guaranteed. The device should not be operated at these limits. The parametric values defined in the "Electrical Characteristics" table are not guaranteed at the absolute maximum ratings. The "Recommended Operating Conditions" table will define the conditions for actual device operation.*

RECOMMENDED OPERATING CONDITIONS

Symbol	Parameter		DM54S195 Min	DM54S195 Nom	DM54S195 Max	DM74S195 Min	DM74S195 Nom	DM74S195 Max	Units
V_{CC}	Supply Voltage		4.5	5	5.5	4.75	5	5.25	V
V_{IH}	High Level Input Voltage		2			2			V
V_{IL}	Low Level Input Voltage				0.8			0.8	V
I_{OH}	High Level Output Current				−1			−1	mA
I_{OL}	Low Level Output Current				20			20	mA
f_{CLK}	Clock Frequency (Note 1)		0	105	70	0	105	70	MHz
f_{CLK}	Clock Frequency (Note 2)		0	90	60	0	90	60	MHz
t_W	Pulse Width (Note 3)	Clock	7			7			ns
		Clear	12			12			
t_{SU}	Setup Time (Note 3)	Shift/Load	11			11			ns
		Data	5			5			
t_H	Data Hold Time (Note 3)		3			3			ns
t_{REL}	Shift/Load Release Time (Note 3)		6			6			ns
	Clear Release Time (Note 3)		9			9			
T_A	Free Air Operating Temperature		−55		125	0		70	°C

Note 1: C_L = 15 pF, R_L = 280Ω, T_A = 25°C and V_{CC} = 5V.

Note 2: C_L = 50 pF, R_L = 280Ω, T_A = 25°C and V_{CC} = 5V.

Note 3: T_A = 25°C and V_{CC} = 5V.

ELECTRICAL CHARACTERISTICS over recommended operating free air temperature (unless otherwise noted)

Symbol	Parameter	Conditions		Min	Typ (Note 4)	Max	Units
V_I	Input Clamp Voltage	V_{CC} = Min, I_I = −18 mA				−1.2	V
V_{OH}	High Level Output Voltage	V_{CC} = Min, I_{OH} = Max V_{IL} = Max, V_{IH} = Min	DM54	2.5	3.4		V
			DM74	2.7	3.4		
V_{OL}	Low Level Output Voltage	V_{CC} = Min, I_{OL} = Max V_{IH} = Min, V_{IL} = Max				0.5	V
I_I	Input Current @ Max Input Voltage	V_{CC} = Max, V_I = 5.5V				1	mA
I_{IH}	High Level Input Current	V_{CC} = Max, V_I = 2.7V				50	μA
I_{IL}	Low Level Input Current	V_{CC} = Max, V_I = 0.5V				−2	mA
I_{OS}	Short Circuit Output Current	V_{CC} = Max (Note 5)	DM54	−40		−100	mA
			DM74	−40		−100	
I_{CC}	Supply Current	V_{CC} = Max (Note 6)			70	109	mA

Note 4: All typicals are at V_{CC} = 5V, T_A = 25°C.

Note 5: Not more than one output should be shorted at a time, and the duration should not exceed one second.

Note 6: With all inputs open, SHIFT/LOAD grounded, and 4.5V applied to the J, K, and data inputs, I_{CC} is measured by applying a momentary ground, then 4.5V to the CLEAR and then applying a momentary ground then 4.5V to the CLOCK.

SWITCHING CHARACTERISTICS at V_{CC} = 5V and T_A = 25°C

Symbol	Parameter	From (Input) To (Output)	R_L = 280Ω, C_L = 15 pF Min	R_L = 280Ω, C_L = 15 pF Max	R_L = 280Ω, C_L = 50 pF Min	R_L = 280Ω, C_L = 50 pF Max	Units
f_{MAX}	Maximum Clock Frequency		70		60		MHz
t_{PLH}	Propagation Delay Time Low to High Level Output	Clock to Any Q		12		15	ns
t_{PHL}	Propagation Delay Time High to Low Level Output	Clock to Any Q		16.5		20	ns
t_{PHL}	Propagation Delay Time High to Low Level Output	Clear to Any Q		18.5		23	ns

FUNCTION TABLE

Inputs									Outputs				
Clear	Shift/ Load	Clock	Serial		Parallel				Q_A	Q_B	Q_C	Q_D	$\overline{Q}_D$
			J	$\overline{K}$	A	B	C	D					
L	X	X	X	X	X	X	X	X	L	L	L	L	H
H	L	↑	X	X	a	b	c	d	a	b	c	d	$\overline{d}$
H	H	L	X	X	X	X	X	X	Q_{A0}	Q_{B0}	Q_{C0}	Q_{D0}	$\overline{Q}_{D0}$
H	H	↑	L	H	X	X	X	X	Q_{A0}	Q_{A0}	Q_{Bn}	Q_{Cn}	$\overline{Q}_{Cn}$
H	H	↑	L	L	X	X	X	X	L	Q_{An}	Q_{Bn}	Q_{Cn}	$\overline{Q}_{Cn}$
H	H	↑	H	H	X	X	X	X	H	Q_{An}	Q_{Bn}	Q_{Cn}	$\overline{Q}_{Cn}$
H	H	↑	H	L	X	X	X	X	$\overline{Q}_{An}$	Q_{An}	Q_{Bn}	Q_{Cn}	$\overline{Q}_{Cn}$

H = High Level (steady state), L = Low Level (steady state), X = Don't Care (any input, including transitions)

↑ = Transition from low to high level

a, b, c, d = The level of steady state input at A, B, C, or D, respectively.

Q_{A0}, Q_{B0}, Q_{C0}, Q_{D0} = The level of Q_A, Q_B, Q_C, or Q_D, respectively, before the indicated steady state input conditions were established.

Q_{An}, Q_{Bn}, Q_{Cn} = The level of Q_A, Q_B, Q_C, respectively, before the most recent transition of the clock.

CONNECTION DIAGRAM

Dual-In-Line Package

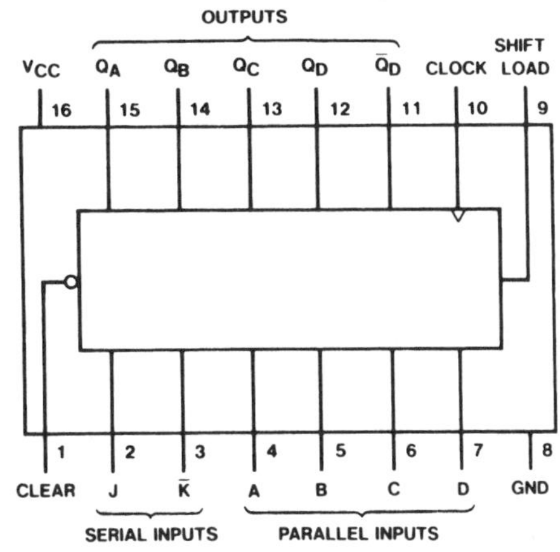

Order Number DM54S195J or DM74S195N
See NS Package Number J16A or N16E

National DM54S240/DM74S240, DM54S241/DM74S241, DM54S244/DM74S244 Octal Tri-state® Buffers/Line Drivers/Line Receivers

- Tri-state outputs drive bus lies directly
- Pnp inputs reduce dc loading on bus lines
- Hysteresis at data inputs improves noise margins
- Typical I_{OL} (sink current):
  - 54S 48 mA
  - 74S 64 mA
- Typical I_{OH} (source current):
  - 54S −12 mA
  - 74S −15 mA
- Typical propagation delay times:
  - Inverting 4.5 ns
  - Noninverting 6 ns
- Typical enable/disable times: 9 ns
- Typical power dissipation (enabled)
  - Inverting 450 mW
  - Noninverting 538 mW

ABSOLUTE MAXIMUM RATINGS (Note)

If Military/Aerospace specified devices are required, please contact the National Semiconductor Sales Office/Distributors for availability and specifications.

Supply Voltage	7V
Input Voltage	5.5V
Operating Free Air Temperature Range	
DM54S	−55°C to +125°C
DM74S	0°C to +70°C
Storage Temperature Range	−65°C to +150°C

Note: *The "Absolute Maximum Ratings" are those values beyond which the safety of the device cannot be guaranteed. The device should not be operated at these limits. The parametric values defined in the "Electrical Characteristics" table are not guaranteed at the absolute maximum ratings. The "Recommended Operating Conditions" table will define the conditions for actual device operation.*

RECOMMENDED OPERATING CONDITIONS

Symbol	Parameter	DM54S			DM74S			Units
		Min	Typ	Max	Min	Typ	Max	
V_{CC}	Supply Voltage	4.5	5	5.5	4.75	5	5.5	V
V_{IH}	High Level Input Voltage	2			2			V
V_{IL}	Low Level Input Voltage			0.8			0.8	V
I_{OH}	High Level Output Current			−12			−15	mA
I_{OL}	Low Level Output Current			48			64	mA
T_A	Free Air Operating Temperature	−55		125	0		70	°C

ELECTRICAL CHARACTERISTICS over recommended operating free-air temperature range (unless otherwise noted)

Symbol	Parameter	Conditions		Min	Typ (Note 1)	Max	Units
V_I	Input Clamp Voltage	V_{CC} = Min, I_I = −18 mA				−1.2	V
H_{ys}	Hysteresis ($V_{T+} - V_{T-}$) (Data Inputs Only)	V_{CC} = Min		0.2	0.4		V
V_{OH}	High Level Output Voltage	V_{CC} = 4.75V, V_{IH} = 2V, V_{IL} = 0.8V, I_{OH} = −1 mA	DM74	2.7			V
		V_{CC} = Min, V_{IH} = 2V, V_{IL} = 0.8V, I_{OH} = −3 mA		2.4	3.4		V
		V_{CC} = Min, V_{IH} = 2V, V_{IL} = 0.5V, I_{OH} = Max		2			V
V_{OL}	Low Level Output Voltage	V_{CC} = Min, I_{OL} = Max, V_{IL} = 0.8V, V_{IH} = 2V	DM54			0.55	V
			DM74			0.55	V
I_{OZH}	Off-State Output Current, High Level Voltage Applied	V_{CC} = Max, V_{IL} = 0.8V, V_{IH} = 2V	V_O = 2.4V			50	μA
I_{OZL}	Off-State Output Current, Low Level Voltage Applied	V_{CC} = Max, V_{IL} = 0.8V, V_{IH} = 2V	V_O = 0.5V			−50	μA
I_I	Input Current at Maximum Input Voltage	V_{CC} = Max, V_I = 5.5V				1	mA
I_{IH}	High Level Input Current	V_{CC} = Max, V_I = 2.7V				50	μA
I_{IL}	Low Level Input Current	V_{CC} = Max, V_I = 0.5V	Any A			−400	μA
			Any G			−2	mA
I_{OS}	Short Circuit Output Current	V_{CC} = Max (Note 2)		−50		−225	mA
I_{CC}	Supply Current	Outputs High	DM54S240		80	123	mA
			DM74S240		80	135	mA
			DM54S241, 244		95	147	mA
			DM74S241, 244		95	160	mA
		Outputs Low	DM54S240		100	145	mA
			DM74S240		100	150	mA
			DM54S241, 244		120	170	mA
			DM74S241, 244		120	180	mA
		Outputs Disabled	DM54S240		100	145	mA
			DM74S240		100	150	mA
			DM54S241, 244		120	170	mA
			DM74S241, 244		120	180	mA

Note 1: All typical values are at V_{CC} = 5V, T_A = 25°C.

Note 2: Not more than one output should be shorted at a time and duration should not exceed one second.

SWITCHING CHARACTERISTICS V_{CC} = 5V, T_A = 25°C

Symbol	Parameter	Conditions		Min	Max	Units
t_{PLH}	Propagation Delay Time Low to High Level Output	C_L = 45 pF R_L = 90Ω	DM54/74S240	2	7	ns
			DM54/74S241, 244	2	9	
t_{PHL}	Propagation Delay Time High to Low Level Output	C_L = 45 pF R_L = 90Ω	DM54/74S240	2	7	ns
			DM54/74S241, 244	2	9	
t_{PZL}	Output Enable Time to Low Level	C_L = 45 pF R_L = 90Ω	DM54/74S240	3	15	ns
			DM54/74S241, 244	3	15	
t_{PZH}	Output Enable Time to High Level	C_L = 45 pF R_L = 90Ω	DM54/74S240	2	10	ns
			DM54/74S241, 244	3	12	
t_{PLZ}	Output Disable Time from Low Level	C_L = 5 pF R_L = 90Ω	DM54/74S240	4	15	ns
			DM54/74S241, 244	2	15	
t_{PHZ}	Output Disable Time from High Level	C_L = 5 pF R_L = 90Ω	DM54/74S240	2	9	ns
			DM54/74S241, 244	2	9	
t_{PLH}	Propagation Delay Time Low to High Level Output	C_L = 150 pF R_L = 90Ω	DM54/74S240	3	10	ns
			DM54/74S241, 244	4	12	
t_{PHL}	Propagation Delay Time High to Low Level Output	C_L = 150 pF R_L = 90Ω	DM54/74S240	3	10	ns
			DM54/74S241, 244	4	12	
t_{PZL}	Output Enable Time to Low Level	C_L = 150 pF R_L = 90Ω	DM54/74S240	6	21	ns
			DM54/74S241, 244	6	21	
t_{PZH}	Output Enable Time to High Level	C_L = 150 pF R_L = 90Ω	DM54/74S240	4	12	ns
			DM54/74S241, 244	4	15	

CONNECTION DIAGRAMS

Dual-In-Line Package

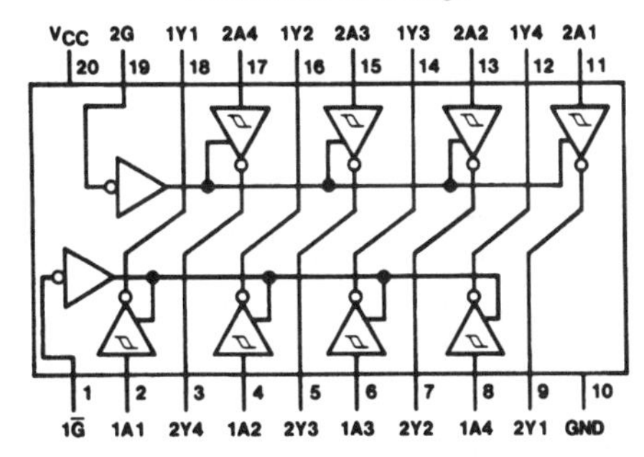

Order Number DM54S240J, DM74S240WM or DM74S240N See NS Package Number J20A, M20B or N20A

Dual-In-Line Package

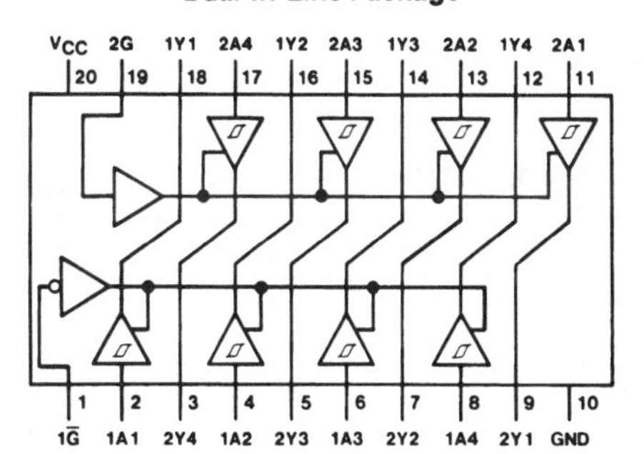

Order Number DM54S241J or DM74S241N See NS Package Number J20A or N20A

Dual-In-Line Package

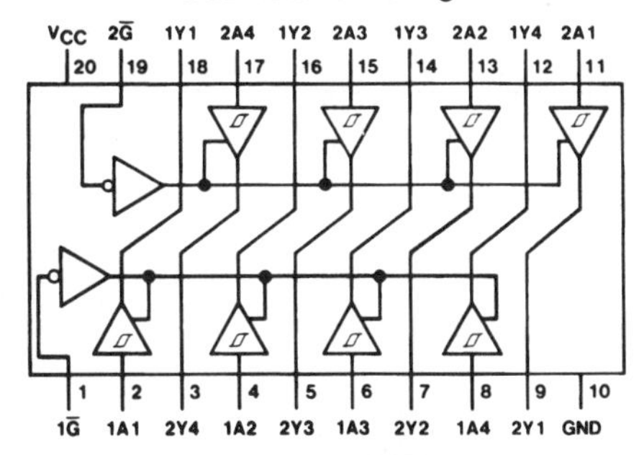

Order Number DM54S244J, DM74S244WM or DM74S244N See NS Package Number J20A, M20B or N20A

Rohm
BA6580DK
Floppy Disk Driver

- Integrates floppy disk drive read, write, and control circuitry on a single chip
- Operates on a single +5V supply. During data write, the device can supply +5V or +12V to the drive heads.
- Read preamplifier's differential voltage gain is selectable from 100 and 200 with a gain-selecting pin.
- Peak shift in the read circuit is less than 1% at an input voltage range of 0.5mV p-p to 10mV p-p with no adjustment
- Tri-state read data output available
- Write current is presettable in a range from 1 to 15mA by an external resistor. Internal voltage regulator ensures stable write current against supply voltage fluctuation and temperature variation.
- Write current correction feature for inner tracks is included. The correcting current can be set by the user with an external resistor.
- Internal supply-voltage drop detector inhibits illegal write at power on, or in the event of supply voltage drop.
- Capable of driving 3-, 3.5-, 5.25-, and 8-inch drives
- Comes in a 44-pin QFP package (lead pitch 0.8mm)

APPLICATION

- Floppy disk drives

The BA6580DK is a floppy disk read/write driver that can operate on a single 5V power supply. The device is applicable to all types of floppy disk drives: 3-, 3.5-, 5.25-, and 8-inch drives. When used with the BU9500K floppy disk drive controller, the BA6580DK significantly reduces external component requirements and achieves compact, low-profile drive design.

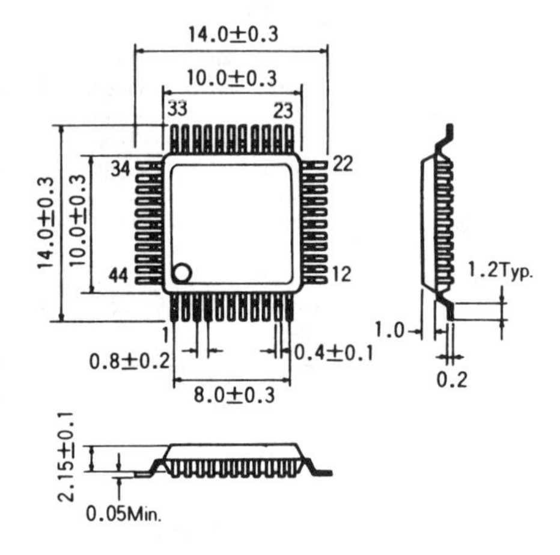

ABSOLUTE MAXIMUM RATINGS (Ta=25°C)

Parameter	Symbol	Limits	Unit	Conditions
Supply voltage	$V_{CC1,3}$	+7	V	—
Supply voltage	V_{CC2}	+16	V	—
Operating temperature range	Topr	0~+70	°C	—
Storage temperature range	Tstg	−55~+125	°C	—
Digital line input voltage	Vi	−0.5~+7.0	V	Applies to $\overline{\text{WC CONT}}$, $\overline{\text{W DATA}}$, $\overline{\text{W GATE}}$, $\overline{\text{E GATE}}$, $\overline{\text{SIDE1}}$, $\overline{\text{OC}}$, $\overline{\text{TDF CONT}}$
RW pin voltage	V_{RW}	25	V	Applies to RW01, RW02, RW11, RE12, WR DUMP pins under Write condition
$\overline{\text{PD}}$ output voltage	V_{PD}	16	V	Applies to $\overline{\text{POWER DOWN}}$ pin
Read data output voltage	V_{RD}	5.5	V	Applies to READ DATA pin
Common drive current	I_{COM}	130	mA	Applies to COM0, COM1 pins
Erase drive current	I_{ER}	130	mA	Applies to EO0, EO1 pins
EO pin voltage	V_{ER}	16	V	—

ELECTRICAL CHARACTERISTICS (Unless otherwise specified, $Ta=25°C$, $V_{CC1}=V_{CC2}=V_{CC3}=5V$)

	Parameter	Symbol	Limits			Unit	Conditions
			Min.	Typ.	Max.		
	Supply voltage range 1	V_{CC1}	4.4	5.0	6.0	V	5.5V max., when $V_{CC2}=12V$
	Supply voltage range 2	V_{CC2}	4.4	5.0	6.0	V	With 5V head
	Supply voltage range 2	V_{CC2}	10.8	12.0	13.2	V	With 12V head
	Supply voltage range 3	V_{CC3}	4.4	5.0	6.0	V	5.5V max.,when $V_{CC2}=12V$
	Current consumption 1	I_{CC1R}	—	34	46	mA	During read
		I_{CC1W}	—	34	46	mA	During write
	Current consumption 2	I_{CC2R}	—	0.17	0.23	mA	During read $V_{CC}=5V$
		I_{CC2W}	—	18	24	mA	During write $V_{CC2}=5V$, $I_W=5mA$, $I_E=0mA$
		I_{CC12R}	—	1.1	1.5	mA	During read $V_{CC2}=12V$
		I_{CC12W}	—	19	26	mA	During write $V_{CC2}=12V$, $I_W=5mA$, $I_E=0mA$
	Current consumption 3	I_{CC3R}	—	14	19	mA	During read
		I_{CC3W}	—	17	23	mA	During write
Preamplifier	Differential voltage gain	A_{VD}	140	200	260	V/V	0.1 μF between pins 7 and 9
	Common-mode input voltage range	V_{ICM}	—	2	—	V	
	Differential input voltage range	V_{ID}	0.5	—	30	mV_{PP}	PS≦1%
	Differential output voltage range	V_{OD}	3.0	—	—	V_{PP}	RL=1.2kΩ, THD=5%
	Output distortion	THD	—	—	5	%	RL=1.2kΩ, Vin=11mV_{P-P}, f=100kHz Between pins 7 and 9, frequency band 10kHz to 1MHz
Peak detector	Peak shift	PS	—	—	1	%	f=125kHz, Vin=0.15 ~ 1V_{P-P}
Pulse shaping	Output high voltage	V_{OH}	2.7	—	—	V	$V_{CC}=4.75V$, IOH= −0.4mA
	Output low voltage	V_{OL}	—	—	0.5	V	$V_{CC}=4.75V$, IOL=2mA
	Output rise time	t_{TLH}	—	—	100	nS	RL=620Ω pull-up
	Output fall time	t_{THL}	—	—	25	nS	RL=620Ω pull-up
	Timing range 1	t_1	0.5	—	4	μS	Time domain filter
	Timing range 2	t_2	0.15	—	2	μS	Read data output
	Timing accuracy 1	Et_1	−5	0	+15	%	$R_{12}=12k\Omega$, $C_8=100pF$
	Timing accuracy 2	Et_2	−15	0	+15	%	$R_{11}=10k\Omega$, $C_7=50pF$
Common Drivers	Output voltage during write selection	V_{CMWR}	4.3	4.5	—	V	$I_{COM}=115mA$
	Output voltage during non-write selection	V_{CMWO}	—	0	—	V	
	Output voltage during read selection	V_{CMRD}	—	2.0	—	V	
	Output voltage during non-read selection	V_{CMRO}	—	0	—	V	
	Output current range	I_{COM}	—	—	115	mA	
Erase Drivers	Output low voltage	V_{OLE}	—	—	0.5	V	$I_{ER}=100mA$
	Output leakage current	I_{LKGE}	—	—	100	μA	
	Erase current range	I_{ER}	—	—	100	mA	
Write Drivers	Write current set-up accuracy	AC_{IW}	−10	—	+10	%	
	Dependence of write current on supply voltage	PS_{IW}	—	±1	—	%/V	
	Dependense of write current on temperature	TC_{IW}	—	±0.05	—	%/°C	
	Write current matching	ΔI_W	−1	—	+1	%	
	Write current set-up range	IWR	1	—	15	mA	
	Leakage current at off	I_{LKGW}	—	—	10	μA	$V_{RW}=20V$
	Write data minimum pulse width	t_{WD}	120	—	—	ns	
Control logic*1	Input high voltage	V_{IH}	2	—	—	V	
	Input low voltage	V_{IL}	—	—	0.8	V	
Control logic (Schmitt)*2	Forward threshold voltage	V_{T+}	2	—	—	V	
	Reverse threshold voltage	V_{T-}	—	—	0.8	V	
	Hysteresis voltage	$V_{T+}-V_{T-}$	0.15	—	—	V	

ELECTRICAL CHARACTERISTICS (Unless otherwise specifed, Ta=25°C, $V_{CC1}=V_{CC2}=V_{CC3}=5V$)

	Parameter	Symbol	Limits			Unit	Conditions
			Min.	Typ.	Max.		
*1	Input current	I_{IH}	—	—	20	μA	$V_{IH}=2.7V$
*2		I_{IL}	—	—	−0.2	mA	$V_{IL}=0.4V$
	Supply falling detector voltage	V_{CC1}	3.5	3.9	4.2	V	
	Supply falling detector voltage	V_{CC2}	3.5	3.9	4.2	V	

*1 Applies to $\overline{OC}$, $\overline{TDF\ CONT}$ pins
*2 Applies to $\overline{WC\ CONT}$, $\overline{W\ DATA}$, $\overline{W\ GATE}$, $\overline{E\ GATE}$, $\overline{SIDE1}$ pins

APPLICATION EXAMPLE

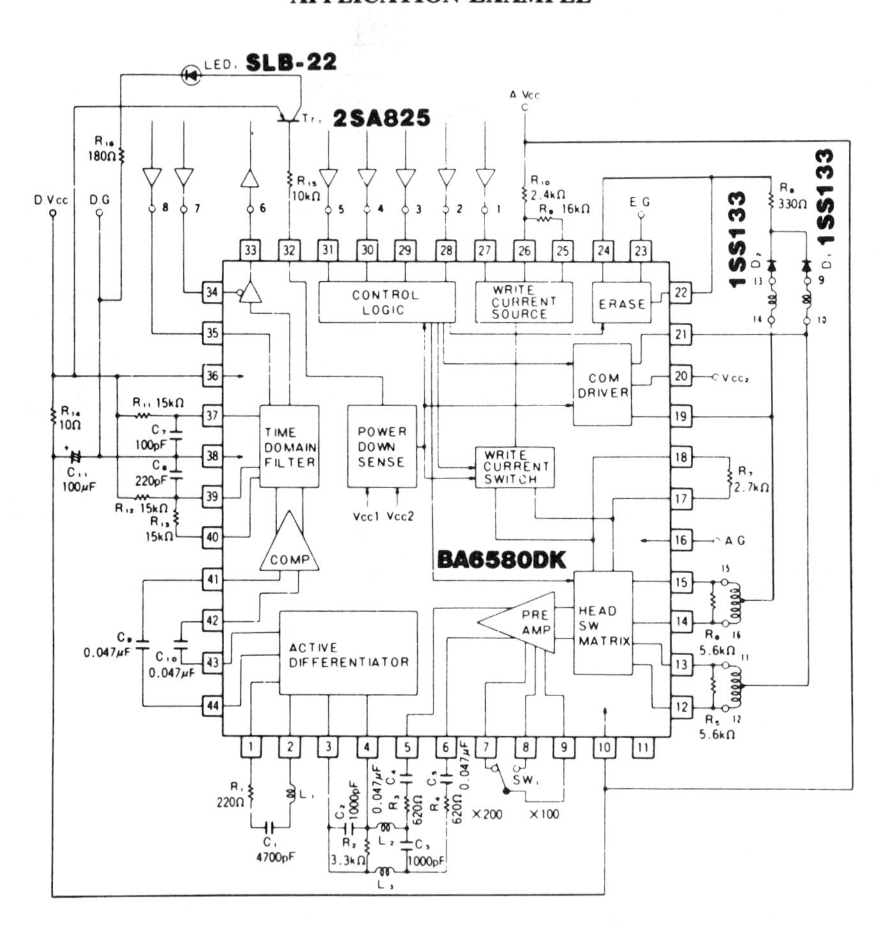

L_1, L_2, L_3: 560 μH (RC-875-561K) SUMIDA DK
SW, (Read preamplifier, gain selecting SW)

Head dump resistor

	SIDE 0	SIDE 1
During read	R_5	R_6
During write	$R_5//R_7$	$R_6//R_7$

BLOCK DIAGRAM

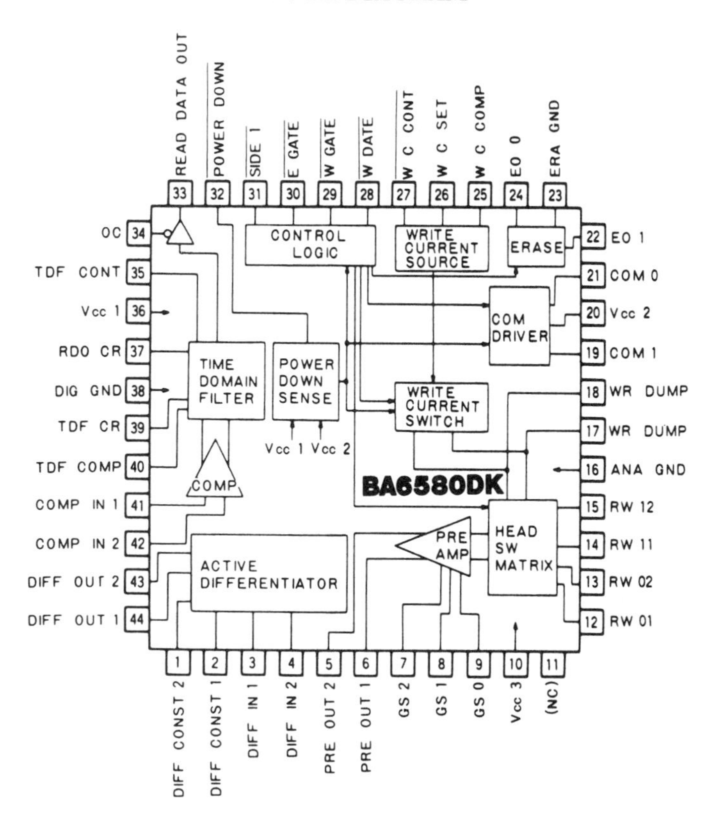

Rohm
BU9500K
Floppy Disk Drive Controller

- Contains stepping motor control, spindle motor on/off control, write control, head control, LED control, and read control logic
- Single 5V power supply, with supply voltage tolerance of +20%-10%
- Different stepping motor excitation modes available: 2-phase excitation, 1/2-phase excitation, etc.
- Stepping motor power control output
- Internal track counter that provides a write current switching or filter switching signal output
- Power-on recalibrate capability
- Erase time switchable for different drive types
- Internal head load drive power save logic
- Different in-use indicator driving modes available
- Drives the motor during chucking
- Internal disk change feature
- Internal ready timing generator

APPLICATION

- Floppy disk drives

The BU9500K is a monolithic floppy disk drive controller that can operate on a single 5V power supply.

In conjunction with the BA6580DK floppy disk driver, the BU9500K can be applied to any type of floppy disk drive.

ABSOLUTE MAXIMUM RATINGS (Ta=25°C)

Parameter	Symbol	Limits	Unit
Supply voltage	V_{DD}	7.0	V
Power dissipation	Pd	500*	mW
Operating temperature range	Topr	−25~75	°C
Storage temperature range	Tstg	−55~125	°C
Input voltage	V_{IN}	$V_{SS}-0.5 \sim V_{DD}+0.5$	V

*Derating is done at 5mW/°C for operation above Ta=25°C.

RECOMMENDED OPERATING CONDITIONS

Parameter	Symbol	Min.	Typ	Max.	Unit	Conditions
Supply voltage	V_{DD}	—	5.0	—	V	—

ELECTRICAL CHARACTERISTICS (Ta=25°C, V_{DD}=5V)

Parameter	Symbol	Min.	Typ	Max.	Unit	Conditions
Index timing range 1	T_1	9.0	10.1	11.0	μsec	C=0.01μF, R=1.0kΩ *1
Index timing range 2	T_2	3.0	3.7	—	msec	C=0.1μF, R=50kΩ *1
Oscillator duty ratio	A_{CLK}	30	52.5	70	%	C=20pF ±30% R=5.1kΩ ±10% *1
Oscillation frequency	f_{CLK}	0.98f	f*3	1.02f	MHz	C=20pF ±30% R=5.1kΩ ±10% *1
Element delay	T_{D1}	0	0	10	nsec	*2
Input high voltage 1	V_{I1H}	2.0	—	—	V	TTL input
Input low voltage 1	V_{I1L}	—	—	0.8	V	TTL input
Input high voltage 2	V_{I2H}	2.75	—	—	V	Comparator input
Input low voltage 2	V_{I2L}	—	—	2.25	V	Comparator input
Input high voltage 3	V_T^+	3.0	3.5	4.0	V	Schmitt trigger input $V_3=V_T^+-V_T^-$
Input low voltage 3	V_T^-	1.2	1.7	2.2	V	Schmitt trigger input $V_3=V_T^+-V_T^-$
Hysteresis voltage	V_3	0.8	1.8	2.8	V	
DLS output on voltage	V_{ON}	0	0.21	1.0	V	I_{ON}=5mA
DLS output leakage current	I_L	—	—	5	μA	V_{DD}=6V, V_O=6V
Output high current	I_{OH}	−0.5	−1.79	—	mA	V_{OH}=3.5V
Output low current	I_{OL}	1.5	11.5	—	mA	V_{OL}=0.4V
Current consumption 1	I_{DD1}	—	5.8	10	mA	at oscillating NO LOAD *1
Current consumption 2	I_{DD2}	—	6.7	11	mA	at oscillating READ *1

*1 Measured data is attached for the thermal characteristic when V_{DD} = 4.5 ~ 6.0V
*2 Repeatability dispersin of the delay time at the input line.
*3 Ceralock oscillation frequency, f = 3.0MHz ~ 5.0MHz at Ta = 252°C, V_{DD} = 5V

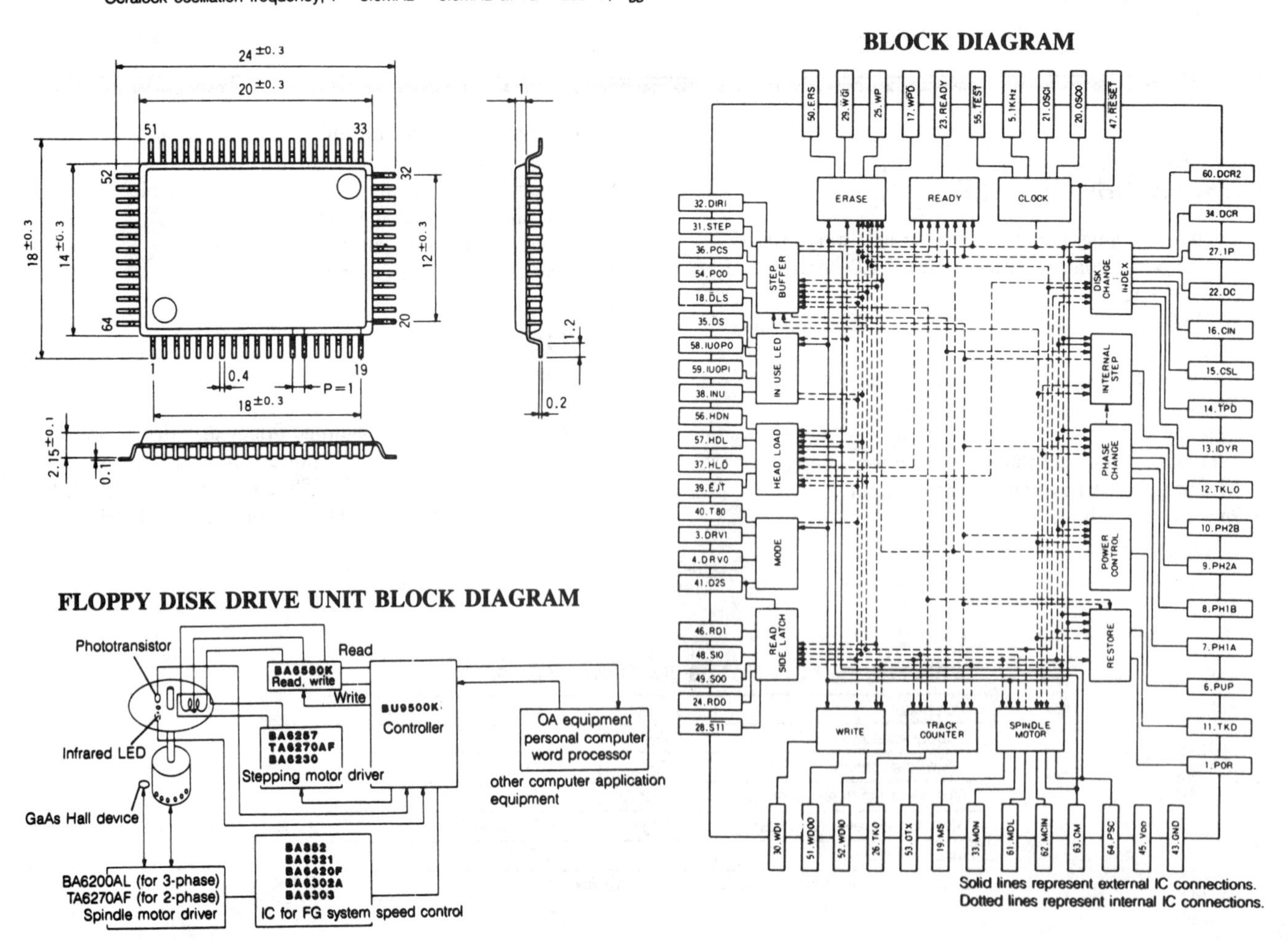

BASIC APPLICATION CIRCUIT

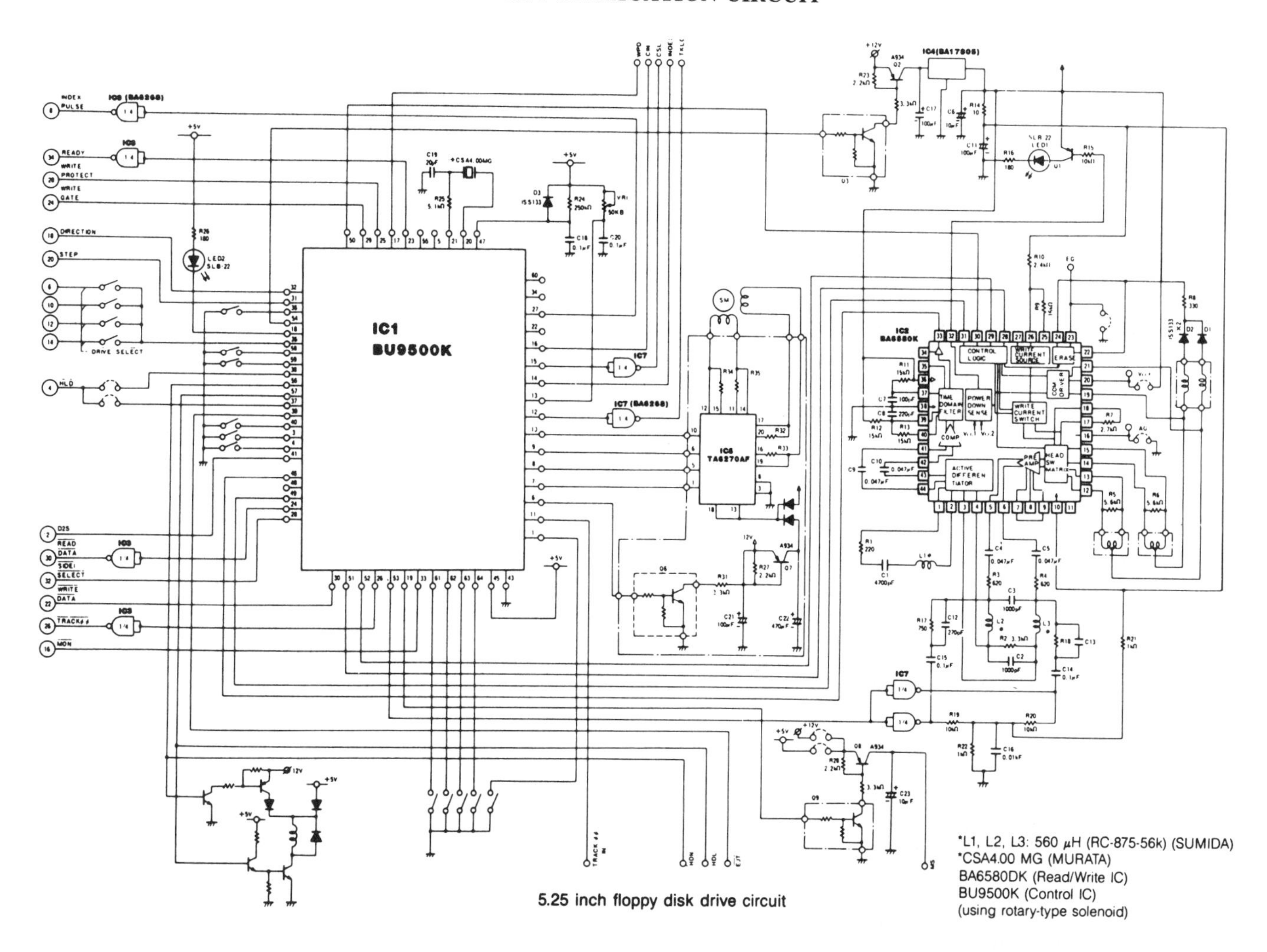

5.25 inch floppy disk drive circuit

Rohm TA6270F Transistor Array

- Low $V_{CE(sat)}$
- Allows easy construction of motor-driving bridges
- Allows compact system design with high reliability
- Comes in a compact 20-pin MF package

APPLICATIONS

- Motor drivers
- Solenoid drivers

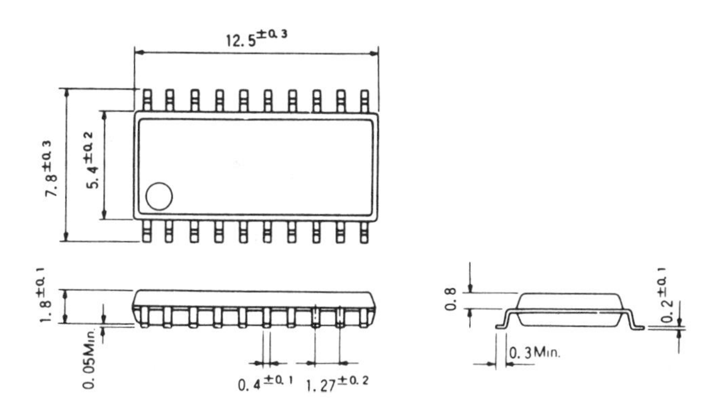

BLOCK DIAGRAM

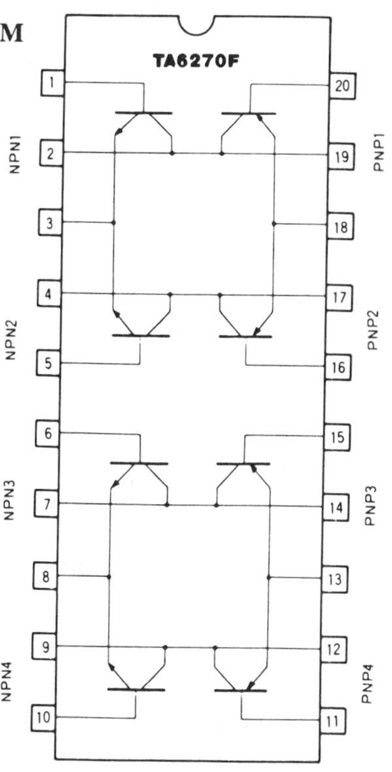

ABSOLUTE MAXIMUM RATINGS (Ta=25°C)

Parameter	Symbol	Limits	Unit
Supply voltage	V_{CEO}	25	V
Power dissipation	Pd	800*	mW
Operating temperature range	Topr	−25~75	°C
Storage temperature range	Tstg	−55~150	°C
Emitter-base voltage	V_{EBO}	5	V
Collector current	I_C	1.5	A

*Package power dissipation of the assembled PC board.

ELECTRICAL CHARACTERISTICS (Ta=25°C)

Parameter	Symbol	Min.	Typ	Max.	Unit	Conditions
Collector, emitter breakdown voltage	V_{CEO}	25	—	—	V	I_C=1mA
Collector-base breakdown voltage	V_{CBO}	25	—	—	V	I_C=50μA
Emitter-base breakdown voltage	V_{EBO}	5	—	—	V	I_E=50μA
Collector cutoff voltage	I_{CBO}	—	—	1	μA	V_{CB}=20V
Emitter cutoff voltage	I_{EBO}	—	—	1	μA	V_{EB}=4V
Collector, emitter saturation voltage (NPN)	$V_{CE\ (sat)}$	—	0.17	—	V	I_C/I_B=0.5A/5mA
Collector, emitter saturation voltage (PNP)	$V_{CE\ (sat)}$	—	0.2	—	V	I_C/I_B=0.5A/5mA
DC current amplifier (NPN)	h_{FE}	—	400	—	—	V_{CE}/I_C=3V/100mA
DC current amplifier (PNP)	h_{FE}	—	200	—	—	V_{CE}/I_C=3V/100mA

Note: PNP is the default characteristic.

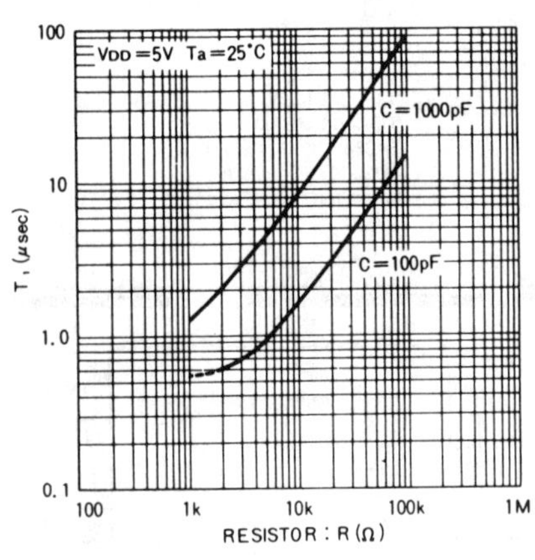

T₁-R and C (TDFR timing range) characteristics

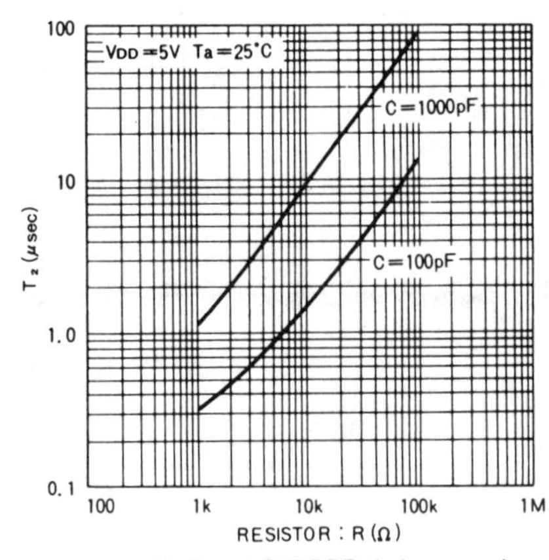

T₂-R and C (RDPR timing range) characteristics

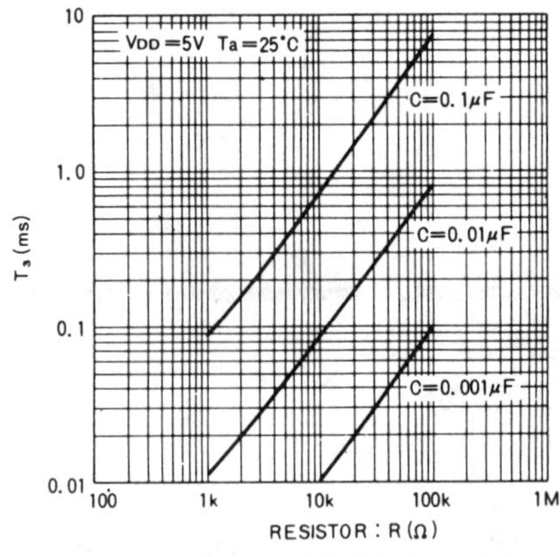

T₃-R and C (IDYR timing range) characteristics

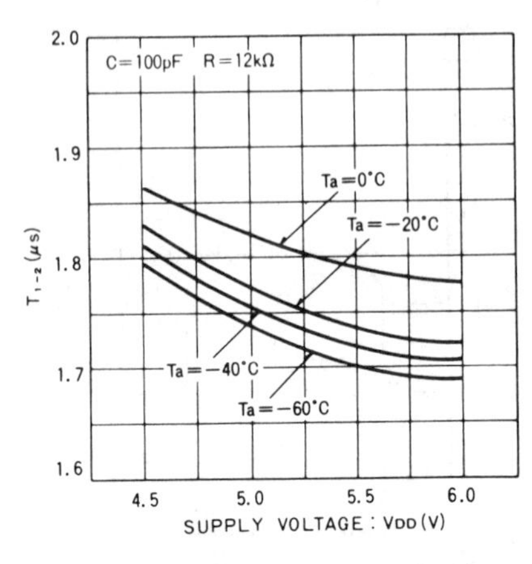

Timing range 1-2 vs. supply voltage

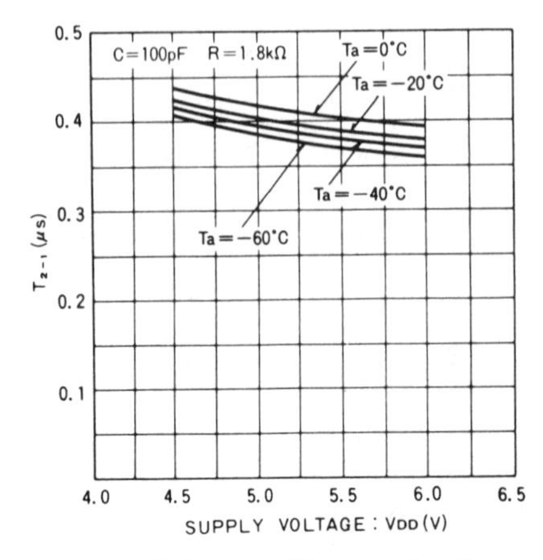

Timing range 2-1 vs. supply voltage

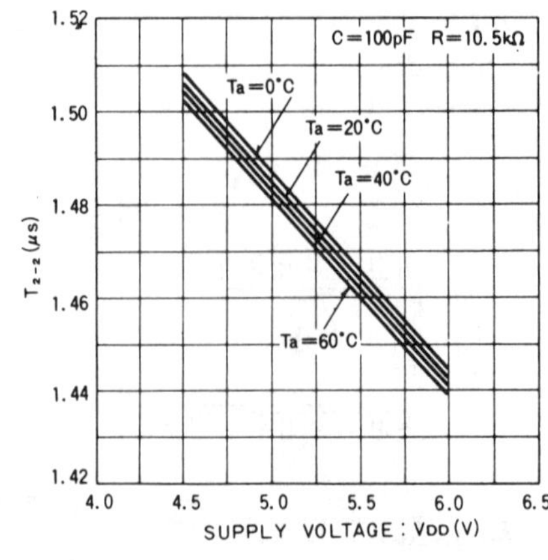

Timing range 2-2 vs. supply voltage

Rohm
CMOS Gate Array BU1200 Series
Single-Chip Microprocessor Series

CMOS GATE ARRAYS

BU1200 Series (Single 5V Power Supply)

☆Under development

Type	No. of gates (dual input)	No. of bonding pads	No. of I/O cells	Propagation delay (internal gate)	Package				I/O level	Toggle frequency
					DIP	Shrink DIP	QFP	MF		
BU1205	156	24	22	3ns typ. (Fo=3 Mean wiring length=3mm)	16,18,20,22,24	22,42	32	18,20,22,24,28	CMOS/TTL	50MHz
BU1206	288	32	30		18,28	42	32	28,40		
BU1201	460	40	38		24,28,40	30,42	32,44	28,40		
BU1202	793	52	50		24,28,40	42	44,64	28,40		
BU1203	1548	72	70		28,40	42	44,64,☆80	40		
BU1204	3025	100	98		40		64,☆80,☆100			

BU1200L Series (Single 3V Power Supply)

☆Under development

Type	No. of gates (dual input)	No. of bonding pads	No. of I/O cells	Propagation delay (internal gate)	Package				I/O level	Toggle frequency
					DIP	Shrink DIP	QFP	MF		
BU1205L	156	24	22	6ns typ. (Fo=3 Mean wiring length=3mm)	16,18,20,22,24	22,42	32	18,20,22,24,28	CMOS	25MHz
BU1206L	288	32	30		18,28	42	32	28,40		
BU1201L	460	40	38		24,28,40	30,42	32,44	28,40		
BU1202L	793	52	50		24,28,40	42	44,64	28,40		
BU1203L	1548	72	70		28,40	42	44,64,☆80	40		
BU1204L	3025	100	98		40		64,☆80,☆100			

SINGLE-CHIP MICROPROCESSOR SERIES

Type	Description and features	Package Type/number of pins	Architecture
BU2400(L)	• Single chip 4 bit microprocessor. • Internal 1K byte programmable ROM. • Internal 32-byte OPLA. • Internal 64 × 4 bit RAM. • Two 4 bit input ports (one with latch mode, the other without latch mode). • 16 independent output ports and a parallel 8-bit output port. • Single +5V (+3V) power supply. • CMOS technology for low-power design. • Clock frequency: 400kHz to 4MHz (100kHz to 1MHz). *Values given in parentheses are for the BU2400L.	DIP40	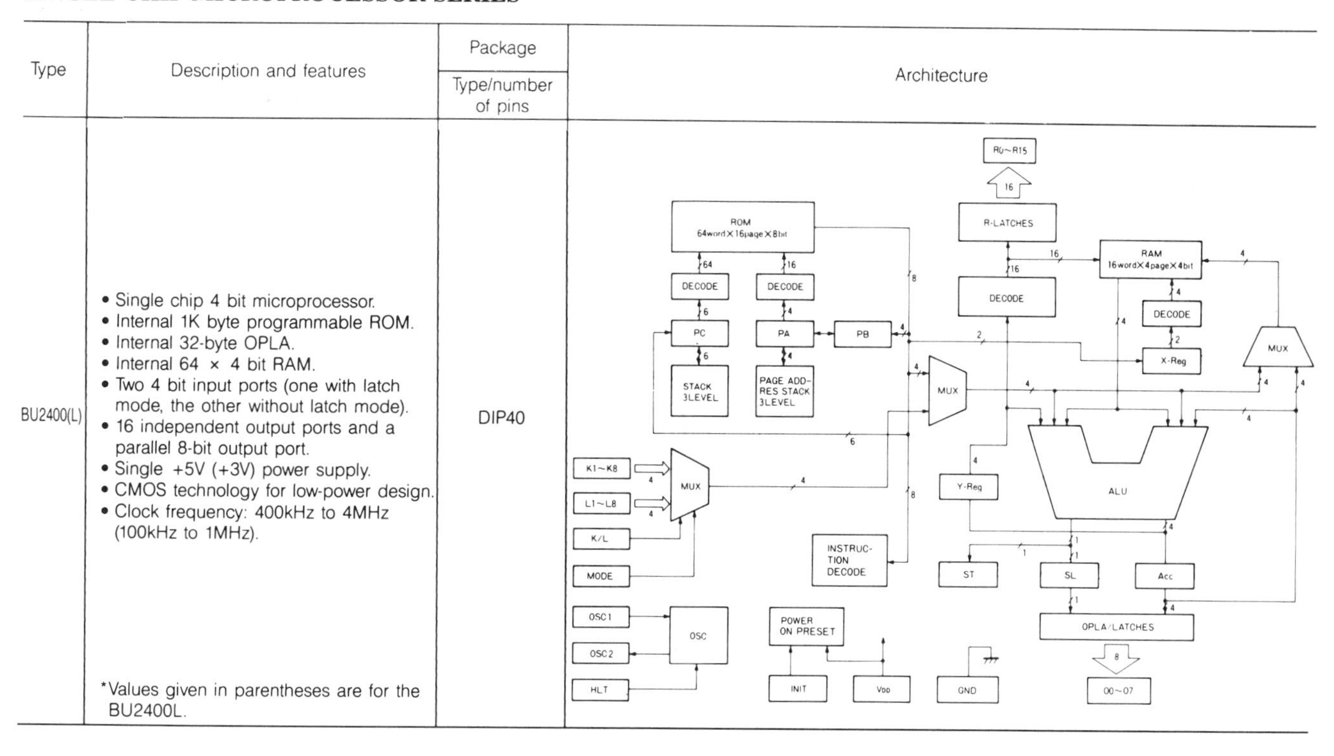

Type	Description and features	Package Type/number of pins	Architecture
BU2405(L)	• Single chip 4 bit microprocessor with high cost effectiveness, suited for use in compact logical processors and small-scale control systems. • Forty-two basic machine-language instructions available. • Instruction cycle time: 1.5μs at 1MHz clock. • Memory capacity: ROM .... 512 × 8 bits RAM .... 16 × 4 bits • Eight input ports (also serve as output ports). • Ten output ports (of which 8 ports also serve as input ports). • Subroutine nesting: 2 levels. • Internal clock generator. • Internal standby feature. • Single +5V (+3V) power supply. • CMOS technology for low-power design. • Available in a 16-pin DIP for compact system design. *Values given in parentheses are for the BU2405L.	DIP16	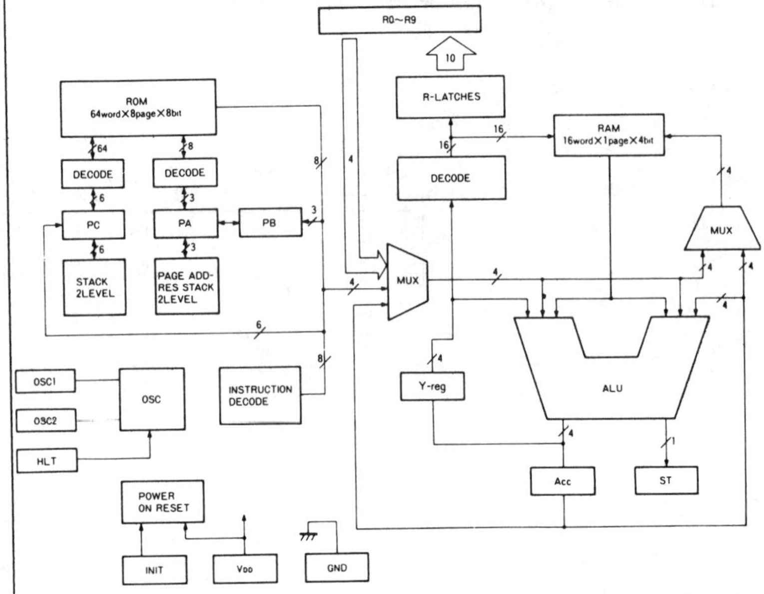
BU13870	• Single chip 8 bit microprocessor. • Internal mask programmable ROM with up to 2K bytes of capacity. • Internal 64 byte scratch pad registers. • 32 bit bidirectional TTL I/O ports (8 bits × 4 ports). • Internal programmable timer can be used in three modes: Interval timer mode Pulsewidth measurement mode Event counter mode • External interrupt feature. • Clock source is selectable from crystal resonator, RC network, and external clock. • Single 5V power supply (V_{DD}=5V ±10%). *CMOS type is under development.	DIP40	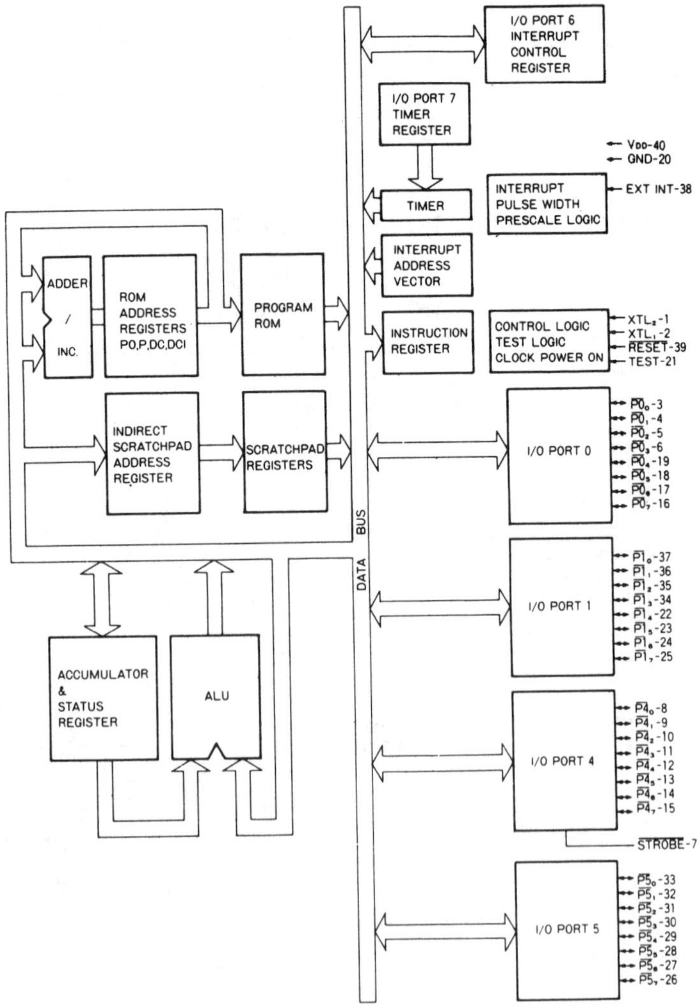

Rohm
BU18400 Series (Z80® Series)
General-Purpose Microprocessors

Z80® is a registered trademark of Zilog, Inc. in the U.S.

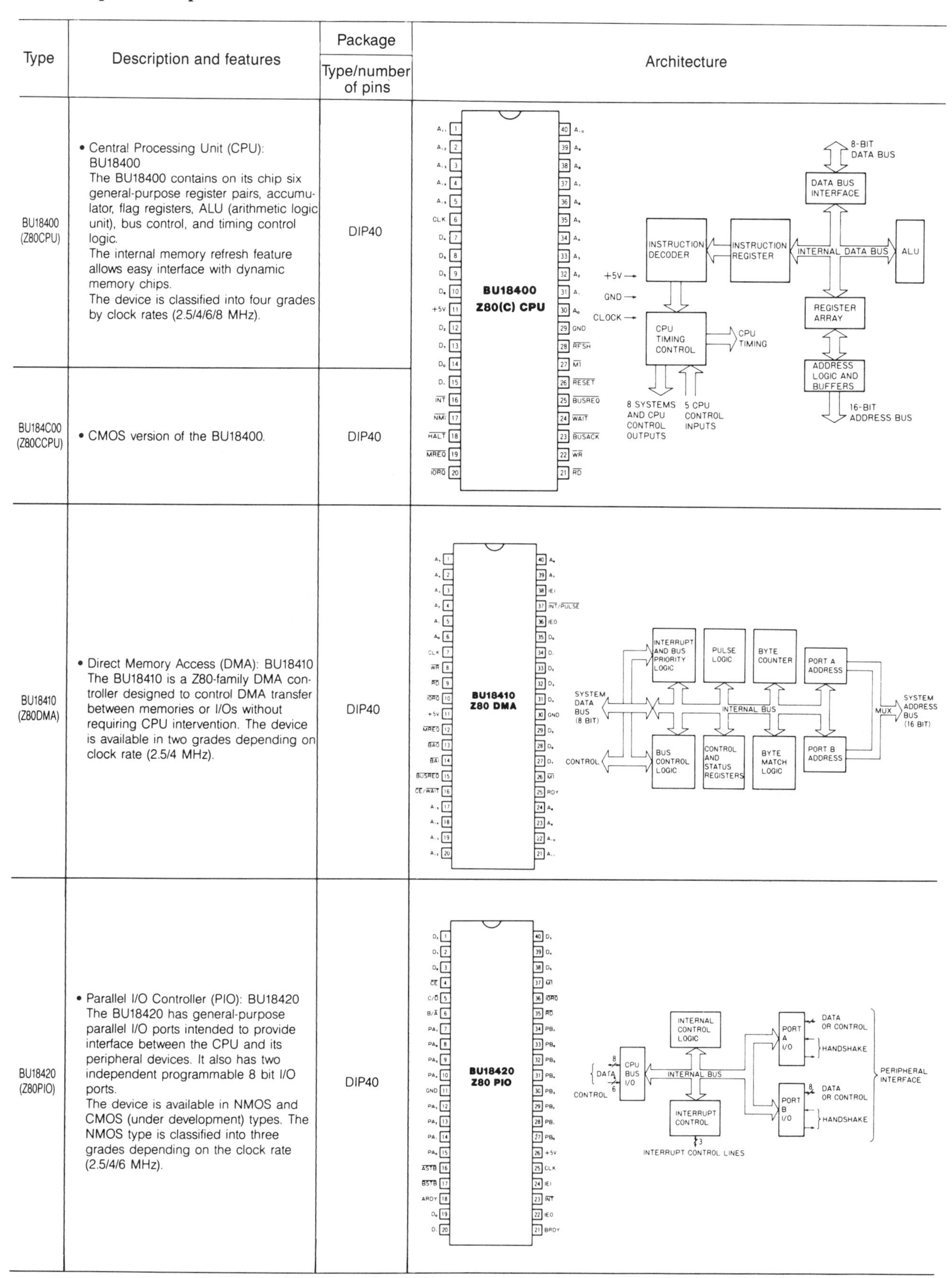

Type	Description and features	Package Type/number of pins	Architecture
BU18400 (Z80CPU)	• Central Processing Unit (CPU): BU18400 The BU18400 contains on its chip six general-purpose register pairs, accumulator, flag registers, ALU (arithmetic logic unit), bus control, and timing control logic. The internal memory refresh feature allows easy interface with dynamic memory chips. The device is classified into four grades by clock rates (2.5/4/6/8 MHz).	DIP40	
BU184C00 (Z80CCPU)	• CMOS version of the BU18400.	DIP40	
BU18410 (Z80DMA)	• Direct Memory Access (DMA): BU18410 The BU18410 is a Z80-family DMA controller designed to control DMA transfer between memories or I/Os without requiring CPU intervention. The device is available in two grades depending on clock rate (2.5/4 MHz).	DIP40	
BU18420 (Z80PIO)	• Parallel I/O Controller (PIO): BU18420 The BU18420 has general-purpose parallel I/O ports intended to provide interface between the CPU and its peripheral devices. It also has two independent programmable 8 bit I/O ports. The device is available in NMOS and CMOS (under development) types. The NMOS type is classified into three grades depending on the clock rate (2.5/4/6 MHz).	DIP40	

Type	Description and features	Package Type/number of pins	Architecture
BU18430 (Z80CTC)	• Counter/Timer Circuit (CTC): BU18430 The BU18430 contains four independent readable counter/timers and a prescaler which is selectable from 16/256. The device is available in NMOS and CMOS (under development) types. The NMOS type is classified into three grades depending on the clock rate (2.5/4/6 MHz).	DIP28	
BU18440 (Z80SIO-0) BU18441 (Z80SIO-1) BU18442 (Z80SIO-2)	• Serial I/O Controller (SIO): BU18440/18441/18442 The BU1844X is a serial I/O controller which can be applied to the various serial data communications required for microcomputer systems. The device is available in three grades depending on the clock rate (2.5/4/6 MHz).	DIP40	

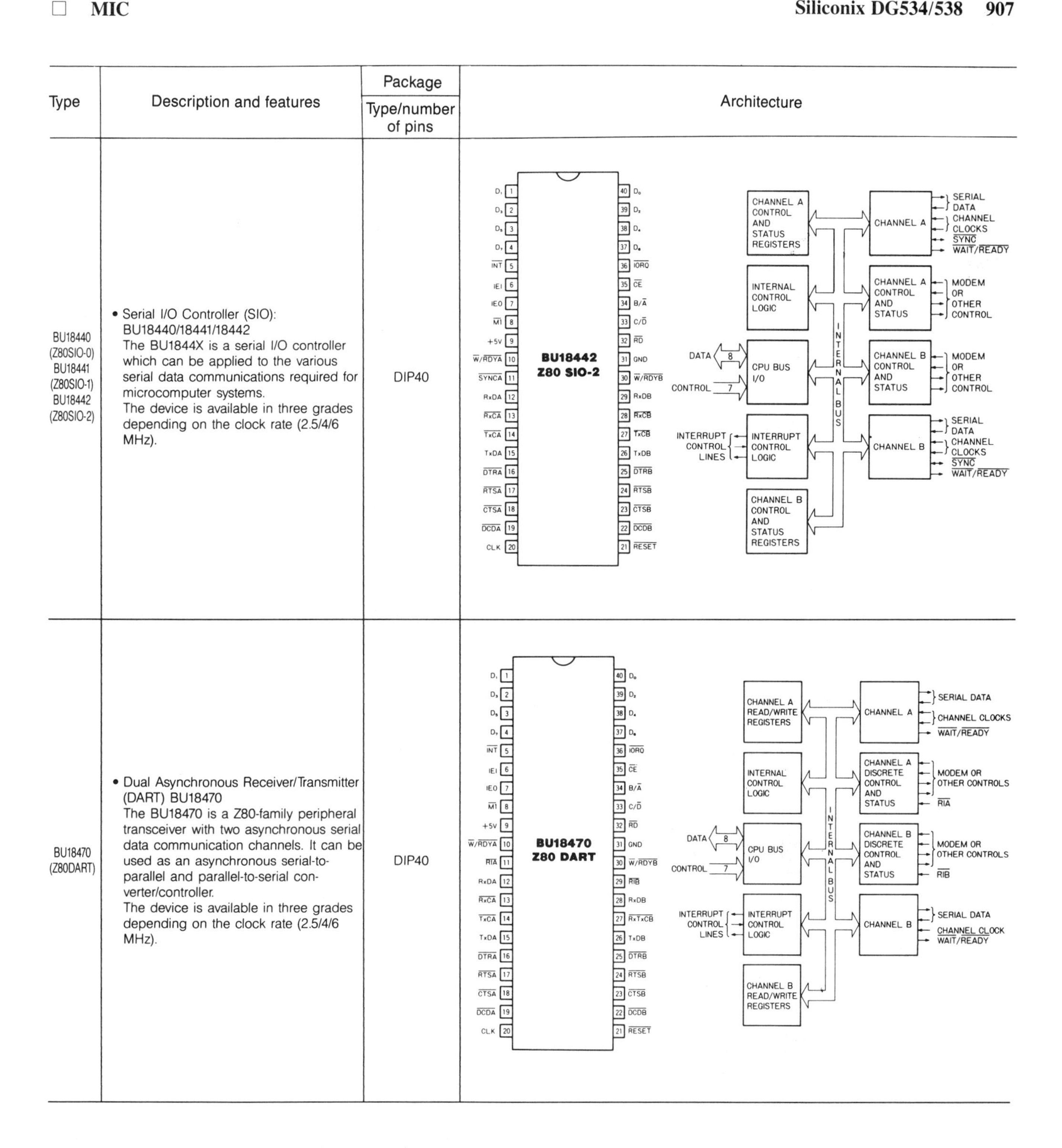

Type	Description and features	Package Type/number of pins	Architecture
BU18440 (Z80SIO-0) BU18441 (Z80SIO-1) BU18442 (Z80SIO-2)	• Serial I/O Controller (SIO): BU18440/18441/18442 The BU1844X is a serial I/O controller which can be applied to the various serial data communications required for microcomputer systems. The device is available in three grades depending on the clock rate (2.5/4/6 MHz).	DIP40	BU18442 Z80 SIO-2
BU18470 (Z80DART)	• Dual Asynchronous Receiver/Transmitter (DART) BU18470 The BU18470 is a Z80-family peripheral transceiver with two asynchronous serial data communication channels. It can be used as an asynchronous serial-to-parallel and parallel-to-serial converter/controller. The device is available in three grades depending on the clock rate (2.5/4/6 MHz).	DIP40	BU18470 Z80 DART

This information is reproduced with permission of Siliconix, Inc.

Siliconix DG534/538

Dual 4-Channel/8-Channel Wideband/Video Multiplexers

- Wide bandwidth (500 MHz)
- Very low crosstalk (−97 dB @ 5 MHz)
- On-board TTL-compatible latches with readback
- Optional negative supply input
- Low $r_{DS(ON)}$ (90 Ω max)
- Single-ended or differential operation
- Improved system bandwidth
- Improved channel off-isolation
- Simplified logic interfacing
- Allows bipolar signal swings
- Reduced insertion loss
- Allows differential signal switching

APPLICATIONS

- Wideband signal routing and multiplexing
- HDTV systems

- μP-controlled systems
- Direct-coupled systems
- ATE systems
- Infrared imaging

The DG534 is a digitally selectable 4-channel or dual 2-channel multiplexer. The DG538 is an 8-channel or dual 4-channel multiplexer. These analog multiplexers/demultiplexers are designed for wideband operation. On-chip TTL-compatible address decoding logic and latches with data readback are included to simplify the interface to a microprocessor data bus. The low on-resistance and low capacitance of these devices make them ideal for wideband data multiplexing and video and audio signal routing in channel selectors and crosspoint arrays.

An optional negative supply pin allows the handling of bipolar signals without dc biasing.

The DG534/538 are built on a D/CMOS process that combines n-channel DMOS switching FETs with low-power CMOS control logic, drivers and latches. The low-capacitance DMOS FETs are in a "T" configuration to achieve extremely high levels of off isolation. Crosstalk is reduced to −97 dB at 5 MHz by including a ground line between each adjacent signal path.

The DG534/DG538 are available in plastic DIP and PLCC packages for operation over the industrial D-suffix (−40 to 85 °C) temperature range. The side-braze DIP is available for military A-suffix (−55 to 125 °C) temperature range operation.

ABSOLUTE MAXIMUM RATINGS

Parameter		Rating
V+ to GND		-0.3 V to +21 V
V+ to V-		-0.3 V to +21 V
V- to GND		-10 V to +0.3 V
V_L		0 V to (V+) + 0.3 V
Digital Inputs		(V-) -0.3 V to (V+) + 0.3 V or 20 mA, whichever occurs first
V_S, V_D		(V-) -0.3 V to (V-) + 21V or 20 mA, whichever occurs first
Current (any terminal) Continuous		20 mA
Current(S or D) Pulsed I ms 10% Duty		40 mA
Storage Temperature	(A Suffix)	-65 to 150°C
	(D Suffix)	-65 to 125°C
Operating Temperature	(A Suffix)	-55 to 150°C
	(D Suffix)	-40 to 85°C

Power Dissipation (Package)

Package	Power
Plastic DIP**	625 mW
Side Braze DIP***	1200 mW
Quad J Lead Plastic****	450 mW

*All leads welded or soldered to PC board.
**Derate 8.3 mW/°C above 75°C.
***Derate 16 mW/°C above 75°C.
****Derate 6 mW/°C above 75°C.

PIN CONFIGURATIONS

Dual-In-Line Package
Top View

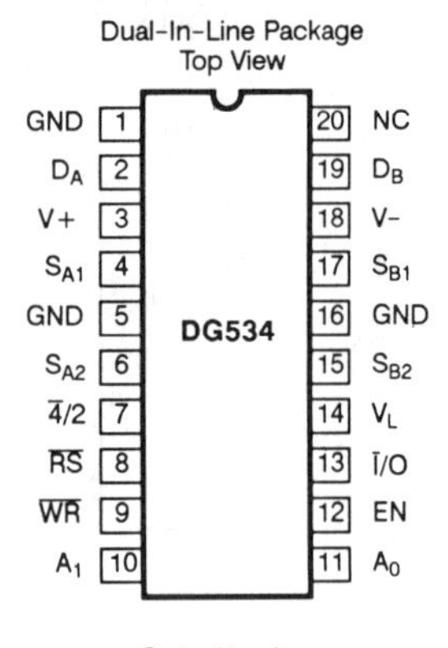

Order Numbers:
Side Braze: DG534AP, DG534AP/883
Plastic: DG534DJ

PLCC Package
Top View

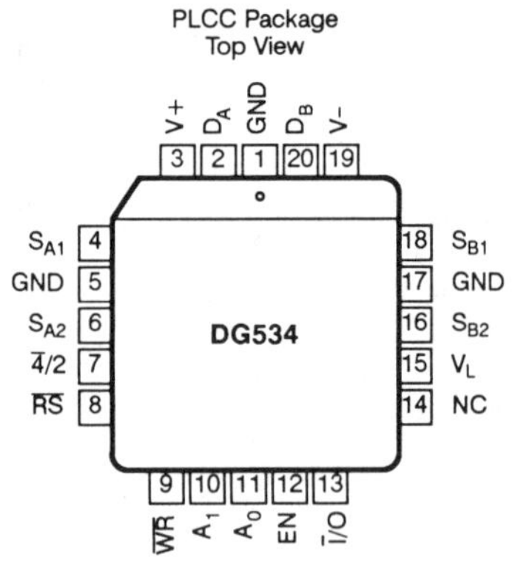

Order Number:
DG534DN

Dual-In-Line Package
Top View

Pin	Name	Pin	Name
1	GND	28	D_B
2	D_A	27	V-
3	V+	26	S_{B1}
4	S_{A1}	25	GND
5	GND	24	S_{B2}
6	S_{A2}	23	GND
7	GND	22	S_{B3}
8	S_{A3}	21	GND
9	GND	20	S_{B4}
10	S_{A4}	19	V_L
11	$\overline{8}$/4	18	$\overline{I}$/O
12	$\overline{RS}$	17	EN
13	$\overline{WR}$	16	A_0
14	A_2	15	A_1

DG538

Order Numbers:
Side Braze: DG538AP, DG538AP/883
Plastic: DG538DJ

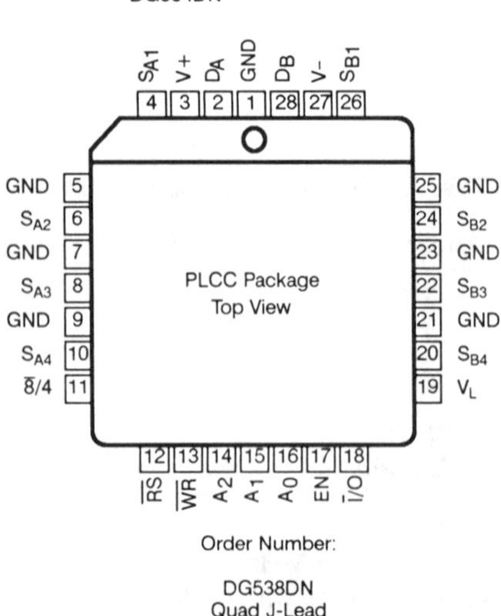

Order Number:
DG538DN
Quad J-Lead
Plastic Chip Carrier

FUNCTIONAL BLOCK DIAGRAMS AND TRUTH TABLES

All input switches are "T" switches.

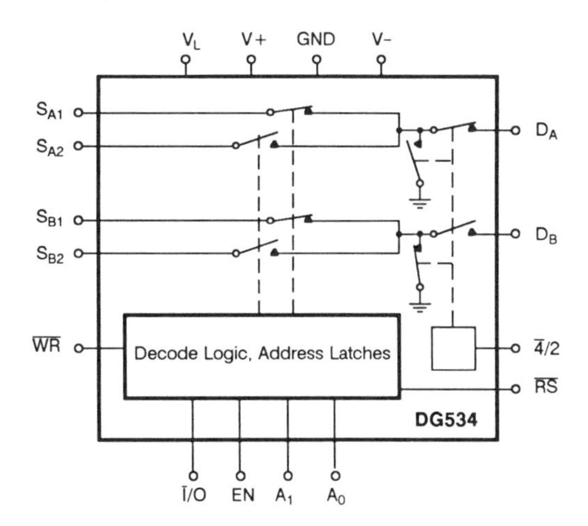

$\bar{I}/O$	A_0	A_1	EN	$\overline{WR}$	$\overline{RS}$	$\overline{4}/2$*	ON Switch		
X	X	X	X	↑	1	1	Maintains previous state		
X	X	X	X	X	0	X	None (latches cleared)		
X	X	X	0	0	1	X	None		Latches Transparent
0	0	0	1	0	1	0	S_{A1}	DA & DB may be connected externally	
0	0	1	1	0	1	0	S_{A2}		
0	1	0	1	0	1	0	S_{B1}		
0	1	1	1	0	1	0	S_{B2}		
0	X	0	1	0	1	1	S_{B1} & S_{B1}		
0	X	1	1	0	1	1	S_{B2} & S_{B2}		
1	Note 2		1	1	1	Note 1			

Logic "1": $V_{AH} \geq 2.0$ V
Logic "0": $V_{AL} \geq 8.0$ V

*D_A and D_B must be connected together externally if single-ended operation is desired.

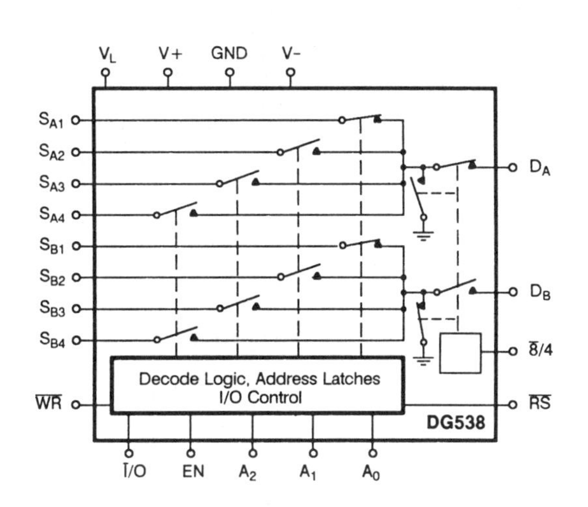

$\bar{I}/O$	A_0	A_1	A_0	EN	$\overline{WR}$	$\overline{RS}$	$\overline{8}/4$*	ON Switch		
X	X	X	X	X	↑	1	1	Maintains previous state		
X	X	X	X	X	X	0	X	None (latches cleared)		
X	X	X	X	0	0	1	X	None		Latches Transparent
0	0	0	0	1	0	1	0	S_{A1}	DA & DB should be connected externally	
0	0	0	1	1	0	1	0	S_{A2}		
0	0	1	0	1	0	1	0	S_{A3}		
0	0	1	1	1	0	1	0	S_{A4}		
0	1	0	0	1	0	1	0	S_{B1}		
0	1	0	1	1	0	1	0	S_{B2}		
0	1	1	0	1	0	1	0	S_{B3}		
0	1	1	1	1	0	1	0	S_{B4}		
0	X	0	0	1	0	1	1	S_{B1} & S_{B1}		
0	X	0	1	1	0	1	1	S_{B2} & S_{B2}		
0	X	1	0	1	0	1	1	S_{B3} & S_{B3}		
0	X	1	1	1	0	1	1	S_{B4} & S_{B4}		
1	Note 2			1	1	1	Note 1			

*D_A and D_B must be connected externally if single-ended operation is desired.

NOTES:
1. $\overline{4}/2$ or $\overline{8}/4$ can be either H or L but should not change during these operations.
2. With $\bar{I}/O$ high, A_n pins become outputs and reflect the contents of the latches. See timing diagrams for more detail.

TYPICAL CHARACTERISTICS

Supply Currents vs. Temperature

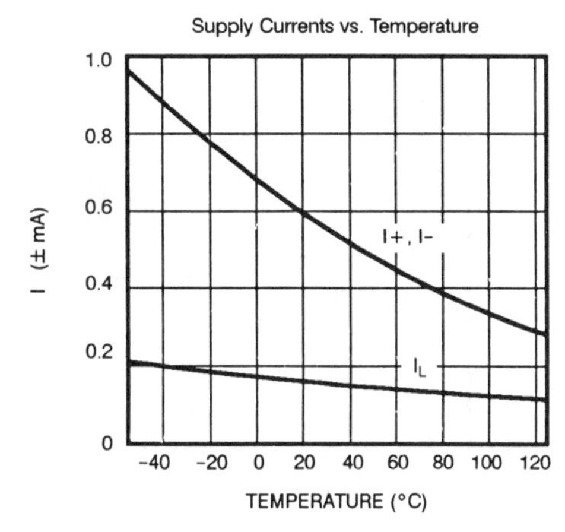

Leakage vs. Temperature

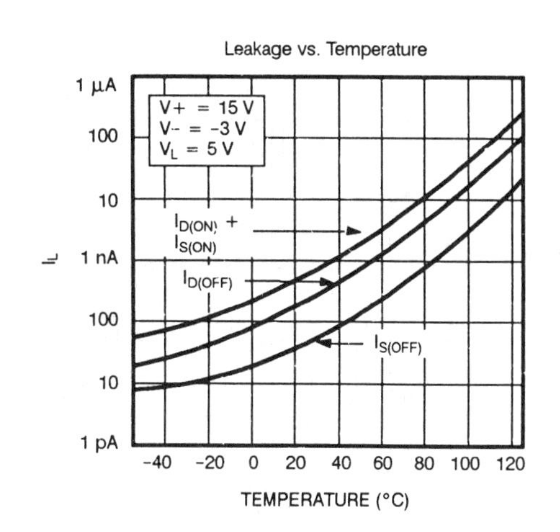

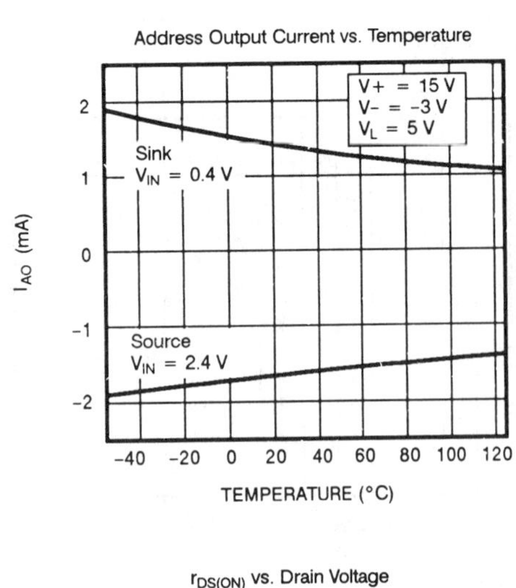
Address Output Current vs. Temperature
V+ = 15 V
V- = -3 V
V_L = 5 V
Sink
V_{IN} = 0.4 V
Source
V_{IN} = 2.4 V
I_{AO} (mA)
TEMPERATURE (°C)

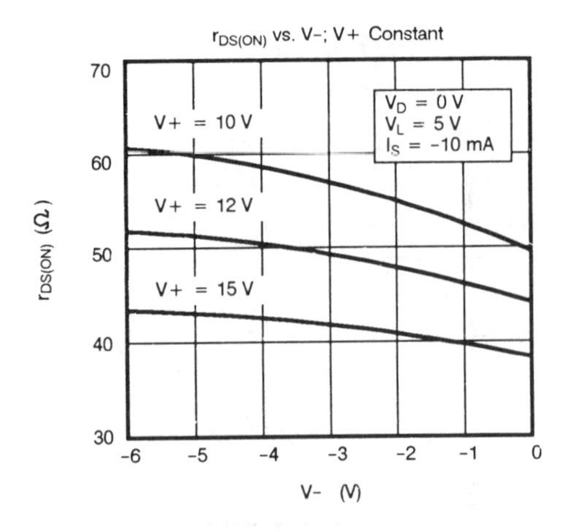
$r_{DS(ON)}$ vs. V−; V+ Constant
V_D = 0 V
V_L = 5 V
I_S = −10 mA
V+ = 10 V
V+ = 12 V
V+ = 15 V
$r_{DS(ON)}$ (Ω)
V− (V)

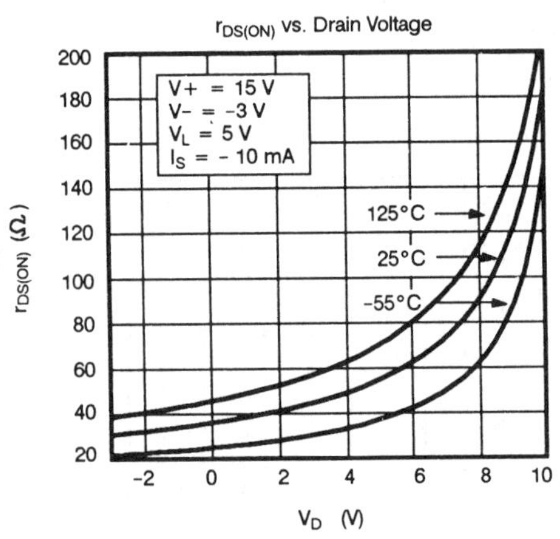
$r_{DS(ON)}$ vs. Drain Voltage
V+ = 15 V
V− = −3 V
V_L = 5 V
I_S = − 10 mA
125°C
25°C
−55°C
$r_{DS(ON)}$ (Ω)
V_D (V)

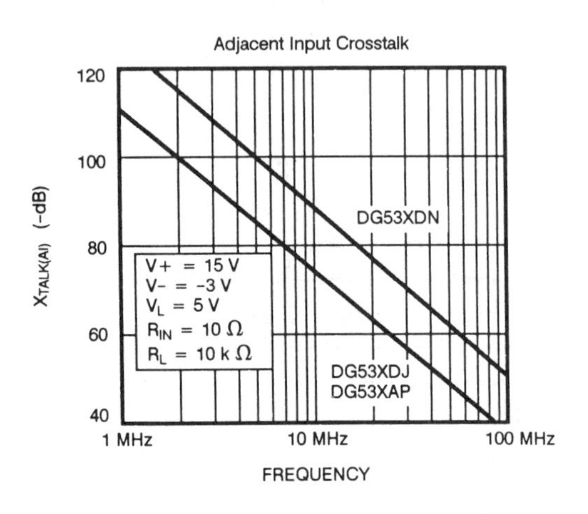
Adjacent Input Crosstalk
DG53XDN
V+ = 15 V
V− = −3 V
V_L = 5 V
R_{IN} = 10 Ω
R_L = 10 k Ω
DG53XDJ
DG53XAP
$X_{TALK(AI)}$ (−dB)
1 MHz
10 MHz
100 MHz
FREQUENCY

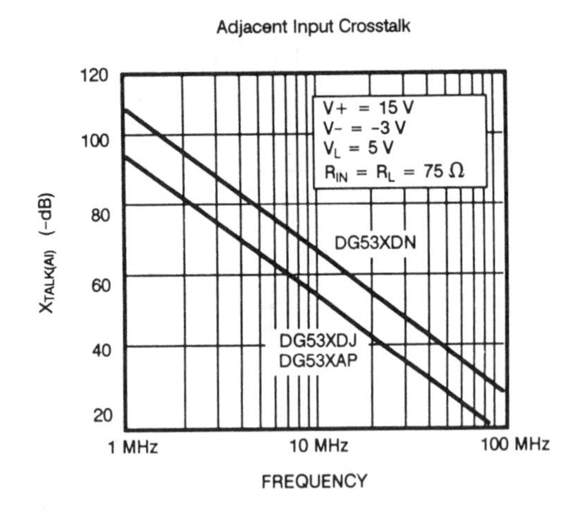
Adjacent Input Crosstalk
V+ = 15 V
V− = −3 V
V_L = 5 V
R_{IN} = R_L = 75 Ω
DG53XDN
DG53XDJ
DG53XAP
$X_{TALK(AI)}$ (−dB)
1 MHz
10 MHz
100 MHz
FREQUENCY

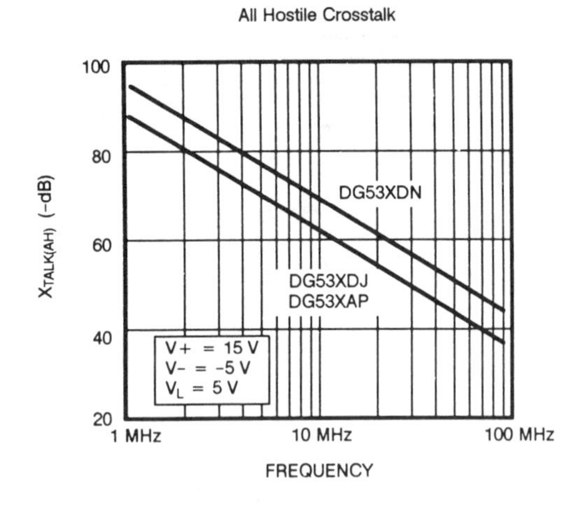
All Hostile Crosstalk
DG53XDN
DG53XDJ
DG53XAP
V+ = 15 V
V− = −5 V
V_L = 5 V
$X_{TALK(AH)}$ (−dB)
1 MHz
10 MHz
100 MHz
FREQUENCY

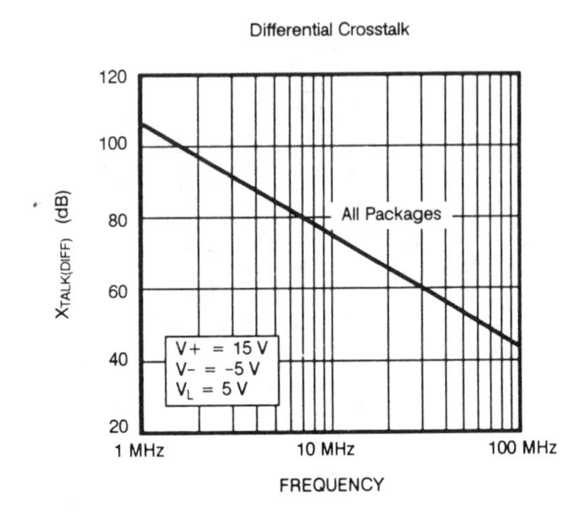
Differential Crosstalk
All Packages
V+ = 15 V
V− = −5 V
V_L = 5 V
$X_{TALK(DIFF)}$ (dB)
1 MHz
10 MHz
100 MHz
FREQUENCY

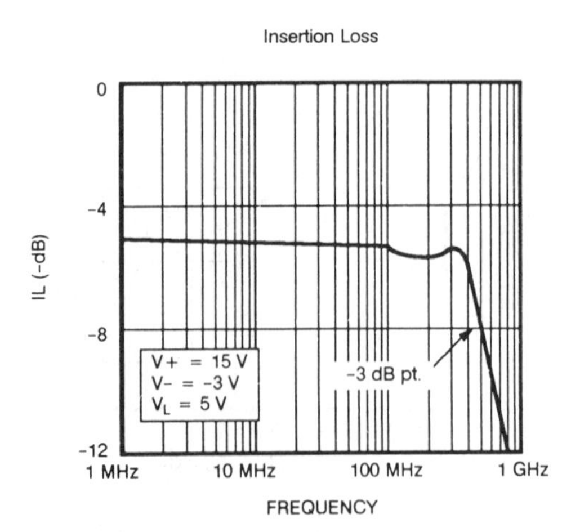
Insertion Loss
V+ = 15 V
V− = −3 V
V_L = 5 V
−3 dB pt.
IL (−dB)
1 MHz
10 MHz
100 MHz
1 GHz
FREQUENCY

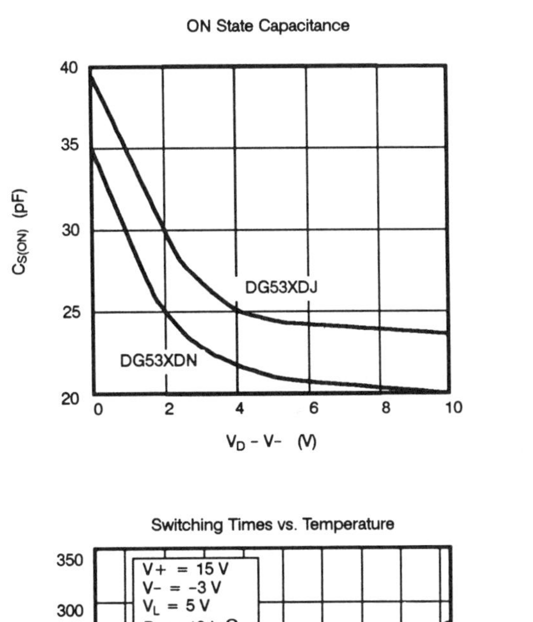

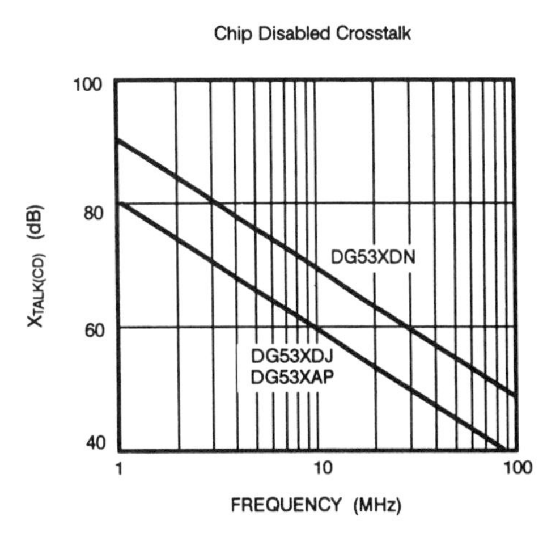

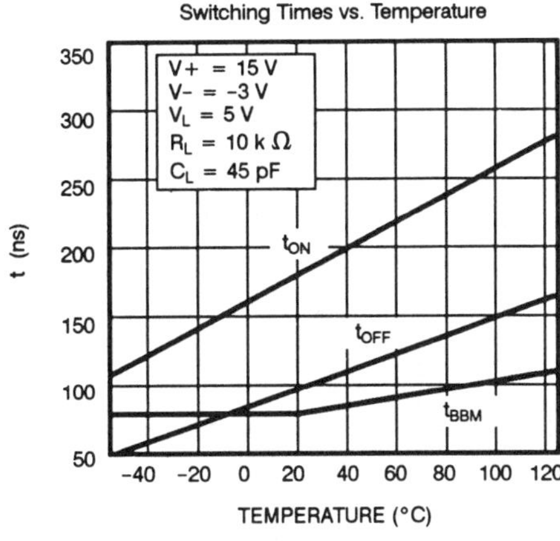

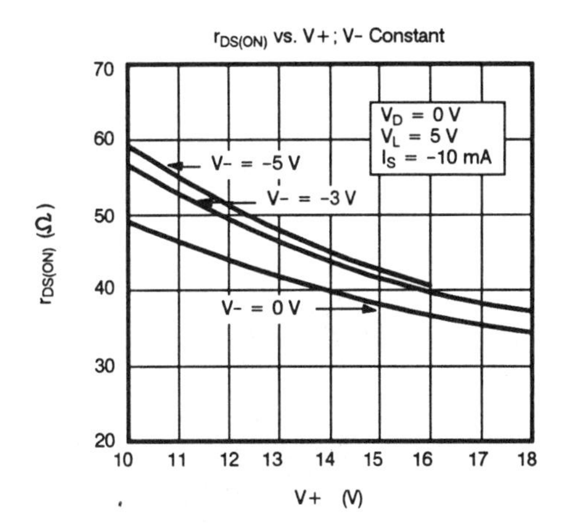

INPUT TIMING REQUIREMENTS

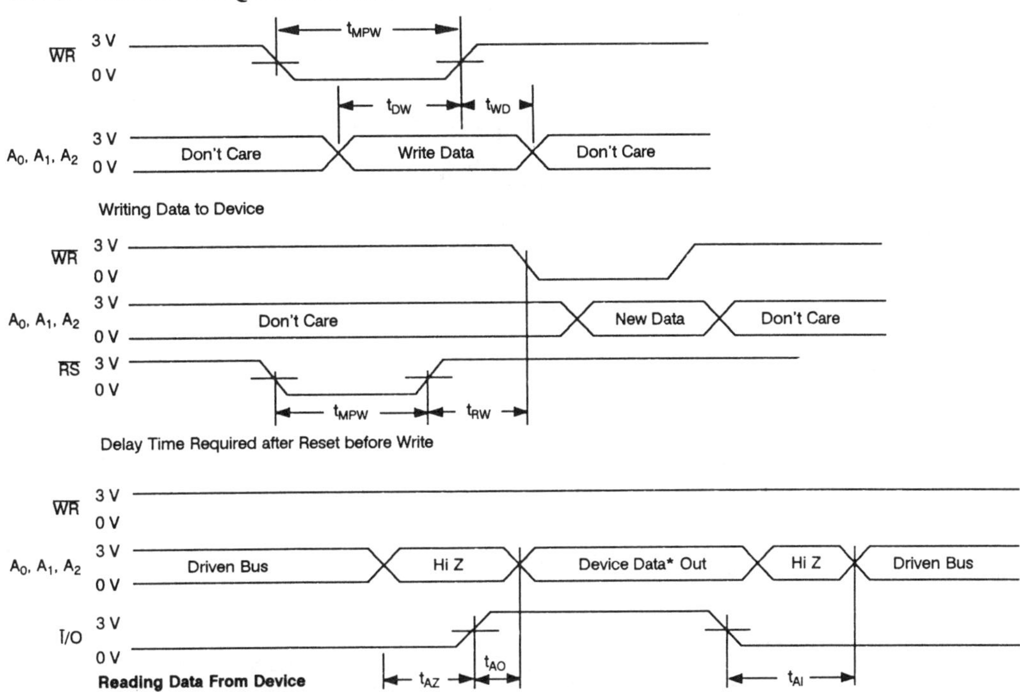

*Enable must be latched "High" to read back data, otherwise BUS is high Z. V− ≤ −3 V required for readback functionality.

CONTROL CIRCUITRY

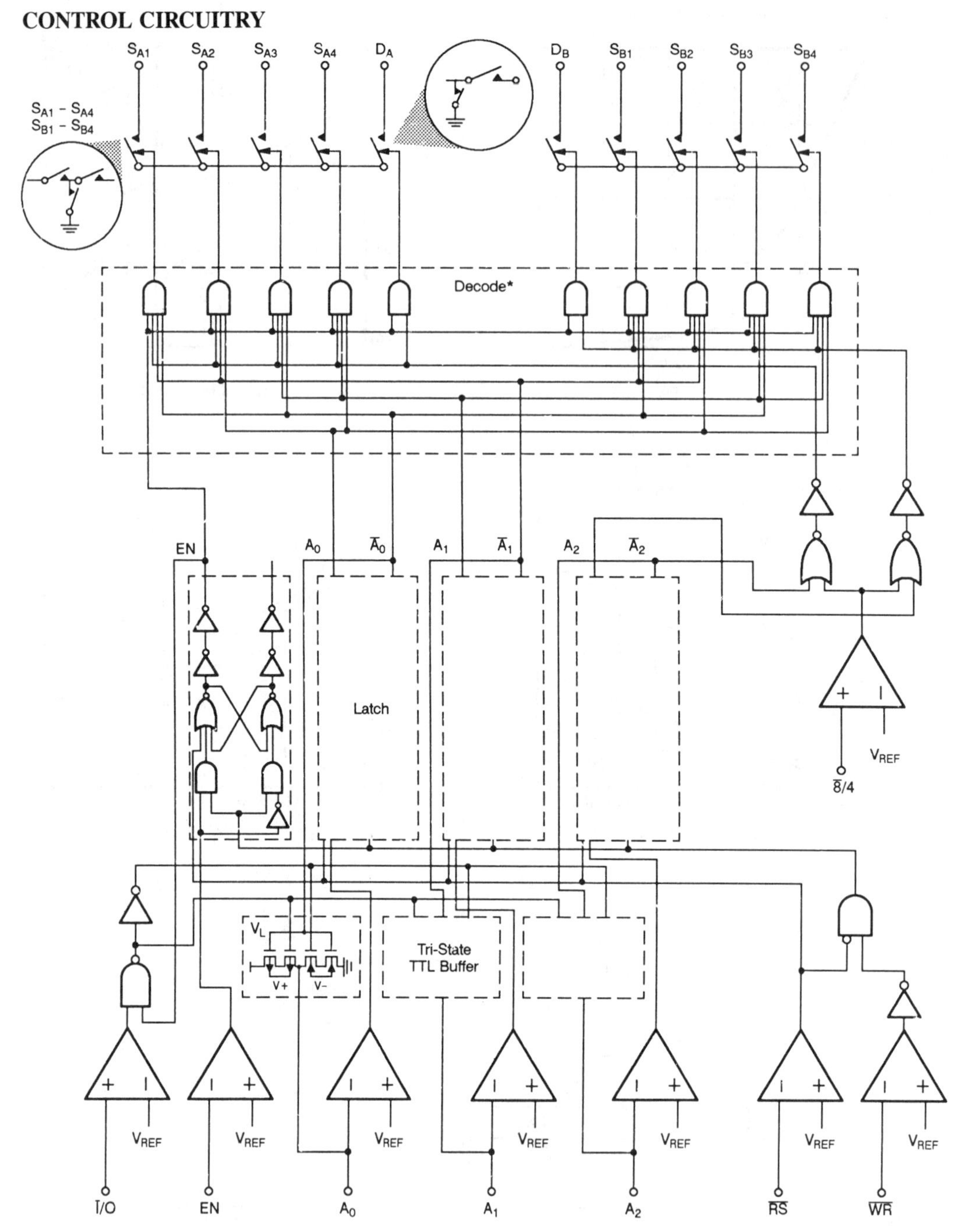

Note: V_{REF} is internally generated from V_L.

* Decode section includes delay circuitry in AND gating to ensure proper break-before-make operation.

OPERATING VOLTAGE RANGE

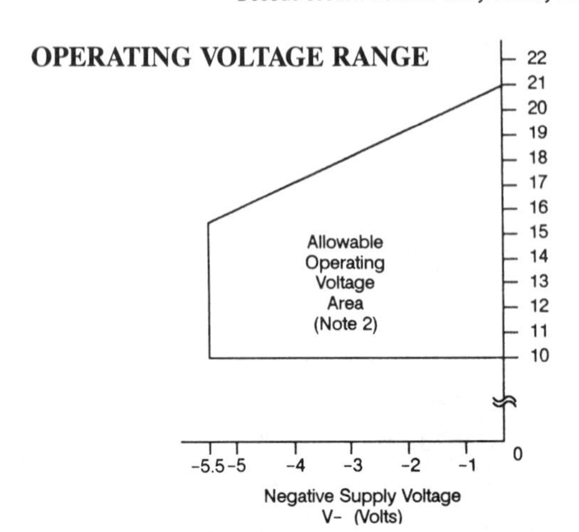

Positive Supply Voltage V+ (Volts)

Notes:

1. Both V+ and V− must have decoupling capacitors mounted as close as possible to the device pins. Typical decoupling capacitors would be 10 µF tantalum bead in parallel with 100 nF ceramic disc.
2. Production tested with V+ = 15 V and V− = −3.0 V.
3. At V_L = 5 V ±10%, 0.8/2.0 TTL compatibility is maintained over the entire operating voltage range.

Siliconix AN88-2
Microprocessor-Compatible Multiplexers

For many new communications systems, such as ISDN (Integrated Services Digital Network), cable TV, and local area networks (LANs), traditional switching techniques have become inadequate. To meet the demands of these applications, semiconductor switching devices must now handle wider bandwidths and offer more on-chip features to achieve low chip-count solutions. Higher integration, smaller packages, easier device paralleling/combining, and improved dynamic performance are essential features for designing large-capacity switching systems.

Analog video information is frequently digitized for processing in frame grabbers, TV standard converter (e.g., NTSC to PAL), time-base correction, special effects, or merely as a means of reducing noise levels and enhancing resolution. However, the price paid for the advantages of the digital technique is the substantially wider bandwidth occupied by the digitized signal. Thus, in a typical 8-bit conversion, the sampling rate must be at least three times the chrominance subcarrier frequency of 4.43 MHz: that is, 13.3 MHz. Thus, the bit rate is $8 \times 13.3 = 106.44$ Mbps. This bandwidth requirement precludes the use of a majority of components and switching techniques commonly employed in video systems.

Video switching applications, such as high-definition TV, digital video equipment, and broadcast studio switches, have forced improvements in semiconductor switch performance. The Siliconix DG534 and DG538 are members of a fast-growing family of multiplexers/demultiplexer devices with performance characteristics optimized for wideband switching applications. This application note presents the benefits of the DG534 and DG538 in a diverse range of wideband switching applications, high-lighting the devices' performance features and providing useful circuit-design techniques.

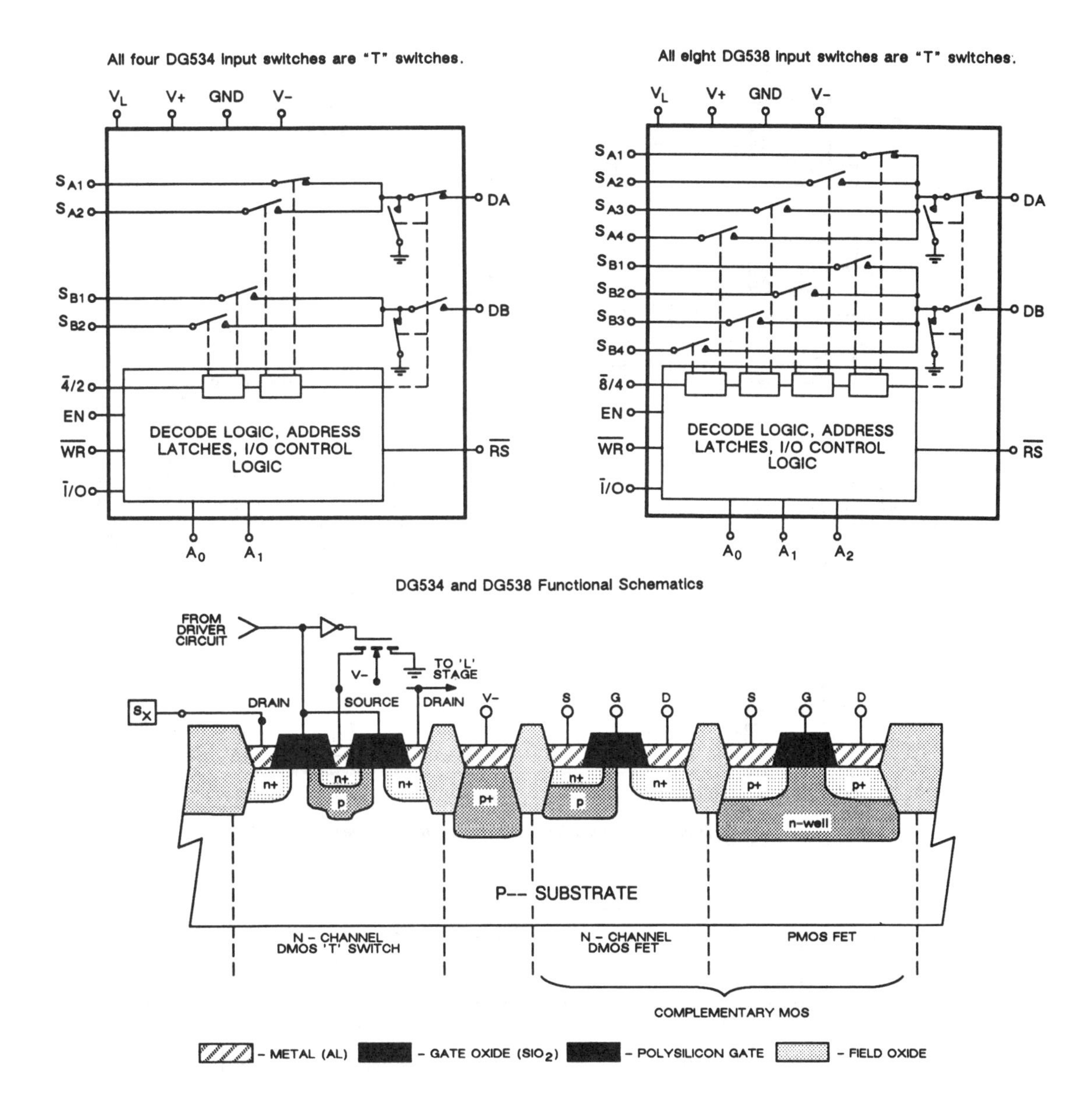

DG534 and DG538 Functional Schematics

Cross Section of the D/CMOS Process

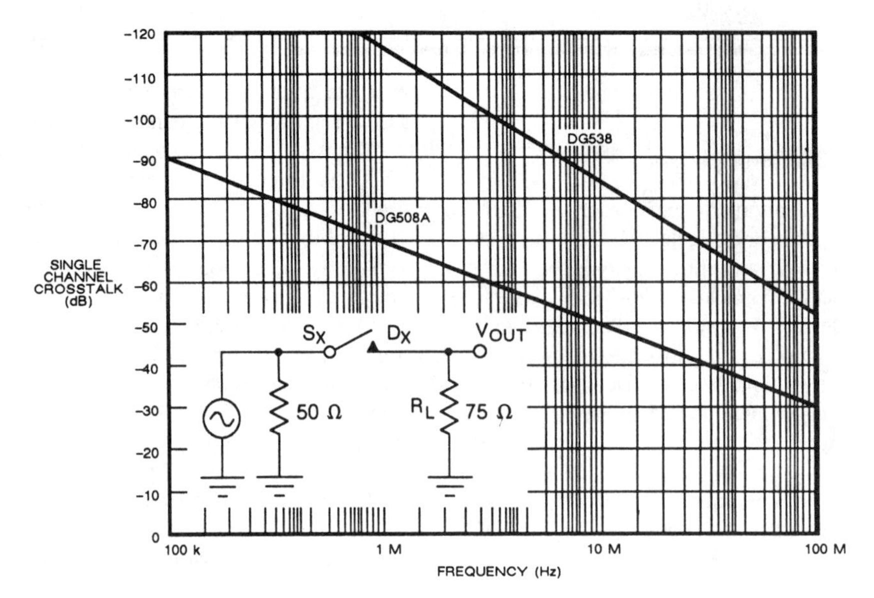

DG538/DG508A Single-Channel Crosstalk vs. Frequency

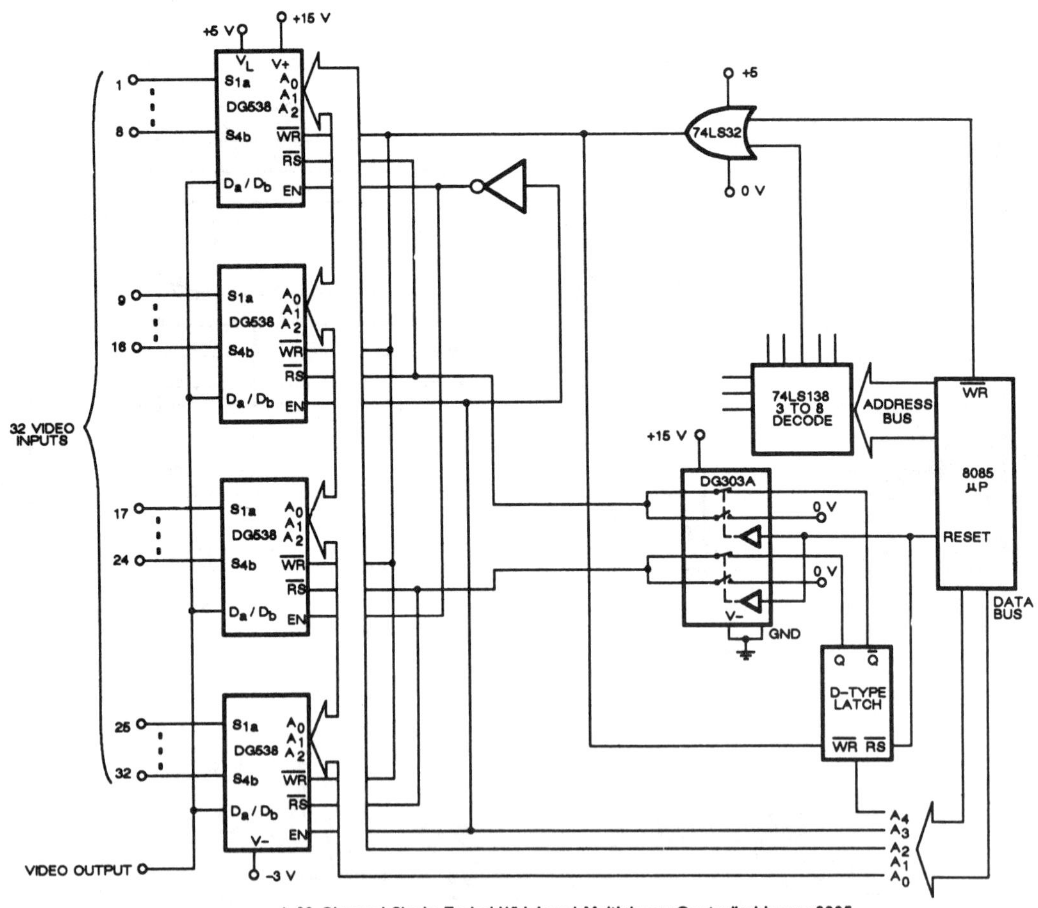

A 32-Channel Single-Ended Wideband Multiplexer Controlled by an 8085

Microprocessor Timing Compatibility

PARAMETER			DG538/4	8085A	8085A-2	Z80	6800
t_{WW}	Width of Control Low ($\overline{WR}$)	(ns)	200 (min)	400 (min)	230 (min)	360 (min)	470 (min)
t_{DW}	Data Valid to Trailing Edge of Write	(ns)	100 (min)	420 (min)	230 (min)	200 (min)	575 (min)
t_{WD}	Data Valid to Trailing Edge of Write	(ns)	50 (min)	100 (min)	60 (min)	100 (min)	150 (min)

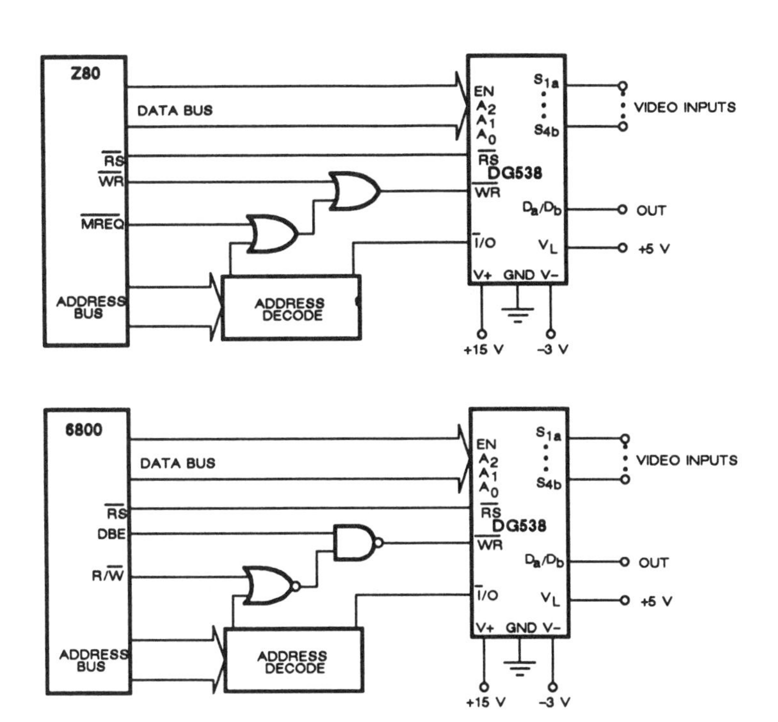

Other Microprocessor Interfaces

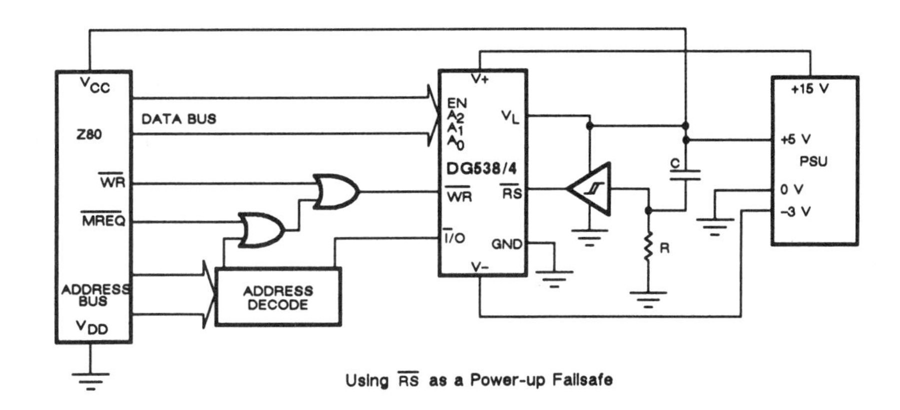

Using $\overline{RS}$ as a Power-up Failsafe

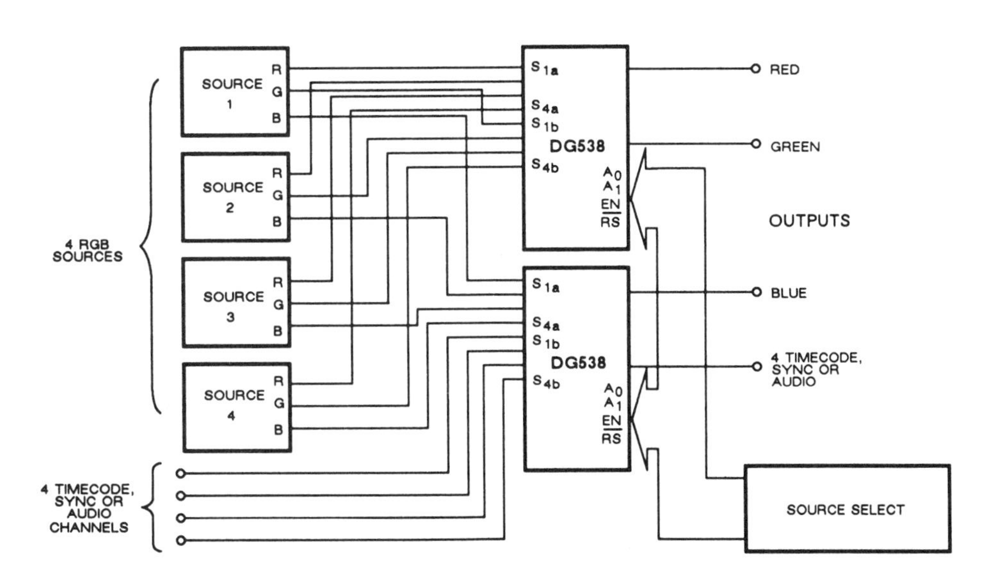

An RGB Plus Timecode, Sync, or Audio Switching System

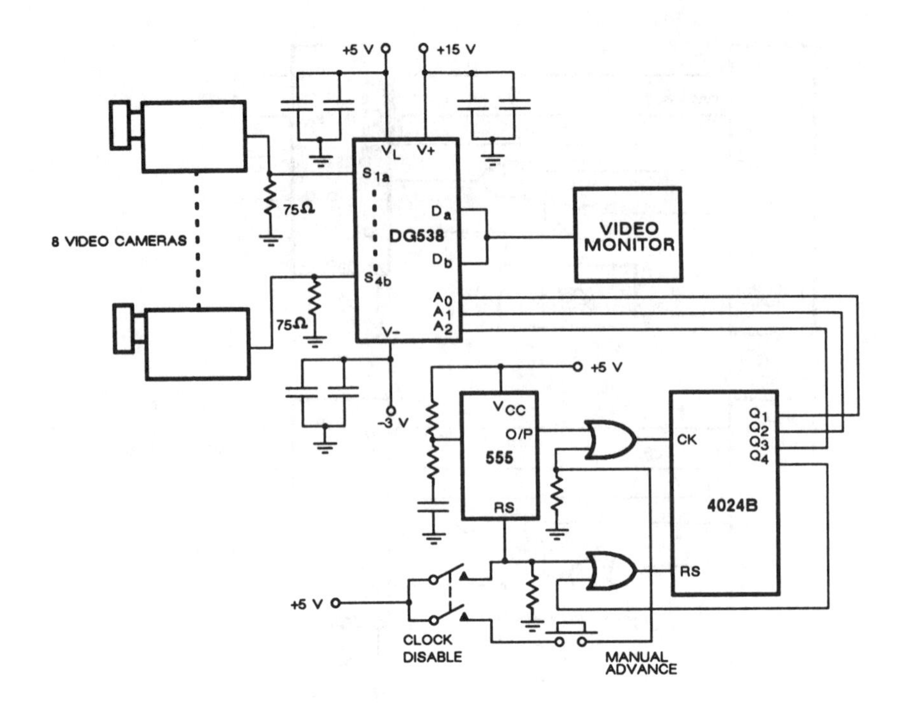

Basic Closed-Circuit TV System

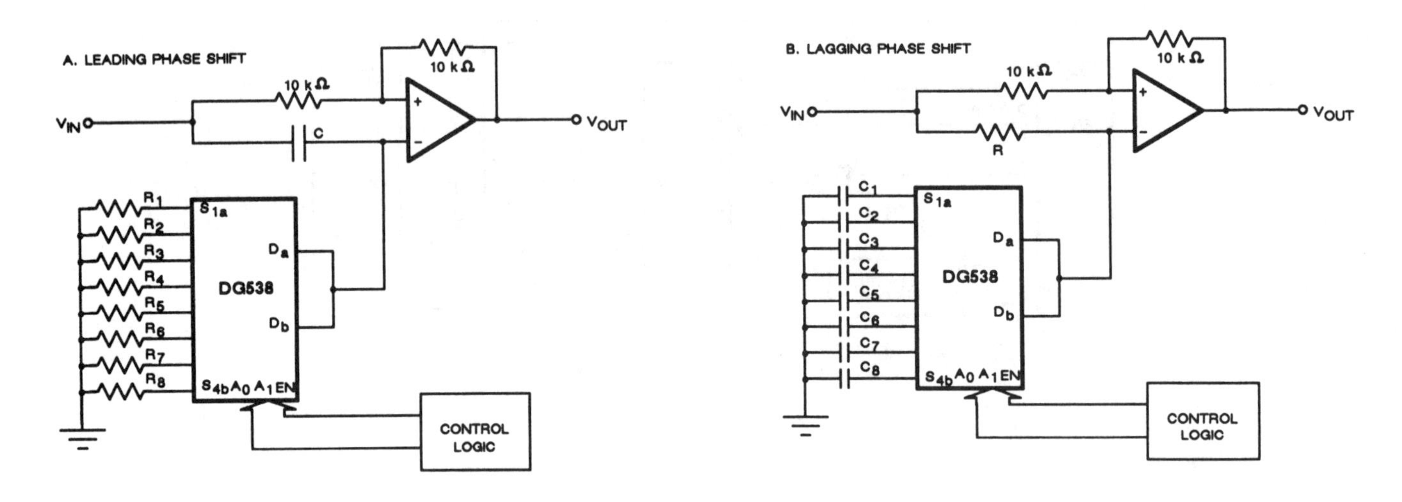

A Digitally Controlled Phase Shifter

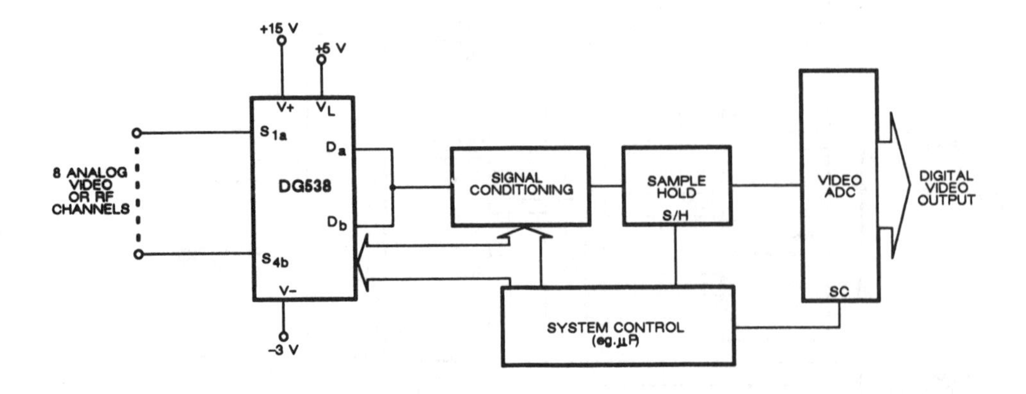

A Basic Multichannel Video/rf Processing Circuit

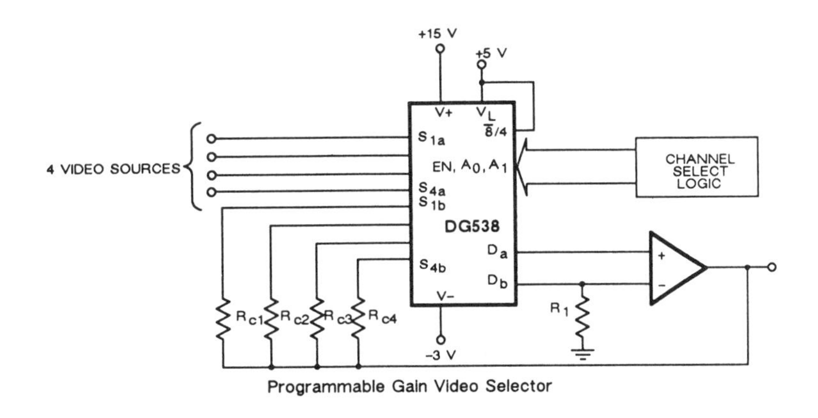

Programmable Gain Video Selector

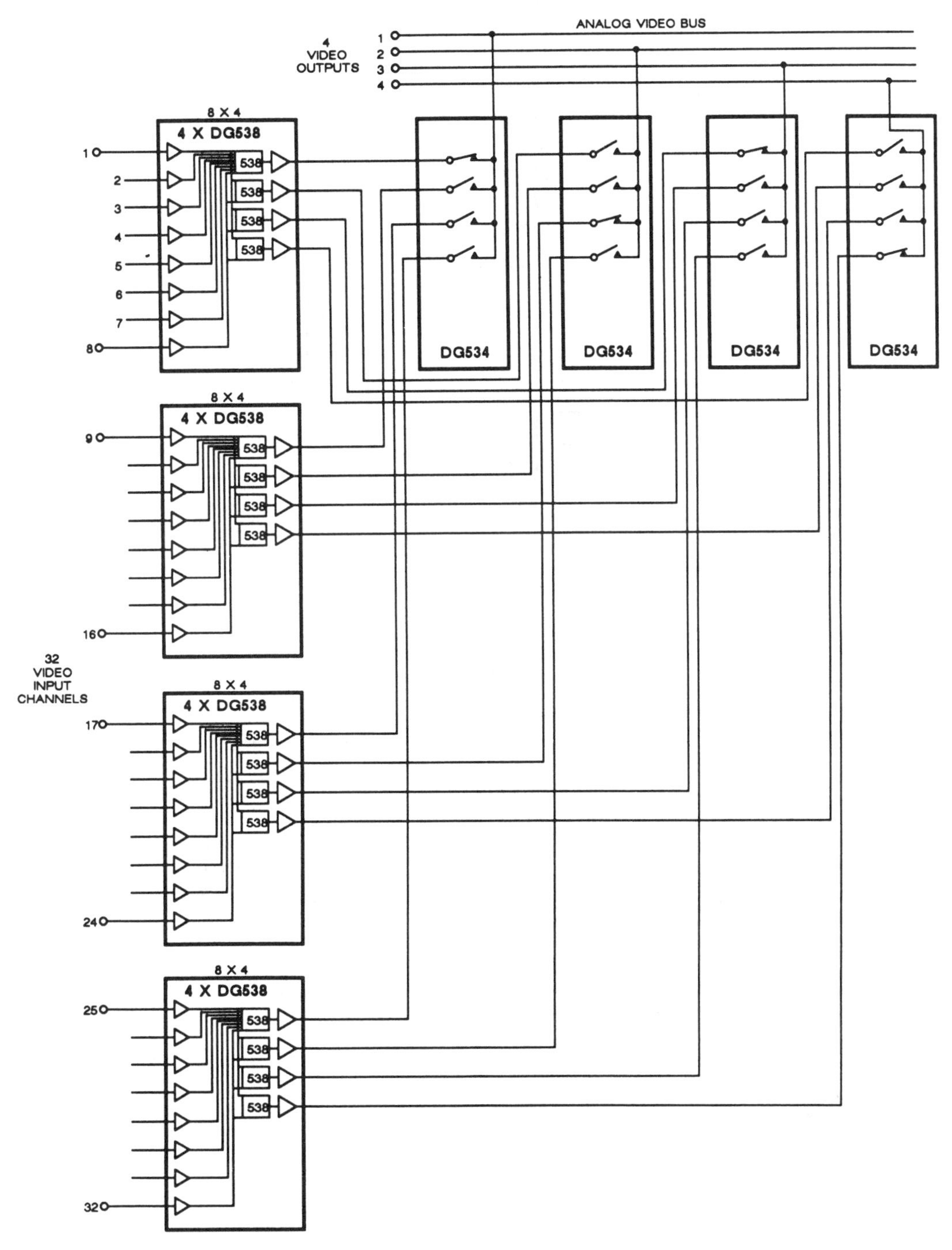

32 X 4 Crosspoint with Bus Isolation

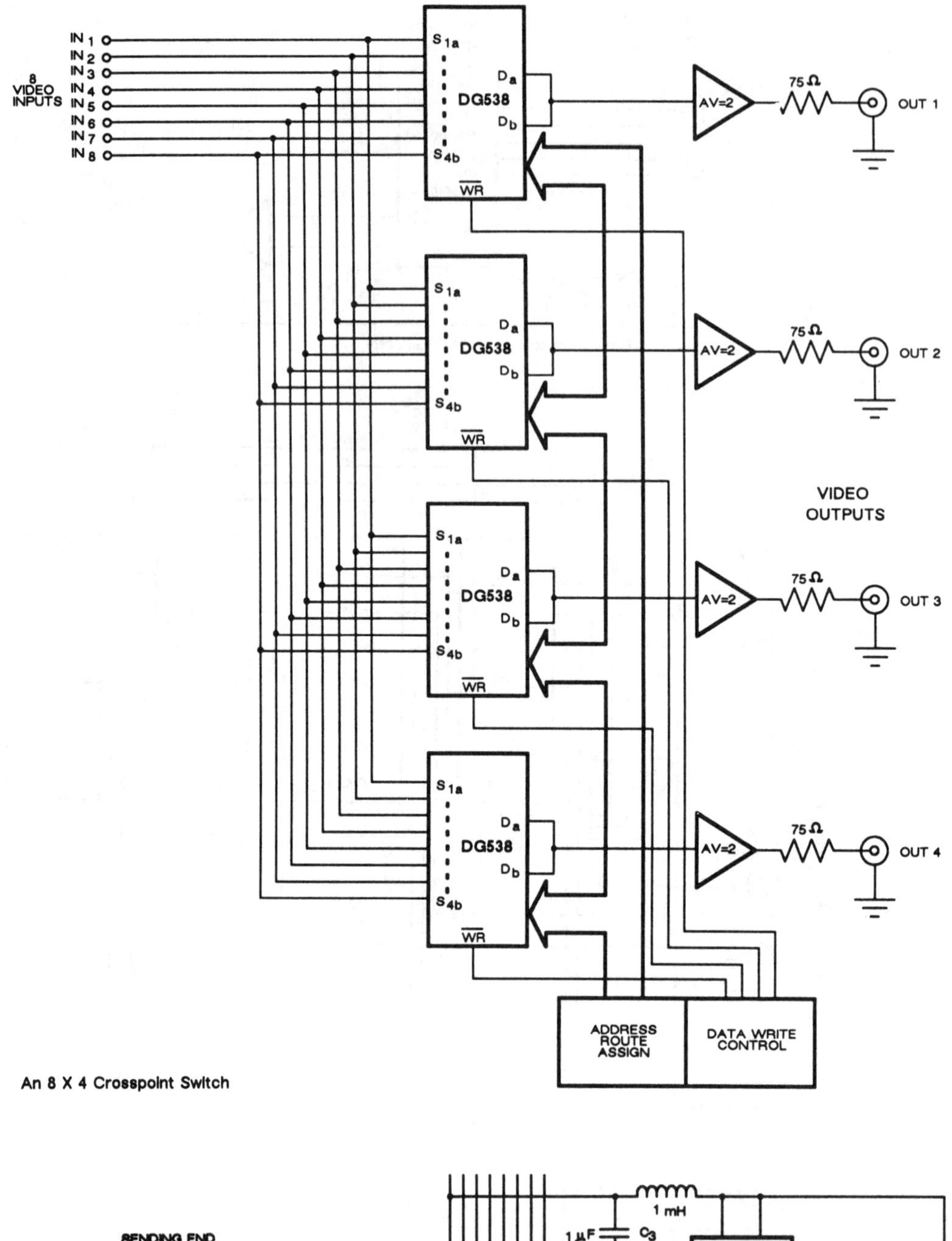

An 8 X 4 Crosspoint Switch

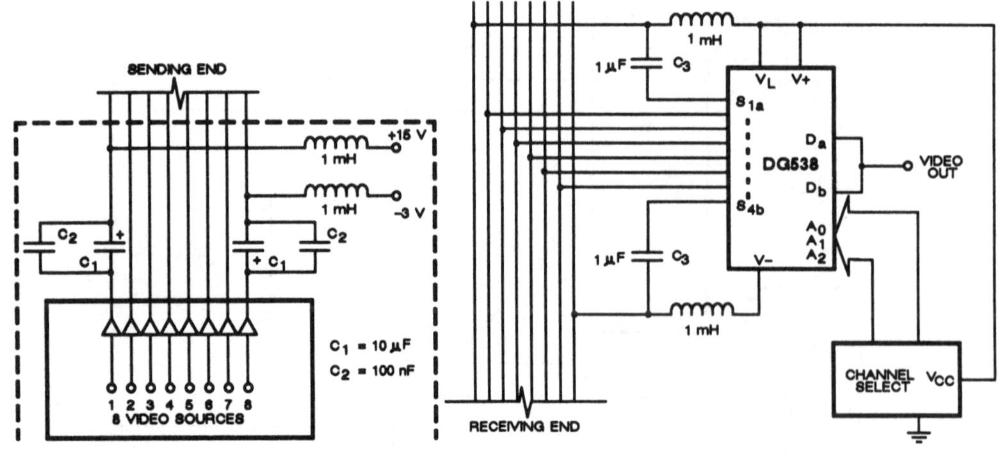

Phantom-Powered Remote Video Switch

Siliconix DG535/536

16-Channel Wideband/Video Multiplexers

- −100 dB crosstalk at 5 MHz
- 300 MHz bandwidth
- 4 pF (max) input and 12 pF output capacitance
- Low power (75 μW)
- $r_{DS(ON)}$ (90 Ω max)
- μP interface latches
- Improved off isolation
- Reduced insertion loss at high frequencies
- Reduced input buffer requirements
- Minimizes system power
- Reduced noise
- Simplifies bus interface

APPLICATIONS

- Video switching/routing
- High speed data routing
- Wideband signal multiplexing
- Precision data acquisition
- Crosspoint arrays
- FLIR systems

The DG535/536 are 16-channel multiplexers designed for routing one of 16 wideband analog or digital input signals to a single output. The feature low input and output capacitance, low on resistance, and n-channel DMOS "T" switches, resulting in wide bandwidth, low crosstalk, and high "off" isolation. In the on state, the switches pass signals in either direction, allowing the DG535/536 to be used as multiplexers or demultiplexers.

On-chip data latches and decode logic simplify microprocessor interface. Chip select and enable inputs simplify addressing in large matrices. Single-supply operation and a low 75 μW power dissipation allows operation in battery-powered systems and vastly reduces power supply requirements.

These devices are built on a D/CMOS process that creates low-capacitance DMOS FETs on the same substrate with dense, high-speed low-power CMOS.

The DG535 is available in the plastic 28-lead DIP for the industrial D-suffix (−40 to 85°C), and the 28-lead side-braze DIP for military A-suffix (−55 to 125°C) temperature range operation.

The DG536 is available in surface-mount plastic PLCC-44 for industrial D-suffix (−40 to 85°C), and the Cerquad hermetic package for military A-suffix (−55 to 125°C) temperature ranges.

ABSOLUTE MAXIMUM RATINGS

V+ to GND .... -0.3 V to +18 V
Digital Inputs .... (GND - 0.3 V) to (V+ plus 2 V) or 20 mA, whichever occurs first
V_S, V_D .... (GND - 0.3 V) to V+ plus 2 V) or 20 mA, whichever occurs first
Current (any terminal) Continuous .... 20 mA
Current (S or D) Pulsed 1 ms 10% duty cycle .... 40 mA
Storage Temperature (A Suffix) .... -65 to 150°C
(D Suffix) .... -65 to 125°C
Operating Temperature (A Suffix) .... -55 to 125°C
(D Suffix) .... -40 to 85°C

Power Dissipation (Package)*
44-Pin J Lead Cerquad** .... 825 mW
44-Pin J Lead Plastic*** .... 450 mW
28-Pin Plastic DIP**** .... 625 mW
28-Pin Sidebraze DIP***** .... 1200 mW

*All leads welded or soldered to PC board.
**Derate 11 mW/°C above 75°C
***Derate 6 mW/°C above 75°C
****Derate 8.6 mW/°C above 75°C
*****Derate 16 mW/°C above 75°C

PIN CONFIGURATIONS

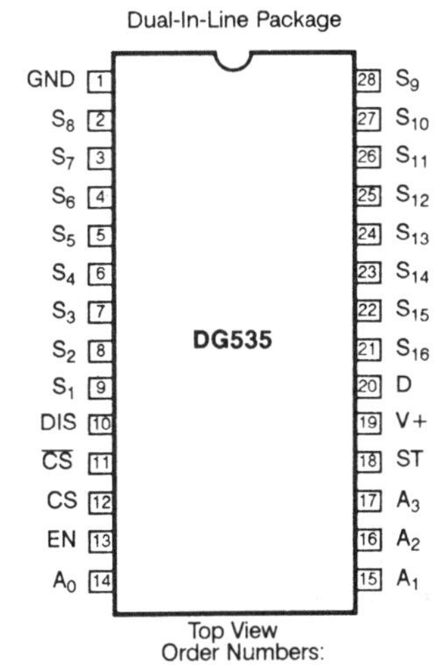

Top View
Order Numbers:
Side Braze: DG535AP, DG535AP/883
Plastic: DG535DJ

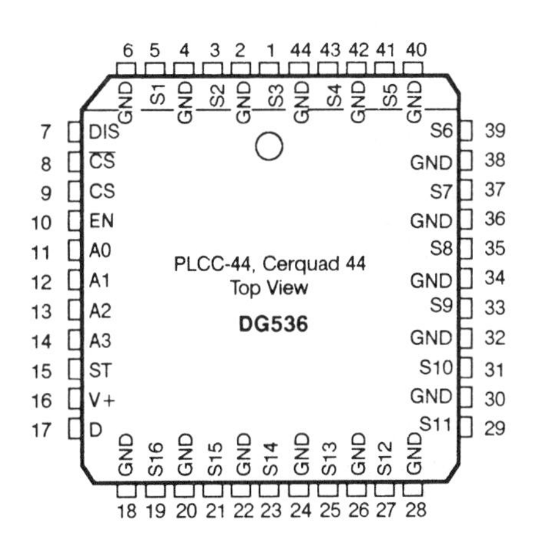

Order Numbers:
Plastic: DG536DN
Cerquad: DG536AM/883

FUNCTIONAL BLOCK DIAGRAMS AND TRUTH TABLES

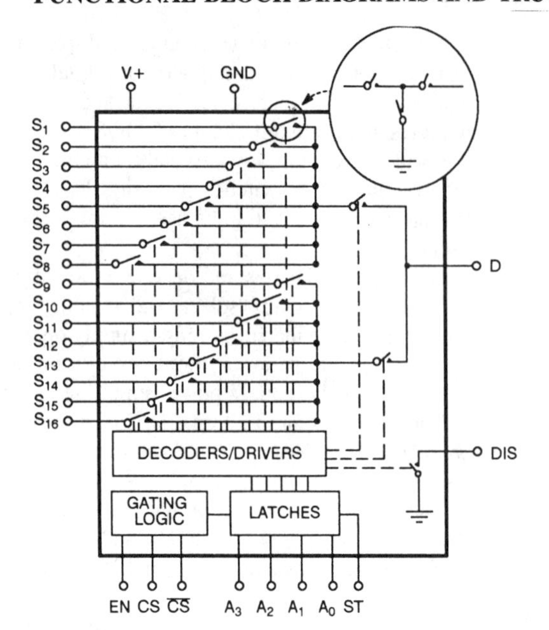

EN	CS	$\overline{CS}$	ST	A_3	A_2	A_1	A_0	Channel Selected	Disable
0	X	X	1	X	X	X	X	NONE	HIGH Z
X	0	X							
X	X	1							
1	1	0	1	0	0	0	0	S1	LOW Z
				0	0	0	1	S2	
				0	0	1	0	S3	
				0	0	1	1	S4	
				0	1	0	0	S5	
				0	1	0	1	S6	
				0	1	1	0	S7	
				0	1	1	1	S8	
				1	0	0	0	S9	
				1	0	0	1	S10	
				1	0	1	0	S11	
				1	0	1	1	S12	
				1	1	0	0	S13	
				1	1	0	1	S14	
				1	1	1	0	S15	
				1	1	1	1	S16	
X	X	X	0	X	X	X	X	Maintains previous switch condition	HIGH Z or LOW Z

Logic "1": $V_{AH} \geq 10.5$ V
Logic "0" $V_{AL} \leq 4.5$ V

1. LOW Z, HIGH Z = Impedance of Disable Output to GND. Disable output sinks current when any channel is selected.
2. Strobe input (ST) is level triggered.

TYPICAL CHARACTERISTICS

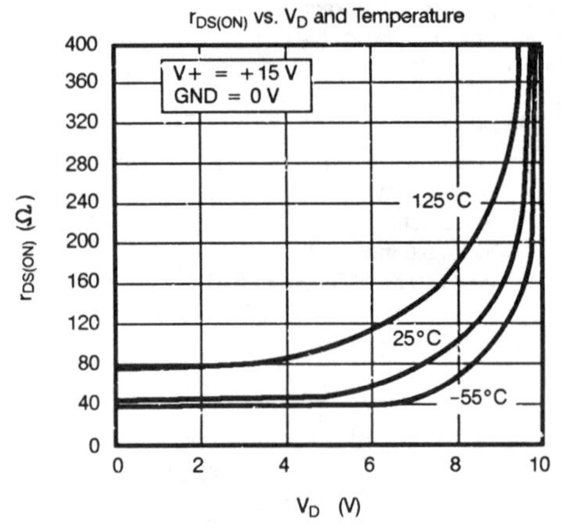

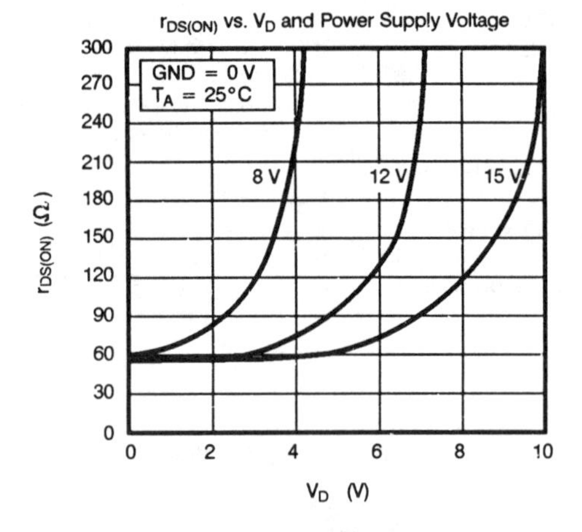

Logic Input Switching Threshold vs. Supply Voltage (V+)

GND = 0 V
T_A = 25°C

V_{TH} (V)

V+ (V)

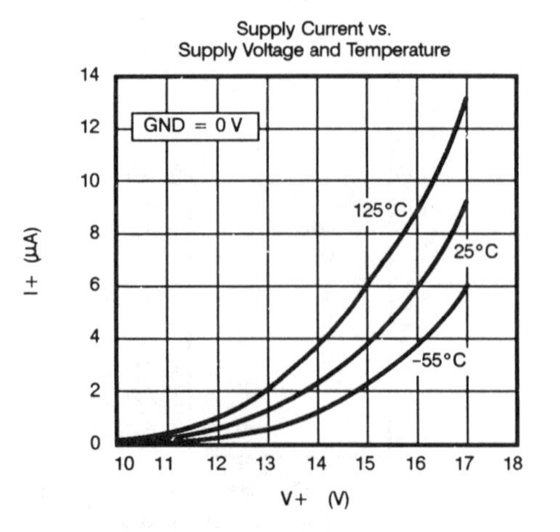

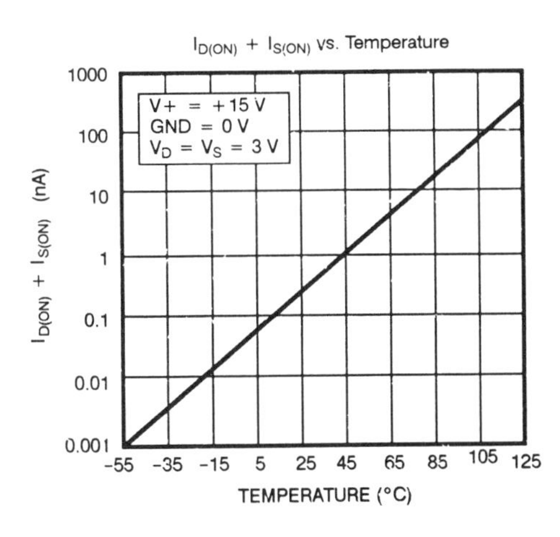
$I_{D(ON)} + I_{S(ON)}$ vs. Temperature
V+ = +15 V
GND = 0 V
$V_D = V_S$ = 3 V
$I_{D(ON)} + I_{S(ON)}$ (nA)
TEMPERATURE (°C)

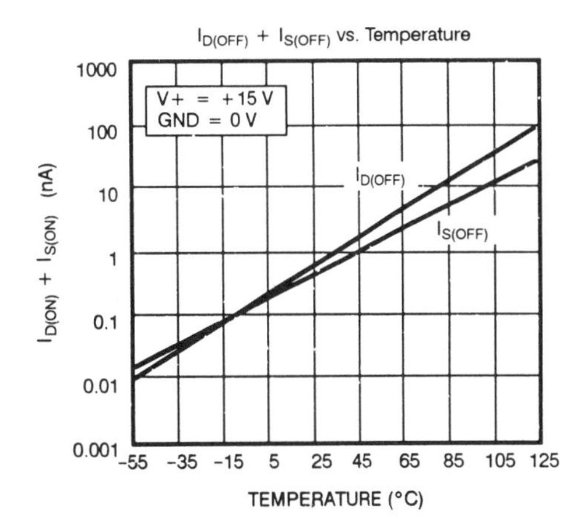
$I_{D(OFF)} + I_{S(OFF)}$ vs. Temperature
V+ = +15 V
GND = 0 V
$I_{D(OFF)}$
$I_{S(OFF)}$
$I_{D(ON)} + I_{S(ON)}$ (nA)
TEMPERATURE (°C)

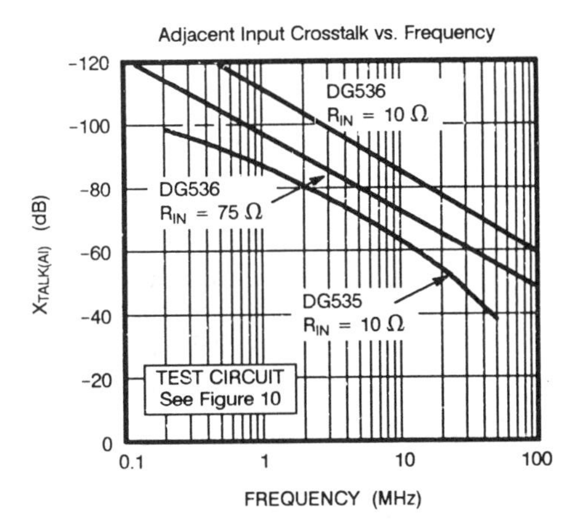
Adjacent Input Crosstalk vs. Frequency
DG536
R_{IN} = 10 Ω
DG536
R_{IN} = 75 Ω
DG535
R_{IN} = 10 Ω
TEST CIRCUIT
See Figure 10
$X_{TALK(AI)}$ (dB)
FREQUENCY (MHz)

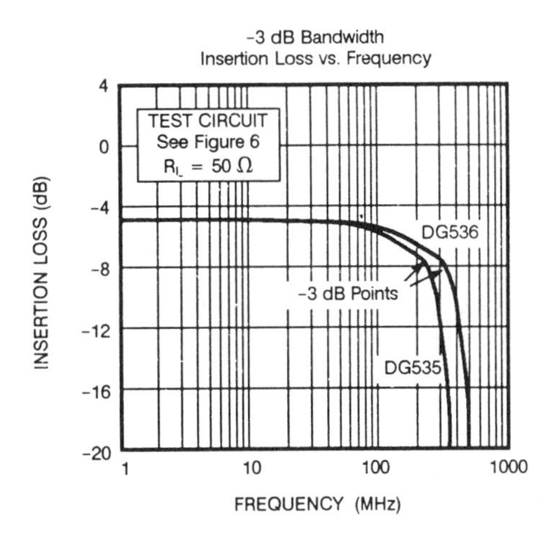
-3 dB Bandwidth
Insertion Loss vs. Frequency
TEST CIRCUIT
See Figure 6
R_L = 50 Ω
DG536
-3 dB Points
DG535
INSERTION LOSS (dB)
FREQUENCY (MHz)

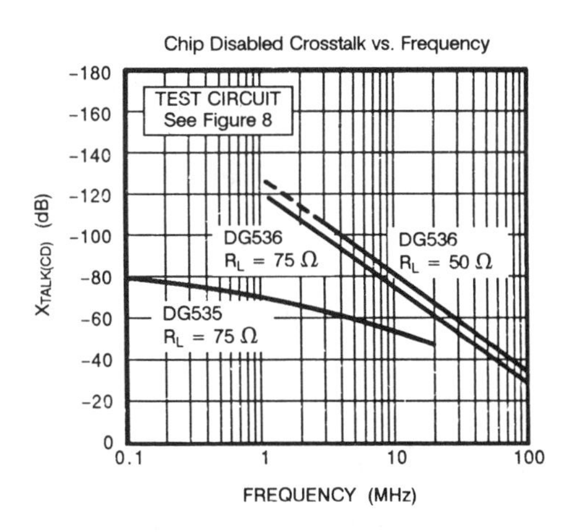
Chip Disabled Crosstalk vs. Frequency
TEST CIRCUIT
See Figure 8
DG536
R_L = 75 Ω
DG536
R_L = 50 Ω
DG535
R_L = 75 Ω
$X_{TALK(CD)}$ (dB)
FREQUENCY (MHz)

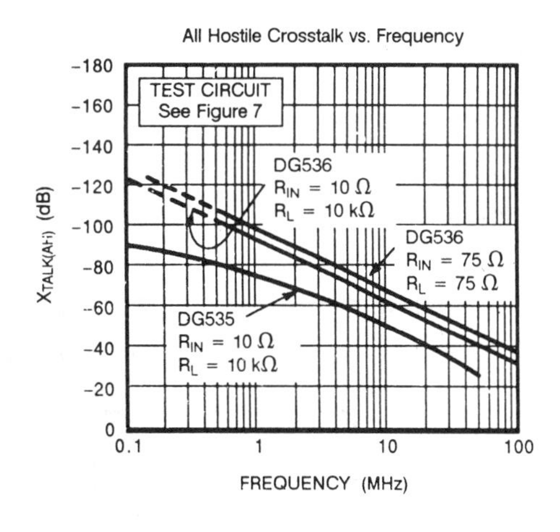
All Hostile Crosstalk vs. Frequency
TEST CIRCUIT
See Figure 7
DG536
R_{IN} = 10 Ω
R_L = 10 kΩ
DG536
R_{IN} = 75 Ω
R_L = 75 Ω
DG535
R_{IN} = 10 Ω
R_L = 10 kΩ
$X_{TALK(AH)}$ (dB)
FREQUENCY (MHz)

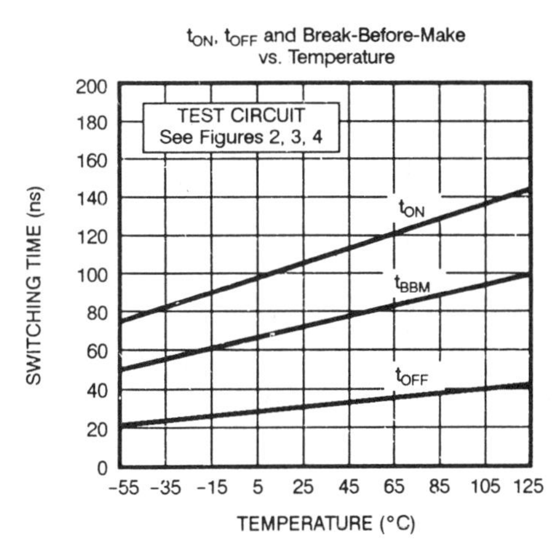
t_{ON}, t_{OFF} and Break-Before-Make
vs. Temperature
TEST CIRCUIT
See Figures 2, 3, 4
t_{ON}
t_{BBM}
t_{OFF}
SWITCHING TIME (ns)
TEMPERATURE (°C)

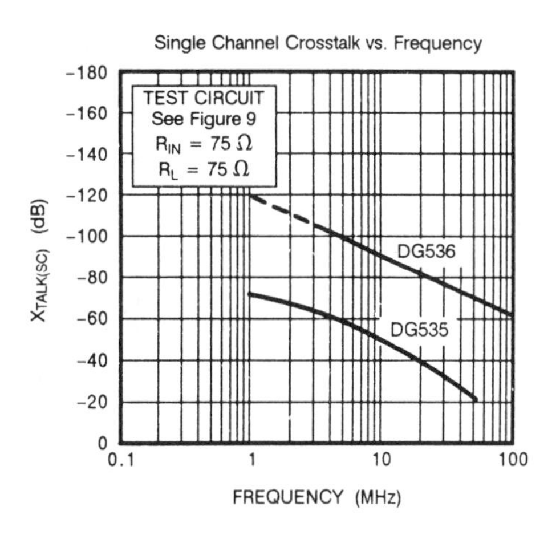
Single Channel Crosstalk vs. Frequency
TEST CIRCUIT
See Figure 9
R_{IN} = 75 Ω
R_L = 75 Ω
DG536
DG535
$X_{TALK(SC)}$ (dB)
FREQUENCY (MHz)

INPUT TIMING REQUIREMENTS

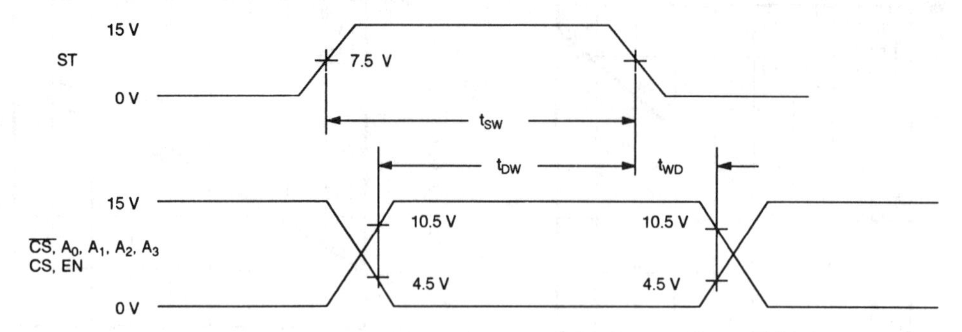

Siliconix DG884

8 × 4 Wideband/Video Crosspoint Array Preliminary Information

- 8 inputs, 4 outputs
- Wide bandwidth (300 MHz)
- Very low crosstalk (−85 dB @ 5 MHz)
- On-board TTL-compatible latches with readback
- Optional negative supply input
- Low $r_{DS(ON)}$ (90 Ω) max
- ESD protection > ±4000 V
- Reduced board space
- Improved system bandwidth
- Improved channel off-isolation
- Simplified logic interfacing
- Allows bipolar signal swings
- Reduced insertion loss
- High reliability

APPLICATIONS

- Wideband signal routing and multiplexing
- High-end video systems
- μP-controlled systems
- Direct-coupled systems
- ATE systems
- Digital video routing

The DG884 is a digitally selectable 8×4 crosspoint array designed for wideband operation. On-chip TTL-compatible crosspoint selection logic and latches with data readback are included to simplify the interface to a microprocessor data bus. The low on-resistance and low capacitance of the DG884 make it ideal for wideband data multiplexing and video and audio signal routing in analog and/or digital video systems. An optional negative-supply pin allows the handling of bipolar signals without dc biasing.

The DG884 is built on a D/CMOS process that combines n-channel DMOS switching FETs with low-power CMOs control logic, drivers, and latches. The low-capacitance DMOS FETs are in a "T" configuration to achieve extremely high levels of off isolation. Crosstalk is reduced to −85 dB at 5 MHz by including a ground line between adjacent signal input pins.

The DG884 is available in the 44-pin plastic and ceramic-leaded chip carrier for operation over the industrial D-suffix (−40 to 85 °C) temperature range, and for military A-suffix (−55 to 125 °C) operation.

FUNCTIONAL BLOCK DIAGRAM

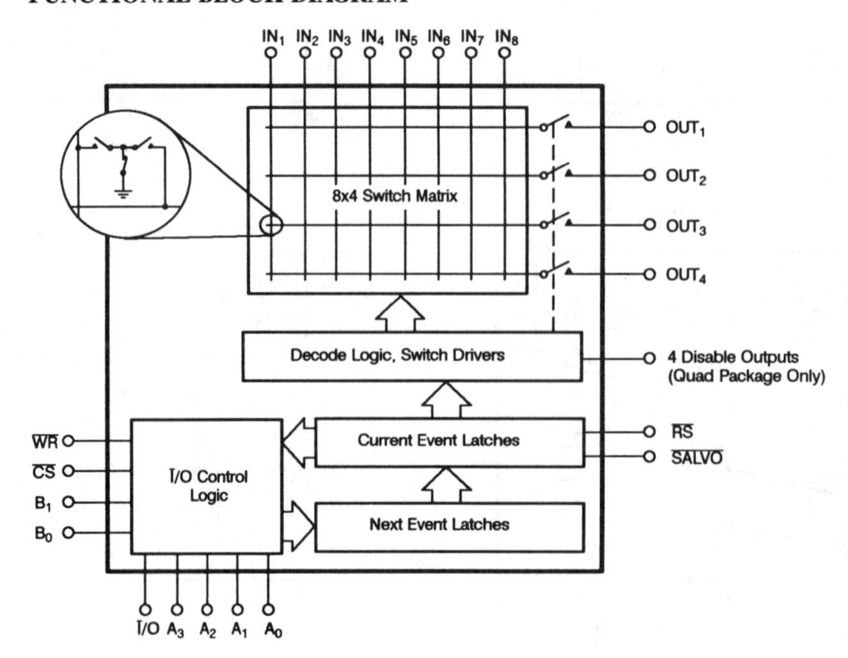

ABSOLUTE MAXIMUM RATINGS

V+ to GND . . . . -0.3 V to 21 V

V+ to V- . . . . -0.3 V to 21 V

V- to GND . . . . -10 V to 0.3 V

Digital Inputs . . . . (V-) - 0.3 V to (V+) + 0.3 V
or 20 mA, whichever occurs first

V_S, V_D . . . . (V-) - 0.3 V to (V+) + 14 V
or 20 mA, whichever occurs first

CURRENT (any terminal) Continuous . . . . 20 mA

CURRENT (S or D) Pulsed 1 ms 10% duty . . . . 40 mA

Storage Temperature (A Suffix) . . . . -65 to 150°C
(D Suffix) . . . . -65 to 125°C

Operating Temperature (A Suffix) . . . . -55 to 125°C
(D Suffix) . . . . -40 to 85°C

Power Dissipation (Package)*
44-Pin Quad J Lead Plastic *** . . . . 450 mW
44-Pin Quad J Lead Hermetic** . . . . 1200 mW

*All Leads welded or soldered to PC board
**Derate 16 mW/°C above 75°C
***Derate 6 mW/°C above 75°C

PIN CONFIGURATIONS

Quad J-Lead Packages

Order Numbers:
PLCC-44: DG884DN
CLCC-44: DG884AM/883

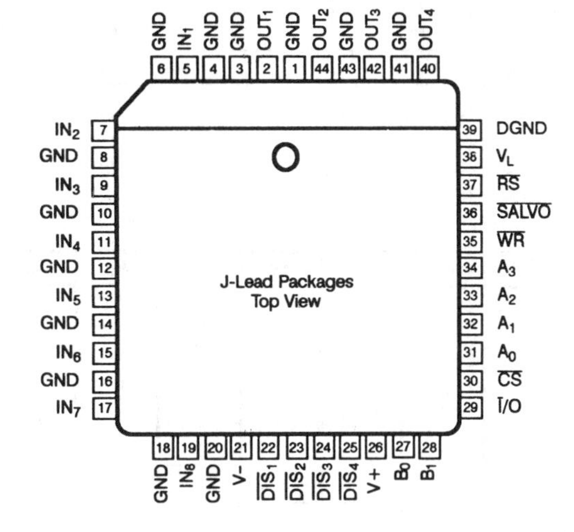

TYPICAL CHARACTERISTICS

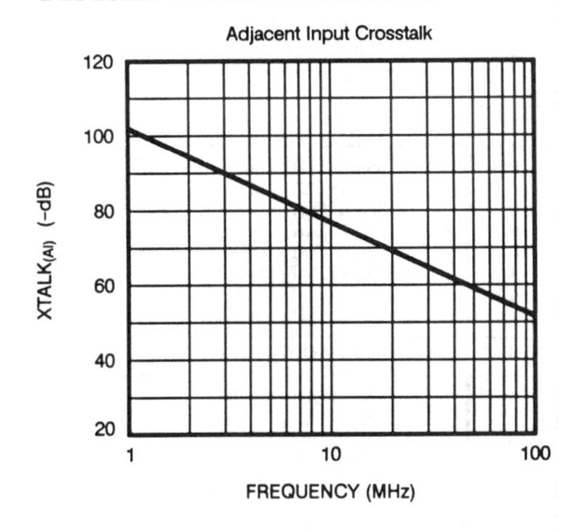

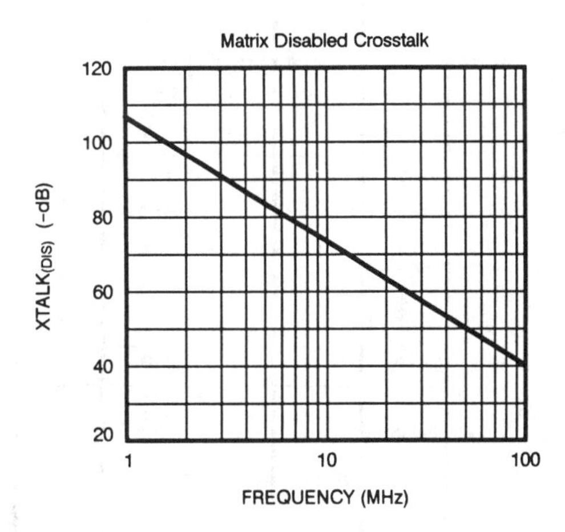

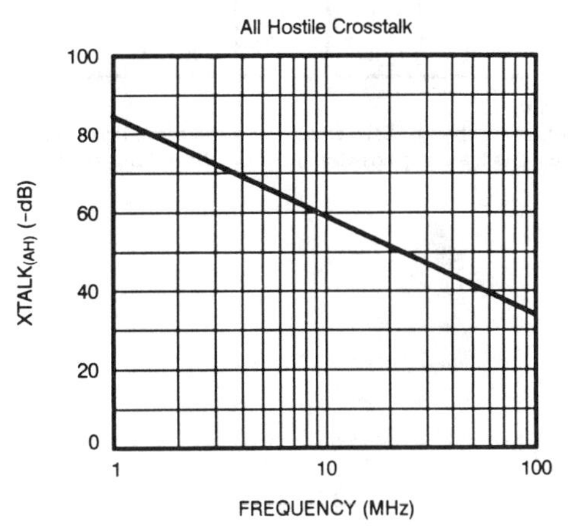

TRUTH TABLES

$\overline{RS}$	I/O	$\overline{CS}$	$\overline{WR}$	$\overline{SALVO}$	ACTIONS
1	0	1	↓	1	No Change to next event latches
1	0	0	↓	1	Next event latches loaded as defined in table above
1	0	0	0	1	Next event latches are transparent.
1	0	0	↑	1	Next event data latched
1	0	X	1	↓	Data in all next event latches is simultaneously loaded into the current event latches, i.e., all new crosspoint addresses change simultaneously when $\overline{SALVO}$ goes low.
1	0	X	1	↑	Current event data latched in
1	0	0	X	0	Current event latches are transparent
1	0	0	0	0	Both next and current event latches are transparent
1	1	1	1	1	A_0, A_1, A_2, A_3 - High impedance
1	1	0	1	1	A_0, A_1, A_2, A_3 become outputs and reflect the contents of the current event latches. B_0, B_1 determine which current event latches are being read
0	X	X	1	1	All crosspoints opened (but data in next event latch is preserved)

$\overline{WR}$	B_1	B_0	A_3	A_2	A_1	A_0	NEXT EVENT LATCHES
0	0	0	1	0	0	0	IN_1 to OUT_1 Loaded
				0	0	1	IN_2 to OUT_1 Loaded
				0	1	0	IN_3 to OUT_1 Loaded
				0	1	1	IN_4 to OUT_1 Loaded
				1	0	0	IN_5 to OUT_1 Loaded
				1	0	1	IN_6 to OUT_1 Loaded
				1	1	0	IN_7 to OUT_1 Loaded
				1	1	1	IN_8 to OUT_1 Loaded
			0	X	X	X	Turn Off OUT_1 Loaded
	0	1	1	0	0	0	IN_1 to OUT_2 Loaded
				0	0	1	IN_2 to OUT_2 Loaded
				0	1	0	IN_3 to OUT_2 Loaded
				0	1	1	IN_4 to OUT_2 Loaded
				1	0	0	IN_5 to OUT_2 Loaded
				1	0	1	IN_6 to OUT_2 Loaded
				1	1	0	IN_7 to OUT_2 Loaded
				1	1	1	IN_8 to OUT_2 Loaded
			0	X	X	X	Turn Off OUT_2 Loaded
	1	0	1	0	0	0	IN_1 to OUT_3 Loaded
				0	0	1	IN_2 to OUT_3 Loaded
				0	1	0	IN_3 to OUT_3 Loaded
				0	1	1	IN_4 to OUT_3 Loaded
				1	0	0	IN_5 to OUT_3 Loaded
				1	0	1	IN_6 to OUT_3 Loaded
				1	1	0	IN_7 to OUT_3 Loaded
				1	1	1	IN_8 to OUT_3 Loaded
			0	X	X	X	Turn Off OUT_3 Loaded
	1	1	1	0	0	0	IN_1 to OUT_4 Loaded
				0	0	1	IN_2 to OUT_4 Loaded
				0	1	0	IN_3 to OUT_4 Loaded
				0	1	1	IN_4 to OUT_4 Loaded
				1	0	0	IN_5 to OUT_4 Loaded
				1	0	1	IN_6 to OUT_4 Loaded
				1	1	0	IN_7 to OUT_4 Loaded
				1	1	1	IN_8 to OUT_4 Loaded
			0	X	X	X	Turn Off OUT_4 Loaded

ALL OTHER STATES ARE NOT RECOMMENDED

NOTE: When $\overline{WR}$ = 0 next event latches are transparent. Each crosspoint is addressed individually, e.g., to connect IN_1 to OUT_1 thru OUT_4 requires A_0, A_1, A_2 = 0 to be latched with each combination of B_0, B_1. When $\overline{RS}$ = 0, all four $\overline{DIS}$ outputs pull low simutaneously.

DIE TOPOGRAPHY

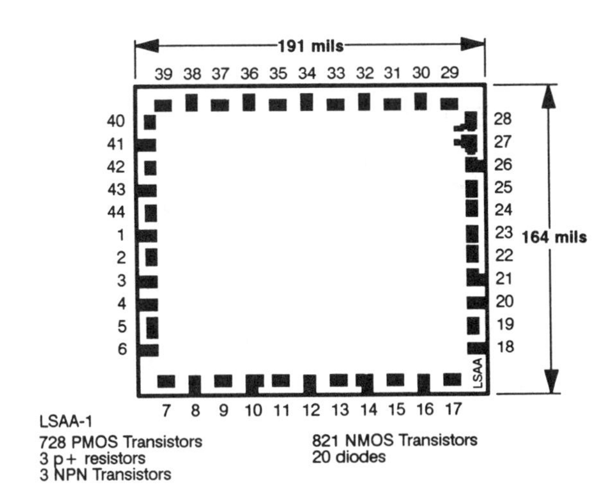

Pad No.	Function	Pad No.	Function
1	GND	23	$\overline{DIS}_2$
2	OUT_1	24	$\overline{DIS}_3$
3	GND	25	$\overline{DIS}_4$
4	GND	26	V+
5	IN_1	27	B_0
6	GND	28	B_1
7	IN_2	29	$\overline{I}/O$
8	GND	30	$\overline{CS}$
9	IN_3	31	A_0
10	GND	32	A_1
11	IN_4	33	A_2
12	GND	34	A_3
13	IN_5	35	$\overline{WR}$
14	GND	36	$\overline{SALVO}$
15	IN_6	37	$\overline{RS}$
16	GND	38	V_L
17	IN_7	39	DGND
18	GND	40	OUT_4
19	IN_8	41	GND
20	GND	42	OUT_3
21	V– (Substrate)	43	GND
22	$\overline{DIS}_1$	44	OUT_2

TIMING REQUIREMENTS

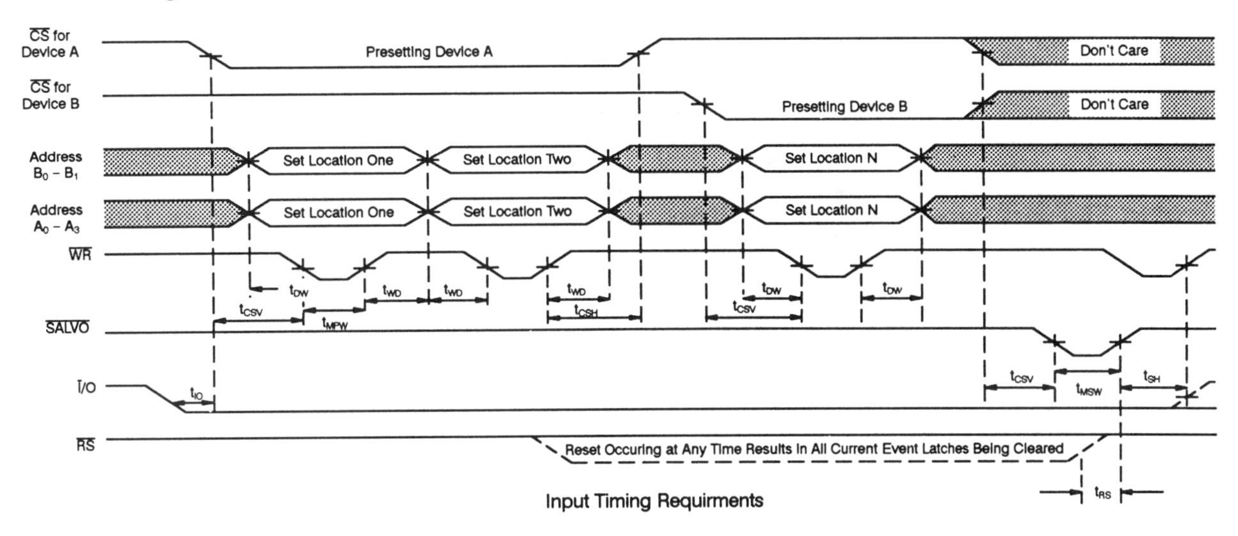

Input Timing Requirments

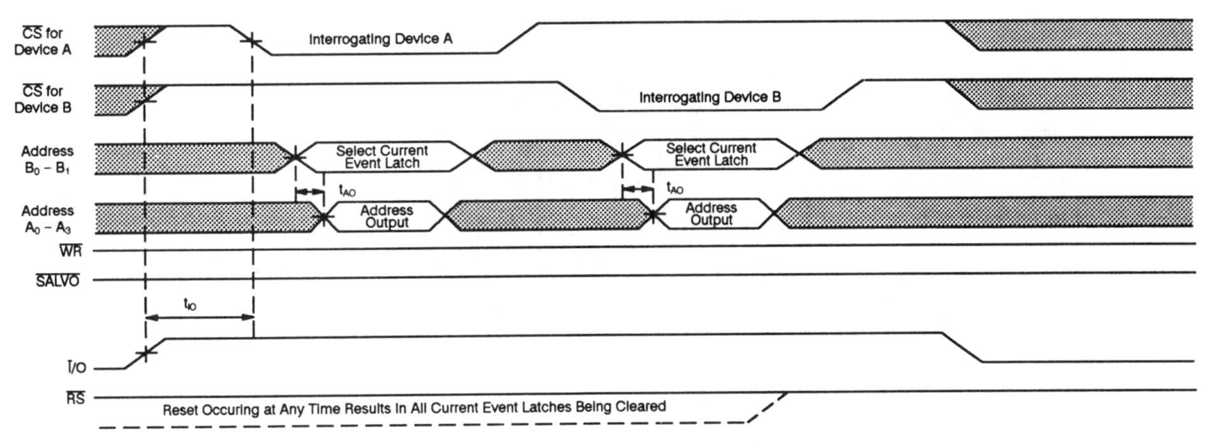

Output Timing Requirments

ON-STATE AND OFF-STATE CAPACITANCES

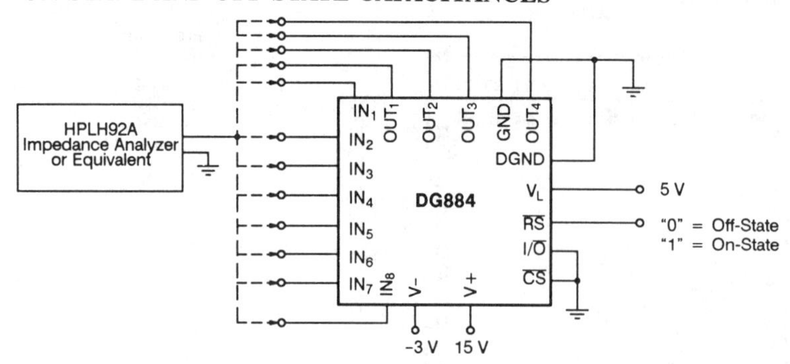

OPERATING VOLTAGE RANGE

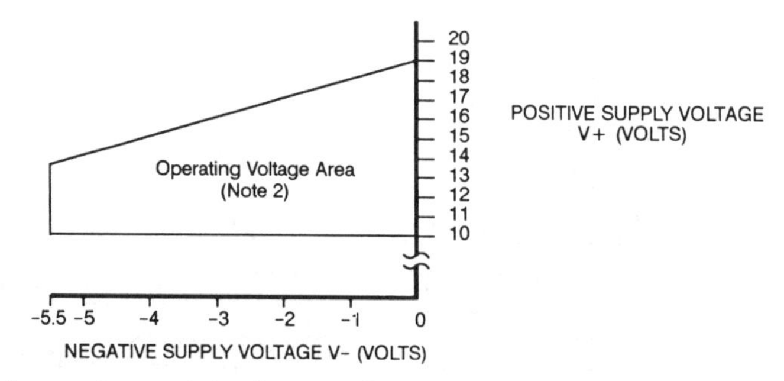

Note:

1. Both V+ and V- must have decoupling capacitors mounted as close as possible to the device pins. Typical decoupling capacitors would be 10 μF tantalum bead in parallel with 0.1 μF ceramic disc.
2. Production tested with V+ = 15 V and V- = -3.0 V. Readback functions for V- $\leq$ -3 V only.
3. At V_L = 5 V ± 10%, 0.8/2.0 V TTL compatibility is maintained over the entire operating voltage range.

APPLICATIONS

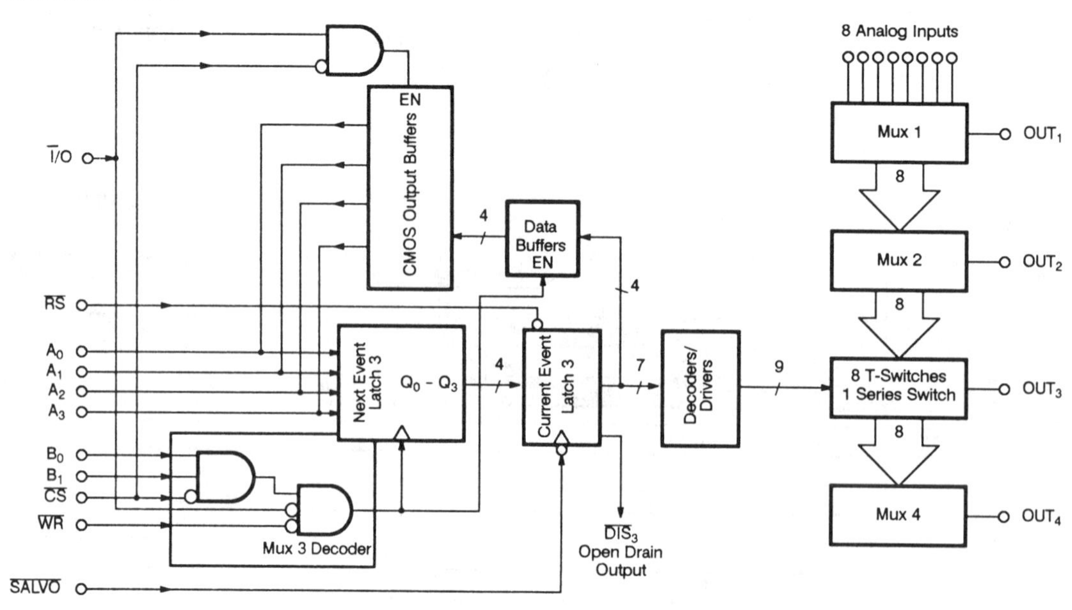

One of Four Blocks of Logic/Latches Shown

Control Circuitry

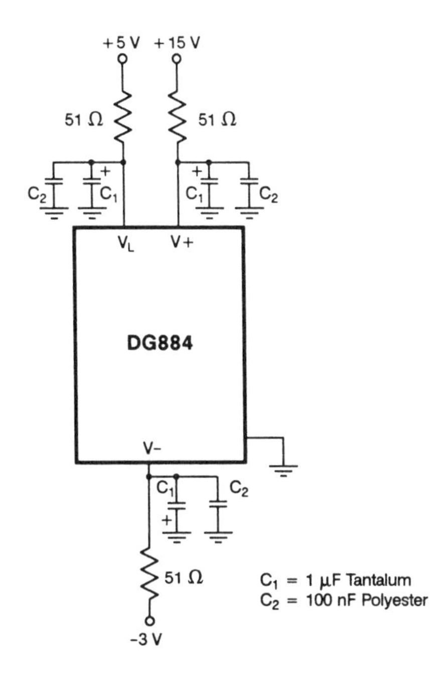

DG884 Power Supply Decoupling

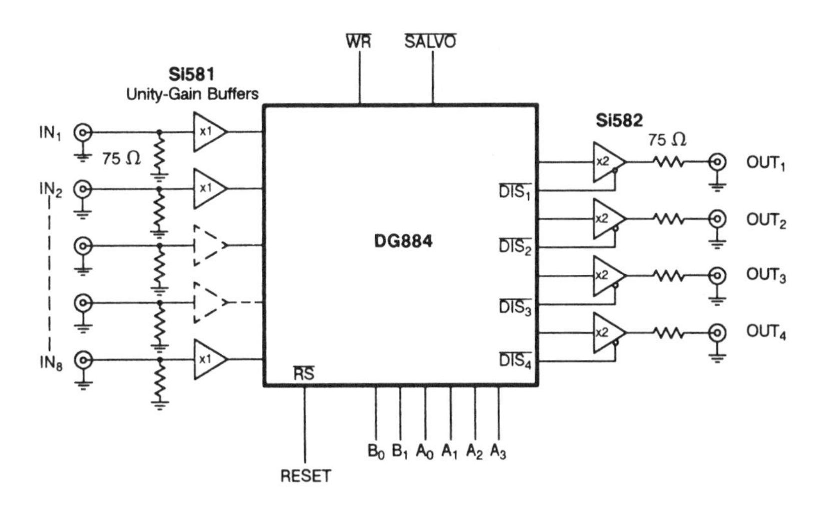

Fully Buffered 8 x 4 Crosspoint. The $\overline{DIS}$ Outputs Are Used to Power Down the Si582 Amplifiers

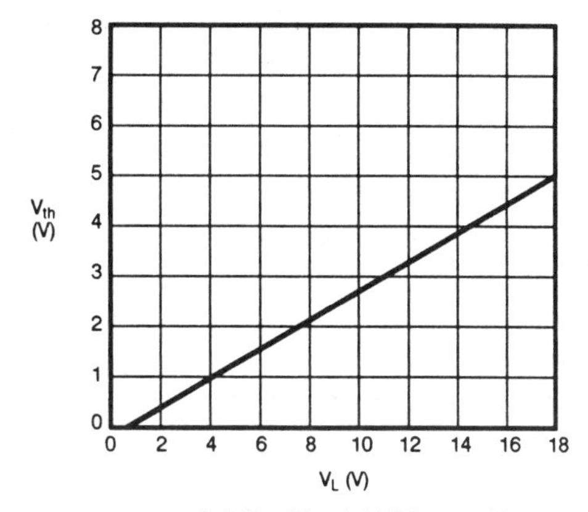

Switching Threshold Voltage vs. V_L

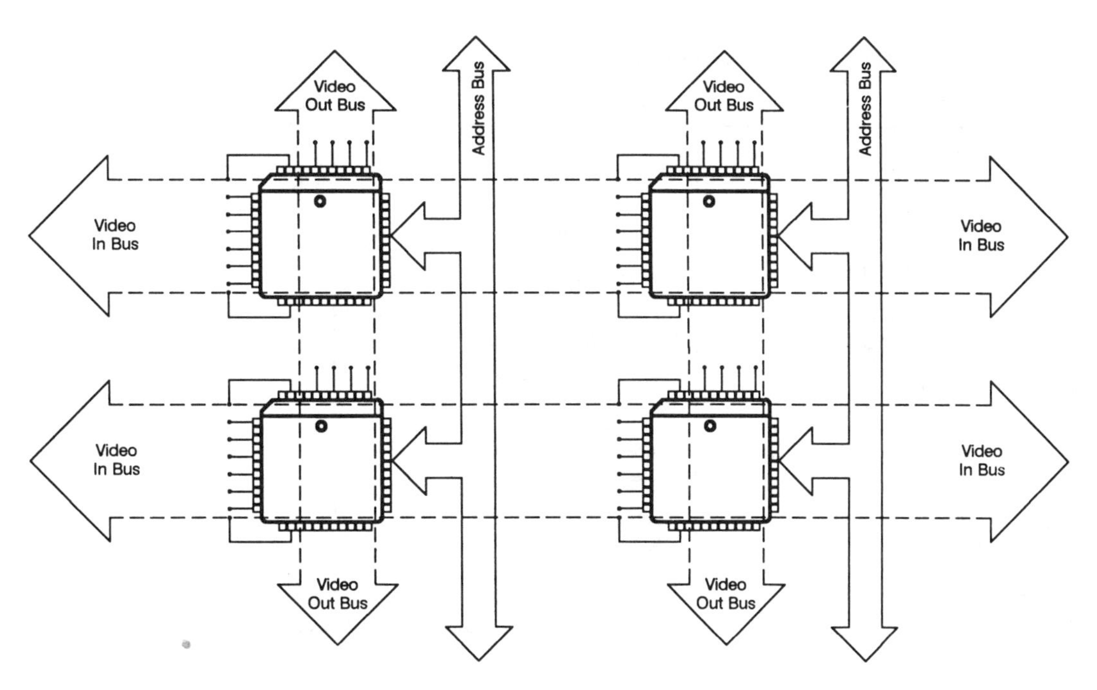

16 x 8 Expandable Crosspoint Matrix Using DG884

No responsibility is assumed by Silicon Systems for use of this product nor for any infringements of patents and trademarks or other rights of third parties resulting from its use. No license is granted under any patents, patent rights or trademarks of Silicon Systems. Silicon Systems reserves the right to make changes in specifications at any time without notice. Accordingly, the reader is cautioned to verify that the data sheet is current before placing orders.

- No adjustments or trims needed for external components
- Programmable data rate, up to 1 Mbit/s
- Internal crystal-controlled oscillator
- Selectable-write precompensation intervals
- Programmable-write clock
- DRQ (Data DMA Request) delay function
- Low-power CMOS, +5V operation
- 28-pin PDIP and 28-pin PLCC

Silicon Systems
SSI 34D441
Data Synchronizer and Write Precompensator

- Ideal for operation with NECμPD765A/μPD7265
- Fast-acquisition analog PLL for precise read-data synchronization

ELECTRICAL CHARACTERISTICS
ABSOLUTE MAXIMUM RATINGS

PARAMETER	RATING	UNIT
Storage Temperature	-40 to +120	°C
Ambient Operating Temperature, TA	0 to +70	°C
Supply Voltages, VPD, VPA	-0.5 to +7.0	VDC
Voltage Applied to Logic inputs	-0.5 to +7.0	VDC
Voltage Supplied to Logic Outputs	-0.5 to +5.5	VDC
Maximum Power Dissipation	750	mW

RECOMMENDED OPERATING CONDITIONS

PARAMETER	CONDITIONS	MIN	NOM	MAX	UNIT
Ambient Temperature, TA		0		70	°C
Power Supply Voltage, VPD, VPA		4.75	5	5.25	VDC
High Level Input Voltage, VIH	Power supply = 4.75V	2.0			V
Input Current High, IIH	Power supply =4.75V VIH = 2.4V			20	µA
Low Level Input Voltage, VIL	Power supply = 4.75V			0.8	V
Input Current Low, IIH	Power supply = 5.25V VIL = 0.4V			-20	µA
High Level Output Voltage, VOH	Power supply = 4.75V IOH=4 mA	2.4			V
Low Level Output Voltage All others, VOL	Power supply = 4.75V IOL = 8 mA			0.4	V
Short Circuit Output Current WDD only IOS (to positive supply)	Power supply = 5.25V	20		150	mA

DC CHARACTERISTICS

(Unless otherwise specified, power supplies = 4.75V to 5.25V, TA = 0 to 70°C, R_{IREF} = 57.6 kΩ ± 1%, R_{PDGAIN} = 39 kΩ ± 5%, XTAL = 16 MHz crystal in series resonance.)

PARAMETER	CONDITIONS	MIN	NOM	MAX	UNIT
Supply Current Analog, IVPA	Power supply = 5.25V 51 MHz data rate			10	mA
Supply Current Digital, IVPD	Power supply = 5.25V 1 MHz data rate			6	mA
Short Circuit Output Current (to ground) All others, IOS	Power supply = 5.25V	30		100	mA

DYNAMIC CHARACTERISTICS AND TIMING (Load Capacitance = 50 pF)

DATA DETECTION CHARACTERISTICS

PARAMETER	CONDITIONS	MIN	NOM	MAX	UNIT
TRDDW RDTA pulse width		25			ns
TRDWP RDW period	R1 = 0, R2 = 1		8		µs
	R1 = 0, R2 = 0		4		µs
	R1 = 1, R2 = 0		2		µs
	R1 = 1, R2 = 1		1		µs

PARAMETER	CONDITIONS	MIN	NOM	MAX	UNIT
TRDWW RDW pulse width high or low	Same R1, R2 as above		$\frac{TRDWP}{2}$		µs
TRDW RDD pulse width		62.5		187.5	ns
TRDDD Propagation Delay from RDW transition to RDD positive edge	Same R1, R2 as above	0.025	$\frac{TRDWP}{4}$		µs

DRQ CHARACTERISTICS

PARAMETER	CONDITIONS	MIN	NOM	MAX	UNIT
TDLY Propagation delay from DRQ positive edge to DRQD positive edge	SCLK = 1	0.75		1.0	µs
	SCLK = 0	1.50		2.0	µs
TDRLL Propagation delay from DRQ negative edge to DRQD negative edge				50	ns

CRYSTAL CHARACTERISTICS

PARAMETER	CONDITIONS	MIN	NOM	MAX	UNIT
TXTALP Crystal oscillator frequency period			62.5		ns

WRITE-DATA CLOCK FREQUENCIES

R2	R1	WCLK	DATA RATE	SCLK	CLK
1	0	250 kHz	125 kHz	1	8 MHz
0	0	500 kHz	250 kHz	0	4 MHz
0	1	1 MHz	500 kHz		
1	1	2 MHz	1 MHz		

PRECOMPENSATION DESCRIPTION

PC2	PC1	PRECOMPENSATION INTERNAL	PS0	PS1	SHIFT
0	0	±62.5 ns	0	1	Normal (no shift)
1	0	±125 ns	0	1	Late (delay)
0	1	±187.5 ns	1	0	Early (advance)
1	1	±250 ns	1	1	Invalid (no Shift)

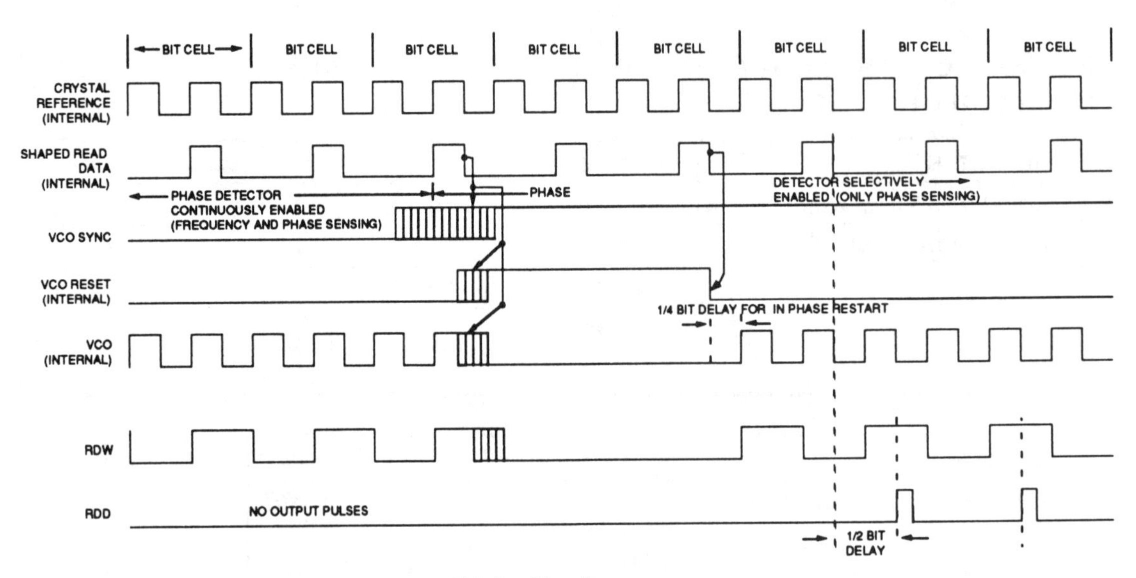

PLL Locking Sequence

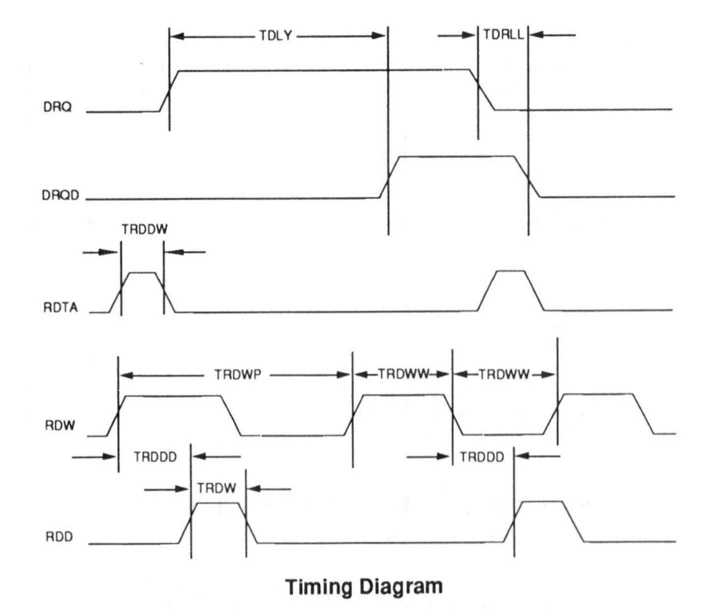

Timing Diagram

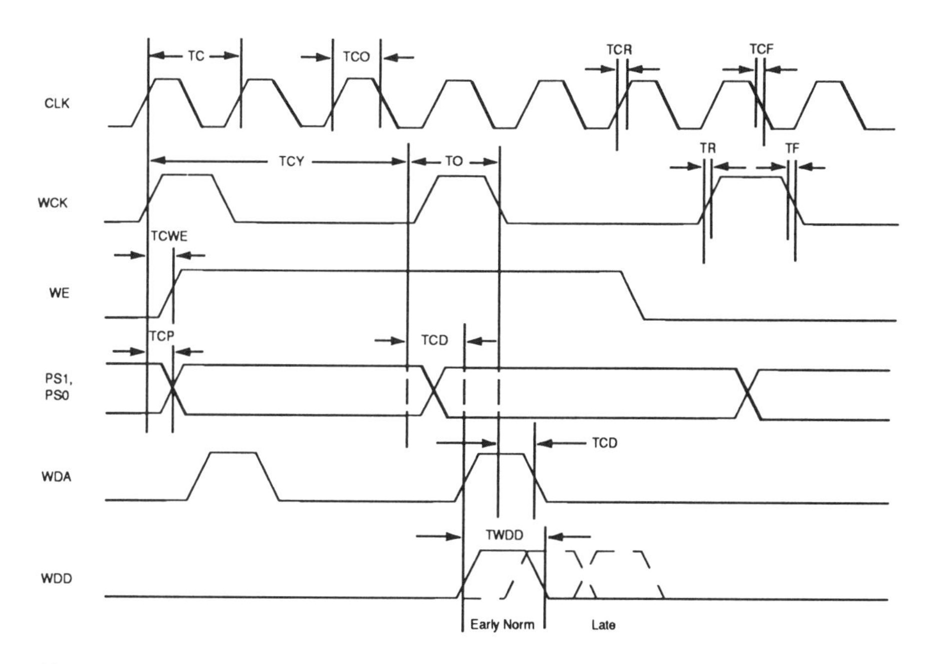

Note 1: Not to scale.
Note 2: Read data waveforms do not show delay between RDIA and RDW.
Note 3: Write data waveforms do not show WDD precompensation by exact amounts specified.
Note 4: WCLK and CLK have their rising edges synchronized.

Switching Characteristics

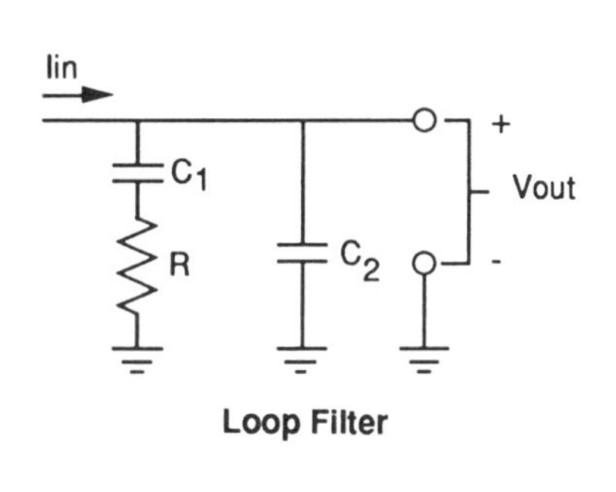

Loop Filter

DATA RATE	LOCK TIME	LOOP FILTER
125 kHz	192 µs	R = 10 kΩ, C_1 = 6800 pF C2 = 360 pF
250 kHz	96 µs	R = 10 kΩ, C_1 = 3300 pF C_2 = 180 pF
500 kHz	46 µs	R = 11 kΩ, C_1 = 1500 pF C_2 = 82 pF
1 MHz	24 µs	R = 13 kΩ, C_1 = 680 pF C_2 = 39 pF

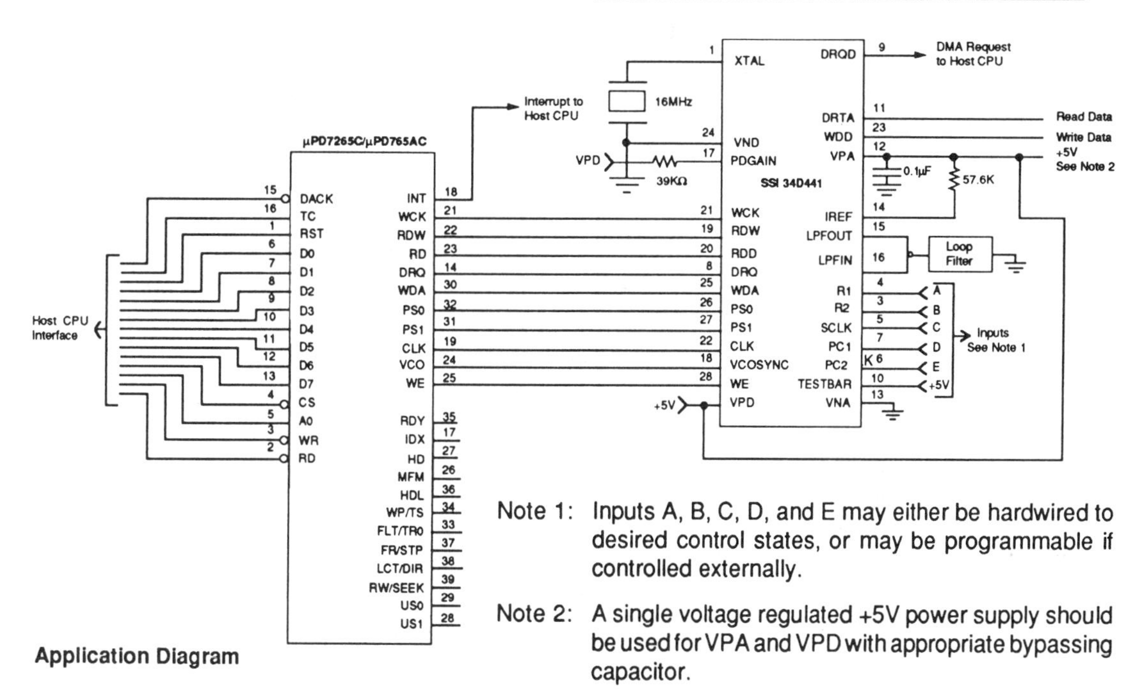

Note 1: Inputs A, B, C, D, and E may either be hardwired to desired control states, or may be programmable if controlled externally.

Note 2: A single voltage regulated +5V power supply should be used for VPA and VPD with appropriate bypassing capacitor.

Application Diagram

PACKAGE PIN DESIGNATIONS

(Top View)

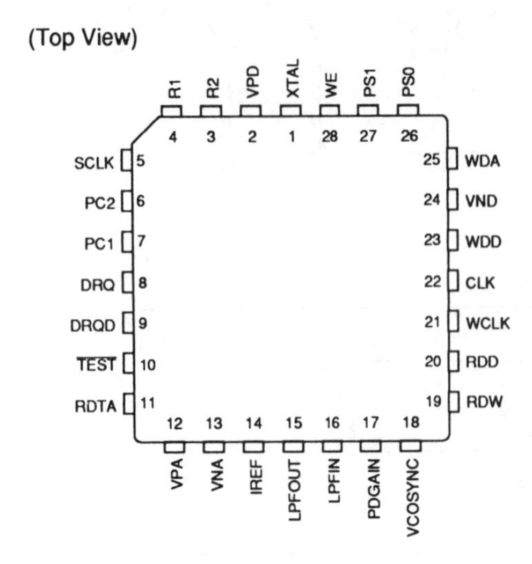

28-lead PLCC
PLCC pinouts are the same as the 28-pin DIP

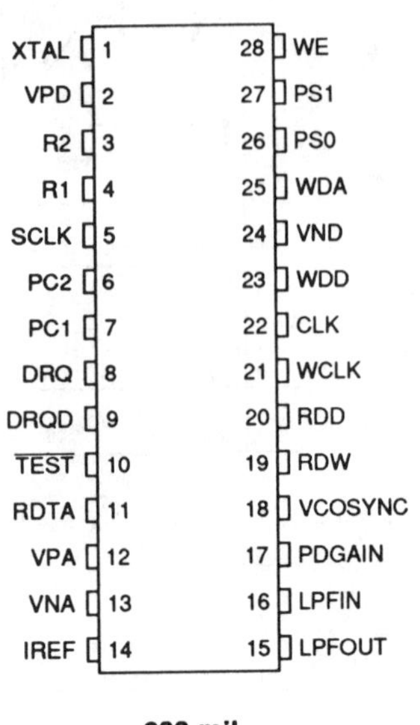

600-mil
28-pin DIP

THERMAL CHARACTERISTICS: θja

28-lead PLCC	55°C/W
28-pin PDIP	65°C/W

BLOCK DIAGRAM

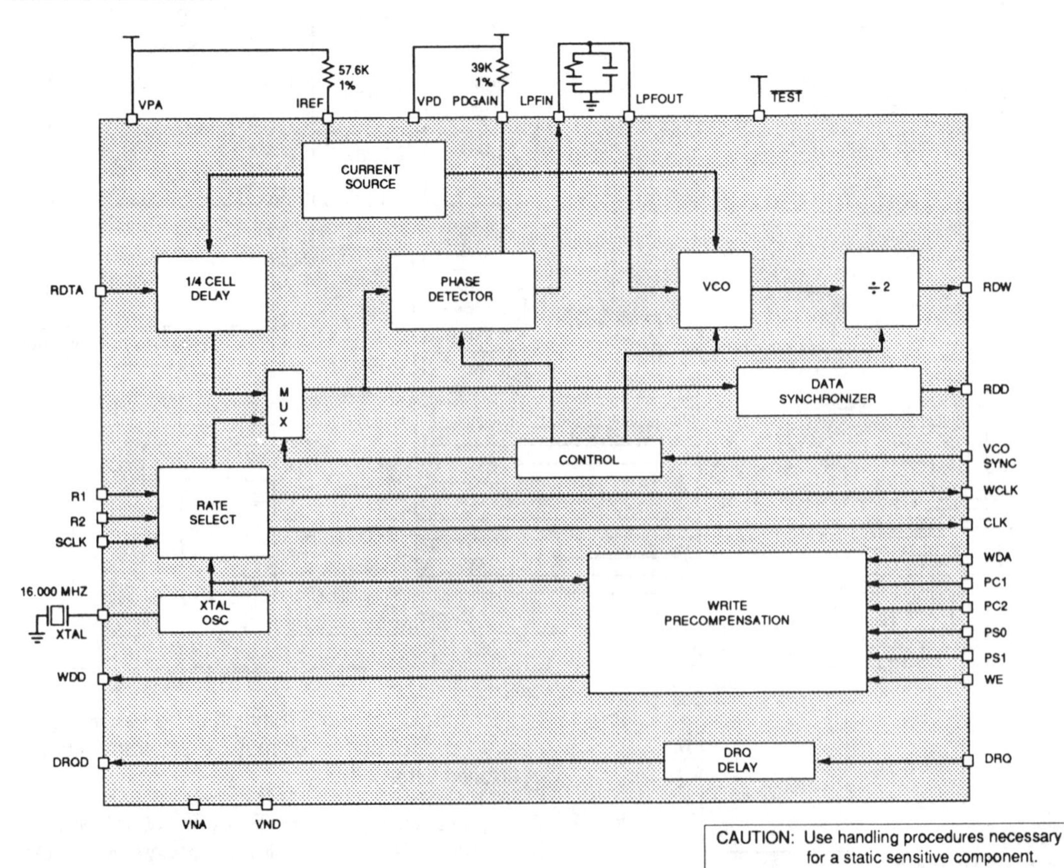

CAUTION: Use handling procedures necessary for a static sensitive component.

Silicon Systems SSI 34P570
2-Channel Floppy Disk Read/Write Device

- Single-chip read/write amplifier and read data processing function
- Compatible with 8″, 5¼″ and 3½″ drives
- Internal write and erase current sources, externally set
- Control signals are TTL compatible
- Schmitt trigger inputs for higher noise immunity on bussed control signals
- TTL-selectable write current boost
- Operate on +12V and +5V power supplies
- High gain, low noise, low peak shift (0.3% typical) read processing circuits

ELECTRICAL CHARACTERISTIC Unless otherwise specified, 4.75 V ≤V_{CC} ≤5.25 V; 11.4 V ≤V_{DD}≤12.6; 0°C≤T_A≤70°C; R_W = 430 Ω; R_{ED} = 62 kΩ; C_E = 0.012 µF; R_{EH} = 62 kΩ; R_{EC} = 220 Ω

ABSOLUTE MAXIMUM RATINGS (Operating above absolute maximum ratings may damage the device.)

PARAMETER	RATING	UNIT
5 V Supply Voltage, V_{CC}	7	V
12 V Supply Voltage, V_{DD}	14	V
Storage Temperature	65 to +130	°C
Junction Operating Temperature	130	°C
Logic Input Voltage	-0.5 V to 7.0 V	dc
Lead Temperature (Soldering, 10 sec.)	260	°C
Power Dissipation	800	mW

POWER SUPPLY CURRENTS

PARAMETER	CONDITIONS	MIN	NOM	MAX	UNIT
I_{CC} - 5 V Supply Current	Read Mode			35	mA
	Write Mode			38	mA
I_{DD} - 12 V Supply Current	Read Mode			26	mA
	Write Mode (excluding Write & Erase currents)			24	mA

LOGIC SIGNALS - READ/$\overline{WRITE}$ (R/$\overline{W}$), CURRENT BOOST (CB)

PARAMETER	CONDITIONS	MIN	NOM	MAX	UNIT
Input Low Voltage (V_{IL})				0.8	V
Input Low Current (I_{IL})	V_{IL} = 0.4 V			-0.4	mA
Input High Voltage (V_{IH})		2.0			V
Input High Current (I_{IH})	V_{IH} = 2.4 V			20	µA

LOGIC SIGNALS - WRITE DATA INPUT (WDI), HEAD SELECT (HSO/$\overline{HS1}$)

PARAMETER	CONDITIONS	MIN	NOM	MAX	UNIT
Threshold Voltage, V_T+ Positive - going		1.4		1.9	V
Threshold Voltage, V_T- Negative - going		0.6		1.1	V
Hysteresis, V_T+ to V_T-		0.4			V
Input High Current, I_{IH}	V_{IH} = 2.4V			20	µA
Input Low Current, I_{IL}	V_{IL} = 0.4V			-0.4	mA

CENTER TAP VOLTAGE REFERENCE

PARAMETER	CONDITIONS	MIN	NOM	MAX	UNIT
Output Voltage (V_{CT})	I_{WC} + I_E = 3 mA to 60 mA	V_{DD} -1.5		V_{DD} - .5	V
V_{CC} Turn-Off Threshold	(See Note 1)	4.0			V
V_{DD} Turn-Off Threshold	(See Note 1)	9.6			V
V_{CT} Disabled Voltage				1.0	V

NOTE1: Voltage below which center tap voltage reference is disabled.

BLOCK DIAGRAM

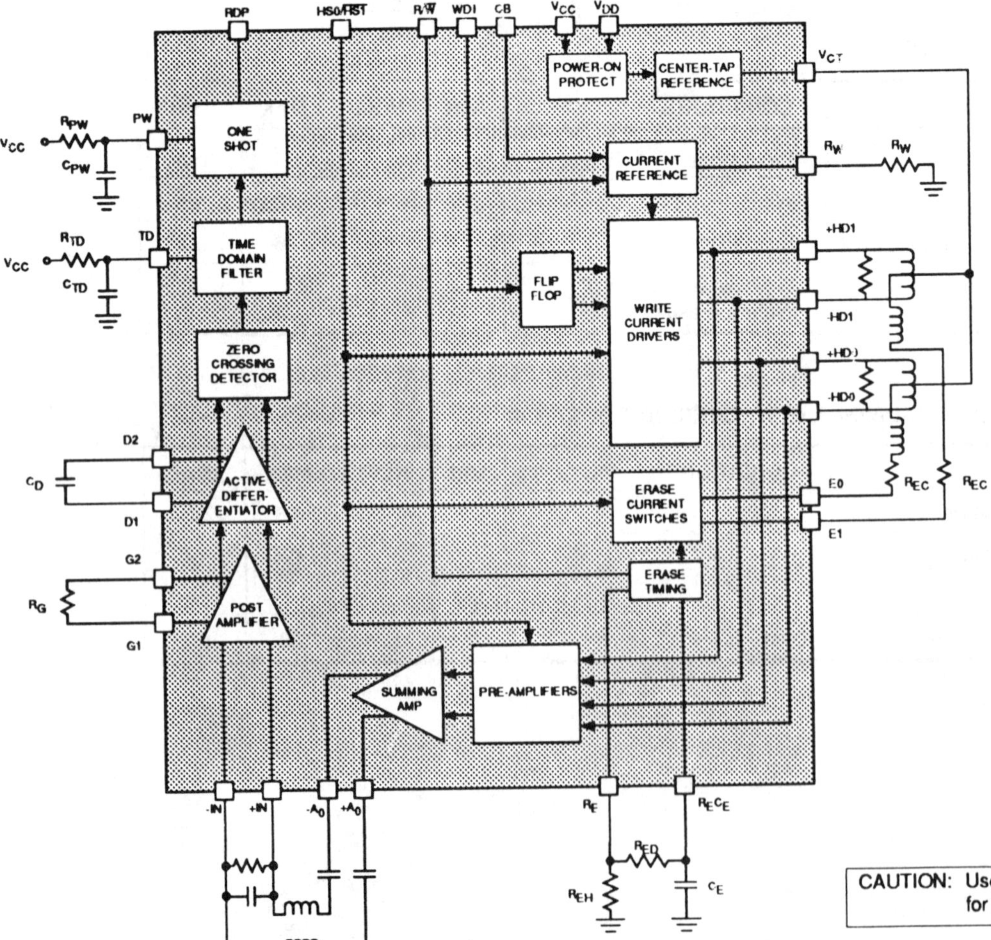

CAUTION: Use handling procedures necessary for a static sensitive component.

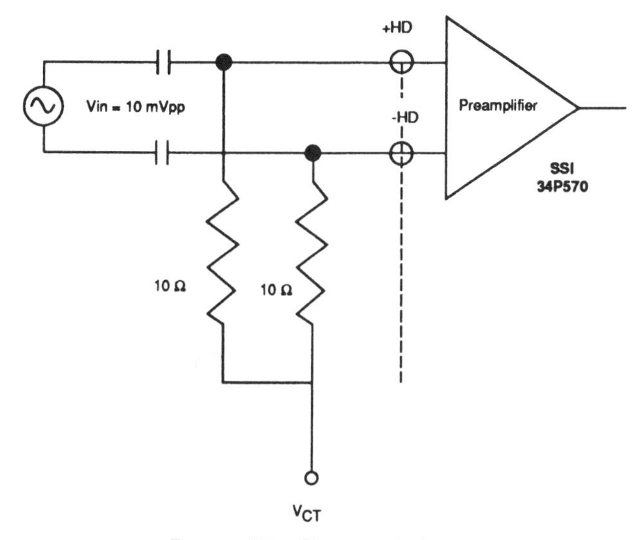

Preamplifier Characteristics

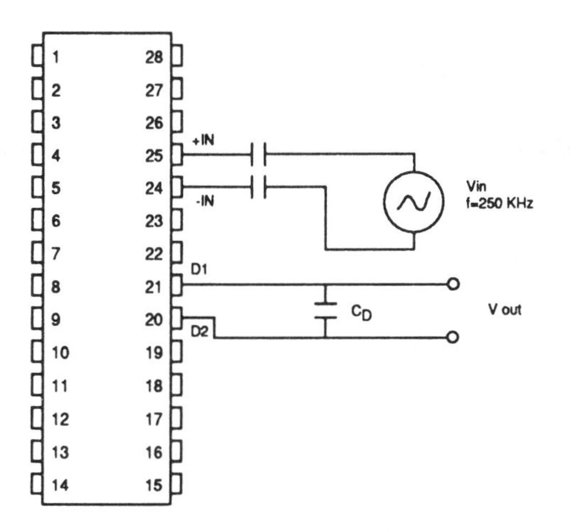

Postamplifier Differential Output Voltage Swing and Voltage Gain

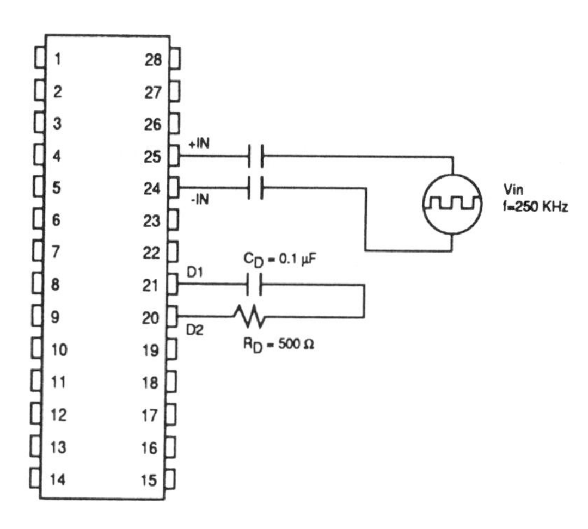

Postamplifier Threshold Differential Input Voltage

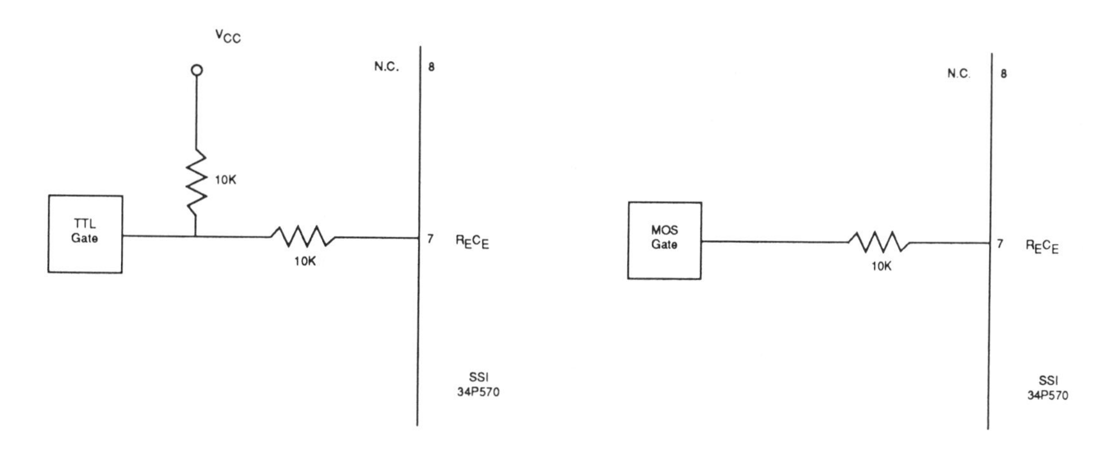

Output HI = Erase Coil Active

External Erase Control Connections

PACKAGE PIN DESIGNATIONS
(TOP VIEW)

Signal	Pin	Pin	Signal
+HD0	1	28	V_{CC}
-HD0	2	27	$+A_0$
+HD1	3	26	$-A_0$
-HD1	4	25	+IN
CB	5	24	-IN
R_W	6	23	G1
R_EC_E	7	22	G2
R_E	8	21	D1
V_{DD}	9	20	D2
V_{CT}	10	19	R/$\overline{W}$
HS0/$\overline{HS1}$	11	18	RDP
WDI	12	17	PW
E1	13	16	TD
GND	14	15	E0

28-Pin DIP

THERMAL CHARACTERISTICS: Øja

28-Pin	DIP	55°C/W
28-Pin	PLCC	65°C/W

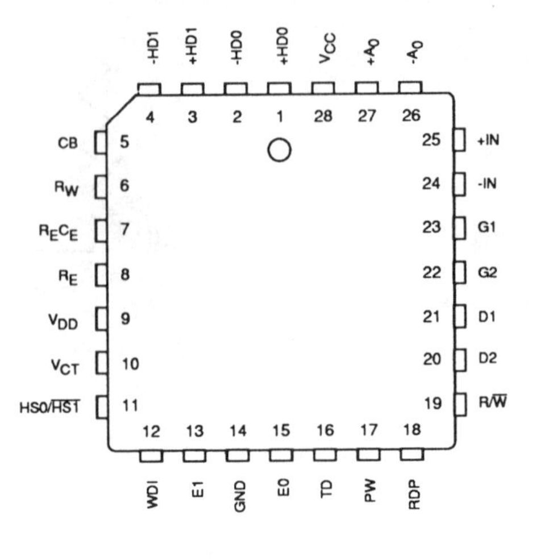

28-Pin PLCC

Silicon Systems SSI 34R575
2 or 4-Channel Floppy Disk Read/Write Device

- Operates on +5V, +12V power supplies
- Two- or four-channel capability
- TTL-compatible control inputs
- Read/Write functions on one-chip
- Internal center tap voltage source
- Supports all disk sizes
- Applicable to tape systems

HEAD SELECT LOGIC

4 - CHANNELS		
HS1	HS0	HEAD
0	0	0
0	1	1
1	0	2
1	1	3

2 - CHANNELS	
HS1	HEAD
0	0
1	1

ELECTRICAL CHARACTERISTICS

ABSOLUTE MAXIMUM RATINGS
(Operating above absolute maximum ratings may damage the device.)

PARAMETER		RATING	UNIT
DC Supply Voltage:	Vcc	6.0	V
	Vdd	14.0	V
Write Current		10	mA
Head Port Voltage		18.0	V
Digital Input Voltages:	DX, DY, HS0, HS1, WD	-0.3 to + 10	V
	$\overline{EG}$, $\overline{WG}$	-0.3 to V_{cc} + 0.3	V
DX, DY Output Current		-5	mA
VCT Output Current		-10	mA
Storage Temperature Range		-65 to + 150	°C
Junction Temperature		125	°C
Lead Temperature (Soldering, 10 sec.)		260	°C

RECOMMENDED OPERATING CONDITIONS (0°C<Ta<50°C, 4.7 V<Vcc<5.3 V, 11 V<VDD<13 V)

PARAMETER	CONDITIONS	MIN	NOM	MAX	UNIT
Vcc Supply Current					
Read mode	Vcc MAX			15	mA
Write mode	Vcc MAX			35	mA
VDD Supply Current					
Read mode	VDD MAX			25	mA
Write mode	VDD MAX			15	mA
Write Current			5.5		mA

ERASE OUTPUT

PARAMETER	CONDITIONS	MIN	NOM	MAX	UNIT
Erase On Voltage	IE = 80 mA	0.7		1.3	VDC
Erase Off Leakage				100	µA

LOGIC SIGNALS

PARAMETER	CONDITIONS	MIN	NOM	MAX	UNIT
Head Select (HS0, HS1) and Write Data (WD)					
Low Level Voltage		-0.3		0.8	VDC
High Level Voltage		2.0		6.0	VDC
Low Level Current	VIN = 0 volts	-1.6			mA
High Level Current	VIN = 2.7 volts			40	µA
$\overline{\text{WRITE GATE}}$ ($\overline{\text{WG}}$) and $\overline{\text{ERASE GATE}}$ ($\overline{\text{EG}}$)					
Low Level Voltage		-0.3		0.81	VDC
High Level Input Current		-300			µA
Low Level Current	VIN = 0 volts	-2.0			mA

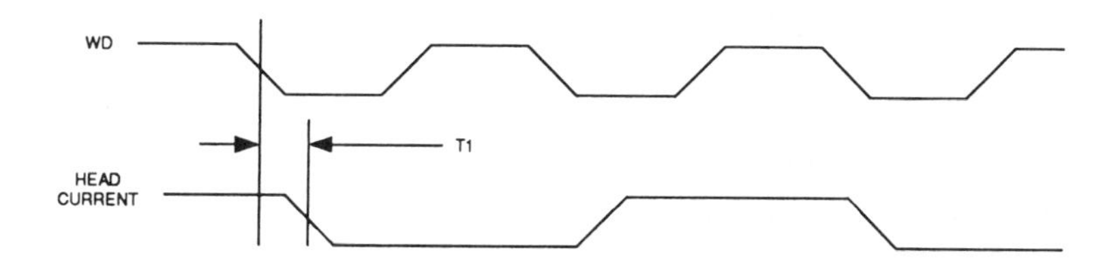

Head Current Switching Delay

BLOCK DIAGRAM

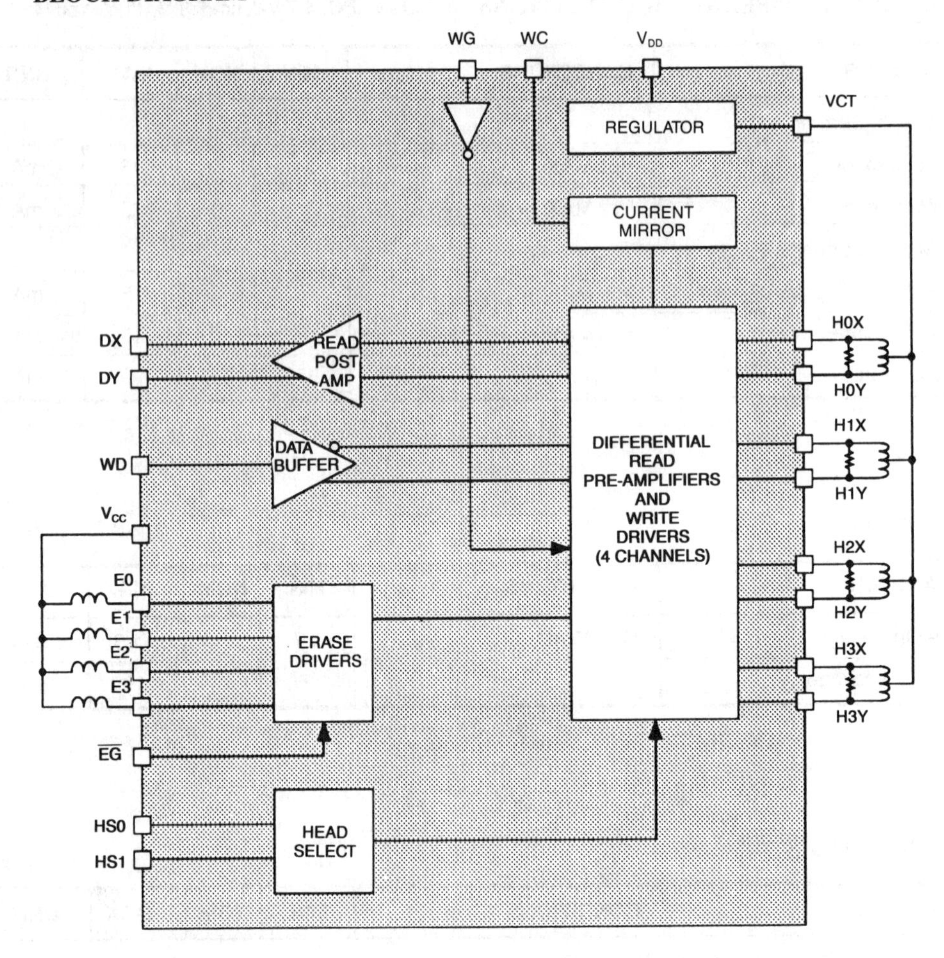

PACKAGE PIN DESIGNATIONS

(TOP VIEW)

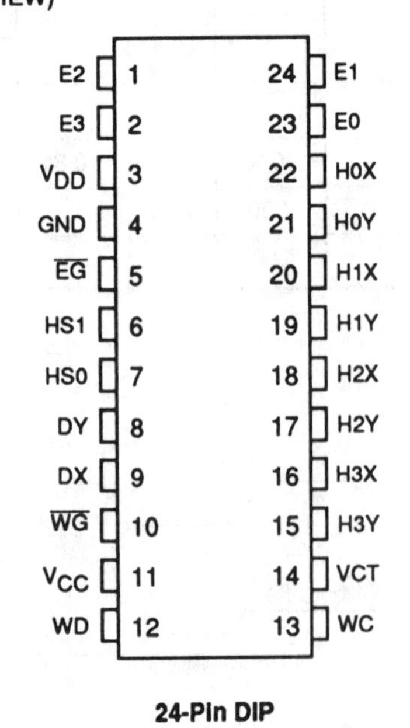

24-Pin DIP

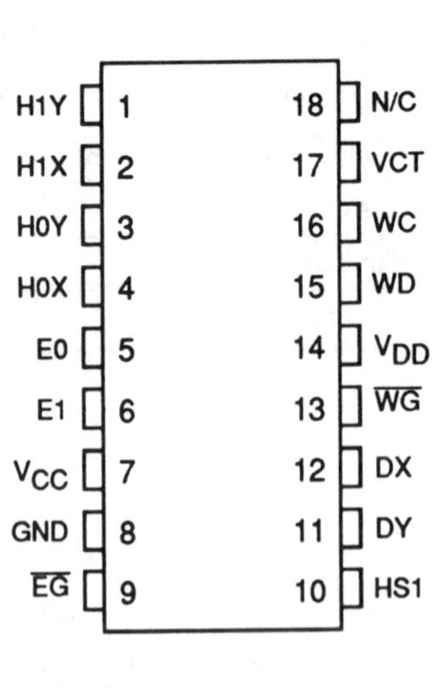

18-Pin DIP

Silicon Systems
SSI 34B580
Port Expander
Floppy Disk Drive

- Reduces package count in flexible disk drive systems
- Replaces bus interface and combinational logic devices between the SSI 34P570 on-board microprocessor and mechanical interfaces
- Surface mount available for further real-estate reduction
- Provides drive capability for mechanical and system interfaces

ABSOLUTE MAXIMUM RATINGS (All voltages referred to GND)

Operation above absolute maximum ratings may permanently damage the device.

PARAMETER	RATING	UNIT
DC Supply	+ 7	VDC
Voltage Range (any pin to GND)	-0.4 to + 7	VDC
Power Dissipation	700	mW
Storage Temperature	-40 to + 125	°C
Lead Temperature (10 sec soldering)	260	°C

ELECTRICAL CHARACTERISTICS

(Unless otherwise specified, 4.75 ≤ Vcc ≤ 5.25 VDC; 0°C < Ta < 70°C)

PARAMETER	CONDITIONS	MIN	NOM	MAX	UNIT
Totem pole outputs (P20 - P23, $\overline{INTR}$, T1)					
Output High Voltage	IOH = -400 A	2.5			V
Output Low Voltage	IOL = 2 mA			0.5	V
Open collector outputs ($\overline{RD\ DATA\ OUT}$, $\overline{INDEX}$, $\overline{WGATE}$, $\overline{TRACK0}$,$\overline{READY}$, $\overline{WR\ PROT}$)					
Output High Current	VOH = 5.25 V			250	µA
Output Low Voltage	IOL = 48 mA			0.5 V	V
Inputs (P20 - P23, PROG, RD DATA IN)					
Input High Voltage		2.0			V
Input Low Voltage				0.8	V
Input Low Current	VIL = 0.5 V			-0.8	mA
Input High Current	VIH = 2.4 V			40	µA
Input Current	Vin = 7.0 V			0.1	mA
Schmitt - Trigger Inputs ($\overline{WGATE\ IN}$, $\overline{MOTOR\ ON}$, $\overline{DIR}$, $\overline{DS}$, $\overline{STEP}$)					
Threshold Voltage	Positive Going, Vcc = 5.0 V	1.3		2.0	V
	Negative Going, Vcc = 5.0 V	0.6		1.1	V
Hysteresis	Vcc = 5.0 V	0.4			V
Input High Current	VIH = 2.4 V			40	µA
Input Low Current	VIL = 0.5 V			-0.4	mA
Input Current	VIN = 7.0 V			0.1	mA

High Impedance Inputs with Hysteresis (WR PROT SENSOR, TRACK 0 SENSOR, INDEX SENSOR)

PARAMETER	CONDITION	MIN	NOM	MAX	UNIT
Input High Voltage				2.0	V
Input Low Voltage		0.8			V
Hysteresis		0.2			V
Input Current	Vin = 0 to Vcc			-0.25	mA

TIMING CHARACTERISTICS

(Unless otherwise specified; Ta = 25°C; 4.75V ≤ Vcc≤ 5.25V; CL = 15 pf.)

PARAMETER	CONDITION	MIN	NOM	MAX	UNIT
Propagation Delay Time	RD DATA IN to $\overline{\text{RD DATA OUT}}$			35	ns
	$\overline{\text{DS}}$ to $\overline{\text{WGATE}}$, $\overline{\text{TRACK 0 READY}}$ $\overline{\text{WR PROT}}$,$\overline{\text{RD DATA}}$, $\overline{\text{INDEX}}$			80	ns
	PROG to $\overline{\text{INTR}}$, $\overline{\text{WGATE}}$, $\overline{\text{TRACK 0}}$ (Rising edge) $\overline{\text{READY}}$, $\overline{\text{WR PROT}}$			100	ns
	WR PROT to $\overline{\text{WGATE}}$, $\overline{\text{WR PROT SENSOR}}$			250	ns
	$\overline{\text{WGATE}}$ IN to $\overline{\text{WGATE}}$			80	ns
	$\overline{\text{STEP}}$ to T1, P20			80	ns
	TRACK 0 SENSOR WR PROT SENSOR to Port 2 INDEX SENSOR			250	ns
	$\overline{\text{MOTOR ON}}$ $\overline{\text{WGATE IN}}$ to Port 2 $\overline{\text{DS}}$			80	ns
Data Setup Time	$\overline{\text{DIR}}$ to $\overline{\text{STEP}}$	50			ns
Data Hold Time	$\overline{\text{DIR}}$ to $\overline{\text{STEP}}$	0			ns
Delay Accuracy (Pin 13)	Td = 0.59 Rd x Cd RD = 3.9 K to 10 K CD = 75 pF to 300 pF	0.8TD		1.2TD	sec
Pulse Width Accuracy (Pin 14)	Tw = 0.59 Rw x Cw Rw = 3.9 K to 10 K Cw = 75 pF to 300 pF	0.8Tw		1.2Tw	sec

PORT 2 (P20 - P23) TIMING (Timing Referenced to PROG signal)

SYMBOL	DESCRIPTION	MIN	NOM	MAX	UNIT
TSA	Addr. setup time	100			ns
THA	Addr. hold time	80			ns
TSD	Data-in setup time	100			ns
THD	Data-in hold time	80			ns
TACC	Data-out access time			700	ns
TDR	Data-out release time			200	ns
TPW	PROG pulse width	1500			ns

INPUT TO PORT2		READ FROM PORT 2				4-BIT
OP Code P22	Addr. P20	P23	P22	P21	P20	Input Port
0	0	$\overline{\text{DS}}$	Index Sensor Latch	WR Sensor	Track 0 Sensor	B
0	1	$\overline{\text{DS}}$	$\overline{\text{WGATEIN}}$	$\overline{\text{MOTORON}}$	$\overline{\text{DIR}}$	A

TABLE 1: Read Mode

INPUT TO PORT2		DATA PROCESSED FROM PORT 2				
OP Code P22	Addr. P20	$\overline{\text{WGATE}}$	$\overline{\text{TRACK0}}$	$\overline{\text{READY}}$	$\overline{\text{INTR}}$	Index Latch Reset
1	0	Z	$\overline{(\text{P22}\bullet\overline{\text{DS}})}$	$\overline{(\text{P21}\bullet\overline{\text{DS}})}$		P20
1	1				See Text	

Where Z = (P23 • WR PROT SENSOR) + $(\overline{\text{DS} \bullet \text{WGATEIN}})$

TABLE 2: Write Mode

PACKAGE PIN DESIGNATIONS

(TOP VIEW)

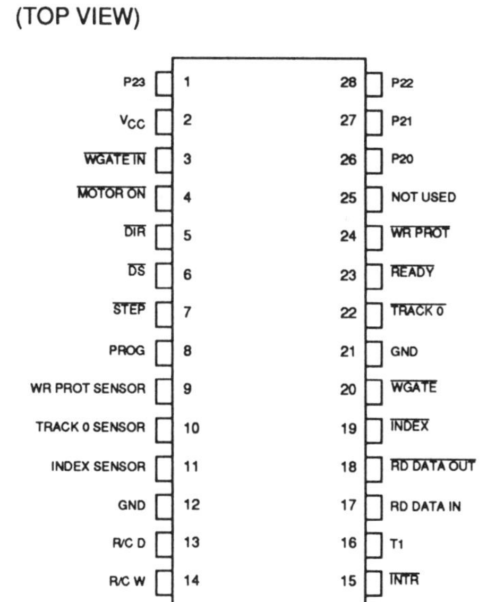

28-Pin DIP

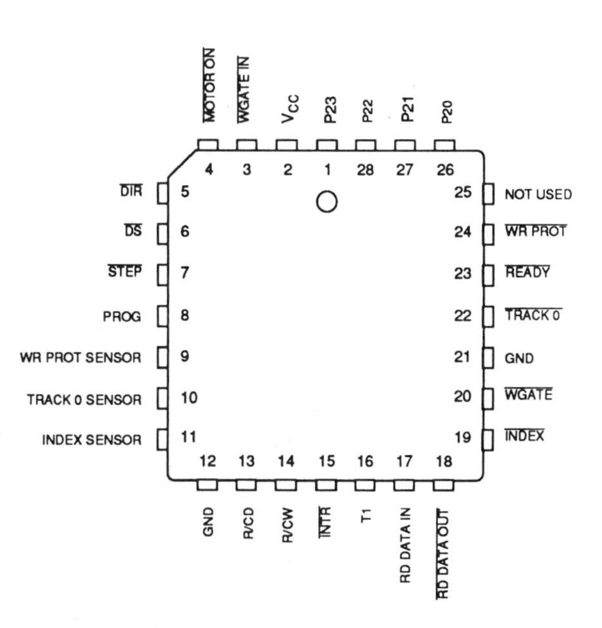

28-Pin PLCC

BLOCK DIAGRAM

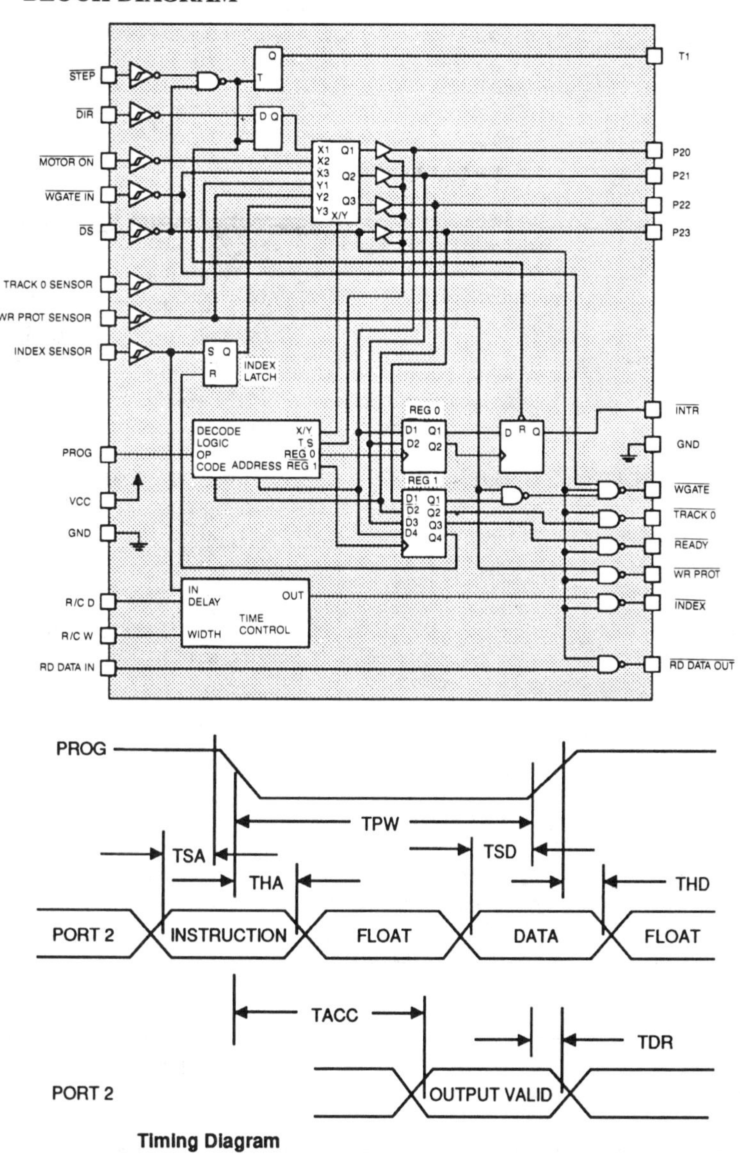

Timing Diagram

Typical Application

Silicon Systems SSI 32C260
PC AT/XT Combo Controller
Advance Information

- PC AT/XT bus interface
  - Single-chip PC AT/XT controller
  - Supports ST506/412, ST412HP, EDSI, and SMD disk interfaces
  - Direct bus interface logic with on-chip 24-mA drivers
  - Logic for daisy-chaining 2 embedded controller drives on a PC AT
  - Supports 15 Mbit/s concurrent disk transfer on a 12-MHz PC AT without wait states
- Buffer manager
  - Supports buffer memory throughput to 6 MB/s
  - Direct buffer memory addressing up to 64-kB static RAM
  - Dual-port circular buffer control
- Storage controller
  - NRZ data rate up to 15 Mbit/s
  - Selectable 16-bit CRC or 56-bit ECC polynomial with fast hardware-correction circuitry
  - Support sector level defect management
  - Support 1:1 interleaved operation
- Microprocessor interface
  - Supports both Intel 8051 and Motorola 68HC11 family of microprocessors
  - Interrupt or polled microprocessor interface
- Others
  - Low-power CMOS technology
  - Plug-and-play compatible with Cirrus CL-SH 260 chip
  - Available in 84-pin PLCC or 100-pin QFP

ABSOLUTE MAXIMUM RATINGS

Maximum limits indicate where a permanent device damage occurs. Continuous operation at these limits is not intended and should be limited to those conditions specified in the DC operating characteristics

PARAMETER	RATING	UNITS
Power Supply Voltage, VCC	7	V
Ambient Temperature	0 to 70	°C
Storage Temperature	-65 to 150	°C
Power Dissipation	750	mW
Input, Output pins	-0.5 to VCC+0.5	V

ELECTRICAL CHARACTERISTICS

PARAMETER	CONDITION	MIN	NOM	MAX	UNITS
VCC Power Supply Voltage		4.75		5.25	V
VIL Input Low Voltage		-0.5		0.8	V
VIH Input High Voltage		2.0		VCC+0.5	V
VOL Output Low Voltage	All pins except PC interface, IOL = 2 mA			0.4	
VOL Output Low Voltage	PC interface pins, IOL = 24 mA			0.5	V
VOH Output High Voltage	IOH = -400 μA			2.4	V
ICC Supply Current				50	mA
ICCS Supply Current Standby	All Inputs at GND or VCC	250			μA
IL Input Leakage Current	0 < VIN < VCC	-10		10	μA
CIN Input Capacitance				10	pF
COUT Output Capacitance				10	pF

BLOCK DIAGRAM

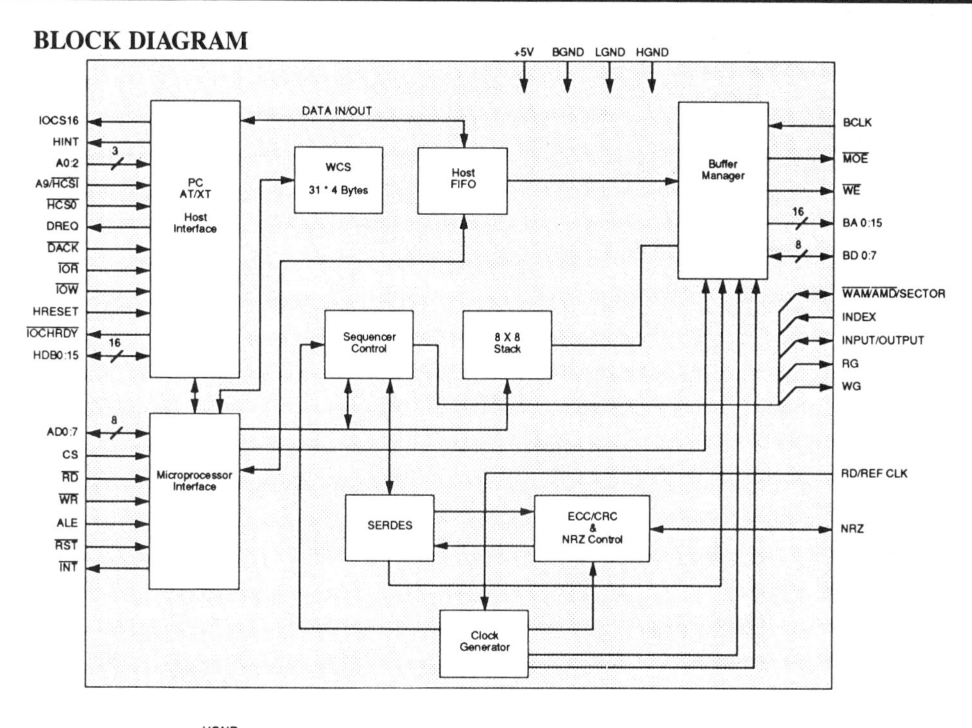

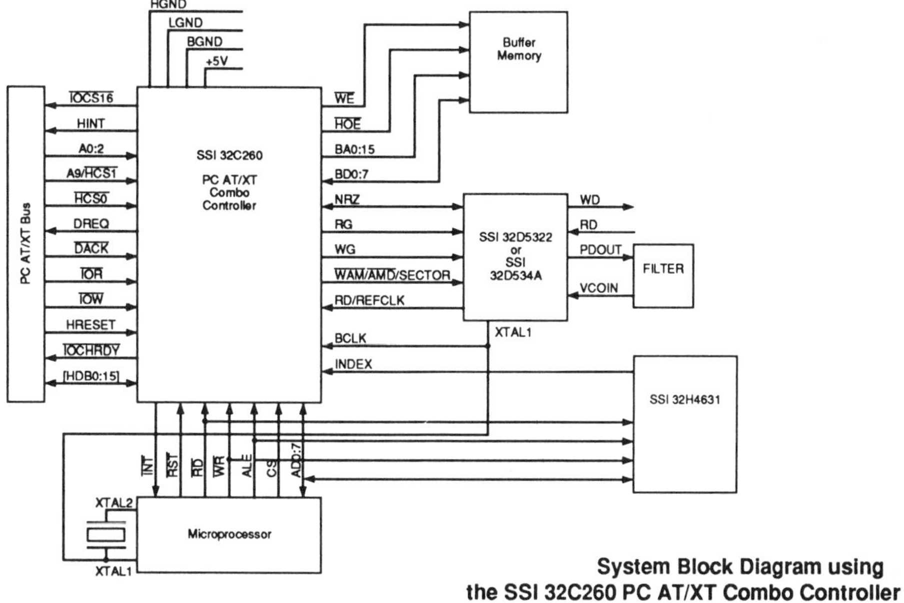

System Block Diagram using
the SSI 32C260 PC AT/XT Combo Controller

PACKAGE PIN DESIGNATIONS

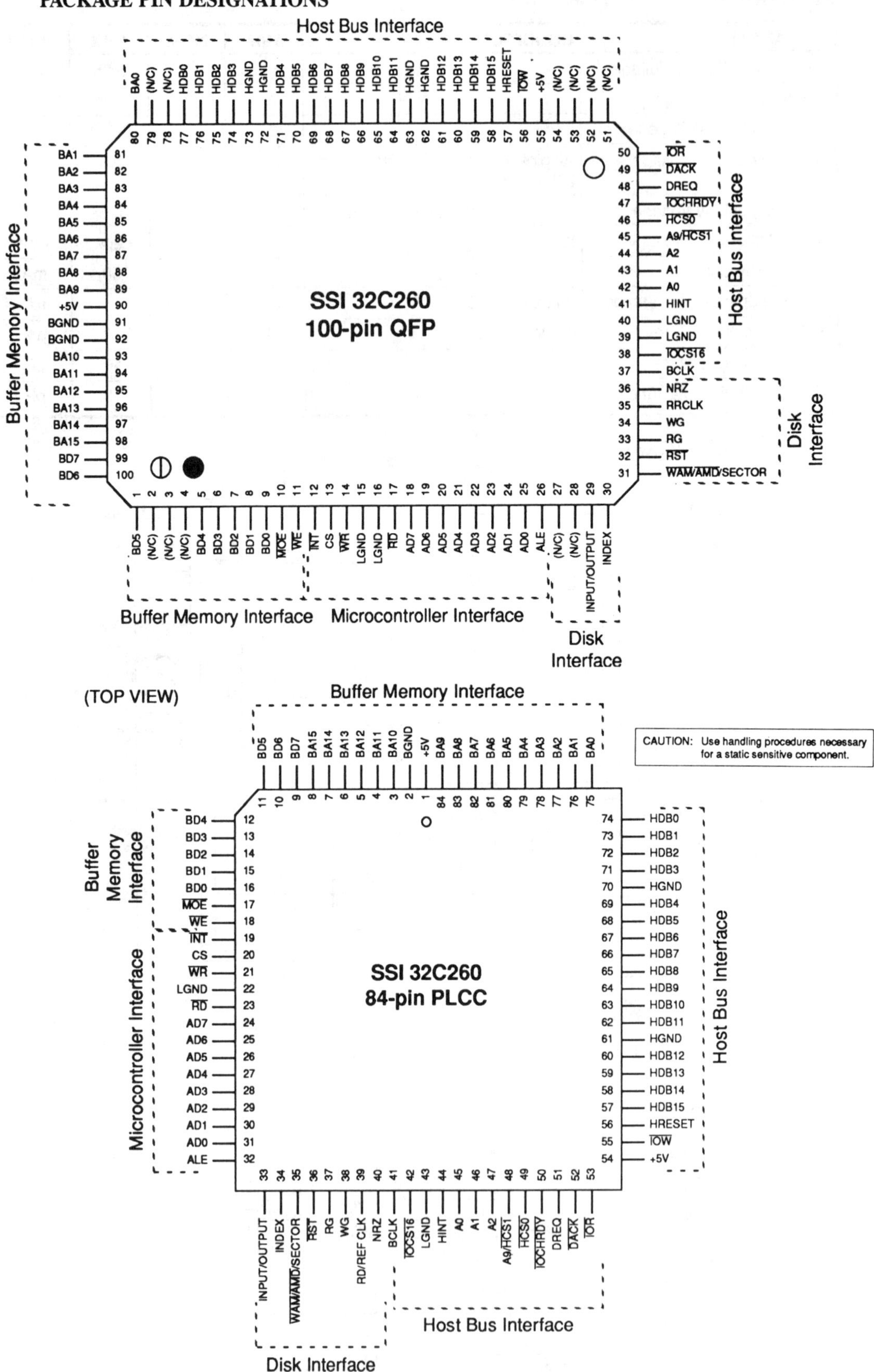

Advance Information: Indicates a product still in the design cycle, and any specifications are based on design goals only. Do not use for final design.

No responsibility is assumed by Silicon Systems for use of this product nor for any infringements of patents and trademarks or other rights of third parties resulting from its use. No license is granted under any patents, patent rights or trademarks of Silicon Systems. Silicon Systems reserves the right to make changes in specifications at any time without notice. Accordingly, the reader is cautioned to verify that the data sheet is current before placing orders.

Silicon Systems
SSI 32C452
Storage Controller
Preliminary Data

- Supports ST506/412, ST412HP, SA100, SMD, ESDI and custom interfaces
- Operates with 16 MHz microprocessors
- Internal RAM-based control sequencer
- Internal user programmable ECC to 32 bits
- Non-interleaved data transfer to 20 Mbit/s
- Hard or soft sector formats
- Programmable sector lengths up to a full track
- High performance, low power CMOS device
- Plug and software compatible with AIC-010F storage controller
- Single 5 volt supply
- Available in 44-pin PLCC or 40-pin-DIP package

ELECTRICAL SPECIFICATIONS

ABSOLUTE MAXIMUM RATINGS

PARAMETER	RATING	UNIT
Ambient Temperature Under Bias	0 to 70	°C
Storage Temperature	-65 to 150	°C
Voltage On Any Pin With Respect To Ground	GND -0.5 or VCC + 0.5	V
Power Supply Voltage	7.0	V
Max Current Injection	25	mA

NOTE: Stress above those listed under Absolute Maximum Ratings may cause permanent damage to the device. This is a stress rating only and functional operation of the device at these or any conditions above those indicated in the operational sections of this specification is not implied. Exposure to absolute maximum rating conditions for extended periods may affect device reliability.

RECOMMENDED OPERATING CONDITIONS

PARAMETER	CONDITIONS	MIN	NOM	MAX	UNIT
VCC Supply Voltage		4.75		5.25	V
TA Operating Free Air Temp.		0		70	°C
Input Low Voltage		0		0.4	V
Input High Voltage		2.4		VCC	V

DC CHARACTERISTICS

TA = 0°C to 70°C, VCC = 5V ± 5%, unless otherwise specified.

PARAMETER	CONDITIONS	MIN	NOM	MAX	UNIT
VIL Input Low Voltage		-0.5		0.8	V
VIH Input High Voltage		2.0		VCC + .5	V
VOL Output Low Voltage	IOL = 4 mA for RG and WG			0.45	V
	IOL = 2 mA all others				
VOH Output High Voltage	IOH = 400 mA			2.4	V
ICCS Supply Current Standby	Inputs at GND or VCC			25	mA
ICC Supply Current				85	mA
Power Dissipation				500	mW
IL Input Leakage	0V < Vin < VCC	-10		10	µA
IOL Output Leakage	0.45V < Vout < VCC	-10		10	µA
Cin Input Capacitance				10	pF
Cout Output Capacitance				10	pF

MICROPROCESSOR INTERFACE REGISTERS

DLR 4DH Read only

DATA LATCH REGISTER
When a microprocessor read from location 70H is detected, the data on the buffer memory bus (D0-D7) is latched by the SSI 32C452 into the DATA LATCH REGISTER. When the microprocessor accesses DLR this data is placed on the address/data bus (AD0-AD7).

SPECIAL ADDRESS DECODES 50H-51H Read/Write

Special decodes
Microprocessor accesses to these locations will cause the address/data bus (AD0-AD7) and the buffer data bus (D0-D7) to be bridged together internally (see external register description).

BUFACC 70H Read/Write

BUFFER ACCESS
Microprocessor accesses to this location cause the address/data bus (AD0-AD7) and the buffer data bus (D0-D7) to be bridged together internally. If a read cycle is performed, the data present will be latched into register DLR as well.

TEST REGISTERS

These registers may not be accessed while the sequencer is running.

TEST0 49H Read only

TEST REGISTER 0
Access to the Next Address field of the current sequencer instruction.

TEST1 4AH Read only

TEST REGISTER 1
Access to the Control field of the current sequencer instruction.

TEST2 4BH Read only

TEST REGISTER 2
Access to the Count/Data Type field of the current sequencer instruction.

TEST3 4CH Read only

TEST REGISTER 3
Access to the Data field of the current sequencer instruction.

EXTERNAL REGISTERS (FOR REFERENCE ONLY)

HOSTL 50H Read/Write

HOST BUS (LOWER BYTE)
External hardware may be used to connect the lower byte of the host bus to the buffer memory when this address is accessed.

HOSTH 51H Read/Write

HOST BUS (UPPER BYTE)
External hardware may be used to connect the upper byte of the host bus to the buffer memory when this address is accessed.

GPREG0 6EH Read/Write

GENERAL PURPOSE REGISTER 0
Systems which need extra I/O on the microprocessor data bus may take advantage of the strobes available on pins GPIO0 (write) and GPIO1 (read) to add an expansion port at this address.

GPREG1 6FH Read/Write

GENERAL PURPOSE REGISTER 1
Systems which need extra I/O on the microprocessor data bus may take advantage of the strobes available on pins GPIO2 (write) and GPIO3 (read) to add an expansion port at this address.

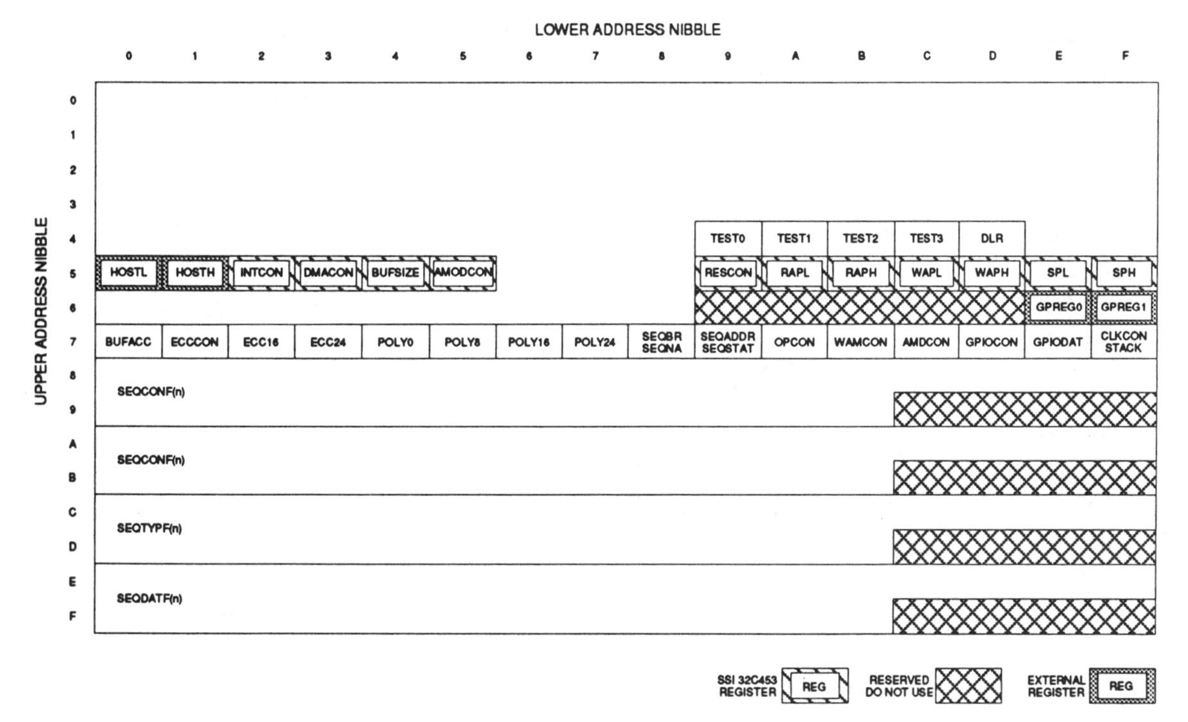

Register Address Map

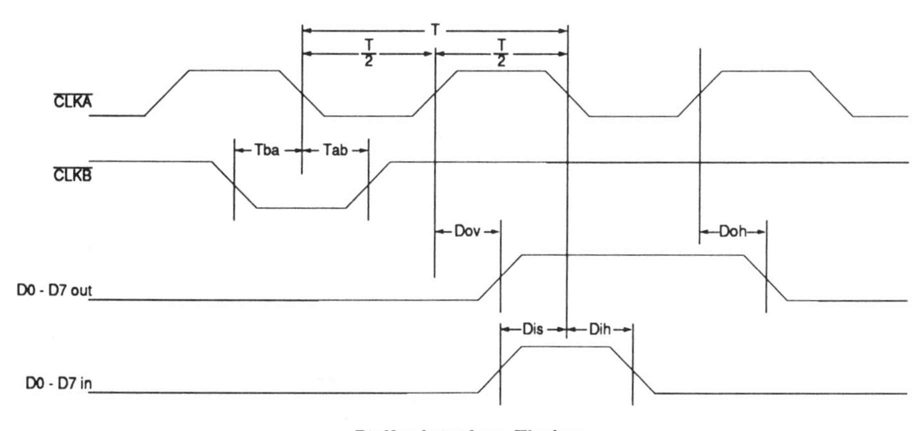

Buffer Interface Timing

REGISTER	ADDRESS	D7	D6	D5	D4	D3	D2	D1	D0	READ/ WRITE
TEST0	49H	SEQUENCER NEXT ADDRESS FIELD								R
TEST1	4AH	SEQUENCER CONTROL FIELD								R
TEST2	4BH	SEQUENCER COUNT/DATA TYPE FIELD								R
TEST3	4CH	SEQUENCER DATA FIELD								R
DLR	4DH	DATA LATCH REGISTER								R
BUFACC	70H	BUFFER MEMORY BYTE								R/W
ECCCON	71H	LEN1	LEN0	RESET	SECTBR	CLRECC	FEEDINH	ECCSHIFT	ECCIN	R/W
ECC16	72H	ECC23	ECC22	ECC21	ECC20	ECC19	ECC18	ECC17	ECC0/16	R
ECC24	73H	ECC31	ECC30	ECC29	ECC28	ECC27	ECC26	ECC25	ECC24	R
POLY0	74H	F7	F6	F5	F4	F3	F2	F1	F0	R/W
POLY8	75H	F15	F14	F13	F12	F11	F10	F9	F8	R/W
POLY16	76H	F23	F22	F21	F20	F19	F18	F17	F16	R/W
POLY24	77H	UNUSED	F30	F29	F28	F27	F26	F25	F24	R/W
SEQBR	78H	UNUSED			BRADR4	BRADR3	BRADR2	BRADR1	BRADR0	W
SEQNA	78H	TEST POINTS			NADR4	NADR3	NADR2	NADR1	NADR0	R
SEQADDR	79H	UNUSED			STADR4	STADR3	STADR2	STADR1	STADR0	W
SEQSTAT	79H	AMACTIVE	DATATRANS	BRACTIVE	STOPPED	UNUSED	ECCERR	COMPLO	COMPEQ	R
OPCON	7AH	CARRYINH	UNUSED	TRANSINH	SEARCHOP	SYNDET	NRZDAT	SECTORP	INDEXP	R/W
WAMCON	7BH	AM7 - AM0								R/W
AMDCON	7CH	AMD7 - AMD0								R/W
GPIOCON	7DH	RGFSEL	WGFSEL	RGESEL	WGESEL	GPDIR3	GPDIR2	GPDIR1	GPDIR0	R/W
GPIODAT	7EH	UNUSED		OUT	INP	GP3	GP2	GP1	GP0	R/W
CLKCON	7FH	CLKF2	CLKF1	UNUSED	CLKFO	CLKINH	SYN2	SYN1	SYN0	W
STACK	7FH	TOP OF STACK								R
SEQADDRF	80H	BRCON2	BRCON1	BRCON0	NEXT4	NEXT3	NEXT2	NEXT1	NEXT0	R/W
	9BH									
SEQCONF	A0H	SETWG	SETRG	RESWG	STACKEN	NRZINH	OUTPIN	COMPEN	DATEN	R/W
	BBH									
SEQTYPF	C0H	CNT7/ DTYP2	CNT6/ DTYP1	CNT5/ DTYP0	CNT4	CNT3	CNT2	CNT1	CNTO	R/W
	DBH									
SEQDATF		DATA FIELD								R/W

Register Bit Map

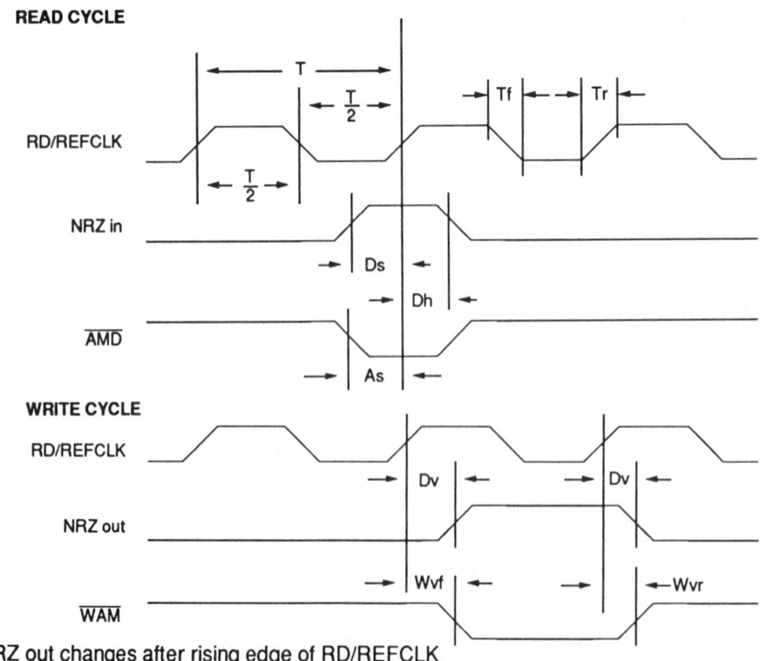

NOTE: NRZ out changes after rising edge of RD/REFCLK

Peripheral Device Interface Timing

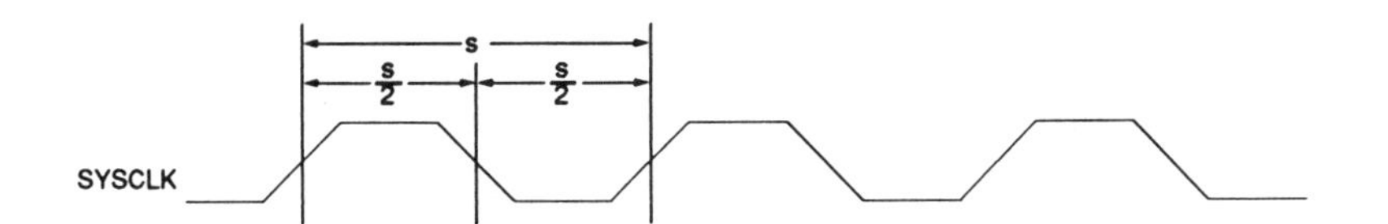

REGISTER WRITE OPERATION

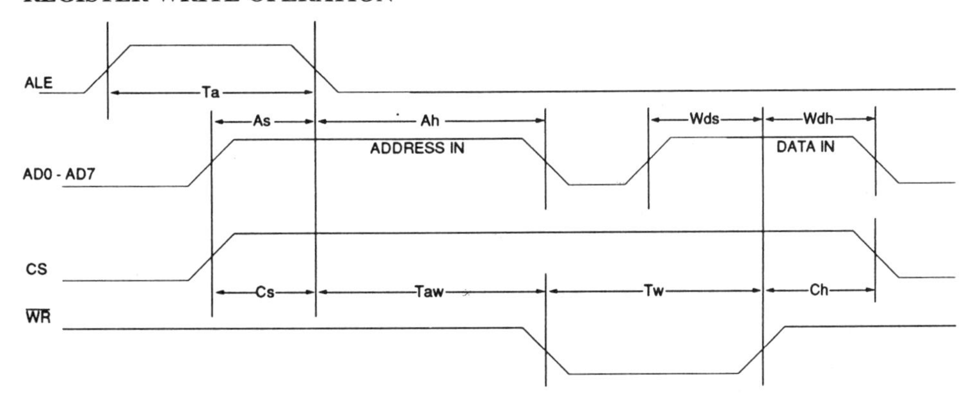

REGISTER READ OPERATION

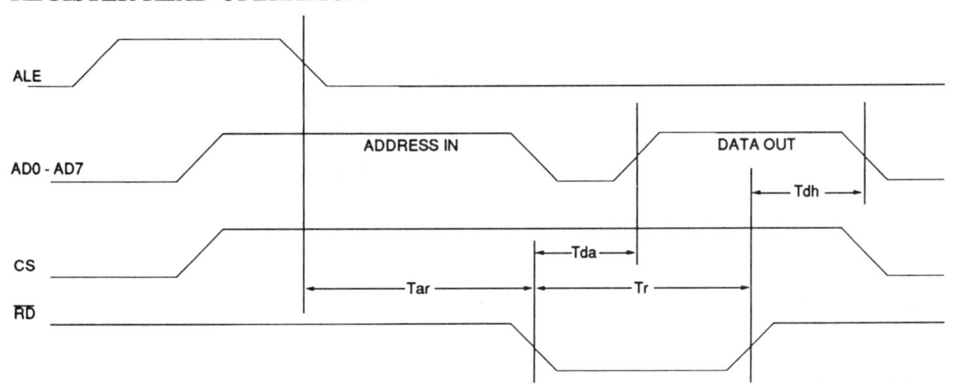

Microprocessor Interface Timing

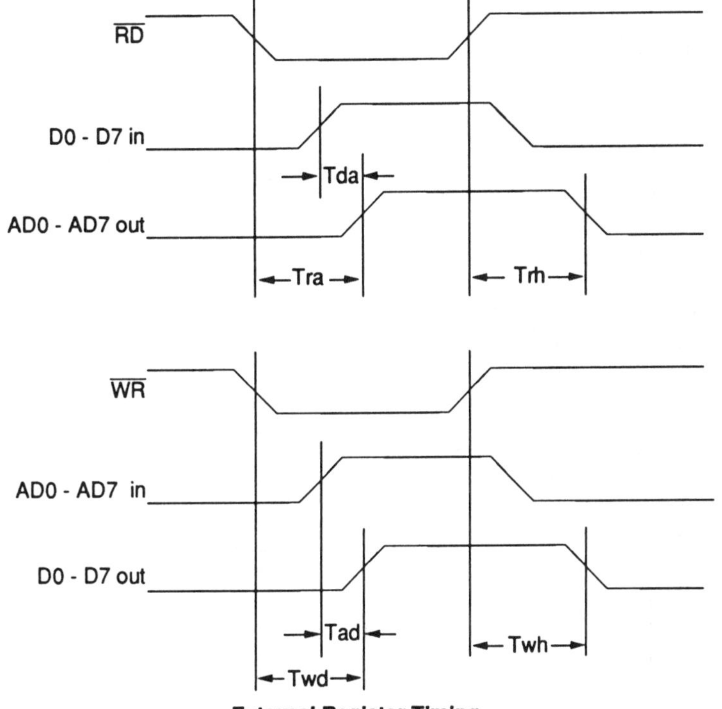

External Register Timing

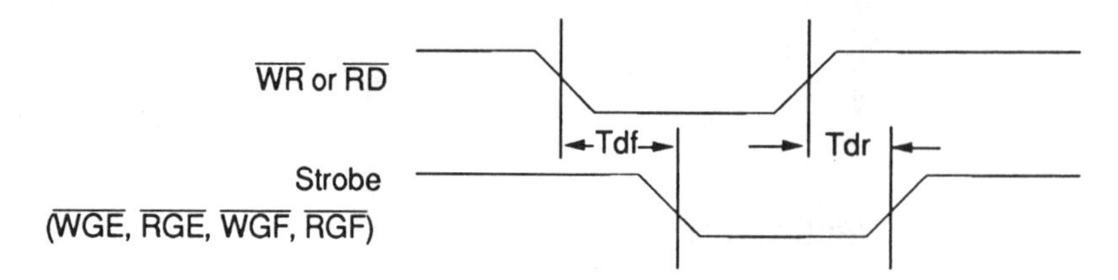

NOTE: The strobe signals are the alternate mode signals of the GPIO pins

Address Decode 6E and 6F Timing

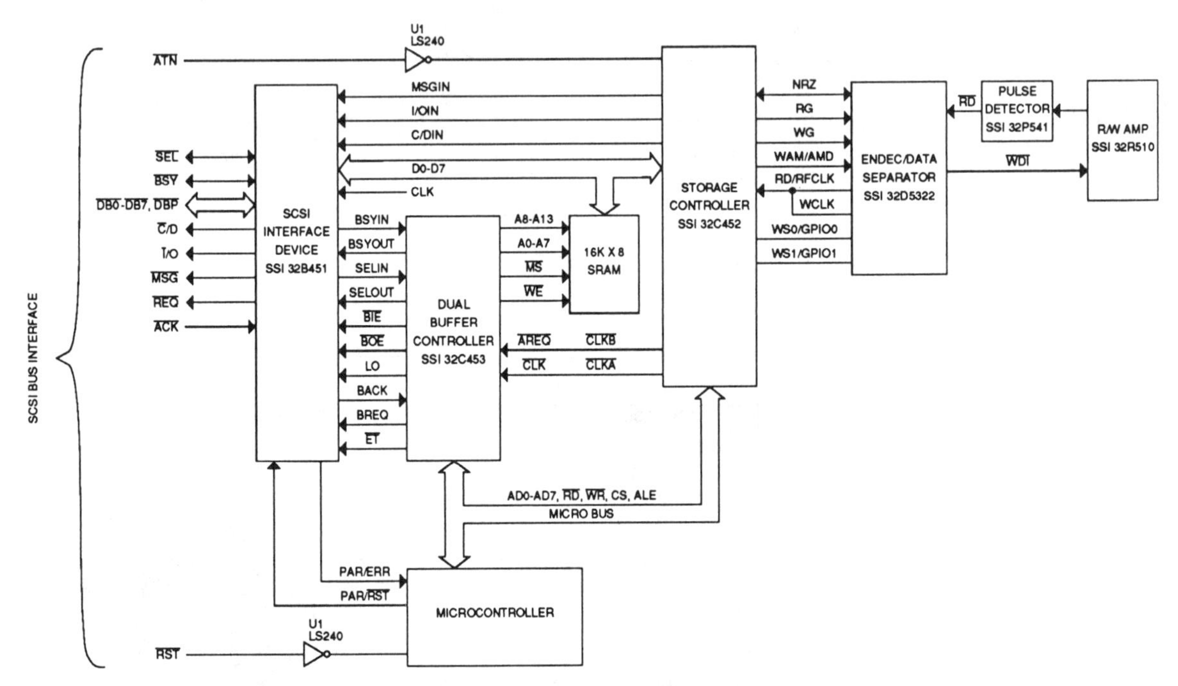

Partial Schematic for SCSI Implementation with Arbitration Support using Silicon Systems Microperipheral Devices

PACKAGE PIN DESIGNATIONS

(TOP VIEW)

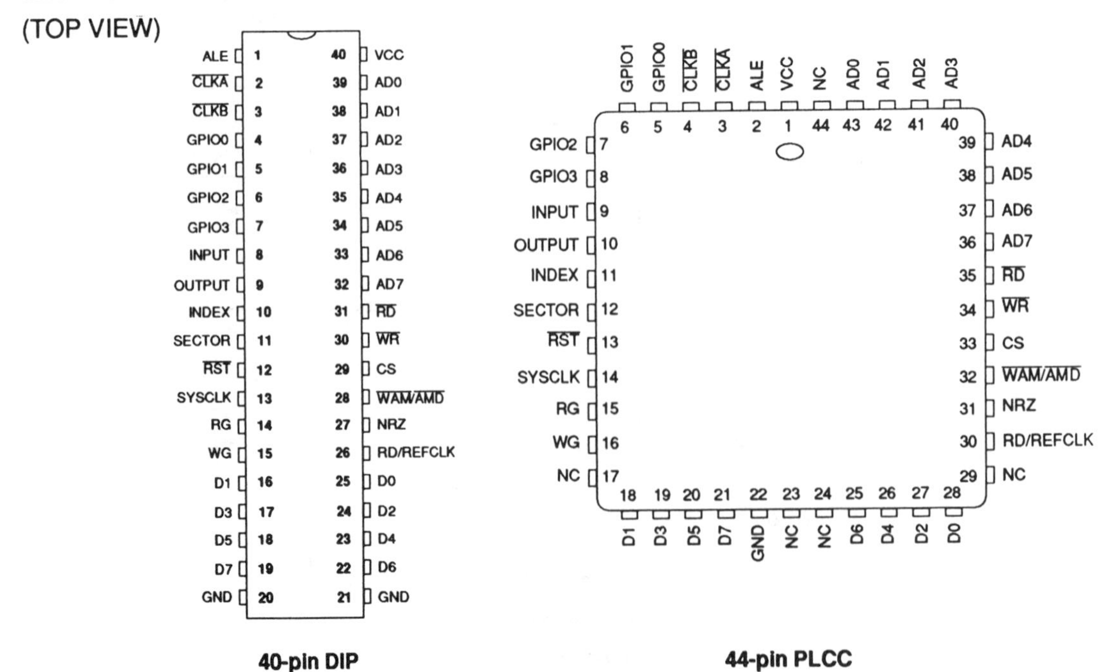

40-pin DIP

44-pin PLCC

BLOCK DIAGRAM

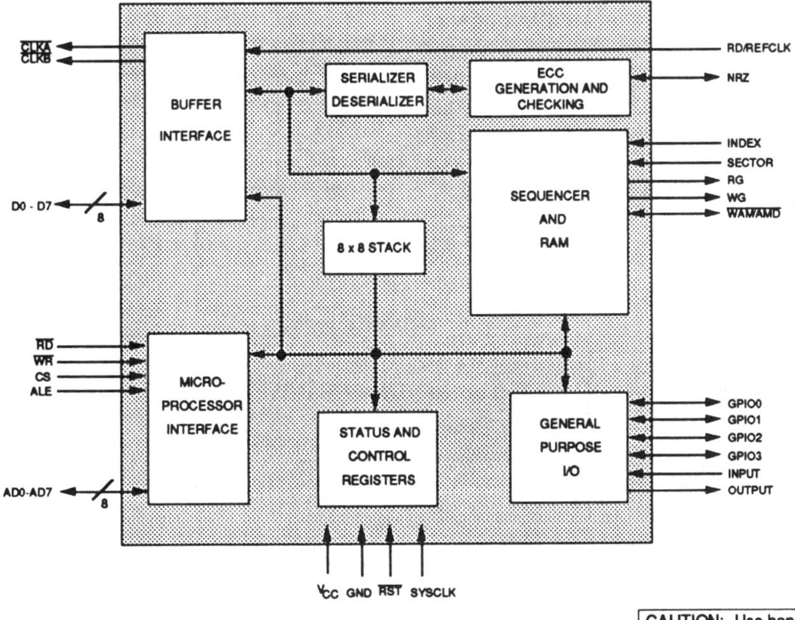

CAUTION: Use handling procedures necessary for a static sensitive components

Preliminary Data: Indicates a product not completely released to production. The specifications are based on preliminary evaluations and are not guaranteed. Small quantities are available, and Silicon Systems should be consulted for current information.

Silicon Systems SSI 32H101
Differential Amplifier

- Very narrow gain range
- 30-MHz bandwidth
- Electrically characterized at two power supply voltages: IBM Model 3340 compatible (8.3V) and standard OEM industry compatible (10V)
- Mechanically compatible with Model 3348-type head arm assembly
- SSI 32H1012 available to operate with a 12V power supply
- Packages include 8-pin DIP or SON

The SSI 32H101 is a two stage differential amplifier applicable for use as a preamplifier for the magnetic servo head circuit of Winchester technology disk drives.

ELECTRICAL CHARACTERISTICS

T_A = 25 °C, (V_{CC}-V_{EE}) = 8.3 to 10V ±10% (12V ±10% for 101-2)

ABSOLUTE MAXIMUM RATINGS

Operation above absolute maximum ratings may permanently damage the device.

PARAMETER	RATING	UNIT
Power Supply Voltage (V_{CC} - V_{EE})	12	V
SSI 32H1012	14	V
Differential Input Voltage	±1	V
Storage Temperature Range	-65 to 150	°C
Operating Temperature Range	0 to 70	°C

DC ELECTRICAL CHARACTERISTICS

PARAMETER	CONDITIONS	MIN	NOM	MAX	UNITS
Gain (differential)	R_P = 130Ω	77	93	110	
Bandwidth (3dB)	V_{IN} = 2 mVpp	10	20		MHz
Input Resistance		750		1200	Ω
Input Capacitance			3		pF
Input Dynamic Range (Differential)	R_L = 130Ω	3			mVpp
Power Supply Current	(V_{CC} - V_{EE}) = 9.15V		26	35	mA
	(V_{CC} - V_{EE}) = 11V		30	40	mA
	(V_{CC} - V_{EE}) = 13.2V (32H101A-2)		35	45	mA
Output Offset (Differential)	R_S = 0, R_L = 130Ω			600	mV
Equivalent Input Noise	R_S = 0, R_L = 130Ω, BW = 4 MHz		8	14	µV
PSRR, Input Referred	R_S = 0, f ≤ 5 MHz	50	65		dB
Gain Sensitivity (Supply)	Δ (V_{CC} - V_{EE}) = ±10%, R_L = 130Ω		±1.3		%
Gain Sensitivity (Temp.)	T_A = 25 °C to 70 °C, R_L = 130Ω		-0.2		%/°C
CMRR, Input Referred	f ≤ 5 MHz	55	70		dB

RECOMMENDED OPERATING CONDITIONS

PARAMETER	CONDITIONS	MIN	NOM	MAX	UNITS
Supply Voltage (V_{CC} - V_{EE})		7.45	8.3	9.15	V
		9.0	10.0	11.0	V
	32H1012 only	10.8	12.0	13.2	V
Input Signal V_{IN}			2		mVpp
Ambient Temp. T_A		0		70	C

BLOCK DIAGRAM

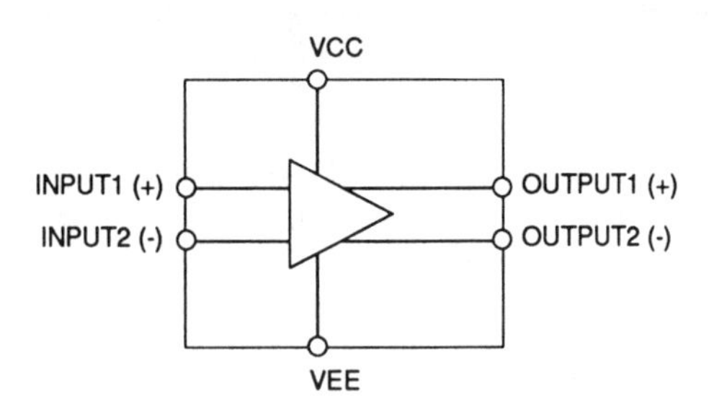

PIN DIAGRAM

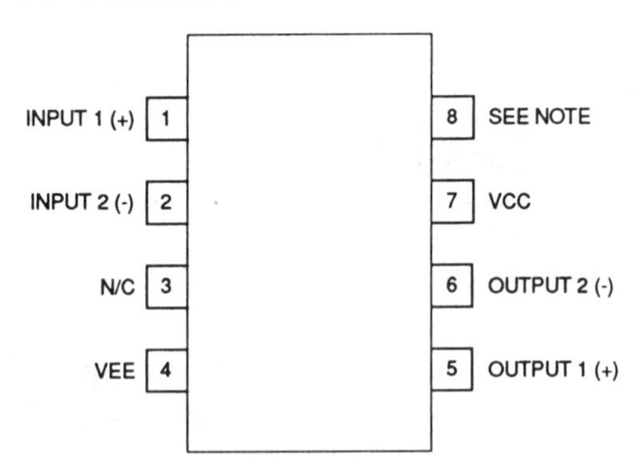

8-Pin PDIP, SON

Note : Pin must be left open and not connected to any circuit etch.

CAUTION: Use handling procedures necessary for a static sensitive component.

APPLICATIONS INFORMATION

CONNECTION DIAGRAM

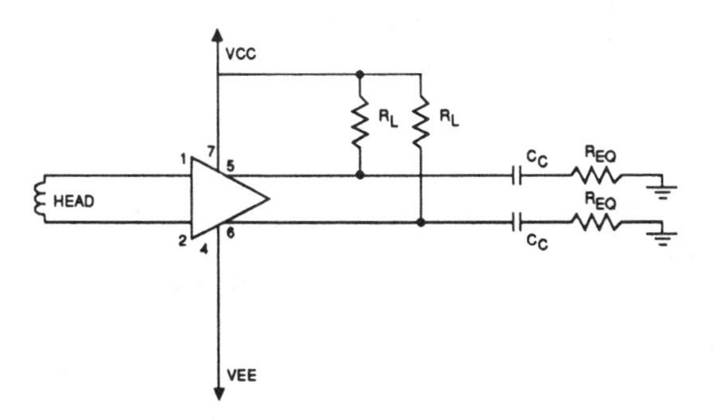

RECOMMENDED LOAD CONDITIONS

1. Input must be AC coupled
2. Cc's are AC coupling capacitors
3. RL's are DC bias and termination resistors (recommended 130Ω)
4. REQ represents equivalent load resistance
5. For gain calculations $R_P = \frac{R_L \cdot R_{EQ}}{R_L + R_{EQ}}$
6. Differential gain = 0.72 RP (± 18%) (RP in Ω)
7. Ceramic capacitors (0.1 μF) are recommended for good power supply noise filtering

---

Silicon Systems SSI 32H116A
Differential Amplifier

- Narrow gain range
- 50-MHz bandwidth
- IBM 3370/3380-compatible performance
- Operates on either IBM-compatible voltages (8.3V) or OEM-compatible (10V)
- Packages include 8-pin CERDIP, plastic DIP or SON, and custom 10-pin flatpack
- SSI 32H1162 available to operate with a 12V power supply

The SSI 32H116A is a high-performance differential applicable for use as a preamplifier for the magnetic servo thin film head in Winchester disk drives.

ELECTRICAL CHARACTERISTICS

Tj = 15 °C to 125 °C, (VCC-VEE) = 7.9V to 10.5V (to 13.2V for 32H1162)

ABSOLUTE MAXIMUM RATINGS

PARAMETER	RATING	UNIT
Power Supply Voltage (VCC-VEE)	12	V
SSI 32H1162	14	V
Differential Input Voltage	±1	V
Storage Temperature Range	-65 to 150	°C
Operating Ambient Temperature (TA)	15 to 60	°C
Operating Junction Temperature (TJ)	15 to 125	°C
Output Voltage	VCC -2.0 to VCC +0.4	V

RECOMMENDED OPERATING CONDITIONS

PARAMETER	CONDITIONS	MIN	NOM	MAX	UNIT
Supply Voltage (VCC-VEE)		7.45	8.3	9.15	V
		9.0	10.0	11.0	V
	SSI 32H1162 only	10.8	12.0	13.2	V
Input Signal Vin			1		mVpp
Ambient Temp TA		+15		+65	°C

DC ELECTRICAL CHARACTERISTICS

(Unless otherwise specified, recommended operating conditions apply.)

PARAMETER	CONDITIONS	MIN	NOM	MAX	UNIT
Gain (Differential)	Vin = 1mVpp, T_A = 25 °C, f = 1 MHz	200	250	310	mV/mV
Bandwidth (3dB)	Vin = 1mVpp, C_L = 15 pF	20	50		MHz
Gain Sensitivity (Supply)				1.0	%/V
Gain Sensitivity (Temp.)	15 °C < T_A < 55 °C		-0.16		%/C
Input Noise Voltage	Input Referred, Rs = 0		0.7	0.94	$nV/\sqrt{Hz}$
Input Capacitance (Differential)	Vin = 0, f = 5 MHz		40	60	pF
Input Resistance (Differential)			200		Ω
Common Mode Rejection Ratio Input Referred	Vin = 100 mVpp, f = 1 MHz	60	70		dB
Power Supply Rejection Ratio Input Referred	VEE + 100 mVpp, f = 1 MHz	46	52		dB
Input Dynamic Range (Differential)	AC input voltage where gain falls to 90% of its small signal gain value, f = 5MHz	±0.75			mV
Output Offset Voltage (Differential)	Vin = 0	-400		+400	mV
Output Voltage (Common Mode)	Inputs shorted together and Outputs shorted together	VCC-0.45	VCC-0.6	VCC-1.0	V
Single Ended Output Capacitance				10	pF
Power Supply Current	VCC-VEE = 9.15V		28	40	mA
	VCC-VEE = 11V		29	42	mA
	VCC-VEE = 13.2V, 32H1162 only		39	50	mA
Input DC Voltage	Common Mode		VEE +2.6		V
Input Resistance	Common Mode		80		Ω

PACKAGE PIN DESIGNATIONS

(TOP VIEW)

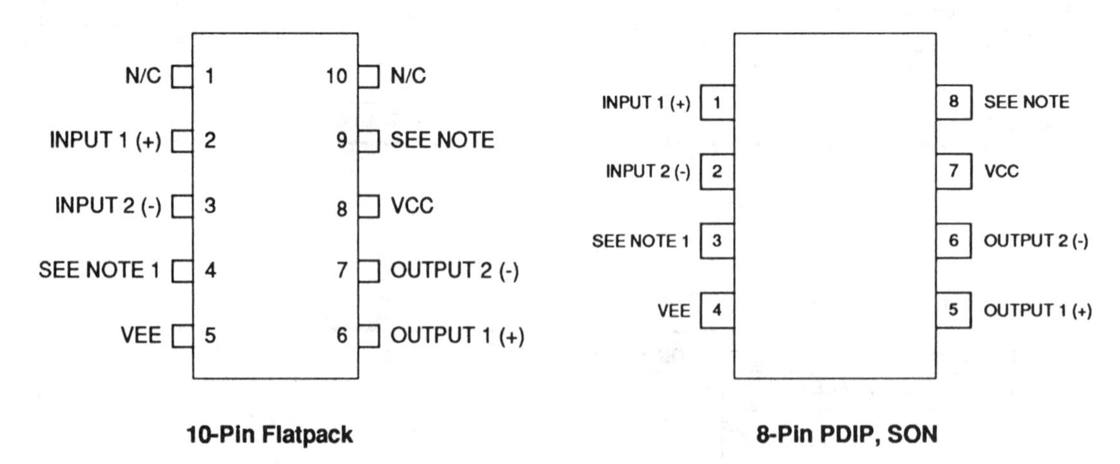

10-Pin Flatpack

8-Pin PDIP, SON

NOTE : Pin must be left open and not connected to any circuit etch.

CAUTION: Use handling procedures necessary for a static sensitive component.

BLOCK DIAGRAM

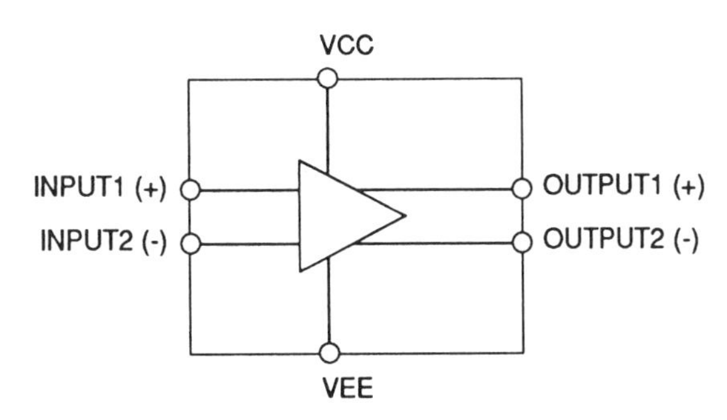

APPLICATIONS INFORMATION

CONNECTION DIAGRAM

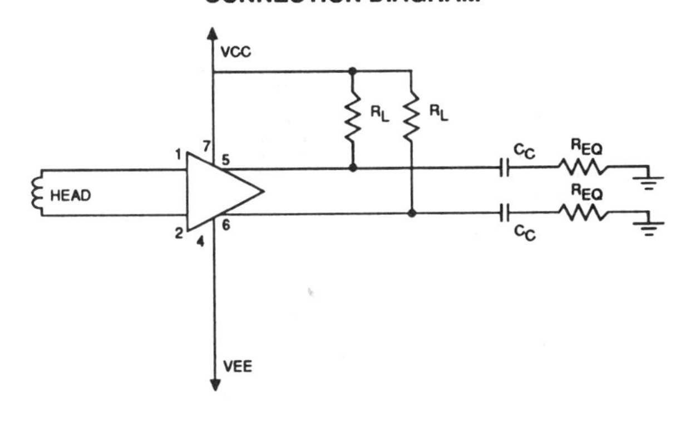

RECOMMENDED LOAD CONDITIONS

1. Input is directly coupled to the head
2. Cc's are AC coupling capacitors
3. RL's are DC bias and termination resistors, 100Ω recommended
4. REQ. represents equivalent load resistance
5. Ceramic capacitors (0.1 µF) are recommended for good power supply noise filtering

Silicon Systems
SSI 32H4631
Hybrid Servo and Spindle Motor Controller
Advance Information

- Head positioning
  - Embedded servo control with digital timing generator and ABCD burst sampling
  - Hybrid servo features with N/Q interface and track counter
  - Quadrature peak detection with sample/hold circuits
  - 250-kHz 8-bit A/D and D/A converters
  - Write protection of embedded servo frame
  - H-Bridge predriver compatible with +5V and +12V designs
- Spindle motor control
  - Sensorless motor commutation with precise speed regulation
  - Compatibility with Bipolar and Unipolar, Delta, and Star motors
  - Adjustable commutation delay for optimum motor operation
  - Microprocessor speed lock status
  - 3-Phase predriver compatible with +5V and +12V designs
- Internal registers addressed by Motorola or Intel µP interface
- Voltage fault protection, including R/$\overline{W}$ guarding, retract, and brake
- Low power, +5V-only operation, including programmable power-down modes
- Available in 100-pin QFP package

The SSI 32H4631 combines the head positioning and spindle motor control electronics along with voltage fault protection and a flexible microprocessor interface into a high-performance low-power CMOS integrated circuit. The spindle motor controller utilizes a sensor-less technique for motor commutation and provides precise motor speed regulation. The voice coil motor is controlled by a hybrid servo system that utilizes embedded servo control and an external dedicated servo demodulator like the SSI 32H6210, all under microprocessor direction. The SSI 32H4631 motor predrivers are compatible with external bridge output structures and can be configured with high-efficiency power FETs in order to maximize the power to the spindle and the voice coil motors. The 32H4631 requires only a +5V power supply, but is compatible with both +5V and +12V motor power drivers. The 32H4631 is available in a 100-pin QFP package.

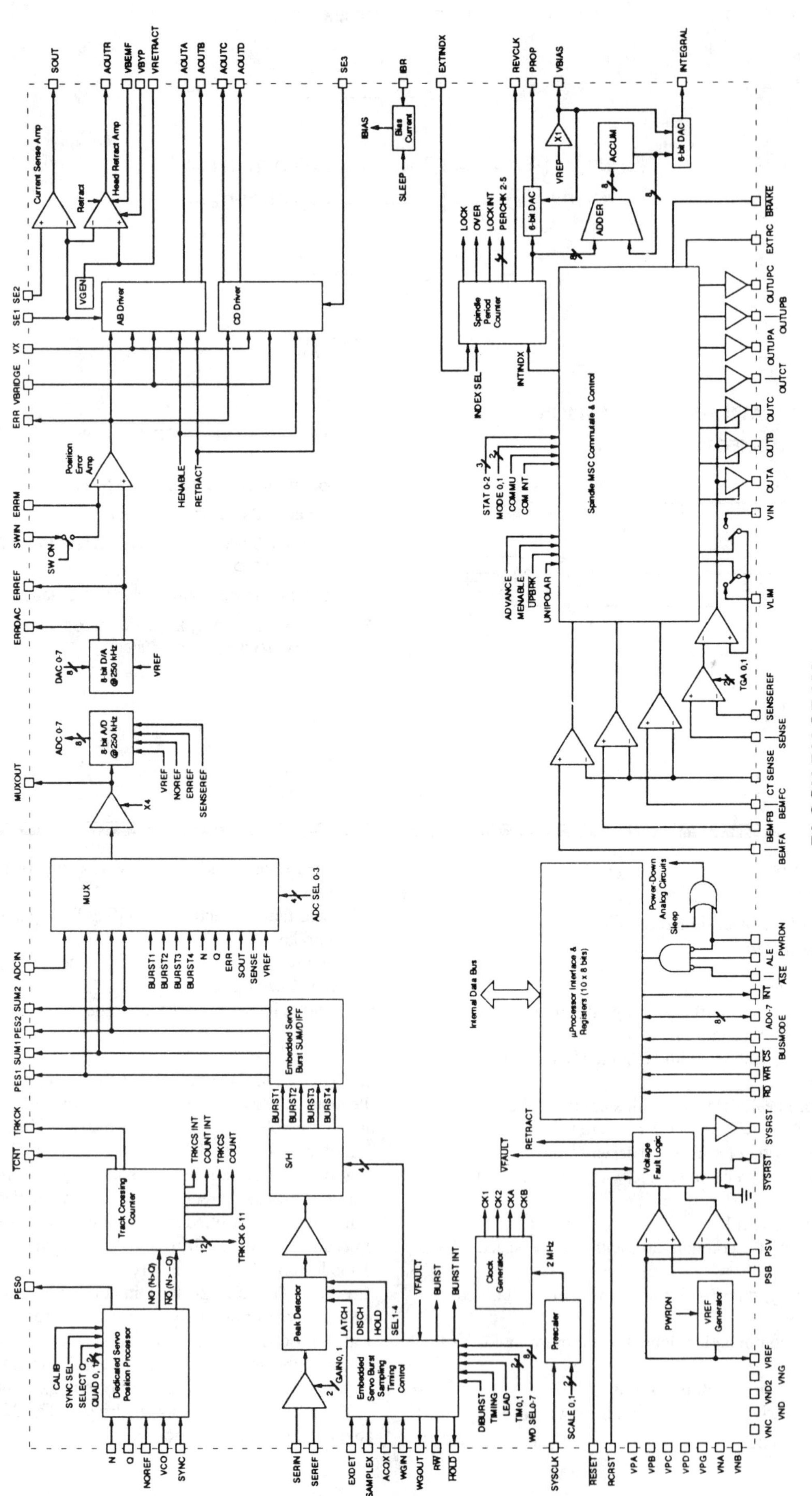

BLOCK DIAGRAM
SSI 32H4631

Silicon Systems SSI 32H6220
Servo Controller
Preliminary Data

- Servo control for Winchester disk drives with dedicated surface head positioning systems
- Accepts standard normal and quadrature position information
- 500-kHz maximum servo frame rate
- Microprocessor bus interface compatible with 16-MHz 8051
- Seek and track modes
- Separate position and velocity error outputs
- Programmable-velocity profile and loop gains
- Internal offset cancellation capability
- Track crossing output clock
- Low-power CMOS design
- Available in a 44-pin PLCC package

ELECTRICAL SPECIFICATIONS

ABSOLUTE MAXIMUM RATINGS

(Maximum limits indicate where permanent device damage occurs. Continous operation at these limits is not intended and should be limited to those conditions specified in the DC operating characteristics.)

PARAMETER	CONDITIONS	MIN	TYP	MAX	UNITS
VPA		0		14	V
Voltage on any pin		0		VPA+0.1V	V
Storage Temp.		-45		165	°C
Solder Temp.	10 sec duration			260	°C

RECOMMENDED OPERATION CONDITIONS (Unless otherwise noted, the following conditions are valid throughout this document.)

PARAMETER	CONDITIONS	MIN	TYP	MAX	UNITS
VPA, VPD		10.8		13.2	V
VDD		4.5		5.5	V
VREF		5.1	5.4	5.7	V
Operating temp.		0		70	°C
RBIAS, bias resistor to AGND		22.3	22.6	22.9	kΩ
Resistive loading (FP1, FV1, PE, FV4, VE)	About VREF	5			kΩ
Capacitive loading (FP1, FV1, PE, FV4, VE)				40	pF

DC CHARACTERISTICS

PARAMETER	CONDITIONS	MIN	TYP	MAX	UNITS
IVP Total VPA and VPD current				40	mA
IDD VDD current				10	mA
IREF VREF current				3	mA

DIGITAL I/O

PARAMETER	CONDITIONS	MIN	TYP	MAX	UNITS
Digital Inputs					
VIH	|IIH| < 10μA	2			V
VIL (Except Reset)	|IIL| < 10μA			0.7	V
VIL Reset Pin	|IIL| < 100μA			0.7	V
Digital Outputs (AD0-AD7, T/$\overline{S}$)					
VOH	|IOH| < 40μA	2.4			V
VOL	|IOL| < 1.6mA			0.4	V
Open Drain Digital Outputs ($\overline{INT}$, $\overline{OFFTRK}$)					
VOL	|IOL| < 1.6mA			0.4	V
Off leakage	VOH = VPD			10	μA

BLOCK DIAGRAM

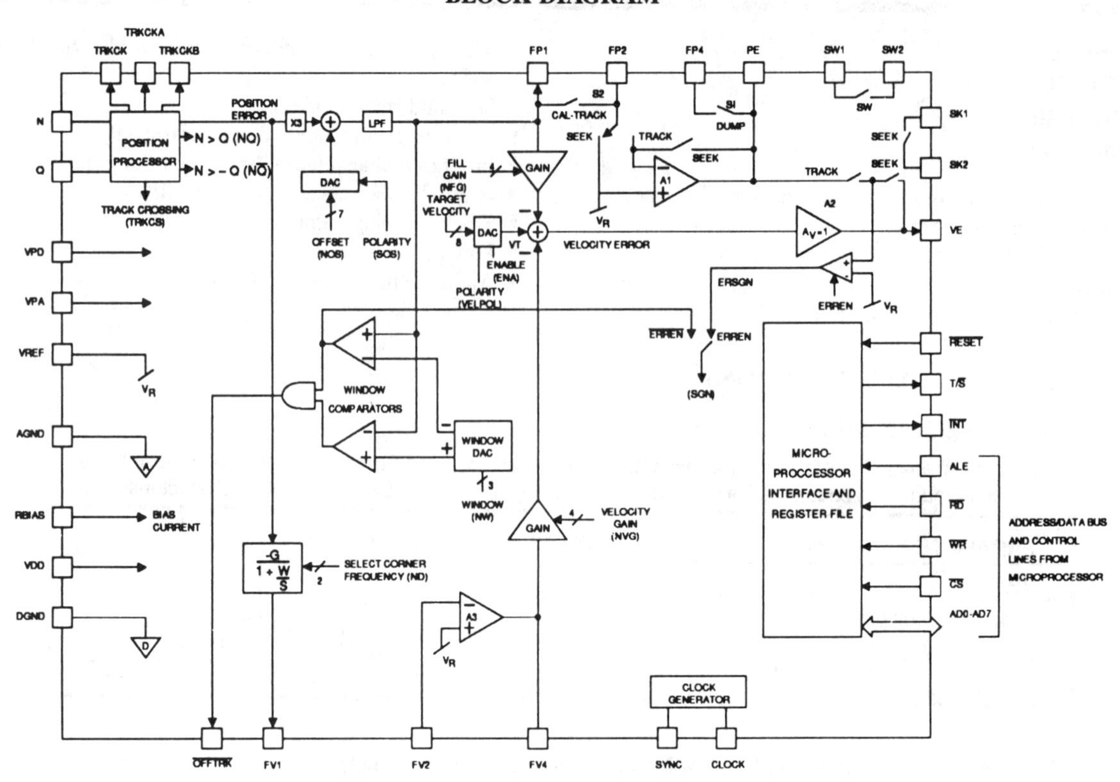

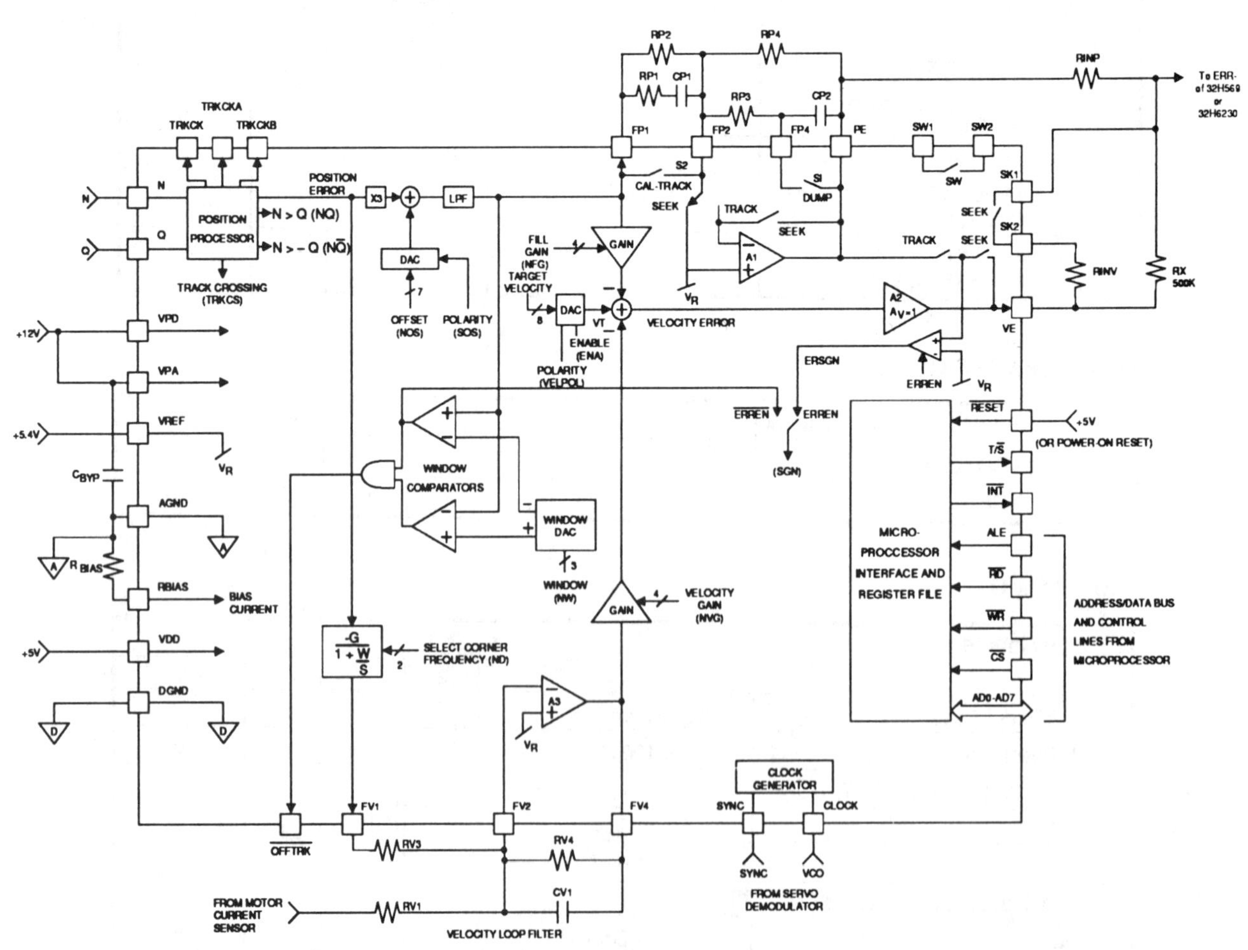

SSI 32H6220 Typical Application

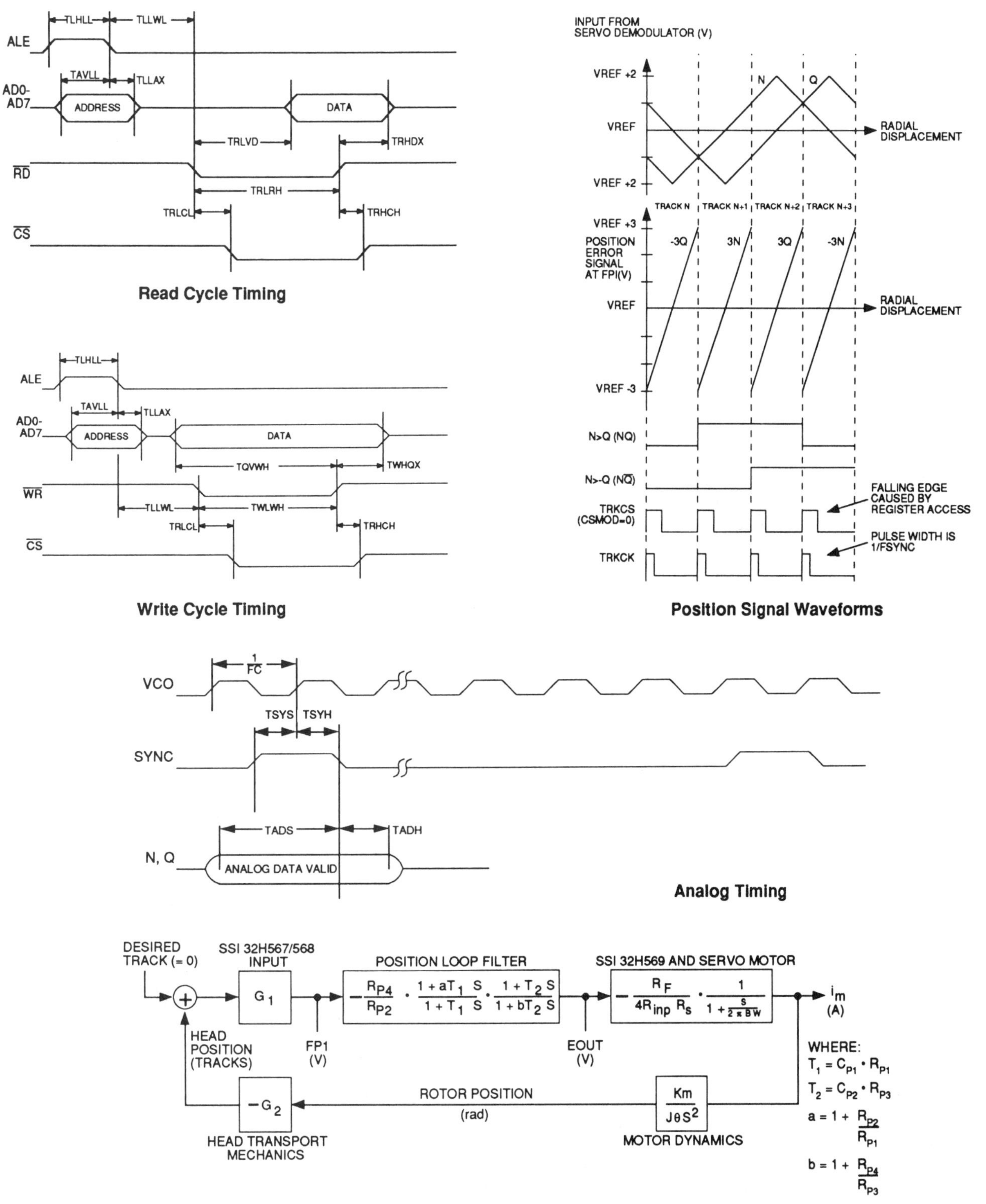

Read Cycle Timing

Write Cycle Timing

Position Signal Waveforms

Analog Timing

System Transfer Function in Track Mode

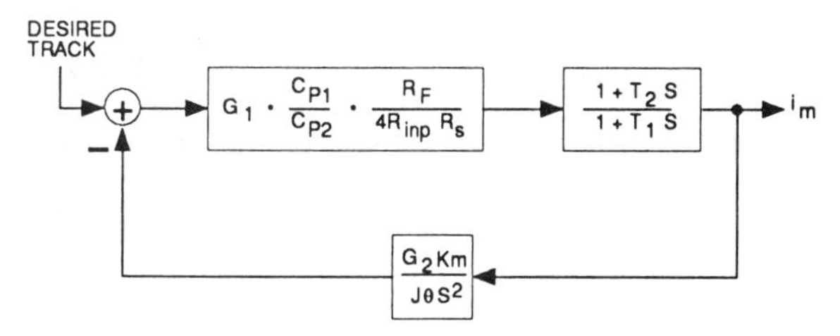

Simplified Track Mode Transfer Function

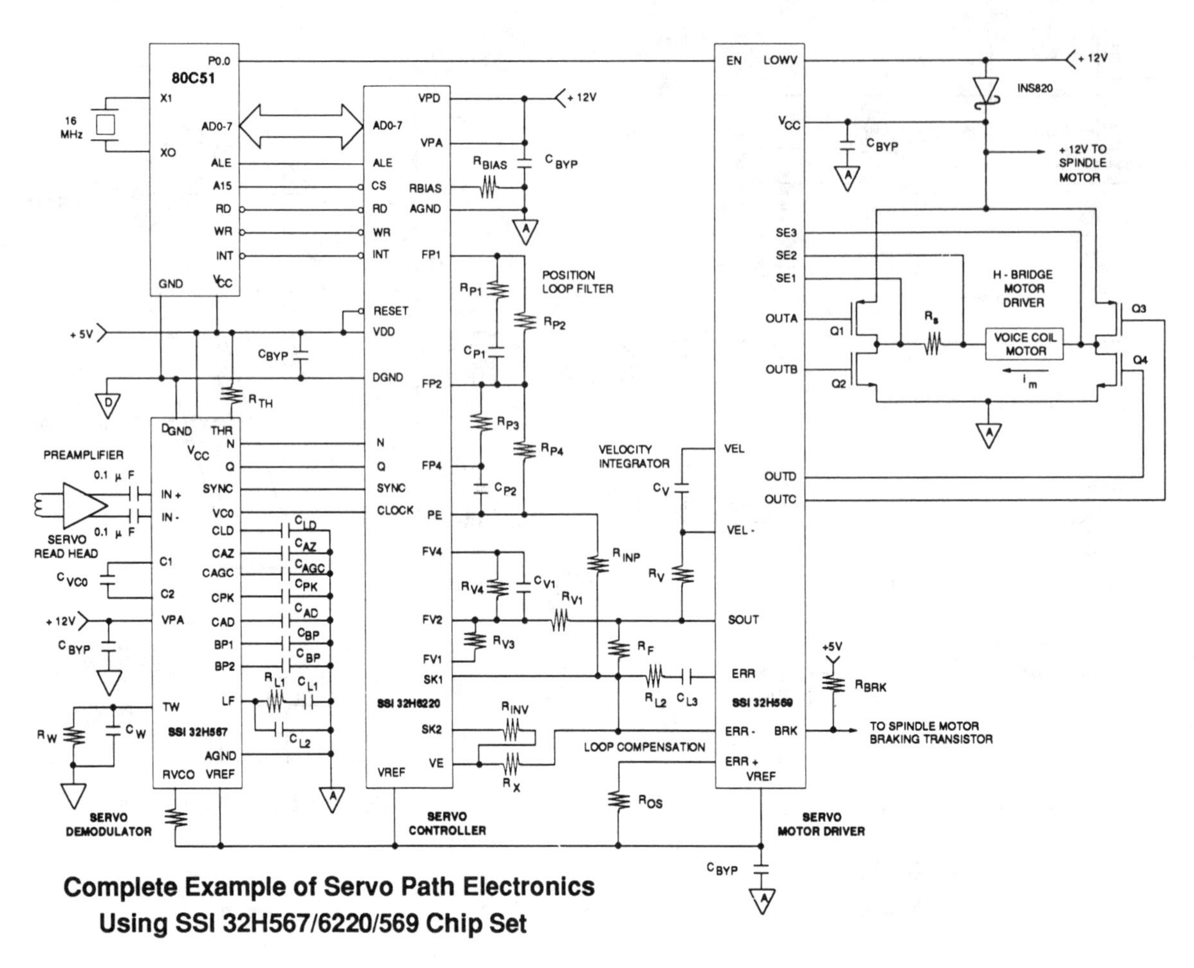

Complete Example of Servo Path Electronics Using SSI 32H567/6220/569 Chip Set

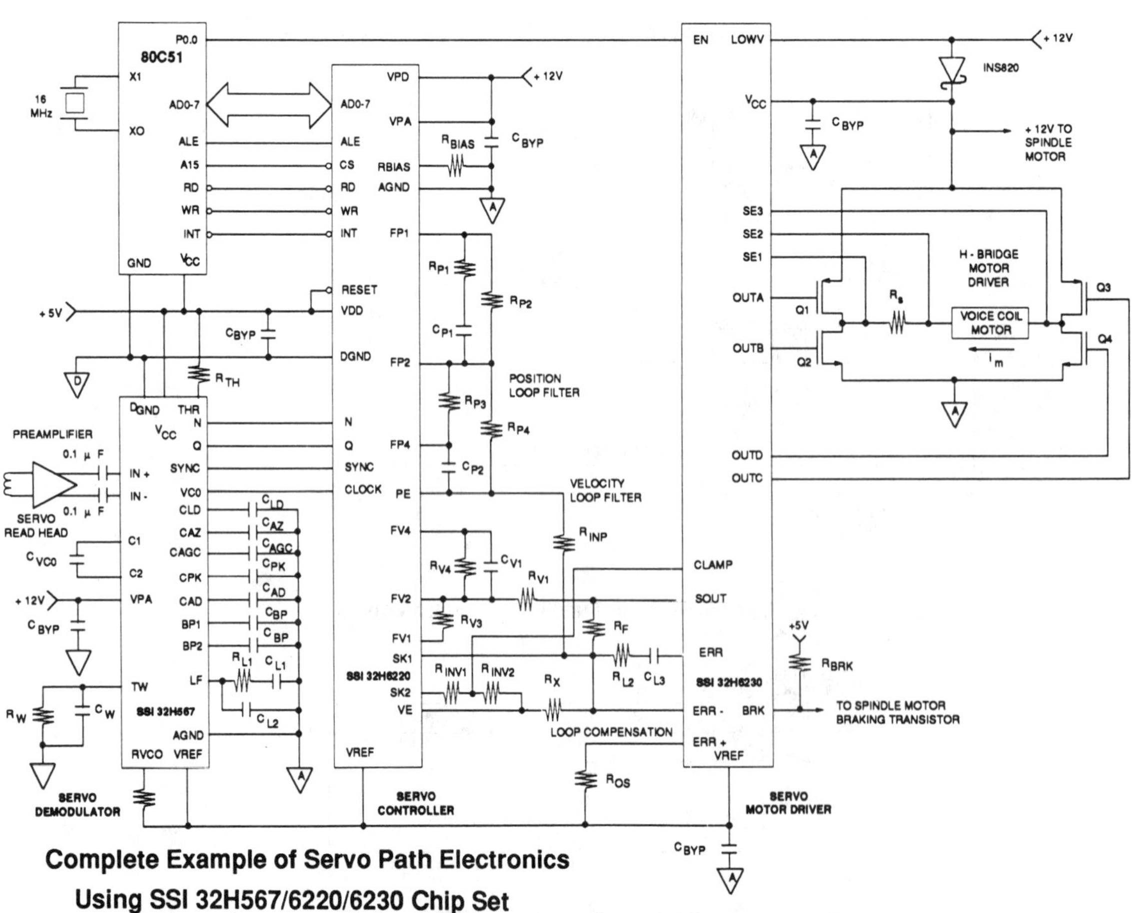

Complete Example of Servo Path Electronics Using SSI 32H567/6220/6230 Chip Set

PACKAGE PIN DESIGNATIONS

(TOP VIEW)

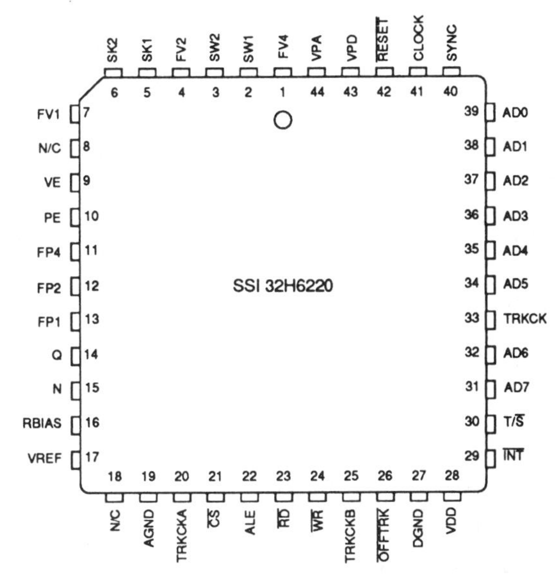

CAUTION: Use handling procedures necessary for a static sensitive component.

44-Pin PLCC

Preliminary Data: Indicates a product not completely released to production. The specifications are based on preliminary evaluations and are not guaranteed. Small quantities are available, and Silicon Systems should be consulted for current information.

Silicon Systems SSI 35P550

4-Channel Magnetic Tape Read Device

- 4-Channel multiplexer with differential-input preamplifiers
- Postamplifier has component-adjustable and programmable gain
- On-chip signal-level detector with programmable threshold and adjustable delay
- Data detection circuit includes spurious signal rejection (adjustable time domain filter) and provides an adjustable uniform data pulse output
- Available in 40-pin DIP or 44-pin PLCC plastic packages

ELECTRICAL CHARACTERISTICS

ABSOLUTE MAXIMUM RATINGS

Operation above absolute maximum ratings may permanently damage the device.

PARAMETER	RATING	UNIT
Storage Temperature	-65 to +150	°C
Junction Operating Temperature	0 to 130	°C
Supply Voltage, VCC1	-0.5 to +6.0	VDC
Supply Voltage, VCC2	-0.5 to +14.0	VDC
Voltage Applied to Logic Inputs	-0.5 to VCC1 +0.5	VDC
Voltage Applied to OFF Logic Outputs	-0.5 to VCC1 +0.5	VDC
Current Into ON Logic Outputs	5.0	mA
Lead Temperature (soldering, 10 sec)	+260	°C

DC CHARACTERISTICS

(Unless otherwise specified, VCC1 = 4.75V to 5.25V, VCC2 = 11.4V to 12.6V, Ta = 0 to 70 °C.)

CHARACTERISTICS	CONDITIONS	MIN	NOM	MAX	UNITS
Input Current Logical Inputs HIGH	Vih = VCC1			100	µA
Input Current Logical Inputs LOW	Vil = 0V			-400	µA
Output Voltage Delay Comparator OFF	Ioh = -400 µA	2.4			V
Output Voltage Delay Comparator ON	Iol = 2.0 mA			0.5	V
Data Pulse Inactive Level Output Voltage	Ioh = -400 µA	2.4			V
Data Pulse Active Level Output Voltage	Iol = 2.0 mA			0.5	V
VCC1 Power Supply Current	No Head Inputs			30	mA
VCC2 Power Supply Current	No Head Inputs			62	mA
NOTE: Characteristic applies to Inputs S0, S1, G0, G1, G2, T0, T1					

PREAMPLIFIER AND MULTIPLEXER CHARACTERSITICS

Output Load = 2 kΩ line-line, Channel Select Signals (S0,S1): VON = 2V Min., VOFF = 0.8V Max.

CHARACTERISTICS	CONDITIONS	MIN	NOM	MAX	UNITS
Differential Voltage Gain	Vin = 4 mVpp @ 100 kHz ref. to CT VOLT	80		120	V/V
Gain Flatness	Vin = 4 mVpp DC to 0.5 MHz ref. to CT VOLT	±0.5			dB
Bandwidth, -1 dB	Vin = 4 mVpp	1.5			MHz
Bandwidth, -3 dB	Vin = 4 mVpp	3.0			MHz
Differential Input Impedance	Vin = 4 mVpp @ 100 kHz ref to CT VOLT	10			kΩ
Common-Mode Rejection Ratio	Vin = 300 mVpp @ 500 kHz Inputs shorted to CT VOLT	50			dB
Power Supply Rejection Ratio	Δ VCC = 300 mVpp @ 500 kHz Inputs shorted to CT VOLT	50			dB
Channel Isolation	Unselected Vin = 100 mVpp @ 2 MHz. Selected Channel inputs connected to CT VOLT	60			dB
Total Harmonic Distortion	Vin = 0.5 to 6.0 mV pp @ 500 kHz			2	%
Equivalent Input Noise	Power BW = 10 kHz to 1MHz Inputs shorted to CT VOLT			10	µVrms
Small Signal Single-Ended Output Res.	Io = 1 mApp @ 100 kHz			35	Ω
Maximum Diff. Output Voltage	Freq = 100 kHz THD ≤ 5%	3			Vpp
Output Offset Voltage	Inputs shorted to CT VOLT Volt Load = Open Circuit			±1.0	V
Common-Mode Output Voltage	Inputs shorted to CT VOLT Volt Load = Open Circuit	2.68		3.5	V
Center Tap Voltage, CT VOLT			3.0		V

SIGNAL LEVEL DETECT CIRCUITS CHARACTERISTICS

Level Comparator Inputs connected in parallel with Differentiator Inputs. Vin (Level Comp) = 100 kHz sine wave, ac-coupled. RDS1 = 5 kΩ; RDS2, CDS = open

CHARACTERISTICS	CONDITIONS	MIN	NOM	MAX	UNITS
Level Comparator Input Thresholds, Single-Ended, Each Input	T0 VT0 = 0.8V VT1 = 0.8V Vo pulse value < 0.5V at MAX LIMIT, >VCC1 - 0.5V at MIN LIMIT	30		70	mV pk
	T1 VT0 = 2.0V VT1 = 0.8V Vo pulse Value <0.5V at MAX LIMIT, >VCC1 - 0.5V at MIN LIMIT	97		153	mV pk
	T2 VT0 = 0.8V VT1 = 2.0V Vo pulse value <0.5V at MAX LIMIT, >VCC1 - 0.5V at MIN LIMIT	138		202	mV pk
	T3 VT0 = 2.0V VT1 = 2.0V Vo pulse value <0.5V at MAX LIMIT, >VCC1 - 0.5V at MIN LIMIT	210		290	mV pk
Level Comparator Diff. Input Resistance	Vin = 5 Vpp @ 100 kHz	5			kΩ
Level Comparator Off Output Leakage	Vo = VCC1			25	μA
Level Comparator ON Output Voltage	VT0 = 0.8V VT1 = 0.8V Vin = ±140 mV diff. dc Io = 2.0 mA			0.25	V
Delay Comparator Upper Threshold Voltage	Vo > 2.4V	.65VCC1		.75VCC1	V
Delay Comparator Lower Threshold Voltage	Vo < 0.5V	.25VCC1		.35VCC1	V
Delay Comparator Input Current	0V < Vin < VCC1			25	μA

DATA DETECTION CIRCUIT CHARACTERISTICS

Vin = 1.0Vpp diff. square wave, Tr, Tf < 20 ns, dc-coupled (for biasing).
RD = 2.5 kΩ; CD = 0.1 μF; RTD = 7.8 kΩ; CTD = 200 pF; RDP = 3.9 kΩ;
CDP = 100 pF. Data Pulse load = 2.5 kΩ to VCC1 plus 20 pF or less to PWR GND.

CHARACTERISTICS	CONDITIONS	MIN	NOM	MAX	UNITS
Differentiator Maximum Differential Input Voltage	Vin = 100 kHz sine wave, dc-coupled. < 5% THD in voltage across CD. CD = 620 pF RD = 0	5.0			Vpp
Differentiator Input Impedance	Vin = 4Vpp diff., 100 kHz sine wave. CD = 620 pF RD = 0	10			kΩ
Differentiator Threshold Differential Input Voltage	Vin = 100 kHz square wave, Tr, Tf , 0.4 μs, no overshoot. Data Pulse from each Vin transition.			300	mVpp
Data Pulse Width Accuracy	TDP = .59 RDP X CDP, RDP = .85 TDP 3.9 kΩ to10 kΩ, CDP = 75 pF to 300 pF. Width measured at 1.5V amplitude	.85TDP		1.15TDP	sec

DATA DETECTION CIRCUIT CHARACTERISTICS

CHARACTERISTICS	CONDITIONS	MIN	NOM	MAX	UNITS
Time Domain Filter Delay Accuracy	TTD = 0.59 RTD X CTD + 50 ns, RTD = 3.9 kΩ to 10 kΩ, CTD =100 pF to 750 pF Delay measured from 50% input amplitude to 1.5V Data Pulse amplitude	.85TTD		1.15TTD	sec
Data Pulse Width Drift from + 25 °C value	Width measure from 1.5V amplitude			±5.0	%
Time Domain Filter Delay Drift from +25 °C value	Delay measured from 50% Input amplitude to 1.5V Data Pulse amplitude			±5.0	%
Note: Differentiating network impedance should be chosen such that 1 mA peak current flows at maximum signal level and frequency.					

POSTAMPLIFIER CHARACTERISTICS

Output Load = 2.5 kΩ + 0.1 μF line-line, Vin = 100 mVpp, 100 kHz sine wave, dc-coupled (to provide proper biasing). CG = 0.1 μF, RG = 0.

CHARACTERISTICS	CONDITIONS	MIN	MAX	UNITS
Differential Voltage Gain	A0 VG0 = 0.8V VG1 = 0.8V VG2 = 0.8V	A7 - 14.75	A7 - 13.25	dB
	A1 VG0 = 2.0V VG1 = 0.8V VG2 = 0.8V	A7 - 12.75	A7 - 11.25	dB
	A2 VG0 = 0.8V VG1 = 2.0V VG2 = 0.8V	A7 - 10.75	A7 - 9.25	dB
	A3 VG0 = 2.0V VG1 = 2.0V VG2 = 0.8V	A7 - 8.75	A7 - 7.25	dB
	A4 VG0 = 0.8V VG1 = 0.8V VG2 = 2.0V	A7 - 6.75	A7 - 5.25	dB
	A5 VG0 = 2.0V VG1 = 0.8V VG2 = 2.0V	A7 - 4.75	A7 - 3.25	dB
	A6 VG0 = 0.8V VG1 = 2.0V VG2 = 2.0V	A7 - 2.75	A7 - 1.25	dB
	A7 VG0 = 2.0V VG1 = 2.0V VG2 = 2.0V	32	36	dB
	ARG VG0 = 2.0V VG1 = 2.0V VG2 = 2.0V when RG = 2.5 kΩ	A7 - 7.5	A7 - 4.5	dB
Differential Input Impedance	VG0 = 2.0V VG1 = 2.0V VG2 = 2.0V	10		kΩ
Bandwidth, 1dB	VG0 = 2.0V VG1 = 2.0V VG2 = 2.0V	1.5		MHz
Bandwidth, 3dB	VG0 = 2.0V VG1 = 2.0V VG2 = 2.0V	3.0		MHz
Maximum Diff. Output Voltage	VG0 = 0.8V VG1 - 0.8V VG2 = 0.8V VIN = 100 kHz sine wave THD < 5%	5		Vpp
Small Signal Single-Ended Output Res	VG0 = 2.0V VG1 = 2.0V VG2 = 2.0V VIN = 0V Io = 1 mApp, 100 kHz		35	Ω
Input Bias Offset Voltage Range	VG0 = 0.8V VG1 = 0.8V VG2 = 0.8V THD < 2.0%		±1.0	V
Input Bias Common-Mode Voltage Range	VG0 = 0.8V VG1 = 0.8V VG2 = 0.8V THD < 2.0%	2.68	3.5	V

BLOCK DIAGRAM

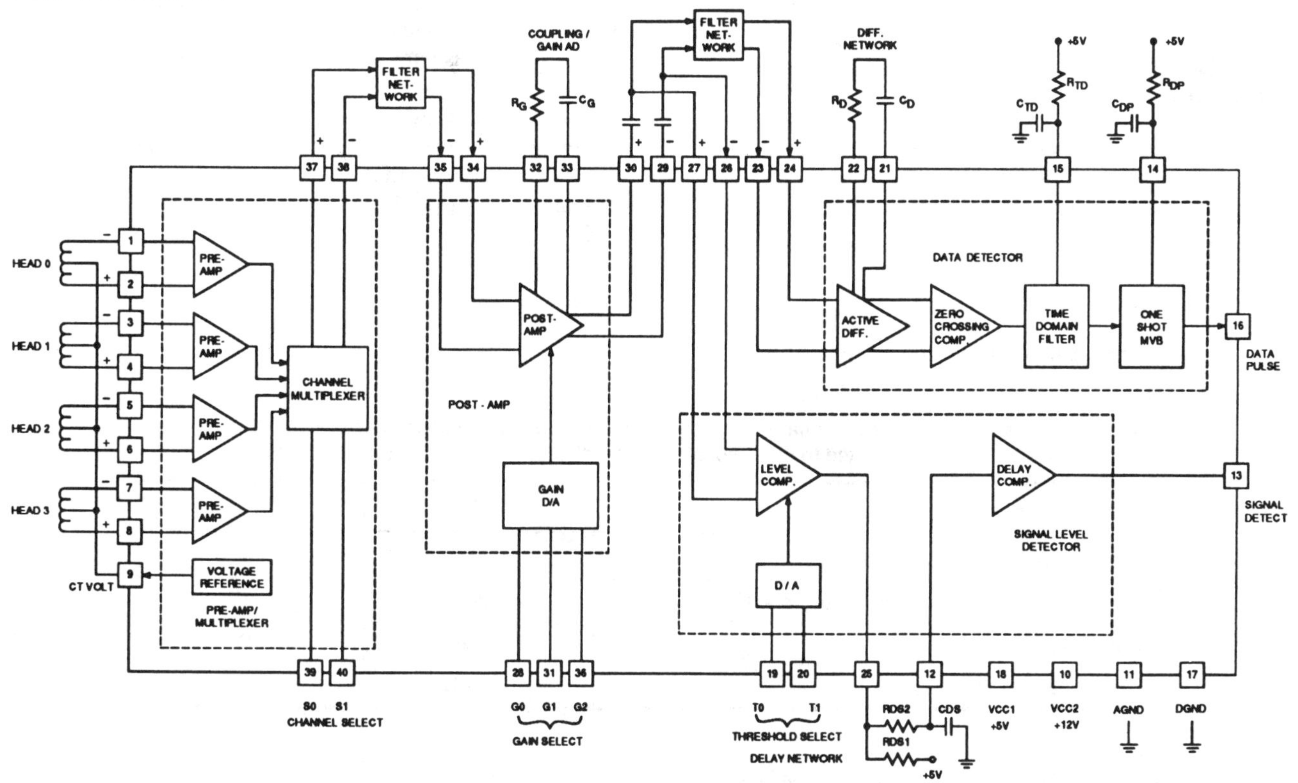

Note: Shown with typical external circuitry.
Pin #s refer to PLCC pinout.

CAUTION: Use handling procedures necessary for a static sensitive component.

THERMAL CHARACTERISTICS: Θja

40-pin PDIP	45°C/W
44-pin PLCC	60°C/W

PACKAGE PIN DESIGNATIONS

(TOP VIEW)

Signal	Pin	Pin	Signal
IN0 -	1	40	S1
IN0 +	2	39	S0
IN1 -	3	38	PREOUT -
IN1 +	4	37	PREOUT +
IN2 -	5	36	G2
IN2 +	6	35	PSTIN -
IN3 -	7	34	PSTIN +
IN3 +	8	33	GAIN 2
CT VOLT	9	32	GAIN 1
VCC2	10	31	G1
AGND	11	30	PSTOUT +
DEL IN	12	29	PSTOUT -
SIGNAL DETECT	13	28	G0
DWP	14	27	LEV +
TDF	15	26	LEV -
DATA PULSE	16	25	LEV OUT
DGND	17	24	DIF +
VCC1	18	23	DIF -
T0	19	22	CAP 2
T1	20	21	CAP 1

40-Pin DIP

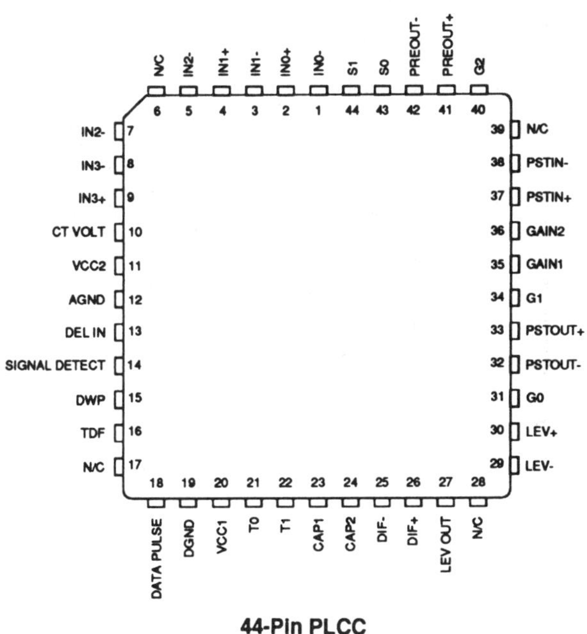

44-Pin PLCC

Silicon Systems
SSI 32B451
SCSI Controller

- Supports asynchronous data transfer up to 1.5 Mbytes/sec
- Supports target role in SCSI applications
- Includes high-current drivers and Schmitt-trigger receivers for direct connection to the SCSI bus
- Full hardware compliance to ANSI X3T9.2 Rev. 17B specification as a target peripheral adapter
- Contains circuitry to support SCSI arbitration, (re)selection and parity features
- Complements the SSI 32C453 buffer controller
- Plug compatible with AIC 500L
- Available in 44-pin PLCC
- Single +5V supply

ELECTRICAL SPECIFICATIONS
ABSOLUTE MAXIMUM RATINGS

(Maximum limits indicate where permanent device damage occurs. Continuous operation at these limits is not intended and should be limited to those conditions specified in the DC Operating Characteristics.)

PARAMETER	RATING	UNIT
VCC with respect to VSS (GND)	+7	V
Max. voltage on any pin with respect to VSS	-0.5 to +7	V
Operating temperature	0 to 70	°C
Storage temperature	-55 to +125	°C

DC OPERATING CHARACTERISTICS

(Ta = 0 to 70°C, VCC = +5V ± 5%, VSS = 0V)

PARAMETER		CONDITION	MIN	MAX	UNITS
IIL	Input Leakage (BREQ, LO, $\overline{BOE}$, $\overline{BIE}$, $\overline{ET}$ SELOUT, BSYOUT, CDIN, I/OIN, MSGIN, PAR/$\overline{RST}$, CLK, $\overline{ACK}$)	0 < Vin < VCC	-10	+10	µA
IOL	SCSI Output Leakage ($\overline{SEL}$, $\overline{BSY}$, $\overline{DB0}$-$\overline{DB7}$, $\overline{DBP}$, $\overline{MSG}$, $\overline{C}$/D, $\overline{I}$/O)	0.5 < Vout < VCC	-50	+50	µA
IOL	D0-D7	0.45 < Vout < VCC	-10	+10	µA
VIL	Input Low Voltage		0	0.8	V
VIH	Input High Voltage		2.0		V
VOH	Output High Voltage	IOH = -400 µA	2.4		V
VOL	SCSI Output Low Voltage	IOL = 48 mA		0.5	V
VOL	All others	IOL = 2 mA		0.4	V
	Power Dissipation			500	mW
Vhsy	Hysteresis Voltage (all SCSI signals)		200		mV
Iccs	Standby Current	Ta = 70°C		600	µA
Icc	Supply Current	Ta = 70°C		30	mA
Cin	Input Capacitance			15	pF

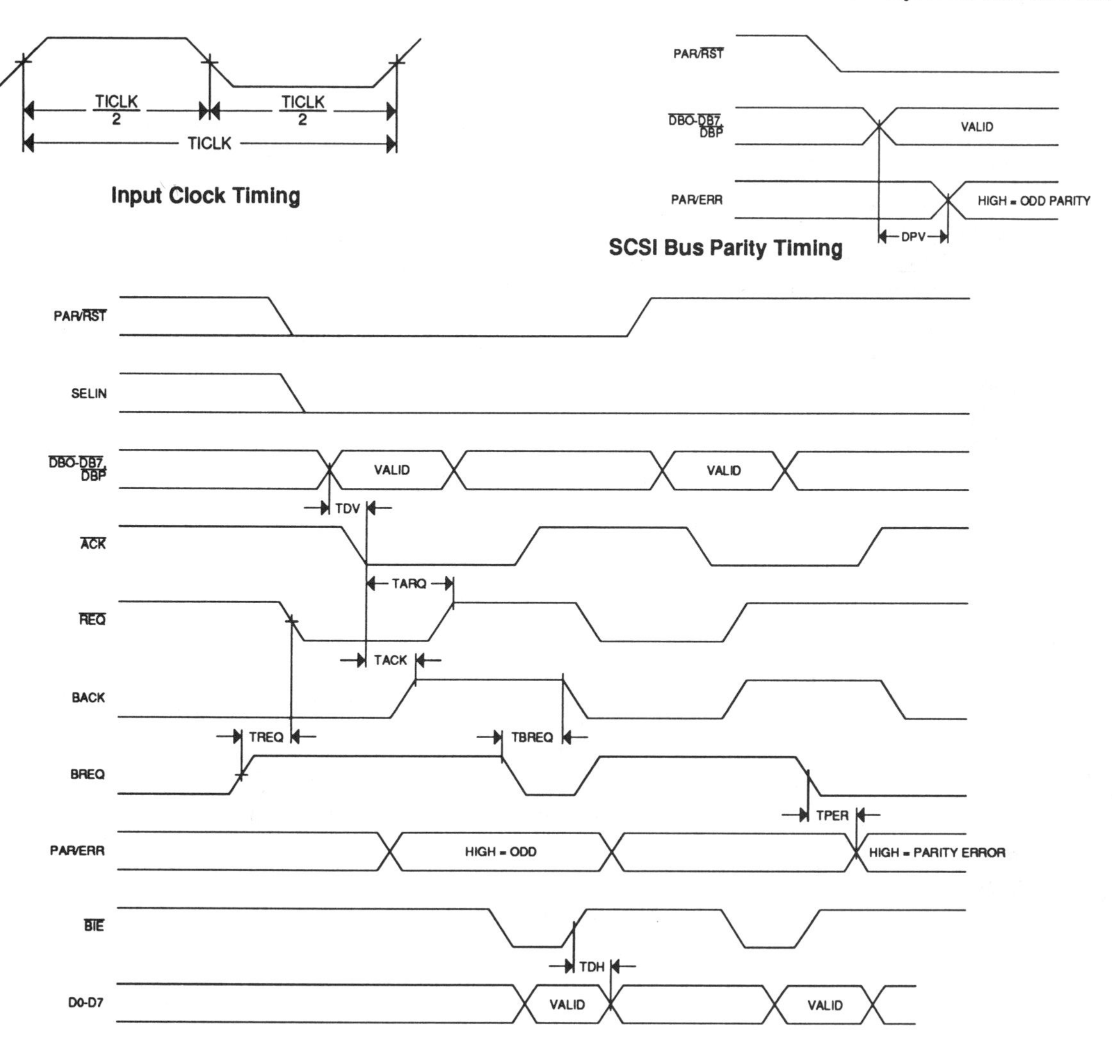

Input Clock Timing

SCSI Bus Parity Timing

Note: Data from host - SCSI initiator - is transferred to buffer RAM.

Write Operation Timing

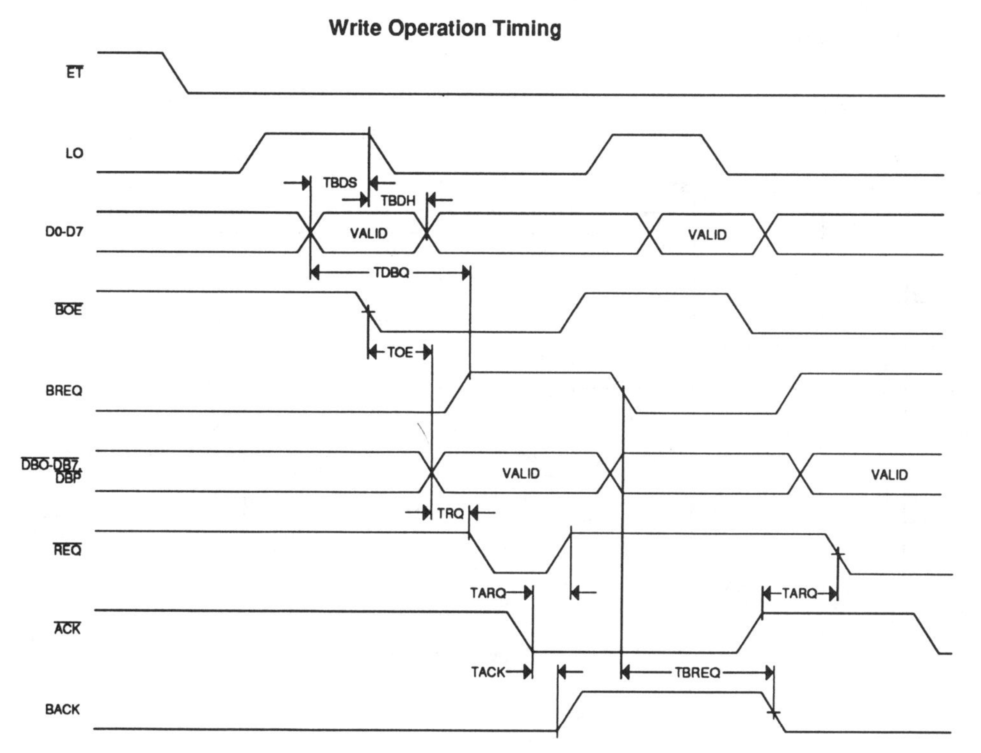

Note: Data read from peripheral drive is transferred to the host - SCSI initiator

Read Operation Timing

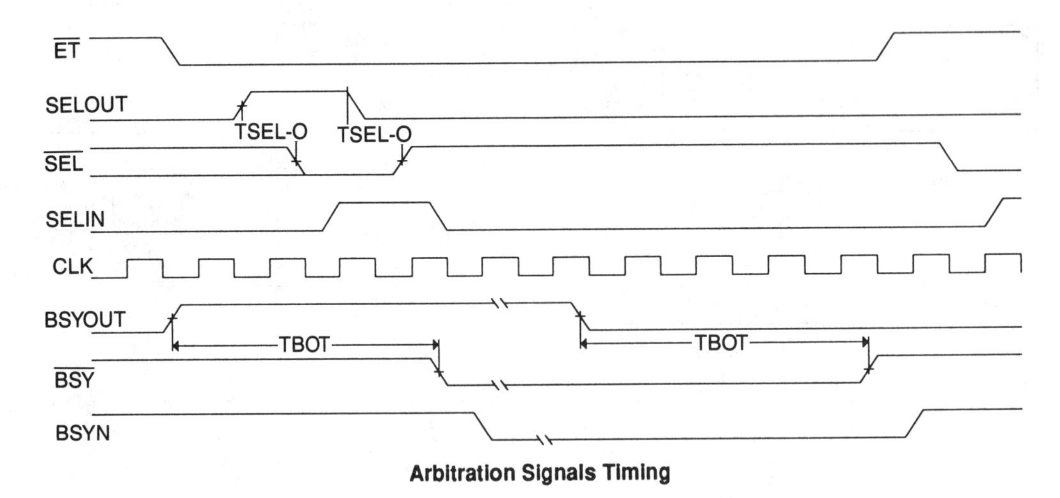

Arbitration Signals Timing

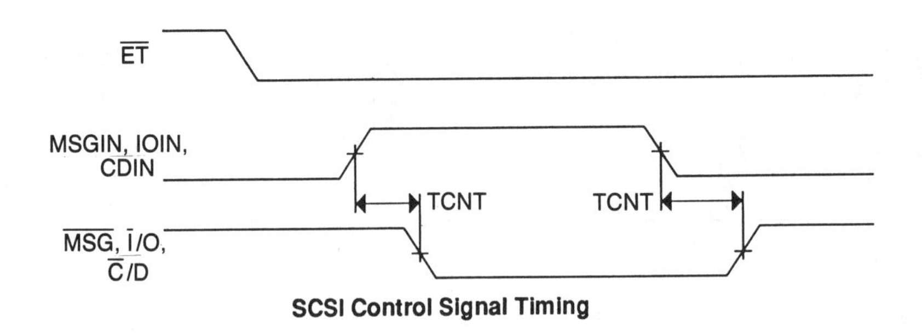

SCSI Control Signal Timing

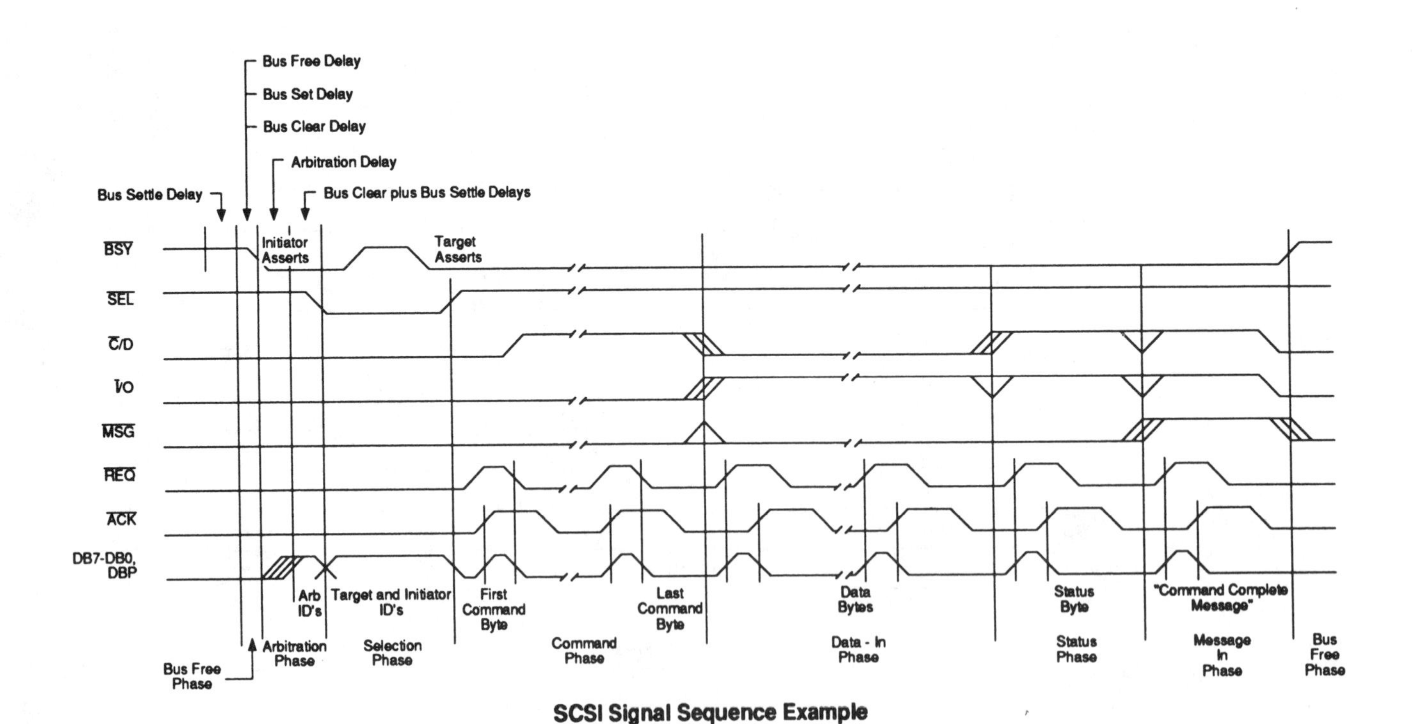

SCSI Signal Sequence Example

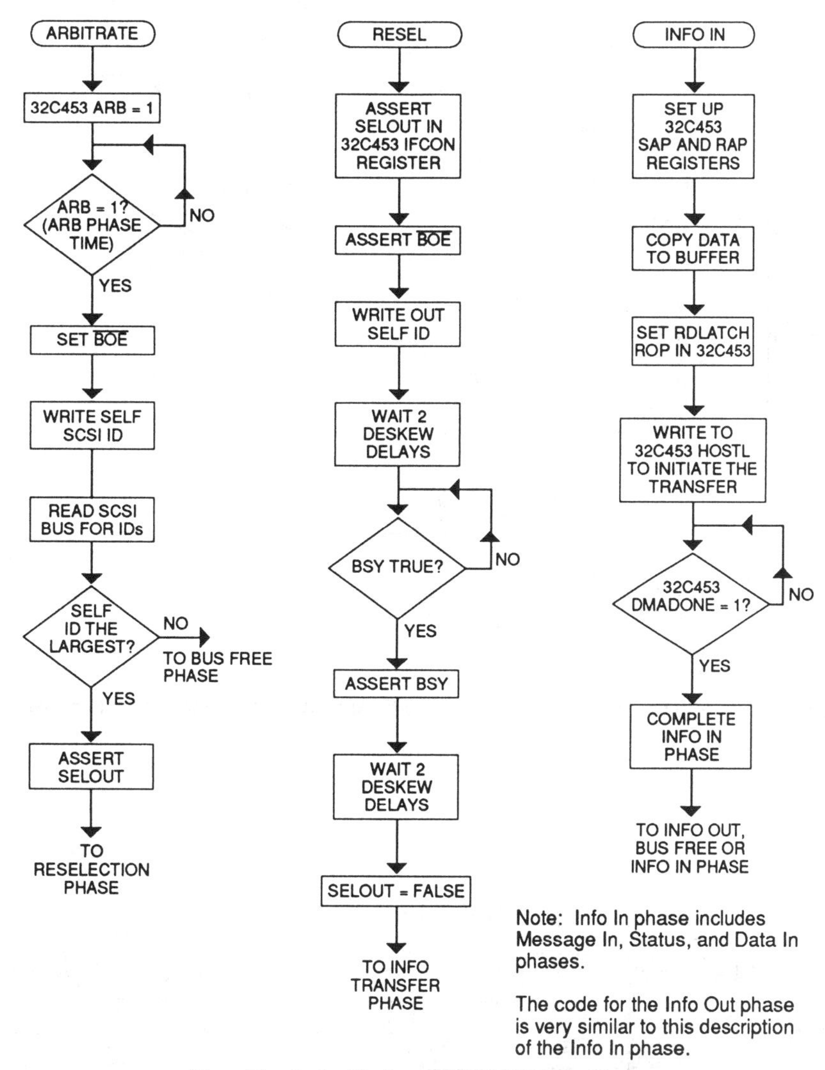

Note: Info In phase includes Message In, Status, and Data In phases.

The code for the Info Out phase is very similar to this description of the Info In phase.

Flow Charts for Various SSI 32B451 Routines

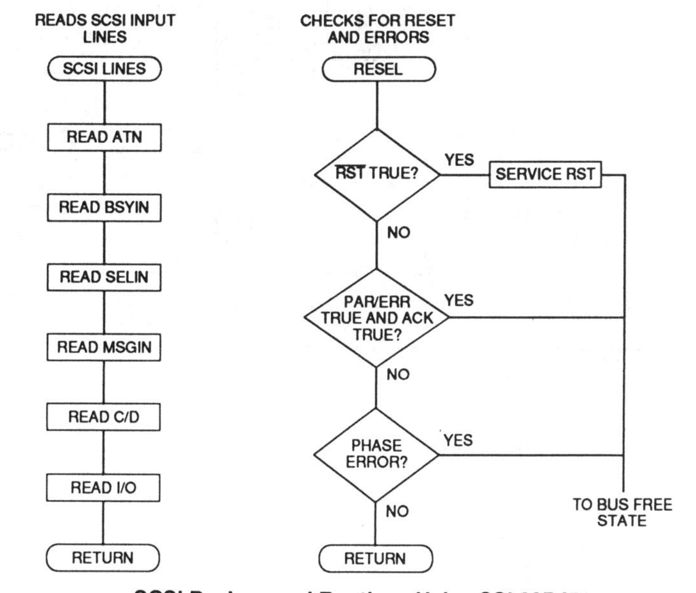

SCSI Background Routines Using SSI 32B451

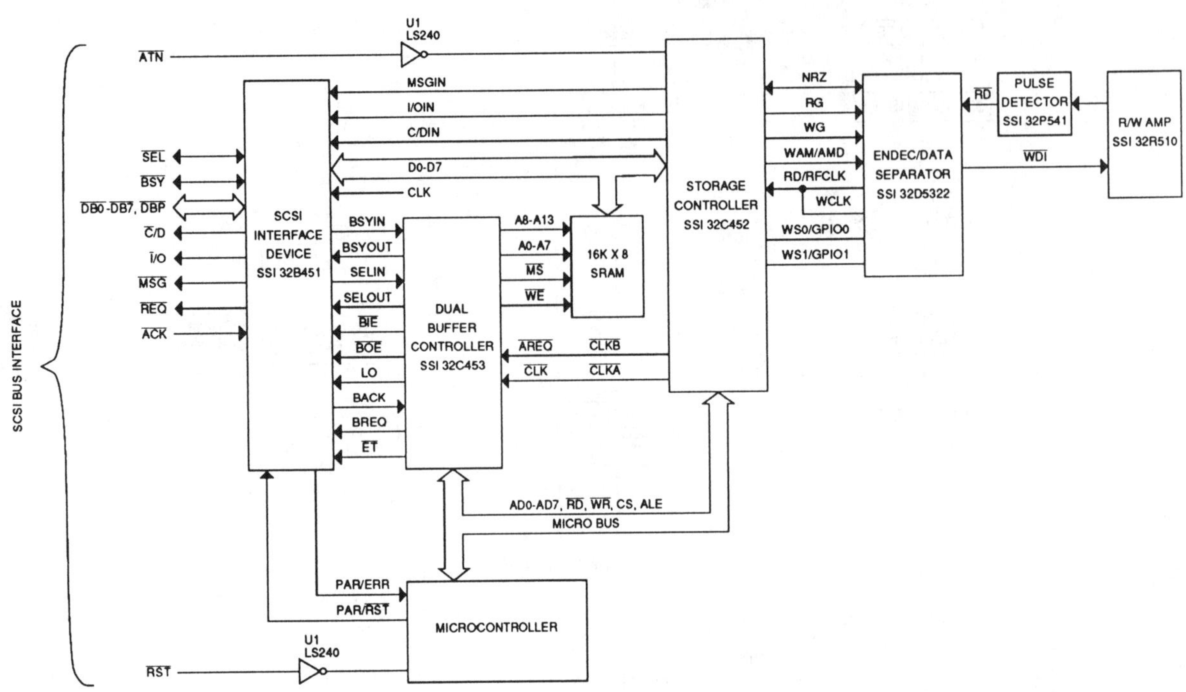

Partial Schematic for SCSI Implementation with Arbitration Support Using SSi Devices

BLOCK DIAGRAM

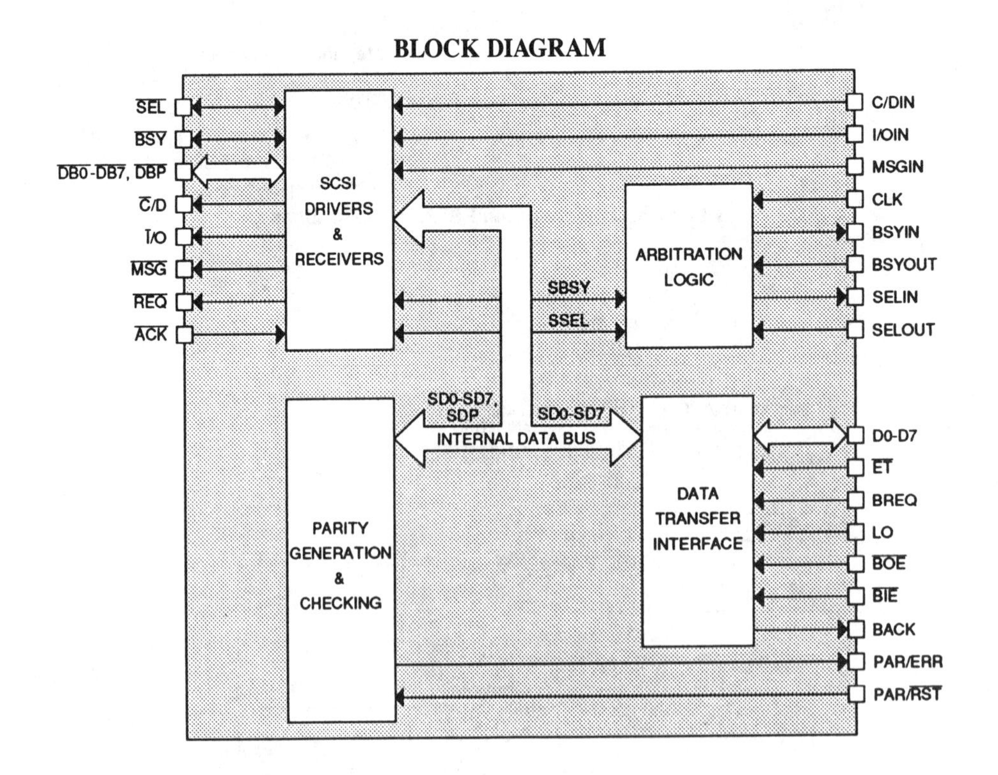

PACKAGE PIN DESIGNATIONS

(TOP VIEW)

CAUTION: Use handling procedures necessary for a static sensitive component.

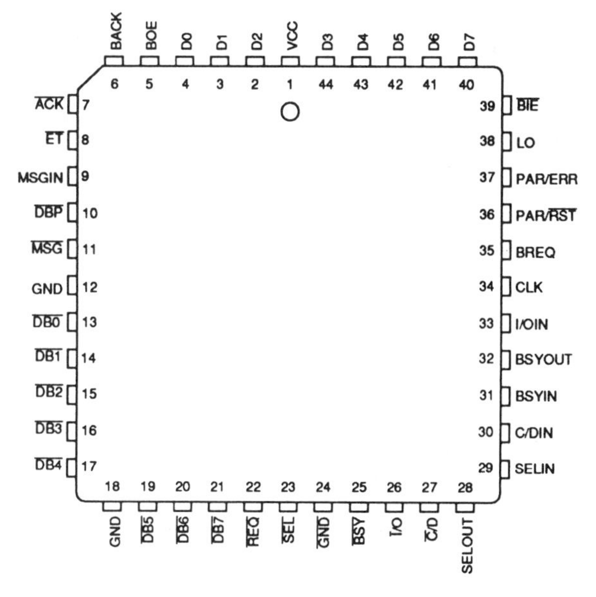

44-pin PLCC

Silicon Systems SSI 73K212/K212L Bell 212A/103 Single-Chip Modem

- One-chip Bell 212A and 103 standard compatible modem data pump
- Full-duplex operation at 0-300 bit/s (FSK) or 1200 bit/s (DPSK)
- Pin and software compatible with other SSI K-Series 1-chip modems
- Interfaces directly with standard microprocessors (8048, 80C51 typical)
- Serial (22-pin DIP) or parallel (28-pin DIP) microprocessor bus for control
- Serial port for data transfer
- Both synchronous and asynchronous modes of operation
- Call progress, carrier, precise answer tone, and long loop detectors
- DTMF generators
- Test modes available: ALB, DL, RDL, mark, space, alternating bit patterns
- Precise automatic gain control allows 45 dB dynamic range
- Space-efficient 22- or 28-pin DIP packages
- CMOS technology for low power consumption using 30 mW @ 5V or 180 mW @ 12V
- Single +5V (73K212L) or +12V (73K212) versions

ELECTRICAL SPECIFICATIONS

ABSOLUTE MAXIMUM RATINGS

PARAMETER	RATING	UNIT
VDD Supply Voltage	14	V
Storage Temperature	-65 to 150	°C
Soldering Temperature (10 sec.)	260	°C
Applied Voltage	-0.3 to VDD+0.3	V

Note: All inputs and outputs are protected from static charge using built-in, industry standard protection devices and all outputs are short-circuit protected.

RECOMMENDED OPERATING CONDITIONS

PARAMETER	CONDITIONS	MIN	NOM	MAX	UNITS
VDD Supply Voltage		4.5	5	5.5	V
TA, Operating Free-Air Temperature		-40		+85	°C
Clock Variation	(11.0592 MHz) Crystal or external clock	-0.01		+0.01	%
External Components (Refer to Application section for placement.)					
VREF Bypass capacitor	(External to GND)	0.1			µF
Bias setting resistor	(Placed between VDD and ISET pins)	1.8	2	2.2	MΩ
ISET Bypass capacitor	(ISET pin to GND)	0.1			µF
VDD Bypass capacitor 1	(External to GND)	0.1			µF
VDD Bypass capacitor 2	(External to GND)	22			µF
XTL1 Load Capacitor	Depends on crystal characteristics;			40	pF
XTL2 Load Capacitor	from pin to GND			20	

REGISTER ADDRESS TABLE

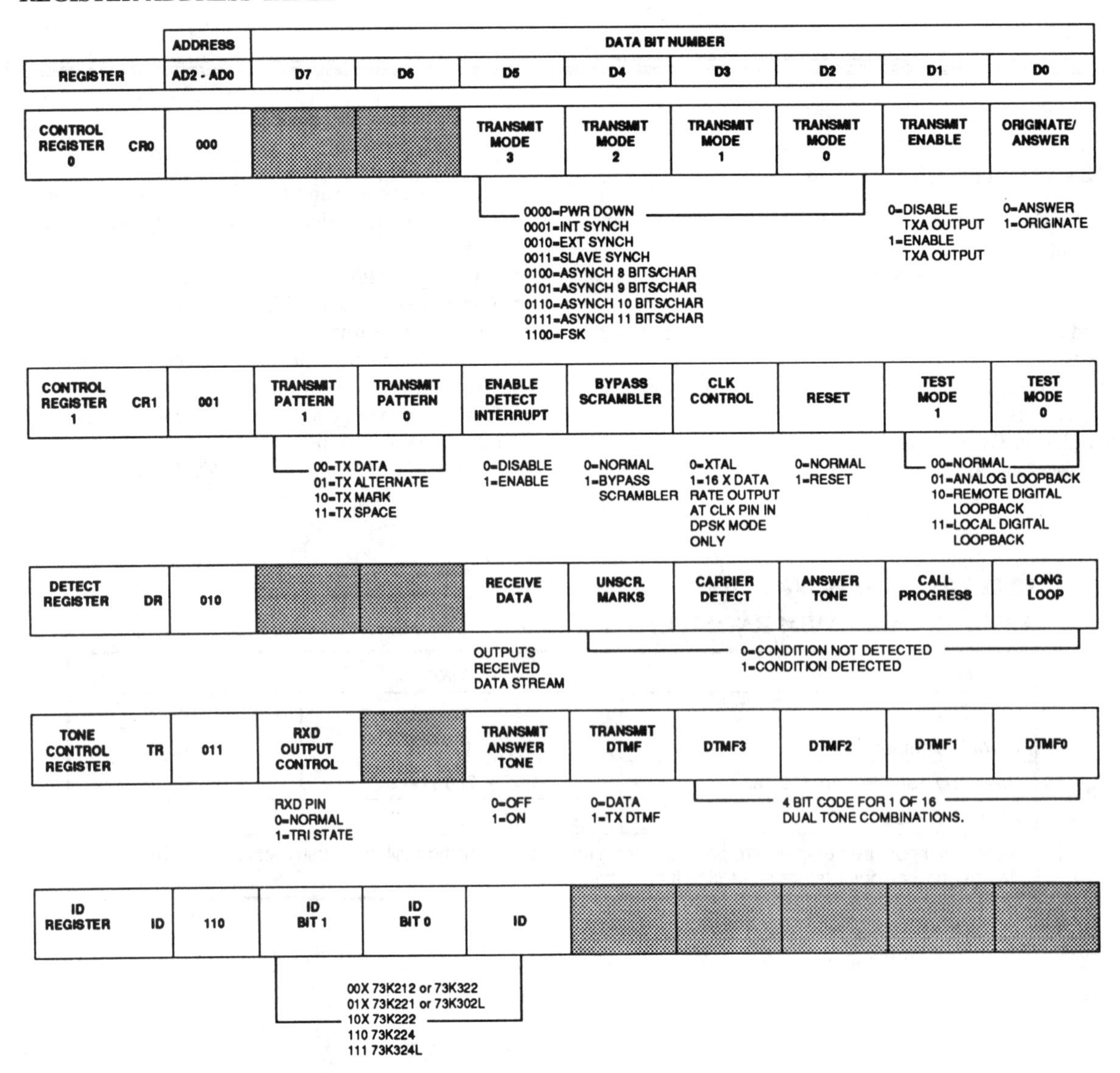

REGISTER	ADDRESS AD2 - AD0	D7	D6	D5	D4	D3	D2	D1	D0
CONTROL REGISTER 0 CR0	000			TRANSMIT MODE 3	TRANSMIT MODE 2	TRANSMIT MODE 1	TRANSMIT MODE 0	TRANSMIT ENABLE	ORIGINATE/ ANSWER
CONTROL REGISTER 1 CR1	001	TRANSMIT PATTERN 1	TRANSMIT PATTERN 0	ENABLE DETECT INTERRUPT	BYPASS SCRAMBLER	CLK CONTROL	RESET	TEST MODE 1	TEST MODE 0
DETECT REGISTER DR	010			RECEIVE DATA	UNSCR. MARKS	CARRIER DETECT	ANSWER TONE	CALL PROGRESS	LONG LOOP
TONE CONTROL REGISTER TR	011	RXD OUTPUT CONTROL		TRANSMIT ANSWER TONE	TRANSMIT DTMF	DTMF3	DTMF2	DTMF1	DTMF0
ID REGISTER ID	110	ID BIT 1	ID BIT 0	ID					

CR0:
- TRANSMIT MODE (D5–D2): 0000=PWR DOWN; 0001=INT SYNCH; 0010=EXT SYNCH; 0011=SLAVE SYNCH; 0100=ASYNCH 8 BITS/CHAR; 0101=ASYNCH 9 BITS/CHAR; 0110=ASYNCH 10 BITS/CHAR; 0111=ASYNCH 11 BITS/CHAR; 1100=FSK
- TRANSMIT ENABLE (D1): 0=DISABLE TXA OUTPUT; 1=ENABLE TXA OUTPUT
- ORIGINATE/ANSWER (D0): 0=ANSWER; 1=ORIGINATE

CR1:
- TRANSMIT PATTERN (D7–D6): 00=TX DATA; 01=TX ALTERNATE; 10=TX MARK; 11=TX SPACE
- ENABLE DETECT INTERRUPT (D5): 0=DISABLE; 1=ENABLE
- BYPASS SCRAMBLER (D4): 0=NORMAL; 1=BYPASS SCRAMBLER
- CLK CONTROL (D3): 0=XTAL; 1=16 X DATA RATE OUTPUT AT CLK PIN IN DPSK MODE ONLY
- RESET (D2): 0=NORMAL; 1=RESET
- TEST MODE (D1–D0): 00=NORMAL; 01=ANALOG LOOPBACK; 10=REMOTE DIGITAL LOOPBACK; 11=LOCAL DIGITAL LOOPBACK

DR:
- RECEIVE DATA (D5): OUTPUTS RECEIVED DATA STREAM
- D4–D0: 0=CONDITION NOT DETECTED; 1=CONDITION DETECTED

TR:
- RXD OUTPUT CONTROL (D7): RXD PIN 0=NORMAL; 1=TRI STATE
- TRANSMIT ANSWER TONE (D5): 0=OFF; 1=ON
- TRANSMIT DTMF (D4): 0=DATA; 1=TX DTMF
- DTMF3–DTMF0 (D3–D0): 4 BIT CODE FOR 1 OF 16 DUAL TONE COMBINATIONS.

ID:
- 00X 73K212 or 73K322
- 01X 73K221 or 73K302L
- 10X 73K222
- 110 73K224
- 111 73K324L

REGISTER BIT SUMMARY

REGISTER	ADDRESS AD2 - AD0	D7	D6	D5	D4	D3	D2	D1	D0
CONTROL REGISTER 0 CR0	000			TRANSMIT MODE 3	TRANSMIT MODE 2	TRANSMIT MODE 1	TRANSMIT MODE 0	TRANSMIT ENABLE	ANSWER/ ORIGINATE
CONTROL REGISTER 1 CR1	001	TRANSMIT PATTERN 1	TRANSMIT PATTERN 0	ENABLE DETECT INTERRUPT	BYPASS SCRAMBLER	CLK CONTROL	RESET	TEST MODE 1	TEST MODE 0
DETECT REGISTER DR	010			RECEIVE DATA	UNSCR. MARKS	CARRIER DETECT	ANSWER TONE	CALL PROGRESS	LONG LOOP
TONE CONTROL REGISTER TR	011	RXD OUTPUT CONTROL		TRANSMIT ANSWER TONE	TRANSMIT DTMF	DTMF3	DTMF2	DTMF1	DTMF0
CONTROL REGISTER 2 CR2	100				THESE REGISTER LOCATIONS ARE RESERVED FOR				
CONTROL REGISTER 3 CR3	101				USE WITH OTHER K-SERIES FAMILY MEMBERS				
ID REGISTER ID	110	0	0						

NOTE: When a register containing reserved control bits is written into, the reserved bits must be programmed as 0's.

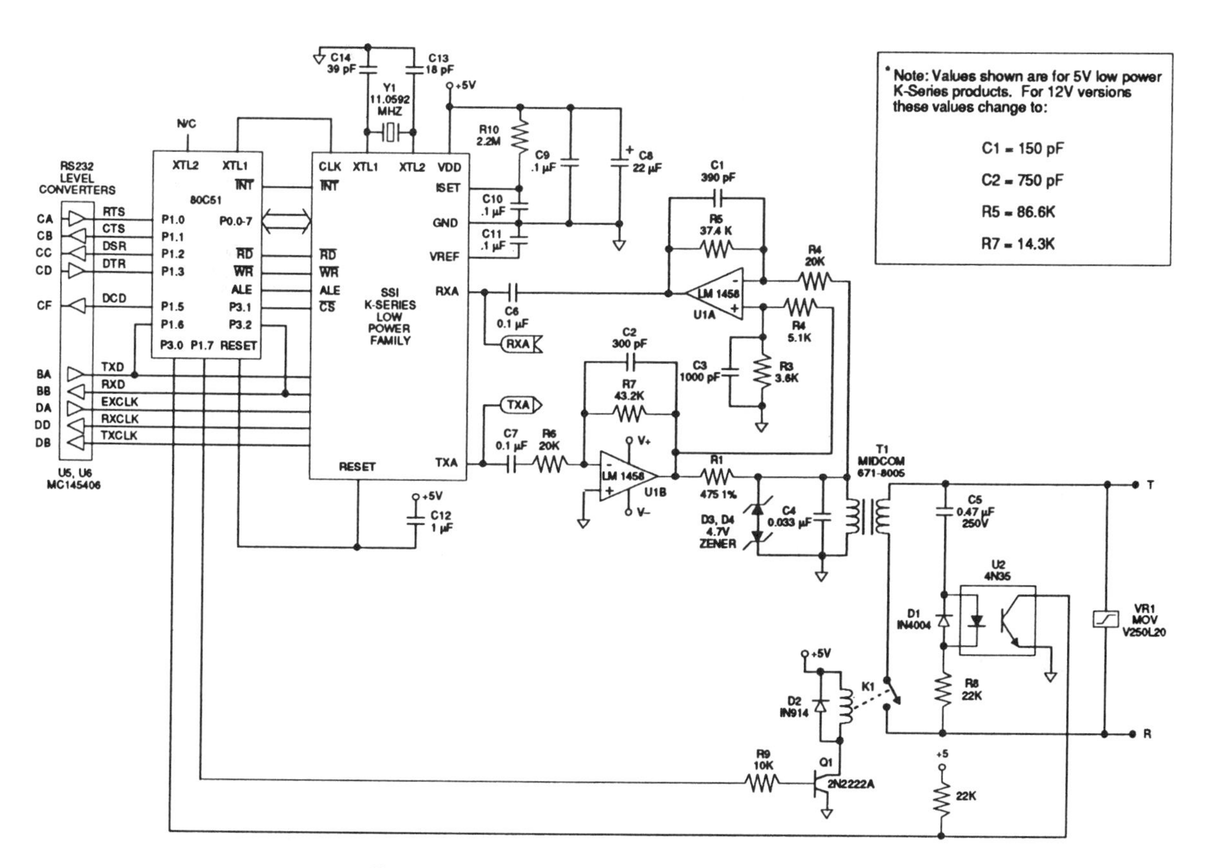

Basic Box Modem with Dual-Supply Hybrid

* Note: Op-amp U1 must be rated for single 5V operation. R10 & R11 values depend on Op-amp used.

Single 5V Hybrid Version

PACKAGE PIN DESIGNATIONS

(TOP VIEW)

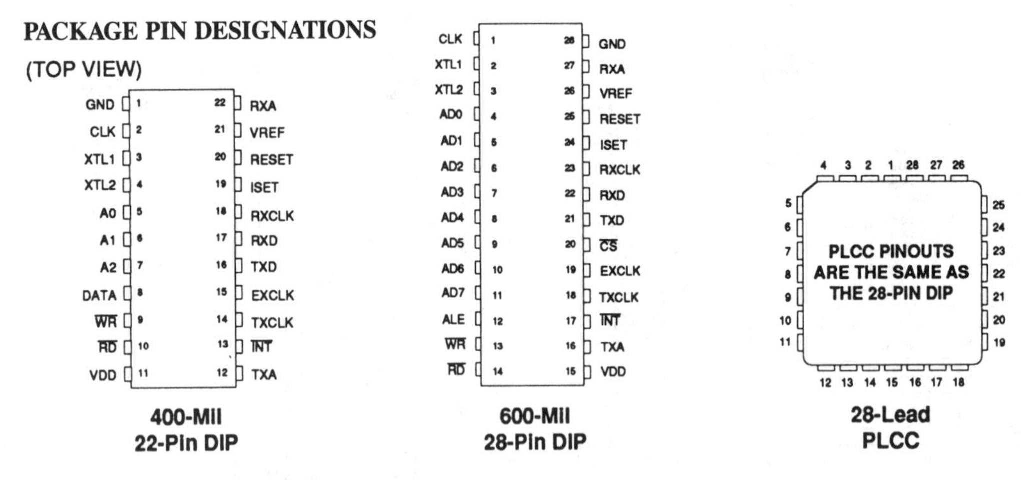

400-Mil 22-Pin DIP

600-Mil 28-Pin DIP

28-Lead PLCC

BLOCK DIAGRAM

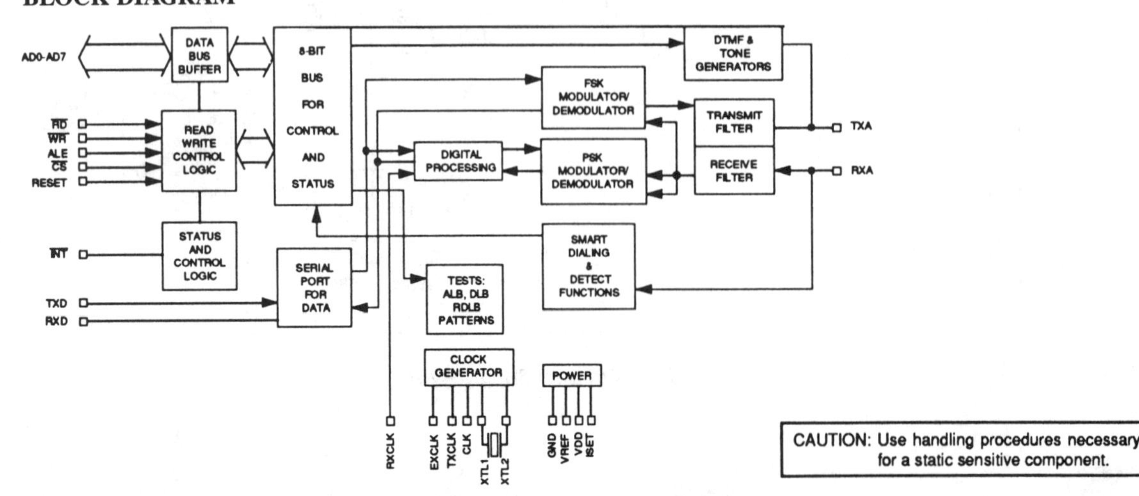

CAUTION: Use handling procedures necessary for a static sensitive component.

Silicon Systems
SSI 73K221/K221L
CCITT V.22, V.21 Single-Chip Modem

- One-chip CCITT V.22 and V.21 standard compatible modem data pump
- Full-duplex operation at 0-300 bit/s (FSK) or 600 and 1200 bit/s (DPSK)
- Pin and software compatible with other SSI K-series 1-chip modems
- Interfaces directly with standard microprocessors (8048, 80C51 typical)
- Serial (22-pin DIP) or parallel (28-pin DIP) microprocessor bus for control
- Serial port for data transfer
- Both synchronous and asynchronous modes of operation
- Call progress, carrier, precise answer tone (2100 Hz), and long loop detectors
- DTMF and 550 (or 1800) Hz guard tone generators
- Test modes available: ALB, DL, RDL, mark, space, alternating bit patterns
- Precise automatic gain control allows 45 dB dynamic range
- Space-efficient 22- or 28-pin DIP packages
- CMOS technology for low power consumption using 30 mW @ 5V or 180 mW @ 12V
- Single +5V (73K221L) or +12V (73K221) versions

ELECTRICAL SPECIFICATIONS

ABSOLUTE MAXIMUM RATINGS

PARAMETER	RATING	UNIT
VDD Supply Voltage	14	V
Storage Temperature	-65 to 150	°C
Soldering Temperature (10 sec.)	260	°C
Applied Voltage	-0.3 to VDD+0.3	V
Note: All inputs and outputs are protected from static charge using built-in, industry standard protection devices and all outputs are short-circuit protected.		

RECOMMENDED OPERATING CONDITIONS

PARAMETER	CONDITIONS	MIN	NOM	MAX	UNITS
VDD Supply voltage		4.5	5	5.5	V
TA, Operating Free-Air Temperature		-40		+85	°C
Clock Variation	(11.0592 MHz) Crystal or external clock	-0.01		+0.01	%
External Components (Refer to Application section for placement.)					
VREF Bypass Capacitor	(External to GND)	0.1			µF
Bias setting resistor	(Placed between VDD and ISET pins)	1.8	2	2.2	MΩ
ISET Bypass Capacitor	(ISET pin to GND)	0.1			µF
VDD Bypass Capacitor 1	(External to GND)	0.1			µF
VDD Bypass Capacitor 2	(External to GND)	22			µF
XTL1 Load Capacitor	Depends on crystal characteristics;			40	pF
XTL2 Load Capacitor	from pin to GND			20	

REGISTER BIT SUMMARY

REGISTER	ADDRESS AD2 - AD0	D7	D6	D5	D4	D3	D2	D1	D0
		DATA BIT NUMBER							
CONTROL REGISTER 0 CR0	000	MODULATION OPTION		TRANSMIT MODE 3	TRANSMIT MODE 2	TRANSMIT MODE 1	TRANSMIT MODE 0	TRANSMIT ENABLE	ANSWER/ ORIGINATE
CONTROL REGISTER 1 CR1	001	TRANSMIT PATTERN 1	TRANSMIT PATTERN 0	ENABLE DETECT INTERRUPT	BYPASS SCRAMBLER	CLK CONTROL	RESET	TEST MODE 1	TEST MODE 0
DETECT REGISTER DR	010			RECEIVE DATA	UNSCR. MARKS	CARRIER DETECT	ANSWER TONE	CALL PROGRESS	LONG LOOP
TONE CONTROL REGISTER TR	011	RXD OUTPUT CONTROL	TRANSMIT GUARD TONE	TRANSMIT ANSWER TONE	TRANSMIT DTMF	DTMF3	DTMF2	DTMF1/ OVERSPEED	DTMF0/ GUARD/
CONTROL REGISTER 2 CR2	100	THESE REGISTER LOCATIONS ARE RESERVED FOR							
CONTROL REGISTER 3 CR3	101	USE WITH OTHER K-SERIES FAMILY MEMBERS							
ID REGISTER ID	110	0	1						

NOTE: When a register containing reserved control bits is written into, the reserved bits must be programmed as 0's.

REGISTER ADDRESS TABLE

REGISTER	ADDRESS AD2 - AD0	D7	D6	D5	D4	D3	D2	D1	D0
		DATA BIT NUMBER							
CONTROL REGISTER 0 CR0	000	MODULATION OPTION		TRANSMIT MODE 3	TRANSMIT MODE 2	TRANSMIT MODE 1	TRANSMIT MODE 0	TRANSMIT ENABLE	ORIGINATE/ ANSWER
		0=1200 BIT/S DPSK 1=600 BIT/S DPSK		D5–D2: 0000=PWR DOWN 0001=INT SYNCH 0010=EXT SYNCH 0011=SLAVE SYNCH 0100=ASYNCH 8 BITS/CHAR 0101=ASYNCH 9 BITS/CHAR 0110=ASYNCH 10 BITS/CHAR 0111=ASYNCH 11 BITS/CHAR 1100=FSK				0=DISABLE TXA OUTPUT 1=ENABLE TXA OUTPUT	0=ANSWER 1=ORIGINATE
CONTROL REGISTER 1 CR1	001	TRANSMIT PATTERN 1	TRANSMIT PATTERN 0	ENABLE DETECT INTERRUPT	BYPASS SCRAMBLER	CLK CONTROL	RESET	TEST MODE 1	TEST MODE 0
		D7–D6: 00=TX DATA 01=TX ALTERNATE 10=TX MARK 11=TX SPACE		0=DISABLE 1=ENABLE	0=NORMAL 1=BYPASS SCRAMBLER	0=XTAL 1=16 X DATA RATE OUTPUT AT CLK PIN IN DPSK MODE ONLY	0=NORMAL 1=RESET	D1–D0: 00=NORMAL 01=ANALOG LOOPBACK 10=REMOTE DIGITAL LOOPBACK 11=LOCAL DIGITAL LOOPBACK	
DETECT REGISTER DR	010			RECEIVE DATA	UNSCR. MARKS	CARRIER DETECT	ANSWER TONE	CALL PROGRESS	LONG LOOP
				OUTPUTS RECEIVED DATA STREAM	D4–D0: 0=CONDITION NOT DETECTED 1=CONDITION DETECTED				
TONE CONTROL REGISTER TR	011	RXD OUTPUT CONTROL	TRANSMIT GUARD/ TONE	TRANSMIT ANSWER TONE	TRANSMIT DTMF	DTMF3	DTMF2	DTMF1/ OVERSPEED	DTMF0/ GUARD/ TONE
		RXD PIN 0=NORMAL 1=TRI STATE	0=OFF 1=ON	0=OFF 1=ON	0=DATA 1=TX DTMF	D3–D0: 4 BIT CODE FOR 1 OF 16 DUAL TONE COMBINATIONS.			0=1800 Hz G.T. 1=550 Hz G.T.
ID REGISTER ID	110	ID	ID	ID					
		D7–D5: 00X 73K212 or 73K322 01X 73K221 or 73K302L 10X 73K222 or 73K321L 110 73K224L 111 73K324L							

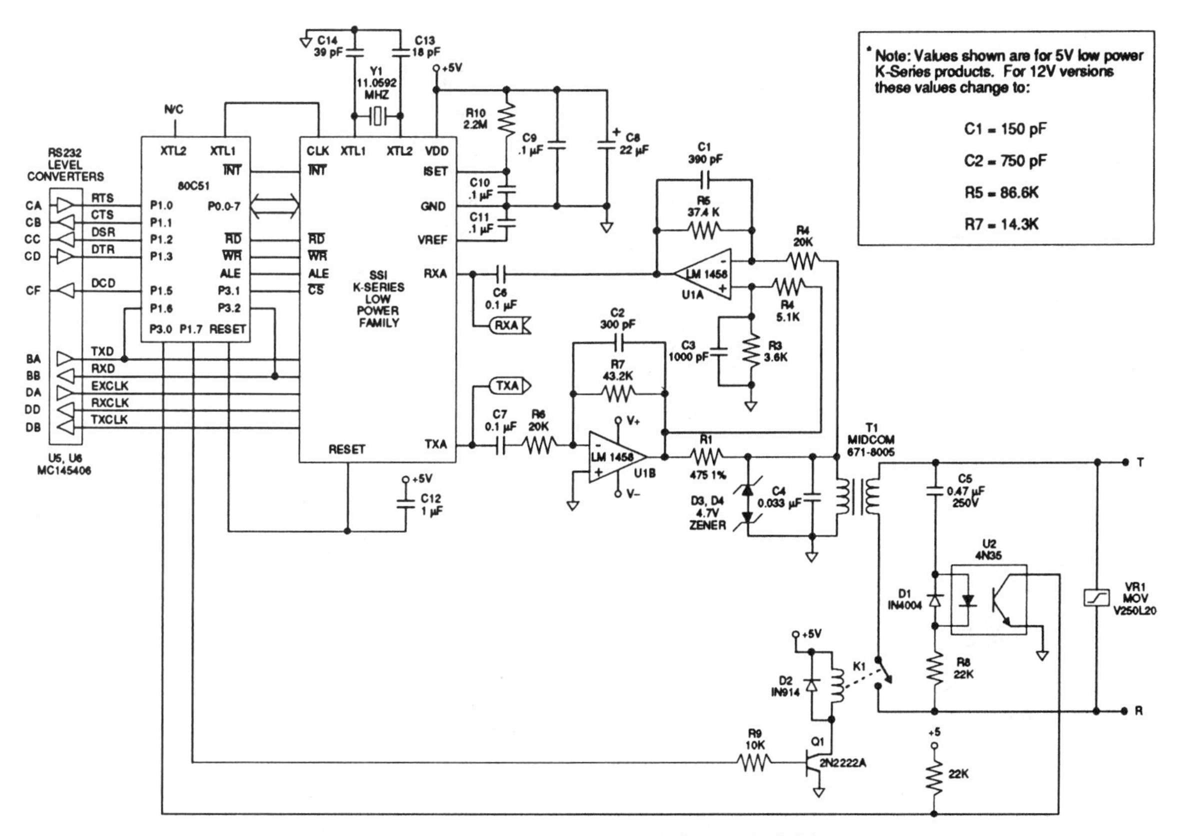

Basic Box Modem with Dual-Supply Hybrid

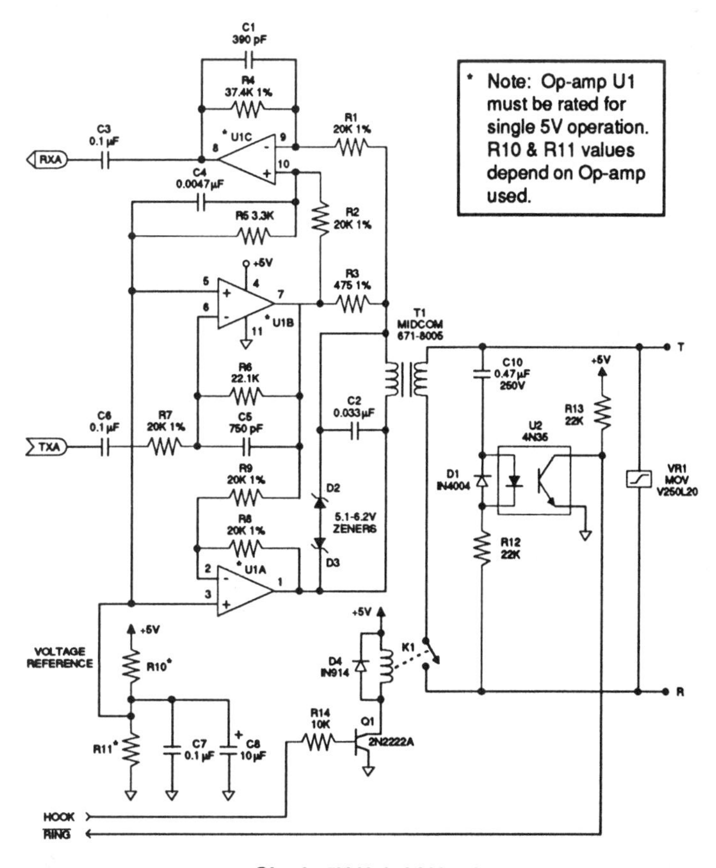

Single 5V Hybrid Version

BLOCK DIAGRAM

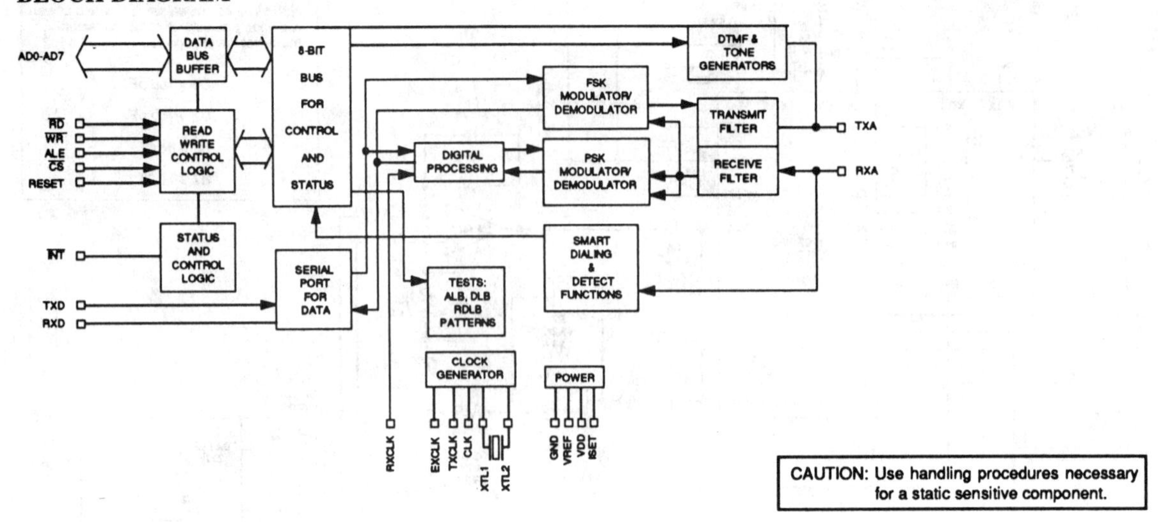

PACKAGE PIN DESIGNATIONS

(TOP VIEW)

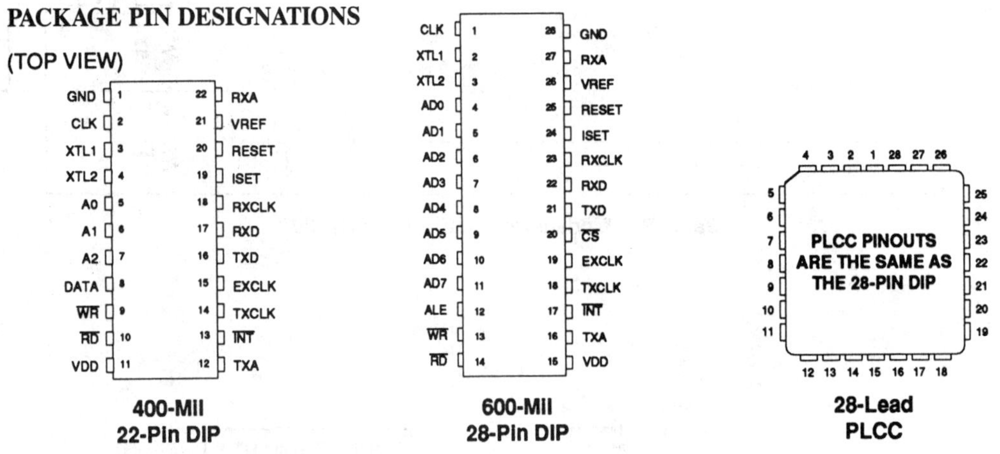

400-Mil 22-Pin DIP

600-Mil 28-Pin DIP

28-Lead PLCC

SECTION 7

POWER SUPPLIES AND TEST EQUIPMENT

This information is reproduced with permission of Analog Devices, Inc.

Analog Devices AD536A Integrated Circuit True rms-to-dc Converter

- True rms-to-dc conversion
- Laser-trimmed to high accuracy
  - 0.2% max error (AD536AK)
  - 0.5% max error (AD536AJ)
- Wide-response capability:
  - Computes rms of ac and dc signals
  - 450-kHz bandwidth: $V_{rms} > 100$ mV
  - 2-MHz bandwidth: $V_{rms} > 1$V
  - Signal crest factor of 7 for 1% error
- dB output with 60 dB range
- Low power: 1.2 mA quiescent current
- Single or dual supply operation
- Monolithic integrated circuit
- −55°C to +125° C operation (AD536AS)
- Low cost

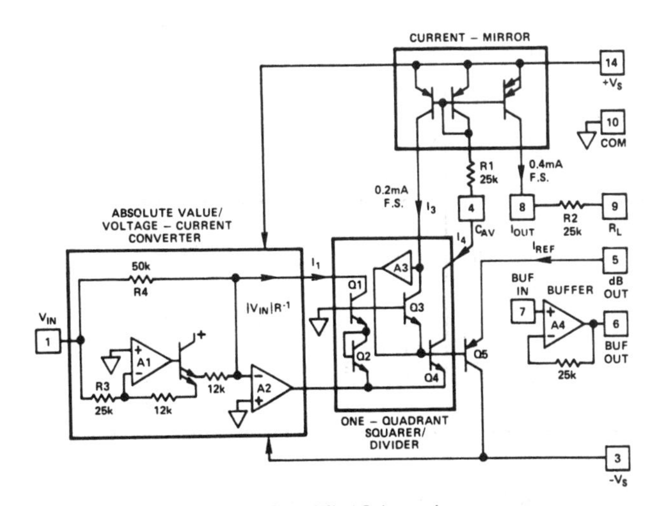

Simplified Schematic

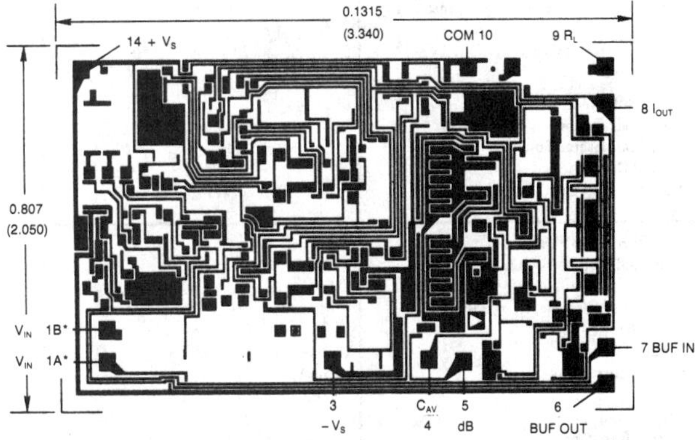

PAD NUMBERS CORRESPOND TO PIN NUMBERS FOR THE TO-116 14-PIN CERAMIC DIP PACKAGE

NOTE
*BOTH PADS SHOWN MUST BE CONNECTED TO V_{IN}.
THE AD536A IS AVAILABLE IN LASER-TRIMMED CHIP FORM.
CONSULT ANALOG DEVICES' CATALOG FOR SPECIFICATIONS AND APPLICATION DETAILS.

Chip Dimensions and Pad Layout. Dimensions shown in inches and (mm).

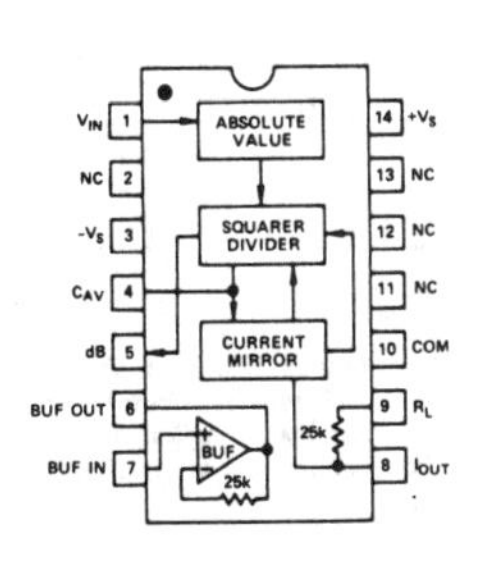

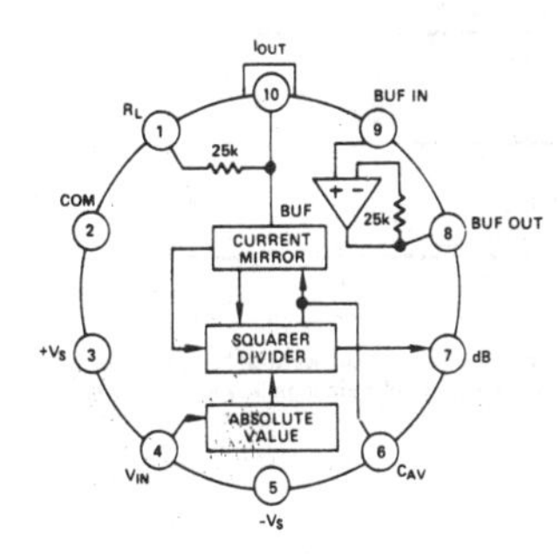

AD536A Pin Connections and Functional Diagram

SPECIFICATIONS (@ +25°C, and ±15V dc unless otherwise noted)

Model	AD536AJ Min	AD536AJ Typ	AD536AJ Max	AD536AK Min	AD536AK Typ	AD536AK Max	AD536AS Min	AD536AS Typ	AD536AS Max	Units
TRANSFER FUNCTION		$V_{OUT} = \sqrt{avg.(V_{IN})^2}$			$V_{OUT} = \sqrt{avg.(V_{IN})^2}$			$V_{OUT} = \sqrt{avg.(V_{IN})^2}$		
CONVERSION ACCURACY										
Total Error, Internal Trim[1] (Figure 1)			**±5 ±0.5**			**±2±0.2**			**±5 ±0.5**	mV ± % of Reading
vs. Temperature, T_{min} to +70°C			±0.1 ±0.01			±0.05 ±0.005			**±0.1 ±0.005**	mV ± % of Reading/°C
+70°C to +125°C									**±0.3 ±0.005**	mV ± % of Reading/°C
vs. Supply Voltage		±0.1 ±0.01			±0.1 ±0.01			±0.1 ±0.01		mV ± % of Reading/V
dc Reversal Error		±0.2			±0.1			±0.2		± % of Reading
Total Error, External Trim[1] (Figure 2)		±3 ±0.3			±2 ±0.1			±3 ±0.3		mV ± % of Reading
ERROR VS. CREST FACTOR[2]										
Crest Factor 1 to 2		Specified Accuracy			Specified Accuracy			Specified Accuracy		
Crest Factor = 3		−0.1			−0.1			−0.1		% of Reading
Crest Factor = 7		−1.0			−1.0			−1.0		% of Reading
FREQUENCY RESPONSE[3]										
Bandwidth for 1% additional error (0.09dB)										
V_{IN} = 10mV		5			5			5		kHz
V_{IN} = 100mV		45			45			45		kHz
V_{IN} = 1V		120			120			120		kHz
±3dB Bandwidth										
V_{IN} = 10mV		90			90			90		kHz
V_{IN} = 100mV		450			450			450		kHz
V_{IN} = 1V		2.3			2.3			2.3		MHz
AVERAGING TIME CONSTANT (Figure 5)		25			25			25		ms/μF CAV
INPUT CHARACTERISTICS										
Signal Range, ±15V Supplies										
Continuous rms Level		0 to 7			0 to 7			0 to 7		V rms
Peak Transient Input			±20			±20			±20	V peak
Continuous rms Level, ±5V Supplies		0 to 2			0 to 2			0 to 2		V rms
Peak Transient Input, ±5V Supplies			±7			±7			±7	V peak
Maximum Continuous Nondestructive Input Level (All Supply Voltages)			±25			±25			±25	V peak
Input Resistance	13.33	16.67	20	13.33	16.67	20	13.33	16.67	20	kΩ
Input Offset Voltage		0.8	±2		0.5	±1		0.8	±2	mV
OUTPUT CHARACTERISTICS										
Offset Voltage, V_{IN} = COM (Figure 1)		±1	**±2**		±0.5	**±1**			**±2**	mV
vs. Temperature		±0.1			±0.1				**±0.2**	mV/°C
vs. Supply Voltage		±0.1			±0.1			±0.2		mV/V
Voltage Swing, ±15V Supplies	**0 to +11**	+12.5		**0 to +11**	+12.5		**0 to +11**	+12.5		V
±5V Supply	0 to +2			0 to +2			0 to +2			V
dB OUTPUT (Figure 12)										
Error, V_{IN} 7mV to 7V rms, 0dB = 1V rms		±0.4	**±0.6**		±0.2	**±0.3**		±0.5	**±0.6**	dB
Scale Factor		−3			−3			−3		mV/dB
Scale Factor TC (Uncompensated, see Figure 12 for Temperature Compensation)		−0.033			−0.033			−0.033		dB/°C
		+0.33			+0.33			+0.33		% of Reading/°C
I_{REF} for 0dB = 1V rms	5	20	**80**	5	20	**80**	5	20	**80**	μA
I_{REF} Range	1		100	1		100	1		100	μA
I_{OUT} TERMINAL										
I_{OUT} Scale Factor		40			40			40		μA/V rms
I_{OUT} Scale Factor Tolerance		±10	±20		±10	±20		±10	±20	%
Output Resistance	20	25	30	20	25	30	20	25	30	kΩ
Voltage Compliance		$-V_S$ to $(+V_S - 2.5V)$			$-V_S$ to $(+V_S - 2.5)$			$-V_S$ to $(+V_S - 2.5V)$		V
BUFFER AMPLIFIER										
Input and Output Voltage Range	$-V_S$ to $(+V_S - 2.5V)$			$-V_S$ to $(+V_S - 2.5V)$			$-V_S$ to $(+V_S - 2.5V)$			V
Input Offset Voltage, R_S = 25k		±0.5	**±4**		±0.5	**±4**		±0.5	**±4**	mV
Input Bias Current		20	**60**		20	**60**		20	**60**	nA
Input Resistance		10^8			10^8			10^8		Ω
Output Current	(+5mA, −130μA)			(+5mA, −130μA)			(+5mA, −130μA)			
Short Circuit Current		20			20			20		mA
Output Resistance			0.5			0.5			0.5	Ω
Small Signal Bandwidth		1			1			1		MHz
Slew Rate[4]		5			5			5		V/μs
POWER SUPPLY										
Voltage Rated Performance		±15			±15			±15		V
Dual Supply	±3.0		±18	±3.0		±18	±3.0		±18	V
Single Supply	+5		+36	+5		+36	+5		+36	V
Quiescent Current										
Total V_S, 5V to 36V, T_{min} to T_{max}		1.2	**2**		1.2	**2**		1.2	**2**	mA
TEMPERATURE RANGE										
Rated Performance	0		+70	0		+70	−55		+125	°C
Storage	−55		+150	−55		+150	−55		+150	°C
PACKAGE OPTIONS										
Ceramic DIP (D-14)		AD536AJD			AD536AKD			AD536ASD		
Metal Can TO-100 (H-10A)		AD536AJH			AD536AKH			AD536ASH		

NOTES

[1] Accuracy is specified for 0 to 7V rms, dc or 1kHz sinewave input with the AD536A connected as in the figure referenced.

[2] Error vs. crest factor is specified as an additional error for 1V rms rectangular pulse input, pulse width = 200μs.

[3] Input voltages are expressed in volts rms, and error is percent of reading.

[4] With 2k external pulldown resistor.

Specifications shown in boldface are tested on all production units at final electrical test. Results from those tests are used to calculate outgoing quality levels. All min and max specifications are guaranteed, although only those shown in boldface are tested on all production units.

Specifications subject to change without notice.

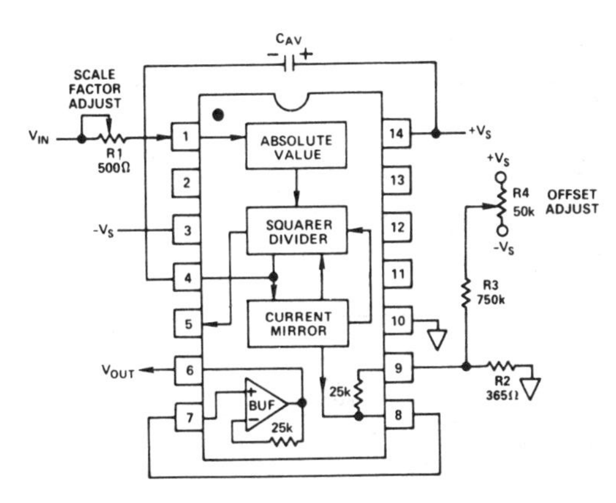

Optional External Gain and Output Offset Trims

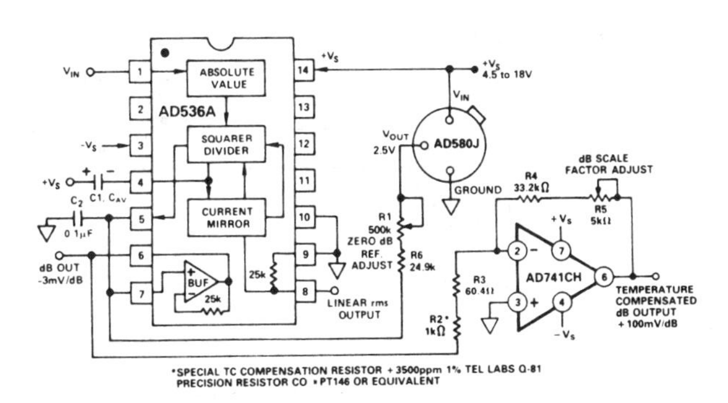

Temperature Compensated dB Output Circuit

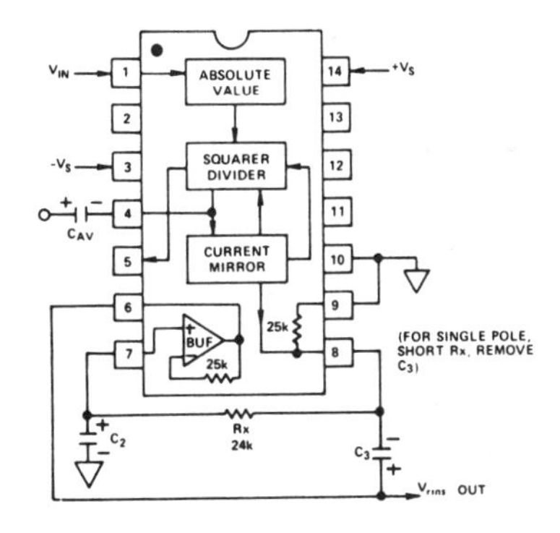

2 Pole "Post" Filter

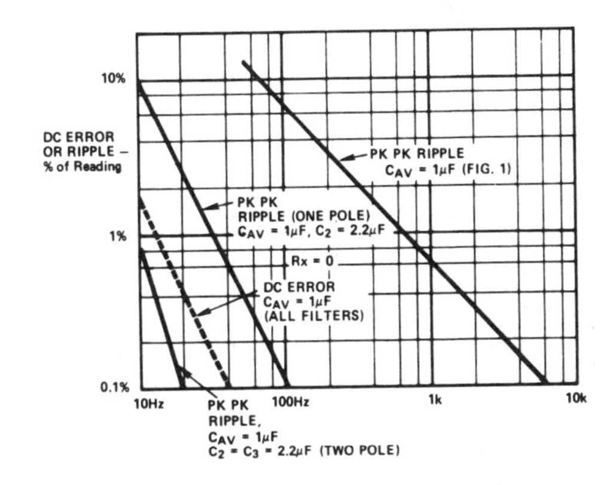

Performance Features of Various Filter Types

Analog Devices AD636
Low-Level True rms-to-dc Converter

- True rms-to-dc conversion
- 200 mV full scale
- Laser-trimmed to high accuracy
  - ○ 0.3% max error (AD636K)
  - ○ 0.6% max error (AD636J)
- Wide response capability:
  - ○ Computes rms of ac and dc signals
  - ○ 1 MHz −3 dB bandwidth: V_{rms} > 100 mV
  - ○ Signal-crest factor of 6 for 0.5% error
- dB output with 50 dB range
- Low power: 800 μA quiescent current
- Single or dual supply operation
- Low cost
- Available in chip form

AD636 FUNCTIONAL BLOCK DIAGRAM

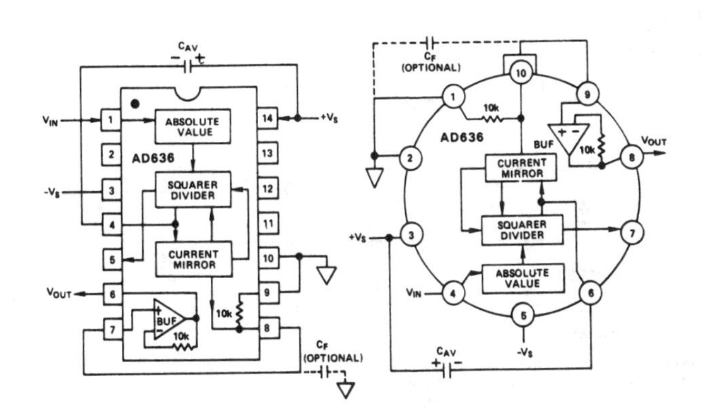

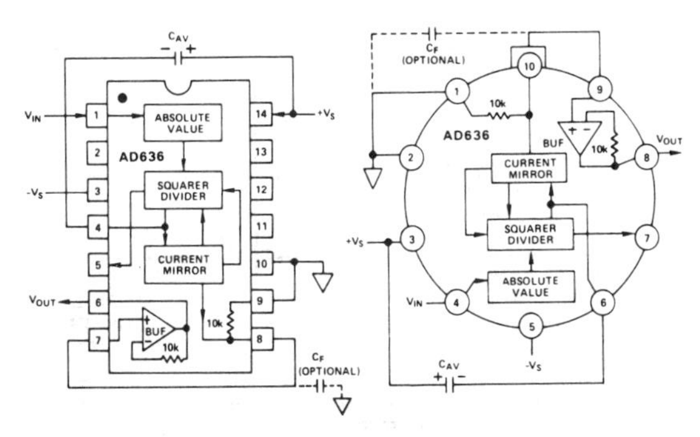

Standard rms Connection

SPECIFICATIONS (@ +25°C, and $+V_S = +3V$, $-V_S = -5V$ unless otherwise noted)

Model	AD636J Min	AD636J Typ	AD636J Max	AD636K Min	AD636K Typ	AD636K Max	Units
TRANSFER FUNCTION		$V_{OUT} = \sqrt{avg.(V_{IN})^2}$			$V_{OUT} = \sqrt{avg.(V_{IN})^2}$		
CONVERSION ACCURACY							
Total Error, Internal Trim[1,2]			**±0.5 ±0.6**			**±0.2±0.3**	mV ± % of Reading
vs. Temperature, 0 to +70°C			±0.1 ±0.01			±0.1 ±0.005	mV ± % of Reading/°C
vs. Supply Voltage		±0.1 ±0.01			±0.1 ±0.01		mV ± % of Reading/V
dc Reversal Error at 200mV		±0.2			±0.1		% of Reading
Total Error, External Trim[1]		±0.3 ±0.1			±0.1 ±0.1		mV ± % of Reading
ERROR VS. CREST FACTOR[3]							
Crest Factor 1 to 2		Specified Accuracy			Specified Accuracy		
Crest Factor = 3		−0.2			−0.2		% of Reading
Crest Factor = 6		−0.5			−0.5		% of Reading
AVERAGING TIME CONSTANT		25			25		ms/μF CAV
INPUT CHARACTERISTICS							
Signal Range, All Supplies							
Continuous rms Level		0 to 200			0 to 200		mV rms
Peak Transient Inputs							
+3V, −5V Supply			±2.8			±2.8	V pk
±2.5V Supply			±2.0			±2.0	V pk
±5V Supply			±5.0			±5.0	V pk
Maximum Continuous Non-Destructive Input Level (All Supply Voltages)			±12			±12	V pk
Input Resistance	5.33	6.67	8	5.33	6.67	8	kΩ
Input Offset Voltage			±0.5			±0.2	mV
FREQUENCY RESPONSE[2,4]							
Bandwidth for 1% additional error (0.09dB)							
V_{IN} = 10mV		14			14		kHz
V_{IN} = 100mV		90			90		kHz
V_{IN} = 200mV		130			130		kHz
±3dB Bandwidth							
V_{IN} = 10mV		100			100		kHz
V_{IN} = 100mV		900			900		MHz
V_{IN} = 200mV		1.5			1.5		MHz
OUTPUT CHARACTERISTICS[2]							
Offset Voltage, V_{IN} = COM			**±0.5**			**±0.2**	mV
vs. Temperature		±10			±10		μV/°C
vs. Supply		±0.1			±0.1		mV/V
Voltage Swing							
+3V, −5V Supply	0 to +1.0			0 to +1.0			V
±5V to ±16.5V Supply	**0 to +1.0**	+1.4		**0 to +1.0**	+1.4		V
Output Impedance	8	10	12	8	10	12	kΩ
dB OUTPUT							
Error, V_{IN} = 7mV to 300mV rms		±0.3	**±0.5**		±0.1	**±0.2**	dB
Scale Factor		−3.0			−3.0		mV/dB
Scale Factor Temperature Coefficient		+0.33			+0.33		% of Reading/°C
		−0.033			−0.033		dB/°C
I_{REF} for 0dB = 1V rms	2	4	8	2	4	8	μA
I_{REF} Range	1		50	1		50	μA
I_{OUT} TERMINAL							
I_{OUT} Scale Factor		100			100		μA/V rms
I_{OUT} Scale Factor Tolerance	−20	±10	+20	−20	±10	+20	%
Output Resistance	8	10	12	8	10	12	kΩ
Voltage Compliance		$-V_S$ to ($+V_S$ −2V)			$-V_S$ to ($+V_S$ −2V)		V
BUFFER AMPLIFIER							
Input and Output Voltage Range	$-V_S$ to ($+V_S$ −2V)			$-V_S$ to ($+V_S$ −2V)			V
Input Offset Voltage, R_S = 10k		±0.8	**±2**		±0.5	**±1**	mV
Input Bias Current		20	**60**		20	**60**	nA
Input Resistance		10^8			10^8		Ω
Output Current	(+5mA, −130μA)			(+5mA, −130μA)			
Short Circuit Current		20			20		mA
Small Signal Bandwidth		1			1		MHz
Slew Rate[5]		5			5		V/μs
POWER SUPPLY							
Voltage, Rated Performance		+3, −5			+3, −5		V
Dual Suply	+2, −2.5		±16.5	+2, −2.5		±16.5	V
Single Suply	+5		+24	+5		+24	V
Quiescent Current[6]		0.80	**1.00**		0.80	**1.00**	mA
TEMPERATURE RANGE							
Rated Performance	0		+70	0		+70	°C
Storage	−55		+150	−55		+150	°C
PACKAGE OPTIONS							
TO-116 (D-14)		AD636JD			AD636KD		
TL-100 (H-10A)		AD636JH			AD636KH		

NOTES
[1]Accuracy specified for 0 to 200mV rms, dc or 1kHz sinewave input. Accuracy is degraded at higher rms signal levels.
[2]Measured at pin 8 of DIP (I_{OUT}), with pin 9 tied to common.
[3]Error vs. crest factor is specified as additional error for a 200mV rms rectangular pulse train, pulse width = 200μs.
[4]Input voltages are expressed in volts rms.
[5]With 10kΩ pull down resistor from pin 6 (BUF OUT) to $-V_S$.
[6]With BUF input tied to Common.

Specifications subject to change without notice.

Specifications shown in boldface are tested on all production units at final electrical test. Results from those tests are used to calculate outgoing quality levels. All min and max specifications are guaranteed, although only those shown in boldface are tested on all production units.

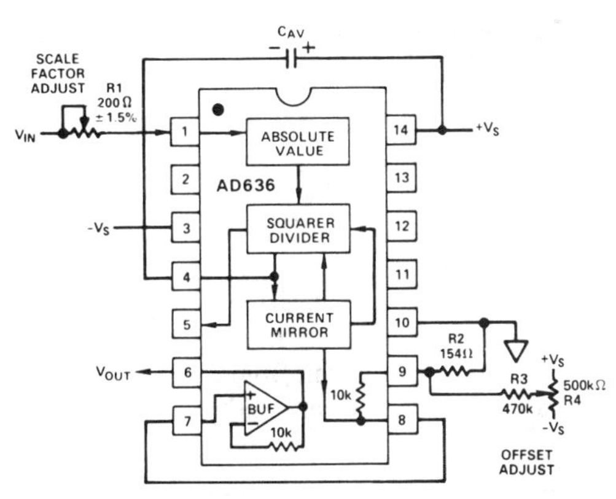

Optional External Gain and Output Offset Trims

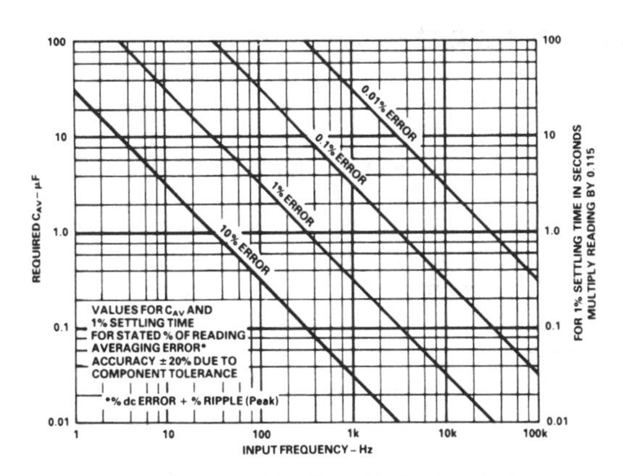

Error/Settling Time Graph for Use with the Standard rms Connection

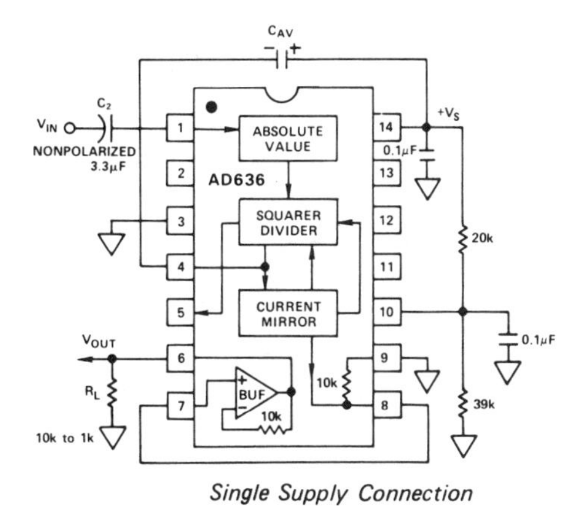

Single Supply Connection

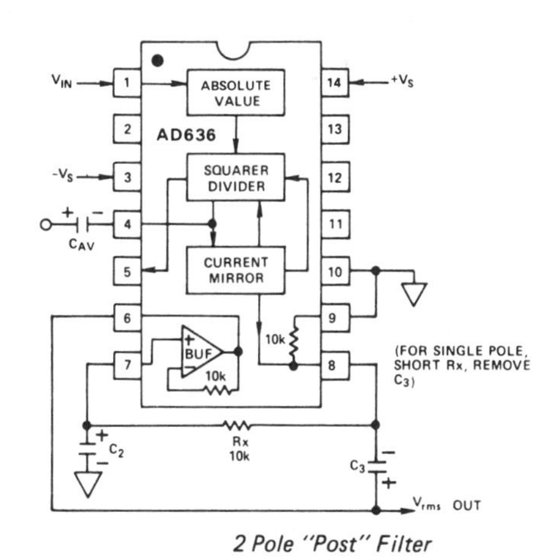

2 Pole "Post" Filter

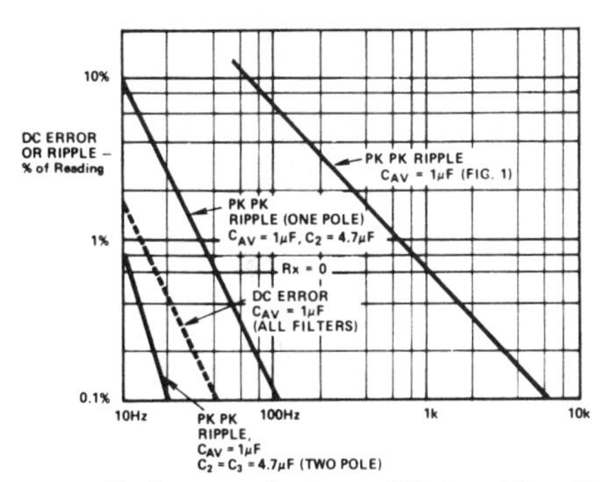

Performance Features of Various Filter Types

Simplified Schematic

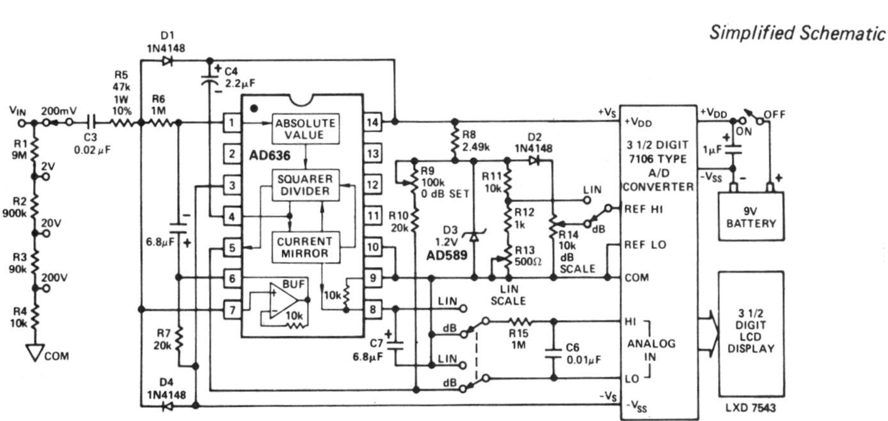

A Portable, High Z Input, rms DPM and dB Meter Circuit

Analog Devices AD637

High-Precision, Wideband rms-to-dc Converter

- High Accuracy
  - 0.2% max nonlinearity, 0 to 2V rms input
  - 0.10% additional error to crest factor of 3
- Wide bandwidth
  - 8 MHz at 2V rms input
  - 600 kHz at 100 mV rms
- Computes:
  - True rms
  - Square
  - Mean square
  - Absolute value
- dB output (60 dB range)
- Chip select-power down feature allows:
  - Analog "3-state" operation
  - Quiescent current reduction from 2.2 mA to 350 μA
  - Side-brazed DIP or low-cost cerdip

AD637 FUNCTIONAL BLOCK DIAGRAM

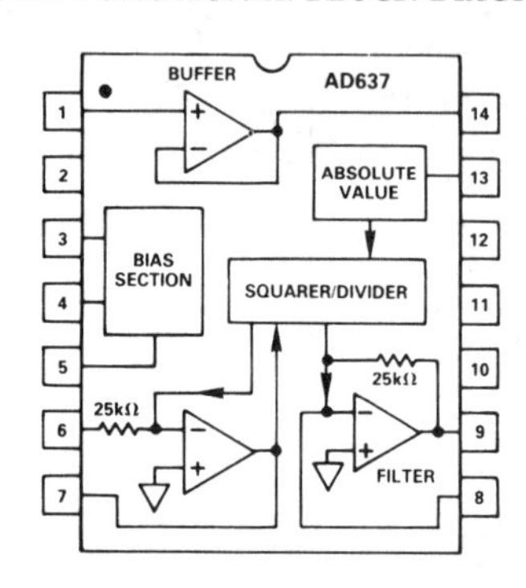

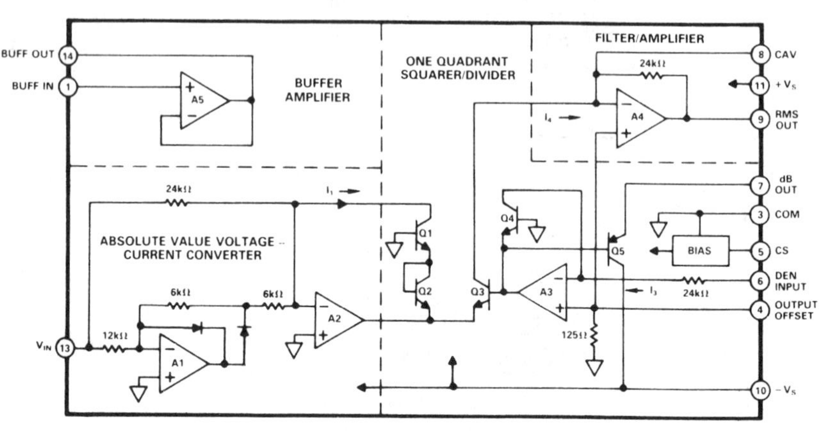

Simplified Schematic

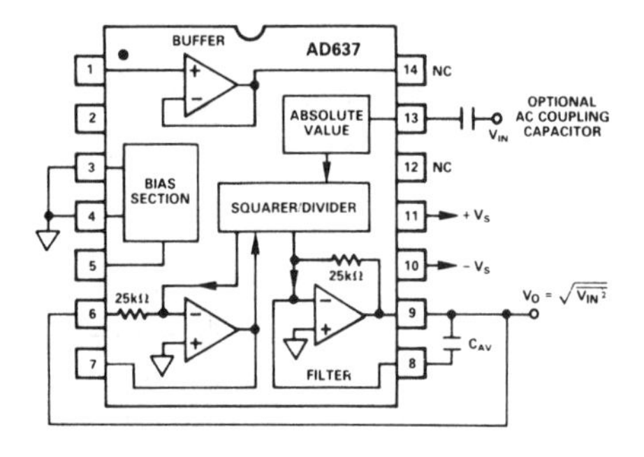

Standard rms Connection

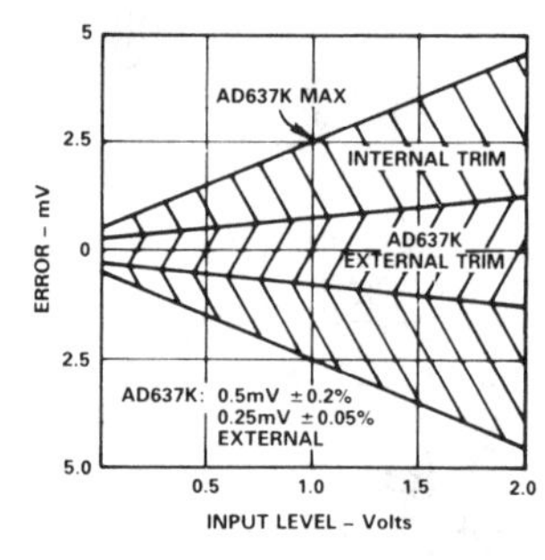

Max Total Error vs. Input Level AD637K Internal and External Trims

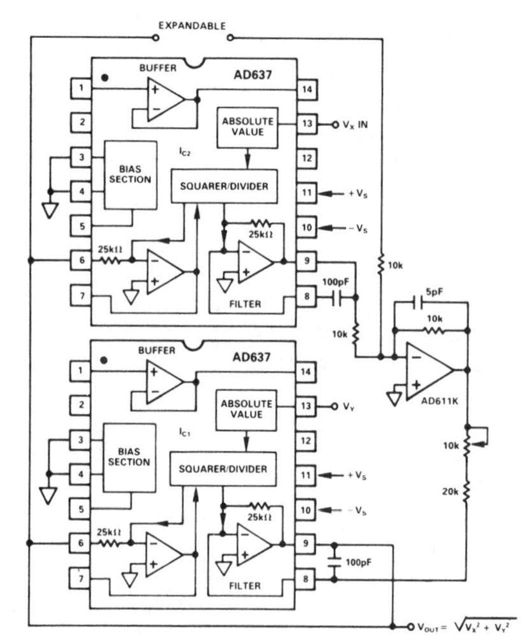

AD637 Vector Sum Configuration

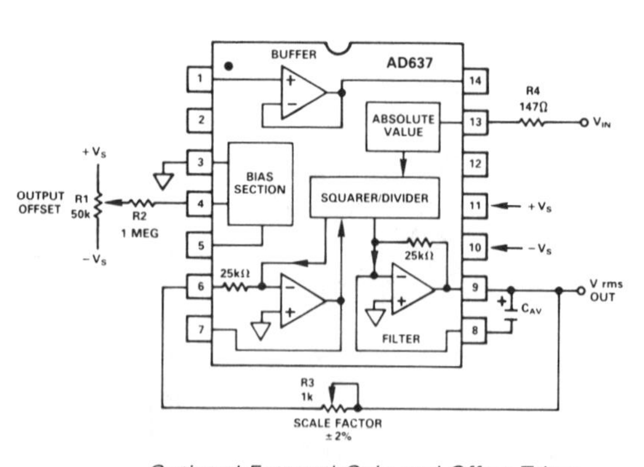

Optional External Gain and Offset Trims

SPECIFICATIONS (@ +25°C, and ±15V dc unless otherwise noted)

Model	AD637AJ Min	AD637AJ Typ	AD637AJ Max	AD637AK Min	AD637AK Typ	AD637AK Max	AD637AS Min	AD637AS Typ	AD637AS Max	Units
TRANSFER FUNCTION		$V_{OUT} = \sqrt{avg.(V_{IN})^2}$			$V_{OUT} = \sqrt{avg.(V_{IN})^2}$			$V_{OUT} = \sqrt{avg.(V_{IN})^2}$		
CONVERSION ACCURACY										
Total Error, Internal Trim[1] (Fig. 2)			**±1 ±0.5**			**±0.5 ±0.2**			**±1 ±0.5**	mV ± % of Reading
T_{min} to T_{max}			**±3.0 ±0.6**			**±2.0 ±0.3**			**±6 ±0.7**	mV ± % of Reading
vs. Supply +		30	**150**		30	**150**		30	**150**	µV/V
vs. Supply −		100	**300**		100	**300**		100	**300**	µV/V
dc Reversal Error at 2V			**0.25**			**0.1**			**0.25**	% of Reading
Nonlinearity 2V Full Scale[2]			**0.04**			**0.02**			**0.04**	% of FSR
Nonlinearity 7V Full Scale			**0.05**			**0.05**			**0.05**	% of FSR
Total Error, External Trim		±0.5 ±0.1			±0.25 ±0.05			±0.5 ±0.1		mV ± % of Reading
ERROR VS. CREST FACTOR[3]										
Crest Factor 1 to 2		Specified Accuracy			Specified Accuracy			Specified Accuracy		
Crest Factor = 3		±0.1			±0.1			±0.1		% of Reading
Crest Factor = 10		±1.0			±1.0			±1.0		% of Reading
AVERAGING TIME CONSTANT		25			25			25		ms/µF CAV
INPUT CHARACTERISTICS										
Signal Range, ±15V Supply										
Continuous rms Level		0 to 7			0 to 7			0 to 7		V rms
Peak Transient Input			±15			±15			±15	V p-p
Signal Range, ±5V Supply										
Continuous rms Level		0 to 4			0 to 4			0 to 4		V rms
Peak Transient Input			±6			±6			±6	V p-p
Maximum Continuous Non-Destructive Input Level (All Supply Voltages)			±15			±15			±15	V p-p
Input Resistance	6.4	8	9.6	6.4	8	9.6	6.4	8	9.6	kΩ
Input Offset Voltage			±0.5			±0.2			±0.5	mV
FREQUENCY RESPONSE[4]										
Bandwidth for 1% additional error (0.09dB)										
V_{IN} = 20mV		11			11			11		kHz
V_{IN} = 200mV		66			66			66		kHz
V_{IN} = 2V		200			200			200		kHz
±3dB Bandwidth										
V_{IN} = 20mV		150			150			150		kHz
V_{IN} = 200mV		1			1			1		MHz
V_{IN} = 2V		8			8			8		MHz
OUTPUT CHARACTERISTICS										
Offset Voltage			**±1**			**±0.5**			**±1**	**mV**
vs. Temperature		±0.05	**±0.089**		±0.04	**±0.056**		±0.04	**±0.07**	**mV/°C**
Voltage Swing, ±15V Supply, 2kΩ Load	**0 to +12.0**	+13.5		**0 to +12.0**	+13.5		**0 to +12.0**	+13.5		**V**
Voltage Swing, ±3V Supply, 2kΩ Load	**0 to +2**	+2.2		**0 to +2**	+2.2		**0 to +2**	+2.2		**V**
Output Current	**6**			**6**			**6**			**mA**
Short Circuit Current		20			20			20		mA
Resistance, Chip Select "High"		0.5			0.5			0.5		MΩ
Resistance, Chip Select "Low"		100			100			100		kΩ
dB OUTPUT										
Error, V_{IN} 7mV to 7V rms, 0dB = 1V rms		±1			±1			±1		dB
Scale Factor		−3			−3			−3		mV/dB
Scale Factor Temperature Coefficient		+0.33			+0.33			+0.33		% of Reading/°C
		−0.033			−0.033			−0.033		dB/°C
I_{REF} for 0dB = 1V rms	**5**	20	**80**	**5**	20	**80**	**5**	20	**80**	**µA**
I_{REF} Range	1		100	1		100	1		100	µA
BUFFER AMPLIFIER										
Input and Output Voltage Range	$-V_S$ to ($+V_S$ − 2.5V)			$-V_S$ to ($+V_S$ − 2.5V)			$-V_S$ to ($+V_S$ − 2.5V)			V
Input Offset Voltage		±0.8	**±2**		±0.5	**±1**		±0.8	**±2**	mV
Input Current		±2	**±10**		±2	**±5**		±2	**±10**	nA
Input Resistance		10^8			10^8			10^8		Ω
Output Current	(+5mA, −130µA)			(+5mA, −130µA)			(+5mA, −130µA)			
Short Circuit Current		20			20			20		mA
Small Signal Bandwidth		1			1			1		MHz
Slew Rate[5]		5			5			5		V/µs
DENOMINATOR INPUT										
Input Range		0 to +10			0 to +10			0 to +10		V
Input Resistance	20	25	30	20	25	30	20	25	30	kΩ
Offset Voltage		±0.2	±0.5		±0.2	±0.5		±0.2	±0.5	mV
CHIP SELECT PROVISION (CS)										
rms "ON" Level		Open or +2.4V < V_C < $+V_S$			Open or +2.4V < V_C < $+V_S$			Open or +2.4V < V_C < $+V_S$		
rms "OFF" Level		V_C < +0.2V			V_C < +0.2V			V_C < +0.2V		
I_{OUT} of Chip Select										
CS "LOW"		10			10			10		µA
CS "HIGH"		Zero			Zero			Zero		
On Time Constant		10µs + ((25kΩ) × C_{AV})			10µs + ((25kΩ) × C_{AV})			10µs + ((25kΩ) × C_{AV})		
Off Time Constant		10µs + ((25kΩ) × C_{AV})			10µs + ((25kΩ) × C_{AV})			10µs + ((25kΩ) × C_{AV})		
POWER SUPPLY										
Operating Voltage Range	**±3.0**		**±18**	**±3.0**		**±18**	**±3.0**		**±18**	**V**
Quiescent Current		2.2	**3**		2.2	**3**		2.2	**3**	**mA**
Standby Current		350	**450**		350	**450**		350	**450**	**µA**
PACKAGE OPTIONS										
TO-116-(D-14)		AD637AJD			AD637AKD			AD637ASD		
Cerdip (Q-14)		AD637AJQ			AD637AKQ			N/A		

NOTES
[1]Accuracy specified 0-7V rms dc with AD637 connected as shown in Figure 2.
[2]Nonlinearity is defined as the maximum deviation from the straight line connecting the readings at 10mV and 2V.
[3]Error vs. crest factor is specified as additional error for 1V rms.
[4]Input voltages are expressed in volts rms. % are in % of reading.
[5]With external 2kΩ pull down resistor tied to $-V_S$.

Specifications subject to change without notice.

Specifications shown in boldface are tested on all production units at final electrical test. Results from those tests are used to calculate outgoing quality levels. All min and max specifications are guaranteed, although only those shown in boldface are tested on all production units.

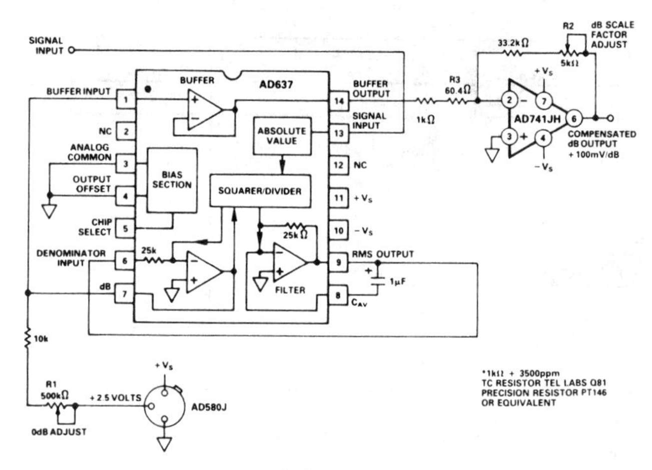

dB Connection

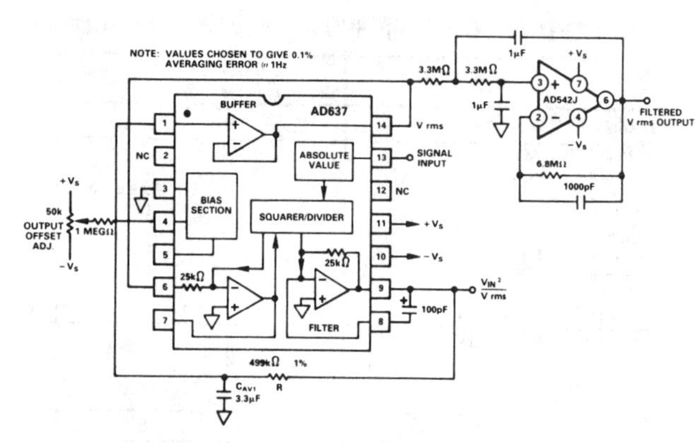

AD637 as a Low Frequency rms Converter

This information is reproduced with permission of GEC Plessey Semiconductors.

Plessey SL9901
50-MHz Transimpedance Amplifier Preliminary Information

- High sensitivity
- 50-MHz bandwidth (100 Mbits/NRZ data rate)
- Wide dynamic range
- 5V supply
- Usable in systems with 10^{-9} BER at −36 dBm average optical power

APPLICATIONS

- Fiber-optic data links
- Nucleonics
- Instrumentation
- Current/voltage conversion

ABSOLUTE MAXIMUM RATINGS

Voltages are with respect to the negative power supply (V_{EE})

Exceeding these ratings may cause permanent damage. Functional operation under these conditions is not implied.

Positive supply voltage, V_{CC}	0V to +7V
Input voltage (device sourcing current), V_I	0V
Input current (device sinking current), I_I	1mA
Output voltage (device sinking current), V_O	0V to V_{CC} -2V
Output current (device sourcing current), I_O	10mA
Storage temperature, T_{ST}	-55°C to +150°C
Relative humidity at 85°C, R_H	85%
Chip temperature - plastic package (Note 1), T_C	145°C

NOTE
1. Thermal conductivities from chip to case and to ambient are given in Electrical Characteristics.

ELECTRICAL CHARACTERISTICS

Test Conditions - Voltages are with respect to the negative power supply (V_{EE})

Characteristic	Symbol	Value			Units
		Min.	Typ.(1)	Max.	
Positive supply voltage	V_{CC}	4.0	5	5.5	V
Ambient temperature	T_{amb}	-40		+85	°C
Input current (RMS) (Note 2)	I_I	0.3		10	μA
Output load	R_O			250	Ω

NOTES
1. Typical figures are for design aid only. They are not guaranteed and not subject to production testing.
2. The device is guaranteed to operate at up to 30 milliamps (RMS), but above 10 milliamps the output may be clipped with associated loss of gain and the mid-points of the output may be distorted by up to 3ns.

Characteristics - Voltages are with respect to the negative power supply (V_{EE})

Characteristic	Symbol	Value			Units	Conditions
		Min.	Typ.(1)	Max.		
Power dissipation	P_D			100	mW	
Thermal resistance: chip to case	θ_{CC}		57	69	°C/W	
Thermal resistance: chip to ambient	θ_{CA}		163	196	°C/W	
Supply current	I_{CC}	9	12	16	mA	Output unloaded
Input bias voltage	V_{IB}		1.5	1.65	V	25°C
Input bias variation	$\delta V_{IB}/\delta T$		1.7	4	mV/°C	
Input impedance	Z_I		400		Ω	
Input current at clipping	I_{IC}	10			μA	
Input current noise (Note 2)	N_I		3.5	4	pA/√Hz	
Input voltage noise (Note 2)	N_V		3200	4000	pV/√Hz	
Transimpedance gain	G_T	30	40		kΩ	
Gain roll-off	G_R	6			dB/Oct.	f > 100MHz
3dB bandwidth (see Fig.3)	f_B	50			MHz	5pF on input
		26			MHz	30pF on input
Output impedance	Z_O		50	100	Ω	
Pin capacitance	C_P		1.5	2	pF	Pin to supplies

NOTES
1. Typical figures are for design aid only. They are not guaranteed and not subject to production testing.
2. The typical input voltage noise would create an equivalent input noise current of 0.12 microamps averaged over DC to 160MHz when the path is completed through an external 5pF input capacitance. The typical current noise source would create an equivalent input noise current of 0.04 microamps over DC to 160MHz.

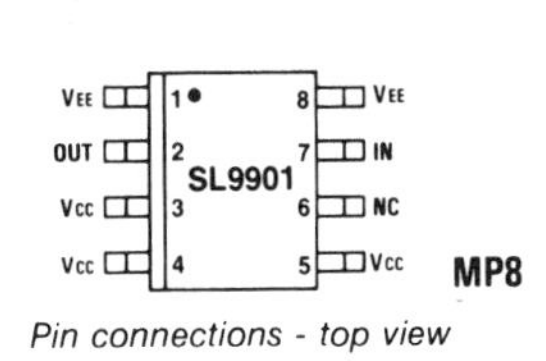

Pin connections - top view

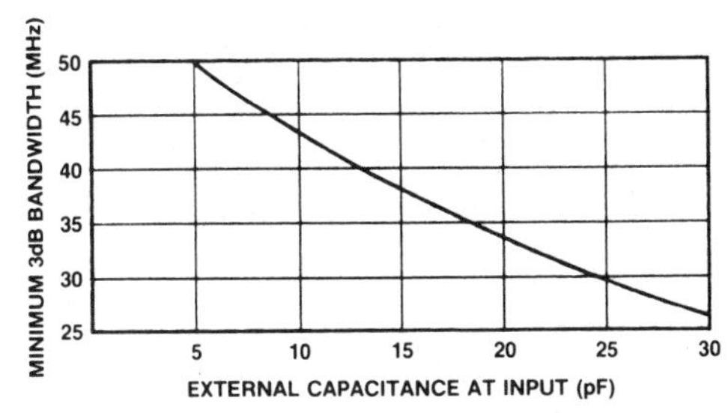

Minimum 3dB bandwidth against external capacitance at input

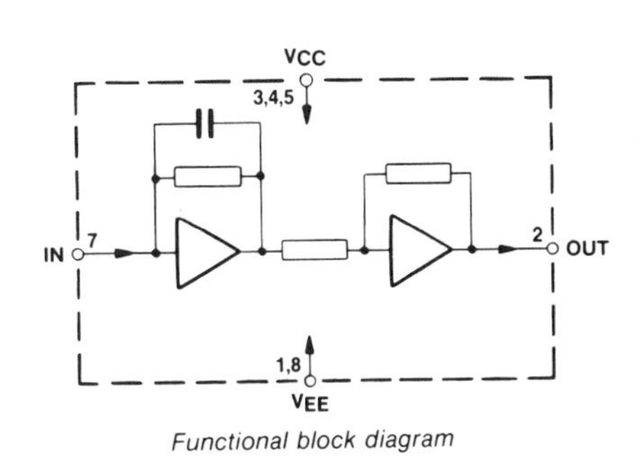

Functional block diagram

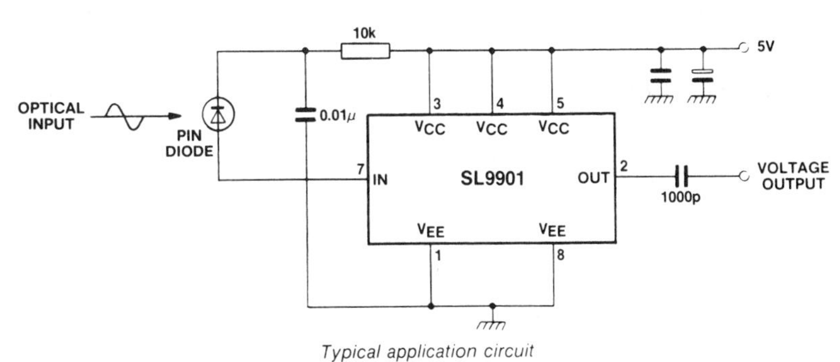

Typical application circuit

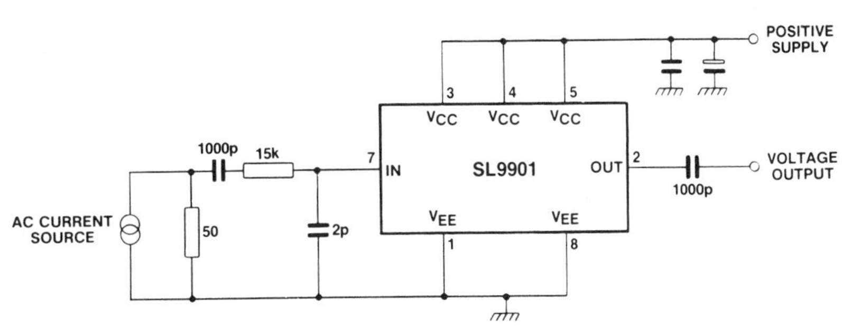

NOTE
For currents in the range 10μA to 30μA (RMS) a DC offset is added to prevent the input from sourcing current. This gives a better approximation to normal use

Test circuit

Plessey
SL441C
Zero Voltage Switch

ELECTRICAL CHARACTERISTICS

Test conditions (unless otherwise stated):
T_{AMB} = 25°C
All voltages measured with respect to common (pin 1)

Characteristics	Value			Units
	Min.	Typ.	Max.	
Shunt regulating voltage pin 3 @ 16mA		14.7		V
Shunt regulating voltage pin 3 @ 16mA @ 75°C			16	V
Supply voltage trip level pin 3		12.2		V
Supply current (less I_4AV, I_5) (see Note 1)			7.5	mA
Regulated voltage pin 5	8.0	8.5	9.0	V
Regulated voltage temperature coefficient pin 5	−1		+1	mV/°C
Triac gate drive pin 4 (See Note 2)				
Open circuit ON voltage		8.5		V
Open circuit OFF voltage			0.1	V
Output current into 2V drain	100	130		mA
Output current into 4V drain	65	80		mA
Output current into short circuit			200	mA
Internal drain resistance		800		Ω
Control input pin 8				
Bias current			1	µA
Hysteresis		20		mV
Sensor malfunction circuit operates at	150	200	250	mV
Input working voltage range	0		12	V
Internal reference voltage (Ramp start) (See Note 3)	4.0	4.25	4.5	V
Internal reference voltage (Ramp finish) (See Note 3)		4.35		V
Peak-to-peak amplitude of ramp	70	100	130	mV
Pin 6 output impedance (R6) (See Note 2)	21.5	27	32.5	kΩ
Maximum ripple voltage pin 3			1	V_{P-P}

NOTES
1. The supply current is 0.45 x (RMS current fed into pin 2). I_5 is the current drained from pin 5 externally. I_{4AV} is the average triac gate current supplied each mains cycle.
2. Triac firing pulse. t_p Pulse width = 0.69 R6C_D microseconds typical
t_f Pulse finish = 1.09 R6C_D microseconds minimum after zero voltage point R6 in kohms. C_D in nF. See Application circuit
t_p Nominal (C_D = 2.7nF) = 50 microseconds
t_f Minimum (C_D = 2.7nF) = 63 microseconds
3. Ramp period = 0.85 ± 0.15 x R_TC_T sec. See Application circuit. The actual value of R_t must lie between 500kohms and 3Mohms.

ABSOLUTE MAXIMUM RATINGS

Voltages
Voltage on pin 8 V_{8-1} Max. 12V
Voltage on pin 4 V_{4-1} Max. 10V

Currents
Supply current (pin 2) Peak value $\pm I_{2M}$ 50mA.
Non-repetitive peak current (tp ≤250µs) $\pm I_{2SM}$ 200mA.
Output current (pin 5) Max. 5mA Short circuit protected.
Output current (pin 4) average value $I_{4(AV)}$ Max 5mA Short circuit protected.

Temperature
Operating ambient temperature T_{AMB} −10°C to +75°C
Storage temperature T_{STG} −30°C to +125°C

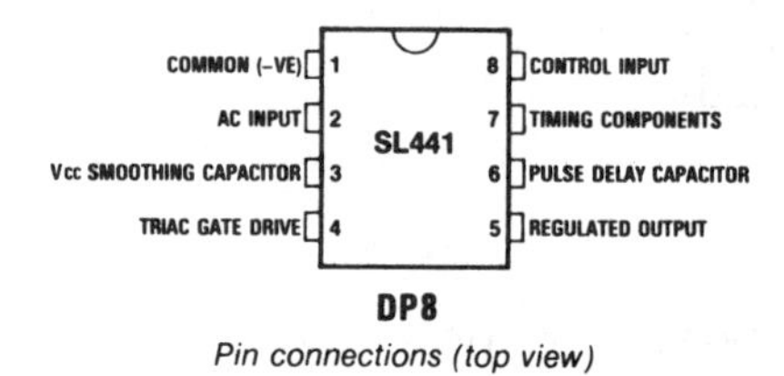

Pin connections (top view)

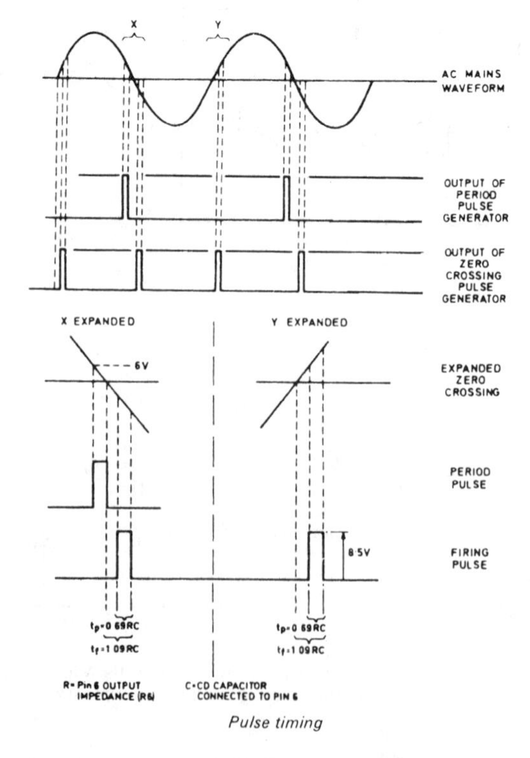

Pulse timing

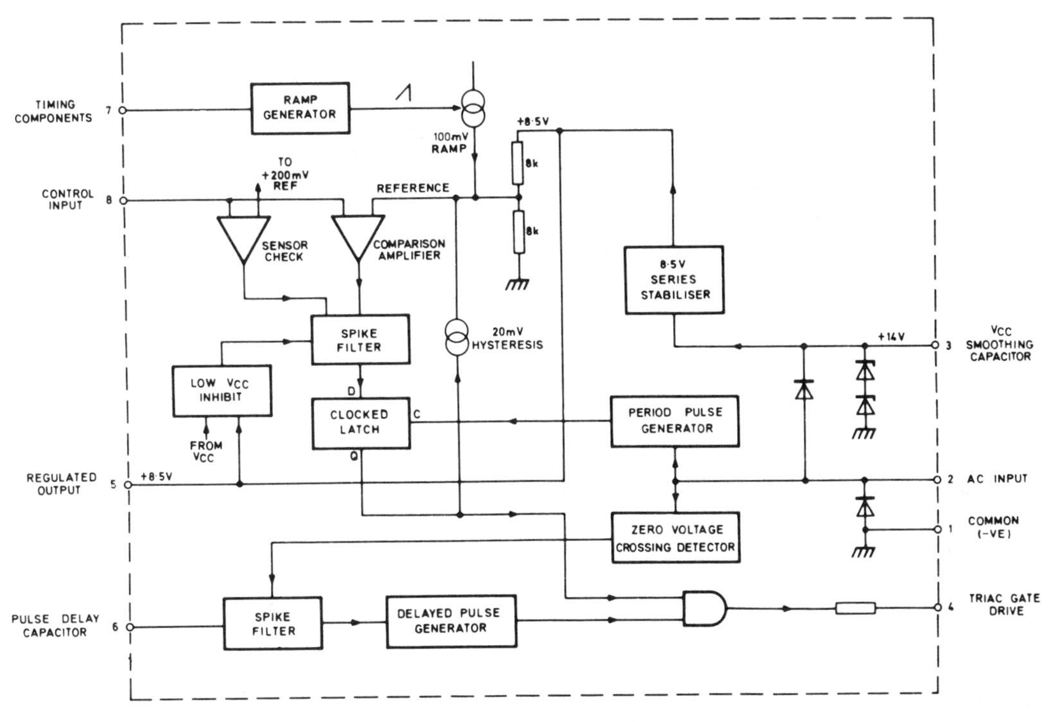

Block schematic of SL441C

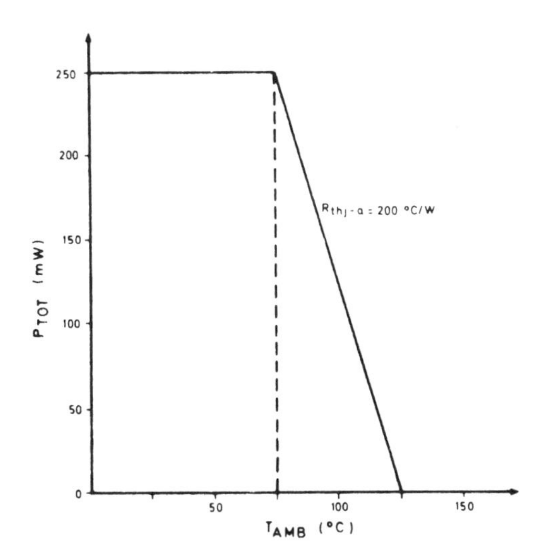

Power dissipation

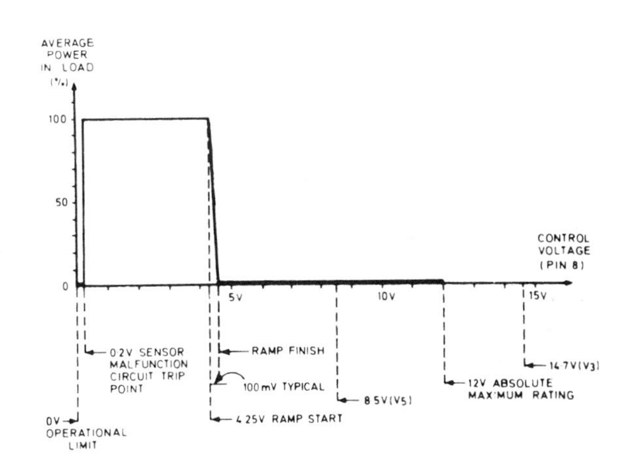

Control characteristic of pin 8

An Electric Thermostat for Room Heaters Using the SL441C

The circuit has a sensitivity of nominally 110 mV/°C. The width of the proportional control band is nominally 1.0 °C and offers a good compromise between temperature stability and regulation performance.

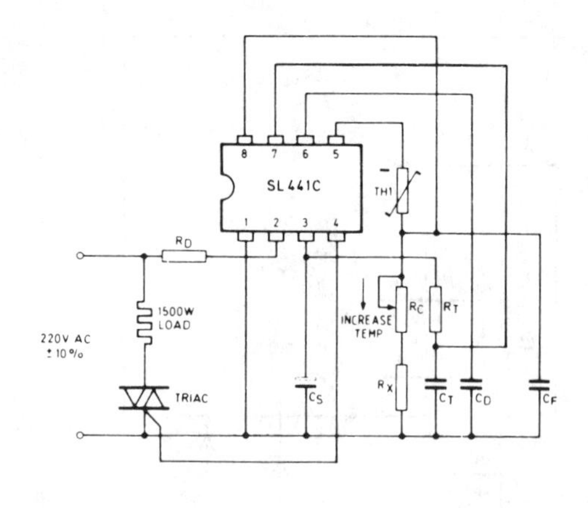

SUGGESTED VALUES

R_D	8.2kΩ 5% 6W (less in air stream)
R_C	22kΩ linear control potentiometer
R_t	2.7MΩ } Ramp period = 11 secs.*
C_t	4.7μF 10V TANT }
C_f	47nF (filter if required)
C_S	150μF 16V
C_D	2.2nF ±10%
Triac	TAG 250-400

TH1 NTC thermistor. R_{25} = 10k Ω
B = 4200. e.g. ITT KQ103

R_x 6.2k ±5% fixed resistor or 10k Ω preset.
See control characteristics Figs. 7 and 8

*If on-off control is required, omit R_T and C_T and link pin 7 to pin 1.

*Application circuit for proportional control system.**

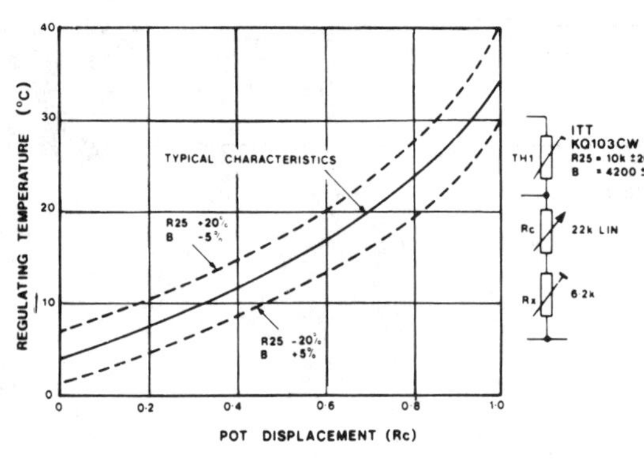

Control characteristics of electronic thermostat (mechanical calibration)

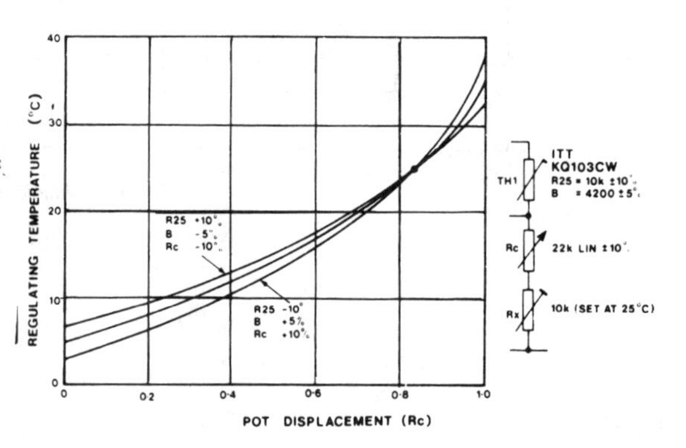

Control characteristics of electronic thermostat (electrical calibration)

Plessey SL443A Zero Voltage Switch

APPLICATIONS

- Cooker hotplates
- Powerful hairdryers

ABSOLUTE MAXIMUM RATINGS

Voltages

Voltage on pin 8,	V_{8-1}	Max	10v
Voltage on pin 4,	V_{4-1}	Max	10v

Currents

Supply current, pin 2 peak value	$\pm I_{2M}$	Max	50mA
Non-repetitive peak current ($t_P \geqslant 250\mu S$)	$\pm I_{2SM}$	Max	200mA
Output current, pin 5	I_5	Short circuit protected	
Output current, pin 4, average value	I_4 (AV)	Max	10mA
		Short circuit protected	

Temperatures

Operating ambient temperature	T_{amb}	—10 to 75°C
Storage temperature	T_{STG}	—55 to +125°C

Power Dissipation

ELECTRICAL CHARACTERISTICS

Test conditions (unless otherwise stated):
T_{amb} = 25 °C,
All voltages measured with respect to common (pin 1)

Characteristic	Value			Units	Conditions
	Min.	Typ.	Max.		
Shunt regulating voltage pin 3		14.7		V	I_3 = 16mA
Shunt regulating voltage pin 3			16	V	I_3 = 16mA, T_{amb} = +75 °C
Supply voltage trip level pin 3		12.2		V	
• Supply current (less I_4 AV, 2 × I_5) See Note 1			7.2	mA	
Potentiometer supply pin 5, V_5	6.8	7.0	7.6	V	
Potentiometer resistance range	18		140	kΩ	
Triac gate drive pin 4					
Open circuit ON voltage		8.5		V	
Open circuit OFF voltage			0.1	V	
Output current into 2V drain	80	100		mA	
Output current into 4V drain	50	70		mA	
Output current into short circuit			200	mA	
Internal drain resistance		800		Ω	
Control input pin 8					
Bias current			1	μA	
Internal reference – ramp start	0.3	0.5	0.7		
– ramp finish	V_5– 0.5	V_5–0.3	V_5– 0.1		
★ Period of ramp generator – T	27	30	33	s	(R_P = 100K, C_t = 0.68μ) (RMS mains voltage=220v)
Pin 6 output impedance R6	21.5	27	32.5	kΩ	

● The supply current is 0.45 x (RMS current fed into Pin 2)

★ Period of ramp = T = 2 x C_T x R_P x (RMS mains voltage) seconds

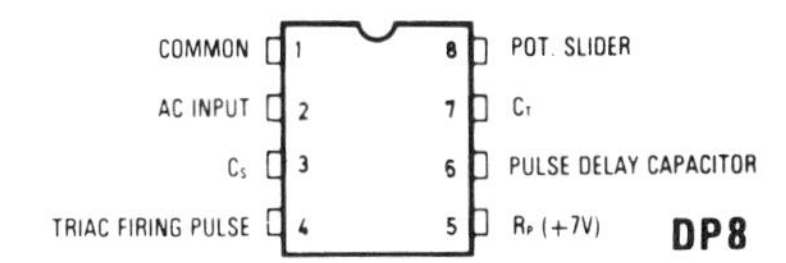

Pin connections - top view

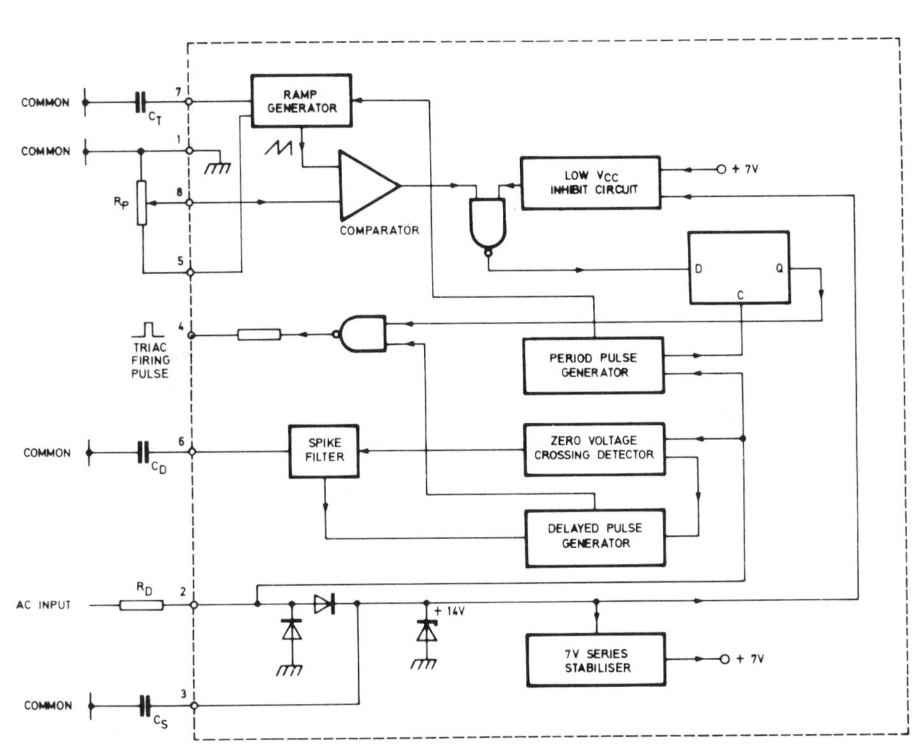

SL443A block diagram

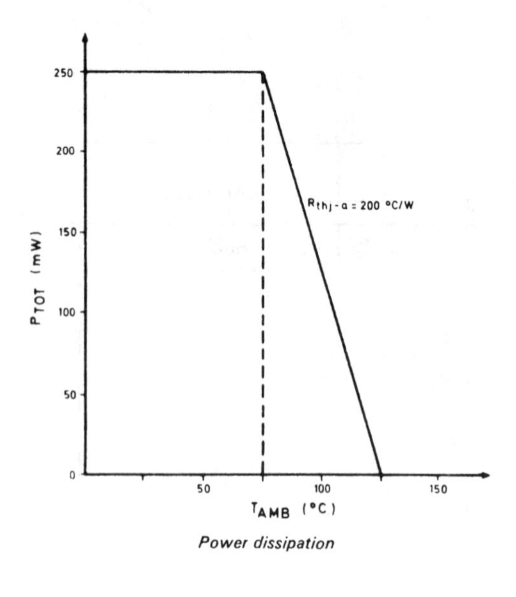

Power dissipation

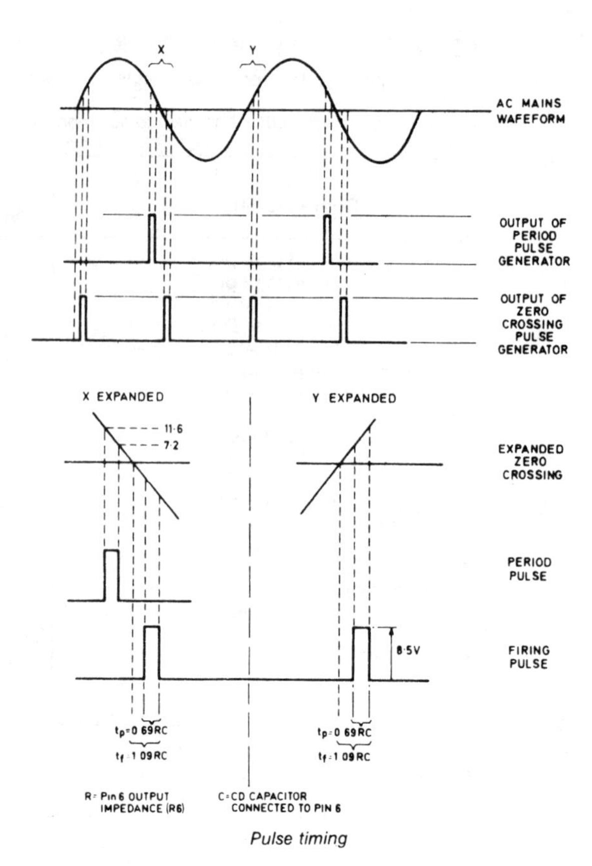

Pulse timing

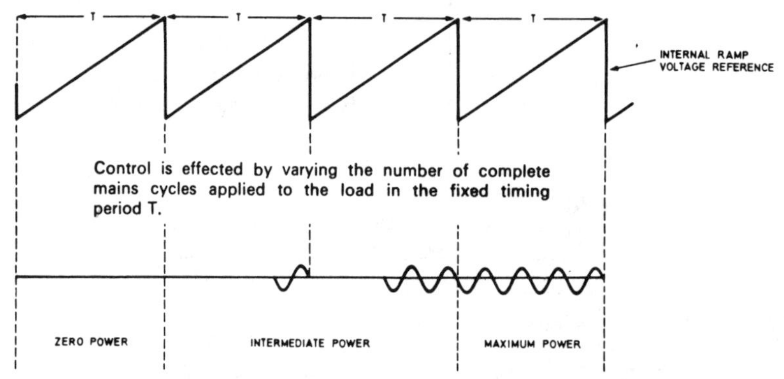

Method of control

SL443A APPLICATION CIRCUITS

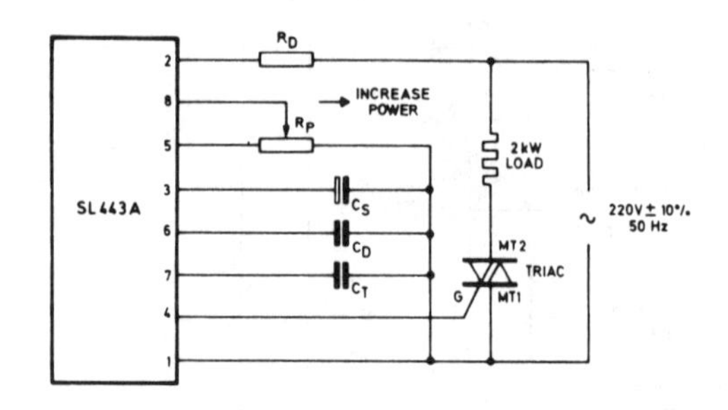

Component values

R_D 8.2kΩ 5% 7W
R_P 100kΩ (Control characteristic of linear potentiometer is shown in Fig.3)
C_S 220μF 16V
C_T 0.47μF (Ramp period = 20 seconds nominal)
C_D 1.5nF ± 10%
TRIAC TAG.255 - 400

Cooker hotplate control

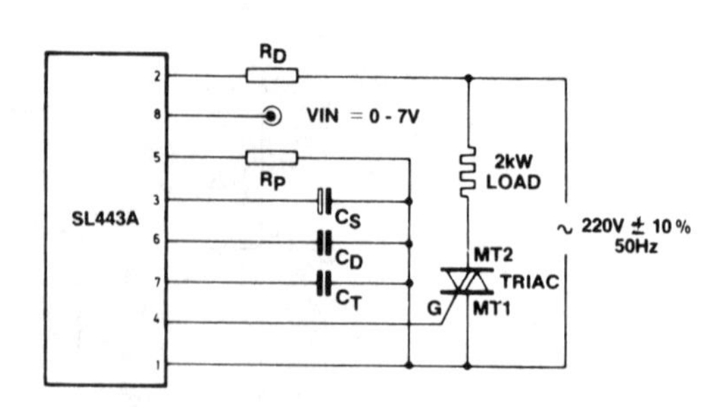

Component values

R_D 8.2kΩ 5% 7W
R_P 100kΩ
C_S 220μF 16V
C_T 0.47μF (Ramp period = 20 seconds nominal)
C_D 1.5nF ±10%
TRIAC TAG.255 – 400

Voltage control

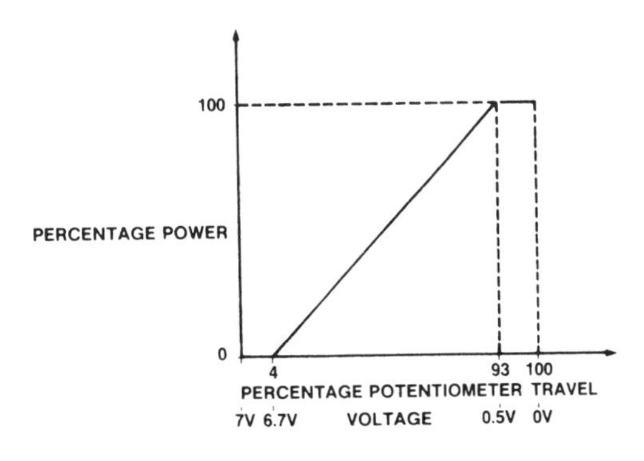

This characteristic applies to a linear potentiometer. Different control characteristics are easily obtained by using a non-linear potentiometer and/or offset resistors in the potentiometer circuit.

Output power v potentiometer displacement or voltage on pin 8

Plessey SL446A Zero Voltage Switch

APPLICATIONS

- Pan temperature control
- Water heaters
- Refrigerators
- Panel heaters

ABSOLUTE MAXIMUM RATINGS

Voltages

Voltage on pins 7,8 (V_{7-1}; V_{8-1})	V_3 (14V)
Voltage on pin 4 (V_{4-1})	10V

Currents

Supply current (pin 2):	
Peak value $\pm I_{2M}$	50mA
Non-repetitive peak current ($t_P < 250\mu s$) $\pm I_{2SM}$	200mA
Output current (pin 5) (I_5)	10mA
Output current (pin 4), average value I_4 (AV)	10mA

Temperatures

Operating ambient temperature T_{AMB}	-10°C to 75°C
Storage temperature T_{STG}	-55°C to +125°C

Power dissipation See Fig.3

ELECTRICAL CHARACTERISTICS

Test Conditions (unless otherwise stated)
$T_{AMB} = 25°C$,
All voltages measured with respect to common (pin 1)

Characteristic	Value			Units	Conditions
	Min.	Typ.	Max.		
Shunt regulating voltage pin 3		14.7		V	$I_3 = 16mA$
Shunt regulating voltage pin 3			16	V	$I_3 = 16mA$, $T_{amb} = +75°C$
Supply voltage trip level pin 3		12.2		V	
*Supply current (less I_4 AV, I_5)		7		mA	
Regulated voltage pin 5	8.0	8.5	9.0	V	
Regulated voltage temperature coefficient pin 5	−1		+1	mV/°C	
Triac gate drive pin 4					
Open circuit ON voltage		8.5		V	
Open circuit OFF voltage			0.1	V	
Output current into 2V drain	80	100		mA	
Output current into 4V drain	50	70		mA	
Output current into short circuit			200	mA	
Internal drain resistance		800		Ω	
Servo Amplifier input bias current			2	μA	
Servo Amplifier hysteresis	20	25	35	mV	
Servo Amplifier input offset voltage	−15	0	+15	mV	
Servo Amplifier input working voltage range	0		10	V	
Pin 6 output impedance R6	21.5	27	32.5	kΩ	
Maximum ripple voltage on supply pin 3			1	Vp-p	

★ The supply current is 0.45 x (RMS current fed into Pin 2)

COMMON (– VE) 1 | 8 SERVO AMPLIFIER +VE I/P
AC INPUT 2 | 7 SERVO AMPLIFIER – VE I/P
POSITIVE LINE (V_{CC}) 3 | 6 PULSE DELAY CAPACITOR
TRIAC GATE DRIVE 4 | 5 REGULATED OUTPUT

DP8

Pin connections - top view

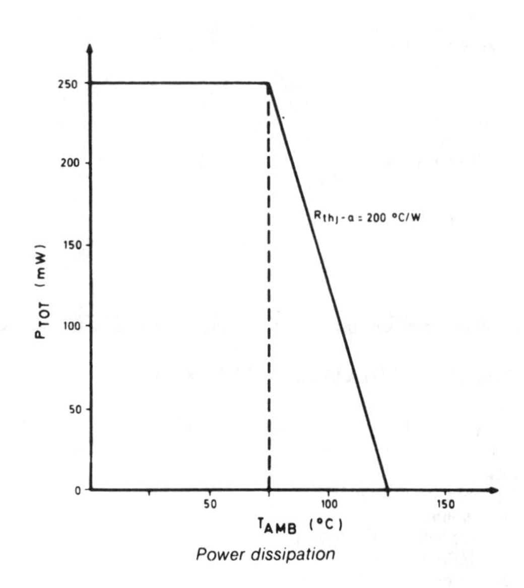

Power dissipation

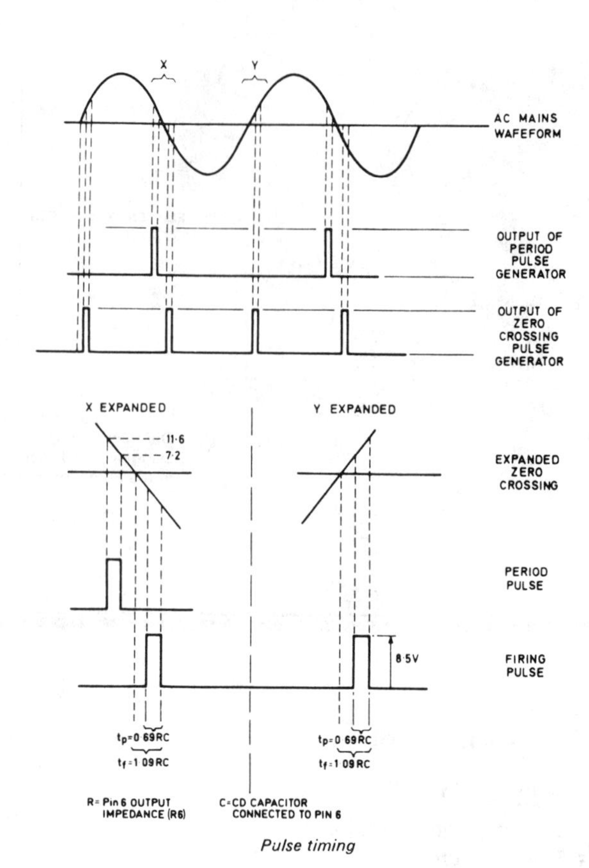

Pulse timing

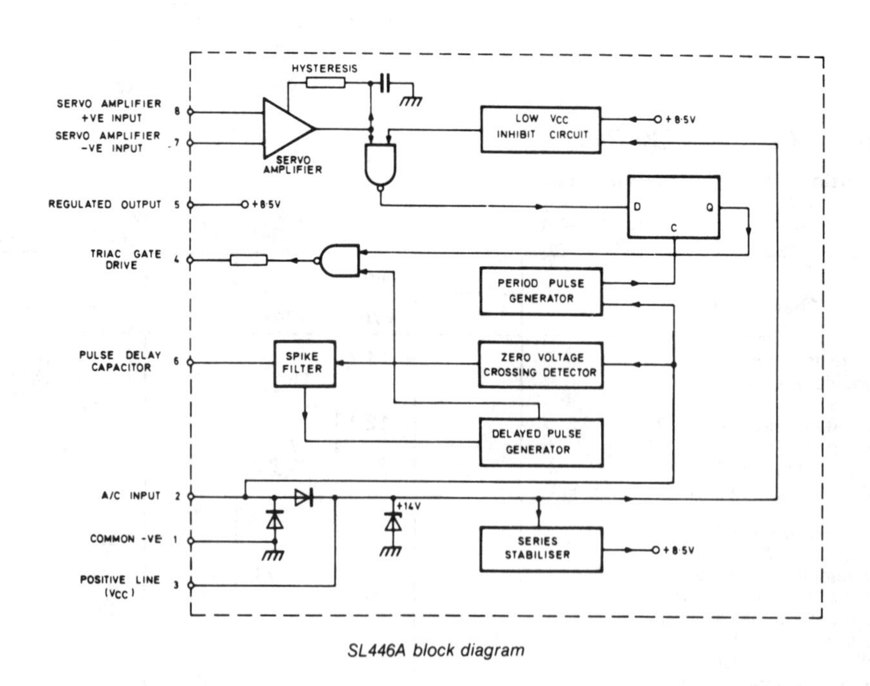

SL446A block diagram

SYSTEM DESIGN WITH THE TDA2090

See section CON for specifications of this IC.

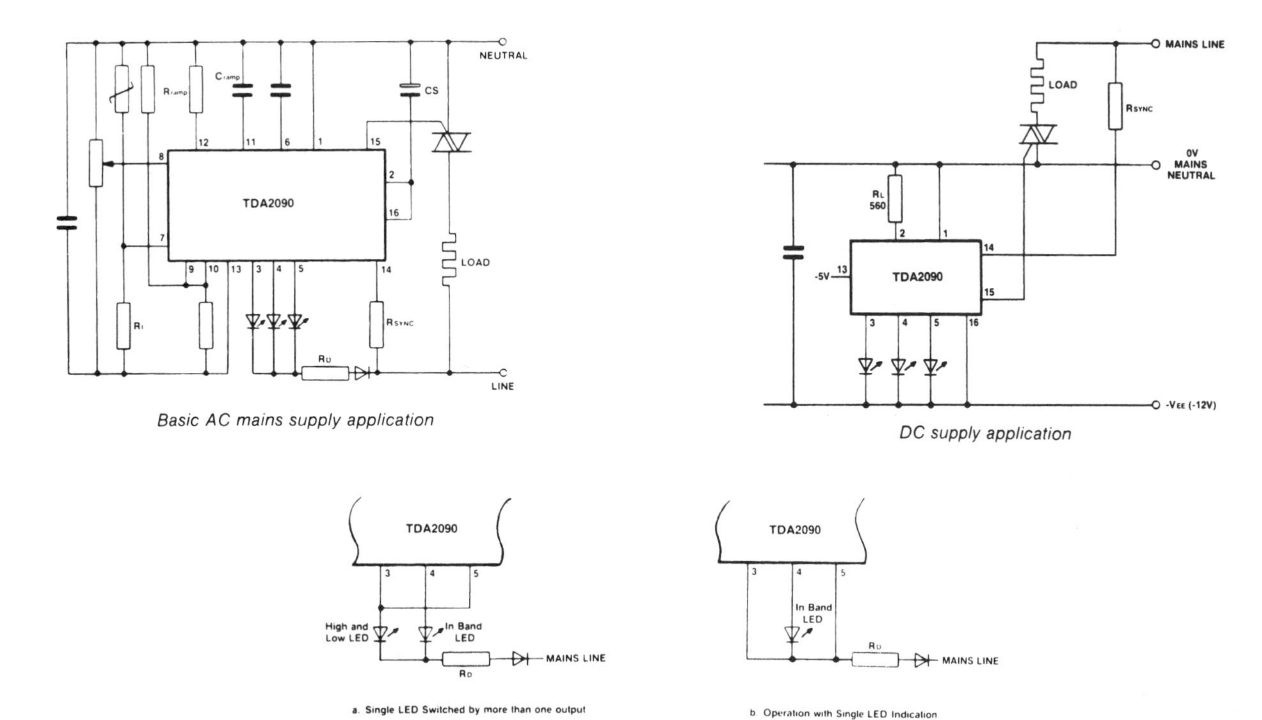

Basic AC mains supply application

DC supply application

Alternative LED connections

SYSTEM DESIGN WITH THE ZN411

See section CON for specifications of this IC.

Component Function	Circuit Ref.	Typical Value	Relevant Section(s)
Error amplifier gain resistor	R_1	infinity	5
Tacho integrator resistor	R_2	180kΩ	4
Current limit resistor	R_3	0.1Ω 2.5W	9
Triac gate current limit resistor	R_4	zero	8
Current sync sense resistor	R_5	270kΩ	7
Voltage sync sense resistor	R_6	330kΩ	7
Supply filter resistor	R_7	430Ω	2,8
Supply dropper resistor	R_8	5.6kΩ 4W	2
Tacho integrator capacitor	C_1	0.1μF	4,5
Tacho output filter capacitor	C_2	47nF	4,5
Dynamic response capacitor	C_3	0.22μF	5
Soft start ramp capacitor	C_4	10μF	3,5,9,10
Tacho monostable capacitor	C_5	6800pF	4
Timing ramp capacitor	C_6	0.1μF	6
Current limit integrator capacitor	C_7	0.47μF	9
Supply decoupling capacitor	C_8	22μF 6.3V	2,3
Supply filter capacitor	C_9	68μF 16V	2
Speed control potentiometer	RV_1	100kΩ	3

Typical external component values

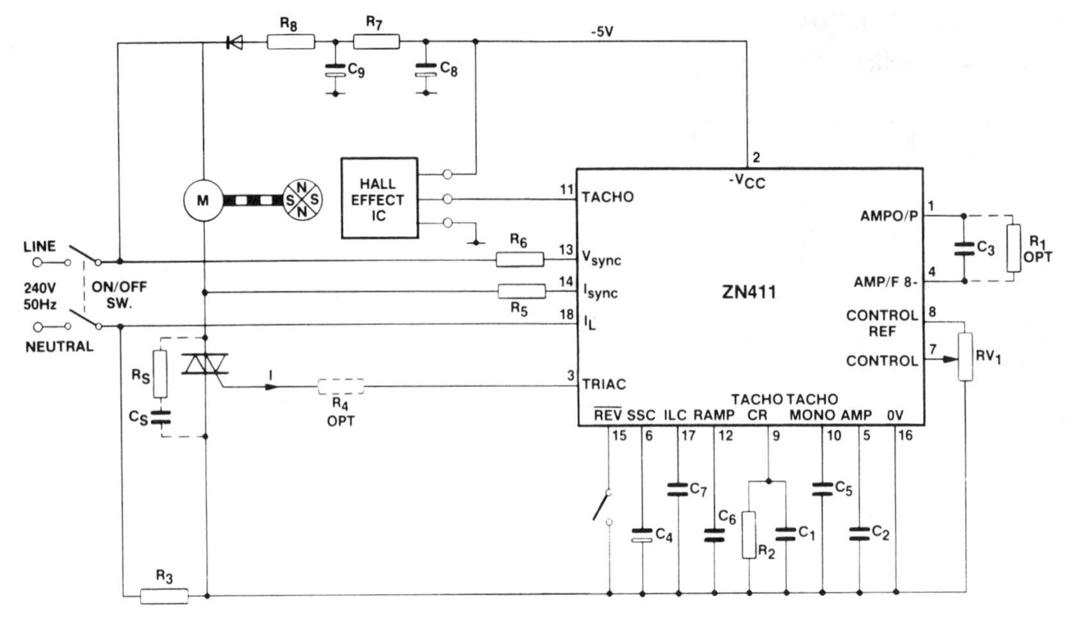

Reference system circuit diagram

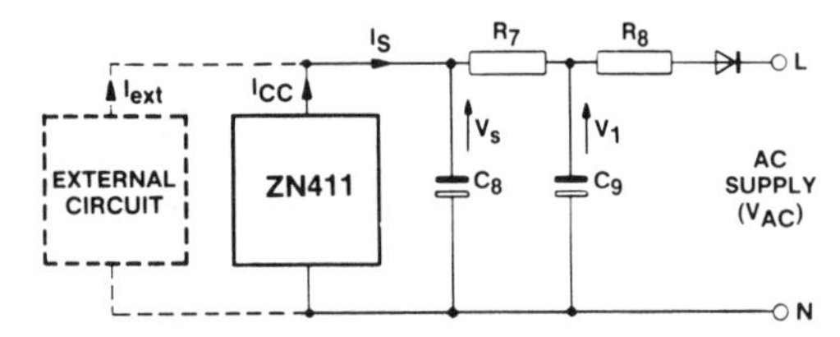

Power supply using diode and dropper resistor

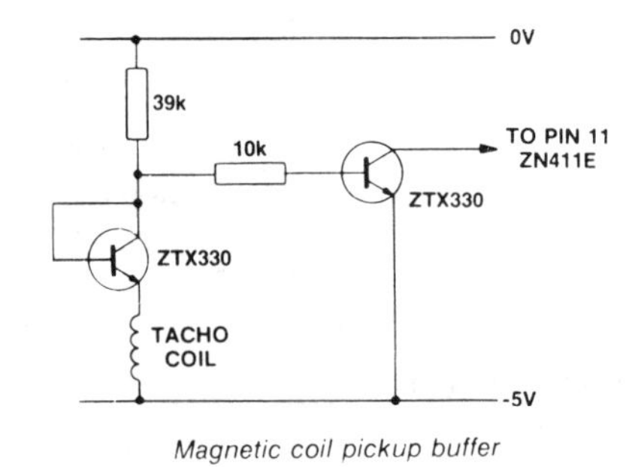

Magnetic coil pickup buffer

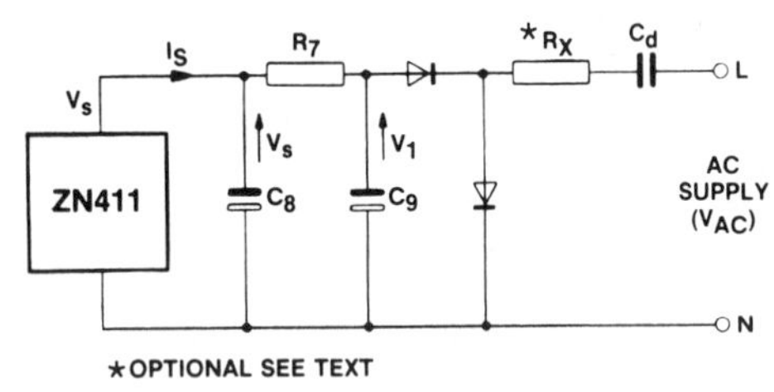

Power supply using reactive dropper circuit

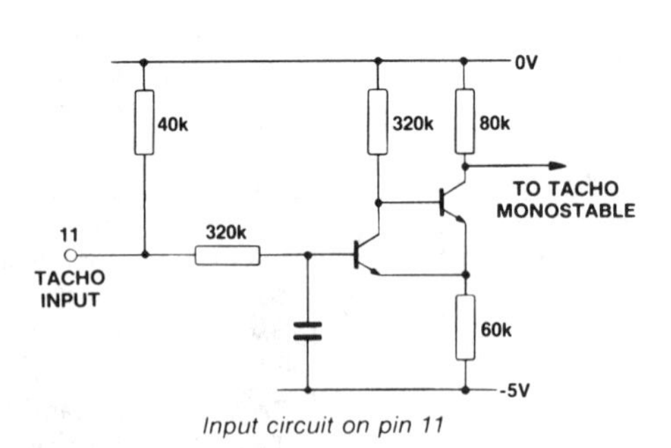

Input circuit on pin 11

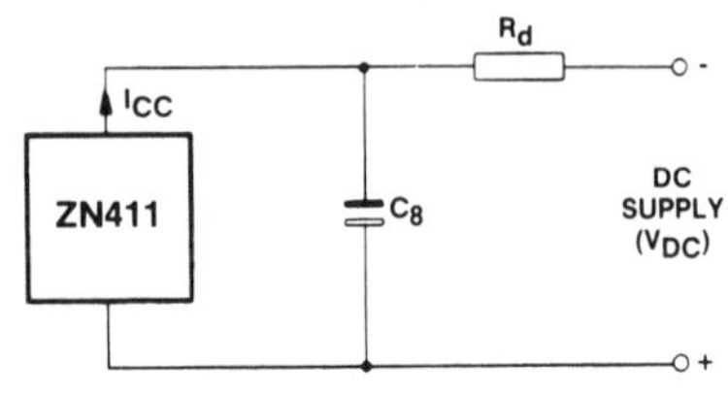

Operation from stabilised or unstabilised DC power supplies

SYSTEM DESIGN WITH THE ZN1060E

See section CON for specifications of this IC.

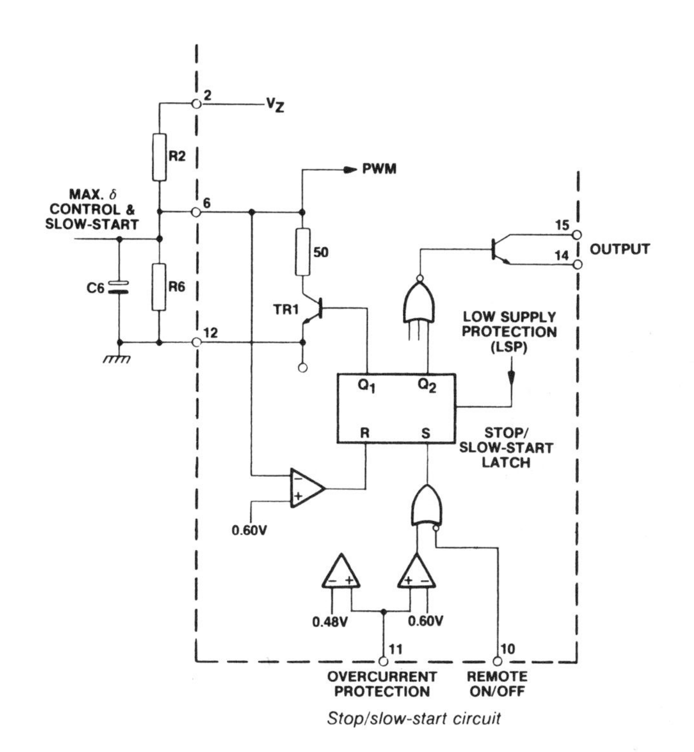

Stop/slow-start circuit

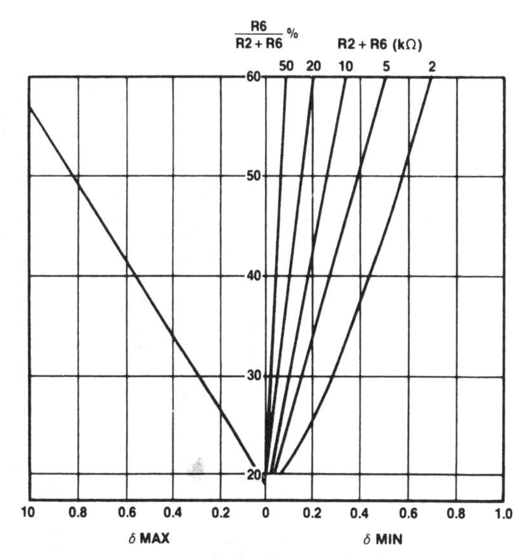

Typical end stop characteristic

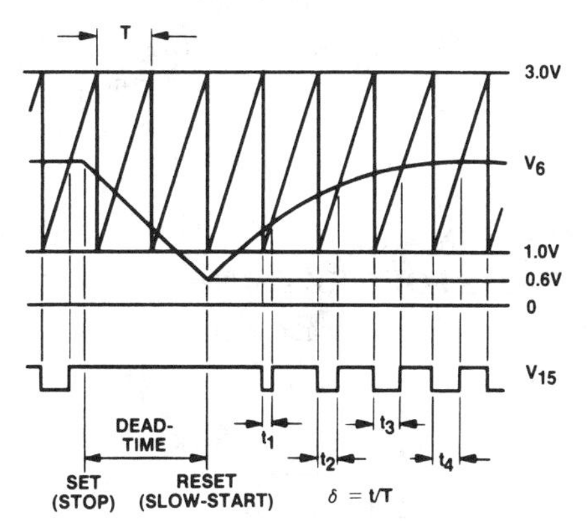

Waveforms associated with the stop/slow-start circuit

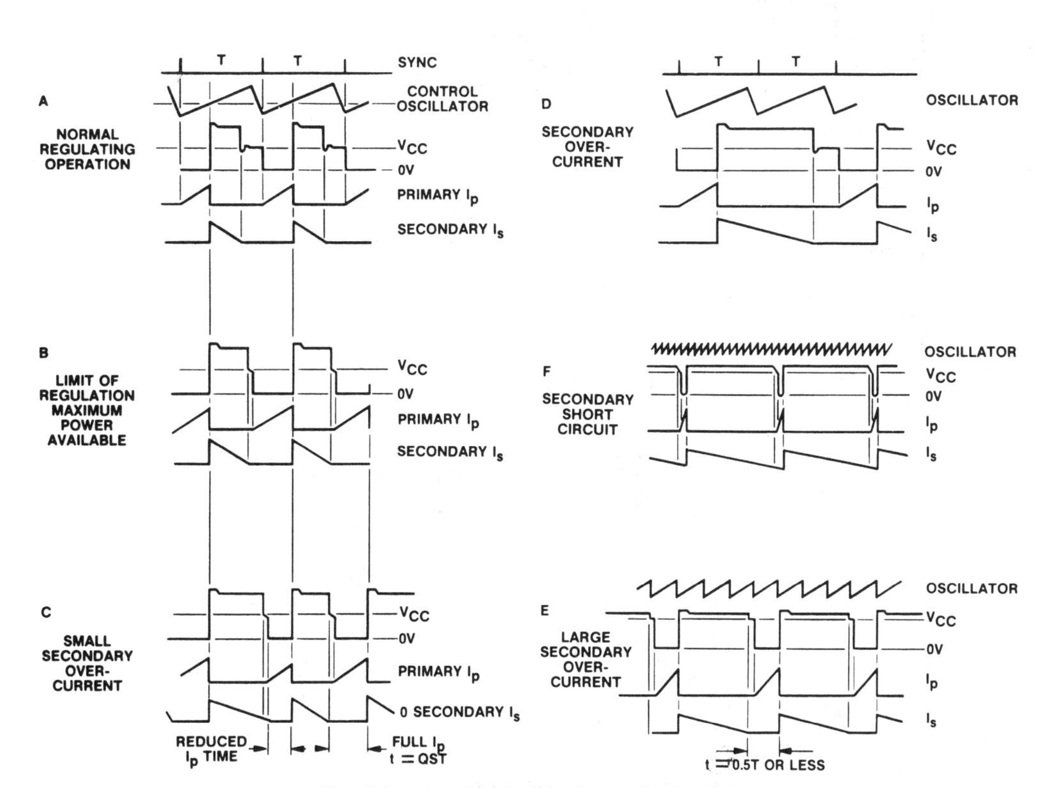

Complete energy transfer flyback operating waveforms

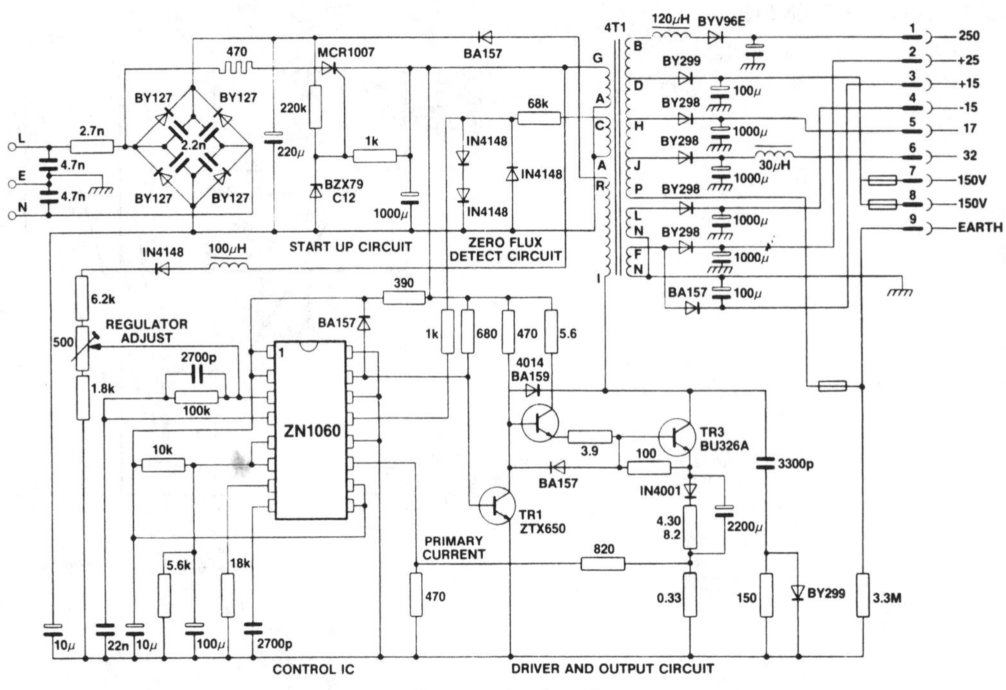

Power supply schematic

SYSTEM DESIGN AND APPLICATIONS FOR THE ZN1066

See section CON for specifications of this IC.

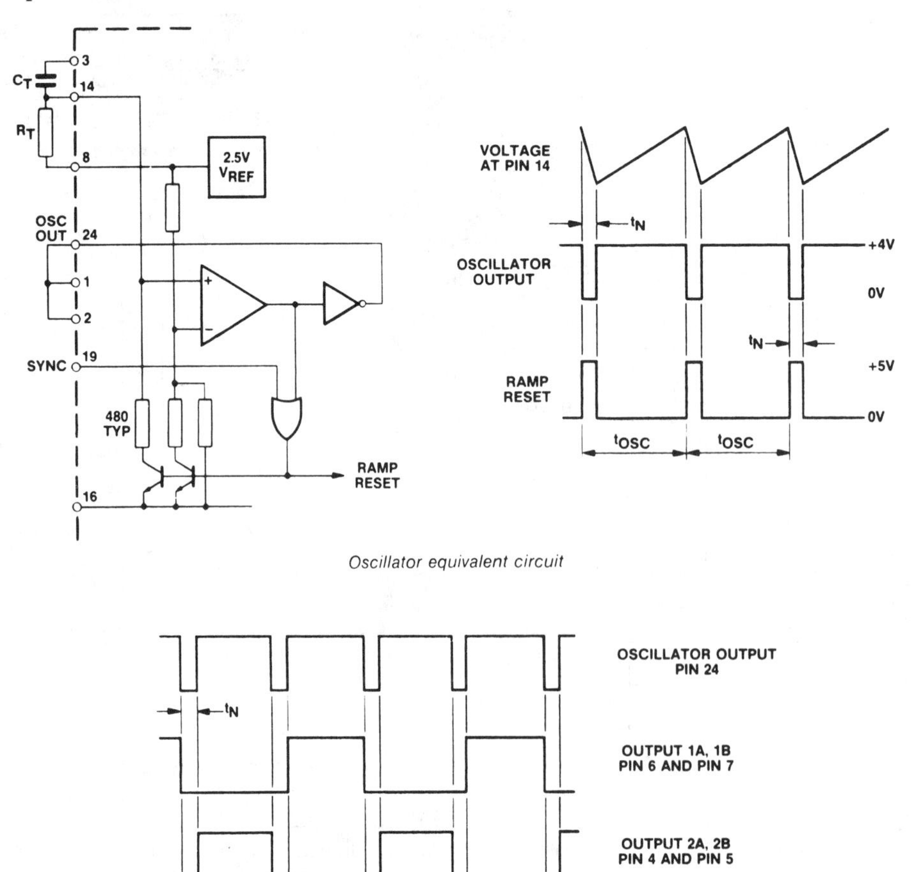

Oscillator equivalent circuit

Relationship between oscillator and output waveforms

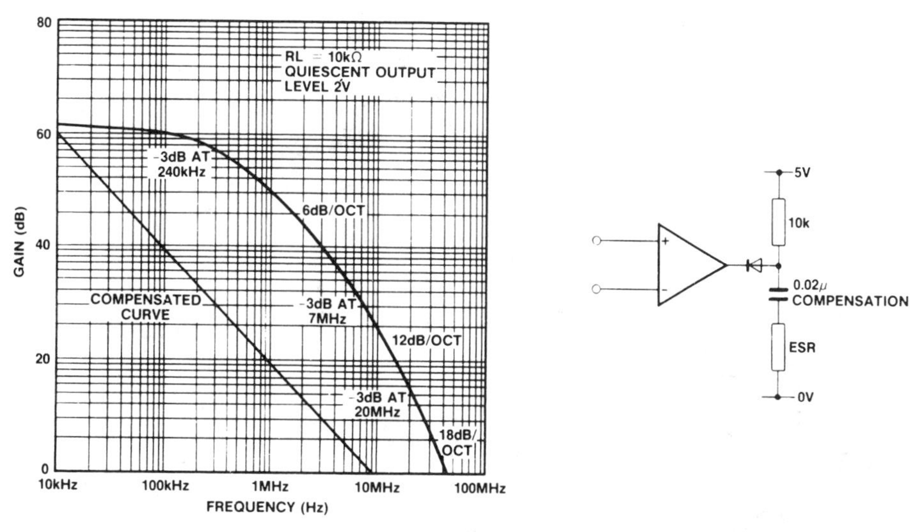

Amplifier open loop gain as a function of frequency

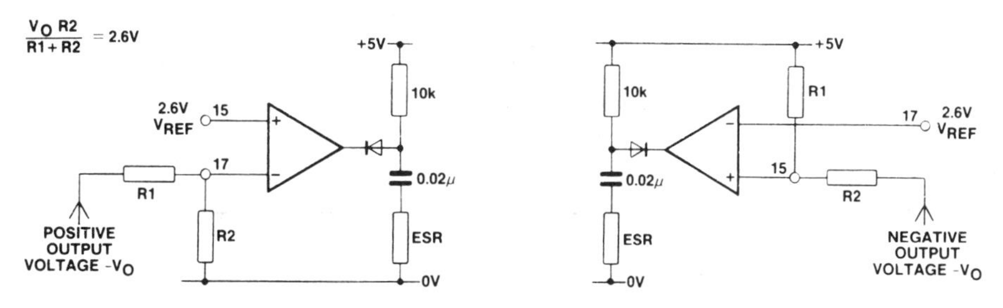

Voltage control biasing networks

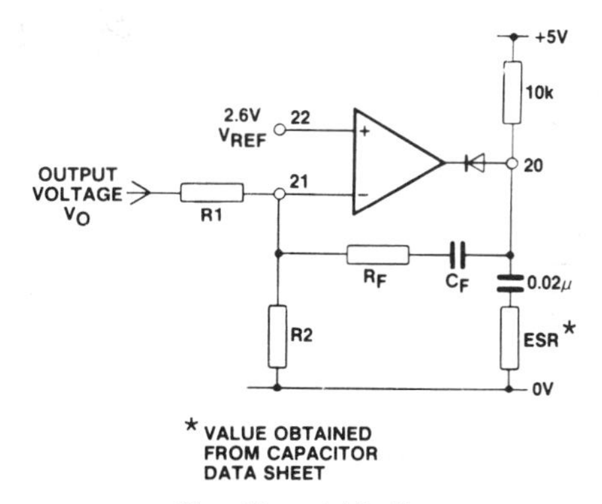

Closed loop stabilisation

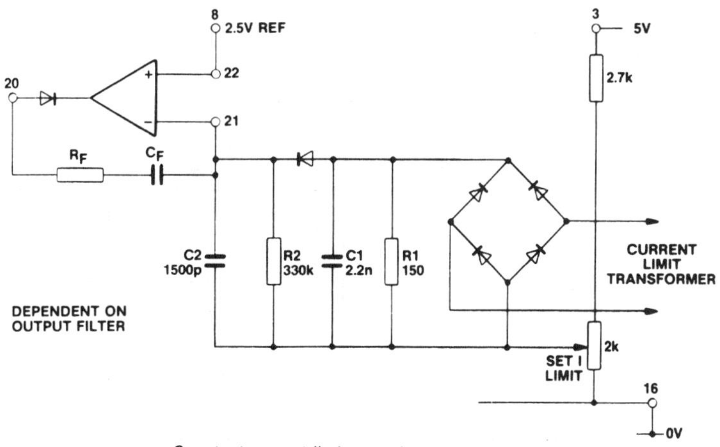

Constant current limit - sensing primary current

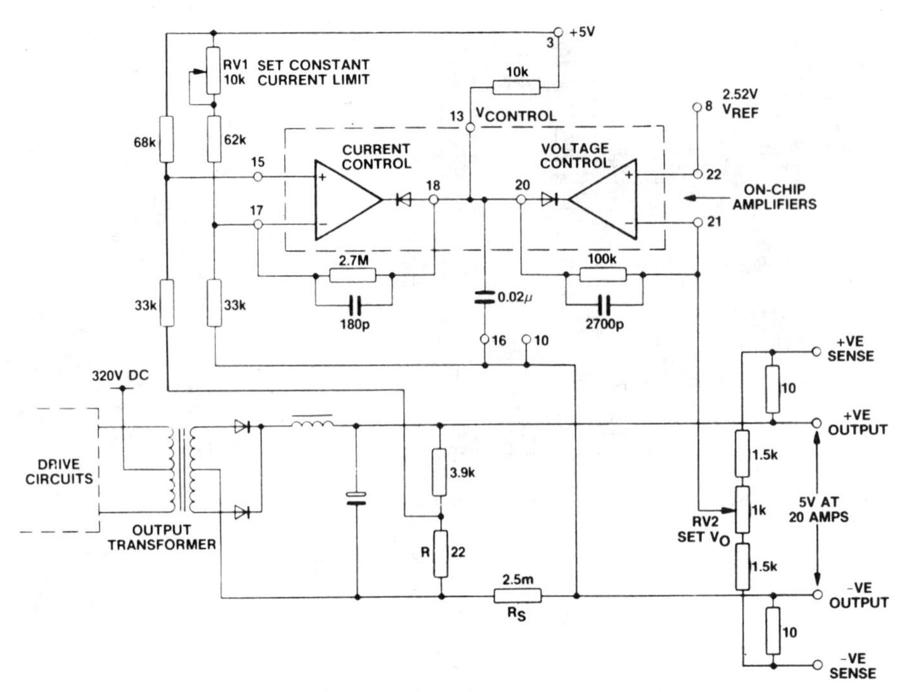

Current and voltage control circuit

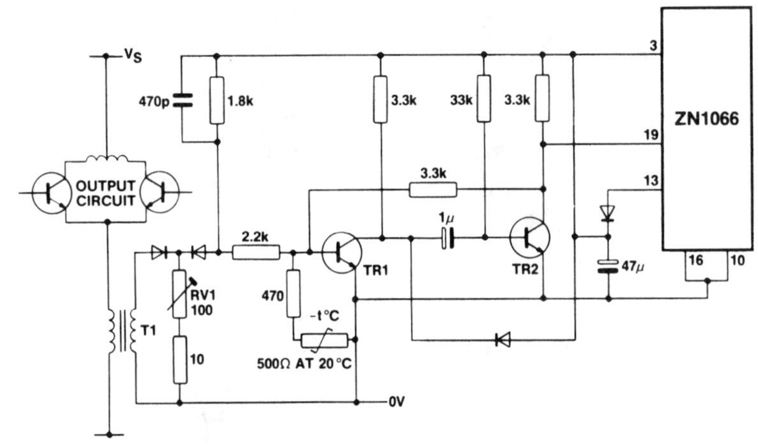

Current limit - primary side

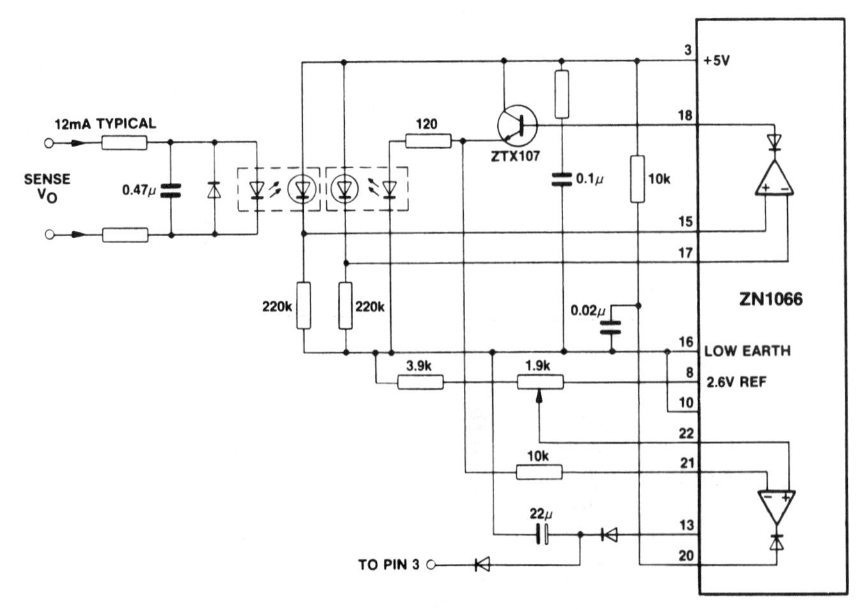

Opto-isolator control circuit

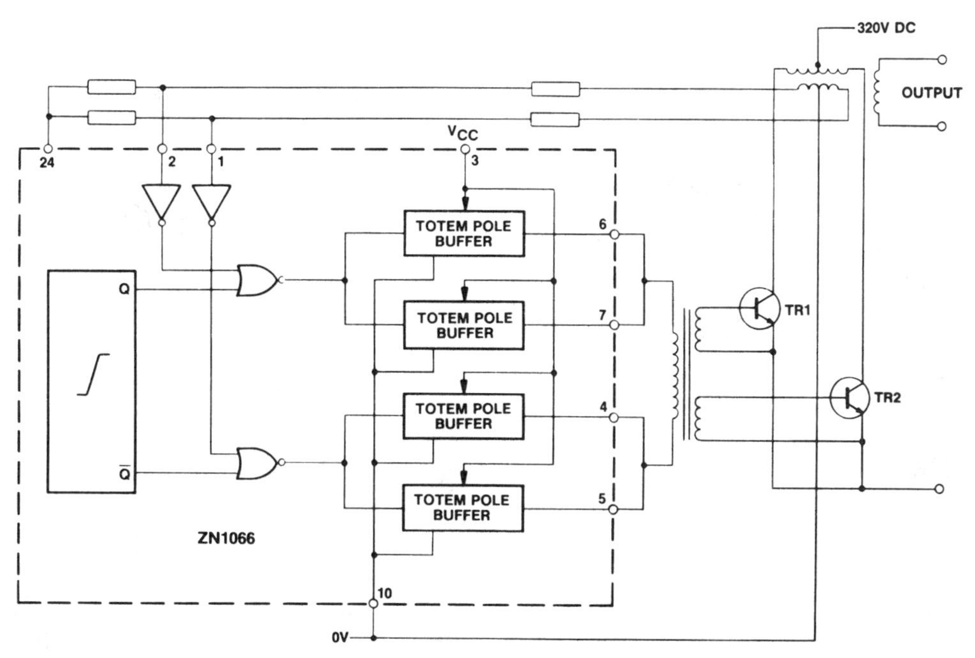

Automatic overlap control

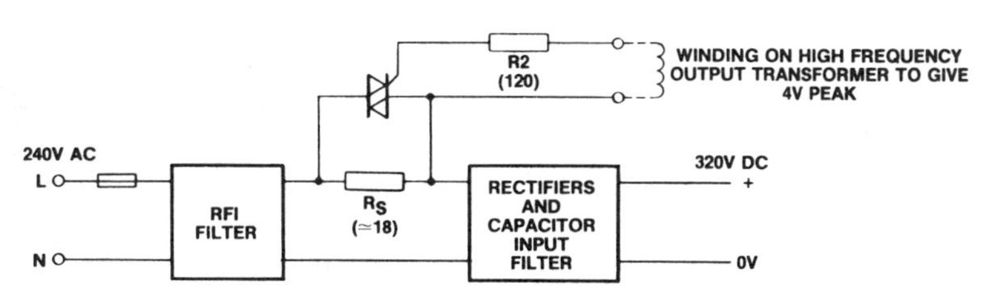

Inrush current limiting

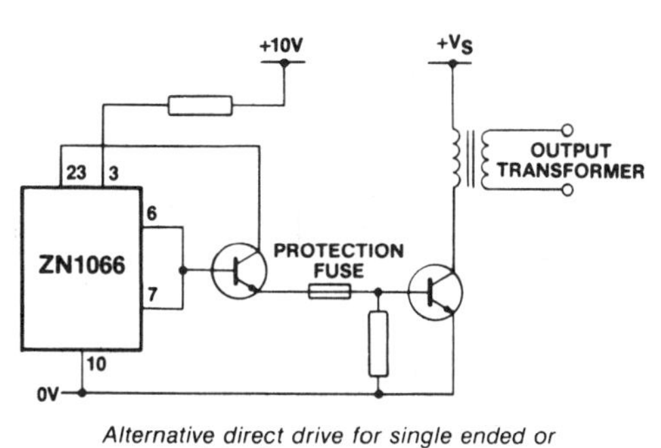

Alternative direct drive for single ended or flyback converters

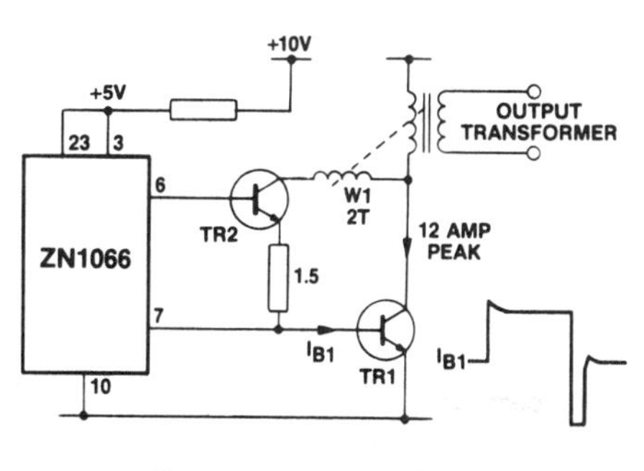

Winding W1 provides approximately 3V to the collector of TR2 thus ensuring that TR1 can fully saturate.

Direct drive for high power single ended converters

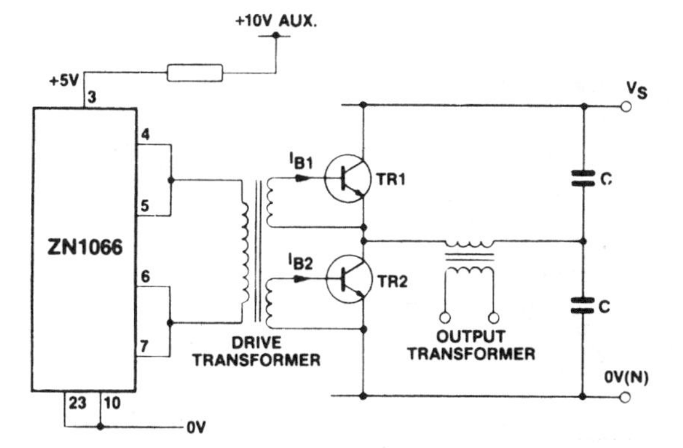

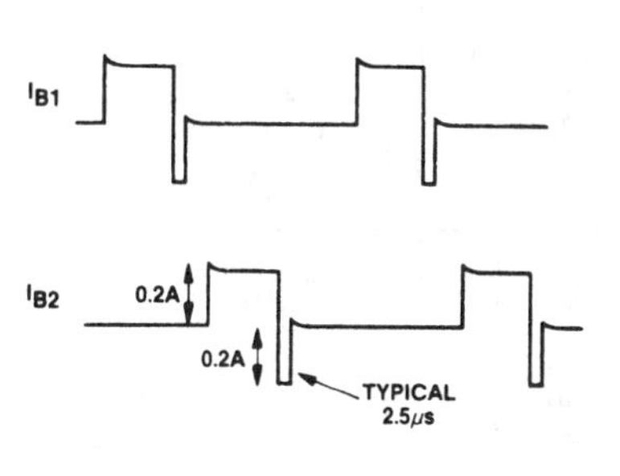

Transformer coupled half bridge push-pull circuit

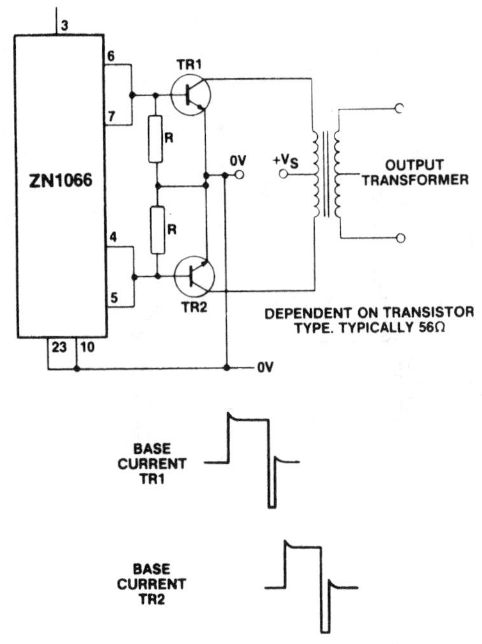

Typical direct coupled push-pull circuit

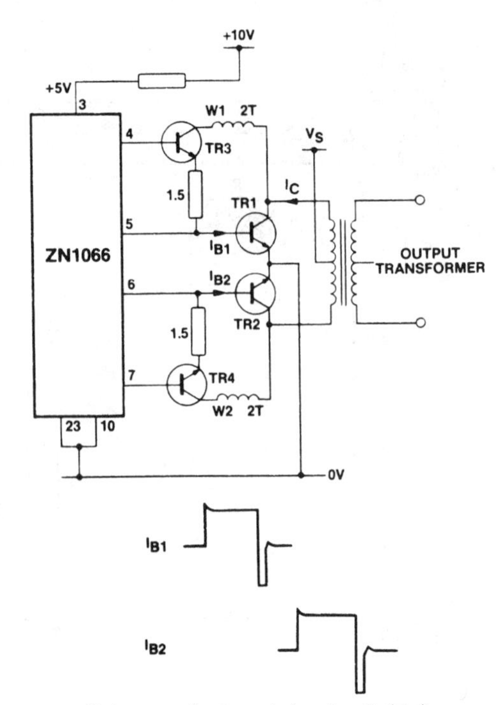

High power direct coupled push-pull circuit

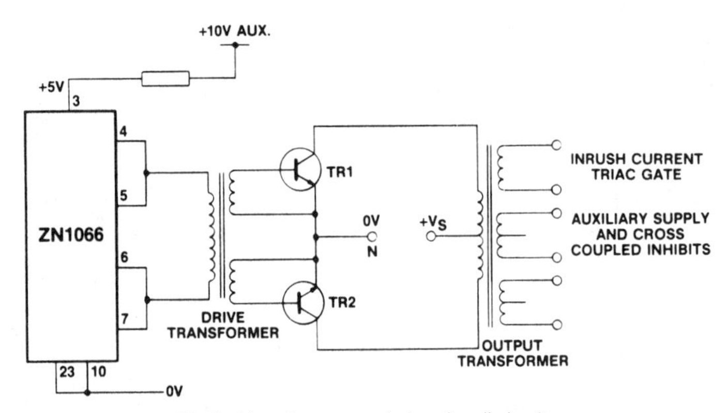

Typical transformer coupled push-pull circuit

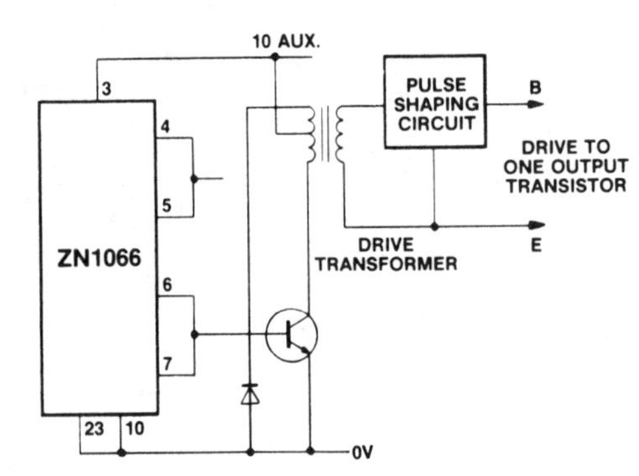

Typical high power high input voltage interface circuit

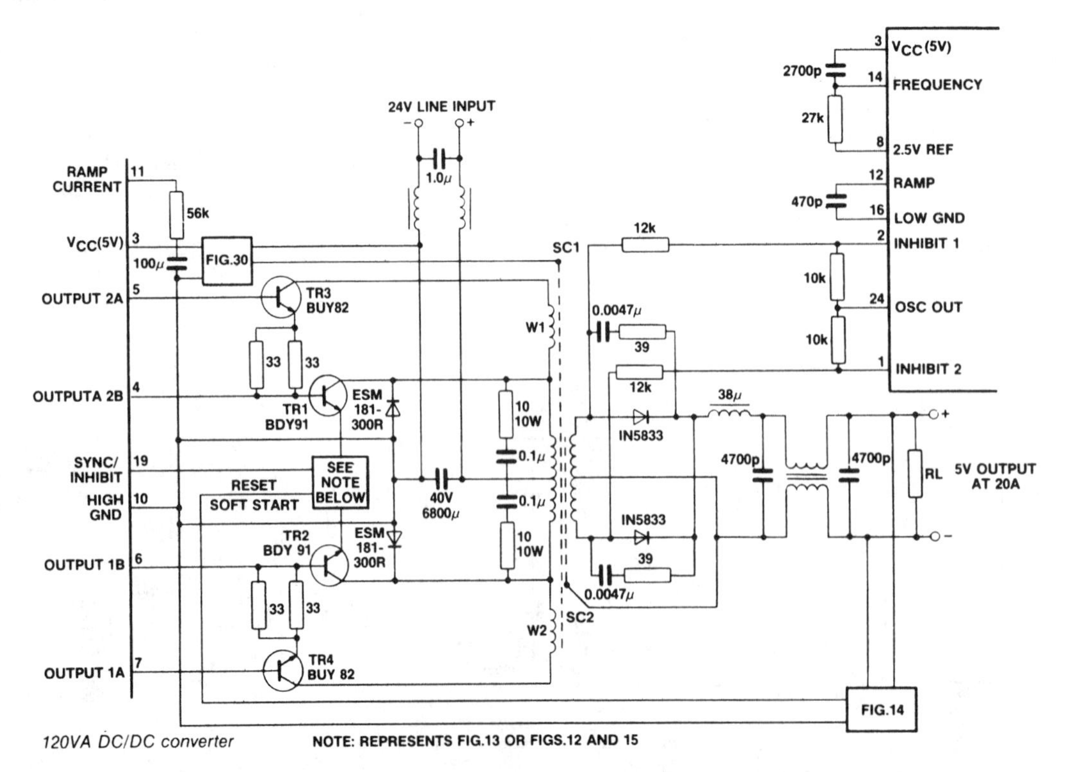

120VA DC/DC converter

NOTE: REPRESENTS FIG.13 OR FIGS.12 AND 15

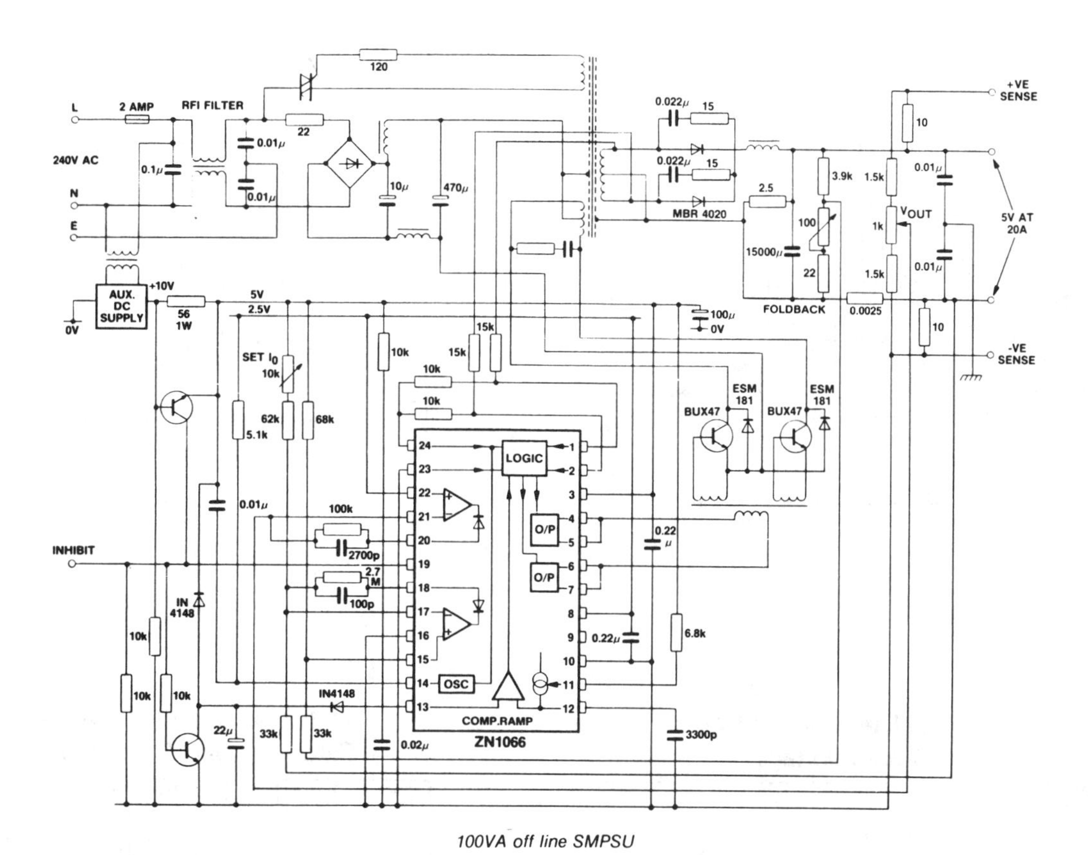

100VA off line SMPSU

Reprinted with permission from Raytheon Company, Semiconductor Division, 1989 Linear Integrated Circuits Data Book.

Raytheon LM1851

Ground Fault Interrupter

- Direct interface to SCR
- Adjustable fault current threshold
- Adjustable fault current integration time
- Compiles with U.S. UL943 standard
- Operates under line reversal; both load vs. line and hot vs. neutral
- Detects grounded neutral faults
- Internal shunt regulator (26V)
- Small outline (SO-8) package available

DESCRIPTION

The LM1851 is a controller for ac outlet ground fault interrupters. These devices detect hazardous grounding conditions (example: a pool of water and electrical equipment connected to opposite phases of the ac line) in consumer and industrial environments. The output of the IC triggers an external SCR, which in turns opens a relay circuit breaker to prevent a harmful or lethal shock.

The information contained herein has been carefully compiled; however, it shall not by implication or otherwise become part of the terms and conditions of any subsequent sale. Raytheon's liability shall be determined solely by its standard terms and conditions of sale. No representation as to application or use or that the circuits licensed or free from patent infringement is intended or implied. Raytheon reserves the right to change the circuitry and other data at any time without notice and assumes no liability for inadvertent errors.

ABSOLUTE MAXIMUM RATINGS

Shunt Current .... 19 mA
Power Dissipation .... 570 mW
Operating Temperature
Range .... -40°C to +70°C
Storage Temperature
Range .... -65°C to +150°C
Lead Soldering Temperature
(SO-8, 10 sec) .... +260°C
Lead Soldering Temperature
(DIP, 60 sec) .... +300°C

THERMAL CHARACTERISTICS

	8-Lead Plastic DIP	8-Lead Small Outline
Max. Junction Temp.	+125°C	+125°C
Max. P_D $T_A < 50°C$	468 mW	300 mW
Therm. Res θ_{JC}	—	—
Therm. Res θ_{JA}	160°C/W	240°C/W
For $T_A < 50°C$ Derate at	6.25 mW/°C	4.17 mW/°C

TYPICAL PERFORMANCE CHARACTERISTICS ($T_A = +25°C$)

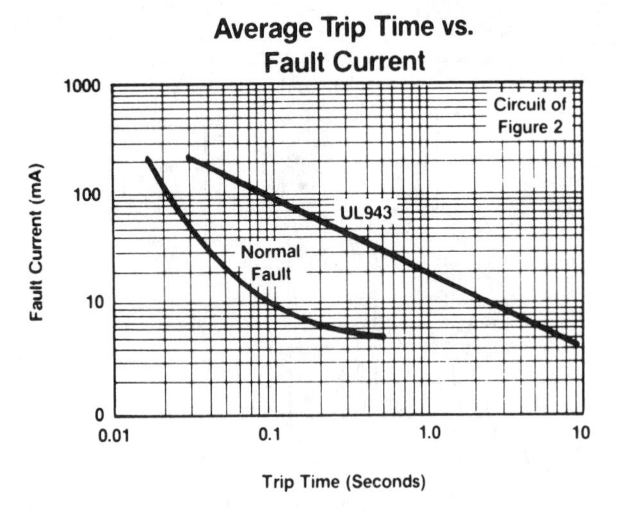

Normal Fault Current Threshold vs. R_{SET}

$R_{SET} = \frac{7V}{I_F(rms)^* \times (0.91)}$

Sense Transformer 1000:1

Fault Current on Line [mA (rms)]

100 / 10 / 1

100k / 1M / 10M

*See Block Diagram R_{SET} (Ω)

Output Drive Current vs. Output Voltage

Output Drive Current @ Pin 1 (μA)

1400 / 1200 / 1000 / 800 / 600 / 400 / 200 / 0

0 / 5 / 10 / 15 / 20 / 25 / 30 / 35

Output Voltage @ V_{PIN1} (V)

31V, 5 mA, 8, 1 mA, 1, A, V_{PIN1}, 4

Pin 1 Saturation Voltage vs. External Load Current, I_L

Pin 1 Saturation Voltage (V)

10 / 1 / 0.1 / 0.01

0.1 / 1 / 10 / 100

External Load Current (mA)

31V, 5 mA, I_L, 8, 1 mA, 1, V, 4

MASK PATTERN

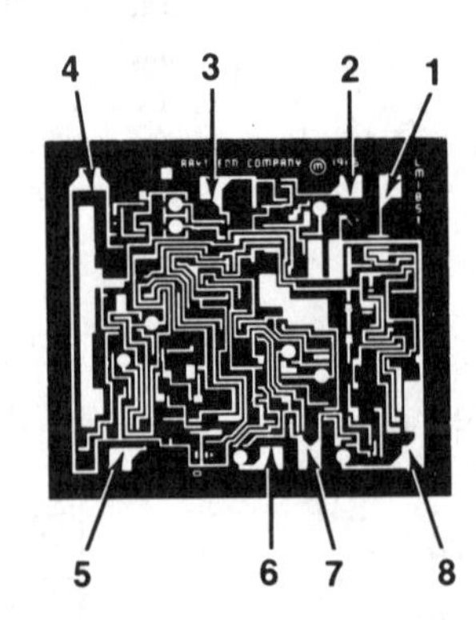

Die Size: 69 x 59 mils
Min. Pad Dimensions: 4 x 4 mils

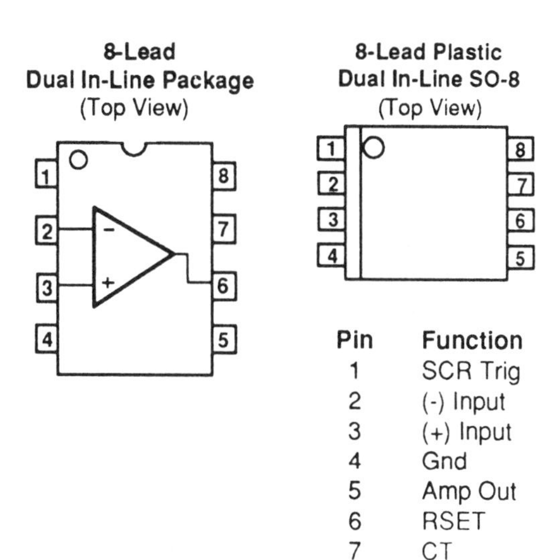

Pin	Function
1	SCR Trig
2	(-) Input
3	(+) Input
4	Gnd
5	Amp Out
6	RSET
7	CT
8	$+V_S$

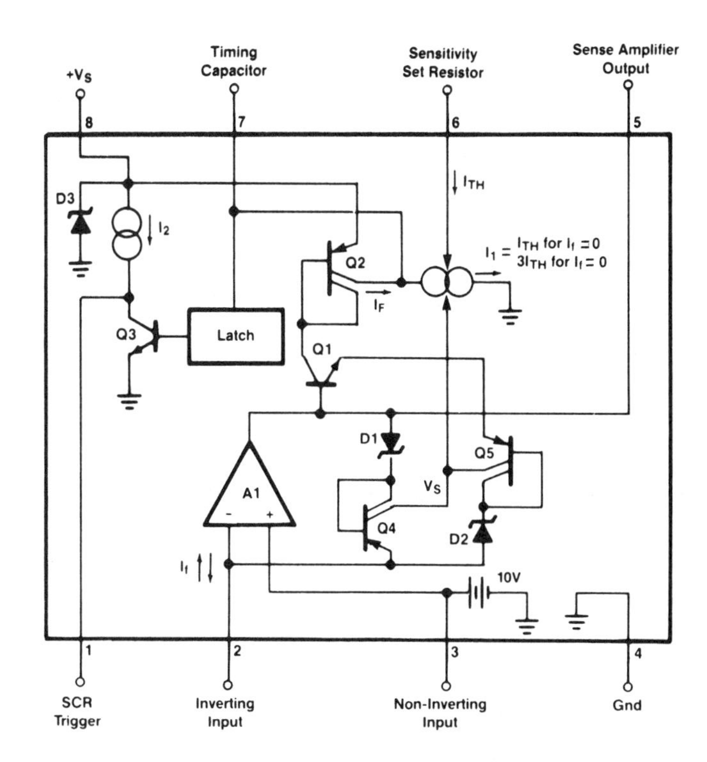

120 Hz NEUTRAL TRANSFORMER APPLICATION

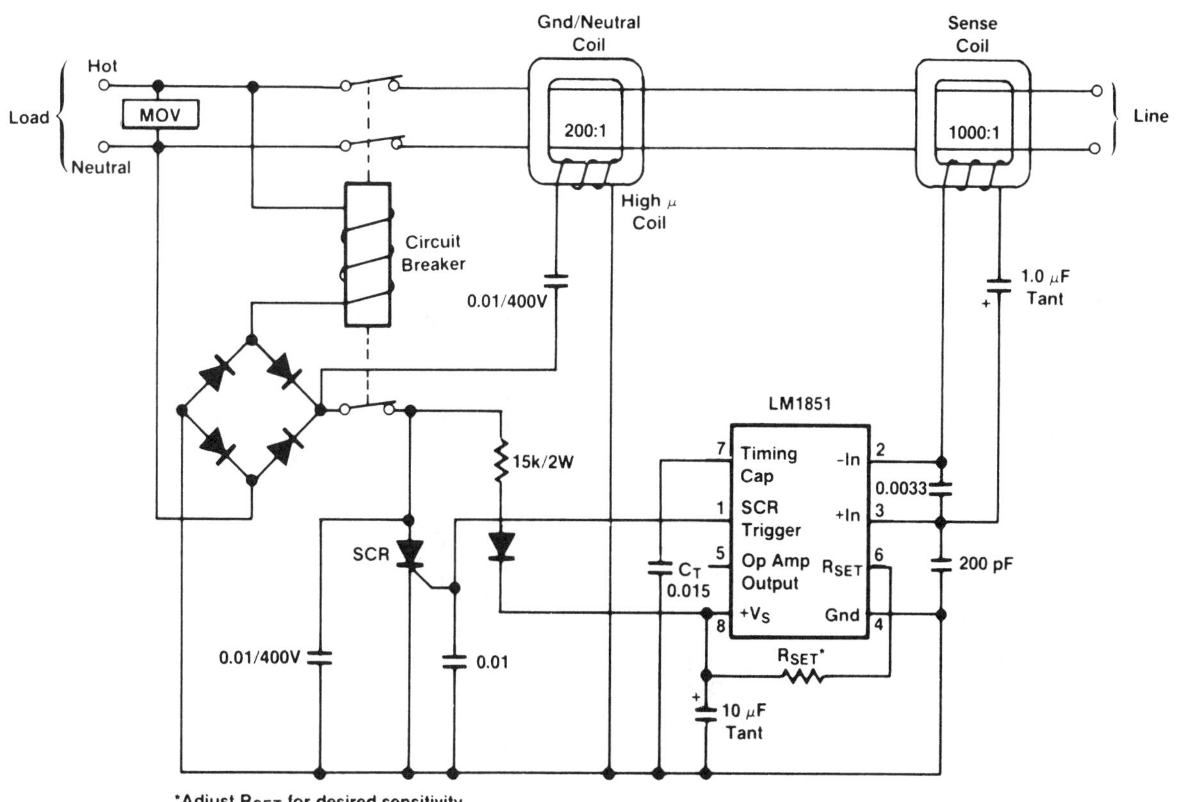

*Adjust R_{SET} for desired sensitivity

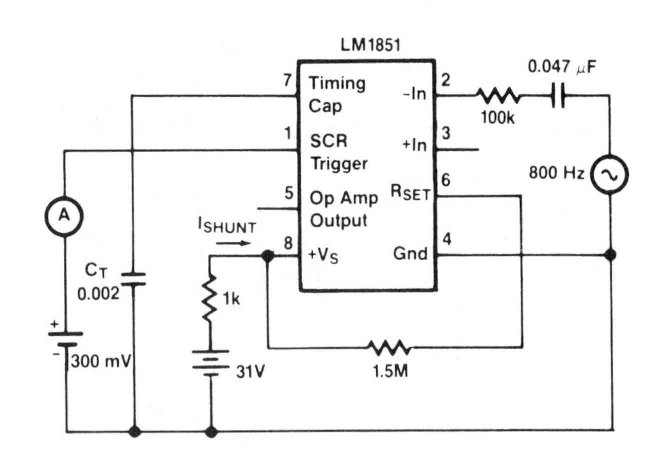

Normal Fault Sensitivity Test Circuit

8-LEAD PLASTIC DUAL IN-LINE SMALL OUTLINE

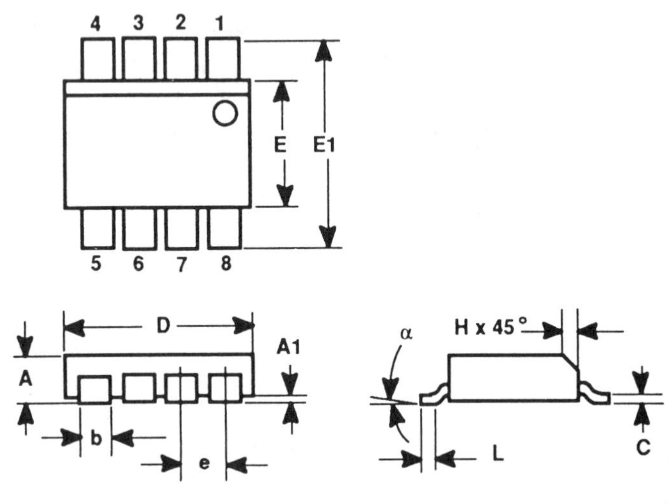

Dimension	Inches		Millimeters	
	Min	Max	Min	Max
A	.053	.069	1.35	1.75
A1	.004	.008	.10	.20
b	.014	.018	.350	.450
C	.007	.009	.19	.22
D	.188	.197	4.80	5.00
E	.150	.158	3.80	4.00
e	.050 BSC		1.27BSC	
E1	.228	.244	5.80	6.20
L	.020	.045	.508	1.143
X	0°	8°	0°	8°
h	.01	.02	.25	.50

Note: C Dimension does not include Hot Solder Dip thickness.

Dimensions conform to JEDEC specification MS-102-AA for SC packages.

8-LEAD CERAMIC DUAL IN-LINE PACKAGE

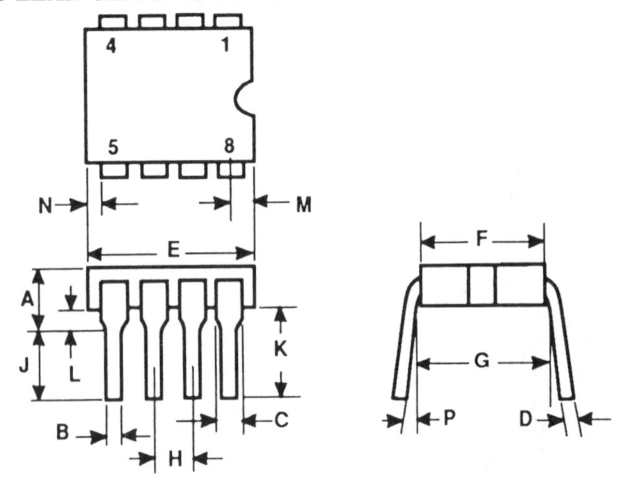

Dimension	Inches		Millimeters	
	Min	Max	Min	Max
A	.115	.125	2.92	3.17
B	.015	.021	0.38	0.53
C	.030	.070	0.76	1.78
D	.010	.015	0.25	0.38
E	.360	.400	9.14	10.16
F	.240	.260	6.09	6.60
G	.290	.310	7.37	7.87
H	.090	.110	2.29	2.79
J	.120	.135	3.05	3.43
K	.140	.165	3.56	4.18
L	.020	.030	0.51	0.75
M	.025	.050	0.64	1.27
N	.005		0.13	
P	0°	15°	0°	15°

SCHEMATIC DIAGRAM

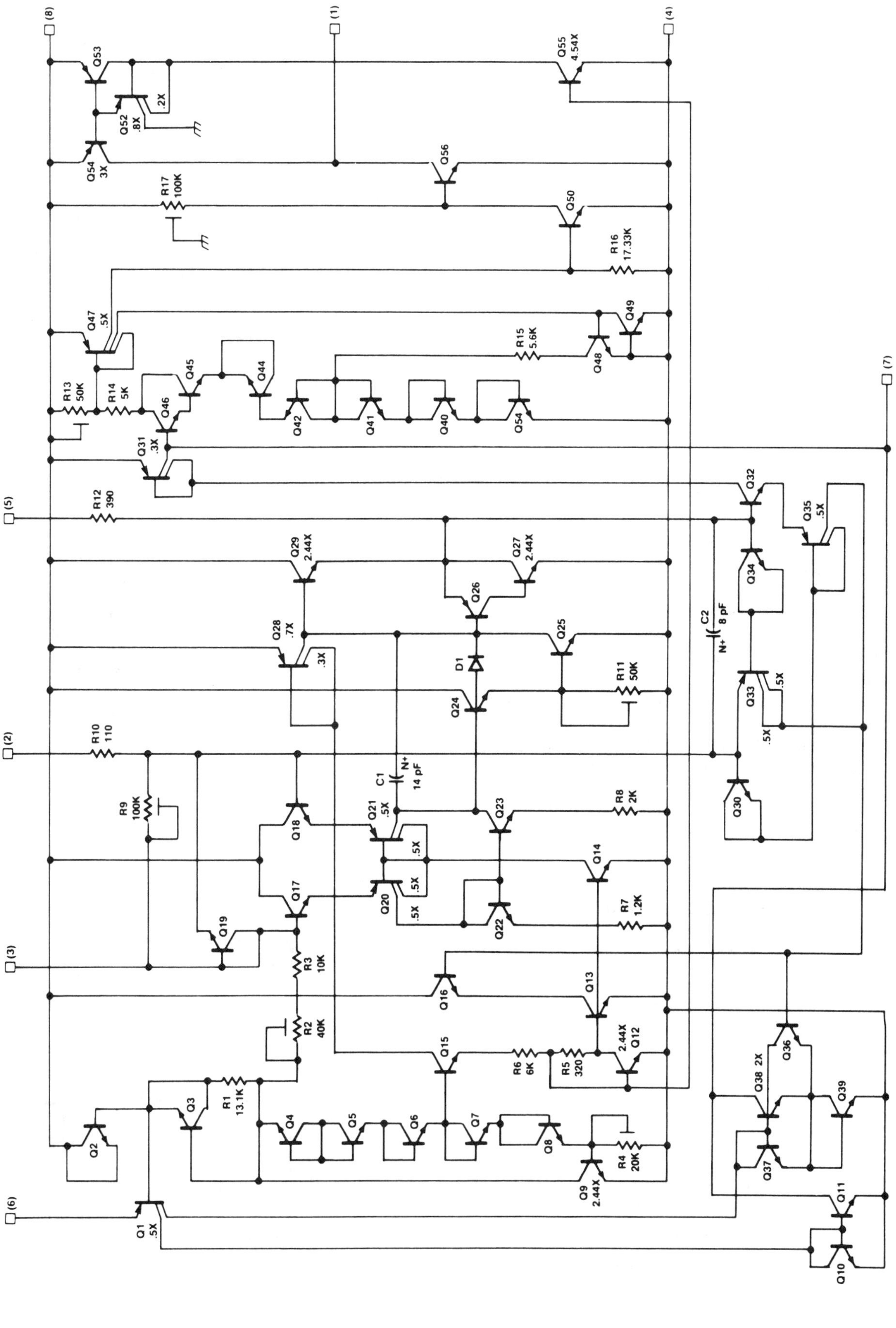

Raytheon RC4292

Negative Switch Mode Power Supply Controller

- Converts a negative voltage into positive and/or negative voltages
- Wide application voltage range: −20V minimum, −120V maximum
- High efficiency: 70% typical
- Adjustable output voltage
- Accurate oscillator frequency: ±10%
- Wide frequency range: 20 kHz to 100 kHz
- Bandgap voltage reference: 50 ppm/°C
- Good line regulation: 0.1%/V
- PWM feedback circuitry
- Short circuit protection
- Soft start
- 8-lead mini DIP
- Load power up to 10W
- Undervoltage lockout

APPLICATIONS

- Small power supplies
- Local on-card regulators
- Telephone peripheral equipment
- Converts −48V off-hook voltage
- Battery-operated equipment

Description of Functional Blocks

1. **Oscillator** The oscillator generates a time base for the V_{DRIVE} pulse. The frequency of oscillation is set by a low-value external capacitor connected between C_X and ground.
2. **Error Amplifier** Here the feedback signals $+V_{FB}$ and $-V_{FB}$ are compared, and the error amplifier output generates an error signal proportional to the input difference.
3. **Current Comparator** This circuit compares the output of the error amplifier to a signal proportional to the current in the primary of the transformer (derived via a low-value resistor in series with the transformer). If the I_{FB} signal is greater than the error signal, the control F/F is reset and the external power transistor will turn off.
4. **Control F/F** The control F/F ensures that the external power transistor receives only one pulse for each oscillator cycle.
5. **Output Driver** This circuit amplifies the control F/F output and provides a fast switching signal to drive the gate of an external power MOSFET or BJT.
6. **Voltage Reference** Generates a voltage reference (−0.5V) for the voltage feedback and also generates internal bias currents.
7. **Shunt Regulator** The shunt regulator acts like a zener diode to clamp the voltage across the 4292, thus regulating the supply voltage applied to the 4292 to within safe limits.

ABSOLUTE MAXIMUM RATINGS

Internal Power Dissipation	750 mW
Storage Temperature Range	-65°C to +150°C
Lead Soldering Temperature (60 Sec)	+300°C
Output Drive Current	750 mA
Shunt Current	20 mA
V_{DRIVE} Voltage	-17V to +1.0V

THERMAL CHARACTERISTICS

	8-Lead Plastic DIP	8-Lead Ceramic DIP
Max. Junction Temp.	+125°C	+175°C
Max. P_D $T_A < 50°C$	468mW	833mW
Therm. Res. θ_{JC}	—	45°C/W
Therm. Res. θ_{JA}	160°C/W	150°C/W
For $T_A > 50°C$ Derate at	6.25mW per °C	8.33mW per °C

8-Lead Dual In-Line Package
(Top View)

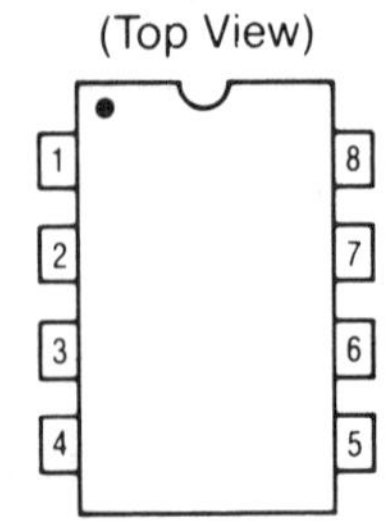

Pin	Function
1	CX — Timing Capacitor
2	-VFB — Error Amp Input
3	+VFB — Error Amp Input
4	V_{REF} — Reference Voltage
5	$-V_S$ — Negative Shunt
6	V_{DRIVE} — Output
7	I_{FB} — Current Limit Feedback
8	GND — Ground

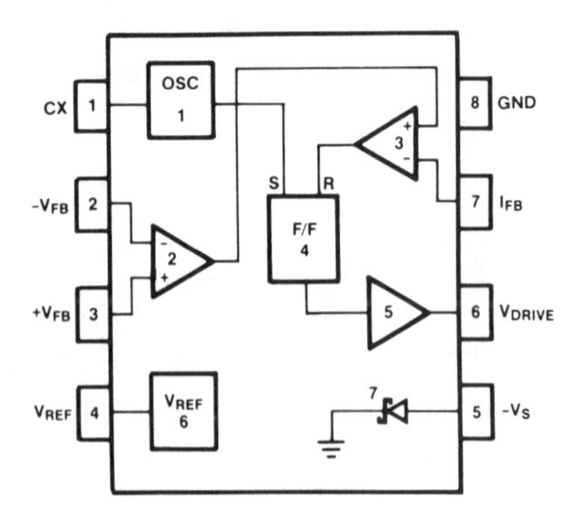

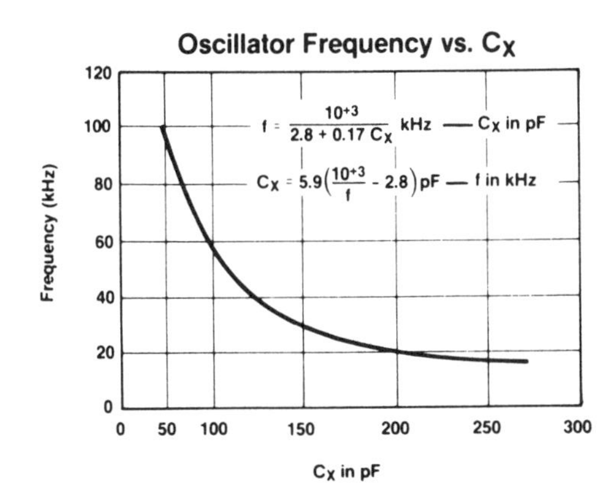
Oscillator Frequency vs. C_X
Frequency (kHz)
$f = \frac{10^{+3}}{2.8 + 0.17\,C_X}$ kHz — C_X in pF
$C_X = 5.9\left(\frac{10^{+3}}{f} - 2.8\right)$ pF — f in kHz
C_X in pF

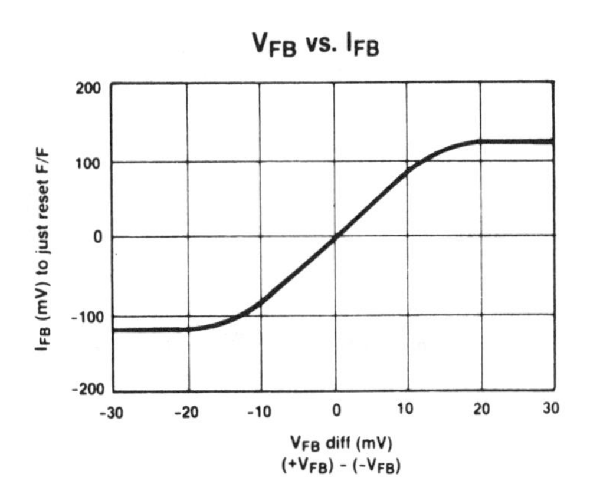
V_{FB} vs. I_{FB}
I_{FB} (mV) to just reset F/F
V_{FB} diff (mV)
$(+V_{FB}) - (-V_{FB})$

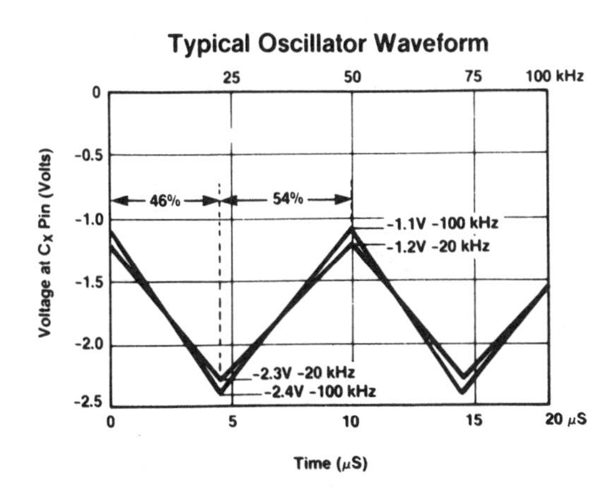
Typical Oscillator Waveform
100 kHz
Voltage at C_X Pin (Volts)
46%
54%
-1.1V -100 kHz
-1.2V -20 kHz
-2.3V -20 kHz
-2.4V -100 kHz
Time (μS)

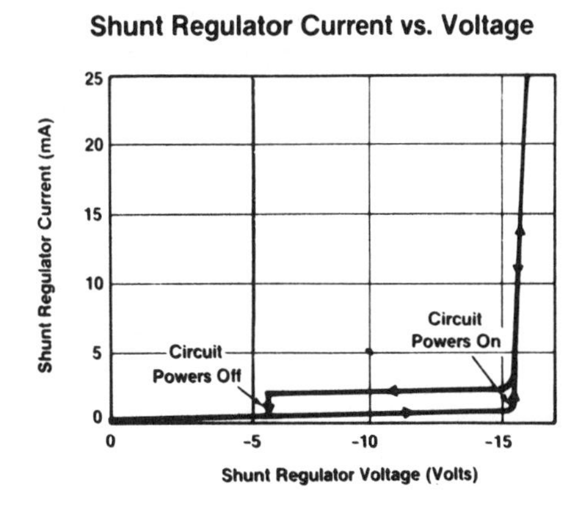
Shunt Regulator Current vs. Voltage
Shunt Regulator Current (mA)
Circuit Powers Off
Circuit Powers On
Shunt Regulator Voltage (Volts)

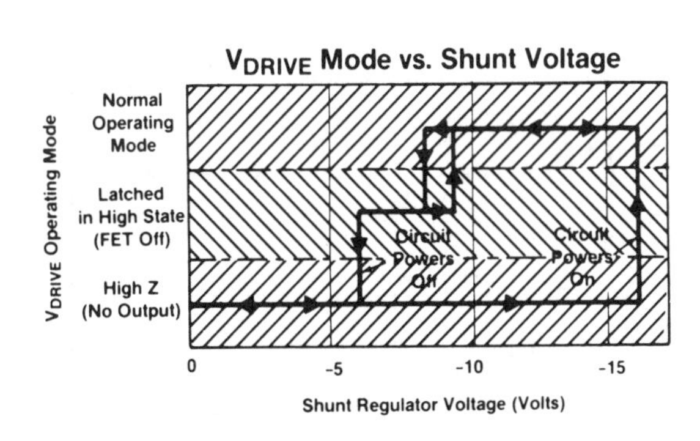
V_{DRIVE} Mode vs. Shunt Voltage
V_{DRIVE} Operating Mode
Normal Operating Mode
Latched in High State (FET Off)
High Z (No Output)
Circuit Powers Off
Circuit Powers On
Shunt Regulator Voltage (Volts)

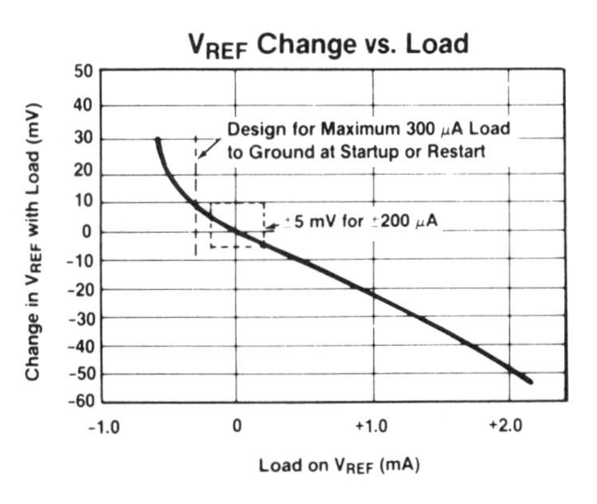
V_{REF} Change vs. Load
Change in V_{REF} with Load (mV)
Design for Maximum 300 μA Load to Ground at Startup or Restart
5 mV for ±200 μA
Load on V_{REF} (mA)

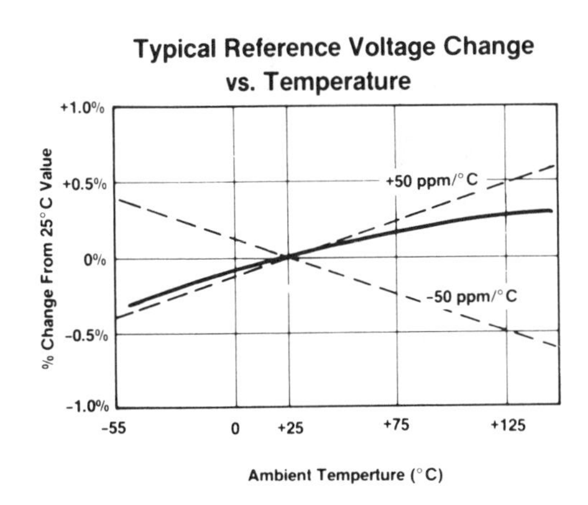
Typical Reference Voltage Change vs. Temperature
% Change From 25°C Value
+50 ppm/°C
-50 ppm/°C
Ambient Temperture (°C)

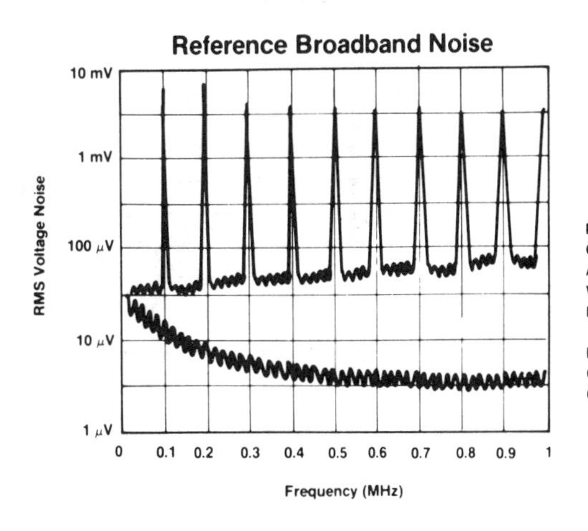
Reference Broadband Noise
RMS Voltage Noise
Frequency (MHz)
Reference Voltage Output in an Application Circuit With Load Oscillator Frequency 100 kHz
Reference Voltage Output With Oscillator Held Off

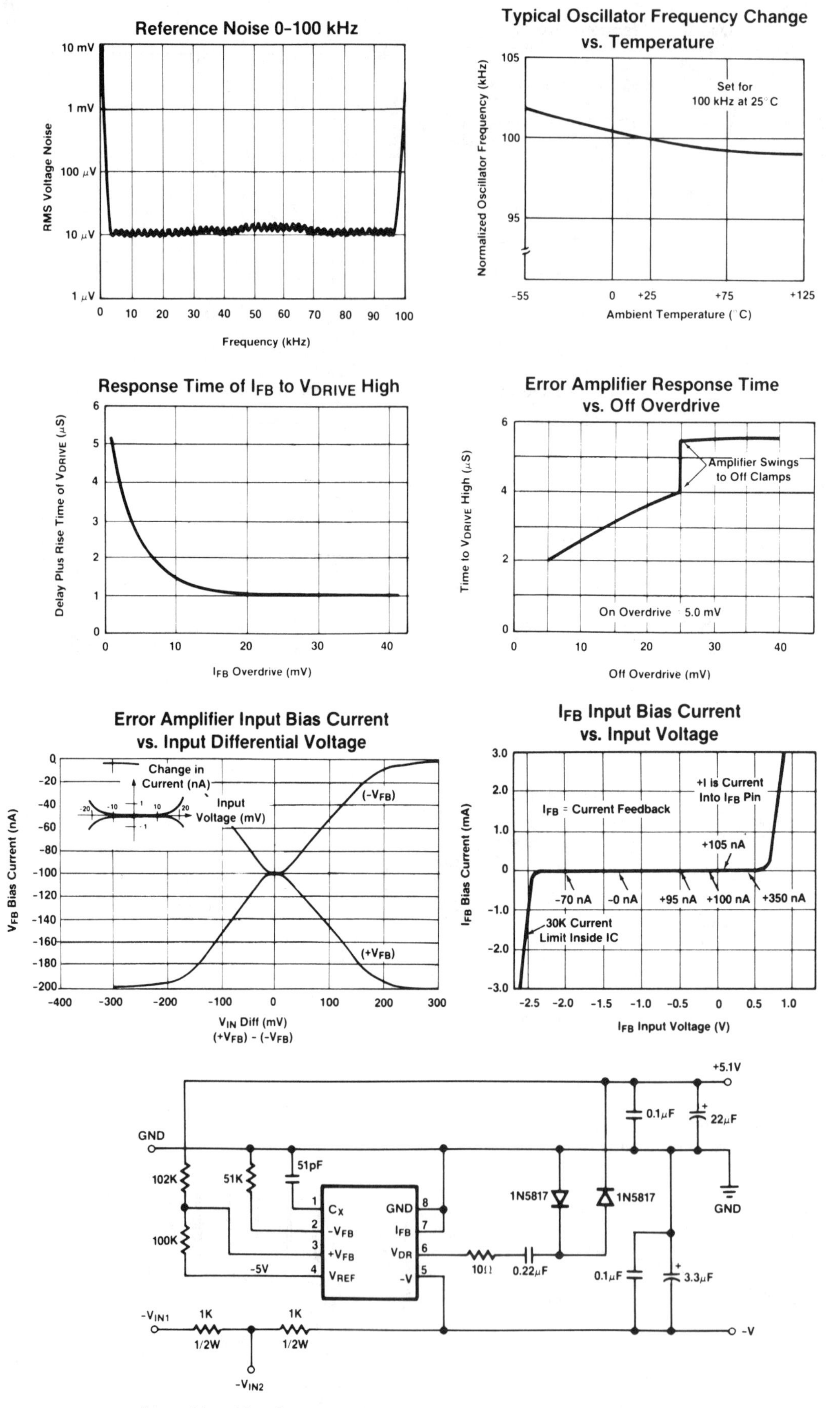

Note: Use $-V_{IN2}$ for V_{IN} = -20V to -35V

Low-Power Switched-Capacitor Regulator

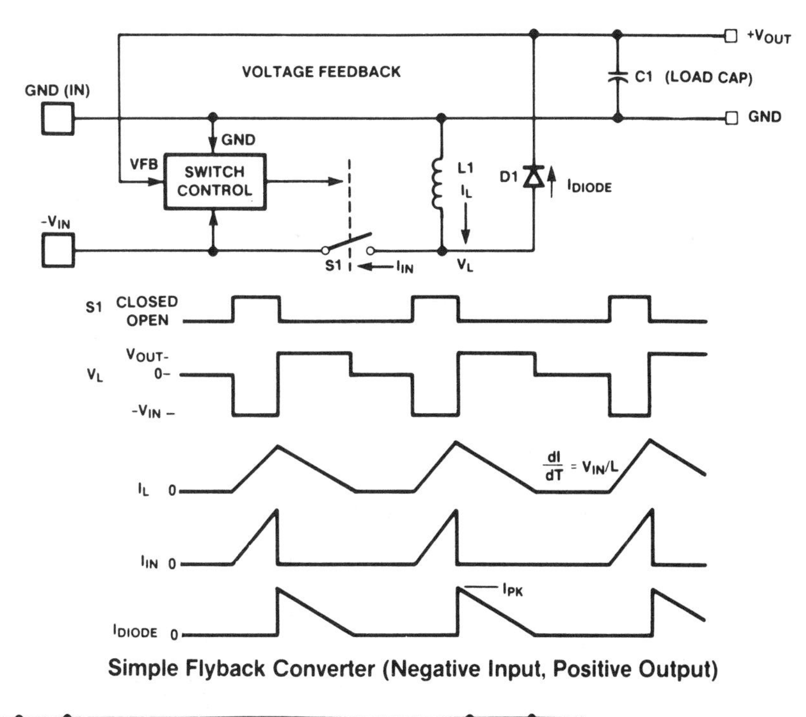

Simple Flyback Converter (Negative Input, Positive Output)

M1 IRF9633
D1,D2 1N5818
D3 1N4148
T1 AIE 315-0906*

*AIE Magnetics, Inc. (615) 244-9024
**To reduce high-frequency noise

Dual-Output PBX Telephone Application (V_{OUT} = +5V and -5.5V)

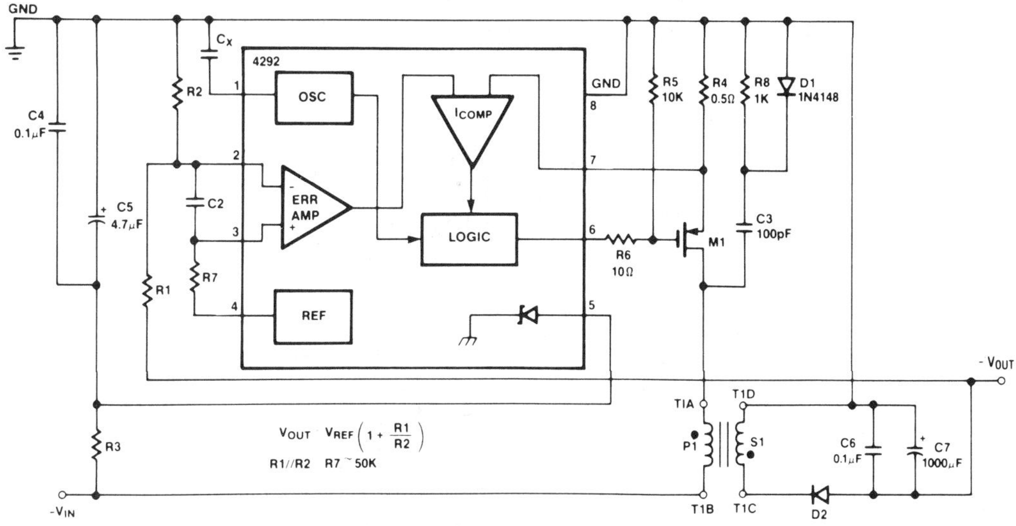

Negative Input, Negative Output Regulator With Transformer

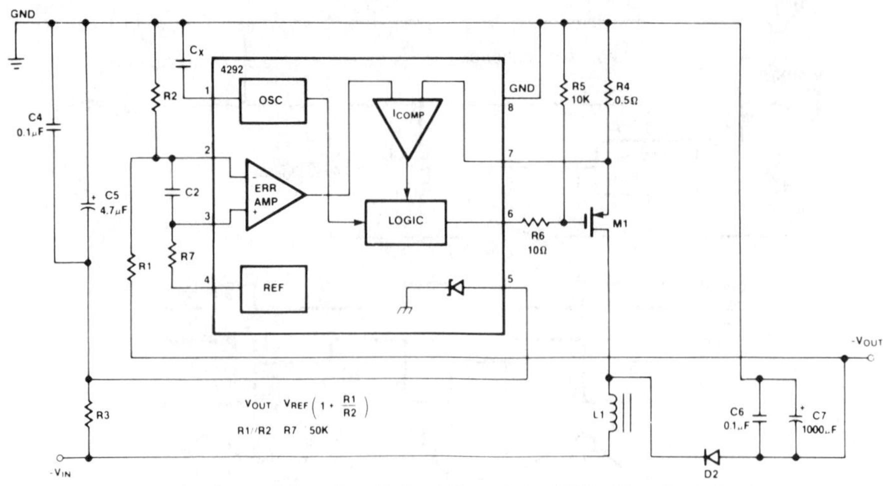

Negative Input, Negative Output Regulator With Simple Inductor

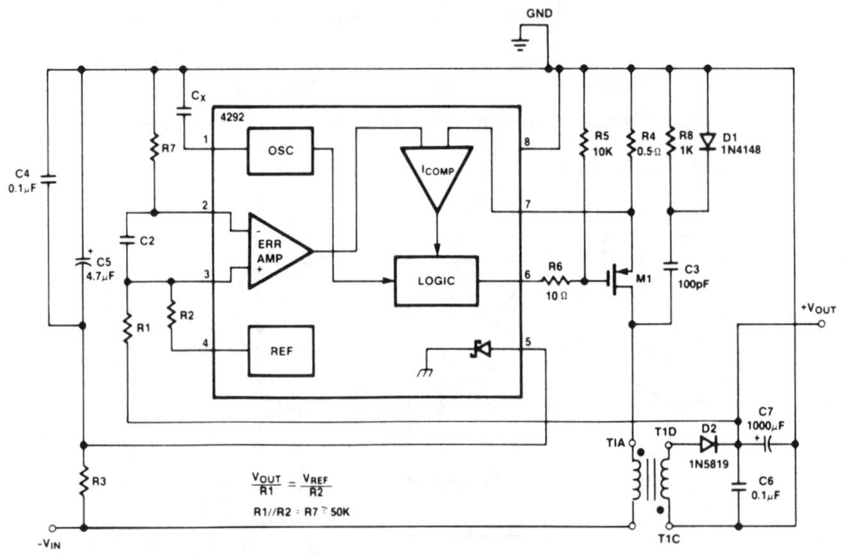

Complete 4292 Application Circuit (Negative Input, Positive Output)

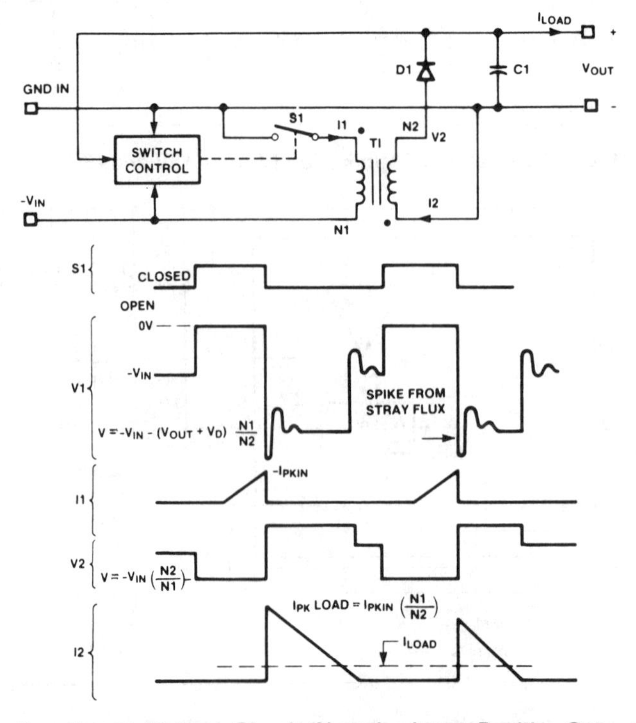

Transformer Flyback Circuit (Negative Input, Positive Output)

Raytheon
RC4194
Dual Tracking Voltage Regulator

- Simultaneously adjustable outputs with one resistor to ±42V
- Load current—±200 mA with 0.04% load regulation
- Internal thermal shutdown at T_j = +175°C
- External balance for $\pm V_o$ unbalancing
- 3W power dissipation

The RM4194 and RC4194 are dual-polarity tracking regulators designed to provide balanced or unbalanced positive and negative output voltages at currents to 200 mA. A single external resistor adjustment can be used to change both outputs between the limits of ±50 mV and ±42V.

These devices are designed for local "on-card" regulation, eliminating distribution problems associated with single-point regulation. To simplify application, the regulators require a minimum number of external parts.

The device is available in three package types to accommodate various power requirements. The TK (TO-66) power package can dissipate up to 3W at T_A = +25°C. The dc 14-pin dual-in-line will dissipate up to 625 mW.

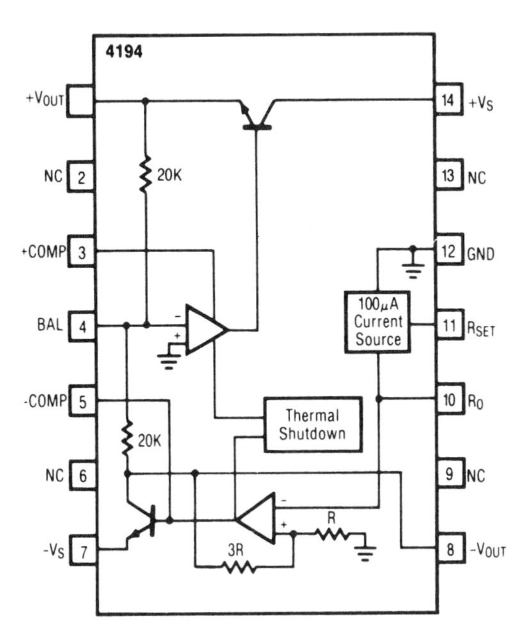

SCHEMATIC DIAGRAM

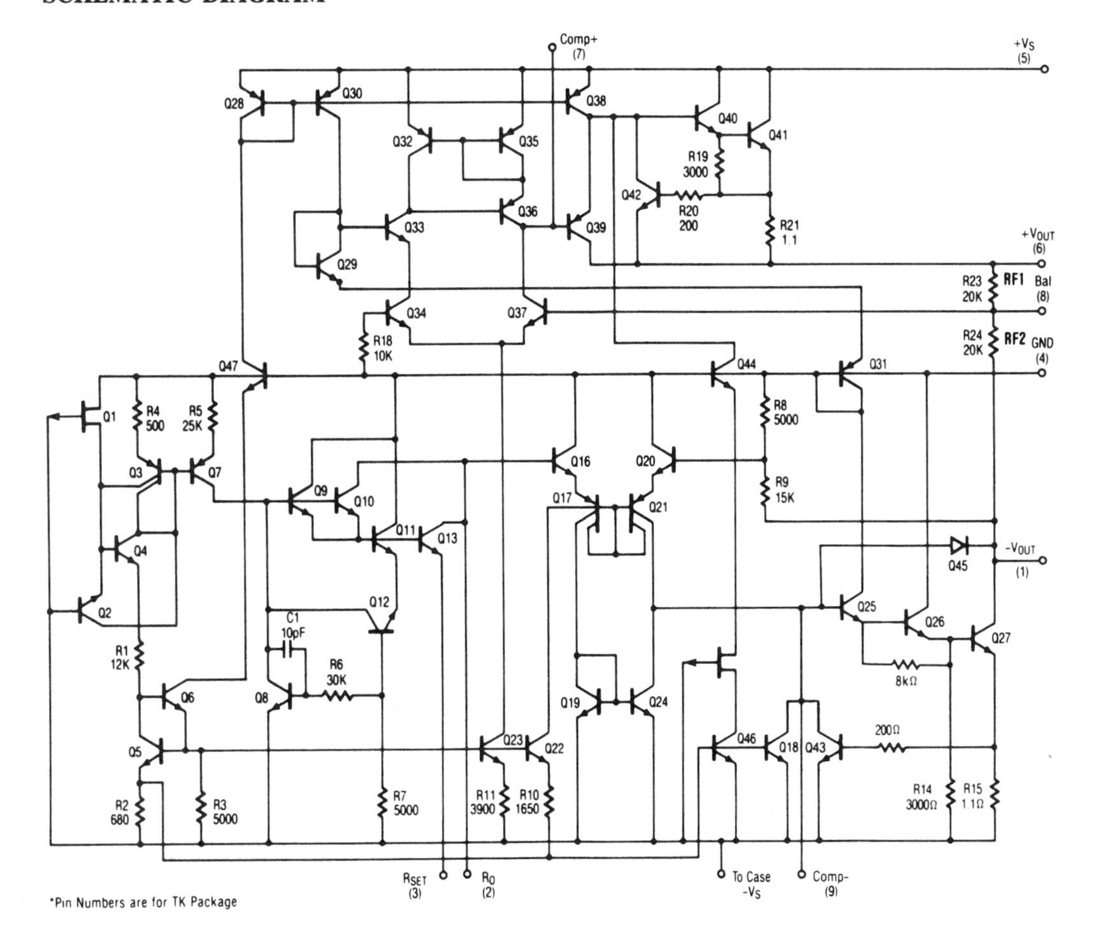

ELECTRICAL CHARACTERISTICS ($\pm 5 \leq V_{OUT} \leq V_{MAX}$; $I_L = \pm 1mA$; RM4194: $-55°C \leq T_j \leq +125°C$; RC4194: $0°C \leq T_j \leq +70°C$ unless otherwise specified)

Parameters	Test Conditions	RC/RM4194 Min	Typ	Max	Units
Line Regulation	$\Delta V_S = 0.1V_{IN}$		0.04	0.1	$\%V_{OUT}$
Load Regulation Output Voltage Drift With Temperature	4194TK: $I_L < 200mA$, 4194DC: $I_L < 100mA$ $\pm V_S = \pm(V_O + 5)V$		0.002	0.004	$\%V_O \times I_L$ (mA)
Positive Output			0.002	0.015	%/°C
Negative Output			0.003	0.015	%/°C
Supply Current[1] (Positive)	$V_S = \pm V_{MAX}$, $V_O = 0V$, $I_L = 0mA$		+0.8	+2.5	mA
Supply Current[1] (Negative)	$V_S = \pm V_{MAX}$, $V_O = 0V$, $I_L = 0mA$		-1.8	-4.0	mA
Supply Voltage	RM4194	±9.5		±45	V
	RC4194	±9.5		±35	
Output Voltage Scale Factor	$R_{SET} = 71.5k\Omega$, $T_j = +25°C$	2.38	2.5	2.62	kΩ/V
Output Voltage Range	RM4194: $R_{SET} = 71.5k\Omega$	0.05		±42	V
	RC4194: $R_{SET} = 71.5k\Omega$	0.05		±32	
Output Voltage Tracking			0.4	2.0	%
Ripple Rejection	f = 120Hz, $T_j = +25°C$		70		dB
Input-Output Voltage Differential	$I_L = 50mA$, $T_j = +25°C$	3.0			V
Short Circuit Current	$V_S = \pm 30V$, $T_j = +25°C$		300		mA
Output Noise Voltage	$C_L = 4.7\mu F$, $V_O = \pm 15V$ f = 10Hz to 100kHz		250		μV_{RMS}
Internal Thermal Shutdown			175		°C

Notes: 1. The current drain will increase by $50\mu A/V_{OUT}$ on positive side and $100\mu A/V_{OUT}$ on negative side.
2. The specifications above apply for the given junction temperatures since pulse test conditions are used.

ORDERING INFORMATION

Part Number	Package	Operating Temperature Range
RC4194N	N	0°C to +70°C
RC4194D	D	0°C to +70°C
RC4194K	K	0°C to +70°C
RM4194D	D	-55°C to +125°C
RM4194D/883B	D	-55°C to +125°C
RM4194K	K	-55°C to +125°C

Notes:
/883B suffix denotes Mil-Std-883, Level B processing
N = 14-lead plastic DIP
D = 14- lead ceramic DIP
K = 9-lead TO-66
Contact a Raytheon sales office or representative for ordering information on special package/temperature range combinations.

MASK PATTERN

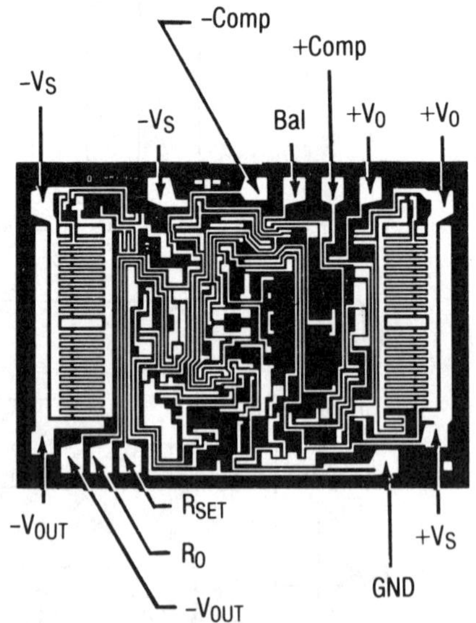

Die Size: 99 x 69 mils
Min. Pad Dimensions: 4 x 4 mils

ABSOLUTE MAXIMUM RATINGS

Supply Voltage
RC4194 ....±35V
RM4194 ....±45V
Supply Input to Output Voltage Differential
RC4194 ....±35V
RM4194 ....±45V
Load Current
N Package ....100 mA
D Package ....150 mA
K Package ....250 mA
Operating Junction Temperature Range
RC4194 ....0°C to +125°C
RM4194 ....-55°C to +150°C
Storage Temperature
Range ....-65°C to +150°C
Lead Soldering Temperature
(60 sec) ....+300°C

9-Lead
TK (TO-66) Package
(Top View)

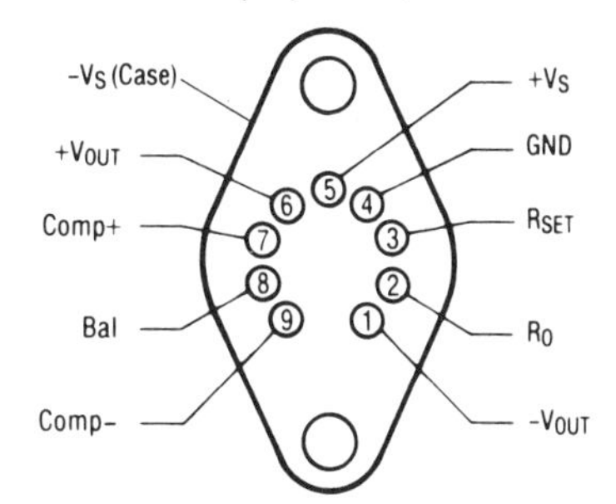

14-Lead
DB and DC Dual In-Line Package
(Top View)

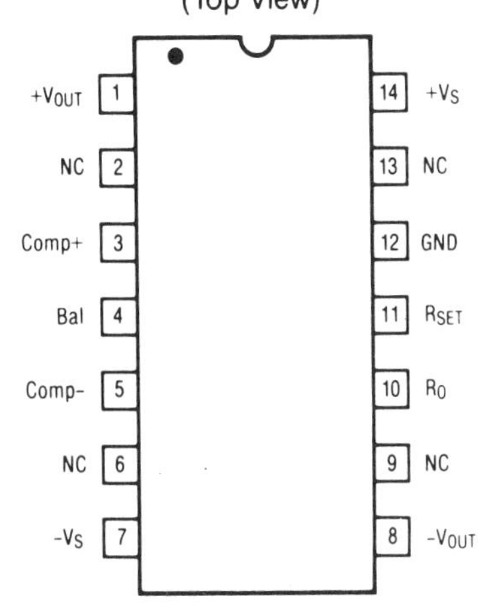

Unbalanced Output Voltage — Comparator Application

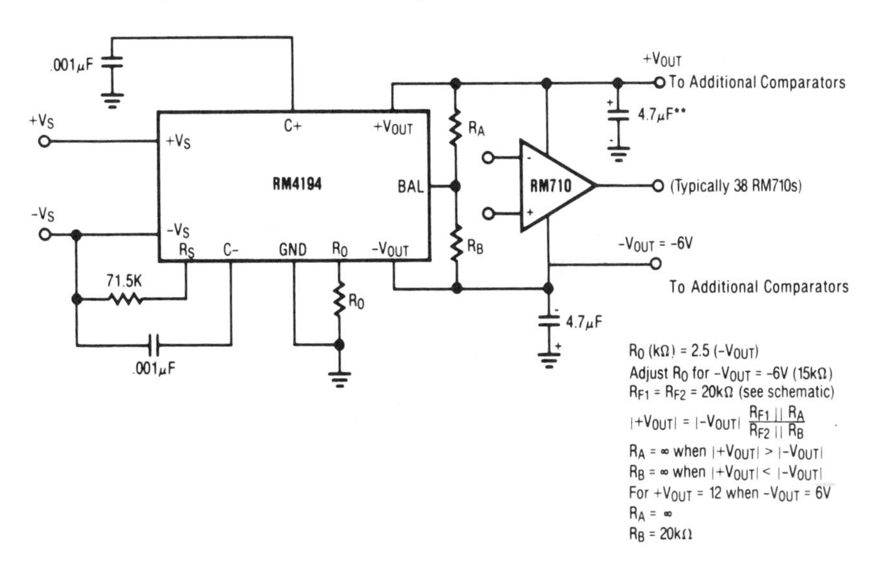

High Output Application

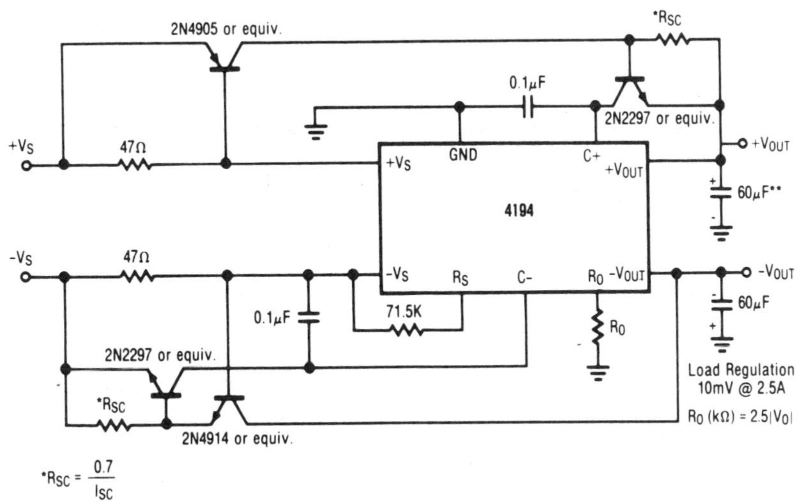

$^*R_{SC} = \frac{0.7}{I_{SC}}$

Note: Compensation and bypass capacitor connections should be close as possible to the 4194.

**Optional usage — not as critical as $-V_O$ bypass capacitors.

Balanced Output Voltage — Op Amp Application

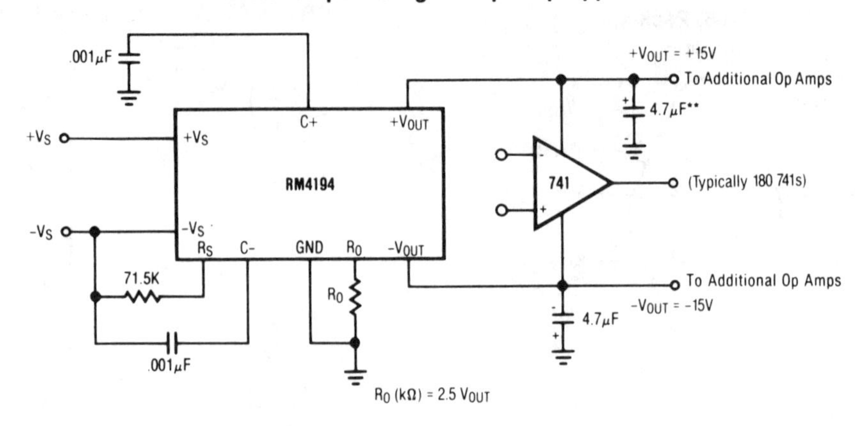

Digitally Controlled Dual 200mA Voltage Regulator

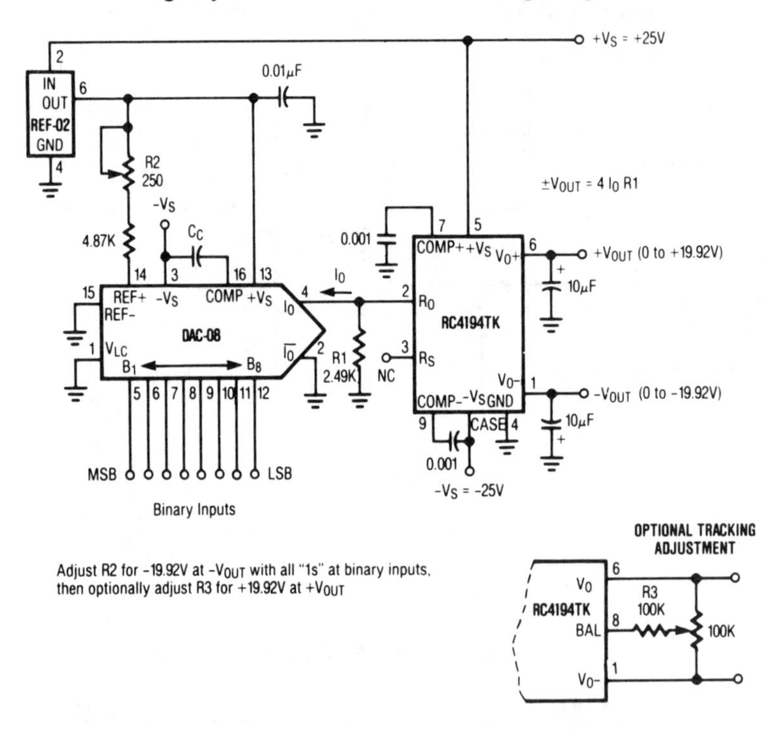

Raytheon RC4195

Fixed ±15V Dual Tracking Voltage Regulator

- ±15V operational amplifier at reduced cost and component density
- Thermal shutdown at $T_j = +175\,°C$ in addition to short circuit protection
- Output currents to 100 mA
- Can be used as single output regulator with up to +50V output
- Available in TO-66, TO-99, and 8-pin plastic mini-DIP
- No external frequency compensation required

The RC/RM4195 is a dual-polarity tracking regulator designed to provide balanced positive and negative 15V output voltages at currents up to 100 mA. This device is designed for local "on-card" regulation, eliminating distribution problems associated with single-point regulation. The regulator is intended for ease of application. Only two external components are required for operation (two 10 μF bypass capacitors).

The device is available in three package types to accommodate various applications requiring economy, high power dissipation, and reduced component density.

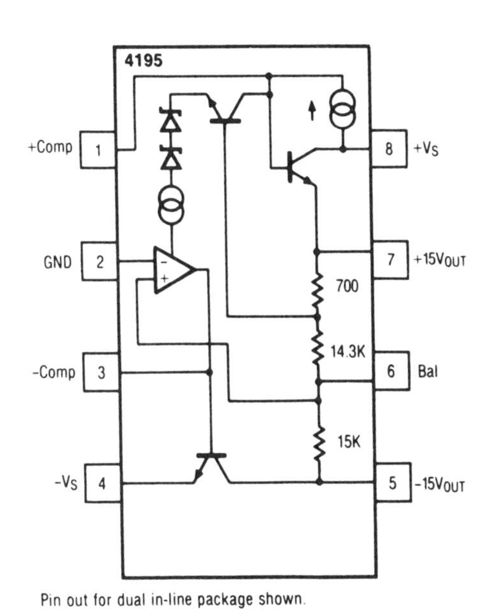

Pin out for dual in-line package shown.

MASK PATTERN

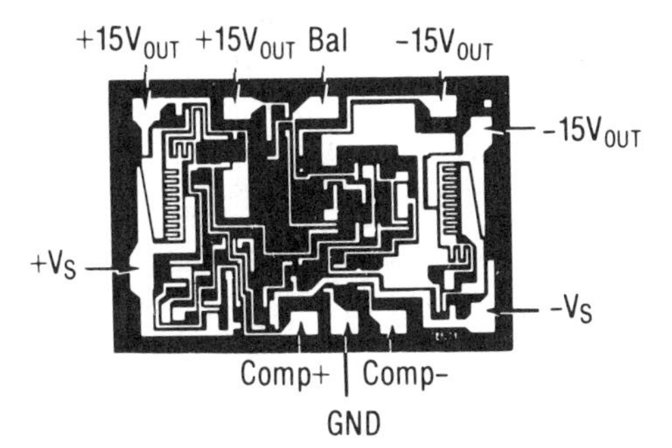

Die Size: 49 x 74 mils
Min. Pad Dimensions: 4 x 4 mils

ABSOLUTE MAXIMUM RATINGS

Supply Voltage ($\pm V_S$) to Ground .... ±30V
Load Current
 TK Package .... 150 mA
 T and N Package .... 100 mA
Storage Temperature
 Range .... -65°C to +150°C
Operating Junction Temperature Range
 RM4195 .... -55°C to +150°C
 RC4195 .... 0°C to +125°C
Lead Soldering Temperature
 (DIP, LCC, TO-99; 60 sec) .... +300°C

THERMAL CHARACTERISTICS

	8-Lead Plastic DIP	8-Lead TO-99 Metal Can	9-Lead TO-66 Metal Can	8-Lead Ceramic DIP
Max. Junction Temp.	125°C	175°C	150°C	175°C
Max. P_D T_A <50°C	468mW	658mW	2381mW	833mW
Therm. Res θ_{JC}	—	50°C/W	7°C/W	45°C/W
Therm. Res. θ_{JA}	160°C/W	190°C/W	42°C/W	150°C/W
For T_A >50°C Derate at	6.25 mW/°C	5.26 mW/°C	23.81 mW/°C	8.33 mW/°C

ELECTRICAL CHARACTERISTICS ($I_L = \pm 1mA$, $V_S = \pm 20V$, $C_L = 10\mu F$
RM4195: $-55°C \leq T_j \leq +125°C$; RC4195: $0°C \leq T_j \leq +70°C$ unless otherwise specified)[1]

Parameters	Test Conditions	RC/RM4195 Min	Typ	Max	Units
Line Regulation	$V_S = \pm 18V$ to $\pm 30V$		2	20	mV
Load Regulation	I_L = 1mA to 100mA		5	30	mV
Output Voltage Drift With Temperature			0.005	0.015	%/°C
Supply Current	$V_S = \pm 30V$, I_L = 0mA		±1.5	±4.0	mA
Supply Voltage		18		30	V
Output Voltage	T_j = +25°C	14.5	15.0	15.5	V
Output Voltage Tracking			±50	±300	mV
Ripple Rejection	f = 120Hz, T_j = +25°C		75		dB
Input-Output Voltage Differential	I_L = 50mA	3			V
Short Circuit Current	T_j = +25°C		220		mA
Output Voltage Noise	T_j = +25°C, f = 100Hz to 10kHz		60		μV_{RMS}
Internal Thermal Shutdown			175		°C

Notes: 1. The specifications above apply for the given junction temperature since pulse test conditions are used.

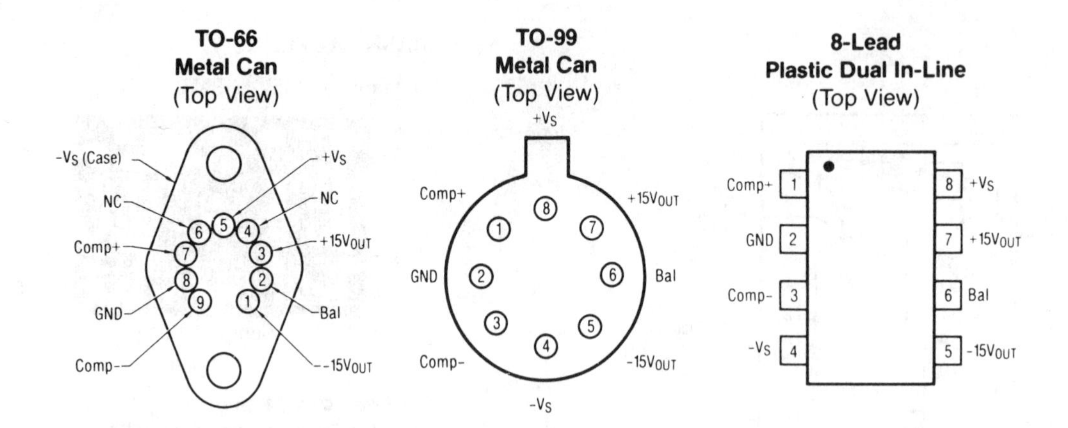

SCHEMATIC DIAGRAM

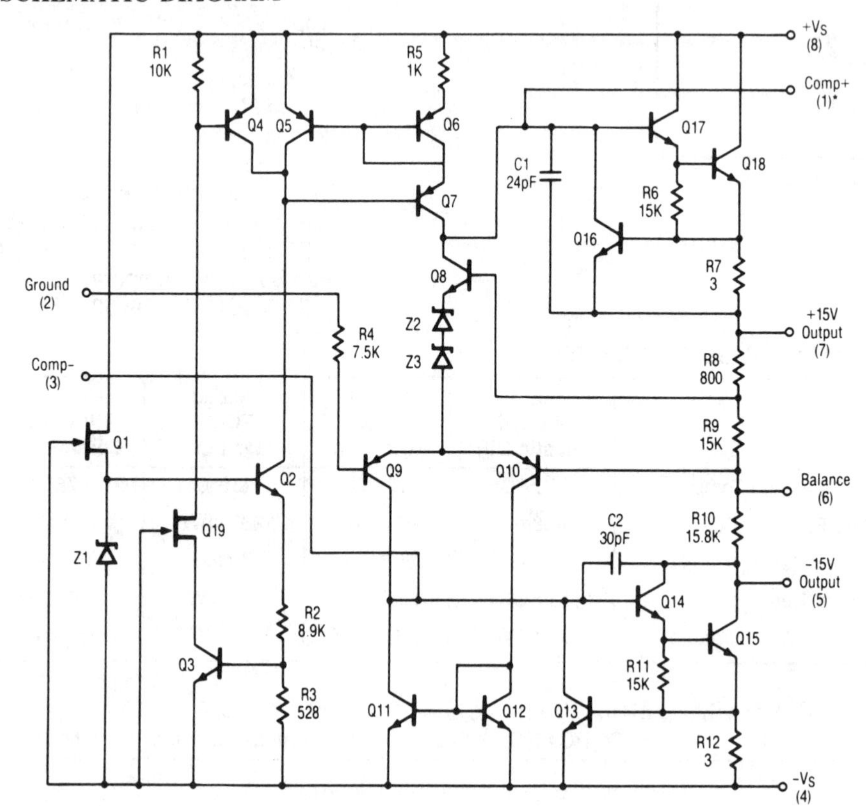

*Pin numbers are for 8-pin packages

Balanced Output ($V_O = \pm 15V$)

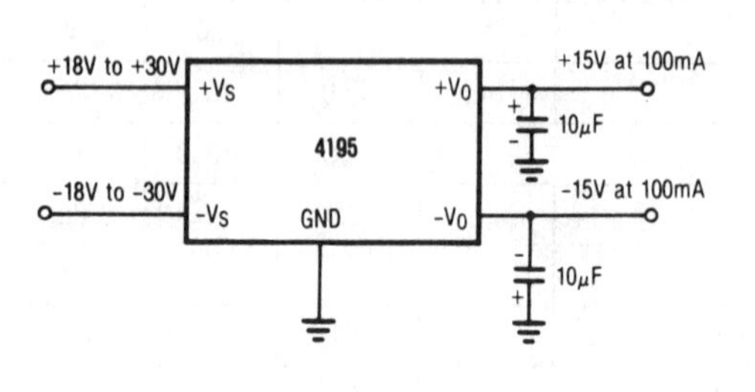

Positive Single Supply ($+15V < V_O < +50V$)

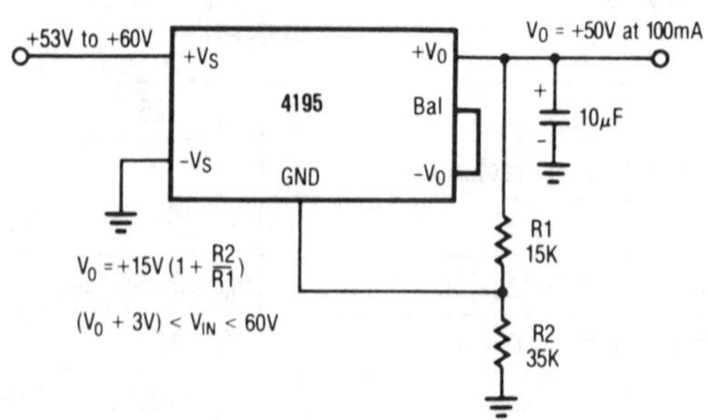

High Output Current

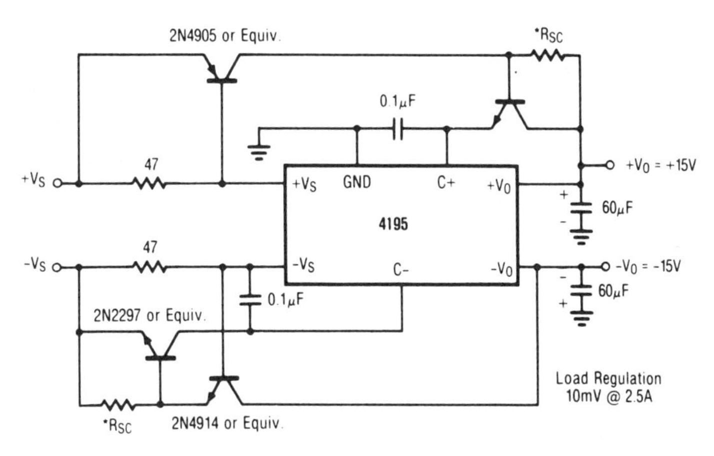

Raytheon RC4191/2/3

Micropower Switching Regulators

- High efficiency: 80% typical
- Low quiescent current: 135 μA
- Adjustable output 2.5V to 30V
- High switch current: 150 mA
- Bandgap reference: 1.31V
- Remote shutdown capability
- Low-battery detection circuitry
- Low component count

Raytheon's micropower switching regulators, 4191, 92 and 93, are the industry's first monolithic low power switching regulators available in an 8-lead mini-DIP, and designed specifically for battery-operated instruments. They each contain a 1.3V temperature-compensated bandgap reference, adjustable free-running oscillator, voltage comparator, low-battery detection circuitry, and a 150 mA switch transistor with all of the functions required to make a complete low power switching regulator.

These regulators can achieve up to 80% efficiency in most applications while being able to operate over a wide input supply voltage range, 2.4V to 30V, at a very low quiescent drain of 135 μA.

The 4191/92/93s have a free-running oscillator that provides the drive signal for the on-chip 150 mA switch transistor. The 100 Hz to 75 kHz oscillator frequency is determined by an external capacitor on pin 2.

These universal regulators can be used as a building block in three basic applications: step-up, step-down, and inverting.

In their most popular configuration step-up, these regulators will reduce the number of battery cells required for a given output voltage, and will maintain that voltage as the batteries decay, thus making available more board space and lengthening battery life.

A practical example of these advantages would be an instrument designed to operate from a nominal 9V supply voltage. If this instrument was powered with just 7 components (a steering diode, an inductor, two resistors, a capacitor, a 4191, 92, or 93 and a 9V battery), it would receive a continuous 9V until the battery had decayed to a terminal voltage of 2.4V. If board space is at a premium, then the designer could remove the 9V source and replace it with a single 3V Ni Cad battery without making any other adjustment(s) to the circuitry or affecting the overall operation of the instrument.

The 4191/92/93 series of micropower switching regulators consists of three devices each with slightly different specifications. The 4191 has a 1.5% maximum output voltage tolerance, 0.2% maximum line regulation, and operation to 30V. The 4192 has a 3.0% maximum output voltage tolerance, 0.5% maximum line regulation, and operation to 30V. The 4193 has a 5.0% maximum output voltage tolerance, 0.5% maximum line regulation, and operation to 24V. Other specifications are identical for the 4191, 4192, and 4193. Each type is available in commercial, industrial, and military temperature ranges, and in plastic and ceramic DIPs.

With some optional external components, the application can be designed to signal a display when the battery has decayed below a predetermined level, or designed to signal a display at one level and then shut itself off after the battery decays to a second level.

CONNECTION INFORMATION

8-Lead Dual In-Line Package (Top View)

1 8
2 7
3 6
4 5

Pin	Function
1	Low Battery (Set) Resistor (LBR)
2	Timing Capacitor (C_X)
3	External Inductor (L_X)
4	Ground
5	+Supply Voltage ($+V_S$)
6	Reference Set Current (I_C)
7	Feedback Voltage (V_{FB})
8	Low Battery Detector Output (LBD)

ORDERING INFORMATION

Part Number	Package	Operating Temperature Range
RC4191N	N	0°C to +70°C
RC4192N	N	0°C to +70°C
RC4193N	N	0°C to +70°C
RV4191N	N	-25°C to +85°C
RV4192N	N	-25°C to +85°C
RV4193N	N	-25°C to +85°C
RM4191D	D	-55°C to +125°C
RM4192D	D	-55°C to +125°C
RM4193D	D	-55°C to +125°C
RM4191D/883B	D	-55°C to +125°C
RM4192D/883B	D	-55°C to +125°C
RM4193D/883B	D	-55°C to +125°C

Notes:
/883B suffix denotes Mil-Std-883, Level B processing
N = 8-lead plastic DIP
D = 8 lead ceramic DIP
Contact a Raytheon sales office or representative for ordering information on special package/temperature range combinations.

MASK PATTERN

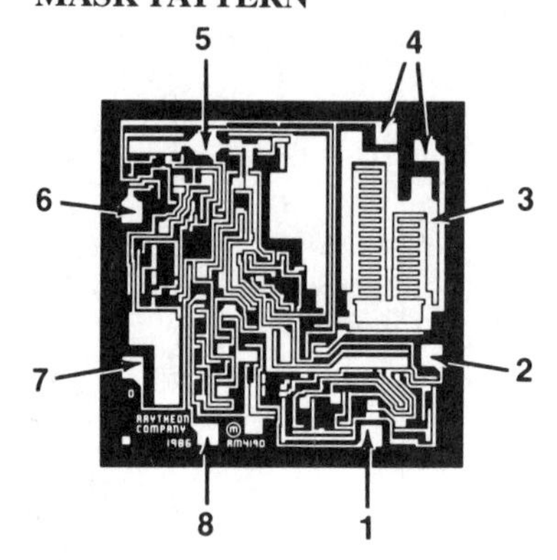

Die Size: 66 x 67 mils
Min. Pad Dimensions: 4 x 4 mils

ELECTRICAL CHARACTERISTICS (V_S = +6.0V, I_C = 5.0μA, and T_A = +25°C unless otherwise noted)

Parameters	Symbol	Conditions	4191			4192			4193			Units
			Min	Typ	Max	Min	Typ	Max	Min	Typ	Max	
Supply Voltage	$+V_S$		2.4		30	2.4		30	2.4		24	V
Reference Voltage (Internal)	V_{REF}		1.29	1.31	1.33	1.27	1.31	1.35	1.24	1.31	1.38	V
Switch Current	I_{SW}	V_3 = 400mV	75	150		75	150		75	150		mA
Supply Current	I_S	Measure at Pin 5 $I_3 = 0$		135	200		135	200		135	200	μA
Efficiency	ef			80			80			80		%
Line Regulation		$0.5V_O < V_S < V_O$		0.08	0.2		0.08	0.5		0.08	0.5	% V_O
Load Regulation	L_I	$V_S = +0.5V_O$, P_L = 150mW		0.2	0.5		0.2	0.5		0.2	0.5	% V_O
Operating Frequency Range	F_O		0.1	25	75	0.1	25	75	0.1	25	75	kHz
Reference Set Current	I_C		1.0	5.0	50	1.0	5.0	50	1.0	5.0	50	μA
Switch Current	V_3	V_3 = 1.0V	100			100			100			mA
Switch Leakage Current	I_{CO}	V_3 = 24V		0.01	5.0		0.01	5.0		0.01	5.0	μA
Supply Current (Disabled)	I_{SO}	$I_C < 0.01μA$		0.1	5.0		0.1	5.0		0.1	5.0	μA
Low Battery Bias Current	I_1	V_1 = 1.2V		0.7			0.7			0.7		μA
Capacitor Charging Current	I_{CX}			5.0			5.0			5.0		μA
Capacitor Threshold Voltage +	$+V_{THX}$			1.78			1.78			1.78		V
Capacitor Threshold Voltage –	$-V_{THX}$			0.62			0.62			0.62		V
Feedback Input Current	I_{FB}	V_7 = 1.3V		0.1			0.1			0.1		μA
Low Battery Output Current	I_{LBD}	V_8 = 0.4V, V_1 = 1.1V	250	600		250	600		250	600		μA

ELECTRICAL CHARACTERISTICS

(V_S = +6.0V, I_C = 5.0μA, unless otherwise noted, over the full operating temperature range)

Parameters	Symbol	Conditions	4191 Min	4191 Typ	4191 Max	4192 Min	4192 Typ	4192 Max	4193 Min	4193 Typ	4193 Max	Units
Supply Voltage	$+V_S$		3.5		24	3.5		24	3.5		24	V
Reference Voltage (Internal)	V_{REF}		1.25	1.31	1.37	1.23	1.31	1.39	1.20	1.31	1.42	V
Quiescent Current	I_S	Measure at Pin 5 $I_3 = 0$		225	300		225	300		225	300	μA
Line Regulation		$0.5V_O < +V_S < V_O$		0.2	0.5		0.5	1.0		0.5	1.0	% V_O
Load Regulation	L_I	$+V_S = 0.5V_O$, P_L = 150mW		0.5	1.0		0.5	1.0		0.5	1.0	% V_O
Reference Set Current	I_C		1.0	5.0	50	1.0	5.0	50	1.0	5.0	50	μA
Switch Leakage Current	I_{CO}	V_3 = 24V			30			30			30	μA
Supply Current (Disabled)	I_{SO}	$I_C < 0.01\mu A$			30			30			30	μA
Low Battery Output Current	I_{LBD}	V_8 = 0.4V, V_1 = 1.1V	250	600		250	600		250	600		μA

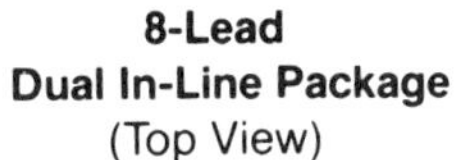

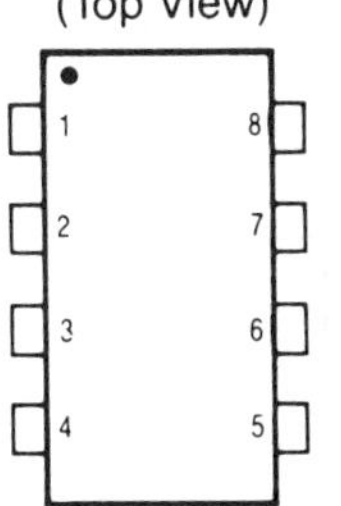

Pin	Function
1	Low Battery (Set) Resistor (LBR)
2	Timing Capacitor (C_X)
3	External Inductor (L_X)
4	Ground
5	+Supply Voltage ($+V_S$)
6	Reference Set Current (I_C)
7	Feedback Voltage (V_{FB})
8	Low Battery Detector Output (LBD)

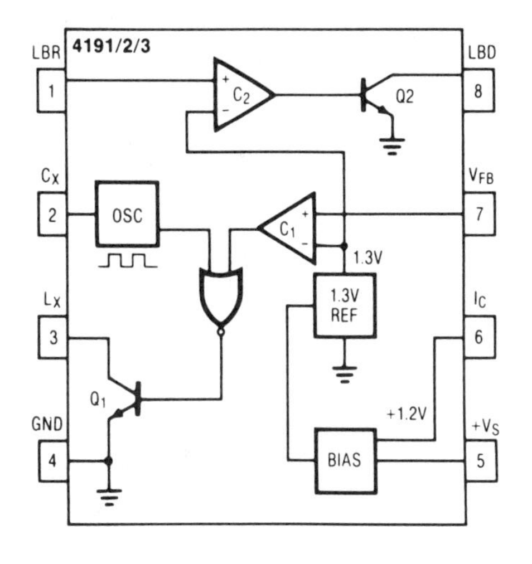

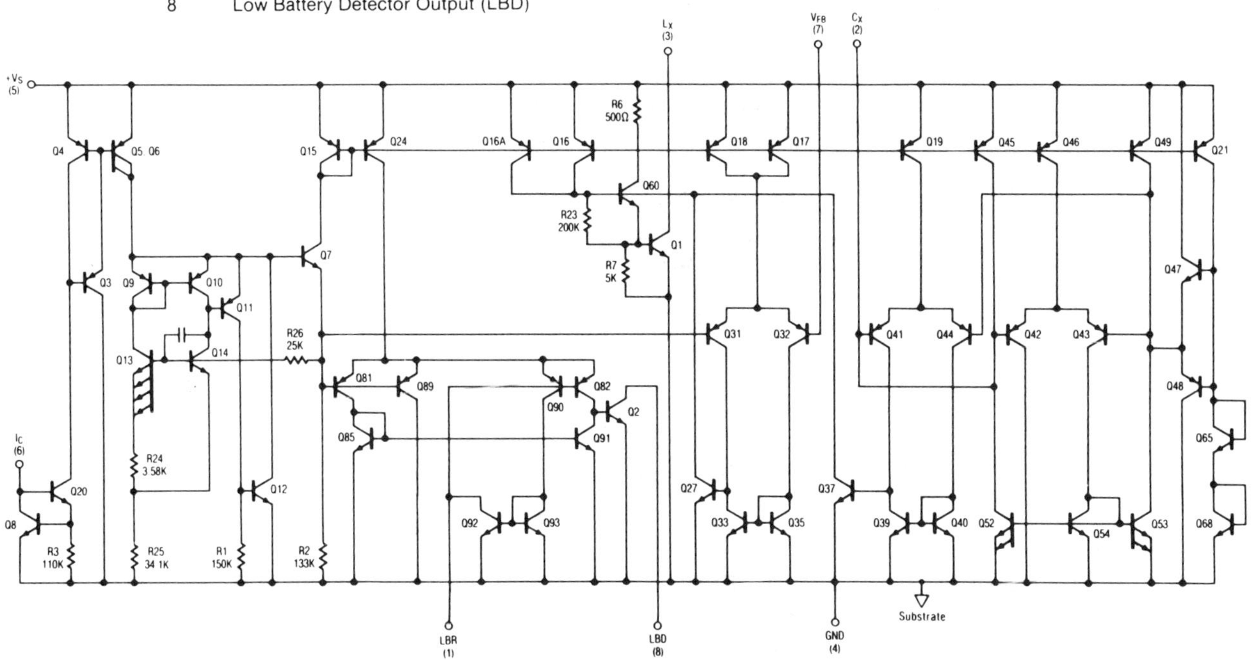

Minimum Supply Voltage vs. Temperature

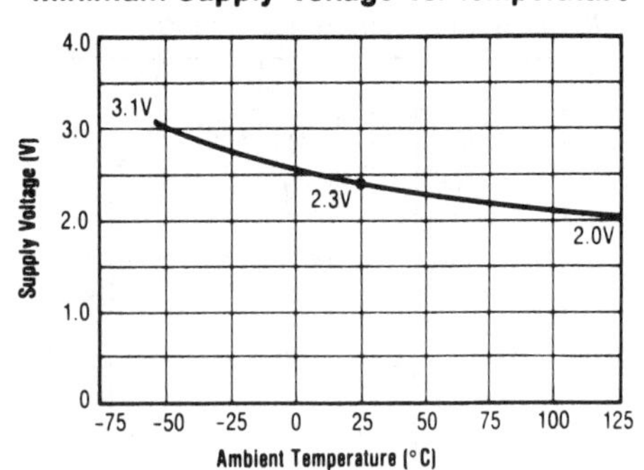

Quiescent Current vs. Temperature

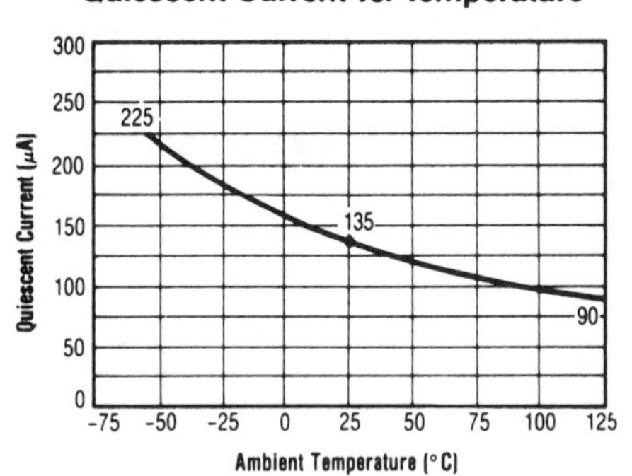

Reference Voltage vs. Temperature

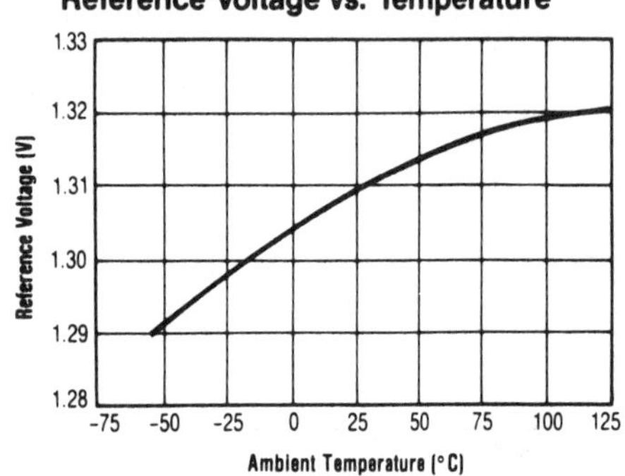

Oscillator Frequency vs. Temperature

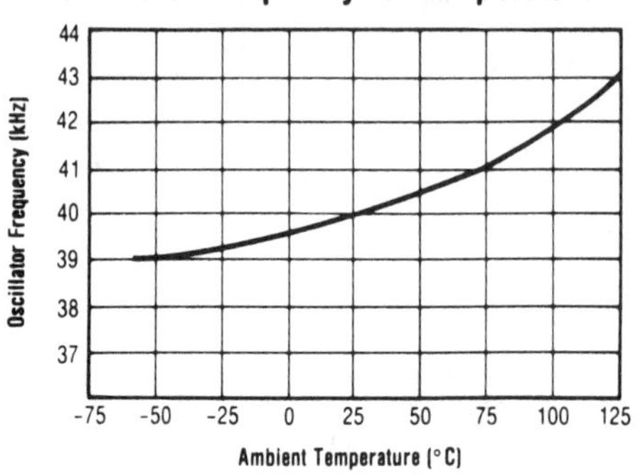

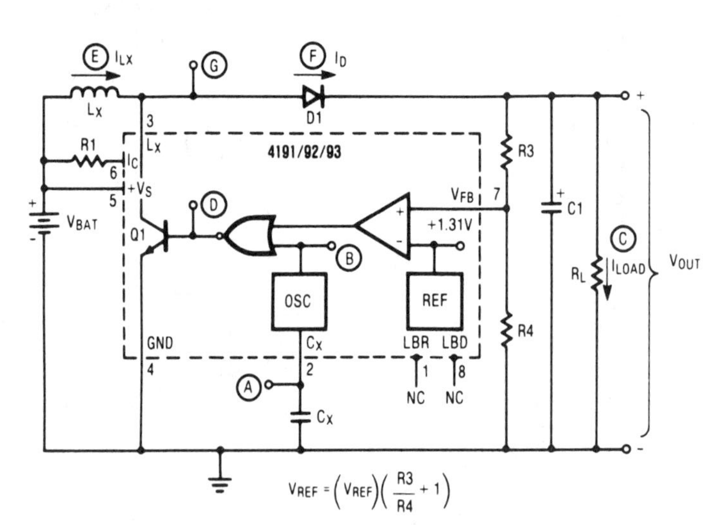

Minimum Step-Up Application

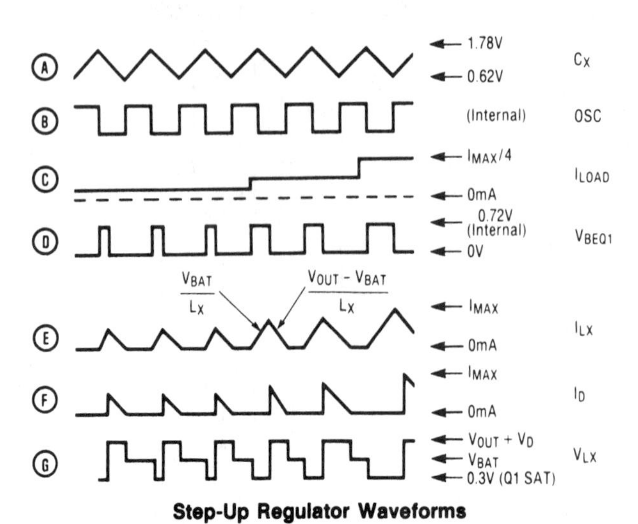

Step-Up Regulator Waveforms

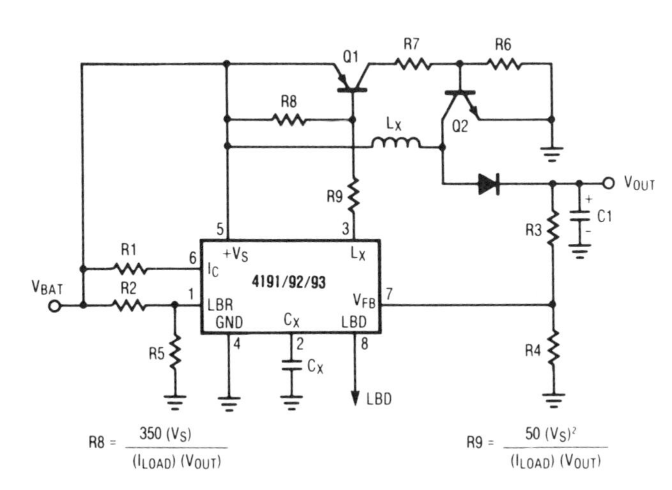

Step-Up Switching Regulator (High Current)

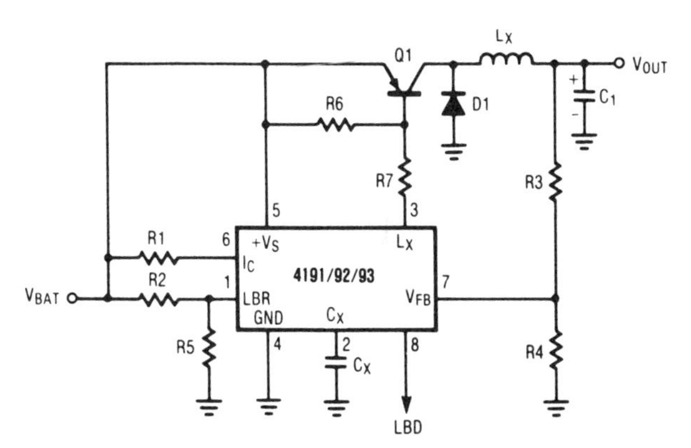

Stepping Down an Input Voltage Greater Than 30V

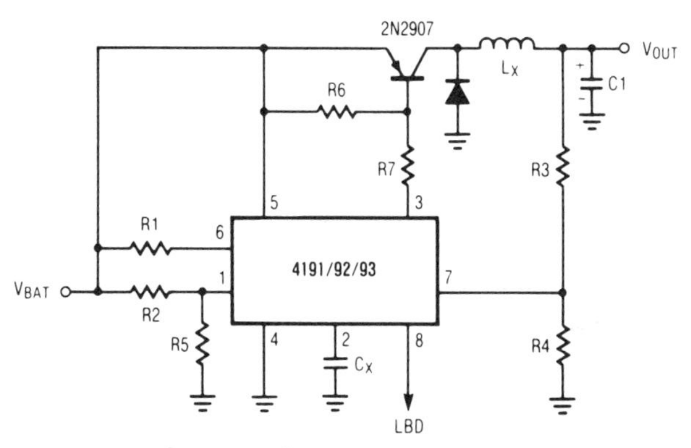

Complete Step-Down Regulator

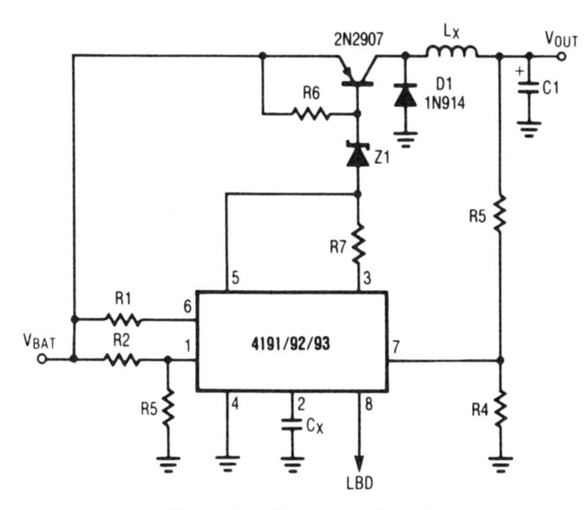

Stepping Down an Input Voltage Greater Than 30V (High Current)

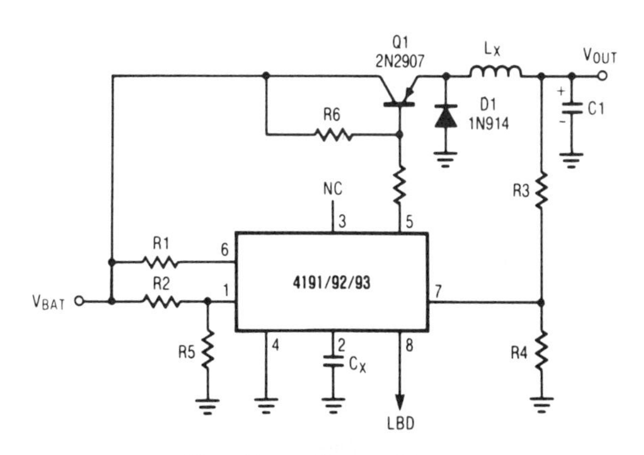

Step-Down Current Saver

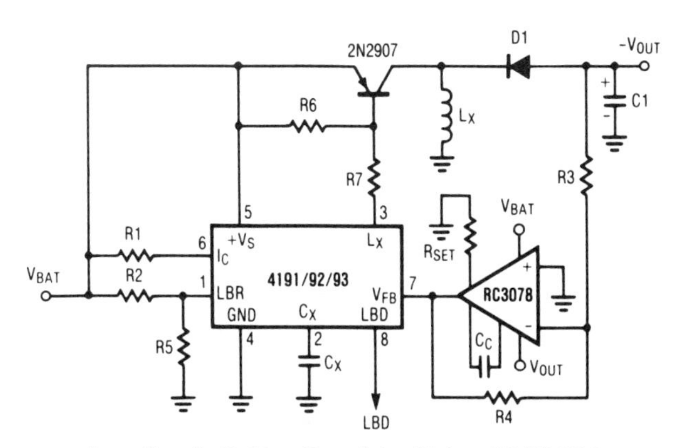

Inverting Switching Regulator Using 4191/92/93

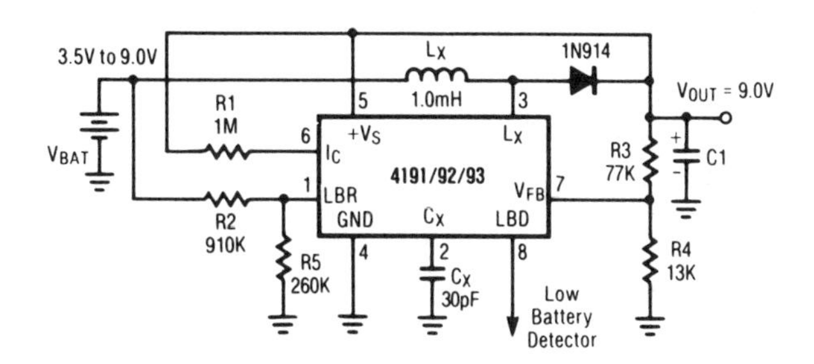

Bootstrapped Operation

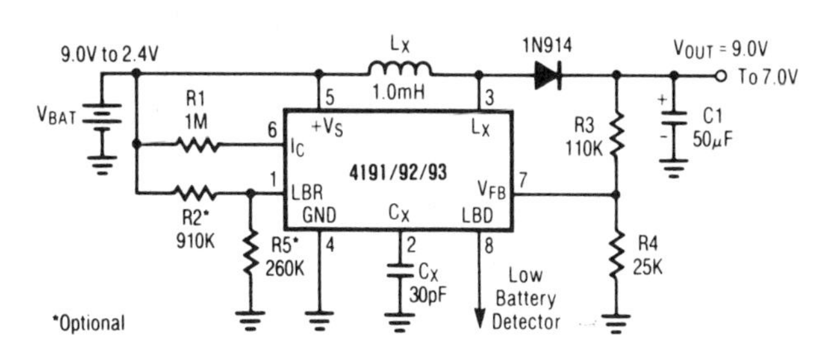

Typical Application: 9.0V Battery Life Extender

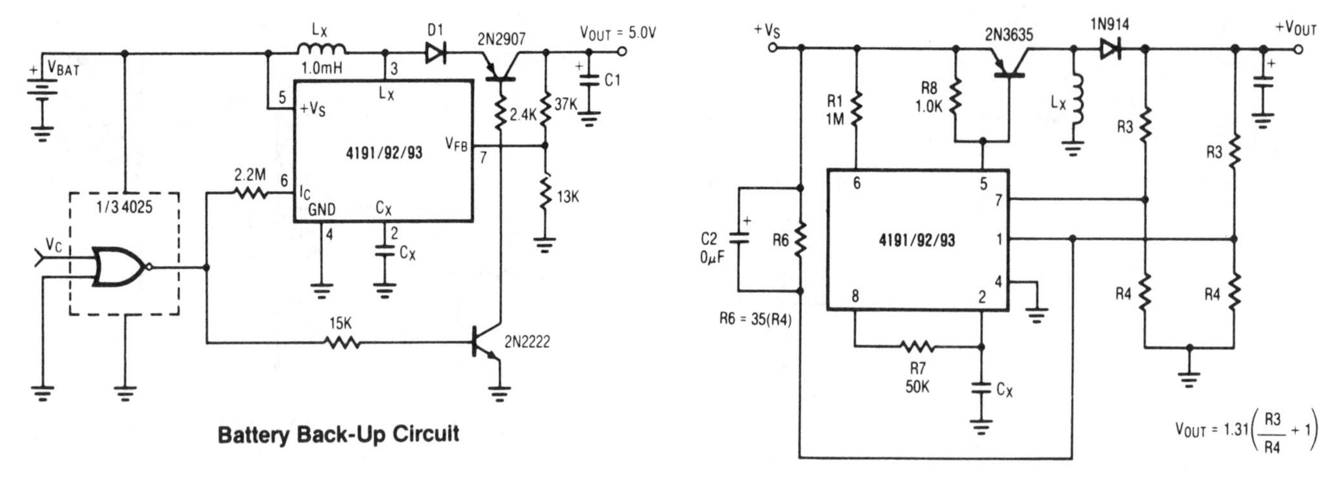

Battery Back-Up Circuit

Step-Down Regulator With Short Circuit Protection

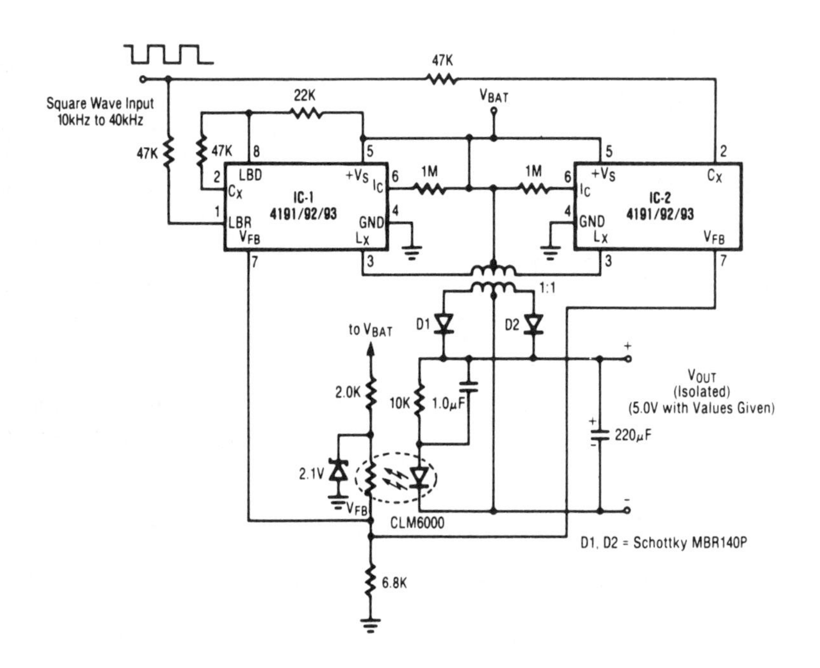

Isolated Push-Pull Transformer Coupled Regulator

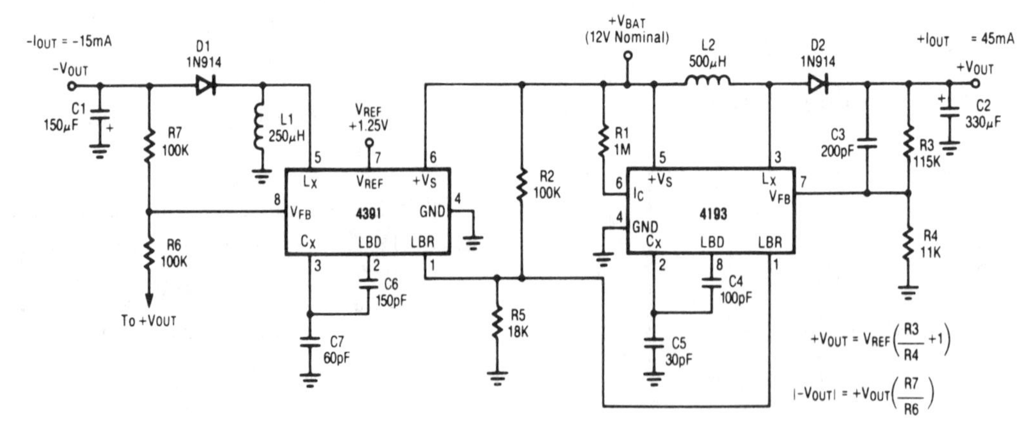

RC4391/RC4193 Power Supply (±15.0V With Values Given)

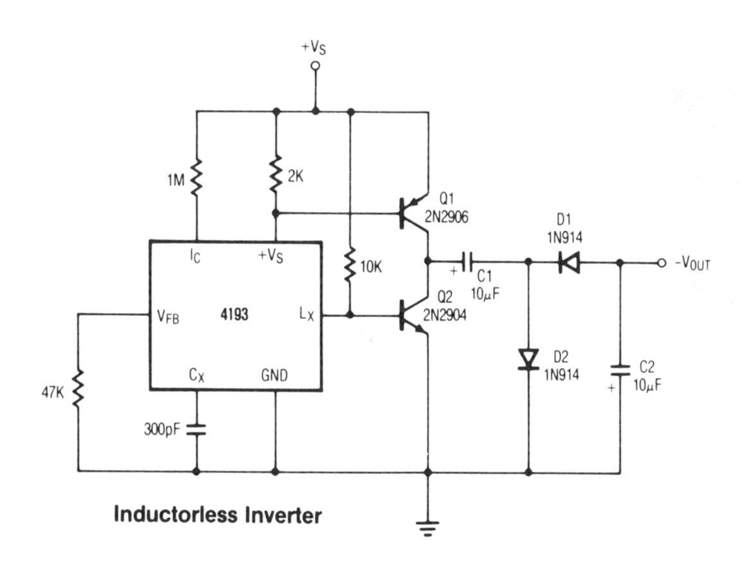

Inductorless Inverter

Raytheon RC4391

Inverting Switching Regulator

- Converts a positive voltage into a negative voltage
- Specifically designed for low-power applications, including batteries
- Adjustable output voltage
- High switch current capability
- Low quiescent supply current: 175 µA typical
- Eight pin mini dual-in-line package
- Low battery detection circuitry
- High efficiency: 70% typical

Raytheon's micropower inverting switching regulator, RC/RM4391, is a monolithic low-power switching regulator specifically designed for low-power inverting applications. The RC4391 contains an internal 1.25V bandgap voltage reference, switch transistor, comparator, free-running oscillator, and low-battery-detection circuitry. These components are interconnected to minimize the number of external components required in typical inverting applications. The RC4391 requires an inductor, diode, timing capacitor, and an R2/R1 network to achieve a negative output voltage. The RC4391 allows the designer flexibility in designing unconventional applications, such as replacing the internal bandgap reference with an external or system reference, or using the low battery detection comparator and transistor as voltage level detectors or for signal generation.

A typical application would combine the RC4391 with the RC4193 micropower switching regulator to convert a single input voltage into a ±12V or ±15V op-amp power supply. The single voltage can be from a battery, bridge rectifier or existing +0.5V bus on card for this application.

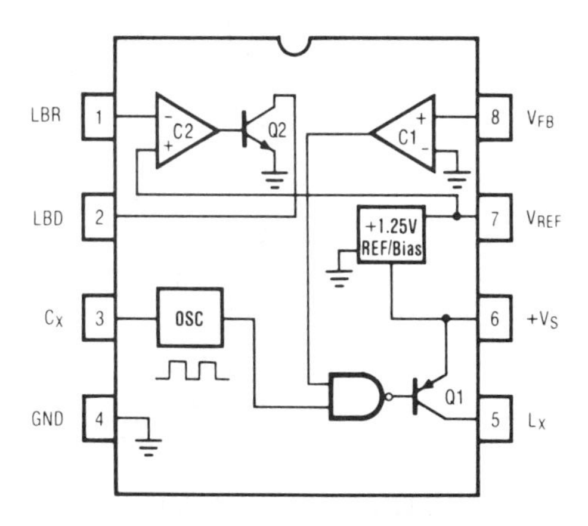

ABSOLUTE MAXIMUM RATINGS

Internal Power Dissipation .... 500 mW
Supply Voltage*
(Pin 6 to Pin 4 or
Pin 6 to Pin 5) .... +30V
Storage Temperature
Range .... -65°C to +150°C
Operating Temperature Range
RM4391 .... -55°C to +125°C
RV4391 .... -25°C to +85°C
RC4391 .... 0°C to +70°C
Switch Current (I_{MAX}) .... 375 mA peak

*The maximum allowable supply voltage (+V_S) in inverting applications will be reduced by the value of the negative output voltage, unless an external power transistor is used in place of Q1.

ORDERING INFORMATION

Part Number	Package	Operating Temperature Range
RC4391N RC4391M	N M	0°C to +70°C 0°C to +70°C
RV4391N	N	-25° C to +85°C
RM4391D RM4391D/883B	D D	-55°C to +125°C -55°C to +125°C

Notes:
/883B suffix denotes Mil-Std-883, Level B processing
N = 8-lead plastic DIP
D = 8 lead ceramic DIP
M = 8-lead plastic SOIC
Contact a Raytheon sales office or representative for ordering information on special package/temperature range combinations.

MASK PATTERN

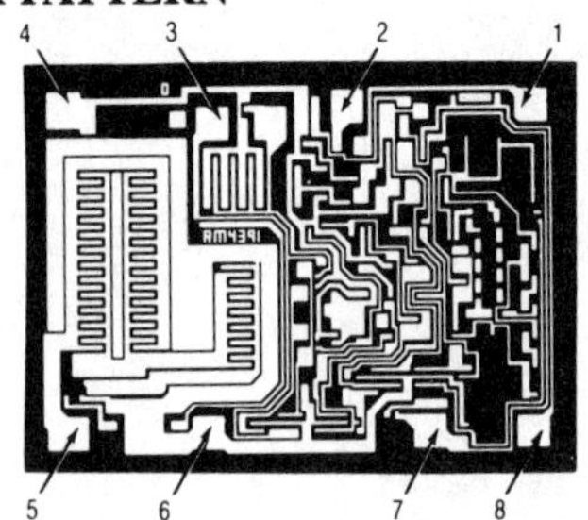

Die Size: 69 x 50 mils
Min. Pad Dimensions: 4 x 4 mils

THERMAL CHARACTERISTICS

	8-Lead Plastic DIP	8-Lead Ceramic DIP	8-Lead Small Outline Plastic SO-8
Max. Junction Temp.	125°C	175°C	125°C
Max. P_D T_A <50°C	468 mW	833 mW	300 mW
Therm. Res θ_{JC}	—	45°C/W	—
Therm. Res. θ_{JA}	160°C/W	150°C/W	240°C/W
For T_A >50°C Derate at	6.25 mW/°C	8.33 mW/°C	4.17 mW/°C

8-Lead
Dual In-Line Package
(Top View)

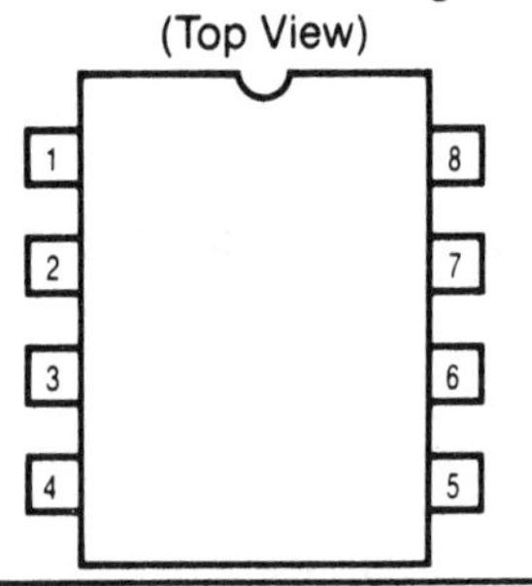

Pin	Function
1	Low Battery Resistor (LBR)
2	Low Battery Detector (LBD)
3	Timing Capacitor (C_X)
4	Ground
5	External Inductor (L_X)
6	+Supply Voltage ($+V_S$)
7	+1.25V Reference Voltage (V_{REF})
8	Feedback Voltage (V_{FB})

Cautionary Note: Care must be taken not to exceed the maximum current rating of the switch transistor. Select the inductor value and timing capacitor value carefully, and when prototyping, start with low input voltages first.

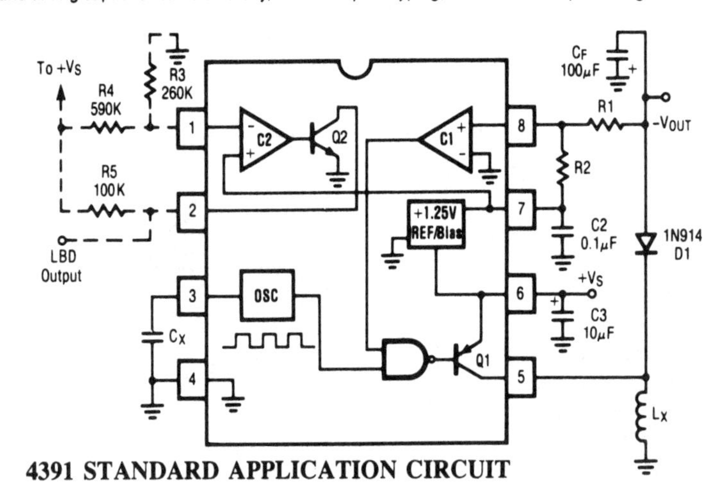

Parts List	-5.0V Output	-15V Output
R1 =	300kΩ	900kΩ
R2 =	75kΩ	75kΩ
Cx =	150pF	150pF
Lx =	1.0mH Dale TE3 Q4 TA	

- - - - = **Optional**

For high performance applications, use larger values C1, C3, and power Schottky for D1.

4391 STANDARD APPLICATION CIRCUIT

ELECTRICAL CHARACTERISTICS

V_S = +6.0V, T_A = +25° C unless otherwise noted)

Parameter	Symbol	Condition	RC4391 Min	RC4391 Typ	RC4391 Max	Units
Supply Current	I_S	V_S = +4.0V, No External Loads		170	250	μA
		V_S = +25V, No External Loads		300	500	
Output Voltage	V_{OUT}	$V_{OUT\ nom}$ = −5.0V	-5.35	-5.0	-4.65	V
		$V_{OUT\ nom}$ = −15V	−15.85	−15	−14.15	
Line Regulation		$V_{OUT\ nom}$ = −5.0V, C_X = 150pF, V_S = +5.8V to +15V		1.5	3.0	% V_{OUT}
		$V_{OUT\ nom}$ = −15V, C_X = 150pF, V_S = +5.8V to +15V		1.0	2.0	

ELECTRICAL CHARACTERISTICS

(V_S = +6.0V over the full operating temperature range unless otherwise noted)

Parameter	Symbol	Condition	RC4391 Min	RC4391 Typ	RC4391 Max	Units
Load Regulation		$V_{OUT\ nom}$ = -5.0V, C_X = 350pF, V_S = +4.5V, P_{LOAD} = 0mW to 75mW		0.2	0.4	% V_{OUT}
		$V_{OUT\ nom}$ = −15V, C_X = 350pF, V_S = +4.5V, P_{LOAD} = 0mW to 75mW		0.07	0.14	
Reference Voltage	V_{REF}		1.18	1.25	1.32	V
Switch Current	I_{SW}	Pin 5 = 5.5V	75	100		mA
Switch Leakage Current	I_{CO}	Pin 5 = −24V		0.01	5.0	μA
Timing Pin Current	I_{CX}	Pin 3 = 0V	6.0	10	14	μA
LBD Leakage Current		Pin 1 = 1.5V, Pin 2 = 6.0V		0.01	5.0	μA
LBD on Current		Pin 1 = 1.1V, Pin 2 = 0.4V	210	600		μA
LBR Bias Current		Pin 1 =1.5V		0.7		μA

ELECTRICAL CHARACTERISTICS

(V_S = +6.0V, T_A = +25° C unless otherwise noted)

Parameter	Symbol	Condition	RC4391 Min	RC4391 Typ	RC4391 Max	Units
Supply Voltage	$+V_S$		+4.0		+30−|V_{OUT}|	V
Supply Current	I_{IN}	V_S = +25V		300	500	μA
Reference Voltage	V_{REF}		1.13	1.25	1.36	V
Output Voltage	V_{OUT}	$V_{OUT\ nom}$ = −5.0V	−5.5	−5.0	−4.5	V
		$V_{OUT\ nom}$ = −15V	−16.5	−15	−13.5	
Line Regulation		$V_{OUT\ nom}$ = −5.0V, C_X = 150pF, V_S = +5.8V to +15V		2.0	4.0	% V_{OUT}
		$V_{OUT\ nom}$ = −15V, C_X =150pF, V_S = +5.8V to +15V		1.5	3.0	
Load Regulation		$V_{OUT\ nom}$ = −5.0V, C_X = 350pF, V_S = +4.5V, P_{LOAD} = 0mW to 50mW		0.2	0.5	% V_{OUT}
		$V_{OUT\ nom}$ = −15V, C_X = 350pF, V_S = +4.5V, P_{LOAD} = 0mW to 50mW		0.2	0.6	
Switch Leakage Current	I_{CO}	Pin 5 = −24V		0.1	30	μA

Power vs. Ripple

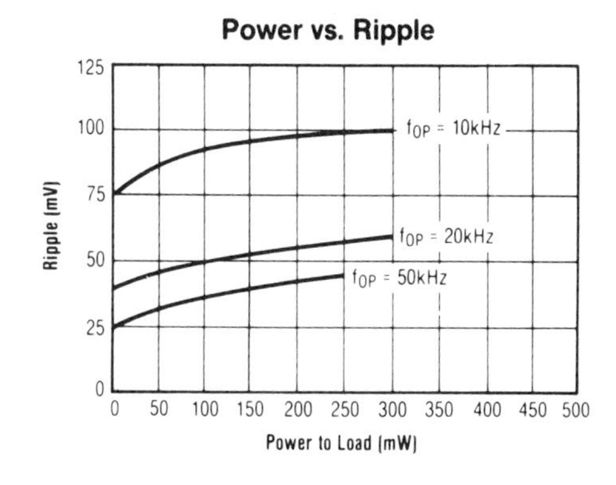

Power vs. Duty Cycle

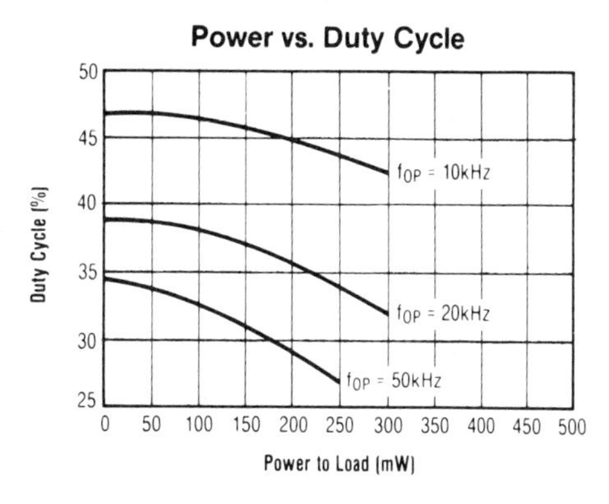

Power vs. Efficiency

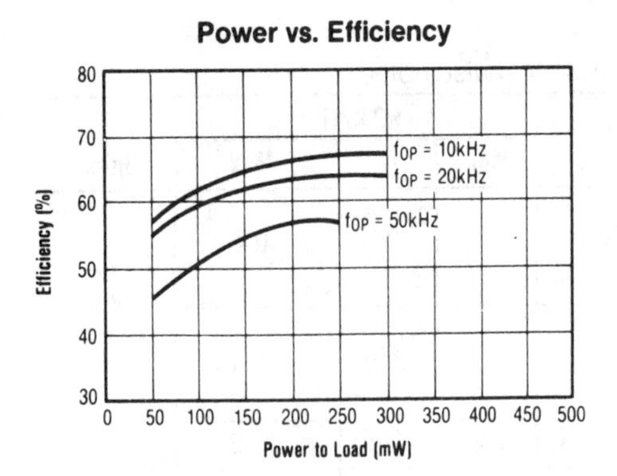

Reference Voltage vs. Temperature

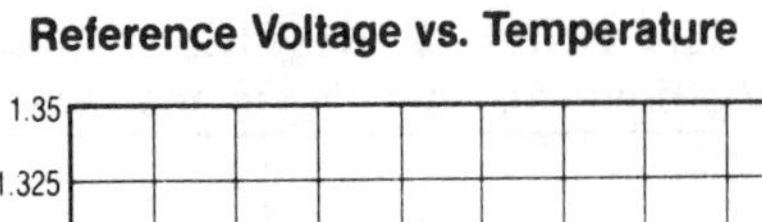
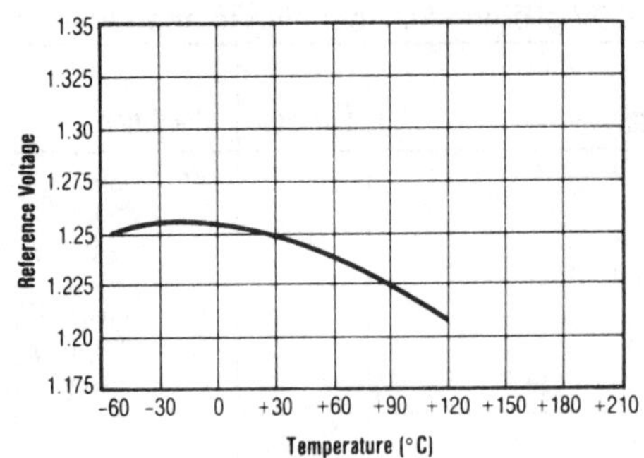

Minimum Supply Voltage vs. Temperature

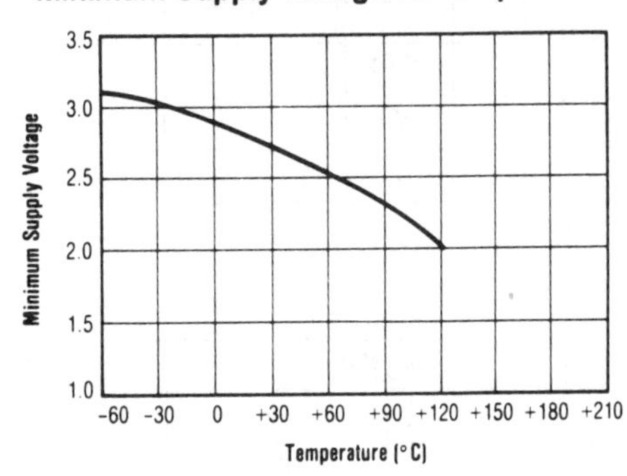

Supply Current vs. Temperature

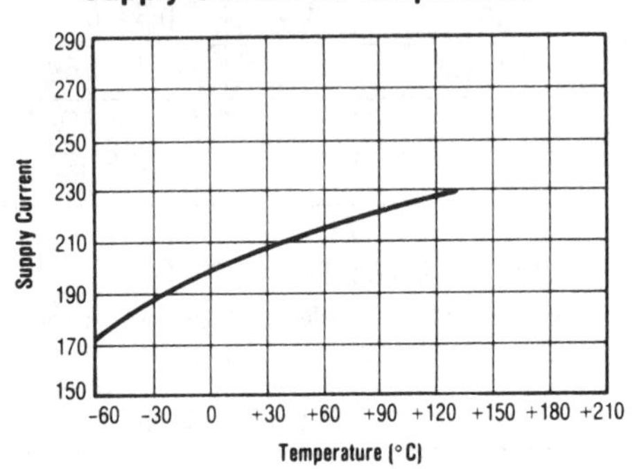

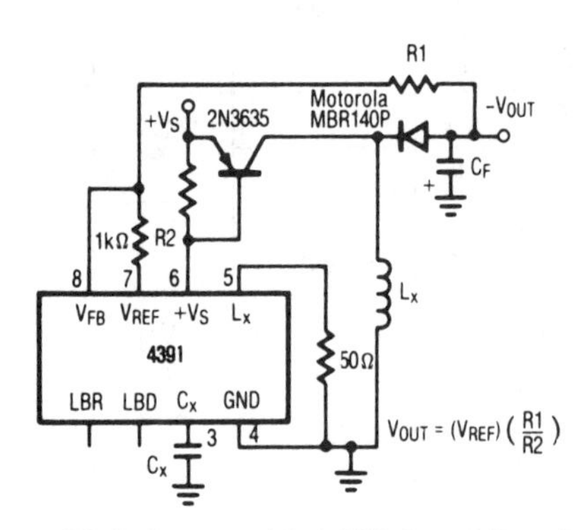

High Current, High Efficiency Supply

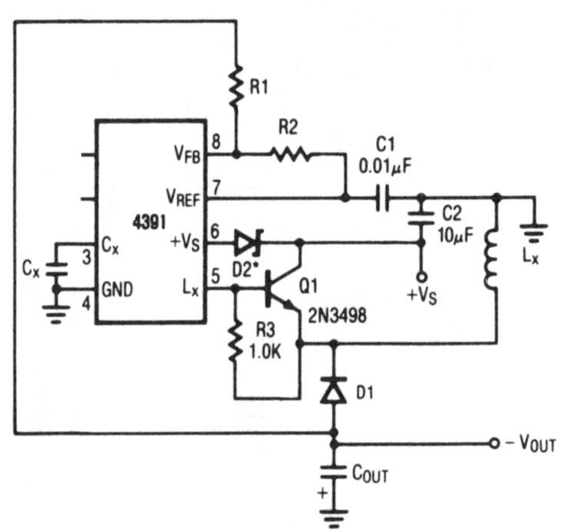

High Voltage/High Power Application Circuit

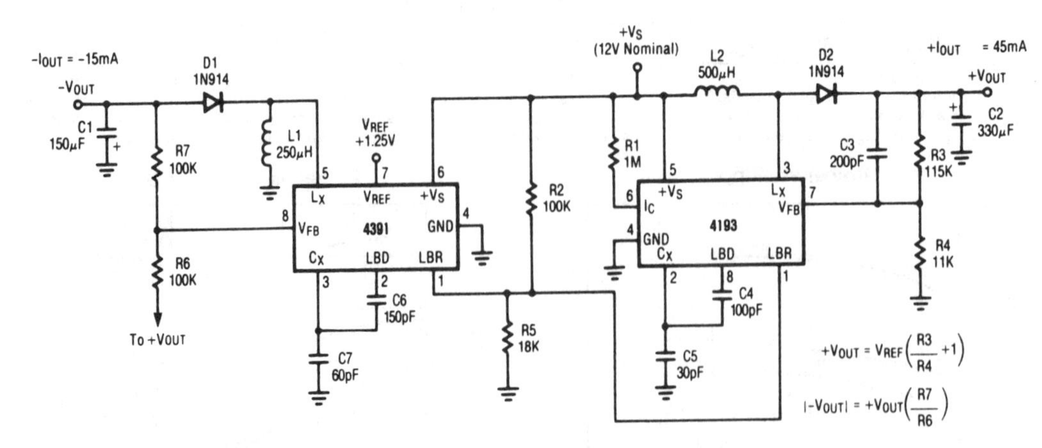

RC4391/RC4193 Power Supply (±15.0V With Values Given)

Raytheon
RC4190
Micropower Switching Regulator

- High efficiency: 85% typical
- Low quiescent current: 215 μA
- Adjustable output: 1.3V to 30V
- High switch current: 150 mA
- Bandgap reference: 1.31V
- Accurate oscillator frequency: ±10%
- Remote shutdown capability
- Low battery detection circuitry
- Low component count
- 8-lead packages including small outline (SO-8)

ABSOLUTE MAXIMUM RATINGS

Supply Voltage (Without External Transistor)
RM4190 .... +30V
RC4190 .... +24V
Storage Temperature
Range .... -65°C to +150°C
Operating Temperature Range
RM4190 .... -55°C to +125°C
RC4190 .... 0°C to +70°C
Switch Current .... 375 mA Peak

CONNECTION INFORMATION

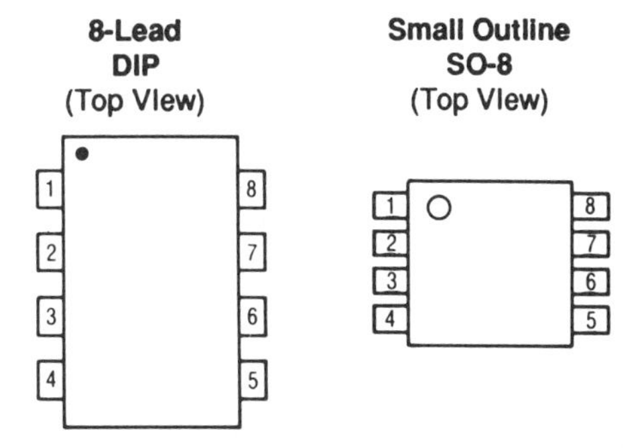

Pin	Function
1	Low Battery (Set) Resistor (LBR)
2	Timing Capacitor (C_X)
3	Ground
4	External Inductor (L_X)
5	+Supply Voltage ($+V_S$)
6	Reference Set Current (I_C)
7	Feedback Voltage (V_{FB})
8	Low Battery Detector Output (LBD)

ORDERING INFORMATION

Part Number	Package	Operating Temperature Range
RC4190M	M	0°C to +70°C
RC4190N	N	0°C to +70°C
RM4190D	D	-55°C to +125°C
RM4190D/883B	D	-55°C to +125°C

Notes:
/883B suffix denotes Mil-Std-883, Level B processing
N = 8-lead plastic DIP
D = 8 lead ceramic DIP
M = 8-lead plastic SOIC
Contact a Raytheon sales office or representative for ordering information on special package/temperature range combinations.

MASK PATTERN

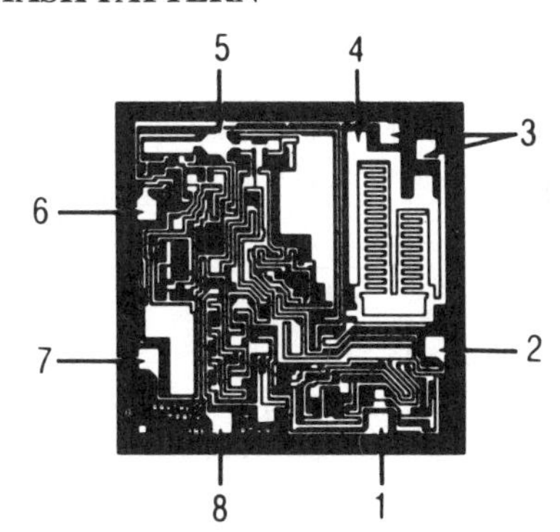

Die Size: 66 x 67 mils
Min. Pad Dimensions: 4 x 4 mils

FUNCTIONAL BLOCK DIAGRAM

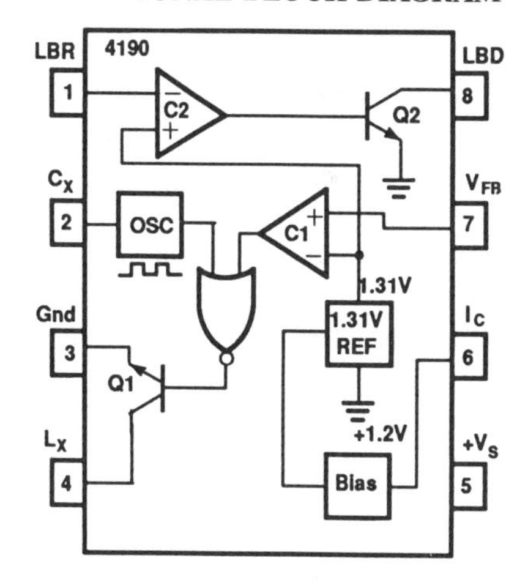

THERMAL CHARACTERISTICS

	8-Lead Plastic DIP	8-Lead Ceramic DIP	Small Outline SO-8
Max. Junction Temp.	+125°C	+175°C	+125°C
Max. P_D T_A <50°C	468 mW	833 mW	240 mW
Therm. Res θ_{JC}	—	45°C/W	—
Therm. Res. θ_{JA}	160°C/W	150°C/W	240°C/W
For T_A >50°C Derate at	6.25 mW/°C	8.33 mW/°C	4.17 mW/°C

ELECTRICAL CHARACTERISTICS

(+V_S = +6.0V, I_C = 5.0 μA, and T_A = +25°C unless otherwise noted)

Parameters	Symbol	Conditions	RM4190			RC4190			Units
			Min	Typ	Max	Min	Typ	Max	
Supply Voltage	$+V_S$		2.2		30	2.2		24	V
Reference Voltage (Internal)	V_{REF}		1.29	1.31	1.33	1.24	1.31	1.38	V
Switch Current	I_{SW}	V_4 = 400 mV	100	200		100	200		mA
Supply Current	I_S	Measure at Pin 5 $I_4 = 0$		215	300		215	300	μA
Efficiency	ef			85			85		%
Line Regulation		$0.5\ V_O < V_S < V_O$		0.04	0.2		0.04	0.5	% V_O
Load Regulation	L_l	$V_S = +0.5\ V_O$, P_L = 150 mW		0.2	0.5		0.2	0.5	% V_O
Operating Frequency Range[1]	F_O		0.1	25	75	0.1	25	75	kHz
Reference Set Current	I_C		1.0	5.0	50	1.0	5.0	50	μA
Switch Leakage Current	I_{CO}	V_4 = 24V		0.01	5.0		0.01	5.0	μA
Supply Current (Disabled)	I_{SO}	$V_C \leq$ 200 mV		0.1	5.0		0.1	5.0	μA
Low Battery Bias Current	I_1	V_1 = 1.2V		0.7			0.7		μA
Capacitor Charging Current	I_{CX}			8.6			8.6		μA
Oscillator Frequency Tolerance				±10			±10		%
Capacitor Threshold Voltage +	$+V_{THX}$			1.4			1.4		V
Capacitor Threshold Voltage –	$-V_{THX}$			0.5			0.5		V
Feedback Input Current	I_{FB}	V_7 = 1.3V		0.1			0.1		μA
Low Battery Output Current	I_{LBD}	V_8 = 0.4V, V_1 = 1.1V	500	1500		500	1500		μA

Note 1. Guaranteed by design.

ELECTRICAL CHARACTERISTICS

($+V_S = +6.0V$, $I_C = 5.0\ \mu A$ unless otherwise noted, over the full operating temperature range)

Parameters	Symbol	Conditions	RM4190			RC4190			Units
			Min	Typ	Max	Min	Typ	Max	
Supply Voltage	$+V_S$		2.6		30	2.6		24	V
Reference Voltage (Internal)	V_{REF}		1.25	1.31	1.37	1.20	1.31	1.42	V
Supply Current	I_S	Measure at Pin 5 $I_4 = 0$		235	350		235	350	μA
Line Regulation		$0.5\ V_O < V_S < V_O$		0.2	0.5		0.5	1.0	% V_O
Load Regulation	L_I	$V_S = +0.5\ V_O$, $P_L = 150$ mW		0.5	1.0		0.5	1.0	% V_O
Reference Set Current	I_C		1.0	5.0	50	1.0	5.0	50	μA
Switch Leakage Current	I_{CO}	$V_4 = 24V$			30			30	μA
Supply Current (Disabled)	I_{SO}	$V_C \leq 200$ mV			30			30	μA
Low Battery Output Current	I_{LBD}	$V_8 = 0.4V$, $V_1 = 1.1V$	500	1200		500	1200		μA
Oscillator Frequency Temperature Drift				±200			±200		ppm/°C

TYPICAL PERFORMANCE CHARACTERISTICS

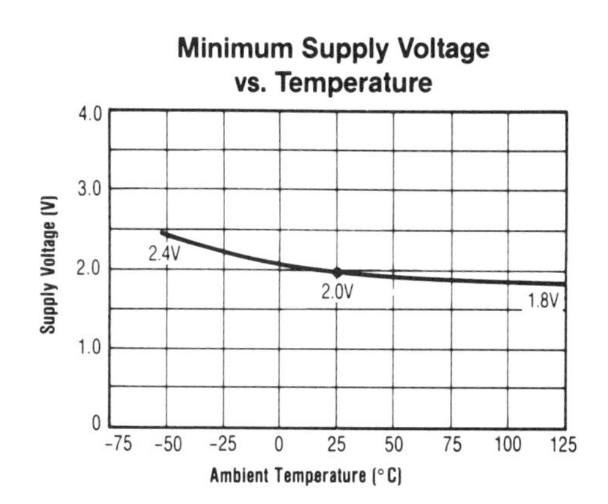

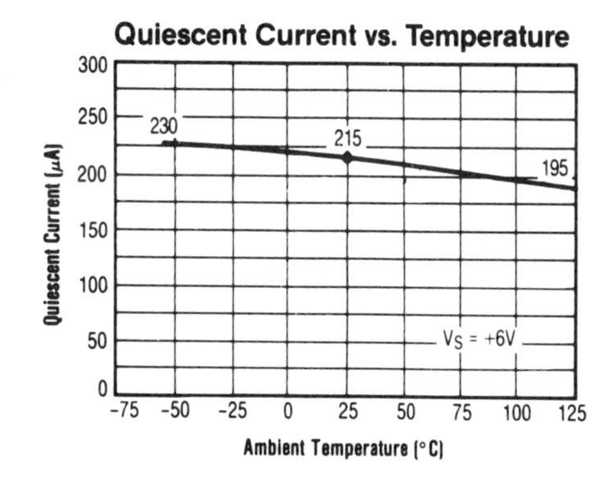

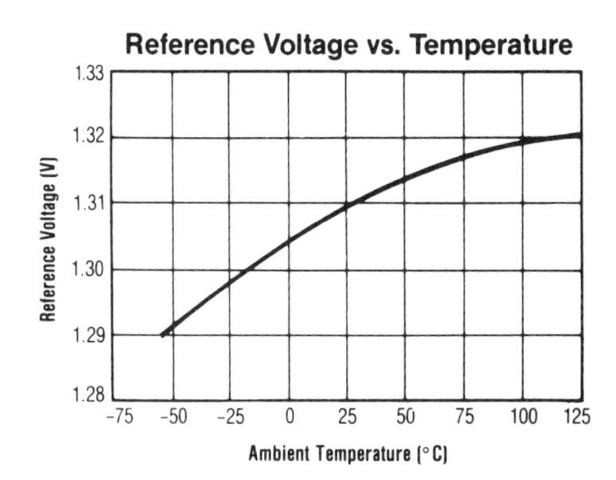

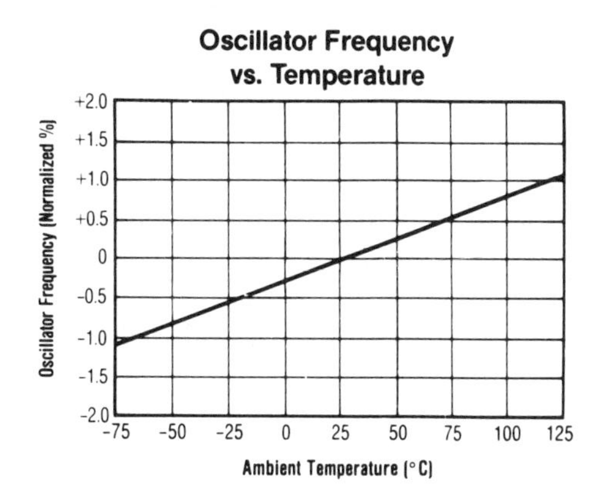

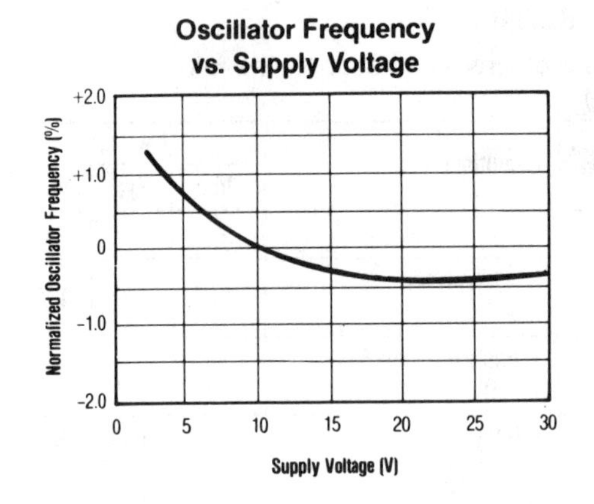

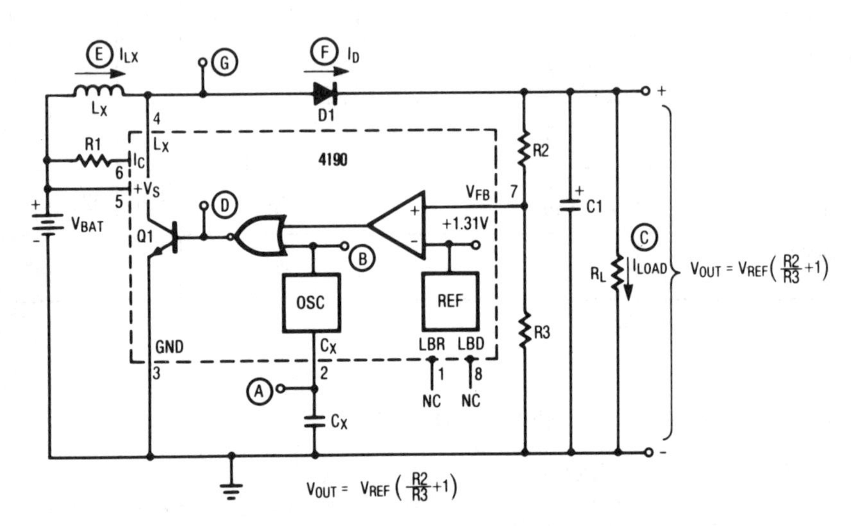

Minimum Step-Up Application

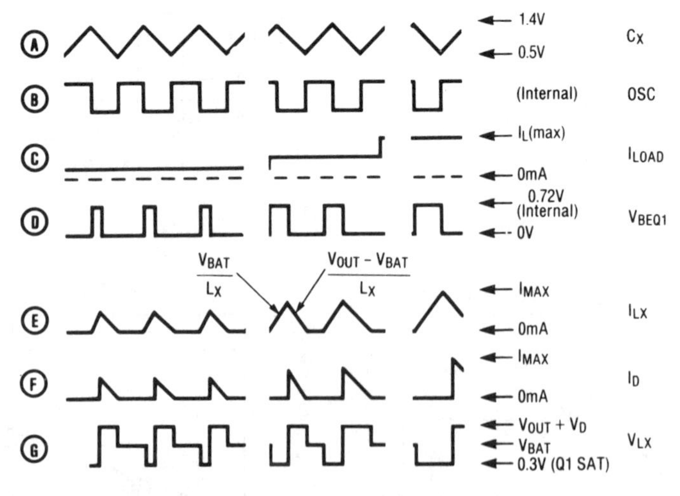

Step-Up Regulator Waveforms

*May not be required

$R5 \approx \frac{50V_S}{I_{MAX}}$ $R4 = 10R5$

High Power Step-Up Application (up to 10W)

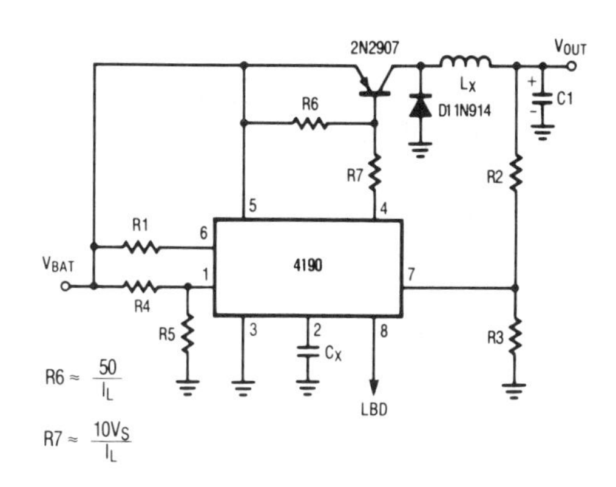

Complete Step-Down Regulator

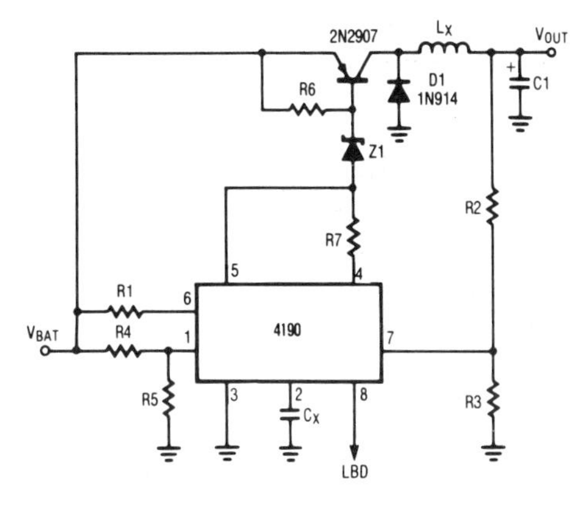

Stepping Down An Input Voltage Greater Than 30V

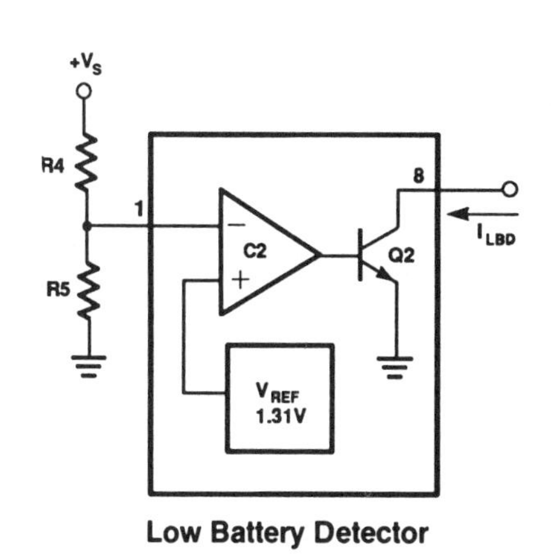

Low Battery Detector

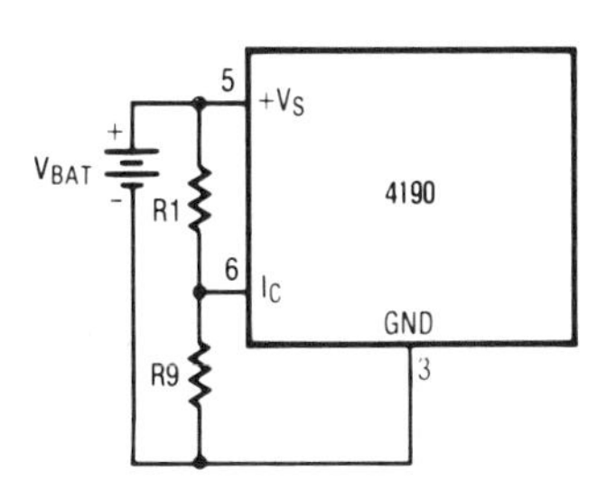

Simple Automatic Shutdown

Battery Back-Up Circuit

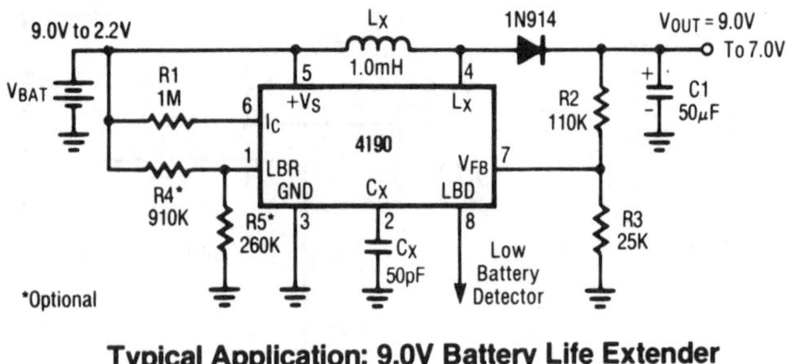

Typical Application: 9.0V Battery Life Extender

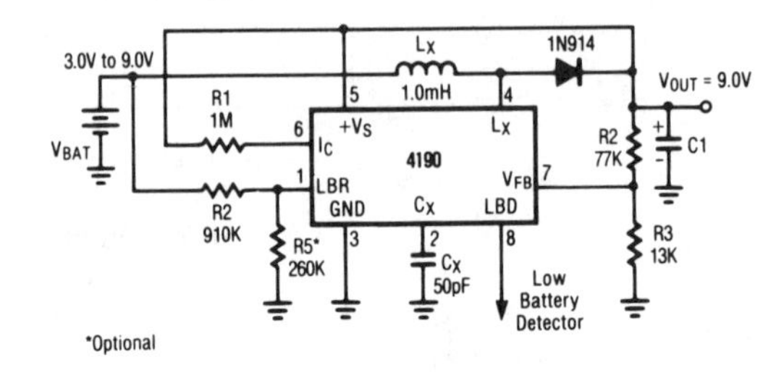

Bootstrapped Operation

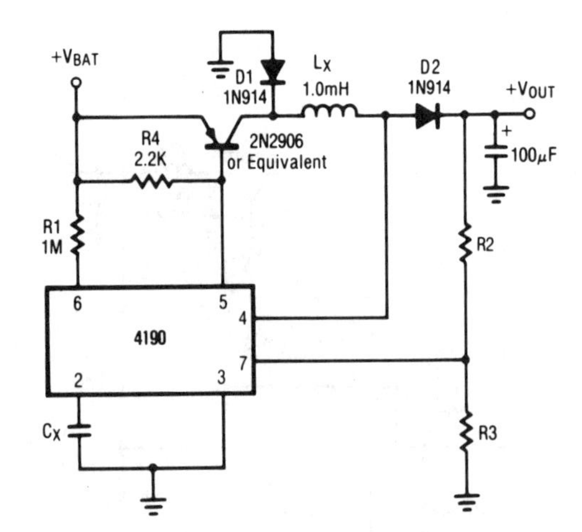

Buck Boost Circuit (V_{BAT} > or < V_{OUT})

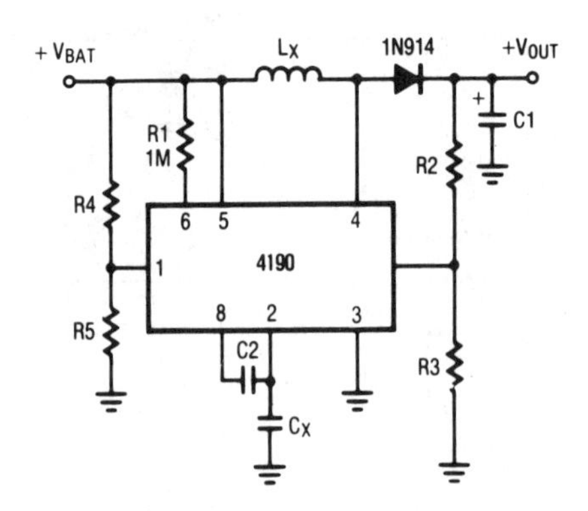

Step-Up Regulator With Voltage-Dependent Oscillator

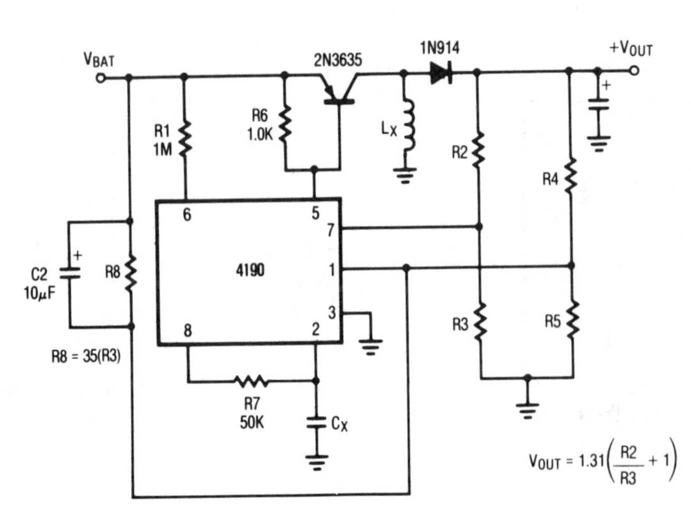

$$V_{OUT} = 1.31\left(\frac{R2}{R3} + 1\right)$$

Step-Down Regulator With Short Circuit Protection

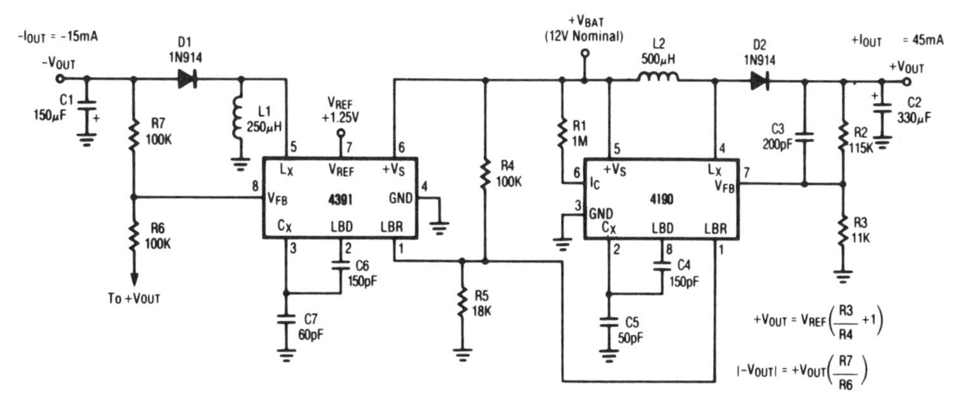

RC4190/4391 Power Supply (±15V With Values Given)

SCHEMATIC DIAGRAM

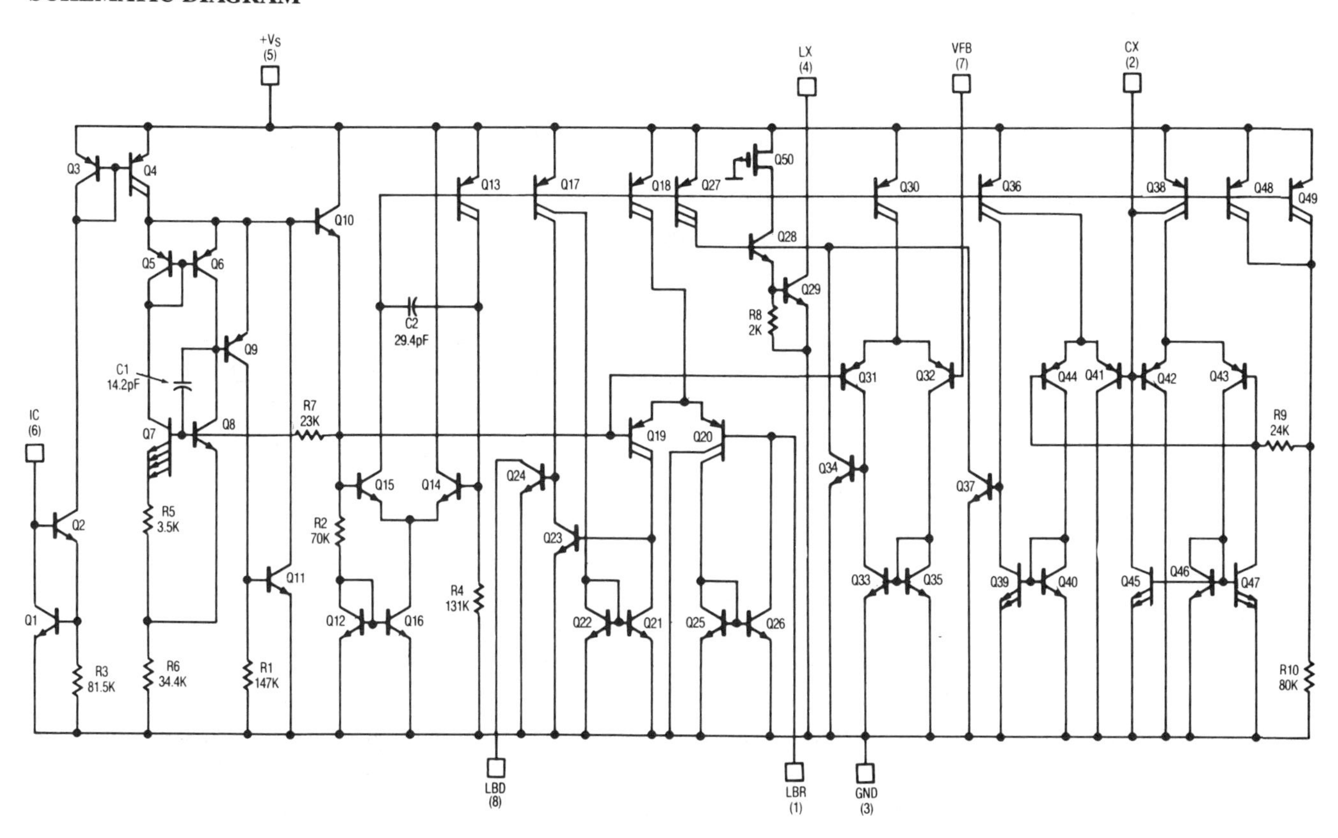

Raytheon RV4143, 4144 Ground Fault Interrupters

- Direct interface to SCR
- Supply voltage derived from ac line—26V shunt
- Adjustable sensitivity
- Grounded neutral fault detection
- Meets U.L. 943 standards

ABSOLUTE MAXIMUM RATINGS

Supply Current 18mA
Internal Power Dissipation 500mW
Storage Temperature
Range .... −65° C to +150° C
Operating Temperature
Range .... −35° C to +80° C
Lead Soldering Temperature
(60 Sec) .... +300° C

THERMAL CHARACTERISTICS

	8-Lead Plastic DIP
Max. Junction Temp.	125° C
Max. P_D $T_A < 50°C$	468mW
Therm. Res. θ_{JC}	—
Therm. Res. θ_{JA}	160° C/W
For $T_A > 50°C$ Derate at	6.25mW per °C

ELECTRICAL CHARACTERISTICS (I_S = 5mA and T_A = +25° C)

Parameters	Test Conditions	Min	Typ	Max	Units
Shunt Regulator					
Zener Shunt Voltage	Pin 6	25	26	29.2	V
Reference Voltage	Pin 3	12.5	13	14.6	V
Op Amp					
Input Offset Voltage	Pin 2 to Pin 3	-3	±1	+3	mV
Output Voltage Swing	Pin 7 to Pin 3	±11	±13.5		V
AC Output Voltage	A_V = 500, f_{IN} = 50kHz V_{IN} = 1mV_{RMS}	50		180	mV_{RMS}
Resistors					
R3		3.8	4.7	5.7	kΩ
R1 RV4143		0.8	1.0	1.2	kΩ
R1 RV4144		0.6	0.75	0.9	kΩ
R2 RV4143		8.0	10.0	12.0	kΩ
R2 RV4144		2.0	2.5	3.0	kΩ
SCR Trigger					
V_{OH}	Across 4.7kΩ	1.5	2.8	6	V
V_{OL}	Across 4.7kΩ		.001	.01	V
Shunt Regulator					
Zener Shunt Voltage	Pin 6	24	26	30	V
Reference Voltage	Pin 3	12	13	15	V
Op Amp					
Input Offset Voltage	Pin 2 to Pin 3	-6	±2	+6	mV
Output Voltage Swing	Pin 7 to Pin 3	±10.5	±13		V
AC Output Voltage	A_V = 500, f_{IN} = 50kHz V_{IN} = 1mV_{RMS}	50		200	mV_{RMS}
Resistors					
R3		3.3	4.7	6.1	kΩ
R1 RV4143		0.7	1.0	1.3	kΩ
R1 RV4144		0.52	0.75	0.98	kΩ
R2 RV4143		7.0	10	13.0	kΩ
R2 RV4144		1.75	2.5	3.25	kΩ
SCR Trigger					
V_{OH}	Across 4.7kΩ	1.3	2.8	5	V
V_{OL}	Across 4.7kΩ		.003	.05	V

CONNECTION INFORMATION

8-Lead Dual In-Line Package (Top View)

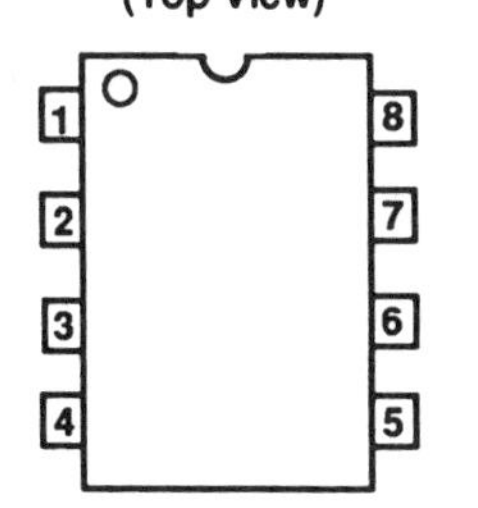

Pin	Function
1	VFB
2	+Input
3	V_{REF} (+13V)
4	Ground
5	SCR Trigger
6	$+V_S$
7	Op Amp Output
8	NC

FUNCTIONAL BLOCK DIAGRAM

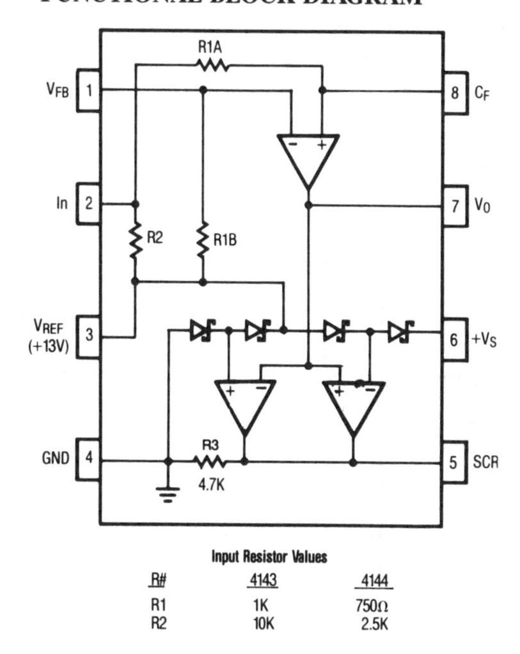

Input Resistor Values

R#	4143	4144
R1	1K	750Ω
R2	10K	2.5K

MASK PATTERN

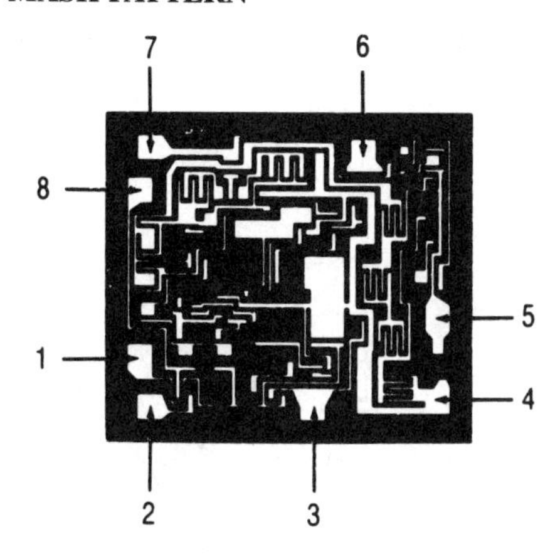

Die Size: 56 x 49 mils
Min. Pad Dimensions: 4 x 4 mils

SCHEMATIC DIAGRAM

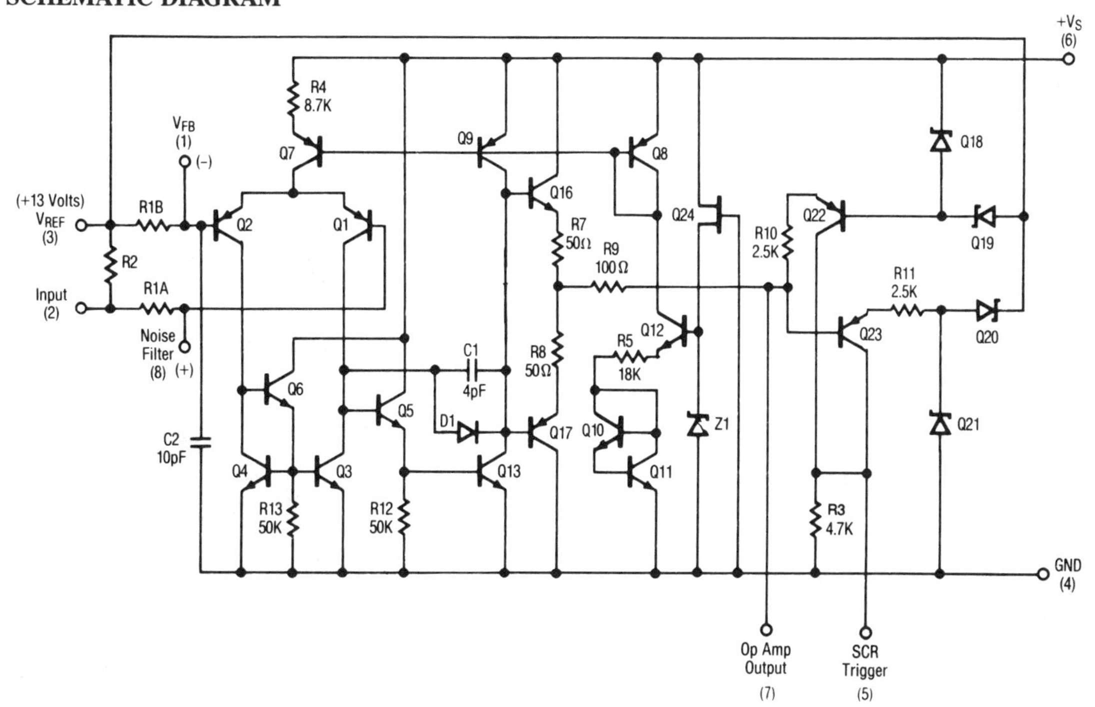

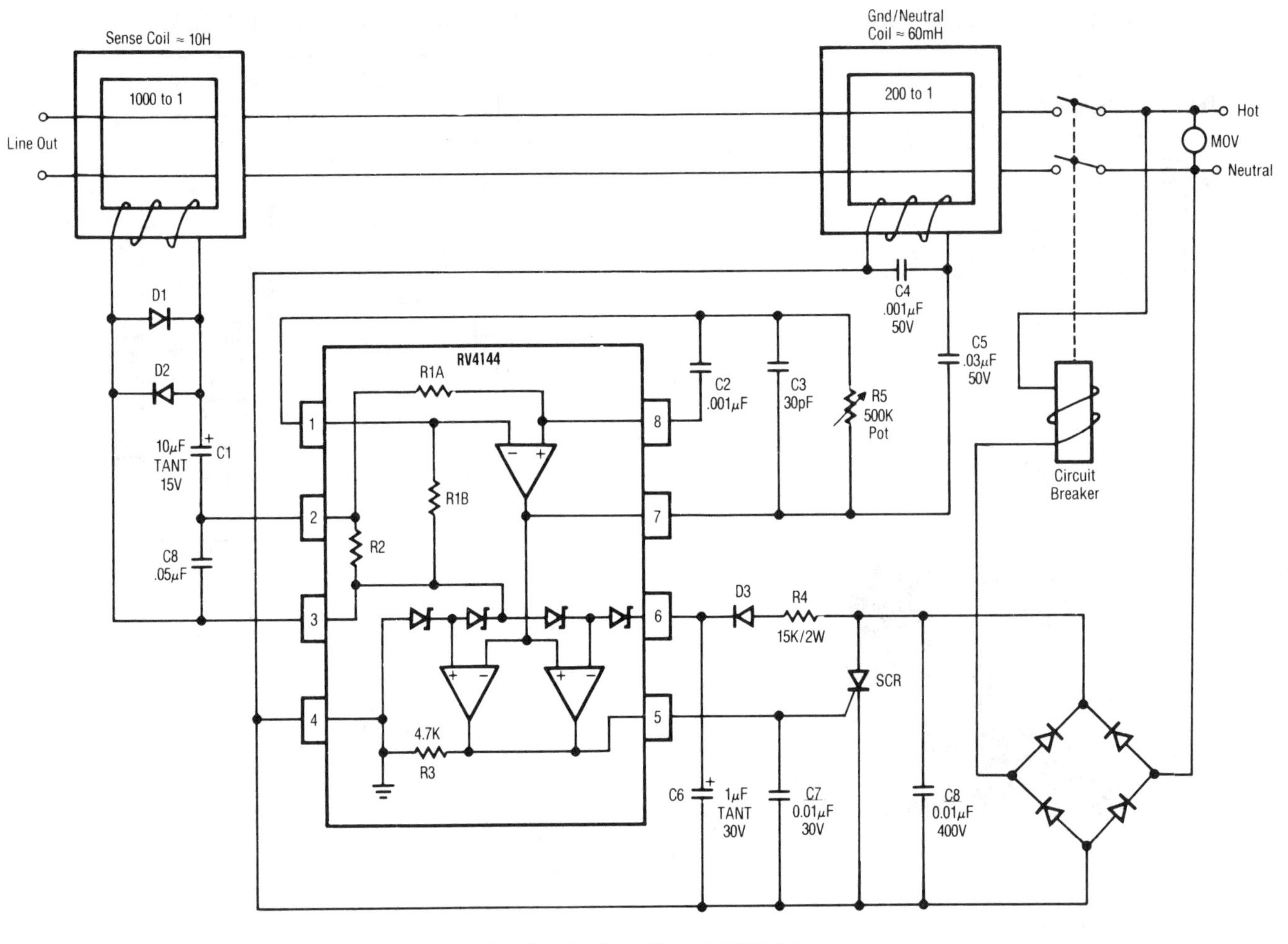

GFI Application Circuit (RV4144)

Raytheon RV4145

Low-Power Ground-Fault Interrupter

- No potentiometer required
- Direct interface to SCR
- Supply voltage derived from ac line—26V shunt
- Adjustable sensitivity
- Grounded neutral fault detection
- Meets U.L. 943 standards
- 450 μA quiescent current
- Ideal for 120V or 220V systems

THERMAL CHARACTERISTICS

	8-Lead Plastic SOIC	8-Lead Plastic DIP
Max. Junction Temp.	+125°C	+125°C
Max. P_D T_A <50°C	300 mW	468 mW
Therm. Res θ_{JC}	—	—
Therm. Res. θ_{JA}	240°C/W	160°C/W
For T_A >50°C Derate at	4.1 mW per °C	6.25 mW per °C

ABSOLUTE MAXIMUM RATINGS

Supply Current .... 18 mA
Internal Power Dissipation .... 500 mW
Storage Temperature
Range .... -65°C to +150°C
Operating Temperature
Range .... -35°C to +80°C
Lead Soldering Temperature
(60 Sec, DIP) .... +300°C
(10 Sec, SO) .... +260°C

MASK PATTERN

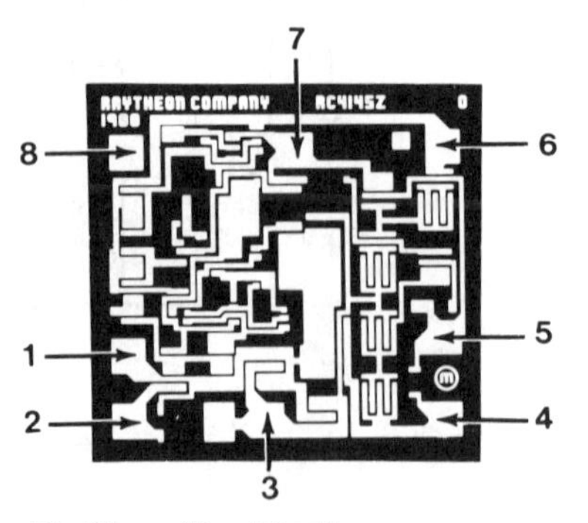

Die Size: 49 x 46 mils
Min. Pad Dimensions: 4 x 4 mils

ELECTRICAL CHARACTERISTICS (I_S = 1.5 mA and T_A = +25°C)

Parameters	Test Conditions	Min	Typ	Max	Units
Shunt Regulator					
Zener Voltage (V_S)	Pin 6 to Pin 4	25	26	29.2	V
Reference Voltage	Pin 3 to Pin 4	12.5	13	14.6	V
Quiescent Current (I_S)	$+V_S$ = 24V		450	750	µA
Operational Amplifier					
Offset Voltage	Pin 2 to Pin 3	-3.0	0.5	+3.0	mV
+Output Voltage Swing	Pin 7 to Pin 3	6.8	7.2	8.1	V
-Output Voltage Swing	Pin 7 to Pin 3	-9.5	-11.2	-13.5	V
+Output Source Current	Pin 7 to Pin 3		650		µA
-Output Sink Current	Pin 7 to Pin 3		1.0		mA
Gain Bandwidth Product	f = 50 kHz	1.0	1.8		MHz
Detector Reference Voltage	Pin 7 to Pin 3	6.8	7.2	8.1	±V
Resistors	I_S = 0 mA				
R1	Pin 2 to Pin 3		10		kΩ
R2	Pin 1 to Pin 3		10		kΩ
R5	Pin 5 to Pin 4	4.0	4.7	5.4	kΩ
SCR Trigger Voltage	Pin 5 to Pin 4				
Detector On		1.5	2.8		V
Detector Off		0	1	10	mV

ELECTRICAL CHARACTERISTICS (I_S = 1.5 mA, < over the specified temperature range)

Parameters	Test Conditions	Min	Typ	Max	Units
Shunt Regulator					
Zener Voltage (V_S)	Pin 6 to Pin 4	24	26	30	V
Reference Voltage	Pin 3 to Pin 4	12	13	15	V
Quiescent Current (I_S)	$+V_S$ = 23V		500		µA
Operational Amplifier					
Offset Voltage	Pin 2 to Pin 3	-5.0	0.5	+5.0	mV
+Output Voltage Swing	Pin 7 to Pin 3	6.5	7.2	8.3	V
-Output Voltage Swing	Pin 7 to Pin 3	-9	-11.2	-14	V
Gain Bandwidth Product	f = 50 kHz		1.8		MHz
Detector Reference Voltage	Pin 7 to Pin 3	6.5	7.2	8.3	±V
Resistors	I_S = 0 mA				
R1	Pin 2 to Pin 3		10		kΩ
R2	Pin 1 to Pin 3		10		kΩ
R5	Pin 5 to Pin 4	3.8	4.7	5.6	kΩ
SCR Trigger Voltage	Pin 5 to Pin 4				
Detector On		1.3	2.8		V
Detector Off		0	3	50	mV

FUNCTIONAL BLOCK DIAGRAM

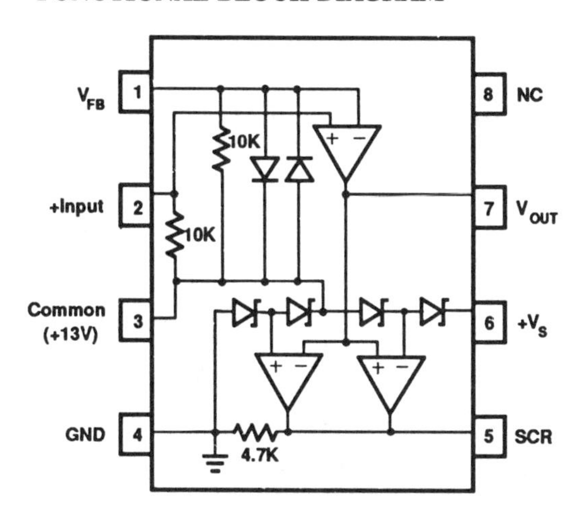

CONNECTION INFORMATION

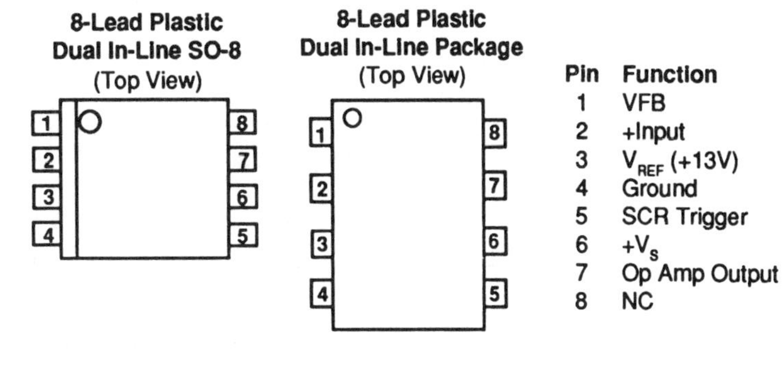

SCHEMATIC DIAGRAM

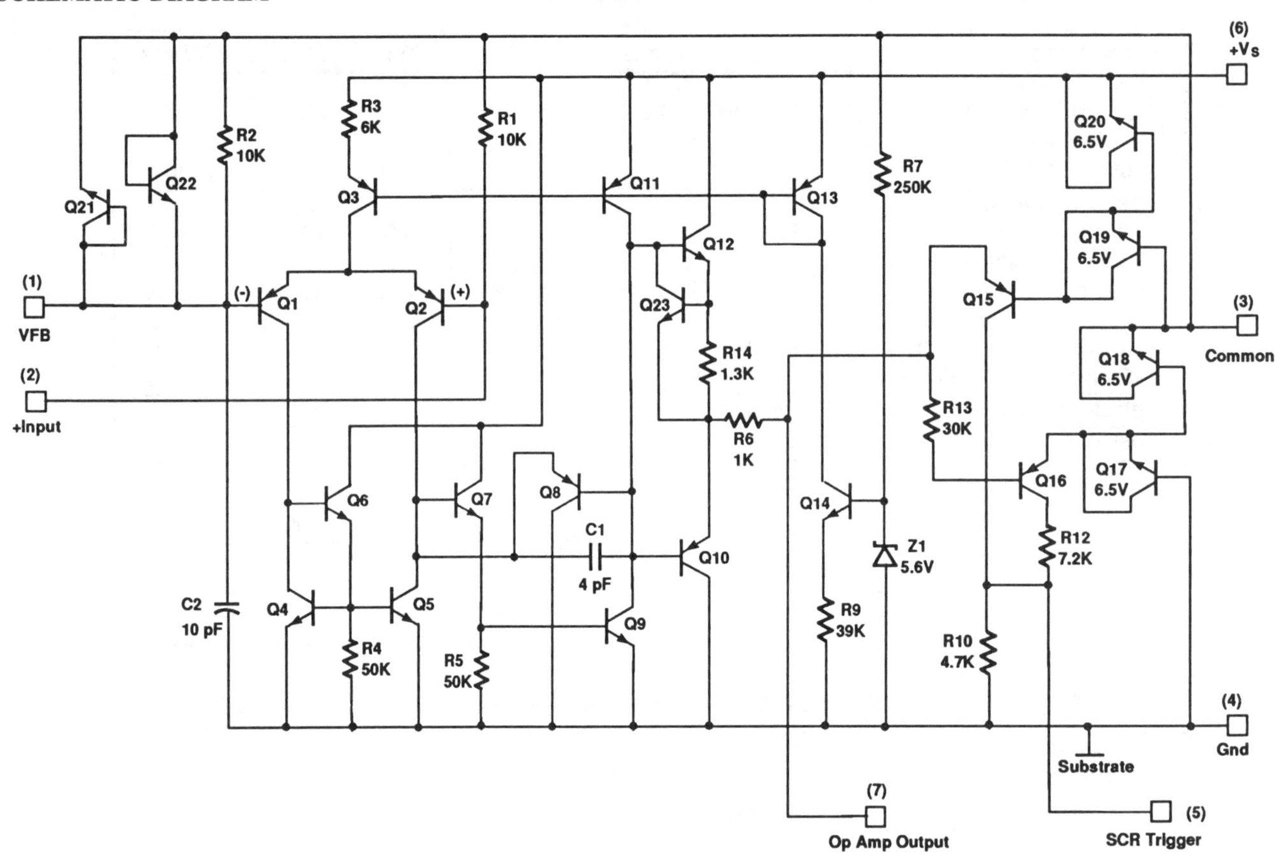

GFI APPLICATION CIRCUIT

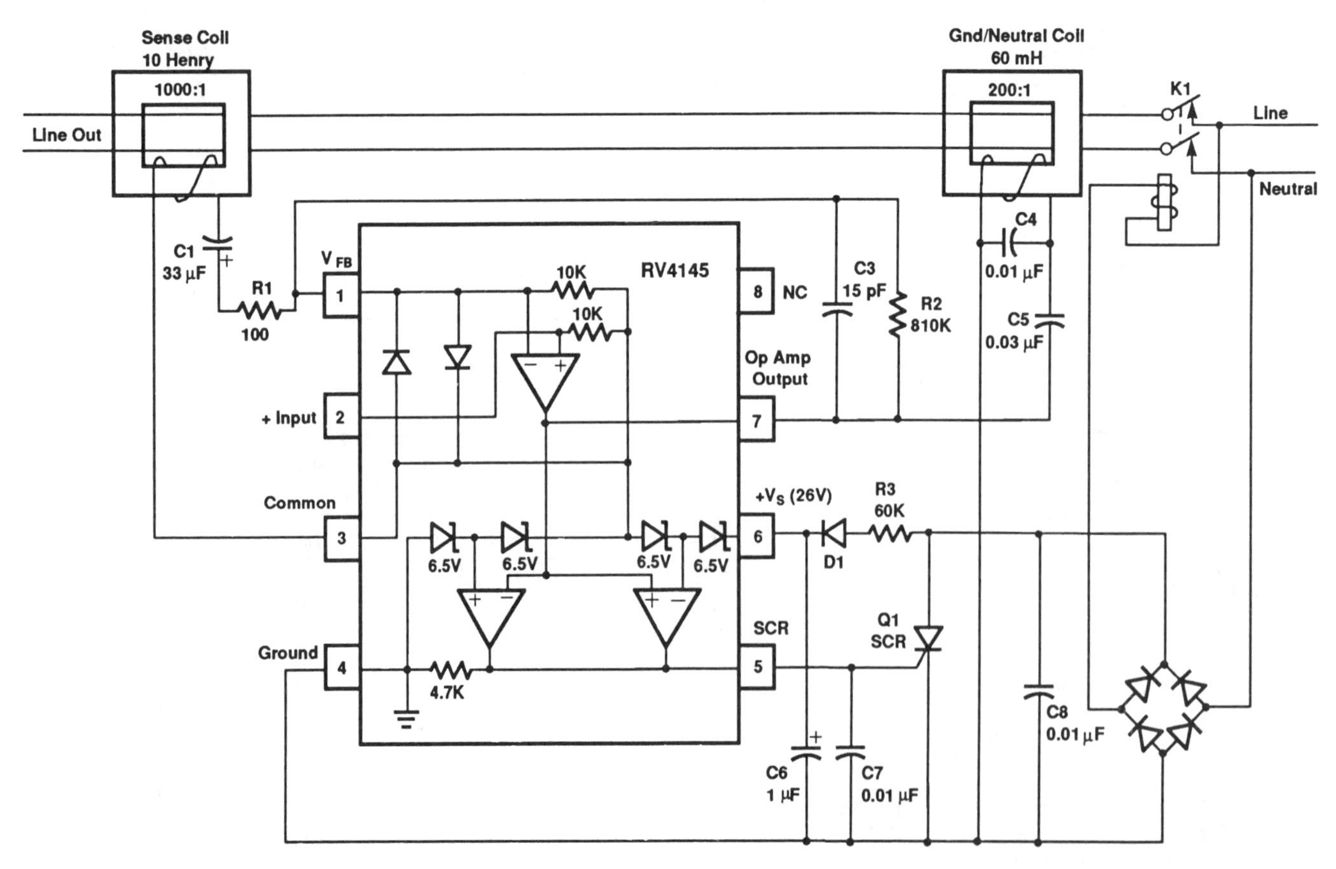

Raytheon
REF-01
+10V Precision Voltage References

- +10V output: ±0.3%
- Adjustable: ±3%
- Excellent temperature stability: 3 ppm/°C
- Low noise: 20 $\mu V_{p\text{-}p}$
- Wide input voltage range: +12V to +40V
- No external components
- Short-circuit proof
- Low power consumption: 15 mW

CONNECTION INFORMATION

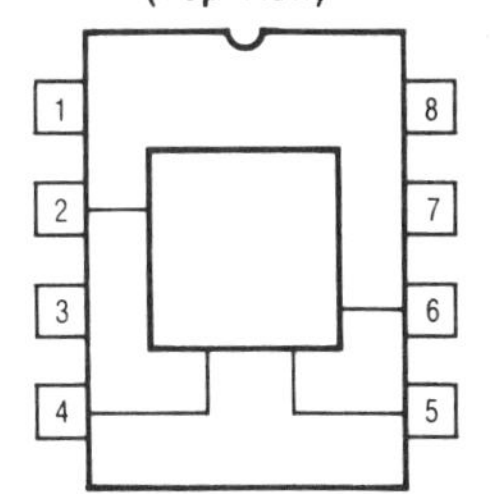

Pin	Function
1	NC
2	$+V_S$
3	NC
4	Ground
5	Adjust
6	Output
7	NC
8	NC

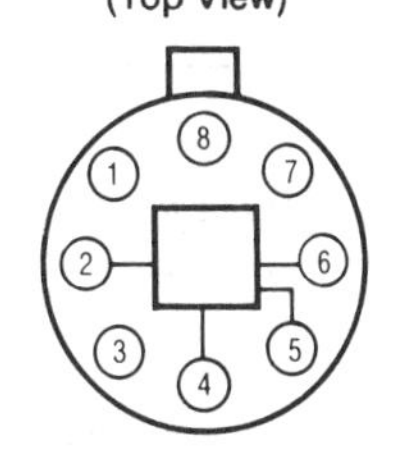

ORDERING INFORMATION

Part Number	Package	Operating Temperature Range
REF-01CD	D	0°C to +70°C
REF-01DD	D	0°C to +70°C
REF-01ED	D	0°C to +70°C
REF-01HD	D	0°C to +70°C
REF-01CN	N	0°C to +70°C
REF-01DN	N	0°C to +70°C
REF-01EN	N	0°C to +70°C
REF-01HN	N	0°C to +70°C
REF-01CT	T	0°C to +70°C
REF-01DT	T	0°C to +70°C
REF-01ET	T	0°C to +70°C
REF-01HT	T	0°C to +70°C
REF-01AD	D	-55°C to +125°C
REF-01AD/883B	D	-55°C to +125°C
REF-01D	D	-55°C to +125°C
REF-01D/883B	D	-55°C to +125°C
REF-01AT	T	-55°C to +125°C
REF-01AT/883B	T	-55°C to +125°C
REF-01T	T	-55°C to +125°C
REF-01T/883B	T	-55°C to +125°C

Notes:
/883B suffix denotes Mil-Std-883, Level B processing
N = 8-lead plastic DIP
D = 8-lead ceramic DIP
T = 8-lead metal can (TO-99)
Contact a Raytheon sales office or representative for ordering information on special package/temperature range combinations.

MASK PATTERN

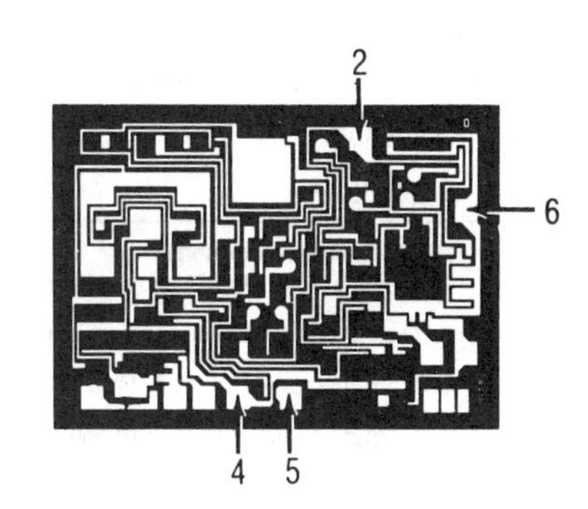

Die Size: 55 x 79 mils
Min. Pad Dimensions: 4 x 4 mils

ABSOLUTE MAXIMUM RATINGS

Supply Voltage
- REF-01A, E, H Grades .... +40V
- REF-01C, D Grades .... +30V

Internal Power Dissipation .... 500 mW
Output Short Circuit Duration .... Indefinite
Storage Temperature
- Range .... -65°C to +150°C

Operating Temperature Range
- REF-01A, -01 .... -55°C to +125°C
- REF-01E,H,C,D .... 0°C to +70°C

Lead Soldering Temperature
- (60 Sec) .... +300°C

THERMAL CHARACTERISTICS

	8-Lead Ceramic DIP	8-Lead TO-99 Metal Can	8-Lead Plastic DIP
Max. Junction Temp.	+175°C	+175°C	+125°C
Max. P_D T_A <50°C	833 mW	658 mW	468 mW
Therm. Res θ_{JC}	45°C/W	50°C/W	—
Therm. Res. θ_{JA}	150°C/W	190°C/W	160°C/W
For T_A >50°C Derate at	8.33 mW/°C	5.26 mW/°C	6.25 mW/°C

ELECTRICAL CHARACTERISTICS (V_S = +15V and T_A = +25°C unless otherwise noted)

Parameters	Test Conditions	REF-01A/E			REF-01/H			Units
		Min	Typ	Max	Min	Typ	Max	
Output Voltage	I_L = 0mA	9.97	10.00	10.03	9.95	10.00	10.05	V
Output Adjustment Range	R_P = 10k Ω	±3.0	±3.3		±3.0	±3.3		%
Output Voltage Noise[1]	0.1Hz to 10Hz		20	30		20	30	μV_{p-p}
Supply Voltage		12		40	12		40	V
Line Regulation[2]	V_S = +13V to +33V		0.006	0.010		0.006	0.010	%/V
Load Regulation[2]	I_L = 0mA to 10mA		0.005	0.008		0.006	0.010	%/mA
Turn-on Settling Time	To ±0.1% of Final Value		5.0			5.0		μS
Supply Current	No Load		1.0	1.4		1.0	1.4	mA
Load Current		10	21		10	21		mA
Sink Current		-0.3	-0.5		-0.3	-0.5		mA
Short Circuit Current	V_O = 0		30			30		mA

ELECTRICAL CHARACTERISTICS (V_S = +15V and −55°C ≤ T_A ≤ +125°C unless otherwise noted)

Parameters	Test Conditions	REF-01A			REF-01			Units
		Min	Typ	Max	Min	Typ	Max	
Output Voltage Change With Temperature[3 4]	Over Temp. Range		0.06	0.15		0.18	0.45	%
Output Voltage Temperature Coefficient[5]	Over Temp. Range		3.0	8.5		10	25	ppm/°C
Change in V_{OUT} Temperature Coefficient With Output Adjustment	R_P = 10k Ω		0.7			0.7		ppm/%
Line Regulation[2]	V_S = +13V to +33V		0.009	0.015		0.009	0.015	%/V
Load Regulation[2]	I_O = 0mA to 8mA		0.007	0.012		0.007	0.012	%/mA

Notes:
1. Guaranteed by design.
2. Line and load regulation specifications include the effects of self heating.
3. Output voltage change with temperature = $\frac{V_{MAX} - V_{MIN}}{10V} \times 100\%$
4. Output voltage change with temperature specification applies untrimmed, or trimmed to +10V.
5. Output voltage temperature coefficient = $\frac{\text{Output voltage change with temperature}}{180°C}$

ELECTRICAL CHARACTERISTICS (V_S = +15V and T_A = +25°C unless otherwise noted)

Parameters	Test Conditions	REF-01C			REF-01D			Units
		Min	Typ	Max	Min	Typ	Max	
Output Voltage	I_L = 0mA	9.90	10.00	10.10	9.850	10.00	10.150	V
Output Adjustment Range	R_P = 10kΩ	±2.7	±3.3		±2.0	±3.3		%
Output Voltage Noise[1]	0.1Hz to 10Hz		25	35		25		μV_{p-p}
Supply Voltage		12		30	12		30	V
Line Regulation[2]	V_S = +13V to +33V		0.009	0.015		0.012	0.04	%/V
Load Regulation[2]	I_L = 0mA to 8mA		0.006	0.015				%/mA
	I_L = 0mA to 4mA		0.006	0.015		0.009	0.04	
Turn-on Settling Time	To ±0.1% of Final Value		5.0			5.0		μS
Supply Current	No Load		1.0	1.6		1.0	2.0	mA
Load Current		8.0	21		8.0	21		mA
Sink Current		-0.2	-0.5		-0.2	-0.5		mA
Short Circuit Current	V_O = 0		30			30		mA

ELECTRICAL CHARACTERISTICS

(V_S = +15V, 0°C ≤ T_A ≤ +70°C, and I_O = 0 unless otherwise noted)

		REF-01E			REF-01H			
Parameters	**Test Conditions**	**Min**	**Typ**	**Max**	**Min**	**Typ**	**Max**	**Units**
Output Voltage Change With Temperature[3 4]	Over Temp. Range		0.02	0.06		0.07	0.17	%
Output Voltage Temperature Coefficient[5]	Over Temp. Range		3.0	8.5		10	25	ppm/°C
Change in V_{OUT} Temperature Coefficient With Output Adjustment	R_P = 10kΩ		0.7			0.7		ppm/%
Line Regulation[2]	V_S = +13V to +33V		0.007	0.012		0.007	0.012	%/V
Load Regulation[2]	I_L = 0mA to 8mA		0.006	0.010		0.007	0.012	%/mA

ELECTRICAL CHARACTERISTICS

(V_S = +15V, 0°C ≤ T_A ≤ +70°C, and I_O = 0 unless otherwise noted)

		REF-01C			REF-01D			
Parameters	**Test Conditions**	**Min**	**Typ**	**Max**	**Min**	**Typ**	**Max**	**Units**
Output Voltage Change With Temperature[3 4]	Over Temp. Range		0.14	0.45		0.49	1.7	%
Output Voltage Temperature Coefficient[5]	Over Temp. Range		20	65		70	250	ppm/°C
Change in V_{OUT} Temperature Coefficient With Output Adjustment	R_P = 10kΩ		0.7			0.7		ppm/%
Line Regulation[2]	V_S = +13V to +30V		0.011	0.018		0.020	0.025	%/V
Load Regulation[2]	I_O = 0mA to 5mA		0.008	0.018		0.020	0.025	%/mA

Notes:
1. Guaranteed by design.
2. Line and load regulation specifications include the effects of self heating.
3. Output voltage change with temperature = $\frac{V_{MAX} - V_{MIN}}{10V} \times 100\%$
4. Output voltage change with temperature specification applies untrimmed, or trimmed to +10V.
5. Output voltage temperature coefficient = $\frac{\text{Output voltage change with temperature}}{70°C}$

TYPICAL PERFORMANCE CHARACTERISTICS

Maximum Load Current vs. Differential Input Voltage

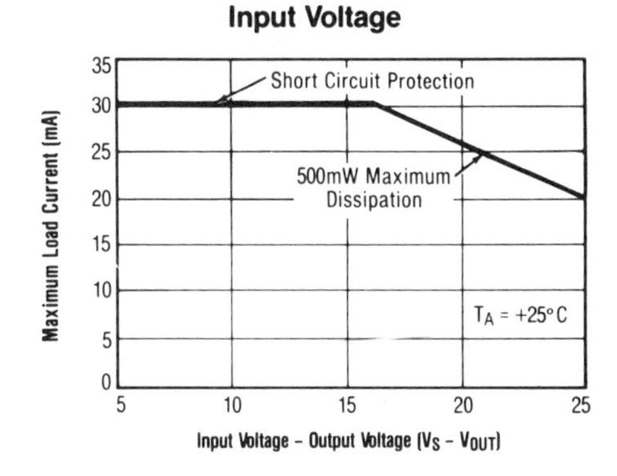

Normalized Load Regulation (ΔIL = 10mA) vs. Temperature

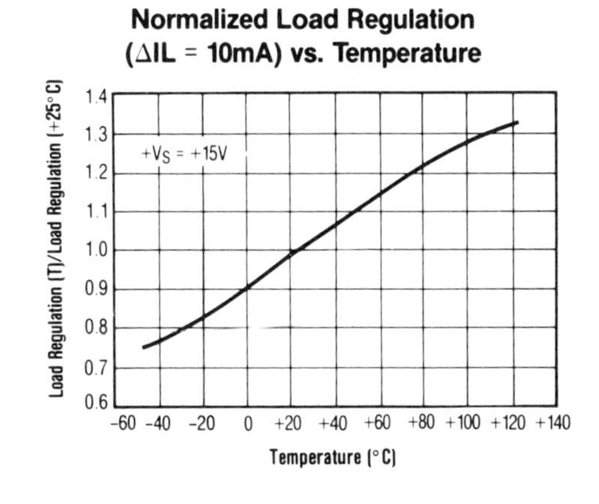

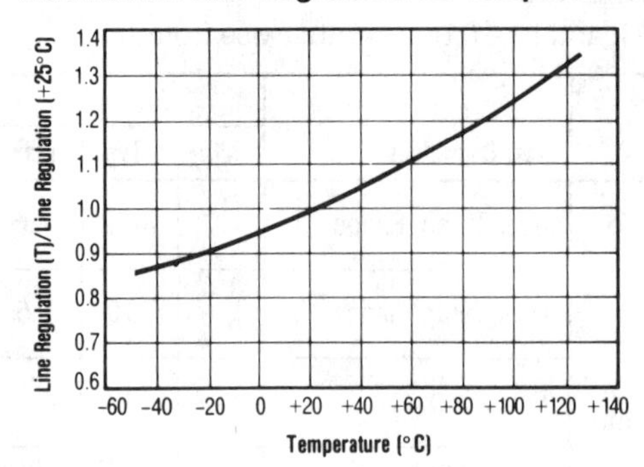

SIMPLIFIED SCHEMATIC DIAGRAM

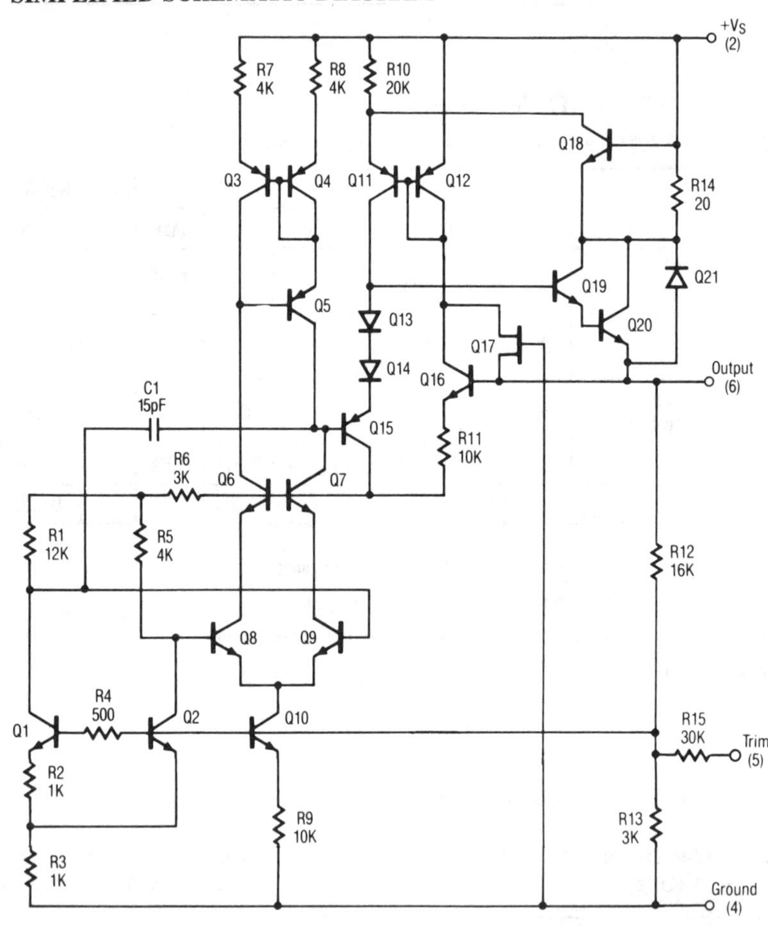

TYPICAL APPLICATIONS

Current Source

+15V

2

+Vs

6

Vo

REF-01

Trim

5

GND

4

R

I_{OUT}

Voltage Compliance: -25V to +3V

$I_{OUT} = \frac{V_{OUT}}{R} + 1mA$

Current Sink

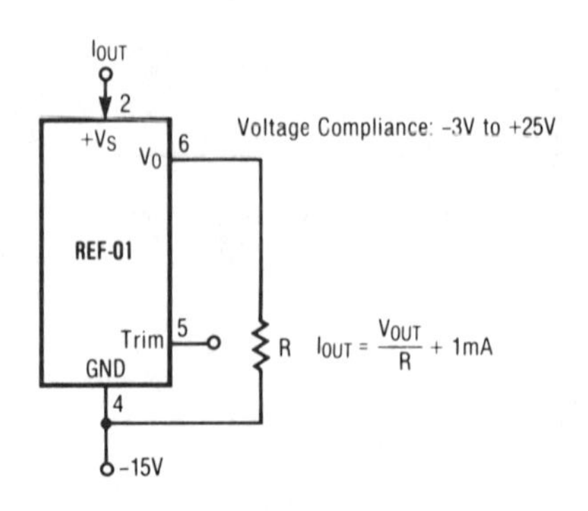

Raytheon REF-02

+5V Precision Voltage References

- +5V output: ±0.3%
- Adjustable: ±3%
- Excellent temperature stability: 3 ppm/°C
- Low noise: 10 $\mu V_{p\text{-}p}$
- Wide input voltage range: +7V to +40V
- No external components
- Short-circuit proof
- Low power consumption: 10 mW

ORDERING INFORMATION

Part Number	Package	Operating Temperature Range
REF-02CD	D	0°C to +70°C
REF-02DD	D	0°C to +70°C
REF-02ED	D	0°C to +70°C
REF-02HD	D	0°C to +70°C
REF-02CN	N	0°C to +70°C
REF-02DN	N	0°C to +70°C
REF-02EN	N	0°C to +70°C
REF-02HN	N	0°C to +70°C
REF-02CT	T	0°C to +70°C
REF-02DT	T	0°C to +70°C
REF-02ET	T	0°C to +70°C
REF-02HT	T	0°C to +70°C
REF-02AD	D	-55°C to +125°C
REF-02AD/883B	D	-55°C to +125°C
REF-02D	D	-55°C to +125°C
REF-02D/883B	D	-55°C to +125°C
REF-02AT	T	-55°C to +125°C
REF-02AT/883B	T	-55°C to +125°C
REF-02T	T	-55°C to +125°C
REF-02T/883B	T	-55°C to +125°C

Notes:
/883B suffix denotes Mil-Std-883, Level B processing
N = 8-lead plastic DIP
D = 8-lead ceramic DIP
T = 8-lead metal can (TO-99)
Contact a Raytheon sales office or representative for ordering information on special package/temperature range combinations.

THERMAL CHARACTERISTICS

	8-Lead Ceramic DIP	8-Lead TO-99 Metal Can	8-Lead Plastic DIP
Max. Junction Temp.	+175°C	+175°C	+125°C
Max. P_D T_A <50°C	833 mW	658 mW	468 mW
Therm. Res θ_{JC}	45°C/W	50°C/W	—
Therm. Res. θ_{JA}	150°C/W	190°C/W	160°C/W
For T_A >50°C Derate at	8.33 mW/°C	5.26 mW/°C	6.25 mW/°C

ABSOLUTE MAXIMUM RATINGS

Supply Voltage
REF-02A, E, H Grades ....+40V
REF-02C, D Grades ....+30V
Internal Power Dissipation ....500 mW
Output Short Circuit Duration ....Indefinite
Storage Temperature
Range ....-65°C to +150°C
Operating Temperature Range
REF-02A, -01 ....-55°C to +125°C
REF-02E,H,C,D ....0°C to +70°C
Lead Soldering Temperature
(60 Sec) ....+300°C

CONNECTION INFORMATION

8-Lead Dual In-Line Package (Top View)

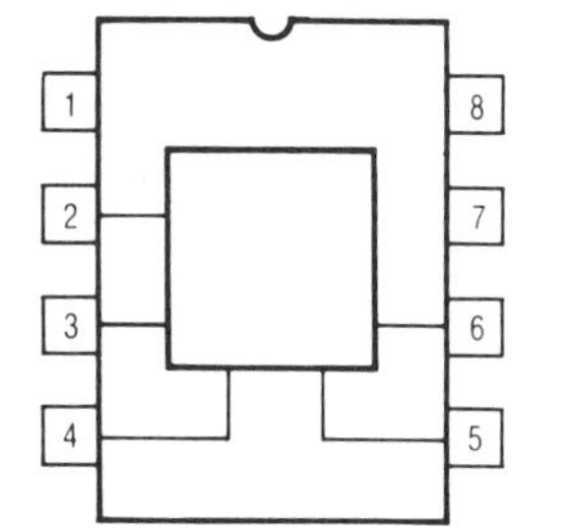

Pin	Function
1	NC
2	$+V_S$
3	Tempco
4	Ground
5	Adjust
6	Output
7	NC
8	NC

8-Lead TO-99 Metal Can (Top View)

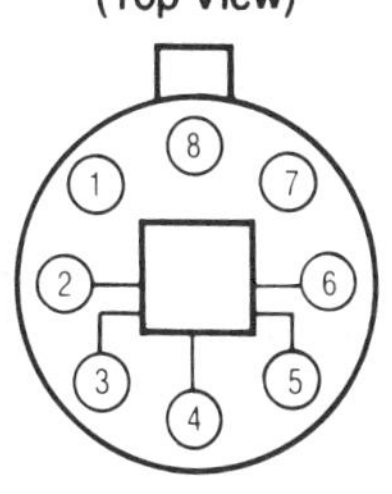

MASK PATTERN

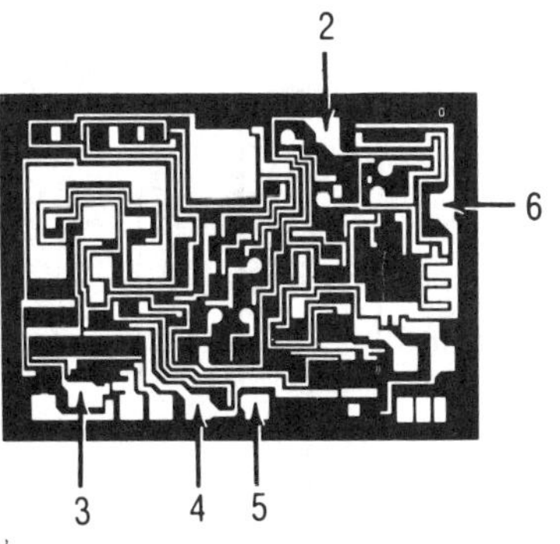

Die Size: 79 x 55 mils
Min. Pad Dimensions: 4 x 4 mils

ELECTRICAL CHARACTERISTICS (V_S = +15V and T_A = +25°C unless otherwise noted)

Parameters	Test Conditions	REF-02A/E			REF-02/H			Units
		Min	Typ	Max	Min	Typ	Max	
Output Voltage	I_L = 0mA	4.985	5.000	5.015	4.975	5.000	5.025	V
Output Adjustment Range	R_P = 10kΩ	±3.0	±6.0		±3.0	±6.0		%
Output Voltage Noise[1]	0.1Hz to 10Hz		10	15		10	15	μV_{p-p}
Supply Voltage		7		40	7		40	V
Line Regulation[2]	V_S = +8V to +33V		0.006	0.010		0.006	0.010	%/V
Load Regulation[2]	I_L = 0mA to 10mA		0.005	0.010		0.006	0.010	%/mA
Turn-on Settling Time	To ±0.1% of Final Value		5.0			5.0		µS
Supply Current	No Load		1.0	1.4		1.0	1.4	mA
Load Current		10	21		10	21		mA
Sink Current		-0.3	-0.5		-0.3	-0.5		mA
Short Circuit Current	V_O = 0		30			30		mA
Tempco Voltage Output[6]			630			630		mV

ELECTRICAL CHARACTERISTICS (V_S = +15V and −55°C ≤ T_A ≤ +125°C unless otherwise noted)

Parameters	Test Conditions	REF-02A			REF-02			Units
		Min	Typ	Max	Min	Typ	Max	
Output Voltage Change With Temperature[3 4]	Over Temp. Range		0.06	0.15		0.18	0.45	%
Output Voltage Temperature Coefficient[5]	Over Temp. Range		3.0	8.5		10	25	ppm/°C
Change in V_{OUT} Temperature Coefficient With Output Adjustment	R_P = 10kΩ		0.7			0.7		ppm/%
Line Regulation[2]	V_S = +8V to +33V		0.009	0.015		0.009	0.015	%/V
Load Regulation[2]	I_O = 0mA to 8mA		0.007	0.012		0.007	0.012	%/mA
Tempco Voltage Output Temperature Coefficient[6]			2.1			2.1		mV/°C

Notes:
1. Guaranteed by design.
2. Line and load regulation specifications include the effects of self heating.
3. Output voltage change with temperature = $\frac{V_{MAX} - V_{MIN}}{5V} \times 100\%$
4. Output voltage change with temperature specification applies untrimmed, or trimmed to +5V.
5. Output voltage temperature coefficient = $\frac{\text{Output voltage change with temperature}}{180°C}$
6. Limit current in or out of pin 3 to 50nA and limit capacitance on pin 3 to 30pF.

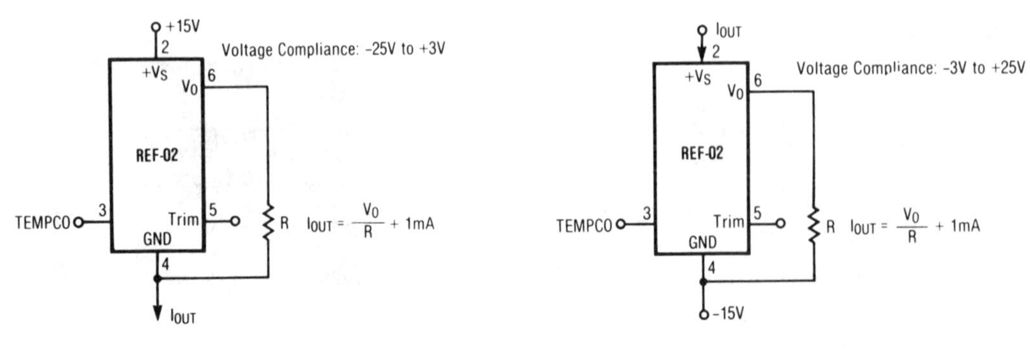

Current Source **Current Sink**

ELECTRICAL CHARACTERISTICS (V_S = +15V and T_A = +25°C unless otherwise noted)

Parameters	Test Conditions	REF-02C			REF-02D			Units
		Min	Typ	Max	Min	Typ	Max	
Output Voltage	I_L = 0mA	4.950	5.000	5.050	4.900	5.000	5.100	V
Output Adjustment Range	R_P = 10kΩ	±2.7	±6.0		±2.0	±6.0		%
Output Voltage Noise[1]	0.1Hz to 10Hz		12	18		12		μV_{p-p}
Supply Voltage		7.0		30	7.0		30	V
Line Regulation[2]	V_S = +8V to +33V		0.009	0.015		0.012	0.04	%/V
Load Regulation[2]	I_L = 0mA to 8mA		0.006	0.015				%/mA
	I_L = 0mA to 4mA					0.009	0.04	
Turn-on Settling Time	To ±0.1% of Final Value		5.0			5.0		µS
Supply Current	No Load		1.0	1.6		1.0	2.0	mA
Load Current		8.0	21		8.0	21		mA
Sink Current		-0.2	-0.5		-0.2	-0.5		mA
Short Circuit Current	V_O = 0		30			30		mA
Tempco Voltage Output[6]			630			630		mV

ELECTRICAL CHARACTERISTICS

(V_S = +15V, 0°C ≤ T_A ≤ +70°C and I_O = 0 unless otherwise noted)

Parameters	Test Conditions	REF-02E			REF-02H			Units
		Min	Typ	Max	Min	Typ	Max	
Output Voltage Change With Temperature[3 4]	Over Temp. Range		0.02	0.06		0.07	0.17	%
Output Voltage Temperature Coefficient[5]	Over Temp. Range		3.0	8.5		10	25	ppm/°C
Change in V_{OUT} Temperature Coefficient With Output Adjustment	R_P = 10kΩ		0.7			0.7		ppm/%
Line Regulation[2]	V_S = +8V to +33V		0.007	0.012		0.007	0.012	%/V
Load Regulation[2]	I_L = 0mA to 8mA		0.006	0.010		0.007	0.012	%/mA
Tempco Voltage Output Temperature Coefficient[6]			2.1			2.1		mV/°C

Notes:
1. Guaranteed by design.
2. Line and load regulation specifications include the effects of self heating.
3. Output voltage change with temperature = $\frac{V_{MAX} - V_{MIN}}{5V} \times 100\%$
4. Output voltage change with temperature specification applies untrimmed, or trimmed to +10V.
5. Output voltage temperature coefficient = $\frac{\text{Output voltage change with temperature}}{70°C}$
6. Limit current in or out of pin 3 to 50nA and limit capacitance on pin 3 to 30pF.

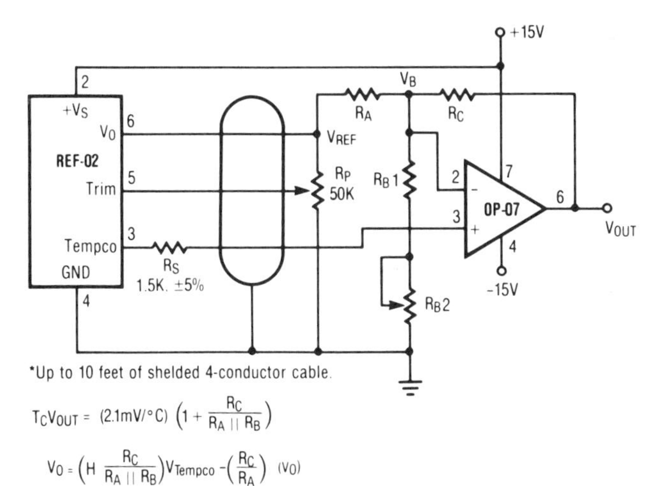

Resistor Values

TCV_{OUT} Slope(s)	10mV/°C	100mV/°C	10mV/°F
Temperature Range	-55°C to +125°C	-55°C to +125°C	-65°F to +257°F
Output Voltage Range	-0.55V to +1.25V	-5.5V to +12.5V	-0.67V to +2.57V
Zero Scale	0V at 0°C	0V at 0°C	0V at 0°F
R_A (±1% Resistor)	9.09KΩ	15KΩ	8.25KΩ
R_{B1} (±1% Resistor)	1.5KΩ	1.82KΩ	1.0KΩ
R_{B2} (Potentiometer)	200Ω	500Ω	200Ω
R_C (±1% Resistor)	5.11KΩ	84.5KΩ	7.5KΩ

Precision Electronic Thermometer

ELECTRICAL CHARACTERISTICS

(V_S = +15V, 0°C ≤ T_A ≤ +70°C and I_O = 0 unless otherwise noted)

Parameters	Test Conditions	REF-02C			REF-02D			Units
		Min	Typ	Max	Min	Typ	Max	
Output Voltage Change With Temperature[3 4]	Over Temp. Range		0.14	0.45		0.49	1.7	%
Output Voltage Temperature Coefficient[5]	Over Temp. Range		20	65		70	250	ppm/°C
Change in V_{OUT} Temperature Coefficient With Output Adjustment	R_P = 10kΩ		0.7			0.7		ppm/%
Line Regulation[2]	V_S = +8V to +33V		0.011	0.018		0.020	0.025	%/V
Load Regulation[2]	I_O = 0mA to 5mA		0.008	0.018		0.020	0.025	%/mA
Tempco Voltage Output Temperature Coefficient[6]			2.1			2.1		mV/°C

Notes:
1. Guaranteed by design.
2. Line and load regulation specifications include the effects of self heating.
3. Output voltage change with temperature = $\frac{V_{MAX} - V_{MIN}}{5V} \times 100\%$
4. Output voltage change with temperature specification applies untrimmed, or trimmed to +5V.
5. Output voltage temperature coefficient = $\frac{\text{Output voltage change with temperature}}{70°C}$
6. Limit current in or out of pin 3 to 50nA and limit capacitance on pin 3 to 30pF.

TYPICAL PERFORMANCE CHARACTERISTICS

Maximum Load Current vs. Differential Input Voltage

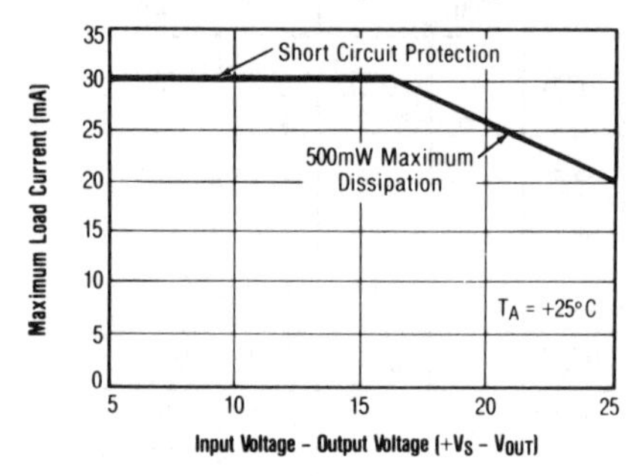

Normalized Load Regulation (ΔIL = 10mA) vs. Temperature

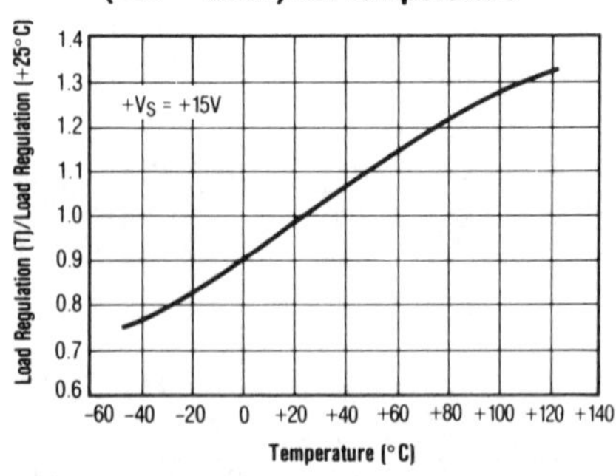

Normalized Line Regulation vs. Temperature

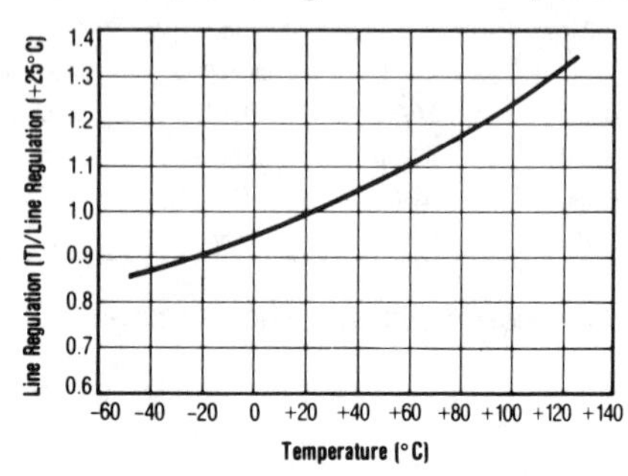

SIMPLIFIED SCHEMATIC DIAGRAM

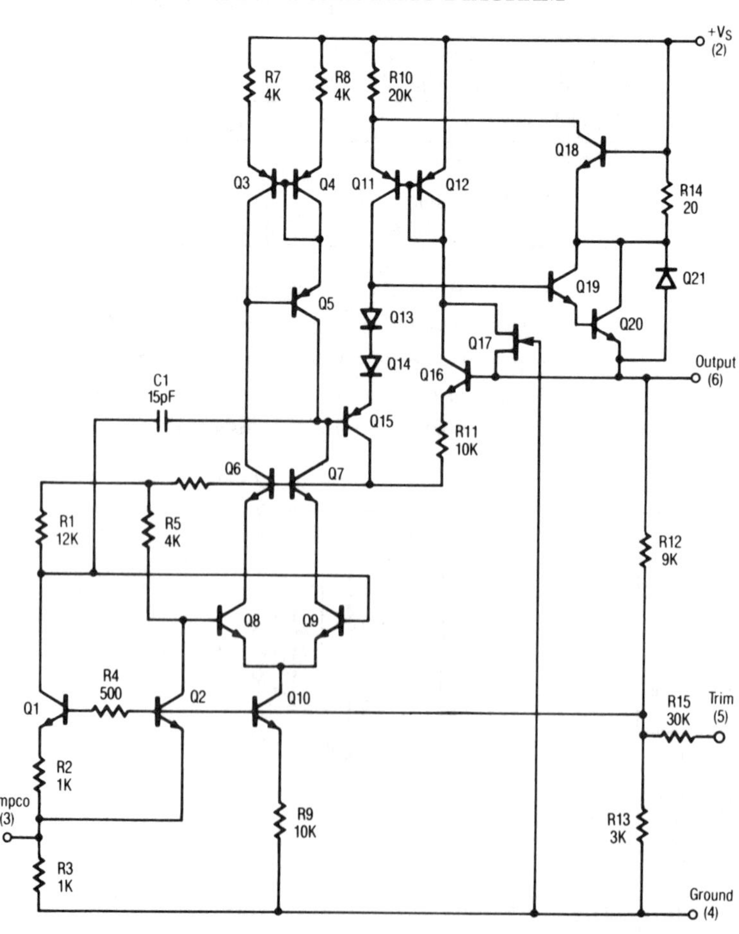

This information is reproduced with permission of Rohm Corporation, Irvine, CA.

Rohm
BA704, BA707
Low-Voltage Regulators

- Wide input voltage range
- Excellent line regulation
- Excellent load regulation
- Excellent temperature stability
- Requires no external components
- In a compact TO-92 package

APPLICATIONS
- Reference voltage supply for cameras
- Instrumentation equipment, etc.

The BA704 and BA707 are three-terminal voltage regulators specifically designed for low supply voltages. The BA704 has a rated output voltage of 2.65V, and the BA707 3.3V.

With special design consideration for input/output and temperature characteristics, the devices have applications in cameras, reference voltage supply for instrumentation equipment, and other low supply-voltage circuits operating under harsh environmental conditions.

ABSOLUTE MAXIMUM RATINGS (Ta=25°C)

Parameter	Symbol	Limits	Unit
Input voltage	V_{IN}	12	V
Power dissipation	Pd	250*	mW
Operating temperature range	Topr	−20 ~ 60	°C
Storage temperature range	Tstg	−55 ~ 125	°C
Load current	I_L	10	mA

*Derating is done at 2.5mW/°C for operation above Ta=25°C.

ELECTRICAL CHARACTERISTICS (Ta = 25°C)
BA704

Parameter	Symbol	Min.	Typ.	Max.	Unit	Conditions
Input voltage	V_{IN}	3.3	—	10.0	V	—
No-load input current	I_{CC}	—	1.5	2.5	mA	V_{IN}=5.5V, I_{OL}=0mA
Output voltage	V_{OUT}	2.40	2.65	2.90	V	V_{IN}=5.5V, I_{OL}=5mA
Output voltage load stability	ΔVo/Io	—	−8	−15	mV	V_{IN}=5.5V, I_{OL}=0 ~ 5mA
Output voltage input stability	ΔVo/Vi	—	5	30	mV	V_{IN}=3.6 ~ 9.0 V, I_{OL}=5mA
Output voltage input stability	ΔVo/Vi	—	—	20	mV	V_{IN}=3.3 ~ 3.6 V, I_{OL}=5mA
Output voltage temperature stability	ΔVo/T	—	±0.3	±1.0	mV/°C	V_{IN}=5.5V, I_{OL}=5mA

BA707

Parameter	Symbol	Min.	Typ.	Max.	Unit	Conditions
Input voltage	V_{IN}	4.3	—	10.0	V	—
No-load input current	I_{CC}	—	1.8	3.0	mA	V_{IN}=5.5V, I_{OL}=0mA
Output voltage	V_{OUT}	3.0	3.3	3.6	V	V_{IN}=5.5V, I_{OL}=5mA
Output voltage load stability	ΔVo/Io	—	−10	−20	mV	V_{IN}=5.5V, I_{OL}=0 ~ 5mA
Output voltage input stability	ΔVo/Vi	—	5	35	mV	V_{IN}=4.3 ~ 9.0 V, I_{OL}=5mA
Output voltage temperature stability	ΔVo/T	—	±0.3	±1.0	mV/°C	V_{IN}=5.5V, I_{OL}=5mA

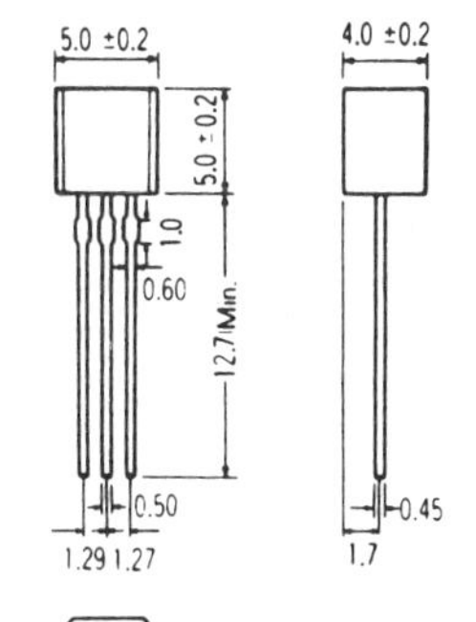

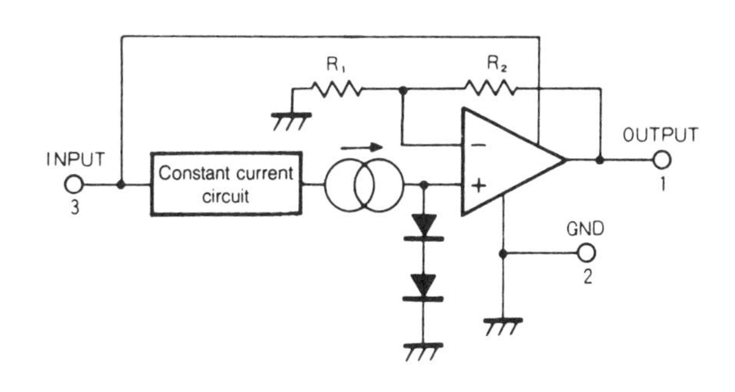

ELECTRICAL CHARACTERISTIC CURVES

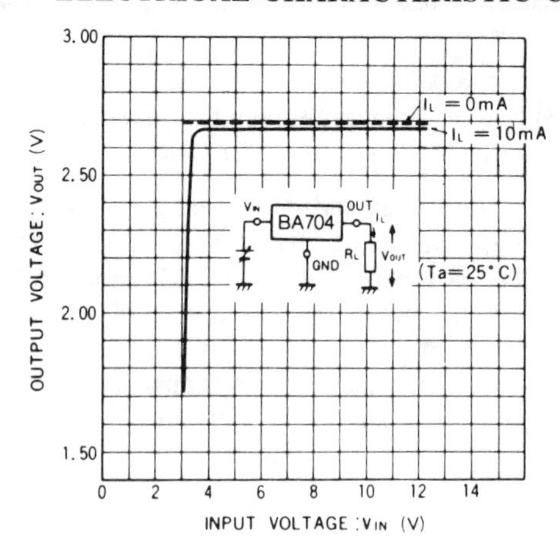

Input/output characteristics

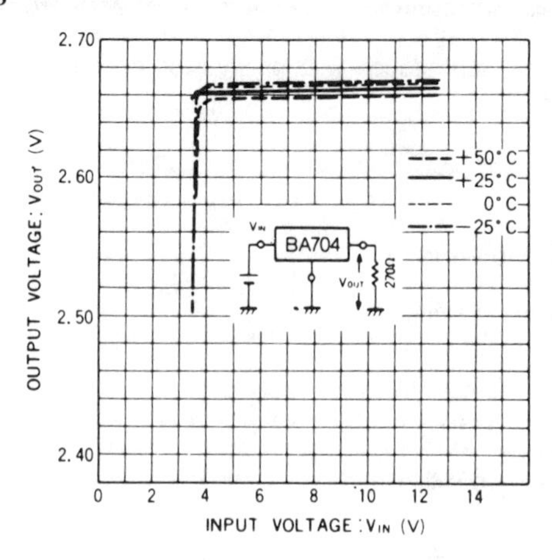

Input/output temperature characteristics

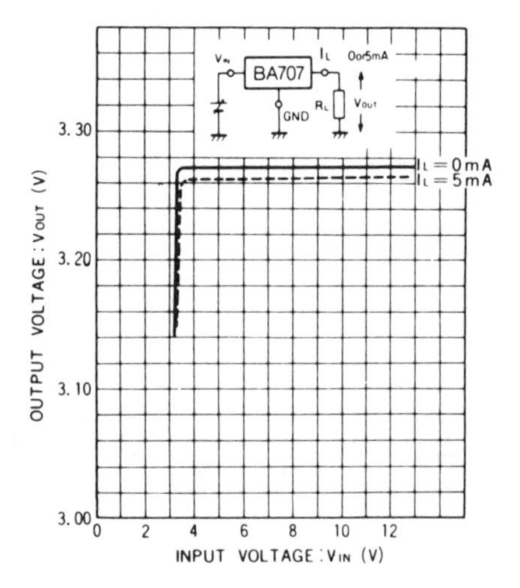

Input/output characteristics

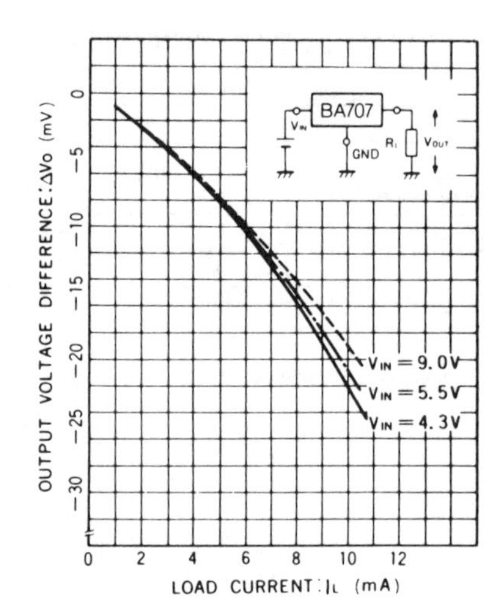

Output voltage difference vs. load current

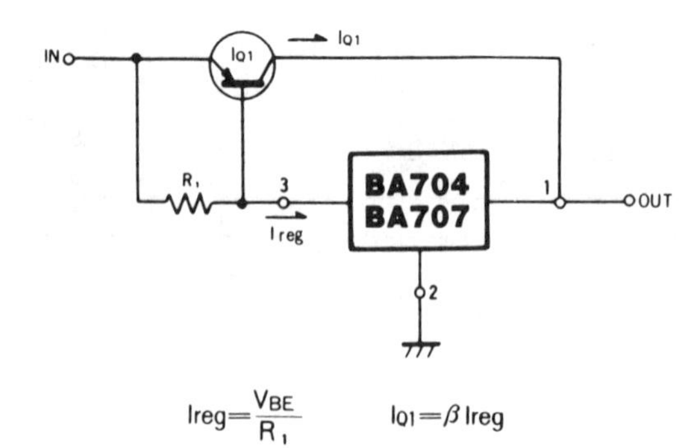

$$I_{reg} = \frac{V_{BE}}{R_1} \qquad I_{Q1} = \beta I_{reg}$$

While the BA704 and BA707 alone can only supply a load current of around 10 mA, they can provide a larger load current when used with a discrete PNP transistor.

Large current circuit

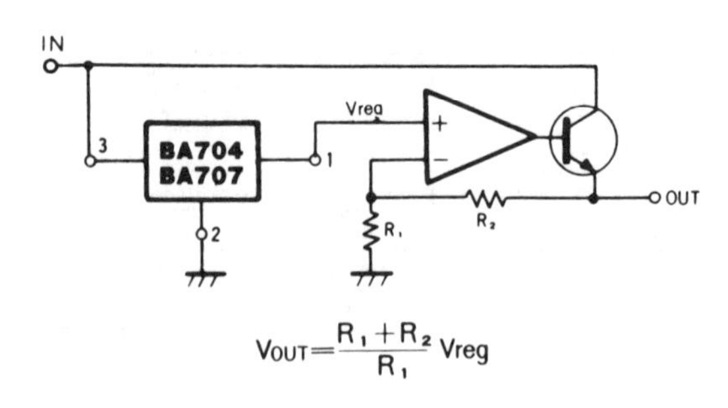

$$V_{OUT} = \frac{R_1 + R_2}{R_1} V_{reg}$$

This figure shows another application example for the BA704 and BA707 as a variable power supply. The output voltage $V_{OUT} = V_{reg} \times (R_1 + R_2)/R_1$ Vreg: BA704 and BA707's output voltage.

Output voltage variable circuit

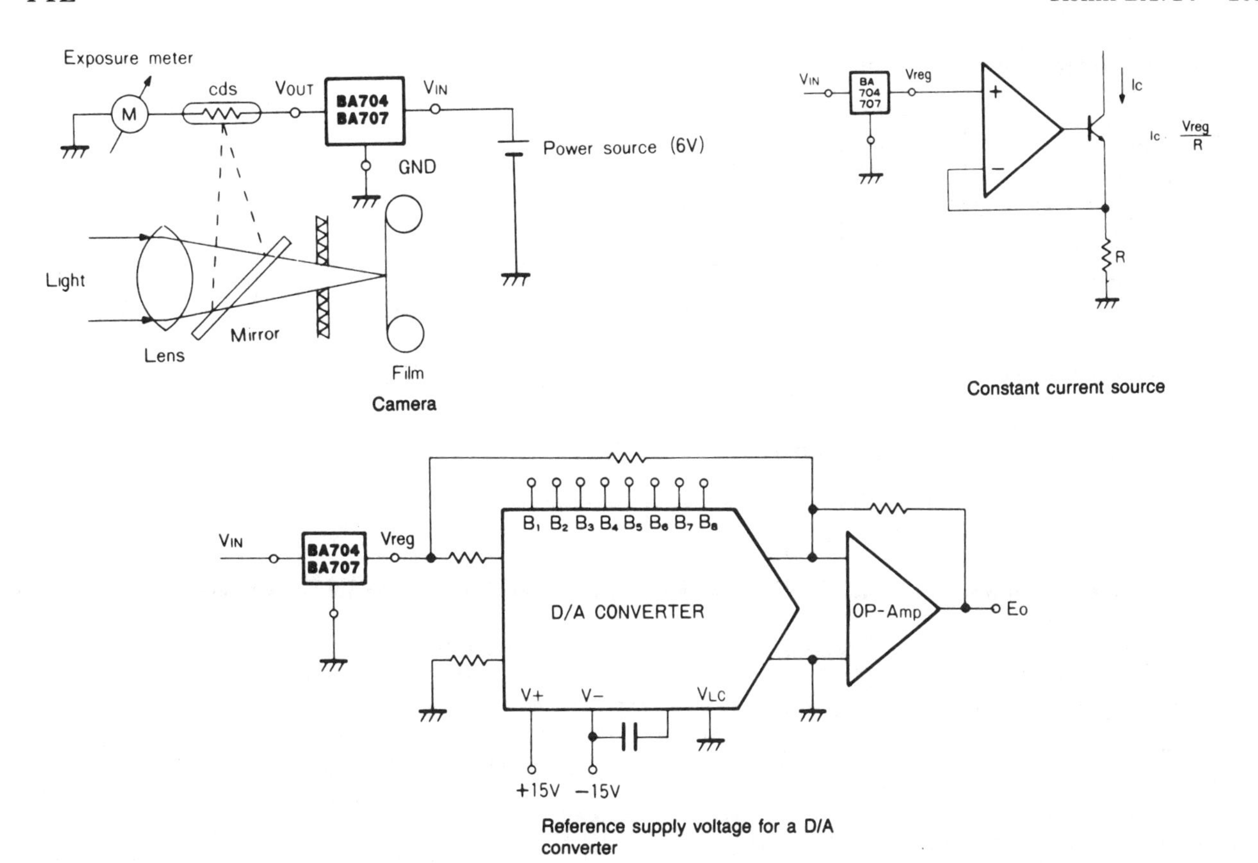

Camera

Constant current source

Reference supply voltage for a D/A converter

Rohm BA714
3-Pin Voltage Regulator (3.3V)

- Compact design
- Output overload protection

The BA714 is a 3-pin voltage regulator designed for supply-voltage stabilization.

The rated output voltage is 3.3V. With a negative temperature characteristic, the device is ideally suited as a regulated power source for liquid crystal displays.

APPLICATIONS

- Cameras
- Reference voltage supply for instrumentation equipment

ABSOLUTE MAXIMUM RATINGS (Ta=25°C)

Parameter	Symbol	Limits	Unit
Supply voltage	V_{CC}	12	V
Power dissipation	Pd	300*	mW
Operating temperature range	Topr	−25~75	°C
Storage temperature range	Tstg	−55~125	°C
Output current	I_{OUT}	300	μA

*Derating is done at 6mW/°C for operation above Ta=25°C.

ELECTRICAL CHARACTERISTICS (Ta=25°C, V_{CC}=6.0V)

Parameter	Symbol	Min.	Typ.	Max.	Unit	Conditions
Output voltage	V_{OUT}	3.05	3.30	3.55	V	V_{CC}=4~7V, I_{OUT}=50~250μA*
40°C output voltage	HV_{OUT}	2.85	—	3.30	V	V_{CC}=4~7V, I_{OUT}=50~250μA*
0°C output voltage	CV_{OUT}	3.30	—	3.95	V	V_{CC}=4~7V, I_{OUT}=50~250μA*
Reactive voltage	V_C	—	0.1	0.3	V	I_{OUT}=250μA
Reactive current	I_{CC}	—	20	40	μA	$R_L=\infty$
Output voltage-to-supply voltage ratio	V_{OUT}/V_{CC}	—	2	—	mV/V	—
Output short-circuit current	I_{OS}	—	1.5	10	mA	—

*The output voltage values at 40°C and 0°C are guaranteed values.

CIRCUIT DIAGRAM

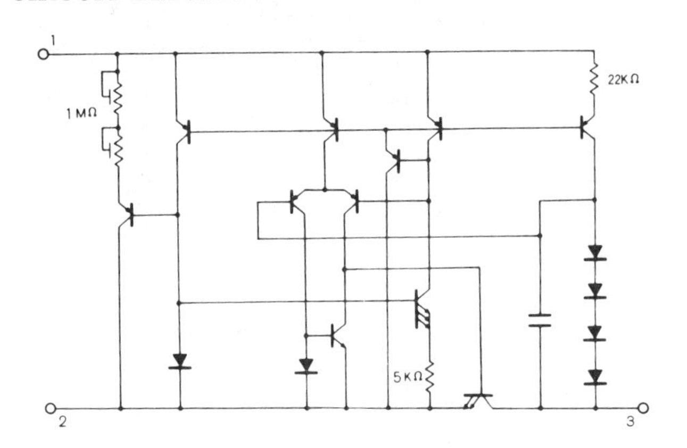

(Unit: mm)

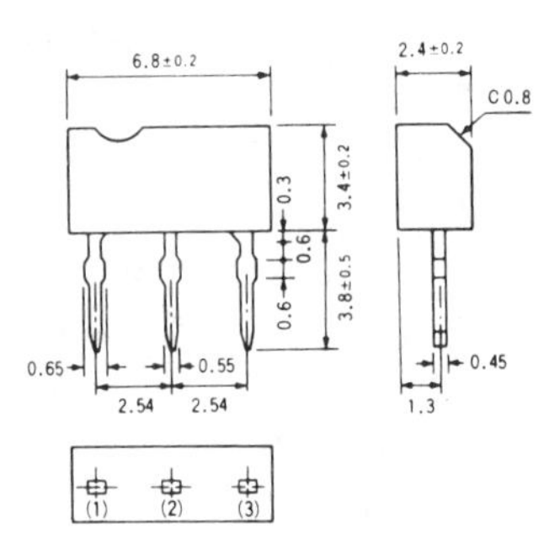

Rohm
BA17805, BA178MO5
3-Pin Voltage Regulators (5.0 V)

- Internal overcurrent protection and thermal protection circuits
- Excellent ripple regulation
- Provides stable +5V voltage supply with no external components required
- Both the 1 A (AB17805) and 0.5 A (BA178MO5) voltage regulators are available in the TO-220 package to ensure a wide application range
- Pin-compatible with equivalent regulators

APPLICATIONS

- Regulated 5V power supply

The BA17805 and BA178MO5 are 3-pin voltage regulators with a rated output voltage of +5V.

When an unregulated dc voltage exceeding 7V is applied to the input, the devices provide a regulated dc voltage of 5V at the output. Featuring excellent ripple regulation characteristics, the devices provide a stable dc voltage supply.

ABSOLUTE MAXIMUM RATINGS (Ta=25°C)

Parameter	Symbol	Limits	Unit
Supply voltage	Vin	35	V
Power dissipation	Pd	2	W
Operating temperature range	Topr	−25 ~ 75	°C
Storage temperature range	Tstg	−55 ~ 125	°C

ELECTRICAL CHARACTERISTICS (Unless otherwise specified, Ta=25°C, V_{IN}=10V, I_O=500mA)

Parameter	Symbol	Min.	Typ.	Max.	Unit	Conditions
Output voltage	V_O	—	5.0	—	V	—
Input regulation	I.R.	—	3	—	mV	V_{IN}=7 ~ 25V
Ripple regulation	R.R.	—	78	—	dB	V_{IN}=8 ~ 18V, f=120Hz
Load regulation	L.R.	—	100	—	mV	I_O=5mA ~ 1.5A
Output voltage temperature coefficient	T_{CVO}	—	−1.1	—	mV/°C	I_O=5mA, Ta=0 ~ 125°C

ELECTRICAL CHARACTERISTICS (Unless otherwise specified, Ta=25°C, V_{IN}=10V, I_O=350mA)

Parameter	Symbol	Min.	Typ.	Max.	Unit	Conditions
Output voltage	V_O	—	5.0	—	V	—
Input regulation	I.R.	—	3	—	mV	V_{IN}=7 ~ 25V, I_O=200mA
Ripple regulation	R.R.	—	80	—	dB	V_{IN}=8 ~ 18V, f=120Hz, I_O=30mA
Load regulation	L.R.	—	30	—	mV	I_O=5 ~ 500mA
Output voltage temperature coefficient	T_{CVO}	—	−1.0	—	mV/°C	I_O=5mA, Ta=0 ~ 125°C

(Unit: mm)

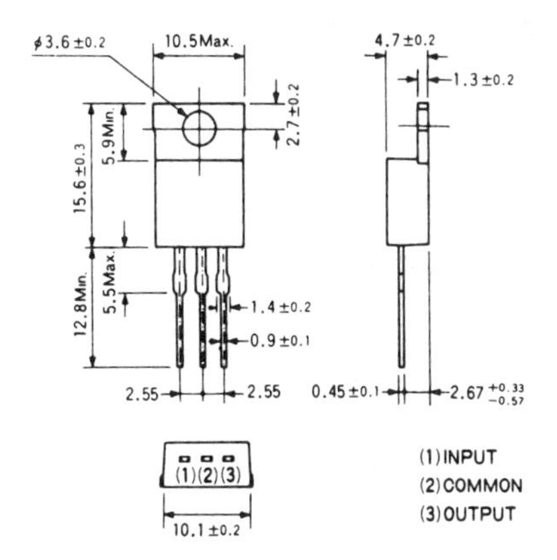

STANDARD APPLICATION

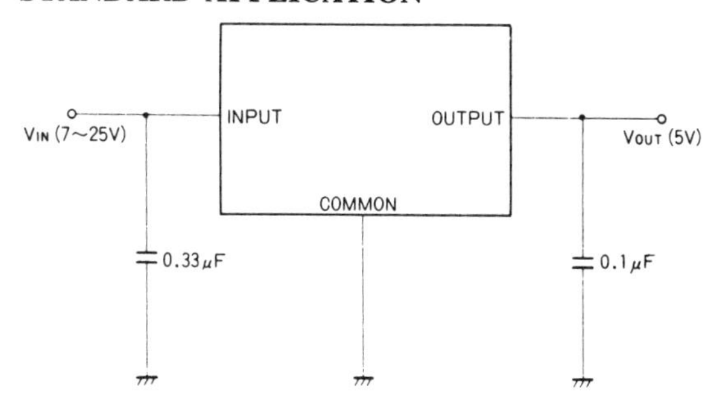

Rohm
BA6121
4-Output Switching Regulator

- Four switching regulator blocks implemented on a single chip
- Power-save pin enables only one regulator output
- External reference voltage control can make the device a variable voltage supply
- The drivability control pin allows the user to set the drain current at pin 1 to enable driving at a minimum power

The BA6121 is a monolithic switching regulator specifically designed for use in VCRs. The device contains four PWM (pulse-width modulated) switching regulator blocks with REF control, drivability control, and feedback-limiting pins. The device also contains an error input pin for each block.

With four blocks of switching regulators and a power-save pin to reduce current consumption, the device is particularly suitable for compact power supply design in portable VCRs.

ABSOLUTE MAXIMUM RATINGS (Ta=25°C)

Parameter	Symbol	Limits	Unit
Supply voltage	V_{CC}	18	V
Power dissipation	Pd	400*	mW
Operating temperature range	Topr	−20~75	°C
Storage temperature range	Tstg	−55~125	°C
Maximum sink current, pin 11	I_{11}	30	mA
Maximum sink current, pin 8	I_8	10	mA
Maximum sink current, pin 14	I_{14}	20	mA
Maximum drain current, pin 1	I_1	1	mA

*Derating is done at 4mW/°C for operation above Ta=25°C.

ELECTRICAL CHARACTERISTICS (Unless otherwise specified, Ta=25°C, V_{CC}=12V)

Parameter	Symbol	Min.	Typ.	Max.	Unit	Conditions
Supply voltage range	V_{CC}	8	—	16	V	—
Quiescent current	I_Q	8.0	11.6	15.0	mA	—
Triangular wave oscillation frequency	fs	25	39	57	kHz	C_T=1500pF
Drain current, pin 15	I_{15}	—	30	—	μA	—
Drain current, pin 2	I_2	—	30	—	μA	—
Drain current, pin 4	I_4	—	200	—	μA	—
Drain current, pin 13	I_{13}	—	0.3	—	mA	13pin=L_O
Sink current, pin 9	I_9	—	—	5	μA	V_9=3.1V
Sink current, pin 7	I_7	—	—	5	μA	V_7=3.1V
Sink current, pin 16	I_{16}	—	—	5	μA	V_{16}=3.1V
Sink current, pin 5	I_5	—	—	5	μA	V_5=3.1V

ELECTRICAL CHARACTERISTIC CURVES

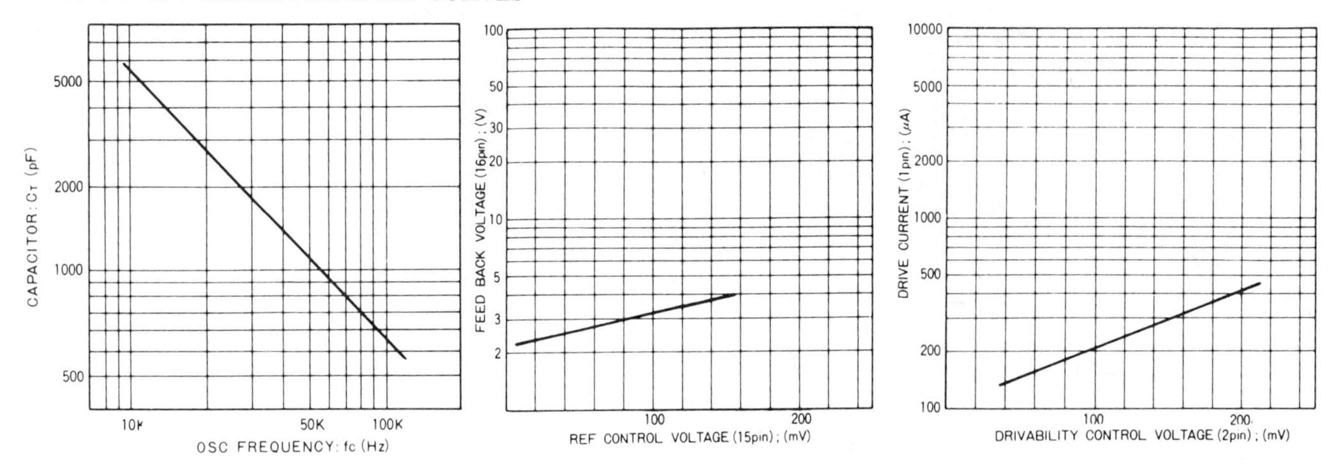

Capacitor vs. OSC frequency

Feed back voltage vs. REF control voltage

Drive current vs. drivability control voltage

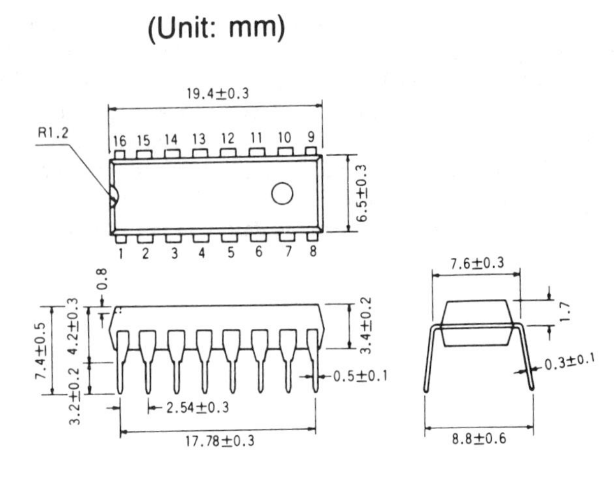

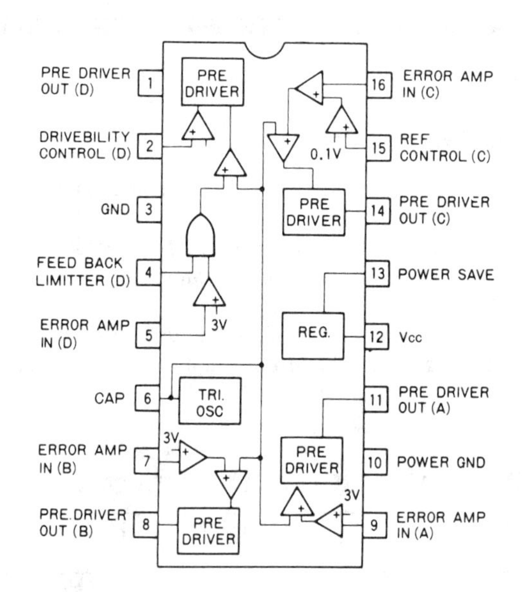

APPLICATION EXAMPLE

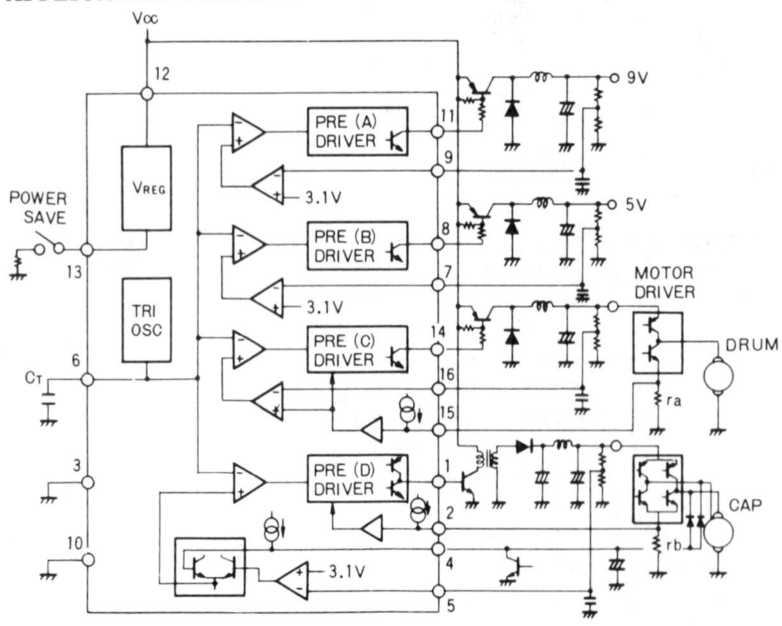

Rohm
BA6122A, BA6122AF
Switching Regulators

- The 9V stop feature of the 9V regulator can make only the 5V regulator enabled
- Direct feedback can drastically reduce ripple
- V_{REG} output is available
- The driver's output duty ratio is variable from 0 to 100% so that the driver output can be completely turned on or off
- Accurate oscillation frequency with stable start-up and temperature characteristics
- Error amplifiers with internal phase compensation
- High conversion efficiency

APPLICATIONS

- Power supply for VCRs
- Power supply for general-purpose equipment

The BA6122 and BA6122AF are dual monolithic PWM (pulse-width modulated) switching regulators. Each regulator block contains a 5V reference-voltage supply, a sawtooth wave oscillator, an error amplifier, a comparator, and a driver. The 9V stop feature that shuts off the 9V regulator block, and the ripple regulation for use with car batteries make this device best suited as a 5V and 9V power supply for portable VCRs.

The devices are available in mini-flat packages as well as LF packages.

PRECAUTION

The BA6122A and BA6122AF are available in different packages with different pin configurations.

ABSOLUTE MAXIMUM RATINGS (Ta=25°C)

Parameter	Symbol	Limits	Unit
Supply voltage	V_{CC}	18	V
Power dissipation	Pd	340*	mW
Operating temperature range	Topr	−10~60	°C
Storage temperature range	Tstg	−55~125	°C
V_{REG} drain current	$I_D(V_{REG})$	5	mA
Sink current, pin 8	I_8	20	mA
Sink current, pin 10	I_{10}	10	mA

*Derating is done at 3.4mW/°C for operation above Ta=25°C.

ELECTRICAL CHARACTERISTICS (Unless otherwise specified, Ta=25°C, V_{CC}=12V)

Parameter	Symbol	Min.	Typ.	Max.	Unit	Conditions
Supply voltage range	V_{CC}	8	—	16	V	$4.2V < V_{REG} < 5.4V$
Quiescent current	I_O	—	5.5	8.0	mA	—
V_{REG} output voltage	V_{REG}	4.65	4.8	4.95	V	—
Input regulation 9V output	$V_{O9\text{-}G}$	−0.1	—	+0.1	V	$10V \leqq V_{CC} \leqq 16V$
5V output voltage	V_{O5}	4.7	5.0	5.3	V	—
Input regulation 5V output	$V_{O5\text{-}G}$	−0.1	—	+0.1	V	$10V \leqq V_{CC} \leqq 16V$
Simple efficiency	I_O	—	—	3.75	mA	I_{O9}=400mA, I_{O5}=50mA
Oscillation frequency	f	36	41	46	kHz	—
Oscillation frequency input regulation	Δf-R	—	−0.3	—	%	R_A=36kΩ, R_B=12kΩ, C_T=1000pF $10V \leqq V_{CC} \leqq 16V$
Reference oscillation output duty	D	14	20	26	%	—
Reference oscillation output voltage-high	H	3.7	—	—	V	—
Reference oscillation output duty	$\bar{D}$	74	80	86	%	—
Reference oscillation output voltage	$\bar{H}$	3.7	—	—	V	—
9V stop input high voltage	V_H	1.5	—	—	V	$V_{O9} < 0.4V$
9V stop input low voltage	V_L	—	—	0.9	V	$8.3V < V_{O9} < 9.7V$

·(Unit: mm)

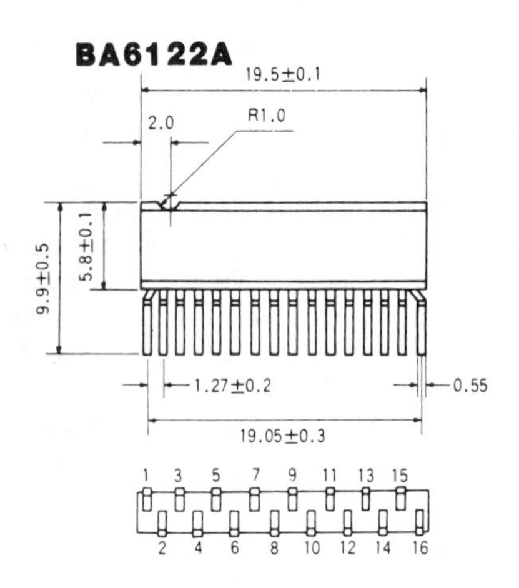

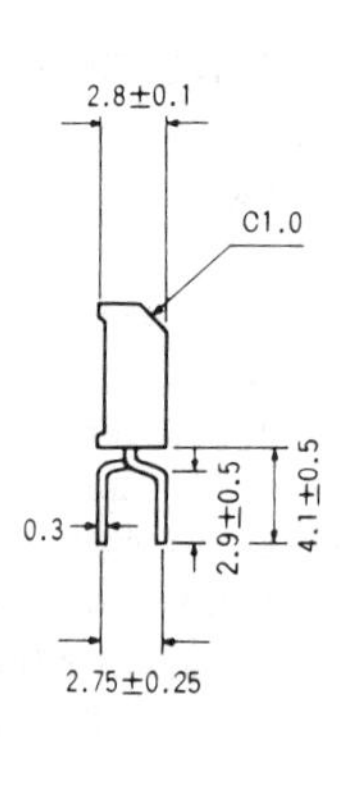

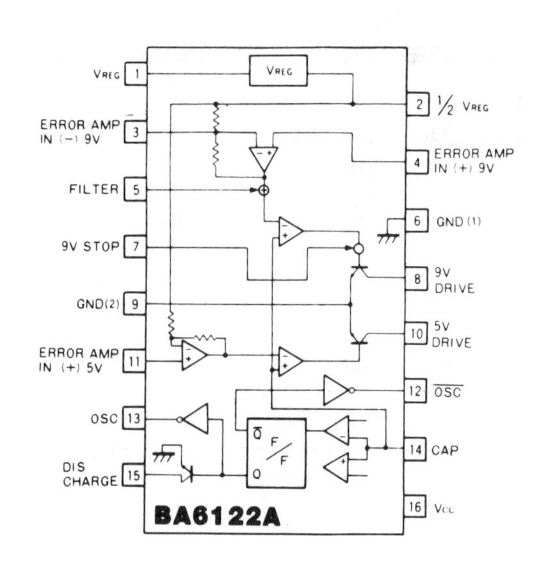

ELECTRICAL CHARACTERISTIC CURVES

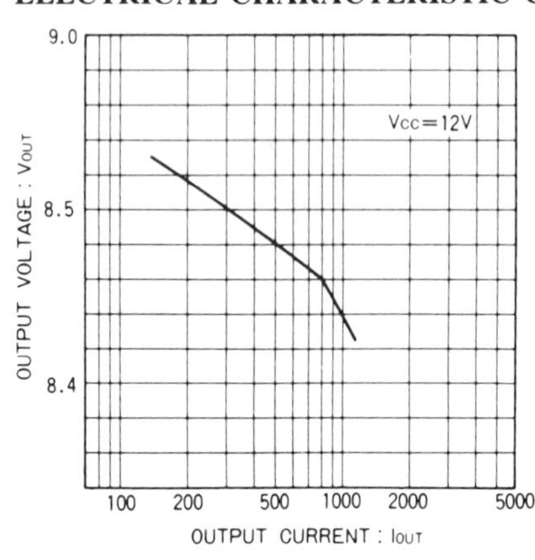

Output voltage vs. output current

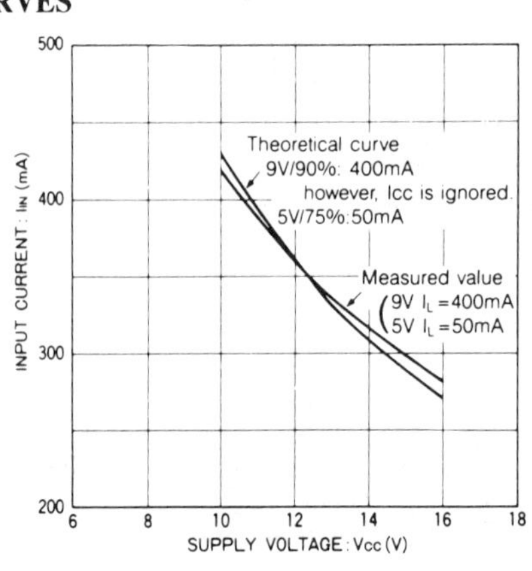

Input current vs. supply voltage

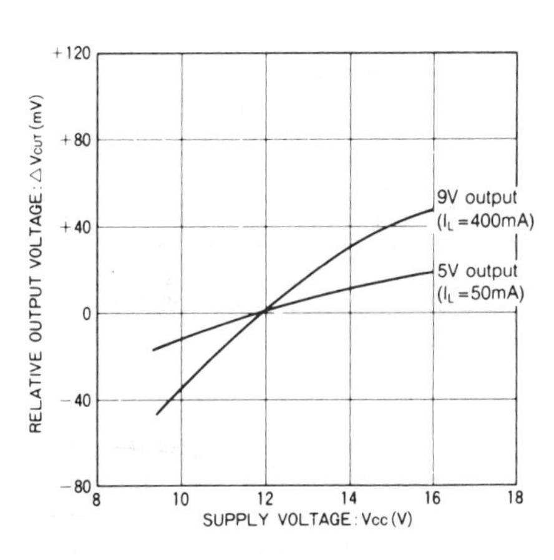

Relative output voltage vs. supply voltage

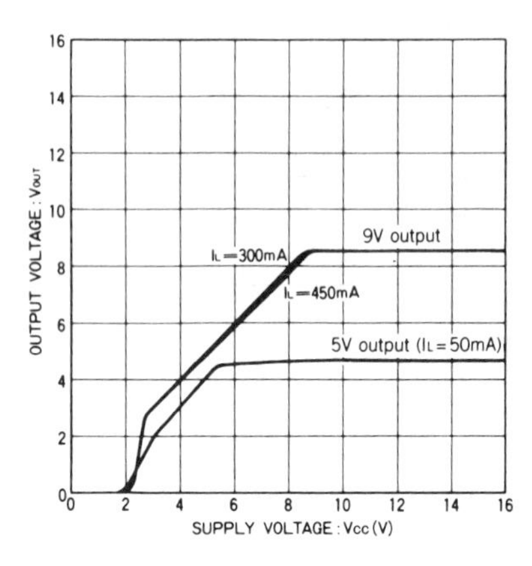

Output voltage vs. supply voltage

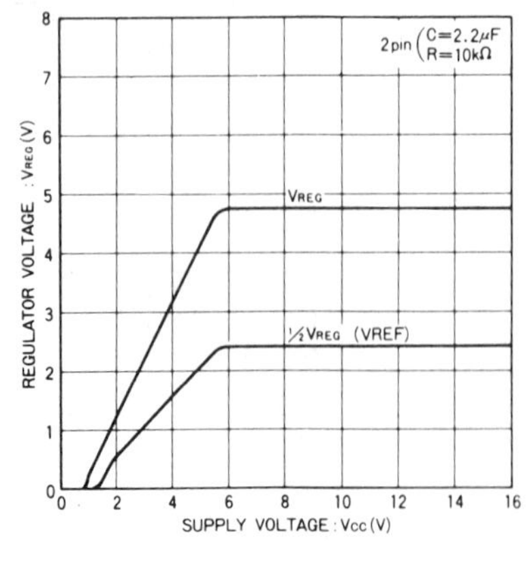

Regulator voltage vs. supply voltage

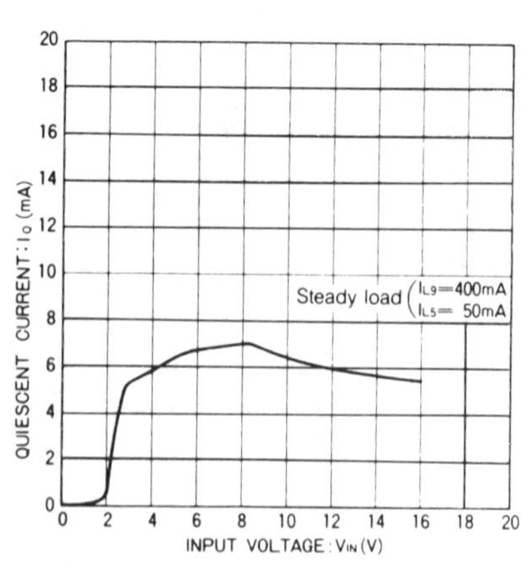

Quiescent current vs. input voltage

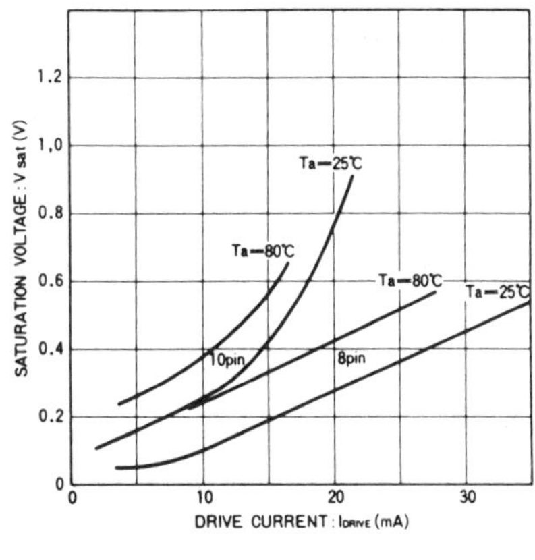

Saturation voltage vs. drive current

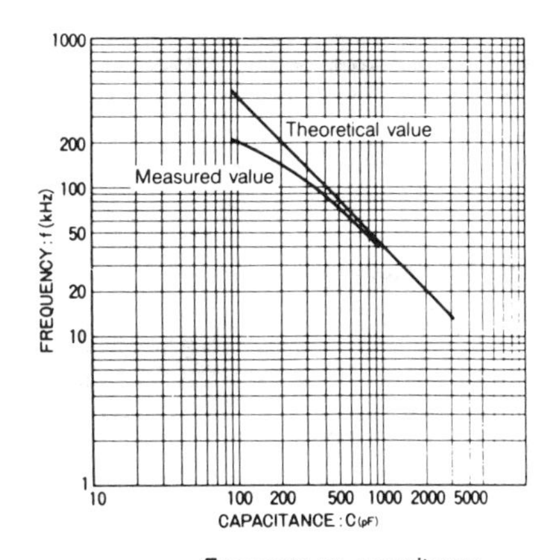

Frequency vs. capacitance

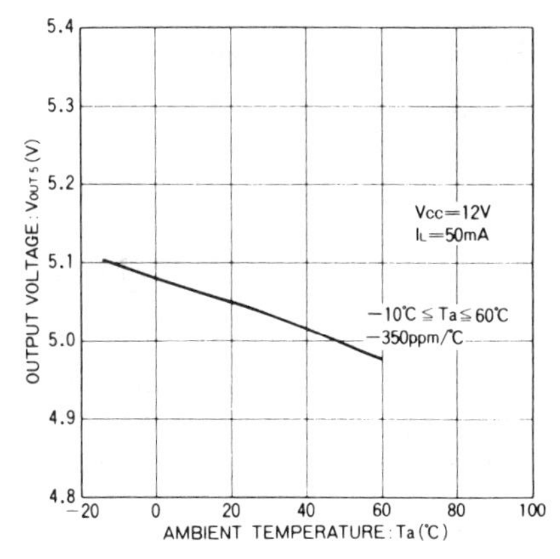

Output voltage vs. ambient temperature

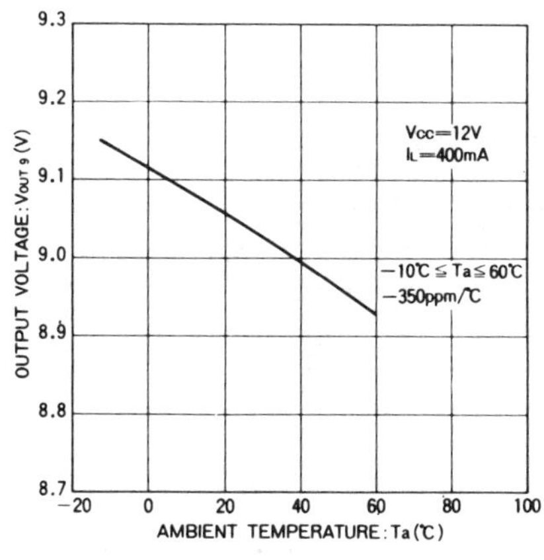

Output voltage vs. ambient temperature

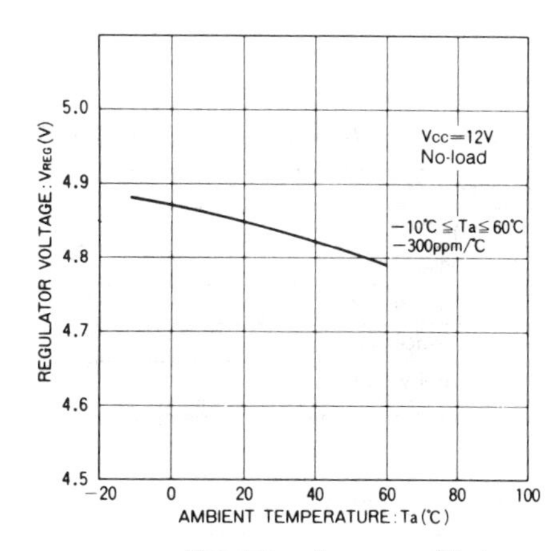

Regulator voltage vs. ambient temperature

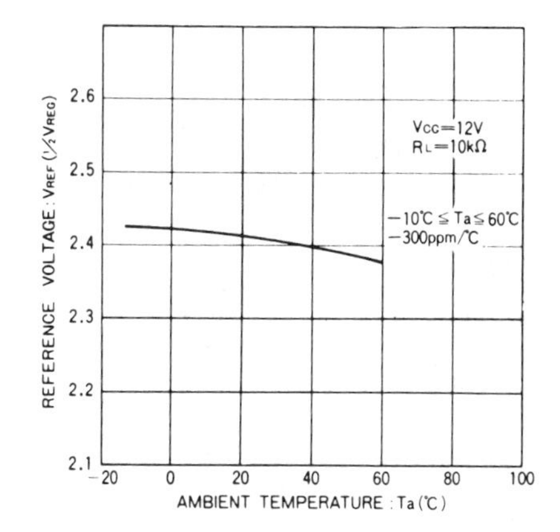

Reference voltage vs. ambient temperature

APPLICATION EXAMPLE

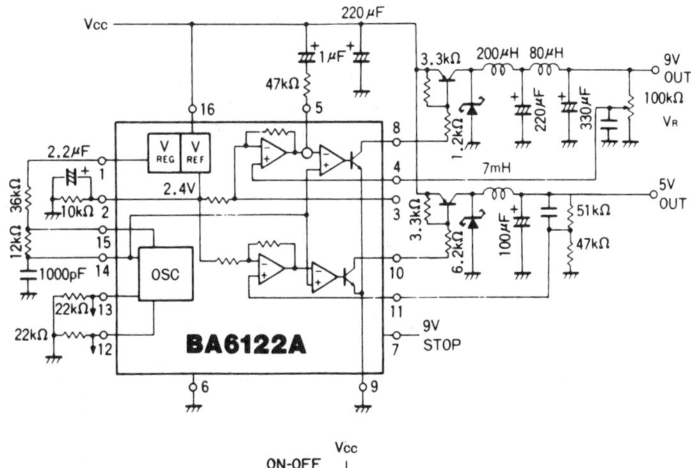

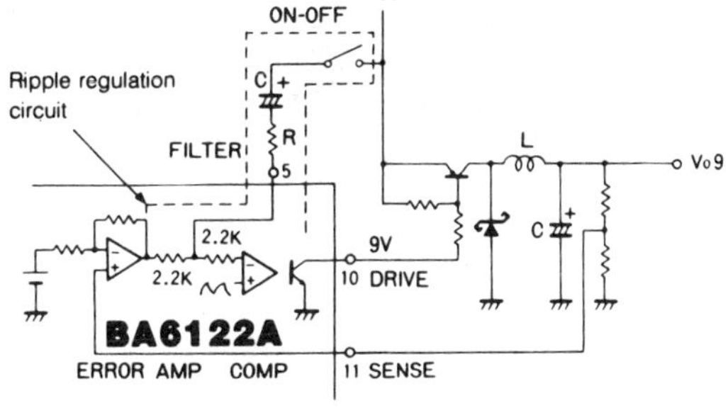

Ripple regulation peripheral circuit

Rohm
BA6132F
5-Output Switching Regulator

- All control circuitry required for the switching regulators is internally implemented
- High-efficiency PWM-type circuitry
- Output on/off control capability (only one block)
- Low current consumption (7 mA typ)
- Reference can be externally controlled

The BA6132F is a monolithic switching regulator which contains five independent switching regulator blocks.

(Unit: mm)

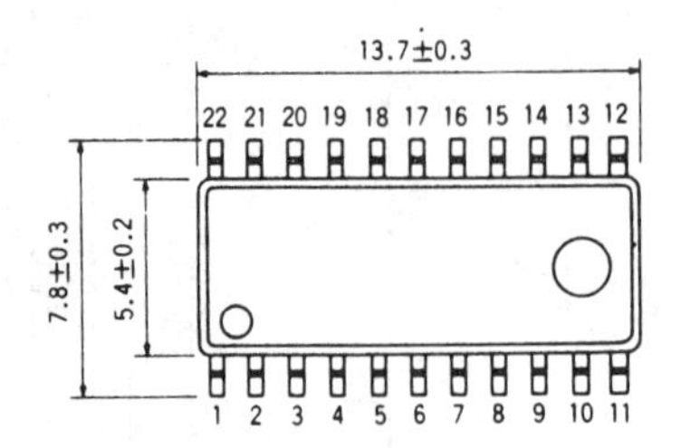

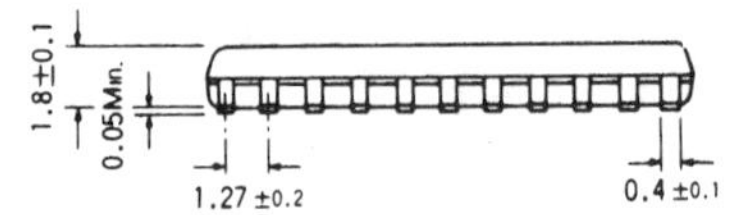

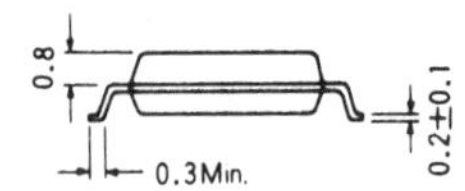

BLOCK DIAGRAM

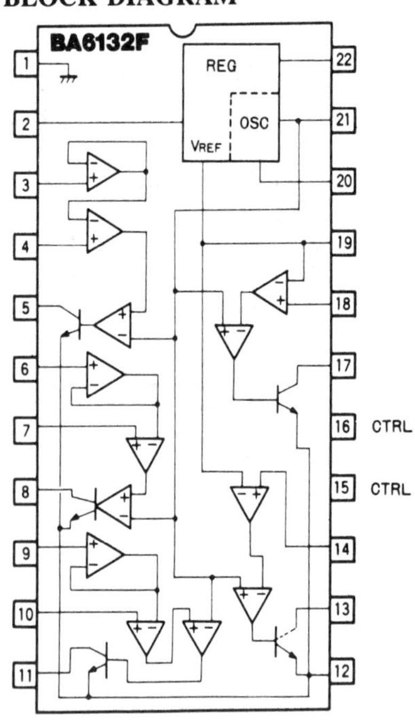

ABSOLUTE MAXIMUM RATINGS (Ta=25°C)

Parameter	Symbol	Limits	Unit
Supply voltage	V_{CC}	18	V
Power dissipation	Pd	600*	mW
Operating temperature range	Topr	−20~75	°C
Storage temperature range	Tstg	−55~125	°C
REF control voltage	Vrc	5	V

*Derating is done at 6mW/°C for operation above Ta=25°C.

CONTROL MODE

Control pin / Output pin	Pin 5(M3)	Pin 8(M2)	Pin 11(M1)	Pin 13(9V)	Pin 17(5V)
Pin 15;L, pin 16;L	ON	ON	ON	ON	ON
Pin 15;L, pin 16;H	OFF	OFF	OFF	ON	ON
Pin 15;H, pin 16;L	OFF	OFF	OFF	OFF	ON
Pin 15;H, pin 16;H	OFF	OFF	OFF	OFF	ON

Blocks 1 to 3 allow external control of the error amplifier's reference voltage, providing a wide output control range. The reference input pins for blocks 4 and 5 are internally connected to V_{REG} (pin 19).

ELECTRICAL CHARACTERISTICS (Unless otherwise specified, Ta=25°C, Vcc=12V)

Parameter	Symbol	Min.	Typ.	Max.	Unit	Conditions
Circuit current	I_{CC}	—	7	—	mA	Rated load
Oscillation frequency temperature regulation	$f_{OSC}/\Delta T$	—	±3	—	%	Ta = −10~60°C
Maximum oscillation frequency	fmax	—	100	—	kHz	—
Pins 5, 8, 11 drive current	$I_{5,8,11}$	—	20	—	mA	Maximum drive current
Pin 13 drive current	I_{13}	—	30	—	mA	Maximum drive current
Pin 17 drive current	I_{17}	—	10	—	mA	Maximum drive current
Control voltage-high	HCTRL	—	2.7~7	—	V	—
Control voltage-low	HCTRL	—	0~0.9	—	V	—

APPLICATION EXAMPLE

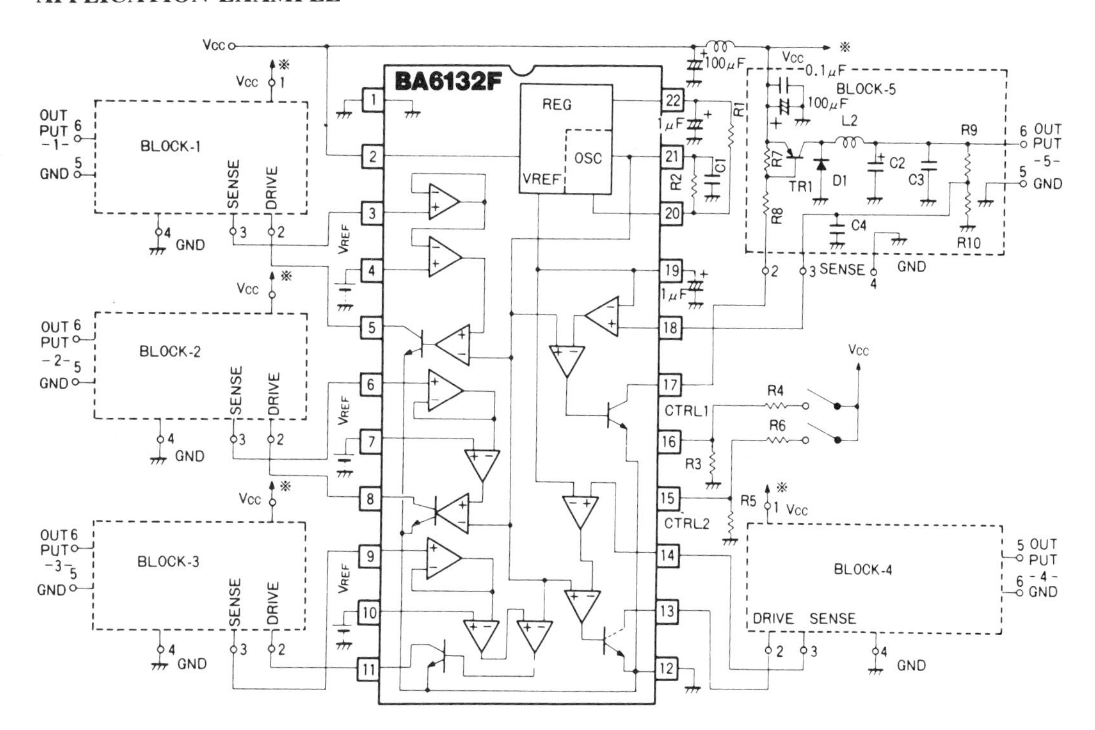

Rohm BA6149LS

6-Output Switching Regulator

- All the control circuitry required for switching regulators is internally implemented
- High-efficiency PWM-type circuitry
- High-accuracy stable triangular-wave oscillator
- Six output voltages available
- Error amplifier gain can be set (two blocks)
- Output on/off control capability (except for 5V)
- Reference can be externally controlled

The BA6149LS is a monolithic switching regulator that contains six independent switching regulator blocks. The outputs of these blocks, except for pin 5, can be turned on and off by pin 19. If a voltage of more than 3.5V is applied to pin 19, all outputs other than pin 5 are turned off (pin 5 is always active).

Blocks 1 to 3 allow external control of the error amplifier's reference voltage, providing a wide output control range. The reference input pins for blocks 4 and 5 are internally connected to V_{REG}.

ABSOLUTE MAXIMUM RATINGS (Ta=25°C)

Parameter	Symbol	Limits	Unit
Supply voltage	V_{CC}	20	V
Power dissipation	Pd	500*	mW
Operating temperature range	Topr	−10~70	°C
Storage temperature range	Tstg	−25~125	°C

*Derating is done at 5mW/°C for operation above Ta=25°C.

BLOCK DIAGRAM

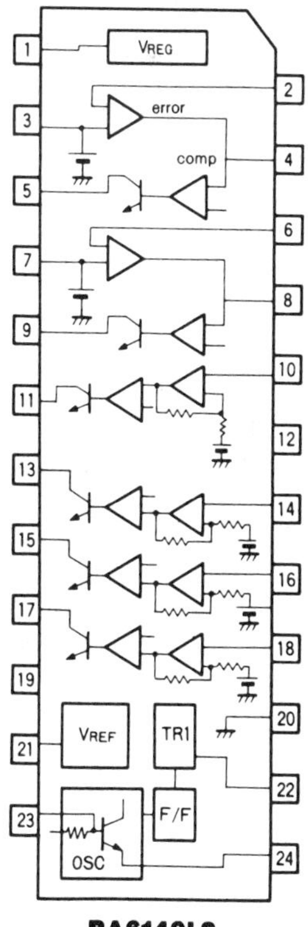

BA6149LS

ELECTRICAL CHARACTERISTICS(Unless otherwise specified, Ta=25°C, V_{CC}=12V)

Parameter	Unit	Min.	Typ.	Max.	Unit	Conditions
Circuit current	I_{CC}	—	7	—	mA	—
Reference voltage	Vref	—	2.53	—	V	—
Triangular wave oscillation frequency	fT	—	101.88	—	kHz	fo=815kHz
5V supply output voltage	V_{05}	—	5	—	V	I_L=227mA
9V supply output voltage	V_{09}	—	9.15	—	V	I_L=100mA
M1 supply output voltage	V_{CY}	—	5	—	V	I_L=100mA
M2 supply output voltage	V_{CA}	—	3.5	—	V	I_L=50mA
M3 supply output voltage	V_{SR}	—	3.5	—	V	I_L=55mA
M4 supply output voltage	V_{TR}	—	3.5	—	V	I_L=200mA
M1 input regulation	V_{r1}	—	80	—	mV	I_L=100mA. V_{CC}=10~16V
M2 input regulation	V_{r2}	—	60	—	mV	I_L=50mA. V_{CC}=10~16V
M3 input regulation	V_{r3}	—	60	—	mV	I_L=55mA. V_{CC}=10~16V
M4 input regulation	V_{r4}	—	60	—	mV	I_L=200mA. V_{CC}=10~16V

APPLICATION EXAMPLE

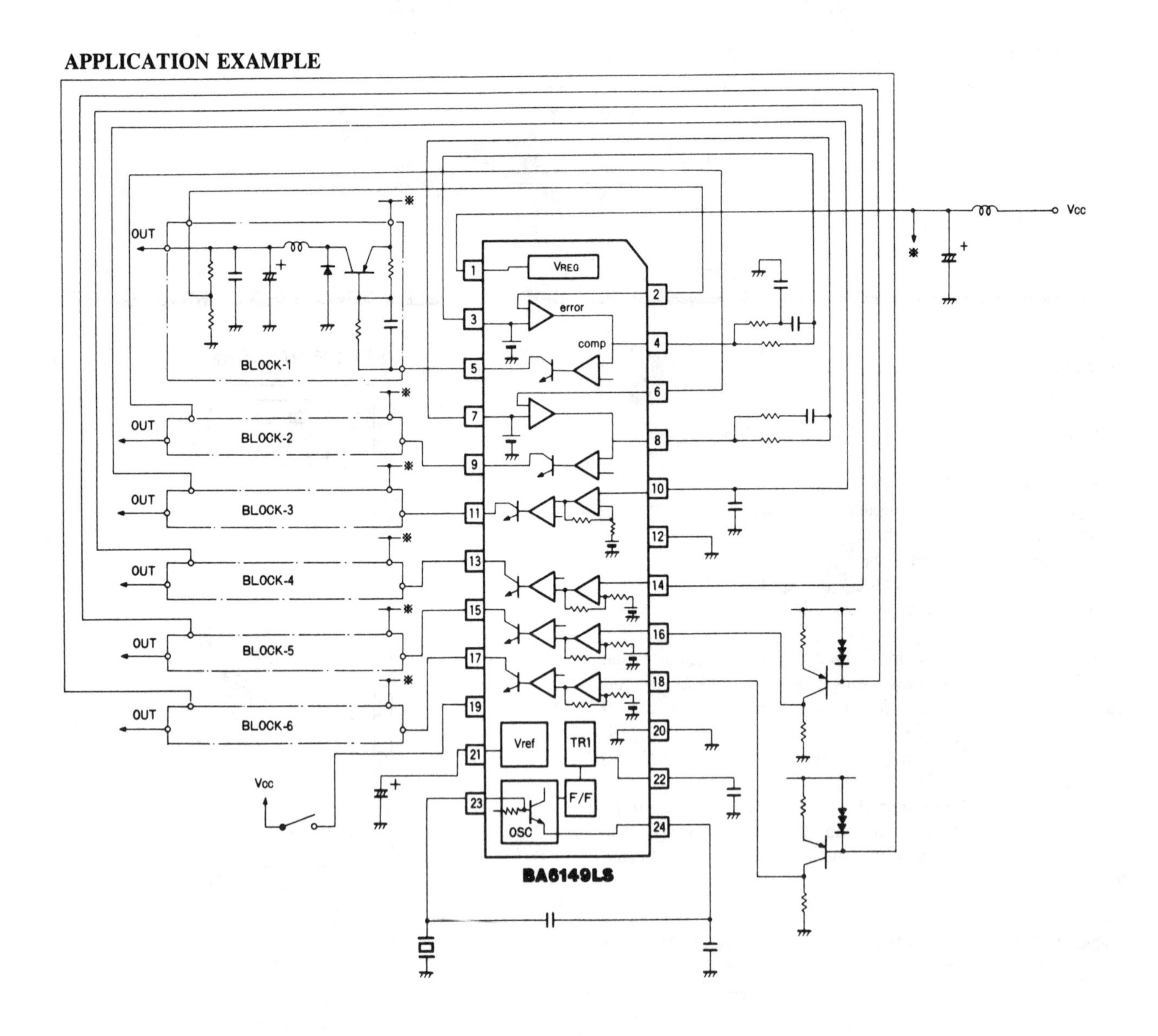

(Unit: mm)

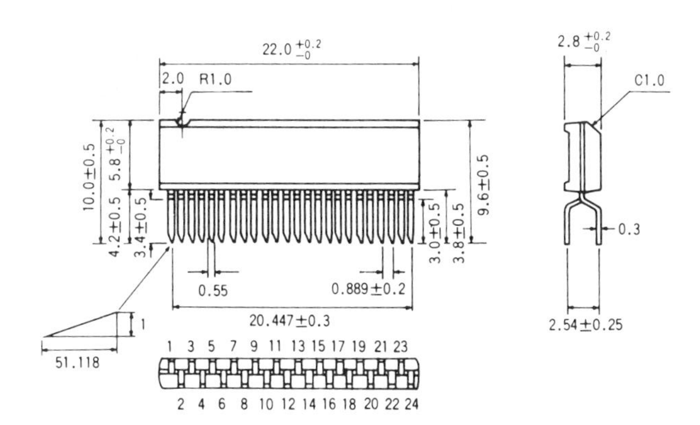

Rohm BA6161M

dc-to-dc converter for electronic tuners

- Low output sensitivity to input voltage variation
- Internal temperature compensation
- Minimum external components required
- Available in compact 4-pin LF-M package

APPLICATIONS

- Electronic tuners for TV sets
- Other applications requiring high dc voltage (30 to 45V) from a battery or other dc power supply.

The BA6161M is a monolithic dc-to-dc converter designed for use with electronic tuners. It has a rated output voltage of 33.3V.

Precautions

1. If you want an output voltage higher than the rated output voltage (33.3V), use a variable resistor with good temperature characteristics (RVG 6PO2-104M by MURATA or equivalent) as shown in the application example. In this case, the voltage at pin 2 should not exceed 42V.
2. The inductor (4.7 mH, RCP095-472K by SUMIDA or equivalent) connected across pins 2 and 3 should have as low a dc resistance as possible.

(Unit: mm)

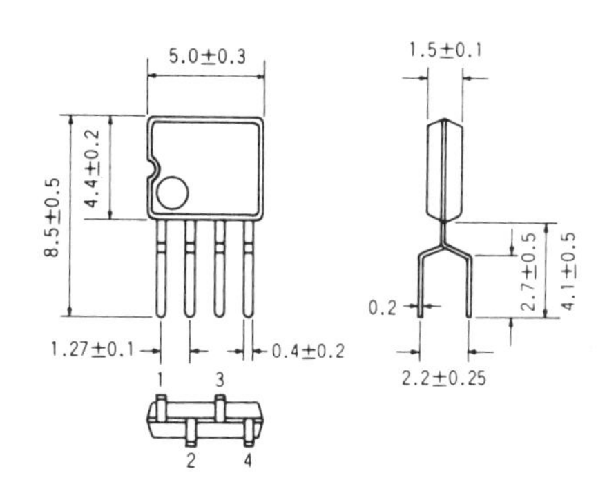

ABSOLUTE MAXIMUM RATINGS (Ta=25°C)

Parameter	Symbol	Limits	Unit
Input voltage	V_{IN}	18.0	V
Power dissipation	Pd	350*	mW
Operating temperature range	Topr	−20~75	°C
Storage temperature range	Tstg	−55~125	°C

*Derating is done at 3.5mW/°C for operation above Ta=25°C.

ELECTRICAL CHARACTERISTICS (Unless otherwise specified, Ta=25°C, V_{IN}=9.0V)

Parameter	Symbol	Min.	Typ.	Max.	Unit	Conditions
Current consumption	I_{IN}	—	11	15	mA	I_{OUT}=1mA, V_{CC}=9V
Input voltage range	V_{IN}	3.0	—	16	V	$I_{OUT} \leqq$ 0.5mA, $V_1=V_{OUT}$
Output voltage	V_{OUT}	30.0	—	35.0	V	I_{OUT}=1mA, $V_1=V_{OUT}$
Supply voltage regulation	$\Delta V_O/\Delta V_{CC}$	—	—	50	mV	I_{OUT}=1mA, ΔV_{CC}=7~11V
Temperature regulation	$\Delta V_O/\Delta T_a$	—	±1.0	—	mV/°C	I_{OUT}=1mA, ΔT_a=−20~75°C
Output current	I_{OUT}	—	—	3.0	mA	V_{CC}>9V, ΔV_0<50mV
Maximum impressed voltage, pin 2	$V_{2\,MAX}$	—	—	42	Vp-p	Maximum impressed voltage at blocking oscillation
Oscillation frequency	f	—	100	—	kHz	I_{OUT}=1mA, L=4.7mH

ELECTRICAL CHARACTERISTIC CURVES

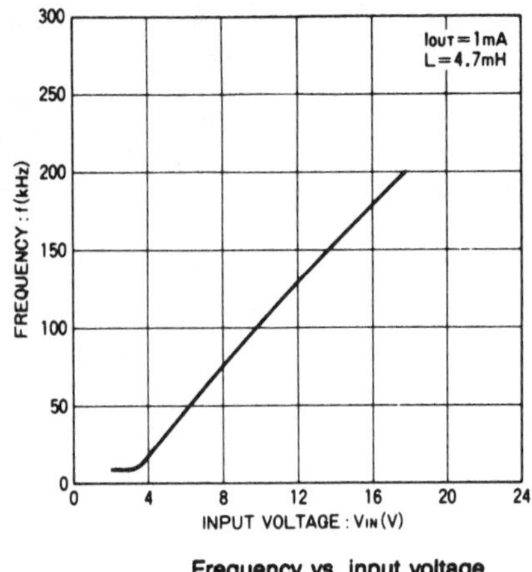

Frequency vs. input voltage

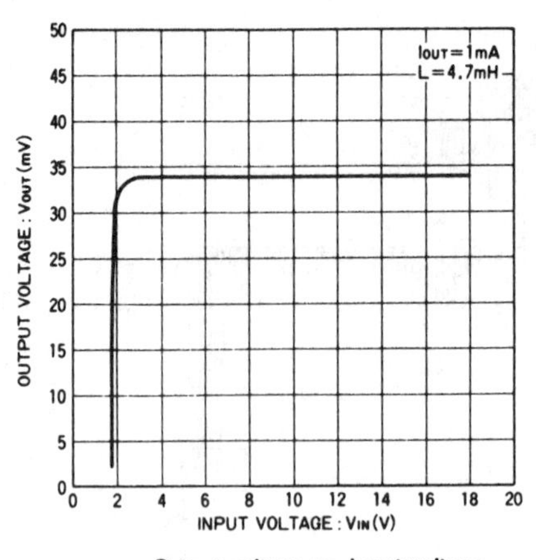

Output voltage vs. input voltage

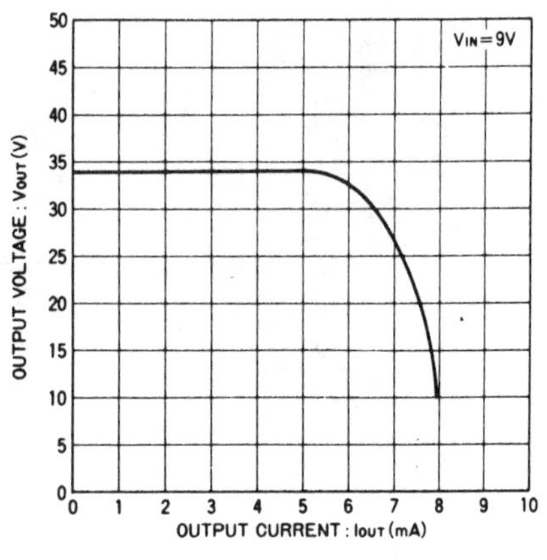

Output voltage vs. output current

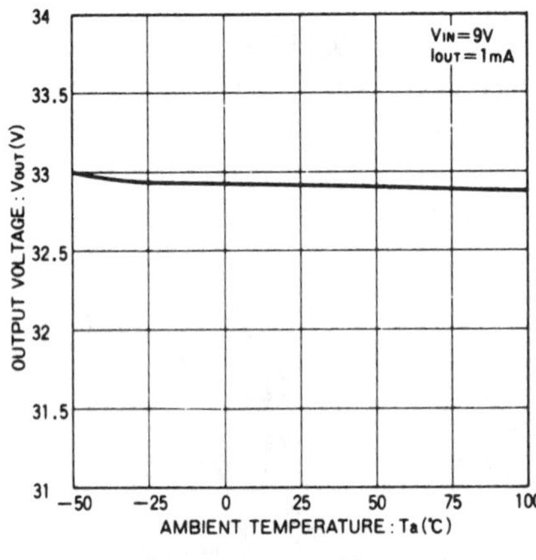

Output voltage vs. ambient temperature

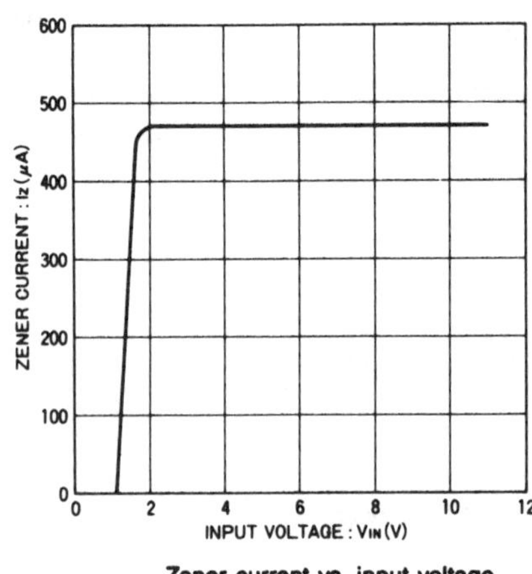

Zener current vs. input voltage

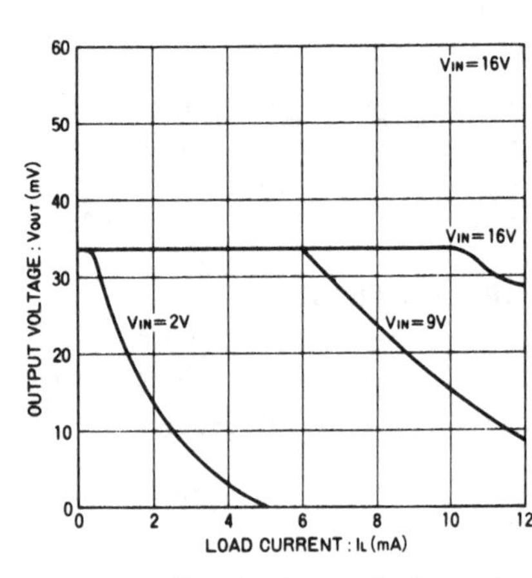

Output voltage vs. load current

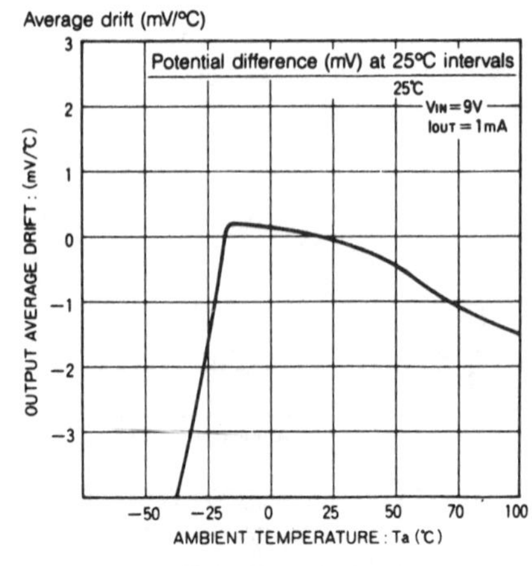

Output average drift vs. ambient temperature

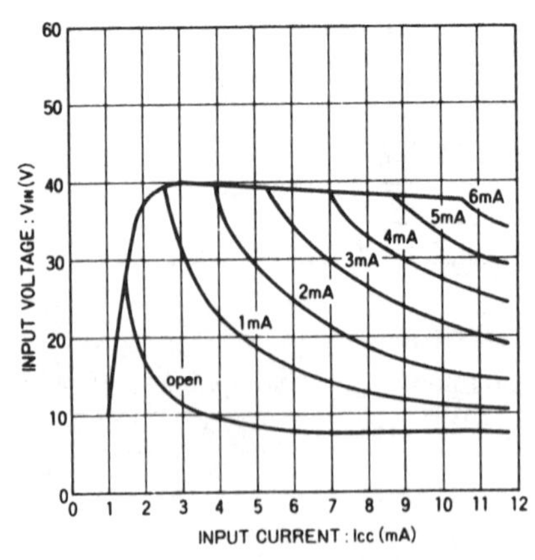

Input voltage vs. input current

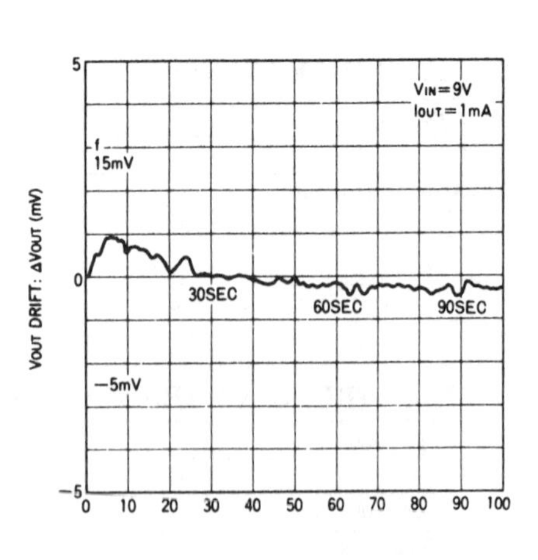

Output voltage drift

BLOCK DIAGRAM

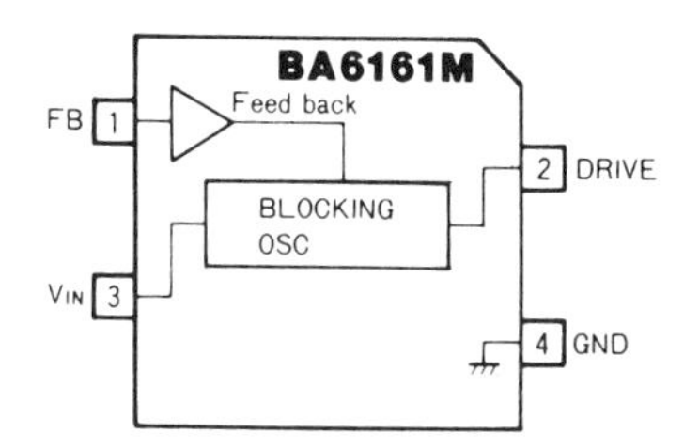

APPLICATION EXAMPLE

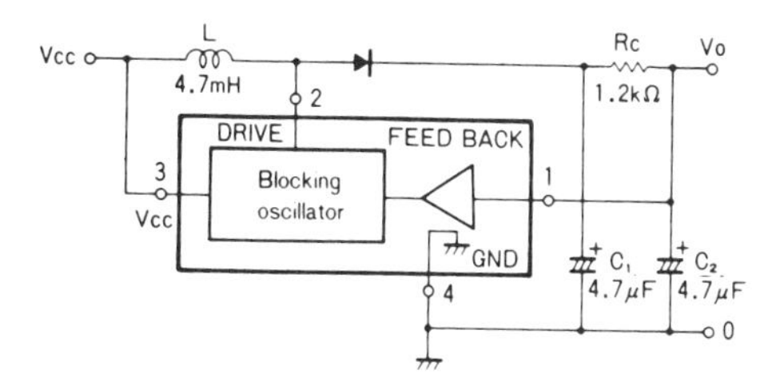

Rohm
BA9700F
Switching Regulator

- Output mode selectable (from step-up, step-down, inverting, etc.)
- Low current consumption (2.0 mA typ.)
- Wide oscillation frequency range
- Internal 2.6V reference voltage source
- Low input voltage capability (down to 3.6V)
- Internal dead-time control, which prevents overcurrent in the event of an output short-circuit
- Power switch input pin can be used to shut down all the internal circuit functions
- 1/2 V_{ref} pin is convenient for biasing the error amplifier
- Available in 14-pin MF package for compact system design

APPLICATIONS

- Products requiring a high supply voltage from low input voltage (3.6V min.)
- Products requiring low supply voltage from high input voltage

The BA9700F is a PWM (pulse-width modulation)-type monolithic switching regulator containing all the basic regulator circuitry. The device consists of a reference voltage source, error amplifier, triangular wave oscillator, PWM comparator, output transistor, and I/O buffers (using JFETs).

ELECTRICAL CHARACTERISTICS (Unless otherwise specified, Ta=25°C, V_{CC}=5.0V)

Parameter	Symbol	Min.	Typ.	Max.	Unit	Conditions
[Reference voltage block]						
Output voltage	Vref	—	2.6	—	V	I_{OR}=1mA
Line regulation	L_{INE}	—	3.0	—	mV	V_{CC}=3.5 ~ 24V
Load regulation	L_{OAD}	—	1.0	—	mV	I_{OR}=0.1 ~ 1mA
Short-circuit output current	I_{OSC}	—	20.0	—	mA	Vref=0V
[Triangular wave oscillator block]						
Oscillation frequency	f_{OSC}	—	240	—	kHz	Ct=330pF, Rt=10kΩ
[Error amplifier block]						
Input offset voltage	V_{IO}	−6	—	+6	mV	—
Open loop gain	A_Y	—	80	—	dB	—
Unity gain bandwidth	G_B	—	1.5	—	MHz	—
Common-mode rejection	CMRR	—	80	—	dB	—
[PWM comparator block]						
Input bias current	I_{IB}	—	180	—	nA	—
Input threshold voltage	V_{t0}	—	1.90	—	V	Duty cycle 0%
Input threshold voltage	V_{t100}	—	1.32	—	V	Duty cycle 100%
Maximum duty cycle		0	—	100	%	—
[Output block and entire device]						
Output saturation voltage	Vsat	—	1.5	—	V	—
Standby current	I_{CCS}	—	2.0	—	mA	When output is off

DIMENSIONS (UNIT: MM)

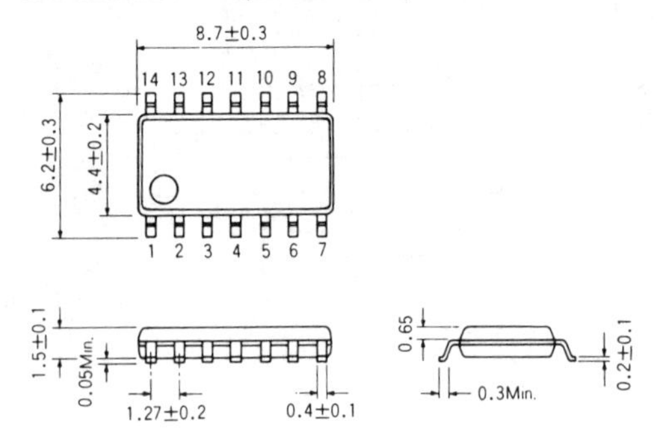

BLOCK DIAGRAM

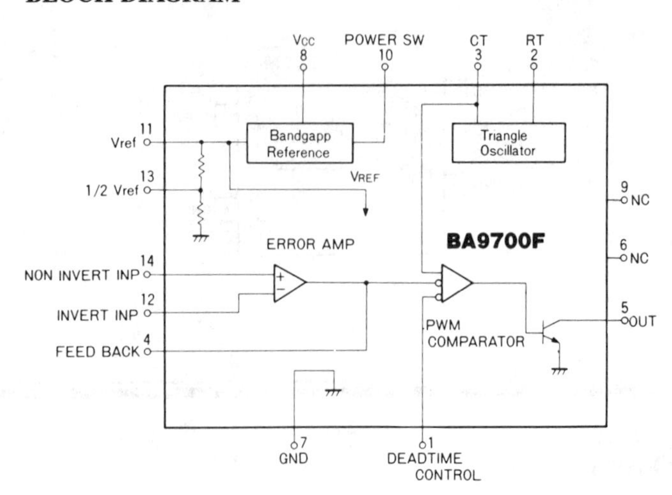

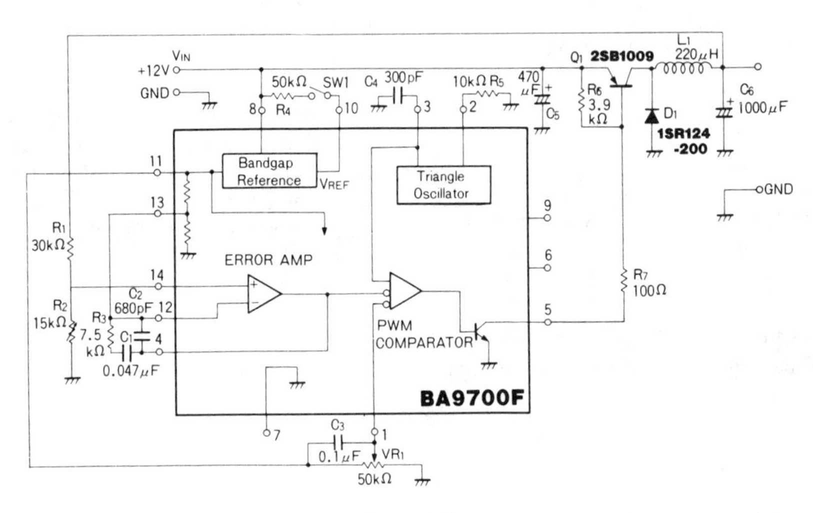

STEP DOWN CONVERTER

This information is reproduced with permission of Siliconix, Inc.

Siliconix Si7660

Monolithic CMOS Voltage Converter

- Improved-no external diode required
- Conversion of 5V logic supply to ±5V supplies
- 99.7% typical open circuit voltage conversion efficiency
- 95% typical power efficiency
- Operating voltage range of 1.5 to 10V
- Requires only 2 capacitors
- Inexpensive negative supply from positive supply
- Easy to use
- Minimum parts count
- Small size
- No diode drop at output

APPLICATIONS

- On-board negative supply for dynamic RAMs
- Localized μ-processor (8080-type) negative supplies
- Inexpensive negative supplies for analog switches
- Data acquisition systems

The Siliconix Si7660 is a monolithic CMOS power supply circuit that offers unique performance advantages over previously available devices. The Si7660 performs a supply voltage conversion from positive to negative for an input range of +1.5 to +10V, resulting in a complementary output voltage of −1.5 to −10V with the addition of only two capacitors.

Typical applications for the Si7660 are data acquisition- and microprocessor-based systems where a +5V supply is available for the digital functions, and an additional −5V supply is required for the analog functions. The Si7660 is also ideally suited for providing low current, −5V body bias supply for dynamic RAMs.

Contained on the chip is a voltage regulator, an RC oscillator, a voltage-level translator, four power MOS switches, and a logic network. This logic network senses the most negative voltage in the device and ensures that the output n-channel switch substrates are not forward-biased. The epitaxial layer prevents latch up.

The oscillator, when unloaded, oscillates at a nominal frequency of 12 kHz for an input supply voltage of 1.5 to 10V. The

OSC terminal can be connected to an external capacitor to lower the frequency or it can be driven by an external clock.

The LV terminal can be tied to ground to bypass the internal regulator and improve low-voltage (LV) operation. At high voltages (+3.5 to 10V), the LV pin should be left disconnected.

Packaging for this device includes the plastic miniDIP and SO options. Performance grades include industrial D-suffix (−40 to 85 °C), and commercial C-suffix (0 to 70 °C) temperature ranges.

ABSOLUTE MAXIMUM RATINGS

Supply Voltage .... 11 V
Oscillator Input Voltage . -0.3 V to (V+) +0.3 V, for V+ < 5.5 V
(V+) -5.5 V to (V+) +0.3 V, for V+ > 5.5 V
LV .... No connection for V+ > 3.5 V
Storage Temperature (C & D Suffix) .... -65 to 125°C
Operating Temperature (C Suffix) .... 0 to 70°C
(D Suffix) .... -40 to 85°C

Power Dissipation:*
8-Pin Plastic DIP** .... 300 mW
SO-8** .... 300 mW

*All leads welded or soldered to PC board.
**Derate 10 mW/°C above 75°C.

SPECIFICATIONS[a]

PARAMETER	SYMBOL	TEST CONDITIONS Unless Otherwise Specified V+ = 5 V, C_{OSC} = 0[g]	TEMP[e]	TYP[d]	C, D SUFFIX MIN[b]	C, D SUFFIX MAX[b]	UNIT
INPUT							
Supply Voltage Range LOW	$V+_L$	R_L = 10 kΩ, LV = GND	Full		1.5	3.5	V
Supply Voltage Range HIGH	$V+_H$	R_L = 10 kΩ, LV = NC	Full		3	10	V
Supply Current	I+	R_L = ∞ , LV = NC	Full	100		250	μA
OUTPUT							
Output Source Resistance	R_{OUT}	V+ = 5 V, LV = OPEN I_O = 20 mA	Room Full	55		100 120	Ω
		V+ = 2 V, LV = GND I_O = 3 mA	Full			300	Ω
Power Conversion Efficiency	PE_1	R_L = 5 kΩ	Room	98	95		%
Voltage Conversion Efficiency	$V_{OUT}E_1$	R_L = ∞	Room	99.9	99		%
DYNAMIC							
Oscillator Frequency[g]	f_{OSC}		Room	12			kHz
Oscillator Impedance	Z_{OSC}	V+ = 2 V, LV = GND	Room	1			MΩ
		V+ = 5 V	Room	100			kΩ

NOTES:
a. Refer to PROCESS OPTION FLOWCHART for additional information.
b. The algebraic convention whereby the most negative value is a minimum and the most positive a maximum, is used in this data sheet.
c. Guaranteed by design, not subject to production test.
d. Typical values are for DESIGN AID ONLY, not guaranteed nor subject to production testing.
e. Room = 25°C, Full = as determined by the operating temperature suffix.
g. For C_{OSC} > 1000 pF, C_1 and C_2 should be increased to 100 μF. C_1 = Pump Capacitor, C_2 = Reservoir Capacitor.

PIN CONFIGURATION

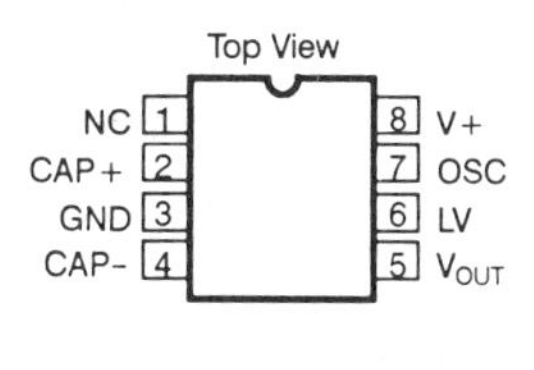

Order Number:
Plastic: Si7660CJ

SO Package
(Same pinout as DIP)

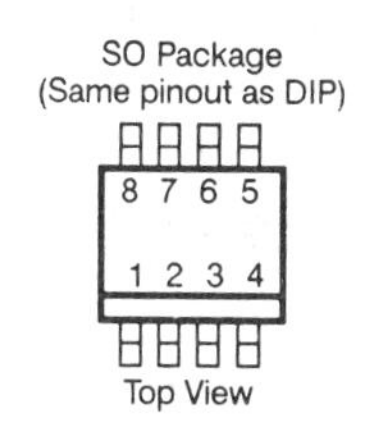

Order Number:
Si7660DY

TYPICAL CHARACTERISTICS

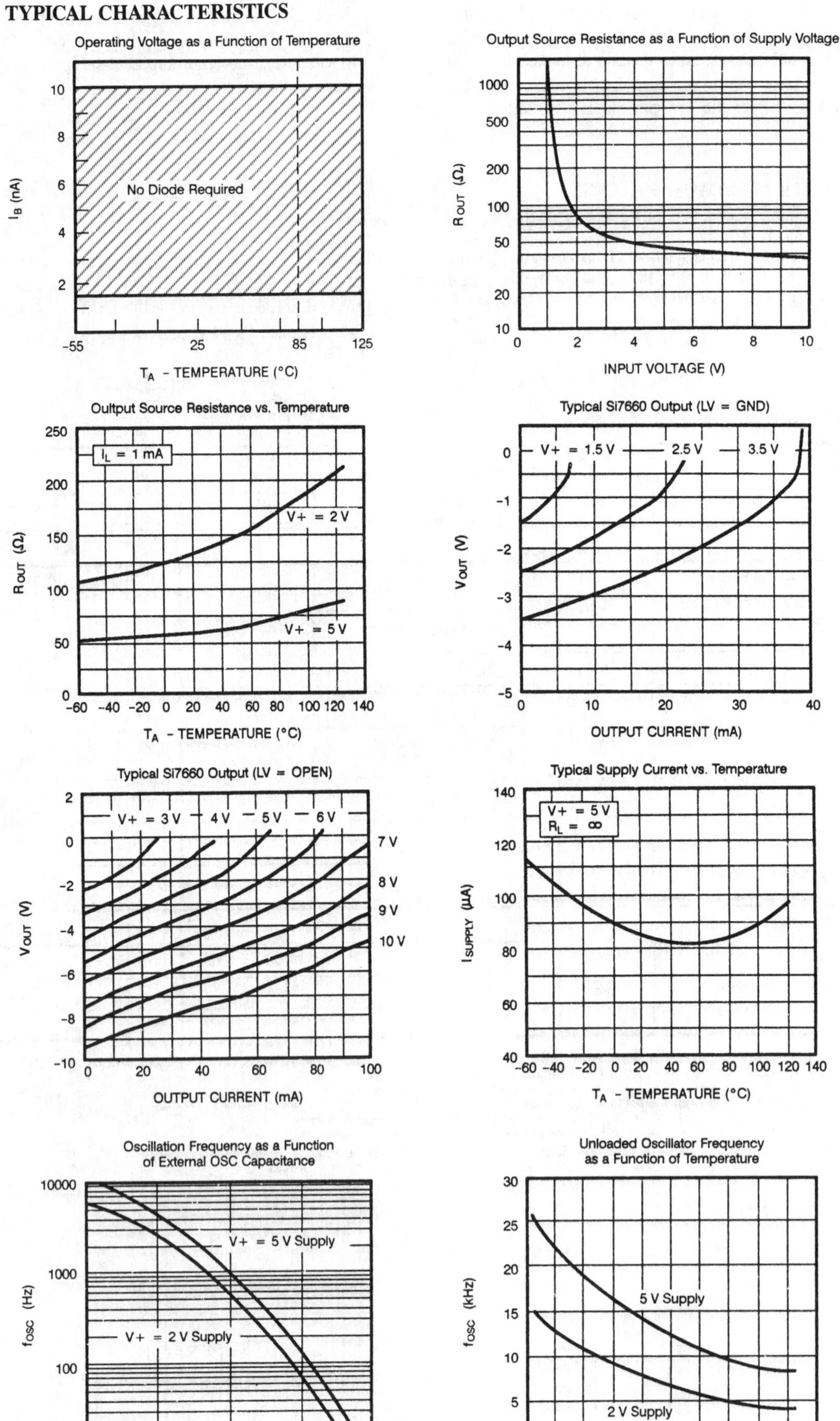

FUNCTIONAL BLOCK DIAGRAM

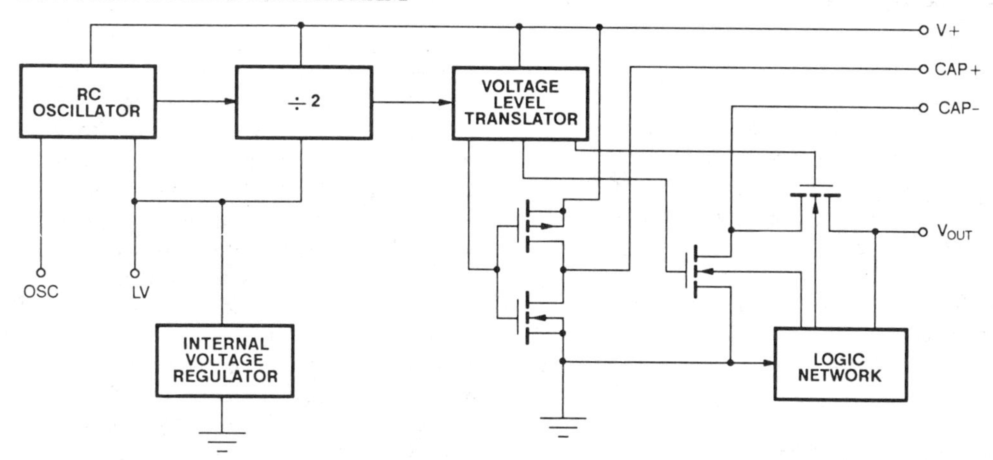

APPLICATIONS

The Siliconix Si7660 is a voltage source, not a current source. Therefore, any heavy load current will either greatly reduce the output voltage (possibly out of the desired range) or cause the device to go into power shutdown. To avoid problems, keep the voltage conversion concept in mind.

The Si7660 is intended for use as a voltage inverter. However, with a few added components, the inverter circuit can be rearranged to provide many different voltage levels. Some of the possibilities include voltage inversion, voltage multiplication, and even simultaneous inversion and multiplication.

In many applications, a low current negative-supply made with an Si7660 would do just as well as a full conventional negative supply or dc-to-dc converter module. Some examples are negative power supplies for microprocessors, dynamic RAMs, or data-acquisition systems.

If the output ripple of the Si7660 is too great for a particular application, the value of the pump and reservoir capacitors can be increased to reduce this effect. However, it is important to note that increasing the capacitor size can lead to surge currents at turn-on. If the current is too great, the power dissipation of the device can be exceeded, causing destruction of the device. The maximum recommended capacitor size is 1000 μF.

The previous version of the Si7660 required a diode in series with pin 5 when operating above 6.5V. The improved Si7660 does not require this diode. The improved version will work in existing circuits that have the diode.

Also shown is a circuit that will produce two output voltages utilizing both of the Si7660 features (i.e., inversion and doubling). The combined output current must be limited so the maximum device dissipation is not exceeded.

Two Si7660's can be paralleled to reduce the effective output resistance of the converter. The output voltage at a given current is increased because the voltage drop is halved when the devices are connected.

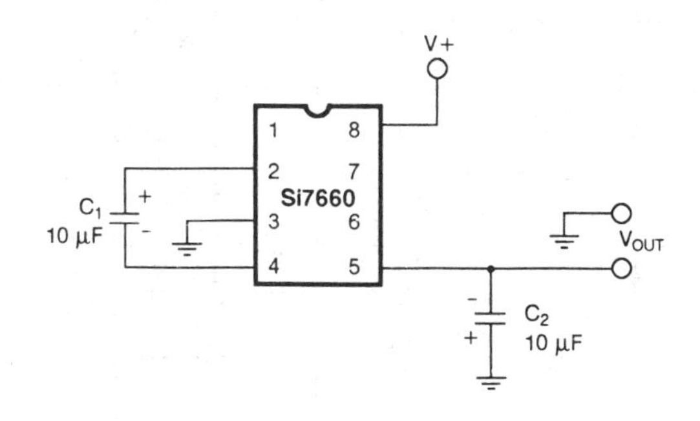

Basic Voltage Inverter Circuit

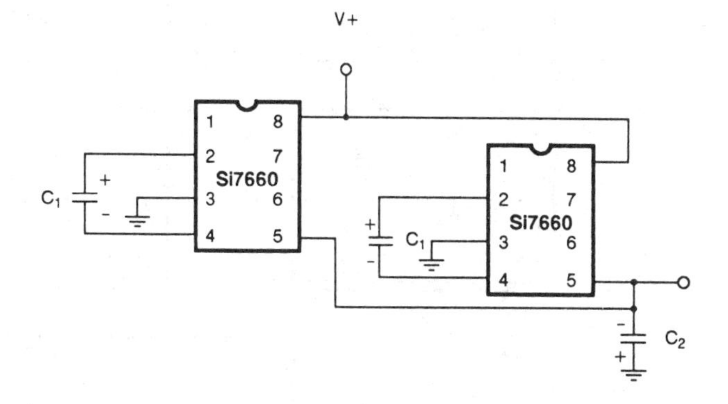

Paralleling Two Si7660s to Reduce the Effective Output Resistance

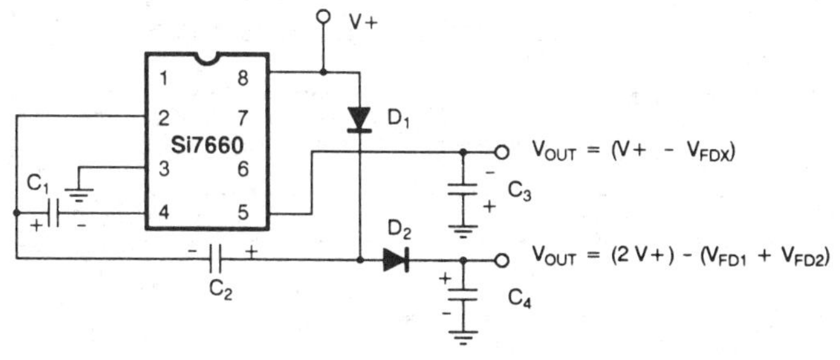

Combination Inverter/Multiplier Circuit

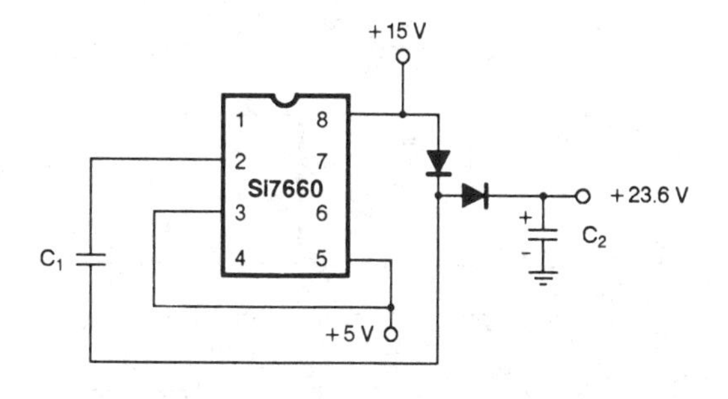

Creating +23.6 V from +15V and +5 V

Siliconix Si7661

Monolithic CMOS Voltage Converter

- Conversion of 4.5 to 20V to −4.5 to −20V supplies
- Voltage multiplication ($V_{OUT} = (-)nV_{IN}$)
- 99.7% typical open-circuit voltage-conversion efficiency
- 95% typical power efficiency
- Inexpensive negative supply generation
- East to use, requires only 2 external capacitors
- Minimum parts count
- Small size

APPLICATIONS

- On-board negative supply for dynamic RAMs
- Localized μ-processor (8080-type) negative supplies
- Inexpensive negative supplies for analog switches
- Data-acquisition systems
- Up to −20V for op amps and other linear circuits

The Siliconix Si7661 is a monolithic CMOS power supply circuit that offers unique performance advantages over previously available devices. The Si7661 performs a supply voltage conversion from positive to negative for an input range of +4.5 to +20V, resulting in a complementary output voltage of −4.5 to −20V with the addition of only two capacitors.

Typical applications for the Si7661 are data acquisition- and microprocessor-based systems, where a +4.5 to +20V supply is available for the digital functions, and an additional −5 to −20V supply is required for analog devices, such as op amps. The Si7661 is also ideally suited for providing low current, −5V body bias supply for dynamic RAMs.

Contained on the chip is a voltage regulator, an RC oscillator, a voltage-level translator, four power MOS switches, and a logic network. This logic network senses the most negative voltage in the device and ensures that the output n-channel switch substrates are not forward-biased. An epitaxial layer prevents latchup.

The oscillator, when unloaded, oscillates at a nominal frequency of 10 kHz for an input supply voltage of 4.5 to 20V. The OSC terminal can be connected to an external capacitor to lower the frequency or it can be driven by an external clock.

The LV terminal can be tied to ground to bypass the internal regulator and improve low-voltage (LV) operation. At high voltages (+8 to +20V), the LV pin should be left disconnected. Packaging for this device is the 8-pin MiniDIP.

ABSOLUTE MAXIMUM RATINGS

Supply Voltage .... 22 V
Oscillator Input Voltage ... −0.3 V to (V+) +0.3 V, for V+ <8 V
(V+) −8 V to (V+) +0.2 V, for V+ >8 V
LV .... No connection for V+ >9 V
Storage Temperature (C Suffix) .... −65 to 125°C
Operating Temperature (C Suffix) .... 0 to 70°C

Power Dissipation:*
8-Pin Plastic DIP** .... 500 mW

*All leads welded or soldered to PC board.
**Derate 6.6 mW/°C above 25°C.

SPECIFICATIONS[a]

PARAMETER	SYMBOL	TEST CONDITIONS Unless Otherwise Specified $C_{OSC} = 0$[g]	TEMP[e]	TYP[d]	C SUFFIX MIN[b]	C SUFFIX MAX[b]	UNIT
INPUT							
Supply Voltage Range (LV)	$V+_{LV}$	R_L = 10 kΩ, LV = 0 V	Full		4.5	9	V
Supply Voltage Range	V+	R_L = 10 kΩ, LV = OPEN	Full		8	20	
Supply Current	I+	V+ = 4.5 V, R_L = ∞, LV = 0 V	Room	100		500	μA
		V+ = 15 V, R_L = ∞, LV = OPEN	Room	0.7		2	mA
OUTPUT							
Output Source Resistance	R_{OUT}	V+ = 4.5 V, LV = 0 V, I_O = 3 mA	Room	75			Ω
		V+ = 15 V, LV = OPEN, I_O = 20 mA	Room Full	55		100 120	
Power Conversion Efficiency	PE_1	V+ = 15 V, R_L = 2 kΩ	Room	92			%
Voltage Conversion Efficiency	$V_{OUT}E_1$	V+ = 15 V, R_L = ∞	Room	99.7	97		
DYNAMIC							
Oscillator Frequency[g]	f_{OSC}	V+ = 15 V	Room	10			kHz
Oscillator Impedance	Z_{OSC}	V+ = 4.5 V, LV = 0 V	Room	1			MΩ
		V+ = 15 V	Room	100			kΩ

NOTES:
a. Refer to PROCESS OPTION FLOWCHART for additional information.
b. The algebraic convention whereby the most negative value is a minimum and the most positive a maximum, is used in this data sheet.
c. Guaranteed by design, not subject to production test.
d. Typical values are for DESIGN AID ONLY, not guaranteed nor subject to production testing.
e. Room = 25°C, Cold and Hot = as determined by the operating temperature suffix.
g. For C_{OSC} > 1000 pF, C_1 and C_2 should be increased to 100 μF. C_1 = Pump Capacitor, C_2 = Reservoir Capacitor.

TYPICAL CHARACTERISTICS

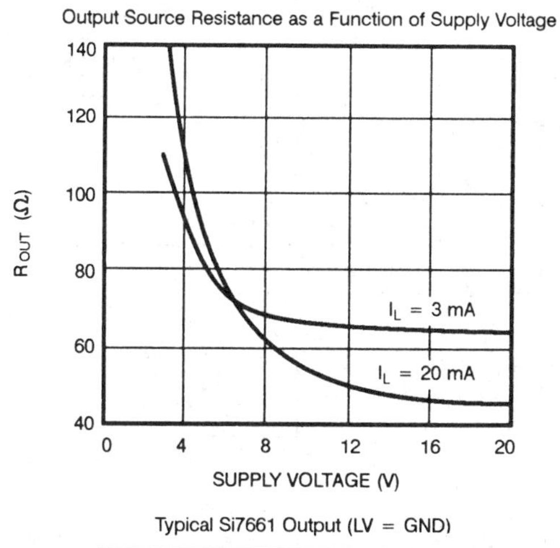

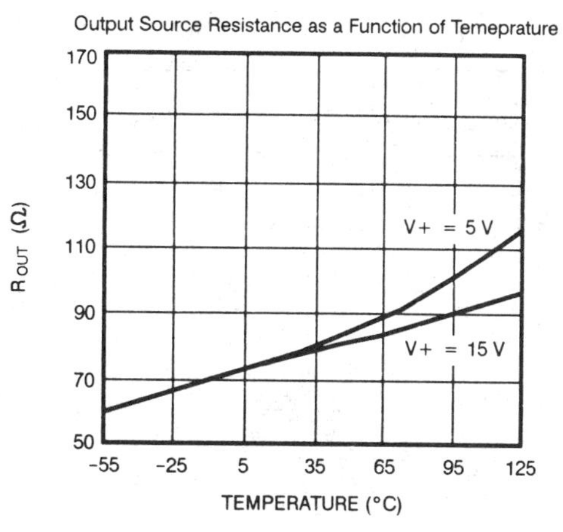

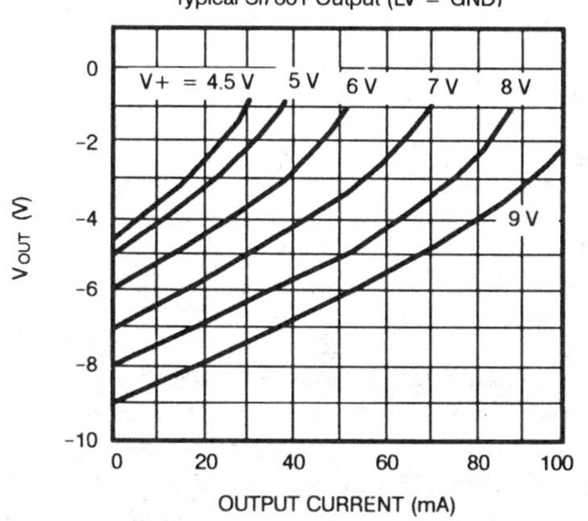

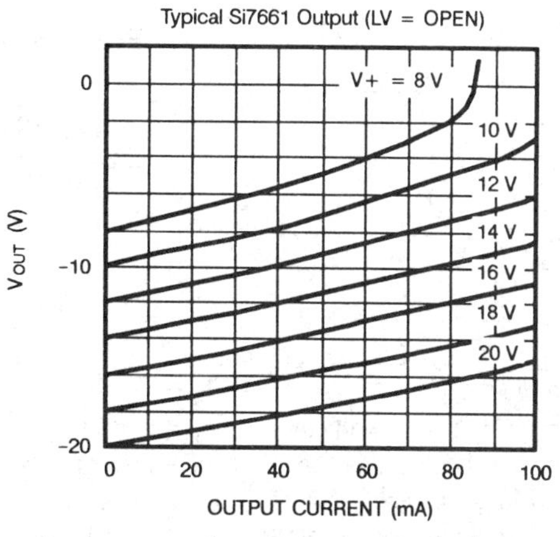

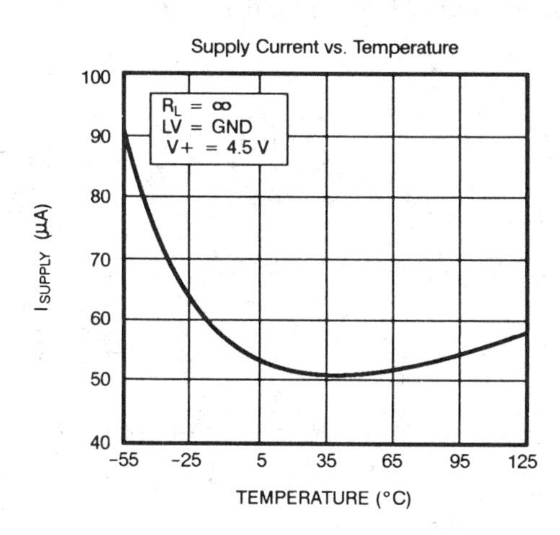

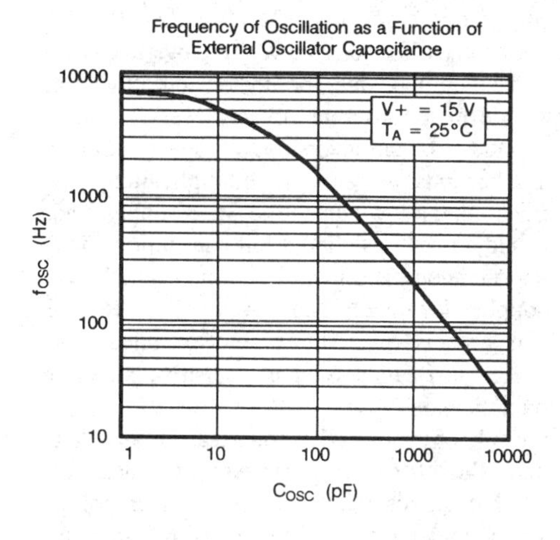

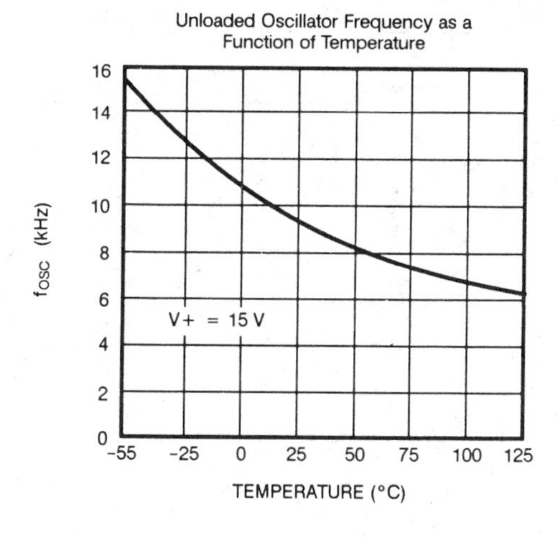

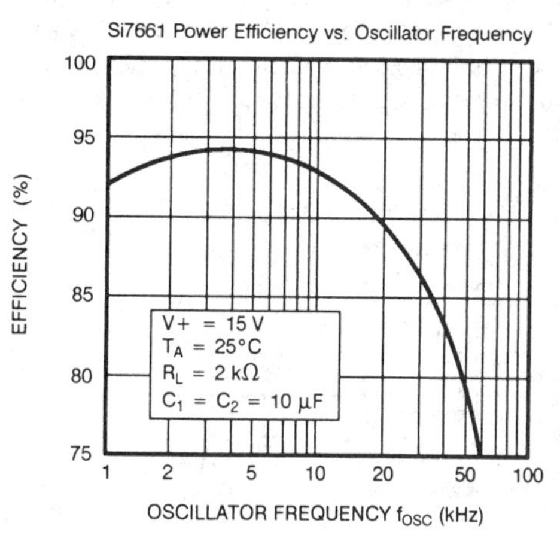

FUNCTIONAL BLOCK DIAGRAM

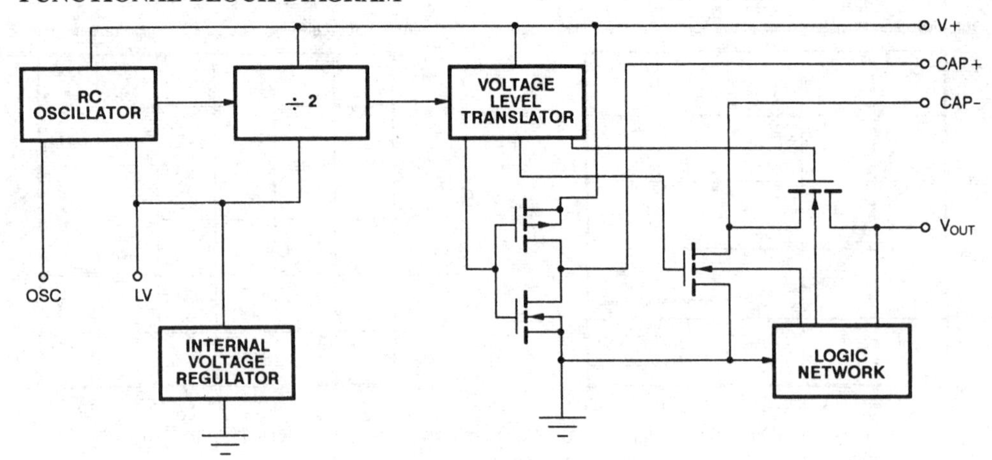

PIN CONFIGURATION

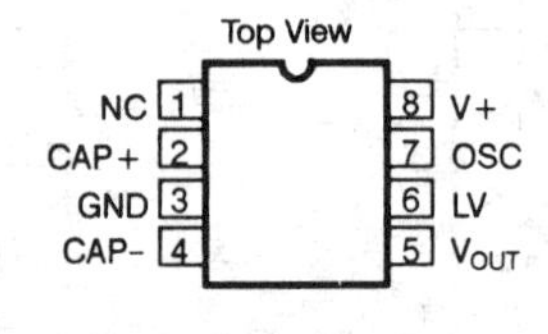

Order Number:
Plastic: Si7661CJ

APPLICATIONS

The Siliconix Si7661 is a voltage source, not a current source. Therefore, any heavy load current will either greatly reduce the output voltage (possibly out of the desired range) or will cause the device to go into power shutdown. To avoid problems, keep the voltage conversion concept in mind.

The SI7661 is intended for use as a voltage inverter. However, with a few added components, the inverter circuit can be rearranged to provide many different voltage levels. Some of the possibilities include voltage inversion, voltage multiplication, and even simultaneous inversion and multiplication.

In many applications, a low current negative supply made with an Si7661 would do just as well as a full conventional negative supply or dc-to-dc converter module. Some examples are negative power supplies for microprocessors, dynamic RAMs, or data-acquisition systems. In addition, the extended input voltage range of the Si7661 can be used as a negative generator for most op-amp applications.

If the output ripple of the Si7661 is too great for a particular application, the value of the pump and reservoir capacitors can be increased to reduce this effect. However, it is important to note that increasing the capacitor size can lead to surge currents at turn-on. If the current is too great, the power dissipation of the device can be exceeded, causing destruction of the device. The maximum recommended capacitor size is 1000 μF.

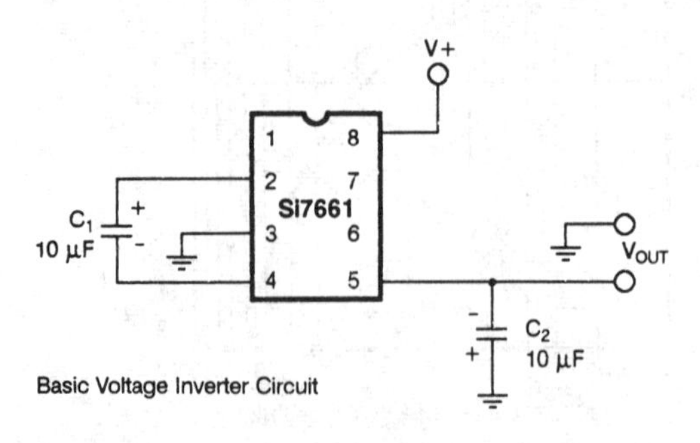

Basic Voltage Inverter Circuit

When an external clock is used to drive the Si7661, a 1 kΩ resistor should be used between the clock source and the OSC input (Pin 7). Also shown is a regulator that will operate with much less than 1V drop between V+ and V_{OUT} at large output currents. Most three terminal voltage regulators would exhibit a drop of a volt or more under these conditions.

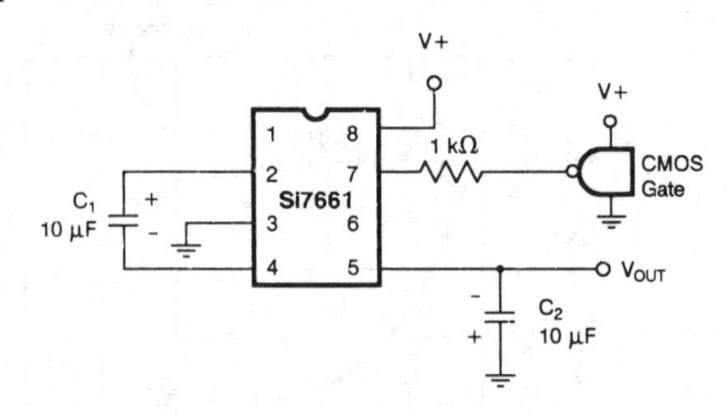

Driving the Si7661 with an External Clock

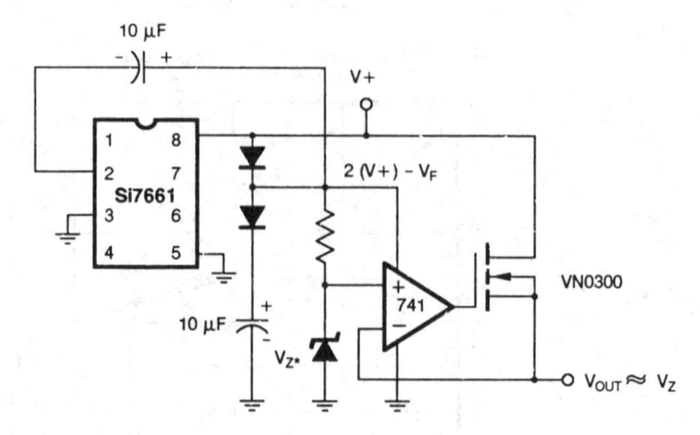

*The Zener voltage sets the output of the regulator.

Low Loss Regulator Circuit

Siliconix
Theory and Applications of the Si7660 and Si7661 Voltage Converters

by Doyle L. Slack.
Revised February 1990.

Many times a simple digital circuit design can be greatly complicated by the needs of just one or two of the onboard devices. For example, analog devices often used along with digital circuits (such as op amps and data-acquisition systems) are notorious for negative voltage requirements of −5, −10, or −15V when everything else in the circuit needs only positive voltages. Until recently, the only answer was to either buy a dc-to-dc converter module (expensive) or redesign the power supply to generate the negative voltages (expensive and wasteful in parts count and space). This application note represents the best alternative to this problem: the Si7660 and Si7661 monolithic voltage converters. With the Si7660 and Si7661, negative voltages from 1.5 to 20V can be generated from a positive supply with minimum parts count and minimum cost.

THEORY OF OPERATION

The basic theory behind the Si7660 and Si7661 is based on the ideal voltage doubler shown in Figure 1. Capacitor C_1 is the pump capacitor, and C_2 is the reservoir capacitor. The pairs of switches (S_1 with S_3, and S_2 with S_4) are driven by an oscillator/toggle circuit, providing charge and transfer cycles of equal length.

During the charge cycle, S_1 and S_3 are closed, and current flows into C_1, charging it to the value of V_{IN}. The oscillator toggle then changes state, and the transfer cycle begins. S_1 and S_3 are opened while S_2 and S_4 are closed, allowing C_1 to dump charge into C_2 until the potential across them has equalized. The oscillator/toggle then switches again, and the process starts over.

For no load conditions, the voltage inversion will be virtually perfect because the amount of charge that must be transferred from C_1 to C_2 will be limited to losses due to leakage from C_2 and any parasitic capacitances. As the load increases, C_1 must transfer more and more charge to make up for the depletion of C_2 as it supplies current to the output during the charge cycle. This action causes the output voltage to drop, making the circuit appear to be a perfect inverter in series with an output resistor that varies in magnitude with the input voltage. Figure 2 shows this concept in a two port diagram of the device, and Figure 3 illustrates the typical output characteristics of both devices configured in the inverter mode.

CIRCUIT OPERATION

With the Si7660 and Si7661, the only parts of the doubler not included inside the package are the pump and reservoir capacitors. The internal switches are made with p-channel and n-channel MOSFETs. The main difference between the Si7660 and Si7661 is the breakdown voltage of the MOSFETs, which in turn dictates the maximum input voltage. Also, the design of the Si7661 offers a much greater resistance to device latchup, discussed later. Because the internal sections of the two devices are very similar, description of the operation of the Si7660 and Si7661 is combined. The internal sections of the circuit are the oscillator, divider, regulator, level translator, and substrate logic. Figure 4 shows a block diagram of the internal sections of the inverter circuit.

The oscillator supplies the signal to the divider section, which in turn drives the rest of the circuit. The OSC input has an input impedance of approximately 1 MΩ. This allows the internal oscillator to be overridden by an external clock or to be slowed down by the addition of an external capacitor.

The internal regulator is a series voltage regulator with a zener reference to ensure that low-voltage components of the circuit are provided with no more than 5V when the input voltage is greater than 5V. It also provides current limiting for the oscillator and divider circuits. When the input voltage is less than 3.5V for the Si7660 (less than 9.0V for the Si7661), the LV pin is grounded, bypassing the internal regulator. However, when the Si7660 is operated above 3.5V, the LV pin must be left open to provide latchup protection. For the Si7661, LV should be left open above 9.0V for proper operation.

The divide-by-two counter provides complementary outputs. Q and $\overline{Q}$ drive the inputs of the level translators, which in turn provide the necessary switching voltages to drive the MOSFET switches. The built-in delay of the translators guarantees that break-before-make action occurs.

The substrate logic network ensures that two things happen. First, it makes sure that the substrate-source/drain junctions at Q_3 and Q_4 are never forward-biased, and that the on-resistance of each of the output transistors will be as low as possible for all operating conditions. Second, the network determines the most negative voltage in the device and uses it to supply power to the level translator.

LATCHUP

Because of the basic internal four-layer geometry of CMOS devices, an SCR action can sometimes occur. Figure 5 depicts the SCR structure. This SCR action can, under certain conditions, cause the Si7660 device to latch up. As Figure 5 shows, the source of the n-channel device becomes the SCR cathode; the source of the p-channel is the anode; and either drain can act as a gate. Since an SCR does not trigger until certain conditions occur, it can sometimes cause no problems at all, yet sometimes it might be fatal to the CMOS device.

The intrinsic SCR needs three conditions to cause latchup; the betas of two parasitic transistors must be greater than one; the current flowing through the channels of the devices must be greater than the holding current of the SCR; a pulse must be applied to one of the gates to trigger the SCR action. The trigger can come from several sources: the power-up sequence might cause problems if the SCR gate receives power before the other terminals. Another possible trigger source is a high slew rate across the intrinsic SCR. When the SCR is triggered, the CMOS devices are suddenly shorted out, and the output impedance of the device becomes very low.

Q_4 of the Si7660 can sometimes experience the conditions to cause SCR latchup when operating at the upper end of the input voltage range. The nearby p-channel substrate logic transistors from the complementary part of the intrinsic SCR. When the SCR action does occur, the circuit suddenly appears to be a short-circuit between V_{IN} and V_{OUT}. The reservoir capacitor (C_2) rapidly discharges through this path. After C_2 has discharged, the current through Q_4 drops below the SCR holding value, and the circuit resets. If the conditions that originally caused the SCR action remain present, the device will latch up repeatedly. If the circuit input is not current limited, this action can sometimes dissipate too much power through Q_4 and the substrate logic, resulting in damage to the device.

To prevent damage to the Si7660 when conditions for latchup occur, older versions required a diode in series with the V_{OUT} pin to block the discharge of C_2 and keep the current below the holding value of the SCR. This diode was used whenever the input voltage could exceed 6.5V at room temperature. Figure 6 shows the operating range of the improved Si7660.

The Si7661 is a higher voltage device than the Si7660, and is designed on a different process. This high-voltage silicon-gate process reduces the parasitic betas in Q_4 to a value that makes it extremely difficult to produce the conditions for latchup. Because of this, the series diode is not needed for proper operation.

GENERAL APPLICATIONS

The Si7660 and Si7661 are intended for use as voltage inverters. However, with a few added components, the inverter circuit can be rearranged to provide many different voltage levels. In some configurations, they can even provide more than one voltage output at the same time. The possibilities include voltage inversion, voltage multiplication, and even simultaneous inversion and multiplication.

BASIC VOLTAGE INVERSION

With no load, the output voltage magnitude of the basic voltage inverter circuit shown in Figure 7 will typically be within 0.1% of the V+ (input voltage) magnitude for the Si7660 and within 0.3% of the input voltage magnitude for Si7661. As the load current increases, the output will drop as shown in Figure 8(a). The effective output resistance will vary with input voltage as given in Figure 8(b). Once the load current reaches its limit (30-40 mA for the 5V case), the inverter can no longer regulate the voltage properly and shuts down to protect itself from extreme power dissipation.

Caution: At higher input voltages (for either device), the output maximum limit can cause the power dissipation to exceed the maximum rating of the package (especially plastic). Always calculate the maximum power dissipation for your design.

The output ripple of the inverter is a function of the oscillator frequency as well as the size of the pump and reservoir capacitors. The nominal oscillator frequency is 12 kHz for the Si7660 and 10 kHz for the Si7661. Because the output ripple is important in some linear applications where supply noise is critical, Table 1 provides ripple values for different pump and reservoir capacitor values.

It is important to note that increasing the capacitor size can lead to other difficulties. The main problem is that the large capacitors might draw excessive amounts of current at turn-on. If the current is too great, the power dissipation of the device can be exceeded, causing destruction of the converter. Even when the device is running, the charge transfer under heavy loads can push the switches to their limits.

As stated before, the LV pin shorts out the internal regulator at low voltages when it is tied to ground. The LV pin should be grounded for operation below 3.5V for the Si7660 and 9.0V for the Si7661. However, it is necessary to leave the LV pin floating for high-voltage operation, as shown in Figure 9. Failure to do so could permanently damage the device. Figure 10 shows the inverter configured for low-voltage operation.

VOLTAGE MULTIPLICATION

Figure 11 is the schematic diagram of the voltage doubler. This circuit requires only two additional diodes and will provide positive voltage multiplication at the expense of the voltage drops of the two diodes in series with the output. This means the positive multiplier will not be able to provide the near-perfect output function like the basic inverter circuit does. The output voltage of the multiplier will be:

$$V_{OUT} = 2\ (V+)\ -2\ V_{diode}$$

The circuit of Figure 11 can also be used as a negative-to-positive voltage converter. To do so, set pin 8 to ground and pins 3 and 5 to the negative input voltage, V−. The output voltage will then be:

$$V_{OUT} = 1\ V-1\ -2V_{diode}$$

SIMULTANEOUS INVERSION AND MULTIPLICATION

The circuit shown in Figure 12 will provide both positive multiplication and inversion at the same time. The output voltages will be the same as those given in the equations. This configuration is limited by the load current that can be drawn out of either output before the circuit overloads.

PARALLEL CONNECTION

Although a single Si7660 and Si7661 cannot supply very large amounts of current, higher currents can be provided when several devices are connected in parallel. As shown in Figure 13, two or more inverter circuits can be paralleled to provide a lower output resistance, providing a smaller output voltage drop for a given current. This circuit will also expand the operating output current ranges slightly. Each device must have its own pump capacitor, but the reservoir capacitor is shared between all of the devices.

When two or more devices are paralleled, the output noise (ripple) will contain components not only at frequencies of each of the oscillators, but also at sum and difference frequencies as a result of a mixing action at the inverter outputs. If such noise cannot be tolerated, the OSC pin of one of the devices can be driven by an exclusive NOR gate that compares the oscillator frequencies of the two devices, as shown in Figure 14. This forces the two devices to alternate their charge and transfer cycles, which will not only reduce output noise, but will also maximize efficiency.

SERIES CONNECTION

When high-voltage inversion is desired, inverter circuits can be placed in series to produce voltage outside of the operating range of a single Si7660 or Si7661. Figure 15 shows two inverters cascaded to double the input voltage magnitude, while inverting the voltage at the same time.

When cascading devices, however, the power dissipation of each device must be considered. As each new stage is added, the previous stages will be subjected to more and more load current, from the quiescent current of the new stage and the multiplying action of the load current through each of the stages, as shown in Figure 16. As the number of cascaded devices increases, the effective output resistance also increases, which severely reduces the output voltage for a given current level, when compared to a single inverter. This effect can be reduced by paralleling devices in the first stages, though the cost in parts increases twofold for every added stage.

CHANGING THE OSCILLATOR FREQUENCY

The typical oscillator frequencies were given in the description of the basic inverter circuit. However, Figure 17(a) shows that the maximum power efficiency is not achieved at the typical oscillator frequency. If maximum power efficiency is desired, an external capacitor can be connected between the OSC pin and ground. Figure 17(b) illustrates the effect of added capacitance on the oscillator frequency. A resistor connected from the OSC pin to V+ can be used to increase the oscillator frequency. This will reduce ripple amplitude at the expense of reduced efficiency. Values above 2 MΩ are usually adequate.

If synchronization with an external driver or clock is needed, the OSC pin can be driven either by a TTL or CMOS logic gate. Figure 18 provides the proper circuits for interfacing to either logic standard. Note that the TTL interface can only be directly connected to the OSC pin if the circuit is using a 5V supply. If the input voltage is other than 5V, some type of buffer circuit will be required. The charge/transfer transitions will occur on each rising edge of the clock.

SPECIFIC APPLICATIONS

When looking at possible applications for the Si7660 and S7661, it must be remembered that these devices are voltage sources, not current sources. Therefore, any heavy load will either greatly reduce output voltage (possibly out of the desired range) or cause the device to go into power shutdown. If the concept of voltage conversion is kept in mind, many problems will be avoided.

In many places, a low-current negative supply made with an Si7660 or Si7661 would do just as well as a full conventional negative supply or dc-to-dc converter module. Some examples of possible uses are power sources for operational amplifiers, dynamic RAMs, microprocessors, and data-conversion products. Several examples of these systems follow:

MEMORIES

Several different memory manufacturers produce 16 K×1 dynamic RAM's that have a need for a −5V low-current supply to provide substrate biasing. The National MM5290, AMD AM9016, and THOMSON MK4116 all use this type of arrangement. Table 2 lists the −5V supply-current requirements for each of these devices.

The only constraint in using the Si7660 or Si7661 for this application is when calculating the voltage fluctuations that will occur when a location is read from or written to. Make sure that the absolute maximum current is considered so that the negative supply for the dynamic RAM will not be pulled down more than 5% (below 4.75V) during a memory read or write. Even with the maximum current taken into account, the Si7660 or Si7661 could easily provide the negative supply voltage supply for an entire 16 K × 8 dynamic memory bank.

OP AMPS

Operational amplifiers are one of the most commonly used integrated circuits and often use negative supply voltages. Although some op amps can supply high-current loads, more often the current requirements involved are well within the capabilities of the Si7661.

Figure 19 shows the Si7661 supplying the negative voltage to a 741 op amp configured as an inverting amplifier. As the current drain through the negative supply terminal of the op amp increases, the output voltage of the Si7661 will decrease. However, this will not affect the output capability of the op amp at its rated output current. Figure 20 illustrates this with a photograph of the input and output of the circuit in Figure 19 when a 1 kΩ load was placed on the amplifier output. The output was undistorted to 26 V peak-to-peak.

The output ripple of the Si7661 must be taken into consideration when using it as an op amp supply. Some op amps do not have adequate power supply rejection to withstand the ripple noise level of the Si7661. The pump and reservoir capacitors can be chosen to minimize this noise condition (see Table 1). The ripple should be measured at the maximum negative supply current (i.e., rated load) to determine if the Si7661 can be used to supply the op amp.

ANALOG SWITCHES

Although in most cases the Si7660 or Si7661 cannot supply sufficient current for analog switch applications, there are some exceptions. For example, Figure 21 is the schematic diagram of a circuit that was used to interface a Northern Telecom telephone set to an ICOM 2AT2-meter amateur radio transceiver. New designs should use the silicon-gate DG402 SPDT switch.

The analog switch provides isolation for the microphone and speaker connections of the transceiver, because the telephone set uses a single path for both transmission and reception. The telephone was operated at 12 V for direct interface to the DG305A, and the supply current from the Si7661 was <1 mA.

MICROPROCESSORS

Some of the older standard microprocessors need a negative supply for a substrate biasing. The Intel 8080 microprocessor is a good example of this. It is an inexpensive 8-bit CPU that has many different support chips available. To provide the negative supply voltage (−5 V), a basic inverter circuit (such as in Figure 7) using an Si7660 is connected to pin 11 of the microprocessor. The 8080 negative supply draws a maximum current of 1 mA, which will not pull down the supply voltage to any great degree.

REGULAR CIRCUITS

This section covers some of the possible methods for using the Si7660 or Si7661 in constant-voltage output circuits over a given output current range. For low-current inverter applications, the circuit shown in Figure 22 can be used. The output impedance of the circuit can be as low as 5 Ω with regulation up to approximately 20 mA. Note that if converters are paralleled on the output of this circuit, they should be synchronized to minimize output voltage fluctuations and output noise.

Another regulator application uses the Si7660 or Si7661 in a positive voltage regulator. Conventional three-terminal voltage regulators have a voltage drop of greater than 1 volt between the input and output when operating with fairly heavy load currents. The circuit given in Figure 23 uses an Si7660 or Si7661 voltage converter to double the voltage, which is then regulated by the op-amp and FET. This configuration allows regulation without the voltage drop, as long as the input voltage does not drop below the Zener voltage plus the product of I_D times $r_{DS(ON)}$.

Therefore, as long as the input voltage does not drop below 5.26 V, the input is guaranteed to be regulated as close to the zener voltage as can be attained by the common-mode offset voltage of the op-amp. By selecting the correct Zener diode, this circuit can supply more than 100 mA and can be adjusted for varying voltage outputs, up to the input voltage limit of the voltage converter.

CONCLUSION

The Si7660 and Si7661 are inexpensive alternatives to full negative supplies in many different low-cost applications. Although they are designed for generation of negative voltages, many different voltage levels can be generated with a few additional parts. The examples given here are only a few of the many possible applications that could utilize the benefits of reduced board space and cost that the Si7660 and Si7661 provide.

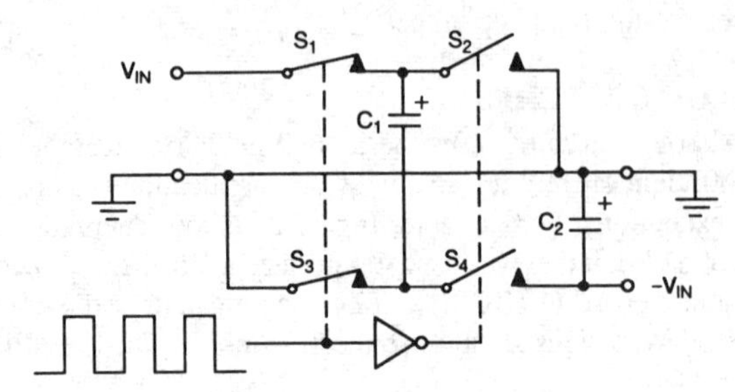

Fig. 1 The Ideal Voltage Doubler

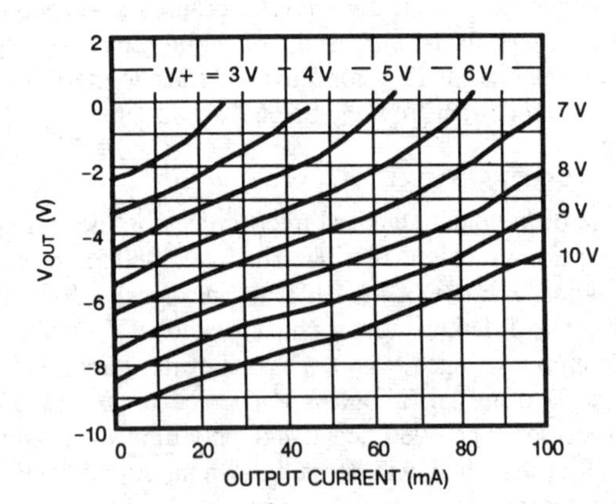

Fig. 3 Output Characteristic of the Si7660 Voltage Converter.

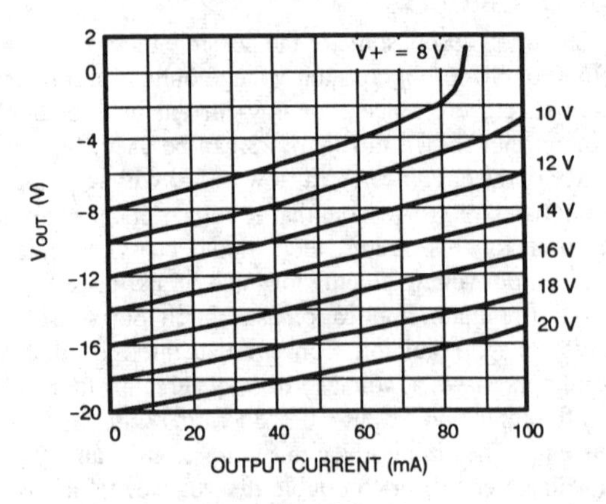

Output Characteristic of the Si7661 Voltage Converter.

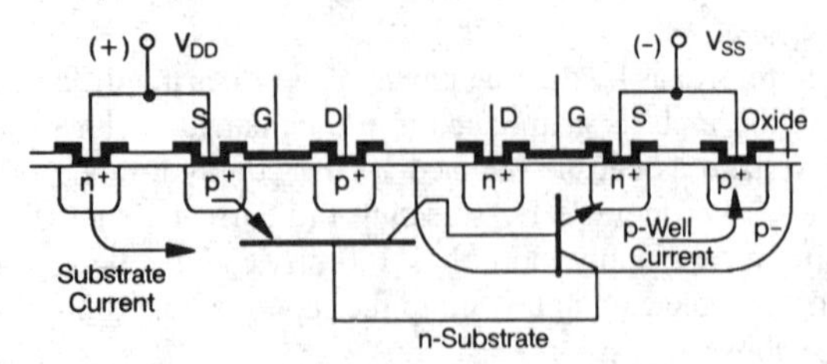

Fig. 5 Intrinsic SCR Superimposed on a CMOS Gate Structure

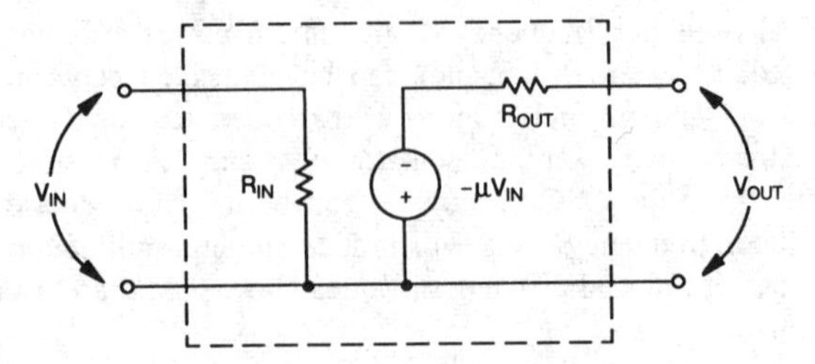

Fig. 2 Two-Port Diagram of the Voltage Converter Circuit

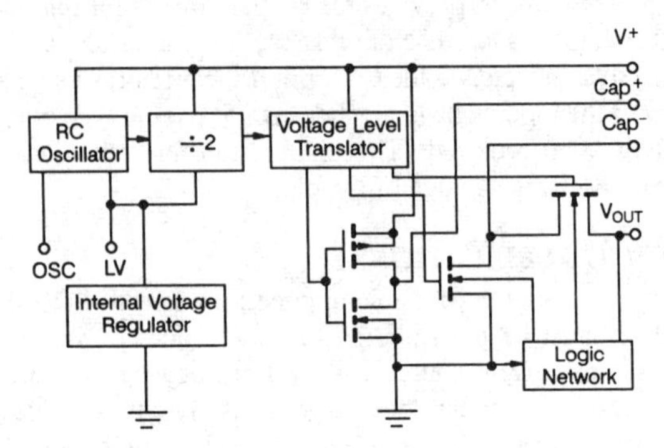

Fig. 4 Block Diagram of the Voltage Converter Circuit

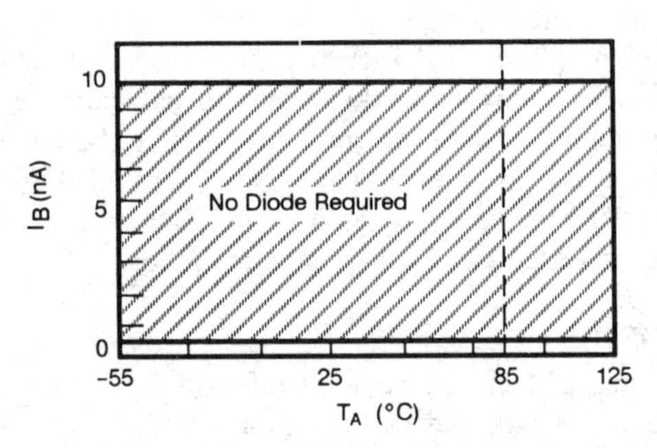

Fig. 6 Range of Input Voltage and Operating Temperature for the Si7660

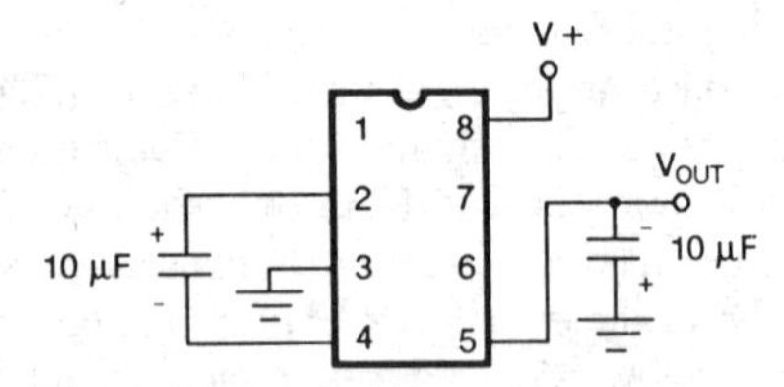

Fig. 7 Schematic Diagram of the Basic Inverter Circuit

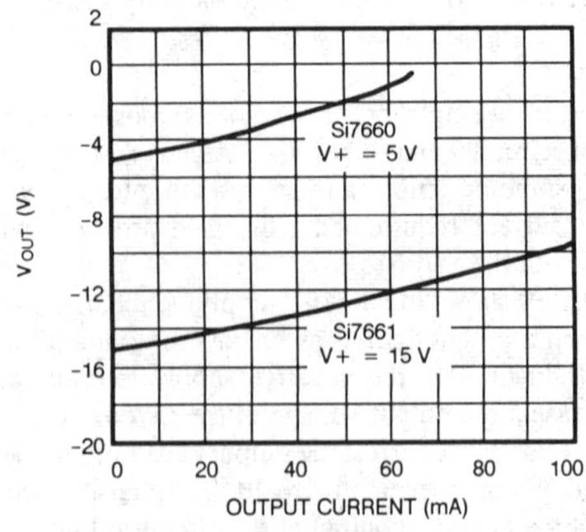

Fig. 8(a) Variation of Output Voltage as a Function of Output Current

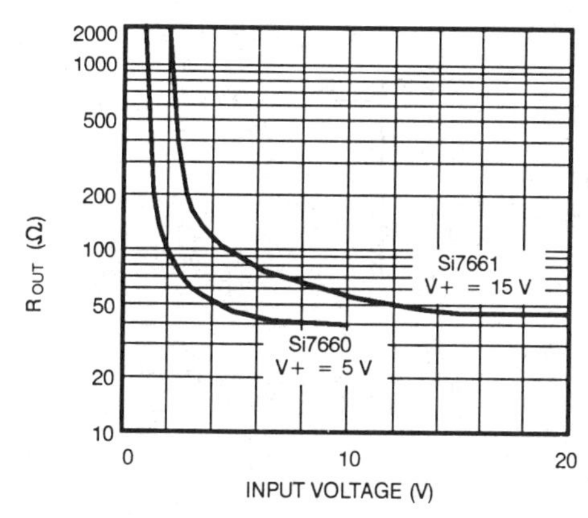

Fig. 8(b) Variation of Output Resistance as a Function of Input Voltage.

Table 1 Effect of Varying the Pump and Reservoir Capacitor Size on Output Ripple Noise.

	Capacitors (μF)	V_{OUT} (V)	VR (mV_{p-p})
Si7660	10	-3.838	150
Inverter Mode	22	-3.862	75
(see Figure 7)	47	-3.873	30
V+ = +5 V	100	-3.874	26
I_{OUT} = 10 mA	470	-3.879	10
	1000	-3.880	5
Si7661	10	-13.849	175
Inverter Mode	22	-13.872	80
(see Figure 7)	47	-13.882	38
V+ = +15 V	100	-13.883	29
I_{OUT} = 10 mA	470	-13.885	10
	1000	-13.890	5

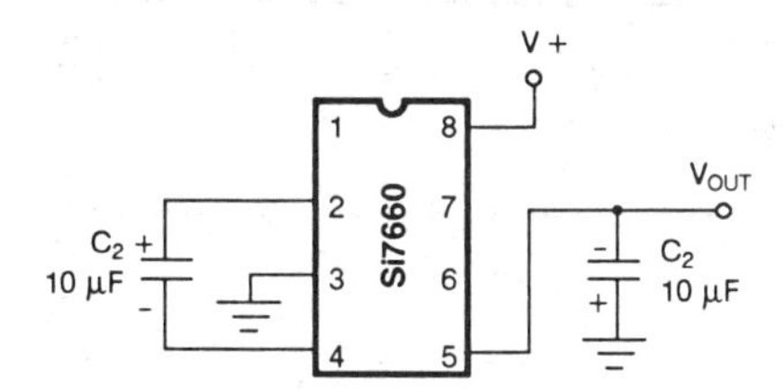

Fig. 9 Inverter Circuit Connections for High Voltage Operation

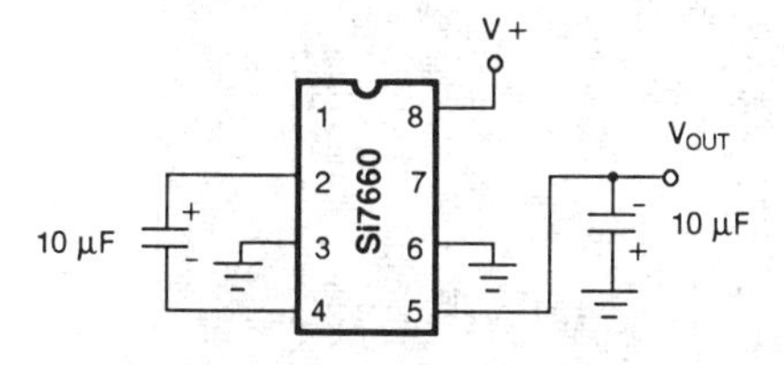

Fig. 10 Inverter Circuit Connections for Low Voltage Operation

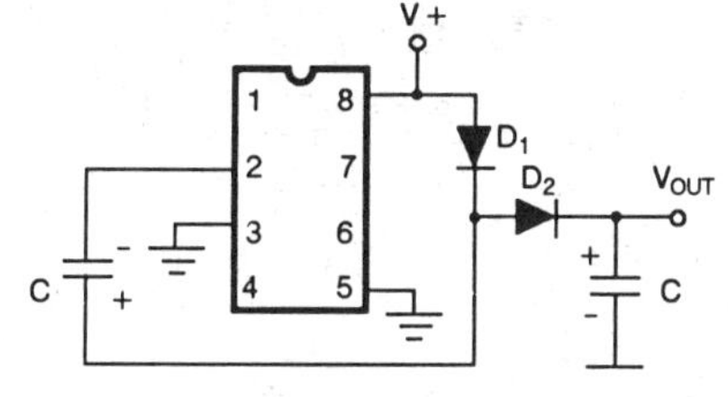

Fig. 11 Voltage Doubler Schematic Diagram

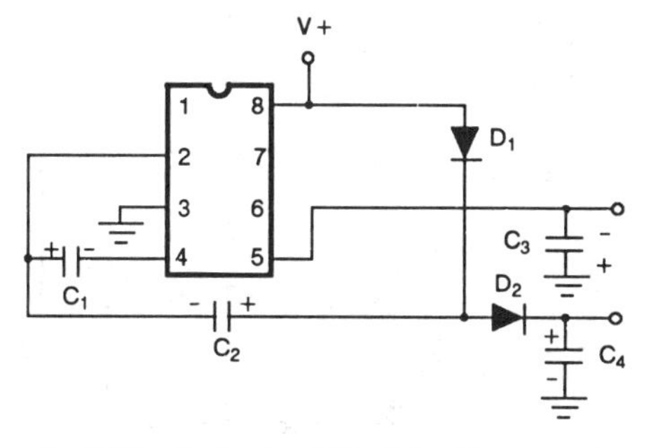

Fig. 12 Combination Inverter/Multiplier Circuit

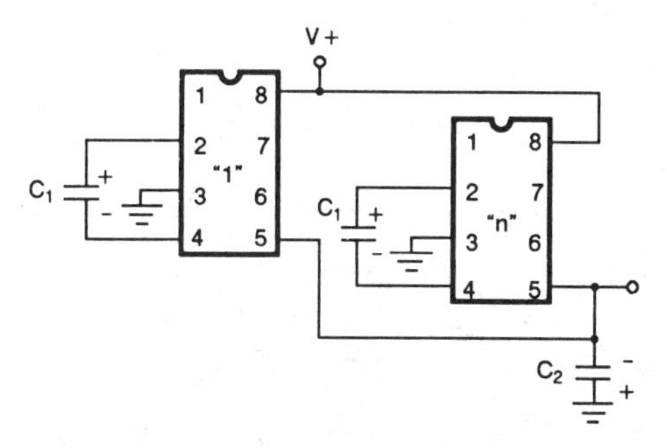

Fig. 13 Paralleling Multiple Voltage Converters for Increased Current Capability

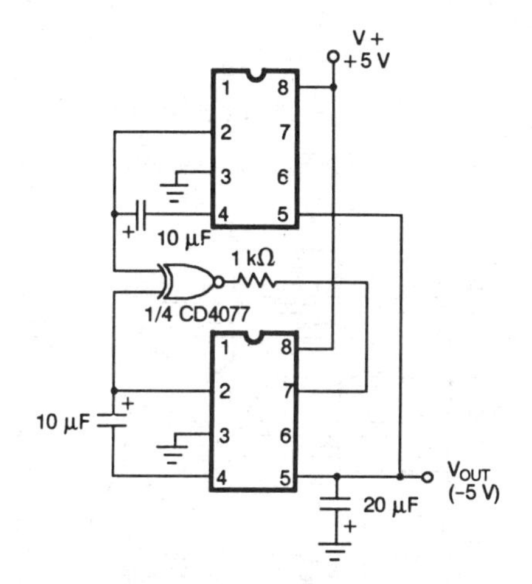

Fig. 14 Synchronizing Two Si7660's or Si7661's with a Single Exclusive NOR Gate

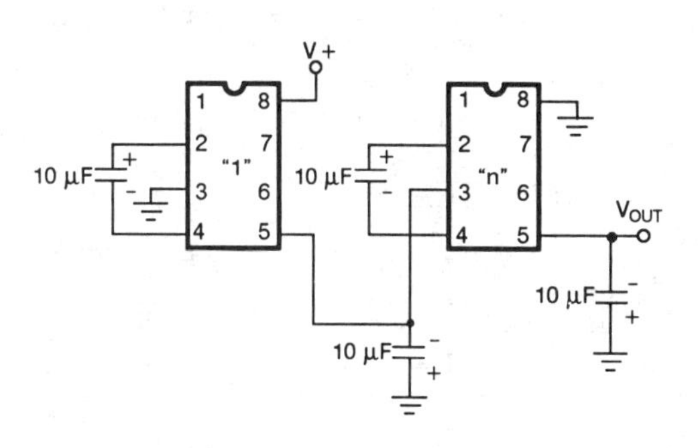

Fig. 15 Cascading Devices for Greater Output Voltage Range

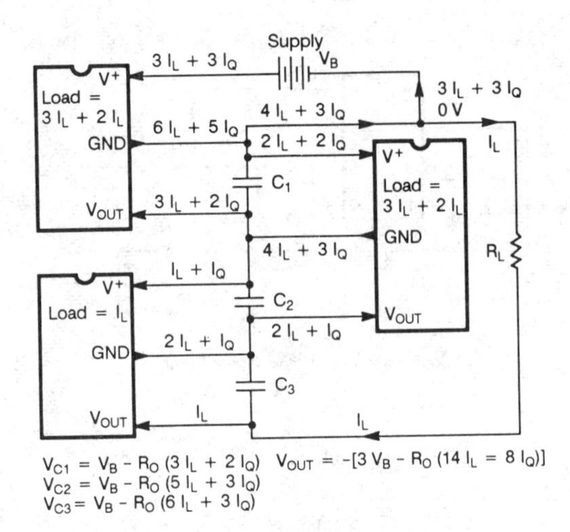

Fig. 16 Current Model of Cascaded Voltage Converters

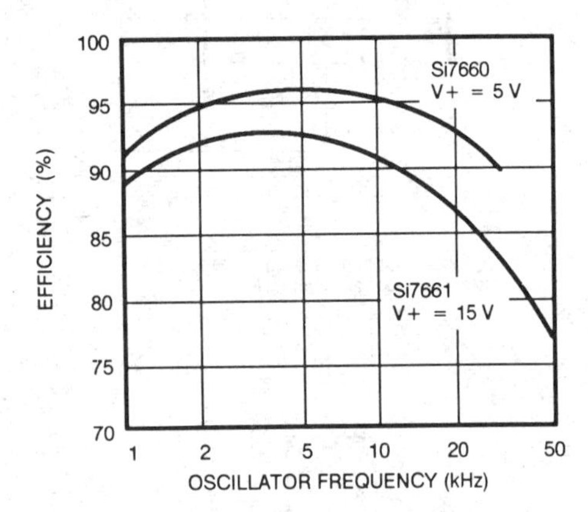

Fig. 17(a) Graph of Efficiency Versus Oscillator Frequency

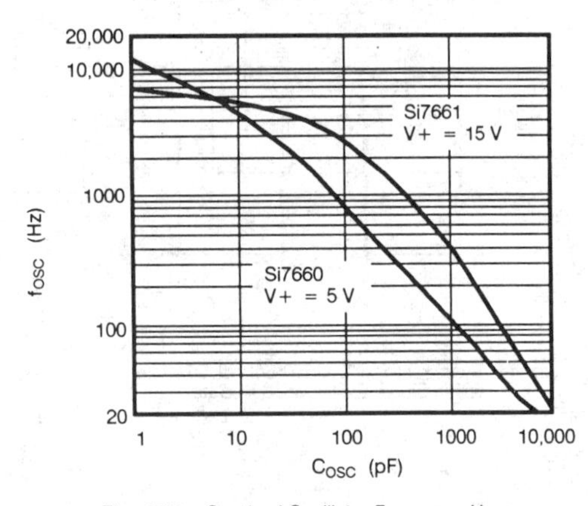

Fig. 17(b) Graph of Oscillator Frequency Versus Added Capacitance

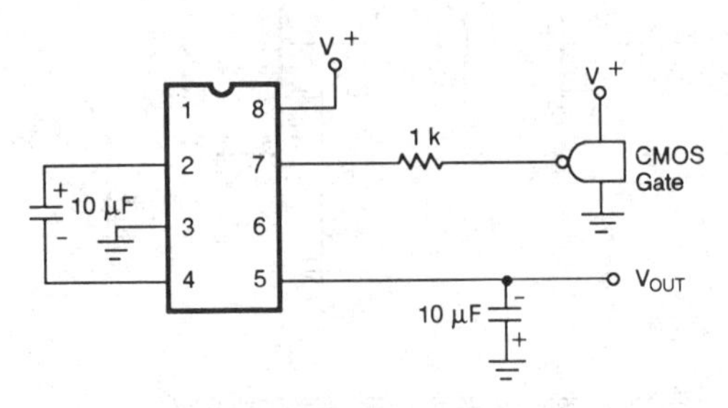

Fig. 18(a) CMOS Drive Circuit for the Si7660 or Si7661

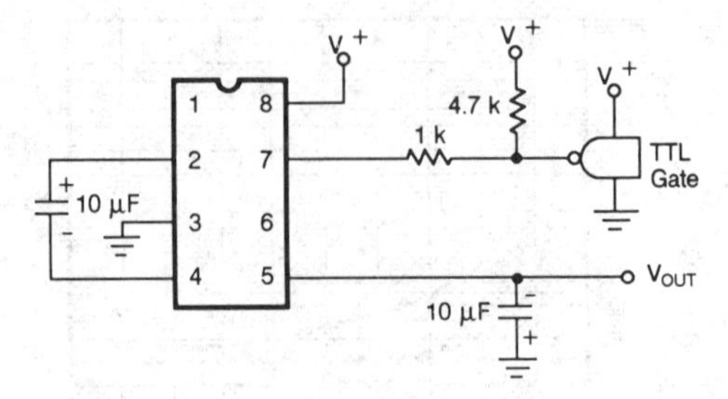

Fig. 18(b) TTL Drive Circuit for the Si7660 Si7661 (5 Volt Input Only)

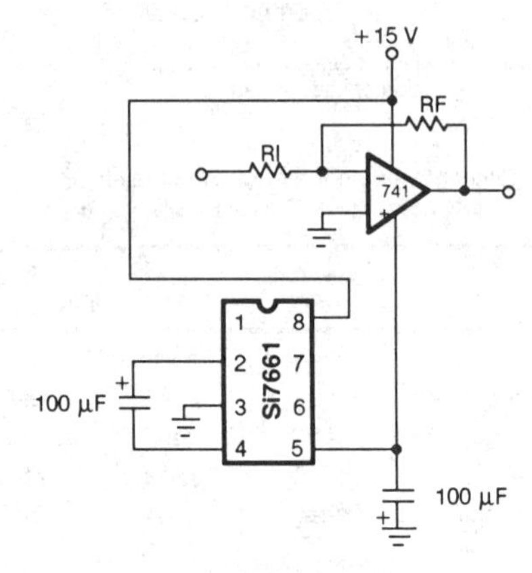

Fig. 19 Using the Si7661 to Generate the Negative Rail for a 741 Op Amp

Table 2 Current Requirements of Several Different Dynamic RAMs.

Device	Operating Current (μA)	Standby Current (μA)	Refresh Current (μA)
MM5290 (0 to 70 °C)	200	100	200
AM9016 (0 to 70 °C)	200	100	200
AM9016 (-55 to 85 °C)	400	200	400
MK4116 (0 to 70 °C)	200	100	200

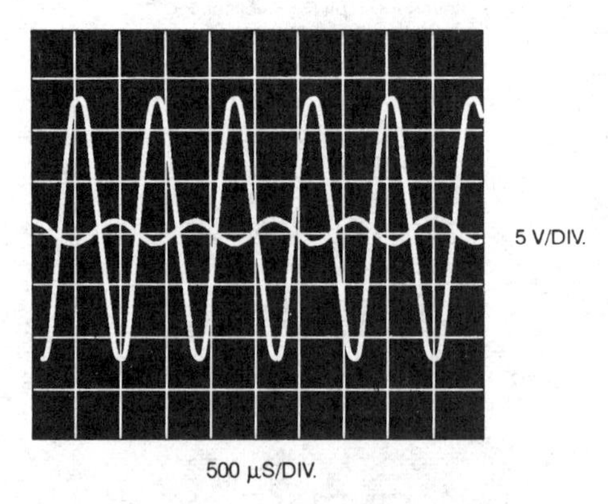

Fig. 20 Input and Output of the 741 Inverting Amplifier at Maximum Undistorted Output

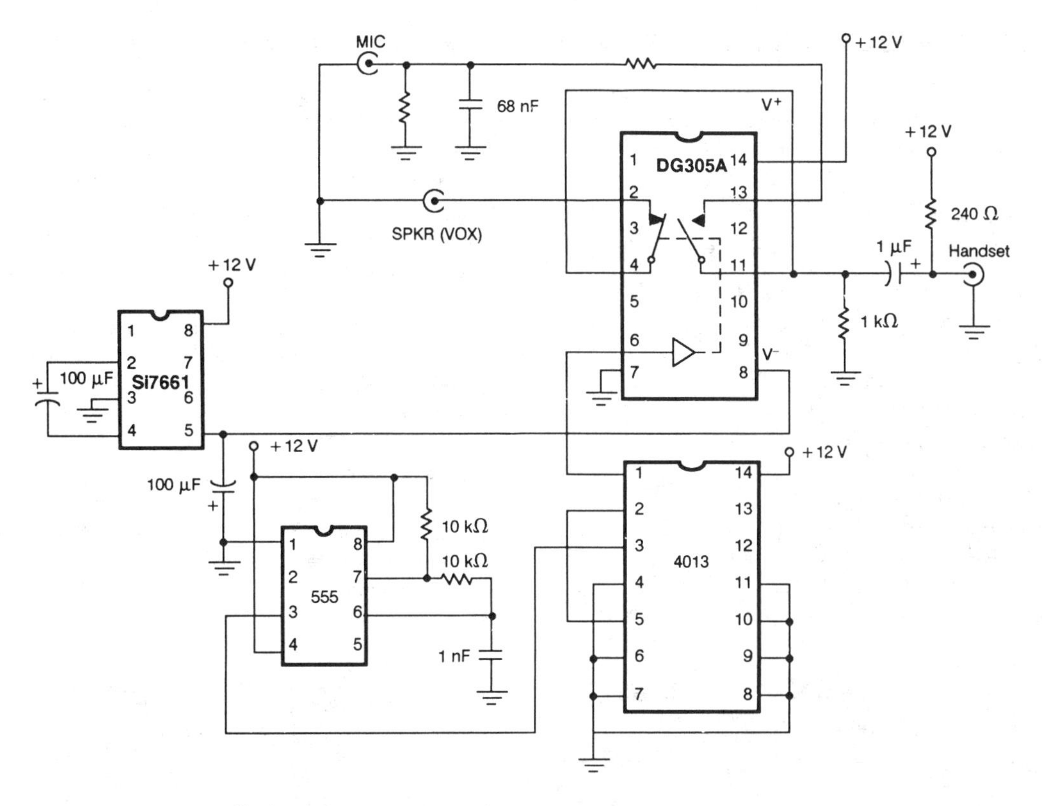

Fig. 21 Using the Si7661 to Supply a Low-current Analog Switch Current

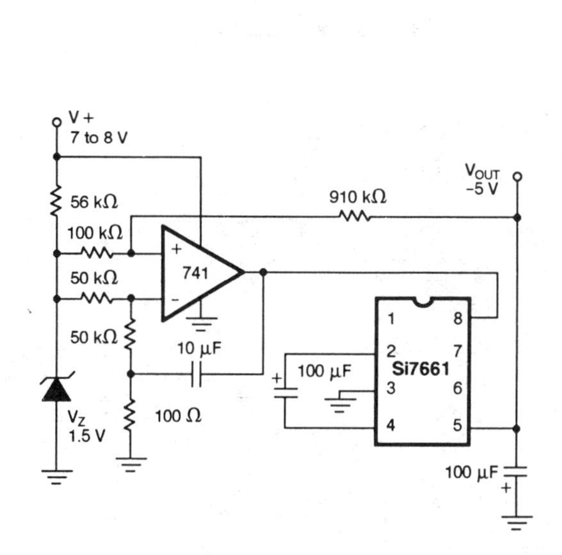

Fig. 22 Low Current Inverting Regulator Circuit

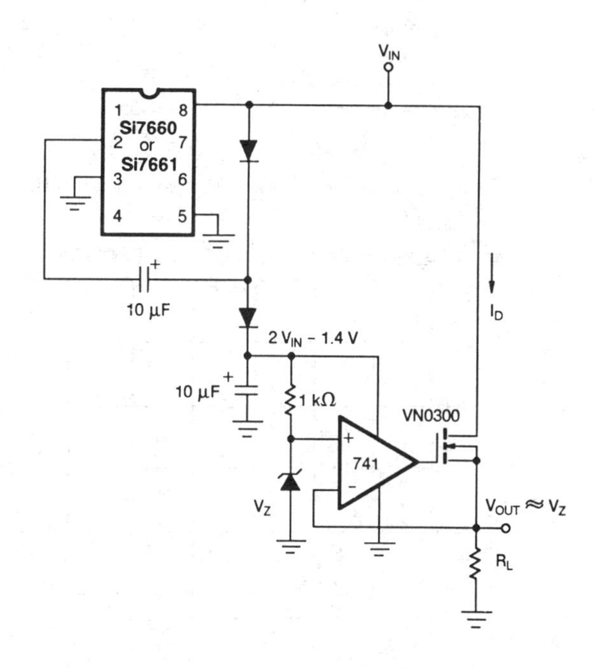

For I_D = 50 mA:
$V_{IN} > V_Z = (I_D \times r_{DS(ON)})$
$V_{IN} > 5.2\ V + (50\ mA \times 1.2\ \Omega)$
$V_{IN} > 5.26\ V$

Fig. 23 Schematic Diagram of the Positive Regulator Circuit

Sprague ULN-8163A, ULN-8163R

Switched-Mode Power-Supply Control Circuits

- Supply range for 4.5V to 14V; internal shunt regulator for higher-voltage source operation
- Low standby current
- 300-kHz sawtooth generator
- Improved feed-forward control (4:1 Range)
- Precision current-limit threshold
- Precision (1%) bandgap voltage reference
- Improved stability over temperature
- Direct pulse-width modulator access
- External TTL-compatible synchronization
- Precision over-voltage threshold
- TTL-compatible shutdown

The ULN-8163A and ULN-8163R are switched-mode power supply control circuits featuring low-voltage operation, precision reference, and protective features. Both have a temperature-compensated bandgap reference, an internal error amplifier, wide-range feedforward capability, a high frequency 300-kHz sawtooth waveform generator, a pulse-width modulator, a variety of protection circuitry, and a 200 mA output driver.

Low-voltage operation and low quiescent current drain make them suitable for automotive and other general SMPS applications, such as dc-to-dc converters operating directly from 5V to 12V supplies and off-line primary-side control.

The ULN-8163A is supplied in a 16-pin dual-in-line package with a copper-lead frame for enhanced power dissipation ratings for operation over a temperature range of −20 °C to +85 °C. The ULN-8163R is furnished in a 16-pin hermetically sealed glass/ceramic package, which will withstand severe environmental contamination.

ABSOLUTE MAXIMUM RATINGS at $T_A = +25°C$

Supply Voltage, V_S	(See Note)
Supply Current, I_{REG}	30 mA
Output Current, I_C (peak)	200 mA
(continuous)	100 mA
Reference Output Current, I_{REF}	10 mA
Logic Input Voltage, V_9, V_{10}	8.0 V
Package Power Dissipation, P_D	
(ULN-8163A)	2.1 W*
(ULN-8163R)	1.7 W*
Operating Temperature Range, T_A	−20°C to +85°C
Storage Temperature Range, T_S	−65°C to +150°C

*Derate linearity to 0 W at T_A = +150°C

NOTE: Maximum allowable supply voltage is dependent on value of external current limiting resistor: 14 V at 0 Ω.

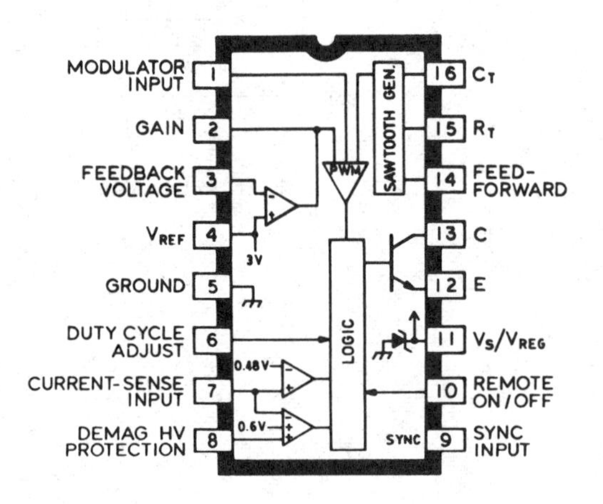

ALLOWABLE AVERAGE PACKAGE POWER DISSIPATION AS A FUNCTION OF TEMPERATURE

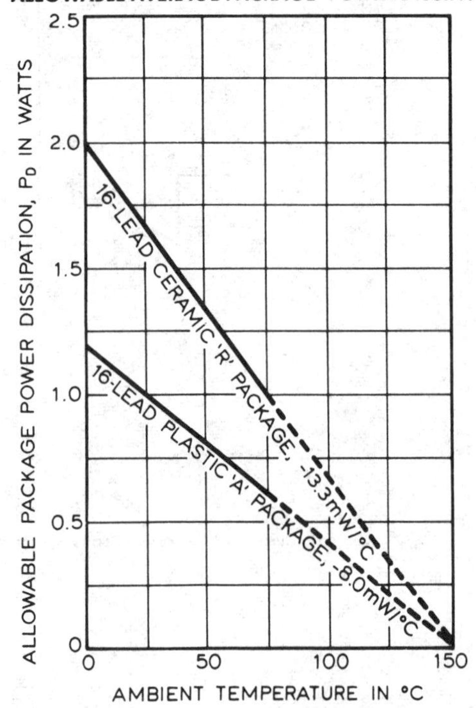

FUNCTIONAL BLOCK DIAGRAM

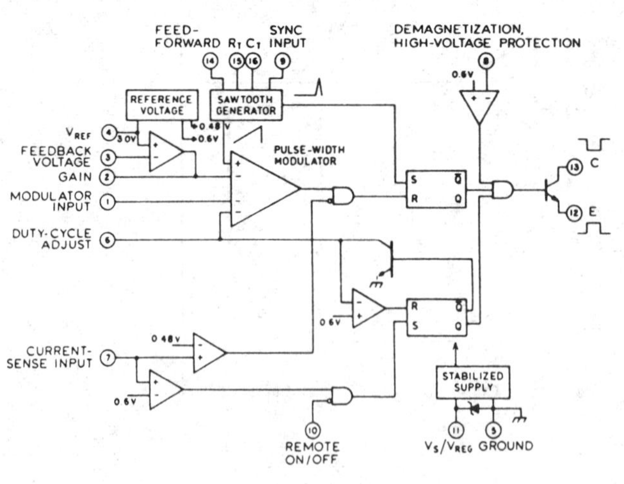

ELECTRICAL CHARACTERISTICS at T_A = 25°C, V_S = 12 V, f_o = 40 kHz (unless otherwise specified).

Characteristic	Test Pin	Test Conditions	Limits			Units
			Min.	Typ.	Max.	
Supply Clamp Voltage	11	I_S = 10 mA	14	—	18	V
	11	I_S = 30 mA	15	—	19	V
Supply Current	11	$V_{12} = V_{13} = 0$	2.0	5.5	7.0	mA
REFERENCE SECTION						
Internal Reference, V_{REF}	4	T_A = +25°C	2.97	3.00	3.03	V
	4	Over Operating Temp. Range†	2.94	—	3.06	V
Temperature Coefficient of V_{REF}	4		—	±100	—	ppm/°C
Line Regulation	4	6 V < V_S < 12 V	—	1.0	3.0	mV/V
Load Regulation	4	0 < I_{REF} < 5 mA	—	3.0	10	mV
OSCILLATOR SECTION						
Min. Oscillator Frequency*	15, 16	R_T = 5 MΩ, C_T = 15000 pF	—	—	50	Hz
Max. Oscillator Frequency	15, 16	R_T = 2.3 kΩ, C_T = 500 pF	200	300	—	kHz
Initial Oscillator Accuracy	15, 16	R_T = 5 kΩ	—	±2.0	—	%
Voltage Stability	15, 16	5 V < V_S < 12 V	—	0.2	0.5	kHz/V
Temperature Stability			—	30	100	ppm/°C
MODULATOR/COMPARATOR SECTION						
Modulator Input Current	1	V_1 = 1.0 V	—	−1.0	−3.0	μA
Maximum Duty Cycle	12	V_6 < 3.0 V	95	—	99	%
Minimum Duty Cycle•	12	V_6 > 0.9 V	—	—	0	%
Duty Cycle Accuracy	12	V_6 = 2.0 V	44	47	50	%
Propagation Delay	6-12		—	200	—	ns
Input Current, Duty Cycle Control	6		—	−1.0	−3.0	μA
PROTECTIVE FUNCTIONS						
Under-Voltage Lockout	11-12		3.80	4.0	4.25	V
Start Threshold	11-12		4.25	4.5	4.75	V
Over-Voltage Threshold	8-12		570	600	630	mV
Over-Voltage Delay	8-12		—	200	500	ns
Over-Voltage Input Current	8		—	2.0	5.0	μA
EXTERNAL SYNCHRONIZATION						
Sync Input OFF Voltage	9		0	—	0.8	V
Sync Input ON Voltage	9		2.0	—	—	V
Sync Input Current	9	V_9 = 0 V, T_A = +25°C	—	−85	−125	μA
	9	V_9 = 0 V, Over Operating Temp. Range†	—	—	−125	μA
REMOTE						
Remote OFF Voltage	10		0	—	0.8	V
Remote ON Voltage	10		2.0	—	—	V
Remote Input Current	10	T_A = +25°C	—	−85	−125	μA
	10	Over Operating Temp. Range†	—	—	−125	μA
CURRENT LIMITING						
Input Current	7	V_7 < 450 mV	—	−5.0	−20	μA
Inhibit Delay*	7	One Pulse, 20% Overdrive at I_C = 40 mA	—	400	600	ns
Trip Levels	7	Shutdown/Slow Start	570	600	630	mV
	7	Current Limit	455	480	505	mV
Shutdown/Current Limit Ratio	—		1.15	1.25	1.40	—
ERROR AMPLIFIER						
Error-Amplifier Gain	3-2	Open Loop	60	66	—	dB
Error-Amplifier Feedback Resistance	2		10	—	—	kΩ
Small-Signal Bandwidth	3-2		700	—	—	kHz
Input Offset Voltage	3		−10	—	10	mV
Input Current	3		—	0.1	1.0	μA
Power Supply Rejection			60	70	—	dB
OUTPUT STAGE						
Output-Saturation Voltage	13	$V_{CE(SAT)}$ at I_C = 10 mA, V_E = 0	—	—	750	mV
		$V_{CE(SAT)}$ at I_C = 100 mA, V_E = 0	—	—	1.0	V
Output Voltage	12		—	—	30	V
Output Source Compliance	11	5 V < V_S < 14 V	V_S − 3	—	—	V

Note: Negative current is defined as coming out of (sourcing) the specified device pin.
†These parameters, although guaranteed over the operating temperature range, are tested at T_A = +25°C only.
*These parameters are tested to a lot sample plan only.
•Any output other than zero is not allowed.

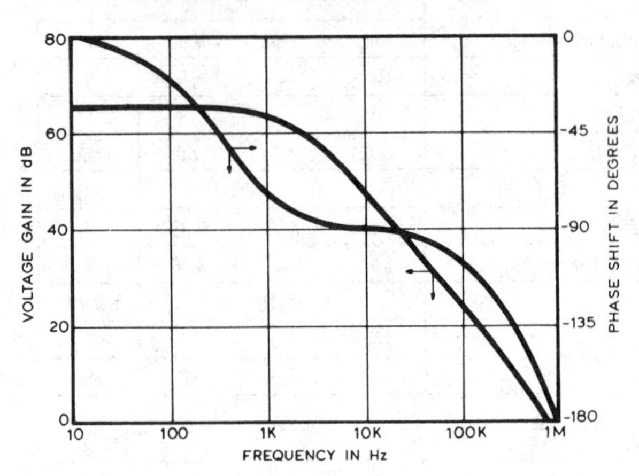

FREQUENCY AS A FUNCTION OF R_T

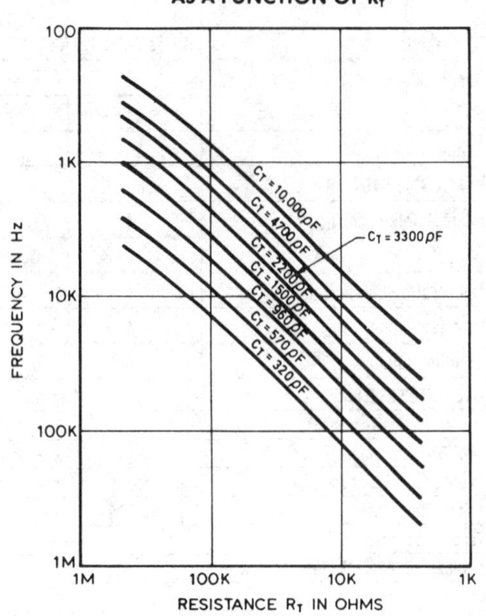

REFERENCE VOLTAGE AS A FUNCTION OF SUPPLY VOLTAGE AND TEMPERATURE

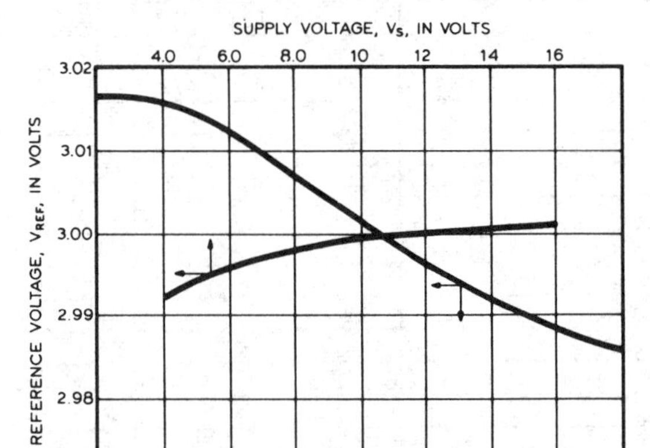

MAXIMUM DUTY CYCLE AS A FUNCTION OF FEED FORWARD VOLTAGE

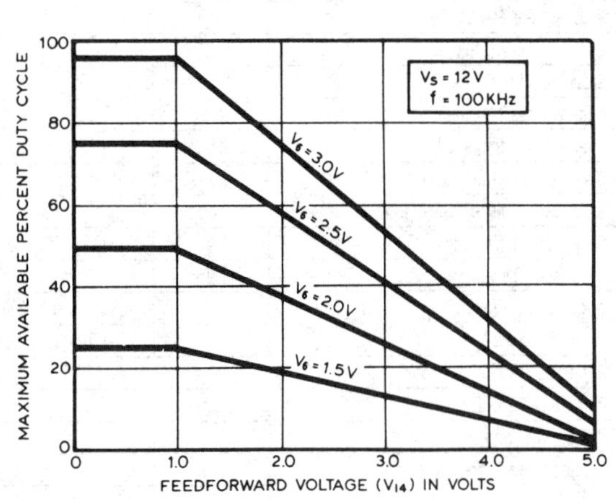

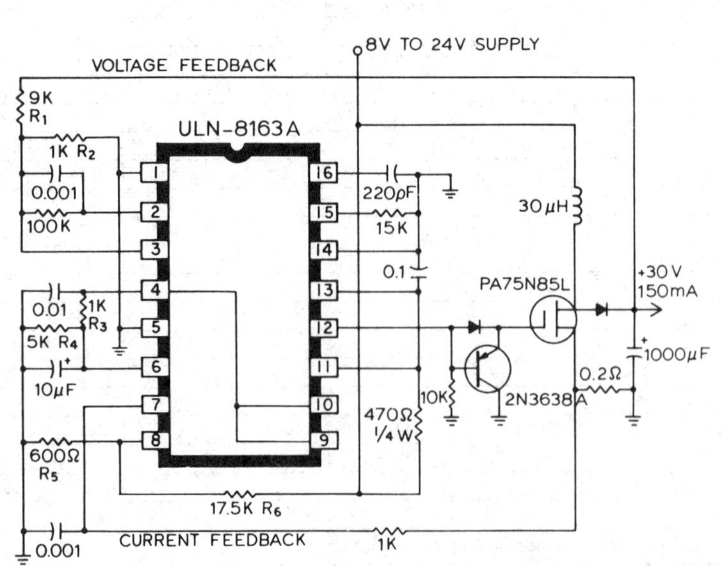

BOOST CONVERTER

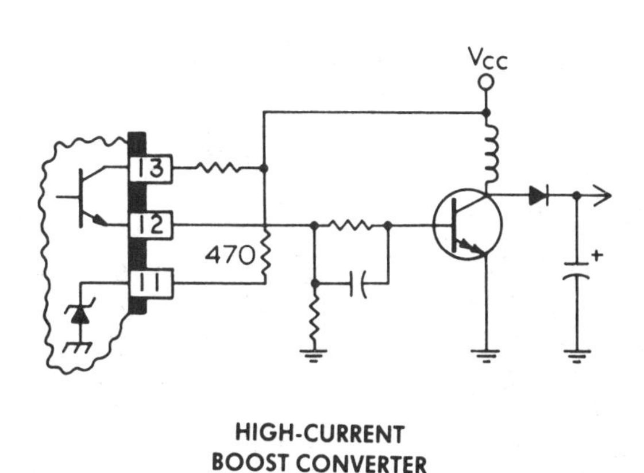

HIGH-CURRENT BOOST CONVERTER

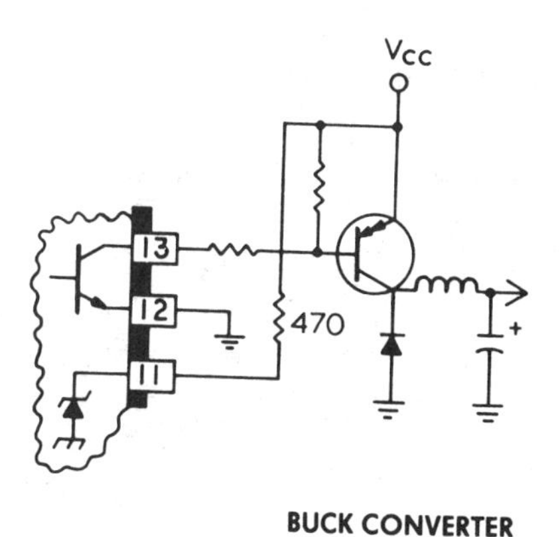

BUCK CONVERTER

Sprague UDN-2998W

Dual Full-Bridge Motor Driver

- ±3 A peak output current
- Output current to 50V
- Integral output-suppression diodes
- Output current sensing
- TTL/CMOS-compatible inputs
- Internal thermal shutdown circuitry
- Crossover-current protected

As an interface between low-level logic and solenoids, brushless dc motors, or stepper motors, the UDN-2998W dual full-bridge driver will operate inductive loads up to 50V with continuous output currents of up 2A per bridge or peak (start-up) currents to 3A. The control inputs are compatible with TTL, DTL, and 5V CMOS, logic. Except for a common supply voltage and thermal shutdown, the two drivers in each package are completely independent.

For external PWM control, an output enable for each bridge circuit is provided and the sink driver emitters are pinned out for connection to external current-sensing resistors. The chopper drive mode is characterized by low power dissipation levels and maximum efficiency. A phase input to each bridge determines load-current direction.

Extensive circuit protection is provided on-chip. Both ground-clamp and flyback diodes for each bridge are provided. A thermal shutdown circuit disables the load drive if the chip temperature rating (package power dissipation) is exceeded. Internally-generated delays provide crossover-current protection.

The UDN-2998W is packaged in a 12-pin single in-line power tab package for high power capabilities. Driving either of the bridges at the full 2A dc rating requires the use of an external heatsink. The tab is a ground potential and needs no insulation.

A similar dual full-bridge driver for use with continuous load currents to ±50 mA is the UDN-2993B.

ABSOLUTE MAXIMUM RATINGS

at $T_{TAB} \leq +70°C$

Parameter	Rating
Supply Voltage, V_{BB}	50 V
Output Current, I_{OUT} (DC)	±2 A
(Peak)	±3 A
Sink Driver Emitter Voltage, V_E	1.5 V
Logic Input Voltage Range, V_{PHASE} or V_{ENABLE}	−0.3 V to 15 V
Package Power Dissipation, P_D	See Graph
Operating Temperature Range, T_A	−20°C to +85°C
Storage Temperature Range, T_S	−55°C to +150°C

NOTE: Output current rating may be limited by chopping frequency, ambient temperature, air flow, or heat sinking. Under any set of conditions, do not exceed the specified current rating or a junction temperature of +150°C.

FUNCTIONAL BLOCK DIAGRAM

(ONE OF TWO DRIVERS)

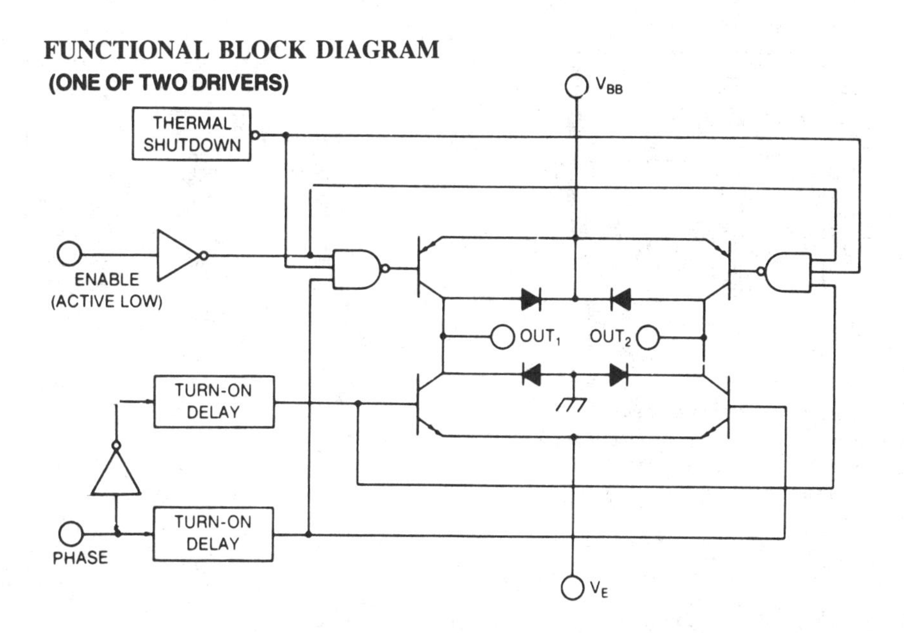

ELECTRICAL CHARACTERISTICS at T_A = +25°C, $T_{TAB} \leq$ +70°C, V_{BB} = 50 V

Characteristic	Symbol	Test Conditions	Limits Min.	Typ.	Max.	Units
Output Drivers						
Operating Voltage Range	V_{BB}		10	—	50	V
Output Leakage Current	I_{CEX}	V_{OUT} = 50 V, V_{ENABLE} = 2.0 V, Note 2	—	<5.0	50	μA
		V_{OUT} = 0, V_{ENABLE} = 2.0 V, Note 2	—	<-5.0	−50	μA
Output Saturation Voltage	$V_{CE(SAT)}$	I_{OUT} = 1 A, Sink Driver	—	1.2	1.4	V
		I_{OUT} = 2 A, Sink Driver	—	1.7	1.9	V
		I_{OUT} = −1 A, Source Driver	—	1.7	1.9	V
		I_{OUT} = −2 A, Source Driver	—	2.0	2.2	V
Output Sustaining Voltage	$V_{CE(sus)}$	I_{OUT} = ±2 A, L = 3.5 mH, Note 2	50	—	—	V
Source Driver Rise Time	t_r	I_{OUT} = −2 A	—	500	—	ns
Source Driver Fall Time	t_f	I_{OUT} = −2 A	—	750	—	ns
Deadtime	t_d	I_{OUT} = ±2 A	—	2.5	—	μs
Clamp Diode Leakage Current	I_R	V_R = 50 V	—	<5.0	50	μA
Clamp Diode Forward Voltage	V_F	I_F = 2 A	—	1.5	2.0	V
Supply Current	I_{BB}	$V_{ENABLE(1)}$ = $V_{ENABLE(2)}$ = 0.8 V	—	25	30	mA
		$V_{ENABLE(1)}$ = $V_{ENABLE(2)}$ = 2.0 V	—	20	25	mA
Control Logic (PHASE or ENABLE)						
Logic Input Voltage	$V_{IN(0)}$		0.8	—	—	V
	$V_{IN(1)}$		—	—	2.0	V
Logic Input Current	$I_{IN(0)}$	V_{PHASE} or V_{ENABLE} = 0.8 V	—	−5.0	−25	μA
	$I_{IN(1)}$	V_{PHASE} or V_{ENABLE} = 2.0 V	—	<1.0	10	μA
Turn-On Delay Time	t_{pd0}	ENABLE Input to Source Drivers	—	0.4	1.0	μs
Turn-Off Delay Time	t_{pd1}	ENABLE Input to Source Drivers	—	2.0	4.0	μs

NOTES:
1. Each driver is tested separately.
2. Test is performed with V_{PHASE} = 0.8 V and then repeated for V_{PHASE} = 2.0 V.
3. Negative current is defined as coming out of (sourcing) the specified device pin.

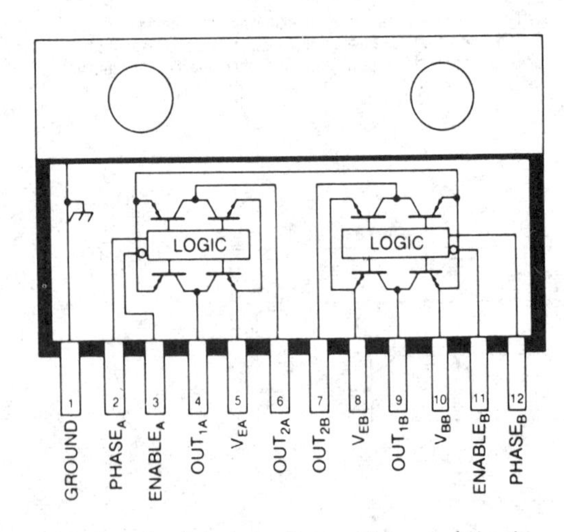

To maintain isolation between integrated circuit components and to provide for normal transistor operation, the ground tab must be connected to the most negative point in the external circuit.

TRUTH TABLE

Enable Input	Phase Input	Output 1	Output 2
Low	High	High	Low
Low	Low	Low	High
High	High	Open	Low
High	Low	Low	Open

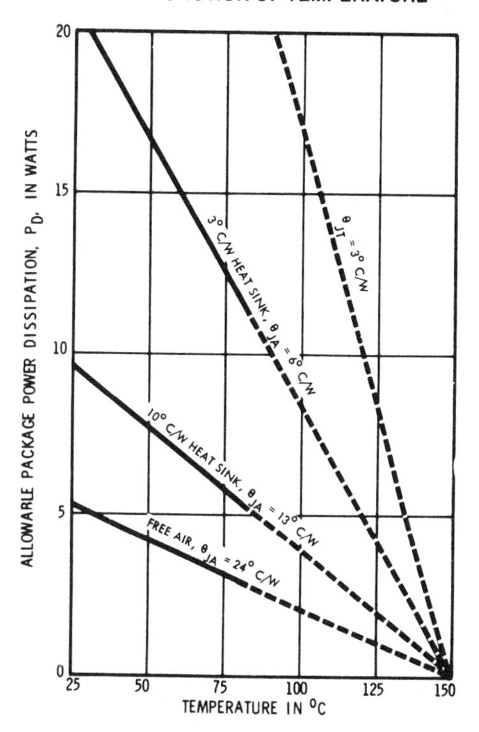

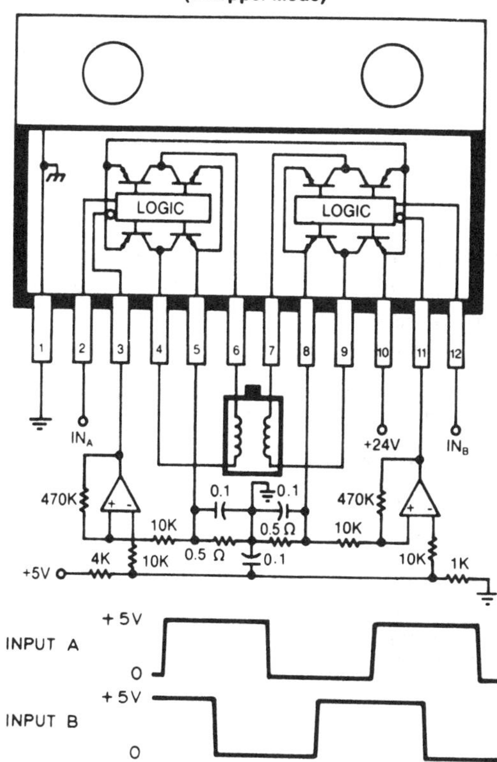

Sprague ULN-3751Z Power Operational Amplifier

- ±3V to ±13V operation
- High output swing
- Peak output current to ±3.5A
- Low input offset
- 90 dB typical open-loop gain
- Internal thermal shutdown
- High common-mode input range
- Unity gain stable
- Pin compatible with L165, L465, SG1173

As a combination general-purpose operational amplifier and power booster, the ULN-3751Z integrated circuit simplifies circuit design, reduces component count, and enhances system reliability.

The power op amp features high-impedance differential inputs, a unity-gain stable amplifier that needs no external compensation, and a high-current power output. Typical applications include use as a voice-coil motor driver, linear servo amplifier, power oscillator, bipolar voltage regulator, and audio power driver.

The ULN-3751Z demands up to ±3.5A of output current. It is furnished in a modified 5-lead JEDEC-style TO-220 plastic package. Lead forming for vertical or horizontal mounting is available (ULN-3751ZV or ULN-3715ZH). The heatsink tab is at substrate potential and must be insulated from ground when the device is used with a split supply.

This power op amp operates over a recommended supply voltage range of ±3V to ±13V. Dual power op amps are available as the ULN-3755B (16-pin DIP) and the high-power ULN-3755W (12-pin SIP). Both of these devices include output current sensing and a voltage boost connection to maximum output voltage swing to ±20V supplies at up to ±3.5A peak output current.

ABSOLUTE MAXIMUM RATINGS
at $T_A = +25°C$

Supply Voltage Differential ($+V_S$ to $-V_S$)	28 V
Peak Output Current, I_{OUT}	±3.5 A
Input Voltage Range, V_{IN}	$+V_S$ to $-V_S - 0.3$ V
Package Power Dissipation, P_D	See Graph
Operating Temperature Range, T_A	0°C to +70°C
Storage Temperature Range, T_S	−40°C to +150°C

ELECTRICAL CHARACTERISTICS at T_A = +25°C, T_{TAB} ≤ +70°C, V_S = ±6.0 V (unless otherwise noted)

Characteristic	Test Conditions	Limits Min.	Typ.	Max.	Unit
Functional Supply Voltage Range	$+V_S$ to $-V_S$	6.0	—	26	V
Quiescent Supply Current		—	40	60	mA
Input Bias Current	$V_{IN} = 0$, $I_{OUT} = 0$	—	−60	−1000	nA
Input Offset Voltage	$V_{IN} = 0$, $I_{OUT} = 0$	—	±2.0	±10	mV
Input Offset Current	$V_{IN} = 0$, $I_{OUT} = 0$	—	10	100	nA
Input Noise Voltage†	BW = 40 Hz to 15 kHz	—	4.0	—	μV
Input Noise Current†	BW = 40 Hz to 15 kHz	—	60	—	pA
Crossover Distortion†	P_{OUT} = 50 mW, R_L = 4 Ω	—	<0.05	—	%
Common Mode Rejection	ΔV_{CM} = 2 V	60	85	—	dB
Input Common Mode Range†	Positive	—	$+V_S - 2$ V	—	V
	Negative	—	$-V_S - 0.3$ V	—	V
Open-Loop Voltage Gain	f = 0	80	90	—	dB
Slew Rate	$V_{IN} = V_{OUT}$ = 6 Vpp, $R_L = \infty$	1.0	2.3	—	V/μs
Gain-Bandwidth Product†	A_V = 40 dB	—	3.5	—	MHz
Output Voltage Swing	I_{out} = 1.0 A	4.5	4.7	—	V
	I_{OUT} = −1 A	−4.5	−4.7	—	V
Supply Voltage Rejection	$+V_S$, ΔV = 1 V	60	85	—	dB
	$-V_S$, ΔV = 1 V	60	80	—	dB
Thermal Shutdown Temp.†		—	160	—	°C

*This parameter is tested to a lot sample plan only.
†Typical values given for circuit design information only.

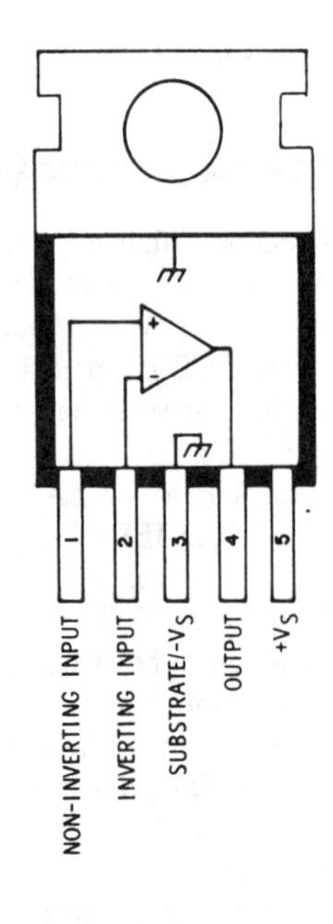

ALLOWABLE POWER DISSIPATION AS A FUNCTION OF AMBIENT TEMPERATURE

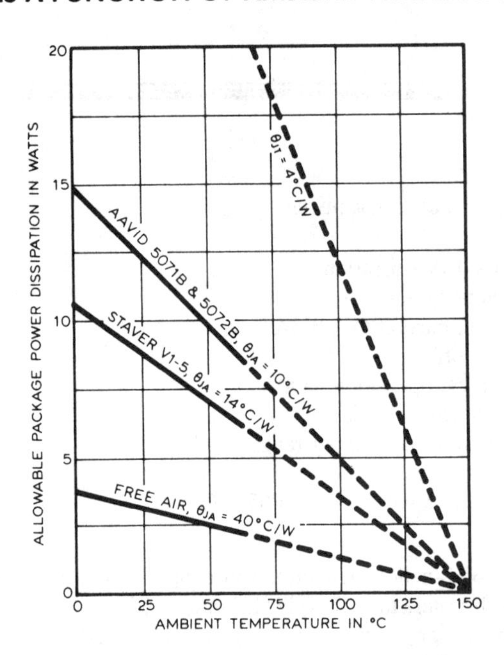

WIEN BRIDGE OSCILLATOR/MOTOR DRIVER

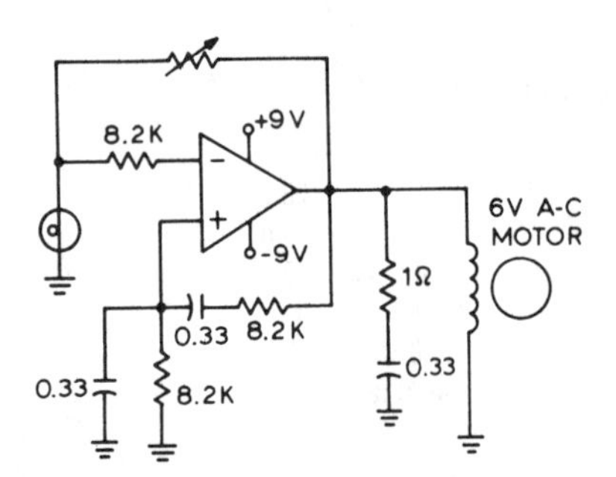

VIDEO MONITOR VERTICAL DEFLECTION MAP

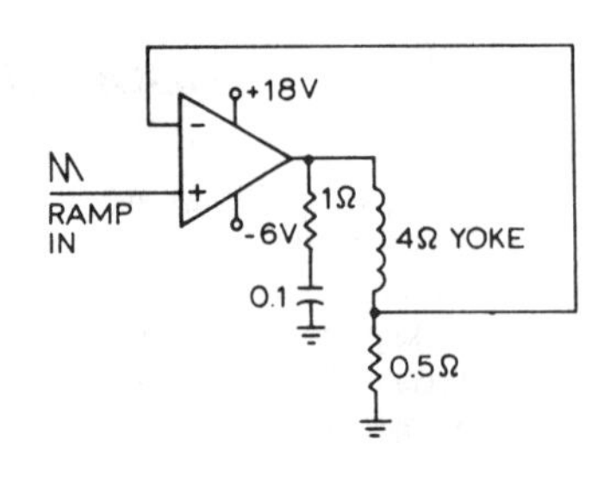

UNITY GAIN VOLTAGE FOLLOWER

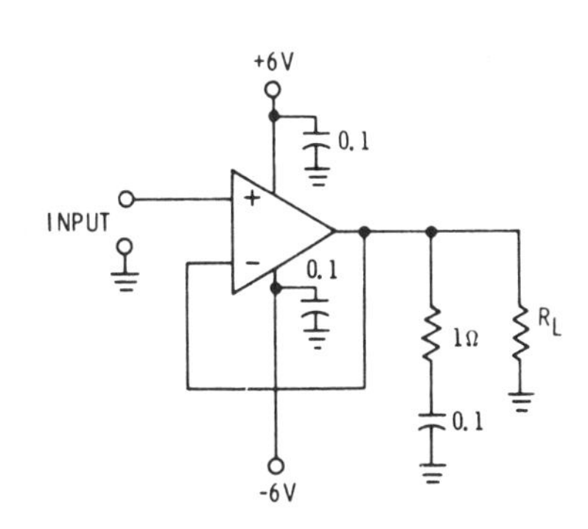

NON-INVERTING POWER AMPLIFIER

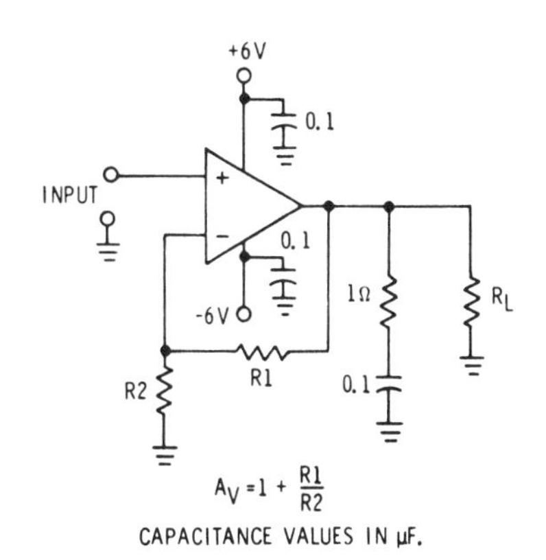

LINEAR VOLTAGE REGULATOR

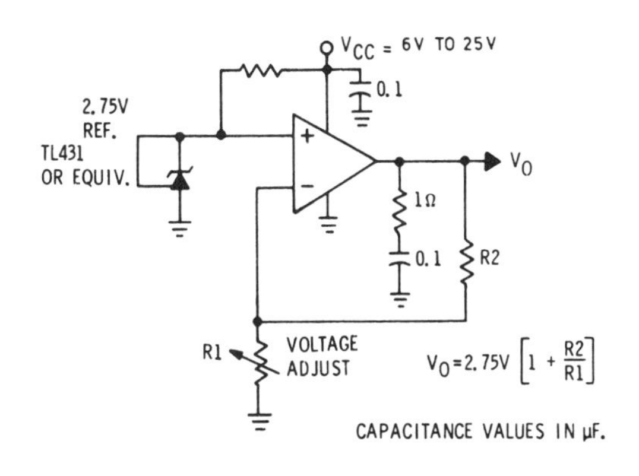

HIGH-POWER LINEAR REGULATOR
(Short-Circuit Protected)

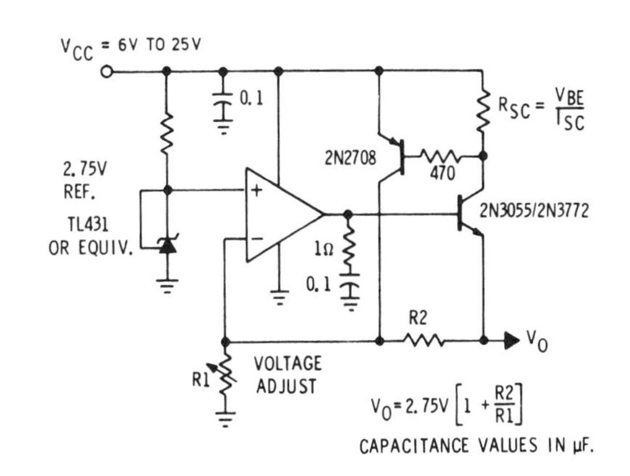

SINGLE-ENDED POSITION SERVO WITH SENSE POTENTIOMETER

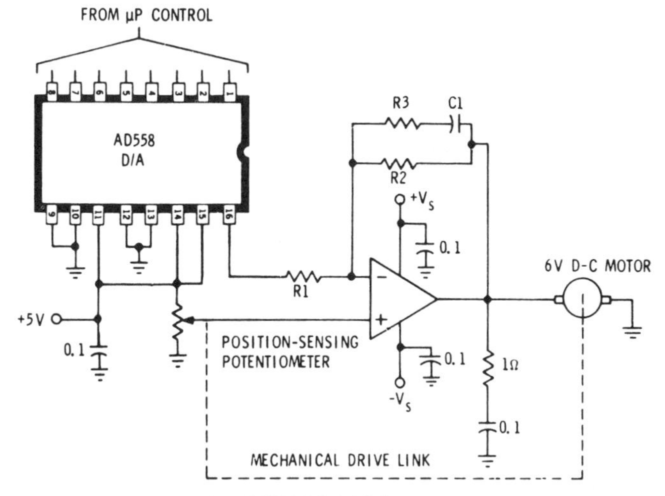

SIMPLIFIED SERVO APPLICATION WITH CONTROL TRANSFORMERS

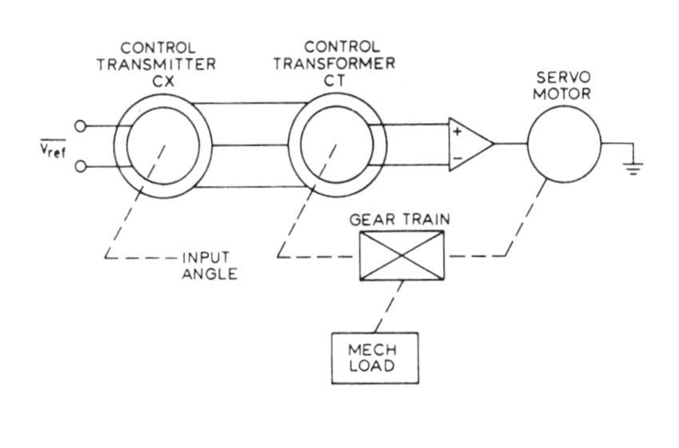

Sprague ULN-3753B, ULN-3753W Dual Power Operational Amplifiers

- Operating supply range ±3V to ±20V
- Output current to ±3.5A peak
- Output-current limiting
- Output-current sensing
- High output-voltage swing
- Low crossover distortion
- Low input offset voltage
- Externally compensated
- High open-loop gain
- Output protection diodes
- Thermal shutdown protection
- Excellent supply and common-mode rejection
- Single or dual-in-line power packages

APPLICATIONS

- Dual half-bridge and full-bridge motor drives
  Linear servo motors
  Voice coil motors
  ac and dc motors
  Microstepping applications
- Power transconducting amplifier
- Audio power amplifier, stereo, or BTL
- Power oscillator/amplifier
- Dual bipolar voltage regulator

High-current linear servo loads, such as voice coil motors used in disc-drive applications, are ideal applications for the ULN-3753B and ULN-3753W dual high-power operational amplifiers. Their building-block design concept also makes them ideal for a wide variety of other motor-drive applications, audio power amplifiers, power oscillators, and linear voltage regulators. External compensation permits user adjustment of bandwidth and phase margin at any gain level.

The ULN-3753B is furnished in a 16-pin dual-in-line package with copper heatsink contact tabs. For higher power requirements, the ULN-3753W is supplied in a 12-pin single in-line power tab package.

The inputs are designed to allow a wide common-mode range from the negative supply, (or ground in single supply applications) to within approximately 2V of the positive supply. Common-mode and power-supply rejection are in excess of 60 dB. The amplifiers' wide output swing is completed by current sensing, which is referenced to the negative supply and allows for feedback, as required to produce a transconductance characteristic.

The ULN-3753B (batwing DIP) can typically dissipate 6W at a tab temperature of 70°C. The lead configuration enables easy attachment of a heatsink while fitting a standard socket or printed wiring board layout. The ULN-3753W (SIP) can safely dissipate significantly higher power levels with appropriate heatsinking. With either package configuration, the heatsink is at the negative supply, or at ground in a single-ended application.

ELECTRICAL CHARACTERISTICS at T_A = +25°C, T_{TAB} ≤ +70°C, V_S = ±6V, C_C = 0, each amplifier tested separately (unless otherwise specified)

Characteristic	Test Conditions	Limits			
		Min.	Typ.	Max.	Units
Functional Supply Voltage Range	$+V_S$ to $-V_S$	6.0	—	40	V
Quiescent Supply Current		—	90	150	mA
Input Bias Current	V_{OUT} = 0	—	−80	−1000	nA
Input Offset Voltage	V_{OUT} = 0, I_{OUT} = 0	—	±1.0	+10	mV
Input Offset Volt. TC†	Over Op. Temp. Range	—	−15	—	μV/°C
Input Offset Current	V_{OUT} = 0, I_{OUT} = 0	—	10	100	nA
Input Noise Voltage†	BW = 40 Hz to 15 kHz	—	4.0	—	μV
Input Noise Current†	BW = 40 Hz to 15 kHz	—	60	—	pA
Crossover Distortion†	P_{OUT} = 50 mW, R_L = 4Ω	—	0.2	—	%
Common Mode Rejection	V_{CM} = 3V	60	85	—	dB
Input Common Mode Range*	V_S = +6V	−6.3	—	+4.0	V
	V_S = +15V	−15.3	—	+13	V
Open Loop Voltage Gain	f = 0	80	100	—	dB
Slew Rate	V_{IN} = V_{OUT} = 6 Vpp	5.0	10	—	V/μs
Gain-Bandwidth Product†	A_V = 40 dB	—	3.0	—	MHz
Channel Separation†	I_{OUT} = 100 mA, f = 1 kHz	—	60	—	dB
Output Voltage Swing	I_{OUT} = 1 A	9.0	9.5	—	Vpp
Supply Voltage Rejection	$+V_S$, ΔV = 1 V	60	85	—	dB
	$-V_S$, ΔV = 1 V	60	80	—	dB
Thermal Resistance, Θ_{JT}.	ULN-3753B	—	—	15	°C/W
	ULN-3753W	—	—	3.0	°C/W
Thermal Shutdown Temp.†		—	165	—	°C

*This parameter is tested to a lot sample plan only.

†Typical values given for circuit design information only.

ABSOLUTE MAXIMUM RATINGS

Supply Voltage Differential ($+V_S$ to $-V_S$)	40 V
Peak Supply Voltage (50 ms)	45 V
Continuous Output Current, I_{OUT} ($V_S = \pm 15$ V)	±2.0 A
($V_S = \pm 6$ V)	±2.5 A
Peak Output Current, I_{OUT} (50 ms)	±3.5 A
Package Power Dissipation, P_D	See Graphs
Operating Temperature Range, T_A	−20°C to +85°C
Storage Temperature Range, T_S	−55°C to +125°C

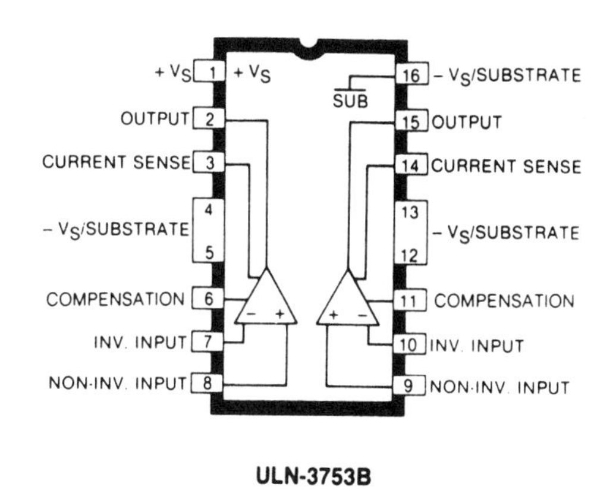

ULN-3753B

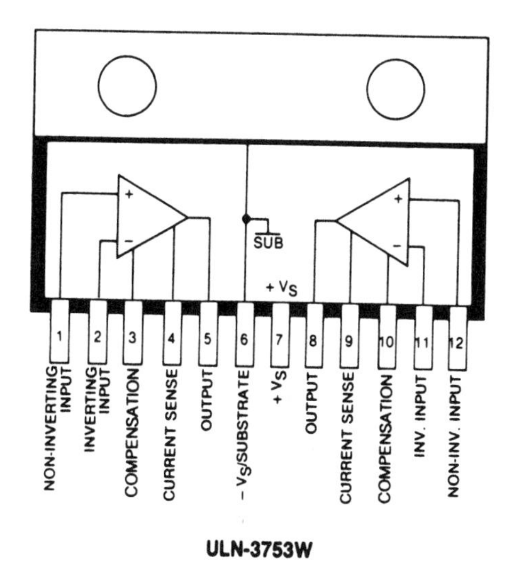

ULN-3753W

NON-INVERTING AMPLIFIER

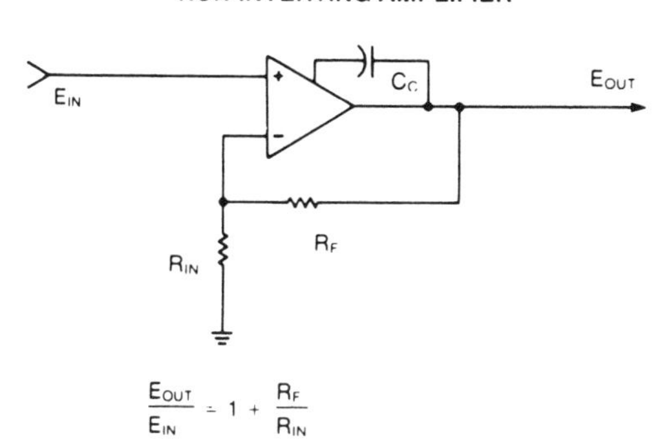

$$\frac{E_{OUT}}{E_{IN}} = 1 + \frac{R_F}{R_{IN}}$$

IF $R_F = 0$ or $R_{IN} = \infty$, $E_{OUT} = E_{IN}$

INVERTING AMPLIFIER

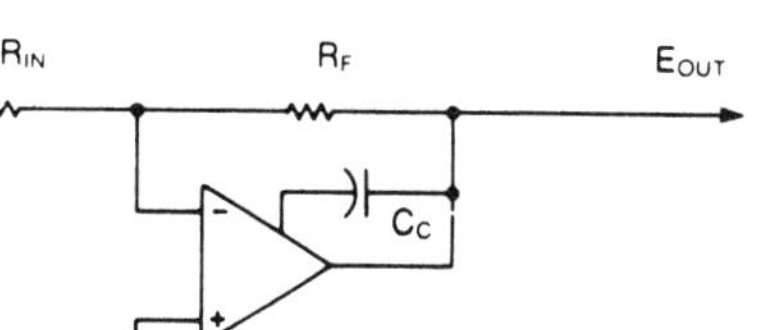

$$\frac{E_{OUT}}{E_{IN}} = -\frac{R_F}{R_{IN}}$$

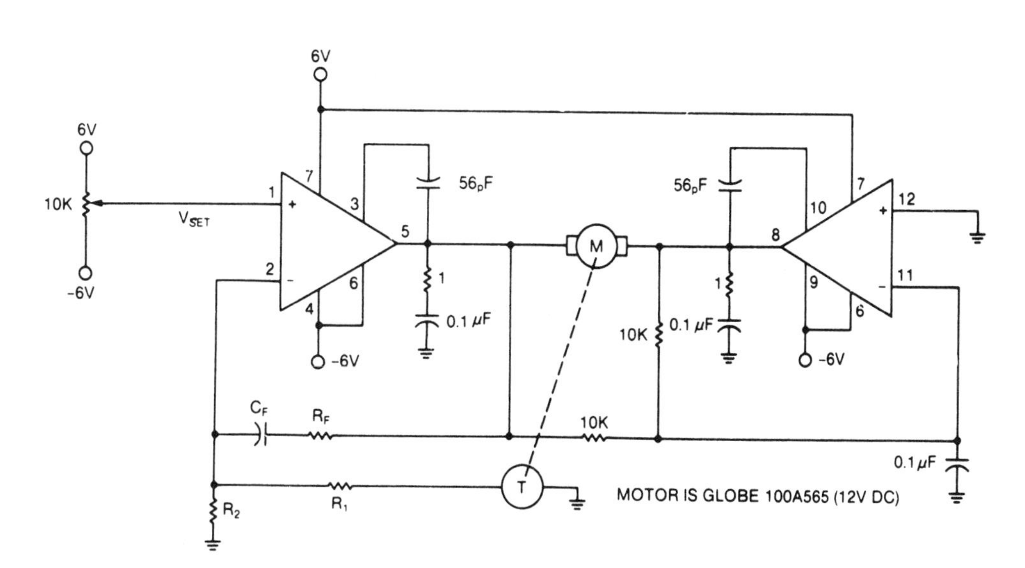

DC MOTOR SPEED CONTROL

ALLOWABLE AVERAGE PACKAGE POWER DISSIPATION AS A FUNCTION OF TEMPERATURE

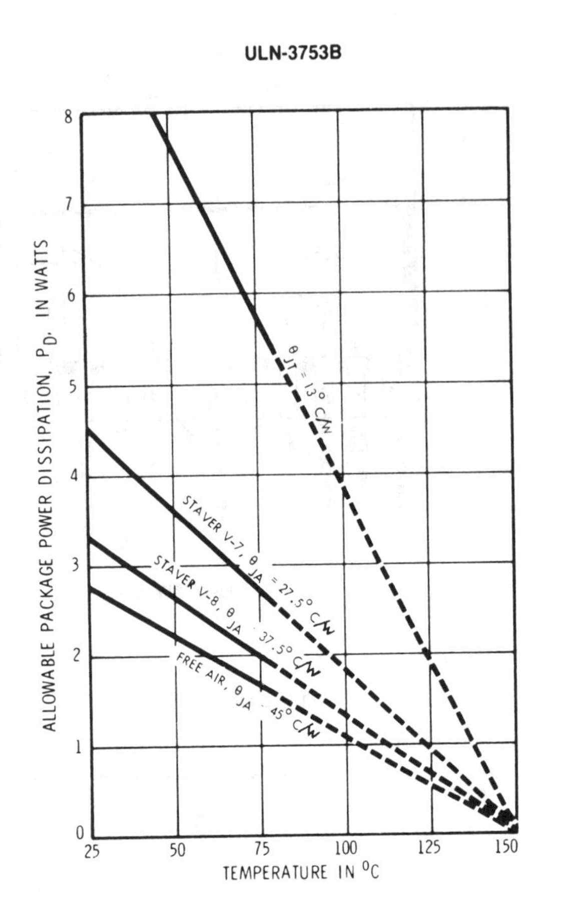

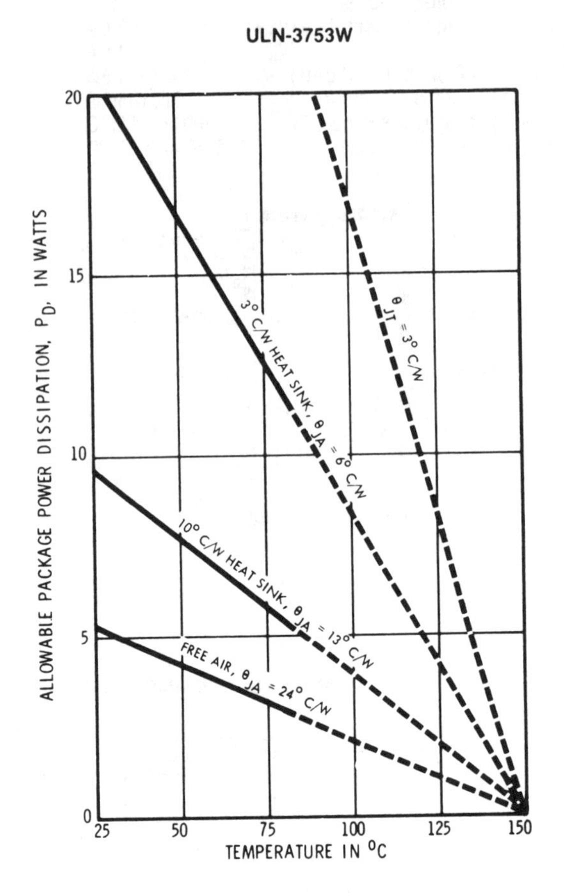

TYPICAL CHARACTERISTICS

OPEN-LOOP VOLTAGE GAIN AS A FUNCTION OF TEMPERATURE

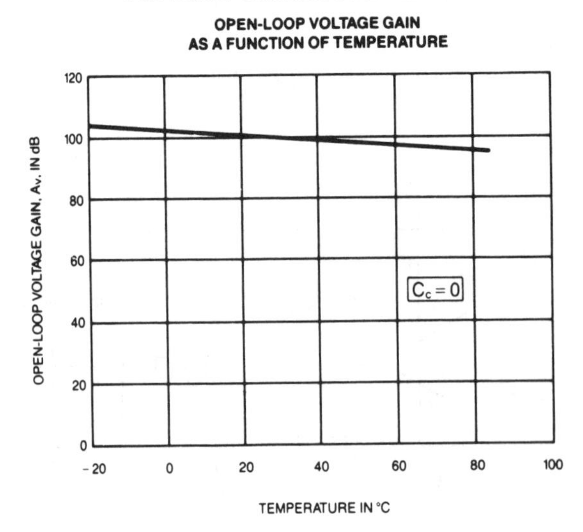

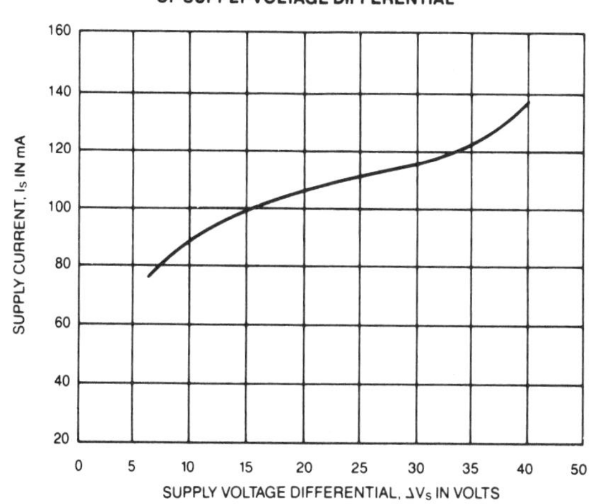
SUPPLY CURRENT AS A FUNCTION
OF SUPPLY VOLTAGE DIFFERENTIAL
SUPPLY CURRENT, I_S IN mA
160
140
120
100
80
60
40
20
0
5
10
15
20
25
30
35
40
50
SUPPLY VOLTAGE DIFFERENTIAL, ΔV_S IN VOLTS

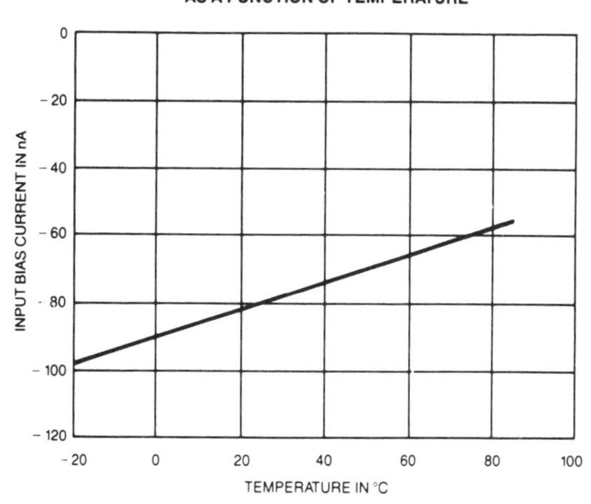
INPUT BIAS CURRENT
AS A FUNCTION OF TEMPERATURE
INPUT BIAS CURRENT IN nA
0
-20
-40
-60
-80
-100
-120
-20
0
20
40
60
80
100
TEMPERATURE IN °C

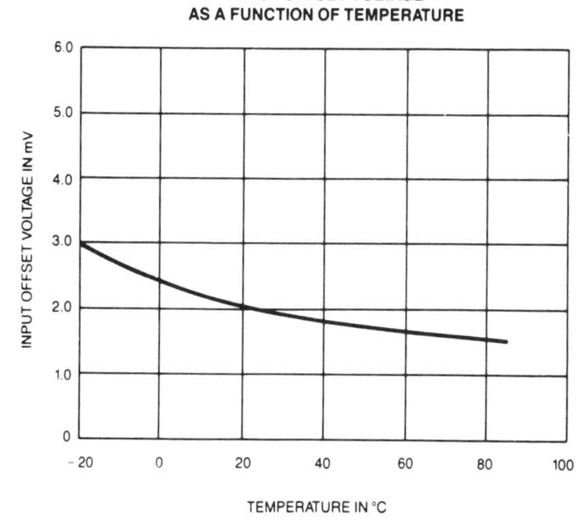
INPUT OFFSET VOLTAGE
AS A FUNCTION OF TEMPERATURE
INPUT OFFSET VOLTAGE IN mV
6.0
5.0
4.0
3.0
2.0
1.0
0
-20
0
20
40
60
80
100
TEMPERATURE IN °C

APPLICATIONS

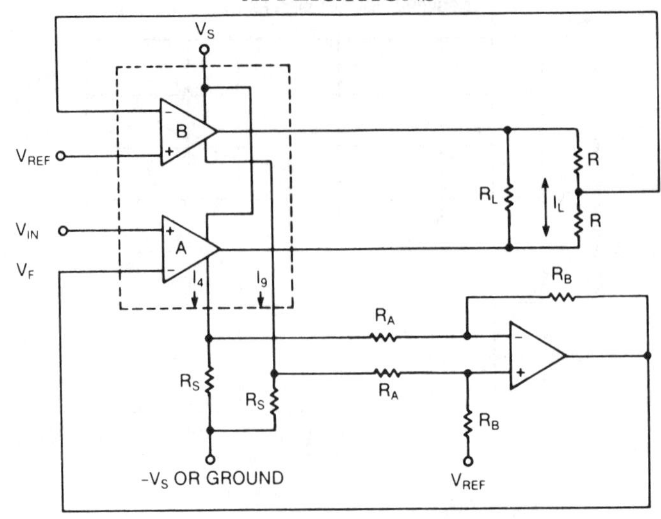

CURRENT-SENSE TRANSCONDUCTANCE AMPLIFIER

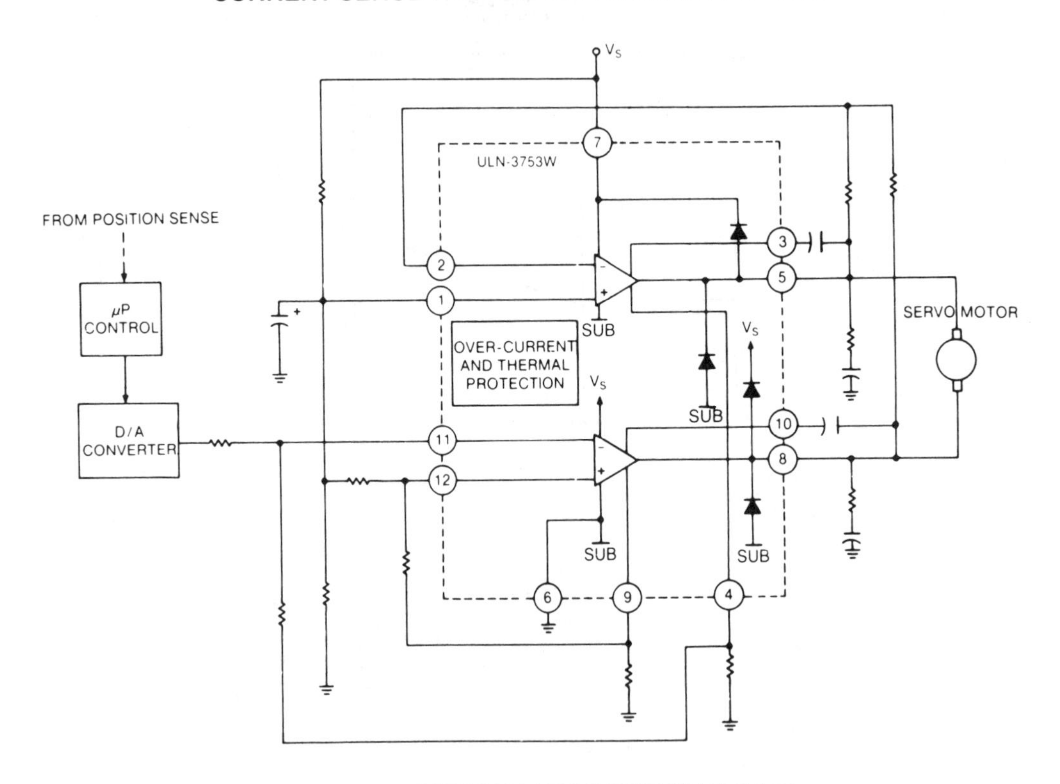

DIGITALLY CONTROLLED POSITION SERVO

TWO-PHASE 60 Hz AC MOTOR DRIVER

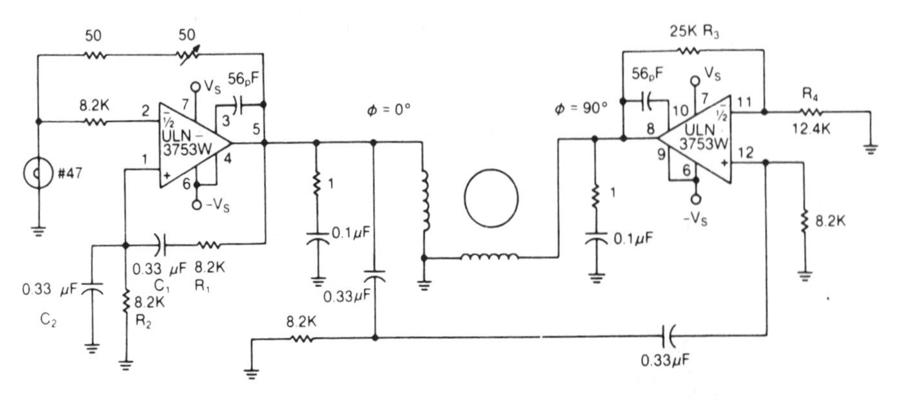

THREE-PHASE 400 Hz AC MOTOR DRIVER

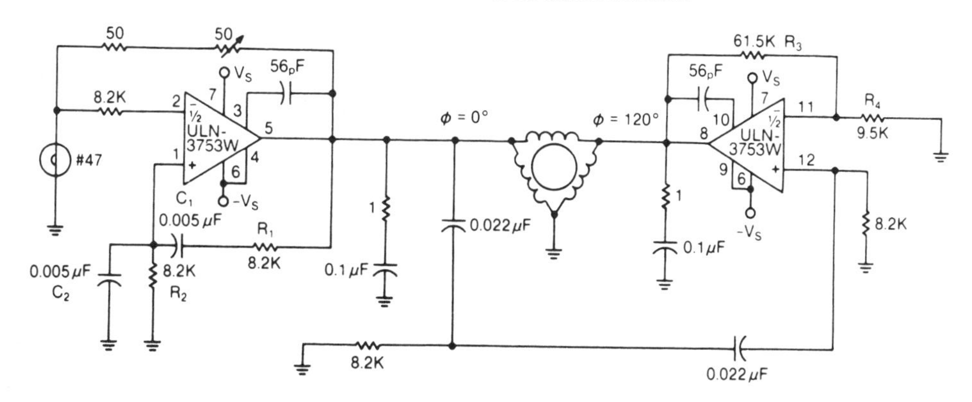

Sprague ULN-3755B, ULN-3755W Dual Power Operational Amplifiers

- Operating supply range ±3 to ±20 volts
- Output current to ±3.5A peak
- Output current limiting
- Output current sensing
- High output-voltage swing
- Low crossover distortion
- Low input offset voltage
- Unity-gain stable
- High open-loop gain
- Output protection diodes
- Thermal shutdown protection
- Excellent supply and common-mode rejection
- Single- or dual-in-line power packages

APPLICATIONS

- Dual half-bridge and full-bridge motor drivers
  Linear servo motors
  Voice coil motors
  ac and dc motors
  Microstepping applications
- Power transconductance amplifier
- Audio power amplifier
  Stereo or BTL
- Power oscillator/amplifier
- Dual bipolar voltage regulator

Consisting of two high-power operational-amplifier circuits in a single-in-line power-tab package or a batwing dual-in-line package, the ULN-3755B and ULN-3755W are specifically designed to drive high-current linear servo loads, such as voice coil motors used in disc-drive applications. Their building-block design concept also makes them ideal for a wide variety of other motor-drive applications, for use as audio power amplifiers, power oscillators, and linear voltage regulators. Low crossover distortion eliminates servo hunting under null conditions and is required for most audio applications.

The ULN-3755B is furnished in a 16-pin dual-in-line package with copper heatsink contact tabs. For higher power requirements, the ULN-3755W is supplied in a 12-pin single-in-line power tab package.

The inputs are designed to allow a wide common mode range from the negative supply, (or ground in single-supply applications) to within approximately two volts of the positive supply. Common-mode and power-supply rejection are in excess of 60 dB. The amplifiers' wide output swing is complemented by current sensing, which is referenced to the negative supply and allows for feedback as required to produce a transconductance characteristic.

Separate supply pins are provided for the low-level input and high-level output circuits to allow voltage boost or bootstrapping to maximize output swing.

The ULN-3755B (batwing DIP) can typically dissipate 6W at a tab temperature of +70°C. The lead configuration enables easy attachment of a heatsink while fitting a standard socket or printed wiring board layout. The ULN-3755W (SIP) can safely dissipate significantly higher power levels with appropriate heatsinking. With either package configuration, the heatsink is at the negative supply, or ground in a single-ended application.

ABSOLUTE MAXIMUM RATINGS

Supply Voltage Differential ($+V_S$ to $-V_S$)	40 V
Peak Supply Voltage (50 ms)	45 V
Continuous Ouput Current	
I_{OUT} ($V_S = \pm 15$ V)	±2.0 A
($V_S = \pm 6$ V)	±2.5 A
Peak Output Current, I_{OUT} (50 ms)	±3.5 A
Package Power Dissipation, P_D	See Graphs
Operating Temperature Range, T_A	−20°C to +85°C
Storage Temperature Range, T_S	−55°C to +125°C

ELECTRICAL CHARACTERISTICS at T_A = +25°C, T_{TAB} ≤ +70°C, V_S = ±6.0 V, V_{BOOST} = +9.0 V, each amplifier tested separately (unless otherwise specified)

Characteristic	Test Conditions	Limits			
		Min.	Typ.	Max.	Unit
Functional Supply Voltage Range	$+V_S$ to $-V_S$	6.0	—	40	V
Quiescent Supply Current	I_{BOOST} (Each Amp.)	—	7.0	10	mA
	$+I_S$ (Total)	—	75	130	mA
Input Bias Current	V_{OUT} = 0	—	−80	−1000	nA
Input Offset Voltage	V_{OUT} = 0, I_{OUT} = 0	—	±1.0	±10	mV
Input Offset Volt. TC†	Over Op. Temp. Range	—	−15	—	μV/°C
Input Offset Current	V_{OUT} = 0, I_{OUT} = 0	—	10	100	nA
Input Noise Voltage†	BW = 40 Hz to 15 kHz	—	4.0	—	μV
Input Noise Current†	BW = 40 Hz to 15 kHz	—	60	—	pA
Crossover Distortion†	P_{OUT} = 50 mW, R_L = 4Ω	—	0.2	—	%
Common Mode Rejection	ΔV_{CM} = 3 V	60	85	—	dB
Input Common Mode Range*	V_S = ±6 V	−6.3	—	+4.0	V
	V_S = ±15 V	−15.3	—	+13	V
Open-Loop Voltage Gain	f = 0	80	100	—	dB
Slew Rate	V_{IN} = V_{OUT} = 6 Vpp	0.5	1.0	—	V/μs
Gain-Bandwidth Product†	A_V = 40 dB	—	800	—	kHz
Channel Separation†	I_{out} = 100 mA, f = 1 kHz	—	60	—	dB
Output Voltage Swing	I_{OUT} = 1 A, V_{BOOST} = +6 V	9.0	9.5	—	Vpp
	I_{OUT} = 1 A, V_{BOOST} = +9 V	9.5	10.1	—	Vpp
Supply Voltage Rejection	$+V_S$, ΔV = 1 V	60	85	—	dB
	$-V_S$, ΔV = 1 V	60	80	—	dB
Thermal Resistance, Θ_{JT}*	ULN-3755B	—	—	15	°C/W
	ULN-3755W	—	—	3.0	°C/W
Thermal Shutdown Temp.†		—	165	—	°C

*This parameter is tested to a lot sample plan only.
†Typical values given for circuit design information only.

ALLOWABLE AVERAGE PACKAGE POWER DISSIPATION AS A FUNCTION OF TEMPERATURE

ULN-3755B

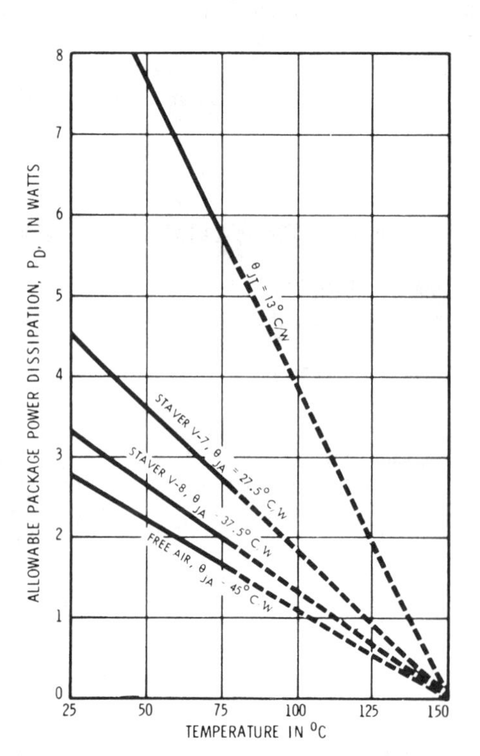

ULN-3755W

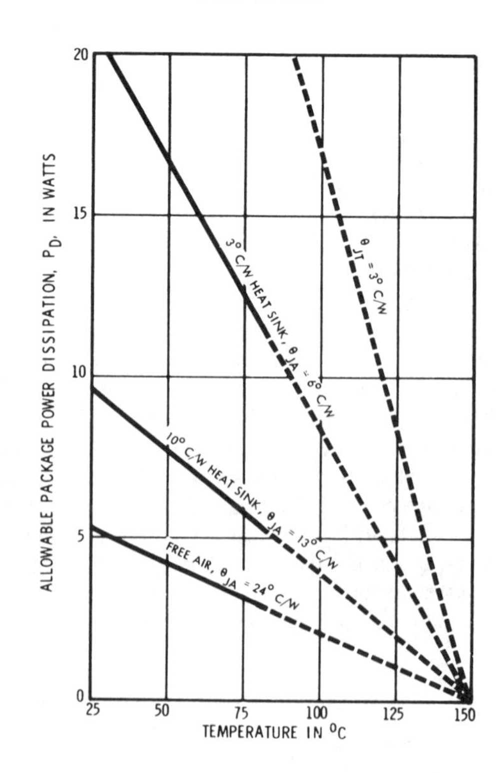

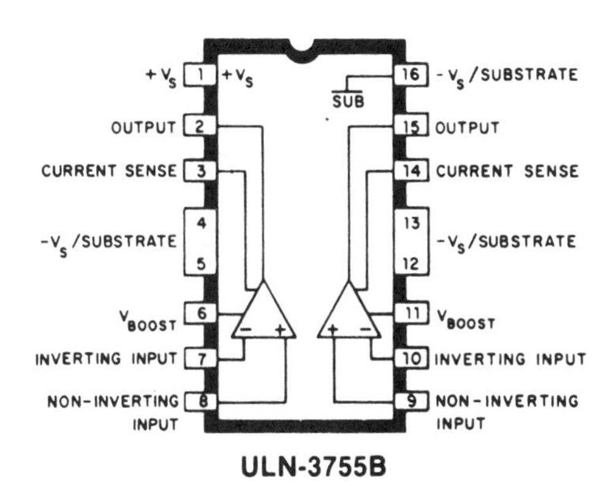

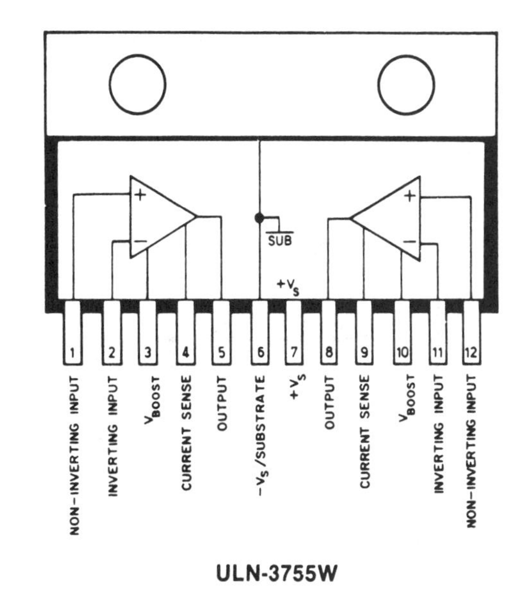

NON-INVERTING AMPLIFIER

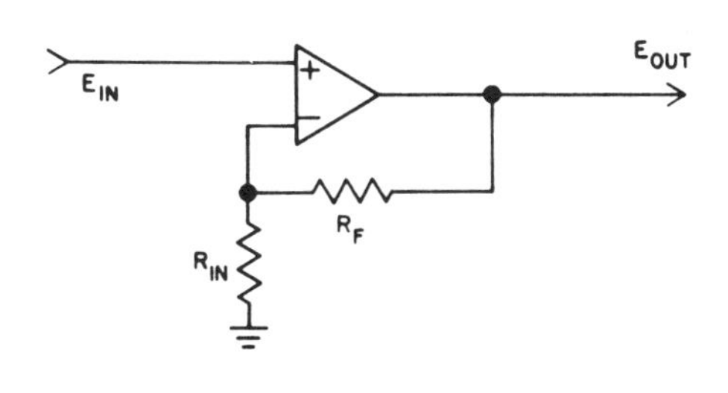

$$\frac{E_{OUT}}{E_{IN}} = 1 + \frac{R_F}{R_{IN}}$$

IF $R_F = 0$ or $R_{IN} = \infty$ THEN $E_{OUT} = E_{IN}$

INVERTING AMPLIFIER

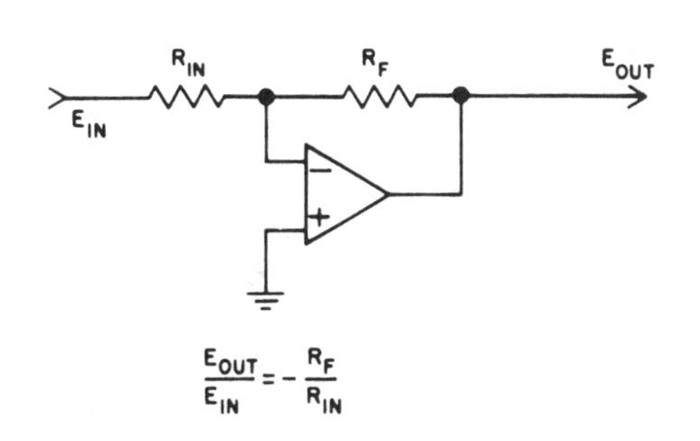

INPUT OFFSET VOLTAGE AS A FUNCTION OF TEMPERATURE

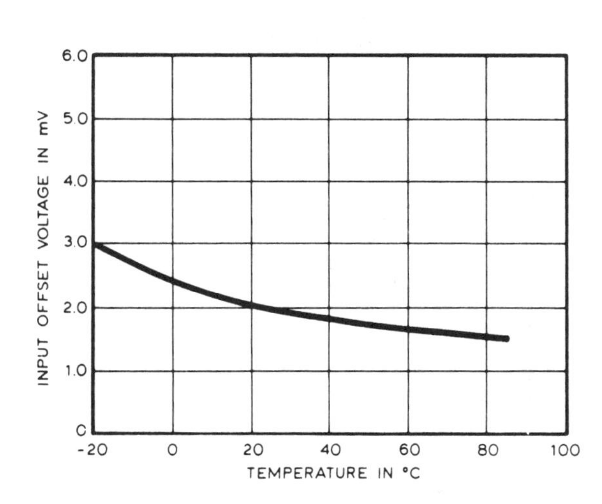

INPUT BIAS CURRENT AS A FUNCTION OF TEMPERATURE

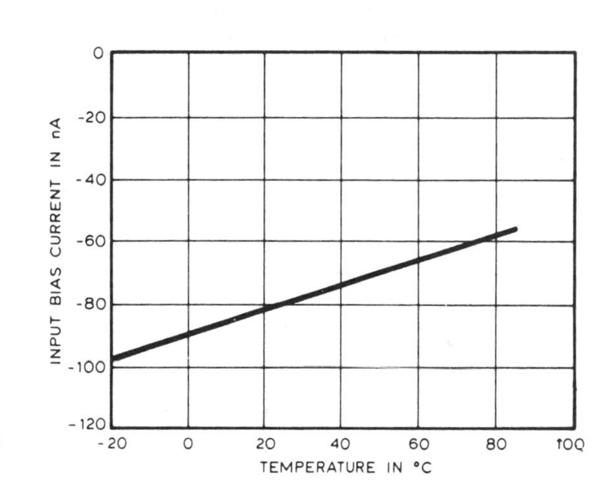

OPEN-LOOP VOLTAGE GAIN AND PHASE AS A FUNCTION OF FREQUENCY

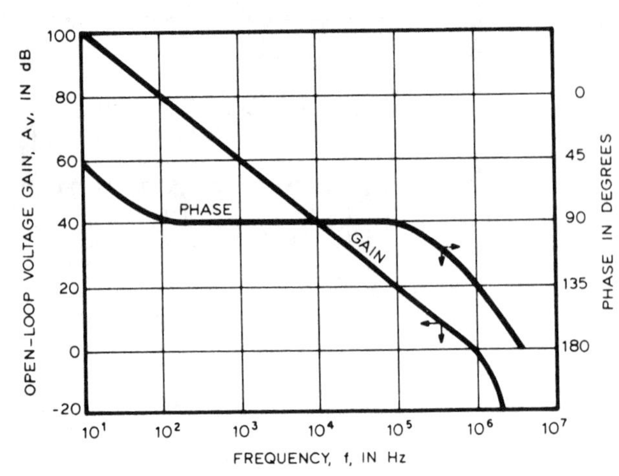

OPEN-LOOP VOLTAGE GAIN AS A FUNCTION OF TEMPERATURE

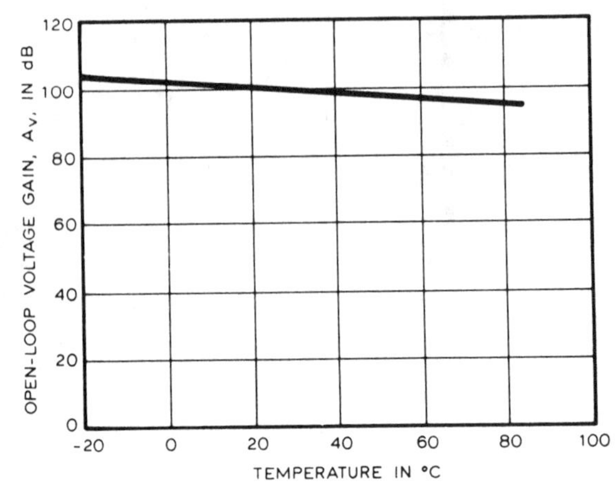

OUTPUT VOLTAGE SWING AS A FUNCTION OF OUTPUT CURRENT

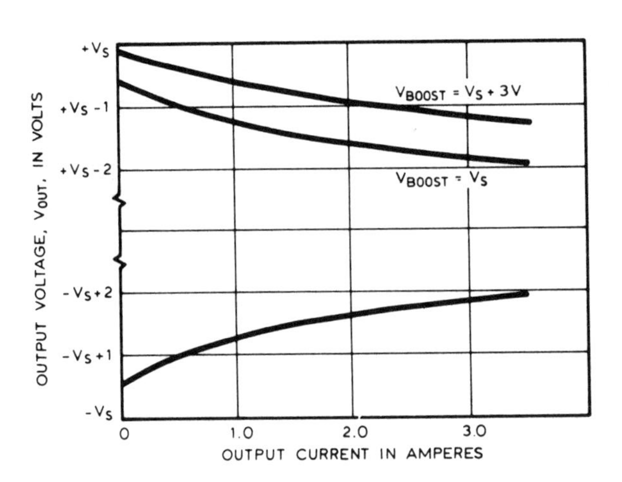

SUPPLY CURRENT AS A FUNCTION OF SUPPLY VOLTAGE

SUPPLY CURRENT, I_S, IN mA

SUPPLY VOLTAGE DIFFERENTIAL, ΔV_S, IN VOLTS

BOOST CURRENT AS A FUNCTION OF BOOST VOLTAGE

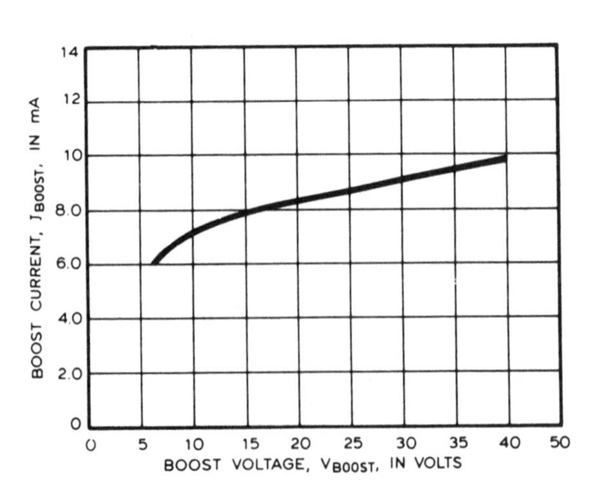

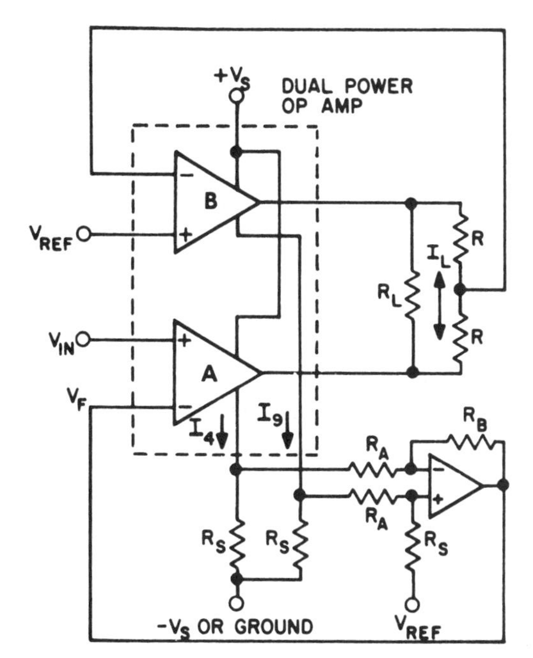

CURRENT-SENSE TRANSCONDUCTANCE AMPLIFIER

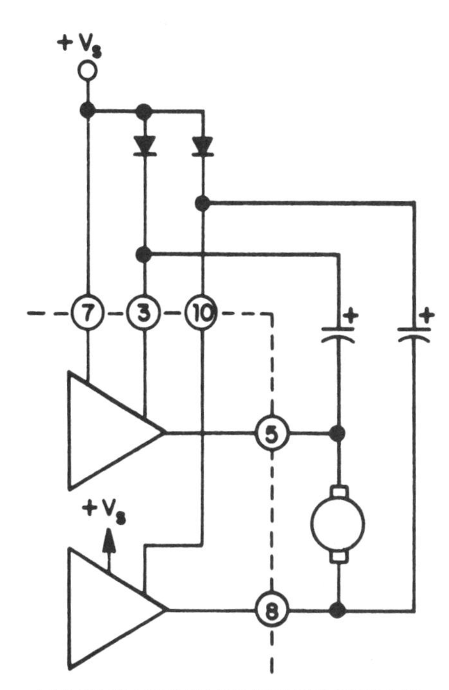

VOLTAGE DOUBLER BOOST SUPPLY

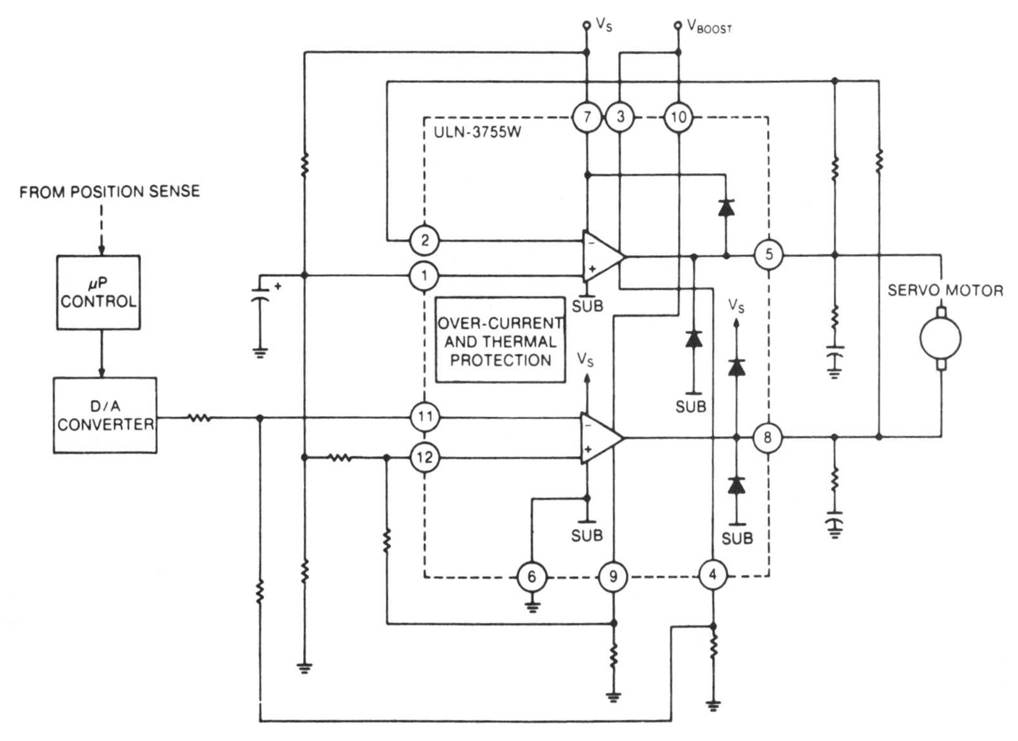

DIGITALLY CONTROLLED POSITION SERVO

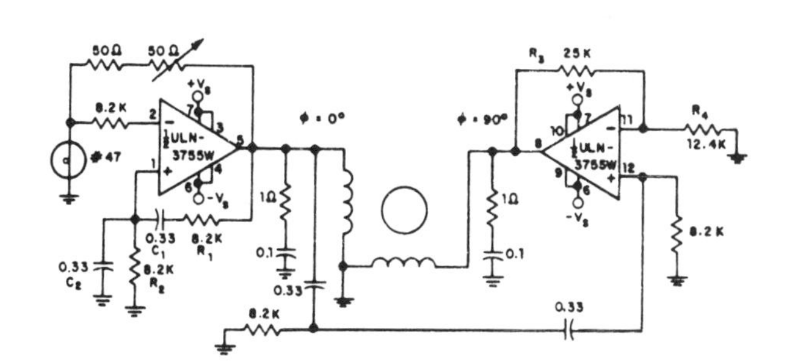

TWO-PHASE 60 HZ AC MOTOR DRIVER

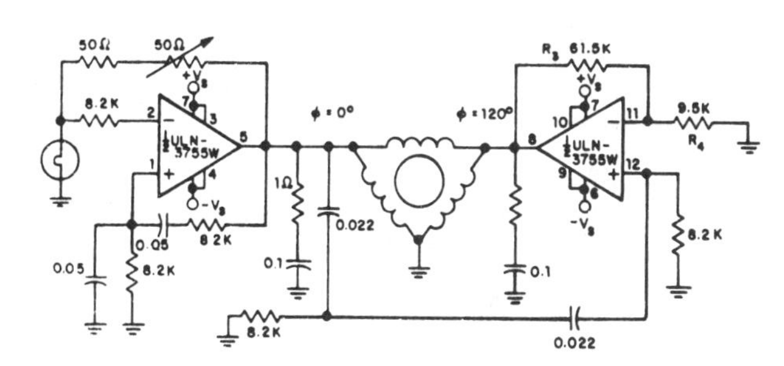

THREE-PHASE 400 HZ AC MOTOR DRIVER

Sprague UDN-5725M
Dual Peripheral/Power Drive

- DTL/TTL/PMOS/CMOS compatible
- Low input current
- Continuous output current to 1A
- 70V output standoff voltage
- Low supply-current requirement

The UDN-5725M power drive combines NAND and NOR logic gates in a configuration particularly useful with small brushless dc motor drivers. The integrated circuit includes high-current saturated output transistors and transient-suppression diodes. It can be used in many applications beyond the capabilities of standard logic buffers. With inputs tied together, one of two loads is energized by a single input signal.

Additional applications include driving peripheral loads, such as solenoids, light-emitting diodes, memories, heaters, and incandescent lamps with peak load currents of up to 1.2A.

Each of the output transistors is capable of sinking 800 mA continuously at 55 °C, or 650 mA at 85 °C. In the off state, the drivers will withstand at least 70V.

The UDN-5725M is supplied in a miniature 8-pin dual-in-line plastic package with a copper lead frame for superior-package power-dissipation ratings.

For application requiring output currents of up to 700 mA, Series UDN-5740M is recommended.

RECOMMENDED OPERATING CONDITIONS

Operating Condition	Min.	Nom.	Max.	Units
Supply Voltage, V_{CC}	4.75	12	14	V
Output Current, I_{ON}	—	—	650	mA
Operating Temperature Range	0	+25	+85	°C

ELECTRICAL CHARACTERISTICS over recommended operating temperature range (unless otherwise noted)

		Test Conditions					Limits				
Characteristic	Symbol	Temp.	V_{CC}	Enable Input	Other Inputs	Output	Min.	Typ.	Max.	Units	Notes
Output Reverse Current	I_{CEX}	—	4.75	0.8 V	2.0 V	70 V	—	—	100	μA	—
			4.75	0.8 V	0.8 V	70 V	—	—	100	μA	—
Output Voltage	$V_{CE(sat)}$	—	14	2.0 V	2.0 V	0.6 A	—	0.4	0.6	V	—
			14	2.0 V	2.0 V	0.8 A	—	0.7	1.0	V	—
			14	2.0 V	2.0 V	1.0 A	—	0.9	1.2	V	3
			14	2.0 V	0.8 V	0.6 A	—	0.4	0.6	V	—
			14	2.0 V	0.8 V	0.8 A	—	0.7	1.0	V	—
			14	2.0 V	0.8 V	1.0 A	—	0.9	1.2	V	3
	$V_{CE(sus)}$	+25°C	14		0 V	0.8 A	50	—	—	V	3,4
			14		2.0 V	0.8 A	50	—	—	V	3,4
Input Voltage	$V_{IN(1)}$	—	—	—	—	—	2.0	—	—	V	—
	$V_{IN(0)}$	—	—	—	—	—	—	—	0.8	V	—
Input Current	$I_{IN(0)}$	—	12.6	0.4 V	30 V	—	—	5.0	25	μA	1
	$I_{IN(1)}$	—	12.6	30 V	0 V	—	—	5.0	25	μA	1
Enable Input Current	$I_{IN(0)}$	—	12.6	0.4 V	30 V	—	—	10	50	μA	—
	$I_{IN(1)}$	—	12.6	30 V	0 V	—	—	10	50	μA	—
Input Clamp Volt.	V_{CLAMP}	—	4.75	−12 mA	—	—	—	—	−1.5	V	—
Diode Leakage Current	I_R	+25°C	5.0	0 V	0 V	Open	—	—	100	μA	2
Diode Forward Voltage	V_F	+25°C	5.0	0 V	0 V	0.6 A	—	1.5	2.0	V	—
			5.0	0 V	0 V	1.0 A	—	1.9	2.5	V	3
Supply Current (Total Package)	$I_{CC(1)}$	+25°C	12.6	0 V	0 V	—	—	3.9	5.0	mA	—
			12.6	0 V	2.0 V	—	—	3.9	5.0	mA	—
	$I_{CC(0)}$	+25°C	12.6	2.0 V	0 V	—	—	22	30	mA	—
			12.6	2.0 V	2.0 V	—	—	22	30	mA	—

NOTES:
1. Except ENABLE input, each input tested separately.
2. Diode leakage current measured at V_R = 70 V.
3. Pulse Test.
4. L_L = 3 mH.

ABSOLUTE MAXIMUM RATINGS

at T_A = +25°C

Output Off-State Voltage, V_{OFF}	70 V
Output On-State Sink Current, I_{ON} (continuous)	1.0 A†
(peak)	1.2 A
Logic Supply Voltage, V_{CC}	16 V
Input Voltage, V_{IN}	30 V
Suppression Diode Off-State Voltage, V_{OFF}	70 V
Suppression Diode On-State Current, I_{ON}	1.0 A
Allowable Package Power Dissipation, P_D	1.5 W*
Operating Free-Air Temperature Range, T_A	−20°C to +85°C
Storage Temperature Range, T_S	−55°C to +150°C

*Derate at the rate of 12.5 mW/°C above T_A = +25°C.
†Limited by P_D.

SWITCHING CHARACTERISTICS at T_A = +25°C, V_{CC} = 5.0 V

Characteristic	Symbol	Test Conditions	Limits Min.	Limits Max.	Limits Units	Notes
Turn-On Delay Time	$t_{pd\,0}$	V_S = 30 V, R_L = 100 (10 W), C_L = 15 pF	—	500	ns	1, 2
Turn-Off Delay Time	$t_{pd\,1}$	V_S = 30 V, R_L = 100 (10 W), C_L = 15 pF	—	750	ns	1, 2

NOTES: 1. Capacitance value specified includes probe and test fixture capacitance.
2. Voltage values shown in test circuit waveforms are with respect to network ground.

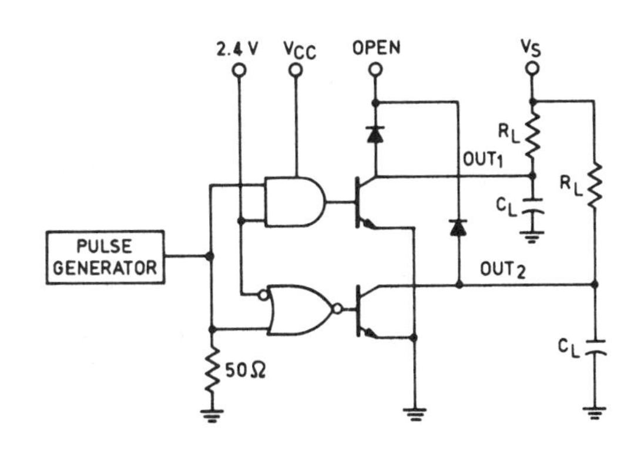

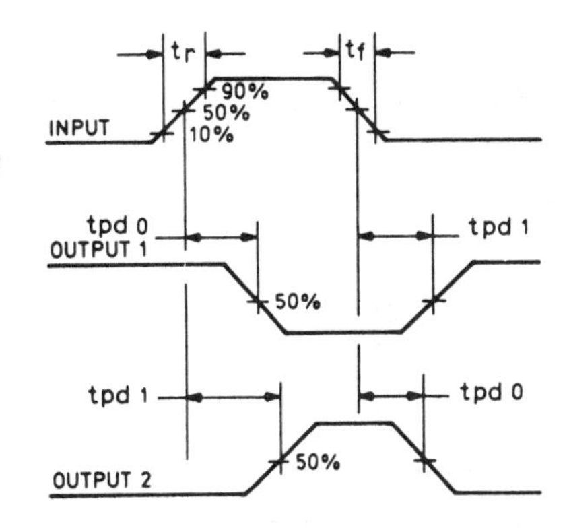

ALLOWABLE AVERAGE PACKAGE POWER DISSIPATION AS A FUNCTION OF TEMPERATURE

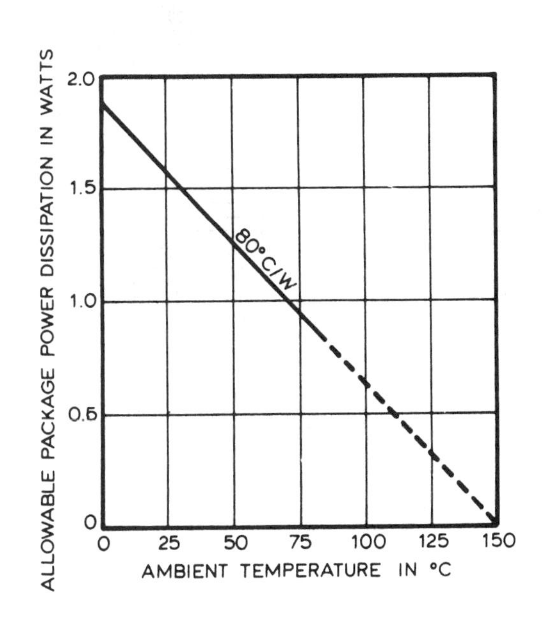

INPUT-PULSE CHARACTERISTICS

$V_{IN(0)}$ = 0 V	t_f = 7 ns	t_p = 1 μs
$V_{IN(1)}$ = 3.5 V	t_r = 14 ns	PRR = 500kHz

TYPICAL APPLICATION

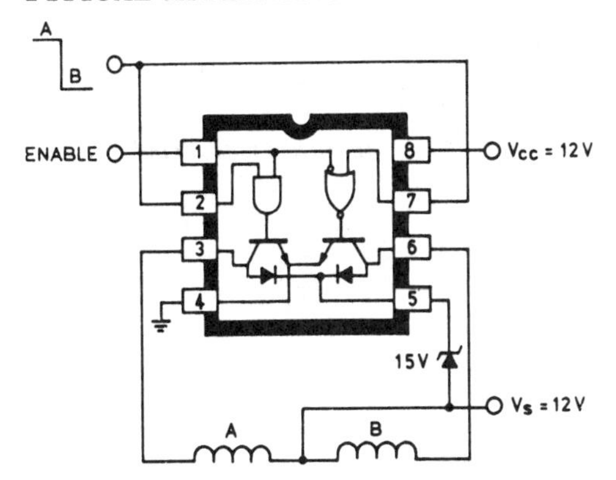

Sprague UDN-7078W
Quad High-Current Darlington Switch

- Output voltage to 90V
- 90V sustaining voltage
- Output current to 3A
- TTL-, DTL-, or CMOS-compatible inputs
- Internal transient-suppression diodes
- Plastic single-in-line package
- Heatsink tab

This quad Darlington array is designed to interface between low-level logic and peripheral power devices, such as solenoids, motors, incandescent lamps, heaters, and similar loads up to 270W per channel. The integrated circuit contains internal transient-suppression diodes that enable use with inductive loads. The input logic is compatible with most TTL, DTL, LSTTL, and 5V CMOS logic. The Darlington array is rated for operation to 90V and is recommended for operation with load currents of 3A or less.

For maximum power-handling capability, the device is supplied in a 12-pin single-in-line plastic package with an integral power tab. The tab is at ground potential and needs no insulation. External heatsinks are usually required for proper operation of this device.

Similar quad high-current Darlington switches, for operation with supply voltages to 50V or 80V (35V or 50V sustaining), are UDN-2878W and UDN-2879W.

ABSOLUTE MAXIMUM RATINGS
at T_{TAB} < +70°C

Output Voltage, V_{CE}	90V
Min. Sustaining Voltage, $V_{CE(sus)}$	90V
Output Current, I_C	3.0A
Supply Voltage, V_S	10V
Input Voltage, V_{IN}	15V
Total Package Power Dissipation, P_D	See Graph
Operating Temperature Range, T_A	−20°C to +85°C
Storage Temperature Range, T_S	−55°C to +150°C

ELECTRICAL CHARACTERISTICS at T_A = +25°C, T_{TAB} < +70°C, V_S = 5 V

Characteristic	Symbol	Test Conditions	Limits Min.	Limits Max.	Units
Output Leakage Current	I_{CEX}	V_{CE} = 90V	—	100	μA
Output Sustaining Voltage	$V_{CE(sus)}$	I_C = 2.5 A	90	—	V
Output Saturation Voltage	$V_{CE(SAT)}$	I_C = 0.5 A, V_{IN} = 2.75V	—	1.1	V
		I_C = 1.0 A, V_{IN} = 2.75V	—	1.3	V
		I_C = 2.0 A, V_{IN} = 2.75V	—	1.6	V
		I_C = 2.5 A, V_{IN} = 2.75V	—	1.9	V
		I_C = 3.0 A, V_{IN} = 2.75V	—	2.2	V
Input Voltage	$V_{IN(ON)}$	I_C = 3.0 A	2.75	—	V
	$V_{IN(OFF)}$		—	0.8	V
Input Current	$I_{IN(0)}$	V_{IN} = 0.8V	—	25	μA
	$I_{IN(1)}$	V_{IN} = 2.75V	—	550	μA
		V_{IN} = 3.75V	—	1.0	mA
Supply Current per Driver	I_S	I_C = 500 mA	—	6.0	mA
Clamp Diode Leakage Current	I_R	V_R = 90V	—	50	μA
Clamp Diode Forward Voltage	V_F	I_F = 2.5 A	—	2.5	V
		I_F = 3.0 A	—	3.0	V
Turn-On Delay	t_{PLH}	0.5 E_{in} to 0.5 E_{out}	—	1.0	μs
Turn-Off Delay	t_{PHL}	0.5 E_{in} to 0.5 E_{out}, I_C = 3.0 A	—	1.5	μs

CAUTION: High-current tests are pulse tests or require heat sinking.

PARTIAL SCHEMATIC
One of 4 Drivers

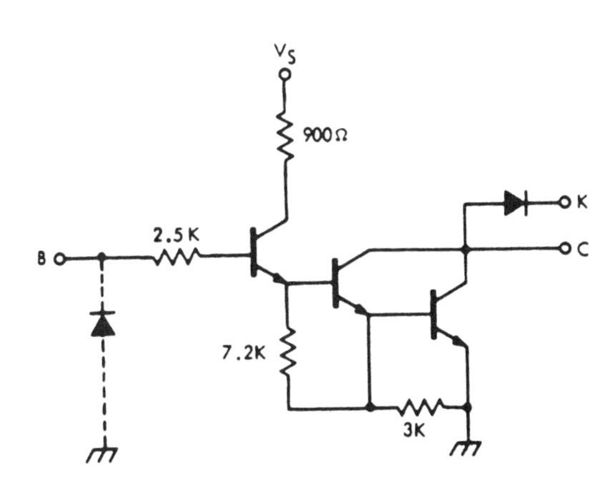

ALLOWABLE AVERAGE PACKAGE POWER DISSIPATION AS A FUNCTION OF TEMPERATURE

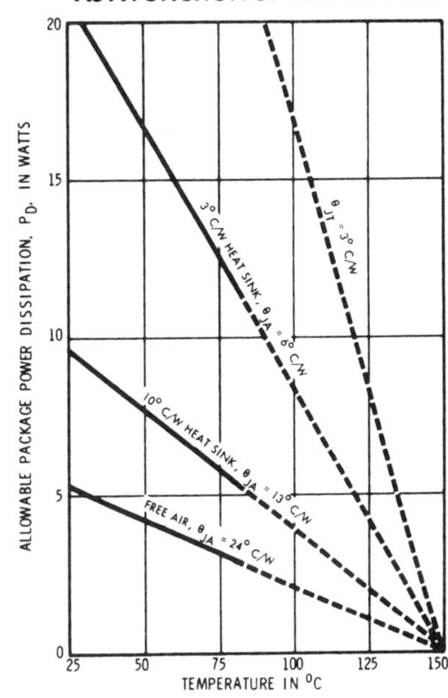

TYPICAL APPLICATION

PRINT-HAMMER DRIVER

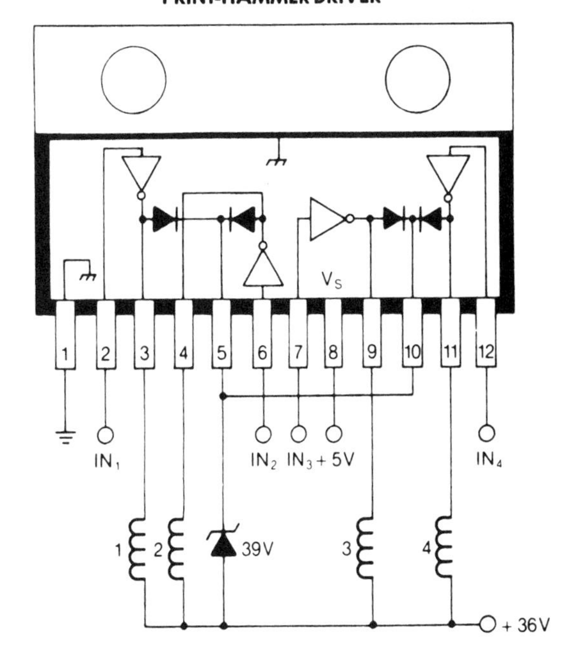

TEST FIGURE
$V_{CE(sat)}$

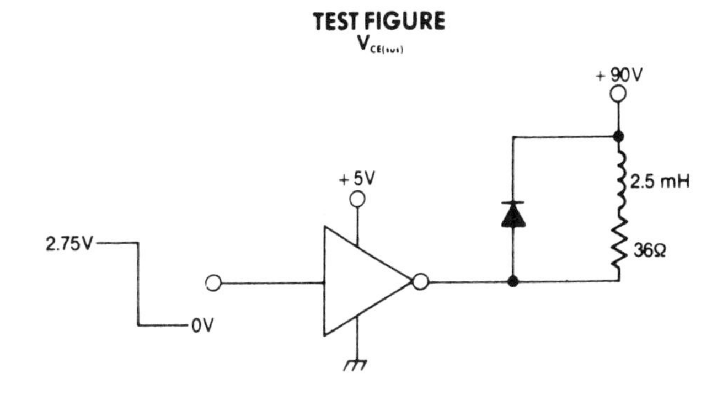

Sprague UDN-2596A through UDN-2599A
8-Channel Saturated Sink Drivers

- Low output on voltages
- Up to 1.0A sink capability
- 50V Min. output breakdown
- Output transient-suppression diodes
- Output pull-down for fast turn-off
- TTL- and CMOS-compatible inputs

Low output saturation voltages at high load currents are provided by UDN-2596A through UDN-2599A sink driver ICs. These devices can be used as interface buffers between standard low-power digital logic (particularly MOS) and high-power loads (such as relays, solenoids, stepping motors, and LED or incandescent displays). The eight saturated sink drivers in each device feature high-voltage, high-current open-collector outputs. Transient-suppression clamp diodes and a minimum 35V output sustaining voltage allow their use with many inductive loads.

The saturated (nonDarlington) npn outputs provide low collector-emitter voltage drops as well as improved turn-off times as a result of an active pull-down function within the output predrive section. The UDN-2596A and UDN-2598A are for use with output loads to 500 mA while the UDN-2597A and UDN-2599A are for use with loads to 1A. Adjacent outputs can be paralleled for higher load currents.

Inputs require very low input current and are activated by a low-logic level consistent with the much greater sinking capability associated with NMOS, CMOS, and TTL logic. The UDN-2596A and UDN-2597A are rated for use with 5V logic levels, and the UDN-2598A and UDN-2599A are for use with 10V to 12V logic levels.

All devices are furnished in 20-pin DIP packages with copper leadframes for improved thermal characteristics.

ABSOLUTE MAXIMUM RATINGS
at $T_A = +25°C$

Output Voltage, V_{CE}	50V
Output Current, I_{OUT}	
(UDN-2596/98A)	500mA
(UDN-2597/99A)	1.0A
Supply Voltage, V_{CC}	
(UDN-2596/97A)	7.0V
(UDN-2598/99A)	15V
Input Voltage, V_{IN}	
(UDN-2596/97A)	7.0V
(UDN-2598/99A)	15V
Package Power Dissipation, P_D	2.27W*
Operating Temperature Range, T_A	−20°C to +85°C
Storage Temperature Range, T_S	−65°C to +150°C

*Derate at the rate of 18.2mW/°C above T_A = 25°C.

RECOMMENDED OPERATING CONDITIONS

Type Number	Logic	I_{OUT}
UDN-2596A	5.0V	300mA
UDN-2597A	5.0V	750mA
UDN-2598A	10-12V	300mA
UDN-2599A	10-12V	750mA

Note: Pins 2 and 12 must both be connected to power ground.

ELECTRICAL CHARACTERISTICS at $T_A = +25°C$, V_{CC} = 5.0V (UDN-2596/97A) or 12V (UDN-2598/99A)

Characteristics	Symbol	Applicable Devices*	Test Conditions	Limits Min.	Limits Max.	Units
Output Leakage Current	I_{CEX}	All	V_{OUT} = 50V, V_{IN} = 2.4V	—	10	µA
Output Sustaining Voltage	$V_{CE(sus)}$	2596/98	I_{OUT} = 300mA, L = 2mH	35	—	V
		2597/99	I_{OUT} = 750mA, L = 2mH	35	—	V
Output Saturation Voltage	$V_{CE(SAT)}$	2596/98	I_{OUT} = 300mA	—	0.5	V
		2597/99	I_{OUT} = 750mA	—	1.0	V
Clamp Diode Leakage Current	I_R	All	V_R = 50V	—	10	µA
Clamp Diode Forward Voltage	V_F	2596/98	I_F = 300mA	—	1.8	V
		2597/99	I_F = 750mA	—	1.8	V
Logic Input Current	$I_{IN(0)}$	2596/97	V_{IN} = 0.8V	—	−15	µA
		2598/99	V_{IN} = 0.8V	—	−50	µA
	$I_{IN(1)}$	2596/97	V_{IN} = 2.4V	—	10	µA
		2598/99	V_{IN} = 12V	—	10	µA
Supply Current (per driver)	$I_{CC(ON)}$	2596/98	V_{IN} = 0.8V	—	6.0	mA
		2597/99	V_{IN} = 0.8V	—	22	mA
	$I_{CC(OFF)}$	2596/97	V_{IN} = 2.4V	—	1.3	mA
		2598/99	V_{IN} = 2.4V	—	2.0	mA
Turn-On Delay	t_{pd0}	All	0.5E_{IN} to 0.5E_{OUT}	—	3.0	µs
Turn-Off Delay	t_{pd1}	All	0.5E_{IN} to 0.5E_{OUT}	—	2.0	µs

*Complete part number includes prefix UDN- and suffix A, e.g. UDN-2596A.

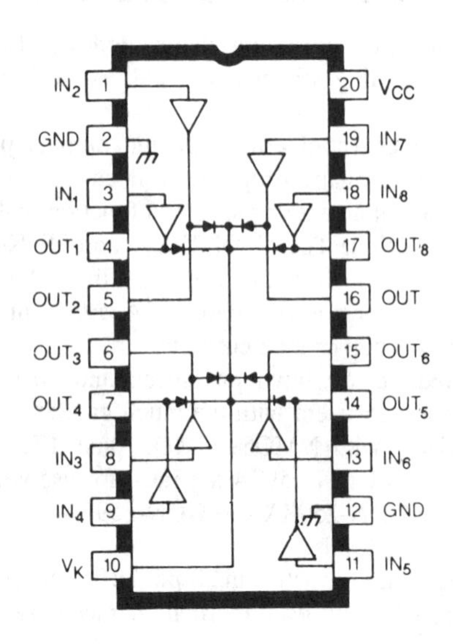

ONE OF EIGHT DRIVERS

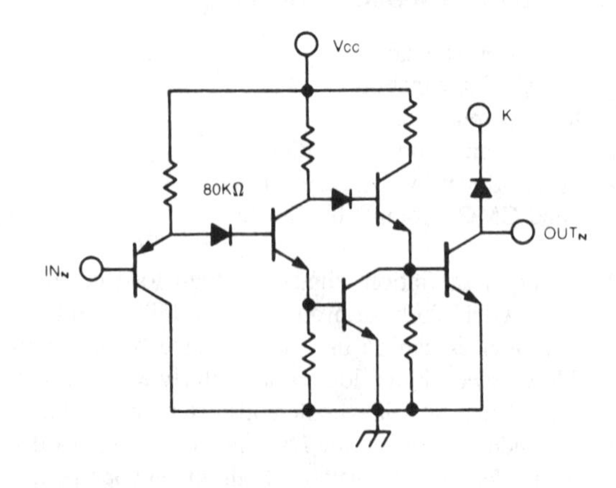

TYPICAL APPLICATION

DUAL STEPPER MOTOR DRIVE

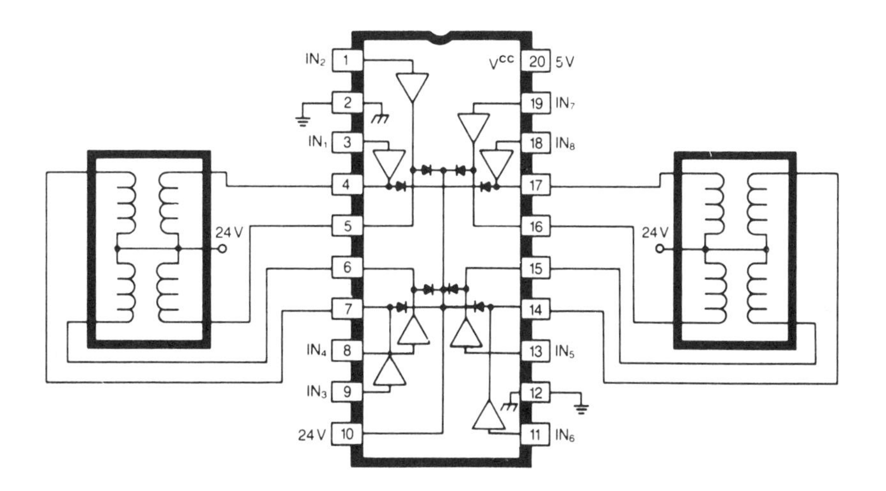

Sprague ULN-8130A

Quad Voltage and Line Monitor

- 10 to 35V operation
- Low standby current
- Reference trimmed to 1%
- Monitors 4 separate dc levels
- Seperate under-voltage comparators
- Fixed under-voltage threshold
- Line sense input
- Pull-up clamped outputs
- Programmable output delays
- V_S under-voltage lockout

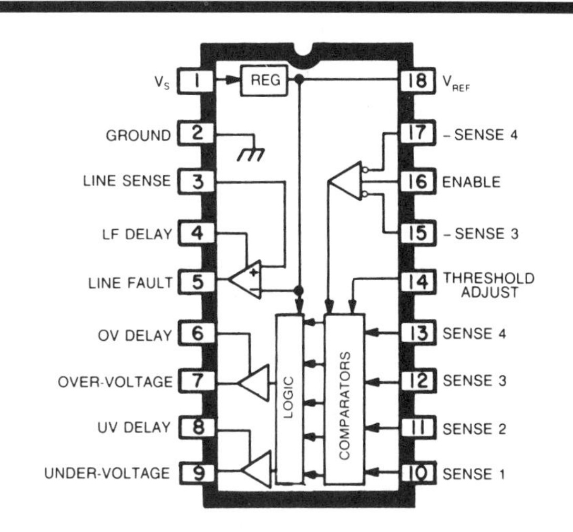

ABSOLUTE MAXIMUM RATINGS at $T_A = +25°C$

Supply Voltage, V_{CC}	35 V
Power Dissipation, P_D	2.3 W*
Operating Temperature, T_A	0°C to +70°C
Storage Temperature, T_S	−65°C to +150°C
Junction Temperature, T_J	+150°C

*Derate at the rate of 18.2 mW/°C above $T_A = 25°C$

NEGATIVE SENSE MONITORING

SENSE 3 and 4 Only

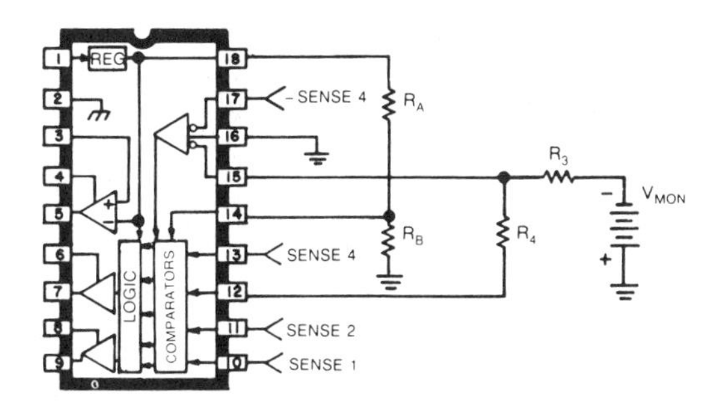

ELECTRICAL CHARACTERISTICS at $T_A = +25°C$, $V_S = 15$ V

Characteristic	Test Pin	Test Conditions	Limits Min.	Limits Max.	Units
Functional V_S Range	1		10	35	V
Quiescent Current	1	$V_S = 35$ V, $V_{16} = V_{18}$, No Fault	—	15	mA
REFERENCE VOLTAGE SECTION					
Reference Voltage	18	No Load, $T_A = +25°C$	2.47	2.53	V
		No Load, Change Over Temp.	—	25	mV
Load Regulation	18	$I_{REF} = 0$ to 10 mA	—	20	mV
Line Regulation	18	$V_S = 10$ to 35 V	—	10	mV
Ripple Rejection	18	f = 120 Hz	60	—	dB
Short-Circuit Current Protection	18		—	40	mA
COMPARATOR SECTION					
Under-Voltage Trip Points	10-13*	$T_A = +25°C$	2.47	2.53	V
		Over Temperature	2.46	2.54	V
Under-Voltage Trip Hysteresis	10-13*	Over Temperature	10	25	mV
Over-Voltage Trip Points	10-13*	$V_{14} = 0$	3.08	3.17	V
Over-Voltage Trip Hysteresis	14	$V_{14} = 0$ to 2.5 V, Over Temp.	10	25	mV
Line Monitor Trip Threshold	3		2.40	2.54	V
Under-Voltage Lockout Enable	1	V_S Decreasing	8.5	—	V
Under-Voltage Lockout Disable	1	V_S Increasing	—	10.5	V
Input Bias Current	3, 10, 11	$V_{IN} = 2.0$ V	—	−6.0	μA
		$V_{IN} = 3.0$ V	—	6.0	μA
	14	$V_{IN} = 0$	—	−50	μA
	15, 17	$V_{IN} = -2.0$ V, $V_{16} = 0$ V	—	−2.0	μA
OUTPUT DRIVERS					
Output Saturation Voltage	5, 9	$I_{SINK} = 5.0$ mA	—	0.5	V
	7	$I_{SINK} = 10$ mA	—	0.5	V
	5, 7, 9	$I_{SOURCE} = 500$ μA	4.0	5.25	V
Output Leakage Current	5, 7, 9	$V_{OUT} = 35$ V	—	50	μA
Line Fault Delay Current Source	4	$V_4 = 2.0$ V	160	350	μA
Line Fault Delay Current Sink	4	$V_4 = 2.0$ V	3.2	7.0	mA
Over-Voltage Delay Current Source	6	$V_6 = 2.0$ V	160	300	μA
Under-Voltage Delay Current Source	8	$V_8 = 2.0$ V	35	75	μA

*All inputs connected to 2.75 V except input being tested.

LINE SENSE AND POSITIVE SUPPLY MONITORING

(SENSE 1, 2, 3, and 4)

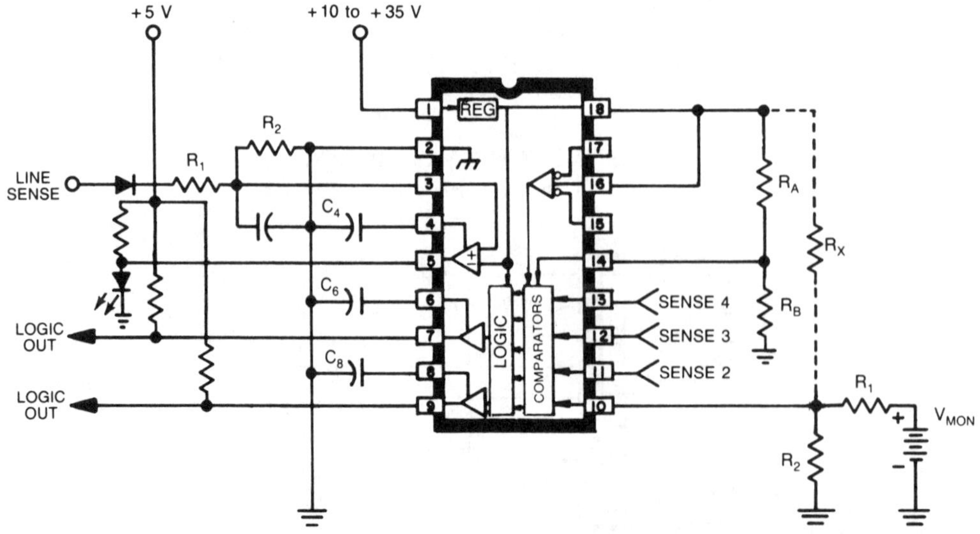

FUNCTIONAL BLOCK DIAGRAM

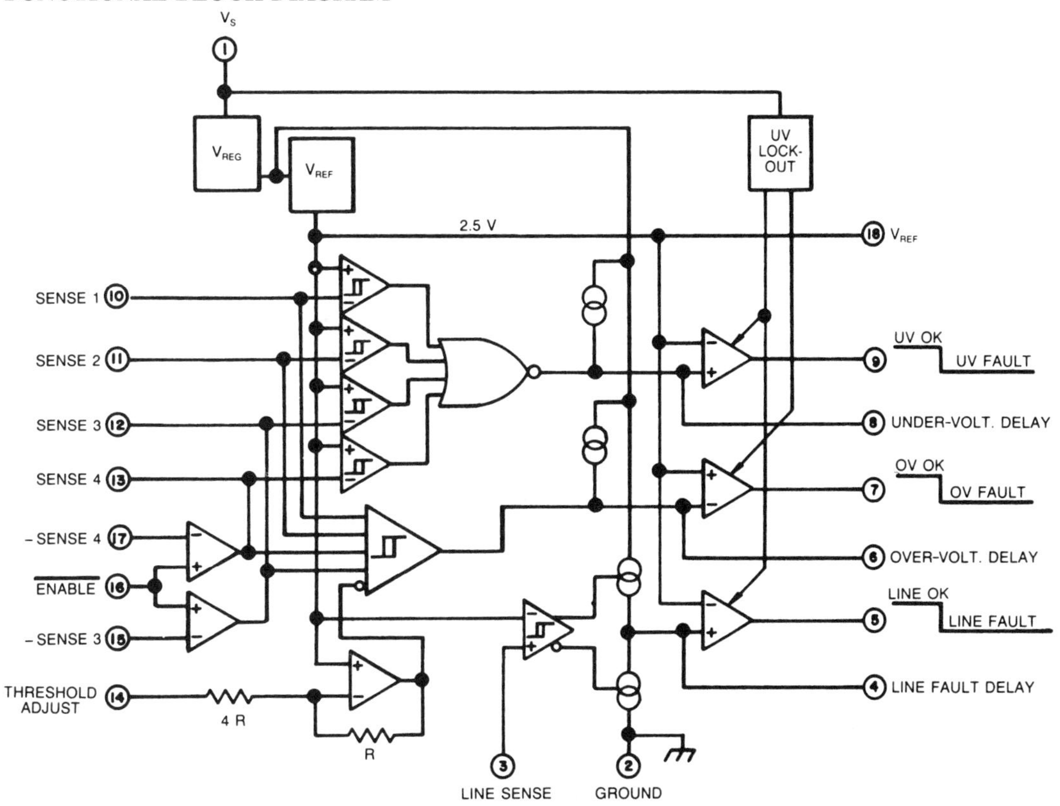

Sprague ULN-8131A
Quad Voltage and Line Monitor

- Reference trimmed to 1%
- Monitors four dc supplies
- 10 to 35V operation
- Low standby current
- Separate undervoltage comparators
- Fixed undervoltage threshold
- Programmable overvoltage threshold
- Line-sense input
- Full-up clamped outputs
- Programmable output delays
- V_S undervoltage lockout

ABSOLUTE MAXIMUM RATINGS

Supply Voltage, V_{CC} 35V
Power Dissipation, P_D 2.3W*
Operating Temperature, T_A .... 0°C to +70°C
Storage Temperature, T_S .... −65°C to +150°C
Junction Temperature, T_J .... +150°C

*Derate at the rate of 18.2 mW/°C above T_A = 25°C

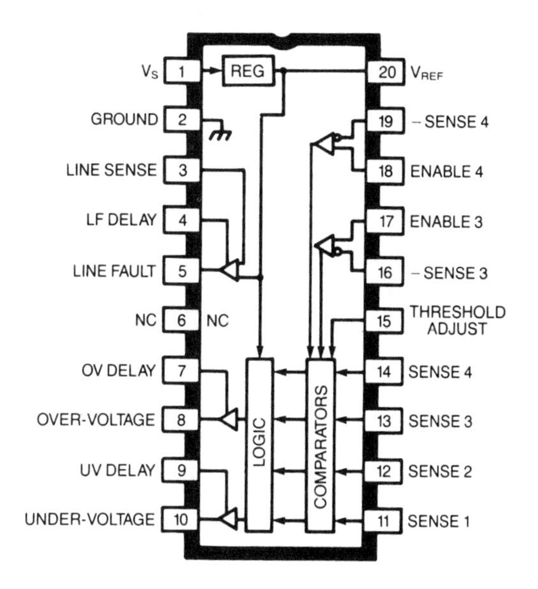

ELECTRICAL CHARACTERISTICS at $T_A = +25°C, V_S = 15\,V$

Characteristic	Test Pin	Test Conditions	Limits Min.	Limits Max.	Units
Functional V_S Range	1		10	35	V
Quiescent Current	1	$V_S = 35V, V_{17} = V_{18} = V_{20}$ No Fault	—	15	mA
REFERENCE VOLTAGE SECTION					
Reference Voltage	20	No Load, $T_A = +25°C$	2.47	2.53	V
		No Load, Change Over Temp.	—	25	mV
Load Regulation	20	$I_{REF} = 0$ to 10mA	—	20	mV
Line Regulation	20	$V_S = 10$ to 35V	—	10	mV
Ripple Rejection	20	f = 120Hz	60	—	dB
Short-Circuit Current Protection	20		—	40	mA
COMPARATOR SECTION					
Under-Voltage Trip Points	11-14*	$T_A = +25°C$	2.47	2.53	V
		Over Temperature	2.46	2.54	V
Under-Voltage Trip Hysteresis	11-14*	Over Temperature	10	25	mV
Over-Voltage Trip Points	11-14*	$V_{15} = 0$	3.08	3.17	V
Over-Voltage Trip Hysteresis	15	$V_{15} = 0$ to 2.5V, Over Temp.	10	25	mV
Line Monitor Trip Threshold	3		2.40	2.54	V
Under-Voltage Lockout Enable	1	V_S Decreasing	8.5	—	V
Under-Voltage Lockout Disable	1	V_S Increasing	—	10.5	V
Input Bias Current	3, 11, 12	$V_{IN} = 2.0V$	—	−6.0	μA
		$V_{IN} = 3.0V$	—	6.0	μA
	15	$V_{IN} = 0$	—	−50	μA
	16, 19	$V_{IN} = -2.0V, V_{17} = V_{18} = 0V$	—	−2.0	μA
OUTPUT DRIVERS					
Output Saturation Voltage	5, 10	$I_{SINK} = 5.0mA$	—	0.5	V
	8	$I_{SINK} = 10mA$	—	0.5	V
	5, 8, 10	$I_{SOURCE} = 500\mu A$	4.0	5.25	V
Output Leakage Current	5, 8, 10	$V_{OUT} = 35V$	—	50	μA
Line Fault Delay Current Source	4	$V_4 = 2.0V$	160	350	μA
Line Fault Delay Current Sink	4	$V_4 = 2.0V$	3.2	7.0	mA
Over-Voltage Delay Current Source	7	$V_7 = 2.0V$	160	300	μA
Under-Voltage Delay Current Source	9	$V_9 = 2.0V$	35	75	μA

**All inputs connected to 2.75 V except input being tested.*

NEGATIVE SENSE MONITORING

SENSE 3 and 4 Only

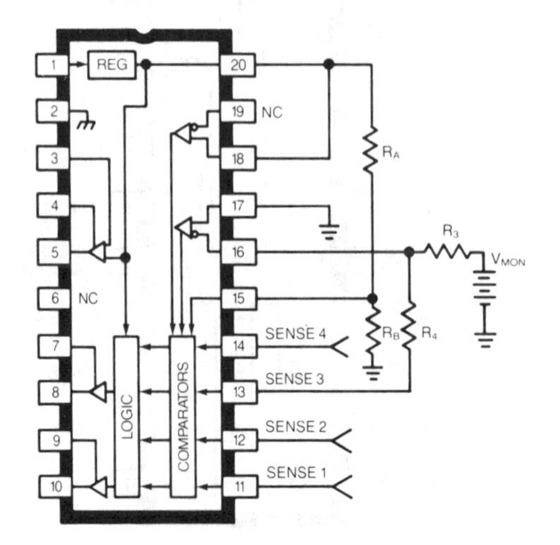

FUNCTION BLOCK DIAGRAM

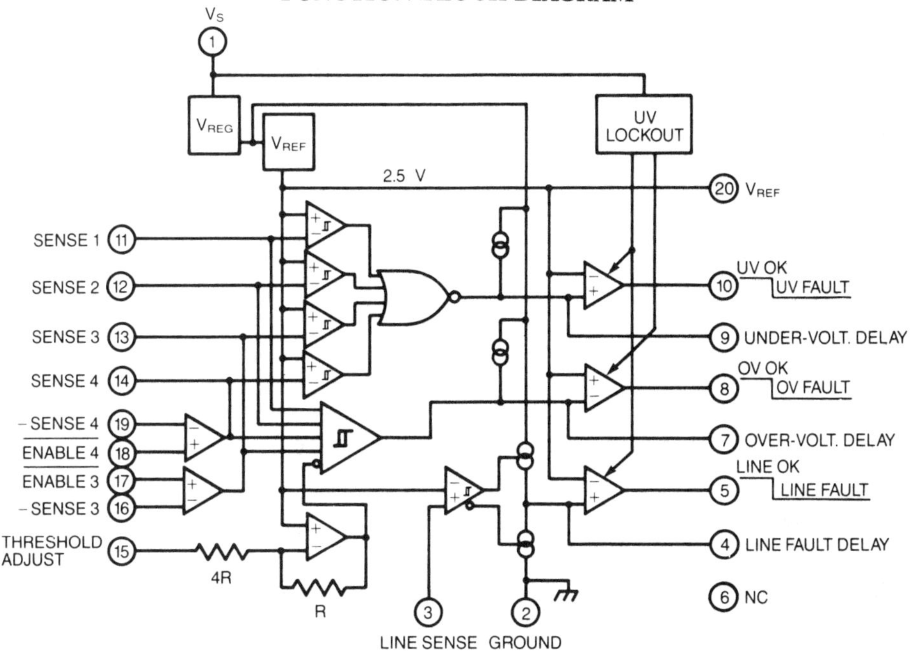

LINE SENSE AND POSITIVE SUPPLY MONITORING

(SENSE 1, 2, 3, and 4)

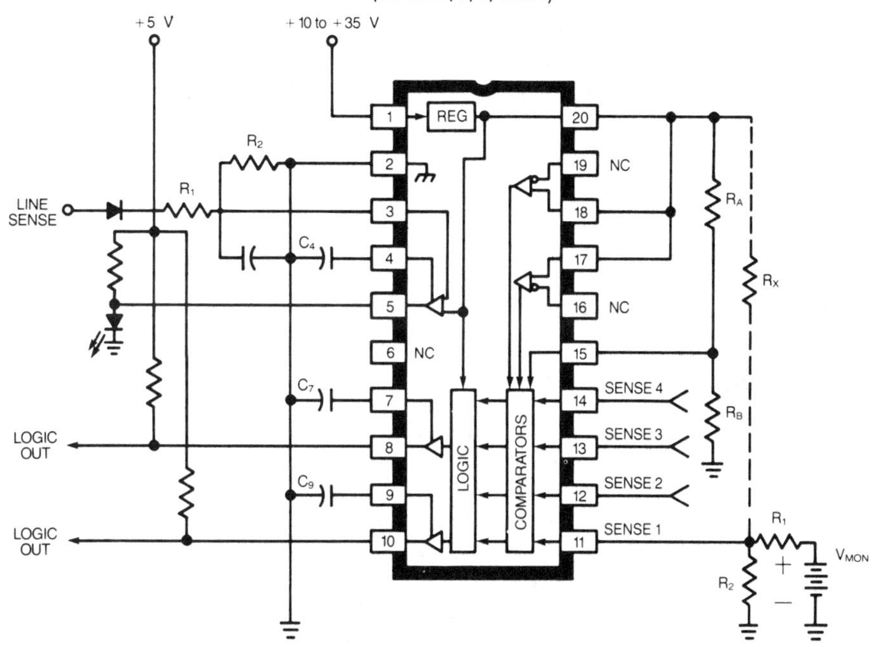

Sprague UCN-5818AF, UCN-5818EPF BiMOS II

32-Bit Serial-Input, Latched-Source Drivers With Active DMOS Pulldowns

- 60 or 80V source outputs
- High-speed source drivers
- Active DMOS pull-downs
- Low-output saturation voltages
- Low-power CMOS logic and latches
- 3.3-MHz minimum data input rate
- Reduced supply current requirements
- Improved replacements for SN75518N/FN

ABSOLUTE MAXIMUM RATINGS

at T_A = 201°C

Logic Supply Voltage, V_{DD}	15 V
Driver Supply Voltage, V_{BB}	60 V
(Suffix "-1")	80 V
Continuous Output Current, I_{OUT}	−40 to +15 mA
Input Voltage Range, V_{IN}	−0.3 V to V_{DD} +0.3 V
Package Power Dissipation, P_D (UCN-5818AF)	2.8 W*
(UCN-5818EPF)	2.0 W†
Operating Temperature Range, T_A	−20°C to +85°C
Storage Temperature Range, T_S	−55°C to +125°C

*Derate at rate of 28 mW/°C above T_A = +25°C
†Derate at rate of 20 mW/°C above T_A = +25°C

ELECTRICAL CHARACTERISTICS at T_A = +25°C, V_{BB} = 60 V (UCN-5818AF/EPF) or 80 V (suffix '-1') unless otherwise noted

Characteristic	Symbol	Test Conditions	Limits at V_{DD} = 5 V Min.	Typ.	Max.	Limits at V_{DD} = 12 V Min.	Typ.	Max.	Units
Output Leakage Current	I_{CEX}	V_{OUT} = 0 V, T_A = +70°C	—	−5.0	−15	—	−5.0	−15	μA
Output Voltage	$V_{OUT(1)}$	I_{OUT} = −25 mA, V_{BB} = 60 V	58	58.5	—	58	58.5	—	V
		I_{OUT} = −25 mA, V_{BB} = 80 V*	78	78.5	—	78	78.5	—	V
	$V_{OUT(0)}$	I_{OUT} = 1 mA	—	2.0	3.0	—	—	—	V
		I_{OUT} = 2 mA	—	—	—	—	2.0	3.0	V
Output Pull-Down Current	$I_{OUT(0)}$	V_{OUT} = 5 V to V_{BB}	2.0	3.5	—	—	—	—	mA
		V_{OUT} = 20 V to V_{BB}	—	—	—	8.0	13	—	mA
Input Voltage	$V_{IN(1)}$		3.5	—	5.3	10.5	—	12.3	V
	$V_{IN(0)}$		−0.3	—	+0.8	−0.3	—	+0.8	V
Input Current	$I_{IN(1)}$	V_{IN} = V_{DD}	—	0.05	0.5	—	0.1	1.0	μA
	$I_{IN(0)}$	V_{IN} = 0.8 V	—	−0.05	−0.5	—	−1.0	−1.0	μA
Serial Data Output Voltage	$V_{OUT(1)}$	I_{OUT} = −200 μA	4.5	4.7	—	11.7	11.8	—	V
	$V_{OUT(0)}$	I_{OUT} = 200 μA	—	200	250	—	100	200	mV
Maximum Clock Frequency	f_{clk}		3.3	5.0	—	—	7.5	—	MHz
Supply Current	$I_{DD(1)}$	All Outputs High	—	100	200	—	200	400	μA
	$I_{DD(0)}$	All Outputs Low	—	100	200	—	200	400	μA
	$I_{BB(1)}$	Outputs High, No Load	—	1.5	3.0	—	1.5	3.0	mA
	$I_{BB(0)}$	Outputs Low	—	10	100	—	10	100	μA
Blanking to Output Delay	t_{PHL}	C_L = 30 pF	—	300	550	—	125	150	ns
	t_{PLH}	C_L = 30 pF	—	250	450	—	170	200	ns
Output Fall Time	t_f	C_L = 30 pF	—	1000	1250	—	250	300	ns
Output Rise Time	t_r	C_L = 30 pF	—	150	170	—	150	170	ns

Negative current is defined as coming out of (sourcing) the specified device pin.
*UCN-5818AF-1 and UCN--5818EPF-1 only.

FUNCTIONAL BLOCK DIAGRAM

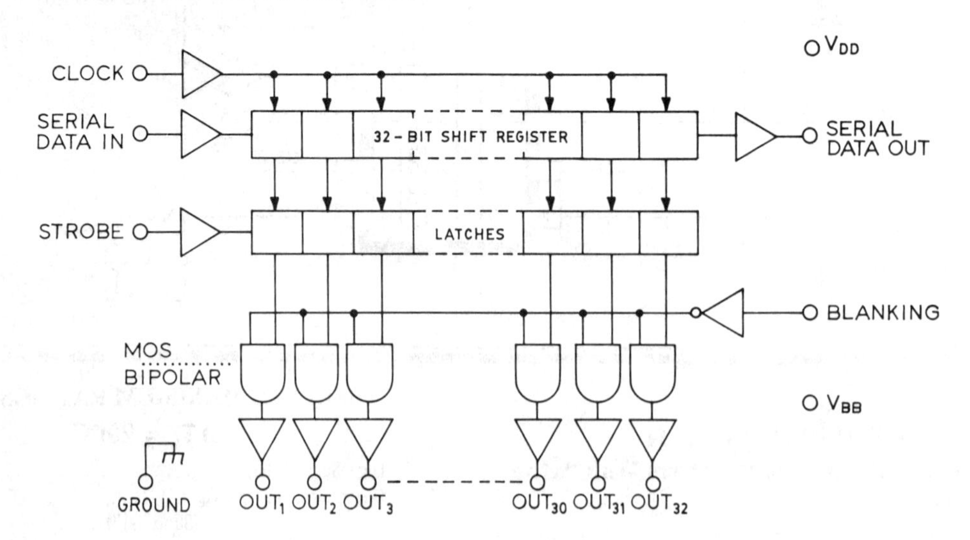

UCN-5818AF

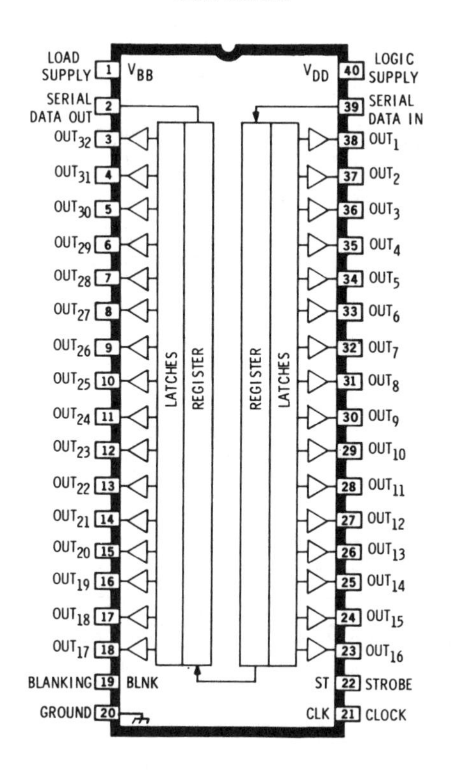

UCN-5818EPF

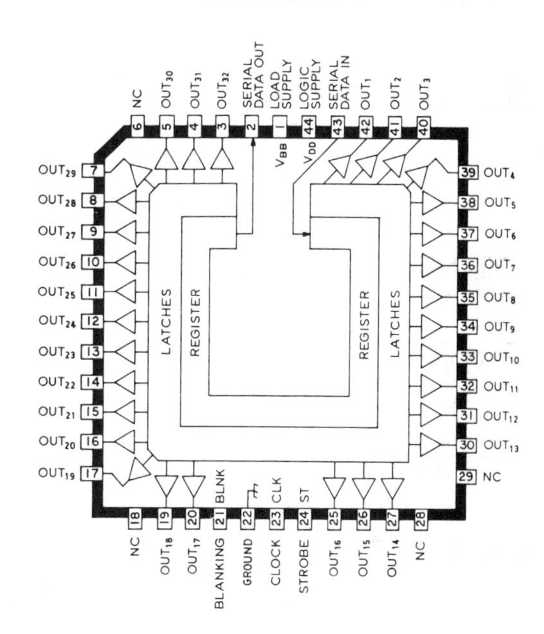

TYPICAL INPUT CIRCUIT

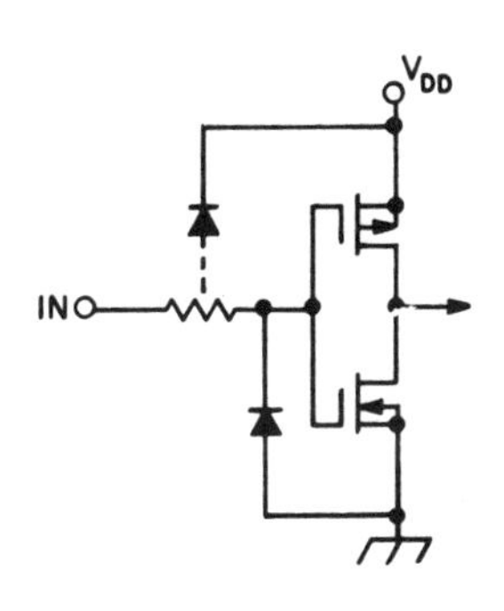

TYPICAL OUTPUT DRIVER

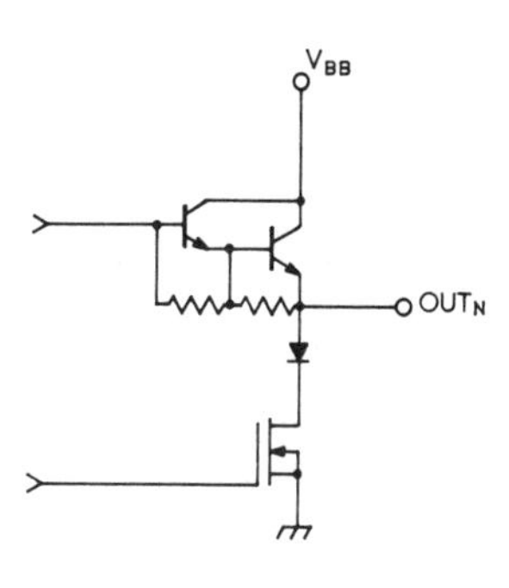

Sprague
Series UCN-5820A
BiMOS II 8-Bit Serial-Input Latched Drivers

- 3.3-MHz minimum data input rate
- CMOS, PMOS, NMOS, TTL compatible
- Internal pull-down resistors
- Low-Power CMOS logic and latches
- High-voltage current-sink outputs
- 16-pin dual-in-line plastic package

ABSOLUTE MAXIMUM RATINGS
at 25°C Free-Air Temperature
and V_{SS} = 0 V

Output Voltage, V_{OUT} (UCN-5821A)	50 V
(UCN-5822A)	80 V
(UCN-5823A)	100 V
Logic Supply Voltage, V_{DD}	15 V
Input Voltage Range, V_{IN}	−0.3 V to V_{DD} +0.3 V
Continuous Output Current, I_{OUT}	500 mA
Package Power Dissipation, P_D	1.67 W*
Operating Temperature Range, T_A	−20°C to +85°C
Storage Temperature Range, T_S	−55°C to +125°C

*Derate at the rate of 16.7 mW/°C above T_A = +25°C

ELECTRICAL CHARACTERISTICS at T_A = +25°C, V_{DD} = 5 V, V_{SS} = 0 V (unless otherwise specified)

Characteristic	Symbol	Applicable Devices	Test Conditions	Limits Min.	Limits Max.	Units
Output Leakage Current	I_{CEX}	UCN-5821A	V_{OUT} = 50 V	—	50	μA
			V_{OUT} = 50 V, = +70°C	—	100	μA
		UCN-5822A	V_{OUT} = 80 V	—	50	μA
			V_{OUT} = 80 V, T_A = +70°C	—	100	μA
		UCN-5823A	V_{OUT} = 100 V	—	50	μA
			V_{OUT} = 100 V, T_A = +70°C	—	100	μA
Collector-Emitter Saturation Voltage	$V_{CE(SAT)}$	ALL	I_{OUT} = 100 mA	—	1.1	V
			I_{OUT} = 200 mA	—	1.3	V
			I_{OUT} = 350 mA, V_{DD} = 7.0 V	—	1.6	V
Input Voltage	$V_{IN(0)}$	ALL		—	0.8	V
	$V_{IN(1)}$	ALL	V_{DD} = 12 V	10.5	—	V
			V_{DD} = 10 V	8.5	—	V
			V_{DD} = 5.0 V	3.5	—	V
Input Resistance	R_{IN}	ALL	V_{DD} = 12 V	50	—	kΩ
			V_{DD} = 10 V	50	—	kΩ
			V_{DD} = 5.0 V	50	—	kΩ
Supply Current	$I_{DD(ON)}$	ALL	One Driver ON, V_{DD} = 12 V	—	4.5	mA
			One Driver ON, V_{DD} = 10 V	—	3.9	mA
			One Driver ON, V_{DD} = 5.0 V	—	2.4	mA
	$I_{DD(OFF)}$	ALL	V_{DD} = 5.0 V, All Drivers OFF, All Inputs = 0 V	—	1.6	mA
			V_{DD} = 12 V, All Drivers OFF, All Inputs = 0 V	—	2.9	mA

Caution: Sprague CMOS devices have input-static protection but are susceptible to damage when exposed to extremely high static electrical charges.

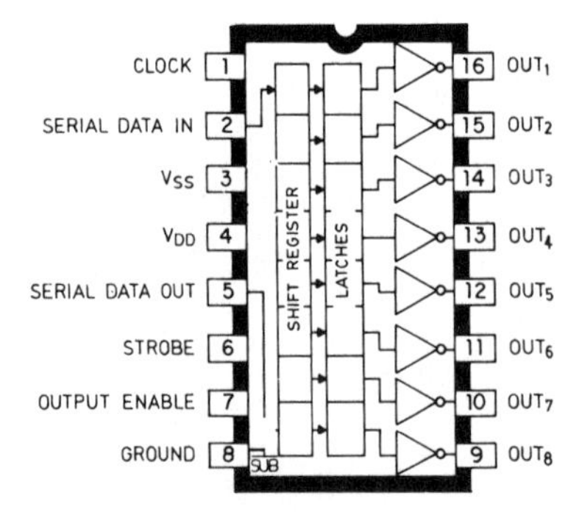

FUNCTIONAL BLOCK DIAGRAM

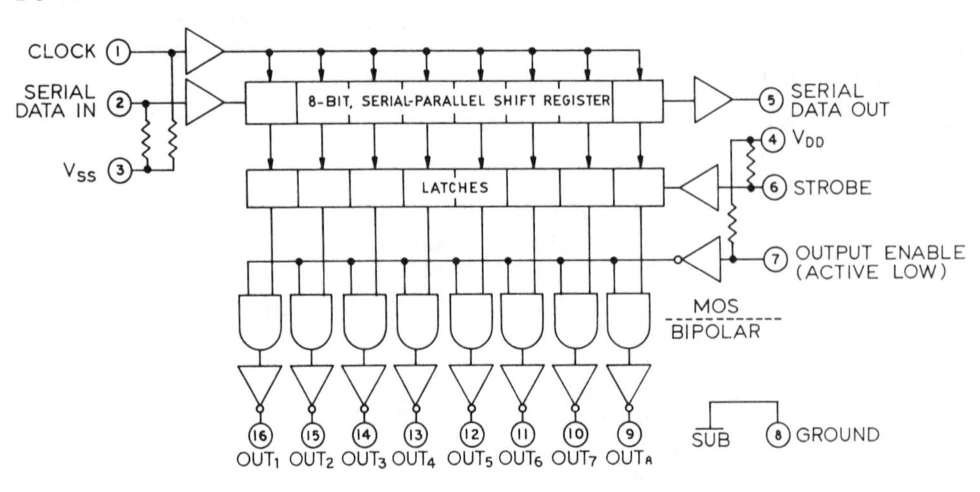

TYPICAL INPUT CIRCUITS

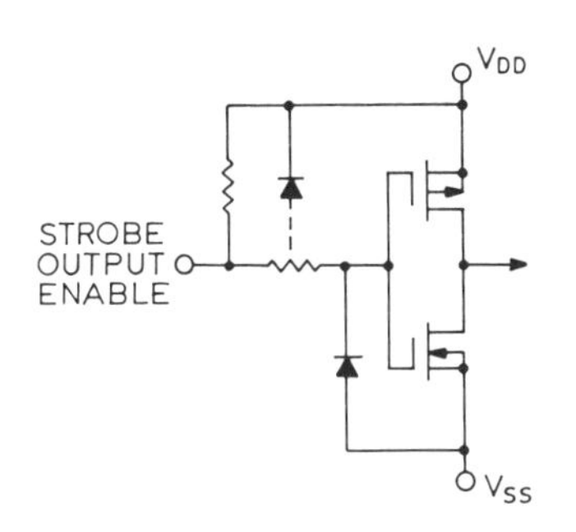

TYPICAL OUTPUT DRIVER

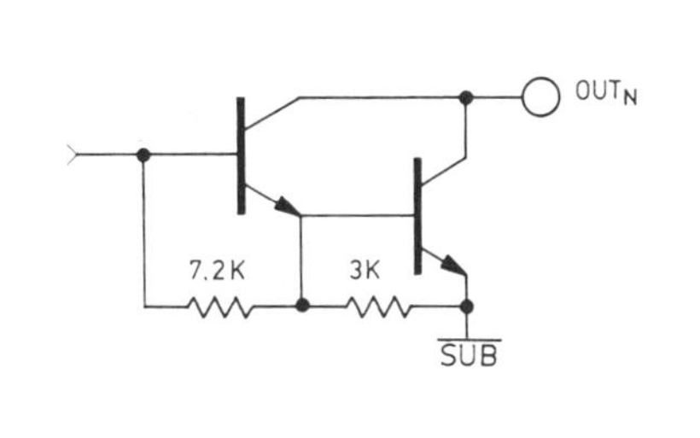

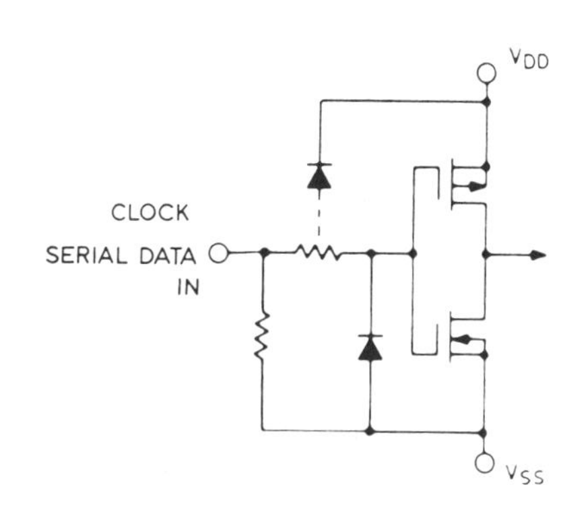

Sprague
UCN-5825B, UCN-5826B
BiMOS II High-Current, Serial-Input Latched Drivers

- 2A open collector outputs
- 60 or 80V minimum output breakdown
- 35 or 60V sustaining voltage
- Output-transient protection
- Low-power CMOS logic and latches
- Typical data input rate > 5 MHz
- Internal pull-down resistors
- CMOS- PMOS- NMOS- TTL-compatible inputs
- Internal thermal shutdown circuitry

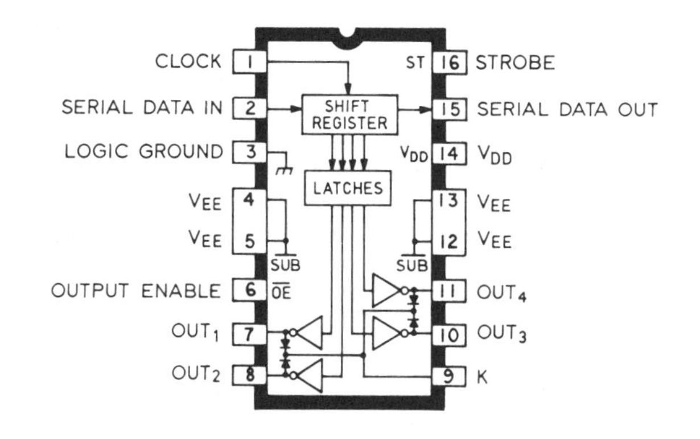

ABSOLUTE MAXIMUM RATINGS

at +25°C Free-Air Temperature

Output Voltage, V_{CE}	
(UCN-5825B)	60 V
(UCN-5826B)	80 V
Output Voltage, $V_{CE(sus)}$	
(UCN-5825B)	35 V*
(UCN-5826B)	60 V*
Logic Supply Voltage Range, V_{DD}	4.5 V to 15 V
V_{DD} with reference to V_{EE}	25 V
Emitter Supply Voltage, V_{EE}	−20 V
Input Voltage Range, V_{IN}	−0.3 V to V_{DD} + 0.3 V
Continuous Output Current, I_{OUT}	2 A
Allowable Package Power Dissipation, P_D	See Graph
Operating Temperature Range, T_A	−20°C to +85°C
Storage Temperature Range, T_S	−55°C to +125°C

*For inductive load applications: The sum of the load supply voltage and clamping voltage(s).

Note: Output-current rating may be limited by duty cycle, ambient temperature, heat sinking, and a number of outputs conducting. Under any combination of conditions, do not exceed the specified maximum current rating and a junction temperature of +125°C.

Caution: Sprague CMOS devices have input-static protection but are susceptible to damage when exposed to extremely high static electrical charges.

ALLOWABLE POWER DISSIPATION AS A FUNCTION OF AMBIENT TEMPERATURE

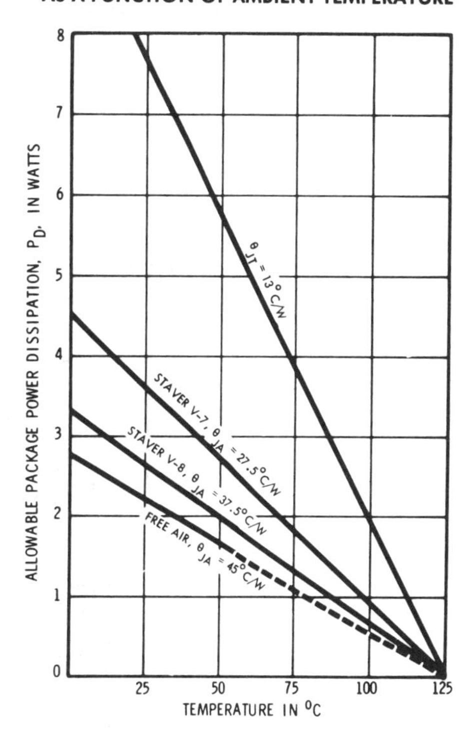

ELECTRICAL CHARACTERISTICS

at T_A = +25°C, V_{CC} = 60 V, V_{DD} = 5 V to 12 V, V_{EE} = 0 V (unless otherwise noted)

Characteristic	Symbol	Applicable Devices	Test Conditions	Limits Min.	Limits Max.	Units
Output Leakage Current	I_{CEX}	UCN-5825B	T_A = +25°C	—	100	μA
			T_A = +70°C	—	500	μA
		UCN-5826B	V_{CC} = 80 V, T_A = +25°C	—	100	μA
			V_{CC} = 80 V, T_A = +70°C	—	500	μA
Output Saturation Voltage	$V_{CE(SAT)}$	Both	I_{OUT} = 1.75 A	—	1.75	V
Output Sustaining Voltage	$V_{CE(SAT)}$	UCN-5825B	I_{OUT} = 1.75 A, L = 2 mH	35	—	V
		UCN-5826B	I_{OUT} = 1.75 A, L = 2 mH	60	—	V
Clamp Diode Leakage Current	I_R	UCN-5825B	V_R = 60 V	—	100	μA
		UCN-5826B	V_R = 80 V	—	100	μA
Clamp Diode Forward Voltage	V_F	Both	I_F = 1.75 A	—	2.0	V
Input Voltage	$V_{IN(1)}$	Both	V_{DD} = 5.0 V	3.5	5.3	V
			V_{DD} = 12 V	10.5	12.3	V
	$V_{IN(0)}$	Both	V_{DD} = 5 V to 12 V	−0.3	+0.8	V
Input Resistance	R_{IN}	Both	V_{DD} = 5.0 V	100	—	kΩ
			V_{DD} = 12 V	50	—	kΩ
Serial Data Output Resistance	R_{OUT}	Both	V_{DD} = 5.0 V	—	20	kΩ
			V_{DD} = 12 V	—	6.0	kΩ
Supply Current	I_{DD}	Both	All outputs OFF	—	3.0	mA
			All outputs ON	—	20	mA
Maximum Clock Frequency	f_C	Both		3.3	—	MHz
Turn-ON Delay	t_{PHL}	Both	0.5 E_{OE} to 0.5 E_{out}	—	1.0	μs
Turn-OFF Delay	t_{PLH}	Both	0.5 E_{OE} to 0.5 E_{out}	—	2.0	μs
Propagation Delay	t_{PD}	Both	0.5 E_{OE} to 0.5 E_{out}	—	100	ns

FUNCTIONAL BLOCK DIAGRAM

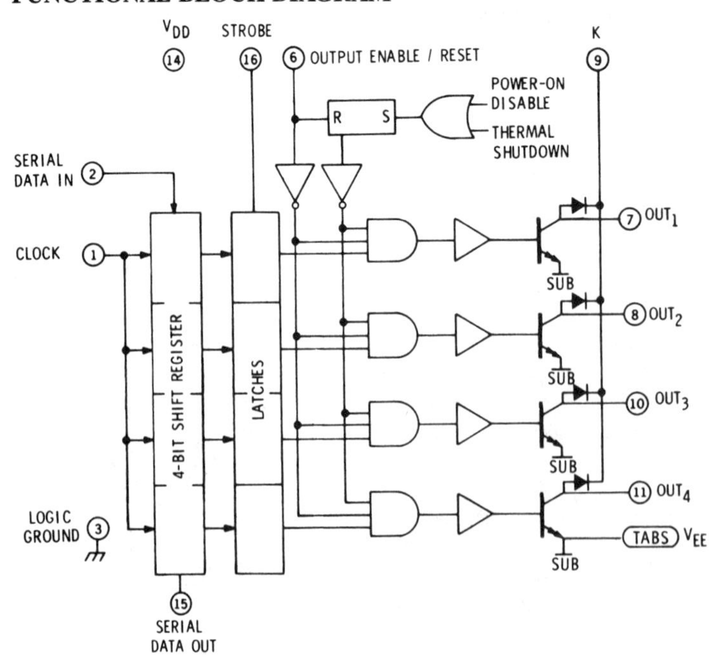

TYPICAL INPUT CIRCUIT

TYPICAL OUTPUT DRIVER

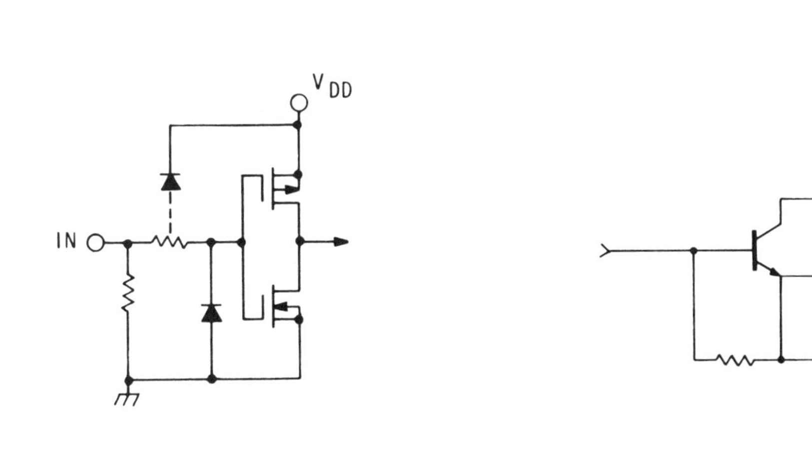

TYPICAL APPLICATION

MULTIPLEXED INCANDESCENT LAMP DRIVE

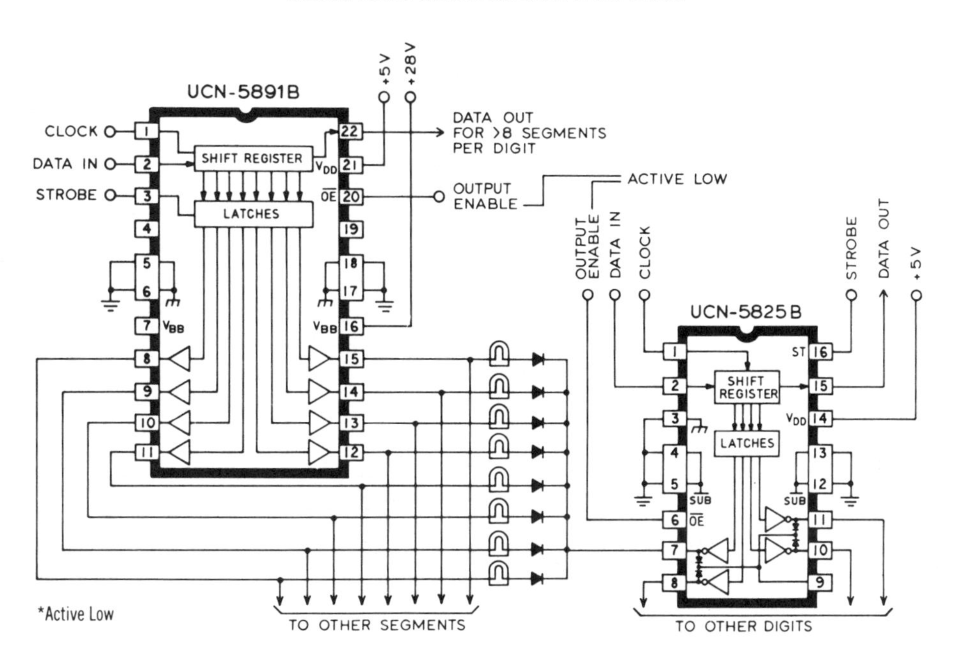

*Active Low

TYPICAL APPLICATION

BILEVEL HAMMER DRIVE

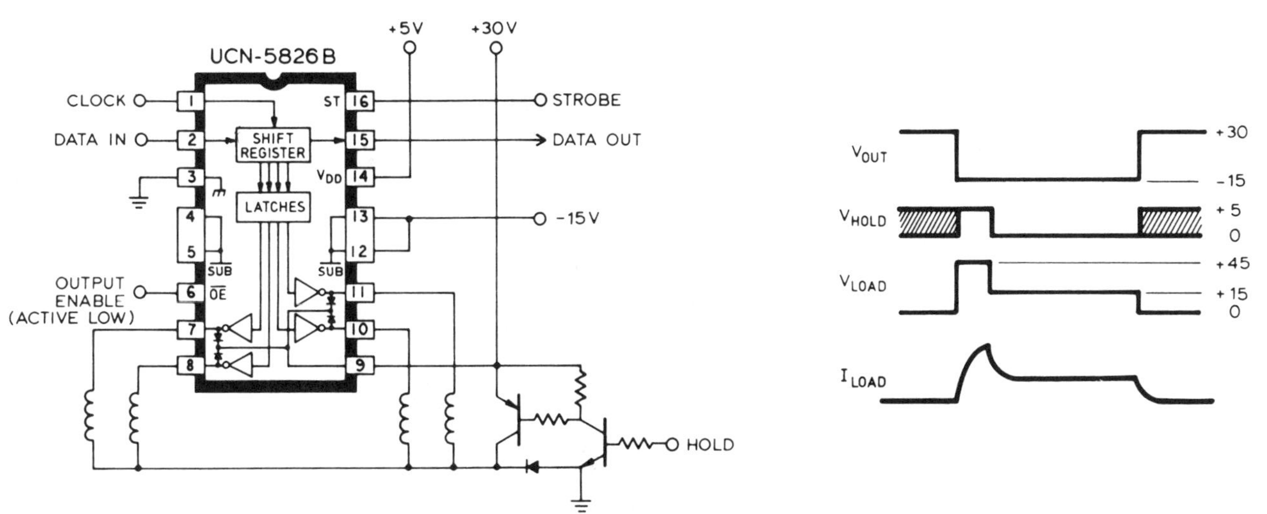

*Active Low

Sprague
UCN-5832A, UCN-5832C
BiMOS II 32-bit Serial-Input, Latched Drivers

- 5-MHz typical data input rate
- Low-power CMOS logic and latches
- 40V current-sink outputs
- Low-saturation voltage

ABSOLUTE MAXIMUM RATINGS
at +25°C Free-Air Temperature

Output Voltage, V_{OUT} .... 40 V
Logic Supply Voltage, V_{DD} .... 15 V
Input Voltage Range, V_{IN} .... −0.3 V to V_{DD} + 0.3 V
Continuous Output Current, I_{OUT} .... 150 mA
Package Power Dissipation, P_D (UCN-5832A) .... 2.8 W*
Operating Temperature Range, T_A .... −20°C to +85°C
Storage Temperature Range, T_S .... −55°C to +125°C

*Derate at the rate of 28 mW/°C above T_A = +25°C

Caution: Sprague CMOS devices have input-static protection but are susceptible to damage when exposed to extremely high static electrical charges.

ELECTRICAL CHARACTERISTICS at T_A = +25°C, V_{DD} = 5 V (unless otherwise noted)

Characteristic	Symbol	Test Conditions	Limits Min.	Limits Max.	Units
Output Leakage Current	I_{CEX}	V_{OUT} = 40 V, T_A = 70°C	—	10	µA
Collector-Emitter Saturation Voltage	$V_{CE(SAT)}$	I_{OUT} = 50 mA	—	275	mV
		I_{OUT} = 100 mA	250	550	mV
Input Voltage	$V_{IN(1)}$		3.5	5.3	V
	$V_{IN(0)}$		−0.3	+0.8	V
Input Current	$I_{IN(1)}$	V_{IN} = 3.5 V	—	1.0	µA
	$I_{IN(0)}$	V_{IN} = 0.8 V	—	−1.0	µA
Input Impedance	Z_{IN}	V_{IN} = 3.5 V	3.5	—	MΩ
Serial Data/Output Resistance	R_{OUT}		—	20	kΩ
Supply Current	I_{DD}	One output ON, I_{OUT} = 100 mA	—	5.0	mA
		All outputs OFF	—	50	µA
Output Rise Time	t_r	I_{OUT} = 100 mA, 10% to 90%	—	1.0	µs
Output Fall Time	t_f	I_{OUT} = 100 mA, 90% to 10%	—	1.0	µs

NOTE: Positive (negative) current is defined as going into (coming out of) the specified device pin.

FUNCTIONAL BLOCK DIAGRAM

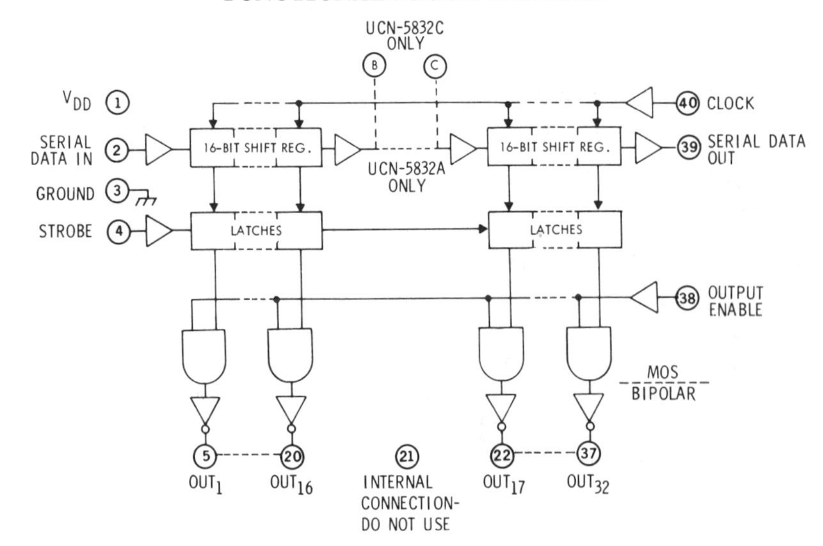

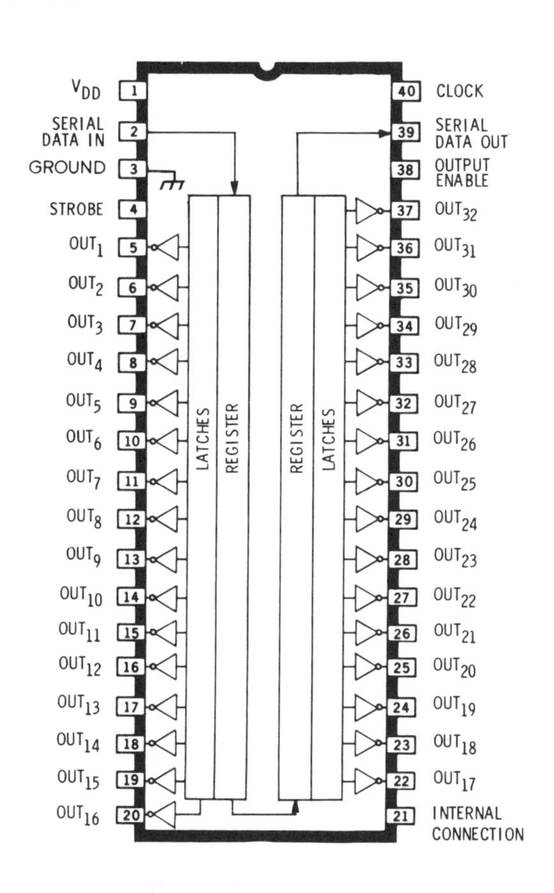

TYPICAL INPUT CIRCUIT

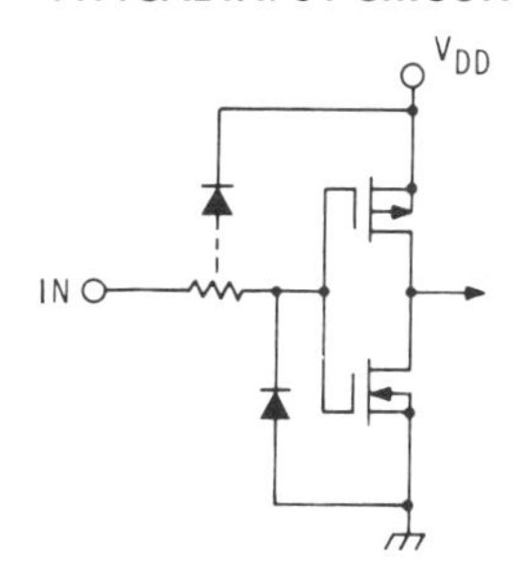

TYPICAL OUTPUT DRIVER

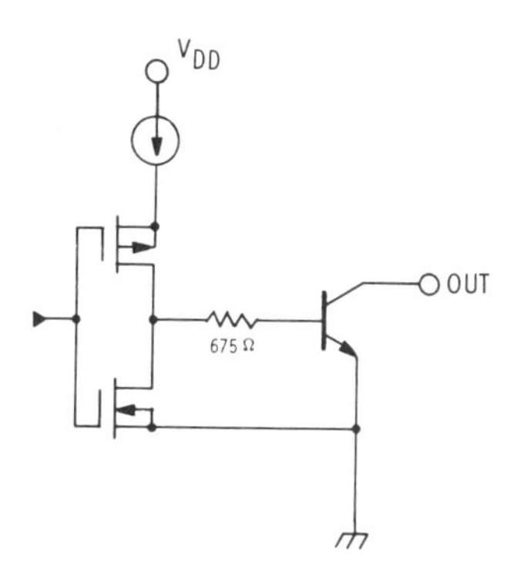

TRUTH TABLE

Serial Data Input	Clock Input	Shift Register Contents I_1 I_2 I_3 ... I_{N-1} I_N	Serial Data Output	Strobe Input	Latch Contents I_1 I_2 I_3 ... I_{N-1} I_N	Output Enable Input	Output Contents I_1 I_2 I_3 ... I_{N-1} I_N
H	↑	H R_1 R_2 ... R_{N-2} R_{N-1}	R_{N-1}				
L	↑	L R_1 R_2 ... R_{N-2} R_{N-1}	R_{N-1}				
X	↓	R_1 R_2 R_3 ... R_{N-1} R_N	R_N				
		X X X ... X X	X	L	R_1 R_2 R_3 ... R_{N-1} R_N		
		P_1 P_2 P_3 ... P_{N-1} P_N	P_N	H	P_1 P_2 P_3 ... P_{N-1} P_N	H	P_1 P_2 P_3 ... P_{N-1} P_N
					X X X ... X X	L	H H H ... H H

L = Low Logic Level
H = High Logic Level
X = Irrelevant
P = Present State
R = Previous State

Sprague UCN-5851A/EP and UCN-5852A/EP

BiMOS II 32-Channel, Serial-Input Drivers

- DMOS outputs
- Output breakdown > 225V
- Sink up to 120 mA
- Low-power CMOS inputs and logic
- 6-MHz data-input rate
- Refresh and output-enable functions
- Replaces SN75551 and SN75552

ABSOLUTE MAXIMUM RATINGS

Voltage Measurements
Referenced to Substrate

Supply Voltage, V_{DD} .... 15 V
Output Voltage, V_{OUT} .... 225 V
Input Voltage, V_{IN} .... −0.3 V to V_{DD} +0.3 V
Output Current, I_{OUT} .... 120 mA
Total Substrate Current, I_{SUB} .... 1.5 A
Package Power Dissipation, P_D .... See Graph
Operating Temperature Range, T_A .. −20°C to +85°C
Storage Temperature Range, T_S ... −55°C to +125°C

NOTE: Output current rating may be limited by duty cycle and ambient temperature (see graphs). Under any set of conditions, do not exceed the specified maximum current ratings or a junction temperature of +125°C.

ELECTRICAL CHARACTERISTICS

at T_A = 0°C to +70°C, V_{DD} = 12 V, V_{SUB} = 0 (unless otherwise specified)

Characteristic	Symbol	Test Conditions	Limits Min.	Limits Typ.	Limits Max.	Units
Functional Supply Voltage Range	V_{DD}		4.5	12	15	V
Output Leakage Current	I_{OUT}	V_{OUT} = 225 V	—	—	10	μA
Output Voltage	$V_{OUT(0)}$	I_{OUT} = 20 mA, V_{DD} = 5 V	—	8.0	10	V
		I_{OUT} = 100 mA, V_{DD} = 12 V	—	15	30	V
Output Clamp Diode Voltage	V_F	I_F = 100 mA	—	1.8	2.5	V
Serial Data Output Voltage	$V_{OUT(1)}$	I_{OUT} = −100 μA	10.5	—	—	V
	$V_{OUT(0)}$	I_{OUT} = 100 μA	—	—	0.8	V
Input Voltage	$V_{IN(1)}$	V_{DD} = 5.0 V	3.5	—	5.3	V
		V_{DD} = 12 V	10.5	—	12.3	V
	$V_{IN(0)}$		−0.3	—	0.8	V
Input Current	$I_{IN(1)}$	V_{IN} = 12 V	—	—	1.0	μA
	$I_{IN(0)}$	V_{IN} = 0	—	—	−1.0	μA
Maximum Clock Frequency	f_{clk}	V_{DD} = 5.0 V	3.3	—	—	MHz
		V_{DD} = 12 V	—	7.5	—	MHz
Supply Current	I_{DD}	f_{clk} = 0	—	—	500	μA
Output Enable to Output Delay	t_{PHL}	C_L = 10 pF	—	200	500	ns
	t_{PLH}	C_L = 10 pF	—	250	500	ns
Output Fall Time	t_f	C_L = 10 pF	—	80	200	ns
Output Rise Time	t_r	C_L = 10 pF	—	300	500	ns

FUNCTIONAL BLOCK DIAGRAM

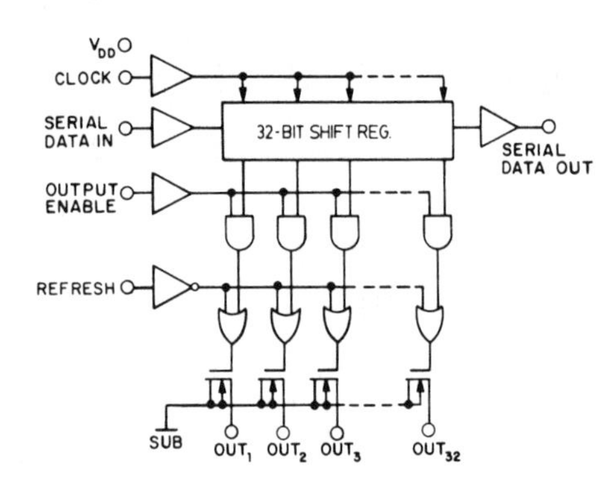

ALLOWABLE POWER DISSIPATION AS A FUNCTION OF AMBIENT TEMPERATURE

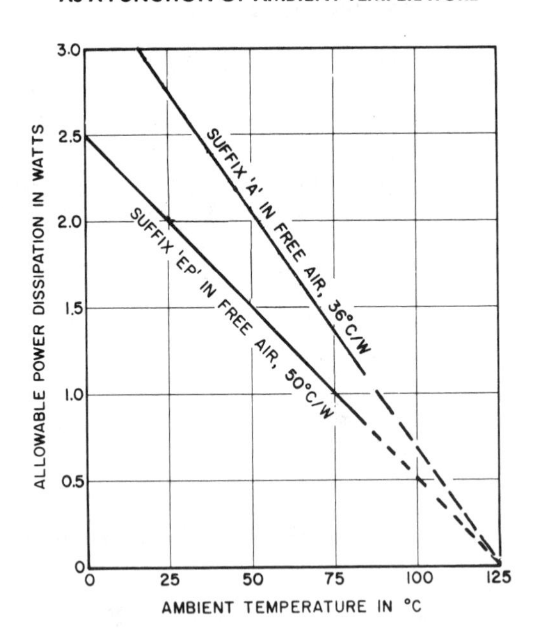

TYPICAL INPUT CIRCUIT

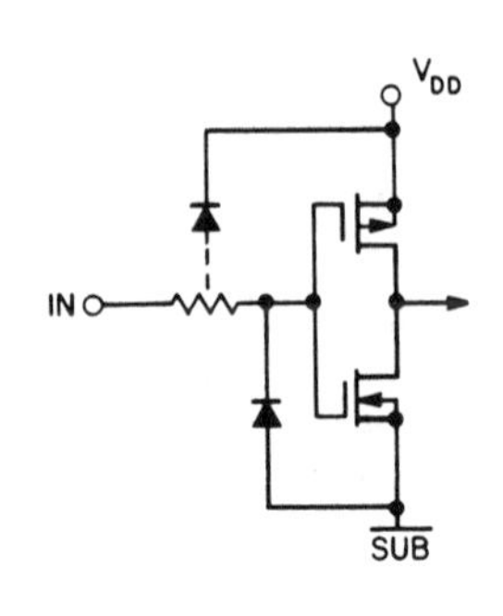

TYPICAL OUTPUT DRIVER

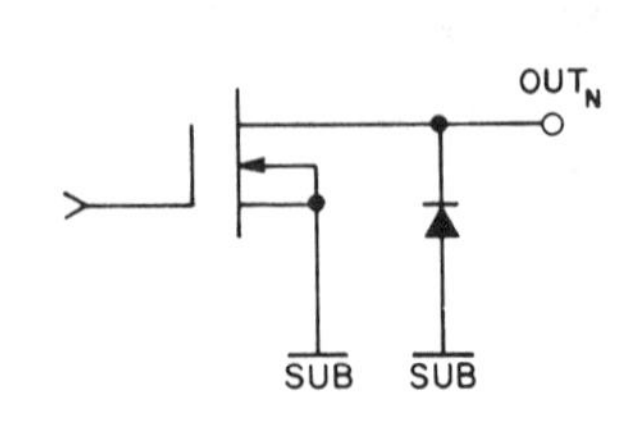

OUTPUT CURRENT AS A FUNCTION OF DUTY CYCLE

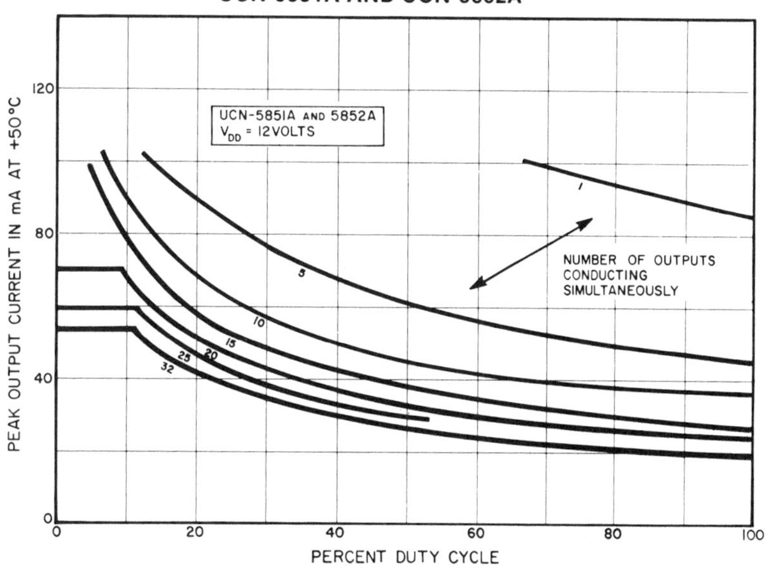

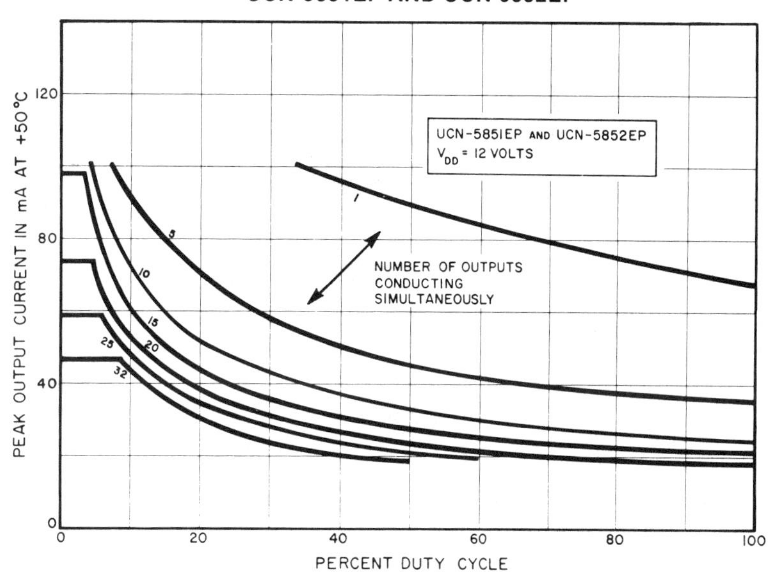

APPLICATIONS

SIMPLIFIED CELL DRIVER

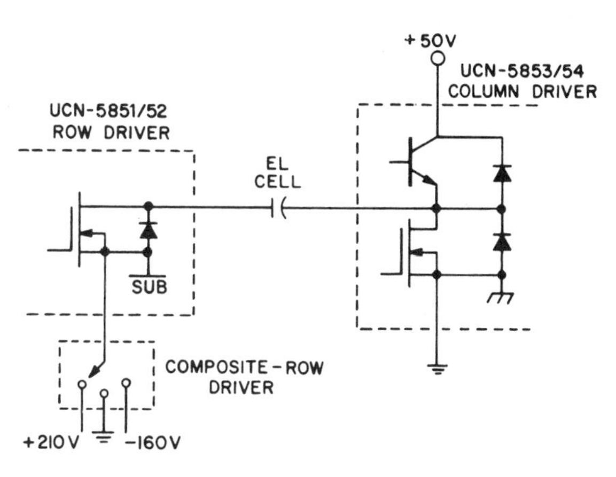

ELECTROLUMINESCENT DISPLAY

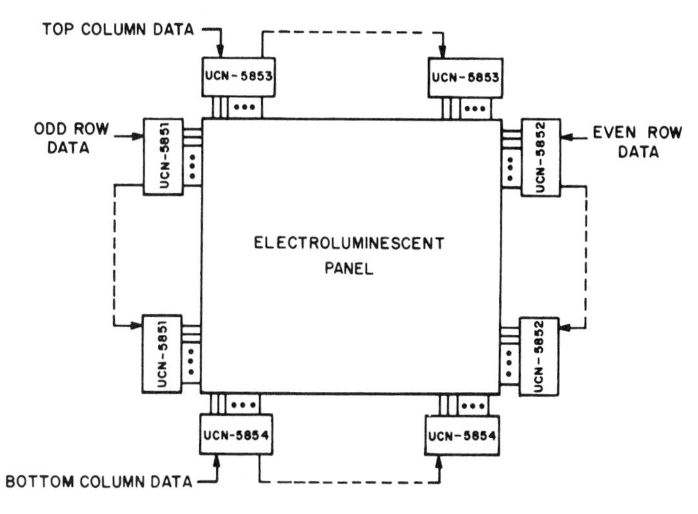

Sprague
UCN-5853A/EP, UCN-5854A/EP
BiMOS II 32-Channel Serial-Input Latched Drivers

- Totem-pole outputs
- High-output breakdown
- Sink or source up to 25 mA
- Low-power CMOS inputs and logic
- 7.5-MHz data input rate
- Strobe and output-enable functions
- Replaces SN75553 and SN75554

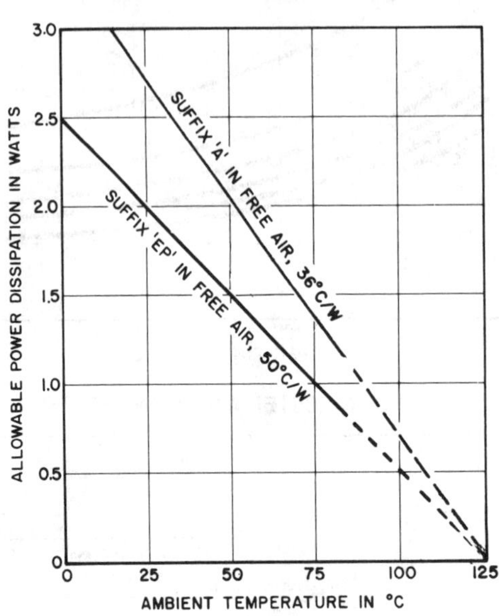

ABSOLUTE MAXIMUM RATINGS

Supply Voltage, V_{DD}	15 V
Supply Voltage, V_{BB}	
(UCN-5853/54A, UCN-5853/54EP)	60 V
(UCN-5853/54A-1, UCN-5853/54EP-1)	80 V
Input Voltage Range, V_{IN}	-0.3 V to V_{DD} $+0.3$ V
Output Current, I_{OUT}	± 25 mA
Total Ground Current, I_{GND}	700 mA
Package Power Dissipation, P_D	See Graph
Operating Temperature Range, T_A	-20°C to $+85$°C
Storage Temperature Range, T_S	-55°C to $+125$°C

FUNCTIONAL BLOCK DIAGRAM

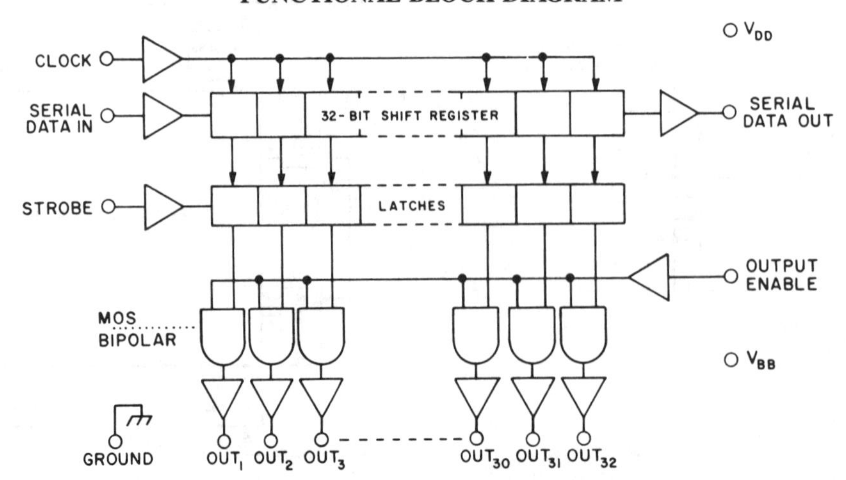

ELECTRICAL CHARACTERISTICS at T_A = 0°C to +70°C, V_{DD} = 12 V, V_{BB} = 60 V (UCN-5853A/EP, UCN-5854A/EP) or V_{BB} = 80 V (Suffix '-1') unless otherwise specified

Characteristic	Symbol	Test Conditions	Limits			Units
			Min.	Typ.	Max.	
Logic Supply Voltage Range	V_{DD}		4.5	12	15	V
Output Voltage	$V_{OUT(1)}$	I_{OUT} = −20 mA, V_{BB} = 60 V	57.5	—	—	V
		I_{OUT} = −20 mA, V_{BB} = 80 V*	77.5	—	—	V
	$V_{OUT(0)}$	I_{OUT} = 20 mA	—	6.0	10	V
Output Clamp Diode Voltage	V_F	I_F = 20 mA	—	2.0	2.5	V
Serial Output Voltage	$V_{OUT(1)}$	I_{OUT} = −100 μA	10.5	—	—	V
	$V_{OUT(0)}$	I_{OUT} = 100 μA	—	—	0.8	V
Input Voltage	$V_{IN(1)}$	V_{DD} = 10.8 V	10.0	—	11.1	V
		V_{DD} = 15 V	14.2	—	15.3	V
	$V_{ON(0)}$	V_{DD} = 10.8 V	−0.3	—	0.8	V
		V_{DD} = 15 V	−0.3	—	0.8	V
Input Current	$I_{IN(1)}$	V_{IN} = 12 V	—	—	1.0	μA
	$I_{IN(0)}$	V_{IN} = 0	—	—	−1.0	μA
Maximum Clock Frequency	f_{clk}	V_{DD} = 5.0 V	3.3	—	—	MHz
		V_{DD} = 12 V	—	7.5	—	MHz
Supply Current	I_{DD}	f_{clk} = 0, Outputs Low	—	—	0.5	mA
		f_{clk} = 0, Outputs High	—	3.0	5.0	mA
	I_{BB}	Outputs High, No Load	—	2.5	3.5	mA
		Outputs Low	—	—	0.5	mA
Output Enable to Ouput Delay	t_{PHL}	C_L = 10 pF	—	200	500	ns
	t_{PHL}	C_L = 10 pF	—	250	500	ns
Output Fall Time	t_f	C_L = 10 pF	—	80	200	ns
Output Rise Time	t_r	C_L = 10 pF	—	300	500	ns

*UCN-5853A/EP-1 and UCN-5854A/EP-1 only.

TYPICAL INPUT CIRCUIT **TYPICAL OUTPUT DRIVER**

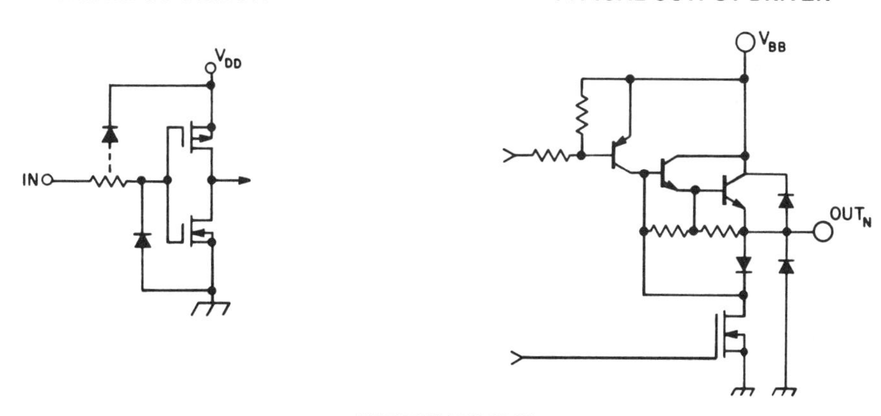

APPLICATIONS

SIMPLIFIED CELL DRIVER

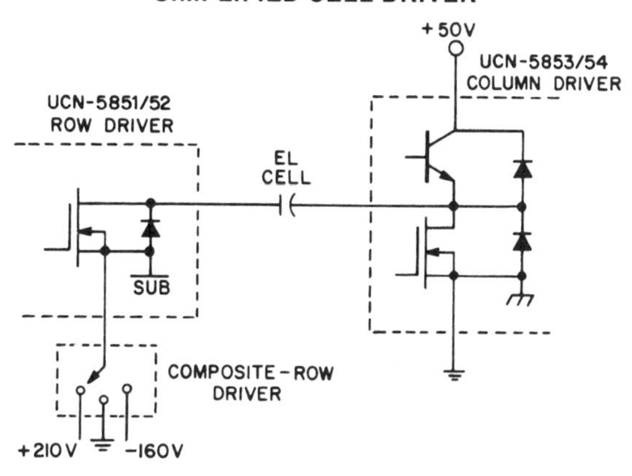

ELECTROLUMINESCENT DISPLAY

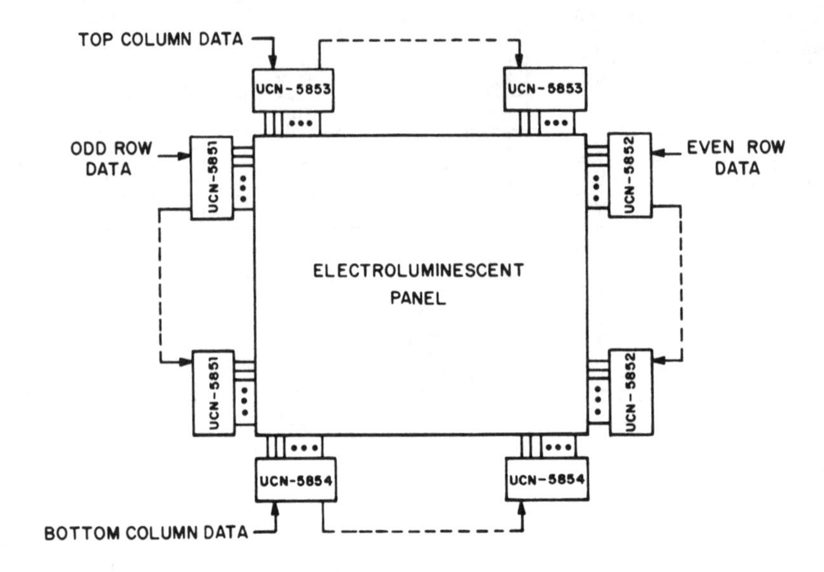

Sprague UCN-5890A/B, UCN-5891A/B BiMOS II
8-Bit Serial-Input Latched Source Drivers

- 50 or 80V source outputs
- Output current to 500 mA
- Output transient-suppression diodes
- 3.3-MHz minimum data-input rate
- Low-power CMOS logic and latches

ABSOLUTE MAXIMUM RATINGS
at T_A = +25°C

Output Voltage, V_{OUT} (UCN-5890A/B) . . . . 80 V
(UCN-5890A/B-2) . . . . 50 V
(UCN-5891A/B) . . . . 50 V
Logic Supply Voltage Range, V_{DD} . . . . 4.5 V to 15 V
Driver Supply Voltage Range, V_{BB}
(UCN-5890A/B) . . . . 20 V to 80 V
(UCN-5890A/B-2) . . . . 20 V to 50 V
(UCN-5891A/B) . . . . 5.0 to 50 V
Input Voltage Range, V_{IN} . . . . −0.3 V to V_{DD} + 0.3 V
Continuous Output Current, I_{OUT} . . . . −500 mA
Allowable Package Power Dissipation, P_D . . . . See Graph
Operating Temperature Range, T_A . . . . −20°C to +85°C
Storage Temperature Range, T_S . . . . −55°C to +125°C

Caution: Sprague Electric CMOS devices have input static protection, but are susceptible to damage when exposed to extremely high static electrical charges.

Number of Outputs ON at I_{OUT} = −200 mA	Max. Allowable Duty Cycle at T_A of					
	50°C	60°C	70°C	50°C	60°C	70°C
	Package "A"			Package "B"		
8	40%	34%	28%	53%	46%	39%
7	45%	39%	33%	60%	52%	44%
6	53%	46%	39%	70%	61%	51%
5	63%	55%	46%	84%	73%	62%
4	79%	68%	58%	100%	91%	77%
3	100%	91%	77%	100%	100%	100%
2	100%	100%	100%	100%	100%	100%
1	100%	100%	100%	100%	100%	100%

Also see Allowable Output Current graphs

ALLOWABLE AVERAGE POWER DISSIPATION AS A FUNCTION OF TEMPERATURE

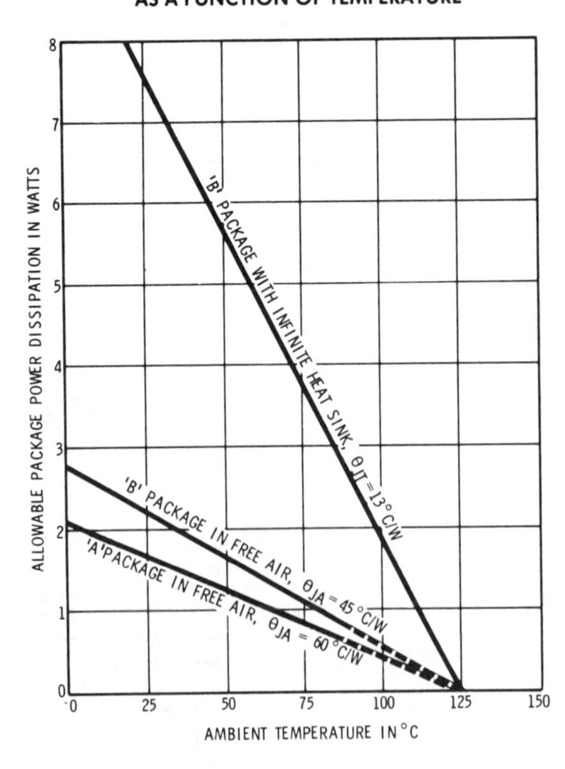

GROUND 1
CLOCK 2
SERIAL DATA IN 3
STROBE 4
OUT₁ 5
OUT₂ 6
OUT₃ 7
OUT₄ 8
16 SERIAL DATA OUT
15 VDD
14 OUTPUT ENABLE
13 VBB
12 OUT 8
11 OUT 7
10 OUT 6
9 OUT 5
SHIFT REGISTER
LATCHES

UCN-5890A
UCN-5891A

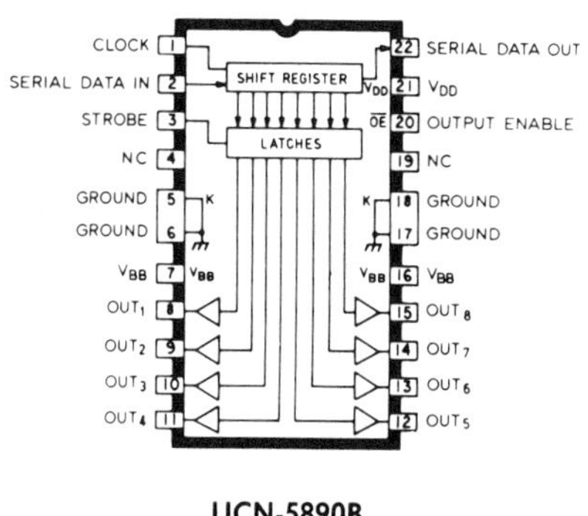

UCN-5890B
UCN-5891B

ELECTRICAL CHARACTERISTICS at $T_A = +25°C$, $V_{BB} = 80$ V (UCN-5890A/B) or 50 V (UCN-5890A/B-2 & UCN-5891A/B), $V_{DD} = 5$ V to 12 V (unless otherwise noted)

Characteristic	Symbol	V_{BB}	Test Conditions	Limits Min.	Limits Max.	Units
Output Leakage Current	I_{CEX}	Max.	$T_A = +25°C$	—	−50	μA
			$T_A = +70°C$	—	−100	μA
Output Saturation Voltage	$V_{CE(SAT)}$	50 V	$I_{OUT} = -100$ mA	—	1.8	V
			$I_{OUT} = -225$ mA	—	1.9	V
			$I_{OUT} = -350$ mA	—	2.0	V
Output Sustaining Voltage	$V_{CE(sus)}$	Max.	$I_{OUT} = -350$ mA, L = 2 mH, UCN-5890A/B-2 & UCN-5891A/B	35	—	V
			$I_{OUT} = -350$ mA, L = 2 mH, UCN-5890A & UCN-5890B only	50	—	V
Input Voltage	$V_{IN(1)}$	50 V	$V_{DD} = 5.0$ V	3.5	5.3	V
			$V_{DD} = 12$ V	10.5	12.3	V
	$V_{IN(0)}$	50 V	$V_{DD} = 5$ V to 12 V	−0.3	+0.8	V
Input Current	$I_{IN(1)}$	50 V	$V_{DD} = V_{IN} = 5.0$ V	—	50	μA
			$V_{DD} = V_{IN} = 12$ V	—	240	μA
Input Impedance	Z_{IN}	50 V	$V_{DD} = 5.0$ V	100	—	kΩ
			$V_{DD} = 12$ V	50	—	kΩ
Clock Frequency	f_c	50 V		3.3	—	MHz
Serial Data Output Resistance	R_{OUT}	50 V	$V_{DD} = 5.0$ V	—	20	kΩ
			$V_{DD} = 12$ V	—	6.0	kΩ
Turn-ON Delay	t_{PLH}	50 V	Output Enable to Output, $I_{OUT} = -350$ mA	—	2.0	μs
Turn-OFF Delay	t_{PHL}	50 V	Output Enable to Output, $I_{OUT} = -350$ mA	—	10	μs
Supply Current	I_{BB}	50 V	All outputs ON, All outputs open	—	10	mA
			All outputs OFF	—	200	μA
	I_{DD}	50 V	$V_{DD} = 5$ V, All outputs OFF, Inputs = 0 V	—	100	μA
			$V_{DD} = 12$ V, All outputs OFF, Inputs = 0 V	—	200	μA
			$V_{DD} = 5$ V, One output ON, All inputs = 0 V	—	1.0	mA
			$V_{DD} = 12$ V, One output ON, All inputs = 0 V	—	3.0	mA
Diode Leakage Current	I_R	Max.	$T_A = +25°C$	—	50	μA
			$T_A = +70°C$	—	100	μA
Diode Forward Voltage	V_F	Open	$I_F = 350$ mA	—	2.0	V

NOTE: Positive (negative) current is defined as going into (coming out of) the specified device pin.

FUNCTIONAL BLOCK DIAGRAM

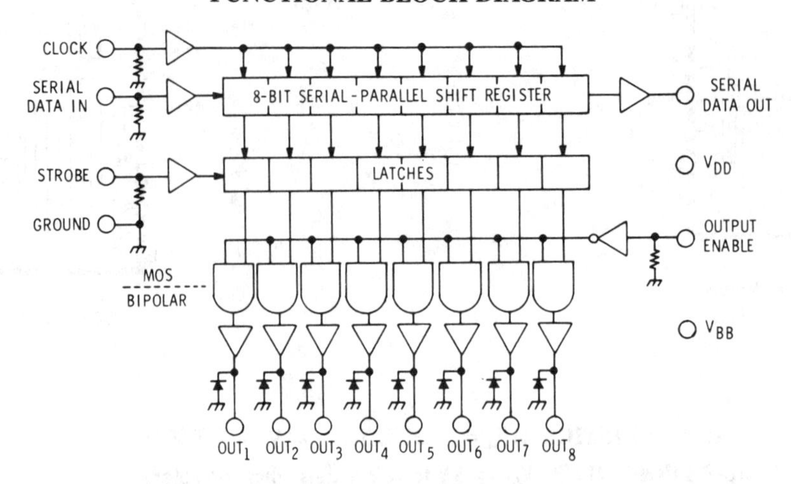

ALLOWABLE OUTPUT CURRENT AS A FUNCTION OF DUTY CYCLE
at +25°C Free-Air Temperature

UCN-5890A AND UCN-5891A

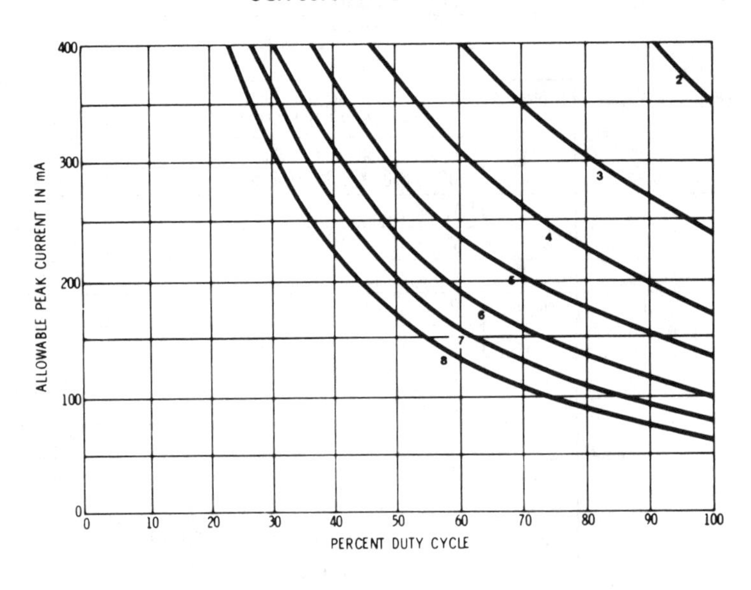

UCN-5890B AND UCN-5891B

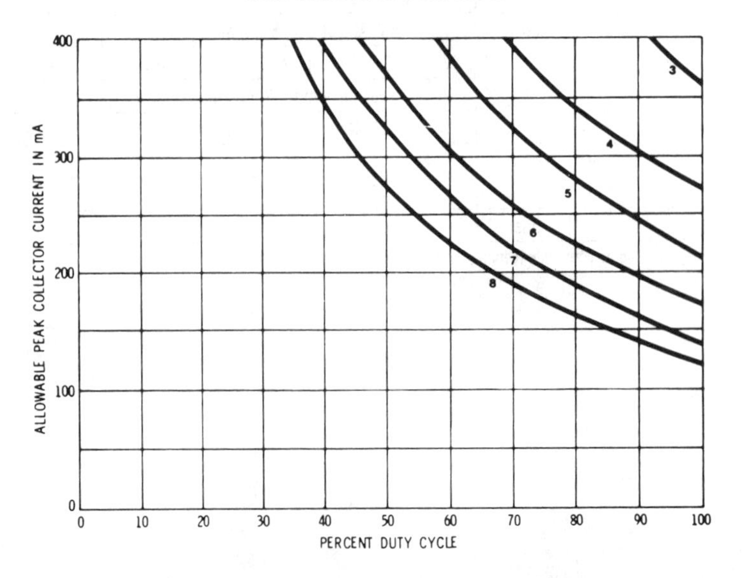

TYPICAL INPUT CIRCUIT

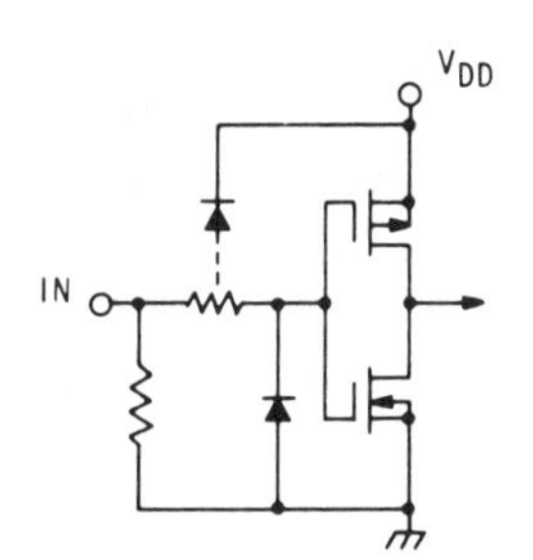

TYPICAL OUTPUT DRIVER

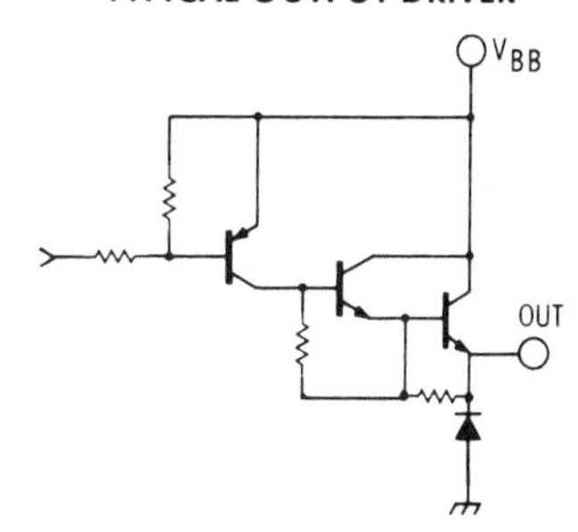

TYPICAL APPLICATIONS

SOLENOID OR RELAY DRIVER

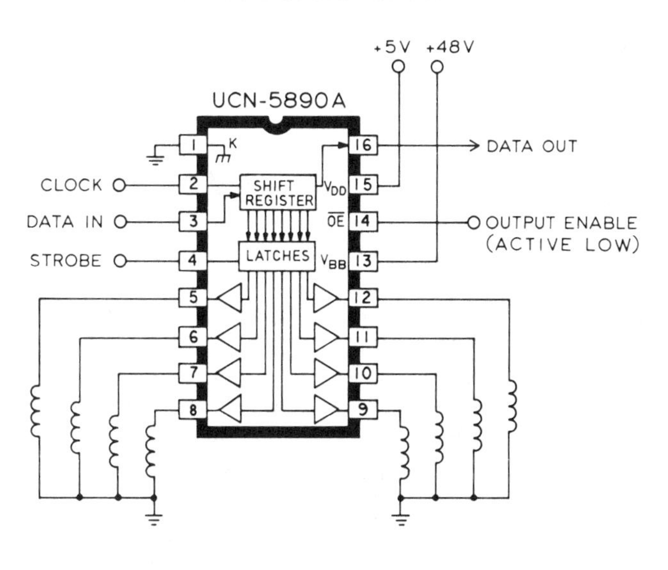

MULTIPLEXED INCANDESCENT LAMP DRIVER

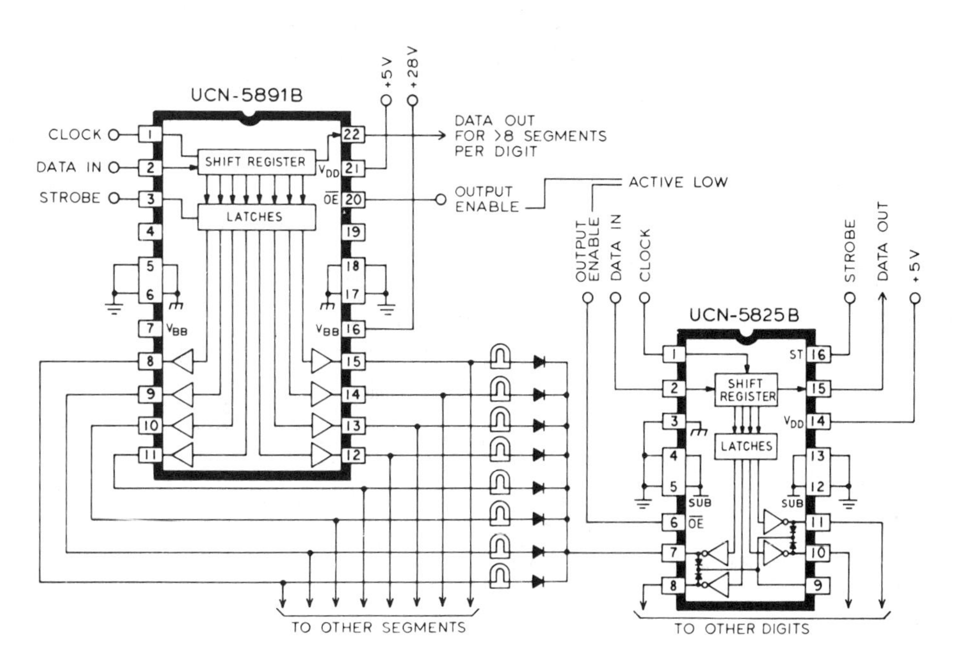

Sprague
UCN-5895A UCN-5895A-2 BiMOS II
8-Bit, Serial-Input Latched Source Drivers

- Low output-saturation voltage
- Source outputs to 50V
- Output current to −250 mA
- 3.3-MHz minimum data-input rate
- Low-power CMOS logic and latches

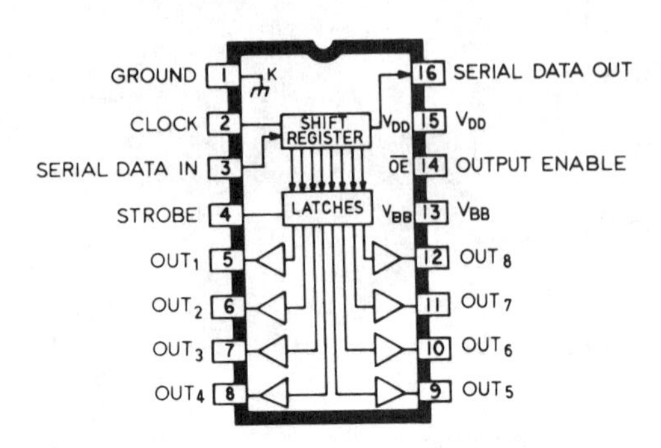

ABSOLUTE MAXIMUM RATINGS
at T_A = +25°C

Output Voltage, V_{OUT} (UCN-5895A) . . . . 50 V
(UCN-5895A-2) . . . . 25 V
Logic Supply Voltage Range, V_{DD} . . . . 4.5 V to 12 V
Driver Supply Voltage Range, V_{BB}
(UCN-5895A) . . . . 5.0 V to 50 V
(UCN-5895A-2) . . . . 5.0 V to 25 V
Input Voltage Range, V_{IN} . . . . −0.3 V to V_{DD} + 0.3 V
Continuous Output Current, I_{OUT} . . . . −250 mA
Allowable Package Power Dissipation, P_D . . . . 1.67 W*
Operating Temperature Range, T_A . . . . −20°C to +85°C
Storage Temperature Range, T_S . . . . −55°C to +125°C
*Derate at the rate of 16.67 mW/°C above T_A = +25°C.

Caution: Sprague Electric CMOS devices have input static protection, but are susceptible to damage when exposed to extremely high static electrical charges.

ELECTRICAL CHARACTERISTICS at T_A = +25°C, V_{BB} = 25 V, V_{DD} = 5 V to 12 V (unless otherwise noted)

Characteristic	Symbol	Test Conditions	Limits Min.	Limits Max.	Units
Output Leakage Current	I_{OUT}	T_A = +25°C	—	−50	µA
		T_A = +70°C	—	−100	µA
Output Saturation Voltage	$V_{CE(SAT)}$	I_{OUT} = −60 mA	—	1.1	V
		I_{OUT} = −120 mA	—	1.2	V
Output Sustaining Voltage	$V_{CE(sus)}$	I_{OUT} = −120 mA, L = 2 mH, UCN-5895A only	35	—	V
		I_{OUT} = −120 mA, L = 2 mH, UCN-5895A-2 only	15	—	V
Input Voltage	$V_{IN(1)}$	V_{DD} = 5.0 V	3.5	5.3	V
		V_{DD} = 12 V	10.5	12.3	V
	$V_{IN(0)}$	V_{DD} = 5 V to 12 V	−0.3	+0.8	V
Input Current	$I_{IN(1)}$	V_{DD} = V_{IN} = 5.0 V	—	50	µA
		V_{DD} = V_{IN} = 12 V	—	240	µA
Input Impedance	Z_{IN}	V_{DD} = 5.0 V	100	—	kΩ
		V_{DD} = 12 V	50	—	kΩ
Clock Frequency	f_c		3.3	—	MHz
Serial Data-Output Resistance	R_{OUT}	V_{DD} = 5.0 V	—	20	kΩ
		V_{DD} = 12 V	—	6.0	kΩ
Turn-ON Delay	t_{PLH}	Output Enable to Output, I_{OUT} = −120 mA	—	2.0	µs
Turn-OFF Delay	t_{PHL}	Output Enable to Output, I_{OUT} = −120 mA	—	10	µs
Supply Current	I_{BB}	All outputs ON, All outputs open	—	10	mA
		All outputs OFF	—	200	µA
	I_{DD}	V_{DD} = 5 V, All outputs OFF, Inputs = 0 V	—	100	µA
		V_{DD} = 12 V, All outputs OFF, Inputs = 0 V	—	200	µA
		V_{DD} = 5 V, One output ON, All inputs = 0 V	—	1.0	mA
		V_{DD} = 12 V, One output ON, All inputs = 0 V	—	3.0	mA
Diode Leakage Current	I_R	V_R = 25 V, T_A = +25°C	—	50	µA
		V_R = 25 V, T_A = +70°C	—	100	µA
Diode Forward Voltage	V_F	I_F = 120 mA	—	2.0	V

NOTE: Positive (negative) current is defined as going into (coming out of) the specified device pin.

FUNCTIONAL BLOCK DIAGRAM

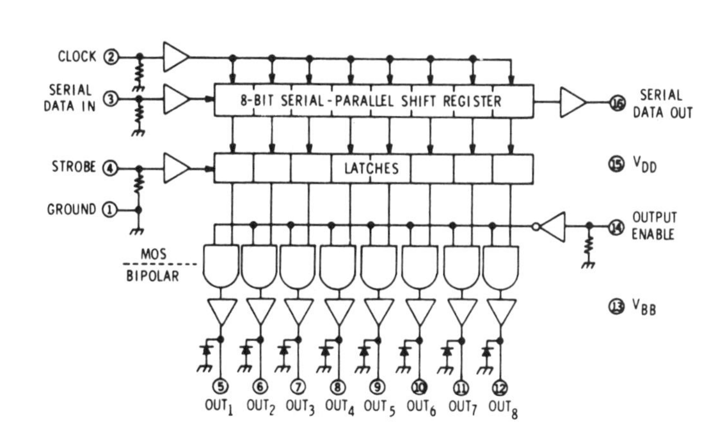

TYPICAL INPUT CIRCUIT

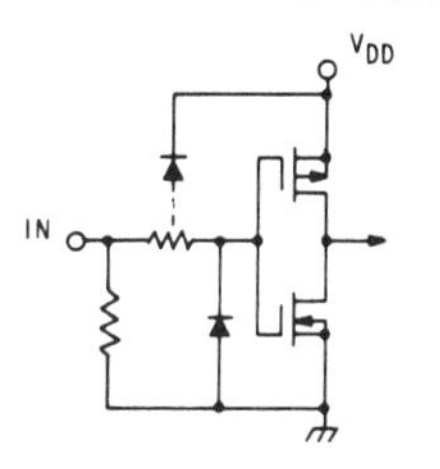

TYPICAL OUTPUT DRIVER

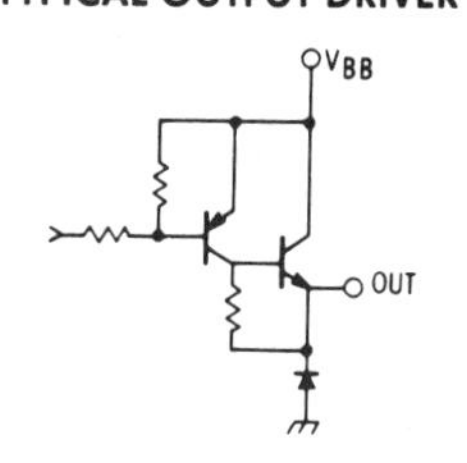

TYPICAL APPLICATION

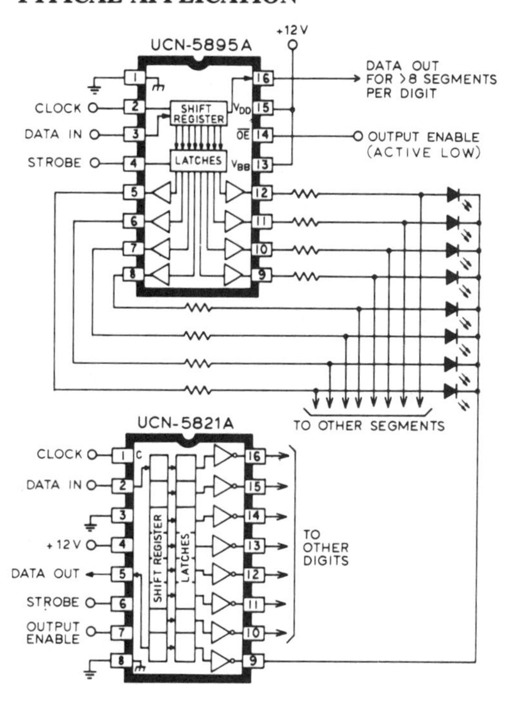

INDEX

B

C

D

E

F

M

N

O

S

Other Bestsellers of Related Interest

HOW TO GET MORE MILES PER GALLON IN THE 1990s
—Robert Sikorsky

This new edition of a bestseller features a wealth of commonsense tips and techniques for improving gas mileage by as much as 100 percent. Sikorsky details specific gas-saving strategies that will greatly reduce aerodynamic drag and increase engine efficiency. New to this edition is coverage of the latest fuel-conserving automotive equipment, fuel additives, engine treatments, lubricants, and maintenance procedures that can help save energy. 184 pages, 39 illustrations. Book No. 3793, $7.95 paperback, $16.95 hardcover

WIRE AND CABLE FOR ELECTRONICS: A User's Handbook—Neil Sclater

In *Wire and Cable for Electronics*, you'll find the most up-to-date information available on the properties of conductors and insulation, system design and assembly techniques, and guidelines for using the latest wire and cable equipment. This easy-to-use reference provides wire tables, unit conversion tables, engineering design formulas, and comparative data on insulation materials. You'll also find the important new regulations imposed by the 1990 National Electrical Code. 248 pages, 200 illustrations. Book No. 3787, $29.95 hardcover only

UPGRADE YOUR IBM® COMPATIBLE AND SAVE A BUNDLE—2nd Edition—Aubrey Pilgrim

Praise for the first edition . . .

"Every aspect is covered . . . liberally and clearly illustrated. . . . invaluable." ***—PC*** magazine

Find valuable advice on adding the newest high-quality, low-cost hardware to your PC. It offers informative how-to's for replacing motherboards with 80286, 80386, and 80486 boards; adding new floppy and hard disk drives; replacing old BIOS chips, installing chips and memory boards; and plugging in internal modems and VGA, fax, and network boards. Reviews of the latest monitors, printers, and other peripherals are also included. 264 pages, 60 illustrations. Book No. 3828, $19.95 paperback, $29.95 hardcover

BUILD YOUR OWN POSTSCRIPT® LASER PRINTER AND SAVE A BUNDLE—Horace W. LaBadie, Jr.

If you can plug in boards and handle a soldering iron, you can build your own high-powered, high-quality laser printer—at about half the cost of the retail model. LaBadie shows you the best way to collect the parts and assemble a low-cost machine that works as well as an off-the-shelf printer. He explains how to convert a stock Canon CX or SX laser engine to full PostScript capabilities. 176 pages, 124 illustrations. Book No. 3738, $16.95 paperback, $26.95 hardcover

TROUBLESHOOTING AND REPAIRING PERSONAL COMPUTERS—2nd Edition—Art Margolis

This all-in-one volume presents the theory and practical techniques necessary to service Apple II, Macintosh, Amiga, Commodore, and IBM PC circuitry and components. Margolis has added new information on the Intel 80386 and 80486 chips, laptops, video displays, disk drives, and power supplies. Fully illustrated coverage ranges from disassembly, maintenance, and testing through performing complicated board-level surgery. 696 pages, 460 illustrations. Book No. 3677, $23.95 paperback, $35.95 hardcover

THE ELECTRONICS WORKBENCH: Tools, Testers, and Tips for the Hobbyist—Delton T. Horn

Not only does Delton T. Horn offer important information about arranging the workbench, construction tips, and electronic safety, but he also gives you guidelines for using a wide variety of equipment. He takes an in-depth look at each major category of test instrument, explaining the characteristics and power capabilities of various models and giving you a reliable set of guidelines with which to choose the right equipment. 264 pages, 125 illustrations. Book No. 3672, $18.95 paperback, $28.95 hardcover

DAT: The Complete Guide to Digital Audio Tape
—Delton T. Horn

This book explains what DAT is, how it works, and how it differs from competing analog and compact disc technology. After a brief overview of basic analog and digital recording concepts, you will find in-depth coverage of DAT techniques and equipment, manufacturer's information on available (and soon to be available) systems, and what the future holds for DAT technology. You'll get a strong foundation in the theory and practice of DAT that you can use to improve all your recording. 264 pages, 98 illustrations. Book No. 3670, $12.95 paperback, $23.95 hardcover

RIP-OFF TIP-OFFS: Winning the Auto Repair Game
—Robert Sikorsky

Don't get ripped off when you take your car for repairs. This book gives you the ammunition to stop repair scams before they start. Sikorsky exposes popular tactics used by cheats and describes how to ensure a fair deal. If you have been ripped off, he tells you how to complain effectively—both to get your money back and to put the charlatans out of business for good. But most importantly, Sikorsky tells how to avoid getting burned in the first place by learning how your car works and by keeping it in good condition. 140 pages, 29 illustrations. Book No. 3572, $9.95 paperback, $16.95 hardcover

ELECTRONIC SIGNALS AND SYSTEMS: Television, Stereo, Satellite TV, and Automotive—Stan Prentiss

This handbook presents a detailed study of signal analysis as it applies to the operation and signal-generating capabilities of today's most advanced electronic devices. It explains the composition and use of a wide variety of test instruments, transmission media, satellite systems, stereo broadcast and reception facilities, antennas, television equipment, and even automotive electrical systems. You'll find coverage of C- and Ku-band video, satellite master TV systems, high-definition television, C-QUAM® AM stereo transmission and reception, and more. 328 pages, 186 illustrations. Book No. 3557, $19.95 paperback, $29.95 hardcover

THE MASTER IC COOKBOOK—2nd Edition
—Clayton L. Hallmark and Delton T. Horn

"A wide range of popular experimenter/hobbyist linear ICs is given in this encyclopedic book." ***—Popular Electronics,*** on the first edition

Find complete information on memories, audio amplifiers, RF amplifiers, and related devices, in addition to other sections on TTL, CMOS, special-purpose CMOS, and other linear devices. Circuits come complete with pinouts, specifications, and a concise description of the IC and its applications. 576 pages, 390 illustrations. Book No. 3550, $22.95 paperback, $34.95 hardcover

DESIGN & BUILD ELECTRONIC POWER SUPPLIES
—Irving M. Gottlieb

This guide brings you up to date on today's most advanced power supply circuits, components, and measurement procedures. It emphasizes practical applications over mathematical theory while focusing on the devices and techniques used with the newer types of switch-mode power supplies. Each section offers on-the-job tips and real-world examples to help you create top-quality electronic power supplies. 176 pages, 98 illustrations. Book No. 3540, $17.95 paperback, $26.95 hardcover

GORDON McCOMB'S TIPS & TECHNIQUES FOR THE ELECTRONICS HOBBYIST—Gordon McComb

This volume covers every facet of your electronics hobby from setting up a shop to making essential equipment. This single, concise handbook will answer almost all of your questions. You'll find general information on electronics practice, important formulas, tips on how to identify components, and more. You can use it as a source for ideas, as a textbook on electronics techniques and procedures, and a databook on electronics formulas, functions, and components. 288 pages, 307 illustrations. Book No. 3485, $17.95 paperback, $27.95 hardcover

MASTERING IC ELECTRONICS—Joseph J. Carr

If you have a basic understanding of integrated circuitry but you have not yet reached an advanced level of expertise, this book is for you. It gives you all the data you need to become well versed in the theory and operation of IC electronics. This book's unique blend of principles and practice provides you with a full working knowledge of component specifications, design standards, and applications for all types of modern ICs. 416 pages, illustrated. Book No. 3660, $19.95 paperback, $32.95 hardcover

BUILD YOUR OWN 80486 PC AND SAVE A BUNDLE
—Aubrey Pilgrim

With inexpensive third-party components and clear, step-by-step photos and assembly instructions—and without any soldering, wiring, or electronic test instruments—you can assemble a 486. This book discusses boards, monitors, hard drives, cables, printers, modems, faxes, UPSs, memory, floppy disks, and more. It includes parts lists, mail order addresses, safety precautions, troubleshooting tips, and a glossary of terms. 240 pages, 62 illustrations. Book No. 3628, $16.95 paperback, $26.95 hardcover

TV REPAIR FOR BEGINNERS—4th Edition
—George Zwick and Homer L. Davidson

Discover how to troubleshoot and repair all kinds of TV malfunctions. You'll look at X-ray protection circuits, scan-derived voltages, saw filter networks, high-voltage shutdowns, chopper power supplies, sandcastle generators, switched-mode power supplies, stereo sound, on-screen display, and surface-mounted components. 360 pages, 303 illustrations. Book No. 3627, $19.95 paperback, $29.95 hardcover

BEGINNER'S GUIDE TO READING SCHEMATICS
—2nd Edition—Robert J. Traister and Anna L. Lisk

Here's a practical guide to schematic diagrams that make even a complex electronic circuit as easy to understand as an ordinary highway map. Never again will you be intimidated by confusing lines and symbols when you open up the back of your TV to see why the sound or picture won't work. With this book, you'll discover how simple electronic circuit repair can be. 140 pages, 131 illustrations. Book No. 3632, $10.95 paperback, $18.95 hardcover

BUILD YOUR OWN MACINTOSH® AND SAVE A BUNDLE—Bob Brant

Building your own Mac is easy and much less expensive than buying one off the shelf. Just assemble economical, readily available components for your own catalog-part Macintosh, or "Cat Mac," with the help of this book. You can also upgrade a Mac you already own, add more memory, a hard disk, a bigger display, or an accelerator card to your existing systems—all at the best price. 240 pages, 113 illustrations. Book No. 3656, $17.95 paperback only